BERNE & LEVY

PHYSIOLOGY

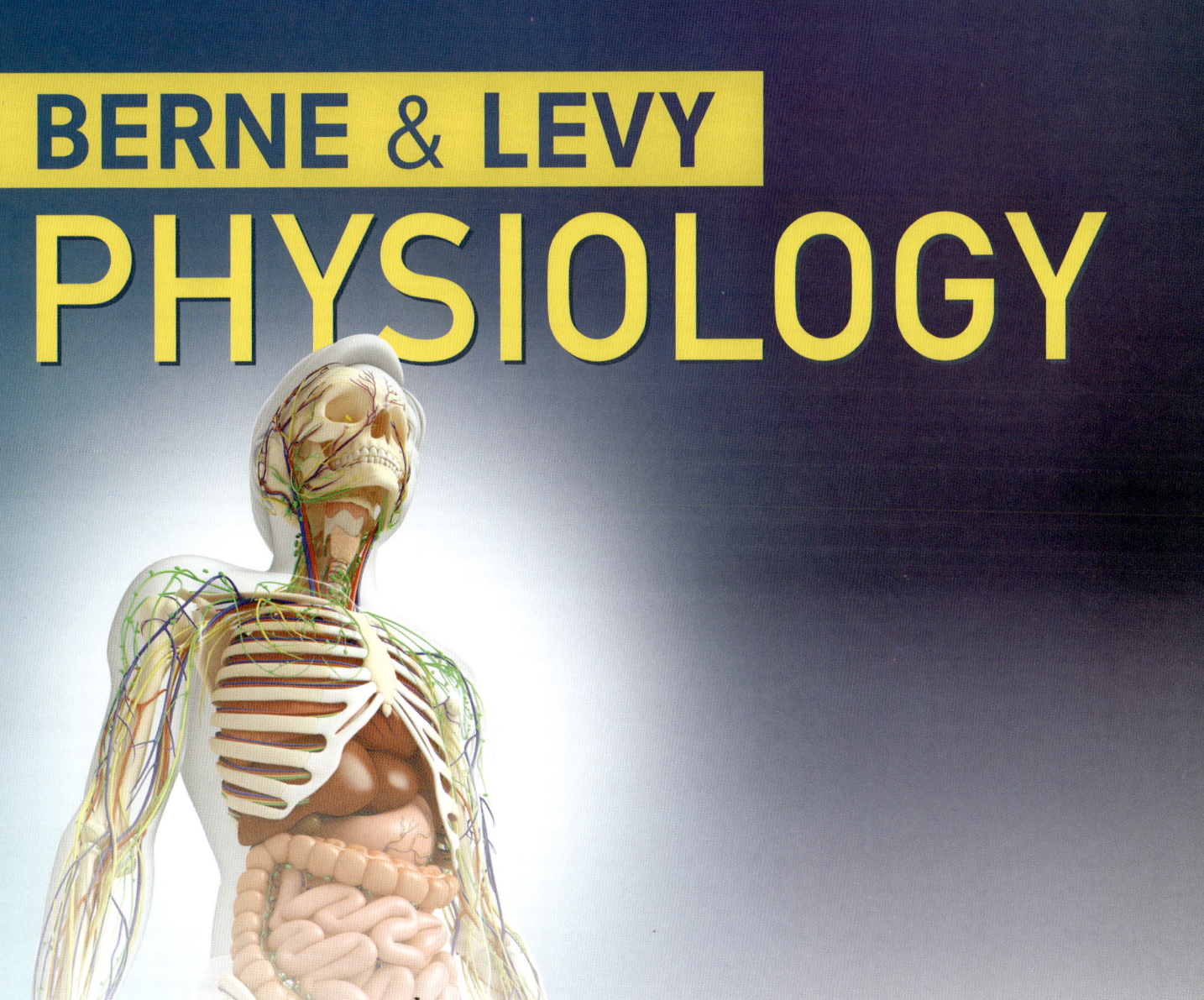

Seventh Edition

BERNE & LEVY
PHYSIOLOGY

Editors

Bruce M. Koeppen, MD, PhD

Dean
Frank H. Netter MD School of Medicine
Quinnipiac University
Hamden, Connecticut

Bruce A. Stanton, PhD

Andrew C. Vail Professor
Microbiology, Immunology, and Physiology
Director of the Lung Biology Center
Geisel School of Medicine at Dartmouth
Hanover, New Hampshire

ELSEVIER

ELSEVIER

1600 John F. Kennedy Blvd.
Ste 1800
Philadelphia, PA 19103-2899

BERNE AND LEVY PHYSIOLOGY, SEVENTH EDITION   ISBN: 978-0-323-39394-2
INTERNATIONAL EDITION   ISBN: 978-0-323-44338-8

Notices

Knowledge and best practice in this field are constantly changing. As new research and experience broaden
our understanding, changes in research methods, professional practices, or medical treatment may become
necessary.

Practitioners and researchers must always rely on their own experience and knowledge in evaluating and
using any information, methods, compounds, or experiments described herein. In using such information or
methods they should be mindful of their own safety and the safety of others, including parties for whom they
have a professional responsibility.

With respect to any drug or pharmaceutical products identified, readers are advised to check the most
current information provided (i) on procedures featured or (ii) by the manufacturer of each product to be
administered, to verify the recommended dose or formula, the method and duration of administration, and
contraindications. It is the responsibility of practitioners, relying on their own experience and knowledge of
their patients, to make diagnoses, to determine dosages and the best treatment for each individual patient,
and to take all appropriate safety precautions.

To the fullest extent of the law, neither the Publisher nor the authors, contributors, or editors, assume any
liability for any injury and/or damage to persons or property as a matter of products liability, negligence or
otherwise, or from any use or operation of any methods, products, instructions, or ideas contained in the
material herein.

Previous editions copyrighted 2010, 2008, 2004, 1998, 1993, 1988, and 1983.

Library of Congress Cataloging-in-Publication Data

Names: Koeppen, Bruce M., editor. | Stanton, Bruce A., editor.
Title: Berne & Levy physiology / editors, Bruce M. Koeppen, Bruce A. Stanton.
Other titles: Berne and Levy physiology | Physiology
Description: Seventh edition. | Philadelphia, PA : Elsevier, [2018] | Includes index.
Identifiers: LCCN 2016039642| ISBN 9780323393942 (hardcover) | ISBN 9780323443388 (international
    edition : hardcover)
Subjects: | MESH: Physiological Phenomena
Classification: LCC QP34.5 | NLM QT 104 | DDC 612–dc23 LC record available at
    https://lccn.loc.gov/2016039642

Executive Content Strategist: Elyse O'Grady
Senior Content Development Specialist: Margaret Nelson
Publishing Services Manager: Patricia Tannian
Senior Project Manager: Cindy Thoms
Book Designer: Patrick Ferguson

Printed in China

Last digit is the print number: 9   8   7   6   5   4   3   2   1

Working together
to grow libraries in
developing countries

www.elsevier.com • www.bookaid.org

This seventh edition of Physiology *is dedicated to the many students who have used this textbook to learn and understand the function of the human body.*

Bruce M. Koeppen, MD, PhD
Bruce A. Stanton, PhD

Section Authors

Kim E. Barrett, PhD
Distinguished Professor of Medicine
University of California, San Diego, School of Medicine
La Jolla, California
Section 6: Gastrointestinal Physiology

Michelle M. Cloutier, MD
Professor
Department of Pediatrics
University of Connecticut School of Medicine
Farmington, Connecticut
and
Director
Asthma Center
Connecticut Children's Medical Center
Hartford, Connecticut
Section 5: The Respiratory System

John R. Harrison, PhD
Associate Professor
Department of Craniofacial Sciences
University of Connecticut Health Center
Farmington, Connecticut
Section 8: The Endocrine and Reproductive Systems

Bruce M. Koeppen, MD, PhD
Dean
Frank H. Netter MD School of Medicine
Quinnipiac University
Hamden, Connecticut
Section 1: Cellular Physiology
Section 7: The Renal System

Eric J. Lang, MD, PhD
Associate Professor
Department of Neuroscience and Physiology
New York University School of Medicine
New York, New York
Section 2: The Nervous System

Achilles J. Pappano, PhD
Professor Emeritus
Department of Cell Biology
Calhoun Cardiology Center
University of Connecticut Health Center
Farmington, Connecticut
Section 4: The Cardiovascular System

Helen E. Raybould, PhD
Professor
Department of Anatomy, Physiology, and Cell Biology
University of California-Davis
School of Veterinary Medicine
Davis, California
Section 6: Gastrointestinal Physiology

Kalman Rubinson, PhD
Emeritus Professor
Department of Neuroscience and Physiology
New York University School of Medicine
New York, New York
Section 2: The Nervous System

Bruce A. Stanton, PhD
Andrew C. Vail Professor
Microbiology, Immunology, and Physiology
Director of the Lung Biology Center
Geisel School of Medicine at Dartmouth
Hanover, New Hampshire
Section 1: Cellular Physiology
Section 7: The Renal System

Roger S. Thrall, PhD
Professor Emeritus
Immunology and Medicine
University of Connecticut Health Center
Farmington, Connecticut
and
Director of Clinical Research
Department of Research
Hospital for Special Care
New Britain, Connecticut
Section 5: The Respiratory System

James M. Watras, PhD
Associate Professor
Department of Cell Biology
University of Connecticut Health Center
Farmington, Connecticut
Section 3: Muscle

Bruce A. White, PhD
Professor
Department of Cell Biology
University of Connecticut Health Center
Farmington, Connecticut
Section 8: The Endocrine and Reproductive Systems

Withrow Gil Wier, PhD
Professor
Department of Physiology
University of Maryland, Baltimore
Baltimore, Maryland
Section 4: The Cardiovascular System

Board of Reviewers

We wish to express our appreciation to all of our colleagues and students who have provided constructive criticism during the revision of this book:

Hannah Carey, PhD
University of Wisconsin, Madison
School of Veterinary Medicine
Madison, Wisconsin
Section 6: Gastrointestinal Physiology

Nathan Davis, PhD
Professor of Medical Sciences
Frank H. Netter MD School of Medicine
Quinnipiac University
Hamden, Connecticut
Section 8: The Endocrine and Reproductive Systems

L. Lee Hamm, MD
Senior Vice President and Dean
Tulane University School of Medicine
New Orleans, Louisiana
Chapter 37: Role of the Kidneys in the Regulation of Acid-Base Balance

Douglas McHugh, PhD
Associate Professor of Medical Sciences
Frank H. Netter MD School of Medicine
Quinnipiac University
Hamden, Connecticut
Section 1: Cellular Physiology

Orson Moe, MD
The Charles Pak Distinguished Chair in Mineral Metabolism
Donald W. Seldin Professorship in Clinical Investigation
University of Texas Southwestern Medical Center
Dallas, Texas
Section 7: The Renal System

R. Brooks Robey, MD, FASN FAHA
Associate Chief of Staff for Research
Chief of Nephrology at the White River Junction VA Medical Center
Geisel School of Medicine at Dartmouth
Hanover, New Hampshire
Section 7: The Renal System

Marion Siegman, PhD
Professor and Chair
Department of Molecular Physiology and Biophysics
Sidney Kimmel Medical College at Thomas Jefferson University
Philadelphia, Pennsylvania
Chapter 14: Smooth Muscle

Travis Solomon, MD, PhD
School of Medicine
University of Missouri
Kansas City, Missouri
Section 6: Gastrointestinal Physiology

Nancy Wills, PhD
Emeritus Professor of Medical Sciences
Frank H. Netter MD School of Medicine
Quinnipiac University
Hamden, Connecticut
Section 1: Cellular Physiology

Preface

We are pleased that the following section authors have continued as members of the seventh edition team: Drs. Kalman Rubinson and Eric Lang (nervous system), Dr. James Watras (muscle), Dr. Achilles Pappano (cardiovascular system), Drs. Michelle Cloutier and Roger Thrall (respiratory system), Drs. Kim Barrett and Helen Raybould (gastrointestinal system), and Dr. Bruce White (endocrine and reproductive systems). We also welcome the following authors: Dr. Withrow Gil Wier (cardiovascular system), and Dr. John Harrison (endocrine and reproduction systems).

As in the previous editions of this textbook, we have attempted to emphasize broad concepts and to minimize the compilation of isolated facts. Each chapter has been written to make the text as lucid, accurate, and current as possible. We have included both clinical and molecular information in each section, as feedback on these features has indicated that this information serves to provide clinical context and new insights into physiologic phenomena at the cellular and molecular levels. New to this edition is a list of sources that the reader can consult for further information on the topics covered in each chapter. We hope that you find this a valuable addition to the book.

The human body consists of billions of cells that are organized into tissues (e.g., muscle, epithelia, and nervous tissue) and organ systems (e.g., nervous, cardiovascular, respiratory, renal, gastrointestinal, endocrine, and reproductive). For these tissues and organ systems to function properly and thus allow humans to live and carry out daily activities, several general conditions must be met. First and foremost, the cells within the body must survive. Survival requires adequate cellular energy supplies, maintenance of an appropriate intracellular milieu, and defense against a hostile external environment. Once cell survival is ensured, the cell can then perform its designated or specialized function (e.g., contraction by skeletal muscle cells). Ultimately, the function of cells, tissues, and organs must be coordinated and regulated. All of these functions are the essence of the discipline of physiology and are presented throughout this book. What follows is a brief introduction to these general concepts.

Cells need a constant supply of energy. This energy is derived from the hydrolysis of **adenosine triphosphate** (**ATP**). If not replenished, the cellular ATP supply would be depleted in most cells in less than 1 minute. Thus, ATP must be continuously synthesized. This in turn requires a steady supply of cellular fuels. However, the cellular fuels (e.g., glucose, fatty acids, and ketoacids) are present in the blood at levels that can support cellular metabolism only for a few minutes. The blood levels of these cellular fuels are maintained through the ingestion of precursors (i.e., carbohydrates, proteins, and fats). In addition, these fuels can be stored and then mobilized when ingestion of the precursors is not possible. The storage forms of these fuels are triglycerides (stored in adipose tissue), glycogen (stored in the liver and skeletal muscle), and protein. The maintenance of adequate levels of cellular fuels in the blood is a complex process involving the following tissues, organs, and organ systems:

- *Liver*: Converts precursors into fuel storage forms (e.g., glucose → glycogen) when food is ingested, and converts storage forms to cellular fuels during fasting (e.g., glycogen → glucose and amino acids → glucose).
- *Skeletal muscle*: Like the liver, stores fuel (glycogen and protein) and converts glycogen and protein to fuels (e.g., glucose) or fuel intermediates (e.g., protein → amino acids) during fasting.
- *Gastrointestinal tract*: Digests and absorbs fuel precursors.
- *Adipose tissue*: Stores fuel during feeding (e.g., fatty acids → triglycerides) and releases the fuels during fasting.
- *Cardiovascular system*: Delivers the fuels to the cells and to and from their storage sites.
- *Endocrine system*: Maintains the blood levels of the cellular fuels by controlling and regulating their storage and their release from storage (e.g., insulin and glucagons).
- *Nervous system*: Monitors oxygen levels and nutrient content in the blood and, in response, modulates the cardiovascular, pulmonary, and endocrine systems and induces feeding and drinking behaviors.

In addition to energy metabolism, the cells of the body must maintain a relatively constant intracellular environment to survive. This includes the uptake of fuels needed to produce ATP, the export from the cell of cellular wastes, the maintenance of an appropriate intracellular ionic environment, the establishment of a resting membrane potential, and the maintenance of a constant cellular volume. All of these functions are carried out by specific membrane transport proteins.

The composition of the extracellular fluid (ECF) that bathes the cells must also be maintained relatively constant. In addition, the volume and temperature of the ECF must be regulated. Epithelial cells in the lungs, gastrointestinal tract, and kidneys are responsible for maintaining the volume and composition of the ECF, while the skin plays a major role in temperature regulation. On a daily basis, H_2O and food are ingested, and essential components are absorbed across the epithelial cells of the gastrointestinal tract. This daily intake of solutes and water must be matched by excretion from the body, thus maintaining **steady-state balance**. The kidneys are critically involved in the maintenance of steady-state balance for water and many components of the ECF (e.g., Na^+, K^+, HCO_3^-, pH, Ca^{++}, organic solutes). The lungs ensure an adequate supply of O_2 to "burn" the cellular fuels for the production of ATP and excrete the major waste product of this process (i.e., CO_2). Because CO_2 can affect the pH of the ECF, the lungs work with the kidneys to maintain ECF pH.

Because humans inhabit many different environments and often move between environments, the body must be able to rapidly adapt to the challenges imposed by changes in ambient temperature and availability of food and water. Such adaptation requires coordination of the function of cells in different tissues and organs as well as their regulation. The nervous and endocrine systems coordinate and regulate cell, tissue, and organ function. The regulation of function can occur rapidly (seconds to minutes), as is the case for levels of cellular fuels in the blood, or over much longer periods of time (days to weeks), as is the case for acclimatization when an individual moves from a cool to a hot environment or changes from a high-salt to a low-salt diet.

The function of the human body represents complex processes at multiple levels. This book explains what is currently known about these processes. Although the emphasis is on the normal function of the human body, discussion of disease and abnormal function is also appropriate, as these often illustrate physiologic processes and principles at the extremes.

The authors for each section have presented what they believe to be the most likely mechanisms responsible for the phenomena under consideration. We have adopted this compromise to achieve brevity, clarity, and simplicity.

Bruce M. Koeppen, MD, PhD
Bruce A. Stanton, PhD

Contents

SECTION 1

Cellular Physiology

BRUCE M. KOEPPEN AND BRUCE A. STANTON

1

Principles of Cell and Membrane Function

LEARNING OBJECTIVES

Upon completion of this chapter, the student should be able to answer the following questions:

1. What organelles are found in a typical eukaryotic cell, and what is their function?
2. What is the composition of the plasma membrane?
3. What are the major classes of membrane transport proteins, and how do they transport biologically important molecules and ions across the plasma membrane?
4. What is the electrochemical gradient, and how it is used to determine whether the transport of a molecule or ion across the plasma membrane is active or passive?
5. What are the driving forces for movement of water across cell membrane and the capillary wall?

In addition, the student should be able to define and understand the following properties of physiologically important solutions and fluids:

- Molarity and equivalence
- Osmotic pressure
- Osmolarity and osmolality
- Oncotic pressure
- Tonicity

The human body is composed of billions of cells. Although cells can perform different functions, they share certain common elements. This chapter provides an overview of these common elements and focuses on the important function of the transport of molecules and water into and out of the cell across its plasma membrane.

Overview of Eukaryotic Cells

Eukaryotic cells are distinguished from prokaryotic cells by the presence of a membrane-delimited nucleus. With the exception of mature human red blood cells and cells within the lens of the eye, all cells within the human body contain a nucleus. The cell is therefore effectively divided into two compartments: the nucleus and the cytoplasm. The cytoplasm is an aqueous solution containing numerous organic molecules, ions, cytoskeletal elements, and a number of organelles. Many of the organelles are membrane-enclosed

compartments that carry out specific cellular function. An idealized eukaryotic cell is depicted in Fig. 1.1, and the primary function of some components and compartments of the cell are summarized in Table 1-1. Readers who desire a more in-depth presentation of this material are encouraged to consult one of the many textbooks on cell and molecular biology that are currently available.

The Plasma Membrane

The cells within the body are surrounded by a plasma membrane that separates the intracellular contents from the extracellular environment. Because of the properties of this membrane and, in particular, the presence of specific membrane proteins, the plasma membrane is involved in a number of important cellular functions, including the following:

- Selective transport of molecules into and out of the cell. A function carried out by membrane transport proteins.
- Cell recognition through the use of cell surface antigens.
- Cell communication through neurotransmitter and hormone receptors and through signal transduction pathways.
- Tissue organization, such as temporary and permanent cell junctions, and interaction with the extracellular matrix, with the use of a variety of cell adhesion molecules.
- Membrane-dependent enzymatic activity.
- Determination of cell shape by linkage of the cytoskeleton to the plasma membrane.

In this chapter, the structure and function of the plasma membrane of eukaryotic cells are considered. More specifically, the chapter focuses on the transport of molecules and water across the plasma membrane. Only the principles of membrane transport are presented here. Additional details that relate to specific cells are presented in the various sections and chapters of this book.

Structure and Composition

The plasma membrane of eukaryotic cells consists of a 5-nm-thick lipid bilayer with associated proteins (Fig. 1.2). Some of the membrane-associated proteins are integrated into the lipid bilayer; others are more loosely attached to the

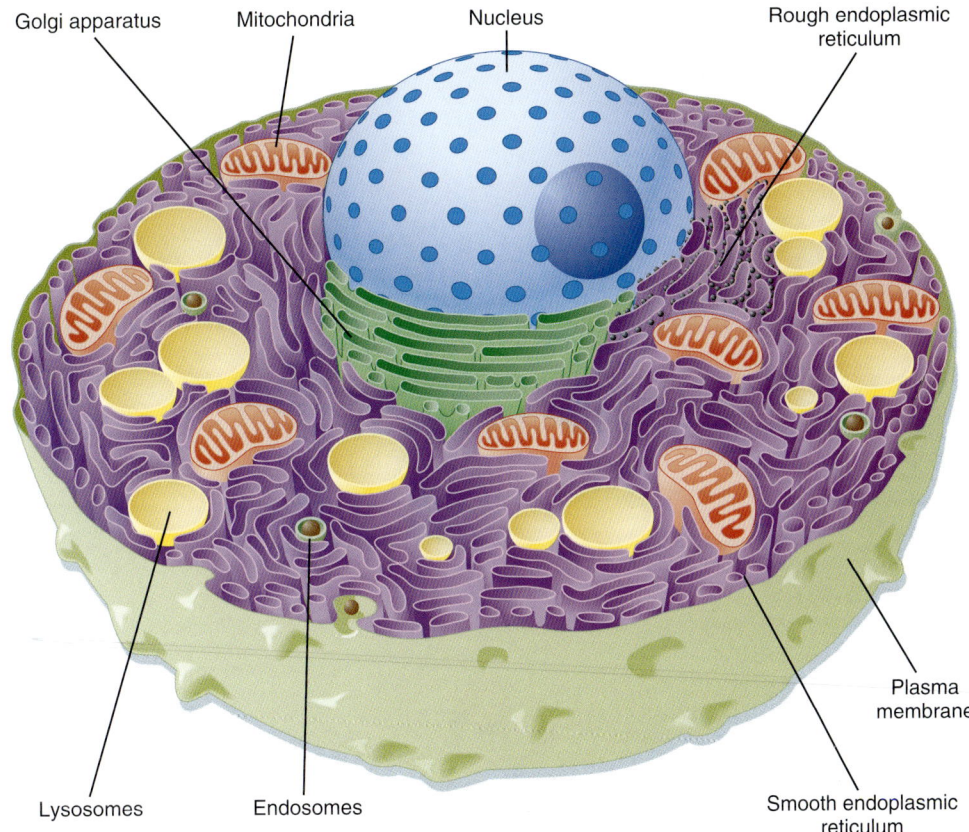

Golgi apparatus   Mitochondria   Nucleus   Rough endoplasmic reticulum

Plasma membrane

Lysosomes   Endosomes   Smooth endoplasmic reticulum

• **Fig. 1.1** Schematic drawing of a eukaryotic cell. The top portion of the cell is omitted to illustrate the nucleus and various intracellular organelles. See text for details.

TABLE 1.1	Primary Functions of Some Eukaryotic Cellular Components and Compartments
Component	**Primary Function**
Cytosol	Metabolism, protein synthesis (free ribosomes)
Cytoskeleton	Cell shape and movement, intracellular transport
Nucleus	Genome (22 autosomes and 2 sex chromosomes), DNA and RNA synthesis
Mitochondria	ATP synthesis by oxidative phosphorylation, Ca^{2+} storage
Smooth endoplasmic reticulum	Synthesis of lipids, Ca^{2+} storage
Free ribosomes	Translation of mRNA into cytosolic proteins
Rough endoplasmic reticulum	Translation of mRNA into membrane associated proteins or for secretion out of the cell
Lysosome	Intracellular degradation
Endosome	Cellular uptake of cholesterol, removal of receptors from the plasma membrane, uptake of small molecules and water into the cell, internalization of large particles (e.g., bacteria, cell debris)
Golgi apparatus	Modification, sorting, and packaging of proteins and lipids for delivery to other organelles within the cell or for secretion out of the cell
Proteosome	Degradation of intracellular proteins
Peroxisome	Detoxification of substances

ATP, adenosine triphosphate; mRNA, messenger RNA.

inner or outer surfaces of the membrane, often by binding to the integral membrane proteins.

Membrane Lipids

The major lipids of the plasma membrane are **phospholipids** and **phosphoglycerides.** Phospholipids are amphipathic molecules that contain a charged (or polar) hydrophilic head and two (nonpolar) hydrophobic fatty acyl chains (Fig. 1.3). The amphipathic nature of the phospholipid molecule is critical for the formation of the bilayer: The hydrophobic fatty acyl chains form the core of the bilayer, and the polar head groups are exposed on the surface.

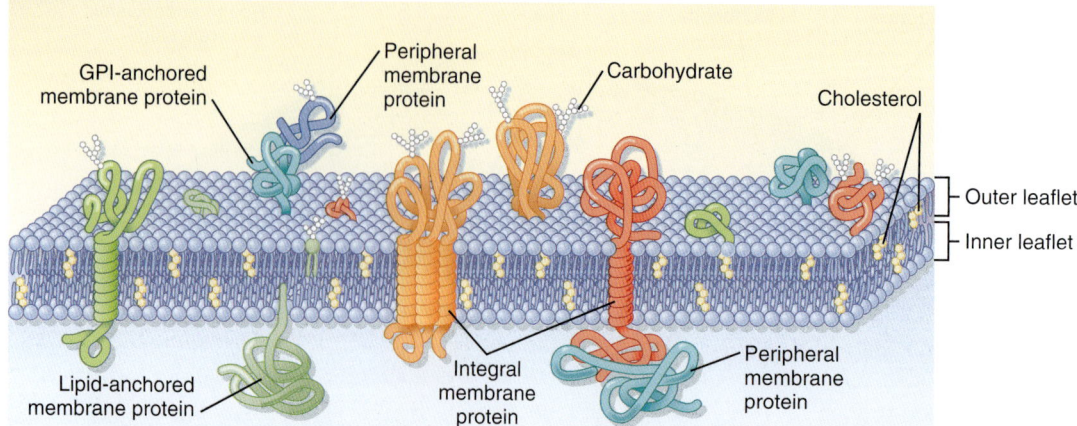

• **Fig. 1.2** Schematic diagram of the cell plasma membrane. Not shown are lipid rafts. See text for details. (Modified from Cooper GM. *The Cell—A Molecular Approach.* 2nd ed. Washington, DC: Sinauer; 2000, Fig. 12.3.)

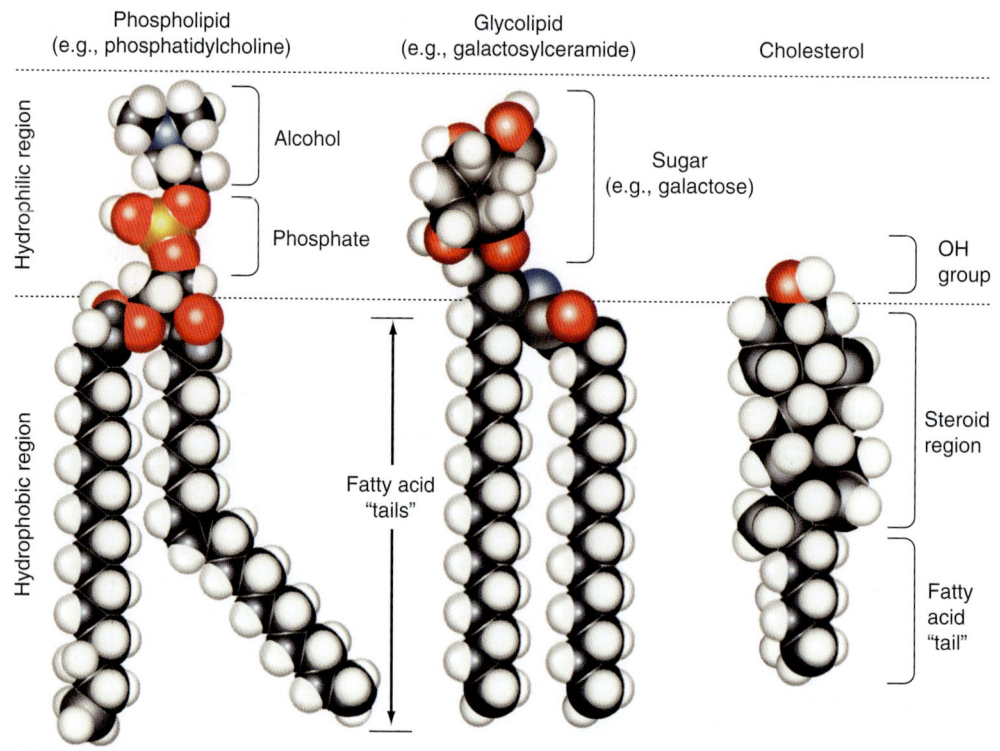

• **Fig. 1.3** Models of the major classes of plasma membrane lipids, depicting the hydrophilic and hydrophobic regions of the molecules. The molecules are arranged as they exist in one leaflet of the bilayer. The opposing leaflet is not shown. One of the fatty acyl chains in the phospholipid molecule is unsaturated. The presence of this double bond produces a "kink" in the fatty acyl chain, which prevents tight packing of membrane lipids and increases membrane fluidity. (Modified from Hansen JT, Koeppen BM: *Netter's Atlas of Human Physiology.* Teterboro, NJ: Icon Learning Systems; 2002.)

The majority of membrane phospholipids have a glycerol "backbone" to which are attached the fatty acyl chains, and an alcohol is linked to glycerol via a phosphate group. The common alcohols are choline, ethanolamine, serine, inositol, and glycerol. Another important phospholipid, sphingomyelin, has the amino alcohol sphingosine as its "backbone" instead of glycerol. Table 1-2 lists these common phospholipids. The fatty acyl chains are usually 14 to 20 carbons in length and may be saturated or unsaturated (i.e., contain one or more double bonds).

The phospholipid composition of the membrane varies among different cell types and even between the bilayer leaflets. For example, in the erythrocyte plasma membrane, phosphatidylcholine and sphingomyelin are found predominantly in the outer leaflet of the membrane, whereas phosphatidylethanolamine, phosphatidylserine,

TABLE 1.2	Plasma Membrane Lipids	
Phospholipid	**Primary Location in Membrane**
Phosphatidylcholine	Outer leaflet
Sphingomyelin	Outer leaflet
Phospatidylethanolamine	Inner leaflet
Phosphatidylserine	Inner leaflet
Phosphatidylinositol*	Inner leaflet

*Involved in signal transduction.

and phosphatidylinositol are found in the inner leaflet. As described in detail in Chapter 3, phosphatidylinositol plays an important role in signal transduction, and its location in the inner leaflet of the membrane facilitates this signaling role.

The sterol molecule **cholesterol** is also a critical component of the bilayer (see Fig. 1.3). It is found in both leaflets and serves to stabilize the membrane at normal body temperature (37°C). As much as 50% of the lipids found in the membrane can be cholesterol. A minor lipid component of the plasma membrane is **glycolipids.** These lipids, as their name indicates, consist of two fatty acyl chains linked to polar head groups that consist of carbohydrates (see Fig. 1.3). As discussed in the section on membrane proteins, one glycolipid, glycosylphosphatylinositol (GPI), plays an important role in anchoring proteins to the outer leaflet of the membrane. Both cholesterol and glycolipids, like the phospholipids, are amphipathic, and they are oriented with their polar groups on the outer surface of the leaflet in which they are located. Their hydrophobic portion is thus located within the interior of the bilayer.

The lipid bilayer is not a static structure. The lipids and associated proteins can diffuse within the plane of the membrane. The fluidity of the membrane is determined by temperature and by its lipid composition. As temperature increases, the fluidity of the membrane increases. The presence of unsaturated fatty acyl chains in the phospholipids and glycolipids also increases membrane fluidity. If a fatty acyl chain is unsaturated, the presence of a double bond introduces a "kink" in the molecule (see Fig. 1.3). This kink prevents the molecule from associating closely with surrounding lipids, and, as a result, membrane fluidity is increased. Although the lipid bilayer is "fluid," movement of proteins in the membrane can be constrained or limited. For example, membrane proteins can be anchored to components of the intracellular cytoskeleton, which limits their movement. Membrane domains can also be isolated from one another. An important example of this can be found in epithelial tissues. Junctional complexes (e.g., tight junctions) separate the plasma membrane of epithelial cells into two domains: apical and basolateral (see Chapter 2). The targeted localization of membrane proteins into one or other of these domains allows epithelial cells to carry

out vectorial transport of substances from one side of the epithelium to the opposite side. The ability to carry out vectorial transport is crucial for the functioning of several organ systems (e.g., the gastrointestinal tract and kidneys). In addition, some regions of the membrane contain lipids (e.g., sphingomyelin and cholesterol) that aggregate into what are called **lipid rafts.** These lipid rafts often have an association with specific proteins, which diffuse in the plane of the membrane as a discrete unit. Lipid rafts appear to serve a number of functions. One important function of these rafts is to segregate signaling molecules.

Membrane Proteins

As much as 50% of the plasma membrane is composed of proteins. These membrane proteins are classified as integral, lipid-anchored, or peripheral.

Integral membrane proteins are imbedded in the lipid bilayer, where hydrophobic amino acid residues are associated with the hydrophobic fatty acyl chains of the membrane lipids. Many integral membrane proteins span the bilayer; such proteins are termed **transmembrane proteins.** Transmembrane proteins have both hydrophobic and hydrophilic regions. The hydrophobic region, often in the form of an α helix, spans the membrane. Hydrophilic amino acid residues are then exposed to the aqueous environment on either side of the membrane. Transmembrane proteins may pass through the membrane multiple times.

 AT THE CELLULAR LEVEL

There is a superfamily of membrane proteins that serve as receptors for many hormones, neurotransmitters, and numerous drugs. These receptors are coupled to heterotrimeric G proteins and are termed *G protein–coupled receptors* (see Chapter 3). These proteins span the membrane with seven α-helical domains. The binding site of each ligand is either on the extracellular portion of the protein (large ligands) or in the membrane-spanning portion (small ligands), whereas the cytoplasmic portion binds to the G protein. This superfamily of membrane proteins makes up the third largest family of genes in humans. Nearly half of all nonantibiotic prescription drugs are targeted toward G protein–coupled receptors.

A protein can also be attached to the membrane via **lipid anchors.** The protein is covalently attached to a lipid molecule, which is then embedded in one leaflet of the bilayer. Glycosylphosphatidylinositol (GPI) anchors proteins to the outer leaflet of the membrane. Proteins can be attached to the inner leaflet via their amino-terminus by fatty acids (e.g., myristate or palmitate) or via their carboxyl-terminus by prenyl anchors (e.g., farnesyl or geranylgeranyl).

Peripheral proteins may be associated with the polar head groups of the membrane lipids, but they more commonly bind to integral or lipid-anchored proteins.

In many cells, some of the outer leaflet lipids, as well as many of the proteins exposed on the outer surface of the membrane, are glycosylated (i.e., have short chains of sugars, called *oligosaccharides,* attached to them). Collectively, these glycolipids and glycoproteins form what is called the glycocalyx. Depending on the cell these glycolipids and glycoproteins may be involved in cell recognition (e.g., cell surface antigens) and formation of cell-cell interactions (e.g., attachment of neutrophils to vascular endothelial cells).

Membrane Transport

Although plasma membrane proteins perform many important cellular functions, as noted previously, the remainder of this chapter focuses on one group of plasma membrane proteins: the membrane transport proteins, or transporters. It has been estimated that approximately 10% of human genes ($\approx$2000) code for transporters. They are also targets for numerous drugs.

The normal function of cells requires the continuous movement of water and solutes into and out of the cell. The intracellular and extracellular fluids are composed primarily of H_2O, in which solutes (e.g., ions, glucose, amino acids) are dissolved. The plasma membrane, with its hydrophobic core, is an effective barrier to the movement of virtually all of these biologically important solutes. It also restricts the movement of water across the membrane. The presence of specific membrane transporters in the membrane is responsible for the movement of these solutes and water across the membrane.

Membrane Transport Proteins

Membrane transporters have been classified in several different ways. In this chapter, the transporters are divided into four general groups: water channels, ion channels, solute carriers, and adenosine triphosphate (ATP)–dependent transporters. Table 1-3 lists these groups of membrane transporters, their modes of transport, and estimates of the rates at which they transport molecules or ions across the membrane.

Water Channels

Water channels, or **aquaporins (AQPs),** are the main routes for water movement into and out of the cell. They are

widely distributed throughout the body (e.g., the brain, lungs, kidneys, salivary glands, gastrointestinal tract, and liver). Cells express different AQP isoforms, and some cells even express multiple isoforms. For example, cells in the collecting ducts of the kidneys express AQP3 and AQP4 in their basolateral membrane and AQP2 in their apical membrane. Moreover, the abundance of AQP2 in the apical membrane is regulated by antidiuretic hormone (also called arginine vasopressin), which is crucial for the ability of the kidneys to concentrate the urine (see Chapter 35).

Although all AQP isoforms allow the passive movement of H_2O across the membrane, some isoforms also provide a pathway for other molecules such as glycerol, urea, mannitol, purines, pyrimidines, CO_2, and NH_3 to cross the membrane. Because glycerol was one of the first molecules identified as crossing the membrane via some AQPs, this group of AQPs is collectively called *aquaglyceroporins* (see also Chapter 34). Regulation of the amount of H_2O that can enter or leave the cell via AQPs occurs primarily by altering the number of AQPs in the membrane.

AT THE CELLULAR LEVEL

Each AQP molecule consists of six membrane-spanning domains and a central water-transporting pore. Four AQP monomers assemble to form a homotetramer in the plasma membrane, with each monomer functioning as a water channel.

Ion Channels

Ion channels are found in all cells, and are especially important for the function of excitable cells (e.g., neurons and muscle cells). Ion channels are classified by their selectivity, conductance and mechanism of channel gating (i.e., opening and closing). *Selectivity* is defined as the nature of the ions that pass through the channel. At one extreme, ion channels can be highly selective, in that they allow only a specific ion through. At the other extreme, they may be nonselective, allowing all or a group of cations or anions through. *Channel conductance* refers to the number of ions that pass through the channel and is typically expressed in picosiemens (pS). The range of conductance is considerable: Some channels have a conductance of only 1 to 2 pS, whereas others have a conductance of more than 100 pS. For some channels, the conductance varies, depending on the direction in which the ion is moving. For example, if the channel has a larger conductance when ions are moving into the cell than when they are moving out of the cell, the channel is said to be an *inward rectifier.* Moreover, ion channels fluctuate between an open state or a closed state, a process called *gating* (Fig. 1.4). Factors that can control gating include membrane voltage, extracellular agonists or antagonists (e.g., acetylcholine is an extracellular agonist that controls the gating of a cation-selective channel in the motor end plate of skeletal muscle cells; see Chapter 6), intracellular messengers (e.g., Ca^{++}, ATP, cyclic guanosine

TABLE 1.3	Major Classes of Plasma Membrane Transporters	
Class	Transport Mode	Transport Rate
Pore*	Open (not gated)	Up to 10^9 molecules/sec
Channel	Gated	10^6-10^8 molecules/sec
Solute carrier	Cycle	10^2-10^4 molecules/sec
ATP-dependent	Cycle	10^2-10^4 molecules/sec

*Examples include porins that are found in the outer membrane of mitochondria, and water channels (i.e., aquaporins) that function as a pore.
ATP, adenosine triphosphate.

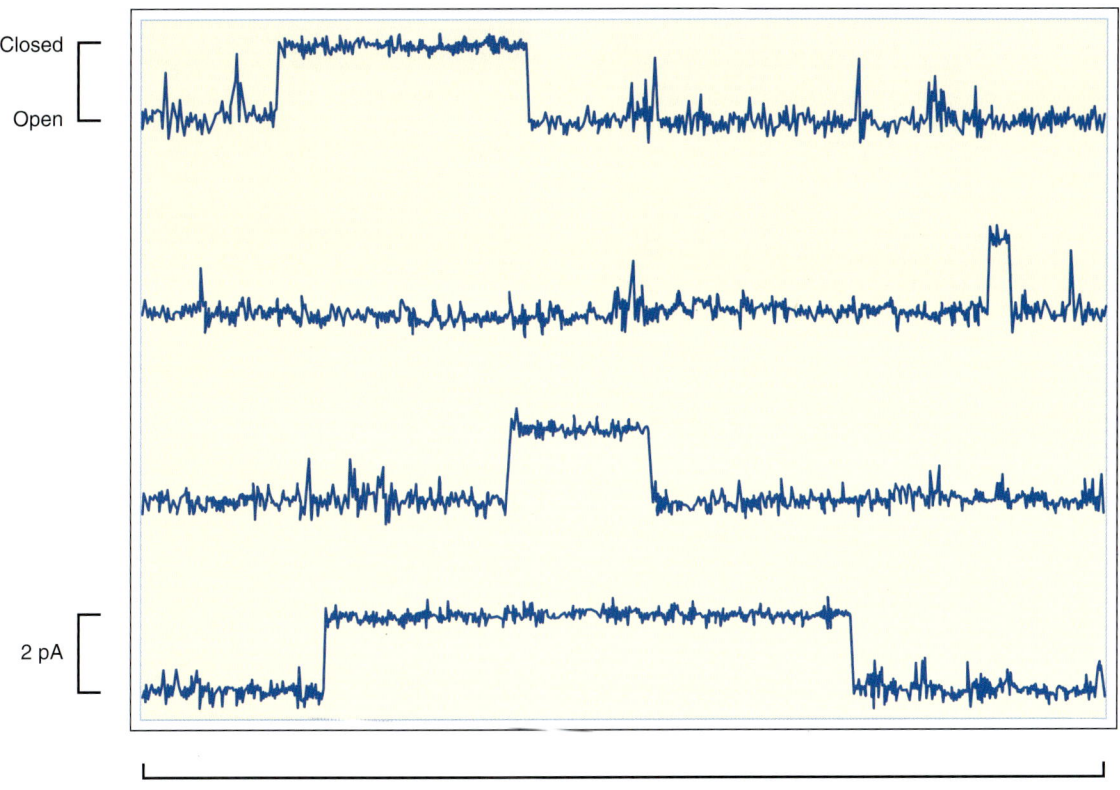

Closed

Open

2 pA

1 second

• **Fig. 1.4** Recording of current flow through a single ion channel. The channel spontaneously fluctuates between an open state and a closed state. The amplitude of the current is approximately 2 pA (2×10^{-12} amps); that is, 12.5 million ions/second cross the membrane.

monophosphate), and mechanical stretch of the plasma membrane. Ion channels can be regulated by a change in the number of channels in the membrane or by gating of the channels.

Solute Carriers

Solute carriers (denoted *SLCs* by the HUGO Gene Nomenclature Committee) represent a large group of membrane transporters categorized into more than 50 families; almost 400 specific transporters have been identified to date. These carriers can be divided into three groups according to their mode of transport. One group, **uniporters** (or **facilitated transporters**), transports a single molecule across the membrane. The transporter that brings glucose into the cell (glucose transporter 1 [GLUT-1], or SLC2A1) is an important member of this group. The second group, **symporters** (or **cotransporters**), couples the movement of two or more molecules/ions across the membrane. As the name implies, the molecules/ions are transported in the same direction. The $Na^+,K^+,2Cl^-$ (NKCC) symporter found in the kidney (NKCC2, or SLC12A1), which is crucial for diluting and concentrating the urine (see Chapter 34), is a member of this group. The third group, **antiporters** (or **exchange transporters**), also couples the movement of two or more molecules/ions across the membrane; in this case, however, the molecules/ions are transported in opposite directions. The Na^+-H^+ antiporter is a member of this group

of solute carriers. One isoform of this antiporter (NHE-1, or SLC9A1) is found in all cells and plays an important role in regulating intracellular pH.

Adenosine Triphosphate–Dependent Transporters

The ATP-dependent transporters, as their name implies, use the energy in ATP to drive the movement of molecules/ions across the membrane. There are two groups of ATP-dependent transporters: the **ATPase ion transporters** and the **ATP-binding cassette (ABC) transporters.** The ATPase ion transporters are subdivided into P-type ATPases and V-type ATPases.[a] The P-type ATPases are phosphorylated during the transport cycle. Na^+,K^+-ATPase is an important example of a P-type ATPase. With the hydrolysis of each ATP molecule, it transports three Na^+ ions out of the cell and two K^+ ions into the cell. Na^+,K^+-ATPase is present in all cells and plays a critical role in establishing cellular ion and electrical gradients, as well as maintaining cell volume (see Chapter 2).

V-type H^+-ATPases are found in the membranes of several intracellular organelles (e.g., endosomes, lysosomes); as a result, they are also referred to as *vacuolar H^+-ATPases.* The

---

[a]Another type of ATPases, F-type ATPases, is found in the mitochondria, and they are responsible for ATP synthesis. They are not considered in this chapter.

AT THE CELLULAR LEVEL

Na⁺,K⁺-ATPase (also called the *Na⁺,K⁺-pump* or just the *Na⁺-pump*) is found in all cells and is responsible for establishing the gradients of Na⁺ and K⁺ across the plasma membrane. These gradients in turn provide energy for several essential cell functions (see Chapter 2). Na⁺,K⁺-ATPase is composed of three subunits (α, β, and γ), and the protein exists in the membrane with a stoichiometric composition of 1α, 1β, 1γ. The α subunit contains binding sites for Na⁺,K⁺ and ATP. It is also the subunit that binds cardiac glycosides (e.g., ouabain), which specifically inhibit the enzyme. It has a transmembrane domain and three intracellular domains: phosphorylation (P-domain), nucleotide binding (N-domain), and actuator (A-domain). Although the α subunit is the functional subunit of the enzyme (i.e., it hydrolyzes ATP, binds Na⁺ and K⁺, and translocates them across the membrane), it cannot function without the β subunit. The β subunit is responsible for targeting the α subunit to the membrane and also appears to modulate the affinity of the Na⁺,K⁺-ATPase for Na⁺ and K⁺. The α and β subunits can carry out Na⁺ and K⁺ transport in the absence of the γ subunit. However, the γ subunit appears to play a regulatory role. The γ subunit is a member of a family of proteins called **FXYD proteins** (so named for the FXYD amino acid sequence found in the protein).

IN THE CLINIC

Cystic fibrosis is an autosomal recessive disease characterized by chronic lung infections, pancreatic insufficiency, and infertility in boys and men. Death usually occurs because of respiratory failure. It is most prevalent in white people and is the most common lethal genetic disease in this population, occurring in 1 per 3000 live births. It is a result of mutations in a gene on chromosome 7 that codes for an ABC transporter. To date, more than 1000 mutations in the gene have been identified. The most common mutation is a deletion of a phenylalanine at position 508 (F508del). Because of this deletion, degradation of the protein by the endoplasmic reticulum in enhanced, and, as a result, the transporter does not reach the plasma membrane. This transporter, called **cystic fibrosis transmembrane conductance regulator (CFTR),** normally functions as a Cl⁻ channel and also regulates other membrane transporters (e.g., the epithelial Na⁺ channel [ENaC]). Thus in individuals with cystic fibrosis, epithelial transport is defective, which is responsible for the pathophysiologic process. For example, in patients not affected by cystic fibrosis, the epithelial cells that line the airway of the lung are covered with a layer of mucus that entraps inhaled particulates and bacteria. Cilia on the epithelial cells then transport the entrapped material out of the lung, a process termed *mucociliary transport* (see Chapter 26 for more details). In patients with cystic fibrosis, the inability to secrete Cl⁻, Na⁺, and H₂O results in an increase in the viscosity of the airway surface mucus; thus the cilia cannot transport the entrapped bacteria and other pathogens out of the lung. This in turn leads to recurrent and chronic lung infections. The inflammatory process that accompanies these infections ultimately destroys the lung tissue, causing respiratory failure and death. In 2015, the U.S. Food and Drug Administration approved lumacaftor/ivacaftor (Orkambi), a drug that increases the amount of F508del CFTR in the plasma membrane of lung epithelial cells.

H⁺-ATPase in the plasma membrane plays an important role in urinary acidification (see Chapter 37).

ABC transporters represent a large group of membrane transporters. They are found in both prokaryotic and eukaryotic cells, and they have amino acid domains that bind ATP (i.e., ABC domains). Seven subgroups of ABC transporters in humans and more than 40 specific transporters have been identified to date. They transport a diverse group of molecules/ions, including Cl⁻, cholesterol, bile acids, drugs, iron, and organic anions.

Because biologically important molecules enter and leave cells through membrane transporters, membrane transport is specific and regulated. Although some membrane transporters are ubiquitously expressed in all cells (e.g., Na⁺,K⁺-ATPase), the expression of many other transporters is limited to specific cell types. This specificity of expression tailors the function of the cell to the organ system in which it is located (e.g., the sodium-glucose–linked transporters SGLT-1 and SGLT-2 in the epithelial cells of the intestines and renal proximal tubules). In addition, the amount of a molecule being transported across the membrane can be regulated. Such regulation can take place through altering the number of transporters in the membrane or altering the rate or kinetics of individual transporters (e.g., the time an ion channel stays in the open versus closed state), or both.

Vesicular Transport

Solute and water can be brought into the cell through a process of **endocytosis** and released from the cell through the process of **exocytosis.** Endocytosis is the process whereby a piece of the plasma membrane pinches off and

is internalized into the cell interior, and exocytosis is the process whereby vesicles inside the cell fuse with the plasma membrane. In both of these processes, the integrity of the plasma membrane is maintained, and the vesicles allow for the transfer of the contents among cellular compartments. In some cells (e.g., the epithelial cells lining the gastro-intestinal tract), endocytosis across one membrane of the cell is followed by exocytosis across the opposite membrane. This allows the transport of substances inside the vesicles across the epithelium, a process termed **transcytosis.**

Endocytosis occurs in three mechanisms. The first is **pinocytosis,** which consists of the nonspecific uptake of small molecules and water into the cell. Pinocytosis is a prominent feature of the endothelial cells that line capillaries and is responsible for a portion of the fluid exchange that occurs across these vessels. The second form of endocytosis, **phagocytosis,** allows for the cellular internalization of large particles (e.g., bacteria, cell debris). This process is an important characteristic of cells in the immune system (e.g., neutrophils and macrophages). Often, but not always, phagocytosis is a receptor-mediated process. For example,

AT THE CELLULAR LEVEL

Proteins within the plasma membrane of cells are constantly being removed and replaced with newly synthesized proteins. As a result, membrane proteins are constantly being replaced. One mechanism by which membrane proteins are "tagged" for replacement is by the attachment of ubiquitin to the cytoplasmic portion of the protein. Ubiquitin is a 76–amino acid protein that is covalently attached to the membrane protein (usually to lysine) by a class of enzymes called *ubiquitin protein ligases.* One important group of these ligases is the developmentally downregulated protein 4 (Nedd4)/Nedd4-like family. Once a membrane protein is ubiquitinated, it undergoes endocytosis and is degraded either by lysosomes or by the proteosome. Cells also contain deubiquitinating enzymes (DUBs). Thus the amount of time a protein stays in the plasma membrane depends on the rate that ubiquitin groups are added by the ligases versus the rate that they are removed by the DUBs. For example, Na^+ reabsorption by the collecting ducts of the kidneys is stimulated by the adrenal hormone aldosterone (see Chapters 34 and 35). One of the actions of aldosterone is to inhibit Nedd4-2. This prevents ubiquitination of ENaC in the apical membrane of epithelial cells. Thus the channels are retained for a longer period of time in the membrane, and as a result, more Na^+ enters the cell and is thereby reabsorbed.

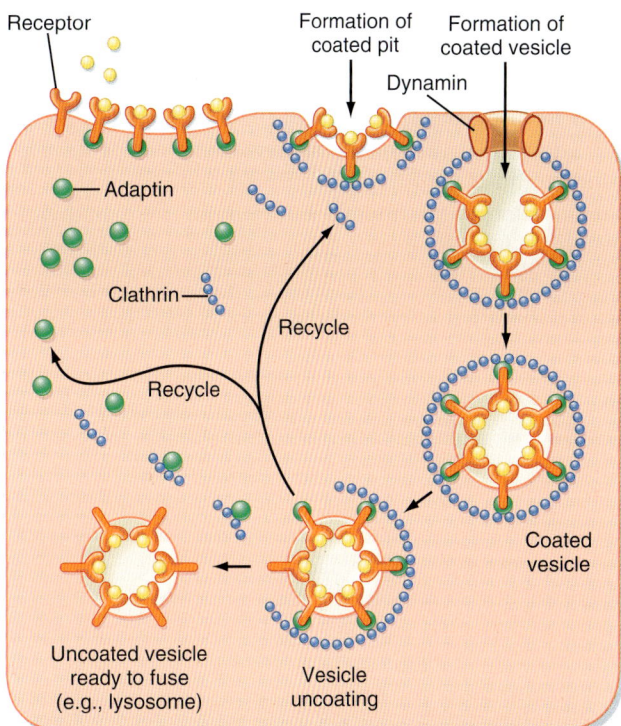

• **Fig. 1.5** Receptor-mediated endocytosis. Receptors on the surface of the cell bind the ligand. A clathrin-coated pit is formed with adaptin linking the receptor molecules to clathrin. Dynamin, a guanosine triphosphatase (GTPase), assists in separation of the endocytic vesicle from the membrane. Once inside the cell, the clathrin and adaptin molecules dissociate and are recycled. The uncoated vesicle is then ready to fuse with other organelles in the cell (e.g., lysosomes). (Adapted from Ross MH, Pawlina W: *Histology.* 5th ed. Baltimore: Lippincott Williams & Wilkins; 2006.)

macrophages have receptors on their surface that bind the Fc portion of immunoglobulins. When bacteria invade the body they are often coated with antibody, a process called opsonization. These bacteria then attach to the membrane of macrophages via the fragment crystallizable (Fc) portion of the immunoglobulin, undergo phagocytosis, and are destroyed inside the cell. The third mechanism of endocytosis is **receptor-mediated endocytosis,** which allows the uptake of specific molecules into the cell. In this form of endocytosis, molecules bind to receptors on the surface of the cell. Endocytosis involves a number of accessory proteins, including adaptin, clathrin, and the GTPase dynamin (Fig. 1.5).

Exocytosis can be either constitutive or regulated. Constitutive secretion occurs, for example, in plasma cells that are secreting immunoglobulin or in fibroblasts secreting collagen. Regulated secretion occurs in endocrine cells, neurons, and exocrine glandular cells (e.g., pancreatic acinar cells). In these cells, the secretory product (e.g., hormone, neurotransmitter, or digestive enzyme), after synthesis and processing in the rough endoplasmic reticulum and Golgi apparatus, is stored in the cytoplasm in secretory granules until an appropriate signal for secretion is received. These signals may be hormonal or neural. Once the cell receives the appropriate stimulus, the secretory vesicle fuses with the plasma membrane and releases its contents into the extracellular fluid. Fusion of the vesicle with the membrane is mediated by a number of accessory proteins. One important group is the SNARE (**s**oluble *N*-ethylmaleimide sensitive fusion protein [**N**SF] **a**ttachment protein **re**ceptors) proteins. These membrane proteins help target the secretory

IN THE CLINIC

Cholesterol is an important component of cells (e.g., it is a key component of membranes). However, most cells are unable to synthesize cholesterol and therefore must obtain it from the blood. Normally, cholesterol is ingested in the diet, and it is transported through the blood in association with lipoproteins. Low-density lipoproteins (LDLs) in the blood carry cholesterol to cells, where they bind to LDL receptors in the plasma membrane. After the receptors bind LDL, they collect into "coated pits" and undergo endocytosis as clathrin-coated vesicles. Once inside the cell, the endosomes release LDL and then recycle the LDL receptors back to the cell surface. Inside the cell, LDL is then degraded in lysosomes, and the cholesterol is made available to the cell. Defects in the LDL receptor prevent cellular uptake of LDL. Individuals with this defect have elevated levels of blood LDL, often called "bad cholesterol," because it is associated with the development of cholesterol-containing plaques in the smooth muscle layer of arteries. This process, atherosclerosis, is associated with an increased risk for heart attacks as a result of occlusion of the coronary arteries.

vesicle to the plasma membrane. The process of secretion is usually triggered by an increase in the concentration of intracellular Ca^{++} ($[Ca^{++}]$). However, two notable exceptions to this general rule exist: (1) Renin secretion by the juxtaglomerular cells of the kidney occurs with a decrease in intracellular Ca^{++} (see Chapters 34 and 35), as does (2) the secretion of parathyroid hormone by the parathyroid gland (see Chapter 40).

Basic Principles of Solute and Water Transport

As already noted, the plasma membrane, with its hydrophobic core, is an effective barrier to the movement of virtually all biologically important molecules into or out of the cell. Thus membrane transport proteins provide the pathway that allows transport to occur into and out of cells. However, the presence of a pathway is not sufficient for transport to occur; an appropriate driving force is also required. In this section, the basic principles of diffusion, active and passive transport, and osmosis are presented. These topics are discussed in greater depth, as appropriate, in the other sections of the book.

Diffusion

Diffusion is the process by which molecules move spontaneously from an area of high concentration to one of low concentration. Thus wherever a concentration gradient exists, diffusion of molecules from the region of high concentration to the region of low concentration dissipates the gradient (as discussed later, the establishment of concentration gradients for molecules requires the expenditure of energy). Diffusion is a random process driven by the thermal motion of the molecules. **Fick's first law of diffusion** quantifies the rate at which a molecule diffuses from point A to point B:

Equation 1.1

$$J = -DA\frac{\Delta C}{\Delta X}$$

where
J = the flux or rate of diffusion per unit time
D = the diffusion coefficient
A = area across which the diffusion is occurring
ΔC = the concentration difference between point A and B
ΔX = the distance along which diffusion is occurring

The diffusion coefficient takes into account the thermal energy of the molecule, its size, and the viscosity of the medium through which diffusion is taking place. For spherical molecules, D is approximated by the **Stokes-Einstein equation:**

Equation 1.2

$$D = \frac{kT}{6\pi r\eta}$$

where
k = Boltzmann's constant
T = temperature in degrees Kelvin

r = radius of the molecule
η = viscosity of the medium

According to eqs. 1.1 and 1.2, the rate of diffusion will be faster for small molecules than for large molecules. In addition, diffusion rates are high at elevated temperatures, in the presence of large concentration gradients, and when diffusion occurs in a low-viscosity medium. With all other variables held constant, the rate of diffusion is linearly related to the concentration gradient.

Fick's equation can also be applied to the diffusion of molecules across a barrier, such as a lipid bilayer. When applied to the diffusion of a molecule across a bilayer, the diffusion coefficient (D) incorporates the properties of the bilayer and especially the ability of the molecule to diffuse through the bilayer. To quantify the interaction of the molecule with the bilayer, the term *partition coefficient* (β) is used. For a molecule that "dissolves" equally in the fluid bathing the lipid bilayer (e.g., water) and in the lipid bilayer, $\beta = 1$. If the molecule dissolves more easily in the lipid bilayer, $\beta > 1$; and if it dissolves less easily in the lipid bilayer, $\beta < 1$. For a simple lipid bilayer, the more lipid soluble the molecule is, the larger the partition coefficient is, and thus the diffusion coefficient—therefore the rate of diffusion of the molecule across the bilayer—is greater. In this situation, ΔC represents the concentration difference across the membrane, A is the membrane area, and ΔX is the thickness of the membrane.

Another useful equation for quantitating the diffusion of molecules across the plasma membrane (or any membrane) is as follows:

Equation 1.3

$$J = -P(C_i - C_o),$$

where
J = the flux or rate of diffusion across the membrane
P = is the permeability coefficient
C_i = concentration of the molecule inside the cell
C_o = the concentration of the molecule outside the cell

This equation is derived from Fick's equation (eq. 1.1). P incorporates D, ΔX, A, and the partition coefficient (β). P is expressed in units of velocity (e.g., centimeters per second), and C the units of moles/cm^3. Thus the units of flux are moles per square centimeter per second ($mol/cm^2/sec$). Values for P can be obtained experimentally for any molecule and bilayer.

As noted, the phospholipid portion of the plasma membrane represents an effective barrier to many biologically important molecules. Consequently, diffusion through the lipid phase of the plasma membrane is not an efficient process for movement of these molecules across the membrane. It has been estimated that for a cell 20 µm in diameter, with a plasma membrane composed only of phospholipids, dissipation of a urea gradient imposed across the membrane would take approximately 8 minutes. Similar gradients for glucose and amino acids would take approximately 14 hours to dissipate, whereas ion gradients would take years to dissipate.

As noted previously, the vast majority of biologically important molecules cross cell membranes via specific membrane transporters, rather than by diffusing through the lipid portion of the membrane. Nevertheless, eq. 1.3 can be and has been used to quantitate the diffusion of molecules across many biological membranes. When this is done, the value of the permeability coefficient (P) reflects the properties of the pathway (e.g., membrane transporter or, in some cases, multiple transporters) that the molecule uses to cross the membrane.

Despite the limitations of using diffusion to describe and understand the transport of molecules across cell membranes, it is also important for understanding gas exchange in the lungs (see Chapter 24), the movement of molecules through the cytoplasm of the cell, and the movement of molecules between cells in the extracellular fluid. For example, one of the physiological responses of skeletal muscle to exercise is the recruitment or opening of capillaries that are not perfused at rest. This opening of previously closed capillaries increases capillary density and thereby reduces the diffusion distance between the capillary and the muscle fiber so that oxygen and cellular fuels (e.g., fatty acids and glucose) can be delivered more quickly to the contracting muscle fiber. In resting muscle, the average distance of a muscle fiber from a capillary is estimated to be 40 μm. However, with exercise, this distance decreases to 20 μm or less.

Electrochemical Gradient

The **electrochemical gradient** (also called the **electrochemical potential difference**) is used to quantitate the driving force acting on a molecule to cause it to move across a membrane. The electrochemical gradient for any molecule ($\Delta\mu_x$) is calculated as follows:

Equation 1.4

$$\Delta\mu_x = RT \ln \frac{[X]_i}{[X]_o} + z_x F V_m,$$

where
R = the gas constant
T = temperature in degrees Kelvin
Ln = natural logarithm
$[X]_i$ = the concentration of X inside the cell
$[X]_o$ = the concentration of X outside the cell
z_x = the valence of charged molecules
F = the Faraday constant
V_m = the membrane potential ($V_m = V_i - V_o$)[b]

The electrochemical gradient is a measure of the free energy available to carry out the useful work of transporting the molecule across the membrane. It has two components: One component represents the energy in the concentration gradient for X across the membrane (**chemical potential**

difference). The second component (**electrical potential difference**) represents the energy associated with moving charged molecules (e.g., ions) across the membrane when a membrane potential exits (i.e., $V_m \neq 0$ mV). Thus for the movement of glucose across a membrane, only the concentrations of glucose inside and outside of the cell need to be considered (Fig. 1.6A). However, the movement of K^+ across the membrane, for example, would be determined both from the K^+ concentrations inside and outside of the cell and from the membrane voltage (see Fig. 1.6B).

Eq. 1.4 can be used to derive the **Nernst equation** for the situation in which a molecule is at equilibrium across the membrane (i.e., $\Delta\mu = 0$):

Equation 1.5a

$$0 = RT \ln \frac{[X]_i}{[X]_o} + z_x F V_m$$

$$-RT \ln \frac{[X]_i}{[X]_o} = z_x F V_m$$

$$V_m = -\frac{RT}{z_x F} \ln \frac{[X]_i}{[X]_o}$$

Alternatively

Equation 1.5b

$$V_m = \frac{RT}{z_x F} \ln \frac{[X]_i}{[X]_o}$$

The value of V_m calculated with the Nernst equation represents the equilibrium condition and is referred to as the **Nernst equilibrium potential** (E_x, the V_m at which there is no net transport of the molecule across the membrane). It should be apparent that the Nernst equilibrium potential quantitates the energy in a concentration gradient and expresses that energy in millivolts. For example, for the cell depicted in Fig. 1.6B, the energy in the K^+ gradient (derived from the Nernst equilibrium potential for K^+ [E_{K^+}]) is proportional to 90.8 mV (causing K^+ to move out of the cell). This is opposite to, and of greater magnitude than, the energy in the membrane voltage ($V_m = -60$ mV), which causes K^+ to enter the cell. As a result, the electrochemical gradient is such that the net movement of K^+ across the membrane will be out of the cell. Another way to state this is that the net driving force for K^+ ($V_m - E_{K^+}$) is 30.8 mV (driving K^+ out of the cell). This is described in more detail in Chapter 2.

The Nernst equation, at 37°C, can be written as follows by replacing the natural logarithm function with the base 10 logarithm function:

Equation 1.6a

$$E_x = -\frac{61.5 \, mV}{z_x} \log \frac{[X]_i}{[X]_o}$$

or

Equation 1.6b

$$E_x = \frac{61.5 \, mV}{z_x} \log \frac{[X]_o}{[X]_i}$$

---

[b]By convention, membrane voltages are determined and reported with regard to the exterior of the cell. In a typical cell, the resting membrane potential (V_m) is negative. Positive V_m values can be observed in some excitable cells at the peak of an action potential.

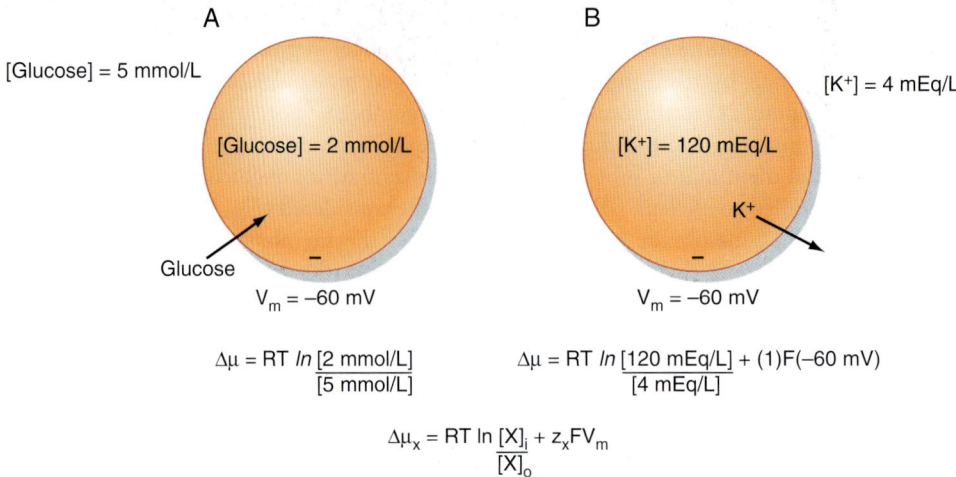

$$\Delta\mu = RT \; ln \frac{[2 \; mmol/L]}{[5 \; mmol/L]} \qquad \Delta\mu = RT \; ln \frac{[120 \; mEq/L]}{[4 \; mEq/L]} + (1)F(-60 \; mV)$$

$$\Delta\mu_x = RT \; ln \frac{[X]_i}{[X]_o} + z_x FV_m$$

• **Fig. 1.6** Electrochemical gradients and cellular transport of molecules. **A,** Because glucose is uncharged, the electrochemical gradient is determined solely by the concentration gradient for glucose across the cell membrane. As shown, the glucose concentration gradient would be expected to drive glucose into the cell. **B,** Because K^+ is charged, the electrochemical gradient is determined by both the concentration gradient and the membrane voltage (V_m). The Nernst equilibrium potential for K^+ (E_{K^+}), calculated with eq. 1.5a, is −90.8 mV ($E_{K^+} = V_m$ at equilibrium). The energy in the concentration gradient, which drives K^+ out of the cell, is thus proportional to +90.8 mV. The membrane voltage of −60 mV drives K^+ into the cell. Thus the electrochemical gradient, or net driving force, is 2.97 kJ/mol (equivalent to 30.8 mV), which drives K^+ out of the cell.

These are the most common forms of the Nernst equation in use. In these equations, it is apparent that for a univalent ion (e.g., Na^+, K^+, Cl^-), a 10-fold concentration gradient across the membrane is equivalent in energy to an electrical potential difference of 61.5 mV (at 37°C), and a 100-fold gradient is equivalent to an electrical potential difference of 123 mV. Similarly, for a divalent ion (e.g., Ca^{++}), a 10-fold concentration gradient is equivalent to a 30.7-mV electrical potential difference, because z in eqs. 1.6a and 1.6b is equal to 2.

Active and Passive Transport

When the net movement of a molecule across a membrane occurs in the direction predicted by the electrochemical gradient, that movement is termed **passive transport.** Thus for the examples given in Fig. 1.6, the movement of glucose into the cell and the movement of K^+ out of the cell would be considered passive transport. Transport that is passive is sometimes referred to as either "downhill transport" or "transport with the electrochemical gradient." In contrast, if the net movement of a molecule across the membrane is opposite to that predicted by the electrochemical gradient, that movement is termed **active transport,** a process that requires the input of energy (e.g., ATP). Active transport is sometimes referred to as either "uphill transport" or "transport against the electrochemical gradient."

In the various classes of plasma membrane transport proteins, the movement of H_2O through water channels is a passive process (see later discussion), as is the movement of ions through ion channels and the transport of molecules via uniporters (e.g., transport of glucose via GLUT-1). The

ATPase-dependent transporters can use the energy in ATP to drive active transport of molecules (e.g., Na^+,K^+-ATPase, H^+-ATPase, or ABC transporters). Because the transport is directly coupled to the hydrolysis of ATP, it is referred to as **primary active transport.** Solute carriers that couple movement of two or more molecules (e.g., $3Na^+$,Ca^{++} antiporter) often transport one or more molecules (one Ca^{++} molecule in this example) against their respective electrochemical gradient through the use of the energy in the electrochemical gradient of the other molecule or molecules (three Na^+ in this example). When this occurs, the molecule or molecules transported against their electrochemical gradient are said to be transported by **secondary active transport** mechanisms (Fig. 1.7).

Osmosis and Osmotic Pressure

The movement of water across cell membranes occurs by the process of **osmosis.** The movement of water is passive, with the driving force for this movement being the osmotic pressure difference across the cell membrane. Fig. 1.8 illustrates the concept of osmosis and the measurement of the osmotic pressure of a solution.

Osmotic pressure is determined by the number of solute molecules dissolved in the solution. It is not dependent on such factors as the size of the molecules, their mass, or their chemical nature (e.g., valence). Osmotic pressure (π), measured in atmospheres (atm), is calculated by **van't Hoff's law** as follows:

Equation 1.7

$$\pi = nCRT,$$

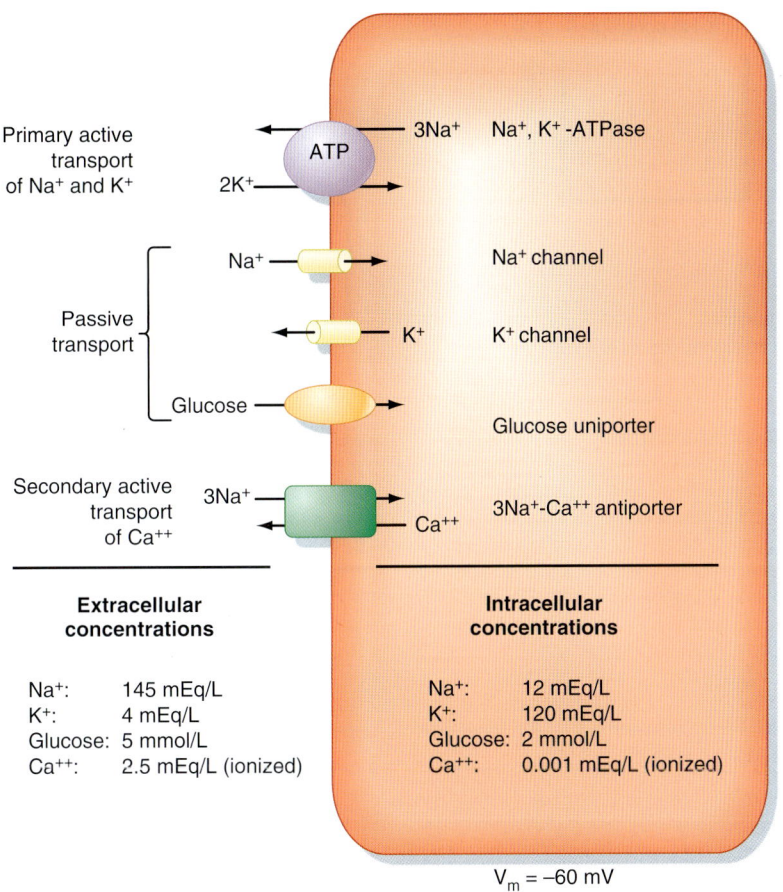

Primary active transport of Na⁺ and K⁺

ATP

3Na⁺   Na⁺, K⁺ -ATPase

2K⁺

Passive transport

Na⁺   Na⁺ channel

K⁺   K⁺ channel

Glucose   Glucose uniporter

Secondary active transport of Ca⁺⁺

3Na⁺   3Na⁺-Ca⁺⁺ antiporter

Ca⁺⁺

Extracellular concentrations

Na⁺:	145 mEq/L
K⁺:	4 mEq/L
Glucose:	5 mmol/L
Ca⁺⁺:	2.5 mEq/L (ionized)

Intracellular concentrations

Na⁺:	12 mEq/L
K⁺:	120 mEq/L
Glucose:	2 mmol/L
Ca⁺⁺:	0.001 mEq/L (ionized)

$V_m = -60$ mV

• **Fig. 1.7** Examples of several membrane transporters, illustrating primary active, passive, and secondary active transport. See text for details. ATP, adenosine triphosphate.

🩺 IN THE CLINIC

Glucose is transported by the epithelial cells that line the gastrointestinal tract (small intestine), and by cells that form the proximal tubules of the kidneys. In the gastrointestinal tract, the glucose is absorbed from ingested food. In the kidney, the proximal tubule reabsorbs the glucose that was filtered across the glomerular capillaries and thereby prevents it from being lost in the urine. The uptake of glucose into the epithelial cell from the lumen of the small intestine and from the lumen of the proximal tubule is a secondary active process involving the sodium-glucose–linked transporters SGLT-1 and SGLT-2. SGLT-2 transports one glucose molecule with one Na⁺ ion, and the energy in the electrochemical gradient for Na⁺ (into the cell) drives the secondary active uptake of glucose. According to the following equation, for calculating the electrochemical gradient, and if the membrane potential (V_m) is –60 mV and there is a 10-fold [Na⁺] gradient across the membrane, an approximate 100-fold glucose gradient could be generated by SGLT-2:

$$\frac{[Glucose]_i}{[Glucose]_o} = \frac{[Na^+]_o}{[Na^+]_i} \times 10^{-V_m/61.5\,mV}$$

Thus, if the intracellular glucose concentration was 2 mmol/L, the cell could lower the extracellular glucose concentration to approximately 0.02 mmol/L. However, by increasing the number of Na⁺ ions transported with glucose from one to two, SGLT-1 can generate a nearly 10,000-fold glucose gradient:

$$\frac{[Glucose]_i}{[Glucose]_o} = \left(\frac{[Na^+]_o}{[Na^+]_i}\right)^2 \times 10^{-2V_m/61.5\,mV}$$

Again, if the intracellular glucose concentration is 2 mmol/L, SGLT-1 could remove virtually all glucose from either the lumen of the small intestine or the lumen of the proximal tubule (i.e., the luminal glucose concentration ≅ 0.0002 mmol/L).

---

where

n = number of dissociable particles per molecule

C = total solute concentration

R = gas constant

T = temperature in degrees Kelvin

For a molecule that does not dissociate in water, such as glucose or urea, a solution containing 1 mmol/L of these molecules at 37°C can exert an osmotic pressure of 2.54 × 10⁻² atm, as calculated with eq. 1.7 and the following values:

n = 1

C = 0.001 mol/L

R = 0.082 atm L/mol K

T = 310 °K

Initial condition   Equilibrium condition

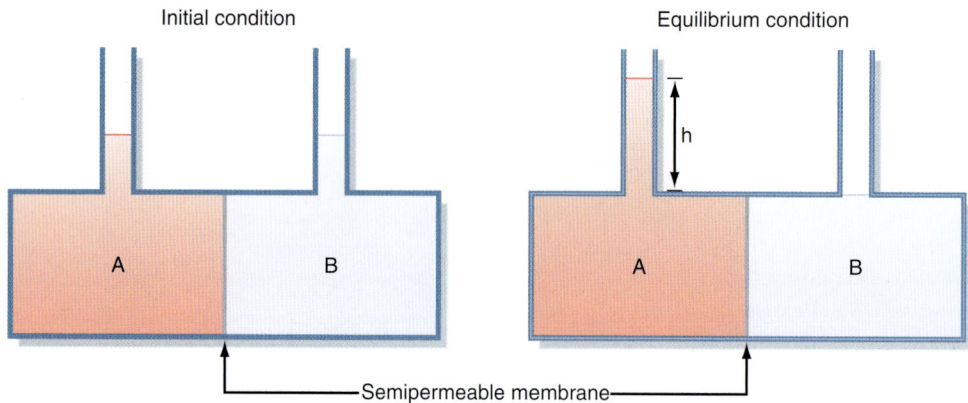

• **Fig. 1.8** Schematic representation of osmotic water movement and the generation of an osmotic pressure. Compartment A and compartment B are separated by a semipermeable membrane (i.e., the membrane is highly permeable by water but impermeable by solute). Compartment A contains a solute, whereas compartment B contains only distilled water. Over time, water moves by osmosis from compartment B to compartment A. (Note: This water movement is driven by the concentration gradient for water. Because of the presence of solute particles in compartment A, the concentration of water in compartment A is less than that in compartment B. Consequently, water moves across the semipermeable membrane from compartment B to compartment A down its concentration gradient.) This causes the level of fluid to be raised in compartment A and lowered in compartment B. At equilibrium, the hydrostatic pressure exerted by the column of water (h) stops the net movement of water from compartment B to A. Thus at equilibrium, the hydrostatic pressure is equal and opposite to the osmotic pressure exerted by the solute particles in compartment A. (Redrawn from Koeppen BM, Stanton BA. *Renal Physiology.* 4th ed. St. Louis: Mosby; 2006.)

Because 1 atm equals 760 mm Hg at sea level, π for this solution can also be expressed as 19.3 mm Hg. Alternatively, osmotic pressure is expressed in terms of osmolarity (see the following section). Regardless of the molecule, a solution containing 1 mmol/L of the molecule therefore exerts an osmotic pressure proportional to 1 mOsm/L.

For molecules that dissociate in a solution, n of eq. 1.7 will have a value other than 1. For example, a 150-mmol/L solution of NaCl has an osmolarity of approximately 300 mOsm/L because each molecule of NaCl dissociates into a Na^+ and a Cl^- ion (i.e., n = 2).[c] If dissociation of a molecule into its component ions is not complete, n will not be an integer. Accordingly, osmolarity for any solution can be calculated as follows:

Equation 1.8

$$Osmolarity = concentration \times number$$
$$of\ dissociable\ particles$$

$$mOsm/L = mmol/L \times number\ of\ particles/mole$$

Osmolarity Versus Osmolality

The terms *osmolarity* and *osmolality* are frequently confused and incorrectly interchanged. *Osmolarity* refers to the osmotic pressure generated by the dissolved solute molecules in 1 L of solvent, whereas osmolality is the number of molecules dissolved in 1 kg of solvent. For a dilute solution, the difference between osmolarity and osmolality is

insignificant. Measurements of osmolarity are temperature dependent because the volume of the solvent varies with temperature (i.e., the volume is larger at higher temperatures). In contrast, osmolality, which is based on the mass of the solvent, is temperature independent. For this reason, *osmolality* is the preferred term for biologic systems and is used throughout this book. Because the solvent in biological solutions and bodily fluids is water, and because of the dilute nature of biological solutions and bodily solutions, osmolalities are expressed as milliosmoles per kilogram of water (mOsm/kg H_2O).

Tonicity

The tonicity of a solution is related to the effect of the solution on the volume of a cell. Solutions that do not change the volume of a cell are said to be **isotonic**. A **hypotonic** solution causes a cell to swell, whereas a **hypertonic** solution causes a cell to shrink. Although related to osmolality, tonicity also accounts for the ability of the molecules in solution to cross the cell membrane.

Consider two solutions: a 300-mmol/L solution of sucrose and a 300-mmol/L solution of urea. Both solutions have an osmolality of 300 mOsm/kg H_2O and therefore are said to be **isosmotic** (i.e., they have the same osmolality). When red blood cells—which for the purpose of this illustration also have an intracellular fluid osmolality of 300 mOsm/kg H_2O—are placed in the two solutions, those in the sucrose solution maintain their normal volume, whereas those placed in urea swell and eventually burst. Thus the sucrose solution is isotonic and the urea solution is hypotonic. The differential effect of these solutions on

---

[c]NaCl does not completely dissociate in water. The value for n is 1.88 rather than 2. However, for simplicity, the value of 2 is most often used.

red blood cell volume is related to the permeability of the red blood cell plasma membrane to sucrose and urea. The red blood cell membrane contains uniporters for urea. Thus urea easily crosses the cell membrane (i.e., the cell is permeable by urea), driven by the concentration gradient (i.e., extracellular urea concentration > intracellular urea concentration). In contrast, the red blood cell membrane does not contain sucrose transporters, and sucrose cannot enter the cell (i.e., the cell is impermeable by sucrose).

To exert an osmotic pressure across a membrane, a molecule must not cross the membrane. Because the red blood cell membrane is impermeable by sucrose, it exerts an osmotic pressure equal and opposite to the osmotic pressure generated by the contents within the red blood cell (in this case, 300 mOsm/kg H_2O). In contrast, urea is readily able to cross the red blood cell membrane, and it cannot exert an osmotic pressure to balance that generated by the intracellular solutes of the red blood cell. Consequently, sucrose is termed an **effective osmole,** whereas urea is an **ineffective osmole.**

To take into account the effect of a molecule's ability to permeate the membrane on osmotic pressure, it is necessary to rewrite eq. 1.7 as follows:

Equation 1.9

$$\Pi_e = \sigma(nCRT),$$

where σ is the **reflection coefficient** (or **osmotic coefficient**) and is a measure of the relative ability of the molecule to cross the cell membrane, and Π_e is the "effective osmotic pressure."

For a molecule that can freely cross the cell membrane, such as urea in the preceding example, $\sigma = 0$, and no effective osmotic pressure is exerted (e.g., urea is an ineffective osmole for red blood cells). In contrast, $\sigma = 1$ for a solute that cannot cross the cell membrane (in the preceding example, sucrose). Such a substance is said to be an effective osmole. Many molecules are neither completely able nor completely unable to cross cell membranes (i.e., $0 < \sigma < 1$) and generate an osmotic pressure that is only a fraction of what is expected from the molecules' concentration in solution.

Oncotic Pressure

Oncotic pressure is the osmotic pressure generated by large molecules (especially proteins) in solution. As illustrated in Fig. 1.9, the magnitude of the osmotic pressure generated by a solution of protein does not conform to van't Hoff's law. The cause of this anomalous relationship between protein concentration and osmotic pressure is not completely understood, but it appears to be related to the size and shape of the protein molecule. For example, the correlation to van't Hoff's law is more precise with small, globular proteins than with larger protein molecules.

The oncotic pressure exerted by proteins in human plasma has a normal value of approximately 26 to 28 mm Hg. Although this pressure appears to be small in relation to osmotic pressure (28 mm Hg ≅ 1.4 mOsm/kg H_2O), it

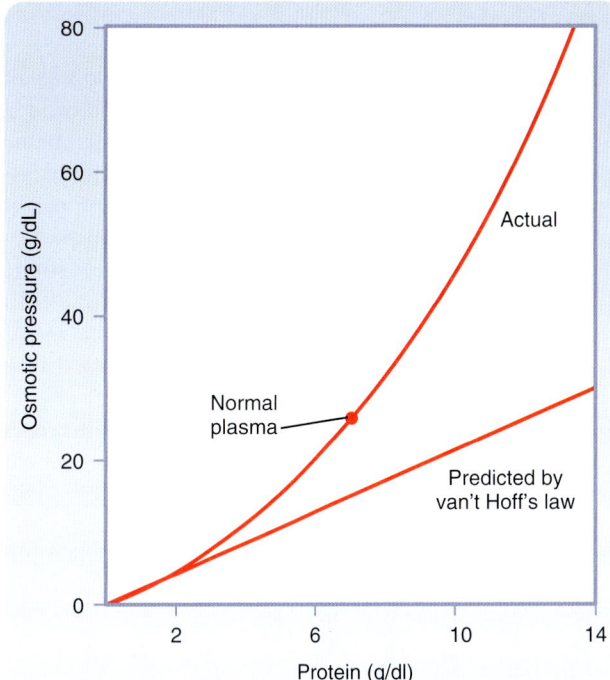

• **Fig. 1.9** Relationship between the concentration of plasma proteins in solution and the osmotic pressure (oncotic pressure) they generate. Protein concentration is expressed in grams per deciliter. Normal plasma protein concentration is indicated. Note how the actual pressure generated exceeds that predicted by van't Hoff's law.

is an important force involved in fluid movement across capillaries (see Chapter 17).

Specific Gravity

The total concentration of all molecules in a solution can also be measured as specific gravity. Specific gravity is defined as the weight of a volume of solution divided by the weight of an equal volume of distilled water. Thus the specific gravity of distilled water is 1. Because biological fluids contain a number of different molecules, their specific gravities are greater than 1. For example, normal human plasma has a specific gravity in the range of 1.008 to 1.010.

 IN THE CLINIC

The specific gravity of urine is sometimes measured in clinical settings and used to assess the urine concentrating ability of the kidneys. The specific gravity of urine varies in proportion to its osmolality. However, because specific gravity depends both on the number of molecules and on their weight, the relationship between specific gravity and osmolality is not always predictable. For example, in patients who have received an injection of radiocontrast dye (molecular weight > 500 g/mole) for x-ray studies, values of urine specific gravity can be high (1.040 to 1.050), even though the urine osmolality is similar to that of plasma (e.g., 300 mOsm/kg H_2O).

Key Points

- The plasma membrane is a lipid bilayer composed of phospholipids and cholesterol, into which are embedded a wide range of proteins. One class of these membrane proteins (membrane transport proteins or transporters) is involved in the selective and regulated transport of molecules into and out of the cell. These transporters include water channels (aquaporins), ion channels, solute carriers, and ATP-dependent transporters.

- The movement of molecules across the plasma membrane through ion channels and solute carriers is driven by chemical concentration gradients and, chemical concentration gradients and electrical potential differences (charged molecules only). The electrochemical gradient is used to quantitate this driving force. ATP-dependent transporters use the energy in ATP to transport molecules across the membrane and often establish the chemical and electrical gradients that then drive the transport of other molecules through channels and by the solute carriers. Water movement through aquaporins is driven by an osmotic pressure difference across the membrane.

- Transport across the membrane is classified as passive or active. Passive transport is the movement of molecules as expected from the electrochemical gradient for that molecule. Active transport represents transport against the electrochemical gradient. Active transport is further divided into primary active and secondary active transport. Primary active transport is directly coupled to the hydrolysis of ATP (e.g., ATP-dependent transporters). Secondary active transport occurs with coupled solute carriers, for which passive movement of one or more molecules drives the active transport of other molecules (e.g., Na^+-glucose symporter, Na^+-H^+ antiporter).

Additional Readings

Alberts B, et al. *Essential Cell Biology*. 4th ed. New York: Garland Science; 2014.

Altenberg GA, Ruess L. Mechanisms of water transport across cell membranes and epithelia. In: Alpern R, Moe O, Kaplan M, eds. *Seldin and Giebisch's The Kidney—Physiology and Pathophysiology*. 5th ed. New York: Academic Press; 2013.

Hediger MA, et al. The ABCs of membrane transporters in health and disease (SLC series): introduction. *Mol Aspects Med*. 2013; 34:95-107.

Pawlina W. *Histology: A Text and Atlas with Correlated Cell and Molecular Biology*. 7th ed. Alphen aan den Rijn, The Netherlands: Wolters Kluwer; 2016.

Rojeck A, et al. A current view of mammalian aquaglyceroporins. *Annu Rev Physiol*. 2008;70:301-327.

Ruess L, Altenberg GA. Mechanisms of ion transport across cell membranes. In: Alpern R, Moe O, Kaplan M, eds. *Seldin and Giebisch's The Kidney—Physiology and Pathophysiology*. 5th ed. New York: Academic Press; 2013.

2

Homeostasis: Volume and Composition of Body Fluid Compartments

LEARNING OBJECTIVES

Upon completion of this chapter, the student should be able to answer the following questions:

1. What is steady-state balance, and, with water balance as an example, what are the elements needed to achieve steady-state balance?
2. What are the volumes of the body fluid compartments, and how do they change under various conditions?
3. How do the body fluid compartments differ with regard to their composition?
4. What determines the resting membrane potential of cells?
5. How do cells regulate their volume in isotonic, hypotonic, and hypertonic solutions?
6. What are the structural features of epithelial cells, how do they carry out vectorial transport, and what are the general mechanisms by which transport is regulated?

Normal cellular function requires that the intracellular composition—with regard to ions, small molecules, water, pH, and a host of other substances—be maintained within a narrow range. This is accomplished by the transport of many substances and water into and out of the cell via membrane transport proteins, as described in Chapter 1. In addition, each day, food and water are ingested, and waste products are excreted from the body. In a healthy individual, these processes occur without significant changes in either the volume of the body fluid compartments or their composition. The maintenance of constant volume and composition of the body fluid compartments (and their temperature in warm-blooded animals and humans) is termed **homeostasis.** The human body has multiple systems designed to achieve homeostasis, the details of which are explained in the various chapters of this book. In this chapter, the basic principles that underlie the maintenance of homeostasis are outlined. In addition, the volume and composition of the various body fluid compartments are defined.

Concept of Steady-State Balance

The human body is an "open system," which means that substances are added to the body each day and, similarly, substances are lost from the body each day. The amounts added to or lost from the body can vary widely, depending on the environment, access to food and water, disease processes, and even cultural norms. In such an open system, homeostasis occurs through the process of **steady-state balance.**

To illustrate the concept of steady-state balance, consider a river on which a dam is built to create a synthetic lake. Each day, water enters the lake from the various streams and rivers that feed it. In addition, water is added by underground springs, rain, and snow. At the same time, water is lost through the spillways of the dam and by the process of evaporation. For the level of the lake to remain constant (i.e., steady-state balance), the rate at which water is added, regardless of source, must be exactly matched by the amount of water lost, again regardless of route. Because the addition of water is not easily controlled and the loss by evaporation cannot be controlled, the only way to maintain a constant level of the lake is to regulate the amount that is lost through the spillways.

To understand steady-state balance as it applies to the human body, the following key concepts are important.

1. There must be a "set point" so that deviations from this baseline can be monitored (e.g., the level of the lake in the preceding example, or setting the temperature in a room by adjusting the thermostat).
2. The sensor or sensors that monitor deviations from the set point must generate "effector signals" that can lead to changes in either input or output, or both, to maintain the desired set point (e.g., electrical signals to adjust the spillway in the dam analogy, or electrical signals sent to either the furnace or air conditioner to maintain the proper room temperature).
3. "Effector organs" must respond in an appropriate way to the effector signals generated by the set point monitor (i.e., the spillway gates must operate, and the furnace or air conditioner must turn on).
4. The sensitivity of the system (i.e., how much of a deviation from the set point is tolerated) depends on several factors, including the nature of the sensor (i.e., how much of a deviation from the set point is needed for the sensor to detect the deviation), the time necessary for generation of the effector signals, and how rapidly the effector organs respond to the effector signals.

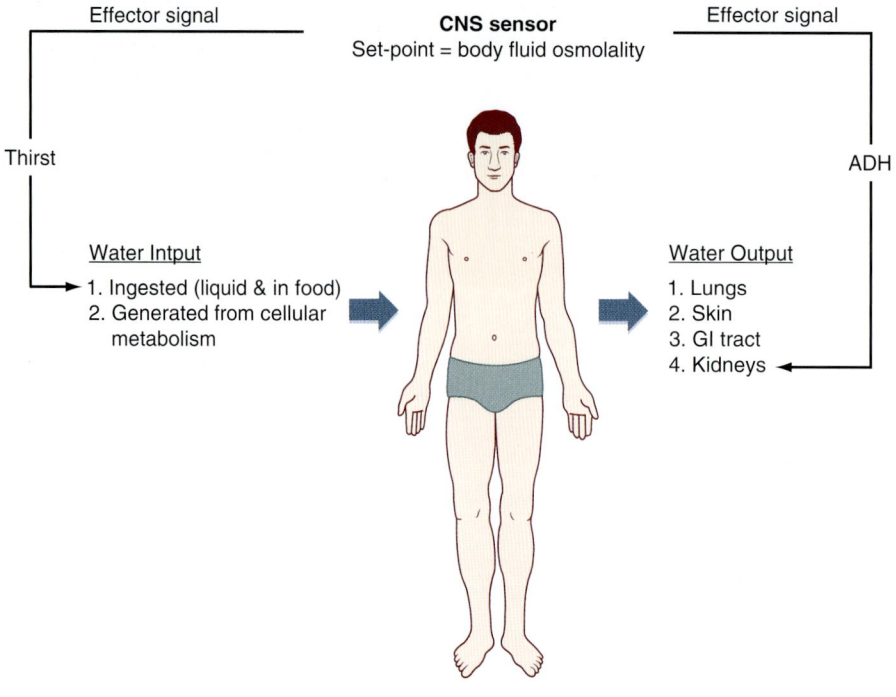

• Fig. 2.1 Whole-Body Steady-State Water Balance. See text for details. ADH, antidiuretic hormone (also called arginine vasopressin); CNS, central nervous system; GI, gastrointestinal.

It is important to recognize that deviations from steady-state balance do occur. When input is greater than output, a state of **positive balance** exists. When input is less than output, a state of **negative balance** exists. Although transient periods of imbalance can be tolerated, prolonged states of positive or negative balance are generally incompatible with life.

Fig. 2.1 illustrates several important concepts for the maintenance of steady-state water balance (details related to the maintenance of steady-state water balance are presented in Chapter 35). As depicted in Fig. 2.1, there are multiple inputs and outputs of water, many of which can vary but nevertheless cannot be regulated. For example, the amount of water lost through the lungs depends on the humidity of the air and the rate of respiration (e.g., low humidity and rapid breathing increase water loss from the lungs). Similarly, the amount of water lost as sweat varies according to ambient temperature and physical activity. Finally, water loss via the gastrointestinal tract can increase from a normal level of 100 to 200 mL/day to many liters with acute diarrhea. Of these inputs and outputs, the only two that can be regulated are increased ingestion of water in response to thirst and alterations in urine output by the kidneys (see Chapter 35).

Water balance determines the osmolality of the body fluids. Cells within the hypothalamus of the brain monitor body fluid osmolality for deviations from the set point (normal range: 280-295 mOsm/kg H_2O). When deviations are sensed, two effector signals are generated. One is neural and relates to the individual's sensation of thirst. The other is hormonal (antidiuretic hormone, also called *arginine*

vasopressin), which regulates the amount of water excreted by the kidneys. With appropriate responses to these two signals, water input, water output, or both are adjusted to maintain balance and thereby keep body fluid osmolality at the set point.

Volumes and Composition of Body Fluid Compartments

Unicellular organisms maintain their volume and composition through exchanges with the environment they inhabit (e.g., sea water). The billions of cells that constitute the human body must maintain their volume and composition as well, but their task is much more difficult. This challenge, as well as its solution, was first articulated by the French physiologist Claude Bernard (1813-1878). He recognized that although cells within the body cannot maintain their volume and composition through exchanges with the environment, they can do so through exchanges with the fluid environment that surrounds them (i.e., the extracellular fluid). Bernard referred to the extracellular fluid as the *milieu intérieur* ("the environment within"). He also recognized that the organ systems of the body are designed and function to maintain a constant milieu interieur or a "constant internal environment." This in turn allows all cells to maintain their volume and composition through exchanges with the extracellular fluid as a result of membrane transport (see Chapter 1).

Transport by the epithelial cells of the gastrointestinal tract, kidneys, and lungs are the body's interface with the external environment and control both the intake and excretion of numerous substances, as well as water. The cardiovascular system delivers nutrients to and removes waste products from the cells and tissues and keeps the extracellular fluid well mixed. Finally, the nervous and endocrine systems provide regulation and integration of these important functions.

To provide background for the study of all organ systems, this chapter presents an overview of the normal volume and composition of the body fluid compartments and describes how cells maintain their intracellular composition and volume. Included is a presentation on how cells generate and maintain a membrane potential, which is fundamental for understanding the function of excitable cells (e.g., neurons and muscle cells). Finally, because epithelial cells are so central to the process of regulating the volume and composition of the body fluids, the principles of solute and water transport by epithelial cells are also reviewed.

Definition and Volumes of Body Fluid Compartments

Water makes up approximately 60% of the body's weight; variability among individuals is a function of the amount of adipose tissue. Because the water content of adipose tissue is lower than that of other tissue, increased amounts of adipose tissue reduce the fraction of water in the total body weight. The percentage of body weight attributed to water also varies with age. In newborns, it is approximately 75%. This decreases to the adult value of 60% by the age of 1 year.

As illustrated in Fig. 2.2, **total body water** is distributed between two major compartments, which are divided by the cell membrane.[a] The **intracellular fluid (ICF)** compartment is the larger compartment, and contains approximately two thirds of the total body water. The remaining third is contained in the **extracellular fluid (ECF)** compartment. Expressed as percentages of body weight, the volumes of total body water, ICF, and ECF are as follows:

$$\text{Total body water} = 0.6 \times (\text{body weight})$$
$$\text{ICF} = 0.4 \times (\text{body weight})$$
$$\text{ECF} = 0.2 \times (\text{body weight})$$

The ECF compartment is further subdivided into interstitial fluid and plasma. The ECF also includes fluid contained within bone and dense connective tissue, as well as the cerebrospinal fluid. The interstitial fluid surrounds the cells in the various tissues of the body and makes up three fourths of the ECF volume. Plasma is contained within the vascular compartment and represents the remaining fourth of the ECF. In some pathological conditions, additional fluid may accumulate in what is referred to as a *third space*. Third-space collections of fluid are part of the ECF; an

---

[a]In these and all subsequent calculations, it is assumed that 1 L of fluid (e.g., ICF and ECF) has a mass of 1 kg. Although, 1 L of the ICF and ECF has a mass of slightly more than 1 kg, this simplification allows conversion from measurements of body weight to volume of body fluids.

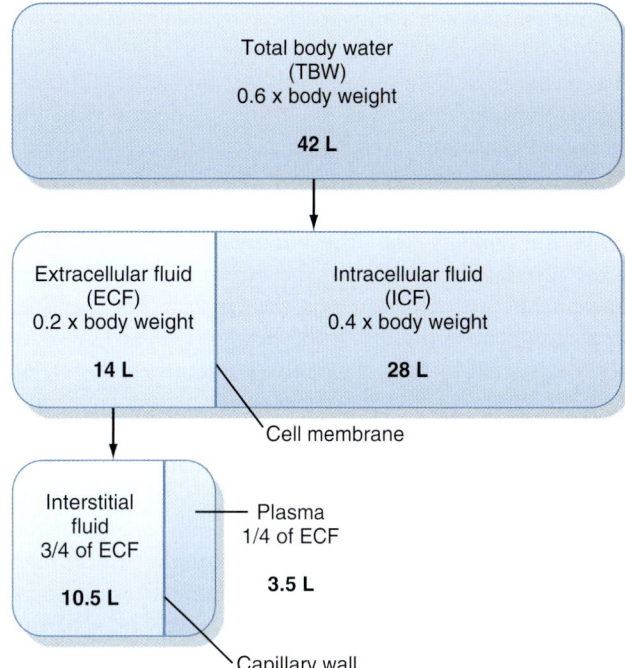

• **Fig. 2.2** Relationship Between the Volumes of the Various Body Fluid Compartments. The actual values shown are for an individual weighing 70 kg. (Modified from Levy MN, Koeppen BM, Stanton BA. *Berne & Levy's Principles of Physiology.* 4th ed. St. Louis: Mosby; 2006.)

example is the accumulation of fluid in the peritoneal cavity (**ascites**) of individuals with liver disease.

Movement of Water Between Body Fluid Compartments

As depicted in Fig. 2.2, water moves between the ICF and ECF compartments across the plasma membranes of cells, and it moves between the vascular (plasma) and interstitial compartments across capillary walls. The pathways and driving forces for this water movement are different across cell membranes, in comparison to the capillary walls.

Movement of water between the ICF and ECF compartments, across cell membranes, occurs through aquaporins expressed in the plasma membrane (see Chapter 1). The driving force for this water movement is an osmotic pressure difference. The osmotic pressure of both the ICF and ECF is determined by the molecules/ions present in these fluids. For simplicity, these can be divided into (1) molecules of low molecular weight (e.g., glucose) and ions (e.g., Na^+) and (2) macromolecules (e.g., proteins). The osmotic pressures of both the ICF and ECF are in the range of 280 to 295 mOsm/kg H_2O. For the ECF, the low-molecular-weight molecules and ions account for nearly all of this pressure because the osmotic pressure contributed by proteins is only 1 to 2 mOsm/kg H_2O. The molecules/ions contributing to the osmotic pressure within the cell are less well understood, but they also include low-molecular-weight molecules (e.g., glucose), ions (e.g., Na^+), and macromolecules (e.g., proteins). The fact that cell volume remains constant when ECF osmolality is constant means that the osmotic pressure inside the cells is equal to that of the ECF. If an osmotic pressure difference did exist, the cells would either swell or shrink, as described in the section "Nonisotonic Cell Volume Regulation."

Movement of water between the vascular (plasma) compartment and the interstitial fluid compartment occurs across the capillary wall. The amount of water that moves across the capillary wall, and the mechanism of the water movement varies depending on the capillary. For example, in the capillary sinusoids of the liver, endothelial cells are often separated by large gaps (discontinuous capillary). As a result, water and all components of the plasma (and some cellular elements) can pass easily across the wall. Other capillaries are lined by endothelial cells that contain fenestrations that are up to 80 to 100 nm in diameter (e.g., in the kidneys). These fenestrations allow all components of the plasma (only cellular elements of blood cannot pass through the fenestrations) to move across the capillary wall. Some capillaries (e.g., in the brain) form a relatively tight barrier to water and small molecules and ions, and water movement occurs through small pores on the endothelial cell surface or through clefts between adjacent endothelial cells. These pores and clefts allow water and molecules smaller than 4 nm to pass. In addition, a small amount of water traverses the capillary wall via pinocytosis by endothelial cells.

The driving forces for fluid (water) movement across the capillary wall are hydrostatic pressure and oncotic pressure (i.e., osmotic pressure generated by proteins). Collectively, these are called the *Starling forces*. Capillary fluid movement is discussed in detail in Chapter 17; in brief, hydrostatic pressure within the capillary (as a result of the pumping of the heart and the effect of gravity on the column of blood in the vessels feeding a capillary) is a force that causes fluid to move out of the capillary. Hydrostatic pressure in the surrounding interstitial tissue opposes the effect of the capillary hydrostatic pressure. The oncotic pressure of the plasma in the capillary tends to draw fluid from the interstitium into the capillary. The oncotic pressure of the interstitial fluid opposes this. Thus the amount of fluid moving across the wall of the capillary is determined as follows:

Equation 2.1

$$\text{Fluid flow}\,(Q_f) = K_f\left[(P_c - P_i) - (\pi_c - \pi_i)\right]$$

or

$$\text{Fluid flow}\,(Q_f) = K_f\left[(P_c + \pi_i) - (P_i + \pi_c)\right]$$

where

Q_f = fluid movement

K_f = filtration constant (measure of surface area + intrinsic permeability)

P_c = capillary hydrostatic pressure

P_i = interstitial fluid hydrostatic pressure

π_c = plasma oncotic pressure

π_i = interstitial fluid oncotic pressure.

Depending on the magnitude of these forces, fluid may move out of the capillary or into the capillary.

The compositions of the various body fluid compartments differ; however, as described later, the osmolalities of the fluid within these compartments are essentially identical.[b] Thus the compartments are in "osmotic equilibrium." In addition, any change in the osmolality of one compartment quickly causes water to redistribute across all compartments, which brings them back into osmotic equilibrium. Because of this rapid redistribution of water, measuring the osmolality of plasma or serum,[c] which is easy to do, reveals the osmolality of the other body fluid compartments (i.e., interstitial fluid and intracellular fluid).

As described later, Na^+ is a major constituent of the ECF. Because of its high concentration in comparison with other molecules and ions, Na^+ (and its attendant anions, primarily Cl^- and HCO_3^-) is the major determinant of the osmolality of this compartment. Accordingly, it is possible to obtain an approximate estimate of the ECF osmolality by simply doubling the sodium concentration $[Na^+]$. For example, if a blood sample is obtained from an individual, and the $[Na^+]$ of the serum is 145 mEq/L, its osmolality can be estimated as follows:

Equation 2.2

$$\text{Plasma Osmolality} = 2(\text{serum}\,[Na^+]) = 290\,\text{mOsm/kg}\,H_2O$$

In contrast to water, the movement of ions across cell membranes is more variable from cell to cell and depends on the presence of specific membrane transport proteins (see the section "Composition of Body Fluid Compartments"). Consequently, in trying to understand the physiology of fluid shifts between body fluid compartments, it can be assumed that while water moves freely between the compartments, there is little net movement of solutes. For most situations, this is a reasonable assumption.

To illustrate the physiologic characteristics of fluid shifts, consider what happens when solutions containing various amounts of NaCl are added to the ECF.[d]

Example 1: Addition of Isotonic Sodium Chloride to the Extracellular Fluid

Addition of an isotonic NaCl solution (e.g., intravenous infusion of 0.9% NaCl: osmolality ≈ 290 mOsm/kg H_2O)[e]

[b]Some exceptions do exist. The cerebrospinal fluid is part of the ECF, but its osmolality is slightly higher than that of the ECF elsewhere in the body. Also, regions within the kidney can have osmolalities that are either less than or greater than that of the ECF. However, these volumes are small (≈150 mL) in comparison with the total volume of the ECF (≥12 L).

[c]Serum is derived from clotted blood. Thus serum differs from plasma by the absence of clotting factors. With regard to osmolality and the concentrations of other molecules and ions, the osmolality and concentrations in plasma and serum are virtually identical.

[d]Fluids are usually administered intravenously. When electrolyte solutions are infused by this route, equilibration between plasma and interstitial fluid is rapid (i.e., minutes) because of the high permeability of many capillary walls for water and electrolytes. Thus these fluids are essentially added to the entire ECF.

[e]A 0.9% NaCl solution (0.9 g NaCl/100 mL) contains 154 mmol/L of NaCl. Because NaCl does not dissociate completely in solution (i.e., 1.88 Osm/mol), the osmolality of this solution is 290 mOsm/kg H_2O, which is very similar to that of normal ECF.

IN THE CLINIC

In some clinical situations, it is possible to obtain a more accurate estimate of the serum osmolality, and thus the osmolalities of the ECF and ICF, by also considering the osmoles contributed by glucose and urea, as these are the next most abundant solutes in the ECF (the other components of the ECF contribute only a few additional milliosmoles). Accordingly, serum osmolality can be estimated as follows:

$$\text{Serum osmolality} = 2(\text{serum [Na}^+]) + \frac{[\text{glucose}]}{18} + \frac{[\text{urea}]}{2.8}$$

The glucose and urea concentrations are expressed in units of milligrams per deciliter (dividing by 18 for glucose and 2.8 for urea* allows conversion from the units of milligrams per deciliter to millimoles per liter and thus to milliosmoles per kilogram of H_2O). This estimation of serum osmolality is especially useful in treating patients who have an elevated serum glucose concentration secondary to diabetes mellitus, and in patients with chronic renal failure, whose serum urea concentration is elevated because of reduced renal excretion.

As discussed in Chapter 1, the ability of a substance to cause water to move across the plasma membrane of a cell depends on whether the substance itself crosses the membrane. Recall Eq. 1.9:

$$\Pi_e = \sigma(nCRT)$$

where Π_e = the effective osmotic pressure and σ = the reflection coefficient for the substance. For many cells, glucose and urea cross the cell membrane. Although they contribute to serum osmolality, as measured by a laboratory osmometer where all molecules are "effective osmoles," they are ineffective osmoles for water movement across many, but not all, cell membranes. In contrast, Na^+ is an "effective osmole" for water movement across the plasma membrane of virtually all cells. Eq. 2.2 gives the best estimate of the effective osmolality of the serum.

*The urea concentration in plasma is measured as the nitrogen in the urea molecule, or blood urea nitrogen (BUN).

IN THE CLINIC

Neurosurgical procedures and cerebrovascular accidents (strokes) often result in the accumulation of interstitial fluid in the brain (i.e., edema) and swelling of the neurons. Because the brain is enclosed within the skull, edema can raise intracranial pressure and thereby disrupt neuronal function, which leads to coma and death. The blood-brain barrier, which separates the cerebrospinal fluid and brain interstitial fluid from blood, can be permeated freely by water but not by most other substances. As a result, excess fluid in brain tissue can be removed by imposing an osmotic gradient across the blood-brain barrier. Mannitol can be used for this purpose. Mannitol is a sugar (molecular weight, 182 g/mol) that does not readily cross the blood-brain barrier and membranes of cells (neurons and other cells in the body). Therefore, mannitol is an effective osmole, and intravenous infusion results in the movement of interstitial fluid out of the brain by osmosis.

the ICF. After osmotic equilibration, the osmolalities of the ICF and ECF are again equal but lower than before the infusion, and the volume of each compartment is increased. The increase in ECF volume is greater than the increase in ICF volume.

Example 3: Addition of Hypertonic Sodium Chloride to the Extracellular Fluid

Addition of a hypertonic NaCl solution to the ECF (e.g., intravenous infusion of 3% NaCl: osmolality ≅ 1000 mOsm/kg H_2O) increases the osmolality of this compartment, which results in the movement of water out of cells. After osmotic equilibration, the osmolalities of the ECF and ICF are again equal but higher than before the infusion. The volume of the ECF is increased, whereas that of the ICF is decreased.

Composition of Body Fluid Compartments

The compositions of the ECF and ICF differ considerably. The ICF has significantly more proteins and macromolecules than the ECF. There are also differences in the concentrations of many ions. The composition of the ICF is maintained by the action of a number of specific cell membrane transport proteins. Principal among these transporters is the Na^+,K^+-adenosine triphosphatase (Na^+,K^+-ATPase), which converts the energy in ATP into ion and electrical gradients, which can in turn be used to drive the transport of other ions and molecules by means of ion channels and solute carriers (e.g., symporters and antiporters).

The compositions of the plasma and interstitial fluid compartments of the ECF are similar because those compartments are separated only by the capillary endothelium, a barrier that ions and small molecules can permeate. The major difference between the interstitial fluid and plasma is that the latter contains significantly more protein. Although this differential concentration of protein can affect the

to the ECF increases the volume of this compartment by the volume of fluid administered. Because this fluid has the same osmolality as does the ECF, and therefore the ICF, there is no driving force for fluid movement between these compartments, and the volume of the ICF remains unchanged. Although Na^+ can cross cell membranes, it is effectively restricted to the ECF by the activity of the Na^+,K^+-ATPase, which is present in the plasma membrane of all cells (see the section "Ionic Composition of Cells"). Therefore, there is no net movement of the infused isotonic NaCl solution into cells.

Example 2: Addition of Hypotonic Sodium Chloride to the Extracellular Fluid

Addition of a hypotonic NaCl solution to the ECF (e.g., intravenous infusion of 0.45% NaCl; osmolality ≅ 145 mOsm/kg H_2O) decreases the osmolality of this fluid compartment, which results in the movement of water into

🩺 IN THE CLINIC

Fluid and electrolyte disorders are observed commonly in clinical practice (e.g., in patients with vomiting or diarrhea, or both). In most instances, these disorders are self-limited, and correction of the disorder occurs without need for intervention. However, more severe or prolonged disorders may necessitate fluid replacement therapy. Such therapy may be administered orally, with special electrolyte solutions, or intravenously, with fluid.

Intravenous solutions are available in many formulations. The type of fluid administered to a particular patient is dictated by the patient's need. For example, if an increase in the patient's vascular volume is necessary, a solution containing substances that do not readily cross the capillary wall is infused (e.g., 5% protein or dextran solutions). The oncotic pressure generated by the albumin molecules causes fluid to be retained in the vascular compartment, which expands its volume. Expansion of the ECF is accomplished most often with isotonic saline solutions (e.g., 0.9% NaCl or lactated Ringer solution). As already noted, administration of an isotonic NaCl solution does not result in the development of an osmotic pressure gradient across the plasma membrane of cells. Therefore, the entire volume of the infused solution remains in the ECF.

Patients whose body fluids are hyperosmotic need hypotonic solutions. These solutions may be hypotonic NaCl (e.g., 0.45% NaCl) or 5% dextrose in water (D_5W). Administration of the D_5W solution is equivalent to the infusion of distilled water because the dextrose is metabolized to CO_2 and water. Administration of these fluids increases the volumes of both the ICF and ECF. In addition, patients whose body fluids are hypotonic need hypertonic solutions. These are typically NaCl-containing solutions (e.g., 3% or 5% NaCl). These solutions expand the volume of the ECF but decrease the volume of the ICF. Other constituents, such as electrolytes (e.g., K^+) or drugs, can be added to intravenous solutions to tailor the therapy to the patient's fluid, electrolyte, and metabolic needs.

TABLE 2.1	Ionic Composition of a Typical Cell	
Ion	**Extracellular Fluid**	**Intracellular Fluid**
Na^+	135-147 mEq/L	10-15 mEq/L
K^+	3.5-5.0 mEq/L	120-150 mEq/L
Cl^-	95-105 mEq/L	20-30 mEq/L
HCO_3^-	22-28 mEq/L	12-16 mEq/L
*Ca^{++}	2.1-2.8 (total) mmol/L	
	1.1-1.4 (ionized) mmol/L	$\approx 10^{-7}$ M (ionized) mmol/L
*Pi	1.0-1.4 (total) mmol/L	
	0.5-0.7 (ionized) mmol/L	0.5-0.7 (ionized) mmol/L

*Ca^{++} and Pi ($H_2PO_4^-$/HPO_4^{-2}) are bound to proteins and other organic molecules. In addition, large amounts of Ca^{++} can be sequestered within cells. Large amounts of Pi are present in cells as part of organic molecules, such as adenosine triphosphate (ATP).

distribution of cations and anions between these two compartments by the Gibbs-Donnan effect (see the section "Isotonic Cell Volume Regulation" for details), this effect is small, and the ionic compositions of the interstitial fluid and plasma can be considered to be identical.

Maintenance of Cellular Homeostasis

Normal cellular function requires that the ionic composition of the ICF be tightly controlled. For example, the activity of some enzymes is pH dependent; therefore, intracellular pH must be regulated. In addition, the intracellular composition of other electrolytes is similarly held within a narrow range. This is necessary for the establishment of the membrane potential, a cell property especially important for the normal function of excitable cells (e.g., neurons and muscle cells) and for intracellular signaling (e.g., intracellular [Ca^{++}]; see Chapter 3 for details). Finally, the volume of cells must be maintained because shrinking or swelling of cells can lead to cell damage or death. The regulation of intracellular

composition and cell volume is accomplished through the activity of specific transporters in the plasma membrane of the cells. This section is a review of the mechanisms by which cells maintain their intracellular ionic environment and their membrane potential and by which they control their volume.

Ionic Composition of Cells

The intracellular ionic composition of cells varies from tissue to tissue. For example, the intracellular composition of neurons is different from that of muscle cells, both of which differ from that of blood cells. Nevertheless, there are similar patterns, and these are presented in Table 2-1. In comparison with the ECF, the ICF is characterized by a low [Na^+] and a high [K^+]. This is the result of the activity of the Na^+,K^+-ATPase, which transports 3 Na^+ ions out of the cell and 2 K^+ ions into the cell for each ATP molecule hydrolyzed. As discussed later in this chapter, the activity of the Na^+,K^+-ATPase not only is important for establishing the cellular Na^+ and K^+ gradients but also is involved in determining, indirectly, the cellular gradients for many other ions and molecules. Of importance is that the cellular K^+ gradient generated by the activity of the Na^+,K^+-ATPase is a major determinant of the membrane voltage because of the leak of K^+ out of the cell through K^+-selective channels (see the section "Membrane Potential"). Thus the Na^+,K^+-ATPase converts the energy in ATP into ion gradients (i.e., Na^+ and K^+), and a voltage gradient (i.e., membrane voltage).

The Na^+,K^+-ATPase–generated ion and electrical gradients are used to drive the transport of other ions and molecules into or out of the cell (Fig. 2.3). For example, as described in Chapter 1, a number of solute carriers couple the transport of Na^+ to that of other ions or molecules. The Na^+-glucose and Na^+–amino acid symporters use the energy in the Na^+ electrochemical gradient, directed to bring Na^+

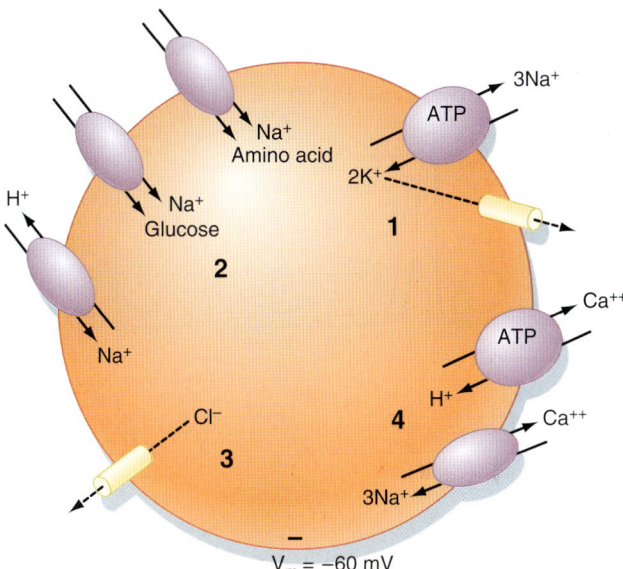

• Fig. 2.3 Cell Model Depicting How Cellular Gradients and the Membrane Potential (V_m) Are Established. (1) The Na$^+$,K$^+$-ATPase decreases the intracellular [Na$^+$] and increases the intracellular [K$^+$]. Some K$^+$ exits the cell via K$^+$-selective channels and generates the V_m (cell's interior is electrically negative). (2) The energy in the Na$^+$ electrochemical gradient drives the transport of other ions and molecules through the use of various solute carriers. (3) The V_m drives Cl$^-$ out of the cell via Cl$^-$-selective channels. (4) The Ca^{++}-ATPase and the 3Na$^+$-Ca^{++} antiporters maintain the low intracellular [Ca^{++}].

into the cell, to drive the secondary active cellular uptake of glucose and amino acids. Similarly, the inwardly directed Na$^+$ gradient drives the secondary active extrusion of H$^+$ from the cell and thus contributes to the maintenance of intracellular pH. The 3Na$^+$-Ca^{++} antiporter, along with the plasma membrane Ca^{++}-ATPase, extrudes Ca^{++} from the cell and thus contributes to the maintenance of a low intracellular [Ca^{++}].[f] In addition, the membrane voltage drives Cl$^-$ out of the cell through Cl$^-$-selective channels, thus lowering the intracellular concentration below that of the ECF.

Membrane Potential

As described previously, the Na$^+$,K$^+$-ATPase and K$^+$-selective channels in the plasma membrane are important determinants of the membrane potential (V_m) of the cell. For all cells within the body, the resting V_m is oriented with the interior of the cell electrically negative in relation to the ECF. However, the magnitude of the V_m can vary widely.

To understand what determines the magnitude of the V_m, it is important to recognize that any transporter that transfers charge across the membrane has the potential to influence the V_m. Such transporters are said to be

electrogenic. As might be expected, the contribution of various electrogenic transporters to the V_m is highly variable from cell to cell. For example, the Na$^+$,K$^+$-ATPase channel transports three Na$^+$ and two K$^+$ ions and thus transfers one net positive charge across the membrane. However, the direct contribution of the Na$^+$,K$^+$-ATPase to the V_m of most cells is only a few millivolts at the most. Similarly, the contribution of other electrogenic transporters, such as the 3Na$^+$-Ca^{++} antiporter and the Na$^+$-glucose symporter is minimal. The major determinants of the V_m are ion channels. The type (e.g., selectivity), number, and activity (e.g., gating) of these channels determine the magnitude of the V_m. As described in Chapter 5, rapid changes in ion channel activity underlies the action potential in neurons and other excitable cells, such as those of skeletal and cardiac muscle (see Chapters 12 and 13).

As ions move across the membrane through a channel, they generate a current. As described in Chapter 1, this current can be measured, even at the level of a single channel. By convention, the current generated by the movement of cations into the cell, or the movement of anions out of the cell, is defined as negative current. Conversely, the movement of cations out of the cell, or the movement of anions into the cell, is defined as positive current. Also by convention, the magnitude of the V_m is expressed in relation to the outside of the cell; thus for a cell with a V_m of −80 mV, the interior of the cell is electrically negative in relation to the outside of the cell.

The current carried by ions moving through a channel depends on the driving force for that ion and on the conductance of the channel. As described in Chapter 1, the driving force is determined by the energy in the concentration gradient for the ion across the membrane (E_i), as calculated by the Nernst equation (Eq. 1.5a) and the V_m:

Equation 2.3
$$\text{Driving force} = V_m - E_i.$$

Thus as defined by **Ohm's law,** the ion current through the channel (I_i) is determined as follows:

Equation 2.4
$$I_i = (V_m - E_i) \times g_i$$

where g_i is the conductance of the channel. For a cell, the conductance of the membrane to a particular ion (G_i) is determined by the number of ion channels in the membrane and by the amount of time each channel is in the open state.

As illustrated in Fig. 2.4, the V_m is the voltage at which there is no net ion flow into or out of the cell. Thus for a cell that has ion channels selective for Na$^+$, K$^+$, and Cl$^-$,

Equation 2.5
$$I_{Na^+} + I_{K^+} + I_{Cl^-} = 0$$

or

Equation 2.6
$$[(V_m - E_{Na^+}) \times G_{Na^+}] + [(V_m - E_{K^+}) \times G_{K^+}] + [(V_m - E_{Cl^-}) \times G_{Cl^-}] = 0.$$

---

[f]In muscle cells, in which contraction is regulated by the intracellular [Ca^{++}], the maintenance of a low intracellular [Ca^{++}] during the relaxed state involves not only the activity of the plasma membrane 3Na$^+$-Ca^{++} antiporter and the Ca^{++}-ATPase but also a Ca^{++}-ATPase molecule located in the smooth endoplasmic reticulum (see Chapters 12 to 14).

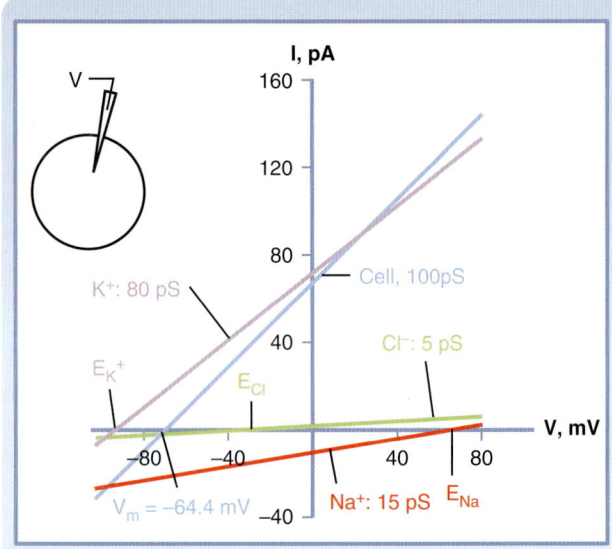

	Intracellular [ ] mEq/L	Extracellular [ ] mEq/L	Nernst potential E_i, mV
Na⁺	12	145	66.6
K⁺	120	4	−90.8
Cl⁻	30	105	−33.5

• **Fig. 2.4** Current-Voltage Relationship of a Hypothetical Cell Containing Na⁺-, K⁺-, and Cl⁻-Selective Channels. Membrane currents are plotted over a range of membrane voltages (i.e., current-voltage relationships). Each ion current is calculated with the use of Ohm's law, the Nernst equilibrium potential for the ion (E_{Cl}, E_K, and E_{Na}), and the membrane conductance for the ion. The current-voltage relationship for the whole cell is also shown. Total cell current (I_{cell}) was calculated with the chord conductance equation (see Eq. 2.7). Because 80% of cell conductance is due to K⁺, the resting membrane voltage (V_m) of −64.4 mV is near to that of the Nernst equilibrium potential for K⁺.

Solving for V_m yields

Equation 2.7

$$V_m = E_{Na^+}\frac{G_{Na^+}}{\Sigma G} + E_{K^+}\frac{G_{K^+}}{\Sigma G} + E_{Cl^-}\frac{G_{Cl^-}}{\Sigma G}$$

where $\Sigma G = G_{Na^+} + G_{K^+} + G_{Cl^-}$.

   Inspection of Eq. 2.7, which is often called the **chord conductance equation,** reveals that the V_m will be near to the Nernst equilibrium potential of the ion to which the membrane has the highest conductance. In Fig. 2.4, 80% of the membrane conductance is attributable to K⁺; as a result, V_m is near to the Nernst equilibrium potential for K⁺ (E_{K^+}). For most cells at rest, the membrane has a high conductance to K⁺, and thus the V_m approximates E_{K^+}. Moreover, the V_m is greatly influenced by the magnitude of E_{K^+}, which in turn is greatly influenced by changes in the [K⁺] of the ECF. For example, if the intracellular [K⁺] is 120 mEq/L and the extracellular [K⁺] is 4 mEq/L, E_{K^+} has a value of −90.8 mV. If the extracellular [K⁺] is increased to 7 mEq/L, E_{K^+} would be −79.9 mV. This change in E_{K^+} **depolarizes** the V_m (i.e., V_m is less negative). Conversely, if the extracellular [K⁺] is decreased to 2 mEq/L, E_{K^+} becomes −109.4 mV, and the V_m **hyperpolarizes** (i.e., V_m is more negative).

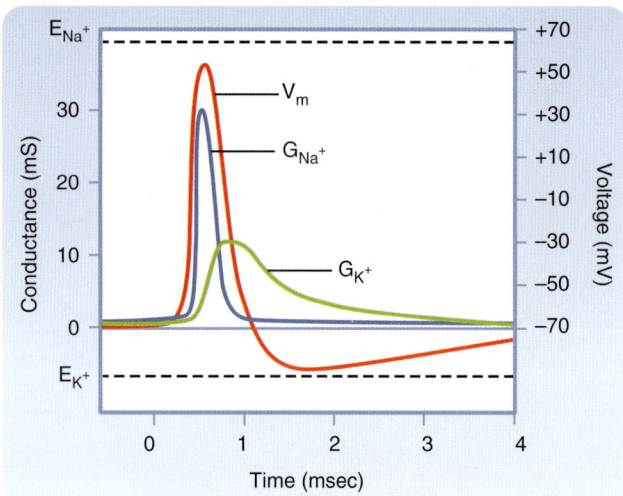

• **Fig. 2.5** Nerve Action Potential Showing the Changes in Na⁺ and K⁺ Conductances (G_{Na^+} and G_{K^+}, Respectively) and the Membrane Potential (V_m). At rest, the membrane has a high K⁺ conductance, and V_m is near the Nernst equilibrium potential for K⁺ (E_{K^+}). With the initiation of the action potential, there is a large increase in the Na⁺ conductance of the membrane, and the V_m approaches the Nernst equilibrium potential for Na⁺ (E_{Na^+}). The increase in Na⁺ conductance is transient, and the K⁺ conductance then increases above its value before the action potential. This hyperpolarizes the cell as V_m approaches E_{K^+}. As the K⁺ conductance returns to its baseline value, V_m returns to its resting value of −70 mV. (Modified from Levy MN, Koeppen BM, Stanton BA. *Berne & Levy's Principles of Physiology.* 4th ed. St. Louis: Mosby; 2006.)

🩺 IN THE CLINIC

Changes in the extracellular [K⁺] can have important effects on excitable cells, especially those of the heart. A decrease in extracellular [K⁺] **(hypokalemia)** hyperpolarizes the V_m of cardiac myocytes and, in so doing, makes initiating an action potential more difficult, because a larger depolarizing current is needed to reach threshold (see Chapter 16). If severe, hypokalemia can lead to cardiac arrhythmias, and eventually the heart can stop contracting **(asystole).** An increase in the extracellular [K⁺] **(hyperkalemia)** can be equally deleterious to cardiac function. With hyperkalemia, the V_m is depolarized, and it is easier to initiate an action potential. However, once the action potential fires the channels become inactivated, and are unable to initiate another action potential, until they are reactivated by normal repolarization of the V_m. Because the V_m is depolarized in hyperkalemia, the channels stay in an inactivated state. Thus depolarization of the V_m with hyperkalemia can lead to cardiac arrhythmias and loss of cardiac muscle contraction.

   Eq. 2.7 also defines the limits for the membrane potential. In the example depicted in Fig. 2.4, it is apparent that the V_m cannot be more negative than E_{K^+} (−90.8 mV), as would be the case if the membrane were only conductive to K⁺. Conversely, the V_m could not be more positive than E_{Na^+} (66.6 mV); such a condition would be met if the membrane were conductive only to Na⁺. The dependence of the V_m on the conductance of the membrane to specific ions is the basis by which action potentials in excitable cells are generated (Fig. 2.5). As noted previously, in all excitable cells, the membrane at rest is conductive predominantly to K^+, and thus V_m is near E_{K^+}. When an action potential is initiated, Na^+-channels open and the membrane is now conductive predominantly to Na^+. As a result, V_m now approaches E_{Na^+}. The generation of action potentials is discussed in more detail in Chapter 5.

✇ AT THE CELLULAR LEVEL

The establishment of the V_m requires the separation of charge across the plasma membrane. However, the number of ions that must move across the membrane is a tiny fraction of the total number of ions in the cell. For example, consider a spherical cell with a diameter of 20 μm and a V_m of −80 mV. Furthermore, assume that this V_m of −80 mV is the result of the diffusion of K^+ out of the cell and that the intracellular $[K^+]$ is 120 mmol/L. The amount of K^+ that would have to diffuse out of the cell to establish the V_m of −80 mV is then calculated as follows:

First the charge separation across the membrane needs to be calculated. This is done with the knowledge that the plasma membrane behaves electrically like a capacitor, the capacitance (C) of which is approximately 1 μF/cm², and

$$C = \frac{Q}{V}$$

where Q = charge and is expressed in units of coulombs. If the surface area of the cell is $4\pi r^2$ or 1.26×10^{-5} cm², the capacitance of the cell is calculated as follows:

$$1 \times 10^{-6} \text{ F/cm}^2 \times 1.26 \times 10^{-5} \text{ cm}^2 = 1.26 \times 10^{-11} \text{ F.}$$

Thus the charge separation across the membrane is calculated as follows:

$$Q = C \times V_m = 1.26 \times 10^{-11} \text{ F} \times 0.08 \text{ volts}$$
$$= 1.01 \times 10^{-12} \text{ coulombs.}$$

Because 1 mole of K^+ contains 96,480 coulombs, the amount of K^+ that had to diffuse across the membrane to establish the V_m of −80 mV is calculated as follows:

$$\frac{1.01 \times 10^{-12} \text{ coulombs}}{96,480 \text{ coulombs/mole}} = 1.05 \times 10^{-17} \text{ mole of } K^+$$

With a cell volume of 4.19×10^{-12} L (volume = $4\pi r^3/3$) and an intracellular $[K^+]$ of 120 mmol/L, the total intracellular K^+ content is

$$4.19 \times 10^{-12} \times 0.12 \text{ mol/L} = 5.03 \times 10^{-13} \text{ moles}$$

Therefore, the diffusion of 1.05×10^{-17} moles of K^+ out of the cell represents only a 0.002% change in the intracellular K^+ content:

$$\frac{1.05 \times 10^{-13} \text{ moles}}{5.03 \times 10^{-13} \text{ moles}} \approx 0.002\%$$

Thus the intracellular $[K^+]$ of the cell is not appreciably altered by the diffusion of K^+ out of the cell.

Regulation of Cell Volume

As already noted, changes in cell volume can lead to cell damage and death. Cells have developed mechanisms to regulate their volume. Most cells are highly permeable by water because of the presence of aquaporins in their plasma membranes. As discussed in Chapter 1, osmotic pressure gradients across the cell membrane that are generated by effective osmoles cause water to move either into or out of the cell, which result in changes in cell volume. Thus cells swell when placed in hypotonic solutions and shrink when placed in hypertonic solutions (see the section "Nonisotonic Cell Volume Regulation"). However, even when a cell is placed in an isotonic solution, the maintenance of cell volume is an active process requiring the expenditure of ATP and specifically the activity of the Na^+,K^+-ATPase.

Isotonic Cell Volume Regulation

The importance of the Na^+,K^+-ATPase in isotonic cell volume regulation can be appreciated by the observation that red blood cells swell when chilled (i.e., reduced ATP synthesis) or when the Na^+,K^+-ATPase is inhibited by cardiac glycosides (e.g., ouabain, digoxin [Lanoxin]). The necessity for energy expenditure to maintain cell volume in an isotonic solution is the result of the effect of intracellular proteins on the distribution of ions across the plasma membrane: the so-called **Gibbs-Donnan effect** (Fig. 2.6).

The Gibbs-Donnan effect occurs when a membrane separating two solutions can be permeated by some but not all of the molecules in solution. As noted previously, this effect accounts for the small differences in the ionic compositions of the plasma and the interstitial fluid. In this case, the capillary endothelium represents the membrane, and the plasma proteins are the molecules whose ability to permeate across the capillary is restricted. For cells, the membrane is the plasma membrane, and the impermeant molecules are the intracellular proteins and organic molecules.

As depicted in Fig. 2.6, the presence of impermeant molecules (e.g., protein) in one compartment results over time in the accumulation of permeant molecules/ions in the same compartment. This increases the number of osmotically active particles in the compartment containing the impermeant anions, which in turn increases the osmotic pressure, and water thereby enters that compartment. For cells, the Gibbs-Donnan effect would increase the number of osmotically active particles in the cell, and result in cell swelling. However, the activity of the Na^+,K^+-ATPase counteracts the Gibbs-Donnan effect by actively extruding cations (three Na^+ ions are extruded, whereas two K^+ ions are brought into the cell). In addition, the K^+ gradient established by the Na^+,K^+-ATPase allows for the development of the V_m (in which the cell's interior is electrically negative), that in turn drives Cl^- and other anions out of the cell. Thus through the activity of the Na^+,K^+-ATPase, the number of intracellular osmotically active particles is

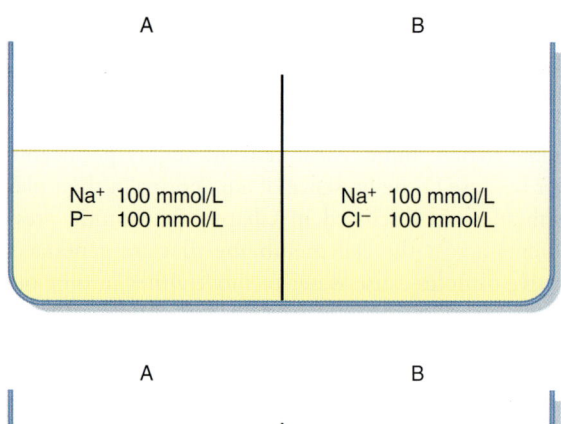

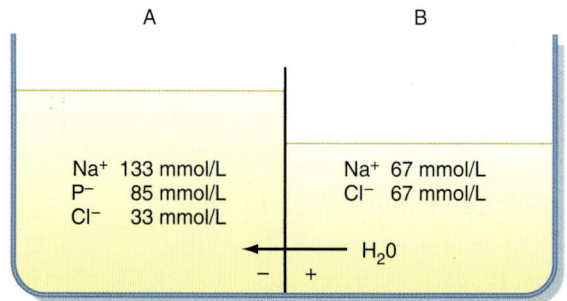

• **Fig. 2.6** The Gibbs-Donnan Effect. **Top,** Two solutions are separated by a membrane that is permeable by Na^+, Cl^-, and H_2O but not permeable by protein (P^-). The osmolality of solution A is identical to that of solution B. **Bottom,** Cl^- diffuses from compartment B to compartment A down its concentration gradient. This causes compartment A to become electrically negative with regard to compartment B. The membrane voltage then drives the diffusion of Na^+ from compartment B to compartment A. The accumulation of additional Na^+ and Cl^- in compartment A increases its osmolality and causes water to flow from compartment B to compartment A (Note: the increase volume of compartment A results in a lower [P^-]). If the container containing the two solutions were sealed at the top so that water could not move from compartment B to compartment A, the pressure in compartment A would increase as the number of osmotically active particles increases in that compartment.

reduced from what would be caused by the Gibbs-Donnan effect, and cell volume is maintained in isotonic solutions.

Nonisotonic Cell Volume Regulation

Most cells throughout the body are bathed with isotonic ECF, the composition of which is tightly regulated (see Chapter 35). However, certain regions within the body are not isotonic (e.g., the medulla of the kidney), and with disorders of water balance, the ECF can become either hypotonic or hypertonic. When this occurs, cells either swell or shrink. Cell swelling or shrinkage can result in cell damage or death, but many cells have mechanisms that limit the degree to which the cell volume changes. These mechanisms are particularly important for neurons, in which swelling within the confined space of the skull can lead to serious neurological damage.

In general, when a cell is exposed to nonisotonic ECF, volume-regulatory responses are activated within seconds to minutes to restore cell volume (Fig. 2.7). With cell swelling, a regulatory volume decrease response transports osmotically active particles (osmolytes) out of the cell, reducing

the intracellular osmotic pressure and thereby restoring cell volume to normal. Conversely with cell shrinking a regulatory volume increase response transports osmolytes into the cell, raising the intracellular osmotic pressure and thereby restoring cell volume to normal. These osmolytes include ions and organic molecules such as polyols (sorbitol and myo-inositol), methylamines (glycerophosphorylcholine and betaine), and some amino acids (taurine, glutamate, and β-alanine). If the cell is exposed to the nonisotonic ECF for an extended period of time, the cell alters the intracellular levels of the organic osmolytes through metabolic processes.

The regulatory volume increase response results in the rapid uptake of NaCl and a number of organic osmolytes. To increase cell volume there is an activation of the Na^+-H^+ antiporter (NHE-1), the $1Na^+,1K^+,2Cl^-$ symporter (NKCC-1), and a number of cation-selective channels, which together bring NaCl into the cell. The Na^+,K^+-ATPase then extrudes the Na^+ in exchange for K^+, so that ultimately the KCl content of the cell is increased. Several organic osmolyte transporters are also activated to increase cell volume. These include a $3Na^+,1Cl^-$-taurine symporter, a $3Na^+,2Cl^-$-betaine symporter, a $2Na^+$–myo-inositol symporter, and a Na^+–amino acid symporter. These transporters use the energy in the Na^+ and Cl^- gradients to drive the secondary active uptake of these organic osmolytes into cells.

The regulatory volume decrease response results in the loss of KCl and organic osmolytes from the cell. The loss of KCl occurs through the activation of a wide range of K^+-selective, Cl^--selective, and anion-selective channels (the specific channels involved vary depending on the cell), as well as through activation of K^+-Cl^- symporters. Some of the organic osmolytes appear to leave the cell via anion channels (e.g., volume-sensitive organic osmolyte-anion channels).

Several mechanisms are involved in activation of these various transporters during the volume regulatory

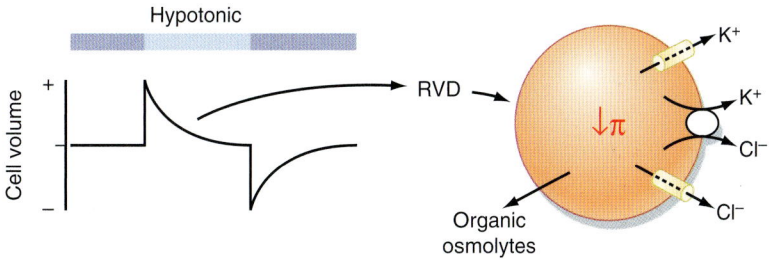

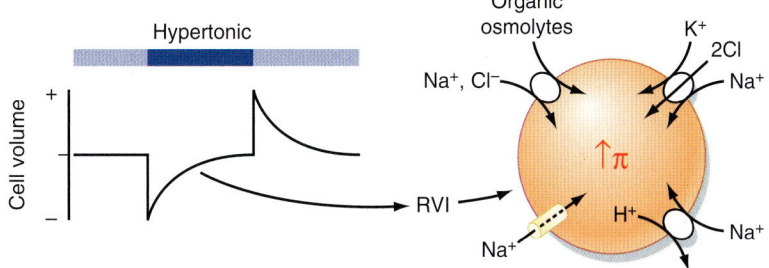

• **Fig. 2.7** Volume Regulation of Cells in Hypotonic and Hypertonic Media. **Top,** When cells are exposed to a hypotonic medium, they swell and then undergo a volume-regulatory decrease (RVD). The RVD involves loss of KCl and organic osmolytes from the cell. The decrease in cellular KCl and organic osmolytes causes intracellular osmotic pressure to decrease, water leaves the cell, and the cell returns to nearly its original volume. **Bottom,** When cells are exposed to a hypertonic medium, they shrink and then undergo a volume-regulatory increase (RVI). During the RVI, NaCl and organic osmolytes enter the cell. The increase in the activity of Na^+,K^+-ATPase (not depicted) enhances the exchange Na^+ for K^+ so that the K^+ (and Cl^-) content of the cell is increased. The increase in cellular KCl, along with a rise in intracellular organic osmolytes, increases intracellular osmotic pressure, which brings water back into the cell, and the cell volume returns to nearly its original volume. π, the oncotic pressure inside the cell.

responses. Changes in cell volume appear to monitored by the cytoskeleton, by changes in macromolecular crowding and ionic strength of the cytoplasm, and by channels whose gating is influenced, either directly or indirectly, by stretch of the plasma membrane (e.g., stretch-activated cation channels). A number of second messenger systems may also be involved in these responses (e.g., intracellular [Ca^{++}], calmodulin, protein kinase A, and protein kinase C), but the precise mechanisms have not been defined completely.

Principles of Epithelial Transport

Epithelial cells are arranged in sheets and provide the interface between the external world and the internal environment (i.e., ECF) of the body. Depending on their location, epithelial cells serve many important functions, such as establishing a barrier to microorganisms (lungs, gastrointestinal tract, and skin), prevention of the loss of water from the body (skin), and maintenance of a constant internal environment (lungs, gastrointestinal tract, and kidneys). This latter function is a result of the ability of epithelial cells to carry out regulated vectorial transport (i.e., transport from one side of the epithelial cell sheet to the opposite side). In this section, the principles of epithelial transport are reviewed. The transport functions of specific epithelial cells are discussed in the appropriate chapters throughout this book.

Epithelial Structure

Fig. 2.8 shows a schematic representation of an epithelial cell. The free surface of the epithelial layer is referred to as the *apical membrane.* It is in contact with the external environment (e.g., air within the alveoli and larger airways of the lungs and the contents of the gastrointestinal tract) or with extracellular fluids (e.g., glomerular filtrate in the nephrons of the kidneys and the secretions of the ducts of the pancreas or sweat glands). The basal side of the epithelium rests on a basal lamina, which is secreted by the epithelial cells, and this in turn is attached to the underlying connective tissue.

Epithelial cells are connected to one another and to the underlying connective tissue by a number of specialized junctions (see Fig. 2.8). The **adhering junction, desmosomes,** and **hemidesmosomes** provide mechanical adhesion by linking together the cytoskeleton of adjacent cells (adhering junction and desmosome) or to the underlying connective tissue (hemidesmosome). The **gap junction** and **tight junction** play important physiological roles.

Gap junctions provide low-resistance connections between cells.[g] The functional unit of the gap junction is the **connexon.** The connexon is composed of six integral

---

[g]Gap junctions are not limited to epithelial cells. A number of other cells also have gap junctions (e.g., cardiac myocytes and smooth muscle cells).

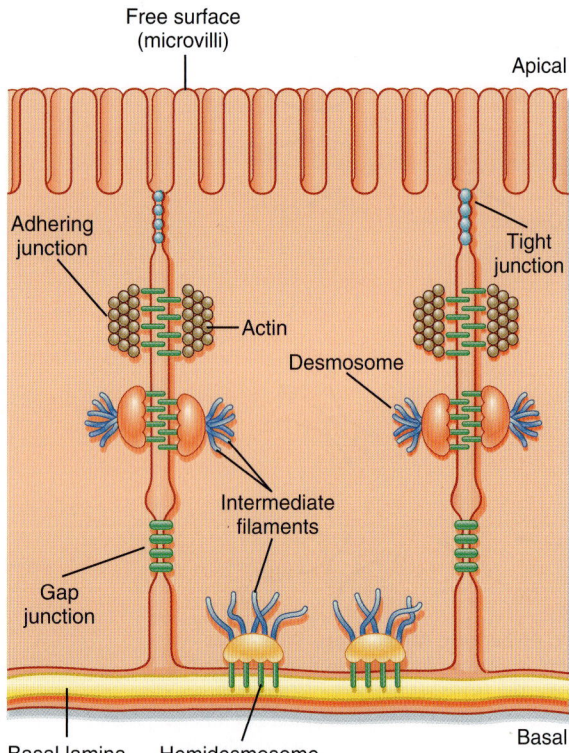

• **Fig. 2.8** Schematic of an Epithelial Cell, Illustrating the Various Adhering Junctions. The tight junction separates the apical membrane from basolateral membrane (see text for details).

membrane protein subunits called **connexins.** A connexon in one cell is aligned with the connexon in the adjacent cell, forming a channel. The channel may be gated, and when it is open, it allows the movement of ions and small molecules between cells. Because of their low electrical resistance, they effectively couple electrically one cell to the adjacent cell.

The tight junction serves two main functions. It divides the cell into two membrane domains (apical and basolateral) and, in so doing, restricts the movement of membrane lipids and proteins between these two domains. This so-called fence function allows epithelial cells to carry out vectorial transport from one surface of the cell to the opposite surface by segregating membrane transporters to one or other of the membrane domains. They also serve as a pathway for the movement of water, ions, and small molecules across the epithelium. This pathway between the cells is referred to as the **paracellular pathway,** as opposed to the **transcellular pathway** through the cells.

The apical surface of epithelial cells may have specific structural features. One such feature is **microvilli** (Fig. 2.9A). Microvilli are small (typically 1 to 3 μm in length), nonmotile projections of the apical plasma membrane that serve to increase surface area. They are commonly located on cells that must transport large quantities of ions, water, and molecules (e.g., epithelial cells lining the small intestine and cells of the renal proximal tubule). The core of the microvilli is composed of actin filaments and a number of accessory proteins. This actin core is connected to the cytoskeleton of the cell via the terminal web (a network of actin fibers at

AT THE CELLULAR LEVEL

Epithelial cell tight junctions (also called **zonula occludens**) are composed of several integral membrane proteins, including **occludins, claudins,** and several members of the immunoglobulin superfamily (e.g., the **junctional adhesion molecule [JAM]**). Occludins and claudins are transmembrane proteins that span the membrane of one cell and link to the extracellular portion of the same molecule in the adjacent cell. Cytoplasmic linker proteins (e.g., tight junction protein [ZO-1, ZO-2, and ZO-3]) then link the membrane spanning proteins to the cytoskeleton of the cell.

Of these junctional proteins, claudins appear to be important in determining the permeability characteristics of the tight junction, especially with regard to cations and anions. To date, 27 mammalian claudin genes have been identified, and 26 are found in the human genome (the gene for claudin 13 is not found in humans). Certain claudins serve as barrier proteins that restrict the movement of ions through the tight junction, whereas others form a "pore" that facilitates the movement of ions through the junction. Thus the permeability characteristics of the tight junction of an epithelium are determined by the complement of claudins expressed by the cell. For example, the proximal tubule of the kidney is termed a "leaky" epithelium, in which water and solutes (e.g., Na^+) move through the junction. Claudin 4 and claudin 10 are expressed in the tight junction of proximal tubule cells. In contrast, the collecting duct of the kidney is considered a "tight" epithelium, with restricted movement of ions through the tight junction. Collecting duct cells express claudins 3, 4, 7, 8, 10, and 18.

The function of claudins can be regulated at several levels, including gene expression, posttranslational modification, interactions with cytoplasmic scaffolding proteins, and interactions with other claudins in the same membrane (*cis*-interaction), as well as with claudins of adjacent cells (*trans*-interaction). The mineralocorticoid hormone aldosterone stimulates Na^+ reabsorption by distal segments of the renal nephron (see Chapters 34 and 35). In addition to the hormone's effect on Na^+ transporters in the cell, aldosterone also upregulates expression of claudin 8 in the tight junction. The increased expression of claudin 8 reduces the ability of Na^+ to permeate the tight junction, which then reduces the backwards leak of Na^+ from the interstitium into the tubule lumen, thereby allowing more efficient Na^+ reabsorption by the epithelium.

IN THE CLINIC

Mutations in the gene that codes for claudin 16 result in the autosomal recessive condition know as *familial hypomagnesemia, hypercalcuria, and nephrocalcinosis* (FHHNC). Claudin 16 is found in the tight junction of the thick ascending portion of Henle's loop in the kidneys and serves as a route for the paracellular reabsorption of Ca^{++} and Mg^{++} from the tubular fluid. Individuals with FHHNC lack functional copies of claudin 16, and reabsorption of these divalent ions is thus reduced, which leads to hypomagnesemia, hypercalcuria, and nephrocalcinosis.

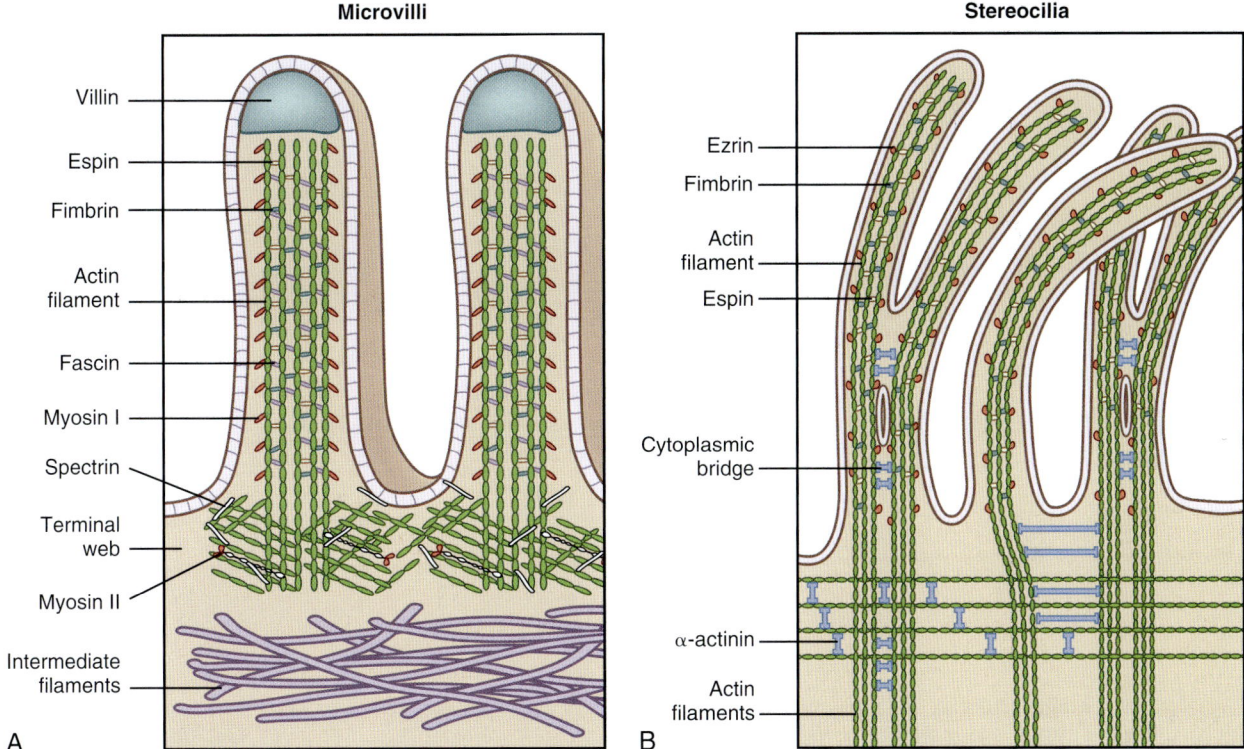

• Fig. 2.9 Illustration of Apical Membrane Specializations of Epithelial Cells (Not Drawn to Scale). **A,** Microvilli 1 to 3 μm in length serve to increase the surface area of the apical membrane (e.g., those of the epithelial cells of the small intestine). **B,** Stereocilia can be up to 120 μm in length (e.g., those of the epididymis of the male reproductive tract). Both microvilli and stereocilia have a core structure composed primarily of actin, with a number of associated proteins. Both are nonmotile. (Redrawn from Pawlina, W. *Histology: A Text and Atlas, with Correlated Cell and Molecular Biology.* 7th ed. Philadelphia: Wolters Kluwer Health, 2016.)

the base of the microvilli) and provides structural support for the microvilli. Another surface feature is stereocilia (see Fig. 2.9B). Stereocilia are long (up to 120 μm), nonmotile membrane projections that, like microvilli, increase the surface area of the apical membrane. They are found in the epididymis of the testis and in the "hair cells" of the inner ear. Their core also contains actin filaments and accessory proteins.

A third apical membrane feature is **cilia** (Fig. 2.10). Cilia may be either motile (called *secondary cilia*) or nonmotile (called *primary cilia*). The motile cilia contain a microtubule core arranged in a characteristic "9+2" pattern (nine pairs of microtubules around the circumference of the cilium, and one pair of microtubules in the center). Dynein is the molecular motor that drives the movement of the cilium. Motile cilia are characteristic features of the epithelial cells that line the respiratory tract. They pulsate in a synchronized manner and serve to transport mucus and inhaled particulates out of the lung, a process termed **mucociliary transport** (see Chapter 26). Nonmotile cilia serve as mechanoreceptors and are involved in determining left-right asymmetry of organs during embryological development, as well as sensing the flow rate of fluid in the nephron of the kidneys (see Chapter 33). Only a single

nonmotile cilium is found in the apical membrane of cells. Nonmotile cilia have a microtubule core ("9+0" arrangement) and lack a motor protein.

As noted previously, the tight junction effectively divides the plasma membrane of an epithelial cell into two domains: an apical surface and a basolateral surface. The basolateral membrane of many epithelial cells is folded or invaginated. This is especially so for epithelial cells that have high transport rates. These invaginations serve to increase the membrane surface area to accommodate the large number of membrane transporters (e.g., Na^+,K^+-ATPase) needed in the membrane.

Vectorial Transport

Because the tight junction divides the plasma membrane into two domains (i.e., apical and basolateral), epithelial cells are capable of vectorial transport, whereby an ion or molecule can be transported from one side of the epithelial sheet to the opposite side (Fig. 2.11). The accomplishment of vectorial transport requires that specific membrane transport proteins be targeted to and remain in one or the other of the membrane domains. In the example shown in Fig. 2.11, the Na^+ channel is present only in the apical

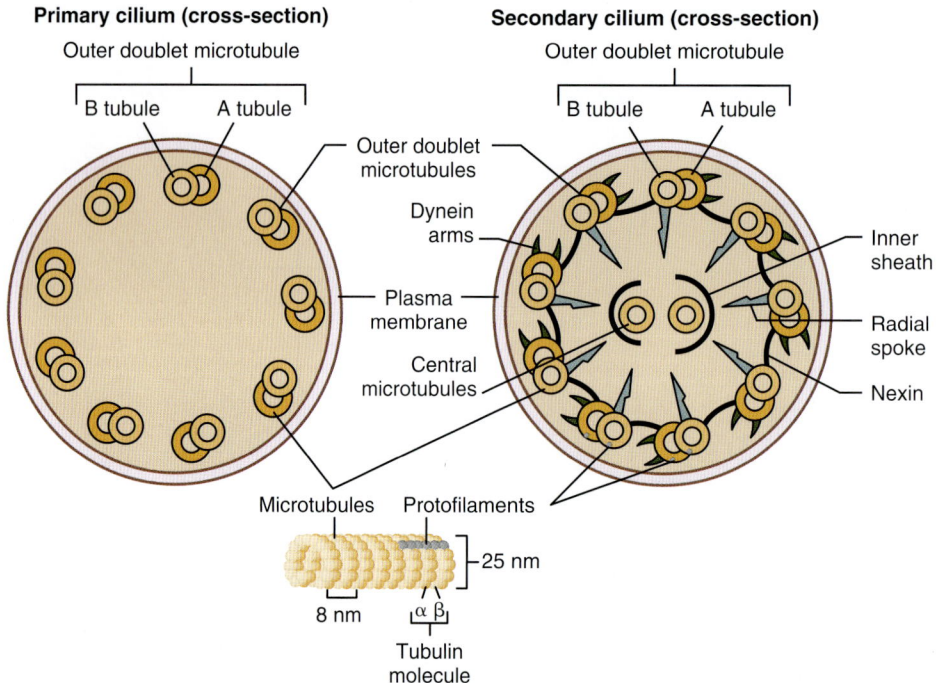

Primary cilium (cross-section)
Outer doublet microtubule
B tubule · A tubule

Secondary cilium (cross-section)
Outer doublet microtubule
B tubule · A tubule

Outer doublet microtubules
Dynein arms
Plasma membrane
Central microtubules

Inner sheath
Radial spoke
Nexin

Microtubules · Protofilaments
25 nm
8 nm
α β
Tubulin molecule

• **Fig. 2.10** Cilia are apical membrane specializations of some epithelial cells. Cilia are 5 to 10 μm in length and contain arrays of microtubules, as depicted in these cross-section diagrams. **Left,** The primary cilium has nine peripheral microtubule arrays. It is nonmotile and serves as a mechanoreceptor (e.g., cells of the renal collecting duct). Cells that have a primary cilium have only a single cilium. **Right,** The secondary cilium has a central pair of microtubules in addition to the nine peripheral microtubule arrays. Also in the secondary cilium, the motor protein dynein is associated with the microtubule arrays and therefore is motile. A single cell can have thousands of secondary cilia on its apical surface (e.g., epithelial cells of the respiratory tract). (Redrawn from Rodat-Despoix L, Delmas P. Ciliary functions in the nephron. *Pflugers Archiv.* 2009;458:179.)

membrane, whereas the Na^+,K^+-ATPase and the K^+ channels are confined to the basolateral membrane. The operation of the Na^+,K^+-ATPase channel and the leakage of K^+ out of the cell across the basolateral membrane sets up a large electrochemical gradient for Na^+ to enter the cell across the apical membrane through the Na^+ channel (intracellular $[Na^+]$ < extracellular $[Na^+]$, and V_m which is oriented with the cell's interior electrically negative with respect to the cell's exterior). The Na^+ is then pumped out of the cell by the Na^+,K^+-ATPase, and vectorial transport from the apical side of the epithelium to the basolateral side of the epithelium occurs. Transport from the apical side to the basolateral side of an epithelium is termed either **absorption** or **reabsorption:** For example, the uptake of nutrients from the lumen of the gastrointestinal tract is termed *absorption,* whereas the transport of NaCl and water from the lumen of the renal nephrons is termed *reabsorption.* Transport from the basolateral side of the epithelium to the apical side is termed **secretion.**

As noted previously, the Na^+,K^+-ATPase and K^+-selective channels play an important role in establishing cellular ion gradients for Na^+ and K^+ and in generating the V_m. In all epithelial cells except the choroid plexus and retinal pigment epithelium,[h] the Na^+,K^+-ATPase channel is located in the basolateral membrane of the cell. Numerous K^+-selective channels are in epithelial cells and may be located in either membrane domain. Through the establishment of these chemical and voltage gradients, the transport of other ions and solutes can be driven (e.g., Na^+-glucose symporter, Na^+-H^+ antiporter, $1Na^+,1K^+,2Cl^-$ symporter, $1Na^+$-$3HCO_3^-$ symporter). The direction of transepithelial transport (reabsorption or secretion) depends simply on which membrane domain the transporters are located. Because of the dependence on the Na^+,K^+-ATPase, epithelial transport requires the expenditure of energy. Other ATP-dependent transporters, such as the H^+-ATPase, H^+,K^+-ATPase, and a host of ABC transporters—such as P-glycoprotein (PGP) and multidrug resistance-associated protein 2 (MRP2), which transport xenobiotics (drugs), and cystic fibrosis transmembrane conductance regulator (CFTR), which transports Cl^-—are involved in epithelial transport.

[h]The choroid plexus is located in the ventricles of the brain and secretes the cerebrospinal fluid. The Na^+,K^+-ATPase channel is located in the apical membrane of these cells.

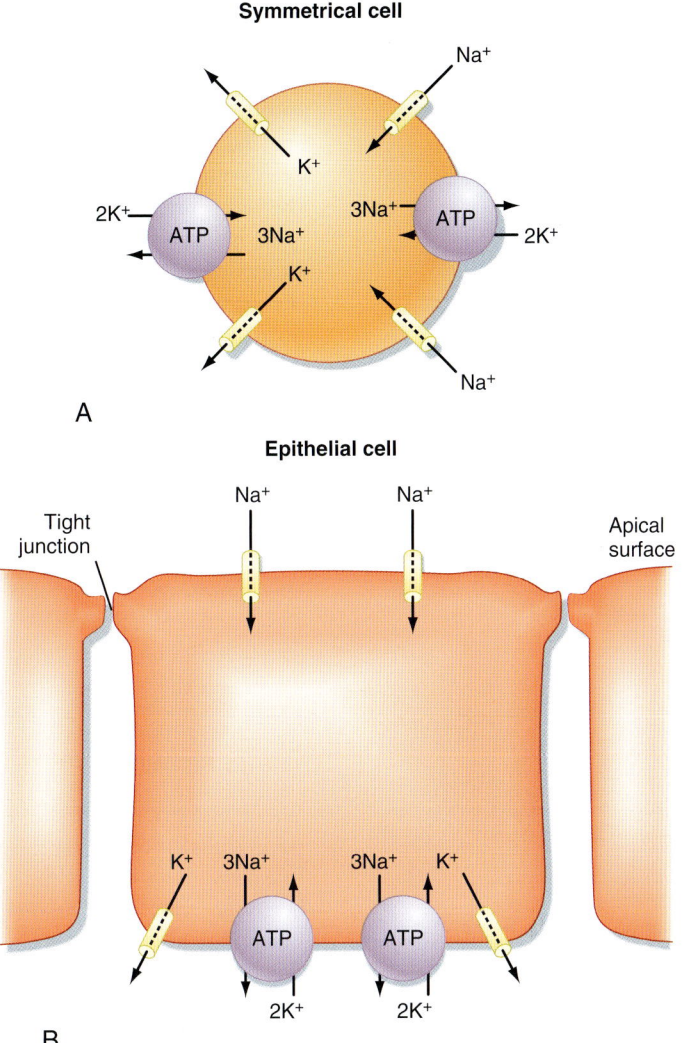

Symmetrical cell

Epithelial cell

• **Fig. 2.11** In symmetrical cells (**A;** e.g., red blood cells), membrane transport proteins are distributed over the entire surface of the cell. Epithelial cells **(B),** in contrast, are asymmetrical and target various membrane transport proteins to either the apical or the basolateral membrane. When the transporters are confined to a membrane domain, vectorial transport can occur. In the cell depicted, Na⁺ is transported from the apical surface to the basolateral surface. ATP, adenosine triphosphate.

Solutes and water can be transported across an epithelium by traversing both the apical and basolateral membranes **(transcellular transport)** or by moving between the cells across the tight junction **(paracellular transport).** Solute transport via the transcellular route is a two-step process, in which the solute molecule is transported across both the apical and basolateral membrane. Uptake into the cell, or transport out of the cell, may be either a passive or an active process. Typically, one of the steps is passive, and the other is active. For the example shown in Fig. 2.11*B,* the uptake of Na^+ into the cell across the apical membrane through the Na^+-selective channel is passive and driven by the electrochemical gradient for Na^+. The exit of Na^+ from the cell across the basolateral membrane is primary active transport via the Na^+,K^+-ATPase channel. Because

a transepithelial gradient for Na^+ can be generated by this process (i.e., the $[Na^+]$ in the apical compartment can be reduced below that of the basolateral compartment, the overall process of transepithelial Na^+ transport is said to be active). Any solute that is actively transported across an epithelium must be transported via the transcellular pathway.

Depending on the epithelium, the paracellular pathway is an important route for transepithelial transport of solute and water. As noted, the permeability characteristics of the paracellular pathway are determined, in large part, by the specific claudins that are expressed by the cell. Thus the tight junction can have low permeability for solutes, water, or both, or it can have a high permeability. For epithelia in which there are high rates of transepithelial transport,

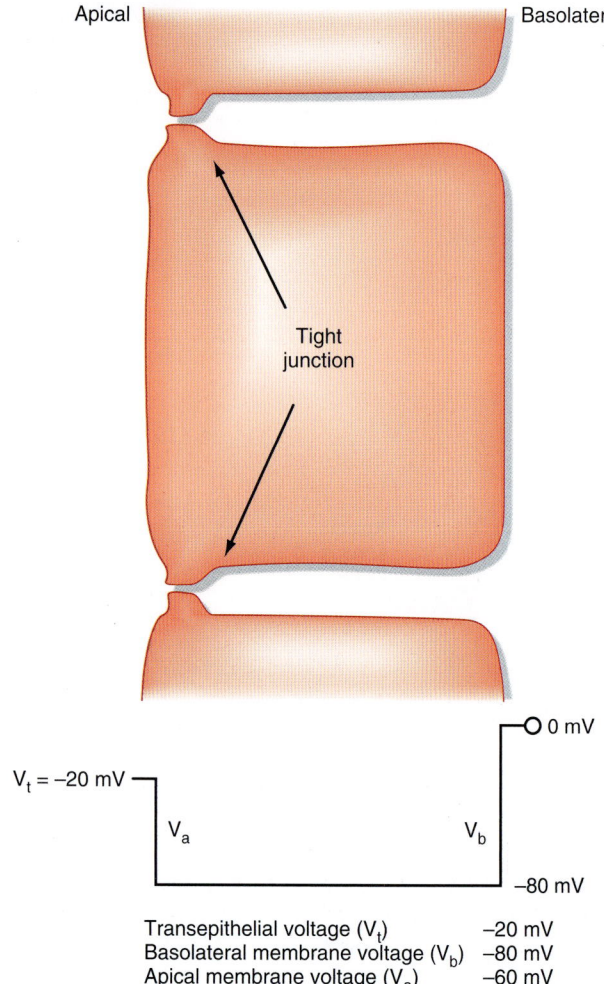

Transepithelial voltage (V_t)   −20 mV
Basolateral membrane voltage (V_b)   −80 mV
Apical membrane voltage (V_a)   −60 mV

• **Fig. 2.12** The Electrical Profile Across an Epithelial Cell. The magnitude of the membrane voltages, and the transepithelial voltage are determined by the various membrane transport proteins in the apical and basolateral membranes. The transepithelial voltage is equal to the sum of the apical and basolateral membrane voltages (see text for details).

the tight junctions typically have a high permeability (i.e., are leaky). Examples of such epithelia include the proximal tubule of the renal nephron and the early segments of the small intestine (e.g., duodenum and jejunum). If the epithelium must establish large transepithelial gradients for solutes, water, or both, the tight junctions typically have low permeability (i.e., are tight). Examples of this type of epithelium include the collecting duct of the renal nephron, the urinary bladder, and the terminal portion of the colon. In addition, the tight junction may be selective for certain solutes (e.g., cation versus anion selective).

All solute transport that occurs through the paracellular pathway is passive in nature. The two driving forces for this transport are the transepithelial concentration gradient for the solute and, if the solute is charged, the transepithelial voltage (Fig. 2.12). The transepithelial voltage may be oriented with the apical surface electrically negative in relation

to the basolateral surface as shown in Fig. 2.12, or it may be oriented with the apical surface electrically positive in relation to the basolateral surface. The polarity and magnitude of the transepithelial voltage is determined by the specific membrane transporters in the apical and basolateral membranes, as well as by the permeability characteristics of the tight junction.

It is important to recognize that transcellular transport processes set up the transepithelial chemical and voltage gradients, which in turn can drive paracellular transport. This is illustrated in Fig. 2.13 for an epithelium that reabsorbs NaCl and for an epithelium that secretes NaCl. In both epithelia, the transepithelial voltage is oriented with the apical surface electrically negative in relation to the basolateral surface. For the NaCl-reabsorbing epithelium, the transepithelial voltage is generated by the active, transcellular reabsorption of Na^+. This voltage in turn drives Cl^- reabsorption through the paracellular pathway. In contrast, for the NaCl-secreting epithelium, the transepithelial voltage is generated by the active transcellular secretion of Cl^-. Na^+ is then secreted passively via the paracellular pathway, driven by the negative transepithelial voltage.

Transepithelial Water Movement

Water movement across epithelia is passive and driven by transepithelial osmotic pressure gradients. Water movement can occur by a transcellular route involving aquaporins in both the apical and basolateral membranes.[i] In addition, water may also move through the paracellular pathway. In the NaCl-reabsorbing epithelium depicted in Fig. 2.13*A*, the reabsorption of NaCl from the apical compartment lowers the osmotic pressure in that compartment, whereas the addition of NaCl to the basolateral compartment raises the osmotic pressure in that compartment. As a result, a transepithelial osmotic pressure gradient is established that drives the movement of water from the apical to the basolateral compartment (i.e., reabsorption). The opposite occurs with NaCl-secreting epithelia (see Fig. 2.13*B*), in which the transepithelial secretion of NaCl establishes a transepithelial osmotic pressure gradient that drives water secretion.

In some epithelia (e.g., proximal tubule of the renal nephron), the movement of water across the epithelium via the paracellular pathway can drive the movement of additional solute. This process is termed **solvent drag** and reflects the fact that solutes dissolved in the water will traverse the tight junction with the water.

As is the case with the establishment of transepithelial concentration and voltage gradients, the establishment of transepithelial osmotic pressure gradients requires transcellular transport of solutes by the epithelial cells.

---

[i]Different aquaporin isoforms are often expressed in the apical and basolateral membrane. In addition, multiple isoforms may be expressed in one or more of the membrane domains.

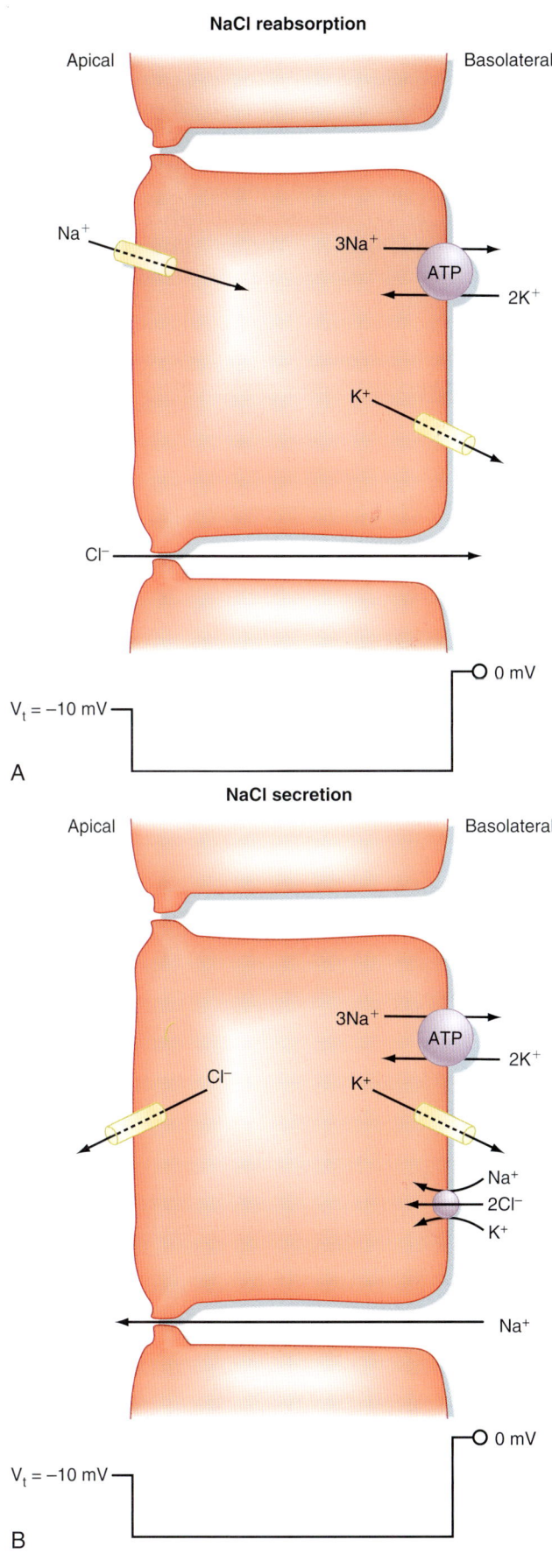

NaCl reabsorption

Apical | Basolateral

Na$^+$

3Na$^+$

ATP

2K$^+$

K$^+$

Cl$^-$

0 mV

$V_t = -10$ mV

A

NaCl secretion

Apical | Basolateral

3Na$^+$

ATP

2K$^+$

Cl$^-$

K$^+$

Na$^+$
2Cl$^-$
K$^+$

Na$^+$

0 mV

$V_t = -10$ mV

B

• **Fig. 2.13** The Role of the Paracellular Pathway in Epithelial Transport. **A,** Na$^+$ transport through the cell generates a transepithelial voltage that then drives the passive movement of Cl$^-$ through the tight junction. NaCl reabsorption results. **B,** Cl$^-$ transport through the cell generates a transepithelial voltage that then drives the passive transport of Na$^+$ through the tight junction. NaCl secretion results.

Regulation of Epithelial Transport

Epithelial transport must be regulated to meet the homeostatic needs of the individual. Depending on the epithelium, this regulation involves neural or hormonal mechanisms, or both. For example, the enteric nervous system of the gastrointestinal tract regulates solute and water transport by the epithelial cells that line the intestine and colon. Similarly, the sympathetic nervous system regulates transport by the epithelial cells of the renal nephron. Aldosterone, a steroid hormone produced by the adrenal cortex (see Chapter 43), is an example of a hormone that stimulates NaCl transport by the epithelial cells of the colon, renal nephron, and sweat ducts. Epithelial cell transport can also be regulated by locally produced and locally acting substances, a process termed **paracrine regulation.** The stimulation of HCl secretion in the stomach by histamine is an example of this process. Cells that are located near the epithelial cells of the stomach release histamine, which acts on the HCl-secreting cells of the stomach (parietal cells) and stimulates them to secrete HCl.

When acted upon by a regulatory signal, the epithelial cell may respond in several different ways, including:

- Retrieval of transporters from the membrane, by endocytosis, or insertion of transporters into the membrane from an intracellular vesicular pool, by a process called *exocytosis*
- Change in activity of membrane transporters (e.g., channel gating)
- Synthesis of specific transporters, and their insertion into the membrane

The first two mechanisms can occur quite rapidly (seconds to minutes), but the synthesis of transporters takes additional time (minutes to days).

Key Concepts

- The body maintains steady-state balance for water and a number of important solutes. This occurs when input into the body equals output from the body. For each solute and water, there is a normal set point. Deviations from this set point are monitored (i.e., when input ≠ output), and effector mechanisms are activated that restore balance. This balance is achieved by adjustment of either intake or excretion of water and solutes. Thereafter, input and output are again equal to maintain balance.
- The Na^+,K^+-ATPase and K^+-selective channels are critically important in establishing and maintaining the intracellular composition, the membrane potential (V_m), and cell volume. Na^+,K^+-ATPase converts the energy in ATP into potential energy of ion gradients and the membrane potential. The ion and electrical gradients created by this process are then used to drive the transport of other ions and other molecules, especially by solute carriers (i.e., symporters and antiporters).
- Epithelial cells constitute the interface between the external world and the internal environment of the body. Vectorial transport of solutes and water across epithelia helps maintain steady-state balance for water and a number of important solutes. Because the external environment constantly changes, and because dietary intake of food and water is highly variable, transport by epithelia is regulated to meet the homeostatic needs of the individual.

Additional Readings

Altenberg GA, Ruess L. Mechanisms of water transport across cell membranes and epithelia. In: Alpern R, Moe O, Kaplan M, eds. *Seldin and Giebisch's The Kidney—Physiology and Pathophysiology*. 5th ed. New York: Academic Press; 2013.

Günzel D, Yu ASL. Claudins and the modulation of tight junction permeability. *Physiol Rev*. 2013;93:525-569.

Hoffman EK, et al. Physiology of cell volume regulation in vertebrates. *Physiol Rev*. 2009;89:193-277.

Lang F. Cell volume control. In: Alpern R, Moe O, Kaplan M, eds. *Seldin and Giebisch's The Kidney—Physiology and Pathophysiology*. 5th ed. New York: Academic Press; 2013.

Pawlina W. *Histology: A Text and Atlas, with Correlated Cell and Molecular Biology*. 7th ed. Philadelphia: Wolters Kluwer Health; 2016.

Pedersen SF, Kapus A, Hoffmann EK. Osmosensory mechanisms in cellular and systemic volume regulation. *J Am Soc Nephrol*. 2011;22:1587-1597.

Sackin H, Palmer LG. Electrophysiological analysis of transepithelial transport. In: Alpern R, Moe O, Kaplan M, eds. *Seldin and Giebisch's The Kidney—Physiology and Pathophysiology*. 5th ed. New York: Academic Press; 2013.

3

Signal Transduction, Membrane Receptors, Second Messengers, and Regulation of Gene Expression

LEARNING OBJECTIVES

Upon completion of this chapter, the student should be able to answer the following questions:

1. How do cells communicate with each other?
2. What are the four classes of receptors, and what signal transduction pathways are associated with each class of receptors?
3. How do steroid and thyroid hormones, cyclic adenosine monophosphate, and receptor tyrosine kinases regulate gene expression?

The human body is composed of billions of cells, each with a distinct function. However, the function of cells is tightly coordinated and integrated by external chemical signals, including hormones, neurotransmitters, growth factors, odorants, and products of cellular metabolism that serve as chemical messengers and provide cell-to-cell communication. Mechanical and thermal stimuli and light are physical external signals that also coordinate cellular function. Chemical and physical messengers interact with receptors located in the plasma membrane, cytoplasm, and nucleus. Interaction of these messengers with receptors initiates a cascade of signaling events that mediate the response to each stimulus. These signaling pathways ensure that the cellular response to external messengers is specific, amplified, tightly regulated, and coordinated. This chapter provides an overview of how cells communicate via external messengers and a discussion of the signaling pathways that process external information into a highly coordinated cellular response. In subsequent chapters, details on signaling pathways in the nervous system, muscular system, cardiovascular system, respiratory system, gastrointestinal system, renal system, and endocrine system are discussed in greater detail.

IN THE CLINIC

The significance of signaling pathways in medicine is illustrated by the following short list of popular drugs that act by regulating signaling pathways. Details on these pathways are presented later in this and other chapters.

- **Aspirin,** the first pharmaceutical (1899), inhibits cyclooxygenase-1 (COX1) and cyclooxygenase-2 (COX2) and therefore is antithrombotic (i.e., reduces the formation of blood clots).
- **β-Adrenergic receptor agonists and antagonists** are used to treat a variety of medical conditions. β_1-Agonists increase cardiac contractility and heart rate in patients with low blood pressure. β_2-Agonists dilate bronchi and are used to treat asthma and chronic obstructive lung disease. In contrast, β-adrenergic antagonists are used to treat hypertension, angina, cardiac arrhythmias, and congestive heart failure (see Chapter 18).
- **Fluoxetine (Prozac)** is an antidepressant medication that inhibits reuptake of the neurotransmitter serotonin into the presynaptic cell, which results in enhanced activation of serotonin receptors (see Chapter 6).
- Several monoclonal antibodies are used to treat cancer caused by the activation of growth factor receptors in cancer cells. For example, **trastuzumab (Herceptin)** is a monoclonal antibody used to treat metastatic breast cancer in women who overexpress **HER2/neu,** a member of the family of epidermal growth factor (EGF) receptors, which stimulate cell growth and differentiation. **Cetuximab (Erbitux)** and **bevacizumab (Avastin)** are monoclonal antibodies that are used to treat metastatic colorectal cancer and cancers of the head and neck. These antibodies bind to and inhibit the EGF receptor and thereby inhibit EGF-induced cell growth in cancer cells.
- Drugs that inhibit cyclic guanosine monophosphate (cGMP)–specific phosphodiesterase type 5, such as **sildenafil (Viagra), tadalafil (Cialis),** and **vardenafil (Levitra),** prolong the vasodilatory effects of nitric oxide and are used to treat erectile dysfunction and pulmonary arterial hypertension (see Chapter 17).

Cell-to-Cell Communication

An overview of how cells communicate with each other is presented in Fig. 3.1. Cells communicate by releasing extracellular signaling molecules (e.g., **hormones and neurotransmitters**) that bind to **receptor** proteins located in the plasma membrane, cytoplasm, or nucleus. This signal is transduced into the activation, or inactivation, of one or more intracellular messengers by interacting with receptors. Receptors interact with a variety of intracellular signaling proteins, including **kinases, phosphatases,** and guanosine triphosphate (GTP)–binding proteins **(G proteins)**. These signaling proteins interact with and regulate the activity of target proteins and thereby modulate cellular function. Target proteins include, but are not limited to, ion channels and other transport proteins, metabolic enzymes, cytoskeletal proteins, gene regulatory proteins, and cell cycle proteins that regulate cell growth and division. Signaling pathways are characterized by (1) multiple, hierarchical steps; (2) amplification of the signal-receptor binding event, which magnifies the response; (3) activation of multiple pathways and regulation of multiple cellular functions; and (4) antagonism by constitutive and regulated feedback mechanisms, which minimize the response and provide tight regulatory control over these signaling pathways. A brief description of how cells communicate follows. Readers who desire a more in-depth presentation of this material are encouraged to consult one of the many cellular and molecular biology textbooks currently available.

Cells in higher animals release into the extracellular space hundreds of chemicals, including (1) **peptides and proteins** (e.g., insulin); **(2) amines** (e.g., epinephrine and norepinephrine); (3) **steroid hormones** (e.g., aldosterone, estrogen); and (4) **small molecules,** including amino acids, nucleotides, ions (e.g., Ca^{++}), and gases, such as nitric oxide and carbon dioxide. Secretion of signaling molecules is

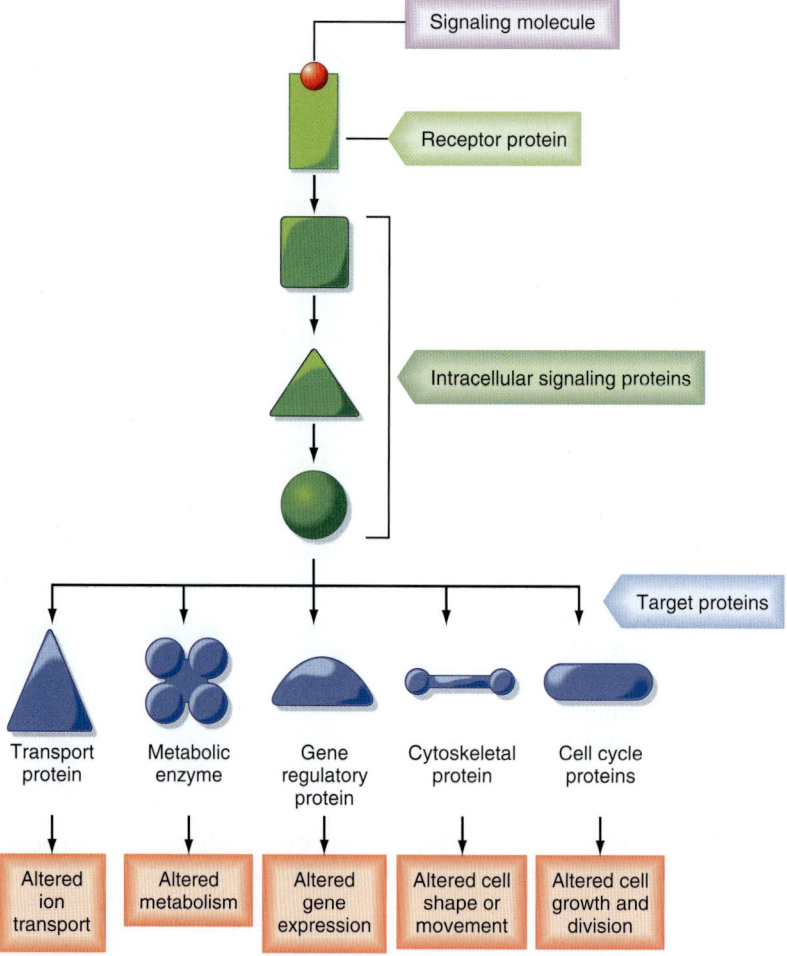

• **Fig. 3.1** An Overview of How Cells Communicate. A signaling molecule (i.e., hormone or neuro-transmitter) binds to a receptor, which may be in the plasma membrane, cytosol, or nucleus. Binding of ligand to a receptor activates intracellular signaling proteins, which interact with and regulate the activity of one or more target proteins to change cellular function. Signaling molecules regulate cell growth, division, and differentiation and influence cellular metabolism. In addition, they modulate the intracellular ionic composition by regulating the activity of ion channels and transport proteins. Signaling molecules also control cytoskeleton-associated events, including cell shape, division, and migration and cell-to-cell and cell-to-matrix adhesion. (Redrawn from Alberts B, et al: *Molecular Biology of the Cell*. 6th ed. New York: Garland Science; 2015.)

cell-type specific. For example, beta cells in the pancreas release insulin, which stimulates glucose uptake into cells. The ability of a cell to respond to a specific signaling molecule depends on the expression of receptors that bind the signaling molecule with high affinity and specificity. Receptors are located in the plasma membrane, the cytosol, and the nucleus (Fig. 3.2).

Signaling molecules can act over long or short distances and can require cell-to-cell contact or very close cellular proximity (Fig. 3.3). **Contact-dependent signaling,** in which a membrane-bound signaling molecule of one cell binds directly to a plasma membrane receptor of another cell, is important during development, in immune responses, and in cancer (see Fig. 3.3A). Molecules that are released and act locally are called **paracrine** (see Fig. 3.3B) or **autocrine** (see Fig. 3.3C) **hormones.** Paracrine signals are released by one type of cell and act on another type; they are usually taken up by target cells or rapidly degraded (within minutes) by enzymes. For example, enterochromaffin-like cells in the stomach secrete histamine, which stimulates the production of acid by neighboring parietal cells (see Chapter 27 for details). Autocrine signaling involves the release of a molecule that affects the same cell or other cells of the same type (e.g., cancer cells). In **synaptic signaling** (see Fig. 3.3D), neurons transmit electrical signals along their axons and release neurotransmitters at synapses that affect the function of other neurons or cells that are distant from the neuron cell body. The close physical relationship between the nerve terminal and the target cell ensures that the neurotransmitter is delivered to a specific cell. Details on synaptic signaling are discussed in Chapter 6. **Endocrine** signals are hormones that are secreted into the blood and are widely dispersed in the body (see Fig. 3.3E). Details on endocrine signaling are discussed in Chapter 38.

In addition to paracrine, autocrine, endocrine, and synaptic signaling, cell-to-cell communication also occurs via **gap junctions** that form between adjacent cells (see Chapter 2). Gap junctions are specialized junctions that allow intracellular signaling molecules, generally less than 1200 D in size, to diffuse from the cytoplasm of one cell to an adjacent cell. The permeability of gap junctions is regulated by cytosolic $[Ca^{++}]$, $[H^+]$, and cyclic adenosine monophosphate (cAMP) and by the membrane potential. Gap junctions also allow cells to be electrically coupled, which is vitally important for the coordinated activity of cardiac and smooth muscle cells (see Chapters 13 and 14).

The speed of a response to an extracellular signal depends on the mechanism of delivery. Endocrine signals are relatively slow (seconds to minutes) because time is required for diffusion and blood flow to the target cell, whereas synaptic signaling is extremely fast (milliseconds). If the response involves changes in the activity of proteins in the cell, the response may occur in milliseconds to seconds. However, if the response involves changes in gene expression and the de novo synthesis of proteins, the response may take hours to occur, and a maximal response may take days. For example, the stimulatory effect of aldosterone on sodium transport by the kidneys requires days to develop fully (see Chapter 35).

The response to a particular signaling molecule also depends on the ability of the molecule to reach a particular cell, on expression of the cognate receptor (i.e., receptors that recognize a particular signaling molecule or ligand with a high degree of specificity), and on the cytoplasmic signaling molecules that interact with the receptor. Thus signaling molecules frequently have many different effects that are dependent on the cell type. For example, the neurotransmitter acetylcholine stimulates contraction of skeletal muscle but decreases the force of contraction in heart muscle. This is because skeletal muscle and heart cells express different acetylcholine receptors.[a]

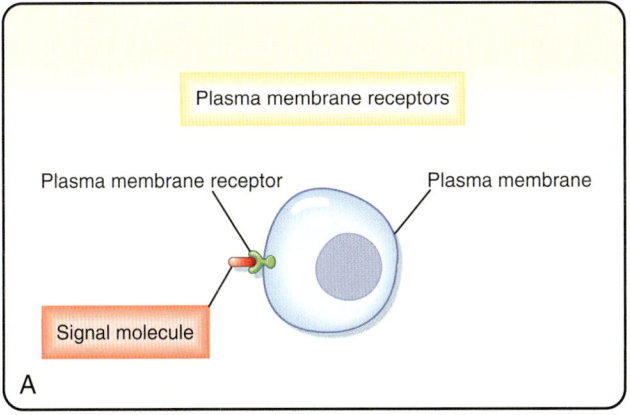

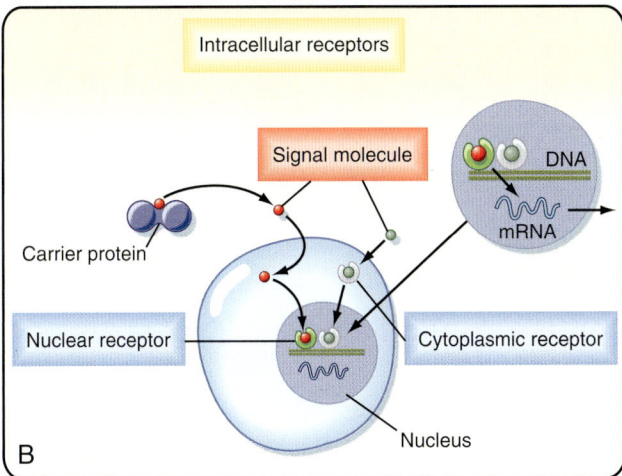

• **Fig. 3.2** Signaling molecules, especially ones that are hydrophilic and cannot cross the plasma membrane, bind directly to their cognate receptors in the plasma membrane **(A).** Other signaling molecules—including steroid hormones, triiodothyronines, retinoic acids, and vitamin D—bind to carrier proteins in blood and readily diffuse across the plasma membrane, where they bind to cognate nuclear receptors in the cytosol or nucleus **(B).** Still other signaling molecules, including nitric oxide, can diffuse without carrier proteins and cross the membrane to act on intracellular protein targets **(B).** Both classes of receptors, when ligand bound, regulate gene transcription. mRNA, messenger RNA. (Redrawn from Alberts B, et al: *Molecular Biology of the Cell.* 6th ed. New York: Garland Science; 2015.)

---

[a]The acetylcholine receptor in skeletal muscle is termed *nicotinic* because nicotine can mimic this action of the neurotransmitter. In contrast, the acetylcholine receptor in cardiac muscle is termed *muscarinic* because this effect is mimicked by muscarine, an alkaloid derived from the mushroom *Amanita muscaria.*

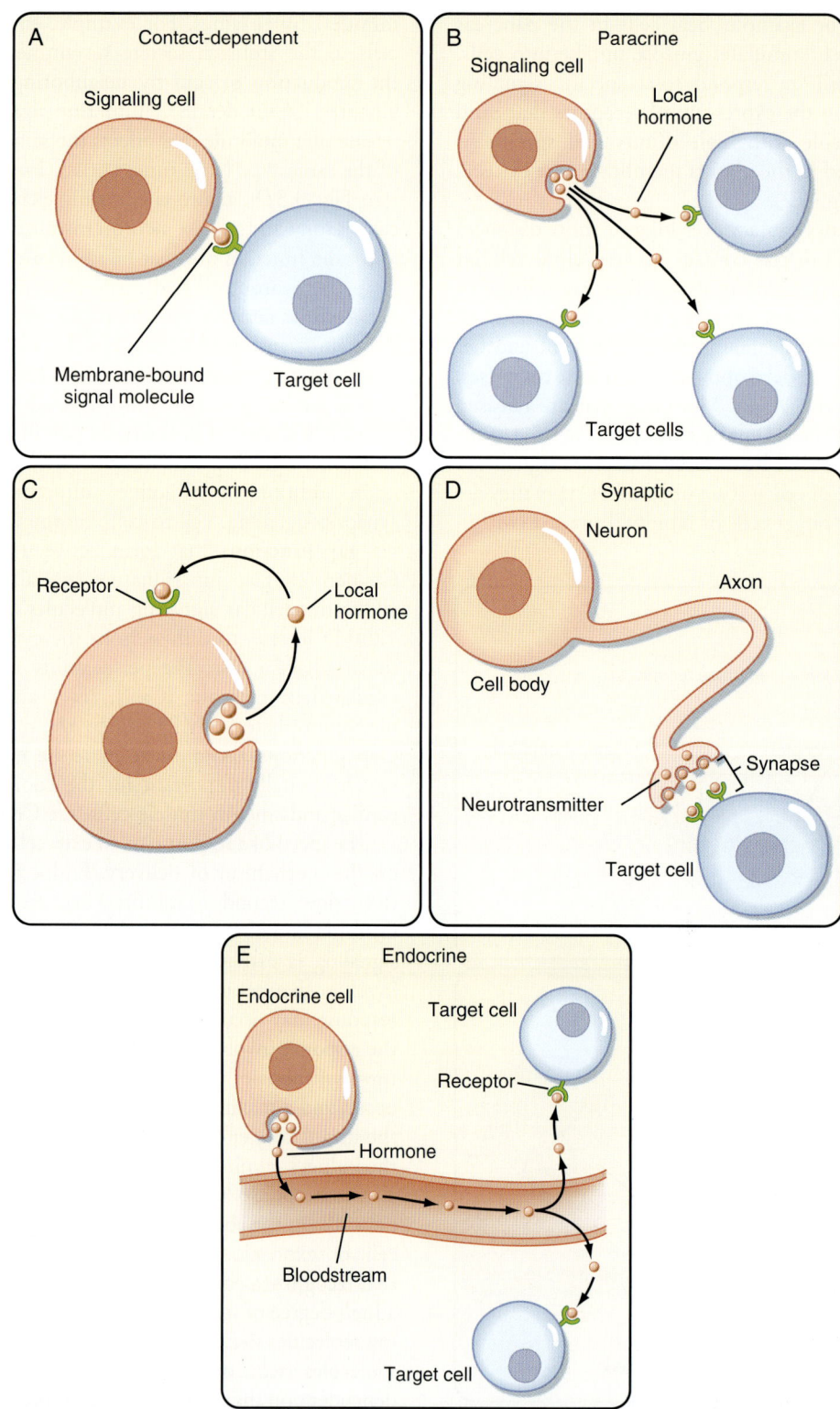

• **Fig. 3.3** Cell-to-cell communication is mediated by five basic mechanisms: contact-dependent **(A),** paracrine **(B),** autocrine **(C),** synaptic **(D),** and endocrine signaling **(E).** These mechanisms are described in detail in the text. (Redrawn from Alberts B, et al: *Molecular Biology of the Cell.* 6th ed. New York: Garland Science; 2015.)

TABLE 3.1 Classes of Membrane Receptors

Receptor Class	Ligand	Signal Transduction Pathway/Target
Ligand-gated ion channels	**Extracellular ligand:**	**Membrane currents:**
	GABA	**Cl^-**
	ACh (muscle)	**Na^+, K^+, Ca^{++}**
	ATP	**Ca^{++}, Na^+, K^+**
	Glutamate: NMDA	**Na^+, K^+, Ca^{++}**
	Intracellular ligand:	
	cAMP (olfaction)	**K^+**
	cGMP (vision)	**Na^+, K^+**
	InsP3	**Ca^{++}**
G protein–coupled receptors	**Neurotransmitters (ACh)**	βγ Subunits activate ion channels
	Peptides (PTH, oxytocin)	α Subunit activates enzymes:
	Odorants	Cyclases that generate cAMP, cGMP, phospholipases that
	Cytokines, lipids	generate InsP3 and diacylglycerol, and phospholipases
		that generate arachidonic acid and its metabolites.
		Monomeric G proteins
Enzyme-linked receptors	**ANP**	**Receptor guanylyl cyclase**
	TGF-β	**Receptor serine/threonine kinase**
	Insulin, EGF	**Receptor tyrosine kinase**
	Interleukin-6, erythropoietin	**Tyrosine kinase–associated receptor**
Nuclear receptors	**Steroid hormones:**	**Bind to regulatory sequences in DNA and increase or**
	Mineralocorticoids	**decrease gene transcription**
	Glucocorticoids	
	Androgens	
	Estrogens	
	Progestins	
	Miscellaneous hormones:	**Bind to regulatory sequences in DNA and increase or**
	Thyroid	**decrease gene transcription**
	Vitamin D	
	Retinoic acid	
	Prostaglandins	

ACh, acetylcholine; ANP, atrial natriuretic peptide; ATP, adenosine triphosphate; cAMP, cyclic adenosine monophosphate; cGMP, cyclic guanosine monophosphate; EGF, epidermal growth factor; GABA, gamma-aminobutyric acid; InsP3, inositol 1,4,5-triphosphate; NMDA, *N*-methyl-D-aspartate; PTH, parathyroid hormone; TGF, transforming growth factor.

Receptors

All signaling molecules bind to specific receptors that act as signal transducers, thereby converting a ligand-receptor binding event into intracellular signals that affect cellular function. Receptors can be divided into four basic classes on the basis of their structure and mechanism of action: (1) **ligand-gated ion channels,** (2) **G protein–coupled receptors (GPCRs),** (3) **enzyme-linked receptors,** and (4) **nuclear receptors** (Table 3.1; Figs. 3.4 and 3.5).

Ligand-gated ion channels mediate direct and rapid synaptic signaling between electrically excitable cells (see Fig. 3.4A). Neurotransmitters bind to receptors and either open or close ion channels, thereby changing the ionic permeability of the plasma membrane and altering the membrane potential. For examples and more details, see Chapter 6.

GPCRs regulate the activity of other proteins, such as enzymes and ion channels (see Fig. 3.4B). In the example in Fig. 3.4B, the interaction between the receptor and the

target protein is mediated by heterotrimeric G proteins, which are composed of α, β, and γ subunits. Stimulation of G proteins by ligand-bound receptors activates or inhibits downstream target proteins that regulate signaling pathways if the target protein is an enzyme or changes membrane ion permeability if the target protein is an ion channel.

Enzyme-linked receptors either function as enzymes or are associated with and regulate enzymes (see Fig. 3.4C). Most enzyme-linked receptors are protein kinases or are associated with protein kinases, and ligand binding causes the kinases to phosphorylate a specific subset of proteins on specific amino acids, which in turn activates or inhibits protein activity.

Nuclear receptors are small hydrophobic molecules, including steroid hormones, thyroid hormones, retinoids, and vitamin D, that have a long biological half-life (hours to days), diffuse across the plasma membrane, and bind to nuclear receptors or to cytoplasmic receptors that, once bound to their ligand, translocate to the nucleus (see Fig. 3.5). Some nuclear receptors, such as those that bind cortisol

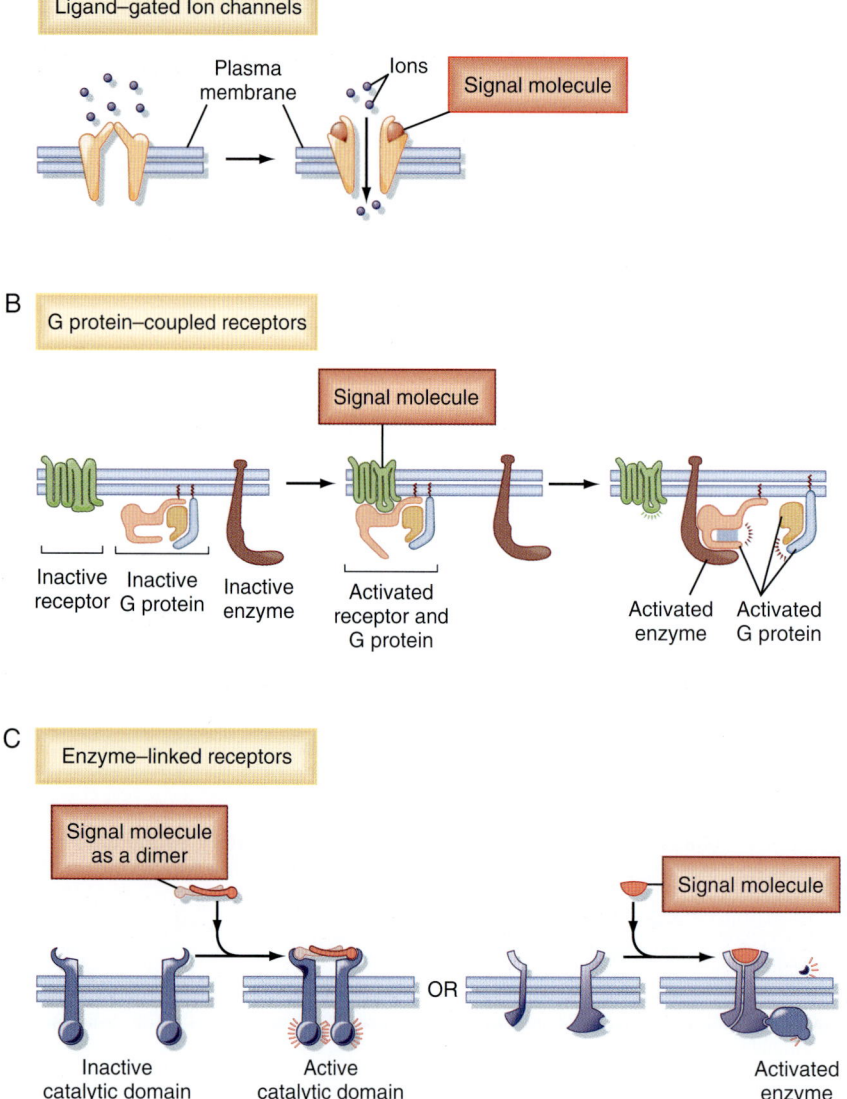

A   Ligand–gated Ion channels

Plasma membrane

Ions

Signal molecule

B   G protein–coupled receptors

Signal molecule

Inactive receptor   Inactive G protein   Inactive enzyme

Activated receptor and G protein

Activated enzyme   Activated G protein

C   Enzyme–linked receptors

Signal molecule as a dimer

Signal molecule

Inactive catalytic domain   Active catalytic domain   OR   Activated enzyme

• **Fig. 3.4** Three of the Four Classes of Plasma Membrane Receptors. See text for details. (Redrawn from Alberts B, et al: *Molecular Biology of the Cell.* 6th ed. New York: Garland Science; 2015.)

and aldosterone, are located in the cytosol and enter the nucleus after binding to hormone, whereas other receptors, including the thyroid hormone receptor, are located in the nucleus. In both cases, inactive receptors are bound to inhibitory proteins, and binding of hormone results in dissociation of the inhibitory complex. Hormone binding causes the receptor to bind coactivator proteins that activate gene transcription. Once activated, the hormone-receptor complex regulates the transcription of specific genes. Activation of specific genes usually occurs in two steps: an early primary response (≈30 minutes), which activates genes that stimulate other genes to produce a delayed (hours to days) secondary response (see Fig. 3.5). Each hormone elicits a specific response that is based on cellular expression of the cognate receptor, as well as on cell type–specific expression of gene regulatory proteins that interact with the activated receptor to regulate the transcription of a specific set of genes (see Chapter 38 for more details). In addition to

steroid receptors that regulate gene expression, evidence also suggests the existence of membrane and juxtamembrane steroid receptors that mediate the rapid, nongenomic effects of steroid hormones.

Some membrane proteins do not fit the classic definition of receptors, but they subserve a receptor-like function in that they recognize extracellular signals and transduce the signals into an intracellular second messenger that has a biological effect. For example, on activation by a ligand, some membrane proteins undergo **regulated intramembrane proteolysis (RIP),** which elaborates a cytosolic peptide fragment that enters the nucleus and regulates gene expression (Fig. 3.6). In this signaling pathway, binding of ligand to a plasma membrane receptor leads to ectodomain shedding, facilitated by members of the metalloproteinase-disintegrin family, and produces a carboxy-terminal fragment that is the substrate for γ-secretase. γ-Secretase induces RIP, thereby causing the release of an intracellular domain of

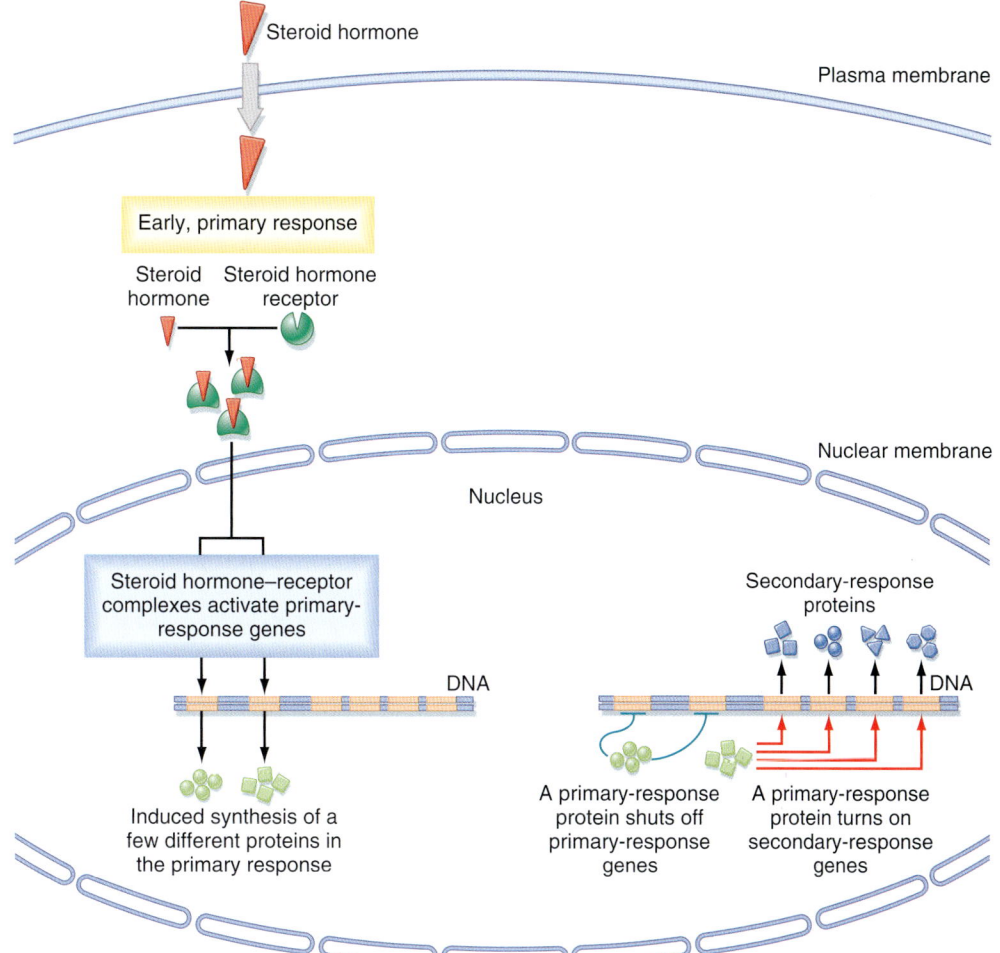

• **Fig. 3.5** Steroid Hormones Stimulate the Transcription of Early-Response Genes and Late-Response Genes. See text for details. (Redrawn from Alberts B, et al: *Molecular Biology of Cell.* 6th ed. New York: Garland Science; 2015.)

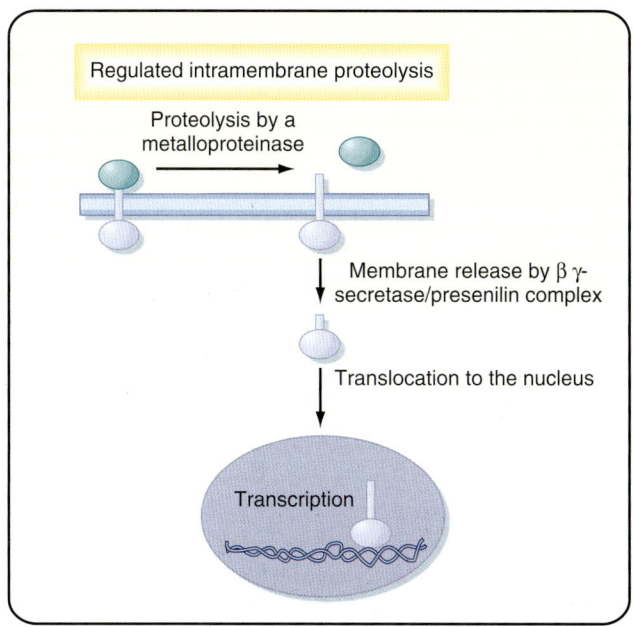

• **Fig. 3.6** Regulated Intramembrane Proteolysis. See text for details. (Redrawn from Alberts B, et al: *Molecular Biology of the Cell.* 6th ed. New York: Garland Science; 2015.)

the protein that enters the nucleus and regulates transcription (see Fig. 3.6). The best characterized example of RIP is the sterol regulatory element–binding protein (SREBP), a transmembrane protein expressed in the membrane of the endoplasmic reticulum. When cellular cholesterol levels are low, SREBP undergoes RIP, and the proteolytically cleaved fragment is translocated into the nucleus, where it transcriptionally activates genes that promote cholesterol biosynthesis.

Receptors and Signal Transduction Pathways

When hormones bind to plasma membrane receptors, signals are relayed to effector proteins via intracellular signaling pathways. When hormones bind to nuclear or cytosolic receptors, they relay signals primarily through regulation of gene expression. Signaling pathways can amplify and integrate signals but can also downregulate and desensitize signals, reducing or terminating the response, even in the continued presence of hormone.

IN THE CLINIC

Alzheimer's disease, a progressive neurodegenerative brain disease characterized by the formation of amyloid plaques, affects approximately 44 million people worldwide. In Alzheimer's disease, regulated intramembrane proteolysis of amyloid β-protein precursor (APP) causes the accumulation of amyloid β-protein (Aβ), which forms amyloid plaques that contribute to the pathogenesis of Alzheimer's disease. APP is a type I transmembrane protein (i.e., its spans the membrane only once). After ectodomain shedding, its sequential proteolysis by β-secretase and γ-secretase produces the Aβ40 and Aβ42 peptides that are normally produced throughout life but accumulate in individuals with Alzheimer's disease. Missense mutations in presenilins, proteins that regulate γ-secretase protease activity, enhance the production of Aβ42, which is more hydrophobic and prone to aggregation into amyloid fibrils than is the more abundant Aβ40 protein.

Intracellular signaling molecules—so-called second messengers (the first messenger of the signal is the ligand that binds to the receptor)—include small molecules such as cAMP, cGMP, Ca^{++}, and diacylglycerol. Signaling pathways often include dozens of small molecules that form complicated networks within the cell (Fig. 3.7). Some proteins in the intracellular signaling pathways relay the signal by passing the message directly to another protein (e.g., by phosphorylating a target, or by binding and causing an allosteric change). Such intracellular signaling proteins act as **reversible molecular switches:** When a signal is received, they switch from an inactive to an active form or vice versa, until another signaling molecule reverses the process. This principle of reversibility is central to many signaling pathways. In many cases, activation is achieved by reversing inhibition: For example, the thyroid hormone receptor is bound to an inhibitory protein in the absence of signal.

Signaling complexes, composed of multiple proteins that interact physically, enhance the speed, efficiency, and specificity of signaling. Many proteins, usually enzymes or ion channels, transduce the signal into a different chemical form and simultaneously amplify the signal either by producing large amounts of additional signaling molecules or by activating a large number of downstream signaling proteins. For example, adenylyl cyclase, the enzyme that makes cAMP, transduces a signal (receptor activation of G proteins) and amplifies the signal by generating large amounts of cAMP. Other types of signaling proteins include those that integrate multiple signals. Other proteins carry the signal from one region of the cell to another: for example, by translocating from the cytosol to the nucleus.

Cells can adjust rapidly to changing signals. Cells can respond quickly and in a graded manner to increasing concentrations of hormone, and the effect of a signaling molecule can be either long- or short-lived. Cells can also adjust their sensitivity to a signal by **desensitization,** whereby prolonged exposure to a hormone decreases the cell's response over time. Desensitization is a reversible process that can involve a reduction in the number of receptors expressed in the plasma membrane, inactivation of receptors, or changes in signaling proteins that mediate the downstream effect of the receptors. Homologous desensitization involves a reduction in the response only to the signaling molecule that caused the response (e.g., opioid dependence and tolerance), whereas heterologous desensitization is when one ligand desensitizes the response to another ligand.

Table 3.1 summarizes the four general classes of receptors and provides a few examples of the signal transduction pathways associated with each class of receptors.

Ligand-Gated Ion Channel Signal Transduction Pathways

This class of receptors transduces a chemical signal into an electrical signal, which elicits a response. For example, the ryanodine receptor, located in the membrane of the sarcoplasmic reticulum of skeletal muscle, is activated by Ca^{++}, caffeine, adenosine triphosphate (ATP), or metabolites of arachidonic acid to release Ca^{++} into the cytosol, which facilitates muscle contraction (see Chapter 12 for details). In glutamergic synapses in which high levels of prior synaptic activity have led to partial membrane depolarization, activation of the *N*-methyl-D-aspartate receptor by glutamate stimulates Ca^{++} influx important for synaptic plasticity.

G Protein–Coupled Signal Transduction Pathways

There are two classes of **GTP-binding proteins** (i.e., GTPases, which are named for their ability to hydrolyze GTP to guanosine diphosphate [GDP] and an inorganic phosphate): low-molecular-weight, **monomeric G proteins** and **heterotrimeric G proteins** composed of α, β, and γ subunits. GTP binding activates, whereas hydrolysis of GTP to GDP inactivates, GTP-binding proteins (Fig. 3.8*A*). All GTPases are controlled by regulatory proteins, including **GTPase-activating proteins,** which induce the hydrolysis of GTP to GDP and thus inactivate the GTPase, and **guanine nucleotide exchange factors (GEFs)** that causes the GTPase to release GDP, which is rapidly replaced by GTP, thereby activating the GTPase (see Fig. 3.8*B*).

Monomeric G proteins are composed of a single 20- to 40-kDa protein and can be membrane bound because of the addition of lipids posttranslationally. Monomeric G proteins have been classified into five families (Ras, Rho, Rab, Ran, and Arf), play a central role in many enzyme-linked receptor pathways, and regulate gene expression and cell proliferation, differentiation, and survival. Rho GTPases regulate actin cytoskeletal organization, cell cycle progression, and gene expression. The Rab GTPases regulate intravesicular transport and trafficking of proteins between organelles in the secretory and endocytic pathways. Ran GTPases regulate nucleocytoplasmic transport of RNA and

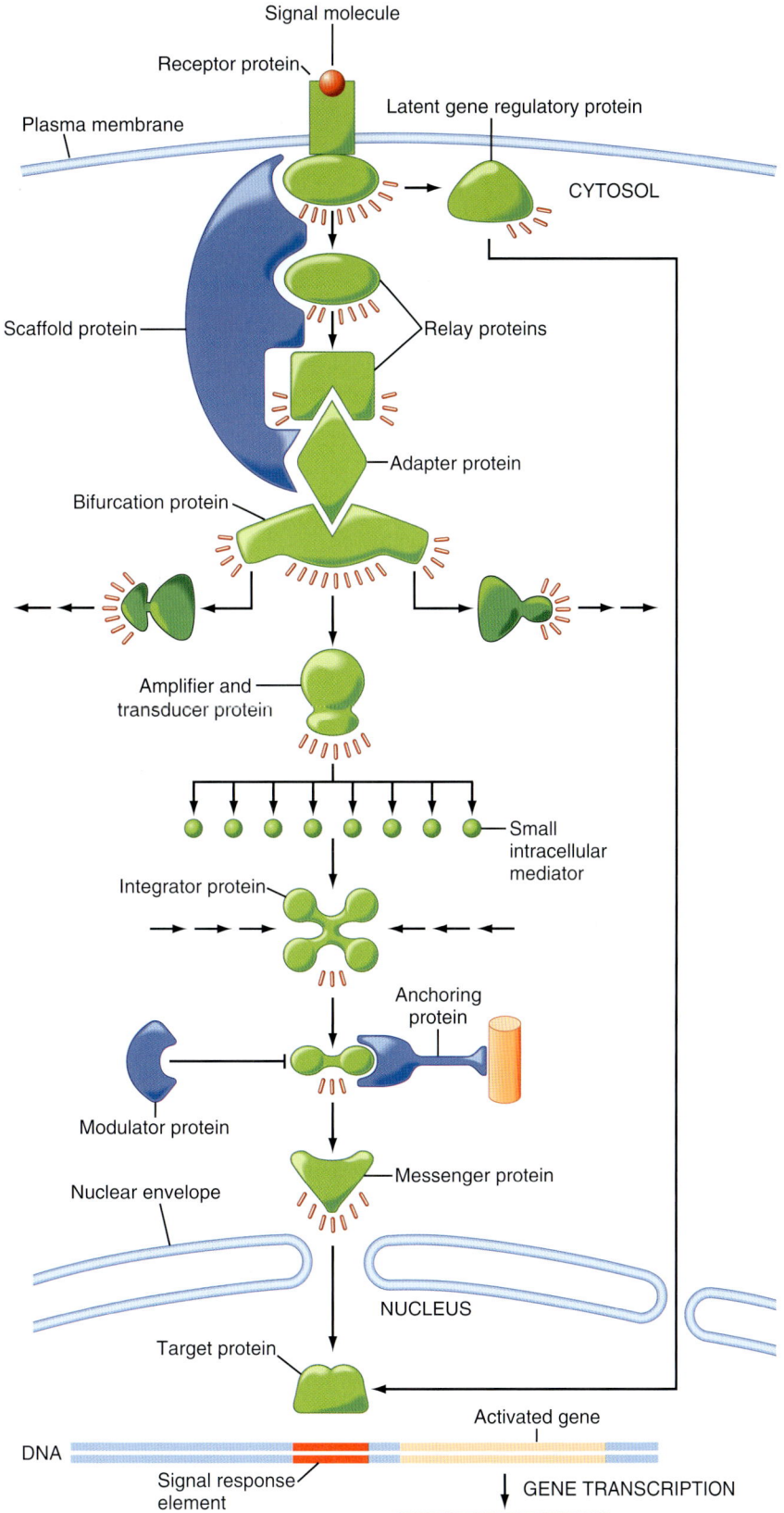

• **Fig. 3.7** Illustration of How Intracellular Signals Are Amplified and Integrated. Signaling pathways often include dozens of proteins and small molecules that form complicated networks within the cell. Some signaling proteins relay the signal by passing the message to another protein. Many proteins amplify the signal either by producing large amounts of additional signaling molecules or by activating a large number of downstream signaling proteins. Other proteins carry the signal from one region of the cell to another. See text for more details. (Redrawn from Alberts B, et al: *Molecular Biology of the Cell.* 6th ed. New York: Garland Science; 2015.).

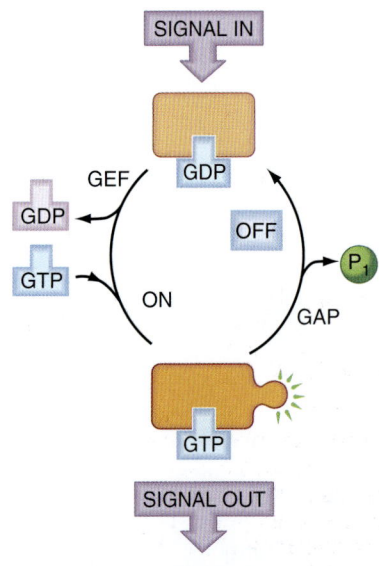

• **Fig. 3.8** **GTP-Binding Proteins.** GTP binding activates whereas hydrolysis of GTP to GDP inactivates GTP-binding proteins **(A).** All GTPase are controlled by regulatory proteins, including GTPase-activating proteins (GAP), which induce the hydrolysis of GTP to GDP, thus inactivating the GTPase, and guanine nucleotide exchange factors, (GEF) that which cause the GTPase to release GDP, which is rapidly replaced by GTP, thereby activating the GTPase **(B).** (Redrawn from Kantrowitx ER, Lipscomb WN. Escherichia coli aspartate transcarbamoylase: the molecular basis for a concerted allosteric transition. *Trends Biochem Sci.* 1990;15:53-59.)

proteins. Ras GTPases are involved in many signaling pathways that control cell division, proliferation and death. Arf GTPases, like Rab GTPases, regulate vesicular transport.

Heterotrimeric G proteins couple to more than 1000 different receptors and thereby mediate the cellular response to an incredibly diverse set of signaling molecules, including hormones, neurotransmitters, peptides, and odorants. Like monomeric G proteins, they can be membrane bound because of the addition of lipids posttranslationally. Heterotrimeric complexes are composed of three subunits: α, β, and γ. There exist 16 α subunits, 5 β subunits, and 11 γ subunits, which can assemble into hundreds of different combinations and thereby interact with a diverse number of receptors and effectors. The assembly of subunits and the association with receptors and effectors depend on the cell type.

An overview of heterotrimeric G protein activation is illustrated in Fig. 3.9. In the absence of ligand, these G proteins are inactive and form a heterotrimeric complex in which GDP binds to the α subunit. Binding of a signal

molecule to an inactive GPCR induces a conformational change in the G protein that promotes the release of GDP and the subsequent binding of GTP to the α subunit. Binding of GTP to the α subunit stimulates dissociation of the α subunit from the heterotrimeric complex and results in release of the α subunit from the $\beta\gamma$ dimer, each of which can interact with and regulate downstream effectors such as adenylyl cyclase and phospholipases (see Fig. 3.9). Activation of downstream effectors by the α subunit and $\beta\gamma$ dimer is terminated when the α subunit hydrolyzes the bound GTP to GDP and inorganic phosphate (P_i). The α subunit bound to GDP associates with the $\beta\gamma$ dimer and terminates the activation of effectors.

Another way to attenuate or terminate signaling through a GPCR involves desensitization and endocytic removal of receptors from the plasma membrane. Binding of hormone to a GPCR increases the ability of **GPCR kinases** to phosphorylate the intracellular domain of GPCRs, which recruits proteins called β-**arrestins** to bind to the GPCR. β-Arrestins inactivate the receptor and promote endocytic

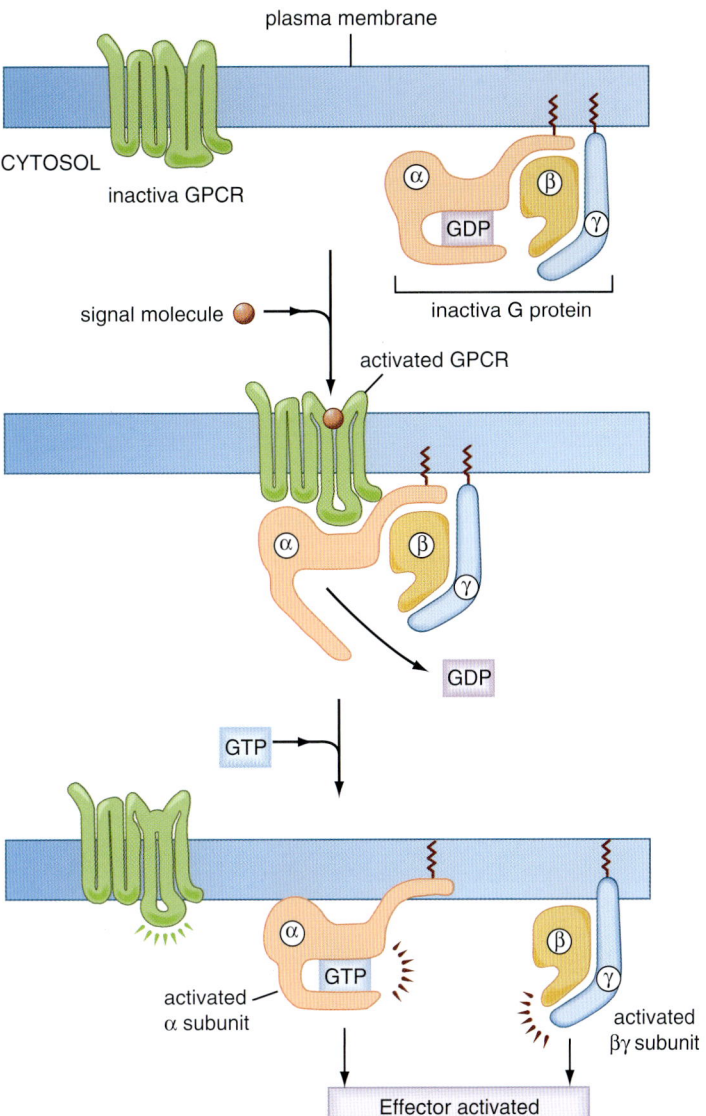

• **Fig. 3.9** Activation of a G Protein–Coupled Receptor (GPCR) and Effector Activation. In the absence of ligand, heterotrimeric G proteins are in an inactive state because GDP binds to the α subunit. Binding of a signal molecule to an inactive G protein–coupled receptor (GPCR) induces a conformational change in the G protein that promotes the release of GDP and the subsequent binding of GTP to the α subunit. Binding of GTP to the α subunit stimulates dissociation of the α subunit from the heterotrimeric complex and results in release of the α subunit from the βγ dimer, each of which can interact with and regulate downstream effectors. (Redrawn from Alberts B, et al: *Molecular Biology of the Cell.* 6th ed. New York: Garland Science; 2015.)

removal of the GPCR from the plasma membrane. GPCR kinase/β-arrestin inactivation with endocytosis of GPCRs is an important mechanism whereby cells downregulate (desensitize) a response during prolonged exposure to elevated hormone levels. One of the major benefits of β blockers, given for congestive heart failure, is that they reverse chronic desensitization and the associated recovery of adrenergic responsiveness.

Activated G protein α subunits couple to a variety of effector proteins, including adenylyl cyclase, **phosphodiesterases,** and **phospholipases** (A₂, C, and D). A very common downstream effector of heterotrimeric G proteins is adenylyl cyclase, which facilitates the conversion of ATP

to cAMP (Fig. 3.10). When a signal molecule binds to a GPCR that interacts with a G protein composed of an α subunit of the α_s class, adenylyl cyclase is activated, which causes an increase in cAMP levels and, as a result, activation of **protein kinase A (PKA).** By phosphorylating specific serine and threonine residues on downstream effector proteins, PKA regulates effector protein activity. In contrast, when a ligand binds to a receptor that interacts with a G protein composed of a α subunit of the α_i class, adenylyl cyclase is inhibited, which causes reductions in cAMP levels and, consequently, in PKA activity.

Some effector proteins, such as ion-gated channels, are also regulated directly by cAMP. cAMP is degraded to AMP

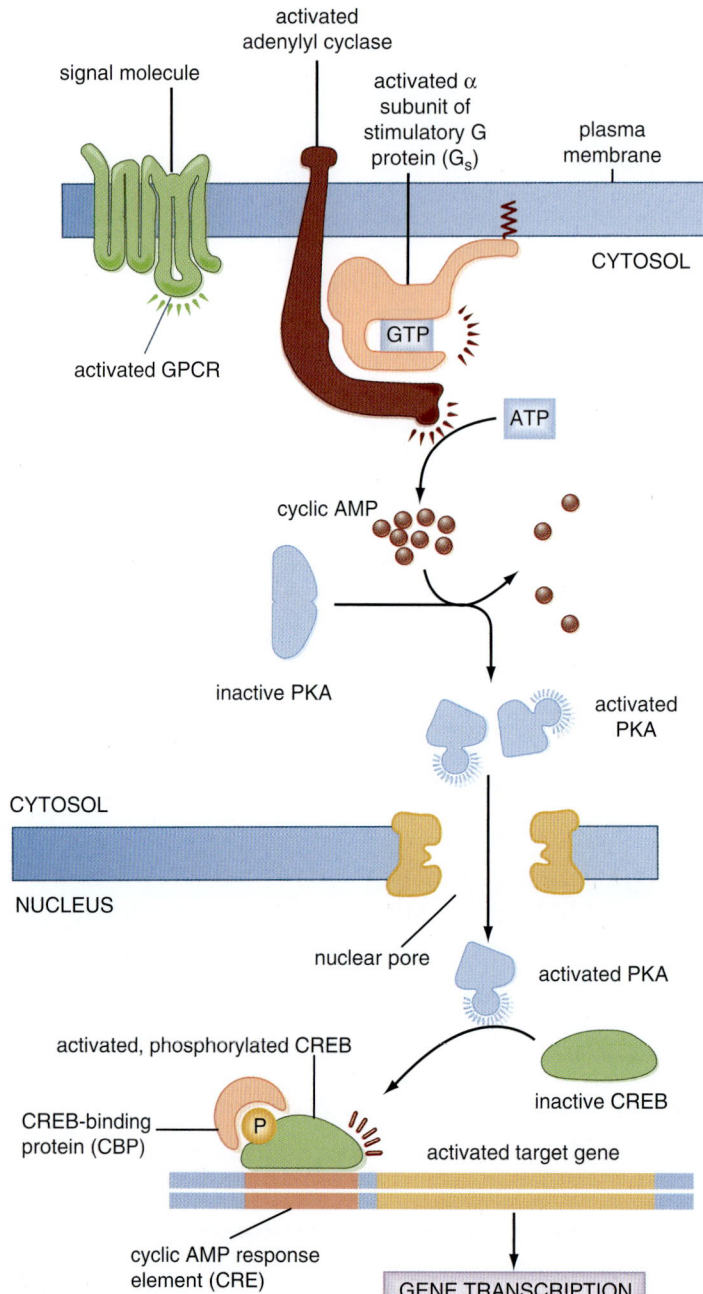

• **Fig. 3.10** GPCR Stimulation of Adenylyl Cyclase, cAMP, and Protein Kinase A. Binding of a signal molecule to a GPCR mediates Gs stimulation of adenylyl cyclase, which increases cytosolic cAMP, which in turn activates protein kinase A (PKA). Activated PKA phosphorylates a number of target proteins to elicit many effects. PKA also enters the nucleus where it phosphorylates CREB (cyclic adenosine monophosphate [cyclic AMP] response element–binding protein). Phosphorylated CREB recruits coactivator CBP, which stimulates gene transcription. (Redrawn from Alberts B, et al: *Molecular Biology of the Cell.* 6th ed. New York: Garland Science; 2015.)

by cAMP phosphodiesterases, which are inhibited by caffeine and other methylxanthines. Thus by interfering with a constitutive "off" signal, caffeine can prolong a cellular response mediated by cAMP and PKA. Because these effects target existing proteins, they can be extremely rapid (e.g., adrenaline response). In addition to cytoplasmic signaling, the catalytic subunit of PKA can enter the nucleus of cells and phosphorylate and activate the transcription factor **cAMP response element–binding (CREB) protein** (see Fig. 3.10). Phospho-CREB protein increases the transcription of many genes, which can in turn produce a distinct set of responses with much slower kinetics. Hence, cAMP has many cellular effects, including direct and indirect effects mediated by PKA.

IN THE CLINIC

Heterotrimeric G proteins also regulate phototransduction. In rod cells in the eye, absorption of light by rhodopsin activates the G protein transducin, which via the α_t subunit activates cGMP phosphodiesterase. Activation of this phosphodiesterase lowers the concentration of cGMP and thereby closes a cGMP-activated cation channel. The ensuing change in cation channel activity alters the membrane voltage. The exquisite sensitivity of rods to light—rods can detect a single photon of light—is due to the abundance of rhodopsin in rods and amplification of the signal (a photon of light) by the G protein–cGMP phosphodiesterase–cGMP channel signaling pathway (see Chapter 8 for more details).

Heterotrimeric G proteins also regulate **phospholipases,** a family of enzymes that modulate a variety of signaling pathways. Ligands that activate receptors that are coupled to the α_q subunit stimulate phospholipase C, an enzyme that converts phosphatidylinositol 4,5-biphosphate to inositol 1,4,5-triphosphate (InsP3) and diacylglycerol. InsP3 is a second messenger that diffuses to the endoplasmic reticulum, where it activates a ligand-activated Ca^{++} channel to release Ca^{++} into the cytosol, whereas diacylglycerol activates protein kinase C, which phosphorylates effector proteins. As noted earlier, both Ca^{++} and protein kinase C influence effector proteins, as well as other signaling pathways, to elicit responses.

Ligand binding to GPCRs can also activate **phospholipase A₂,** an enzyme that releases arachidonic acid from membrane phospholipids **Arachidonic acid,** which can also be released from diacylglycerol via an indirect pathway, can be released from cells and thereby regulate neighboring cells or stimulate inflammation. It can also be retained within cells, where it is incorporated into the plasma membrane or is metabolized in the cytosol to form intracellular second messengers that affect the activity of enzymes and ion channels. In one pathway, cytosolic **cyclooxygenases** facilitate the metabolism of arachidonic acid to prostaglandins, thromboxanes, and prostacyclins. Prostaglandins mediate aggregation of platelets, cause constriction of the airways, and induce inflammation. Thromboxanes also induce platelet aggregation and constrict blood vessels, whereas prostacyclin inhibits platelet aggregation and causes dilation of blood vessels. In a second pathway of arachidonic acid metabolism, the enzyme 5-lipoxygenase initiates the conversion of arachidonic acid to **leukotrienes,** which participate in allergic and inflammatory responses, including those causing asthma, rheumatoid arthritis, and inflammatory bowel disease. The third pathway of arachidonic acid metabolism is initiated by epoxygenase, an enzyme that facilitates the generation of hydroxyeicosatetraenoic acid (HETE) and cis-epoxyeicosatrienoic acid (cis-EET). HETE and cis-EET and their metabolites increase release of Ca^{++} from the endoplasmic reticulum, stimulate cell proliferation, and regulate inflammatory responses.

Ca^{++} is also an intracellular messenger that elicits cellular effects via Ca^{++}-binding proteins, most notably **calmodulin** (CaM). When Ca^{++} binds to CaM, its conformation is altered, and the structural change in CaM allows it to bind to and regulate other signaling proteins, including cAMP phosphodiesterase, an enzyme that degrades cAMP to AMP, which is inactive and unable to activate PKA. By binding to **CaM-dependent kinases,** CaM also phosphorylates specific serine and threonine residues in many proteins, including myosin light-chain kinase, which facilitates smooth muscle contraction (see Chapter 14).

Protein Phosphatases and Phosphodiesterases Counteract the Activation of Cyclic Nucleotide Kinases

There are two ways to terminate a signal initiated by cAMP and cGMP: enhancing degradation of these cyclic nucleotides by phosphodiesterases and dephosphorylation of effectors by protein **phosphatases.** Phosphodiesterases facilitate the breakdown of cAMP and cGMP to AMP and GMP, respectively, and are activated by ligand activation of GPCRs. Phosphatases dephosphorylate effector proteins that were phosphorylated by kinases such as PKA. The balance between kinase-mediated phosphorylation and phosphatase-mediated dephosphorylation allows rapid and exquisite regulation of the phosphorylated state and thus the activity of signaling proteins.

Enzyme Receptor–Linked Signal Transduction Pathways

There are several classes of receptors that have enzymatic activity or are intimately associated with proteins that have enzymatic activity. Four of these classes are discussed next, including receptors that mediate the cellular responses to atrial natriuretic peptide (ANP) and nitric oxide (**guanylyl cyclase receptors);** transforming growth factor-β (TGF-β; **threonine/serine kinase receptors**); EGF, platelet-derived growth factor (PDGF), and insulin (**tyrosine kinase receptors);** and interleukins (**tyrosine kinase–associated receptors**).

IN THE CLINIC

There are two isoforms of **cyclooxygenase: COX1** and **COX2.** When activated in endothelial cells, COX1 facilitates the production of prostacyclins, which inhibit blood clots (**thrombin**). In vascular smooth muscle cells and platelets, COX1 facilitates the production of thromboxane A_2, which is prothrombotic. Thus cardiovascular health depends in part on the balance between prostacyclins and thromboxane A_2, generated by distinct cell types. Low doses of **aspirin,** a **nonsteroidal antiinflammatory drug (NSAID),** reduce thromboxane A_2 production by platelets with little effect on endothelial prostacyclin production. Thus low-dose aspirin is antithrombotic (i.e., reduces blood clots). COX2 is activated by inflammatory stimuli. Thus the ability of NSAIDs (e.g., aspirin, ibuprofen, naproxen, acetaminophen, indomethacin) to suppress the inflammatory response is due to inhibition of COX2. Both COX1 and COX2 facilitate the production of prostanoids that protect the stomach. Several lines of evidence suggest that both COX1 and COX2 must be inhibited to elicit damage to the gastrointestinal tract. Consequently, the negative effects of NSAIDs on the gastric mucosa (e.g., increased incidence of gastrointestinal bleeding) are most likely due to inhibition of COX1 and COX2 by these nonselective COX inhibitors.

Selective COX2 inhibitors (e.g., celecoxib, rofecoxib) are very effective in selectively inhibiting COX2 and are used extensively to reduce the inflammatory response. Because COX2 inhibitors are thought to lack the negative effects elicited by NSAIDs on the gastrointestinal tract, their use has increased dramatically. However, in 2005, the U.S. Food and Drug Administration (FDA) announced that selective COX2 inhibitors, as well as nonselective NSAIDs, increase the risk for heart attacks and strokes and required that COX2-selective and nonselective NSAIDs carry a warning label on product packaging that highlighted the potential for the increased risk for adverse cardiovascular events and stroke. In addition, although much evidence suggests that COX2-selective inhibitors do not cause gastrointestinal bleeding, in 2005 the FDA also required the pharmaceutical industry to add to the warning label on COX2-selective drugs a caution about the potential for increased risk for gastrointestinal bleeding. In 2015, the FDA strengthened warnings that both COX2-selective and COX2-nonselective NSAIDs increase the risk of heart attacks and strokes.[b]

[b]See U.S. Food and Drug Administration. *FDA Drug Safety Communication: FDA Strengthens Warning That Non-aspirin Nonsteroidal Anti-inflammatory Drugs (NSAIDs) Can Cause Heart Attacks or Strokes.* <http://www.fda.gov/Drugs/DrugSafety/ucm451800.htm>; 2015 Accessed 25.07.16.

AT THE CELLULAR LEVEL

Ras GTPases, monomeric G proteins, are involved in many signaling pathways that control cell division, proliferation, and death. Many mutations of proteins in the Ras signaling pathway are oncogenic (cancer causing) or inactivate tumor suppressors. Mutations in *Ras* genes that inhibit GTPase activity, as well as overexpression of Ras proteins as a result of transcriptional activation, lead to continuous cell proliferation, a major step in the development of cancer in many organs, including the pancreas, colon, and lungs. In addition, mutations in and overexpression of GEFs, which facilitate exchange of GTP for GDP, and GTPase-activating proteins, which accelerate GTP hydrolysis, may also be oncogenic (see Fig. 3.8*B*).

nitric oxide, which increases cGMP and thereby relaxes smooth muscle in coronary arteries, it has long been used to treat **angina pectoris** (i.e., chest pain caused by inadequate blood flow to heart muscle; see Chapter 17).

Threonine/Serine Kinase Receptors

The TGF-β receptor is a threonine/serine kinase that has two subunits. Binding of TGF-β to the type II subunit induces it to phosphorylate the type I subunit on specific serine and threonine residues, which in turn, phosphorylates other downstream effector proteins on serine and threonine residues and thereby elicits cellular responses, including cell growth, cell differentiation, and apoptosis.

Tyrosine Kinase Receptors

There are two classes of tyrosine kinase receptors. Nerve growth factor (NGF) receptors are typical examples of one class. Ligand binding to two NGF receptors facilitates their dimerization and thus enables the cytoplasmic tyrosine kinase domain of each monomer to phosphorylate and activate the other monomer. Once the other monomer is phosphorylated, the cytoplasmic domains can recruit GEFs such as growth factor-receptor bound protein 2 to the plasma membrane, which in turn activates Ras and downstream kinases that regulate gene transcription programs important for cell survival and proliferation.

Activation of the insulin receptor (which is tetrameric and composed of two α and two β subunits) by insulin is an example of the other type of tyrosine kinase receptor. Binding of insulin to the α subunits produces a conformational change that facilitates interaction between the two α and β pairs. Binding of insulin to its receptor causes autophosphorylation of tyrosine residues in the catalytic domains of the β subunits, and the activated receptor then phosphorylates cytoplasmic proteins to initiate its cellular effects, including stimulating the absorption of glucose from the blood into skeletal muscle and fat tissue.

Tyrosine Kinase–Associated Receptors

The tyrosine kinase–associated receptors have no intrinsic kinase activity but associate with proteins that do have

Guanylyl Cyclase Receptors

ANP binds to the extracellular domain of the plasma membrane receptor guanylyl cyclase and induces a conformational change in the receptor that causes receptor dimerization and activation of guanylyl cyclase, which metabolizes GTP to cGMP. cGMP activates **cGMP-dependent protein kinase,** which phosphorylates proteins on specific serine and threonine residues. In the kidney, ANP inhibits reabsorption of sodium and water by the collecting duct (see Chapter 35).

Nitric oxide activates a soluble receptor guanylyl cyclase that converts GTP to cGMP, which relaxes smooth muscle. Because nitroglycerin increases blood concentrations of tyrosine kinase activity, including tyrosine kinases of the Src family and Janus family. Receptors in this class bind several cytokines, including interleukin-6, a proinflammatory cytokine that is necessary for resistance to bacterial infections, and erythropoietin, which stimulates the production of red blood cells. Tyrosine kinase–associated receptor subunits assemble into homodimers ($\alpha\alpha$), heterodimers ($\alpha\beta$), or heterotrimers ($\alpha\beta\gamma$) when ligands bind. Subunit assembly enhances the binding of tyrosine kinases, which induces kinase activity and thereby phosphorylates tyrosine residues on the kinases, as well as on the receptor. Most polypeptide growth factors bind to tyrosine-kinase-associated receptors.

Regulation of Gene Expression by Signal Transduction Pathways

Steroid and thyroid hormones, cAMP, and receptor tyrosine kinases are transcription factors that regulate gene expression and thereby participate in signal transduction pathways. This section discusses the regulation of gene expression by steroid and thyroid hormones, cAMP, and receptor tyrosine kinases.

Nuclear Receptor Signal Transduction Pathways

The family of nuclear receptors includes more than 30 genes and has been divided into two subfamilies on the basis of structure and mechanism of action: (1) steroid hormone receptors and (2) receptors that bind retinoic acid, thyroid hormones (iodothyronines), and vitamin D. When ligands bind to these receptors, the ligand-receptor complex activates transcription factors that bind to DNA and regulate the expression of genes (see Figs. 3.2B, 3.5, and 3.7).

The location of nuclear receptors varies. Glucocorticoid and mineralocorticoid receptors are located in the cytoplasm, where they interact with chaperones (i.e., heat shock proteins; see Fig. 3.2B). Binding of hormone to these receptors results in a conformational change that causes chaperones to dissociate from the receptor, thereby revealing a nuclear localization motif that facilitates translocation of the hormone-bound receptor complex to the nucleus. Estrogen and progesterone receptors are located primarily in the nucleus, and thyroid hormone and retinoic acid receptors are located in the nucleus bound to DNA.

When activated by hormone binding, nuclear receptors bind to specific DNA sequences in the regulatory regions of responsive genes called **hormone response elements.** Ligand-receptor binding to DNA causes a conformational change in DNA that initiates transcription. Nuclear receptors also regulate gene expression by acting as transcriptional repressors. For example, glucocorticoids suppress the **transcription activator protein-1 (AP-1)** and **nuclear factor κB,** which stimulate the expression of genes that cause inflammation. By this mechanism glucocorticoids reduce inflammation.

Cell-Surface Signal Transduction Pathways Control Gene Expression

As noted previously, cAMP is an important second messenger. In addition to its importance in activating PKA, which phosphorylates specific serine and threonine residues on proteins, cAMP stimulates the transcription of many genes, including those that code for hormones, including somatostatin, glucagon, and vasoactive intestinal polypeptide (see Fig. 3.10). Many genes activated by cAMP have a **cAMP response element (CRE)** in their DNA. Increases in cAMP stimulate PKA, which not only acts in the cytoplasm but also can translocate to the nucleus, where it phosphorylates **CREB** and thereby increases its affinity for **CREB-binding protein (CBP).** The CREB-CBP complex activates transcription. The response is terminated when PKA phosphorylates a phosphatase that dephosphorylates CREB (see Fig. 3.10).

Many growth factors, including EGF, PDGF, NGF, and insulin, bind to and activate enzyme-linked receptors that have tyrosine kinase activity. Activation of tyrosine kinases initiates a cascade of events that enhance the activity of the small GTP-binding protein Ras, which in a series of steps and intermediary proteins phosphorylates the **mitogen-activated protein kinase,** which then translocates to the nucleus and stimulates transcription of genes that stimulate cell growth.

Tyrosine kinase–associated receptors, as noted earlier, are activated by a variety of hormones, including cytokines, growth hormone, and interferon. Although these receptors do not have tyrosine kinase activity, they are associated with **Janus family proteins,** which do have tyrosine kinase activity. Once activated, hormone tyrosine kinase–associated receptors activate Janus family protein, which phosphorylates latent transcription factors called **signal transducers and activators of transcription (STATs).** When phosphorylated on tyrosine residues, STATs dimerize and then enter the nucleus and regulate transcription.

Key Points

1. The function of cells is tightly coordinated and integrated by external chemical signals, including hormones, neurotransmitters, growth factors, odorants, and products of cellular metabolism that serve as chemical messengers and provide cell-to-cell communication. Chemical and physical signals interact with receptors located in the plasma membrane, cytoplasm, and nucleus. Interaction of these signals with receptors initiates a cascade of events that mediate the response to each stimulus.

These pathways ensure that the cellular response to external signals is specific, amplified, tightly regulated, and coordinated.

2. There are two classes of GTP-binding proteins: monomeric G proteins and heterotrimeric G proteins composed of α, β, and γ subunits. Monomeric G proteins regulate actin cytoskeleton organization, cell cycle progression, intracellular vesicular transport, and gene expression. Heterotrimeric G proteins regulate ion channels, adenylyl cyclase and the cAMP-PKA signaling pathway, phosphodiesterases (which also regulate cAMP and cGMP signaling pathways), and phospholipases, which regulate the production of prostaglandins, prostacyclins, and thromboxanes.

3. There are four subtypes of enzyme-linked receptors that mediate the cellular response to a wide variety of signals, including ANP, nitric oxide, TGF-β, PDGF, insulin, and interleukins.

4. There are two types of nuclear receptors: (1) one type that in the absence of ligand is located in the cytoplasm and when bound to ligand translocates to the nucleus and (2) another type that permanently resides in the nucleus. Both classes of receptors regulate gene transcription.

Additional Readings

Journal Articles

Cheung E, Kraus WL. Genomic analyses of hormone signaling and gene regulation. *Annu Rev Physiol*. 2010;72:191-218.

Huang P, Chandra V, Rastinejad F. Structural overview of the nuclear receptor superfamily: insights into physiology and therapeutics. *Annu Rev Physiol*. 2010;72:247-272.

Levin ER. Extranuclear steroid receptors are essential for steroid hormone actions. *Annu Rev Med*. 2015;66:271-280.

Riccardi D, Kemp P. The calcium-sensing receptor beyond extracellular calcium homeostasis: conception, development, adult physiology, and disease. *Annu Rev Physiol*. 2012;74:271-297.

Wu H. Higher-order assemblies in a new paradigm of signal transduction. *Cell*. 2013;153:287-292.

Book Chapters

Cantley L. Signal transduction. In: Boron WF, Boulpaep EL, eds. *Medical Physiology*. 3rd ed. Philadelphia: Elsevier; 2016 [Chapter 3].

Caplan MJ. Functional organization of the cell. In: Boron WF, Boulpaep EL, eds. *Medical Physiology*. 3rd ed. Philadelphia: Elsevier; 2016 [Chapter 2].

Heald R. Cell signaling. In: Alberts B, Johnson A, Lewis J, et al., eds. *Molecular Biology of the Cell*. 6th ed. New York: Garland Science; 2015 [Chapter 15].

Igarashi P. Regulation of gene expression. In: Boron WF, Boulpaep EL, eds. *Medical Physiology*. 3rd ed. Philadelphia: Elsevier; 2016 [Chapter 4].

The Nervous System

ERIC J. LANG AND KALMAN RUBINSON

4

The Nervous System: Introduction to Cells and Systems

LEARNING OBJECTIVES

Upon completion of this chapter the student should be able to answer the following questions:

1. What are the major cell types of the central and peripheral nervous systems?
2. What are the major components of a neuron, and what are their functional roles?
3. What are the functional roles of the major glial cell types?
4. What are the main divisions of the central nervous system?
5. How and where is the cerebrospinal fluid formed, and how does it circulate and exit the ventricular system?
6. How is axon transport related to the response of the axon to transection?

The nervous system is a communications and control network that allows an organism to interact rapidly and adaptively with its environment, where environment includes both the external environment (the world outside the body) and the internal environment (the components and cavities of the body). To carry out its function the nervous system takes in sensory information from a variety of sources using specialized sensors (receptors), integrates this information with previously obtained information stored as memories and with the intrinsic goals and drives of the organism that have been embedded in its nervous system through evolution, decides on a course of action, and then issues commands to the effector organs (muscles and glands) to execute the chosen behavioral response.

Moreover, almost all behavioral responses require the coordination of many body parts. For example, even a simple reaching movement of the arm may require coactivation of axial muscles and possibly muscles in the lower extremity to maintain posture and balance, which themselves may be monitored by up to three different sensory systems (vision, vestibular, and proprioceptive) whose information has to be integrated. Furthermore, movements can alter the internal environment and thus can require compensatory changes in heart and breathing rates, blood vessel diameters, and other internal processes. All these variables are monitored and controlled by various specialized subsystems of the nervous system, all of which must work together for the organism to perform movements and more generally to survive. The succeeding chapters will describe these major subsystems individually; however, it should be remembered that in reality their activity is integrated to generate normal behavior.

To begin, it is useful to divide the nervous system into central and peripheral parts. The *central nervous system* (CNS) consists of the brain and spinal cord. The *peripheral nervous system* (PNS) consists of nerves and ganglia (small groups of neurons) that innervate all parts of the body and provide an interface between the environment and the CNS. The transition between the CNS and PNS occurs on the dorsal and ventral rootlets near to where they emerge from the spinal cord and on the cranial nerve fibers near to where they arise from the brain.

Cellular Components of the Nervous System

The nervous system is made up of cells, connective tissue, and blood vessels. The major cell types are **neurons** (nerve cells) and **glia** (neuroglia = "nerve glue"). In its most general form a neuron's function can be defined as generation of signals (to be sent to other neurons or effector cells [e.g., muscle cells]) based on an integration of its own electrical properties with electrochemical signals from other neurons. The points where specific neuron-to-neuron communication occurs are known as *synapses,* and the process of synaptic transmission is critical to neuronal function (see Chapter 6). Neuroglia, or just glia, traditionally have been characterized as supportive cells that sustain neurons both metabolically and physically, isolate individual neurons from each other, and help maintain the internal milieu of the nervous system; however, it is now known that they also have important roles in shaping the flow of activity through the nervous system.

Neurons

The typical neuron consists of three main cellular compartments: a *cell body* (also referred to as a *perikaryon* or

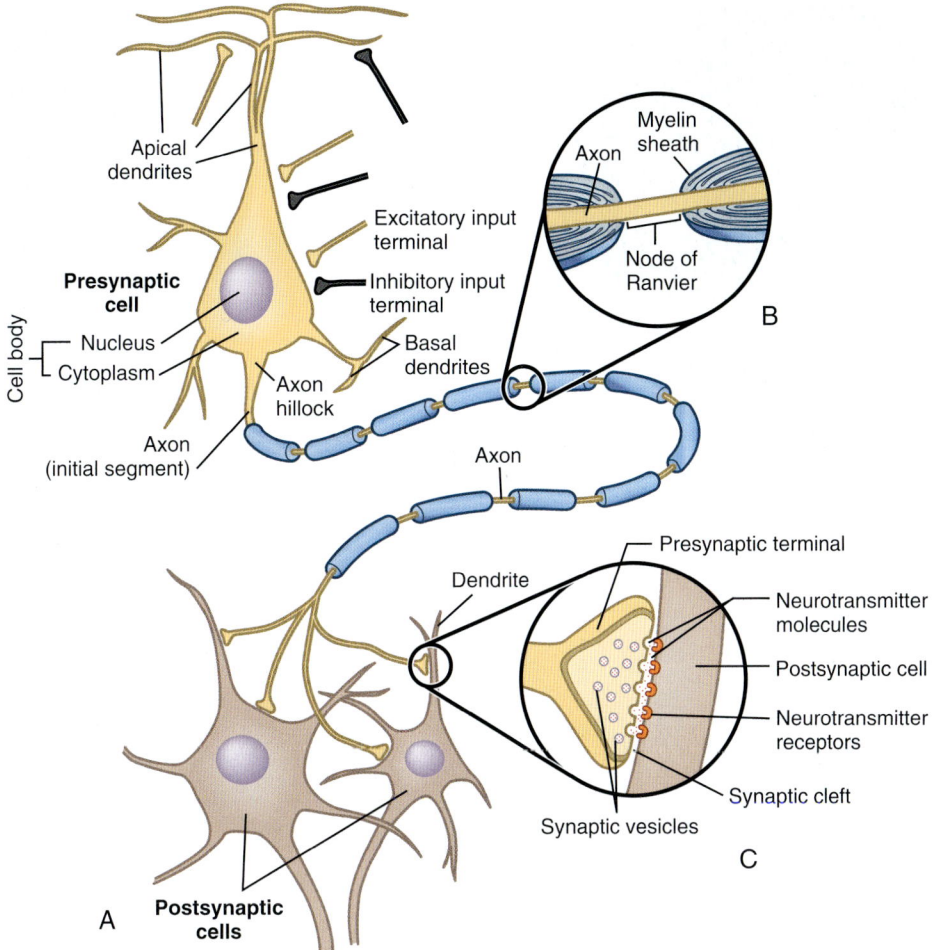

• **Fig. 4.1** Schematic diagram of an idealized neuron and its major components and connections. A, Afferent input from axons of other cells terminates in synapses on the dendrites and cell body. The initial segment of the axon attaches at the axon hillock. This axon is myelinated, as indicated by the blue structures that encapsulate segments of the axon. The axon terminates on two postsynaptic neurons by forming synaptic terminals. B, Nodes of Ranvier are the gaps between the myelin segments where the axon membrane is exposed to the extracellular space. C, Higher-magnification view of synapse. (Redrawn from Blumenfeld H. *Neuroanatomy Through Clinical Cases.* 2nd ed. Sunderland, MA: Sinauer Associates; 2010.)

soma), a variable number of processes that extend from the soma called *dendrites,* and an *axon* (Fig. 4.1). A tremendous number of morphological variants of this basic template exist, including cases where dendrites or an axon may be absent (Fig. 4.2). These variations do not occur randomly but rather relate to the distinct functional properties of each neuronal class. Indeed, neurons with similar morphologies often characterize specific regions of the CNS and reflect the distinct neuronal processing performed in each CNS region.

The cell body is the main genetic and metabolic center of the neuron. Correspondingly it contains the nucleus and nucleolus of the cell and also possesses a well-developed biosynthetic apparatus for manufacturing membrane constituents, synthetic enzymes, and other chemical substances needed for the specialized functions of nerve cells. The neuronal biosynthetic apparatus includes *Nissl bodies,* which are stacks of rough endoplasmic reticulum, and a prominent *Golgi apparatus.* The soma also contains numerous mitochondria and cytoskeletal elements, including neurofilaments and microtubules.

The cell body is also a region in which the neuron receives synaptic input (i.e., electrical and chemical signals from other neurons). Although quantitatively the synaptic input to the soma is usually much less than that to dendrites, it often differs qualitatively from dendritic inputs, and by virtue of the closeness of the soma to the axon, inputs to the soma can override those to the dendrites (see Chapter 6).

Dendrites are tapering and branching extensions of the soma and are the main direct recipients of signals from other neurons. They can be thought of as a way to expand and specialize the surface area of a neuron, and indeed, they may account for more than 90% of the surface area available for synaptic contact (soma plus dendrites). Dendrites can be divided into primary dendrites (those that extend directly from the soma) and higher-order dendrites (daughter branches extending from a more proximal branch, in

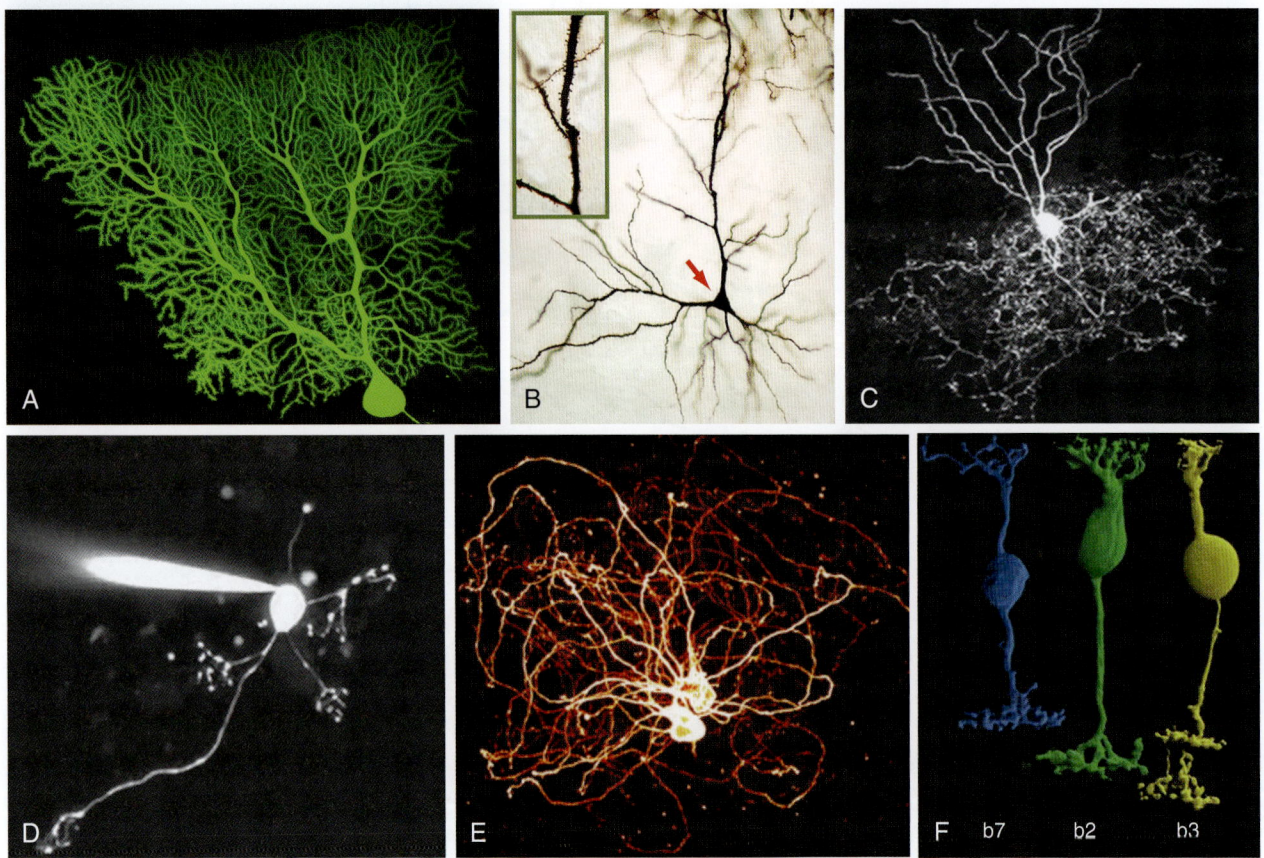

Fig. 4.2 A, Purkinje cell. B, Pyramidal cell. C, Golgi cell. D, Granule cell. E, Inferior olive cells. F, Bipolar cells. (A, Courtesy of Boris Barbour. B, Courtesy of T.F. Fletcher, from http://vanat.cvm.umn.edu/ neurHistAtls/pages/neuron3.html. C, Figure was provided by Court Hull and Wade Regehr, Department of Neurobiology, Harvard Medical School. D, From Delvendahl I et al. *Front Cell Neurosci* 2015;9:93, Fig. 1A. E, From Mathy A, Clark BA. In: Manto M et al [eds]. *Handbook of the Cerebellum and Cerebellar Disorders.* Dordrecht, Netherlands: Springer Science+Business Media Dordrecht; 2013. F, From Li W, DeVries SH. *Nat Neurosci* 2006;9:669-675, Fig. 2.)

which *proximal* refers to closeness to the soma). The main cytoplasmic organelles in dendrites are microtubules and neurofilaments; however, the primary dendrites can also contain Nissl bodies and parts of the Golgi apparatus.

A neuron's set of dendrites is termed its *dendritic tree.* Dendritic trees differ tremendously between different types of neurons in terms of the size, number, and spatial organization of the dendrites. A dendritic tree can consist of just a few unbranched dendrites or of many highly ramified dendrites. Individual dendrites can be longer than 1 mm or only 10 to 20 μm in length. Another major morphological variation is whether or not a dendrite has spines, which are small mushroom- or lollipop-shaped protrusions from the main dendrite. Spines are sites specialized for synaptic contact (usually, but not always) from excitatory inputs. The shape and size of the dendritic tree, as well as the population and distribution of channels in the dendritic membrane, are all important determinants of how the synaptic input will affect the neuron (see Chapter 6).

The axon is an extension of the cell that conveys the output of the cell to other neurons or, in the case of a motor neuron, to muscle cells as well. In general, each neuron has only one axon, and it is usually of uniform diameter. The length and diameter of axons vary with the neuronal type. Some axons do not extend much beyond the length of the dendrites, whereas others may be a meter or more long. Axons may have orthogonal branches en passant, but they often end in a spray of branches called a *terminal arborization* (represented by the four terminal branches and their synaptic terminals in Fig. 4.1A). The size, shape, and organization of the terminal arborization determine which other cells it will contact. The first part of the axon is known as the **initial segment** and arises from the soma (or sometimes from a proximal dendrite) in a specialized region called the *axon hillock.* The axon differs from the soma and proximal dendrites in that it lacks rough endoplasmic reticulum, free ribosomes, and a Golgi apparatus. The initial segment is usually the site where action potentials (spikes) that are propagated down the axon are initiated (see Chapter 5). An axon may terminate in a synapse and/or it may make synapses along its length. Synapses will be described in detail in Chapter 6.

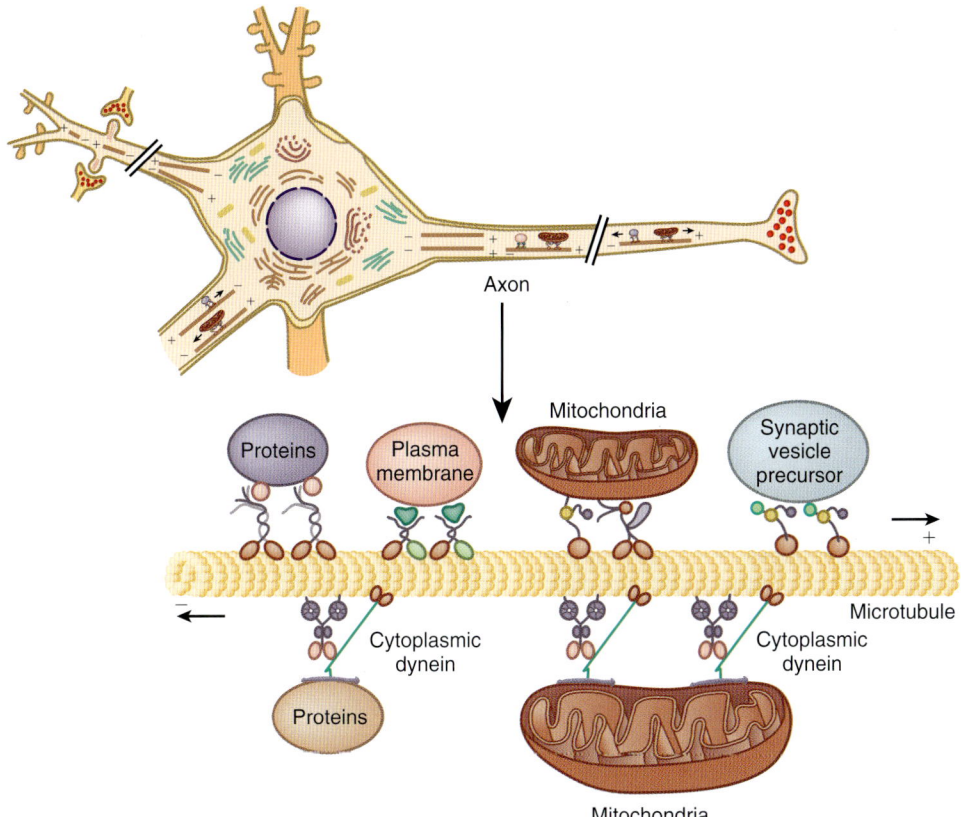

• Fig. 4.3 Axonal transport. Schematic of neuron and enlargement of axonal transport mechanism. Axonal transport depends on movement of material along transport filaments such as microtubules. Transported components attach to transport filaments by means of cross-bridges. Different objects are transported anterogradely (from cell body to axon terminal) and others retrogradely (toward the cell body). The direction of transport—retrograde and anterograde—is determined by specific proteins such as dynein and kinesin, respectively.

Neurons are special because of their ability to control and respond to electricity. Moreover, the response and control mechanisms of each part of a neuron are distinct from those in other parts. This intraneuronal specialization is a consequence of the particular morphology and the ion channel composition of each part of the neuron. For example, dendrites have ligand gated ion channels that allow neurons to respond to chemicals released by other neurons, and their characteristic branching pattern allows for integration of multiple input signals. In contrast the axon typically has a long length and high concentration of voltage-gated channels that allows it to convey electrical signals (action potentials) rapidly over long distances without alteration.

Axonal Transport

Because the soma is the metabolic engine of the neuron, substances needed to support axonal and synaptic function are synthesized there. These substances must be distributed to replenish secreted or inactivated materials along the axon and especially to the presynaptic terminals. Most axons are too long to allow efficient movement of substances from the soma to the synaptic endings by simple diffusion. Thus special axonal transport mechanisms have evolved

to accomplish this task (Fig. 4.3). A consequence of this metabolic dependency is that axons degenerate when disconnected from the cell body, a fact that has been used by scientists tracing out neuronal pathways; they would cut an axonal pathway and then determine where the degenerating axons distal to the cut projected to.

Several types of axonal transport exist. Membrane-bound organelles and mitochondria are transported relatively rapidly by fast axonal transport. Substances that are dissolved in cytoplasm (e.g., proteins) are moved by slow axonal transport. In mammals, fast axonal transport proceeds as rapidly as 400 mm/day, whereas slow axonal transport occurs at about 1 mm/day. Synaptic vesicles, which travel by fast axonal transport, can travel from the soma of a motor neuron in the spinal cord to a neuromuscular junction in a person's foot in about 2.5 days. In comparison the movement of some soluble proteins over the same distance can take nearly 3 years.

Axonal transport requires metabolic energy and involves calcium ions. Microtubules provide a system of guidewires along which membrane-bound organelles move (see Fig. 4.3). Organelles attach to microtubules through a linkage similar to that between the thick and thin filaments of skeletal muscle fibers. Ca^{++} triggers movement of the organelles along the microtubules. Special microtubule-associated motor proteins called *kinesin* and *dynein* are required for axonal transport.

Axonal transport occurs in both directions. Transport from the soma toward the axonal terminals is called *antero-grade axonal transport.* This process involves kinesin, and it allows replenishment of synaptic vesicles and enzymes responsible for the synthesis of neurotransmitters in synaptic terminals. Transport in the opposite direction, which is driven by dynein, is called *retrograde axonal transport.* This process returns recycled synaptic vesicle membrane to the soma for lysosomal degradation.

IN THE CLINIC

Certain viruses and toxins can be conveyed by axonal transport along peripheral nerves. For example, herpes zoster, the virus of chickenpox, invades dorsal root ganglion cells. The virus may be harbored by these neurons for many years. However, eventually the virus may become active because of a change in immune status. The virus may then be transported along the sensory axons to the skin, causing shingles, a very painful disease. Another example is the axonal transport of tetanus toxin. *Clostridium tetani* bacteria may grow in a dirty wound, and if the person had not been vaccinated against tetanus toxin, the toxin can be transported retrogradely in the axons of motor neurons. The toxin can escape into the extracellular space of the spinal cord ventral horn and block the synaptic receptors for inhibitory amino acids. This process can result in tetanic convulsions.

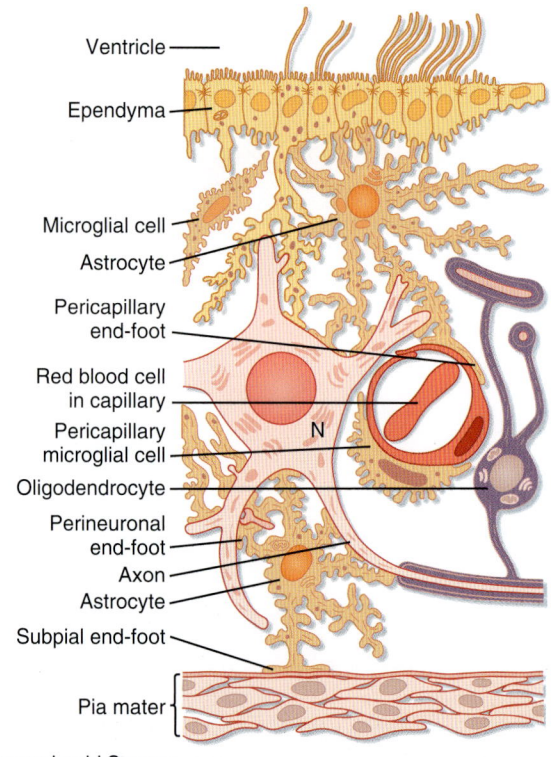

Ventricle
Ependyma
Microglial cell
Astrocyte
Pericapillary end-foot
Red blood cell in capillary
Pericapillary microglial cell
N
Oligodendrocyte
Perineuronal end-foot
Axon
Astrocyte
Subpial end-foot
Pia mater
Sub-arachnoid Space

• **Fig. 4.4** Schematic representation of cellular elements in the CNS. Two astrocytes are shown ending on a soma and dendrites of a neuron. Astrocytes also contact the pial surface or capillaries or both. An oligodendrocyte provides the myelin sheaths for axons. Also shown are microglia and ependymal cells. N, neuron. (Redrawn from Williams PL, Warwick R. *Functional Neuroanatomy of Man.* Edinburgh: Churchill Livingstone; 1975.)

Glia

The major nonneuronal cellular elements of the nervous system are the glia (Fig. 4.4). Glial cells in the human CNS outnumber neurons by an order of magnitude: there are about 10^{13} glia and 10^{12} neurons. Glial cells in the CNS include astrocytes, oligodendrocytes, microglia, and ependymal cells (see Fig. 4.4); in the PNS the glial cells are Schwann cells and satellite cells. Traditionally, glial cells were thought of as supportive cells, and consistent with that conception, their functions include regulation of the microenvironment and myelination of axons. Glial cells are now also recognized to be important determinants of the flow of signals through neuronal circuits, and related to that they act to modulate synaptic and nonsynaptic transmission and have important roles in synaptogenesis and maintenance.

Astrocytes (named for their star shape) help regulate the microenvironment of the CNS, both under normal conditions and in response to damage to the nervous system. Astrocytes have a cell body from which several main branches arise. Through repeated branching these main processes give rise to hundreds to thousands of branchlets. Astrocyte processes contact neurons and surround synaptic endings, isolating them from adjacent synapses and the general extracellular space. Astrocytes also have foot processes that contact the capillaries and connective tissue at the surface of the CNS, the pia mater (see Fig. 4.4). These foot processes may help mediate the entry of substances into the CNS. Astrocytes can actively take up K^+ ions and neurotransmitter substances, which they metabolize, biodegrade, or recycle. Thus astrocytes serve to buffer the extracellular environment of neurons with respect to both ions and neurotransmitters. The cytoplasm of astrocytes contains glial filaments that provide mechanical support for CNS tissue. After injury the astrocytes undergo a variety of changes to become reactive astrocytes. One example is a class of reactive astrocytes that act to form a glial scar around an area of focal damage, which segregates the damaged tissue and thereby allows inflammatory processes to act selectively at the site of damage, minimizing the impact on surrounding normal tissue. Astrocytes can also affect the properties of synaptic transmission, which is discussed in Chapter 6.

Oligodendrocytes and **Schwann cells** are critical for the function of axons. Many axons are surrounded by a myelin sheath, which is a spiral multilayered wrapping of glial cell membrane (Fig. 4.5A–B). In the CNS, myelin is formed by the oligodendrocytes, whereas in the PNS Schwann cells form myelin. Myelin increases the speed of action potential conduction, in part by restricting the

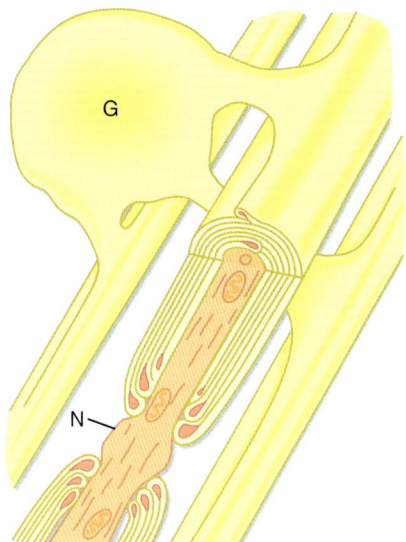

• **Fig. 4.5** Axonal/glial associations. A, Myelinated axons in the CNS. A single oligodendrocyte (G) emits several processes, each of which winds in a spiral fashion around an axon to form a myelin segment. The axon is shown in cutaway. The myelin from a single oligodendrocyte ends before the next wrapping from another oligodendrocyte. The bare axon between the myelinated segments is the node of Ranvier (N). B, Electron micrograph of myelinated axon in the PNS shown in cross section. The axon (Ax) is seen at the center within a sheath consisting of multiple wrappings of the Schwann cell's cytoplasmic membrane. The Schwann cell soma (SC) is at the upper right. mes, mesaxon, internal and external. C, Electron micrograph of unmyelinated axons in PNS. Nine axons *(asterisks)* cut in cross section are seen embedded in a Schwann cell whose nucleus is at the center (N). At lower right a portion of a myelinated axon is visible. (B, From Peters A et al. *The Fine Structure of the Nervous System.* New York: Oxford University Press; 1991, Fig. 6.5. C, From Pannese E. *Neurocytology.* 2nd ed. Basel, Switzerland: Springer International; 2015.)

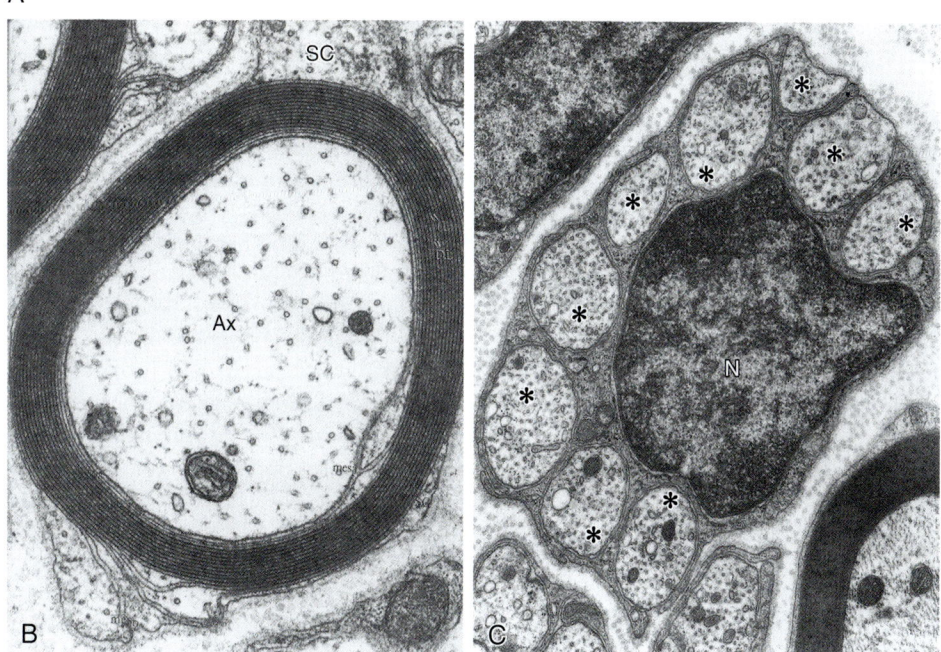

⚛ AT THE CELLULAR LEVEL

Astrocytes are coupled to each other by gap junctions such that they form a syncytium through which small molecules and ions can redistribute along their concentration gradients or by current flow. When normal neural activity gives rise to a local increase in extracellular [K^+], this coupled network can enable spatial redistribution of K^+ over a wide area via current flow in many astrocytes.

Under conditions of hypoxia, such as might be associated with ischemia secondary to blockage of an artery (i.e., a stroke), [K^+] in the extracellular space of a brain region can increase by a factor of as much as 20. This will depolarize neurons and synaptic terminals and result in release of transmitters such as glutamate, which will cause further release of K^+ from neurons. The additional release only exacerbates the problem and can lead to neuronal death. Under such conditions, local astroglia will probably take up the excess K^+ by K^+-Cl^- symport rather than by spatial buffering, because the elevation in extracellular [K^+] tends to be widespread rather than local.

flow of ionic current to small unmyelinated portions of the axon between adjacent glial cells, the nodes of Ranvier (see Chapter 5). Although both act to increase the speed of conduction, there are several important differences in the relationship between axons and either oligodendrocytes or Schwann cells. One major difference is that a single oligodendrocyte typically helps myelinate multiple axons in the CNS, whereas each Schwann cell helps myelinate only a single axon in the PNS. A second difference is that in the CNS, unmyelinated axons are bare, whereas in the PNS, unmyelinated axons are not. Rather, they are surrounded by Schwann cell processes; the Schwann cell, however, does not form a multilayered covering (i.e., myelin), but instead extends processes that surround parts of several axons (the Schwann cell with its set of unymyelinated axons is called a *Remak bundle*) (see Fig. 4.5C).

Satellite cells encapsulate dorsal root and cranial nerve ganglion cells and regulate their microenvironment in a fashion similar to that of astrocytes.

Microglia are derived from erythromyeloid stem cells that migrate into the CNS early in development. They play an important role in immune responses within the CNS. When the CNS is damaged, microglia help remove the cellular products of the damage by phagocytosis. They are assisted by other glia and by other phagocytes that invade the CNS from the circulation. In addition to their role in immune responses, recent evidence suggests they are also active in healthy brain tissue and may have important roles in normal brain development and function, including pruning of excess synapses that are formed during development and synaptic plasticity.

Ependymal cells form the epithelium lining the ventricular spaces of the brain, which contain (cerebrospinal fluid) CSF. CSF is secreted in large part by specialized ependymal cells of the choroid plexuses located in the ventricular system. Many substances diffuse readily across the ependyma, which lies between the extracellular space of the brain and the CSF.

 IN THE CLINIC

Most neurons in the adult nervous system are postmitotic cells (although some stem cells may also remain in certain sites in the brain). Many glial precursor cells are present in the adult brain, and they can still divide and differentiate. Thus the cellular elements that give rise to most intrinsic brain tumors in the adult brain are the glial cells. For example, brain tumors can be derived from astrocytes (which vary in malignancy from the slowly growing astrocytoma to the rapidly fatal glioblastoma multiforme), from oligodendroglia (oligodendroglioma), or from ependymal cells (ependymoma). Meningeal cells can also give rise to slowly growing tumors (meningiomas) that compress brain tissue, as can Schwann cells (e.g., acoustic schwannomas, which are tumors formed by Schwann cells of the eighth cranial nerve). In the brain of infants, neurons that are still dividing can sometimes give rise to neuroblastomas (e.g., of the roof of the fourth ventricle) or retinoblastomas (in the eye).

The Peripheral Nervous System

The PNS provides an interface between the environment and the CNS, both for sensory information flowing to the CNS and for motor commands issued from the CNS. It includes sensory (or primary afferent) neurons, somatic motor neurons, and autonomic motor neurons.

Sensory pathways into the nervous system start with a receptor, which may simply be a specialized part of an axon in the PNS or may include additional cells. Each sensory receptor is organized so that it transduces a specific type of energy into an electrical signal. Thus they can be classified in terms of the type of energy they transduce (e.g., photoreceptors transduce light, mechanoreceptors transduce displacement and force). They may also be classified according to the source of the input (e.g., exteroceptors signal external events, proprioceptors signal the state of a body part such as the angle of elbow, and interoceptors signal the distension of the gut).

The transduction process leads to an electrical response in the primary afferent called a *receptor potential*, which triggers action potentials in the primary afferent fibers innervating the receptor. These action potentials contain information about the sensory stimulus that is conveyed to the CNS via the primary afferent.

Somatic and autonomic motor neurons convey signals from the CNS to their respective effector targets. The somatic motor neurons innervate the skeletal muscles throughout the body. Their cell bodies lie in the ventral horn (or equivalent brainstem nuclei) and project out of the CNS via a ventral root or cranial nerve. The details of their relationship to muscles is covered in Chapter 9. The autonomic motor pathway is responsible for controlling the functioning of organs, smooth muscle, and glands. It is actually a two-neuron pathway, and its properties are covered in Chapter 11.

The Central Nervous System

The CNS is built from the cellular elements just described and includes the spinal cord and brain (Fig. 4.6A). These cellular elements are connected in a variety of complex ways to form the subsystems that underlie the multitude of functions performed by the CNS. The physiology of these systems is covered in Chapters 7 through 11; however, a basic knowledge of CNS anatomy is needed to understand systems physiology and will be briefly discussed here.

Regions of the CNS containing high concentrations of axon pathways (and very few neurons) are called **white matter** because the axonal myelin sheaths of the axons are highly refractive to light. Regions containing high concentrations of neurons and dendrites are by contrast called **gray matter.** Note that axons are also present in gray matter. These axons may be related to local processing (i.e., either originating from local neurons or terminating on them) or may be fibers of passage. Thus effects of damage to an area may reflect either loss of local function or disconnection

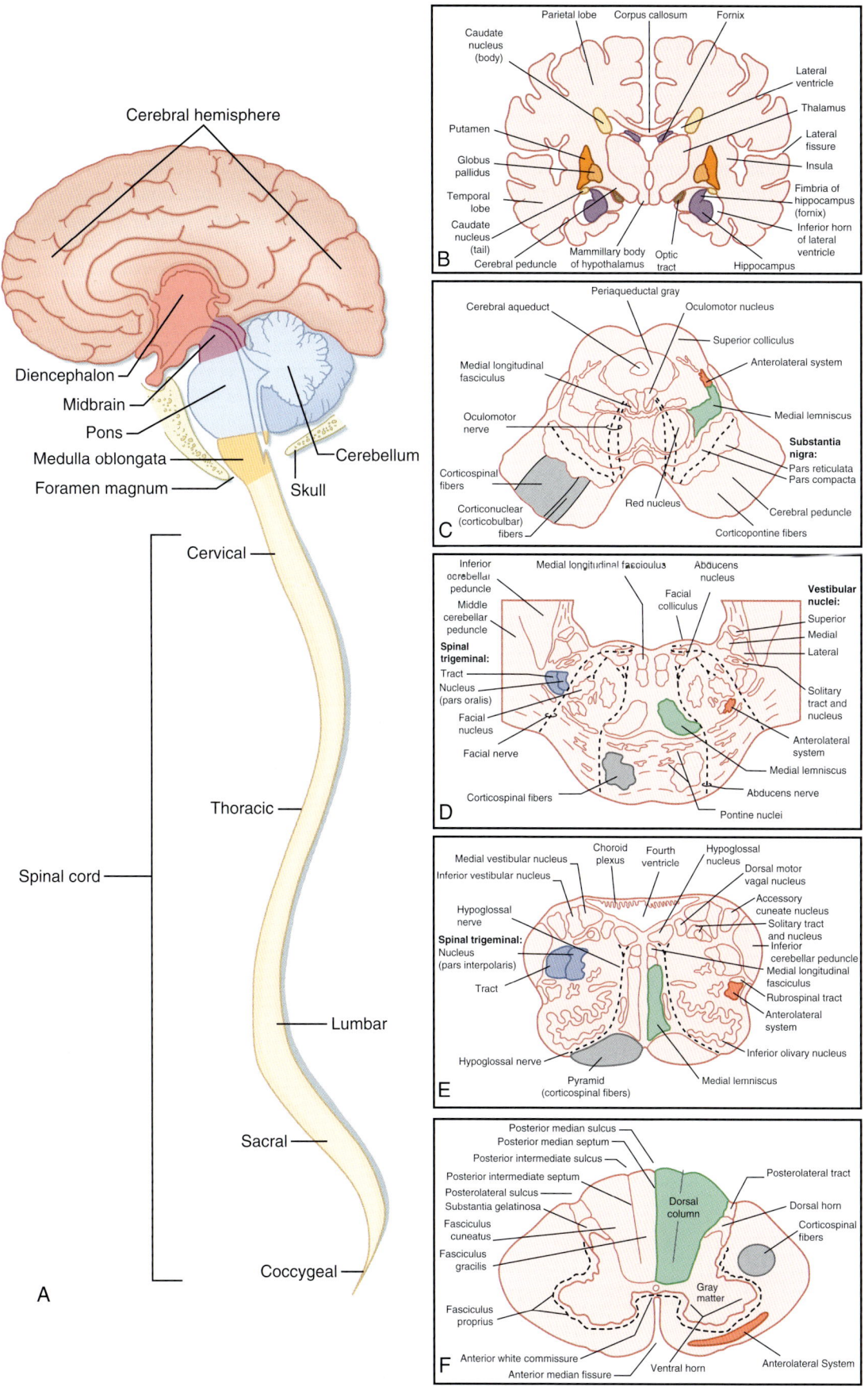

• Fig. 4.6 A, Schematic of the major components of the CNS as shown in a longitudinal midline view. B–F, Representative sections through the brain and spinal cord, with the major landmarks labeled. B, Cerebrum and thalamus; C, midbrain; D, pons; E, medulla; F, cervical spinal cord. Note that many pathways (e.g., corticospinal fibers) cross sides (decussate) as they travel through the CNS, but these descussations are not indicated in the figure (see Chapters 7 and 9 for details on the motor and sensory pathway crossings). (A, From Haines DE [ed]. *Fundamental Neuroscience for Basic and Clinical Applications.* 3rd ed. Philadelphia: Churchill Livingstone; 2006.)

of remote regions that had been linked by fibers of passage through the area that was damaged.

In the CNS, axons often travel in bundles or tracts. The names applied to tracts usually describe their origin and termination. For example, the spinocerebellar tracts convey information from the spinal cord to the cerebellum. The term **pathway** is similar to tract but is generally used to suggest a particular function (e.g., the auditory pathway: a series of neuron-to-neuron links across several synapses that convey and process auditory information).

Gray matter exists in two main configurations in the CNS. A **nucleus** is a group of neurons in the CNS (in the PNS such a grouping is called a **ganglion**). Examples include the thalamic, cerebellar, and cranial nerve nuclei. A **cortex** is neurons that are organized into layers and usually found on the surface of the CNS. The most prominent are the cerebral and cerebellar cortices, which cover the surface of the cerebral hemispheres and the cerebellum, respectively (Fig. 4.7).

In most nuclei and cortices, one can classify neurons into two broad categories: projection cells and local interneurons. Projection cells are neurons that send their axon to another region and thus are the origins of the various tracts of the nervous system. In contrast, local interneurons have axons that terminate in the same neural structure as their cell of origin and are involved with local computations rather than conveying signals from one region to another. These categories are not exclusive; many neurons have axons that both give off local branches and project to one or more distant regions.

Regional Anatomy of the CNS

The **spinal cord** can be subdivided into a series of regions (see Fig. 4.6A), each composed of a number of segments named for the vertebrae where their nerve roots enter or leave: 8 cervical, 12 thoracic, 5 lumbar, 5 sacral, and 1 coccygeal. Each portion maintains its tubular appearance, although its lumen, the spinal canal, may not remain patent. Within the gray matter the dorsal horn receives and processes sensory information from the dorsal roots, whereas the ventral horn is primarily a motor structure and contains the motor neurons whose axons project out via the ventral roots (Fig. 4.8).

The surrounding white matter consists of many tracts interconnecting spinal cord levels and for communication with the brain. Three major ones are the lateral corticospinal tract (motor), spinothalamic tract/anterolateral system (sensory), and dorsal column-medial lemniscus pathway (sensory) (see Fig. 4.6F).

The brainstem consists of the **medulla, pons,** and **midbrain** (Fig. 4.9; also see Fig. 4.6). In addition to the longitudinal pathways interconnecting with the spinal cord, the brainstem contains nuclei and many additional pathways that vary by level. These structures have many functions, some of which are analogous to those of the spinal cord (e.g., conveying basic sensory information and motor commands) and others related to a variety of other brain functions, such as cardiac control and state of consciousness. The brainstem also receives input and sends motor output via cranial nerves (Table 4.1).

The **cerebellum** sits dorsal to the pons and medulla. It receives inputs from spinal cord, brainstem, and cerebral cortex and projects back to many of these same structures. The cerebellum is critical for motor coordination but is increasingly recognized as having key roles in other cognitive functions.

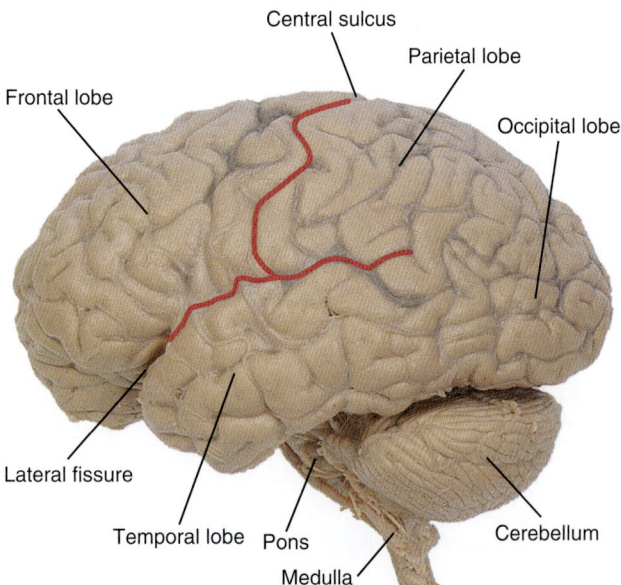

• **Fig. 4.7** Lateral view of the human brain showing the left cerebral hemisphere, cerebellum, pons, and medulla. Note the division of the lobes of the cerebrum (frontal, parietal, occipital, and temporal) and the two major fissures (lateral and central). (From Nolte J, Angevine J. *The Human Brain in Photographs and Diagrams.* 2nd ed. St Louis: Mosby; 2000.)

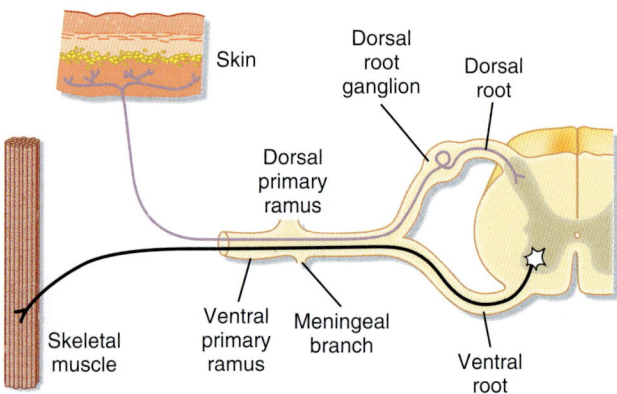

• **Fig. 4.8** Diagram of the spinal cord, spinal roots, and spinal nerve. The spinal nerve begins where the dorsal and ventral roots fuse, and has multiple branches (rami), the first few of which are represented. A primary afferent neuron is shown with its cell body in the dorsal root ganglion and its central and peripheral processes distributed, respectively, to the spinal cord gray matter and to a sensory receptor in the skin. An α motor neuron is shown to have its cell body in the spinal cord gray matter and to project its axon out the ventral root to innervate a skeletal muscle fiber.

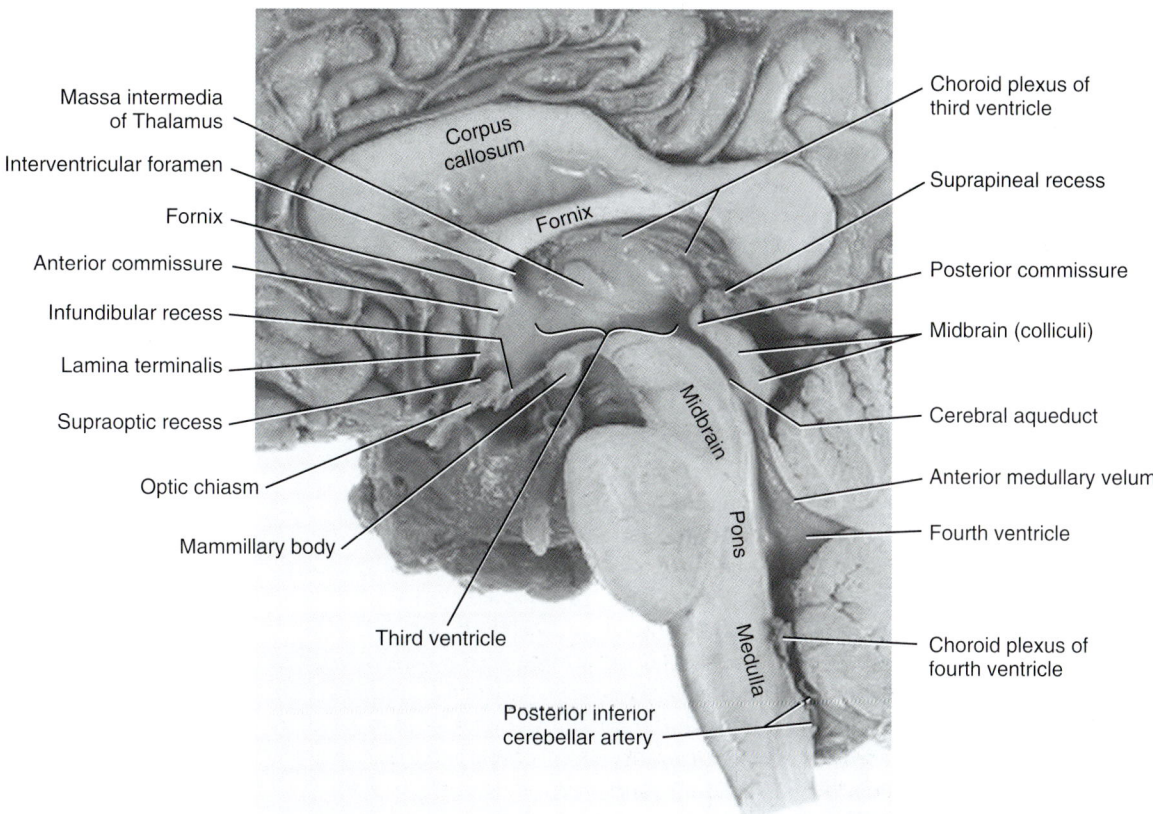

Massa intermedia of Thalamus
Interventricular foramen
Fornix
Anterior commissure
Infundibular recess
Lamina terminalis
Supraoptic recess
Optic chiasm
Mammillary body
Corpus callosum
Fornix
Third ventricle
Midbrain
Pons
Medulla
Posterior inferior cerebellar artery
Choroid plexus of third ventricle
Suprapineal recess
Posterior commissure
Midbrain (colliculi)
Cerebral aqueduct
Anterior medullary velum
Fourth ventricle
Choroid plexus of fourth ventricle

• **Fig. 4.9** Midsagittal view of the brain showing the third and fourth ventricles, the cerebral aqueduct of the midbrain, and the choroid plexus. The CSF formed by the choroid plexus in the lateral ventricles enters this circulation via the interventricular foramen. Note also the location of the corpus callosum and other commissures. (From Haines DE [ed]: *Fundamental Neuroscience for Basic and Clinical Applications.* 3rd ed. Philadelphia: Churchill Livingstone; 2006.)

TABLE 4.1	Parts and Functions of the Central Nervous System	
Region	**Nerves (Input/Output)**	**General Functions of the Region**
Spinal cord	Dorsal/ventral roots	Sensory input, reflex circuits, somatic and autonomic motor output
Medulla	Cranial nerves VIII–XII	Cardiovascular and respiratory control, auditory and vestibular input, brainstem reflexes
Pons	Cranial nerves V–VIII	Respiratory/urinary control, control of eye movement, facial sensation/motor control
Cerebellum	Cranial nerve VIII	Motor coordination, motor learning, equilibrium
Midbrain	Cranial nerves III–IV	Acoustic relay and mapping, control of the eye (including movement, lens and pupillary reflexes), pain modulation
Thalamus	Cranial nerve II	Sensory and motor relay to the cerebral cortex, regulation of cortical activation, visual input
Hypothalamus		Autonomic and endocrine control, motivated behavior
Basal ganglia		Shape patterns of thalamocortical motor inhibition
Cerebral cortex	Cranial nerve I	Sensory perception, cognition, learning and memory, motor planning and voluntary movement, language

The **thalamus** sits at the upper end of the brainstem and is enclosed by the **cerebrum** with which it is highly interconnected (see Fig. 4.6B). With a few exceptions, ascending information first reaches the thalamus, which conveys it to the cerebral cortex. These structures play a major role in many functions, including conscious awareness, volition, memory, and language. In addition to the cortex, the cerebrum contains a group of deep nuclei, the **basal ganglia,** that are interconnected with the cortex and thalamus and whose function will be described in Chapter 9.

The major functions of the different parts of the CNS are listed in Table 4.1.

Cerebrospinal Fluid

CSF fills the ventricular system, a series of interconnected spaces within the brain, and the subarachnoid space directly surrounding the brain. The intraventricular CSF reflects the composition of the brain's extracellular space via free exchange across the ependyma, and the brain "floats" in the subarachnoid CSF to minimize the effect of external mechanical forces. The volume of CSF within the cerebral ventricles is approximately 30 mL, and that in the subarachnoid space is about 125 mL. Because about 0.35 mL of CSF is produced each minute, CSF is turned over more than three times daily.

CSF is a filtrate of capillary blood formed largely by the choroid plexuses, which comprise pia mater, invaginating capillaries, and ependymal cells specialized for transport. The choroid plexuses are located in the lateral, third, and fourth ventricles (see Fig. 4.9). The lateral ventricles are situated within the two cerebral hemispheres. They each connect with the third ventricle through one of the interventricular foramina (of Monro). The third ventricle lies in the midline between the diencephalon on the two sides. The cerebral aqueduct (of Sylvius) traverses the midbrain and connects the third ventricle with the fourth ventricle. The fourth ventricle is a space defined by the pons and medulla below and the cerebellum above. The central canal of the spinal cord continues caudally from the fourth ventricle, although in adult humans the canal is not fully patent and continues to close with age.

CSF escapes from the ventricular system through three apertures or foramina (a medial foramen of Magendie and two lateral foramina of Luschka) located in the roof of the fourth ventricle. After leaving the ventricular system, CSF circulates through the subarachnoid space that surrounds the brain and spinal cord. Regions where these spaces are expanded are called *subarachnoid cisterns.* An example is the lumbar cistern, which surrounds the lumbar and sacral spinal roots below the level of termination of the spinal cord. The lumbar cistern is the target for lumbar puncture, a clinical procedure to sample CSF. A large part of CSF is removed by bulk flow through the valvular arachnoid granulations into the dural venous sinuses in the cranium.

Because the extracellular fluid within the CNS communicates with the CSF, the composition of the CSF is a useful indicator of the composition of the extracellular environment of neurons in the brain and spinal cord. The main constituents of CSF in the lumbar cistern are listed in Table 4.2. For comparison the concentrations of the same constituents in blood are also given. CSF has a lower concentration of K^+, glucose, and protein but a greater concentration of Na^+ and Cl^- than blood does. Furthermore, CSF contains practically no blood cells. The increased concentration of Na^+ and Cl^- enables CSF to be isotonic to blood.

The pressure in the CSF column is about 120 to 180 mm H_2O when a person is recumbent. The rate at which CSF is formed is relatively independent of the pressure in the

TABLE 4.2	Constituents of Cerebrospinal Fluid and Blood	
Constituent	Lumbar CSF	Blood
Na^+ (mEq/L)	148	136–145
K^+ (mEq/L)	2.9	3.5–5
Cl^- (mEq/L)	120–130	100–106
Glucose (mg/dL)	50–75	70–100
Protein (mg/dL)	15–45	6.8×10^3
pH	7.3	7.4

From Willis WD, Grossman RG. *Medical Neurobiology.* 3rd ed. St Louis: Mosby; 1981.

ventricles and subarachnoid space, as well as systemic blood pressure. However, the absorption rate of CSF is a direct function of CSF pressure.

 IN THE CLINIC

Obstruction of the circulation of CSF leads to increased CSF pressure and hydrocephalus, an abnormal accumulation of fluid in the cranium. In hydrocephalus the ventricles become distended, and if the increase in pressure is sustained, brain substance is lost. When the obstruction is within the ventricular system or in the foramina of the fourth ventricle, the condition is called a *noncommunicating hydrocephalus.* If the obstruction is in the subarachnoid space or the arachnoid villi, it is known as a *communicating hydrocephalus.*

The Blood-Brain Barrier

The local environment of most CNS neurons is controlled such that neurons are normally protected from extreme variations in the composition of the extracellular fluid that bathes them. Part of this control is provided by the presence of a blood-brain barrier (other mechanisms are the buffering functions of glia, regulation of CNS circulation, and exchange of substances between the CSF and extracellular fluid of the CNS). Movement of large molecules and highly charged ions from blood into the brain and spinal cord is severely restricted. The restriction is at least partly due to the barrier action of the capillary endothelial cells of the CNS and the tight junctions between them. Astrocytes may also help limit the movement of certain substances. For example, astrocytes can take up potassium ions and thus regulate $[K^+]$ in the extracellular space. Some pharmaceutical agents, such as penicillin, are removed from the CNS by transport mechanisms.

Nervous Tissue Reactions to Injury

Injury to nervous tissue elicits responses by neurons and glia. Severe injury causes cell death. Except in specific instances, once a neuron is lost it cannot be replaced because, in general,

IN THE CLINIC

The blood-brain barrier can be disrupted by pathology of the brain. For example, brain tumors may allow substances that are otherwise excluded to enter the brain from the circulation. Radiologists can exploit this by introducing a substance into the circulation that normally cannot penetrate the blood-brain barrier. If the substance can be imaged, its leakage into the region occupied by the brain tumor can be used to demonstrate the distribution of the tumor.

neurons are postmitotic cells. In animals, two exceptions are olfactory bulb and hippocampal neurons; however, in humans, only for the hippocampus has evidence been found for significant levels of neurogenesis in the adult CNS.

Degeneration

When an axon is transected, the soma of the neuron may show chromatolysis, or "axonal reaction." Normally, Nissl bodies stain well with basic aniline dyes, which attach to the RNA of ribosomes (Fig. 4.10A). After injury to the axon (see Fig. 4.10B), the neuron attempts to repair the axon by making new structural proteins, and the cisterns of the rough endoplasmic reticulum become distended with the products of protein synthesis. The ribosomes appear to be disorganized, and the Nissl bodies are stained weakly by basic aniline dyes. This process, called *chromatolysis,* alters the staining pattern (see Fig. 4.10C). In addition, the soma may swell and become rounded, and the nucleus may assume an eccentric position. These morphological changes reflect the cytological processes that accompany increased protein synthesis.

Because it cannot synthesize new protein, the axon distal to the transection dies (see Fig. 4.10C). Within a few days the axon and all the associated synaptic endings disintegrate. If the axon had been a myelinated axon in the CNS, the myelin sheath would also fragment and eventually be removed by phagocytosis. However, in the PNS the Schwann cells that had formed the myelin sheath remain viable, and in fact they undergo cell division. This sequence of events was originally described by Waller and is called *wallerian degeneration.*

If the axons that provide the sole or predominant synaptic input to a neuron or to an effector cell are interrupted, the postsynaptic cell may undergo transneuronal degeneration

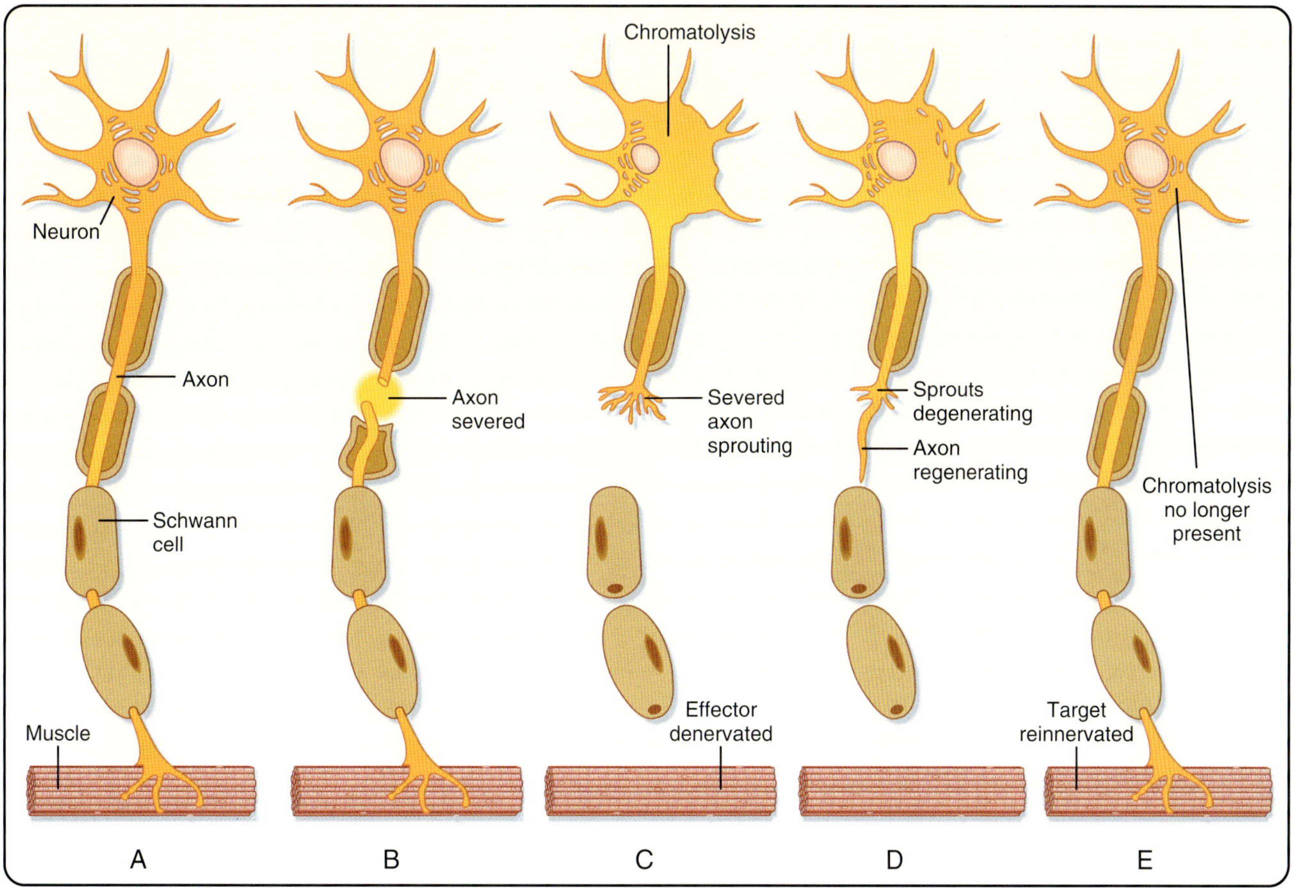

• **Fig. 4.10** A, Normal motor neuron innervating a skeletal muscle fiber. B, A motor axon has been severed, and the motor neuron is undergoing chromatolysis. C, This is associated in time with sprouting and, in D, with regeneration of the axon. The excess sprouts degenerate. E, When the target cell is reinnervated, chromatolysis is no longer present.

and even death. The best known example of this is atrophy of skeletal muscle fibers after their innervation by motor neurons has been interrupted. However, if only one or a few of the innervating axons are removed, the other surviving axons may sprout additional terminals, thereby taking up the synaptic space of the damaged axons and increasing their influence on the postsynaptic cell.

Regeneration

In the PNS, after an axon is lost through injury, many neurons can regenerate a new axon. The proximal stump of the damaged axon develops sprouts (see Fig. 4.10C), these sprouts elongate, and they grow along the path of the original nerve if this route is available (see Fig. 4.10D). The Schwann cells in the distal stump of the nerve not only survive the wallerian degeneration but also proliferate and form rows along the course previously taken by the axons. Growth cones of the sprouting axons find their way along these rows of Schwann cells, and they may eventually reinnervate the original peripheral target structures (see Fig. 4.10E). The Schwann cells then remyelinate the axons. The rate of regeneration is limited by the rate of slow axonal transport to about 1 mm/day.

In the CNS, transected axons also sprout. However, proper guidance for the sprouts is lacking, in part because the oligodendroglia do not form a path along which the sprouts can grow. This limitation may be a consequence of the fact that a single oligodendroglial cell myelinates many central axons, whereas a single Schwann cell provides myelin for only a single axon in the periphery. In addition, different chemical signals may affect peripheral and central attempts at regeneration differently. Other obstacles to successful CNS regeneration include formation of a glial scar by astrocytes and lack of trophic influences that guided axonal trajectories during development.

Key Concepts

1. The functions of the nervous system include excitability, sensory detection, information processing, and behavior.
2. The CNS includes the spinal cord and brain. The brain includes the medulla, pons, cerebellum, midbrain, thalamus, hypothalamus, basal ganglia, and cerebral cortex.
3. The neuron is the functional unit of the nervous system. Neurons have three major compartments: the dendrites, cell body, and axon. The first two receive and integrate signals, and the axon conveys the output signals of the neuron to other cells.
4. The PNS includes primary afferent neurons and the sensory receptors they innervate, the axons of somatic motor neurons, and autonomic neurons.
5. Information is conveyed through neural circuits by action potentials in the axons of neurons and by synaptic transmission between axons and the dendrites and somas of other neurons or between axons and effector cells.
6. Different types of neurons are specialized as a consequence of their individual morphology and the ion channel distribution in the cell membrane of their soma, dendrites, and axons.
7. Sensory receptors include exteroceptors, interoceptors, and proprioceptors. *Stimuli* are environmental events that excite sensory receptors, *responses* are the effects of stimuli, and *sensory transduction* is the process by which stimuli are detected by transforming their energy into electrical signals.
8. Sensory receptors can be classified in terms of the type of energy they transduce or by the source of the input. Central pathways are usually named by their origin and termination or for the type of information conveyed.
9. Chemical substances are distributed along the axons by fast or slow axonal transport. The direction of axonal transport may be anterograde or retrograde.
10. Glial cells include astrocytes (regulate the CNS microenvironment), oligodendroglia (form CNS myelin), Schwann cells (form PNS myelin), ependymal cells (line the ventricles), and microglia (CNS macrophages). Myelin sheaths increase the conduction velocity of axons.
11. The choroid plexus forms CSF. CSF differs from blood in having a lower concentration of K^+, glucose, and protein and a higher concentration of Na^+ and Cl^-; CSF normally lacks blood cells.
12. The extracellular fluid composition of the CNS is regulated by CSF, the blood-brain barrier, and astrocytes.
13. Damage to the axon of a neuron causes an axonal reaction (chromatolysis) in the cell body and wallerian degeneration of the axon distal to the injury. Regeneration of PNS axons is more likely than regeneration of CNS axons.

Additional Reading

Kiernan JA, Rajakumar N. *Barr's The Human Nervous System: An Anatomical Viewpoint*. 10th ed. Philadelphia: Lippincott Williams & Wilkins; 2013.

Squire L, Berg D. *Fundamental Neuroscience*. 4th ed. Waltham, MA: Academic Press; 2012.

Vanderah T, Gould D. *Nolte's The Human Brain: An Introduction to its Functional Anatomy*. 7th ed. St. Louis: Mosby; 2015.

5

Generation and Conduction of Action Potentials

LEARNING OBJECTIVES

Upon completion of this chapter, the student should be able to answer the following questions:

1. How is a nerve membrane's response to small-amplitude stimuli like a passive electric circuit comprising batteries, resistors, and capacitors?
2. What factors determine the time and length constants of a nerve membrane? How do these constants shape the electric responses of the nerve membrane?
3. How does an action potential differ from the subthreshold responses of a membrane (i.e., the passive and local responses)?
4. What is the sequence of conductances that underlies the action potential?
5. How are the responses of Na^+ and K^+ channels to membrane depolarization similar? How does the presence of the Na^+ channel inactivation gate cause the responses to differ?
6. How do the gating properties of Na^+ and K^+ channels relate to the absolute and relative refractory periods of the action potential?
7. How is the action potential propagated without decrement? What are the factors that determine its propagation velocity?
8. What are the structural properties of myelin that underlie its ability to increase conduction velocity?
9. Given the all-or-none nature of action potentials, how are the characteristics of different stimuli distinguished by the central nervous system?

An **action potential** is a rapid, all-or-none change in the membrane potential, followed by a return to the resting membrane potential. This chapter describes how action potentials are generated by voltage-dependent ion channels in the plasma membrane and propagated with the same shape and size along the length of an axon. The influences of axon geometry, ion channel distribution, and myelin are discussed and explained. The ways in which information is encoded by the frequency and pattern of action potentials in individual cells and in groups of nerve cells are also described. Finally, because the nervous system provides important information about the external world

through specific sensory receptors, general principles of sensory transduction and coding are introduced. More detailed information about these sensory mechanisms and systems is provided in other chapters.

Membrane Potentials

Observations on Membrane Potentials

When a microelectrode (tip diameter <0.5 μm) is inserted through the plasma membrane of a neuron, a difference in potential is observed between the tip of the microelectrode inside the cell and an electrode placed outside the cell. The internal electrode is approximately 70 mV negative with regard to the external electrode, and this difference is referred to as the **resting membrane potential** or, simply, the *resting potential* (see Chapter 1 for details on the basis of the resting potential). (By convention, membrane potentials are expressed as the intracellular potential minus the extracellular potential.) Neurons have a resting potential that typically is around −70 mV.

One of the signature features of neurons is their ability to change their membrane potential rapidly from rest in response to an appropriate stimulus. Two such classes of responses are action potentials and synaptic potentials, which are described in this chapter and the next, respectively. Current knowledge about the ionic mechanisms of action potentials comes from experiments with many species. One of the most studied is the squid because the large diameter (up to 0.5 mm) of the squid giant axon makes it an excellent model for electrophysiological research with intracellular electrodes.

The Passive Response

To understand how an action potential is generated and why it is needed, it is necessary to understand the passive electrical properties of the nerve cell membrane. The term *passive properties* refers to the fact that components of the cell membrane behave very similarly to some of the passive elements of electric circuits, including batteries, resistors, and capacitors. This is very useful because the properties of these elements are well understood. In particular, a piece of membrane containing ion channels responds to changes in voltage across it much as a circuit containing a resistor and capacitor in parallel (parallel RC circuit) would: The ion channels correspond to the resistor, and the lipid bilayer acts as a capacitor. When a battery is *first* connected across the two terminals of a parallel RC circuit, all of the current flows through the branch of the circuit with the capacitor, causing the voltage across it to begin changing (recall that for a capacitor, I $\tilde{\alpha}$ dV/dt). Over time, however, the current flow through the capacitor decreases, whereas that through the resistor increases. As this happens, the rate of voltage change across the capacitor (and resistor) slows, and

the voltage approaches a steady-state value. This change in voltage has an exponential time course whose specific characteristics depend on the resistance (R) and capacitance (C) of the resistor and capacitor. Moreover, a time constant, τ, for this circuit can be defined by the equation τ = R * C, and it equals the time it takes for the voltage to rise (or fall) exponentially by approximately 63% of the difference between its initial and final values.

With regard to how an axon actually responds to electrical stimulation, Fig. 5.1 illustrates the results of an experiment in which the membrane potential of an axon is altered by passing rectangular pulses of **depolarizing** (upward-going

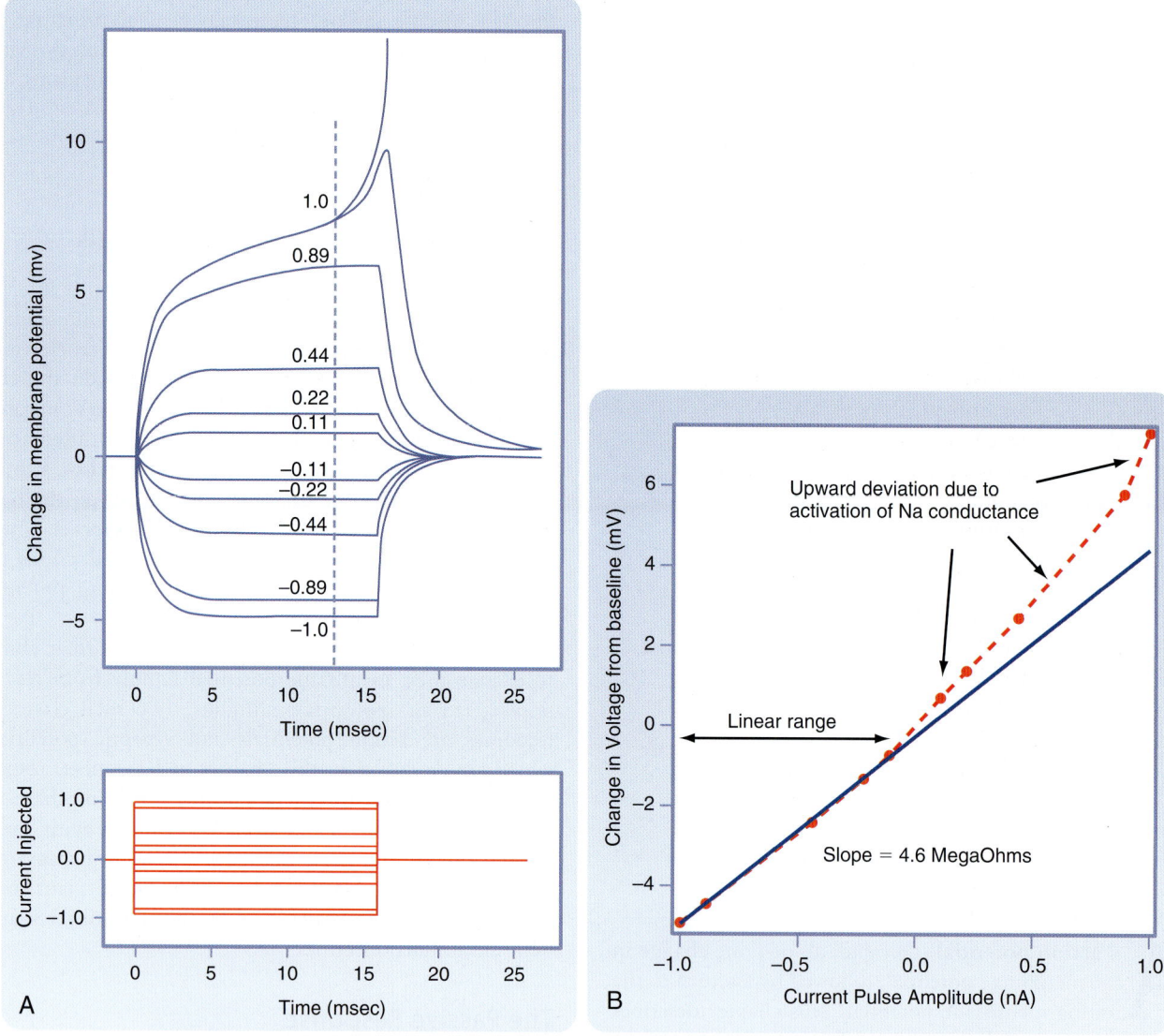

• **Fig. 5.1 (A)** Voltage responses of an axon to rectangular pulses of hyperpolarizing current (negative numbers) or depolarizing current (positive numbers) as injected and recorded from an intracellular electrode. The changes in transmembrane potential are mirror images of the small amplitude pulses. At the threshold level (current = 1.0), there is a 50:50 chance of returning to resting potential or of generating an action potential. For clarity, only the rising phase of the action potential is shown. **B,** Current-voltage (I-V) plot derived from data in A. Current pulse amplitude is plotted on the x-axis, and voltage response (measured at dotted line) is plotted on the y-axis. Note the deviation from linearity with large depolarizations, which is due to activation of voltage-gated conductances. (Redrawn from Hodgkin AL, Rushton WAH: The electrical constants of a crustacean nerve fibre. *Proc R Soc Lond B Biol Sci.* 1946;133:444-479.)

pulses) or **hyperpolarizing** (downward-going pulses) current across its cell membrane. The injection of positive charge is depolarizing because it makes the cell less negative (i.e., decreases the potential difference across the cell membrane). Conversely, the injection of negative charge makes the membrane potential more negative, and this change in potential is called *hyperpolarization*. The larger the current that is injected, the greater the change in the membrane potential will be. The responses to hyperpolarizing and small-amplitude depolarizing current pulses (see Fig. 5.1*A*) all have the same fundamental shape because of the passive properties of the membrane. In contrast, the shapes of the responses to the larger depolarizing stimulus pulses differ from those to hyperpolarizing and small-amplitude depolarizing current pulses because the larger stimuli activate nonpassive elements in the membrane.

For the responses to hyperpolarizing current pulses, once a long enough time has elapsed from the start of the current pulse to allow the membrane voltage to plateau (essentially several times τ), virtually all of the injected current is flowing through the membrane resistance. If the difference between the initial and steady-state voltages is plotted against the amplitude of the current pulse (see Fig. 5.1*B*), a linear relationship is observed for the hyperpolarizing pulses, which is exactly what is expected from Ohm's law (V = I * R) for current flowing through a resistor. The slope of this line (ΔV/ΔI) is referred to as the **input resistance** of the cell (**R_{in}**) and is determined experimentally, exactly as just described. R_{in} is related to the **membrane resistance (r_m)** of the cell, but the exact relationship depends on the geometry of the cell and is complex in most cases.

Next, note that although the current is injected as rectangular pulses, with vertical rising and falling edges, the shape of the membrane voltage responses just after the starts and ends of the pulses have slower rises and falls. Moreover, with regard to only the responses to hyperpolarizing and small-amplitude depolarizing current pulses (see Fig. 5.1*A*), the fall and rise in the membrane voltage have exponential shapes. This shows that the membrane is responding to these current pulses as a parallel RC circuit would; that is, the stimulus causes no change in membrane resistance or **capacitance (c_m)**, and thus the time course of the rise and fall in voltage is the same in all cases because it is governed by the same membrane time constant (τ).

The relationships between voltage and current just described show that within a certain range of stimulation, the cell membrane in one region of the axon can be modeled by a passive RC circuit. However, this model circuit, with only a single resistor and capacitor, takes no account of the fact that axons are spatially extended structures and that because of this, the resistance of the intracellular space is a significant factor in how electrical events in one region affect other regions. That is, if axons had no intracellular resistance, their intracellular space would be isoelectric, and voltage changes, like those just described, across one part of the axonal membrane would occur across all regions instantaneously. In this case, there would be no need for a

special mechanism (i.e., the action potential) to propagate signals actively down the axon. In actuality, axons (and neurons in general) are spatially extended structures with significant resistance to current flow between different regions (this is one reason the relationship of R_{in} and r_m is complicated). Therefore, it is important to understand how current injected at one point along the axon affects the membrane potential at other points because this both helps explain why action potentials are needed and helps explain some of their characteristics.

When current pulses that elicit only passive responses are passed across the plasma membrane, the size of the change in potential recorded depends on the distance of the recording electrode from the point of passage of the current (Fig. 5.2). The closer the recording electrode is to the site of current passage, the larger and steeper the change in potential is. The magnitude of the change in potential decreases exponentially with distance from the

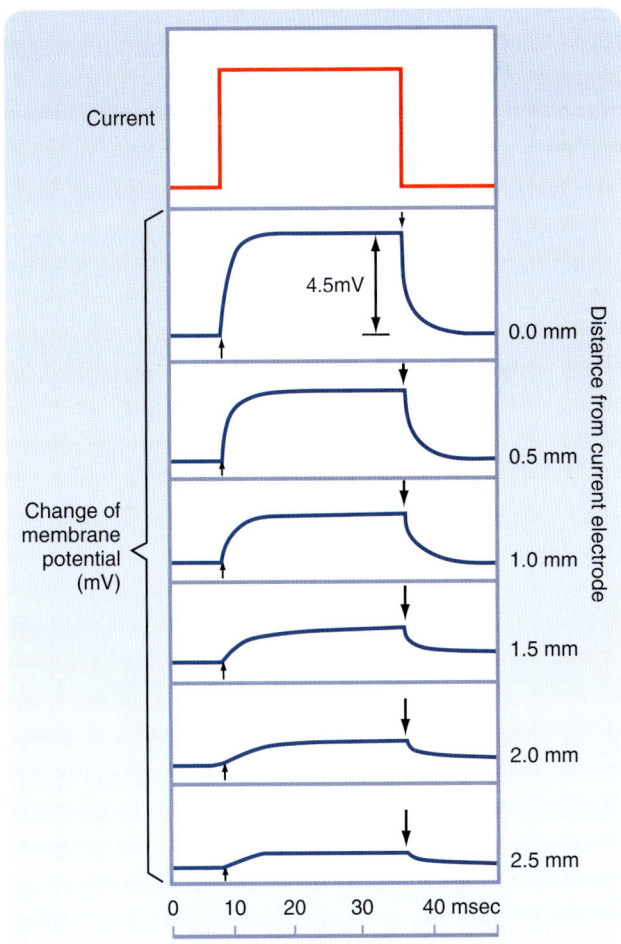

• **Fig. 5.2** Responses of an axon of a shore crab to a subthreshold rectangular current pulse by an extracellular electrode applied closely to its surface and located at different distances from the current-passing electrode. As the recording electrode is moved farther from the point of stimulation, the response of the membrane potential is slower and smaller. (Redrawn from Hodgkin AL, Rushton WAH: The electrical constants of a crustacean nerve fibre. *Proc R Soc Lond B Biol Sci.* 1946;133:444-479.)

site of passage of the current, and the change in potential is said to reflect **passive** or **electrotonic conduction.** Such passively conducted changes in potential do not spread very far along the membrane before they become insignificant. As shown in Fig. 5.2, an electrotonically conducted signal dies away over a distance of a few millimeters. The distance over which the change in potential decreases to 1/e (37%) of its maximal value is called the **length constant** or **space constant** (where e is the base of natural logarithms and is equal to 2.7182). A length constant of 1 to 3 mm is typical for mammalian axons, which can be more than a meter long, which makes obvious the need for a mechanism to propagate information about electrical events generated at the soma to the far end of the axon.

The length constant can be related to the electrical properties of the axon according to cable theory because nerve fibers have many of the properties of an electrical cable. In a perfect cable, the insulation surrounding the core conductor prevents all loss of current to the surrounding medium, so that a signal is transmitted along the cable with undiminished strength. If an unmyelinated nerve fiber (discussed later) is compared to an electrical cable, the plasma membrane equates to the insulation and the cytoplasm as the core conductor, but the plasma membrane is not a perfect insulator. Thus the spread of signals depends on the ratio of the membrane resistance to the **axial resistance of the axonal cytoplasm (r_a).** When the ratio of r_m to r_a is high, less current is lost across the plasma membrane per unit of axonal length, the axon can function better as a cable, and the distance that a signal can be conveyed electrotonically without significant decrement is longer. A useful analogy is to think of the axon as a garden hose with holes poked in it. The more holes there are in the hose, the more water leaks out along its length (analogous to more loss of current when r_m is low) and the less water is delivered to its nozzle.

According to cable theory, the length constant can be related to axonal resistance and is equal to $\sqrt{r_m/r_a}$. This relationship can be used to determine how changes in axonal diameter affect the length constant and, hence, how the decay of electrotonic potentials varies. An increase in the diameter of the axon reduces both r_a and r_m. However, r_m is inversely proportional to diameter (because it is related to the circumference of the axon), whereas r_a varies inversely to the diameter squared (because it is related to the cross-sectional area of the axon). Thus r_a decreases more rapidly than r_m does as axonal diameter increases, and the length constant therefore increases (Fig. 5.3).

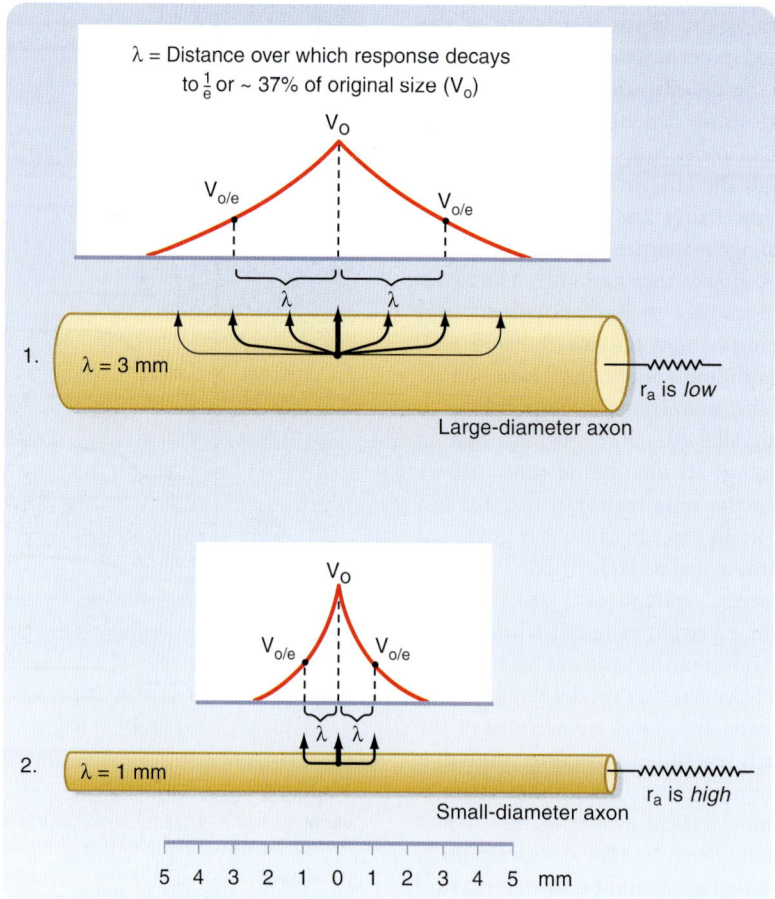

• **Fig. 5.3** Comparison of the Length Constant to Axon Diameter. Note that the increase in diameter is associated with a decrease in axial resistance of the axonal cytoplasm (r_a) and an increase in the length constant (λ). (Redrawn from Blankenship J. *Neurophysiology.* Philadelphia: Mosby; 2002.)

In sum, in the passive domain, the membrane response to electrical stimuli is essentially identical to that of a circuit composed of passive electrical elements, and it can thus be characterized by the length and time constants of the membrane, which will determine how far and how rapidly electrical signals at one point in the cell spread to other parts.

The Local Response

With regard to the experiment shown in Fig. 5.1, if larger depolarizing current pulses are injected, the voltage response of the membrane no longer resembles that of a passive RC circuit. This is most easily observed with pulses that elicit depolarizations either just below or to the **threshold membrane potential** for an action potential but fail to evoke an action potential (tracings 0.89 and 1.0; the threshold membrane potential can be defined as the voltage at which the probability of evoking an action potential is 50%). In these cases, the voltage response shape is altered from that of the passive responses because the stimulus has changed the membrane potential sufficiently to cause the opening of significant numbers of voltage-sensitive Na^+ channels (described later).

Also, note the upward deviation from linearity for the corresponding points in the I-V curve (see Fig. 5.1B). Opening of these voltage-sensitive channels changes the membrane's resistance and allows Na^+ to enter more easily, driven by its electrochemical gradient. This entry of positive charge (Na^+ current) enhances the depolarization by adding to the current pulse delivered by the electrode. The resulting depolarization is called a **local response.** The local response results from active changes in membrane properties (specifically, its Na^+ conductance), whereas in a passive electrotonic response, the conductance to various ions remains constant. Nevertheless, the local response is not self-regenerating but, again, decreases in amplitude with distance. The change in membrane properties is insufficient for what is needed to generate an action potential.

Suprathreshold Response: The Action Potential

Local responses will increase in size as the amplitude of the depolarizing current pulse is increased, until the threshold membrane potential is reached, at which point a different sort of response, the **action potential** (or **spike**), can occur. The threshold value is typically near −55 mV. Normally, when the membrane potential exceeds this value, an action potential is always triggered.

Fig. 5.4 shows the typical shape of an action potential. When the membrane is depolarized past threshold, the depolarization becomes explosive and overshoots in such a way that the membrane potential reverses from negative to positive and approaches, but does not reach, the Nernst equilibrium potential for Na^+ (E_{Na}; see Chapter 2). The membrane potential then returns toward the resting membrane potential (**repolarizes**) almost as rapidly as it was depolarized, and in general, it hyperpolarizes beyond its resting potential (the **afterhyperpolarization**). The main phase of the action potential (from the onset to the return to the resting potential) typically has a duration of 1 to 2 msec, but the afterhyperpolarization, it can persist

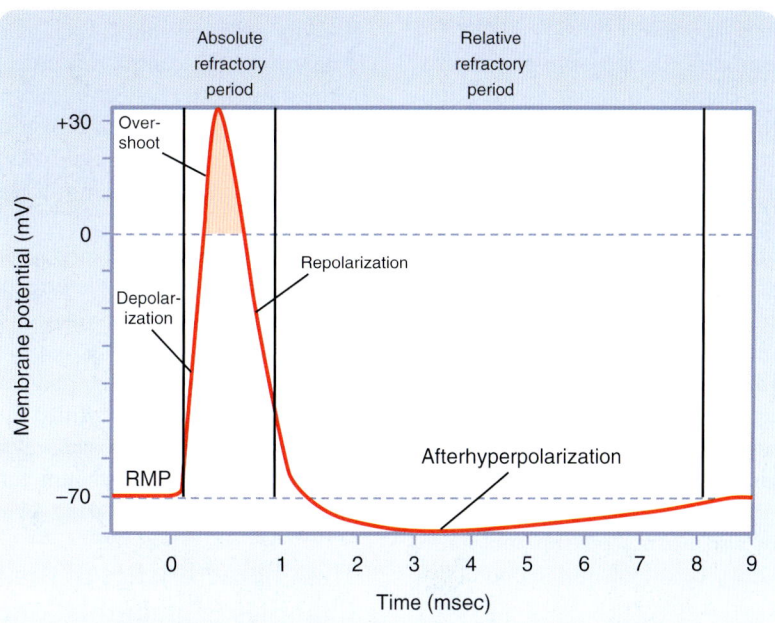

• **Fig. 5.4** Components of the Action Potential With Regard to Time and Voltage. Markers indicate the absolute and relative refractory periods. Note that the time scale for the first few milliseconds has been expanded for clarity. RMP, resting membrane potential. (Redrawn from Blankenship J. *Neurophysiology.* Philadelphia: Mosby; 2002.)

from a few to 100 msec, depending on the particular type of neuron.

The action potential differs from the subthreshold and passive responses in three important ways: (1) It is a response of much larger amplitude, in which the polarity of the membrane potential actually overshoots 0 mV (the cell interior becomes positive in relation to the exterior). (2) The action potential is generally propagated down the entire length of the axon without decrement (i.e., it maintains its size and shape because it is regenerated as it travels along the axon). (3) It is an **all-or-none response,** which means that a stimulus normally either produces a full-sized action potential or fails to produce one. This all-or-none nature is in contrast both to the graded nature of the passive and local responses described previously and to synaptic responses (see Chapter 6).

Ionic Basis of Action Potentials

Recall that the resting membrane potential is determined primarily by the weighted average of the Nernst potentials for Na^+ (E_{Na}) and K^+ (E_K), as defined by the chord conductance equation (see Chapter 2). The weighting factors are the conductance ($g = 1/resistance$) to each ion. At rest the conductance to K^+ (g_K) is high in relation to that for Na^+ (g_{Na}), and so the resting membrane potential (V_r) is closer to E_K ($V_r \cong -70$ mV). If, however, the relative conductances to these ions were to change, this would cause a corresponding change in the membrane potential. For example, an increase in g_K would hyperpolarize the membrane, whereas a decrease in g_K would depolarize the membrane because E_K is approximately -100 mV. Conversely, an increase in g_{Na} would depolarize the membrane and, if of sufficient magnitude, even a lead to reversal in membrane polarity because E_{Na} is approximately $+65$ mV.

An axonic action potential is, in fact, the result of a rapid sequence of transient changes in g_{Na} or g_K, or both. In all axons there is a brief rise in g_{Na}, followed by a decline back to baseline levels. In some axons, this change in g_{Na} occurs against a fixed resting g_K (because of leak channels, which are not voltage-gated; discussed later). In many other cases, however, both g_{Na} and g_K change. Thus as with the resting membrane potential, the action potential depends on the opposing tendencies of (1) the Na^+ gradient to bring the resting membrane potential toward the Nernst potential for Na^+ and (2) the K^+ gradient to bring the resting membrane potential toward the Nernst potential for K^+; but in contrast to when the neuron is at rest, the g_K/g_{Na} ratio is not constant but is changing continuously. One additional difference is that because the membrane potential is changing, a capacitative current also exists, and this must also be taken into account to describe the membrane potential quantitatively during an action potential (as a corollary, note that the chord conductance equation is valid only when the membrane potential is constant because then there is no capacitative current).

The early phase of the action potential (the positive deflection of the membrane potential toward E_{Na}) is a result of a rapid increase in g_{Na} and thus of the Na^+ current (I_{Na}). These changes cause the membrane potential to move toward the equilibrium potential for Na^+. The peak of the action potential does not reach E_{Na} because the rise in g_{Na} is not infinite (i.e., the g_K/g_{Na} ratio does not fall to zero).

Because of the nature of the underlying Na^+ channels (described later), the rise in g_{Na} with depolarization is transient. Moreover, in many cases, the depolarization leads to a rise in g_K. These two factors cause the g_K/g_{Na} ratio to stop falling and start increasing; as a result, the membrane potential is driven back toward E_K and thus repolarizes toward its resting value. In cases in which the repolarization involves a rise in g_K, the membrane potential hyperpolarizes temporarily beyond its normal rest value (if g_K does not change, the drop in g_{Na} causes the membrane simply to return to its resting potential). This afterhyperpolarization occurs because g_K remains elevated for a period of time after the action potential. As g_K returns to its baseline level, the membrane potential returns to its rest value.

These changes in conductance can be explained by the properties of Na^+ and K^+ ion channels, which are described next.

Ion Channels and Gates

Early studies of the mechanism underlying action potentials indicated that ion currents pass through separate Na^+ and K^+ channels, each with distinct characteristics, in the cell membrane. Subsequent research has supported this interpretation. The amino acid sequences of the channel proteins and many of the functional and structural characteristics of the channels are now known in detail.

The structure of a voltage-gated Na^+ channel (Fig. 5.5) consists of four α subunits and two β subunits. The α subunit has four repeated motifs each of six transmembrane helices that surround a central ion pore. The pore walls are partly formed by the six helices in each motif. Most voltage-gated K^+ channels are composed of four separate subunits, each consisting of a polypeptide with six membrane-spanning segments, similar to the motifs that make up the α subunit of the Na^+ channel.

An important characteristic of some channels, such as those that underlie the action potential, is that they are gated by the membrane voltage. These voltage-gated channels sense the potential across the membrane and then act to either open or close the pore according to the membrane potential. The gates are formed by groups of charged amino acid residues, and the voltage dependence of the Na^+ and K^+ channel gates can account for the complex changes in g_{Na} and g_K that occur during an action potential.

The Characteristics of the Na⁺ and K⁺ Channels Explain the Conductance Changes During the Action Potential

Use of standard intracellular recording along with voltage-clamp techniques enabled investigators to characterize the underlying ionic currents and conductance changes

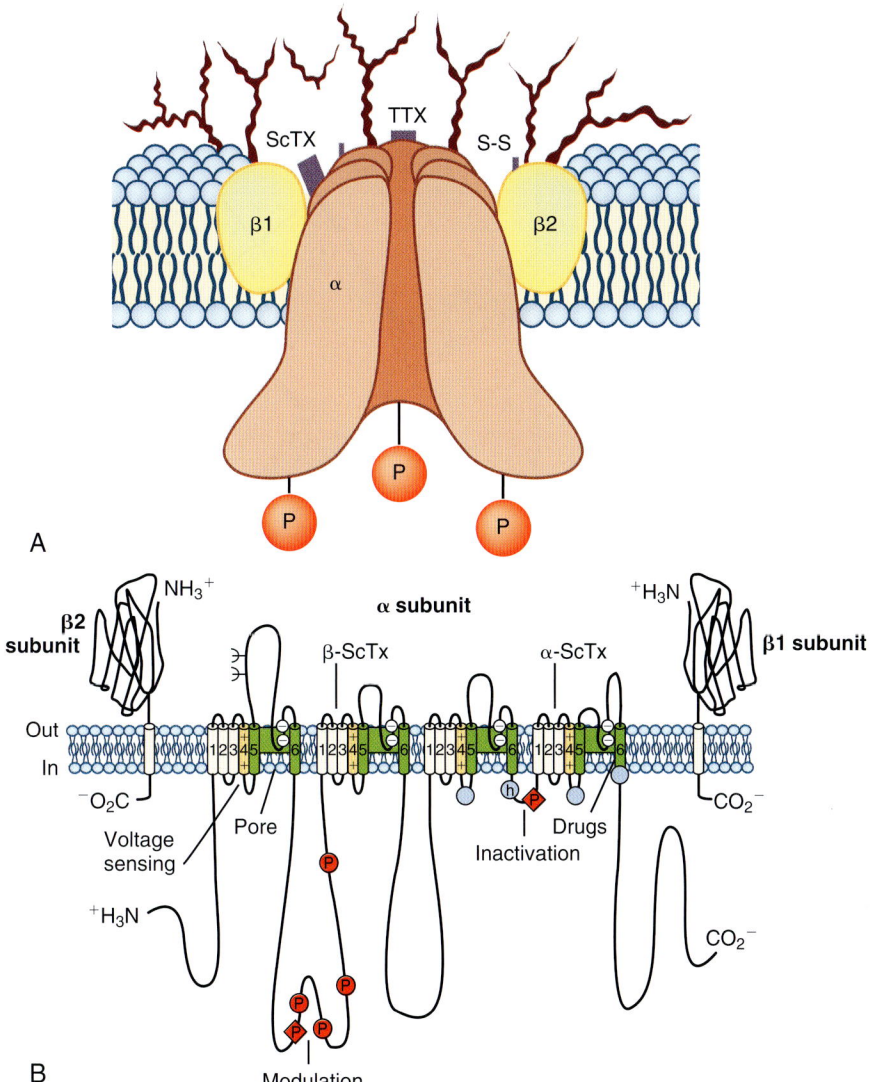

• **Fig. 5.5** A Model of the Voltage-Gated Na⁺ Channel. **A,** The large red elements represent the four α subunits and the two yellow elements are β subunits with the receptor sites for α scorpion toxin (ScTX) and tetrodotoxin (TTX) indicated. **B,** β1 and β2 subunits flanking an α subunit are shown with their transmembrane helices. (Redrawn from Catterall WA: Structure and function of voltage-gated sodium channels at atomic resolution. *Exp Physiol.* 2014;99:35-51.)

associated with the action potential. Detailed statistical analyses of these recordings also allowed remarkable inferences to be made about the nature of the channels that passed these Na⁺ and K⁺ currents. The development of the **patch recording,** however, enabled direct observation of the behavior of individual channels. In this technique, a specially shaped microelectrode is placed against the surface of a cell, and suction is applied to the microelectrode. As a result, a high-resistance seal is formed between the membrane and the tip of the microelectrode (Fig. 5.6*A*), which allows recording of the activity of whatever channels happen to be in the patch of membrane that is inside the seal. Under ideal conditions only one or a few ion channels of a single type are present in the membrane patch.

Patch recordings show that many ion channels flip spontaneously between open and closed conductance states as if they have gates that open and close the entrance to their pore. In the case of voltage-gated channels, the gate is sensitive to the voltage across the membrane, and thus the time a gate spends in each state is a probabilistic function of the membrane potential. A patch recording of a K⁺ channel demonstrates this probabilistic behavior (see Fig. 5.6*B*). As the membrane potential is clamped to more depolarized levels, the channel spends more time in its open state, which reflects the voltage dependence of the probability that the channel will open (see Fig. 5.6*C*). Also, the amplitude of the current in the open state increases with the level of depolarization; this is because the driving force for K⁺ is greater at more depolarized levels (i.e., the membrane potential is farther from the K⁺ Nernst potential).

The behavior of a Na⁺ channel is more complex than that of the K⁺ channel. Like the K⁺ channel, it has a

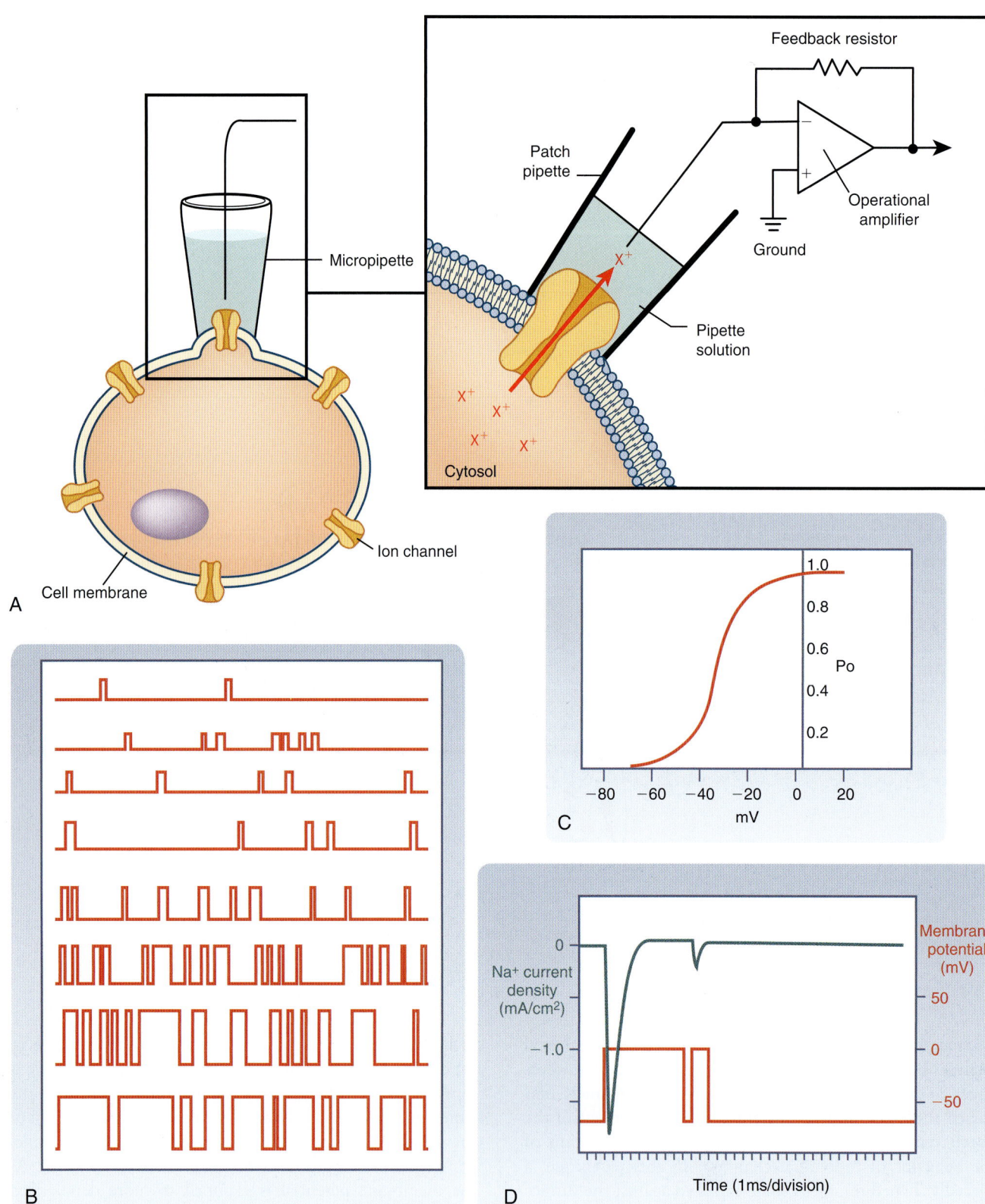

A

B

C

D

• Fig. 5.6 A, A micropipette is applied to the cell membrane and sufficient suction is applied to electrically isolate a single channel at the tip. An amplifier records the current that passes through the channel. **B,** Each line shows the current passed through a K$^+$ channel as it opens spontaneously. Note that as the transmembrane voltage is progressively depolarized (from top to bottom), both the probability that the channel will open and the amplitude of the current are increased. **C,** A graph of the probability that the channel will open versus membrane depolarization. **D,** A graph of transmembrane voltage (lower tracing, right-sided scale) and the current density from a population of Na$^+$ channels (upper tracing, left-sided scale) isolated in a patch similar to that in **A** (except that it contains several Na$^+$ channels). Initially, at resting potential, there is no current flow. With depolarization to 0mV, there is an inward Na$^+$ current that is curtailed even while the depolarization continues. This is due to the closing of the channels' inactivation gates. After a brief return to resting potential, another depolarization to 0mV evokes inward current flow, but it is smaller and briefer because there has not been enough time for most of the slow inactivation gates to reopen. (**B, C,** and **D** are redrawn from http://www.physiologymodels.info/electrophysiology.)

voltage-sensitive gate (activation gate) whose probability of being open increases with depolarization. However, unlike K$^+$ channels, with maintained depolarization, Na$^+$ channels open only at the onset of the depolarization and then remain closed. This suggests that Na$^+$ channels have a second gate (inactivation gate) whose probability of being open goes down as the membrane is depolarized. Thus any Na$^+$ current conducted by these channels will be transient (see Fig. 5.6D), because the same stimulus (depolarization) increases both the probabilities that the activation gate will open and that the inactivation gate will close. Note that the Na$^+$ channel thus has two closed states, one in which the activation gate is closed, and the channel is said to be "closed," and one in which the inactivation gate is closed, and the channel is referred to as "inactivated." The reason the channel is called "inactivated" when the second gate closes is that once this gate closes, it will remain so until the membrane is repolarized. The activation gate, in contrast, can open and close at all membrane potentials, just with differing probabilities.

With the knowledge of the Na$^+$ and K$^+$ channel gating behavior just discussed, we can understand how the action potential is generated by the interaction of these channels (in the following, we assume that both g_{Na} and g_K change during the action potential). As stated previously, the action potential starts with a rapid increase in Na$^+$ conductance (g_{Na}; Fig. 5.7). This increase in Na$^+$ conductance reflects the opening of many Na$^+$ channels in response to the depolarization. The open channels allow the influx of Na$^+$ ions, and the effect of this current is to depolarize the membrane further. Note that this is a positive feedback loop, which accounts for the explosive nature of the action potential: the Na$^+$ current depolarizes the membrane, which causes more Na$^+$ channels to open, which in turn increases the Na$^+$ current. In sum, the voltage-dependent opening of Na$^+$ channels and the depolarizing action of the Na$^+$ current account for the rising phase of the action potential.

The end of the rising phase and the subsequent falling (repolarization) phase of the action potential is the result of two processes: a reduction in g_{Na} and an increase in g_K. The rise in g_K is simply a consequence of membrane depolarization, which increases the probability that the K$^+$ channel will be open. The decrease in g_{Na} results from two

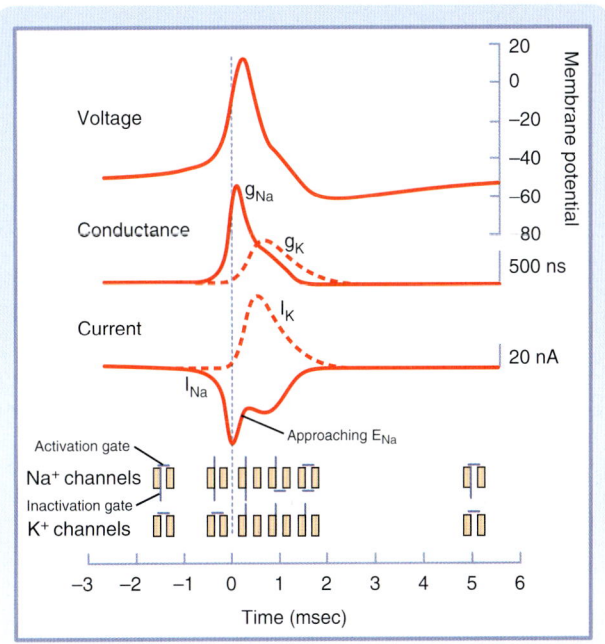

• Fig. 5.7 The Action Potential and the Conductance and Currents That Underlie the Action Potential With Regard to Time. Note that the increased conductance for Na$^+$ (g_{Na}), as well as its inward flow, is associated with the rising phase of the action potential, whereas the slower increase in conductance for K$^+$ (g_K), as well as its outward flow, is associated with repolarization of the membrane and with afterhyperpolarization. The reduction in the Na$^+$ current (I_{Na}) before the peak of the action potential (even though g_{Na} is still high) is due to inactivation of the Na$^+$ channels. (Redrawn from Squires LR, et al: *Fundamental Neuroscience.* 2nd ed. San Diego, CA: Academic Press; 2002.)

factors. First, Na$^+$ channels are inactivated as a result of the closing of the inactivation gate with depolarization. Unlike the activation gate, which can flip between states even when the membrane is depolarized, the inactivation gate, once closed, remains closed until significant repolarization occurs. Second, as the g_K/g_{Na} ratio increases (as a result of both inactivation of Na$^+$ channels and opening of K$^+$ channels), the membrane begins to repolarize, and this repolarization acts to shut the activation gate of the Na$^+$ channel. The closure of both voltage-gated Na$^+$ and K$^+$ channels during the falling phase brings the membrane back to its resting state. If only Na$^+$ channels had opened during the action potential (as is the case for some axons), the membrane would simply return to its rest potential. If voltage-gated K⁺ channels also had opened during the action potential, an afterhyperpolarization would be present because these K⁺ channels close slowly in response to hyperpolarization.

AT THE CELLULAR LEVEL

Knowledge of the molecular structure of channels has increased the understanding of the basis of their properties. For example, most channels are highly selective for a particular ion. First, if the channel walls are lined with either positive or negative charges, then either cations or anions can be excluded; however, most channels are also differentially permeable by different ions of the same charge. This further selectivity appears to be the result of requiring ions to become dehydrated as they pass through the narrowest part of a channel, known as the **selectivity filter.** Ions in solution are hydrated (are surrounded by a shell of H_2O molecules), and the radius of this hydration shell is different for each type of ion. In Na^+ and K^+ channels, to make dehydration energetically possible, the pore of the channel is lined with negatively polarized amino acid substituents of a particular geometry, and these substituents substitute for the water molecules. Such substitution, however, requires close matching of the filter's size to the ion's hydration shell. Because each ion has a different-sized shell, a particular channel will best allow passage of one particular ionic species.

AT THE CELLULAR LEVEL

Tetrodotoxin (TTX), one of the most potent poisons known, specifically blocks the Na^+ channel. TTX binds to the extracellular side of the sodium channel (see Fig. 5.5A). **Tetraethylammonium (TEA⁺),** another poison, blocks K^+ channels. TEA⁺ enters the K^+ channel from the cytoplasmic side and blocks the channel because TEA⁺ is unable to pass through it. The ovaries of certain species of puffer fish, also known as blowfish, contain TTX. Raw puffer fish is a highly prized delicacy in Japan. Connoisseurs of puffer fish enjoy the tingling numbness of the lips caused by the minuscule quantities of TTX present in the flesh. Sushi chefs who are trained to remove the ovaries safely are licensed by the government to prepare puffer fish. Despite these precautions, several people die each year as a result of eating improperly prepared puffer fish.

Saxitoxin is another blocker of Na^+ channels that is produced by the reddish dinoflagellates that are responsible for so-called red tides. Shellfish eat the dinoflagellates and concentrate saxitoxin in their tissues. A person who eats these shellfish may experience life-threatening paralysis within 30 minutes after the meal.

Accommodation

When a nerve is depolarized very slowly, the normal threshold may be passed without the firing of an action potential; this phenomenon is called **accommodation.** Both Na^+ and K^+ channels are involved in accommodation. In response to membrane depolarization, g_{Na} first increases and then,

a short time later, decreases. This is due to the opening of the activation gates and closing of the inactivation gates of the Na^+ channels. Normally, membrane depolarization to threshold or beyond triggers an action potential; however, the explosive depolarization of the action potential can occur only if a critical number of Na^+ channels are recruited. Thus if a cell is slowly depolarized, Na^+ channels can become inactivated even without the occurrence of an action potential, and the pool of available noninactivated Na^+ channels (i.e., channels in the closed state) can be reduced to the point at which a stimulus to may not be able to recruit a sufficient number of Na^+ channels to generate an action potential. An additional factor in causing accommodation is that K^+ channels open slowly in response to the depolarization. The increased g_K tends to oppose depolarization of the membrane, which makes it even less likely to fire an action potential.

Refractory Periods

When a cell is refractory, it is either completely unable to fire an action potential or it requires a much stronger stimulation than usual. During much of the action potential, the cell is completely refractory because it will not fire another action potential no matter how strongly it is stimulated. This **absolute refractory period** (see Fig. 5.4) occurs when a large fraction of the Na^+ channels are inactivated and therefore cannot be reopened until the membrane is repolarized. During this period, the critical number of Na^+ channels required to produce an action potential cannot be recruited.

IN THE CLINIC

In an inherited disorder called **primary hyperkalemic paralysis,** patients have episodes of painful spontaneous muscle contractions followed by periods of paralysis of the affected muscles. These symptoms are accompanied by elevated [K⁺] in plasma and extracellular fluid. Some patients with this disorder have mutations of voltage-gated Na^+ channels that result in a decreased rate of voltage inactivation. This results in longer-lasting action potentials in skeletal muscle cells and increased efflux of K^+ during each action potential, which can raise extracellular [K⁺].

The elevation in extracellular [K⁺] causes depolarization of skeletal muscle cells. Initially, the depolarization brings muscle cells closer to threshold, and so spontaneous action potentials and contractions are more likely to occur. As depolarization of the cells becomes more marked, the cells become refractory because increasing numbers of Na^+ channels become inactivated. Consequently, the cells become unable to fire action potentials and are not able to contract in response to action potentials in their motor axons.

During the latter part of the action potential, and during the afterhyperpolarization period, the cell is able to fire a second action potential, but a stimulus stronger than normal is required. This period is called the **relative refractory period.** Early in the relative refractory period, before the membrane potential has returned to the resting potential level, some Na⁺ channels are still voltage inactivated, but there are enough in the closed state (and therefore have the potential to open when the membrane is depolarized) to support the generation of an action potential if they are stimulated to open. However, a stimulus stronger than normal is necessary to recruit the critical number of Na⁺ channels needed to trigger an action potential (i.e., the reduction in the total number of available Na⁺ channels is countered by increasing the probability of opening). Throughout the relative refractory period, conductance to K⁺ is elevated, which opposes depolarization of the membrane. This increase in K⁺ conductance continues throughout the afterhyperpolarization and accounts for most of the duration of the relative refractory period.

Conduction of Action Potentials

Fundamental to nervous system function is the transmission of information along neuronal pathways. To accomplish this, neurons generate action potentials that propagate down the length of their axon without decrement in size in order to trigger neurotransmitter release from the presynaptic terminals. How action potentials propagate down an axon and how the characteristics of the axon affect this propagation are discussed in this section. How they trigger transmitter release is covered in Chapter 6.

Action Potential as a Self-Reinforcing Signal

Passive conduction will not transport a signal from one end of an axon to the other unless the axon is very short (i.e., on the order of its length constant) because passively conducted signals decrease in size rapidly with distance from their origin. Neurons with such short axons exist; for example, in the retina of the eye, the distance from one neuron to the next is so small that electrotonic (passive) conduction is sufficient. However, in most cases, axons are many times longer than their length constant. In fact, they can be up to 1 m or more in length (e.g., those of motor neurons) and thus hundreds of times their length constant. Nonetheless, if researchers were to record from points along a typical axon, they would find that as the action potential arrives at successive points traveling along the axon, its shape and size remain constant. This is because the action potential regenerates itself as it is conducted along the fiber and thus is said to be actively **propagated.**

Fig. 5.8 shows how in a local response the current that flows in through one part of the membrane acts to depolarize the neighboring membrane. The same thing happens when the Na⁺ channels are opened by an action potential at one site along the axon, except that in this case, the current will be large enough to depolarize the areas on either side past threshold and thus generate action potentials in these neighboring areas. The inward Na⁺ current in these areas can then provide the current to depolarize their neighbors past threshold so that they in turn generate action potentials,

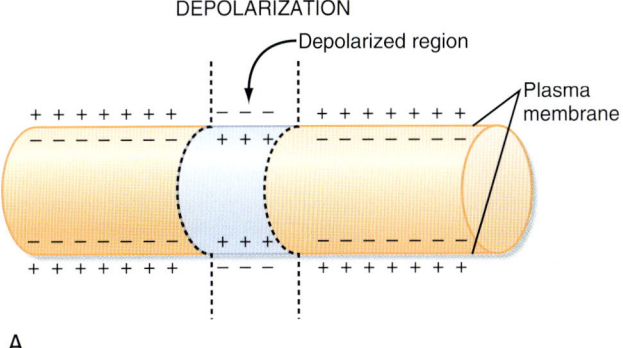

DEPOLARIZATION

A

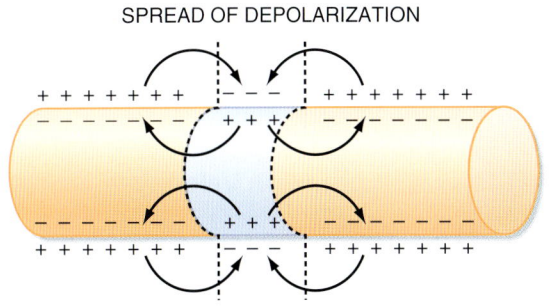

SPREAD OF DEPOLARIZATION

B

• **Fig. 5.8** Mechanism of Electrotonic Spread of Depolarization. **A,** The reversal of membrane polarity that occurs with local depolarization. **B,** The local currents that flow to depolarize adjacent areas of the membrane and allow conduction of the depolarization.

and so on. In short, action potential propagation along an axon involves recurring cycles of depolarization to provide sufficient local current flow for generation of an action potential in an adjacent region of the cell membrane. Thus the action potential is said to be propagated down the axon, with "new" action potentials being generated along its length. In this way, the action potential can propagate down the entire length of the axon while retaining the same size and shape.

Normally, action potentials are first generated at the axon's initial segment (i.e., where the axon is attached to the neuron cell body or proximal dendrite) and then conducted to the terminal end. The reason for this is that the initial segment has a very high density of voltage-gated Na⁺ channels, and thus it has a lower threshold for spiking than does the soma or dendrites. However, axons are not inherently unidirectional conductors. For example, as implied by the local circuits shown in Fig. 5.8, an action potential generated by a depolarization in the middle of an axon is conducted in both directions from its initiation site simultaneously.

Why does a spike that starts at the initial segment not propagate in both directions? In fact, it does. In addition to propagating down the axon, the current flowing from the initial segment back to the soma can cause a spike to be generated in the soma because the soma also has voltage-gated Na⁺ channels. Additional "backpropagating" spikes from the axon do not occur, however, nor does the somatic spike cause the initial segment of the axon to fire a second time (and thereby send another spike down the axon and start a repeating cycle). This does not happen because the refractory period of the membrane makes any area that has already spiked unable to fire a second spike for a short time. Thus for a spike that started at the initial segment and has begun traveling down the axon, current flowing in at the site of the spike depolarizes the membrane on both sides of that site. However, the side closer to the cell body, which has recently fired a spike, cannot respond to this depolarization because its Na^+ channels are still inactivated. By the time Na^+ channels are de-inactivated (have returned to their closed state and would be able to open) the depolarization of membrane at that site has ended (because the action potential lasts only for ≈ 1 msec). Thus the inactivation gate of the Na^+ channel not only helps determine the duration of the action potential but it also is responsible for its singular and unidirectional propagation from its origin at the initial segment.

Action Potential Conduction Velocity Is Correlated With Axon Diameter

The speed of conduction in a nerve fiber is determined by the electrical properties of the cytoplasm and the plasma membrane that surrounds the fiber, as well as by its geometry. In nonmyelinated fibers, conduction velocity is proportional to the square root of the cross-sectional diameter (Fig. 5.9). This effect is related to the changes in r_a and r_m with diameter. As the diameter of a fiber increases, r_a decreases with the square of the diameter, and r_m increases only linearly with diameter; as a result, resistance to current

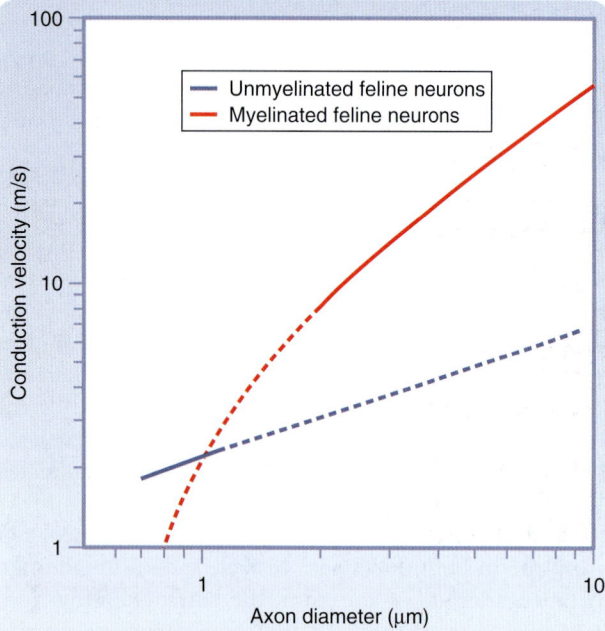

• **Fig. 5.9** Conduction velocities of unmyelinated *(blue)* and myelinated *(red)* feline axons as functions of axon diameter. *Solid lines* represent measured data. *Dotted lines* represent extrapolations that show the advantage of myelination over simply increasing axon diameter as a mechanism for increased conduction velocity. (From Schmidt-Nielsen K. *Animal Physiology: Adaptation and Environment.* 5th ed. Cambridge, UK: Cambridge University Press; 1997.)

flow down the axon decreases more than it does to current flow across the membrane. This increases the length constant (see Fig. 5.3), which means that a greater amount of the current entering at one site is delivered to neighboring regions of the axon, which brings those regions to threshold more quickly, and thus the action potential is conducted faster along fibers with large diameters.

However, increasing the diameter also increases the surface area of the plasma membrane over which inner negative and outer positive charges are held to each other. Discharging this increased capacitance tends to slow conduction and mitigate the increase in conduction velocity gained by increasing diameter.

Myelination Greatly Increases Conduction Velocity

In vertebrates, many nerve fibers are coated with **myelin,** and such fibers are said to be *myelinated.* Myelin consists of the plasma membranes of **Schwann cells** (in the peripheral nervous system) or **oligodendroglia** (in the central nervous system [CNS]), which wrap around and insulate the nerve fiber (Fig. 5.10*A* and *B*). The myelin sheath consists of several to more than 100 layers of glial cell plasma membrane. Gaps about 1-2 μm wide, known as **nodes of Ranvier,** separate the contribution of one Schwann cell (or oligodendrocyte) from that of another. For all but the axons of smallest diameter, a myelinated axon has much greater conduction velocity than does an unmyelinated fiber of the same caliber because the myelin sheath increases the effective membrane resistance of the axon, decreases the capacitance of the axon membrane, and limits the generation of action potentials to the nodes of Ranvier. In short, myelination greatly alters the electrical properties of the axon.

Because the many wrappings of membrane around the axon increase the effective membrane resistance r_m/r_a and the length constant are much greater. The increased membrane resistance means that less current is lost through the membrane per length of axon, and thus the amplitude of a conducted signal decreases less with distance along the axon and needs to be regenerated (by opening of Na^+ channels) less often.

In addition, the thicker myelin-wrapped membrane results in a much larger separation of charges across it than exists across the bare membrane of an axon, so that the charges across it are much less tightly bound to each other. This is analogous to when the plates of a capacitor are moved apart and reduce its capacitance. Because the effect of membrane capacitance is to slow the rate at which the membrane potential can be changed, the reduced capacitance of myelinated axons means that the depolarization occurs more rapidly. For all these reasons, conduction velocity is greatly increased by myelination, and the current generated at one node of Ranvier is conducted at great speed to the next (see Fig. 5.10).

In myelinated axons, the Na^+ channels that bring about generation of an action potential are highly concentrated

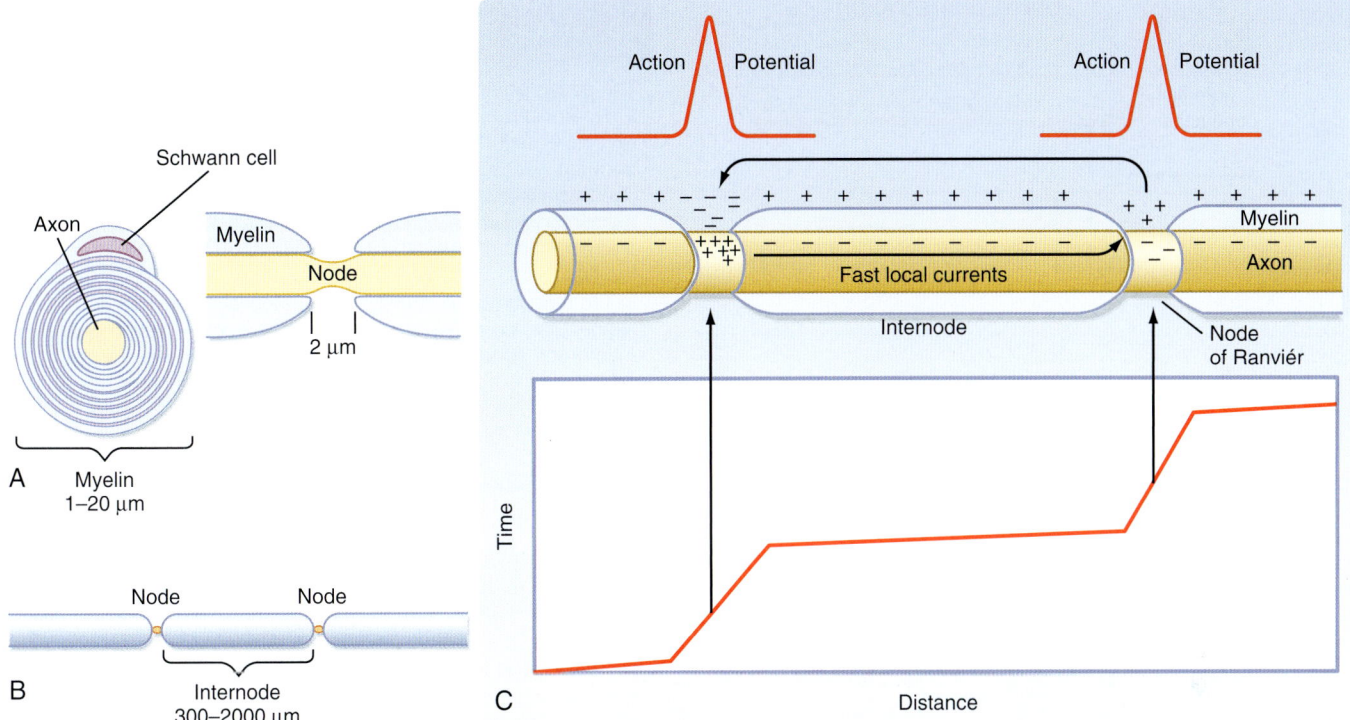

• **Fig. 5.10 A,** Schematic illustrations, in cross section and longitudinal section through a node of Ranvier, of a Schwann cell wrapped around an axon to form myelin. Note that the axon is exposed to the extracellular space only at the node of Ranvier. **B,** View of two nodes and the intervening internode of myelin. **C,** Saltatory conduction in a myelinated axon with a plot of the action potential location along the axon (x-axis) versus time (y-axis). Note the short time taken for the action potential to traverse the large distance between nodes (*shallow sloped lines* on the plot) because of the high resistance and low capacitance of the internodal region. In contrast, the action potential slows as it crosses each node (*steep sloped line segments*). (**B,** Redrawn from Squires LR et al: *Fundamental Neuroscience.* 2nd ed. San Diego, CA: Academic Press; 2002. **C,** Redrawn from Blankenship J: *Neurophysiology.* Philadelphia: Mosby; 2002.)

at the nodes of Ranvier and are not found between them. Thus the action potential is regenerated only at the nodes of Ranvier (0.3-2 mm apart) rather than being regenerated continuously along the fiber, as is the case in an unmyelinated fiber. Resistance to the flow of ions across the many layers that make up the myelin sheath is so high that transmembrane currents are largely restricted to the short stretches of naked plasma membrane that are present at the nodes of Ranvier (see Fig. 5.10C). Therefore, the action potential is regenerated at each successive node. The local currents entering the node are almost entirely conducted from one node to the next node, bringing each node to threshold in about 20 μsec. Thus the action potential appears to "jump" from one node of Ranvier to the next, and the process is called **saltatory** (from the Latin word *saltare,* "to leap") **conduction** (Fig. 5.11).

Functional Consequences of Myelination

The functional consequences of myelination can be highlighted by a comparison of squid and mammalian axons. Although human nerve fibers are much smaller in diameter than squid giant axons, human axons conduct at comparable or even faster speeds because of myelination. The unmyelinated squid giant axon has a 500-μm diameter and a conduction velocity of about 20 m/sec. In mammals, axon diameters range from about 0.2 to 20 μm, and all fibers with diameters larger than 1-2 μm are myelinated. An unmyelinated mammalian nerve fiber, which has a diameter of less than 1 to 2 μm, has a conduction velocity of less than 2 m/sec (see Fig. 5.9), as expected because of its smaller diameter in comparison to the squid giant axon. In contrast, a 10-μm myelinated mammalian fiber has a conduction velocity in the range of 50 m/sec, more than twice that of the 500-μm squid giant axon, despite being 1/50 of its diameter. Thus the high conduction velocity with far narrower axons achieved by myelination allows a tremendous increase in neuronal connectivity without enormously expanding the volume of the CNS. This is certainly one factor that enabled the evolution of mammalian nervous systems with their huge numbers of neurons that are able to generate everything from fast reflexes to efficient and complex mental processing.

Sensory Transduction

To receive information about the world, the CNS contains a wide variety of sensory receptors, each of which is specialized

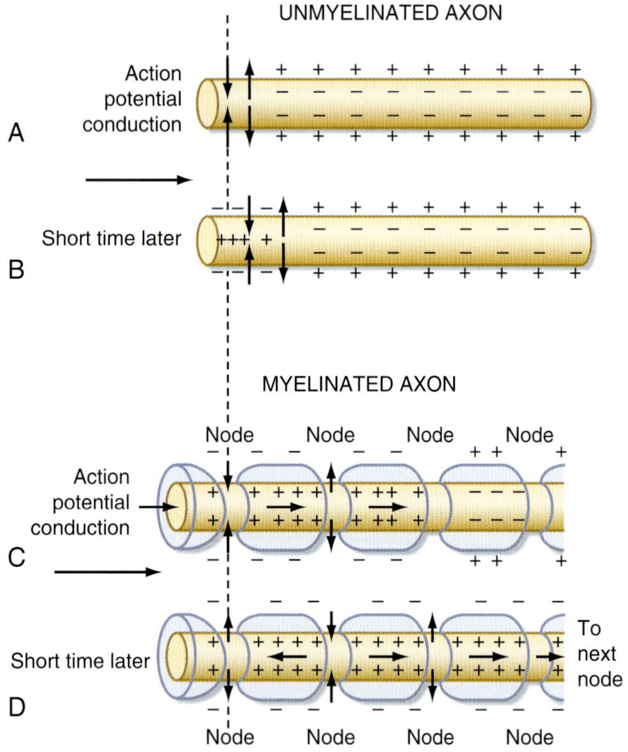

UNMYELINATED AXON

A

Action
potential
conduction

B

Short time later

MYELINATED AXON

Node   Node   Node   Node

C

Action
potential
conduction

D

Short time later

To
next
node

Node   Node   Node   Node

• **Fig. 5.11** Comparison of Action Potential Conduction in an Unmyelinated Axon and in a Myelinated Axon. At the initial time (**A** and **C**), an action potential is being generated at the left side of each axon. Note that the inward current in the unmyelinated axon (**A**) is depolarizing an adjacent portion, whereas the inward current in the myelinated axon (**C**) is depolarizing all of the membrane to the next node. At the second instant in time (**B** and **D**), the action potential in the unmyelinated axon (**B**) has been generated in the adjacent portion, whereas the action potential in the myelinated axon (**D**) has been generated at subsequent nodes and is already depolarizing the last node to the right. (Redrawn from Castro A, et al: *Neuroscience: An Outline Approach.* Philadelphia: Mosby; 2002.)

IN THE CLINIC

In some diseases known as **demyelinating disorders,** the myelin sheath deteriorates. In **multiple sclerosis,** scattered progressive demyelination of axons in the CNS results in loss of motor control and sensory deficits. The neuropathy common in severe cases of diabetes mellitus is caused by the demyelination of peripheral axons. When myelin is lost, the length constant becomes much shorter. Hence, the action potential loses amplitude as it is electrotonically conducted from one node of Ranvier to the next. If demyelination is sufficiently severe, the action potential may arrive at the next node of Ranvier with insufficient strength to fire an action potential at that node, leading to propagation failure.

to detect a particular type of energy (**stimulus).** When a stimulus activates a sensory receptor, it initiates a process called **sensory transduction** by which information about the stimulus (e.g., its intensity and duration) is converted into local electrical signals. These local signals are called

AT THE CELLULAR LEVEL

The action potentials of myelinated axons may not have a hyperpolarizing afterpotential or an extended relative refractory period because their K^+ channels are displaced from the nodes into the partly exposed flanking paranodes. That increases the rate at which these fast-conducting axons can fire. Myelinated axons are also more metabolically efficient than unmyelinated axons. Na^+,K^+-ATPase extrudes the Na^+ that enters the cell and causes the K^+ that leaves the cell to reaccumulate during action potentials. In myelinated axons, ionic currents are restricted to the small fraction of the membrane surface at the nodes of Ranvier. For this reason, far fewer ions traverse a unit length of fiber membrane, and much less ion pumping—and energy expenditure—is necessary to maintain the gradients.

IN THE CLINIC

Investigators can record an action potential with a microelectrode without penetrating the axon by placing two spaced electrodes on its surface and comparing the electrical charge at each point. An electrode located where there is an action potential would yield a somewhat negative signal in comparison to an electrode where there is no action potential. As the action potential is conducted to the second electrode, the polarity of the recording reverses. This technique is used clinically to assess nerve function. Peripheral nerves and many central pathways consist of a population of axons of various diameters (Fig. 5.12); some of the axons are myelinated, and some are not. Consequently, action potentials travel at different velocities in the individual axons. As a result, a recording from such a nerve with external electrodes does not show a single synchronous peak but a series of peaks that vary in time (which reflects the conduction velocity of groups of axons) and in magnitude (which reflects the number of axons in each velocity group). This is called a **compound action potential**. The clinical value of such a recording is its ability, in certain disease states, to reveal the dysfunction of a particular group of axons associated with specific functions, as well as the noninvasive nature of the technique because it can be performed with skin surface electrodes (Table 5-1).

receptor or **generator potentials.** The receptor potentials can then be transformed into patterns of action potentials that are conducted over one or more axons into the CNS. In order for this to happen, the stimulus must produce receptor potentials that are large enough to change the spiking levels of one or more primary afferent fibers that are connected to the receptor. Weaker intensities of stimulation can produce subthreshold receptor potentials, but such stimuli do not change the activity of central sensory neurons and thus are not detected. Thus s**timulus threshold** is defined as the weakest stimulus that can be reliably detected.

Environmental events that evoke sensory transduction can be mechanical, thermal, chemical, or other forms of energy. However, the types of information used by a

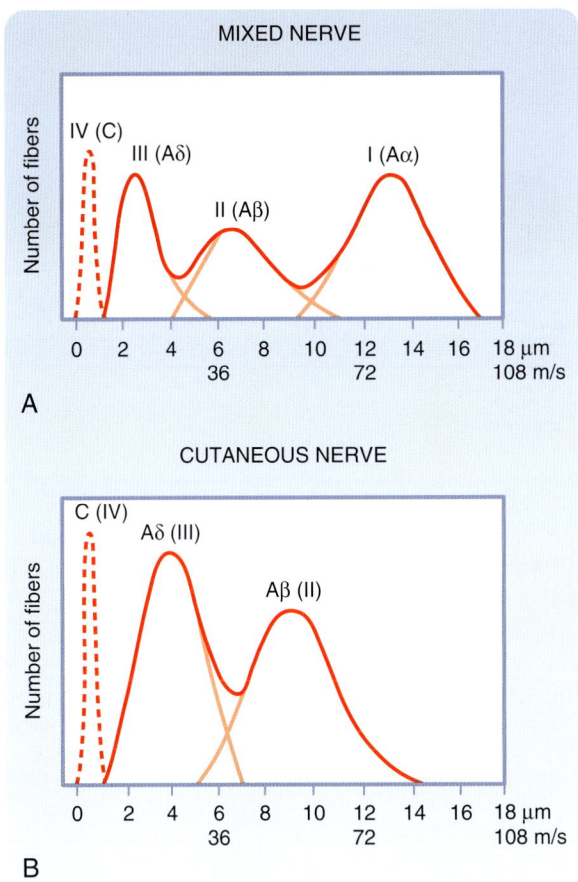

MIXED NERVE

CUTANEOUS NERVE

• **Fig. 5.12** The distribution of axons, by size and conduction velocity, in a mixed (muscle) nerve **(A)** and a cutaneous nerve **(B)**. Note the increased number of small-diameter fibers and the absence of Aα fibers in the cutaneous nerve. (From Haines DE [ed]. *Fundamental Neuroscience for Basic and Clinical Applications.* 3rd ed. Philadelphia: Churchill Livingstone; 2006.)

particular organism depend on its set of sensory receptors. For example, humans cannot sense electrical or magnetic fields, but other animals can sense such stimuli. In particular, many fish have electroreceptors, and various fish and birds use the earth's magnetic field to orient themselves during migration.

The transduction process varies with the type of environmental stimulus being detected. Fig. 5.13 shows three examples of how stimuli can alter the membrane properties of the specific sensory receptors that transduce such stimuli (further details for each of these examples are given in other chapters). Fig. 5.13*A* illustrates how a **chemoreceptor,** such as that used for taste and smell, might respond when a chemical stimulant reacts with receptor molecules on the plasma membrane of the sensory receptor. Binding of the chemical stimulant to the receptor molecule opens an ion channel, which enables the influx of an ionic current that depolarizes the sensory receptor cell. (This is similar to what is described for ligand-gated channels in Chapter 6.) In Fig. 5.13*B*, the ion channel of a **mechanoreceptor,** such as those in the skin, opens in response to the application of a mechanical force along the membrane, and this allows an influx of current to depolarize the sensory receptor. In Fig. 5.13*C*, the ion channel of a retinal **photoreceptor** cell (so-called because it responds to light) is open in the dark and closed when a photon is absorbed by pigment on an internal disc membrane. In this case, an influx of current occurs in the dark; the current ceases when light is applied. When the current stops, the photoreceptor hyperpolarizes. (Because capture of the photon is distant from the ion channel that it influences, this process must involve an intracellular "second messenger" mechanism.)

The nature of the receptor also can vary. In the simplest situation, a receptor is just a specialized portion of an axon,

TABLE 5.1	**Correlation of Axon Groups, as Revealed by Compound Action Potential Recordings, With Their Functional Properties**			
Electrophysiological Classification of Peripheral Nerves	Classification of Only Afferent Fibers (Class/Group)	Fiber Diameter (μm)	Conduction Velocity (m/sec)	Receptor Supplied
Sensory Fiber Type				
Aα	Ia	13-20	80-120	Primary muscle spindles
Aβ	Ib and II	6-12	35-75	Golgi tendon organ, secondary muscle spindles, skin mechanoreceptors
Aδ	III	1-5	5-30	Skin mechanoreceptors, thermal receptors, nociceptors
C	IV	0.2-1.5	0.5-2	Skin mechanoreceptors, thermal receptors, nociceptors
Motor Fiber Type				
Aα	N/A	8-13	44-78	Extrafusal skeletal muscle fibers
Aγ	N/A	2-8	12-48	Intrafusal muscle fibers
B	N/A	1-3	6-18	Preganglionic autonomic fibers
C	N/A	0.2-2	0.5-2	Postganglionic autonomic fibers

From Haines DE (ed). *Fundamental Neuroscience for Basic and Clinical Applications.* 3rd ed. Philadelphia: Churchill Livingstone; 2006.
N/A, not applicable.

CHEMORECEPTOR

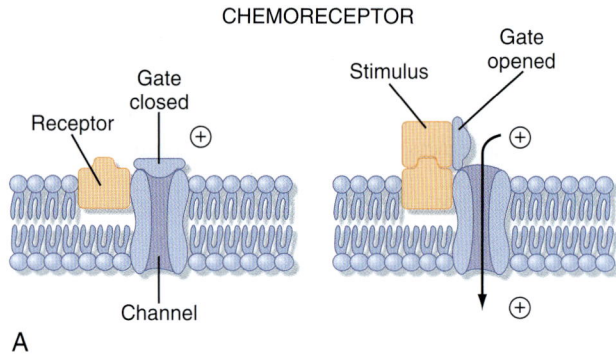

A

MECHANORECEPTOR

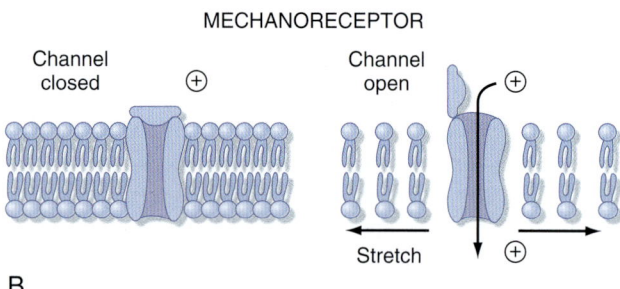

B

PHOTORECEPTOR

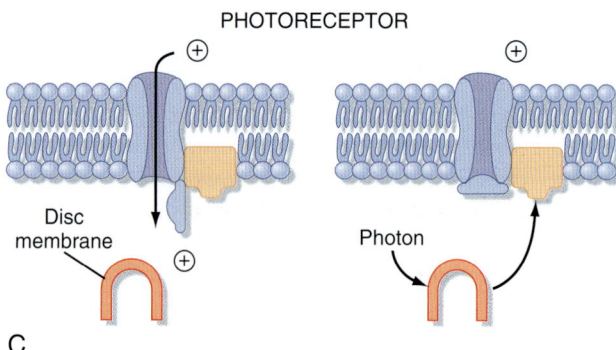

C

• **Fig. 5.13** Models of Transducer Mechanisms in Three Types of Receptors. **A,** Chemoreceptor. **B,** Mechanoreceptor. **C,** Photoreceptor.

in which case the transduction of a stimulus into receptor potential and the translation of this potential into a spike train all take place in the same cell. For example, a mechanical stimulus, such as pressure on the skin of a finger, can distort the membrane of an axon that forms part of a mechanoreceptor, as shown in Fig. 5.14A. This distortion causes inward current flow at the end of the axon and longitudinal and outward current flow along the neighboring parts of the axon. The outward current produces a depolarization (the receptor potential) that might exceed the threshold for an action potential (see Fig. 5.14B). If so, one or more action potentials are evoked and then travel along this primary afferent fiber to the CNS and thereby convey information about the mechanical stimulus.

In many other cases, the receptor is composed of more than one cell. In this situation, transduction occurs in one cell, but spikes are generated in other cells that are

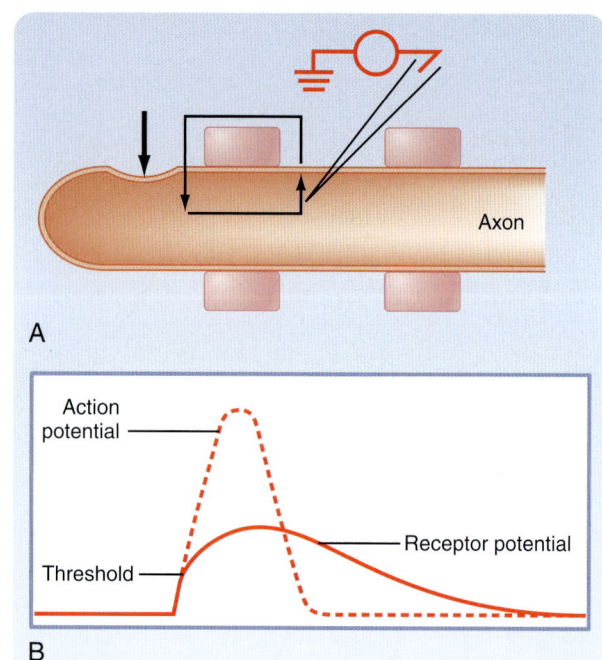

A

B

• **Fig. 5.14 A,** Current flow *(thin arrows)* produced by stimulation *(thick arrow)* of a mechanoreceptor at the tip of an axon. An intracellular recording electrode is placed at the first node of Ranvier. **B,** The receptor potential produced by the current and an action potential that would be superimposed on the receptor potential if it were to exceed threshold at the first node of Ranvier.

synaptically connected to it (see Chapter 6). For example, in the **cochlea,** the primary afferent fibers get synaptic input from mechanoreceptive **hair cells.** Sensory transduction in such sense organs can be more complex in this arrangement. In photoreceptors, moreover, the receptor potential is hyperpolarizing, as mentioned earlier, and interruption of the dark current is the signal event. Information about each of these mechanisms is discussed in Chapter 8.

Although the mechanisms of sensory transduction vary between stimulus types, the end result is typically a receptor potential in either the receptor cell or the primary afferent neuron (i.e., the first neuron in a sensory pathway) that has a synapse with the receptor cell.

Receptive Fields

The relationship between the location of a stimulus and activation of particular sensory neurons is a major theme in the field of sensory physiology. The **receptive field** of a sensory neuron is the region that, when stimulated, affects the activity of that neuron. For example, a sensory receptor might be activated by indentation of only a small area of skin. That area is the **excitatory receptive field** of the sensory receptor. Moreover, a neuron in the CNS might have a receptive field several times as large as that of a sensory receptor because it may receive information from many sensory receptors, each with a slightly different receptive field. The receptive field of that CNS neuron is thus the sum of the receptive fields of the sensory receptors that influence it. The location of the receptive field is determined by the location of the sensory transduction apparatus responsible for signaling information about the stimulus to the sensory neuron.

In general, sensory receptive fields are excitatory. However, a central sensory neuron can have either an excitatory or an inhibitory receptive field or, indeed, a complex receptive field that includes areas that excite it and areas that inhibit it. Examples of such complex receptive fields are discussed in Chapters 7 and 8.

Coding of Information by Action Potentials

Central to CNS function is the transmission of information between neurons. This is accomplished primarily through action potentials, which propagate down the axon to the presynaptic terminals and cause neurotransmitter release, signaling the postsynaptic cells. As already explained, the regenerative nature of action potentials allows them to carry signals regardless of the length of the axon, whereas local signals, such as receptor or synaptic potentials (see Chapter 6), decay with distance and are therefore not suitable for this purpose. The tradeoff, however, is that the all-or-none nature of action potentials means that their shape and size do not generally convey information in the way gradations of local potentials do. Instead, the variations in the rate or timing of action potentials appear to be used primarily as the "codes" for transmission of information between neurons.

Rate coding refers to information being coded in the firing rate of a neuron, where *firing rate* is defined as the number of spikes fired per unit time, usually expressed as spikes/second, also called hertz (Hz). For example, the force of a mechanical stimulus to the skin can be encoded in the firing rate of the primary afferent neuron that innervates the skin; the greater the force applied to the skin, the larger the resulting receptor potential in the primary afferent neuron will be and, as a consequence, the faster the rate of action potentials triggered by the receptor potential will be. Research has shown many neurons employ rate coding in the sense that the firing rate of a neuron shows a consistent relationship to particular parameters of sensory stimuli, upcoming movements, or other aspects of behavior.

The amount of information in such rate codes is constrained by several factors. One factor is a neuron's range of firing rates. The upper limit of this range is set by the maximal frequency that a neuron can fire action potentials, which is determined by the duration of the absolute and relative refractory periods (see Fig. 5.4) and rarely exceeds 1000 Hz. The lower limit of the firing range is, of course, 0 Hz, as neurons cannot fire at negative rates. To avoid this problem, many neurons have spontaneous activity levels. These can be quite high (e.g., some Purkinje cells fire spontaneously at 100 Hz) and let a cell either increase or decrease its activity over a similar range in response to inputs. A second constraining factor is the variability of neuron's firing rate, which determines the resolution of the neuron's information coding.

Timing, or **temporal coding,** refers to spike codes in which the specific timing of spikes rather than the overall firing rate encodes information. One often-studied version of temporal coding is the synchronization of spikes across neurons. Synchronization of neuronal spiking has been shown to occur in a number of brain regions and has been related to function in a number of instances. An advantage of temporal coding is that it can convey information more quickly than can rate coding, inasmuch as it does not require averaging, which takes time. Moreover, rate coding and temporal coding are not mutually exclusive, inasmuch as overall firing rates can be varied while synchronous events are superimposed. Such multiplexing of codes may increase the information transmission capacity of neuronal pathways.

Sensory Coding

Sensory neurons encode information about stimuli. In the process of sensory transduction, one or more aspects of the stimulus must be encoded in a way that can be interpreted by the CNS. The encoded information is an abstraction based on (1) which sensory receptors are activated, (2) the responses of sensory receptors to the stimulus, and (3) information processing in the sensory pathway. Some stimulus parameters that can be encoded include **sensory modality, location, intensity, frequency,** and **duration.** Other aspects of stimuli that are encoded are described in relation to particular sensory systems in later chapters.

A **sensory modality** is a class of sensation. For example, sustained mechanical stimuli applied to the skin result in sensations of touch or pressure, and transient mechanical stimuli may evoke sensations of flutter or vibration. Other cutaneous modalities include cold, warmth, and pain. Vision, audition, taste, and smell are examples of noncutaneous sensory modalities. The specific sensory receptors define the normal energy associated with the modality of a sensory pathway. For example, the visual pathway includes photoreceptors, neurons in the retina, the lateral geniculate nucleus of the thalamus, and the visual areas of the cerebral cortex (see Chapter 8). The normal means of activating the visual pathway is light striking the retina. However, mechanical stimulation (e.g., pressure on the eyeball) or electrical stimulation of neurons in the visual pathway also produce a visual sensation. Thus neurons of the visual system can be regarded as a **labeled line,** which, when activated by whatever means, results in a visual sensation.

The **location** of a stimulus is signaled by activation of the particular population of sensory neurons whose receptive fields are affected by the stimulus. The information may be encoded in the CNS as a neural map. For example, a **somatotopic map** is formed by arrays of neurons in the somatosensory cortex that receive information from corresponding locations on the body surface (see Chapter 7). In the visual system, points on the retina are represented by neuronal arrays that form **retinotopic maps** (see Chapter 8).

Intensity may be encoded in a number of ways. Because action potentials have a uniform magnitude, some sensory neurons encode intensity by their frequency of discharge (rate coding). The relationship between stimulus intensity and response can be plotted as a stimulus-response function. For many sensory neurons, the stimulus-response function approximates an exponential curve with an exponent that can be less than, equal to, or greater than 1. Stimulus-response functions with fractional exponents characterize many **mechanoreceptors.** **Thermoreceptors,** which detect changes in temperature, have linear stimulus-response curves (exponent of 1). **Nociceptors,** which detect painful stimuli, may have linear or positively accelerating stimulus-response functions (i.e., the exponent for these curves is 1 or greater). The positively accelerating stimulus-response functions of nociceptors help explain the urgency that is experienced as the pain sensation increases.

Another way in which stimulus intensity is encoded is according to the number of sensory receptors that are activated. A stimulus at the threshold for perception may activate only one or only a few primary afferent neurons of an appropriate class, whereas a strong stimulus of the same type may recruit many similar receptors. Central sensory neurons that receive input from sensory receptors of this particular class would be more powerfully affected as more primary afferent neurons discharge. Greater activity in central sensory neurons may be perceived as a stronger stimulus.

Stimuli of different intensities may also activate different sets of sensory receptors. The limit of a neuron's firing rate of action potentials can also limit its range of response to a stimulus. However, mechanoreceptors with different thresholds can overcome this problem: Those with low thresholds can signal over a range of low input intensities, whereas others with higher thresholds can signal higher input intensities. Together they allow fine resolution over an extended range of intensities. In addition, still higher intensities might recruit nociceptors, and that will also change the perceived quality of the stimulus.

Stimulus **frequency** can sometimes be encoded by action potentials whose interspike intervals correspond exactly to the intervals between stimuli (e.g., at intervals corresponding

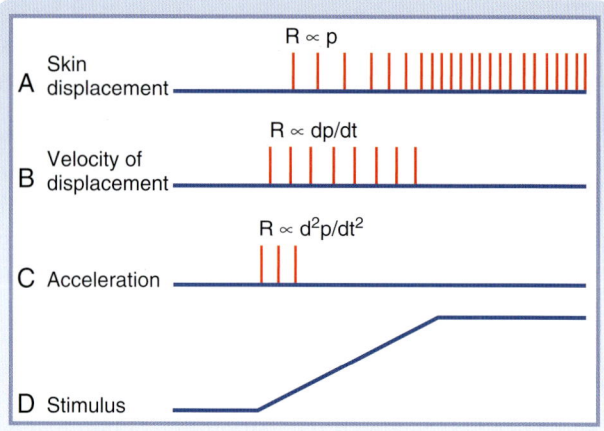

• **Fig. 5.15** Responses of Slowly and Rapidly Adapting Mechanoreceptors to Displacement of the Skin. **A** to **C** are the discharges of primary afferent fibers during a ramp-and-hold stimulus shown in **D**. **A,** The response of a slowly adapting receptor that signals the magnitude and duration of displacement. **B,** The response of a rapidly adapting receptor whose output signals the velocity of displacement. **C,** The response of a different rapidly adapting receptor that responds to acceleration. p, displacement; R, response; t, time.

to that of a low-frequency vibration). However, this mechanism is limited by the firing rate limits of neurons as discussed earlier. When higher frequencies need to be encoded (e.g., the auditory system, which in humans is capable of detecting frequencies up to 20,000 Hz; see Chapter 8), other strategies are needed. Other candidate codes depend on the spatiotemporal patterns of firing across populations of neurons.

The **duration** and the onset and offset of events are encoded by different populations of sensory neurons. For example, slowly adapting receptors in the skin produce a repetitive discharge throughout a prolonged stimulus. However, rapidly adapting receptors produce spikes at the onset (or offset) of the same stimulus. Fig. 5.15 shows the responses of three types of receptors to the slow deflection of the skin, which is depicted in the graph at the bottom of the figure. The functional implication is that different temporal features of a stimulus can be signaled by receptors with different adaptation rates.

Key Concepts

1. Ion channels are integral membrane proteins that have ion-selective pores. An ion channel typically has two states: high conductance (open) and zero conductance (closed). Different regions of an ion channel protein act as gates to open and close the channel. The channel flips spontaneously between the open and closed states.

2. For a voltage-dependent channel, the fraction of time that the channel spends in the open state is a function of the transmembrane potential difference.

3. The action potential is generated by the rapid opening and subsequent voltage inactivation of voltage-dependent Na^+ channels and by the delayed opening and closing of voltage-dependent K^+ channels.

4. The absolute and relative refractory periods result from voltage inactivation of Na^+ channels and the delayed closure of K^+ channels in response to membrane repolarization. These refractory periods limit the firing rate of action potentials.

5. Subthreshold signals and action potentials are conducted along the length of a cell by local circuit currents. Subthreshold signals are conducted only electrotonically, and thus decrease with distance.

6. The action potential is propagated rather than merely conducted; it is regenerated as it moves along the axon. In this way, an action potential retains the same size and shape as it travels along the axon.

7. A large-diameter axon has greater propagation velocity because increased axon diameter lowers axial resistance and allows greater amounts of current to flow farther down the axon.

8. Myelination dramatically increases the conduction velocity of a nerve axon because myelin increases membrane resistance and lowers membrane capacitance. Myelination allows an action potential to be conducted very rapidly from one node of Ranvier to the next. This makes the action potential appear to jump from node to node in a form of conduction called *saltatory conduction.*

9. A receptor responds preferentially to a particular form of stimulus energy. Its receptive field is that part of a sensory domain in which energy can affect the receptor.

10. Receptor potentials are the result of transduction of sensory stimuli. These potentials reflect the specific parameters of the stimulus and, if they exceed threshold, alter the action potential firing patterns of the afferent neurons.

Additional Readings

Fain GL. *Molecular and Cellular Physiology of Neurons.* 2nd ed. Cambridge, MA: Harvard University Press; 2014.

Hille B. *Ion Channels of Excitable Membranes.* 3rd ed. Sunderland, MA: Sinauer Associates; 2001.

Hodgkin AL, Huxley AF. A quantitative description of membrane current and its application to conduction and excitation in nerve. *J Physiol.* 1952;117:500-544.

Johnston D, Wu SM-S. *Foundations of Cellular Neurophysiology.* Cambridge, MA: MIT Press; 1994.

Sakmann B, Neher E. *Single-Channel Recording.* 2nd ed. Philadelphia: Springer; 1995.

6

Synaptic Transmission

LEARNING OBJECTIVES

Upon completion of this chapter the student should be able to answer the following questions:

1. What are the characteristics of electrical synapses?
2. What are the specializations found in the presynaptic and postsynaptic elements of a chemical synapse?
3. What sequence of events connect the arrival of the action potential at the presynaptic terminal to the entry of calcium?
4. What sequence of events connect the entry of calcium at the presynaptic terminal to release of neurotransmitter?
5. What is the quantal hypothesis of synaptic transmission, and how does the presence of miniature end plate potentials support this hypothesis?
6. Why is the reversal potential of a typical EPSP near 0 mV?
7. What distinguishes EPSPs and IPSPs in terms of underlying ionic conductances, effect on membrane potential, and neuronal firing probability?
8. How does an IPSP still inhibit a neuron when its reversal potential is equal to or more positive than the neuron's resting potential?
9. What are the mechanisms by which synaptic effects can change over time?
10. What are the criteria for determining a substance is a neurotransmitter, and what are the major excitatory and inhibitory neurotransmitters?
11. What are the major classes of neurotransmitter receptors?

Synaptic transmission is the major process by which electrical signals are transferred between cells within the nervous system (or between neurons and muscle cells or sensory receptors). Within the nervous system, synaptic transmission is usually conceived of as an interaction between two neurons that occurs in a point-to-point manner at specialized junctions called *synapses*. Two main classes of synapses are distinguished: electrical and chemical. However, as the list of chemical neurotransmitters has grown and as understanding of their mechanisms of action has increased, the definition and conception of what constitutes synaptic transmission has had to be refined and expanded. We no longer think of synaptic transmission as a process that involves only neurons, but now realize that glia form an important element of the synapse and that signaling occurs between neurons and glia. Moreover, in some cases, neurotransmitter released at a synapse will act over a widespread territory rather than just at the synapse from which it is released. Thus, we must either generalize the definition of synaptic transmission or consider classically defined synaptic transmission as but one of several mechanisms by which cells in the nervous system communicate with each other. In this chapter we first describe the classic conception of synaptic transmission (electrical and chemical) and then introduce some of the nontraditional neurotransmitters and discuss how they have forced modifications in our conception of chemical communication between cells in the nervous system.

Electrical Synapses

Although their existence in the mammalian central nervous system (CNS) has been known for a long time, electrical synapses, or gap junctions, between neurons were thought to be of relatively little importance for the functioning of the adult mammalian CNS. Only recently has it become apparent that these synapses are quite common and that they may underlie important neuronal functions.

An electrical synapse is effectively a low-resistance pathway between cells that allows current to flow directly from one cell to another and, more generally, allows the exchange of small molecules between cells. Electrical synapses are present in the CNS of animals from invertebrates to mammals. They are present between glial cells as well as between neurons. Electrical coupling of neurons has been demonstrated for most brain regions, including the inferior olive, cerebellum, spinal cord, neocortex, thalamus, hippocampus, olfactory bulb, retina, and striatum.

A gap junction is the morphological correlate of an electrical synapse (see also Chapter 1). These junctions are plaque-like structures in which the plasma membranes of the coupled cells become closely apposed (the intercellular space narrows to ≈ 3 nm) and filled with electron-dense material (Fig. 6.1). Freeze-fracture electron micrographs of gap junctions display regular arrays of intramembrane particles that correspond to proteins that form the intercellular channels connecting the cells. The typical channel diameter is large (1 to 2 nm), thus making it permeable not only to ions but also other small molecules up to approximately 1 kDa in size.

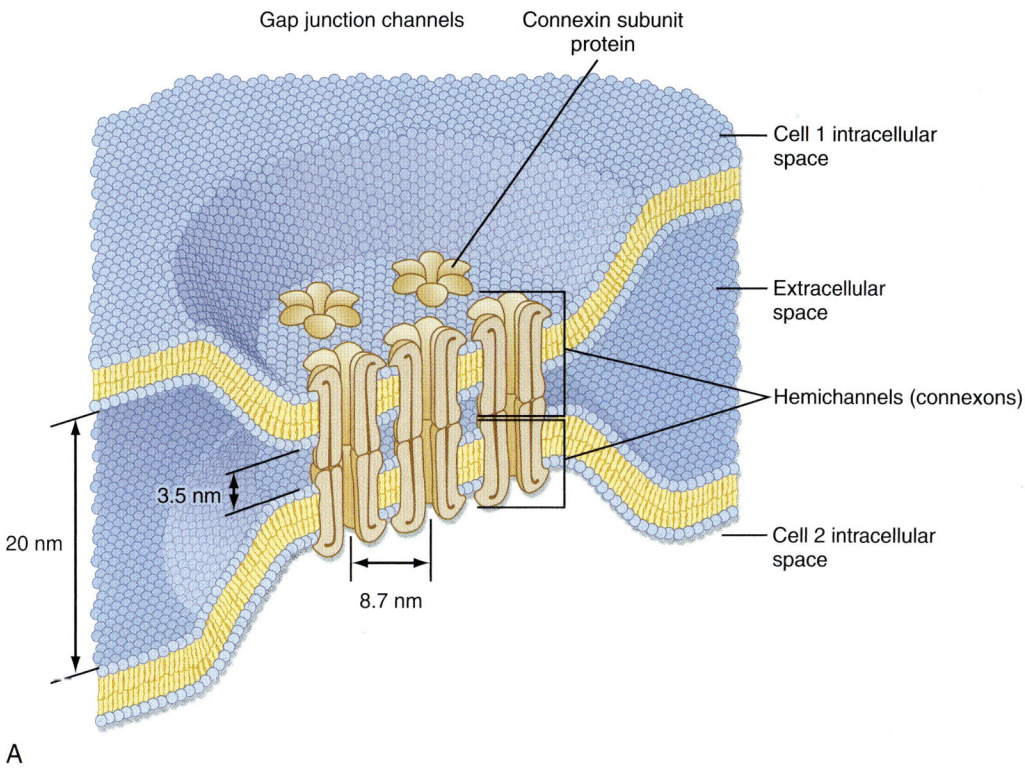

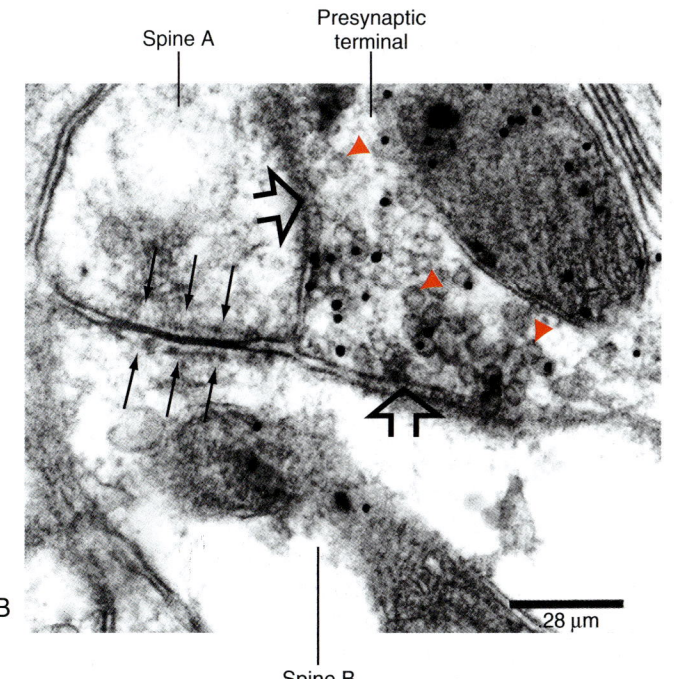

• **Fig. 6.1** Gap junction structure. **A,** Schematic view of the gap junction showing narrowing of the intercellular space to 3.5 nm at the junction. The gap junction has multiple channels, with each channel formed by two connexon hemichannels. Each connexon in turn comprises six connexin subunits. **B,** Electron micrograph of part of a complex synaptic arrangement called a *glomerulus* that is found in the inferior olive and some other CNS regions. Two dendritic spines are coupled by a gap junction *(small black arrows).* An axon terminal packed with synaptic vesicles fills the upper right part of the panel. Large arrowheads point to the electron-dense material that marks the active zones. Black dots are immunogold labeling for GABA, thus identifying this terminal as GABAergic. Red arrowheads point to synaptic vesicles. (From De Zeeuw CI et al. *J Neurosci* 1996;16:3420. Copyright 1996 by the Society for Neuroscience.)

AT THE CELLULAR LEVEL

Each gap junction channel is formed by two hemichannels (called *connexons*), one contributed by each cell. Each connexon, in turn, is a hexamer of connexin protein subunits, which are encoded for by a gene family of at least 21 different members in mammals. (A second family of proteins that form gap junctions, the pannexins, has also been identified.) Gap junctions formed by different connexins have distinct biophysical properties (gating and conductance) and cellular distributions. Although at least 10 connexin types are expressed in the CNS, connexin 36 (connexins are named according to their molecular weight; thus, the number refers to the approximate molecular weight of the connexin in kilodaltons) is the major neuronal connexin in the adult CNS. Other connexin types found in the CNS form gap junctions between glial cells or are primarily expressed transiently during development.

Electrical synapses are fast (essentially no synaptic delay) and bidirectional (i.e., current generated in either cell can flow across the gap junction to influence the other cell). In addition they act as **low-pass filters.** That is, slow electrical events are much more readily transmitted than are fast signals such as action potentials. One important role for neuronal gap junctions appears to be synchronization of network activity. For example, the activity of inferior olivary neurons is normally synchronized but becomes uncorrelated when pharmacological blockers of gap junctions are injected into the inferior olive. It also appears that the patterns of electrical coupling by gap junctions may be highly specific. For example, neocortical interneurons almost exclusively couple to interneurons of the same type. This specific gap junction–coupling pattern suggests that multiple, independent, electrically coupled networks of interneurons may coexist across the neocortex.

Finally, although electrical synapses are generally regarded as relatively simple and static in comparison to chemical synapses, they may actually be fairly dynamic entities. For example, the properties of electrical synapses can be modulated by several factors, including voltage, intracellular pH, and [Ca^{++}]. Moreover, they are subject to regulation by G protein–coupled receptors, and connexins (the protein subunits that form a gap junction, see At The Cellular Level) contain sites for phosphorylation. These factors can change the coupling between cells by causing changes in single-channel conductance, formation of new gap junctions, or removal of existing ones.

Chemical Synapses

Chemical synaptic transmission was first demonstrated between the vagus nerve and the heart by a simple experiment by Otto Loewi. The vagus nerve of a frog was stimulated to slow the heart rate down while the solution perfusing the heart was collected. This solution was then used to perfuse a second heart, whose beating then also slowed, demonstrating that the vagal nerve stimulation had caused a chemical to be released into the solution. The chemical responsible was found to be acetylcholine, which we now know is also a neurotransmitter at the neuromuscular junction and at other synapses in the peripheral and central nervous systems.

Unlike the situation at electrical synapses, at chemical synapses there is no direct communication between the cytoplasm of the two cells. Instead the cell membranes are separated by a synaptic cleft of some 20 nm, and interaction between the cells occurs via chemical intermediaries known as **neurotransmitters.** Chemical synapses are generally unidirectional, and thus one can refer to the presynaptic and postsynaptic elements that are diagrammed in Fig. 6.2. The **presynaptic** element is often the terminal portion of an axon and is packed with small vesicles whose exact shape and size vary with the neurotransmitter they contain. In addition, the presynaptic membrane apposed to the **postsynaptic** element has regions, known as **active zones**, of electron-dense material that corresponds to the proteins involved in transmitter release (see Fig. 6.1B). Moreover, mitochondria and rough endoplasmic reticulum are typically found in the presynaptic terminal. The postsynaptic membrane is also characterized by electron-dense material, which in this case corresponds to the receptors for the neurotransmitter.

Chemical synapses occur between different parts of neurons. Traditionally, focus has been placed on synapses formed by an axon onto the dendrites or soma of a second cell (**axodendritic** or **axosomatic synapses**), and our description will be based primarily on such synapses. However, there are many additional types of chemical synapses, such as **axoaxonic** (axon to axon), **dendrodendritic** (dendrite to dendrite), and **dendrosomatic** (dendrite to soma). Furthermore, complex synaptic arrangements are possible, such as mixed synapses, in which cells form both electrical and chemical synapses with each other; serial synapses, in which an axoaxonic synapse is made onto the axon terminal and influences the efficacy of that terminal's synapse with yet a third element; and reciprocal synapses, in which both cells can release transmitter to influence the other. Fig. 6.1B shows a complex synaptic arrangement called a *glomerulus* that involves both chemical and electrical synapses among the participating elements.

Much of what we know about chemical synapses comes from the study of two classic preparations, the frog neuromuscular junction (the synapse from a motor neuron onto a muscle fiber) and the squid giant synapse (the synapse from a second-order neuron onto third-order neurons that innervate the muscle of the squid's mantle; i.e., the motor neurons whose axons were used to characterize the conductances underlying the action potential [see Chapter 5]). The principles governing transmission at these synapses mostly apply to synapses within the mammalian CNS as well, at least with regard to synapses using what are called the "classic" neurotransmitters (see the section Neurotransmitters). Thus, much of the following discussion will be based on results from these two preparations; however, some differences in CNS synapses will also be pointed out.

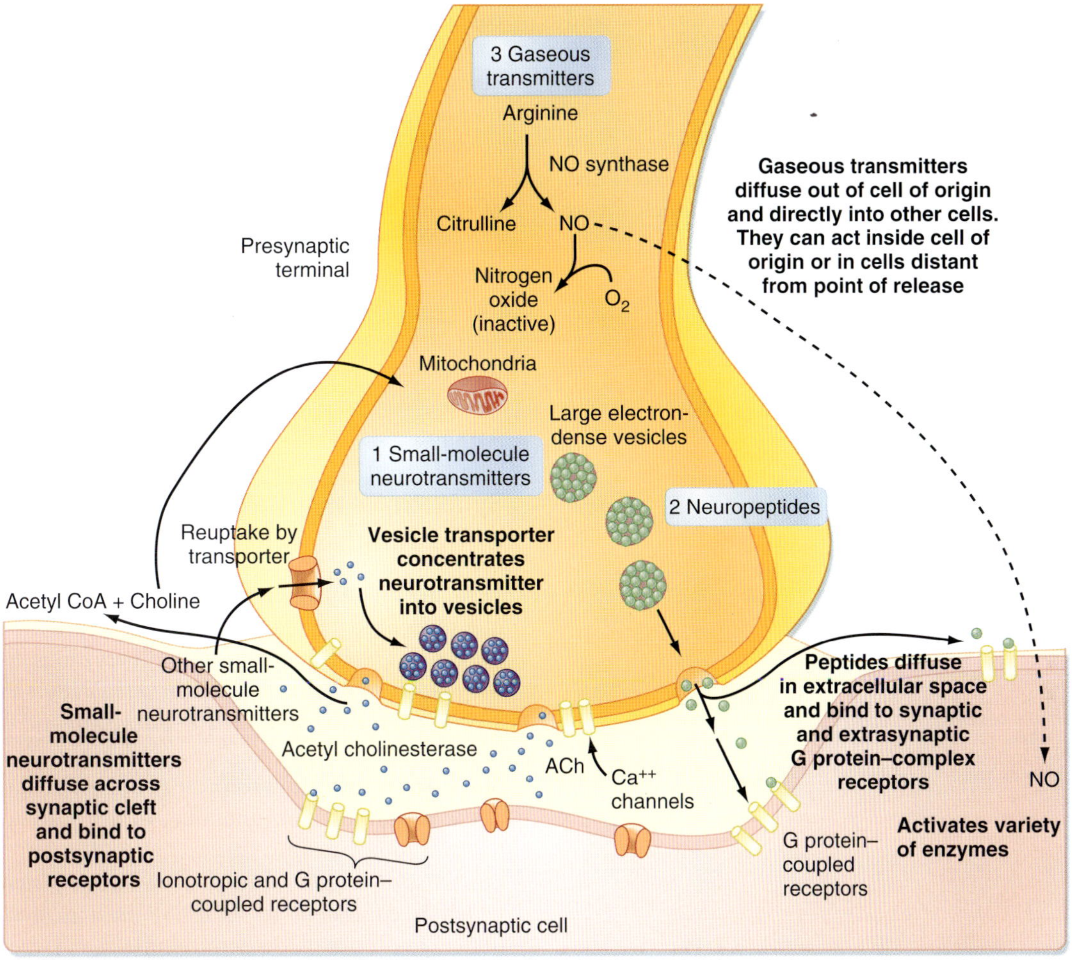

• Fig. 6.2 Schematic of a chemical synaptic terminal releasing all three main classes of neurotransmitter. For each, the mechanisms of release, sites of action, and mechanisms for termination of activity are shown. Real synapses release transmitter from one or more classes.

Synaptic transmission at a chemical synapse may be summarized as follows. Synaptic transmission is initiated by arrival of the action potential at the presynaptic terminal. The action potential depolarizes the terminal, which causes Ca^{++} channels to open. The subsequent rise in $[Ca^{++}]$ within the terminal triggers the fusion of vesicles containing neurotransmitter with the plasma membrane. The transmitter is then expelled into the synaptic cleft, diffuses across it, and binds to specific receptors on the postsynaptic membrane. Binding of transmitter to receptors then causes the opening (or less often, the closing) of ion channels in the postsynaptic membrane, which in turn results in changes in the potential and resistance of the postsynaptic membrane that alter the excitability of the cell. The changes in membrane potential of the postsynaptic cell are termed **excitatory** and **inhibitory postsynaptic potentials** (**EPSPs** and **IPSPs**) (Fig. 6.3), depending on whether they increase or decrease, respectively, the cell's excitability, which can be defined as its probability of firing action potentials. The transmitter acts for only a very short time (milliseconds) because reuptake and degradation mechanisms rapidly clear the transmitter from the synaptic cleft.

The succeeding sections will amplify specific points of this summary. However, it is worth mentioning at this point that some of the nonclassic types of neurotransmitters (e.g., neuropeptides and gaseous neurotransmitters such as nitric oxide) and the discovery of **metabotropic receptors** have required modifications of several aspects of this basic conception. (Whereas an **ionotropic receptor** usually contains the ion channel as an integral part of itself, a metabotropic receptor does not contain an ion channel but instead is coupled to a G protein that initiates second messenger cascades that can ultimately affect ion channels.) Some of the differences between classic and peptide transmitters are listed in Table 6.1. More details on the properties of peptide and gaseous transmitters are provided in the relevant parts of the Neurotransmitters section of this chapter, and metabotropic receptors are covered in the Receptors section.

Calcium Entry Is the Signal for Transmitter Release

Depolarization of the presynaptic membrane by the action potential causes voltage-gated Ca^{++} channels to open, which makes it possible for Ca^{++} to flow into the terminal and trigger the release of transmitter. However, Ca^{++} will enter the terminal only if there is a favorable electrochemical gradient to do so. Recall that it is the combination of the concentration and voltage gradients that determines the direction of ion flow through open channels. Extracellular [Ca^{++}] is high relative to intracellular [Ca^{++}], which favors entry into the terminal; however, during the peak of the action potential, the membrane potential is positive, and the voltage gradient opposes the entry of Ca^{++} because of its positive charge. Thus, at the peak of the action potential, relatively little Ca^{++} enters the terminal because although the membrane is highly permeable to Ca^{++}, the overall driving force is small. In fact, by using a voltage clamp, one can experimentally make the membrane potential positive and

equal to the Nernst equilibrium potential for Ca^{++}. If this is done, no Ca^{++} will enter the terminal despite Ca^{++} channels being open, and as a result no transmitter is released and no postsynaptic response is observed. This voltage is known as the **suppression potential.** If the membrane potential is rapidly made negative again (because of either the end of the action potential or by adjusting the voltage clamp), Ca^{++} rushes into the terminal as a result of the large driving force (which arises instantaneously on repolarization) and the high membrane permeability to Ca^{++} (which remains high because it takes the Ca^{++} channels several milliseconds to close in response to the new membrane potential), thereby resulting in release of transmitter and a postsynaptic response (Fig. 6.4).

Synaptic Vesicles and the Quantal Nature of Transmitter Release

How neurotransmitter is stored and how it is released are questions fundamental to synaptic transmission. Answering these questions began with two observations. The first was the discovery of small round or irregularly shaped organelles known as *synaptic vesicles* in presynaptic terminals by electron microscopy (see Figs. 6.1B and 6.2). The second observation came from recordings of postsynaptic responses at the neuromuscular junction. Normally an action potential in a motor neuron causes a large depolarization in the postsynaptic muscle, termed an **end plate potential (EPP),** which is equivalent to an EPSP in a neuron. However, under conditions of low extracellular [Ca^{++}], the EPP amplitude is reduced (because the presynaptic Ca^{++} current is reduced, leading to a smaller rise in intracellular [Ca^{++}], and transmitter release is proportional to [Ca^{++}]). In this condition, the EPP is seen to fluctuate among discrete values (Fig. 6.5). Moreover, small spontaneous depolarizations of the postsynaptic membrane, termed **miniature end plate potentials (mEPPs),** are observable. The amplitude of the mEPP (≤ 1 mV) corresponds to that of the smallest EPP evoked under low [Ca^{++}], and the amplitudes of other EPPs were shown to be integral multiples of the mEPP amplitude; thus, it was proposed that each mEPP corresponded

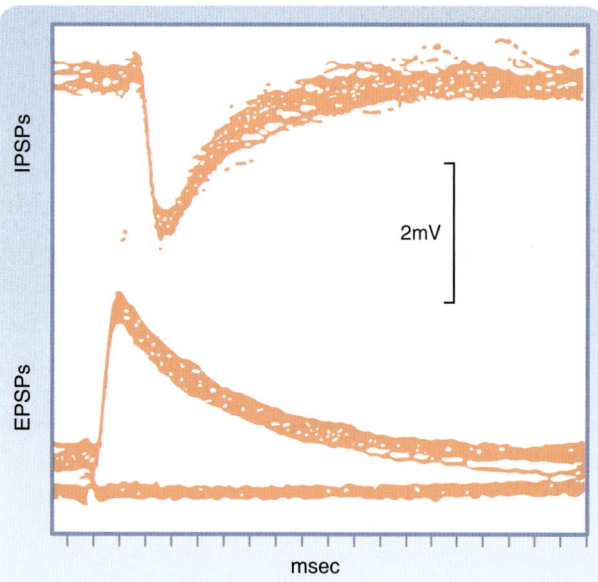

• **Fig. 6.3** IPSPs and EPSPs recorded with a microelectrode in a cat spinal motor neuron in response to stimulation of appropriate peripheral afferent fibers. Forty traces are superimposed. Note that these IPSPs are hyperpolarizing, but in some cases IPSPs can be depolarizing—see text for an explanation. (Redrawn from Curtis DR, Eccles JC. *J Physiol* 1959;145:529.)

TABLE 6.1	Distinctions Between Classic Nonpeptide Neurotransmitters and Peptide Neurotransmitters	
Nonpeptide Transmitters	**Peptide Transmitters**
Synthesized and packaged in the nerve terminal	Synthesized and packaged in the cell body; transported to the nerve terminal by fast axonal transport
Synthesized in active form	Active peptide formed when it is cleaved from a much larger polypeptide that contains several neuropeptides
Usually present in small clear vesicles	Usually present in large electron-dense vesicles
Released into a synaptic cleft	May be released some distance from the postsynaptic cell There may be no well-defined synaptic structure
Action of many terminated because of uptake by presynaptic terminals via Na$^+$-powered active transport	Action terminated by proteolysis or by the peptide diffusing away
Typically, action has short latency and short duration (msec)	Action may have long latency and may persist for many seconds

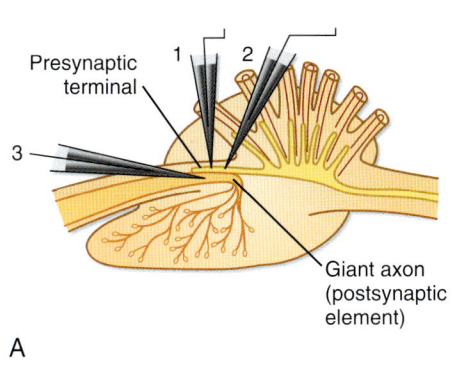

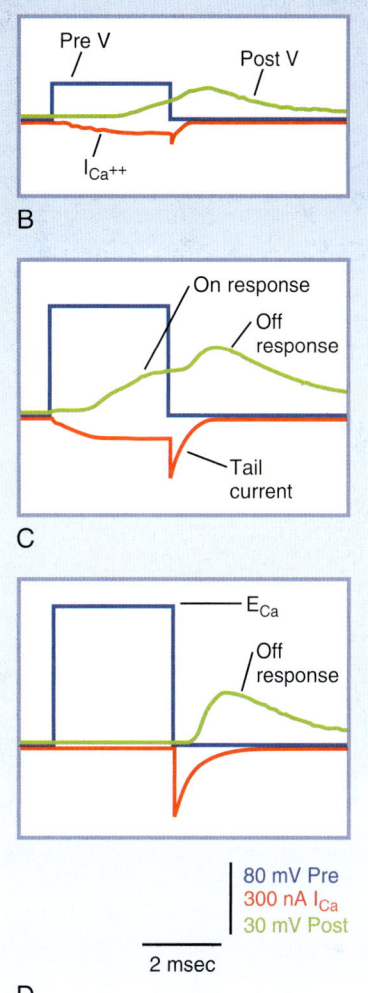

• **Fig. 6.4** Presynaptic Ca^{++} current and its relationship to the postsynaptic response. **A,** Schematic of a squid giant synapse preparation. Electrodes 1 and 2 are used to voltage-clamp the presynaptic terminal and record its voltage and current. (Note that tetrodotoxin and tetraethyl ammonium were present to block Na^{+} and K^{+} conductance to isolate the Ca^{++} conductance.) Electrode 3 records the membrane potential of the postsynaptic axon. The presynaptic terminal was voltage-clamped to increasingly more depolarized levels *(blue traces)*. With a small depolarization **(B),** a small Ca^{++} current starts shortly after the voltage step, continues to grow for the duration of the step (on current), and then decays exponentially after its termination (off or tail current). A larger voltage step **(C)** increases both the on and the off components of the Ca^{++} current, and now distinct on and off responses are observed in the postsynaptic response. **D,** The voltage step is to the Nernst potential for Ca^{++}, so there is no Ca^{++} current during the step, but a large tail current and off response are observed. (Based on data of Llinas R et al. *Biophys J* 1981;33:323.)

to the release of transmitter from a single vesicle and that EPPs represented the combined simultaneous release of transmitter from many vesicles.

This linking of mEPPs and vesicles implies that each mEPP is caused by the action of many molecules of neurotransmitter binding to postsynaptic receptors. The alternative that each mEPP could be caused by a single transmitter molecule binding to and opening a single postsynaptic receptor was rejected, in part because responses smaller in amplitude than mEPPs could be generated experimentally by directly applying dilute solutions of acetylcholine to the muscle. In fact, mEPPs were calculated to be caused by the action of approximately 10,000 molecules, which corresponds well to estimates of the number of neurotransmitter molecules contained within a single vesicle.

Many additional studies have confirmed the vesicle hypothesis of neurotransmitter release. For example, biochemical studies have shown that neurotransmitter is concentrated in vesicles, and fusion of vesicles to the plasma membrane and their depletion in the terminal cytoplasm after action potentials have been shown with electron microscopic techniques.

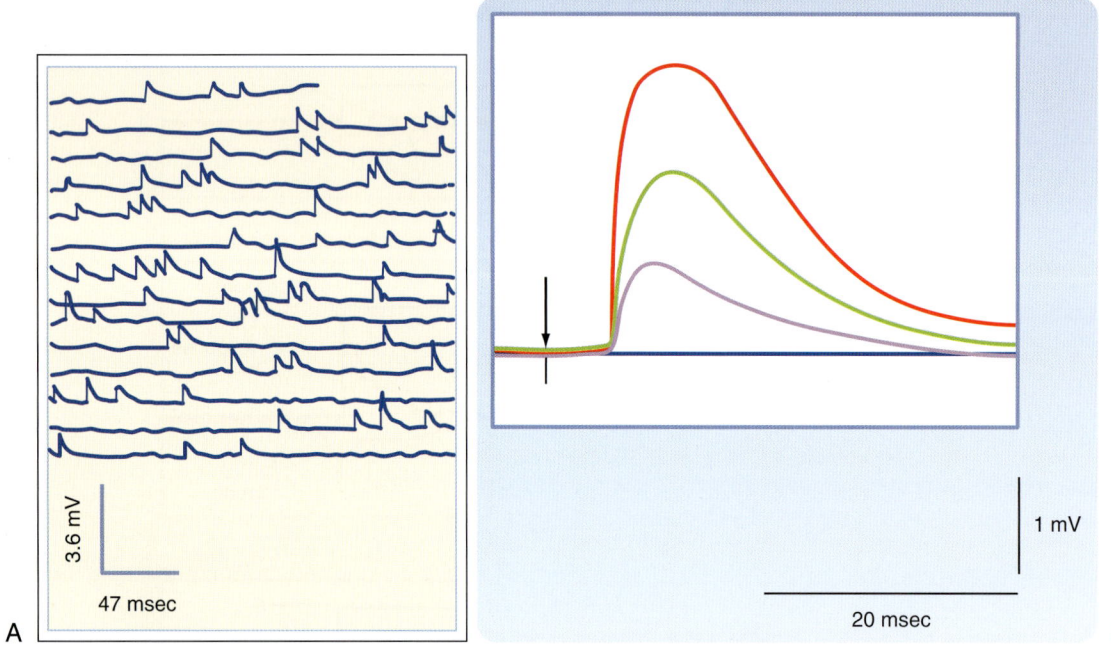

A

B

• **Fig. 6.5 A,** Spontaneous mEPPs recorded at a neuromuscular junction in a fiber of frog extensor digitorum longus. **B,** EPPs evoked by nerve stimulation under low-[Ca^{++}] conditions, which reduce the probability of transmitter release. The small-amplitude EPPs evoked under these conditions vary in amplitude in a step-like manner, where the size of the step is equal to the smallest EPP, which in turn equals the size of the mEPPs. (Note that in these conditions the stimulus often fails to evoke any response, as indicated by a flat response.) (**A,** Data from Fatt P, Katz B. *Nature* 1950;166:597; **B,** data from Fatt P, Katz B. *J Physiol* 1952;117:109.)

Molecular Apparatus Underlying Vesicular Release

The small vesicles that contain nonpeptide neurotransmitters can fuse with the presynaptic membrane only at specific sites called *active zones.* To become competent to fuse with the presynaptic membrane at an active zone, a small vesicle must first dock at the active zone and then undergo a priming process. Once primed the vesicle can fuse and release its transmitter into the synaptic cleft in response to an increase in local cytoplasmic [Ca^{++}]. On the order of 25 proteins may play roles in docking, priming, and fusion. Some of these proteins are cytosolic, whereas others are proteins associated with the vesicle membrane or the presynaptic plasma membrane. The functions of most of these proteins are incompletely understood; however, knowledge of the molecular details of transmitter release has increased dramatically in recent years.

As with other exocytotic processes, neurotransmitter release involves **SM** (sec1/Munc18-like) and **SNARE** (**s**oluble **N**-ethyl maleimide-sensitive factor **a**ttachment protein **r**eceptor) proteins: v-SNARES in the vesicle membrane and t-SNARES in the (target) presynaptic plasma membrane. Zipper-like interactions between **synaptobrevin** (a v-SNARE) and **syntaxin** and **SNAP-25** (both are t-SNARES) with the assistance of SM proteins bring the vesicle membrane and the presynaptic plasma membrane close together before fusion. The SNARE proteins are targets for various **botulinum toxins,** which disrupt synaptic transmission, thus demonstrating their critical role in this process. Nevertheless, they do not bind Ca^{++}, so another protein must be the Ca^{++} sensor that triggers the actual fusion event. Evidence indicates that a **synaptotagmin** protein is almost certainly the Ca^{++} sensor and, even more specifically, that the second of its two cytoplasmic domains contains the Ca^{++} binding site. Interestingly, synaptotagmins differ in their kinetics, and brain regions vary as to which synaptotagmin family member acts as the Ca^{++} sensor for vesicular fusion. Thus, differential expression of synaptotagmin genes in neurons may be a mechanism to adapt the kinetics of vesicle release and thereby tailor the specific characteristics of synaptic transmission to the functional needs of each CNS region.

Calcium channels are located in the active zone membrane at sites adjacent to the docked vesicles. When they open, a small area of high [Ca^{++}], a microdomain is created at the active zone. This local high concentration (which lasts for less than a millisecond), allows the rapid binding of Ca^{++} to synaptotagmin, triggering the fusion of a docked vesicle and allowing release of its neurotransmitter. Despite the multiple steps involved, the process of vesicular release at a synapse is extremely rapid because of the close proximity of the molecular apparatuses involved to each other. Indeed, the time from Ca^{++} influx to vesicle fusion is about 0.2 msec.

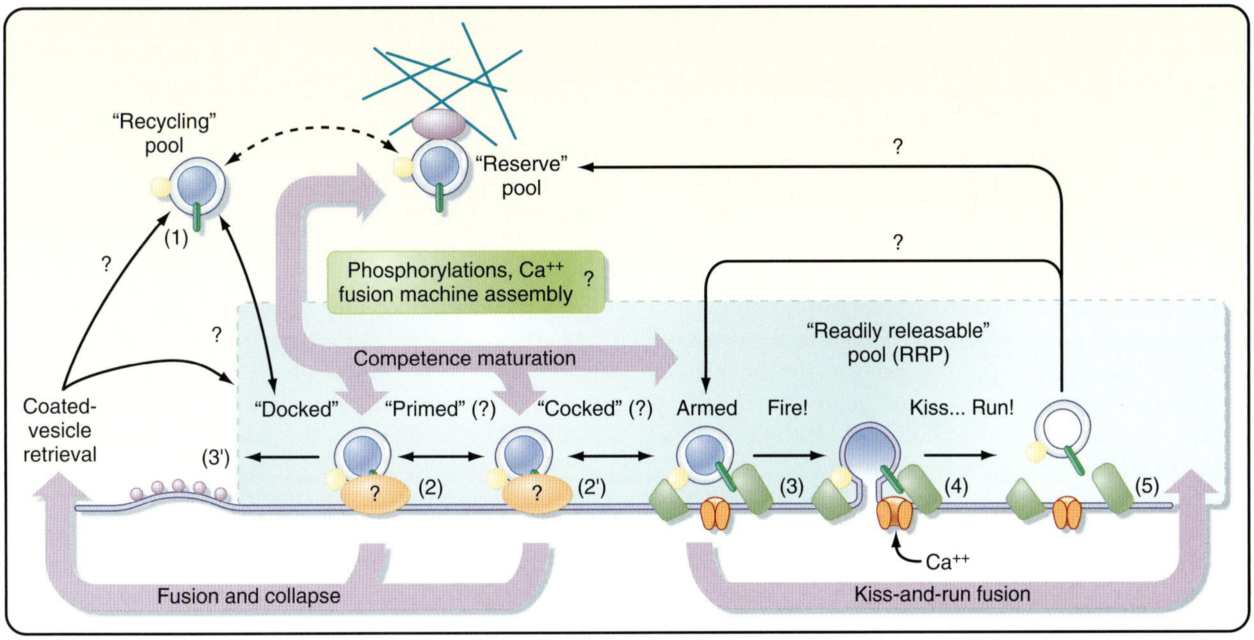

• **Fig. 6.6** Vesicle recycling pathways. Synaptic vesicles have been thought to fuse with the membrane while emptying their contents and then be recycled by forming clathrin-coated pits that are endocytosed to form coated vesicles (1 → [2 or 2′] → 3′ → 1). An alternative pathway that may allow more rapid recycling of vesicles has been proposed. This pathway, called "kiss and run," involves only transient fusion of the vesicle to the presynaptic membrane to form a pore through which the vesicle contents may be emptied, followed by detachment of the vesicle from the membrane (1 → 2 → 3 → 4 → 5 → 1). (Redrawn from Valtorta F, Meldolesi J, Fesce R. *Trends Cell Biol* 2001;11:324.)

Synaptic Vesicles Are Recycled

During synaptic transmission, vesicles must fuse with the plasma membrane to release their contents into the synaptic cleft. However, there must be a reverse process; otherwise, not only would it be hard to sustain the vesicle population, but the presynaptic membrane's surface area would also grow with each bout of synaptic transmission, and its molecular content and functionality would likewise change (because, as just discussed, the protein content of the vesicle membrane is distinct from that of the terminal membrane).

There appear to be two distinct mechanisms by which vesicles are retrieved after release of their neurotransmitter content (Fig. 6.6). One mechanism is the endocytotic pathway commonly found in most cell types. Coated pits are formed in the plasma membrane, which then pinch off to form coated vesicles within the cytoplasm of the presynaptic terminal. These vesicles then lose their coat and undergo further transformations (i.e., acquire the correct complement of membrane proteins and be refilled with neurotransmitter) to become once again synaptic vesicles ready for release.

Evidence for a second, more rapid recycling mechanism has been obtained (see Fig. 6.6). It involves transient fusion of the vesicle to the synaptic membrane and has been called "kiss and run." In this case, fusion of the vesicle with the synaptic membrane leads to the formation of a pore through which the transmitter is expelled, but there is no wholesale collapse of the vesicle into the membrane. Instead, the

duration of the fusion is very brief, after which the vesicle detaches from the plasma membrane and reseals itself. Thus, the vesicle membrane retains its molecular identity. Its contents can then simply be replenished, thereby making the vesicle ready for use again.

The relative importance of these two mechanisms is still being debated. However, at central synapses, which tend to be small and contain relatively few vesicles in comparison to the neuromuscular junction, the rapid time course of the kiss-and-run mechanism may help avoid the problem of vesicle depletion and the consequent failure of synaptic transmission during periods of high activity (many neurons in the CNS can show firing rates of several hundred hertz, and a few types of neuron can fire at rates of ≈ 1000 Hz).

Postsynaptic Potentials

Following vesicle fusion the neurotransmitter molecules are released and diffuse across the synaptic cleft (a very rapid process) and bind to receptors on the postsynaptic membrane. This binding leads to the opening (or less often the closing) of ion channels. These channels are termed **ligand-gated** because their opening and closing are primarily controlled by the binding of neurotransmitter. This mechanism can be contrasted with that of the **voltage-gated** channels underlying the action potential, whose opening and closing are determined by the membrane potential. However, there are some channels, most notably the NMDA (*N*-methyl-D-aspartate) channel, that are both ligand and voltage gated.

It is also worth noting here that what follows in this section refers to what happens when neurotransmitter binds to receptors in which the ion channel is part of the receptor itself. These receptors are referred to as **ionotropic** receptors and underlie what is now called "fast" synaptic transmission. There is also "slow" synaptic transmission, mediated by what are called **metabotropic** receptors, in which the receptor and ion channel are not part of the same molecule, and binding of neurotransmitter to the receptor initiates biochemical cascades that lead to postsynaptic potentials with slow onsets (see the section Receptors for details). Despite the differing time courses, many of the same basic principles apply to both types of postsynaptic potential.

EPSPs. As stated earlier, the binding of neurotransmitter generally changes the membrane potential of the postsynaptic cell, and these changes are referred to as *EPSPs* when they increase the excitability of the neuron and *IPSPs* when they inhibit the neuron from firing action potentials. EPSPs are always depolarizing potentials, and IPSPs are usually hyperpolarizing.

Once a ligand-gated channel is open, the direction of current flow through it is determined by the electrochemical driving force for the permeant ion(s). It turns out that the pores of most channels that underlie EPSPs are relatively large and therefore allow passage of most cations with similar ease. As an example, consider the acetylcholine-gated channel that is opened at the neuromuscular junction. Na^+ and K^+ are the major cations present (Na^+ extracellularly and K^+ intracellularly); therefore, the net current through the channel is approximately the sum of the Na^+ and K^+ currents ($I_{net} = I_{Na} + I_K$). Recall that the current through a channel from a particular ion is dependent on two factors: the conductance of the channel to the ion and the driving force on the ion. This relationship is expressed by the equation

Equation 6.1

$$I_x = g_x \times (V_m - E_x)$$

where g_x is the conductance of the channel to ion x, V_m is the membrane potential, and E_x is the Nernst equilibrium potential for ion x. In this case g_x is similar for Na^+ and K^+, so the main determinant of net current is the relative driving forces ($V_m - E_x$). If the membrane is at its resting potential (typically around −70 mV), there is a strong driving force ($V_m - E_{Na}$) for Na^+ to enter the cell because this potential is far from the Na^+ Nernst potential (about +55 mV), whereas there is only a small driving force for K^+ to leave the cell because V_m is close to the K^+ Nernst potential (about −90 mV). Thus, if acetylcholine-gated channels open when the membrane is at its resting potential, a large inward Na^+ current and a small outward K^+ current will flow through the acetylcholine channel, thereby resulting in a net inward current, which acts to depolarize the membrane.

The net inward current that results from opening such channels is called the **excitatory postsynaptic current (EPSC).** Fig. 6.7A contrasts the time course of the EPSC and the resulting EPSP for fast synaptic transmission. The EPSC is much shorter (≈1 to 2 msec in duration) and corresponds to the time the channels are actually open. The short duration of the EPSC is due to the fact that the released neurotransmitter remains in the synaptic cleft for only a short while before being either enzymatically degraded or taken up by either glia or neurons. Binding and unbinding of a neurotransmitter to its receptor take place rapidly, so once its concentration falls in the cleft, the postsynaptic receptor channels rapidly close as well and terminate the EPSC. Note how the end of the EPSC corresponds to the peak of the EPSP, which is followed by a long tail. The duration of the tail and the rate of the decay in EPSP amplitude reflect the passive membrane properties of the cell (i.e., its RC properties) (see Chapter 5). In slow synaptic transmission, the duration of the EPSP reflects the activation and deactivation of biochemical processes more than the membrane properties. The long duration of even fast EPSPs (relative to EPSCs and action potentials) is functionally important because it allows EPSPs to overlap and thereby summate. Such summation is central to the integrative properties of neurons (see the next section, Synaptic Integration).

Normally an EPSP depolarizes the membrane, and if this depolarization reaches threshold, an action potential is generated. However, consider what happens if the channels underlying the action potential are blocked and the membrane of the postsynaptic cell is experimentally depolarized by injecting current through an intracellular electrode. Because the membrane potential is now more positive, the driving force for Na^+ is decreased and that for K^+ increased. If the synapse is activated at this point, the net current through the receptor channel (the EPSC) will be smaller because of changes in the relative driving force. This implies that if the membrane potential is depolarized enough, there will be a point at which the Na^+ and K^+ currents through the channel are equal and opposite, and thus there is no net current and no EPSP. If the membrane is depolarized beyond this point, there is a net outward current through the receptor channels and the membrane will hyperpolarize (i.e., the EPSP will be negative). Thus, the potential at which there is no EPSP (or EPSC) is known as the **reversal potential.** For excitatory synapses, the reversal potential is usually around 0 mV (±10 mV), depending on the synapse (see Fig. 6.7B–C).

It is worth noting that a reversal potential is a key criterion for demonstrating the chemical-gated as opposed to the voltage-gated nature of a synaptic response because currents through voltage-gated channels do not reverse, except at the Nernst potential of the ion for which they are selective (and then only if the channel is open at that potential). Consequently, beyond a certain membrane potential, no current will flow through voltage-gated channels because they will be closed. In contrast, ligand-gated channels can be opened at any membrane potential and thus can always have a net current flow through them, except at one specific voltage, the reversal potential.

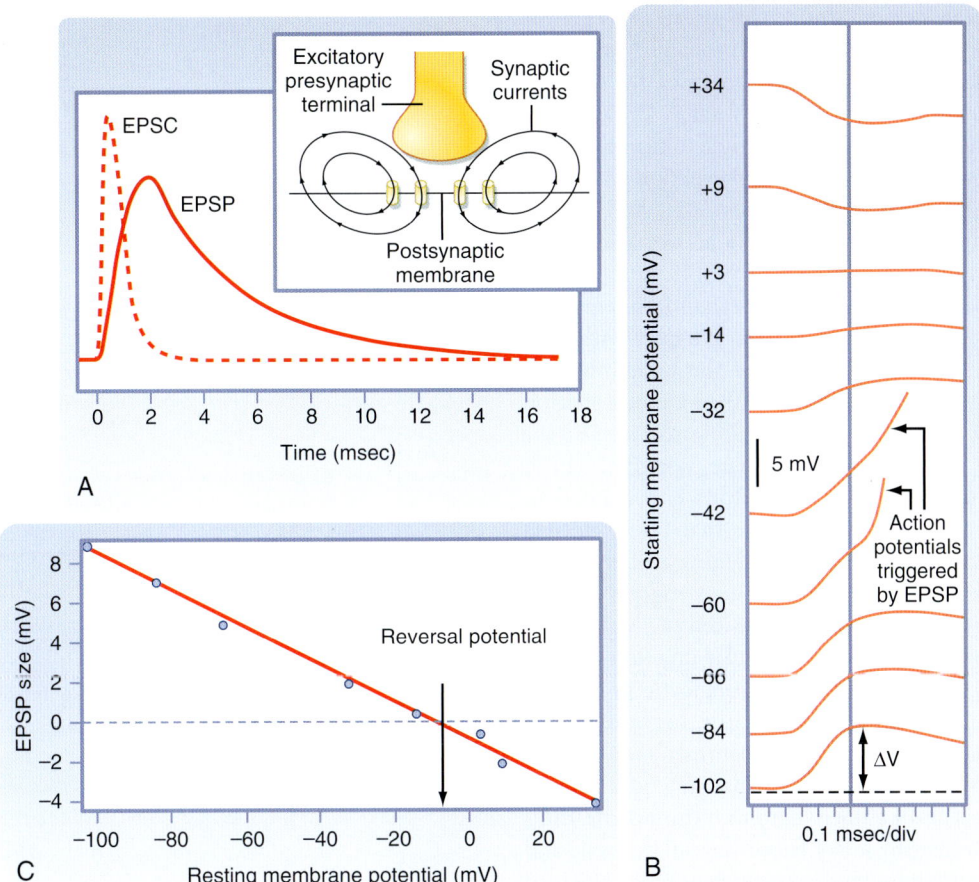

• Fig. 6.7 Properties of EPSPs. **A,** Time course of a fast EPSP compared with that of the underlying EPSC. In many cases, such as this one, the EPSC is much shorter than the EPSP; however, sometimes the EPSC can have a fairly extensive tail. **B,** Intracellularly recorded EPSPs at different levels of depolarization. EPSPs were evoked in motor neurons by stimulation of Ia afferents. The number to the left of each trace indicates the membrane potential induced by injection of current through the electrode. At initial membrane potentials of −42 and −60 mV, the EPSP triggered an action potential. At more depolarized levels, Na^+ channels are inactivated, so no spike occurs. **C,** To determine the EPSP reversal potential, the initial membrane potential is plotted against the size of the EPSP (ΔV). This EPSP reversed at −7 mV. (**A,** Data from Curtis DR, Eccles JC. *J Physiol* 1959;145:529; **B,** data from Coombs JS et al. *J Physiol* 1955;130:374.)

IPSPs. Like EPSPs, IPSPs are triggered by the binding of neurotransmitter to receptors on the postsynaptic membrane and typically involve an increase in membrane permeability as a result of the opening of ligand-gated channels. They differ in that IPSP channels are permeable to only a single ionic species, either Cl^- or K^+. Thus, IPSPs will have a reversal potential equal to the Nernst potential of the ion carrying the underlying current. Typically the Nernst potential for these ions is somewhat negative relative to the resting potential, so when IPSP channels open, there is an outward flow of current through them that results in hyperpolarization of the membrane (see Fig. 6.3).

However, in some cells, activation of an inhibitory synapse may produce no change in potential (if the membrane potential equals the Nernst potential for Cl^- or K^+) or may actually result in a small depolarization. Nevertheless, in both these cases, the reversal potential for the IPSP is still negative with regard to the threshold for eliciting an action potential (otherwise it would increase the probability of the cell spiking and by definition be an EPSP). It may seem counterintuitive that something that depolarizes the membrane can still be considered inhibitory, but if it decreases the probability of spiking, then it is indeed inhibitory (a further explanation is given in the Synaptic Integration section).

In sum, starting from the resting membrane potential, EPSPs are always depolarizing, IPSPs can be either depolarizing or hyperpolarizing, and a hyperpolarizing potential is always an IPSP. Thus, the key distinction between inhibitory and excitatory synapses (and IPSPs and EPSPs) is how they affect the probability of the cell firing an action potential: EPSPs increase the probability, whereas IPSPs decrease the probability.

Safety factor. Synapses between cells vary in strength and thus in the size of the PSP generated in the postsynaptic cell. Many factors determine synaptic strength, including the size and number of synaptic contacts between two cells, its activity level and past history, and the probability of vesicle fusion for the synapse. For excitatory synapses, the strength of the synapse may be quantified by what is known as its **safety factor** (the ratio of EPSP amplitude to the amplitude needed to reach the threshold to trigger an action potential). Most synapses have low safety factors (<1), and thus it takes the summed EPSPs of multiple active synapses to trigger an action potential in the postsynaptic neuron. This summation process is at the core of synaptic integration, which is taken up in the next section. Synapses with high (>1) safety factors exist, however, and the neuromuscular junction is one prominent example. When a motor neuron action potential triggers release of neurotransmitter at the neuromuscular junction, an end plate potential (EPP; the equivalent of an EPSP in a neuron) is generated in the muscle fiber. The EPP is so large that under normal circumstances it depolarizes the sarcolemma well above the action potential threshold and thus always triggers a spike, leading to contraction of the muscle cell. A high safety factor makes sense for the neuromuscular junction because each muscle cell is contacted by only a single motor neuron, and if that motor neuron is firing, the nervous system has basically made the decision to contract that muscle. In certain diseases of the neuromuscular junction, such as myasthenia gravis and Lambert-Eaton syndrome, the EPPs are reduced such that the safety factor can fall below 1, and thus the EPPs sometimes fail to trigger action potentials in the muscle fibers, leading to weakness.

Synaptic Integration

The overall effect of a particular synapse is dependent on its location. To understand this concept fully, we must first recall that action potentials are typically generated at the initial segment of the cell because it has the highest density of voltage-gated Na^+ channels and therefore the lowest threshold for initiation of a spike. Thus, it is the summed amplitudes of the synaptic potentials at this point, the initial segment, that is critical for the decision to spike. EPSPs generated by synapses close to the initial segment (i.e., synapses onto the soma or proximal dendrites) will result in a larger depolarization at the initial segment than will EPSPs generated by synapses on distal dendrites (Fig. 6.8A single action potential in axon 2 versus 1). This is because the cell membrane is leaky and synaptic currents are generated locally at the synapse, so even if two synapses generate a local EPSC of the same size, less of the initial current will arrive at the initial segment from the more distal synapse than from the more proximal one, thereby resulting in the generation of a smaller EPSP at the initial segment by the distal synapse (see discussion of length constant in Chapter 5). Thus, the synapse's spatial location in the dendritic tree is an important determinant of its efficacy. However, as already mentioned, EPSPs generated by most CNS synapses, even those in favorable positions (i.e., close to the initial segment), are too small by themselves to reach

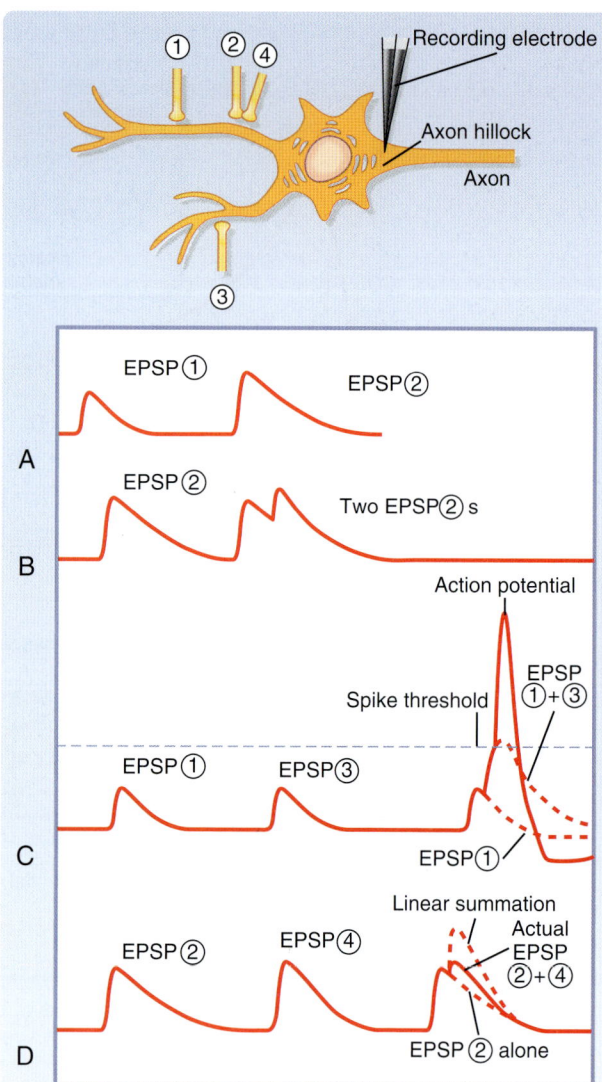

• **Fig. 6.8** Synaptic integration of EPSPs recorded at the axon hillock adjacent to the initial segment. **A,** Comparison of EPSPs evoked by proximal versus distal synapses (2 versus 1). **B,** Temporal summation. EPSPs in response to two spikes in the same axon occurring in rapid succession (axon 2). **C,** Spatial summation. Responses evoked by synapses that are electrically distant from each other (1 and 3). **D,** Sublinear summation of two synapses located near each other because of shunting (2 and 4).

the spiking threshold in the postsynaptic cell, as illustrated in Fig. 6.8A, where an action potential in either axon 1 (distal) or 2 (proximal) both produce EPSPs that are too small to trigger a spike. Thus, generally the summed EPSPs from multiple synapses are required to reach threshold and trigger a spike.

The requirement for multiple EPSPs to summate in order to trigger a spike is what makes the relatively long duration of EPSPs so important. **Temporal summation** refers to the fact that EPSPs that are separated by a latency less than their duration can sum. This is illustrated in Fig. 6.8B, where the same synapse is activated multiple times in rapid succession (axons can fire action potentials at rates well over 100 Hz); in this situation, successive EPSPs will be less than 10 msec apart and therefore overlap and sum. Note the higher amplitude of the second peak.

Spatial summation refers to the fact that synaptic potentials generated by different synapses can interact. For example, in Fig. 6.8, suppose axon 1 and 3 each fire an action potential but at widely separated times. Each produces an EPSP that depolarizes the cell but is too small to reach threshold (see Fig. 6.8C, EPSP1, EPSP3). Instead, if both axons fire within a short enough time of each other, their effect can be additive, as shown in Fig. 6.8C *(EPSP 1+3)*. The combined EPSP amplitude may then reach threshold and lead to spiking of the cell. If the EPSPs generated by axons 1 and 3 were simultaneous, then we would have an example of pure spatial summation. In the example shown, however, the times of the two EPSPs were slightly separated, thus we have both spatial and temporal summation present. The fact that EPSPs have a long time course (when compared with action potentials or the underlying EPSCs) facilitates both types of synaptic integration.

In the foregoing example, the combined EPSP was approximately the linear summation of the two individual EPSPs evoked by action potentials in axons 1 and 3. This is the case when two synapses are far apart. If the two synapses are close together, such as for axons 2 and 4 (see Fig. 6.8D), the summation becomes less than linear because of what is known as a **shunting effect.** That is, when synapse 2 is active, channels are opened in the cell membrane, which means that it is more leaky. Therefore, when synapse 4 is also active, more of its EPSC will be lost (shunted) through the dendritic membrane, and less current will be left to travel down the dendrite to the initial segment. The result is that synapse 4 causes a smaller EPSP at the initial segment than it would have generated in isolation. Nevertheless, the combined EPSP is still larger than an EPSP caused by either synapse 2 or 4 alone.

Where do IPSPs fit into synaptic integration? In many cases one can think of them as negative EPSPs. Thus, whereas EPSPs add together to help bring the membrane potential up to and beyond the spiking threshold, IPSPs subtract from the membrane potential to make it more negative and therefore further from threshold. In deciding whether to spike, a cell adds up the ongoing EPSPs and subtracts the IPSPS to determine whether the sum reaches threshold. As with an EPSP, the efficacy of an IPSP varies with its location.

In addition to subtracting algebraically from the membrane potential, IPSPs exert an inhibitory action via the shunting mechanism, just as was described earlier for EPSPs. That is, while the IPSP channels are open, they make the membrane more leaky (i.e., lower its resistance) and thereby reduce the size of EPSPs, thus making them less effective. This shunting mechanism explains how IPSPs that do not change the membrane potential—or even those that slightly depolarize it—can still decrease the excitability of the cell. An alternative way to look at this effect is to view each synapse as a device that tries to bring the membrane potential to its own equilibrium potential. Because this potential is below the action potential threshold in the case of IPSPs, IPSPs make it harder for the cell to spike.

Thus far the interaction of synaptic potentials has been presented under the assumption that the postsynaptic cell membrane is passive (i.e., it acts as though it were simply resistors and capacitors in parallel with each other). However, it is clear that the dendrites and somas of most, if not all, neurons contain active elements (i.e., gated channels) that can amplify and alter EPSPs and IPSPs. For example, a distal EPSP can have a larger-than-expected effect if the EPSP activates dendritically-located voltage-gated Na^+ or Ca^{++} channels that boost its amplitude or even generate propagated dendritic action potentials. Another example is Ca^{++}-activated K^+ channels that are present in the dendrites of some neurons. These channels are activated by the influx of Ca^{++} either through synaptic channels or via dendritic voltage-gated Ca^{++} channels opened by EPSPs and can cause long-lasting hyperpolarizations that effectively make the cell inexcitable for tens to hundreds of milliseconds. As a final example, there are some Ca^{++} channels that underlie a low-threshold Ca^{++} spike. These channels are normally inactive at resting membrane potentials, but the hyperpolarization that results from a large IPSP can de-inactivate them and allow them to open (and produce a spike) after termination of the IPSP. In this case "inhibition" actually increases the cell's excitability. In sum, synaptic integration is a highly complex, nonlinear process. Nevertheless, the basic principles just described remain at its core.

Modulation of Synaptic Activity

Integration of synaptic input by a postsynaptic neuron, as described in the previous section, represents one aspect of the dynamic nature of synaptic transmission. A second aspect is that the strength of individual synapses can vary as a function of their use or activity. That is, a synapse's current functional state reflects, to some extent, its history.

Activation of a synapse typically produces a response in the postsynaptic cell (i.e., a postsynaptic potential) that will be roughly the same each time, assuming the postsynaptic cell is in a similar state. Certain patterns of synaptic activation, however, result in changes in the response to subsequent activation of the synapse. Such use-related changes may remain for short (milliseconds) or long (minutes to days) durations and may be either a potentiation or suppression of the synapse's strength. These changes probably underlie cognitive abilities such as learning and memory. Thus, the processes by which activity results in changes in a synapse's efficacy are a critical feature of synaptic transmission.

Paired-Pulse Facilitation

When a presynaptic axon is stimulated twice in rapid succession, it is often found that the postsynaptic potential evoked by the second stimulus are larger in amplitude than the one evoked by the first (Fig. 6.9). This increase

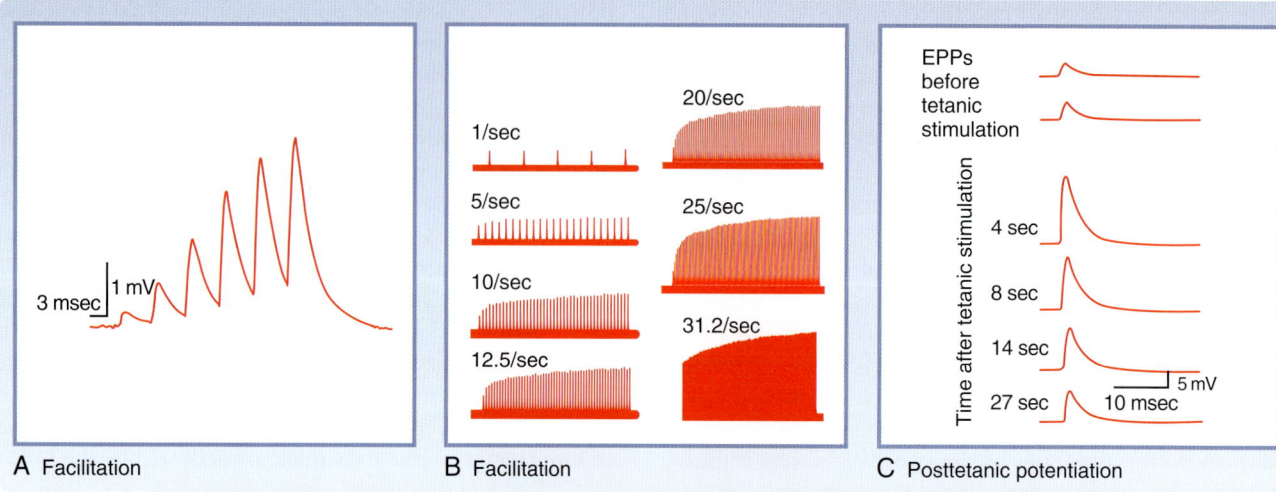

• **Fig. 6.9 A,** Facilitation at a neuromuscular junction. EPPs at a neuromuscular junction in toad sartorius muscle were elicited by successive action potentials in the motor axon. Neuromuscular transmission was depressed by 5 mM Mg^{++} and 2.1 mM curare so that action potentials did not occur. **B,** EPPs at a frog neuromuscular junction elicited by repetitively stimulating the motor axon at different frequencies. Note that facilitation failed to occur at the lowest frequency of stimulation (1/sec) and that the degree of facilitation increased with increasing frequency of stimulation in the range of frequency used. Neuromuscular transmission was inhibited by bathing the preparation in 12 to 20 mM Mg^{++}. **C,** Posttetanic potentiation at a frog neuromuscular junction. The top two traces indicate control EPPs in response to single action potentials in the motor axon. Subsequent traces indicate EPPs in response to single action potentials after tetanic stimulation (50 impulses/sec for 20 seconds) of the motor neuron. The time interval between the end of tetanic stimulation and the single action potential is shown on each trace. The muscle was treated with tetrodotoxin to prevent generation of action potentials. (**A,** Redrawn from Belnave RJ, Gage PW. *J Physiol* 1977;266:435; **B,** redrawn from Magelby KL. *J Physiol* 1973;234:327; **C,** redrawn from Weinrich D. *J Physiol* 1971;212:431.)

is known as **paired-pulse facilitation (PPF).** Note PPF is distinct from temporal summation, in which two EPSPs overlap and sum to a larger response; with PPF the second EPSP itself is greater in size. If one plots the relative size of the two postsynaptic potentials (PSPs) as a function of the time between two stimuli, the amount of increase in the second PSP will be seen to depend on the time interval. Maximal facilitation occurs at around 20 msec, followed by a gradual reduction in facilitation as the interstimulus interval continues to increase; with intervals of several hundred milliseconds, the two PSPs are equal in amplitude and no facilitation is observed. Thus, PPF is a relatively rapid and short-lasting change in synaptic efficacy.

Posttetanic Potentiation

Posttetanic potentiation (PTP) is similar to PPF; however, in this case the responses are compared before and after stimulation of the presynaptic neuron tetanically (tens to hundreds of stimuli at a high frequency). Such a tetanic stimulus train causes an increase in synaptic efficacy (see Fig. 6.9C). PTP, like PPF, is an enhancement of the postsynaptic response, but it lasts longer: tens of seconds to several minutes after the cessation of tetanic stimulation.

Numerous experiments have shown that PPF and PTP are the result of changes in the presynaptic terminal and do not generally involve a change in the sensitivity of the postsynaptic cell to transmitter. Rather, the repeated stimulation leads to an increased number of quanta of transmitter being released. This increase is thought to be due to residual amounts of Ca^{++} that remain in the presynaptic terminal after each stimulus and help potentiate subsequent release of transmitter. However, the exact mechanism or mechanisms by which this residual Ca^{++} enhances release is not yet clear. The residual Ca^{++} does not, however, appear to act simply by binding to the same sites as the Ca^{++} that enters at the active zone and directly triggers vesicle fusion in response to the action potential.

Synaptic Depression

Use of a synapse can also lead to a short-term depression in its efficacy. Most commonly, the postsynaptic cell at such a fatigued or depressed synapse responds normally to transmitter applied from a micropipette; hence, as was the case for PPF and PTP, the change is presynaptic. In general, the depression is thought to reflect depletion of the number of releasable presynaptic vesicles. Thus, short-term depression of synaptic transmission is most often and most easily seen at synapses in which the probability of release after a single stimulus is high and under conditions that favor release (i.e., high [Ca^{++}]). A postsynaptically related cause of synaptic depression can be desensitization of the receptors in the postsynaptic membrane.

Both potentiation and depressive processes can occur at the same synapse. So in general, the type of modulation observed will depend on which process dominates. This in turn can reflect stimulus parameters, local ionic conditions, and the properties of the synapse. In particular, synapses have different baseline probabilities for releasing vesicles. Synapses with a high release probability will be more likely to show poststimulus depression, whereas those with low release probability are less likely to deplete their vesicle store and thus can be facilitated more easily. Sometimes mixed responses can occur. For example, during a tetanic stimulus train a synapse may show a depressed response, but after the train the synapse can show posttetanic facilitation once the vesicles are recycled.

Presynaptic Receptors Can Modulate Transmitter Release

Just as the postsynaptic membrane contains receptors for neurotransmitters, so does the presynaptic membrane. When these presynaptic receptors bind neurotransmitter, they cause events that can modulate subsequent release of transmitter by the terminal. There are several sources of transmitter that bind to presynaptic receptors: it can be the transmitter released by the terminal itself (i.e., self-modulation, in which case the receptors are referred to as *autoreceptors*), it can be released by another presynaptic terminal that synapses onto the terminal (a serial synapse), or it can be a nonsynaptically acting neurotransmitter (see the section Neurotransmitters).

Presynaptic receptors can be either ionotropic or metabotropic. In the latter case, recall that their action will be relatively slow in onset and long in duration and the effect will depend on the specific second messenger cascades that are activated. Such cascades can ultimately regulate presynaptic voltage-gated Ca^{++} and K^+ channels and other presynaptic proteins and thereby alter the probability of vesicle release.

In contrast, activation of presynaptic ionotropic receptors will directly alter the electrical properties of the presynaptic terminal and cause rapid transient (millisecond time scale) changes in the probability of vesicle release (although they too can have much longer lasting effects). Binding of an ionotropic receptor will open channels in the presynaptic terminal and thereby alter the amount of transmitter released by an action potential.

Presynaptic inhibition refers to occasions when binding of presynaptic receptors leads to a decrease in release of transmitter, and it can be the result of one or more mechanisms (Fig. 6.10). First, opening of channels decreases membrane resistance and creates a current shunt. The shunt acts to divert the current associated with the action potential from the active zone membrane and thereby lessens the depolarization of the active zone, which results in less activation of Ca^{++} channels, less Ca^{++} entry, and less release of transmitter. A second mechanism is the change in membrane potential caused by the opening of presynaptic

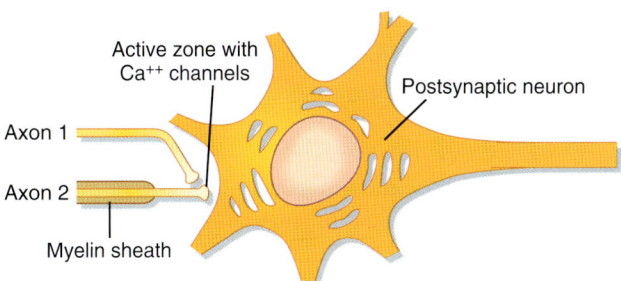

• **Fig. 6.10** Presynaptic inhibition. Active regeneration of action potentials in axon 2 ends at the last node. The action potential is then passively conducted into the terminal. Axon 1 makes an axoaxonic synapse with axon 2. Activation of this synapse reduces conduction of the action potential in axon 2 to the active zone of its synaptic terminal by mechanisms described in the text. This reduces the opening of voltage-gated Ca^{++} channels and therefore release of neurotransmitter.

ionotropic channels. If a small depolarization is the result, there will be inactivation of voltage-gated Na^+ channels and thereby lessening of the action potential–associated current and transmitter release. Presynaptic γ-aminobutyric acid A receptors ($GABA_A$) occur in the spinal cord and mediate presynaptic inhibition by these mechanisms. They control Cl^- channels. Generally, opening of Cl^- channels generates a hyperpolarization. However, in the presynaptic terminal, the $[Cl^-]$ gradient is such that Cl^- flows out of the cell and generates a small depolarization. This depolarization is small enough that it does not cause significant opening of voltage-gated Ca^{++} channels; otherwise, it would increase release of transmitter (presynaptic facilitation). In fact, there are other receptors that control cation channels and create large depolarizations, thereby increasing the release of transmitter. In addition, presynaptic nicotinic acetylcholine receptors control a cation channel that is permeable to Ca^{++}. By allowing additional entry of Ca^{++}, these receptors increase the release of transmitter from the terminal.

Long-Term Changes in Synaptic Strength

Repetitive stimulation of certain synapses in the brain can also produce more persistent changes in the efficacy of transmission at these synapses, a process called **long-term potentiation** or **long-term depression.** Such changes can persist for days to weeks and are believed to be involved in the storage of memories.

The increased synaptic efficacy that occurs in long-term potentiation probably involves both presynaptic (greater transmitter release) and postsynaptic (greater sensitivity to transmitter) changes, in contrast to the short-term changes that involve changes only in presynaptic function. Entry of calcium into the postsynaptic region is an early step required for initiating the changes that result in long-term enhancement of the response of the postsynaptic cell to neurotransmitter. Entry of calcium occurs through NMDA and some AMPA (α-amino-3-hydroxy-5-methyl-4-isoxazole propionic acid) receptors (classes of glutamate receptors; see the section Receptors). Entry of Ca^{++} is believed to activate

Ca^{++}-calmodulin kinase II, a multifunctional protein kinase that is present in very high concentrations in postsynaptic densities. In the presence of high [Ca^{++}], this kinase can phosphorylate itself and thereby become active. Calcium-calmodulin kinase II is believed to phosphorylate proteins that are essential for the induction of long-term potentiation. Long-term potentiation may also have an anatomic component. After appropriate stimulation of a presynaptic pathway, the number of dendritic spines and the number of synapses on the dendrites of postsynaptic neurons may increase rapidly. Changes in the presynaptic nerve terminal may also contribute to long-term potentiation. The postsynaptic neuron may release a signal (nitric oxide has been suggested) that enhances release of transmitter by the presynaptic nerve terminal.

Neurotransmitters

Neurotransmitters are the substances that mediate chemical signaling between neurons. For a substance to be considered a neurotransmitter, it must meet several generally recognized criteria. First, the substance must be demonstrated to be present in the presynaptic terminal, and the cell must be able to synthesize the substance. It should be released on depolarization of the terminal. Finally, there should be specific receptors for it on the postsynaptic membrane. This last criterion is certainly true for substances that act as synaptic transmitters, but if we want to be inclusive and include substances that act over widespread territories rather than just at a single synapse, the last criterion needs to be relaxed to include situations in which receptors are located at sites outside the synapse. *Neurotransmission* has been suggested as a general term to describe both synaptic and nonsynaptic signaling between cells.

More than 100 substances have been identified as potential neurotransmitters because they have met some (hence the "potential" qualifier) or all of these criteria. These substances can be subdivided into three major categories: small-molecule transmitters, peptides, and gaseous transmitters. The small-molecule neurotransmitters may be further subdivided into acetylcholine, amino acids, biogenic amines, and purines. The first three groups of small-molecule transmitters contain what are considered the classic neurotransmitters. Remaining transmitters are substances that are more recent additions to the list of neurotransmitters, although many of them have been known as biologically important molecules in other contexts for a long time.

Small-Molecule Neurotransmitters
Acetylcholine

In the peripheral nervous system, acetylcholine is the transmitter at neuromuscular junctions, at sympathetic and parasympathetic ganglia, and of the postganglionic fibers from all parasympathetic ganglia and a few sympathetic ganglia. It is also a transmitter within the CNS, most prominently of neurons in some brainstem nuclei, in several parts of the basal forebrain (septal nuclei and nucleus basalis) and basal ganglia, and in the spinal cord (e.g., motor neuron axon collaterals). Cholinergic neurons from the basal forebrain areas project diffusely throughout the neocortex and to the hippocampus and amygdala, and they have been implicated in memory functions. Indeed, degeneration of these cells occurs in Alzheimer's disease, a form of dementia in which memory function is gradually and progressively lost.

IN THE CLINIC

A number of drugs known as **anticholinesterases** interfere with acetylcholinesterase and thereby enhance the action of acetylcholine by prolonging its presence at its synapses. Such drugs include insecticides and chemical warfare agents, as well as some therapeutic drugs, such as those used to treat **myasthenia gravis.** Myasthenia gravis is an autoimmune disease in which antibodies bind to acetylcholine receptors at the neuromuscular junction, thereby disrupting their functionality and causing them to be more rapidly degraded. This reduction in receptors leads to severe weakness and ultimately paralysis. The weakness is characterized by rapid tiring of the muscle with repeated use. Rapid tiring occurs because the number of presynaptic vesicles available for release drops during the high-frequency train of motor neuron action potentials that generates such contractions. Normally, because of the high safety factor of the neuromuscular junction, smaller but still suprathreshold EPPs would still be generated and maintain muscle contraction during repeated use. In people with myasthenia gravis, the safety factor is so reduced by the loss of acetylcholine receptors that the decrease in release of acetylcholine with repeated activity leads to EPPs that fail to trigger spikes, and thus muscular contraction fails. Standard treatments include anticholinesterases, which allow a greater concentration of acetylcholine to partially overcome the deficit caused by the reduced number of functional postsynaptic receptors, and immunosuppressive therapies and plasma exchange, which reduce levels of autoantibodies against the acetylcholine receptor. These therapies are all relatively nonspecific and can therefore have many side effects. Potential future therapies are being developed and include inducing tolerance to the acetylcholine receptor and selective destruction of the B cells that make antibodies against the receptor.

Acetylcholine is synthesized from acetyl coenzyme A and choline by the enzyme choline acetyltransferase, which is located in the cytoplasm of cholinergic presynaptic terminals. After synthesis, acetylcholine is concentrated in vesicles. After release, the action of acetylcholine is terminated by the enzyme acetylcholinesterase, which is highly concentrated in the synaptic cleft. Acetylcholinesterase hydrolyzes acetylcholine into acetate and choline. The choline is then taken up by an Na$^+$ symporter in the presynaptic membrane for the resynthesis of acetylcholine. The extracellular enzymatic degradation of acetylcholine is unusual for a neurotransmitter inasmuch as the synaptic action of other classic neurotransmitters is terminated via reuptake by a series of specialized transporter proteins.

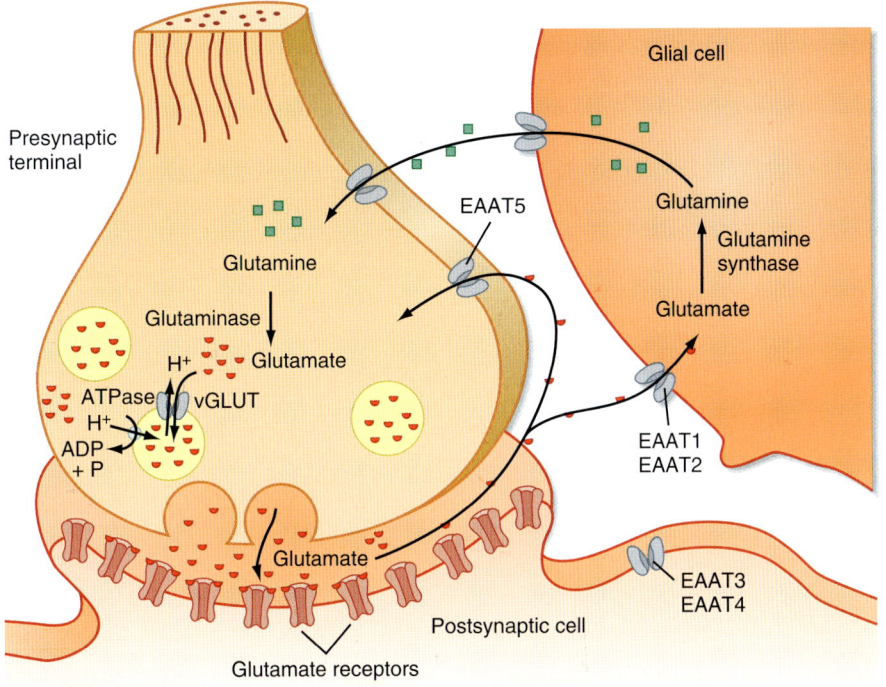

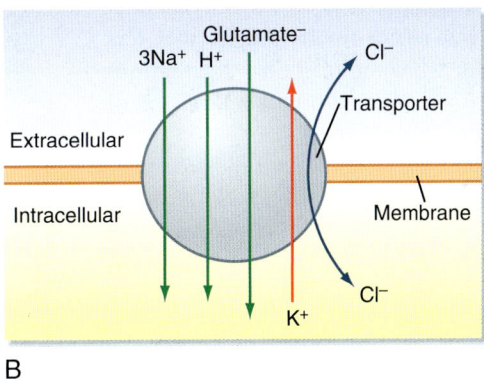

• **Fig. 6.11** Glutamate transport cycle. **A,** Schematic shows the fate of glutamate released from a presyn-aptic terminal. Distinct glutamate transporters exist on the presynaptic and postsynaptic cell membranes for reuptake. In addition, glial cells take up glutamate and convert it to glutamine. The glutamine is then released and taken into the presynaptic terminal, where it is converted back to glutamate before being repackaged into synaptic vesicles. **B,** Schematic of transporter showing direction of ion flow associated with the movement of glutamate across the membrane.

Amino Acids

A variety of amino acids function as neurotransmitters. The three most important are glutamate, glycine, and GABA.

Glutamate is the neurotransmitter at the overwhelming majority of excitatory synapses throughout the CNS. Despite its ubiquity, it was initially difficult to identify specific neurons as glutamatergic because glutamate is present in all cells; it has a key role in multiple metabolic pathways, and it is a precursor to GABA, the major inhibitory neurotransmitter. Nevertheless, experimental results have now clearly established glutamate as the major excitatory CNS neurotransmitter. When applied to cells, it causes

depolarization and is released from neurons, and specific receptors and transporters for it have been identified.

In addition to being the main excitatory neurotransmitter, glutamate is a potent neurotoxin at high concentrations. Thus, strict limitation of glutamate's activity after its release from the presynaptic terminal is necessary, not only to allow normal synaptic transmission but also to prevent cell death. This task is accomplished by specialized membrane transporter proteins (Fig. 6.11 and At the Cellular Level box).

GABA and **glycine** act as inhibitory neurotransmitters. GABA is the major inhibitory transmitter throughout the nervous system. GABA is produced from glutamate by a specific enzyme (glutamic acid decarboxylase) that is present only in neurons that use GABA as a transmitter. Thus, experimentally, it is possible to identify cells as inhibitory GABAergic neurons by using antibodies to this enzyme to mark them (immunolabeling, see Fig. 6.1B). Many local interneurons are GABAergic. In addition, several brain regions contain large numbers of GABAergic projection neurons. The most notable are the spiny neurons of the striatum and the Purkinje cells of the cerebellar cortex. The inhibitory nature of Purkinje cells was especially surprising because they represent the entire output of the cerebellar cortex, and thus cerebellar cortical activity basically functions to suppress the activity of its downstream targets (cerebellar and vestibular nuclei).

Glycine functions as an inhibitory neurotransmitter in a much more restricted territory. Glycinergic synapses are predominantly found in the spinal cord, where they represent approximately half of the inhibitory synapses. They are likewise present in the lower brainstem, cerebellum, and retina in significant numbers. Interestingly, glycine also has another synaptic function. At excitatory NMDA-type glutamate receptors, glycine must also be bound for the ion channel to open. Thus, it acts as a cotransmitter at these synapses. It was generally thought that under physiological conditions the extracellular glycine concentration was high enough that the glycine binding sites of the NMDA channel were always saturated, but recent results suggest that this may not always be true, which implies that fluctuations in glycine levels may also be an important modulator of NMDA-mediated synaptic transmission.

After GABA and glycine are released from the presynaptic terminal, they are taken back up into the nerve terminal and neighboring glia by high-affinity Na^+-Cl^-–coupled membrane transporters. These Na^+-Cl^- transporters are part of what is called the *solute carrier 6* (SLC6) family of transporters that also includes those for the biogenic amine neurotransmitters, but it is distinct from those for glutamate. Transport of the neurotransmitter into the cell is accomplished by symport with two Na^+ and one Cl^- ion. Four GABA transporter (*GAT1, GAT2, GAT3,* and *BGT1*) genes have been identified; however, *GAT1* and *GAT3* are the ones that are highly expressed in the CNS. Depending on region and species, they may be expressed in neurons and/or in glia. There are two main glycine transporters, GlyT1 and GlyT2. GlyT1 is found predominantly on astrocytes and is present throughout the CNS. In contrast, GlyT2 is located on glycinergic nerve terminals and is largely restricted to the spinal cord, brainstem, and cerebellum.

Biogenic Amines

Many of the neurotransmitters in this category may be familiar because they have roles outside the nervous system, often as hormones. Among the amines known to act as neurotransmitters are **dopamine, norepinephrine** (noradrenaline), **epinephrine** (adrenaline), **serotonin** (5-hydroxytryptamine [5-HT]), and **histamine.** Dopamine, norepinephrine, and epinephrine are catecholamines, and they share a common biosynthetic pathway that starts with the amino acid tyrosine. Tyrosine is converted to L-dopa by the enzyme tyrosine hydroxylase. L-Dopa is then converted to dopamine by dopa-decarboxylase. In dopaminergic neurons, the pathway stops here. In noradrenergic neurons, another enzyme, dopamine β-hydroxylase, converts dopamine to norepinephrine. Epinephrine is obtained by adding a methyl group to norepinephrine via phenylethanolamine-*N*-methyl transferase. In serotoninergic neurons, serotonin is synthesized from the essential amino acid tryptophan. Tryptophan is first converted to 5-hydroxytryptophan by tryptophan 5-hydroxylase, which is then converted to serotonin by aromatic L-amino acid decarboxylase. Finally, in histaminergic neurons the conversion of histidine to histamine is catalyzed by histidine decarboxylase.

Removal of synaptically released biogenic amines is generally accomplished by reuptake into glia and neurons via transporters belonging to the Na^+-Cl^-–dependent transporter family. The catecholamines are then degraded by two enzymes, monoamine oxidase and catechol *O*-methyltransferase.

🧬 AT THE CELLULAR LEVEL

At least five transporters (EAAT1 to EAAT5, in which *EAAT* stands for *excitatory amino acid transporter*) that carry glutamate across the plasma membrane have been identified. They are all part of the Na^+-K^+–dependent family of transporters. Inward movement of each glutamate molecule is driven by the cotransport of three Na^+ ions and one H^+ ion and the countertransport of one K^+ ion out of the cell (see Fig. 6.11B). In addition, the transporter has Cl^- conductance, although passage of Cl^- ions is not stoichiometrically linked to glutamate transport. Glutamate transporters are found on both neurons and glia. However, the transporters differ in their regional and cellular distribution and in their pharmacological and biophysical properties. For example, EAAT2 is found on glia and is generally responsible for more than 90% of glutamate uptake from the extracellular space. The glutamate taken up into glial cells by EAAT2 is eventually returned to the presynaptic terminal by the glutamate-glutamine cycle (see Fig. 6.11). Inside glial cells, glutamate is converted to glutamine. Glutamine is then transported out of the glial cell and back into the presynaptic terminal, where it is subsequently converted back to glutamate. Glutamate inside the presynaptic terminal is packaged into synaptic vesicles by a second set of glutamate transporters known as **vGLUTs (vesicular glutamate transporters)**, which are present in the membrane of glutamatergic vesicles. Transport of glutamate into synaptic vesicles by vGLUT is driven by the countertransport of H^+ ions, the electrochemical gradient for which having been established by an H^+-ATPase in the vesicle membrane.

Within the CNS, nerve cells that use biogenic amines as neurotransmitters are primarily found within one of a few brainstem nuclei, most of which project rather diffusely throughout large areas of the brain. Noradrenergic neurons are primarily found in the locus coeruleus and nucleus subcoeruleus, which are located near each other in the dorsal part of the rostral pons. The neurons of the locus coeruleus project throughout the entire brain. Targets of the nucleus subcoeruleus are more limited but still widespread and include the pons, medulla, and spinal cord. (Norepinephrine is also important in the peripheral nervous system because it is used by postganglionic sympathetic cells.) Serotoninergic fibers arise from a series of nuclei located at the midline of the brainstem, known as the *raphe nuclei.* Similar to the noradrenergic fibers, serotoninergic fibers are distributed throughout most of the brain and spinal cord. Dopaminergic fibers arise from two main brainstem regions: the substantia nigra pars compacta, which projects to the striatum, and the ventral tegmental area, which projects more widely to the neocortex and subcortical areas, including the nucleus accumbens. Histaminergic neurons are located within the tuberomammillary nucleus of the hypothalamus and project diffusely throughout the CNS. Finally, adrenergic neurons are relatively few in number when compared with the other biogenic amine transmitters. They have cell bodies localized to small cell groups in the rostral medulla. The largest group, termed C1, has projections to the locus coeruleus and down to the thoracic and lumbar levels of the spinal cord, where they terminate in the autonomic nuclei of the intermediolateral and intermediomedial cell columns. Thus, these neurons are important for autonomic functions, particularly vasomotor ones, such as control of arterial pressure.

The diffuse nature of the projection pattern of most of the amine systems is mirrored in their proposed functions. Activity in the different aminergic systems is believed to be important in setting global brain states. For example, these systems are involved in setting the level of arousal (sleep, waking), attention, and mood. Their involvement in pathways connected with the hypothalamus and other autonomic centers also indicates that they have important homeostatic functions. The role of dopamine in balancing the flow of activity through the basal ganglia pathways and how its loss leads to the motor symptoms observed in Parkinson's disease are described in Chapter 9.

Purines

ATP has the potential to act as a transmitter or cotransmitter at synapses in the peripheral and central nervous systems. ATP is found in all synaptic vesicles and thus is co-released during synaptic transmission. ATP has its own receptors, which like standard neurotransmitters, are coupled to ion channels, but it can also modify the action of other neurotransmitters with which it is co-released, including norepinephrine, serotonin, glutamate, dopamine, and GABA. Glial cells may also release ATP after certain types of stimulation. Once released, ATP is broken down by ATPases and 5-nucleotidase to **adenosine,** which can be taken up again by the presynaptic terminal.

Peptides

Peptide neurotransmitters consist of chains of between 3 and about 40 amino acids. Studies of neuropeptides

IN THE CLINIC

Hyperactivity of dopaminergic synapses may be involved in some forms of psychosis. **Chlorpromazine** and related antipsychotic drugs inhibit dopamine receptors on postsynaptic membranes and thus diminish the effects of dopamine released from presynaptic nerve terminals. Overdoses of such antipsychotic drugs can produce a temporary parkinsonian-like state.

focused on the hypothalamus for many years. However, it is now clear that neuropeptides are released by neurons and act on receptors throughout the CNS and thus are a fundamental mechanism of neurotransmission throughout the CNS. To date, more than 100 neuropeptides have been identified. They can be classified into several functional groups, as shown in Box 6.1, which lists some of the known neuropeptides. It is now clear that many neurons that release classic neurotransmitters also release neuropeptides. As detailed later, understanding the interaction between coexisting classic and peptide transmitters has become an important area of research. In addition to being co-released with another transmitter, neuropeptides can also function as the sole or primary neurotransmitter at a synapse.

In some ways neuropeptides are like the classic neurotransmitters: they are packaged into synaptic vesicles, their release is dependent on Ca^{++}, and they bind to specific receptors on target neurons. However, there are also significant differences, ones that have led to alternative names for the intercellular communication mediated by neuropeptides, such as nonsynaptic, parasynaptic, and volume transmission. Table 6.1 summarizes some of these differences between classic and peptide neurotransmitters.

Unlike classic neurotransmitters, which are synthesized at the presynaptic terminal, neuropeptides are synthesized at the cell body and then transported to the terminal (see Fig. 6.2). Neuropeptides are packaged into large electron-dense vesicles that are scattered throughout the presynaptic terminal rather than in small electron-lucent vesicles docked at the active zone, where small-molecule transmitters are stored. (In neurons that make multiple neuropeptides, the various peptides are co-stored in the same vesicles.) Neuropeptide receptors are not confined to the synaptic region, and in general, peptide action is not limited by reuptake mechanisms.

Each of these differences has functional implications. For example, the separate storage of peptide and nonpeptide transmitters immediately raises the question of whether the two transmitters are co-released or differentially released in response to particular stimulation patterns.

In fact, differential release of peptide and classic transmitters from the same cell has been demonstrated for several types of neurons and is probably a result of the differences in vesicle storage described earlier. Because of their proximity to the active zones, nonpeptide vesicles can be released rapidly (<1 msec) in response to single action potentials

Hypothalamic Hormones

Corticotropin-releasing hormone (CRH)
Growth hormone–releasing hormone (GHRH)
Luteinizing hormone–releasing hormone (LHRH)
Oxytocin
Somatostatin
Thyrotropin-releasing hormone (TRH)
Vasopressin

NPY-Related Peptides

Neuropeptide Y

Opioid Peptides

Dynorphin
Methionine enkephalin
Leucine enkephalin

Tachykinins

Neurokinin α
Neurokinin β
Neuropeptide K
Substance P

VIP-Glucagon Family

Glucagon-like peptide 1
Peptide histidine-leucine
Pituitary adenylyl cyclase–activating peptide (PACAP)
Vasoactive intestinal polypeptide (VIP)

Others

Adrenocorticotropic hormone (ACTH)
Brain natriuretic peptide
Cholecystokinin (CCK)
Galanin
Hypocretins/orexins
Neurotensin
Motilin
Insulin
α-Melanocyte–stimulating hormone (α-MSH)
Neurotensin
Prolactin-releasing peptide
Secretoneurin
Urocortin

---

as a result of localized influx of Ca^{++}. Thus, low-frequency stimulation of the cell causes just the release of nonpeptide transmitter. In contrast, with higher-frequency stimulation of the presynaptic neuron, there is a more global increase in $[Ca^{++}]$ throughout the nerve terminal that leads to release of neuropeptide as well as neurotransmitter.

When neuropeptides are co-released with other transmitters, they may act synergistically or antagonistically. For example, in the spinal cord, **tachykinins** and **calcitonin gene–related peptide (CGRP)** act synergistically with glutamate and with **substance P** to enhance the action of serotonin. Conversely, tachykinins and CGRP antagonize norepinephrine's action at other synapses. The interactions, however, are not simply a one-to-one synergism or antagonism at a particular synapse because of the differing temporal and spatial profiles of the action of peptides versus classic transmitters. In particular, the slower release and lack of rapid reuptake mean that neuropeptides can act for long durations, diffuse over a region of brain tissue, and affect all cells in that region (that have the appropriate receptors) rather than just acting at the specific synapse at which it was released. In fact, studies have shown that there is often a spatial mismatch between the presynaptic terminals that contain a particular neuropeptide and the sites of the receptors for that peptide. In sum, peptides released from a particular synapse probably affect the local neuronal population as a whole, whereas the co-released classic transmitters act in more of a point-to-point manner.

Opioid Peptides

Opiates are drugs derived from the juice of the opium poppy. Compounds that are not derived from the opium poppy but that exert direct effects by binding to opiate receptors are called **opioids** and form a clinically and functionally important class of neuropeptides. Operationally, opioids are defined as compounds whose effects are stereospecifically antagonized by a morphine derivative called *naloxone.*

The three major classes of endogenous opioid peptides in mammals are **enkephalins, endorphins,** and **dynorphins.** Enkephalins are the simplest opioids; they are pentapeptides. Dynorphins and endorphins are somewhat longer peptides that contain one or the other of the enkephalin sequences at their N-terminal ends.

Opioid peptides are widely distributed in neurons of the CNS and intrinsic neurons of the gastrointestinal tract. The endorphins are discretely localized in particular structures of the CNS, whereas the enkephalins and dynorphins are more widely distributed. Opioids inhibit neurons in the brain involved in the perception of pain. Indeed, opioid peptides are among the most potent analgesic (pain-relieving) compounds known, and opiates are used therapeutically as powerful analgesics. They exert their analgesic effect by binding to specific opiate receptors.

Substance P

Substance P is a peptide consisting of 11 amino acids. It is present in specific neurons in the brain, in primary sensory neurons, and in plexus neurons in the wall of the gastrointestinal tract. The wall of the gastrointestinal tract is richly innervated with neurons that form networks or plexuses (see also Chapter 33). The intrinsic plexuses of the gastrointestinal tract exert primary control over its motor and secretory activities. These enteric neurons contain many of the neuropeptides, including substance P, that are found in the brain and spinal column. Substance P is involved in pain transmission and has a powerful effect on smooth muscle.

Substance P is probably the transmitter used at synapses made by primary sensory neurons (their cell bodies are in the dorsal root ganglia) with spinal interneurons in the dorsal horn of the spinal column, and thus it is an example of a peptide acting as a primary transmitter at a synapse.

Enkephalins act to decrease the release of substance P at these synapses and thereby inhibit the pathway for pain sensation at the first synapse in the pathway.

Gas Neurotransmitters

This is the newest category of neurotransmitter to be defined and stretches the usual definition of synaptic transmission even further than neuropeptides do. Gas neurotransmitters are neither packaged into synaptic vesicles nor released by exocytosis. Instead, gas neurotransmitters are highly permeant and simply diffuse from synaptic terminals to neighboring cells after synthesis, their synthesis being triggered by depolarization of the nerve terminal (the influx of Ca^{++} activates synthetic enzymes). Moreover, there are no specific reuptake mechanisms, nor do they undergo enzymatic destruction, so their action appears to be ended by diffusion or binding to superoxide anions or various scavenger proteins. Both **nitric oxide (NO)** and **carbon monoxide (CO)** are examples of gaseous neurotransmitters. NO is a transmitter at synapses between inhibitory motor neurons of the enteric nervous system and gastrointestinal smooth muscle cells (see Chapter 33). NO also functions as a neurotransmitter in the CNS. The enzyme NO synthase catalyzes the production of NO as a product of the oxidation of arginine to citrulline. This enzyme is stimulated by an increase in cytosolic $[Ca^{++}]$.

In addition to serving as a neurotransmitter, NO functions as a cellular signal transduction molecule both in neurons and in nonneuronal cells (e.g., vascular smooth muscle; see Chapter 14). One way that NO functions as a signal transduction molecule is by regulating **guanylyl cyclase,** the enzyme that produces **cGMP** from **GTP.** NO binds to a heme group in soluble guanylyl cyclase and potently stimulates the enzyme. Stimulation of this enzyme leads to an elevation in cGMP in the target cell. The cGMP can then influence multiple cellular processes.

Neurotransmitter Receptors

The multitude of neurotransmitters used in the nervous system provides it with a specific and flexible interneuronal communications system. These characteristics are even further enhanced by the variety of receptors for each neurotransmitter. Receptors for a particular neurotransmitter were traditionally distinguished primarily by pharmacological differences in their sensitivity to particular agonists and antagonists. For example, acetylcholine receptors were split into **muscarinic** and **nicotinic** classes, depending on whether they bind muscarine or nicotine. Similarly, glutamate receptors were split into three main groups according to their sensitivity to the agonists NMDA, kainic acid, or AMPA. Though useful, this classification scheme has several limitations: some receptors fail to be activated by agonists, and it fails to disclose all the various receptor subtypes for a particular transmitter. Over the past 15 years or so, molecular biological approaches have been used to identify and sequence the receptor genes for many of the known neurotransmitters. It is thought that we now have a relatively complete catalog of the genes for these receptors. What this work has revealed is that there is a tremendous diversity of actual and potential receptor subtypes that are or could be used by the nervous system. Moreover, knowledge of the gene sequences has enabled an understanding of the relationship of different receptor proteins to each other and to other important proteins. This knowledge, combined with the results of biochemical, crystallographic, and other types of studies, has led to a much deeper understanding of the structural and functional workings of receptor proteins. In particular, various receptors can be grouped into families based on gene sequences, and members of each family share various structural and functional features.

Neurotransmitter receptors are members of one of two large groups or families of proteins: ligand-gated ion channels, also known as *ionotropic receptors,* and G protein–coupled receptors, also referred to as *metabotropic receptors* (Fig. 6.12A–B). Almost all classic neurotransmitters and neuropeptides have at least one metabotropic-type receptor. Many of the classic neurotransmitters also have at least one ionotropic receptor. Ionotropic receptors are protein complexes that both have an extracellular binding site for the transmitter and form an ion channel (pore) through the cell membrane. The receptor is made up of several protein subunits, usually three to five, each of which typically has a series of membrane-spanning domains, some of which contribute to the wall of the ion channel. Binding of the neurotransmitter alters (usually increases) the probability of the ion channel being in the open state and thus typically results in postsynaptic events that are rapid in both onset and decay, with a duration of several milliseconds. Ionotropic receptors underlie fast synaptic EPSPs and IPSPs, as described earlier.

AT THE CELLULAR LEVEL

The ionotropic receptors can be divided into several superfamilies. Members of the cys-loop superfamily have peptide subunits that have an N-terminal extracellular domain that contains a loop delimited by cysteine residues. This family includes the ionotropic receptors for acetylcholine, serotonin, GABA, and glycine. In addition to their family-defining cysteine loop, these receptors share the following common features: they are pentamers, with each peptide subunit having four transmembrane domains; the neurotransmitter binds to the N-terminal domain; and the second transmembrane domains are thought to form the wall of the ion pore.

Ionotropic glutamate and ATP receptors form two other ionotropic receptor superfamilies; the details for each are given in the later corresponding sections. Transient receptor potential (TRP) channels, which are important for transduction of pain and thermal sensations, form yet another family (see Chapter 7).

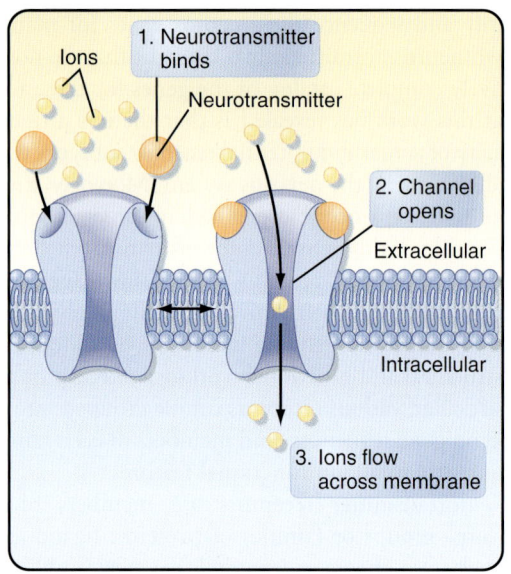

A Ligand-gated ion channels (ionotropic)

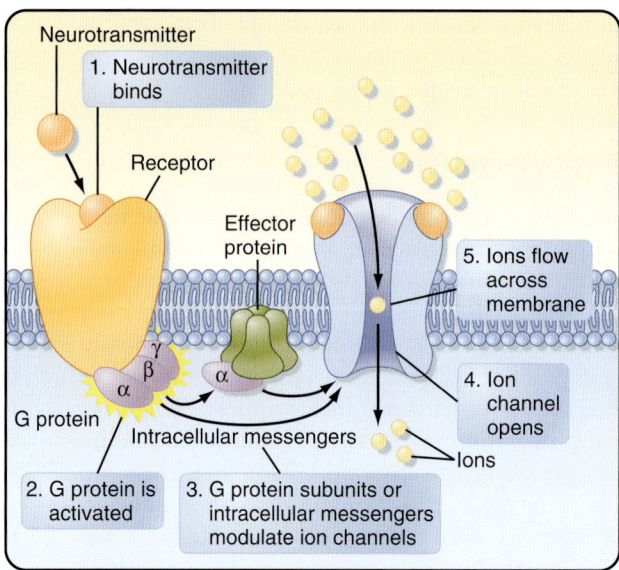

B G protein–coupled receptors (metabotropic)

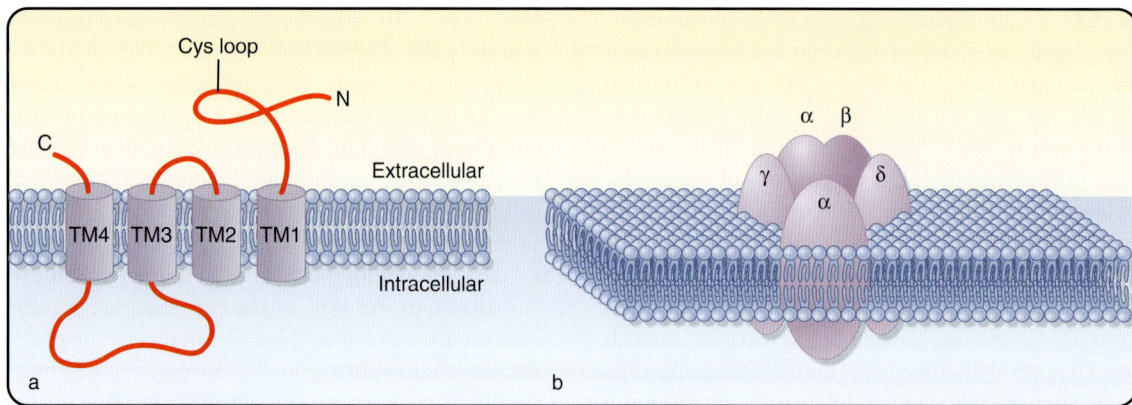

C Cys-loop family channels

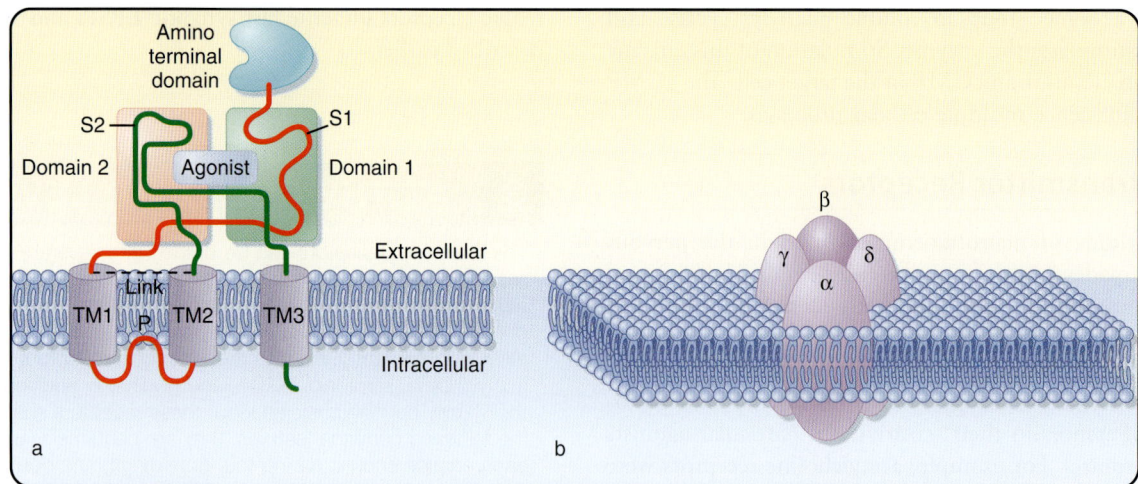

D Glutamate channels

• **Fig. 6.12** Neurotransmitter receptors. The basic structure and mechanism of action are shown for ligand-gated ion channels (ionotropic receptors) **(A)** and G protein–coupled (metabotropic) receptors **(B)**. Detailed structures of cys-loop and glutamate ionotropic receptors are shown in **C** and **D,** respectively. Cys-loop receptors include ionotropic receptors for GABA, glycine, serotonin, and acetylcholine. Note the differing membrane topologies of the individual subunits of these two classes of receptors: four transmembrane domains for cys-loop receptors and three plus a pore loop for glutamate receptors. Pore loops form the internal wall of the glutamate channel, whereas transmembrane domain 2 forms the internal wall of cys-loop receptors. (**A** and **B,** From Purves D, Augustine GJ, Fitzpatrick D. *Neuroscience.* 2nd ed. Sunderland, MA: Sinauer Associates; 2001.)

Metabotropic receptors are not ion channels. Instead, they are protein monomers that have an extracellular binding site for a particular transmitter and an intracellular site for binding a G protein. Binding of the receptor leads to activation of a G protein, which is the first step in a signal transduction cascade that alters the function of an ion channel in the postsynaptic membrane. In contrast to ionotropic receptors, metabotropic receptors mediate post-synaptic phenomena that have a slow onset and that may persist from hundreds of milliseconds to minutes. Because of the various biochemical cascades they initiate, they have great potential to cause changes in the neuron beyond just generating a postsynaptic potential.

Acetylcholine Receptors

Acetylcholine receptors were originally classified on a pharmacological basis (being sensitive to nicotine or mus-carine) into two major groups. This grouping corresponds to groupings based on structural and molecular biological studies. Nicotinic receptors are members of the ionotropic cys-loop family, and muscarinic receptors are part of the metabotropic family of receptor proteins.

The nicotinic receptors mediate synaptic transmission at the neuromuscular junction as described earlier; however, nicotinic receptors are also present within the CNS. The nicotinic receptor contains a relatively nonselective cationic channel, so binding of acetylcholine produces an EPSP. Being members of the cys-loop family, acetylcholine receptors are pentamers constructed from a series of subunit types called α, β, γ, δ, ϵ, some of which contain multiple members. At the neuromuscular junction the channel is constructed from 2α, β, δ, ϵ, whereas in the CNS the composition is typically 3α, 2β. Furthermore, the junctional receptors all use the α_1 subunit, whereas centrally located receptors use one of the α subunits between α_2 and α_{10}. As noted, the differing subunits result in receptors with differing pharmacological sensitivities and channel kinetics and selectivity.

There are five known muscarinic subtypes of acetylcho-line receptors (M_1 to M_5). All are metabotropic receptors; however, they are coupled to different G proteins and can thus have distinct effects on the cell. M_1, M_3, and M_5 are coupled to pertussis toxin–insensitive G proteins, whereas M_2 and M_4 are coupled to pertussis toxin–sensitive G pro-teins. Each set of G proteins is coupled to different enzymes and second messenger pathways (see Chapter 3 for details of these pathways).

Inhibitory Amino Acid Receptors: GABA and Glycine

As noted, the most common inhibitory synapses in the CNS use either glycine or GABA as their transmitter. Glycine-mediated inhibitory synapses are common in the spinal cord, whereas GABAergic synapses make up the majority of inhibitory synapses in the brain.

Both glycine and GABA have ionotropic receptors that are members of the cys-loop family, thus sharing a number of characteristics as already described. In addition, each of these receptors has a Cl^- channel, which opens while the receptor portion is bound. Therefore, the probability of these channels opening and the average time a channel stays open are controlled by the concentration of the neurotrans-mitter for which the receptor is specific.

Glycine receptors are pentamers and may be heteromers of α and β subunits ($3:2$ ratio) or homomers. Interestingly, the molecular composition appears to be related to its cel-lular location, with heteromers located postsynaptically and homomers located extrasynaptically. The β subunit seems to bind to an intracellular scaffold protein called *gephyrin* that appears to help localize receptors to the postsynaptic site. The α subunit contains the glycine binding site, and there are four genes coding for distinct α subunits (and splice variants of each). Each variant results in a receptor having distinct conductance, kinetics, agonist and antagonist affin-ity, and modulatory sites. Intriguingly, subunit variants are differentially expressed during development and in different brain regions.

GABA has two separate ionotropic receptors ($GABA_A$ and $GABA_C$) coded for by distinct sets of genes. Like glycine receptors, both control a Cl^- channel. $GABA_A$ receptors are heteromers generated from seven classes of subunits, three of which have multiple members. The most common configuration is α_1, β_2, γ_2 in a $2:2:1$ stoichiometry, which may account for 80% of the receptors; however, many other heteromers are found in the brain. As with glycine, different subunits confer distinct properties on the receptor. For example, $GABA_A$ receptors are the targets of two major classes of drugs: benzodiazepines and barbiturates. Benzo-diazepines (e.g., diazepam) are widely used antianxiety and relaxant drugs. Barbiturates are used as sedatives and anti-convulsants. Both classes of drugs bind to distinct sites on the α subunits of $GABA_A$ receptors and enhance opening of the receptors' Cl^- channels in response to GABA. The seda-tive and anticonvulsant actions of benzodiazepines appear to be mediated by receptors with the α_1 subunit, whereas the anxiolytic effects reflect binding to receptors with the α_2 subunit. $GABA_C$ receptors are structurally similar to $GABA_A$ receptors but have a distinct pharmacological profile (e.g., they are not affected by benzodiazepines) and are coded for by a separate set of genes (ρ_1, ρ_2, and ρ_3).

The $GABA_B$ receptor is a metabotropic receptor. Binding of GABA to this receptor activates a heterotrimeric GTP-binding protein (G protein; see Chapter 3), which leads to activation of K^+ channels and hence hyperpolarization of the postsynaptic cell, as well as inhibition of Ca^{++} channels (when located presynaptically) and thus a reduction in release of transmitter.

Excitatory Amino Acid Receptors: Glutamate

Glutamate has both ionotropic and metabotropic recep-tors. Based on pharmacological properties and subunit composition, several distinct ionotropic receptor subtypes are recognized: AMPA, kainate, and NMDA. Overall, there are 18 known genes that code for glutamate subunits for the ionotropic glutamate receptors. The genes are divided into several families (AMPA, kainate, NMDA, and δ) that essentially correspond to the pharmacological subtypes of receptors. Each glutamate receptor is a tetramer. Thus, there is a certain correspondence between the genes and the receptor types that are formed. For example, AMPA receptors are formed from GluR1 to GluR4 subunits, kainate receptors require either KA1 or KA2 and GluR5 to GluR7 subunits, and NMDA receptors all have NR1 subunits plus some combination of NR2 and NR3 subunits. As was mentioned for the other receptors, the receptor properties vary with subunit composition. Ionotropic glutamate receptors are excitatory and contain a cationic-selective channel. Thus, all the channels are permeable to Na^+ and K^+, but only a subset allow Ca^{++} to pass.

AMPA and kainate receptors behave as classic ligand-gated channels as already discussed; on binding of glutamate to the receptor, the channel opens and allows current to flow, thereby generating an EPSP. NMDA channels are different. First, they require binding of both glutamate and glycine to open. Second, they display voltage sensitivity as a result of Mg^{++} blockade of the channel. That is, at resting (or more negative) membrane potentials, a Mg^{++} ion blocks the entrance to the channel so that even when glutamate and glycine are bound, no current flows through the channel. However, if the cell is depolarized (either experimentally by injection of current through an electrode or by other EPSPs), the Mg^{++} block is relieved and current can flow through the channel. A further interesting feature of NMDA channels is that they are generally permeable to Ca^{++}, which can act as a second messenger. The combination of voltage sensitivity and Ca^{++} permeability of the NMDA channels has led to hypotheses concerning their role in learning and memory-related functions (see Chapter 10).

Eight genes coding for metabotropic glutamate receptors have been identified and classified into three groups. Group I receptors are found postsynaptically, whereas groups II and III are found presynaptically. These receptors generate slow EPSPs, but probably at least as importantly, they trigger second messenger cascades (see Chapter 3).

Purine (ATP) Receptors

Purines have two receptor families: an ionotropic (P2X) and a metabotropic (P2Y) family. There are seven identified P2X subunit types that form channels, and they represent their own superfamily of ligand-gated channels. Each subunit has only two transmembrane domains, with the loop between these two domains located extracellularly and containing the ATP binding site. The receptors are heterotrimers or homotrimers or hexamers. In general, these receptors form a cationic channel that is permeable to Na^+, K^+, and Ca^{++}. The distribution of subunits in the brain varies significantly, with some subunits having a widespread distribution ($P2X_2$)

and others being quite limited ($P2X_3$ is present mostly on cells involved in pain-related pathways).

Metabotropic purine receptors are coded for by 10 genes, but only 6 are expressed in the human CNS. They have the typical features of G protein–coupled receptors and are known to activate K^+ currents and modulate both NMDA and voltage-gated Ca^{++} currents. An interesting localization distinction between P2X and P2Y receptors is that although both are present on neurons, the latter dominates on astrocytes.

Finally, in addition to the P2X and P2Y receptors, which respond to ATP, there are adenosine receptors that respond to the adenosine that is released after the enzymatic breakdown of ATP. These receptors are located presynaptically and act to inhibit synaptic transmission by inhibiting influx of Ca^{++}.

Biogenic Amine Receptors: Serotonin, Dopamine, Noradrenaline, Adrenaline, Histamine

With the exception of one class of serotonin receptors ($5-HT_3$) that are part of the cys-loop ionotropic family, receptors for the various biogenic amines are all metabotropic-type receptors. Thus, these neurotransmitters tend to act on relatively long time scales by generating slow synaptic potentials and by initiating second messenger cascades. Agonists and blockers of many of these receptors are important clinical tools for treating various neurological and psychiatric disorders. The role of different dopamine receptors in basal ganglia disorders will be covered in the motor systems (see Chapter 9).

Neuropeptide Receptors

As is the case with the biogenic amines, receptors for the various peptides are essentially all of the metabotropic type and are coupled to G proteins that mediate effects via second messenger cascades. It is worth mentioning again that studies consistently show a mismatch between the locations of terminals containing a particular peptide and the receptors for it. Thus, these receptors are often activated by neurotransmitter diffusing through the extracellular space rather than at synapses. This implies that these receptors will experience much lower concentrations of agonist, and indeed, they are very sensitive to their agonists.

Gas Neurotransmitter Receptors

Unlike the other neurotransmitters that were covered, NO and CO do not bind to receptors. One way they do affect cell activity is to activate enzymes involved in second messenger cascades, such as guanylyl cyclase. In addition, NO has been shown to modify the activity of other proteins, such as NMDA receptors and the Na^+,K^+-ATPase pump, by nitrosylating them.

Key Concepts

1. Both electrical and chemical synapses are important means of cellular communication in the mammalian nervous system.

2. Electrical synapses directly connect the cytosol of two neurons and allow rapid bidirectional current flow between neurons. They act as low-pass filters.

3. Gap junctions are the morphological correlate of electrical synapses. Gap junctions contain channels formed by hemichannels called *connexons*. Connexons are formed by proteins called *connexins*.

4. Standard chemical synaptic transmission involves the release of transmitter from a presynaptic terminal, diffusion of transmitter across a synaptic cleft, and binding of the transmitter to receptors on the apposed postsynaptic membrane.

5. Entry of calcium into the presynaptic terminal triggers the release of neurotransmitter. Release of neurotransmitter is quantal, as first demonstrated by the recording of mEPPs at the frog neuromuscular junction.

6. Transmitter is packaged into synaptic vesicles in the presynaptic terminal. The vesicles are the quantal elements. That is, the release of transmitter from one vesicle causes a mEPP at the neuromuscular junction or, equivalently, one mPSP at a central synapse.

7. Many proteins are involved in priming, docking, and fusion of synaptic vesicles. Synaptotagmin is the Ca^{++} sensor for triggering vesicle fusion.

8. Excitatory and inhibitory synapses increase or decrease, respectively, the probability that the postsynaptic neuron will spike.

9. The reversal potential is the membrane potential at which net current flow through a ligand-gated channel reverses. Excitatory synapses generate depolarizing potentials (EPSPs) that have reversal potentials positive to the spike threshold, most often as a result of the opening of nonselective cation channels.

10. Inhibitory synapses generate IPSPs that have reversal potentials more negative than the spike threshold but not necessarily negative to the resting potential. Inhibitory synapses can decrease spike probability by two mechanisms: hyperpolarization of the membrane and a decrease in the input resistance of the neuron, leading to a shunt of synaptic currents.

11. The process by which a neuron decides to fire an action potential as a result of its inputs is referred to as *synaptic integration*. The summation of EPSPs and IPSPs can be highly nonlinear and depends on many factors, including the geometry of the dendritic tree, location of the synaptic inputs relative to the initial segment, and the passive (RC) and active membrane properties of the cell.

12. The efficacy of synaptic transmission depends on the timing and frequency of action potentials in the presynaptic neuron. Facilitation, posttetanic potentiation, and long-term potentiation are examples of increased efficacy of synaptic transmission in response to previous multiple stimulations of a synapse. Long-term depression is an example of reduced efficacy resulting from previous activation of the synapse.

13. The nervous system uses hundreds of neurotransmitters. Neurotransmitters can be subdivided into a few broad functional classes: small-molecule transmitters (acetylcholine, amino acids, biogenic amines, and purines), peptides, and gases (CO, NO). The action of a neurotransmitter depends on its postsynaptic receptors. Most nongaseous transmitters have both ionotropic and metabotropic receptors.

14. Small-molecule transmitters act locally, mainly across a single synapse, and their duration of action is limited by reuptake and enzymatic degradation. Peptides can diffuse from their presynaptic release site and thus have the potential to affect all cells within a local region. Gaseous transmitters are free to diffuse from their release site.

15. Ionotropic receptors contain an ion channel whose state (open versus closed) is gated by the binding of neurotransmitter to the receptor. Metabotropic receptors activate second messengers on binding neurotransmitter.

16. Many synapses can release multiple types of transmitters, and which ones they release depends on the activity pattern of the terminal. Co-released transmitters may function independently or act synergistically or antagonistically.

Additional Reading

Alabi AA, Tsien RW. Perspectives on kiss-and-run: role in exocytosis, endocytosis, and neurotransmission. *Annu Rev Physiol.* 2013;75:393-422.

Fain GL. *Molecular and Cellular Physiology of Neurons.* 2nd ed. Cambridge, MA: Harvard University Press; 2014.

Hille B. *Ion Channels of Excitable Membranes.* 3rd ed. Sunderland, MA: Sinauer Associates; 2001.

7

The Somatosensory System

LEARNING OBJECTIVES

Upon completion of this chapter the student should be able to answer the following questions:

1. What are the major modalities of somatosensory information, and what are the corresponding pathways that convey each from the periphery to the primary somatosensory cortex?
2. What body regions and categories of information are the exteroceptive, proprioceptive, and enteroceptive divisions of the somatosensory system associated with?
3. What are the main receptors for fine/discriminatory touch sensations?
4. What types of somatosensory information does the cerebellum receive?
5. What are the main receptors for pain and temperature touch sensations?
6. What is the phenomenon of referred pain?
7. What proteins are involved in transducing different categories of somatosensory information?
8. How do descending pathways act to regulate the flow of activity in ascending somatosensory pathways?

The somatosensory system provides information to the central nervous system (CNS) about the state of the body and its contact with the world. It does so by using a variety of sensory receptors that transduce mechanical (pressure, stretch, and vibrations) and thermal energies into electrical signals. These electrical signals are called *generator* or *receptor potentials* and occur in the distal ends of axons of first-order somatosensory neurons, where they trigger action potential trains that reflect information about the characteristics of the stimulus. The cell bodies of these neurons are located in dorsal root (Fig. 7.1A; see Fig. 4.8) and cranial nerve ganglia.

Each ganglion cell gives off an axon that after a short distance divides into a peripheral process and a central process. The peripheral processes of the ganglion cells coalesce to form peripheral nerves. A purely sensory nerve will have only axons from such ganglion cells; however, mixed nerves, which innervate muscles, will contain both afferent (sensory) fibers and efferent (motor) fibers. At the target organ the peripheral process of an afferent axon divides repeatedly, with each terminal branch ending as a sensory receptor. In most cases the free nerve ending by itself forms a functional receptor, but in some the nerve ending is encapsulated by accessory cells and the entire structure (axon terminal plus accessory cells) forms the receptor.

The central axonal process of the ganglion cell either enters the spinal cord via a dorsal root or enters the brainstem via a cranial nerve. A central process typically gives rise to numerous branches that may synapse with a variety of cell types, including second-order neurons of the somatosensory pathways. The terminal location of these central branches varies depending on the type of information being transmitted. Some terminate at or near the segmental level of entry, whereas others project to brainstem nuclei.

Second-order neurons that are part of the pathway for the perception of somatosensory information project to specific thalamic nuclei where the third-order neurons reside. These neurons in turn project to the primary somatosensory cortex (S-I). Within the cortex, somatosensory information is processed in S-I and in numerous higher-order cortical areas. Somatosensory information is also transmitted by other second-order neurons to the cerebellum for use in its motor coordination function.

The organization of the somatosensory system is quite distinct from that of the other senses, which has both experimental and clinical implications. In particular, other sensory systems have their receptors localized to a single organ, where they are present at high density (e.g., the eye for the visual system). In contrast, somatosensory receptors are distributed throughout the body (and head).

Subdivisions of the Somatosensory System

The somatosensory system receives three broad categories of information based on the distribution of its receptors. Its **exteroceptive** division is responsible for providing information about contact of the skin with objects in the external world, and a variety of cutaneous mechanoceptive, nociceptive (pain), and thermal receptors are used for this purpose. Understanding this division will be the main focus of this chapter. The **proprioceptive** component provides information about body and limb position and movement and relies primarily on receptors found in joints, muscles, and tendons. The ascending central pathways that originate

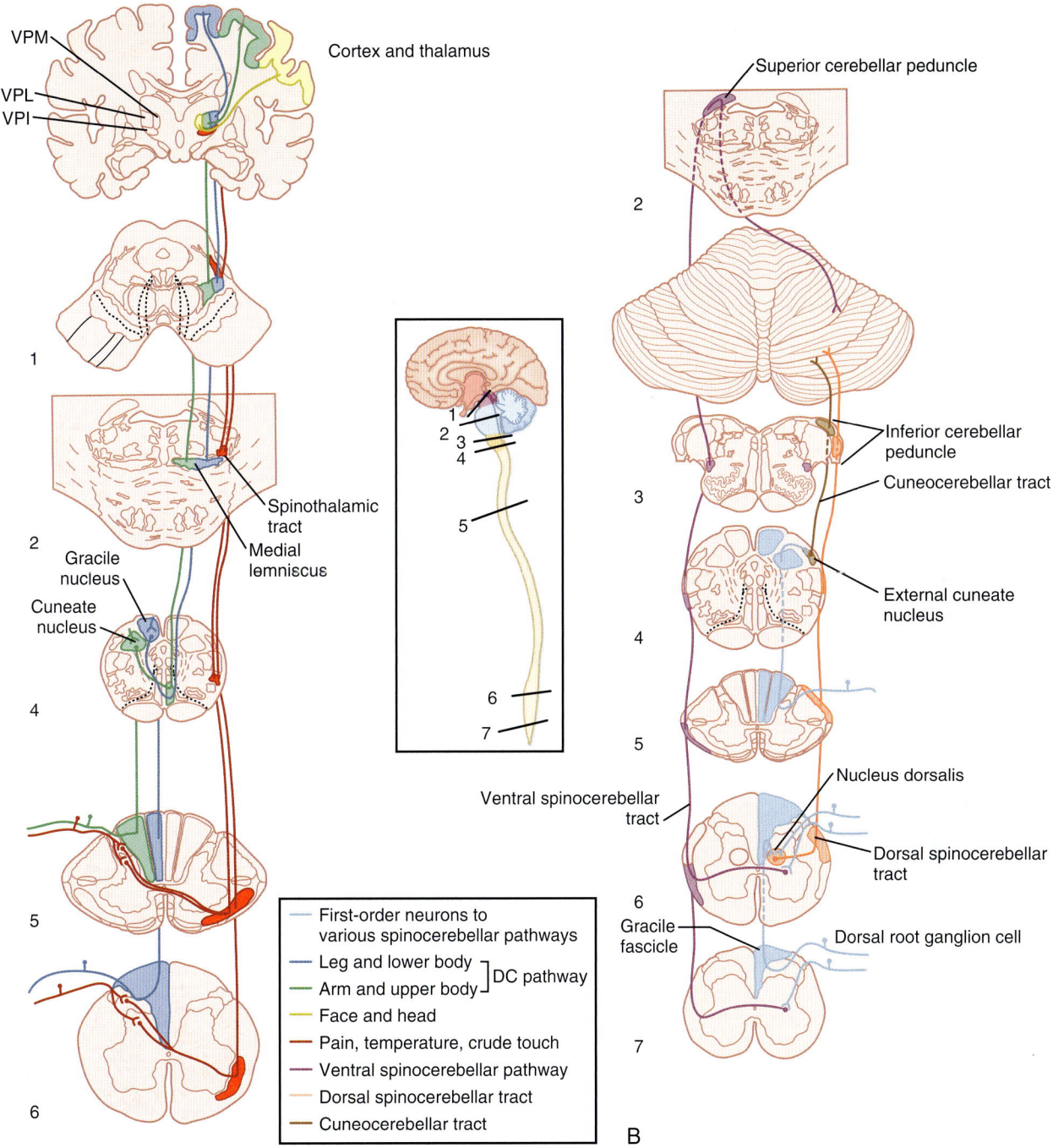

VPM

VPL
VPI

Cortex and thalamus

Superior cerebellar peduncle

1

2

Gracile
nucleus

Cuneate
nucleus

4

Spinothalamic
tract

Medial
lemniscus

Inferior cerebellar
peduncle

Cuneocerebellar tract

External cuneate
nucleus

5

6

A

Nucleus dorsalis

Ventral spinocerebellar
tract

Dorsal spinocerebellar
tract

Gracile
fascicle

Dorsal root ganglion cell

First-order neurons to
various spinocerebellar pathways

Leg and lower body
Arm and upper body ⎤ DC pathway
Face and head
Pain, temperature, crude touch
Ventral spinocerebellar pathway
Dorsal spinocerebellar tract
Cuneocerebellar tract

B

• **Fig. 7.1** Ascending somatosensory pathways from the body. A, First-, second-, and third-order neurons are shown for the two main pathways conveying cutaneous information from the body to the cerebral cortex: the dorsal column/medial lemniscal and the spinothalamic pathways. Note that the axon of the second-order neuron crosses the midline in both cases, so sensory information from one side of the body is transmitted to the opposite side of the brain, but the levels in the neuraxis at which this takes place are distinct for each pathway. Homologous central pathways for the head originate in the trigeminal nucleus and are described in text, but they are not illustrated for clarity. B, Major spinocerebellar pathways carrying tactile and proprioceptive information to the cerebellum from the upper and lower parts of the body. Again, pathways from the head originate in the trigeminal nuclei but are not shown for clarity. A midsagittal view of the nervous system shows the levels of the spinal and brainstem cross sections in panels A and B.

with them and that underlie conscious and unconscious proprioceptive functions will be covered in this chapter. However, because these receptors also initiate pathways that are intimately involved in the control of movement, they will be discussed again in Chapter 9. Finally, the **enteroceptive** division has receptors for monitoring the internal state of the body and includes mechanoreceptors that detect distention of the gut or fullness of the bladder. Aspects of enteroceptive division are also covered in Chapter 11 as they are related to autonomic functions.

The somatosensory pathways can also be classified by the type of information they carry. Two broad functional categories are recognized, each of which subsumes several somatosensory submodalities. **Fine discriminatory touch** sensations include light touch, pressure, vibration, flutter (low-frequency vibration), and stretch or tension. The second major functional group of sensations is that of **pain and temperature.** Submodalities here include both noxious and innocuous cold and warm sensations and mechanical and chemical pain. Itch is also closely related to pain and appears to be carried by particular fibers associated with the pain system.

IN THE CLINIC

The sensory functions of various cutaneous sensory receptors have been studied in human subjects with a technique known as **microneurography,** in which a fine metal microelectrode is inserted into a nerve trunk in the arm or leg to record the action potentials from single sensory axons. When a recording can be made from a single sensory axon, the receptive field of the fiber is mapped. Most of the various types of sensory receptors that have been studied in experimental animals have also been found in humans with this technique.

After the receptive field of a sensory axon has been characterized, the electrode can be used to stimulate the same sensory axon. In these experiments the subject is asked to locate the perceived receptive field of the sensory axon, which turns out to be identical to the mapped receptive field.

Of great importance experimentally, the afferent fibers that convey these somatosensory submodalities to the CNS are different sizes. Recall that the compound action potential recorded from a peripheral nerve (see Chapter 5, Table 5.1) consists of a series of peaks, thus implying that the diameters of axons in a nerve are grouped rather than being uniformly distributed. Information about tactile sensations is carried primarily by large-diameter myelinated fibers in the Aβ class, whereas pain and temperature information travels via small-diameter, lightly myelinated (Aδ) and unmyelinated (C) fibers. It is possible to block or stimulate selectively a class of axons of particular size, thereby allowing study of the different somatosensory submodalities in isolation.

Discriminatory Touch and Proprioception

Innervation of the Skin

Low-Threshold Mechanoreceptors

The skin is an important sensory organ and not surprisingly is richly innervated with a variety of afferents. We first consider the afferent types related to fine or discriminatory touch sensations. These afferents are related to what are called **low-threshold mechanoreceptors.** Nociceptor and thermoceptor innervation will be considered separately in a later section of this chapter.

To study the responsiveness of tactile receptors, a small-diameter rod or wire is used to press on a localized region of skin. With this technique, two basic types of responses may be seen when recording sensory afferent fibers: fast-adapting (FA) and slow-adapting (SA) responses (Fig. 7.2). They are present in similar quantities. FA fibers will show a short burst of action potentials when the rod first pushes down on the skin, but then they will cease firing despite continued application of the rod. They may also burst at the cessation of the stimulus (i.e., when the rod is lifted off). In contrast, SA units will start firing action potentials (or increase their firing rate) at the onset of the stimulus and continue to fire until the stimulus ends.

Both the FA and SA afferent classes can be subdivided on the basis of other aspects of their receptive fields, where **receptive field** is defined as the region of skin from which stimuli can evoke a response (i.e., change the firing of the afferent axon). Type 1 units have small receptive fields with well-defined borders. Particularly for glabrous skin (i.e., hairless skin, such as on the palms of the hands and soles of the feet), the receptive field has a circular or ovoid shape within which there is relatively uniform and high sensitivity to stimuli that decreases sharply at the border (Fig. 7.3). Type 1 units, particularly SA1 units, respond best to edges. That is, a larger response is elicited from them when the edge of a stimulus cuts through their receptive field than when the entire receptive field is indented by the stimulus.

Type 2 units have wider receptive fields with poorly defined borders and only a single point of maximal sensitivity, from which there is a gradual reduction in sensitivity with distance (see Fig. 7.3). For comparison a type 1 unit's receptive field typically will cover approximately four papillary ridges in the fingertip, whereas a type 2 unit will have a receptive field that covers most or all of a finger.

Receptive Field Properties

Thus, four main classes of low-threshold mechanosensitive afferents have been identified physiologically (FA1, FA2, SA1, and SA2). Peripherally these axons may terminate either as free nerve endings, associated with a hair follicle, or within a specialized receptor structure made up of supporting cells.

For glabrous skin the four afferent classes have been associated with four specific types of histologically identified receptor structures whose locations and physical

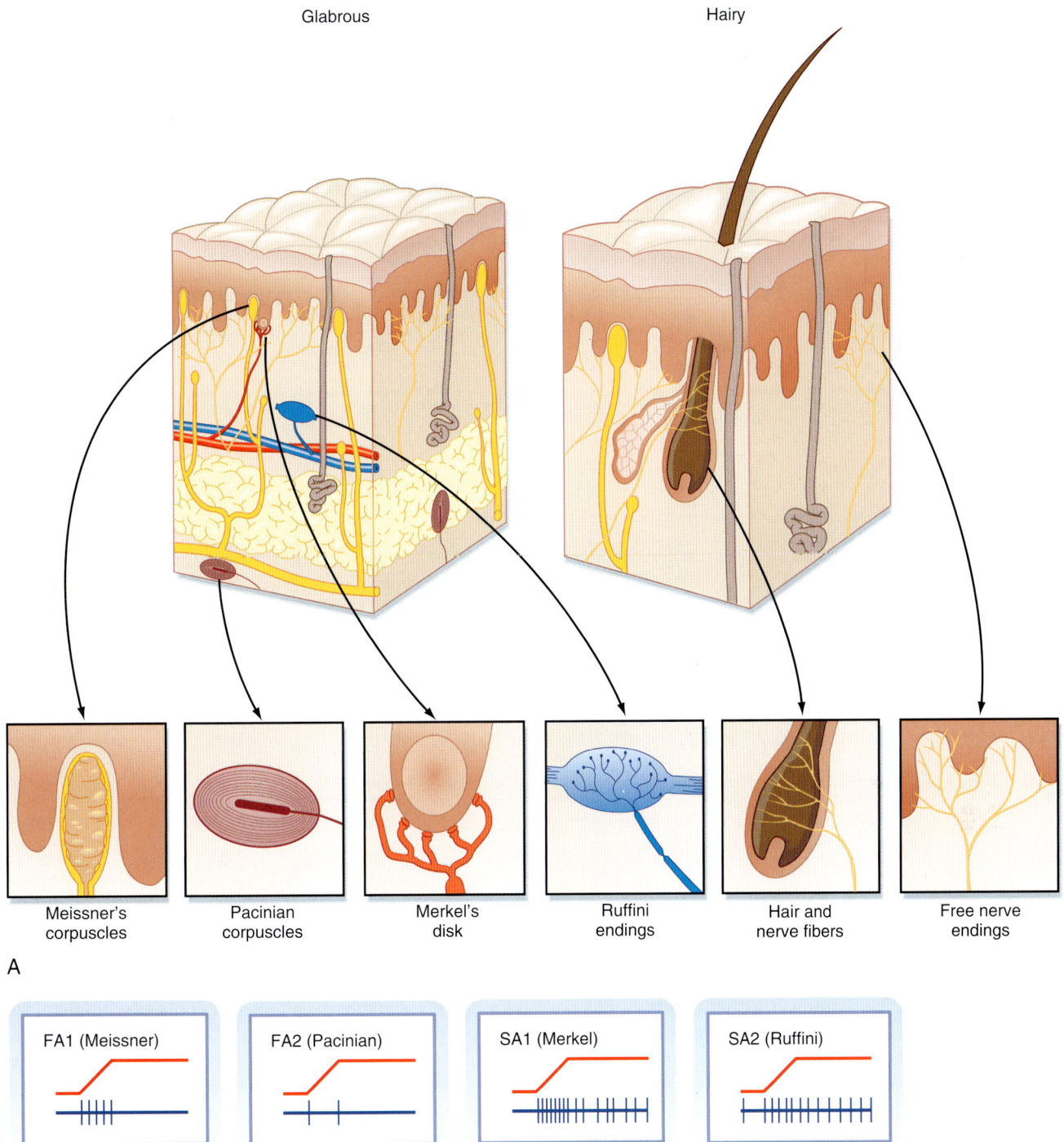

Glabrous   Hairy

Meissner's   Pacinian   Merkel's   Ruffini   Hair and   Free nerve
corpuscles   corpuscles   disk   endings   nerve fibers   endings

A

FA1 (Meissner)   FA2 (Pacinian)   SA1 (Merkel)   SA2 (Ruffini)

B

• **Fig. 7.2** Cutaneous mechanoreceptors and the response patterns of associated afferent fibers. A, Schematic views of glabrous (hairless) and hairy skin showing the arrangement of the various major mechanoreceptors. B, Firing patterns of the different cutaneous low-threshold mechanosensitive afferent fibers that innervate the various encapsulated receptors of the skin. (Traces in B are based on data from Johansson RS, Vallbo ÅB. *Trends Neurosci* 1983;6:27.)

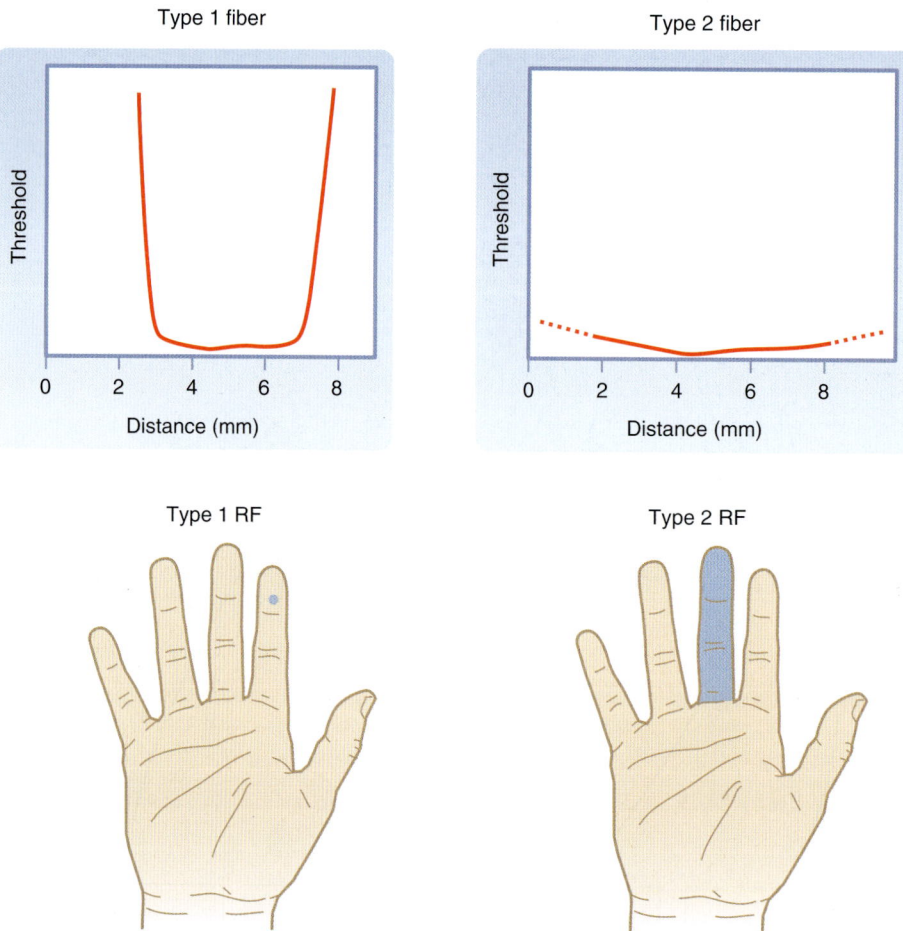

• **Fig. 7.3** Receptive field characteristics for type 1 and type 2 sensory afferents. Plots in the top row show the threshold level of force needed to evoke a response as a function of the distance across the receptive field. Receptive field size is shown on the hand below each plot. (Data from Johansson RS, Vallbo ÅB. *Trends Neurosci* 1983;6:27.)

characteristics help explain the firing properties of these sensory afferents. FA1 afferents terminate in **Meissner's corpuscles,** whereas SA1 afferents terminate in **Merkel's disks.** In both cases the receptor is located relatively superficially, either in the basal epidermis (Merkel) or just below the epidermis (Meissner) (see Fig. 7.2). These receptors are small and oriented to detect stimuli pressing down on the skin surface just above them, thus allowing SA1 and FA1 afferents to have small receptive fields. For glabrous skin, SA2 afferents terminate in **Ruffini endings** and FA2 afferents end in **Pacinian corpuscles.** Both of these receptors lie deeper in the dermis and connective tissue and therefore are sensitive to stimuli applied over much larger territory. The capsules of both Pacinian and Meissner receptors act to filter out slowly changing or steady stimuli, thus making these afferents selectively sensitive to changing stimuli.

For hairy skin, the relationship between receptors and afferent classes is similar to that of glabrous skin. SA1 and SA2 fibers connect to Merkel and Ruffini endings, the same as for glabrous skin. Pacinian corpuscles also underlie the properties of FA2 afferents; however, they are not found in hairy skin but instead are located in deep tissues surrounding

muscles and blood vessels. There is not an exact analogue to the FA1 afferents. Rather there are **hair units,** which are afferents whose free endings wrap around hair follicles (see Fig. 7.2). Each such hair unit will connect with about 20 hairs to produce a large ovoid or irregularly shaped receptive field. These units are extremely sensitive to movement of even a single hair. There are also **field units** that respond to touch of the skin, but unlike FA1 units, they have large receptive fields.

Several psychophysical and neural coding questions can be related to the receptive field properties and sensitivities of the various categories of afferents. For example, is the threshold of perception of tactile stimuli due to the sensitivity of the peripheral receptors or to central processes? In fact, by using microneurography it is possible to show that a single spike in an FA1 afferent from the finger can be perceived, thus indicating that the receptors limit the sensitivity; however, for other skin regions, perception is more dependent on central factors such as attention.

An important behavioral and clinical measure of somatosensory function is spatial acuity or two-point discrimination. Clinically a doctor will apply two needle-like points simultaneously to the skin of a patient. The patient will generally perceive the points as two distinct stimuli as long as they are farther apart than some threshold distance, which varies across the body. The best discrimination (shortest threshold distance) is at the fingertips. Type 1 units underlie spatial acuity, which is not surprising given the smaller receptive fields of type 1 units than type 2 units. Moreover, the threshold distance for a region of skin is most closely related to its density of type 1 units because these units have similarly sized receptive fields throughout the glabrous skin, but their density falls off from fingertip to palm to forearm, and this fall-off correlates with the rise in threshold distance. Note that this variation in innervation density also matches the overall sensitivity of different skin regions to cutaneous stimuli.

The relationship of the firing rates in the various afferent classes to perceived stimulus quality is another important issue that has been addressed with microneurographic techniques. When a single SA fiber is stimulated with brief current pulses such that each pulse triggers a spike, a sensation of steady pressure is felt at the receptive field area of that fiber. As pulse frequency is increased a concurrent increase in pressure is perceived. Thus the firing rate in SA fibers codes for the force of the tactile stimulus. As another example, when an FA fiber is repetitively stimulated, a sensation of tapping results first, and as the frequency of the stimulus is increased, the sensation turns to one of vibration. Interestingly, in neither case does the stimulus change its qualitative character—for example, to a feeling of pain—so long as the stimulus activates only a particular fiber class. This is evidence that pain is a distinct submodality that uses a set of fibers distinct from those used by low-threshold mechanoreceptors.

These findings illustrate an important principle of sensory systems called **labeled line.** The idea is that the quality (i.e., modality) of a particular sensation results from the fact that it is conveyed to the CNS by a specific set of afferents that have a distinct set of targets in the nervous system. Alterations in activity in these afferents will therefore change only quantitative aspects of the sensation. As will be seen in more detail later, the various somatosensory submodalities (i.e., information arising from FA and SA mechanoreceptors, proprioceptors, and nociceptors) appear to use relatively separate dedicated cell populations even at relatively high levels of the CNS such as the thalamus and primary somatosensory cortex.

Innervation of the Body

Axons of the peripheral nervous system (PNS) enter or leave the CNS through the **spinal roots** (or through cranial nerves). The dorsal root of a given spinal segment is composed entirely of the central processes of its associated dorsal root ganglion cells. The ventral root consists chiefly of motor axons, including α and γ motor neuron axons (see Chapter 9), and at certain segmental levels, autonomic preganglionic axons (see Chapter 11).

The pattern of innervation is determined during embryological development. In adults a given dorsal root ganglion supplies a specific cutaneous region called a **dermatome.** Many dermatomes become distorted during development, chiefly because of rotation of the upper and lower extremities as they are formed, but also because humans maintain an upright posture. However, the sequence of dermatomes can readily be understood if depicted on the body of a person in a quadrupedal position (Fig. 7.4).

Although a dermatome receives its densest innervation from the corresponding spinal cord segment, collaterals of afferent fibers from the adjacent spinal segments also supply the dermatome. Thus transection of a single dorsal root causes little sensory loss in the corresponding dermatome. Anesthesia of any given dermatome requires interruption of several adjacent dorsal roots.

IN THE CLINIC

A common disease that illustrates the dermatomal organization of the dorsal roots is **shingles.** Shingles is the result of reactivation of the herpes zoster virus, which typically causes chickenpox during the initial infection. During the initial infection the virus infects dorsal root (and cranial nerve) ganglion cells, where it can remain latent for years to decades. When the virus reactivates, the cells of that particular dorsal root ganglion become infected, and the virus travels along the peripheral axon branches and gives rise to a painful or itchy rash that is confined to one side of the body (ends at the midline) in a dermatomal, belt-like distribution or to the distribution of a cranial nerve.

Within the dorsal roots, fibers are not randomly distributed. Rather the large myelinated primary afferent fibers assume a medial position in the dorsal root, whereas the small myelinated and unmyelinated fibers are more lateral. The large medially placed afferent fibers enter the dorsal column, where they bifurcate into rostrally and caudally directed branches. These branches give off collaterals that terminate in several neighboring segments. The rostral branch also ascends to the medulla as part of the **dorsal column–medial lemniscus pathway.** The axonal branches that terminate locally in the spinal cord gray matter transmit sensory information to neurons in the dorsal horn and also provide the afferent limb of reflex pathways (see Chapter 9).

Innervation of the Face

The arrangement of primary afferent fibers that supply the face is comparable to that of fibers that supply the body and is provided for primarily by fibers of the **trigeminal nerve.** Peripheral processes of neurons in the trigeminal ganglion (also called the *gasserian* or *semilunar ganglion*) pass through the ophthalmic, maxillary, and mandibular divisions of the trigeminal nerve to innervate dermatome-like regions of the face. These fibers carry both tactile information and pain and temperature information. The trigeminal nerve

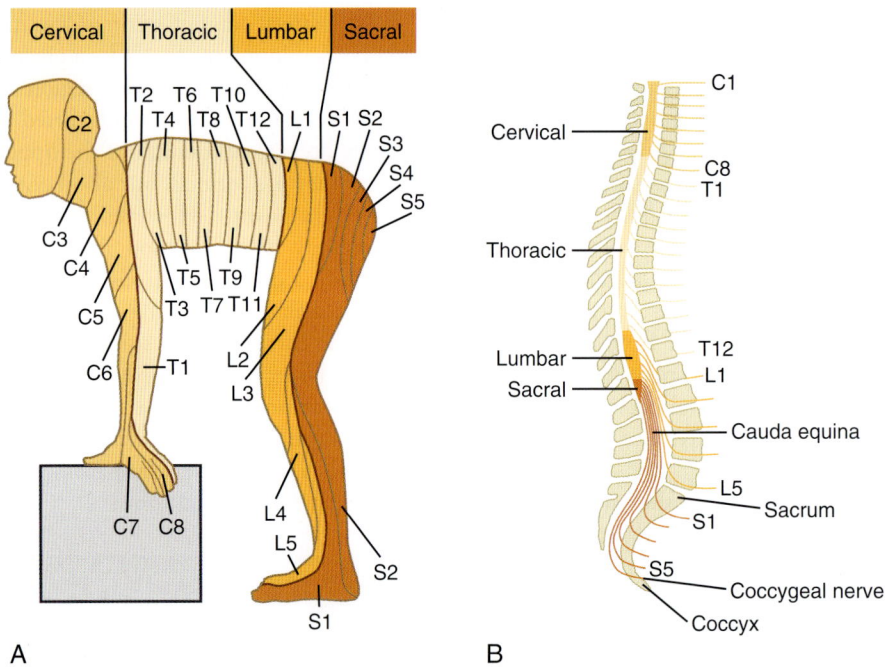

• **Fig. 7.4** A, Dermatomes represented on a drawing of a person assuming a quadrupedal position. Note nerve C1 generally has little or no sensory component, and the unlabeled portion of the head and the face are innervated by sensory fibers of the cranial nerves, primarily the trigeminal nerve. B, Sagittal view of the spinal cord showing the origin of nerves corresponding to each of the dermatomes shown in A.

AT THE CELLULAR LEVEL

The trigeminal nuclear complex consists of four main divisions, three of which are sensory. The three sensory divisions (from rostral to caudal) are the **mesencephalic, chief** (or **main**) **sensory,** and **descending** (or **spinal**) **trigeminal nuclei.** The latter two are typical sensory nuclei in that the cell bodies contained in them are second-order neurons. The mesencephalic nucleus actually contains first-order neurons and thus is analogous to a dorsal root ganglion. The fourth division of the trigeminal complex is the motor nucleus of the trigeminal nerve, whose motor neurons project to skeletal muscles of the head via the trigeminal nerve (see Chapter 4, Fig. 4.6D–E).

also innervates the teeth, the oral and nasal cavities, and the cranial dura mater.

The central processes of trigeminal ganglion cells enter the brainstem at the midpontine level, which also corresponds to the level of the chief sensory trigeminal nucleus (nucleus of cranial nerve V). Some axons terminate in this nucleus (primarily large-caliber axons carrying the information needed for fine discriminative touch), whereas others (intermediate- and small-caliber axons that carry information about touch as well as pain and temperature) form the descending trigeminal tract, which descends through the medulla just lateral to the descending trigeminal nucleus. As the tract descends, axons peel off and synapse in this nucleus.

Proprioceptive information is also conveyed via the trigeminal nerve; however, in this unique case the cell bodies of the first-order fibers are located within the CNS in the mesencephalic portion of the trigeminal nucleus. The central processes of these neurons terminate in the motor trigeminal nucleus (to subserve segmental reflexes equivalent to the segmental spinal cord reflexes [see Chapter 9]), the reticular formation, and the chief sensory trigeminal nucleus.

Central Somatosensory Pathways for Discriminatory Touch and Proprioception

As may already be clear, information related to the different somatosensory submodalities travels to a large extent into the CNS via distinct sets of axons and targets different structures in the spinal cord and brainstem. Within the CNS this segregation continues as the information travels via separate pathways up the spinal cord and brainstem. For example, from the body, fine discriminatory touch information is conveyed by the dorsal column–medial lemniscus pathway, whereas pain, temperature, and crude touch information is conveyed by the **anterolateral system.**

Proprioceptive information is transmitted by yet another route that partially overlaps with the dorsal column–medial lemniscal pathway. Note, however, that this functional segregation is not absolute, so, for example, there can be some recovery of discriminative touch ability after a lesion of the dorsal columns. The anterolateral system will be discussed in the section on pain because it is the critical pathway for that information. Here, the central pathways for discriminatory touch and proprioception are considered in detail.

Dorsal Column–Medial Lemniscus Pathway

This pathway is shown in its entirety in Fig. 7.1A. The dorsal columns are formed by ascending branches of the large myelinated axons of dorsal root ganglion cells (the first-order neurons). These axons enter at each spinal segmental level and travel rostrally up to the caudal medulla to synapse in one of the dorsal column nuclei: the nucleus gracilis, which receives information from the lower part of the body and leg, and the nucleus cuneatus, which receives information from the upper part of the body and arm. Note that across the dorsal columns and across the dorsal column nuclei there is a somatotopic representation of the body, with the legs represented most medially, followed by the trunk and then the upper limb. This somatotopy is a consequence of newly entering afferents being added to the lateral border of the dorsal funiculus as the spinal cord is ascended. Such somatotopic maps are present at all levels in the somatosensory system, at least through the primary sensory cortices.

The dorsal column nuclei are located in the medulla and contain the second-order neurons of the pathway for discriminatory touch sensation. These cells respond similarly to the primary afferent fibers that synapse on them (see the earlier description of afferent types). The main differences between the responses of dorsal column neurons and primary afferent neurons are as follows: (1) dorsal column neurons have larger receptive fields because multiple primary afferent fibers synapse on a given dorsal column neuron, (2) dorsal column neurons sometimes respond to more than one class of sensory receptor because of the convergence of several different types of primary afferent fibers on the second-order neurons, and (3) dorsal column neurons often have inhibitory receptive fields that are mediated through local interneurons.

The axons of dorsal column nuclear projection neurons exit the nuclei and are referred to as the *internal arcuate fibers* as they sweep ventrally and then medially to cross the midline at the same medullary level as the nuclei. Immediately after crossing the midline, these fibers form the medial lemniscus (see Chapter 4, Fig. 4.6C–E), which projects rostrally to the thalamus. Knowledge of the level of this decussation is clinically important because damage to the dorsal column–medial lemniscal pathway below this level, which includes all of the spinal cord, will produce loss of fine somatosensory discriminatory abilities on the same, or ipsilateral, side of the lesion, whereas lesions above this level will produce contralateral deficits. Moreover, because there is a clear somatotopic arrangement of fibers in the medial lemniscus, localized lesions cause selective loss of fine-touch sensations limited to specific body regions.

The third-order neurons of the pathway are located in the **ventral posterior lateral (VPL) nucleus of the thalamus** and project to somatosensory areas of the cerebral cortex (Fig. 7.5).

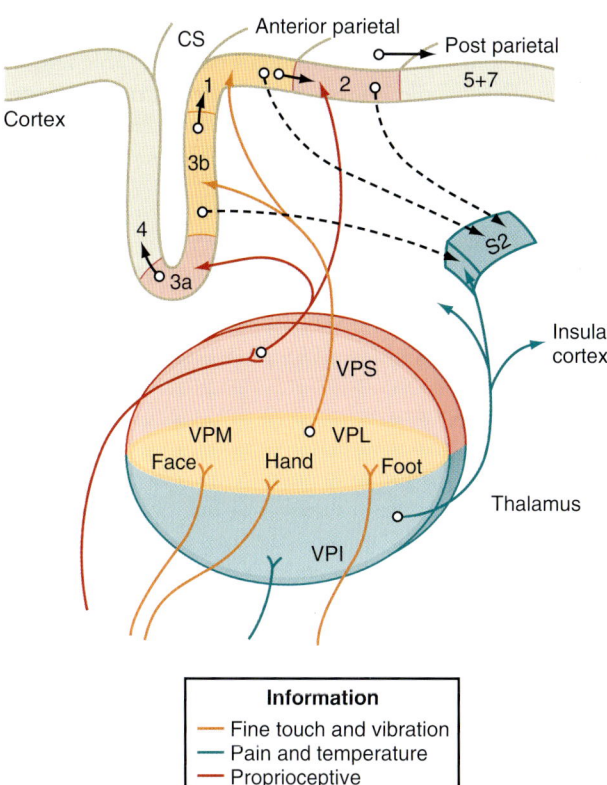

• **Fig. 7.5** Diagram of connections from the somatosensory receiving nuclei of the thalamus to the somatosensory cortex of the parietal lobe. Note the parallel flow of different types of somatosensory information through the thalamus and onto the cortex. CS, central sulcus; S1 and S2 are primary and secondary somatosensory areas, respectively. Note: collectively, areas 3a, 3b, 1, and 2 are referred to as *S1*.

The dorsal column–medial lemniscus pathway conveys information about fine-touch and vibratory sensations. This information is critical for many of the discriminatory tactile abilities we have. For example, spatial acuity is lowered by damage to this pathway, and the ability to identify objects by their shape and texture can be lost by damage to this pathway. Clinically, one may test for impaired graphesthesia, or the ability to recognize letters or numbers traced on the skin, or for loss of the ability to tell the direction of a line drawn across the skin. Importantly, some tactile function remains even after complete loss of the dorsal columns, and awareness and localization of nonnoxious tactile stimuli can still occur. Thus at least some of the information carried by the dorsal column pathway is also conveyed by additional ascending pathways. In contrast to the severe deficits in discriminatory touch sensation, cutaneous pain and temperature sensations are unaffected by lesions of the dorsal columns. However, visceral pain is substantially diminished by damage to the dorsal columns.

Trigeminal Pathway for Fine-Touch Sensation From the Face

Primary afferent fibers that supply the face, teeth, oral and nasal cavities, and cranial meninges synapse in several brainstem nuclei, including the main sensory nucleus and the descending nucleus of the trigeminal nerve.

The pathway through the main sensory nucleus resembles the dorsal column–medial lemniscus pathway. This sensory nucleus relays tactile information to the contralateral **ventral posterior medial (VPM) thalamic nucleus** by way of the **trigeminothalamic tract.** Third-order neurons in the VPM nucleus project to the facial area of the somatosensory cortex.

Spinocerebellar and Proprioceptive Pathways

Proprioceptors provide information about the positions and movement of parts of the body. In addition to being used for local reflexes (see Chapter 9), this information has two main targets, the cerebellum and the cerebral cortex. The cerebellum uses this information for its motor coordination functions. The information sent to the cerebral cortex is the basis for conscious awareness of our body parts (e.g., position of our hand), which is referred to as *kinesthesia.*

The major pathways by which somatosensory information is brought to the cerebellum are shown in Fig. 7.1B. These pathways carry both cutaneous and proprioceptive information to the cerebellum. For the trunk and lower limb, the pathway starts with dorsal root ganglion cells whose axons synapse in **Clarke's column** (nucleus dorsalis). The cells of Clarke's column send their axons into the ipsilateral lateral funiculus to form the **dorsal spinocerebellar tract,** which enters the cerebellum via the inferior cerebellar peduncle. The **ventral spinocerebellar tract** also provides somatosensory input from the lower limb to the cerebellum. Note the double decussation of the ventral spinocerebellar pathway (one decussation at the spinal cord levels and a second one in the cerebellar white matter). This double crossing highlights the general rule that each half of the cerebellum is functionally related to the ipsilateral side of the body.

To provide proprioceptive information from the lower limb to the cerebral cortex, the main axons of the dorsal spinocerebellar tract give off a branch in the medulla that terminates in nucleus z, which is just rostral to the nucleus gracilis. The axons of cells from **nucleus z** then form part of the internal arcuate fibers and medial lemniscus and ascend to the VPL nucleus of the thalamus.

The ascending somatosensory pathways to the cerebellum for the upper limb are simpler than those from the lower limb (see Fig. 7.1B). The route to the cerebellum starts with dorsal root ganglion fibers from the cervical spinal levels that ascend in the cuneate fasciculus to the **external cuneate nucleus.** The axons of the external cuneate nucleus then form the cuneocerebellar tract, which enters the cerebellum via its inferior peduncle.

The route to the cerebral cortex for proprioceptive information from the upper limb is identical to that for discriminative touch: the dorsal column–medial lemniscal pathway, with a synapse in the cuneate nucleus and then in the VPL nucleus of the thalamus.

For the head, proprioceptive input is carried by cells of the mesencephalic nucleus of the trigeminal nerve. Recall that the neurons in this nucleus are actually the cell bodies of the primary afferents that innervate stretch receptors in the muscles of mastication (muscles that move the jaw) and in other muscles of the head. The central processes of these neurons project to the trigeminal motor nucleus for local reflexes or to the nearby reticular formation. Axons from these reticular formation neurons join the trigeminothalamic tract, which terminates in the VPM of the thalamus. There are also trigeminocerebellar pathways for conveying somatosensory (tactile and proprioceptive) information from the head to the cerebellum.

Thalamic and Cortical Somatosensory Areas

Thalamus

The ventroposterior nuclear complex of the thalamus represents the main termination site for ascending somatosensory information in the diencephalon. It consists of two major nuclei, the VPL and VPM, and a smaller nucleus called the *ventral posterior inferior* (VPI) (see Fig. 7.5). The medial lemniscus forms the main input to the VPL nucleus, and the equivalent trigeminothalamic tract from the main sensory nucleus of the trigeminal nerve forms the main input to the VPM nucleus. These nuclei also receive input conveying pain and temperature information from the spinothalamic or equivalent trigeminothalamic tracts, respectively. The VPI nucleus receives input from the spinothalamic tract. In addition, the spinothalamic tract terminates in parts of the posterior nuclear complex and several other thalamic nuclei.

Single-unit recordings from the ventroposterior complex of nuclei have shown that the responses of many of the neurons in these nuclei to stimuli resemble those of first- and second-order neurons in the ascending tracts. The receptive fields of thalamic cells are small but somewhat larger than those of primary afferent fibers. Moreover, the responses may be dominated by a particular type of sensory receptor. For example, VPL and VPM nuclei have cells whose receptive fields typically reflect input either from one type of cutaneous receptor (FA or SA) or from proprioceptive receptors, as expected from their medial lemniscal input. In contrast, cells of the VPI and posterior nuclei show responses to activation of nociceptors, the main input to the spinothalamic pathway.

Thalamic neurons often have inhibitory as well as excitatory receptive fields. The inhibition may actually take place in the dorsal column nuclei or in the dorsal horn of the spinal cord. However, inhibitory circuits are also situated within the thalamus. For example, the VPL and VPM nuclei contain GABAergic inhibitory interneurons (in primates but not in rodents), and GABAergic inhibitory interneurons in the **reticular nucleus of the thalamus** project into the VPL and VPM nuclei. (The reticular nucleus of the thalamus is a thin shell of neurons that surrounds much of the thalamus. Thalamocortical and corticothalamic neurons both send axon collaterals to the reticular nucleus, whose neurons then respectively complete feedback and/ or feed-forward inhibitory circuits with neurons in other thalamic nuclei.)

One difference between neurons in the VPL and VPM nuclei and sensory neurons at lower levels of the somatosensory system is that thalamic neuron excitability depends on the stage of the sleep/wake cycle and on the presence or absence of anesthesia. During a state of drowsiness or during barbiturate anesthesia, thalamic neurons tend to undergo an alternating sequence of excitatory and inhibitory postsynaptic potentials. The alternating bursts of discharges in turn intermittently excite neurons in the cerebral cortex. Such patterns of excitation and inhibition result in an α rhythm or in spindling on the electroencephalogram. This alternation of excitatory and inhibitory postsynaptic potentials during these two states may reflect the level of excitation of thalamic neurons by excitatory amino acids that act at NMDA and non-NMDA receptors. It may also reflect inhibition of the thalamic neurons by recurrent pathways through the reticular nucleus.

Thalamic neuron receptive fields are on the side of the body contralateral to the neuron, and the receptive field locations vary systematically across the ventroposterior nuclear complex. That is, the VPL and VPM nuclei are somatotopically organized such that the lower limb is represented most laterally and the upper limb most medially in the VPL nucleus, and the head is represented even more medially in the VPM nucleus. Moreover, the fact that thalamic neurons often receive input from only one class of receptor suggests that there are multiple somatotopic maps laid out across the ventroposterior nuclear complex. That is, there appear to be separate somatotopic maps for SA, FA, and proprioceptive and pain sensations laid out across the ventroposterior nuclear complex.

These maps are not randomly interspersed. As already mentioned, pain sensation is largely mapped across the VPI nucleus. In addition, the cutaneous receptors appear to drive cells located in a central "core" region of the VPL-VPM complex, whereas proprioceptive information is directed to cells that form a "shell" (VPS) around this core. This parallel flow of information into thalamus and then onto the cortex is diagrammed in Fig. 7.5.

The spinothalamic tract also projects to other thalamic regions, including the posterior nucleus and the central lateral nucleus of the intralaminar complex of the thalamus. The intralaminar nuclei of the thalamus are not somatotopically organized, and they project diffusely to the cerebral cortex as well as to the basal ganglia (see Chapter 9). The projection of the central lateral nucleus to the S-I cortex may be involved in arousal of this part of the cortex and in selective attention.

Somatosensory Cortex

Third-order sensory neurons in the thalamus project to the somatosensory cortex. The details of this projection pattern are shown in Fig. 7.5. The main somatosensory receiving areas of the cortex are called the *S-I* and *S-II areas.* The S-I cortex (or primary somatosensory cortex) is located on the postcentral gyrus, and the S-II cortex (secondary somatosensory cortex) is in the superior bank of the lateral fissure (see Fig. 7.5).

As previously discussed, the S-I cortex, like the somatosensory thalamus, has a somatotopic organization. The S-II cortex also contains a somatotopic map, as do several other less understood areas of the cortex. In the S-I cortex the face is represented in the lateral part of the postcentral gyrus, above the lateral fissure. The hand and the rest of the upper extremity are represented in the dorsolateral part of the postcentral gyrus, and the lower extremity on the medial surface of the hemisphere. A map of the surface of the body and face of a human on the postcentral gyrus is called a **sensory homunculus.** The map is distorted because the volume of neural tissue devoted to a body region is proportional to the density of its innervation. Thus in humans, the perioral area, the thumb, and other digits take up a disproportionately large expanse of cortex relative to their size.

The sensory homunculus is an expression of place coding of somatosensory information. A locus in the S-I cortex encodes the location of a somatosensory stimulus on the surface of the body or face. For example, the brain knows that a certain part of the body has been stimulated because certain neurons in the postcentral gyrus are activated.

The S-I cortex has several morphological and functional subdivisions, and each subdivision has a somatotopic map. These subdivisions were originally described by Brodmann, and they were based on the arrangements of neurons in the various layers of the cortex, as seen in Nissl-stained preparations. The subdivisions are therefore known as *Brodmann areas 3a, 3b, 1,* and *2* (see Chapter 10). Cutaneous input dominates in areas 3b and 1, whereas muscle and joint input (proprioceptive) dominates in areas 3a and 2. Thus separate cortical zones are specialized for processing tactile and proprioceptive information (see Fig. 7.5).

Within any particular area of the S-I cortex, all the neurons along a line perpendicular to the cortical surface have similar response properties and receptive fields. The S-I cortex is thus said to have a *columnar organization.* A comparable columnar organization has also been demonstrated for other primary sensory receiving areas, including the primary visual and auditory cortices (see Chapter 8). Nearby cortical columns in the S-I cortex may process information for different sensory modalities. For example, the cutaneous information that reaches one cortical column in area 3b may come from FA mechanoreceptors, whereas the information that reaches a neighboring column might originate from SA mechanoreceptors.

Besides being responsible for the initial processing of somatosensory information, the S-I cortex also begins higher-order processing such as feature extraction. For example, certain neurons in area 1 respond preferentially to a stimulus that moves in one direction across the receptive field but not in the opposite direction (Fig. 7.6). Such

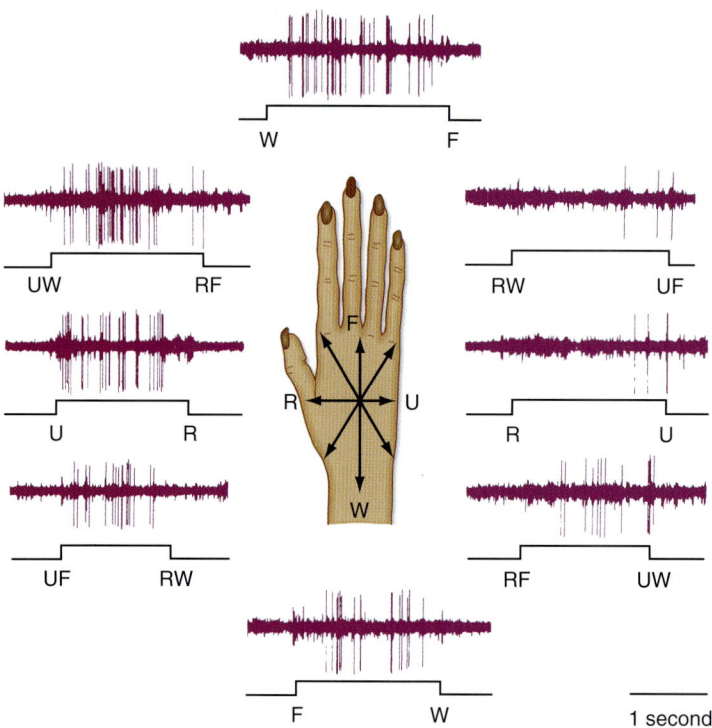

• **Fig. 7.6** Feature extraction by cortical neurons. The responses were recorded from a neuron in the somatosensory cortex of a monkey. The direction of a stimulus was varied, as shown by the arrows in the drawing. Note that the responses were greatest when the stimulus moved in the direction from UW to RF and least from RW to UF. F, fingers; R, radial side; U, ulnar side; W, wrist. (From Costanzo RM, Gardner EP. *J Neurophysiol* 1980;43:1319.)

neurons presumably contribute to the perceptual ability to recognize the direction of an applied stimulus and could help detect slippage of an object being grasped by the hand.

Effects of Lesions of the Somatosensory Cortex

A lesion of the S-I cortex in humans produces sensory changes similar to those produced by a lesion of the somatosensory thalamus. However, usually only a part of the cortex is involved, and thus the sensory loss may be confined, for example, to the face or to the leg, depending on the location of the lesion with respect to the sensory homunculus. The sensory modalities most affected are discriminative touch and position sense. Graphesthesia and stereognosis (i.e., the ability to recognize objects such as coins and keys as they are handled) are particularly disturbed. Pain and thermal sensation may be relatively unaffected, although loss of pain sensation may follow cortical lesions. Conversely, cortical lesions can result in a central pain state that resembles thalamic pain (see "Effects of Interruption of the Spinothalamic Tract and Lesions of the Thalamus on Somatosensory Sensation").

Pain and Temperature Sensation

The sensations of pain and temperature are related and often grouped together because they are mediated by overlapping sets of receptors and are conveyed by the same types of fibers in the PNS and the same pathways in the CNS. One consequence of these labeled lines is that pain sensations in particular are not due to stronger activation of touch pathways, as might naively be thought. This difference is borne out experimentally because if SA afferents, for example, are stimulated more and more frequently, the sensation of tactile pressure becomes stronger but not painful.

Nociceptors and Primary Afferents

The axons that carry painful and thermal sensations are members of the relatively slowly conducting Aδ and C classes. However, not all Aδ and C axons carry pain and temperature information; some respond to light touch in a manner similar to what was described for low-threshold mechanoreceptors.

Unlike the case for low-threshold mechanoreceptors in which morphologically distinct receptors correspond to response properties, the Aδ and C axons conveying pain and temperature information appear to originate mostly as "free nerve endings." (This description is not entirely accurate; the endings are mostly but not entirely covered by Schwann cells.) Despite the lack of distinct morphological specialization associated with their endings, Aδ and C axons constitute a heterogeneous population that is differentially sensitive to a variety of tissue-damaging or thermal stimuli (or both). This ability to sense tissue-damaging stimuli (mechanical, thermal, or chemical) is mediated by what are called **nociceptors.** These receptors share some features with low-threshold mechanoreceptors but are distinct in many ways, such as the ability to become sensitized (see later). Indeed, there appear to be a significant number of C fibers that are silent or unresponsive to any stimuli until first sensitized.

The first functional distinction that may be made in the pain system is between Aδ and C axons. Aδ axons conduct signals faster than C fibers do and are thought to underlie what is called **first pain,** whereas C fibers are responsible for **second pain.** Thus after a damaging stimulus, one first feels an initial sharp, pricking, highly localized sensation (first pain) followed by a duller, more diffuse, burning sensation (second pain). Experiments in which Aδ or C fibers were selectively activated demonstrated that activity in Aδ fibers produces sensations similar to first pain and that activity in C fibers produces second pain–like sensations.

Each fiber class in turn forms a heterogeneous group with regard to sensitivity to stimuli. Thus afferents are classified according to both size and their sensitivity to mechanical, thermal, and chemical stimuli. Fibers may have a low or high threshold to mechanical stimulation or be completely insensitive to it. Thermal sensitivity has been classified as responsiveness to warmth, noxious heat, cool, and noxious cold. Note that 43°C and 15°C are the approximate limits above and below which, respectively, thermal stimuli are sensed as painful. Chemical sensitivity to a variety of irritating compounds has been tested, including capsaicin (found in chili peppers), mustard oil, and acids.

Afferent fibers may be sensitive to one or more types of stimuli and have been named accordingly. For example, C fibers sensitive only to high-intensity (damaging) mechanical stimuli are called *C mechanosensitive fibers,* whereas those sensitive to heat and mechanical stimuli are labeled *C mechanoheat-sensitive fibers* (also called *polymodal fibers*). Other identified fiber types include Aδ and C cold-sensitive, Aδ mechanosensitive, and mechanoheat-sensitive fibers. Thus there is quite a variety of afferent types; however, the most common afferent type is the C polymodal fiber, which accounts for nearly half of the cutaneous C fibers. Surprisingly the second most common type is the mechanoheat-insensitive afferent (i.e., an afferent that is not sensitive to noxious stimuli until sensitized—see later).

Spinal Cord Gray Matter and Trigeminal Nucleus

The central portion of the Aδ and C axons carrying pain and temperature information from the body terminates in the dorsal horn of the spinal cord. The Aδ fibers target lamina I (dorsalmost part of dorsal horn), V (base of dorsal horn), and X (area surrounding central canal) of the gray matter, whereas the C fibers terminate in lamina I and II. The distinct termination patterns of the Aδ and C fibers in the spinal cord suggest that the messages they are carrying to the CNS are kept separated, and this is consistent with our ability to feel two distinct types of pain.

The primary afferent termination patterns in the spinal cord are also important because they may help determine the possible interactions pain fibers can have with other afferents and with descending control systems (see later). Indeed, the **gate control theory of pain** refers to the phenomenon that

innocuous stimuli, such as rubbing a hurt area, can block or reduce painful sensations. Such stimulation activates the large-diameter (Aα and Aβ) fibers, and their activity leads to release of GABA and other neurotransmitters by interneurons within the dorsal horn. GABA then acts by both presynaptic and postsynaptic mechanisms to shut down the activity of spinothalamic tract cells. Presynaptically, GABA activates both GABA$_A$ and GABA$_B$ receptors, which leads to partial depolarization of the presynaptic terminal and blocking of Ca^{++} channels, respectively. Both actions will decrease release of transmitter by the afferent terminal and thereby lessen excitation of the tract cell (see Chapter 6 section on presynaptic inhibition).

Nociceptive and thermoreceptive information that originates from regions of the head is processed in a fashion similar to that for the trunk and limbs. The primary afferent fibers of nociceptors and thermoreceptors in the head enter the brainstem through the trigeminal nerve (some also enter through the facial, glossopharyngeal, and vagus nerves). Of note, the trigeminal distribution includes both tooth and headache pain. These fibers then descend through the brainstem as far as the upper cervical spinal cord via the descending tract of the trigeminal nerve. Some mechanoreceptive afferent fibers also join the descending tract of the trigeminal nerve. Axons in the descending tract synapse on second-order neurons in the descending nucleus of the trigeminal nerve.

 IN THE CLINIC

Elderly people are sometimes susceptible to a condition of chronic pain known as **trigeminal neuralgia.** People with this condition experience spontaneous episodes of severe, often lancinating pain in the distribution of one or more branches of the trigeminal nerve. Frequently the pain is triggered by weak mechanical stimulation in the same region. A major contributing factor to this painful state appears to be mechanical damage to the trigeminal ganglion by an artery that impinges on the ganglion. Surgical displacement of the artery can often resolve the condition.

Central Pain Pathways

The central pain pathways include the spinothalamic, spinoreticular, and spinomesencephalic tracts. The **spinothalamic tract** is the most important sensory pathway for somatic pain and thermal sensations from the body (see Fig. 7.1A). It also contributes to tactile sensation. The spinothalamic tract originates from second-order neurons located in the spinal cord (primarily laminae I and IV to VI). The axons of these cells cross to the opposite side of the cord at or near to their level of origin. They then ascend to the brain in the ventral part of the lateral funiculus and subsequently through the brainstem to the thalamus, where they terminate on third-order neurons (thalamocortical neurons), as described earlier. Spinothalamic cells conveying pain and temperature target the VPI portion of the ventroposterior complex (although some also end in the VPL), the posterior nucleus, and the intralaminar nuclei of the thalamus. Nociceptive signals are then forwarded to several cortical areas, including not only the somatosensory cortex but also cortical areas that are involved in affective responses, such as the cingulate gyrus and the insula, which have limbic system functions (see Fig. 7.5).

Most spinothalamic tract cells receive excitatory input from nociceptors in the skin, but many can also be excited by noxious stimulation of muscle, joints, or viscera. Few receive input only from viscera. Effective cutaneous stimuli include noxious mechanical, thermal (hot or cold), and chemical stimuli. Thus different spinothalamic tract cells respond in a manner appropriate for signaling noxious, thermal, or mechanical events.

Some nociceptive spinothalamic tract cells receive convergent excitatory input from several different classes of cutaneous sensory receptors. For example, a given spinothalamic neuron may be activated weakly by tactile stimuli but more powerfully by noxious stimuli (Fig. 7.7A). Such neurons are called *wide–dynamic range cells* or *multireceptive cells* because they are activated by stimuli with a wide range of intensities. Wide–dynamic range neurons signal mainly noxious events; weak responses to tactile stimuli appear to be ignored by the higher centers. However, in certain pathological conditions, these neurons may be sufficiently activated by tactile stimuli to evoke a sensation of pain, possibly as a result of activity in sensitized afferents that were previously silent. This would explain some pain states in which activation of mechanoreceptors causes pain (mechanical allodynia). Other spinothalamic tract cells are activated only by noxious stimuli. Such neurons are often called *high-threshold* or *nociceptive-specific cells* (see Fig. 7.7B).

Because cells signaling visceral input also typically convey information from cutaneous receptors, the brain may misidentify the source of the pain. This phenomenon is called **referred pain.** A typical example is when the heart muscle becomes ischemic and pain is felt in the chest wall and left arm.

Neurotransmitters released by **nociceptive afferents** that activate spinothalamic tract cells include the excitatory amino acid glutamate and any of several peptides, such as SP, CGRP, and vasoactive intestinal polypeptide (VIP). Glutamate appears to act as a fast transmitter by its action on non-NMDA excitatory amino acid receptors. However, with repetitive stimulation, glutamate can also act through NMDA receptors. Peptides appear to act as neuromodulators. For example, through a combined action with an excitatory amino acid such as glutamate, SP can produce a long-lasting increase in the responses of spinothalamic tract cells; this enhanced responsiveness is called **central sensitization.** CGRP seems to increase release of SP and prolong the action of SP by inhibiting its enzymatic degradation.

Spinothalamic tract cells often have inhibitory receptive fields. Inhibition may result from weak mechanical stimuli, but usually the most effective inhibitory stimuli

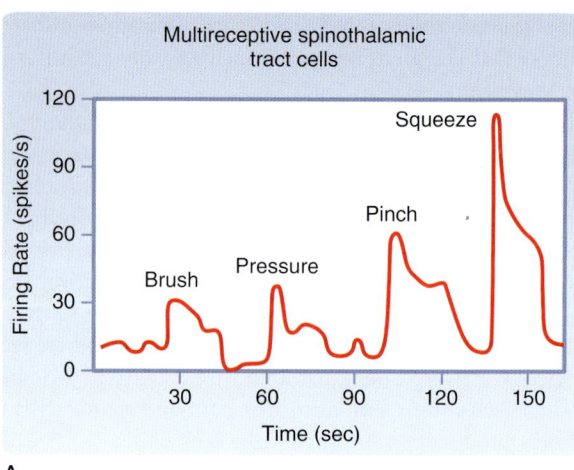

A

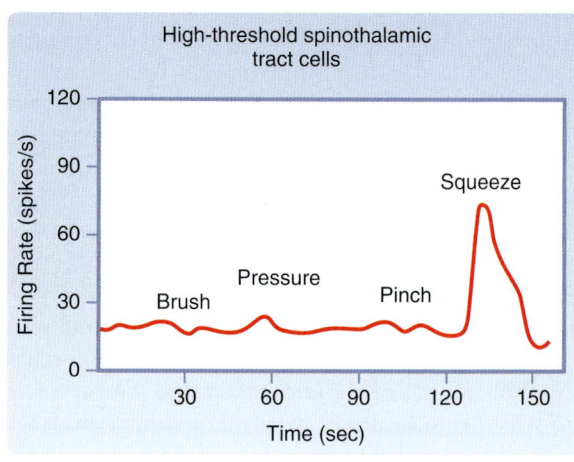

B

• **Fig. 7.7** A, Responses of a wide–dynamic range or multireceptive spinothalamic tract cell. B, Responses of a high-threshold spinothalamic tract cell. Graphs show responses to graded intensities of mechanical stimulation. Brush stimulus is a camel's hair brush repeatedly stroked across the receptive field. Pressure is applied by attachment of an arterial clip to the skin. This is a marginally painful stimulus to a human. Pinch is achieved by attachment of a stiff arterial clip to the skin and is distinctly painful. Squeeze is applied by compressing a fold of skin with forceps and is damaging to the skin.

are noxious ones. The nociceptive inhibitory receptive fields may be very large and may include most of the body and face. Such receptive fields may account for the ability of various physical manipulations, including transcutaneous electrical nerve stimulation and acupuncture, to suppress pain. Neurotransmitters that can inhibit spinothalamic tract cells include the inhibitory amino acids GABA and glycine, as well as monoamines and the endogenous opioid peptides.

Spinoreticular tract neurons frequently have large, sometimes bilateral receptive fields, and effective stimuli include noxious ones. These dorsal horn neurons target multiple regions in the medullary and pontine reticular formation. The reticular formation, which projects to the intralaminar complex of the thalamus and thereby to wide areas of the cerebral cortex, is involved in attentional mechanisms and arousal (see Chapter 10). The reticular formation also gives rise to descending reticulospinal projections, which contribute to the descending systems that control transmission of pain.

Many cells of the **spinomesencephalic tract** respond to noxious stimuli, and the receptive fields may be small or large. The terminations of this tract are in several midbrain nuclei, including the periaqueductal gray, which is an important component of the endogenous analgesia system. Motivational responses may also result from activation of the periaqueductal gray matter. For example, stimulation in the periaqueductal gray matter can cause vocalization and aversive behavior. Information from the midbrain is relayed not only to the thalamus but also to the amygdala. This provides one of several pathways by which noxious stimuli can trigger emotional responses.

Pain and temperature information originating from the face and head is conveyed along analogous ascending central pathways, as is such information from the body. Neurons in the descending trigeminal nucleus transmit pain and temperature information to specific nuclei (VPM, VPI) of the contralateral thalamus via the ventral trigeminothalamic tract, which runs in close association with the medial lemniscus. The descending nucleus also projects to the intralaminar complex and other thalamic nuclei in a fashion similar to that of the spinothalamic tract. The thalamic nuclei in turn project to the somatosensory cerebral cortex for sensory discrimination of pain and temperature and to other cortical regions responsible for motivational-affective responses.

Effects of Interruption of the Spinothalamic Tract and Lesions of the Thalamus on Somatosensory Sensation

When the spinothalamic tract and accompanying ventral spinal cord pathways are interrupted, both the sensory-discriminative and the motivational-affective components of pain are lost on the contralateral side of the body. This result motivated development of the surgical procedure known as *anterolateral cordotomy,* which was used to treat pain in many individuals, especially those suffering from cancer. This operation is now used infrequently because of improvements in drug therapy and because pain often returns months to years after an initially successful cordotomy. Return of pain may reflect either extension of the disease or development of a central pain state. In addition to loss of pain sensation, anterolateral cordotomy produces loss of cold and warmth sensation on the contralateral side of the body. Careful testing may reveal a minimal tactile deficit as well, but the intact sensory pathways of the dorsal part of the spinal cord provide sufficient tactile information that any loss caused by interruption of the spinothalamic tract is insignificant.

Destruction of the VPL or VPM nuclei diminishes sensation on the contralateral side of the body or face. The sensory qualities that are lost reflect those that are transmitted mainly by the dorsal column–medial lemniscus pathway and its trigeminal equivalent. The sensory-discriminative component of pain sensation is also lost. However, the motivational-affective component of pain is still present if the medial thalamus is intact. Presumably, pain persists because of the spinothalamic and spinoreticulothalamic projections to this part of the thalamus. In some individuals a lesion of the somatosensory thalamus results in a central pain state known as *thalamic pain.* Patients with thalamic pain report that even the slightest touch feels painful, although the intensity of the touch is lower than the threshold of any pain receptor. It is thought that their pain sensitivity is due to a post-lesion sprouting of low-threshold dorsal column system fibers that synapse onto surviving thalamic neurons that normally mediate only pain. Pain that is indistinguishable from thalamic pain can also be produced by lesions in the brainstem or cortex.

Neuropathic Pain

Pain sometimes occurs in the absence of nociceptor stimulation. This type of pain is most likely to occur after damage to peripheral nerves or to parts of the CNS that are involved in transmitting nociceptive information. Pain caused by damage to neural structures is called *neuropathic pain.* Neuropathic pain states include peripheral neuropathic pain, which may follow damage to a peripheral nerve, and central neuropathic pain, which sometimes occurs after damage to CNS structures.

Examples of pain secondary to damage to a peripheral nerve are causalgia and phantom limb pain. Causalgia may develop after traumatic damage to a peripheral nerve. Even though evoked pain is reduced, severe pain may develop in the area innervated by the damaged nerve. This pain may be very difficult to treat, even with strong analgesic drugs. The pain is caused in part by spontaneous activity that develops in dorsal root ganglion cells; such activity may be attributed to upregulation of Na^+ channels. In some cases the pain seems to be maintained by sympathetic neural activity, because a sympathetic nerve block may alleviate the pain. Sympathetic involvement may relate to the sprouting of damaged sympathic postganglionic axons into the dorsal root ganglia, and it may be accompanied by upregulation of adrenoreceptors in primary afferent neurons. Phantom limb pain follows traumatic amputation in some individuals. Such phantom pain is clearly not caused by activation of nociceptors in the area in which pain is felt, because these receptors are no longer present.

Lesions of the thalamus or at other levels of the spinothalamocortical pathway may cause central pain, which is a severe spontaneous pain. However, interruption of the nociceptive pathway by the same lesion may simultaneously prevent or reduce the pain evoked by peripheral stimulation. The mechanism of such trauma-induced pain caused by neural damage is poorly understood. The pain appears to depend on changes in the activity and response properties of more distant neurons in the nociceptive system.

Transduction in the Somatosensory System

Unraveling the transduction processes for the somatosensory system has proven difficult because of a variety of factors. The heart of the transduction process occurs at the specialized endings of the peripheral branch of the axon of the sensory neuron (dorsal root ganglia and trigeminal ganglia cells). These endings have channels that are activated by mechanical, thermal, and/or chemical stimuli and that allow the current flow that underlies the local generator potential (see Chapter 5, Fig. 5.13). The generator potential then triggers spikes in the afferent fiber (see Chapter 5, Fig. 5.14). However, the fact is that somatosensory axons are found throughout the body at low densities, making purification of the proteins difficult (e.g., compare with the retina, in which photoreceptors are packed at high density onto a small surface).

Complicating matters, the generator potential can be modified by a host of voltage-gated channels, both excitatory (e.g., Na^+ and Ca^{++} channels) and inhibitory (K^+ channels). Moreover, as described earlier, in many cases the axon terminal is encapsulated (e.g., Pacinian corpuscles), and the properties of this accessory structure modify the transduction process. They may do so passively, as a result of the mechanical characteristics of the capsule, or actively by having the accessory cells release transmitter in response to a stimulus. An example of the latter is Merkel cells, where recent evidence suggests the Merkel cells and the axon innervating them both have mechanosensitive channels and the Merkel cells release neurotransmitter.

Despite these complications, using a number of approaches, our knowledge of somatosensory transduction processes has begun to increase rapidly.

Mechanotransduction

Until recently, ASIC (acid-sensing ion channel) proteins, which belong to the DEG/ENaC family, had been thought to be the channel proteins underlying the cutaneous mechanoactivated currents, because homologs of these proteins underlie the touch sensation in invertebrates such as *Caenorhabditis elegans,* and because some ASIC proteins are highly expressed in dorsal root ganglion cells. However, the mechanoactivated currents in dorsal root ganglion cells are not affected by knockdown of the ASIC proteins, indicating that they are not the channel underlying cutaneous mechanotransduction. Nevertheless, touch and pain sensitivity is altered in such knockdown mutants, so they may still play a modulatory role in the transduction process.

Currently, Piezo2 is thought to be channel protein underlying the transduction for cutaneous mechanical rapidly adapting responses, because it forms a nonselective cation pore that opens in response to mechanical stimuli. Moreover, the activation and inactivation kinetics of this channel are consistent with its causing the rapidly adapting mechanoactivated current, and it is blocked by agents

(gadolinium, ruthenium red) that block these rapidly adapting currents. Furthermore, transfection of ganglion cells with small interfering (si)RNA for Piezo2 blocks the rapid mechanoactivated current in these cells. Lastly, messenger (m)RNA expression of Piezo2 has been shown in dorsal root ganglion cells.

Both low-threshold (light touch) and nociceptive-type ganglion cells have been found to express Piezo2, indicating that it plays a role in both innocuous and painful touch sensation. The Piezo proteins (Piezo1 and Piezo2) are both found in a variety of organs and thus may underlie visceral sensation as well. Piezo2 has been found in the main proprioceptors (muscle spindles and Golgi tendon organs [see Chapter 9]) and appears to be the main mechanotransducer protein there as well.

Thermal Transduction

The receptor that binds capsaicin (the molecule in chili peppers responsible for their spiciness) has been identified, and either it or one of a family of related proteins has been found to be expressed in populations of dorsal root ganglion cells. These proteins belong to the **TRP (transient receptor potential)** protein family and are currently the most likely candidates for being the transducers of thermal sensations.

It is important to note that many ion channels (and other proteins, e.g., enzymes) are sensitive to temperature; however, in the case of TRP channels, temperature is acting directly as the gating mechanism. The temperatures at which specific TRP channels are active are indicated by arrows in Fig. 7.8 (bottom), in which the direction of each

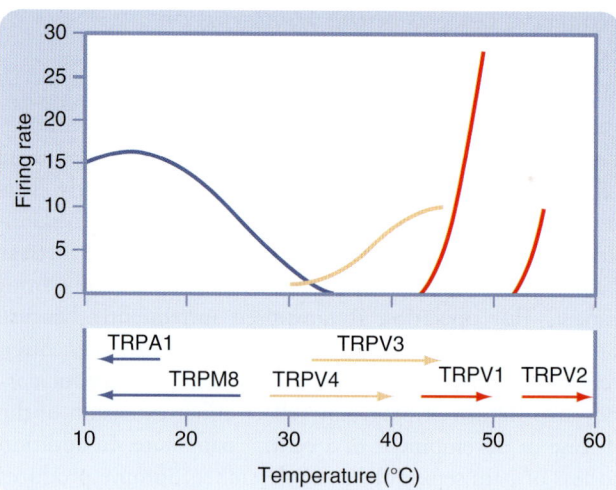

• **Fig. 7.8** Temperature dependence of firing rates in different thermosensitive afferents. Below the firing rate curves are shown the ranges over which the different TRP channels are activated. The direction of increasing activation is indicated by an arrow in each case. Note how in some cases the range over which an afferent is active (top graphs) corresponds well to the activation range of a single TRP channel, thus suggesting that the afferent would need to express only a single type of channel. In other cases the active range of the afferent suggests that multiple TRP channels would be needed to underlie the complete responsiveness of the afferent.

arrow indicates which temperatures cause greater activation. For comparison, Fig. 7.8 also plots the firing rates of several thermosensitive fibers as a function of temperature. Note how the response ranges of the afferents largely match up with those of individual heat-sensitive channels. The cold fibers, however, show firing over a wider range than any one TRP channel does. One possible explanation for this discrepancy is that dorsal root ganglion cells may express multiple classes of TRP, which would enable them to respond over a wider range of physiological temperatures.

AT THE CELLULAR LEVEL

Members of the TRP protein family were first identified in *Drosophila* and were found to be part of the phototransduction process in *Drosophila* photoreceptors. Thus the name *(TRP)* refers to the fact that a mutation in the gene leads to a transient depolarizing response to a light stimulus instead of the normal sustained response. On the basis of sequence homology, a number of genes encoding TRP proteins have been found in mammals (27 in humans alone), which are currently divided into seven subfamilies. TRP channels are cation permeable and have a structure similar to voltage-gated K^+ channels. They are homotetramers or heterotetramers. Each subunit has six transmembrane domains. TRP proteins appear to have a variety of functions (e.g., phototransduction, chemotransduction, and mechanotransduction) and are expressed in a number of cell types. Those listed in Table 7.1 appear to act as temperature sensors with distinct thermal sensitivities that span the range of physiologically relevant temperatures.

Modulation of the Transduction Process

As with the low-threshold mechanoreceptors for innocuous touch sensations, activation of the various nociceptor

TABLE 7.1   TRP Family Proteins Involved in Thermal Transduction

Receptor Protein	Threshold or Temperature Range for Activation (°C)	Other Characteristics
TRPV1	>42	Activated by capsaicin
TRPV2	>52	
TRPV3	34-38	Activated by camphor
TRPV4	27-40	
TRPM8	<25	Activated by menthol
TRPA1	<18	Activated by mustard oil

The fourth letter in the name identifies the subfamily and was chosen because of the first member of the subfamily identified: V, vanilloid; M, melastatin; A, ankyrin-like. Each of the proteins listed is expressed in at least some dorsal root ganglion cells, but they are also expressed in other cell types.

transduction proteins leads to a generator potential that causes spiking of the afferent, which transmits information to the CNS. In addition, activation of nociceptors also leads to local release of various chemical compounds, including **tachykinins (substance P [SP])** and **calcitonin gene–related protein (CGRP).** These substances and others released from the damaged cells cause neurogenic inflammation (edema and redness of the surrounding skin).

In addition to causing a local reaction, these substances may serve to activate the insensitive or silent nociceptors mentioned earlier, such that they can henceforth respond to any subsequent damaging stimuli. Sensitization of silent nociceptors has been suggested to underlie *allodynia* (elicitation of painful sensations by stimuli that were innocuous before an injury) and *hyperalgesia* (increase in the level of pain felt to already painful stimuli). Thermal hyperalgesia is due to sensitization of TRPV1 channels. The Piezo2 current is increased by substances known to cause mechanical hyperalgesia and allodynia, suggesting changes in this current underlie these phenomena.

Centrifugal Control of Somatosensation

Sensory experience is not just the passive detection of environmental events. Instead, it more often depends on exploration of the environment. Tactile cues are sought by moving the hand over a surface. Visual cues result from scanning targets with the eyes. Thus sensory information is often received as a result of activity in the motor system. Furthermore, transmission in pathways to the sensory centers of the brain is regulated by descending control systems. These systems allow the brain to control its input by filtering the incoming sensory messages. Important information can be attended to and unimportant information can be ignored.

The tactile and proprioceptive somatosensory pathways are regulated by descending pathways that originate in the S-I and motor regions of the cerebral cortex. For example, cortical projections to the dorsal column nuclei help control the sensory input that is transmitted by the dorsal column–medial lemniscus pathway.

Of particular interest is the descending control system that regulates transmission of nociceptive information. This system presumably suppresses excessive pain under certain circumstances. For example, it is well known that soldiers on the battlefield, accident victims, and athletes in competition often feel little or no pain at the time a wound occurs or a bone is broken. At a later time, pain may develop and become severe. Although the descending regulatory system that controls pain is part of a more general centrifugal control system that modulates all forms of sensation, the pain control system is so important medically that it is distinguished as a special system called the **endogenous analgesia system.**

Several centers in the brainstem and pathways descending from these centers contribute to the endogenous analgesia system. For example, stimulation in the midbrain

• **Fig. 7.9 A,** Models of descending control of ascending pain pathways. A, Schematic showing the ascending nociceptive pathway that conveys pain information to the brain, represented by the spinothalamic tract (STT), and the descending pathway that modulates (gates) the flow of information into it. The descending pathway starts in the periaqueductal gray (PAG), which projects via the rostroventromedial (RVM) medulla to the dorsal horn. Activation of the PAG or RVM blocks transmission of pain information at the spinal cord level. B, The standard model of descending control of pain perception. Tonically active GABAergic interneurons in the PAG and RVM inhibit the projection cells (note that the PAG has local GABAergic interneurons but also receives GABAergic terminals from other brain regions, and both may be involved in tonically inhibiting the PAG projection cells). Opioids appear to inhibit the GABAergic interneurons, thus disinhibiting the projection neurons, whose increased activity leads to blocking the activation of STT cells by nociceptive dorsal root ganglion (DRG) cells. C, Parallel inhibition-excitation model. Some studies suggest both excitatory and inhibitory cells project from the PAG and the RVM. In this model, gating of transmission of pain information would be due to a balance of the excitatory and inhibitory activity in the descending pathways. Note that the excitatory and inhibitory pathways are shown on opposite sides just for clarity. D, Possible presynaptic and postsynaptic sites of action of enkephalin (Enk). Many of the terminals from the descending pathways for pain modulation release both GABA and enkephalin (EnK), which appears to have both pre- and postsynaptic sites of action. The presynaptic action might prevent release of substance P (Sub P.) from nociceptors. (A-C, Redrawn from Lau BK, Vaughan CW. *Curr Opin Neurobiol* 2014;29:159. D, Redrawn from Henry JL. In: Porter R, O'Connor M [eds]: Ciba Foundation Symposium 91. London: Pitman; 1982.)

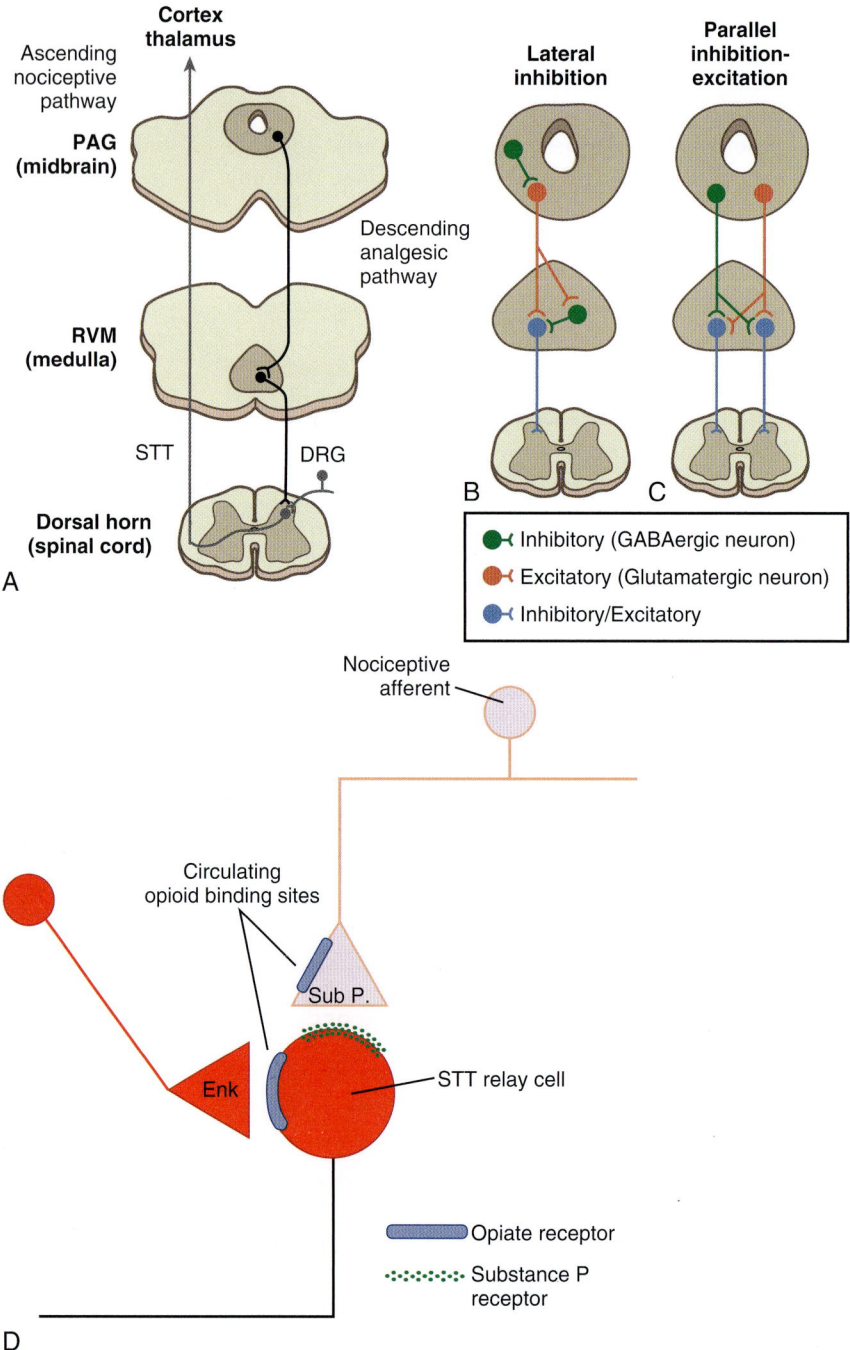

periaqueductal gray, locus coeruleus, or medullary raphe nuclei inhibits nociceptive neurons at the spinal cord and brainstem level, including spinothalamic and trigeminothalamic tract cells (Fig. 7.9A-C). Other inhibitory pathways originate in the sensorimotor cortex, hypothalamus, and reticular formation.

The endogenous analgesia system can be subdivided into two components: one component uses endogenous **opioid** peptides as neurotransmitters and the other does not. Endogenous opioids are neuropeptides that activate one of several types of opiate receptors. Some of the endogenous opioids include enkephalin, dynorphin, and β-endorphin. Opiate analgesia can generally be prevented or reversed by the narcotic antagonist naloxone. Therefore naloxone is frequently used to determine whether analgesia is mediated by an opioid mechanism.

The opioid-mediated endogenous analgesia system can be activated by exogenous administration of morphine or other opiate drugs. Thus one of the oldest medical treatments of pain depends on the triggering of a sensory control system. Opiates typically inhibit neural activity in nociceptive pathways. Two sites of action have been proposed for opiate inhibition, presynaptic and postsynaptic (see Fig. 7.9D). The presynaptic action of opiates on nociceptive afferent terminals is thought to prevent release of excitatory transmitters such as SP. The postsynaptic action of opiates produces an inhibitory postsynaptic potential. How can an inhibitory neurotransmitter activate descending pathways? One hypothesis is that the descending analgesia system is under tonic inhibitory control by inhibitory interneurons in both the midbrain and medulla. The action of opiates would inhibit the inhibitory interneurons and thereby disinhibit the descending analgesia pathways.

Some endogenous analgesia pathways operate by neurotransmitters other than opioids and thus are unaffected by naloxone. One way of engaging a nonopioid analgesia pathway is through certain forms of stress. The analgesia thus produced is a form of stress-induced analgesia.

Many neurons in the raphe nuclei use serotonin as a neurotransmitter. Serotonin can inhibit nociceptive neurons and presumably plays an important role in the endogenous analgesia system. Other brainstem neurons release catecholamines, such as norepinephrine and epinephrine, in the spinal cord. These catecholamines also inhibit nociceptive neurons; therefore catecholaminergic neurons may contribute to the endogenous analgesia system. Furthermore, these monoamine neurotransmitters interact with endogenous opioids. Undoubtedly, many other substances are involved in the analgesia system. In addition, there is evidence for the existence of endogenous opiate antagonists that can prevent opiate analgesia.

Key Concepts

1. Sensory neurons have cell bodies in sensory nerve ganglia: (1) dorsal root ganglia for neurons innervating the body and (2) cranial nerve ganglia for neurons innervating the face, oral and nasal cavities, and dura, except for proprioceptive neurons, which are in the trigeminal mesencephalic nucleus. They connect peripherally to a sensory receptor and centrally to second-order neurons in the spinal cord or brainstem.
2. Skin contains low-threshold mechanoreceptors, thermoreceptors, and nociceptors. Muscle, joints, and viscera have mechanoreceptors and nociceptors. Low-threshold mechanoreceptors may be rapidly or slowly adapting. Thermoreceptors include cold and warm receptors. Aδ and C nociceptors detect noxious mechanical, thermal, and chemical stimuli and may be sensitized by release of chemical substances from damaged cells. Peripheral release of substances, such as peptides, from nociceptors themselves may contribute to inflammation.
3. Large primary afferent fibers enter the dorsal funiculus through the medial part of the dorsal root; collaterals synapse in the deep dorsal horn, intermediate zone, and ventral horn. Small primary afferent fibers enter the spinal cord through the lateral part of the dorsal root; collaterals synapse in the dorsal horn.
4. Ascending branches of large primary afferent fibers synapse on second-order neurons in the dorsal column nuclei. These second-order neurons project via the medial lemniscus to the contralateral thalamus and synapse on third-order neurons of the VPL nucleus. The equivalent trigeminal pathway is relayed by the main sensory nucleus to the contralateral VPM nucleus.
5. The dorsal column spinal cord pathways signal the sensations of flutter-vibration, touch-pressure, and proprioception. They also contribute to visceral sensation, including visceral pain.
6. The spinothalamic tract includes nociceptive, thermoreceptive, and tactile neurons; its cells of origin are mostly in the dorsal horn, and the axons cross, ascend in the ventrolateral funiculus, and synapse in the VPL, VPI, and posterior and intralaminar nuclei of the thalamus. The equivalent trigeminal pathway is relayed by the descending trigeminal nucleus and projects to the contralateral VPM and intralaminar nuclei.
7. The spinothalamic relay in the VPL, VPM, and VPI nuclei helps account for the sensory-discriminative aspects of pain. Parallel nociceptive pathways in the ventrolateral funiculus are the spinoreticular and spinomesencephalic tracts; these tracts and the spinothalamic projection to the medial thalamus contribute to the motivational-affective aspects of pain.
8. Referred pain is explained by convergent input to spinothalamic tract cells from the body wall and from viscera.
9. The VPL and VPM nuclei are somatotopically organized and contain inhibitory circuits. These nuclei contain multiple somatotopic maps, one for each somatosensory submodality. The somatosensory cortex includes the S-I and S-II regions; these regions are also somatotopically organized.
10. The S-I cortex contains columns of neurons with similar receptive fields and response properties. Some S-I neurons are involved in feature extraction.
11. Transmission in somatosensory pathways is regulated by descending control systems. The endogenous analgesia system regulates nociceptive transmission, and it uses transmitters such as endogenous opioid peptides, norepinephrine, and serotonin.

Additional Readings

Coste B, Xiao B, Santos JS, et al. Piezo proteins are pore-forming subunits of mechanically activated channels. *Nature*. 2012;483:176-181.

Lau BK, Vaughan CW. Descending modulation of pain: the GABA disinhibition hypothesis of analgesia. *Curr Opin Neurobiol*. 2014;29:159-164.

Squire L, Berg D, Bloom FE, et al., eds. *Fundamental Neuroscience*. 4th ed. Waltham, MA: Academic Press; 2013.

8

The Special Senses

LEARNING OBJECTIVES

Objectives Heading

Upon completion of this chapter, the student should be able to answer the following questions:

1. What is the dark current, and how does the absorption of a photon change it?
2. What are the synaptic pathways for the central and surround portions of the receptive field of an on-center bipolar cell? Of an off-center bipolar cell?
3. What are the receptive field properties of simple and complex cells in the visual cortex?
4. What is the frequency theory of sound encoding? Why is the place theory also required?
5. What are the stimuli that are normally transduced by the hair cells in the semicircular canals and otolith organs?
6. What are the functional consequences of the differing numbers of different receptor molecules between olfactory and gustatory receptor cells?

The evolution of vertebrates shows a trend called **cephalization** in which special sensory organs develop in the heads of animals, along with the corresponding development of the brain. These special sensory systems, which include the visual, auditory, vestibular, olfactory, and gustatory systems, detect and analyze light, sound, and chemical signals in the environment, as well as signal the position and movement of the head. The stimuli transduced by these systems are most familiar to humans when they provide conscious awareness of the environment, but they are equally important as the sensory basis for reflexive and subconscious behavior.

The Visual System

Vision is one of the most important special senses in humans and, along with audition, is the basis for most human communication. The visual system detects electromagnetic waves between 400 and 750 nm long as **visible light**, which enters the eye and impinges on **photoreceptors** in a specialized sensory epithelium, the **retina.**

The photoreceptors, rods and cones, can distinguish two aspects of light: its **brightness** (or luminance) and its **wavelength** (or color). **Rods** have high sensitivity for detecting low light intensities but do not provide well-defined visual images, nor do they contribute to color vision. Rods operate best under conditions of reduced lighting **(scotopic vision). Cones,** in contrast, are not as sensitive to light as rods are and thus operate best under daylight conditions **(photopic vision).** Cones are responsible for high visual acuity and color vision.

The retina is an outgrowth of the thalamus. Thus information processing within the retina is performed by **retinal neurons,** and the output signals are carried to the brain by the axons of **retinal ganglion cells** in the **optic nerves.** There is a partial crossing of these axons in the **optic chiasm** that causes all input from one side of the visual space to pass to the opposite side of the brain. Posterior to the optic chiasm, the axons of retinal ganglion cells form the **optic tracts** and synapse in nuclei of the brain. The main visual pathway in humans targets the **lateral geniculate nucleus (LGN)** of the thalamus and this nucleus, through the visual radiations, projects the visual information to the visual cortex. Other visual pathways project to the **superior colliculus, pretectum,** and **hypothalamus,** structures that participate in orientation of the eyes, control of pupil size, and circadian rhythms, respectively.

Structure of the Eye

The wall of the eye is composed of three concentric layers (Fig. 8.1). The outer layer, or the fibrous coat, includes the transparent **cornea,** with its epithelium, and the opaque **sclera.** The middle layer, or vascular coat, includes the iris and the choroid. The **iris** contains both radially and circularly oriented smooth muscle fibers, which make up the pupillary dilator and constricter muscles, respectively. The **choroid** is rich in blood vessels that support the outer layers of the retina, and it also contains pigment. The innermost layer of the eye, the retina, is embryologically derived from the diencephalon and therefore is part of the central nervous system (CNS). The functional part of the retina covers the entire posterior aspect of the eye except for the optic nerve head, or **optic disc,** which is where the optic nerve axons leave the retina. Because there are no receptors at this location, it is often referred to as the anatomical "blind spot" (see Fig. 8.1).

A number of functions of the eyes are under muscular control. Externally attached extraocular muscles aim the eyes toward an appropriate visual target (see Chapter 9). These muscles are innervated by the **oculomotor nerve**

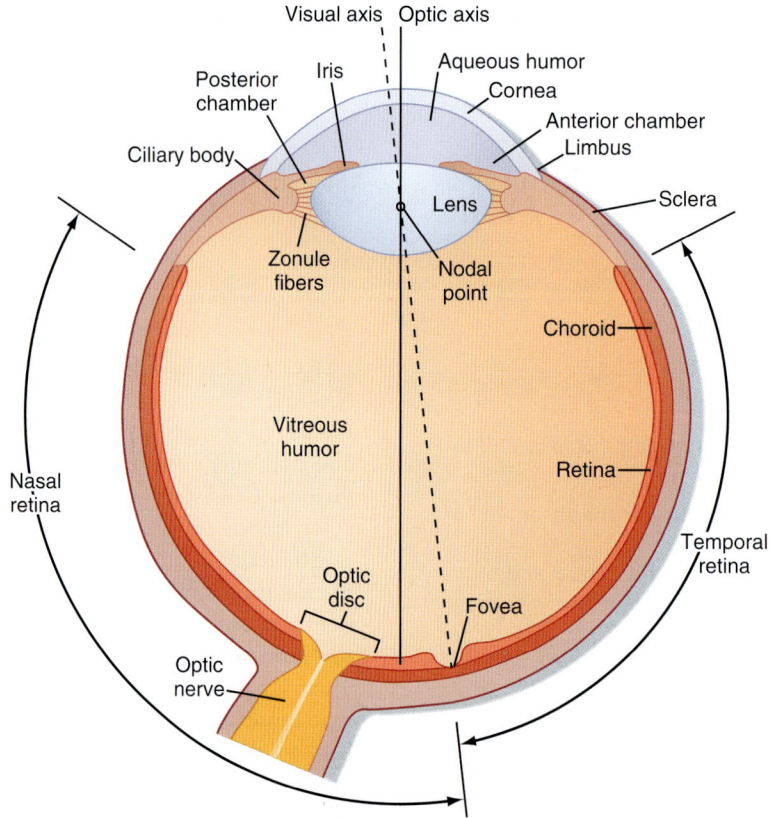

• **Fig. 8.1** Illustration of a view of a horizontal section of the right eye. (Redrawn from Wall GL. *The Vertebrate Eye and Its Adaptive Radiation*. Bloomfield Hills, MI: Cranbrook Institute of Science; 1942.)

(cranial nerve [CN] III), the trochlear nerve (CN IV), and the abducens nerve (CN VI). Several muscles are also found within the eye (intraocular muscles). The **muscles in the ciliary body** control lens shape and thereby the focus of images on the retina. The **pupillary dilator** and **sphincter** muscles in the iris control the amount of light entering the eye, in a way similar to that of the diaphragm of a camera. The dilator is activated by the sympathetic nervous system, whereas the sphincter and ciliary muscles are controlled by the parasympathetic nervous system (through the oculomotor nerve; see Chapter 11).

Light enters the eye through the cornea and passes through a series of transparent fluids and structures that are collectively called the **dioptric media.** These fluids and structures consist of the cornea, aqueous humor, lens, and vitreous humor (see Fig. 8.1). The aqueous humor (located in the **anterior** and **posterior chambers**) and the vitreous humor (located in the space behind the lens) help maintain the shape of the eye.

Although the geometrical optic axis of the human eye passes through the nodal point of the lens and reaches the retina at a point between the fovea and the optic disc (see Fig. 8.1), the eyes are oriented by the oculomotor system to a point, called the **fixation point,** on the visual target. Light from the fixation point passes through the nodal point of the lens and is focused on the **fovea.** Light from the remainder of the visual target falls on the retina surrounding the fovea.

Normally, light from a visual target is focused sharply on the retina by the cornea and lens, which bend or refract the light. The cornea is the major refractive element of the eye, with a refractive power of 43 diopters[a] (D). However, unlike the cornea, the lens can change shape and vary its refractive power between 13 and 26 D. Thus the lens is responsible for adjusting the optical focus of the eye. **Suspensory ligaments** (or **zonule fibers**) attach to the wall of the eye at the ciliary body (see Fig. 8.1) and hold the lens in place. When the muscles in the ciliary body are relaxed, the tension exerted by the suspensory ligaments flattens the lens. When the ciliary muscles contract, the tension on the suspensory ligaments is reduced; this process allows the somewhat elastic lens to assume a more spherical shape. The ciliary muscles are activated by the parasympathetic nervous system via the oculomotor nerve.

In this way, the lens allows the eye to focus on, or accommodate to, either near or distant objects. For instance, when light from a distant visual target enters a normal eye (one with a relaxed ciliary muscle), the target image is in focus on the retina. However, if the eye is directed at a nearby visual target, the light is initially focused behind the retina (i.e., the image at the retina is blurred) until accommodation

---

[a]A diopter is a unit of measurement of optical power that is equal to the reciprocal of the focal length measured in meters. Thus it is a unit of reciprocal length, and a 2-D lens would bring parallel rays of light into focus at a distance of 0.5 m.

occurs; that is, until the ciliary muscle contracts, causing the lens to become more spherical, the increased convexity causes the lens to refract the light waves more strongly, bringing the image into focus on the retina.

Proper imaging of light on the retina depends not only on the lens and cornea but also on the iris, which adjusts the amount of light that can enter the eye through the pupil. In this regard, the pupil is analogous to the aperture in a camera, which also controls the depth of field of the image and the amount of spherical aberration produced by the lens. When the pupil is constricted, the depth of field is increased, and the light is directed through the central part of the lens, where spherical aberration is minimal. Pupillary constriction occurs reflexively when the eye accommodates for near vision or adapts to bright light, or both. Thus when a person reads or does other fine visual work, the quality of the image is improved by adequate light.

Retina

Layers of the Retina

The 10 layers of the retina are shown in Fig. 8.2. The outermost portion is the **pigmented epithelium** (layer 1), which is just inside the choroid. The pigment cells have

 IN THE CLINIC

As an individual ages, the elasticity of the lens gradually declines. As a result, accommodation of the lens for near vision becomes progressively less effective, a condition called **presbyopia.** A young person can change the power of the lens by as much as 14 D. However, by the time that a person reaches 40 years of age, the amount of accommodation halves, and after 50 years, it can decrease to 2 D or less. Presbyopia can be corrected by convex lenses.

Defects in focus can also be caused by a discrepancy between the size of the eye and the refractive power of the dioptric media. For example, in **myopia** (near-sightedness), the images of distant objects are focused in front of the retina. Concave lenses correct this problem. Conversely, in **hypermetropia** (far-sightedness), the images of distant objects are focused behind the retina; this problem can be corrected with convex lenses. In **astigmatism,** an asymmetry exists in the radii of curvature of different meridians of the cornea or lens (or sometimes of the retina). Astigmatism can often be corrected with lenses that possess complementary radii of curvature.

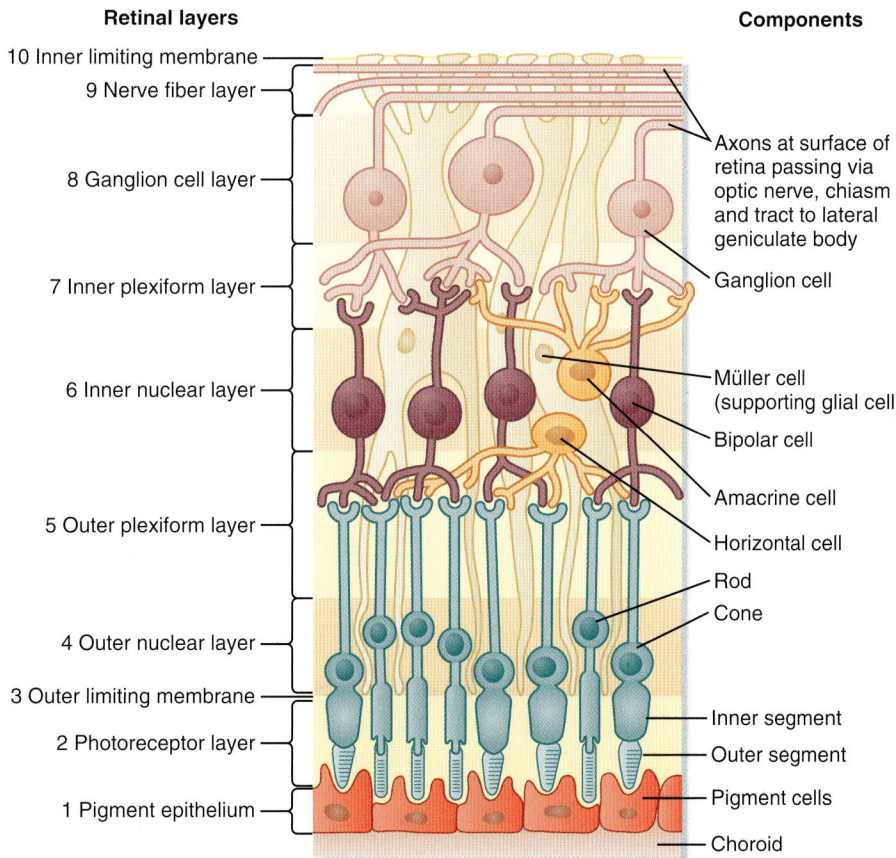

Retinal layers

10 Inner limiting membrane
9 Nerve fiber layer
8 Ganglion cell layer
7 Inner plexiform layer
6 Inner nuclear layer
5 Outer plexiform layer
4 Outer nuclear layer
3 Outer limiting membrane
2 Photoreceptor layer
1 Pigment epithelium

Components

Axons at surface of retina passing via optic nerve, chiasm and tract to lateral geniculate body
Ganglion cell
Müller cell (supporting glial cell)
Bipolar cell
Amacrine cell
Horizontal cell
Rod
Cone
Inner segment
Outer segment
Pigment cells
Choroid

Fig. 8.2 Layers of the Retina. Light impinging on the retina would be coming from the top of the figure, and would pass through all the superficial layers to reach the photoreceptor rods and cones.

tentacle-like processes that extend into the **photoreceptor layer** (layer 2) and surround the outer segments of the rods and cones. These processes prevent transverse scatter of light between photoreceptors. In addition, they serve a mechanical function in maintaining contact between layers 1 and 2 so that the pigmented epithelium can (1) provide nutrients and remove waste from the photoreceptors; (2) phagocytose the ends of the outer segments of the rods, which are continuously shed; and (3) reconvert metabolized visual pigment into a form that can be reused after it is transported back to the photoreceptors.

Retinal glial cells, known as **Müller cells,** play an important role in maintaining the internal geometry of the retina. Müller cells are oriented radially, parallel to the light path through the retina. The outer ends of Müller cells form tight junctions with the inner segments of the photoreceptors, and these numerous connections have the appearance of a continuous layer, the **outer limiting membrane** (layer 3 of the retina).

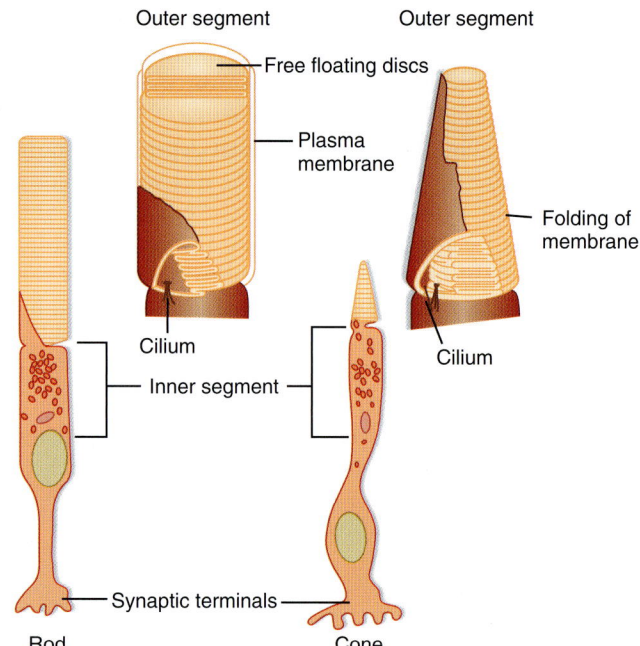

• **Fig. 8.3** Rods and Cones. The drawings at the *bottom* show the general features of a rod and a cone. The *insets* show the outer segments.

 IN THE CLINIC

The junction between layers 1 and 2 of the retina in adults represents the surface of contact between the anterior and posterior walls of the embryonic optic cup during development and is structurally weak. Retinal detachment is separation at this surface and can cause loss of vision because of displacement of the retina from the focal plane of the eye. It can also lead to the death of photoreceptor cells, which are maintained by the blood supply of the choroid (the photoreceptor layer itself is avascular). Deterioration of the pigmented epithelium can also result in macular degeneration, a critical loss of high-acuity central and color vision that does not affect peripheral vision.

Inside the external limiting membrane is the **outer nuclear layer** (layer 4) that contains the cell bodies and nuclei of the rods and cones. The **outer plexiform layer** (layer 5) contains synapses between the photoreceptors and retinal interneurons, including bipolar cells and horizontal cells, whose cell bodies are found in the **inner nuclear layer** (layer 6). This layer also contains the cell bodies of other retinal interneurons (the amacrine and interplexiform cells) and the Müller cells.

The **inner plexiform layer** (layer 7) contains synapses between the retinal neurons of the inner nuclear layer, including the bipolar and amacrine cells, and the ganglion cells, whose cell bodies lie in the **ganglion cell layer** (layer 8). As previously mentioned, the ganglion cells are the output cells of the retina; it is their axons that transmit visual information to the brain. These axons form the **optic fiber layer** (layer 9), pass along the inner surface of the retina while avoiding the fovea, and enter the optic disc, where they leave the eye as the optic nerve. The portions of the ganglion cell axons that are in the optic fiber layer remain unmyelinated, but they become myelinated after they reach the optic disc. The lack of myelin where the

axons cross the retina helps permit light to pass through the inner retina with minimal distortion.

The innermost layer of the retina is the **inner limiting membrane** (layer 10) formed by the end-feet of Müller cells.

Structure of Photoreceptors: Rods and Cones

Each rod or cone photoreceptor cell is composed of a cell body (in layer 4), an inner and an outer segment that extend into layer 2, and a set of synaptic terminals that synapse in layer 5 onto other retinal cells (Fig. 8.3). The outer segments of cones are not as long as those of rods, and they contain stacks of disc membranes formed by infoldings of the plasma membrane. The outer segments of rods are longer, and they contain stacks of membrane discs that float freely in the outer segment, having completely disconnected from the plasma membrane when formed at the base. Both sets of discs are rich in visual pigment molecules, but rods have a greater visual pigment density, which partly accounts for their greater sensitivity to light. A single photon can elicit a rod response, whereas several hundred photons may be required for a cone response.

The outer segments of the photoreceptors are connected by a modified cilium to the inner segments, which contain a number of organelles, including numerous mitochondria. The inner segments are the sites where the visual pigment is synthesized before it is incorporated into the membranes of the outer segment. In rods, the pigment is inserted into new membranous discs, which are then displaced distally until they are eventually shed at the apex of the outer segment, where they undergo phagocytosis by cells of the pigmented

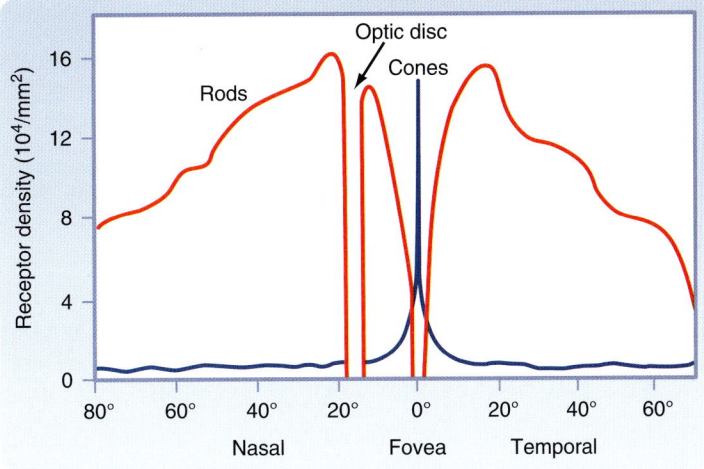

• **Fig. 8.4** Graph of a Plot of the Density of Cones and Rods as a Function of Retinal Eccentricity From the Fovea. Note that cone density peaks at the fovea, rod density peaks at about 20 degrees eccentricity, and no photoreceptors are found at the optic disc, where the ganglion cell axons leave to form the optic nerve. (Data from Cornsweet TN. *Visual Perception.* New York: Academic Press; 1970.)

epithelium. This process determines the rod-like shape of the outer segments of rods. In cones, the visual pigment is inserted randomly into the membranous folds of the outer segment, and shedding, comparable to that seen in rods, does not take place.

Regional Variations in the Retina

The **macula lutea** is the area of central vision and is characterized by a slight thickening and a pale color. The thickness is due to the high concentration of photoreceptors and interneurons, which are needed for high-resolution vision. It is pale because both optic nerve fibers and blood vessels are routed around it.

The fovea, which is a depression in the macula lutea, is the region of the retina with the very highest visual resolution and, as noted previously, the light from the fixation point is focused on the fovea. (A major function of eye movements is to bring objects of interest into view on the fovea.) The retinal layers in the foveal region are unusual because several of them appear to be pushed aside into the surrounding macula. Consequently, light can reach the foveal photoreceptors without having to pass through the inner layers of the retina, and both image distortion and light loss are minimized. The fovea has cones with unusually long and thin outer segments, which allows for high packing density. In fact, cone density is maximal in the fovea, providing for high visual resolution, as well as high quality of the image (Fig. 8.4).

The optic disc, where ganglion cell axons leave the retina, lacks photoreceptors and therefore lacks photosensitivity. Thus it is a so-called blind spot in the visual surface of the retina (see Figs. 8.4 and 8.9). A person is normally unaware of the blind spot because the corresponding part of the visual field can be seen by the contralateral eye and because of the psychological process in which incomplete visual images tend to be completed perceptually.

 IN THE CLINIC

As mentioned, the axons of retinal ganglion cells cross the retina in the optic fiber layer (layer 9) to enter the optic nerve at the optic disc. These axons in the optic fiber layer pass around the macula and fovea, as do the blood vessels that supply the inner layers of the retina. The optic disc can be visualized on physical examination with an **ophthalmoscope.** The normal optic disc has a slight depression in its center. Changes in the appearance of the optic disc are important clinically. For example, the depression may be exaggerated by loss of ganglion cell axons **(optic atrophy),** or the optic disc may protrude into the vitreous space because of edema **(papilledema)** that results from increased intracranial pressure.

Visual Transduction

To be detected by the retina, light energy must be absorbed. This is the job primarily of the rods and cones (a small class of ganglion cells are also photosensitive) and is accomplished by visual pigment molecules located in their outer segments. For both rods and cones, the pigment molecule consists of a chromophore, 11-*cis* retinal, bound to an opsin protein. The visual pigment found in the outer segments of rods is **rhodopsin,** or visual purple (so named because it has a purple appearance when light has been absorbed). It absorbs light best at a wavelength of 500 nm. Three variants of visual pigment, resulting from the binding of different opsins to retinal, are found in cones (in most species, each cone expresses one of the three cone pigments). The cone pigments absorb best at 419 nm (blue), at 533 nm (green), and at 564 nm (red). However, the absorption spectrum of these visual pigments is broad so that they overlap considerably (Fig. 8.5).

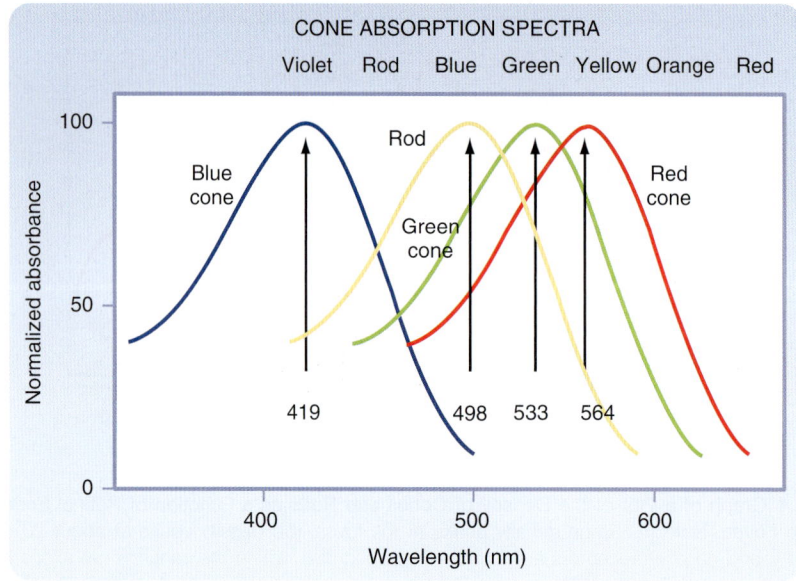

• **Fig. 8.5** The Spectral Sensitivity of the Three Types of Cone Pigments and of the Rod Pigment (Rhodopsin) in the Human Retina. Note that the curves overlap and that the so-called blue and red cones actually absorb maximally in the violet and yellow ranges, respectively. (Data from Squire LR, et al [eds]. *Fundamental Neuroscience.* San Diego, CA: Academic Press; 2002.)

Despite the differences in spectral sensitivity, the transduction process is similar in rods and cones. The absorption of a photon by a visual pigment molecule leads to the isomerization of 11-*cis* retinal to all-*trans* retinal, release of the bond with the opsin, and conversion of retinal to retinol. These changes trigger a second messenger cascade that leads to a change in the electrical activity of the rod or cone (discussed later in this section).

The separation of all-*trans* retinal from opsin also causes both the loss of its ability to absorb light and bleaching (i.e., the visual pigment loses its color). In both rods and cones, regeneration of the visual pigment molecule is a multistep process: the all-*trans* retinal is transported to the retinal pigmented cell layer, where it is reduced to retinol, isomerized, and esterified back to 11-*cis* retinal. It is then transported back to the photoreceptor layer, taken up by outer segments, and recombined with opsin to regenerate the visual pigment molecule, which can again absorb light. There is evidence that cones also use a second pathway to regenerate visual pigment. This pathway is much more rapid and involves transport of the retinal molecule to and from the Müller cells (see Fig. 8.2) rather than the pigmented epithelial cells. The potential importance of this more rapid pathway is discussed in the section "Visual Adaptation."

Ultimately, the transduction process triggered by absorption of photons causes the photoreceptor to hyperpolarize. To understand this action and its consequences fully, it is necessary to know the baseline state of the photoreceptor in the dark (i.e., before it absorbs a photon). In darkness, photoreceptors are slightly depolarized (≈ -40 mV) in relation to most neurons because cyclic guanosine monophosphate (cGMP)–gated cation channels in their outer segments are open (Fig. 8.6A). These channels allow a steady influx of Na$^+$ and Ca^{2+}. The resulting current is known as the **dark current,** and the depolarization it causes leads to the tonic release of the neurotransmitter glutamate at the photoreceptor's synapses.

When light is absorbed in a rod (an equivalent sequence happens in cones), photoisomerization of rhodopsin activates a G protein called **transducin** (see Fig. 8.6B). This G protein, in turn, activates **cyclic guanosine monophosphate phosphodiesterase,** which is associated with the rhodopsin-containing discs, hydrolyzes cGMP to 5′-GMP, and lowers the cGMP concentration in the rod cytoplasm. The reduction in cGMP leads to closing of the cGMP-gated cation channels, hyperpolarization of the rod cell membrane, and a reduction in the release of neurotransmitters. Thus cGMP acts as a "second messenger" to translate the absorption of a photon by rhodopsin into a change in membrane potential.

In sum, in all photoreceptors (cones undergo a process analogous to that described for rod transduction), capture of light energy leads to (1) hyperpolarization of the photoreceptor and (2) a reduction in the release of neurotransmitters. Because of the very short distance between the site of transduction and the synapse, the modulation of neurotransmitter release is accomplished without the generation of an action potential.

Visual Adaptation

Adaptation refers to the ability of the retina to adjust its sensitivity according to ambient light. This ability allows the retina to operate efficiently over a wide range of lighting conditions, and it reflects a switching between the use of the cone and rod systems for bright- and low-light conditions, respectively.

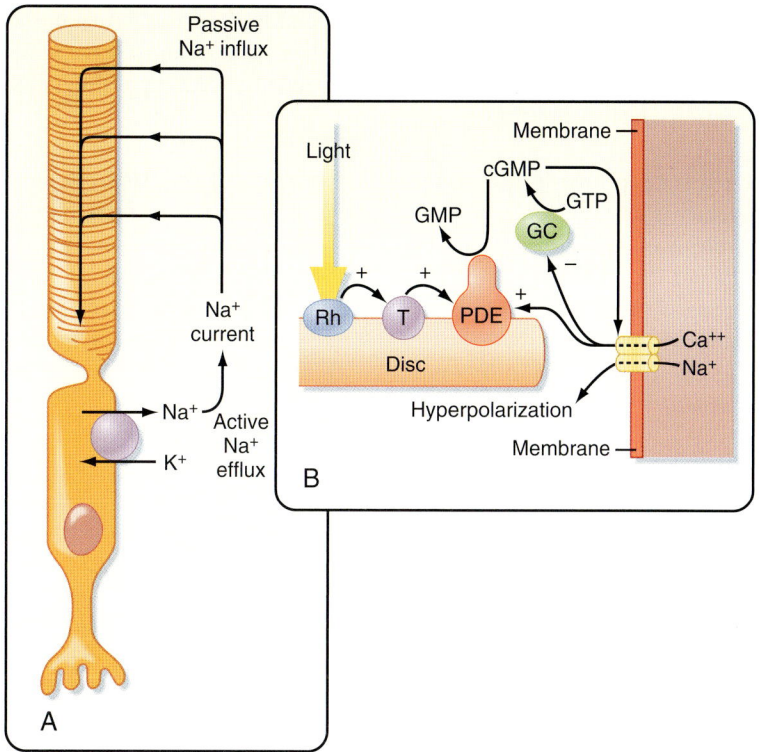

• Fig. 8.6 A, Drawing of a rod with the flow of current in the dark. With the assistance of the Na⁺,K⁺ pump, the rod is kept depolarized. **B,** Sequence of the second messenger events that follow the absorption of light through the reduction of cGMP. Because cGMP maintains open Na⁺ channels in the dark, the results of light absorption are the closing of the Na⁺ channels and hyperpolarization of the rod. cGMP, cyclic guanosine monophosphate; GC, guanylate cyclase; GTP, guanosine triphosphate; PDE, phosphodiesterase; Rh, rhodopsin; T, transducin.

AT THE CELLULAR LEVEL

Rhodopsin contains a chromophore called *retinal,* which is the aldehyde of **retinol,** or vitamin A. Retinol is derived from carotenoids, such as β-carotene, the orange pigment found in carrots. Like other vitamins, retinol cannot be synthesized by humans; instead, it is derived from food sources. Individuals with a severe vitamin A deficiency suffer from "night blindness," a condition in which vision is defective in low-light situations.

The extraordinary sensitivity of rods, which can signal the capture of a single photon, is enhanced by an amplification mechanism in which photoactivation of only one rhodopsin molecule can activate hundreds of transducin molecules. In addition, each phosphodiesterase molecule hydrolyzes thousands of cGMP molecules per second. Similar events occur in cones, but the membrane hyperpolarization occurs much more quickly than in rods and requires thousands of photons.

Light Adaptation

As described previously, absorption of a photon causes 11-*cis* retinal to be converted to all-*trans* retinal, which then splits from the opsin (bleaching). The visual pigments in rods and cones are bleached at a similar rate; however, regeneration of the visual pigment occurs much more rapidly in cones than in rods. This difference is, at least in part, due to the cones' ability to utilize a second pathway for regeneration (see previous section). This more rapid regeneration of visual pigment prevents cones from becoming unresponsive in bright light conditions. In contrast, the slowness of the regeneration of rhodopsin molecules means that at light levels not much above those found in evening hours, essentially all of the rhodopsin molecules are bleached. Thus in bright-light conditions, only the cone system is functioning, and the retina is said to be **light-adapted.**

When entering a darkened movie theater, a person can observe evidence of the existing light adaptation (decreased light sensitivity in association with the reduced amount of rhodopsin) in the inability to see the empty seats (or much else). The gradual return of the ability to see the seats while the person remains in the theater reflects the slow regeneration of rhodopsin and recovery of function of the rod system, a process known as **dark adaptation.**

Dark Adaptation

This process refers to the gradual increase in light sensitivity of the retina when in low-light conditions. Rods adapt to darkness slowly as their rhodopsin levels are restored, and, indeed, it may take more than 30 minutes for the retina to become fully dark-adapted. In contrast, cones adapt rapidly to darkness, but their adapted threshold is relatively high, and so they do not function when the ambient light level is low. Thus within 10 minutes in a dark room, rod vision is more sensitive than cone vision and becomes the main system for seeing.

In sum, in the dark-adapted state, primarily rod vision is operative, and thus visual acuity is low and colors are not distinguished (this is called *scotopic vision*). However, when light levels are higher (e.g., when the movie is projected) and cone function resumes (this is called *photopic vision*), visual acuity and color vision are restored. There is an intermediate range of light levels at which rod and cones are both functional *(mesopic vision)*.

Color Vision

The visual pigments in the cone outer segments contain different opsins. As a result of these differences, the three types of cones absorb light best at different wavelengths. Although the cone pigments have maximum efficiency closer to violet, green, and yellow wavelengths, they are referred to as blue, green, and red pigments, respectively (see Fig. 8.5). The differences in the cone absorption spectra underlie humans' ability to see colors, as opposed to only shades of gray.

According to the **trichromacy theory,** the differences in absorption efficiency of the cone visual pigments are presumed to account for color vision because a suitable mixture of three colors can produce any other color. However, a neural mechanism must also exist for the analysis of color brightness because the amount of light absorbed by a visual pigment, as well as the subsequent response of the cell, depends on both the wavelength and the intensity of the light (see Fig. 8.5). Two or three of the cone pigments may absorb a particular wavelength of light, but the amount absorbed by each differs according to its efficiency at that wavelength. If the intensity of the light is increased (or decreased), all will absorb more (or less), but the ratio of absorption among them will remain constant. Consequently, there must be a neural mechanism to compare the absorption of light of different wavelengths by the different types of cones for the visual system to distinguish different colors. At least two different kinds of cones are required for color vision. The presence of three kinds decreases the ambiguity in distinguishing colors when all three absorb light, and it ensures that at least two types of cones will absorb most wavelengths of visible light.

The **opponent process theory** is based on observations that certain pairs of colors seem to activate opposing neural processes. Green and red are opposed, as are yellow and blue, as well as black and white. For example, if a gray area is surrounded by a green ring, the gray area appears to acquire a reddish color. Furthermore, a greenish red or a bluish yellow color does not exist. These observations are supported by findings that neurons activated by green wavelengths are inhibited by red wavelengths. Similarly, neurons excited by blue wavelengths may be inhibited by yellow wavelengths. Neurons with these characteristics are present both in the retina and at higher levels of the visual pathway and seem to serve to increase the ability to see the contrast between opposing colors.

Retinal Circuitry

A diagram of the basic circuitry of the retina is shown in Fig. 8.7. Several features of this circuitry are noteworthy. Input to the retina is provided by light striking the photoreceptors. The output is carried by axons of the retinal ganglion cells to the brain. Information is processed within the retina by the interneurons. The most direct pathway through the retina is from a photoreceptor to a bipolar cell and then to a ganglion cell (see Fig. 8.7). More indirect pathways that provide for intraretinal signal processing involve photoreceptors, bipolar cells, amacrine cells, and ganglion cells, as well as horizontal cells to provide lateral interactions between adjacent pathways.

 IN THE CLINIC

Observations on color blindness are consistent with the trichromacy theory. In color blindness, a genetic defect (sex-linked recessive), one or more cone mechanisms are lost. People with normal color vision are trichromats because they have three cone mechanisms. Individuals who lack one of the cone mechanisms are called dichromats. When the long-wavelength cone mechanism is absent, the resulting condition is called protanopia; absence of the medium-wavelength system causes deuteranopia; and absence of the short-wavelength system causes tritanopia. Monochromats lack two or more cone mechanisms.

Contrasts in Rod and Cone Pathway Functions

Rod and cone pathways have several important functional differences in their phototransduction mechanisms and their retinal circuitry. As described previously, rods have more visual pigment and a better signal amplification system than cones do, and there are many more rods than cones. Thus rods function better in dim light (scotopic vision), and loss of rod function results in night blindness. In addition, all rods contain the same visual pigment, so they cannot signal color differences. Furthermore, many rods converge onto individual bipolar cells and the results are very large receptive fields and low spatial resolution. Finally, in bright light, most rhodopsin is bleached, so that rods no longer function under photopic conditions.

Cones have a higher threshold to light and thus are not activated in dim light after dark adaptation. However, they operate very well in daylight. They provide high-resolution vision because only a few cones converge onto individual bipolar cells in cone pathways. Moreover, no convergence occurs in the fovea, where the cones make one-to-one connections to bipolar cells. As a result of the reduced convergence, cone pathways have very small receptive fields and can resolve stimuli that originate from sources very close to each other. Cones also respond to sequential stimuli with good temporal resolution. Finally, cones have three

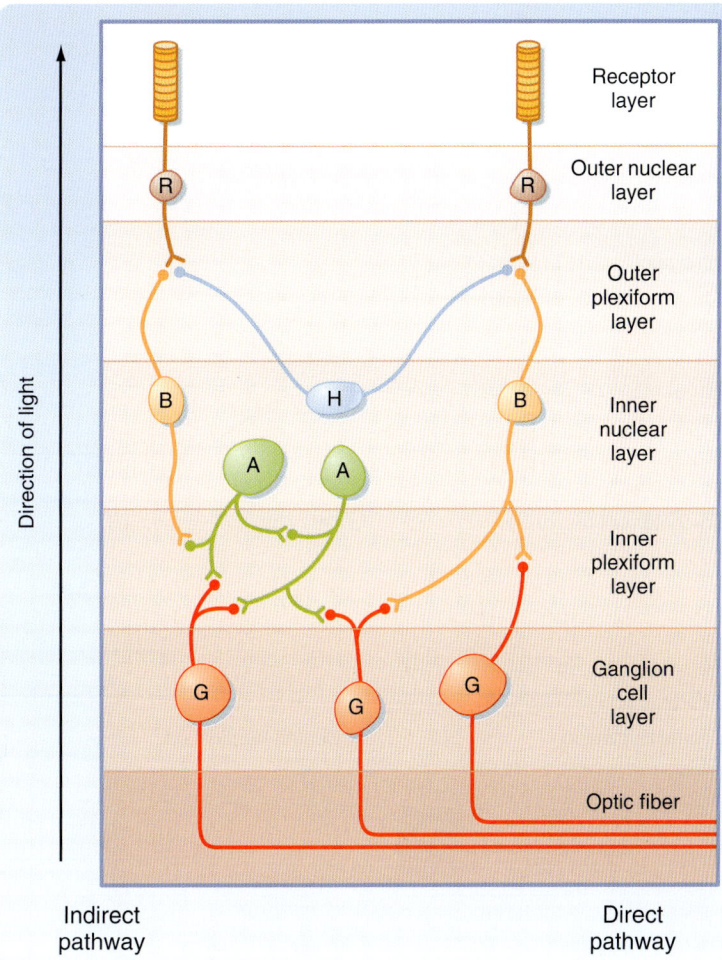

• **Fig. 8.7** Basic Retinal Circuitry. The *arrow* at the left indicates the direction of light through the retina. Photoreceptors (R) synapse on the dendrites of bipolar cells (B) and horizontal cells (H) in the outer plexiform layer. The horizontal cells make reciprocal synaptic connections with photoreceptor cells and are electrically coupled to other horizontal cells. Bipolar cells reach synapse on the dendrites of ganglion cells (G) and on the processes of amacrine cells (A) in the inner plexiform layer. Amacrine cells connect with ganglion cells and other amacrine cells.

different visual pigments and therefore provide for color vision. Loss of cone function results in functional blindness; rod vision is not sufficient for normal visual requirements.

Synaptic Interactions and Receptive Field Organization

The receptive field of an individual photoreceptor is circular. Light in the receptive field hyperpolarizes the photoreceptor cell and cause it to release less neurotransmitter. The receptive fields of photoreceptors and retinal interneurons determine the receptive fields of the retinal ganglion cells onto which their activity converges. The characteristics of the receptive fields of retinal ganglion cells constitute an important step in visual information processing because all the information about visual events that is conveyed to the brain is contained in ganglion cell activity.

The bipolar cell, which receives input from a photoreceptor, can have either of two types of receptive fields, as shown in Fig. 8.8. Both are described as having a center-surround

organization in which the light that strikes the central region of the receptive field either excites or inhibits the cell, whereas the light that strikes a region that surrounds the central portion has the converse effect. The receptive field with a centrally located excitatory region surrounded by an inhibitory annulus is called an **on-center, off-surround receptive field** (see Fig. 8.8A). Bipolar cells with such a receptive field are described as "on" bipolar cells. The other type of receptive field has an **off-center, on-surround** arrangement, which characterizes "off" bipolar cells (see Fig. 8.8F).

The center response of a bipolar cell receptive field is due to only the photoreceptors that directly synapse with the bipolar cell. Photoreceptor cells respond to light with hyperpolarization and a decrease in glutamate release and respond to the removal of light with depolarization and increased glutamate release. This implies that the difference in the center responses of "on" and "off" bipolar cells lies in their response to glutamate. In fact, off-center bipolar

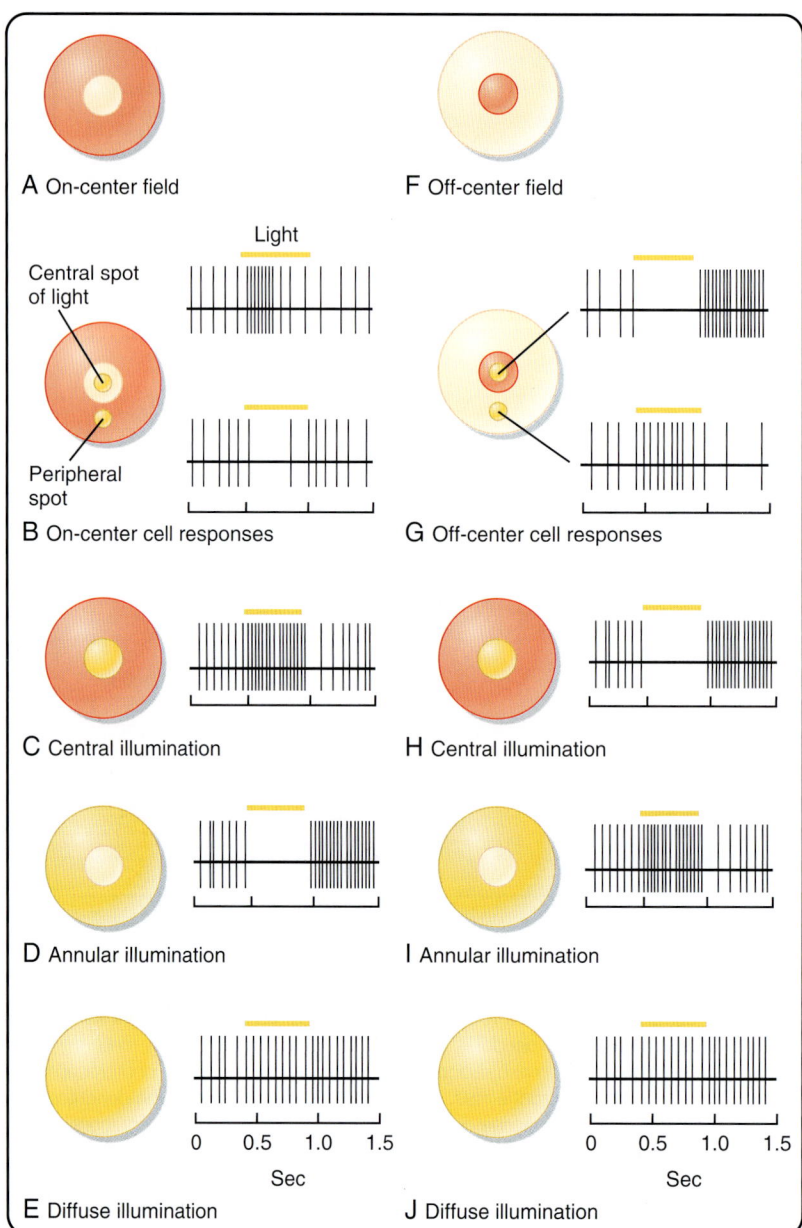

• **Fig. 8.8** The receptive fields of on-center **(A)** and off-center **(F)** bipolar cells and, below them, the receptive fields of ganglion cells **B** through **E** and **G** through **J** to which they are connected. Ganglion cell responses to central spots (*upper recording*) and peripheral spots (*lower recording*) are shown in **B** and **G**. Also shown are responses to central (**C** and **H**), surround (**D** and **G**), and diffuse whole-field (**E** and **J**) illumination in their receptive fields. The ganglion cells and the on-center and off-center bipolar cells providing input to these ganglion cells have similar receptive fields, but whereas ganglion cells increase or decrease their spike frequency, bipolar cells depolarize or hyperpolarize, without generating action potentials. (Redrawn from Squire LR et al [eds]. *Fundamental Neuroscience*. San Diego, CA: Academic Press; 2002.)

cells have ionotropic glutamate receptor channels that open in response to glutamate, and thus they are excited by the removal of light stimuli from the center of their receptive field. In contrast, on-center bipolar cells have metabotropic glutamate receptors that close their channels in response to glutamate. They are depolarized by light on the center of their receptive field, because the reduced release of glutamate by the photoreceptors results in more open metabotropic channels. Thus on-center bipolar cells are excited by light stimulation of the center of their receptive fields.

The antagonistic surround response of bipolar cells is due to photoreceptors that surround those that synapse directly on them. These photoreceptors (which also connect directly with their own bipolar cells) synapse with horizontal cells that participate in complex triadic synapses with many photoreceptors and bipolar cells. The pathway through the horizontal cells results in a response that is opposite in sign to that produced directly by the photoreceptors that mediate the center response. The reason for this is that horizontal cells are depolarized by glutamate released from photoreceptors and thus, like "off" bipolar cells, are hyperpolarized in the light. Moreover, because they are electrically coupled to each other by gap junctions, they have very large receptive fields. Darkness in the periphery of a bipolar cell's receptive field (such as an annulus that does not affect the photoreceptors to which it is directly connected) causes neighboring photoreceptors and horizontal cells to depolarize. The depolarized horizontal cells release gamma-aminobutyric acid (GABA) onto central (and peripheral) photoreceptor terminals, reducing their release of glutamate. Thus when darkness surrounds central illumination, there is increased excitation of **on-center** bipolar cells. There is a complementary effect on **off-center** bipolar cells when a bright annulus surrounds a central dark spot (see Fig. 8.8).

Bipolar cells may not respond to large or diffuse areas of illumination, covering both the receptors that are responsible for the center response and those that cause the surround response because of their opposing actions. Thus bipolar cells may not signal changes in the intensity of light that strikes a large area of the retina. On the other hand, a small spot of light moving across the receptive field may sequentially and dramatically alter the activity of the bipolar cell as the light crosses the receptive field from the surround portion to the center and then back again to the surround portion. This demonstrates that bipolar cells respond best to the local contrast of stimuli and function as contrast detectors.

Amacrine cells receive input from different combinations of on-center and off-center bipolar cells. Thus their receptive fields are mixtures of on-center and off-center regions. There are many different types of amacrine cells, and they may use at least eight different neurotransmitters. Accordingly, the contributions of amacrine cells to visual processing are complex.

Ganglion cells may receive dominant input from bipolar cells, dominant input from amacrine cells, or mixed input from amacrine and bipolar cells. When amacrine cell input dominates, the receptive fields of ganglion cells tend to be diffuse, and they are either excitatory or inhibitory. Most ganglion cells, however, are dominated by bipolar cell input and have a center-surround organization, similar to that of the bipolar cells that connect to them (see Fig. 8.8).

The distances between retinal components are short. Hence, modulation of transmitter release by changes in transmembrane potential and the resulting postsynaptic potentials are sufficient for most of the activity in retinal circuits, and action potentials are not required except for ganglion cells and some amacrine cells, which generate action potentials. It is unclear why amacrine cells have action potentials, but ganglion cells must generate them to transmit information over the relatively long distance from the retina to the brain.

P, M, and W Cells

Experiments have shown that in primates, retinal ganglion cells can be subdivided into three general types called **P cells, M cells,** and **W cells.** P and M cells are fairly homogeneous groups, whereas W cells are heterogeneous. P cells are so named because they project to the parvocellular layers of the LGN of the thalamus, whereas M cells project to the magnocellular layers of the LGN. P and M cells have center-surround receptive fields; hence, they are presumably controlled by bipolar cells. W cells have large, diffuse receptive fields and slowly conducting axons. They are probably influenced chiefly through amacrine cell pathways, but less is known about them than about M and P cells.

Several of the physiological differences among these cell types correspond to morphological differences (Table 8.1). For example, P cells have small receptive fields (which corresponds to smaller dendritic trees) and more slowly conducting axons than M cells do. In addition, P cells show a linear response in their receptive field; that is, they respond with a sustained, tonic discharge of action potentials in response to maintained light but do not signal shifts in the pattern of illumination as long as the overall level of illumination is constant. Thus a small object entering a P cell's central receptive field will change the cell's firing, but

TABLE 8.1 Properties of Retinal Ganglion Cells

Properties	P Cells	M Cells	W Cells
Cell body and axon	Medium sized	Large	Small
Dendritic tree	Restricted	Extensive	Extensive
Receptive field			
Size	Small	Medium	Large
Organization	Center-surround	Center-surround	Diffuse
			Poorly responsive
Adaptation	Tonic	Phasic	
Linearity	Linear	Nonlinear	
Wavelength	Sensitive	Insensitive	Insensitive
Luminance	Insensitive	Sensitive	Sensitive

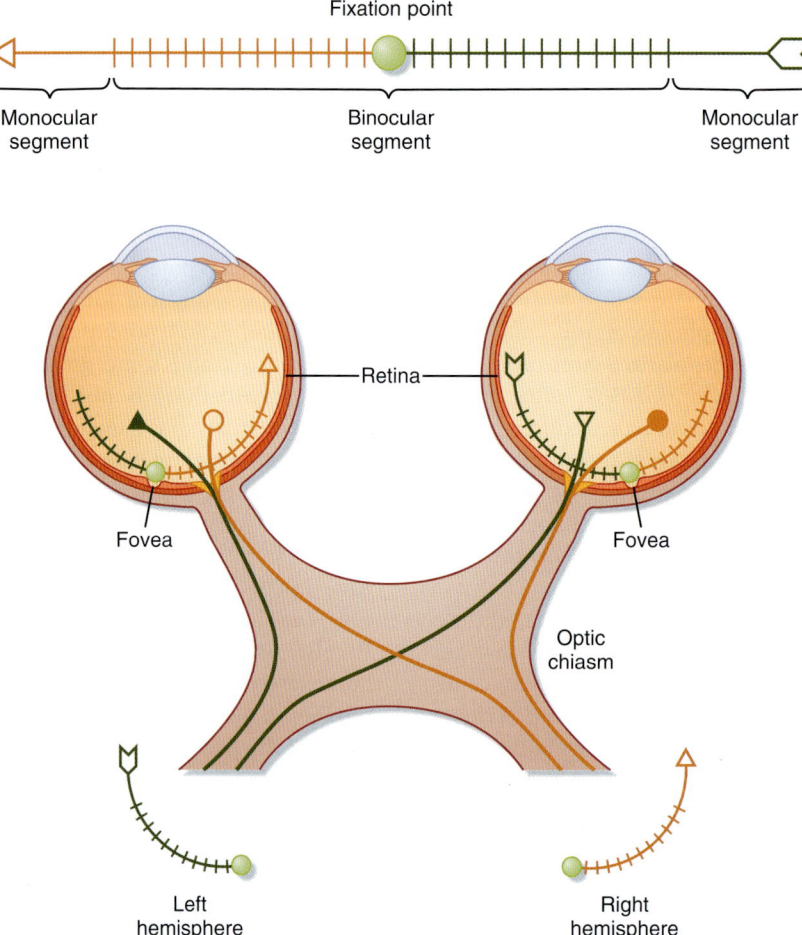

• **Fig. 8.9** Relationships among a visual target (*long arrow,* **top**), images on the retinas of the two eyes **(middle),** and projections of the ganglion cells carrying visual information about these images **(bottom).** The target image is so large that it extends into the monocular segments of the eyes, where one side of it is seen by only the ipsilateral eye. Note how the axons are sorted in the chiasm so that all information about the left visual field of both eyes is conveyed to the right side of the brain and all information about the right visual field is conveyed to the left side.

continued object movement within the field will not be signaled. P cells respond differently to different wavelengths of light. Because there are blue, green, and red cones, many combinations of color properties are possible, but in fact P cells have been shown to have opposing responses only to red and green or only to blue and yellow (a combination of red and green). These mechanisms can greatly reduce the ambiguity of color detection caused by the overlap in cone color sensitivity and may provide a substrate for the opponency process observations.

M cells, on the other hand, respond with phasic bursts of action potentials to the redistribution of light, such as would be caused by the movement of an object within their large receptive fields. M cells are not sensitive to differences in wavelength but are more sensitive to luminance than P cells are.

Thus the output of the retina consists primarily of ganglion cell axons from (1) sustained, linear P cells with small receptive fields that convey information about color, form, and fine details and (2) phasic, nonlinear M cells

with larger receptive fields that convey information about illumination and movement. Both exist in on-center and off-center varieties (Fig. 8.8).

The Visual Pathway

Retinal ganglion cells transmit information to the brain by way of the optic nerve, optic chiasm, and optic tract. Fig. 8.9 shows the relationships among a visual target, the retinal images of the target in the two eyes, and the projections of retinal ganglion cells to the two hemispheres of the brain. The eyes and the optic nerves, chiasm, and tract are viewed from above.

The visual target, an arrow, is in the visual fields of both eyes (see Fig. 8.9) and, in this case, is so long that it extends into the monocular segments of each retina (i.e., one end of the target can be seen by only one eye and the other end by only the other eye). The shaded circle at the center of the target represents the fixation point. The image of the target on the retinas is reversed by the lens system. The left half of the visual target is imaged on the nasal retina of the left eye and the temporal retina of the right eye. Thus the left visual field is seen by the left nasal retina and the right temporal retina. Similarly, the right half of the visual target is imaged on and seen by the left temporal retina and the right nasal retina. The lens system also causes an inversion in the vertical axis, with the upper visual field imaged on the lower retina and vice versa.

The axons of retinal ganglion cells may or may not cross in the optic chiasm, depending on the location of the ganglion cell in the retina (see Fig. 8.9). Axons from the temporal portion of each retina pass through the optic nerve, the lateral side of the optic chiasm, and the ipsilateral optic tract and terminate ipsilaterally in the brain. Axons from the nasal portion of each retina pass through the optic nerve, cross to the opposite side in the optic chiasm, and then pass through the contralateral optic tract to end in the contralateral side of the brain. As a result of this arrangement, objects in the left field of vision are represented in the right side of the brain, and those in the right field of vision are represented in the left side of the brain.

Lateral Geniculate Nucleus

Retinal ganglion cell axons can synapse in several parts of the brain, but the main target for vision is the **lateral geniculate nucleus (LGN)** of the thalamus. There is a point-to-point projection from the retina to the LGN. The LGN thus has a retinotopic map. Cells that represent a particular retinal location are aligned along projection lines that can be drawn across the layers of the LGN.

The projection from each eye is distributed to three of the layers of the LGN, one of the magnocellular layers (layers 1 and 2 receive M cell input) and two of the parvocellular layers (layers 3 to 6 receive P cell input). Color-coded ganglion cells project to groups of cells between the major layers, the intralaminar zones. Thus the properties of LGN neurons are very similar to those of retinal ganglion cells. For example, LGN neurons can be classified as P or M cells, and they have on-center or off-center receptive fields.

The LGN also receives input from the visual areas of the cerebral cortex, the thalamic reticular nucleus, and several nuclei of the brainstem reticular formation. The activity of LGN projection neurons is inhibited by interneurons both in the LGN and in the thalamic reticular nucleus. These cells use GABA as their inhibitory neurotransmitter. In addition, the activity of LGN neurons is influenced by corticofugal pathways and by brainstem neurons that transmit signals via monoamine neurotransmitters. These control systems filter visual information and may be important for selective attention.

Striate Cortex

The LGN projects to the **primary visual cortex** or **striate cortex** by way of the **visual radiations.** The visual radiation fibers carrying information derived from the lower half of the appropriate hemiretinas (and therefore the contralateral upper visual field) project to the **lingual gyrus,** which lies

IN THE CLINIC

Interruption of the visual pathway at any level causes a defect in the appropriate part of the visual field (see Fig. 8.9). For example, a tiny lesion in the retina would result in a blind spot **(scotoma)** in that eye, whereas a similar lesion in the striate cortex would produce corresponding scotomas in both eyes. Interruption of the optic nerve on one side produces blindness in that eye. Damage to the optic nerve fibers as they cross in the optic chiasm results in loss of vision in both temporal fields of vision; this condition is known as **bitemporal hemianopsia** and occurs because the crossing fibers originate from ganglion cells in the nasal halves of each retina. A lesion of the entire optic tract, LGN, visual radiation, or visual cortex on one side causes **homonymous hemianopsia,** which is loss of vision in the entire contralateral visual field. Partial lesions result in partial visual field defects. For example, a lesion in the lingual gyrus causes an upper **homonymous quadrantanopsia,** which in this case is loss of vision in the contralateral, upper visual field.

on the medial surface of the occipital lobe, just below the calcarine sulcus. Axons in the visual radiation that represent the contralateral lower visual field project to the adjacent **cuneus gyrus,** which lies just above the calcarine sulcus. Together, the portions of these two gyri that line and border the calcarine sulcus constitute the primary visual cortex (or Brodmann area 17; Fig. 8.10).

Like the LGN, the striate cortex contains a retinotopic map. The representation of the macula occupies the most posterior and largest part of both gyri, and progressively more peripheral areas of the retina are projected to more anterior parts of these gyri. Overall, there is an orderly mapping of retinal loci across the surface of the striate cortex (see Fig. 8.10).

The geniculostriate pathway ends chiefly in layer 4 of the striate cortex (Fig. 8.11), whereas the projection from the intralaminar LGN terminates in so-called blobs in layers 2 and 3. Similarly, axons that represent one eye or the other terminate within layer 4C in alternate adjacent patches that define **ocular dominance columns.** Cortical neurons in such a column respond preferentially to input from one eye. Near the border between two ocular dominance columns, neurons respond about equally to input from the two eyes.

The receptive fields of neurons in the striate cortex, aside from the monocular cells in layer 4C, are more complex than those of LGN neurons. Neurons in other layers may be binocular and respond to stimulation of both eyes, although the input from one eye often dominates (see Chapter 10). In addition, cortical neurons outside layer 4C often show **orientation selectivity** (i.e., they respond best when the stimulus, such as a bar or an edge, is oriented and positioned in a particular way; Fig. 8.12). These "simple cells" appear to be responding as though they received input from cells whose concentric center-surround receptive fields were arranged in such a way that their "on" centers were aligned in a row flanked by antagonistic regions. "Complex" cortical

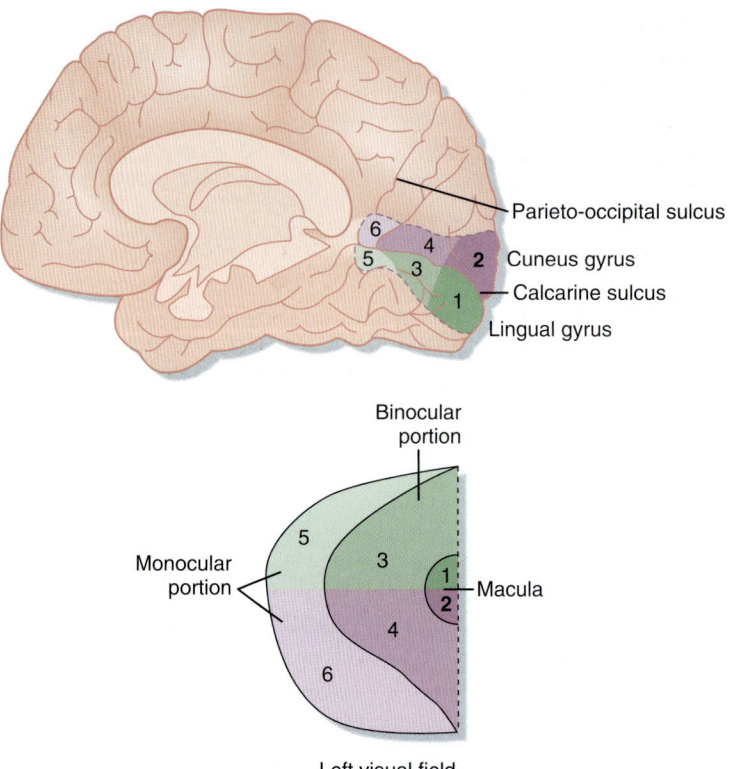

Left visual field

• **Fig. 8.10** The left visual field is relayed (via the LGN and visual radiation) to the primary visual cortex of the right hemisphere, as a point-to-point retinotopic map. The representation of each part of visual space is proportional to the number of afferent axons with receptive fields in that part of space. As a result, the area of macular representation (near the occipital pole) is larger than that for the rest of the binocular and monocular fields. Note that the lower half of the field is represented in the cuneus gyrus above the calcarine sulcus and the upper half of the field in the lingual gyrus below the sulcus. (Redrawn from Purves D, et al [eds]. *Neuroscience.* 3rd ed. Sunderland, MA: Sinauer; 2004.)

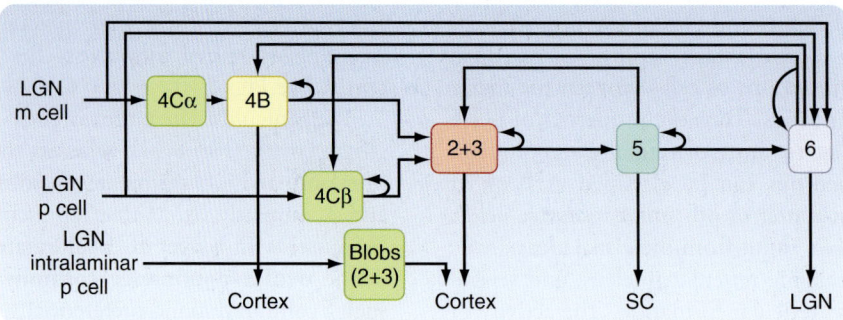

• **Fig. 8.11** Diagram of visual information flow into the visual cortex from the LGN and its projection to the extrastriate cortex, to the superior colliculus (SC), and back to the LGN. M, magnocellular path; P, parvocellular path. (Redrawn from Squire LR, et al [eds]. *Fundamental Neuroscience.* San Diego, CA: Academic Press; 2002.)

neurons are similar to "simple" cells in that they are orientation specific, but instead of having flanking excitatory and inhibitory zones, they respond best to a particular stimulus orientation anywhere in their receptive field. They may also display **direction selectivity;** that is, they may respond when the stimulus is moved in one direction but not when it is moved in the opposite direction (see Fig. 8.12). The receptive field of a "complex" cell may be thought of as a composite of adjacent "simple" cells with similar orientation selectivity. Because such neurons in a particular zone of the cortex all tend to have the same orientation selectivity, they are considered to form an **orientation column** (Fig. 8.13).

As already discussed, color vision may depend on the presence in the retina of three different types of cones, as well as neurons in the visual pathway that show spectral opposition. Retinal ganglion cells, LGN neurons, and some P cells display spectral opponent properties. The spectral opponent neurons in the striate cortex are found in cortical blobs, and these show double-opponency, in which both the center and the surround portions respond antagonistically

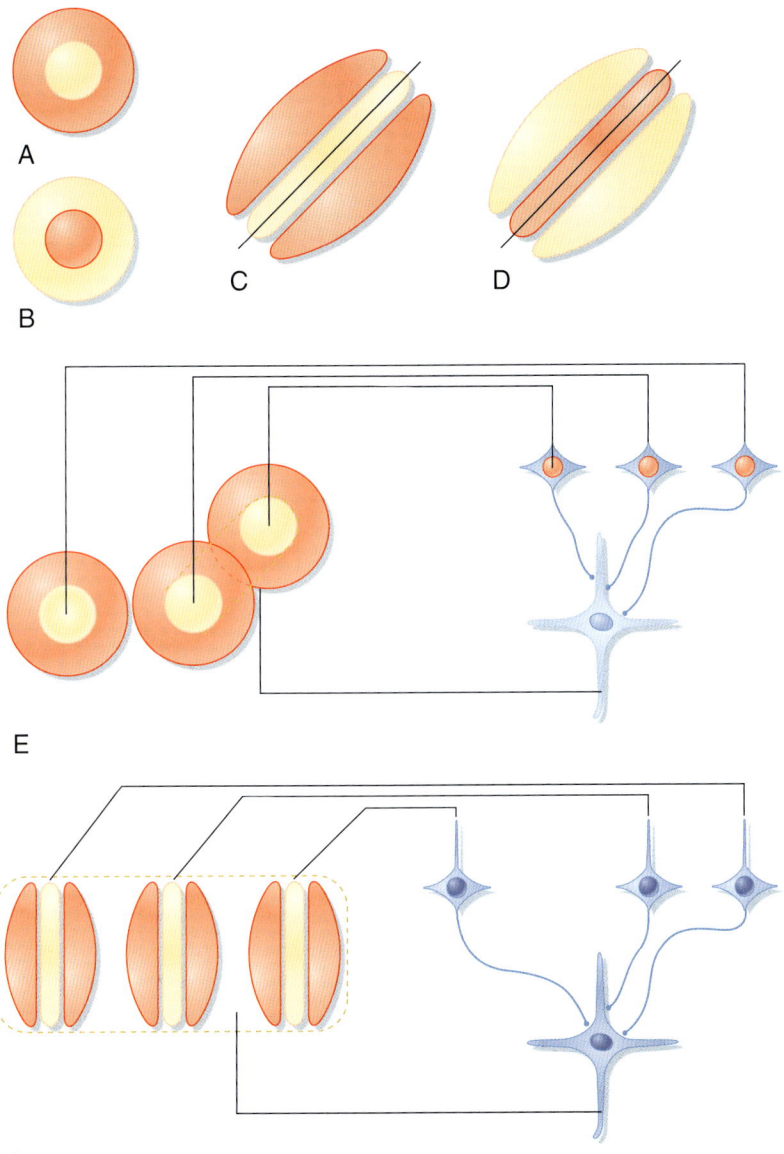

• **Fig. 8.12** Simple and Complex Receptive Fields in the Visual Cortex Can Be Generated From Multiple Inputs With Concentric Fields. **A** and **B** represent on-center and off-center input, respectively, from the retina. If three on-center cells **(A)** with adjacent receptive fields converged onto one cortical neuron **(E)**, that neuron, a simple cell, would respond best to a long bar stimulus at a specific location and orientation **(C)**. For three off-center inputs **(B)**, the resulting receptive field is shown in **D**. The convergence of multiple simple cells onto another cortical neuron **(F)** would result in a complex cell that responds best to a bar stimulus with a vertical orientation that can be placed anywhere within its receptive field. (Redrawn from Squire LR, et al [eds]. *Fundamental Neuroscience*. San Diego, CA: Academic Press; 2002.)

to two colors. Such a cell, whose center responds to red but not green (R^+G^-) and whose surround portion responds to green but not red (R^-G^+), is shown in Fig. 8.13*A*. The relationships between the ocular dominance and orientation columns and the cortical color blobs are shown in Fig. 8.13*B*.

Extrastriate Visual Cortex

In animal studies, at least 25 different visual areas have been identified in the cerebral cortex, in addition to the striate cortex (Brodmann area 17, or V1). The extrastriate

areas include several parallel pathways of visual processing. The P pathway originates with P cells and functions in the recognition of form and color. Structures in the P pathway include LGN layers 3 to 6, layer 4Cβ of the striate cortex, V4 (Brodmann area 19), and several areas in the inferotemporal region (Fig. 8.14). Processing of form includes recognition of complex visual patterns, such as faces. Color information is processed separately from form. The M pathway originates with M cells and functions in motion detection and control of eye movement. Cortical structures in the M pathway include layers 4B and 4Cα of the striate cortex and areas MT (medial temporal) and

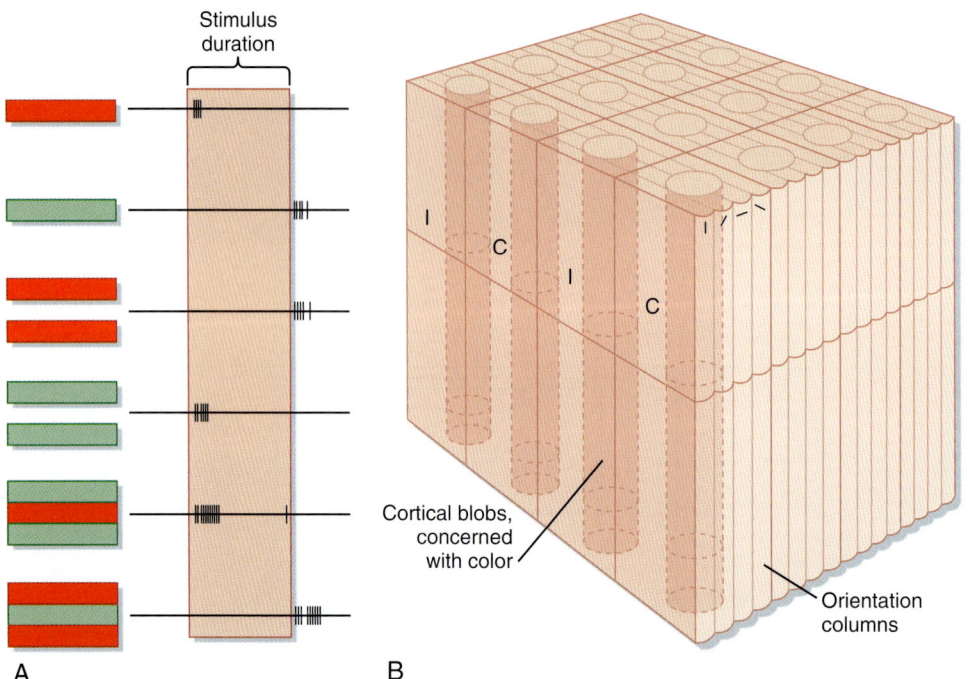

• **Fig. 8.13 A,** The receptive field and responses of a double-opponent (R⁺G⁻/R⁻G⁺) neuron in a blob of the striate cortex as it responds to various combinations of red and green bars. The best "on" response is to a red bar flanked by two green bars. **B,** Diagram of the columnar arrangement of the visual cortex. Ocular dominance columns are indicated by I (for ipsilateral) and C (for contralateral). Orientation columns are indicated by the smaller columns marked with short bars at varying angles. The cortical blobs contain neurons like that in **A** and have spectral opponent-receptive fields.

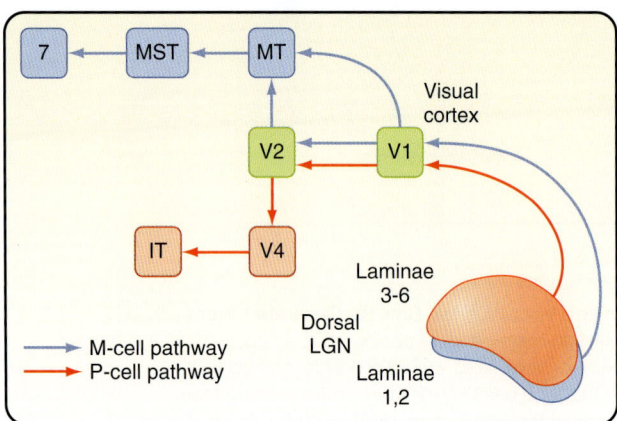

• **Fig. 8.14** Distribution of P and M cell influences on different areas of the visual cortex. IT, inferotemporal area; MST, medial superior temporal area; MT, medial temporal area; V1, striate cortex; V2 and V4, higher order visual areas.

MST (medial superior temporal) on the lateral aspect of the temporal lobe, as well as Brodmann area 7a of the parietal lobe (see Fig. 8.14).

Both P and M pathways contribute to depth perception or stereopsis, which is dependent on slight differences in the retinal images formed in the two eyes. Stereopsis is useful only for relatively nearby objects. However, in such cases, these disparities provide visual cues about depth. Interestingly, the anatomy of the visual pathways indicates that depth perception must be a cortical function because

it depends on convergent input from the two eyes and the left/right eye inputs are parallel but segregated in the LGN and in Layer 4 of striate cortex.

The separation of M and P pathways from the retina through the thalamus and all the cortical regions raises the issue of how all the parts are combined to account for the clear, coherent images of events, objects, and persons that humans perceive. It seems unlikely that all the components that represent a percept, such as parts of a face and whether that face belongs to a familiar person, are somehow converged onto a single neuron that recognizes it. Rather, complex percepts probably arise from the coordinated activity of large sets of neurons across multiple regions of the CNS. The process by which a "binding" of such disparate neuronal elements into a percept is unclear, but one working hypothesis is that it may be accomplished by the temporal synchronization of many anatomically distributed neural events.

⚕ IN THE CLINIC

Lesions of the extrastriate visual cortex can produce various deficits. Bilateral lesions of the inferotemporal cortex can result in cortical color blindness **(achromatopsia)** or in an inability to recognize faces, even of close members of the family **(prosopagnosia).** A lesion in area MT or MST can interfere with motion detection and eye movements.

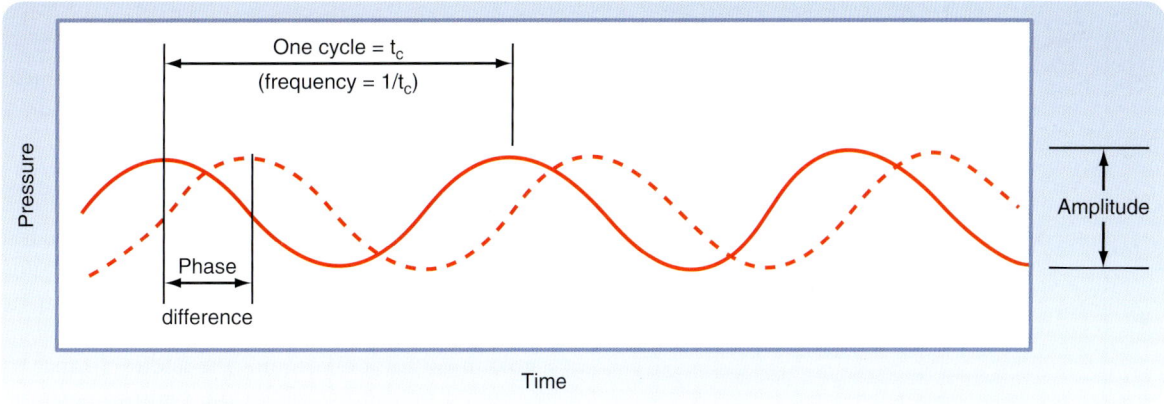

• Fig. 8.15 Two pure tones are shown by the *solid* and *dashed lines*. Frequency is determined from the wavelength as indicated. Amplitude is the peak-to-peak change in sound pressure. The two tones have the same frequency and amplitude but differ in phase.

Other Visual Pathways

The **superior colliculus** of the midbrain is a layered structure that is important for certain types of eye movements (see Chapter 9). The three most superficial layers are involved exclusively in visual processing, whereas the deeper layers receive multimodal input from the somatosensory and auditory systems, as well as the visual system, particularly from cortical areas involved in eye movement.

Another retinal projection is to the **pretectum,** which bilaterally activates parasympathetic preganglionic neurons in the **Edinger-Westphal nucleus** that cause pupillary constriction in the pupillary light reflex. The pretectal areas are also interconnected through the posterior commissure, and thus the reflex causes both ipsilateral (direct) and contralateral (consensual) pupillary constriction when a light is shown in one eye.

The visual pathways also include connections to nuclei that serve functions other than vision. For example, a retinal projection to the **suprachiasmatic nucleus** of the hypothalamus controls circadian rhythmicity (see Chapter 37).

The Auditory and Vestibular Systems

The peripheral parts of the auditory and vestibular systems share components of the bony and membranous labyrinths, use hair cells as mechanical transducers, and transmit information to the CNS through the vestibulocochlear nerve (CN VIII). However, the CNS processing and sensory functions of the auditory and vestibular systems are distinct. The function of the auditory system is to transduce sound. This allows us to recognize environmental cues and to communicate with other organisms. The most complex auditory functions are those involved in language. The function of the vestibular system is to provide the CNS with information related to the position and movements of the head in space. Control of eye movement by the vestibular system is discussed in Chapter 9.

Audition

Sound

Sound is produced by compression and decompression waves in air or in other elastic media, such as water. Sound frequency is measured in cycles per second, or **hertz (Hz).** Each pure tone results from a sinusoidal wave at a particular frequency and is characterized not only by its frequency but also, instantaneously, by its amplitude and phase (Fig. 8.15). Most naturally occurring sound, however, is a mixture of pure tones. **Noise** is unwanted sound and may have any composition of pure tones. Sound propagates at about 335 m/sec in air. The waves are associated with certain pressure changes, called *sound pressure.* The unit of sound pressure is Newtons per meter squared (N/m^2), but sound pressure is more commonly expressed as the **sound pressure level (SPL).** The unit of SPL is the **decibel (dB):**

Equation 8.1

$$SPL = 20 \log P/P_R$$

where P is sound pressure and P_R is a reference pressure (0.0002 dyne/cm^2, the absolute threshold for human hearing at 1000 Hz). A sound with intensity 10 times greater would be 20 dB; one 100 times greater would be 40 dB.

The normal young human ear is sensitive to pure tones with frequencies that range between about 20 and 20,000 Hz. The threshold for detection of a pure tone varies with its frequency (Fig. 8.16). The lowest thresholds for human hearing are, for pure tones, approximately 3000 Hz. The threshold at these frequencies is approximately −3 to −5 dB, in comparison with the reference 0 dB at 1000 Hz. In reference to this scale, normal speech has an intensity of about 65 dB, and its main frequencies fall in the range of 300 to 3500 Hz. Sounds that exceed 100 dB can damage the peripheral auditory apparatus, and those higher than 120 dB can cause pain and permanent damage. As people age, their thresholds at high frequencies rise, thereby reducing their ability to hear such tones, a condition called **presbycusis.**

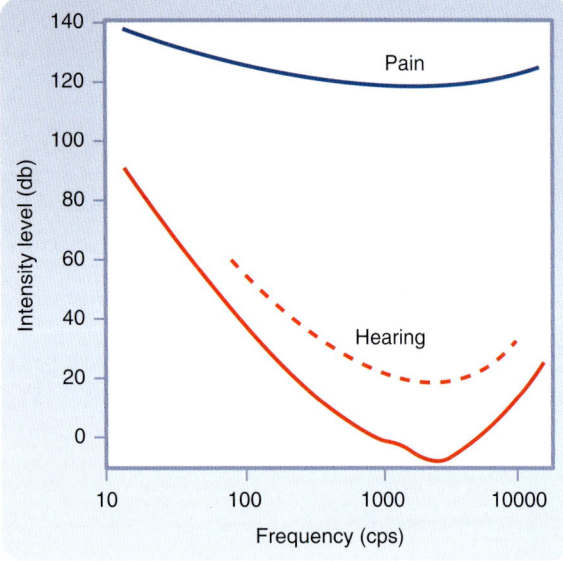

• **Fig. 8.16** Sound Threshold Intensities at Different Frequencies. The *bottom curve* indicates the absolute intensity needed to detect a sound. The *dashed curve* represents the threshold for functional hearing. The *top curve* indicates levels at which sound is painful and damaging.

The Ear

The peripheral auditory apparatus is the ear, which can be subdivided into the external ear, the middle ear, and the inner ear (Fig. 8.17).

External Ear

The external ear includes the pinna and the external auditory meatus (auditory canal). The auditory canal contains glands that secrete **cerumen,** a waxy protective substance. The pinna helps direct sounds into the auditory canal and plays a role in sound localization. The auditory canal transmits the sound pressure waves to the tympanic membrane. In humans, the auditory canal has a resonant frequency of about 3500 Hz, and this resonance contributes to the low perceptual threshold for sounds in that range.

Middle Ear

The external ear is separated from the middle ear by the **tympanic membrane** (see Fig. 8.17*A*). The middle ear contains air. Three ossicles are present and serve to link the tympanic membrane to the oval window of the inner ear.

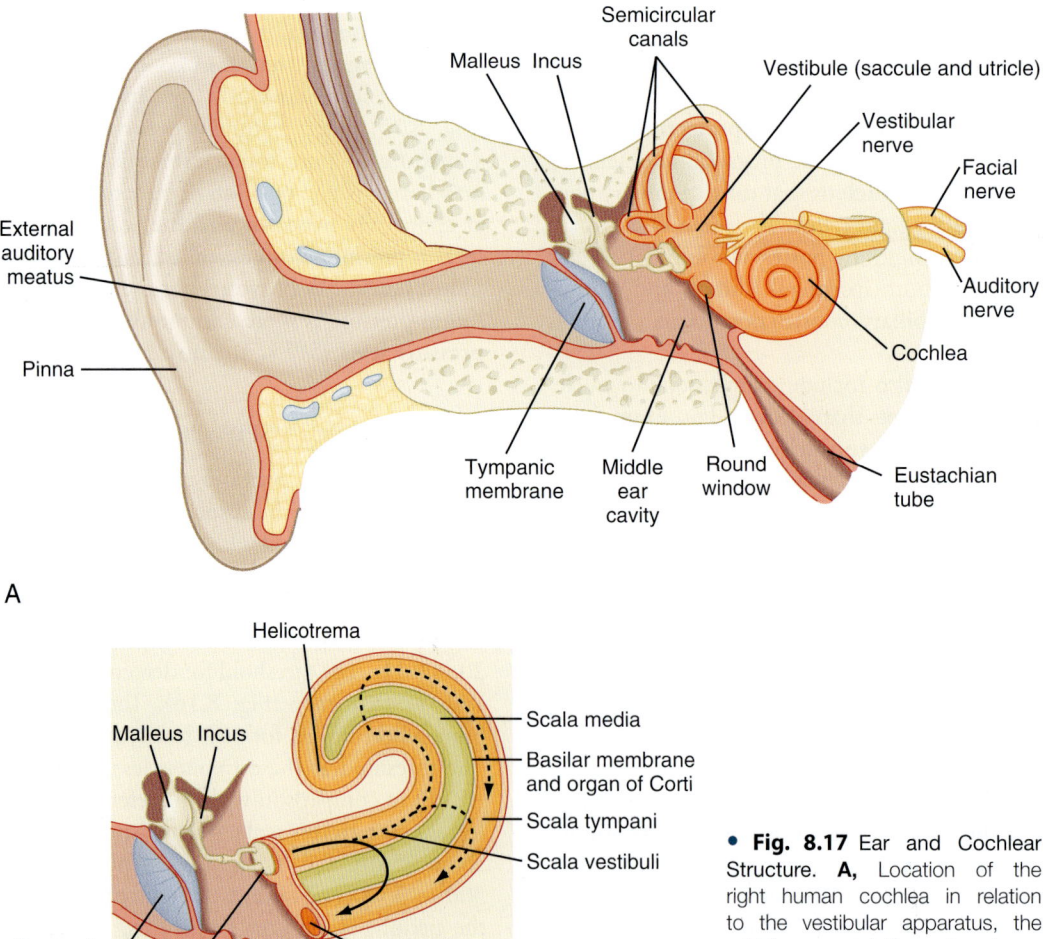

• **Fig. 8.17** Ear and Cochlear Structure. **A,** Location of the right human cochlea in relation to the vestibular apparatus, the middle ear, and the external ear. **B,** Relationships between the outer, middle, and inner ear spaces; the cochlea is depicted unrolled for clarity.

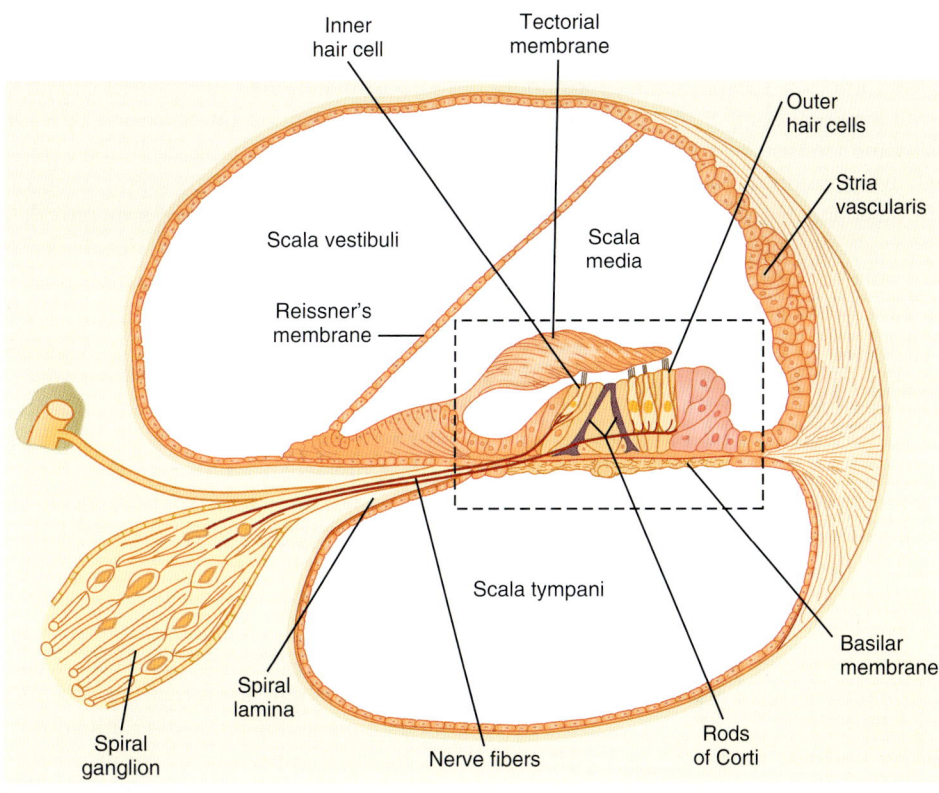

C

● **Fig. 8.17, cont'd C,** Drawing of a cross-section through the cochlea. The organ of Corti (see Fig. 8.18A and B) is outlined.

Adjacent to the oval window is the round window, another membrane-covered opening between the middle ear and inner ear (see Fig. 8.17*A* and *B*).

The ossicles include the **malleus,** the **incus,** and the **stapes.** The stapes has a footplate that inserts into the oval window. Behind the oval window is a fluid-filled component of the inner ear, the **vestibule.** It is continuous with a tubular structure known as the **scala vestibuli.** Inward movement of the tympanic membrane by a sound pressure wave causes the chain of ossicles to push the footplate of the stapes into the oval window (see Fig. 8.17*B*). This movement of the stapes footplate in turn displaces the fluid within the scala vestibuli. The pressure wave that ensues within the fluid is transmitted through the **basilar membrane** of the **cochlea** to the **scala tympani** (described later), and it causes the round window to bulge into the middle ear.

The tympanic membrane and the chain of ossicles serve as an impedance-matching device. The ear must detect sound waves traveling in air, but the neural transduction mechanism depends on movement in the fluid-filled cochlea, where acoustic impedance is much higher than that of air. Therefore, without a special device for impedance matching, most sound reaching the ear would simply be reflected, as are voices from shore when a person is swimming under water. Impedance matching in the ear depends on (1) the ratio of the surface area of the large tympanic membrane to that of the smaller oval window and (2) the mechanical

advantage of the lever system formed by the ossicles. This impedance matching is sufficient to increase the efficiency of energy transfer by nearly 30 dB in the range of hearing from 300 to 3500 Hz.

 IN THE CLINIC

The middle ear also serves other functions. Two muscles are found in the middle ear: the tensor tympani attached to the malleus and the stapedius attached to the stapes. When these muscles contract, they damp movements of the ossicles and decrease the sensitivity of the acoustic apparatus. This action can protect the acoustic apparatus against damaging sounds that can be anticipated. However, a sudden explosion can still damage the acoustic apparatus because reflex contraction of the middle ear muscles does not occur quickly enough. The chamber of the middle ear connects to the pharynx through the eustachian tube. Pressure differences between the external ear and middle ear can be equalized through this passage. If fluid collects in the middle ear, as during an infection, the eustachian tube may become blocked. The resulting pressure difference between the external ear and middle ear can produce painful displacement of the tympanic membrane and, in extreme cases, cause rupture of the tympanic membrane. Unequalized pressure changes as a result of flying or diving can also cause discomfort.

Inner Ear

The inner ear includes the bony and membranous labyrinths. The bony labyrinth is a complex but continuous series of spaces in the temporal bone of the skull, whereas the membranous labyrinth consists of a series of soft tissue spaces and channels lying inside the bony labyrinth. The cochlea and the vestibular apparatus are formed from these structures.

The cochlea is a spiral-shaped organ (see Fig. 8.17A and B). In humans, the spiral consists of 2¾ turns from a broad base to a narrow apex, although its internal lumen is small at the base and wide at the top. The apex of the cochlea faces laterally (see Fig. 8.17A). The bony labyrinth component of the cochlea is subdivided into several chambers. The vestibule is the space facing the oval window (see Fig. 8.17A). Continuous with the vestibule is the scala vestibuli, the spiral-shaped chamber that extends to the apex of the cochlea, where it meets and merges with the **scala tympani** at the **helicotrema.** The scala tympani is another spiral-shaped space that winds back down the cochlea and ends at the round window (see Fig. 8.17B). Separating the two, except at the helicotrema, is the scala media enclosed in the membranous labyrinth.

The **scala media,** or **cochlear duct** (see Fig. 8.17B and C), is a membrane-bound spiral tube that extends along the cochlea, between the scala vestibuli and scala tympani. One wall of the scala media is formed by the **basilar membrane,** another by **Reissner's membrane,** and the third by the **stria vascularis** (see Fig. 8.17C).

The spaces within the cochlea are filled with fluid. The fluid in the bony labyrinth, including the scala vestibuli and scala tympani, is **perilymph,** which closely resembles cerebrospinal fluid. The fluid in the membranous labyrinth, including the scala media, is endolymph, which is very different from perilymph. **Endolymph,** generated by the **stria vascularis,** contains high [K^+] (about 145 mM) and low [Na^+] (about 2 mm) and has a high positive potential (about +80 mV) with regard to the perilymph. As a result, a very large potential gradient (about 140 mV) exists across the membranes of the hair cell cilia that extend into the endolymph. (These hair cells, which are the sensory receptors for sound, are discussed in more detail later.)

The neural apparatus responsible for transduction of sound is the **organ of Corti** (see Fig. 8.17C), which is located within the cochlear duct. It lies on the basilar membrane and consists of several components, including three rows of **outer hair cells,** a single row of **inner hair cells,** a gelatinous **tectorial membrane,** and a number of types of supporting cells. The organ of Corti in humans contains 15,000 outer and 3500 inner hair cells. The **rods of Corti** help provide a rigid scaffold. Located on the apical surface of the hair cells are stereocilia, which can be described as nonmotile cilia that contact the tectorial membrane.

The organ of Corti is innervated by nerve fibers of the cochlear division of the vestibulocochlear nerve (CN VIII). The 32,000 auditory afferent fibers in humans originate in sensory ganglion cells in the **spiral ganglion.** These nerve fibers penetrate the organ of Corti and terminate at the bases of the hair cells (Fig. 8.18; see also Fig. 8.17C). Approximately 90% of the fibers end on inner hair cells, and the remainder end on outer hair cells. Thus approximately 10 afferent fibers supply each inner hair cell, whereas other afferent fibers diverge to supply about five outer hair cells each. The inner hair cells clearly provide most of the neural information about acoustic signals that the CNS processes for hearing. The sensory function of the outer hair cells is less clear.

In addition to afferent fibers, the organ of Corti is supplied by efferent fibers, most of which terminate on the outer hair cells. These cochlear efferent fibers originate in the superior olivary nucleus of the brainstem and are often called **olivocochlear fibers.** The length of the outer hair cells varies; this characteristic suggests that changes in outer hair cell length may affect the sensitivity, or "tuning," of the inner hair cells. The cochlear efferent fibers may control outer hair cell length. Such a mechanism could conceivably influence the sensitivity of the cochlea and the way that the brain recognizes sound. Other efferent fibers that end on cochlear afferent fibers may be inhibitory, and they may help improve frequency discrimination.

Sound is transduced by the organ of Corti. Sound waves that reach the ear cause the tympanic membrane to oscillate, and these oscillations are transmitted to the scala vestibuli by the ossicles. This creates a pressure difference between the scala vestibuli and the scala tympani (see Fig. 8.17B) that serves to displace the basilar membrane and, with it, the organ of Corti (see Fig. 8.18A and B). Because of the shear forces set up by the relative displacement of the basilar and tectorial membranes, the stereocilia of the hair cells bend. Upward displacement bends the stereocilia toward the tallest cilium, which depolarizes the hair cells; downward deflection bends the stereocilia in the opposite direction, which hyperpolarizes the hair cells.

🩺 IN THE CLINIC

A common cause of deafness is the destruction of hair cells by loud sounds. Hair cells can be destroyed, for example, by exposure to industrial noise or by listening to loud music. Typically, hair cells in certain parts of the cochlea are selectively damaged by exposure to high levels of sound at particular frequencies (as predicted by the place theory), and thus hearing may be lost over a discrete frequency range. Presbycusis, or the loss of high-frequency hearing with age, is probably increased by the loss of hair cells as a result of long-term noise exposure in urban environments.

Sound Transduction

In view of the wide range of frequencies and amplitudes of sound stimuli, it is no surprise that hair cell transduction must provide for a fast response. The fast response to deflection of the cilia is based on direct opening of ion channels by so-called tip links that connect the tip of each stereocilium with the shaft of the next taller one (see

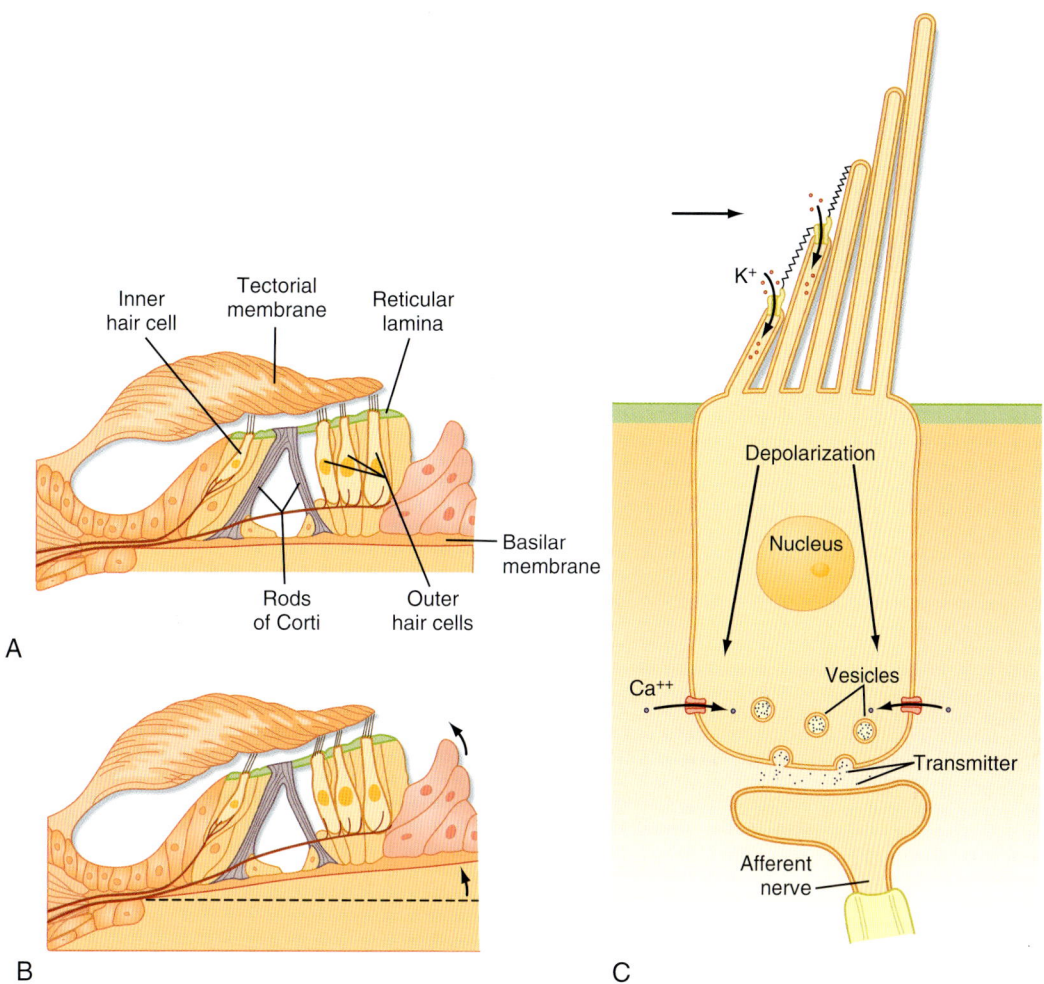

• Fig. 8.18 Detail of the organ of Corti at rest **(A)** and with upward movement of the basilar membrane **(B).** The upward movement causes the stereocilia to bend because of shear forces produced by relative displacement of the hair cells and the tectorial membrane. **C,** Diagram of a hair cell with tip link connections between the hair cell cilia to show how shear forces open mechanoreceptor channels and depolarize the hair cell.

Fig. 8.18C). With deflection, the tip links are subjected to a lever action that transiently opens the channels, allows the entry of K$^+$ (because of the high [K$^+$] and high potential in endolymph), and depolarizes the hair cell. Several mechanisms have been proposed to account for the equally important rapid adaptation necessary for a high-frequency response. A "spring" response by the tip links would allow the attachment point of the tip link to be moved along the stereocilium's shaft to reset the mechanical leverage of the tip link. In addition, it has been observed that Ca^{++} can enter and bind to the open channel, change it to require greater opening force, and thereby reduce the statistical probability of opening.

The potential gradient that induces movement of ions into hair cells includes both the resting potential of the hair cells and the positive potential of the endolymph. As noted previously, the total gradient across the apical membrane of hair cells is about 140 mV. Therefore, a change in K$^+$ conductance in the apical membranes of hair cells results in a rapid current flow that produces the **receptor potential** in

these cells. This current flow can be recorded extracellularly as a **cochlear microphonic potential,** an oscillatory event that has the same frequency as the acoustic stimulus. The cochlear microphonic potential represents the sum of the receptor potentials of a number of hair cells.

Hair cells, like retinal photoreceptors, release an excitatory neurotransmitter (probably glutamate) when depolarized. The neurotransmitter produces an excitatory postsynaptic potential (EPSP) in the cochlear afferent nerve fibers with which the hair cell synapses. In summary, sound is transduced when oscillatory movements of the basilar membrane cause transient changes in the transmembrane voltage of the hair cells and, finally, the generation of action potentials in cochlear afferent nerve fibers. The activity of a large number of cochlear afferent fibers in the auditory nerve can be recorded extracellularly as a compound action potential.

On the basis of differences in width and tension, investigators originally concluded that different parts of the basilar membrane have different resonant frequencies.

• **Fig. 8.19** Different Frequencies of Sound Result in Different Amplitudes of Displacement at Different Sites Along the Organ of Corti. **A,** Traveling wave produced in the basilar membrane by a sound of 200 Hz. The curves at a, b, c, and d represent displacement of the basilar membrane at different times, and the *dashed line* is the envelope formed by the peaks of the wave at different times. Maximum deflection occurs at about 29 mm from the oval window. **B,** Envelopes of traveling waves produced by several frequencies of sound. Note that the maximum displacement varies with frequency and is closest to the stapes when the frequency is highest. (Redrawn from von Bekesy G. *Experiments in Hearing.* New York: McGraw-Hill; 1960.)

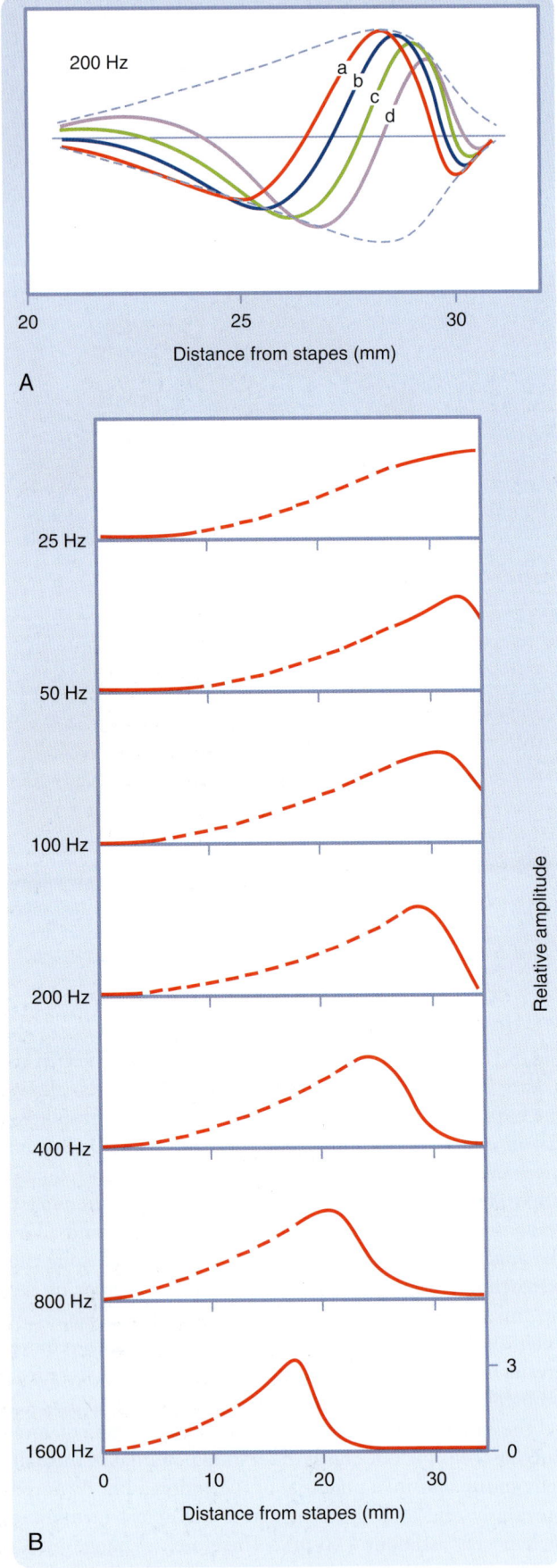

For example, the basilar membrane is about 100 μm wide at the base and 500 μm wide at the apex. It also has higher tension at the base. Thus, the investigators predicted that the base would vibrate at higher frequencies than would the apex, as do the shorter strings of musical instruments. Experiments have shown that the basilar membrane moves as a whole in traveling waves (Fig. 8.19), but displacement of the basilar membrane is maximal nearer the base of the cochlea during high-frequency tones and maximal nearer the apex during low-frequency tones.

In effect, the basilar membrane serves as a frequency analyzer; it distributes the stimulus along the organ of Corti, and different hair cells respond differentially to particular frequencies of sound. This is the basis of the **place theory of hearing.** In addition, hair cells located at different places along the organ of Corti may be tuned to different frequencies because of variations in their stereocilia and biophysical properties. As a result of these factors, the basilar membrane and organ of Corti have a so-called tonotopic map (Fig. 8.20).

Cochlear Nerve Fibers

Neurotransmitter release by hair cells in the organ of Corti can evoke action potentials in the primary afferent fibers of the cochlear nerve. Afferent fibers in the vestibulocochlear nerve (CN VIII) are bipolar cells with a myelin sheath around the cell bodies, as well as around the axons. The cell bodies are in the spiral ganglion, their peripheral processes synapse at the base of hair cells, and their central processes synapse in the cochlear nuclei of the brainstem.

Characteristic Frequencies

A cochlear afferent fiber discharges maximally when stimulated by a particular sound frequency called its **characteristic frequency.** The characteristic frequency can be determined from a tuning curve for the fiber (Fig. 8.21). A **tuning curve** is a plot of the threshold for activation of the nerve fiber by different sound frequencies. The major factor that influences the activity of individual afferent fibers is the location along the basilar membrane of the hair cells that they innervate. The location of those hair cells is important because for any given sound frequency, there is a site of maximum displacement of the basilar membrane as the pressure wave travels along its length (see Fig. 8.19). Typically, tuning curves are sharp near the characteristic

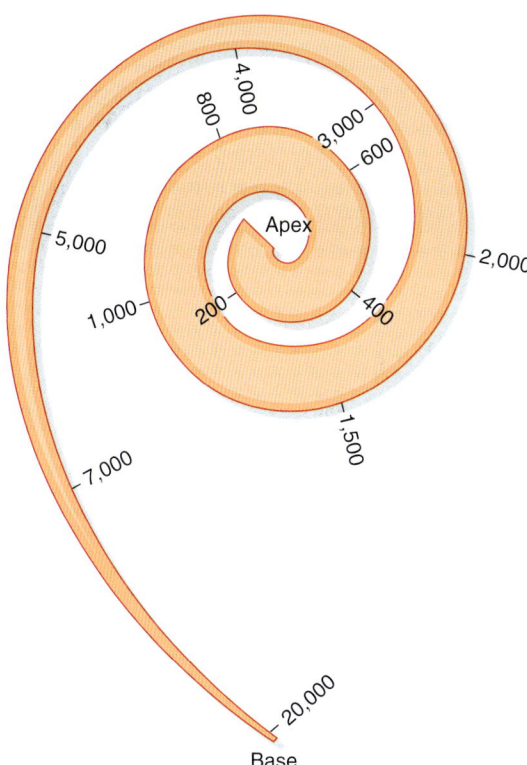

• **Fig. 8.20** The tonotopic map of the cochlea. (Redrawn from Stuhlman O. *An Introduction to Biophysics.* New York: John Wiley & Sons; 1943.)

frequency, but they broaden at high sound pressure levels. Tuning curves can have excitatory and inhibitory areas (see Fig. 8.21*A*). The sharpness of the excitatory regions may reflect inhibitory processes.

Encoding

The different features of an acoustic stimulus are encoded in the discharges of cochlear nerve fibers. Duration is signaled by the duration of activity; intensity is signaled both by the amount of neural activity and by the number of fibers that discharge. For low-frequency sounds, the frequency is signaled by the tendency of an afferent fiber to discharge in phase with the stimulus (**phase locking;** see Fig. 8.22*A*). If the tone is much more than 1 kHz, a single fiber cannot discharge with every cycle, but phase locking can also occur for sounds with periods shorter than the absolute refractory period of the afferent fiber. This allows the CNS to detect higher frequency information from the activity of a population of afferent fibers, each of which discharges in phase with the stimulus and which, as a group, signal the frequency of the stimulus (see Fig. 8.22*B*). This observation is the basis of the **frequency theory of hearing.**

For still higher frequencies (>5000 Hz), the place theory dominates: the CNS interprets sounds that activate afferent fibers supplying hair cells near the base of the cochlea as being of high frequency. Thus both the place and the frequency theories are necessary to explain the frequency coding of sound (**duplex theory**) across the entire range from 20 to 20,000 Hz.

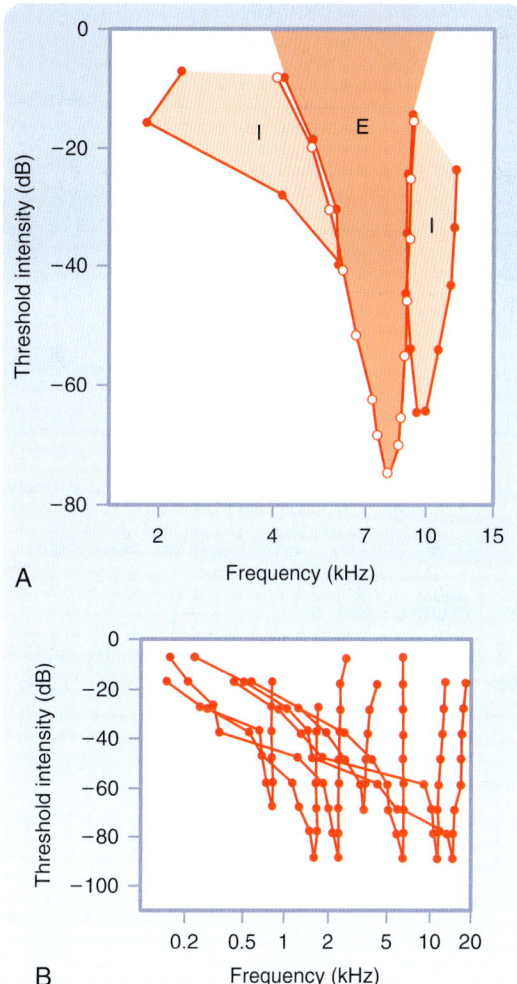

• **Fig. 8.21** Tuning Curves of Neurons in the Auditory System. Tuning curves can be considered as receptive field plots. **A,** Tuning curve with central excitatory frequencies (E) and flanking inhibitory frequencies (I). **B,** Tuning curves for cochlear nerve fibers. (**A,** Redrawn from Arthur RM, et al. *J Physiol [Lond]* 1971;212:593. **B,** Redrawn from Katsui Y. In: Rosenblith WA [ed]. *Sensory Communication.* Cambridge, MA: MIT Press; 1961.)

⚕ IN THE CLINIC

An important, although relatively uncommon, condition that can interrupt the function of cochlear nerve fibers is an **acoustic neuroma,** a tumor of Schwann cells of the vestibulocochlear nerve (CN VIII). As the tumor grows, irritation of cochlear nerve fibers may cause a ringing sound in the affected ear **(tinnitus).** Eventually, conduction in cochlear nerve fibers is blocked, and the ear becomes deaf. The tumor may be operable while still small; therefore, early diagnosis is important. If the tumor is allowed to enlarge substantially, it could interrupt the entire vestibulocochlear nerve and cause vestibular as well as auditory difficulties. It could also impinge on or distort neighboring cranial nerves (e.g., V, VII, IX, and X), and it could produce cerebellar signs by compressing the cerebellar peduncles.

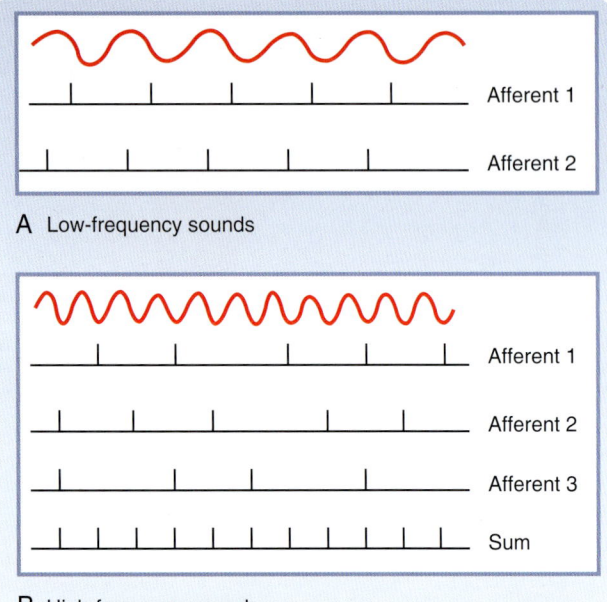

A Low-frequency sounds

B High-frequency sounds

• **Fig. 8.22 A,** At low frequencies, individual auditory afferent fibers can respond at each cycle to the signal frequency. **B,** At higher frequencies, each afferent fiber generates an action potential only at certain cycles, limited by its maximum firing frequency. However, the overall population of afferent fibers can still signal stimulus frequency by their aggregate firing frequency.

Central Auditory Pathway

Cochlear afferent fibers synapse on neurons of the dorsal and ventral cochlear nuclei. The neurons in these nuclei have axons that contribute to the central auditory pathways. Some of the axons from the cochlear nuclei cross to the contralateral side and ascend in the **lateral lemniscus,** the main ascending auditory tract. Others connect with various ipsilateral or contralateral nuclei, such as the **superior olivary nuclei,** which project through the ipsilateral and contralateral lateral lemnisci. Each lateral lemniscus ends in an **inferior colliculus.** Neurons of the inferior colliculus project to the **medial geniculate nucleus** of the thalamus, which gives rise to the auditory radiation. The auditory radiation ends in the **primary auditory cortex** (Brodmann areas 41 and 42), located on the superior surface of the temporal lobe.

The input from each ear is bilaterally represented in the ascending auditory system pathway at the level of the lateral lemniscus and above. Thus the representation of auditory space is complex, even at the brainstem level. Consequently, unilateral deafness may occur with isolated lesions of the cochlear nuclei or more peripheral structures. Central lesions do not cause unilateral deafness, although they may interfere with overall sensitivity to speech or with sound localization.

Functional Organization of the Central Auditory System

Receptive Fields and Tonotopic Maps

The responses of neurons in several structures that belong to the auditory system can be described by **tuning curves**

(see Fig. 8.21*B*). By plotting the distribution of the characteristic frequencies of neurons within a nucleus or in the auditory cortex, a **tonotopic map** may be revealed in which neurons are ordered according to their "best" frequencies. Tonotopic maps have been found in the cochlear nuclei, superior olivary complex, inferior colliculus, medial geniculate nucleus, and auditory cortex. A given auditory structure may, in fact, contain several tonotopic maps.

Binaural Interactions

Most auditory neurons at levels above the cochlear nuclei respond to stimulation of either ear (i.e., they have **binaural receptive fields**). Binaural receptive fields contribute to sound localization. A human can distinguish sounds originating from sources separated by as little as 1 degree. The auditory system uses several clues to judge the origin of sounds, including differences in the time (or phase) of arrival of the sound at the two ears and differences in sound intensity on the two sides of the head.

For example, neurons in the medial superior olivary nucleus have medial and lateral dendrites. The synapses on the medial dendrites are largely excitatory, and they originate from the contralateral ventral cochlear nucleus. Those on the lateral dendrites are mostly inhibitory and come from the ipsilateral ventral cochlear nucleus. Differences in the phase of the sound reaching the two ears affect the strength and timing of the excitation and inhibition reaching a particular medial superior olivary neuron. The lateral superior olivary nucleus processes differences in the sound intensity that reaches the two ears to provide information about the source of the sound. The activity of superior olivary neurons can provide information about sound localization.

Cortical Organization

Several features of the primary auditory cortex resemble those of other primary sensory areas. Not only are sensory maps—in this case, tonotopic maps—present in the auditory cortex but also this cortical region performs feature extraction. Neurons in the primary auditory cortex form **isofrequency columns** (in which the neurons in the column have the same characteristic frequency), and they also form alternating columns, known as summation and suppression columns. Neurons in **summation columns** are more responsive to binaural than to monaural input. Neurons in **suppression columns** are less responsive to binaural than to monaural stimulation, and, accordingly, the response to one ear is dominant. Some neurons are selective for the direction of frequency change.

Bilateral lesions of the auditory cortex have some effect on the ability to distinguish the frequency or intensity of different sounds, and they reduce the abilities to localize sound and to understand speech. Unilateral lesions, however, have little effect, especially if the nondominant (for language) hemisphere is involved (see Chapter 10). Evidently, frequency discrimination depends on activity at lower levels of the auditory pathway, possibly the inferior colliculus.

IN THE CLINIC

Two simple tests are often used clinically to distinguish the most important types of deafness: **conduction loss** and **sensorineural loss.** Conduction hearing loss occurs in disorders of the external ear (e.g., ear canal blocked by cerumen) or middle ear (e.g., rupture of the eardrum). Sensorineural hearing loss reflects disorders of the inner ear, the cochlear nerve, or central connections.

The **Weber test** is used to evaluate the magnitude of conduction hearing loss. In this test, the base of a vibrating tuning fork is placed against the middle of the person's forehead, and the person is asked to localize the sound. Normally, the sound is not localized to a particular ear. However, if the person has conductive hearing loss (e.g., due to a punctured tympanic membrane, fluid in the middle ear, otosclerosis, or loss of continuity of the ossicular chain), the sound is localized to the deaf ear because it is conducted to the cochlea through bone. The sound is also conducted to the cochlea of the undamaged ear, but bone-conducted sound does not activate the organ of Corti as well as does sound conducted normally through the tympanic membrane and ossicle chain. One reason why the sound in the Weber test is not localized to the normal ear may be that hearing in the normal ear is inhibited by the ambient sound level **(auditory masking).** Conversely, in people with sensorineural hearing loss (e.g., due to damage to the organ of Corti, the cochlear nerve, or the cochlear nuclei), the sound is localized to the normal side.

In the **Rinne test,** a vibrating tuning fork is placed against the bone behind the person's ear, and the person is asked to indicate when the sound dies out. The tuning fork is then held near the external auditory meatus of that ear. In people with normal hearing, the sound is again heard because the sound is more effectively transmitted to the cochlea in air (i.e., air conduction > bone conduction). If the conduction mechanism is damaged, the sound is not heard when the tuning fork is held near the external auditory meatus. Bone conduction in this case is better than air conduction. If the hearing loss is sensorineural, the sound is heard again when the tuning fork is placed by the external auditory meatus because with sensorineural hearing loss, the inner ear and cochlear nerve are less able to transmit impulses regardless of whether the sound vibrations reach the cochlea via air or bone. Thus because air conduction is more effective than bone conduction, the bone conduction pattern seen with sensorineural hearing loss is the same as in a normal ear.

The Vestibular System

The vestibular system detects angular and linear accelerations of the head. Signals from the vestibular system allow the body to make adjustments in posture that maintain balance and trigger head and eye movements to stabilize the visual image on the retina. The following description of the vestibular system emphasizes the sensory aspects of vestibular function, and it introduces the central vestibular pathways. The role of the vestibular apparatus in motor control is discussed in Chapter 9.

The Vestibular Apparatus

Structure of the Vestibular Labyrinth

The vestibular apparatus, like the cochlea, consists of a component of the membranous labyrinth located within the bony labyrinth. The vestibular apparatus on each side is composed of three **semicircular canals** and two **otolith organs** (Fig. 8.23; see also Fig. 8.17A). These structures contain endolymph and are surrounded by perilymph. The semicircular canals are named the **horizontal, anterior,** and **posterior** canals. The otolith organs are the **utricle** and the **saccule** (together indicated as "Vestibule" in Fig. 8.17A). Each semicircular canal has a swelling called an **ampulla** at the point where it joins the utricle. The saccule connects with the cochlea, through which endolymph (produced by the stria vascularis of the cochlea) can reach the vestibular apparatus.

The three semicircular canals on one side are matched with corresponding coplanar semicircular canals on the other side. The horizontal canals on each side of the head correspond, as do the anterior canal on one side and the posterior canal on the other side (see Fig. 8.23B). This arrangement allows the sensory epithelia, in corresponding pairs of canals on the two sides, to cooperate in sensing acceleration of the head about three nearly orthogonal axes in space. The horizontal canals are not truly horizontal; rather, they lie in the horizontal plane if the head is tilted down 30 degrees in relation to the horizon.

The ampulla of each of the semicircular canals contains a sensory epithelium called a **crista ampullaris,** or **ampullary crest** (Fig. 8.24). An ampullary crest consists of a ridge, transverse to the long axis of the canal, that is covered by epithelium containing vestibular hair cells. These hair cells are innervated by primary afferent fibers of the vestibular nerve, which is a subdivision of the vestibulocochlear nerve (CN VIII).

Like cochlear hair cells, each vestibular hair cell contains a set of stereocilia on its apical surface. However, unlike cochlear hair cells, vestibular hair cells also contain a large single kinocilium. The cilia on ampullary hair cells are embedded in a gelatinous structure called the *cupula.* The cupula and the crista occlude the lumen of the ampulla completely. Movement of endolymph, produced by angular acceleration of the head about an axis perpendicular to the plane of the canal, deflects the cupula and consequently bends the cilia on the hair cells. The cupula has the same specific gravity as endolymph, and thus it is unaffected by linear acceleratory forces, such as gravity.

The sensory epithelia of the otolith organs are called the **macula utriculi** and the **macula sacculi** (Fig. 8.25). The utricle is oriented nearly horizontally; the saccule is oriented vertically. Their hair cells are embedded in the epithelium that overlies each macula. As in the ampullary crests, the stereocilia and kinocilia of the macula project into a gelatinous mass. However, the gelatinous mass in the macula contains numerous **otoliths** ("ear stones") composed of calcium carbonate crystals. Together, the gelatinous mass and its otoliths are known as an **otolithic membrane.** The otoliths increase the specific gravity of the otolithic membrane to about twice that of the endolymph. Hence, the otolithic membrane tends to move when subjected to acceleration, whether linear (such as that produced by gravity) or angular, particularly when the center of rotation is outside the head.

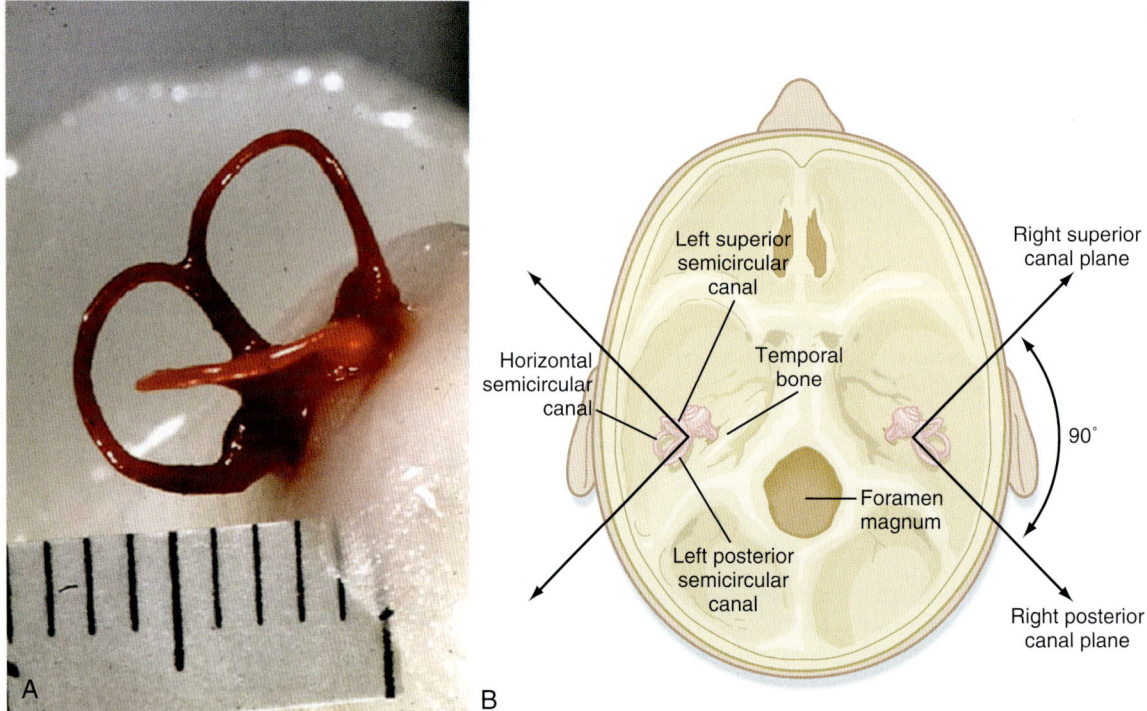

• **Fig. 8.23 A,** Lateral view of the right semicircular canals of a rhesus monkey that were dissected after being filled with plastic. Note the ampullae associated with each canal. Scale is in millimeters. **B,** Overhead view of the base of the skull showing the orientation of structures of the inner ear. Coplanar pairs of semicircular canals include the horizontal canals, as well as the anterior and contralateral posterior canals. (**A,** Courtesy of Dr. John Simpson, New York University School of Medicine. **B,** Redrawn from Haines DE [ed]. *Fundamental Neuroscience for Basic and Clinical Applications.* 3rd ed. Philadelphia: Churchill Livingstone; 2006.)

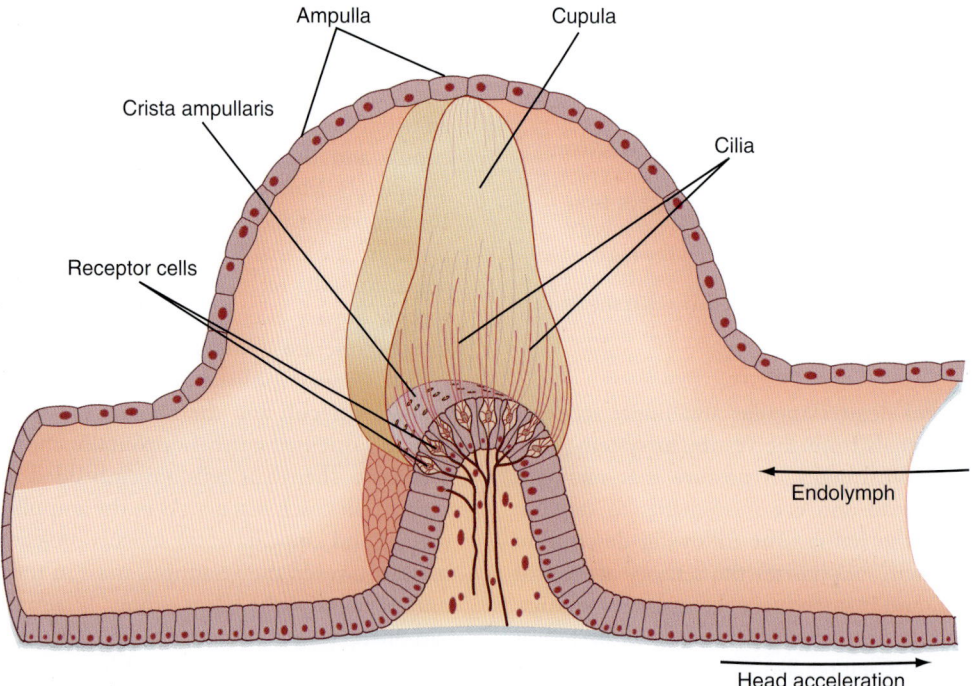

• **Fig. 8.24** Drawing of an Ampullary Crest Inside an Ampulla. The stereocilia and the kinocilium of each hair cell extend into the cupula, which extends across the entire cross-section of the ampulla. Head movement (acceleration) to the right would result in endolymph pressure to the left and deflection of the cupula to the left.

Innervation of Sensory Epithelia of the Vestibular Apparatus

The cell bodies of the primary afferent fibers of the vestibular nerve are located in the Scarpa ganglion. The neurons are bipolar, and their cell bodies, as well as axons, are myelinated. Peripherally, the vestibular nerve gives off separate branches to each of the vestibular epithelia; centrally, it accompanies the cochlear and facial nerves as they enter the internal auditory meatus of the skull.

Vestibular Transduction

Like cochlear hair cells, vestibular hair cells are functionally polarized, and the transduction mechanism is presumed to be similar. When the stereocilia are bent toward the longest cilium (in this case, the **kinocilium**), conductance of the apical membrane increases for cations and, because of the high K^+ concentration of the endolymph, K^+ enters, and the vestibular hair cell is depolarized (Fig. 8.26). Conversely, when the cilia are bent away from the kinocilium, the hair cell is hyperpolarized. The hair cell releases an excitatory neurotransmitter (either glutamate or aspartate) tonically, so that the afferent fiber on which it synapses has a resting discharge. When the hair cell is depolarized, more neurotransmitter is released, and the discharge rate of the afferent fiber increases. Conversely, when the hair cell is hyperpolarized, less neurotransmitter is released, and the firing rate of the afferent fiber slows.

Semicircular Canals

Angular accelerations of the head produce minute movement of the endolymph in relation to the head (Fig. 8.27). This happens because the inertia of the endolymph causes it to resist the initial acceleration of the membranous labyrinth. This lag pushes on the cupula, causes the cilia to bend, and consequently changes the discharge rates of the vestibular afferent fibers. All the cilia in a given ampullary crest are oriented in the same direction. In the horizontal canal, the cilia are oriented toward the utricle, and in the other ampullae, they are oriented away from the utricle.

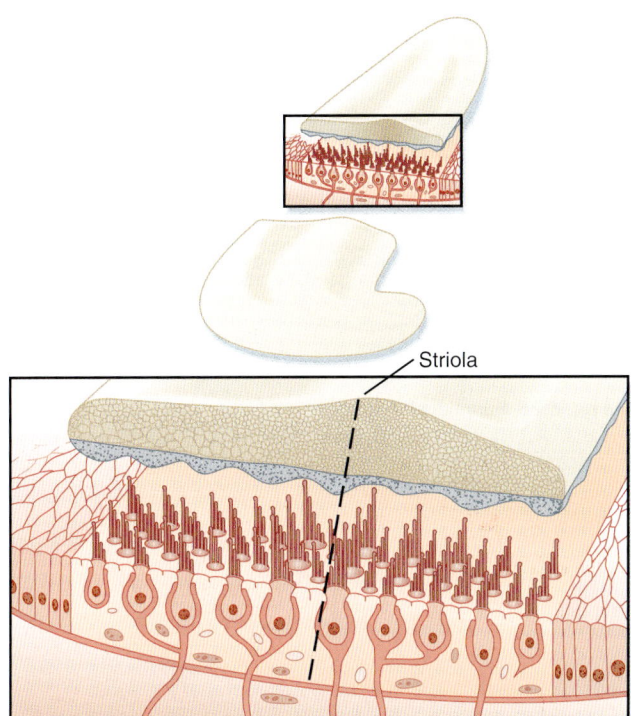

• **Fig. 8.25** Structure of one of the otolith organs, the saccule. Note the orderly variation in kinocilium orientation, as well as their mirror symmetry with regard to the striola. (Redrawn from Lindeman HH. *Adv Otorhinolaryngol* 1973;20:405.)

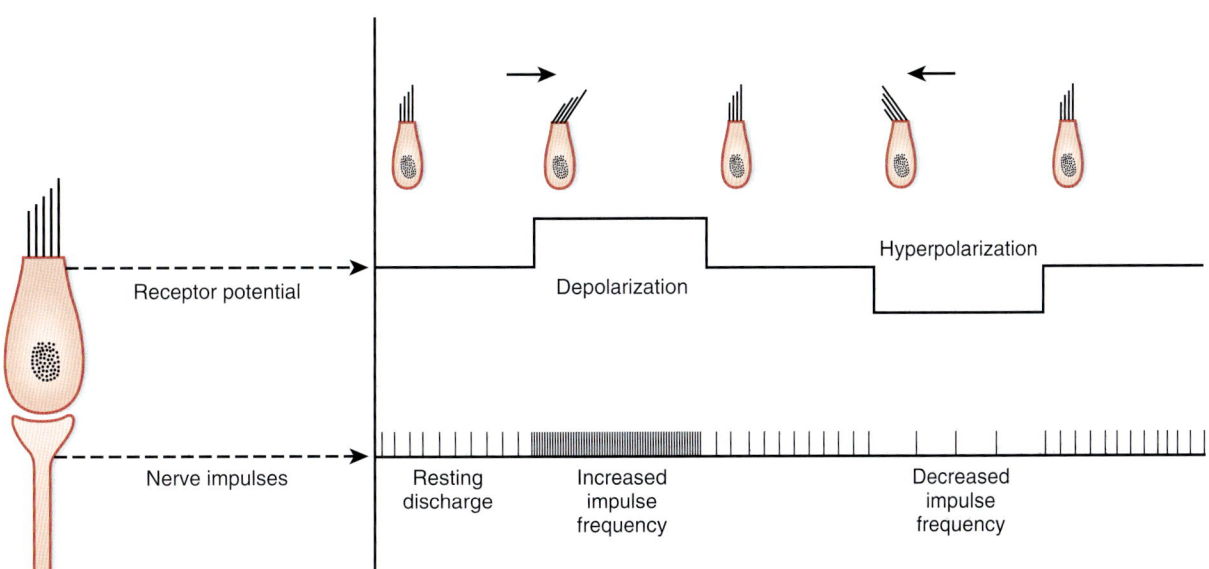

• **Fig. 8.26** Functional Polarization of Vestibular Hair Cells. When the stereocilia are bent toward the kinocilium, the hair cell is depolarized, and the afferent fiber is excited. When the stereocilia are bent away from the kinocilium, the hair cell is hyperpolarized, and the afferent discharge slows or stops. (Redrawn from Kandel ER, Schwartz JH. *Principles of Neural Science.* New York: Elsevier; 1981.)

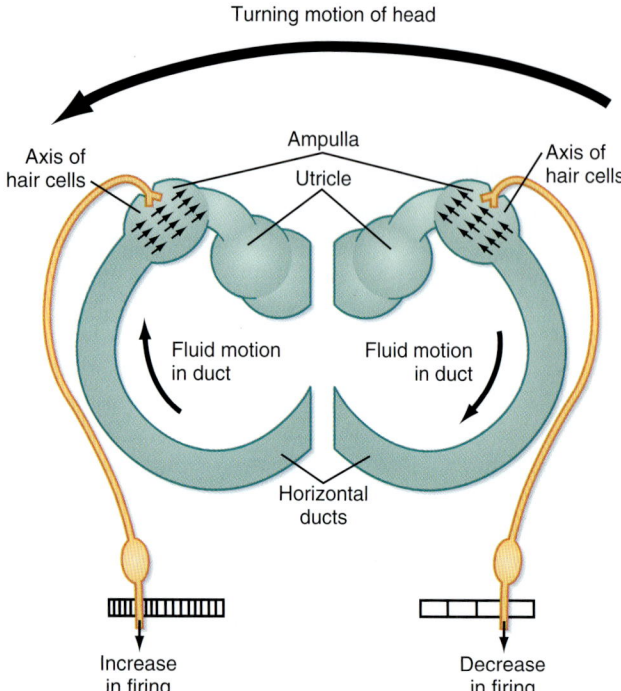

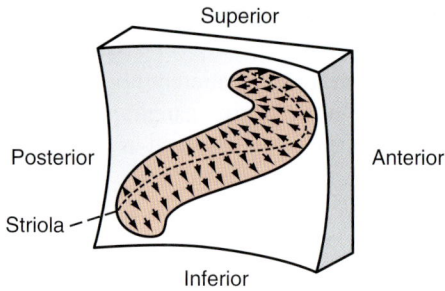

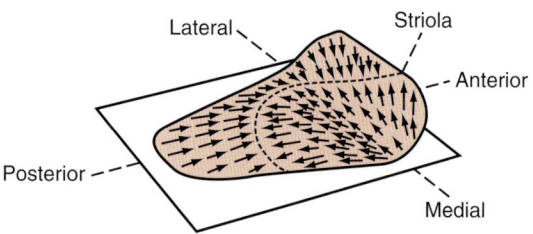

• **Fig. 8.27** Effect of Leftward Head Movement on the Activity of Vestibular Afferent Fibers Supplying Hair Cells in the Horizontal Semicircular Canals. The perspective in the figure is from above the head looking down. In this case, head rotation (acceleration) to the left causes endolymph pressure to the right and the result is increased output from the left canal and decreased output from the right canal. *Small arrows* in the ampulla indicate the functional polarity of the hair cells. The *large curved arrow* at the top indicates movement of the head; *smaller curved arrows* indicate relative movement of the endolymph.

The way in which angular acceleration of the head affects the discharge of vestibular afferent fibers is exemplified by the activity that originates from the horizontal canals. Fig. 8.27 shows the horizontal canals and utricle, as seen from above as the head is rotated (accelerated) to the left. As acceleration to the left begins, the inertia of the endolymph in the horizontal canals increases pressure toward the right side. This causes the cilia to bend on hair cells of the ampulla of the left horizontal canal toward the utricle and bends the cilia of the right canal away from the utricle. These actions increase the firing rate in the afferent fibers on the left and decrease the firing rate of the afferent fibers on the right. Once the head is moving at a constant velocity of rotation (i.e., no acceleration), there would be no force on either cupula, and therefore the hair cells of both canals would be firing as they do at rest. However, when the indicated rotation is stopped, the inertia of the endolymph creates a force on both cupulas, but in the direction opposite to that caused by the original acceleration. This results in an increase in the discharge rate of afferent fibers on the right side and a decrease in the discharge rate on the left. This postrotatory effect is of functional and clinical significance.

Otolith Organs

Unlike the hair cells in the ampullary crests, not all the hair cells in the otolith organs are oriented in the same

• **Fig. 8.28** Functional Polarization of Hair Cells in the Otolith Organs. **A,** The saccule. **B,** The utricle. The striola in each case is indicated by the *dotted line.* (Redrawn from Spoendlin HH. In: Wolfson RJ [ed]. *The Vestibular System and Its Diseases.* Philadelphia: University of Pennsylvania Press; 1966.)

🩺 IN THE CLINIC

Irritation of the vestibular labyrinth, as in **Meniere's disease,** can result in rhythmic conjugate deviations of the eyes, followed by quick return saccades. This condition is known as **nystagmus** (see Chapter 9). These eye movements are accompanied by a sense of **vertigo** and often **nausea.** The brain interprets a difference in the input from the two sides of the vestibular system as head motion. Irritation (or destruction) of one labyrinth produces an asymmetry of input that results in abnormal eye movement and associated psychological effects.

direction. Instead, they are oriented in relation to a ridge, called the **striola,** along the otolith organ (see Fig. 8.25). In the utricle, the hair cells on either side of the striola are polarized toward the striola, whereas in the saccule they are polarized away from the striola. Because the striola in each otolith organ is curved, there are hair cells with all orientations in the plane of the organ (Fig. 8.28). In any particular orientation of the head, the cilia of the hair cells are bent to varying extents according to their orientation in relation to the gravitational vector. This results in a particular pattern of input from the otolith organs to the CNS. When the head is tilted to a new position, the orientation of the otolithic membranes in relation to the gravitational vector changes, and so the cilia of the hair cells are bent in a new way. This change in the bending of the cilia of the hair cells changes the pattern of input from the otolith organs to the CNS and creates the sensation of movement, as well as possibly triggering various reflexes. Similarly, a linear acceleration caused by other forces, such as might occur in a fall or the angular acceleration when a car turns around a curve (angular accelerations have linear centripetal and instantaneous tangential components), also affects output from the otolith organs.

Central Vestibular Pathways

The vestibular afferent fibers project to the brainstem through the vestibular nerve. As previously mentioned, the cell bodies of these afferent fibers are located in the Scarpa ganglion. These primary afferent fibers terminate in the **vestibular nuclei** (see Fig. 4.6D to E), which are located in the rostral medulla and caudal pons and in specific regions of the **cerebellum**, most prominently in the **nodulus.**

The vestibular nuclei give rise to various projections, including projections through the **medial longitudinal fasciculus** (see Fig. 4.6C to E) to the oculomotor nuclei. Therefore, it is not surprising that the vestibular nuclei exert powerful control over eye movements (the **vestibule-ocular reflex**). Other projections give rise to the **lateral** and **medial vestibulospinal tracts,** which, respectively, provide for the activation of trunk and neck muscles and thereby contribute to equilibrium and to head movements **(vestibulocolic reflex).** There are vestibular pathways to the cerebellum, the reticular formation, and the contra-lateral vestibular complex, as well as to the thalamus. The pathway to the thalamus, via a projection to the cerebral cortex, mediates conscious sensation of vestibular activity. Vestibular reflexes and clinical tests of vestibular function are described in Chapter 9.

The Chemical Senses

The senses of **gustation** (taste) and **olfaction** (smell) help detect chemical stimuli that are present either in food and drink or in the air. In the evolution of humans, these chemical senses apparently did not have the survival value of some of the other senses, but they contribute consider-ably to quality of life and food selection, and they are important stimulants of digestion. In other animals, the chemical senses have greater survival value, and their activa-tion evokes a number of social behaviors, including mating, territoriality, and feeding.

Taste

The stimuli that we commonly know as tastes are actually mixtures of five elementary taste qualities: salty, sweet, sour, bitter, and umami.[b] Taste stimuli that are particularly effec-tive in eliciting these sensations are, respectively, sodium chloride, sucrose, hydrochloric acid, quinine, and mono-sodium glutamate. Umami has been described as having a proteinaceous, meaty character.

Taste Receptors

The sensation of taste depends on the activation of che-moreceptors located in taste buds. A taste bud consists of a group of 50 to 150 receptor cells, as well as supporting cells and basal cells (Fig. 8.29A). The chemoreceptor cells synapse at their bases with primary afferent nerve fibers, and their apices have microvilli that extend toward a taste pore. Chemoreceptor cells live only about 10 days. They are continuously replaced by new chemoreceptor cells that differentiate from basal cells located near the base of the taste bud.

Chemoreceptor molecules, each specialized for one type of taste stimulus, sit on the microvilli of chemoreceptor cells and detect molecules that diffuse into the taste pore from the overlying mucus of the tongue, part of which originates from glands adjacent to the taste buds. Some stimuli can pass directly into the cell to depolarize it (Na^+ for salty and H^+ for sour) or open cation channels to generate a receptor potential (also salty and sour), whereas others (sucrose, quinine, and glutamate for sweet, bitter, and umami) activate a second messenger that can either open cation channels or directly activate intracellular Ca^{++} stores (see Fig. 8.29B). In each case, depolarization of the receptor results in the release of an excitatory neurotransmitter and, consequently, action potentials in the primary afferent nerve fiber that are transmitted to the CNS.

Coding of taste, however, is not based entirely on the selectivity of the chemoreceptors for the different primary qualities because each cell responds to a range of stimuli, although most intensely to one. Because most natural tastes have chemicals that effect responses from a number of che-moreceptors, recognition of taste quality appears to depend on the patterned input from a population of chemorecep-tors, each responding differentially to the components of the stimulus. The intensity of the stimulus is reflected in the total amount of activity evoked.

Distribution and Innervation of Taste Buds

Taste buds are located on different types of taste papillae found on the tongue, palate, pharynx, and larynx. Types of taste papillae include **fungiform** and **foliate papillae** on the anterior and lateral aspects, respectively, of the tongue and **circumvallate papillae** on the base of the tongue (see Fig. 8.29C). The circumvallate papillae may contain several hundred taste buds. The tongue in humans may have several thousand taste buds. The sensitivity of different regions of the tongue for different taste qualities varies slightly because taste buds responding to each type of taste are widely distributed. The taste buds are innervated by three cranial nerves. The chorda tympani branch of the **facial nerve** (CN VII) innervates taste buds on the anterior two thirds of the tongue, and the **glossopharyn-geal nerve** (CN IX) innervates taste buds on the posterior third of the tongue (see Fig. 8.29C). The **vagus nerve** (CN X) innervates a few taste buds in the larynx and upper esophagus.

---

[b]The existence of a sixth, taste, fat (free fatty acids), is currently being debated.

Central Taste Pathways

The cell bodies of taste fibers in cranial nerves VII, IX, and X are located in the **geniculate, petrosal,** and **nodose ganglia,** respectively. The central processes of the afferent fibers enter the medulla, join the solitary tract, and synapse in the **nucleus of the solitary tract** (see Fig. 4.6D to E). In some animals, including several rodent species, the second-order taste neurons of the solitary nucleus project rostrally to the ipsilateral parabrachial nucleus. The parabrachial nucleus then projects to the small-celled (parvocellular) part of the **ventroposterior medial** (VPMpc) nucleus of the thalamus. In monkeys, the solitary nucleus projects directly to the VPMpc nucleus. The VPMpc nucleus is connected to two different gustatory areas of the cerebral cortex: one

in the face area of the S1 cortex and the other in the insula. An unusual feature of the central gustatory pathway is that it is predominantly an uncrossed pathway (unlike the central somatosensory pathways, which are predominantly crossed).

Olfaction

The sense of smell is much better developed in some animals **(macrosmatic animals)** than in humans. The ability of dogs to track other animals on the basis of odor is legendary, as is the use of **pheromones** by insects to attract mates. However, olfaction contributes to humans' emotional life, and odors can effectively conjure up memories. It also helps

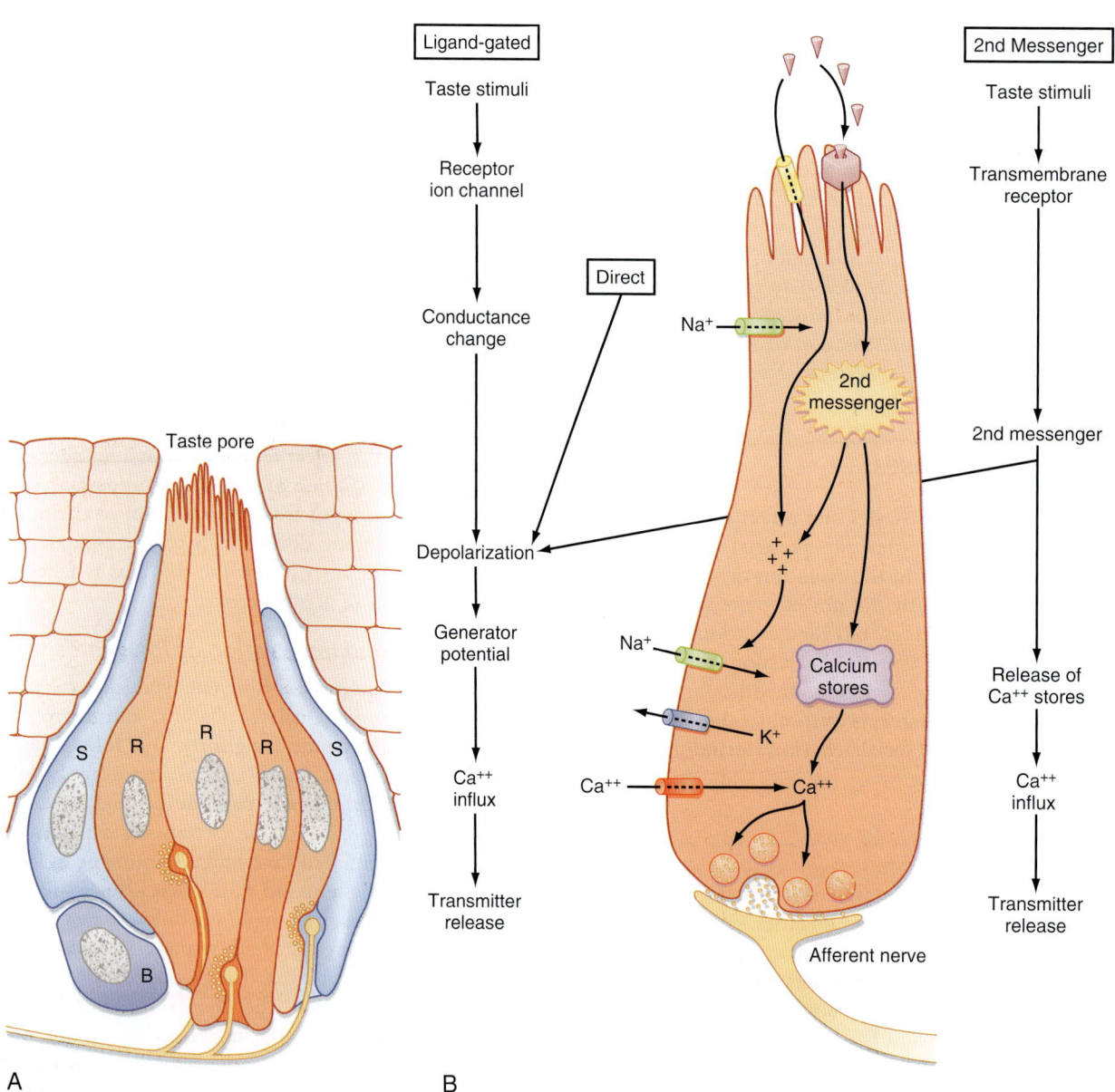

• **Fig. 8.29 A,** A taste bud is shown with the taste pore at the top and its innervation below. B, basal cell; R, ciliated taste receptor cells; S, supporting cells. **B,** Taste receptor cell showing second messenger, ligand-gated, and direct depolarization resulting in depolarization of the cell.

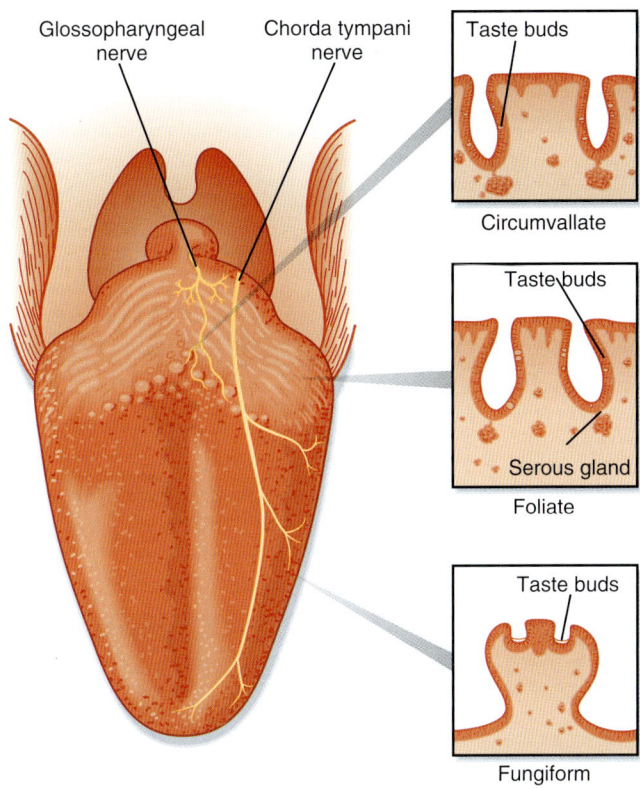

Glossopharyngeal nerve

Chorda tympani nerve

Taste buds

Circumvallate

Taste buds

Serous gland

Foliate

Taste buds

Fungiform

C

• **Fig. 8.29, cont'd C,** Distribution of the taste buds on the tongue and their innervation. (Redrawn from Squire LR, et al [eds]. *Fundamental Neuroscience.* San Diego, CA: Academic Press; 2002.)

people avoid consuming spoiled food and detect dangerous situations. For example, an unpleasant odorant is added to odorless, colorless natural gas so that people can easily detect a leak.

Odor has more primary qualities than taste does.[c] As many as 1000 different odor receptors are coded in the human genome, and although only approximately 350 types are functional, they represent the largest population of G protein–coupled receptors in the genome. The olfactory mucosa also contains somatosensory receptors of the trigeminal nerve. When performing clinical tests of olfaction, clinicians must avoid activating these somatosensory receptors with thermal or noxious stimuli, such as the ammonia used in "smelling salts."

Olfactory Receptors

The olfactory chemoreceptor cells are located in the **olfactory mucosa,** a specialized part of the nasopharynx. Olfactory chemoreceptors are bipolar nerve cells (Fig. 8.30). The

nonmotile cilia on the apical surface of these cells contain chemoreceptors that detect odorant chemicals dissolved in the overlying mucus layer. From its opposite side, the cell gives off an unmyelinated axon that joins other **olfactory nerve filaments** and penetrates the base of the skull through openings in the **cribriform plate** of the ethmoid bone. These olfactory nerves synapse in the **olfactory bulb,** a portion of the cerebral hemisphere of the brain located at the base of the cranial cavity, just below the frontal lobe (Fig. 8.31).

Humans have about 10 million olfactory chemoreceptors. Like taste cells, olfactory chemoreceptors have a short life span (about 60 days), and they are also continuously replaced. However, olfactory receptor cells are true neurons and, as such, are the only neurons that are continuously regenerated throughout life.

The olfactory mucosa is exposed to odorant molecules by ventilatory air currents or from the oral cavity during feeding. Sniffing increases the influx of odorants. The odorants are temporarily bound in mucus to an olfactory binding protein that is secreted by a gland in the nasal cavity.

Olfactory coding resembles taste coding in that most natural odors are complex and consist of many molecules that excite a wide variety of olfactory chemoreceptors. Coding for a particular perceived odor depends on the responses of many olfactory chemoreceptors, and the strength of the odorant is represented by the overall amount of afferent neural activity.

Central Pathways

The initial synapse of the olfactory pathway is located in the olfactory bulb, which is a specialized portion of the cerebral cortex and located on the underside of the frontal lobe. It contains **mitral cells,** interneurons (**granule cells; periglomerular cells),** and distinct synaptic clusters (**glomeruli;** see Fig. 8.31) in which theses interact with olfactory afferent fibers. As the olfactory afferent fibers reach the olfactory bulb from the olfactory mucosa, they branch as they approach an olfactory glomerulus to synapse on the dendrites of mitral cells. Each glomerulus is the target of thousands of olfactory afferent fibers, but all the afferent fibers to a single glomerulus convey input from the one type of olfactory receptor. This is all the more remarkable because olfactory receptor cells are being regenerated continuously and new axons must therefore navigate their way to a correct glomerulus.

The granule and periglomerular cells are inhibitory interneurons. They form **dendrodendritic reciprocal synapses** with the dendrites of mitral cells. Activity in a mitral cell depolarizes these inhibitory cells, and they in turn inhibit the original and adjacent glomeruli. Because each glomerulus is specialized by being the target of afferent fibers for a unique combination of odor qualities, this appears to be a way of enhancing stimulus contrast, much the way horizontal cells do in the retina. In addition, it provides a mechanism for adaptation to continuous stimulation.

---

[c]The conscious perception of **flavor,** particularly of foods, is the result of both olfactory and gustatory input based on directly inhaled odor, taste from the food as it is macerated in the mouth, and retronasal odor from the volatile molecules that are released by maceration and pass up into the nasal cavity from the pharynx.

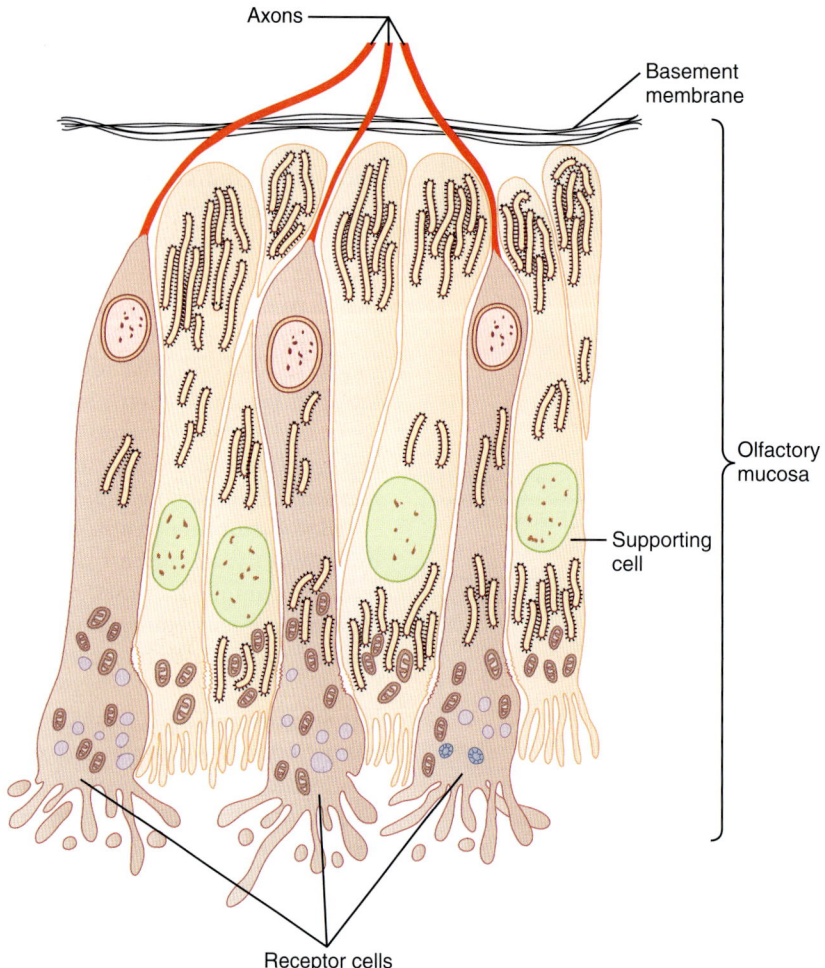

• **Fig. 8.30** Olfactory chemoreceptors and supporting cells. (Redrawn from de Lorenzo AJD. In: Zotterman Y [ed]. *Olfaction and Taste.* Elmsford, NY: Pergamon; 1963.)

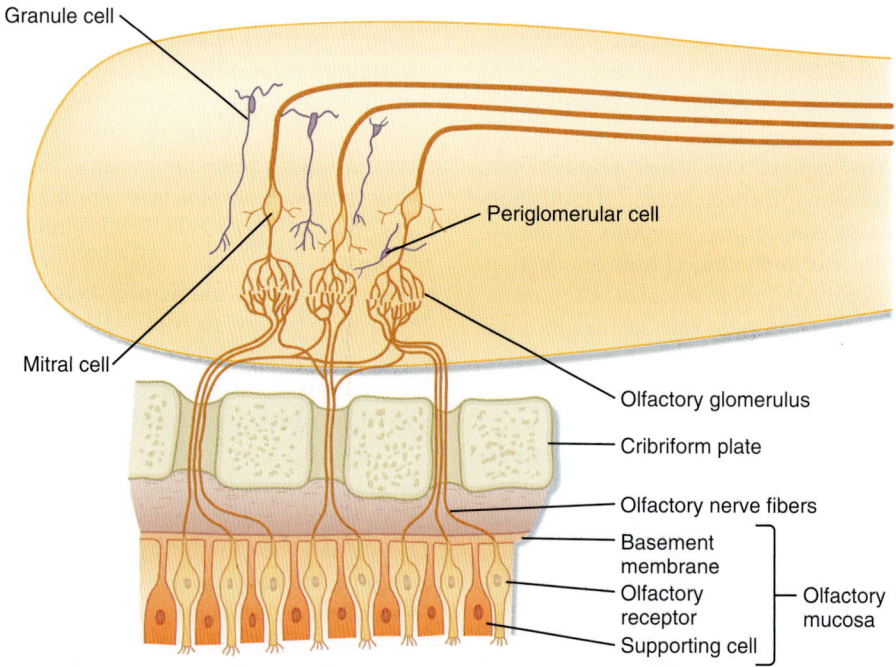

• **Fig. 8.31** Drawing of a Sagittal Section Through an Olfactory Bulb, Showing Terminations of the Olfactory Chemoreceptor Cells in the Olfactory Glomeruli and the Intrinsic Neurons of the Olfactory Bulb. The axons of the mitral cells are shown exiting in the olfactory tract to the right. (Modified from House EL, Pansky B. *A Functional Approach to Neuroanatomy.* 2nd ed. New York: McGraw-Hill; 1967.)

The axons of mitral cells leave the olfactory bulb and enter the olfactory tracts. From here, the olfactory connections become highly complex. Within the olfactory tracts is a nucleus, called the **anterior olfactory nucleus,** that receives input from the olfactory bulb and projects to the contralateral olfactory bulb through the **anterior commissure.** As each olfactory tract approaches the base of the brain, it splits into the **lateral** and **medial olfactory striae.** Axons of the lateral olfactory stria synapse in the primary olfactory cortex, which includes the **prepiriform cortex** (and, in many animals, the piriform lobe). The medial olfactory stria includes projections to the **amygdala,** as well as to the basal forebrain. These structures are portions of, or directly connected to, the limbic system (see Chapter 10).

Of note is that the olfactory pathway is the only sensory system that does not have an obligatory synaptic relay in the thalamus before signals reach the cortex. However, olfactory information does reach the mediodorsal nucleus of the thalamus, and it is then transmitted to the prefrontal and orbitofrontal cortex. The functional roles of olfaction, in addition to the conscious perception of odor, include providing much of the subtleties of taste by enhancing the

IN THE CLINIC

Olfaction is not generally examined in a routine neurological examination. However, smell can be tested by having the patient inhale and identify an odorant. One nostril should be examined at a time while the other nostril is occluded. Strong odorants, such as ammonia, should be avoided because they also activate trigeminal nerve fibers. Smell sensation can be lost **(anosmia)** after a basal skull fracture or after damage to one or both olfactory bulbs or tracts by a tumor (such as an **olfactory groove meningioma).** Concussions can cause anosmia because the sudden movement of the brain inside the skull can shear the small unmyelinated olfactory nerve fibers. An aura of a disagreeable odor, often the smell of burning rubber, occurs during **uncinate seizures,** which are epileptic seizures that originate in the medial temporal lobe.

narrow range of gustatory receptors with the wide repertoire of olfactory receptors. In addition, via its intimate connections with limbic and, by extension, hypothalamic structures, it provides input to subconscious mechanisms related to emotions, memory, and sexual behavior.

Key Points

1. Light enters the eye through the cornea and lens and is focused on the retina, which lines the back of the eye. The cornea is the most powerful refractive surface, but the lens has a variable power that allows images of near objects to be focused on the retina. The iris regulates depth of field and the amount of illumination that enters the eye.

2. The outer segments of the photoreceptor cells transduce light. Photoreceptors synapse on retinal bipolar cells, which in turn synapse on other interneurons and on ganglion cells. The ganglion cells project to the brain through the optic nerve. The optic disc, where the optic nerve leaves the retina, contains no photoreceptors and is therefore a blind spot. The portion of the retina with the highest degree of spatial resolution is the fovea and the surrounding macula.

3. Rod photoreceptors have high sensitivity, do not discriminate among colors, and function best under low light levels. Cone photoreceptors have lower sensitivity but higher spatial resolution. Color vision relies on the three types of cones that have different spectral sensitivities.

4. Bipolar cells and many ganglion cells have concentric receptive fields with an on-center/off-surround or off-center/on-surround organization. Horizontal cells mediate this center-surround antagonism. Photoreceptor, bipolar, and horizontal cells respond to stimulation by modulating their membrane potential and their release of neurotransmitters,

but ganglion cells respond by generating action potentials.

5. The axons of ganglion cells in the temporal retina project to the brain ipsilaterally; those in the nasal retina cross in the optic chiasm. Because the lens inverts the image that falls on the retina, each side of the visual field is projected to the contralateral side of the brain for both eyes. In the lateral geniculate nucleus (LGN) of the thalamus, the input from each eye terminates in separate layers, and the M ganglion cells (sensitive to movement) and P ganglion cells (sensitive to detail and color) project to separate layers as well.

6. The LGN projects to primary visual (striate) cortex via the visual radiation and terminates largely in layer 4, where there is an orderly retinotopic map. Within the map, information from each eye maps to alternating adjacent points to create ocular dominance columns that extend vertically in the cortex. Striate cortical neurons outside of layer 4 respond best to bar or edge stimuli oriented in a particular way. Cells that "prefer" a particular stimulus orientation are grouped in orientation columns.

7. The extrastriate visual areas have different functions. Some in the inferotemporal cortex are influenced chiefly by P cells, and they function in form detection, color vision, and face discrimination. M cells influence regions of the middle temporal and parietal cortex, which function in motion detection and the control of eye movements.

8. A pure tone is characterized in terms of its amplitude, frequency, and phase. Natural sounds are combinations of pure tones. Sound pressure is measured in decibels (dB), in relation to a reference level.

9. The pinna and auditory canal convey airborne sound waves to the tympanic membrane. The three small bones (ossicles) of the middle ear transmit the vibrations of the tympanic membrane to the oval window of the fluid-filled inner ear. Hearing is most sensitive at about 3000 Hz because of the dimensions of the auditory canal and the mechanics of the ossicles.

10. The cochlea of the inner ear has three main compartments: the scala vestibuli, the scala tympani, and the intervening scala media (cochlear duct). The cochlear duct is bounded on one side by the basilar membrane, on which lies the organ of Corti, the sound transduction mechanism.

11. When the basilar membrane oscillates in response to pressure waves introduced into the scala vestibuli at the oval window, the stereocilia of the hair cells of the organ of Corti are subjected to shear forces, which open mechanoreceptor K^+ channels. This results in a membrane conductance change that modulates the release of neurotransmitters on to cochlear nerve fibers.

12. High-frequency sounds best activate the hair cells near the base of the cochlea, and low-frequency sounds activate cells near the apex. Such a tonotopic organization is also present in central auditory structures, including the cochlear nuclei, superior olivary complex, inferior colliculus, medial geniculate nucleus, and primary auditory cortex.

13. Auditory processing at many sites in the central auditory pathway contributes to sound localization, frequency and intensity analysis, and speech recognition.

14. The vestibular apparatus is part of the inner ear. It includes three semicircular canals (horizontal, anterior, and posterior) and two otolith organs (utricle and saccule) on each side. These transduce, respectively, angular and linear accelerations of the head. The three semicircular canals are mutually orthogonal, so they can resolve angular acceleration of the head about any axis of rotation.

15. In each semicircular canal, there are sensory hair cells whose cilia extend into a cupula, which blocks the cross-section of the endolymph-filled canal. Angular head acceleration displaces the endolymph and the cupula, bending the cilia. If the stereocilia bend toward the kinocilium, the hair cell is depolarized, which causes an increase in the firing rate in the afferent fiber.

16. In the otolith organs, the cilia project into an otolithic membrane. Acceleration of the head, as with linear movement, or change in position in relation to gravity displaces the otolithic membrane (because of the mass of the otoliths) and changes the firing patterns of the hair cells, depending on their orientation.

17. Central vestibular pathways include afferent connections to the vestibular nuclei and the cerebellum. Activation of the vestibular afferent fibers is detected by the brain as head acceleration or position change and is relayed via the vestibular nuclei to pathways that mediate compensatory eye movements, neck movements, and adjustments to posture.

18. Taste buds contain chemoreceptor cells arranged around a taste pore. Taste buds are located on several kinds of papillae on the tongue and in the pharynx and larynx. Five types of taste-receptor cells detect the five elementary qualities of taste: salty, sweet, sour, bitter, and umami. Complex flavors are signaled by the patterned activity of multiple classes of taste receptor and by central correlation with accompanying olfactory input.

19. Afferent taste fibers synapse in the nucleus of the solitary tract. The thalamic relay is via a part of the ventroposterior medial nucleus to the taste-receiving areas located in the S1 cortex and the insula.

20. Odors are detected by olfactory chemoreceptor cells, which are continuously regenerated in the olfactory mucosa. These cells are true neurons that are endowed with a wide array of G protein–coupled receptors that enable the detection of hundreds of odor molecules.

21. Individual olfactory axons project to olfactory glomeruli, specific for each stimulus type, in the olfactory bulb. They synapse on the dendrites of mitral cells, which have reciprocal synapse with inhibitory interneurons. This synaptic organization in the glomerulus underlies stimulus adaptation and contrast enhancement.

Additional Reading

Squire L, Berg D. *Fundamental Neuroscience.* 4th ed. New York: Academic Press; 2012.

9

Organization of Motor Function

LEARNING OBJECTIVES

Upon completion of this chapter, the student should be able to answer the following questions:

1. What is a motor neuron, and how are α and γ motor neurons different?
2. What is a motor unit? How does the "size principle" apply to the orderly recruitment of motor units?
3. What is a reflex, and why are reflexes useful for clinical and scientific understanding?
4. What information about the state of the muscle is sensed by the muscle spindles, and what afferent fibers convey this information to the central nervous system (CNS)?
5. How do γ motor neurons modulate the responses of the muscle spindle?
6. What are the pathways and functions of the basic spinal reflexes?
7. What is a central pattern generator, and what types of movements can it be used for?
8. What distinguishes the pathways of the medial and lateral descending pathways in motor control?
9. What is decerebrate rigidity, and what are its implications for the control of muscle tone?
10. What distinguishes the cortical motor areas from each other?
11. What motor parameters are coded for in the activity of neurons in motor cortex?
12. How does the organization of the mossy and olivocerebellar (climbing) fiber afferent systems to the cerebellum differ in their origins, topography, and synaptic connections.
13. What is the geometric relationship between the major cellular elements of the cerebellar cortex?
14. What are simple and complex spikes in Purkinje cells?
15. What are the direct and indirect pathways in the basal ganglia, and how does their activity influence movement?
16. How is the balance of activity between the direct and indirect pathways altered in Parkinson's disease and Huntington's disease?
17. How do the vestibuloocular and optokinetic reflexes act to stabilize gaze? How do they complement each other?
18. What are the roles of saccades and smooth pursuit movements in visual tracking?
19. What is nystagmus, and what types of sensory stimulation can drive nystagmus in a normal individual?
20. What is the somatotopic organization of the different CNS regions involved in motor control.

Movements are the major way in which humans interact with the world. Most activities—including running, reaching, eating, talking, writing, and reading—ultimately involve motor acts. Thus motor control is a major task of the central nervous system (CNS), and from an evolutionary perspective, it is probably the reason that nervous systems first arose. Not surprisingly, a large amount of the CNS is devoted to motor control, which can be defined as the generation of signals to coordinate contraction of the musculature of the body and head, either to maintain a posture or to make a movement (transition between two postures).

Because large amounts of the nervous system are involved in motor control, damage or diseases of the nervous system often result in motor abnormalities. Conversely, particular motor symptoms help determine the location of the damaged or malfunctioning region; thus assessment of motor function is an important clinical tool.

In this chapter, each major CNS area involved in motor control is described, starting with the spinal cord and continuing with the brainstem, cerebral cortex, cerebellum, and basal ganglia. Eye movements are discussed at the end of the chapter because of their importance and the specialized circuits involved in their generation. Each CNS area is described separately; however, CNS regions do not function in isolation, and most movements result from the coordinated action of multiple brain regions. For example, even spinal reflexes, which are mediated by local circuits in the spinal cord, can be modified by descending motor commands, and virtually all voluntary movements, which arise from cerebral activity, are ultimately generated by activation of the spinal cord circuitry (or analogous brainstem nuclei for muscles in the head and face).

Principles of Spinal Cord Organization

The spinal cord has a cylindrical shape in which the white matter is located superficially and the gray matter is found deep to the white matter shell. The gray matter forms a continuous column that runs the length of the cord. However, the nerve roots that enter and exit the spinal cord bundle into discrete nerves, which form the basis for naming the

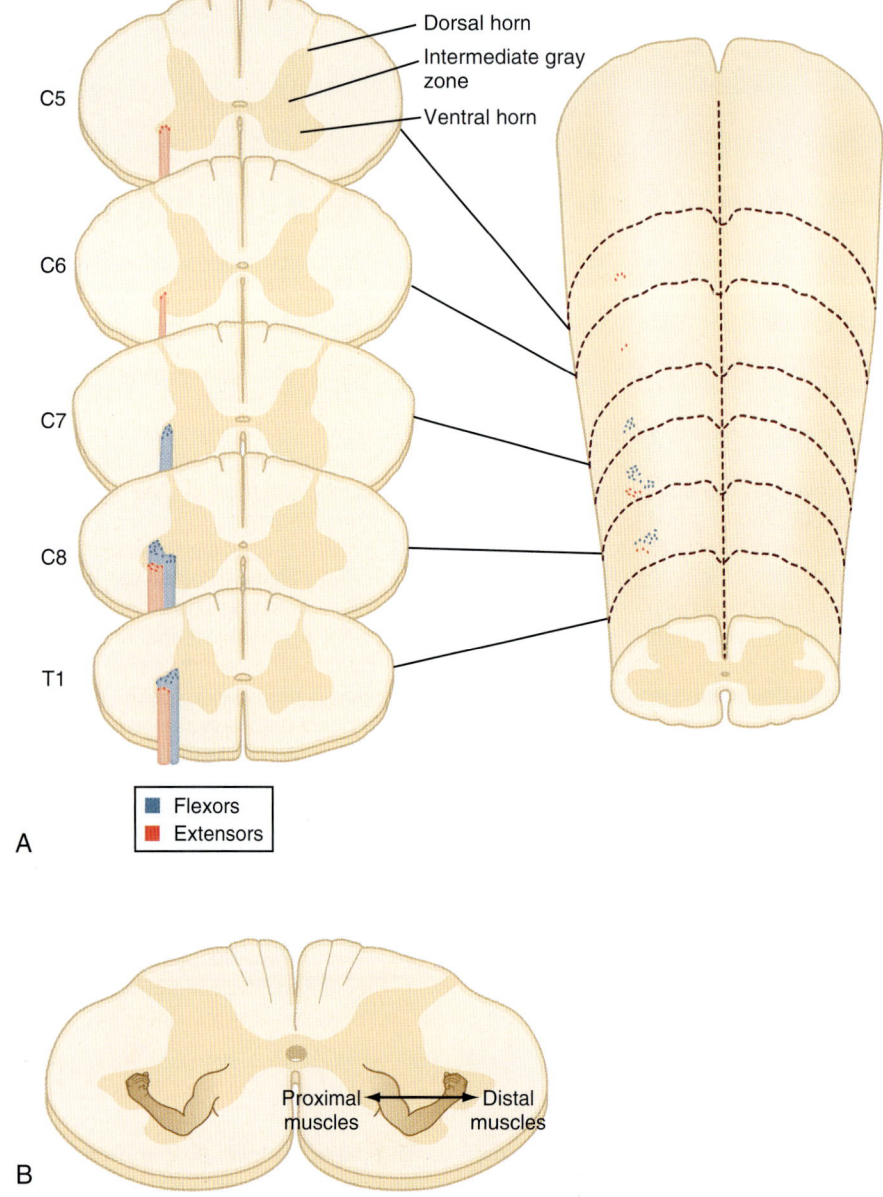

specific levels ("segments") of the spinal cord (8 cervical, 12 thoracic, 5 lumbar, 5 sacral, and 1 coccygeal). When viewed in cross-section, the gray matter column typically has an "H" or butterfly shape. The "butterfly wings" are divided into dorsal and ventral horns that are separated by an intermediate zone (Fig. 9.1) (at some spinal cord levels, a small lateral horn is also present; see Chapter 11). The ventral horn is where motor neurons reside, and thus it has primarily a motor function. Correspondingly, it is the main target of descending motor pathways from the brain. In contrast, the dorsal horn is the major recipient of incoming sensory information and the main source of ascending sensory pathways (e.g., the spinothalamic tract; see Chapter 7). The description of the motor function of the spinal cord therefore begins with the properties of motor neurons and their organization in the ventral horn. Similar organizational principles hold for cranial nerve nuclei that are involved in controlling the musculature of the head and face (e.g., the facial and motor trigeminal nuclei).

Somatic Motor Neurons

A motor neuron is a neuron that projects to muscle cells. Because motor neurons represent the only route for CNS activity to control muscle activity, motor neurons have been termed the **final common pathway.** There are both somatic motor neurons and autonomic motor neurons. However, in this chapter, the term *motor neuron* refers only to somatic motor neurons; autonomic motor neurons are discussed in Chapter 11. Somatic motor neurons innervate the skeletal (striated) muscles of the body. Two main classes are distinguished on the basis of their axonal diameters: α and γ **motor neurons.**

α Motor Neurons

The α motor neurons are large, multipolar neurons that range in size up to 70 μm in diameter (see Fig. 4.10*A*). Their axons leave the spinal cord through the ventral roots and the brainstem via several cranial nerves and are distributed to the appropriate skeletal muscles via peripheral nerves. The α motor neuron axon also projects to other neurons by giving off collateral axons before leaving the CNS. The main axon terminates by synapsing onto the extrafusal muscle fibers. These synapses are called *neuromuscular junctions* or *end plates*. Extrafusal fibers are large muscle fibers that make up the bulk of a skeletal muscle and essentially generate its contractile force (a muscle also contains intrafusal fibers whose functions are detailed later in this chapter; also see Chapter 12).

A key functional aspect of the motor neuron projection pattern out of the CNS is that each neuron's axon innervates only one muscle, but it branches to innervate multiple fibers within that muscle. Moreover, each extrafusal muscle fiber in mammals is supplied by only one α motor neuron. Thus a **motor unit** can be defined as an α motor neuron and all of the skeletal muscle fibers that its axon supplies. The motor unit can be regarded as the basic unit of movement because when an α motor neuron discharges under normal circumstances, all of the muscle fibers of the motor unit contract. That is, the safety factor (see Chapter 6) of the neuromuscular junction is greater than 1, and so each action potential in the motor neuron axon triggers an action potential in every muscle fiber of the motor unit.

Of importance is that the average size of the motor unit (i.e., the number of muscle fibers innervated by an axon) varies between muscles, depending on how fine a control of the muscle is required. For finely controlled muscles, such as the eye muscles, an α motor neuron may supply only a few muscle fibers. However, in a proximal limb muscle, such as the quadriceps femoris, a single α motor neuron may innervate thousands of muscle fibers.

The muscle fibers that belong to a given motor unit are called a *muscle unit*. All the muscle fibers in a muscle unit are of the same histochemical type (i.e., they are all either slow-twitch [type I] or fast-twitch [type IIA or IIB] fibers). For an in-depth description of muscle fiber types, see Chapter 12. Of importance in this chapter is that a number of physiological properties are correlated with this histochemical classification scheme. In particular, slow-twitch fibers, which contract and relax slowly as implied by their name, also generate low-force levels but essentially never fatigue. In contrast, the fast-twitch fiber contract and relax rapidly, generate higher levels of force, and fatigue at varying rates.

The first motor units to be activated in many cases, either by voluntary effort or during reflex action or just to maintain posture, are those with the smallest motor axons. These motor units contain slow-twitch fibers and thus generate the smallest contractile force, allowing the initial contraction to be finely graded. These units tend to be active much of the time, if not continuously, and so their lack of fatigability makes good functional sense.

As more motor units are recruited for a motor act, motor neurons with progressively larger axons become involved, and these axons synapse onto the fast-twitch fibers. Thus they generate progressively larger amounts of tension. The most powerful such motor units are typically recruited only for tasks requiring large amounts of force (e.g., sprinting, jumping, and lifting a heavy weight), tasks that people can perform for only for short periods of time.

The orderly recruitment of motor units helps the CNS generate a large range of forces and also maintain relatively precise control at the different force levels. This recruitment pattern is called the **size principle** because the motor units are recruited in order of motor neuron axon size. The size principle depends on the fact that small motor neurons are activated more easily than are large motor neurons. Recall that if an excitatory synapse is active, it opens channels in the postsynaptic membrane and causes an excitatory postsynaptic current. The same size excitatory postsynaptic current generates a larger potential change at the initial segment of an axon of a small motor neuron than it does at a larger motor neuron, simply as a consequence of Ohm's law ($V = IR$) and the fact that smaller motor neurons have higher membrane resistance than larger motor neurons do. Thus because excitatory postsynaptic potentials in the CNS are small and need to summate to reach threshold for triggering spikes, as the level of synaptic bombardment rises from zero, the resulting depolarization will reach spiking threshold in smaller motor neurons first. As the size principle is usually obeyed, this assumption generally appears to hold; however, there can be exceptions, and in these cases, the descending motor pathways presumably must provide differing levels of synaptic drive to the different-sized motor neurons.

γ Motor Neurons

The γ motor neurons are smaller than α motor neurons; they have a soma diameter of about 35 μm. The γ motor neurons that project to a particular muscle are located in the same regions of the ventral horn as the α motor neurons that supply that muscle. γ Motor neurons do not supply extrafusal muscle fibers; instead, they synapse on specialized striated muscle fibers called **intrafusal muscle fibers,** which traverse receptors called *muscle spindles* that are embedded in skeletal muscles. The function of γ motor neurons is to regulate the sensitivity of these receptors (discussed later).

Topographic Organization of Motor Neurons in the Ventral Horn

The spatial distribution of motor neurons in the spinal cord is highly organized. (This is also true in the cranial nerve nuclei.) A given skeletal muscle is supplied by a group

IN THE CLINIC

A clinically useful way to monitor the activity of motor units is **electromyography.** An electrode is placed within a skeletal muscle to record the summed action potentials of the skeletal muscle fibers of a muscle unit. If no spontaneous activity is noted, the patient is asked to contract the muscle voluntarily to increase the activity of motor units in the muscle. As the force of voluntary contraction increases, more motor units are recruited. In addition to the recruitment of more motor neurons, contractile strength increases with increases in the rate of discharge of the active α motor neurons. Electromyography is used for various purposes. For example, the conduction velocity of motor axons can be estimated as the difference in latency of motor unit potentials when a peripheral nerve is stimulated at two sites separated by a known distance. Another use is to observe fibrillation potentials that occur when muscle fibers are denervated. Fibrillation potentials are spontaneously occurring action potentials in single muscle fibers. These spontaneous potentials contrast with motor unit potentials, which are larger and have a longer duration because they represent the action potentials in a set of muscle fibers that belong to a motor unit.

of α motor neurons, called a **motor nucleus,** located in the ventral horn. Each such motor nucleus takes the form of a rostrocaudally running column that can span several spinal cord levels (see Fig. 9.1*A*). Motor neurons that supply the axial musculature collectively form a column of cells that extends the length of the spinal cord. In the cervical and lumbosacral enlargements, these cells are located in the most medial part of the ventral horn; at other levels, they essentially form the entire ventral horn. The motor neurons innervating the limb muscles are in the cervical and lumbosacral enlargements, where they form columns that are lateral to those for the axial muscles. Motor neurons to muscles of the distal part of the limb are located most laterally, whereas those that innervate more proximal muscles are located more medially (Fig. 9.1*B*). Also, motor neurons to flexors are dorsal to those that innervate extensors. Note that the α and γ motor neurons to a given muscle are found intermixed within the same motor neuron column.

The interneurons that connect with the motor neurons in the enlargements are also similarly topographically organized. In general, interneurons that supply the limb muscles are located mainly in the lateral parts of the deep dorsal horn and the intermediate region between the dorsal and ventral horns. Those that supply the axial muscles, however, are located in the medial part of the ventral horn. All of these interneurons receive synaptic connections from primary afferent fibers and from the axons of pathways that descend from the brain, and thus they are part of both spinal reflex arcs and descending motor control pathways.

An important aspect of interneuronal systems is that the laterally placed interneurons project ipsilaterally to motor neurons that supply the distal or the proximal limb muscles, whereas the medial interneurons project bilaterally. This

arrangement of the lateral interneurons allows the limbs to be controlled independently. In contrast, the bilateral arrangement of the medial interneurons allows bilateral control of motor neurons to the axial muscles to provide postural support to the trunk and neck.

Spinal Reflexes

Although motor neurons are the final common pathway from the CNS to muscles and thus shape how neuronal activity is transformed into muscular contraction, each motor neuron directly acts on only a single muscle. Normal movements (or postures), however, are rarely, if ever, caused by the isolated contraction of an individual muscle. Rather, they reflect the coordinated activity of large groups of muscles. For example, elbow flexion involves an initial burst of activity in flexor muscles, such as the biceps, and in relaxation of extensor muscles, such as the triceps. This activity is then succeeded by a burst of activity of the triceps and then a second burst of activity in the biceps to stop the flexion movement at the desired position. Furthermore, other muscles are also activated during the elbow flexion to maintain overall balance and posture.

As the elbow flexion example shows, different roles are played by each muscle during a movement. The muscle that initiates, and is the prime cause of, the movement is called the *agonist.* Muscles that act similarly to the agonist are called *synergists,* whereas muscles whose activity opposes the action of the agonist are *antagonists.* In addition, muscles can act as *fixators* to immobilize a joint and in postural roles. Moreover, the relationship two muscles have to each other may depend on the specific movement being performed. For example, during elbow flexion, the triceps acts an antagonist to the biceps. In contrast, during supination of the forearm without rotation occurring about the elbow, the biceps (which also acts to supinate the forearm) is again an agonist, but the role of the triceps is that of an elbow fixator.

Thus motor control requires flexibly linking (and unlinking) the activity of groups of motor neurons that connect to different muscles. The circuits of the spinal cord are a major mechanism used by the CNS for this aspect of motor control. Indeed, descending pathways from the brain target primarily the interneurons of the spinal cord, although there are some descending axons that synapse directly onto motor neurons.

Spinal cord circuitry has several levels of organization. The most basic is the segmental level: that is, a circuit that is largely confined to a single or several neighboring segments and that is repeated again at many levels. The basic spinal reflexes covered below (i.e., the myotatic, inverse myotatic, and flexion reflexes) are mediated by such circuits. Superimposed on this segmental organization is the propriospinal system, which is a series of neurons whose axons run up and down the spinal cord to interconnect the different levels of the cord. This system allows the coordination of activity at different spinal levels, which is important for behavior involving both the forelimbs and the hind limbs, such as locomotion. Finally, there are descending motor pathways that interact with these spinal circuits. These motor pathways carry signals related to voluntary movement, but they are also important for the more automatically (or nonconsciously) controlled aspects of motor function, such as the setting of muscle tone (the resting resistance of muscles to changes in length).

Spinal cord circuits are thus involved in all movements made by the body, but they have been most extensively studied with the use of reflexes. A **reflex** is a relatively predictable, involuntary, and stereotyped response to an eliciting stimulus. Because of these properties, spinal reflexes have been used to identify and classify spinal cord neurons, determine their connectivity, and study their response properties. Thus knowledge of spinal reflexes is essential for understanding spinal cord function.

The basic circuit that underlies a reflex is called a **reflex arc.** A reflex arc can be divided into three parts: an afferent limb (sensory receptors and axons) that carries information to the CNS, a central component (synapses and interneurons within the CNS), and an efferent limb (motor neurons) that causes the motor response. The knee jerk response to a physician's tapping on the patellar tendon with a reflex hammer is a common example of a spinal reflex and illustrates the various components of the definition. The tap on the tendon actually causes a brief stretching of the quadriceps muscle (eliciting stimulus) and thus activates sensory receptors (group Ia fibers in muscle spindles). Activation of sensory receptors causes an excitatory signal to be sent to the spinal cord to activate motor neurons that go back to the quadriceps and cause it to contract, which results in a kick (stereotyped response). The person feels the kicking motion, but it is involuntary, and the person has no sense of having generated it. In this case, the afferent limb is represented by the group Ia fibers and the efferent limb by the motor neurons. The central portion of this arc is minimal (a synapse from the group Ia afferent fibers onto the motor neurons), but in most reflexes, it is more complex and can involve multiple types of interneurons.

It is the predictable linking of stimulus and response that makes reflexes useful tools both for clinicians and for neuroscientists trying to understand spinal cord function. However, one danger to avoid is thinking that a particular neuron's function is solely participation in a particular reflex because these same neurons are the targets of descending motor pathways and are involved in generating voluntary movement. Indeed, many of these neurons are active even when the afferent leg of their reflex arc is silent. One such example is the interneurons of the flexion reflex arc that are also part of the central pattern generator for locomotion.

In the next several sections, three well-known spinal reflexes are discussed in detail because they illustrate important aspects of spinal cord circuitry and function and because of their behavioral and clinical importance. However, many additional reflexes mediated by spinal circuits exist (e.g., see micturition reflex; Fig. 11.3).

The Myotatic or Stretch Reflex

The stretch reflex, as implied by its name, is a group of motor responses elicited by stretch of a muscle. The knee jerk reflex described previously is a well-known example. The stretch reflex is crucial for the maintenance of posture and helps overcome unexpected impediments during a voluntary movement. Changes in the stretch reflex are involved in actions commanded by the brain, and pathological alterations in this reflex are important signs of neurological disease. The phasic stretch reflex occurs in response to rapid, transient stretches of the muscle, such as those elicited by a physician's use of a reflex hammer or by an unexpected impediment to an ongoing movement. The tonic stretch reflex occurs in response to a slower or steady stretch applied to the muscle. The receptor responsible for initiating a stretch reflex is the muscle spindle. Muscle spindles are found in almost all skeletal muscles and are particularly concentrated in muscles that exert fine motor control (e.g., the small muscles of the hand and eye). Thus this reflex circuit essentially is a universal mechanism for helping govern muscle activity.

Structure of the Muscle Spindle

As its name implies, a muscle spindle is a spindle or fusiform-shaped organ composed of a bundle of specialized muscle fibers richly innervated both by sensory axons and by motor axons (Fig. 9.2). A muscle spindle is about 100 µm in diameter and up to 10 mm long. The innervated part of the muscle spindle is encased in a connective tissue capsule. Muscle spindles lie between regular muscle fibers and are typically located near the tendinous insertion of the muscle. The ends of the spindle are attached to the connective tissue within the muscle (endomysium). The key point is that muscle spindles are connected in parallel with the regular muscle fibers and thus are able to sense changes in the length of the muscle.

The muscle fibers within the spindle are called **intrafusal fibers,** to distinguish them from the regular or extrafusal fibers that make up the bulk of the muscle. Individual intrafusal fibers are much narrower than extrafusal fibers and do not run the length of the muscle. Thus they are too weak to contribute significantly to muscle tension or to cause changes in the overall length of the muscle directly by their contraction.

Morphologically, two types of intrafusal muscle fibers are found within muscle spindles: **nuclear bag** and **nuclear chain fibers** (see Fig. 9.2*B*). These names are derived from the arrangement of nuclei in the fibers. (Muscle fibers are formed by the fusion of many individual myoblasts during development; thus mature muscle cells are multinucleate.) Nuclear bag fibers are larger than nuclear chain fibers, and their nuclei are bunched together like a bag of oranges in the central, or equatorial, region of the fiber. In nuclear chain fibers, the nuclei are arranged in a row. Functionally, nuclear bag fibers are divided into two types: bag1 and bag2. As detailed later, bag2 fibers are functionally similar to chain fibers.

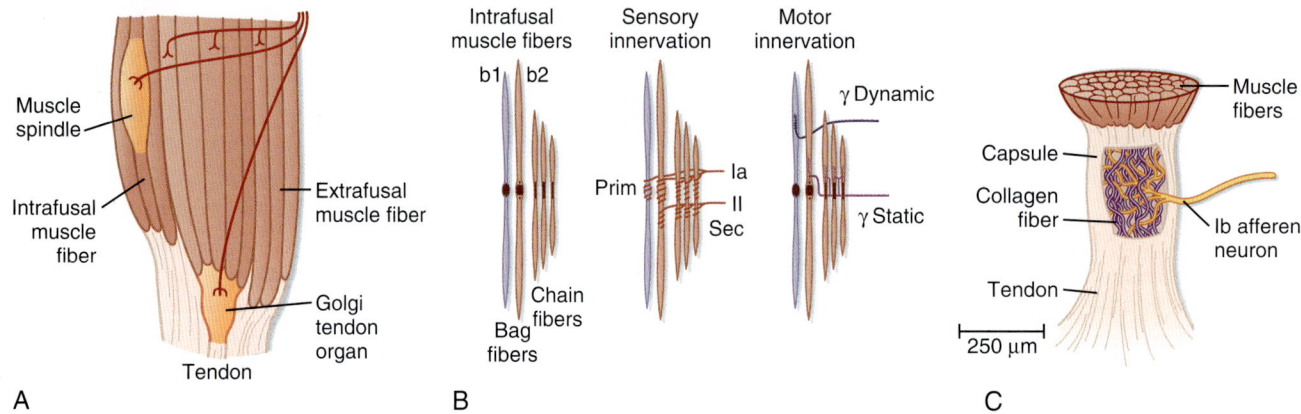

• **Fig. 9.2** Muscle Proprioceptors. Skeletal muscles contain sensory receptors embedded within the muscle (spindles) and within their tendons (Golgi tendon organs). **A,** Schematic view of a muscle, showing the arrangement of a spindle in parallel with extrafusal muscle fibers and a tendon organ in series with muscle fibers. **B,** Structure and innervation (motor and sensory) of a muscle spindle. **C,** Structure and innervation of a tendon organ.

The neural innervation of an intrafusal fiber differs significantly from that of an extrafusal fiber, which is innervated by a single motor neuron. Intrafusal fibers are multiply innervated and receive both sensory and motor innervation. The sensory innervation typically includes a single group Ia afferent fiber and a variable number of group II afferent fibers (see Fig. 9.2*B*). Group Ia fibers belong to the class of sensory nerve fibers with the largest diameters and conduct at 80 to 120 m/sec; group II fibers are intermediate in size and conduct at 35 to 75 m/sec. A group Ia afferent fiber forms a spiral-shaped termination, referred to as a *primary ending,* on each of the intrafusal muscle fibers in the spindle. Thus primary endings are found on both types of nuclear bag fibers and on nuclear chain fibers. The group II afferent fiber forms a secondary type ending on nuclear chain and bag2 fibers, but not on bag1 fibers. The primary and secondary endings have mechanosensitive channels that are sensitive to the level of tension on the intrafusal muscle fiber.

The motor supply to a muscle spindle consists of two types of γ motor axons (see Fig. 9.2*B*). Dynamic γ motor axons end on nuclear bag1 fibers, and static γ motor axons end on nuclear chain and bag2 fibers.

Muscle Spindles Detect Changes in Muscle Length

Muscle spindles respond to changes in muscle length because they lie in parallel with the extrafusal fibers and therefore are also stretched or shortened along with the extrafusal fibers. Because intrafusal fibers, like all muscle fibers, display spring-like properties, a change in their length changes the tension that they are under, and this change is sensed by mechanoreceptors of the group Ia and group II spindle afferent fibers. The nonselective cation channel Piezo2 has been identified as the principal transduction channel that allows spindle sensory afferent fibers to sense changes in mechanical stress that occur when a muscle changes length.

Fig. 9.3 shows the changes in activity of the afferent fibers of a muscle spindle when the muscle is stretched. It

is clear that group Ia and group II fibers respond differently to stretch. Group Ia fibers are sensitive both to the amount of muscle stretch and to its rate, whereas group II fibers respond chiefly to the amount of stretch. Thus when a muscle is stretched to a new longer length, group II firing increases in proportion to the amount of stretch (see Fig. 9.3, *left*), and when the muscle is allowed to shorten, its firing rate decreases proportionately (see Fig. 9.3, *right*). Group Ia fibers show this same **static-type response,** and thus under steady-state conditions (i.e., constant muscle length), their firing rate reflects the amount of muscle stretch, similar to that of group II fibers.

While muscle length is changing, however, group Ia firing also reflects the rate of stretch or shortening that the muscle is undergoing. Its activity overshoots during muscle stretch and undershoots (and possibly ceases) during muscle shortening. These are called **dynamic responses.** This dynamic sensitivity also means that the activity of group Ia fibers is much more sensitive to transient and oscillatory stretches, such as shown in the middle diagrams of Fig. 9.3. In particular, the tap profile is what occurs when a reflex hammer is used to hit the muscle tendon and thereby cause a brief stretching of the attached muscle. The change in muscle length is too brief for significant changes in group II firing to occur, but because the magnitude of the rate of change (slopes of the tap profile) is so high with this stimulus, large dynamic responses are elicited in the group Ia fibers. Thus the functionality of reflex arcs involving group Ia afferent fibers is what is being assessed when a reflex hammer is used to tap on tendons.

γ Motor Neurons Adjust the Sensitivity of the Spindle

Up to this point, we have described only how muscle spindles behave when there are no changes in γ motor neuron activity. The efferent innervation of muscle spindles is extremely important, however, because it determines the sensitivity of muscle spindles to stretch. For example, in Fig. 9.4*A,* the activity of a muscle spindle afferent fiber is shown during

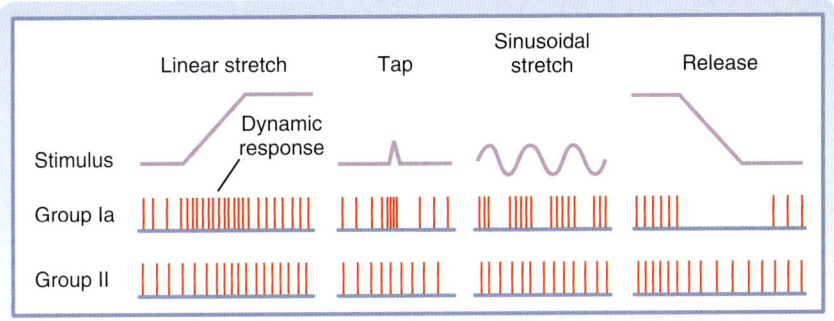

• **Fig. 9.3** Responses of a Primary Ending (Group Ia) and a Secondary Ending (Group II) to Changes in Muscle Length. Note the difference in dynamic and static responsiveness of these endings. The waveforms at the top represent the changes in muscle length. The middle and bottom rows show the discharges of a group Ia fiber and a group II fiber, respectively, during the various changes in muscle length.

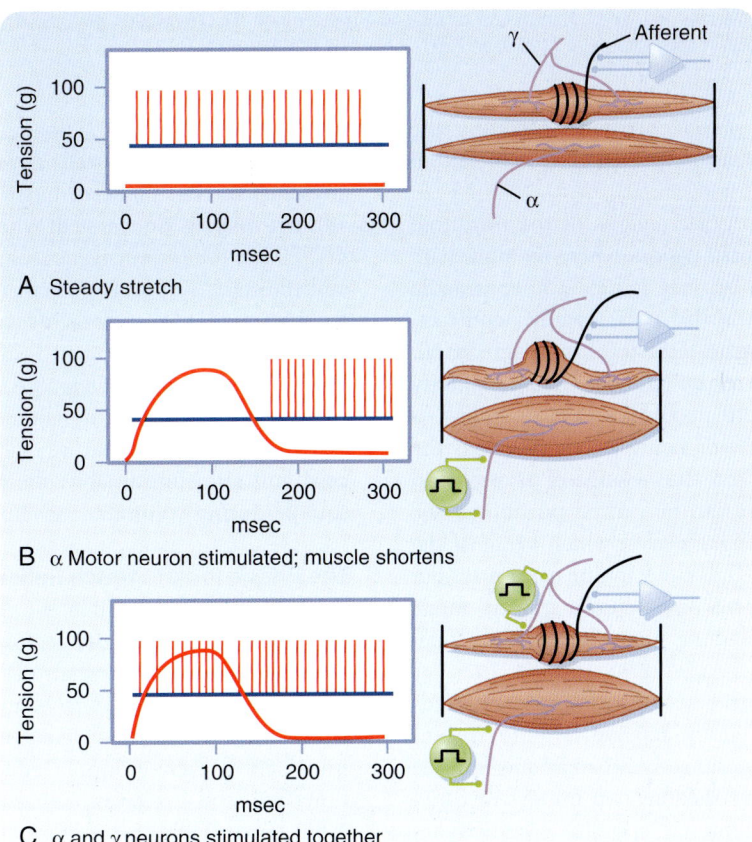

• **Fig. 9.4** The Activity of γ Motor Neurons Can Counteract the Effects of Unloading on the Discharge of a Muscle Spindle Afferent Fiber. **A,** The activity of a muscle spindle afferent fiber during steady stretch. **B,** Stimulation of α motor neuron at 0 msec causes contraction of the extrafusal fibers, which leads to muscle shortening and increased muscle tension but unloading of the tension across the muscle spindle, which in turn induces the afferent fiber to stop firing. Upon relaxation, the muscle returns to its original length, and tension is restored on the intrafusal fibers, causing the return of activity in the group Ia afferent fiber. **C,** Coactivation of α and γ motor neurons causes shortening of both extrafusal and intrafusal fibers. Thus there is no unloading of the spindle, and the afferent fiber maintains its spontaneous activity. (Redrawn from Kuffler SW, Nicholls JG. *From Neuron to Brain.* Sunderland, MA: Sinauer; 1976.)

a steady stretch. If only the extrafusal muscle fibers were to contract (this can be done experimentally by selective stimulation of α motor neurons; see Fig. 9.4*B*), the muscle spindle would be unloaded by the resultant shortening of the muscle. If this happens, the muscle spindle afferent fiber

may stop discharging and become insensitive to further decreases in muscle length. However, the unloading of the spindle can be prevented if α and γ motor neurons are stimulated simultaneously. Such combined stimulation causes the intrafusal muscle fibers of the spindle to shorten

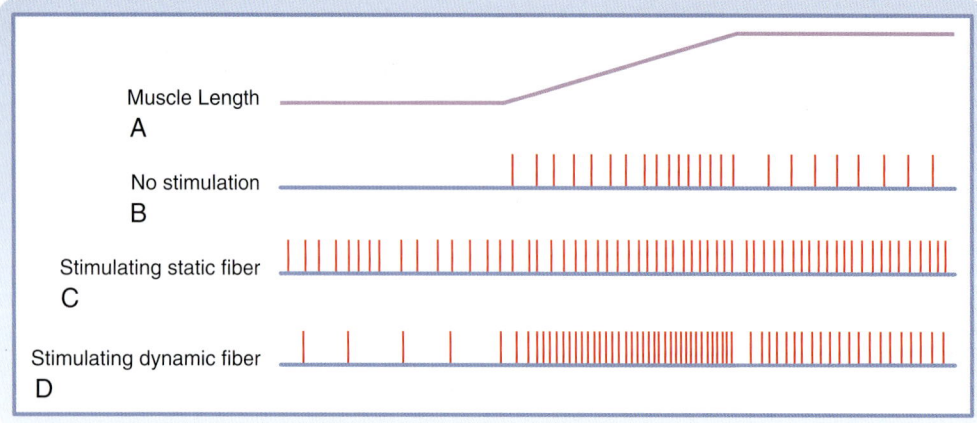

- **Fig. 9.5** Effects of Static and Dynamic γ Motor Neurons on the Responses of a Primary Ending to Muscle Stretch. **A,** The time course of the stretch. **B,** The discharge of group Ia fibers in the absence of γ motor neuron activity. **C,** Stimulation of a static γ motor axon. **D,** Stimulation of a dynamic γ motor axon. (Redrawn from Crowe A, Matthews PBC. *J Physiol.* 1964;174:109.)

along with the extrafusal muscle fibers, maintaining the baseline tension on the equatorial portion of the intrafusal fibers (see Fig. 9.4*C*).

Note that only the two polar regions of the intrafusal muscle contract; the equatorial region, where the nuclei are located, does not contract because it has little contractile protein. Nevertheless, when the polar regions contract, the equatorial region elongates and regains its sensitivity. Conversely, when a muscle relaxes (α motor neuron activity drops) and thus elongates (if its ends are being pulled), a concurrent decrease in γ motor neuron activity allows the intrafusal fibers to relax (and thus elongate) as well and thereby prevent the tension on the central portion of the intrafusal fiber from reaching a level at which firing of the afferent fibers is saturated. Thus the γ motor neuron system allows the muscle spindle to operate over a wide range of muscle lengths while retaining high sensitivity to small changes in length.

For voluntary movements, descending motor commands from the brain in fact typically activate α and γ motor neurons simultaneously, presumably to maintain spindle sensitivity as just described. This has two important functions. By maintaining the muscle spindle's sensitivity as the muscle changes length, the spindle remains capable of sensing, and signaling to the CNS, any disturbances to the ongoing movement that cause an unexpected stretch of the muscle, and this in turn allows the CNS to initiate both reflex (see next section) and voluntary corrections. Second, if the spindle were to become unloaded during the movement, this would oppose the intended movement by decreasing the excitatory drive, via the group Ia reflex arc (see next section), to the α motor neurons driving the agonist muscles.

As mentioned earlier, there are two types of γ motor neurons: dynamic and static (see Fig. 9.2). This allows the CNS to have very precise control over the sensitivity of the muscle spindle. Dynamic γ motor axons end on nuclear bag1 fibers, and static γ motor axons synapse on

nuclear chain and bag2 fibers. Thus when a dynamic γ motor neuron is activated, the response of the group Ia afferent fiber is enhanced, but the activity of the group II afferent fibers is unchanged; when a static γ motor neuron discharges, the responsiveness of the group II afferent fibers and the static responsiveness of the group Ia afferent fibers are increased. The effects of stimulating the static and dynamic fibers on a group Ia afferent fiber's response to stretch are illustrated in Fig. 9.5. Descending pathways can preferentially influence dynamic or static γ motor neurons and thereby alter the nature of reflex activity in the spinal cord and also, presumably, the functioning of the muscle spindle during voluntary movements.

The Phasic (or Ia) Stretch Reflex

The reflex arc responsible for the phasic stretch reflex is depicted in Fig. 9.6; the rectus femoris muscle serves as an example. A rapid stretch of the rectus femoris muscle strongly activates the group Ia fibers of the muscle spindles, which then convey this signal into the spinal cord. In the spinal cord, each group Ia afferent fiber branches many times to form excitatory synapses directly (monosynaptically) on virtually all α motor neurons that supply the same (also known as the *homonymous*) muscle and with many α motor neurons that innervate synergists, such as the vastus intermedius muscle in this case, which also acts to extend the leg at the knee. If the excitation is powerful enough, the motor neurons discharge and cause a contraction of the muscle. Note that the group Ia fibers do not contact the γ motor neurons, possibly to avoid a positive-feedback loop situation. This selective targeting of α motor neurons is exceptional in that most other reflex and descending pathways target both α and γ motor neurons.

Other branches of group Ia fibers end on a variety of interneurons; however, one type, the reciprocal Ia inhibitory interneuron (black cell in Fig. 9.6), is particularly important with regard to the stretch reflex. These interneurons are identifiable because they are the only inhibitory interneurons

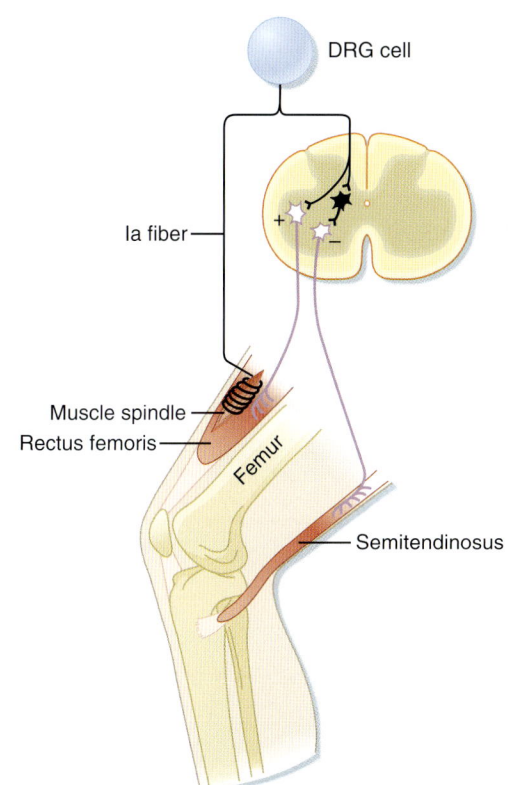

● Fig. 9.6 Reflex Arc of the Stretch Reflex. The pathway back to the rectus femoris in this arc contains a single synapse within the central nervous system; hence, it is a monosynaptic reflex. The interneuron, shown in *black,* is a group Ia inhibitory interneuron. DRG, dorsal root ganglion.

that receive input from both the group Ia afferent fibers and Renshaw cells (see Fig. 9.12). They end on α motor neurons that innervate the antagonist muscles—in this case, the hamstring muscles, including the semitendinosus muscle—which act to flex the knee. Other branches of the group Ia afferent fibers synapse with yet other neurons that originate ascending pathways that provide various parts of the brain (particularly the cerebellum and cerebral cortex) with information about the state of the muscle.

The organization of the stretch reflex arc guarantees that one set of α motor neurons is activated and the opposing set is inhibited. This arrangement is known as **reciprocal innervation.** Although many reflexes involve such reciprocal innervation, this type of innervation is not the only possible organization of a motor control system; descending motor pathways can override such patterns.

The stretch reflex is quite powerful, in large part because of its monosynaptic nature. The power of this reflex also derives from the essentially maximal convergence and divergence that exist in this pathway, which is not apparent from the circuit diagrams, such as Fig. 9.6, that are typically used to illustrate reflex pathways. That is, each group Ia fiber contacts virtually all homonymous α motor neurons, and each such α motor neuron receives input from every spindle in that muscle. Although its monosynaptic nature makes the group Ia reflex rapid and powerful, it also means that there is relatively little opportunity for direct control of

activity flow through its reflex arc. The CNS overcomes this problem by controlling muscle spindle sensitivity via the γ motor neuron system as described previously.

The Tonic Stretch Reflex

The tonic stretch reflex can be elicited by passive bending of a joint. This reflex circuit includes both group Ia and group II afferent fibers from muscle spindles. Group II fibers make monosynaptic excitatory connections with α motor neurons, but they also excite them through disynaptic and polysynaptic pathways. Normally, there is ongoing activity in the group Ia and group II afferent fibers that helps maintain a baseline rate of firing of α motor neurons; therefore, the tonic stretch reflex contributes to muscle tone. Its activity also contributes to the ability to maintain a posture. For example, if the knee of a soldier standing at attention begins to flex because of fatigue, the quadriceps muscle is stretched, a tonic stretch reflex is elicited, and the quadriceps contracts more, thereby opposing the flexion and restoring the posture.

The foregoing discussion suggests that stretch reflexes can act like a negative-feedback system to control muscle length. By following the stretch reflex arc, it is possible to see that changes in its activity act to oppose changes in muscle length from a particular equilibrium point. For example, if the muscle's length is increased, there will be an increase in firing by group Ia and group II fibers, which excites homonymous α motor neurons and leads to contraction of the muscle and reversal of the stretch. Similarly, passive shortening of the muscle unloads the spindles and leads to a decrease in the excitatory drive to the motor neurons and thus relaxation of the muscle. So how are humans able to rotate their joints? It is partly because the γ motor neurons are coactivated during a movement and thereby shift the equilibrium point of the spindle and partly because the gain or strength of the reflex is low enough that other input to the motor neuron can override the stretch reflex.

Inverse Myotatic or Group Ib Reflex

The inverse myotatic reflex acts to oppose changes in the level of force in the muscle. Just as the stretch reflex can be thought of as a feedback system to regulate muscle length, the inverse myotatic, or group Ib, reflex can be thought of as a feedback system to help maintain force levels in a muscle. With the upper part of the leg as an example, the group Ib reflex arc is depicted in Fig. 9.7.

The arc starts with the Golgi tendon organ receptor, which senses the tension in the muscle. Golgi tendon organs are located at the junction of the tendon and the muscle fibers and thus lie in series with the muscle fibers, in contrast to the parallel arrangement of the muscle spindles (see Fig. 9.2). Golgi tendon organs have a diameter of about 100 μm and a length of about 1 mm. A Golgi tendon organ is innervated by the terminals of group Ib afferent fibers. These terminals wrap about bundles of collagen fibers in the tendon of a muscle (or in tendinous inscriptions within the muscle).

⚕ IN THE CLINIC

Hyperactive stretch reflexes can lead to tremors and clonus, which are types of involuntary rhythmic movements. Although the negative-feedback action of the stretch reflex can help stabilize the limb at a particular position, if an external perturbation to the limb occurs, the conduction delay between the initiating stimulus (muscle stretch) and the response (muscle contraction) can cause the stretch reflex circuit to be a source of instability that leads to rhythmic movements. Specifically, clonus is elicited by a sustained stretch of a muscle in a person who has spinal cord damage. Normally, an imposed sustained stretch on a muscle elicits an increase in group Ia and group II fiber activity, which after a delay causes a contraction in the muscle that opposes the stretch but does not completely return the muscle to its initial length because the gain of the stretch reflex is much less than 1.*

 This partial compensation, in turn, leads to a decrease in group Ia and group II fiber activity, which causes the limb to lengthen again, but not fully. This lengthening once again increases group Ia and group II fiber activity, and so on. The delay is essential in setting up this oscillation because it causes the feedback signal to continue even after the muscle has compensated and thus results in an overcompensation that leads to the next overcorrection. However, because the reflex gain is normally much less than 1, this oscillation normally dies out quickly (the overcompensations decrease in amplitude rapidly), and the muscle comes to rest at an intermediate length. In contrast, when descending motor pathways are damaged, the resulting changes in spinal cord connectivity and increases in neuronal excitability result in a hyperactive reflex (which is equivalent to raising the gain of the stretch reflex close to 1). In this case, the successive overcompensations are much larger, and an overt but transient oscillation can be observed (clonus). If the gain equals 1, the clonus does not die out but rather persists for as long as the initial stretch stimulus is maintained.

*In general, *gain* of a system is defined as its output for a given input. In this case, the input to the system is the imposed stretch, and the output is movement caused by the stretch reflex–evoked contraction.

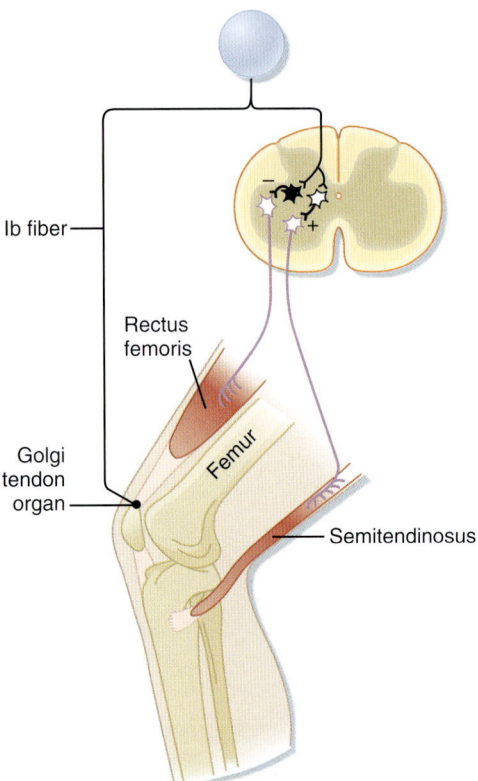

• **Fig. 9.7** Reflex Arc of the Inverse Myotatic Reflex. The interneurons include both excitatory (*white*) and inhibitory (*black*) interneurons. This is an example of a disynaptic reflex.

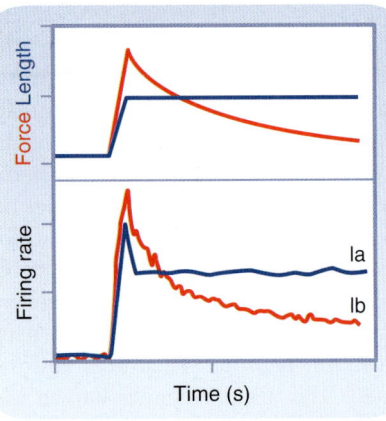

• **Fig. 9.8** Changes in Group Ia and Group Ib Firing Rates When Muscle Is Stretched to a New Length. After a transient burst, the firing rate of the group Ia fiber remains constant at a new higher level that is proportional to the increase in length (compare *blue lines in upper and lower graphs*). In contrast, the group Ib fiber shows an initial rapid increase in firing followed by a slow decrease back toward its original level (*lower graph, red line*) and has a firing profile that matches the tension level in the muscle caused by the stretch (*upper graph, red line*).

Because of their in-series relationship to the muscle, Golgi tendon organs can be activated either by muscle stretch or by muscle contraction. In both cases, however, the actual stimulus sensed by the Golgi tendon organ is the force that develops in the tendon to which it is linked. Thus the response to stretch is the result of the spring-like nature of the muscle (i.e., by Hooke's law, the force on a spring is proportional to how much it is stretched).

To distinguish between the responsiveness of the muscle spindles and Golgi tendon organs, the firing patterns of group Ia and group Ib fibers can be compared when a muscle is stretched and then held at a longer length (Fig. 9.8). The firing rate of the group Ia fibers maintains its increase until the stretch is reversed. In contrast, the group Ib fiber shows an initial large increase in firing, reflecting the increased tension on the muscle caused by the stretch, but then shows a gradual return toward its initial firing rate as the tension on the muscle is lowered because of cross-bridge recycling and the resultant lengthening of the

sarcomeres. Therefore, Golgi tendon organs signal force, whereas spindles signal muscle length. Further evidence of this distinction is that group Ib firing is correlated with force level during isometric contraction even though muscle length and therefore group Ia activity are unchanged.

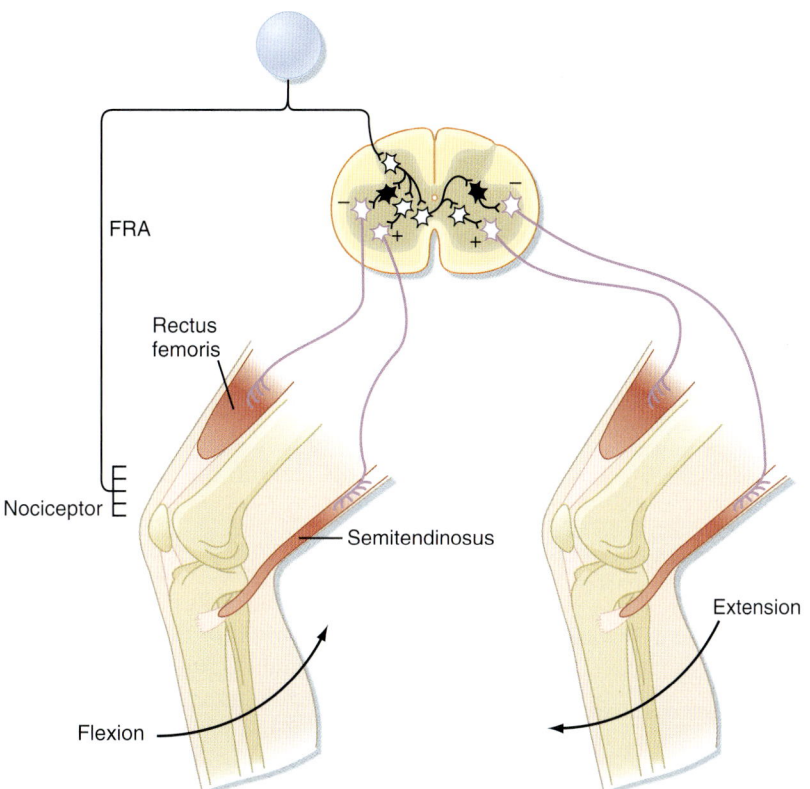

• **Fig. 9.9** The Reflex Arc of the Flexion Reflex. *Black* interneurons are inhibitory, and *white* ones are excitatory. FRA, flexion reflex afferent fiber.

The group Ib afferent fibers branch as they enter the spinal cord and end on interneurons. There are no monosynaptic connections to α motor neurons. Rather, the group Ib afferent fibers synapse onto two classes of interneurons: interneurons that inhibit α motor neurons that supply the homonymous muscle (in this case the rectus femoris muscle) and excitatory interneurons that activate α motor neurons to the antagonist (the semitendinosus muscle). Because there are two synapses in series in the CNS, this is a disynaptic reflex arc. Because of these connections, group Ib fiber activity should have the opposite action of the group Ia stretch reflex during passive stretch of the muscle, which explains the group Ib reflex's other name, the inverse myotatic reflex.

Functionally, however, the two reflex arcs can act synergistically, as the following example shows. Recall that the Golgi tendon organs monitor force levels across the tendon that they supply. If during maintained posture (such as standing at attention) knee extensors (such as the rectus femoris muscle) begin to fatigue, the force pulling on the patellar tendon declines. The decline in force reduces the activity of Golgi tendon organs in this tendon. Because the group Ib reflex normally inhibits the α motor neurons to the rectus femoris muscle, reduced activity of the Golgi tendon organs enhances the excitability of (i.e., disinhibits) the α motor neurons and thereby helps reverse the decrease in force caused by the fatigue. Simultaneously, bending of the knee stretches the knee extensors and activates the

afferent fibers from the muscle spindles, which then excite the same α motor neurons. Thus coordinated action of afferent fibers from both the muscle spindle and Golgi tendon organ help oppose the decrease in contraction of the rectus femoris muscle due to fatigue and thereby work together to maintain the standing posture.

Flexion Reflexes and Locomotion

The flexion reflex starts with activation of one or more of a variety of sensory receptors, including nociceptors, whose signals can be carried to the spinal cord via a variety of afferent fibers, including group II and group III fibers, collectively called the **flexion reflex afferent (FRA)** fibers. In flexion reflexes, afferent volleys (1) cause excitatory interneurons to activate the α motor neurons that supply the flexor muscles in the ipsilateral limb and (2) cause inhibitory interneurons to inhibit the α motor neurons that supply the antagonistic extensor muscles (Fig. 9.9). This pattern of activity causes one or more joints in the stimulated limb to flex. In addition, commissural interneurons evoke the opposite pattern of activity in the contralateral side of the spinal cord (see Fig. 9.9), which results in extension of the opposite limb, the **crossed extension reflex.** For lower limbs in humans (or for both forelimbs and hind limbs in quadrupeds), the crossed extension part of the reflex helps in maintaining balance by enabling the contralateral limb to be able to support the additional load that is transferred to it when the flexed limb is lifted.

IN THE CLINIC

After damage to the descending motor pathways, hyperactive stretch reflexes may result in spasticity, in which there is large resistance to passive rotation of the limbs. In this condition, it may be possible to demonstrate what is called the **clasp-knife reflex.** When spasticity is present, attempts to rotate a limb about a joint initially meet high resistance. However, if the applied force is increased, there comes a point at which the resistance suddenly dissipates and the limb rotates easily. This change in resistance is caused by reflex inhibition. The group Ib reflex arc suggests that rising activity in this pathway could underlie the sudden release of resistance, and indeed, the clasp-knife reflex was once attributed to the activation of Golgi tendon organs when these receptors were thought to have a high threshold to muscle stretch. However, the tendon organs have since been shown to be activated at very low levels of force and are no longer thought to cause the clasp-knife reflex. It is now thought that this reflex is caused by the activation of other high-threshold muscle receptors that supply the fascia around the muscle. Signals from these receptors cause the activation of interneurons that lead to inhibition of the homonymous motor neurons.

Because flexion typically brings the affected limb in closer to the body and away from a painful stimulus, flexion reflexes are a type of withdrawal reflex. In Fig. 9.9, the neural circuit of the flexion reflex is shown for neurons that affect only the knee joint. Actually, however, considerable divergence of the primary afferent and interneuronal pathways occurs in the flexion reflex. In fact, all the major joints of a limb (e.g., hip, knee, and ankle) may be involved in a strong flexor withdrawal reflex. Details of the flexor withdrawal reflex vary, depending on the nature and location of the stimulus.

The interneurons subserving flexion reflexes also appear to be part of the **central pattern generator (CPG)** for generating locomotion and thus are an example of how reflex circuits are used for multiple purposes. A CPG is a set of neurons and circuits capable of generating the rhythmic activity that underlies motor acts, even in the absence of sensory input. For example, activation of the FRA interneurons leads to a pattern of flexor excitation and extensor inhibition on one side and the converse pattern on the opposite side, and if the FRA interneurons on each side of the spinal cord alternated in being active, a stepping pattern would emerge. That is, walking motion could result from alternately activating the FRA interneurons on each side. Note that such a rhythmic activity pattern in the FRA circuits need not be dependent on activity from the FRA fibers themselves (e.g., they could be activated by descending pathways from the brain).

To show that these circuits are actually involved in generating the locomotion rhythm, spinal cord preparations were made that showed spontaneous locomotion (i.e., if the

brainstem is transected and weight is supported, the spinal cord circuits can generate activity that causes the limbs to perform a normal locomotion sequence). In one such preparation, the electromyographic signals from the flexors and extensors of a limb were recorded, and the FRA fibers then stimulated to demonstrate the effect on locomotion rhythm (Fig. 9.10). Before any stimulus, a spontaneous alternating pattern of flexor and extensor electromyographic (EMG) activity exists. If the FRA fibers were not involved in the locomotion circuit, or at least were not a critical part of the circuits responsible for generating the rhythm (see Fig. 9.10B), the stimulus would be expected to produce only a transient response (i.e., a single burst in the flexor EMG record and brief inhibition of activity in the extensor EMG record) but to have no long-term effect on ongoing EMG pattern. Such a transient response is observed (see Fig. 9.10A; EMG records just after the stimulus). However, the stimulus also causes a permanent, approximately 180-degree phase shift in locomotor rhythm, as can shown from a comparison of the times of contractions before and after the stimulus. The dashed vertical lines indicate the times at which a flexor EMG response would be expected if the stimulus had produced no phase shift from the EMG activity pattern. Before the stimulus, each vertical line is aligned with the onset of a flexor EMG burst, whereas after the stimulus, each vertical line occurs at the end of the flexor burst. Therefore, the stimulus affected the locomotor CPG itself and the FRA interneurons are a critical part of this CPG (see Fig. 9.10C).

A second important point illustrated by this experiment is that the locomotion CPG (and CPGs in general) can be influenced by strong afferent fiber activity. The afferent fiber's influence ensures that the pattern generator adapts to changes in the terrain as locomotion proceeds. Such changes may occur rapidly during running, and locomotion must then be adjusted to ensure proper coordination.

Determining Spinal Cord Organization Through the Use of Reflexes

Convergence and divergence are important aspects of reflex pathways and of neuronal circuits in general. Several examples of these phenomena have been described in the previous discussion of the reflexes. Reflexes can be used to identify and characterize these phenomena in the spinal cord. For example, convergent input can be demonstrated through the phenomenon of **spatial facilitation,** which is illustrated in Fig. 9.11.

In this example, a monosynaptic reflex is elicited by electrical stimulation of the group Ia fibers in each of two nerves (see Fig. 9.11A). The reflex response is characterized by a recording of the discharges of α motor axons from the appropriate ventral root (as a compound action potential). When nerve A is stimulated, a small compound action potential is recorded as reflex A. Similarly, when nerve B is stimulated, reflex B is recorded. Fig. 9.11B depicts the motor neurons contained within the motor nucleus.

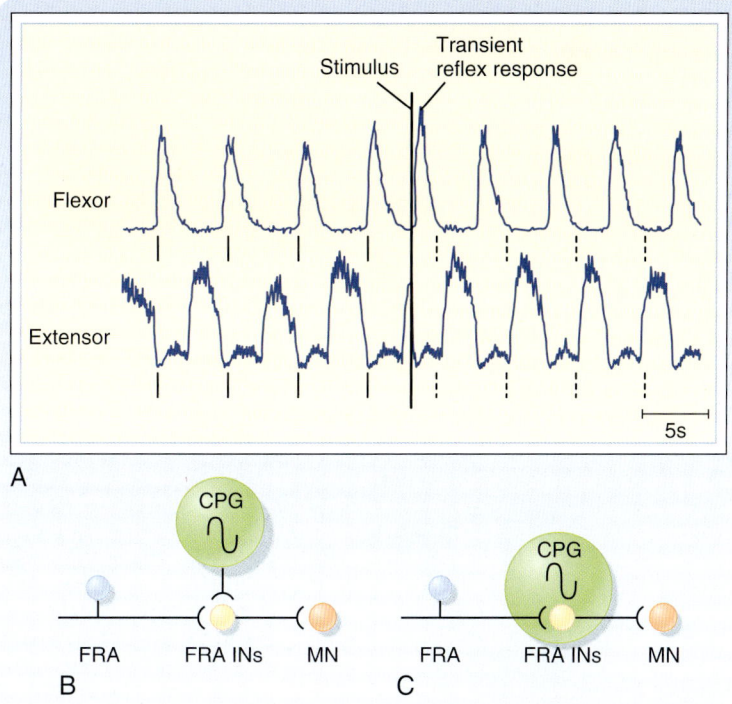

• Fig. 9.10 Phase Reset of Locomotion Rhythm by Flexion Reflex Afferent (FRA) Fiber Stimulation Helps Identify Neuronal Components of the Underlying Central Pattern Generator (CPG). **A,** EMG records from knee flexor and extensor muscles. Note the rhythmic alternating pattern before application of the stimulus. The *solid vertical lines* below each trace indicate the times at which flexor contraction is initiated. The *dashed vertical lines* indicate the times at which flexor contraction would have been initiated if the stimulus caused no lasting effect on the rhythmic pattern. **B** and **C,** Two possible models for the CPG underlying the locomotor rhythm depicted in **A. B,** The FRA interneurons (INs) in the CPG are not shown. **C,** The FRA interneurons are shown. The data shown in **A** support the model shown in **C.** MN, motoneuron. (Data from Hultborn H, et al. *Ann N Y Acad Sci.* 1998;860:70.)

The α motor neurons in the discharge zones are activated above threshold when each nerve branch is stimulated separately. Thus a distinct pair of α motor neurons spike when each nerve is stimulated alone. In addition, each of these motor neuron pairs is surrounded by a subliminal fringe of eight additional motor neurons that are excited but not sufficiently to trigger spikes. When the two nerves are stimulated at the same time, a much larger reflex discharge is recorded (compare R_A and R_B with $R_A + R_B$ recordings at the right of Fig. 9.11B). As the figure demonstrates, this reflex represents the discharge of seven α motor neurons: the four that spiked after the singular stimulation of each nerve (two per nerve) and three additional α motor neurons (located in the facilitation zone) that are made to discharge only when the two nerves are stimulated simultaneously because they lie in the subliminal fringe for both nerves.

A similar effect could be elicited by repetitive stimulation of one of the nerves, provided that the stimuli occur close enough together that some of the excitatory effect of the first volley still persists after the second volley arrives. This effect is called **temporal summation.** Both spatial summation and temporal summation depend on the properties of the excitatory postsynaptic potentials evoked in α motor neurons by the group Ia afferent fibers (see Fig. 6.8).

Convergence can also lead to inhibitory interactions between stimuli, a phenomenon called **occlusion.** If a volley in one of the two nerves in Fig. 9.11 reaches the motor nucleus at a time when the motor neurons are highly excitable, the reflex discharge is relatively large (see Fig. 9.11C). A similar volley in the other nerve might also produce a large reflex response. However, when the two nerves are excited simultaneously, the reflex can be less than the sum of the two independently evoked reflexes if the cells reaching threshold to activation of either of the two nerves alone overlap significantly. In this case, each afferent nerve activates 7 α motor neurons, but the volleys in the two nerves together cause only 12 α motor neurons to discharge because two motor neurons lie in the individual discharge zones of both afferent nerves.

The phenomena of spatial and temporal summation and occlusion can also be used to demonstrate interactions between spinal cord neurons and the various reflex circuits. To start, a monosynaptic reflex discharge can be evoked by stimulation of the group Ia afferent fibers in a muscle nerve. This is a test of the reflex excitability of a population of α motor neurons. The discharges of either extensor or flexor α motor neurons can be recorded if the proper muscle nerve to be stimulated is chosen. Other kinds of afferent fibers

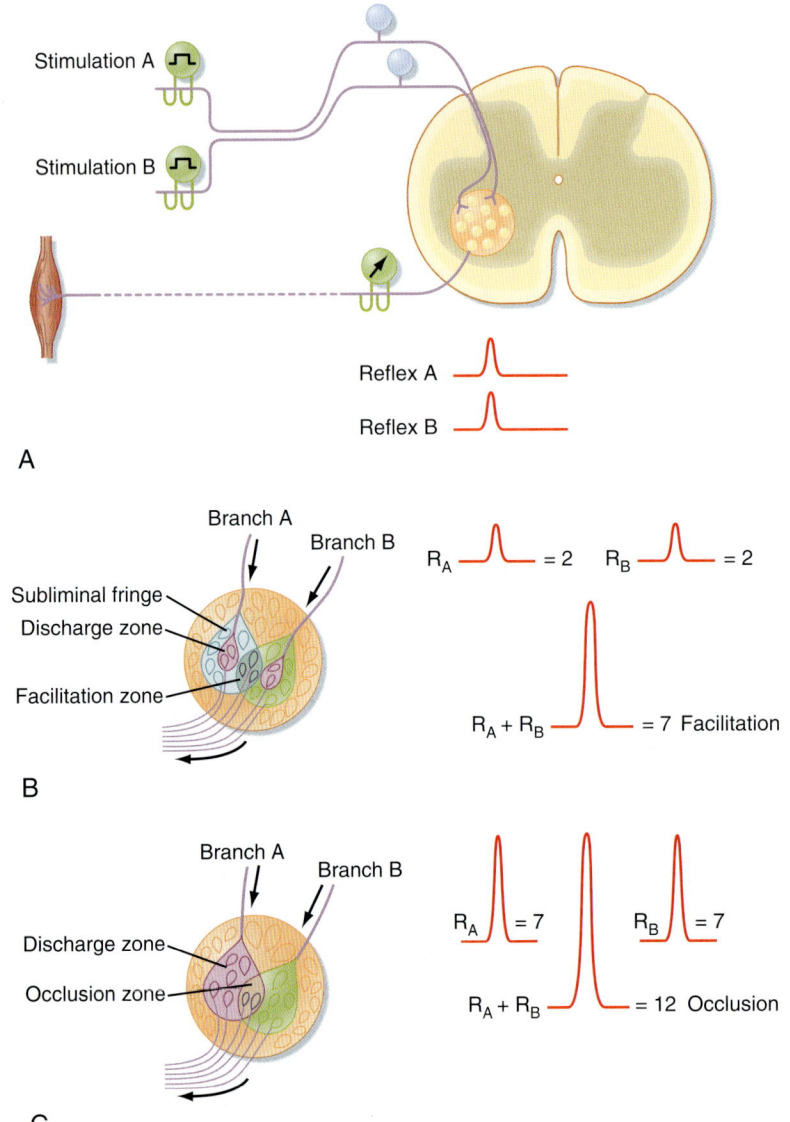

Stimulation A

Stimulation B

Reflex A

Reflex B

A

Branch A
Branch B

Subliminal fringe
Discharge zone

Facilitation zone

R_A = 2 R_B = 2

$R_A + R_B$ = 7 Facilitation

B

Branch A
Branch B

Discharge zone

Occlusion zone

R_A = 7 R_B = 7

$R_A + R_B$ = 12 Occlusion

C

• **Fig. 9.11** Spatial Facilitation. **A,** Arrangement for using electrically evoked afferent volleys and recordings from motor axons in a ventral root to study reflexes. **B,** Experiment in which combined stimulation of afferent fibers in two muscle nerves resulted in spatial summation (R_A and R_B). The discharge zones *(pink areas)* enclose α motor neurons that are activated above threshold when each nerve branch is stimulated separately. In **C,** the combined volleys caused occlusion. (Redrawn from Eyzaguirre C, Fidone SJ. *Physiology of the Nervous System.* 2nd ed. Chicago: Mosby–Year Book; 1975.)

are then stimulated along with the homonymous group Ia afferent fibers from the muscle to demonstrate whether the response to the group Ia stimulation changes. For example, stimulation of group Ia afferent fibers in the nerve to the antagonist muscles produces inhibition of the response to the homonymous group Ia stimulation (which is mediated by the reciprocal group Ia inhibitory interneuron described previously).

As another example, if the small afferent fibers of a cutaneous nerve are stimulated to evoke a flexion reflex, the responses to group Ia stimulation of the α motor neurons that innervate the extensor muscles are inhibited (and those of α motor neurons that innervate flexor muscles are potentiated).

As a final example, stimulation of a ventral root causes inhibition of group Ia responses and inhibits the reciprocal group Ia inhibition. Because the ventral root contains only motor neuron axons, this result implies the presence of axon collaterals that excite inhibitory interneurons that feed back onto the same motor neuron population (Fig. 9.12). These interneurons are named **Renshaw cells**. Because ventral root stimulation also inhibits the group Ia inhibition of antagonist motor neurons, but no other classes of interneurons, the reciprocal group Ia interneurons are uniquely inhibited by ventral root stimulation (and activated by group Ia stimulation). Experiments like these described have been used to provide a detailed knowledge of the circuitry of the spinal cord.

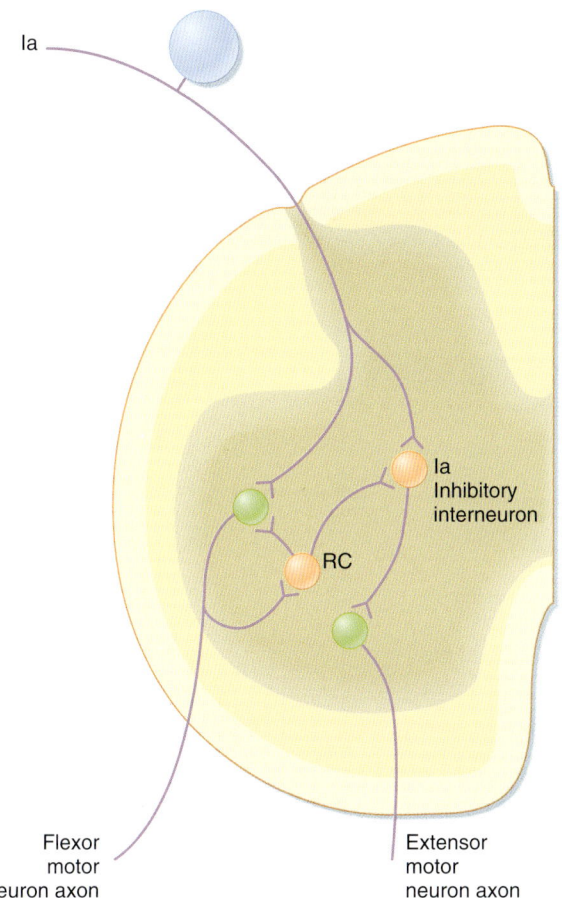

la

la
Inhibitory
interneuron

RC

Flexor
motor
neuron axon

Extensor
motor
neuron axon

• **Fig. 9.12** Renshaw Cell (RC) Connections With Motor Neurons and Group Ia Inhibitory Interneurons. The circuits shown mediate group Ia reciprocal inhibition of antagonist muscles (in this case, an extensor) and inhibition of this reciprocal inhibition by Renshaw cells. Note that equivalent numbers of Renshaw cells and group Ia inhibitory interneurons are associated with extensor motor neurons and group Ia input from spindles in extensor muscles, but they are not shown for simplicity. *Orange cells* are inhibitory, and *blue* and *green cells* are excitatory.

Descending Motor Pathways

Classification of Descending Motor Pathways

Descending motor pathways were traditionally subdivided into **pyramidal** and **extrapyramidal pathways.** This terminology reflects a clinical dichotomy between pyramidal tract disease and extrapyramidal disease. In pyramidal tract disease, the **corticospinal** (pyramidal) tract is interrupted. The signs of this disease were originally attributed to the loss of function of the pyramidal tract (so named because the corticospinal tract passes through the medullary pyramid). However, in many cases of pyramidal tract disease, the functions of other pathways are also altered, and most signs of pyramidal tract disease (see the later section "Motor Deficits Caused by Lesions of Descending Motor Pathways") are apparently not caused solely by loss of the corticospinal tract but also reflect damage to additional motor pathways. The

term *extrapyramidal* is even more problematic. Thus this classification system is not used in this book.

Another way of classifying the motor pathways is based on their sites of termination in the spinal cord and the consequent differences in their roles in the control of movement and posture. The **lateral pathways** terminate in the lateral portions of the spinal cord's gray matter (Fig. 9.13). The lateral pathways can excite motor neurons directly, although interneurons are their main target. They influence reflex arcs that control fine movement of the distal ends of limbs, as well as those that activate supporting musculature in the proximal ends of limbs. The **medial pathways** end in the medial ventral horn on the medial group of interneurons (see Fig. 9.13). These interneurons connect bilaterally with motor neurons that control the axial musculature and thereby contribute to balance and posture. They also contribute to the control of proximal limb muscles. In this book, the terms *lateral* and *medial* are used to classify the descending motor pathways. However, even this terminology is not perfect, partly because although motor neuron cell bodies form localized columns, motor neuron dendritic trees are rather large and typically span most of the ventral horn. Thus any motor neuron can potentially receive input from so-called medial or lateral system pathways.

The Lateral System

Lateral Corticospinal and Corticobulbar Tracts

The corticospinal and corticobulbar tracts originate from a wide region of the cerebral cortex. This region includes the primary motor, premotor, supplementary, and cingulate motor areas of the frontal lobe and the somatosensory cortex of the parietal lobe. The cells of origin of these tracts include both large and small pyramidal cells of layer V of the cortex, including the **giant pyramidal cells of Betz.** Although Betz cells are a defining feature of the primary motor cortex, they represent a small minority (<5%) of the cells that contribute to these tracts, in part because they are found only in the primary motor cortex, and even there they represent a minority of the cells contributing to the tracts. These tracts leave the cortex and enter the internal capsule, then traverse the midbrain in the cerebral peduncle, pass through the basilar pons, and emerge to form the pyramids on the ventral surface of the medulla (see Fig. 9.13*A*). The corticobulbar axons leave the tract as it descends in the brainstem and terminate in the motor nuclei of the various cranial nerves. The corticospinal fibers continue caudally, and in the most caudal region of the medulla, about 90% of them cross to the opposite side. They then descend in the contralateral lateral funiculus as the lateral corticospinal tract. The lateral corticospinal axons terminate at all spinal cord levels, primarily on interneurons, but also on motor neurons. The remaining uncrossed axons continue caudally in the ventral funiculus on the same side as the ventral corticospinal tract, which belongs to the medial system. Many of these fibers ultimately decussate (cross) at the spinal cord level at which they terminate.

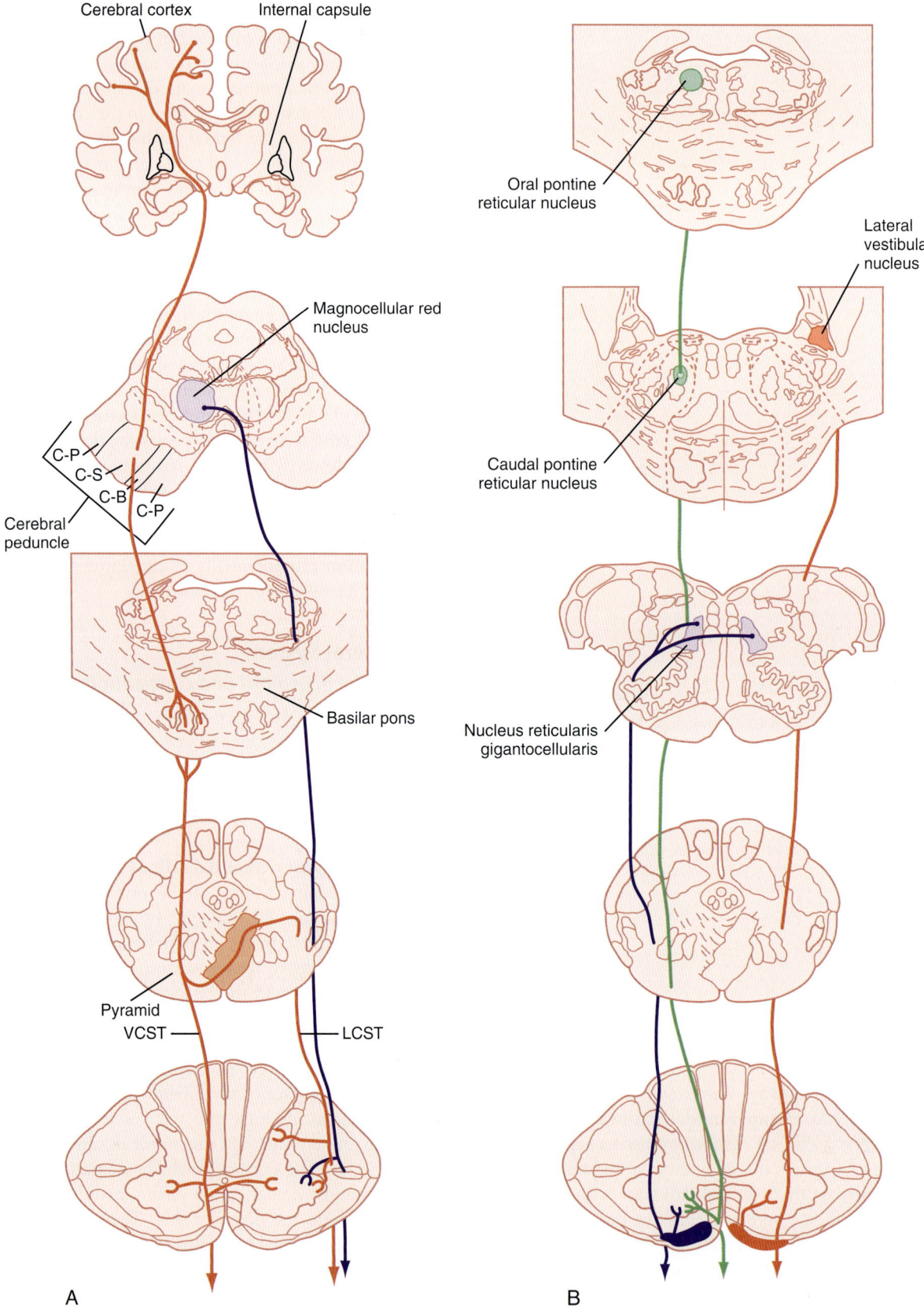

Cerebral cortex   Internal capsule

Oral pontine reticular nucleus

Lateral vestibular nucleus

Magnocellular red nucleus

Caudal pontine reticular nucleus

C-P
C-S
C-B
C-P
Cerebral peduncle

Basilar pons

Nucleus reticularis gigantocellularis

Pyramid
VCST   LCST

A   B

• **Fig. 9.13** Descending Motor Pathways. Major pathways connecting the cortical and brainstem motor areas to the spinal cord are shown. **A,** Lateral system pathways, corticospinal *(red)* and rubrospinal *(blue)* pathways. Note that the ventral corticospinal pathway is part of the medial system but is shown in **A** for simplicity. **B,** Medial system pathways, medullary *(blue)* and pontine *(green)* reticulospinal and lateral vestibulospinal *(red)* pathways. C-B, corticobulbar; C-P, corticopontine; C-S, corticospinal; LCST, lateral corticospinal tract; VCST, ventral corticospinal tract.

The lateral corticospinal tract is a relatively minor tract in lower mammals but is quantitatively and functionally very important in primates, particularly in humans, in which it contains more than 1 million axons. This number still represents a relatively small proportion of the outflow from the cortex because there are approximately 20 million axons in the cerebral peduncles. Nevertheless, the corticospinal pathway is critical for the fine independent control of finger movement, inasmuch as isolated lesions of the corticospinal tract typically lead to a permanent loss of this ability, even though other movement abilities are often recovered with such lesions. Indeed, in primates, corticospinal synapses directly onto motor neurons are particularly prevalent for the motor neurons controlling finger muscles and are probably the basis of the ability to make independent, finely controlled finger movements.

The corticobulbar tract, which projects to the cranial nerve motor nuclei, has subdivisions that are comparable with the lateral and ventral corticospinal tracts. For example, part of the corticobulbar tract ends contralaterally in the portion of the facial nucleus that supplies muscles of the lower part of the face and in the hypoglossal nucleus. This component of the corticobulbar tract is organized like the lateral corticospinal tract. The remainder of the corticobulbar tract ends bilaterally.

Rubrospinal Tract

The rubrospinal tract originates in the magnocellular portion of the red nucleus, which is located in the midbrain tegmentum. These fibers decussate in the midbrain, descend through the pons and medulla, and then take up a position just ventral to the lateral corticospinal tract in the spinal cord. They preferentially affect motor neurons controlling distal musculature, as do the corticospinal fibers. Red nucleus neurons receive input from the cerebellum and from the motor cortex; thus making this an area of integration of activity from these two motor systems.

The Medial System

The ventral corticospinal tract and much of the corticobulbar tract can be regarded as medial system pathways. These tracts end on the medial group of interneurons in the spinal cord and on equivalent neurons in the brainstem. The axial muscles are controlled by these pathways. These muscles often contract bilaterally to provide postural support or some other bilateral function, such as swallowing or wrinkling of the brow.

Other medial system pathways originate in the brainstem. These include the pontine and medullary reticulospinal tracts, the lateral and medial vestibulospinal tracts, and the tectospinal tract.

Pontine and Medullary Reticulospinal Tracts

The cells that give rise to the pontine reticulospinal tract are in the medial pontine reticular formation. The tract descends in the ventral funiculus, and it ends on the ipsilateral medial group of interneurons. Its function is to excite motor neurons to the proximal extensor muscles to support posture.

The medullary reticulospinal tracts arise from neurons of the medial medulla, particularly those of the gigantocellularis reticular nucleus. The tracts descend bilaterally in the ventral lateral funiculus, and they end mainly on interneurons associated with cell groups of medial motor neurons. The function of the pathway is mainly inhibitory.

Lateral and Medial Vestibulospinal Tracts

The lateral vestibulospinal tract originates in the lateral vestibular nucleus, also known as *Deiter's nucleus*. This tract descends ipsilaterally through the ventral funiculus of the spinal cord and ends on interneurons associated with the medial motor neuron groups. The lateral vestibulospinal tract excites motor neurons that supply extensor muscles of the proximal part of the limb that are important for postural control. In addition, this pathway inhibits flexor motor neurons because it also excites the reciprocal group Ia interneurons that receive group Ia input from extensor muscles, which in turn inhibit flexor motor neurons. The excitatory input to the lateral vestibular nucleus is from both the semicircular canals and the otolith organs, whereas the inhibitory input is from the Purkinje cells of the anterior vermis region of the cerebellar cortex. An important function of the lateral vestibulospinal tract is to assist in postural adjustments after angular and linear accelerations of the head.

The medial vestibulospinal tract originates from the medial vestibular nucleus. This tract descends in the ventral funiculus of the spinal cord to the cervical and midthoracic levels, and it ends on the medial group of interneurons. Sensory input to the medial vestibular nucleus from the labyrinth is chiefly from the semicircular canals. This pathway thus mediates adjustments in head position in response to angular acceleration of the head.

The Tectospinal Tract

The tectospinal tract originates in the deep layers of the superior colliculus. The axons cross to the contralateral side, just below the periaqueductal gray matter. They then descend in the ventral funiculus of the spinal cord to terminate on the medial group of interneurons in the upper cervical spinal cord. The tectospinal tract regulates head movement in response to visual, auditory, and somatic stimuli.

Monoaminergic Pathways

In addition to the lateral and medial systems, less specifically organized systems descend from the brainstem to the spinal cord. These include several pathways in which monoamines serve as synaptic transmitters.

The locus coeruleus and the nucleus subcoeruleus are nuclei located in the rostral pons, and they are composed of norepinephrine-containing neurons. These nuclei project widely throughout the CNS, and their projection to the spinal cord travels in the lateral funiculus. Their terminals are on interneurons and motor neurons. The dominant effect of the pathway is inhibitory.

The raphe nuclei of the medulla also project widely throughout the CNS and give rise to several raphe-spinal pathways. With regard to motor function, the ventral horn projection may enhance motor activity.

In general, the monoaminergic pathways act to alter the responsiveness of spinal cord circuits, including the reflex arcs. In this way, they induce widespread changes in excitability rather than discrete movements or specific changes in behavior.

Motor Deficits Caused by Lesions of Descending Motor Pathways

A common cause of motor impairment in humans is interruption of the cerebral cortical efferent fibers in the internal capsule; such interruptions occur in capsular strokes. The resulting disorder is often termed a **pyramidal tract syndrome,** or **upper motor neuron disease,** although these names are misnomers. Motor changes characteristic of this disorder include (1) increased phasic and tonic stretch reflexes (spasticity); (2) weakness, usually of the distal muscles, especially the finger muscles; (3) pathological reflexes, including the **sign of Babinski** (dorsiflexion of the big toe and fanning of the other toes when the sole of the foot is stroked); and (4) a reduction in superficial reflexes, such as the abdominal and cremasteric reflexes. Of importance is that if only the corticospinal tract is interrupted, as can occur with a lesion of the medullary pyramid, most of these signs are much reduced or absent. In this situation, the most prominent deficits are weakness of the distal muscles, especially those of the fingers, and a Babinski sign. Spasticity does not occur; instead, muscle tone may actually decrease. Evidently, the presence of spasticity requires the disordered function of other pathways, such as the reticulospinal tracts, as would occur after loss of the descending cortical influence to the brainstem nuclei of origin of these tracts.

The effects of interruption of the medial system pathways are quite different from those produced by corticospinal tract lesions. The main deficits associated with medial system interruption are an initial reduction in the tone of postural muscles and loss of righting reflexes. Long-term effects include locomotor impairment and frequent falling. However, manual manipulation of objects is perfectly normal.

The Decerebrate Preparation

The decerebrate preparation has been useful for experimentally investigating how various descending pathways interact with the spinal cord circuitry. Surgical decerebration is achieved either by transection of the midbrain, often at an intercollicular level, or by occlusion of the blood vessels feeding this area. In the latter case, a lesion also occurs in the anterior vermis of the cerebellum, an important distinction.

With the intercollicular transection, some descending pathways, such as those originating in the cerebral cortex, are interrupted, whereas others, such as those originating in the brainstem, remain intact.

However, remember that the corticospinal tract is only one component of the cortical descending fibers. Many other cortical fibers project to locations throughout the brainstem, including the nuclei of origin for the medial descending pathways. Loss of these cortical control systems results in altered activity in the intact descending pathways. As a result, affected animals show hypertonia and suppression of some spinal reflexes, such as the flexion reflex, and exaggeration of others, such as the stretch reflex; this condition is called **decerebrate rigidity.** Decerebrate animals maintain a posture that has been called **exaggerated standing.** Human patients with brainstem damage may also develop a decerebrate state that has many of the same reflex features as animal preparations. The prognosis in such patients is poor if signs of decerebration appear.

Loss of descending control on the reticular formation results in increased activity in the pontine reticulospinal pathway and decreased activity in the medullary reticulospinal pathway. Such increase and decrease in activity, respectively, produces increased excitation and decreased inhibition (disinhibition) of the motor neurons, which explains the observed rigidity. Interestingly, this hypertonia can be relieved by cutting the dorsal roots, which indicates that the reticulospinal tracts have a major effect on γ motor neurons. This is because γ motor neuron activity alters muscle stiffness by increasing muscle spindle sensitivity and thereby causes increased activity in the group Ia and group II afferent fibers, which travel through the dorsal roots into the spinal cord in order to innervate the α motor neurons.

When vessel occlusion is used to generate the decerebrate state, the lateral vestibulospinal tract becomes hyperactive because of damage to Purkinje cells in the anterior vermis of the cerebellum, which provide the major inhibitory projection to the lateral vestibular nucleus. This hypertonia is actually not lost after transection of the dorsal roots, which implies that the lateral vestibulospinal tract is acting to a significant extent directly on α motor neurons (either monosynaptically or via interneurons).

Brainstem Control of Posture and Movement

The importance of motor control pathways that originate in the brainstem is evident from observations of the extensor hypertonus and increased phasic stretch reflexes that occur in decerebrate animals. Particular brainstem systems have been identified as influencing posture and locomotion. Brainstem circuits are also critically involved in the control of eye movement; these circuits are discussed in a separate section at the end of the chapter.

Postural Reflexes

Several reflex mechanisms are evoked when the head is moved or the neck is bent. There are three types of postural reflexes: vestibular reflexes, tonic neck reflexes, and righting reflexes. The sensory receptors responsible for these reflexes include the vestibular apparatus (see Chapter 8), which is stimulated by head movement, and stretch receptors in the neck.

The **vestibular reflexes** constitute one class of postural reflex. Rotation of the head activates sensory receptors of the semicircular canals (see Chapter 8). In addition to generating eye movement, the sensory input to the vestibular nuclei results in postural adjustments. Such adjustments are mediated by commands transmitted to the spinal cord through the lateral and medial vestibulospinal tracts and the reticulospinal tracts. The lateral vestibulospinal tract activates extensor muscles that support posture. For instance, if the head is rotated to the left, postural support is increased on the left side. This increased support prevents the person from falling to the left as the head rotation continues. A person who has any disease that eliminates labyrinthine function in the left ear tends to fall to the left. Conversely, a person with a disease that irritates (stimulates) the left labyrinth tends to fall to the right. The medial vestibulospinal tract causes contractions of neck muscles that oppose the induced movement (**vestibulocollic reflex).**

Tilting the head also changes the linear acceleration on individual hair cells of the otolith organs of the vestibular apparatus. The resulting changes in hair cell activity can produce eye movement and postural adjustment. For example, when a quadruped, such as a cat, tilts the head and body forward (without bending the neck and consequently without evoking the tonic neck reflexes), the result is extension of the forelimbs and flexion of the hind limbs. This vestibular action tends to restore the body toward its original orientation. Conversely, if the quadruped tilts the head and body backward (without bending the neck), the forelimbs flex and the hind limbs extend. Otolithic organs also contribute to the **vestibular placing reaction.** If an animal, such as a cat, is dropped, stimulation of the utricles leads to extension of the forelimbs in preparation for landing.

The **tonic neck reflexes** are another type of positional reflex. These reflexes are activated by the muscle spindles found in neck muscles. These muscles contain the largest concentration of muscle spindles of any muscle in the body. If the neck is bent (without tilting of the head), the neck muscle spindles evoke tonic neck reflexes without interference from the vestibular system. When the neck is extended, the forelimbs extend and the hind limbs flex. The opposite effects occur when the neck is flexed. Note that these effects are opposite those evoked by the vestibular system. Furthermore, if the neck is bent to the left, the extensor muscles in the limbs on the left contract more, and the flexor muscles in the limbs on the right side relax.

The third class of postural reflex is the **righting reflexes.** These reflexes tend to restore an altered position of the head and body toward normal. The receptors responsible for righting reflexes include the vestibular apparatus, the neck stretch receptors, and mechanoreceptors of the body wall.

Brainstem Control of Locomotion

The spinal cord contains neural circuits that serve as **central pattern generators** for locomotion, as discussed earlier. These CPG circuits produce very regular rhythmic output that characterizes stereotyped behavior, such as walking. The irregularities of real-world environments, however, often require modification of this stereotyped output (e.g., if you are walking and see a hole in the floor where you are about to step, you can extend the forward swing of your leg past the hole onto solid ground beyond it).

Such modifications can be the result of sensory input to the spinal cord, as shown in Fig. 9.10, in which stimulation of FRA fibers in a peripheral nerve caused a phase shift in the locomotor pattern. They can also be the result of descending commands along the motor pathways discussed earlier. In this case, sensory data (e.g., visual) can be used by the brain to make anticipatory modifications in CPG activity so that potential obstacles can be avoided. In addition, people can voluntarily control activation, or shutdown, of the CPG (i.e., deciding consciously when to start and stop walking). Such voluntary regulation of spinal CPGs originates in the cerebral cortex; however, much of the cortical influence on locomotion appears to be mediated via projections to brainstem regions known as *locomotor regions.* A locomotor region can be defined as a brain area that, when stimulated, leads to sustained locomotion.

There are several such locomotor regions in the brainstem, and they are located at different levels ranging from the subthalamus to the medulla and are connected with each other. The best known is the **midbrain locomotor region,** which is thought to organize commands to initiate locomotion. It is located in the midbrain at the level of the inferior colliculus. Voluntary activity that originates in the motor cortex can trigger locomotion by the action of corticobulbar fibers projecting to the midbrain locomotor region. The commands are relayed through the reticular formation and then to the spinal cord via the reticulospinal tracts.

Motor Control by the Cerebral Cortex

Thus far in this chapter, emphasis has been on reflexes and relatively automatic types of movement. We now discuss the neural basis for more complex, goal-directed voluntary movement. Such movement often varies when repeated and is frequently initiated as a result of cognitive processes rather than in direct response to an external stimulus. Thus it requires the participation of motor areas of the cerebral cortex.

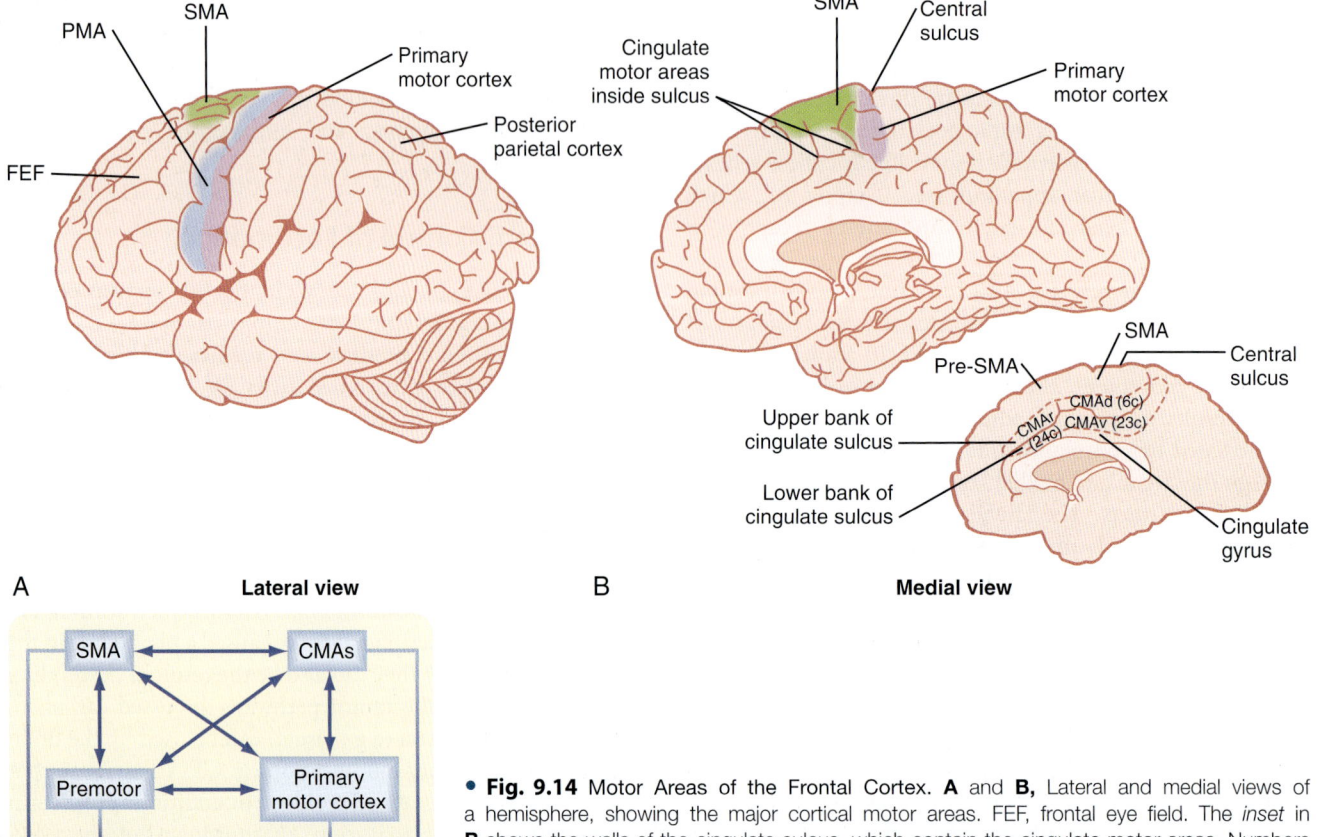

• **Fig. 9.14** Motor Areas of the Frontal Cortex. **A** and **B,** Lateral and medial views of a hemisphere, showing the major cortical motor areas. FEF, frontal eye field. The *inset* in **B** shows the walls of the cingulate sulcus, which contain the cingulate motor areas. Numbers in parentheses are Brodmann area numbers for the cingulate motor areas (CMAs). **C,** Diagram showing interconnections of the motor areas. PMA, premotor area; SMA, supplementary motor area.

First, consider what is necessary to generate a voluntary movement. For example, to make a reaching movement with your arm, you must first identify the target (or goal) and locate it in external space. Next, a limb trajectory must be determined on the basis of an internal representation of your arm and, in particular, your hand in relation to the target. Finally, a set of forces necessary to generate the desired trajectory must be computed. This process is often thought of as a series of transformations between coordinate systems. For example, the location of a visually identified target is measured in a retinotopic space, but its location is perceived in an external or world space (i.e., the position of a nonmoving target is perceived as stable, even when the eye, and thus the target's image on the retina, changes). Next, calculation of a trajectory would involve a body- or hand-centered system, and finally, forces must ultimately be computed in a muscle-based reference frame.

These steps form a linear sequence, and traditionally it was thought that a hierarchy of motor areas carried out the successive steps. For example, the target of the movement was thought to be identified by pooling of sensory information in the **posterior parietal cerebral cortex** (Fig. 9.14*A*). This information would then be transmitted to the supplementary motor and premotor areas, where a motor plan would be developed and then forwarded to the primary motor cortex, whose activity would be related to the final execution stage (e.g., generation of appropriate force levels). The motor cortex would then transmit commands, via the descending pathways discussed earlier, to the spinal cord and brainstem motor nuclei.

Although there is significant evidence in support of this hierarchical view of the generation of voluntary movement by the cortical motor system, more recent results have suggested a different conception: namely, that the various motor areas should be thought of as forming a parallel distributed network rather than a strict hierarchy (see Fig. 9.14*C*). For example, each cortical motor area makes its own significant contribution to the descending motor pathways; the primary motor cortex contributes only approximately half the fibers in the corticospinal tract that arise from the frontal lobe. Moreover, the various motor areas are all bidirectionally connected to each other, and results of the single-unit recording studies described later suggest that each of the areas plays a role in several of the stages of planning and executing a movement. This debate forms one of the themes of the following discussion because in its various guises, the distributed network versus hierarchical organization debate has been ongoing for decades and will probably continue for some time.

Cortical Motor Areas

The motor areas in the cerebral cortex were originally defined on the basis of experiments in which electrical stimuli applied to the cortex evoked discrete, contralateral movement. Movement, however, can also be evoked when other cortical areas are stimulated more intensely. Thus motor areas are defined as those from which movement can be evoked by the lowest stimulus intensity. On the basis of these stimulation studies, the effects produced by lesions, results of anatomical experiments, electrophysiological recordings, and modern imaging studies in humans, many "motor" areas of the cerebral cortex have been recognized (see Fig. 9.14), including the **primary motor cortex** in the precentral gyrus, the **premotor area** just rostral to the primary motor cortex, the **supplementary motor cortex** on the medial aspect of the hemisphere, and three **cingulate motor areas** located on the walls of the cingulate sulcus in the frontal lobe. There are also cortical regions in other lobes whose activity is related specifically to eye movement (see the section "Eye Movement").

Somatotopic Organization of Cortical Motor Areas

Primary Motor Cortex

The primary motor cortex (or just motor cortex) can be defined as the region of cortex from which movements are elicited with the least amount of electrical stimulation. It is essentially congruent with the Brodmann cytoarchitectonic area 4 (see Fig. 10.3). In humans it is located on the parts of the precentral gyrus that form the rostral wall of the central sulcus and the caudal half of the apex of the gyrus. On the basis of initial mapping studies, which were done with surface stimulation, the motor cortex was described as having a topographic organization that parallels that of the somatosensory cortex. The face, body, and upper limb were represented on the lateral surface with the face located inferiorly, near the lateral fissure, the torso most superiorly, and the lower extremity mostly on the medial aspect of the hemisphere. This somatotopic organization is often represented as a figurine or in a graphic form called a **motor homunculus** (Fig. 9.15B). The distortion of the various body parts in the homunculus indicates approximately how much of the cortex is devoted to their motor control. This simple homunculus was likened to a piano keyboard and fit well with traditional conceptions of the motor cortex being the final cortical stage and acting as a relay for sending motor commands to the spinal cord.

Beginning in the 1960s and 1970s, investigators in mapping studies began using microelectrodes inserted to the deep, or output, layers of the cortex to apply stimuli. With this technique, called **intracortical microstimulation,** much lower stimulus intensities could be used to evoke movements and thus allowed higher resolution mapping of the motor cortex, which revealed a much more complex topography than was previously imagined (see

Fig. 9.15C). Movement about each joint was found to be evoked by many noncontiguous columns throughout wide regions of the motor cortex. Thus cell columns related to movement about a particular joint are actually interspersed among columns that control movement about many other joints. In sum, although the motor cortex may have large subdivisions corresponding to a limb or the head, within each such area there is a complex intermingling of cell columns that control the muscles within that body part.

Such mixing of cell columns makes functional sense because most movement requires the coordinated action of muscles throughout a limb and most connectivity in the cortex is localized (i.e., axon collaterals that connect different cell columns are primarily confined to a 1- to 3-mm region surrounding the column from which they originate). Thus when multiple cell columns that control movement about a joint are present and intermixed with columns controlling movement about other joints, multijoint movement can be generated as a whole.

Although the topographic map of the motor cortex is in part anatomically determined by the topography of the corticospinal pathway, it is also a dynamic map. Axon collaterals link the different cell columns, so that activity in one column could potentially lead to movement about multiple joints. In fact, this can happen, but these intercolumnar connections are modulated by inhibitory interneurons that transmit or secrete gamma-aminobutyric acid (GABA). This was shown by locally blocking GABA in one region of the motor cortex and then stimulating the neighboring region. Before the block, stimuli evoked contractions of one set of muscles, but once inhibition was blocked, contractions were also evoked in muscles controlled by the region that was no longer inhibited (Fig. 9.16). Functional connections between cell columns can be controlled on a millisecond time scale, and depending on their state, the motor cortex map can be radically changed. Longer term plastic changes are also known to occur; for example, the use (or disuse) of a body part can affect the size of its somatotopic representation.

Supplementary Motor Area

The supplementary motor area (SMA) is located mainly on the medial surface of the hemisphere, just anterior to the primary motor cortex, and corresponds to the medial portion of Brodmann area 6 (see Fig. 9.14). It is subdivided into two regions: the more posterior part is referred to as the *SMA proper* (or just *SMA*), and the anterior portion is called the *pre-SMA*. The SMA proper is similar to the other motor areas already listed: it contains a complete somatotopic map, it contributes to the corticospinal tract, and it is interconnected with the other motor areas. In contrast, the pre-SMA is not strongly connected with the other motor areas and spinal cord but rather is connected to the prefrontal cortex.

The results of stimulation studies show that as in the motor cortex, there is a complete somatotopic map in the SMA. Stimulation of the SMA can evoke isolated movement

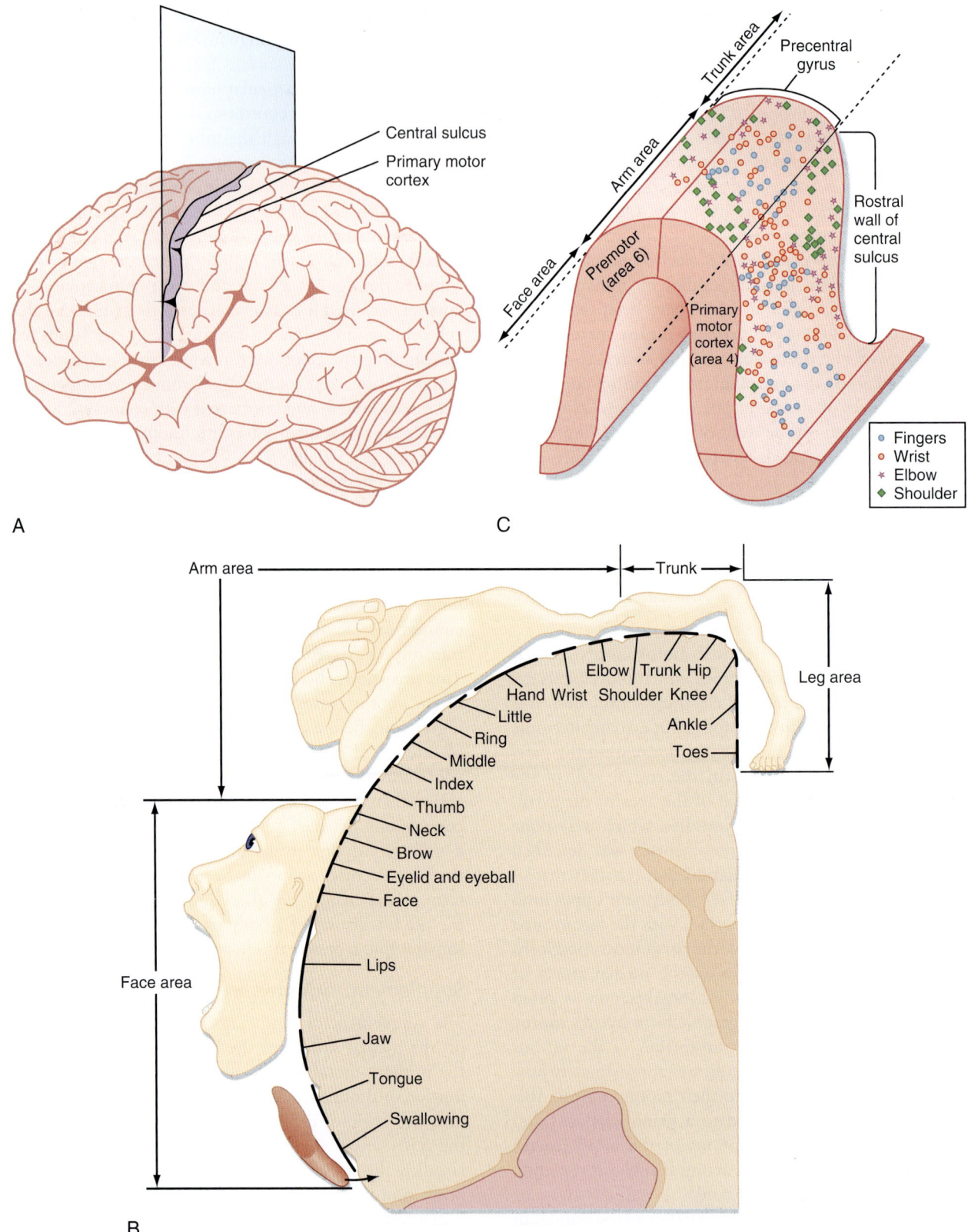

A

B

C

• **Fig. 9.15** Traditional and Modern Views of Motor Cortex Musculotopic Organization. **A,** Lateral view of the cerebrum, showing a plane of section through the precentral gyrus (primary motor cortex) that corresponds to the section shown in **B. B,** Classic view of motor cortex musculotopy ("muscular homunculus"). **C,** Modern view of motor cortex organization in which each body part is represented multiple times across several discrete regions.

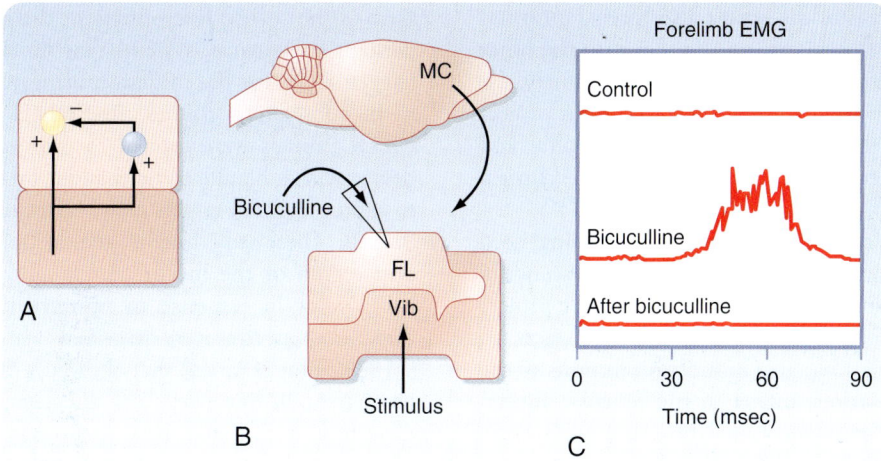

• **Fig. 9.16** Dynamic Nature of a Motor Cortex Musculotopic Map. Inhibitory GABAergic interneurons play an important role in shaping motor responses to stimulation of each region of the motor cortex. **A,** Schematic view of excitatory connections between two regions of primary motor cortex and local inhibitory neurons within a single region. **B,** Schematic view of a rat brain, indicating motor cortex (MC) regions where electrical stimuli were applied to evoke movements (Vib region) and bicuculline was applied to block GABAergic synapses (in the FL region). FL, forelimb; Vib, vibrissa. **C,** Forelimb EMG records showing response to stimulation of the Vib region before and during the application of bicuculline and after it was washed out. Note that Vib stimulation evoked vibrissae movement in all conditions but evoked forelimb movement only when inhibitory interneurons in FL were blocked. (Data from Jacobs K, Donoghue J. *Science.* 1991;251:944.)

about single joints, similar to that after stimulation of the motor cortex, but stimulation must be of higher intensity and longer duration; moreover, the evoked movements are often more complex than those evoked by stimulation of the motor cortex. However, longer duration stimulation of the primary motor cortex can also evoke complex, apparently purposeful movement sequences, and so the distinction is not absolute. In addition, stimulation of the SMA can produce vocalization or complex postural movements, but it can also have the opposite result: namely, a temporary arrest of movement or speech. Removal of the supplementary motor cortex retards movement of the opposite extremities and may result in forced grasping movements with the contralateral hand.

Premotor Area

This area lies rostral to the primary motor cortex and is contained in Brodmann area 6 on the lateral surface of the brain (see Fig. 9.14). It can be distinguished from the primary motor cortex by the higher stimulus intensities needed to evoke movement. The premotor area has been divided into two functionally distinct subdivisions: dorsal and ventral. Like the motor cortex, both subdivisions are somatotopically organized and both contribute to the corticospinal tract. The dorsal division (PMd) contains a relatively complete map representing the leg, trunk, arm, and face. In contrast, the somatotopic map of the ventral division (PMv) is mostly limited to the arm and face, with only a small leg representation. Thus the PMv appears to be specialized for control of upper limb and head movement. A second difference between the subdivisions is that the PMd contains a

large representation of the proximal muscles, whereas the PMv has a large representation of the distal muscles.

Cingulate Motor Areas

These motor areas are located within the cingulate sulcus at approximately the same anterior-posterior level as the SMA. There are three cingulate motor areas (dorsal, ventral, and rostral; see Fig. 9.14B). Each contains a somatotopic map and contributes to the corticospinal tract. Microstimulation in these areas evokes movement similar to that evoked by motor cortex stimulation, except that, again, higher stimulus intensities are needed. Single-cell recordings during movements have shown that the spontaneous activity of neurons in the cingulate motor areas is related to the preparation and execution of movements.

Connections of the Cortical Motor Areas

The motor areas of the cortex receive input from a number of sources, cortical and subcortical; however, the largest source of synapses in an area is the area itself: specifically, the local intrinsic connections. Moreover, all the motor areas described earlier are bidirectionally connected to each other with high topographic specificity (see Fig. 9.14C). For example, the arm regions of the primary motor cortex and the cingulate motor areas project to each other. Ascending pathways relay sensory information to the thalamus. This information can reach the motor cortex directly from the thalamus or indirectly by way of the somatosensory cortex. Both somatosensory information and visual information are conveyed to the motor areas from the posterior parietal cortex. The motor areas of the cortex also receive information through circuits that interconnect them with the other major brain regions involved in motor control: namely, the cerebellum and basal ganglia. These two structures project to distinct parts of the thalamus (the ventral lateral and ventral anterior nuclei), which then project to the cortical motor areas.

The output of the cortical motor areas to the spinal cord and brainstem is conducted through several descending pathways. These pathways include not only direct projections through the corticospinal and corticobulbar tracts (to the cranial nerve nuclei) but also indirect projections to the red nucleus and to various nuclei in the reticular formation. Descending projections from these brainstem sites were reviewed in the section "Descending Motor Pathways." Control of head and neck muscles is mediated by projections to the various cranial nerve nuclei. The motor regions also project to the cerebellum and basal ganglia, thus completing neuroanatomical loops with these structures. The major connection to the cerebellum is via the corticopontine projections to the basilar pontine nuclei, which in turn project to the cerebellum. In addition, the cortical motor areas project, mostly via disynaptic pathways that synapse in the midbrain, to the inferior olivary nucleus, another important precerebellar area. The cortical motor regions project directly to the striatum of the basal ganglia. Finally, there are major projections to the thalamus by which the cortex regulates the information that it receives.

Activity of Motor Cortex Cells

The role of individual motor cortex neurons in the control of movement has been extensively investigated in trained monkeys. In these experiments, discharges from a neuron in the primary motor cortex are recorded during the execution of a previously learned simple movement, such as wrist flexion, made immediately in response to a sensory cue (Fig. 9.17). Motor cortex neurons were found to change their firing rates before initiation of the movement, and the onset of this change was correlated with the reaction time (i.e., the time from the cue to onset of the movement). Moreover, in this task, the change in firing of motor cortex neurons was often correlated with the contractile force of the muscle that generates the movement and with the rate of change in force rather than with the position of the joint. These findings suggest that these cells are involved in the final stages of planning and executing movements, which is consistent with the hierarchical view of the cortical motor areas.

However, even in these early experiments, the firing rates of some motor cortex cells appeared to relate to earlier planning stages. Moreover, even when a monkey was trained to withhold the movement for a certain period after the cue, the firing rates of motor cortex neurons still changed despite the absence of any movement. Such "set-related" activity has been amply confirmed in a variety of other tasks and suggests that motor cortex activity may be involved in the earlier planning stages along with activity in other motor

areas of the cortex. It also suggests the possibility that other, perhaps subcortical, systems may be needed to generate a trigger signal for the initiation of movement.

In subsequent studies, researchers have used tasks in which animals were trained to move a manipulandum (a device with a handle to hold and a small circle on the end) to capture lighted targets on a surface in front of them (Fig. 9.18A). These experiments demonstrated that cells in the arm region of the motor cortex showed changes in their firing rates in response to movement in many different directions and thus were described as broadly tuned (see Fig. 9.18B). That is, a cell that showed a maximal increase for movement in one particular direction, called its preferred direction, would also show somewhat smaller increases or even decreases for movement in other directions (see Fig. 9.18C). Moreover, the preferred directions of the different cells were uniformly distributed across all 360 degrees of possible movement directions.

These results implied that a particular cell is probably involved in most arm movements, but they also raised the issue of how precise movements could be made with such broadly tuned cells. It was suggested that although changes in the activity of individual cells could not precisely predict or specify the direction of the upcoming movement, the net activity of the population could. To test this idea, models were made in which the activity of each cell is represented as a vector (see Fig. 9.18D). The direction of each cell vector is determined by the preferred direction of the cell, and the magnitude of the vector for a particular movement is proportional to the firing rate of the cell during the time preceding the movement. The individual cell vectors (see Fig. 9.18D, black lines) from hundreds of cells can then be vectorially summed to get a resultant or population vector (see Fig. 9.18D, red lines) that accurately predicts the upcoming movement.

One of the difficulties in assessing the relationship between firing of cortical cells and various movement parameters, such as force, velocity, displacement, and target location, is that these parameters are normally correlated with each other. Therefore, variations of the tasks described earlier have been used to decorrelate these various parameters (e.g., using weights to vary the force needed to make a movement without changing the displacement, as illustrated in Fig. 9.17A, or rotating the starting position of the wrist so that different muscles are required to generate the same trajectory in external space). The results of these experiments showed that the activity of motor cortex cells may be related to each of the various motor planning stages. Furthermore, the activity of a single cell may be correlated with one parameter initially and then switch as the time for onset of movement approaches.

Activity in Other Cortical Motor Areas

Activity in the premotor and supplementary motor areas is in many ways similar to that in the primary motor cortex. Cells in these areas show activity related to upcoming

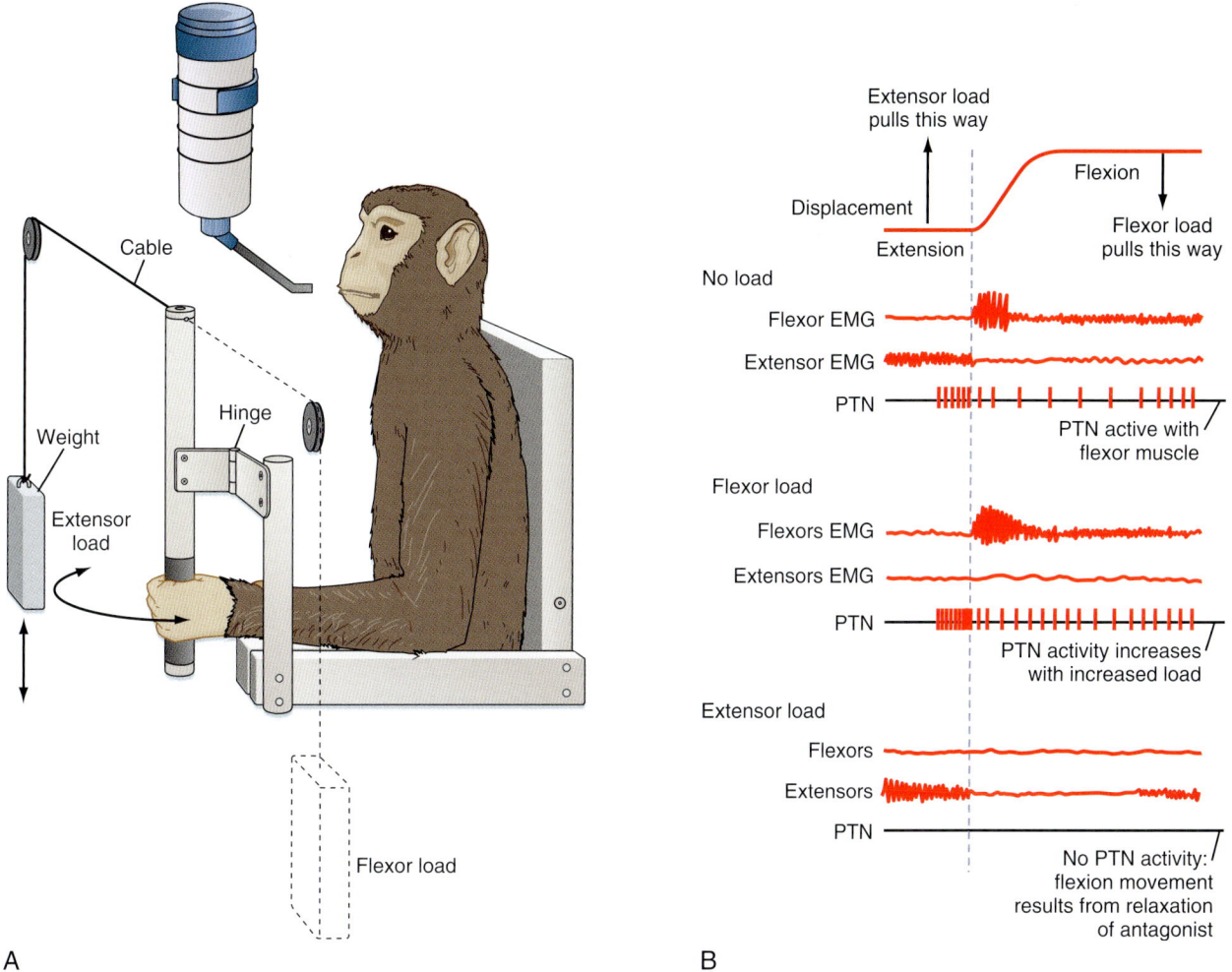

• Fig. 9.17 A, Experimental arrangement for recording from a corticospinal neuron while a monkey performs trained wrist movements. A stimulation electrode is used to elicit antidromic spikes that are used to identify the motor cortex neuron specifically as a pyramidal tract neuron. Stimuli are not applied while the monkey is performing movements. **B,** The pyramidal tract neuron (PTN) discharges before the onset of movement or EMG activity when flexors need to generate force (no load and flexor load conditions). Moreover, the firing rate is correlated with the level of flexor force that is needed. In the extensor load condition, flexors do not need to contract to generate movement, and thus there is no activity in this PTN. The top trace shows wrist movement, which is essentially identical for all three experimental conditions. Thus this cell's activity involves encoding force magnitude and direction, but not displacement. (Figure based on work of Evarts and colleagues.)

movement, and the activity is correlated with movement parameters, such as displacement, force, and target location, just as primary motor cortex activity can be, which is consistent with the distributed network view of the cortical motor areas. There do, however, appear to be some real differences between the areas as well, although these differences may be more quantitative than qualitative. For example, the percentage of cells in the premotor and supplementary motor areas that show activity related to earlier motor planning stages is higher than that of such cells in the primary motor cortex. In addition, the premotor and supplementary motor areas can be distinguished from each other by the apparently greater involvement of the premotor area in movements made to external cues (such as in the task shown in Fig. 9.18) and the greater involvement of the supplementary motor area in movements made in response to internal cues

(i.e., self-initiated). Research has also revealed that each of these areas is functionally heterogeneous and can therefore be further subdivided; however, such details are beyond the scope of this discussion.

Motor Control by the Cerebellum

Overview of the Role of the Cerebellum in Motor Control

In the early 1900s, scientists showed that damage to the cerebellum led to deficits in motor coordination. That is, damage or loss of the cerebellum does not lead to paralysis, loss of sensation, or an inability to understand the nature of a task; rather, it leads to an inability to perform movements well. Yet it has been hard to define the precise role or

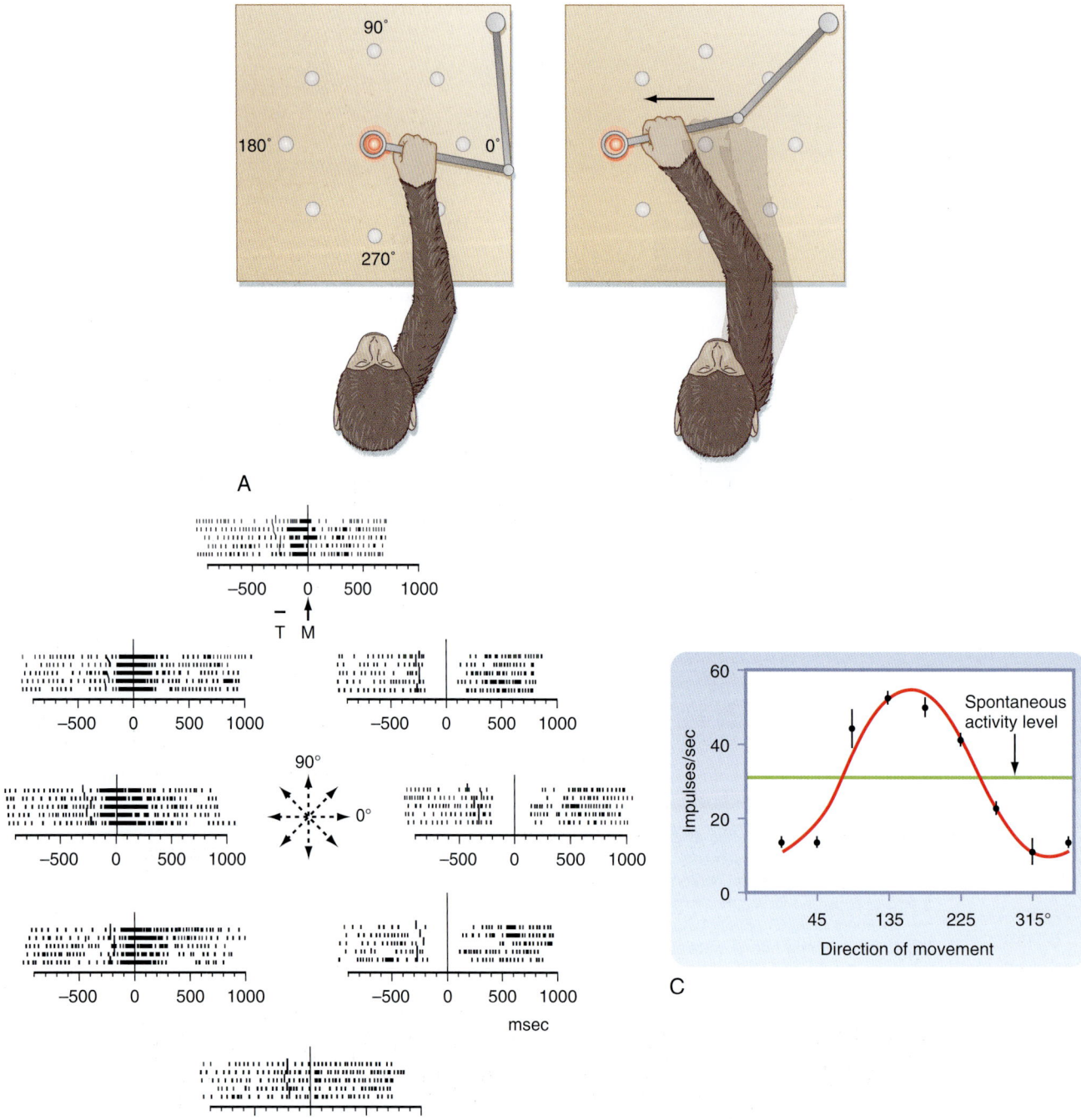

• **Fig. 9.18 A,** Experimental setup in which a monkey holds onto the arm of the apparatus and captures light spots with the distal end of the arm. The monkey first captures the central light spot and then captures whichever of the surrounding targets that becomes illuminated. **B,** Raster plots showing the activity of one motor cortex cell during movement in eight different directions. T indicates the time at which the target light turns on, whereas M indicates the time at the onset of movement, which is at the center of each raster. Each mark on a raster represents a spike of a motor cortex cell, and each row of marks shows the cell's activity during one trial. **C,** Cosine function was fit to the firing level as a function of the direction of movement. The horizontal bar indicates the average spontaneous firing rate in the absence of an upcoming movement. Note that for most directions, the activity in the periods just before and during movement changed significantly from baseline.

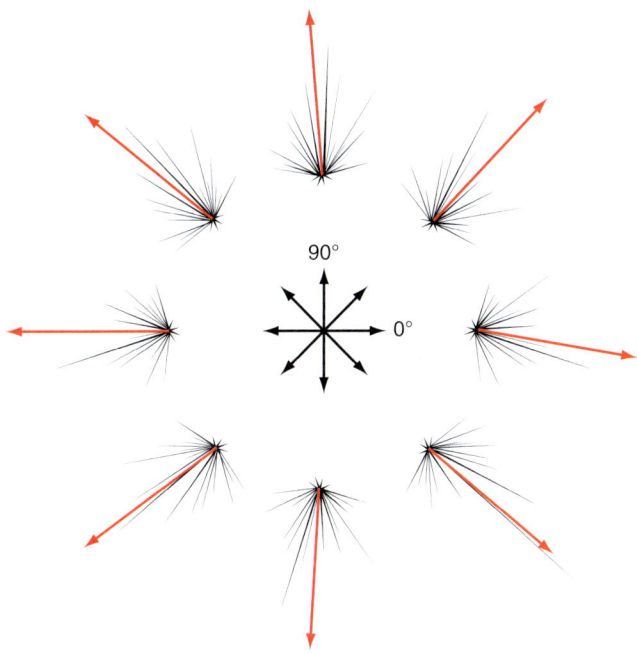

D

• **Fig. 9.18, cont'd D,** Vector model of population activity in the motor cortex. *Black lines* represent individual cell vectors. When all of them are summed for a particular direction of movement, the resulting population vector *(red)* points in essentially the direction of the upcoming movement. (**B** and **C,** Modified from Georgopoulos AP, et al. *J Neurosci.* 1982;2:1527. **D,** Modified from Georgopoulos AP, et al. In Massion J, et al [eds]. *Experimental Brain Research Series,* vol. 7: *Neuronal Coding of Motor Performance.* Berlin: Springer-Verlag; 1983.)

roles of the cerebellum in generating movement, although, paradoxically, scientists have more detailed knowledge of its deceptively simple anatomic and physiological organization than of any other CNS region. The cerebellum is proposed to play a critical role in the learning and execution of both voluntary and certain reflex movements. However, hypotheses about these roles face significant challenges that prevent their complete acceptance. In this section, the behavioral effects of damaging the cerebellum are discussed, followed by a description of its connectivity, both intrinsic and with the rest of the CNS, and then finally a discussion of its activity.

Behavioral Consequences of Cerebellar Damage

Damage to one side of the cerebellum impairs motor function on the ipsilateral side of the body. This reflects a double crossing of most cerebellum-related output as it travels to the motor neurons. The first crossing typically occurs in the cerebellar efferent pathway, whereas the second crossing takes place in the descending motor pathways. For example, the cerebellum projects to the contralateral motor cortex, via the thalamus, and the corticospinal pathway recrosses the midline at the lower medulla.

The specific motor deficits that result from cerebellar lesions depend on which functional component of the

cerebellum is most affected. If the flocculonodular lobe is damaged, the motor disorders resemble those produced by a lesion of the vestibular apparatus; such disorders include difficulty in balance and in gait and often nystagmus. If the vermis is affected, the motor disturbance affects the trunk, and if the intermediate region or hemisphere is involved, motor disorders occur in the limbs. The part of the limbs affected depends on the site of damage; hemispheric lesions affect the distal muscles more than paravermal lesions do.

Types of motor dysfunction in cerebellar disease include disorders of coordination, equilibrium, and muscle tone. Incoordination is called **ataxia** and is often manifested as **dysmetria,** a condition in which errors in the direction and force of movement prevent a limb from being moved smoothly to a desired position. Ataxia may also be manifested as **dysdiadochokinesia,** in which rapid alternating of supination and pronation of the arm is difficult to execute. When more complicated movement is attempted, **decomposition of movement** occurs, in which the movement is accomplished in a series of discrete steps rather than as a smooth sequence. An **intention tremor** appears when the subject is asked to touch a target; the affected hand (or foot) develops a tremor that increases in magnitude as the target is approached. When equilibrium is disturbed, impaired balance may be seen, and the individual tends to fall toward the affected side and may walk with a wide-based stance (gait ataxia). Speech may be slow and slurred; such a defect is called **scanning speech.** Muscle tone may be diminished **(hypotonia),** except for lesions of the anterior vermis (see earlier section on decerebrate rigidity); the diminished tone may be associated with a **pendular knee jerk.** This can be demonstrated by eliciting a phasic stretch reflex of the quadriceps muscle by striking the patellar tendon. The leg continues to swing back and forth because of the hypotonia, in contrast to the highly damped oscillation in a normal person.

These disorders reflect, in part, abnormal timing of muscle contractions. Normally, limb movements involve precisely timed EMG bursts in both agonist and antagonist muscles. There is an initial agonist burst followed by a burst in the antagonist and, finally, a second agonist burst. With cerebellar damage, the relative timing of these bursts is abnormal (Fig. 9.19).

Cerebellar Organization

The cerebellum ("little brain") is located in the posterior fossa of the cranium, just below the occipital lobe, and is connected to the brainstem via three cerebellar peduncles (superior, middle, and inferior). From the outer surface, only the cortex is visible. Deep to the cortex is the white matter of the cerebellum, and buried within the white matter are the four cerebellar nuclei: proceeding medially to laterally, the fastigial, globose, emboliform, and dentate nuclei. The middle two nuclei are often grouped together and referred to as the *interpositus nucleus.* For the most part, cerebellar afferent fibers to the cortex and nuclei enter the cerebellum

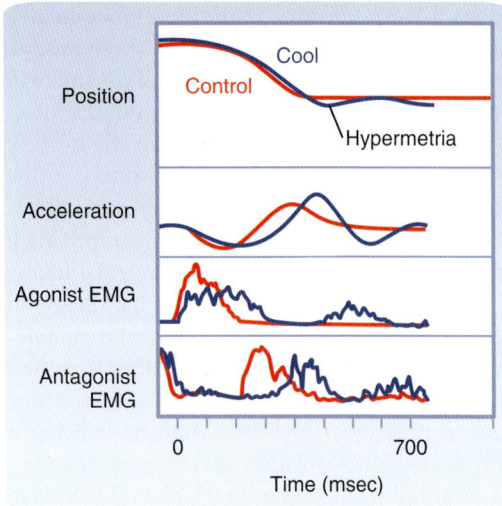

• **Fig. 9.19** Disruption of Cerebellar Activity Alters the Timing of EMG Responses During Movement. The cerebellar nuclei were cooled to block their functioning temporarily while monkeys performed movements about their elbow. Loss of cerebellar activity disrupts the relative timing of agonist and antagonist EMG bursts. This leads to abnormal acceleration of the limb and a movement trajectory that overshoots the target position (hypermetria). (Data from Flament D, Hore J. *J Neurophysiol.* 1986;55:1221.)

via the inferior and middle peduncles, and efferent fibers from the cerebellar nuclei leave via the superior peduncle.

The cerebellar cortex is subdivided into three rostrocaudally arranged lobes: the **anterior lobe,** the **posterior lobe,** and the **flocculonodular lobe** (Fig. 9.20*A*). The cerebellar lobes are separated by two major fissures, the **primary fissure** and the **posterolateral fissure,** and each lobe is made up of one or more **lobules.** Each lobule of the cerebellar cortex is composed of a series of transverse folds called **folia.**

The cerebellar cortex has also been divided into longitudinal compartments (see Fig. 9.20*B* and *C*). Initially, the cerebellar cortex was divided into three such compartments: the vermis, which lies on the midline; the paravermis, which lies adjacent to both sides of the vermis; and the lateral hemispheres. These regions have now been subdivided into many further compartments on the basis of **myeloarchitectonics** (patterns of axonal bundles in the white matter) and the expression patterns of specific molecules, such as aldolase C. Although the functional significance of these compartments is not fully known, the topography of cerebellar afferent fibers, particularly the olivocerebellar system, is precisely aligned with them, and the receptive field properties of cerebellar Purkinje cells also tend to follow this organizational scheme.

Cerebellar Cortex
Afferent Systems

There are two major classes of cerebellar afferent systems: mossy fibers and **olivocerebellar fibers. Mossy fibers** are named for their distinctive appearance in the cerebellar

cortex: as a mossy fiber courses through the granule layer, on occasion it swells and sends out a bunch of short twisted branchlets. These entities are called *rosettes* and are points of synaptic contact between these fibers and neurons in the granule cell layer. Mossy fibers arise from many sources, including the spinal cord (the spinocerebellar pathways), dorsal column nuclei, trigeminal nucleus, nuclei in the reticular formation, primary vestibular afferent fibers, vestibular nuclei, cerebellar nuclei, and the basilar pontine nuclei. The details of specific mossy fiber projection patterns are beyond the scope of this chapter; however, several general points are worth noting:

1. Mossy fibers are excitatory.
2. They convey exteroceptive and proprioceptive information from the body and head and form at least two somatotopic maps of the body across the cerebellar cortex. However, like those of the motor cortex, these maps are fractured in the sense that contiguous body regions are not necessarily represented on contiguous areas of the cerebellar cortex; rather, the maps are complicated mosaics.
3. Mossy fibers conveying vestibular information are restricted to the flocculonodular lobe and regions of the vermis. As a result, the flocculonodular lobe and regions of the vermis are sometimes referred to as the *vestibulocerebellum.* However, these same regions also receive a variety of other information (e.g., visual, neck, oculomotor), and so their function is not exclusively vestibular.
4. The largest sources of mossy fibers are the basilar pontine nuclei, which serve to relay information from areas throughout much of the cerebral cortex.
5. Mossy fibers enter the cerebellum via all three cerebellar peduncles and provide collateral fibers to the cerebellar nuclei before heading up to the cortex. In sum, via the mossy fiber system, the cerebellum receives a wide variety of sensory information, as well as descending motor-related and cognitive-related activity.

In contrast to the diverse origins of mossy fibers, olivocerebellar fibers all originate from a single nucleus: the inferior olivary nucleus, which is located in the rostral medulla, just dorsal and lateral to the pyramids. Almost all the olivary neurons are projection cells whose axons leave the nucleus without giving off collaterals and then cross the brainstem to enter the cerebellum primarily via the inferior cerebellar peduncle. Like mossy fibers, olivocerebellar axons are excitatory and send collaterals to the cerebellar nuclei as they ascend through the cerebellar white matter to the cortex. In the cerebellar cortex, olivocerebellar axons may synapse with basket, stellate, and Golgi cells, but they form a special synaptic arrangement with Purkinje cells. Each Purkinje cell receives input from only a single climbing fiber, which "climbs" up its proximal dendrites and makes hundreds of excitatory synapses. (The terminal portion of the olivocerebellar axon is referred to as a climbing fiber.) Conversely, each olivary axon branches to form about 10 to 15 climbing fibers.

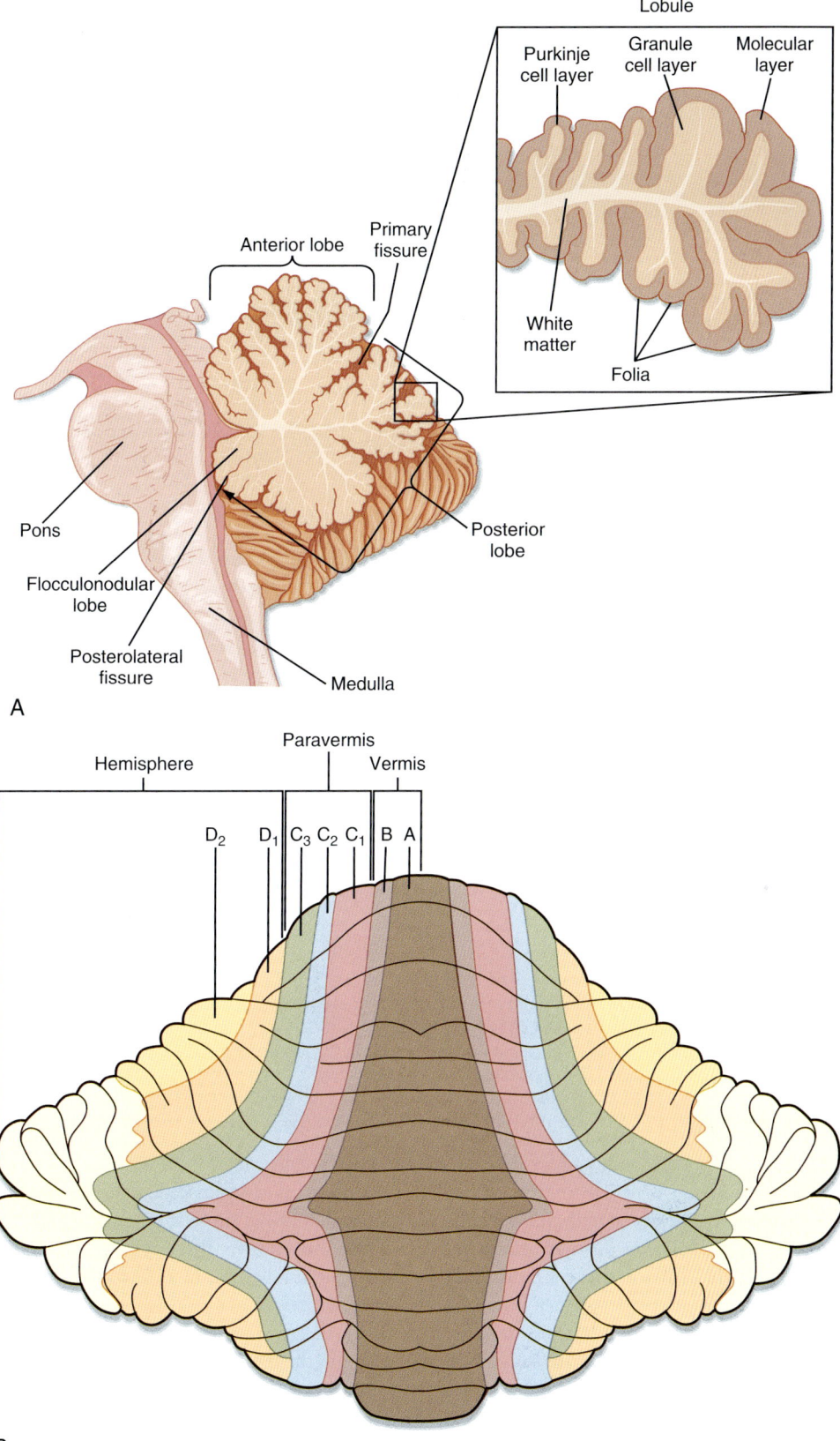

A

B

• Fig. 9.20 Anatomic Divisions of the Cerebellum. **A,** Schematic midsagittal view of the folding of the cortex into lobe, lobules, and folia. **B,** Schematic view of an unfolded ferret cerebellar cortex to illustrate earlier compartmentation schemes for subdividing the cerebellar cortex into three (vermis, paravermis, and hemisphere) and then seven longitudinally running zones (A, B, C_1, C_2, C_3, D_1, D_2). The *light yellow* portion of each hemisphere indicates an area for which no data were available.

Continued

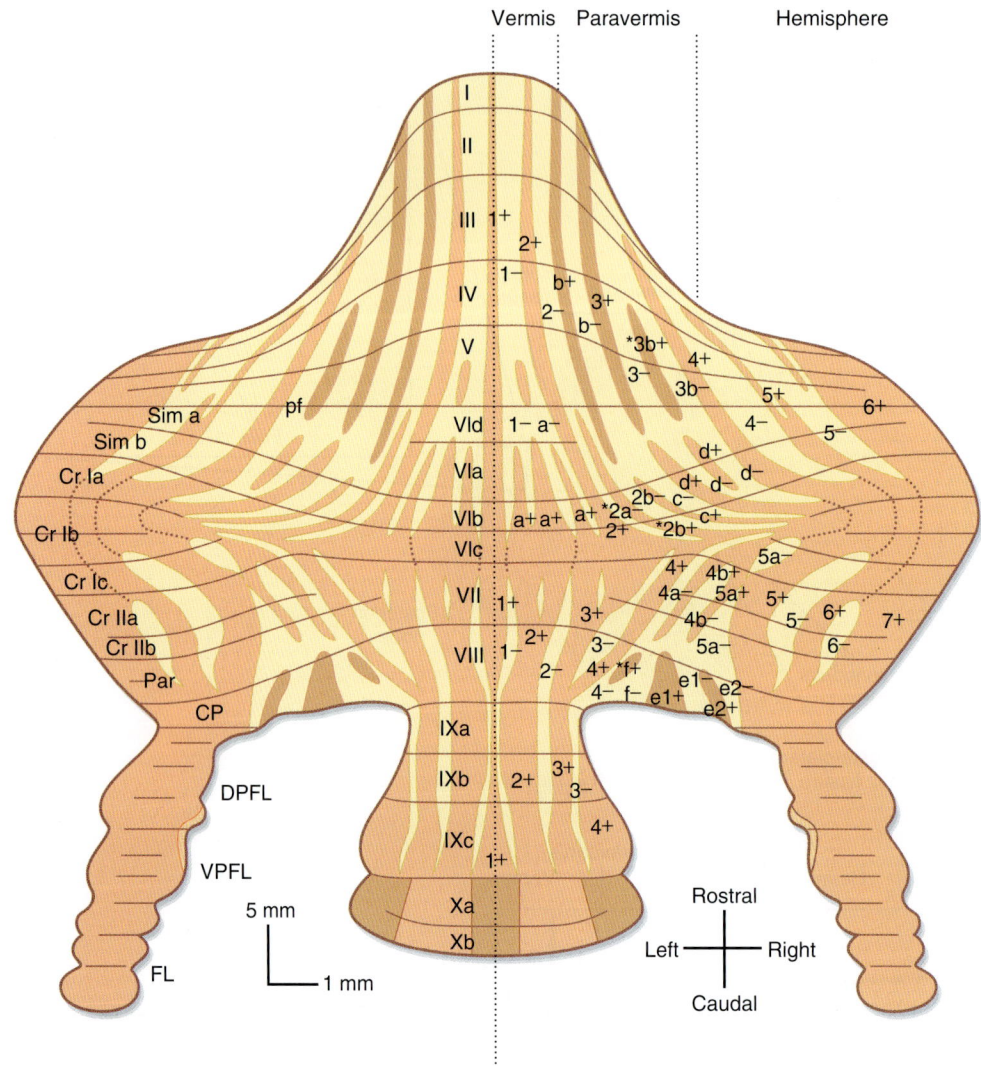

C

• **Fig. 9.20, cont'd** **C,** Schematic view of an unfolded rat cerebellum, showing its subdivision into more than 20 compartments, according to staining for molecular markers: in this case, zebrin II (aldolase C). *Letters and numbers on the right half* of the cerebellum indicate the zebrin compartment number. *Roman numerals down the center* indicate cerebellar lobules. *Names on left hemisphere* indicate names of cerebellar lobules. CP, copula pyramis; Cr, crus; DPFL, dorsal paraflocculus; FL, flocculus; Par, paramedian; pf, primary fissure; Sim, simplex; VPFL, ventral paraflocculus. (**B,** Modified from Voogd J. In Llinás RR (ed). *Neurobiology of Cerebellar Evolution and Development.* Chicago: American Medical Association; 1969. **C,** Courtesy of Dr. Izumi Sugihara.)

The inferior olivary nucleus is a distinctive brain region for several reasons. As already noted, its neurons are virtually all projection cells, and so there is little local chemical synaptic interaction between the cells. Instead, olivary neurons are electrically coupled to each other by gap junctions. In fact, the inferior olivary nucleus has the highest density of neuronal gap junctions in the CNS. This allows olivary neurons to have synchronized activity that gets transmitted to the cerebellum. Afferent fibers to the inferior olivary nucleus may be divided into two main classes: those transmitting excitatory input, which arises from many regions throughout the CNS, and those transmitting inhibitory GABAergic input from the cerebellar nuclei and a few brainstem nuclei. Although these afferent fibers can modulate the firing rates of olivary neurons (as is typical in most brain regions), the membrane properties of olivary neurons limit this modulation to a range of a few hertz and endows these neurons with the potential to be intrinsic oscillators. Instead of just modulating firing rates, olivary afferent activity also acts to modify the effectiveness of the electrical coupling between olivary neurons and thus changes the patterns of synchronous activity delivered to the cerebellum. Afferent activity may also modulate expression of the oscillatory potential of olivary neurons. Thus the inferior olivary nucleus appears to be organized to generate patterns of synchronous activity across the cerebellar cortex. The functional significance of these patterns remains controversial. One hypothesis is that they provide a gating signal for synchronizing motor commands to various muscle combinations.

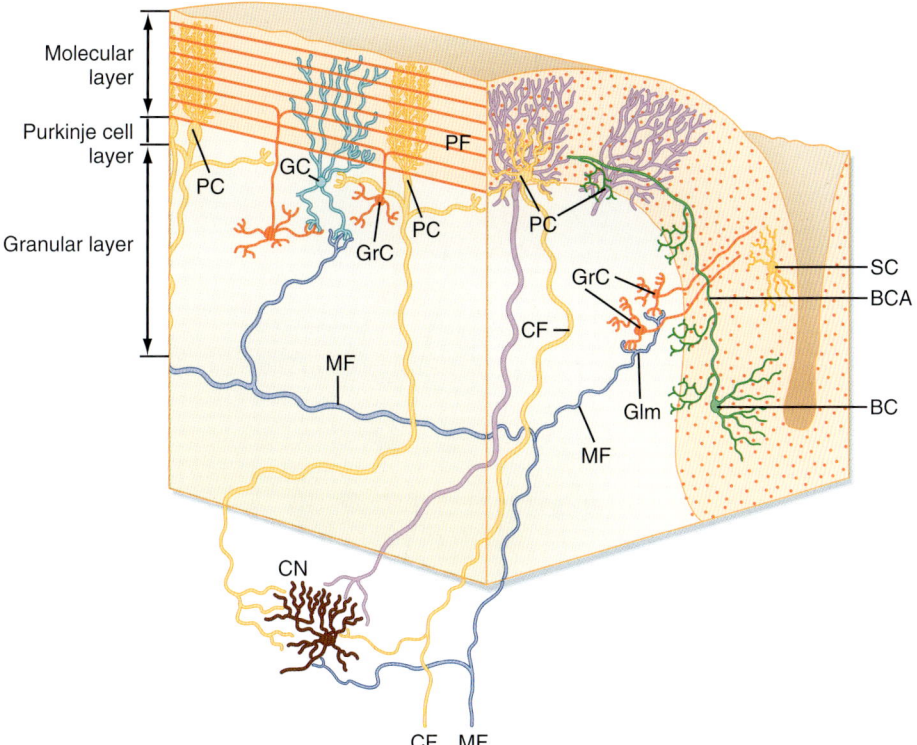

• **Fig. 9.21** Three-Dimensional View of the Cerebellar Cortex, Showing Some of the Cerebellar Neurons. The cut face at the left is along the long axis of the folium; the cut face at the right is at right angles to the long axis. BC, basket cell; BCA, basket cell axon; CF, climbing (olivocerebellar) fiber; CN, cerebellar nuclear cell; GC, Golgi cell; Glm, glomerulus; GrC, granule cell; MF, mossy fiber; PC, Purkinje cell; PF, parallel fiber; SC, stellate cell.

Cellular Elements and Efferent Fibers of the Cortex

Despite its enormous expansion throughout vertebrate evolution, the basic anatomical organization of the cerebellar cortex has remained nearly invariant. The circuitry is also among the most regular and stereotyped of any brain region. The cerebellar cortex contains eight different neuronal types: Purkinje cells, Golgi cells, granule cells, Lugaro cells, basket cells, stellate cells, unipolar brush cells, and candelabrum cells. These cells are found in all regions of the cerebellar cortex, with the exception of unipolar brush cells, which are limited mainly to cerebellar areas receiving vestibular input. These eight cell types are distributed among the three layers that make up the cerebellar cortex of higher vertebrates (Fig. 9.21). The outer or superficial layer is the **molecular layer; stellate** and **basket cells** are found there. The deepest layer is the **granule cell layer;** this layer has the highest cellular density in the CNS and contains granule, **Golgi,** and **unipolar brush cells.** Separating the molecular and granule cell layers is the **Purkinje cell layer,** formed by Purkinje cell somata, which are arranged as a one-cell-thick sheet of cells. **Candelabrum cells** are also located in this layer. **Lugaro cells** are situated slightly deeper at the upper border of the granule cell layer.

The sole efferent fiber from the cortex is the Purkinje cell axon, which also has local collaterals and is GABAergic and inhibitory. Thus the remaining seven cell types are local interneurons. Of these, the stellate, basket, Golgi, Lugaro, and candelabrum cells are also inhibitory GABAergic neurons, whereas the granule and unipolar brush cells are excitatory.

Microcircuitry of the Cortex

The dendrites, axons, and patterns of synaptic connections of most neurons within the cerebellar cortex are organized with regard to the transverse (short) and longitudinal (long) axes of the folium (Fig. 9.21). In the vermis, where the folia run perpendicular to the sagittal plane, these axes lie in the sagittal and coronal planes, respectively. In the hemispheres, where the folia are oriented at various angles with regard to the sagittal plane, this correspondence is lost, and the local axes of the folia must then serve as the reference axes.

The Purkinje cell dendritic tree is the largest in the CNS. It extends from the Purkinje cell layer through the molecular layer to the surface of the cerebellar cortex and for several hundred microns along the transverse axis of the folium but for only 30 to 40 μm in the longitudinal direction. Thus it is like a flat pancake that lies in a plane parallel to the transverse axis of the folium. Accordingly, a set of Purkinje cell dendritic trees can be thought of as a stack of pancakes, with the stack running along the longitudinal axis of the folium.

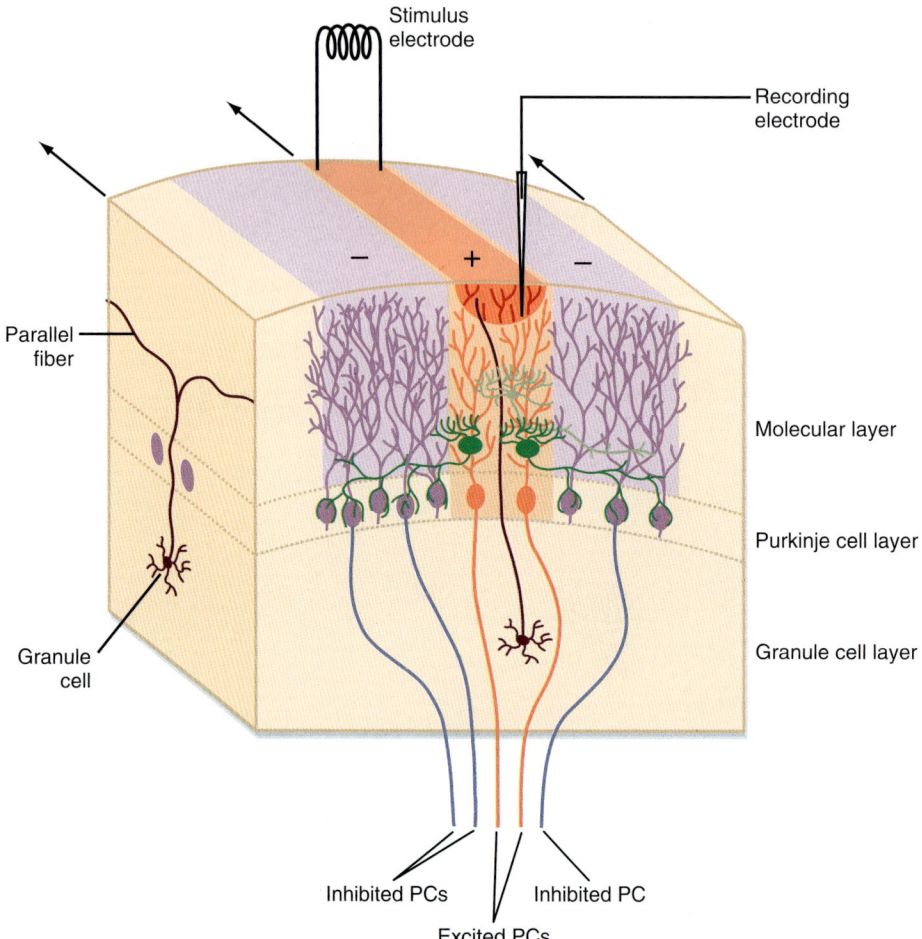

• **Fig. 9.22** Functional Connectivity of the Cerebellar Cortex. Because of the geometric organization of the cerebellar cortical circuits, the functional connectivity of the cellular elements can be determined electrophysiologically. The figure depicts a classic paradigm in which stimulation of the cerebellar cortex activates a beam of parallel fibers *(orange).* Recordings from the stellate and basket cells *(green cells)* and Purkinje cells (PCs; *orange cells)* in line with this beam show that they are excited by the parallel fibers. In contrast, Purkinje cells flanking the beam receive only inhibition *(purple areas)* as a result of the perpendicular spatial relationship of the parallel fibers and the stellate and basket cell axons.

The dendritic trees of the molecular layer interneurons (stellate and basket cells) are oriented in a manner similar to that of the Purkinje cell dendritic tree, although they are much less extensive. The axons of stellate and basket cells run transversely across the folium and form synapses with Purkinje cells. Stellate and basket cells synapse onto Purkinje cell dendrites. In addition, basket cells make synapses on the Purkinje cell soma and form a basket-like structure around the base of the soma, which gives the basket cell its name.

Granule cells are small neurons with four to five short unbranched dendrites, each ending in a claw-like expansion that synapses with a mossy fiber rosette and with terminals from Golgi cell axons in a complex arrangement known as a *glomerulus.* The axons of granule cells ascend through the Purkinje cell layer to the molecular layer, where they bifurcate and form parallel fibers. The parallel fibers run parallel to the cerebellar surface along the longitudinal axis of the folium (perpendicular to the planes of the Purkinje, stellate, and basket cell dendritic trees) and form excitatory synapses with the dendrites of the Purkinje, Golgi, stellate, and basket cells.

The orthogonal relationship between the parallel fibers and the dendritic trees of the Purkinje cells and molecular layer interneurons (basket and stellate cells) has significant functional consequences. This arrangement allows maximal convergence and divergence to occur. A single parallel fiber, which can be up to 6 mm long, passes through more than 100 Purkinje cell dendritic trees (and also interneuron dendrites); however, it has the chance to make only one or two synapses with any particular cell because it crosses through the short dimension of the dendritic tree. Conversely, a given Purkinje cell receives synapses from on the order of 100,000 parallel fibers. Thus a beam of parallel fibers can be excited experimentally, which excites a row of Purkinje cells and interneurons that are in line with this beam (Fig. 9.22). In addition, because the axons of the interneurons run perpendicular to the parallel fibers, this beam of excitation is flanked by inhibition. Although this classic electrophysiological experiment clearly demonstrates the functional connectivity of the cerebellar cortex, whether such beams of excitation occur normally remains a controversial question.

The Golgi cells are inhibitory interneurons in the granule cell layer. The geometric organization of their axonal and dendritic arbors is an exception to the orthogonal and planar organization of the cortex in that their dendrites and axons carve out approximately conical territories: like two cones, tip to tip, in which the soma is at the point where the two cone tips meet. The dendritic tree forms the upper cone, which often extends into the molecular layer, and the axon forms the lower one. Golgi cells are excited by mossy and olivocerebellar fibers and by granule cell axons (parallel fibers) and inhibited by basket, stellate, and Purkinje cell axon collaterals. They in turn inhibit granule cells. Thus they participate both in feedback (when excited by parallel fibers) and in feedforward (when excited by mossy fibers) inhibitory loops that control activity in the mossy fiber–parallel fiber pathway to the Purkinje cell.

Lugaro cells have fusiform somata from which emerge two relatively unbranched dendrites, one from each side, that run along the transverse axis of the folium for several hundred microns, usually just under the Purkinje cell layer. Purkinje cell axon collaterals provide the main input to these neurons, and granule cell axons add minor input. The axon terminates mainly in the molecular layer on basket, stellate, and possibly Purkinje cells. Thus these cells appear to sample the activity of Purkinje cells and provide both positive-feedback signals (they inhibit the interneurons that inhibit Purkinje cells) and negative-feedback signals (they directly inhibit the Purkinje cell).

Unipolar brush cells have only a single dendrite that ends as a tight bunch of branchlets that resemble a brush. These cells receive excitatory input from mossy fibers and inhibitory input from Golgi cells. It is thought that they synapse with granule and Golgi cells, which would make these cells an excitatory feedforward link in the mossy fiber–parallel fiber pathway.

Candelabrum cells are GABAergic cells located in the Purkinje layer. Their dendrites and axons terminate in the molecular layer, where the axonal arborization pattern resembles a candelabrum.

Cerebellar Nuclei

The cerebellar nuclei are the main targets of the cerebellar cortex. This projection is topographically organized in such a way that each longitudinal strip of cortex targets a specific region of the cerebellar nuclei. The gross pattern is that the vermis projects to the fastigial and vestibular nuclei, the paravermal region projects to the interpositus, and the lateral hemisphere projects to the dentate nucleus.

The cerebellar nuclear neurons in turn provide the output from the cerebellum to the rest of the brain (with the primary exception of Purkinje cells that project to the vestibular nuclei). In discussing the output of the cerebellar nuclei, it is useful to group the nuclear cells according to whether they are GABAergic because the GABAergic cells project back to the inferior olivary nucleus and form a negative-feedback loop to one of the cerebellum's principal afferent sources. Of importance is that GABAergic cells project to the specific part of the inferior olivary nucleus from which they receive input and from which their overlying longitudinal strip of cortex receives climbing fibers. Thus the cerebellar cortex, cerebellar nuclei, and inferior olivary nucleus are functionally organized as a series of closed loops. The non-GABAergic, excitatory nuclear cells project to a variety of targets from the spinal cord to the thalamus. In general, each nucleus gives rise to crossed ascending and descending projections that leave the cerebellum via the superior cerebellar peduncle. The fastigial nucleus also gives rise to significant uncrossed fibers, as well as a second crossed projection called the *uncinate gyrus,* or *hook bundle,* that leaves via the inferior cerebellar peduncle.

Although there are differences in the specific targets of each nucleus, in general, the ascending cerebellar projections target midbrain structures, such as the red nucleus and superior colliculus, and the ventral lateral nucleus of the thalamus, which connects to the primary motor cortex and thereby links the cerebellum to motor areas of the cerebrum. (The cerebral motor areas are likewise linked to the cerebellum by multiple pathways, including ones that relay in the basilar pons and inferior olivary nucleus.) It should also be mentioned that ascending cerebellar projections, particularly from the dentate nucleus, also target nonmotor regions of the cerebrum, particularly in the frontal lobe. The descending fibers target mainly the basilar pontine nuclei, inferior olivary nucleus, and several reticular nuclei. Lastly, a small cerebellospinal pathway arises principally from the fastigial nucleus. In addition, the fastigial nucleus has significant projections to the vestibular nuclei.

Activity of Purkinje Cells in the Cerebellar Cortex in the Context of Motor Coordination

Mossy fiber input to the cerebellar cortex, via their excitation of granule cells, causes a Purkinje cell to discharge single action potentials, referred to as *simple spikes* (Fig. 9.23). The spontaneous simple spike firing rate of a Purkinje cell typically is between 20 and 100 Hz but can be modulated over a much wider range (from 0 to >200 Hz), depending on the relative balance of excitation from parallel fiber input and inhibition from cerebellar cortex interneurons. Thus this activity reflects the state of the cerebellar cortex. Interestingly, evidence from studies performed in the early 2000s indicates that the spontaneous levels of simple spike activity vary systematically across the cerebellar cortex: Firing rates in zebrin-negative regions average twice those of the zebrin-positive regions. The full significance of this discovery is not known, but it suggests that despite the anatomical uniformity of the cerebellar cortical circuits, they might be functionally quite distinct.

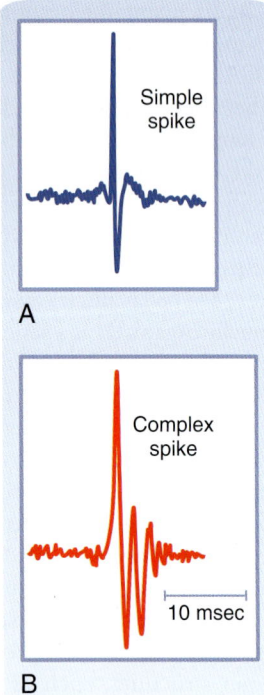

A — Simple spike

B — Complex spike

10 msec

• **Fig. 9.23** Responses of a Purkinje Cell to Excitatory Input, Recorded Extracellularly. **A,** Granule cells, via their ascending axons and parallel fibers, excite Purkinje cells and trigger simple spikes. **B,** Climbing fiber activity leads to high-frequency (≈500-Hz) bursts of spikes known as *complex spikes* in Purkinje cells. Note that the spikes following the initial one are smaller and referred to as spikelets.

In contrast, a climbing fiber discharge causes a high-frequency burst of action potentials, called a *complex spike* (see Fig. 9.23), in an all-or-none manner because of the massive excitation that is provided by the single climbing fiber that a Purkinje cell receives. This excitation is so powerful that there is essentially a one-to-one relationship between climbing fiber discharge and a complex spike. Thus complex spikes essentially override what is happening at the cortex level and reflect the state of the inferior olivary nucleus. The average firing rate of a spontaneous complex spike is only about 1 Hz.

Because the climbing fibers generate complex spikes at such a low frequency, they do not substantially change the average firing rates of Purkinje cells, and as a consequence, it is commonly argued that they have no direct role in shaping the output of the cerebellar cortex and are therefore not involved in ongoing motor control. Instead, it is commonly thought that their function is to alter the responsiveness of Purkinje cells to parallel fiber input. In particular, under certain circumstances, complex spike activity produces a prolonged depression in synaptic efficacy of parallel fibers, termed *long-term depression* (**LTD**). This phenomenon is the proposed mechanism by which climbing fibers act in motor-learning hypotheses. According to typical hypotheses about it, the parallel fiber system and hence simple spikes are involved in generating ongoing movement, and when there is a mismatch between the intended and actual movement, this error activates the inferior olivary nucleus and complex spikes result, which then lead to LTD of the active parallel fiber synapses. This adjustment in synaptic weight changes the motor output in the future. If this change results in a properly executed movement, activation of the inferior olivary nucleus does not occur, and the motor program is unchanged, but if there is still an error, the olivocerebellar system triggers additional complex spikes that cause further changes in synaptic efficacy, and so on. Major challenges to this view are that motor learning can occur when LTD is chemically blocked and that learned behavior can remain after removal of portions of the cerebellum in which the memory is supposedly stored.

An alternative view is that the olivocerebellar system is directly involved in motor control (note that this does not preclude a role in motor learning as well) and, in particular, helps in the timing of motor commands. This view follows from the types of motor deficits observed in cerebellar damage and accounts for the special properties of the inferior olivary nucleus mentioned earlier: namely, that it can generate rhythmic, synchronous complex spike discharges across populations of Purkinje cells. These complex spikes would then produce synchronized inhibitory postsynaptic currents (IPSPs) on cerebellar nuclear neurons as a result of the convergence present in the Purkinje cell axon to the cerebellar nuclear projection. Because of the membrane properties of cerebellar nuclear neurons, these synchronized IPSPs could have a qualitatively different effect on nuclear cell firing than would the IPSPs caused by simple spikes, which are more numerous but largely asynchronous. Specifically, they could trigger large precisely timed changes in the nuclear cell activity that would then be transmitted to other motor systems as a gating signal. In fact, voluntary movements appear to be composed of a series of periodic accelerations that reflect a central oscillatory process. However, determining whether the olivocerebellar system helps time motor commands requires further evidence.

Motor Control by the Basal Ganglia

The basal ganglia are deep nuclei in the cerebrum. Like the cerebellum, a major function of the basal ganglia is to regulate motor activity and we will focus on that role in this section. However, it is worth noting that they, like the cerebellum, also contribute to affective and cognitive functions. To understand basal ganglia function in motor control, the following discussion is organized around two major themes: (1) the connections between basal ganglia and cortex form loops through which activity flows and (2) there are two functionally distinct pathways through the basal ganglia, the direct and indirect pathways.

Organization of the Basal Ganglia and Related Nuclei

The basal ganglia include the **caudate nucleus,** the **putamen,** and the **globus pallidus** (Fig. 9.24). The term **striatum,** derived from the striated appearance of these

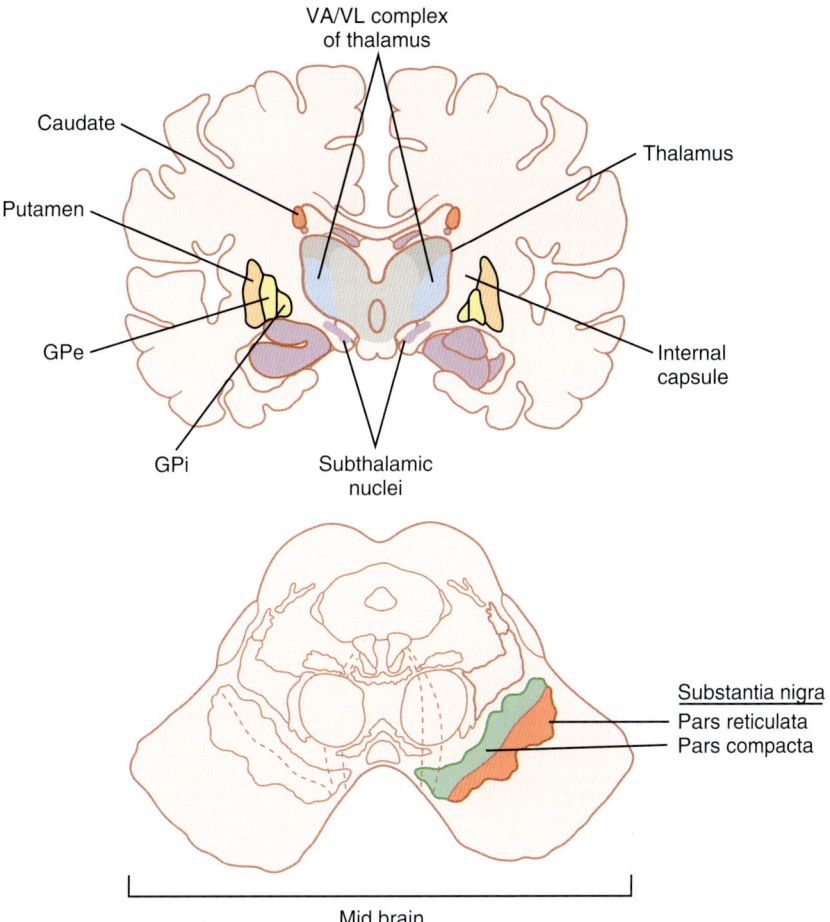

VA/VL complex
of thalamus

Caudate

Putamen

Thalamus

GPe

Internal
capsule

GPi

Subthalamic
nuclei

Substantia nigra
Pars reticulata
Pars compacta

Mid brain

• **Fig. 9.24** Components of Basal Ganglia and Other Closely Associated Brain Regions. The main components of the basal ganglia are the caudate nucleus, putamen, globus pallidus, and substantia nigra pars reticulata. Major portions of the basal ganglia connect with motor areas in the frontal cortex, via the ventral anterior and ventral lateral thalamic nuclei, and with the superior colliculus. Input from the substantia nigra pars compacta is critical for normal basal ganglia function. GPe, external segment of globus pallidus; GPi, internal segment of globus pallidus.

nuclei, refers only to the caudate nucleus and putamen. The striations are produced by the fiber bundles formed by the anterior limb of the internal capsule as it separates the caudate nucleus and putamen. The globus pallidus typically has two parts: an **external segment** and an **internal segment.** The combination of putamen and globus pallidus is often referred to as the **lentiform nucleus.**

Associated with the basal ganglia are several thalamic nuclei. These include the **ventral anterior (VA)** and **ventral lateral (VL) nuclei** and several components of the intralaminar complex. Other associated nuclei are the **subthalamic nucleus** of the diencephalon and the **substantia nigra** of the midbrain (see Fig. 9.24). The substantia nigra ("black substance") derives its name from its content of melanin pigment. Many of the neurons in the **pars compacta** of this nucleus contain melanin, a byproduct of dopamine synthesis. The other subdivision of the substantia nigra is the **pars reticulata.** This structure can be regarded as an extension of the internal segment of the globus pallidus because these nuclei have an identical origin and analogous connections.

Connections and Operation of the Basal Ganglia

With the exception of the primary visual and auditory cortices, most regions of the cerebral cortex project topographically to the striatum. The corticostriatal projection arises from neurons in layer V of the cortex. Glutamate appears to be the excitatory neurotransmitter of these neurons. The striatum then influences neurons in the nuclei of the thalamus by two pathways: direct and indirect (Fig. 9.25A). The thalamic neurons in turn excite neurons of the cerebral cortex thereby forming closed loops with most of the cortex. Several distinct loops have been identified on the basis of cortical regions and function; however, here we will focus on the motor-related loops as a model for basal ganglia operation (Fig. 9.25A).

Direct Pathway

The overall action of the direct pathway through the basal ganglia to motor areas of the cortex is to enhance motor activity. In the direct pathway, the striatum projects to

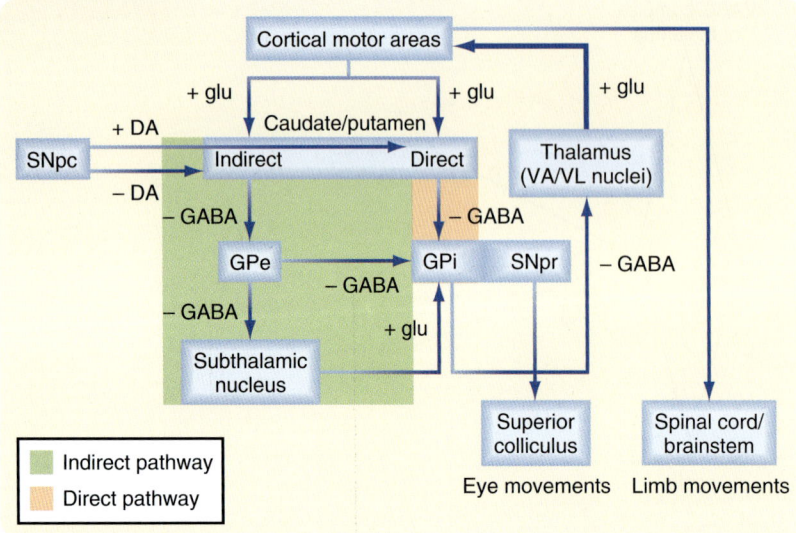

A

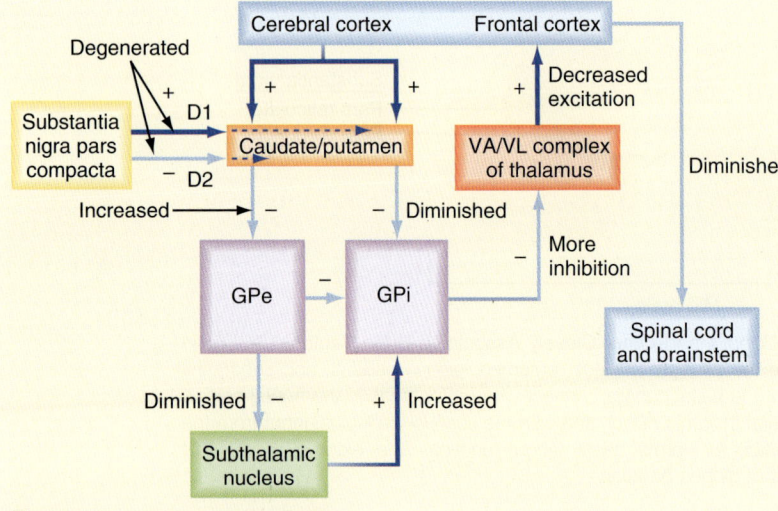

B Parkinson's disease (hypokinetic)

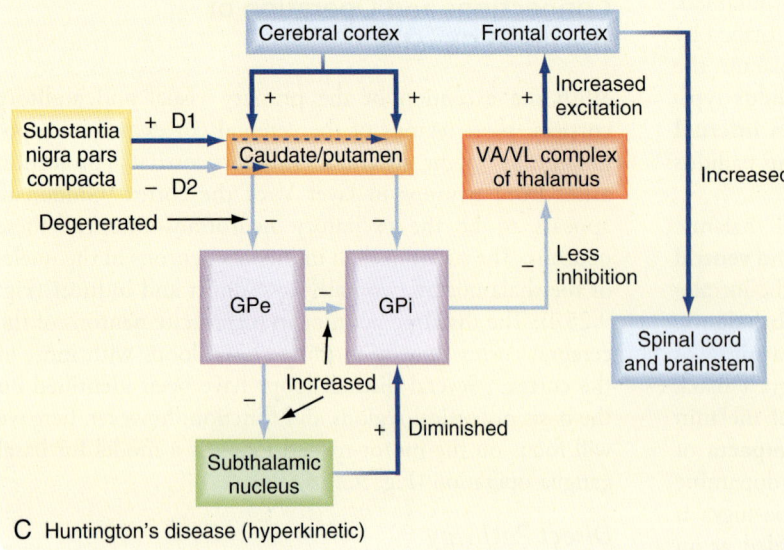

C Huntington's disease (hyperkinetic)

• **Fig. 9.25** Functional Connectivity of the Basal Ganglia for Motor Control. **A,** Connections between various basal ganglia components and other associated motor areas. The excitatory cortical input to the caudate and putamen influences output from the GPi and the substantia nigra pars reticulata (SNpr) via a direct and an indirect pathway. The two inhibitory steps in the indirect pathway mean that activity through this pathway has an effect on basal ganglia output to the thalamus and superior colliculus opposite that of the direct pathway. Dopamine (DA) is a neuromodulator that acts on D_1 and D_2 receptors on striatal neurons that participate in the direct and indirect pathways, respectively. **B,** Changes in activity flow that occur in Parkinson's disease, in which the substantia nigra pars compacta (SNpc) degenerates. **C,** Changes in activity flow in Huntington's disease, in which inhibitory control of the GPe is lost. *Plus symbols* (+) and *minus symbols* (–), respectively, indicate the excitatory or inhibitory nature of a synaptic connection. glu, glutamate; GPe, external globus pallidus; GPi, internal globus pallidus; VA/VL, ventral anterior/ventral lateral nuclei of the thalamus.

the internal segment of the globus pallidus (GPi). This projection is inhibitory, and the main transmitter is GABA. The GPi projects to the VA and VL nuclei of the thalamus. These connections also function with GABA and are inhibitory. The VA and VL nuclei send excitatory connections to the prefrontal, premotor, and supplementary motor cortex. This input to the cortex influences motor planning, and it also affects the discharge of corticospinal and corticobulbar neurons.

The direct pathway appears to function as follows: Neurons in the striatum have little background activity, but during movement they are activated by their input from the cortex. In contrast, neurons in the GPi have a high level of background activity. When the striatum is activated, its inhibitory projections to the globus pallidus slow the activity of pallidal neurons. However, the pallidal neurons themselves are inhibitory, and they normally provide tonic inhibition of neurons in the VA and VL nuclei of the thalamus. Therefore, activation of the striatum causes **disinhibition** of neurons of the VA and VL nuclei. When disinhibited, the VA/VL neurons increase their firing rates, exciting their target neurons in the motor areas of the cerebral cortex. Because the motor areas evoke movement by activating α and γ motor neurons in the spinal cord and brainstem, the basal ganglia can regulate movement by enhancing the activity of neurons in the motor cortex.

Indirect Pathway

The overall effect of the indirect pathway is to reduce the activity of neurons in motor areas of the cerebral cortex. The indirect pathway involves inhibitory connections from the striatum to the external segment of the globus pallidus (GPe), which in turn sends an inhibitory projection to the subthalamic nucleus and to the GPi. The subthalamic nucleus then sends an excitatory projection back to the GPi (see Fig. 9.25A).

In this pathway, pallidal neurons in the external segment are inhibited by the GABA released from striatal terminals in the globus pallidus. The GPe normally releases GABA in the subthalamic nucleus and thereby inhibits the subthalamic neurons. Therefore, striatal inhibition of the GPe results in the disinhibition of neurons of the subthalamic nucleus. The subthalamic neurons are normally active, and they excite neurons in the GPi by releasing glutamate. When the neurons of the subthalamic nucleus become more active because of disinhibition, they release more glutamate in the GPi. This transmitter excites neurons in the GPi and consequently activates inhibitory projections that affect the VA and VL thalamic nuclei. The activity of the thalamic neurons consequently decreases, as does the activity of the cortical neurons that they influence.

The direct and indirect pathways thus have opposing actions; an increase in the activity of either one of these pathways might lead to an imbalance in motor control. Such imbalances, which are typical of basal ganglion diseases, may alter the motor output of the cortex.

Actions of Neurons in the Pars Compacta of the Substantia Nigra on the Striatum

Dopamine is the neurotransmitter used by neurons of the substantia nigra pars compacta. In the nigrostriatal pathway, release of dopamine has an overall excitatory action on the direct pathway and an inhibitory action on the indirect pathway. This is, however, a modulatory type of effect; that is, dopamine is apparently causing its action not by triggering spikes directly but rather by altering the striatal cells' response to other transmitters. The different actions on the direct and indirect pathways result from the expression of different types of dopamine receptors (D_1 and D_2) by the spiny projection cells of the striatum that contribute to the direct and indirect pathways. D_1 receptors are found on striatal cells that form the direct pathway by projecting to the GPi, whereas D_2 receptors are found on striatal cells that participate in the indirect pathway and project to the GPe. The overall consequence of dopamine release in both cases is facilitation of activity in the motor areas of the cerebral cortex.

Subdivision of the Striatum Into Striosomes and Matrix

On the basis of the associated neurotransmitters, the striatum has been subdivided into zones called **striosomes** and **matrix.** The cortical projections related to motor control end in the matrix area. The limbic system projects to the striosomes. Striosomes are thought to synapse in the pars compacta of the substantia nigra and to influence the dopaminergic nigrostriatal pathway.

Role of the Basal Ganglia in Motor Control

The basal ganglia influence the cortical motor areas. Therefore, the basal ganglia have an important influence on the lateral system of motor pathways. Such an influence is consistent with some of the movement disorders observed in diseases of the basal ganglia. However, the basal ganglia must additionally regulate the medial motor pathways because diseases of the basal ganglia can also affect the posture and tone of proximal muscles.

The deficits seen in the various basal ganglia diseases include abnormal movement (**dyskinesia**), increased muscle tone (**cogwheel rigidity**), and slowness in initiating movement (**bradykinesia**). Abnormal movement includes tremor, **athetosis, chorea, ballism,** and **dystonia.** The **tremor** of basal ganglion disease is a 3-Hz "pill-rolling" tremor that occurs when the limb is at rest. Athetosis consists of slow, writhing movement of the distal parts of the limbs, whereas chorea is characterized by rapid, flicking movement of the extremities and facial muscles. Ballism is associated with violent, flailing movement of the limbs (ballistic movement). Finally, dystonic movements are slow involuntary movements that may cause distorted body postures.

Parkinson's disease is a common disorder characterized by tremor, rigidity, and bradykinesia. This disease is caused by loss of neurons in the pars compacta of the substantia nigra. Consequently, the striatum suffers a severe loss of dopamine. Neurons of the locus coeruleus and the raphe nuclei, as well as other monoaminergic nuclei, are also lost. The loss of dopamine diminishes the activity of the direct pathway and increases the activity of the indirect pathway (see Fig. 9.25B). The net effect is an increase in the activity of neurons in the internal segment of the globus pallidus. This results in greater inhibition of neurons in the VA and VL nuclei and less pronounced activation of the motor cortical areas. The consequence is slowed movement (bradykinesia).

Before the dopaminergic neurons are completely lost, administration of levodopa can relieve some of the motor deficits in Parkinson's disease. Levodopa is a precursor of dopamine, and it can cross the blood-brain barrier. Currently, the possibility of transplanting dopamine-synthesizing neurons into the striatum is being explored. Future research will no doubt focus on the potential for human embryonic stem cells to play such a therapeutic role.

Another basal ganglia disturbance is Huntington's disease, which results from a genetic defect that involves an autosomal dominant gene. This defect leads to the preferential loss of striatal GABAergic and cholinergic neurons that project to the GPe as part of the indirect pathway (and also degeneration of the cerebral cortex, with resultant dementia). Loss of inhibition of the GPe presumably leads to diminished activity of neurons in the subthalamic nucleus (see Fig. 9.25C). Hence, the excitation of neurons of the GPi would be reduced. This would disinhibit neurons in the VA and VL nuclei. The resulting enhancement of activity in neurons in the motor areas of the cerebral cortex may help explain the choreiform movements of Huntington's disease. The rigidity in Parkinson's disease may in a sense be the opposite of chorea because overtreatment of patients with Parkinson's disease with levodopa can result in chorea.

Hemiballism is caused by a lesion of the subthalamic nucleus on one side of the brain. In this disorder, involuntary, violent flailing movements of the limbs may occur on the side of the body contralateral to the lesion. Because the subthalamic nucleus excites neurons of the GPi, a lesion of the subthalamic nucleus would reduce the activity of these pallidal neurons. Therefore, neurons in the VA and VL nuclei of the thalamus would be less inhibited, and the activity of neurons in the motor cortex would be increased.

In all these basal ganglia disorders, the motor dysfunction is contralateral to the diseased component. This is understandable because the main final output of the basal ganglia to the body is mediated by the corticospinal tract.

Eye Movement

Eye movement has a number of features that distinguish it from other motor behavior. In comparison with the movement that limbs, with their multiple joints and muscles, can perform, eye movement is relatively simple. For example, each eye is controlled by only three agonist-antagonist muscle pairs: the medial and lateral recti, the superior and inferior recti, and the superior and inferior oblique muscles. These muscles allow the eye to rotate about three axes. Assuming that the head is in an upright position, the axes are the vertical axis, a horizontal axis that runs left to right, and the torsional axis (which is directed along the axis of sight). The medial and lateral recti control movement about the vertical axis; the other four muscles generate movement about the horizontal and torsional axes. Another simplifying feature is that there are no external loads for which to be compensated. Furthermore, eye movement appears to be separable into a few distinct types, with each type being controlled by its own specialized circuitry. Thus eye movement offers a number of advantages as a model system for studying motor control. Moreover, deficits in eye movement provide important clinical clues to the diagnosis of neurological problems. We first review the different eye movement types and then discuss the neural circuitry underlying their generation.

Types of Eye Movement
Vestibuloocular Reflex

Eye movement probably first evolved to hold the image of the external world still. (In contrast, limb movements evolved to generate changes in the position of the limb with regard to the external world.) The reason is that visual acuity degrades rapidly when there is eye movement in relation to the external world (i.e., the visual scene slips across the retina). A major cause of such slippage is movement of the head. The vestibuloocular reflex (VOR) is one of the main mechanisms by which head movement is compensated in order to maintain stability of the visual scene on the retina.

To maintain a stable visual scene on the retina, the VOR produces movements of the eyes that are equal and opposite the movement of the head. This reflex is initiated by stimulation of the receptors (hair cells) in the vestibular system (see Chapter 8). Recall that the vestibular organs are sensitive to head acceleration, not visual cues, and thus the VOR occurs in both the light and dark. Functionally, it is what is called an *open-loop system* in that it generates an output (eye movement) in response to a stimulus (head acceleration), but its immediate behavior is not regulated by feedback about the success or failure of its output. It is worth noting, however, that in the light at least, any failure by the VOR to match eye and head rotation results in what is called *retinal slip* (i.e., slip of the visual image across the retina), and this error signal can be fed back to the VOR circuits by other neuronal pathways and over time can lead to adjustments in the strength of the VOR to eliminate the error. This adaptation of the VOR is a major model for studying plasticity in the brain.

As stated, acceleration signals initiate the VOR. The output of the VOR, however, must be a change in eye position in the orbit. Thus the problem to be solved by the nervous system is to translate the acceleration signals sensed by the vestibular organs into correct positional signals for the eyes. Mathematically, this can be thought of as a double integration. The first integration occurs in the vestibular receptor apparatus because although the hair cells respond to head acceleration, the signals in the vestibular afferent fibers are proportional to head velocity (at least for most stimuli that are encountered physiologically). The second integration, from velocity to position, occurs in the CNS in circuits described later.

The head can move in six different ways, often referred to as *six degrees of freedom:* three translational and three rotational. To compensate for these different types of movement, there are both translational and angular VORs, as well as separate subsystems for handling movement about different directions (e.g., rotation about a vertical or a horizontal axis).

Optokinetic Reflex

The optokinetic reflex (OKR) is a second mechanism by which the CNS stabilizes the visual scene on the retina, and it often works in conjunction with the VOR. Whereas the VOR is activated only by head motion, the OKR is activated by movement of the visual scene, whether caused by motion of the scene itself or by head motion. Specifically, the sensory stimulus for this reflex is slip of the visual scene on the retina as detected by motion-sensitive retinal ganglion cells. An example of the former occurs when you are sitting in a train and a train on the adjacent track begins moving: Your eyes rotate to keep the image of the neighboring car stable. This often leads to a sensation that you are moving (this is not entirely surprising because OKR circuits feed into the same circuits as used by the vestibular system).

The OKR can work in conjunction with the VOR to stabilize the visual image and is particularly important for maintaining a stable image when head movements are slow because the VOR works poorly in these conditions.

Saccades

In animals whose eyes have a fovea, it becomes particularly advantageous to be able to move the eye in relation to the world (i.e., the main visual scene) so that objects of importance can be focused onto the fovea and scrutinized with this high-resolution part of the retina. Two classes of eye movement underlie this ability: saccadic and smooth pursuit. Very rapid discrete movements that bring a particular region of the visual world onto the fovea are called *saccades.* For example, to read this sentence, you are making a series of saccades to bring successive words onto your fovea to be read. However, even in animals that lack a fovea, the eyes make saccades, and thus saccades may also be used to rapidly scan the visual environment.

Saccades are extremely rapid eye movements. In humans, eye velocity during a saccade can reach 800 degrees/second, in comparison with movement velocity of less than 10 degrees/second generated in response to typical VOR and OKR stimuli (velocities of up to ≈120 degrees/second can be produced by OKR stimuli in humans; however, they are still much slower than the maximal saccade velocities). Saccades can be made voluntarily or reflexively. Moreover, although they are usually made in response to visual targets, they can also be made toward auditory or other sensory cues, in the dark, or toward memorized targets.

Interestingly, visual processing appears to be suppressed just before and during saccades, particularly in the magnocellular visual pathway that is concerned with visual motion. This phenomenon is known as *saccadic suppression* and may function to prevent sensations of sudden, rapid movement of the visual world that would result during a saccade in the absence of such suppression. The mechanisms underlying saccadic suppression are not fully known, but in areas of the cortex related to visual processing, the responsiveness of the cells to visual stimuli is reduced and altered during saccades.

Smooth Pursuit

Once a saccade has brought a moving object of interest onto the fovea, the smooth pursuit system allows the person to keep it stable on the fovea despite its continued motion. This ability appears to be limited to primates and allows prolonged continuous observation of a moving object. Note that in some respects, smooth pursuit might seem similar to the OKR; in fact, there may not be an absolute difference because as the target size grows, the distinction between target and background is lost; however, for small moving targets, smooth pursuit requires suppression of the OKR. You can see the effect of this suppression by moving your finger back and forth in front of this text while tracking it with your eyes. Your finger will be in focus, but the words on this page will be part of the background scene and will become illegible as they slip along your retina.

Nystagmus

When there is a prolonged OKR or VOR stimulus (e.g., if you keep turning in one direction), these reflexes will initially counterrotate the eyes in an attempt to maintain a stable image on the retina, as described earlier. However, with a prolonged stimulus, the eyes will reach their mechanical limit, no further compensation will be possible, and the image will begin to slip on the retina. To avoid this situation, a fast saccade-like movement of the eyes occurs in the opposite direction, essentially resetting the eyes to begin viewing the visual scene again. Then the slow OKR- or VOR-induced counterrotation will start anew. This alternation of slow and fast movement in opposite directions is nystagmus and can be recorded on a nystagmogram (Fig. 9.26). Thus nystagmus can be defined as oscillatory or rhythmic movements of the eye in which there is a fast phase and a slow phase. The nystagmus is named according to the direction of the fast phase because the fast phase is more easily observed.

In addition to being induced physiologically by VOR or OKR stimuli, nystagmus can result from damage to the

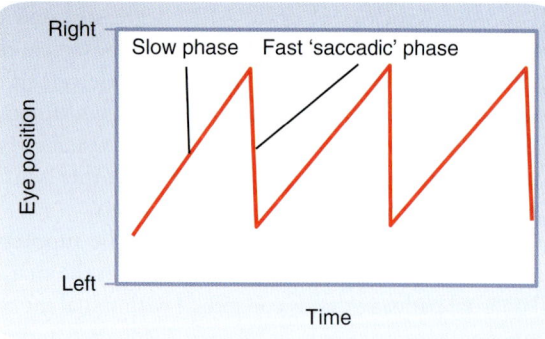

• **Fig. 9.26** Nystagmogram Showing Eye Movements That Occur During Nystagmus. The plot shows a left nystagmus because the fast phase is directed toward the left (downward on the graph).

vestibular circuits, either in the periphery (e.g., cranial nerve VIII) or centrally (e.g., vestibular nuclei), and can be an informative diagnostic symptom.

Vergence

Conjugate eye movement is movement of both eyes in the same direction and in an equal amount. Such coordination allows a target to be maintained on both foveae during eye movement and is necessary to maintain binocular vision without diplopia (double vision). However, when objects are close (<30 m), maintaining a target on both foveae requires non-identical movements of the two eyes. Such disjunctive, or vergence, movements are also necessary for fixation of both eyes on objects that are approaching or receding. Stimuli that trigger vergence movements are diplopia and blurry images. It should be noted that when tracking an approaching object in addition to convergence movements, the lens accommodates for near vision, and pupillary constriction occurs.

Neural Circuitry and Activity Underlying Eye Movement

Motor Neurons of the Extraocular Muscles

Three cranial nerve nuclei supply the extraocular muscles: oculomotor, trochlear, and abducens nuclei. These three nuclei are sometimes referred to collectively as the *oculomotor nuclei;* however, the context (the specific nucleus or all three) should be clear. Motor neurons for the ipsilateral medial and inferior recti, ipsilateral inferior oblique, and contralateral superior rectus muscles reside in the oculomotor nucleus; those for the contralateral superior oblique muscle reside in the trochlear nucleus; and those for the ipsilateral lateral rectus muscle are located in the abducens nucleus. These motor neurons form some of the smallest motor units (1:10 nerve-to-muscle ratio), which is consistent with the very fine control needed for precise eye movement.

An important point regarding motor neurons innervating the extraocular muscles is that most have spontaneous activity when the eye is in the primary position (looking straight ahead), and their firing rate is correlated with eye position and velocity. This spontaneous activity allows the antagonist muscle pairs to act in a push-pull manner, which increases the responsiveness of the system. That is, as motor neurons innervating one muscle are activated and cause increased contraction, those innervating its antagonist are inhibited, which leads to relaxation.

In addition to motor neurons, the abducens nuclei have internuclear neurons. These neurons project, via the medial longitudinal fasciculus, to medial rectus motor neurons in the contralateral oculomotor nucleus. As described later, this projection facilitates the coordinated action of the medial and lateral recti muscles that is needed for conjugate movements, such as those that occur in the VOR.

Circuits Underlying the Vestibuloocular Reflex

The VOR acts to counter head motion by causing rotation of the eyes in the opposite direction. There are separate circuits for rotational and translational movement of the head. The sensors for the former are the semicircular canals, and the sensors for the latter are the otoliths (the utricle and saccule). The circuits for the angular VOR are more straightforward (but still complex), and this section focuses on these pathways to illustrate how this reflex works; however, the basic scheme is the same: Vestibular afferent fibers go to vestibular nuclei, the vestibular nuclei in turn project to the various oculomotor nuclei, and motor neurons in the oculomotor nuclei give rise to axons that innervate the extraocular muscles. What varies are the specific vestibular and oculomotor nuclei that are involved.

With regard to the angular VOR pathways, the pathway for generating horizontal eye movement originates in the horizontal canals, and the analogous one for vertical movement originates in the anterior and posterior canals. Fig. 9.27*A* shows the basic circuit for the horizontal VOR. Note that only the major central circuits originating in the left horizontal canal and vestibular nuclei are shown; however, mirror image pathways arise from the right canal and vestibular nuclei. Vestibular afferent fibers involved in the horizontal VOR pathway synapse primarily in the medial vestibular nucleus, which projects to the abducens nucleus bilaterally; inhibitory neurons project ipsilaterally, and excitatory ones project contralaterally. Control of the medial rectus muscle is achieved by abducens internuclear neurons that project from the abducens to the part of the oculomotor nucleus controlling the medial rectus muscle. Note the double decussation of this pathway results in aligning of the responses of functional synergists (e.g., the left medial rectus with the right lateral rectus).

The vertical VOR pathway involves primarily the superior vestibular nucleus, which has direct bilateral projections to the oculomotor nucleus.

Consider what happens in the horizontal canal pathway when there is head rotation to the left, as shown in Fig. 9.27*B*. Leftward head rotation would cause the visual image to slip to the right. However, compensation by the VOR would be triggered by depolarization of the hair cells of the left canal in response to the angular acceleration

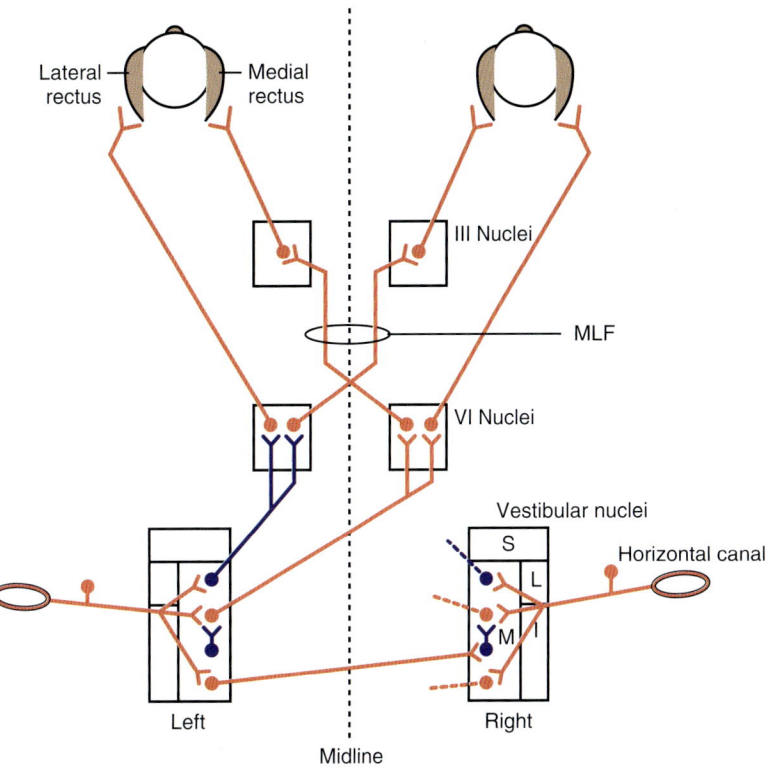

• **Fig. 9.27** Circuits Underlying the Horizontal Vestibuloocular Reflex (VOR). **A,** The vestibular nuclei receive excitatory input from the afferent fibers of the horizontal canal and project to the abducens (cranial nerve VI) nucleus. This nucleus innervates the lateral rectus muscle and projects to the contralateral oculomotor (cranial nerve III) nucleus, which controls the medial rectus muscle. Excitatory neurons are shown in *red;* inhibitory ones, in *blue.* Note that only the major pathways originating in the left vestibular nuclei are shown. For clarity, only the beginnings of mirror image pathways from the right vestibular nuclei are shown *(dotted lines).* **B,** Flow of activity in the VOR circuitry induced by leftward head rotation. Increased axonal thickness indicates increased activity; thinner axons indicate decreased activity in comparison with levels at rest **(A).** Note that leftward rotation causes both an increase in activity of the left vestibular afferent fibers and a decrease in activity of the right ones. MLF, medial longitudinal fasciculus; vestibular nuclei: I, inferior; L, lateral; M, medial; S, superior.

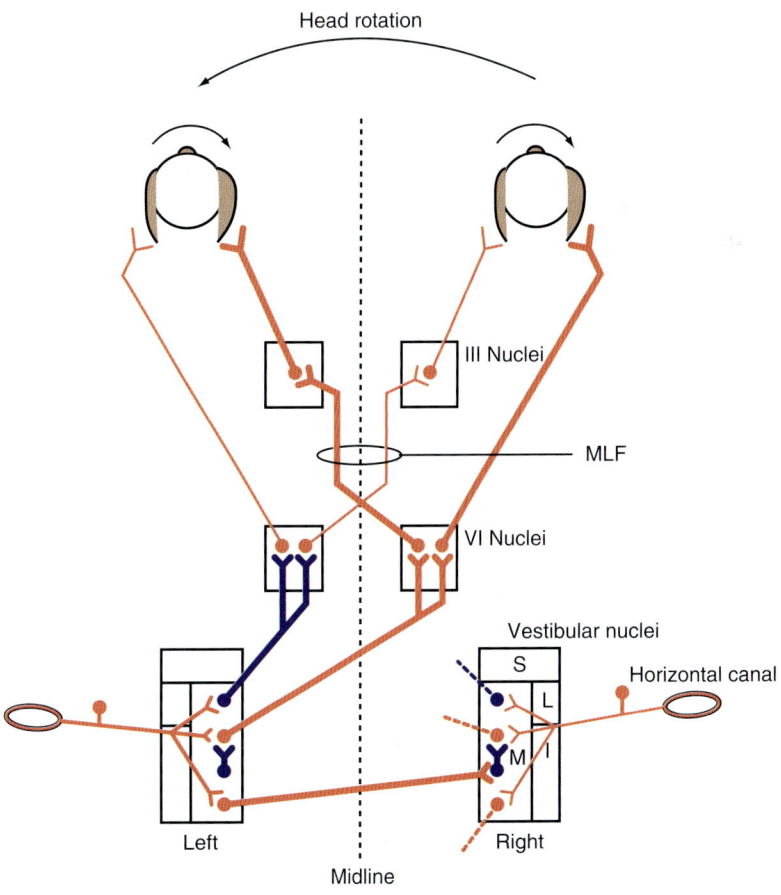

(see Fig. 8.27). The depolarized hair cells cause increased activity in the left vestibular afferent fibers and thereby excite neurons of the left medial vestibular nucleus. These include excitatory neurons that project to the contralateral abducens nucleus and synapse with both motor neurons and internuclear neurons. Excitation of the motor neurons leads to contraction of the right lateral rectus muscle and rotation of the right eye to the right, whereas excitation of the internuclear neurons of the right abducens nucleus leads to excitation of the medial rectus motor neurons in the left oculomotor nucleus, thus causing the left eye to rotate to the right as well.

Along the pathway starting with the inhibitory vestibular neurons that project from the left medial vestibular nucleus to the ipsilateral abducens nucleus, the activity of these cells leads to inhibition of motor neurons to the left lateral rectus muscle and motor neurons to the right medial rectus muscle (the latter via internuclear neurons to the right oculomotor nucleus). Consequently, these muscles relax, thereby facilitating rotation of the eyes to the right. Thus the eye is being pulled by the increased tension of one set of muscles and "pushed" by the release of tension in the antagonist set of muscles.

Note again that the mirror image pathways originating from the right canal have been left out of Fig. 9.27 for clarity, but the changes in activity through them with leftward head rotation would be exactly the opposite, and thus they would function synergistically with those that are shown. As an exercise, work out the resulting changes in activity through these circuits. Remember that leftward head rotation hyperpolarizes the hair cells of the right canal, thereby leading to a decrease in right vestibular afferent activity and disfacilitation of the right vestibular nuclear neurons.

Now, consider the commissural fibers that connect the two medial vestibular nuclei are excitatory but end on local inhibitory interneurons of the contralateral vestibular nucleus and thus inhibit the projection neurons of that nucleus. This pathway reinforces the actions of the contralateral vestibular afferent fibers on their target vestibular nuclear neurons. In the aforementioned example, commissural cells in the left vestibular nucleus are activated and therefore cause active inhibition of the right medial vestibular nuclei projection neurons, which reinforces the disfacilitation caused by the decrease in right afferent activity. In fact, this commissural pathway is powerful enough to modulate the activity of the contralateral vestibular nuclei even after unilateral labyrinthectomy, which destroys the direct vestibular afferent input to these nuclei.

Of importance is that superimposed on the brainstem circuits is the cerebellum. Parts of the vermis and flocculonodular lobe receive primary vestibular afferent fibers or secondary vestibular afferent fibers (axons of the vestibular nuclear neurons), or both, and in turn project back to the vestibular nuclei directly and via a disynaptic pathway involving the fastigial nucleus. The exact role of these cerebellar circuits in generating the VOR is much debated, but they are critical inasmuch as damage to them leads to

abnormal eye movement, such as spontaneous nystagmus, and other symptoms of vestibular dysfunction.

IN THE CLINIC

When a labyrinth is irritated in one ear, as in **Meniere's disease,** or when a labyrinth is rendered nonfunctional, as may happen as a result of head trauma or disease of the labyrinth, the signals transmitted through the VOR pathways from the two sides become unbalanced. Vestibular nystagmus can then result. For example, irritation of the labyrinth of the left ear can increase the discharge of afferent fibers that supply the left horizontal semicircular duct. The signal produced resembles that normally generated when the head is rotated to the left. Because the stimulus is ongoing, a left nystagmus results, with a slow phase to the right (caused by the VOR pathway) and a fast phase to the left. Destruction of the labyrinth in the right ear produces effects similar to those induced by irritation of the left labyrinth. Interestingly, the nystagmus is temporary, which shows the ability of these circuits to adapt over time.

Circuits Underlying the Optokinetic Reflex

The stimulus eliciting the OKR is visual (retinal slip), and so photoreceptors are the start of the reflex arc. Key brainstem centers for this reflex lie in the tegmentum and pretectal region of the rostral midbrain. They are the nucleus of the optic tract (NOT) and a group of nuclei collectively known as the accessory optic nuclei (AON). Direction-selective, motion-sensitive retinal ganglion cells are a major afferent source carrying visual information to these nuclei. In addition, input comes from primary and higher order visual cortical areas in the occipital and temporal lobes. These latter afferent sources are particularly important in primates and humans. Cells of the NOT and AON have large receptive fields, and their responses are selective for the direction and speed of movement of the visual scene. Of interest is that the preferred directions of movement of the NOT/AON cells correspond closely to motion caused by rotation about axes perpendicular to the semicircular canals, thereby facilitating coordination of the VOR and OKR to produce stable retinal images.

The efferent connections of these nuclei are numerous and complex and not fully understood. There are polysynaptic pathways to the oculomotor and abducens nuclei and monosynaptic input to the vestibular nuclei, which allow interaction with the VOR. There are projections to various precerebellar nuclei, including the inferior olivary nucleus and basilar pontine nuclei. These pathways then loop through the flocculus and back to the vestibular nuclei. In sum, via several pathways operating in parallel, activity ultimately arrives at the various oculomotor nuclei whose motor neurons are activated, and proper counterrotation of the eyes results.

Circuits Underlying Saccades

Saccades are generated in response to activity in the superior colliculus or the cerebral cortex (frontal eye fields and

IN THE CLINIC

Clinical testing of labyrinthine function is done either by rotating the patient in a Bárány chair to activate the labyrinths in both ears or by introducing cold or warm water into the external auditory canal of one ear **(caloric test).** When a person is rotated in a Bárány chair, nystagmus develops during the rotation. The direction of the fast phase of the nystagmus is in the same direction as the rotation. When the rotation of the chair is halted, nystagmus in the opposite direction develops (postrotatory nystagmus) because stopping a rotation has the same effect as accelerating in the opposite direction.

The caloric test is more useful because it can distinguish between malfunction of the labyrinths on the two sides. The neck is bent backward about 60 degrees so that the two horizontal canals are essentially vertical. If warm water is introduced into the left ear, the endolymph in the outer portion of the loop of the left semicircular canal tends to rise as the specific gravity of the endolymph decreases because of heating. This sets up a convection flow of endolymph, and as a result, the kinocilia of the left ampullary crest hair cells are deflected toward the utricle, as if the head had rotated to the left; the discharge of the afferent fibers that supply this canal increases; and nystagmus occurs with the fast phase toward the left. The nystagmus produces a sensation that the environment is spinning to the right, and the person tends to fall to the right. The opposite effects are produced if cold water is placed in the ear. A mnemonic expression that can help in remembering the direction of the nystagmus in the caloric test is COWS ("cold opposite, warm same"). In other words, cold water in one ear results in a fast phase of nystagmus toward the opposite side, and warm water causes a fast phase toward the same side.

posterior parietal areas). Activity in the superior colliculus is related to computation of the direction and amplitude of the saccade. Indeed, the deep layers of the superior colliculus contain a topographic motor map of saccade locations. From the superior colliculus, information is forwarded to distinct sites for control of horizontal and vertical saccades, referred to as the *horizontal* and *vertical gaze centers,* respectively. The horizontal gaze center consists of neurons in the paramedian pontine reticular formation, in the vicinity of the abducens nucleus (Fig. 9.28A). The vertical gaze center is located in the reticular formation of the midbrain: specifically, the rostral interstitial nucleus of the medial longitudinal fasciculus and the interstitial nucleus of Cajal. Because the circuitry and operation of the horizontal gaze center are better understood than those of the vertical gaze center, it is discussed here in detail. However, cells showing analogous activity patterns have been described in the vertical gaze center.

Fig. 9.28A is an overview of the neural circuitry by which saccades are generated, and Fig. 9.28B shows the activity of certain types of neurons found in the gaze center that are responsible for horizontal saccades. Each horizontal gaze center has excitatory burst neurons that project to motor neurons in the ipsilateral abducens nucleus and to the internuclear neurons (which excites medial rectus motor

neurons in the contralateral oculomotor nucleus). It also has inhibitory burst neurons that inhibit the contralateral abducens. These burst neurons are capable of extremely high bursts of spikes (up to 1000 Hz). Moreover, the gaze center has neurons showing tonic activity and burst-tonic activity.

Normally, both inhibitory and excitatory burst neurons are inhibited by omnipause neurons located in the nucleus of the dorsal raphe. When a saccade is to be made, activity from the frontal eye fields or the superior colliculus, or both, leads to inhibition of the omnipause cells and excitation of the burst cells on the contralateral side. The resulting high-frequency bursts in the excitatory burst neurons provide a powerful drive to motor neurons of the ipsilateral lateral rectus and contralateral medial rectus muscles (see Fig. 9.28A); at the same time, inhibitory burst neurons enable relaxation of the antagonists. The initial bursts of these neurons allow strong contraction of the appropriate extraocular muscles, which overcomes the viscosity of the extraocular muscle and enables rapid movement to occur.

Circuits Underlying Smooth Pursuit

Smooth pursuit involves tracking a moving target with the eyes (Fig. 9.29). Visual information about target velocity is processed in a series of cortical areas, including the visual cortex in the occipital lobe, several temporal lobe areas, and the frontal eye fields. In the past, the frontal eye fields were thought to be related only to control of saccades, but more recent evidence has shown that there are distinct regions within the frontal eye fields dedicated to either saccade production or smooth pursuit. Indeed, there may be two distinct cortical networks, each specialized for one of these types of eye movement. Cortical activity from multiple cortical areas is fed to the cerebellum via parts of the pontine nuclei and nucleus reticularis tegmenti pontis. Specific areas in the cerebellum—namely, parts of the posterior lobe vermis, the flocculus, and the paraflocculus—receive this input, and they in turn project to the vestibular nuclei. From the vestibular nuclei, activity can then be forwarded to the oculomotor, abducens, and trochlear nuclei, as was described for the VOR earlier.

Circuits Underlying Vergence

The neural circuits underlying vergence movements are not well known. There are premotor neurons (neurons that feed onto motor neurons) located in the brainstem areas surrounding the various oculomotor nuclei. In some cortical visual areas and the frontal eye fields, there are neurons whose activity is related to the disparity of the image on the two retinas or to the variation of the image during vergence movements. How vergence signals in these cortical areas feed into the brainstem premotor neurons is not clear. The cerebellum also appears to play a role in vergence movements because cerebellar lesions impair this type of eye movement. Note that lesions of the medial longitudinal fasciculus that result in a loss of the VOR do not compromise vergence.

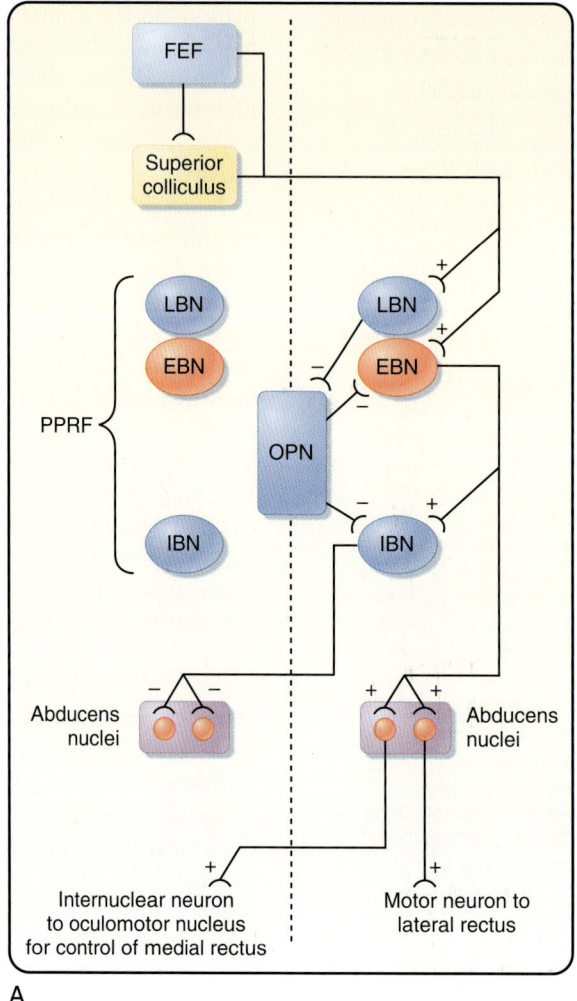

A

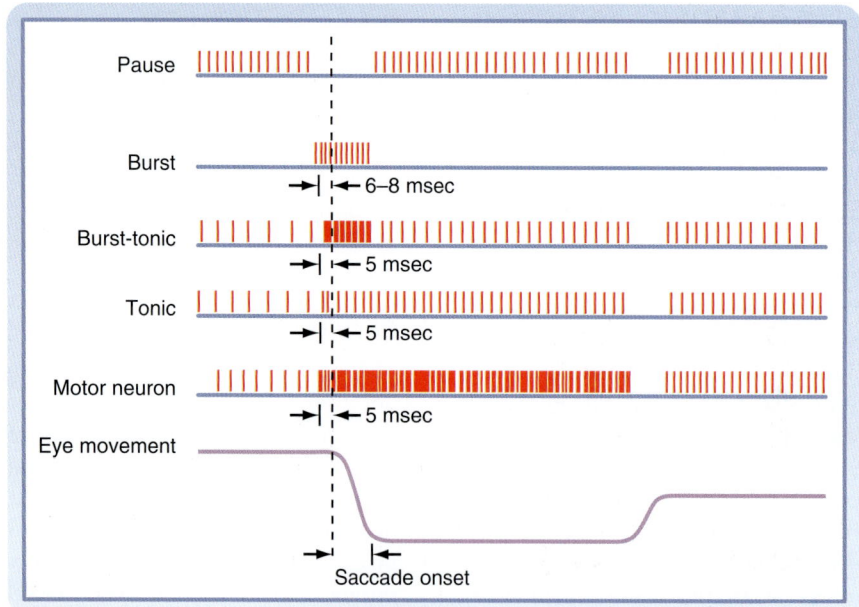

B

• **Fig. 9.28** Horizontal Saccade Pathways. **A,** Circuit diagram of the major pathways. EBN, excitatory burst neuron; FEF, frontal eye field; IBN, inhibitory burst neuron; LBN, long lead burst neuron; OPN, omnipause neuron; PPRF, paramedian pontine reticular formation. **B,** Firing patterns of some of the neurons involved in making saccades. Excitation of burst neurons of the right horizontal gaze center causes abducens motor neurons on the right and medial rectus motor neurons on the left to be activated. The ascending pathway to the oculomotor nucleus is through the medial longitudinal fasciculus. The left horizontal gaze center is simultaneously inhibited.

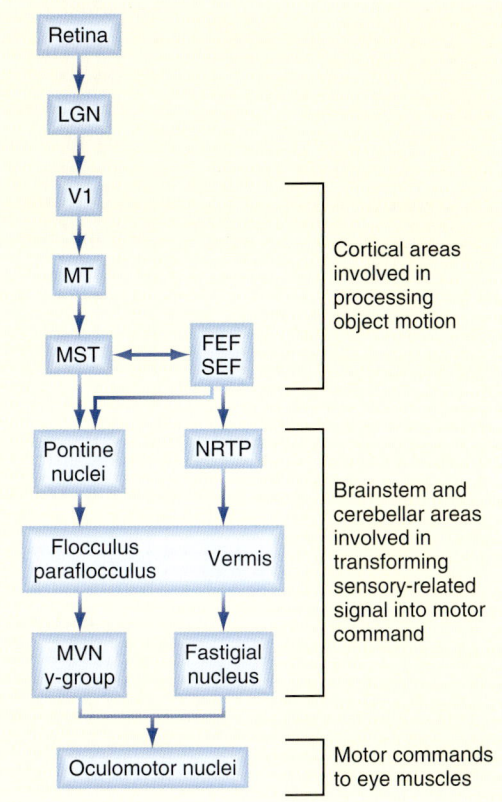

• Fig. 9.29 Smooth Pursuit Pathways. The stimulus for smooth pursuit eye movement is a moving visual target. This causes activity to flow through the circuitry diagrammed in the figure and leads to maintenance of the fovea on the target. FEF, frontal eye field; LGN, lateral geniculate nucleus; MST and MT, higher order visual association areas; MVN, medial vestibular nucleus; NRTP, nucleus reticularis tegmenti pontis; SEF, supplementary eye field; V1, primary visual cortex.

Key Points

1. The extrafusal skeletal muscle fibers are innervated by α motor neurons. A motor unit consists of a single α motor neuron and all the muscle fibers with which it synapses. Motor unit size varies greatly among muscles; small motor units allow finer control of muscle force.

2. The size principle refers to the orderly recruitment of α motor neurons according to their size, from smallest to largest. Because smaller motor neurons connect to weaker motor units, the relative fineness of motor control is similar for weak and strong contractions.

3. A reflex is a simple, stereotyped motor response to a stimulus. A reflex arc includes the afferent fibers, interneurons, and motor neurons responsible for the reflex.

4. Muscle spindles are complex sensory receptors found in skeletal muscle. They lie parallel to extrafusal muscle fibers, and they contain nuclear bag and nuclear chain intrafusal muscle fibers. By being in parallel to the main muscle, the spindle can detect changes in muscle length.

5. Group Ia afferent fibers form primary endings on nuclear bag1, bag2, and chain fibers, and group II fibers form secondary endings on nuclear chain and bag2 fibers.

6. Primary endings demonstrate both static and dynamic responses that signal muscle length and rate of change in muscle length. Secondary endings demonstrate only static responses and signal only muscle length.

7. The intrafusal muscle fibers associated with muscle spindles are innervated by γ motor neurons. Contraction of intrafusal fibers does not directly cause significant changes in muscle tension or length; however, when the level of tension in these fibers is adjusted, γ motor neurons influence the sensitivity of the muscle spindle to stretch.

8. Golgi tendon organs are located in the tendons of muscles and are thus arranged in series with the muscle. They are supplied by group Ib afferent fibers. Their in-series relationship means that tendon organs can detect the force level generated by the muscle,

whether it is due to passive stretch or to active contraction of the muscle.

9. The phasic stretch (or myotatic) reflex includes (1) a monosynaptic excitatory pathway from group Ia afferent fibers in muscle spindles to α motor neurons that supply the same and synergistic muscles and (2) a disynaptic inhibitory pathway to antagonistic motor neurons.

10. The inverse myotatic reflex is evoked by Golgi tendon organs. Afferent volleys in group Ib fibers from a given muscle cause disynaptic inhibition of α motor neurons to the same muscle, and they excite α motor neurons to antagonist muscles.

11. The flexion reflex is an important protective response because it acts to withdraw a limb from damaging stimuli. The reflex is evoked by volleys in afferent fibers that supply various receptors, particularly nociceptors. Via polysynaptic pathways, these volleys cause excitation of flexor motor neurons and inhibition of extensor motor neurons ipsilaterally. Concurrently, the opposite pattern of action (inhibition of flexor and excitation of extensor motor neurons) occurs contralaterally and is referred to as the *crossed extension reflex.*

12. Descending pathways can be subdivided into (1) a lateral system, which ends on motor neurons to limb muscles and on the lateral group of interneurons, and (2) a medial system, which ends on the medial group of interneurons.

13. The lateral system includes the lateral corticospinal tract and part of the corticobulbar tract. These pathways influence the contralateral motor neurons that supply the musculature of the limbs, especially of the digits, and the muscles of the lower part of the face and the tongue.

14. The medial system includes the ventral corticospinal, lateral and medial vestibulospinal, reticulospinal, and tectospinal tracts. These pathways affect mainly posture and provide the motor background for movement of the limbs and digits.

15. Locomotion is triggered by commands relayed through the midbrain locomotor center. However, central pattern generators formed by spinal cord circuits and influenced by afferent input provide for the detailed organization of locomotor activity.

16. Voluntary movements depend on interactions among motor areas of the cerebral cortex, the cerebellum, and the basal ganglia.

17. Motor areas of the cerebral cortex are arranged as a parallel distributed network, in which each contributes to the various descending motor pathways. The areas primarily involved in body and head movement include the primary motor cortex, the premotor area, the supplementary motor cortex, and the cingulate motor areas. The frontal eye fields are important for eye movement and help initiate voluntary saccades.

18. Individual corticospinal neurons discharge before voluntary contractions of related muscles occur. The discharges are typically related to contractile force rather than to joint position. However, the activity of an individual neuron may encode different parameters of a movement at different times in relation to the execution of that movement.

19. The population activity of motor cortex neurons can be used to predict the direction of upcoming movements.

20. The cerebellum influences the rate, range, force, and direction of movements. It also influences muscle tone and posture, as well as eye movement and balance.

21. The intrinsic circuitry of the cerebellum is remarkably uniform. Differences in function of different parts of the cerebellum arise largely as a result of differing afferent sources and efferent targets.

22. Anatomical and physiological techniques have shown that the cerebellar cortex may be divided into many functionally distinct, longitudinally running compartments.

23. Most of the input to the cerebellum is through pathways that end as mossy fibers. Mossy fibers excite granule cells, which in turn can evoke single action potentials, called *simple spikes,* in Purkinje cells, whose axons form the only output pathway from the cerebellar cortex.

24. The projections of the inferior olivary nucleus to the cerebellum end as climbing fibers and are the only source of them. Each Purkinje cell receives massive input from just one climbing fiber. As a result, each climbing fiber discharge produces a high-frequency burst of several action potentials, known as a *complex spike,* in the Purkinje cell.

25. Although complex spike activity is relatively rare in comparison to simple spike activity, complex spikes are precisely synchronized across populations of Purkinje cells, and because of the convergence of these cells onto cerebellar nuclear neurons, this synchronization may allow complex spike activity to significantly affect cerebellar output. Synchronization of complex spikes is the result of electrical coupling of inferior olivary neurons by gap junctions.

26. The basal ganglia include several deep telencephalic nuclei (including the caudate nucleus, putamen, and globus pallidus). The basal ganglia interact with the cerebral cortex, subthalamic nucleus, substantia nigra, and thalamus.

27. Activity transmitted from the cerebral cortex through the basal ganglia can either facilitate or inhibit the thalamic neurons that project to motor areas of the cortex, depending on the balance between direct and indirect basal ganglia pathways. When there is an imbalance of these two pathways, hyperkinetic or hypokinetic disorders occur.

28. Some types of eye movement help stabilize the view of the visual world. This is critical because visual acuity drops dramatically when the visual world moves, or slips, across the retina. Vestibuloocular and optokinetic movements help stabilize the visual world on the retina by compensating for movement of the head or external world (or both). Smooth pursuit movements allow tracking of a visual target so that it remains centered on the foveae.

29. Saccades act to move a specific part of the visual scene to the fovea, the retinal area of highest acuity, for detailed inspection.

30. There are specialized circuits and areas in the brainstem for control of vertical and horizontal eye movements. These areas are used both by the cortex (when voluntary eye movements are made) and by the sensory input that initiates reflexive eye movement.

Additional Readings

Eccles JC, Ito M, Szentagothai J. *The Cerebellum as a Neuronal Machine*. New York: Springer; 2013. (Original work published 1967.)

Shadmehr R, Wise SP. *The Computational Neurobiology of Reaching and Pointing: A Foundation for Motor Learning*. Cambridge, MA: Bradford Books; 2004.

Shepherd GM. Spinal cord. In: *Synaptic Organization of the Brain*. 5th ed. Oxford, UK: Oxford University Press; 2003: Chapter 3.

Shepherd GM. Cerebellum. In: *Synaptic Organization of the Brain*. 5th ed. Oxford, UK: Oxford University Press; 2003: Chapter 7.

Shepherd GM. Basal ganglia. In: *Synaptic Organization of the Brain*. 5th ed. Oxford, UK: Oxford University Press; 2003: Chapter 9.

Xiao J, Cerminara NL, Kotsurovskyy Y, et al. Systematic regional variations in Purkinje cell spiking patterns. *PLoS ONE*. 2014;9: e105633.

10

Integrative Functions of the Nervous System

LEARNING OBJECTIVES

Upon completion of this chapter, the student should be able to answer the following questions:

1. What is the basic layering pattern of the neocortex, and how do cortical inputs and outputs align with this layering pattern? What is the functional significance of the variation in the layering pattern between cortical areas?
2. What are the major functions of each of the lobes of the cerebrum?
3. How does the electroencephalogram (EEG) reflect cortical activity? What are evoked potentials?
4. How does cerebral dominance correlate with language and hand preference?
5. What is aphasia, and what is compromised in the different types of aphasia?
6. How do synaptic and cellular processes support learning and memory? How is memory distributed in the brain?
7. What role does plasticity play in neural development and in response to damage of the nervous system?

In earlier chapters, the interaction of the nervous system with the body and the outside world was discussed in terms of the transduction and analysis of sensory events, the organization of motor function, and relatively simple central processes that link them, such as reflexes (e.g., the stretch reflex and the vestibuloocular reflex). The nervous system has other capabilities, so-called integrative or higher functions, that are less directly tied to specific sensory modalities or motor behavior. These functions, in particular, require interactions between different parts of the cerebral cortex, and, as is being increasingly recognized, between the cerebral cortex and other parts of the brain. The neural basis for some of these higher functions is discussed in this chapter. Because these functions (as well as sensory perception and voluntary motor function) are so highly dependent on the cerebral cortex, its basic organization is described first.

The Cerebral Cortex

The human cerebral cortex occupies a volume of about 600 cm³ and has a surface area of 2500 cm². The surface of the cortex is highly convoluted and folded into ridges known as **gyri.** Gyri are separated by grooves called **sulci** (if shallow) or **fissures** (if deep; see Fig. 4.7). This folding greatly increases the surface area of cortex that can be fit into the limited and fixed volume within the skull. Indeed, most of the cortex cannot be seen from the brain surface because of this folding.

The cerebral cortex can be divided into the left and right hemispheres and subdivided into a number of lobes (Fig. 10.1; see also Fig. 4.7), including the **frontal, parietal, temporal,** and **occipital lobes.** The frontal and parietal lobes are separated by the central sulcus; both are separated from the temporal lobe by the **lateral fissure.** The occipital and parietal lobes are separated (on the medial surface of the hemisphere) by the **parieto-occipital fissure** (see Fig. 10.1). Buried within the lateral fissure is another lobe, the **insula** (see Fig. 4.6A). A group of structures that make up the **limbic lobe** is on the medial aspect of the hemisphere, and its largest part, the **hippocampal formation,** is folded into the **parahippocampal gyrus** of the temporal lobe and cannot be seen from the surface of the brain.

Activity in the two hemispheres of the cerebral cortex is coordinated by interconnections through the cerebral commissures. The bulk of the cortex is connected through the massive **corpus callosum** (see Figs. 4.9, 10.1), and parts of the temporal lobes connect through the anterior commissure.

There are three types of cerebral cortex: **neocortex, archicortex,** and **paleocortex.** The neocortex has six cortical layers (Fig. 10.2). In contrast, the archicortex has only three layers, and the paleocortex has four to five layers. In humans, approximately 90% of the cerebral cortex is neocortex.

The Neocortex
Neuronal Cell Types in the Neocortex

A number of different neuronal cell types in the neocortex have been described (see Fig. 10.2). **Pyramidal cells** are the most abundant cell type and account for approximately 75% of neocortical neurons. **Stellate cells** and various other types of nonpyramidal neurons make up the balance. Pyramidal cells have a large triangular cell body, a long

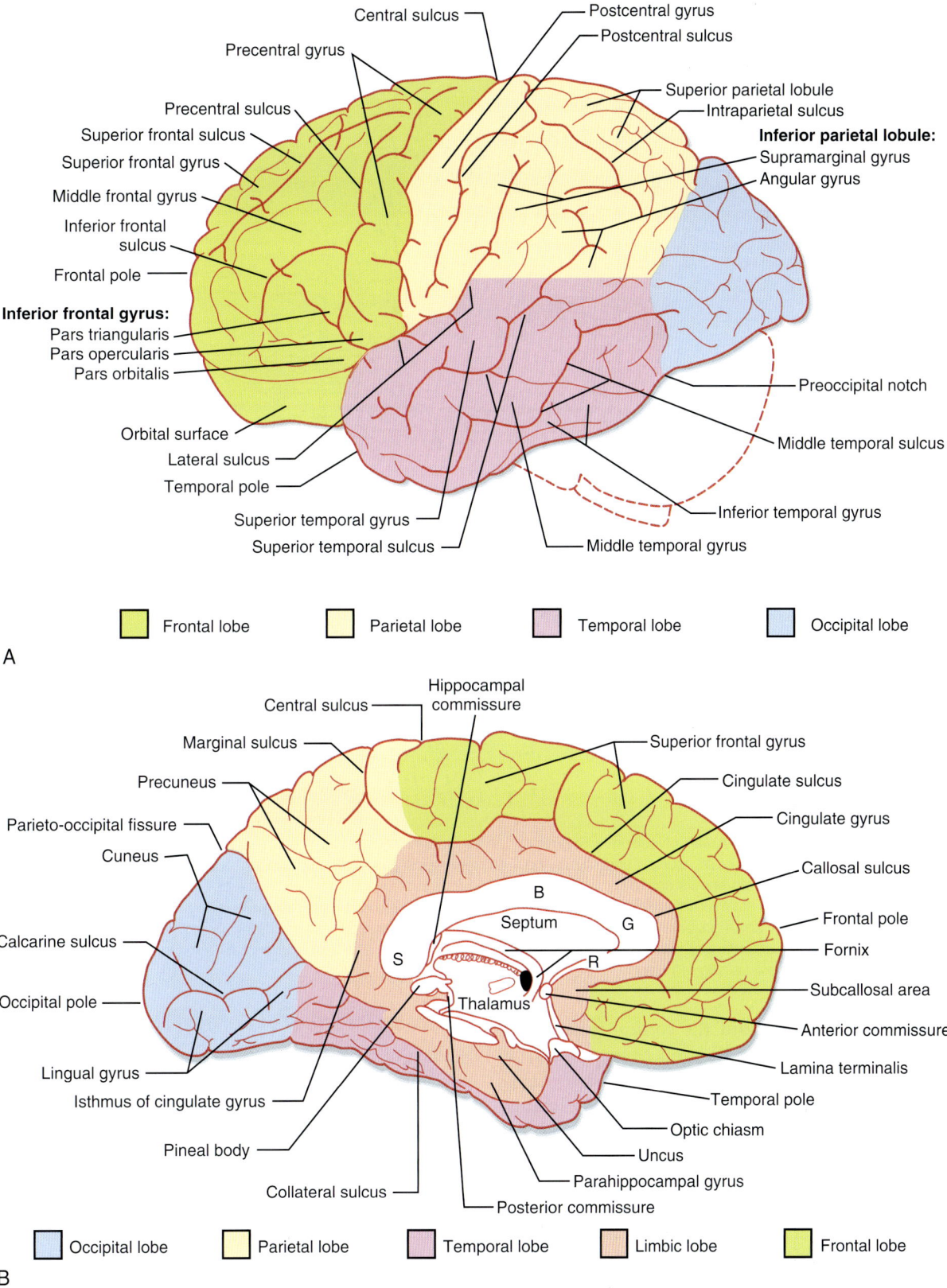

• Fig. 10.1 Lateral **(A)** and medial **(B)** illustrations of the left hemisphere of the human cerebrum with the major features labeled and the lobes indicated by color. R, G, B, and S indicate, respectively, the rostrum, genu, body, and splenium of the corpus callosum. (From Haines DE [ed]. *Fundamental Neuroscience for Basic and Clinical Applications.* 3rd ed. Philadelphia: Churchill Livingstone; 2006.)

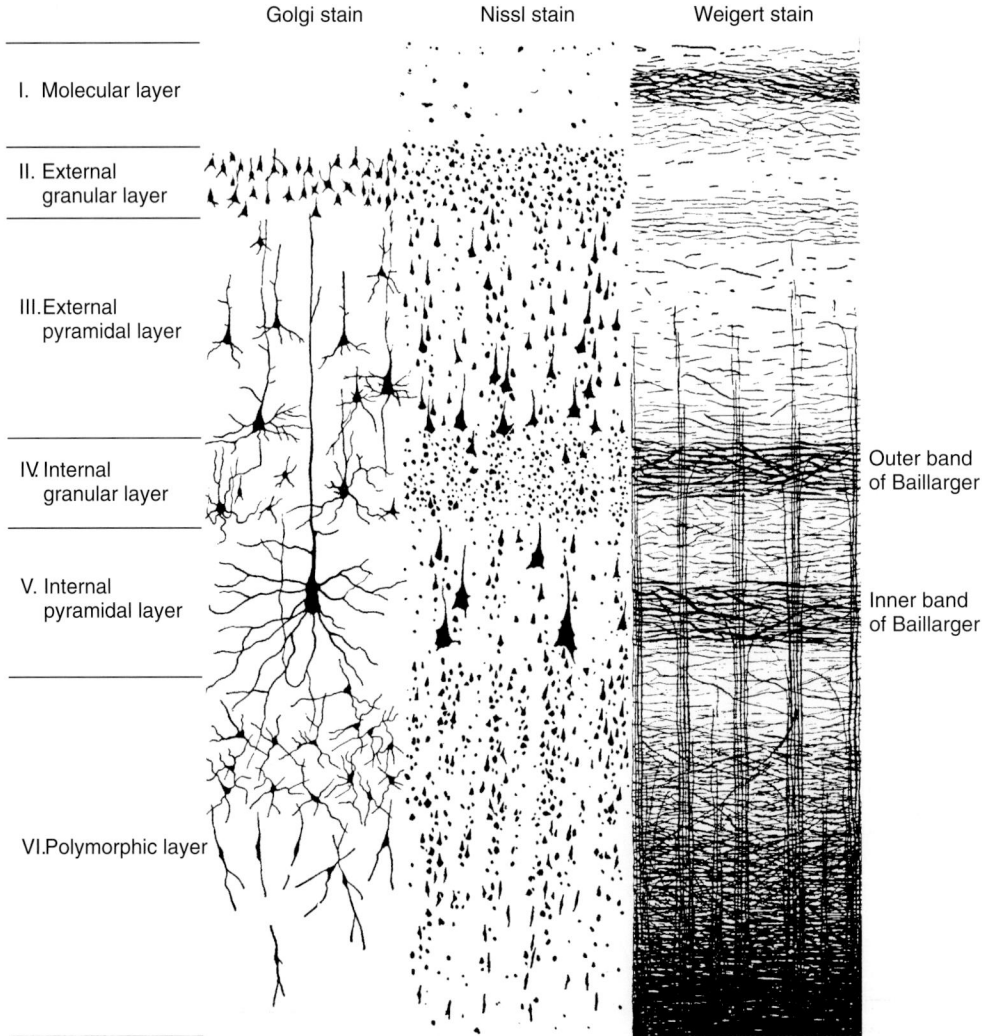

Golgi stain   Nissl stain   Weigert stain

I. Molecular layer

II. External granular layer

III. External pyramidal layer

IV. Internal granular layer

V. Internal pyramidal layer

VI. Polymorphic layer

Outer band of Baillarger

Inner band of Baillarger

• **Fig. 10.2** **An Area of Neocortex Stained by Three Different Methods.** The Nissl stain *(center)* shows the cell bodies of all neurons and reveals how different types are distributed among the six layers. The Golgi stain *(left)* shows only a sample of the neuronal population but reveals details of their dendrites. The Weigert stain for myelin *(right)* demonstrates vertically oriented bundles of axons entering and leaving the cortex and horizontally coursing fibers that interconnect neurons within a layer. (From Brodmann K. *Vergleichende Lokalisationslehre der Grosshirnrinde in ihren prinzipien Dargestellt auf Grund des Zellenbaues.* Leipzig: JA Barth; 1909.)

apical dendrite directed toward the cortical surface, and several basal dendrites. The cell's axon emerges from the body opposite the apical dendrite, and those from the larger pyramidal cells project into the subcortical white matter. The axon may give off collateral branches as it descends through the cortex. The neurotransmitter of pyramidal cells is an excitatory amino acid (glutamate or aspartate). Stellate cells, often called **granule cells,** are interneurons with local connections. They have a small soma and numerous branched dendrites, although many have an apical dendrite and thus look like small pyramidal cells. Some are excitatory interneurons; these cells are abundant in layer IV of the cortex (described in the next section). Their axons remain in the same cortical region, and many ascend into the upper cortical layers. Some stellate cells are inhibitory interneurons whose neurotransmitter is gamma-aminobutyric acid (GABA).

Cytoarchitecture of Cortical Layers

Each of the six layers of the neocortex has a characteristic cellular content (see Fig. 10.2). Layer I (molecular layer) has few neuronal cell bodies and contains mostly axon terminals synapsing on apical dendrites. Layer II (external granular layer) contains mostly stellate cells. Layer III (external pyramidal layer) consists mostly of small pyramidal cells. Layer IV (internal granular layer) contains mostly stellate cells and a dense matrix of axons. Layer V (internal pyramidal layer) is dominated by large pyramidal cells, the main source of cortical efferents to most subcortical regions. Layer VI (multiform layer) contains pyramidal, fusiform, and other types of cells.

Cortical Afferent and Efferent Fibers

Most input to the cortex from other regions of the central nervous system (CNS) is relayed by neurons in the thalamus, as described in earlier chapters for sensory and motor pathways. The projections from the thalamus to the cortex are a significant component of cortical organization that is observed clearly in the layering pattern. Thalamocortical fibers from thalamic nuclei that have specific (topographically mapped) cortical projections end chiefly in layer IV but also in layers III and VI. Neurons in other thalamic nuclei (particularly those relaying input from the brainstem reticular formation) project diffusely and terminate in layers I and VI to modulate cortical activity globally, perhaps in conjunction with changes in state (e.g., sleep or waking).

In addition to subcortical inputs, every region of the cortex receives input from other cortical regions. There are some large fiber bundles that connect widely separated cortical regions, and commissural fibers connect corresponding regions in each hemisphere (these projections terminate in layers I and VI), but in relative terms, the largest source of synapses in a cortical region is local, either from within the region itself or from its neighbors.

The cortical efferent axons originate from pyramidal cells. The smaller pyramidal cells of layers II and III project to adjacent cortical areas directly and to contralateral regions via the corpus callosum. The larger pyramidal cells of layer V project in many pathways to synaptic targets in the spinal cord, brainstem, striatum, and thalamus. The pyramidal cells of layer VI form corticothalamic projections that target the same thalamic nuclei that are their input creating circuits of reciprocal thalamocortical and corticothalamic interconnections. In addition, intrathalamic connections serve to associate activity in different cortical regions.

The specific patterns of input from the thalamus have another influence on cortical organization. As discussed in the sensory and motor systems, the topographic mapping of cortical input defines a **columnar organization.** A column is a narrow, vertically oriented (from the white matter to the cortical surface) region in which the neurons have correlated activity because of shared input from the thalamus. Within a column, there is a great richness in vertical interconnections and fewer lateral interconnections (to cells in neighboring columns), which suggests that columns may be the functional unit of the cortex. Despite their relative paucity, however, the lateral interconnections can exert powerful actions, as shown by inhibitory interconnection between regions within motor cortex (see Fig. 9.16). Interestingly, the columnar organization can greatly influenced by functional interactions, as well as by genetics. (See the section "Neural Plasticity.")

Regional Variations in Neocortical Structure

The architecture of the neocortex varies regionally, which presumably reflects the functional specialization of cortical areas. Different aspects of this variation are the bases of several methods for subdividing the cortex into discrete areas. The most widely used method is **cytoarchitectonics,** in which variations in cell density and structure are used; however, myeloarchitectonics (variations in axon density and size) and chemoarchitectonics (expression of molecular markers) are also used. Although several cytoarchitectonic maps of the cortex have been devised, the one by Korbinian Brodmann is most commonly used. In this map, the cortex is divided into 52 discrete areas (Fig. 10.3), numbered in the order that Brodmann studied them. Areas commonly referred to include **Brodmann areas 3, 1,** and **2** (the primary somatosensory cortex located on the postcentral gyrus); **area 4** (the primary motor cortex located on the precentral gyrus); **area 6** (the premotor and supplementary motor cortex); **areas 41** and **42** (the primary auditory cortex on the superior temporal gyrus); and **area 17** (the primary visual cortex, mostly on the medial surface of the occipital lobe). Subsequent studies confirmed that Brodmann areas are distinctive with regard to their cytoarchitecture, interconnections, and functions, but more recent work has shown that there is some plasticity both in the size of the areas and in their internal organization (see the section "Neural Plasticity").

Although cytoarchitectonic maps, like Brodmann's, give the impression of sharp boundaries between contiguous areas, the variation between many of the defined cortical areas is actually fairly subtle and, rather than sharing a well-defined border, most neighboring regions may gradually transition into one another. Nevertheless, some areas have quite distinct cortical characteristics, particularly the primary sensory and motor cortices. For example, the primary and premotor areas are referred to as **agranular cortex,** because no clear layer IV is present in these areas. Moreover, among the motor areas, the primary motor cortex is distinguished by the presence of large layer V pyramidal cells, the largest of which are called **Betz cells.** These enormous cells have axons that contribute to the corticospinal tracts and whose soma size (diameter > 150 μm) is necessary for the metabolic maintenance of so much axoplasm. Note that despite being the histological criterion for identifying primary motor cortex, Betz cell axons account for less than 5% of all corticospinal fibers.

In contrast to the motor areas, the primary sensory cortices (e.g., somatosensory, auditory, and visual) typically have a very prominent layer IV (internal granular layer), which is dominated by stellate cells (see Fig. 10.2), and therefore they are classified as **granular cortices.** Indeed, the primary visual cortex is also known as the **striate cortex** because of a particularly prominent horizontal sheet of myelinated axons in layer IV known as the **stripe of Gennari.** In a sense, the terms *granular* and *agranular* are inaccurate because all cortical areas have similar percentages of pyramidal calls (≈75%) and nonpyramidal cells (25%). Nevertheless, the key idea is that the grouping of the cell types into layers varies dramatically between the frontal motor areas, where the nonpyramidal neurons do not form a distinct internal granular layer, and the primary sensory cortices, where they do.

Archicortex and Paleocortex

About 10% of the human cerebral cortex is archicortex and paleocortex. The archicortex has a three-layered structure;

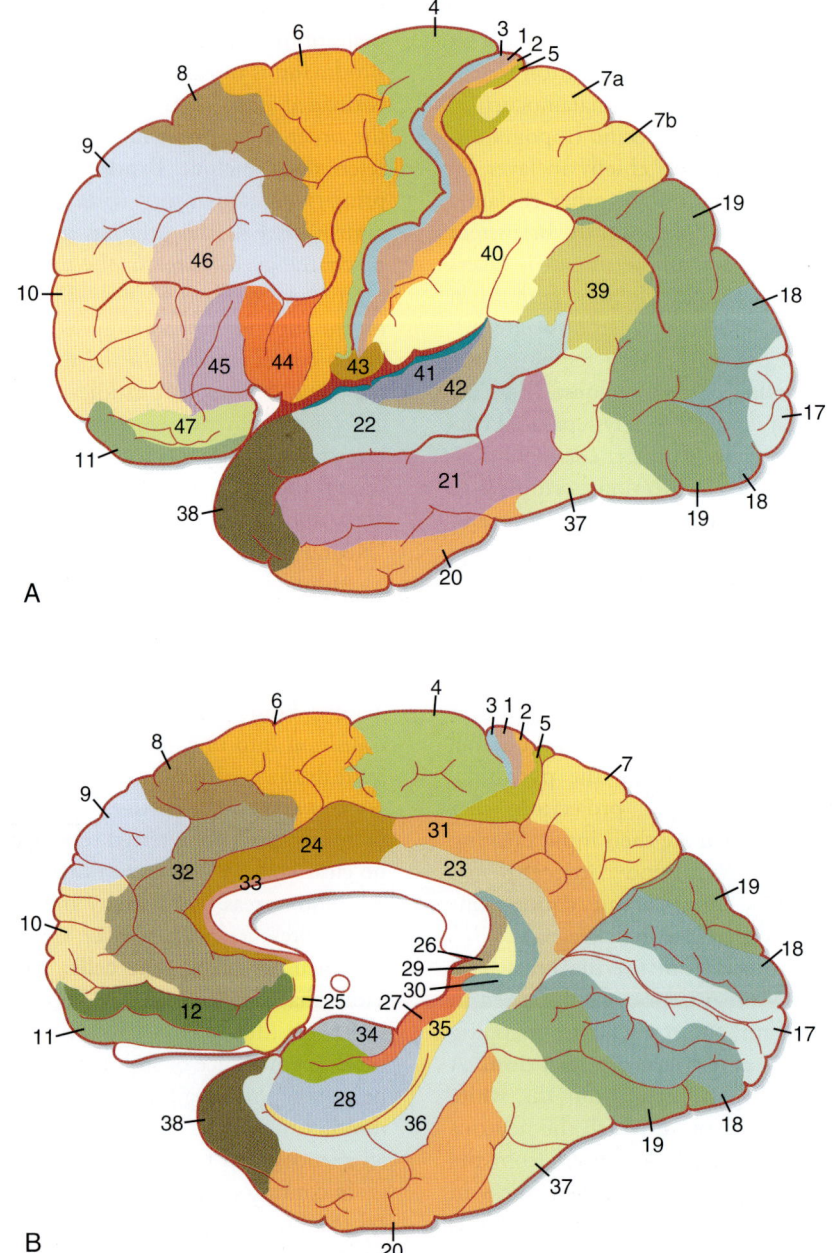

Fig. 10.3 Brodmann's areas in the human cerebral cortex. (Redrawn from Crosby EC, et al. *Correlative Anatomy of the Nervous System.* New York: Macmillan; 1962.)

the paleocortex has four to five layers. The paleocortex is located at the border between the archicortex and neocortex.

In humans, the hippocampal formation is part of the archicortex. It is folded into the temporal lobe and can be viewed only when the brain is dissected. The hippocampal cortex has three layers: the molecular, pyramidal cell, and polymorphic layers. They resemble layers I, V, and VI of the neocortex. The white matter covering the hippocampus is called the **alveus,** which contains hippocampal afferent and efferent fibers. The efferent axons coalesce to form the fornix (Fig. 10.4).

Functions of the Lobes of the Cerebral Cortex

There is no exact correspondence between the folds (lobes and gyri) of the cerebral cortex and function; nevertheless, some with the individual lobes of the cerebral hemispheres have a general association with function that helps clarify cortical organization.

Frontal Lobe

One of the main functions of the **frontal lobe** is motor behavior. As discussed in Chapter 9, the motor, premotor, cingulate motor, and supplementary motor areas are located

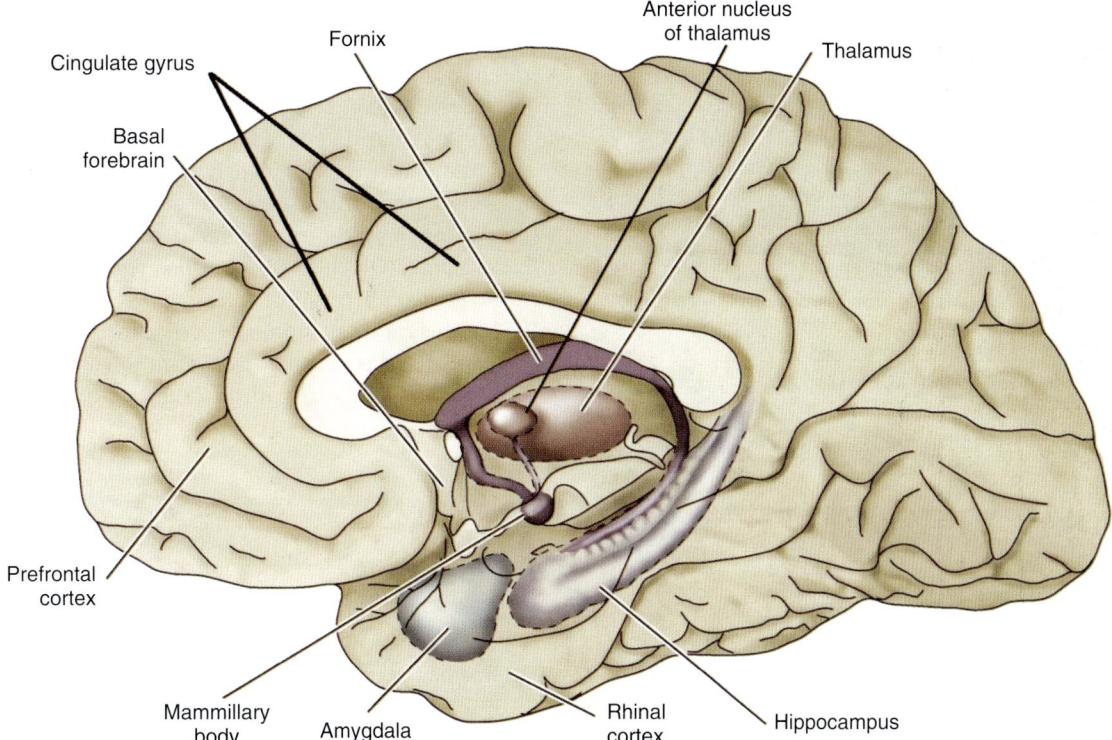

• Fig. 10.4 The hippocampus and the amygdala are located on the medial aspect of the temporal lobe. The fornix, the major output pathway from the hippocampus, projects to the mammillary body, which in turn connects to the anterior nucleus of the thalamus via the mammillothalamic tract. Also illustrated are the cingulate gyrus, the basal forebrain area (septal nuclei, bed nucleus of the stria terminalis, nucleus accumbens), and the prefrontal cortex. (From Purves D. Sleep and wakefulness. In: Purves D, et al [eds]. *Neuroscience.* 3rd ed. Sunderland, MA: Sinauer; 2004.)

in the frontal lobe, as is the frontal eye field. These areas are crucial for planning and executing motor behavior. **Broca's area,** essential for the generation of speech, is located in the inferior frontal gyrus of the dominant hemisphere for human language (almost always the left hemisphere, as explained later). In addition, the more anterior prefrontal cortex in plays a major role in personality and emotional behavior.

Bilateral lesions of the prefrontal cortex may be produced either by disease or by a surgical frontal lobotomy. Such lesions produce deficits in attention, difficulty in planning and problem solving, and inappropriate social behavior. Aggressive behavior is also lessened and the motivational-affective component of pain is reduced although pain sensation remains. Frontal lobotomies are rarely performed today because modern drug therapies provide more focal and effective management for mental illness and chronic pain.

Parietal Lobe

The **parietal lobe** contains the **somatosensory cortex** (see Chapter 7) and the adjacent **parietal association cortex.** The parietal association cortex gets information from somatosensory, visual, and auditory cortices and is involved in the processing, perception, and integration of sensory information. Connections with the frontal lobe allow somatosensory information to aid in voluntary motor activity. Somatosensory, visual, and auditory information can also be transferred to language centers, such as **Wernicke's area,** as described later. Lesions in the left parietal lobe can result in Gerstmann's syndrome, which includes a person's inability to name his or her fingers (or those of another) and a loss of the ability to perform numerical calculations. The right parietal lobe is involved in determining spatial context. Localized lesions can result in the **neglect syndrome,** in which the patient seems unaware of the left side of his or her body and of persons, objects, and events on his or her left. (See an "In the Clinic" box later in this chapter.)

Occipital Lobe

The major function of the **occipital lobe** is visual processing and perception (see Chapter 8). The primary visual cortex (Brodmann area 17) lines the calcarine sulcus and is flanked by secondary (Brodmann area 18) and tertiary (Brodmann area 19) visual cortices. Lesions of these areas in the cuneus gyrus result in blindness in the lower contralateral visual field; those in the lingual gyrus result in blindness in the upper contralateral visual field. Connections to the frontal eye fields affect direction of gaze, and projections to the midbrain assists in the control of convergent eye movements, pupillary constriction, and accommodation, all of which occur when the eyes adjust for near vision.

Temporal Lobe

The **temporal lobe** has many different functions, including the processing and perception of sounds and vestibular information and higher order visual processing (see Chapter 8). For example, the infratemporal cortex, on its inferior surface, is involved in the recognition of faces. In addition, Meyer's loop, which forms part of the optic pathway, passes through the temporal lobe. As a result, temporal lobe lesions can damage vision in upper part of the visual field. Similarly, a portion of Wernicke's area, essential for the understanding of language, lies in the posterior region of the temporal lobe.

The limbic system dominates the medial temporal lobe (see Fig. 10.4), and it participates in emotional behavior and in learning and memory (see the "Learning and Memory" section). The limbic system helps control emotional behavior, in part by an influence on the hypothalamus via the Papez circuit. This circuit projects from the cingulate gyrus to the entorhinal cortex and hippocampus, and from there, via the fornix, to the mammillary bodies in the hypothalamus. The mammillothalamic tract then connects the hypothalamus with the anterior thalamic nuclei, which project back to the cingulate gyrus (see Fig. 10.4). In addition, the hippocampus and amygdala are connected to the prefrontal cortex, the basal forebrain, and the anterior cingulate cortex.

Bilateral temporal lobe lesions can produce Klüver-Bucy syndrome, which is characterized by loss of the ability to recognize the meaning of objects from visual cues (visual agnosia); a tendency to examine all objects, even dangerous ones, orally; attention to irrelevant stimuli; hypersexuality; a change in dietary habits; and decreased emotionality.

Although this syndrome was originally described as following large lesions of most or all of the temporal lobe, more recent studies have highlighted the role of the amygdala. The amygdala conditions the association of fear with painful stimuli and may trigger, via connections to the medial frontal cortex and anterior cingulate gyrus, emotional or avoidance responses when these stimuli recur. In addition, the amygdala projects to the **nucleus accumbens,** a region of the basal ganglia which has been called a "reward center." The nucleus accumbens signals pleasurable events in response to dopaminergic input from the ventral tegmental area of the brainstem and is also the most common site of the plaques and tangles associated with Alzheimer's disease.

The Electrical Activity of the Cortex

An **electroencephalogram (EEG)** is a recording of the neuronal electrical activity of the cerebral cortex made by electrodes placed on the skull. EEG waves normally reflect the summed extracellular currents that result from the generation of synaptic potentials in the pyramidal cells and are thus a type of **field potential.** Because the currents generated by a single cell are too small to be detected as discrete events by an electrode on the skull (to record the activity of a single neuron, a microelectrode must be placed within microns of the neuron), the EEG waves reflect the combined activity of many pyramidal cells. Moreover, for the activity of a group of neurons to generate an event detectable on EEG, they must be oriented so that their individual currents summate to produce a detectable field. The arrangement of pyramidal neurons, with their apical dendrites aligned in parallel to form a dipole sheet, is particularly favorable for generating large field potentials. One pole of this sheet is oriented toward the cortical surface and the other toward the subcortical white matter. Thus the currents generated

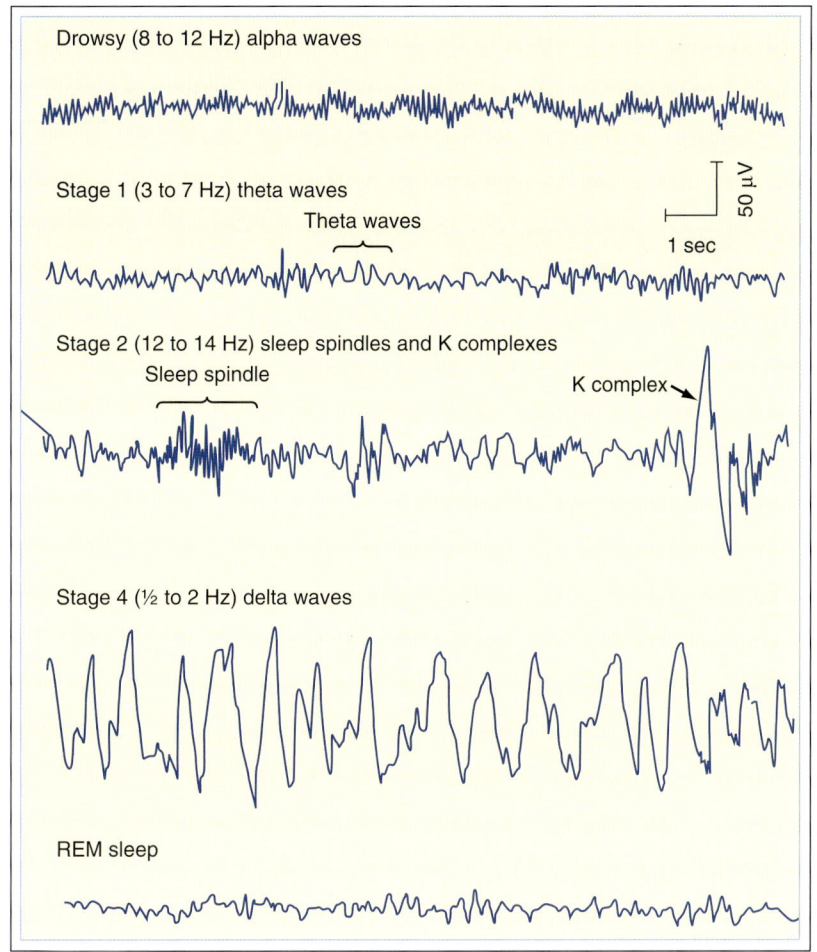

Drowsy (8 to 12 Hz) alpha waves

Stage 1 (3 to 7 Hz) theta waves

Theta waves

50 µV

1 sec

Stage 2 (12 to 14 Hz) sleep spindles and K complexes

Sleep spindle

K complex

Stage 4 (½ to 2 Hz) delta waves

REM sleep

• **Fig. 10.5** Electroencephalographic tracings during drowsiness; stages 1, 2, and 4 of slow-wave (non–rapid eye movement [non-REM]) sleep; and REM sleep. (Modified from Shepherd GM. *Neurobiology.* London: Oxford University Press; 1983.)

by cortical pyramidal cells have a similar orientation and can therefore summate to produce a field potential that can be detected. The need for summation also explains why EEG signals reflect primarily synaptic potentials rather than action potentials: electrical events must overlap in time in order to sum, and synaptic potentials have much longer durations than do action potentials.

The sign of an EEG wave can be positive or negative, but its direction alone does not indicate whether pyramidal cells are being excited or inhibited. For instance, a negative EEG potential may be generated at the surface of the skull (or cortex) by excitation of apical dendrites or by inhibition near the somas. Conversely, a positive EEG wave can be produced by inhibition of apical dendrites or by excitation near the somas.

A normal EEG tracing consists of waves of various frequencies. The dominant frequencies depend on several factors, including the state of wakefulness, the age of the subject, the location of the recording electrodes, and the absence or presence of drugs or disease. When a normal awake adult is relaxed with the eyes closed, the dominant frequencies of the EEG recorded over the parietal and occipital lobes are about 8 to 12 Hz, the **alpha rhythm.** If the subject is asked to open the eyes, the wave becomes less synchronized, and the dominant frequency increases to 13 to 30 Hz, which is called the **beta rhythm.** The **delta** (0.5 to 2 Hz) and **theta** (3 to 7 Hz) **rhythms** are observed during sleep (see the following discussion; Fig. 10.5). Also, brief EEG waves do exist and, because of their shape, are sometimes referred to as **spikes,** but this does not imply that they are associated with action potentials.

Evoked Potentials

An EEG change that can be elicited by a stimulus is called a **cortical evoked potential.** A cortical evoked potential is best recorded from the part of the skull located over the cortical area being activated. For example, a visual stimulus results in an evoked potential that can be recorded best over the occipital bone, whereas a somatosensory evoked potential is recorded most effectively near the junction of the frontal and parietal bones. Evoked potentials reflect the activity in large numbers of cortical neurons. They may also reflect activity in subcortical structures.

Evoked potentials are small in comparison with the size of the EEG waves. However, their appearance can be enhanced by a process called **signal averaging.** In this process, the stimulation is repeated, and the EEGs recorded during each trial are electronically averaged. With each repetition of the stimulus, the evoked potential occurs at a fixed time after the stimulus. When the records are averaged, the components of the EEG that have a random temporal association with the stimulus cancel each other, whereas the evoked potentials sum.

IN THE CLINIC

Evoked potentials are used clinically to assess the integrity of a sensory pathway, at least to the level of the primary sensory receiving area. These potentials can be recorded in comatose individuals, as well as in infants too young to undergo a sensory examination. The initial parts of the auditory evoked potential actually reflect activity in the brainstem; therefore, this evoked potential can be used to assess the function of brainstem structures.

Sleep-Wake Cycle

Sleep and wakefulness are among the many functions of the body that show **circadian** (about 1-day) periodicity. Characteristic changes in the EEG can be correlated with changes in the behavioral state during the sleep-wake cycle. **Beta wave** activity dominates in an awake, aroused individual. The EEG is said to be **desynchronized;** it displays low-voltage, high-frequency activity. In relaxed individuals with their eyes closed, the EEG is dominated by **alpha waves** (see Fig. 10.5). A person falling asleep passes sequentially through four stages of **slow-wave sleep** (called stages 1 through 4) over a period of 30 to 45 minutes (see Fig. 10.5). In stage 1, alpha waves are interspersed with lower frequency waves called **theta waves.** In stage 2, the waves slow further, but the slow-wave activity is interrupted by **sleep spindles,** which are bursts of activity at 12 to 14 Hz, and by large **K complexes** (large, slow potentials). Stage 3 sleep is associated with **delta waves** and with occasional sleep spindles. Stage 4 is characterized by delta waves without spindles.

During slow-wave sleep, the muscles of the body relax, but the posture is adjusted intermittently. The heart rate and blood pressure decrease, and gastrointestinal motility increases. The ease with which individuals can be awakened decreases progressively as they pass through these sleep stages. As individuals awaken, they pass through the sleep stages in reverse order.

About every 90 minutes, slow-wave sleep changes to a different form of sleep, called **rapid eye movement (REM)** sleep. In REM sleep, the EEG again becomes desynchronized. The low-voltage, fast activity of REM sleep resembles that seen in the EEG from an aroused subject (see Fig. 10.5, bottom trace). Because of the similarity of the EEG to that of an awake individual and the difficulty awaking the person, the term **paradoxical sleep** characterizes this type of sleep. Muscle tone is completely lost, but phasic contractions occur in a number of muscles, most notably the eye muscles. The resulting rapid eye movements are basis of the name for this type of sleep. Many autonomic changes also take place. Temperature regulation is lost, and meiosis occurs. Penile erection may occur during this type of sleep. Heart rate, blood pressure, and respiration change intermittently. Several episodes of REM sleep occur each night. Although it is difficult to arouse a person from REM sleep, internal arousal is common. Most dreaming occurs during REM sleep.

IN THE CLINIC

The sleep-wake cycle has an endogenous periodicity of about 25 hours, but it normally becomes entrained to the day-night cycle. The source of circadian periodicity appears to be the suprachiasmatic nucleus of the hypothalamus. This nucleus receives projections from the retina, and its neurons seem to form a biological clock that adapts to the light-dark cycle. However, the entrainment can be disrupted when the subject is isolated from the environment or changes time zones (jet lag). Destruction of the suprachiasmatic nucleus disrupts a number of biological rhythms, including the sleep-wake cycle.

The proportion of slow-wave (non-REM) sleep to REM sleep varies with age. Newborns spend about half of their sleep time in REM sleep, whereas elderly people have little REM sleep. About 20% to 25% of the sleep of young adults is REM sleep.

The mechanism of sleep is incompletely understood. Stimulation in the brainstem in a large region known as the **reticular activating system** causes arousal and low-voltage, fast EEG activity. Sleep was once thought to be caused by a reduced level of activity in the reticular activating system. However, substantial data, including the observations that anesthesia of the lower brainstem results in arousal and that stimulation in the medulla near the nucleus of the solitary tract can induce sleep, suggest that sleep is an active process. Investigators have tried to find a relationship between sleep mechanisms and brainstem networks in which particular neurotransmitters, including serotonin, norepinephrine, and acetylcholine, are used; manipulations of the levels of these transmitters in the brain can affect the sleep-wake cycle. However, a detailed neurochemical explanation of the neural mechanisms of sleep is not yet available.

Similarly, the purpose of sleep is still unclear. However, it must have a high value because so much of life is spent in sleep and because lack of sleep can be debilitating. Medically important disorders of the sleep-wake cycle include insomnia, bed-wetting, sleepwalking, sleep apnea, and narcolepsy.

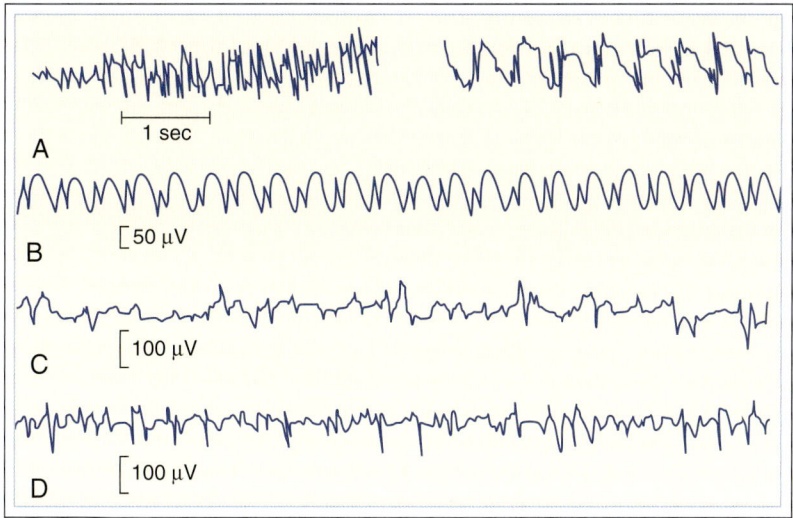

• **Fig. 10.6** Electroencephalographic (EEG) Abnormalities in Several Forms of Epilepsy. **A,** EEG tracings during the tonic *(left)* and clonic *(right)* phases of a tonic-clonic (grand mal) seizure. **B,** Spike and wave components of an absence (petit mal) seizure. **C,** EEG tracing in a person with temporal lobe epilepsy. **D,** EEG tracing of a focal seizure. (Redrawn from Eyzaguirre C, Fidone SJ. *Physiology of the Nervous System.* 2nd ed. St. Louis: Mosby; 1975.)

 ## IN THE CLINIC

The EEG becomes abnormal in a variety of pathological circumstances. For example, during coma, the EEG is dominated by delta activity. **Brain death** is defined by a maintained flat EEG wave.

Epilepsy commonly causes, and can be diagnosed by, specific EEG abnormalities. There are many forms of epilepsy, and examples of EEG patterns from some of these types of epilepsy are shown in Figure 10.6. Epileptic seizures can be either partial or generalized.

One form of partial seizures originates in the motor cortex and results in localized contractions of contralateral muscles. The contractions may then spread to other muscles; such spread follows the somatotopic sequence of the motor cortex (see Chapter 9). This stereotypical progression is called a **Jacksonian march.** Complex partial seizures (which may occur in **psychomotor epilepsy**) originate in the limbic structures of the temporal lobe and result in illusions and semipurposeful motor activity. During and between focal seizures, scalp recordings may reveal EEG spikes (see Fig. 10.6C and D).

Generalized seizures involve wide areas of the brain and loss of consciousness. Two major types are *petit mal* and

grand mal **seizures.** In petit mal epilepsy, consciousness is lost transiently (typically for less than 15 seconds), and the EEG displays **spike and wave activity** (see Fig. 10.6B). In grand mal seizures, consciousness is lost for a longer period, and the affected individual may fall if standing when the seizure starts. The seizure begins with a generalized increase in muscle tone **(tonic phase),** followed by a series of jerky movements **(clonic phase).** The bowel and bladder may be evacuated. The EEG shows widely distributed seizure activity (see Fig. 10.6A).

EEG spikes that occur between full-blown seizures are called **interictal spikes.** Similar events can be studied experimentally. These spikes arise from abrupt, long-lasting depolarizations, called **depolarization shifts,** that trigger repetitive action potentials in cortical neurons. These depolarization shifts may reflect several changes in epileptic foci. Such changes include regenerative Ca^{++}-mediated dendritic action potentials in cortical neurons and a reduction in inhibitory interactions in cortical circuits. Electrical field potentials and the release of K^+ and excitatory amino acids from hyperactive neurons may also contribute to the increased cortical excitability.

Cerebral Dominance and Language

Although right-handedness represents a sensorimotor dominance of the left hemisphere and left-handedness represents a sensorimotor dominance of the right hemisphere, **cerebral dominance** is assigned to the hemisphere in which language is to communicate; in humans, the left hemisphere is the **dominant hemisphere** in more than 90% of both right- and left-handed people. This dominance

has been demonstrated (1) by the effects of lesions of the left hemisphere that produce deficits in language function **(aphasia)** and (2) by the transient aphasia (inability to speak or write) that results when a short-acting anesthetic is introduced into the left carotid artery. Lesions of the nondominant hemisphere and injection of anesthetic into it do not usually affect language substantially.

Several areas in the left hemisphere are involved in language. **Wernicke's area** is a large area in the posterior part of the superior temporal gyrus, extending from behind the auditory cortex into the parietal lobe. Another important language area, **Broca's area,** is in the posterior part of the inferior frontal gyrus, close to the face representation of the motor cortex. Damage to Wernicke's area results in **receptive aphasia,** in which the person has difficulty comprehending spoken *and* written language; however, speech production remains fluent, if meaningless. Conversely, a lesion in Broca's area causes **expressive aphasia,** in which individuals have difficulty in generating speech and writing, although they can understand language relatively well.

The terms *sensory aphasia* and *motor aphasia* are often interchanged with *receptive aphasia* and *expressive aphasia,* respectively. The former terms, however, are misleading: A person with receptive aphasia may not have auditory or visual impairment, and one with expressive aphasia may have normal motor control of the muscles responsible for speech or writing. Aphasia does not depend on a deficit of sensation or of motor skill; rather, it is an inability to decode language-encoded sensory information into concepts or to encode concepts into language. However, lesions in the dominant hemisphere may be large enough to result in mixed forms of aphasia, as well as sensory changes or paralysis of some of the muscles used to express language. For example, the latter situation could occur with a lesion of the face representation portion of the motor cortex that results in an inability to manipulate the motor apparatus needed for speaking (vocal cords, jaws, tongue, lips) and would be manifest as unclear speech because of dysarthria, a mechanical deficit. An affected individual would, however, be able to write if the motor cortex serving the upper limb were unaffected.

Interhemispheric Communication and the Corpus Callosum

The two cerebral hemispheres can function somewhat independently, as in the control of one hand. However, information must be transferred between the hemispheres to coordinate activity on the two sides of the body. Much of that information is transmitted through the corpus callosum, although some is transmitted through other commissures (e.g., the anterior commissure or the hippocampal commissure).

The importance of the corpus callosum for interhemispheric transfer of information is illustrated in Figure 10.7*A.* An animal with an intact optic chiasm and corpus callosum and with the left eye closed learns a visual discrimination task (see Fig. 10.7*A*). The information is transmitted to both hemispheres through bilateral connections made by the optic chiasm or through the corpus callosum, or both. When the animal is tested with the left eye open and the right eye closed (see Fig. 10.7*A, center*), the task can still be performed because both hemispheres have learned the task. If the optic chiasm is transected before the animal is trained, the result is the same (see Fig. 10.7*B*). Information

is presumably transferred between the two hemispheres through the corpus callosum. This finding can be confirmed by cutting both the optic chiasm and the corpus callosum before training (see Fig. 10.7*C*). Then the information is not transferred, and each hemisphere must learn the task independently.

A similar experiment was conducted in human patients who had undergone surgical transection of the corpus callosum to prevent the interhemispheric spread of epilepsy (Fig. 10.8). The optic chiasm remained intact, but visual information was directed to one or the other hemisphere by the patient's fixing vision on the central point of the screen. A picture or name of an object was then flashed to one side of the fixation point, so that visual information about the picture reached only the contralateral hemisphere. An opening beneath the screen allowed the patient to manipulate objects that could not be seen. The objects included those shown in the projected pictures. Normal individuals would be able to locate the correct object with either hand. However, patients with a transected corpus callosum could locate the correct object only with the hand ipsilateral to the projected image (contralateral to the hemisphere that received the visual information). For the hand to explore and recognize the correct object, the visual information must have access to the somatosensory and motor areas of the cortex. With the corpus callosum cut, the visual and motor areas are interconnected only on the same side of the brain.

Another test was to ask the patient to verbally identify what object was seen in the picture. The patient would make a correct verbal response to a picture that was projected to the right of the fixation point because the visual information reached only the left (language-dominant) hemisphere. However, the patient could not verbally identify a picture that was presented to the left hemifield because visual information reached only the right hemisphere.

Similar observations can be made in patients with a transected corpus callosum when different forms of stimuli are used. For example, when such patients are given a verbal command to raise the right arm, they do so without difficulty. The language centers in the left hemisphere send signals to the ipsilateral motor areas, and these signals produce the movement of the right arm. However, these patients cannot respond to a command to raise the left arm. The language areas on the left side cannot influence the motor areas on the right unless the corpus callosum is intact. Somatosensory stimuli applied to the right side of the body can be described by patients with a transected corpus callosum, but these patients cannot describe the same stimuli applied to the left side of the body. Information that reaches the right somatosensory areas of the cortex cannot reach the language centers if the corpus callosum has been cut.

In addition to language, other differences in the functional capabilities of the two hemispheres can be compared by exploring the performance of individuals with a transected corpus callosum. Such patients solve

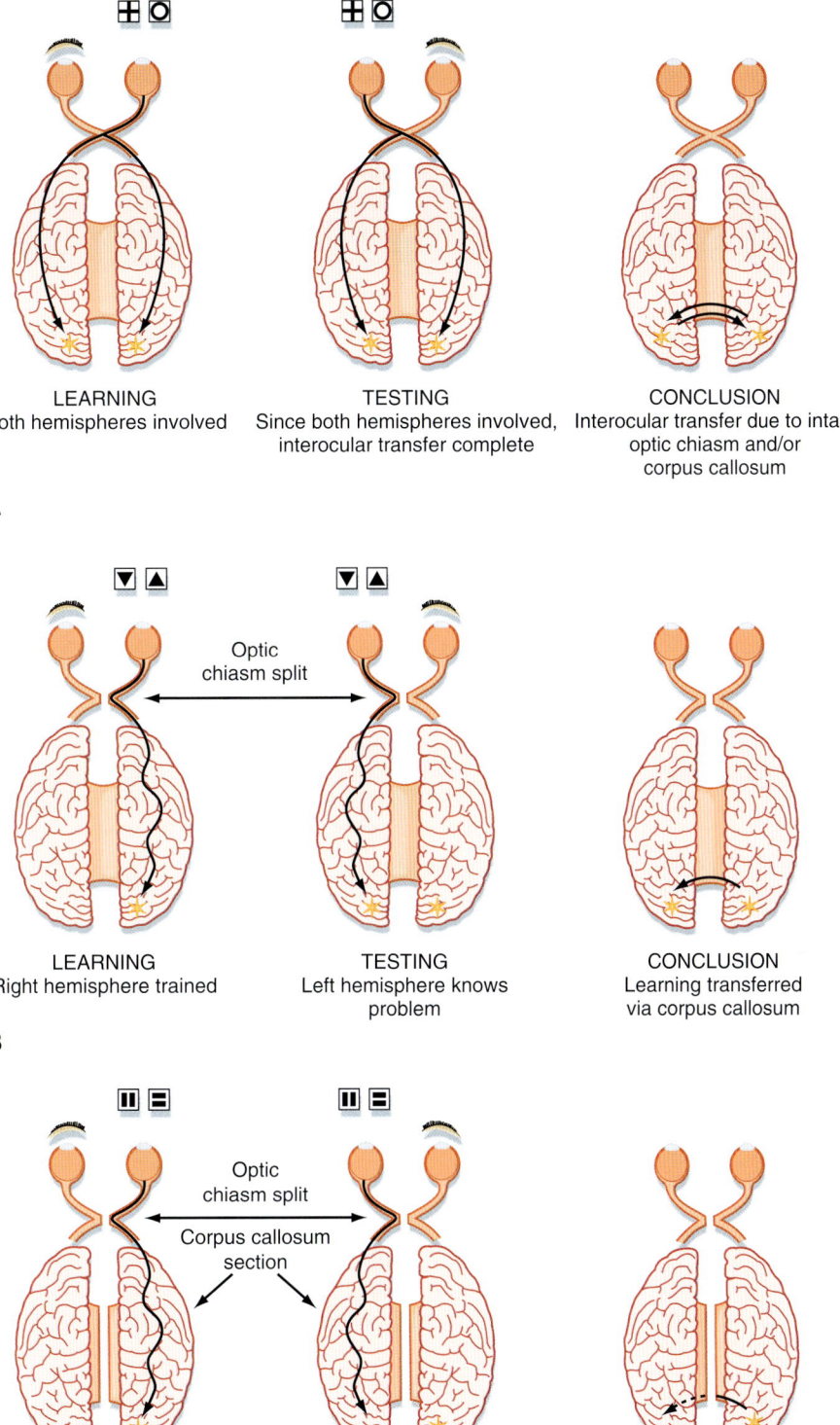

LEARNING
Both hemispheres involved

TESTING
Since both hemispheres involved, interocular transfer complete

CONCLUSION
Interocular transfer due to intact optic chiasm and/or corpus callosum

A

Optic chiasm split

LEARNING
Right hemisphere trained

TESTING
Left hemisphere knows problem

CONCLUSION
Learning transferred via corpus callosum

B

Optic chiasm split

Corpus callosum section

LEARNING
Right hemisphere trained

TESTING
Left hemisphere does not know problem

CONCLUSION
Transfer pathway was blocked

C

• **Fig. 10.7** Role of the Corpus Callosum in the Interhemispheric Transfer of Visual Information When Learning Involves One Eye. **A,** Discrimination depends on distinguishing between a cross and a circle. **B,** Discrimination is between triangles oriented with the apex up or down. **C,** Discrimination is between vertical and horizontal bars.

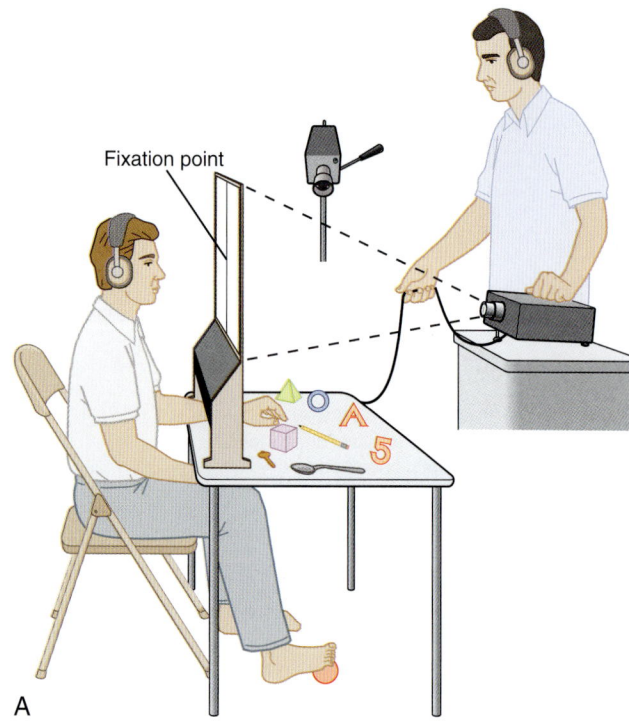

A

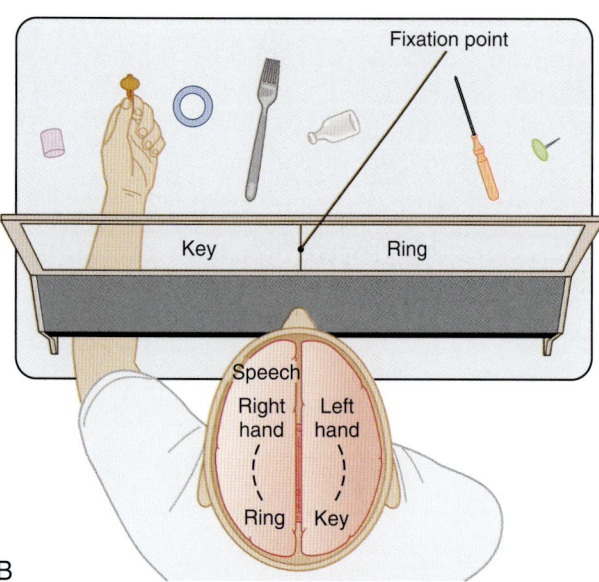

B

• **Fig. 10.8** Illustration of Tests in a Patient With a Transected Corpus Callosum. **A,** The patient fixes on a point on a rear projection screen, and pictures are projected to either side of the fixation point. The hand can palpate objects that correspond to the projected pictures, but these objects cannot be seen. **B,** Response by the left hand to a picture of a key in the left field of view. However, the verbal response is that the patient sees a picture of a ring. (Redrawn from Sperry RW. In: Schmitt FO, Worden FG [eds]. *The Neurosciences: Third Study Program.* Cambridge, MA: MIT Press; 1974.)

three-dimensional puzzles better with the right than with the left hemisphere, which suggests that the right hemisphere has specialized functions for spatial tasks. Other functions that seem to be more associated with the right than the left hemisphere are facial expression, body language, and speech intonation (Fig. 10.9). Patients with a transected corpus callosum lack normal interhemispheric coordination. When they are dressing, for example, one hand may button a shirt while the other tries to unbutton it. Observation of these patients indicates that the two hemispheres can operate quite independently when they are no longer interconnected. However, one hemisphere can express itself with language, whereas the other communicates only nonverbally.

IN THE CLINIC

One of the more striking examples of interhemispheric differences is the phenomenon of **"cortical neglect,"** which is a consequence of a lesion in the parietal cortex of the nondominant (usually right) hemisphere. In such cases, the patient ignores objects and individuals in the left visual field, draws objects that are incomplete on the left, denies the existence of his or her left arm and leg, and fails to dress the left side of his or her body. The patient also denies having any such difficulties **(anosognosia).** Although the patient may respond to touch and pinprick on the left side of the body, he or she cannot identify objects placed in the left hand. The lesion is adjacent to the first somatosensory (SI) cortex, as well as the visual association cortex, and it suggests that this region plays a special role in the perception of body image and immediate extrapersonal space. Similar lesions on the dominant side result only in loss of some higher order somesthesias, such as **agraphesthesia** (inability to identify characters drawn on the palm) and **astereognosis** (inability to identify an object only by touch).

Learning and Memory

Major functions of the higher levels of the nervous system are learning and memory. *Learning* is a neural mechanism by which the organism's behavior changes as a result of experience. *Memory* is the storage mechanism for what is learned.

The neural circuitry involved in memory and learning in mammals is complex and difficult to study. Alternative approaches are animal studies (especially in the simpler nervous systems of invertebrates), analysis of the functional consequences of lesions, and anatomical/physiological studies at the cellular and pathway level. For example, in the marine mollusk *Aplysia,* it has been possible to isolate a connection between a single sensory neuron and a motor neuron, which shows aspects of **habituation** (learning not to respond to repetitions of an insignificant stimulus), **sensitization** (increased responsiveness to innocuous stimuli that follow the presentation of a strong or noxious stimulus), and even **associative conditioning** (learning to respond to a previously insignificant event after it has been

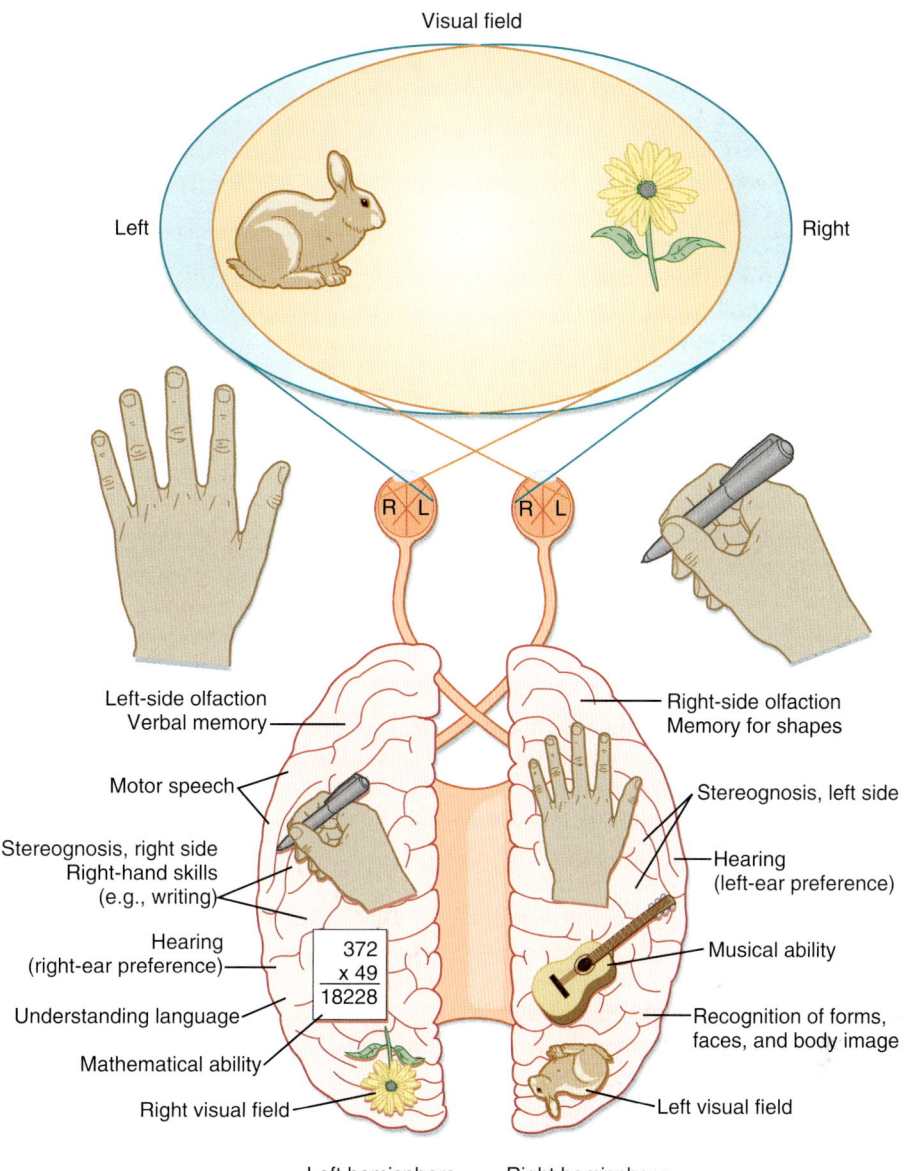

• **Fig. 10.9** Schematic illustration of the functional specializations of the left and right hemispheres, as determined in patients after section of the corpus callosum. (Modified from Siegel A, Sapru HN. *Essential Neuroscience.* 5th ed. Philadelphia: Lippincott Williams & Wilkins; 2005.)

paired with a significant one). In the case of habituation, the amount of transmitter released in successive responses gradually diminishes. The change involves an alteration in the Ca^{++} current that triggers release of neurotransmitter. The cause of this change is inactivation of presynaptic Ca^{++} channels by repeated action potentials. Long-term habituation can also be produced. In this case, the numbers of synaptic endings and active zones in the remaining terminals decreases.

Long-Term Potentiation

Additional models of learning, provided by synaptic phenomena, are called **long-term potentiation (LTP)** and **long-term depression (LTD).** LTD has been studied most extensively in the cerebellum (see Chapter 9), but it also occurs in the hippocampus and in other regions of the CNS.

LTP has been studied most intensively in slices of the hippocampus in vitro, but it has also been studied in the neocortex, cerebellum, and other parts of the CNS. Repetitive activation of an afferent pathway to the hippocampus or repetitive activation of one of the intrinsic connections increases the responses of pyramidal cells. The increased responses (the LTP) last for hours in vitro (and even days to weeks in vivo). The forms of LTP differ, depending on the particular synaptic system. The mechanism of the enhanced synaptic efficacy seems to involve both presynaptic and postsynaptic events. The neurotransmitters involved in LTP include excitatory amino acids that act on *N*-methyl-D-aspartate (NMDA) receptors, the responses of which are associated with an influx of Ca^{++} into the postsynaptic neuron. Second messenger pathways (including G proteins, Ca^{++}/calmodulin-dependent kinase II, protein kinase G, and protein kinase C) are also involved, and these kinases cause protein phosphorylation and changes in the responsiveness of neurotransmitter receptors. A retrograde messenger, perhaps nitric oxide (or carbon monoxide), may be released from postsynaptic neurons to act on presynaptic endings in such a way that transmitter release is enhanced. Immediate-early genes are also activated during LTP. Hence, changes in gene expression may also be involved.

Memory

With regard to the stages of memory storage, a distinction between **short-term memory** and **long-term memory** is useful. Recent events appear to be stored in short-term memory by ongoing neural activity because short-term memory persists for only minutes. Short-term memory is used, for instance, to remember page numbers in a book after looking them up in the index. Long-term memory can be subdivided into an intermediate form, which can be disrupted, and a long-lasting form, which is difficult to disrupt. Memory loss, or **amnesia,** can be caused by a loss of memory information per se, or it can result from interference with the mechanism for accessing the information. Long-term memory probably involves structural changes because it can remain intact even after events that disrupt short-term memory.

The temporal lobes appear to be particularly important for memory because bilateral removal of the hippocampal formation can severely and permanently disrupt recent memory. Short-term and long-term memories are unaffected, but new long-term memories can no longer be established. Thus patients with such amnesia remember events before their surgery but fail to recall new events, even with multiple exposures, and must be reintroduced repeatedly to people they meet after the surgery. This loss of **declarative memory** involves the conscious recall of personal events, places, and general history. Such patients, however, can still learn some tasks because they retain **procedural memory,** which involves associational and motor skills. If such patients are given a complex task to perform (e.g., mirror writing), they not only improve during the first training session but also perform better on subsequent days despite their denial of having any earlier experience with that task. The cerebral structures involved in procedural memory are not yet defined.

Neural Plasticity

Plasticity most commonly refers to the ability of the CNS to change its connectivity. Such changes can occur in various contexts, including learning and memory (see previous discussion), damage, and development. Damage to the CNS can induce remodeling of neural pathways and thereby alter behavior. Plasticity is greatest in the developing brain, but

AT THE CELLULAR LEVEL

Cellular studies of the hippocampus and the entorhinal cortex (which is adjacent and parallel to the hippocampus) have demonstrated the existence of "place cells" that fire when the subject enters a specific place in a test environment. Although there are many place cells, they are not distributed in an orderly manner that would resemble a topographic map. Place cells appear in very young animals as soon as they are able to explore. More recent studies reveal the presence of "grid cells," which also respond to specific sites but are distributed in hexagonal arrays that resembles an orderly map of the environmental space in the posterior entorhinal cortex. Although this cognitive map is fixed for any context, changes to the environment or removal of the person to a new test environment causes the generation of a new and appropriate map of grid cells.

Because the entorhinal cortex is a major source of input to the hippocampus, it is interesting that studies of London's taxi drivers—who must demonstrate an extremely detailed knowledge of the city's streets and most efficient routes before being licensed—indicate that the posterior hippocampus in trained and experienced drivers is larger than that in beginners or the general population. "Getting lost," a common complaint associated with amnesia, may be due to the loss of spatial memory skills.

some degree of plasticity remains in the adult brain, as evidenced by responses to certain manipulations, such as lesions of the brain, sensory deprivation, or even experience.

The capability for developmental plasticity may be maximal for some neural systems at a time referred to as the **critical period.** For example, it is possible to alter some connections formed in the visual pathways during their development by preventing one eye from providing input, but only during a specific critical period early in development. In such visually deprived animals, the visual connections become abnormal (Fig. 10.10), and restoration of normal visual input after the critical period does not undo the abnormality, nor does it restore functional vision from the deprived eye. In contrast, similar visual deprivation later in life does not result in abnormal connections. The plastic changes seen in such experiments may reflect a competition for synaptic connections, whereby the less functional connections are pruned away. Research has shown a correlation between the form of a gene that modulates the efficacy of synaptic pruning and the probability of schizophrenia.

Plastic changes can also occur after injury to the brain in adults. Sprouting of new axons does occur in the damaged CNS; however, the sprouts do not necessarily restore normal function, and many neural pathways do not appear to produce sprouts. Additional knowledge concerning neural plasticity in the adult CNS is vital if medical therapy is to be improved for many diseases of the CNS and after neural trauma. Research is currently being conducted to explore

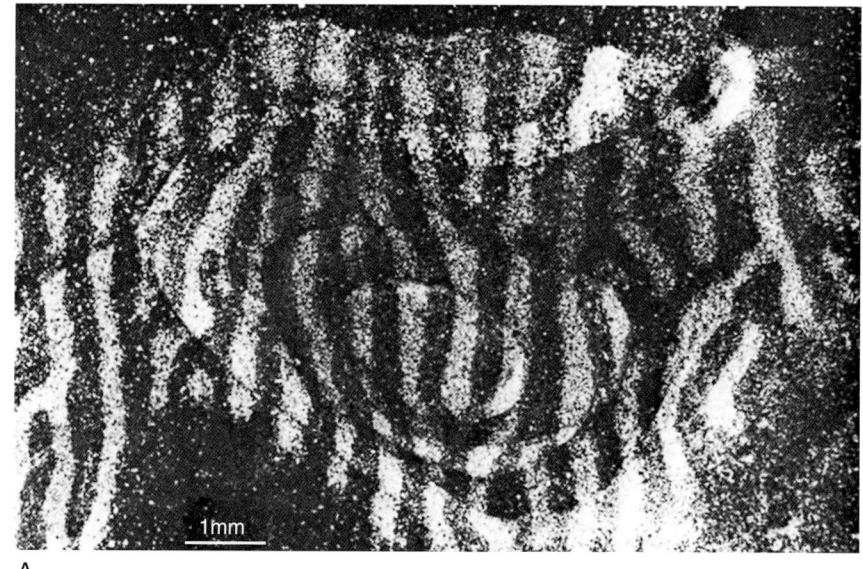

A

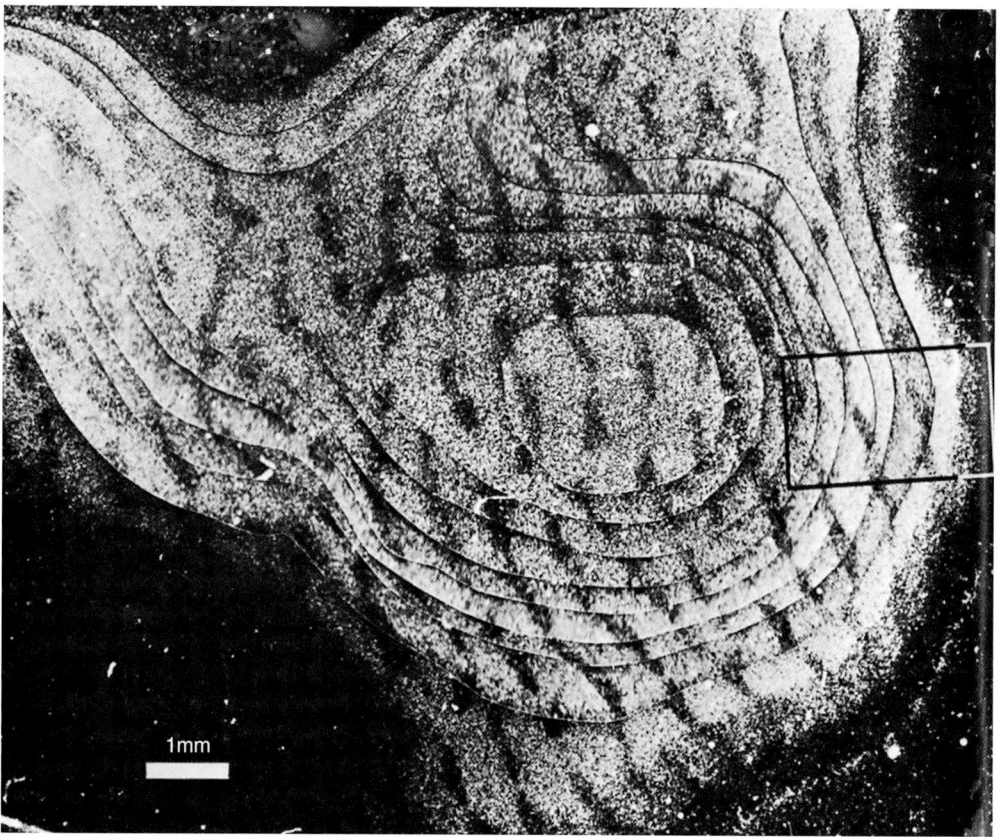

B

• **Fig. 10.10** Plasticity in the Visual Pathway as a Result of Sensory Deprivation During Development. The ocular dominance columns are demonstrated by autoradiography after injection of a radioactive tracer into one eye. The tracer is transported to the lateral geniculate nucleus and then transneurally transported to the striate cortex. The cortex is labeled in bands that alternate with unlabeled bands whose input is from the uninjected eye. **A,** Normal pattern. **B,** Changed pattern in an animal raised with monocular visual deprivation. The injection was made into the nondeprived eye, and the ocular dominance columns for this eye were clearly expanded. Other experiments showed that the ocular dominance columns for the deprived eye contracted. (**A,** From Hubel DH, Wiesel TN. *Proc R Soc Lond B*. 1977;198:1. **B,** From LeVay S, et al. *J Comp Neurol*. 1980;191:1.)

the potential of human embryonic stem cells for restoring CNS function.

Phantom limb sensation is an example of neural plasticity in adults. A patient whose limb has been amputated often perceives sensations on the missing limb when stimulated elsewhere on the body. Functional imaging studies suggest that this is a result of the spread of connections from the surrounding cortical territories into the cortical region that had served the amputated limb.

🩺 IN THE CLINIC

It was traditional policy to delay corrective surgery for a child born with a congenital cataract until the child was older and more able to cope with the stress of surgery. However, if the correction is deferred until after the "critical period," full recovery of function is unlikely. Similarly, children born with **amblyopia,** a condition characterized by strabismus (cross-eye) because of relative weakness of one of the extraocular muscles, tend to use the unaffected eye in preference. In both cases, early surgery is now common practice so that the cortical circuitry can be correctly sculpted by balanced input from the two eyes.

Such remapping can also occur after surgical amputation of the second and third digits of the hand. Before surgery, each of the digits was represented in discrete and somatotopically organized areas of the postcentral gyrus (SI cortex). After surgery, the area that represented the amputated digits is now mapped with an enlarged representation of the adjacent digits (Fig. 10.11). Conversely, individuals born with syndactyly (fusion of two or more digits of the hand) have a single or mostly overlapping representation of these digits in the SI cortex. After corrective surgery, the independent digits come to have distinctive representations. Even more remarkable is that monkeys that were trained on a sensory discrimination task requiring repeated daily use of their fingertips showed cortical differences after training. Not only were the SI cortical territories of their fingertips larger than before training but also the number of cortically recorded receptive fields on the fingertips was likewise increased.

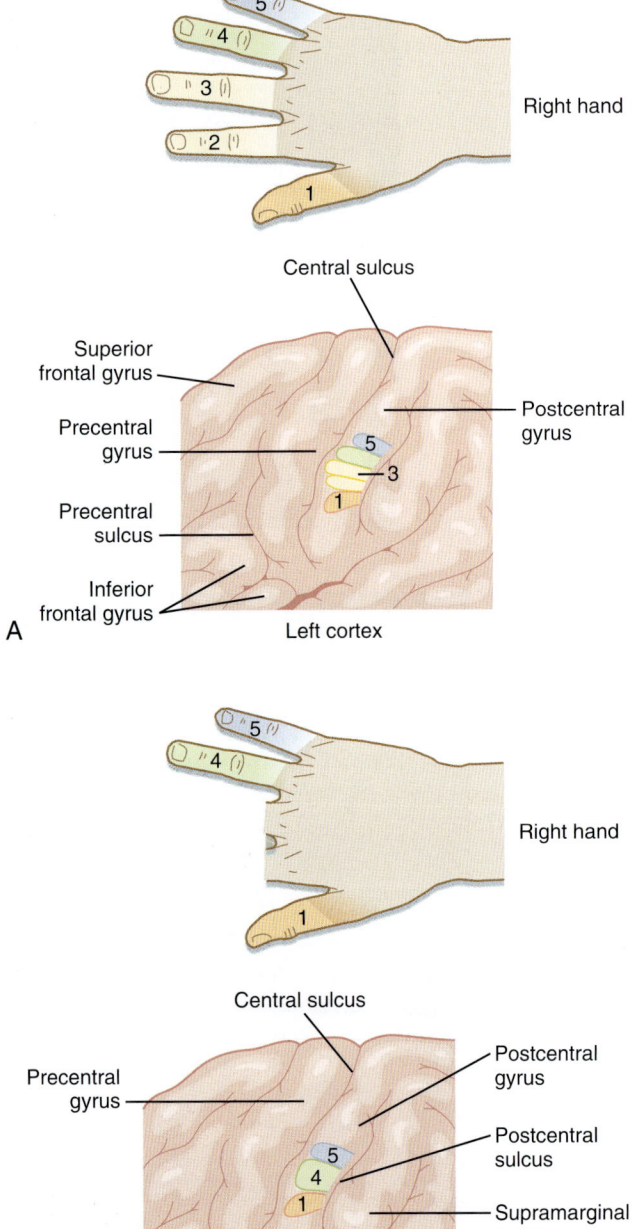

• **Fig. 10.11** Representation of the digit region of the left first somatosensory (SI) cortex **(A)** and reorganization of this representation **(B)** after amputation of the second and third digits. (From Haines DE [ed]. *Fundamental Neuroscience for Basic and Clinical Applications.* 3rd ed. Philadelphia: Churchill Livingstone; 2006.)

Key Points

1. The cerebral cortex can be divided into lobes on the basis of the pattern of gyri and sulci. Each lobe has distinctive functions, as shown by the effects of lesions. The left cerebral hemisphere is dominant for language in most individuals. Wernicke's area (in the posterior temporal lobe) is responsible for the understanding of language, and Broca's area (in the inferior frontal lobe) is responsible for its expression.

2. The neocortex contains pyramidal cells and several kinds of interneurons. Specific thalamocortical afferent fibers terminate mainly in layer IV of the

neocortex; diffuse thalamocortical afferent fibers synapse in layers I and VI. Axons from pyramidal cells in layer V are the major source of output to subcortical targets, including the spinal cord, brainstem, striatum, and thalamus.

3. The cortical structure varies in different regions. Brodmann's designations reflect these variations in cortical structure and correspond to functionally discrete areas.

4. The EEG reflects electrical fields generated by the activity of pyramidal and varies with the state of the sleep-wake cycle, disease, and other factors. Cortical evoked potentials are stimulus-triggered changes in the EEG and are useful clinical data about sensory transmission.

5. EEG patterns during sleep are divided into slow-wave and REM forms. Slow-wave sleep progresses through stages 1 through 4, each with a characteristic EEG pattern. Most dreams occur in REM sleep. Sleep is produced actively by a brainstem mechanism,

and its circadian rhythmicity is controlled by the suprachiasmatic nucleus.

6. Information is transferred between the two hemispheres primarily through the corpus callosum. The right hemisphere is more capable than the left in spatial tasks, facial expression, body language, and speech intonation. The left hemisphere is specialized for the understanding and generation of language, for logic, and for mathematical computation.

7. Learning and memory can be studied on the cellular level, in invertebrates, and in higher animals. Memory includes short-term (lasting minutes), recent, and long-term storage processes and a retrieval mechanism. The hippocampal formation is important for storing declarative and spatial memory.

8. Lesion studies and behavioral studies indicate that plasticity occurs in the brain throughout life. However, there appears to be more plasticity early in life, and synaptic competition in "critical periods" is important for the establishment of neural circuitry.

Additional Reading

Squire L, Berg D. *Fundamental Neuroscience*. 4th ed. New York: Academic Press; 2012.

11

The Autonomic Nervous System and Its Central Control

LEARNING OBJECTIVES

Objectives Heading

Upon completion of this chapter, the student should be able to answer the following questions:

1. What are the similarities and differences in the general organizations of the parasympathetic and sympathetic systems?
2. What are the respective actions of the parasympathetic and sympathetic innervation of the eye, and what symptoms arise when the parasympathetic or sympathetic innervation is lost?
3. What are the changes in the balance of parasympathetic and sympathetic activity to the bladder that occur during micturation?
4. What is meant by a "servomechanism"?
5. What are the specific feedback loops that regulate body temperature, feeding and body weight, and water intake?
6. What is the role of the hypothalamus in each of these feedback loops?

The main function of the **autonomic nervous system** is to assist the body in maintaining a constant internal environment **(homeostasis).** When internal stimuli signal that regulation of the body's environment is required, the central nervous system (CNS) and its autonomic outflow issue commands that lead to compensatory actions. For example, a sudden increase in systemic blood pressure activates the baroreceptors, which in turn modify the activity of the autonomic nervous system so that the blood pressure is lowered toward its previous level (see Chapter 17).

The autonomic nervous system has both sensory and motor divisions. The motor division is further divided into the **sympathetic** and **parasympathetic divisions.** Because much of the autonomic nervous system's actions relate to control of the viscera, it is sometimes called the **visceral nervous system.**

In service of its homeostatic function, the autonomic nervous system mediates visceral reflexes (e.g., the gastro-colic reflex, where stomach distention triggers peristalsis in the intestines) and provides sensory information to the CNS for the perception of the state of our viscera, a percept known to anyone who has eaten too much at a meal. More generally, activation of autonomic receptors can evoke a variety of sensory experiences such as pain, hunger, thirst, nausea, and a sense of visceral distention; these perceptions can then lead to compensatory voluntary behaviors that assist in maintaining homeostasis.

In addition to its central role in homeostasis, the autonomic nervous system also participates in appropriate and coordinated responses to external stimuli that are required for the optimal functioning of the somatic nervous system in performing voluntary behaviors. For example, the autonomic nervous system helps regulate pupil size in response to different intensities of ambient light, thus helping the visual system to operate over a large range of light intensity.

In this chapter, the **enteric nervous system** is also considered part of the autonomic nervous system, although it is sometimes considered a separate entity (see also Chapter 33). In addition, because the autonomic nervous system is under CNS control, the central components of the autonomic nervous system are discussed in this chapter. These central components include the hypothalamus and higher levels of the limbic system, which are associated with emotions (see Chapter 10) and with many visceral types of behavior (e.g., feeding, drinking, thermoregulation, reproduction, defense, and aggression) that have survival value.

Organization of the Autonomic Nervous System

The sensory autonomic neurons are located in the dorsal root ganglia and in the cranial nerve ganglia. Like the other neurons of the dorsal root ganglia, they are pseudounipolar cells with a peripheral axonal branch extended to one of the viscera and a central branch that enters the CNS. With regard to autonomic motor output, both the sympathetic and parasympathetic nervous systems use a two-neuron motor pathway, which consists of a preganglionic neuron, whose cell body is located in the CNS, and a postganglionic neuron, whose cell body is located in one of the autonomic ganglia (Figs. 11.1 and 11.2). The targets of this motor

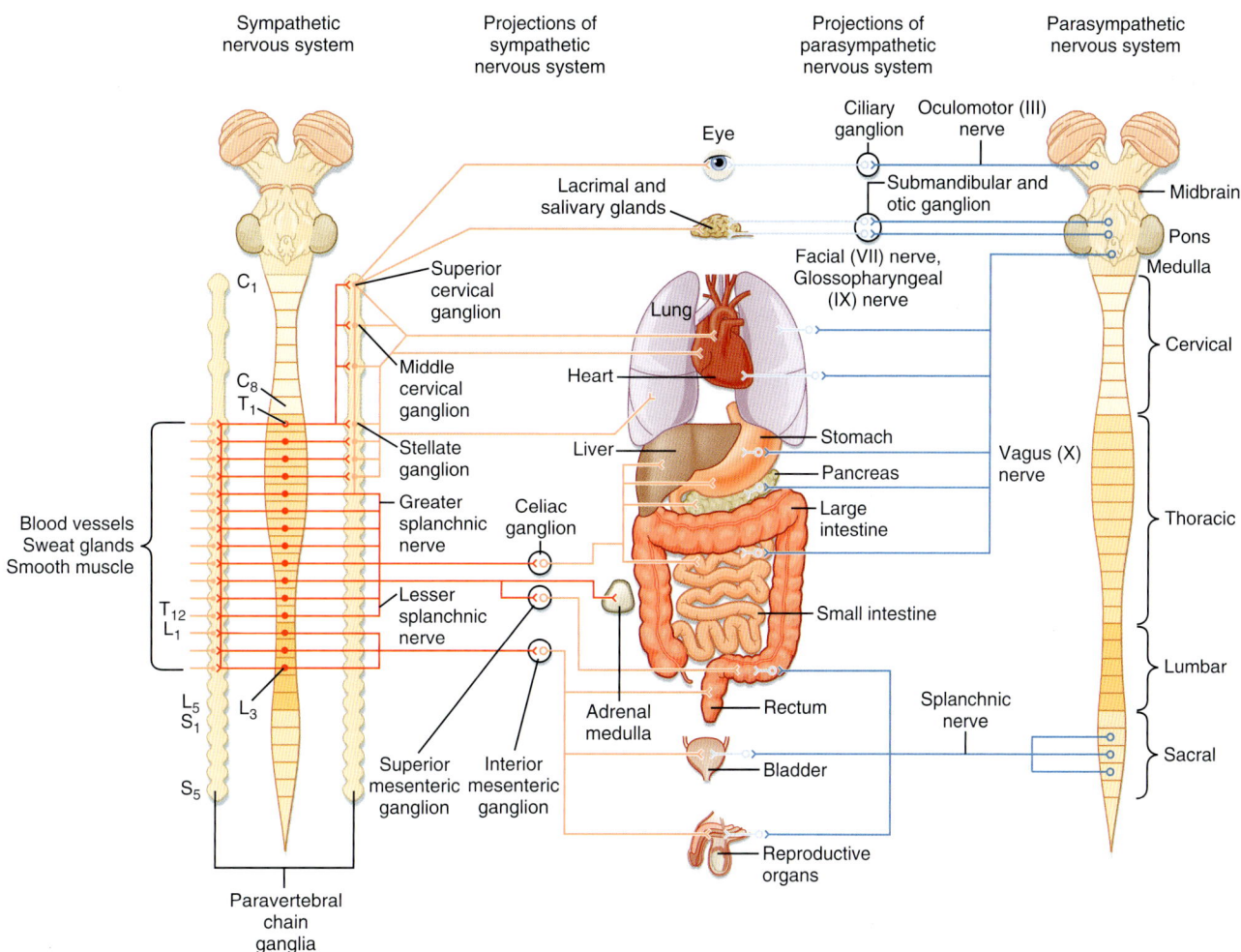

• **Fig. 11.1** Schematic Illustration of the Sympathetic and Parasympathetic Pathways. Sympathetic pathways are shown in *red* and parasympathetic pathways in *blue*. Preganglionic neurons are shown in *darker shades,* and postganglionic neurons, in *lighter shades.*

pathway are smooth muscle, cardiac muscle, and glands. The enteric nervous system includes the neurons and nerve fibers in the myenteric and submucosal plexuses, which are located in the wall of the gastrointestinal tract.

Control of the sympathetic and parasympathetic nervous systems of many organs is often antagonistic. To highlight this contrast, the sympathetic and parasympathetic systems are sometimes referred to as the "fight or flight" and the "rest and digest" systems, respectively. Indeed, the fight-or-flight response to a threat to the organism reflects an intense activation of the sympathetic nervous system, which leads to a variety of responses, including increased heart rate and blood pressure, redistribution of blood to the muscles, decreased peristalsis and gastrointestinal secretions, pupil dilation, and sweating.

However, under most conditions, the two parts of the autonomic control system work in a coordinated manner—sometimes acting reciprocally and sometimes synergistically—to regulate visceral function. Furthermore, not all visceral structures are innervated by both systems. For example, the smooth muscles and glands in the skin and most of the blood vessels in the body receive sympathetic

innervation exclusively; only a small fraction of the blood vessels have parasympathetic innervation. Indeed, the parasympathetic nervous system innervates not the body wall but only structures in the head and in the thoracic, abdominal, and pelvic cavities.

The Sympathetic Nervous System

The sympathetic preganglionic neurons are located in the thoracic and upper lumbar segments of the spinal cord. For this reason, the sympathetic nervous system is sometimes referred to as the **thoracolumbar division** of the autonomic nervous system. Specifically, sympathetic preganglionic neurons are concentrated in the **intermediolateral cell column** (lateral horn) in the thoracic and upper lumbar segments of the spinal cord (see Fig. 11.2). Some neurons may also be found in the C8 segment. In addition to the intermediolateral cell column, groups of sympathetic preganglionic neurons are found in other locations, including the lateral funiculus, the intermediate gray matter, and the gray matter dorsal to the central canal. Sympathetic postganglionic neurons are generally found in the paravertebral

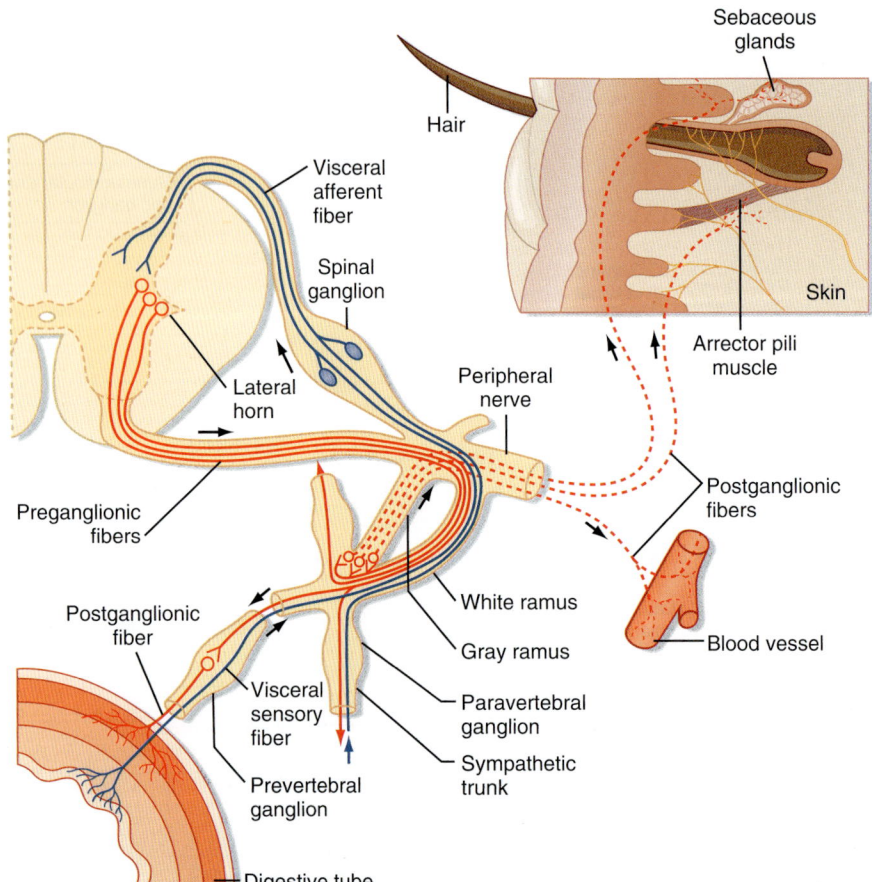

• Fig. 11.2 Details of the Sympathetic Pathway at a Thoracic Spinal Segment. Autonomic sensory fibers are represented by *blue lines;* sympathetic fibers, by *red lines;* preganglionic axons, by *solid lines,* and postganglionic axons, by *dashed lines.* (Redrawn from Parent A, Carpenter MB. *Carpenter's Human Neuroanatomy.* 9th ed. Philadelphia: Williams & Wilkins; 1996:295.)

or prevertebral ganglia. The paravertebral ganglia form two sets of ganglia, each lateral to one side of the vertebral column. The individual ganglia on each side are linked by longitudinally running axons that form a sympathetic trunk (see Figs. 11.1 and 11.2). Prevertebral ganglia are located in the abdominal cavity and include the Celiac and Superior and Inferior Mesenteric Ganglia (see Fig. 11.1). Thus paravertebral and prevertebral ganglia are located at some distance from their target organs.

The axons of preganglionic neurons are often small, myelinated nerve fibers known as *B fibers* (see Table 5.1). However, some are unmyelinated *C fibers.* They leave the spinal cord in the ventral root and enter the paravertebral ganglion at the same segmental level through a white communicating ramus. White rami are found only from the levels of T1 to L2. The preganglionic axon may synapse on postganglionic neurons in the ganglion at its level of entry; may travel rostrally or caudally within the sympathetic trunk and give off collaterals to the ganglia that it passes; or may pass through the ganglion, exit the sympathetic trunk, and enter a splanchnic nerve to travel to a prevertebral ganglion (see Figs. 11.1 and 11.2). Splanchnic nerves innervate the viscera; they contain both visceral afferents and autonomic motor fibers (sympathetic or parasympathetic).

Postganglionic neurons whose somata lie in paravertebral ganglia generally send their axons through a gray communicating ramus to enter a spinal nerve (see Fig. 11.2). Each of the 31 pairs of spinal nerves has a gray ramus. Postganglionic axons are distributed through the peripheral nerves to effectors, such as piloerector muscles, blood vessels, and sweat glands, located in the skin, muscle, and joints. Postganglionic axons are generally unmyelinated (C fibers), although some exceptions exist. The names white and gray rami reflect the relative contents of myelinated and unmyelinated axons in these rami.

Preganglionic axons in a splanchnic nerve often travel to a prevertebral ganglion and synapse, or they may pass through the ganglion and an autonomic plexus and end in a more distant ganglion. Some preganglionic axons pass through a splanchnic nerve and end directly on cells of the adrenal medulla, which are equivalent to postganglionic cells.

The sympathetic chain extends from the cervical to the coccygeal levels of the spinal cord. This arrangement serves as a distribution system that enables preganglionic neurons, which are limited to the thoracic and upper lumbar segments, to activate postganglionic neurons that innervate all body segments. However, there are fewer paravertebral ganglia than there are spinal segments because some of the segmental ganglia fuse during development. For example, the superior cervical sympathetic ganglion represents the fused ganglia of C1 through C4; the middle cervical sympathetic ganglion is the fused ganglia of C5 and C6; and the inferior cervical sympathetic ganglion is a combination of the ganglia at C7 and C8. The term **stellate ganglion** refers to fusion of the inferior cervical sympathetic ganglion with the ganglion of T1. The superior cervical sympathetic ganglion provides postganglionic innervation to the head and neck, and the middle cervical and stellate ganglia innervate the heart, lungs, and bronchi.

In general, the sympathetic preganglionic neurons are distributed to ipsilateral ganglia and thus control autonomic function on the same side of the body. Important exceptions are the sympathetic innervation of the intestines and the pelvic viscera, which are both bilateral. As with motor neurons to skeletal muscle, sympathetic preganglionic neurons that control a particular organ are spread over several segments. For example, the sympathetic preganglionic neurons that control sympathetic functions in the head and neck region are distributed at levels C8 to T5, whereas those that control the adrenal gland are distributed at levels T4 to T12.

The Parasympathetic Nervous System

The parasympathetic preganglionic neurons are found in several of the cranial nerve nuclei of brainstem and in the sacral spinal cord (S3-S4) gray matter (see Fig. 11.1). Hence, this part of the autonomic nervous system is sometimes called the **craniosacral division.** The cranial nerve nuclei that contain parasympathetic preganglionic neurons are the **Edinger-Westphal nucleus** (cranial nerve III), the **superior** (cranial nerve VII) and **inferior** (cranial nerve IX) **salivatory nuclei,** and the **dorsal motor nucleus of the vagus** and **nucleus ambiguus** (cranial nerve X). Postganglionic parasympathetic cells are located in cranial ganglia, including the **ciliary ganglion** (preganglionic input is from the Edinger-Westphal nucleus), the **pterygopalatine** and **submandibular ganglia** (input is from the superior salivatory nucleus), and the **otic ganglion** (input is from the inferior salivatory nucleus). The ciliary ganglion innervates the pupillary sphincter and ciliary muscles in the eye. The pterygopalatine ganglion supplies the lacrimal gland, as well as glands in the nasal and oral pharynx. The submandibular ganglion projects to the submandibular and sublingual salivary glands and to glands in the oral cavity. The otic ganglion innervates the parotid salivary gland and glands in the mouth.

Other parasympathetic postganglionic neurons are located near or in the walls of visceral organs in the thoracic, abdominal, and pelvic cavities. Neurons of the enteric plexus include cells that can also be considered parasympathetic postganglionic neurons. All of these cells receive input from the vagus or pelvic nerves. The vagus nerves innervate the heart, lungs, bronchi, liver, pancreas, and gastrointestinal

tract from the esophagus to the splenic flexure of the colon. The remainder of the colon and rectum, as well as the urinary bladder and reproductive organs, is supplied by sacral parasympathetic preganglionic neurons that travel through the pelvic nerves to postganglionic neurons in the pelvic ganglia.

The parasympathetic preganglionic neurons that project to the viscera of the thorax and part of the abdomen are located in the dorsal motor nucleus of the vagus (see Fig. 4.6E) and the nucleus ambiguus. The dorsal motor nucleus is largely **secretomotor** (it activates glands), whereas the nucleus ambiguus is **visceromotor** (it modifies the activity of cardiac muscle). The dorsal motor nucleus supplies visceral organs in the neck (pharynx, larynx), thoracic cavity (trachea, bronchi, lungs, heart, and esophagus), and abdominal cavity (including much of the gastrointestinal tract, liver, and pancreas). Electrical stimulation of the dorsal motor nucleus results in gastric acid secretion, as well as secretion of insulin and glucagon by the pancreas. Although projections to the heart have been described, their function is uncertain. The nucleus ambiguus contains two groups of neurons: (1) a dorsal group (**branchiomotor**) that activates striated muscle in the soft palate, pharynx, larynx, and esophagus and (2) a ventrolateral group that innervates and slows the heart (see also Chapter 18).

Visceral Afferent Fibers

The visceral motor fibers in the autonomic nerves are accompanied by visceral afferent fibers. Most of these afferent fibers supply information that originates from sensory receptors in the viscera. The activity of these sensory receptors only rarely reaches the level of consciousness; however, these receptors initiate the afferent limb of reflex arcs. Both viscerovisceral and viscerosomatic reflexes are elicited by these afferent fibers. Even though these visceral reflexes generally operate at a subconscious level, they are very important for homeostatic regulation and adjustment to external stimuli.

The fast-acting neurotransmitters released by visceral afferent fibers are not well documented, although many of these neurons release an excitatory amino acid transmitter such as glutamate. However, visceral afferent fibers also contain many neuropeptides or combinations of neuropeptides, including angiotensin II, arginine vasopressin, bombesin, calcitonin gene–related peptide, cholecystokinin, galanin, substance P, enkephalin, oxytocin, somatostatin, and vasoactive intestinal polypeptide.

Visceral afferent fibers that can mediate conscious sensation include nociceptors that travel in sympathetic nerves, such as the splanchnic nerves. Visceral pain is caused by excessive distention of hollow viscera, contraction against an obstruction, or ischemia. The origin of visceral pain is often difficult to identify because of the diffuse nature of the pain and its tendency to be referred to somatic structures (see Chapter 7). Visceral nociceptors in sympathetic nerves reach the spinal cord via the sympathetic chain, white rami, and dorsal roots. The terminals of nociceptive afferent fibers project to the dorsal horn and to the region surrounding the central canal. They activate not only local interneurons, which participate in reflex arcs, but also projection cells, which include spinothalamic tract cells that signal pain to the brain.

A major visceral nociceptive pathway from the pelvis involves a relay in the gray matter of the lumbosacral spinal cord. These neurons send axons into the fasciculus gracilis that terminate in the nucleus gracilis. Thus the dorsal columns not only contain primary afferents for somatic sensation (their main component) but also second-order neurons of the visceral pain pathway (recall that second-order axons for somatic pain travel in the lateral funiculus as part of the spinothalamic tract). Visceral nociceptive signals are then transmitted to the ventral posterior lateral nucleus of the thalamus and presumably from there to the cerebral cortex. Interruption of this pathway accounts for the beneficial effects of surgically induced lesions of the dorsal column at lower thoracic levels to relieve pain produced by cancer of the pelvic organs.

Other visceral afferent fibers travel in parasympathetic nerves. These fibers are generally involved in reflexes rather than sensation (except for taste afferent fibers; see Chapter 8). For example, the baroreceptor afferent fibers that innervate the carotid sinus are in the glossopharyngeal nerve. They enter the brainstem, pass through the solitary tract, and terminate in the nucleus of the solitary tract (see Fig. 4.6E). These neurons connect with interneurons in the brainstem reticular formation. The interneurons, in turn, project to the autonomic preganglionic neurons that control heart rate and blood pressure (see Chapter 18).

The nucleus of the solitary tract receives information from all visceral organs, except those in the pelvis. This nucleus is subdivided into several areas that receive information from specific visceral organs.

The Enteric Nervous System

The enteric nervous system, which is located in the wall of the gastrointestinal tract, contains about 100 million neurons. The enteric nervous system is subdivided into the myenteric plexus, which lies between the longitudinal and circular muscle layers of the gut, and the submucosal plexus, which lies in the submucosa of the gut. The neurons of the myenteric plexus primarily control gastrointestinal motility (see Chapter 27), whereas those in the submucosal plexus primarily regulate body fluid homeostasis (see Chapter 35).

The types of neurons found in the myenteric plexus include not only excitatory and inhibitory motor neurons (which can be considered parasympathetic postganglionic neurons) but also interneurons and primary afferent neurons. Afferent neurons supply mechanoreceptors within the wall of the gastrointestinal tract. These mechanoreceptors are the beginning of the afferent limb of reflex arcs within the enteric plexus. Local excitatory and inhibitory interneurons participate in these reflexes, and the output is sent through

the motor neurons to smooth muscle cells. Excitatory motor neurons release acetylcholine and substance P; inhibitory motor neurons release dynorphin and vasoactive intestinal polypeptide. The circuitry of the enteric plexus is so extensive that it can coordinate the movements of an intestine that has been completely removed from the body. However, normal function requires innervation by the autonomic preganglionic neurons and regulation by the CNS.

Activity in the enteric nervous system is modulated by the sympathetic nervous system. Sympathetic postganglionic neurons that contain norepinephrine inhibit intestinal motility, those that contain norepinephrine and neuropeptide Y regulate blood flow, and those that contain norepinephrine and somatostatin control intestinal secretion. Feedback is provided by intestinofugal neurons that project back from the myenteric plexus to the sympathetic ganglia.

The submucosal plexus regulates ion and water transport across the intestinal epithelium and glandular secretion. It also communicates with the myenteric plexus to ensure coordination of the functions of the two components of the enteric nervous system. The neurons and neural circuits of the submucosal plexus are not as well understood as those of the myenteric plexus, but many of the neurons contain neuropeptides, and the neural networks are well organized.

Autonomic Ganglia

The main type of neuron in autonomic ganglia is the post-ganglionic neuron. These cells receive synaptic connections from preganglionic neurons, and they project to autonomic effector cells. However, many autonomic ganglia also contain interneurons. These interneurons process information within the autonomic ganglia; the enteric plexus can be regarded as an elaborate example of this kind of processing. One type of interneuron found in some autonomic ganglia contains a high concentration of catecholamines; hence, these interneurons have been called **small, intensely fluorescent (SIF) cells.** SIF cells are believed to be inhibitory.

Neurotransmitters

Neurotransmitters in Autonomic Ganglia

The classic neurotransmitter of autonomic ganglia, whether sympathetic or parasympathetic, is acetylcholine. The two classes of acetylcholine receptors in autonomic ganglia are **nicotinic** and **muscarinic receptors,** so named because of their responses to the plant alkaloids **nicotine** and **muscarine.** Nicotinic acetylcholine receptors can be blocked by such agents as **curare** or **hexamethonium,** and muscarinic receptors can be blocked by **atropine.** Nicotinic receptors in autonomic ganglia differ somewhat from those on skeletal muscle cells.

Nicotinic and muscarinic receptors both mediate excitatory postsynaptic potentials (EPSPs), but these potentials have different time courses. Stimulation of preganglionic neurons elicits a fast EPSP, followed by a slow EPSP. The fast EPSP results from activation of nicotinic receptors, which cause ion channels to open. The slow EPSP is mediated by muscarinic receptors (primarily the M_2 receptor; see Chapter 6) that inhibit the **M current,** a current produced by potassium conductance.

Neurons in autonomic ganglia also release neuropeptides that act as neuromodulators. Besides acetylcholine, sympathetic preganglionic neurons may release enkephalin, substance P, luteinizing hormone–releasing hormone, neurotensin, or somatostatin.

Catecholamines such as norepinephrine and dopamine serve as the neurotransmitters of SIF cells in autonomic ganglia.

Neurotransmitters Between Postganglionic Neurons and Autonomic Effectors

Sympathetic Postganglionic Neurons

Sympathetic postganglionic neurons typically release norepinephrine, which excites some effector cells but inhibits others. The receptors on target cells may be either α- or β-adrenergic receptors. These receptors are further subdivided into α_1, α_2, β_1, β_2, and β_3 receptor types on the basis of pharmacological and genetic features. The distribution of these types of receptors and the actions that they mediate when activated by sympathetic postganglionic neurons are listed for various target organs in Table 11.1.

α_1 receptors are located postsynaptically, but α_2 receptors may be either presynaptic or postsynaptic. Receptors located presynaptically are generally called **autoreceptors;** they usually inhibit release of transmitter. The effects of agents that excite α_1 or α_2 receptors can be distinguished through the use of antagonists to block these receptors specifically. For example, prazosin is a selective α_1-adrenergic antagonist, and yohimbine is a selective α_2-adrenergic antagonist. The effects of α_1 receptors are mediated by activation of the inositol triphosphate/diacylglycerol second messenger system (see Chapter 3). In contrast, α_2 receptors decrease the rate of synthesis of cyclic adenosine monophosphate (cAMP) through action on a G protein.

β receptors were originally classified on the basis of the ability of antagonists to block them, but this has been supplemented by genetic studies. The β_1 and β_2 proteins have been much more extensively studied than has β_3, but it is thought that the proteins that make up all three types of β receptors are similar, with seven membrane-spanning regions connected by intracellular and extracellular domains (see Chapter 3). Agonist drugs that work on β receptors activate a G protein that stimulates adenylyl cyclase to increase the cAMP concentration. This action is terminated by the buildup of guanosine diphosphate.

TABLE 11.1 Responses of Effector Organs to Autonomic Nerve Impulses

Effector Organs	Receptor Type	Adrenergic Impulses,[a] Responses[b]	Cholinergic Impulses,[a] Responses[b]
Eye			
Radial muscle, iris	α	Contraction (mydriasis) ++	—
Sphincter muscle, iris	α	—	Contraction (miosis) +++
Ciliary muscle	β	Relaxation for far vision +	Contraction for near vision +++
Heart			
Sinoatrial node	β_1	Increase in heart rate ++	Decrease in heart rate; vagal arrest +++
Atria	β_1	Increase in contractility and conduction velocity ++	Decrease in contractility and (usually) increase in conduction velocity ++
Atrioventricular (AV) node	β_1	Increase in automaticity and conduction velocity ++	Decrease in conduction velocity; AV block +++
His-Purkinje system	β_1	Increase in automaticity and conduction velocity +++	Little effect
Ventricles	β_1	Increase in contractility, conduction velocity, automaticity, and rate of idioventricular pacemakers +++	Slight decrease in contractility
Arterioles			
Coronary	α, β_2	Constriction +; dilation[c] ++	Dilation +
Skin and mucosa	α	Constriction +++	Dilation[d]
Skeletal muscle	α, β_2	Constriction ++; dilation[c,e] ++	Dilation[f] +
Cerebral	α	Constriction (slight)	Dilation[d]
Pulmonary	α, β_2	Constriction +; dilation[c]	Dilation[d]
Abdominal viscera, renal	α, β_2	Constriction +++; dilation[e] +	—
Salivary glands	α	Constriction +++	Dilation ++
Veins (Systemic)	α, β_2	Constriction ++; dilation ++	—

Continued

TABLE 11.1	Responses of Effector Organs to Autonomic Nerve Impulses—cont'd			
Effector Organs	**Receptor Type**	**Adrenergic Impulses,[a] Responses[b]**	**Cholinergic Impulses,[a] Responses[b]**	
Lungs				
Bronchial muscle	β_2	Relaxation +	Contraction ++	
Bronchial glands	?	Inhibition (?)	Stimulation +++	
Stomach				
Motility and tone	α_2, β_2	Decrease (usually)[g] +	Increase +++	
Sphincters	α	Contraction (usually) +	Relaxation (usually) +	
Secretion		Inhibition (?)	Stimulation +++	
Intestine				
Motility and tone	α_2, β_2	Decrease[g] +	Increase +++	
Sphincters	α	Contraction (usually) +	Relaxation (usually) +	
Secretion		Inhibition (?)	Stimulation +++	
Gallbladder and Ducts		Relaxation +	Contraction +	
Kidney	β_2	Renin secretion ++	—	
Urinary Bladder				
Detrusor	β	Relaxation (usually) +	Contraction +++	
Trigone and sphincter	α	Contraction +++	Relaxation ++	
Ureter				
Motility and tone	α	Increase (usually)	Increase (?)	
Uterus	α, β_2	Pregnant: contraction (α); nonpregnant: relaxation (β)	Variable[h]	
Sex Organs, Male	α	Ejaculation +++	Erection +++	
Skin				
Pilomotor muscles	α	Contraction ++	—	
Sweat glands	α	Localized secretion[i] +	Generalized secretion +++	
Spleen Capsule	α, β_2	Contraction +++; relaxation +	—	
Adrenal Medulla		—	Secretion of epinephrine and norepinephrine	
Liver	α, β_2	Glycogenolysis, gluconeogenesis[j] +++	Glycogen synthesis +	
Pancreas				
Acini	α	Decreased secretion +	Secretion ++	
Islets (beta cells)	α	Decreased secretion +++	—	
	β_2	Increased secretion +	—	
Fat cells	α, β_1	Lipolysis[j] +++	—	
Salivary glands	α	K$^+$ and water secretion +	K$^+$ and water secretion +++	
	β	Amylase secretion +	—	
Lacrimal glands		—	Secretion +++	
Nasopharyngeal glands		—	Secretion +++	
Pineal gland	β	Melatonin synthesis	—	

From Goodman LS, Gilman A. *The Pharmacological Basis of Therapeutics*. 6th ed. New York: Macmillan; 1980.

[a]A long dash (—) signifies no known functional innervation.

[b]Responses are designated + to +++ to provide an approximate indication of the importance of adrenergic and cholinergic nerve activity in control of the various organs and functions listed.

[c]Dilation predominates in situ because of metabolic autoregulatory phenomena.

[d]Cholinergic vasodilation at these sites is of questionable physiological significance.

[e]Over the usual concentration range of physiologically released circulating epinephrine, a β receptor response (vasodilation) predominates in blood vessels of skeletal muscle and the liver, and an α receptor response (vasoconstriction) predominates in blood vessels of other abdominal viscera. The renal and mesenteric vessels also contain specific dopaminergic receptors, activation of which causes dilation, but their physiological significance has not been established.

[f]The sympathetic cholinergic system causes vasodilation in skeletal muscle, but this is not involved in most physiological responses.

[g]It has been proposed that adrenergic fibers terminate at inhibitory β receptors on smooth muscle fibers and at inhibitory α receptors on parasympathetic cholinergic (excitatory) ganglion cells of the Auerbach plexus.

[h]Depends on the stage of the menstrual cycle, the amount of circulating estrogen and progesterone, and other factors.

[i]Palms of the hands and some other sites ("adrenergic sweating").

[j]There is significant variation among species in the type of receptor that mediates certain metabolic responses.

β receptor activity is controlled in a number of ways. It can be antagonized by the action of α_1 receptors. The β receptors can also be desensitized by phosphorylation with prolonged exposure to agonists. Regulation of β receptor numbers represents a third control mechanism. For example, β receptor numbers can be decreased by being internalized. Alternatively, β receptor numbers can be increased (upregulated) in certain circumstances: for example, after denervation. Note that the number of α receptors is likewise regulated.

In addition to releasing norepinephrine, sympathetic postganglionic neurons release neuropeptides such as somatostatin and neuropeptide Y. For example, cells that release both norepinephrine and somatostatin supply the mucosa of the gastrointestinal tract, and cells that release both norepinephrine and neuropeptide Y innervate blood vessels in the gut and the limb. Another chemical mediator in sympathetic postganglionic neurons is adenosine triphosphate (ATP).

The endocrine cells of the adrenal medulla are similar in many ways to sympathetic postganglionic neurons (see also Chapter 43). They receive input from sympathetic preganglionic neurons, are excited by acetylcholine, and release catecholamines. However, the cells of the adrenal medulla differ from sympathetic postganglionic neurons in that they release catecholamines into the circulation rather than into a synapse. Moreover, the main catecholamine released is epinephrine, not norepinephrine. In humans, 80% of the catecholamine released by the adrenal medulla is epinephrine, and 20% is norepinephrine.

Some sympathetic postganglionic neurons release acetylcholine rather than norepinephrine as their neurotransmitter. For example, sympathetic postganglionic neurons that innervate eccrine sweat glands are cholinergic. The acetylcholine receptors involved are muscarinic, and they are therefore blocked by atropine. Similarly, some blood vessels are innervated by cholinergic sympathetic postganglionic neurons. In addition to releasing acetylcholine, the postganglionic neurons that supply the sweat glands also release neuropeptides, including calcitonin gene–related peptide and vasoactive intestinal polypeptide.

Parasympathetic Postganglionic Neurons

The neurotransmitter released by parasympathetic postganglionic neurons is acetylcholine. The effects of these neurons on various target organs are listed in Table 11.1. Parasympathetic postganglionic actions are mediated by muscarinic receptors. On the basis of binding studies, the action of selective antagonists, and molecular cloning, five types of muscarinic receptors have been identified (see Chapter 6). Activation of M_1 receptors enhances the secretion of gastric acid in the stomach. M_2 receptors are the most abundant receptor type in smooth muscle, including smooth muscle in the intestines, uterus, trachea, and bladder. In addition, they are present in autonomic ganglia and in the heart, where they exert negative chronotropic and inotropic actions (see Chapter 18). M_3 receptors are also present in the smooth

muscle of a variety of organs, and although they are less abundant than M_2 receptors, normal contractile patterns appear to require an interaction between the two types of receptors. M_4 receptors, like M_2 receptors, are present in autonomic ganglia and thus play a role in synaptic transmission at these sites. M_5 receptors are present in the sphincter muscle of the pupil, in the esophagus, and in the parotid gland, as well as in cerebral blood vessels.

Muscarinic receptors, like adrenergic receptors, have diverse actions. Some of their effects are mediated by specific second messenger systems. For example, cardiac M_2 muscarinic receptors may act by way of the inositol triphosphate system, and they may also inhibit adenylyl cyclase and thus cAMP synthesis. Muscarinic receptors also open or close ion channels, particularly K^+ or Ca^{++} channels. This action on ion channels is likely to occur through activation of G proteins. A third action of muscarinic receptors is to relax vascular smooth muscle by an effect on endothelial cells, which produce endothelium-derived relaxing factor (EDRF). EDRF is actually nitric oxide, a gas released when arginine is converted to citrulline by nitric oxide synthase (see Chapter 18). Nitric oxide relaxes vascular smooth muscle by stimulating guanylate cyclase and thereby increasing levels of cyclic guanosine monophosphate (cGMP), which in turn activates a cGMP-dependent protein kinase (see Chapter 3). The number of muscarinic receptors is regulated, and exposure to muscarinic agonists decreases the number of receptors by internalization of the receptors.

IN THE CLINIC

Chagas disease is the result of infection by the parasite *Trypanosoma cruzi*. About 18 million people are infected worldwide, and approximately 50,000 die each year as a result of complications from the disease. The most serious forms involve enlargement of the esophagus, colon, and heart. Loss of parasympathetic control is a significant component of the initial stages of the disease; shortly after the initial infection, the parasympathetic neurons innervating the heart, esophagus, and colon are destroyed, which leads to arrhythmias (and potentially sudden death) and aperistalsis. Chronic cardiomyopathy (malfunction of the heart muscle) that can lead to death occurs in approximately 30% of those infected. Although the pathogenesis of the cardiomyopathy is not fully understood, one leading idea involves autoimmunity. Antibodies against the parasitic antigens have been found to bind to the β-adrenergic and M_2 acetylcholine receptors in the heart. These antibodies not only trigger autoimmune responses that destroy heart muscle but also act as agonists at these receptors and cause inappropriate responses of the cardiovascular system to changing external demands.

Central Control of Autonomic Function

The discharges of autonomic preganglionic neurons are controlled by pathways that synapse on autonomic preganglionic neurons. The pathways that influence autonomic activity include spinal cord and brainstem reflex pathways, as well as descending control systems originating at higher levels of the nervous system, such as the hypothalamus.

Examples of Autonomic Control of Particular Organs

Pupil

The dilator and constrictor muscles of the iris, which are under the control of sympathetic and parasympathetic fibers, respectively, determine the size of the pupil. Activation of sympathetic innervation of the eye, via thoracic white rami and sympathetic trunk ganglia, dilates the pupil, which occurs during emotional excitement and also in response to painful stimulation. The neurotransmitter at the sympathetic postganglionic synapses is norepinephrine, and it acts at α receptors.

The parasympathetic nervous system exerts an action on pupillary size opposite that of the sympathetic nervous system. Whereas the sympathetic system elicits pupillary dilation, the parasympathetic system constricts the pupil. The preganglionic parasympathetic nerves that innervate the pupillary constrictor are in the Edinger-Westphal nucleus, which is in the midbrain, and travel in cranial nerve III, and so damage to this nerve can lead to a dilated pupil (mydriasis).

IN THE CLINIC

Sympathetic control of the pupil is sometimes affected by disease. For example, interruption of the sympathetic innervation of the head and neck results in **Horner's syndrome.** This syndrome is characterized by the triad of miosis (abnormal pupillary constriction), ptosis (caused by paralysis of the superior tarsal muscle), and anhydrosis (loss of sweating) on the face. **Enophthalmos** (retraction of the eye into the orbit) also occurs in some animals (rats, cats, and dogs, among others), but in humans no true enophthalmos occurs; however, there is an apparent enophthalmos, an illusion created by partial closure of the eyelid from the ptosis. Horner's syndrome can be produced by a lesion that (1) destroys the sympathetic preganglionic neurons in the upper thoracic spinal cord, (2) interrupts the cervical sympathetic chain, or (3) damages the lower brainstem in the region of the reticular formation, through which pathways descend to the spinal cord to activate sympathetic preganglionic neurons. In the last case, there is also a loss of sweating on the side of the body ipsilateral to the lesion.

Pupil size is reduced by the **pupillary light reflex** and during accommodation for near vision. In the pupillary light reflex, light that strikes the retina is processed by retinal circuits that excite W-type retinal ganglion cells (see Chapter 8). These cells respond to diffuse illumination. The axons of some of the W-type cells project through the optic nerve and tract to the pretectal area, where they synapse in the olivary pretectal nucleus. This nucleus contains neurons that also respond to diffuse illumination. Activity of neurons of the olivary pretectal nucleus causes pupillary constriction by means of bilateral connections with parasympathetic preganglionic neurons in the Edinger-Westphal nuclei. The reflex results in contraction of the pupillary sphincter muscles in both eyes, even when light is shone into only one eye.

The **accommodation response,** which is important for focusing on near objects, involves pupillary constriction, increasing the curvature of the lens, and convergence of the eyes. This response is triggered by information from M cells of the retina that is transmitted to the striate cortex through the geniculostriate visual pathway (see Chapter 8). The specific stimuli that trigger accommodation are thought to be a blurred retinal image and disparity of the image between the two eyes. After the information is processed in the visual cortex, signals are transmitted directly or indirectly to the middle temporal cortex, where they activate neurons in a visual area known as MT. Area MT neurons transmit signals to the midbrain that activate parasympathetic preganglionic neurons in the Edinger-Westphal nuclei, which results in pupillary constriction. At the same time, signals are transmitted to the ciliary muscle that cause it to contract. The ciliary muscle contraction allows the lens to round up and increase its refractile power. (Convergence is a somatic response mediated by neurons in the oculomotor [cranial nerve III] nucleus of the midbrain.)

IN THE CLINIC

The pupillary light reflex is sometimes absent in patients with tertiary (advanced) syphilis, which affects the CNS (i.e., in the form of tabes dorsalis). Although the pupil fails to respond to light, it has a normal accommodation response. This condition is known as the **Argyll Robertson pupil.** The exact mechanism is controversial. One explanation rests on the fact that some optic tract fibers project to the pretectal area in the midbrain. These fibers can be damaged in syphilitic meningitis, possibly by the presence of spirochetes in the subarachnoid space. Note that the pretectal area projects to the Edinger-Westphal nucleus, also in the midbrain, whose cells originate the parasympathetic innervation of the eye, which controls the pupillary sphincter muscle. Although input to the olivary pretectal nucleus is interrupted, the optic tract fibers projecting to the lateral geniculate nucleus are not destroyed, and thus vision is maintained, as is pupillary constriction during accommodation.

Urinary Bladder

The urinary bladder is controlled by reflex pathways in the spinal cord and also by a supraspinal center (Fig. 11.3). The sympathetic innervation originates from preganglionic sympathetic neurons in the upper lumbar segments of the spinal cord. Postganglionic sympathetic axons act to inhibit the smooth muscle (**detrusor muscle**) throughout the body of the bladder, and they also act to excite the smooth muscle of the trigone region and the internal urethral sphincter.

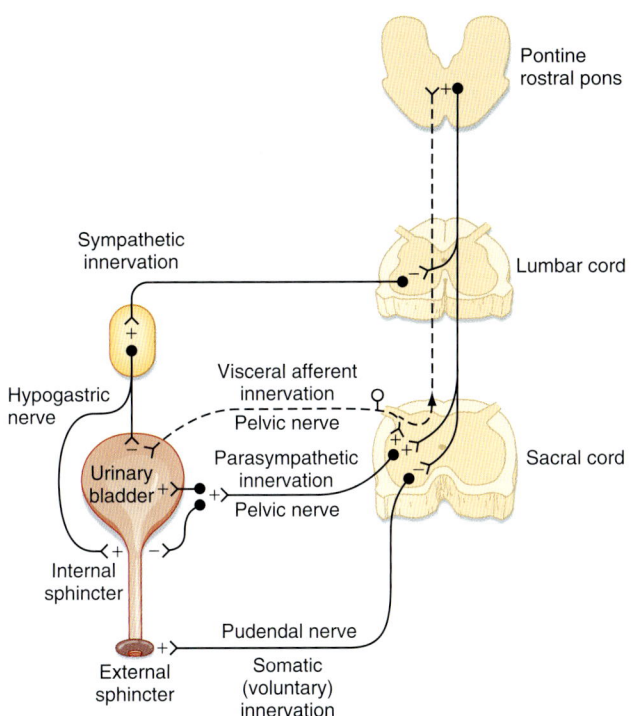

Pontine
rostral pons

Lumbar cord

Sympathetic
innervation

Hypogastric
nerve

Visceral afferent
innervation
Pelvic nerve

Parasympathetic
innervation

Sacral cord

Urinary
bladder

Pelvic nerve

Internal
sphincter

Pudendal nerve

External
sphincter

Somatic
(voluntary)
innervation

• **Fig. 11.3** Illustration of Descending and Efferent Pathways for Reflexes That Control the Urinary Bladder. For clarity, only some of the major involved pathways are shown. (Redrawn from de Groat WC, Booth AM. In Dyck PJ, et al [eds]. *Peripheral Neuropathy.* 2nd ed. Philadelphia: WB Saunders; 1984.)

The detrusor muscle is tonically inhibited during filling of the bladder, and such inhibition prevents urine from being voided. Inhibition of the detrusor muscle is mediated by the action of norepinephrine on β receptors, whereas excitation of the trigone and internal urethral sphincter is elicited by the action of norepinephrine on α receptors.

The external sphincter of the urethra also helps control voiding. This sphincter is a striated muscle, and it is innervated by motor axons in the pudendal nerves, which are somatic nerves. The motor neurons are located in the **Onuf nucleus,** in the ventral horn of the sacral spinal cord.

The parasympathetic preganglionic neurons that control the bladder are located in the sacral spinal cord (the S2 and S3 or S3 and S4 segments). These cholinergic neurons project through the pelvic nerves and are distributed to ganglia in the pelvic plexus and the bladder wall. Postganglionic parasympathetic neurons in the bladder wall innervate the detrusor muscle, as well as the trigone and sphincter. The parasympathetic activity contracts the detrusor muscle and relaxes the trigone and internal sphincter. These actions result in **micturition,** or urination. Some of the postganglionic neurons are cholinergic and others are purinergic (they release ATP).

Micturition is normally controlled by the **micturition reflex** (see Fig. 11.3). Mechanoreceptors in the bladder wall are excited by both stretch and contraction of the muscles in the bladder wall. Thus as urine accumulates and distends the bladder, the mechanoreceptor afferents

begin to discharge. The pressure in the urinary bladder is low during filling (5 to 10 cm H_2O), but it increases abruptly when micturition begins. Micturition can be triggered either reflexively or voluntarily. In reflex micturition, bladder afferent fibers excite neurons that project to the brainstem and activate the micturition center in the rostral pons (**Barrington's nucleus**). The descending projections also inhibit sympathetic preganglionic neurons that prevent voiding. When a sufficient level of activity occurs in this ascending pathway, micturition is triggered by the micturition center. Commands reach the sacral spinal cord through a reticulospinal pathway. Activity in the sympathetic projection to the bladder is inhibited, and the parasympathetic projections to the bladder are activated. Contraction of muscle in the wall of the bladder causes a vigorous discharge of the mechanoreceptors that supply the bladder wall and thereby further activates the supraspinal loop. The result is complete emptying of the bladder.

A spinal reflex pathway also exists for micturition. This pathway is operational in newborn infants. However, with maturation, the supraspinal control pathways take on a dominant role in triggering micturition. After spinal cord injury, human adults lose bladder control during the period of spinal shock (urinary incontinence). As the spinal cord recovers from spinal shock, some degree of bladder function is recovered because of enhancement of the spinal cord micturition reflex. However, the bladder has increased muscle tone and fails to empty completely. These circumstances frequently lead to urinary infections.

Autonomic Centers in the Brain

Influence over autonomic output is maintained by autonomic centers, which consist of local networks of neurons, in a variety of brain regions. The micturition center in the pons, which was just discussed, is one example. Many other autonomic centers with diverse functions exist. Vasomotor and vasodilator centers are in the medulla, and respiratory centers are in the medulla and pons. Perhaps the greatest concentration of autonomic centers is found in the hypothalamus.

The Hypothalamus and Preoptic Area

The hypothalamus is part of the diencephalon. Some of the nuclei of the hypothalamus are shown in Fig. 11.4. Located anteriorly from the hypothalamus are telencephalic structures: the preoptic region and septum, both of which help regulate autonomic function. Important fiber tracts that course through the hypothalamus are the **fornix,** the **medial forebrain bundle,** and the **mammillothalamic tract.** The fornix is used as a landmark to divide the hypothalamus into medial and lateral zones.

The hypothalamus has many functions; see Chapter 41 for a discussion of hypothalamic control of endocrine function. Its control of autonomic function is emphasized here. In its control of autonomic function, the hypothalamus functions much like a control system that is termed, in

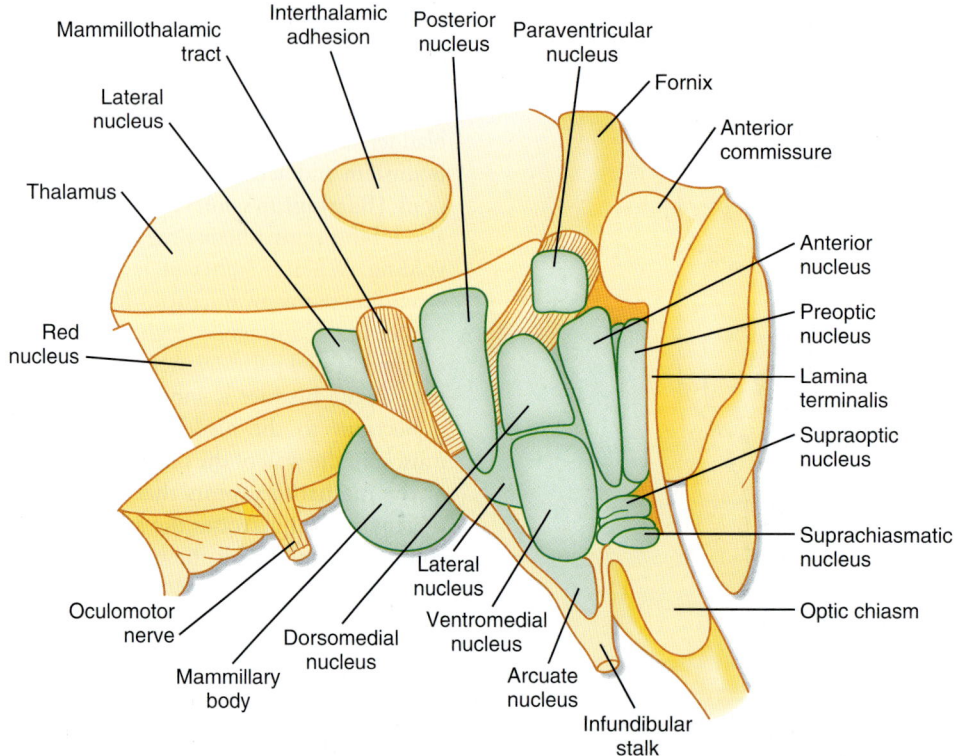

• **Fig. 11.4** Illustration of Main Nuclei of the Hypothalamus, Viewed From the Third Ventricle. Anterior is to the right. (Redrawn from Nauta WJH, Haymaker W. *The Hypothalamus.* Springfield, IL: Charles C Thomas; 1969.)

engineering, a *servomechanism:* that is, a system in which a particular physiological parameter is controlled through the use of negative feedback loops to maintain the parameter at a particular set point or value. The following examples illustrate this principle for body temperature, body weight and adiposity, and water intake.

Temperature Regulation

Homeothermic animals maintain a relatively constant core body temperature in situations of fluctuating environmental temperatures and differing levels of bodily activity that cause endogenous heat production. This ability rests on information from three main groups of thermoreceptors located in the skin, CNS, and viscera.

Information about the external temperature is provided by thermoreceptors in the skin. Core body temperature is monitored by central thermoreceptive neurons in the preoptic area (and possibly the spinal cord), which monitor the temperature of local blood. Thermoreceptors in the viscera monitor the temperature in these organs. All of these receptors provide temperature information to the preoptic area (pathways described later), along with parts of the hypothalamus, in which this information is used to keep core body temperature constant. Thus the preoptic area and hypothalamus act together as a servomechanism with a set point at the normal body temperature.

Although the signals from each of these sources are integrated, their relative importance may shift, depending on the situation. Changes in environmental temperature evoke more rapid and much larger changes in the temperature of the skin than in the body core, and so cutaneous receptors are probably the initial and most often used mechanism for compensating for external changes in temperature. Central thermoreceptors are more important for situations with internal causes of temperature change, such as during exercise, or in which external changes temperature are so severe or prolonged that core body temperature starts to change despite the signals from peripheral thermoreceptors. Last, alteration of body temperature by ingestion of hot or cold food or liquids is detected by the visceral thermoreceptors.

Error signals (i.e., cooling and warming of the body), which represent a deviation from the set point of the servomechanism, evoke responses that tend to restore body temperature toward the set point. These responses are mediated by the autonomic, somatic, and endocrine systems.

Situations involving cooling, for example, trigger a variety of responses that increase heat production (thermogenesis) and minimize heat loss. Heat production is increased by mechanisms that include **shivering thermogenesis** (asynchronous contractions of skeletal muscle that increases heat production) and **brown adipose tissue (BAT) thermogenesis** (in BAT thermogenesis, oxidative phosphorylation is uncoupled from ATP synthesis, which allows the energy released by the reaction to be dissipated as heat instead), and increased thyroid hormone levels lead to increased metabolism. Heat loss is reduced by cutaneous vasoconstriction and by piloerection. Piloerection is effective in animals with fur but not in humans; in the latter, the result is only goose bumps. In addition, tachycardia occurs, which may help provide metabolites to be used in thermogenesis to the thermogenic tissues (fat and muscle) and help distribute the heat generated throughout the body. Finally, the perception of being cold influences the decision to initiate voluntary behaviors: in this case, possibly putting on a jacket.

Warming the body generally causes changes in the opposite direction. The activity of the thyroid gland diminishes, which leads to reduced metabolic activity and less heat production. Heat loss is increased by sweating, salivation (in some animals but not humans), and cutaneous vasodilation (because of decreased sympathetic activity). However, again tachycardia occurs, this time presumably to allow optimal perfusion of the cutaneous circulation for heat dissipation.

Early studies identified the preoptic region and anterior hypothalamus as a heat loss center and the posterior hypothalamus as a heat conservation center. For example, lesions in the preoptic region prevent sweating and cutaneous vasodilation, and if an individual with a lesion in this region is placed in a warm environment, **hyperthermia** occurs. Conversely, electrical stimulation of the heat loss center causes cutaneous vasodilation and inhibits shivering. In contrast, lesions in the area dorsolateral to the mammillary body interfere with heat production and conservation and can cause **hypothermia** when the person is in a cold environment. Electrical stimulation in this region of the brain evokes shivering.

Many details of the circuitry and physiologic processes underlying the temperature regulation responses are now known, and they indicate that the preoptic area and **dorsal medial hypothalamic nucleus** are key components in the regulation of body temperature. The preoptic area in particular appears to be the target of the various sources of sensory information. Cutaneous temperature information is conveyed by thermosensitive primary afferents that synapse in the dorsal horn of the spinal cord onto neurons that project up to and excite the **parabrachial nucleus** in the caudal midbrain. Information from visceral afferents is relayed by the solitary nucleus to the parabrachial nucleus as well. The parabrachial neurons in turn excite neurons within a specific part of the preoptic area, the **median nucleus** (of the preoptic area). Many median nucleus neurons are also sensitive to local changes in blood temperature and contain prostaglandin E2 (PGE2) receptor 3 (EP3), which mediates fever responses (see the following "In the Clinic" box).

Thus the median preoptic nucleus is a key component of the thermoregulatory control system in which information from the various types of thermoreceptors is integrated. Output from this nucleus is directed to the neighboring medial preoptic area, which projects to regions of the rostral medulla, both directly and via the dorsal medial hypothalamic nucleus. The rostral medulla has spinally projecting neurons that project to the lateral horn of the spinal cord, where the preganglionic sympathetic neurons

are located, and the activity of these neurons regulates BAT thermogenesis and modulates cutaneous vasomotor tone. The rostral medulla also projects to the ventral horn of the spinal cord, which contains the somatic motor neurons that contract skeletal muscle and thus mediate the shivering response.

 IN THE CLINIC

Fever, which accompanies some infections, can be thought of as an elevation of the set point for body temperature. This elevation can be caused by the release of **pyrogens** by microorganisms or by cells mediating the inflammatory response. The pyrogen's effect to raise the set point is mediated primarily by the action of prostaglandin PGE2's binding to EP3 receptors on neurons in the preoptic area. PGE2 is released by peripheral tissues and by blood vessels supplying the preoptic area. The binding of PGE2 to EP3 receptors causes a reduction in the activity of preoptic neurons. This reduction in neuronal activity leads to increased heat production through shivering and BAT thermogenesis and to heat conservation by cutaneous vasoconstriction, the combined effect of which is to raise body temperature. Evidence of this mechanism of fever production includes studies in which injection of PGE2 into the preoptic area induced fever and others in which selective deletion of the EP3 receptor from preoptic neurons abolished the ability of PGE2 injections to induce fever.

Regulation of Feeding and Body Weight

Energy homeostasis is crucial for the survival of the animal. The challenge is that most cells need a continuous supply of nutrients to function, but most animals do not constantly eat; instead, they have periodic meals. Thus to achieve energy homeostasis, feeding behavior is controlled by many factors, which operate both on a short-term basis to control ingestion and on a long-term basis to control body weight in order to ensure sufficient energy stores. Both hedonic and homeostatic factors are involved; however, in this chapter, the focus is on the latter because of the central role that the hypothalamus plays in energy homeostasis.

In the short run, eating is controlled by a number of mechanisms. First, the stomach wall has stretch receptors that signal distention as food fills the stomach. These signals are conveyed by the afferents of the vagus nerve to the solitary nucleus in the medulla. From there the information is relayed to several brain areas, including the hypothalamus, either directly or via a relay in the parabrachial nucleus, to organize autonomic responses to the ingested material, and the thalamus and cortex, for conscious awareness of the fullness of the stomach. In the hypothalamus, the paraventricular, dorsomedial, and arcuate nuclei and the lateral hypothalamus are the major targets of these signals.

Sensory afferents also sense the concentrations of glucose and lipids in the intestines and the hepatic portal circulation and send this information to the solitary nucleus and, from there, to the hypothalamus, in a manner similar to that described for the stretch receptors. In addition, the stomach and gut release a number of hormones in response to feeding, including cholecystokinin, peptide YY, glucagon-like peptide-1 (GLP-1), and ghrelin. Hypothalamic cells have receptors for many of these hormones and can be influenced directly by them. In addition, cells in other brain areas have receptors for these hormones and thus may provide an indirect pathway to the hypothalamus. One such region is the **area postrema,** which is just dorsal to the solitary nucleus and projects to it. The area postrema is not protected by the blood-brain barrier (it is one of the circumventricular organs), and its neurons respond to cholecystokinin and GLP-1, which leads to decreased food intake.

Control of body weight over the long run is influenced by many factors and involves the interaction of the nervous and endocrine systems. In this section, the focus is on the role of hypothalamus and its control of the autonomic nervous system, which provides another example of how the hypothalamus is part of a servomechanism. In this case, adiposity is the controlled parameter. For further details on the endocrine system's role, see Chapters 38 and 40.

Early studies in which researchers used lesions and electrical stimulation provided evidence that the ventromedial and ventrolateral hypothalamus are involved in energy homeostasis. A lesion in the ventromedial region causes an increase in food intake (hyperphagia) that results in obesity, whereas electrical stimulation of the same region decreases feeding behavior. These lesions were also shown to alter autonomic activity, increasing parasympathetic and decreasing sympathetic tone, both of which lead to high blood insulin levels, which in turn promote energy conservation and storage (see Chapter 38). These observations led to the idea that the ventromedial hypothalamus contains a satiety center. However, an alternative interpretation is that the primary controlled variable may not be simply eating behavior per se but rather body weight and, even more specifically, body fat levels (i.e., adiposity). Thus modulation of feeding behavior may be just one of several actions used to defend a body weight set point. Evidence for this is that although lesions cause an initial period of dynamic weight gain in which hyperphagia is present, this is followed by a static period in which the higher weight is maintained without hyperphagia. Moreover, animals with a lesion in the ventromedial hypothalamus that are fed a fixed (normal) amount of food to prevent hyperphagia nonetheless become obese, which implies changes in the regulation of other metabolic processes. Last, lesions of the ventromedial hypothalamus have been shown to alter levels of energy expenditure.

In contrast to lesions of the ventromedial hypothalamus, those of the lateral hypothalamus suppress food intake (hypophagia) and lead to a decrease in body weight; indeed, animals can starve to death after such lesions. Conversely, electrical stimulation of the medial forebrain bundle in the lateral hypothalamus evokes exploratory behavior and eating, if food is present. This stimulation also provides a dopamine-dependent reward that mediates the incentive effects of natural rewards (food, sex) as well as the rewarding effects of most drugs of abuse. These observations led to the view that the lateral hypothalamus contains a feeding center. This interpretation, however, is complicated by the fact that the dopaminergic axons of substantia nigra neurons pass just lateral to the lateral hypothalamus on their way to the striatum, and so loss or stimulation of these fibers could account for the effects produced in these experiments. However, lateral hypothalamic neurons have been found to synthesize peptides, such as orexin, that affect feeding behavior, and so the lateral hypothalamus probably does play a role in energy homeostasis.

In newer studies, investigators have identified a number of hormones and neuropeptides involved in feeding and control of body weight, and many of the interactions between the endocrine and nervous systems that underlie energy homeostasis have been clarified.

In normal individuals, blood **insulin** levels are correlated with adiposity (in addition to varying acutely with blood levels of glucose and other substances). Similarly, the level of the protein **leptin,** a hormone released by adipocytes (primarily those forming white adipose tissue), is correlated with adiposity. High levels of leptin inhibit food intake and stimulate catabolic processes, including loss of fat tissue, whereas low leptin levels trigger the reverse actions. Similarly, high insulin levels promote energy storage processes.

The ability of leptin and insulin to regulate body weight has been linked to their actions on the hypothalamus, particularly the arcuate nucleus, whose neurons express receptors for both hormones (see also Chapter 40). Two major classes of arcuate nucleus neurons that respond to leptin and insulin have been identified. Neurons that express proopiomelanocortin (POMC) and cocaine- and amphetamine-related transcript (CART) are stimulated by leptin and insulin, and their activity leads to increased catabolism. In contrast, the activity of a second group of neurons, those that express neuropeptide Y (NPY) and agouti-related peptide (AgRP), triggers anabolic processes but is inhibited by leptin and insulin. Thus increased body fat levels lead to high leptin and insulin levels, which in turn both (1) increase the activity of POMC- and CART-expressing neurons, leading to increased catabolism, and (2) decrease the activity of NPY- and AgRP-expressing neurons, leading to decreased anabolism; both of which act to return body fat levels to their set point. Lowering body fat levels would result in a sequence of events opposite to that just described to increase body fat levels to their original level or set point.

The efferent limb that mediates the actions of these sets of arcuate neurons is not fully worked out. However, the arcuate nucleus projection to the **paraventricular hypothalamic nucleus** appears to be an important step in the pathway. Paraventricular neurons contain oxytocin. Many of them project to the posterior pituitary gland and are involved in lactation and uterine contractions during labor (see Chapter 43). However, the paraventricular neurons involved in body weight regulation are a distinct subset of neurons that project down to the brainstem and spinal cord, where they probably synapse with autonomic and preautonomic nuclei that control parasympathetic vagal fibers to the pancreas, which act to stimulate insulin release, and sympathetic fibers, which act to inhibit its release.

Regulation of Water Intake

Water intake also depends on a servomechanism. Fluid intake is influenced by blood osmolality and volume (Fig. 11.5).

With water deprivation, the extracellular fluid becomes hyperosmotic, which in turn causes the intracellular fluid to become hyperosmotic. The brain contains neurons that serve as osmoreceptors for detection of increases in the osmotic pressure of extracellular fluid (see also Chapter 35). The osmoreceptors appear to be located in the organum

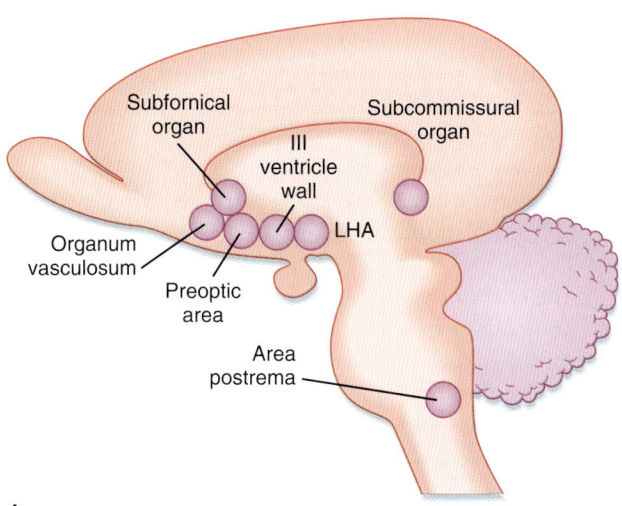

A

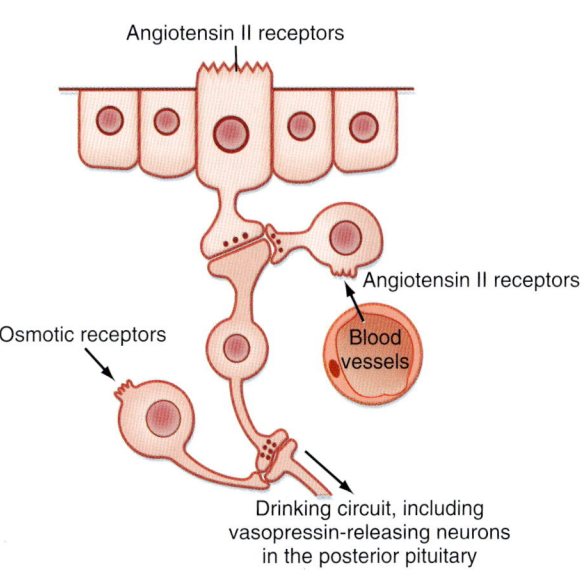

B

• **Fig. 11.5 A,** Structures thought to play a role in the regulation of water intake in rats. LHA, lateral hypothalamic area. **B,** Neural circuits that signal changes in blood osmolality and volume. (**A,** Redrawn from Shepherd GM. *Neurobiology.* New York: Oxford University Press; 1983.)

vasculosum of the lamina terminalis, which is a circumventricular organ. Circumventricular organs surround the cerebral ventricles and lack a blood-brain barrier. The subfornical organ and the organum vasculosum are involved in thirst.

Water deprivation also causes a decrease in blood volume, which is sensed by receptors in the low-pressure side of the vasculature, including the right atrium of the heart (see also Chapter 17). In addition, decreased blood volume triggers the release of renin by the kidneys. Renin breaks down angiotensinogen into angiotensin I, which is then hydrolyzed to angiotensin II (see Chapter 34). This peptide stimulates drinking by an action on angiotensin II receptors in another of the circumventricular organs, the subfornical organ. Angiotensin II also causes vasoconstriction and release of aldosterone and antidiuretic hormone (ADH).

Insufficient water intake is usually a greater problem than excessive water intake. However, when more water is taken in than required, it is easily eliminated by inhibition of the release of ADH from neurons in the supraoptic nucleus at their terminals in the posterior pituitary gland (see Chapter 41). As mentioned previously, signals that inhibit release of ADH include increased blood volume and decreased osmolality of extracellular fluid. Other areas of the hypothalamus, particularly the preoptic region and lateral hypothalamus, help regulate water intake, as do several structures outside the hypothalamus.

Other Autonomic Control Structures

Several regions of the forebrain other than the hypothalamus also play a role in autonomic control. These regions include the central nucleus of the amygdala and the bed nucleus of the stria terminalis, as well as a number of areas of the cerebral cortex. Information reaches these higher autonomic centers from viscera through an ascending system that involves the nucleus of the solitary tract, the parabrachial nucleus, the periaqueductal gray matter, and the hypothalamus. Descending pathways that help control autonomic activity originate in such structures as the paraventricular nucleus of the hypothalamus, noradrenergic cell group A5, the rostral ventrolateral medulla, and the raphe nuclei and adjacent structures of the ventromedial medulla.

Neural Influences on the Immune System

Environmental stress can cause immunosuppression, in which the number of helper T cells and the activity of natural killer cells are reduced. Immunosuppression can even be the result of classical conditioning. One mechanism for such an effect involves the release of corticotropin-releasing factor from the hypothalamus. Corticotropin-releasing factor causes the release of adrenocorticotropic hormone (ACTH) from the pituitary gland; release of ACTH stimulates the secretion of adrenal corticosteroids, which cause immunosuppression (see Chapter 43). Other mechanisms include direct neural actions on lymphoid tissue. The immune system may also influence neural activity.

Key Points

1. The autonomic nervous system controls smooth muscle, cardiac muscle, and glands. It helps maintain homeostasis and coordinates responses to external stimuli. It has sensory and motor components, and the motor component consists of sympathetic and parasympathetic divisions. The enteric nervous system is often considered as part of the autonomic nervous system but is concerned specifically with control of the gastrointestinal tract.

2. Autonomic motor pathways have preganglionic and postganglionic neurons. Preganglionic neurons reside in the CNS, whereas postganglionic neurons lie in peripheral ganglia. Sympathetic preganglionic neurons are located in the thoracolumbar region of the spinal cord, and sympathetic postganglionic neurons are located in paravertebral and prevertebral ganglia. Parasympathetic preganglionic neurons are located in cranial nerve nuclei or in the sacral portion of the spinal cord. Parasympathetic postganglionic neurons reside in ganglia located in or near the target organs.

3. Autonomic afferent fibers innervate sensory receptors in the viscera. Most function to activate reflexes; for some, activation also leads to sensations that are experienced consciously.

4. The enteric nervous system includes the myenteric and submucosal plexuses in the wall of the gastrointestinal tract. The myenteric plexus regulates motility, and the submucosal plexus regulates ion and water transport and secretion.

5. Neurotransmitters at the synapses of preganglionic neurons in autonomic ganglia include acetylcholine (acting at both nicotinic and muscarinic receptors) and a number of neuropeptides. Interneurons in the ganglia release catecholamines. Norepinephrine (acting on adrenergic receptors) is the neurotransmitter generally released by sympathetic postganglionic neurons; neuropeptides are also released. Sympathetic postganglionic neurons that supply sweat glands release acetylcholine. Parasympathetic postganglionic neurons release acetylcholine (acting on muscarinic receptors).

6. The pupil is controlled reciprocally by the sympathetic and parasympathetic nervous systems. Sympathetic activity causes pupillary dilation (mydriasis); parasympathetic activity causes pupillary constriction (meiosis).

7. Emptying of the urinary bladder depends on parasympathetic outflow during the micturition reflex. Sympathetic constriction of the internal sphincter of the urethra prevents voiding. The micturition reflex is triggered by stretch receptors, and it is controlled in normal adults by a micturition center in the pons.

8. The hypothalamus contains many nuclei that have a variety of functions related to regulation of basic bodily functions, including body temperature, body weight, and fluid intake.

9. The goal of hypothalamic function is to maintain homeostasis of critical physiological parameters by acting as a servomechanism. The hypothalamus receives information about specific physiological parameters and uses this information to maintain each of these parameters at a specific set point. It does so via multiple mechanisms. This chapter illustrates how it maintains homeostasis via its control of the autonomic system.

Additional Readings

Nakamura K. Central circuitries for body temperature regulation and fever. *Am J Physiol Regul Integr Comp Physiol.* 2011;301: R1207-R1228.

Saper CB, Chou TC, Elmquist JK. The need to feed: homeostatic and hedonic control of eating. *Neuron.* 2002;36:199-211.

Squire L, Berg D. *Fundamental Neuroscience.* 4th ed. New York: Academic Press; 2012.

Muscle

JAMES M. WATRAS

12

Skeletal Muscle Physiology

LEARNING OBJECTIVES

Upon completion of this chapter, the student should be able to answer the following questions:

1. Describe the organization of skeletal muscle, including the structural features/proteins within the skeletal muscle fiber that link the contractile elements to the extracellular matrix and bone to effect movement. While describing the various linkages, identify congenital conditions that commonly affect particular structures and how they might contribute to a myopathy.
2. Describe the molecular mechanisms by which an action potential in the α motor neuron in the ventral horn of the spinal column can lead to contraction of a skeletal muscle.
3. Describe the mechanisms by which the force of skeletal muscle contraction increases.
4. Compare skeletal muscle fiber types in terms of recruitment pattern, metabolic characteristics, contractile characteristics, and thus their suitability for various types of activity.
5. Discuss the signaling pathways that contribute to the expression of the slow-twitch muscle phenotype versus the fast-twitch muscle phenotype.
6. Describe general signaling pathways that contribute to hypertrophy or atrophy.
7. Discuss mechanisms underlying the development of muscular fatigue.
8. Describe mechanisms underlying the monosynaptic reflex.
9. Discuss the length-tension curves and force-velocity curves for skeletal muscle, including the molecular bases of both curves.

Skeletal Muscle Physiology

Muscle cells are highly specialized for the conversion of chemical energy to mechanical energy. Specifically, muscle cells use the energy in adenosine triphosphate (ATP) to generate force or do work. Because work can take many forms (such as locomotion, pumping blood, or peristalsis), several types of muscle have evolved. The three basic types of muscle are **skeletal muscle, cardiac muscle,** and **smooth muscle.**

Skeletal muscle acts on the skeleton. In limbs, for example, skeletal muscle spans a joint, thereby allowing a lever action. Skeletal muscle is under voluntary control (i.e., controlled by the central nervous system) and plays a key role in numerous activities such as maintenance of posture, locomotion, speech, and respiration. When viewed under the microscope, skeletal muscle exhibits transverse striations (at intervals of 2 to 3 µm) that result from the highly organized arrangement of actin and myosin molecules within the skeletal muscle cells. Thus skeletal muscle is classified as a **striated muscle.** The heart is composed of cardiac muscle, and although it is also a striated muscle, it is an involuntary muscle (i.e., controlled by an intrinsic pacemaker and modulated by the autonomic nervous system). Smooth muscle (which lacks the striations evident in skeletal and cardiac muscle) is an involuntary muscle typically found lining hollow organs such as the intestine and blood vessels. In all three muscle types, force is generated by the interaction of actin and myosin molecules, a process that requires transient elevation of intracellular $[Ca^{++}]$.

In this chapter, attention is directed at the molecular mechanisms underlying contraction of skeletal muscle. Mechanisms for regulating the force of contraction are also addressed. To put this information into perspective, it is important to first examine the basic organization of skeletal muscle.

Organization of Skeletal Muscle

Fig. 12.1 illustrates skeletal muscles spanning the elbow joint. The muscles are attached to bone on either side of the joint. The point of attachment closest to the spine (proximal) is called the **origin,** whereas the point of attachment on the far side of the joint (distal) is called the **insertion.** These points of attachment occur through **tendons** (connective tissue) at the end of the muscle. Note that the point of insertion is close to the elbow joint, which enables a broad range of motion. Also note that the joint is spanned by a **flexor** muscle on one side and an **extensor** muscle on the opposite side of the joint. Thus contraction of the flexor muscle (see the biceps muscle in Fig. 12.1) results in a decrease in the angle of the elbow joint (bringing the forearm closer to the shoulder), whereas contraction of the extensor muscle (see the triceps muscle in Fig. 12.1) results in the reverse motion (extending the arm).

The basic structure of skeletal muscle is shown in Fig. 12.2. Each muscle is composed of numerous cells called

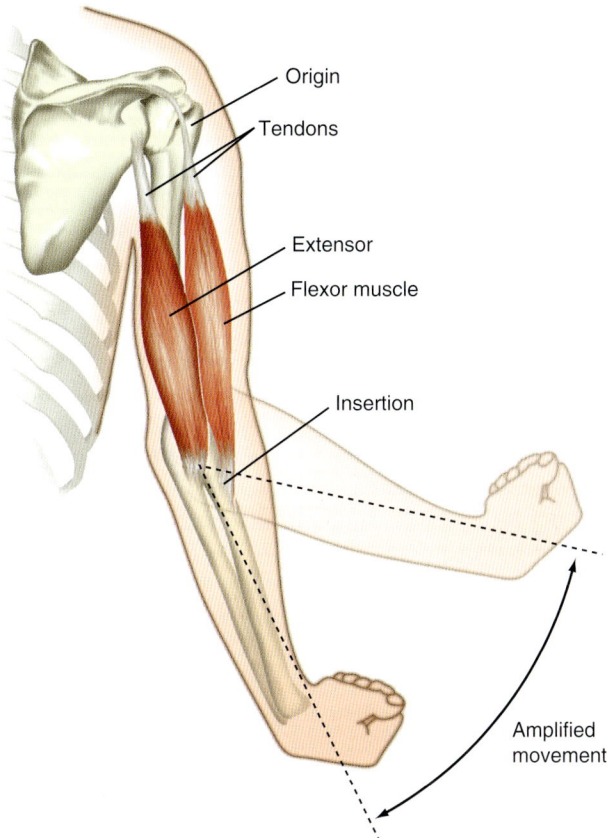

• Fig. 12.1. Skeletal Muscle Attaches to the Skeleton by Way of Tendons and Typically Spans a Joint. The proximal and distal points of attachment of the tendon are termed *origin* and *insertion,* respectively. Note that the insertion is close to the joint, which allows a broad range of motion. Also note that skeletal muscles span both sides of the joint, which allows both flexion and extension of the forearm.

Labels in figure: Origin; Tendons; Extensor; Flexor muscle; Insertion; Amplified movement

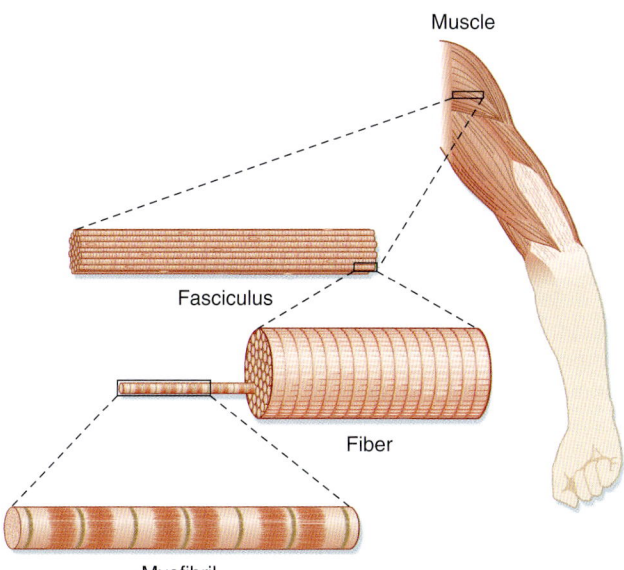

Labels in figure: Muscle; Fasciculus; Fiber; Myofibril

• Fig. 12.2. Skeletal muscle is composed of bundles of muscle fibers; each such bundle is called a *fasciculus.* A muscle fiber represents an individual muscle cell and contains bundles of myofibrils. The striations are due to the arrangement of thick and thin filaments. See text for details. (Redrawn from Bloom W, Fawcett DW. *A Textbook of Histology.* 10th ed. Philadelphia: Saunders; 1975.)

muscle fibers. A connective tissue layer called the **endomysium** surrounds each of these fibers. Individual muscle fibers are then grouped together into **fascicles,** which are surrounded by another connective tissue layer called the **perimysium.** Within the perimysium are the blood vessels and nerves that supply the individual muscle fibers. The fascicles are joined together to form the muscle. The connective tissue sheath that surrounds the muscle is called the **epimysium.** At the ends of the muscle, the connective tissue layers come together to form a tendon, which attaches the muscle to the skeleton. The **myotendinous junction** is a specialized region of the tendon where the ends of the muscle fibers interdigitate with the tendon for the transmission of the force of contraction of the muscle to the tendon to effect movement of the skeleton (discussed later in this section). The tendon and the connective tissue layers are composed mainly of elastin and collagen fibers, and thus they also contribute to passive tension of muscle and prevent damage to the muscle fibers as a result of overstretching or contraction.

Individual skeletal muscle cells are narrow ($\approx$10 to 80 μm in diameter), but they can be extremely long (up to 25 cm in length). Each skeletal muscle fiber contains bundles of filaments, called **myofibrils,** running along the axis of the cell. The gross striation pattern of the cell results from a repeating pattern in the myofibrils. Specifically, it is the regular arrangement of the thick and thin filaments within these myofibrils, coupled with the highly organized alignment of adjacent myofibrils, that gives rise to the striated appearance of skeletal muscle. Striations can be observed in intact muscle fibers and in the underlying myofibrils.

A myofibril can be subdivided longitudinally into **sarcomeres** (Fig. 12.3). The sarcomere is demarcated by two dark lines called **Z lines** and represents a repeating contractile unit in skeletal muscle. The average length of a sarcomere is 2 μm. On either side of the Z line is a light band (**I band**) that contains thin filaments composed primarily of the protein **actin.** The area between two I bands within a sarcomere is the **A band,** which contains thick filaments composed primarily of the protein **myosin.** The thin actin filaments extend from the Z line toward the center of the sarcomere and overlap a portion of the thick filaments. The dark area at the end of the A band represents this region of overlap between thick and thin filaments. A light area in the center of the sarcomere is called the **H band.** This area represents the portion of the A band that contains myosin thick filaments but no thin actin filaments. Thus thin actin filaments extend from the Z line to the edge of the H band and overlap a portion of the thick filament in the A band. A dark line called the **M line** is evident in the center of the sarcomere and includes proteins that appear to be critical for organization and alignment of the thick filaments in the sarcomere.

As illustrated in Fig. 12.3, each myofibril in a muscle fiber is surrounded by **sarcoplasmic reticulum (SR).** The SR is an intracellular membrane network that plays a critical

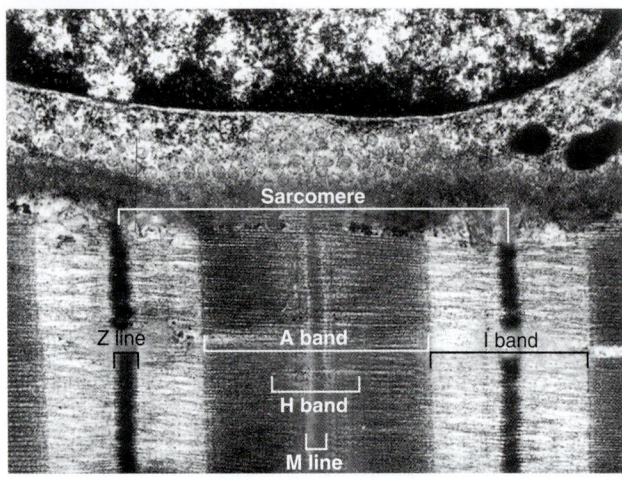

A

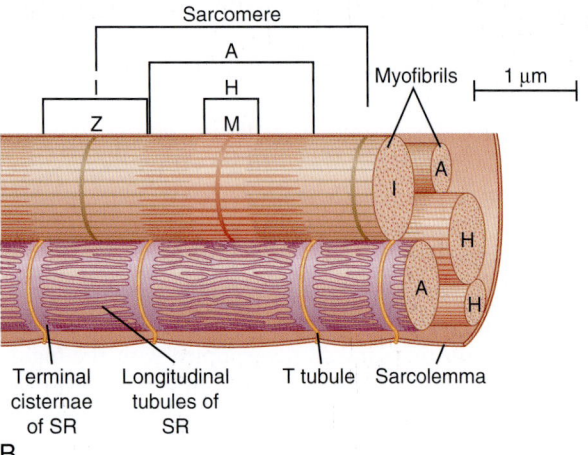

B

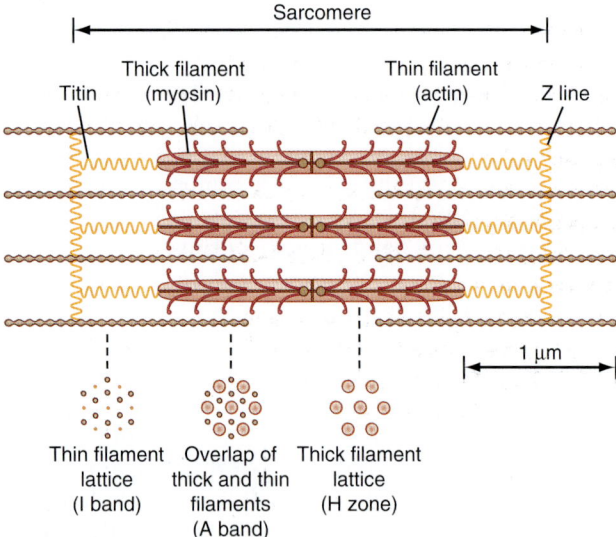

C

• **Fig. 12.3.** **A,** Myofibrils are arranged in parallel within a muscle fiber. **B,** Each fibril is surrounded by sarcoplasmic reticulum (SR). Terminal cisternae of the SR are closely associated with T tubules and form a triad at the junction of the I and A bands. The Z lines define the boundary of the sarcomere. The striations are formed by overlap of the contractile proteins. Three bands can be observed: the A band, I band, and H band. An M line is visible in the middle of the H band. **C,** Organization of the proteins within a single sarcomere. The cross-sectional arrangement of the proteins is also illustrated.

role in the regulation of intracellular [Ca^{++}]. Invaginations of the sarcolemma, called **T tubules,** pass into the muscle fiber near the ends of the A band (i.e., close to the SR). The SR and the T tubules, however, are distinct membrane systems: The SR is an intracellular network, whereas the T tubules are in contact with the extracellular space. A gap ($\approx$15 nm in width) separates the T tubules from the SR. The portion of the SR nearest the T tubules is called the **terminal cisternae,** and it is the site of Ca^{++} release, which is critical for contraction of skeletal muscle (see the section "Excitation-Contraction Coupling"). The longitudinal portions of the SR are continuous with the terminal cisternae and extend along the length of the sarcomere. This portion of the SR contains a high density of Ca^{++} pump protein (i.e., **SERCA:** **S**arcoplasmic **E**ndoplasmic **R**eticulum **C**a^{++}-**A**TPase), which is critical for reaccumulation of Ca^{++} in the SR and hence for relaxation of the muscle.

The thick and thin filaments are highly organized in the sarcomere of myofibrils (see Fig. 12.3). As mentioned, thin actin filaments extend from the Z line toward the center of the sarcomere, whereas thick myosin filaments are centrally located and overlap a portion of the opposing thin actin filaments. The thick and thin filaments are oriented in such a way that in the region of overlap within the sarcomere, each thick myosin filament is surrounded by a hexagonal array of thin actin filaments. The Ca^{++}-dependent interaction of the thick myosin filaments and the thin actin filaments generate the force of contraction after stimulation of the muscle (see the section "Actin-Myosin Interaction: Cross-Bridge Formation").

The thin filament is formed by aggregation of actin molecules (termed **globular actin** or **G-actin**) into a two-stranded helical filament called **filamentous actin** or **F-actin**. The elongated cytoskeletal protein **nebulin** extends along the length of the thin filament and may participate in regulation of the length of the thin filament. Dimers of the protein **tropomyosin** extend over the entire actin filament and cover myosin binding sites on the actin molecules. Each tropomyosin dimer extends across seven actin molecules, with sequential tropomyosin dimers arranged in a head-to-tail configuration. A **troponin complex** consisting of three subunits (**troponin T, troponin I,** and **troponin C**) is present on each tropomyosin dimer and influences the position of the tropomyosin molecule on the actin filament and hence the ability of tropomyosin to inhibit binding of myosin to the actin filament at low cytosolic Ca concentrations (see the section "Actin-Myosin Interaction: Cross-Bridge Formation"). Binding of cytosolic Ca^{++} to troponin C promotes the movement of tropomyosin on the actin filament that exposes myosin binding sites on actin, thereby facilitating actin-myosin interaction, and hence contraction (see the section "Actin-Myosin Interaction: Cross-Bridge Formation"). Additional proteins associated with the thin filament include **tropomodulin, α-actinin,** and **CapZ protein.** Tropomodulin is located at the end of the thin filament, toward the center of the sarcomere, and may participate in setting the length of the thin filament.

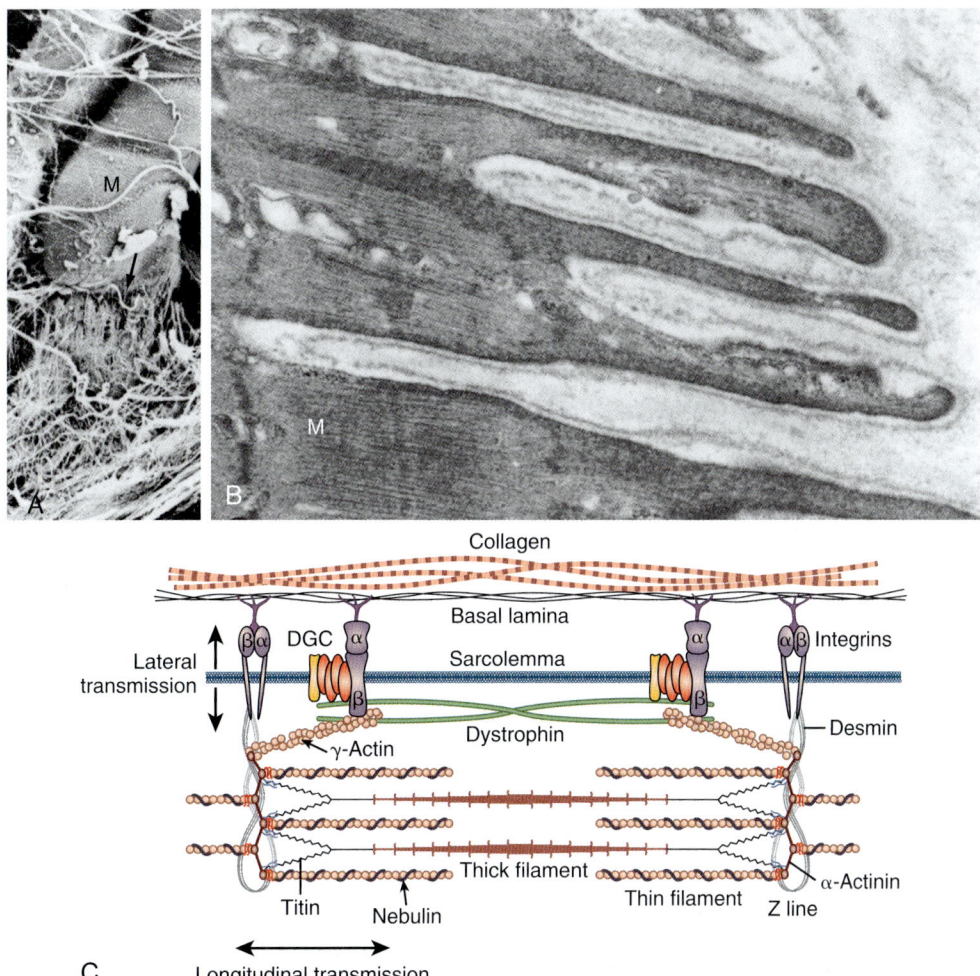

• Fig. 12.4. The force of contraction of the muscle fiber is transmitted both longitudinally to the tendon (at the myotendinous junction) and laterally to adjacent extracellular connective tissue (at costameres). The force of contraction is transmitted from the end of the muscle fiber (M) to the tendon by connections with numerous collagen fibers (**A,** tip of *arrow*). Folds in the sarcolemma at the end of the muscle fiber (**B**) result in an interdigitation of the muscle fiber with the tendon, and represents the myotendinous junction. Costameres are located on the sides of the muscle fibers, and represent the bridges between the Z lines in the subsarcolemmal myofibrils and the extracellular connective tissue (**C**). Costameres facilitate the lateral transmission of force of contraction, which helps stabilize the sarcolemma. DGC, dystrophin-associated glycoprotein complex. (**A** and **B,** From Tidball JG. Myotendinous junction: morphological changes and mechanical failure associated with muscle cell atrophy. *Exp Molec Pathol.* 1984;40:1-12. **C,** From Hughes D, Wallace M, Baar K. Effects of aging, exercise, and disease on force transfer in skeletal muscle. *Am J Physiol Endocrin Metab.* 2015;309:E1-E10.)

CapZ protein and α-actinin serve to anchor the thin filament to the Z line.

The thick myosin filaments are tethered to the Z lines by a cytoskeletal protein called **titin.** Titin is a very large, elastic protein (molecular weight, >3000 kDa) that extends from the Z line to the center of the sarcomere and appears to be important for organization and alignment of the thick filaments in the sarcomere. Some forms of muscular dystrophy have been attributed to defects in titin (i.e., titinopathies).

The cytoskeleton (including the intermediate filament protein desmin) participates in the highly organized alignment of sarcomeres. Desmin extends from the Z lines of adjacent sarcomeres to the integrin protein complexes on the sarcolemma and thus participates in both the alignment of sarcomeres across muscles and the lateral transmission of force (described later in this section).

The force of contraction is transmitted both longitudinally to the tendon (via myotendinous junctions) and laterally to connective tissue adjacent to the muscle fibers (via costameres). The myotendinous junction represents a specialized region where the muscle fiber connects to the tendon (Fig. 12.4A and B). Folding of the sarcolemma at the myotendinous junction results in an interdigitation of the tendon with the end of the muscle fiber, which increases the contact area between the muscle fiber and the connective tissue and hence reduces the force per unit area at the end of the muscle fiber. Some forms of muscular dystrophy are associated with decreased folding at the myotendinous junction. Proteins involved in longitudinal transmission of force at the myotendinous junction include the talin-vinculin-integrin protein complexes and the dystrophin-glycoprotein complexes. The myotendinous junction also contains Z disk proteins that have been implicated in signaling.

Lateral transmission of the force of contraction involves costameres, which link the Z lines of subsarcolemmal sarcomeres to extracellular matrix through a series of proteins (see Fig. 12.4C). This is particularly important for muscle fibers that do not extend the entire length of the muscle. The lateral transmission of force is also thought to stabilize the sarcolemma and to protect it from damage during contraction. The myotendinous junction and the costameres also contain signaling molecules.

⚕ AT THE CELLULAR LEVEL

The muscular dystrophies constitute a group of genetically determined degenerative disorders. **Duchenne's muscular dystrophy** (described by G.B. Duchenne in 1861) is the most common of the muscular dystrophies and affects 1 per 3500 boys (3 to 5 years of age). Severe muscle wasting occurs, and most affected patients are wheelchair bound by the age of 12; many die of respiratory failure in adulthood (30 to 40 years of age). Duchenne's muscular dystrophy is an X-linked recessive disease that has been linked to a defect in the dystrophin gene that leads to a deficiency of the dystrophin protein in skeletal muscle, brain, retina, and smooth muscle. **Dystrophin** is a large (427-kDa) protein that is present in low abundance (0.025%) in skeletal muscle. It is localized on the intracellular surface of the sarcolemma in association with several integral membrane glycoproteins (forming a dystrophin-glycoprotein complex; Fig. 12.5). This dystrophin-glycoprotein complex provides a structural link between the subsarcolemmal cytoskeleton of the muscle cell and the extracellular matrix and appears to stabilize the sarcolemma and hence prevent contraction-induced injury (rupture). The dystrophin-glycoprotein complex may also serve as a scaffold for cell signaling cascades. The enzyme nitric oxide synthase is present in the dystrophin-glycoprotein complex.

Although defects in the dystrophin-glycoprotein complex are involved in many forms of muscular dystrophy, some forms of muscular dystrophy that involve other mechanisms have been identified. Specifically, a defect in sarcolemma repair (attributed to loss/mutation of the protein dysferlin) appears to underlie at least one form of muscular dystrophy (**limb-girdle muscular dystrophy 2B,** associated with muscle wasting in the pelvic region). Defects in the protein titin (titinopathies) have been implicated in other forms of muscular dystrophy (e.g., **limb-girdle muscular dystrophy 2J** and **tibial muscular dystrophy**). Mutations in the protease **calpain 3** (resulting in loss of protease activity) have also been implicated in some types of muscular dystrophy (e.g., limb-girdle muscular dystrophy 2A), apparently secondary to apoptosis.

Organization of the thick filament is shown in Fig. 12.6. **Myosin** is a large protein (≈480 kDa) that consists of six different polypeptides with one pair of large heavy chains ($\approx$200 kDa) and two pairs of light chains ($\approx$20 kDa). The heavy chains are wound together in an α-helical configuration to form a long rod-like segment, and the N-terminal portions of each heavy chain form a large globular head. The head region extends away from the thick filament toward the actin thin filament and is the portion of the molecule that can bind to actin. Myosin is also able to hydrolyze ATP, and ATPase activity is located in the globular head as well. Two pairs of light chains are associated with the globular head. One of these pairs of light chains, termed *essential light chains,* is crucial for the ATPase activity of myosin. The other pair of light chains, called *regulatory light chains,* can be phosphorylated by Ca++/calmodulin-dependent myosin light chain protein kinase, which can influence the interaction of myosin with actin (see the section "Skeletal Muscle Types"). Thus myosin ATPase activity occurs in the globular head of myosin and requires the presence of light chains (namely, the "essential" light chains).

Myosin filaments form by a tail-to-tail association of myosin molecules, which results in a bipolar arrangement of the thick filament. The thick filament then extends on either side of the central bare zone by a head-to-tail association of myosin molecules, thus maintaining the filament's bipolar organization centered on the M line. Such a bipolar arrangement is critical for drawing the Z lines together (i.e., shortening the length of the sarcomere) during contraction. The mechanisms controlling this highly organized structure of the myosin thick filament are not clear, although the cytoskeletal protein titin is thought to participate in the formation of a scaffold for organization and alignment of the thick filament in the sarcomere. Additional proteins found in the thick filaments (e.g., **myomesin** and **C protein**) may also participate in the bipolar organization or packing of the thick filament (or both).

Control of Skeletal Muscle Activity

Motor Nerves and Motor Units

Skeletal muscle is controlled by the central nervous system. Specifically, each skeletal muscle is innervated by an α **motor neuron.** The cell bodies of α motor neurons are located in the ventral horn of the spinal cord (Fig. 12.7; see also Chapter 9). The motor axons exit via the ventral roots and reach the muscle through mixed peripheral nerves. The motor nerves branch in the muscle, and each branch innervates a single muscle fiber. The specialized cholinergic synapse that forms the **neuromuscular junction** and the neuromuscular transmission process that generates an action potential in the muscle fiber are described in Chapter 6.

A **motor unit** consists of the motor nerve and all the muscle fibers innervated by the nerve. The motor unit is the functional contractile unit because all the muscle cells within a motor unit contract synchronously when the motor nerve fires. The size of motor units within a muscle varies, depending on the function of the muscle.

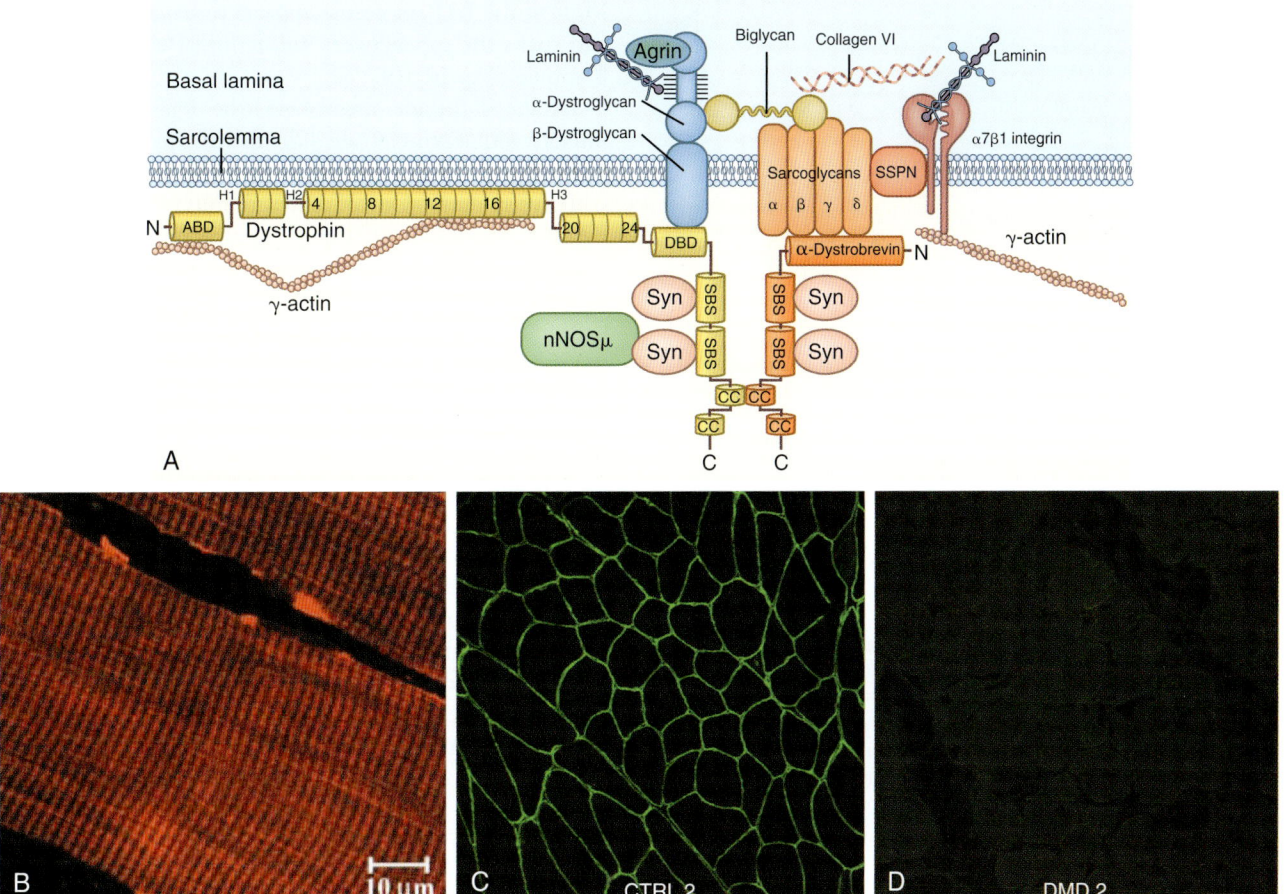

● **Fig. 12.5.** **A,** Organization of the dystrophin-glycoprotein complex in skeletal muscle. The dystrophin-glycoprotein complex provides a structural link between the cytoskeleton of the muscle cell and the extracellular matrix, which appears to stabilize the sarcolemma and hence prevents contraction-induced injury (rupture). Duchenne's muscular dystrophy is associated with loss of dystrophin. Numbers in dystrophin indicate hinge regions (e.g., H1, H2) and spectrin-like repeat domains (e.g., 4, 8, 12). ABD, actin-binding domain; C, carboxy terminus; CC, coiled-coil domain; DBD, dystroglycan-binding domain; N, amino terminus; nNOSμ, neuronal nitric oxide synthase μ; SBS, syntrophin-binding site; SSPN, sarcospan; Syn, syntrophin. Electron micrographs of a longitudinal view **(B)** and a cross-sectional view **(C)** show the distribution of dystrophin in skeletal muscle of a normal patient (CTRL). Another cross-sectional view **(D)** shows the loss of dystrophin from skeletal muscle in a patient with Duchenne's muscular dystrophy (DMD). (**A,** From Allen D, Whitehead N, Froehner S. Absence of dystrophin disrupts skeletal muscle signaling: roles of Ca^{2+}, reactive oxygen species, and nitric oxide in the development of muscular dystrophy. *Physiol Rev.* 2016;96:253-305. **B,** From Anastasi G, et al. Costameric proteins in human skeletal muscle during muscular inactivity. *J Anat.* 2008;213:284-295. **C** and **D,** From Beekman C, et al. A sensitive, reproducible and objective immunofluorescence analysis method of dystrophin in individual fibers in samples from patients with Duchenne muscular dystrophy. *PLoS One.* 2014;9[9]:e107494.)

Activation of motor units with a small number of fibers facilitates fine motor control. Activation of varying numbers of motor units within a muscle is one way in which the tension developed by a muscle can be controlled (see "Recruitment" in the section "Modulation of the Force of Contraction").

The neuromuscular junction formed by the α motor neuron is called an **end plate** (see Chapter 6 for details). Acetylcholine released from the α motor neuron at the neuromuscular junction initiates an action potential in the muscle fiber that rapidly spreads along its length. The duration of the action potential in skeletal muscle is less than 5 msec. The duration of the action potential in cardiac muscle, in contrast, is approximately 200 msec. The short duration of the skeletal muscle action potential allows very rapid contractions of the fiber and provides yet another mechanism by which the force of contraction can be increased. Increasing tension by repetitive stimulation of the muscle is called *tetany* (see the section "Modulation of the Force of Contraction").

Excitation-Contraction Coupling

When an action potential is transmitted along the sarcolemma of the muscle fiber and then down the T tubules, Ca^{++} is released from the terminal cisternae SR into the

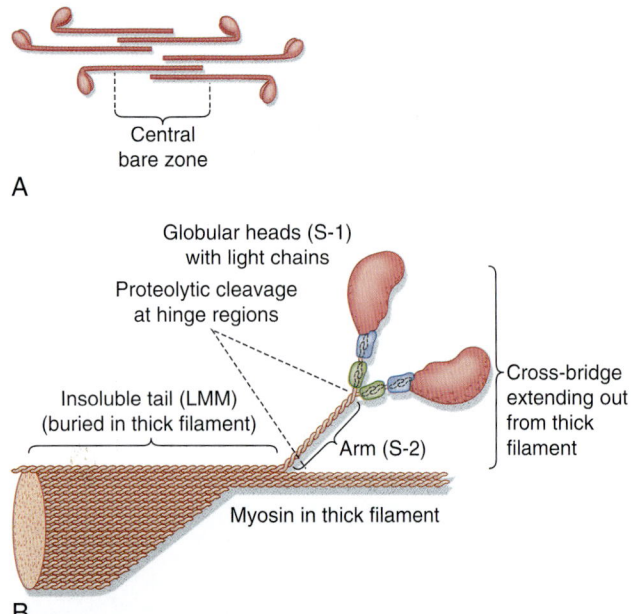

Central bare zone

A

Globular heads (S-1) with light chains

Proteolytic cleavage at hinge regions

Insoluble tail (LMM) (buried in thick filament)

Arm (S-2)

Cross-bridge extending out from thick filament

Myosin in thick filament

B

• **Fig. 12.6.** Organization of a Thick Filament. A thick filament is formed by the polymerization of myosin molecules in a tail-to-tail configuration extending from the center of the sarcomere **(A).** An individual myosin molecule has a tail region and a cross-bridge region. The cross-bridge region is composed of an arm and globular heads **(B).** The globular heads contain light chains that are important for the function of myosin ATPase activity. LMM, light meromyosin; S-1 and S-2, myosin subfragments 1 and 2.

myoplasm. This release causes intracellular [Ca^{++}] to rise, which in turn promotes actin-myosin interaction and contraction. The time course for the increase in intracellular [Ca^{++}] in relation to the action potential and development of force is shown in Fig. 12.8. The action potential is extremely short-lived (≈ 5 msec). The elevation in intracellular [Ca^{++}] begins slightly after the action potential and peaks at approximately 20 msec. This increase in intracellular [Ca^{++}] initiates a contraction called a *twitch*.

The mechanism underlying the elevation in intracellular [Ca^{++}] involves an interaction between protein in the T tubule and the adjacent terminal cisternae of the SR. As previously described (see Fig. 12.3), the T tubule represents an invagination of the sarcolemma that extends into the muscle fiber and forms a close association with two terminal cisternae of the SR. The association of a T tubule with two opposing terminal cisternae is called a **triad.** Although there is a gap (≈ 15 nm in width) between the T tubule and the terminal cisternae, proteins bridge this gap. On the basis of their appearance on electron micrographs, these bridging proteins are called **feet** (Fig. 12.9). These feet are the Ca^{++} release channels in the membrane of the terminal cisternae that are responsible for the elevation in intracellular [Ca^{++}] in response to the action potential. Because this channel binds the drug **ryanodine,** it is commonly called the **ryanodine receptor (RYR).** RYR is a large protein (≈ 500 kDa) that exists as a homotetramer. Only a small portion of the RYR molecule is actually embedded in the SR membrane.

Most of the RYR molecule appears to be in the myoplasm and spans the gap between the terminal cisternae and the T tubule (Fig. 12.10).

At the T tubule membrane, the RYR is thought to interact with a protein called the **dihydropyridine receptor (DHPR).** DHPR is an L-type voltage-gated Ca^{++} channel with five subunits. One of these subunits binds the dihydropyridine class of channel-blocking drugs and appears to be crucial for the ability of the action potential in the T tubule to induce release of Ca^{++} from the SR. However, influx of Ca^{++} into the cell through the DHPR is not needed for the initiation of Ca^{++} release from the SR. In fact, skeletal muscle is able to contract in the absence of extracellular Ca^{++} or with a mutated DHPR that does not conduct Ca^{++}. Instead, release of Ca^{++} from the terminal cisternae of the SR is thought to result from a conformational change in the DHPR as the action potential passes down the T tubule, and this conformational change in the DHPR, by means of a protein-protein interaction, opens the RYR and releases Ca^{++} into the myoplasm.

Structural analysis, including the use of freeze-fracture techniques, provides evidence for a close physical association of DHPR and RYR (see Fig. 12.9). DHPR in the T tubule membrane appears to reside directly opposite the four corners of the underlying homotetrameric RYR channel in the SR membrane. Studies have shown that the SH3-cysteine rich domain 3 protein **Stac3** is also critical for coupling of the DHPR to the RYR during excitation-contraction coupling in skeletal muscle SR. Stac3 is not present in cardiac muscle, which relies on Ca^{++} influx through the sarcolemma to initiate Ca^{++} release from the RYR instead of the direct coupling of the DHPR and the RYR in skeletal muscle.

Other proteins that reside near the RYR include **calsequestrin, triadin,** and **junctin** (see Fig. 12.10). Calsequestrin is a low-affinity Ca^{++}-binding protein that is present in the lumen of the terminal cisternae. It allows Ca^{++} to be "stored" at high concentration and thereby establishes a favorable concentration gradient that facilitates the efflux of Ca^{++} from the SR into the myoplasm when the RYR opens. Triadin and junctin are in the terminal cisternae membrane and bind both RYR and calsequestrin; they could anchor calsequestrin near the RYR and thereby increase Ca^{++} buffering capacity at the site of Ca^{++} release. **Histidine-rich calcium-binding protein** is another low-affinity Ca^{++}-binding protein in the SR lumen, although it is less abundant than calsequestrin. It appears to bind triadin in a Ca^{++}-dependent manner, which raises the possibility that it has a role more important than serving simply as a Ca^{++} buffer. There is also evidence for the presence of **st**ore-**o**perated **Ca e**ntry (**SOCE**) in skeletal muscle (e.g., via the Orai/Stim1 complex) during tetany. Inhibition of Ca influx did not affect excitation-contraction coupling but did reduce maximal tetanic tension at high rates of electrical stimulation, which suggests that there may be some extrusion of intracellular Ca during tetany, which is compensated by Ca influx to maintain maximal tetanic tension.

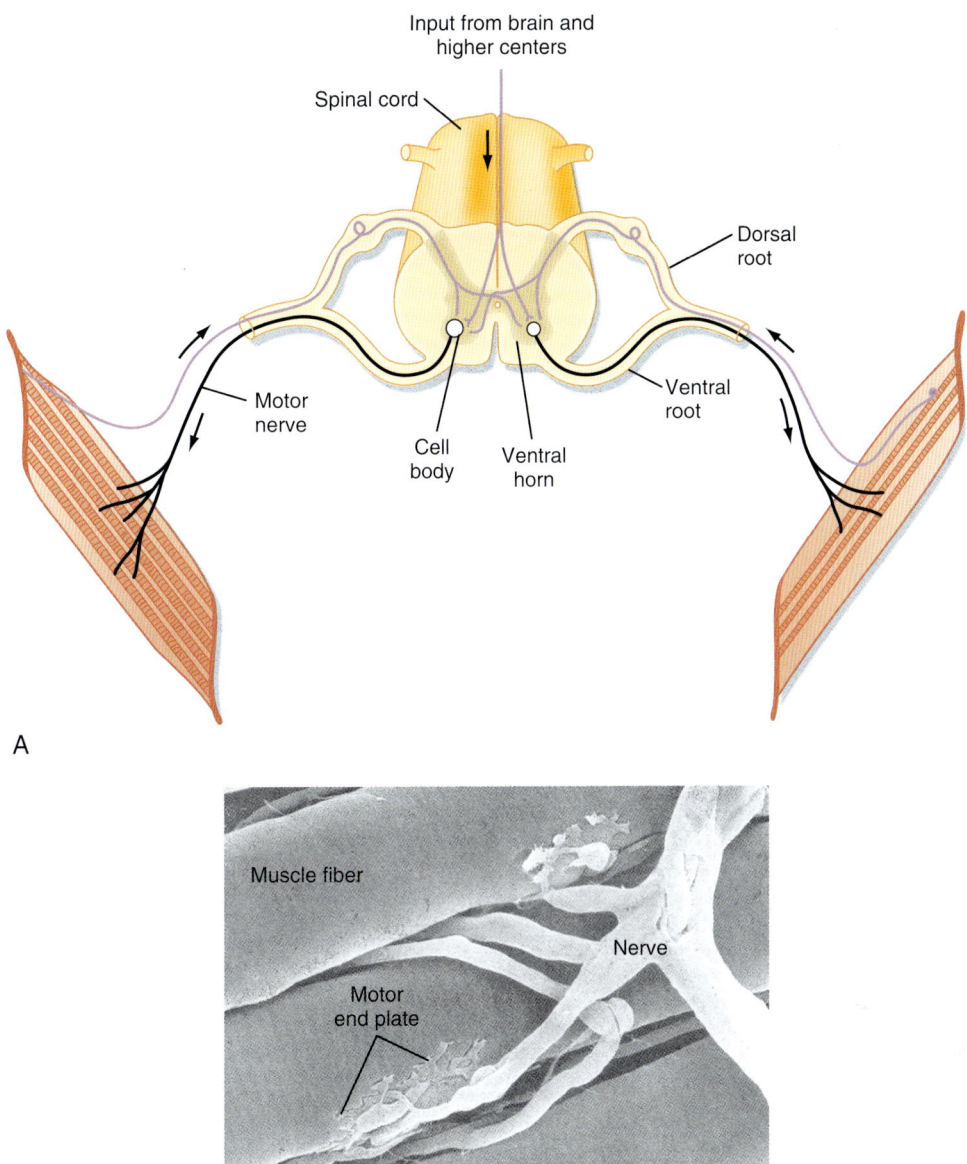

• **Fig. 12.7.** Skeletal Muscle Is Voluntary Muscle Controlled by the Central Nervous System, With Efferent Signals (i.e., action potentials) Passing Through an α Motor Neuron to Muscle Fibers. Each motor neuron may innervate many muscle fibers within a muscle, although each muscle fiber is innervated by only one motor neuron **(A).** A scanning electron micrograph **(B)** shows innervation of several muscle fibers by a single motor neuron. (**B,** From Bloom W, Fawcett DW: *A Textbook of Physiology.* 12th ed. New York: Chapman & Hall; 1994.)

AT THE CELLULAR LEVEL

A variety of mutational studies have been conducted to ascertain the region of the DHPR that is critical for opening of the RYR. One possible site of interaction (depicted in Fig. 12.10) is the myoplasmic loop between transmembrane domains II and III in the α_1 subunit of the DHPR. The voltage-sensing region of the DHPR involved in intramembranous charge movement is thought to reside in the S_4 transmembrane segments of the α_1 subunit. Genetic mutations in the RYR or DHPR, or in both, have been associated with pathological disturbances in myoplasmic [Ca^{++}]. Such disturbances include malignant hyperthermia and central core disease, as described later. These mutations are typically observed in the myoplasmic portion of the RYR, although mutations have also been observed in a myoplasmic loop in the DHPR. Results of studies indicate that the protein Stac3 is also important for the coupling of the DHPR to the RYR in skeletal muscle and that a mutation in Stac3 that is present in the rare congenital disorder Native American myopathy results in impaired excitation-contraction coupling in skeletal muscle.

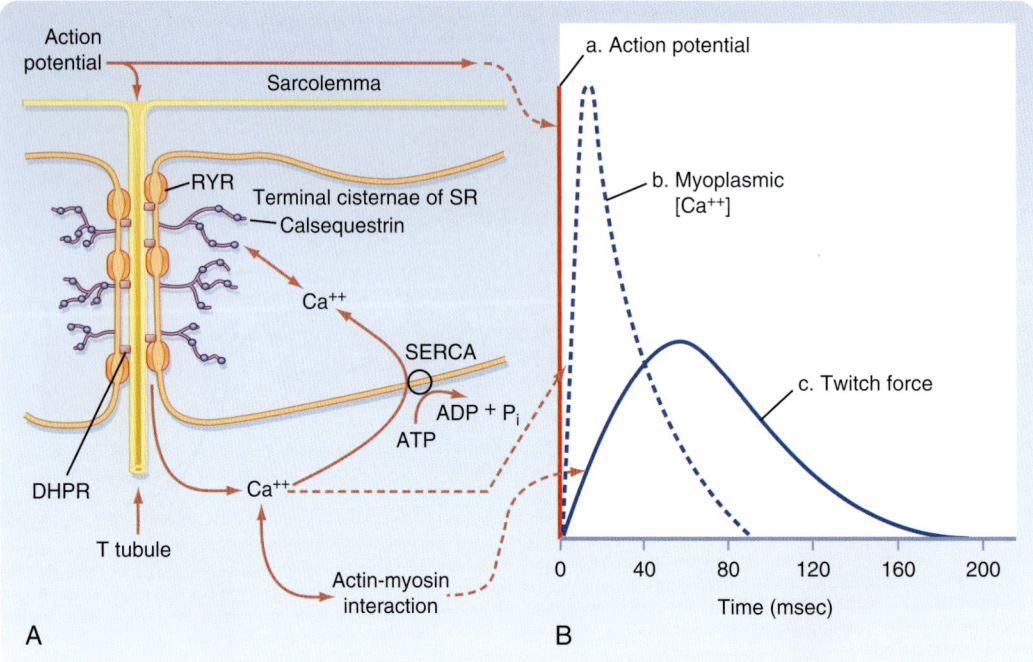

• **Fig. 12.8. A,** Stimulation of a skeletal muscle fiber initiates an action potential in the muscle that travels down the T tubule and induces release of Ca++ from the terminal cisternae of the sarcoplasmic reticulum (SR). The rise in intracellular [Ca++] causes a contraction. As Ca++ is pumped back into the SR by sarcoplasmic endoplasmic reticulum Ca++-ATPase (SERCA), relaxation occurs. DHPR, dihydropyridine receptor; Pi, inorganic phosphate; RYR, ryanodine receptor. **B,** Time courses of the action potential, myoplasmic Ca++ transient, and force of the twitch contraction.

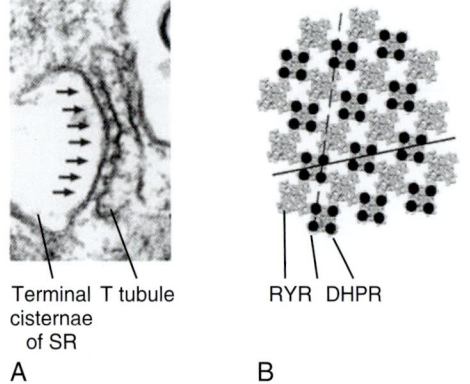

• **Fig. 12.9. A,** Electron micrograph of a triad illustrating the "feet" *(arrows)* between the T tubule and the sarcoplasmic reticulum (SR), which are thought to be the ryanodine receptors (RYRs) in the SR. **B,** Each RYR in the SR is associated with four dihydropyridine receptors (DHPRs) in the T tubule. (From Protasi F, et al. RYR1 and RYR3 have different roles in the assembly of calcium release units of skeletal muscle. *Biophys J.* 2000;79:2494.)

Relaxation of skeletal muscle occurs as intracellular Ca++ is resequestered by the SR. Uptake of Ca++ into the SR is due to the action of a Ca++ pump (i.e., Ca++-ATPase). This pump is not unique to skeletal muscle; it is found in all cells in association with the endoplasmic reticulum. Accordingly, it is named **SERCA,** which stands for **s**arcoplasmic **e**ndoplasmic **r**eticulum **c**alcium **A**TPase. SERCA is the most abundant protein in the SR of skeletal muscle, and it

is distributed throughout the longitudinal tubules and the terminal cisternae as well. It transports two molecules of Ca++ into its lumen for each molecule of ATP hydrolyzed.* Thus the Ca++ transient seen during a twitch contraction (see Fig. 12.8) reflects release of Ca++ from the terminal cisternae via the RYR and reuptake primarily into the longitudinal portion of the SR by SERCA. The low-affinity Ca++-binding protein **sarcalumenin** is present throughout the longitudinal tubules of the SR and nonjunctional regions of the terminal cisternae and is thought to be involved in the transfer of Ca++ from sites of Ca++ uptake in the longitudinal tubules to sites of Ca++ release in the terminal cisternae. Results of studies suggest that sarcalumenin increases Ca++ uptake by SERCA, at least in part by buffering luminal Ca++ near the pump.

The endogenous micropeptides phospholamban, sarcolipin, and myoregulin have been shown to regulate the activity of SERCA by decreasing the Ca sensitivity of Ca uptake (Fig. 12.11). Protein kinase A–dependent phosphorylation of phospholamban in slow-twitch skeletal muscle has been reported to increase Ca transport in the SR, similar to the effect of phospholamban phosphorylation in the heart. Phospholamban and sarcolipin are present in slow-twitch muscle, whereas myoregulin is present in both fast- and slow-twitch muscle.

---

*During the transport of Ca++, SERCA exchanges two Ca++ ions for two H+ ions (i.e., H+ is pumped out of the SR).

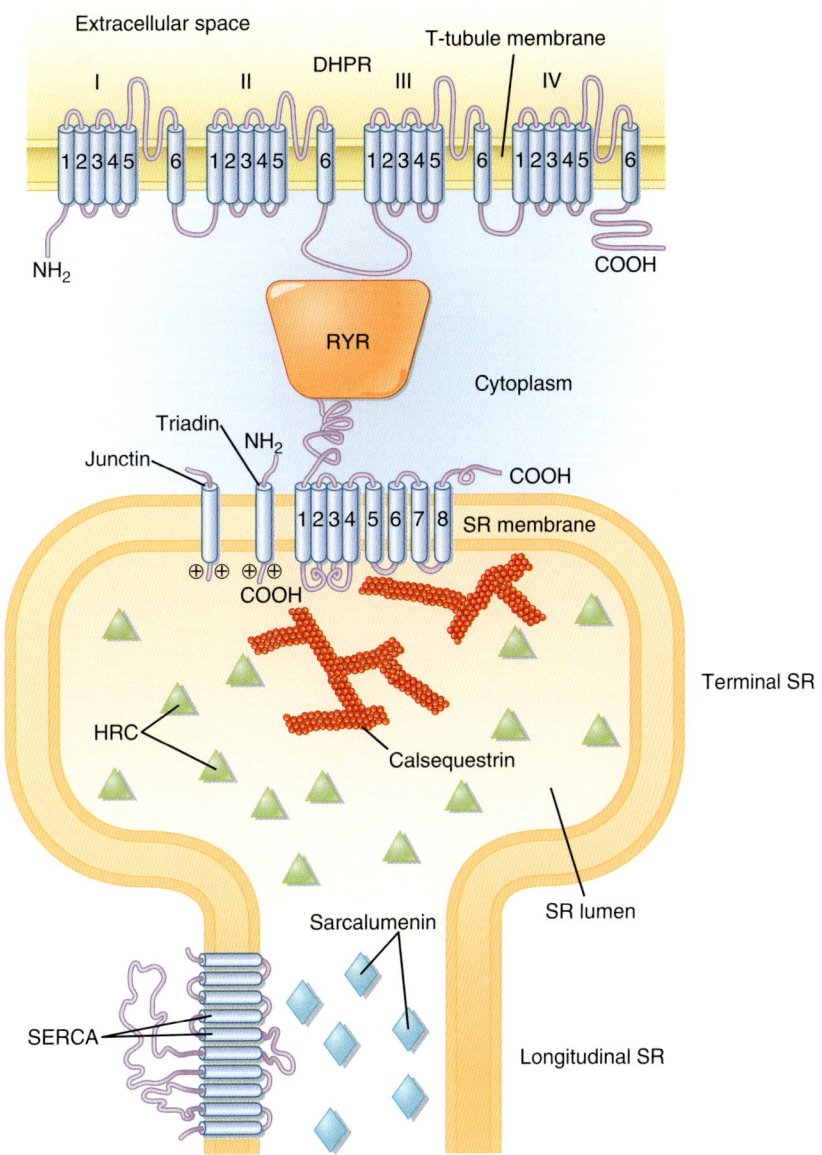

• **Fig. 12.10.** Molecular Structure and Relationships Between the Dihydropyridine Receptor (DHPR) in the T Tubule Membrane and the Ryanodine Receptor (RYR) in the Sarcoplasmic Reticulum (SR) Membrane. Triadin is an associated SR protein that may participate in the interaction of RYR and DHPR. Calsequestrin is a low-affinity Ca^{++}-binding protein that helps accumulate Ca^{++} in the terminal cisternae. See text for details. COOH, carboxylic acid. (From Rossi AE, Dirksen RT. Sarcoplasmic reticulum: the dynamic calcium governor of muscle. *Muscle Nerve.* 2006;33:715.)

Actin-Myosin Interaction: Cross-Bridge Formation

As noted, contraction of skeletal muscle requires an increase in intracellular [Ca^{++}]. Moreover, the process of contraction is regulated by the thin filament. As shown in Fig. 12.12, contractile force (i.e., tension) increases in a sigmoidal manner as intracellular [Ca^{++}] is elevated above 0.1 μM, with half-maximal force occurring at less than 1 μM Ca^{++}. The mechanism by which Ca^{++} promotes this increase in tension is as follows: Ca^{++} released from the SR binds to troponin C. Once bound with Ca^{++}, troponin C facilitates

movement of the associated tropomyosin molecule toward the cleft of the actin filament. This movement of tropomyosin exposes myosin binding sites on the actin filament and allows a cross-bridge to form and thereby generate tension (see section "Cross-Bridge Cycling: Sarcomere Shortening"). Troponin C has four Ca^{++} binding sites. Two of these sites have high affinity for Ca^{++} but also bind Mg^{++} at rest. These sites seem to be involved in controlling and enhancing the interaction between the troponin I and troponin T subunits. The other two binding sites have lower affinity and bind Ca^{++} as its concentration rises after release from the SR. Binding of myosin to the actin

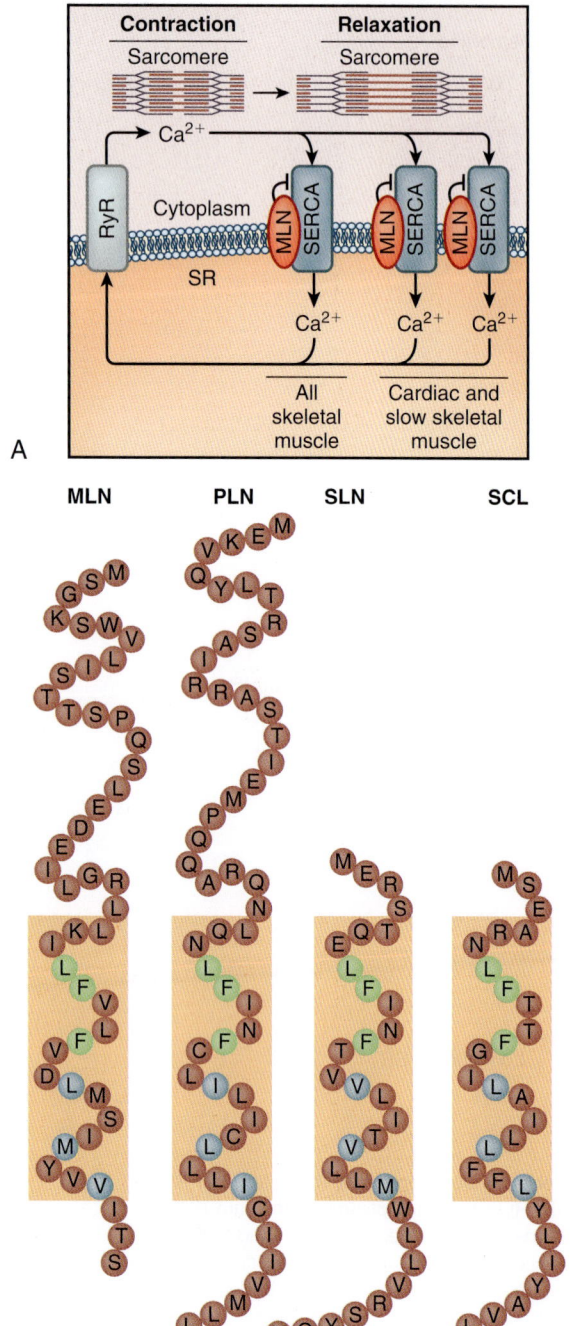

A

B   Vertebrate   Invertebrate

• **Fig. 12.11.** Identification of three endogenous micropeptides that inhibit sarcoplasmic endoplasmic reticulum Ca⁺⁺-ATPase (SERCA; **A**), and their predicted organization in the SR membrane **(B).** MLN, myoregulin; PLN, phospholamban; SCL, sarcolamban; SLN, sarcolipin. (Modified from Anderson DM, et al. A micropeptide encoded by a putative long noncoding RNA regulates muscle performance. *Cell.* 2015;160:595-606.)

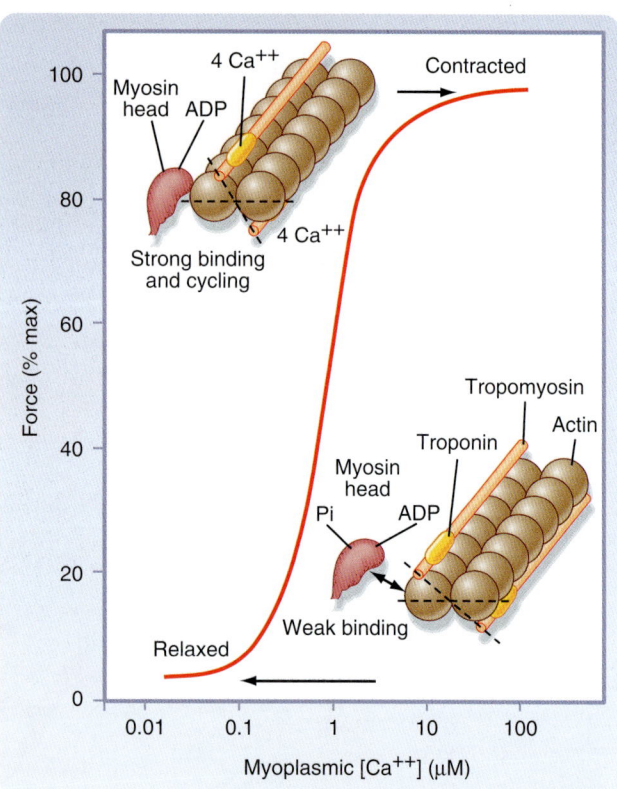

• **Fig. 12.12.** The contractile force of skeletal muscle increases in a Ca⁺⁺-dependent manner as a result of binding of Ca⁺⁺ to troponin C and the subsequent movement of tropomyosin away from myosin binding sites on the underlying actin molecules. See text for details. (Redrawn from Hartshorne DJ. In Lapedes DN. ed. *Yearbook of Science and Technology.* New York: McGraw-Hill; 1976.)

filaments appears to cause a further shift in tropomyosin. Although a given tropomyosin molecule extends over seven actin molecules, it is hypothesized that the strong binding of myosin to actin results in movement of an adjacent tropomyosin molecule, perhaps exposing myosin binding sites on as many as 14 actin molecules. This ability of one tropomyosin molecule to influence the movement of another may be a consequence of the close proximity of adjacent tropomyosin molecules.

Cross-Bridge Cycling: Sarcomere Shortening

Once myosin and actin are bound, ATP-dependent conformational changes in the myosin molecule result in movement of the actin filaments toward the center of the sarcomere. Such movement shortens the length of the sarcomere and thereby contracts the muscle fiber. The mechanism by which myosin produces force and shortens the sarcomere is thought to involve four basic steps that are collectively termed the *cross-bridge cycle* (labeled *a* to *d* in Fig. 12.13). In the resting state, myosin is thought to have partially hydrolyzed ATP (state *a*). When Ca⁺⁺ is released from the terminal cisternae of the SR, it binds to troponin C, which in turn promotes movement of tropomyosin on the actin filament in such a way that myosin-binding sites on actin are exposed. This then allows the "energized"

IN THE CLINIC

Genetic diseases that cause disturbances in Ca⁺⁺ homeostasis in skeletal muscle include **malignant hyperthermia, central core disease,** and **Brody's disease.** Malignant hyperthermia is an autosomal dominant trait that has life-threatening consequences in certain surgical instances. Anesthetics such as halothane or ether and the muscle relaxant succinylcholine can produce uncontrolled release of Ca⁺⁺ from the SR, thereby resulting in skeletal muscle rigidity, tachycardia, hyperventilation, and hyperthermia. This condition is lethal if not treated immediately. There are currently a series of tests (involving contractile responses of muscle biopsy specimens) to assess whether a patient has malignant hyperthermia. The incidence of this condition is approximately 1 per 15,000 children and 1 per 50,000 adults treated with anesthetics. Malignant hyperthermia is the result of a defect in the SR Ca⁺⁺ release channel (RYR), which becomes activated in the presence of the aforementioned anesthetics, causes the release of Ca⁺⁺ into the myoplasm, and hence prolongs muscle contraction (rigidity). The defect in the RYR is not restricted to a single locus. In some cases, malignant hyperthermia has been linked to a defect in the DHPR of the T tubule.

Central core disease is a rare autosomal dominant trait that results in muscle weakness, loss of mitochondria in the core of skeletal muscle fibers, and some disintegration of contractile filaments. It is often closely associated with malignant hyperthermia, and so patients with central core disease are treated as though they are susceptible to malignant hyperthermia in surgical situations. It is hypothesized that central cores devoid of mitochondria represent areas of elevated intracellular Ca⁺⁺ secondary to a mutation in the RYR. The loss of mitochondria is thought to occur when they take up the elevated Ca⁺⁺, which leads to mitochondrial Ca⁺⁺ overload.

Brody's disease is characterized by painless muscle cramping and impaired muscle relaxation during exercise. While an affected person runs upstairs, for example, muscles may stiffen and temporarily cannot be used. This relaxation

abnormality is seen in muscles of the legs, arms, and eyelid, and the response is worsened in cold weather. Brody's disease can be either autosomal recessive or autosomal dominant and may involve mutations in up to three genes; however, it is rare (affecting 1 per 10,000,000 births). It appears to result from decreased activity of the SERCA1 Ca⁺⁺ pump found in fast-twitch skeletal muscle (see the section "Skeletal Muscle Types"). The decreased activity of SERCA1 has been associated with mutations in the gene that encodes SERCA1, although another accessory factor may contribute to the decreased SR Ca⁺⁺ uptake in the fast-twitch skeletal muscle of individuals with Brody's disease.

Myotonia congenita is also associated with prolonged muscle contractions (painless cramping) after voluntary contractions, as a result of mutations in the *CLCN1* gene, which encodes the chloride voltage-gated channel 1 in skeletal muscle sarcolemma and T tubules. Chloride conductance in the skeletal muscle is important for repolarization and stabilization of the membrane potential, and so the reduced chloride conductance in skeletal muscles of individuals with myotonia congenita results in hyperexcitability of the muscle fiber. Voluntary contraction may therefore be followed by a series of action potentials (afterdepolarizations) in the muscle that result in prolonged contractions (i.e., cramping). Epinephrine (e.g., during stressful situations) often worsens the condition, as shown in myotonic ("fainting") goats. Muscle stiffness can be relieved by repeated contractions (i.e., the warm-up phenomenon), although the mechanism underlying the warm-up phenomenon is not known. Mutations in the *CLCN1* gene in myotonia congenita may be transmitted in either an autosomal recessive manner (as in Becker's disease, one type of myotonia congenita) or an autosomal dominant manner (as in Thomsen's disease, the other type of myotonia congenita). The prevalence of myotonia congenita is approximately 1 per 100,000 worldwide; the incidence is higher (≈1:10,000) in northern Scandinavia.

myosin head to bind to the underlying actin (state *b*). Myosin next undergoes a conformational change termed "ratchet action" that pulls the actin filament toward the center of the sarcomere (state *c*). Myosin releases adenosine diphosphate (ADP) and inorganic phosphate during the transition to state *c*. Binding of ATP to myosin decreases the affinity of myosin for actin, thereby resulting in the release of myosin from the actin filament (state *d*). Myosin then partially hydrolyzes the ATP, and part of the energy in the ATP is used to recock the head and return to the resting state.

If intracellular [Ca⁺⁺] is still elevated, myosin undergoes another cross-bridge cycle and produces further contraction of the muscle. The ratchet action of the cross-bridge is capable of moving the thin filament approximately 10 nm. The cycle continues until the SERCA pumps Ca⁺⁺ back into the SR. As [Ca⁺⁺] falls, Ca⁺⁺ dissociates from troponin C, and the troponin-tropomyosin complex moves and blocks the myosin binding sites on the actin filament. If the supply of ATP is exhausted, as occurs with death, the cycle stops

in state *c* with the formation of permanent actin-myosin complexes (i.e., the rigor state). In this state, the muscle is rigid, and the condition is termed **rigor mortis.**

As already noted, formation of the thick filaments involves the association of myosin molecules in a tail-to-tail configuration to produce a bipolar orientation (see Fig. 12.6). Such a bipolar orientation allows myosin to pull the actin filaments toward the center of the sarcomere during the cross-bridge cycle. The myosin molecules are also oriented in a helical array in the thick filament in such a way that cross-bridges extend toward each of the six thin filaments surrounding the thick filament (see Fig. 12.3). These myosin projections/cross-bridges can be seen on electron micrographs of skeletal muscle and appear to extend perpendicular from the thick filaments at rest. In the contracted state, the myosin cross-bridges slant toward the center of the sarcomere, which is consistent with the ratchet action of the myosin head.

The cross-bridge cycling mechanism just described is called the **sliding filament theory** because the myosin

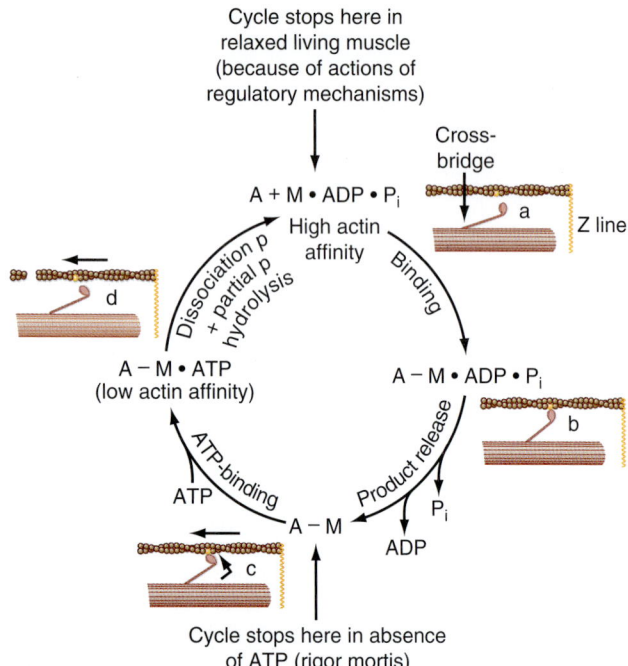

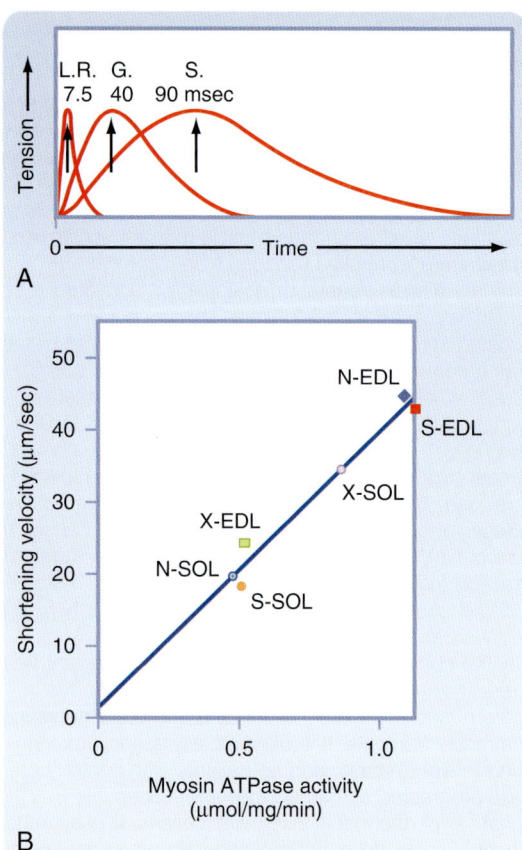

• **Fig. 12.13.** Cross-Bridge Cycle. In the relaxed state (state *a*), ATP is partially hydrolyzed (M • ADP • P$_i$). In the presence of elevated myoplasmic Ca^{++} (state *b*), myosin (M) binds to actin (A). Hydrolysis of ATP is completed (state *c*) and causes a conformational change in the myosin molecule that pulls the actin filament toward the center of the sarcomere. A new ATP molecule binds to myosin and causes release of the cross-bridge (state *d*). Partial hydrolysis of the newly bound ATP recocks the myosin head, which is now ready to bind again and again. If myoplasmic [Ca^{++}] is still elevated, the cycle repeats. If myoplasmic [Ca^{++}] is low, relaxation results.

• **Fig. 12.14. A,** Muscles vary in terms of the speed of contraction. G, gastrocnemius muscle of the leg; LR, lateral rectus muscle of the eye; S, soleus muscle of the leg. **B,** The speed of shortening is correlated with myosin ATPase activity. N-SOL, normal soleus muscle (slow twitch); N-EDL, normal extensor digitorum longus muscle (fast twitch); S-EDL, self-innervated extensor digitorum longus muscle (EDL motor nerve transected and resutured); S-SOL, self-innervated soleus muscle (soleus motor nerve transected and resutured); X-EDL, cross-innervated extensor digitorum longus muscle (EDL innervated by soleus motor nerve); X-SOL, cross-innervated SOL muscle (soleus innervated by EDL motor nerve). (**A,** From Montcastle V [ed]. *Medical Physiology.* 12th ed. St. Louis: Mosby; 1974. **B,** From Bárány M, Close RI. *J Physiol.* 1971;213:455.)

cross-bridge is pulling the actin thin filament toward the center of the sarcomere, which results in an apparent "sliding" of the thin filament past the thick filament. There is, however, uncertainty about how many myosin molecules contribute to the generation of force and whether both myosin heads in a given myosin molecule are involved. It has been calculated that there may be 600 myosin heads per thick filament, with a stoichiometry of 1 myosin head per 1.8 actin molecules. As a result of steric considerations, it is unlikely that all myosin heads can interact with actin, and calculations suggest that even during maximal force generation, only 20% to 40% of the myosin heads bind to actin.

The conversion of chemical energy (i.e., ATP) to mechanical energy by muscle is highly efficient. In isolated muscle preparations, maximum mechanical efficiency ($\approx$65% efficiency) is obtained at a submaximal force of 30% maximal tension. In humans performing steady-state ergometer exercise, mechanical efficiencies range from 40% to 57%.

Skeletal Muscle Types

Skeletal muscle fibers can be classified into two main groups according to the speed of contraction: fast-twitch

and slow-twitch muscle fibers. As shown in Fig. 12.14A, the lateral rectus of the eye contracts very quickly in response to an action potential, reaching peak tension within 8 msec, and then relaxes quickly, which results in a short duration of contraction. The soleus muscle of the leg, in contrast, requires 90 msec to reach peak tension in response to an action potential, and then it relaxes slowly. The gastrocnemius muscle requires an intermediate time to reach peak tension (40 msec) because of the presence of both fast-twitch and slow-twitch muscle fibers in this muscle.

The difference in speed of contraction between fast-twitch and slow-twitch muscles is correlated with myosin ATPase activity (see Fig. 12.14B), which in turn reflects the type of myosin present in the muscle fiber (Fig. 12.15; Table 12.1). Thus fast-twitch muscle fibers contain myosin isoforms that hydrolyze ATP quickly, whereas slow-twitch muscle fibers contain myosin isoforms that hydrolyze ATP

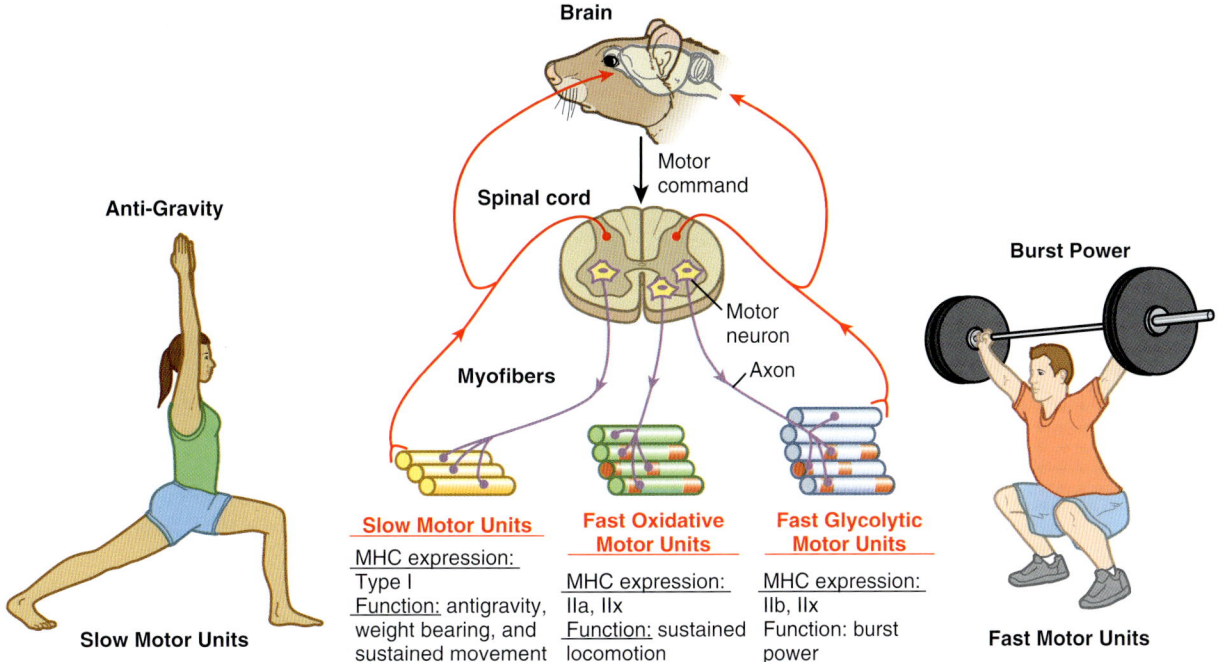

- **Fig. 12.15.** Comparison of three basic motor unit phenotypes in skeletal muscle of extremities and trunk. MHC, myosin heavy chain. (Redrawn from Baldwin K, Haddad F, Pandorf C, et al. Alterations in muscle mass and contractile phenotype in response to unloading models: role of transcriptional/pretranslational mechanisms. *Front Physiol.* 2013;4:284.)

TABLE 12.1   Basic Classification of Skeletal Muscle Fiber Types

Classification Parameters	Type I: Slow Oxidative	Type IIa: Fast Oxidative	Type IIb: Fast Glycolytic
Myosin isoenzyme ATPase rate and gene classification	Slow type I	Fast type IIa, IIx	Fast type IIb*, IIx
Sarcoplasmic reticular Ca++-pumping capacity	Moderate	High	High
Diameter (diffusion distance)	Moderate	Moderate	Large
Oxidative capacity: mitochondrial content, capillary density, myoglobin	High	Moderate	Low
Glycolytic capacity	Moderate	High	High

*Human "fast glycolytic" skeletal muscle fibers typically express the type IIx myosin isoenzyme instead of the type IIb myosin isoenzyme. In contrast, rodents express one type of slow-twitch muscle fiber (type I) and three types of fast-twitch muscle fibers (expressing type IIa, type IIx, or type IIb myosin isoenzymes); the type IIx muscle fibers express metabolic properties intermediate between those of type IIa and type IIb muscle fibers. The type IIx myosin isoenzyme may be coexpressed with type IIa myosin isoenzyme in "fast oxidative" muscle fibers. In text, the simple designation *type II fiber* refers to fast glycolytic muscle fibers.

slowly. These two types of myosin isoforms have the same basic structure described previously, with two heavy chains and two pairs of light chains, although they differ in amino acid composition.

It is very difficult to convert a slow-twitch muscle fiber into a fast-twitch fiber, although it can be accomplished by cross-innervation, which involves surgically interconnecting two motor neurons. As shown in Fig. 12.14B, when the soleus muscle and extensor digitorum longus muscle underwent cross-innervation, so that contraction of the soleus muscle was controlled by the extensor digitorum motor neuron (and vice versa), the speed of the contraction and

the myosin ATPase activity of the soleus muscle increased (labeled *X-SOL* in Fig. 12.14), whereas the extensor digitorum longus exhibited a decrease in shortening velocity and myosin ATPase activity (labeled *X-EDL*). Thus the motor innervation of the muscle fiber plays an important role in determining which type of myosin isoform is expressed in the muscle fiber. Further study showed that the intracellular Ca concentration in the muscle (secondary to differences in the activity pattern of the motor neuron) was an important determinant of whether the muscle fiber expressed the slow myosin isoform or the fast myosin isoform (see the section "Growth and Development").

The myosin isoforms expressed in skeletal muscle can be distinguished on the basis of myosin heavy chain composition (see Fig. 12.15 and Table 12.1). Slow-twitch muscle fibers express type I myosin heavy chain, whereas fast-twitch skeletal muscle fibers could contain type IIa, type IIx, or type IIb myosin heavy chains. Some fast-twitch muscle fibers may contain a mixture of type II myosin isoforms.

Slow-twitch skeletal muscles are also characterized by a high oxidative capacity (see Table 12.1), which in combination with the low myosin ATPase activity contribute to the fatigue resistance of slow-twitch muscle fibers. The oxidative capacity of the fast-twitch muscle fiber ranges from relatively high (in muscle fibers expressing type IIa myosin heavy chain) to low (in muscle fibers expressing type IIb myosin heavy chains). The low oxidative capacity of fast type IIb muscle fibers, coupled with the high myosin ATPase activity, increases the susceptibility of these muscle fibers to fatigue. Humans, however, rarely express the type IIb myosin heavy chain. Type IIb myosins are expressed in small animals such as rabbit and rats.

Motor units are generally composed of only one type of muscle fiber (Table 12.2; see also Fig. 12.15), unless the muscle fibers are undergoing a transition. Conditions that may trigger a change in the type of myosin expressed in a muscle fiber within a motor unit include chronic conditions such as microgravity (in space flight), denervation, and chronic unloading, which are associated with severe atrophy and promote the gradual transition from the expression of slow muscle myosin (type I) in the fiber to the expression of fast muscle myosin (types IIa and IIx).

An important function of slow motor units is in the maintenance of posture (see Fig. 12.15). The low ATPase activity of myosin in slow motor units, coupled with their high oxidative capacity, facilitates the ability of these slow motor units to maintain posture at low energy cost and thus resist fatigue. The smaller diameter of slow muscle fibers,

and the higher capillary density in slow muscle, also helps slow muscle resist fatigue.

Fast muscle, in contrast, is recruited for activities that require faster movements, more force, or both (see Fig. 12.15). Weightlifting, for example, can require a lot of power for short duration. In order to meet the demands for more force, additional motor units are recruited. In comparison with slow motor units, the fast motor units typically contain more muscle fibers (see Table 12.2). Fast muscle fibers also have a larger diameter than do slow muscle fibers. Thus recruitment of fast motor units can help meet the increased demands of burst activities such as weightlifting. The high myosin ATPase activity in fast muscle fibers and the increase in diffusion distance (resulting from the large diameter of the fast muscle fibers), however, increase the susceptibility of fast muscle fibers to fatigue.

Additional differences between fast and slow muscles include the following:

1. The motor neuron in slow muscle is more easily excited than that in fast muscle, and so slow muscles are typically recruited first. As stated previously, the high oxidative capacity of slow muscle, coupled with the low myosin ATPase activity, helps reduce the susceptibility to fatigue in slow muscle.

2. The neuromuscular junction of fast muscle differs from that in slow muscle in terms of acetylcholine vesicle content, the amount of acetylcholine released, the density of nicotinic acetylcholine receptors, the acetylcholine esterase activity, and Na channel density, all of which endow the fast muscle with a higher safety factor for initiation of an action potential. During repetitive stimulation, however, the safety factor in fast muscle drops quickly (faster than that seen in slow muscle).

3. The SR is more highly developed in fast muscle than in slow muscle, with higher levels of RYR, SERCA, lumenal Ca, and a higher DHPR/RYR ratio, all of which promote the development of a larger, faster intracellular Ca transient in fast muscle, which is important for quick, forceful contraction.

In addition to the differences between fast and slow fibers just noted, other muscle proteins are also expressed in a fiber type–specific manner. Such proteins include the three troponin subunits, tropomyosin, and C protein. The differential expression of troponin and tropomyosin isoforms influences the dependency of contraction on Ca^{++}. Slow fibers begin to develop tension at lower $[Ca^{++}]$ than fast fibers do. This difference in sensitivity to Ca^{++} is related in part to the fact that the troponin C isoform in slow fibers has only a single low-affinity Ca^{++}-binding site, whereas the troponin C of fast fibers has two low-affinity binding sites. Changes in the dependence of contraction on Ca^{++}, however, are not restricted to differences in the troponin C isoforms. Differences in troponin T and tropomyosin isoforms are also found. Thus regulation of the dependence of contraction on Ca^{++} is complex and involves contributions from multiple proteins on the thin filament. Phosphorylation of the regulatory light chain of myosin by Ca^{++}/

TABLE 12.2	Properties of Motor Units	
	Motor Unit Classification	
Characteristics	Type I	Type II
Properties of Nerve		
Cell diameter	Small	Large
Conduction velocity	Fast	Very fast
Excitability	High	Low
Properties of Muscle Cells		
Number of fibers	Few	Many
Fiber diameter	Moderate	Large
Force of unit	Low	High
Metabolic profile	Oxidative	Glycolytic
Contraction velocity	Moderate	Fast
Fatigability	Low	High

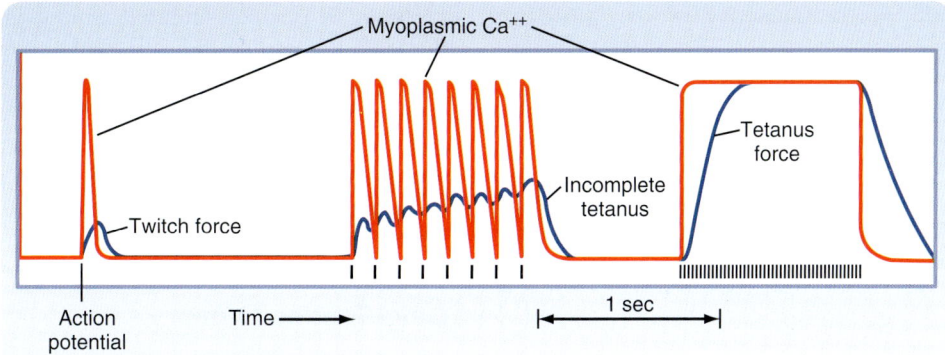

• **Fig. 12.16.** Increasing the Frequency of Electrical Stimulation of Skeletal Muscle Results in an Increase in the Force of Contraction. This is attributable to prolongation of the intracellular Ca^{++} transient and is termed *tetany*. Incomplete tetany results from initiation of another intracellular Ca^{++} transient before the muscle has completely relaxed. Thus there is a summation of twitch forces. See text for details.

calmodulin-dependent myosin light chain kinase, however, can increase Ca^{++} sensitivity of contraction, particularly in fast muscle fibers (partly because of the reported higher activity of myosin light chain kinase in fast muscle fibers).

Modulation of the Force of Contraction

Recruitment

A simple means of increasing the force of contraction of a muscle is to recruit more muscle fibers. Because all the muscle fibers within a motor unit are activated simultaneously, a muscle recruits more muscle fibers by recruiting more motor units. As already noted, muscle fibers can be classified as fast-twitch or slow-twitch. The type of fiber is determined by its innervation. Because all fibers in a motor unit are innervated by a single α motor neuron, all fibers within a motor unit are of the same type. Slow-twitch motor units tend to be small (100 to 500 muscle fibers) and are innervated by an α motor neuron that is easily excited (see Table 12.2). Fast-twitch motor units, in contrast, tend to be large (containing 1000 to 2000 muscle fibers) and are innervated by α motor neurons that are more difficult to excite. Thus slow-twitch motor units tend to be recruited first. As more and more force is needed, fast-twitch motor units are recruited. The advantage of such a recruitment strategy is that the first muscle fibers recruited are those that have high resistance to fatigue. Moreover, the small size of slow-twitch motor units allows fine motor control at low levels of force. The process of increasing the force of contraction by recruiting additional motor units is termed **spatial summation** because forces from muscle fibers are being "summed" within a larger area of the muscle. This is in contrast to **temporal summation,** which is discussed later.

Tetany

Action potentials in skeletal muscles are quite uniform and lead to the release of a reproducible pulse of Ca^{++} from the

SR (Fig. 12.16). A single action potential releases sufficient Ca^{++} to cause a twitch contraction. However, the duration of this contraction is very short because Ca^{++} is very rapidly pumped back into the SR. If the muscle is stimulated a second time before it is fully relaxed, the force of contraction increases (see Fig. 12.16, middle). Thus twitch forces are amplified as stimulus frequency increases. At a high level of stimulation, intracellular $[Ca^{++}]$ increases and is maintained throughout the period of stimulation (see Fig. 12.16, right), and the amount of force developed greatly exceeds that observed during a twitch. The response is termed *tetany*. At intermediate stimulus frequency, intracellular $[Ca^{++}]$ returns to baseline just before the next stimulus. However, there is a gradual rise in force (see Fig. 12.16, middle). This phenomenon is termed *incomplete tetany*. In both cases, the increased frequency of stimulation is said to produce a fusion of twitches.

The low force generation during a twitch, in comparison with that during tetany, may be due to the presence of a series elastic component in the muscle. Specifically, when the muscle is stretched a small amount shortly after initiation of the action potential, the muscle generates a twitch force that approximates the maximal tetanic force. This result, coupled with the observation that the size of the intracellular Ca^{++} transient during a twitch contraction is comparable with that during tetany, suggests that enough Ca^{++} is released into the myoplasm during a twitch to allow the actin-myosin interactions to produce maximal tension. However, the duration of the intracellular Ca^{++} transient during a twitch is sufficiently short that the contractile elements may not have enough time to fully stretch the series elastic components in the fiber and muscle. As a result, the measured tension is submaximal.

An increase in the duration of the intracellular Ca^{++} transient, as occurs with tetany, provides the muscle with sufficient time to completely stretch the series elastic component and thereby results in expression of the full contractile force of the actin-myosin interactions (i.e., maximal tension). Partial stretching of the series elastic component (as might be expected during a single twitch), followed

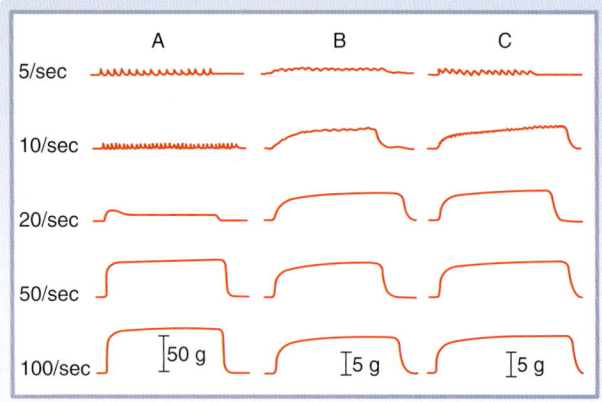

• **Fig. 12.17.** Slow-Twitch Muscles Exhibit Tetany at a Lower Stimulation Frequency Than Do Fast-Twitch Muscles. **A,** Fast-twitch motor unit in the gastrocnemius muscle. **B,** Slow-twitch motor unit in the gastrocnemius muscle. **C,** Slow-twitch muscle unit in the soleus muscle. The motor units were stimulated at the frequencies indicated on the left. The calibration bar for tension (in grams) generated during concentration is indicated by the *vertical brackets* under the curves. Note the large force generated by the fast-twitch motor unit **(A)**. (From Montcastle V [ed]. *Medical Physiology*. 12th ed. St. Louis: Mosby: 1974.)

by restimulation of the muscle before complete relaxation, however, would be expected to yield an intermediate level of tension, similar to that seen with incomplete tetany. The location of the series elastic component in skeletal muscle is not known. One potential source is the myosin molecule itself. In addition, it is likely that there are other sources of the series elastic component, such as the connective tissue and titin.

The stimulus frequency needed to produce tetany depends on whether the motor unit consists of slow or fast fibers (Fig. 12.17). Slow fibers can be tetanized at lower frequencies than can fast fibers. The ability of slow-twitch muscle to tetanize at lower stimulation frequencies reflects, at least in part, the longer duration of contraction seen in slow fibers. As also illustrated in Fig. 12.17, fast fibers develop a larger maximal force than slow fibers do because fast fibers are larger in diameter than slow fibers and there are more fibers in fast motor units than in slow motor units.

Modulation of Force by Reflex Arcs

Stretch Reflex

Skeletal muscles contain sensory fibers (**muscle spindles;** also called **intrafusal fibers**) that run parallel to the skeletal muscle fibers. The muscle spindles assess the degree of stretch of the muscle, as well as the speed of contraction. In the stretch reflex, rapid stretching of the muscle (e.g., tapping the tendon) lengthens the spindles in the muscle and results in an increased frequency of action potentials in the afferent sensory neurons of the spindle. These afferent fibers in turn excite the α motor neurons in the spinal cord that innervate the stretched muscle. The result is that the

reflex arc is a stretch-induced contraction of the muscle that does not require input from high centers in the brain. As the muscle shortens, efferent output is also sent to the spindle, which thereby takes the slack out of the spindle and ensures its ability to respond to stretch at all muscle lengths. By their action, muscle spindles provide feedback to the muscle in terms of its length and thus help maintain a joint at a given angle.

Golgi Tendon Organ

Golgi tendon organs are located in the tendons of muscles and provide feedback regarding contraction of the muscle. The main component of the tendon organ is an elongated fascicle of collagen bundles that is in series with the muscle fibers and can respond to contractions of individual muscle fibers. A given tendon organ may attach to several fast-twitch or slow-twitch muscle fibers (or both) and sends impulses through type Ib afferent nerve fibers in response to muscle contraction. The type Ib afferent impulses enter the spinal cord, which can promote inhibition of α motor neurons to the contracting (and synergistic) muscles while promoting excitation of α motor neurons to antagonistic muscles. The inhibitory actions are mediated through interneurons in the cord that release an inhibitory transmitter to the α motor neuron and create an inhibitory postsynaptic potential. The type Ib afferent impulses are also sent to higher centers of the brain (including the motor cortex and cerebellum). It is hypothesized that feedback from the tendon organs in response to muscle contraction may smooth the progression of muscle contraction by limiting the recruitment of additional motor units. Of interest is that the response of the tendon organ is not linearly related to force; rather, it drops off at higher levels of force, which may facilitate the recruitment of motor units at higher levels of effort.

Skeletal Muscle Tone

The skeletal system supports the body in an erect posture with the expenditure of relatively little energy. Nonetheless, even at rest, muscles normally exhibit some level of contractile activity. Isolated (i.e., denervated) unstimulated muscles are in a relaxed state and are said to be *flaccid*. However, relaxed muscles in the body are comparatively firm. This firmness, or tone, is caused by low levels of contractile activity in some of the motor units and is driven by reflex arcs from the muscle spindles. Interruption of the reflex arc by sectioning of the sensory afferent fibers abolishes this resting muscle tone. The tone in skeletal muscle is distinct from the "tone" in smooth muscle (see Chapter 14).

Energy Sources During Contraction

Adenosine Triphosphate

Muscle cells convert chemical energy to mechanical energy. ATP is the energy source used for this conversion. The ATP pool in skeletal muscle is small and capable of supporting only a few contractions if not replenished. This pool, however, is continually replenished during contraction, as described later, so that even when the muscle fatigues, ATP stores are only modestly decreased.

Creatine Phosphate

Muscle cells contain creatine phosphate, which is used to convert ADP to ATP and thus replenish the ATP store during muscle contraction. The creatine phosphate store represents the immediate high-energy source for replenishing the ATP supply in skeletal muscle, especially during intense exercise. The enzyme **creatine phosphokinase** catalyzes the reaction:

$$ADP + creatine\ phosphate \rightarrow ATP + creatine$$

Although much of the creatine phosphokinase is present in the myoplasm, a small amount is located in the thick filament (near the M line). The creatine phosphokinase in the thick filament may participate in the rapid resynthesis of ATP near the myosin heads during muscle contraction. The phosphate store created, however, is only about five times the size of the ATP store and thus cannot support prolonged periods of contraction (less than a minute of maximal muscle activity). Skeletal muscle fatigue during intense exercise is associated with depletion of the creatine phosphate store, although as described subsequently, this does not necessarily imply that the fatigue is caused by depletion of the creatine phosphate store. Because the creatine phosphokinase–catalyzed reaction is reversible, the muscle cell replenishes the creatine phosphate pool during recovery from fatigue by using ATP synthesized through oxidative phosphorylation.

Carbohydrates

Muscle cells contain glycogen, which can be metabolized during muscle contraction to provide glucose for oxidative phosphorylation and glycolysis, both of which generate ATP to replenish the ATP store. Muscle cells can also take up glucose from blood, a process that is stimulated by insulin (see Chapter 39). The cytosolic enzyme phosphorylase releases glucose 1-phosphate residues from glycogen, which are then metabolized by a combination of glycolysis (in the cytosol) and oxidative phosphorylation (in the mitochondria) to yield the equivalent of 37 mol of ATP per mole of glucose 1-phosphate. Blood glucose yields 36 mol of ATP per mole of glucose because 1 ATP is used to phosphorylate glucose at the start of glycolysis. These ATP yields, however, are dependent on an adequate oxygen supply. Under anaerobic conditions, in contrast, metabolism of glycogen and glucose yields only 3 and 2 mol of ATP per mole of glucose 1-phosphate and glucose, respectively (along with 2 mol of lactate). As discussed later, muscle fatigue during prolonged exercise is associated with depletion of glycogen stores in the muscle.

Fatty Acids and Triglycerides

Fatty acids represent an important source of energy for muscle cells during prolonged exercise. Muscle cells contain fatty acids but can also take up fatty acids from blood. In addition, muscle cells can store triglycerides, which can be hydrolyzed when needed to produce fatty acids. The fatty acids are subjected to β oxidation within the mitochondria. For fatty acids to enter the mitochondria, however, they are converted to acylcarnitine in the cytosol and then transported into the mitochondria, where they are converted to acyl coenzyme A (CoA). Within the mitochondria, the acyl CoA is subjected to β oxidation and yields acetyl CoA, which then enters the citric acid cycle and ultimately produces ATP.

Oxygen Debt

If the energy demands of exercise cannot be met by oxidative phosphorylation, an **oxygen debt** is incurred. After completion of exercise, respiration remains above the resting level in order to "repay" this oxygen debt. The extra oxygen consumption during this recovery phase is used to restore metabolite levels (such as creatine phosphate and ATP) and to metabolize the lactate generated by glycolysis. The increased cardiac and respiratory work during recovery also contributes to the increased oxygen consumption seen at this time and explains why more oxygen has to be "repaid" than was "borrowed." Some oxygen debt occurs even with low levels of exercise because slow oxidative motor units consume considerable ATP, derived from creatine phosphate or glycolysis, before oxidative metabolism can increase ATP production to meet steady-state requirements. The oxygen debt is much greater with strenuous exercise, when fast glycolytic motor units are used (Fig. 12.18). The oxygen debt is approximately equal to the energy consumed during exercise minus that supplied by oxidative metabolism (i.e., the dark- and light-colored areas in Fig. 12.18 are approximately equal). As indicated earlier, the additional oxygen used during recovery from exercise represents the energy requirements for restoring normal cellular metabolite levels.

Fatigue

The ability of muscle to meet energy needs is a major determinant of the duration of the exercise. However, fatigue is not the result of depletion of energy stores. Instead, metabolic byproducts seem to be important factors in the onset of fatigue. Fatigue may potentially occur at any of the points involved in muscle contraction, from the brain to the muscle cells, as well as in the cardiovascular and respiratory systems that maintain energy supplies (i.e., fatty acids and glucose) and oxygen delivery to the exercising muscle.

Several factors have been implicated in **muscle fatigue.** During brief periods of tetany, the oxygen supply to the muscle is adequate as long as the circulation is intact. However, the force/stress generated during these brief

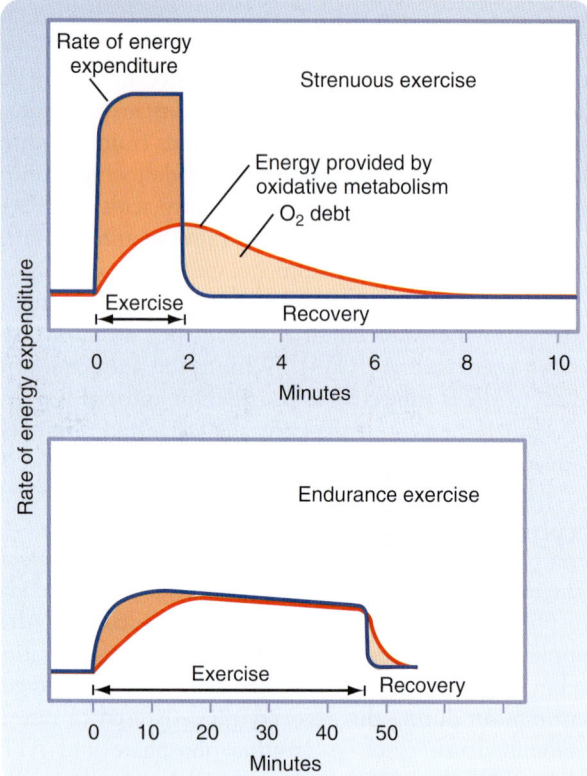

• **Fig. 12.18.** An Oxygen Debt Is Incurred by the Exercising of Muscle When the Rate of Energy Expenditure Exceeds the Rate of Energy Production by Oxidative Metabolism. **Upper panel,** Energy expenditure during strenuous exercise. **Lower panel,** Energy expenditure during endurance exercise. See text for details.

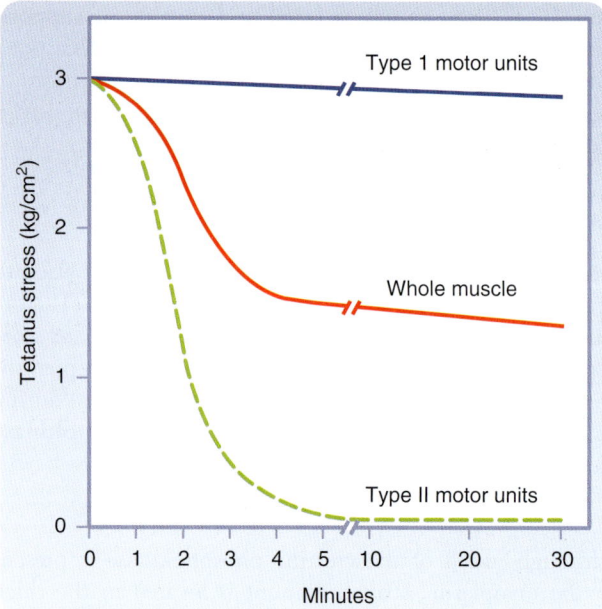

• **Fig. 12.19.** A series of brief tetanic stimulations of skeletal muscle result in a rapid decrease in force (tetanic stress, exemplified by the "Whole muscle" line in plot) that is attributable to fatigue of fast-twitch (type II) motor units in the muscle. Under these conditions, however, slow-twitch (type I) motor units are resistant to fatigue.

tetanic periods decays rapidly to a level that can be maintained for long periods (Fig. 12.19). This decay represents the rapid and almost total failure of the fast motor units. The decline in force/stress is paralleled by depletion of glycogen and creatine phosphate stores and the accumulation of lactic acid. Of importance is that the decline in force/stress occurs when the ATP pool is not greatly reduced, so that the muscle fibers do not go into rigor. In contrast, the slow motor units are able to meet the energy demands of fibers under this condition, and they do not exhibit significant fatigue, even after many hours. Evidently, some factor associated with energy metabolism can inhibit contraction (e.g., in the fast fibers), but this factor has not been clearly identified.

During intense exercise, accumulation of inorganic phosphate (P_i) and lactic acid in the myoplasm accounts for muscle fatigue. The accumulation of lactic acid, to levels as high as 15 to 26 mmol/L, decreases myoplasmic pH (from ≈7 to ≈6.2) and inhibits actin-myosin interactions. This decrease in pH reduces the sensitivity of the actin-myosin interaction to Ca^{++} by altering Ca^{++} binding to troponin C and by decreasing the maximum number of actin-myosin interactions. P_i has also been implicated as an important factor in the development of fatigue during intense exercise, inasmuch as phosphate concentrations can increase from approximately 2 mmol/L at rest to nearly 40 mmol/L

in working muscle. Such an elevation in $[P_i]$ can reduce tension by at least the following three different mechanisms: (1) inhibition of Ca^{++} release from the SR, (2) decrease in the sensitivity of contraction to Ca^{++}, and (3) alteration in actin-myosin binding. A number of other factors, including glycogen depletion from a specialized compartment, a localized increase in [ADP], extracellular elevation of $[K^+]$, and generation of oxygen free radicals, have also been implicated in various forms of exercise-induced muscle fatigue. Finally, the central nervous system contributes to fatigue, especially in how fatigue is perceived by the individual.

Regardless of whether the muscle is fatigued as a consequence of high-intensity exercise or prolonged exercise, the myoplasmic ATP level does not decrease substantially. In view of the reliance of all cells on the availability of ATP to maintain viability, fatigue has been described as a protective mechanism to minimize the risk of muscle cell injury or death. Consequently, it is likely that skeletal muscle cells have developed redundant systems to ensure that ATP levels do not drop to dangerously low levels and hence risk the viability of the cell.

Most persons tire and cease exercise long before the motor unit fatigues. General physical fatigue may be defined as a homeostatic disturbance produced by work. The basis for the perceived discomfort (or even pain) probably involves many factors. These factors may include a decrease in plasma glucose levels and accumulation of metabolites. Motor system function in the central nervous system is not impaired. Highly motivated and trained athletes can withstand the discomfort of fatigue and may exercise to the point at which some motor unit fatigue occurs. Part of the enhanced performance observed after training involves motivational factors.

Growth and Development

Skeletal muscle fibers differentiate before they are innervated, and some neuromuscular junctions are formed well after birth. Before innervation, the muscle fibers physiologically resemble slow (type I) cells. **Acetylcholine receptors** are distributed throughout the sarcolemma of these uninnervated cells and are supersensitive to that neurotransmitter. An end plate is formed when the first growing nerve terminal establishes contact with a muscle cell. The cell forms no further association with nerves, and receptors to acetylcholine become concentrated in the end plate membranes. Cells innervated by a small motor neuron form slow (type I) oxidative motor units. Fibers innervated by large motor nerves develop all the characteristics of fast (type II) motor units. Innervation produces major cellular changes, including synthesis of the fast and slow myosin isoforms, which replace embryonic or neonatal variants. Thus muscle fiber type is determined by the nerves that innervate the fiber.

An increase in muscle strength and size occurs during maturation. As the skeleton grows, the muscle cells lengthen. Lengthening is accomplished by the formation of additional sarcomeres at the ends of the muscle cells (Fig. 12.20), a process that is reversible. For example, the length of a cell decreases when terminal sarcomeres are eliminated, which can occur when a limb is immobilized with the muscle in a shortened position or when improper setting of a fracture causes shortening of the limb segment. Changes in muscle length affect the velocity and extent of shortening but do not influence the amount of force that can be generated by the muscle. The gradual increase in strength and diameter of a muscle during growth is achieved mainly by hypertrophy. Doubling the myofibrillar diameter by adding more sarcomeres in parallel (**hypertrophy,** for example) may double the amount of force generated but has no effect on the maximal velocity of shortening. Resistance exercise can promote hypertrophy by activation of the Akt-mTOR signaling pathway and simultaneous inhibition of the forkhead box O protein (FoxO)–atrogene pathway, which results in an increase in net protein synthesis (Fig. 12.21A). The increase in intracellular Ca and decrease in ATP concentration can also stimulate mitochondrial biogenesis and expression of a slow muscle phenotype through activation of peroxisome proliferator-activated receptor γ coactivator 1α (PGC-1α) signaling (see Fig. 12.21B).

Skeletal muscles have a limited ability to form new fibers (**hyperplasia**). These new fibers result from differentiation of satellite cells that are present in the tissues. However, major cellular destruction leads to replacement by scar tissue.

Muscles not only must be used to maintain normal growth and development but must also experience loading.

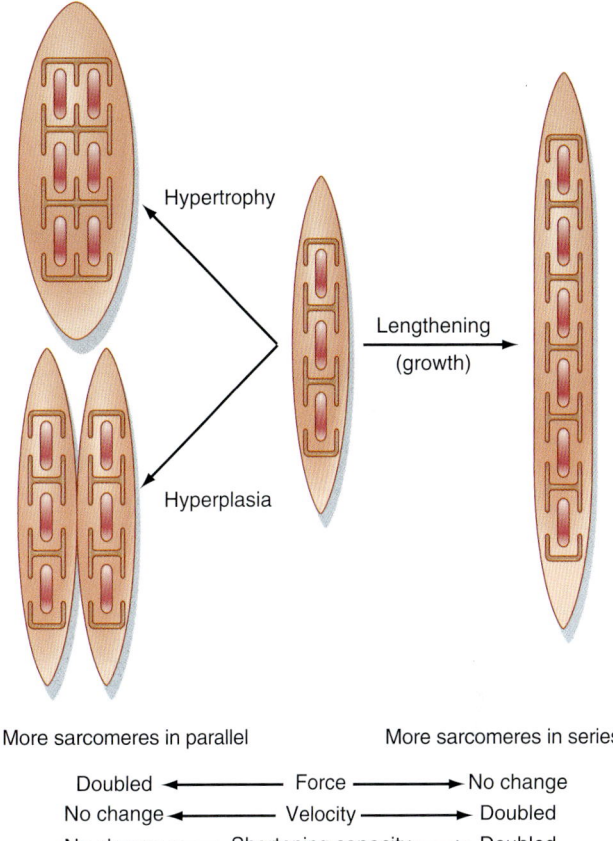

More sarcomeres in parallel   **More sarcomeres in series**

Doubled	◄──	Force	──►	No change
No change	◄──	Velocity	──►	Doubled
No change	◄──	Shortening capacity	──►	Doubled

• **Fig. 12.20.** Effects of Growth on the Mechanical Output of a Muscle Cell. Typically, skeletal muscle cell growth involves either lengthening (adding more sarcomeres to the ends of the muscle fibers) or increasing muscle fiber diameter (hypertrophy as a result of the addition of more myofilaments/myofibrils in parallel within the muscle fiber). The formation of new muscle fibers is called *muscle hyperplasia*, and it is infrequent in skeletal muscle.

Muscles immobilized in a cast lose mass. In addition, space flight exposes astronauts to a microgravity environment that mechanically unloads their muscles. Such unloading leads to rapid loss of muscle mass (i.e., **atrophy**) and weakness. Disuse atrophy appears to involve both inhibition of protein synthesis and stimulation of protein degradation (with net activation of the FoxO-atrogene pathway). Other categories of skeletal muscle atrophy include sarcopenia (which is atrophy associated with the aging process) and cachexia (which is atrophy associated with an illness).

Muscles that frequently contract to support the body typically have a high number of slow (type I) oxidative motor units. These slow motor units atrophy more rapidly than the fast (type II) motor units during prolonged periods of unloading. This atrophy of slow motor units is associated with a decrease in maximal tetanic force but also an increase in maximal shortening velocity. The increase in velocity is correlated with expression of the fast myosin isoform in these fibers. An important aspect of space medicine is the design of exercise programs that minimize such phenotypic changes during prolonged space flight.

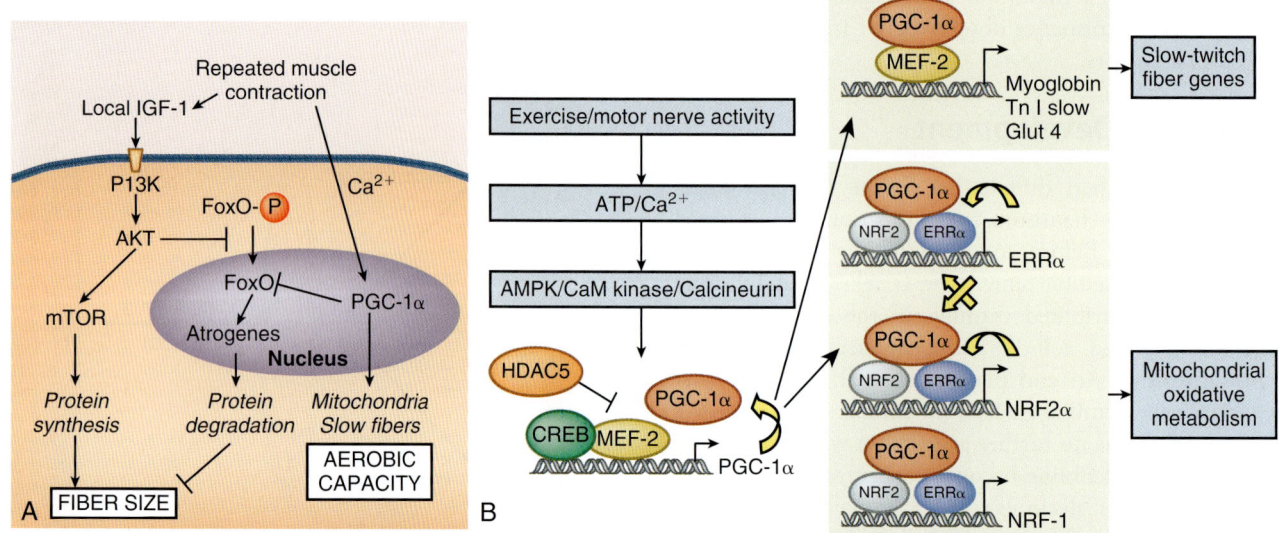

• **Fig. 12.21.** Basic molecular signaling pathways involved in contributing to net protein synthesis **(A)** and mitochondrial biogenesis **(B)** in skeletal muscle in response to exercise. AKT, a serine/threonine-specific protein kinase (protein kinase B); AMPK, 5′ adenosine monophosphate–activated protein kinase; ATP, adenosine triphosphate; CaM kinase, Ca^{++}/calmodulin-dependent protein kinase; CREB, cAMP response element binding protein; ERRα, estrogen-related receptor-α; FoxO, forkhead box O class of transcription factors; FoxO-P, phosphorylated FoxO; Glut-4, glucose transporter type 4; HDAC5, histone deacetylase 5; IGF-1, insulin-like growth factor 1; MEF2, myocyte enhancer factor 2; mTOR, mammalian target of rapamycin; Nrf1, Nrf2, and Nrf2α, nuclear respiratory factors 1, 2, and 2α; PGC-1α, peroxisome proliferator-activated receptor γ coactivator 1α; PI3K, phosphatidylinositol-3-kinase; Tn I, troponin I. (**A,** From Sandri M, et al. PGC-1alpha protects skeletal muscle from atrophy by suppressing FoxO3 action and atrophy-specific gene transcription. *PNAS.* 2006;103:16260-16265. **B,** From Lin J, Handschin C, Spiegelman BM. Metabolic control through the PGC-1 family of transcription coactivators. *Cell Metab.* 2005;1:361-370.)

Testosterone is a major factor responsible for the greater muscle mass in men because it has both myotrophic action and androgenic (masculinization) effects (see Chapter 44). A variety of synthetic molecules, called *anabolic steroids,* have been designed to enhance muscle growth while minimizing their androgenic action. These drugs are widely used by bodybuilders and athletes in sports in which strength is important. The doses are typically 10- to 50-fold greater than might be prescribed therapeutically for individuals with impaired hormone production. Unfortunately, none of these compounds lack androgenic effects. Hence, at the doses used, they induce serious hormone disturbances, including depression of testosterone production. A major issue is whether these drugs do in fact increase muscle and athletic performance in individuals with normal circulating levels of testosterone. Although they have been used since the 1950s, the scientific facts remain uncertain, and most experimental studies in animals have not demonstrated any significant effects on muscle development. Reports of studies in humans remain controversial. Proponents claim increases in strength that provide advantages in world-class performance. Critics argue that these increases are largely placebo effects associated with expectations and motivational factors. The public debate on abuse of anabolic steroids has led to their designation as controlled substances, along with opiates, amphetamines, and barbiturates.

Denervation, Reinnervation, and Cross-Innervation

As already noted, innervation is crucial for the skeletal muscle phenotype. If the motor nerve is cut, muscle fasciculation occurs. **Fasciculation** is characterized by small, irregular contractions caused by release of acetylcholine from the terminals of the degenerating distal portion of the axon. Several days after denervation, muscle fibrillation begins. **Fibrillation** is characterized by spontaneous, repetitive contractions. At this time, the cholinergic receptors have spread out over the entire cell membrane, in effect reverting to their preinnervation embryonic arrangement. The muscle fibrillations reflect supersensitivity to acetylcholine. Affected muscles also atrophy, with a decrease in the size of the muscle and its cells. Atrophy is progressive in humans, with degeneration of some cells 3 or 4 months after denervation. Most of the muscle fibers are replaced by fat and connective tissue after 1 to 2 years. These changes can be reversed if reinnervation occurs within a few months. Reinnervation is normally achieved by growth of the peripheral stump of motor nerve axons along the old nerve sheath.

Reinnervation of formerly fast (type II) fibers by a small motor axon causes that cell to redifferentiate into a slow (type I) fiber, and vice versa. This suggests that large and small motor nerves differ qualitatively and that the nerves have a specific "trophic" effect on the muscle fibers. This "trophic" effect reflects the rate of fiber stimulation. For example, stimulation via electrodes implanted in the muscle can lessen denervation atrophy. More strikingly, chronic low-frequency stimulation of fast motor units causes these units to be converted to slow units. Some conversion toward a typical fast-fiber phenotype can occur when the frequency of contraction in slow units is greatly decreased by reducing the excitatory input. Excitatory input can be reduced by sectioning the appropriate spinal or dorsal root or by severing the tendon, which functionally inactivates peripheral mechanoreceptors.

The frequency of contraction determines fiber development and phenotype through changes in gene expression and protein synthesis. Fibers that undergo frequent contractile activity form many mitochondria and synthesize the slow isoform of myosin. Fibers innervated by large, less excitable axons contract infrequently. Such relatively inactive fibers typically form few mitochondria, have large concentrations of glycolytic enzymes and synthesize the fast isoform of myosin.

AT THE CELLULAR LEVEL

The transcription factor **nuclear factor from activated T cells (NFAT)** has been implicated in this transition from fast-twitch to slow-twitch muscle (Fig. 12.22A). Specifically, it appears that stimulation of adult fast-twitch muscle cells at a frequency consistent with slow-twitch muscle cells can activate the Ca^{++}-dependent phosphatase calcineurin, which in turn can dephosphorylate NFAT and result in translocation of NFAT from the myoplasm to the nucleus, followed by the transcription of slow-twitch muscle genes (and inhibition of fast-twitch muscle genes). In accordance with this mechanism, expression of constitutively active NFAT in fast-twitch muscle promotes the expression of slow-twitch myosin while inhibiting the expression of fast-twitch myosin. The transcription factor **myocyte enhancing factor 2 (MEF2)** has also been implicated in this transition from fast-twitch to slow-twitch muscle (see Fig. 12.22B). Activation of MEF2 is thought to result from Ca^{++}/calmodulin-dependent phosphorylation of an inhibitor of MEF2: namely, histone deacetylase (HDAC).

Intracellular $[Ca^{++}]$ appears to play an important role in expression of the slow myosin isoform. Slow-twitch muscle fibers have a higher resting level of intracellular Ca^{++} than do fast-twitch muscle fibers. In addition, chronic electrical stimulation of fast-twitch muscle is accompanied by a 2.5-fold increase in resting myoplasmic $[Ca^{++}]$ that precedes the increased expression of slow-twitch myosin and decreased expression of fast-twitch myosin. Similarly, chronic elevation of intracellular Ca^{++} (approximately fivefold) in muscle cells expressing fast-twitch myosin induces a change in gene expression from the fast muscle myosin isoform to the slow myosin isoform within 8 days. An increase in citrate synthetase activity (an indicator of oxidative capacity) and a decrease in lactate dehydrogenase activity (an indicator of glycolytic capacity) accompany this Ca^{++}-dependent

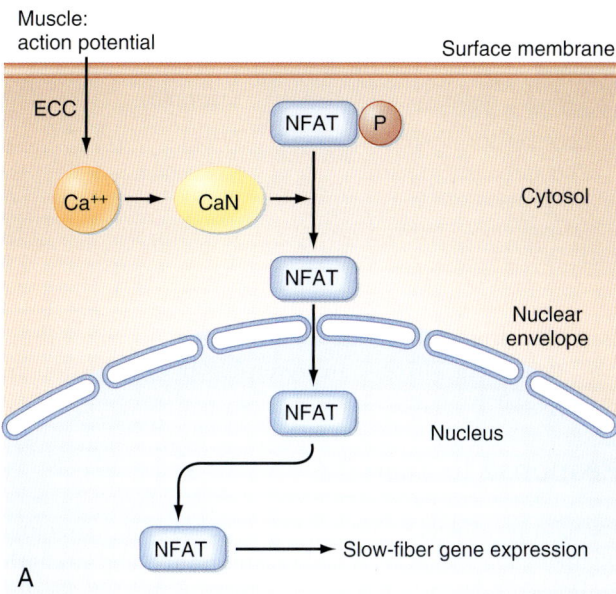

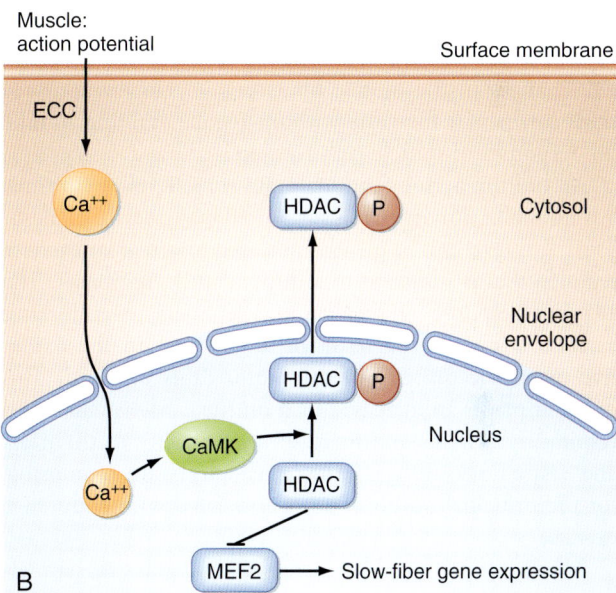

• **Fig. 12.22.** Molecular Signaling Pathways Contributing to the Transition From Fast-Twitch Muscle to Slow-Twitch Muscle. Chronic electrical stimulation of a fast-twitch muscle in a pattern consistent with a slow-twitch muscle results in development of the slow-twitch muscle phenotype because of dephosphorylation of the transcription factor nuclear factor from activated T cells (NFAT) by the Ca^{++}/calmodulin-dependent protein phosphatase calcineurin (CaN); this in turn results in nuclear translocation of NFAT and expression of slow-twitch muscle fiber genes **(A)**. Activation of the transcription factor myocyte enhancer factor 2 (MEF2) also appears to contribute to this fiber type transition **(B)**, in which activation of MEF2 involves Ca^{++}/calmodulin-dependent phosphorylation of an inhibitor, histone deacetylase (HDAC). CaMK, Ca^{++}/calmodulin-dependent protein kinase; ECC, excitation-contraction coupling; P, phosphorylation of HDAC. (From Liu Y, et al:. Signaling pathways in activity-dependent fiber type plasticity in adult skeletal muscle. *J Muscle Res Cell Motil.* 2005;26:13-21.)

TABLE 12.3	Effects of Exercise		
Type of Training	Example	Major Adaptive Response
Learning/coordination skills	Typing	Increased rate and accuracy of motor units (central nervous system)
Endurance (submaximal, sustained efforts)	Marathon running	Increased oxidative capacity in all involved motor units, with limited cellular hypertrophy
Strength (brief, maximal efforts)	Weightlifting	Hypertrophy and enhanced glycolytic capacity of the motor units used

transition from fast-twitch to slow-twitch myosin. These Ca^{++}-dependent changes are reversible by a reduction of intracellular $[Ca^{++}]$.

Response to Exercise

Exercise physiologists identify three categories of training regimens and responses: **learning, endurance, and strength training** (Table 12.3). Typically, most athletic endeavors involve elements of all three. The learning aspect of training involves motivational factors, as well as neuromuscular coordination. This aspect of training does not involve adaptive changes in the muscle fibers per se. However, motor skills can persist for years without regular training, unlike the responses of muscle cells to exercise.

All healthy persons can maintain some level of continuous muscular activity that is supported by oxidative metabolism. This level can be greatly increased by a regular exercise regimen that is sufficient to induce adaptive responses. The adaptive response of skeletal muscle fibers to endurance exercise is mainly the result of an increase in the oxidative metabolic capacity of the motor units involved. This demand places an increased load on the cardiovascular and respiratory systems and increases the capacity of the heart and respiratory muscles. The latter effects are responsible for the principal health benefits associated with endurance exercise.

Muscle strength can be increased by regular massive efforts that involve most motor units. Such efforts recruit fast glycolytic motor units, as well as slow oxidative motor units. During these efforts, blood supply to the working muscles may be interrupted as tissue pressures rise above intravascular pressure. The reduced blood flow limits the duration of the contraction. Regular maximal-strength exercise, such as weightlifting, induces the synthesis of more myofibrils and hence hypertrophy of the active muscle cells. The increased stress also induces the growth of tendons and bones.

Endurance exercise does not cause fast motor units to become slow, nor does maximal muscular effort produce a shift from slow to fast motor units. Thus any practical exercise regimen, when superimposed on normal daily activities, probably does not alter muscle fiber phenotype.

Delayed-Onset Muscle Soreness

Activities such as hiking or, in particular, downhill running, in which contracting muscles are stretched and lengthened too vigorously, are followed by more pain and stiffness than after comparable exercise that does not involve vigorous muscle stretching and lengthening (e.g., cycling). The resultant dull, aching pain develops slowly and reaches its peak within 24 to 48 hours. The pain is associated with reduced range of motion, stiffness, and weakness of the affected muscles. The prime factors that cause the pain are swelling and inflammation from injury to muscle cells, most commonly near the myotendinous junction. Type II motor units are affected more than type I motor units because the maximal force is highest in large cells, in which the loads imposed are approximately 60% greater than the maximal force that the cells can develop. Recovery is slow and depends on regeneration of the injured sarcomeres.

Biophysical Properties of Skeletal Muscle

The molecular mechanisms of muscle contraction described earlier underlie and are responsible for the biophysical properties of muscle. Historically, these biophysical properties were well described before elucidation of the molecular mechanisms of contraction. They remain important ways of describing muscle function.

Length-Tension Relationship

When muscles contract, they generate force (often measured as tension or stress) and decrease in length. In examination of the biophysical properties of muscle, one of these parameters is usually held constant, and the other is measured after an experimental maneuver. Accordingly, an **isometric contraction** is one in which muscle length is held constant, and the force generated during the contraction is then measured. An **isotonic contraction** is one in which the force (or tone) is held constant, and the change in length of the muscle is then measured.

When a muscle at rest is stretched, it resists stretch by a force that increases slowly at first and then more rapidly as the extent of stretch increases (Fig. 12.23). This purely passive property is due to the elasticity of the muscle tissue. If the muscle is stimulated to contract at these various lengths, a different relationship is obtained. Specifically, contractile force increases as muscle length is increased up to a point (designated L_O to indicate optimal length). As the muscle is stretched beyond L_O, contractile force decreases. This length-tension curve is consistent with the sliding filament theory, described previously. At a very

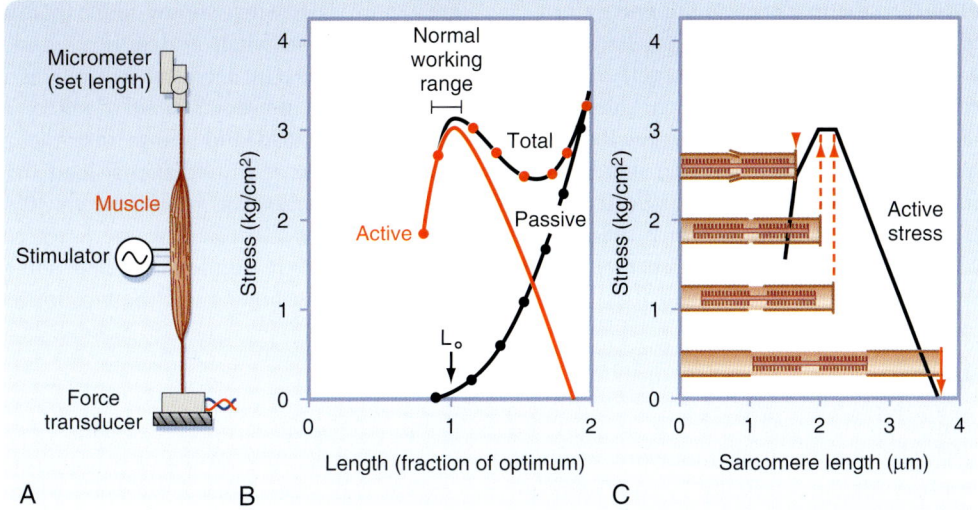

Fig. 12.23. Length-Tension Relationship in Skeletal Muscle. **A,** Experimental setup in which maximal isometric tetanic tension is measured at various muscle lengths. **B,** How active tension was calculated at various muscle lengths (i.e., by subtracting passive tension from total tension at each muscle length). **C,** Plot of active tension as a function of muscle length, with the predicted overlap of thick and thin filaments at selected points.

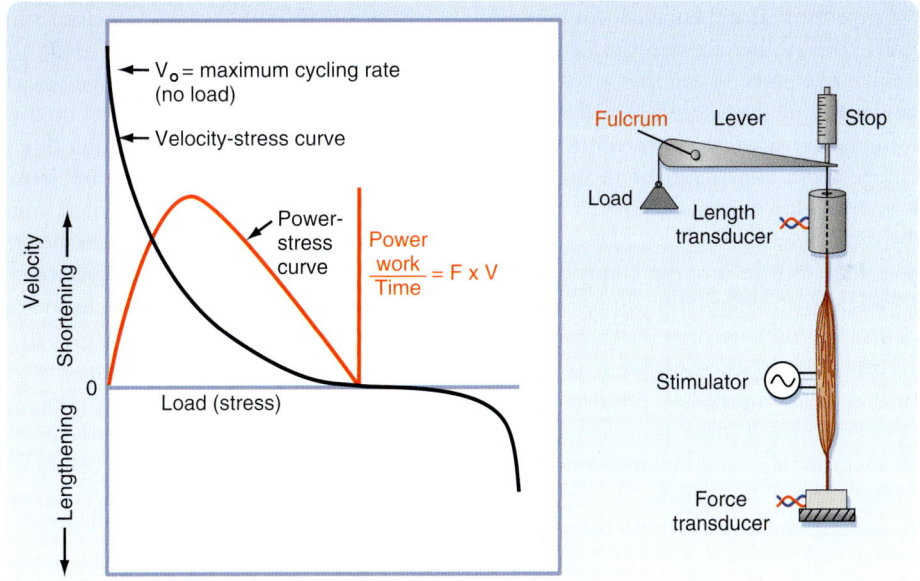

Fig. 12.24. Force-Velocity Relationship of Skeletal Muscle. The experimental setup is shown on the right. The initial muscle length was kept constant, but the amount of weight that the muscle had to lift during tetanic stimulation varied. While these various amounts of weight were lifted, muscle-shortening velocity was measured. See text for details. F, force; V, velocity.

long sarcomere length (3.7 µm), actin filaments no longer overlap with myosin filaments, and so there is no contraction. As muscle length is decreased toward L_O, the amount of overlap increases, and contractile force progressively increases. As sarcomere length decreases below 2 µm, the thin filaments collide in the middle of the sarcomere, the actin-myosin interaction is disturbed, and hence contractile force decreases. For construction of the length-tension curves, muscles were maintained at a given length, and then contractile force was measured (i.e., isometric contraction).

Thus the length-tension relationship supports the sliding filament theory of muscle contraction.

Force-Velocity Relationship

The velocity at which a muscle shortens is strongly dependent on the amount of force that the muscle must develop (Fig. 12.24). In the absence of any load, the shortening velocity of the muscle is maximal (denoted as V_0). V_0 corresponds to the maximal cycling rate of the cross-bridges (i.e., it is proportional to the maximal rate of energy turnover [ATPase activity] by myosin). Thus V_0 for fast-twitch muscle is higher than that for slow-twitch muscle. Increasing the load decreases the velocity of muscle shortening until, at maximal load, the muscle cannot lift the load and hence cannot shorten (zero velocity). Further increases in load result in stretching the muscle (negative velocity). The maximal isometric tension (i.e., force at which shortening velocity is zero) is proportional to the number of active cross-bridges between actin and myosin, and it is usually

greater for fast-twitch motor units (because of the larger diameter of fast-twitch muscle fibers and greater number of muscle fibers in a typical fast-twitch motor unit). In Fig. 12.24, the power-stress curve reflects the rate of work done at each load and shows that the maximal rate of work was done at a submaximal load (namely, when the force of contraction was approximately 30% of the maximal tetanic tension). To calculate the latter curve, the x- and y-coordinates were simply multiplied, and then the product was plotted as a function of the x-coordinate.

Key Points

1. Skeletal muscle is composed of numerous muscle cells (muscle fibers) that are typically 10 to 80 μm in diameter and up to 25 cm in length. The appearance of striations in skeletal muscle is due to the highly organized arrangement of thick and thin filaments in the myofibrils of skeletal muscle fibers. The sarcomere is a contractile unit in skeletal muscle. Each sarcomere is approximately 2 μm in length at rest and is bounded by two Z lines. Sarcomeres are arranged in series along the length of the myofibril. Thin filaments, containing actin, extend from the Z line toward the center of the sarcomere. Thick filaments, containing myosin, are positioned in the center of the sarcomere and overlap the actin thin filaments. Muscle contraction results from the Ca^{++}-dependent interaction of myosin and actin, in which myosin pulls the thin filaments toward the center of the sarcomere.

2. Contraction of skeletal muscle is under control of the central nervous system (i.e., voluntary). Motor centers in the brain control the activity of α motor neurons in the ventral horns of the spinal cord. These α motor neurons, in turn, synapse on skeletal muscle fibers. Whereas each skeletal muscle fiber is innervated by only one motor neuron, a motor neuron innervates several muscle fibers within the muscle. A *motor unit* refers to all the muscle fibers innervated by a single motor neuron.

3. The motor neuron initiates contraction of skeletal muscle by producing an action potential in the muscle fiber. As the action potential passes down the T tubules of the muscle fiber, dihydropyridine receptors (DHPRs) in the T tubules undergo conformational changes that result in the opening of neighboring SR Ca^{++} channels called ryanodine receptors (RYRs), which then release Ca^{++} to the myoplasm from the SR. The increase in myoplasmic Ca^{++} promotes muscle contraction by exposing myosin-binding sites on the actin thin filaments (a process that involves binding of Ca^{++} to troponin C, followed by movement of tropomyosin toward the groove in the thin filament). Myosin cross-bridges then appear to undergo a ratchet action, with the thin filaments pulled toward the center of the sarcomere and contracting the skeletal muscle

fiber. Relaxation of the muscle follows as myoplasmic Ca^{++} is resequestered by Ca^{++}-ATPase (SERCA) in the SR.

4. The force of contraction can be increased by the activation of more motor neurons (i.e., recruiting of more muscle fibers) or by an increase in the frequency of action potentials in the muscle fiber, which produces tetany. The increase in force during tetanic contractions is due to prolonged elevation of intracellular $[Ca^{++}]$.

5. The two basic types of skeletal muscle fibers are distinguished on the basis of their speed of contraction (i.e., fast-twitch versus slow-twitch). The difference in speed of contraction is attributed to the expression of different myosin isoforms that differ in myosin ATPase activity. In addition to the difference in myosin ATPase activity, fast- and slow-twitch muscles also differ in metabolic activity, fiber diameter, motor unit size, sensitivity to tetany, and recruitment pattern.

6. Typically, slow-twitch muscles are recruited before fast-twitch muscle fibers because of the greater excitability of motor neurons innervating slow-twitch muscles. The high oxidative capacity of slow-twitch muscle fibers supports sustained contractile activity. Fast-twitch muscle fibers, in contrast, tend to be large and typically have low oxidative capacity and high glycolytic capacity. The fast-twitch motor units are thus best suited for short periods of activity when high levels of force are required.

7. Fast-twitch muscle fibers can be converted to slow-twitch muscle fibers (and vice versa), depending on the stimulation pattern. Chronic electrical stimulation of a fast-twitch muscle results in the expression of slow-twitch myosin and decreased expression of fast-twitch myosin, along with an increase in oxidative capacity. The mechanism or mechanisms underlying this change in gene expression are unknown, but the change appears to be secondary to an elevation in resting intracellular $[Ca^{++}]$. The Ca^{++}-dependent phosphatase calcineurin and the transcription factor NFAT have been implicated in this transition from the fast-twitch to the slow-twitch phenotype. Ca^{++}/calmodulin-dependent kinase and the transcription factor MEF2 may also participate in the phenotype transition.

8. Skeletal muscle fibers atrophy after denervation. Muscle fibers depend on the activity of their motor nerves for maintenance of the differentiated phenotype. Reinnervation by axon growth along the original nerve sheath can reverse these changes. Skeletal muscle has a limited capacity to replace cells lost as a result of trauma or disease. Inhibition of the PI3K/Akt signaling pathways and activation of the FoxO pathway appears to contribute to the decreased rate of protein synthesis and the increased rate of protein degradation (respectively) observed during disuse atrophy. The increased protein degradation during atrophy is attributed to increases in both protease activity (e.g., activation of caspase 3) and ubiquitination (through elevated levels of ubiquitin ligases).

9. Skeletal muscle exhibits considerable phenotypic plasticity. Normal growth is associated with cellular hypertrophy, caused by the addition of more myofibrils and more sarcomeres at the ends of the cell to match skeletal growth. Strength training induces cellular hypertrophy, whereas endurance training increases the oxidative capacity of all involved motor units. Training regimens cannot alter fiber type or the expression of myosin isoforms.

10. Muscle fatigue during exercise is not due to depletion of ATP. The mechanism or mechanisms underlying exercise-induced fatigue are not known, although the accumulation of various metabolic products (lactate, P_i, ADP) has been implicated. In view of the importance of preventing depletion of myoplasmic ATP, which would affect the viability of the cell, it is likely that multiple mechanisms may have been developed to induce fatigue and hence lower the rate of ATP hydrolysis before the individual risks injury or death of the skeletal muscle cell.

11. When the energy demands of an exercising muscle cannot be met by oxidative metabolism, an oxygen debt is incurred. Increased breathing during the recovery period after exercise reflects this oxygen debt. The greater the reliance on anaerobic metabolism to meet the energy requirements of muscle contraction, the greater the oxygen debt.

Additional Readings

Allen DG, Whitehead NP, Froehner SC. Absence of dystrophin disrupts skeletal muscle signaling: roles of Ca^{2+}, reactive oxygen species, and nitric oxide in the development of muscular dystrophy. *Physiol Rev.* 2016;96(1):253-305.

Anderson DM, Anderson KM, Chang CL, et al. A micropeptide encoded by a putative long noncoding RNA regulates muscle performance. *Cell.* 2015;160(4):595-606.

Baldwin KM, Haddad F, Pandorf CE, Roy RR, Edgerton VR. Alterations in muscle mass and contractile phenotype in response to unloading models: role of transcriptional/pretranslational mechanisms. *Front Physiol.* 2013;4:284.

Burr AR, Molkentin JD. Genetic evidence in the mouse solidifies the calcium hypothesis of myofiber death in muscular dystrophy. *Cell Death Differ.* 2015;22:1402-1412.

Gordon AM, Homsher E, Regnier M. Regulation of contraction in striated muscle. *Physiol Rev.* 2000;80(2):853-924.

Imbrici P, Altamura C, Pessia M, Mantegazza R, Desaphy JF, Camerino DC. ClC-1 chloride channels: state-of-the-art research and future challenges. *Front Cell Neurosci.* 2015;9:156.

Joyner MJ, Casey DP. Regulation of increased blood flow (hyperemia) to muscles during exercise: a hierarchy of competing physiological needs. *Physiol Rev.* 2015;95(2):549-601.

Nelson CR, Fitts RH. Effects of low cell pH and elevated inorganic phosphate on the pCa-force relationship in single muscle fibers at near-physiological temperatures. *Am J Physiol Cell Physiol.* 2014;306(7):C670-C678.

Sandri M. Protein breakdown in cancer cachexia. *Semin Cell Dev Biol.* 2016;54:11-19.

Schiaffino S, Reggiani C. Fiber types in mammalian skeletal muscles. *Physiol Rev.* 2011;91(4):1447-1531.

13

Cardiac Muscle

LEARNING OBJECTIVES

Upon completion of this chapter, the student should be able to answer the following questions:

1. Describe the organization of cardiac muscle and how it meets the demands of the organ.
2. Describe the molecular mechanisms involved in excitation-contraction coupling in cardiac muscle and its suitability for this organ.
3. Describe the molecular mechanisms that lead to an increase in the force of contraction of the heart.
4. Discuss the length-tension relationship and the force-velocity curve for cardiac muscle, including the molecular basis for both curves.

If the student has already completed Chapter 12 on skeletal muscle, the student will be able to compare cardiac and skeletal muscle for each of the learning objectives just listed.

The function of the heart is to pump blood through the circulatory system, and this is accomplished by the highly organized contraction of cardiac muscle cells. Specifically, the cardiac muscle cells are connected together to form an electrical syncytium, with tight electrical and mechanical connections between adjacent cardiac muscle cells. An action potential initiated in a specialized region of the heart (e.g., the sinoatrial node) is therefore able to pass quickly throughout the heart to facilitate synchronized contraction of the cardiac muscle cells, which is important for the pumping action of the heart. Likewise, refilling of the heart requires synchronized relaxation of the heart; abnormal relaxation often results in pathological conditions.

This chapter begins with a description of the organization of cardiac muscle cells within the heart, including discussion of the tight electrical and mechanical connections. The mechanisms that underlie contraction, relaxation, and regulation of the force of contraction of cardiac muscle cells are also addressed. Although cardiac muscle and skeletal muscle are both striated muscles, they are significantly different in terms of organization, electrical and mechanical coupling, excitation-contraction coupling, and mechanisms to regulate the force of contraction. These differences are also highlighted.

Basic Organization of Cardiac Muscle Cells

Cardiac muscle cells are much smaller than skeletal muscle cells. Typically, cardiac muscle cells measure 10 μm in diameter and approximately 100 μm in length. As shown in Fig. 13.1*A,* cardiac cells are connected to each other through **intercalated disks,** which include a combination of mechanical junctions and electrical connections. The mechanical connections, which keep the cells from pulling apart when contracting, include the **fascia adherens** and **desmosomes. Gap junctions** between cardiac muscle cells, on the other hand, provide electrical connections between cells to allow propagation of the action potential throughout the heart. Thus the arrangement of cardiac muscle cells within the heart is said to form an electrical and mechanical syncytium that allows a single action potential (generated within the sinoatrial node) to pass throughout the heart so that the heart can contract in a synchronous, wave-like manner. Blood vessels course through the myocardium.

The basic organization of thick and thin filaments in cardiac muscle cells is comparable with that in skeletal muscle (see Chapter 12). Electron microscopy reveals repeating light and dark bands that represent I bands and A bands, respectively (see Fig. 13.1*B* and Chapter 12, Fig. 12.3). Thus cardiac muscle is classified as a striated muscle. The Z line transects the I band and represents the point of attachment of the thin filaments. The region between two adjacent Z lines represents the sarcomere, which is the contractile unit of the muscle cell. The thin filaments are composed of actin, tropomyosin, and troponin and extend into the A band. The A band is composed of thick filaments, along with some overlap of thin filaments. The thick filaments are composed of myosin and extend from the center of the sarcomere toward the Z lines.

Myosin filaments are formed by a tail-to-tail association of myosin molecules in the center of the sarcomere, followed by a head-to-tail association as the thick filament extends toward the Z lines. Thus the myosin filament is polarized and poised for pulling the actin filaments toward the center of the sarcomere. A cross-section view of the sarcomere near the end of the A band shows that each thick filament is surrounded by six thin filaments, and each thin filament receives cross-bridge attachments from three thick filaments. This complex array of thick and thin filaments is characteristic of both cardiac and skeletal muscle and helps

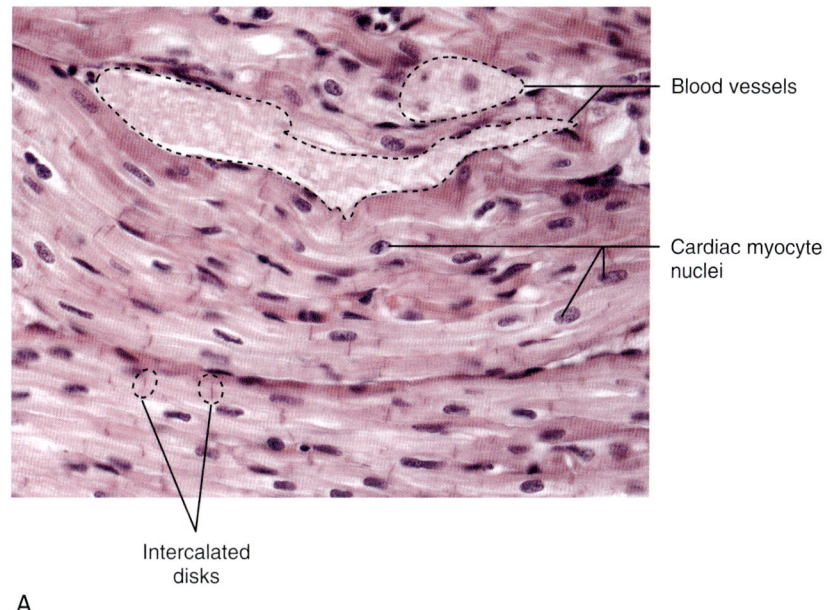

Blood vessels

Cardiac myocyte
nuclei

Intercalated
disks

A

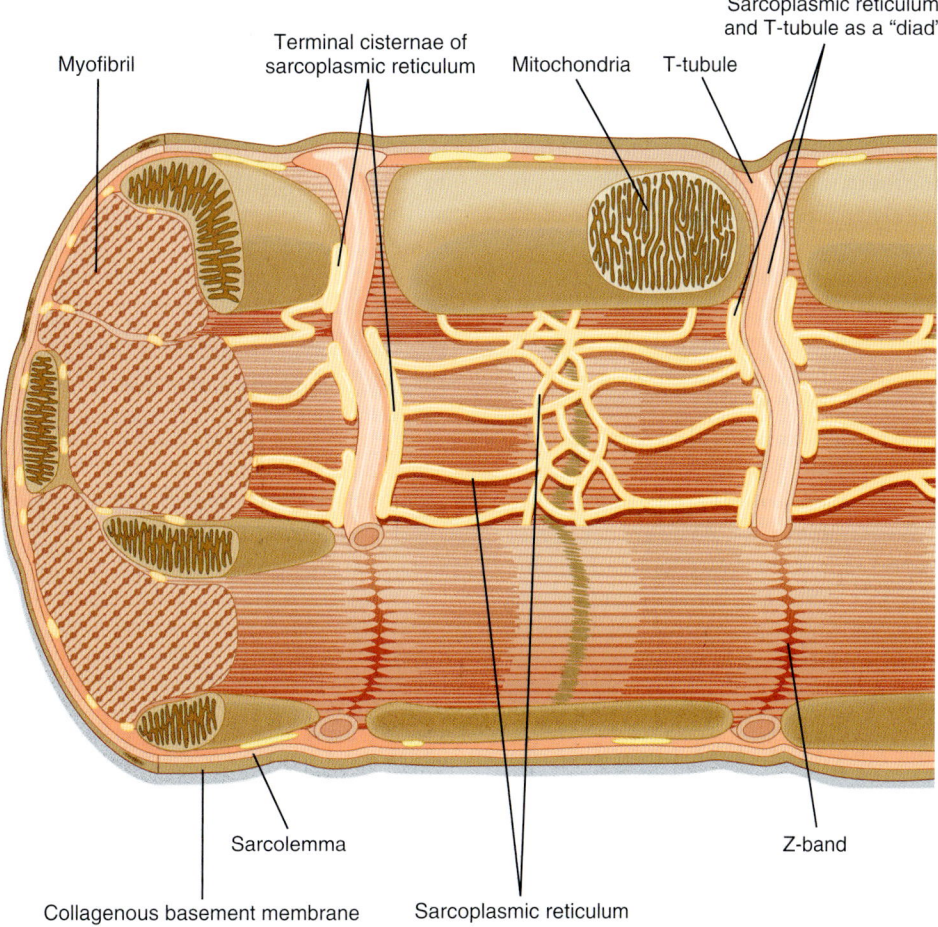

Myofibril

Terminal cisternae of
sarcoplasmic reticulum

Mitochondria

T-tubule

Sarcoplasmic reticulum
and T-tubule as a "diad"

Sarcolemma

Collagenous basement membrane

Sarcoplasmic reticulum

Z-band

B

• **Fig. 13.1 A,** Photomicrograph of cardiac muscle cells (magnification, ×210). Intercalated disks at either end of a muscle cell are identified in the lower left portion of the micrograph. The intercalated disk physically connects adjacent myocytes and, because of the presence of gap junctions, electrically couples the cells as well so that the muscle functions as an electrical and mechanical syncytium. **B,** Schematic representation of the organization of a sarcomere within a cardiac muscle cell. (**A,** From Telser A. *Elsevier's Integrated Histology.* St. Louis: Mosby; 2007. **B,** Redrawn from Fawcett D, McNutt NS. The ultrastructure of the cat myocardium. I. Ventricular papillary muscle. *J Cell Biol.* 1969;42:1-45.)

stabilize the filaments during muscle contraction (see Fig. 12.3B, for the hexagonal array of thick and thin filaments in the sarcomere of striated muscle).

Several proteins may contribute to the organization of the thick and thin filaments, including meromyosin and C protein (in the center of the sarcomere), which appear to serve as a scaffold for organization of the thick filaments. Similarly, nebulin extends along the length of the actin filament and may serve as a scaffold for the thin filament. The actin filament is anchored to the Z line by α-actinin, whereas the protein tropomodulin resides at the end of the actin filament and regulates the length of the thin filament. These proteins are present in both cardiac and skeletal muscle cells.

The thick filaments are tethered to the Z lines by a large elastic protein called **titin.** Although titin was postulated to tether myosin to the Z lines and thus prevent overstretching of the sarcomere, there is evidence indicating that titin may participate in cell signaling (perhaps by acting as a stretch sensor and thus modulating protein synthesis in response to stress). Such signaling by titin has been observed in both cardiac and skeletal muscle cells. Moreover, genetic defects in titin result in atrophy of both cardiac and skeletal muscle cells and may contribute to both cardiac dysfunction and skeletal muscle dystrophies (termed **titinopathies**). Titin is also thought to contribute to the ability of cardiac muscle to increase force upon stretch (discussed in the later section "Stretch.").

Although both cardiac muscle and skeletal muscle contain an abundance of connective tissue, there is more connective tissue in the heart. The abundance of connective tissue in the heart helps prevent muscle rupture (as in skeletal muscle), but it also prevents overstretching of the heart. Length-tension analysis of cardiac muscle, for example, shows a dramatic increase in passive tension as cardiac muscle is stretched beyond its resting length. Skeletal muscle, in contrast, tolerates a much greater degree of stretch before passive tension increases to a comparable level. The reason for this difference between cardiac and skeletal muscle is not known, although one possibility is that stretch of skeletal muscle is typically limited by the range of motion of the joint, which in turn is limited by the ligaments/connective tissue surrounding the joint.

The heart, on the other hand, appears to rely on the abundance of connective tissue around cardiac muscle cells to prevent overstretching during periods of increased venous return. During intense exercise, for example, venous return may increase fivefold. However, the heart is capable of pumping this extra volume of blood into the arterial system with only minor changes in the ventricular volume of the heart (i.e., end-diastolic volume increases less than 20%). Although the abundance of connective tissue in the heart limits stretch of the heart during these periods of increased venous return, additional regulatory mechanisms help the heart pump the extra blood that it receives (as discussed in the section "Stretch"). Conversely, if the heart were to be overstretched, the contractile ability of cardiac

muscle cells would be expected to decrease (because of decreased overlap of the thick and thin filaments), which would result in insufficient pumping, increased venous pressure, and perhaps pulmonary edema.

Within cardiac muscle cells, myofibrils are surrounded by the **sarcoplasmic reticulum (SR),** an internal network of membranes (see Fig. 13.1B). This is similar to the SR in skeletal muscle except that the SR in the heart is less dense and not as well developed. Terminal regions of the SR abut the **T tubule** or lie just below the **sarcolemma** (or both) and play a key role in the elevation of intracellular [Ca⁺⁺] during an action potential. The mechanism by which an action potential initiates release of Ca⁺⁺ in the heart, however, differs significantly from that in skeletal muscle (as discussed in the section "Excitation-Contraction Coupling"). The heart contains an abundance of mitochondria; up to 30% of the volume of the heart is occupied by these organelles. The high density of mitochondria provides the heart with great oxidative capacity, more so than is typical in skeletal muscle.

The sarcolemma of cardiac muscle also contains invaginations **(T tubules)** comparable to those seen in skeletal muscle. In cardiac muscle, however, T tubules are positioned at the Z lines, whereas in mammalian skeletal muscle, T tubules are positioned at the ends of the I bands. In cardiac muscle, the connections between the T tubules and the SR are fewer than, and not as well developed as, those in skeletal muscle.

 ## AT THE CELLULAR LEVEL

Familial cardiomyopathic hypertrophy (FCH) occurs in approximately 0.2% of the general population but is a leading cause of sudden death in otherwise healthy adults. It has been linked to genetic defects in a variety of proteins in cardiac sarcomeres, including myosin, troponin, tropomyosin, and myosin-binding protein C, a structural protein located in the middle of the A band of the sarcomere. FCH is an autosomal dominant disease, and transgenic studies indicate that expression of only a small amount of the mutated protein can result in development of the cardiomyopathic phenotype. Moreover, mutation of a single amino acid in the myosin molecule is sufficient to produce cardiomyopathic hypertrophy. The pathogenesis of FCH, however, is variable, even within a family with a single gene defect, in terms of both onset and severity; this variability suggests the presence of modifying loci.

Control of Cardiac Muscle Activity

Cardiac muscle is an involuntary muscle with an intrinsic pacemaker. The pacemaker represents a specialized cell (located in the **sinoatrial node** of the right atrium) that is able to undergo spontaneous depolarization and generate action potentials. Of importance is that although several cells in the heart are able to depolarize spontaneously, the fastest spontaneous depolarizations occur in cells in the sinoatrial node. Moreover, once a given cell spontaneously depolarizes and fires an action potential, this action potential is then propagated throughout the heart (by specialized conduction pathways and cell-to-cell contact). Thus depolarization from only one cell is needed to initiate a wave of contraction in the heart (i.e., a heartbeat). The mechanisms underlying this spontaneous depolarization are discussed in depth in Chapter 16.

As shown in Fig. 16.17, once an action potential is initiated in the sinoatrial node, it is propagated between atrial cells via gap junctions, as well as through specialized conduction fibers in the atria. The action potential can pass throughout the atria within approximately 70 msec. For the action potential to reach the ventricles, it must pass through the **atrioventricular node,** after which the action potential passes throughout the ventricle via specialized conduction pathways (the **bundle of His** and the **Purkinje system**) and gap junctions in the intercalated disks of adjacent cardiac myocytes. The action potential can pass through the entire heart within 220 msec after initiation in the sinoatrial node. Because contraction of a cardiac muscle cell typically lasts 300 msec, this rapid conduction promotes nearly synchronous contraction of heart muscle cells. This is a very different scenario from that of skeletal muscle, in which cells are grouped into motor units that are recruited independently as the force of contraction is increased.

Excitation-Contraction Coupling

Blood and extracellular fluids typically contain 1 to 2 mmol/L of free Ca^{++}, and it has been known since the days of the physiologist Sidney Ringer (ca. 1882) that the heart requires extracellular Ca^{++} to contract. Thus an isolated heart typically continues to beat when perfused with a warm (37°C), oxygenated, physiological salt solution that contains approximately 2 mmol/L Ca^{++} (e.g., Tyrode's solution), but it stops beating in the absence of extracellular Ca^{++}. This cessation of contractions in Ca^{++}-deficient media is also observed in hearts that are electrically stimulated, which further demonstrates the importance of extracellular Ca^{++} for contraction of cardiac muscle. This situation is quite different from that of skeletal muscle, which can contract in the total absence of extracellular Ca^{++}.

Action potentials in cardiac muscle are prolonged, lasting 150 to 300 msec (Fig. 13.2 inset), which is substantially longer than the action potentials in skeletal muscle ($\approx$5 msec). The long duration of the action potential in cardiac muscle is due to a slow inward Ca^{++} current through a **voltage-gated L-type calcium channel** in the sarcolemma. The amount of Ca^{++} coming into the cardiac muscle cell is relatively small and serves as a trigger for release of Ca^{++} from the SR. In the absence of extracellular Ca^{++}, an action potential can still be initiated in cardiac muscle, although it

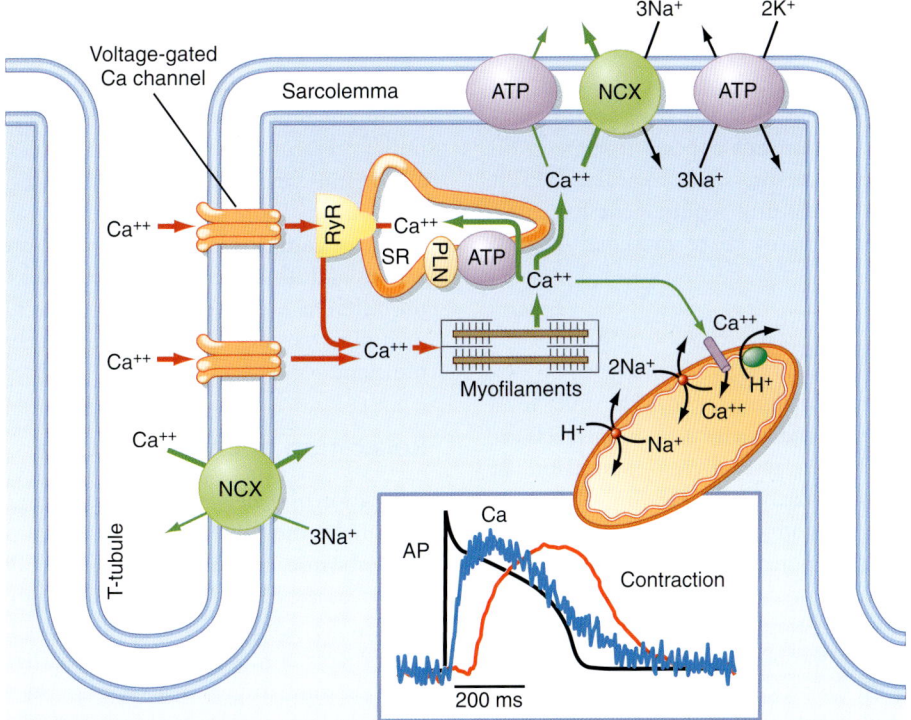

• **Fig. 13.2** Excitation-contraction coupling in the heart requires Ca^{++} influx through L-type calcium channels in the sarcolemma and T tubules. See text for details. *Inset* shows time course of action potential (AP), intracellular Ca transient (Ca), and contraction. ATP, adenosine triphosphate; NCX, sarcolemmal 3Na⁺-Ca⁺⁺ antiporter; PLN, phospholamban; RYR, ryanodine receptor. (Modified from Bers DM. Cardiac excitation-contraction coupling. *Nature.* 2002;415:198-205. *Inset* modified from Mountcastle VB: Medical Physiology, 13 ed. St Louis, Mosby, 1974; Brooks CM, Hoffman BF, Suckling EE, Orias O: Excitability of the Heart. New York, Grune & Stratton, 1955.)

is considerably shorter in duration and unable to initiate a contraction. Thus influx of Ca^{++} during the action potential is crucial for triggering release of Ca^{++} from the SR and thus initiating contraction.

The L-type calcium channel is composed of five subunits (α_1, α_2, β, γ, and δ). The α_1 subunit is also called the **dihydropyridine receptor (DHPR)** because it binds the dihydropyridine class of calcium channel–blocking drugs (e.g., nitrendipine and nimodipine). Although this channel complex is present in both skeletal and cardiac muscle, it serves very different functions in the two muscle types (discussed in the next paragraph).

In each cardiac muscle sarcomere, terminal regions of the SR abut T tubules and the sarcolemma (see Figs. 13.1B, 13.2). These junctional regions of the SR are enriched in **ryanodine receptors** (**RYRs;** a Ca^{++} release channel in the SR). The RYR is a Ca^{++}-gated calcium channel, and so influx of Ca^{++} during an action potential is able to initiate release of Ca^{++} from the SR in cardiac muscle. The amount of Ca^{++} released into the cytosol from the SR is much greater than that entering the cytosol from the sarcolemma, although release of Ca^{++} from the SR does not occur without this entry of "trigger" Ca^{++}. In contrast, in skeletal muscle, release of Ca^{++} from the SR does not involve entry of Ca^{++} across the sarcolemma but instead results from a voltage-induced conformational change in the DHPR. Thus excitation-contraction coupling in cardiac muscle is termed **electrochemical coupling** (involving Ca^{++}-induced release of Ca^{++}), whereas excitation-contraction coupling in skeletal muscle is termed **electromechanical coupling** (involving direct interactions between the DHPR in the T tubule and the RYR in the SR). The basis for this difference in Ca^{++} release mechanisms appears to depend on the DHPR isoform because expression of cardiac DHPR in skeletal muscle cells results in a requirement for extracellular Ca^{++} for contraction of these modified skeletal muscle cells.

Contraction Mechanism

As in skeletal muscle, contraction of cardiac muscle is regulated by thin filaments, and an elevation in intracellular $[Ca^{++}]$ is necessary to promote actin-myosin interaction. At low (<50 nmol/L) intracellular $[Ca^{++}]$, binding of myosin to actin is blocked by tropomyosin. As cytosolic $[Ca^{++}]$ increases during an action potential, however, binding of Ca^{++} to troponin C results in a conformational change in the troponin/tropomyosin complex in which tropomyosin slips into the groove of the actin filament and exposes myosin binding sites on the actin filament. As long as cytosolic $[Ca^{++}]$ remains elevated, and hence myosin binding sites are exposed, myosin will bind to actin, undergo a ratchet action, and contract the cardiac muscle cell. Note that because myosin binding sites on actin are blocked at low $[Ca^{++}]$ and exposed during a rise in intracellular $[Ca^{++}]$, contraction of cardiac muscle is termed *thin filament regulated*. This is identical to the situation in skeletal muscle;

in smooth muscle, in contrast, contraction is thick filament regulated (see Chapter 14).

During a rise in intracellular $[Ca^{++}]$ and exposure of myosin-binding sites on actin, the myosin cross-bridges undergo a series of steps that result in contraction of the cardiac muscle cell. At rest, the myosin molecules are energized in that they have partially hydrolyzed adenosine triphosphate (ATP) to "cock the head" and are thus ready to interact with actin. An elevation in intracellular $[Ca^{++}]$ then exposes myosin-binding sites on actin and thus allows myosin to bind actin (step 1). The bound myosin subsequently undergoes a powerstroke in which the actin filament is pulled toward the center of the sarcomere (step 2). Adenosine diphosphate (ADP) and inorganic phosphate (P_i) are released from the myosin head during this step as the energy from ATP is used to contract the muscle. The myosin head moves approximately 70 nm during each ratchet action (cross-bridge cycle). Binding of ATP to myosin decreases the affinity of myosin for actin and thus allows myosin to release from actin (step 3). Myosin then partially hydrolyzes the bound ATP to reenergize ("cock") the head (step 4) and ready the cross-bridge for another cycle. This four-step cycle is identical to that described for skeletal muscle (see Chapter 12, Fig. 12.13).

Cardiac muscle and skeletal muscle differ, however, in the level of intracellular $[Ca^{++}]$ attained after an action potential and hence in the number of actin-myosin interactions. In skeletal muscle, intracellular $[Ca^{++}]$ rises and the number of actin-myosin interactions is high after an action potential. In cardiac muscle, the rise in intracellular $[Ca^{++}]$ can be regulated, which affords the heart an important means of modulating the force of contraction without recruiting more muscle cells or undergoing tetany. Recall that in the heart, all the muscle cells are activated during a contraction, and so recruiting more muscle cells is not an option. Moreover, tetany of cardiac muscle cells would prevent any pumping action and thus be fatal. Consequently, the heart relies on different means of increasing the force of contraction, including varying the amplitude of the intracellular Ca^{++} transient.

Relaxation of Cardiac Muscle

Relaxation of skeletal muscle simply requires reaccumulation of Ca^{++} by the SR through the action of the **sarcoplasmic endoplasmic reticulum calcium-ATPase (SERCA),** also known as the **SR Ca^{++} pump.** Although SERCA plays a key role in the decrease in cytosolic $[Ca^{++}]$ in cardiac muscle, the process is more complex than that in skeletal muscle because some trigger Ca^{++} enters the cardiac muscle cell through the sarcolemmal calcium channels during each action potential. A mechanism must therefore exist to extrude this trigger Ca^{++}; otherwise, the amount of Ca^{++} in the SR would continuously increase, and Ca^{++} overload would result. In particular, some Ca^{++} is extruded from the cardiac muscle cell though the sarcolemmal **3Na$^+$-Ca^{++} antiporter** and a **sarcolemmal Ca^{++} pump** (Fig. 13.2). The extracellular [Ca^{++}] is in the millimolar range, whereas the amount of intracellular [Ca^{++}] is submicromolar, and so extrusion of Ca^{++} is accomplished against a large chemical gradient. Similarly, [Na$^+$] is considerably higher in the extracellular media than within the cell. The antiporter uses the Na$^+$ gradient across the cell to power the uphill movement of Ca^{++} out of the cell. Because three Na$^+$ ions enter the cell in exchange for one Ca^{++} ion, the 3Na$^+$-Ca^{++} antiporter is electrogenic and creates a depolarizing current. The sarcolemmal Ca^{++} pump, on the other hand, uses the energy in ATP to extrude Ca^{++} from the cell. Both extrusion mechanisms and SERCA thus contribute to the relaxation of cardiac muscle by decreasing cytosolic [Ca^{++}].

Although the interaction of actin and myosin requires a relatively small increase in free intracellular [Ca^{++}], the abundance of Ca^{++}-binding proteins in the myoplasm necessitates a much larger increase in total intracellular [Ca^{++}]. The resting intracellular [Ca^{++}] is approximately 50 to 100 nmol/L; half-maximal force of contraction requires approximately 600 nmol/L of free Ca^{++}. However, because of Ca^{++}-binding proteins such as parvalbumin and troponin C, the total myoplasmic concentration must increase by 70 µmol/L. As already noted, much of this increase in total myoplasmic [Ca^{++}] occurs through release of Ca^{++} from the SR. In a number of species, including rabbits, dogs, cats, guinea pigs, and humans, uptake and release of Ca^{++} by the SR account for approximately 70% of the intracellular Ca^{++} transient. Thus up to 30% of the rise in intracellular [Ca^{++}] may be attributable to influx of Ca^{++} through voltage-gated calcium channels in the sarcolemma, and the 3Na$^+$-Ca^{++} antiporter contributes significantly to Ca^{++} extrusion during relaxation.

The sarcolemmal Ca^{++} pump is in lower abundance than the 3Na$^+$-Ca^{++} antiporter but has a higher affinity for Ca^{++} and thus may contribute more to the regulation of resting intracellular [Ca^{++}] (see Fig. 13.2). The relative contribution of the Ca^{++} extrusion mechanisms, however, varies between species. For example, rat and mouse myocytes rely primarily on Ca^{++} reuptake by the SR (i.e., the SR accounts for 92% of Ca^{++} transport).

Regulation of the Force of Contraction

Intracellular Calcium

Because the heart represents an electrical syncytium, in which all the cardiac muscle cells contract during a single beat, it is not possible to increase the force of contraction by recruiting more muscle cells. Moreover, tetany of the heart would be lethal because it would defeat the critical pumping action of the heart. The heart has therefore developed alternative strategies to increase the force of contraction. The long duration of the action potential found in cardiac muscle, which is due to activation of the voltage-gated L-type calcium channel, results in a long refractory period, which in turn prevents tetany. Modulation of Ca^{++} influx through L-type calcium channels during an action potential,

however, provides the heart with a mechanism to alter cytosolic [Ca^{++}] and hence the force of contraction.

A simple means of modulating the force of contraction of cardiac muscle cells in vitro is to vary extracellular [Ca^{++}]. As noted previously, contraction of the heart requires extracellular Ca^{++}. Decreasing extracellular [Ca^{++}] from a normal range of 1 to 2 mmol/L to 0.5 mmol/L, for example, reduces the force of the contraction. This reduction in force of contraction is not associated with a change in the duration of the contraction because the kinetic characteristics of Ca^{++} sequestration by the SR and Ca^{++} extrusion have not been modified. Although this approach of varying extracellular [Ca^{++}] to alter the force of contraction is demonstrable in vitro, it is not a common means of modulating the force of cardiac contraction in vivo.

In vivo, an increase in the size of the intracellular Ca^{++} transient and hence the force of contraction occurs in response to sympathetic stimulation (see the section "β-Adrenergic Agonists" and also Chapter 18). Sympathetic stimulation often occurs during periods of excitement or fright and involves activation of β-adrenergic receptors on the heart by norepinephrine (released from nerve terminals in the heart) or epinephrine (released from the adrenal medulla into the bloodstream). As shown in Fig. 13.3, the β-adrenergic agonist isoproterenol results in a dramatic increase in the size of the intracellular Ca^{++} transient and, consequently, a more forceful contraction. An increase in the force of contraction is termed **positive inotropy.** Typically, the rate of relaxation accompanying this β-adrenergic stimulation also increases, which results in a shorter contraction. The increase in the rate of muscle relaxation is termed **positive lusitropy.** The frequency of contractions of the heart also increases with β-adrenergic stimulation and is termed **positive chronotropy.** Thus β-adrenergic stimulation of the heart produces stronger, briefer, and more frequent contractions.

β-Adrenergic Agonists

The sympathetic nervous system is stimulated when a human or an animal becomes excited, and it is said to prepare the individual for "fight or flight." In the case of the heart, increased levels of the adrenal medullary hormone **epinephrine** or the sympathetic neurotransmitter **norepinephrine** activate β-adrenergic receptors on the cardiac muscle cells, which in turn activates **adenylate cyclase,** increases **cyclic adenosine monophosphate (cAMP),** and thus promotes cAMP-dependent phosphorylation of numerous proteins in cardiac muscle cells (Fig. 13.4).

Both voltage-gated L-type calcium channels (responsible for the trigger Ca^{++}) and a protein associated with SERCA, called **phospholamban,** are phosphorylated by cAMP-dependent protein kinase. The combined action of these phosphorylations increases the amount of Ca^{++} in the SR. Specifically, phosphorylation of the sarcolemmal calcium channel causes more trigger Ca^{++} to enter the cell, and phosphorylation of phospholamban increases the activity

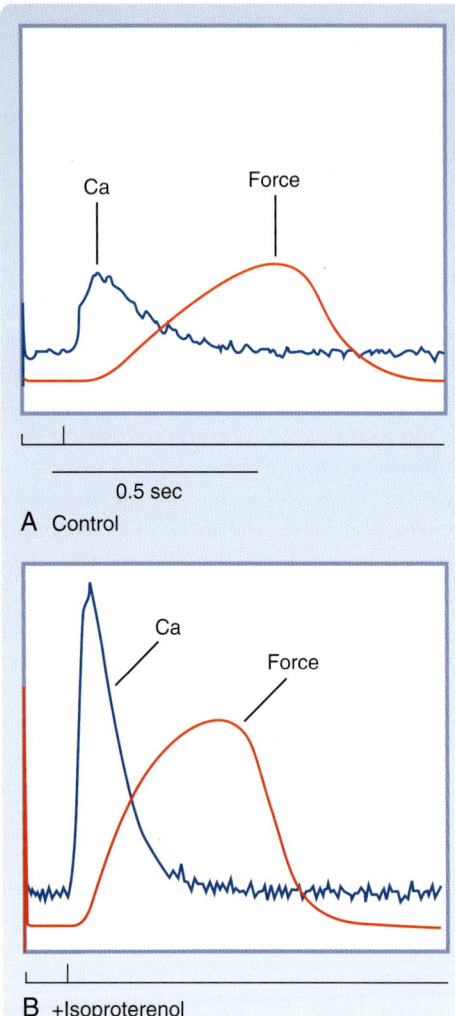

• **Fig. 13.3** Stimulation of β-adrenergic receptors in the heart increases the force of contraction. Electrical stimulation of myocardium results in a transient rise in intracellular [Ca^{++}] and production of force **(A)**. Isoproterenol (a β-adrenergic receptor agonist) increases the amplitude of the intracellular Ca^{++} transient and hence the amount of force generated **(B)**.

of SERCA, thereby allowing the SR to accumulate more Ca^{++} before it is extruded by the 3Na^{+}-Ca^{++} antiporter and the sarcolemmal Ca^{++} pump. The net result is that the SR releases more Ca^{++} into the cytosol during the next action potential, which promotes more actin-myosin interactions and hence greater force of contraction (see Fig. 13.3). The increased activity of SERCA after sympathetic stimulation also results in a shortened contraction because of the rapid reaccumulation of Ca^{++} by the SR. This in turn allows the heart to increase its rate of relaxation. An additional consequence of sympathetic stimulation is an increase in heart rate through a direct effect on the pacemaker cells (see Chapter 18).

Additional proteins and some micropeptides also appear to be associated with SERCA and influence SR calcium transport. This includes the 34–amino acid peptide **dwarf open reading frame (DWORF),** which increases SERCA

calcium affinity (Fig. 13.5), apparently by displacing phospholamban. DWORF was identified in putative non-coding RNA.

AT THE CELLULAR LEVEL

The mechanisms underlying the response of the heart to β-adrenergic stimulation are complex and involve cAMP-dependent phosphorylation of several proteins. An **A kinase anchoring protein (AKAP)** has been shown to be closely associated with the L-type calcium channel in the heart, thereby positioning **cAMP-dependent protein kinase** close to the channel and facilitating cAMP-dependent phosphorylation of this channel during sympathetic stimulation. How these cAMP-dependent phosphorylations increase the amplitude of the intracellular Ca^{++} transient and, in so doing, result in a more forceful, briefer cardiac contraction is discussed in general terms later (see also Chapter 18).

AT THE CELLULAR LEVEL

Mutations in the cardiac ryanodine receptor (RYR2) have been associated with cardiac arrhythmias. Specifically, catecholaminergic polymorphic ventricular tachycardia (CPVT) is an inherited autosomal dominant disease that is typically manifested during childhood as an exercise-induced tachycardia that can progress to arrhythmias during exercise (or stress) and result in sudden death. Approximately 40% of patients with CPVT exhibit a defect in RYR2 that has been associated with increased release of Ca^{++} from the SR. The mutation in RYR2 may involve substitution of a highly conserved amino acid, which differs from malignant hyperthermia, in which splicing errors or deletions within RYR have been reported. It is hypothesized that during periods of exercise or stress, increased levels of intracellular Ca^{++} (because of the combined effects of β-adrenergic stimulation and increased activity of the mutated RYR2) promote the development of delayed afterdepolarizations and hence arrhythmias. Elevation of intracellular [Ca^{++}] during diastole is thought to promote the development of delayed afterdepolarizations through activation of the 3Na^{+}-Ca^{++} antiporter, wherein Ca^{++} extrusion during diastole results in a net inward current sufficient to depolarize the cell to the threshold for an action potential. Treatment of CPVT involves antiadrenergic therapy (with β-adrenergic antagonists) or (for unresponsive patients) an implanted defibrillator.

Stretch

Stretching of the heart increases the force of contraction both in vivo and in vitro and is an intrinsic mechanism for regulating contractile force. In contrast, skeletal muscle typically exhibits maximal tension at resting length. Stretching of the heart in vivo occurs during times of increased venous return of blood to the heart (e.g., during exercise or when the heart rate is slowed, or both). The **Frank-Starling law of the heart** refers to this ability of the heart to increase its force of contraction when stretched, which occurs at

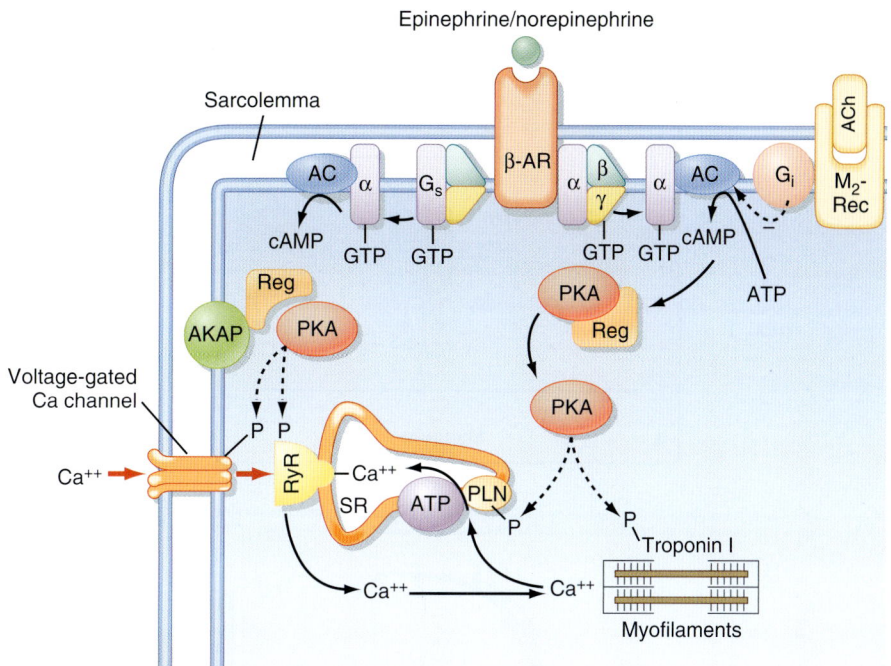

Epinephrine/norepinephrine

• **Fig. 13.4** Sympathetic stimulation of the heart results in an increase in cytosolic cyclic adenosine monophosphate (cAMP) and hence phosphorylation of several proteins by protein kinase A (PKA). An A kinase anchor protein (AKAP) adjacent to the L-type calcium channel facilitates phosphorylation of this channel and possibly nearby sarcoplasmic reticulum calcium channels. Other proteins phosphorylated by PKA include phospholamban (PLN) and troponin I. Muscarinic agonists (e.g., acetylcholine [ACh]), on the other hand, inhibit this sympathetic cascade by inhibiting the production of cAMP by adenylate cyclase (AC). β-AR, β-adrenergic receptor; ATP, adenosine triphosphate; G_i, inhibitory G protein; M_2Rec, muscarinic acetylcholine M_2 receptor; Reg, regenerating gene receptor. (Redrawn from Bers DM. Cardiac excitation-contraction coupling. *Nature* 2002;415:198.)

times of increased venous return (Fig. 13.6*A*; also see Chapter 16).

The importance of this mechanism is that it helps the heart pump whatever volume of blood it receives. Thus when the heart receives a lot of blood, the ventricles are stretched, and the force of contraction is increased, which ensures ejection of this extra volume of blood. Stretching of cardiac muscle also increases passive tension, which helps prevent overstretching of the heart. This passive resistance in the heart is greater than that in skeletal muscle and is attributed to both extracellular matrix (connective tissue) and intracellular elastic proteins (**titin**).

This stretch-induced increase in force of contraction of cardiac muscle occurs over a narrow range of sarcomere lengths (ca. 1.6-2.3 um), resulting in a steep length-dependent activation of contraction. This ascending limb of the length-tension relationship in cardiac muscle is much steeper than that seen in skeletal muscle. It is important to note that this stretch-induced increase in force can occur within a singe heart beat.

The mechanism(s) underlying this stretch-induced increase in force of contraction of cardiac muscle is controversial, but appears to involve changes in the overlap of the thick and thin filaments as well as stretch-induced increase in the Ca^{++} sensitivity of contraction (Fig. 13.6B). In rat ventricular trabecular muscle, approximately 60%

of the stretch-induced increase in force of contraction has been attributed to an increase in Ca^{++} sensitivity, whereas the remaining 40% of the stretch-induced increase in force of contraction has been attributed to changes in the overlap of the thick and thin filaments. The changes in myofilament overlap, however, are less likely to contribute to the continued increase in force of cardiac contraction as sarcomere length increases from 2.0um to 2.3 um, as this region is thought to represent a region of optimal overlap of myofilaments (and represents a plateau in the length tension relationship in skeletal muscle).

The mechanism(s) contributing to the stretch-induced increase in Ca^{++} sensitivity of cardiac contraction are not clear, but appears to involve the intracellular elastic protein titin, as well as regulatory proteins (e.g., troponinC).

Cardiac Muscle Metabolism

As in skeletal muscle, myosin uses the energy in ATP to generate force, so the ATP pool, which is small, must be continually replenished. Typically, this replenishment of ATP pools is accomplished by aerobic metabolism, including the oxidation of fats and carbohydrates. During times of ischemia, the **creatine phosphate** pool, which converts ADP to ATP, may decrease. As in skeletal muscle, the creatine phosphate pool is small.

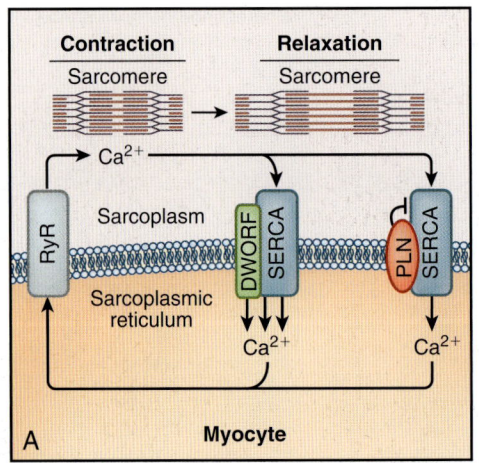

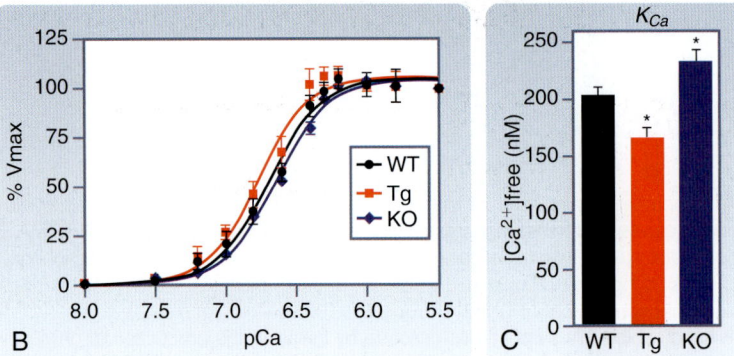

• **Fig. 13.5** A small (34–amino acid) peptide encoded by a long noncoding RNA appears to improve sarcoplasmic reticulum function and myocyte performance by countering the inhibitory effect of phospholamban (PLN) (as depicted in the working model shown in panel A). The ability of DWORF to increase cardiac SR Ca uptake is shown in panels B and C, where overexpression of DWORF in cardiomyocytes (labeled Tg) increased the Ca sensitivity of SR Ca transport, whereas knockout of DWORF (labeled KO) had the reverse effect. DWORF, dwarf open reading frame; KO, knockout; SERCA, sarcoplasmic endoplasmic reticulum calcium-ATPase; Tg, transgenic; WT, wild-type. (From Nelson BR, Makarewich CA, Anderson DM, et al. A peptide encoded by a transcript annotated as long noncoding RNA enhances SERCA activity in muscle. *Science.* 2016;351:271-275.)

When cardiac muscle is completely deprived of O_2 because of occlusion of a coronary vessel (i.e., stopped-flow ischemia), contractions quickly cease (within 30 seconds). This is not due to depletion of either ATP or creatine phosphate because these levels decline more slowly. Even after 10 minutes of stopped-flow ischemia, when creatine phosphate levels are near zero and only 20% of the ATP remains, reperfusion can restore these energy stores, as well as contractile ability. However, prolonging the stopped-flow ischemia for 20 minutes results in further drops in ATP, so that reperfusion has considerably less effect, with only limited restoration of ATP and creatine phosphate levels or contractile activity.

Cardiac Muscle Hypertrophy

Exercise such as endurance running can increase the size of the heart as a result of hypertrophy of individual cardiac muscle cells. Concomitant with this enlarged so-called athlete's heart is improved cardiac performance, as assessed by an increase in stroke volume, increased oxygen consumption, and preserved relaxation. Thus the athlete's heart represents an example of "physiological hypertrophy," with beneficial contractile effects.

In contrast, if exposed to chronic pressure overload, the heart may undergo either **concentric left ventricular hypertrophy** or **dilated left ventricular hypertrophy,** which causes impairment of function. Details regarding the morphological, functional, and mechanistic differences between these various types of hypertrophy can be found elsewhere in this textbook (see Chapter 18).

Concentric hypertrophy is characterized by thickening of the left ventricular wall and represents a compensatory hypertrophy to the increased load. Dilated hypertrophy is characterized by increased ventricular volume (end-diastolic volume). Both concentric/compensatory left ventricular hypertrophy and dilated left ventricular hypertrophy have been shown to exhibit decreased contractile response to β-adrenergic stimulation, which limits the contractile reserve. In dilated left ventricular hypertrophy, normal

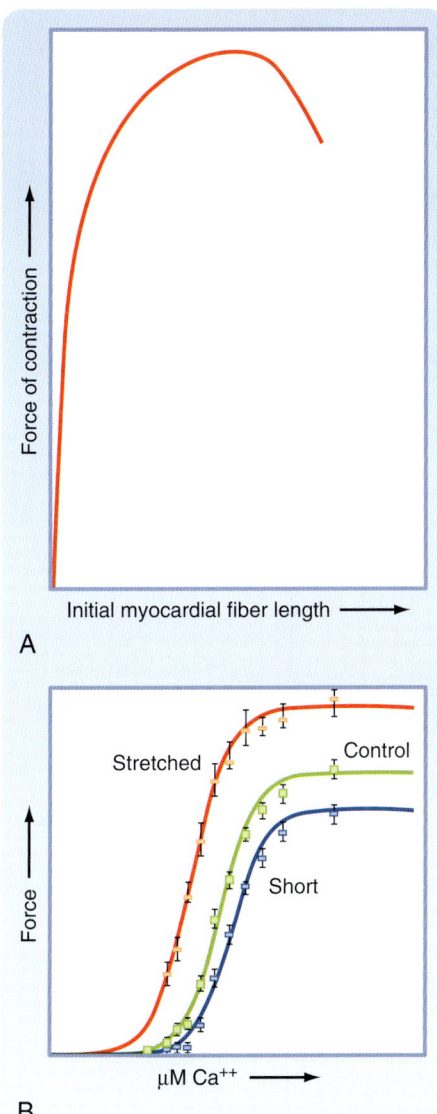

A

B

• **Fig. 13.6** Stretching of the heart increases the force of contraction **(A).** This is attributable to both an increase in the maximal force of contraction and an increase in the sensitivity of contraction to Ca^{++} **(B).** It reflects an intrinsic regulatory process referred to as the *Frank-Starling law of the heart.* (**B**, Redrawn from Dobesh D, Konhilas J, de Tombe P. Cooperative activation in cardiac muscle: impact of sarcomere length. *Am J Physiol Heart Circ Physiol.* 2002;282:H1055-H1062.)

contractile function, along with the Frank-Starling response, may also be impaired.

The cellular and molecular mechanisms underlying the development of cardiac hypertrophy are not clear, although an elevation in intracellular $[Ca^{++}]$ has been implicated.

The link or links between cardiac hypertrophy, decreased cardiac performance, and impaired β-adrenergic response during chronic pressure overload are unclear. Decreased cardiac performance has been attributed to dysregulation of intracellular $[Ca^{++}]$. Alterations in the level, activity, and phosphorylation status of a variety of proteins, including L-type calcium channels, phospholamban, SERCA, and RYR, have all been implicated in the Ca^{++}

dysregulation associated with a failing heart (pathological hypertrophy).

A microRNA (miR-222) has been shown to be important for cardiac growth in response to exercise. It also appeared to inhibit maladaptive remodeling of the heart after ischemia/reperfusion injury.

AT THE CELLULAR LEVEL

A modest elevation in intracellular $[Ca^{++}]$ (as a result of increased contractile activity, for example), has been proposed to activate a Ca^{++}/calmodulin-dependent protein phosphatase **(calcineurin)** that can dephosphorylate the transcription factor **nuclear factor of activated T cells (NFAT),** thereby facilitating translocation of NFAT to the nucleus and ultimately promoting protein synthesis and thus hypertrophy (Fig. 13.7). Activation of Ca^{++}/calmodulin-dependent protein kinase has also been implicated in activation of the transcription factor **myocyte enhancer factor 2 (MEF2)** by promoting the dissociation (nuclear export) of an inhibitor of MEF2 (namely, **histone deacetylase [HDAC]).**

The impaired β-adrenergic response of cardiac muscle after chronic pressure overload involves, at least in part, a decrease in β-adrenergic receptors because of internalization. Both **phosphatidylinositol-3-kinase (PI3K)** and **β-adrenergic receptor kinase 1** have been implicated in the internalization of β-adrenergic receptors.

AT THE CELLULAR LEVEL

High blood pressure, defects in heart valves, and ventricular walls weakened as a result of myocardial infarction can all lead to heart failure, a leading cause of death. Heart failure may be seen with thickening of the walls of the ventricle or with dilation (i.e., increased volume) of the ventricles.

Results of studies suggest that dilated cardiomyopathy can be prevented in an animal model by downregulating phospholamban. The mechanism underlying this preventive effect of phospholamban downregulation is thought to involve an increase in Ca^{++} uptake activity in the SR because phospholamban typically inhibits SERCA. Increased activity of SERCA would facilitate relaxation of the heart as a result of rapid Ca^{++} uptake by the SR. In addition, the force of contraction is increased because more Ca^{++} is available for release. Increased Ca^{++} uptake by the SR may also decrease activation of the Ca^{++}-dependent phosphatases that have been implicated in the development of cardiac hypertrophy.

There is evidence that cardiac hypertrophy may not be associated with some functional impairments. Intermittent aortic constrictions, for example, result in decreased β-adrenergic signaling, decreased capillary density, and decreased SERCA2 levels, without evidence of hypertrophy. Activation of PI3K appears to be involved in this response.

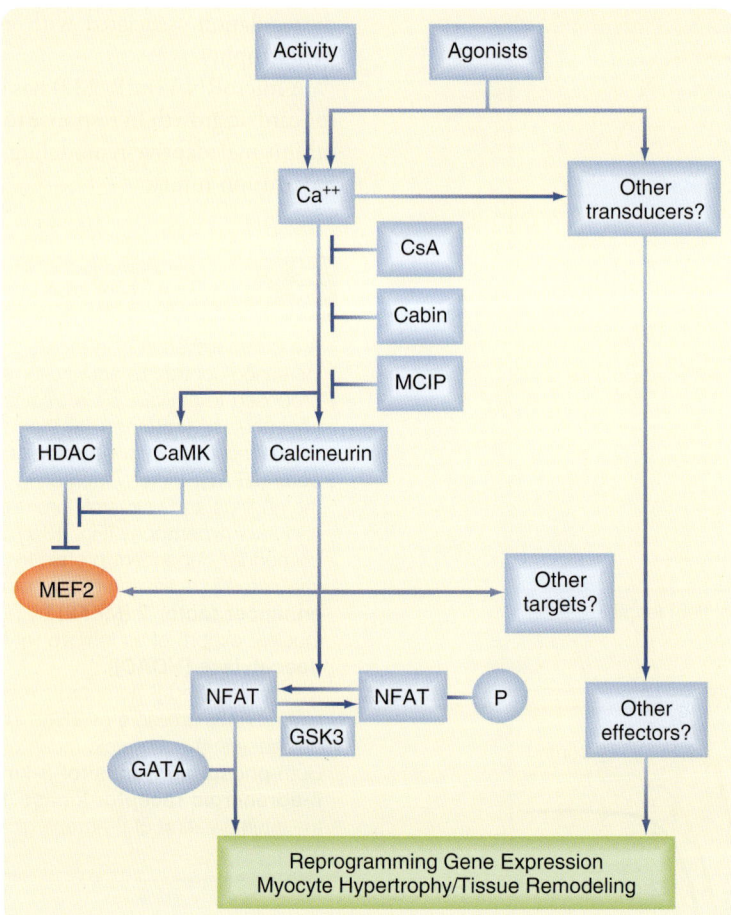

• **Fig. 13.7** Calcium-dependent activation of calcineurin and calmodulin-dependent protein kinase have been implicated in the development of cardiac hypertrophy and involve activation of the following transcription factors: nuclear factor of activated T cells (NFAT), transcription factor binding to DNA sequence GATA (GATA), and myocyte enhancer factor 2 (MEF2). Cabin, calcineurin-binding protein/ inhibitor; CaMK, Ca^{++}/calmodulin-dependent protein kinase; CsA, cyclosporine; GSK3, glycogen synthase kinase 3; HDAC, histone deacetylase; MCIP, modulatory calcineurin-interacting protein. (Redrawn from Olson EN, Williams RS. Calcineurin signaling and muscle remodeling. *Cell.* 2000;101:689-692.)

Key Points

1. Cardiac muscle is an involuntary, striated muscle. Cardiac muscle cells are relatively small (10 μm × 100 μm) and form an electrical syncytium with tight electrical and mechanical connections between adjacent cardiac muscle cells. Action potentials are initiated in the sinoatrial node and spread quickly throughout the heart to allow synchronous contraction, a feature important for the pumping action of the heart.

2. Contraction of cardiac muscle involves the Ca^{++}-dependent interaction of actin and myosin filaments, as in skeletal muscle. However, unlike skeletal muscle, cardiac muscle requires an influx of extracellular Ca^{++}. Specifically, the influx of Ca^{++} during an action potential triggers release of Ca^{++} from the SR, which then promotes actin-myosin interaction and contraction.

3. Relaxation of cardiac muscle involves reaccumulation of Ca^{++} by the SR and extrusion of Ca^{++} from the cell via the $3Na^+$-Ca^{++} antiporter and the sarcolemmal Ca^{++} pump. The Ca^{++} pump of the SR is associated with numerous proteins (forming a regulosome), including some endogenous micropeptide inhibitors and activators.

4. The force of contraction of cardiac muscle is increased by stretch (Frank-Starling Law of the Heart) and by sympathetic stimulation. Skeletal muscle, in contrast, increases force by recruiting more muscle fibers or by tetany.

5. Hypertrophy of the heart can occur in response to exercise, chronic pressure overload, or genetic mutations. The cardiac hypertrophy resulting from exercise is typically beneficial, with improved cardiac performance, increased oxygen consumption, and

normal relaxation. Chronic pressure overload, on the other hand, can result in cardiac hypertrophy that is initially associated with a decreased β-adrenergic response but may progress to dilated cardiac hypertrophy, characterized by decreased contractile ability. Genetic mutations resulting in cardiac hypertrophy include familial hypertrophic cardiomyopathy, in which a mutation in a single intracellular protein may alter contractile function and promote a hypertrophic response. Researchers have identified a microRNA that appears to contribute to the exercise-induced hypertrophy of the heart and to inhibit maladaptide remodeling after ischemia/reperfusion injury.

Additional Readings

Anderson DM, Anderson KM, Chang CL, et al. A micropeptide encoded by a putative long noncoding RNA regulates muscle performance. *Cell.* 2015;160:595-606.

Endoh M. Cardiac Ca^{2+} signaling and Ca^{2+} sensitizers. *Circ J.* 2008;72:1915-1925.

Haghighi K, Bidwell P, Kranias EG. Phospholamban interactome in cardiac contractility and survival: a new vision of an old friend. *J Mol Cell Cardiol.* 2014;77:160-167.

Hidalgo C, Granzier H. Tuning the molecular giant titin through phosphorylation: role in health and disease. *Trends Cardiovasc Med.* 2013;23:165-171.

Kobirumaki-Shimozawa F, Inoue T, Shintani SA, et al. Cardiac thin filament regulation and the Frank–Starling mechanism. *J Physiol Sci.* 2014;64:221-232.

Marks AR. Calcium cycling proteins and heart failure: mechanisms and therapeutics. *J Clin Invest.* 2013;123(1):46-52.

Tao L, Bei Y, Zhang H, Xiao J, Li X. Exercise for the heart: signaling pathways. *Oncotarget.* 2015;6:20773-20784.

Williams GS, Boyman L, Lederer WJ. Mitochondrial calcium and the regulation of metabolism in the heart. *J Mol Cell Cardiol.* 2015;78:35-45.

14

Smooth Muscle

LEARNING OBJECTIVES

Upon completion of this chapter the student should be able to answer the following questions:

1. Describe the organization of smooth muscle in various tissues, and how it meets the demands of each tissue/organ.
2. Discuss the mechanisms that promote contraction and relaxation of smooth muscle in the vasculature and various organs.
3. Describe the autoregulatory mechanism by which an artery can maintain relatively constant blood flow to a tissue over a broad range of perfusion pressures.
4. Describe the basis and utility of a transition from phasic contraction to tonic contraction.
5. Discuss the length-tension curves and force-velocity curves for smooth muscle, and the molecular basis for each of these curves.

If the student has already completed Chapters 12 and 13 on skeletal muscle and cardiac muscle, comparison of all three tissues for each of the learning objectives listed should be possible.

Nonstriated, or smooth, muscle cells are a major component of hollow organs such as the alimentary canal, airways, vasculature, and urogenital tract. Contraction of smooth muscle serves to alter the dimensions of the organ, which may result in either propelling the contents of the organ (as in peristalsis of the intestine) or increasing the resistance to flow (as in vasoconstriction). The basic mechanism underlying contraction of smooth muscle involves an interaction of myosin with actin (as in striated muscle), although there are some important differences. Specifically, contraction of smooth muscle is thick-filament regulated and requires an alteration in myosin before it can interact with actin, whereas contraction of striated muscle is thin-filament regulated and requires movement of the troponin-tropomyosin complex on the actin filament before myosin can bind to actin. Smooth muscle can contract in response to either electrical or hormonal signals and exhibits the ability to remain contracted for extended periods at low levels of energy consumption, which is important for functions such as maintaining vascular tone and hence blood pressure. Thus regulation of contraction of smooth muscle is complex, sometimes involving multiple intracellular signaling cascades. In the present chapter, effort is made

to identify mechanisms underlying this diverse regulation of smooth muscle contraction and, when appropriate, compare these regulatory mechanisms with those observed in striated muscle. Alterations in smooth muscle function/regulation that have been implicated in various pathological conditions are also discussed.

Overview of Smooth Muscle

Types of Smooth Muscle

Smooth muscle has been subdivided into two groups: **single unit** and **multiunit.** In single-unit smooth muscle the smooth muscle cells are electrically coupled such that electrical stimulation of one cell is followed by stimulation of adjacent smooth muscle cells. This results in a wave of contraction, as in peristalsis. Moreover, this wave of electrical activity, and hence contraction, in single-unit smooth muscle may be initiated by a pacemaker cell (i.e., a smooth muscle cell that exhibits spontaneous depolarization). In contrast, multiunit smooth muscle cells are not electrically coupled, so stimulation of one cell does not necessarily result in activation of adjacent smooth muscle cells. Examples of multiunit smooth muscle include the vas deferens of the male genital tract and the iris of the eye. Smooth muscle, however, is even more diverse, with the single-unit and multiunit classifications representing ends of a spectrum. The terms *single-unit* and *multiunit* represent an oversimplification, however; many smooth muscles are modulated by a combination of neural elements with at least some degree of cell-to-cell coupling and locally produced activators or inhibitors that also promote a somewhat coordinated response of smooth muscles.

A second consideration when discussing types of smooth muscle is the activity pattern (Fig. 14.1). In some organs the smooth muscle cells contract rhythmically or intermittently, whereas in other organs the smooth muscle cells are continuously active and maintain a level of "tone." Smooth muscle exhibiting rhythmic or intermittent activity is termed **phasic smooth muscle** and includes smooth muscles in the walls of the gastrointestinal (GI) and urogenital tracts. Such phasic smooth muscle corresponds to the single-unit category described earlier because the smooth muscle cells contract in response to action potentials that propagate from cell to cell. Smooth muscle that is continuously active, on the other hand, is termed **tonic smooth muscle.**

Vascular smooth muscle, respiratory smooth muscle, and some sphincters are continuously active. The continuous partial activation of tonic smooth muscle is not associated with action potentials, although it is proportional to membrane potential. Tonic smooth muscle would thus correspond to the multiunit smooth muscle described earlier. Phasic and tonic contractions of smooth muscle result from interactions of actin and myosin filaments, although as discussed later in this chapter, there is a change in cross-bridge cycling kinetics during tonic contraction, such that the smooth muscle can maintain force at low energy cost.

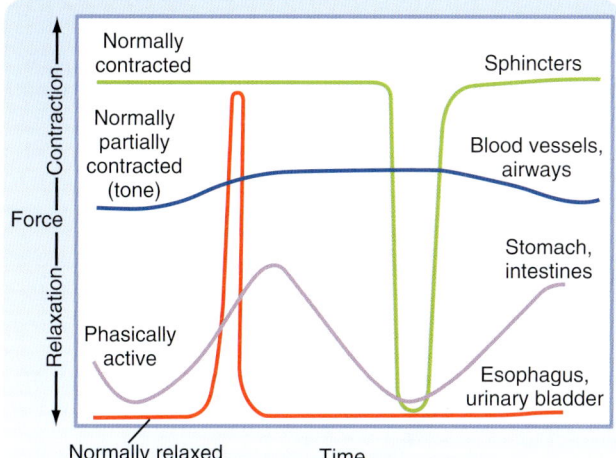

• **Fig. 14.1** Some contractile activity patterns exhibited by smooth muscles. Tonic smooth muscles are normally contracted and generate a variable steady-state force. Examples are sphincters, blood vessels, and airways. Phasic smooth muscles commonly exhibit rhythmic contractions (e.g., peristalsis in the GI tract) but may contract intermittently during physiological activities under voluntary control (e.g., voiding of urine from urinary bladder, swallowing).

Structure of Smooth Muscle Cells

Smooth muscle cells typically form layers around hollow organs (Fig. 14.2). Blood vessels and airways exhibit a simple tubular structure in which the smooth muscle cells are arranged circumferentially, so contraction reduces the diameter of the tube. This contraction increases resistance to the flow of blood or air but has little effect on the length of the organ. Smooth muscle cell organization is more complex in the GI tract. Layers of smooth muscle in both circumferential and longitudinal orientations provide the mechanical action for mixing food and also propelling the luminal contents from the mouth to the anus. Coordination between these layers depends on a complex system of autonomic nerves linked by plexuses. These plexuses are located between the two muscle layers. The smooth muscle in the walls of saccular structures such as the urinary bladder or rectum allows the organ to increase in size with accumulation of urine or feces. The varied arrangement of cells in the walls of these organs contributes to their ability to reduce internal volume to almost zero during urination or defecation. Smooth muscle cells in hollow organs occur in a spectrum of forms, depending on their function and mechanical loads.

In all hollow organs the smooth muscle is separated from the contents of the organ by other cellular elements, which may be as simple as vascular endothelium or as complex as the mucosa of the digestive tract. The walls of hollow organs also contain large amounts of connective tissue that bear an increasing share of the wall stress as organ volume increases.

The following sections describe the structural components that enable smooth muscle to set or alter hollow organ volume. These components include contractile and regulatory proteins, force-transmitting systems such as the

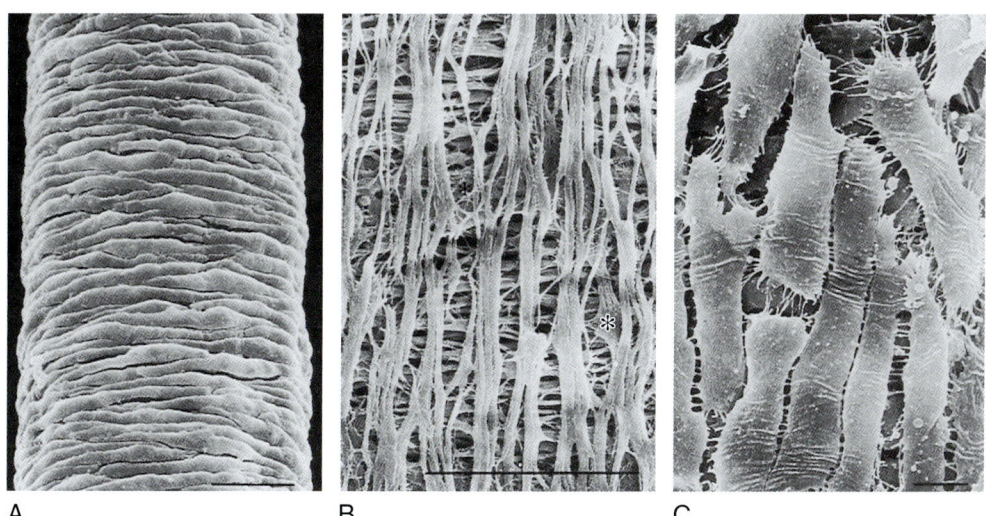

A   B   C

• **Fig. 14.2** Scanning electron micrographs of smooth muscle. A, Muscular arteriole with fusiform smooth muscle cells in a circular orientation (*bar,* 20 µm). B, Superimposed images of circular *(below)* and longitudinal *(above)* layers of intestinal smooth muscle sandwiching neural components of the myenteric plexus *(asterisk)* (*bar,* 50 mm). C, Rectangular smooth muscle cells with thin projections to adjacent cells in a small testicular duct (*bar,* 5 µm). (From Motta PM [ed]. *Ultrastructure of Smooth Muscle.* Norwell, MA: Kluwer Academic; 1990.)

cytoskeleton, linkages between cells and the extracellular matrix, and membrane systems that transduce extracellular signals into changes in myoplasmic [Ca++].

Cell-to-Cell Contact

A variety of specialized contacts exists between smooth muscle cells. Such contacts allow mechanical linkage and communication between cells. In contrast to skeletal muscle cells, which are normally attached at either end to a tendon, smooth (and cardiac) muscle cells are connected to each other. Because smooth muscle cells are anatomically arranged in series, they not only must be mechanically linked but must also be activated simultaneously and to the same degree. This mechanical and functional linkage is crucial to smooth muscle function. If such linkage did

not exist, contraction in one region would simply stretch another region without a substantial decrease in radius or increase in pressure. The mechanical connections are provided by attachments to sheaths of connective tissue and by specific junctions between muscle cells.

Several types of junctions are found in smooth muscle. Functional linkage of the cells is provided by **gap junctions** (Fig. 14.3A). Gap junctions form low-resistance pathways between cells (see Chapter 2). They also allow chemical communication by diffusion of low-molecular-weight compounds. In certain tissues, such as the outer longitudinal layer of smooth muscle in the intestine, large numbers of such junctions exist. Action potentials are readily propagated from cell to cell through such tissues.

Adherens junctions (also called *dense plaques* or *attachment plaques*) provide mechanical linkage between smooth

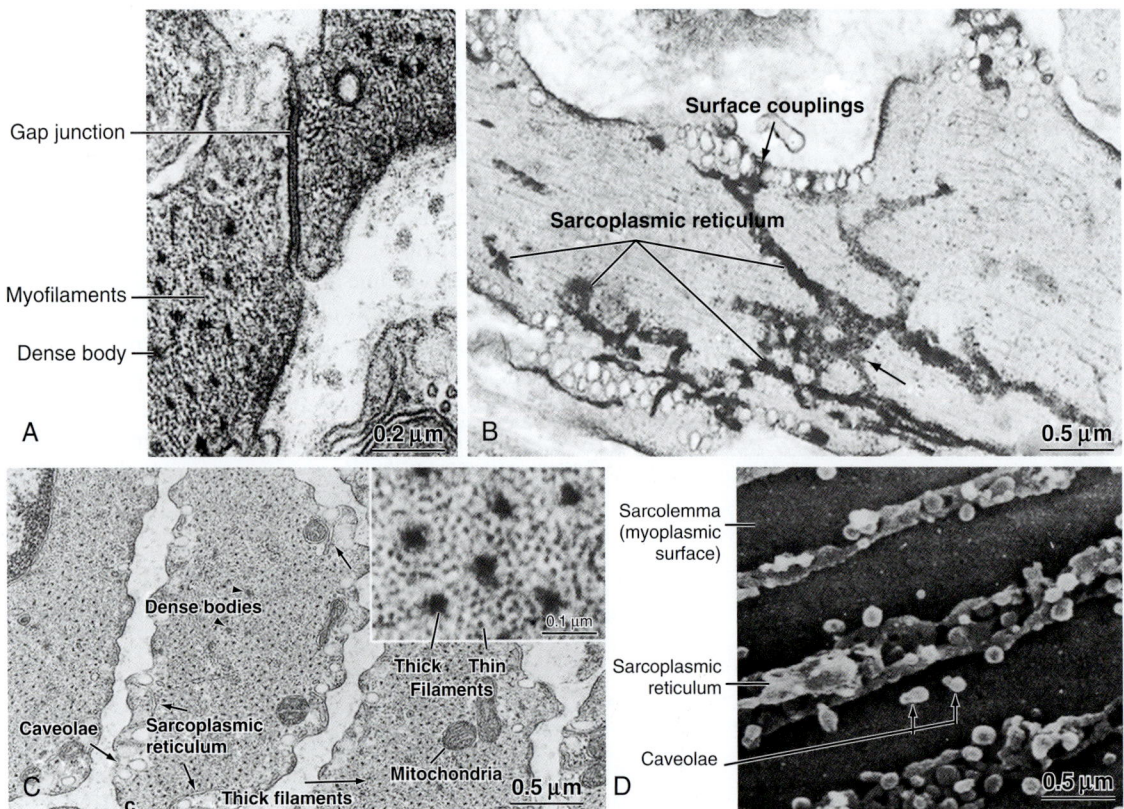

• **Fig. 14.3** Junctions, membranes, and myofilaments, in smooth muscle. A, Transmission electron micrograph of Gap junction between intestinal smooth muscle cells. B, Longitudinal view of a pulmonary artery smooth muscle cell. The sarcoplasmic reticulum is stained with osmium ferricyanide and appears to form a continuous network throughout the cell consisting of tubules, fenestrated sheets (*long arrows*), and surface couplings at the cell membrane (*short arrows*). C, Transverse section of a bundle of venous smooth muscle cells illustrating the regular spacing of thick filaments (*long line*) and the relatively large number of surrounding thin (actin) filaments (*inset*). Dense bodies (*arrowheads*) are sites of attachment for the thin actin filaments and equivalent to the Z lines of striated muscles. Elements of sarcoplasmic reticulum (*short line*) occur at the periphery of these cells. D, Scanning electron micrograph of the inner surface of the sarcolemma of an intestinal smooth muscle cell. Longitudinal rows of caveolae project into the myoplasm (*small, light-colored spheres*), surrounded by darker elements of the tubular sarcoplasmic reticulum. The attachments of thin filaments to the sarcolemma between the rows of membrane elements were removed during preparation of the specimen. (A and D, From Motta PM [ed]. *Ultrastructure of Smooth Muscle*. Norwell, MA: Kluwer Academic; 1990. B and C, From Somlyo AP, Somlyo AV. Smooth muscle structure and function. In: Fozzard HA et al. [eds]. *The Heart and Cardiovascular System*. 2nd ed. New York: Raven Press; 1992.)

muscle cells. The adherens junction appears as thickened regions of opposing cell membranes that are separated by a small gap ($\approx$60 nm) containing dense granular material. Thin filaments extend into the adherens junction to allow the contractile force generated in one smooth muscle cell to be transmitted to adjacent smooth muscle cells.

Cells and Membranes

Embryonic smooth muscle cells do not fuse, and each differentiated cell has a single centrally located nucleus. Though dwarfed by skeletal muscle cells, smooth muscle cells are nevertheless quite large (typically 40–600 μm long). These cells are 2 to 10 μm in diameter in the region of the nucleus, and most taper toward their ends. Contracting cells become quite distorted as a result of the force exerted on the cell by attachments to other cells or to the extracellular matrix, and cross sections of these cells are often very irregular.

Smooth muscle cells lack T tubules, the invaginations of the skeletal muscle sarcolemma that provide electrical links to the sarcoplasmic reticulum (SR). However, the sarcolemma of smooth muscle has longitudinal rows of tiny sac-like inpocketings called *caveolae* (see Fig. 14.3B–D). Caveolae increase the surface-to-volume ratio of the cells and are often closely apposed to the underlying SR. A gap of approximately 15 nm has been observed between the caveolae and the underlying SR, comparable to the gap between the T tubules and terminal SR in skeletal muscle. Moreover, "Ca^{++} sparks" and a variety of Ca^{++}-handling proteins have been observed in the vicinity of caveolae, thus raising the possibility that the caveolae and the underlying SR may contribute to regulation of intracellular [Ca^{++}] in smooth muscle. The voltage-gated L-type Ca^{++} channel and the 3Na$^+$-1Ca^{++} antiporter, for example, are associated with caveolae. The proteins caveolin and cholesterol are both critical for the formation of caveolae, and it is hypothesized that the caveolae reflect a specialized region of the sarcolemma that may also contain various signaling molecules in addition to the Ca^{++} signaling mentioned earlier.

Smooth muscle also has an intracellular membrane network of SR that serves as an intracellular reservoir for Ca^{++} (see Fig. 14.3B–D). Calcium can be released from the SR into the myoplasm when stimulatory neurotransmitters, hormones, or drugs bind to receptors on the sarcolemma. Importantly, intracellular Ca^{++} channels in the SR of smooth muscle include the **ryanodine receptor (RYR),** which is similar to that found in skeletal muscle SR, and the **inositol 1,4,5-trisphosphate (InsP3)**-gated Ca^{++} channel. The RYR is typically activated by a rise in intracellular [Ca^{++}] (i.e., Ca^{++}-induced release of Ca^{++} in response to an influx of Ca^{++} through the sarcolemma). The InsP3-gated Ca^{++} channel is activated by InsP3, which is produced when a hormone or hormones bind to various Ca^{++}-mobilizing receptors on the sarcolemma. Intracellular [Ca^{++}] is lowered through the action of an **SR Ca^{++}-ATPase (SERCA)** and extrusion of Ca^{++} from the cell via a 3Na$^+$-1Ca^{++} antiporter and a sarcolemmal Ca^{++}-ATPase. The amount of SR in smooth muscle cells varies from 2% to 6% of cell volume and approximates that of skeletal muscle. Chemical signals such as InsP3 or a localized increase in intracellular [Ca^{++}] (e.g., within the gap between the caveolae and SR) functionally link the sarcolemma and the SR.

Smooth muscle cells contain a prominent rough endoplasmic reticulum and Golgi apparatus, which are located centrally at each end of the nucleus. These structures reflect significant protein synthetic and secretory functions. The scattered mitochondria are sufficient for oxidative phosphorylation to generate the increased adenosine triphosphate (ATP) consumed during contraction.

Contractile Apparatus

The thick and thin filaments of smooth muscle cells are about 10,000 times longer than their diameter and are tightly packed. Therefore the probability of observing an intact filament by electron microscopy is extremely low. In contrast to skeletal muscle, which contains a transverse alignment of thick and thin filaments that results in striations, the contractile filaments in smooth muscle are not in uniform transverse alignment, and thus smooth muscle has no striations. The lack of striations in smooth muscle does not imply a lack of order. The thick and thin filaments are organized in contractile units that are analogous to sarcomeres.

The thin filaments of smooth muscle have an actin and tropomyosin composition and structure similar to that in skeletal muscle. However, the cellular content of actin and tropomyosin in smooth muscle is about twice that of striated muscle. Smooth muscle lacks troponin and nebulin but contains two proteins not found in striated muscle: **caldesmon** and **calponin.** The precise roles of these proteins are unknown, but they do not appear to be fundamental to cross-bridge cycling. It has been suggested that both calponin and caldesmon may regulate the contractility of smooth muscle. (in part by inhibiting actomyosin ATPase activity). Most of the myoplasm is filled with thin filaments that are roughly aligned along the long axis of the cell. The myosin content of smooth muscle is only a fourth that of striated muscle. Small groups of three to five thick filaments are aligned and surrounded by many thin filaments. These groups of thick filaments with interdigitating thin filaments are connected to **dense bodies** or **areas** (Fig. 14.4; also see Fig. 14.3A–B) and represent the equivalent of the sarcomere. The contractile apparatus of adjacent cells is mechanically coupled by the links between membrane-dense areas.

Cytoskeleton

The cytoskeleton in smooth muscle cells serves as an attachment point for the thin filaments and permits transmission of force to the ends of the cell. In contrast to skeletal muscle, the contractile apparatus in smooth muscle is not organized into myofibrils, and Z lines are lacking. The functional equivalents of the Z lines in smooth muscle cells

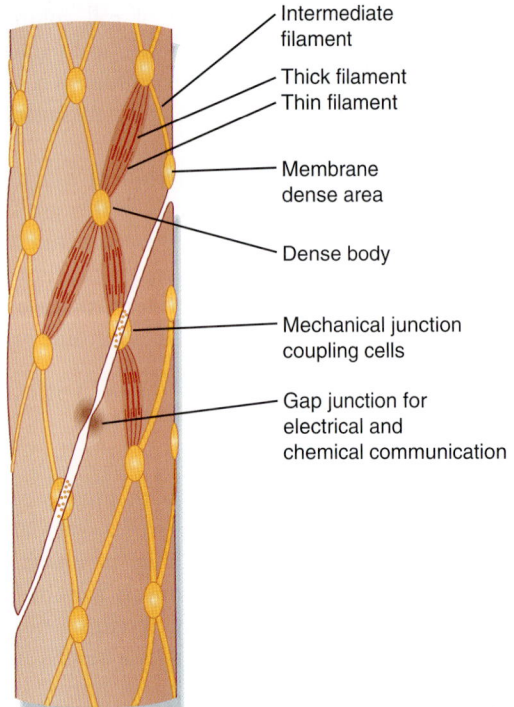

- Intermediate filament
- Thick filament
- Thin filament
- Membrane dense area
- Dense body
- Mechanical junction coupling cells
- Gap junction for electrical and chemical communication

• **Fig. 14.4** Apparent organization of cell-to-cell contacts, cytoskeleton, and myofilaments in smooth muscle cells. Small contractile elements functionally equivalent to a sarcomere underlie the similarities in mechanics between smooth and skeletal muscle. Linkages consisting of specialized junctions or interstitial fibrillar material functionally couple the contractile apparatus of adjacent cells. Dense bodies, the functional equivalent of Z lines in striated muscle, are interconnected by intermediate filaments. The myofilaments are oriented largely in parallel with the longitudinal axis of the cell, though an oblique orientation has been observed in some arteries.

are ellipsoidal dense bodies in the myoplasm and dense areas that form bands along the sarcolemma (see Figs. 14.3A–B and 14.4). These structures serve as attachment points for the thin filaments and contain α-actinin, a protein also found in the Z lines of striated muscle. Intermediate filaments with diameters between those of thin filaments (7 nm) and thick filaments (15 nm) are prominent in smooth muscle. These filaments link the dense bodies and areas into a cytoskeletal network. The intermediate filaments consist of protein polymers of **desmin** or **vimentin.**

Control of Smooth Muscle Activity

The contractile activity of smooth muscle can be controlled by numerous factors, including hormones, autonomic nerves, pacemaker activity, and a variety of drugs. Like skeletal or cardiac muscle, contraction of smooth muscle is dependent on Ca^{++}, and the agents just listed induce smooth muscle contraction by increasing intracellular $[Ca^{++}]$. However, in contrast to skeletal or cardiac muscle, action potentials in smooth muscle are highly variable and not always needed to initiate contraction. Moreover, several agents can increase intracellular $[Ca^{++}]$ and hence contract smooth muscle without changing the membrane potential.

Fig. 14.5 shows various types of action potentials in smooth muscle and the corresponding changes in force. An action potential in smooth muscle can be associated with a slow twitch-like response, and the twitch forces can summate during periods of repetitive action potentials (i.e., similar to tetany in skeletal muscle). Such a pattern of activity is characteristic of single-unit smooth muscle in many viscera.

Periodic oscillations in membrane potential can occur as a result of changes in the activity of Na^+,K^+-ATPase in the sarcolemma. These oscillations in membrane potential can trigger multiple action potentials in the cell. Alternatively the contractile activity of smooth muscle may not be associated with generation of action potentials or even a change in membrane potential. In many smooth muscles the resting membrane potential is sufficiently depolarized (−60 to −40 mV) that a small decrease in membrane potential can significantly inhibit influx of Ca^{++} through voltage-gated Ca^{++} channels in the sarcolemma. By decreasing Ca^{++} influx, the force developed by smooth muscle decreases. Such a graded response to slight changes in the resting membrane potential is common in multiunit smooth muscles that maintain constant tension (e.g., vascular smooth muscle).

Contraction of smooth muscle in response to an agent that does not produce a change in membrane potential is termed **pharmacomechanical coupling** and typically reflects the ability of the agent to increase the level of the intracellular second messenger InsP3. Other agents result in a decrease in tension, also without a change in membrane potential. These agents typically increase levels of the intracellular second messengers cyclic guanosine monophosphate (cGMP) or cyclic adenosine monophosphate (cAMP). The molecular mechanisms by which InsP3, cGMP, cAMP, and Ca^{++} alter the contractile force of smooth muscle are presented later.

Phosphorylation of a myosin light chain is required for the interaction of myosin with actin, and although Ca^{++}-dependent phosphorylation plays a key role in this process, the level of myosin phosphorylation (and hence the degree of contraction) is dependent on the relative activities of both **myosin light-chain kinase** (**MLCK,** which promotes phosphorylation) and **myosin phosphatase** (**MP,** which promotes dephosphorylation). Several agonists/hormones increase the level of myosin light-chain phosphorylation by simultaneously activating MLCK through an increase in intracellular $[Ca^{++}]$ and inhibiting MP through a signaling cascade involving the monomeric G protein **RhoA** and its effector **Rho kinase (ROK).** Moreover, hyperactivity of this RhoA/ROK signaling cascade has been implicated in various pathological conditions such as hypertension and vasospasm (discussed later).

Innervation of Smooth Muscle

Neural regulation of smooth muscle contraction depends on the type of innervation and neurotransmitters released, the proximity of the nerves to the muscle cells, and the

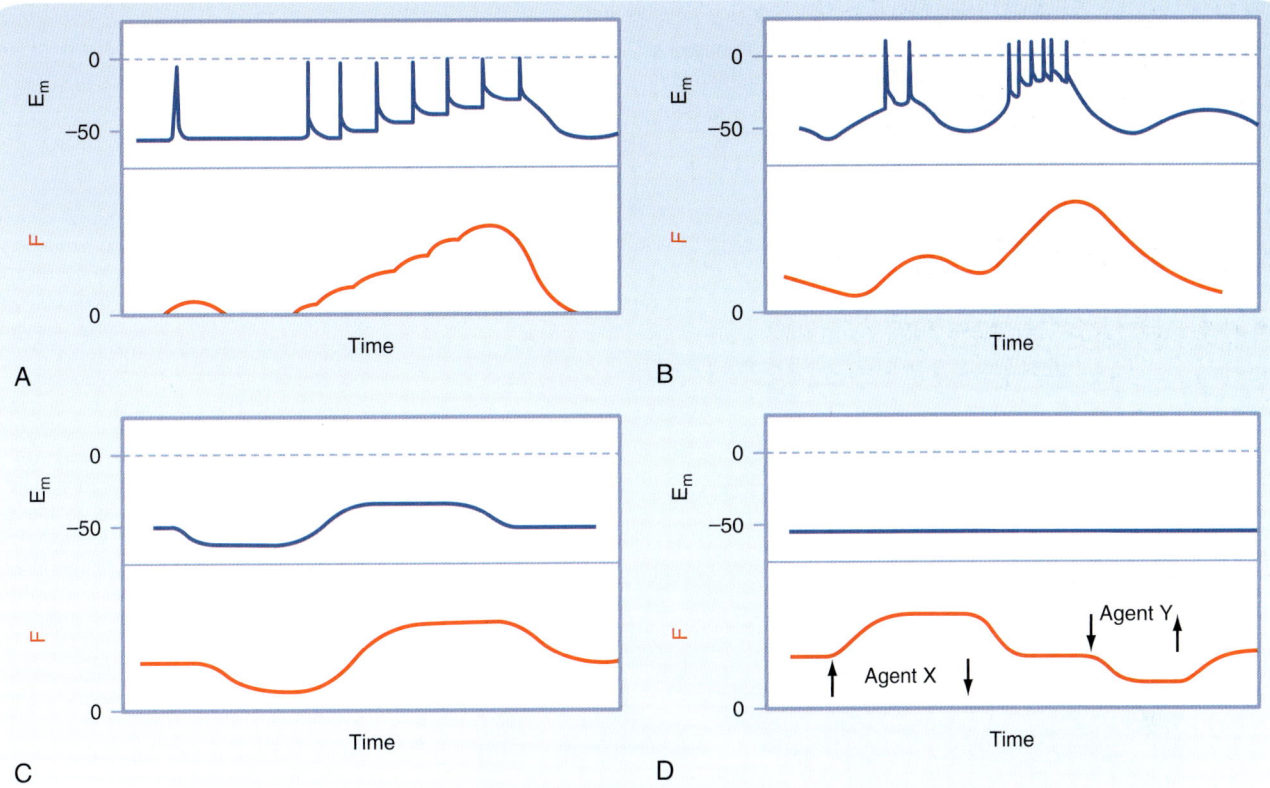

• **Fig. 14.5** Relationships between membrane potential (E_m) and generation of force (F) in different types of smooth muscle. A, Action potentials may be generated and lead to a twitch or larger summed mechanical responses. Action potentials are characteristic of single-unit smooth muscles (many viscera). Gap junctions permit the spread of action potentials throughout the tissue. B, Rhythmic activity produced by slow waves that trigger action potentials. The contractions are generally associated with a burst of action potentials. Slow oscillations in membrane potential usually reflect the activity of electrogenic pumps in the cell membrane. C, Tonic contractile activity may be related to the value of the membrane potential in the absence of action potentials. Graded changes in E_m are common in multiunit smooth muscles (e.g., vascular), where action potentials are not generated and propagated from cell to cell. D, Pharmacomechanical coupling; changes in force produced by the addition or removal (arrows) of drugs or hormones that have no significant effect on membrane potential.

type and distribution of the neurotransmitter receptors on the muscle cell membranes (Fig. 14.6). In general, smooth muscle is innervated by the autonomic nervous system. The smooth muscle in arteries is innervated primarily by sympathetic fibers, whereas the smooth muscle in other tissues can have both sympathetic and parasympathetic innervation. In the GI tract, smooth muscle is innervated by nerve plexuses that make up the enteric nervous system. The smooth muscle cells of some tissues (e.g., uterus) have no innervation.

The neuromuscular junctions and neuromuscular transmission in smooth muscle are functionally comparable to that of skeletal muscle but structurally less complex. The autonomic nerves that supply smooth muscle have a series of swollen areas, or varicosities, that are spaced at intervals along the axon. These varicosities contain vesicles for the neurotransmitter (see Fig. 14.6). The postsynaptic membrane of smooth muscle exhibits little specialization when compared with that of skeletal muscle (see Chapter 6). The synaptic cleft is typically about 80 to 120 nm wide but can

be as narrow as 6 to 20 nm or even greater than 120 nm. In synapses in which a wide synaptic cleft is found, release of neurotransmitter can affect multiple smooth muscle cells. There are a large number of neurotransmitters that affect smooth muscle activity. A partial listing is provided in Table 14.1.

Regulation of Contraction

Contraction of smooth muscle requires phosphorylation of a myosin light chain. Typically this phosphorylation occurs in response to a rise in intracellular [Ca^{++}], either after an action potential or in the presence of a hormone/agonist. As depicted in Fig. 14.7, a rise in intracellular [Ca^{++}] in smooth muscle results in the binding of 4 Ca^{++} ions to the protein calmodulin, and then the Ca^{++}-calmodulin complex activates MLCK, which phosphorylates the regulatory light chain of myosin. This phosphorylation step is critical for the interaction of smooth muscle myosin with actin. In addition to this phosphorylation step in smooth muscle,

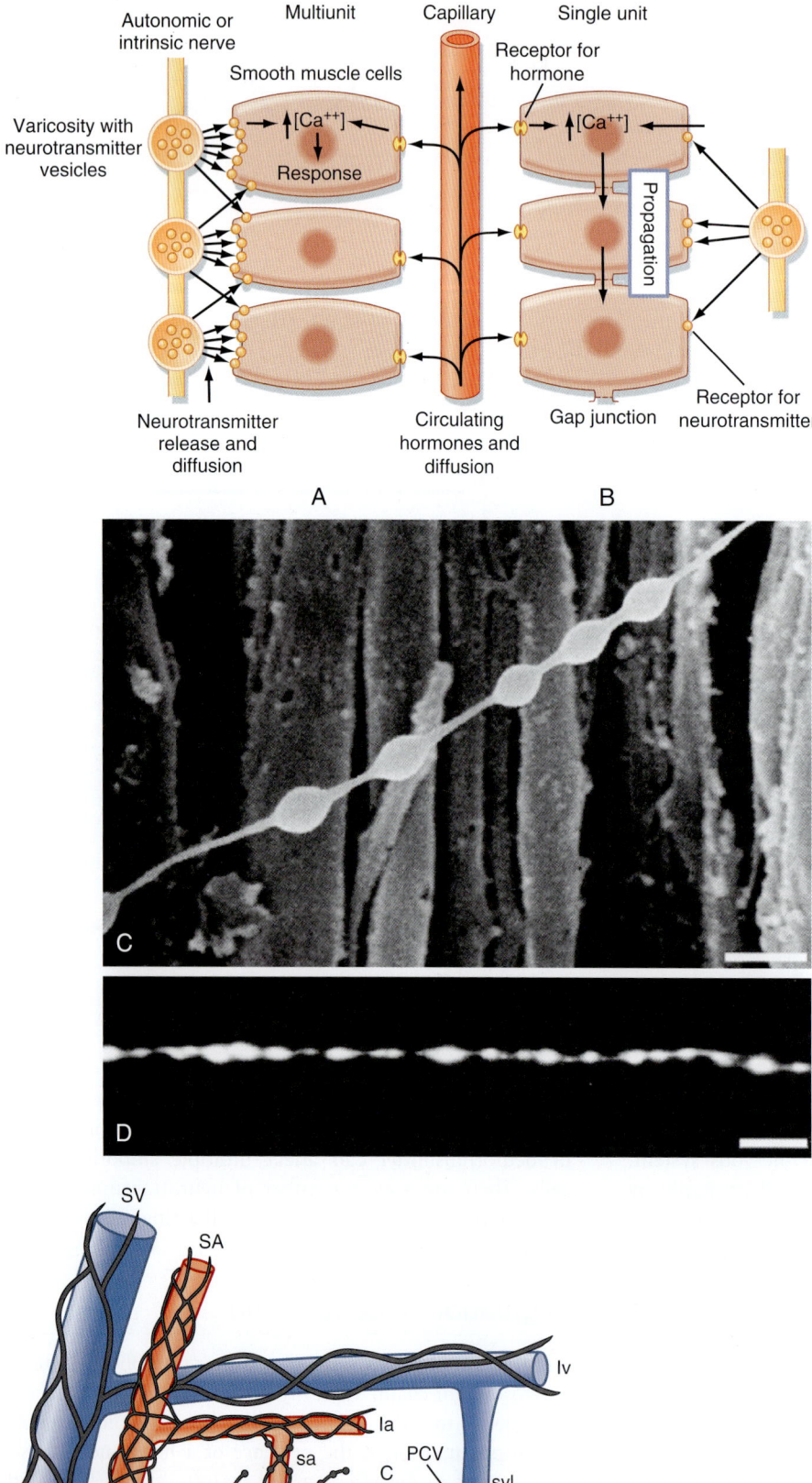

• Fig. 14.6 Control systems of smooth muscle. In both multiunit smooth muscle (A) and single unit smooth muscle (B), neurally released transmitters or circulating or locally generated hormones or signaling molecules can induce contraction or relaxation of the smooth muscle. The combination of a neurotransmitter, hormone, or drug with specific receptors activates contraction by increasing cell Ca^{++}. The response of the cells depends on the concentration of the transmitters or hormones at the cell membrane and the nature of the receptors present. Hormone concentrations depend on diffusion distance, release, reuptake, and catabolism. Consequently, cells lacking close neuromuscular contacts will have a limited response to neural activity unless they are electrically coupled so that depolarization is transmitted from cell to cell. A, Multiunit smooth muscles resemble striated muscles in that there is no electrical coupling and neural regulation is important. B, Single-unit smooth muscles are like cardiac muscle, and electrical activity is propagated throughout the tissue. Most smooth muscles probably lie between the two ends of the single unit–multiunit spectrum. C, Scanning micrograph of varicose nerve lying over rat small intestine scale bar = 3 μm. D, Fluorescent image of catecholamines in single adrenergic axon in guinea pig mesentery. Scale bar = 10 μm. E, Depiction of distribution of sympathetic nerves *(in black)* in vasculature. Sympathetic nerves are associated with small arteries (SA), large arterioles (la), small veins (SV), large venules (lv), small arterioles (sa), and terminal arterioles (TA). Capillaries (c), postcapillary venules (pcv), and small venules (svl) appear to lack sympathetic innervation. (C, From Burnstock G: Autonomic neural control mechanisms. With special reference to the airways. In Kaliner MA, Barnes PJ (Eds.), The Airways. Neural Control in Health and Disease, Marcel Dekker, New York (1988), pp. 1–22. D, From Chamley JH, Mark GE, Campbell GR, Burnstock G. (1972) Sympathetic ganglia in culture. I. Neurons. Z. Zellforsch. Mikrosk. Anat. 135, 287-314.)

TABLE 14.1 Modulation of Smooth Muscle Activity by Neurotransmitters, Hormones, and Local Factors			
Agonist	**Response**	**Receptor**	**Second Messenger**
Norepinephrine and epinephrine from sympathetic stimulation	Contraction[a] (predominant)	α_1-AR	InsP3
	Relaxation[b]	β_2-AR	cAMP
Acetylcholine from parasympathetic stimulation	Contraction[c] (direct)	Muscarinic receptor on SMC	
	Relaxation[c] (indirect)	Muscarinic receptor on EC	
Angiotensin II	Contraction[d]	AT-II receptor	InsP3
Vasopressin	Contraction[d]	Vasopressin receptor	InsP3
Endothelin	Contraction[d]	Endothelin receptor	InsP3
Adenosine	Relaxation[e]	Adenosine receptor	cAMP

[a]The predominant effect of sympathetic stimulation is smooth muscle contraction caused by the abundance of α_1-AR relative to β_2-AR in smooth muscle.
[b]Activation of β_2-AR on smooth muscle modulates the degree of smooth muscle contraction during sympathetic stimulation. Therapeutic β_2-AR agonists are important for the relaxation of bronchial smooth muscle during asthmatic attacks.
[c]Vascular smooth muscles are poorly innervated by the parasympathetic system. During vagal stimulation, however, acetylcholine (ACh) can become elevated in the coronary circulation and result in coronary relaxation (mediated by binding of ACh to endothelial cells). Note that this effect of ACh is indirect because binding of ACh to endothelial cells results in release of the smooth muscle relaxant nitric oxide from the endothelial cells. In regions of the coronary circulation with damaged endothelium, binding of ACh to coronary smooth muscle could promote contraction (vasospasm; direct effect).
[d]A variety of hormones can elevate InsP3 in smooth muscle and thereby result in smooth muscle contraction. Such hormones include angiotensin II, vasopressin, and endothelin, along with the neurotransmitters norepinephrine and acetylcholine. As noted above, however, each hormone/transmitter binds to a specific receptor type.
[e]During periods of intense muscular activity, adenosine can be released from the working muscle, diffuse to the neighboring vasculature, and promote vasodilation. Thus, adenosine is acting as a local factor to increase blood flow to a specific region (i.e., working muscle).
AR, adrenergic receptor; EC, endothelial cell; InsP3, inositol 1,4,5-trisphosphate; SMC, smooth muscle cell.

an ATP molecule is also needed to energize the myosin cross-bridge for the development of force.

IN THE CLINIC

The enteric nervous system controls many aspects of GI function, including motility. Some children are born without enteric nerves in the distal portion of the colon. The absence of nerves is caused by mutant genes that disrupt the signals necessary for the embryonic nerves to migrate to the colon. In these children, normal motility of the colon does not occur and severe constipation results. This condition is called **Hirschsprung's disease.** It can be corrected by surgically removing the portion of the colon that does not contain enteric nerves.

Contraction of smooth muscle is thus said to be *thick-filament regulated,* which contrasts with the *thin-filament regulation* of contraction of striated muscle, where binding of Ca^{++} to troponin exposes myosin binding sites on the actin thin filament. The thick-filament regulation is attributable to expression of a distinct myosin isoform in smooth muscle.

The myosin cross-bridge cycle in smooth muscle is similar to that in striated muscle in that after attachment to the actin filament, the cross-bridge undergoes a ratchet action in which the thin filament is pulled toward the center of the thick filament and force is generated. ADP and P_i are released from the myosin head at this time, thereby allowing ATP to bind. ATP decreases the affinity of myosin for actin, which allows release of myosin from actin. Energy from the

newly bound ATP is then used to produce a conformational change in the myosin head (i.e., recocking the head) so that the cross-bridge is ready for another contraction cycle. The cross-bridge cycle continues as long as the myosin cross-bridge remains phosphorylated. Note that although the four basic steps of the cross-bridge cycle appear to be the same for striated and smooth muscle, the kinetics of cross-bridge cycling is much slower for smooth muscle.

Cross-bridge cycling continues with the hydrolysis of 1 ATP molecule per cycle until myoplasmic $[Ca^{++}]$ falls. With the decrease in $[Ca^{++}]$, MLCK becomes inactive, and the cross-bridges are dephosphorylated by MP (see Fig. 14.7).

As indicated in Fig. 14.4, the thin filaments in smooth muscle are attached to dense bodies, and the myosin thick filaments appear to reside between two dense bodies and overlap a portion of the thin filaments, much like the overlap of thick and thin filaments in the sarcomere of striated muscle. A bipolar arrangement of myosin molecules within the thick filament is thought to allow the myosin cross-bridges to pull the actin filaments toward the center of the thick filament, thus contracting the smooth muscle and hence developing force.

From a structural standpoint, smooth muscle myosin is similar to striated muscle myosin in that they both contain a pair of heavy chains and two pairs of light chains. Despite this similarity, they represent different gene products and thus have different amino acid sequences. As noted, smooth muscle myosin, unlike skeletal muscle myosin, is unable to interact with the actin thin filament unless the regulatory light chain of myosin is phosphorylated. Moreover, the thin filament in smooth muscle lacks troponin, which plays a

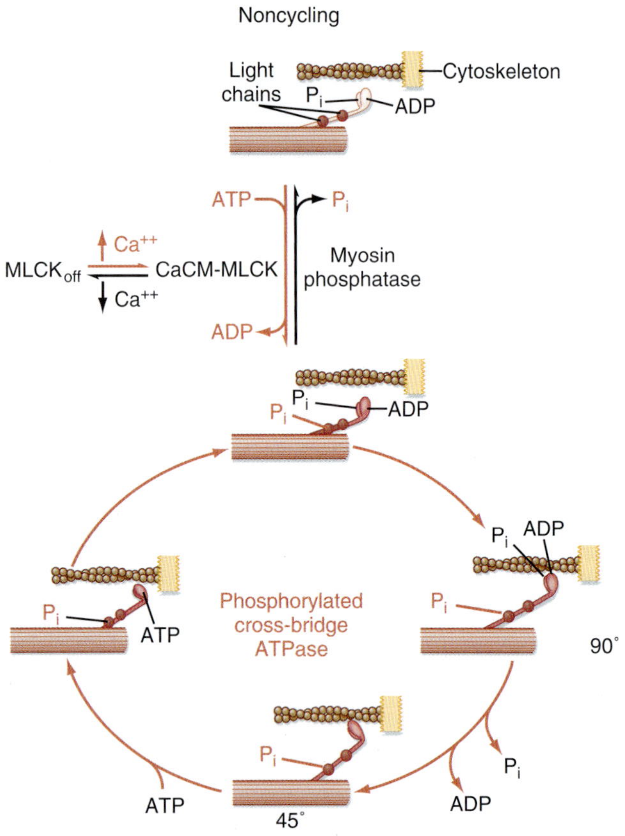

• **Fig. 14.7** Regulation of smooth muscle myosin interactions with actin by Ca^{++}-stimulated phosphorylation. In the relaxed state, cross-bridges are present as a high-energy myosin-ADP-P$_i$ complex in the presence of ATP. Attachment to actin depends on phosphorylation of the cross-bridge by a Ca^{++}-calmodulin–dependent myosin light-chain kinase (MLCK). Phosphorylated cross-bridges cycle until they are dephosphorylated by myosin phosphatase. Note that cross-bridge phosphorylation at a specific site on a myosin regulatory light chain requires ATP in addition to that used in each cyclic interaction with actin.

critical role in the thin-filament regulation of contraction in striated muscle (see Chapter 12).

Although intracellular Ca^{++} is required for smooth muscle contraction, the sensitivity of contraction to Ca^{++} is variable. Several hormones/agonists, for example, increase the force of contraction at a given submaximal intracellular [Ca^{++}], thereby resulting in **Ca^{++} sensitization** (Fig. 14.8). Ca^{++} sensitization is depicted as a leftward shift in the Ca^{++} dependence of smooth muscle contraction (see Fig. 14.8B) and can occur in response to a decrease in the activity of MP activity at a given intracellular [Ca^{++}]. Likewise an increase in MP activity at a given intracellular [Ca^{++}] promotes a rightward shift in the Ca^{++} dependence of smooth muscle contraction, resulting in Ca^{++} desensitization (see Fig. 14.8B). Reciprocal changes in the activities of MP and MLCK can occur, as shown in Fig. 14.8A, where stimulation of the ROK signaling cascade by an agonist can simultaneously inhibit MP and stimulate MLCK at a given intracellular [Ca^{++}], thereby increasing the Ca^{++} sensitivity of contraction (i.e., Ca^{++} sensitization).

AT THE CELLULAR LEVEL

Inhibition of MP underlies the phenomenon of Ca^{++} sensitization that occurs in response to activation of the monomeric G protein RhoA signaling cascade (see Fig. 14.8A). RhoA activates Rho kinase (ROK), which in turn inhibits MP by both direct and indirect mechanisms. Direct inhibition of MP by activated ROK involves ROK phosphorylation of the myosin-binding subunit (MBS) of MP. Indirect inhibition of MP by activated ROK involves phosphorylation of **CPI-17,** an endogenous 17-kDa protein, which then inhibits MP. Hormones/agonists such as catecholamines (acting on α_1-adrenergic receptors), vasopressin, endothelin, angiotensin, and muscarinic agonists increase the sensitivity of smooth muscle contraction to Ca^{++} through activation of RhoA/ROK signaling. ROK can also be activated by arachidonic acid and inhibited by Y-27632, a highly specific inhibitor (see Fig. 14.8A). Though not shown in Fig. 14.8, inactive RhoA is typically located in the cytosol, bound to GDP and an inhibitory protein **(Rho-GDP dissociation inhibitor [GDI]).** Binding of agonist to various G-coupled receptors can activate RhoA by stimulating **guanine nucleotide exchange factor (GEF)** to yield RhoA-GTP, which localizes to the sarcolemma and activates ROK. Conversely, stimulation of MP activity reduces the Ca^{++} sensitivity of contraction of smooth muscle, thereby promoting relaxation (and hence vasodilation).

Hyperactivity of the RhoA/ROK signaling cascade has been implicated in various pathological conditions such as hypertension and vasospasm. Hyperactivity of RhoA/ROK in the vascular smooth muscle of hypertensive animals, for example, was manifested by increased levels of activated RhoA, upregulation of ROK, enhancement of agonist-induced Ca^{++} sensitization of contraction, and a greater reduction in blood pressure by ROK inhibitors as compared with normotensive controls. A similar trend was observed in humans, in that ROK inhibitors decreased forearm vascular resistance in hypertensive patients to a greater extent than in normotensive controls. ROK inhibitors have also been shown to reverse or prevent experimentally induced cerebral vasospasm and coronary vasospasm, as well as the associated upregulation of RhoA/ROK and increased myosin light-chain phosphorylation. Hyperactivity of RhoA/ROK has additionally been implicated in bronchial asthma, erectile dysfunction, and preterm labor, as evidenced by the effects of ROK inhibitors. In addition, ROK inhibitors have decreased vascular smooth muscle proliferation and reduced restenosis after balloon angioplasty in rat carotid artery.

Phasic Versus Tonic Contraction

During a phasic contraction, myoplasmic [Ca^{++}], cross-bridge phosphorylation, and force reach a peak and then return to baseline (Fig. 14.9). In contrast, during a tonic contraction, myoplasmic [Ca^{++}] and cross-bridge phosphorylation decline after an initial spike but do not return to baseline levels. During this later phase, force slowly increases and is sustained at a high level (see Fig. 14.9). This sustained force is maintained with only 20% to 30% of the cross-bridges phosphorylated, and thus ATP utilization is reduced. The term *latch state* refers to this condition of tonic contraction during which force is maintained at low energy expenditure.

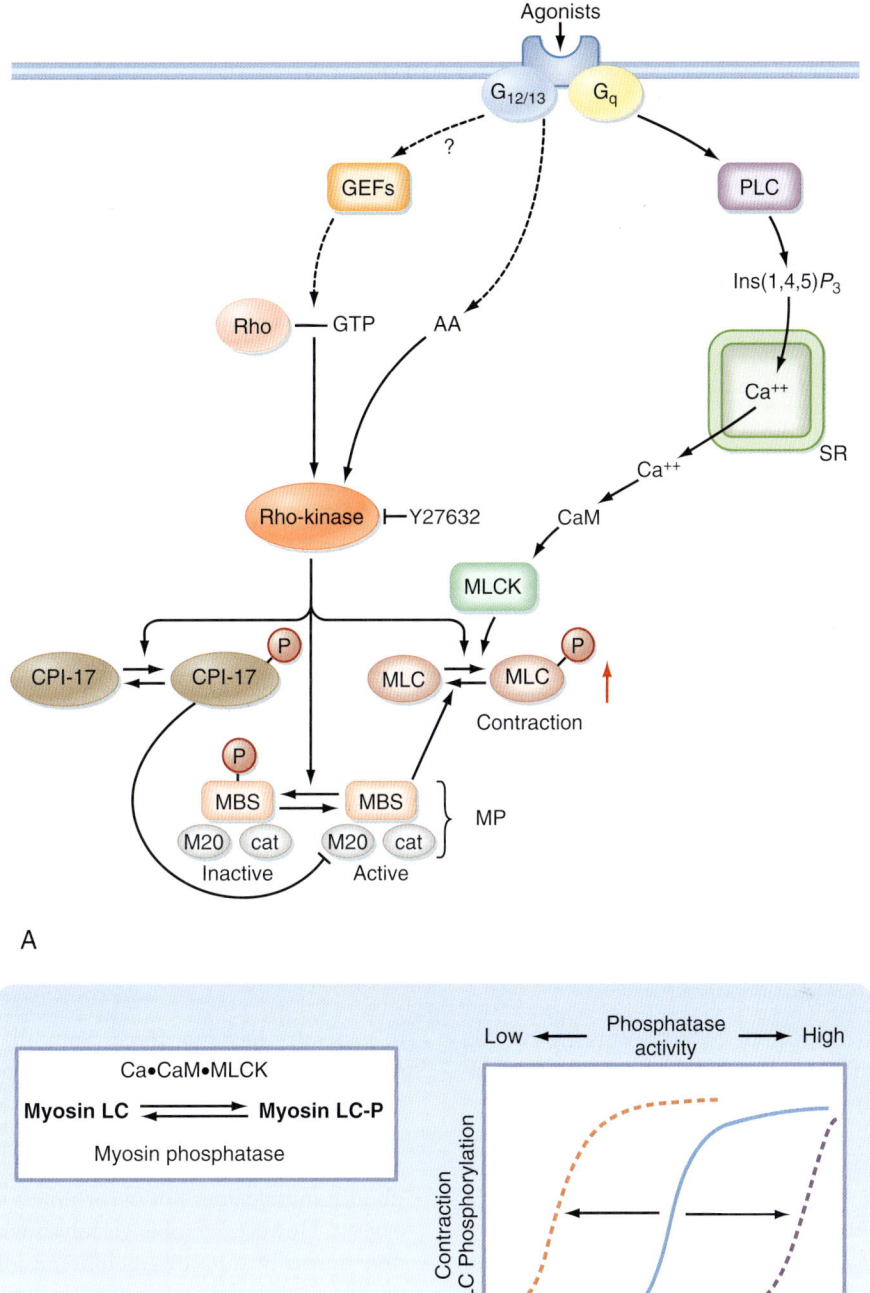

Fig. 14.8 Rho-kinase (ROK) signaling in smooth muscle. A, A variety of agonists of G-coupled receptors simultaneously stimulate InsP3 production and activate RhoA-ROK signaling. InsP3 (Ins[1,4,5]P_3) is produced by phospholipase C (PLC)-mediated hydrolysis of PIP2. InsP3 increases intracellular [Ca^{++}] by opening InsP3-gated Ca^{++} channels in the sarcoplasmic reticulum (SR), thereby resulting in Ca^{++}-calmodulin–dependent activation of myosin light-chain kinase (MLCK) and subsequent phosphorylation of the myosin regulatory light chain (MLC) and promotion of actin-myosin interaction (contraction). Activated RhoA (depicted as Rho-GTP) stimulates ROK, which inhibits myosin phosphatase (MP) by phosphorylating the myosin-binding subunit (MBS) of MP. ROK also inhibits MP indirectly by phosphorylating/activating CPI-17, a 17-kDa inhibitor of MP. The net effect of ROK phosphorylations is a decrease in MP activity, which results in an increased level of myosin light-chain phosphorylation (MLCP) and hence greater force of contraction at a given intracellular [Ca^{++}] (i.e., increased sensitivity of contraction to Ca^{++}). B, Ca^{++} sensitization refers to an increase in the force of contraction at a given intracellular [Ca^{++}] and is depicted as a leftward shift in the Ca^{++} dependence of contraction of smooth muscle. Ca^{++} sensitization can result from a decrease in MP activity and/or an increase in MLCK activity at a given intracellular [Ca^{++}] (as shown in the ROK signaling cascade in panel A). Conversely, Ca^{++} desensitization refers to a decrease in the force of contraction at a given intracellular [Ca^{++}] and is depicted as a rightward shift in the Ca^{++} dependence of contraction. AA, arachidonic acid; CaM, Ca^{++}-calmodulin complex; cat, catalytic subunit of myosin phosphatase; G$_{12/13}$ and G$_q$, heterotrimeric G-proteins; GEF, guanine-nucleotide exchange factor; M20, subunit of myosin phosphatase; MP, myosin phosphatase complex; Y27632, commercial inhibitor of ROK. (From Fukata Y et al. *Trends Pharmacol Sci* 2001;22:32-39.)

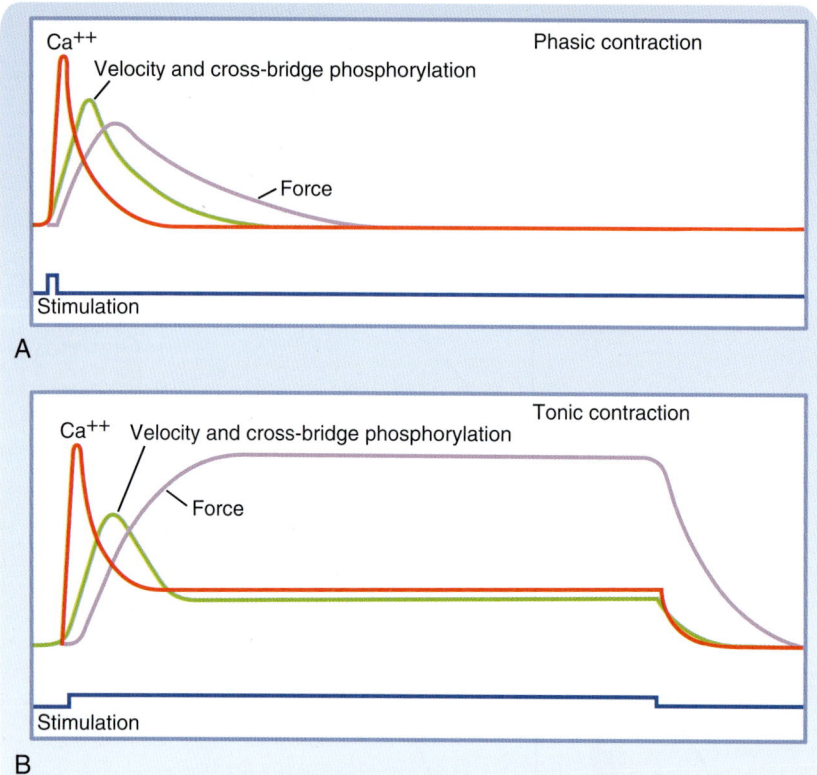

• **Fig. 14.9** Time course of events in cross-bridge activation and contraction in smooth muscle. A, A brief period of stimulation is associated with Ca^{++} mobilization, followed by cross-bridge phosphorylation and cycling to produce a brief phasic, twitch-like contraction. B, In a sustained tonic contraction produced by prolonged stimulation, the Ca^{++} and phosphorylation levels typically fall from an initial peak. Force is maintained during tonic contractions at a reduced $[Ca^{++}]$ (and hence a low level of myosin light-chain phosphorylation), with lower cross-bridge cycling rates manifested by lower shortening velocities and ATP consumption.

The latch state is thought to reflect a slowing of the cross-bridge cycle, so that the myosin heads remain in contact with the actin filament for a longer time, thereby maintaining tension at low energy cost. Note that the intracellular $[Ca^{++}]$ falls to a low level during the tonic phase of contraction, though it is still above the resting/basal $[Ca^{++}]$. The mechanism contributing to the ability of smooth muscle to maintain force at a low intracellular $[Ca^{++}]$ during tonic contraction is thought to involve dephosphorylation of the myosin regulatory light chain while the myosin cross-bridge is attached to the actin filament, resulting in slowing of the rate of dissociation of the myosin from the actin, allowing the myosin to spend more time in an attached, force-generating conformation. Relaxation of smooth muscle following tonic contraction occurs when intracellular $[Ca^{++}]$ decreases to a level that prevents a net phosphorylation of the regulatory light chain by MLCK. It has also been proposed that caldesmon may participate in the transition to the latch state.

Energetics and Metabolism

As already noted, ATP consumption is reduced during the latch state. Under this condition, smooth muscle uses 300-fold less ATP than would be required by skeletal muscle to generate the same force. Smooth muscle, like skeletal muscle, requires ATP for ion transport to maintain the resting membrane potential, sequester Ca^{++} in the SR, and extrude Ca^{++} from the cell. All these metabolic needs are readily met by oxidative phosphorylation. Fatigue of smooth muscle does not occur unless the cell is deprived of oxygen. However, aerobic glycolysis with lactic acid production normally supports membrane ion pumps even when oxygen is plentiful.

Regulation of Myoplasmic Calcium Concentration

The mechanisms that couple activation to contraction in smooth muscle involve two Ca^{++} sources: one involving the sarcolemma and the other involving the SR. The sarcolemma regulates Ca^{++} influx and efflux from the extracellular Ca^{++} pool. The SR membranes determine Ca^{++} movement between the myoplasm and the SR pool. Skeletal muscle contraction does not require extracellular Ca^{++} (see Chapter 12). In contrast, extracellular Ca^{++} is important for smooth muscle contraction. Thus regulation of myoplasmic $[Ca^{++}]$ involves not only the SR but also the sarcolemma (Fig. 14.10). A number of factors can alter the myoplasmic

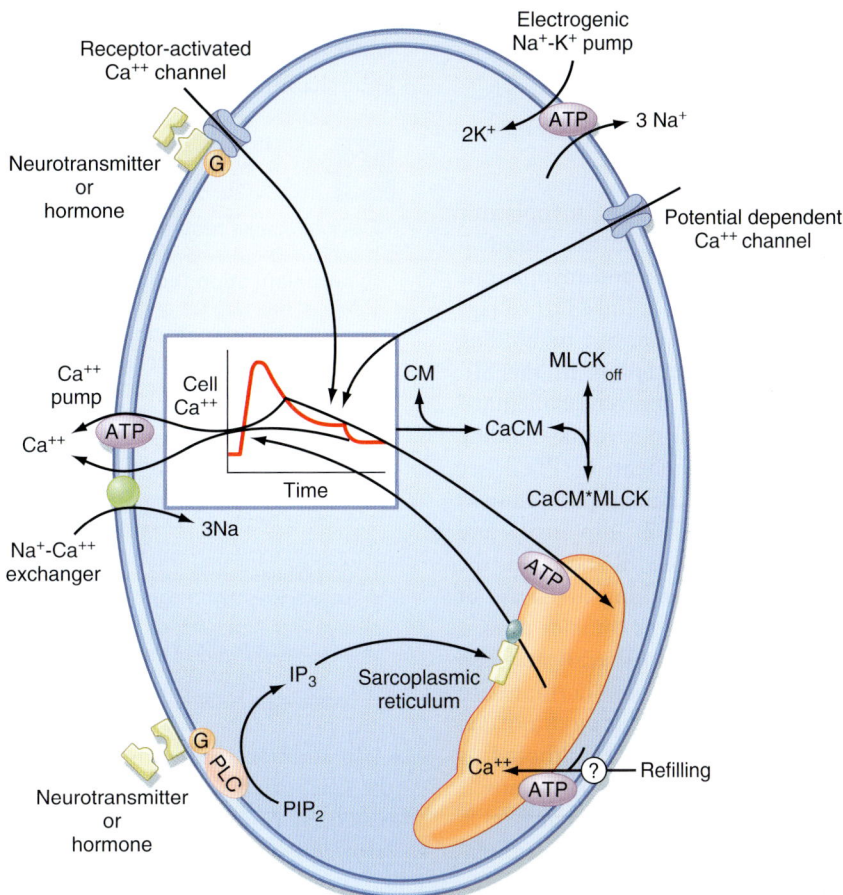

• **Fig. 14.10** Principal mechanisms determining myoplasmic [Ca^{++}] in smooth muscle. Release of calcium from the sarcoplasmic reticulum (SR) is a rapid initial event in activation, whereas both the SR and the sarcolemma participate in the subsequent stimulus-dependent regulation of myoplasmic [Ca^{++}]. The sarcolemma integrates many simultaneous excitatory and inhibitory inputs to govern the cellular response. Higher-order regulatory mechanisms can alter the activity of various pumps, exchangers, or enzymes (the *asterisk* designates well-established instances). The label *ATP* indicates that the process requires ATP hydrolysis; the *question mark* on the sarcolemma refers to the Ca^{++} pathway important for refilling the SR (i.e., store-operated Ca^{++} entry [SOCE]), which appears to involve the interaction of a Ca^{++}-sensitive stromal interaction molecule (STIM) on the SR and the sarcolemmal Ca^{++} channel Orai. CaCM, Ca^{++}-calmodulin complex; G, guanine nucleotide–binding proteins; IP$_3$, inositol 1,4,5-trisphosphate; MLCK, myosin light-chain kinase; PIP$_2$, phosphatidylinositol bisphosphate; PLC, phospholipase C.

 IN THE CLINIC

Inappropriate contraction of smooth muscle is associated with many pathological situations. One example is sustained vasospasm of a cerebral artery that develops several hours after a subarachnoid hemorrhage. It is thought that free radicals generated as a result of the hemorrhage raise myoplasmic [Ca^{++}] in surrounding arterial smooth muscle cells. The rise in myoplasmic [Ca^{++}] activates MLCK, which leads to cross-bridge phosphorylation and contraction. The vasoconstriction deprives other areas of the brain of oxygen and may lead to permanent injury or death of surrounding neurons. For a few days the cerebral artery remains sensitive to vasoactive agents, and therefore treatment with vasodilators may restore flow. An increase in ROK activity and MP phosphorylation has been observed during cerebral vasospasm. Administration of ROK inhibitors promotes relaxation of the vasospasm and decreases the level of myosin light-chain phosphorylation. The smooth muscle cells cease to respond to the vasodilators after several days, and they lose contractile proteins and secrete extracellular collagen. The lumen of the artery remains constricted as a result of structural and mechanical changes that do not involve active contraction.

[Ca^{++}] of smooth muscle. This differs from skeletal muscle, in which action potential–induced release of Ca^{++} from the SR fully activates the contractile apparatus.

Sarcoplasmic Reticulum

The role of smooth muscle SR in regulating myoplasmic [Ca^{++}] is comparable to that of skeletal muscle. Stimulation of the cell opens SR Ca^{++} channels, and myoplasmic [Ca^{++}] increases rapidly. This release is not linked to voltage sensors, as is the case in skeletal muscle, but to binding of the second messenger InsP3 to receptors in the SR. InsP3 is generated by a stimulus that acts on sarcolemmal receptors that are coupled via a guanine nucleotide–binding protein (G protein) to activate phospholipase C (PLC) (see Chapter 3). PLC hydrolyzes the membrane phospholipid phosphatidylinositol bisphosphate (PIP2) into InsP3 and diacylglycerol. InsP3 then diffuses to the SR and opens the InsP3-gated Ca^{++} channel, thereby resulting in release of Ca^{++} from the SR into the myoplasm. This complex process may permit graded release of Ca^{++} from the SR and also enable many different neurotransmitters and hormones to effect smooth muscle contraction. Calcium is reaccumulated by the SR through the activity of the SERCA, although as indicated later, extrusion of Ca^{++} from the smooth muscle cell also contributes to the reduction in myoplasmic [Ca^{++}]. Refilling of the SR with Ca^{++} not only involves reaccumulation of cytosolic Ca^{++} but also depends on the extracellular [Ca^{++}]. The dependence on extracellular [Ca^{++}] is thought to reflect the operation of a **"store-operated"** Ca^{++} channel present in the sarcolemma at points near underlying SR called **junctional SR.**

A variety of hormones and neurotransmitters elevate myoplasmic [Ca^{++}] by stimulating InsP3 production. Vascular smooth muscle, for example, is innervated by sympathetic fibers of the autonomic nervous system. These fibers use norepinephrine as a neurotransmitter, which when released binds to α_1-adrenergic receptors on vascular smooth muscle cells and results in G protein–dependent activation of PLC. Activation of PLC results in production of InsP3, which activates the InsP3-gated Ca^{++} channel in the SR, thereby elevating myoplasmic [Ca^{++}] and causing vasoconstriction. Other agents that promote vasoconstriction by activating the InsP3 cascade include angiotensin II and vasopressin. Development of drugs that block production of angiotensin II (e.g., angiotensin-converting enzyme [ACE] inhibitors) provides a means of promoting vasodilation that is important for individuals with hypertension or congestive heart failure. As mentioned previously, a variety of agents can produce contraction of smooth muscle without altering membrane potential (i.e., pharmacomechanical coupling). Agonist-induced activation of the InsP3 cascade represents an example of pharmacomechanical coupling. Many of the hormones/agonists that activate PLC through G protein–coupled receptors also promote sarcolemmal Ca^{++} influx and activation of RhoA/ROK. The net effect is a rise in intracellular [Ca^{++}], which activates MLCK, concomitant

with a rise in ROK activity, which inhibits MP, both of which act complementarily to increase the level myosin light-chain phosphorylation.

 AT THE CELLULAR LEVEL

Calcium sparks have also been observed to occur in smooth muscle in the presence of an **endothelial-dependent hyperpolarization factor (EDHF)** (Fig. 14.11). Specifically, EDHF appears to be an arachidonic acid metabolite (e.g., **epoxyeicosatrienoic acid [EET]**) that is produced by endothelial cells in response to various stimuli and then released to the underlying vascular smooth muscle. EET has been shown to activate a **transient receptor channel** (e.g., **TRPV4**) in the sarcolemma of smooth muscle that leads to the influx of Ca^{++}, which then opens RYR channels in the SR and results in Ca^{++} sparks. The Ca^{++} sparks in turn activate a large-conductance K$^+$ channel in the sarcolemma (BK$_{Ca}$), and the smooth muscle cell becomes hyperpolarized. Hyperpolarization in turn decreases basal Ca^{++} influx through voltage-gated Ca^{++} channels in the smooth muscle, thereby decreasing intracellular [Ca^{++}] and hence relaxing the smooth muscle, as described earlier.

In addition to the InsP3 receptor, the SR also contains the Ca^{++}-gated Ca^{++} channel, also called the *RYR*, which may be activated during periods of Ca^{++} influx through the sarcolemma. Short-lived spontaneous opening of the RYR

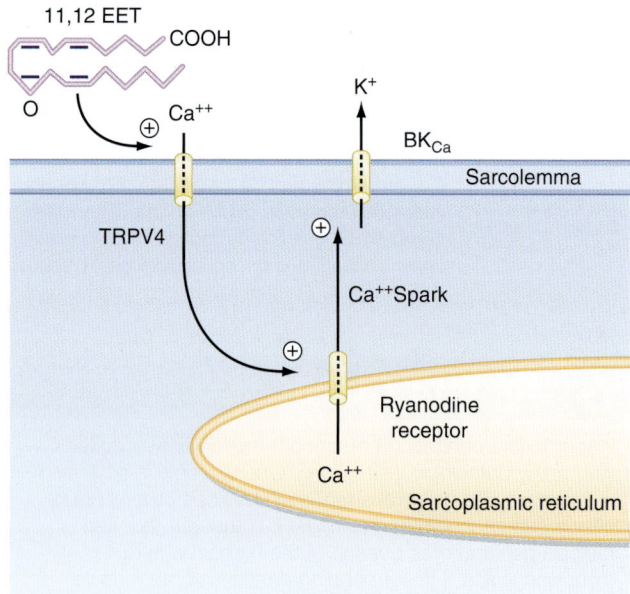

• **Fig. 14.11** An arachidonic acid metabolite (11,12-epoxyeicosatrienoic acid [11,12 EET]) released from endothelial cells can open the transient receptor channel TRPV4 in the underlying smooth muscle to permit the influx of Ca^{++}, which in turn initiates brief openings of the SR ryanodine receptor (Ca^{++} sparks) localized near the sarcolemma. Opening of Ca^{++}-activated K$^+$ channels in the sarcolemma by calcium sparks results in hyperpolarization of the smooth muscle and hence vasodilation. BK$_{Ca}$, large conductance Ca^{++}-activated potassium channel. (From Earley S et al. *Circ Res* 2005;97:1270-1279.)

resulting in localized elevations in myoplasmic [Ca^{++}] occurs in many cells, including smooth muscle. When observed with Ca^{++}-sensitive fluorescent dyes, these spontaneous localized elevations in myoplasmic [Ca^{++}] produce brief light flashes and as a result are named ***Ca^{++} sparks.*** In smooth muscle an increase in cAMP has been associated with an increase in the frequency of Ca^{++} sparks, particularly in situations in which the SR is in close proximity to the sarcolemma (i.e., junctional SR, perhaps near caveolae). An increase in the frequency of these sparks hyperpolarizes vascular smooth muscle by activation of a large-conductance Ca^{++}-gated K$^+$ channel in the sarcolemma. This hyperpolarization then decreases overall myoplasmic [Ca^{++}], and relaxation occurs.

Sarcolemma

Calcium is extruded from the smooth muscle cell by the activity of sarcolemmal Ca^{++}-ATPase and by a 3Na$^+$-1Ca^{++} antiporter (i.e., 3 Na$^+$ ions enter the cell for each Ca^{++} ion extruded). Extrusion of Ca^{++} from the cell competes with sequestration of Ca^{++} in the SR by SERCA and thus reduces the accumulation of Ca^{++} in the SR. When the [Ca^{++}] in the SR decreases, the SR is thought to initiate Ca^{++} influx into the cell through a process called ***store-operated Ca^{++} entry*** **(SOCE)** to facilitate refilling of the SR. Specifically, *stromal interaction molecule 1* **(STIM1)** in the SR is hypothesized to monitor SR [Ca^{++}] and then initiate Ca^{++} influx through the sarcolemmal channel protein **Orai** through a protein–protein interaction. Thus the influx of Ca^{++} during SOCE is thought to occur in the confined space between the caveolae and peripheral SR of smooth muscle. Recent studies raise the possibility that STIM1-Orai–mediated SOCE may also contribute to the rise in intracellular Ca^{++} transient in smooth muscle following α_1-adrenergic receptor stimulation. The Stim1-Orai complex has also been implicated in remodeling of smooth muscle in pathological conditions such as restenosis following balloon angioplasty.

In addition to the stimulatory effects of various agents on sarcolemma Ca^{++} channels and InsP3 cascades, there are several inhibitory factors that lower myoplasmic [Ca^{++}] and thereby relax smooth muscle. For example, the dihydropyridine class of Ca^{++} channel blocking drugs decreases the influx of Ca^{++} through sarcolemmal L-type voltage-gated Ca^{++} channels and reduces vasomotor tone. Similarly, drugs that open K$^+$ channels in the sarcolemma (e.g., hydralazine) promote relaxation (e.g., vasodilation) by hyperpolarizing the membrane potential, which reduces the influx of Ca^{++} through voltage-gated Ca^{++} channels. Conversely, agents that decrease K$^+$ permeability of the sarcolemma may promote vasoconstriction by inducing membrane depolarization, which then increases influx of Ca^{++} through these same voltage-gated Ca^{++} channels. Smooth muscle also contains receptor-activated Ca^{++} channels. Conductance of these receptor-activated Ca^{++} channels is linked to receptor occupancy.

A variety of drugs and hormones relax smooth muscle by increasing the cellular concentrations of cAMP or cGMP.

Nitric oxide (NO) is produced by nerves and vascular endothelial cells and relaxes smooth muscle by increasing cGMP (Fig. 14.12). Acetylcholine released from parasympathetic fibers causes vasodilation in some vascular beds as a result of stimulating the production of NO by vascular endothelial cells. Shear stress and adenosine (e.g., released from exercising muscle) may also promote NO release from vascular endothelial cells. The molecular mechanisms underlying the cGMP-dependent relaxation of vascular smooth muscle are complex and have been reported to involve (1) inhibition of InsP3 production, (2) inhibition of the InsP3 receptor, (3) activation of myosin light-chain phosphatase, and (4) activation of the Ca^{++}-activated K$^+$ channel (BK$_{Ca}$), which promotes hyperpolarization of the cells and thus inhibits Ca^{++} influx through voltage-gated Ca^{++} channels.

Similarly, elevation of cAMP in vascular smooth muscle by activation of β-adrenergic receptors or adenosine receptors promotes vasodilation through multiple mechanisms, including (1) decreased affinity of MLCK for Ca^{++}/calmodulin, (2) decreased cytosolic [Ca^{++}] concentration, and/or (3) increased MP activity. The decreased affinity of MLCK for Ca^{++}/calmodulin following elevation of cAMP involves protein kinase A (PKA)-dependent phosphorylation of MLCK. The ability of PKA to reduce cytosolic [Ca^{++}] is complex and can involve activation of potassium channels (e.g., ATP-dependent potassium channels), resulting in hyperpolarization of the smooth muscle and hence decreased Ca^{++} influx through voltage-gated Ca^{++} channels. cAMP has also been shown to increase the frequency of Ca^{++} sparks in vascular smooth muscle, which as described earlier, hyperpolarizes the membrane potential by activation of Ca^{++}-gated K$^+$ channels, thereby reducing the influx of Ca^{++} through voltage-gated Ca^{++} channels. cAMP can also promote relaxation of smooth muscle by increasing the activity of MP (through either PKA-dependent phosphorylation of a subunit of MP or through inhibition of ROK). Inhibition of ROK can occur through PKA-dependent phosphorylation or through a signaling pathway involving the cAMP-modulated guanine nucleotide exchange factor **Epac** (**e**xchange **p**rotein directly **a**ctivated by **c**AMP)

Relaxation of smooth muscle by elevation of cAMP has afforded asthmatics a means of reversing bronchiolar constriction with use of β$_2$-adrenergic agonists. The local vasodilatory effect of adenosine produced in working muscle during periods of intense exercise has also been attributed at least in part to elevated cAMP levels in vascular smooth muscle secondary to adenosine-induced stimulation of purinergic receptors on the sarcolemma of vascular smooth muscle. Adenosine may also activate a sarcolemmal K$^+$ channel to induce membrane hyperpolarization, which as already noted will decrease the influx of Ca^{++} through voltage-gated Ca^{++} channels and cause vasodilation. Thus regulation of smooth muscle tone may be under the influence of not only the autonomic nervous system and circulating hormones but also neighboring endothelial cells and skeletal muscle cells via diffusible substances such as NO and adenosine.

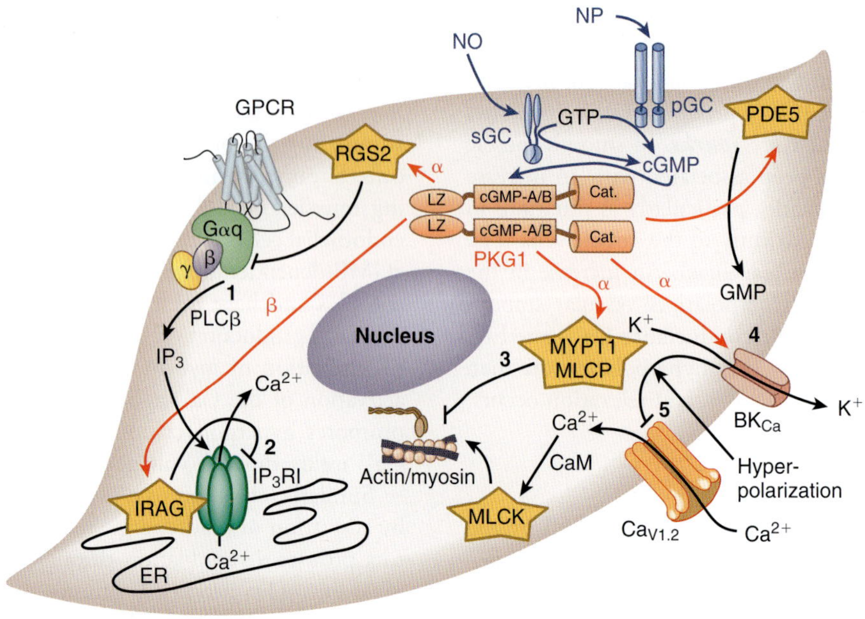

• Fig. 14.12 Signaling mechanisms by which nitric oxide (NO) and nitroprusside (NP) promote relaxation of vascular smooth muscle. Nitric oxide stimulates soluble guanylcyclase (sGC) in the vascular smooth muscle cell, resulting in production of cGMP, activation of protein kinase G (PKG1), which in turn phosphorylates various substrates that (1) inhibit InP3 production, (2) inhibit InsP3-induced Ca++ release, (3) activate myosin light-chain phosphatase, (4) activate the large conductance Ca++-activated K+ channel (BKCa), which hyperpolarizes the cell and thus inhibits the voltage-gated Ca++ channel (Cav1.2) (5). BKCa, calcium-activated maxi-K channel; CaM, calmodulin; Cat, catalytic domain; Cav1.2, L-type calcium channel; ER, endoplasmic reticulum; GPCR, G protein–coupled receptor; IP3RI, inositol 1,4,5-trisphosphate (IP3) receptor I; IRAG, inositol trisphosphate receptor-associated cGMP-kinase substrate; LZ, leucine/isoleucine zipper; MLCK, myosin light-chain kinase; MLCP, myosin light-chain phosphatase; MYPT, regulatory myosin phosphatase targeting subunit 1; PDE5, phosphodiesterase 5; pGC, particulate guanylyl cyclase; PLCβ, phospholipase C beta; RGS2, regulator of G-protein signaling 2; sGC, soluble guanylyl cyclase. (Modified from Schlossmann J, Desch M. *Am J Physiol Heart Circ Physiol* 2011;301:H672–H682.)

Myogenic Response

Blood flow to tissues such as the brain is maintained at a relatively constant flow over a wide range of blood pressures through a process called *autoregulation.* The mechanism underlying autoregulation of blood flow involves the *myogenic response,* wherein an increase in distending pressure in an artery results in vasoconstriction, whereas a decrease in transmural pressure results in a vasodilation (over a given range of pressures). As shown in Fig. 14.13, distention of the resistance artery resulted in an immediate elevation of intracellular [Ca++], followed by vasoconstriction in an effort to maintain relatively constant flow. The mechanism underlying this elevation of intracellular [Ca++] and subsequent vasoconstriction is complex. The rise in intracellular [Ca++] in response to stretch been reported to involve stretch-activated channels, InsP3-signaling, and voltage-gated Ca++ channels.

Development and Hypertrophy

During development and growth, the number of smooth muscle cells increases (Fig. 14.14). Smooth muscle tissue mass also increases if an organ is subjected to a sustained increase in mechanical work. This increase in mass is called **compensatory hypertrophy.** A striking example occurs with arterial smooth muscle cells (i.e., in the tunica media of the artery) in hypertensive patients. The increased mechanical load on the muscle cells appears to be the common factor that induces this hypertrophy. Chromosomal replication can result in significant numbers of polyploid muscle cells. The polyploid cells contain multiple sets of the normal number of chromosomes. They synthesize more contractile proteins and thus increase the size of the cell (see Fig. 14.14).

The myometrium, which is the smooth muscle component of the uterus, undergoes hypertrophy as parturition (birth) approaches. Hormones play an important role in this response. The smooth muscle is quiescent during pregnancy when the hormone progesterone predominates, and few gap junctions that electrically couple the smooth muscle cells are present. At term, under the dominant influence of estrogen, the myometrium undergoes marked hypertrophy. Large numbers of gap junctions form just before birth and convert the myometrium to a single-unit tissue to coordinate contraction during parturition.

Synthetic and Secretory Functions

The growth and development of tissues that contain smooth muscle are associated with increases in the connective tissue

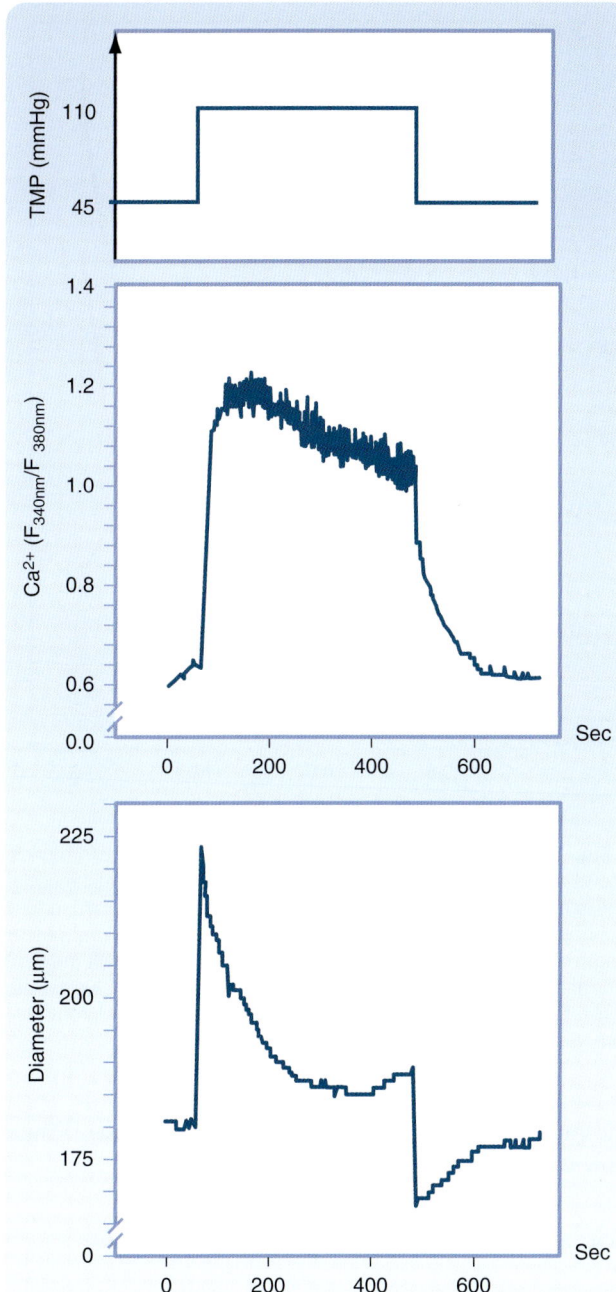

• **Fig. 14.13** Myogenic response in vascular smooth muscle is an autoregulatory response to maintain constant flow to a tissue. An increase in transmural pressure *(TMP; top panel)* elevates intracellular [Ca⁺⁺] *(middle panel)*, resulting in a vasoconstriction *(bottom panel)*. Note that the increase in transmural pressure initially stretches the skeletal muscle resistance artery *(bottom panel)*, but this is quickly followed by vasoconstriction. (From Schubert R et al. *Cardiovasc Res* 2008;77:8-18.)

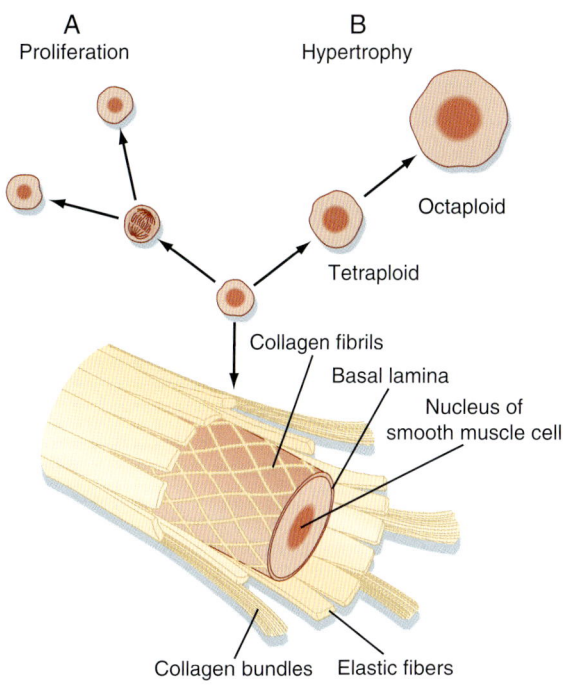

• **Fig. 14.14** Smooth muscle cells carry out many activities. A, They retain the capacity to divide during normal growth or in certain pathological responses such as formation of atherosclerotic plaque. B, Cells may also hypertrophy in response to increased loads. Chromosomal replication not followed by cell division yields cells with a greater content of contractile proteins. C, Smooth muscle cells also synthesize and secrete the constituents of the extracellular matrix.

🩺 IN THE CLINIC

Although smooth muscle is involved in physiological adjustments to exercise, sustained changes in the mechanical loading that induce cellular adaptations are usually the result of a pathological condition (e.g., hypertension). A fairly common example in men is **urinary bladder hypertrophy** caused by benign or cancerous enlargement of the prostate gland, which obstructs the bladder outlet. The clinical result is difficulty urinating, distention of the bladder, and impaired emptying. In this situation the ability of the bladder smooth muscle to contract and develop stress is diminished. The reasons for this remain unexplained, but phenotypic modulation of the smooth muscle cells with altered contractile protein isoform expression and gross anatomic distortion of the bladder wall occurs. Neuromuscular changes also affect myoplasmic Ca⁺⁺ mobilization and cross-bridge phosphorylation. Fortunately, normal structure and function are usually restored after the obstruction is alleviated.

matrix. Smooth muscle cells can synthesize and secrete the materials that make up this matrix, including collagen, elastin, and proteoglycans (see Fig. 14.14). The synthetic and secretory capacities are evident when smooth muscle cells are isolated and placed in tissue culture. The cells rapidly

lose thick myosin filaments and much of the thin-filament lattice, and there is expansion of the rough endoplasmic reticulum and Golgi apparatus. The phenotypically altered cells multiply and lay down connective tissue. This process is reversible, and some degree of redifferentiation with the formation of thick filaments occurs after cell replication ceases. Determinants of the smooth muscle cell phenotype are largely unknown, but hormones and growth factors in blood, as well as mechanical loads on cells, have been implicated in the control of phenotypic modulation.

IN THE CLINIC

Atherosclerosis is a disease characterized by lesions located in the wall of blood vessels. The lesions are induced by disorders that injure the endothelium, such as hypertension, diabetes, and smoking. Three formed elements (monocytes, T lymphocytes, and platelets) that circulate in the bloodstream act on the damaged vascular endothelium. There, they generate chemotactic factors and mitogens that modify the structure of the surrounding smooth muscle cells. The latter lose most of their thick and thin filaments and develop an extensive rough endoplasmic reticulum and Golgi complex. These cells migrate to the subendothelial space (i.e., the tunica media of the artery), proliferate, and participate in formation of the fatty lesions or the fibrous plaques that characterize atherosclerosis. Inhibition or downregulation of Rho kinase (ROK) has been shown to promote regression of atherosclerotic-like lesions in an animal model. The mechanism or mechanisms underlying this beneficial effect of ROK inhibition are unclear but may be related to regulation of both endothelial permeability and monocyte migration by ROK. That is, hyperactivity of ROK has been implicated in various pathological conditions, including increased transendothelial permeability (perhaps secondary to increased actomyosin activity), whereas inhibition of ROK has been shown to decrease transendothelial migration of monocytes and neutrophils.

Biophysical Properties of Smooth Muscle

Length-Tension Relationship

Smooth muscle contains large amounts of connective tissue composed of **extensible elastin fibrils** and **inextensible collagen fibrils.** Because this extracellular matrix can withstand high distending forces or loads, it is responsible for the passive length-tension curve measured in relaxed tissues. This ability of the matrix also limits organ volume.

When lengths are normalized to the optimal length for development of force (i.e., L_0), the **length-tension curves** for smooth and skeletal muscle are very similar (Fig. 14.15; see also Chapter 12). However, the length-tension curves of striated and smooth muscle differ quantitatively. For example, smooth muscle cells shorten more than skeletal muscle cells do. In addition, smooth muscle is characteristically only partially activated, and the peak isometric force attained varies with the stimulus. In skeletal muscle the stimulus (i.e., action potential) always produces a full twitch contraction. Smooth muscle can generate active force comparable to that of skeletal muscle, even though smooth muscle contains only about a fourth as much myosin. This does not imply that the cross-bridges in smooth muscle have greater force-generating capacity. Instead, active cross-bridges in smooth muscle are much more likely to be in

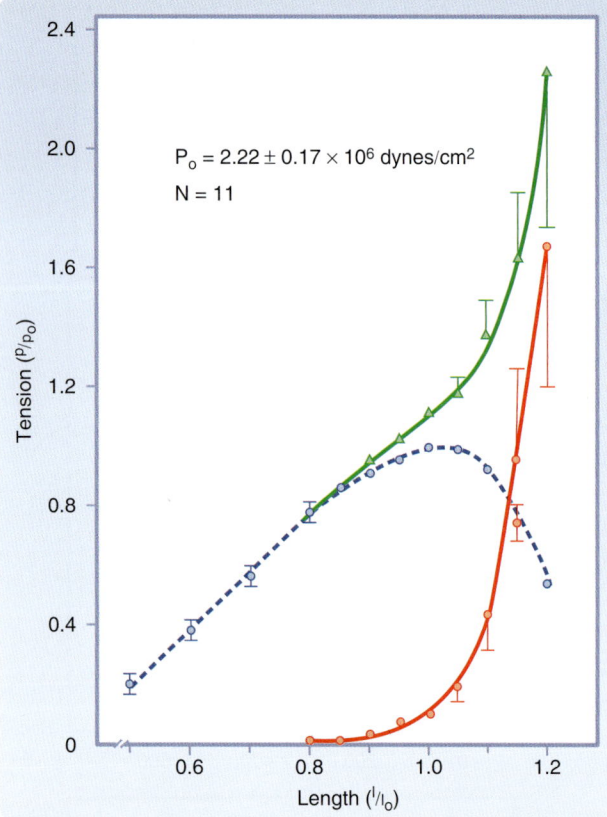

$$P_0 = 2.22 \pm 0.17 \times 10^6 \text{ dynes/cm}^2$$
$$N = 11$$

• **Fig. 14.15** Length dependence of contraction of smooth muscle shows a bell-shaped response similar to that seen in skeletal muscle. The smooth muscle strips were obtained from hog carotid artery. Contraction was induced by potassium depolarization. Optimal muscle length (L_0) was graphically determined as the length at which maximal tension (P_0) was developed. Red circles, passive tension; blue circles, active tension, triangles, total tension; (From Herlihy JT, Murphy RA. *Circ. Res.* 1973;33:275-283.)

the attached force-generating configuration because of their slow cycling kinetics.

Force-Velocity Relationship

Smooth and striated muscles both exhibit a hyperbolic dependence of shortening velocity on load. However, contraction velocities are far slower in smooth muscle than in striated muscle. One factor that underlies these slow velocities is that the myosin isoform in smooth muscle cells has low ATPase activity.

Skeletal muscle cells have a **force-velocity curve** in which shortening velocities are determined only by load and the myosin isoform (see Chapter 12). In contrast, both force and shortening velocity, which reflect the number of cycling cross-bridges and their cycling rates, vary in smooth muscle. When activation of smooth muscle is altered, for example, by different frequencies of nerve stimulation or changing hormone concentrations, a "family" of velocity-stress curves can be derived (Fig. 14.16). This implies that both cross-bridge cycling rates and the number of active cross-bridges in smooth muscle are regulated in some way,

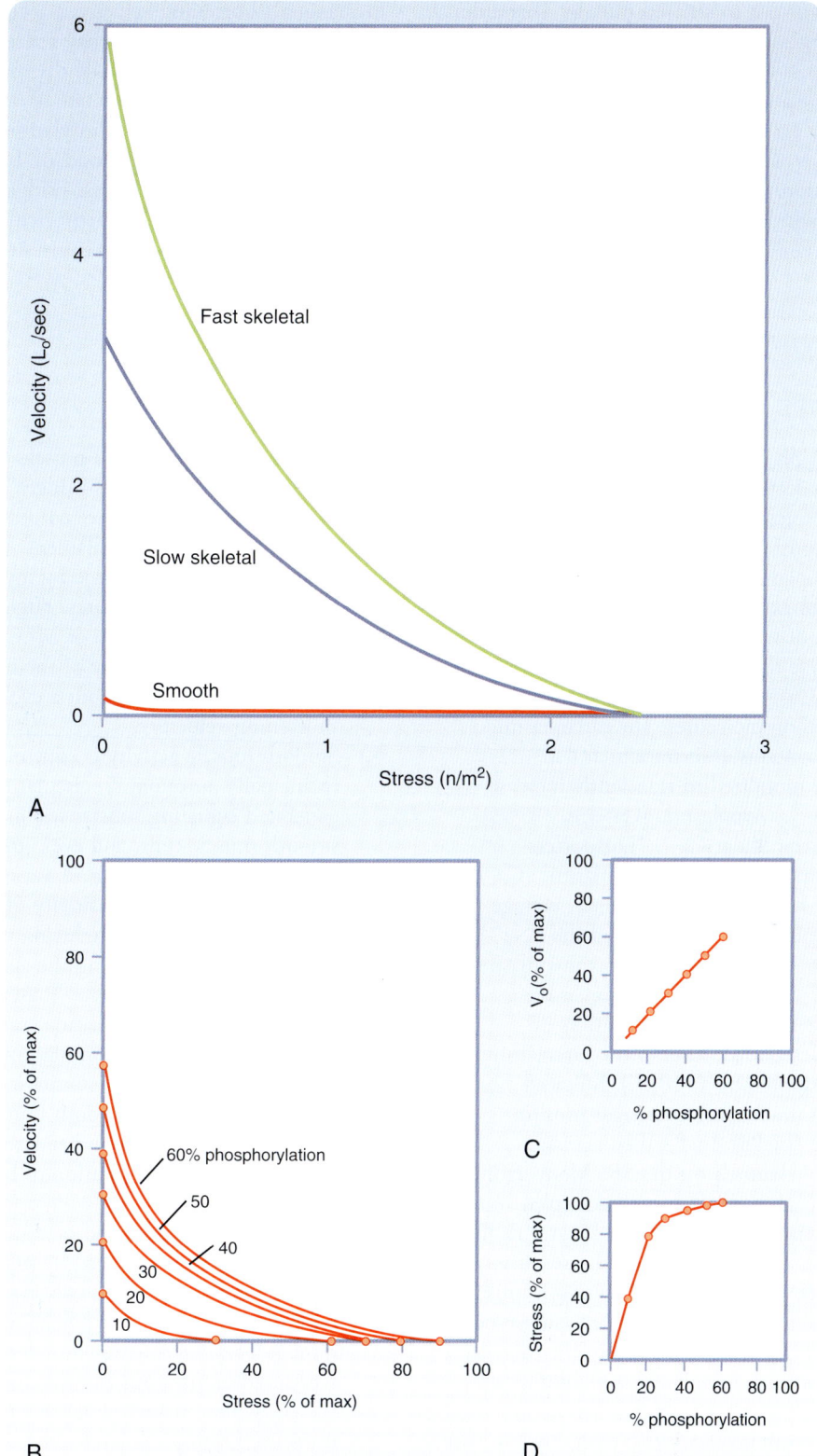

• **Fig. 14.16** A, Force-velocity curves for fast and slow human skeletal muscle cells and smooth muscle. B, Smooth muscles have variable force-velocity relationships that are determined by the level of Ca^{++}-stimulated cross-bridge phosphorylation. C, Maximal shortening velocities with no load (V_o, which represents the intercepts on the ordinate in B) are directly dependent on cross-bridge phosphorylation by MLCK. D, Active force/stress (abscissa intercepts in B) rises rapidly with phosphorylation; near maximal stress, it may be generated with only 20% to 30% of the cross-bridges in the phosphorylated state.

which is in marked contrast to striated muscle. This difference is conferred by a regulatory system that depends on phosphorylation of cross-bridges, which in turn depends on myoplasmic [Ca^{++}]. Because myosin light-chain phosphorylation is required for actin-myosin interaction in smooth muscle, a dependence of maximal force on the degree of myosin phosphorylation is expected (i.e., phosphorylation of more myosin molecules results in more actin-myosin interactions and hence more force generated). The variation in maximal shortening velocity as a function of the degree of myosin phosphorylation may reflect dephosphorylation of the myosin light chain while the myosin is still attached to the actin, thus slowing the rate of detachment (i.e., latch state) at low levels of phosphorylation. At higher levels of phosphorylation, the likelihood of latch states would be reduced and the myosin cross-bridges would be released more quickly from actin, thereby yielding a higher shortening velocity at all loads (see Fig. 14.16B).

Key Concepts

1. Smooth muscle cells are linked by a variety of junctions that serve both mechanical and communication roles. These linkages are essential in cells that must contract uniformly.

2. The sarcolemma plays an important role in Ca^{++} exchange between extracellular fluid and myoplasm. The sarcolemma of smooth muscle contains numerous caveolae that contribute to regulation of intracellular [Ca^{++}] and also appear to serve as a scaffold for signaling molecules. The sarcoplasmic reticulum (SR) contains an intracellular Ca^{++} pool that can be mobilized to transiently increase myoplasmic [Ca^{++}]. Myoplasmic [Ca^{++}] is dependent on extracellular Ca^{++}. Transporters in the sarcolemma that regulate myoplasmic [Ca^{++}] include receptor-mediated Ca^{++} channels, voltage-gated Ca^{++} channels, Ca^{++}-ATPase, and the 3Na$^+$-1Ca^{++} antiporter. The SR also regulates myoplasmic [Ca^{++}]. The Ca^{++} channels in the SR open in response to a chemical. Neurotransmitters or hormones that act via receptors in the sarcolemma can activate phospholipase C (PLC), followed by generation of the second messenger inositol 1,4-triphosphate (InsP3). InsP3 then activates InsP3-gated Ca^{++} channels on the SR. Many agonists that activate PLC through G protein–coupled receptors also activate the RhoA/Rho kinase (ROK) signaling cascade, thereby increasing the sensitivity of smooth muscle contraction to Ca^{++}. Smooth muscle SR also contains Ca^{++}-gated Ca^{++} channels (ryanodine receptor [RYR]). Ca^{++} reaccumulates in the SR via SERCA (sarcoplasmic endoplasmic reticulum Ca^{++} ATPase).

3. Smooth muscles contain contractile units that consist of small groups of thick myosin filaments that interdigitate with large numbers of thin filaments attached to Z-line equivalents termed *dense bodies* or *membrane-dense areas*. No striations are evident. Contraction is caused by a sliding filament–cross-bridge mechanism.

4. Contraction of smooth muscle is dependent on both release of Ca^{++} from the SR and entry of Ca^{++} across the sarcolemma. Smooth muscle lacks troponin. Phosphorylation of cross-bridges by a Ca^{++}-dependent myosin light chain kinase (MLCK) is necessary for attachment to the thin filament. Dephosphorylation of an attached cross-bridge by myosin phosphatase (MP) slows its cycling rates. Higher myoplasmic [Ca^{++}] increases the ratio of MLCK to MP activity, with the result that more of the cross-bridges remain phosphorylated throughout a cycle. This increases shortening velocities.

5. Smooth muscle activity is controlled by nerves (principally autonomic), circulating hormones, locally generated signaling substances, junctions with other smooth muscle cells, and even junctions with other non–smooth muscle cells. A variety of hormones/agonists increase the sensitivity of smooth muscle contraction to Ca^{++} by reducing the activity of MP and reciprocally increasing the activity of MLCK at a given intracellular [Ca^{++}]. Activation of the RhoA/ROK signaling cascade contributes to this inhibition of MP and stimulation of MLCK, and hence to the increase in sensitivity of smooth muscle contraction to Ca^{++}. Smooth muscle also has an intrinsic ability to respond to stretch, which is important for autoregulation of blood flow to various tissues.

6. The response to sustained or tonic stimulation is a rapid contraction followed by sustained maintenance of force with reduced cross-bridge cycling rates and ATP consumption. This behavior, called the *latch state,* is advantageous for muscles that may need to withstand continuous external force, such as blood vessels, which must be able to withstand blood pressure. During the latch state, ATP is consumed at less than 1/300 the rate needed to maintain the same force in skeletal muscle.

7. The length-tension relationships, hyperbolic velocity-load relationships, power output curves, and ability to resist imposed loads are comparable to those of skeletal muscle. Shortening velocities and ATP consumption rates are very low in smooth muscle, in keeping with expression of a myosin isoform with

low activity. Smooth muscles also have the unusual ability to alter velocity-stress relationships, which reflects regulation of both the number of active cross-bridges (determining force) and their average cycling rates for a given load (determining velocity).

8. Smooth muscle is also a synthetic and secretory cell with a major role in formation of the extensive extracellular matrix that surrounds and links the cells. Cellular hypertrophy occurs in response to physiological needs, and smooth muscle cells retain the potential to divide.

Additional Reading

Burnstock G, Ralevic V. Purinergic signaling and blood vessels in health and disease. *Pharmacol Rev.* 2014;66:102-192.

Joyner MJ, Casey DP. Regulation of increased blood flow (hyperemia) to muscles during exercise: a hierarchy of competing physiological needs. *Physiol Rev.* 2015;95:549-601.

Longden TA, et al. Ion channel networks in the control of cerebral blood flow. *J Cereb Blood Flow Metab.* 2016;36:492-512.

Momotani K, Somlyo AV. p63RhoGEF: a new switch for G q-mediated activation of smooth muscle. *Trends Cardiovasc Med.* 2012;22:122-127.

Roberts OL, Dart C. cAMP signalling in the vasculature: the role of Epac (exchange protein directly activated by cAMP). *Biochem Soc Trans.* 2014;42:89-97.

Schlossmann J, Desch M. IRAG and novel PKG targeting in the cardiovascular system. *Am J Physiol Heart Circ Physiol.* 2011;301:H672-H682.

Shimokawa H, et al. RhoA/Rho-Kinase in the Cardiovascular System. *Circ Res.* 2016;118:352-366.

Somlyo AP, Somlyo AV. Ca2+ sensitivity of smooth muscle and nonmuscle myosin II: modulated by G proteins, kinases, and myosin phosphatase. *Physiol Rev.* 2003;83:1325-1358.

Spinelli AM, Trebak M. ORAI channel-mediated Ca2+ signals in vascular and airway smooth muscle. *Am J Physiol Cell Physiol.* 2016;310:C402-C413.

Tinker A, et al. The role of ATP-sensitive potassium channels in cellular function and protection in the cardiovascular system. *Br J Pharmacol.* 2014;171:12-23.

SECTION 4

The Cardiovascular System

ACHILLES J. PAPPANO AND WITHROW GIL WIER

15

Overview of Circulation

LEARNING OBJECTIVES

Upon completion of this chapter, the student should be able to answer the following questions:

1. How does the arrangement of heart and vessels enable unidirectional flow of well-oxygenated blood to the body?
2. What is the advantage of the reciprocal relation between blood flow velocity and vascular cross-sectional area?
3. How do the differing compositions (smooth muscle, fibrous and elastic tissue) of blood vessels contribute to their respective functions?

The circulatory system transports and distributes essential substances to tissues and removes metabolic byproducts. This system also participates in homeostatic mechanisms such as regulation of body temperature, maintenance of fluid balance, and adjustment of O_2 and nutrient supply in various physiological states. The cardiovascular system, which accomplishes these tasks, is composed of a pump (the heart), a series of distributing and collecting tubes (blood vessels), and an extensive system of thin vessels (capillaries) that enable rapid exchange between the tissues and vascular channels. Blood vessels throughout the body are filled with a heterogeneous fluid (blood) that is essential for the transport processes performed by the heart and blood vessels. This chapter is a general, functional overview of the heart and blood vessels, whose functions are analyzed in much greater detail in subsequent chapters.

The Heart

The heart consists of two pumps in series: one pump propels blood through the lungs for exchange of O_2 and CO_2 (the **pulmonary circulation**) and the other pump propels blood to all other tissues of the body (the **systemic circulation**). Flow of blood through the heart is in one direction (unidirectional). Unidirectional flow through the heart is achieved by the appropriate arrangement of flap valves. Although cardiac output is intermittent, continuous flow to body tissues (periphery) occurs by distention of the aorta and its branches during ventricular contraction (**systole**) and by elastic recoil of the walls of the large arteries with forward propulsion of the blood during ventricular relaxation (**diastole**).

The Cardiovascular Circuit

In the normal intact circulation, the total volume of blood is constant, and an increase in the volume of blood in one area must be accompanied by a decrease in another. However, the distribution of blood circulating to the different regions of the body is determined by the output of the left ventricle and by the contractile state of the resistance vessels (arterioles) of these regions. The circulatory system is composed of conduits arranged in series and in parallel (Fig. 15.1). This arrangement, which is discussed in subsequent chapters, has important implications in terms of resistance, flow, and pressure in blood vessels.

Blood entering the right ventricle via the right atrium is pumped through the pulmonary arterial system at a mean pressure about one seventh that in the systemic arteries. The blood then passes through the lung capillaries, in which CO_2 in the blood is released and O_2 is taken up. The O_2-rich blood returns via the pulmonary veins to the left atrium, where it is pumped from the ventricle to the periphery, which thus completes the cycle.

Blood Vessels

Blood moves rapidly through the aorta and its arterial branches. As these branches approach the periphery, the branches narrow, and their walls become thinner. They also change histologically. The aorta is a predominantly elastic structure, but the peripheral arteries become more muscular until, at the arterioles, the muscular layer predominates (Fig. 15.2).

In the large arteries, frictional resistance is relatively small, and pressures are only slightly less than those in the aorta. The small arteries, on the other hand, offer moderate resistance to blood flow. This resistance reaches a maximal level in the arterioles, which are sometimes referred to as the "stopcocks" of the vascular system. Hence, the pressure drop is greatest across the terminal segment of the small arteries and the arterioles (Fig. 15.3). Adjustment in the degree of contraction of the circular muscle of these small vessels allows regulation of tissue blood flow and aids in the control of arterial blood pressure.

In addition to the reduction in pressure along the arterioles, there is a change from pulsatile to steady blood flow (see Fig. 15.3). Pulsatile arterial blood flow, caused by the intermittent ejection of blood from the heart, is damped

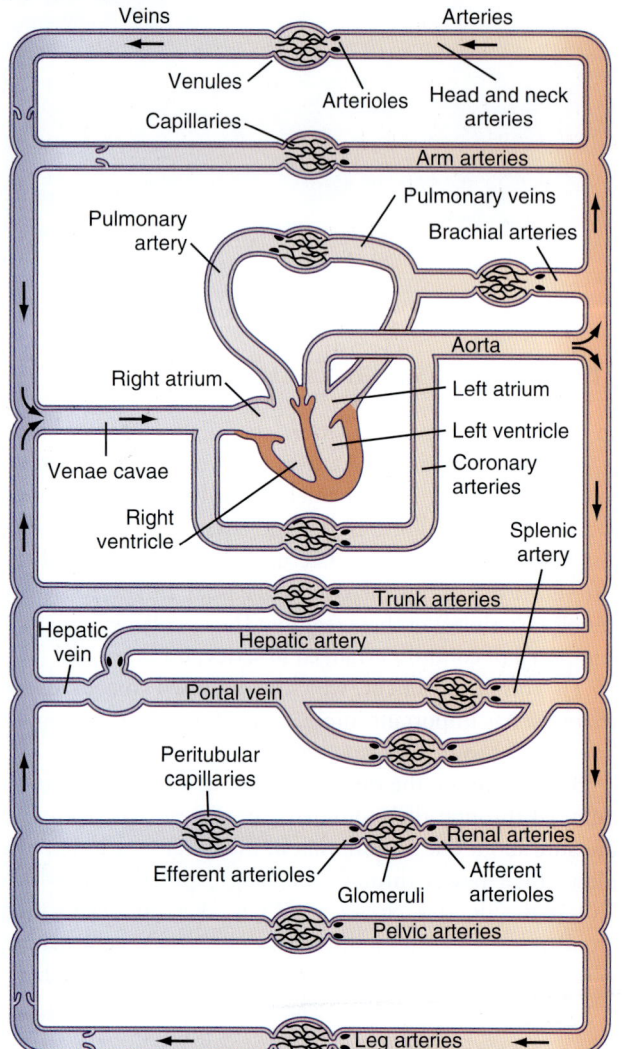

• **Fig. 15.1** Schematic Diagram of the Parallel and Series Arrangement of the Vessels That Constitute the Circulatory System. The capillary beds are represented by *thin lines* connecting the arteries (on the right) with the veins (on the left). The *crescent-shaped thickenings* proximal to the capillary beds represent the arterioles (resistance vessels). (Redrawn from Green HD. In: Glasser O [ed]. *Medical Physics*. Vol 1. Chicago: Year Book; 1944.)

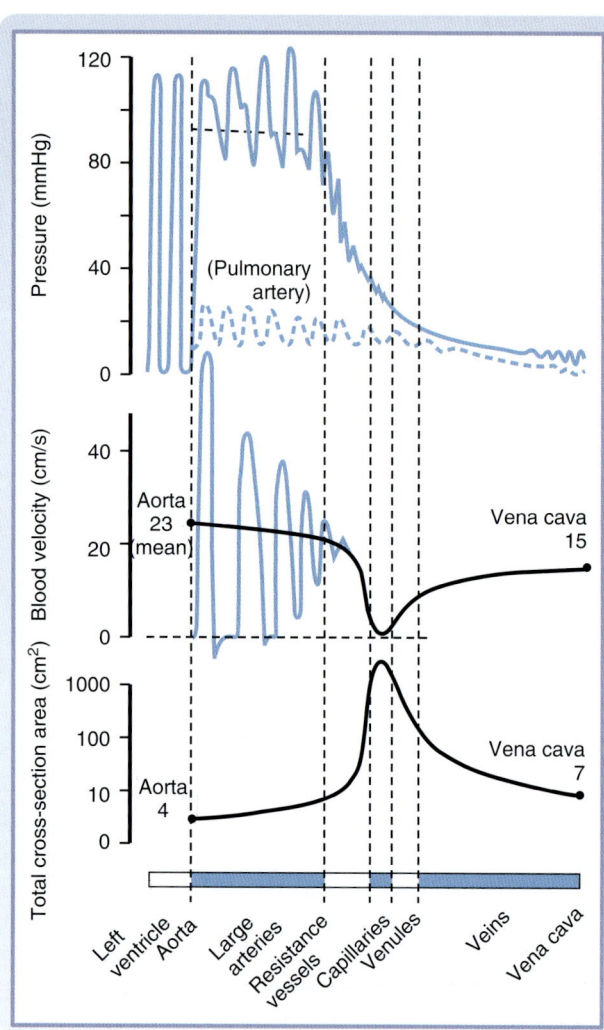

• **Fig. 15.3** Phasic Pressure, Velocity of Flow, and Cross-Sectional Area of the Systemic Circulation. The important features are the major pressure drop across the small arteries and arterioles, the inverse relationship between blood flow velocity and cross-sectional area, and the maximal cross-sectional area and minimal flow rate in the capillaries. (From Levick JR. *An Introduction to Cardiovascular Physiology*. 5th ed. London: Hodder Arnold; 2010.)

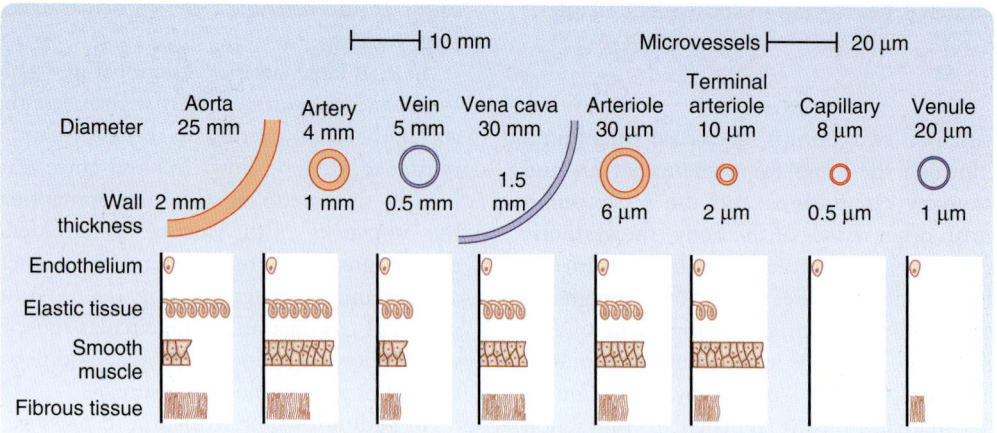

• **Fig. 15.2** Internal Diameter, Wall Thickness, and Relative Amounts of the Principal Components of the Vessel Walls of the Various Blood Vessels That Constitute the Circulatory System. Cross-sections of the vessels are not drawn to scale because of the huge range in size from aorta and venae cavae to capillary. (Redrawn from Burton AC. Relation of structure to function of the tissues of the wall of blood vessels. *Physiol Rev.* 1954;34:619.)

at the capillary level by a combination of two factors: distensibility of the large arteries and frictional resistance in the small arteries and arterioles.

Many capillaries arise from each arteriole. The total cross-sectional area of the capillary bed is very large despite the fact that the cross-sectional area of each capillary is less than that of each arteriole. As a result, blood flow velocity becomes quite slow in the capillaries (see Fig. 15.3), which is analogous to the decrease in velocity of flow in the wide regions of a river. Conditions in the capillaries are ideal for the exchange of diffusible substances between blood and tissue because capillaries consist of short tubes with walls that are only one cell thick and in which flow velocity is low.

On its return to the heart from the capillaries, blood passes through venules and then through veins of increasing size. Pressure within these vessels progressively decreases until the blood reaches the right atrium (see Fig. 15.3). Near the heart, the number of veins decreases, the thickness and composition of the vein walls change (see Fig. 15.2), the total cross-sectional area of the venous channels diminishes, and the velocity of blood flow increases (see Fig. 15.3). The velocity of blood flow and the cross-sectional area at each level of the vasculature are essentially mirror images (see Fig. 15.3).

Data from humans (Table 15.1) indicate that between the aorta and the capillaries, the total cross-sectional area increases about 500-fold. The volume of blood in the systemic vascular system is greatest in the veins and venules (64%). Only 6% of total blood volume exists in the capillaries, and 14% of total blood volume is found in the aorta, arteries, and arterioles. In contrast, blood volume in the pulmonary vascular bed is about equally divided among the arterial, capillary, and venous vessels. The cross-sectional area of the venae cavae is larger than that of the aorta. Therefore, the velocity of flow is slower in the venae cavae than in the aorta (see Fig. 15.3).

TABLE 15.1	Distribution of Blood Volume*	
Location	Absolute Volume (mL)	Relative Volume (%)
Systemic Circulation		
Aorta and large arteries	300	6.0
Small arteries	400	8.0
Capillaries	300	6.0
Small veins	2300	46.0
Large veins	900	18.0
Total	4200	84.0
Pulmonary Circulation		
Arteries	130	2.6
Capillaries	110	2.2
Veins	200	4.0
Total	440	8.8
Heart (End-Diastole)	360	7.2
Total	5000	100

Data from Boron WF, Boulpaep EL. *Medical Physiology*. 2nd ed. Philadelphia: Elsevier Saunders; 2009.
*Values refer to a 70-kg woman.

IN THE CLINIC

In a patient with hyperthyroidism **(Graves' disease),** basal metabolism is elevated and often associated with arteriolar vasodilation. This reduction in arteriolar resistance diminishes the damping effect on pulsatile arterial pressure and is manifested as pulsatile flow in the capillaries, as observed in the fingernail beds of patients with this ailment.

Key Points

1. The circulatory system consists of a pump (the heart), a series of distributing and collecting tubes (blood vessels), and an extensive system of thin vessels (capillaries) that enable rapid exchange of substances between tissues and blood.

2. Pulsatile pressure is progressively damped by the elasticity of the arterial walls and the frictional resistance of the small arteries and arterioles in such a way that capillary blood flow is essentially nonpulsatile. The resistance to blood flow and hence the pressure drop in the arterial system are greatest at the level of the small arteries and arterioles.

3. The velocity of blood flow is inversely related to the cross-sectional area at any point along the vascular system.

Additional Readings

Burton AC. *Physiol Rev*. 1954;34:619.

Green HD. In: Glasser O, ed. *Medical Physics*. Vol. 1. Chicago: Year Bookl; 1944.

Levick JR. *An introduction to cardiovascular physiology*. 5th ed. London: Hodder Arnold; 2010.

16

Elements of Cardiac Function

LEARNING OBJECTIVES

Upon completion of this chapter, the student should be able to answer the following questions:

1. How does the action potential contribute to excitability and contraction in heart muscle?
2. What is automaticity, and how does it differ from excitability? How do derangements of these properties contribute to arrhythmias?
3. What is the structural basis of the electrocardiogram?
4. How are the concepts of preload and afterload, developed for skeletal muscle, applied to the contraction of the heart?
5. What is the role of Ca^{++} in excitation-contraction coupling?
6. What changes occur in atrial and ventricular pressures and volumes during a cardiac cycle, and what is their temporal relation to the electrocardiogram?
7. How does the relation between end-diastolic volume and left ventricular developed pressure define the Frank-Starling law of the heart and regulate the force of cardiac contraction?
8. What is the pressure-volume loop of the left ventricle, and how does it define changes in left ventricular function?
9. How is cardiac metabolism linked to O_2 consumption, and how are these processes affected by changes in cardiac work?

Electrical Properties of the Heart

The cells of the heart, like neurons, are excitable and generate action potentials. These action potentials initiate contraction and thus determine the heart rate. Disorders in electrical activity can induce serious and sometimes lethal disturbances in cardiac rhythm.

In this section, the electrical properties of cardiac cells are described. In addition, how these electrical properties account for the **electrocardiogram (ECG)** is considered. The initiation of contraction as a result of the electrical properties of cardiac cells is considered in a later section.

The Cardiac Action Potential

Fig. 16.1 illustrates action potentials found in different cardiac cells. Two main types of action potentials occur in the heart and are depicted. One type, the fast response,

occurs in normal atrial and ventricular myocytes and in the specialized conducting fibers (Purkinje fibers of the heart) and is divided into five phases. The rapid upstroke of the action potential is designated phase 0. The upstroke is followed by a brief period of partial, early repolarization (phase 1) and then by a plateau (phase 2) that persists for approximately 0.1 to 0.2 second. The membrane then repolarizes (phase 3) until the resting state of polarization (phase 4) is again attained. Final repolarization (phase 3) develops more slowly than depolarization (phase 0). The other type of action potential, the slow response, occurs in the **sinoatrial (SA) node,** which is the natural pacemaker region of the heart, and in the **atrioventricular (AV) node,** which is the specialized tissue that conducts the cardiac impulse from the atria to the ventricles. The slow-response cells lack the early repolarization phase (phase 1).

Other differences between the electrical properties of the fast-response and slow-response cells include the following: The resting membrane potential (phase 4) of the fast-response cells is considerably more negative than that of the slow-response cells. Moreover, the slope of the upstroke (phase 0), the amplitude of the action potential, and the overshoot (membrane voltage positive to 0 mV) are greater in the fast-response cells than in the slow-response cells. The action potential amplitude and the steepness of the upstroke are important determinants of propagation velocity along the myocardial fibers. In slow-response cardiac tissue, the action potential is propagated more slowly and conduction is more likely to be blocked than in fast-response cardiac tissue. Slow conduction and a tendency toward conduction block increase the likelihood of some rhythm disturbances (see the section "Reentry").

The action potential initiates contraction of the myocyte. The relationship between the action potential and contraction of a cardiac myocyte is shown in Fig. 16.2. Rapid depolarization (phase 0) precedes cell shortening, and completion of repolarization occurs just before peak shortening. Relaxation of the muscle takes place mainly during phase 4 of the action potential. The duration of contraction usually parallels the duration of the action potential.

The various phases of the cardiac action potential are associated with changes in cell membrane permeability, mainly by Na^+, K^+, and Ca^{++} ions. Changes in cell membrane permeability alter the rate of movement of these ions across the membrane and thereby change the membrane voltage

(V_m). These changes in permeability are accomplished by the opening and closing of ion channels that are specific for individual ions (see Chapters 1 and 2).

As with all other cells in the body, the concentration of K^+ inside a cardiac muscle cell (intracellular [K^+]) exceeds the concentration outside the cell (extracellular [K^+]). The concentration gradient is reversed for Na^+ and Ca^{++}. Estimates of the extracellular and intracellular concentrations of Na^+, K^+, and Ca^{++} and the Nernst equilibrium potentials (see Chapter 1, Eq. 1.5b) for these ions are compiled in Table 16.1.

Resting Membrane Voltage

The resting cell membrane has a relatively high permeability for K^+; permeability for Na^+ and Ca^{++} is much less. Because of the existing chemical gradient for K^+ and V_m, K^+ tends to diffuse from the inside to the outside of the cell. Any diffusion of K^+ that occurs at the resting membrane potential (i.e., during phase 4) takes place mainly through specific potassium channels. Several types of potassium channels exist in cardiac cell membranes. Opening and closing of some of these channels are regulated by V_m, whereas others are controlled by a chemical signal (e.g., the extracellular acetylcholine concentration). The specific potassium channel through which K^+ passes during phase 4 is a voltage-regulated channel that conducts the **inward rectifying K^+ current (i_{K1})**, which is discussed in more detail later. For now, it is necessary only to know how this current is established.

The dependence of V_m on conductance and the intracellular and extracellular concentrations of K^+, Na^+, and other ions are described by the **chord conductance equation**

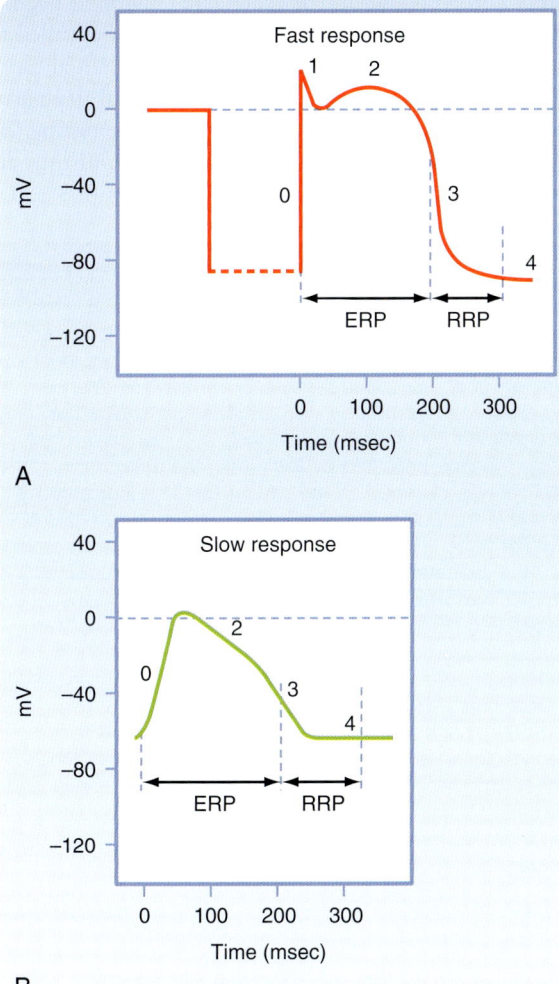

A

B

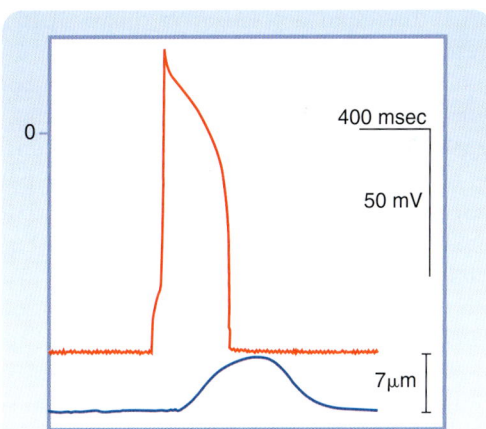

• **Fig. 16.1** Action Potentials of Cardiac Fibers. **A,** Fast-response cardiac fibers. **B,** Slow-response cardiac fibers. The phases of the action potentials are labeled (see text for details), as are the effective refractory period (ERP) and the relative refractory period (RRP). Note that in comparison with fast-response fibers, the resting potential of slow fibers is less negative, the upstroke (phase 0) of the action potential is less steep, the amplitude of the action potential is smaller, phase 1 is absent, and the RRP extends well into phase 4 after the fibers have fully repolarized.

• **Fig. 16.2** Temporal relationship between the changes in transmembrane potential *(top trace)* and cell shortening *(bottom trace)* in a single ventricular myocyte. The value of 0 mV membrane potential is indicated by "0." (From Pappano A. unpublished record, 1995.)

TABLE 16.1	Intracellular and Extracellular Ion Concentrations and Equilibrium Potentials in Cardiac Muscle Cells		
Ion	Extracellular Concentrations (mmol/L)	Intracellular Concentrations (mmol/L)*	Equilibrium Potential (mV)
Na^+	145	10	71
K^+	4	135	−93
Ca^{++}	2	10^{-4}	129

Data from Ten Eick RE, et al. *Prog Cardiovasc Dis.* 1981;24:157.
*The intracellular concentrations are estimates of the free concentrations in cytoplasm.

(see Chapter 2, Eq. 2.7). In a resting cardiac cell, K^+ conductance (g_K) is approximately 100 times greater than Na^+ conductance (g_{Na}). Therefore, V_m is similar to the Nernst equilibrium potential for K^+. As a result, alterations in extracellular $[K^+]$ can significantly change V_m: Hypokalemia causes hyperpolarization, and hyperkalemia causes depolarization. In contrast, because g_{Na} is so small in the resting cell, changes in extracellular $[Na^+]$ do not significantly affect V_m.

IN THE CLINIC

Fast responses of cardiac muscle may change to slow responses in certain pathological conditions. For example, in coronary artery disease, a region of cardiac muscle may be deprived of its normal blood supply. As a result, $[K^+]$ in the interstitial fluid that surrounds the affected muscle cells rises because K^+ is lost from the inadequately perfused (or ischemic) cells. The action potentials in some of these cells may then be converted from fast to slow responses. Conversion from a fast to a slow response as a result of increasing interstitial $[K^+]$ is illustrated later in Fig. 16.13.

Fast-Response Action Potentials
Genesis of the Upstroke (Phase 0)

Any stimulus that abruptly depolarizes V_m to a critical value (the *threshold*) elicits an action potential. The characteristics of fast-response action potentials are shown in Fig. 16.1*A*. The rapid depolarization (phase 0) is related almost exclusively to the influx of Na^+ into the myocyte as a result of a sudden increase in g_{Na}. The action potential amplitude (the membrane potential change during phase 0) is dependent on extracellular $[Na^+]$. When extracellular $[Na^+]$ is decreased, the amplitude of the action potential decreases, and when it is reduced from its normal value of approximately 140 mEq/L to approximately 20 mEq/L, the cell is no longer excitable.

When the resting membrane potential, V_m, is suddenly depolarized from −90 mV to the threshold level of approximately −65 mV, the cell membrane properties change dramatically. Na^+ enters the myocyte through specific fast **voltage-activated sodium channels** that exist in the membrane (see Fig. 5.7). These channels can be blocked by the puffer fish toxin tetrodotoxin. In addition, many drugs used to treat certain cardiac rhythm disturbances (cardiac arrhythmias) act by blocking these fast sodium channels.

The sodium channels open very rapidly, or **activate** (in ≈0.1 msec), thereby resulting in an abrupt increase in g_{Na}. However, once open, the sodium channels **inactivate** (time course ≅ 1 to 2 msec), and g_{Na} rapidly decreases (Fig. 16.3). The sodium channels remain in the inactivated state until the membrane begins to repolarize. With repolarization, the channel transitions to the **closed** state, from which it can then be reopened by another depolarization of V_m to the threshold. These properties of the sodium channel are the basis of the refractory period of the action potential. When the sodium channels are in the inactivated state, they cannot be reopened, and another action potential cannot be generated. During this period, the cell is said to be in the **effective refractory period.** This prevents a sustained, tetanic contraction of cardiac muscle, which would retard ventricular relaxation and therefore interfere with the normal intermittent pumping action of the heart. As the cell repolarizes (phase 3), the inactivated channels begin to transition to the closed state. During this period, called the **relative refractory period,** another action potential can be generated, but it requires a larger-than-normal depolarization of V_m. Only when V_m has returned to the resting level (phase 4) are all the sodium channels closed and thus able to be reactivated by the normal depolarization of V_m.

AT THE CELLULAR LEVEL

An ionic current through one membrane channel can be measured with the patch clamp technique. The individual channels open and close repeatedly in a random manner. This process is illustrated in Fig. 16.4, which shows the current flow through sodium channels in a myocardial cell. Before the time indicated by the arrow, the membrane potential was clamped at −85 mV. At the time indicated by the arrow, the potential was suddenly changed to −45 mV, at which value it was held for the remainder of the recording. Fig. 16.4 indicates that immediately after the membrane potential was made less negative, one sodium channel opened (1.5 pA in amplitude), and then a second sodium channel opened (3 pA total current from both channels), whereupon both channels closed (i.e., the current returned to 0). Both channels closed for approximately 4 or 5 msec, and then both channels opened at the same time. Thereafter, one channel closed and then the second rapidly closed. Finally, after several seconds, one channel opened and then closed. Thereafter, both channels remained closed for the rest of the recording, even though the membrane voltage was held constant at −45 mV.

Genesis of Early Repolarization (Phase 1)

In many cardiac cells that have a prominent plateau, phase 1 is an early, brief period of limited repolarization. This brief repolarization results in the notch between the end of the upstroke and the beginning of the plateau (see Figs. 16.1 and 16.3). Repolarization is brief because of activation of a **transient outward current (i_{to})** carried mainly by K^+. Activation of potassium channels during phase 1 causes a brief efflux of K^+ from the cell because the cell interior is positively charged and intracellular $[K^+]$ greatly exceeds extracellular $[K^+]$ (see Fig. 16.3). The cell is briefly and partially repolarized as a result of this transient efflux of K^+.

The size of the phase 1 notch varies among cardiac cells. It is prominent in myocytes in the epicardial and midmyocardial regions of the left ventricular wall (Fig. 16.5) and in ventricular Purkinje fibers. However, the notch is negligible in myocytes from the endocardial region of the left ventricle (see Fig. 16.5) because the density of i_{to} channels is less in these cells. The notch is also less prominent in the presence

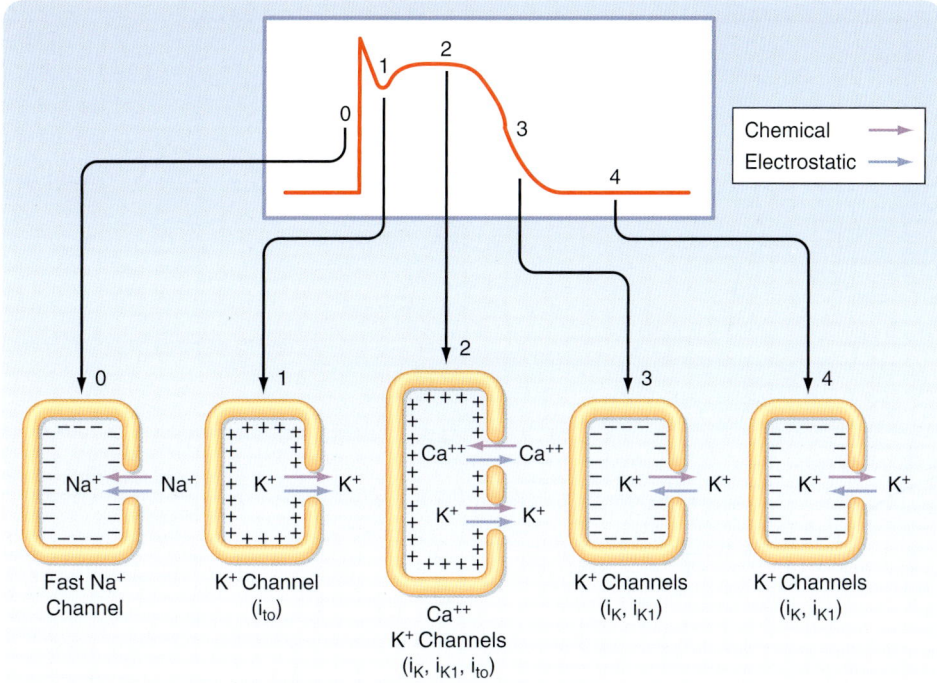

Fig. 16.3 Principal Ionic Currents and Channels That Generate the Various Phases of the Action Potential in a Cardiac Cell. **Phase 0,** The chemical and electrostatic forces both favor the entry of Na^+ into the cell through fast sodium channels to generate the upstroke. **Phase 1,** Both the chemical and electrostatic forces favor the efflux of K^+ through transient outward current (i_{to}) channels to generate early, partial repolarization. **Phase 2,** During the plateau, the net influx of Ca^{++} through calcium channels is balanced by the efflux of K^+ through outward rectifying current (i_K), inward rectifying current (i_{K1}), and i_{to} channels. **Phase 3,** The chemical forces that favor the efflux of K^+ through i_K and i_{K1} channels predominate over the electrostatic forces that favor the influx of K^+ through these same channels. **Phase 4,** The chemical forces that favor the efflux of K^+ through i_K and i_{K1} channels very slightly exceed the electrostatic forces that favor the influx of K^+ through these same channels.

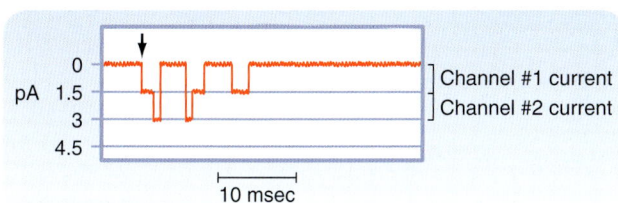

Fig. 16.4 Current (in Picoamperes [pA]) Through Two Individual Sodium Channels in a Cultured Heart Cell, Recorded With the Patch Clamp Technique. Membrane voltage was held at −85 mV, then abruptly changed to −45 mV at the time indicated by the *arrow*, and held at this potential for the remainder of the record. (Redrawn from Cachelin AB, et al. *J Physiol.* 1983;340:389.)

of 4-aminopyridine, which blocks the potassium channels that carry i_{to}.

Genesis of the Plateau (Phase 2)

During the action potential plateau, Ca^{++} enters myocardial cells through calcium channels (described in next paragraph) that activate and inactivate much more slowly than the fast sodium channels do. During the flat portion of phase 2 (see Figs. 16.1 and 16.3), this influx of Ca^{++} is counterbalanced by the efflux of K^+. K^+ exits through channels that conduct mainly the i_{to}, i_K, and i_{K1} currents. The i_{to} current

is responsible for phase 1, as described previously, but it is not completely inactivated until after phase 2 has expired. The i_K and i_{K1} are described later in this section of the chapter.

Ca^{++} enters the cell via voltage-regulated calcium channels, which are activated as V_m becomes progressively less negative during the action potential upstroke. Two types of calcium channels (**L-type** and **T-type**) have been identified in cardiac tissue. Some of their important characteristics are illustrated in Fig. 16.6. L-type channels are so designated because once open, they inactivate slowly (see Fig. 16.6, lower panel) and enable a long-lasting Ca^{++} current. They are the predominant type of calcium channel in the heart, and they are activated during the action potential upstroke when V_m reaches approximately −20 mV. L-type channels are blocked by **calcium channel antagonists** such as verapamil, amlodipine, and diltiazem (Fig. 16.7).

T-type (or "transient") calcium channels are much less abundant in the heart. They are activated at membrane potentials that are more negative (approximately −70 mV) than the potential that activates L-type channels (−20 mV). They also inactivate more quickly than do L-type channels (see Fig. 16.6; compare the upper and lower panels).

Opening of calcium channels results in an increase in Ca^{++} conductance (g_{Ca}) and calcium current (i_{Ca}) soon after

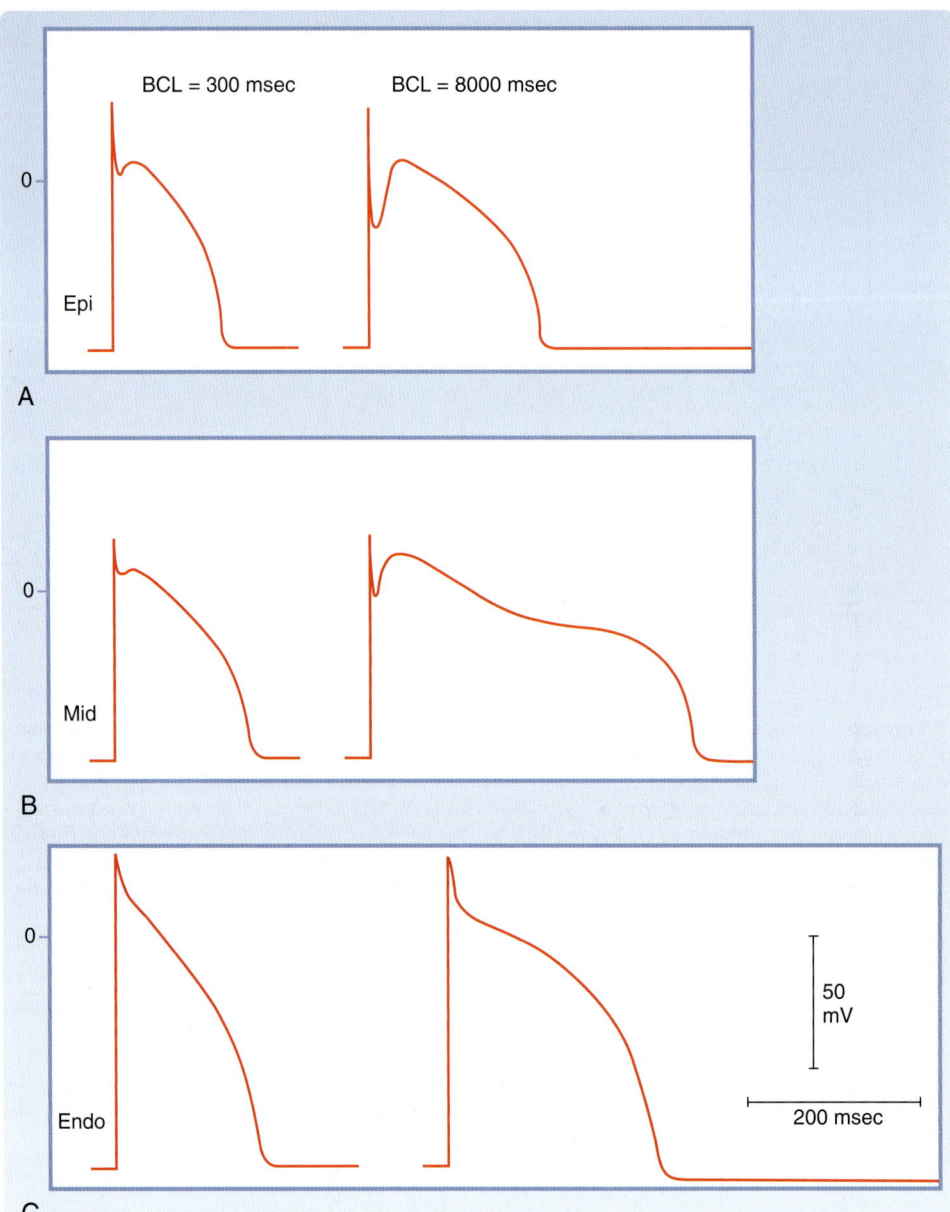

• **Fig. 16.5** Action potentials recorded from the epicardial (Epi) **(A)**, midmyocardial (Mid) **(B)**, and endocardial (Endo) **(C)** regions of the free wall of the left ventricle. The preparations were driven at a basic cycle length (BCL): that is, at interstimulus intervals of 300 and 8000 msec. The value of 0 mV membrane potential is indicated by "0." (From Liu D-W, et al. *Circ Res.* 1993;72:671.)

the action potential upstroke (see Fig. 16.3). Because the intracellular concentration of Ca^{++} is much lower than the extracellular concentration of Ca^{++} (see Table 16.1), the increase in g_{Ca} promotes the influx of Ca^{++} into the cell throughout the plateau. This Ca^{++} influx during the plateau is involved in excitation-contraction coupling, as described in the section "Excitation-Contraction Coupling" (see also Chapter 13).

Various neurotransmitters and drugs may substantially influence g_{Ca}. The adrenergic neurotransmitter norepinephrine, the β-adrenergic receptor agonist isoproterenol, and various other catecholamines enhance g_{Ca}, whereas the

parasympathetic neurotransmitter acetylcholine decreases g_{Ca}. Enhancement of g_{Ca} by catecholamines is the principal mechanism by which they enhance cardiac muscle contractility.

During the plateau (phase 2) of the action potential, the concentration gradient for K^+ across the cell membrane is virtually the same as it is during phase 4; at that time, however, V_m is positive. Therefore, there is a large gradient that favors efflux of K^+ from the cell (see Fig. 16.3). If g_K were the same during the plateau as it is during phase 4, efflux of K^+ during phase 2 would greatly exceed the influx of Ca^{++}, and a sustained plateau could not be achieved.

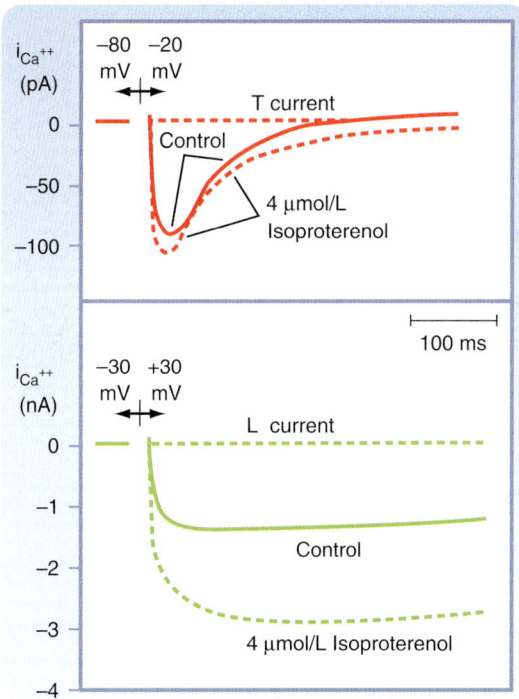

• **Fig. 16.6** Effects of Isoproterenol on the Ca⁺⁺ Currents. The currents were conducted by T-type **(upper panel)** and L-type **(lower panel)** calcium channels in atrial myocytes. **Upper panel,** The membrane potential was changed from −80 to −20 mV. **Lower panel,** The membrane potential was changed from −30 to +30 mV. (Redrawn from Bean BP. *J Gen Physiol.* 1985;86:1.)

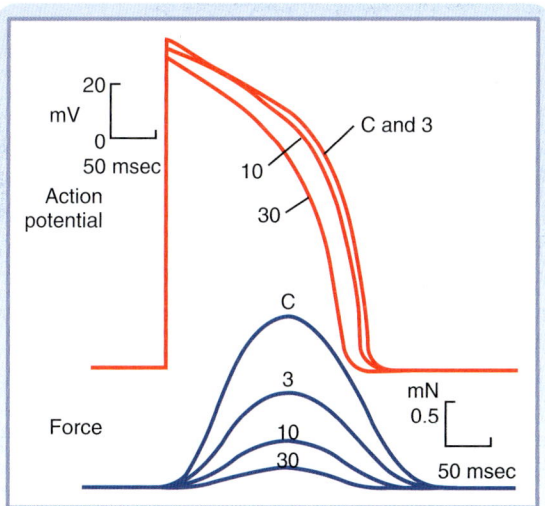

• **Fig. 16.7** Effects of diltiazem, a calcium channel antagonist, on the action potentials (in millivolts [mV]) and isometric contractile forces (in millinewtons [mN]) recorded from a papillary muscle in vitro. The tracings were recorded under control conditions (C) and in the presence of diltiazem in concentrations of 3, 10, and 30 μmol/L. (Redrawn from Hirth C, et al. *J Mol Cell Cardiol.* 1983;15:799.)

However, as V_m approaches and then attains positive values near the peak of the action potential upstroke, g_K suddenly decreases (Fig. 16.8). The diminished K⁺ current associated with the reduction in g_K prevents excessive loss of K⁺ from the cell during the plateau.

AT THE CELLULAR LEVEL

To enhance g_{Ca}, catecholamines first bind to β-adrenergic receptors in the cardiac cell membrane. This interaction stimulates the membrane-bound enzyme adenylate cyclase, which raises the intracellular concentration of cyclic adenosine monophosphate (cAMP; see also Chapter 3). The rise in cAMP activates cAMP-dependent protein kinase, which in turn promotes phosphorylation of L-type calcium channels in the cell membrane and thus augments the influx of Ca⁺⁺ into the cells (see Fig. 16.6). Conversely, acetylcholine interacts with muscarinic receptors in the cell membrane to inhibit adenylate cyclase. In this way, acetylcholine antagonizes the activation of calcium channels and thereby diminishes g_{Ca}.

IN THE CLINIC

Calcium channel antagonists are substances that block calcium channels. Examples include the drugs verapamil, amlodipine, and diltiazem. These drugs decrease g_{Ca} and thereby impede the influx of Ca⁺⁺ into myocardial cells. Calcium channel antagonists decrease the duration of the action potential plateau and diminish the strength of the cardiac contraction (see Fig. 16.7). Calcium channel antagonists also depress the contraction of vascular smooth muscle and thereby induce generalized vasodilation. This diminished vascular resistance reduces the counterforce (afterload) that opposes the propulsion of blood from the ventricles into the arterial system, as explained in Chapter 17. Hence, vasodilator drugs such as the calcium channel antagonists are often referred to as *afterload-reducing drugs.*

The reduction in g_K at both positive and low negative values of V_m is called **inward rectification.** Inward rectification is a characteristic of several K⁺ currents, including i_{K1} (Fig. 16.9). For these channels, large K⁺ currents flow at negative values of V_m (i.e., g_K is large). However, when V_m is near 0 mV or positive, as occurs during the plateau (phase 2), little or no K⁺ current flows (i.e., g_K is small). Thus the substantial g_K that prevails during phase 4 of the cardiac action potential (see Fig. 16.8) is large because of the i_{K1} channels, but current through these channels is greatly diminished during the plateau (see Fig. 16.9).

Other potassium channels play a role in phase 2 of the action potential. These potassium channels are characterized as **delayed i_K** channels. They are closed during phase 4 and are activated very slowly by the potentials that prevail toward the end of phase 0. Hence, activation of these channels tends to increase g_K very gradually during phase 2. These channels play only a minor role during phase 2, but they contribute to the process of final repolarization (phase 3), as described in the section "Genesis of Repolarization (Phase 3)." There exist two types of i_K channels, classified according to their rates of activation: The more slowly activating channel is designated the **i_{Ks} channel,** and the more rapidly activating channel is designated the **i_{Kr} channel** (see Fig. 16.8). The

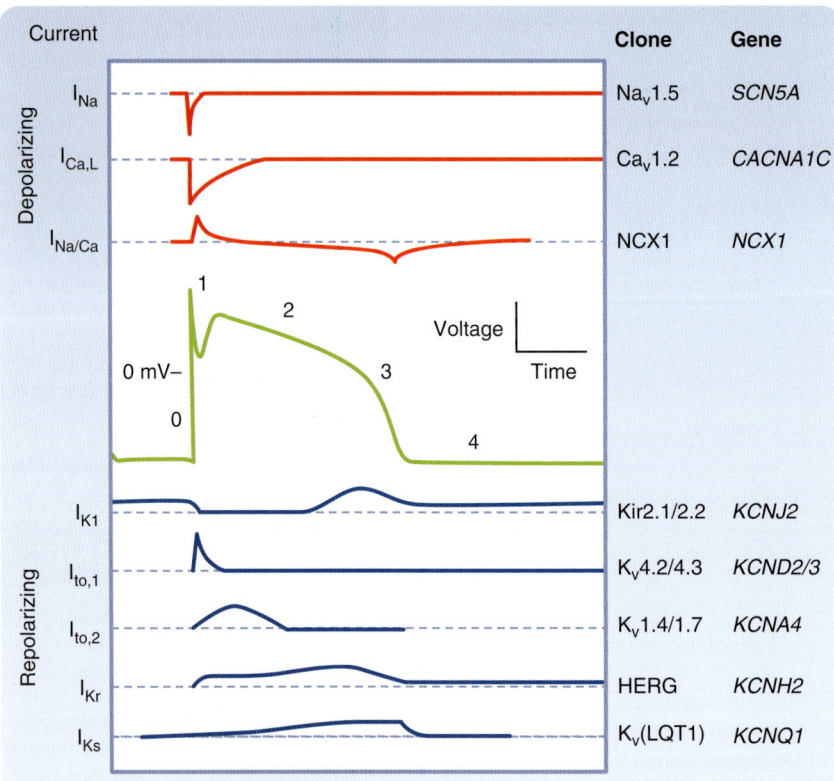

• **Fig. 16.8** Changes in Ion Currents During the Various Phases of the Action Potential in a Fast-Response Cardiac Ventricular Cell. **Upper panels,** Depolarizing ion currents. **Lower panels,** Repolarizing ion currents. The inward currents include the fast Na^+ (i_{Na}) and L-type Ca^{++} ($i_{Ca,L}$) currents. Outward currents are i_{K1}, i_{to}, and the rapid (i_{Kr}) and slow (i_{Ks}) delayed rectifier K^+ currents. The clones and genes for the principal ionic currents are also included. (Redrawn from Tomaselli G, Marbán E. *Cardiovasc Res.* 1999;42:270.)

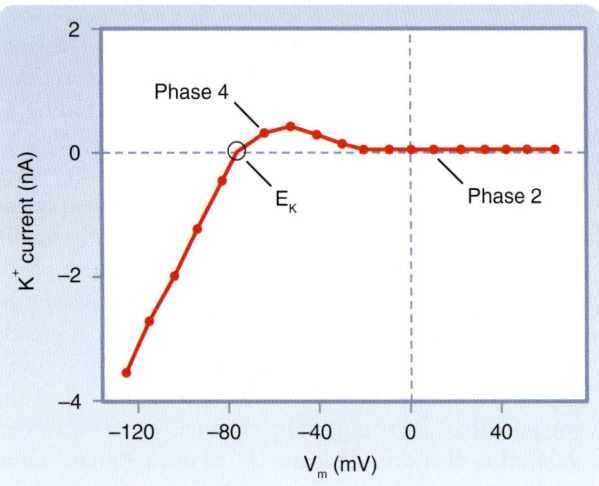

• **Fig. 16.9** Inward Rectified K^+ Currents Recorded From a Ventricular Myocyte When the Potential Was Changed From a Holding Potential of −80 mV to Various Test Potentials. Positive values along the vertical axis represent outward currents; negative values represent inward currents. The point at which the current trace intersects the x-axis is the reversal potential *(open circle);* it denotes the Nernst equilibrium potential (E_K), at which point the chemical and electrostatic forces are equal. V_m, membrane potential. (Redrawn from Giles WR, Imaizumi Y. *J Physiol [Lond].* 1988;405:123.)

duration of the action potential in myocytes in various regions of the ventricular myocardium is determined in part by the relative distributions of these i_{Kr} and i_{Ks} channels.

The action potential plateau persists as long as the efflux of charge carried mainly by K^+ is balanced by the influx of charge carried mainly by Ca^{++}. The effects of altering this balance are exemplified by the action of the calcium channel antagonist diltiazem in a papillary muscle preparation (see Fig. 16.7). With increasing concentrations of diltiazem, the plateau voltage becomes progressively less positive and the plateau duration diminishes. Conversely, administration of certain potassium channel antagonists prolongs the plateau substantially.

Genesis of Final Repolarization (Phase 3)

The process of final repolarization (phase 3) starts at the end of phase 2, when efflux of K^+ from the cardiac cell begins to exceed influx of Ca^{++}. As noted, at least three outward K^+ currents (i_{to}, i_K, and i_{K1}) contribute to the final repolarization (phase 3) of the cardiac cell (see Figs. 16.3 and 16.8).

The i_{to}, i_{Kr}, and i_{Ks} currents help initiate repolarization. These currents are therefore important determinants of the duration of the plateau. For example, the duration of the plateau is substantially less in atrial myocytes than in ventricular myocytes (Fig. 16.10) because the magnitude

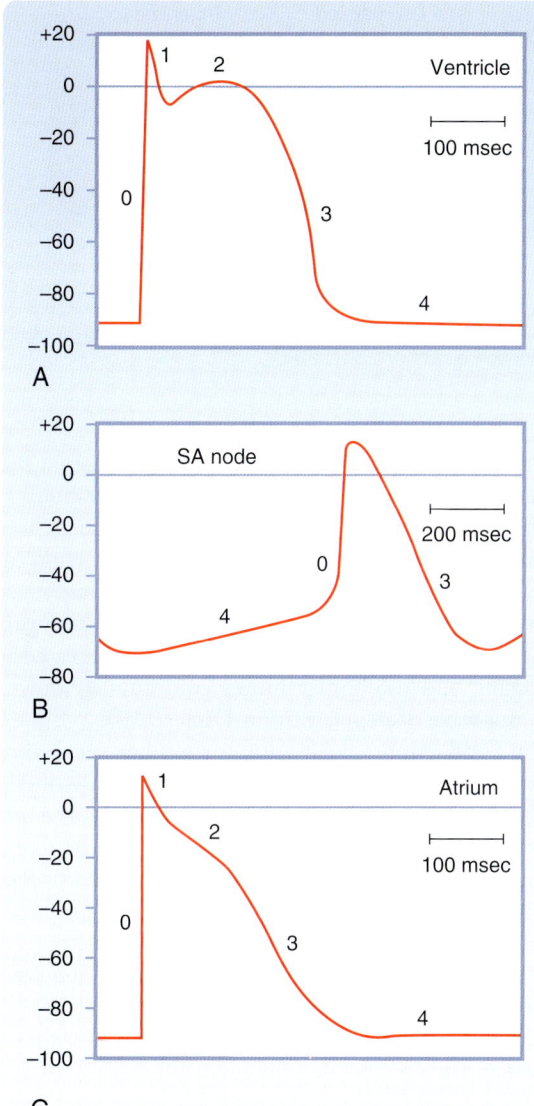

Fig. 16.10 Typical Action Potentials (in Millivolts, y-Axis). These potentials were recorded from cells in the ventricle **(A),** sinoatrial (SA) node **(B),** and atrium **(C).** Note that the time calibration in **B** differs from that in **A** and **C.** Numbers on the curves denote the phases of the action potential. (From Hoffman BF, Cranefield PF. *Electrophysiology of the Heart.* New York: McGraw-Hill; 1960.)

of i_{to} during the plateau is greater in atrial myocites than in ventricular myocytes. As already noted, the duration of the action potential in ventricular myocytes varies considerably with the location of these myocytes in the ventricular walls (see Fig. 16.5). The i_{to} and i_K currents mainly account for these differences. In endocardial myocytes, in which the duration of the action potential is least, the magnitude of i_K is greatest. The converse applies to the midmyocardial myocytes. The magnitude of i_K and the duration of the action potential are intermediate for epicardial myocytes.

The i_{K1} current does not participate in the initiation of repolarization because the conductance of these channels is very small over the range of V_m values that prevail during the plateau. However, the i_{K1} channels contribute substantially to the rate of repolarization once phase 3 has

been initiated. As V_m becomes increasingly negative during phase 3, the conductance of the channels that carry the i_{K1} current progressively increases and thereby accelerates repolarization (see Fig. 16.3).

Restoration of Ionic Concentrations (Phase 4)

The steady inward leak of Na^+ that enters the cell rapidly during phase 0 and more slowly throughout the cardiac cycle would gradually depolarize the resting membrane voltage were it not for **Na^+,K^+-ATPase,** which is located in the cell membrane (see Chapter 1). Similarly, most of the excess Ca^{++} ions that had entered the cell mainly during phase 2 are eliminated principally by a $3Na^+$-Ca^{++} antiporter, which exchanges three Na^+ ions for one Ca^{++} ion. However, some of the Ca^{++} ions are eliminated by an ATP-driven Ca^{++} pump.

 AT THE CELLULAR LEVEL

Cardiac ion channels are connected to cellular proteins to form macromolecular complexes. These complexes are involved in modulating the transport, membrane localization, operation, posttranslational modification and turnover of particular ion channels. The carboxy terminals of ion channels link them with several intracellular proteins such as PDZ (postsynaptic density, disc large, and zonula occludens-1) domain proteins whose binding sites interact with synapse-associated protein (SAP97), syntrophin, and A-kinase anchoring protein (AKAP5) among others. Different macromolecular complexes are found in distinct cellular locations, and this is thought to underlie ion channel distribution. The voltage-sensitive sodium channel (Na$_v$1.5) is linked with syntrophin/dystrophin at lateral cell membranes and with ankyrin B, plakophilin-2, and calmodulin-dependent protein kinase II at the intercalated disk. Also, different ion channels can found in the same complex. Thus Na$_v$1.5 channels and inward-rectifying K (Kir2.1) channels can be connected in a complex, or *channelosome,* with SAP97. This link not only affects the localization but also allows changes in the abundance of Kir2.1 channels to produce reciprocal changes in Na$_v$1.5 abundance; the converse is also observed. Thus the complex contains Kir2.1, which sets the resting membrane potential, and Na$_v$1.5, which accounts for rapid excitation. Colocalization of these two channels therefore exerts a powerful effect on excitability and its regulation under normal and pathological conditions (arrhythmias).

Slow-Response Action Potentials

Fast-response action potentials (see Fig. 16.1*A*) consist of four principal components: an upstroke (phase 0), an early partial repolarization (phase 1), a plateau (phase 2), and a final repolarization (phase 3). However, in the slow-response action potential (see Fig. 16.1*B*), the upstroke is much less steep, early repolarization (phase 1) is absent, the plateau is less prolonged and not as flat, and the transition from the plateau to the final repolarization is less distinct.

Blocking fast sodium channels with tetrodotoxin in a fast-response fiber can generate slow responses under

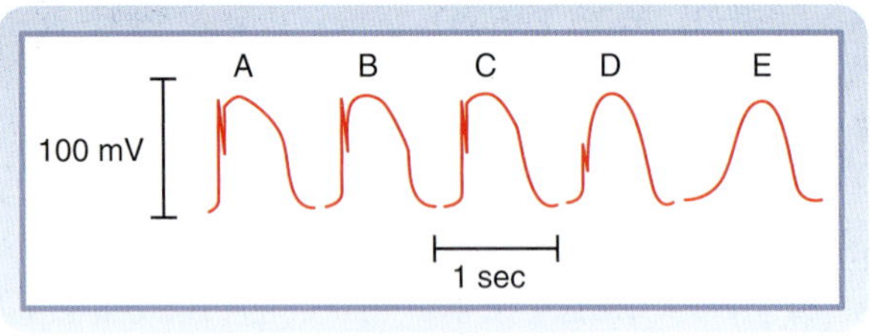

• **Fig. 16.11** Effect of Tetrodotoxin, Which Blocks Fast Sodium Channels, on the Action Potentials (Tracings A to E) Recorded in a Purkinje Fiber. The concentrations of tetrodotoxin were 0 mol/L in tracing A, 3×10^{-8} mol/L in tracing B, 3×10^{-7} mol/L in tracing C, and 3×10^{-6} mol/L in tracings D and E; E was recorded later than D. (Redrawn from Carmeliet E, Vereecke J. *Pflügers Arch.* 1969;313:300.)

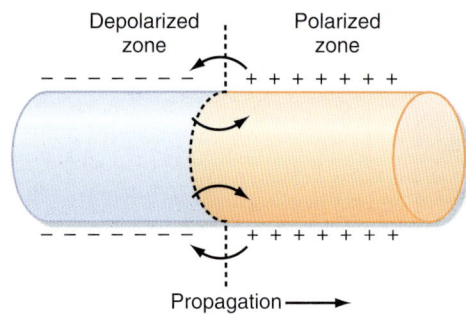

• **Fig. 16.12** The role of local currents in the propagation of a wave of excitation down a cardiac fiber.

appropriate conditions. The Purkinje fiber action potentials shown in Fig. 16.11 clearly exhibit the two response types. In the control tracing (A), the typical fast-response action potential displays a prominent notch, as a result of i_{to}, that separates the upstroke from the plateau. Progressively higher concentrations of tetrodotoxin produce a graded blockade of the fast sodium channels, as demonstrated in tracings B to E. The upstroke and notch are progressively less prominent in tracings B to D. In tracing E, the notch has disappeared, and the upstroke is very gradual; this action potential resembles a typical slow response.

Certain cells in the heart, notably those in the SA and AV nodes, exhibit slow-response action potentials. In these cells, depolarization is achieved mainly by influx of Ca^{++} through L-type calcium channels instead of influx of Na^+ through fast sodium channels. Repolarization is accomplished in these fibers by inactivation of the calcium channels and by the increased K^+ conductance through the i_{K1} and i_K channels (see Fig. 16.3).

Conduction in Cardiac Fibers

An action potential traveling along a cardiac muscle fiber is propagated by local circuit currents, much as it is in nerve and skeletal muscle fibers (see Chapter 5). When the wave of depolarization reaches the end of the cell, the impulse is conducted to adjacent cells through gap junctions (see Chapter 2). Impulses pass more readily along the length of the cell (isotropic) than laterally from cell to cell (anisotropic) because gap junctions are preferentially located at the ends of the cell. Gap junctions are rather nonselective in their permeability by ions and have a low electrical resistance that allows ionic current to pass from one cell to another. The electrical resistance of gap junctions is similar to that of cytoplasm. The flow of charge from cell to cell follows the principles of local circuit currents and therefore allows intercellular propagation of the impulse.

Conduction of the Fast Response

In fast- and slow-response fibers, the characteristics of conduction differ. In fast-response fibers, fast sodium channels are activated when the transmembrane potential

of one region of the fiber suddenly changes from a resting value of approximately −90 mV to the threshold value of approximately −65 mV. The inward Na^+ current then rapidly depolarizes the cell at that site. This portion of the fiber subsequently becomes part of the depolarized zone, and the border is displaced accordingly. The same process then begins at the new border. This process is repeated again and again, and the border moves continuously down the fiber as a wave of depolarization (Fig. 16.12).

The conduction velocity along the fiber varies directly with the action potential amplitude and the rate of change of the potential (dV_m/dt) during phase 0. The action potential amplitude is the potential difference between the fully depolarized and the fully polarized regions of the cell interior. The magnitude of the local current is proportional to this potential difference (see Chapter 5). Because these local currents shift the potential of the resting zone toward the threshold value, they are local stimuli that depolarize the adjacent resting portion of the fiber to its threshold potential. The greater the potential difference between the depolarized and polarized regions (i.e., the greater the action potential amplitude), the more effectively local stimuli can depolarize adjacent parts of the membrane, and the more rapidly the wave of depolarization is propagated down the fiber.

The dV_m/dt during phase 0 is also an important determinant of conduction velocity. If the active portion of the fiber depolarizes gradually, the local currents between the resting region and the neighboring depolarizing region are small. The resting region adjacent to the active zone is depolarized gradually, and more time is therefore required for each new section of the fiber to reach threshold. This allows some sodium channels to inactivate.

The resting membrane potential is another important determinant of conduction velocity. Changes in the resting membrane potential influence both the amplitude of the action potential and dV_m/dt, which in turn alter the conduction velocity (Fig. 16.13). Depolarization of V_m inactivates the fast sodium channels, which in turn decreases the amplitude of the action potential and the dV_m/dt, and as a consequence conduction velocity is slowed. In addition to changes in extracellular $[K^+]$, premature excitation of a cell that has not completely repolarized also results in a

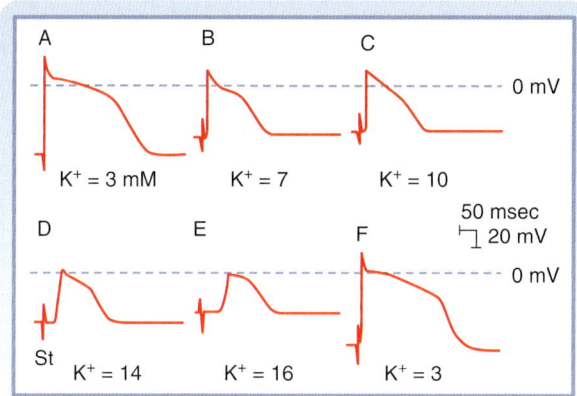

• **Fig. 16.13** Effect of Changes in Extracellular Potassium Concentration (Extracellular [K⁺]) on the Action Potentials Recorded From a Purkinje Fiber. The stimulus artifact (St in tracing D) appears as a biphasic spike to the left of the upstroke of the action potential. The *horizontal dashed lines* near the peaks of the action potentials denote 0 mV. When extracellular [K⁺] is 3 mmol/L ("mM"; tracings A and F), the resting membrane voltage (V_m) is −82 mV, and the slope of phase 0 is steep. At the end of phase 0, the overshoot attains a value of +30 mV. Hence, the action potential amplitude is 112 mV. The distance from the stimulus artifact to the beginning of phase 0 is inversely proportional to the conduction velocity. When extracellular [K⁺] is increased gradually to 16 mmol (tracings B to E), the resting V_m becomes progressively less negative. At the same time, the amplitudes and durations of the action potentials and the steepness of the upstrokes all diminish. As a consequence, conduction velocity decreases progressively. At extracellular [K⁺] levels of 14 and 16 mmol (tracings D and E), the resting V_m attains levels sufficient to inactivate all the fast sodium channels and lead to the characteristic slow-response action potentials. (From Myerburg RJ, Lazzara R. In: Fisch E [ed]. *Complex Electrocardiography*. Philadelphia: FA Davis; 1973.)

decrease in conduction velocity. This too reflects the fact that when V_m is depolarized, more fast sodium channels are inactivated, and thus only a fraction of the sodium channels are available to conduct the inward Na⁺ current during phase 0.

Conduction of the Slow Response

Local circuits (see Fig. 16.12) also propagate the slow response, whose conduction characteristics differ quantitatively from those of the fast response. The threshold potential is approximately −40 mV for the slow response, and conduction is much slower than for the fast response. The conduction velocities of the slow response in the SA and AV nodes are approximately 0.02 to 0.10 m/sec. The fast-response conduction velocities are approximately 0.30 to 1.00 m/sec for myocardial cells and 1.00 to 4.00 m/sec for the specialized conducting (Purkinje) fibers in the ventricles. Slow responses are more readily blocked than are fast responses; that is, conduction ceases before the impulse reaches the end of the myocardial fiber. Also, fast-response fibers can respond at repetition rates that are much faster than those of slow-response fibers.

Most of the experimentally induced changes in transmembrane potential shown in Fig. 16.13 also take place in the cardiac tissue of patients with coronary artery disease. When regional myocardial blood flow is diminished, the supply of O₂ and metabolic substrates delivered to the ischemic tissues is insufficient. The Na⁺,K⁺-ATPase in the membrane of cardiac myocytes requires considerable metabolic energy to maintain the normal transmembrane gradients of Na⁺ and K⁺. When blood flow is inadequate, the activity of Na⁺,K⁺-ATPase is impaired, and the ischemic myocytes gain excess Na⁺ and lose K⁺ to the surrounding interstitial space. As a consequence, [K⁺] in the extracellular fluid surrounding the ischemic myocytes is elevated. Such changes in extracellular [K⁺] may, if sufficiently large, disturb cardiac rhythm and conduction critically (see Fig. 16.13).

Cardiac Excitability

Because of the rapid development of artificial pacemakers and other electrical devices for correcting cardiac rhythm disturbances, detailed knowledge of cardiac excitability is essential. The excitability characteristics of various types of cardiac cells differ considerably, depending on whether the action potentials are fast or slow responses.

Fast Response

Once the fast response has been initiated, the depolarized cell is no longer excitable until it has partially repolarized (see Fig. 16.1*A*). In the fast response, the interval from the beginning of the action potential until the fiber is able to conduct another action potential is called the effective refractory period which extends from the beginning of phase 0 to a point in phase 3 at which repolarization has reached approximately −50 mV. At approximately this value of V_m, many of the fast sodium channels have transitioned from the inactivated state to the closed state. However, the cardiac fiber is not fully excitable until it has been completely repolarized. Before complete repolarization (i.e., during the relative refractory period), an action potential may be evoked only when the stimulus is stronger than a stimulus that could elicit a response during phase 4.

When a fast response is evoked during the relative refractory period of a previous excitation, its characteristics vary with the membrane potential that exists at the time of stimulation (Fig. 16.14). The later in the relative refractory period that the fiber is stimulated, the greater are the increases in the amplitude of the response and the slope of the upstroke because the number of fast sodium channels that have recovered from inactivation increases as repolarization proceeds. As a consequence, propagation velocity also increases the later in the relative refractory period that the fiber is stimulated. Once the fiber is fully repolarized, the response is constant no matter what time in phase 4 the stimulus is applied.

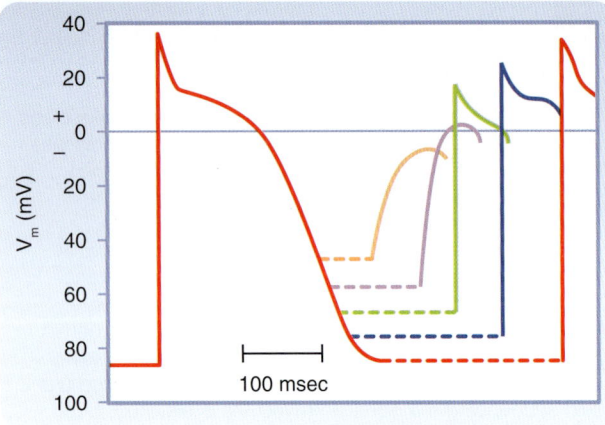

• **Fig. 16.14** Changes in action potential amplitude and upstroke slope as action potentials are initiated at different stages of the relative refractory period of the preceding excitation. (Redrawn from Rosen MR, et al. *Am Heart J.* 1974;88:380.)

Slow Response

In slow-response fibers, the relative refractory period frequently extends well beyond phase 3 (see Fig. 16.1*B*). Even after the cell has completely repolarized, it may be difficult to evoke a propagated response for some time. This characteristic of slow-response fibers is called *postrepolarization refractoriness*.

🩺 IN THE CLINIC

In a patient who has occasional premature atrial depolarizations (see Fig. 16.32), the timing of these early beats may determine their clinical consequence. If they occur late in the relative refractory period of the preceding depolarization or after full repolarization, the premature depolarization is inconsequential. However, if the premature depolarizations originate early in the relative refractory period of the ventricles, conduction of the premature impulse from the site of origin is slow, and hence a cardiac impulse is more likely to reexcite some myocardial region through which it had passed previously (a phenomenon known as *reentry*). If that reentry is irregular (i.e., if ventricular fibrillation ensues), the heart cannot pump effectively, and death may result.

Action potentials evoked early in the relative refractory period are small, and the upstrokes are not very steep (Fig. 16.15). The amplitudes and upstroke slopes progressively improve as action potentials are elicited later in the relative refractory period. Recovery of full excitability is much slower than recovery of the fast response. Impulses that arrive early in the relative refractory period are conducted much more slowly than those that arrive late in that period. The long refractory periods also lead to conduction blocks. Even when slow responses recur at low frequency, the fiber may be able to conduct only a fraction of these impulses; for example, in certain conditions, only alternate impulses may be propagated.

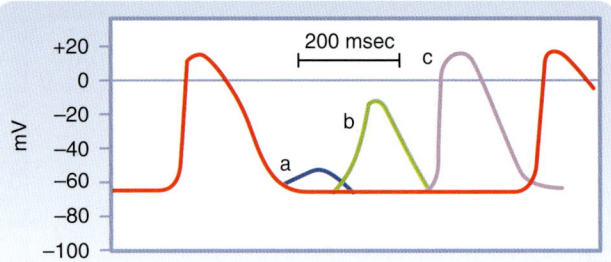

• **Fig. 16.15** Effects of Excitation at Various Times After the Initiation of an Action Potential in a Slow-Response Fiber. In this fiber, excitation very late in phase 3 (or early in phase 4) induces a small, nonpropagated (local) response (wave a). Later in phase 4, a propagated response (wave b) can be elicited, but its amplitude is small and the upstroke is not very steep; this response is conducted very slowly. Still later in phase 4, full excitability is regained, and the response (wave c) displays normal characteristics. (Modified from Singer DH, et al. *Prog Cardiovasc Dis.* 1981;24:97.)

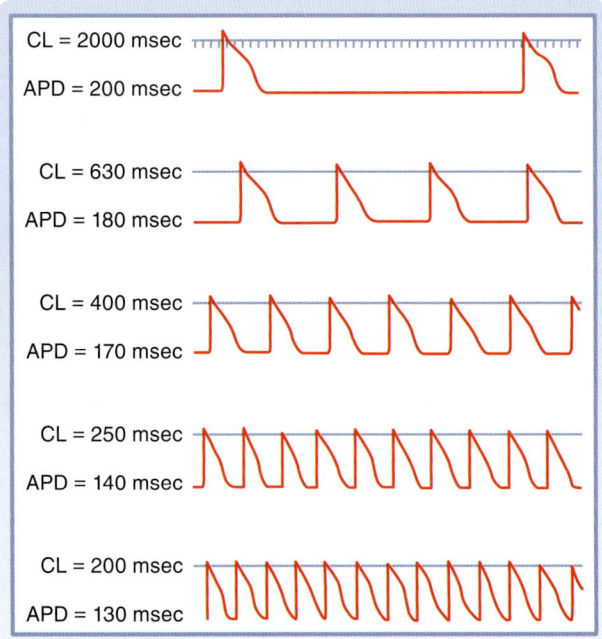

• **Fig. 16.16** Effect of changes in cycle length (CL) on the action potential duration (APD) of Purkinje fibers. (Modified from Singer D, Ten Eick RE. *Am J Cardiol.* 1971;28:381.)

Effects of Cycle Length

Cycle length is the time between successive action potentials. Changes in cycle length alter the duration of the action potential in cardiac cells (Fig. 16.16; also see Fig. 16.5) and thus change their refractory periods. As a consequence, changes in cycle length are often important factors in the initiation or termination of certain arrhythmias (irregular heart rhythms).

The changes in action potential duration produced by stepwise reductions in cycle length from 2000 to 200 msec in a Purkinje fiber are shown in Fig. 16.16. Note that as cycle length diminishes, the duration of the action potential decreases. This direct correlation between action potential duration and cycle length is mediated by changes in g_K that involve at least two types of potassium channels: namely, those that conduct the delayed rectifier K^+ currents (i_{Kr} and i_{Ks}) and those that conduct the transient outward K^+ current (i_{to}).

The i_K current is activated at values of V_m near zero, but the current activates slowly, remains activated for hundreds of milliseconds, and also inactivates very slowly. As a consequence, as the basic cycle length diminishes, each action potential tends to occur earlier in the inactivation period of the i_K current initiated by the preceding action potential. Therefore, the shorter the basic cycle length, the greater the outward K^+ current during phase 2 and hence the shorter the action potential duration.

The i_{to} current also influences the relationship between cycle length and action potential duration. The i_{to} current is also activated at near zero potential, and its magnitude varies inversely with cardiac frequency. Therefore, as cycle length decreases, the consequent increase in the outward K^+ current shortens the plateau.

Natural Excitation of the Heart and the Electrocardiogram

Excitation of the heart normally occurs in an ordered manner, which allows effective pumping of blood. This ordered excitation occurs via the heart's conduction system (Fig. 16.17). The SA node is the pacemaker of the heart and initiates the spread of action potentials throughout the atria. This spread of excitation reaches the AV node, where conduction is slowed so that atrial contraction can occur and the ventricles can be adequately filled. Excitation then spreads rapidly throughout the ventricles via the Purkinje fibers so that the ventricular myocytes contract in a coordinated manner. The properties of each component of the heart's conduction system are described in the next sections.

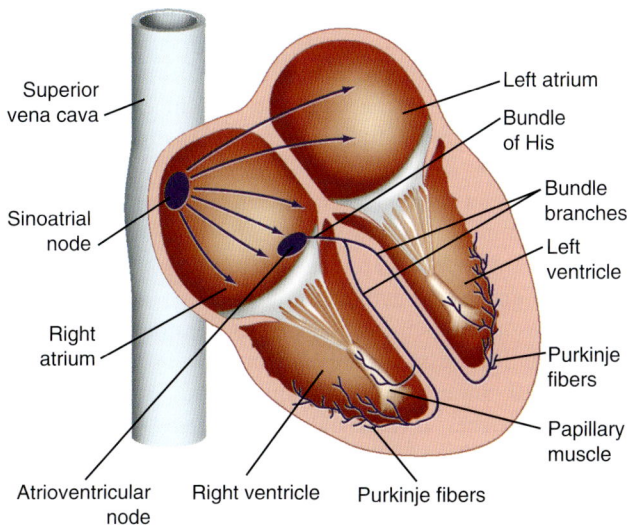

Superior
vena cava

Left atrium

Bundle
of His

Bundle
branches

Sinoatrial
node

Left
ventricle

Right
atrium

Purkinje
fibers

Papillary
muscle

Atrioventricular
node   Right ventricle   Purkinje fibers

• **Fig. 16.17** The cardiac conduction system.

The autonomic nervous system controls various aspects of cardiac function, such as the heart rate and contraction strength. However, cardiac function does not require intact innervation. Indeed, a patient with a cardiac transplant, whose heart is completely denervated, may still adapt well to stressful situations. The ability of a denervated, transplanted heart to adapt to changing conditions lies in certain intrinsic properties of cardiac tissue, especially its automaticity.

The properties of **automaticity** (the ability to initiate its own beat) and **rhythmicity** (the regularity of pacemaking activity) allow a perfused heart to beat even when it is completely removed from the body. The vertebrate heartbeat is myogenic in origin. If the coronary vasculature of an excised heart is artificially perfused with blood or an oxygenated electrolyte solution, rhythmic cardiac contractions may persist for many hours. Some cells in the atria and ventricles can initiate beats; such cells reside mainly in nodal tissues or specialized conducting fibers of the heart.

Sinoatrial Node

As noted, the region of the mammalian heart that ordinarily generates impulses at the greatest frequency is the SA node; it is the main cardiac pacemaker. Detailed mapping of the electrical potentials on the surface of the right atrium has revealed that two or three sites of automaticity, located 1 or 2 cm from the SA node itself, serve along with the SA node as an atrial pacemaker complex. At times, all these loci initiate impulses simultaneously. At other times, the site of earliest excitation shifts from locus to locus, depending on certain conditions, such as the level of autonomic neural activity.

In humans, the SA node is approximately 8 mm long and 2 mm thick, and it lies posteriorly in the groove at the junction between the superior vena cava and the right atrium. The sinus node artery runs lengthwise through the center of the node. The SA node contains two principal cell types: (1) small, round cells that have few organelles and myofibrils and (2) slender, elongated cells that are intermediate in appearance between the round and "ordinary" atrial myocardial cells. The round cells are probably the pacemaker cells; the slender, elongated cells probably conduct the impulses within the node and to the nodal margins.

A typical transmembrane action potential recorded from an SA node cell is depicted in Fig. 16.10B. In comparison with the transmembrane potential recorded from a ventricular myocardial cell (see Fig. 16.10A), the resting potential of the SA node cell is usually less negative, the upstroke of the action potential (phase 0) is less steep, the plateau is not sustained, and repolarization (phase 3) is more gradual. These are characteristic attributes of the slow response. Tetrodotoxin (which blocks the fast Na^+ current) has no influence on the SA nodal action potential because the action potential upstroke is not produced by an inward Na^+ current through fast channels. Thus tetrodotoxin has no effect on cells that exhibit the slow response.

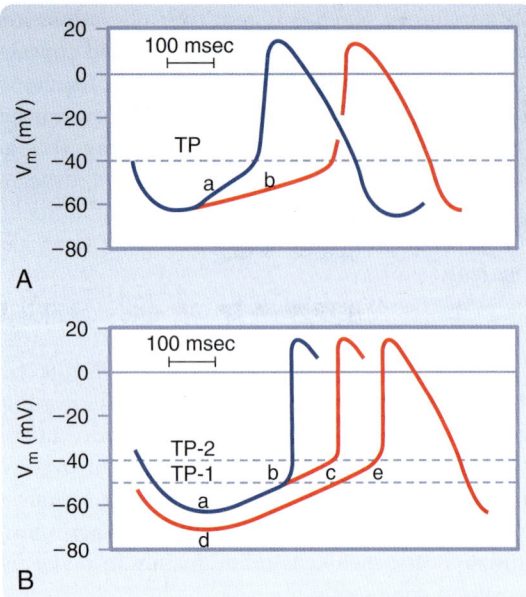

• **Fig. 16.18** Mechanisms Involved in the Changes in Frequency of Pacemaker Firing. **A,** A reduction in the slope (from wave a to wave b) of slow diastolic depolarization diminishes the firing frequency. TP, threshold potential. **B,** An increase in the threshold potential (from TP-1 to TP-2) or an increase in the magnitude of the maximum diastolic potential (from wave segments a to d) also diminishes the firing frequency. (From Hoffman BF, Cranefield PF. *Electrophysiology of the Heart.* New York: McGraw-Hill; 1960.)

The transmembrane potential during phase 4 is much less negative in SA (and AV) nodal automatic cells than in atrial or ventricular myocytes because nodal cells lack the i_{K1} (inward rectifying) type of potassium channel. Thus the ratio of g_K to g_{Na} during phase 4 is much less in nodal cells than in myocytes. Hence, during phase 4, V_m deviates much more from the K^+ Nernst equilibrium potential (E_K) in nodal cells than it does in myocytes.

The principal feature of a pacemaker cell that distinguishes it from the other cells manifests in phase 4. In nonautomatic cells, the potential remains constant during this phase, whereas a pacemaker fiber is characterized by slow diastolic depolarization throughout phase 4. Depolarization proceeds at a steady rate until a threshold is attained, and an action potential is then triggered.

Pacemaker cell frequency may be varied by a change in (1) the rate of depolarization during phase 4, (2) the maximal negativity during phase 4, or (3) the threshold potential (Fig. 16.18). When the rate of slow diastolic depolarization is increased, the threshold potential is attained earlier, and the heart rate increases. A rise in the threshold potential delays the onset of phase 0, and the heart rate is reduced. Similarly, when the maximal negative potential is increased, more time is required to reach the threshold potential, when the slope of phase 4 remains unchanged, and the heart rate therefore diminishes.

Ionic Basis of Automaticity

Several ionic currents contribute to the slow diastolic depolarization that characteristically occurs in the automatic

 IN THE CLINIC

Ordinarily, the frequency of pacemaker firing is controlled by the activity of both divisions of the autonomic nervous system. Increased sympathetic nervous activity, through the release of norepinephrine, raises the heart rate principally by increasing the slope of the slow diastolic depolarization. This mechanism of increasing heart rate functions during physical exertion, anxiety, or certain illnesses such as febrile infectious diseases.

Increased vagal activity, through the release of acetylcholine, diminishes the heart rate by hyperpolarizing the pacemaker cell membrane and reducing the slope of the slow diastolic depolarization. These mechanisms of decreasing the heart rate occur when vagal activity is predominant over sympathetic activity. An extreme example is vasovagal syncope, a brief period of lightheadedness or loss of consciousness caused by an intense burst of vagal activity. This type of syncope is a reflex response to pain or to certain psychological stimuli.

Changes in autonomic neural activity do not usually change the heart rate by altering the threshold level of V_m in the nodal pacemaker cells. However, certain antiarrhythmic drugs, such as quinidine and procainamide, do shift the threshold potential of the automatic cells to less negative values.

cells in the heart. In the pacemaker cells of the SA node, at least three ionic currents mediate the slow diastolic depolarization: (1) an outward K^+ current, i_K; (2) a hyperpolarization-induced inward current, i_f; and (3) an inward Ca^{++} current, I_{Ca} (Fig. 16.19).

The repetitive firing of the pacemaker cell begins with the i_K current. Efflux of K^+ tends to repolarize the cell after the upstroke of the action potential. K^+ continues to move out well beyond the time of maximal repolarization, but its efflux diminishes throughout phase 4 (see Fig. 16.19). As the current diminishes, its opposition to the depolarizing effects of the two inward currents (i_f and i_{Ca}) also gradually decreases. The progressive diastolic depolarization is mediated by the i_f and i_{Ca} currents, which oppose the repolarizing effect of the i_K current.

The inward current i_f is activated near the end of repolarization and is carried mainly by Na^+ through specific channels that differ from the fast sodium channels. The current was dubbed "funny" because its discoverers had not expected to detect an inward Na^+ current in pacemaker cells at the end of repolarization. This current is activated as the membrane potential becomes hyperpolarized beyond −50 mV. The more negative the membrane potential at this time, the greater the activation of i_f.

The second current responsible for diastolic depolarization is the inward rectifying Ca^{++} current, i_{Ca}. This current is activated toward the end of phase 4 as the membrane potential reaches a value of approximately −35 mV (see Fig. 16.19). Once the calcium channels are activated, influx of Ca^{++} into the cell increases. This influx accelerates the rate of diastolic depolarization, which then leads to the action potential upstroke. A decrease in extracellular

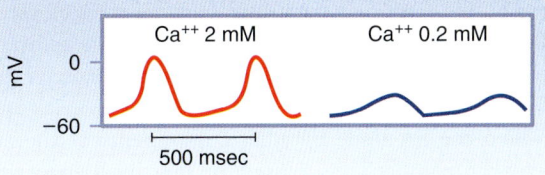

• **Fig. 16.20** Membrane Action Potentials Recorded From an Sino-atrial Node Pacemaker Cell. The concentration of Ca⁺⁺ in the bath was reduced from 2 to 0.2 mmol/L ("mM"). (Modified from Kohlhardt M, et al. *Basic Res Cardiol.* 1976;71:17.)

[Ca⁺⁺] (Fig. 16.20) or the addition of calcium channel antagonists diminishes the amplitude of the action potential and the slope of the slow diastolic depolarization in SA node cells. Research evidence indicates that additional ion currents—including a sustained (background) inward Na^+ current (i_{Na}), the T-type Ca^{++} current, and the Na^+/Ca^{++} exchange current triggered by spontaneous release of Ca^{++} from the sarcoplasmic reticulum—may also be involved in pacemaking. These observations illustrate the many ways to sustain this vital function.[a]

IN THE CLINIC

Regions of the heart other than the SA node may initiate beats in special circumstances. Such sites are called *ectopic foci* or *ectopic pacemakers.* Ectopic foci may become pacemakers when (1) their own rhythmicity becomes enhanced, (2) the rhythmicity of the higher order pacemakers becomes depressed, or (3) all conduction pathways between the ectopic focus and regions with greater rhythmicity become blocked. Ectopic pacemakers may act as a safety mechanism when normal pacemaking centers fail. However, if an ectopic center fires while the normal pacemaking center still functions, the ectopic activity may induce either sporadic rhythm disturbances, such as premature depolarizations, or continuous rhythm disturbances, such as paroxysmal tachycardias (see the section "Ectopic Tachycardias").

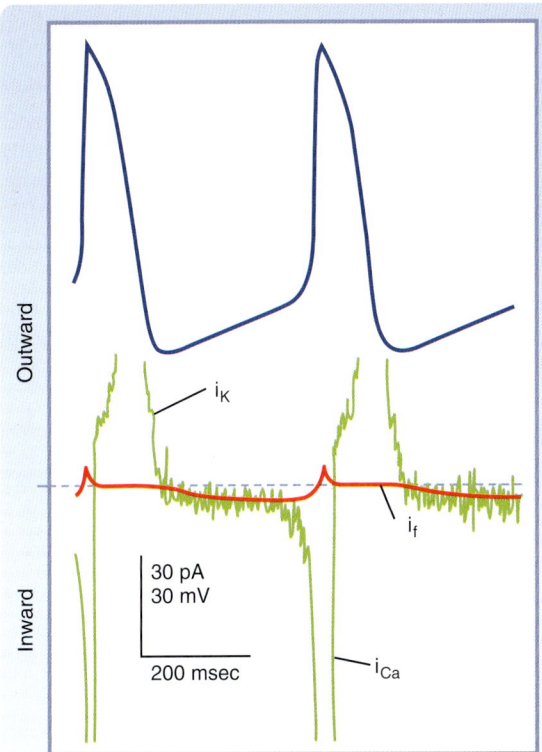

• **Fig. 16.19** The membrane potential changes (*blue trace, top half*) that occur in sinoatrial node cells are produced by three principal currents (*green trace, bottom half):* the inward Ca⁺⁺ current (i_{Ca}); a hyperpolarization-induced inward current (i_f), and an outward K⁺ current (i_K). The thin noisy *green trace* shows net membrane current and the approximate time course of the repolarizing outward K⁺ current (i_K), the hyperpolarization-induced inward current (i_f), and the L-type Ca⁺⁺ current (i_{Ca}). The *thick bold red line* in the current trace indicates the magnitude and direction of estimated I_f. (Redrawn from van Ginneken ACG, Giles W. *J Physiol.* 1991;434:57.)

AT THE CELLULAR LEVEL

The so-called funny current (i_f) in cardiac SA node cells is activated by *hyperpolarization* and gated by *cyclic nucleotides*, and its channel is designated HCN. There are four members of the *HCN* gene family, and such channels are found in central nervous system neurons that generate action potentials repetitively. Transmembrane segment 4 (S₄) of HCN has many positively charged amino acids that act as voltage sensors, which are also present in voltage-gated sodium, potassium, and calcium channels. The dominant channel expressed in the heart is derived from the *HCN4* gene. Mutations in amino acids in S₄ and in the S₄-to-S₅ linker cause marked changes in the voltage dependence of activation, in such a way that greater hyperpolarization is needed to open the channel. This effect is like that of acetylcholine, and it has been predicted that the occurrence of such mutations in the human heart could underlie sinus bradycardia and sick sinus syndrome.

The autonomic neurotransmitters affect automaticity by altering membrane ionic currents. The adrenergic transmitters increase all three currents involved in SA nodal automaticity. To increase the slope of diastolic depolarization, the augmentation of i_f and i_{Ca} by adrenergic transmitters must exceed the enhancement of i_K by these same transmitters.

The hyperpolarization induced by acetylcholine released from vagus nerve endings in the heart is achieved by the activation of specific potassium channels: the

[a]The ionic basis for automaticity in AV node pacemaker cells resembles that in SA node cells. Similar mechanisms also account for automaticity in ventricular Purkinje fibers, except that the fast Na⁺ current rather than i_{Ca} is involved. Also, a voltage- and time-dependent K⁺ current rather than the hyperpolarization-induced inward current i_f has been suggested to mediate the slow diastolic depolarization; however, this remains to be clarified.

acetylcholine-regulated potassium channels (K_{ACh}). Acetylcholine also depresses the i_f and i_{Ca} currents. The autonomic neural effects on cardiac cells are described in greater detail in Chapter 18.

When the SA node or other components of the atrial pacemaker complex are excised or destroyed, pacemaker cells in the AV junction generally take over the pacemaker function for the entire heart. After some time, which may vary from minutes to days, automatic cells in the atria usually become dominant again and resume their pacemaker function. Purkinje fibers in the specialized conduction system of the ventricles also display automaticity. These fibers characteristically fire at a very slow rate. When the AV junction cannot conduct cardiac impulses from the atria to the ventricles, these idioventricular pacemakers in the Purkinje fiber network initiate the ventricular contractions, but at a frequency of only 30 to 40 beats per minute.

IN THE CLINIC

If an ectopic focus in one of the atria suddenly began to fire at a high rate (e.g., 150 impulses per minute) in an individual with a normal heart rate of 70 beats per minute, the ectopic site would become the pacemaker for the entire heart. If that rapid ectopic focus suddenly stopped firing, the SA node would remain briefly quiescent because of overdrive suppression. The interval from the end of the period of overdrive until the SA node resumes firing is called the *sinus node recovery time.* In patients with sick sinus syndrome, the sinus node recovery time is prolonged. The consequent period of asystole (absence of a heartbeat) may cause loss of consciousness.

Overdrive Suppression

The automaticity of pacemaker cells diminishes after these cells have been excited at a high frequency. This phenomenon is known as **overdrive suppression.** Because the intrinsic rhythmicity of the SA node is greater than that of the other latent pacemaking sites in the heart, firing of the SA node tends to suppress the automaticity in other loci.

Overdrive suppression results from the activity of membrane Na^+,K^+-ATPase. A certain amount of Na^+ enters the cardiac cell during each depolarization. The more frequently the cell is depolarized, the more Na^+ enters the cell per minute. At high excitation frequencies, the activity of Na^+,K^+-ATPase increases to extrude this larger amount of Na^+ from the cell. The activity of Na^+,K^+-ATPase hyperpolarizes the cell because three Na^+ ions are extruded by the pump in exchange for two K^+ ions that enter the cell (see Chapter 1). Therefore, slow diastolic depolarization requires more time to reach the firing threshold. In addition, when the overdrive suddenly ceases, the activity of Na^+,K^+-ATPase does not slow instantaneously but temporarily remains overactive. This continued extrusion of Na^+ opposes the gradual depolarization of the pacemaker cell during phase 4, and it temporarily suppresses the cell's intrinsic automaticity.

Atrial Conduction

From the SA node, the cardiac impulse spreads radially throughout the right atrium (see Fig. 16.17) along ordinary atrial myocardial fibers at a conduction velocity of approximately 1 m/second. A special pathway, the anterior interatrial myocardial band (or Bachmann's bundle), conducts the SA node impulse directly to the left atrium. The wave of excitation proceeds inferiorly through the right atrium and ultimately reaches the AV node (see Fig. 16.17), which is normally the sole entry route of the cardiac impulse to the ventricles.

IN THE CLINIC

Some people have accessory AV pathways. Because these pathways often serve as a part of a reentry loop (see the section "Reentry"), they can be associated with serious cardiac rhythm disturbances. Wolff-Parkinson-White syndrome, a congenital disturbance, is the most common clinical disorder in which a bypass tract of myocardial fibers becomes an accessory pathway between the atria and ventricles. Ordinarily, the syndrome causes no functional abnormality. The disturbance is easily detected on an ECG because a portion of the ventricle is excited via the bypass tract before the remainder of the ventricle is excited via the AV node and the His-Purkinje system. This preexcitation appears as a bizarre configuration in the ventricular (QRS) complex of the ECG. On occasion, however, a reentry loop develops in which the atrial impulse travels to the ventricles via one of the two AV pathways (AV node or bypass tract) and then back to the atria through the other of these two pathways. Continuous circling around the loop leads to a very rapid rhythm (supraventricular tachycardia). This rapid rhythm may be incapacitating because it might not allow sufficient time for ventricular filling. Transient block of the AV node by an injection of adenosine intravenously or by a reflexive increase in vagal activity (by pressing on the neck over the carotid sinus region) usually abolishes the tachycardia and restores a normal sinus rhythm.

The atrial plateau (phase 2) is briefer and less developed, and repolarization (phase 3) is slower (see Fig. 16.10) than in a typical ventricular fiber. The action potential duration in atrial myocytes is briefer than that in ventricular myocytes because efflux of K^+ is greater during the plateau in atrial myocytes than in ventricular myocytes. The presence of an ultrarapid K^+ current (I_{Kur}) in atrial myocytes, but not in ventricular myocytes, contributes to greater K^+ efflux and shorter action potentials in atrial myocytes.

Atrioventricular Conduction

The atrial excitation wave reaches the ventricles via the AV node. In adult humans, this node is approximately 15 mm long, 10 mm wide, and 3 mm thick. The node is situated posteriorly on the right side of the interatrial septum near the ostium of the coronary sinus. The AV node contains the same two cell types as the SA node, but the round cells

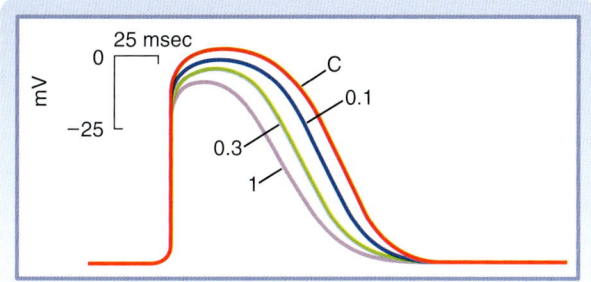

• **Fig. 16.21** Membrane potentials recorded from an atrioventricular node cell under control conditions (C) and in the presence of the calcium channel antagonist diltiazem at concentrations of 0.1, 0.3, and 1 μmol/L. (Redrawn from Hirth C, et al. *J Mol Cell Cardiol.* 1983;15:799.)

in the AV node are less abundant and the elongated cells predominate.

The AV node is made up of three functional regions: (1) the atrionodal (AN) region, or the transitional zone between the atrium and the remainder of the node; (2) the nodal (N) region, or the midportion of the AV node; and (3) the nodal-His (NH) region, or the zone in which nodal fibers gradually merge with the **bundle of His,** which is the upper portion of the specialized conducting system for the ventricles (see Fig. 16.17). Normally, the AV node and the bundle of His are the only pathways along which the cardiac impulse travels from atria to ventricles.

Several features of AV conduction are of physiological and clinical significance. The principal delay in conduction of impulses from the atria to the ventricles occurs in the AN and N regions of the AV node. Conduction velocity is actually less in the N region than in the AN region. However, the path length is substantially greater in the AN region than in the N region. Conduction times through the AN and N regions account for the delay between the start of the P wave (the electrical manifestation of atrial excitation) and the QRS complex (the electrical manifestation of ventricular excitation) on an ECG (see the section "Scalar Electrocardiography"). In terms of function, the delay between atrial and ventricular excitation enables optimal ventricular filling during atrial contraction.

In the N region, slow-response action potentials prevail. The resting potential is approximately −60 mV, the upstroke velocity is low (≈5 V/sec), and the conduction velocity is approximately 0.05 m/sec.[b] Tetrodotoxin, which blocks the fast sodium channels, has virtually no effect on action potentials in this region (or on any other slow-response fibers). Conversely, calcium channel antagonists decrease the amplitude and duration of the action potentials (Fig. 16.21) and depress AV conduction.

Like other slow-response action potentials, the relative refractory period of cells in the N region extends well beyond the period of complete repolarization; that is, these cells display postrepolarization refractoriness (see Fig. 16.15). As the heart rate increases, the time between successive atrial depolarizations is decreased, and conduction through the AV junction slows. Abnormal prolongation of the AV conduction time is called a *first-degree AV block* (see the section "Atrioventricular Conduction Blocks"). Most of the prolongation of AV conduction induced by an increase of atrial frequency takes place in the N region of the AV node.

Impulses tend to be blocked in the AV node at stimulation frequencies that are easily conducted in other regions of the heart. If the atria are depolarized at a high repetition rate, only a fraction (e.g., half) of the atrial impulses might be conducted through the AV junction to the ventricles. The conduction pattern in which only a fraction of the atrial impulses are conducted to the ventricles is called a *second-degree AV block* (see the section "Atrioventricular Conduction Blocks"). This type of block may protect the ventricles from excessive contraction frequencies, wherein the filling time between contractions might be inadequate.

Retrograde conduction can occur through the AV node. However, the conduction time is significantly longer, and the impulse is blocked at lower repetition rates when the impulse is conducted in the retrograde instead of the antegrade direction. In addition, the AV node is a common site for reentry (see the section "Reentry").

As in the SA node, the autonomic nervous system regulates AV conduction. Weak vagal activity may simply prolong the AV conduction time. Thus for any given atrial cycle length, the atrium-to-His or atrium-to-ventricle conduction time is prolonged by vagal stimulation. Stronger vagal activity may cause some or all of the impulses arriving from the atria to be blocked in the node. The conduction pattern in which none of the atrial impulses reaches the ventricles is called a *third degree,* or *complete, AV block* (see the section "Atrioventricular Conduction Blocks"). The vagally induced delay or absence of conduction through the AV node occurs mainly in the N region. This effect of vagal stimulation reflects the action of acetylcholine to hyperpolarize the membrane of the conducting fibers in the N region (Fig. 16.22). The greater the hyperpolarization at the time of arrival of the atrial impulse, the more impaired the AV conduction.

Cardiac sympathetic nerves, in contrast, facilitate AV conduction. They decrease the AV conduction time and enhance the rhythmicity of latent pacemakers in the AV junction. The norepinephrine released at the postganglionic sympathetic nerve terminals increases the amplitude and slope of the upstroke of the AV nodal action potentials, principally in the AN and N regions of the node.

Ventricular Conduction

The bundle of His passes subendocardially down the right side of the interventricular septum for approximately 1 cm

---

[b]The shapes of the action potentials in the AN region are intermediate between those in the N region and those in the atria. Similarly, action potentials in the NH region are transitional between those in the N region and those in the bundle of His.

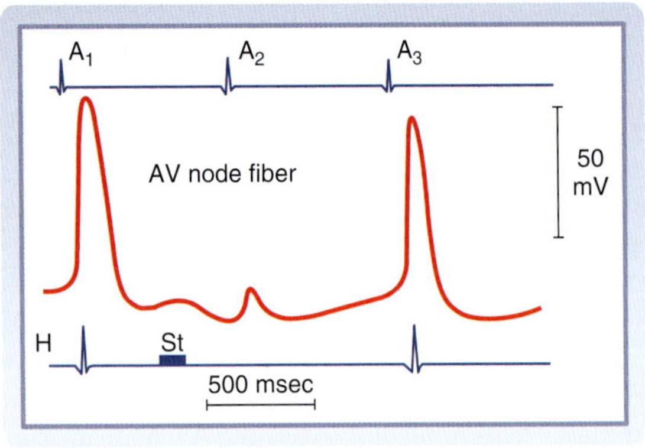

• **Fig. 16.22** Effects of a Brief Vagal Stimulus (St) on the Transmembrane Potential. This response was recorded from an atrioventricular (AV) nodal fiber *(red trace)*. Compound action potentials were recorded from the atrium (A_1, A_2, A_3; *upper blue trace*) and from the His-Purkinje system (H; *lower blue trace*). Note that shortly after vagal stimulation, the membrane of the fiber was hyperpolarized. The atrial excitation (A_2) that arrived at the AV node when the cell was hyperpolarized failed to be conducted, as denoted by the absence of a depolarization in the His electrogram (H). The atrial excitations that preceded (A_1) and followed (A_3) excitation A_2 were conducted to the His bundle region. (Redrawn from Mazgalev T. et al. *Am J Physiol.* 1986;251:H631.)

and then divides into the right and left bundle branches (see Fig. 16.17). The right bundle branch, a direct continuation of the bundle of His, proceeds down the right side of the interventricular septum. The left bundle branch, which is considerably thicker than the right, arises almost perpendicular from the bundle of His and perforates the interventricular septum. On the subendocardial surface of the left side of the interventricular septum, the left bundle branch splits into a thin anterior division and a thick posterior division.

> ### 🩺 IN THE CLINIC
>
> Conduction of impulses in the right or left bundle branch or in either division of the left bundle branch may be impaired. Conduction blocks may develop in one or more of these conduction pathways as a consequence of coronary artery disease or degenerative processes associated with aging, and they give rise to characteristic ECG patterns. Block of either of the main bundle branches is known as *right* or *left bundle branch block.* Block of either division of the left bundle branch is called *left anterior* or *left posterior hemiblock.*

The right bundle branch and the two divisions of the left bundle branch ultimately subdivide into a complex network of conducting fibers, called *Purkinje fibers,* that spread out over the subendocardial surfaces of both ventricles. Purkinje fibers have abundant, linearly arranged sarcomeres, as do myocytes. However, the transverse (T) tubular system, which is well developed in myocytes, is absent in the Purkinje fibers of many species. Purkinje fibers are the broadest cells in the heart: 70 to 80 μm in diameter, in comparison with diameters of 10 to 15 μm for

ventricular myocytes. Partly because of the large diameter of the Purkinje fibers, conduction velocity (1 to 4 m/second) in these fibers exceeds that in any other fiber type within the heart. The increased conduction velocity enables rapid activation of the entire endocardial surface of the ventricles.

The action potentials recorded from Purkinje fibers resemble those of ordinary ventricular myocardial fibers. However, because of the long refractory period of Purkinje fiber action potentials, many premature excitations of the atria are conducted through the AV junction but are then blocked by the Purkinje fibers. Blockade of these atrial excitations prevents premature contraction of the ventricles. This function of protecting the ventricles against the effects of premature atrial depolarization is especially pronounced at slow heart rates because the action potential duration, and hence the effective refractory period, of the Purkinje fibers vary inversely with the heart rate (see Fig. 16.16). At slow heart rates, the effective refractory period of the Purkinje fibers is especially prolonged.[c] In contrast to Purkinje fibers, the effective refractory period of AV node cells does not change appreciably over the normal range of heart rates and actually increases at very rapid heart rates. Therefore, when the atrium is excited at high repetition rates, it is the AV node that normally protects the ventricles from these excessively high frequencies.

The first portions of the ventricles to be excited by impulses arriving from the AV node are the interventricular septum (except the basal portion) and the papillary muscles. The activation wave spreads into the substance of the septum from both its left and right endocardial surfaces. Early contraction of the septum makes it more rigid and allows it to serve as an anchor point for contraction of the remaining ventricular myocardium. Furthermore, early contraction of the papillary muscles prevents eversion of the AV valves into the atria during ventricular systole.

The endocardial surfaces of both ventricles are activated rapidly, but the wave of excitation spreads from endocardium to epicardium at a slower velocity (≈0.3 to 0.4 m/sec). The epicardial surface of the right ventricle is activated earlier than that of the left ventricle because the right ventricular wall is appreciably thinner than the left. In addition, the apical and central epicardial regions of both ventricles are activated somewhat earlier than their respective basal regions. The last portions of the ventricles to be excited are the posterior basal epicardial regions and a small zone in the basal portion of the interventricular septum.

Reentry

The conditions necessary for reentry (defined in the following box) are illustrated in Fig. 16.23. In each of the four panels, a single bundle of cardiac fibers splits into a left branch and a right branch. A connecting bundle runs

---

[c]Similar directional changes in the refractory period also occur in ventricular myocytes in response to changes in heart rate.

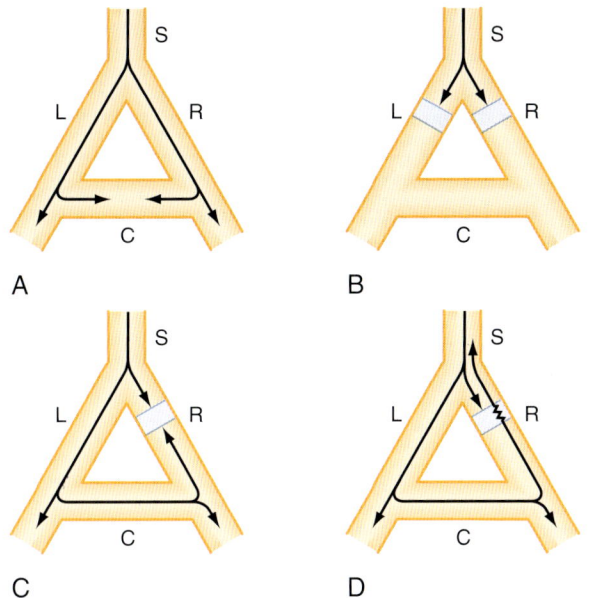

• Fig. 16.23 The Role of Unidirectional Block in Reentry. **A,** An excitation wave traveling down a single bundle (S) of fibers continues down the left (L) and right (R) branches. The depolarization wave enters the connecting branch (C) from both ends and is extinguished at the zone of collision. **B,** The wave is blocked in the L and R branches. **C,** A bidirectional block exists in the R branch. **D,** A unidirectional block exists in the R branch. The antegrade impulse is blocked, but the retrograde impulse is conducted through and reenters the S bundle.

between the two branches. Normally, the impulse moving down the single bundle is conducted along the left and right branches (see Fig. 16.23A). As the impulse reaches the link to the connecting bundle (Fig. 16.23C), it enters from both sides and becomes extinguished at the point of collision. The impulse from the left side cannot proceed because the tissue beyond is absolutely refractory; it has just been depolarized from the other direction. The impulse also cannot pass through the connecting bundle from the right for the same reason.

Fig. 16.23B shows that the impulse cannot complete the circuit if an antegrade block exists in the left and right branches of the fiber bundle. Also, if a bidirectional block exists at any point in the loop (e.g., branch R in Fig. 16.23C), the impulse also cannot reenter.

IN THE CLINIC

In certain conditions, a cardiac impulse may reexcite some myocardial region through which it had passed previously. This phenomenon, known as *reentry,* is responsible for many clinical arrhythmias (disturbances in cardiac rhythm). The reentry may be ordered or random. In ordered reentry, the impulse traverses a fixed anatomical path, whereas in random reentry, the path continues to change.

A necessary condition for reentry is that at some point in the loop the impulse can pass in one direction but not in the other. This phenomenon is called *unidirectional block.* As shown in Fig. 16.23D, the impulse may travel down

the left branch normally but is blocked in the antegrade direction in the right branch because of some pathological change in the myocardial cells in that branch. The impulse that was conducted down the left branch and through the connecting branch of the connecting bundle may then be able to penetrate the depressed region in the right branch from the retrograde direction, even though the antegrade impulse had been blocked previously at this same site. Why is the antegrade impulse blocked but not the retrograde impulse? The reason is that the antegrade impulse arrives at the depressed region in the right branch earlier than the retrograde impulse because the path length of the antegrade impulse is very short, whereas the retrograde impulse traverses a much longer path. Therefore, the antegrade impulse may be blocked simply because it arrives at the depressed region during its effective refractory period. If the retrograde impulse is delayed sufficiently, the refractory period may have ended in the affected region, and the impulse can then be conducted back through this region and return to the single bundle.

Although unidirectional block is a necessary condition for reentry, it alone cannot cause reentry. For reentry to occur, the effective refractory period of the reentered region must be shorter than the conduction time around the loop. In Fig. 16.23D, if the tissue just beyond the depressed zone in the right branch is still refractory from the antegrade depolarization, the retrograde impulse will not be conducted into the single bundle branch. Therefore, the conditions that promote reentry are those that prolong the conduction time or shorten the effective refractory period.

The functional characteristics of the various components of the reentry loops responsible for specific cardiac arrhythmias are diverse. Some loops are large and involve entire specialized conduction bundles, whereas others are microscopic. The loop may include myocardial fibers, specialized conducting fibers, nodal cells, and junctional tissues in almost any conceivable arrangement. In addition, the various cardiac cells in the loop may be normal or abnormal.

The propagation velocity along a multicellular cardiac conduction fiber is normally facilitated by gap junctions that lie between consecutive conducting fibers. Variations in the protein structure of the connexins in the gap junctions can affect the propagation velocity along these fibers. The chemical structure of the specific connexins can vary locally in cardiac tissues and, as a result, can establish local variations in propagation velocity. Such topical variations in velocity might include regions of unidirectional block that induce reentrant rhythm disturbances.

Triggered Activity

Triggered activity is so named because it is always coupled to a preceding action potential. Because reentrant activity is also coupled to a preceding action potential, the arrhythmias induced by triggered activity are usually difficult to distinguish from those induced by reentry. Triggered activity is caused by **afterdepolarizations,** of which two types

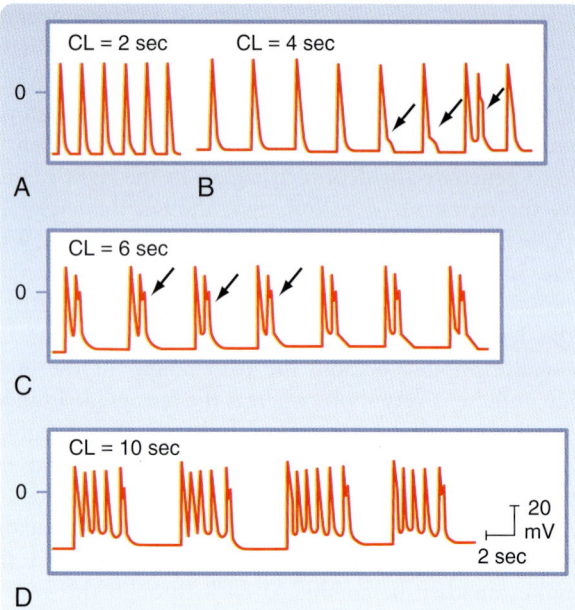

• **Fig. 16.24** Effect of Pacing at Different Cycle Lengths (CL) on Cesium-Induced Early Afterdepolarizations (EADs) in a Purkinje Fiber. **A,** EADs are not evident. **B,** EADs first appear *(arrows)*. The third EAD reaches threshold and triggers an action potential *(third arrow)*. **C,** EADs that appear after each driven depolarization trigger an action potential. **D,** Triggered action potentials occur in salvos. (Modified from Damiano BP, Rosen M. *Circulation.* 1984;69:1013.)

are recognized: **early afterdepolarizations (EADs)** and **delayed afterdepolarizations (DADs).** EADs may appear either at the end of the action potential plateau (phase 2) or approximately midway through repolarization (phase 3), whereas DADs occur near the very end of repolarization or just after full repolarization (phase 4).

Early Afterdepolarizations

EADs are more likely to occur when the prevailing heart rate is slow; a rapid heart rate suppresses EADs (Fig. 16.24). EADs are also more likely to occur in cardiac cells with prolonged action potentials than in cells with shorter action potentials. For example, EADs can be induced more readily in myocytes from the midmyocardial region of the ventricular walls than in myocytes from the endocardial or epicardial regions because of the longer action potential of midmyocardial myocytes (see Fig. 16.5). Certain antiarrhythmic drugs, such as quinidine, prolong the action potential. As a consequence, such drugs increase the likelihood that EADs may occur. Hence, some antiarrhythmic drugs are also proarrhythmic.

The direct correlation between a cell's action potential duration and its susceptibility to EADs is related to the time required for calcium channels in the cell membranes to recover from inactivation. When action potentials are sufficiently prolonged, the calcium channels that were activated at the beginning of the plateau have sufficient time to recover from inactivation and thus may be reactivated before the cell fully repolarizes. This secondary activation could then trigger an EAD.

IN THE CLINIC

The pathological significance of EADs was recognized in connection with the congenital and drug-induced long QT syndrome. As the action potential lengthens, EADs occur and cause triggered automaticity. In the electrocardiogram, this problem is manifested as polymorphic ventricular tachycardia, also called *torsades de pointes*. Episodes of torsades de pointes can be self-limited but may progress to ventricular fibrillation and sudden death. Hypokalemia and bradycardia (see Fig. 16.24) are clinical hallmarks of this condition. Restoring extracellular K+ to normal levels and increasing heart rate are two approaches used to overcome the propensity for development of torsades de pointes.

Delayed Afterdepolarizations

In contrast to EADs, DADs are more likely to occur when the heart rate is high (Fig. 16.25). DADs are associated with elevated intracellular [Ca++]. The amplitudes of DADs are increased by interventions that raise intracellular [Ca++], such as increasing extracellular [Ca++] and administering toxic amounts of digitalis glycosides. The elevated levels of intracellular Ca++ provoke the oscillatory release of Ca++ from the sarcoplasmic reticulum. Hence, in myocardial cells, DADs are accompanied by small rhythmic changes in the force developed. The high intracellular [Ca++] also activates certain membrane channels that allow the passage of Na+ and K+. The net flux of these cations constitutes a transient inward current (i_{ti}), that contributes to the appearance of DADs. The elevated intracellular [Ca++] may also activate the 3Na+-Ca++ antiporter. This electrogenic antiporter, which moves three Na+ ions into the cell for each Ca++ ion that it releases, also creates a net inward cation current that contributes to the appearance of DADs.

Electrocardiography

The **ECG** enables physicians to infer the course of cardiac electrical impulses by recording the variations in electrical potential at various loci on the surface of the body. By analyzing the details of these fluctuations in electrical potential, the physician gains valuable insight into (1) the anatomical orientation of the heart; (2) the relative sizes of its chambers; (3) various disturbances in rhythm and conduction; (4) the extent, location, and progress of ischemic damage to the myocardium; (5) the effects of

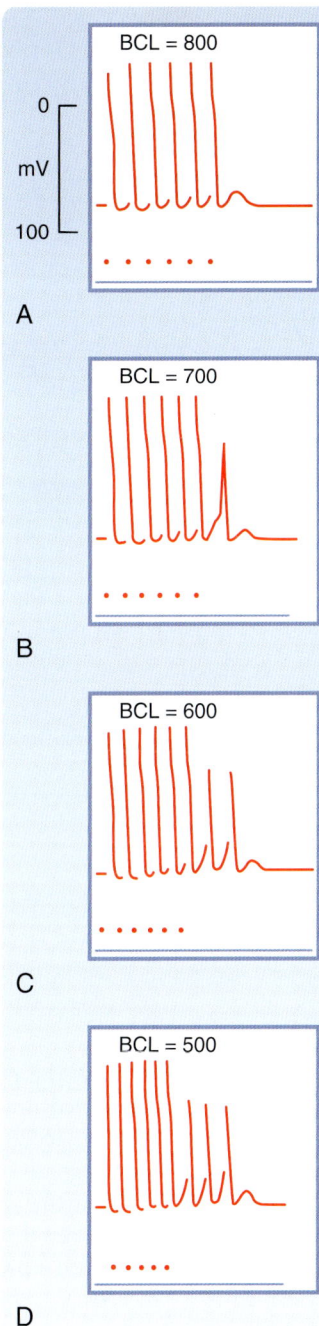

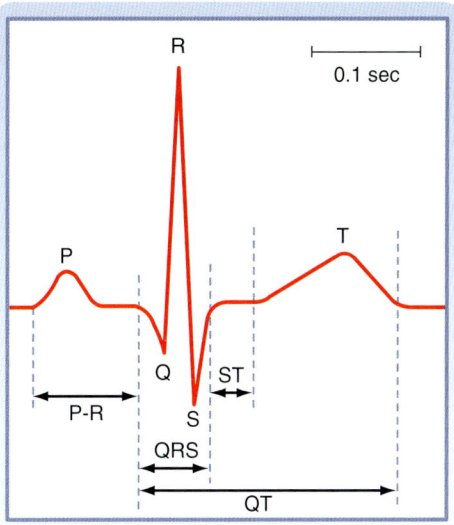

• **Fig. 16.26** Important deflections and intervals of a typical scalar electrocardiogram.

• **Fig. 16.25** Transmembrane Action Potentials Recorded From Purkinje Fibers. Acetylstrophanthidin, a digitalis glycoside, was added to the bath, and sequences of six driven beats (denoted by the *dots*) were each produced at a basic cycle length (BCL) of 800 **(A),** 700 **(B),** 600 **(C),** and 500 **(D)** msec. Note that delayed afterpotentials occurred after the driven beats and that these afterpotentials reached threshold after the last driven beat in **B** to **D.** (From Ferrier GR, et al. *Circ Res.* 1973;32:600.)

altered electrolyte concentrations; and (6) the influence of certain drugs (notably digitalis, antiarrhythmic agents, and calcium channel antagonists). Because electrocardiography is an extensive and complex discipline, only the elementary principles are considered in this section.

Scalar Electrocardiography

In electrocardiography, a lead is the electrical connection from the patient's skin to a recording device (**electrocardiograph**) that measures the electrical activity of the heart. The system of leads used to record routine ECGs is oriented in certain planes of the body. The diverse electrical events that exist in the heart at any moment can be represented by a three-dimensional vector (a quantity with magnitude and direction). A system of recording leads oriented in a given plane detects only the projection of the three-dimensional vector on that plane. The potential difference between two recording electrodes represents the projection of the vector on the line between the two leads. Components of vectors projected on such lines are not vectors but scalar quantities (having magnitude but not direction). Hence, a recording of changes in the difference in potential between two points on the skin surface over time is called a *scalar ECG.*

A scalar ECG detects temporal changes in the electrical potential between some point on the surface of the skin and an indifferent electrode or between pairs of points on the skin surface. The cardiac impulse progresses through the heart in a complex three-dimensional pattern. Hence, the precise configuration of the ECG varies from individual to individual, and in any given individual, the pattern varies with the anatomical location of the leads. The graphic display of the electrical impulse recorded in an ECG is called a *tracing.*

In general, a tracing consists of P, QRS, and T waves (Fig. 16.26). The P wave reflects the spread of depolarization through the atria, the QRS wave (or complex) reflects depolarization of the ventricles, and the T wave represents repolarization of the ventricles (repolarization of the atria occurs, and is therefore masked, during ventricular depolarization). The PR interval (or more precisely, the PQ interval) is a measure of the time from the onset of atrial activation to the onset of ventricular activation; it normally ranges from

0.12 to 0.20 second. A large fraction of this time involves passage of the impulse through the AV conduction system. Pathological prolongations of the PR interval are associated with disturbances in AV conduction. Such disturbances may be produced by inflammatory, circulatory, pharmacological, or neuronal mechanisms.

The configuration and amplitude of the QRS complex vary considerably among individuals. The duration is usually between 0.06 and 0.10 second. An abnormally prolonged QRS complex may indicate a block in the normal conduction pathways through the ventricles (such as a block of the left or right bundle branch). During the ST interval, the entire ventricular myocardium is depolarized. Therefore, the ST segment normally lies on the isoelectric line. Any appreciable deviation of the ST segment from the isoelectric line may indicate ischemic damage to the myocardium. The QT interval, sometimes referred to as the period of "electrical systole" of the ventricles, is closely correlated with the mean action potential duration of the ventricular myocytes. The duration of the QT interval is approximately 0.4 second, but it varies inversely with the heart rate, mainly because the duration of the myocardial cell's action potential varies inversely with the heart rate (see Fig. 16.16).

In most leads, the T wave is deflected in the same direction from the isoelectric line as the major component of the QRS complex, although biphasic (i.e., oppositely directed) T waves are perfectly normal in certain leads. Deviation of the T wave and QRS complex in the same direction from the isoelectric line indicates that the repolarization process is proceeding in a direction counter to that of the depolarization process. T waves that are abnormal either in direction or in amplitude may indicate myocardial damage, electrolyte disturbances, or cardiac hypertrophy.

Standard Limb Leads

The original ECG lead system was devised by Willem Einthoven at the beginning of the 20th century. In this system, the vector sum of all cardiac electrical activity at any moment is called the **resultant cardiac vector.** This directional electrical force is considered to lie in the center of an equilateral triangle whose apices are located in the left and right shoulders and the pubic region (Fig. 16.27). This triangle, called **Einthoven's triangle,** is oriented in the frontal plane of the body. Hence, only the projection of the resultant cardiac vector on the frontal plane is detected by this system of leads. For convenience, the electrodes are connected to the right and left forearms rather than to the corresponding shoulders because the arms represent simple electrical extensions of leads from the shoulders. Similarly, the leg represents an extension of the lead system from the pubis, and thus the third electrode is generally connected to an ankle (usually the left one).

Certain conventions dictate the manner in which these standard limb leads are connected to the electrocardiograph. Lead I records the potential difference between the left arm and the right arm. The connections are such that when the

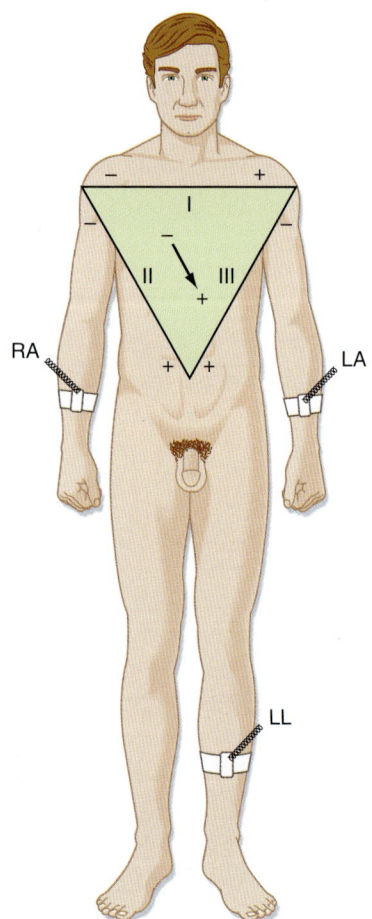

• **Fig. 16.27** Einthoven's triangle, illustrating the electrocardiographic connections for standard limb leads I, II, and III.

potential at the left arm (V_{LA}) exceeds the potential at the right arm (V_{RA}), the tracing is deflected upward from the isoelectric line. In Fig. 16.27 and Fig. 16.28, this arrangement of connections for lead I is designated by a plus sign at the left arm and by a minus sign at the right arm. Lead II records the potential difference between the right arm and the left leg, and the tracing is deflected upward when the potential at the left leg (V_{LL}) exceeds V_{RA}. Finally, lead III registers the potential difference between the left arm and the left leg, and the tracing is deflected upward when V_{LL} exceeds V_{LA}. These connections were chosen arbitrarily so that the QRS complexes are upright in all three standard limb leads in most normal individuals.

If the frontal projection of a resultant cardiac vector at some moment is represented by an arrow (tail negative, head positive), as in Fig. 16.27, the potential difference, V_{LA} − V_{RA}, recorded in lead I is represented by the component of the vector projected along the horizontal line between the left arm and the right arm, also shown in Fig. 16.27. If the vector makes an angle (θ) of 60 degrees with the horizontal line (as in Fig. 16.28A), the deflection recorded in lead I is upward because the positive arrowhead lies closer to the left arm than to the right arm. The deflection in lead II is also upright because the arrowhead lies closer to the left leg than

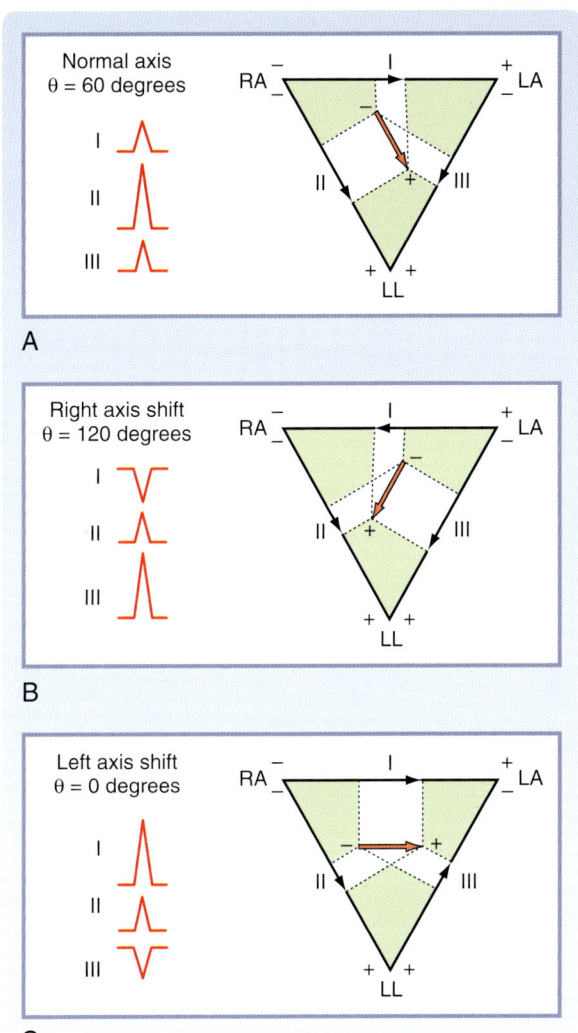

• Fig. 16.28 Magnitude and direction of the QRS complexes in limb leads I, II, and III when the mean electrical axis (θ) is 60 degrees **(A)**, 120 degrees **(B)**, and 0 degrees **(C)**.

to the right arm. The magnitude of the lead II deflection is greater than that in lead I because in this example, the direction of the vector parallels that of lead II; therefore, the magnitude of the projection on lead II exceeds that on lead I. Similarly, in lead III, the deflection is upright and its magnitude equals that in lead I.

If the vector in Fig. 16.27A is the result of electrical events that occur during the peak of the QRS complex, the orientation of this vector is said to represent the mean electrical axis of the heart in the frontal plane. The positive rotatory direction of this axis is assumed to be in the clockwise direction from the horizontal plane (contrary to the usual mathematical convention). In normal individuals, the average mean electrical axis is approximately +60 degrees (as in Fig. 16.28A). Therefore, QRS complexes are usually upright in all three leads and largest in lead II.

If the mean electrical axis shifts substantially to the right (as in Fig. 16.28B, in which θ = 120 degrees), projections of the QRS complexes on the standard leads change

considerably. In this case, the largest upright deflection is in lead III, and the deflection in lead I is inverted because the arrowhead is closer to the right arm than to the left arm. Such a shift is termed *right axis deviation* and occurs with hypertrophy (i.e., increased thickness) of the right ventricle. When the axis shifts to the left, as occurs with hypertrophy of the left ventricle (see Fig. 16.28C, where θ = 0 degrees), the largest upright deflection is in lead I, and the QRS complex in lead III is inverted.

In addition to limb leads I, II, and III, other limb leads that are also oriented in the frontal plane are routinely recorded in patients. These leads are (1) **aVR,** for which the right arm is defined as the positive lead and the middle of the heart is defined as the negative lead (i.e., the left arm and ankle leads are connected together); (2) **aVL,** for which the left arm is the positive lead and the middle of the heart is defined as the negative lead (i.e., the right arm and ankle leads are connected together); and (3) **aVF,** for which the ankle (foot) lead is defined as positive and the middle of the heart is defined as the negative lead (i.e., the two arm leads are connected together). The axes of these leads form angles of +90 degrees for aVF, −30 degrees for aVL, and −150 degrees for aVR (all with respect to the horizontal axis).

Leads can also be applied to the surface of the chest, so-called **precordial leads,** to determine the projections of the cardiac vector on the sagittal and transverse planes of the body. These precordial leads are recorded from six selected points on the anterior and lateral surfaces of the chest in the vicinity of the heart. The leads extend from the right border of the sternum in the fourth intercostal space (lead V_1) to under the left arm (midaxillary line) in the fifth intercostal space (lead V_6). Each precordial lead (V_1 to V_6) is defined as a positive lead, whereas the middle of the heart is defined as the negative lead. Detailed analysis of the ECG, as detected by the various lead systems just described, is beyond the scope of this book. Interested students are referred to textbooks on electrocardiography for more information.

IN THE CLINIC

Changes in the mean electrical axis may occur if the anatomical position of the heart is altered or if the relative mass of the right and left ventricles is abnormal, as it is in certain cardiovascular disturbances. For example, the axis tends to shift toward the left (more horizontal) in short, stocky individuals and toward the right (more vertical) in tall, thin persons. In addition, in left or right ventricular hypertrophy (increased myocardial mass of either ventricle), the axis shifts toward the hypertrophied side.

Arrhythmias

Cardiac arrhythmias are disturbances in either impulse initiation or impulse propagation. Disturbances in impulse initiation include those that arise from the SA node and

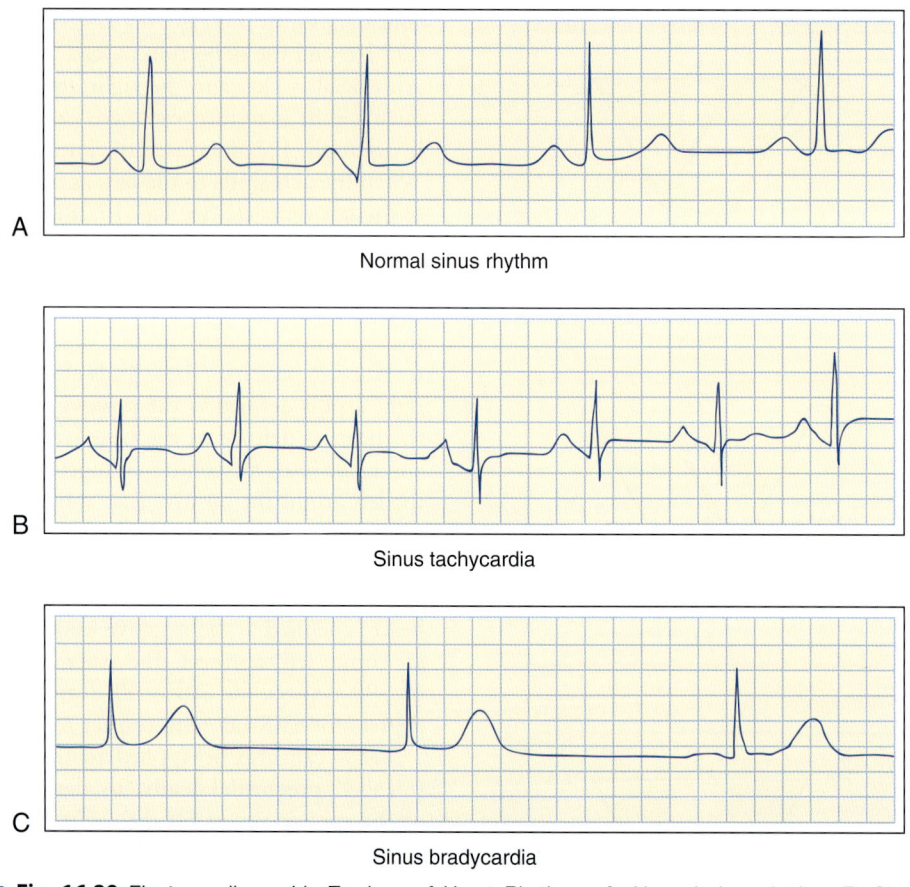

A
Normal sinus rhythm

B
Sinus tachycardia

C
Sinus bradycardia

• **Fig. 16.29** Electrocardiographic Tracings of Heart Rhythms. **A,** Normal sinus rhythm. **B,** Sinus tachycardia. **C,** Sinus bradycardia.

those that originate from various ectopic foci. The principal disturbances in impulse propagation are reentrant rhythms and conduction blocks.

Altered Sinoatrial Rhythms

Mechanisms that vary the firing frequency of cardiac pacemaker cells were described previously. Changes in the firing rate of the SA node are usually produced by cardiac autonomic nerves. When the firing rate of the SA node is decreased, the heart rate also decreases **(bradycardia).** Conversely, increased SA node firing results in elevation of the heart rate **(tachycardia).** Examples of ECGs of sinus tachycardia and sinus bradycardia are shown in Fig. 16.29. The P, QRS, and T deflections are all normal, but duration of the cardiac cycle (the PP interval) is altered. The changes in cardiac frequency are characteristically gradual. A rhythmic variation of the PP interval at the respiratory frequency (i.e., a respiratory sinus arrhythmia) is a normal, common occurrence (see Fig. 18.7).

Atrioventricular Conduction Blocks

Various physiological, pharmacological, and pathological processes can impede transmission of an impulse through the AV node. The site of block can be localized more precisely by an electrocardiogram of the His bundle (Fig. 16.30). To obtain such tracings, an electrode catheter is inserted into a peripheral vein and advanced centrally into the right side of the heart until the electrode lies in the AV junctional region. When the electrode is properly positioned, a distinct deflection (H in Fig. 16.30) is registered as the cardiac impulse passes through the bundle of His. The time intervals required for propagation from the atrium to the bundle of His and from the bundle of His to the ventricles (His-to-ventricle interval) may be measured accurately. Abnormal prolongation of the atrium-to-His or His-to-ventricle interval indicates block above or below the bundle of His, respectively.

Premature Depolarizations

Premature depolarizations occur occasionally in most normal individuals, but they arise more commonly in certain abnormal conditions. Such depolarizations may originate in the atria, AV junction, or ventricles. One type of premature depolarization follows a normally conducted depolarization at a constant time interval (the **coupling interval**). If the normal depolarization is suppressed in some way (e.g., by vagal stimulation), the premature depolarization is also abolished. Such premature depolarizations are called **coupled extrasystoles,** or simply **extrasystoles,** and they

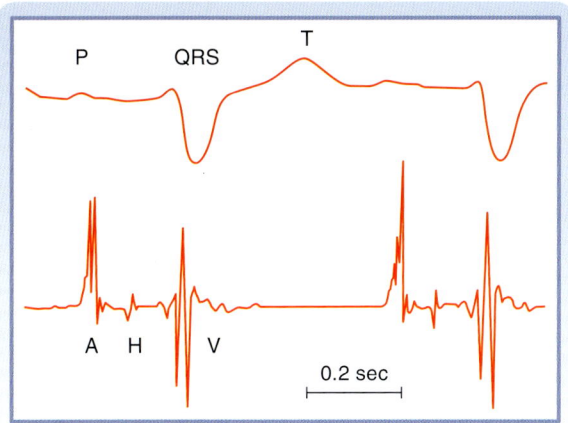

• **Fig. 16.30** Electrocardiogram of the His bundle (**lower tracing, retouched**) and lead II recording of the scalar electrocardiogram (**upper tracing**). The deflection H, which represents conduction of the impulse over the bundle of His, is clearly visible between the atrial (A) and the ventricular (V) deflections. The conduction time from the atria to the bundle of His is denoted by the A-H interval; that from the bundle of His to the ventricles, by the H-V interval. (Courtesy of Dr. J. Edelstein.)

IN THE CLINIC

Three degrees of AV block can be distinguished, as shown in Fig. 16.31. First-degree AV block is characterized by a prolonged PR interval. In most cases of first-degree block, the atrium-to-His interval is prolonged and the His-to-ventricle interval is normal. Hence, the delay in a first-degree AV block is located above the His bundle (i.e., in the AV node). In second-degree AV block, all QRS complexes are preceded by P waves, but not all P waves are followed by QRS complexes. The ratio of P waves to QRS complexes is usually the ratio of two small integers (such as 2:1, 3:1, or 3:2). The site of block may be located above or below the His bundle. A block below the bundle is usually more serious than one above the bundle because the former is more likely to evolve into a third-degree block. An artificial pacemaker is frequently implanted when the block is below the bundle. Third-degree AV block is often referred to as *complete heart block* because the impulse is completely unable to traverse the AV conduction pathway from atria to ventricles. The most common sites of complete block are distal to the bundle of His. In complete heart block, the atrial and ventricular rhythms are entirely independent. Because of the slow ventricular rhythm that results, the volume of blood pumped by the heart is often inadequate, especially during muscular exercise. Third-degree block is frequently associated with syncope (pronounced lightheadedness), which is caused principally by insufficient cerebral blood flow. Third-degree block is one of the most common conditions that necessitate the implantation of artificial pacemakers.

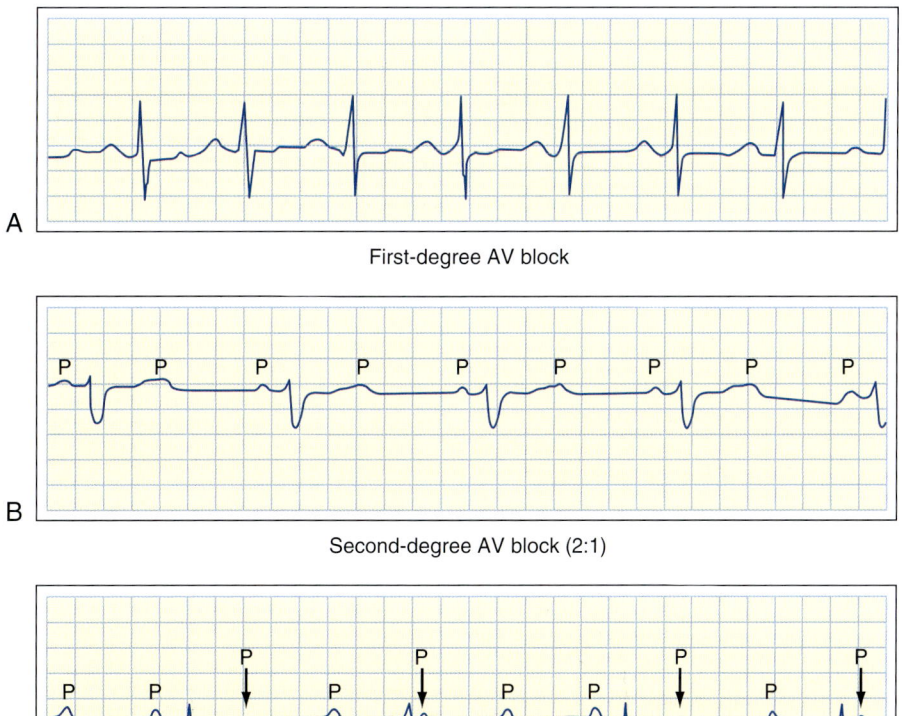

• **Fig. 16.31** Atrioventricular (AV) Blocks. **A,** First-degree block; the PR interval is 0.28 second (normal, <0.20 sec). **B,** Second-degree block (ratio of P waves to QRS complexes, 2:1). **C,** Third-degree block; note the dissociation between the P waves and the QRS complexes.

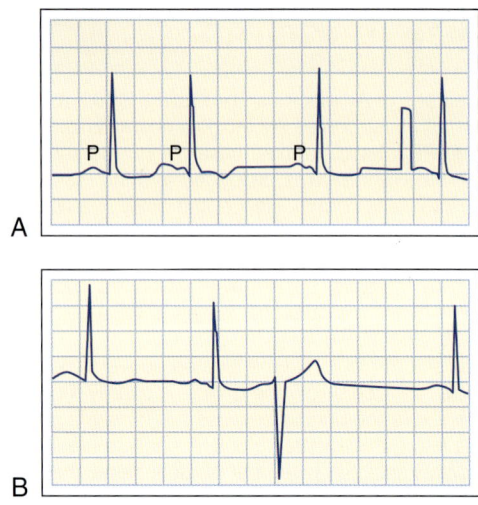

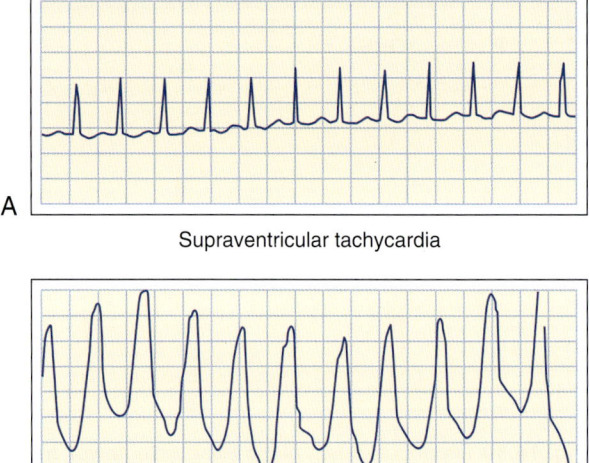

Supraventricular tachycardia

Ventricular tachycardia

• **Fig. 16.32** Premature Atrial Depolarization and Premature Ventricular Depolarization. The premature atrial depolarization (**A;** second beat) is characterized by an inverted P wave (just below the second "P") and normal QRS complexes and T waves. The interval after the premature atrial depolarization is not much longer than the usual interval between beats. The brief rectangular deflection just before the last atrial depolarization is a standardization signal. The premature ventricular depolarization (**B**) is characterized by bizarre, inverted QRS complexes and elevated T waves and is followed by a compensatory pause.

• **Fig. 16.33** Paroxysmal tachycardias. **A,** supraventricular tachycardia; P waves precede each QRS complex. **B,** ventricular tachycardia; P waves not readily observed.

generally reflect a reentry phenomenon. A second type of premature depolarization occurs as the result of enhanced automaticity in some ectopic focus. This ectopic center may fire regularly, and a zone of tissue that conducts unidirectionally may protect this center from being depolarized by the normal cardiac impulse. If this premature depolarization occurs at a regular interval or at an integral multiple of that interval, the disturbance is called **parasystole.**

A tracing of a premature atrial depolarization is shown in Fig. 16.32*A*. With a premature atrial depolarization, the normal interval between beats is shortened. In addition, the configuration of the premature P wave differs from that of the other normal P waves because the course of atrial excitation, which originates at some ectopic focus in the atrium, differs from the normal spread of excitation, which originates at the SA node. The QRS complex of the premature depolarization is generally normal because the ventricular excitation spreads over the usual pathways.

A tracing of a premature ventricular depolarization is shown in Fig. 16.32*B*. Propagation of the impulse is abnormal, and the configuration of the QRS complex and T wave is entirely different from the normal ventricular deflections because the premature excitation originates at some ectopic focus in the ventricles. The time interval between the premature QRS complex and the preceding normal QRS complex is shortened, whereas the interval after the premature QRS complex and the next normal QRS complex is prolonged by a compensatory pause. The interval from the QRS complex just before the premature excitation to the QRS complex just after it is virtually equal to the duration of two normal cardiac cycles.

As noted, a compensatory pause usually follows a premature ventricular depolarization. This pause occurs because the ectopic ventricular impulse does not disturb the natural rhythm of the SA node. There are two possible reasons for this: The ectopic ventricular impulse is not conducted in a retrograde direction through the AV conduction system, or the SA node had already fired at its natural interval before the ectopic impulse could have reached and depolarized it prematurely. Likewise, the SA nodal impulse generated just before or after the ventricular extrasystole generally does not affect the ventricle because the AV junction and perhaps also the ventricles are still refractory from the premature ventricular excitation.

    IN THE CLINIC

Paroxysmal tachycardias that originate either in the atria or in the AV junctional tissues (Fig. 16.33*A*) are usually indistinguishable, and therefore the term *paroxysmal supraventricular tachycardia* refers to both types. In this tachycardia, the impulse often circles a reentry loop that includes atrial and AV junctional tissue. The QRS complexes are frequently normal because ventricular activation proceeds over the usual pathways. As its name implies, paroxysmal ventricular tachycardia originates from an ectopic focus in the ventricles. The ECG is characterized by repeated bizarre QRS complexes that reflect the abnormal intraventricular impulse conduction (see Fig. 16.33*B*). Paroxysmal ventricular tachycardia is much more ominous than supraventricular tachycardia because the former is frequently a precursor of ventricular fibrillation, a lethal arrhythmia described in the next section.

Ectopic Tachycardias

In contrast to the gradual rate changes that characterize sinus tachycardia, tachycardias that originate from an

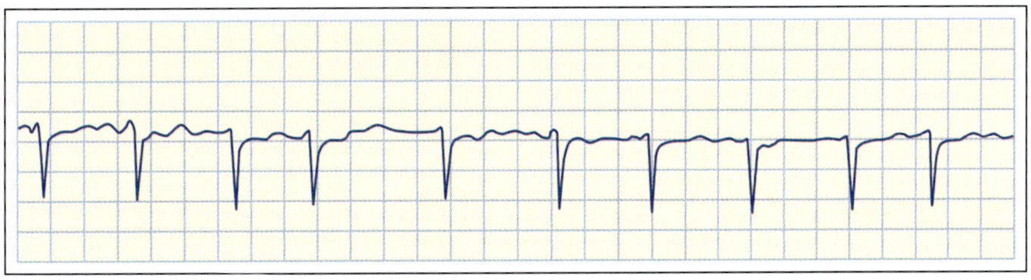

A
Atrial fibrillation

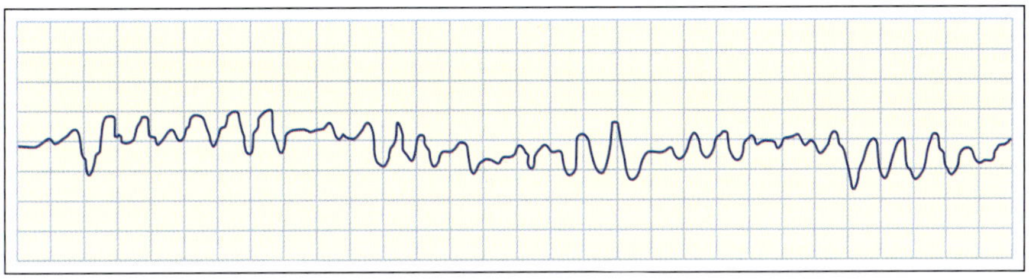

B
Ventricular fibrillation

• **Fig. 16.34** Atrial (**A**) and ventricular (**B**) fibrillation.

ectopic focus typically begin and end abruptly. Such ectopic tachycardias are generally called **paroxysmal tachycardias.** Episodes of paroxysmal tachycardia may persist for only a few beats or for many hours or days, and episodes often recur. Paroxysmal tachycardias may result from (1) rapid firing of an ectopic pacemaker, (2) triggered activity secondary to afterpotentials that reach threshold, or (3) an impulse that circles a reentry loop repetitively.

Fibrillation

Under certain conditions, cardiac muscle undergoes an irregular type of contraction that is ineffectual in propelling blood. Such an arrhythmia is termed **fibrillation,** and the disturbance may involve either the atria or the ventricles. Fibrillation probably represents a reentry phenomenon in which the reentry loop fragments into multiple, irregular circuits.

The electrocardiographic changes in atrial fibrillation are shown in Fig. 16.34*A.* This arrhythmia occurs in various types of chronic heart disease. The atria do not contract and relax sequentially during each cardiac cycle, and thus they do not contribute to ventricular filling. Instead, the atria undergo a continuous, uncoordinated rippling motion. P waves do not appear on the ECG; instead, the tracing shows continuous irregular fluctuations in potential called *f waves.* The AV node is activated at intervals that may vary considerably from cycle to cycle. Hence, no constant interval occurs between successive QRS complexes or between successive ventricular contractions. Because the strength of ventricular contraction depends on the interval between beats (see Chapter 18), the volume and rhythm of the pulse are irregular. In many affected patients, the atrial reentry loop and the pattern of AV conduction are more regular than they are in atrial fibrillation. The rhythm is then referred to as *atrial flutter.*

IN THE CLINIC

Atrial fibrillation and flutter are not usually life-threatening; some people with these disturbances can function normally. However, because the atria do not contract and relax rhythmically, blood clots tend to form in the atria. Such clots, if dislodged, may then travel to the pulmonary or systemic vascular beds. Patients with atrial fibrillation or flutter are generally treated with anticoagulant drugs such as dicumarol to prevent the formation of such clots. Ventricular fibrillation, in contrast, leads to loss of consciousness within a few seconds. The irregular, continuous, uncoordinated twitching of the ventricular muscle fibers pumps no blood. Death ensues unless immediate resuscitation is achieved or the rhythm spontaneously reverts to normal, which rarely occurs. Ventricular fibrillation may supervene when the entire ventricle, or some portion of it, is deprived of its normal blood supply. It may also occur as a result of electrocution or in response to certain drugs and anesthetics. On the ECG (see Fig. 16.34*B*), the fluctuations in potential are highly irregular.

Ventricular fibrillation is often initiated when a premature impulse arrives during the vulnerable period of the cardiac cycle. This period coincides with the downslope of the T wave on the ECG. During this period, the excitability of cardiac cells varies spatially. Some fibers are still in their effective refractory periods, others have almost fully recovered their excitability, and still others are able to conduct impulses, but only at very slow conduction velocities. As a consequence, the action potentials are propagated over the chambers in many irregular wavelets that travel along circuitous paths and at various conduction velocities. As a region of cardiac cells becomes excitable again, it is ultimately activated by one of the wave fronts traveling around the chamber. Hence, the process is self-sustaining.

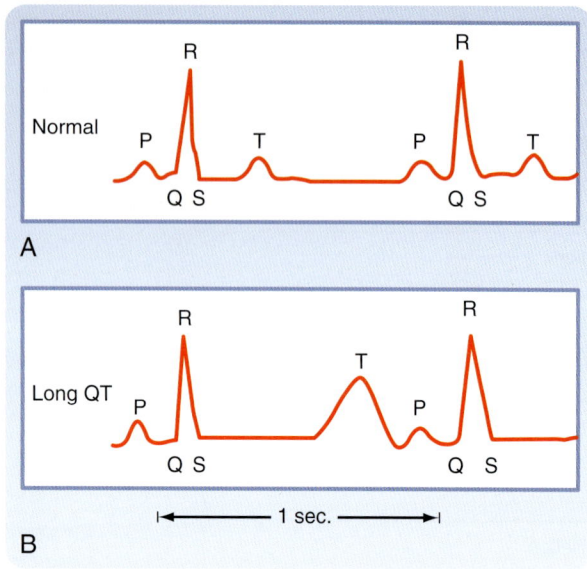

• **Fig. 16.35** Electrocardiograms recorded from a normal subject **(A)** and from a patient with long QT syndrome **(B)**.

AT THE CELLULAR LEVEL

In some individuals, the interval between the QRS complex and the T wave is abnormally prolonged, a condition termed *long QT syndrome* (Fig. 16.35). Several congenital forms of long QT syndrome have been identified in human subjects. Two of the many genes that have been identified as a basis for this syndrome are the *HERG* gene (a K⁺ channel gene), located on chromosome 7, and the *SCN5A* gene (a Na⁺ gene), located on chromosome 3. Patients with congenital forms of long QT syndrome may have periodic episodes of syncope (fainting), and approximately 10% of pediatric subjects with this disorder may die suddenly, without any preceding symptoms. In some individuals with silent ion channel mutations, long QT syndrome may not be evident until a drug is taken that affects the ion channel involved. Many drugs, including several antiarrhythmic agents, have been identified as causing acquired long QT syndrome.

Atrial fibrillation may be changed to a normal sinus rhythm by drugs that prolong the refractory period. As the cardiac impulse completes the reentry loop, it may then encounter refractory myocardial fibers. When atrial fibrillation does not respond adequately to drugs, electrical defibrillation may be used to correct this condition.

Dramatic therapy is required for ventricular fibrillation. Conversion to a normal sinus rhythm is accomplished by means of a strong electrical current that places the entire myocardium briefly in a refractory state. Techniques have been developed to administer the current safely through the intact chest wall. In successful cases, the SA node again takes over the normal pacemaker function for the entire heart.

The Cardiac Pump

The great amount of work performed by the heart over an individual's lifetime is impressive. A useful way to

IN THE CLINIC

Implantable cardioverter-defibrillator (ICD) devices have been developed to prevent death in patients in whom either ventricular fibrillation or paroxysmal ventricular tachycardia has suddenly developed. The former is lethal unless it is treated immediately, and the latter often leads to ventricular fibrillation and sudden death. The ICD device is implanted subcutaneously in the left subclavicular region of the chest wall. Atrial and ventricular leads enable recording of the right atrial and right ventricular electrocardiograms and provide the ability for right atrial or right ventricular pacing, or both. The defibrillation coil in the right ventricle enables the application of a strong electrical current to the ventricle and thereby usually terminates the lethal arrhythmia.

understand how the heart accomplishes its important task is to consider the relationships between the structure and function of its components.

Relationship of Heart Structure to Function
The Myocardial Cell

Myocardial and skeletal muscle cells have many important morphological and functional differences and similarities (see Chapters 12 and 13). Of importance is that both are striated as a result of regular arrangement of the contractile proteins actin and myosin, and generation of force and contraction of muscle fiber occur as a result of their interactions (i.e., sliding filament mechanism).

Skeletal muscle and cardiac muscle show similar length-force relationships. As shown in Chapter 13, this relationship for the heart may be expressed graphically, as in Fig. 16.36, by substituting ventricular systolic pressure for force and end-diastolic ventricular volume for resting myocardial fiber (and hence sarcomere) length. The lower curve in Fig. 16.36 represents the increment in pressure produced by each increment in volume when the heart is in diastole. The upper curve represents the peak force or pressure developed by the ventricle during systole as a function of initial fiber length (or diastolic filling pressure). This curve illustrates the **Frank-Starling law of the heart.**

The pressure-volume curve during diastole is initially quite flat (compliant), which indicates that large increases in volume can be accommodated with only small increases in pressure. In contrast, the development of systolic pressure is considerable at the lower filling pressures. However, the ventricle becomes much less distensible with greater filling, as evidenced by the sharp rise in the diastolic pressure curve at large intraventricular volumes.

In a normal intact heart, peak force may be attained at a filling pressure of approximately 12 mm Hg. At this intraventricular diastolic pressure, which is near the upper limit observed in a normal heart, sarcomere length is near its resting length of 2.2 μm. However, the force causes filling pressures to peak as high as 30 mm Hg. At even higher diastolic pressures (>50 mm Hg), sarcomere length is no

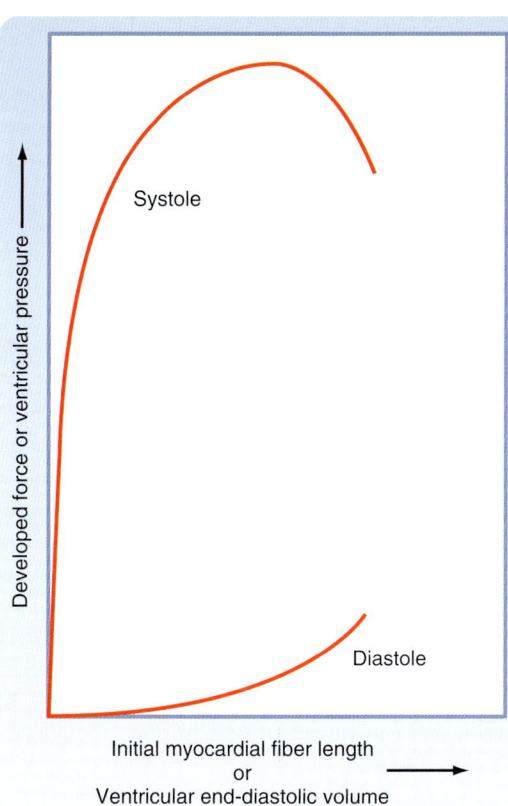

• Fig. 16.36 Relationship of myocardial resting fiber length (sarcomere length) or end-diastolic volume to the force developed or peak systolic ventricular pressure during ventricular contraction in an intact heart. (Redrawn from Patterson SW, et al. *J Physiol.* 1914;48:465.)

greater than 2.6 μm. This ability of the myocardium to resist stretch at high filling pressures probably resides in the noncontractile constituents of the heart tissue (connective tissue), and it may serve as a safety factor against overloading of the heart in diastole. Usually, ventricular diastolic pressure is approximately 0 to 7 mm Hg, and the average diastolic sarcomere length is approximately 2.2 μm. Thus a normal heart operates on the ascending portion of the Frank-Starling curve, as depicted in Fig. 16.36.

Functional Anatomy
Cardiac Muscle

Cardiac muscle functions as a syncytium; that is, a stimulus applied to any part of the cardiac muscle results in contraction of the entire muscle. Gap junctions with high conductance are present in the intercalated disks between adjacent cells and facilitate conduction of the cardiac impulse from one cell to the next.

Cardiac muscle must contract repetitively for a lifetime, and hence it requires a continuous supply of O_2. Cardiac muscle is therefore very rich in mitochondria. The large number of mitochondria (see Fig. 13.1*B*), which have the enzymes necessary for oxidative phosphorylation, allows rapid oxidation of substrates and synthesis of ATP and thus sustains the myocardial energy requirements.

To provide adequate O_2 and substrate for its metabolic machinery, the myocardium is also endowed with a rich capillary supply, approximately one capillary per fiber. Thus diffusion distances are short, and O_2, CO_2, substrates, and waste material can move rapidly between the myocardial cell and capillary. The **transverse (T) tubular system** within myocardial cells participates in this exchange of substances between capillary blood and myocardial cells (as described later, the T tubule system also plays a key role in excitation-contraction coupling). The T tubular system is absent or poorly developed in the atrial cells of many mammals.

Excitation-Contraction Coupling

Results of the earliest studies on isolated hearts indicated that optimal concentrations of Na^+, K^+, and Ca^{++} in extracellular fluid are necessary for contraction of cardiac muscle. Without Na^+, the heart is not excitable and does not beat. As already described, the resting membrane potential is independent of the extracellular [Na^+] gradient across the membrane but very much dependent on extracellular [K^+]. Decreases or increases in extracellular [K^+], especially if they are large or occur quickly, can lead to arrhythmias, loss of excitability of the myocardial cells, and even cardiac arrest. Ca^{++} is also essential for cardiac contraction. Removal of Ca^{++} from the extracellular fluid results in decreased contractile force and eventual arrest in diastole. Conversely, an increase in extracellular [Ca^{++}] enhances contractile force, and very high extracellular [Ca^{++}] induces cardiac arrest in systole (rigor). The free intracellular [Ca^{++}] is the factor principally responsible for the contractile state of the myocardium.

The process by which the action potential of the cardiac myocyte leads to contraction is termed **excitation-contraction coupling** (see also Chapter 13). Cardiac muscle is excited when a wave of excitation spreads rapidly along the myocardial sarcolemma from cell to cell via gap junctions. Excitation also spreads into the interior of the cells via the T tubules, which invaginate the cardiac fibers at the Z lines. Electrical stimulation at the Z line or the application of Ca^{++} to the Z lines in a skinned cardiac fiber (whose sarcolemma is removed) elicits localized contraction of the adjacent myofibrils. During the plateau (phase 2) of the action potential, permeability of the sarcolemma by Ca^{++} increases. Ca^{++} flows down its electrochemical gradient and enters the cell through calcium channels in the sarcolemma and in the T tubules.

During the action potential, Ca^{++} enters the cell via calcium channels (L-type). However, the amount of Ca^{++} that enters the cell interior from the extracellular/interstitial fluid is not sufficient to induce contraction of the myofibrils. Instead, it acts as a trigger (**trigger Ca^{++}**) to release Ca^{++} from the sarcoplasmic reticulum, where the intracellular Ca^{++} is stored (see Fig. 13.2). Ca^{++} leaves the sarcoplasmic reticulum through Ca^{++} release channels, which are called **ryanodine receptors** because the channel protein binds ryanodine avidly. Cytoplasmic [Ca^{++}] increases from a resting level of approximately 10^{-7} mol to levels of approximately 10^{-5} mol during excitation. This Ca^{++} then binds to the protein troponin C. The Ca^{++}–troponin C complex interacts with tropomyosin to unblock active sites between the actin and myosin filaments. This unblocking initiates cross-bridge cycling and hence contraction of the myofibrils.

Mechanisms that raise cytosolic $[Ca^{++}]$ increase the force developed, and those that lower cytosolic $[Ca^{++}]$ decrease the force developed. For example, catecholamines increase the movement of Ca^{++} into the cell by phosphorylation of the sarcolemmal calcium channels via a cAMP-dependent protein kinase (see Fig. 13.4). This in turn causes the release of more Ca^{++} from the sarcoplasmic reticulum, and as a result, contractile force increases. Increasing extracellular $[Ca^{++}]$ increases the amount of Ca^{++} that enters the cell via the calcium channels and thereby increases contractile force as just described. Reducing the Na^+ gradient across the sarcolemma also increases contractile force, an effect mediated by the $3Na^+$-Ca^{++} antiporter that normally extrudes Ca^{++} from the cell (see Fig. 13.2). For example, reducing extracellular $[Na^+]$ causes less Na^+ to enter the cell in exchange for Ca^{++}, which results in an increase in intracellular $[Ca^{++}]$ and thus contractile force. Raising intracellular $[Na^+]$ has a similar effect. In fact, this is the mechanism by which cardiac glycosides increase contractile force. Cardiac glycosides inhibit Na^+,K^+-ATPase and thereby raise intracellular $[Na^+]$ in the cells. The elevated cytosolic $[Na^+]$ reverses the direction of the $3Na^+$-Ca^{++} antiporter, and therefore less Ca^{++} is removed from the cell. The increase in intracellular $[Ca^{++}]$ results in an increase in contractile force. Finally, contractile force is diminished when intracellular $[Ca^{++}]$ is decreased by a reduction in extracellular $[Ca^{++}]$, by an increase in the Na^+ gradient across the sarcolemma, or by the administration of a calcium channel antagonist that prevents Ca^{++} from entering the myocardial cell.

At the end of systole, the influx of Ca^{++} stops, and the sarcoplasmic reticulum is no longer stimulated to release Ca^{++}. In fact, the sarcoplasmic reticulum avidly takes up Ca^{++} by means of Ca^{++}-ATPase. The Ca^{++}-ATPase in the sarcoplasmic reticulum is similar to but distinct from the Ca^{++}-ATPase found in the sarcolemma. Cytosolic $[Ca^{++}]$ is also reduced during diastole through the action of the $3Na^+$-Ca^{++} antiporter in the sarcolemma, as well as by a sarcolemmal Ca^{++}-ATPase (see Fig. 13.2).

Both cardiac contraction and relaxation are accelerated by catecholamines (see Fig. 13.3). When catecholamines bind to their receptor (β_1-adrenoceptor), adenylate cyclase is activated, thereby increasing intracellular cAMP levels, which then leads to activation of cAMP-dependent protein kinase A. Protein kinase A has multiple effects in the cell. As already described, it phosphorylates the calcium channel in the sarcolemma and causes increased entry of Ca^{++} into the cell, thus increasing the force of contraction. In addition, protein kinase A phosphorylates other proteins that facilitate relaxation. One such protein is **phospholamban.** Phospholamban normally inhibits the Ca^{++}-ATPase of the sarcoplasmic reticulum. However, when phospholamban is

phosphorylated, its inhibitory action is reduced, and uptake of Ca^{++} into the sarcoplasmic reticulum is enhanced. The increased activity of the Ca^{++}-ATPase in the sarcoplasmic reticulum decreases intracellular $[Ca^{++}]$, thereby causing relaxation. Protein kinase A also phosphorylates troponin I, which in turn inhibits binding of Ca^{++} by troponin C. As a result, tropomyosin returns to its position of blocking the myosin-binding sites on the actin filaments, and relaxation results.

Myocardial Contractile Machinery and Contractility

Contraction of cardiac muscle is influenced by both **preload** and **afterload** (Fig. 16.37). Preload is the force that stretches the relaxed muscle fibers, thereby increasing their resting length (see Fig. 16.37B). In the left ventricle, for example, the blood filling and thus the stretching of the wall during diastole represents the preload. Afterload is the force added to the muscle (see Fig. 16.37C) against which the contracting muscle must act (see Fig. 16.37D). Again from the perspective of the left ventricle, afterload is the pressure in the aorta that must be overcome by the contracting left ventricular muscle to open the aortic valve and eject the blood.

Preload can be increased by greater filling of the left ventricle during diastole (i.e., increasing **end-diastolic volume**). From lower initial end-diastolic volumes, increments in filling pressure during diastole increase resting fiber length and elicit a greater systolic pressure during the subsequent contraction (see Fig. 16.36). Systolic pressure increases until a maximal systolic pressure is reached at the optimal preload (see Fig. 16.36). If diastolic filling continues beyond this point, the pressure developed will no longer increase. At very high filling pressures, peak pressure development in systole is actually reduced.

At a constant preload, higher systolic pressure can be reached during ventricular contractions by an increase in the afterload (e.g., increasing aortic pressure by restricting the runoff of arterial blood to the periphery). Incremental increases in afterload produce progressively higher peak systolic pressures. However, if the afterload continues to increase, it becomes so great that the ventricle can no longer generate enough force to open the aortic valve. At this point, ventricular systole is totally isometric (i.e., there is no ejection of blood), and thus no change occurs in ventricular volume during systole. The maximal pressure developed by the left ventricle under these conditions is the maximal isometric force that the ventricle is capable of generating at a given preload. At preloads below the optimal filling volume, an increase in preload can yield greater maximal isometric force (see Fig. 16.36).

Preload and afterload depend on certain characteristics of the vascular system and the behavior of the heart. With regard to the vasculature, the degree of venomotor tone and peripheral resistance influences preload and afterload. With regard to the heart, a change in rate or stroke volume can also alter preload and afterload. Hence, cardiac and

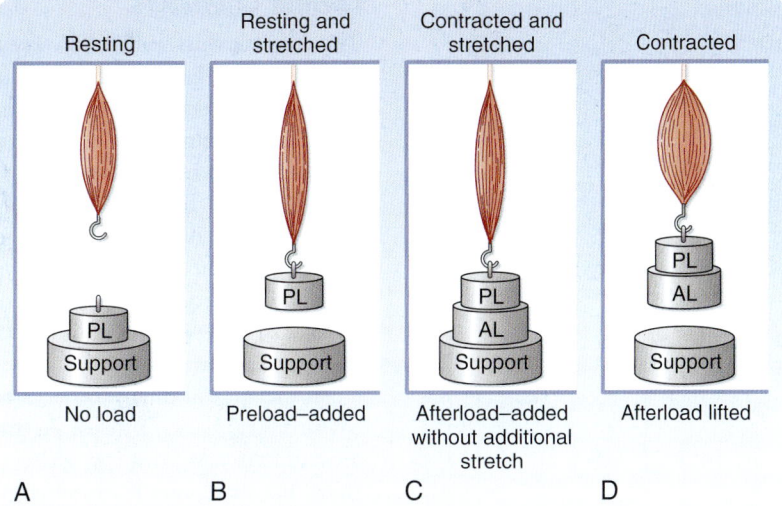

• **Fig. 16.37** Preload and Afterload in a Papillary Muscle. **A,** Resting stage (no load) in the intact heart just before opening of the AV valves. **B,** Preload (PL) in the intact heart at the end of ventricular filling. **C,** Supported preload plus afterload (AL) in the intact heart just before opening of the aortic valve. **D,** Lifting preload plus afterload in the intact heart during ventricular ejection, with a decrease in ventricular volume. PL + AL, total load.

vascular factors interact with each other to affect preload and afterload (see Chapter 19).

Contractility defines cardiac performance at a given preload and afterload. Contractility determines the change in peak isometric force (isovolumic pressure) at a given initial fiber length (end-diastolic volume). Contractility can be augmented by drugs, such as norepinephrine or digitalis, or by an increase in contraction frequency **(tachycardia).** The increase in contractility **(positive inotropic effect)** produced by these interventions is reflected by incremental increases in the force developed and in the velocity of contraction.

Indices of Contractility

A reasonable index of myocardial contractility can be derived from the contour of ventricular pressure curves (Fig. 16.38). A hypodynamic heart is characterized by elevated end-diastolic pressure, slowly rising ventricular pressure, and a somewhat reduced ejection phase (curve C in Fig. 16.38). A hyperdynamic heart (curve B in Fig. 16.38) is characterized by reduced end-diastolic pressure, fast-rising ventricular pressure, and a brief ejection phase. The slope of the ascending limb of the ventricular pressure curve indicates the maximal rate of force development by the ventricle. The maximal rate of change in pressure with time—that is, the **maximum dP/dt**—is illustrated by the tangents to the steepest portion of the ascending limbs of the ventricular pressure curves in Fig. 16.38. The slope of the ascending limb is maximal during the isovolumic phase of systole (see Fig. 16.38). At any given degree of ventricular filling, the slope provides an index of the initial contraction velocity and hence an index of contractility.

Similarly, the contractile state of the myocardium can be calculated from the velocity of blood flow that occurs

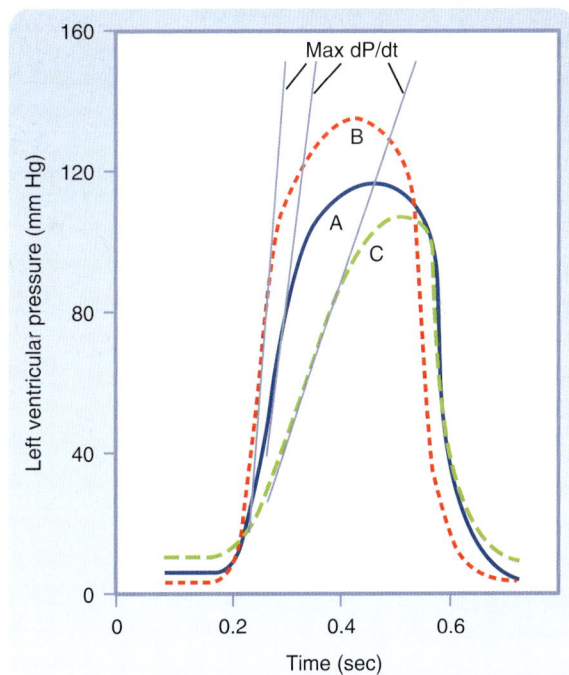

• **Fig. 16.38** Left ventricular pressure curves with tangents drawn to the steepest portions of the ascending limbs to indicate maximal rates of change in pressure with time (dP/dt values). A *(blue curve),* control; B *(dashed red curve),* hyperdynamic heart, as with administration of norepinephrine; C *(green dashed curve),* hypodynamic heart, as in cardiac failure.

initially in the ascending aorta during the cardiac cycle (Fig. 16.39). In addition, the **ejection fraction,** which is the ratio of the volume of blood ejected from the left ventricle per beat **(stroke volume)** to the volume of blood in the left ventricle at the end of diastole (end-diastolic volume), is widely used clinically as an index of contractility.

Cardiac Chambers

The atria are thin-walled, low-pressure chambers that function more as large-reservoir conduits of blood for their respective ventricles than as important pumps for the forward propulsion of blood. The ventricles comprise a continuum of muscle fibers originating from the fibrous skeleton at the base of the heart (chiefly around the aortic orifice). These fibers sweep toward the cardiac apex at the epicardial surface. They pass toward the endocardium and gradually undergo a 180-degree change in direction to lie parallel to the epicardial fibers and to form the endocardium and papillary muscles.

At the apex of the heart, the fibers twist and turn inward to form papillary muscles. At the base of the heart and around the valve orifices, these myocardial fibers form a thick, powerful muscle mass that not only decreases the ventricular circumference to implement the ejection of blood but also narrows the AV valve orifices, which aids in closure of the valve. Ventricular ejection is also accomplished by a decrease in the longitudinal axis as the heart begins to narrow toward the base. The early contraction of the apical part of the ventricles, coupled with the approximation of the ventricular walls, propels the blood toward the ventricular outflow tracts. The right ventricle, which develops a mean pressure that is approximately one seventh that developed by the left ventricle, is considerably thinner than the left ventricle.

Cardiac Valves

The cardiac valve leaflets consist of thin flaps of flexible, tough, endothelium-covered fibrous tissue that are firmly attached at the base to the fibrous valve rings. Movement of the valve leaflets is essentially passive, and the orientation of the cardiac valves is responsible for the unidirectional flow of blood through the heart. There are two types of valves in the heart: **atrioventricular** and **semilunar** (Figs. 16.40 and 16.41).

Atrioventricular Valves

The tricuspid valve, located between the right atrium and the right ventricle, is made up of three cusps, whereas the mitral valve, which lies between the left atrium and the left ventricle, has two cusps. The total area of the cusps of each AV valve is approximately twice that of the respective AV orifice, and so considerable overlap of the leaflets occurs when the valves are in the closed position. Attached to the free edges of these valves are fine, strong ligaments (chordae tendineae cordis) that arise from the powerful papillary muscles of the respective ventricles. These ligaments prevent the valves from becoming everted during ventricular systole.

In a normal heart, the valve leaflets remain relatively close together during ventricular filling. The partial approximation of the valve surfaces during diastole is caused by eddy currents that prevail behind the leaflets and by tension that is exerted by the chordae tendineae cordis and papillary muscles.

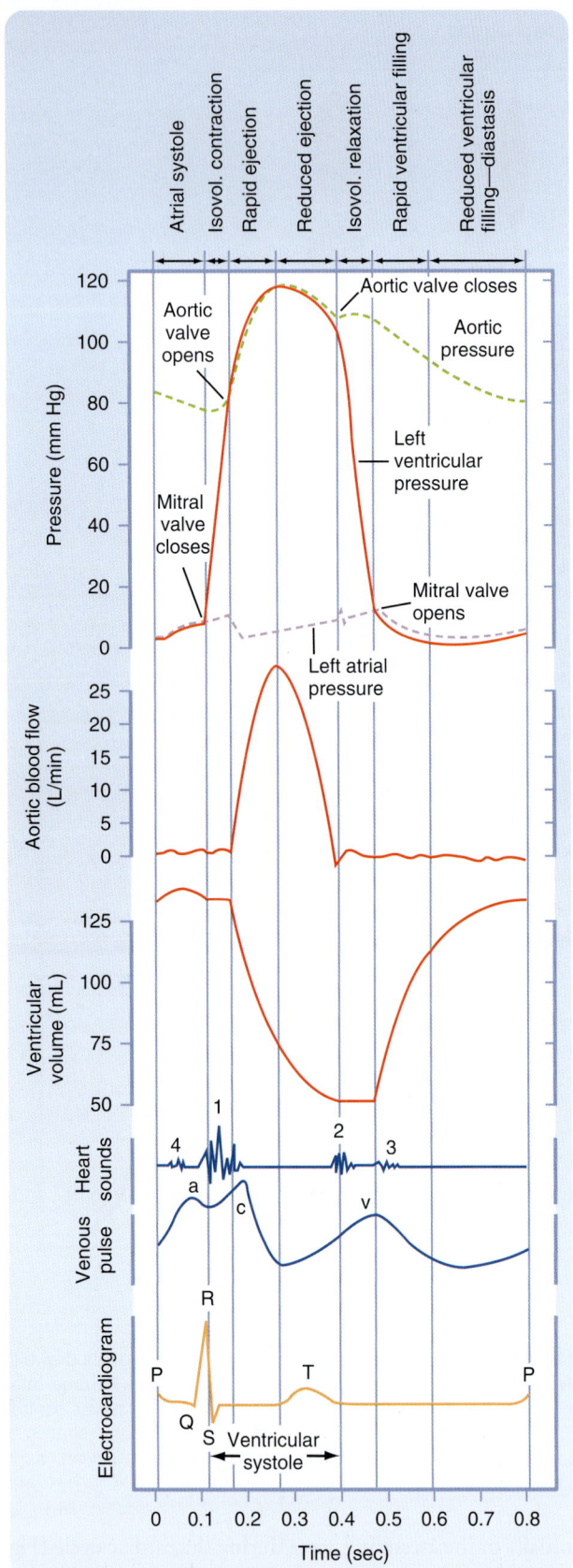

• **Fig. 16.39** Left atrial, aortic, and left ventricular pressure pulses correlated in time with aortic flow, ventricular volume, heart sounds, venous pulse, and the electrocardiogram for a complete cardiac cycle in a human subject. Isovol, isovolumic.

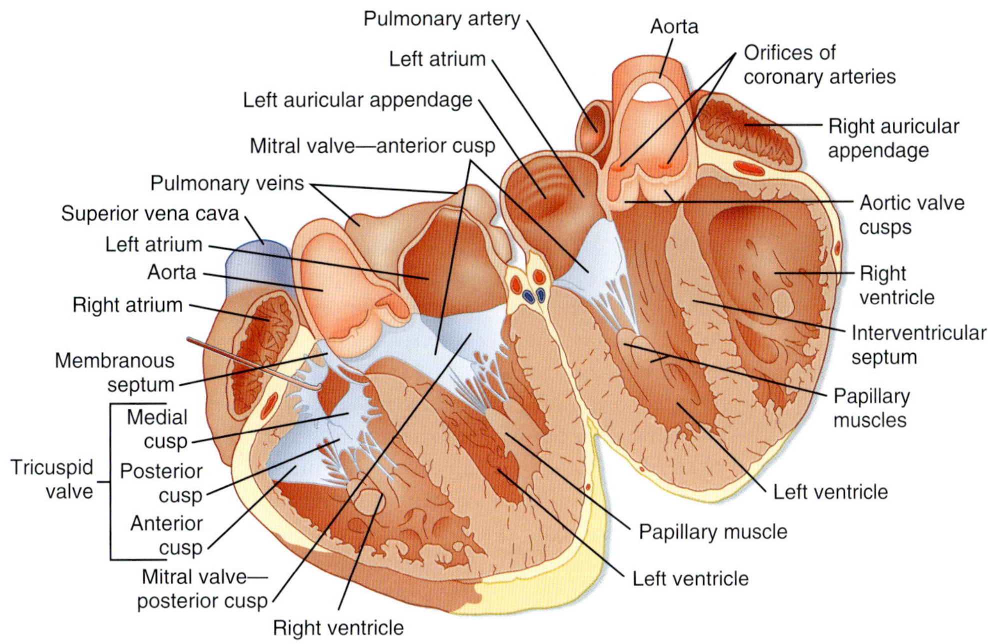

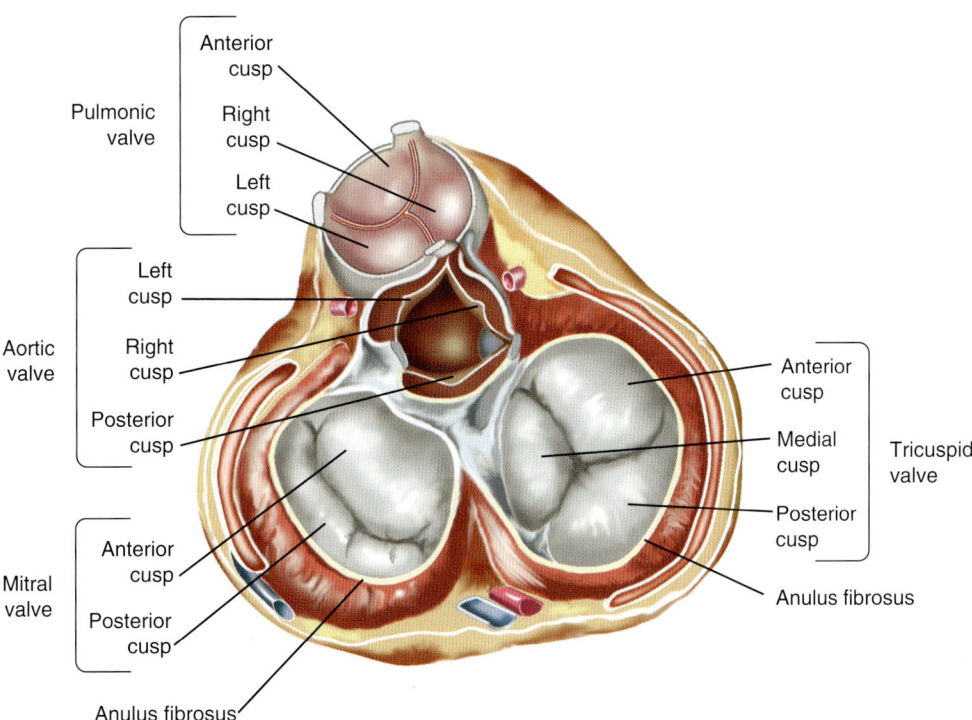

• **Fig. 16.40** Illustration of a heart split perpendicular to the interventricular septum to depict the anatomical relationships of the leaflets of the atrioventricular and aortic valves.

• **Fig. 16.41** Illustration of four cardiac valves as viewed from the base of the heart. Note how the leaflets overlap in the closed valves.

Semilunar Valves

The pulmonic and aortic valves are located between the right ventricle and the pulmonary artery and between the left ventricle and the aorta, respectively. These valves consist of three cup-like cusps that are attached to the valve rings (see Figs. 16.40 and 16.41). At the end of the reduced ejection phase of ventricular systole, blood flow briefly reverses toward the ventricles. This reversal of blood flow snaps the cusps together and prevents regurgitation of blood into the ventricles. During ventricular systole, the cusps do not lie back against the walls of the pulmonary artery and aorta; instead, they float in the bloodstream at a point approximately midway between the vessel walls and their closed position. Behind the semilunar valves are small outpocketings (sinuses of Valsalva) of the pulmonary artery and aorta. In these sinuses, eddy currents develop, which tend to keep the valve cusps away from the vessel walls. Furthermore, the orifices of the right and left coronary arteries are behind the right and the left cusps, respectively, of the aortic valve. Were it not for the presence of the sinuses of Valsalva and the eddy currents developed therein, the coronary ostia could be blocked by the valve cusps, and coronary blood flow would cease.

The Pericardium

The pericardium invests the entire heart and the cardiac portion of the great vessels, and it is reflected onto the cardiac surface as the epicardium. The sac normally contains a small amount of fluid, which provides lubrication for the continuous movement of the enclosed heart. The pericardium is not very distensible; it strongly resists a large, rapid increase in cardiac size and hence prevents sudden overdistention of the chambers of the heart. However, in congenital absence of the pericardium or after its surgical removal, cardiac function is not seriously affected. Nevertheless, with the pericardium intact, an increase in diastolic pressure in one ventricle increases the pressure and decreases the compliance of the other ventricle.

Heart Sounds

Four sounds are usually generated by the heart, but only two are ordinarily audible through a stethoscope. With electronic amplification, the less intense sounds can be detected and recorded graphically as a phonocardiogram. This means of registering faint heart sounds helps delineate the precise timing of the heart sounds in relation to other events in the cardiac cycle.

The first heart sound is initiated at the onset of ventricular systole (Fig. 16.42) and reflects closure of the AV valves. It is the loudest and longest of the heart sounds, has a crescendo-decrescendo quality, and is heard best over the apical region of the heart. The tricuspid valve sounds are heard best in the fifth intercostal space just to the left of the sternum; the mitral sounds are heard best in the fifth intercostal space at the cardiac apex.

The second heart sound, which occurs with abrupt closure of the semilunar valves (see Fig. 16.42), is composed of higher frequency vibrations (higher pitch) and is of shorter duration and lower intensity than is the first heart sound. The portion of the second sound caused by closure of the pulmonic valve is heard best in the second thoracic interspace just to the left of the sternum, whereas that caused by closure of the aortic valve is heard best in the same intercostal space but to the right of the sternum. The aortic valve sound is generally louder than the pulmonic valve, but in cases of pulmonary hypertension, the reverse is true. The nature of the second heart sound changes with respiration. During expiration, a single heart sound is heard that reflects simultaneous closing of the pulmonic and aortic valves. However, during inspiration, closure of the pulmonic valve is delayed, mainly as a result of increased blood flow from an inspiration-induced increase in venous return.[d] With this delayed closure of the pulmonic valve, the second heart sound can be heard as two components; this is termed **physiological splitting** of the second heart sound.

A third heart sound is sometimes heard in children with thin chest walls or in patients with left ventricular failure. It consists of a few low-intensity, low-frequency vibrations heard best in the region of the cardiac apex. The vibrations occur in early diastole and are caused by the abrupt cessation of ventricular distention and by the deceleration of blood entering the ventricles. A fourth, or atrial, sound consists of a few low-frequency oscillations. This sound is occasionally heard in individuals with normal hearts. It is caused by the oscillation of blood and cardiac chambers as a result of atrial contraction.

The Cardiac Cycle

Ventricular Systole

Isovolumic Contraction

The phase between the start of ventricular systole and opening of the semilunar valves (when ventricular pressure rises abruptly) is called the *isovolumic* (literally, "same volume") *contraction period*. This term is appropriate because ventricular volume remains constant during this brief period (see Fig. 16.39). The onset of isovolumic contraction also coincides with the peak of the R wave on an ECG, initiation of the first heart sound, and the earliest rise in ventricular pressure on the ventricular pressure curve after atrial contraction.

Ejection

Opening of the semilunar valves marks the onset of the ventricular ejection phase, which may be subdivided

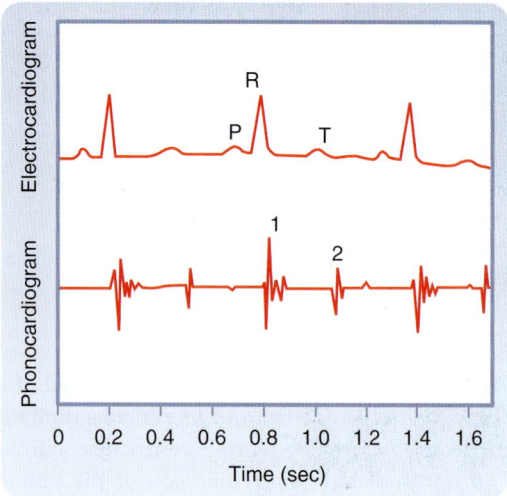

• **Fig. 16.42** The first and second heart sounds of the phonocardiogram *(bottom tracing)* are shown in regard to their relationship to the P, R, and T waves of the electrocardiogram *(top tracing)*.

---

[d]With inspiration, intrathoracic pressure is reduced (see Chapter 21), which then increases venous blood flow to the right atrium.

In overloaded hearts, as in congestive heart failure, when ventricular volume is very large and the ventricular walls are stretched maximally, a third heart sound is often heard. A third heart sound in patients with heart disease is usually a grave sign. When the third and fourth (atrial) sounds are accentuated, as occurs in certain abnormal conditions, triplets of sounds resembling the sound of a galloping horse (called *gallop rhythms*) may occur. Mitral insufficiency and mitral stenosis produce, respectively, systolic and diastolic murmurs that are heard best at the cardiac apex. Aortic insufficiency and aortic stenosis, in contrast, produce, respectively, diastolic and systolic murmurs that are heard best in the second intercostal space just to the right of the sternum. The characteristics of the murmurs serve as an important guide in the diagnosis of valvular disease.

into an earlier, shorter phase (rapid ejection) and a later, longer phase (reduced ejection). The rapid ejection phase is distinguished from the reduced ejection phase by three characteristics: (1) a sharp rise in ventricular and aortic pressure that terminates at peak ventricular and aortic pressure, (2) an abrupt decrease in ventricular volume, and (3) a pronounced increase in aortic blood flow (see Fig. 16.39). The sharp decrease in left atrial pressure at the onset of ventricular ejection results from descent of the base of the heart and consequent stretching of the atria. During the reduced ejection period, runoff of blood from the aorta to the peripheral blood vessels exceeds the rate of ventricular output, and aortic pressure therefore declines. Throughout ventricular systole, the blood returning from the peripheral veins to the atria produces a progressive increase in atrial pressure.

During the rapid ejection period, left ventricular pressure slightly exceeds aortic pressure and aortic blood flow accelerates (continues to increase), whereas during the reduced ventricular ejection phase, aortic pressure is higher and aortic blood flow decelerates. This reversal of the ventricular-aortic pressure gradient in the presence of continuous flow of blood from the left ventricle to the aorta is the result of storage of potential energy in the stretched arterial walls. This stored potential energy causes blood flow from the left ventricle into the aorta to decelerate. The peak of the flow curve coincides with the point at which the left ventricular pressure curve intersects the aortic pressure curve during ejection. Thereafter, flow decelerates (continues to decrease) because the pressure gradient has been reversed.

Fig. 16.39 shows a tracing of a venous pulse curve recorded from a jugular vein. Three waves are apparent. The **a wave** occurs with the rise in pressure caused by atrial contraction. The **c wave** is caused by impact of the common carotid artery with the adjacent jugular vein and, to some extent, by the abrupt closure of the tricuspid valve in early ventricular systole. The **v wave** reflects the rise in pressure associated with atrial filling. Except for the c wave, the venous pulse curve closely resembles the left atrial pressure curve.

At the end of ventricular ejection, a volume of blood approximately equal to that ejected during systole remains in the ventricular cavities. This residual volume is fairly constant in normal hearts. However, residual volume decreases somewhat when the heart rate increases or when peripheral vascular resistance has diminished.

Ventricular Diastole

Isovolumic Relaxation
Closure of the aortic valve produces the characteristic incisura (notch) on the descending limb of the aortic pressure curve, and it also produces the second heart sound (with some vibrations evident on the atrial pressure curve). The incisura marks the end of ventricular systole. The period between closure of the semilunar valves and opening of the AV valves is termed *isovolumic relaxation.* It is characterized by a precipitous fall in ventricular pressure without a change in ventricular volume.

Rapid Filling Phase
The major portion of ventricular filling occurs immediately after the AV valves open. At this point, the blood that returned to the atria during the previous ventricular systole is abruptly released into the relaxing ventricles. This period of ventricular filling is called the *rapid filling phase.* In Fig. 16.39, the onset of the rapid filling phase is indicated by the decrease in left ventricular pressure below left atrial pressure. This pressure reversal opens the mitral valve. The rapid flow of blood from atria to relaxing ventricles produces transient decreases in atrial and ventricular pressures and a sharp increase in ventricular volume.

Diastasis
The rapid ventricular filling phase is followed by a phase of slow ventricular filling called *diastasis.* During diastasis, blood returning from the peripheral veins flows into the right ventricle and blood from the lungs flows into the left ventricle. This small, slow addition to ventricular filling is indicated by gradual rises in atrial, ventricular, and venous pressures and in ventricular volume (see Fig. 16.39).

An increase in myocardial contractility, as produced by catecholamines or by digitalis in a patient with a failing heart, may decrease residual ventricular volume and increase the stroke volume and ejection fraction. In severely hypodynamic and dilated hearts, residual volume can become much greater than stroke volume.

Atrial Systole
The onset of atrial systole occurs soon after the beginning of the P wave (atrial depolarization) of the ECG. The transfer of blood from atrium to ventricle achieved by atrial contraction completes the period of ventricular filling.

Atrial systole is responsible for the small increases in atrial, ventricular, and venous pressure, as well as in ventricular volume (see Fig. 16.39). Throughout ventricular diastole, atrial pressure barely exceeds ventricular pressure. This small pressure difference indicates that the pathway through the open AV valves during ventricular filling has low resistance.

Because there are no valves at the junction of the venae cavae and right atrium or at the junction of the pulmonary veins and left atrium, atrial contraction may force blood in both directions. However, little blood is actually pumped back into the venous tributaries during the brief atrial contraction, mainly because of the inertia of the inflowing blood.

The contribution of atrial contraction to ventricular filling is governed to a great extent by the heart rate and the position of the AV valves. At slow heart rates, filling practically ceases toward the end of diastasis, and atrial contraction contributes little additional filling. During tachycardia, however, diastasis is abbreviated and the atrial contribution can become substantial. Should tachycardia become so severe that the rapid filling phase is attenuated, atrial contraction assumes great importance in rapidly propelling blood into the ventricle during this brief period of the cardiac cycle. If the period of ventricular relaxation is so brief that filling is seriously impaired, even atrial contraction cannot provide adequate ventricular filling. The consequent reduction in cardiac output may result in syncope (fainting).

IN THE CLINIC

Atrial contraction is not essential for ventricular filling, as can be observed in patients with atrial fibrillation or complete heart block. In atrial fibrillation, the atrial myofibers contract in a continuous, uncoordinated manner and therefore cannot pump blood into the ventricles. In complete heart block, the atria and ventricles beat independently of each other. However, ventricular filling may be normal in patients with these two arrhythmias. In certain disease states, the AV valves may be markedly narrowed (stenotic). In such conditions, atrial contraction plays a much more important role in ventricular filling than it does in a normal heart.

Pressure-Volume Relationship

The changes in left ventricular pressure and volume throughout the cardiac cycle are summarized in Fig. 16.43. Diastolic filling starts when the mitral valve opens (point A in Fig. 16.43), and it terminates when the mitral valve closes (point C). The initial decrease in left ventricular pressure (from points A to B), despite the rapid inflow of blood from the left atrium, is attributed to progressive ventricular relaxation and distensibility. During the remainder of diastole (from points B to C), the increase in ventricular pressure reflects ventricular filling and changes in the passive elastic characteristics of the ventricle. Note that only a small increase in pressure accompanies the substantial increase in ventricular volume during diastole (from points B to C).

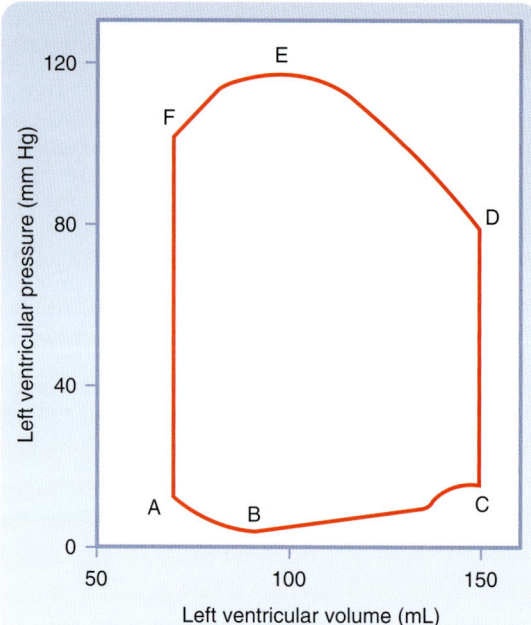

• **Fig. 16.43** Schematized pressure-volume loop of the left ventricle for a single cardiac cycle.

The small pressure increase reflects the compliance of the left ventricle during diastole. The small increase in pressure just before the mitral valve closes (to the left of point C) is caused by the contribution of atrial contraction to ventricular filling. With isovolumic contraction (from points C to D), pressure rises steeply, but ventricular volume does not change because the mitral and aortic valves are both closed. When the aortic valve opens (point D), and during the first (rapid) phase of ejection (from points D to E), the large reduction in volume is associated with a steady increase in ventricular pressure. This reduction in volume is followed by reduced ejection (from points E to F) and a small decrease in ventricular pressure. Closure of the aortic valve (point F) is followed by isovolumic relaxation (from points F to A), which is characterized by a sharp drop in pressure. Ventricular volume does not change during the interval between closing of the aortic valve and opening of the mitral valve (from points F to A) because both the mitral and aortic valves are closed. The mitral valve opens (point A) to complete one cardiac cycle.

Several key cardiovascular system parameters are evident on a left ventricular pressure-volume loop (P-V loop), or may be calculated from it. End-diastolic volume is obtained at mitral valve closure (point C in Fig. 16.43) and the end-systolic volume at mitral valve opening (point A). The stroke volume is then apparent as the "width" of the P-V loop and is calculated as follows:

$$\text{Stroke volume} = \text{end-diastolic volume} - \text{end-systolic volume}$$

The preload of the left ventricle, considered here as left ventricular end-diastolic pressure, is the pressure coordinate when the mitral valve closes (point C in Fig. 16.43). The

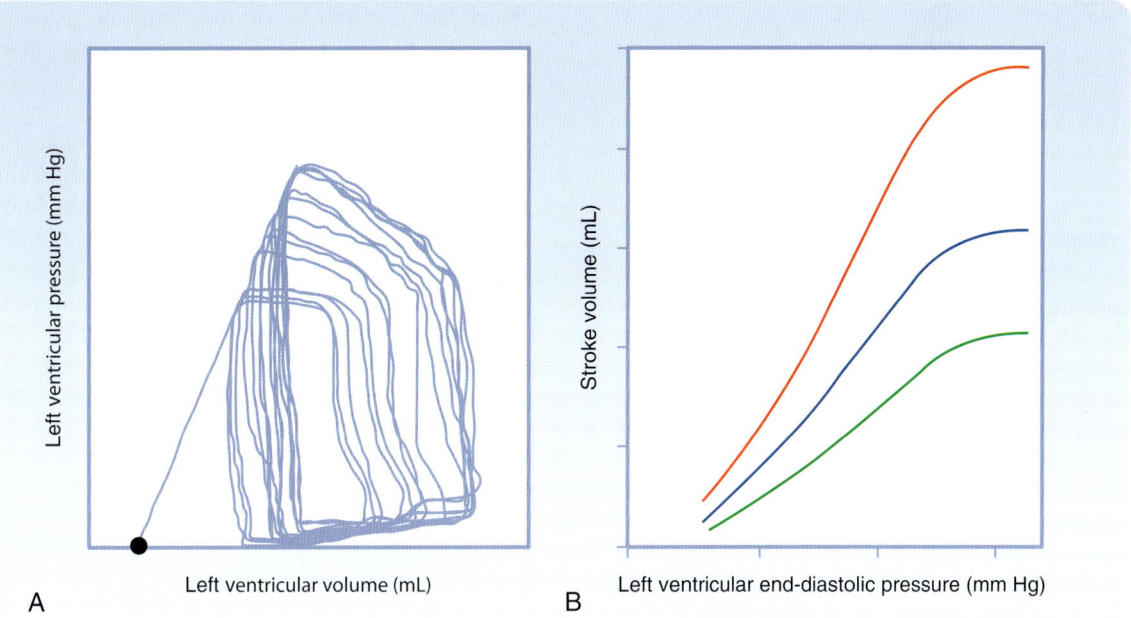

• **Fig. 16.44 A,** Left ventricular pressure-volume loops recorded in a human. The left ventricle was subjected to different preloads by transient occlusion of blood flow in the inferior vena cava. As preload (left ventricular end-diastolic pressure [LVEDP]) was decreased, both end-diastolic volume and end-systolic volume decreased, but end-diastolic volume decreased more, which resulted in decreased stroke volume. **B,** Cardiac function curves when stroke volume is plotted as a function of LVEDP for a normal heart in the basal state *(blue line)*, a heart with increased contractility *(red line)*, and a heart with reduced contractility *(green line)*. At any given preload, stroke volume is higher in hearts with increased contractility and lower in hearts with decreased contractility than stroke volume in the normal basal state. (**A,** From Senzaki H, et al. Single-beat estimation of end-systolic pressure-volume relation in humans. A new method with the potential for noninvasive application. *Circulation.* 1996;94[10]:2497-2506.)

approximate "diastolic" arterial blood pressure may be read when the aortic valve opens (point D), and the approximate "systolic" arterial blood pressure is read during systole (point E). Left ventricular P-V loops recorded from a human (Fig. 16.44A) are similar to the schematized version (see Fig. 16.43). The slope of the end-systolic P-V relation (line extending from the dot in Fig. 16.44A) defines contractility; a steeper slope indicates increased contractility. In this subject, partial occlusion of the inferior vena cava reduced the preload of the left ventricle on successive beats (as the inflow of blood to the left ventricle was reduced, as a consequence of the reduced return of blood to the right ventricle) and the effect was successively smaller stroke volume of the left ventricle.

This illustrates, in humans, the operation of the Frank-Starling law of the heart (Chapter 13), whereby changes in preload (left ventricular volume) change myocardial fiber length, and thus the strength of the subsequent contraction, as well as stroke volume produced. This important phenomenon is characterized by the cardiac function curve (see Fig. 16.44B). As preload increases, stroke volume increases (solid blue line). If contractility of the heart is increased, as in the action of norepinephrine, the slope of the end-systolic P-V relation becomes steeper, and the entire cardiac function curve shifts upward (solid red line), which reflects the fact that the ventricle is now able to produce

a larger stroke volume at a given preload. The increased stroke volume at any given preload is largely produced by a decreased end-systolic volume: Hearts with increased contractility are able to "squeeze down" to a greater extent. Conversely, if the ventricle is damaged, as after cardiac ischemia, or if its contractility is otherwise reduced from normal (as after calcium channel blockade), the cardiac function curve is shifted downward (green line), which reflects a reduced stroke volume at any given preload. The cardiac function curve is also known as *ventricular function curves* or *Starling curves*. To assess the integrated functioning of the cardiovascular system (Chapter 19), cardiac output, rather than stroke volume, is usually measured; preload of the heart is considered to be the filling pressure of the right ventricle, typically measured as mean right atrial pressure ($\bar{P}_{ra}$), or central venous pressure.

Measurement of Cardiac Output

The Fick Principle

In 1870 the German physiologist Adolph Fick contrived the first method for measuring cardiac output in intact animals and people. The basis for this method, called the **Fick principle,** is simply an application of the law of conservation of mass. The principle is derived from the fact that the quantity of O_2 delivered to the pulmonary capillaries via

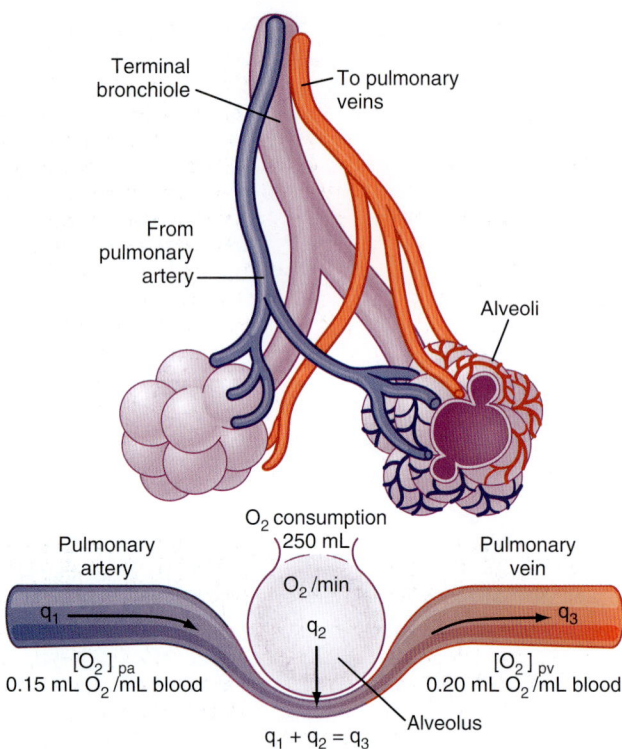

Terminal
bronchiole

To pulmonary
veins

From
pulmonary
artery

Alveoli

Pulmonary
artery

O_2 consumption
250 mL

Pulmonary
vein

O_2/min

q_1

q_2

q_3

$[O_2]_{pa}$
0.15 mL O_2/mL blood

$[O_2]_{pv}$
0.20 mL O_2/mL blood

Alveolus

$q_1 + q_2 = q_3$

• **Fig. 16.45** Schema Illustrating the Fick Principle for Measuring Cardiac Output. The change in color from pulmonary artery to pulmonary vein represents the change in color of the blood as venous blood becomes fully oxygenated.

the pulmonary artery, plus the quantity of O_2 that enters the pulmonary capillaries from the alveoli, must equal the quantity of O_2 that is carried away by the pulmonary veins.

The Fick principle is depicted schematically in Fig. 16.45. The rate of O_2 delivery to the lungs (q_1) equals the O_2 concentration in pulmonary arterial blood ($[O_2]_{pa}$), multiplied by pulmonary arterial blood flow (Q), which equals cardiac output; that is,

Equation 16.1

$$q_1 = Q[O_2]_{pa}$$

If q_2 is the net rate of O_2 uptake by the pulmonary capillaries from the alveoli, then at steady state, q_2 equals the O_2 consumption of the body. The rate at which O_2 is carried away by the pulmonary veins (q_3) equals the O_2 concentration in pulmonary venous blood ($[O_2]_{pv}$), multiplied by total pulmonary venous flow, which is virtually equal to pulmonary arterial blood flow (Q); that is,

Equation 16.2

$$q_3 = Q[O_2]_{pv}$$

From the law of conservation of mass,

Equation 16.3

$$q_1 + q_2 = q_3$$

Therefore,

Equation 16.4

$$Q[O_2]_{pa} + q_2 = Q[O_2]_{pv}$$

When this is solved for cardiac output,

Equation 16.5

$$Q = q_2/([O_2]_{pv} - [O_2]_{pa})$$

Eq. 16.5 is the statement of the Fick principle.

To determine cardiac output by this method, three values must be known: (1) O_2 consumption of the body, (2) the O_2 concentration in pulmonary venous blood ($[O_2]_{pv}$), and (3) the O_2 concentration in pulmonary arterial blood ($[O_2]_{pa}$). O_2 consumption is computed from measurements of the volume and O_2 content of expired air over a given interval. Because the O_2 concentration of peripheral arterial blood is essentially identical to that in the pulmonary veins, $[O_2]_{pv}$ is determined with a sample of peripheral arterial blood withdrawn by needle puncture. The compositions of pulmonary arterial blood and mixed systemic venous blood are virtually identical to one another. Samples for O_2 analysis are obtained from the pulmonary artery or right ventricle through a catheter. A very flexible catheter with a small balloon near the tip can be inserted into a peripheral vein. As the flexible tube is advanced, the flowing blood carries it toward the heart. By monitoring the pressure changes, the physician can advance the catheter tip into the pulmonary artery.

With the values depicted in Fig. 16.45, cardiac output can be calculated as follows: If the O_2 consumption is 250 mL/minute, the arterial (pulmonary venous) O_2 content is 0.20 mL of O_2 per milliliter of blood, and the mixed venous (pulmonary arterial) O_2 content is 0.15 mL of O_2 per milliliter of blood, cardiac output equals 250/(0.20 − 0.15) = 5000 mL/minute.

The Fick principle is also used to estimate the O_2 consumption of organs when blood flow and the O_2 content of arterial and venous blood can be determined. Algebraic rearrangement reveals that O_2 consumption equals blood flow multiplied by the difference in the arteriovenous O_2 concentration. For example, if blood flow through one kidney is 700 mL/minute, the arterial O_2 content is 0.20 mL of O_2 per milliliter of blood, and the renal venous O_2 content is 0.18 mL of O_2 per milliliter of blood, the rate of O_2 consumption by that kidney must be 700 (0.20 − 0.18) = 14 mL of O_2 per minute.

In the clinic, cardiac output is most commonly measured noninvasively with Doppler echocardiography. By this method, the velocity of blood in the ascending aorta is measured. Obtaining the cross-sectional area of the aorta (also measured by echocardiography) allows the volume of blood ejected in a single beat (i.e., stroke volume) to be determined (see Eq. 17.1). Multiplying stroke volume by the heart rate then yields a value for cardiac output in liters per minute.

Cardiac Oxygen Consumption and Work

Consumption of O_2 by the heart depends on the amount and type of activity that the heart performs. Under basal conditions, myocardial O_2 consumption ($\dot{V}_{O_2}$) is approximately 8 to 10 mL/minute/100 g of heart. It can increase severalfold during exercise and decrease moderately in such conditions as hypotension and hypothermia. The O_2 content of cardiac venous blood is normally low ($\approx$5 mL/dL), and the myocardium can receive little additional O_2 by further extraction of O_2 from coronary blood. Therefore, increased O_2 demands of the heart must be met mainly by an increase in coronary blood flow (see Chapter 17). In experiments in which the heartbeat is arrested but coronary perfusion is maintained, O_2 consumption falls to 2 mL/minute/100 g or less, which is still six to seven times greater than the O_2 consumption of resting skeletal muscle.

Cardiac work has external and internal components. Left ventricular work per beat (stroke work) is approximately equal to the product of stroke volume and the mean aortic pressure against which the blood is ejected by the left ventricle. External cardiac work ($\mathbf{W_e}$) may be defined as follows:

Equation 16.6
$$W_e = \int_{t_1}^{t_2} PdV + \rho v^2/2$$

That is, each small increment in volume that is pumped, ($\mathbf{dV}$) is multiplied by the associated pressure ($\mathbf{P}$), and the products ($\mathbf{PdV}$) are integrated over the time interval of interest ($\mathbf{t_2 - t_1}$) to calculate total work. Added to this is kinetic work due to the velocity (v) of blood flow and the density (ρ) of blood. The mean pressure during expulsion is used to simplify this to

Equation 16.7
$$W_e = PV + \rho v^2/2$$

At resting levels of cardiac output, the kinetic energy component is negligible. However, with high cardiac output, as in strenuous exercise, the kinetic energy component can account for up to 50% of total cardiac work.

The internal work (W_i) of the heart can be written as

Equation 16.8
$$W_i = \alpha \int T \, dt$$

where α is a proportionality constant that converts T dt into units of work, T is wall tension, and dt is time. Clinically, $\dot{V}_{O_2}$ and left ventricular power are difficult to measure; however, both are closely related to the systolic pressure–time index, the integral of left ventricular pressure, and time during systole. Such measurements are important because internal work is a large determinant of myocardial O_2 need. An alternative approach to evaluate cardiac work and its relation to O_2 consumption has been developed in which P-V loops (see Fig. 16.44) are examined in conditions with varied preload and afterload; contractility is maintained constant.

Simultaneously halving aortic pressure and doubling cardiac output, or vice versa, result in the same value for cardiac work. However, the O_2 requirements are greater for any given amount of cardiac work when a major proportion of the work is pressure work, as opposed to volume work. An increase in cardiac output at a constant aortic pressure (volume work) is accomplished with only a small increase in left ventricular O_2 consumption, whereas increased

arterial pressure at constant cardiac output (pressure work) is accompanied by a large increase in myocardial O_2 consumption. Thus myocardial O_2 consumption may not be well correlated with overall cardiac work. The magnitude and duration of left ventricular pressure are correlated with left ventricular O_2 consumption. The work of the right ventricle is one-seventh that of the left ventricle because pulmonary vascular resistance is much less than systemic vascular resistance.

Cardiac Efficiency

The efficiency of the heart may be calculated as the ratio of the work accomplished to the total energy used. If the average O_2 consumption is assumed to be 9 mL/minute/100 g for the two ventricles, a 300-g heart will consume 27 mL of O_2 per minute. This value is equivalent to 130 small calories when the respiratory quotient is 0.82. Together, the two ventricles do approximately 8 kg-m of work per minute, which is equivalent to 18.7 small calories. Therefore, the gross efficiency of the heart is approximately 14%:

Equation 16.9
$$18.7/130 \times 100 \cong 14\%$$

 IN THE CLINIC

The greater energy demand of pressure work than of volume work is clinically important, especially in aortic stenosis. In this condition, left ventricular O_2 consumption is increased, mainly because of the high intraventricular pressure developed during systole. However, coronary perfusion pressure (and hence O_2 supply) is either normal or reduced because of the pressure drop across the narrow orifice of the diseased aortic valve.

The actual gross mechanical efficiency of the heart is slightly higher (18%) than the value calculated and is determined through subtracting the O_2 consumption of the nonbeating (asystolic) heart ($\approx$2 mL/minute/100 g) from the total cardiac O_2 consumption in the calculation of efficiency. The efficiency of the heart as a pump is relatively low. During physical exercise, efficiency improves because mean blood pressure does not change appreciably, whereas cardiac output and work increase considerably, without a proportional increase in myocardial O_2 consumption. Of interest is that the chemical efficiency of the heart is rather high, as indicated by the estimate of 60% for the efficiency of generating ATP from oxidative phosphorylation. The energy expended in cardiac metabolism that does not contribute to the propulsion of blood through the body appears in the form of heat. The energy of flowing blood is also dissipated as heat.

Myocardial Adenosine Triphosphate and Its Relation to Mechanical Function

The chemical energy that fuels cardiac contractile work and relaxation is derived from ATP hydrolysis (Fig. 16.46).

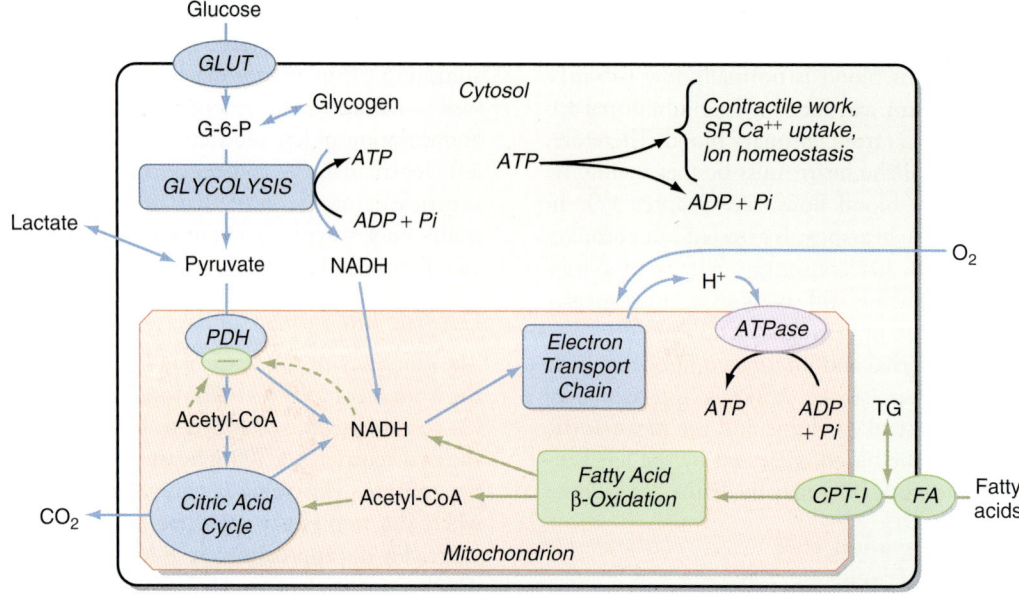

• **Fig. 16.46** Overall Scheme for Production and Utilization of Adenosine Triphosphate (ATP) Within a Cardiac Myocyte. Pathways for utilization of glucose, fatty acids (FA), and lactate are indicated, as are the requirements for O_2 and H^+ by the electron transport chain in mitochondria. ADP, adenosine diphosphate; CoA, coenzyme A; CPT-I, carnitine palmitoyltransferase; G-6-P, glucose-6-phosphate; GLUT, glucose transporter; NADH, nicotinamide adenosine dehydrogenase; PDH, pyruvate dehydrogenase; Pi, inorganic phosphate; SR, sarcoplasmic reticulum; TG, triglyceride.

The healthy heart has a relatively constant level of ATP ($\approx5\mu$mol/g wet weight) despite an extremely high rate of ATP hydrolysis ($\approx0.3\mu$mol/g^{-1}/second^{-1}). The tissue content of ATP is low in relation to the rate of breakdown and production; complete turnover of the myocardial ATP content occurs approximately every 12 seconds in the heart at rest. ATP hydrolysis (see Fig. 16.46) provides energy for contractile work (actin-myosin interaction and cell shortening), pumping Ca^{++} back into the sarcoplasmic reticulum at the end of systole and maintaining normal ion gradients (low Na^+ and high K^+ in the cell). Approximately two thirds of the ATP hydrolyzed by the heart is used to fuel contractile work, and the remaining third is used for ion pumps and "housekeeping" functions such as synthesis of proteins and nucleic acids. The Ca^{++}-ATPase of the sarcoplasmic reticulum is the primary ion pump consuming ATP. This process occurs during the end of systole, when cytosolic Ca^{++} is rapidly sequestered into the sarcoplasmic reticulum to initiate diastolic relaxation. In the healthy heart, the rate of ATP hydrolysis is exquisitely matched to the rate of ATP resynthesis. There is no significant change in the concentration of ATP or ADP even with the onset of heavy exercise.

ATP resynthesis occurs primarily through oxidative phosphorylation in mitochondria (>98%) and, to a small degree, through glycolysis (<2%). Oxidative phosphorylation requires O_2 and H_2. The O_2 is delivered to the myocardium and consumed in the mitochondria to make H_2O, and the H_2 is from the metabolism of carbon fuels (mainly fatty acids, glucose, and lactate) and the generation of the reduced form of nicotinamide adenine dinucleotide (NADH). Pyruvate dehydrogenase (PDH) regulates the oxidation of glucose and lactate. The activity of PDH is inhibited by product inhibition from acetyl–coenzyme A and NADH. Also, PDH activity is inhibited when phosphorylated by PDH kinase and activated when dephosphorylated by PDH phosphatase. ATP is formed from ADP and inorganic phosphate with the use of H^+. Of importance is that the rates of ATP formation and breakdown depend on an adequate delivery of O_2 to the myocardium, which is a function of myocardial blood flow and oxygenation of arterial blood. An increase in ATP breakdown in the myocardium, such as occurs when heart rate, systolic blood pressure, and contractility are increased (as during exercise), necessitates an increase in O_2 delivery to the myocardium so that the mitochondria can generate sufficient ATP by oxidative phosphorylation to meet the demand for ATP. Thus the rate of myocardial O_2 consumption is tightly linked to the work rate (or power) of the myocardium.

Substrate Utilization

The heart is versatile in its use of substrates, and within certain limits, uptake of a particular substrate is directly proportional to its arterial concentration. The use of one substrate by the heart is also influenced by the presence or absence of other substrates. For example, the addition of lactate to the blood that perfuses a heart metabolizing glucose leads to a reduction in glucose uptake and vice versa. At normal blood concentrations, glucose and lactate are consumed at approximately equal rates.

In contrast, uptake of pyruvate is very low, as is its arterial concentration. For glucose, the threshold concentration is approximately 4 mmol/L. Below this blood level, no glucose is taken up by the myocardium. Insulin reduces the glucose threshold and increases the rate of glucose uptake by the heart. Cardiac utilization of lactate occurs in most circumstances; insulin does not affect its uptake by the myocardium. Under hypoxic conditions, glucose utilization is facilitated by an increase in the rate of transport across the myocardial cell wall. However, lactate cannot be metabolized by the hypoxic heart and is produced by the heart under anaerobic conditions. Associated with lactate production by the hypoxic heart is the breakdown of cardiac glycogen.

Of the total cardiac O_2 consumption, only 35% to 40% can be accounted for by the oxidation of carbohydrate. Thus the heart derives the major part of its energy from the oxidation of noncarbohydrate sources: namely, esterified and nonesterified fatty acids, which account for approximately 60% of the myocardial O_2 consumption in people in the postabsorptive state. Various fatty acids have different thresholds for myocardial uptake, but these acids are generally used in direct proportion to their arterial concentration. Ketone bodies, especially acetoacetate, are readily oxidized by the heart, and they are a major source of energy in diabetic acidosis. As is true of carbohydrate substrates, use of a specific noncarbohydrate is influenced by the presence of other substrates, whether noncarbohydrate or carbohydrate. Therefore, within certain limits, the heart preferentially uses the substrate that is available in the largest concentration. The oxidation of amino acids makes a small contribution to myocardial energy expenditure.

Normally, the heart derives its energy by oxidative phosphorylation, in which each mole of glucose yields 36 mol of ATP. However, during hypoxia, glycolysis takes over, and 2 mol of ATP is provided by each mole of glucose; β-oxidation of fatty acids is also curtailed. If hypoxia is prolonged, cellular creatine phosphate and eventually ATP are depleted.

In ischemia, lactic acid accumulates and decreases intracellular pH. This condition inhibits glycolysis, fatty acid use, and protein synthesis, and therefore it results in cellular damage and eventually necrosis of myocardial cells.

Key Points

1. The transmembrane action potentials recorded from cardiac myocytes may contain the following five phases:

   Phase 0: The action potential upstroke is initiated when a suprathreshold stimulus rapidly depolarizes the membrane by activating the fast sodium channels.

   Phase 1: The notch is an early partial repolarization that is achieved by the efflux of K^+ through channels that conduct the transient outward current (i_{to}).

   Phase 2: The plateau represents a balance between the influx of Ca^{++} through calcium channels and the efflux of K^+ through several types of potassium channels.

   Phase 3: Final repolarization is initiated when the efflux of K^+ exceeds the influx of Ca^{++}. The resultant partial repolarization rapidly increases K^+ conductance and restores full repolarization.

   Phase 4: The resting potential of the fully repolarized cell is determined by conductance of the cell membrane to K^+, mainly through i_{K1} channels.

2. Fast-response action potentials are recorded from atrial and ventricular myocardial fibers and from ventricular specialized conducting (Purkinje) fibers. Such an action potential is characterized by a large amplitude, a steep upstroke, and a relatively long plateau. The effective refractory period of fast-response fibers begins at the upstroke of the action potential and persists until midway through phase 3. The fiber is relatively refractory during the remainder of phase 3, and it regains full excitability soon after it is fully repolarized (phase 4).

3. Slow-response action potentials are recorded from normal SA and AV nodal cells and from abnormal myocardial cells that have been partially depolarized. The action potential is characterized by a less negative resting potential, a smaller amplitude, a less steep upstroke, and a shorter plateau than is typical of the fast-response action potential. The upstroke in slow-response fibers is produced by the activation of calcium channels. Slow-response fibers become absolutely refractory at the beginning of the upstroke, and partial excitability may not be regained until very late in phase 3 or until after the fiber is fully repolarized.

4. Normally, the SA node serves as the cardiac pacemaker to initiate the cardiac impulse. This impulse is propagated from the SA node to the atria and ultimately reaches the AV node. After a delay in the AV node, the cardiac impulse is propagated throughout the ventricles. Ectopic foci in the atrium, the AV node, or the His-Purkinje system may initiate propagated cardiac impulses if the normal pacemaker cells in the SA node are suppressed or if the rhythmicity of the ectopic automatic cells is abnormally enhanced.

5. Under certain abnormal conditions, afterdepolarizations may be triggered by an otherwise normal action potential. EADs arise early in phase 3 of a normal action potential. They are more likely to occur when the basic cycle length of the

initiating beats is very long and when the cardiac action potentials are abnormally prolonged. DADs appear late in phase 3 or in phase 4. They are more likely to occur when the basic cycle length of the initiating beats is short and when the cardiac cells are overloaded with Ca^{++}.

6. Reentrant arrhythmias occur when a cardiac impulse traverses a loop of cardiac fibers and reenters previously excited tissue, when the impulse is conducted slowly around the loop, and when the impulse is blocked unidirectionally in some section of the loop.

7. The ECG, which is recorded from the surface of the body, traces the conduction of the cardiac impulse throughout the heart. The ECG may be used to detect and analyze certain cardiac arrhythmias, such as altered SA rhythms, AV conduction blocks, premature depolarizations, ectopic tachycardias, and atrial and ventricular fibrillation.

8. On excitation, voltage-gated calcium channels open to admit extracellular Ca^{++} into the cardiac myocites. The influx of Ca^{++} triggers the release of Ca^{++} from the sarcoplasmic reticulum. The elevated intracellular $[Ca^{++}]$ elicits contraction of the myofilaments. Relaxation is accomplished through the restoration of resting cytosolic $[Ca^{++}]$ by the pumping of Ca^{++} back into the sarcoplasmic reticulum and the exchange of Ca^{++} for extracellular Na^+ across the sarcolemma. Velocity and force of contraction are functions of intracellular $[Ca^{++}]$. Force and velocity are inversely related, and so with no load, velocity is maximal. In an isovolumic contraction, no external shortening occurs.

9. In ventricular contraction, preload is stretch of the fibers by blood during ventricular filling. Afterload is the arterial pressure against which the ventricle ejects the blood. An increase in myocardial fiber length, as occurs with augmented ventricular filling (preload) during diastole, produces a more forceful ventricular contraction. This relationship between fiber length and strength of contraction is known as the Frank-Starling law of the heart.

10. Contractility is an expression of cardiac performance at a given preload and afterload. Contractility can be modulated by the autonomic nervous system.

11. To determine cardiac output, according to the Fick principle, the O_2 consumption of the body (q_2) and the oxygen content of arterial blood ($[O_2]_a$) and mixed venous blood ($[O_2]_v$) are measured. Cardiac output $= q_2/([O_2]_a - [O_2]_v)$. It can also be measured noninvasively by Doppler echocardiography.

12. In the normal heart, the rate of ATP hydrolysis is matched by the rate of ATP synthesis. ATP is generated by oxidative phosphorylation of fatty acids and glucose, a process that requires O_2. When the rate of ATP hydrolysis increases, as during exercise, the consumption of O_2 for oxidative phosphorylation also increases. The rate of O_2 utilization is thereby coupled to the rate of cardiac work. The myocardium functions only aerobically, and in general it uses substrates in proportion to their arterial concentration.

Additional Readings

Abriel H, Rougier J-S, Jalife J. Ion channel macromolecular complexes in cardiomyocytes: Roles in sudden cardiac death. *Circ Res*. 2015;116:1971.

Bers DM, Guo T. Calcium signaling in cardiac ventricular myocytes. *Ann N Y Acad Sci*. 2005;1047:86.

Cannell MB, Kong CH. Local control in cardiac E-C coupling. *J Mol Cell Cardiol*. 2012;52:298.

Gima K, Rudy Y. Ionic current basis of electrocardiographic waveforms. *Circ Res*. 2003;90:889.

Priori SG. The fifteen years of discoveries that shaped molecular electrophysiology: time for appraisal. *Circ Res*. 2010;107:451.

17

Properties of the Vasculature

LEARNING OBJECTIVES

Upon completion of this chapter, the student should be able to answer the following questions:

1. What physical properties of blood vessels and blood determine hemodynamics, and how are they defined by Poiseuille's law?
2. How is arterial compliance related to stroke volume and pulse pressure? How does arterial compliance affect the arterial pulse wave and cardiac work?
3. What are mean, systolic, diastolic, and pulse pressures, and how are they measured?
4. What vessels constitute the microcirculation? How is pulsatile blood flow in large arteries converted into steady flow in the microcirculation?
5. What are the hydrostatic and osmotic factors that underlie Starling's hypothesis for capillary function?
6. How do intrinsic and extrinsic factors modulate peripheral circulation, and how do these factors affect blood flow in particular organs?
7. How does the myogenic hypothesis account for autoregulation of blood flow? What is the effect of tissue metabolism on autoregulation?

The vasculature consists of a closed system of tubes or vessels that distributes blood from the heart to the tissues and returns blood from the tissues to the heart. It can be divided into three components: the **arterial system,** which takes blood from the heart and distributes it to the tissues; the **venous system,** which returns blood from the tissues to the heart; and the **microcirculation,** which separates the arterial and venous systems and is the site where nutrients and cellular waste products are exchanged between blood and tissues. These components of the vasculature are described in this chapter. In addition, the properties of blood flow to specific vascular beds and tissues are considered. As an introduction to this material, the physics of blood/fluid flow through the vasculature (i.e., **hemodynamics**) is reviewed.

Hemodynamics

The physics of fluid flow through rigid tubes provides a basis for understanding the flow of blood through blood vessels, even though the blood vessels are not rigid tubules (i.e., they are distensible) and blood is not a simple homogeneous fluid. Knowledge of these physical principles underlies understanding of the interrelationships among velocity of blood flow, blood pressure, and the dimensions of the various components of the systemic circulation.

Velocity of the Bloodstream

Velocity, as relates to fluid movement, is the distance that a particle of fluid travels with regard to time, and it is expressed in units of distance per unit time (e.g., centimeters per second). Flow, in contrast, is the rate of displacement of a volume of fluid, and it is expressed in units of volume per unit time (e.g., cubic centimeters per second). In a rigid tube, velocity (v) and flow (Q) are related to one another by the cross-sectional area (A) of the tube:

Equation 17.1
$$v = Q/A$$

The interrelationships among velocity, flow, and area are shown in Fig. 17.1. Because conservation of mass requires that the fluid flowing through a rigid tube be constant, the velocity of the fluid varies inversely with the cross-sectional area. Thus fluid flow velocity is greatest in the section of the tube with the smallest cross-sectional area and slowest in the section of the tube with the greatest cross-sectional area.

As shown in Fig. 15.3, velocity decreases progressively as blood traverses the arterial system. In the capillaries, velocity decreases to a minimal value. As the blood then passes centrally through the venous system toward the heart, velocity progressively increases again. The relative velocities in the various components of the circulatory system are related only to the respective cross-sectional areas.

Relationship Between Velocity and Pressure

The total energy in a hydraulic system consists of three components: pressure, gravity, and velocity. The velocity of blood flow can have an important effect on the pressure within the tube. Consider the effect of velocity on pressure in a tube with different cross-sectional areas (Fig. 17.2). In this system, the total energy remains constant. The total pressure within the tube equals the lateral (static) pressure plus the dynamic pressure. The gravitational component can be neglected because the tube is horizontal. The total pressures in segments A, B, and C are equal, provided that

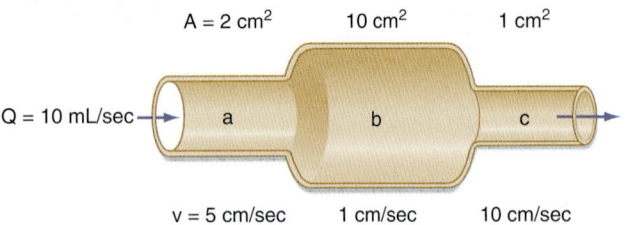

• **Fig. 17.1** As fluid flows through a tube of variable cross-sectional area (A), the linear velocity (v) varies inversely with the cross-sectional area. Q, flow.

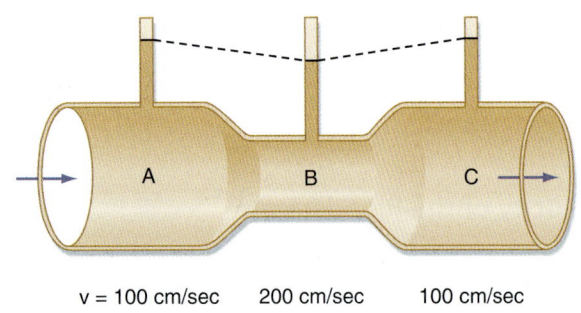

$\rho v^2/2 = 3.8$ mm Hg   15 mm Hg   3.8 mm Hg

• **Fig. 17.2** In a narrow section (B) of a tube, the linear velocity (v) and hence the dynamic component of pressure ($\rho v^2/2$) are greater than in the wide sections (A and C), of the same tube. If the total energy is virtually constant throughout the tube (i.e., if the energy loss because of viscosity is negligible), the lateral pressure in the narrow section is lower than the lateral pressure in the wide sections of the tube (see the heights of the column of fluid above compartments A, B, and C each, which reflect pressure).

the energy loss from viscosity is negligible (i.e., this fluid is an "ideal fluid"). The effect of velocity on the dynamic component (P_{dyn}) can be estimated as follows:

Equation 17.2

$$P_{dyn} = \rho v^2/2$$

where ρ is the density of the fluid (grams per cubic centimeters) and v is velocity (centimeters per second). Assume that the fluid has a density of 1 g/cm³. In section A in Fig. 17.2, the lateral pressure is 100 mm Hg; note that 1 mm Hg equals 1330 dynes/cm². According to Eq. 17.2, $P_{dyn} = 5000$ dynes/cm², or 3.8 mm Hg. In the narrow section B of the tube, where the velocity is twice as high, $P_{dyn} = 20,000$ dynes/cm², or 15 mm Hg. Thus the lateral pressure in section B is 15 mm Hg lower than the total pressure, whereas the lateral pressures in sections A and C are only 3.8 mm Hg lower. In most arterial locations, the dynamic component is a negligible fraction of the total pressure. However, at sites of an arterial constriction or obstruction, the high flow velocity is associated with large kinetic energy, and the dynamic pressure component may therefore increase significantly. Hence, the pressure would be reduced, and perfusion of distal segments would be correspondingly decreased. This example helps explain how pressure changes in a vessel that is narrowed by atherosclerosis or spasm of the blood vessel

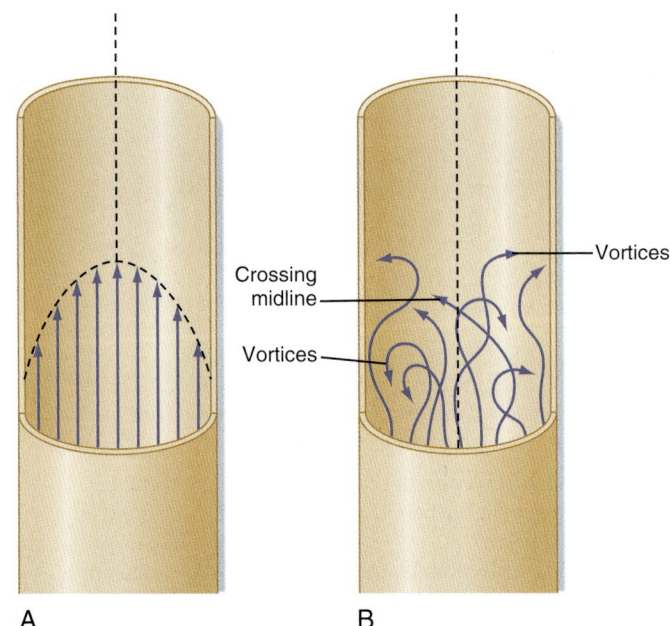

• **Fig. 17.3** Laminar and Turbulent Flow. **A,** When flow is laminar, all elements of the fluid move in streamlines that are parallel to the axis of the tube; the fluid does not move in a radial or circumferential direction. The layer of fluid in contact with the wall is motionless; the fluid that moves along the central axis of the tube has the maximal velocity. **B,** In turbulent flow, the elements of the fluid move irregularly in axial, radial, and circumferential directions. Vortices frequently develop.

wall: that is, in narrowed sections of a tube, the dynamic component increases significantly because the flow velocity is associated with large kinetic energy.

Relationship Between Pressure and Flow

The most fundamental law that governs the flow of fluids through cylindrical tubes was derived empirically by the French physiologist Jean Léonard Marie Poiseuille in the 1840s. He was interested primarily in the physical determinants of blood flow, but he replaced blood with simpler liquids in his measurements of flow through glass capillary tubes. His work was so precise and important that his observations have been designated **Poiseuille's law.**

Poiseuille's Law

Poiseuille's law applies to the steady (i.e., nonpulsatile) laminar flow of newtonian fluids through rigid cylindrical tubes. A newtonian fluid is one whose viscosity remains constant, and laminar flow is the type of motion in which the fluid moves as a series of individual layers, with each layer moving at a velocity different from that of its neighboring layers (Fig. 17.3A). In the case of laminar flow through a tube, the fluid consists of a series of infinitesimally thin concentric tubes sliding past one another, of which the central tube has the highest velocity. The velocities of the concentric laminae decrease parabolically towards the vessel wall. Despite the differences between the vascular system (i.e., flow is pulsatile, the vessels are not rigid cylinders, and blood is not a newtonian fluid), Poiseuille's law does provide valuable insight into the determinants of blood flow through the vascular system. In certain unusual situations, however, flow can become turbulent (see Fig. 17.3B), rather than laminar. Under these conditions, vortices (swirls) are present, and the distribution of flow velocities is chaotic. This condition is described in more detail later in this chapter.

Poiseuille's law describes the laminar flow of fluids through cylindrical tubes in terms of pressure, the dimensions of the tube, and the viscosity of liquid:

Equation 17.3

$$Q = \pi(P_i - P_o)r^4/8\eta l$$

where

Q = flow

$P_i - P_o$ = pressure gradient from the inlet (i) of the tube to the outlet (o)

r = radius of the tube

η = viscosity of the fluid

l = length of the tube

As is clear from the equation, flow through the tube increases as the pressure gradient is increased, and it decreases as either the viscosity of the fluid or the length of the tube increases. The radius of the tube is a critical factor in determining flow because it is raised to the fourth power.

Resistance to Flow

In electrical theory, **Ohm's law** is that the resistance (R) equals the ratio of voltage drop (E) to current flow (I).

Equation 17.4

$$R = E/I$$

Similarly, in fluid mechanics, hydraulic resistance (R) may be defined as the ratio of the pressure drop ($P_i - P_o$) to flow (Q):

Equation 17.5

$$R = (P_i - P_o)/Q$$

For the steady, laminar flow of a newtonian fluid through a cylindrical tube, the physical components of hydraulic resistance may be appreciated by the rearranging of Poiseuille's law to yield the hydraulic resistance equation:

Equation 17.6

$$R = (P_i - P_o)/Q = 8\eta l/\pi r^4$$

Thus when Poiseuille's law applies, the resistance to flow depends on only the dimensions of the tube and the characteristics of the fluid.

The principal determinant of resistance to blood flow through any vessel is the caliber of the vessel because resistance varies inversely as the fourth power of the radius of the tube. In Fig. 17.4, the resistance to flow through small blood vessels is measured, and the resistance per unit length of vessel (R/l) is plotted against the vessel diameter. As shown, resistance is highest in the capillaries (diameter of 7 μm), and it diminishes as the vessels increase in diameter

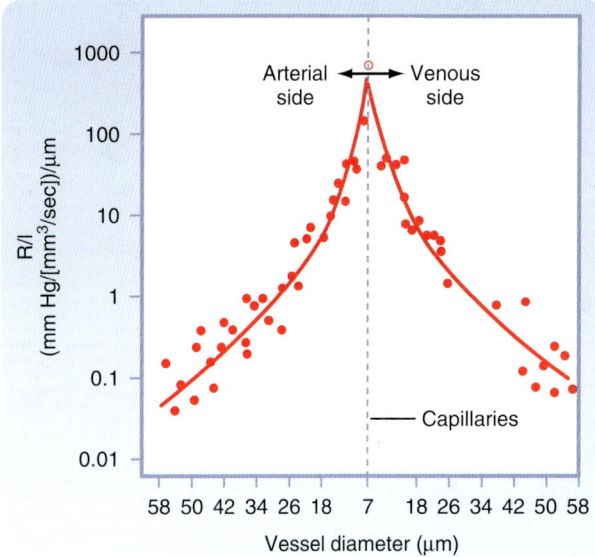

• **Fig. 17.4** Resistance Per Unit Length (R/l) of Individual Small Blood Vessels. The capillaries, with a diameter of 7 μm, are denoted by the *vertical dashed line.* Resistances of arterioles are plotted to the left of the vertical dashed line, and resistances of venules, to the right of the vertical dashed line. For both types of vessels, the resistance per unit length is inversely proportional to the fourth power of the vessel diameter. (Redrawn from Lipowsky HH, et al. *Circ Res.* 1978;43:738.)

on the arterial and venous sides of the capillaries. Values of R/l are virtually inversely proportional to the fourth power of the diameter (or radius) of the larger vessels on both sides of the capillaries.

Changes in vascular resistance occur when the caliber of vessels changes. The most important factor that leads to a change in vessel caliber is contraction of the circular smooth muscle cells in the vessel wall. Changes in internal pressure also alter the caliber of blood vessels and therefore alter the resistance to blood flow through these vessels. Blood vessels are elastic tubes. Hence, the greater the transmural pressure (i.e., the difference between internal and external pressure) across the wall of a vessel, the greater the caliber of the vessel and the less its hydraulic resistance.

It is apparent from Fig. 15.3 that the greatest drop in pressure occurs in the very small arteries and arterioles. However, capillaries, which have a mean diameter of approximately 7 μm, have the greatest resistance to blood flow. Nevertheless, of all the different varieties of blood vessels that lie in series with one another (as in Fig. 15.3), the arterioles, not the capillaries, have the greatest resistance. This seeming paradox is related to the relative numbers of parallel capillaries and parallel arterioles: There are far more capillaries than arterioles in the systemic circulation, and total resistance across the many capillaries arranged in parallel is much less than total resistance across the fewer arterioles arranged in parallel. In addition, arterioles have a thick coat of circularly arranged smooth muscle fibers that can vary the lumen radius. Even small changes in radius alter resistance greatly, as can be seen from the hydraulic

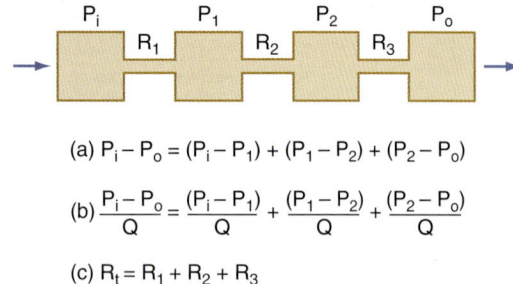

(a) $P_i - P_o = (P_i - P_1) + (P_1 - P_2) + (P_2 - P_o)$

(b) $\dfrac{P_i - P_o}{Q} = \dfrac{(P_i - P_1)}{Q} + \dfrac{(P_1 - P_2)}{Q} + \dfrac{(P_2 - P_o)}{Q}$

(c) $R_t = R_1 + R_2 + R_3$

• **Fig. 17.5** For resistances (R_1, R_2, and R_3) arranged in series, total resistance (R_t) equals the sum of the individual resistances. P, pressure; Q, flow.

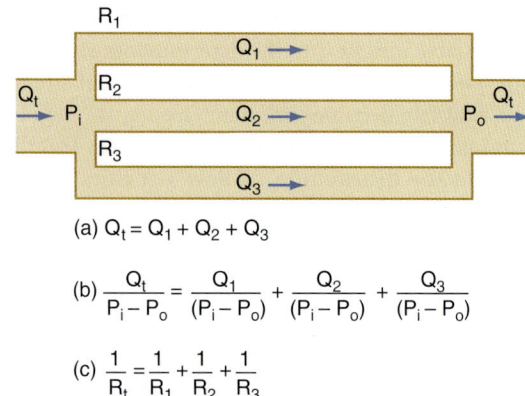

(a) $Q_t = Q_1 + Q_2 + Q_3$

(b) $\dfrac{Q_t}{P_i - P_o} = \dfrac{Q_1}{(P_i - P_o)} + \dfrac{Q_2}{(P_i - P_o)} + \dfrac{Q_3}{(P_i - P_o)}$

(c) $\dfrac{1}{R_t} = \dfrac{1}{R_1} + \dfrac{1}{R_2} + \dfrac{1}{R_3}$

• **Fig. 17.6** For resistances (R_1, R_2, and R_3) arranged in parallel, the reciprocal of the total resistance (R_t) equals the sum of the reciprocals of the individual resistances. P, pressure; Q, flow.

resistance equation (Eq. 17.6), wherein R varies inversely with r^4.

Resistances in Series and in Parallel

In the cardiovascular system, the various types of vessels listed along the horizontal axis in Fig. 15.3 lie in series with one another. The individual members of each category of vessels are ordinarily arranged in parallel with one another (see Fig. 15.1). Thus capillaries are in most instances parallel elements throughout the body, except in the renal vasculature (in which the peritubular capillaries are in series with the glomerular capillaries) and the splanchnic vasculature (in which the intestinal and hepatic capillaries are aligned in series with each other). The total hydraulic resistance of components arranged in series or in parallel can be derived in the same manner as those for analogous combinations of electrical resistance.

Resistance of Vessels in Series

In the system depicted in Fig. 17.5, three hydraulic resistances, R_1, R_2, and R_3, are arranged in series. The pressure drop across the entire system (i.e., the difference between inflow pressure [P_i] and outflow pressure [P_o]) consists of the sum of the pressure drops across each of the individual resistances (equation a in Fig. 17.5). In the steady state, the flow (Q) through any given cross-section must equal the flow through any other cross-section. When each component in equation (a) is divided by Q (equation [b] in Fig. 17.5), it is evident from the definition of resistance (Eq. 17.5) that for resistances in series, the total resistance (R_t) of the entire system equals the sum of the individual resistances; that is,

Equation 17.7
$$R_t = R_1 + R_2 + R_3$$

Resistance of Vessels in Parallel

For resistances in parallel, as illustrated in Fig. 17.6, inflow and outflow pressure is the same for all tubes. In steady state, the total flow (Q_t) through the system equals the sum of the flows through the individual parallel elements (equation [a] in Fig. 17.5). Because the pressure gradient ($P_i - P_o$) is identical for all parallel elements, each term

in equation (a) may be divided by that pressure gradient to yield equation (b). From the definition of resistance, equation (c) in Fig. 17.5 may be derived. According to this equation, for resistances in parallel, the reciprocal of the total resistance (R_t) equals the sum of the reciprocals of the individual resistances; that is,

Equation 17.8
$$1/R_t = (1/R_1) + (1/R_2) + (1/R_3)$$

In a few simple illustrations, some of the fundamental properties of parallel hydraulic systems become apparent. For example, if the resistances of the three parallel elements in Fig. 17.6 were all equal, then

Equation 17.9
$$R_1 = R_2 = R_3$$

Therefore, from Eq. 17.8,

Equation 17.10
$$1/R_t = 3/R_1$$

When the reciprocals of these terms, are equated,

Equation 17.11
$$R_t = R_1/3$$

Thus the total resistance is less than the individual resistances. For any parallel arrangement, the total resistance must be less than that of any individual component. For example, consider a system in which a tube with very high resistance is added in parallel to a low-resistance tube. The total resistance of the system must be less than that of the low-resistance component by itself because the high-resistance component affords an additional pathway, or conductance, for flow of fluid.

Consider the physiological relationship between the **total peripheral resistance (TPR)** of the entire systemic vascular bed and the resistance of one of its components, such as the renal vasculature. TPR is the ratio of the arteriovenous (AV) pressure difference (arterial pressure [P_a] − venous pressure [P_v]) to the flow through the entire systemic vascular bed (i.e., the cardiac output [Q_t]). For example, the renal vascular resistance (R_r) would be the ratio of the same AV pressure difference ($P_a - P_v$) to renal blood flow (Q_r).

In an individual with an P_a of 100 mm Hg, a peripheral P_v of 0 mm Hg, and a cardiac output of 5000 mL/min, TPR is 0.02 mm Hg/mL/minute, or 0.02 **peripheral resistance units (PRUs).** Normally, the rate of blood flow through one kidney would be approximately 600 mL/minute. Renal resistance would therefore be 100 mm Hg ÷ 600 mL/minute, or 0.17 PRUs, which is 8.5 times greater than the TPR. In an organ such as the kidney, which weighs only approximately 1% as much as the whole body, the vascular resistance is much greater than that of the entire systemic circulation. Hence, it is not surprising that the resistance to flow would be greater for a component organ, such as the kidney, than for the entire systemic circulation because the systemic circulation has not only one kidney but also many more alternative pathways for blood to flow.

Laminar and Turbulent Flow

In *laminar* flow (see Fig. 17.3A), a thin layer of fluid in contact with the tube wall adheres to the wall and hence is motionless. The layer of fluid just central to the external lamina must shear against this motionless layer, and therefore that layer moves slowly but with a finite velocity. Similarly, the next more central layer moves still more rapidly; the longitudinal velocity profile is that of a paraboloid (see Fig. 17.3A). The fluid elements in any given lamina remain in that lamina as the fluid moves longitudinally along the tube. The velocity at the center of the stream is maximal and equal to twice the mean velocity of flow across the entire cross-section of the tube.

Irregular motions of the fluid elements may develop in the flow of fluid through a tube; such flow is called *turbulent.* In this condition, fluid elements do not remain confined to a specific laminae; instead, rapid, radial mixing occurs (see Fig. 17.3B). Greater pressure is necessary to force a given flow of fluid through the same tube when the flow is turbulent than when it is laminar. In turbulent flow, the pressure drop is approximately proportional to the square of the flow rate, whereas in laminar flow, the pressure drop is proportional to the first power of the flow rate. Hence, to produce a given flow, a pump such as the heart must do considerably more work if turbulent flow develops.

Whether turbulent or laminar flow exists in a tube under given conditions may be predicted on the basis of a dimensionless number called **Reynold's number (N_R).** This number represents the ratio of inertial to viscous forces. For a fluid flowing through a cylindrical tube,

Equation 17.12

$$N_R = \rho D v / \eta$$

where ρ = fluid density, D = tube diameter, v = mean velocity, and η = viscosity. When N_R is 2000 or less, the flow is usually laminar; when N_R is 3000 or greater, the flow is turbulent; and when N_R is between 2000 and 3000, the flow is transitional between laminar and turbulent. Eq. 17.12 indicates that high fluid densities, small tube diameters,

high flow velocities, and low fluid viscosities predispose to turbulence. In addition to these factors, abrupt variations in tube dimensions or irregularities in the tube walls may produce turbulence.

Shear Stress on the Vessel Wall

As blood flows through a vessel, it exerts a force on the vessel wall parallel to the wall. This force is called a *shear stress* (τ). Shear stress is directly proportional to the flow rate and viscosity of the fluid:

Equation 17.13

$$\tau = 4 \eta Q / \pi r^3$$

 **IN THE CLINIC**

Turbulence is usually accompanied by audible vibrations. Turbulent flow within the cardiovascular system may be detected through a stethoscope during physical examination. When the turbulence occurs in the heart, the resultant sound is termed a *murmur;* when it occurs in a vessel, the sound is termed a *bruit.* In severe anemia, functional cardiac murmurs (murmurs not caused by structural abnormalities) are frequently detectable. The physical bases for such murmurs resides are (1) the reduced viscosity of blood in anemia and (2) the high flow velocities associated with the high cardiac output that usually prevails in anemic patients. Blood clots, or thrombi, are more likely to develop in turbulent flow than in laminar flow. A problem with the use of artificial valves in the surgical treatment of valvular heart disease is that thrombi may occur in association with the prosthetic valve. The thrombi may be dislodged and occlude a crucial blood vessel. It is important to design such valves to avert turbulence and to include anticoagulants as a part of therapy.

 IN THE CLINIC

In certain types of arterial disease, particularly hypertension, the subendothelial layers of vessels tend to degenerate locally, and small regions of the endothelium may lose their normal support. The viscous drag on the arterial wall may cause a tear between a normally supported region and an unsupported region of the endothelial lining. Blood may then flow from the vessel lumen through the rift in the lining and become dissected between the various layers of the artery. Such a lesion is called a *dissecting aneurysm.* It occurs most often in the proximal portions of the aorta and is extremely serious. One reason for its predilection for this site is the high velocity of blood flow, with associated large shear rate values at the endothelial wall. Shear stress at the vessel wall also influences many other vascular functions, such as the permeability of the vessel walls by large molecules, the biochemical activity of endothelial cells, the integrity of the formed elements in blood, and blood coagulation. An increase in shear stress on the endothelial wall is also an effective stimulus for the release of nitric oxide (NO) from vascular endothelial cells; NO is a potent vasodilator (see the section "Microcirculation and Lymphatic System").

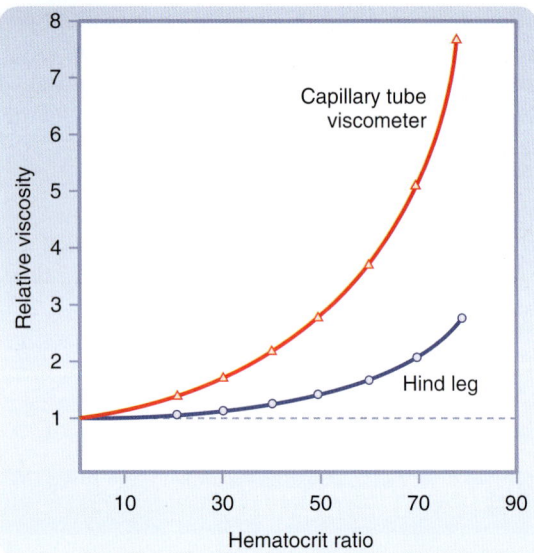

• **Fig. 17.7** The relative viscosity of whole blood increases at a progressively greater rate as the hematocrit ratio increases. For any given hematocrit ratio, the apparent viscosity of blood is lower when measured in a biological viscometer (such as a hind leg blood vessel) than in a conventional capillary tube viscometer. (Redrawn from Levy MN, Share L. *Circ Res.* 1953;1:247.)

Rheologic Properties of Blood

The viscosity of a given newtonian fluid at a specified temperature stays constant over a wide range of tube dimensions and flows. However, for a nonnewtonian fluid such as blood, viscosity may vary considerably as a function of tube dimensions and flows. Therefore, the term *viscosity* does not have a unique meaning for blood. The term *apparent viscosity* is frequently used for the derived value of blood viscosity obtained under the particular conditions of measurement.

Rheologically, blood is a suspension of formed elements, principally erythrocytes, in a relatively homogeneous liquid, the blood plasma. Because blood is a suspension, the apparent viscosity of blood varies as a function of the hematocrit (ratio of the volume of red blood cells to the volume of whole blood). The viscosity of plasma is 1.2 to 1.3 times that of water. The upper curve in Fig. 17.7 shows that the apparent viscosity of blood with a normal hematocrit ratio of 45% is 2.4 times that of plasma.* In severe anemia, blood viscosity is low. As the hematocrit increases, the slope of the curve increases progressively; it is especially steep at the upper range of erythrocyte concentrations (see Fig. 17.7).

For any given hematocrit, the apparent viscosity of blood depends on the dimensions of the tube used in estimating the viscosity. Fig. 17.8 demonstrates that the apparent viscosity of blood diminishes progressively as tube diameter decreases to less than approximately 0.3 mm. The diameters of the blood vessels with the highest resistance, the arterioles, are considerably less than this critical value. This phenomenon

---

*Fig. 17.7 also illustrates that the apparent viscosity of blood, when measured in living tissues, is considerably less than the apparent viscosity of the same blood measured in a conventional capillary tube viscometer.

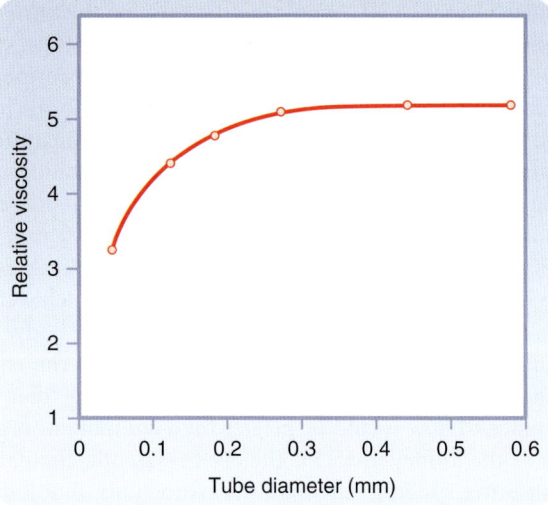

• **Fig. 17.8** The relative viscosity of blood relative to that of water increases as a function of tube diameter up to a diameter of approximately 0.3 mm. (Redrawn from Fåhraeus R, Lindqvist T. *Am J Physiol.* 1931;96:562.)

therefore reduces the resistance to flow in blood vessels that possess the greatest resistance. The influence of tube diameter on apparent viscosity is explained in part by the actual change in blood composition as it flows through small tubes. The composition of blood changes because the red blood cells tend to accumulate in the faster axial stream, whereas plasma tends to flow in the slower marginal layers. Because the axial portions of the bloodstream contain a greater proportion of red cells and this axial portion moves at greater velocity, the red blood cells tend to traverse the tube in less time than plasma does. Furthermore, the hematocrit of the blood contained in small blood vessels is lower than that in blood in large arteries or veins.

The physical forces responsible for the drift of erythrocytes toward the axial stream and away from the vessel walls when blood is flowing at normal rates are not fully understood. One factor is the great flexibility of red blood cells. At low flow rates, like those in the microcirculation, rigid particles do not migrate toward the central axis of a tube, whereas flexible particles do. The concentration of flexible particles near the tube's central axis is enhanced by an increase in the shear rate.

The apparent viscosity of blood diminishes as the shear rate is increased (Fig. 17.9), a phenomenon called *shear thinning*. The greater the amount of flow, the greater the rate that one lamina of fluid shears against an adjacent lamina. The greater tendency for erythrocytes to accumulate in the axial laminae at higher flow rates is partly responsible for this nonnewtonian behavior. However, a more important factor is that at very slow flow rates, the suspended cells tend to form aggregates; such aggregation increases blood viscosity. As flow is increased, this aggregation decreases, and so does the apparent viscosity of blood (see Fig. 17.9).

The tendency for erythrocytes to aggregate at low flow rates depends on the concentration of the larger protein

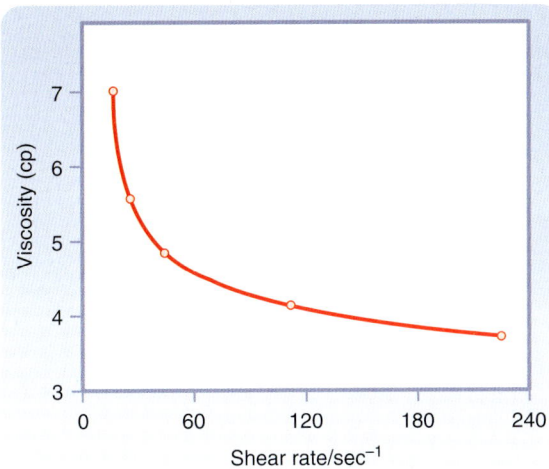

• Fig. 17.9 Decrease in the viscosity of blood (cp, centipoise) at increasing rates of shear (sec⁻¹). The shear rate is the velocity of one layer of fluid in relation to that of the adjacent layers and is directionally related to the rate of flow. (Redrawn from Amin TM, Sirs JA. *Q J Exp Physiol.* 1985;70:37.)

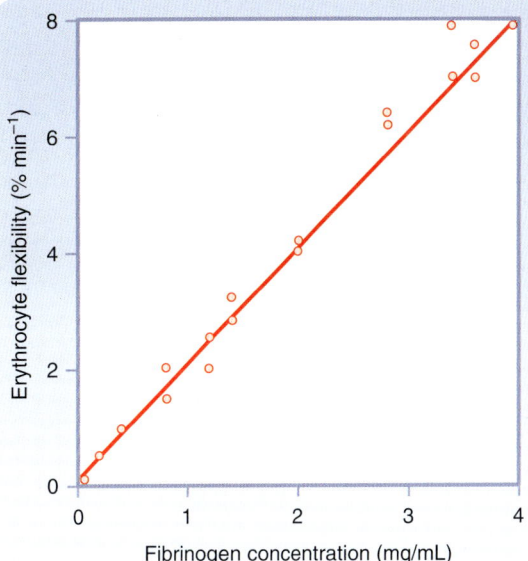

• Fig. 17.10 Effect of the plasma fibrinogen concentration on the flexibility of human erythrocytes. (Redrawn from Amin TM, Sirs JA. *Q J Exp Physiol.* 1985;70:37.)

molecules in plasma, especially fibrinogen. For this reason, changes in blood viscosity with flow rate are much more pronounced when the concentration of fibrinogen is high. In addition, at low flow rates, leukocytes tend to adhere to the endothelial cells of the microvessels and thereby increase the apparent viscosity of the blood.

The deformability of erythrocytes is also a factor in shear thinning, especially when the hematocrit is high. The mean diameter of human red blood cells is approximately 7 μm, but they are able to pass through openings with a diameter of only 3 μm. As blood with densely packed erythrocytes flows at progressively greater rates, the erythrocytes become more and more deformed. Such deformation diminishes the apparent viscosity of blood. The flexibility of human erythrocytes is enhanced as the concentration of fibrinogen in plasma increases (Fig. 17.10). If the red blood cells become hardened, as they are in certain spherocytic anemias, shear thinning may diminish.

The Arterial System

Arterial Elasticity

The systemic and pulmonary arterial systems distribute blood to the capillary beds throughout the body. The arterioles are high-resistance vessels of this system that regulate the distribution of flow to the various capillary beds. The aorta, the pulmonary artery, and their major branches have a large amount of elastin in their walls, which makes these vessels highly distensible (i.e., compliant). This distensibility serves to dampen the pulsatile nature of blood flow that results as the heart pumps blood intermittently. When blood is ejected from the ventricles during systole, these vessels distend, and during diastole, they recoil and propel the blood forward (Fig. 17.11). Thus the intermittent

output of the heart is converted to a steady flow through the capillaries.

The elastic nature of the large arteries also reduces the work of the heart. If these arteries were rigid rather than compliant, the pressure would rise dramatically during systole. This increased pressure would require the ventricles to pump against a large load (i.e., afterload) and thus increase the work of the heart. Instead, as blood is ejected into these vessels, they distend, and the resultant increase in systolic pressure, and thus the work of the heart, are reduced.

 IN THE CLINIC

As people age, the elastin content of the large arteries is reduced and replaced by collagen. This reduces arterial compliance (Fig. 17.12). Thus with age, systolic pressure increases, as does the difference between systolic and diastolic blood pressure, called the *pulse pressure* (described in the next section).

Determinants of Arterial Blood Pressure

Arterial blood pressure is routinely measured in patients, and it provides a useful estimate of their cardiovascular status. Arterial pressure can be defined as **mean arterial pressure ($\bar{P}_a$),** which is the pressure averaged over time, and as **systolic** (maximal) and **diastolic** (minimal) arterial pressure within the cardiac cycle (Fig. 17.13). The difference between systolic and diastolic pressure is termed **pulse pressure.**

The determinants of arterial blood pressure are arbitrarily divided into "physical" and "physiological" factors. The two

COMPLIANT

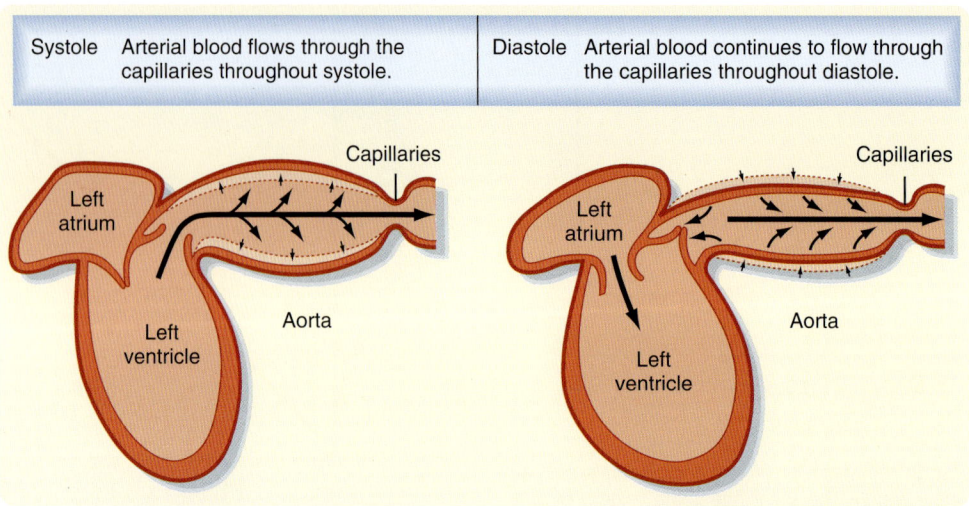

| Systole | Arterial blood flows through the capillaries throughout systole. | Diastole | Arterial blood continues to flow through the capillaries throughout diastole. |

A When the arteries are normally compliant, a substantial fraction of the stroke volume is stored in the arteries during ventricular systole. The arterial walls are stretched.

B During ventricular diastole the previously stretched arteries recoil. The volume of blood that is displaced by the recoil furnishes continuous capillary flow throughout diastole.

RIGID ARTERIES

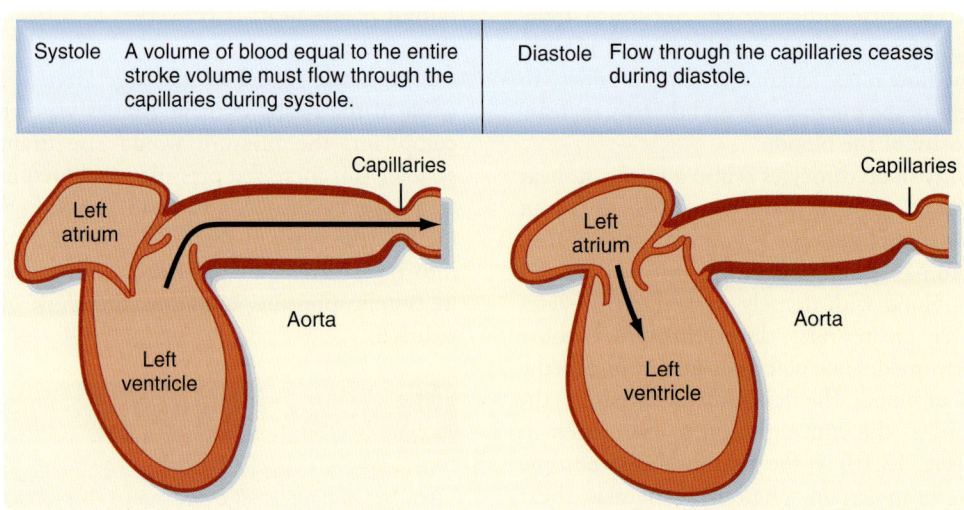

| Systole | A volume of blood equal to the entire stroke volume must flow through the capillaries during systole. | Diastole | Flow through the capillaries ceases during diastole. |

C When the arteries are rigid, virtually none of the stroke volume can be stored in the arteries.

D Rigid arteries cannot recoil appreciably during diastole.

• **Fig. 17.11** When arteries are normally compliant (**A** and **B**), blood flows through the capillaries throughout the cardiac cycle. When the arteries are rigid, blood flows through the capillaries during systole (**C**), but flow ceases during diastole (**D**).

physical factors, or fluid mechanical characteristics, are fluid volume (i.e., blood volume) within the arterial system and the static elastic characteristics (compliance) of the system. The physiological factors are cardiac output (which equals heart rate × stroke volume) and peripheral resistance.

Mean Arterial Pressure

To estimate $\overline{P}_a$ from an arterial blood pressure tracing, the area under the pressure curve is divided by the time interval involved (see Fig. 17.13). Alternatively, $\overline{P}_a$ can be

approximated from the measured values of systolic pressure (P_s) and diastolic pressure (P_d) by means of the following formula:

Equation 17.14
$$\overline{P}_a = P_d + (P_s - P_d/3)$$

Consider that $\overline{P}_a$ depends on only two physical factors: mean blood volume in the arterial system and arterial compliance (Fig. 17.14). Arterial volume (V_a), in turn, depends on the rate of inflow, (Q_h) into the arteries from the heart (cardiac output) and on the rate of outflow (Q_r) from the arteries through the resistance vessels (peripheral runoff). These relationships are expressed mathematically as

Equation 17.15

$$dV_a/dt = Q_h - Q_r$$

where dV_a/dt is the change in arterial blood volume per unit of time. If Q_h exceeds Q_r, arterial volume increases, the arterial walls are stretched further, and pressure rises. The converse happens when Q_r exceeds Q_h. When Q_h equals Q_r, P_a remains constant. Thus increases in cardiac output raise $\overline{P}_a$, as do increases in peripheral resistance. Conversely, decreases in cardiac output or peripheral resistance decrease $\overline{P}_a$.

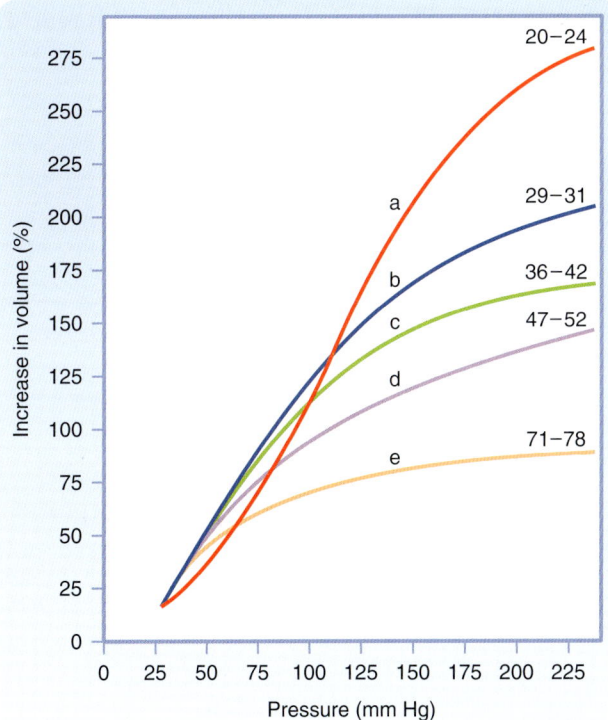

• **Fig. 17.12** Pressure-volume relationships of aortas obtained at autopsy from humans in different age groups (denoted by the numbers at the right end of each of the curves). Note that compliance ($\Delta V/\Delta P$) decreases with age. (Redrawn from Hallock P, Benson IC. *J Clin Invest.* 1937;16:595.)

Arterial Pulse Pressure

Arterial pulse pressure is systolic pressure minus diastolic pressure. It is principally a function of just one physiological factor, stroke volume, which determines the change in arterial blood volume (a physical factor) during ventricular systole. This physical factor, in addition to a second physical factor (arterial compliance), determines the arterial pulse pressure (see Fig. 17.14).

Stroke Volume

As described previously, $\overline{P}_a$ depends on cardiac output and peripheral resistance. During the rapid ejection phase of systole, the volume of blood introduced into the arterial system exceeds the volume that exits the system through the arterioles. Arterial pressure and volume therefore peak; the peak arterial pressure is systolic pressure. During the remainder of the cardiac cycle (i.e., ventricular diastole), cardiac ejection is zero, and peripheral runoff now greatly exceeds cardiac ejection. The resultant decrement in arterial blood volume thus causes pressure to fall to a minimum, which is diastolic pressure. Fig. 17.15 illustrates the effect of stroke volume on pulse pressure when arterial compliance is constant.

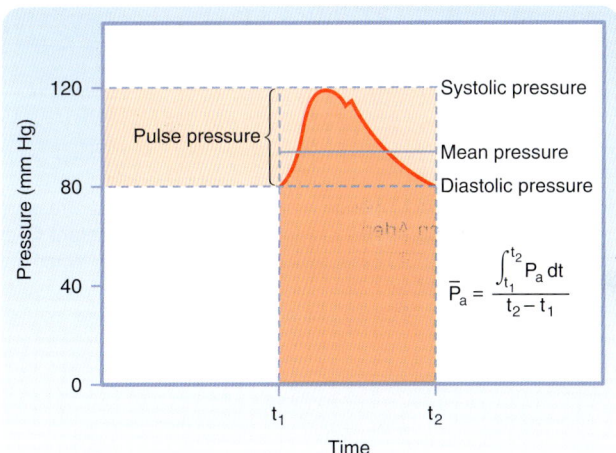

• **Fig. 17.13** Arterial Systolic, Diastolic, Pulse, and Mean Pressure. Mean arterial pressure ($\overline{P}$) represents the area under the arterial pressure curve (*dark red*) divided by the duration of the cardiac cycle ($t_2 - t_1$).

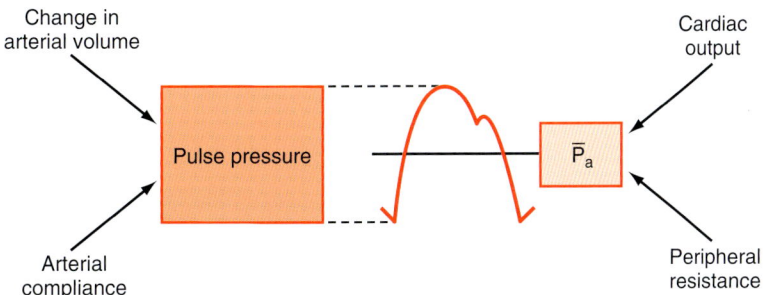

• **Fig. 17.14** The two physical determinants of pulse pressure are arterial compliance (C_a) and the change in arterial volume. The two physiological determinants of mean arterial pressure ($\overline{P}$) are cardiac output and total peripheral resistance.

Arterial Compliance

Arterial compliance (C_a), the ratio of blood volume to mean blood pressure (see Eq. 19.1), also affects pulse pressure. This relationship is illustrated in Fig. 17.16. When cardiac output and TPR are constant, a decrease in arterial compliance results in an increase in pulse pressure. Diminished arterial compliance also imposes a greater workload on the left ventricle (i.e., increased afterload), even if stroke volume, TPR, and $\overline{P}_a$ are equal in the two individuals.

Total Peripheral Resistance and Arterial Diastolic Pressure

As previously discussed, if the heart rate and stroke volume remain constant, an increase in TPR causes $\overline{P}_a$ to increase.

When arterial compliance is constant, an increase in TPR leads to proportional increases in systolic and diastolic pressure so that the pulse pressure is unchanged (Fig. 17.17A). However, arterial compliance is not linear. As $\overline{P}_a$ increases and the artery is stressed, compliance decreases (see Fig. 17.17B). Because of the decrease in arterial compliance with increased P_a, pulse pressure increases when P_a is elevated.

Effect of Arterial Compliance on Myocardial Energy Consumption

The increased cardiac energy requirement imposed by a rigid arterial system is illustrated in Fig. 17.18. In the data depicted in Fig. 17.18, the cardiac output from the left ventricle either was allowed to flow through the natural route (the aorta) or was directed through a stiff plastic tube to the peripheral arteries. In this experiment, the TPR values were virtually identical, regardless of which pathway

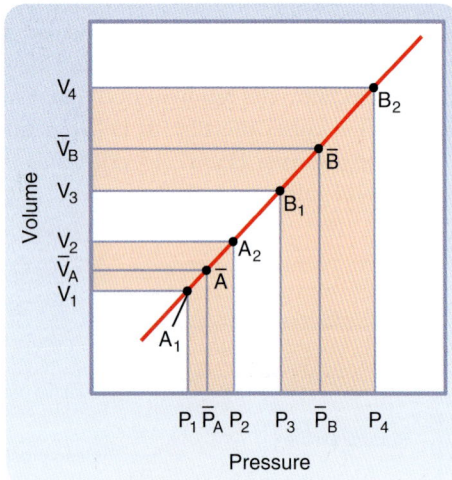

• **Fig. 17.15** Effect of a Change in Stroke Volume on Pulse Pressure in a System in Which Arterial Compliance Remains Constant Over the Prevailing Range of Pressures and Volumes. A larger increment in blood volume, whereby $(V_4 - V_3) > (V_2 - V_1)$, results in greater mean blood pressure ($\overline{P}_B > \overline{P}_A$) and a greater pulse pressure, so that $(P_4 - P_3) > (P_2 - P_1)$.

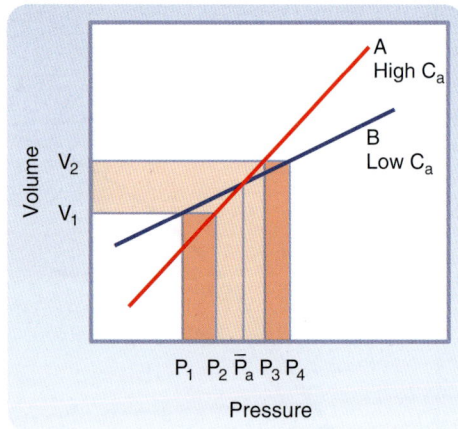

• **Fig. 17.16** For a given volume increment ($V_2 - V_1$), reduced arterial compliance (compliance B [Low C_a] < compliance A [High C_a]) results in increased pulse pressure, whereby $(P_4 - P_1) > (P_3 - P_2)$. $\overline{P}_a$, mean arterial pressure.

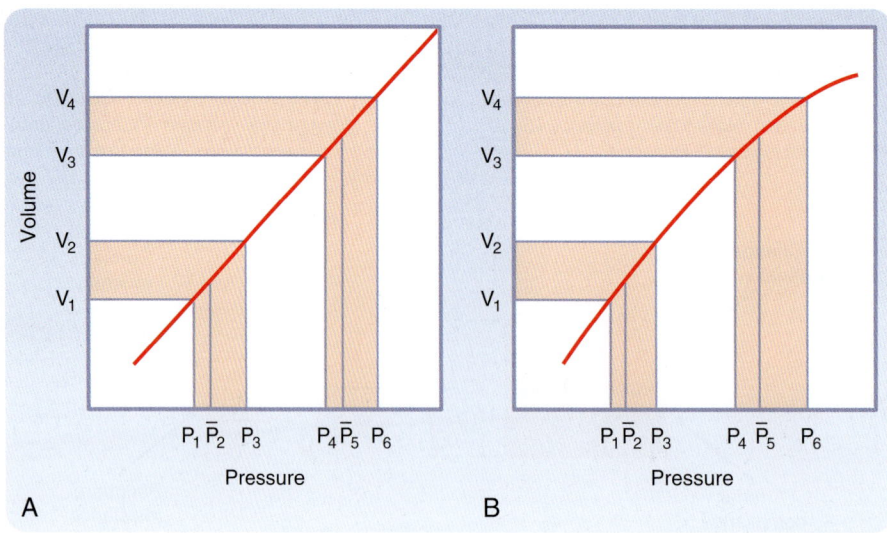

A   B

• **Fig. 17.17** Comparison of the effects of a given change in peripheral resistance on pulse pressure (P) when the pressure-volume curve for the arterial system is either rectilinear **(A)** or curvilinear **(B)**. The increment in arterial volume is the same for both conditions; that is, $(V_4 - V_3) = (V_2 - V_1)$.

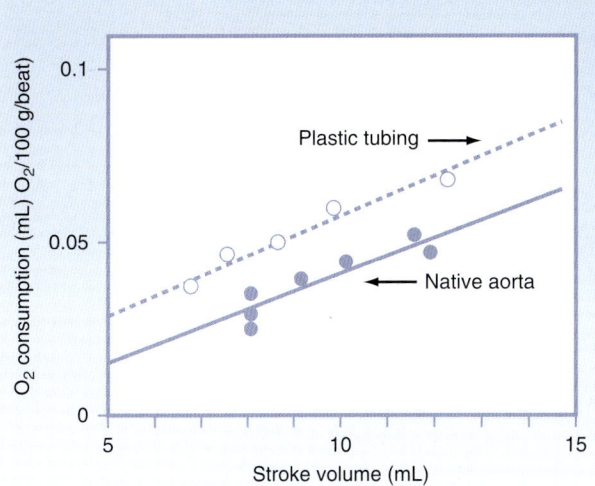

• **Fig. 17.18** The relationship between myocardial oxygen consumption (1 mL/100 g/beat) and stroke volume (in milliliters) in an anesthetized dog whose cardiac output could be pumped by the left ventricle either through the aorta or through a stiff plastic tube to the peripheral arteries. (Modified from Kelly RP, Tunin R, Kass DA. *Circ Res.* 1992;71:490.)

was selected. The results showed that for any given stroke volume, myocardial oxygen consumption was substantially greater when the blood was diverted through the plastic tubing than when it flowed through the aorta. The increased oxygen consumption indicates that the left ventricle has to expend significantly more energy to pump blood through a less compliant conduit than through a more compliant conduit.

 IN THE CLINIC

Arterial pulse pressure provides valuable information about a person's stroke volume, provided that arterial compliance is essentially normal. Patients who have severe congestive heart failure or who have suffered a severe hemorrhage are likely to have a very low arterial pulse pressure because their stroke volumes are abnormally small. Conversely, individuals with large stroke volumes, as in aortic valve regurgitation, are likely to have an increased arterial pulse pressure. Similarly, well-trained athletes at rest tend to have large stroke volumes because their heart rates are usually low. The prolonged ventricular filling times in these individuals induce the ventricles to pump a large stroke volume, and hence their pulse pressure is large.

Peripheral Arterial Pressure Curves

The radial stretch of the ascending aorta brought about by left ventricular ejection initiates a pressure wave that is propagated down the aorta and its branches. The pressure wave travels much faster (≈4 to 12 m/second) than the blood itself does. This pressure wave is the "pulse" that can be detected through palpation of a peripheral artery.

 IN THE CLINIC

In chronic hypertension, a condition characterized by a persistent elevation in TPR, the arterial pressure-volume curve resembles that shown in Fig. 17.17B. Because arteries become substantially less compliant when P_a rises, an increase in TPR causes systolic pressure to be more elevated than diastolic pressure. Diastolic pressure is elevated in such individuals but ordinarily not more than 10 to 40 mm Hg above the average normal level of 80 mm Hg. Not uncommonly, however, systolic pressure is elevated by 50 to 100 mm Hg above the average normal level of 120 mm Hg.

The velocity of the pressure wave varies inversely with arterial compliance. In general, transmission velocity increases with age, which confirms the observation that the arteries become less compliant with advancing age. Velocity also increases progressively as the pulse wave travels from the ascending aorta toward the periphery. This increase in velocity reflects the decrease in vascular compliance in the more distal portions than in the more proximal portions of the arterial system.

The P_a contour becomes distorted as the wave is transmitted down the arterial system. This distortion in the pressure wave contour of the human arterial tree is demonstrated as a function of age and of recording site in Fig. 17.19. Damping of the high-frequency components of the arterial pulse is caused largely by the viscoelastic properties of the arterial walls. The pulse pressure wave travels more rapidly in older people than in the younger people, as a consequence of reduced compliance. Several factors—including wave reflection and resonance, vascular tapering, and pressure-induced changes in transmission velocity—contribute to peaking of the P_a wave.

Blood Pressure Measurement in Humans

Most commonly, blood pressure is estimated indirectly by means of a sphygmomanometer. In hospital intensive care units, needles or catheters may be introduced into the peripheral arteries of patients to measure arterial blood pressure directly by means of strain gauges. When blood pressure readings are taken from the arm, systolic pressure may be estimated by palpation of the radial artery at the wrist (palpatory method). While pressure in the cuff exceeds the systolic level, no pulse is perceived. As pressure falls just below the systolic level (Fig. 17.20A), a spurt of blood passes through the brachial artery under the cuff during the peak of systole, and a slight pulse is felt at the wrist.

The auscultatory method is a more sensitive and therefore more precise technique for measuring systolic pressure, and it also enables diastolic pressure to be estimated. The practitioner listens with a stethoscope applied to the skin of the antecubital space over the brachial artery. While the pressure in the cuff exceeds systolic pressure, the brachial artery is occluded, and no sounds are heard (see Fig. 17.20B). When the inflation pressure falls just below the systolic level (120 mm Hg in Fig. 17.20A), a small spurt of blood escapes the occluding pressure of the cuff, and slight tapping sounds (called *Korotkoff sounds*) are heard with each heartbeat. The pressure at which the first sound is detected

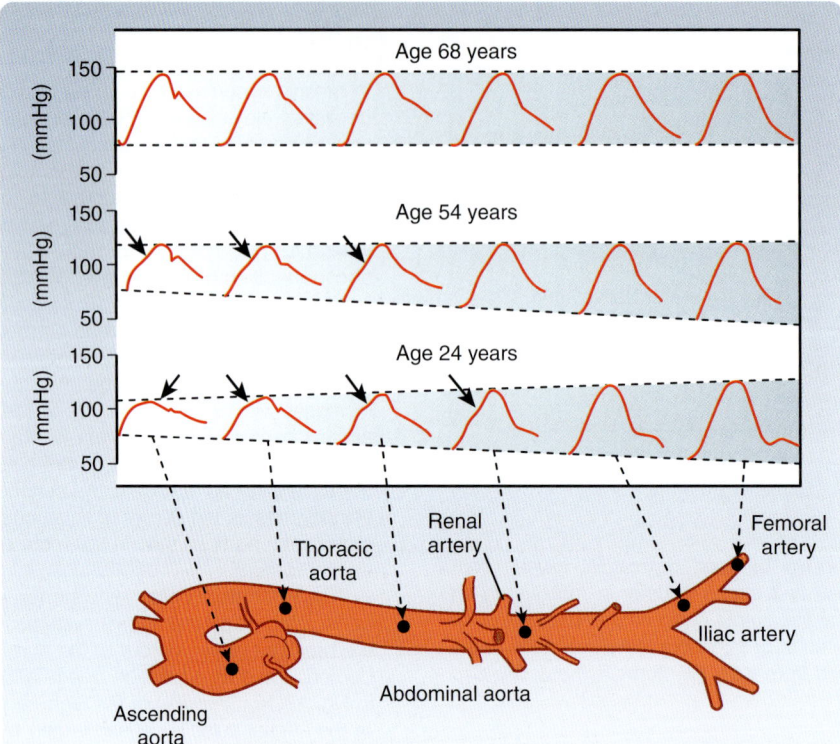

• **Fig. 17.19** Pulse Pressure Curves Recorded From Various Sites in the Arterial Trees of Humans at Different Ages. In the 24-year old, the arterial pulse displays striking changes in the pulse pressure amplitude and contour as it passes down the arterial tree. The pulse pressure wave in the 68-year-old shows little amplification and is relatively unchanged as the pulse travels because there is less wave reflection. (Reproduced by permission of Hodder Education from Nichols WW, O'Rourke M, eds. *McDonald's Blood Flow in Arteries: Theoretical, Experimental and Clinical Principles.* 5th ed. London: Arnold; 2005.)

represents systolic pressure. It usually corresponds closely to the directly measured systolic pressure. As the inflation pressure of the cuff continues to fall, more blood escapes under the cuff per beat and the sounds become louder. When the inflation pressure approaches the diastolic level, the Korotkoff sounds become muffled. When the inflation pressure falls just below the diastolic level (80 mm Hg in Fig. 17.20*A*), the sounds disappear; the pressure reading at this point indicates diastolic pressure. The origin of the Korotkoff sounds is related to the discontinuous spurts of blood that pass under the cuff and meet a static column of blood beyond the cuff; the impact and turbulence generate audible vibrations. Once the inflation pressure is less than diastolic pressure, flow is continuous in the brachial artery, and sounds are no longer heard (see Fig. 17.20*C*).

The Venous System

Capacitance and Resistance

Veins are elements of the circulatory system that return blood to the heart from tissues. Moreover, veins constitute a very large reservoir that contains up to 70% of the blood in the circulation. The reservoir function of veins makes them able to adjust the volume of blood returning to the heart,

IN THE CLINIC

The ankle-brachial index (ABI) is the ratio of systolic blood pressures at the ankle (dorsalis pedis artery) to that in the brachial artery. The ABI, which is obtained by simple measurements, is an indicator of possible peripheral artery disease. The ABI has also been proposed as a predictor of risk for cardiovascular and cerebrovascular disease. People with a normal ABI ratio of 1.1 to 1.4 have a lower incidence of either coronary or cerebrovascular events than do those with a ratio of 0.9 or lower. In addition, as the rate of ABI increases with time, the incidences of cardiovascular morbidity and mortality also increase.

or preload, so that the needs of the body can be matched when cardiac output is altered (see Chapter 19). This high capacitance is an important property of veins.

The *hydrostatic pressure* in postcapillary venules is approximately 20 mm Hg, and it decreases to approximately 0 mm Hg in the thoracic venae cavae and right atrium. Hydrostatic pressure in the thoracic venae cavae and right atrium is also termed *central venous pressure*. Veins are very distensible and have very low resistance to blood flow. Such low resistance allows movement of blood from

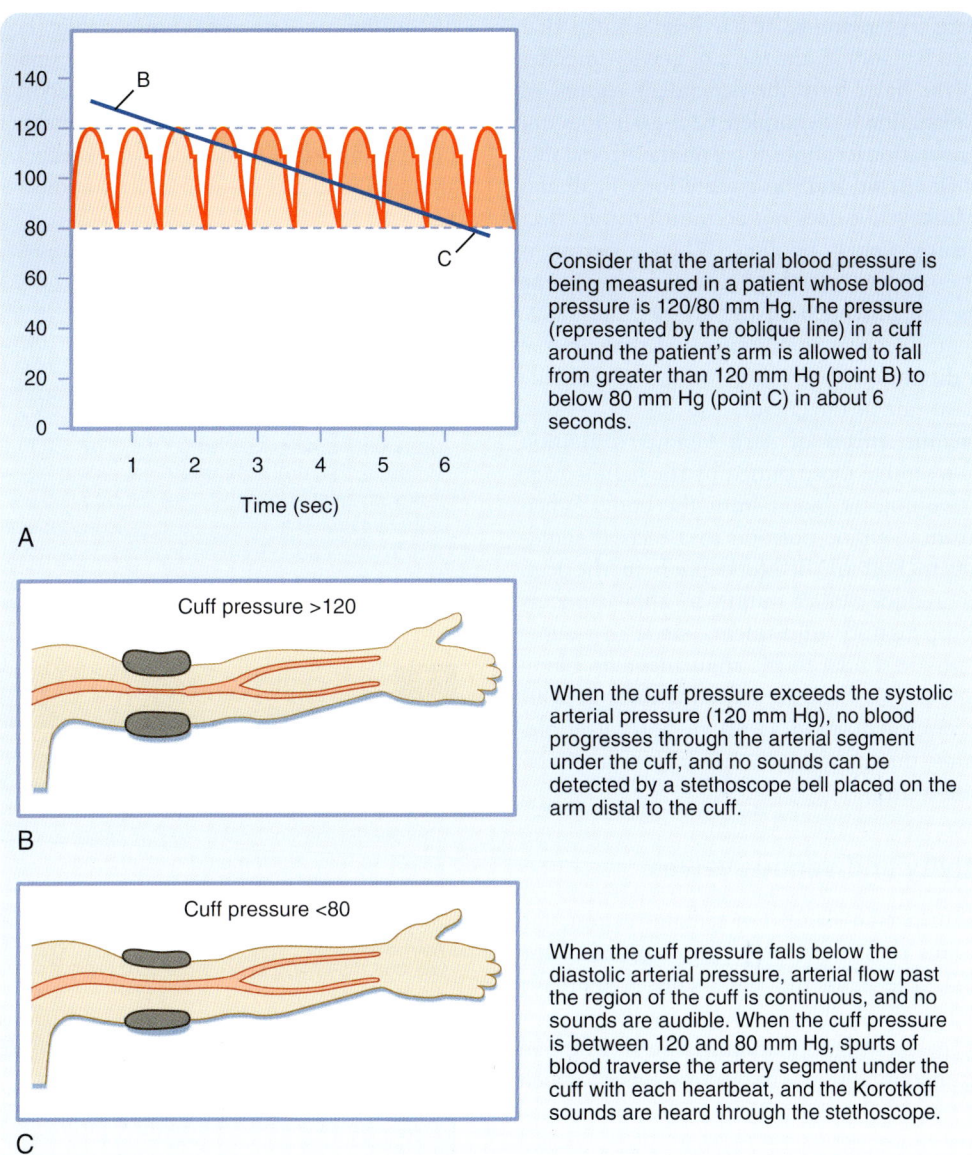

Consider that the arterial blood pressure is being measured in a patient whose blood pressure is 120/80 mm Hg. The pressure (represented by the oblique line) in a cuff around the patient's arm is allowed to fall from greater than 120 mm Hg (point B) to below 80 mm Hg (point C) in about 6 seconds.

When the cuff pressure exceeds the systolic arterial pressure (120 mm Hg), no blood progresses through the arterial segment under the cuff, and no sounds can be detected by a stethoscope bell placed on the arm distal to the cuff.

When the cuff pressure falls below the diastolic arterial pressure, arterial flow past the region of the cuff is continuous, and no sounds are audible. When the cuff pressure is between 120 and 80 mm Hg, spurts of blood traverse the artery segment under the cuff with each heartbeat, and the Korotkoff sounds are heard through the stethoscope.

• **Fig. 17.20 A** to **C,** Measurement of arterial blood pressure with a sphygmomanometer.

peripheral veins to the heart with only small reductions in central venous pressure. Moreover, veins control filtration and absorption by adjusting postcapillary resistance (see the section "Hydrostatic Forces") and assist in the cardiovascular adjustments that accompany changes in body position.

The ability of veins to participate in these various functions depends on their distensibility, or compliance. Venous compliance varies with the position in the body in such a way that veins in the lower limb are less compliant than those at or above the level of the heart. Veins in the lower limbs are also thicker than those in the brain or upper limbs. The compliance of veins, like that of arteries, decreases with age, and the vascular thickening that occurs is accompanied by a reduction in elastin and an increase in collagen content.

Variations in venous return are achieved by adjustments in venomotor tone, respiratory activity (see Chapter 19), and orthostatic stress or gravity.

Gravity

Gravitational forces influence the amount of blood in the venous system and therefore may profoundly affect cardiac output. For example, soldiers standing at attention for a long time may faint because gravity causes blood to pool in the dependent blood vessels, which reduces cardiac output. Warm ambient temperatures interfere with the compensatory vasomotor reactions, and the absence of muscular activity exaggerates these effects. Gravitational effects are amplified in airplane pilots during pullout from dives. The centrifugal force in the footward direction may be several times greater than the force of gravity. Pilots characteristically black out momentarily during the pullout maneuver as blood is drained from the cephalic regions and pooled in the lower parts of the body.

Some explanations have been advanced to explain the gravitationally induced reduction in cardiac output, but they are inaccurate. For example, it has been argued that when an individual is standing, the force of gravity impedes venous return to the heart from the dependent regions of the body. This explanation is incomplete because it does not account for the gravitational counterforce on the arterial side of the same vascular circuit, and this counterforce facilitates venous return. Moreover, it does not account for the effect of gravity in causing venous pooling. When a person is standing upright, gravity causes blood to accumulate in the lower extremities and distend both the arteries and veins. Because venous compliance is so much greater than arterial compliance, this distention occurs more on the venous side than on the arterial side of the circuit.

The hemodynamic effects of such venous distention (venous pooling) resemble those caused by the hemorrhage of an equivalent volume of blood from the body. When an adult shifts from a supine position to a relaxed standing position, 300 to 800 mL of blood pools in the legs. This pooling may reduce cardiac output by approximately 2 L/min. The compensatory adjustments made to assume a standing position are similar to the adjustments to blood loss (see also Chapter 19): There are reflex increases in heart rate and cardiac contractility. In addition, both arterioles and veins constrict; the arterioles are affected to a greater extent than are the veins.

Muscular Activity and Venous Valves

When a recumbent person stands but remains at rest, the pressure in the veins rises in the dependent regions of the body (Fig. 17.21). The P_v in the legs increases gradually and does not reach an equilibrium value until almost 1 minute after the person begins standing. The slowness of this rise in P_v is attributable to the venous valves, which allow flow only toward the heart. When a person stands, the valves prevent blood in the veins from falling toward the feet. Hence, the column of venous blood is supported at numerous levels by these valves. Because of these valves, the venous column can be thought of as consisting of many discontinuous segments. However, blood continues to enter the column from many venules and small tributary veins, and the pressure continues to rise. As soon as the pressure in one segment exceeds that in the segment just above it, the intervening valve is forced open. Ultimately, all the valves are open, and the column is continuous.

 IN THE CLINIC

Some of the drugs used to treat chronic hypertension interfere with the reflex adaptation to standing. Similarly, astronauts exposed to weightlessness lose their adaptations to gravity after a few days in space, and they experience pronounced difficulties when they first return to earth. When such astronauts and other individuals with impaired reflex adaptations stand, their blood pressure may drop substantially. This response is called *orthostatic hypotension,* which may cause lightheadedness or fainting.

 IN THE CLINIC

The superficial veins in the neck ordinarily are partially collapsed when a normal individual is sitting or standing. Venous return from the head is conducted largely through the deeper cervical veins, which are protected from collapse because they are tethered to surrounding structures. When central venous pressure is abnormally elevated, the superficial neck veins are distended, and they do not collapse even when the person sits or stands. Such cervical venous distention is an important clinical sign of congestive heart failure.

 IN THE CLINIC

The auxiliary pumping mechanism generated by skeletal muscle contractions is much less effective in people with varicose veins in their legs. The valves in these defective veins do not function properly, and therefore when the leg muscles contract, the blood in the leg veins is forced in both the retrograde and antegrade directions. Thus when an individual with varicose veins stands or walks, P_v in the ankles and feet is excessively high. The consequent high capillary pressure leads to the accumulation of edematous fluid in the ankles and feet.

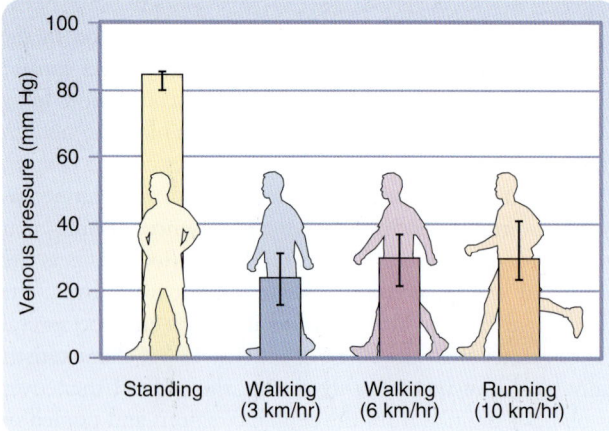

• **Fig. 17.21** Mean pressures (±95% confidence intervals) in the foot veins of human subjects during quiet standing, during walking, and during running. (From Stick C, et al. *J Appl Physiol.* 1992;72:2063.)

Precise measurement reveals that the final level of P_v in the feet during quiet standing is only slightly greater than that in a static column of blood extending from the right atrium to the feet. This finding indicates that the pressure drop caused by blood flow from the foot veins to the right atrium is very small. Because of this very low resistance, all the veins can be viewed as having a common venous compliance in the model of the circulatory system illustrated in Chapter 19. When an individual who has been standing quietly begins to walk, P_v in the legs decreases appreciably (see Fig. 17.21). Because of the intermittent venous compression exerted by the contracting leg muscles, and because of the operation of the venous valves, blood is forced from the veins toward the heart. Hence, muscular contraction lowers the mean P_v in the legs and serves as an auxiliary pump. Furthermore, muscular contraction prevents venous pooling and lowers capillary hydrostatic pressure. In this way, muscular contraction reduces the tendency for edematous fluid to collect in the feet during standing.

Microcirculation and Lymphatic System

The circulatory system supplies the tissues with blood in amounts that meet the body's requirements for O_2 and nutrients. The capillaries, whose walls consist of a single layer of endothelial cells, allow rapid exchange of gases, water, and solutes with interstitial fluid. The muscular arterioles, which are the major resistance vessels, regulate regional blood flow to the capillary beds. Venules and veins serve primarily as collecting channels and storage vessels. The lymphatic system is composed of lymphatic vessels, nodes, and lymphoid tissue. This system collects the fluid and proteins that have escaped from blood and transports them back into the veins for recirculation in blood. In this section, the network of the smallest blood vessels of the body, as well as the lymphatic vessels, are examined in detail.

Microcirculation

The *microcirculation* is defined as the circulation of blood through the smallest vessels of the body: arterioles, capillaries, and venules. Arterioles (5 to 100 µm in diameter) have a thick smooth muscle layer, a thin adventitial layer, and an endothelial lining (see Fig. 15.2). Arterioles give rise directly to capillaries (5 to 10 µm in diameter) or, in some tissues, to metarterioles (10 to 20 µm in diameter), which then give rise to capillaries (Fig. 17.22). Metarterioles can bypass the capillary bed and connect to venules, or they can connect directly to the capillary bed. Arterioles that give rise directly to capillaries regulate flow through these capillaries by constriction or dilation. The capillaries form an interconnecting network of tubes with an average length of 0.5 to 1 mm.

Functional Properties of Capillaries

In metabolically active organs, such as the heart, skeletal muscle, and glands, capillary density is high. In less active tissues, such as subcutaneous tissue or cartilage, capillary density is low. Capillary diameter also varies. Some capillaries have diameters smaller than those of erythrocytes. Passage through these tiny vessels requires the erythrocytes to become temporarily deformed. Fortunately, normal erythrocytes are quite flexible.

Blood flow in capillaries depends chiefly on the contractile state of arterioles. The average velocity of blood flow

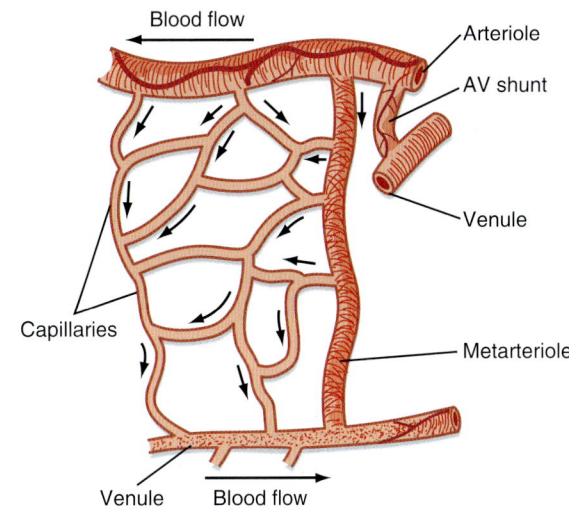

• **Fig. 17.22** Composite Schematic Illustration of the Microcirculation. The *circular structures* on the arteriole and venule represent smooth muscle fibers, and the *branching solid lines* represent sympathetic nerve fibers. The *arrows* indicate the direction of blood flow. AV, arteriovenous.

in capillaries is approximately 1 mm/second; however, it can vary from zero to several millimeters per second in the same vessel within a brief period. These changes in capillary blood flow may be random or rhythmic. The rhythmic oscillatory behavior of capillaries is caused by contraction and relaxation (vasomotion) of the precapillary vessels (i.e., the arterioles and small arteries).

Vasomotion is an intrinsic contractile behavior of vascular smooth muscle and is independent of external input. Changes in transmural pressure (intravascular pressure minus extravascular pressure) also influence the contractile state of precapillary vessels. An increase in transmural pressure, caused either by an increase in P_v or by dilation of arterioles, results in contraction of the terminal arterioles. A decrease in transmural pressure causes precapillary vessel relaxation. Humoral and possibly neural factors also affect vasomotion. For example, when increased transmural pressure causes the precapillary vessels to contract, the contractile response can be overridden and vasomotion abolished. This effect is accomplished by metabolic (humoral) factors when the O_2 supply becomes too low for the requirements of parenchymal tissue, as occurs in skeletal muscle during exercise.

Although a reduction in transmural pressure relaxes the terminal arterioles, blood flow through the capillaries cannot increase if the reduction in intravascular pressure is caused by severe constriction of the upstream microvessels. Large arterioles and metarterioles also exhibit vasomotion. However, their contraction usually does not completely occlude the lumen of the vessel and arrest blood flow, whereas contraction of the terminal arterioles may arrest blood flow. Thus the flow rate in capillaries may be altered by contraction and relaxation of small arteries, arterioles, and metarterioles.

Blood flow through the capillaries has been called *nutritional flow* because it provides for exchange of gases and solutes between blood and tissue. Conversely, blood flow that bypasses the capillaries as it passes from the arterial to the venous side of the circulation via metarterioles has been termed *nonnutritional,* or *shunt, flow* (see Fig. 17.22). In some areas of the body (e.g., fingertips, ears), true AV shunts exist (see Fig. 17.37). However, in many tissues, such as muscle, anatomical shunts are lacking. Even in the absence of these shunts, nonnutritional flow can occur. In tissues with metarterioles, nonnutritional flow may be continuous from arteriole to venule during low metabolic activity, when many precapillary vessels are closed. When metabolic activity increases in these tissues, more precapillary vessels open to allow capillary perfusion.

True capillaries lack smooth muscle and are therefore incapable of active constriction. Nevertheless, the endothelial cells that form the capillary wall contain actin and myosin, and they can alter their shape in response to certain chemical stimuli.

Because of its narrow lumen (i.e., small radius), a thin-walled capillary can withstand high internal pressures without bursting. This property can be explained in terms of the law of Pierre-Simon Laplace:

Equation 17.16

$$T = \Delta Pr$$

where

T = tension in the vessel wall

ΔP = transmural pressure difference

r = radius of the vessel

Laplace's equation applies to very thin-walled vessels, such as capillaries. Wall tension opposes the distending force (ΔPr) that tends to pull apart a theoretical longitudinal slit in the vessel (Fig. 17.23). Transmural pressure in a blood vessel in vivo is essentially equal to intraluminal pressure because extravascular pressure is generally negligible. To calculate wall tension, pressure in mm Hg is converted to dynes per square centimeter according to the equation $P = h\rho g$, where h is the height of an Hg column in centimeters, ρ is the density of Hg in g/cm^3, and g is gravitational acceleration in cm/s^2. For a capillary with a pressure of 25 mm Hg and a radius of 5×10^{-4} cm, the pressure (2.5 cm Hg × 13.6 g/cm^3 × 980 cm/sec^2) is 3.33×10^4

dyne/cm^2. Wall tension is then 16.7 dyne/cm. For an aorta with a pressure of 100 mm Hg and a radius of 1.5 cm, wall tension is 2×10^5 dyne/cm. Thus at the pressures normally found in the aorta and capillaries, the wall tension of the aorta is approximately 12,000 times greater than that of the capillaries. In a person standing quietly, capillary pressure in the feet may reach 100 mm Hg. Even under such conditions, capillary wall tension increases to a value that is still only one three-thousandth of the wall tension in the aorta at the same internal pressure.

The diameter of the resistance vessels (arterioles) is determined from the balance between the contractile force of the vascular smooth muscle and the distending force produced by intraluminal pressure. The greater the contractile activity of the vascular smooth muscle of an arteriole, the smaller its diameter. In small arterioles, contraction can continue to the point at which the vessel is completely occluded. Occlusion is caused by infolding of the endothelium and by trapping of blood cells in the vessel.

With a progressive reduction in intravascular pressure, vessel diameter decreases (as does vessel wall tension, according to the law of Laplace) and blood flow eventually ceases, although pressure within the arteriole is still greater than tissue pressure. The pressure that causes flow to cease has been called the *critical closing pressure,* and its mechanism is still unclear. The critical closing pressure is low when vasomotor activity is reduced by inhibition of sympathetic nerve activity in the vessel and is increased when vasomotor tone is enhanced by activation of the vascular sympathetic nerve fibers.

IN THE CLINIC

If the heart becomes greatly distended with blood during diastole, as may occur with cardiac failure, it functions less efficiently. To eject a given volume of blood per beat, more energy is required (wall tension must be greater) for the distended heart than for a normal undilated heart. The less efficient pumping of a distended heart is an example of Laplace's law, according to which the tension in the wall of a vessel or chamber (in this case, the ventricles) equals transmural pressure (pressure across the wall, or distending pressure) multiplied by the radius of the vessel or chamber. Laplace's relationship ordinarily applies to infinitely thin-walled vessels, but it can be applied to the spherical, dilated heart if correction is made for wall thickness. Under these conditions, the equation is $\sigma = \Delta Pr/2w$, where σ = wall stress, ΔP = transmural pressure difference, r = radius, and w = wall thickness.

Vasoactive Role of the Capillary Endothelium

The endothelium is an important source of substances that cause contraction or relaxation of vascular smooth muscle. One of these substances is **prostacyclin,** or prostaglandin I2 **(PGI2).** PGI2 can relax vascular smooth muscle via an increase in cyclic adenosine monophosphate (cAMP; Fig. 17.24). PGI2 is formed in the endothelium from

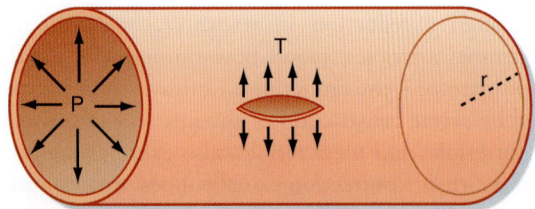

• **Fig. 17.23** Diagram of a Small Blood Vessel to Illustrate the Law of Laplace. T = Pr, where P = intraluminal pressure, r = radius of the vessel, and T = wall tension as the force per unit length tangential to the vessel wall. Wall tension prevents rupture along a theoretical longitudinal slit in the vessel.

- **Fig. 17.24** Endothelium-Mediated and Non–Endothelium-Mediated Vasodilation. Prostacyclin (PGI2) is formed from arachidonic acid (AA) by the action of cyclooxygenase (Cyc Ox) and prostacyclin synthase (PGI2 Syn) in the endothelium and elicits relaxation of the adjacent vascular smooth muscle via increases in cAMP. Stimulation of the endothelial cells with acetylcholine (ACh) or other agents (see text) results in the formation and release of an endothelium-derived relaxing factor identified as nitric oxide (NO). NO stimulates guanylyl cyclase (G Cyc) to increase cGMP in the vascular smooth muscle to produce relaxation. The vasodilator nitroprusside (NP) acts directly on vascular smooth muscle. Substances such as adenosine, H^+, CO_2, and K^+ can arise in the parenchymal tissue and elicit vasodilation by direct action on vascular smooth muscle. ADP, adenosine diphosphate; AMP, adenosine monophosphate; ATP, adenosine triphosphate; cAMP, cyclic adenosine monophosphate; cGMP, cyclic guanosine monophosphate; L-arg, L-arginine.

arachidonic acid, and the process is catalyzed by PGI2 synthase. The mechanism that triggers synthesis of PGI2 is not known. However, PGI2 may be released by an increase in shear stress caused by accelerated blood flow. The primary function of PGI2 is to inhibit platelet adherence to the endothelium and platelet aggregation and thus prevent intravascular clot formation. PGI2 also causes relaxation of vascular smooth muscle.

Of far greater importance in endothelium-mediated vascular dilation is the formation and release of **nitric oxide (NO),** a component of endothelium-derived relaxing factor (see Fig. 17.24). When endothelial cells are stimulated by acetylcholine or other vasodilator agents (e.g., adenosine triphosphate [ATP], bradykinin, serotonin, substance P, histamine), NO is released. These agents do not cause vasodilation in blood vessels lacking the endothelium. NO (synthesized from L-arginine) activates guanylyl cyclase in vascular smooth muscle to increase the concentration of cyclic guanosine monophosphate (cGMP), which produces relaxation by decreasing myofilament sensitivity to $[Ca^{++}]$. Release of NO can be stimulated by the shear stress of blood flow on the endothelium. The drug nitroprusside also increases cGMP by acting directly on vascular smooth

muscle; its action is not endothelium mediated. Vasodilator agents such as adenosine, H^+, CO_2, and K^+ may be released from parenchymal tissue and act locally on resistance vessels (see Fig. 17.24).

Acetylcholine also stimulates the release of an endothelium-dependent hyperpolarizing factor that underlies the relaxation of adjacent smooth muscle. Although arachidonic acid metabolites have been suggested, the factor remains unknown. Moreover, how the factor reaches vascular smooth muscle (diffusion through the extracellular space or passage via myoepithelial junctions) is unclear. Nevertheless, there are diverse ways by which endothelial cells communicate with vascular smooth muscle.

The endothelium can also synthesize **endothelin,** a potent vasoconstrictor peptide. Endothelin affects vascular tone and blood pressure and may be involved in pathological states, including atherosclerosis, pulmonary hypertension, congestive heart failure, and renal failure.

Passive Role of the Capillary Endothelium

Transcapillary Exchange

Solvent and solute move across the capillary endothelial wall by three processes: diffusion, filtration, and pinocytosis. Diffusion is the most important process for transcapillary exchange, and pinocytosis is the least important.

Diffusion. Under normal conditions, only approximately 0.06 mL of water per minute moves across the capillary wall per 100 g of tissue as a result of filtration. In contrast, 300 mL of water per minute per 100 g of tissue moves across the capillary wall by diffusion. Thus diffusion is the key factor in providing exchange of gases, substrates, and waste products between capillaries and tissue cells.

The process of diffusion is described by Fick's law (see also Chapter 1):

Equation 17.17
$$J = -DA(\Delta C/\Delta x)$$

where

J = quantity of a substance moved per unit time
D = free diffusion coefficient for a particular molecule
A = cross-sectional area of the diffusion pathway
ΔC = concentration gradient of the solute
Δx = distance over which diffusion occurs

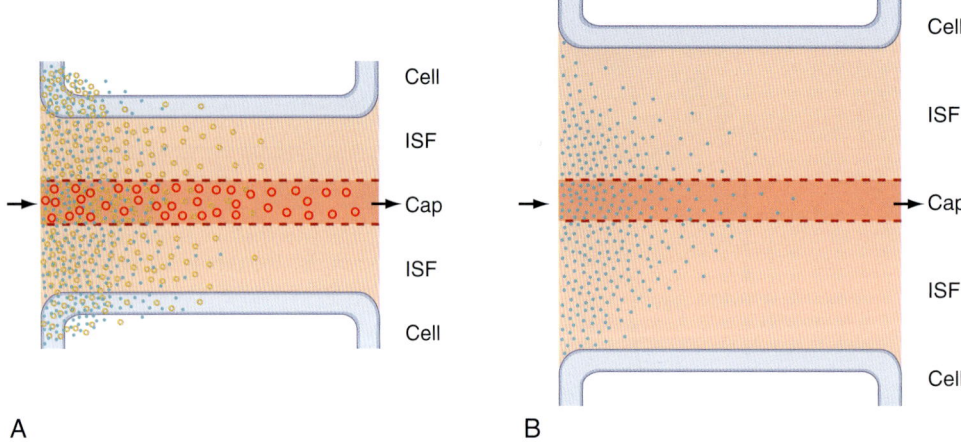

A   B

• **Fig. 17.25** Flow- and Diffusion-Limited Transport From Capillaries (Cap) to Tissue. **A,** Flow-limited transport. The smallest water-soluble inert tracer particles *(blue dots)* reach negligible concentrations after passing only a short distance down the capillary. Larger particles *(brown dots)* with similar properties travel farther along the capillary before reaching an insignificant intracapillary concentration. Both substances cross the interstitial fluid (ISF) and reach the parenchymal tissue (Cell). Because of their size, more of the smaller particles are taken up by the tissue cells. The largest particles *(red circles)* cannot penetrate the capillary pores and hence do not escape from the capillary lumen except by pinocytotic vesicle transport. An increase in the volume of blood flow or an increase in capillary density increases tissue supply of the diffusible solutes. Note that capillary permeability is greater at the venous end of the capillary (also in the venule, not shown) because of the larger number of pores in this region. **B,** Diffusion-limited transport. When the distance between capillaries and parenchymal tissue is large as a result of edema or low capillary density, diffusion becomes a limiting factor in solute transport from capillary to tissue, even at high rates of capillary flow.

For diffusion across a capillary wall, Fick's law can also be expressed as

Equation 17.18

$$J = -PS(C_o - C_i)$$

where

P = capillary permeability by the substance

S = capillary surface area

C_o = concentration of the substance outside the capillary

C_i = concentration of the substance inside the capillary

The PS product provides a convenient expression of available capillary surface area because the intrinsic permeability of the capillary is rarely altered much under physiological conditions. However, in pathological conditions, as with a bee sting, capillary permeability may be altered.

In capillaries, diffusion of lipid-insoluble molecules is restricted to water-filled channels or pores. Movement of solute across the capillary endothelium is complex and involves corrections for attractions between solute and solvent molecules, interactions between solute molecules, pore configuration, and charge on the molecules in relation to charge on the endothelial cells. Such solute motion is not simply a matter of random thermal movement of molecules that seemingly run down a concentration gradient. For small molecules, such as water, NaCl, urea, and glucose, the capillary pores offer little restriction to diffusion (i.e., they have a low reflection coefficient; see the section "Osmotic Forces"). Diffusion of these substances is so rapid that the mean concentration gradient across the capillary endothelium is extremely small. The larger the lipid-insoluble molecules are, the more restricted is their diffusion through

capillaries. Diffusion eventually becomes minimal when the molecular weight of the molecules exceeds approximately 60,000. With small molecules, the only limitation to net movement across the capillary wall is the rate at which blood flow transports the molecules to the capillary. Transport of these molecules is said to be **flow limited.**

With flow-limited small molecules, the concentration of the molecule in blood reaches equilibrium with its concentration in interstitial fluid at a location near the origin of the capillary from its parent arteriole. Its concentration falls to negligible levels near the arterial end of the capillary (Fig. 17.25*A*). If the flow is large, the small molecule can still be present at a distant locus downstream in the capillary. A somewhat larger molecule moves farther along the capillary before it reaches an insignificant concentration in blood. Furthermore, the number of still larger molecules that enter the arterial end of the capillary but cannot pass through the capillary pores equals the number that leaves the venous end of the capillary (see Fig. 17.25*A*).

With large molecules, diffusion across the capillaries becomes the limiting factor (**diffusion limited**); that is, the permeability of a capillary to a large solute molecule limits its transport across the capillary wall. Diffusion of small lipid-insoluble molecules is so rapid that diffusion limits blood-tissue exchange only when distances between capillaries and parenchymal cells are great (e.g., as in tissue edema or very low capillary density; see Fig. 17.25*B*).

Movement of lipid-soluble molecules across the capillary wall is not limited to capillary pores (only ≈0.02% of the capillary surface); it also occurs directly through the lipid membranes of the entire capillary endothelium. Consequently, lipid-soluble molecules move rapidly between blood and tissue. The degree of lipid solubility (oil-to-water partition coefficient) provides a good index of the ease of transfer of lipid molecules through the capillary endothelium.

Both O_2 and CO_2 are lipid soluble, and they readily pass through endothelial cells. Calculations based on (1) the diffusion coefficient for O_2, (2) capillary density and diffusion distances, (3) blood flow, and (4) tissue O_2 consumption indicate that the O_2 supply of normal tissue at rest and during activity is not limited by diffusion or by the number of open capillaries.

Measurements of the partial pressure of O_2 (PO_2) and O_2 saturation of blood in microvessels indicate that in many tissues, O_2 saturation at the entrance of capillaries has decreased to approximately 80% as a result of diffusion of O_2 from arterioles and small arteries. Moreover, CO_2 loading and the resulting intravascular shifts in the oxyhemoglobin dissociation curve occur in the precapillary vessels. Hence, in addition to gas exchange at the capillaries, O_2 and CO_2 pass directly between adjacent arterioles and venules and possibly between arteries and veins (countercurrent exchange). The countercurrent exchange represents a diffusional shunting of gas away from the capillaries; this shunting may limit the supply of O_2 to the tissue at low blood flow rates.

Capillary Filtration. The permeability of the capillary endothelial membrane is not uniform. For example, liver capillaries are quite permeable, and albumin escapes from them at a rate several times greater than that from the less permeable muscle capillaries. Furthermore, permeability is not uniform along the length of the capillary. The venous ends are more permeable than the arterial ends, and permeability is greatest in the venules, a property attributed to the greater number of pores in these regions.

Where does filtration occur? Some water passes through the capillary endothelial cell membranes, but most flows through apertures (pores) in the endothelial walls of the capillaries (Figs. 17.26 and 17.27). The pores in skeletal and cardiac muscle capillaries have diameters of approximately 4 nm. There are clefts between adjacent endothelial cells in cardiac muscle, and the gap at the narrowest point is approximately 4 nm. The clefts (pores) are sparse and represent only approximately 0.02% of the capillary surface area. Pores are absent in cerebral capillaries, where the blood-brain barrier blocks the entry of many small molecules.

In addition to clefts, some of the more porous capillaries (e.g., those in the kidneys and intestines) contain fenestrations 20 to 100 nm wide, whereas other capillaries (e.g., those in the liver) have a discontinuous endothelium (see Fig. 17.27). Fenestrations and discontinuous endothelia allow the passage of molecules that are too large to pass through the intercellular clefts of the endothelium.

The direction and magnitude of water movement across the capillary wall can be estimated as the algebraic sum of the hydrostatic and osmotic pressure that exists across the wall. An increase in intracapillary hydrostatic pressure favors movement of fluid from the vessel interior to the interstitial space, whereas an increase in the concentration of osmotically active particles within vessels favors movement of fluid into the vessels from the interstitial space (Fig. 17.28).

Hydrostatic Forces. Hydrostatic pressure (blood pressure) within capillaries is not constant. Instead, it depends on arterial and venous pressure and on precapillary resistance (in the arterioles) and postcapillary resistance (in the venules and small veins). An increase in arterial or venous pressure elevates capillary hydrostatic pressure, whereas a reduction in arterial or venous pressure has the opposite effect. An increase in arteriolar resistance or closure of arteries reduces capillary pressure, whereas a greater resistance to flow in venules and veins increases capillary pressure.

Hydrostatic pressure is the principal force in capillary filtration. A given change in P_v produces a greater effect on capillary hydrostatic pressure than does the same change in P_a. Approximately 80% of an increase in P_v is transmitted back to the capillaries.

Capillary hydrostatic pressure (P_c) varies from tissue to tissue. Average values, obtained from direct measurements in human skin, are approximately 32 mm Hg at the arterial end of capillaries and approximately 15 mm Hg at the venous end of capillaries at the level of the heart (see Fig. 17.28). As discussed previously, when a person stands, hydrostatic pressure increases in the legs and decreases in the head.

Tissue pressure, or, more specifically, interstitial fluid pressure (P_i) outside the capillaries, opposes capillary filtration. The difference between P_c and P_i constitutes the driving force for filtration. Normally, P_i is close to zero, and so P_c essentially represents the hydrostatic driving force.

Osmotic Forces. The key factor that restrains fluid loss from capillaries is the osmotic pressure of plasma proteins (such as albumin). This osmotic pressure is called *colloid osmotic pressure* or *oncotic pressure* (π_p). The total osmotic pressure of plasma is approximately 6000 mm Hg (reflecting the presence of electrolytes and other small molecules, as well as plasma proteins), whereas oncotic pressure is only approximately 25 mm Hg. This low level of oncotic pressure is an important factor in fluid exchange across the capillary because plasma proteins are essentially confined to the intravascular space, whereas electrolytes are virtually equal in concentration on both sides of the capillary endothelium. The relative permeability of solute by water influences the actual magnitude of osmotic pressure. The **reflection coefficient (σ)** is the relative impediment to the passage of a substance through the capillary membrane. The reflection coefficient of water is 0, and that of albumin (to which the endothelium is essentially impermeable) is 1. Filterable solutes have reflection coefficients between 0 and 1. In addition, different tissues have different reflection coefficients for the same molecule. Hence, movement of a given solute across the endothelial wall varies with the tissue. The actual oncotic pressure of the plasma (π_p) is defined by the following equation (see also Chapter 1):

Equation 17.19

$$\pi_p = \sigma RTC_p$$

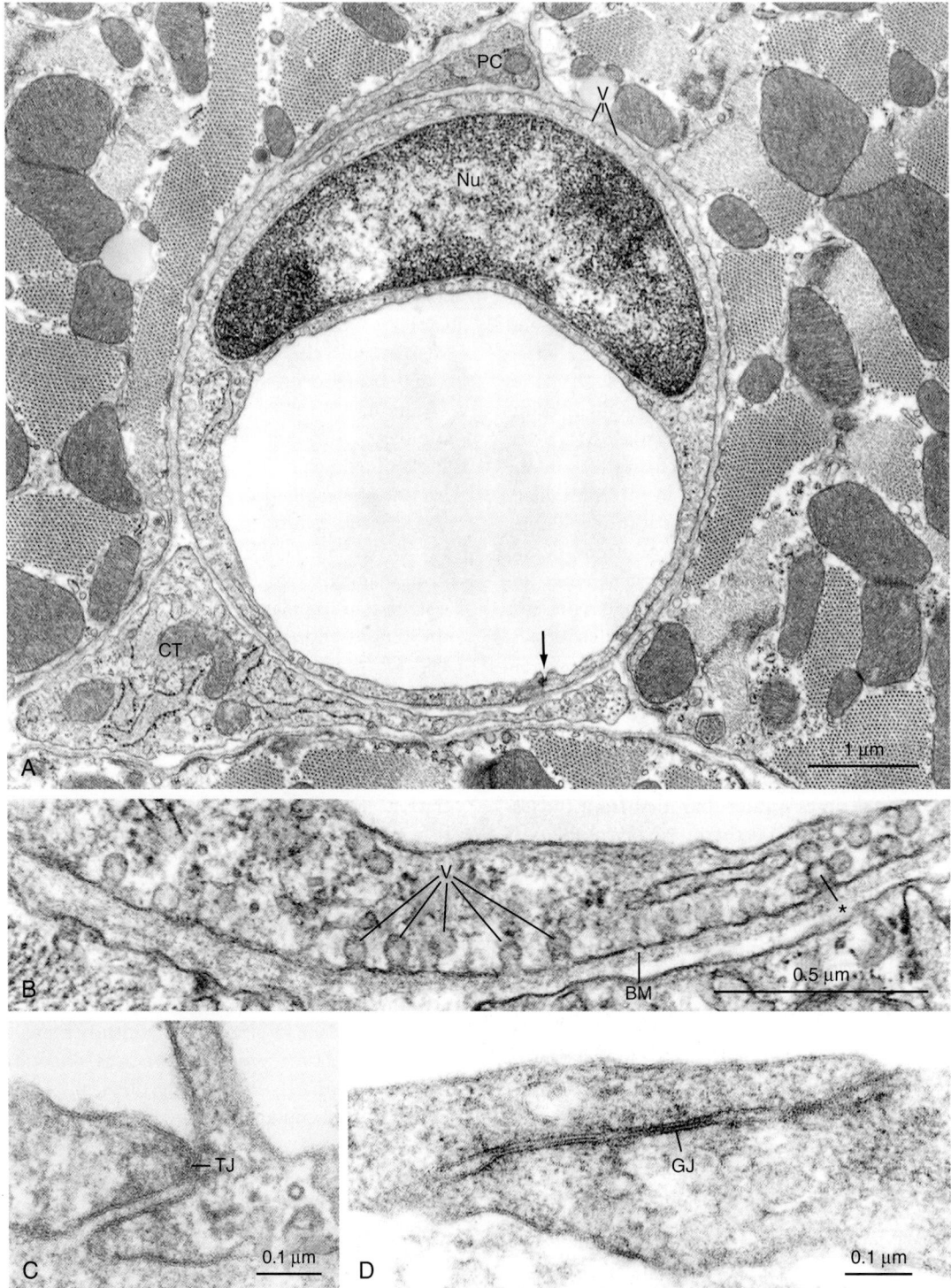

• **Fig. 17.26 A,** Electron micrograph of a cross-section of a capillary in a mouse ventricle. The luminal diameter is approximately 4 μm. In this section, the capillary wall is formed by a single endothelial cell (Nu, endothelial nucleus). The thin pericapillary space is occupied by a pericyte (PC) and a connective tissue (CT) cell ("fibroblast"), which forms a functional complex (arrow) with itself. V, plasmalemmal vesicles. **B,** Detail of the endothelial cell in **A** showing plasmalemmal vesicles (V) attached to the endothelial cell surface. These vesicles are especially prominent in vascular endothelium and are involved in transport of substances across the blood vessel wall. Note the complex alveolar vesicle *(asterisk).* BM, basement membrane. **C,** Junctional complex in a capillary of a mouse heart. "Tight" junctions (TJ) typically form in these small blood vessels and appear to consist of fusions between apposed endothelial cell surface membranes. **D,** Interendothelial junction in a muscular artery of a papillary muscle. Although tight junctions similar to those of capillaries are found in these large blood vessels, extensive junctions that resemble gap junctions in the intercalated disks between myocardial cells often appear in arterial endothelium (example shown at GJ).

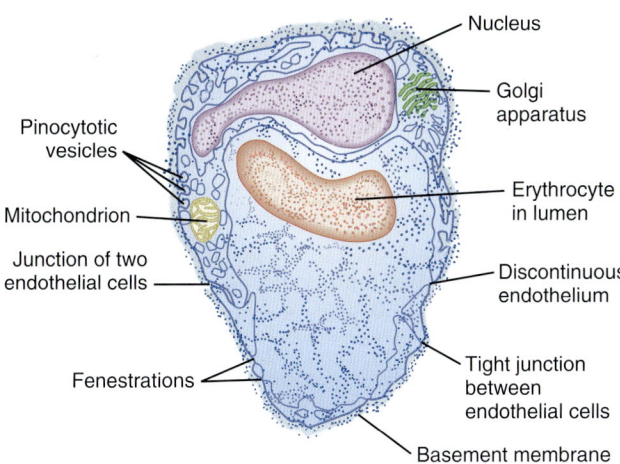

• **Fig. 17.27** Illustration of an electron micrograph of a capillary in cross-section.

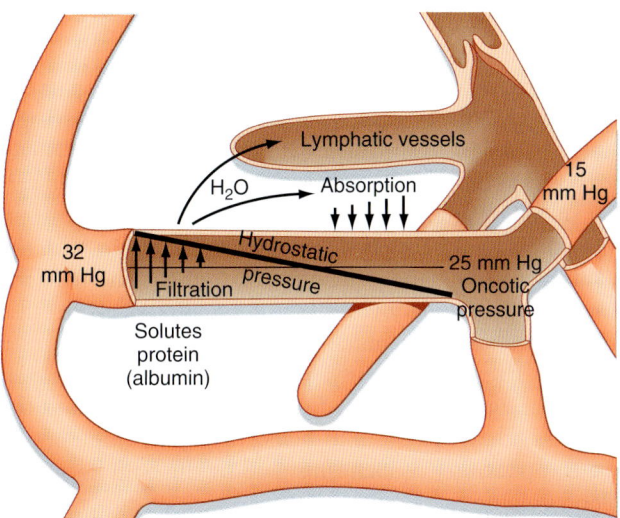

• **Fig. 17.28** Schematic representation of the factors responsible for filtration and absorption across the capillary wall and the formation of lymph.

where
σ = reflection coefficient
R = gas constant
T = temperature in degrees Kelvin
C_p = plasma solute concentration

Albumin is the most important plasma protein that determines oncotic pressure. Its molecular weight is 69,000 D. Albumin exerts an osmotic force greater than can be accounted for solely on the basis of its concentration in plasma. Therefore, it cannot be replaced on a mole-by-mole basis by inert substances of appropriate molecular size, such as dextran. This additional osmotic force becomes disproportionately great at high concentrations of albumin (as in plasma), and this force is weak to absent in dilute solutions of albumin (as in interstitial fluid). The reason for this activity of albumin is its negative charge at normal blood pH and the attraction and retention of cations (principally Na^+) in the vascular compartment (Gibbs-Donnan effect).

IN THE CLINIC

With prolonged standing, particularly when associated with elevation of P_v in the legs (such as that caused by pregnancy and congestive heart failure), filtration across capillaries is greatly enhanced, exceeding the capacity of the lymphatic system to remove the filtrate from the interstitial space and thus leading to edema.

The concentration of plasma proteins may change in several pathological states and thus alter the osmotic force and movement of fluid across the capillary membrane. The plasma protein concentration is increased in conditions of dehydration (e.g., water deprivation, prolonged sweating, severe vomiting, diarrhea). In this condition, less water moves by osmotic force from the tissues to the vascular compartment, thereby decreasing the volume of the interstitial fluid. In contrast, the plasma protein concentration is reduced in some renal diseases because of its loss in urine, and edema may occur.

When capillary injury is extensive, as in severe burns, intravascular fluid and plasma protein leak into the interstitial space in the damaged tissues. The protein that escapes from the vessel lumen increases the oncotic pressure of the interstitial fluid. This greater osmotic force outside the capillaries leads to additional fluid loss and possibly to severe dehydration.

Balance of Hydrostatic and Osmotic Forces. The relationship between hydrostatic pressure and oncotic pressure and the role of these forces in regulating fluid passage across the capillary endothelium were expounded by Frank Starling in 1896. This relationship constitutes Starling's hypothesis. It can be expressed as follows:

Equation 17.20
$$Q_f = k[(P_c - P_i) - (\pi_p - \pi_i)]$$

where
Q_f = fluid movement
k = filtration constant for the capillary membrane
P_c = capillary hydrostatic pressure
P_i = interstitial fluid hydrostatic pressure
π_p = plasma oncotic pressure
π_i = interstitial fluid oncotic pressure

Filtration occurs when the algebraic sum is positive; absorption occurs when it is negative.

Traditionally, filtration was thought to occur at the arterial end of the capillary, and absorption was thought to occur at its venous end because of the gradient of hydrostatic pressure along the capillary. This scheme is true for an idealized capillary (see Fig. 17.28). However, in well-perfused capillaries, arteriolar vasoconstriction can reduce P_c in such a way that absorption at the arteriolar end can occur transiently. With continued vasoconstriction, absorption diminishes with time because P_i increases. In some vascular beds (e.g., the renal glomerulus), hydrostatic pressure in the capillary is high enough to cause filtration along the entire length of the capillary. In other vascular beds (e.g., the intestinal mucosa), the hydrostatic and oncotic forces are such that absorption occurs along the whole capillary.

In the steady-state P_a, P_v, postcapillary resistance, hydrostatic and oncotic pressure of interstitial fluid, and oncotic pressure of plasma are relatively constant. Hence, in the normal state, filtration and absorption across the capillary wall are well balanced. However, a change in precapillary resistance influences fluid movement across the capillary wall. Vasoconstriction reduces net filtration, and vasodilation increases filtration.

IN THE CLINIC

In the lungs, mean capillary hydrostatic pressure is only approximately 8 mm Hg (see Chapter 22). Because plasma oncotic pressure is 25 mm Hg and pressure of the interstitial fluid in the lungs is approximately 15 mm Hg, the net force slightly favors net absorption (i.e., fluid leaves the interstitial space). Despite net absorption, pulmonary lymph is formed. This lymph consists of fluid that is osmotically withdrawn from the capillaries by the small amount of plasma protein that escapes through the capillary endothelium. In pathological conditions, such as left ventricular failure or mitral valve stenosis, pulmonary capillary hydrostatic pressure may exceed plasma oncotic pressure. When this occurs, it may cause pulmonary edema, a condition in which excessive fluid accumulates in the pulmonary interstitium. This fluid accumulation seriously interferes with gas exchange in the lungs.

Capillary Filtration Coefficient. The rate of fluid movement (Q_f) across the capillary membrane depends not only on the algebraic sum of the hydrostatic and osmotic forces across the endothelium (ΔP) but also on the area (A_m) of the capillary wall available for filtration, the distance (Δx) across the capillary wall, the viscosity (η) of the filtrate, and the filtration constant (k) of the membrane. These factors may be expressed as follows:

Equation 17.21
$$Q_f = kA_m\Delta P/\eta\Delta x$$

This expression, which describes the flow of fluid through the membrane pores, is essentially Poiseuille's law for flow through tubes.

Because the thickness of the capillary wall and the viscosity of the filtrate are relatively constant, they can be included in the filtration constant k. If the area of the capillary membrane is not known, the rate of filtration can be expressed per unit weight of tissue. Hence, the equation can be simplified as

Equation 17.22
$$Q_f = k_t\Delta P$$

where k_t is the capillary filtration coefficient for a given tissue and the units for Q_f are milliliters per minute per 100 g of tissue.

In any given tissue, the filtration coefficient per unit area of capillary surface, and hence capillary permeability, is not changed by various physiological conditions, such as arteriolar dilation and capillary distention, or by such adverse conditions as hypoxia, hypercapnia, or reduced pH. When capillaries are injured (as by toxins or severe burns), significant amounts of fluid and protein leak out of the capillaries into the interstitial space. This increase in capillary permeability is reflected by an increase in the filtration coefficient.

Because capillary permeability is constant under normal conditions, the filtration coefficient can be used to determine the relative number of open capillaries (i.e., the capillary surface area available for filtration in tissue). For example, the increased metabolic activity of contracting skeletal muscle relaxes the precapillary resistance vessels and hence opens more capillaries. This process, called **capillary recruitment,** increases the filtering surface area.

Disturbances in Hydrostatic-Osmotic Balance. Relatively small changes in P_a may have little effect on filtration. The change in pressure may be countered by adjustments in precapillary resistance vessels (autoregulation; see Chapter 18) so that hydrostatic pressure remains constant in the open capillaries. However, a severe reduction in $\overline{P}_a$ usually evokes arteriolar constriction mediated by the sympathetic nervous system. This response may occur in hemorrhage, and it is often accompanied by a fall in P_v. These changes reduce capillary hydrostatic pressure. However, the lowering of blood pressure in hemorrhage causes a decrease in blood flow (and hence in O_2 supply) to the tissue, with the result that vasodilator metabolites accumulate and relax the arterioles. Precapillary vessel relaxation also occurs because of the reduced transmural pressure (autoregulation; see Chapter 18). Consequently, absorption predominates over filtration, and fluid moves from the interstitium into the capillary. These responses to hemorrhage constitute one of the compensatory mechanisms used by the body to restore blood volume (see Chapter 19).

An increase in P_v alone, as occurs in the feet when a person stands up, would elevate capillary pressure and enhance filtration. However, the increase in transmural pressure closes precapillary vessels (myogenic mechanism; see Chapter 18), and hence the capillary filtration coefficient actually decreases. This reduction in capillary surface available for filtration prevents large amounts of fluid from leaving the capillaries and entering the interstitial space.

In a healthy individual, the filtration coefficient (k_t) for the whole body is approximately 0.006 mL/minute/100 g of tissue/mm Hg. For a 70-kg man, an elevation in P_v of 10 mm Hg for 10 minutes would increase filtration from capillaries by 420 mL. Edema does not usually occur because the fluid is returned to the vascular compartment by the lymphatic vessels. When edema develops, it usually appears in the dependent parts of the body, where the hydrostatic pressure is greatest, but its location and magnitude are also determined by the type of tissue. Loose tissues, such as the subcutaneous tissue around the eyes or in the scrotum, are more prone than firm tissues, as in a muscle, or encapsulated structures, as in a kidney, to collect larger quantities of interstitial fluid.

Pinocytosis. Some transfer of substances across the capillary wall can occur in tiny pinocytotic vesicles. These vesicles (see Figs. 17.26 and 17.27), formed by the pinching off of the endothelial cell membrane, can take up substances on one side of the capillary wall, move them across the cell by kinetic energy, and deposit their contents on the other side. This process is termed *transcytosis*. The amount of material transported in this way is very small in relation to that moved by diffusion. However, pinocytosis may be responsible for the movement of large (30-nm) lipid-insoluble molecules between blood and interstitial fluid. The number of pinocytotic vesicles in endothelium varies among tissues (amount in muscle > amount in lung > amount in brain), and the number increases from the arterial end to the venous end of the capillary.

Lymphatic System

The terminal vessels of the lymphatic system consist of a widely distributed, closed-end network of highly permeable lymphatic capillaries. These lymphatic capillaries resemble blood capillaries, with two important differences: tight junctions are not present between endothelial cells, and fine filaments anchor lymphatic vessels to the surrounding connective tissue. With muscular contraction, these fine strands pull on the lymphatic vessels to open spaces between the endothelial cells and enable the entrance of protein and large particles into the lymphatic vessels. The lymphatic capillaries drain into larger vessels that finally enter the right and left subclavian veins, where they connect with the respective internal jugular veins.

Only cartilage, bone, epithelia, and tissues of the central nervous system lack lymphatic vessels. These vessels return the plasma capillary filtrate to the circulation. This task is accomplished by means of tissue pressure, and it is facilitated by intermittent skeletal muscle activity, lymphatic vessel contractions, and an extensive system of one-way valves. In this regard, lymphatic vessels resemble veins, although the larger lymphatic vessels do have thinner walls than do the corresponding veins, and they contain only a small amount of elastic tissue and smooth muscle.

The volume of fluid transported through the lymphatic vessels in 24 hours is approximately equal to the body's total plasma volume. The lymphatic vessels return all of the proteins filtered back to the blood; these proteins account for approximately one fourth to half of the circulating plasma proteins in the blood. The lymphatic vessels are the only means by which the protein that leaves the vascular compartment can be returned to blood. Net backward diffusion of protein into the capillaries cannot occur against the large protein concentration gradient. If the protein were not removed by the lymph vessels, it would accumulate in interstitial fluid and act as an oncotic force that draws fluid from the blood capillaries and produces edema.

In addition to returning fluid and protein to the vascular bed, the lymphatic system filters the lymph at the lymph nodes and removes foreign particles such as bacteria. The largest lymphatic vessel, the thoracic duct, not only drains the lower extremities but also returns the protein lost through the permeable liver capillaries. Moreover, the thoracic duct carries substances absorbed from the gastrointestinal tract. The principal substance is fat, in the form of chylomicrons.

Lymph flow varies considerably. The flow from resting skeletal muscle is almost nil, and it increases during exercise in proportion to the degree of muscular activity. It is increased by any mechanism that enhances the rate of blood capillary filtration; such mechanisms include increased capillary pressure or permeability and decreased plasma oncotic pressure. When the volume of interstitial fluid exceeds the drainage capacity of the lymphatic vessels, or when the lymphatic vessels become blocked, interstitial fluid accumulates and gives rise to clinical edema.

Coronary Circulation

Functional Anatomy of Coronary Vessels

The right and left coronary arteries arise at the root of the aorta behind the right and left cusps of the aortic valve, respectively. These arteries provide the entire blood supply to the myocardium. The right coronary artery supplies mainly the right ventricle and atrium. The left coronary artery, which divides near its origin into the anterior descending and the circumflex branches, supplies mainly the left ventricle and atrium. There is some overlap between the regions supplied by the left and right arteries. In humans, the right coronary artery is dominant (supplying most of the myocardium) in approximately 50% of individuals. The left coronary artery is dominant in another 20%, and the flow delivered by each main artery is approximately equal in the remaining 30%. The epicardial distribution of the coronary arteries and veins is illustrated in Fig. 17.29.

Coronary arterial blood passes through the capillary beds; most of it returns to the right atrium through the coronary sinus. Of the coronary arteries, epicardial arteries are largest (2 to 5 mm in diameter), large arterioles are medium in size (1.0 to 0.5 mm in diameter), and small arterioles are smallest (<0.1 mm in diameter). Some of the coronary venous blood reaches the right atrium via the anterior coronary veins. In addition, vascular communications directly link the myocardial vessels with the cardiac chambers; these communications are the **arteriosinusoidal, arterioluminal,** and **thebesian** vessels. The arteriosinusoidal channels consist of small arteries or arterioles that lose their arterial structure as they penetrate the chamber walls, where they divide into irregular, endothelium-lined sinuses. These sinuses anastomose with other sinuses and with capillaries, and they communicate with the cardiac chambers. The arterioluminal vessels are small arteries or arterioles that open directly into the atria and ventricles. The thebesian vessels are small veins that connect capillary beds directly with the cardiac chambers and also communicate with the cardiac veins. All the minute vessels of the myocardium communicate in the form of an extensive

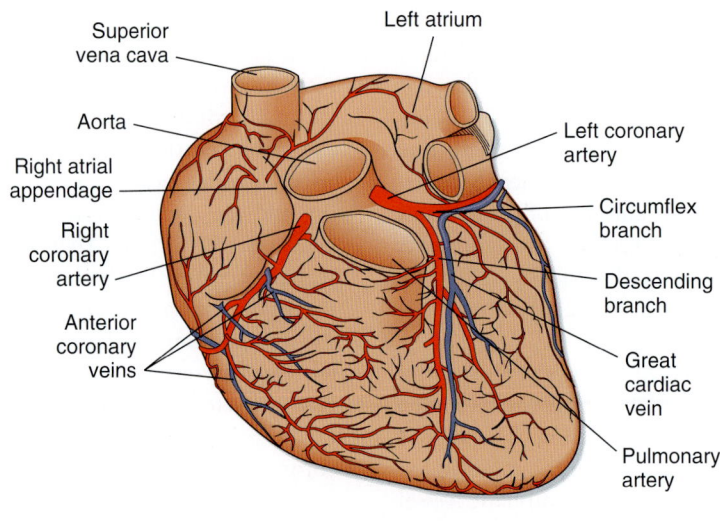

POSTERIOR VIEW

ANTERIOR VIEW

• **Fig. 17.29** Illustrations of the anterior and posterior surfaces of the heart, depicting the location and distribution of the principal coronary vessels.

plexus of subendocardial vessels. However, the myocardium does not receive significant nutritional blood flow directly from the cardiac chambers.

Factors That Influence Coronary Blood Flow
Physical Factors

The primary factor responsible for perfusion of the myocardium is aortic pressure. Changes in aortic pressure generally evoke parallel directional changes in coronary blood flow. This is caused in part by changes in coronary perfusion pressure. However, the major factor in the regulation of coronary blood flow is a change in arteriolar resistance engendered by changes in the metabolic activity of the heart. When the metabolic activity of the heart increases,

coronary resistance decreases; when cardiac metabolism decreases, coronary resistance increases (see Chapter 18).

Blood flow in the heart is autoregulated. If a cannulated coronary artery is perfused by blood from a pressure-controlled reservoir, perfusion pressure can be altered without a change in aortic pressure and cardiac work. The relationship between initial and steady-state blood flow is shown in the experiment depicted in Fig. 17.30. This is an example of autoregulation of blood flow, which is mediated by a myogenic mechanism in large and small arterioles (Chapter 18). The metabolic activity of cardiac muscle in small arterioles and the endothelium modulate autoregulation. The coronary circulation adjusts serial resistances within the microvasculature thereby adapting blood flow to O_2 requirements. Blood pressure is kept within narrow

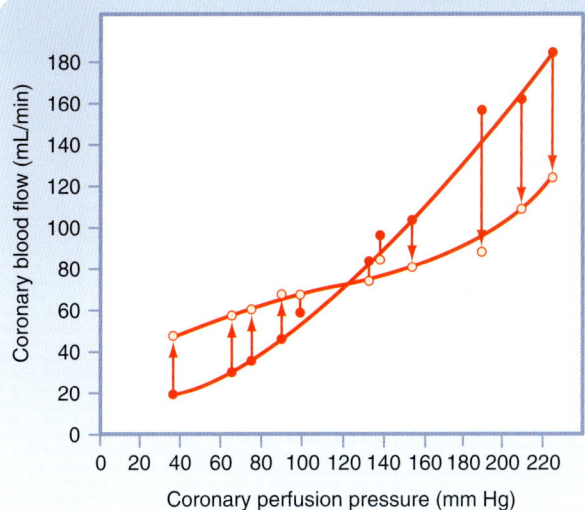

• **Fig. 17.30** Pressure-Flow Relationships in the Coronary Vascular Bed. As aortic pressure was held constant, the cardiac output, heart rate, and coronary artery perfusion pressure were abruptly increased or decreased from the control level, which is indicated by the point at which the two lines cross. The *solid circles* represent the flows that resulted immediately after the change in perfusion pressure; the *open circles* represent the steady-state flows at the new pressures. There is a tendency for flow to return toward the control level (autoregulation of blood flow), and this is most prominent over the intermediate pressure range (≈60 to 180 mm Hg). (From Berne RM, Rubio R. Coronary circulation. In Page E, ed. *Handbook of Physiology: Section 2: The Cardiovascular System: The Heart.* Vol 1. Bethesda, MD: American Physiological Society; 1979.)

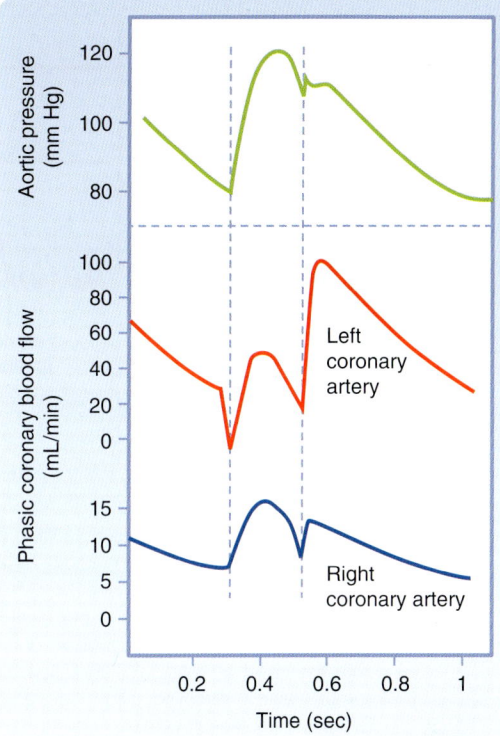

• **Fig. 17.31** Comparison of Phasic Coronary Blood Flow in the Left and Right Coronary Arteries. Extravascular compression is so great during early ventricular systole that the direction of blood flow in the large coronary arteries supplying the left ventricle is briefly reversed. Maximal inflow in the left coronary artery occurs in early diastole, when the ventricles have relaxed and extravascular compression of the coronary vessels is virtually absent. After an initial reversal in early systole, blood flow in the left coronary artery follows the aortic pressure until early diastole, when it rises abruptly and then declines slowly as aortic pressure falls during the remainder of diastole.

limits by baroreceptor reflex mechanisms. Hence, changes in coronary blood flow are caused mainly by changes in the diameter of coronary resistance vessels in response to the metabolic demands of the heart.

In addition to providing the pressure to move blood through the coronary vessels, the heart also affects its blood supply by the squeezing effect (extravascular compression) of the contracting myocardium on its own blood vessels. The patterns of flow in the left and right coronary arteries are shown in Fig. 17.31. In the left ventricle, coronary perfusion pressure is the difference between aortic diastolic pressure and left ventricular end-diastolic pressure.

Left ventricular myocardial pressure (pressure within the wall of the left ventricle) is highest near the endocardium and lowest near the epicardium. This pressure gradient does not normally impair endocardial blood flow because the greater blood flow to the endocardium during diastole compensates for the greater blood flow to the epicardium during systole. Measurements of coronary blood flow indicate that the epicardial and endocardial halves of the left ventricle receive approximately equal blood flow under normal conditions. Because extravascular compression is greatest at the endocardial surface of the ventricle, the equality of epicardial and endocardial blood flow indicates that the tone of the endocardial resistance vessels is less than that of the epicardial vessels.

IN THE CLINIC

The minimal extravascular resistance and absence of left ventricular work during diastole can be used to improve myocardial perfusion in patients with a damaged myocardium and low blood pressure. In a method called *counterpulsation,* an inflatable balloon is inserted into the thoracic aorta through a femoral artery. The balloon is inflated during each ventricular diastole and deflated during each systole. This procedure enhances coronary blood flow during diastole by raising diastolic pressure at a time when coronary extravascular resistance is lowest. Furthermore, it reduces cardiac energy requirements by lowering aortic pressure (afterload) during ventricular ejection.

The flow pattern in the right coronary artery is similar to that in the left coronary artery (see Fig. 17.31). In contrast to the left ventricle, reversal of blood flow does not occur in the right ventricle in early systole because pressure in the thin right ventricle is lower during systole. Hence, systolic blood flow constitutes a much greater proportion of total coronary inflow than it does in the left coronary artery.

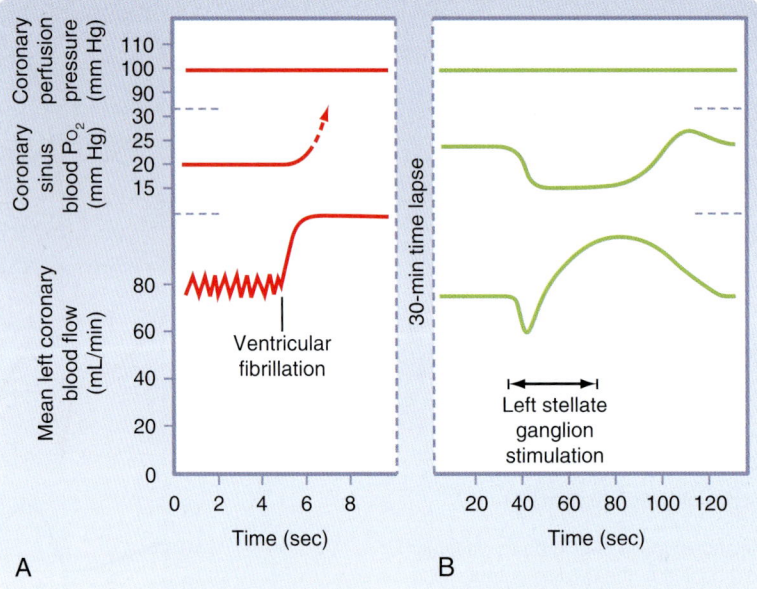

• Fig. 17.32 A, Unmasking of the restricting effect of ventricular systole on mean coronary blood flow by induction of ventricular fibrillation during perfusion of the left coronary artery at constant pressure. With the onset of ventricular fibrillation, coronary blood flow increases abruptly because extravascular compression is removed. Flow then gradually returns toward and often falls below the prefibrillation level. This increase in coronary resistance that occurs despite the removal of extravascular compression demonstrates the heart's ability to adjust its blood flow to meet its energy requirements. **B,** Effect of cardiac sympathetic nerve stimulation on coronary blood flow and on blood O_2 tension (Po_2) in the coronary sinus in a fibrillating heart during perfusion of the left coronary artery at constant pressure. (Berne RM. Unpublished observations.)

The extent to which extravascular compression restricts coronary inflow can be readily observed when the heart is suddenly arrested in diastole or with the induction of ventricular fibrillation. Fig. 17.32A depicts mean left coronary flow when the vessel was perfused with blood at a constant pressure from a reservoir. When ventricular fibrillation was electrically induced, blood flow increased immediately and substantially. A subsequent increase in coronary resistance over a period of many minutes reduced myocardial blood flow to below the level that existed before induction of ventricular fibrillation (see Fig. 17.32B, just before stellate ganglion stimulation).

When diastolic pressure in the coronary arteries is abnormally low (as in severe hypotension, partial coronary artery occlusion, or severe aortic stenosis), the ratio of endocardial to epicardial blood flow falls below a value of 1. This ratio indicates that blood flow to the endocardial regions is more severely impaired than that to the epicardial regions of the ventricle. There is also an increase in the gradient of myocardial lactic acid and myocardial adenosine concentrations from epicardium to endocardium. For this reason, the myocardial damage observed in atherosclerotic heart disease (e.g., after coronary occlusion) is greatest in the inner wall of the left ventricle.

Tachycardia and bradycardia have dual effects on coronary blood flow. A change in heart rate mainly alters diastole. In tachycardia, the proportion of time spent in systole, and consequently the period of restricted inflow,

increases. However, this mechanical effect is overridden by the dilation of coronary resistance vessels associated with the increased metabolic activity of the more rapidly beating heart. With bradycardia, the opposite occurs: Coronary inflow is less restricted (more time spent in diastole), but so are the metabolic (O_2) requirements of the myocardium.

Neural and Neurohumoral Factors

Stimulation of cardiac sympathetic nerves markedly increases coronary blood flow. However, the increase in flow is associated with an increased heart rate and more forceful systole. The stronger contraction and the tachycardia tend to restrict coronary flow. The increase in myocardial metabolic activity, however, tends to dilate coronary resistance vessels. The increase in coronary blood flow evoked by cardiac sympathetic nerve stimulation reflects the sum of these factors. In perfused hearts in which the mechanical effect of extravascular compression is eliminated by cardiac arrest or by ventricular fibrillation, an initial coronary vasoconstriction of the coronary vessels is often observed. After this initial vasoconstriction, the metabolic effect evokes vasodilation (see Fig. 17.32B).

Furthermore, when β-adrenergic receptor blockade eliminates the positive chronotropic and inotropic effects, activation of the cardiac sympathetic nerves increases coronary resistance. These observations indicate that the direct action of the sympathetic nerve fibers on the coronary resistance vessels is vasoconstriction.

Both α-adrenergic receptors (constrictors) and β_2-adrenergic receptors (dilators) are present on the coronary vessels. Coronary resistance vessels also participate in the baroreceptor and chemoreceptor reflexes, and the sympathetic constrictor tone of the coronary arterioles can be modulated by such reflexes. Nevertheless, coronary resistance is predominantly under local nonneural control.

Vagus nerve stimulation causes slight dilation of the coronary resistance vessels, and activation of the carotid and aortic chemoreceptors can cause a slight decrease in coronary resistance via the vagus nerves to the heart. Failure of strong vagal stimulation to increase coronary blood flow is not due to lack of muscarinic receptors on the coronary resistance vessels because intracoronary administration of acetylcholine elicits marked vasodilation. In the human heart, acetylcholine caused vasodilation when administered directly into the left anterior descending coronary artery of subjects with no evidence of coronary artery disease. However, acetylcholine caused vasoconstriction in the coronary artery of subjects whose endothelium had been damaged and rendered dysfunctional by atherosclerosis.

Metabolic Factors

A striking characteristic of the coronary circulation is the close relationship between the level of myocardial metabolic activity and the magnitude of coronary blood flow (Fig. 17.33). This relationship is also found in a denervated heart and in a completely isolated heart, either in the beating state or in the fibrillating state. Ventricles can fibrillate for many hours when the coronary arteries are perfused with arterial blood from some external source. As already noted, a fibrillating heart uses less O_2 than a pumping heart does, and blood flow to the myocardium is reduced accordingly.

The mechanisms that link the cardiac metabolic rate and coronary blood flow remain unsettled. However, it appears that a decrease in the ratio of O_2 supply to O_2 demand releases vasodilator substances from the myocardial cells into the interstitial fluid, where they relax the coronary resistance vessels. Decreases in arterial blood O_2 content or in coronary blood flow and increases in metabolic rate all decrease the O_2 supply/demand ratio (Fig. 17.34). As a consequence, substances are released that dilate the arterioles and thereby

adjust the O_2 supply to the O_2 demand. A decrease in O_2 demand diminishes the release of vasodilators and enables greater expression of basal tone.

Numerous metabolites participate in the vasodilation that accompanies increased cardiac work. Accumulation of vasoactive metabolites can also account for the increase in blood flow that results from a brief period of ischemia (i.e., **reactive hyperemia;** see Chapter 18). The duration of the enhanced coronary flow after release of the briefly occluded vessel is, within certain limits, proportional to the duration of the period of occlusion. Among the factors implicated in reactive hyperemia are ATP-sensitive potassium (K_{ATP}) channels, NO, CO_2, H^+, K^+, hypoxia, H_2O_2, and adenosine.

Of these agents, the key factors appear to be adenosine, NO, opening of the K_{ATP} channels, and H_2O_2. The contributions of each of these agents and their interaction under basal conditions and during increased myocardial

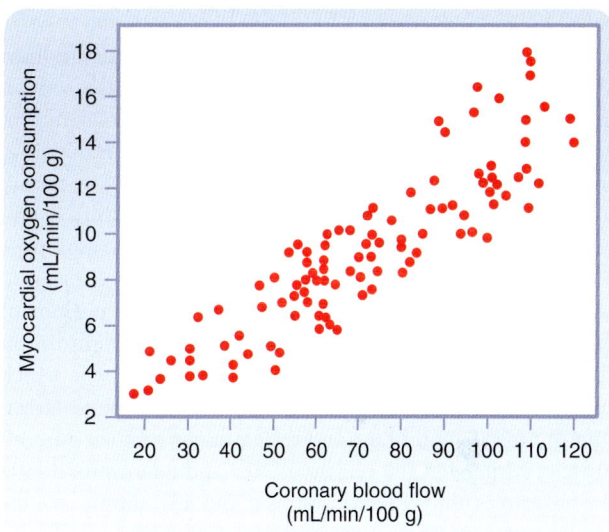

• **Fig. 17.33** Relationship between myocardial O_2 consumption and coronary blood flow during a variety of interventions that increase or decrease the myocardial metabolic rate. (From Berne RM, Rubio R: Coronary circulation. In Page E, ed. *Handbook of Physiology: Section 2: The Cardiovascular System: The Heart.* Vol 1. Bethesda, MD: American Physiological Society; 1979.)

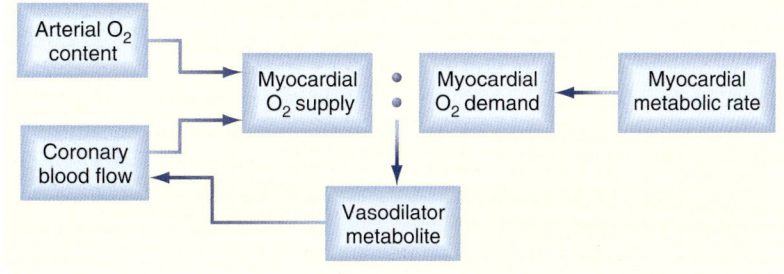

• **Fig. 17.34** Imbalance in the O_2 supply–O_2 demand ratio alters coronary blood flow by the rate of release of a vasodilator metabolite from cardiomyocytes. A decrease in the ratio elicits an increase in vasodilator release, whereas an increase in the ratio has the opposite effect.

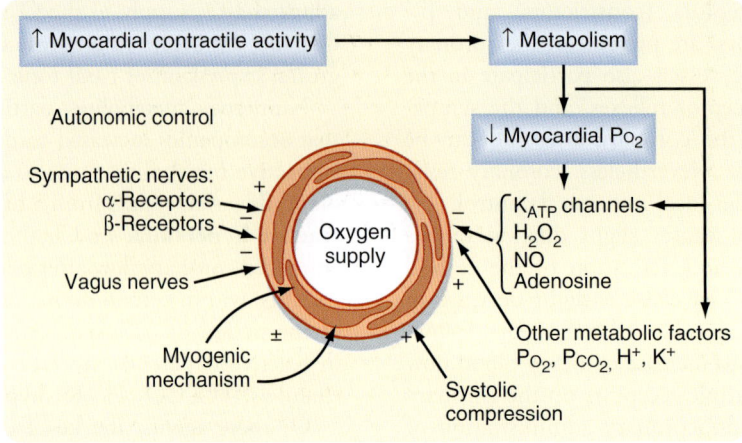

• **Fig. 17.35** Schematic Representation of Factors That Increase (+) or Decrease (−) Coronary Vascular Resistance. Intravascular pressure (arterial blood pressure) stretches the vessel wall. K_{ATP} channels, adenosine triphosphate–sensitive potassium channels; NO, nitric oxide; PCO_2, partial pressure of carbon dioxide; PO_2, partial pressure of oxygen.

activity are complex. A reduction in oxidative metabolism in vascular smooth muscle reduces ATP synthesis, which in turn opens K_{ATP} channels and causes hyperpolarization. This change in potential reduces entry of Ca^{++} and relaxes coronary vascular smooth muscle to increase flow. A reduction in ATP also opens K_{ATP} channels in cardiac muscle and generates an outward current that reduces action potential duration and limits Ca^{++} entry during phase 2 of the action potential. This action may be protective during periods of imbalance between O_2 supply and demand. In addition, as cardiac work increases, H_2O_2 production rises, which activates $K_v1.5$ channels and thereby causes hyperpolarization of muscle membrane and relaxation of vascular smooth muscle. Moreover, the release of NO and adenosine dilates the arterioles and thereby adjusts the O_2 supply to the O_2 demand. At low concentrations, adenosine appears to activate endothelial K_{ATP} channels and to enhance release of NO. Conversely, at higher concentrations, adenosine acts directly on vascular smooth muscle by activating K_{ATP} channels. Decreased O_2 demand would sustain the ATP level, as well as reduce the amount of vasodilator substances released, and allows greater expression of basal tone. If production of all these agents is inhibited, coronary blood flow is reduced, both at rest and during exercise. Furthermore, contractile dysfunction and signs of myocardial ischemia become evident.

According to the adenosine hypothesis, a reduction in myocardial O_2 tension produced by inadequate coronary blood flow, hypoxemia, or increased metabolic activity of the heart leads to release of adenosine from the myocardium. Adenosine enters the interstitial fluid space to reach the coronary resistance vessels and induces vasodilation by activating adenosine receptors. However, it cannot be responsible for the increased coronary flow observed during prolonged enhancement of cardiac metabolic activity because release of adenosine from cardiac muscle is

transitory. Factors that alter coronary vascular resistance are illustrated in Fig. 17.35.

Effects of Diminished Coronary Blood Flow

Most of the O_2 in coronary arterial blood is extracted during one passage through the myocardial capillaries. Thus the supply of O_2 to myocardial cells is **flow limited;** any substantial reduction in coronary blood flow curtails O_2 delivery to the myocardium because O_2 extraction is nearly maximal even when blood flow is normal.

A reduction in coronary flow that is neither too prolonged nor too severe to induce myocardial necrosis can nonetheless cause substantial (but temporary) dysfunction of the heart. A relatively brief period of severe ischemia followed by reperfusion can result in pronounced mechanical dysfunction (myocardial stunning). However, the heart eventually recovers fully from the dysfunction. The pathophysiological basis for myocardial stunning appears to be intracellular Ca^{++} overload, initiated during the period of ischemia, combined with the generation of OH^- and superoxide free radicals early in the period of reperfusion. These changes impair the responsiveness of myofilaments to Ca^{++}.

Coronary Collateral Circulation and Vasodilators

In the normal human heart, there are virtually no functional intercoronary channels. Abrupt occlusion of a coronary artery or one of its branches leads to ischemic necrosis and eventual fibrosis of the areas of myocardium supplied by the occluded vessel. However, if a coronary artery narrows slowly and progressively over a period of days or weeks, collateral vessels develop and may furnish sufficient blood to the ischemic myocardium to prevent or reduce the extent of necrosis. Collateral vessels may develop between branches of occluded and nonoccluded arteries. They originate from preexisting small vessels that undergo proliferative changes of the endothelium and smooth muscle. These changes may occur in response to wall stress and to chemical agents, including vascular endothelial growth factors (VEGFs) released by the ischemic tissue. The VEGFs, of which there are at least five in mammals, are glycoproteins. The VEGFs induce angiogenesis, elicit vasodilation and increase endothelial permeability. By causing vasodilation, VEGFs enable perfusion of more capillaries and increase capillary permeability by opening tight junctions between endothelial cells and by adding fenestrations.

IN THE CLINIC

Myocardial stunning, prolonged ventricular dysfunction without myocardial necrosis, may be evident in patients who have suffered an acute coronary artery occlusion. If the patient is treated sufficiently early by coronary bypass surgery or balloon angioplasty, and if adequate blood flow is restored to the ischemic region, the myocardial cells in this region may recover fully. However, for many days or even weeks, the contractility of the myocardium in the affected region may be grossly subnormal.

Prolonged reductions in coronary blood flow (myocardial ischemia) may critically and permanently impair the mechanical and electrical behavior of the heart. Diminished coronary blood flow as a consequence of coronary artery disease (usually coronary atherosclerosis) is one of the most common causes of serious cardiac disease. The ischemia may be global (affects an entire ventricle) or regional (affects some fraction of the ventricle). The impairment in mechanical contraction of the affected myocardium is produced not only by the diminished delivery of O_2 and metabolic substrates but also by the accumulation of potentially harmful substances (e.g., K^+, lactic acid, H^+) in the cardiac tissues. If the reduction in coronary flow to any region of the heart is sufficiently severe and prolonged, necrosis of the affected cardiac cells results.

Myocardial hibernation describes the phenomenon in which cellular metabolism is downregulated in cells whose function is impaired by inadequate delivery of O_2 and nutrients. Myocardial hibernation occurs mainly in patients with coronary artery disease. The coronary blood flow in such patients is diminished persistently and significantly, and the mechanical function of the heart is impaired. If coronary blood flow is restored to normal by bypass surgery or angioplasty, mechanical function returns to normal.

Cutaneous Circulation

The O_2 and nutrient requirements of the skin are relatively small. Unlike other body tissues, the supply of O_2 and nutrients is not the chief factor in the regulation of cutaneous blood flow. The primary function of the cutaneous circulation is to maintain a constant body temperature. Thus the skin undergoes wide fluctuations in blood flow, depending on whether the body needs to lose or conserve heat. Changes in ambient and internal body temperature

activate mechanisms responsible for alterations in skin blood flow.

IN THE CLINIC

Numerous surgical attempts have been made to enhance the development of coronary collateral vessels. However, the techniques used do not increase the collateral circulation over and above that produced by coronary artery narrowing alone. When discrete occlusions or severe narrowing occurs in coronary arteries, as in coronary atherosclerosis, the lesions can be bypassed with an artery or a vein graft. Frequently, the narrow segment can be dilated by insertion of a balloon-tipped catheter into the diseased vessel via a peripheral artery and then inflation of the balloon. Distention of the vessel by balloon inflation (angioplasty) can produce lasting dilation of a narrowed coronary artery (Fig. 17.36), particularly when a drug-eluting stent (the drugs help prevent restenosis) is inserted during angioplasty.

Many drugs are available for use in patients with coronary artery disease to relieve angina pectoris, the chest pain associated with myocardial ischemia. These compounds include organic nitrates/nitrites, calcium channel antagonists, and β-adrenoceptor antagonists. Organic nitrates/nitrites are metabolized to NO. NO dilates the great veins to reduce venous return (preload), thereby reducing cardiac work (see Chapter 19) and myocardial O_2 requirements. In addition, NO dilates the coronary arteries to increase collateral flow. Of importance is that organic nitrates/nitrites do not interfere with coronary autoregulation. Calcium channel antagonists also cause vasodilation; none selectively dilates the coronary vessels. The β-adrenoceptor antagonists reduce the heart rate to indirectly increase coronary flow and oppose the reflex tachycardia that has been observed with organic nitrates/nitrites.

In patients with marked narrowing of a coronary artery, administration of dipyridamole, a vasodilator, can fully dilate normal vessel branches that are parallel to the narrowed segment and thereby reduce the pressure on the partially occluded vessel. The reduced pressure on the narrowed vessel further compromises blood flow to the ischemic myocardium. This phenomenon, known as *coronary steal,* occurs because dipyridamole acts by blocking the cellular uptake and metabolism of endogenous adenosine. Of note is that dipyridamole interferes with coronary autoregulation.

Regulation of Skin Blood Flow
Neural Factors

The skin contains essentially two types of resistance vessels: arterioles and **arteriovenous anastomoses.** AV anastomoses shunt blood from the arterioles to the venules and venous plexuses; hence, they bypass the capillary bed. Such anastomoses are found in the fingertips, palms of the hands, toes, soles of the feet, ears, nose, and lips. AV anastomoses differ morphologically from arterioles; the anastomoses are either short and straight or long coiled vessels, approximately 20 to 40 μm in luminal diameter, and they have thick muscular walls richly supplied with nerve fibers (Fig. 17.37). These vessels are almost exclusively under sympathetic neural control, and they dilate maximally when their nerve

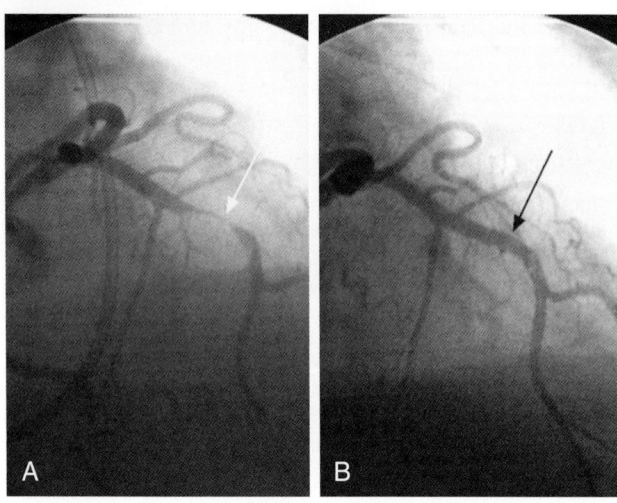

• **Fig. 17.36 A,** Angiogram (with intracoronary radiopaque dye) of marked narrowing of the left anterior descending branch of the left coronary artery *(white arrow).* **B,** The same segment of the coronary artery *(black arrow)* after angioplasty and insertion of a drug-eluting stent. (Courtesy of Dr. Michael Azrin.)

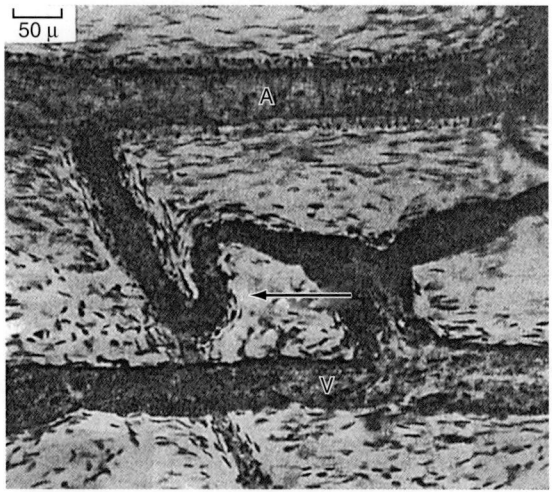

• **Fig. 17.37** Arteriovenous (AV) Anastomosis in the Ear Injected With Berlin Blue Dye. A, artery; V, vein; arrow points to an AV anastomosis. The walls of the AV anastomosis in the fingertips are thicker and more cellular. (From Pritchard MML, Daniel PM. *J Anat.* 1956;90:309.)

supply is interrupted. Conversely, reflex stimulation of the sympathetic fibers to these vessels may constrict them and obliterate the vascular lumen. Although AV anastomoses do not exhibit basal tone, they are highly sensitive to vasoconstrictor agents such as epinephrine and norepinephrine. Furthermore, AV anastomoses are not under metabolic control, and they do not show reactive hyperemia or autoregulation of blood flow. Thus regulation of blood flow through these anastomotic channels is governed principally by the nervous system in response to reflex activation by temperature receptors or from higher centers of the central nervous system.

Most of the resistance vessels in the skin exhibit some basal tone and are under dual control of the sympathetic nervous system and local regulatory factors. However, neural control predominates. Stimulation of sympathetic nerve fibers

induces vasoconstriction, and cutting of the sympathetic nerves induces vasodilation. After chronic denervation of the cutaneous blood vessels, the degree of tone that existed before denervation is gradually regained over a period of several weeks. This restoration of tone is accomplished by an enhancement of basal tone. Denervation of the skin vessels results in enhanced sensitivity to catecholamines in circulation (**denervation hypersensitivity**).

Parasympathetic vasodilator nerve fibers do not innervate cutaneous blood vessels. However, stimulation of the sweat glands, which are innervated by sympathetic cholinergic fibers, dilates the skin resistance vessels. Sweat contains an enzyme that lyses a protein (kallidin) in the tissue fluid to produce bradykinin, a polypeptide with potent vasodilator properties. Bradykinin, formed locally, dilates the arterioles and increases blood flow to the skin.

Certain skin vessels, particularly those in the head, neck, shoulders, and upper part of the chest, are regulated by higher centers in the brain. Blushing, in response to embarrassment or anger, and blanching, in response to fear or anxiety, are examples of cerebral inhibition and stimulation, respectively, of the sympathetic nerve fibers to the affected cutaneous regions.

In contrast to AV anastomoses in the skin, the resistance vessels display autoregulation of blood flow and reactive hyperemia. If the arterial inflow to a limb is stopped briefly by inflation of a blood pressure cuff, the skin becomes bright red below the point of vascular occlusion when the cuff is subsequently deflated. The increased cutaneous blood flow (reactive hyperemia) is also manifested by distention of the superficial veins in the affected extremity.

The Role of Temperature in the Regulation of Skin Blood Flow

The primary function of the skin is to maintain a constant internal environment and protect the body from adverse changes. Ambient temperature is one of the most important external variables with which the body must contend. Exposure to cold elicits a generalized cutaneous vasoconstriction that is especially pronounced in the hands and feet. This response is chiefly mediated by the nervous system. Arrest of the circulation to a hand by a pressure cuff plus immersion of that hand in cold water induces vasoconstriction in the skin of the other extremities that are exposed to room temperature. When the circulation to the chilled hand is not occluded, the reflex-generalized vasoconstriction is caused in part by the cooled blood that returns to the general circulation. This returned blood then stimulates the temperature-regulating center in the anterior hypothalamus, which then activates heat preservation centers in the posterior hypothalamus to evoke cutaneous vasoconstriction.

The skin vessels of the cooled hand also respond directly to cold. Moderate cooling or a brief exposure to severe cold (0°C to 15°C) constricts the resistance and capacitance vessels, including the AV anastomoses. Prolonged exposure to severe cold evokes a secondary vasodilator response. Prompt vasoconstriction and severe pain are elicited by immersion of the hand in ice water. However, this response is soon followed by dilation of the skin vessels, with reddening of the immersed part and alleviation of the pain. With continued immersion of the hand, alternating periods of constriction and dilation occur, but the skin temperature rarely drops as much as it did in response to the initial vasoconstriction. Prolonged severe cold, of course, damages tissue. The rosy faces of people exposed to a cold environment are examples of cold-induced vasodilation. However, blood flow through the skin of the face may be greatly reduced despite the flushed appearance. The red color of the slowly flowing blood is mainly caused by reduced O_2 uptake by the cold skin and the cold-induced shift of the oxyhemoglobin dissociation curve to the left (see Chapter 23).

Direct application of heat to the skin not only dilates the local resistance and capacitance vessels and the AV anastomoses but also reflexively dilates blood vessels in other parts of the body. The local effect is independent of the vascular nerve supply, whereas the reflex vasodilation is a combined response to stimulation of the anterior hypothalamus by the returning warmed blood and stimulation of cutaneous heat receptors in the heated regions of the skin.

The close proximity of the major arteries and veins allows countercurrent heat exchange between them. Cold blood that flows in veins from a cooled hand toward the heart takes up heat from adjacent arteries; this warms the venous blood and cools the arterial blood. Heat exchange takes place in the opposite direction when the extremity is exposed to heat. Thus heat conservation is enhanced during exposure of extremities to cold environments, and heat conservation is minimized during exposure of the extremities to warm environments.

Skin Color: Relationship to Skin Blood Volume, Oxyhemoglobin, and Blood Flow

Skin color is determined mainly by the pigment content. However, the degree of pallor or ruddiness is mainly a function of the amount of blood in the skin, except when the skin is very dark. With little blood in the venous plexus, the skin appears pale, whereas with moderate to large quantities of blood in the venous plexus, the skin displays a color. This color may be red, blue, or some shade between, depending

on the degree of oxygenation of the blood. A combination of vasoconstriction and reduced hemoglobin can impart an ashen gray color to the skin. A combination of venous engorgement and reduced hemoglobin content can impart a dark purple hue.

Skin color provides little information about the rate of cutaneous blood flow. Rapid blood flow may be accompanied by skin pallor when the AV anastomoses are open, and slow blood flow may be associated with skin ruddiness when the skin is exposed to cold.

Skeletal Muscle Circulation

The rate of blood flow in skeletal muscle varies directly with the contractile activity of the tissue and the type of muscle. Blood flow and capillary density are greater in red muscle (slow-twitch muscle with high oxidative capacity) than in white muscle (fast-twitch muscle with low oxidative capacity). In resting muscle, the precapillary arterioles contract and relax intermittently. Thus at any given moment, most of the capillary bed is not perfused, and total blood flow through quiescent skeletal muscle is low (1.4 to 4.5 mL/minute/100 g). During exercise, the resistance vessels relax, and muscle blood flow may increase to 15 to 20 times the resting level, depending on the intensity of the exercise.

Regulation of Skeletal Muscle Blood Flow

Neural and local factors regulate muscle circulation. Physical factors such as P_a, tissue pressure, and blood viscosity influence muscle blood flow. However, another physical factor, the squeezing effect of the active skeletal muscle, affects blood flow in the vessels. With intermittent contractions, inflow is restricted, and as previously described, venous outflow is enhanced. The venous valves prevent backward flow of blood between contractions and thereby aid in the forward propulsion of blood. With strong sustained contractions, as occur during exercise, the vascular bed can be compressed to the point at which blood flow actually ceases temporarily.

Neural Factors

The resistance vessels of muscle possess a high degree of basal tone; they also display tone in response to continuous low-frequency activity in the sympathetic vasoconstrictor nerve fibers. The basal firing frequency of sympathetic vasoconstrictor fibers is only approximately 1 to 2 per second, and maximal vasoconstriction occurs at frequencies of approximately 10 per second.

Vasoconstriction evoked by sympathetic nerve activity is caused by the local release of norepinephrine. Intra-arterially injected norepinephrine elicits only vasoconstriction (α_1-adrenergic receptor). In contrast, low doses of epinephrine produce vasodilation (β_2-adrenergic receptor), whereas large doses cause vasoconstriction.

Baroreceptor reflexes greatly influence the tonic activity of the sympathetic nerves. An increase in carotid sinus pressure causes the muscle vascular bed to dilate, whereas a decrease in carotid sinus pressure elicits vasoconstriction (Fig. 17.38). When sympathetic constrictor tone is high, the decrease in blood flow evoked by common carotid artery occlusion is small, but the increase in flow after the release of occlusion is large. The vasodilation produced by

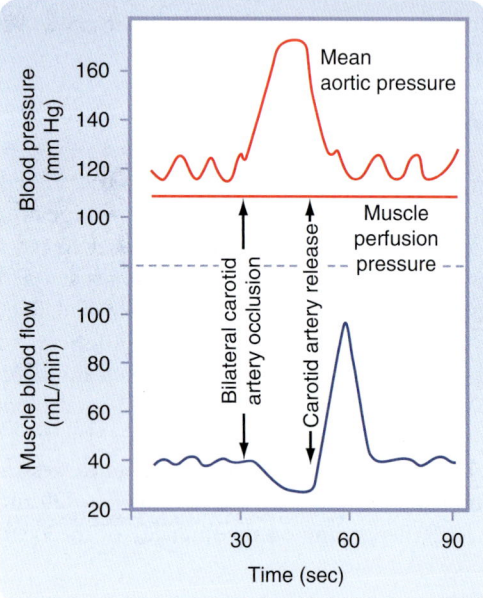

• Fig. 17.38 Evidence of Participation of the Muscle Vascular Bed in Vasoconstriction and Vasodilation Mediated by the Carotid Sinus Baroreceptors After Occlusion and Release of the Common Carotid Artery. In this preparation, the sciatic and femoral nerves constituted the only direct innervation of the hind leg muscle mass. The muscle was perfused with blood at a constant pressure. (Redrawn from Jones RD, Berne RM. *Am J Physiol.* 1963;204:461.)

baroreceptor stimulation is caused by inhibition of sympathetic vasoconstrictor activity.

The resistance vessels in skeletal muscle contribute significantly to maintenance of arterial blood pressure because skeletal muscle constitutes a large fraction of the body's mass, and the muscle vasculature thus constitutes the largest vascular bed. Participation of the skeletal muscle vessels in vascular reflexes is important in maintaining normal arterial blood pressure.

A comparison of the sympathetic neural effects on the blood vessels of muscle and skin is summarized in Fig. 17.39. Note that the lower the basal tone of the skin vessels, the greater their constrictor response; also note the absence of active cutaneous vasodilation.

Local Factors

In active skeletal muscle, blood flow is regulated by metabolic factors. In resting muscle, neural factors predominate, and they superimpose neurogenic tone on basal tone (see Fig. 17.39). Cutting of the sympathetic nerves to muscle abolishes the neural component of vascular tone, and it unmasks the intrinsic basal tone of the blood vessels. The neural and local mechanisms that regulate blood flow oppose each other, and during muscle contraction, the local vasodilator mechanism supervenes. However, during exercise, strong sympathetic nerve stimulation slightly attenuates the vasodilation induced by locally released metabolites.

Cerebral Circulation

Blood reaches the brain through the internal carotid and vertebral arteries. The vertebral arteries join to form the basilar artery, which, in conjunction with branches of the

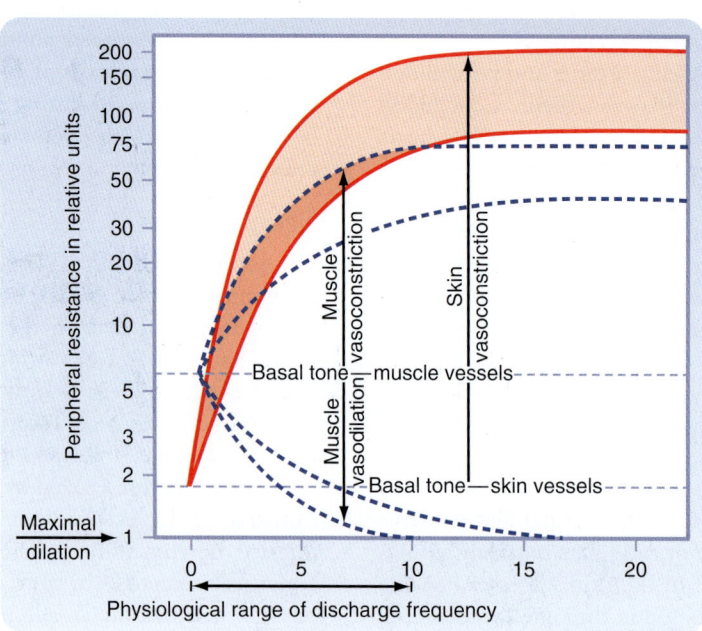

• Fig. 17.39 Basal tone and the range of response of resistance vessels in muscle *(dashed lines)* and skin *(shaded areas)* to stimulation and section of sympathetic nerves. Peripheral resistance is plotted on a logarithmic scale. (Redrawn from Celander O, Folkow B. *Acta Physiol Scand.* 1953;29:241.)

internal carotid arteries, forms the circle of Willis. Arteries on the brain surface differ from those that penetrate the brain parenchyma. Pial arteries and arterioles have an extrinsic innervation (e.g., via superior cervical ganglion, sphenopalatine nerves, trigeminal nerve); parenchymal arterioles have an intrinsic innervation (via cerebral neurons). Pial arteries have more smooth muscle cells than do parenchymal arterioles. Also, pial arteries and arterioles have collateral branches, whereas parenchymal arterioles do not. Therefore, parenchymal arterioles regulate blood flow to discrete cortical regions, and their occlusion can reduce blood flow significantly.

The cerebral circulation is unique because it lies within a rigid structure, the cranium. Any increase in arterial inflow must be associated with a comparable increase in venous outflow because the intracranial contents cannot be compressed. The volume of blood and extravascular fluid can vary considerably in most body tissues. In the brain, however, the volume of blood and extravascular fluid is relatively constant; a change in one of these fluid volumes must be accompanied by a reciprocal change in the other. The rate of cerebral blood flow is maintained within a narrow range; in humans, it averages 55 mL/minute/100 g of brain tissue.

Regulation of Cerebral Blood Flow

At rest, the brain consumes 20% of total body oxygen and 25% of total body glucose. Of all body tissues, the brain is the least tolerant of ischemia. Interruption of cerebral blood flow for as little as 5 seconds results in loss of consciousness. Ischemia that lasts just a few minutes may cause irreversible tissue damage. Fortunately, regulation of the cerebral circulation is primarily under the direction of the brain itself. Local regulatory mechanisms and reflexes that originate in the brain tend to maintain a relatively constant cerebral circulation in the presence of such adverse effects as sympathetic vasomotor nerve activity, circulating humoral vasoactive agents, and changes in arterial blood pressure. Under certain conditions, the brain also regulates its blood flow by initiating changes in systemic blood pressure.

Changes in cerebral blood flow are associated with "functional recruitment" of capillaries. Thus the rate of flow through each capillary is adjusted to meet the needs of the organ. In "capillary recruitment," in contrast, more capillaries are open to accommodate greater blood flow.

The brain has several protective mechanisms that regulate blood flow. These mechanisms include the blood-brain barrier, extrinsic regulation of central cardiovascular centers, intrinsic control (autoregulation) of circulation, and functional hyperemia, in which blood flow increases to a brain region that is active.

Blood-Brain Barrier

The blood-brain barrier regulates ion and nutrient transport between the blood and the brain and also limits the entry of harmful substances from the blood into the brain. The

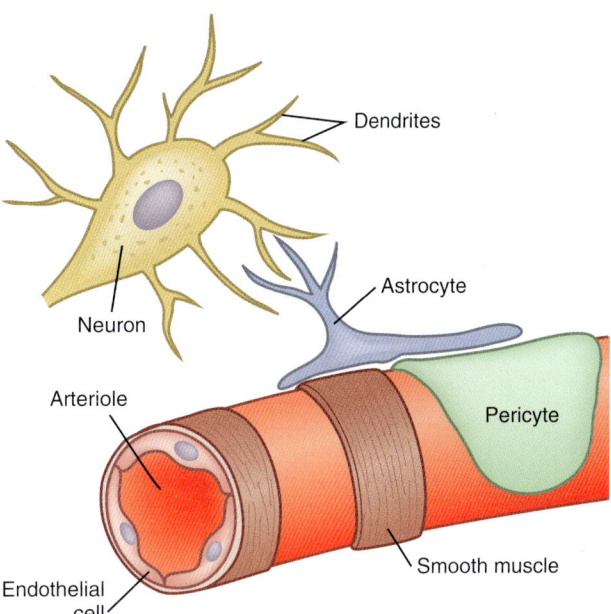

• **Fig. 17.40** Diagram of a Neurovascular Unit With an Astrocyte Linking a Neuron to an Arteriole of the Brain Microcirculation. Arteriolar tone is modulated by vascular smooth muscle and by the action of pericytes. The endothelial cell restricts diffusion of substances by virtue of tight junctions. The neurovascular unit is a component of the blood-brain barrier and also serves as a regulator of blood flow during neuronal activity.

blood-brain barrier includes tight junction proteins (junctional adhesion molecule-1, occludins, claudins), which are connected to the endothelial cell cytoskeleton to form a barrier that opposes paracellular movement of substances from blood to brain. In addition, the blood-brain barrier includes the **neurovascular unit** (microcirculation, pericytes, the extracellular matrix, astrocytes, and neurons; Fig. 17.40). Pericytes regulate blood flow by adjusting vascular diameter, and they secrete angiopoetin, a growth factor that stimulates the expression of occludins in endothelial cells. Occludins are prominently expressed in brain endothelial cells, in contrast to their sparse distribution in nonneural endothelium. The neurovascular unit regulates blood flow and capillary permeability. Thus the neurovascular unit is involved in pathological states, including hypoxia, neurodegenerative diseases, and inflammation, that are characterized by dysfunction of the blood-brain barrier.

Neural Factors

The extrinsic innervation of cerebral (pial) vessels consists of components of the autonomic nervous system. Cervical sympathetic nerve fibers that accompany the internal carotid and vertebral arteries into the cranial cavity innervate the cerebral vessels. In comparison to other vascular beds, sympathetic control of the cerebral vessels is weak, and the contractile state of the cerebrovascular smooth muscle depends primarily on local metabolic factors. The density of α_1-adrenergic receptors is less than in other vascular beds. Cerebral vessels receive parasympathetic fibers from the facial nerve that produce a slight vasodilation

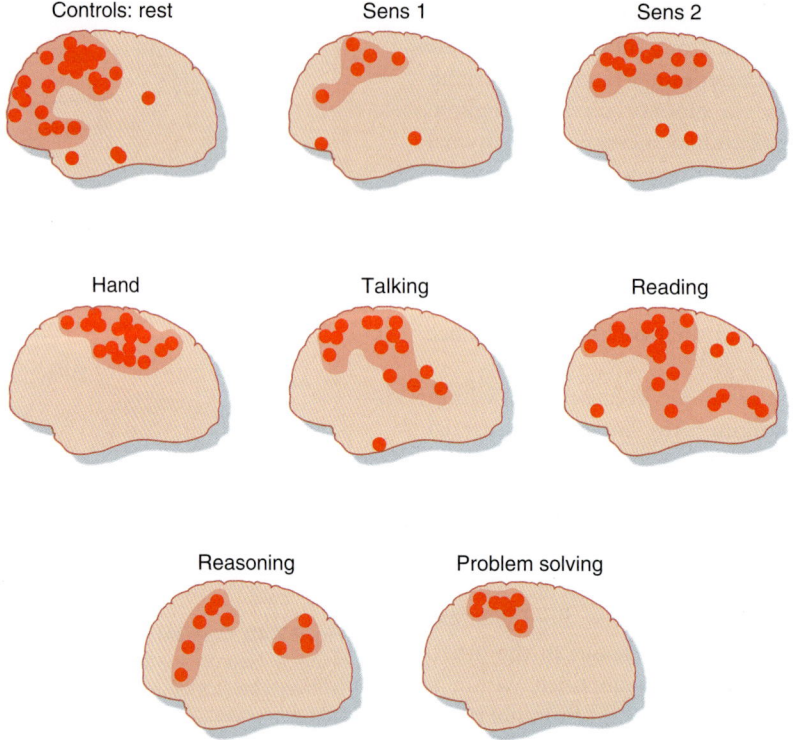

• **Fig. 17.41** Effects of Different Stimuli on Regional Blood Flow in the Contralateral Human Cerebral Cortex. Sens 1, low-intensity electrical stimulation of the hand; Sens 2, high-intensity (painful) electrical stimulation of the hand. Other stimuli are as noted. (Redrawn from Ingvar DH. *Brain Res.* 1976;107:181.)

on stimulation. The sympathetic nervous system exerts the most prominent effect on cerebral blood flow during pathophysiological conditions.

Local Factors

In general, total cerebral blood flow is relatively constant and is autoregulated. Autoregulation of cerebral blood flow involves interplay among myogenic, metabolic, and neural mechanisms much as described for peripheral vessels (see Chapter 18). However, regional blood flow in the brain is associated with regional neural activity. For example, movement of one hand results in increased blood flow only in the hand area of the contralateral sensorimotor and premotor cortex. Talking, reading, and other stimuli to the cerebral cortex are also associated with increased blood flow in the appropriate regions of the contralateral cortex (Fig. 17.41). Glucose uptake also corresponds to regional cortical neuronal activity. Thus when the retina is stimulated by light, uptake of glucose is enhanced in the visual cortex.

The neurovascular unit plays an integral role in the discrete regulation of blood flow. Production of vasoactive compounds couples increased neuronal activity to greater uptake of oxygen and glucose. Within the neurovascular unit, astrocytes link neurons with the microcirculation (see Fig. 17.40). At one pole, astrocytes surround presynaptic and postsynaptic neurons at synapses. At the other pole, astrocytes converge on vascular smooth muscle and endothelial cells of cerebral vessels. When activated by the neurotransmitter glutamate or acetylcholine, astrocytes produce

inositol trisphosphate (IP_3), which causes the release of Ca^{++}, which, in turn, activates large conductance potassium (BK_{Ca}) channels. The released K^+ raises extracellular $[K^+]$ to 8 to 15 mEq/L in the space between the astrocyte and arteriolar smooth muscle. The elevated extracellular $[K^+]$ causes hyperpolarization of smooth muscle by activating Na^+/K^+-ATPase and also by increasing the conductance of inward-rectifying K^+ channels. The hyperpolarization reduces Ca^{++} entry into vascular smooth muscle because the membrane potential is shifted away from the threshold. Hence, the parenchymal arteriole dilates, and blood flow increases.

With regard to K^+, stimuli such as hypoxia, electrical stimulation of the brain, and seizures elicit rapid increases in cerebral blood flow, and they are associated with increases in perivascular K^+. The increments in K^+ are similar to those that produce pial arteriolar dilation when K^+ is applied topically to these vessels. When extracellular K^+ exceeds 15 mEq/L, smooth muscle cells depolarize, and Ca^{++} entry is increased to cause contraction and vasoconstriction. Thus the extracellular $[K^+]$ has a dual effect on smooth muscle function that is derived from its actions on Na^+/K^+-ATPase, K^+ conductance, and the K^+ concentration gradient.

The cerebral vessels are also regulated by CO_2. Increases in arterial blood CO_2 tension ($PaCO_2$) elicit marked cerebral vasodilation; for example, inhalation of 7% CO_2 increases cerebral blood flow twofold. Conversely, decreases in $PaCO_2$, caused by hyperventilation, diminish cerebral blood flow. CO_2 evokes changes in arteriolar resistance by altering perivascular pH. When PaCO$_2$ and the HCO$_3^-$ concentration are independently changed, pial vessel diameter and blood flow are inversely related to pH, regardless of the level of PaCO$_2$ or [HCO$_3^-$]. Acidosis initiates a marked vasodilation of brain arterioles. The vasodilation is mediated by a very localized release of Ca^{++} from the endoplasmic reticulum (Ca^{++} "sparks"). This local Ca^{++} signal activates large conductance BK$_{Ca}$ channels; the ensuing hyperpolarization stabilizes the vascular smooth muscle cell and opposes vasoconstriction.

IN THE CLINIC

Elevation in intracranial pressure, caused by a brain tumor, results in an increase in systemic blood pressure. This response, called *Cushing's phenomenon,* is evoked by ischemic stimulation of vasomotor regions in the medulla. Cushing's phenomenon helps maintain cerebral blood flow in conditions such as expanding intracranial tumors.

Carbon dioxide diffuses into vascular smooth muscle from brain tissue or from the lumen of blood vessels, whereas H$^+$ in blood is prevented from reaching arteriolar smooth muscle by the blood-brain barrier. Hence, cerebral vessels dilate when the [H$^+$] of cerebrospinal fluid is increased, but these vessels dilate only minimally in response to an increase in the [H$^+$] of arterial blood. Chemical regulation of cerebral blood flow by PaCO$_2$ is impaired in humans with endothelial dysfunction (e.g., diabetes, hypertension); the relative roles of H$^+$ and NO in response to changes of PaCO$_2$ are not clear.

Potassium concentration also affects cerebral blood flow. Hypoxia, electrical stimulation of the brain, and seizures elicit rapid increases in cerebral blood flow and in perivascular [K$^+$]. The increases in [K$^+$] are similar in magnitude to those that produce pial arteriolar dilation when K$^+$ is applied topically to these vessels. However, the increase in [K$^+$] is not sustained throughout the period of cerebral stimulation. Thus only the initial increase in cerebral blood flow can be attributed to the release of K$^+$.

Adenosine also has a major effect on cerebral blood flow. Adenosine levels in the brain increase in response to ischemia, hypoxemia, hypotension, hypocapnia, electrical stimulation of the brain, and induced seizures. Topically applied adenosine is a potent dilator of the pial arterioles. Any intervention that either reduces the O$_2$ supply to the brain or increases the O$_2$ requirements of the brain results in the rapid (within 5 seconds) formation of adenosine in cerebral tissue. Unlike the changes in pH or [K$^+$], the adenosine concentration in the brain increases with initiation of the change in O$_2$ supply, and it remains elevated throughout the period of O$_2$ imbalance. The adenosine that is released into cerebrospinal fluid during cerebral ischemia becomes incorporated into adenine nucleotides in cerebral tissue. These local factors, including pH, K$^+$, and adenosine, act in concert to adjust cerebral blood flow to the metabolic activity of the brain. The cerebral circulation displays reactive hyperemia

and excellent autoregulation when arterial blood pressure is between 60 and 160 mm Hg. Mean arterial pressures below 60 mm Hg result in reduced cerebral blood flow and then syncope, whereas mean pressures above 160 mm Hg may lead to increased permeability of the blood-brain barrier and consequently to cerebral edema. Hypercapnia or any other potent vasodilator abolishes autoregulation of cerebral blood flow. Autoregulation of cerebral blood flow is probably mediated by a myogenic mechanism that is modulated by a metabolic component.

Intestinal Circulation

Anatomy

The gastrointestinal tract is supplied by the celiac, superior mesenteric, and inferior mesenteric arteries. The superior mesenteric artery carries more than 10% of the cardiac output. Small mesenteric arteries form an extensive vascular network in the submucosa of the gastrointestinal tract. The arterial branches penetrate the longitudinal and circular muscle layers of the tract, and they give rise to third- and fourth-order arterioles. Some third-order arterioles in the submucosa supply the tips of the villi (Fig. 17.42).

The direction of blood flow in the capillaries and venules in a villus is opposite that in the main arteriole (see Fig. 17.42). This arrangement is a countercurrent exchange system. Effective countercurrent exchange enables diffusion of O$_2$ from arterioles to venules. At low blood flow rates, a substantial portion of the O$_2$ may be shunted from arterioles to venules near the base of the villus. This reduces the O$_2$ supply to the mucosal cells at the tip of the villus. When intestinal blood flow is very low, shunting of O$_2$ is so great that extensive necrosis of the intestinal villi takes place.

Neural Regulation

Neural control of the mesenteric circulation is almost exclusively sympathetic. Increased sympathetic activity, through α_1-adrenergic receptors, constricts the mesenteric arterioles and capacitance vessels. These receptors are preeminent in the mesenteric circulation. However, β_2-adrenergic receptors are also present, and so the agonist isoproterenol causes vasodilation.

In response to aggressive behavior or to artificial stimulation of the hypothalamic "defense" area, pronounced vasoconstriction occurs in the mesenteric vascular bed. This vasoconstriction shifts blood flow from the less important intestinal circulation to the more crucial skeletal muscles, heart, and brain.

Autoregulation

Autoregulation of blood flow is not as well developed in the intestinal circulation as in other vascular beds. The principal mechanism responsible for autoregulation is metabolic, although a myogenic mechanism probably also

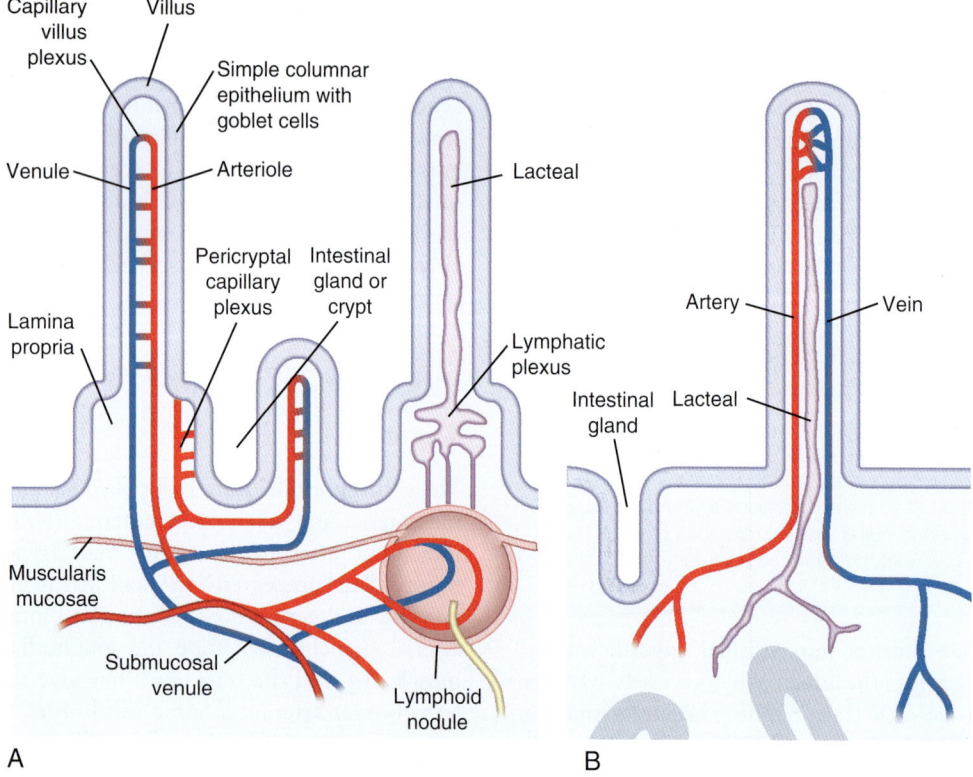

• **Fig. 17.42** Microcirculation Pattern of the Small Intestine. **A,** Capillary plexuses arise from arterioles in the villus and also in the intestinal crypt. Blood leaves the intestinal crypt via venules that enter the portal circulation. **B,** Lymphatic vessels (lacteals) originate within the villus and eventually form a plexus at the base of the villus. (Redrawn from Kierszenbaum A. *Histology and Cell Biology: An Introduction to Pathology.* Philadelphia: Mosby; 2002.)

participates (see Chapter 18). The adenosine concentration in mesenteric venous blood rises fourfold after brief arterial occlusion. It also rises during enhanced metabolic activity of the intestinal mucosa, such as during absorption of food. Adenosine, a potent vasodilator in the mesenteric vascular bed, may be the principal metabolic mediator of autoregulation. However, [K+] and altered plasma osmolality may also contribute to autoregulation.

Oxygen consumption by the small intestine is more rigorously controlled than is blood flow. Experiments have shown that O_2 uptake of the small intestine remains constant when arterial perfusion pressure is varied between 30 and 125 mm Hg.

Functional Hyperemia

Food ingestion increases intestinal blood flow. Secretion of certain gastrointestinal hormones contributes to this hyperemia. Gastrin and cholecystokinin augment intestinal blood flow, and they are secreted when food is ingested. Absorption of food also affects intestinal blood flow. Undigested food has no vasoactive influence, whereas several products of digestion are potent vasodilators. Among the various constituents of chyme, the principal mediators of mesenteric hyperemia are glucose and fatty acids.

Hepatic Circulation

Anatomy

Normally, blood flow to the liver is approximately 25% of cardiac output. Hepatic blood flow is supplied by two sources: the portal vein (≈75%) and the hepatic artery. Because portal venous blood has already passed through the gastrointestinal capillary bed, much of the O_2 of the portal vein blood flow has already been extracted. The hepatic artery delivers the remaining 25% of the blood, which is fully saturated with O_2. Hence, approximately three fourths of the O_2 used by the liver is derived from hepatic arterial blood.

The small branches of the portal vein and hepatic artery give rise to terminal portal venules and hepatic arterioles (Fig. 17.43). These terminal vessels enter the hepatic acinus (the functional unit of the liver) at its center. Blood flows from these terminal vessels into the sinusoid capillaries, which constitute the capillary network of the liver. The sinusoid capillaries radiate toward the periphery of the acinus, where they connect with the terminal hepatic venules. Blood from these terminal venules drains into progressively larger branches of the hepatic veins, which are tributaries of the inferior vena cava.

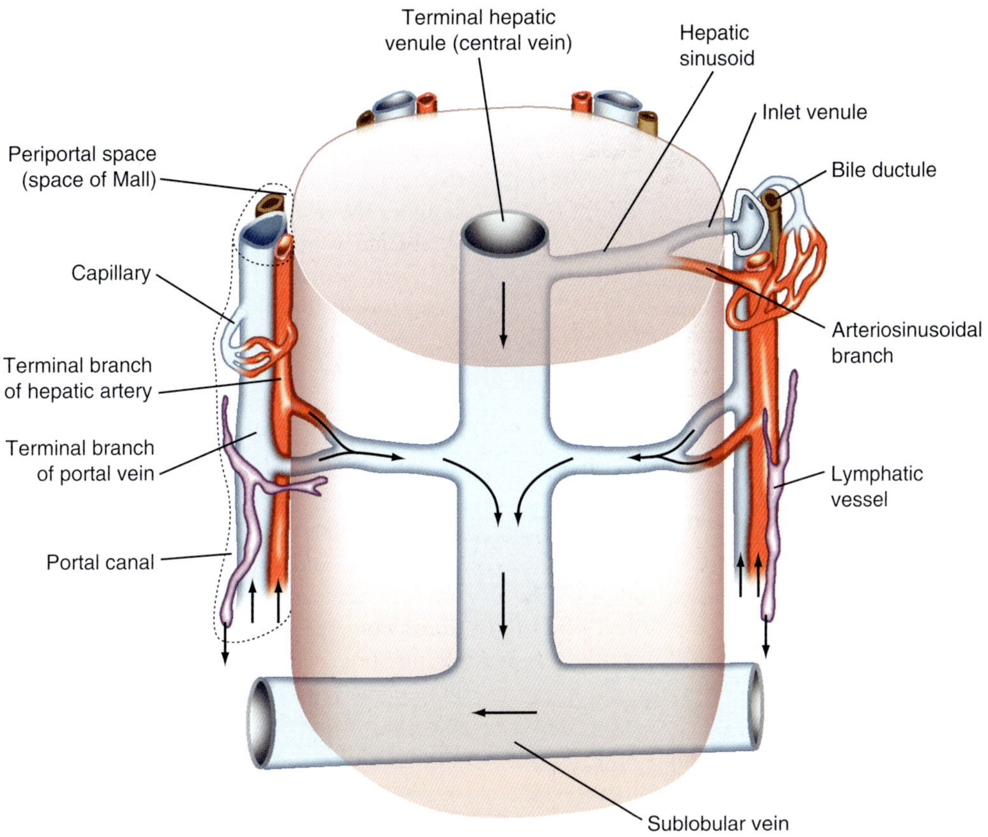

• **Fig. 17.43** Microcirculation of the Hepatic Acinus. *Arrows* indicate the direction of blood flow from the terminal portions of the hepatic artery and portal vein to the sinusoid capillaries. The mixture of arterial and venous blood flows into the central vein and then passes into the sublobular vein. (Redrawn from Ross MH, Pawling W. *Histology: A Text and Atlas: With Correlated Cell and Molecular Biology.* Philadelphia: Lippincott Williams & Wilkins; 2006.)

Hemodynamics

Mean blood pressure in the portal vein is approximately 10 mm Hg, and mean blood pressure in the hepatic artery is approximately 90 mm Hg. The resistance of the vessels upstream to the hepatic sinusoid capillaries is considerably greater than that of the downstream vessels. Consequently, the pressure in the sinusoid capillaries is only 2 or 3 mm Hg greater than that in the hepatic veins and inferior vena cava. The ratio of presinusoidal to postsinusoidal resistance is much greater in the liver than in almost any other vascular bed. Hence, drugs and other interventions that alter presinusoidal resistance usually affect pressure in the sinusoid capillaries and fluid exchange across the sinusoidal wall only slightly. However, changes in hepatic and central venous pressure are transmitted almost quantitatively to the hepatic sinusoid capillaries, and they profoundly affect the transsinusoidal exchange of fluids.

Regulation of Flow

Blood flow in the portal venous and hepatic arterial systems varies reciprocally. When blood flow is curtailed in one system, flow increases in the other but does not fully compensate for the decreased flow in the first system.

The portal venous system is not autoregulated. As portal P_v and flow are raised, resistance either remains constant or decreases. The hepatic arterial system is autoregulated, however, and adenosine may be involved in this adjustment of blood flow.

The liver tends to maintain constant O_2 consumption because O_2 extraction from hepatic blood is very efficient. As the rate of O_2 delivery to the liver varies, the liver compensates by an appropriate change in the fraction of O_2 extracted from blood. Such extraction is facilitated by the distance between the presinusoidal vessels at the acinar center and the postsinusoidal vessels at the periphery of the acinus (see Fig. 17.43). The substantial distance between these types of vessels prevents countercurrent exchange of O_2, in contrast to the countercurrent exchange that occurs in an intestinal villus.

The sympathetic nerves constrict the presinusoidal resistance vessels in the portal venous and hepatic arterial systems. Neural effects on the capacitance vessels are more important, however. The liver contains approximately 15% of the total blood volume of the body. In appropriate conditions, as in response to hemorrhage, approximately half of the hepatic blood volume can be rapidly expelled by constriction of the capacitance vessels (see also Chapter 19). Hence, the liver is an important blood reservoir in humans.

IN THE CLINIC

When central venous pressure is elevated, as in congestive heart failure, large volumes of plasma water diffuse from the liver into the peritoneal cavity; this accumulation of fluid in the abdomen is known as **ascites**. Extensive fibrosis of the liver, as in hepatic cirrhosis, markedly increases hepatic vascular resistance and thereby raises pressure substantially in the portal venous system. The consequent increase in capillary hydrostatic pressure through the splanchnic circulation also leads to extensive fluid transudation into the abdominal cavity. The pressure may likewise rise substantially in other veins that anastomose with the portal vein. For example, the esophageal veins may enlarge considerably to form esophageal varices. These varices may rupture and lead to severe, frequently fatal internal bleeding. To prevent these grave problems associated with elevated portal P_v in cirrhosis of the liver, an anastomosis (portacaval shunt) is often inserted surgically between the portal vein and the inferior vena cava to lower portal P_v.

Fetal Circulation

In Utero

Fetal circulation differs from the circulation in postnatal infants. Of most importance is that the fetal lungs are functionally inactive, and the fetus depends completely on the placenta for O_2 and nutrients. Oxygenated fetal blood from the placenta passes through the umbilical vein to the fetal liver. Approximately half the flow from the placenta passes through the liver, and the remainder bypasses the fetal liver and reaches the inferior vena cava through the **ductus venosus** (Fig. 17.44). Blood from the ductus venosus joins the blood returning from the lower part of the fetal trunk and the extremities in the inferior vena cava. This blood merges with blood from the fetal liver through the hepatic veins.

The streams of blood tend to maintain their characteristics in the inferior vena cava and are divided into two streams of unequal size by the edge of the interatrial septum (crista dividens). The larger stream, which contains mainly blood from the umbilical vein, is shunted from the inferior vena cava to the left atrium through the **foramen ovale** (see Fig. 17.44). The other stream passes into the right atrium, where it merges with blood returning from the upper parts of the fetal body through the superior vena cava and with blood from the myocardium.

Unlike the ventricles in adults, those in a fetus operate essentially in parallel. Only a tenth of right ventricular output passes through the lungs because the pulmonary vascular resistance in the fetus is high. The remainder passes from the fetal pulmonary artery through the **ductus arteriosus** to the aorta at a point distal to the origins of the arteries to the fetal head and upper extremities. Blood flows from the pulmonary artery to the aorta because pulmonary vascular resistance is high, and the diameter of the ductus arteriosus is as large as that of the descending aorta.

The large volume of blood that passes through the foramen ovale into the fetal left atrium is joined by blood returning from the lungs, and it is pumped out by the left ventricle into the aorta. Most of the blood in the ascending aorta goes to the fetal head, upper thorax, and arms; the remainder joins blood from the ductus arteriosus and supplies the rest of the body. The amount of blood pumped by the left ventricle is approximately half that pumped by the right ventricle. The major fraction of the blood that passes down the descending aorta comes from the ductus arteriosus and right ventricle and flows by way of the two umbilical arteries to the placenta.

Oxygen saturation of fetal blood occurs at various loci (see Fig. 17.44). Thus the fetal tissues that receive the most highly saturated blood are the liver, heart, and upper parts of the body, including the head.

At the placenta, the chorionic villi dip into the maternal sinuses, and O_2, CO_2, nutrients, and metabolic waste products are exchanged across the membranes. The barrier to exchange prevents equilibration of O_2 between the two circulations at normal rates of blood flow. Therefore, the PO_2 of the fetal blood that leaves the placenta is very low. Were it not for the fact that fetal hemoglobin has a greater affinity for O_2 than adult hemoglobin does, the fetus would not receive an adequate O_2 supply. The fetal oxyhemoglobin dissociation curve is shifted to the left. Therefore, at equal pressures of O_2, fetal blood carries significantly more O_2 than maternal blood does.

In early gestation, the high glycogen levels that prevail in cardiac myocytes may protect the heart from acute periods of hypoxia. Glycogen levels decrease in late gestation, and they reach adult levels by term.

Circulatory Changes That Occur at Birth

The umbilical vessels have thick muscular walls that react to trauma, tension, sympathomimetic amines, bradykinin, angiotensin, and changes in PO_2. In animals in which the umbilical cord is not tied, hemorrhage of the newborn is minimized by constriction of these large umbilical vessels in response to stretching of the umbilical arteries and by an associated increase in PO_2 in systemic arteries.

Closure of the umbilical vessels increases total peripheral resistance and the arterial blood pressure of the infant. When blood flow through the umbilical vein ceases, the ductus venosus, a thick-walled vessel with a muscular sphincter, closes. The factor that initiates closure of the ductus venosus is unknown.

IN THE CLINIC

If a pregnant woman is subjected to hypoxia, the reduced blood PO_2 in the fetus evokes tachycardia and an increase in blood flow through the umbilical vessels. If the hypoxia persists or if flow through the umbilical vessels is impaired, fetal distress occurs and is manifested initially as bradycardia.

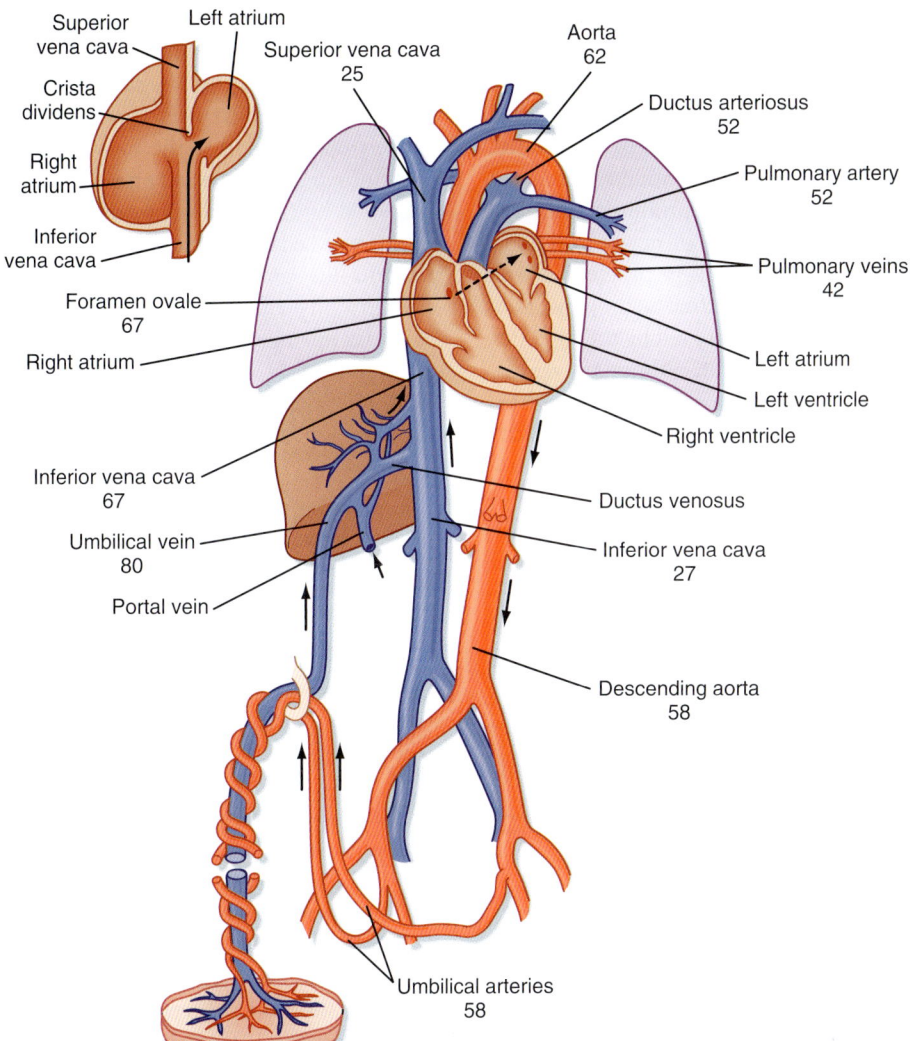

• **Fig. 17.44** Schematic Diagram of the Fetal Circulation. The *numbers* represent the percentage of O_2 saturation of the blood flowing in the indicated blood vessel. Fetal blood that leaves the placenta is 80% saturated, but the saturation of the blood that passes through the foramen ovale is reduced to 67%. This reduction in O_2 saturation is caused by the mixing of saturated blood with desaturated blood returning from the lower part of the fetal body and the liver. Addition of the desaturated blood from the fetal lungs reduces the O_2 saturation of left ventricular blood to 62%, which is the level of saturation of the blood reaching the fetal head and upper extremities. The blood in the right ventricle—which is a mixture of desaturated superior vena caval blood, coronary venous blood, and inferior vena caval blood—is only 52% saturated with O_2. When the major portion of this blood traverses the ductus arteriosus and joins that pumped by the left ventricle, the resulting O_2 saturation of the blood traveling to the lower part of the fetal body and back to the placenta is 58%. The *inset* at upper left illustrates the direction of flow of a major portion of the inferior vena caval blood through the foramen ovale to the left atrium. *Arrows* indicate the directions of flow. (Data from Dawes GS, et al. *J Physiol.* 1954;126:563.)

Immediately after birth, the asphyxia caused by constriction or clamping of the umbilical vessels, together with cooling of the body, activates the respiratory center of the newborn infant. As the lungs fill with air, pulmonary vascular resistance decreases to approximately 10% of the value that existed before lung expansion. This change in vascular resistance is not caused by the presence of O_2 in the lungs because the change is just as great if the lungs are filled with N_2. However, filling the lungs with liquid does not reduce pulmonary vascular resistance.

After birth, left atrial pressure is raised above that in the inferior vena cava and right atrium by (1) the decrease in pulmonary resistance, with the consequent large flow of blood through the lungs to the left atrium; (2) the reduction of flow to the right atrium caused by occlusion of the umbilical vein; and (3) the increased resistance to left ventricular output produced by occlusion of the umbilical arteries. Reversal of the pressure gradient across the atria abruptly closes the valve over the foramen ovale, and the septal leaflets fuse over a period of several days.

The decrease in pulmonary vascular resistance causes the pressure in the pulmonary artery to fall to approximately half its previous level (to ≈35 mm Hg). This change in pressure, coupled with a slight increase in aortic pressure, reverses the flow of blood through the ductus arteriosus. However, within several minutes, the large ductus arteriosus begins to constrict. This constriction produces turbulent flow, which is manifested as a murmur in newborn infants. Constriction of the ductus arteriosus is progressive and usually complete within 1 to 2 days after birth. Closure of the ductus arteriosus appears to be initiated by the high PO_2 of the arterial blood passing through it; pulmonary ventilation with O_2 closes the ductus, whereas ventilation with air low in O_2 opens this shunt vessel. Whether O_2 acts directly on the ductus or through the release of a vasoconstrictor substance is not known.

At birth, the walls of the two ventricles have approximately equal thickness. In addition, the muscle layer of the pulmonary arterioles is thick; this thickness is partly responsible for the high pulmonary vascular resistance of the fetus. After birth, the thickness of the walls of the right ventricle diminishes, as does the muscle layer of the pulmonary arterioles. In contrast, the left ventricular walls become thicker. These changes progress over a period of weeks after birth and reflect the effects of different hemodynamic forces (e.g., vascular resistance) on the two ventricles. Cardiac hypertrophy underlies the increase of heart weight during the normal growth period after birth. The physical demands imposed by the developing cardiovascular system, together with increased levels of soluble factors (e.g., growth hormone, insulin-like growth factor-1), account for physiological hypertrophy by which left ventricular mass more than doubles during the period from birth to early adulthood.

 IN THE CLINIC

The ductus arteriosus occasionally fails to close after birth. In the newborn, this congenital cardiovascular abnormality, called **patent ductus arteriosus,** can sometimes be corrected by the administration of nonsteroidal anti-inflammatory agents such as ibuprofen. If this does not result in closure of the ductus or if the child is older, closure must be achieved surgically.

Key Points

1. The vascular system is composed of two major subdivisions: the systemic circulation and the pulmonary circulation. These subdivisions are in series with each other and are composed of a number of vessel types (e.g., arteries, arterioles, capillaries) that are aligned in series with one another. In general, the vessels of a given type are arranged in parallel with each other.

2. The mean velocity (v) of blood flow in a given type of vessel is directly proportional to the total blood flow being pumped by the heart, and it is inversely proportional to the cross-sectional area of all the parallel vessels of that type.

3. Poiseuille's law characterizes blood flow that is steady and laminar in vessels larger than arterioles. However, blood flow is nonnewtonian in very small blood vessels (i.e., Poiseuille's law is not applicable).

4. Flow tends to become turbulent when (1) flow velocity is high, (2) fluid viscosity is low, (3) fluid density is great, (4) vessel diameter is large, or (5) the wall of the vessel is irregular.

5. Arteries not only conduct blood from the heart to the capillaries but also store some of the ejected blood during each cardiac systole. Hence, blood flow continues through the capillaries during cardiac diastole. Veins return blood to the heart from the capillaries and have a relatively low resistance and high capacitance that enables them to serve as reservoirs for blood.

6. The aging process diminishes compliance of the arteries, as well as of the veins. The less compliant the arteries are, the more work the heart must do to achieve a given cardiac output. The less compliant the veins are, the poorer is their ability to store blood.

7. Mean arterial pressure varies directly with cardiac output and total peripheral resistance. Arterial pulse pressure varies directly with stroke volume but inversely with arterial compliance.

8. Blood flow through capillaries is regulated chiefly by contraction of arterioles (resistance vessels). The capillary endothelium is the source of NO and PGI2, which relax vascular smooth muscles.

9. Water and small solutes move between the vascular and interstitial fluid compartments through capillary pores mainly by diffusion but also by filtration and absorption. Molecules larger than approximately 60 kD are essentially confined to the vascular compartment. Lipid-soluble substances, such as CO_2 and O_2, pass directly through the lipid membranes of the capillary; the rate of transfer is directly proportional to their lipid solubility. Large molecules can move across the capillary wall in vesicles by pinocytosis. The vesicles are formed from the lipid membrane of the capillaries.

10. Capillary filtration and absorption are described by Starling's equation:

$$\text{Fluid movement} = k[(P_c - P_i) - (\pi_p - \pi_i)]$$

Filtration occurs when the algebraic sum of these terms is positive; absorption occurs when it is negative.

11. Fluid and protein that have escaped from blood capillaries enter lymphatic capillaries and are

transported via the lymphatic system back to the blood vascular compartment.

12. Physical factors that influence coronary arterial blood flow are the viscosity of the blood, frictional resistance of the vessel walls, aortic pressure, and extravascular compression of the vessels within the walls of the left ventricle. Left coronary arterial blood flow is restricted during ventricular systole by extravascular compression, and the flow is greatest during diastole, when the intramyocardial vessels are not compressed. Neural regulation of coronary arterial blood flow is much less important than metabolic regulation. Activation of the cardiac sympathetic nerves constricts the coronary resistance vessels. However, the enhanced myocardial metabolism caused by the associated increase in heart rate and contractile force produces vasodilation, which overrides the direct constrictor effect of sympathetic nerve stimulation. Stimulation of the cardiac branches of the vagus nerves causes slight dilation of the coronary arterioles. A striking parallelism exists between metabolic activity of the heart and coronary arterial blood flow. A decrease in O_2 supply or an increase in O_2 demand apparently releases vasodilators that decrease coronary arterial resistance. Of the known factors (CO_2, O_2, H^+, K^+, H_2O_2, adenosine) that can mediate this response, K_{ATP} channels, NO, H_2O_2 and adenosine are the most likely candidates, although CO_2, O_2, and H^+ cannot be ruled out.

13. Most of the resistance vessels in the skin are under dual control of the sympathetic nervous system and local vasodilator metabolites. The AV anastomoses found in the hands, feet, and face, however, are solely under neural control. The main function of skin blood vessels is to aid in the regulation of body temperature by constricting to conserve heat and by dilating to lose heat. Skin blood vessels dilate directly and reflexively in response to heat, and they constrict directly and reflexively in response to cold.

14. Blood flow in skeletal muscle is regulated centrally by sympathetic nerves and locally by the release of vasodilator metabolites. In persons at rest, neural regulation of blood flow is paramount, but it yields to metabolic regulation during muscle contractions (as during exercise).

15. Cerebral blood flow is regulated predominantly by metabolic factors, especially CO_2, K^+, and adenosine. The increased regional cerebral activity produced by stimuli such as touch, pain, hand motion, talking, reading, reasoning, and problem solving are associated with enhanced blood flow in the activated area of the contralateral cerebral cortex. The neurovascular unit (microcirculation, pericytes, the extracellular matrix, astrocytes and neurons), a component of the blood-brain barrier, is thought to link brain activity with increased blood flow and oxygenation.

16. The microcirculation in intestinal villi constitutes a countercurrent exchange system for O_2. Because of the presence of this countercurrent exchange system, the villi are in jeopardy in states of low blood flow. The splanchnic resistance and capacitance vessels are very responsive to changes in sympathetic neural activity.

17. The liver receives approximately 25% of cardiac output; approximately three fourths of this output is from the portal vein and approximately a fourth from the hepatic artery. When flow is diminished in either the portal or hepatic system, flow in the other system usually increases, but not proportionately. The liver tends to maintain constant O_2 consumption, in part because its mechanism for extracting O_2 from blood is so efficient. The liver normally contains approximately 15% of the total blood volume. It serves as an important blood reservoir for the body.

18. In the fetus, a large percentage of right atrial blood passes through the foramen ovale to the left atrium, and a large percentage of pulmonary arterial blood passes through the ductus arteriosus to the aorta. At birth, the umbilical vessels, ductus venosus, and ductus arteriosus close by contraction of their muscle layers. The reduction in pulmonary vascular resistance caused by lung inflation is the main factor that reverses the pressure gradient between the atria and thereby causes the foramen ovale to close.

Additional Readings

Camici PG, d'Amati G, Rimoldi O. Coronary microvascular dysfunction: mechanisms and functional assessment. *Nat Rev Cardiol*. 2015;12:48.

Chiu J-J, Chien S. Effects of disturbed flow on vascular endothelium: pathophysiological basis and clinical perspectives. *Physiol Rev*. 2011;91:327.

Edwards G, Félétou M, Weston AH. Endothelium-derived hyperpolarization factors and associated pathways: a synopsis. *Pflügers Arch*. 2010;459:863.

Kara T, Narkiewicz K, Somers VK. Chemoreflexes—physiology and clinical implications. *Acta Physiol Scand*. 2003;177:377.

Ludmer PL, Selwyn AP, Shook TL, et al. Paradoxical vasoconstriction induced by acetylcholine in atherosclerotic coronary arteries. *N Engl J Med*. 1986;315:1046.

Ohanyan V, Yin L, Bardakjian R, et al. Requisite role of Kv1.5 channels in coronary metabolic dilation. *Circ Res*. 2015;117:612.

Potente M, Gerhardt H, Carmeliet P. Basic and therapeutic aspects of angiogenesis. *Cell*. 2011;146:873.

Resnick N, Yahav H, Shay-Salit A, et al. Fluid shear stress and the vascular endothelium: for better or for worse. *Prog Biophys Mol Biol*. 2003;81:177.

18

Regulation of the Heart and Vasculature

LEARNING OBJECTIVES

Upon completion of this chapter, the student should be able to answer the following questions:

1. How do the parasympathetic and sympathetic nervous systems regulate the functions of heart and vasculature?
2. What factors affect the differential sympathetic regulation of resistance and capacitance vessels?
3. How does the baroreceptor-mediated reflex mimic the operation of skeletal muscle proprioceptor reflex?
4. What are the two major mechanisms, intrinsic to heart muscle, that regulate myocardial performance?
5. What are the major hormones that regulate myocardial performance?
6. How is myocardial performance affected by changes in the arterial blood concentrations of O_2, CO_2, and H^+?
7. What is the myogenic mechanism of vascular smooth muscle, and how does it participate in regulation of tissue blood flow?
8. What are the humoral factors that participate in regulation of blood flow, and what are their actions?

Regulation of Heart Rate and Myocardial Performance

Cardiac output is defined as the quantity of blood pumped by the heart each minute. Cardiac output may be varied by a change in the **heart rate** or the volume of blood ejected from either ventricle with each heartbeat; this volume is called the **stroke volume.** Mathematically, cardiac output (CO) can be expressed as the product of heart rate (HR) and stroke volume (SV):

Equation 18.1
$$CO = HR \times SV$$

Thus to understand how cardiac activity is controlled, consider how the heart rate and stroke volume are regulated. Heart rate is regulated by the activity of the autonomic nervous system to modulate the intrinsic cardiac pacemaker. Stroke volume is determined by myocardial performance

(which is determined by cardiac cell contractility) and by the hemodynamic loads on the heart. All of these determinants are interdependent, inasmuch as a change in one determinant of cardiac output almost invariably alters another.

Nervous Control of the Heart Rate

Although certain local factors, such as temperature changes and stretching of tissue, can affect the heart rate, the autonomic nervous system is the principal means by which the heart rate is controlled.

The average resting heart rate is approximately 70 beats per minute in normal adults, and it is significantly faster in children. During sleep, the heart rate decreases by 10 to 20 beats per minute. It may increase during emotional excitement, and during muscular exercise, it may increase to rates well above 150 beats per minute. In well-trained athletes, the usual resting rate is only approximately 50 beats per minute.

Both divisions of the autonomic nervous system tonically influence the cardiac pacemaker, which is normally the sinoatrial (SA) node. The sympathetic nervous system enhances automaticity, whereas the parasympathetic nervous system inhibits it. Changes in heart rate usually involve a reciprocal action of these two divisions of the autonomic nervous system. Thus the heart rate ordinarily increases with a combined decrease in parasympathetic activity and increase in sympathetic activity; the heart rate decreases with the opposite changes in autonomic neural activity.

Parasympathetic tone usually predominates in healthy, resting individuals. When a resting individual is given atropine, a muscarinic receptor antagonist that blocks parasympathetic effects, the heart rate generally increases substantially. If a resting individual is given propranolol, a β-adrenergic receptor antagonist that blocks sympathetic effects, the heart rate usually decreases only slightly (Fig. 18.1). When both divisions of the autonomic nervous system are blocked, the heart rate of young adults averages approximately 100 beats per minute. The rate that prevails after complete autonomic blockade is called the **intrinsic heart rate.**

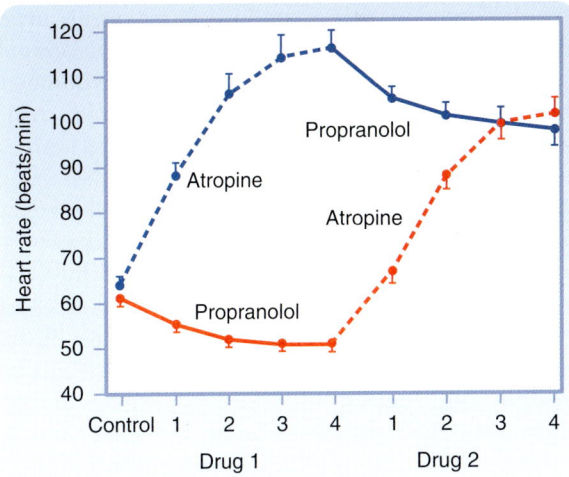

• **Fig. 18.1** Effects of four equal doses of atropine (muscarinic receptor antagonist that blocks parasympathetic effects) and propranolol (β-adrenergic receptor antagonist that blocks sympathetic effects) on the heart rates of 10 healthy young men. In half the trials, atropine was given first *(top curve)*; in the other half, propranolol was given first *(bottom curve)*. (Redrawn from Katona PG, et al. *J Appl Physiol.* 1982;52:1652.)

Parasympathetic Pathways

The cardiac parasympathetic fibers originate in the medulla oblongata, in cells that lie in the dorsal motor nucleus of the vagus nerve or in the nucleus ambiguus (see Chapter 11). In humans, centrifugal vagal fibers pass inferiorly through the neck near the common carotid arteries and then through the mediastinum to synapse with postganglionic vagal cells. These cells are located either on the epicardial surface or within the walls of the heart. Most of the vagal ganglion cells are located in epicardial fat pads near the SA and atrioventricular (AV) nodes.

The right and left vagus nerves are distributed to different cardiac structures. The right vagus nerve affects the SA node predominantly; stimulation of this nerve slows SA nodal firing and can even stop the firing for several seconds. The left vagus nerve mainly inhibits AV conduction tissue to produce various degrees of AV block (see Chapter 16). However, the distribution of the efferent vagal fibers is overlapping in such a way that left vagal stimulation also depresses the SA node and right vagal stimulation impedes AV conduction.

The SA and AV nodes are rich in acetylcholinesterase, an enzyme that rapidly hydrolyzes the neurotransmitter acetylcholine (ACh). The effects of a given vagal stimulus decay very quickly (Fig. 18.2*A*) when vagal stimulation is discontinued because ACh is rapidly destroyed. In addition, vagal effects on SA and AV nodal function have a very short latency (≈50 to 100 msec) because the ACh released quickly activates special ACh-regulated potassium (K$_{ACh}$) channels in the cardiac cells. These channels open quickly because the muscarinic receptor is coupled directly to the K$_{ACh}$ channel by a guanine nucleotide–binding protein. These two features of the vagus nerves—brief latency and rapid

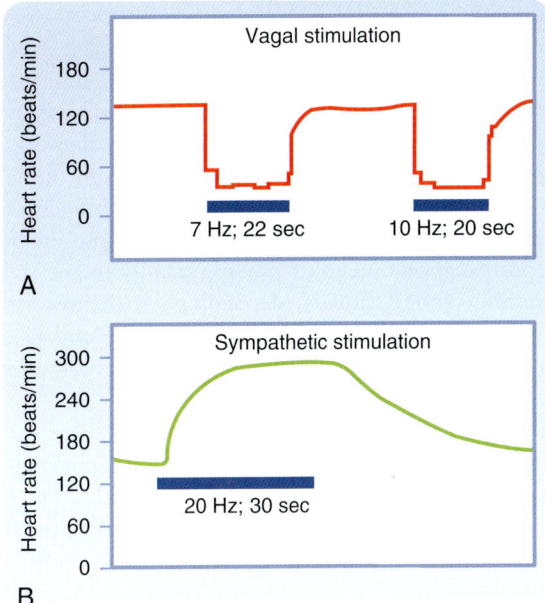

• **Fig. 18.2** Changes in heart rate evoked by stimulation *(horizontal bars)* of the vagus **(A)** and sympathetic nerves **(B).** (Modified from Warner HR, Cox A. *J Appl Physiol.* 1962;17:349.)

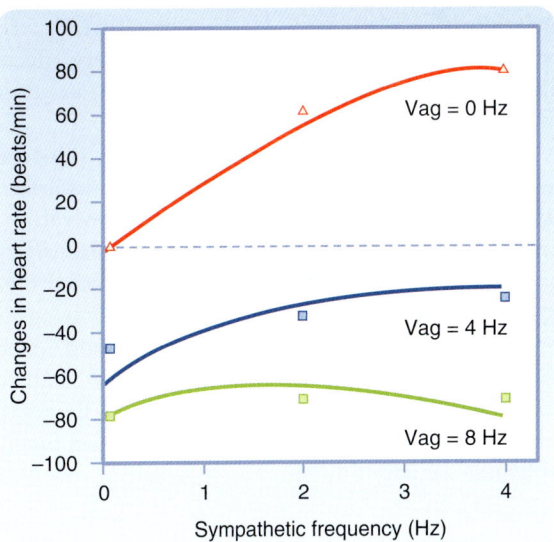

• **Fig. 18.3** Changes in Heart Rate When the Vagus and Cardiac Sympathetic Nerves Are Stimulated Simultaneously. The sympathetic nerves were stimulated at 0, 2, and 4 Hz in the presence of vagal nerve stimulation (Vag) at 0, 4, and 8 Hz. (Modified from Levy MN, Zieske H. *J Appl Physiol.* 1969;27:465.)

decay of the response—enable them to exert beat-by-beat control of SA and AV nodal function.

Parasympathetic influences usually predominate over sympathetic effects at the SA node, as shown in Fig. 18.3. When the frequency of sympathetic stimulation increases from 0 to 4 Hz, the heart rate increases by approximately 80 beats per minute in the absence of vagal nerve stimulation (0 Hz). However, when the vagus nerves are stimulated at 8 Hz, increasing the sympathetic stimulation frequency from 0 to 4 Hz has only a negligible influence on heart rate.

Sympathetic Pathways

The cardiac sympathetic fibers originate in the intermediolateral columns of the upper five or six thoracic segments and the lower one or two cervical segments of the spinal cord (see Chapter 11). These fibers emerge from the spinal column through the white communicating branches and enter the paravertebral chains of ganglia. The preganglionic and postganglionic neurons synapse mainly in the stellate or middle cervical ganglia, depending on the species. In the mediastinum, postganglionic sympathetic and preganglionic parasympathetic fibers join to form a complicated plexus of mixed efferent nerves to the heart.

The postganglionic cardiac sympathetic fibers in this plexus approach the base of the heart along the adventitial surface of the great vessels. From the base of the heart, these fibers are distributed to the various chambers as an extensive epicardial plexus. They then penetrate the myocardium, usually accompanying the coronary vessels.

In contrast to abrupt termination of the response after vagal activity, the effects of sympathetic stimulation decay gradually after stimulation is stopped (see Fig. 18.2B). Nerve terminals take up to 70% of the norepinephrine released during sympathetic stimulation; much of the remainder is carried away by the bloodstream. These processes are slow. Furthermore, the facilitatory effects of sympathetic stimulation on the heart attain steady-state values much more slowly than do the inhibitory effects of vagal stimulation. The onset of the cardiac response to sympathetic stimulation begins slowly for two main reasons. First, norepinephrine appears to be released slowly from the sympathetic nerve terminals. Second, the cardiac effects of the neurally released norepinephrine are mediated mainly by a relatively slow second messenger system involving cyclic adenosine monophosphate (cAMP; see Chapter 3). Hence, sympathetic activity alters the heart rate and AV conduction much more slowly than vagal activity does. Whereas vagal activity can exert beat-by-beat control of cardiac function, sympathetic activity cannot.

Control by Higher Centers

Stimulation of various brain regions can have significant effects on cardiac rate, rhythm, and contractility (see Chapter 11). In the cerebral cortex, centers that regulate cardiac function are located in the anterior half of the brain, principally in the frontal lobe, the orbital cortex, the motor and premotor cortex, the anterior portion of the temporal lobe, the insula, and the cingulate gyrus. Stimulation of the midline, ventral, and medial nuclei of the thalamus elicits tachycardia. Stimulation of the posterior and posterolateral regions of the hypothalamus can also change the heart rate. Stimuli applied to the H2 field of Forel in the posterior hypothalamus evoke various cardiovascular responses, including tachycardia and associated limb movements; these changes resemble those observed during muscular exercise. Undoubtedly, the cortical and hypothalamic centers initiate the cardiac reactions that occur during excitement, anxiety, and other emotional states. The hypothalamic centers also initiate the cardiac response to alterations in environmental temperature. Experimentally induced temperature changes in the preoptic anterior hypothalamus alter the heart rate and peripheral resistance.

Stimulation of the parahypoglossal area of the medulla reciprocally activates cardiac sympathetic pathways and inhibits cardiac parasympathetic pathways. In certain dorsal regions of the medulla, distinct cardiac accelerator sites (increase the heart rate) and augmentor sites (increase cardiac contractility) have been detected in animals with transected vagus nerves. The accelerator regions are more abundant on the right side, whereas the augmentor sites are more prevalent on the left. A similar distribution also exists in the hypothalamus. Therefore, the sympathetic fibers mainly descend ipsilaterally through the brainstem.

IN THE CLINIC

Cortical centers have important effects on autonomic function. The insula exerts distinct regulation of the balance between sympathetic and parasympathetic actions on the cardiovascular system. In patients subjected to electrical stimulation, stimuli applied to the left insular cortex elicit predominantly parasympathetic responses (bradycardia and vasodepression), whereas stimuli applied to the right insular cortex evoke sympathetic actions (tachycardia and vasopression). As predicted, patients with acute, stroke-induced damage of the left insular cortex display increased sympathetic tone and an increased risk of arrhythmias and cardiovascular mortality. When the right insular cortex is acutely involved in the stroke, the incidence of cardiovascular mortality/morbidity is unchanged.

Baroreceptor Reflex

Sudden changes in arterial blood pressure initiate a reflex that evokes an inverse change in heart rate (Fig. 18.4). Baroreceptors located in the aortic arch and carotid sinuses are responsible for this reflex (see the section "Arterial Baroreceptors"). The inverse relationship between heart rate and arterial blood pressure is generally most pronounced over an intermediate range of arterial blood pressures. Below this intermediate range, the heart rate maintains a constant, high value; above this pressure range, the heart rate maintains a constant, low value.

The effects of changes in carotid sinus pressure on the activity in cardiac autonomic nerves are described in Fig. 18.5, which shows that over an intermediate range of carotid sinus pressures (100 to 180 mm Hg), reciprocal changes are evoked in efferent vagal and sympathetic neural activity. Below this range of carotid sinus pressure, sympathetic activity is intense, and vagal activity is virtually absent. Conversely, above the intermediate range of carotid sinus pressure, vagal activity is intense and sympathetic activity is minimal.

Bainbridge Reflex, Atrial Receptors, and Atrial Natriuretic Peptide

In 1915, Francis A. Bainbridge reported that infusing blood or saline into dogs accelerated their heart rate. This increase did not seem to be tied to arterial blood pressure because the heart rate rose regardless of whether arterial blood pressure did or did not change. However, Bainbridge also noted that the heart rate increased whenever central venous pressure rose sufficiently to distend the right side of the heart. This response is termed the **Bainbridge reflex.** Bilateral transection of the vagus nerves abolished this response.

Many investigators have confirmed Bainbridge's observations and have noted that the magnitude and direction of the response depend on the prevailing heart rate. When the heart rate is slow, intravenous infusions of blood or electrolyte solutions usually accelerate the heart. At more rapid heart rates, however, such infusions ordinarily slow the heart. What accounts for these different responses? Increases in blood volume not only evoke Bainbridge reflex but also activate other reflexes (of note, the baroreceptor reflex). These other reflexes tend to elicit opposite changes in heart rate. Therefore, changes in heart rate evoked by an alteration in blood volume are the result of these antagonistic reflex effects (Fig. 18.6). Evidently, the Bainbridge reflex predominates over the baroreceptor reflex when blood volume rises, but the baroreceptor reflex prevails over the Bainbridge reflex when blood volume diminishes.

Both atria have receptors that are affected by changes in blood volume and that influence the heart rate. These receptors are located principally in the venoatrial junctions: in the right atrium at its junctions with the venae cavae and in the left atrium at its junctions with the pulmonary veins. Distention of these atrial receptors sends afferent impulses to the brainstem in the vagus nerves. The efferent impulses are carried from the brainstem to the SA node by fibers from both autonomic divisions.

The cardiac response to these changes in autonomic neural activity is highly selective. Even when the reflex increase in heart rate is large, changes in ventricular contractility

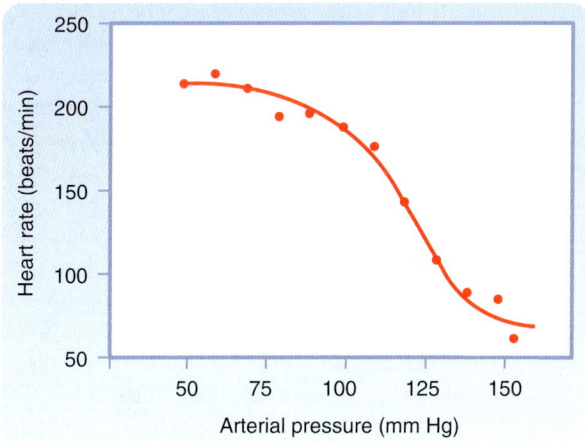

• Fig. 18.4 Experimental data showing that heart rate decreases as arterial pressure is increased. (Adapted from Cornish KG, et al. *Am J Physiol.* 1989;257:r595.)

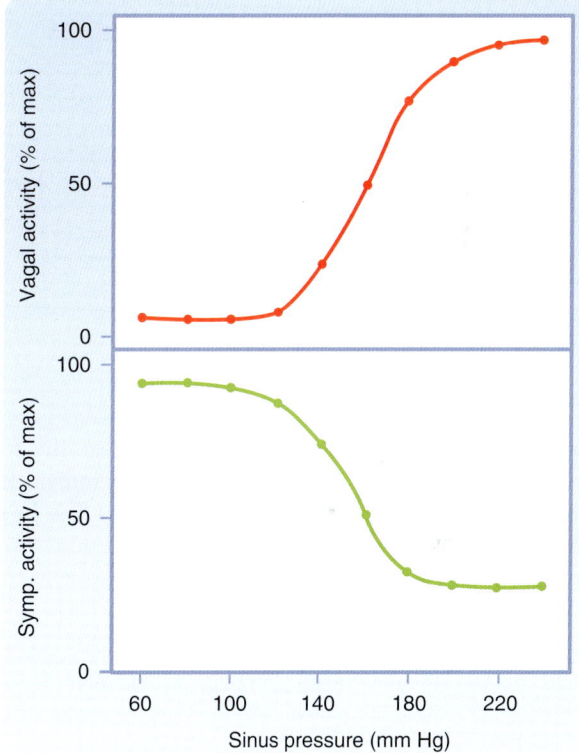

• Fig. 18.5 Experimental data showing that increases in carotid sinus pressure result in decreased sympathetic efferent nerve (Symp) activity and increased cardiac vagal nerve activity. (Adapted from Kollai M, Koizumi K. *Pflügers Arch.* 1989;413:365.)

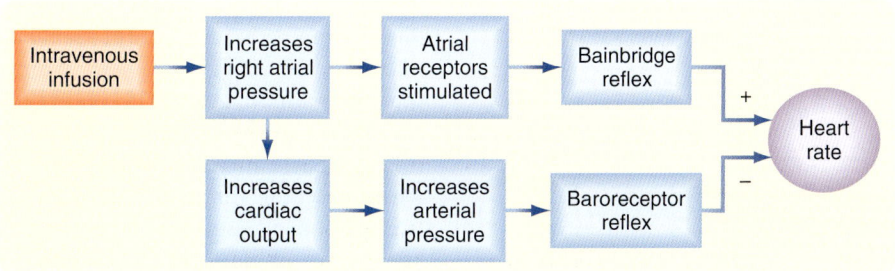

• Fig. 18.6 Intravenous Infusions of Blood or Electrolyte Solutions Tend to Increase the Heart Rate Through the Bainbridge Reflex and to Decrease the Heart Rate Through the Baroreceptor Reflex. The actual change in heart rate induced by such infusions is the result of these two opposing effects.

are generally negligible. Furthermore, the neurally induced increase in heart rate is not usually accompanied by an increase in sympathetic activity in the peripheral arterioles.

Stimulation of the atrial receptors increases not only the heart rate but also urine volume. Reduced activity in the renal sympathetic nerve fibers may partially account for this diuresis. However, the principal mechanism appears to be a neurally mediated reduction in **vasopressin (antidiuretic hormone)** secretion by the posterior pituitary gland (see Chapters 35 and 41). Stretch of the atrial walls also releases **atrial natriuretic peptide (ANP)** from the atria.* ANP, a 28–amino acid peptide, exerts potent diuretic and natriuretic effects on the kidneys (see also Chapter 35) and vasodilator effects on the resistance and capacitance vessels. Thus ANP is an important regulator of blood volume and blood pressure.

 IN THE CLINIC

In congestive heart failure, NaCl and water are retained, mainly because stimulation by the renin-angiotensin system increases the release of aldosterone from the adrenal cortex. The plasma level of ANP is also increased in congestive heart failure. By enhancing the renal excretion of NaCl and water, ANP gradually reduces fluid retention and the consequent elevations in central venous pressure and cardiac preload.

Respiratory Sinus Arrhythmia

Rhythmic variations in heart rate, occurring at the frequency of respiration, are detectable in most individuals and tend to be more pronounced in children. The heart rate typically accelerates during inspiration and decelerates during expiration (Fig. 18.7).

---

*The myocytes of the ventricles secrete a related peptide in response to stretch. This peptide, termed **brain natriuretic peptide (BNP)** because of its initial discovery in the central nervous system, has actions similar to those of ANP (see Chapter 35).

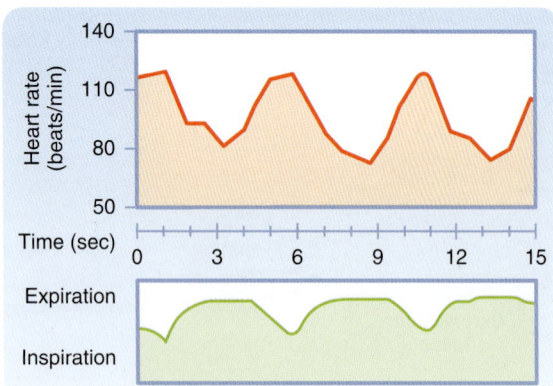

• **Fig. 18.7** Respiratory Sinus Arrhythmia. Note that the heart rate increases during inspiration and decreases during expiration. (Modified from Warner MR, et al. *Am J Physiol.* 1986;251:H1134.)

Recordings from cardiac autonomic nerves reveal that neural activity increases in the sympathetic fibers during inspiration and increases in the vagal fibers during expiration. The heart rate response to cessation of vagal stimulation is very quick because, as already noted, ACh released from the vagus nerves is rapidly hydrolyzed by acetylcholinesterase. This short latency enables the heart rate to vary rhythmically at the respiratory frequency. Conversely, the norepinephrine released periodically at the sympathetic endings is removed very slowly. Therefore, the rhythmic variations in sympathetic activity that accompany inspiration do not induce any appreciable oscillatory changes in heart rate. Thus respiratory sinus arrhythmia is brought about almost entirely by changes in vagal activity. In fact, respiratory sinus arrhythmia is exaggerated when vagal tone is enhanced.

Both reflex and central factors help initiate respiratory sinus arrhythmia (Fig. 18.8). Stretch receptors in the lungs are stimulated during inspiration, and this action leads to a reflex increase in heart rate. The afferent and efferent limbs of this reflex are located in the vagus nerves. Intrathoracic pressure also decreases during inspiration and thereby increases venous return to the right side of the heart (see Chapter 19). The consequent stretch of the right atrium elicits the Bainbridge reflex. After the time delay required for the increased venous return to reach the left side of the heart, left ventricular output increases and raises arterial blood pressure. This rise in blood pressure in turn reduces the heart rate through the baroreceptor reflex.

Central factors are also responsible for respiratory cardiac arrhythmia. The respiratory center in the medulla directly influences the cardiac autonomic centers (see Fig. 18.8). In heart-lung bypass studies, the chest is opened, the lungs are collapsed, venous return is diverted to a pump-oxygenator, and arterial blood pressure is maintained at a constant level. In such studies, rhythmic movement of the rib cage attests to the activity of the medullary respiratory centers, and is often accompanied by rhythmic changes in heart rate at the respiratory frequency. This respiratory cardiac arrhythmia is almost certainly induced by a direct interaction between the respiratory and cardiac centers in the medulla.

Chemoreceptor Reflex

The cardiac response to peripheral chemoreceptor stimulation illustrates the complex interactions that may ensue when one stimulus excites two organ systems simultaneously. Stimulation of carotid chemoreceptors consistently increases ventilatory rate and depth (see Chapter 24), but ordinarily it changes the heart rate only slightly. The magnitude of the ventilatory response determines whether the heart rate increases or decreases as a result of carotid chemoreceptor stimulation. Mild chemoreceptor-induced stimulation of respiration decreases the heart rate moderately; more pronounced stimulation increases the heart rate only slightly. If the pulmonary response to chemoreceptor stimulation is blocked, the heart rate response may be greatly exaggerated, as described later.

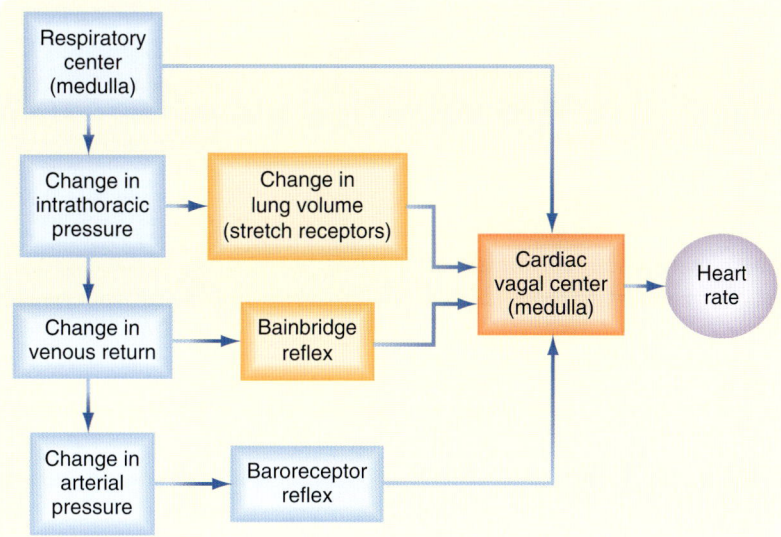

• **Fig. 18.8** Respiratory sinus arrhythmia is generated by a direct interaction between the respiratory and cardiac centers in the medulla, as well as by reflexes that originate from stretch receptors in the lungs, from stretch receptors in the right atrium (the Bainbridge reflex), and from baroreceptors in the carotid sinuses and aortic arch.

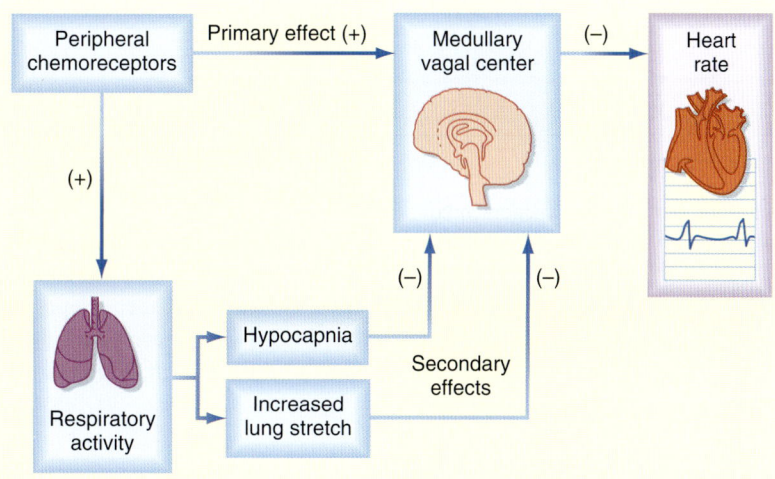

• **Fig. 18.9** The Primary Effect of Stimulation of Peripheral Chemoreceptors on the Heart Rate Is to Excite the Cardiac Vagal Center in the Medulla and Thus to Decrease the Heart Rate. Peripheral chemoreceptor stimulation also excites the respiratory center in the medulla. This effect produces hypocapnia and increases lung inflation, both of which secondarily inhibit the medullary vagal center. Thus these secondary influences attenuate the primary reflex effect of peripheral chemoreceptor stimulation on heart rate.

The cardiac response to peripheral chemoreceptor stimulation is the result of primary and secondary reflex mechanisms (Fig. 18.9). The principal effect of the primary reflex stimulation is to excite the medullary vagal center and thereby decrease the heart rate. The respiratory system mediates secondary reflex effects. The respiratory stimulation by arterial chemoreceptors tends to inhibit the medullary vagal center. This inhibition varies with the level of concomitant stimulation of respiration; small increases in respiration inhibit the vagal center slightly, whereas large increases in ventilation inhibit the vagal center more profoundly.

An example of the primary inhibitory influence is shown in Fig. 18.10. In this example, the lungs are completely collapsed, and blood oxygenation is accomplished with an artificial oxygenator. When the carotid chemoreceptors are stimulated, intense bradycardia and some degree of AV block ensue. Such effects are mediated primarily by efferent vagal fibers.

The pulmonary hyperventilation that is ordinarily evoked by carotid chemoreceptor stimulation influences the heart rate secondarily, both by initiating more pronounced pulmonary inflation reflexes and by producing hypocapnia (see Fig. 18.9). Both influences tend to depress the primary cardiac response to chemoreceptor stimulation and thereby accelerate the heart rate. Hence, when pulmonary hyperventilation is not prevented, the primary and secondary effects neutralize each other, and carotid chemoreceptor stimulation affects the heart rate only moderately.

Ventricular Receptor Reflexes

Sensory receptors located near the endocardial surfaces of the ventricles initiate reflex effects similar to those elicited

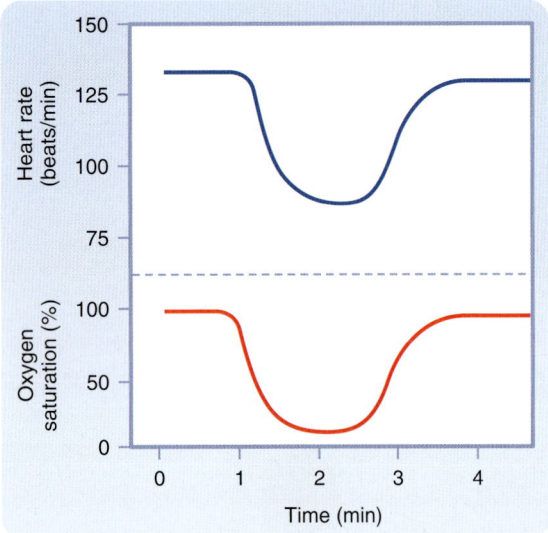

• **Fig. 18.10** Changes in Heart Rate With Carotid Chemoreceptor Stimulation During Total Heart Bypass. The lungs remain deflated, and respiratory gas exchange is accomplished by an artificial oxygenator. The lower tracing represents the oxygen saturation of the blood perfusing the carotid chemoreceptors. The blood perfusing the remainder of the body, including the myocardium, is fully saturated with oxygen. (Modified from Levy MN, et al. *Circ Res.* 1966;18:67.)

 IN THE CLINIC

The electrocardiogram in Fig. 18.11 was recorded from a quadriplegic patient who could not breathe spontaneously and required tracheal intubation and artificial respiration. When the tracheal catheter was briefly disconnected (near the beginning of the top strip in the figure, indicated by the arrow) to allow nursing care, profound bradycardia developed after 9 heart beats. The patient's heart rate was 65 beats per minute just before the tracheal catheter was disconnected. In less than 10 seconds after cessation of artificial respiration, his heart rate dropped to approximately 20 beats per minute. This bradycardia could be prevented by blocking the effects of efferent vagal activity with atropine, and its onset could be delayed considerably by hyperventilation of the patient before the tracheal catheter is disconnected.

by the arterial baroreceptors. Excitation of these endocardial receptors causes the heart rate and peripheral resistance to diminish. Other sensory receptors have been identified in the epicardial regions of the ventricles. Although all these ventricular receptors are excited by various mechanical and chemical stimuli, their exact physiological functions remain unclear.

Regulation of Myocardial Performance

Intrinsic Regulation of Myocardial Performance

As noted previously, the heart can initiate its own beat in the absence of any nervous or hormonal control. The myocardium can also adapt to changing hemodynamic conditions by means of mechanisms that are intrinsic to cardiac muscle itself. For example, racing greyhounds with

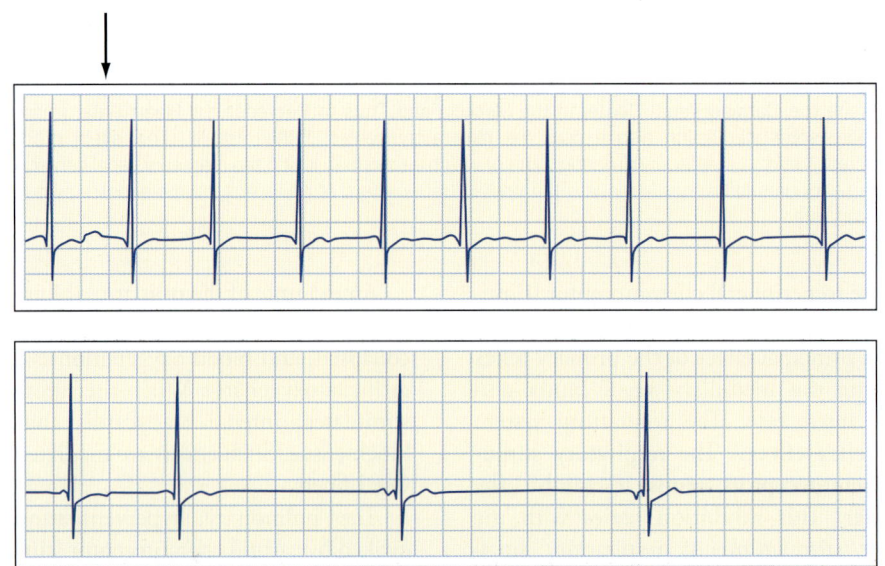

• **Fig. 18.11** Electrocardiogram of a 30-Year-Old Quadriplegic Man Who Could Not Breathe Spontaneously and Required Tracheal Intubation and Artificial Respiration. The two strips are continuous. (Modified from Berk JL, Levy MN. *Eur Surg Res.* 1977;9:75.)

denervated hearts perform almost as well as those with intact innervation. Their maximal running speed decreases by only 5% after complete cardiac denervation. In these dogs, the threefold to fourfold increase in cardiac output during a race is achieved principally by an increase in stroke volume. Normally, the increase in cardiac output with exercise is accompanied by a proportionate increase in heart rate; stroke volume does not change much (see Chapter 19). This adaptation in the denervated heart is not achieved entirely by intrinsic mechanisms; circulating catecholamines undoubtedly contribute. For example, if β-adrenergic receptor antagonists are given to greyhounds with denervated hearts, their racing performance is severely impaired.

 IN THE CLINIC

Ventricular receptors have been implicated in the initiation of **vasovagal syncope,** a feeling of lightheadedness or brief loss of consciousness that may be triggered by psychological or orthostatic stress. The ventricular receptors are believed to be stimulated by reduced ventricular filling volume in combination with vigorous ventricular contraction. In a person standing quietly, ventricular filling is diminished because blood tends to pool in the veins in the abdomen and legs, as explained in Chapter 17. Consequently, the reduction in cardiac output and arterial blood pressure leads to a generalized increase in sympathetic neural activity through the baroreceptor reflex (see Fig. 18.5). The enhanced sympathetic activity to the heart evokes a vigorous ventricular contraction that stimulates the ventricular receptors. Excitation of the ventricular receptors initiates the autonomic neural changes that evoke vasovagal syncope: namely, a combination of profound, vagally mediated bradycardia and generalized arteriolar vasodilation mediated by a reduction in sympathetic neural activity.

 IN THE CLINIC

The heart is partially or completely denervated in various clinical situations: (1) a surgically transplanted heart is totally denervated, although the intrinsic, postganglionic parasympathetic fibers persist; (2) atropine blocks vagal effects on the heart, and propranolol blocks sympathetic β-adrenergic influences; (3) certain drugs, such as reserpine, deplete cardiac norepinephrine stores and thereby restrict or abolish sympathetic control; and (4) in chronic congestive heart failure, cardiac norepinephrine stores are often severely diminished, and any sympathetic influences are attenuated.

Two principal intrinsic mechanisms, the **Frank-Starling mechanism** and **rate-induced regulation,** enable the myocardium to adapt to changes in hemodynamic conditions. The Frank-Starling mechanism **(Frank-Starling law of the heart)** is invoked in response to changes in the resting length of myocardial fibers. Rate-induced regulation is evoked by changes in the frequency of the heartbeat.

Frank-Starling Mechanism

In the 1910s, the German physiologist Otto Frank and the English physiologist Ernest Starling independently studied the response of isolated hearts to changes in preload and afterload (see Chapter 16). When ventricular filling pressure (preload) is increased, ventricular volume increases progressively, and after several beats, becomes constant and larger. At equilibrium, the volume of blood ejected by the ventricles (stroke volume) with each heartbeat increases to equal the greater quantity of venous return to the right atrium.

The increased ventricular volume facilitates ventricular contraction and enables the ventricles to pump a greater stroke volume. This increase in ventricular volume is associated with an increase in length of the individual ventricular cardiac fibers. The increase in fiber length alters cardiac performance mainly by altering the number of myofilament cross-bridges that interact (see Chapter 16). More recent evidence indicates that the principal mechanism involves a stretch-induced change in the sensitivity of cardiac myofilaments to Ca^{++} (see Chapters 13 and 16). There exists an optimal fiber length, however. Excessively high filling pressures that overstretch the myocardial fibers may depress rather than enhance the pumping capacity of the ventricles (see Fig. 16.36).

Starling also showed that isolated heart preparations could adapt to changes in the counterforce to the ventricular ejection of blood during systole (i.e., afterload). As the left ventricle contracts, it does not eject blood into the aorta until the ventricle has developed a pressure that just exceeds the prevailing aortic pressure (see Fig. 16.39). The aortic pressure during ventricular ejection essentially constitutes the left ventricular afterload. In Starling's experiments, arterial pressure was controlled by a hydraulic device in the tubing that led from the ascending aorta to the right atrial blood reservoir. To hold venous return to the right atrium constant, the hydrostatic level of the blood reservoir was maintained. As Starling raised arterial pressure to a new constant level, the left ventricle responded at first to the increased afterload by pumping a diminished stroke volume. Because venous return was held constant, the diminution in stroke volume was accompanied by a rise in ventricular diastolic volume, as well as by an increase in the length of the myocardial fibers. This change in end-diastolic fiber length finally enabled the ventricle to pump a normal stroke volume against the greater peripheral resistance. As mentioned, a change in the number of cross-bridges between the thick and thin filaments probably contributes to this adaptation, but the major factor appears to be a stretch-induced change in the sensitivity of the contractile proteins to Ca^{++}.

Cardiac adaptation to alterations in heart rate also involves changes in ventricular volume. During bradycardia, for example, the increased duration of diastole allows greater ventricular filling. The consequent increase in myocardial fiber length increases stroke volume. Therefore, the reduction in heart rate may be fully compensated by the increase in stroke volume, and cardiac output may therefore remain constant.

When cardiac compensation involves ventricular dilation, the effect of the increased size of the ventricle on the generation of intraventricular pressure must be considered. According to Laplace's relationship (see Chapter 17), if the ventricle enlarges, the force required by each myocardial fiber to generate a given intraventricular systolic pressure must be appreciably greater than that developed by the fibers in a ventricle of normal size. Thus more energy is required for a dilated heart to perform a given amount of external work than for a normal-sized heart to do so. Hence, computation of afterload on contracting myocardial fibers in the walls of the ventricles must account for ventricular dimensions along with intraventricular (and aortic) pressure.

The relatively rigid pericardium that encloses the heart determines the pressure-volume relationship at high levels of pressure and volume. The pericardium limits heart volume even under normal conditions, when an individual is at rest and the heart rate is slow. In patients with **chronic congestive heart failure,** the sustained cardiac dilation and hypertrophy may stretch the pericardium considerably. In such patients, the pericardial limitation of cardiac filling is exerted at pressures and volumes entirely different from those in normal individuals.

To assess changes in ventricular performance, the Frank-Starling mechanism is often represented by a family of **ventricular function curves.** To construct a control ventricular function curve, for example, blood volume is altered over a range of values, and stroke work (i.e., stroke volume × mean arterial pressure) and end-diastolic ventricular pressure are measured at each step. Similar observations are then made during the desired experimental intervention. For example, the ventricular function curve obtained during infusion of norepinephrine lies above and to the left of the control ventricular function curve (Fig. 18.12). Clearly, for a given level of left ventricular end-diastolic pressure (an index of preload), the left ventricle performs more work during the norepinephrine infusion than during control conditions. Hence, the upward and leftward shift of the ventricular function curve signifies improved ventricular contractility. Conversely, a shift downward and to the right indicates impaired contractility and a tendency toward **cardiac failure.**

Balance Between Right and Left Ventricular Output

The Frank-Starling mechanism is well suited to match cardiac output to venous return. Any sudden, excessive output by one ventricle soon causes an increase in venous return to the second ventricle. The consequent increase in diastolic fiber length in the second ventricle augments the output of that ventricle to correspond to the output of its mate. In this way, the Frank-Starling mechanism maintains a precise balance between the output of the right and left ventricles. If the two ventricles were not arranged in series

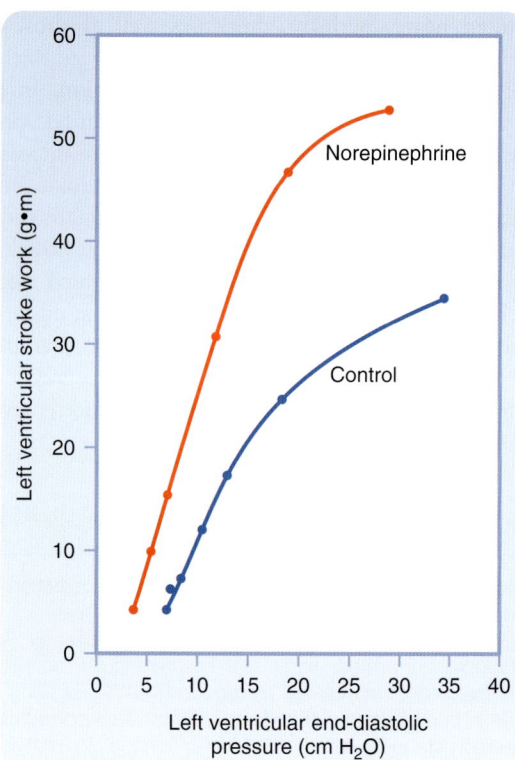

• **Fig. 18.12** A constant infusion of norepinephrine shifts the ventricular function curve up and to the left. This shift signifies an enhancement in ventricular contractility. (Redrawn from Sarnoff SJ, et al. *Circ Res.* 1960;8:1108.)

in a closed circuit, any small but maintained imbalance in output of the two ventricles would be catastrophic.

The curves that relate cardiac output to mean atrial pressure for the two ventricles do not coincide; the curve for the left ventricle usually lies below that for the right ventricle (Fig. 18.13). At equal right and left atrial pressures (points A and B in Fig. 18.13), right ventricular output exceeds left ventricular output. Hence, venous return to the left ventricle (a function of right ventricular output) exceeds left ventricular output, and left ventricular diastolic volume and pressure rise. According to the Frank-Starling mechanism, left ventricular output therefore increases (from point B toward point C in Fig. 18.13). Only when the output of both ventricles is identical (points A and C) is equilibrium reached. Under such conditions, however, left atrial pressure (point C) exceeds right atrial pressure (point A). This is precisely the relationship that ordinarily prevails.

 IN THE CLINIC

The fact that left atrial pressure exceeds right atrial pressure accounts for the observation that in individuals with congenital atrial septal defects in which the two atria communicate with each other via a patent foramen ovale, the direction of shunt flow is usually from left to the right side of the heart.

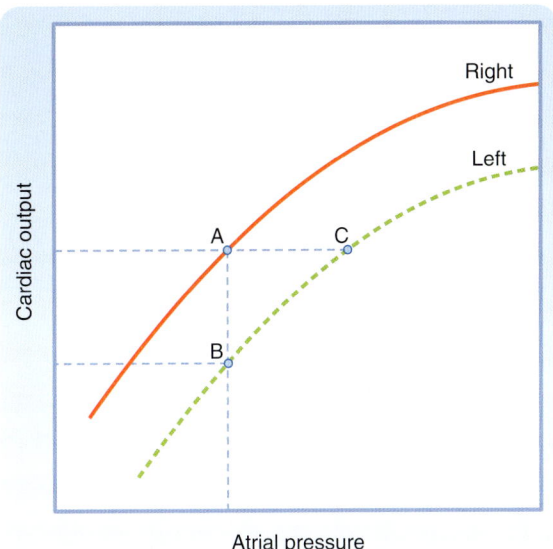

• Fig. 18.13 Relationships Between the Output of the Right and Left Ventricles and Mean Pressure in the Right and Left Atria, Respectively. At any given level of cardiac output, mean left atrial pressure (e.g., point C) exceeds mean right atrial pressure (point A).

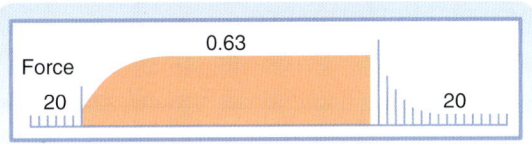

• Fig. 18.14 Changes in development of force (y-axis) in an isolated papillary muscle as the interval between contractions is varied from 20 seconds to 0.63 second and then back to 20 seconds. (Redrawn from Koch-Weser J, Blinks JR. *Pharmacol Rev.* 1963;15:601.)

Rate-Induced Regulation

Myocardial performance is also regulated by changes in the frequency at which the myocardial fibers contract. The effects of changes in contraction frequency on the force developed in an isometrically contracting papillary muscle are shown in Fig. 18.14. Initially, the cardiac muscle is stimulated to contract once every 20 seconds. When the muscle is suddenly made to contract once every 0.63 seconds, the force developed increases progressively over the next several beats. At the new steady state, the force developed is more than five times greater than the force at the larger contraction interval. A return to the larger interval (20 seconds) has the opposite influence on the development of force.

The rise in the force developed when the contraction interval is decreased is caused by a gradual increase in intracellular [Ca^{++}]. Two mechanisms contribute to the rise in intracellular [Ca^{++}]: an increase in the number of depolarizations per minute and an increase in the inward Ca^{++} current per depolarization.

In the first mechanism, Ca^{++} enters the myocardial cell during each action potential plateau (see Chapters 13 and 16). As the interval between beats is diminished, the number of plateaus per minute increases. Although the duration of

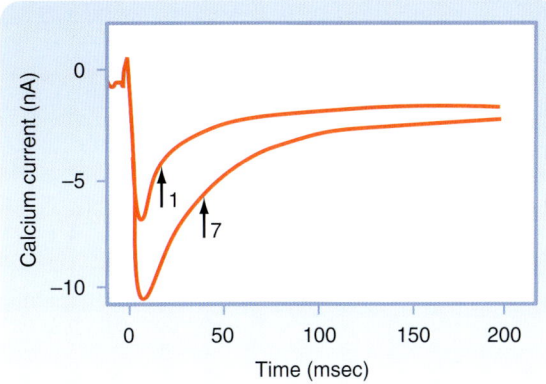

• Fig. 18.15 Calcium Currents Induced in a Myocyte During the First (Labeled *1*) and Seventh (Labeled *7*) Depolarizations in a Consecutive Sequence of Depolarizations. The *arrows* indicate the half-times of inactivation of the calcium current, as obtained from kinetic analysis. By the seventh depolarization, the maximal inward rectifying calcium current had increased by more than 50%, and the half-time of inactivation had increased by 20 msec. (Modified from Lee KS. *Proc Natl Acad Sci U S A.* 1987;84:3941.)

each action potential (and of each plateau) decreases as the interval between beats is reduced, the overriding effect of the increased number of plateaus per minute on the influx of Ca^{++} prevails, and intracellular [Ca^{++}] increases.

In the second mechanism, as the interval between beats is suddenly diminished, the inward rectifying calcium current (i_{Ca}) progressively increases with each successive beat until a new steady state is attained at the new basic cycle length. In an isolated ventricular myocyte, influx of Ca^{++} into the myocyte increases on successive depolarizations (Fig. 18.15). Both the increased magnitude and the slowed inactivation of i_{Ca} result in greater Ca^{++} influx into the myocyte during the later depolarizations than during the first depolarization. This greater Ca^{++} influx strengthens contraction.

Transient changes in the intervals between beats also profoundly affect the strength of contraction. When the left ventricle contracts prematurely (Fig. 18.16, beat A), the premature contraction (extrasystole) itself is weak, whereas contraction B (postextrasystolic contraction) after the compensatory pause is very strong. In the intact circulatory system, this response depends partly on the Frank-Starling mechanism. Inadequate time for ventricular filling just before the premature beat results in the weak premature contraction. Subsequently, the exaggerated degree of filling associated with the long compensatory pause (see Fig. 18.16, beat B) contributes to the vigorous postextrasystolic contraction.

The weakness of the premature beat is directly related to its degree of prematurity: The earlier the premature beat, the weaker its force of contraction. The curve that represents strength of contraction of a premature beat in relation to the coupling interval is called a **mechanical restitution curve.** Fig. 18.17 shows the restitution curve obtained when the coupling intervals of test beats were isolated in an isolated ventricular muscle preparation.

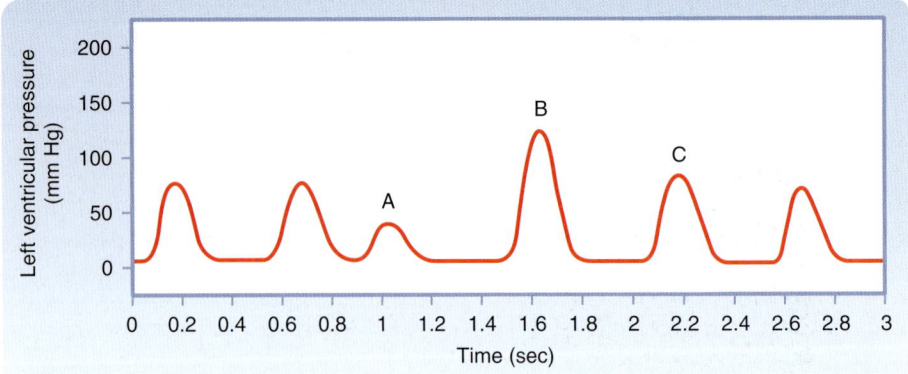

• **Fig. 18.16** In an isovolumic left ventricle preparation, a premature ventricular systolic contraction (beat A) is typically weak, whereas the postextrasystolic contraction (beat B) is characteristically strong, and the enhanced contractility may diminish over a few contractions (e.g., beat C). (From Levy MN. Unpublished tracing.)

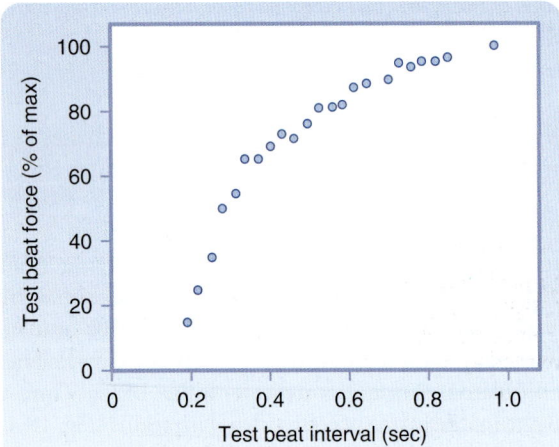

• **Fig. 18.17** Force Generated During Premature Contractions in an Isolated Ventricular Muscle Preparation. The muscle was stimulated to contract once per second. Periodically, the muscle was stimulated prematurely. The scale along the x-axis denotes the time between the driven and the premature beat. The scale along the y-axis denotes the ratio of the contractile force of the premature beat to that of the driven beat. (Modified from Seed WA, Walker JM. *Cardiovasc Res.* 1988;22:303.)

Restitution of the force of contraction depends on the time course of the intracellular circulation of Ca^{++} in cardiac myocytes during contraction and relaxation. During relaxation, the Ca^{++} that dissociates from the contractile proteins is taken up by the sarcoplasmic reticulum for subsequent release. However, there is a lag of approximately 500 to 800 msec before this Ca^{++} is available for release from the sarcoplasmic reticulum in response to the next depolarization. Thus the strength of the premature beat is reduced because the time during the preceding relaxation is insufficient to allow much of the Ca^{++} taken up by the sarcoplasmic reticulum to become available for release during the premature beat. Conversely, the postextrasystolic beat is considerably stronger than normal because more Ca^{++} is released from the sarcoplasmic reticulum as a result of the

relatively large amount of Ca^{++} taken up by it during the time that had elapsed from the end of the last regular beat until the beginning of the postextrasystolic beat.

Extrinsic Regulation of Myocardial Performance

Although a completely isolated heart can adapt well to changes in preload and afterload, various extrinsic factors also influence the heart in an individual. Often, these extrinsic regulatory mechanisms may overwhelm the intrinsic mechanisms. The extrinsic regulatory factors may be subdivided into nervous and chemical components.

Nervous Control

Sympathetic Influences

Sympathetic nervous activity enhances atrial and ventricular contractility. The alterations in ventricular contraction evoked by electrical stimulation of the left stellate ganglion in an isovolumic left ventricle preparation are shown in Fig. 18.18. Note that the duration of systole is reduced and the rate of ventricular relaxation is increased during the early phases of diastole; both these effects assist ventricular filling. For any given cardiac cycle length, the abbreviated systole allows more time for diastole and hence for ventricular filling.

Sympathetic nervous activity also enhances myocardial performance by altering intracellular Ca^{++} dynamics (see Chapter 16). Neurally released norepinephrine or circulating catecholamines interact with β-adrenergic receptors on the cardiac cell membranes (Fig. 18.19). This interaction activates adenylate cyclase, which raises intracellular levels of cAMP (see Chapter 3). Consequently, protein kinases that promote the phosphorylation of various proteins are activated within the myocardial cells. Phosphorylation of phospholamban facilitates reuptake of Ca^{++} by the sarcoplasmic reticulum, and phosphorylation of troponin I reduces the sensitivity of contractile proteins to Ca^{++}. These effects facilitate relaxation and reduce end-diastolic pressure (see Chapter 19). Phosphorylation of specific sarcolemmal proteins also activates calcium channels in the membranes of myocardial cells.

Activation of calcium channels increases the influx of Ca^{++} during the action potential plateau, and more Ca^{++} is released from the sarcoplasmic reticulum in response to each cardiac excitation. The contractile strength of the heart is thereby increased. Fig. 18.20 shows the correlation between the contractile force in a thin strip of ventricular muscle and the free $[Ca^{++}]$ (indicated by the aequorin light signal) in the myoplasm as the concentration of isoproterenol (a β-adrenergic agonist) is increased (see also Fig. 13.3).

The overall effect of increased cardiac sympathetic activity in intact animals can best be appreciated in terms of families of ventricular function curves. When the frequency of electrical stimulation applied to the left stellate ganglion increases, the ventricular function curves shift progressively to the left. The changes parallel those produced by infusions of norepinephrine (see Fig. 18.12). Hence, for any given left

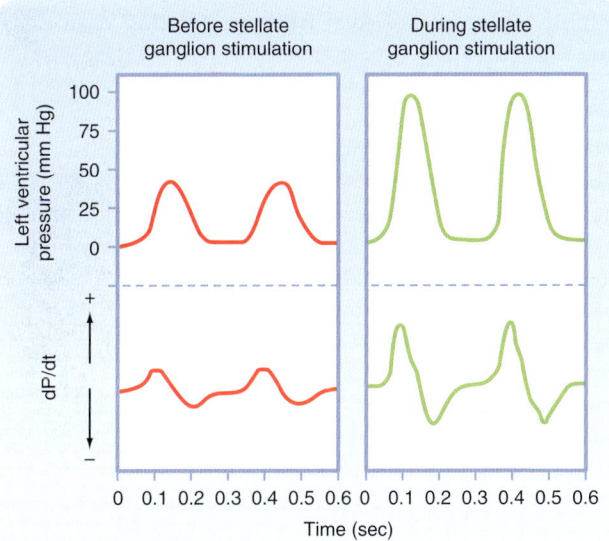

Fig. 18.18 In an isovolumic left ventricle preparation, stimulation of cardiac sympathetic nerves evokes a substantial rise in peak left ventricular pressure and in the maximal rates of rise and fall in intraventricular pressure (dP/dt). (From Levy MN. Unpublished tracing.)

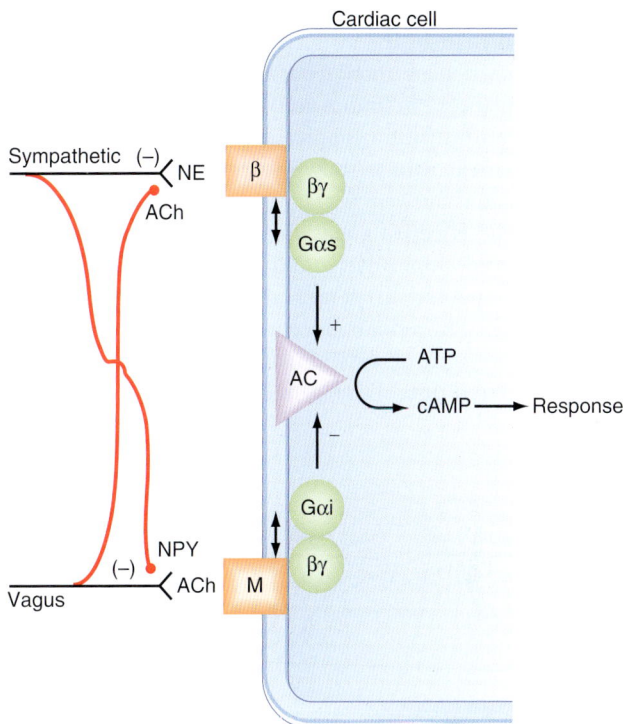

Fig. 18.19 Interneuronal and Intracellular Mechanisms Responsible for Interactions Between the Sympathetic and Parasympathetic Systems in the Neural Control of Cardiac Function. AC, adenylate cyclase; ACh, acetylcholine; ATP, adenosine triphosphate; β, β-adrenergic receptor; βγ, beta/gamma; cAMP, cyclic adenosine monophosphate; Gs and Gi, stimulatory and inhibitory G proteins; M, muscarinic receptor; NE, norepinephrine; NPY, neuropeptide Y. (Modified from Levy MN. In: Kulbertus HE, Franck G, eds. *Neurocardiology.* Mt. Kisco, NY: Futura; 1988.)

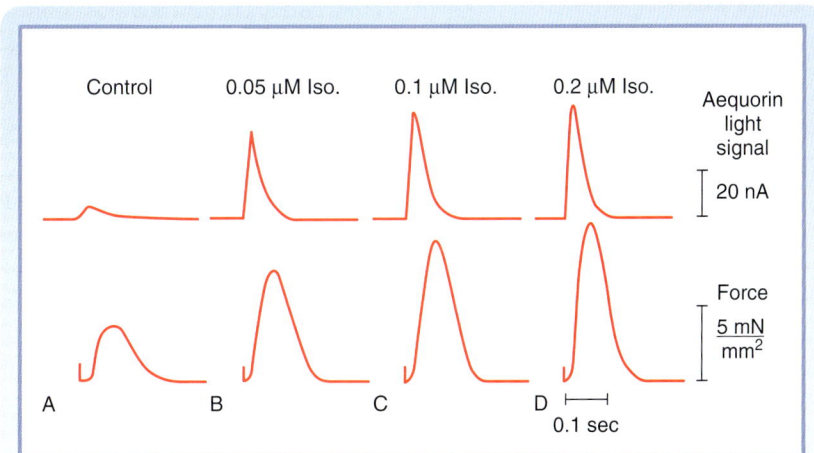

Fig. 18.20 Effects of Various Concentrations of Isoproterenol (Iso) on the Aequorin Light Signal (in Nanoamperes) and Contractile Force (in Millinewtons Per Square Millimeter) in a Rat Ventricular Muscle Injected With Aequorin. The aequorin light signal reflects the instantaneous changes in intracellular $[Ca^{++}]$. μM, micromolar. (Modified from Kurihara S, Konishi M. *Pflügers Arch.* 1987;409:427.)

ventricular end-diastolic pressure, the ventricle can perform more work as sympathetic nervous activity is increased.

Parasympathetic Influences

The vagus nerves inhibit the cardiac pacemaker, atrial myocardium, and AV conduction tissue. The vagus nerves also depress the ventricular myocardium, but the effects are less pronounced than in the atria. In pumping heart preparations, the ventricular function curve shifts to the right during vagal stimulation, an indication of reduced contractility. Vagus nerve stimulation suppresses the contractile force in the left ventricle. This is shown in pressure-volume curves obtained at constant ventricular rate (Fig. 18.21). The negative inotropic effect of vagus nerve stimulation, depicted as the reduced slope of the end-systolic pressure-volume relation, is opposed by a muscarinic receptor antagonist and diminished by a β-adrenoceptor antagonist. The results indicate that vagus nerve stimulation reduces contractility in the heart and does so by at least two pathways.

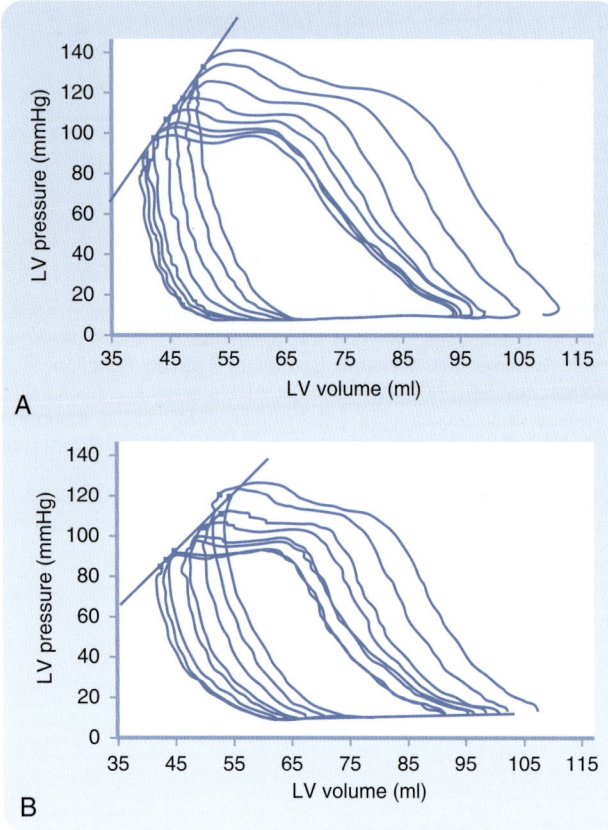

A

B

● **Fig. 18.21** Vagus Nerve Stimulation Reduces Ventricular Contractility. Pressure-volume curves obtained at constant ventricular rate before open heart surgery in human. **A,** Control pressure-volume curves were calculated during occlusion of the inferior vena cava. The end-systolic pressure-volume relation, defined by the slope of the *straight line,* measured approximately 4 mm Hg/mL. **B,** During stimulation of the left vagus nerve, the slope of the end-systolic pressure-volume relation decreased to approximately 3 mm Hg/mL, an indication that contractility had decreased. LV, left ventricular. (Redrawn from Lewis ME, Al-Khalidi AH, Bonser RS, et al. Vagus nerve stimulation decreases left ventricular contractility in vivo in the human and pig heart. *J Physiol.* 2001;534:547.)

Mechanisms for the vagal effects on the ventricular myocardium are shown in Fig. 18.19. ACh released by the vagal nerves interact with muscarinic receptors on the cardiac ventricular cell membrane to inhibit adenylate cyclase and the cAMP/protein kinase A cascade. This direct inhibition diminishes the Ca^{++} conductance of the cell membrane, reduces phosphorylation of the calcium channel, and hence decreases myocardial contractility. The ACh released from vagal nerves can also inhibit norepinephrine release by activating muscarinic receptors on neighboring sympathetic nerves, a mechanism of indirect inhibition. Thus vagal activity can decrease ventricular contractility partly by antagonizing any stimulatory effects that concomitant sympathetic activity may be exerting on ventricular contractility. Similarly, sympathetic nerves release norepinephrine and certain neuropeptides, including neuropeptide Y, which inhibits the release of ACh from neighboring vagal fibers (see Fig. 18.19).

Chemical Control

Adrenomedullary Hormones

The adrenal medulla is essentially a component of the autonomic nervous system (see Chapters 11 and 43). The principal hormone secreted by the adrenal medulla is epinephrine; some norepinephrine is also released. The rate of secretion of these catecholamines by the adrenal medulla is regulated by mechanisms that control the activity of the sympathetic nervous system. Concentrations of catecholamines in blood thus rise under the same conditions that activate the sympathetic nervous system. However, the cardiovascular effects of circulating catecholamines are probably minimal under normal conditions. Moreover, the pronounced changes in myocardial contractility with exercise, for example, are mediated mainly by the norepinephrine released from cardiac sympathetic nerve fibers rather than by the catecholamines released from the adrenal medulla.

Adrenocortical Hormones

How adrenocortical steroids influence myocardial contractility is controversial. Cardiac muscle taken from adrenalectomized animals and placed in a tissue bath is more likely to fatigue in response to stimulation than is cardiac muscle obtained from normal animals. In some species, however, adrenocortical hormones enhance contractility. In addition, the glucocorticoid hydrocortisone potentiates the cardiotonic effects of catecholamines. This potentiation is mediated in part by the ability of adrenocortical steroids to inhibit the extraneuronal catecholamine uptake mechanisms.

Thyroid Hormones

Thyroid hormones enhance myocardial contractility. Rates of ATP hydrolysis and Ca^{++} uptake by the sarcoplasmic reticulum are increased in hyperthyroidism; the opposite effects occur in hypothyroidism. Thyroid hormones increase cardiac protein synthesis, and this response leads to cardiac hypertrophy. These hormones also affect the composition of myosin isoenzymes in cardiac muscle. By increasing isoenzymes with the greatest ATPase activity, thyroid hormones enhance myocardial contractility.

IN THE CLINIC

Cardiovascular problems are common in patients with adrenocortical insufficiency (Addison's disease). Blood volume tends to fall, which may lead to severe hypotension and cardiovascular collapse, the so-called addisonian crisis (see Chapter 43).

AT THE CELLULAR LEVEL

Thyroid hormone exerts its cardiac actions by two paths: genomic and nongenomic. The genomic route involves interaction of thyroxine (T_3) with nuclear receptors that regulate the transcription of T_3-responsive genes. In hyperthyroidism, messenger mRNA for cardiac myocyte proteins involved in regulating intracellular [Ca^{++}] (e.g., sarcoplasmic endoplasmic reticulum calcium-ATPase [SERCA], ryanodine channel) is increased, as are amounts of contractile proteins (e.g., myosin heavy chain, actin, troponin I). Consequently, the rates of contraction and relaxation increase as ATP hydrolysis and O_2 consumption increase. In the hyperthyroid state, the use of ATP is less efficient, and the fractional loss of heat is greater. If untreated, severe hyperthyroidism can result in heart failure.

IN THE CLINIC

Cardiac activity is depressed in patients with inadequate thyroid function (hypothyroidism). The converse is true in patients with overactive thyroid glands (hyperthyroidism). Characteristically, patients with hyperthyroidism exhibit tachycardia, high cardiac output, and arrhythmias such as atrial fibrillation. In such patients, sympathetic neural activity may be increased, or the sensitivity of the heart to such activity may be enhanced. Studies have shown that thyroid hormone increases the density of β-adrenergic receptors in cardiac tissue (see also Chapter 42). In experimental animals, the cardiovascular manifestations of hyperthyroidism may be simulated by the administration of excess thyroxine.

increases the body's metabolic rate, which in turn results in arteriolar vasodilation. The consequent reduction in total peripheral resistance increases cardiac output, as explained in Chapter 19.

Insulin
Insulin has a positive inotropic effect on the heart. The effect of insulin is evident even when hypoglycemia is prevented by glucose infusions and when β-adrenergic receptors are blocked. Indeed, the positive inotropic effect of insulin is potentiated by β-adrenergic receptor antagonists. The enhanced contractility cannot be explained satisfactorily by the concomitant augmentation of glucose transport into myocardial cells.

Glucagon
Glucagon has potent positive inotropic and chronotropic effects on the heart. This endogenous hormone is probably not important in normal regulation of the cardiovascular system, but it has been used clinically to enhance cardiac performance. The effects of glucagon on the heart and certain metabolic effects are similar to those of catecholamines. Both glucagon and catecholamines activate adenylate cyclase to increase myocardial levels of cAMP. The catecholamines activate adenylate cyclase by interacting with β-adrenergic receptors, but glucagon activates this enzyme by a different mechanism. Nevertheless, the rise in cAMP increases influx of Ca^{++} through calcium channels in the sarcolemma and facilitates release and reuptake of Ca^{++} by the sarcoplasmic reticulum, just as catecholamines do.

Anterior Pituitary Hormones
The cardiovascular derangements in hypopituitarism are related principally to the associated deficiencies in adrenocortical and thyroid function. Growth hormone affects the myocardium, at least in combination with thyroxine. In hypophysectomized animals, growth hormone alone has little effect on the depressed heart, whereas thyroxine by itself restores adequate cardiac performance under basal conditions. However, when blood volume or peripheral resistance is increased, thyroxine alone does not restore adequate cardiac function, but the combination of growth hormone and thyroxine reestablishes normal cardiac performance. In certain animal models of heart failure, administration of growth hormone alone increases cardiac output and myocardial contractility.

Blood Gases
Changes in cardiac performance as a result of stimulation of central and peripheral chemoreceptors have been described previously in this chapter. These effects usually predominate. However, O_2 and CO_2 do have direct effects on the myocardium.

Oxygen
Hypoxia has a biphasic effect on myocardial performance. Mild hypoxia stimulates performance, but more severe hypoxia depresses performance because oxidative metabolism is limited.

Carbon Dioxide and Acidosis
An increase in partial pressure of carbon dioxide (PCO_2)—which results in a decrease in pH—has a direct depressant effect on the heart. This effect is mediated by changes in intracellular pH. A reduction in intracellular pH, induced by an increase in PCO_2, diminishes the amount of Ca^{++}

released from the sarcoplasmic reticulum in response to excitation. The diminished pH also decreases the sensitivity of the myofilaments to Ca^{++}. Increases in intracellular pH have the opposite effect; that is, they enhance sensitivity to Ca^{++}.

Regulation of the Peripheral Circulation

The peripheral circulation is essentially under dual control: centrally through the nervous system and locally by conditions in tissues surrounding the blood vessels. Nervous and humoral regulation of vascular smooth muscle is described in Chapter 14 (see Fig. 14.7 and Table 14.1), in which transmitters, hormones and their receptors are discussed. Aspects of local control are discussed in Chapter 17, in which the relative importance of these two control mechanisms is shown to vary in different tissues.

The arterioles are involved in regulating the rate of blood flow throughout the body. These vessels offer the greatest resistance to the flow of blood pumped to the tissues by the heart, and thus these vessels are important in the maintenance of arterial blood pressure. The arteriole walls are composed in large part of smooth muscle fibers that allow the diameter of the vessel lumen to vary (see Fig. 15.2). When this smooth muscle contracts strongly, the endothelial lining folds inward and completely obliterates the vessel lumen. When the smooth muscle is completely relaxed, the vessel lumen is maximally dilated. Some resistance vessels are closed at any given time. In addition, the smooth muscle in these vessels is partially contracted (which accounts for the tone of these vessels). If all the resistance vessels in the body dilated simultaneously, arterial blood pressure would fall precipitously.

Vascular smooth muscle controls total peripheral resistance, arterial and venous tone, and the distribution of blood flow throughout the body. The properties of vascular smooth muscle are discussed in Chapter 14. In the following sections, intrinsic and extrinsic control of vascular smooth muscle tone, and thus perfusion of peripheral tissues, is reviewed.

Intrinsic or Local Control of Peripheral Blood Flow

Autoregulation and Myogenic Regulation

In certain tissues, blood flow is adjusted to the existing metabolic activity of the tissue. Furthermore, when tissue metabolism is steady, changes in perfusion pressure (arterial blood pressure) evoke changes in vascular resistance that tend to maintain a constant blood flow. This myogenic mechanism, which is illustrated graphically in Fig. 18.22, is commonly referred to as **autoregulation of blood flow.** When pressure is abruptly increased or decreased from a control pressure of 100 mm Hg, flow increases or decreases, respectively. However, even with pressure maintained at its new level, blood flow returns toward the control level within 30 to 60 seconds.

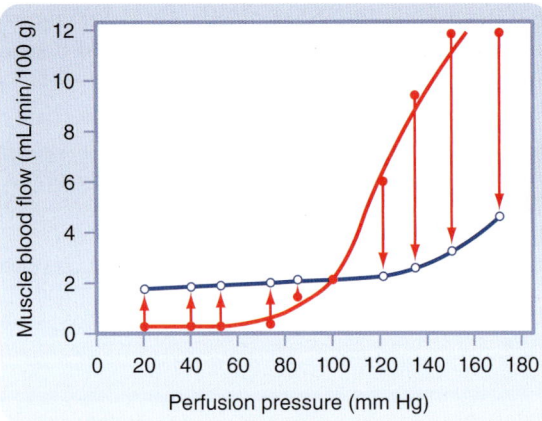

• **Fig. 18.22** Pressure-flow relationship in the vascular bed of the skeletal muscle. *Filled (red) circles* represent the flow rates obtained immediately after abrupt changes in perfusion pressure from the control level (the point where lines cross). *Open (blue) circles* represent the steady-state flow rates obtained at the new perfusion pressure. (Redrawn from Jones RD, Berne RM. *Circ Res.* 1964;14:126.)

Over the pressure range of 20 to 120 mm Hg, the steady-state flow is relatively constant. Calculation of hydraulic resistance (pressure/flow) across the vascular bed during steady-state conditions shows that the resistance vessels constrict with an elevation in perfusion pressure but dilate with a reduction in perfusion pressure. This response to perfusion pressure is independent of the endothelium because it is identical in intact vessels and in vessels that have been stripped of their endothelium. According to the myogenic mechanism, vascular smooth muscle contracts in response to an increase in the pressure difference across the wall of a blood vessel (transmural pressure), and it relaxes in response to a decrease in transmural pressure. The signaling mechanisms that allow distention of a vessel to elicit contraction are unknown. However, because stretch of vascular smooth muscle has been shown to raise intracellular $[Ca^{++}]$, an increase in transmural pressure is believed to activate membrane calcium channels.

🧬 AT THE CELLULAR LEVEL

Transient receptor potential (TRP) channels have been implicated in the myogenic mechanism. These channels are mammalian homologues of a *Drosophila melanogaster* gene that, when mutated, allows only a transient response to a sustained light stimulus. The pressure-induced vasoconstrictive response of an artery (myogenic response) appears to have the following signal path: pressure → increased phospholipase C activity → synthesis of diacylglycerol → activation of TRP channel → smooth muscle depolarization and opening of L-type calcium channels that increase intracellular $[Ca^{++}]$ and muscle tone. This is a means of regulating vascular resistance. Other TRP channel types have been proposed to participate in chronic hypoxic pulmonary hypertension and in the vasoconstriction caused by the α-adrenergic agonist norepinephrine.

In normal subjects, blood pressure is maintained at a fairly constant level via the baroreceptor reflex. Hence, the myogenic mechanism may play little role in regulating blood flow to tissues under normal conditions. However, when a person changes from a lying to a standing position, transmural pressure rises in the lower extremities, and the precapillary vessels constrict in response to this imposed stretch.

Endothelium-Mediated Regulation

As described in Chapter 17, the endothelium lining the vasculature produces a number of substances that can relax (e.g., nitric oxide) or contract (e.g., angiotensin II and endothelin) vascular smooth muscle. Thus the endothelium plays an important role in regulating blood flow to specific vascular beds.

Metabolic Regulation

The metabolic activity of a tissue governs blood flow in that tissue. Any intervention that results in inadequate O_2 supply prompts the formation of vasodilator metabolites that are released from the tissue and act locally to dilate the resistance vessels. When the metabolic rate of the tissue increases, or when O_2 delivery to the tissue decreases, more vasodilator substances are released (see Chapter 17).

Candidate Vasodilator Substances

Potassium, inorganic phosphate ions, and interstitial fluid osmolarity induce vasodilation. During skeletal muscle contraction, both (1) K^+ and phosphate are released and (2) osmolarity is increased. Therefore, these factors may contribute to active hyperemia (increase in blood flow caused by enhanced tissue activity). However, significant increases in the phosphate concentration and in osmolarity are not always observed during muscle contraction, and they may increase blood flow only transiently. Therefore, they probably do not mediate the vasodilation observed during muscular activity.

Potassium is released at the onset of skeletal muscle contraction or with an increase in cardiac muscle activity. Hence, release of K^+ could underlie the initial decrease in vascular resistance observed in response to physical exercise or to increased cardiac work. However, release of K^+ is not sustained, but arteriolar dilation persists throughout the period of enhanced muscle activity. Furthermore, reoxygenated venous blood obtained from active cardiac and skeletal muscles does not elicit vasodilation when the blood is infused into a test vascular bed. It is unlikely that oxygenation of venous blood alters its K^+ or phosphate content or its osmolarity and thereby neutralizes its vasodilator effect. Therefore, some agent other than K^+ must mediate the vasodilation associated with metabolic activity of the tissue.

Adenosine, which contributes to the regulation of coronary blood flow, may also participate in control of the resistance vessels in skeletal muscle. In addition, some prostaglandins may be important vasodilator mediators in certain vascular beds. Many prostaglandins have thus been proposed as mediators of metabolic vasodilation, and the relative contribution of each remains to be determined.

Basal Vessel Tone

Metabolic control of vascular resistance by the release of a vasodilator substance requires the existence of a basal vessel tone. Tonic activity in vascular smooth muscle is readily demonstrable, but in contrast to tone in skeletal muscle, the tone in vascular smooth muscle is independent of the nervous system. Thus some metabolic factor must be responsible for maintaining this tone. The following factors may be involved: (1) the myogenic response to the stretch imposed by blood pressure, (2) the high partial pressure of oxygen in arterial blood (PaO_2), or (3) the presence of Ca^{++}.

Reactive Hyperemia

If arterial inflow to a vascular bed is stopped temporarily, blood flow on release of the occlusion immediately exceeds the flow that prevailed before occlusion, and the flow gradually returns to the control level. This increase in blood flow is called *reactive hyperemia*. This type of event provides evidence for the existence of a local metabolic factor that regulates tissue blood flow.

In the experiment shown in Fig. 18.23, blood flow to the leg was stopped by clamping of the femoral artery for 15, 30, and 60 seconds. Release of the 60-second occlusion resulted in a peak blood flow that was 70% greater than the control flow, and the flow returned to the control level within 110 seconds.

Within limits, peak flow and particularly the duration of reactive hyperemia are proportional to the duration of the occlusion (see Fig. 18.23). If the extremity is exercised

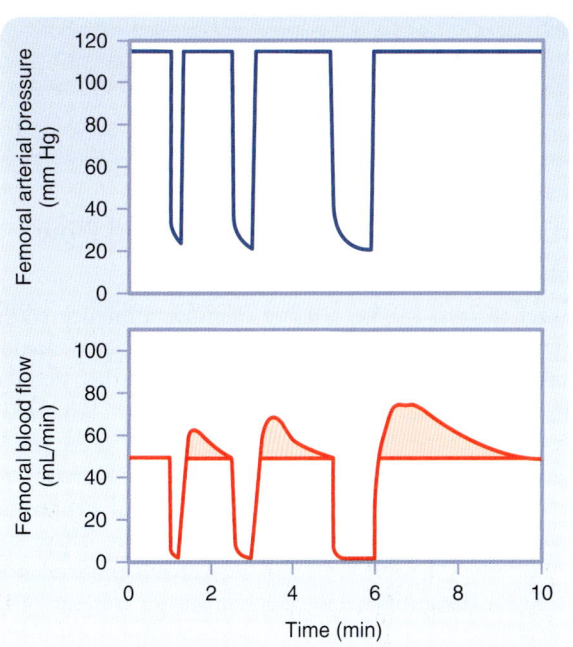

• **Fig. 18.23** Graphs of reactive hyperemia in the hind limb of the leg after 15-, 30-, and 60-second occlusion of the femoral artery. (From Berne RM. Unpublished observations.)

during the occlusion period, reactive hyperemia is increased. These observations and the close relationship between metabolic activity and blood flow in an unoccluded limb are consistent with the notion of a metabolic mechanism in the local regulation of tissue blood flow.

Coordination of Arterial and Arteriolar Dilation

When the vascular smooth muscle of arterioles relaxes in response to vasodilator metabolites whose release is caused by a decrease in the ratio of O_2 supply to O_2 demand of the tissue, resistance may diminish concomitantly in the small upstream arteries that feed these arterioles. The result is blood flow greater than that produced by arteriolar dilation alone. There are two possible mechanisms for this coordination of arterial and arteriolar dilation. First, the vasodilation in the microvessels may be propagated, and when dilation is initiated in the arterioles, it can propagate along the vessels from the arterioles back to the small arteries. Second, the metabolite-mediated dilation of the arterioles accelerates blood flow in the feeder arteries. This greater blood flow velocity increases the shear stress on the arterial endothelium, which in turn can induce flow-mediated vasodilation by release of one or more vasodilators (e.g., nitric oxide, prostacyclin, H_2O_2, epoxyeicosatrienoic acid).

IN THE CLINIC

In the legs, disease of the arterial walls can lead to obstruction of the arteries and symptoms, a condition called *intermittent claudication*. The symptoms consist of leg pain when the person walks or climbs stairs, and the pain is relieved by rest. The disease is called *thromboangiitis obliterans,* and it appears most frequently in men who are smokers. With minimal walking, the resistance vessels become maximally dilated by local release of metabolites; when the O_2 demand of the muscles increases with more rapid walking, blood flow cannot increase sufficiently to meet the muscle needs for O_2, and pain caused by muscle ischemia results.

Extrinsic Control of Peripheral Blood Flow
Sympathetic Neural Vasoconstriction

Several regions in the cerebral medulla influence cardiovascular activity. Stimulation of the dorsal lateral medulla (pressor region) evokes vasoconstriction, cardiac acceleration, and enhanced myocardial contractility. Stimulation of cerebral centers caudal and ventromedial to the pressor region decreases arterial blood pressure. This depressor area exerts its effect by direct inhibition of spinal regions and by inhibition of the medullary pressor region. These areas are not true anatomical centers in which a discrete group of cells is discernible, but they constitute "physiological" centers.

The cerebrospinal vasoconstrictor regions are tonically active. Reflexes or humoral stimuli that enhance this activity increase the frequency of impulses that reach the terminal neural branches to the vessels. A constrictor neurohumor (norepinephrine) is released at the terminals to elicit a constrictive α-adrenergic effect on the resistance vessels. Inhibition of the vasoconstrictor areas diminishes the impulse frequency in the efferent nerve fibers, and vasodilation results. Thus neural regulation of the peripheral circulation is achieved mainly by alteration in the impulse frequency in the sympathetic nerves to the blood vessels. Surgical section of the sympathetic nerves to an extremity abolishes sympathetic vascular tone and thereby increases blood flow to that limb. With time, vascular tone is regained by an increase in basal (intrinsic) tone.

Both the pressor and depressor regions may undergo rhythmic changes in tonic activity that are manifested as oscillations in arterial pressure. Some rhythmic changes **(Traube-Hering waves)** occur at the frequency of respiration and are caused by a cyclic fluctuation in sympathetic impulses to the resistance vessels. Other fluctuations in sympathetic activity **(Mayer waves)** occur at a frequency lower than that of respiration.

Sympathetic Constrictor Influence on Resistance and Capacitance Vessels

Vasoconstrictor fibers of the sympathetic nervous system supply the arteries, arterioles, and veins; the neural influence is much less on larger vessels than on arterioles and small arteries. Capacitance vessels (veins) respond more to sympathetic nerve stimulation than do resistance vessels; the capacitance vessels are maximally constricted at a lower stimulation frequency than the are resistance vessels. However, capacitance vessels lack β-adrenergic receptors, and they respond less to vasodilator metabolites. Norepinephrine is the neurotransmitter released at the sympathetic nerve terminals in blood vessel. Factors such as circulating hormones and particularly locally released substances mediate the release of norepinephrine from the nerve terminals.

The response of the resistance and capacitance vessels to stimulation of sympathetic fibers is illustrated in Fig. 18.24. When arterial pressure is held constant, stimulation of sympathetic fibers reduces blood flow (constriction of resistance vessels) and decreases the blood volume of the tissue (constriction of capacitance vessels). Constriction of the resistance vessels establishes a new equilibrium of the forces responsible for filtration and absorption across the capillary wall (see Eq. 17.20).

In addition to active changes (contraction and relaxation of vascular smooth muscle) in vessel caliber, passive changes are also caused by alterations in intraluminal pressure. An increase in intraluminal pressure distends the vessels, and a decrease reduces the caliber of the vessels as a consequence of elastic recoil of the vessel walls.

At basal vascular tone, approximately a third of the blood volume of a tissue can be mobilized when the sympathetic nerves are stimulated at physiological frequencies. Basal tone is very low in capacitance vessels; if these vessels are denervated experimentally, the increases in volume evoked

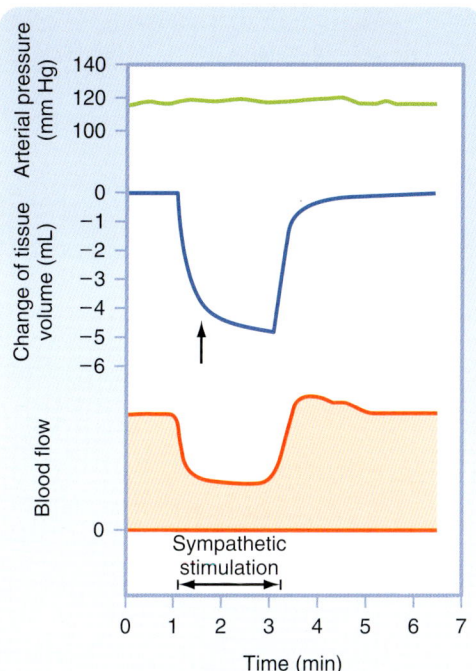

• **Fig. 18.24** Effect of Sympathetic Nerve Stimulation (2 Hz) on Blood Flow and Tissue Volume of the Lower Limb. The *upward arrow* denotes the change in slope of the tissue volume curve at the point at which the decrease in volume caused by emptying of capacitance vessels ceases and loss of extravascular fluid becomes evident. The abrupt decrease in tissue volume is caused by movement of blood out of the capacitance vessels and out of the lower limb. The late, slow, progressive decline in volume (to the right of the arrow) is caused by the movement of extravascular fluid into the capillaries and hence away from the tissue. The loss of tissue fluid results from the lowering of the capillary hydrostatic pressure secondary to constriction of the resistance vessels. (From Mellander S. *Acta Physiol Scand Suppl.* 1960;50[176]:1.)

by maximal doses of ACh are small. Therefore, at basal vascular tone, blood volume is close to the maximal blood volume of the tissue. More blood can be mobilized from the capacitance vessels in the skin than from those in the muscle. This disparity depends in part on the greater sensitivity of the skin vessels to sympathetic stimulation, but it also occurs because basal tone is lower in skin vessels than in muscle vessels. Therefore, in the absence of a neural influence, skin capacitance vessels contain more blood than do muscle capacitance vessels.

Physiological stimuli mobilize blood from capacitance vessels. For example, during physical exercise, activation of sympathetic nerve fibers constricts the peripheral veins and hence augments cardiac filling pressure. In arterial hypotension (as occurs in hemorrhage), the capacitance vessels constrict and thereby correct the decreased central venous pressure associated with blood loss.

Parasympathetic Neural Influence

The efferent fibers of the cranial division of the parasympathetic nervous system innervate the blood vessels of the head and some of the viscera, whereas fibers of the sacral division innervate blood vessels of the genitalia, bladder,

and large bowel. Skeletal muscle and skin do not receive parasympathetic innervation. The effect of cholinergic fibers on total vascular resistance is small because only a small proportion of the resistance vessels of the body receive parasympathetic fibers.

 IN THE CLINIC

In hemorrhagic shock, the resistance vessels constrict and thereby assist in the maintenance of normal arterial blood pressure. With arterial hypotension, the enhanced arteriolar constriction also leads to a small mobilization of blood from the tissue by virtue of recoil of the postarteriolar vessels when intraluminal pressure is reduced. Furthermore, extravascular fluid is mobilized because of greater fluid absorption into the capillaries in response to the lowered capillary hydrostatic pressure.

Stimulation of the parasympathetic fibers to the salivary glands induces marked vasodilation. A vasodilator polypeptide, bradykinin, formed locally from the action of an enzyme on a plasma protein substrate in the glandular lymphatic vessels, mediates this vasodilation. Bradykinin is formed in other exocrine glands, such as the lacrimal and sweat glands. Its presence in sweat may be partly responsible for the dilation of cutaneous blood vessels.

Humoral Factors

Epinephrine and norepinephrine exert a powerful effect on peripheral blood vessels. In skeletal muscle, low concentrations of epinephrine dilate resistance vessels (β_2-adrenergic effect), but high concentrations produce constriction (α_1-adrenergic effect), as noted in Table 14.1. In all vascular beds, the primary effect of norepinephrine is vasoconstriction. When stimulated, the adrenal gland can release epinephrine and norepinephrine into the systemic circulation. However, under physiological conditions, the effect of catecholamine release from the adrenal medulla is less important than norepinephrine release from sympathetic nerve endings.

Local humoral substances have an important role in regulating vessel tone (Table 14.1). Some are released from the endothelium (e.g., nitric oxide, endothelin, thromboxane A_2), whereas others are derived from perivascular tissues (e.g., histamine, adenosine, angiotensin II).

Vascular Reflexes

Areas of the cerebral medulla that mediate sympathetic and vagal effects are under the influence of neural impulses that originate in the baroreceptors, chemoreceptors, hypothalamus, cerebral cortex, and skin. These areas of the medulla are also affected by changes in the blood concentrations of CO_2 and O_2.

Arterial Baroreceptors

The baroreceptors (or pressoreceptors) are stretch receptors located in the carotid sinuses and in the aortic arch (Figs. 18.25 and 18.26). The carotid sinuses are the slightly

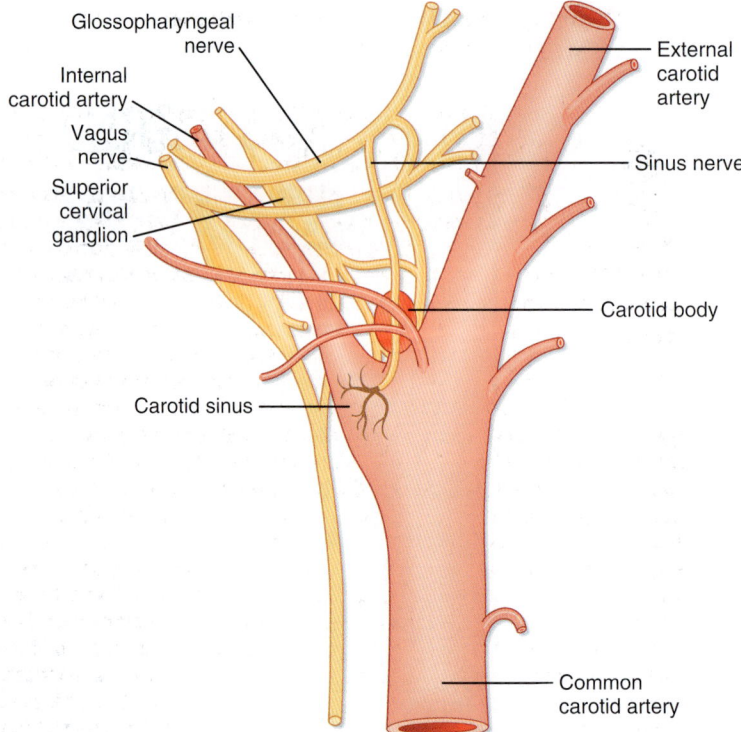

• **Fig. 18.25** Diagrammatic representation of the carotid sinus and carotid body and their innervation. (Redrawn from Adams WE. *The Comparative Morphology of the Carotid Body and Carotid Sinus.* Springfield, IL: Charles C Thomas; 1958.)

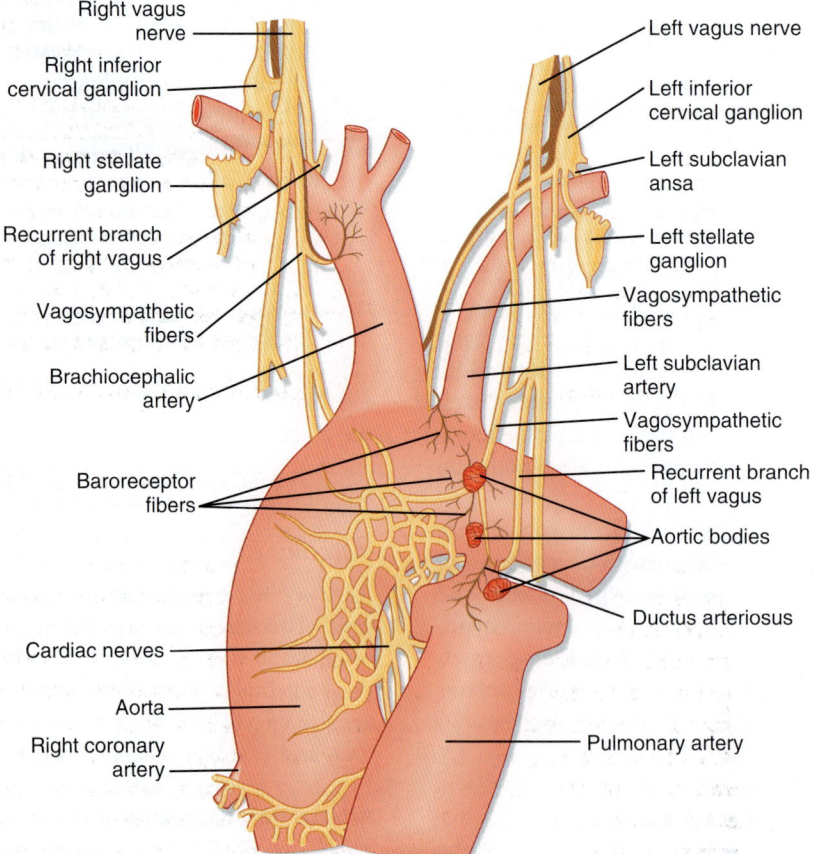

• **Fig. 18.26** Anterior view of the aortic arch showing the innervation of the aortic bodies and baroreceptors. (Modified from Nonidez JF. *Anat Rec.* 1937;69:299.)

widened areas at the origins of the internal carotid arteries. Impulses that arise in the carotid sinus travel up the carotid sinus nerve (nerve of Hering) to the glossopharyngeal nerve (cranial nerve IX) and, via the latter, to the nucleus of the tractus solitarius (NTS) in the medulla. The NTS is the site of the central projections of the chemoreceptors and baroreceptors. Stimulation of the NTS inhibits sympathetic nerve outflow to the peripheral blood vessels (depressor effect), whereas lesions of the NTS produce vasoconstriction (pressor effect). Impulses that arise in the aortic arch baroreceptors reach the NTS via afferent fibers in the vagus nerves.

Baroreceptor nerve terminals in the walls of the carotid sinus and aortic arch respond to the vascular stretch and deformation induced by changes in arterial blood pressure. The frequency of firing of these nerves is enhanced by an increase in arterial blood pressure and diminished by a reduction in arterial blood pressure. An increase in impulse frequency, as occurs with a rise in arterial pressure, inhibits the cerebral vasoconstrictor regions and results in peripheral vasodilation and lowering of arterial blood pressure. Bradycardia brought about by activation of the cardiac branches of the vagus nerves contributes to this lowering of blood pressure.

The carotid sinus baroreceptors are more sensitive than those in the aortic arch. Changes in carotid sinus pressure evoke greater changes in systemic arterial pressure and peripheral resistance than do equivalent changes in aortic arch pressure.

The receptors in the carotid sinus walls respond more to pulsatile pressure than to constant pressure. This is illustrated in Fig. 18.27, which shows that at normal levels of mean arterial blood pressure (≈100 mm Hg), a barrage of impulses from a single fiber of the sinus nerve is initiated in early systole by the pressure rise; only a few spikes occur during late systole and early diastole. At lower arterial pressure, these phasic changes are even more evident, but the overall discharge frequency is reduced. The blood pressure threshold for evoking sinus nerve impulses is approximately 50 mm Hg; maximal sustained firing is reached at approximately 200 mm Hg. Because the baroreceptors adapt, their response at any mean arterial pressure level is greater to a high pulse pressure than to a low pulse pressure.

The increases in resistance that occur in response to reduced pressure in the carotid sinus vary from one peripheral vascular bed to another. These variations allow blood flow to be redistributed. The resistance changes elicited by altering carotid sinus pressure are greatest in the femoral vessels, less in the renal vessels, and least in the mesenteric and celiac vessels.

In addition, the sensitivity of the carotid sinus reflex can be altered. Local application of norepinephrine or stimulation of sympathetic nerve fibers to the carotid sinuses enhances the sensitivity of its receptors in such a way that a given increase in intrasinus pressure produces a greater depressor response. Baroreceptor sensitivity decreases in hypertension because the carotid sinuses become stiffer as a result of the high intra-arterial pressure. Consequently, a given increase in carotid sinus pressure elicits a smaller decrease in systemic arterial pressure than it does at a normal level of blood pressure. Thus the set point of the baroreceptors is raised in hypertension in such a way that the threshold is increased and the pressure receptors are less sensitive to changes in transmural pressure. As would be expected, denervation of the carotid sinus can produce temporary and, in some instances, prolonged hypertension.

The arterial baroreceptors play a key role in short-term adjustments in blood pressure in response to relatively abrupt changes in blood volume, cardiac output, or peripheral resistance (as in exercise). However, long-term control of blood pressure—over a period of days or weeks—is determined by the fluid balance of the individual: namely, the balance between fluid intake and fluid output. By far, the most important organ in the control of body fluid volume, and hence blood pressure, is the kidney (see also Chapter 35).

Cardiopulmonary Baroreceptors

Cardiopulmonary receptors are located in the atria, ventricles, and pulmonary vessels. These baroreceptors are innervated by vagal and sympathetic afferent nerves. Cardiopulmonary reflexes are tonically active and can alter peripheral resistance in response to changes in intracardiac, venous, or pulmonary vascular pressure.

The atria contain two types of cardiopulmonary baroreceptors: those activated by the tension developed during atrial systole (type A receptors) and those activated by stretch of the atria during atrial diastole (type B receptors).

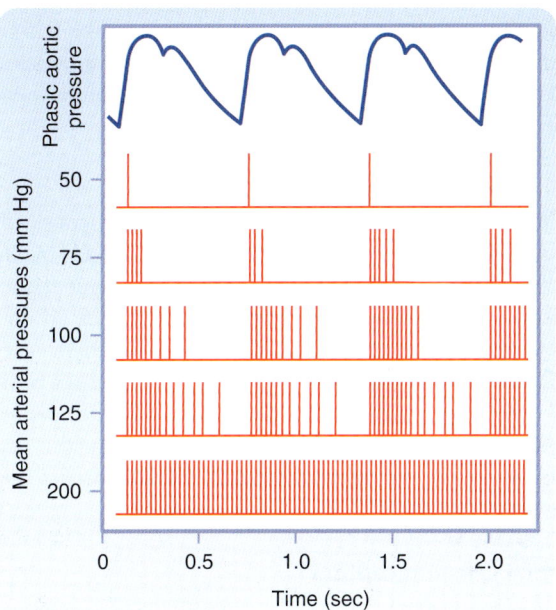

• **Fig. 18.27** Relationship of phasic aortic blood pressure in the firing of a single afferent nerve fiber from the carotid sinus at different levels of mean arterial pressure.

Stimulation of these atrial receptors sends impulses up vagal fibers to the vagal center in the medulla. Consequently, sympathetic activity is decreased to the kidney and increased to the sinus node. These changes in sympathetic activity increase renal blood flow, urine flow, and heart rate.

Activation of the cardiopulmonary receptors can also initiate a reflex that lowers arterial blood pressure by inhibiting the vasoconstrictor center in the cerebral medulla. Stimulation of the cardiopulmonary receptors inhibits release of angiotensin, aldosterone, and vasopressin (antidiuretic hormone); interruption of the reflex pathway has the opposite effects.

The role that activation of these baroreceptors plays in the regulation of blood volume is apparent in the body's responses to hemorrhage. The reduction in blood volume (hypovolemia) enhances sympathetic vasoconstriction in the kidney and increases the secretion of renin, angiotensin, aldosterone, and vasopressin (see also Chapter 35). The renal vasoconstriction (primarily afferent arterioles) reduces glomerular filtration and increases release of renin from the kidney. Renin acts on a plasma substrate to yield angiotensin II, which stimulates aldosterone secretion by the adrenal cortex. The enhanced release of vasopressin decreases renal water excretion, and the release of aldosterone decreases renal NaCl excretion. The kidneys retain salt and water, and hence blood volume increases. Angiotensin II (formed from angiotensin I by angiotensin-converting enzyme) also raises systemic arteriolar tone (see Table 14.1).

IN THE CLINIC

In some individuals, the carotid sinus is abnormally sensitive to external pressure. Hence, tight collars or other forms of external pressure over the region of the carotid sinus may elicit marked hypotension and fainting. Such hypersensitivity is known as the *carotid sinus syndrome*.

Peripheral Chemoreceptors

Peripheral chemoreceptors consist of small, highly vascular bodies in the region of the aortic arch (aortic bodies; see Fig. 18.26) and just medial to the carotid sinuses (carotid bodies; see Fig. 18.25). These vascular bodies are sensitive to changes in the PO_2, PCO_2, and pH of arterial blood. Although they primarily regulate respiration, they also influence the vasomotor regions. A reduction in PaO_2 stimulates the chemoreceptors. The increased activity in afferent nerve fibers from the carotid and aortic bodies stimulates the vasoconstrictor regions and thereby increases the tone of resistance and capacitance vessels.

The chemoreceptors are also stimulated by increased arterial blood PCO_2 ($PaCO_2$) and by reduced pH. However, the reflex effect is small in comparison to the direct effects of hypercapnia (high $PaCO_2$) and acidosis on the vasomotor regions in the medulla. When hypoxia and hypercapnia occur simultaneously, the effects of the chemoreceptors are

greater than the sum of the effects of each of the two stimuli when they act alone.

Chemoreceptors are also located in the heart. These cardiac chemoreceptors are activated by ischemia of cardiac muscle, and they transmit the precordial pain (angina pectoris) associated with an inadequate blood supply to the myocardium.

Hypothalamus

Optimal function of the cardiovascular reflexes requires integrity of the pontine and hypothalamic structures. Furthermore, these structures are responsible for behavioral and emotional control of the cardiovascular system (see also Chapter 11). Stimulation of the anterior hypothalamus produces both a fall in blood pressure and bradycardia, whereas stimulation of the posterolateral region of the hypothalamus increases both blood pressure and the heart rate. The hypothalamus also contains a temperature-regulating center that affects blood vessels in the skin. Stimulation by the application of cold to the skin or by cooling of the blood perfusing the hypothalamus results in constriction of the skin vessels and heat conservation, whereas warm stimuli to the skin result in cutaneous vasodilation and enhanced heat loss.

Cerebrum

The cerebral cortex also affects blood flow distribution in the body. Stimulation of the motor and premotor areas affects blood pressure; usually, a pressor response occurs. However, vasodilation and depressor responses may be evoked, as in blushing or fainting, in response to an emotional stimulus.

Skin and Viscera

Painful stimuli can elicit either pressor or depressor responses, depending on the magnitude and location of the stimulus. Distention of the viscera often evokes a depressor response, whereas painful stimuli to the body surface generally evoke a pressor response.

Pulmonary Reflexes

Inflation of the lungs initiates a reflex that induces systemic vasodilation and a decrease in arterial blood pressure. Conversely, collapse of the lungs evokes systemic vasoconstriction. Afferent fibers that mediate this reflex are in the vagus nerves and possibly also in the sympathetic nerves. Stimulation of these fibers by stretch of the lungs inhibits the vasomotor areas. The magnitude of the depressor response to lung inflation is directly related to the degree of inflation and to the existing level of vasoconstrictor tone (see also Chapter 22).

Central Chemoreceptors

Increases in PCO_2 stimulate chemosensitive regions of the medulla (the central chemoreceptors), and they elicit vasoconstriction and increased peripheral resistance. A reduction in PCO_2 to subnormal levels (in response to hyperventilation) decreases tonic activity in these areas in the medulla and thereby decreases peripheral resistance. The chemosensitive regions are also affected by changes in pH. Lowering of blood pH stimulates these cerebral areas, and a rise in blood pH inhibits them. These effects of changes in PCO_2 and blood pH may operate through changes in cerebrospinal fluid pH, as may also the respiratory center.

PaO_2 has little direct effect on the medullary vasomotor region. The primary effect of hypoxia is mediated by reflexes via the carotid and aortic chemoreceptors. A moderate reduction in PaO_2 stimulates the vasomotor region, but a severe reduction depresses vasomotor activity in the same manner by which other areas of the brain are depressed by very low O_2 tension.

Balance Between Extrinsic and Intrinsic Factors in Regulation of Peripheral Blood Flow

Dual control of peripheral vessels by intrinsic and extrinsic mechanisms evokes a number of important vascular adjustments. Such regulatory mechanisms enable the body to direct blood flow to areas where it is most needed and away from areas that have fewer requirements. In some tissues, the effects of the extrinsic and intrinsic mechanisms are fixed; in other tissues, the ratio is changeable and depends on the state of activity of that tissue.

 IN THE CLINIC

Cerebral ischemia, which may occur because of excessive pressure exerted by an expanding intracranial tumor, results in a marked increase in peripheral vasoconstriction. The stimulation is probably caused by local accumulation of CO_2 and a reduction in O_2 and possibly by excitation of intracranial baroreceptors. With prolonged, severe ischemia, central depression eventually supervenes, and blood pressure falls.

In the brain and heart, which are vital structures with limited tolerance for a reduced blood supply, intrinsic flow-regulating mechanisms are dominant. For instance, massive discharge of the vasoconstrictor region via the sympathetic nerves, which might occur in severe, acute hemorrhage, has negligible effects on the cerebral and cardiac resistance vessels, whereas the cutaneous, renal, and splanchnic blood vessels become greatly constricted.

In the skin, extrinsic vascular control is dominant. The cutaneous vessels not only participate strongly in a general vasoconstrictor discharge but also respond selectively via hypothalamic pathways to subserve the functions of heat loss and heat conservation required for regulation of body temperature. However, intrinsic control can be elicited by local temperature changes that modify or override the central influence on resistance and capacitance vessels (see also Chapter 17).

In skeletal muscle, the extrinsic and intrinsic mechanisms interact. In resting skeletal muscle, neural control (vasoconstrictor tone) is dominant, as can be demonstrated by the large increase in blood flow that occurs immediately after section of the sympathetic nerves to the tissue. After the onset of exercise, the intrinsic flow-regulating mechanism assumes control, and vasodilation occurs in the active muscles because of the local increase in metabolites. Vasoconstriction occurs in the inactive tissues as a manifestation of the general sympathetic discharge. However, constrictor impulses that reach the resistance vessels of the active muscles are overridden by the local metabolic effect. Operation of this dual control mechanism thus increases blood flow where it is required and shunts it away from relatively inactive areas (see also Chapter 17). Similar effects may be achieved in response to an increase in PCO_2. Normally, the hyperventilation associated with exercise keeps PCO_2 at normal levels. However, if PCO_2 is increased, generalized vasoconstriction would occur because CO_2 stimulates the medullary vasoconstrictor region. In active muscles, where $[CO_2]$ would be highest, the smooth muscle of the arterioles would relax in response to the local PCO_2. Factors that affect and are affected by the vasomotor region are summarized in Fig. 18.28.

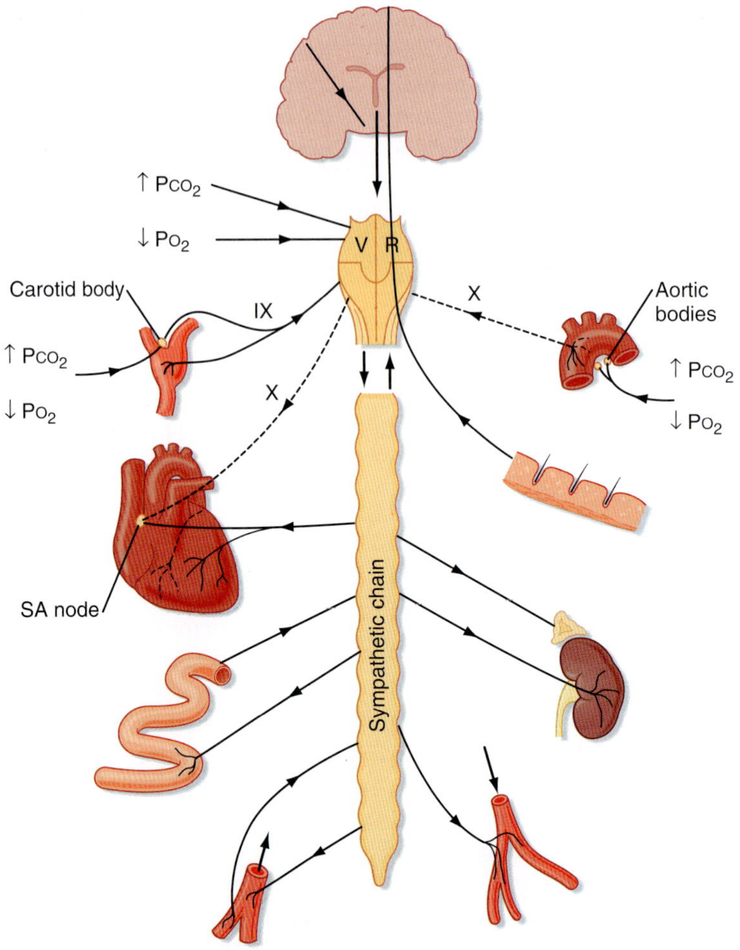

• **Fig. 18.28** Schematic Diagram Illustrating Neural Input and Output of the Vasomotor Region (VR). *Arrows* indicate direction of neural input and output. IX, glossopharyngeal nerve; ↑ P_{CO_2}, increased partial pressure of carbon dioxide; ↓ P_{CO_2}, increased P_{CO_2}; ↑ P_{O_2}, increased partial pressure of oxygen; ↓ P_{O_2}, increased P_{O_2}; SA, sinoatrial; X, vagus nerve.

Key Points

1. Cardiac function is regulated by a number of intrinsic and extrinsic mechanisms. The principal intrinsic mechanisms that regulate myocardial contraction are the Frank-Starling mechanism and rate-induced regulation.

2. The heart rate is regulated mainly by the autonomic nervous system. Sympathetic nervous activity increases the heart rate, whereas parasympathetic (vagal) activity decreases the heart rate. When both systems are active, the vagal effects usually dominate. The autonomic nervous system regulates myocardial performance mainly by varying the Ca^{++} conductance of the cell membrane via the adenylate cyclase system.

3. The reflexes that regulate the heart rate are the baroreceptor, chemoreceptor, pulmonary inflation, atrial receptor (Bainbridge), and ventricular receptor reflexes.

4. Certain hormones—such as epinephrine, adrenocortical steroids, thyroid hormones, insulin, glucagon, and anterior pituitary hormones—regulate myocardial performance. Changes in the arterial blood concentrations of O_2, CO_2, and H^+ alter cardiac function directly and, through the chemoreceptors, indirectly.

5. The arterioles (resistance vessels) regulate blood flow mainly through their downstream capillaries. Smooth muscle, which makes up most of the walls of arterioles, contracts and relaxes in response to neural and humoral stimuli. Neural regulation of blood flow is almost completely accomplished by the sympathetic nervous system. Sympathetic nerves to blood vessels are tonically active; inhibition of the vasoconstrictor center in the medulla reduces peripheral vascular resistance. Stimulation of the sympathetic nerves constricts the resistance and capacitance (veins) vessels. Parasympathetic fibers innervate the head, viscera, and genitalia; they do not innervate the skin and muscle.

6. Autoregulation of blood flow occurs in most tissues. This process is characterized by a constant blood flow in the presence of a change in perfusion pressure. Autoregulation is mediated by a myogenic mechanism whereby an increase in transmural pressure elicits a contraction of vascular smooth muscle and a decrease in transmural pressure elicits a relaxation.

7. The striking parallelism between tissue blood flow and tissue O_2 consumption indicates that blood flow is regulated largely by a metabolic mechanism. A decrease in the ratio of O_2 supply to O_2 demand of a tissue releases vasodilator metabolites that dilate arterioles and thereby enhance the O_2 supply.

8. The baroreceptors in the internal carotid arteries and aorta are tonically active and regulate blood pressure on a moment-to-moment basis. An increase in arterial pressure stretches these receptors to initiate a reflex that inhibits the medullary vasoconstrictor center and induces vasodilation. Conversely, a decrease in arterial pressure disinhibits the vasoconstrictor center and induces vasoconstriction. The baroreceptors in the internal carotid arteries predominate over those in the aorta, and they respond more vigorously to changes in pressure (stretch) than to elevated or reduced nonpulsatile pressure.

9. Peripheral chemoreceptors (carotid and aortic bodies) and central chemoreceptors in the medulla oblongata are stimulated by a decrease in blood PaO_2 and by an increase in blood PCO_2. Stimulation of these chemoreceptors increases the rate and depth of respiration, but it also produces peripheral vasoconstriction. Cardiopulmonary baroreceptors are also present in the cardiac chambers and large pulmonary vessels. They have less influence on blood pressure but do participate in regulation of blood volume.

10. Peripheral resistance and hence blood pressure are affected by stimuli that arise in the skin, viscera, lungs, and brain. The combined effect of neural and local metabolic factors distributes blood to active tissues and diverts it from inactive tissues. In vital structures, such as in the heart and brain and in contracting skeletal muscle, the metabolic factors predominate.

Additional Readings

Fukuda K, Kanazawa H, Aizawa Y, et al. Cardiac innervation and sudden cardiac death. *Circ Res*. 2015;116:2005.

Iancu RV, Jones SW, Harvey RD. Compartmentation of cAMP signaling in cardiac myocytes. *Biophys J*. 2007;92:3317.

Liu Y, Bubolz AH, Mendoza S, et al. H_2O_2 is the transferrable factor mediating flow-induced dilation in human coronary arterioles. *Circ Res*. 2011;108:566.

Llewellyn-Smith IJ, Verberne AJM, eds. *Central Regulation of Autonomic Functions*. New York: Oxford University Press; 2011.

Timmers HJ, Wieling W, Karemaker JM, et al. Cardiovascular responses to stress after carotid baroreceptor denervation in humans. *Ann N Y Acad Sci*. 2004;1018:515.

Zhang P, Mende U. Regulators of G-protein signaling in the heart and their potential as therapeutic targets. *Circ Res*. 2011;109: 320.

19

Integrated Control of the Cardiovascular System

LEARNING OBJECTIVES

Upon completion of this chapter, the student should be able to answer the following questions:

1. What are the four major factors that determine cardiac output? Which two of these factors are said to be "coupling factors," and what is the reason for that description?
2. What is a cardiac function curve, and how is it related to the Frank-Starling mechanism?
3. What is a vascular function curve, and how is it affected by changes in total peripheral resistance, blood volume, and venous tone?
4. Why does the operating point of the cardiovascular system occur at the intersection of the vascular and cardiac function curves?
5. How does evaluation of the cardiac function curve and the vascular function curve enable clinicians to determine the effect of changes in blood volume, vascular tone, and contractility on cardiac output?
6. What mechanisms in the central nervous system, heart, and systemic vasculature allow cardiac output to increase to the necessary levels during vigorous exercise?
7. What are the cardiovascular consequences of hemorrhage, and what are the compensatory mechanisms that tend to restore arterial pressure and cardiac output?

Regulation of Cardiac Output and Blood Pressure

Four factors control cardiac output: heart rate, myocardial contractility, preload, and afterload (Fig. 19.1). Heart rate and myocardial contractility are strictly cardiac factors, although they are controlled by various neural and humoral mechanisms. Preload and afterload are factors that are mutually dependent on function of the heart and the vasculature and are important determinants of cardiac output. Preload and afterload are themselves determined by cardiac output and by certain vascular characteristics. Preload and afterload are called *coupling factors* because they constitute a functional coupling between the heart and blood vessels.

To understand regulation of cardiac output, the nature of the coupling between the heart and the vascular system must be appreciated.

In this chapter, two kinds of graphed curves are used to analyze interactions between the cardiac and vascular components of the circulatory system. The first curve, the **cardiac function curve,** is an expression of the well-known **Frank-Starling relationship,** and it illustrates the dependence of cardiac output on preload (i.e., central venous or right atrial pressure). The cardiac function curve is a characteristic of the heart itself and is usually studied in hearts completely isolated from the rest of the circulation. This curve has already been discussed in detail in Chapter 16. Later in this chapter, this curve is discussed in association with the other characteristic curve, the **vascular function curve,** to analyze interactions between the heart and the vasculature. The vascular function curve defines the dependence of central venous pressure on cardiac output. This relationship depends only on several vascular system characteristics, including peripheral vascular resistance, arterial and venous compliance, and blood volume. The vascular function curve is entirely independent of the characteristics of the heart. Because of this independence, it can be derived experimentally even if a mechanical pump replaces the heart.

Vascular Function Curve

The vascular function curve defines the changes in central venous pressure (P_v) that are caused by changes in cardiac output. In this curve, P_v is the dependent variable (or response), and cardiac output is the independent variable (or stimulus). These variables are opposite those of the cardiac function curve, in which P_v (or preload) is the independent variable and cardiac output is the dependent variable.

The simplified model of the circulation shown in Fig. 19.2 helps explain how cardiac output determines the level of P_v. In this model, all essential components of the cardiovascular system have been lumped into four basic elements. The right and left sides of the heart, as well as the pulmonary vascular bed, constitute a **pump-oxygenator,** much like an artificial heart-lung machine used to perfuse

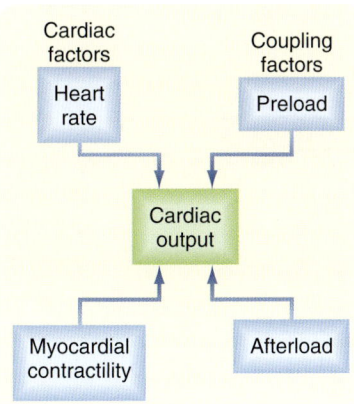

• **Fig. 19.1** The four factors (in *blue squares*) that determine cardiac output.

the body during open heart surgery. The high-resistance microcirculation is designated the **peripheral resistance.** Finally, the compliance of the system is subdivided into **arterial compliance (C_a)** and **venous compliance (C_v).** As defined in Chapter 17, the compliance (C) of a blood vessel is the change in volume (ΔV) that is accommodated in that vessel per unit change in transmural pressure (ΔP); that is,

Equation 19.1
$$C = \Delta V/\Delta P$$

Venous compliance is approximately 20 times greater than arterial compliance. In the example in Fig. 19.2, the ratio of C_v to C_a is set at 19:1 to simplify calculations.*

To show how a change in cardiac output causes an inverse change in P_v, the hypothetical model has certain characteristics that mimic those of an average adult (see Fig. 19.2A). The flow generated by the heart (i.e., cardiac output; Q_h) is 5 L/minute; mean arterial pressure (P_a) is 102 mm Hg; and P_v is 2 mm Hg. Peripheral resistance (R) is the ratio of the arteriovenous pressure difference ($P_a - P_v$) to flow (Q_r) through the resistance vessels; this ratio is equal to 20 mm Hg/L/minute.

An arteriovenous pressure difference of 100 mm Hg is sufficient to force a flow rate (Q_r) of 5 L/minute through a peripheral resistance of 20 mm Hg/L/minute (see Fig. 19.2A). Under equilibrium conditions, this flow rate (Q_r) is precisely equal to the flow rate (Q_h) pumped by the heart. From heartbeat to heartbeat, the volume of blood in the arteries (V_a) and the volume of blood in the veins (V_v) remain constant because the volume of blood transferred from the veins to the arteries by the heart is equal to the volume of blood that flows from the arteries through the resistance vessels and into the veins.

Effects of Cardiac Arrest on Arterial and Venous Pressure

Fig. 19.2B depicts the circulation at the very beginning of an episode of cardiac arrest; that is, $Q_h = 0$. In the instant immediately after arrest of the heart, the volume of blood in the arteries (V_a) and veins (V_v) has not had time to change appreciably. Because arterial pressure and venous pressure depend on V_a and V_v, respectively, these pressures are identical to the respective pressures in Fig. 19.2A (i.e., $P_a = 102$ and $P_v = 2$). This arteriovenous pressure gradient of 100 mm Hg forces a flow rate (Q_r) of 5 L/minute through the peripheral resistance of 20 mm Hg/L/minute. Thus although cardiac output (Q_h) at that point is 0 L/minute, the rate of flow through the microcirculation (Q_r) is 5 L/minute because the potential energy stored in the arteries by the preceding pumping action of the heart causes blood to be transferred from arteries to veins. This transfer occurs initially at the control (steady-state) rate, even though the heart can no longer transfer blood from the veins to the arteries.

As cardiac arrest continues, blood flow through the resistance vessels causes the blood volume in the arteries to decrease progressively and the blood volume in the veins to increase progressively at the same absolute rate. Because the arteries and veins are elastic structures, arterial pressure falls gradually, and the venous pressure rises gradually. This process continues until arterial and venous pressures become equal (see Fig. 19.2C). Once this condition is reached, the rate of flow (Q_r) from the arteries to the veins through the resistance vessels is 0 L/minute, as is Q_h.

When the effects of cardiac arrest reach this equilibrium state (see Fig. 19.2C), the pressure attained in the arteries and veins depends on the relative compliance of these vessels. If arterial compliance (C_a) and venous compliance (C_v) are equal, the decline in P_a is equal to the rise in P_v because the decrease in arterial volume would be equal to the increase in venous volume (according to the principle of conservation of mass). Both P_a and P_v would attain the average of their combined values in Fig. 19.2A; that is, $P_a = P_v = (102 + 2)/2 = 52$ mm Hg. However, C_a and C_v in a living person are not equal. Veins are much more compliant than arteries; the compliance ratio (C_v/C_a) is approximately 19, the ratio assumed for the model in Fig. 19.2. When the effects of cardiac arrest reach equilibrium in an intact subject, the pressure in the arteries and veins is much lower than the average value of 52 mm Hg that occurs when C_a and C_v are equal. Hence, transfer of blood from arteries to veins at equilibrium induces a fall in arterial pressure 19 times greater than the concomitant rise in venous pressure. As Fig. 19.2C shows, P_v would increase by 5 mm Hg (to 7 mm Hg), whereas P_a would fall by 95 (i.e., 19 × 5) mm Hg (to 7 mm Hg). This equilibrium pressure, which prevails in the absence of flow, is referred to as either **mean circulatory pressure** or **static pressure.** The pressure in the static system reflects the total blood volume in the system and the overall compliance of the system.

---

*Thus if it were necessary to add *x* mL of blood to the arterial system to produce a 1–mm Hg increase in arterial pressure, 19*x* mL of blood would need to be added to the venous system to raise venous pressure by the same amount.

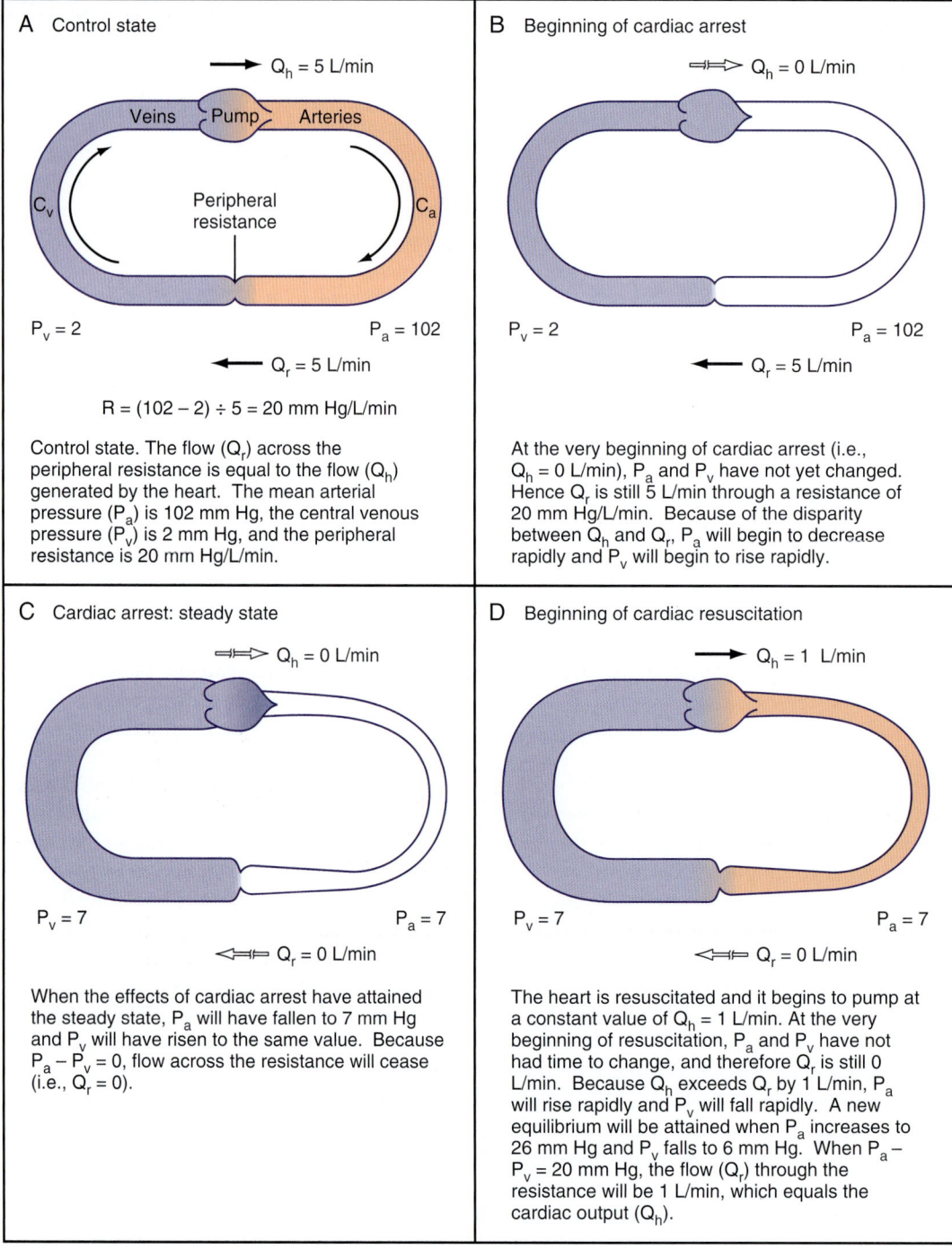

A   Control state

$\longrightarrow Q_h = 5$ L/min

Veins   Pump   Arteries

C_v

Peripheral resistance

C_a

$P_v = 2$

$P_a = 102$

$\longleftarrow Q_r = 5$ L/min

R = (102 − 2) ÷ 5 = 20 mm Hg/L/min

Control state. The flow (Q_r) across the peripheral resistance is equal to the flow (Q_h) generated by the heart. The mean arterial pressure (P_a) is 102 mm Hg, the central venous pressure (P_v) is 2 mm Hg, and the peripheral resistance is 20 mm Hg/L/min.

B   Beginning of cardiac arrest

$Q_h = 0$ L/min

$P_v = 2$

$P_a = 102$

$\longleftarrow Q_r = 5$ L/min

At the very beginning of cardiac arrest (i.e., $Q_h = 0$ L/min), P_a and P_v have not yet changed. Hence Q_r is still 5 L/min through a resistance of 20 mm Hg/L/min. Because of the disparity between Q_h and Q_r, P_a will begin to decrease rapidly and P_v will begin to rise rapidly.

C   Cardiac arrest: steady state

$Q_h = 0$ L/min

$P_v = 7$

$P_a = 7$

$Q_r = 0$ L/min

When the effects of cardiac arrest have attained the steady state, P_a will have fallen to 7 mm Hg and P_v will have risen to the same value. Because $P_a − P_v = 0$, flow across the resistance will cease (i.e., $Q_r = 0$).

D   Beginning of cardiac resuscitation

$\longrightarrow Q_h = 1$ L/min

$P_v = 7$

$P_a = 7$

$Q_r = 0$ L/min

The heart is resuscitated and it begins to pump at a constant value of $Q_h = 1$ L/min. At the very beginning of resuscitation, P_a and P_v have not had time to change, and therefore Q_r is still 0 L/min. Because Q_h exceeds Q_r by 1 L/min, P_a will rise rapidly and P_v will fall rapidly. A new equilibrium will be attained when P_a increases to 26 mm Hg and P_v falls to 6 mm Hg. When $P_a − P_v = 20$ mm Hg, the flow (Q_r) through the resistance will be 1 L/min, which equals the cardiac output (Q_h).

• **Fig. 19.2 A** to **D,** Simplified model of the cardiovascular system, consisting of a pump, arterial compliance (C_a), peripheral resistance, and venous compliance (C_v).

The example of cardiac arrest aids in the understanding of the vascular function curve. The clinician can now begin to assemble a vascular function curve (Fig. 19.3). The independent variable (plotted along the x-axis) is cardiac output, and the dependent variable (plotted along the y-axis) is P_v. Two important points on this curve can be derived from the example in Fig. 19.2. One point (A in Fig. 19.3) represents the control state; that is, when cardiac output is 5 L/minute,

P_v is 2 mm Hg. When the heart is arrested (cardiac output = 0), P_v becomes 7 mm Hg at equilibrium (see Fig. 19.2C); this pressure is the mean circulatory pressure.

The inverse relationship between P_v and cardiac output simply means that when cardiac output is suddenly decreased, the rate at which blood flows from arteries to veins through the capillaries is temporarily greater than the rate at which the heart pumps blood from the veins back

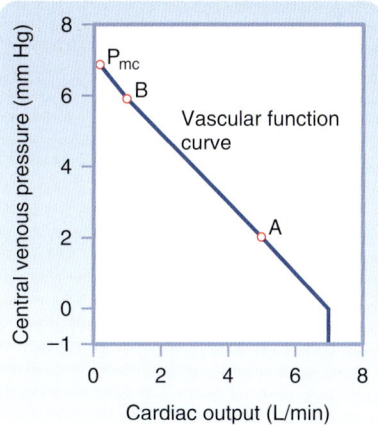

• **Fig. 19.3** Changes in Central Venous Pressure Produced by Changes in Cardiac Output. The mean circulatory (or static) pressure (P_mc) is the equilibrium pressure throughout the cardiovascular system when cardiac output is 0. Points B and A represent the values of venous pressure at cardiac outputs of 1 and 5 L/minute, respectively.

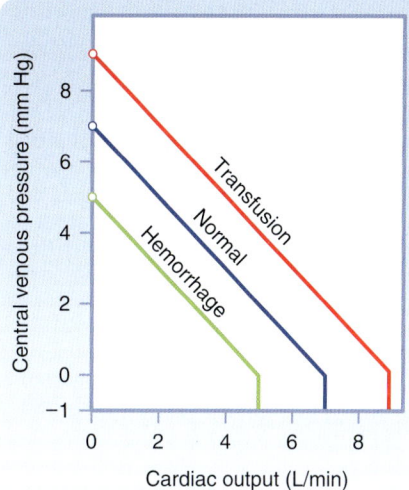

• **Fig. 19.4** Effects of increased blood volume (*transfusion curve*) and decreased blood volume (*hemorrhage curve*) on the vascular function curve. Similar shifts in the vascular function curve can be produced by increases and decreases, respectively, in venomotor tone.

into the arteries. During that transient period, a net volume of blood is transferred from arteries to veins; hence, P_a falls and P_v rises.

Now suppose that cardiac output suddenly increases. This example illustrates how a third point (B in Fig. 19.3) on the vascular function curve is derived. Consider that the arrested heart is suddenly restarted and immediately begins pumping blood from the veins into the arteries at a rate of 1 L/minute (see Fig. 19.2D). When the heart first begins to beat, the arteriovenous pressure gradient is 0, and no blood is transferred from the arteries through the capillaries and into the veins. Thus when beating resumes, blood is depleted from the veins at the rate of 1 L/minute, and arterial blood volume is replenished from venous blood volume at that same absolute rate. Hence, P_v begins to fall and P_a begins to rise. Because of the difference in arterial and venous compliance, P_a rises at a rate 19 times faster than the rate at which P_v falls. The resultant arteriovenous pressure gradient causes blood to flow through the peripheral resistance vessels. If the heart maintains a constant output of 1 L/minute, P_a continues to rise and P_v continues to fall until the pressure gradient becomes 20 mm Hg. This gradient forces a rate of flow of 1 L/minute through a peripheral resistance of 20 mm Hg/L/minute. This gradient is achieved by a 19–mm Hg rise (to 26 mm Hg) in P_a and a 1–mm Hg fall (to 6 mm Hg) in P_v. This equilibrium value of P_v (6 mm Hg) for a cardiac output of 1 L/minute also appears on the vascular function curve of Fig. 19.3 (point B). The 1–mm Hg reduction in P_v reflects a net transfer of blood from the veins to the arteries of the circuit.

The reduction in P_v that can be evoked by a sudden increase in cardiac output is limited. At some critical maximal value of cardiac output, sufficient fluid is transferred from the veins to the arteries of the circuit for P_v to fall below ambient pressure. In a system of very distensible vessels, such as the venous system, the greater external pressure causes the vessels to collapse (see Chapter 17). This

venous collapse impedes venous return to the heart. Hence, it limits the maximal value of cardiac output to 7 L/minute in this example (see Fig. 19.3), regardless of the capabilities of the pump.

Factors That Influence the Vascular Function Curve

Dependence of Venous Pressure on Cardiac Output

According to experimental and clinical observations, changes in cardiac output do indeed evoke the alterations in P_a and P_v that have been predicted by the simplified model in Fig. 19.2.

Blood Volume

The vascular function curve is affected by variations in total blood volume. During circulatory standstill (zero cardiac output), mean circulatory pressure depends only on total vascular compliance and blood volume. For a given vascular compliance, mean circulatory pressure is increased when blood volume is expanded (**hypervolemia**) and is decreased when blood volume is diminished (**hypovolemia**). This relationship is illustrated by the y-axis intercepts in Fig. 19.4, in which mean circulatory pressure is 5 mm Hg after hemorrhage and 9 mm Hg after transfusion, in comparison with a value of 7 mm Hg at normal blood volume (**normovolemia or euvolemia**).

Venomotor Tone

The effects of changes in venomotor tone on the vascular function curve closely resemble those of changes in blood volume. In Fig. 19.4, for example, the transfusion curve could also represent increased venomotor tone, whereas the hemorrhage curve could represent decreased tone. During

IN THE CLINIC

Cardiac output may decrease abruptly when a major coronary artery suddenly becomes occluded. The **acute heart failure** that occurs as a result of **myocardial infarction** (death of myocardial tissue) is usually accompanied by a fall in arterial blood pressure and a rise in P_v. In Fig. 19.4 it is also apparent that the cardiac output at which $P_v = 0$ varies directly with blood volume. Therefore, the maximal value of cardiac output becomes progressively more limited as the total blood volume is reduced. However, the P_v at which the veins collapse (illustrated by the sharp change in slope of the vascular function curve) is not significantly altered by changes in blood volume. This pressure depends only on the ambient pressure surrounding the central veins. Ambient pressure is the pleural pressure in the thorax (see Chapter 21).

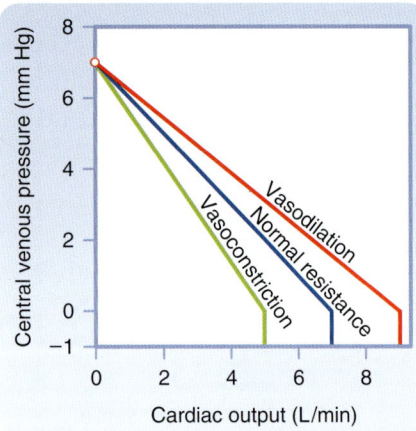

• **Fig. 19.5** Effects of arteriolar dilation and constriction on the vascular function curve.

circulatory standstill, for a given blood volume, the pressure within the vascular system rises as smooth muscle tension exerted within the vascular walls increases (these contractile changes in arteriolar and venous smooth muscle are under nervous and humoral control). The fraction of the blood volume located within the arterioles is very small, whereas the blood volume in the veins is large percentage of total blood volume (see Table 15.1). Thus changes in peripheral resistance (arteriolar tone) have no significant effect on mean circulatory pressure, but changes in venous tone can alter mean circulatory pressure appreciably. Hence, mean circulatory pressure rises with increased venomotor tone and falls with diminished venomotor tone.

In experiments, the mean circulatory pressure attained approximately 1 minute after abrupt circulatory standstill is usually substantially above 7 mm Hg, even when blood volume is normal. The elevation to this pressure level is attributable to the generalized venoconstriction that is caused by cerebral ischemia, activation of chemoreceptors, and reduced excitation of baroreceptors. If resuscitation fails, this reflex response subsides as central nervous activity ceases, and mean circulatory pressure then usually falls to a value close to 7 mm Hg.

Blood Reservoirs

Venoconstriction is considerably greater in certain regions of the body than in others. In effect, vascular beds that undergo significant venoconstriction constitute blood reservoirs. The skin's vascular bed is one of the major blood reservoirs in humans. Blood loss evokes profound subcutaneous venoconstriction, which gives rise to the characteristic pale appearance of the skin in response to hemorrhage. Diversion of blood away from the skin frees several hundred milliliters of blood that can be perfused through more vital regions of the body. The vascular beds of the liver, lungs, and spleen are also important blood reservoirs. In humans, however, the volume changes in the spleen are considerably less extensive (see also the sections "Exercise" and "Hemorrhage").

Peripheral Resistance

The changes in the vascular function curve induced by alterations in arteriolar tone are depicted in Fig. 19.5. The amount of blood in the arterioles is small; these vessels contain only approximately 3% of total blood volume (see Chapter 15). Changes in the contractile state of arterioles do not significantly alter mean circulatory pressure. Thus vascular function curves that represent different peripheral resistances converge at a common point on the y-axis (see Fig. 19.5).

P_v varies inversely with **total peripheral resistance (TPR)** when all other factors remain constant. Physiologically, the relationship between P_v and TPR can be explained as follows: if cardiac output is held constant, a sudden increase in TPR causes a progressively greater volume of blood to be retained in the arterial system. Blood volume in the arterial system continues to increase until P_a rises sufficiently to force a flow of blood equal to cardiac output through the resistance vessels. If total blood volume does not change, this increase in arterial blood volume is accompanied by an equivalent decrease in venous blood volume. Hence, an increase in TPR diminishes P_v proportionately. This relationship between TPR and P_v, together with the inability of peripheral resistance to affect mean circulatory pressure, accounts for the clockwise rotation of the vascular function curves in response to increased arteriolar constriction (see Fig. 19.5). Similarly, arteriolar dilation produces a counterclockwise rotation from the same vertical axis intercept. A higher maximal level of cardiac output is attainable when the arterioles are dilated than when they are constricted (see Fig. 19.5).

Interrelationships Between Cardiac Output and Venous Return

Cardiac output and venous return are tightly linked. Except for small, transient disparities, the heart cannot pump any more blood than is delivered to it through the venous system. Similarly, because the circulatory system is a closed circuit, venous return to the heart must equal cardiac output over any appreciable time interval. The flow around the entire closed circuit depends on the capability of the pump, the characteristics of the circuit, and the total fluid volume of the system.

Thus *cardiac output* and *venous return* are simply two terms for the flow around this closed circuit. Cardiac output is the volume of blood being pumped by the heart per unit time. Venous return is the volume of blood returning to the heart per unit time. At equilibrium, these two volumes are equal. In the following section, certain techniques of circuit analysis are discussed to provide some insight into the control of flow around the circuit.

Relating the Cardiac Function Curve to the Vascular Function Curve

Coupling Between the Heart and the Vasculature

In accordance with the Frank-Starling law of the heart, cardiac output depends closely on right atrial (or central venous) pressure. Furthermore, right atrial pressure is approximately equal to right ventricular end-diastolic pressure because the normal tricuspid valve acts as a low-resistance junction between the right atrium and ventricle. Graphs of cardiac output as a function of P_v are called **cardiac function curves;** extrinsic regulatory influences may be expressed as shifts in such curves.

A typical cardiac function curve is plotted on the same coordinates as those for a normal vascular function curve in Fig. 19.6. The cardiac function curve is plotted according to the usual convention; that is, the independent variable (P_v) is plotted along the x-axis, and the dependent variable (cardiac output) is plotted along the y-axis. In accordance with the Frank-Starling mechanism, the cardiac function curve reveals that a rise in P_v increases cardiac output.

Conversely, the vascular function curve characterizes an inverse relationship between cardiac output and P_v; that is, a rise in cardiac output diminishes P_v. P_v is the dependent variable (or response) and cardiac output is the independent variable (or stimulus) for the vascular function curve. Therefore, to plot a vascular function curve in the conventional manner, P_v should be scaled along the y-axis and cardiac output along the x-axis.

To plot the cardiac and vascular function curves on the same set of axes requires a modification of the plotting convention for one of these curves. The convention for the vascular function curve is violated arbitrarily in this chapter. Note that the vascular function curve in Fig. 19.6 is intended to reflect how P_v (scaled along the x-axis) varies in response to a change in cardiac output (scaled along the y-axis).

When the cardiovascular system is represented by a given pair of cardiac and vascular function curves, the intersection of these two curves defines the **equilibrium point** of that system. The coordinates of this equilibrium point represent

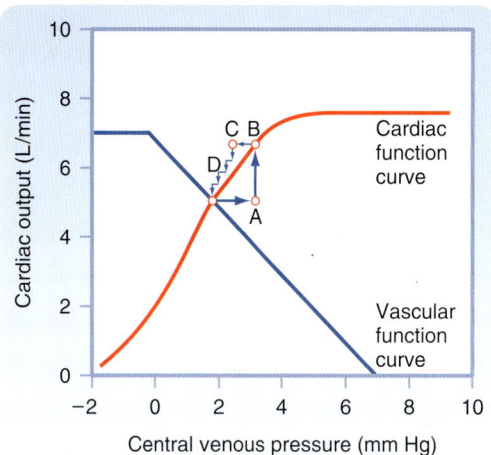

• **Fig. 19.6** Typical Vascular and Cardiac Function Curves Plotted on the Same Coordinate Axes. To plot both curves on the same graph, the x-axis and y-axis for the vascular function curves had to be switched; compare the assignment of axes with those in Figs. 19.3, 19.4, and 19.5. The coordinates of the equilibrium point, at the intersection of the cardiac and vascular function curves, represent the stable values of cardiac output and central venous pressure at which the system tends to operate. Any perturbation (e.g., a sudden increase in venous pressure to point A) institutes a sequence of changes in cardiac output and venous pressure that restore these variables to their equilibrium values.

the values of cardiac output and P_v at which the system tends to operate. Only transient deviations from such values of cardiac output and P_v are possible, as long as the given cardiac and vascular function curves characterize the system accurately.

The tendency to operate about this equilibrium point may best be illustrated by the response to a sudden change. Consider the changes caused by a sudden rise in P_v from the equilibrium point to point A in Fig. 19.6. This change in P_v might be caused by the rapid injection, during ventricular diastole, of a given volume of blood on the venous vessels of the circuit and simultaneous withdrawal of an equal volume from the arterial vessels of the circuit. Thus although P_v rises, total blood volume remains constant.

As defined by the cardiac function curve, this elevated P_v would increase cardiac output (from point A to point B in Fig. 19.6) during the next ventricular systole. The increased cardiac output would then cause the transfer of a net quantity of blood from the veins to the arteries of the circuit, with a consequent reduction in P_v. In one heartbeat, the reduction in P_v would be small (from point B to point C) because the heart would transfer only a fraction of the total venous blood volume to the arteries. As a result of this reduction in P_v, cardiac output during the very next beat diminishes (from point C to point D) by an amount dictated by the cardiac function curve. Because point C is still above the intersection point, the heart pumps blood from the veins to the arteries at a rate greater than that at which blood flows across the peripheral resistance from arteries to veins. Hence, P_v continues to fall. This process continues in diminishing steps until the point of

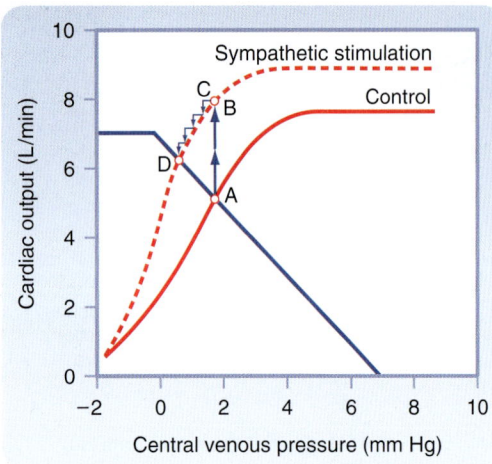

• **Fig. 19.7** Enhancement of myocardial contractility, as by stimulation of cardiac sympathetic nerves, causes the equilibrium values of cardiac output and central venous pressure (P_v) to shift from the intersection (point A) of the control vascular and cardiac function curves (continuous curve) to the intersection (point D) of the same vascular function curve with the cardiac function curve (dashed curve) that represents the response to sympathetic stimulation.

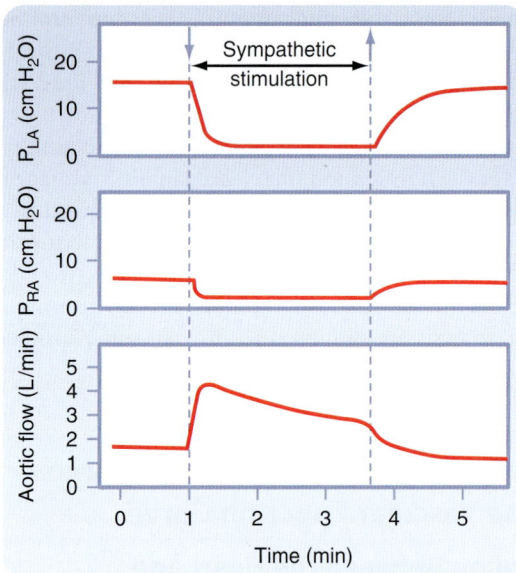

• **Fig. 19.8** During electrical stimulation of the cardiac sympathetic nerve fibers, aortic blood flow (cardiac output) increased, whereas pressures in the left atrium (P_{LA}) and right atrium (P_{RA}) diminished. These data conform to the conclusions derived from Fig. 19.7, in which the equilibrium values of cardiac output and venous pressure are observed to shift from point A to point D (i.e., cardiac output increased, but central venous pressure decreased) during cardiac sympathetic nerve stimulation. (Redrawn from Sarnoff SJ, et al. *Circ Res.* 1960;8:1108.)

intersection is reached. Only one specific combination of cardiac output and venous pressure—the equilibrium point, denoted by the coordinates of the point at which the curves intersect—satisfies the requirements of the cardiac and vascular function curves simultaneously. At the equilibrium point, cardiac output equals venous return, and the system is stable.

Myocardial Contractility

Combinations of cardiac and vascular function curves also help explain the effects of alterations in ventricular contractility on cardiac output and P_v. In Fig. 19.7, the lower cardiac function curve represents the control state, whereas the upper curve reflects the influence of increased myocardial contractility. This pair of curves is analogous to the ventricular function curves shown in Fig. 18.12. The enhanced ventricular contractility represented by the upper curve in Fig. 19.7 can be produced by electrical stimulation of the cardiac sympathetic nerves. When the effects of such neural stimulation are restricted to the heart, the vascular function curve is unaffected. Therefore, only one vascular function curve is needed for this hypothetical intervention (see Fig. 19.7).

During the control state of the model, the equilibrium values for cardiac output and P_v are designated by point A in Fig. 19.7. Cardiac sympathetic nerve stimulation abruptly raises cardiac output to point B because of the enhanced myocardial contractility. However, this high cardiac output causes an increase in the net transfer of blood from the veins to the arteries of the circuit, and as a consequence, P_v subsequently begins to fall (to point C). The reduction in P_v then leads to a small decrease in cardiac output. However, cardiac output is still sufficiently high to effect

the net transfer of blood from the veins to the arteries of the circuit. Thus both P_v and cardiac output continue to fall gradually until a new equilibrium point (point D) is reached. This equilibrium point is located at the intersection of the vascular function curve and the new cardiac function curve. In Fig. 19.7, point D lies above and to the left of the control equilibrium point (point A) and indicates that sympathetic stimulation can evoke greater cardiac output despite the lower level of P_v.

The biological response to enhancement of myocardial contractility is mimicked by the hypothetical change predicted by the model in this chapter. As depicted in Fig. 19.8, sympathetic nerves innervating the heart are stimulated during the time denoted by the double-headed arrow. During neural stimulation, cardiac output (aortic flow) rises quickly to a peak value and then falls gradually to a steady-state value significantly higher than the control level. The increase in aortic flow is accompanied by reductions in right and left atrial pressures.

Blood Volume

Changes in blood volume do not directly affect myocardial contractility, but they do influence the vascular function curve in the manner shown in Fig. 19.4. Thus to understand how changes in blood volume affect cardiac output and P_v, the appropriate cardiac function curve is plotted along with the vascular function curves that represent the control and experimental states (Fig. 19.9). When blood volume

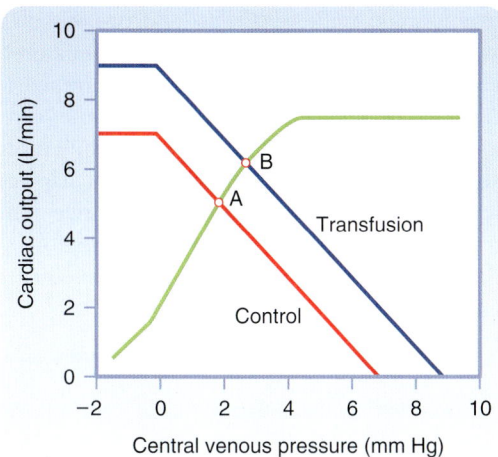

• **Fig. 19.9** After a blood transfusion, the vascular function curve is shifted to the right. Therefore, both cardiac output and venous pressure are increased, as denoted by translocation of the equilibrium point from point A to point B.

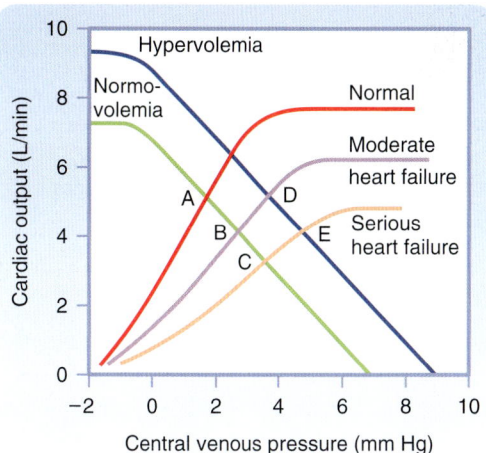

• **Fig. 19.10** Moderate or severe heart failure shifts the cardiac function curves downward and to the right. Before changes in blood volume, cardiac output decreases and central venous pressure rises (from control equilibrium point A to point B or point C). After the increase in blood volume that usually occurs in heart failure, the vascular function curve is shifted to the right. Hence, central venous pressure may be elevated with no reduction in cardiac output (point D) or, in severe heart failure, with some reduction in cardiac output (point E).

is increased by a blood transfusion, the equilibrium point (point B in Fig. 19.9), which denotes the values of cardiac output and P_v after transfusion, lies above and to the right of the control equilibrium point (point A). Thus transfusion increases both cardiac output and P_v. Hemorrhage causes the opposite effect. Mechanistically, the change in ventricular filling pressure (P_v) evoked by a given change in blood volume alters cardiac output by changing the sensitivity of the contractile proteins to the prevailing concentration of intracellular Ca^{++} (see Chapter 18). For reasons explained earlier, pure increases or decreases in venomotor tone elicit responses that are like those evoked by increases or decreases, respectively, in total blood volume.

Peripheral Resistance

Analysis of the effects of changes in peripheral resistance on cardiac output and P_v is complex because both the cardiac and vascular function curves shift. When peripheral resistance increases (Fig. 19.11), the vascular function curve is rotated counterclockwise, but it converges on the same P_v axis intercept as the control curve does. Note that vasoconstriction causes a counterclockwise rotation of the vascular function curve in Fig. 19.11 but a clockwise rotation in Fig. 19.5. The direction of rotation differs because the axes for the vascular function curves were switched in these two figures, as explained earlier. The cardiac function curve in Fig. 19.11 is also shifted downward because at any given P_v, the heart is able to pump less blood against the greater cardiac afterload imposed by the increased peripheral resistance. Because both curves in Fig. 19.11 are displaced downward, the new equilibrium point (point B) is below the control point (point A); that is, an increase in peripheral resistance diminishes cardiac output.

Whether point B falls directly below point A or lies slightly to the right or left of it depends on the magnitude of the shift in each curve. For example, if a given increase

 IN THE CLINIC

Heart failure is a general term that applies to conditions in which the pumping capability of the heart is impaired to the extent that the tissues of the body are not adequately perfused. In heart failure, myocardial contractility is impaired. Heart failure may be acute or chronic. Consequently, in a graph of cardiac and vascular function curves, the cardiac function curve is shifted downward and to the right, as depicted in Fig. 19.10.

Acute heart failure may be caused by toxic concentrations of drugs or by certain pathological conditions such as coronary artery occlusion. In acute heart failure, blood volume does not change immediately. In Fig. 19.10, therefore, the equilibrium point shifts from the intersection (point A) of the normal curves to the intersection (point B or point C) of the normal vascular function curve.

Chronic heart failure may occur in conditions such as essential hypertension or ischemic heart disease. In chronic heart failure, both the cardiac function and vascular function curves shift. The vascular function curve shifts because of an increase in blood volume caused in part by fluid retention by the kidneys. The fluid retention is related to the concomitant reduction in glomerular filtration rate and the decreased renal excretion of NaCl and water (see also Chapter 35). The resultant hypervolemia is reflected by a rightward shift of the vascular function curve, as shown in Fig. 19.10. Hence, with moderate degrees of heart failure, P_v is elevated, but cardiac output may be normal (point D). With more severe degrees of heart failure, P_v is still elevated, but cardiac output is subnormal (point E).

in peripheral resistance shifts the vascular function curve more than it does the cardiac function curve, equilibrium point B is below and to the left of point A; that is, both cardiac output and P_v diminish. Conversely, if the cardiac function curve is displaced more than the vascular function

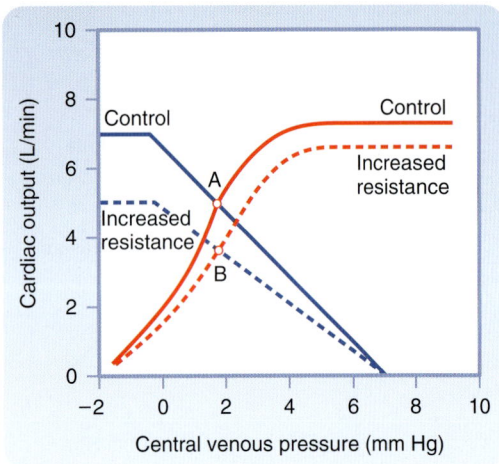

• **Fig. 19.11** An increase in peripheral resistance shifts the cardiac and vascular function curves downward. At equilibrium, cardiac output is less (point B) when peripheral resistance is high than when peripheral resistance is normal (point A).

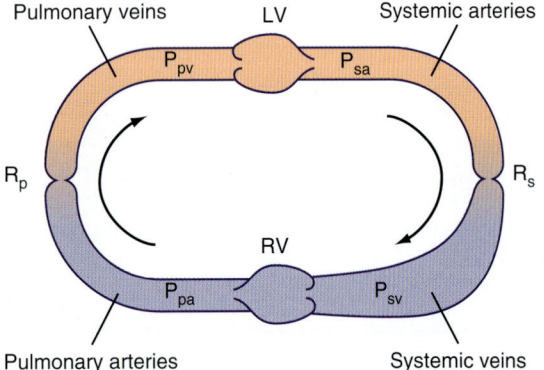

• **Fig. 19.12** Simplified model of the cardiovascular system that consists of the left ventricle (LV) and right ventricle (RV), systemic vascular resistance (R_s) and pulmonary vascular resistance (R_p), systemic arterial and venous compliance, and pulmonary arterial and venous compliance. P_{sa} and P_{sv} are the pressures in the systemic arteries and veins, respectively; P_{pa} and P_{pv} are the pressures in the pulmonary arteries and veins, respectively.

curve, point B falls below and to the right of point A; that is, cardiac output decreases, but P_v rises.

A More Complete Theoretical Model: The Two-Pump System

The preceding discussion shows that the interrelationships between cardiac output and P_v are complex, even in an oversimplified circulation model that includes only one pump and just the systemic circulation. In reality, the cardiovascular system includes the systemic and pulmonary circulations and two pumps: the left and right ventricles. Thus the interrelationships among ventricular output, arterial pressure, and atrial pressure are much more complex.

Fig. 19.12 depicts a more complete (but still oversimplified) cardiovascular system model that has two pumps in

series (the left and right ventricles) and two vascular beds in series (the systemic and pulmonary vasculature). The series arrangement requires that the flow pumped by the two ventricles be virtually equal to each other over any substantial period; otherwise, all the blood would ultimately accumulate in one or the other of the vascular systems. Because the cardiac function curves for the two ventricles differ substantially, the filling (atrial) pressures for the two ventricles must differ appropriately to ensure equal stroke volumes (see Fig. 18.13).

 IN THE CLINIC

Any change in contractility that affects the two ventricles differently alters the distribution of blood volume in the two vascular systems. If a coronary artery to the left ventricle becomes occluded, left ventricular contractility is impaired, and **acute left ventricular failure** ensues. In the instant after occlusion, left atrial pressure does not change, and the left ventricle begins to pump at a diminished rate of flow. If the right ventricle is not affected by the acute coronary artery occlusion, the right ventricle initially continues to pump the normal flow. The disparate right and left ventricular outputs result in a progressive increase in left atrial pressure and a progressive decrease in right atrial pressure. Therefore, left ventricular output increases toward the normal value, and right ventricular output falls below the normal value. This process continues until the outputs of the two ventricles again become equal. At this new equilibrium, the output of the two ventricles is subnormal. The elevation in left atrial pressure is accompanied by an equal elevation in pulmonary venous pressure, which can have serious clinical consequences. The high pulmonary venous pressure can increase lung stiffness and lead to respiratory distress by increasing the mechanical work of pulmonary ventilation (see Chapter 22). Furthermore, the high pulmonary venous pressure causes an elevate the hydrostatic pressure in the pulmonary capillaries and may lead to the transudation of fluid from the pulmonary capillaries to the pulmonary interstitium or into the alveoli (**pulmonary edema**), which may be lethal.

Two basic principles to remember about ventricular function are that (1) the left ventricle pumps blood through the systemic vasculature, and (2) the right ventricle pumps blood through the pulmonary vasculature. However, these principles do not necessarily imply that both ventricles are essential to perfuse the systemic and pulmonary vascular beds adequately. To better understand the relationships between the two ventricles and the two vascular beds, the right ventricular function is examined in more detail as follows.

In the circulatory system model shown in Fig. 19.12, consider the hemodynamic consequences that would occur if the right ventricle suddenly ceased its pump function but instead served merely as a passive, low-resistance conduit between the systemic veins and the pulmonary arteries. Under these conditions, the only functional pump would be the left ventricle, which would then be required to pump blood through both the systemic and pulmonary resistances (for the purposes of this discussion, consider the resistance to the flow of blood through the inactive right ventricle to be negligible).

Normally, pulmonary vascular resistance is approximately 10% as great as systemic vascular resistance. Because the two resistances are in series with one another, total resistance would be 10% greater than systemic resistance alone (see Chapter 17). In a normal cardiovascular system, a 10% increase in systemic vascular resistance would increase P_a (and hence left ventricular afterload) by approximately 10%. This increase would not drastically affect left ventricular function. Under certain conditions, however, this increase in P_a could significantly alter the function of the cardiovascular system. If the 10% increase in total resistance is achieved by adding a small degree of resistance (i.e., pulmonary vascular resistance) to that of the much larger systemic resistance, and if the pulmonary vascular resistance is separated from the systemic resistance by a large degree of compliance (the combined systemic venous and pulmonary arterial compliance), the 10% increase in total resistance could drastically impair operation of the cardiovascular system.

The simulated effects of inactivating the pumping action of the right ventricle in a hydraulic analogue of the circulatory system are shown in Fig. 19.13. In the model, the right and left ventricles generate cardiac outputs that vary directly with their respective filling pressures. Under control conditions (when the right ventricle is functioning normally), the outputs of the left and right ventricles are equal (5 L/minute). The right ventricular pumping action causes the pressure in the pulmonary artery (not shown) to exceed the pressure in the pulmonary veins (P_{pv}) by an amount that forces fluid through the pulmonary vascular resistance at a rate of 5 L/minute. When the right ventricle ceases pumping (arrow 1 in Fig. 19.13), the systemic venous and pulmonary arterial systems, along with the right ventricle itself, become a common passive conduit with a large compliance. When the right ventricle ceases to transfer blood actively from the systemic veins to the pulmonary arteries, pulmonary arterial pressure (P_{pa}) decreases rapidly (not shown) and systemic venous pressure (P_{sv}) rises rapidly to a common value (≈5 mm Hg). At this low pressure, however, fluid flows from the pulmonary arteries to the pulmonary veins at a greatly reduced rate.

At the start of right ventricular arrest, the left ventricle is pumping fluid from the pulmonary veins to the systemic arteries at the control rate of 5 L/minute, which greatly exceeds the rate at which blood returns to the pulmonary veins once the right ventricle ceases to operate. Hence, pulmonary venous pressure (P_{pv}) drops sharply. Because pulmonary venous pressure is the preload for the left ventricle, left ventricular (cardiac) output drops abruptly as well and attains a steady-state value of approximately 2.5 L/minute. This effect in turn leads to a rapid reduction in systemic arterial pressure (P_{sa}). In short, stoppage of right ventricular pumping markedly curtails cardiac output, systemic arterial pressure, and pulmonary venous

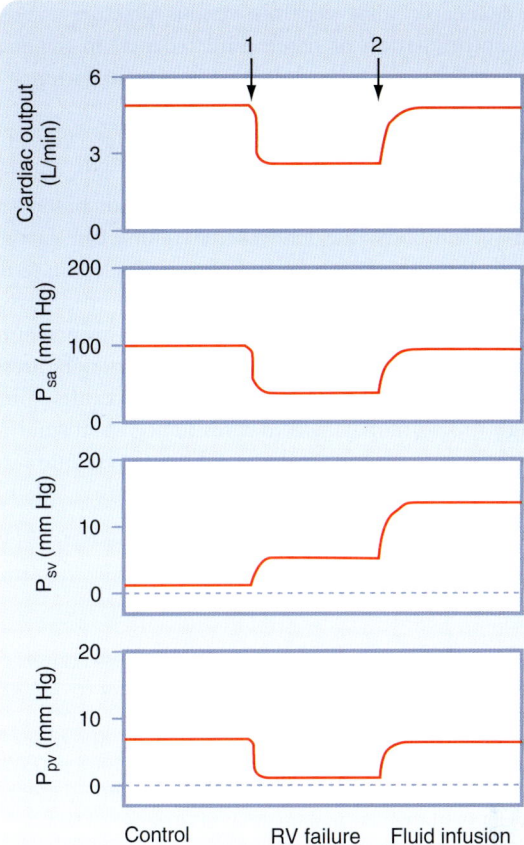

• **Fig. 19.13** Changes in cardiac output, systemic arterial pressure (P_{sa}), systemic venous pressure (P_{sv}), and pulmonary venous pressure (P_{pv}) evoked by simulated right ventricular (RV) failure and by simulated infusion of fluid in the circulatory model shown in Fig. 19.12. At *arrow 1,* the pumping action of the right ventricle was discontinued (simulated RV failure), and the right ventricle served only as a low-resistance conduit. At *arrow 2,* the fluid volume in the system was expanded, and the right ventricle continued to serve only as a conduit. (Modified from Furey SA, et al. *Am Heart J.* 1984;107:404.)

pressure and raises systemic venous pressure moderately (see Fig. 19.13).

Most of the hemodynamic problems induced by inactivation of the right ventricle can be reversed by an increase in the fluid (blood) volume of the system (arrow 2 in Fig. 19.13). If fluid is added until pulmonary venous pressure (left ventricular preload) is raised to its control value, cardiac output and systemic arterial pressure are restored almost to normal, but systemic venous pressure is abnormally elevated. If left ventricular function is normal, adding a normal left ventricular preload evokes normal left ventricular output. The 10% increase in peripheral resistance caused by adding the pulmonary vascular resistance to that of the systemic vascular resistance does not impose a serious burden on left ventricular pumping capacity.

When the right ventricle is inoperative, however, pulmonary blood flow is not normal unless the usual pulmonary arteriovenous pressure gradient (≈10 to 15 mm Hg) prevails. Hence, systemic venous pressure (P_{sv}) must exceed pulmonary venous pressure (P_{pv}) by this amount.

Maintenance of high systemic venous pressure may lead to the accumulation of tissue fluid (edema) in dependent regions of the body, a characteristic finding in patients with right ventricular heart failure.

With this information, the principal function of the right ventricle may be characterized as as follows. From the viewpoint of providing sufficient flow of blood to all tissues in the body, the left ventricle alone can carry out this function. Operation of the two ventricles in series is not essential to provide adequate blood flow to the tissues. The crucial function of the right ventricle is to prevent the rise in systemic venous (and pulmonary arterial) pressure that would be required to force the normal cardiac output through the pulmonary vascular resistance. A normal right ventricle, by preventing an abnormal rise in systemic venous pressure, prevents the development of extensive edema in dependent regions of the body.

 IN THE CLINIC

Clinically, **right ventricular heart failure** may be caused by occlusive disease predominantly of the coronary vessels to the right ventricle. These vessels are affected much less commonly than the vessels to the left ventricle. The major hemodynamic effects of acute right-sided heart failure are pronounced reductions in cardiac output and arterial blood pressure, and the principal treatment is infusion of blood or plasma. Bypass of the right ventricle (by anastomosis of the right atrium to the pulmonary artery) may be performed surgically in patients with certain **congenital cardiac defects,** such as severe narrowing of the tricuspid valve or maldevelopment of the right ventricle. The effects of acute right-sided heart failure or right ventricular bypass are directionally similar to those predicted previously from analysis of the model shown in Fig. 19.13.

Role of the Heart Rate in Control of Cardiac Output

Cardiac output is the product of stroke volume and heart rate. Analysis of the control of cardiac output has thus far been restricted to the control of stroke volume, and the role of heart rate has not been considered. Analysis of the effect of a change in heart rate on cardiac output is complex because a change in heart rate alters the other three factors (preload, afterload, and myocardial contractility) that determine stroke volume (see Fig. 19.1). An increase in heart rate, for example, shortens the duration of diastole. Hence, ventricular filling is diminished; that is, preload is reduced. If an increase in heart rate altered cardiac output, arterial pressure would change; that is, afterload would be altered. A rise in heart rate would increase the net influx of Ca^{++} per minute into myocardial cells (see also Chapter 18), and this influx would enhance myocardial contractility.

The effects of changes in heart rate on cardiac output have been studied extensively, and the results are similar to those shown in Fig. 19.14. As atrial pacing frequency is

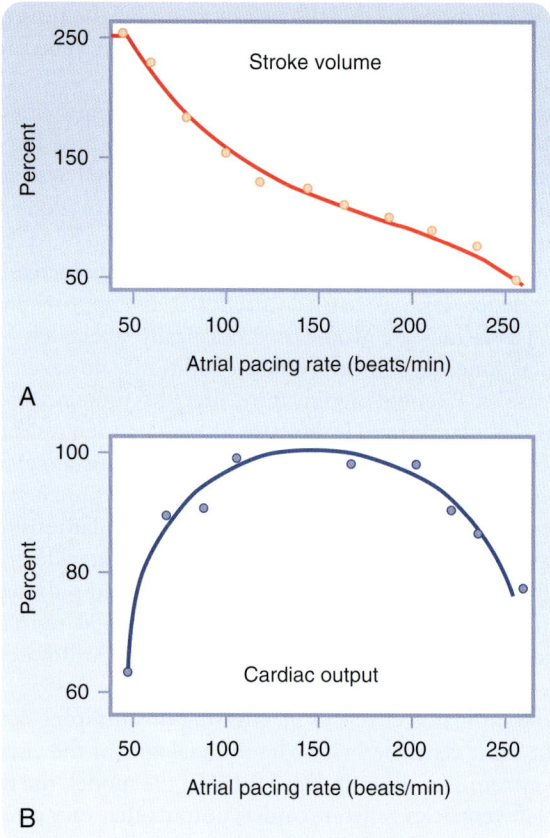

• **Fig. 19.14** Changes in stroke volume **(A)** and cardiac output **(B)** induced by changes in the rate of atrial pacing. (Redrawn from Kumada M, et al. *Jpn J Physiol.* 1967;17:538.)

gradually increased, stroke volume progressively diminishes (see Fig. 19.14A). The decrease in stroke volume is caused by the reduced time for ventricular filling. The change in stroke volume is not inversely proportional to the change in heart rate because the direction of the change in cardiac output (Q_h) is markedly influenced by the actual heart rate (see Fig. 19.14B). For example, as pacing frequency is increased from 50 to 100 beats/minute, the increase in heart rate augments Q_h. Because $Q_h = SV \times HR$, the decrease in stroke volume (SV) over this frequency range must be proportionately less than the increase in heart rate (HR).

Over the frequency range from approximately 100 to 200 beats/minute, however, cardiac output is not affected significantly by changes in pacing frequency (see Fig. 19.14B). Hence, as pacing frequency is increased, the decrease in stroke volume must be approximately equal to the increase in heart rate. In addition, generalized vascular autoregulation tends to keep tissue blood flow constant (see also Chapter 17). This adaptation leads to changes in preload and afterload that also keep cardiac output nearly constant.

Moreover, at excessively high pacing frequencies (above 200 beats/minute; see Fig. 19.14), further increases in heart rate decrease cardiac output. Therefore, the induced decrease in stroke volume must have exceeded the increase in heart rate at this high range of pacing frequencies. At such high pacing frequencies, the ventricular filling time is so severely restricted that compensation is inadequate, and cardiac output decreases sharply. Although the relationship of cardiac output to heart rate is characteristically that of an inverted U in the general population, the relationship varies quantitatively among subjects and among physiological states.

Strong correlations between heart rate and cardiac output must be interpreted cautiously. In people who are exercising, for example, cardiac output and heart rate usually increase proportionately, and stroke volume may remain constant or increase only slightly (see the later section "Exercise"). It is tempting to conclude that the increase in cardiac output during exercise must be caused solely by the observed increase in heart rate. However, Fig. 19.14 shows that over a wide range of heart rates, a change in heart rate may have little influence on cardiac output. The principal increase in cardiac output during exercise must therefore be attributed to other factors. Such ancillary factors include the pronounced reduction in peripheral vascular resistance because of the vasodilation in the active skeletal muscles and the increased contractility of cardiac muscle associated with the generalized increase in sympathetic neural activity. Nevertheless, the increase in heart rate is still an important factor. Abundant data show that if the heart rate cannot increase normally during exercise, the augmentation in cardiac output and the capacity for exercise are severely limited. Because stroke volume changes only slightly during exercise, the increase in heart rate may play an important permissive role in augmenting cardiac output during physical exercise.

IN THE CLINIC

The characteristic relationship between cardiac output and heart rate explains the urgent need of treatment by patients who have excessively slow or excessively fast heart rates. Profound **bradycardia** (slow rate) may occur as a result of a very slow sinus rhythm in patients with **sick sinus syndrome** or as a result of a slow idioventricular rhythm in patients with **complete atrioventricular block.** In either rhythm disturbance, the capacity of the ventricles to fill during prolonged diastole is limited (often by the noncompliant pericardium). Hence, cardiac output usually decreases substantially because the very slow heart rate cannot be counterbalanced by a sufficiently large stroke volume. As a consequence, such bradycardia often necessitates the implantation of an artificial pacemaker. In patients with **supraventricular** or **ventricular tachycardia,** excessively high heart rates frequently necessitate emergency treatment because in such patients, cardiac output may be critically low, and the filling time is so restricted at very high heart rates that even small additional reductions in filling time cause disproportionately severe reductions in filling volume. Slowing the heart rate to a more normal rhythm can generally be accomplished pharmacologically, but electrical cardioversion may be required in emergencies (see Chapter 16).

Ancillary Factors That Affect the Venous System and Cardiac Output

In earlier sections of this chapter, the descriptions of the interrelationships between P_v and cardiac output were simplified by restricting the discussion to the effects evoked by individual variables. However, because the cardiovascular system is regulated by so many feedback control loops, its responses are rarely simple. A change in blood volume, for example, not only affects cardiac output directly through the Frank-Starling mechanism but also triggers reflexes that alter other aspects of cardiac function (such as the heart rate, atrioventricular conduction, and myocardial contractility) and other characteristics of the vascular system (such as peripheral resistance and venomotor tone). Several other factors, especially gravity (see Chapter 17) and respiration, also regulate cardiac output.

Circulatory Effects of Respiratory Activity

The normal, periodic activity of the respiratory muscles causes rhythmic variations in vena caval flow (Fig. 19.15). During respiration, the reduction in intrathoracic pressure is transmitted to the lumens of the thoracic blood vessels. The reduction in P_v during inspiration increases the pressure gradient between extrathoracic and intrathoracic veins. The consequent acceleration in venous return to the right atrium is shown in Fig. 19.15 as an increase in superior vena caval blood flow from 5.2 mL/sec during expiration to 11 mL/sec during inspiration.

The exaggerated reduction in intrathoracic pressure achieved by a strong inspiratory effort against a closed glottis (called **Müller's maneuver**) does not increase venous return proportionately. The extrathoracic veins collapse near their entry into the chest when their internal pressures fall below the ambient level. As the veins collapse, flow into the chest momentarily stops. The cessation of flow causes pressure upstream to rise, which forces the collapsed segment to reopen.

During normal expiration, flow into the central veins decelerates. However, the mean rate of venous return during

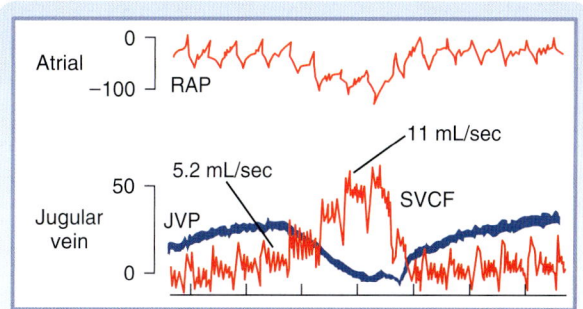

• **Fig. 19.15** During a normal inspiration, intrathoracic pressure, right atrial pressure (RAP), and jugular venous (JVP) pressure decrease, and flow in the superior vena cava (SVCF) increases (from 5.2 to 11 mL/sec). All pressures are given in mm H_2O. Femoral arterial pressure (not shown) did not change substantially during the normal inspiration.

normal respiration exceeds the flow during a brief period of **apnea** (cessation of respiration). Hence, normal inspiration apparently facilitates venous return more than normal expiration impedes it. In part, venous return is facilitated by the valves in the veins of the extremities. These valves prevent any reversal of flow during expiration. Thus the respiratory muscles and venous valves constitute an auxiliary pump for venous return.

 IN THE CLINIC

The dramatic increase in intrathoracic pressure induced by coughing constitutes an auxiliary pumping mechanism for the blood despite its concurrent tendency to impede venous return. Certain diagnostic procedures, such as coronary angiography or electrophysiological testing of cardiac function, increase the risk for ventricular fibrillation; therefore, patients undergoing such procedures are trained to cough rhythmically on command during the study. If ventricular fibrillation does occur, each cough can generate substantial increases in arterial blood pressure, and enough cerebral blood flow may be promoted to sustain consciousness. The cough raises intravascular pressure equally in the intrathoracic arteries and veins. Blood is propelled through the extrathoracic tissues because the increased pressure is transmitted to the extrathoracic arteries but not to the extrathoracic veins because the venous valves prevent backward flow from the intrathoracic to the extrathoracic veins.

In most forms of artificial respiration (mouth-to-mouth resuscitation, mechanical respiration), endotracheal pressure above atmospheric pressure is used to inflate the lungs, and expiration occurs by passive recoil of the thoracic cage (see Chapter 21). Thus lung inflation is accompanied by an appreciable rise in intrathoracic pressure. Vena caval flow decreases sharply during the phase of positive-pressure lung inflation when the endotracheal pressure progressively rises. When negative endotracheal pressure is used to facilitate deflation, vena caval flow accelerates more than when the lungs are allowed to deflate passively.

Sustained expiratory efforts increase intrathoracic pressure and thus impede venous return. Straining against a closed glottis (**Valsalva's maneuver**) regularly occurs during coughing, defecation, and heavy lifting. Intrathoracic pressures in excess of 100 mm Hg have been recorded in trumpet players, and pressures higher than 400 mm Hg have been observed during paroxysms of coughing. Such increases in pressure are transmitted directly to the lumens of the intrathoracic arteries. After coughing stops, arterial blood pressure may fall precipitously because of the preceding impediment to venous return.

Interplay of Central and Peripheral Factors in Control of the Circulation

The primary function of the circulatory system is to deliver the nutrients needed for tissue metabolism and growth and to remove the products of metabolism. The contributions of the components of the cardiovascular system to maintain adequate tissue perfusion under different physiological conditions were discussed previously. In this section, the interrelationships among the various components of the circulatory system are explored. The autonomic nervous system and the baroreceptors and chemoreceptors play key roles in regulating the cardiovascular system. Control of fluid balance by the kidneys, with maintenance of a constant blood volume, is also very important.

In any well-regulated system, one way to evaluate the extent and sensitivity of its regulatory mechanisms is to disturb the system and to observe how it restores the preexisting steady state. Two such disturbances, physical exercise and hemorrhage, are discussed in the following sections to illustrate operation of the various regulatory factors.

Exercise

The cardiovascular adjustments that occur during exercise consist of a combination of neural and local (chemical) factors. Neural factors include (1) central command, (2) reflexes that originate in the contracting muscle, and (3) the baroreceptor reflex. Central command is the cerebrocortical activation of the sympathetic nervous system that produces cardiac acceleration, increased myocardial contractile force, and peripheral vasoconstriction. Reflexes are activated intramuscularly by stimulation of mechanoreceptors (by stretch, tension) and chemoreceptors (by metabolic products) in response to muscle contraction. Impulses from these receptors travel centrally via small myelinated (group III) and unmyelinated (group IV) afferent nerve fibers. Group IV unmyelinated fibers may represent the muscle chemoreceptors, inasmuch as no morphological chemoreceptor has been identified. The central connections of this reflex are unknown, but the efferent limb consists of sympathetic nerve fibers to the heart and peripheral blood vessels. The baroreceptor reflex is described in Chapter 18, and local factors that influence skeletal muscle blood flow (metabolic vasodilators) are described in Chapter 17. Vascular chemoreceptors are important in regulation of the cardiovascular system during exercise. Evidence for this assertion comes from the observations that the $PaCO_2$, the PaO_2, and the pH of arterial blood remain normal during exercise.

Mild to Moderate Exercise

In humans or trained animals, anticipation of physical activity inhibits vagal nerve impulses to the heart and increases sympathetic discharge. The result is an increase in heart rate and myocardial contractility. The tachycardia and enhanced contractility increase cardiac output.

Peripheral Resistance

When cardiac stimulation occurs, the sympathetic nervous system also changes vascular resistance in the periphery. Sympathetic nervous system–mediated vasoconstriction increases vascular resistance and thereby diverts blood away from the skin, kidneys, splanchnic regions, and inactive

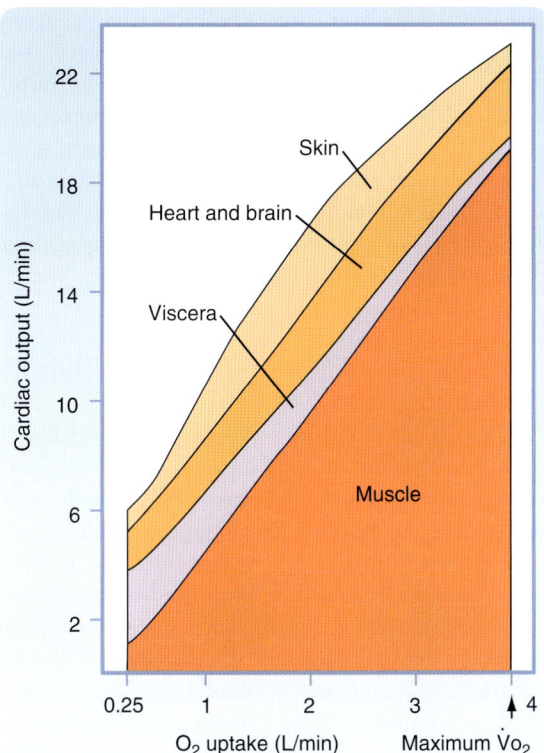

• **Fig. 19.16** Approximate distribution of cardiac output at rest and at different levels of exercise up to the maximal O₂ consumption (V̇o₂) in a normal young man. (Redrawn from Ruch HP, Patton TC. *Physiology and Biophysics.* 12th ed. Philadelphia: Saunders; 1974.)

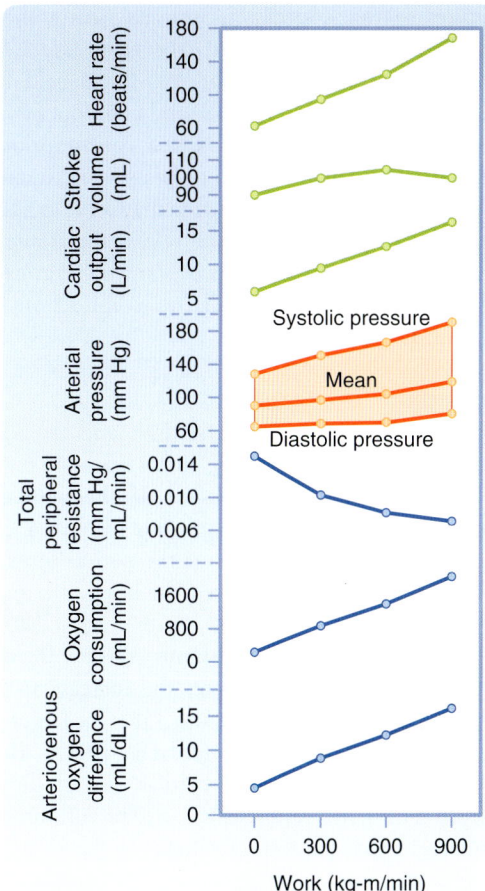

• **Fig. 19.17** Effects of different levels of exercise (i.e., work) on several cardiovascular variables. (Data from Carlsten A, Grimby G. *The Circulatory Response to Muscular Exercise in Man.* Springfield, IL: Charles C Thomas; 1966.)

muscle (Fig. 19.16). This increase in vascular resistance persists throughout the period of exercise.

Cardiac output and blood flow to active muscles increase as the intensity of exercise increases. Blood flow to the myocardium increases, whereas flow to the brain is unchanged. Blood flow in the skin initially decreases during exercise, then increases as body temperature rises with increments in the duration and intensity of exercise, and finally decreases when the skin vessels constrict as total body O₂ consumption nears its maximal value (see Fig. 19.16).

The major circulatory adjustment to prolonged exercise occurs in the vasculature of the active muscles. Local formation of vasoactive metabolites causes marked dilation of the resistance vessels. This dilation progresses with increases in the intensity of exercise. Potassium is one of the vasodilator substances released by the contracting muscle, and this ion may be partly responsible for the initial decrease in vascular resistance in the active muscles. Other contributing factors may be the release of adenosine and a decrease in tissue pH during sustained exercise. The local accumulation of metabolites causes the terminal arterioles to relax, and blood flow through the muscle may increase 15- to 20-fold above the resting level. This metabolic vasodilation of the precapillary vessels in active muscles occurs very soon after the onset of exercise. The decrease in TPR enables the heart to pump more blood at a lesser load, and it pumps more efficiently than if TPR were unchanged (see Chapters 17 and 18).

Marked changes in the capillary circulation also occur during exercise. At rest, only a small percentage of the capillaries are perfused, whereas in actively contracting muscle, all or nearly all of the capillaries contain flowing blood **(capillary recruitment).** The surface area available for exchange of gases, water, and solutes is increased many times. Furthermore, hydrostatic pressure in the capillaries is increased because of relaxation of the resistance vessels. Hence, water and solutes move into the muscle tissue. Tissue pressure rises and remains elevated during exercise as fluid continues to move out of the capillaries; this tissue fluid is carried away by the lymphatic vessels. Lymph flow is increased as a result of the rise in capillary hydrostatic pressure and the massaging effect of the contracting muscles on the valve-containing lymphatic vessels (see Chapter 17).

Contracting muscle avidly extracts O₂ from the perfusing blood and thereby increases the arteriovenous O₂ difference (Fig. 19.17). This release of O₂ from blood is facilitated by the shift in the oxyhemoglobin dissociation curve during exercise. During exercise, the high concentration of CO₂ and the formation of lactic acid cause a reduction in tissue pH. This decrease in pH, in addition to the increase in temperature in the contracting muscle, shifts the oxy-hemoglobin dissociation curve to the right (see Chapter 23). Therefore, at any given PO_2, less O_2 is held by the hemoglobin in the red blood cells, and consequently more O_2 is available for the tissues. Oxygen consumption may increase as much as 60-fold, with only a 15-fold increase in muscle blood flow. Muscle myoglobin may serve as a limited O_2 store during exercise, and it can release the attached O_2 at very low partial pressures. However, myoglobin can also facilitate O_2 transport from capillaries to mitochondria by serving as an O_2 carrier.

Cardiac Output

Because the enhanced sympathetic drive and the reduced parasympathetic inhibition of the sinoatrial node continue during exercise, tachycardia persists. If the workload is moderate and constant, the heart rate reaches a certain level and remains there throughout the period of exercise. However, if the workload increases, the heart rate increases concomitantly until a plateau of approximately 180 beats per minute is reached during strenuous exercise. In contrast to the large increase in heart rate, the increase in stroke volume is only approximately 10% to 35%, the larger values occurring in trained individuals (see Fig. 19.17). In well-trained distance runners, whose cardiac output can reach six to seven times the resting level, stroke volume attains approximately twice the resting value.

IN THE CLINIC

Cardiac muscle size (growth) is directly related to the amount of work that is imposed upon it. During development and in endurance exercise, cardiac growth is achieved at a constant relation between systolic blood pressure and the ratio of wall thickness to ventricular chamber radius. An echocardiographic measurement used to distinguish physiological from pathological hypertrophy is relative wall thickness (ratio of left ventricular wall thickness to chamber radius). In physiological hypertrophy, left ventricular mass and radius increase proportionately so that relative wall thickness does not change significantly. Examples of physiological hypertrophy occur in endurance athletes and in pregnant women, in whom left ventricular enlargement occurs with volume overload at constant relative wall thickness. Physiological hypertrophy is associated with an increased arteriolar diameter in experimental animals. Also, capillary density increases in proportion to the degree of hypertrophy. This is in contrast to the situation in pathological hypertrophy, in which a reduction of capillary density (rarefaction) can occur. Neither myocardial fibrosis nor derangement of muscle fiber orientation is detected in physiological hypertrophy, in contrast to the findings in pathological hypertrophy.

Thus the increase in cardiac output observed during exercise is correlated principally with an increase in heart rate. If the baroreceptors are denervated, the cardiac output and heart rate responses to exercise are small in comparison with those in individuals with normally innervated baroreceptors. However, with total cardiac denervation, exercise still increases cardiac output as much as it does in normal individuals. This increase in cardiac output is achieved chiefly by means of an elevated stroke volume. However, if a β-adrenergic receptor antagonist is given to dogs with denervated hearts, exercise performance is impaired. The β-adrenergic receptor antagonist prevents the cardiac acceleration and enhanced contractility caused by increased amounts of circulating catecholamines. Therefore, the increase in cardiac output necessary for maximal exercise performance is limited.

Venous Return

In addition to the contribution made by sympathetically mediated constriction of the capacitance vessels in both exercising and nonexercising parts of the body, venous return is aided by the auxiliary pumping action of the working skeletal muscles and the muscles of respiration (see also Chapters 21 and 24). The intermittently contracting muscles compress the veins that course through them. Because the venous valves are oriented toward the heart, the contracting muscle pumps blood back toward the right atrium (see Chapter 17). In exercise, the flow of venous blood to the heart is also aided by the deeper and more frequent respirations that increase the pressure gradient between the abdominal and thoracic veins (intrathoracic pressure becomes more negative during exercise).

In humans, blood reservoirs do not contribute much to the circulating blood volume. In fact, blood volume is usually reduced slightly during exercise, as evidenced by a rise in the hematocrit ratio. This decrease in blood volume is caused by water loss externally through sweating and enhanced ventilation and by fluid movement into the contracting muscle. However, fluid loss is counteracted in several ways. Fluid loss from the vascular compartment into the contracting muscles eventually reaches a plateau as interstitial fluid pressure rises and opposes the increased hydrostatic pressure in capillaries of the active muscle. Fluid loss is partially offset by movement of fluid from the splanchnic regions and inactive muscle into the bloodstream. This influx of fluid results from (1) a decrease in hydrostatic pressure in the capillaries of these tissues and (2) an increase in plasma osmolarity because of movement of osmotically active molecules into blood from the contracting muscle. Reduction in urine formation by the kidneys also helps conserve body water.

The large volume of venous blood returning to the heart is so effectively pumped through the lungs and out into the aorta that P_v remains essentially constant. Thus the Frank-Starling mechanism of a greater initial fiber length does not account for the greater stroke volume in moderate exercise. Radiographs of individuals at rest and during exercise reveal a decrease in heart size during exercise. However, during maximal or near-maximal exercise, right atrial pressure and end-diastolic ventricular volume do increase, and the Frank-Starling mechanism contributes to the enhanced stroke volume in very vigorous exercise.

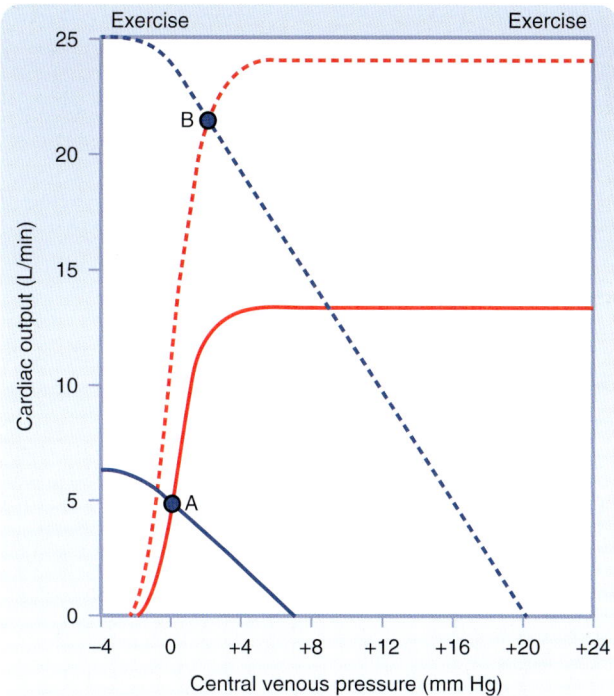

• **Fig. 19.18** Cardiac and vascular function curves are greatly altered during strenuous exercise, which allows cardiac output to increase fourfold to fivefold. The operating point of the cardiovascular system moves from point A to point B. The cardiac function curve during strenuous exercise is the result of increased heart rate, stroke volume, and contractility. The vascular function curve reflects greatly decreased total peripheral resistance and increased mean circulatory pressure. At the new operating point (point B), cardiac output is increased more than fourfold, but filling pressure is increased only slightly.

Coupling Between Heart and Vasculature During Exercise

In an active, healthy (untrained) individual, the mechanisms previously described typically lead to a fourfold to fivefold increase in cardiac output during vigorous exercise (Fig. 19.18). Increased cardiac output is the fundamental means by which more O_2 is delivered to exercising muscles (see Fig. 19.17). The cardiac function curve during exercise reflects increased stroke volume (up to ≈1.5-fold) and heart rate (up to ≈3-fold). The vascular function curve during dynamic exercise reflects a marked decrease in peripheral resistance (changed slope) and an increase in mean circulatory filling pressure (changed intercept), which result from increased venous constriction (tone) in the skeletal muscle "pump," and the respiratory "pump." In these conditions, the cardiovascular system is able to operate at a new point (point B in Fig. 19.18), in which cardiac output is increased while filling pressure is little changed. The graphical analysis (see Fig. 19.18) shows that without the systemic changes in vascular function, even a heart beating strongly and fast would be able to produce only a small increase in cardiac output.

Arterial Pressure

If exercise involves a large proportion of the body musculature, as in running or swimming, the reduction in

total vascular resistance can be considerable. Nevertheless, arterial pressure starts to rise with the onset of exercise, and the increase in blood pressure approximately parallels the severity of the exercise performed (see Fig. 19.17). Therefore, the increase in cardiac output is proportionally greater than the decrease in TPR. The vasoconstriction produced in the inactive tissues by the sympathetic nervous system (and to some extent by the release of catecholamines from the adrenal medulla) is important for maintenance of normal or increased blood pressure. Sympathectomy or drug-induced blockade of the adrenergic sympathetic nerve fibers decreases arterial pressure (hypotension) during exercise.

Sympathetic neural activity also elicits vasoconstriction in active skeletal muscle when additional muscles are recruited. In experiments in which one leg is working at maximal levels and then the other leg starts to work, blood flow decreases in the first working leg. Furthermore, blood levels of norepinephrine rise significantly during exercise, and most of the norepinephrine is released from sympathetic nerves to the active muscles.

As body temperature rises during exercise, the skin vessels dilate in response to thermal stimulation of the heat-regulating center in the hypothalamus, and TPR decreases further. This reduction in TPR would reduce blood pressure were it not for the increased cardiac output and the constriction of arterioles in the renal, splanchnic, and other tissues.

In general, P_a rises during exercise as a result of the increase in cardiac output. However, the effect of enhanced cardiac output is offset by an overall decrease in TPR, and therefore mean blood pressure increases only slightly. Vasoconstriction in the inactive vascular beds helps maintain normal arterial blood pressure for adequate perfusion of the active tissues. The actual P_a attained during exercise thus represents a balance between cardiac output and TPR (see Chapter 17). Systolic pressure usually increases more than diastolic pressure, which results in an increase in pulse pressure (see Fig. 19.17). The larger pulse pressure is primarily attributable to a greater stroke volume, but also to more rapid ejection of blood by the left ventricle and diminished peripheral runoff during the brief ventricular ejection period (see also Chapter 17).

Severe Exercise

During exhaustive exercise, the compensatory mechanisms begin to fail. The heart rate attains a maximal level of approximately 180 beats per minute, and stroke volume reaches a plateau. The heart rate may then decrease, which results in a fall in blood pressure. The exercising individual also frequently becomes dehydrated. Sympathetic vasoconstrictor activity supersedes the vasodilator influence on vessels of the skin so that the rate of heat loss is decreased. Body temperature is normally elevated during exercise. A reduction in heat loss through cutaneous vasoconstriction can lead to very high body temperatures and to acute distress during severe exercise. Tissue pH and blood pH decrease as a result of increased production of lactic acid and CO_2. The reduced pH may be a key factor that determines

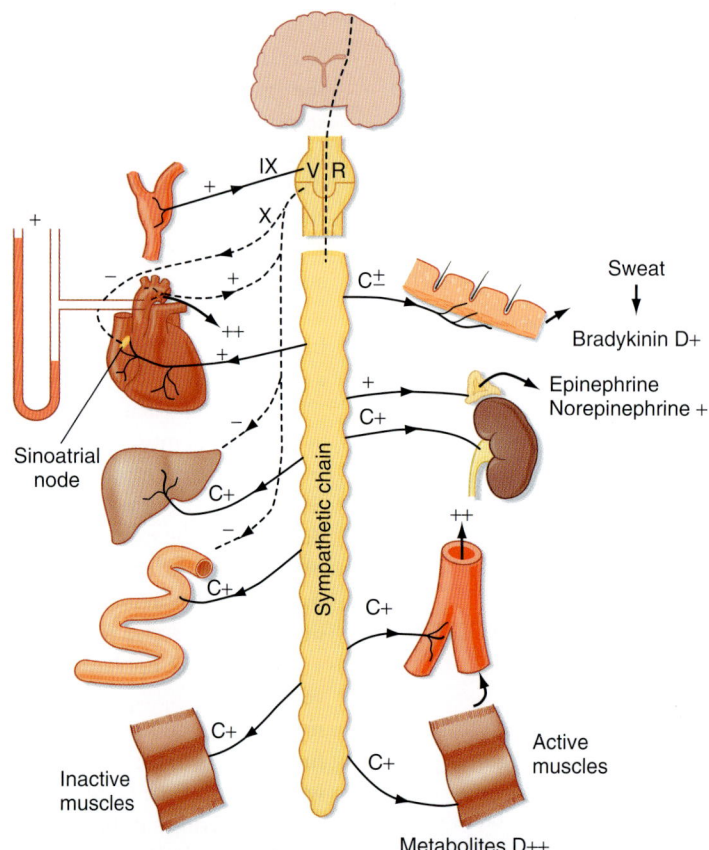

• **Fig. 19.19** Cardiovascular adjustments in exercise. *Plus signs* indicate increased activity, and *minus signs* indicate decreased activity. C, vasoconstrictor activity; D, vasodilator activity; IX, glossopharyngeal nerve; VR, vasomotor region; X, vagus nerve.

the maximal amount of exercise that a given individual can tolerate. Muscle pain, a subjective feeling of exhaustion, and loss of the will to continue determine exercise tolerance. A summary of the neural and local effects of exercise on the cardiovascular system is diagrammed in Fig. 19.19.

Postexercise Recovery

When exercise stops, the heart rate and cardiac output quickly decrease: The sympathetic drive to the heart is essentially removed. In contrast, TPR remains low for some time after the exercise is stopped, presumably because vasodilator metabolites have accumulated in the muscles during the exercise period. As a result of the reduced cardiac output and persistence of vasodilation in the muscles, arterial pressure falls, often below preexercise levels, for brief periods. Blood pressure is then stabilized at normal levels by the baroreceptor reflexes.

Limits of Exercise Performance

The two main factors that limit skeletal muscle performance in humans are the rate of O_2 use by the muscles and the O_2 supply to the muscles. However, O_2 use by muscle is probably not a critical factor. During exercise, maximal O_2 consumption (maximal $\dot{V}O_2$) by a large percentage of the body's muscle mass is unchanged or increases only slightly when additional muscles are activated. In fact, during

exercise of a large muscle mass, as in vigorous bicycling, the addition of bilateral arm exercise without change in the cycling effort produces only a small increase in cardiac output and maximal $\dot{V}O_2$. However, the additional arm exercise decreases blood flow to the legs. This centrally mediated (baroreceptor reflex) vasoconstriction during maximal cardiac output prevents the fall in blood pressure that would otherwise be caused by metabolically induced vasodilation in the active muscle. If use of O_2 by muscles were a significant limiting factor, recruitment of more contracting muscles would entail the use of much more O_2 to meet the enhanced O_2 requirements.

Limitation of the O_2 supply could be caused by inadequate oxygenation of blood in the lungs or limitation of the supply of O_2-laden blood to the muscles. Failure by the lungs to oxygenate blood fully can be ruled out because even with the most strenuous exercise at sea level, arterial blood is fully saturated with O_2. Therefore, O_2 delivery to the active muscles (or blood flow because the arterial blood O_2 content is normal) appears to be the limiting factor in muscle performance. This limitation could be caused by the inability to increase cardiac output beyond a critical level. In turn, this inability is caused by a limitation in stroke volume because the heart rate reaches maximal levels before maximal $\dot{V}O_2$ is reached. Hence, the major factor that limits muscle performance is the pumping capacity of the heart.

Physical Training and Conditioning

The response of the cardiovascular system to regular exercise is to increase its capacity to deliver O_2 to the active muscles and improve the ability of the muscle to use O_2. Maximal $\dot{V}O_2$ varies with the level of physical conditioning. Training progressively increases maximal $\dot{V}O_2$, which reaches a plateau at the highest level of conditioning. Highly trained athletes have a lower resting heart rate, a greater stroke volume, and lower peripheral resistance than they had before training or after deconditioning. The low resting heart rate is caused by a higher vagal tone and a lower sympathetic tone. During exercise, the maximal heart rate of a trained individual is the same as that in an untrained person, but it is attained at a higher level of exercise. A trained person also exhibits low vascular resistance in the muscles. If an individual exercises one leg regularly over an extended period and does not exercise the other leg, vascular resistance is lower and maximal $\dot{V}O_2$ is higher in the "trained" leg than in the "untrained" leg.

Physical conditioning is also associated with greater extraction of O_2 from the blood (greater arteriovenous O_2 difference) by the muscles. With long-term training, capillary density in skeletal muscle increases. Also, an increase in the number of arterioles may account for the decrease in muscle vascular resistance. The number of mitochondria increases, as does the number of oxidative enzymes in mitochondria. In addition, levels of adenosine triphosphatase (ATPase) activity, myoglobin, and enzymes involved in lipid metabolism increase in response to physical conditioning.

 IN THE CLINIC

Endurance training, such as running or swimming, increases left ventricular volume without increasing left ventricular wall thickness. In contrast, strength exercises, such as weightlifting, increase left ventricular wall thickness (hypertrophy) with little effect on ventricular volume. However, this increase in wall thickness is small in relation to that observed in chronic hypertension, in which afterload is persistently elevated because of high peripheral resistance.

Hemorrhage

The cardiovascular system is the system mainly affected in an individual who has lost a large quantity of blood. Arterial systolic, diastolic, and pulse pressures decrease, and the arterial pulse is rapid and feeble. The cutaneous veins collapse, and they fill slowly when compressed centrally. The skin is pale, moist, and slightly cyanotic. Respiration is rapid but may be shallow or deep.

Course of Arterial Blood Pressure Changes

Cardiac output decreases as a result of blood loss. The amount of blood removed when it is donated is approximately 10% of total blood volume; its removal is well tolerated, and mean arterial blood pressure changes little. This is not the case when greater amounts are lost from the

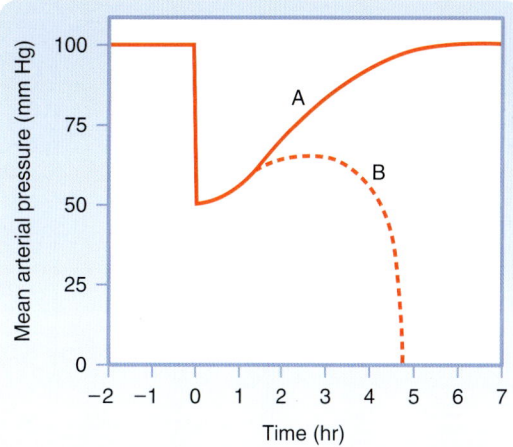

• Fig. 19.20 Changes in Mean Arterial Pressure After Rapid Hemorrhage. At time 0, rapid loss of blood causes a reduction in the mean arterial pressure to 50 mm Hg. After a period in which the pressure returns toward the control level, some individuals continue to improve (curve A) until the control pressure is attained. However, in other individuals, the pressure begins to decline (curve B) until death ensues.

circulation. The changes in P_a evoked by acute hemorrhage are illustrated in Fig. 19.19. If sufficient blood is rapidly withdrawn to decrease P_a to 50 mm Hg, the pressure then tends to rise spontaneously toward the control level over the next 20 or 30 minutes. In some individuals (curve A in Fig. 19.20), this trend continues, and normal pressure is regained within a few hours. In others (curve B in Fig. 19.20), the pressure rises initially after the cessation of hemorrhage. The pressure then begins to decline, and it continues to fall at an accelerating rate until death ensues. This progressive deterioration in cardiovascular function is termed *hemorrhagic shock*. At some time after the hemorrhage, the deterioration in the cardiovascular system becomes irreversible. A lethal outcome in patients with hemorrhagic shock can be prevented only temporarily by any known therapy, including massive transfusions of donor blood.

Compensatory Mechanisms

The changes in arterial pressure immediately after acute blood loss (see Fig. 19.20) indicate that certain compensatory mechanisms must be operative. Any mechanism that raises arterial blood pressure toward normal in response to a reduction in pressure is designated a negative feedback mechanism. This mechanism is termed *negative* because the direction of the secondary change in pressure is opposite the direction of the initiating change after the acute blood loss. The following negative feedback responses are evoked: (1) baroreceptor reflexes, (2) chemoreceptor reflexes, (3) cerebral ischemia responses, (4) reabsorption of tissue fluids, (5) release of endogenous vasoconstrictor substances, and (6) renal conservation of salt and water.

Baroreceptor Reflexes

The reductions in P_a and pulse pressure during hemorrhage decrease stimulation of the baroreceptors in the carotid

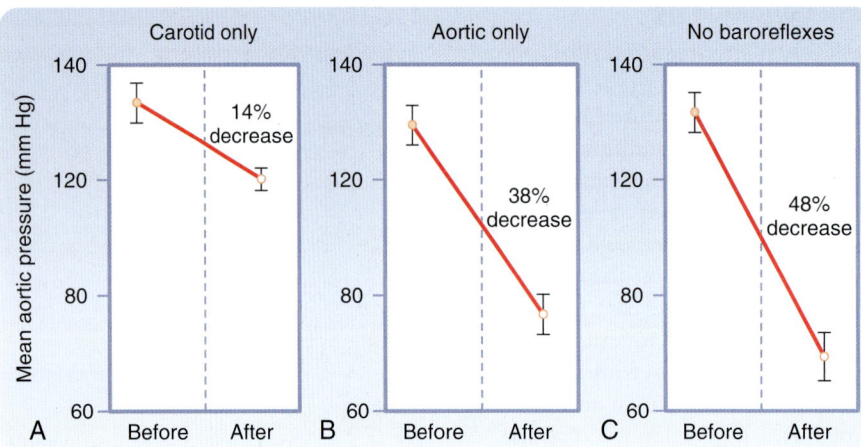

● Fig. 19.21 Changes in Mean Aortic Pressure in Response to an 8% Blood Loss in Three Conditions. **A,** The carotid sinus baroreceptors were intact and the aortic reflexes were interrupted. **B,** The aortic reflexes were intact and the carotid sinus reflexes were interrupted. **C,** All sinoaortic reflexes were abrogated. (Data from Shepherd JT. *Circulation.* 1974;50:418. Derived from the data of Edis AJ. *Am J Physiol.* 1971;221:1352.)

sinuses and aortic arch (see Chapter 18). Several cardiovascular responses are thus evoked, all of which tend to restore arterial pressure to the normal level. Such responses include reduction of vagal tone and enhancement of sympathetic tone, increased heart rate, and enhanced myocardial contractility.

The increased sympathetic tone also produces generalized venoconstriction, which has the same hemodynamic consequences as transfusion of blood (see Fig. 19.9). Sympathetic activation constricts certain blood reservoirs. This vasoconstriction acts as an autotransfusion of blood into the circulation. In humans, the cutaneous, pulmonary, and hepatic branches of the vasculature constitute the principal blood reservoirs.

Generalized arteriolar constriction is a prominent response to the reduced baroreceptor stimulation during hemorrhage. The reflex increase in peripheral resistance minimizes the fall in arterial pressure caused by the reduction in cardiac output. Fig. 19.21 shows the effect of an 8% blood loss on mean aortic pressure. When both vagus nerves were cut to eliminate the influence of the aortic arch baroreceptors and only the carotid sinus baroreceptors were operative (see Fig. 19.21*A*), this hemorrhage decreased mean aortic pressure by 14%. This pressure change did not differ significantly from the decline in pressure (12%) evoked by the same hemorrhage before vagotomy (not shown). When the carotid sinuses were denervated and the aortic baroreceptor reflexes were intact, the 8% blood loss decreased mean aortic pressure by 38% (see Fig. 19.21*B*). Hence, the carotid sinus baroreceptors were more effective than the aortic baroreceptors in attenuating the fall in pressure. However, when both sets of afferent baroreceptor pathways were interrupted (see Fig. 19.21*C*), an 8% blood loss reduced arterial pressure by 48%.

Arteriolar constriction is widespread during hemorrhage but it is by no means uniform. Vasoconstriction is most pronounced in the cutaneous, skeletal muscle, and splanchnic vascular beds, and it is slight or absent in the cerebral and coronary circulations in response to hemorrhage. In many instances, cerebral and coronary vascular resistance is diminished. The reduced cardiac output is redistributed to favor flow through the brain and the heart.

In the early stages of mild to moderate hemorrhage, renal resistance changes only slightly. The tendency for increased sympathetic activity to constrict the renal vessels is counteracted by autoregulatory mechanisms (see Chapters 18 and 35). With more prolonged and severe hemorrhage, however, renal vasoconstriction becomes intense.

The renal and splanchnic vasoconstriction during hemorrhage is least severe in the heart and brain. However, if such constriction persists too long, it may be detrimental. Frequently, patients survive the acute hypotensive period of a prolonged, severe hemorrhage, only to die several days later from the kidney failure that results from renal ischemia. Intestinal ischemia may also have dire effects. For example, intestinal bleeding and extensive sloughing of the mucosa can occur after only a few hours of hemorrhagic hypotension. Furthermore, the diminished splanchnic flow swells the centrilobular cells in the liver. The resulting obstruction of the hepatic sinusoids causes portal venous pressure to rise, and this response intensifies intestinal blood loss.

Chemoreceptor Reflexes

Reductions in arterial pressure below approximately 60 mm Hg do not evoke any additional responses through the baroreceptor reflexes because this pressure level constitutes the threshold for stimulation (see Chapter 18). However, low arterial pressure may stimulate peripheral chemoreceptors because inadequacy of local blood flow leads to hypoxia in the chemoreceptor tissue. Chemoreceptor excitation may then enhance the already existent peripheral vasoconstriction evoked by the baroreceptor reflexes. In addition, respiratory stimulation assists venous return by the auxiliary pumping mechanism described earlier (see also Chapter 24).

Cerebral Ischemia

When arterial pressure falls below approximately 40 mm Hg as a consequence of blood loss, the resulting cerebral ischemia activates the sympathoadrenal system. The discharge by sympathetic nerves is several times greater than the maximal neural activity that occurs when the baroreceptors cease to be stimulated. The vasoconstriction and increase in myocardial contractility may be pronounced. With more severe degrees of cerebral ischemia, however, the vagal centers also become activated. The resulting bradycardia aggravates the hypotension that initiated the cerebral ischemia

Reabsorption of Tissue Fluids

The arterial hypotension, arteriolar constriction, and reduced venous pressure during hemorrhagic hypotension cause the hydrostatic pressure in the capillaries to drop. The balance of these forces promotes the net reabsorption of interstitial fluid into the vascular compartment (see Chapter 17). The rapidity of this response is displayed in Fig. 19.22. When 45% of the estimated blood volume is removed over a 30-minute period, mean arterial blood pressure declines rapidly and then is largely restored to nearly the control level. Plasma colloid osmotic pressure declines markedly during the bleeding and continues to decrease more gradually for several hours. The reduction in colloid osmotic pressure reflects dilution of the blood by tissue fluids that contain little protein.

Considerable quantities of fluid may thus be drawn into the circulation during hemorrhage. Approximately 0.25 mL of fluid per minute per kilogram of body weight may be reabsorbed by the capillaries. Thus approximately 1 L of fluid per hour might be autoinfused from the interstitial spaces into the circulatory system of an average individual after acute blood loss. Substantial quantities of fluid may shift slowly from the intracellular space to the extracellular

space. This fluid exchange is probably mediated by secretion of cortisol from the adrenal cortex in response to hemorrhage. Cortisol is essential for the full restoration of plasma volume after hemorrhage.

Endogenous Vasoconstrictors

The catecholamines epinephrine and norepinephrine are released from the adrenal medulla in response to the same stimuli that evoke widespread sympathetic nervous discharge (see Chapter 43). Blood levels of catecholamines are high during and after hemorrhage. When blood loss is such that arterial pressure is reduced to 40 mm Hg, the level of catecholamines increases as much as 50-fold. Epinephrine comes almost exclusively from the adrenal medulla, whereas norepinephrine is derived from both the adrenal medulla and peripheral sympathetic nerve endings. These humoral substances reinforce the effects of the sympathetic nervous activity listed previously.

Vasopressin (antidiuretic hormone), a potent vasoconstrictor, is secreted by the posterior pituitary gland in response to hemorrhage (see Chapters 35 and 41). The plasma concentration of vasopressin rises progressively as arterial blood pressure diminishes (Fig. 19.23). The receptors responsible for the augmented release of vasopressin are the aortic arch and carotid sinus baroreceptors (high pressure) and stretch receptors in the left atrium (low pressure).

The diminished renal perfusion during hemorrhagic hypotension leads to the secretion of renin from the juxtaglomerular apparatus (see Chapter 35). This enzyme acts on a plasma protein, angiotensinogen, to form the decapeptide angiotensin I, which in turn is cleaved to the active octapeptide angiotensin II by angiotensin-converting enzyme; angiotensin II is a very powerful vasoconstrictor.

Renal Conservation of Salt and Water

Fluid and electrolytes are conserved by the kidneys during hemorrhage in response to various stimuli, including increased secretion of vasopressin, as noted previously (see Fig. 19.23), and increased renal sympathetic nerve activity, which enhances NaCl reabsorption by the nephron (decreased excretion). The lower arterial pressure decreases

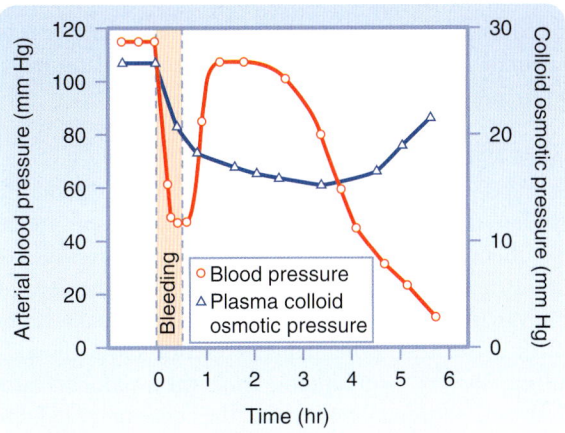

• **Fig. 19.22** Changes in arterial blood pressure and plasma colloid osmotic pressure in response to withdrawal of 45% of the estimated blood volume over a 30-minute period, beginning at time 0. (Redrawn from Zweifach BW. *Anesthesiology.* 1974;41:157.)

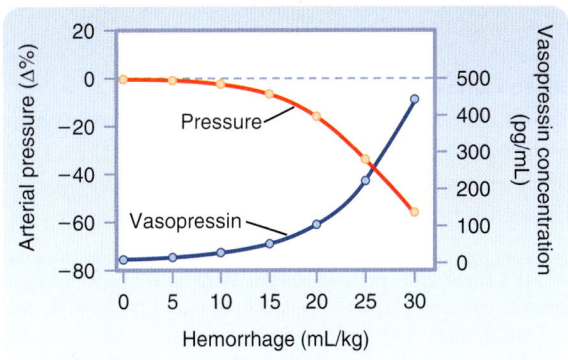

• **Fig. 19.23** Mean percentage changes in arterial blood pressure and plasma vasopressin concentration in response to blood loss. (Redrawn from Shen YT, et al. *Circ Res.* 1991;68:1422.)

the glomerular filtration rate, which also curtails the excretion of water and electrolytes. In addition, the elevated levels of angiotensin II, as described earlier, stimulate the release of aldosterone from the adrenal cortex. Aldosterone, in turn, stimulates reabsorption of NaCl by the nephrons. Thus NaCl and water excretion is decreased (see also Chapter 35).

Decompensatory Mechanisms

In contrast to negative feedback mechanisms, hemorrhage also evokes latent positive feedback mechanisms. These mechanisms exaggerate any primary change initiated by the blood loss. Specifically, positive feedback mechanisms aggravate the hypotension induced by blood loss and tend to initiate vicious cycles, which may lead to death.

Whether a positive feedback mechanism will lead to a vicious cycle depends on the gain of that mechanism. Gain is the ratio of the secondary change evoked by a given mechanism to the initiating change itself. A gain greater than 1 induces a vicious cycle; a gain of less than 1 does not. Consider a positive feedback mechanism with a gain of 2. If P_a were to decrease by 10 mm Hg, a positive feedback mechanism with a gain of 2 would then evoke a secondary reduction in pressure of 20 mm Hg, which in turn would cause a further decrease of 40 mm Hg. Thus each change would induce a subsequent one that is twice as great. Hence, P_a would decline at an ever-increasing rate until death occurred. This process is depicted in curve B in Fig. 19.20.

Conversely, a positive feedback mechanism with a gain of 0.5 also exaggerates any change in P_a, but the change would not necessarily lead to death. If arterial pressure suddenly decreased by 10 mm Hg, a positive feedback mechanism would initiate a secondary, additional fall of 5 mm Hg. This decrease, in turn, would provoke a further decrease of 2.5 mm Hg. The process would continue in ever-diminishing steps until arterial pressure approached an equilibrium value.

Some of the more important positive feedback mechanisms that are evident during hemorrhage include (1) cardiac failure, (2) acidosis, (3) central nervous system depression, (4) aberrations in blood clotting, and (5) depression of the mononuclear phagocytic system (MPS).* These mechanisms are discussed next.

Cardiac Failure

Shifts to the right in ventricular function curves, particularly in the later stages of hemorrhagic shock (Fig. 19.24), provide evidence of a progressive depression in myocardial contractility during hemorrhage.

---

*The MPS (previously called the *reticuloendothelial system*) consists of macrophages that are distributed throughout the body. They are derived from bone marrow and exist for a short period in circulating blood as monocytes. They then migrate into tissues, where they phagocytose foreign material and present antigens to lymphocytes to initiate the adaptive immune response. Cells of the MPS include the Kupffer cells of the liver, alveolar macrophages, microglia, and Langerhans cells.

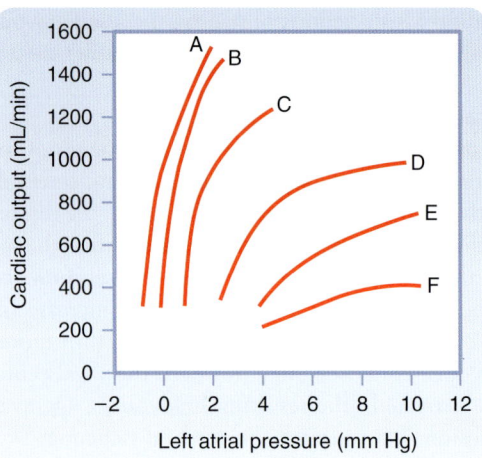

• **Fig. 19.24** Ventricular Function Curves for the Left Ventricle During the Course of Hemorrhagic Shock. *Curve A* represents the control function curve; *curves B* to *F* represent time after the hemorrhage: 117 minutes *(curve B)*, 247 minutes *(curve C)*, 280 minutes *(curve D)*, 295 minutes *(curve E)*, and 310 minutes *(curve F)*. (Redrawn from Crowell JW, Guyton AC. *Am J Physiol*. 1962;203:248.)

The hypotension induced by hemorrhage reduces coronary blood flow and therefore depresses ventricular function. The consequent reduction in cardiac output further reduces arterial pressure, a classic example of a positive feedback mechanism. Furthermore, reduced blood flow to peripheral tissues leads to an accumulation of vasodilator metabolites that decrease peripheral resistance and therefore aggravate the fall in arterial pressure.

Acidosis

The inadequate blood flow during hemorrhage affects the metabolism of all cells. The decreased O_2 delivery to cells accelerates tissue production of lactic acid and other acid metabolites. Moreover, impaired kidney function prevents adequate excretion of the excess H^+, and generalized metabolic acidosis ensues. The resulting depressant effect of acidosis on the heart is a further reduction in tissue perfusion, which aggravates the metabolic acidosis. Acidosis also reduces the reactivity of the heart and resistance vessels to neurally released and circulating catecholamines and thereby intensifies the hypotension.

Central Nervous System Depression

The hypotension in shock reduces cerebral blood flow. Moderate degrees of cerebral ischemia induce pronounced sympathetic nervous stimulation of the heart, arterioles, and veins, as noted earlier. In severe hypotension, however, the cardiovascular centers in the brainstem eventually become depressed because of inadequate cerebral blood flow. The resulting loss of sympathetic tone then reduces cardiac output and peripheral resistance. The consequent reduction in P_a intensifies the inadequate cerebral perfusion.

Endogenous opioids, such as enkephalins and β-endorphin, may be released into the brain substance and into the circulation in response to the same stresses that provoke circulatory shock. Opioids are stored, along with catecholamines, in secretory granules in the adrenal medulla and in sympathetic nerve terminals, and they are released together in response to stress. Similar stimuli cause the release of β-endorphin and adrenocorticotropic hormone from the anterior pituitary gland. Opioids depress the brainstem centers that mediate some of the compensatory autonomic adaptations to blood loss, endotoxemia, and other shock-provoking stress. Conversely, the opioid antagonist naloxone improves cardiovascular function and rates of survival in various forms of shock.

Aberrations in Blood Clotting

The alterations in blood clotting after hemorrhage are typically biphasic. An initial phase of hypercoagulability is followed by a secondary phase of hypocoagulability and fibrinolysis. In the initial phase, platelets and leukocytes adhere to the vascular endothelium, and intravascular clots, or thrombi, develop within a few minutes of the onset of severe hemorrhage. This phenomenon, called *disseminated intravascular coagulation* (DIC), occurs when thrombin is activated and causes widespread deposition of fibrin within narrow and medium-diameter vessels.

The initial phase is further enhanced by the release of thromboxane A_2 from various ischemic tissues. Thromboxane A_2 aggregates platelets. As more platelets aggregate, more thromboxane A_2 is released and more platelets are trapped. This form of positive feedback intensifies and prolongs the clotting tendency. Inflammatory cytokines (interleukin-6, tumor necrosis factor) also contribute to DIC. The rate of mortality from certain standard shock-provoking procedures has been reduced considerably by the administration of anticoagulants such as heparin.

In the later stages of hemorrhagic hypotension, the clotting time is prolonged, and fibrinolysis is prominent. Fibrinolysis occurs when clotting factors and platelets are depleted.

Depression of the Mononuclear Phagocytic System

During the course of hemorrhagic hypotension, MPS function becomes depressed. The phagocytic activity of the MPS is modulated by an opsonic protein. The opsonic activity in plasma diminishes during shock, and this change may account in part for the depression in MPS function. As a result, antibacterial and antitoxin defense mechanisms are impaired. Hypoperfusion also suppresses the barrier function of the adherens junctions and tight junctions in the

intestinal epithelium. Endotoxins from the normal bacterial flora of the intestine constantly enter the circulation. Ordinarily, they are inactivated by the MPS, principally in the liver. Disruption of the intestinal epithelial barrier, together with depression of the MPS, allows these endotoxins to invade the general circulation. Endotoxins produce profound, generalized vasodilation, mainly by inducing the synthesis of an isoform of nitric oxide synthase in the smooth muscle of blood vessels throughout the body. The profound vasodilation aggravates the hemodynamic changes caused by blood loss.

In addition to their role in inactivating endotoxin, macrophages release many of the mediators associated with shock. These mediators include acid hydrolases, neutral proteases, oxygen free radicals, certain coagulation factors, and the following arachidonic acid derivatives: prostaglandins, thromboxanes, and leukotrienes. Macrophages also release certain monokines that modulate temperature regulation, intermediary metabolism, hormone secretion, and the immune system.

Interactions of Positive and Negative Feedback Mechanisms

Hemorrhage provokes a multitude of circulatory and metabolic derangements. Some of these changes are compensatory, and others are decompensatory. Some of these feedback mechanisms possess high gain and others possess low gain. Furthermore, the gain of any specific mechanism varies with the severity of the hemorrhage. For example, with only a slight loss of blood, P_a is maintained within the normal range and the gain of the baroreceptor reflexes is high. With greater losses of blood, when P_a is below 60 mm Hg (i.e., below the threshold for the baroreceptors), further reductions in pressure have no additional influence through the baroreceptor reflexes. Hence, below this critical pressure, the baroreceptor reflex gain is zero or near zero.

In general, with minor degrees of blood loss, the gains of negative feedback mechanisms are high, whereas those of positive feedback mechanisms are low. The opposite is true with more severe hemorrhage. The gains of the various mechanisms add algebraically. Therefore, whether a vicious cycle develops depends on whether the sum of the positive and negative gains exceeds 1. Total gains in excess of 1 are, of course, more likely to occur with severe losses of blood. Therefore, to avert a vicious cycle, serious hemorrhages must be treated quickly and intensively, preferably by whole blood transfusion, before the process becomes irreversible.

Key Points

1. Two important relationships between cardiac output (Q_h) and central venous pressure (P_v) prevail in the cardiovascular system. With regard to the heart, Q_h varies directly with P_v (or preload) over a very wide range of P_v. This relationship is represented by the cardiac function curve, and it expresses the Frank-Starling mechanism. In the vascular system, P_v varies

inversely with Q_h. This relationship is represented by the vascular function curve, and it reflects the fact that as Q_h increases, a greater fraction of the total blood volume resides in the arteries and a smaller volume resides in the veins.

2. The principal cardiac mechanisms that govern cardiac output are the changes in numbers of myocardial

cross-bridges that interact and in the affinity of the contractile proteins for Ca^{++}. The principal factors that govern the vascular function curve are arterial and venous compliance, peripheral vascular resistance, and total blood volume.

3. The equilibrium values of Q_h and P_v that prevail under a given set of conditions are determined by the intersection of the cardiac and vascular function curves. At very low and very high heart rates, the heart is unable to produce adequate Q_h. At very low heart rates, the increase in filling during diastole cannot compensate for the small number of cardiac contractions per minute. At very high heart rates, the large number of contractions per minute cannot compensate for the inadequate filling time.

4. Gravity influences Q_h because the veins are so compliant, and substantial quantities of blood tend to pool in the veins of dependent portions of the body. Respiration changes the pressure gradient between the intrathoracic and extrathoracic veins. Hence, respiration serves as an auxiliary pump, which may affect the mean level of Q_h and induce rhythmic changes in stroke volume during the various phases of the respiratory cycle.

5. In anticipation of exercise, vagus nerve impulses to the heart are inhibited and the sympathetic nervous system is activated by central command. The result is an increase in heart rate, myocardial contractile force, and regional vascular resistance. In addition, vascular resistance increases in the skin, kidneys, splanchnic regions, and inactive muscles and decreases markedly in the active muscles. The overall effect is a pronounced reduction in total peripheral resistance, which, along with the auxiliary pumping action of the contracting skeletal muscles, greatly increases venous return. The increases in heart rate and myocardial contractility, both induced by the activation of cardiac sympathetic nerves, enables the heart to transfer blood to the pulmonary and systemic circulations, thereby increasing cardiac output. Stroke volume increases only slightly. O_2 consumption and blood O_2 extraction increase, and systolic pressure and mean blood pressure increase slightly. As body temperature rises during exercise, the skin blood vessels dilate. However, when the heart rate becomes maximal during severe exercise, the skin vessels constrict. This increases the effective blood volume but causes greater increases in body temperature and a feeling of exhaustion. The limiting factor in exercise performance is delivery of blood to the active muscles.

6. Acute blood loss induces tachycardia, hypotension, generalized arteriolar constriction, and generalized venoconstriction. Acute blood loss invokes a number of negative feedback (compensatory) mechanisms, such as baroreceptor and chemoreceptor reflexes, responses to moderate cerebral ischemia, reabsorption of tissue fluids, release of endogenous vasoconstrictors, and renal conservation of water and electrolytes. Acute blood loss also invokes a number of positive feedback (decompensatory) mechanisms, such as cardiac failure, acidosis, central nervous system depression, aberrations in blood coagulation, and depression of the mononuclear phagocytic system. The outcome of acute blood loss depends on the sum of gains of the positive and negative feedback mechanisms and on the interactions between these mechanisms.

Additional Readings

Eijsvogels TMH, Fernandez AB, Thompson PD. Are there deleterious cardiac effects of acute and chronic endurance exercise? *Physiol Rev.* 2016;96:99.

Fernandes T, Baraúna VG, Negrão CE, et al. Aerobic exercise training promotes physiological cardiac remodeling involving a set of microRNAs. *Am J Physiol.* 2015;309:H543.

Gruen RL, Brohi K, Schreiber M, et al. Haemorrhage control in severely injured patients. *Lancet.* 2012;380:1099.

Laughlin MH, Korthuis RJ, Duncker DJ, et al. Control of blood flow to cardiac and skeletal muscles during exercise. In: Rowell LB, Shepherd JT, eds. *Handbook of Physiology, Section 12: Exercise: Regulation and Integration of Multiple Systems.* Bethesda, MD: Oxford University Press; 1996.

Lavie CJ, Arena R, Swift DL, et al. Exercise and the cardiovascular system: clinical science and cardiovascular outcomes. *Circ Res.* 2015;117:207.

Valparaiso AP, Vicente DA, Bograd BA, et al. Modeling acute traumatic injury. *J Surg Res.* 2015;194:220.

The Respiratory System

MICHELLE M. CLOUTIER AND ROGER S. THRALL

433

20

Introduction to the Respiratory System

LEARNING OBJECTIVES

Upon completion of this chapter, the student should be able to answer the following questions:

1. Explain the anatomical structure/function relationships of the upper and lower components of the respiratory system.
2. Compare and contrast the pulmonary and bronchial circulatory systems.
3. Explain the relationships between innervation and muscles in the control of respiration.
4. Compare and contrast the roles of the conducting airways and components of the respiratory unit.
5. Compare and contrast the effects of stimulation of the parasympathetic and sympathetic nervous systems on respiratory responses.
6 Describe lung development and alveolar repair after injury to regeneration of normal architecture.

The primary function of the lungs is gas exchange, which consists of movement of oxygen (O_2) into the body and removal of carbon dioxide (CO_2). This chapter provides an overview of lung anatomical structure/ function relationships (i.e., upper and lower airways, the two circulatory systems, innervation, and muscles), lung development (at the embryo stage and throughout life), and lung repair. This chapter is designed to provide a broad conceptual understanding of structure/function interactions and is not intended to afford a comprehensive understanding of individual lung structures and anatomy.

Lung Anatomical Structure/Function Relationships

The lungs are contained in a space with a volume of approximately 4 L, but they have a surface area for gas exchange that is the size of a tennis court (≈ 85 m^2). This large surface area is composed of myriads of independently functioning respiratory units. Unlike the heart, but like the kidneys, the lungs demonstrate functional unity; that is, each unit is structurally identical and functions just like every other unit. Because the divisions of the lung and the sites of disease are designated by their anatomical locations (e.g., right upper lobe, left lower lobe), it is essential to

understand pulmonary anatomy in order to comprehend respiratory physiology and pathophysiological alterations in respiratory diseases.

Upper Airways: Nose, Sinuses, and Pharynx

The respiratory system begins at the nose and ends in the most distal **alveolus.** Thus the **nasal cavity,** the **posterior pharynx,** the **glottis** and **vocal cords,** the **trachea,** and all divisions of the **tracheobronchial tree** are included in the respiratory system. The **upper airway** consists of all structures from the nose to the vocal cords, including sinuses and the larynx, whereas the **lower airway** consists of the trachea, airways, and alveoli. The upper airways "condition" inspired air so that by the time air reaches the trachea, inspired air is at body temperature and fully humidified. The nose also functions to filter, entrap, and clear particles larger than 10 µm in size. The interior of the nose is lined by respiratory epithelial cells interspersed with surface secretory cells. These secretory cells produce important immunoglobulins, inflammatory mediators, and interferons, which are the first line of host defense.

The paranasal **sinuses (frontal, maxillary, sphenoid, and ethmoid)** are lined by ciliated epithelial cells and surround the nasal passages (Fig. 20.1*A*). The cilia facilitate the movement of mucus from the upper airways and clear the main nasal passages approximately every 15 minutes. The functions of the sinuses are (1) to lessen the weight of the skull, which makes upright posture easier; (2) to offer resonance to the voice; and (3) to protect the brain from frontal trauma. The fluid covering their surfaces is continually being propelled into the nose. In some sinuses (e.g., the **maxillary sinus**), the opening **(ostium)** is at the upper edge, which makes them particularly susceptible to retention of mucus. The ostia are readily obstructed by nasal edema (swelling), and retention of secretions and secondary infection **(sinusitis)** can result. The volume of the nose in an adult is approximately 20 mL, but its surface area is greatly increased by the **nasal turbinates,** which are a series of three continuous ribbons of tissue that protrude into the nasal cavity (see Fig. 20.1*B*). The nose enables the sense of smell. Neuronal endings in the roof of the nose above the **superior turbinate** carry impulses through the **cribriform plate** to the **olfactory bulb.**

The **pharynx** is divided into three sections: the **nasopharynx, oropharynx,** and **laryngopharynx.** Important

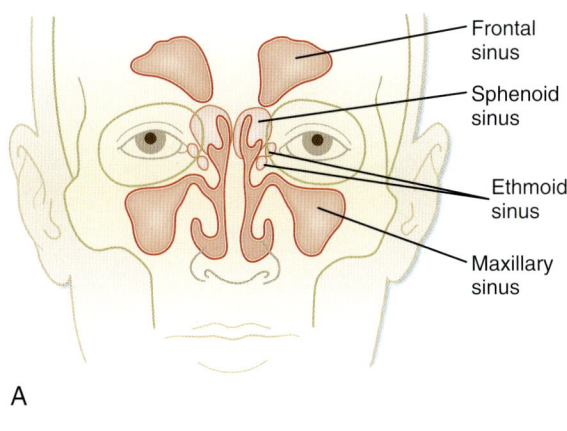

A

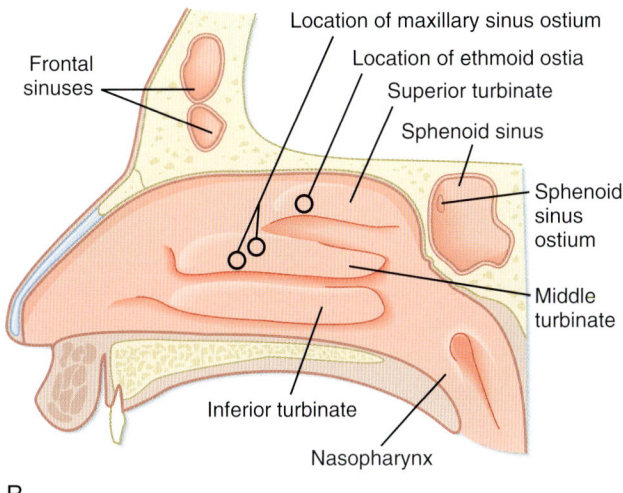

B

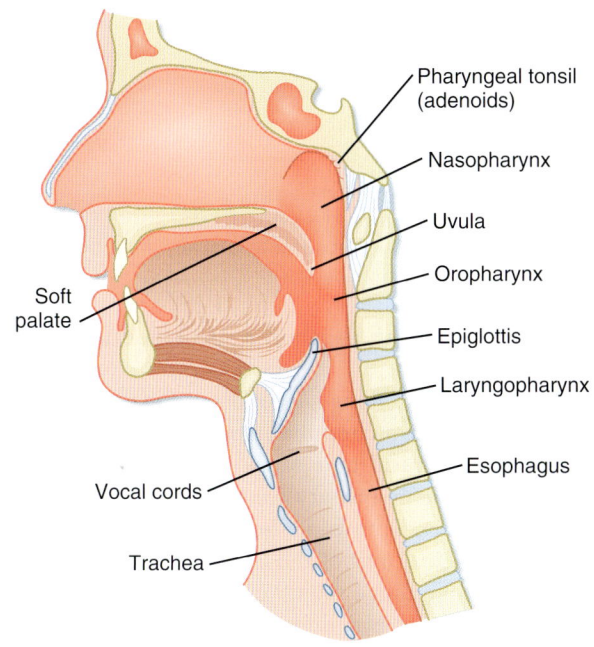

C

• **Fig. 20.1** Illustrations of Upper Airway Anatomy. **A,** Anterior view of the paranasal sinuses. **B,** Lateral view of the nasal passage structures demonstrating the superior, middle, and inferior turbinates and the sinus ostia. **C,** Lateral midsagittal section view of the head and neck, showing the three divisions of the pharynx (nasopharynx, oropharynx, and laryngopharynx) and surrounding upper airway structures.

structures within these regions include the **epiglottis, vocal cords,** and **arytenoid cartilage** which is attached to the vocal cords (see Fig. 20.1*C*). The nasopharynx (2 to 3 cm wide and 3 to 4 cm long) is the most anterior and lies behind the nose. In this region, the nose and mouth are connected via an isthmus (canals) that enables both oral and nasal breathing. Also, the nasopharynx contains small masses of lymphoid tissue (adenoids), also known as *pharyngeal tonsils,* which fight infections. The nasopharynx is connected to the middle ear cavity via the eustachian tubes, which aid in equalizing pressure in the ear to atmospheric pressure; thus they represent a drainage pathway of lymphatic fluid between the throat, nose, and ears. This network of structures provides a means of fighting infections but also is a common location for infections in the head.

The soft palate separates the nasopharynx and the oropharynx, which ends at the epiglottis. The laryngopharynx begins at the epiglottis and ends at the esophagus. Its major role is to help regulate the passage of food into the esophagus and air into the lungs. With some infections, these structures can become **edematous** and contribute significantly to airflow resistance. The epiglottis and arytenoid cartilage (attached to the vocal cords) cover or act as a hood over the vocal cords during swallowing. Thus under normal circumstances, the epiglottis and arytenoid cartilage function to prevent aspiration of food and liquid into the lower respiratory tract. The act of swallowing food after **mastication** (chewing) usually occurs within 2 seconds, and it is closely synchronized with muscle reflexes that coordinate opening and closing of the airway. Hence, air is allowed to enter the lower airways, and food and liquids are kept out. Patients with some neuromuscular diseases have altered muscle reflexes and can lose this coordinated swallowing mechanism. Such patients may become susceptible to aspiration of food and liquid, which poses a risk for **pneumonia.**

Lower Airways: Trachea, Bronchi, Bronchioles, and Respiratory Unit

The right lung, located in the right **hemithorax,** is divided into three lobes (**upper, middle,** and **lower**) by two interlobular fissures (oblique, horizontal), whereas the left lung, located in the left hemithorax, is divided into two lobes (**upper,** including the **lingula,** a tongue-like projection of the anterior aspect of the upper lobe, and **lower**) by an **oblique fissure** (Fig. 20.2). Both the right and left lungs are covered by a thin membrane called the **visceral pleura** and are encased by another membrane called the **parietal pleura.** The interface of these two pleurae allows for smooth gliding of the lung as it expands in the chest and produces a potential space. Air can enter between the visceral and parietal pleurae by trauma, surgery, or rupture of a group of alveoli; the resulting condition is a **pneumothorax.** Fluid can also enter this space and create a **pleural effusion** or, in the case of severe infection, an **empyema.**

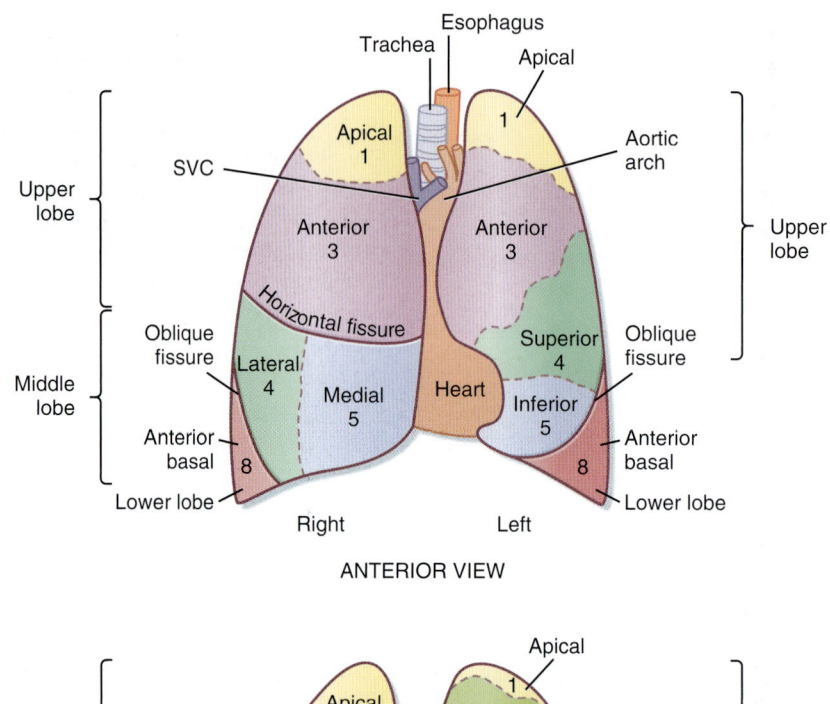

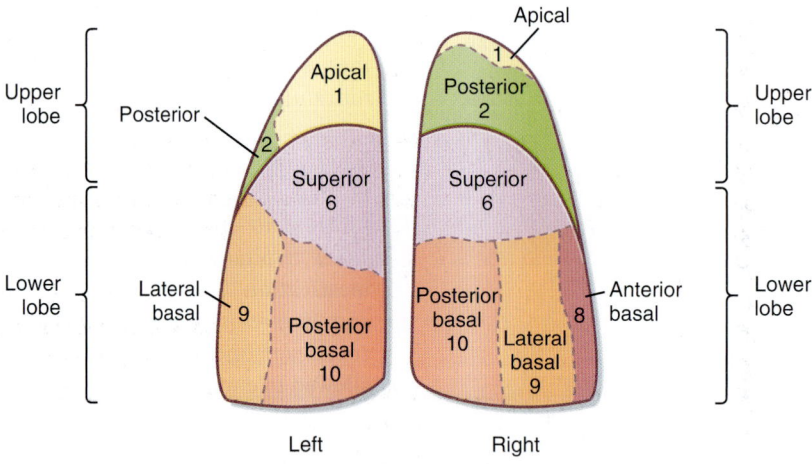

• **Fig. 20.2** Illustrations of the Topography of the Lung With Anterior and Posterior Views, Demonstrating the Lobes, Segments, and Fissures. The fissures (or chasms) demarcate the lobes in each lung. Numbers refer to specific bronchopulmonary segments, as depicted in Fig. 20.3. SVC, superior vena cava.

The **trachea** branches into two main stem bronchi (Fig. 20.3). These main stem bronchi then divide (like the branches of a tree) into lobar bronchi (one for each lobe), which in turn divide into **segmental bronchi** (Fig. 20.4; see also Fig. 20.3) and then into smaller and smaller branches **(bronchioles)** until ending in the **alveolus** (Fig. 20.5). Bronchi and bronchioles differ not only in size but also by the presence of cartilage, the type of epithelium, and their blood supply (Table 20.1). Beyond the segmental bronchi, the airways divide in a dichotomous or asymmetrical branching pattern. Bronchi, distinguished by their size and the presence of cartilage, eventually become **terminal bronchioles,** which are the smallest airways without alveoli. Each branching of an airway results in an increase in the number of airways with smaller diameters; as a result, the total surface area for the next generation of branches increases. Terminal bronchioles terminate in an opening

(duct) to a group of alveoli and are called **respiratory bronchioles.**

The region of the lung supplied by a segmental bronchus is called a **bronchopulmonary segment** and is the functional **anatomical** unit of the lung. Because of their structure, bronchopulmonary segments that have become irreversibly diseased can easily be removed surgically. The basic **physiological** unit of the lung is the gas-exchanging unit **(respiratory unit),** which consists of the respiratory bronchioles, the alveolar ducts, and the alveoli (see Figs. 20.4 and 20.5). The bronchi, which contain cartilage, and the terminal bronchioles (i.e., lacking alveoli), in which cartilage is absent, serve to move gas from the airways to the alveoli and are referred to as the **conducting airways.** This area of the lung (≈150 mL in volume) does not participate in gas exchange and forms the **anatomical dead space.** The respiratory bronchioles with alveoli and the area beginning

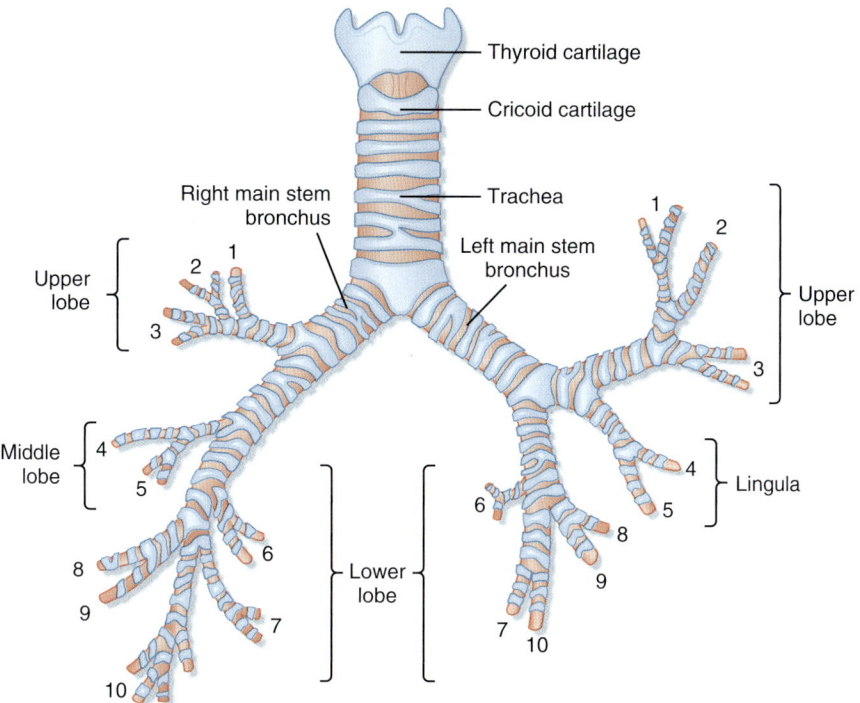

• **Fig. 20.3** Illustration of Bronchopulmonary Segments, Anterior View. The numbers correspond to those in Fig. 20.2: 1, apical; 2, posterior; 3, anterior; 4, lateral (superior); 5, medial (inferior); 6, superior; 7, medial basal; 8, anterior basal; 9, lateral basal; 10, posterior basal. The medial basal regions (7) are located in the upper region of the posterior basal regions (10) in Fig. 20.2.

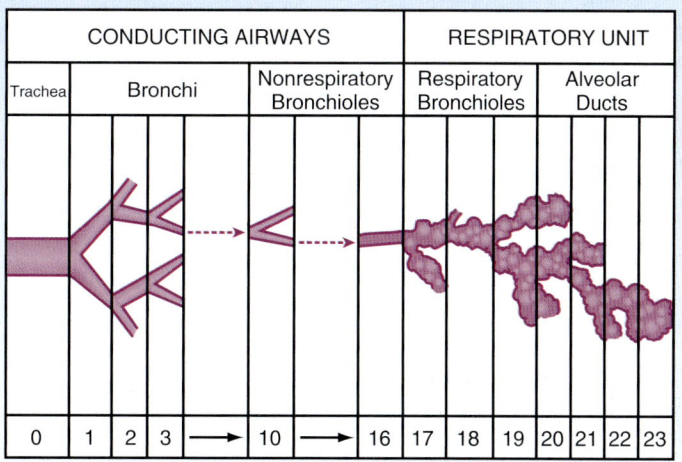

CONDUCTING AIRWAYS			RESPIRATORY UNIT	
Trachea	Bronchi	Nonrespiratory Bronchioles	Respiratory Bronchioles	Alveolar Ducts
0	1 2 3 ⟶ 10 ⟶ 16		17 18 19	20 21 22 23

• **Fig. 20.4** Illustration of Conducting Airways and Alveolar Units of the Lung. The relative size of the alveolar unit is greatly enlarged. Numbers at the bottom indicate the approximate number of generations from trachea to alveoli, which may vary from as few as 10 to as many as 23. (From Weibel ER. *Morphometry of the Human Lung.* Heidelberg: Springer-Verlag; 1963.)

with the respiratory bronchioles to the alveoli are where all gas exchange occurs. This region is only approximately 5 mm long, but it is the largest volume of the lung, at a volume of approximately 2500 mL and with a surface area of 70 m² when the lung and chest wall are at the resting volume (see Table 20.1).

The alveoli are polygonal in shape and approximately 250 μm in diameter. **Alveolar spaces** are responsible for most of the lung's volume; these spaces are divided by tissue

known collectively as the **interstitium.** The interstitium is composed primarily of lung collagen fibers and is a space in which fluid and cells can potentially accumulate. An adult has approximately 5×10^8 alveoli (Fig. 20.6), which are composed of **type I** and **type II** epithelial cells. Under normal conditions, type I and type II cells exist in a 1:1 ratio. The type I cell occupies 96% to 98% of the surface area of the alveolus, and it is the primary site for gas exchange. The thin cytoplasm of type I cells is ideal for optimal gas diffusion. In addition, the basement membrane of type I cells and the capillary endothelium are fused, which minimizes the distance for gas diffusion and thereby facilitates gas exchange.

Type II cells are cuboidal and usually found in the "corners" of the alveolus, where they occupy 2% to 4% of the surface area. During embryonic development, the alveolar epithelium is composed entirely of type II cells, and only very late in gestation do they differentiate into type I cells and form the "normal" alveolar epithelium for optimal gas exchange. Also, type II cells synthesize **pulmonary surfactant** (see Chapter 26), which reduces surface tension (nonsticking properties) in the alveolus and thereby promotes less resistance during inhalation and exhalation. Gas exchange occurs in the alveoli through a dense mesh-like network of capillaries and alveoli called the **alveolar-capillary network.** The barrier between gas in the alveoli and the red blood cell is only 1 to 2 μm thick and consists of type I alveolar epithelial cells, capillary endothelial cells, and their respective basement membranes. O_2 and CO_2 passively diffuse across this barrier into plasma and red blood cells (see Chapter 24). Red blood cells pass through the network in less than 1 second, which is sufficient time for CO_2 and O_2 gas exchange.

Circulatory Systems in the Lung

The circulation to the lung is unique in its duality and ability to accommodate large volumes of blood at low pressure. The lung has two separate blood supplies, one

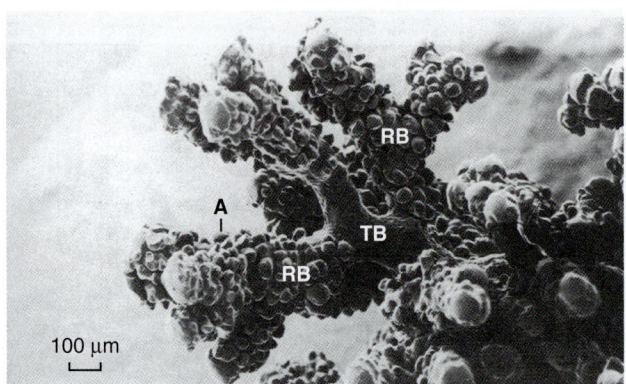

• **Fig. 20.5** The Airway From the Terminal Bronchiole to the Alveolus. Note the absence of alveoli in the terminal bronchiole. A, alveolus; RB, respiratory bronchiole; TB, terminal bronchiole.

IN THE CLINIC

The conducting airways are involved in several major pulmonary diseases collectively referred to as **obstructive pulmonary disease,** including asthma, bronchiolitis, chronic bronchitis, and cystic fibrosis. Obstruction of airflow through the airways is commonly caused by increased amounts of mucus, airway inflammation, and smooth muscle constriction. **Asthma** is a chronic inflammatory disease of the large and small airways, mediated predominantly by lymphocytes and eosinophils, and is associated with increased amounts of mucus in the airways and with reversible constriction of the airway smooth muscle (bronchospasm). **Bronchiolitis** is a disease of the bronchioles that usually occur in young infants and is caused by viruses, primarily **respiratory syncytial virus. Chronic bronchitis,** a disease typically of people who smoke, is associated with a marked increase in mucus-secreting cells in the airways and an increase in mucus production. **Cystic fibrosis** is an autosomal recessive genetic disease caused by mutations in the Cl⁻ ion channel of the cystic fibrosis transmembrane conductance regulator (CFTR). Mutations in the *CFTR* gene cause a reduction in chloride and water secretion into the mucus overlying the epithelia cells, which increases the viscosity of mucus. This situation results in mucus accumulation and chronic pulmonary infections, primarily by *Pseudomonas aeruginosa.*

Not all important obstructive lung diseases involve the airways directly. **Emphysema** is an irreversible, obstructive lung disease common in people who smoke. The pathogenesis involves a progressive destruction of the elastic tissues in the lung with a loss of alveolar/capillary structure. The mechanisms of tissue destruction are unclear but may involve proteolytic enzymes and other toxic compounds in cigarette smoke. Emphysema also occurs in nonsmoking individuals with the genetic disorder α_1-antitrypsin deficiency, caused by the inability to regulate proteolytic enzymes, particularly elastase.

for the uptake of O_2 and removal of CO_2 from the body (pulmonary circulation) and the other to supply O_2 to lung tissue (bronchial circulation; Fig. 20.7).

Pulmonary Circulation

The pulmonary circulation begins in the right atrium of the heart. Deoxygenated blood from the right atrium enters the right ventricle via the tricuspid valve, and it is then pumped under low pressure (9 to 24 mm Hg) into the pulmonary artery (pulmonary trunk), which is approximately 3 cm

TABLE 20.1	Anatomical Characteristics of Bronchi and Bronchioles					
Anatomical Site	Cartilage	Diameter (mm)	Epithelium	Blood Supply	Alveoli	Volume (mL)
Bronchi	Present	>1	Pseudostratified Columnar	Bronchial	Absent	—
Terminal bronchioles	Absent	<1	Cuboidal	Bronchial	Absent	>150
Respiratory bronchioles	Absent	<1	Cuboidal/ alveolar	Pulmonary	Present	2500

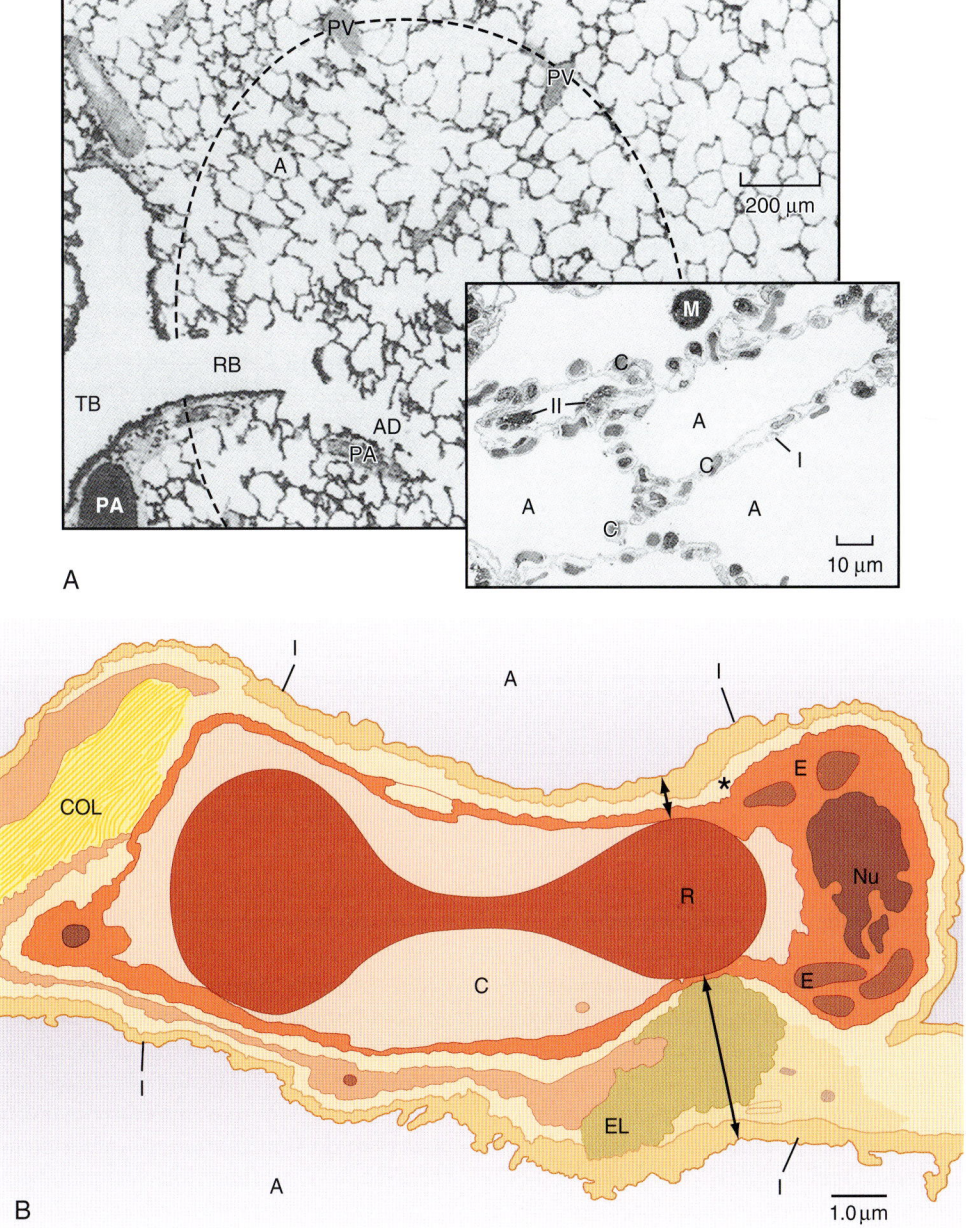

• Fig. 20.6 Alveoli. A. The terminal respiratory unit consists of the alveoli (A) and the alveolar ducts (AD) arising from a respiratory bronchiole (RB). Each unit is approximately spherical, as suggested by the *dashed outline.* Pulmonary venous vessels (PV) have a peripheral location. PA, pulmonary artery; TB, terminal bronchiole. **Inset,** type I and type II alveolar epithelial cells. A large fraction of the alveolar wall consists of capillaries (C) and their contents. **B.** Illustration of a cross-section of an alveolar wall, showing the path of diffusion of O_2 and CO_2. The thin side of the alveolar wall barrier *(short double arrow)* consists of type I epithelium (I), interstitium *(asterisk)* formed by the fused basal laminae of the epithelial and endothelial cells, capillary endothelium (E), plasma in the alveolar capillary (C), and the cytoplasm of the red blood cell (R). The thick side of the gas-exchange barrier *(long double arrow)* has an accumulation of elastin (EL), collagen (COL), and matrix that jointly separate the alveolar epithelium from the alveolar capillary endothelium. Nu, nucleus of the capillary endothelial cell.

in diameter, and branches quickly (5 cm from the right ventricle) into the right and left main pulmonary arteries, which supply blood to the right and left lungs, respectively. The arteries of the pulmonary circulation are the only arteries in the body that carry deoxygenated blood. The deoxygenated blood in the pulmonary arteries passes through a progressively smaller series of branching vessels—arteries (diameter, >500 µm); arterioles (diameter, 10 to 200 µm); and capillaries (diameter, <10 µm)—that end in a complex mesh-like network of capillaries. The sequential branching pattern of the pulmonary arteries follows the pattern of airway branching.

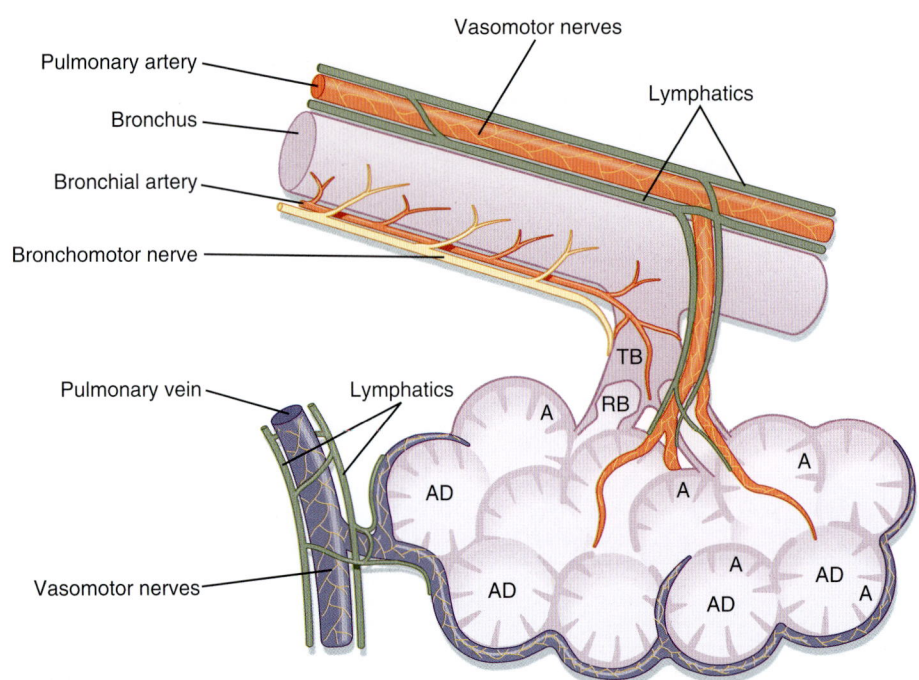

• **Fig. 20.7** Illustration of the anatomical relationship of the pulmonary artery, the bronchial artery, the airways, and the lymphatic vessels. A, alveoli; AD, alveolar ducts; RB, respiratory bronchioles; TB, terminal bronchioles.

The functions of the pulmonary circulatory system are (1) to reoxygenate the blood and release CO_2, (2) to aid in fluid balance in the lung, and (3) to distribute metabolic products to and from the lung. Red blood cells are oxygenated in the capillaries that surround the alveoli, where the pulmonary capillary bed and the alveoli come together in the alveolar wall in a unique configuration for optimal gas exchange. Gas exchange occurs through this alveolar-capillary network (see Chapter 24).

The total blood volume of the pulmonary circulation is approximately 500 mL, which is approximately 10% of the circulating blood volume. Approximately 75 mL of blood is present in the alveolar-capillary network of healthy adults at any one time. The pulmonary capillary bed is the largest vascular bed in the body. It covers a surface area of 70 to 80 m², which is nearly as large as the alveolar surface area. During exercise, the pulmonary capillary blood volume increases from 75 mL to as high as 200 mL because of the recruitment of new capillaries as the result of an increase in pressure and flow. This recruitment of new capillaries is a unique feature of the lungs, and it allows for compensation in periods of stress, as in the case of exercise.

The oxygenated blood leaves the alveolus through a network of small pulmonary venules (15 to 500 μm in diameter) and veins. These small vessels quickly coalesce to form larger pulmonary veins (>500 μm in diameter), through which the oxygenated blood returns to the left atrium of the heart. In contrast to arteries, arterioles, and capillaries, which closely follow the branching patterns of the airways, venules and veins run quite distant from the airways.

Structure of the Pulmonary Circulation

The arteries of the pulmonary circulation have thin walls, with minimal smooth muscle. They are seven times more compliant than systemic vessels, and they are easily distensible. This highly compliant state of the pulmonary arterial vessels requires lower pressure for blood flow through the pulmonary circulation than do the more muscular, noncompliant arterial walls of the systemic circulation. The vessels in the pulmonary circulation, under normal circumstances, are in a dilated state and have larger diameters than do similar arteries in the systemic system. All of these factors contribute to a very compliant, low-resistance circulatory system, which aids in the flow of blood through the pulmonary circulation via the relatively weak pumping action of the right ventricle. This low-resistance, low-work system also explains why the right ventricle is less muscular than the left ventricle. The pressure gradient differential for the pulmonary circulation from the pulmonary artery to the left atrium is only 6 mm Hg (14 mm Hg in the pulmonary artery minus 8 mm Hg in the left atrium). This pressure gradient differential is less than 7% of the pressure gradient differential of 87 mm Hg present in the systemic circulation (90 mm Hg in the aorta minus 3 mm Hg in the right atrium).

Structures of the Extra-Alveolar and Alveolar Vessels and the Pulmonary Microcirculation

Although not well defined anatomically, vessels in the pulmonary circulation can be divided into three categories (extra-alveolar, alveolar, and microcirculation) on the basis of differences in their physiological properties. The extra-alveolar vessels (arteries, arterioles, veins, and venules) are larger than their systemic counterparts. They are not influenced by alveolar pressure changes, but they are affected by intrapleural and interstitial pressure changes. Thus the caliber of extra-alveolar vessels is affected by lung volume and by lung elastin. At high lung volumes, the decrease in pleural pressure increases the caliber of extra-alveolar vessels, whereas at low lung volumes, an increase in pleural pressure decreases vessel caliber. In contrast, alveolar capillaries reside within the interalveolar septa, and they are very sensitive to changes in alveolar pressure but not to changes in pleural or interstitial pressure. Positive-pressure ventilation increases alveolar pressure and compresses these capillaries and thus blocks blood flow. The pulmonary microcirculation comprises the small vessels that participate in liquid and solute exchange in maintenance of fluid balance in the lung.

Structure of the Alveolar-Capillary Network

The sequential branching of the pulmonary arteries culminates in a dense mesh-like network of capillaries that surround alveoli. This alveolar-capillary network is composed of thin epithelial cells of the alveolus and endothelial cells of the vessels and their supportive matrix, and it has an alveolar surface area of approximately 85 m^2 (the approximate size of a tennis court). The structural matrix and the tissue components of this alveolar-capillary network provide the only barrier between gas in the airway and blood in the capillary. The cells of this barrier, which is 1 to 2 μm thick, include type I alveolar epithelial cells, capillary endothelial cells, and their respective basement membranes, which are back to back (see Fig. 20.6B). Surrounded mostly by air, this alveolar-capillary network is an ideal environment for gas exchange. Red blood cells pass through the capillary component of this network in single file in less than 1 second, which is sufficient time for CO_2 and O_2 gas exchange.

In addition to gas exchange, the alveolar-capillary network regulates the amount of fluid within the lung. At the pulmonary capillary level, the balance between hydrostatic and oncotic pressure across the wall of the capillary results in a small net movement of fluid out of the vessels into the interstitial space. The fluid is then removed from the lung interstitium by the lymphatic system and enters the circulation via the vena cava in the area of the lung hilus. In normal adults, an average of 30 mL of fluid per hour is returned to the circulation via this route.

Bronchial Circulation

The bronchial circulation is a distinct system, separate from the pulmonary circulation in the lung, that provides systemic arterial blood to the trachea, upper airways, surface secretory cells, glands, nerves, visceral pleural surfaces, lymph nodes, pulmonary arteries, and pulmonary veins. The bronchial circulation is similar in structure to the systemic circulatory system and perfuses the upper respiratory tract; it does not reach the terminal or respiratory bronchioles or the alveoli. Venous blood from the capillaries of the bronchial circulation flows to the heart through either true bronchial veins or bronchopulmonary veins. True bronchial veins are present in the region of the lung hilus, and blood flows into the azygos, hemiazygos, or intercostal veins before entering the right atrium. The bronchopulmonary veins are formed through a network of tributaries from the bronchial and pulmonary circulatory vessels that anastomose and form vessels with an admixture of blood from both circulatory systems. Blood from these anastomosed vessels returns to the left atrium through pulmonary veins. Approximately two thirds of the total bronchial circulation is returned to the heart via the pulmonary veins and this anastomosis route.

The bronchial circulation receives only approximately 1% of total cardiac output; in comparison, the pulmonary circulation receives almost 100%. In the presence of diseases such as cystic fibrosis, the bronchial arteries, which normally receive only 1% to 2% of cardiac output, increase in size (hypertrophy) and receive as much as 10% to 20% of the cardiac output. The erosion of inflamed tissue into these vessels as a result of bacterial infection is responsible for the **hemoptysis** (coughing up blood) that can occur in this disease.

Innervation

Breathing is automatic and under control of the central nervous system (CNS). The lungs are innervated by the autonomic nervous system of the peripheral nervous system (PNS), which is under CNS control (Fig. 20.8). The autonomic nervous system has four distinct components: **parasympathetic, sympathetic, nonadrenergic noncholinergic inhibitory,** and **nonadrenergic noncholinergic stimulatory.**

Stimulation of the parasympathetic system leads to airway smooth muscle constriction, blood vessel dilation, and increased glandular cell secretion, whereas stimulation of the sympathetic system causes relaxation of the airway smooth muscle, constriction of blood vessels, and inhibition of glandular secretion (see Chapter 26, Fig. 26.1). The functional unit of the autonomic nervous system is composed of preganglionic and postganglionic neurons in the CNS and postganglionic neurons in the ganglia of the specific organ. As with most organ systems, the CNS and PNS work in concert to maintain homeostasis. There is no voluntary motor innervation in the lung, nor are there pain fibers. Pain fibers are found only in the pleura.

The parasympathetic innervation of the lung originates from the medulla in the brainstem (cranial nerve X, the **vagus nerve**). Preganglionic fibers from the vagal nuclei descend in the vagus nerve to ganglia adjacent to airways and blood vessels in the lung. Postganglionic fibers from the ganglia then complete the network by innervating smooth muscle cells, blood vessels, and bronchial epithelial cells (including goblet cells and submucosal glands). In the lungs, both preganglionic and postganglionic fibers

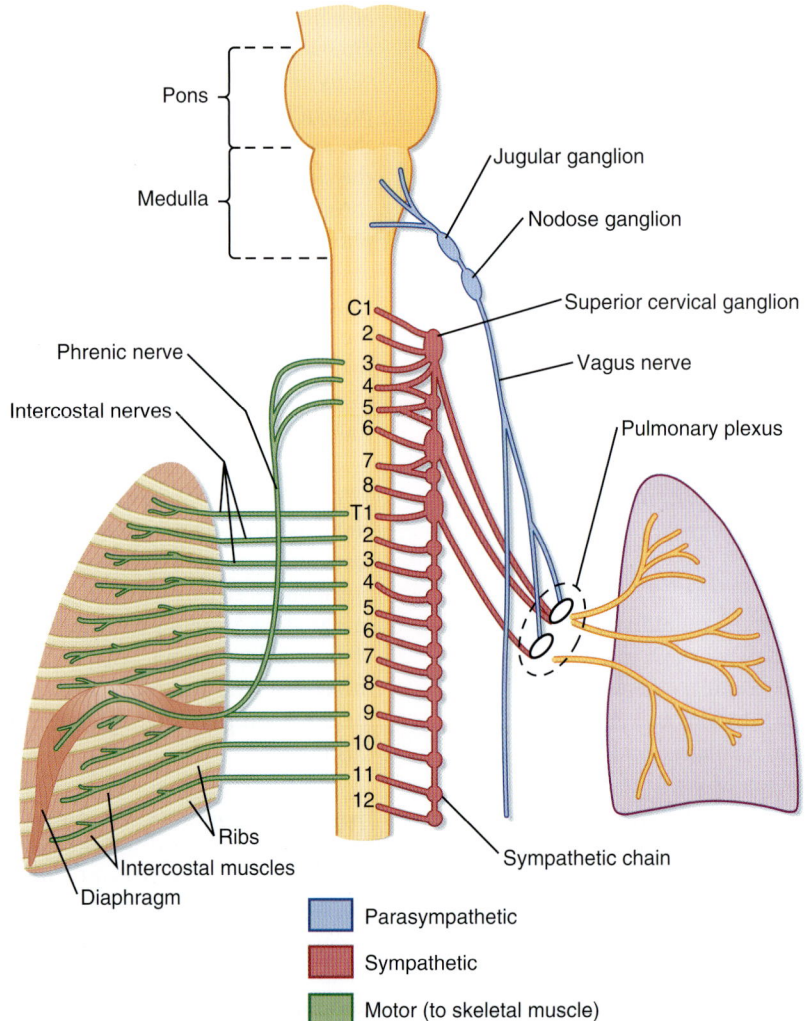

Pons

Medulla

Jugular ganglion

Nodose ganglion

Superior cervical ganglion

Vagus nerve

Pulmonary plexus

Phrenic nerve

Intercostal nerves

C1
2
3
4
5
6
7
8
T1
2
3
4
5
6
7
8
9
10
11
12

Ribs

Intercostal muscles

Diaphragm

Sympathetic chain

Parasympathetic

Sympathetic

Motor (to skeletal muscle)

● **Fig. 20.8** Innervation of the Lung. The autonomic innervation (motor and sensory) of the lung and the somatic (motor) nerve supply to the intercostal muscles and diaphragm are depicted.

contain excitatory (cholinergic) and inhibitory (nonadrenergic) motor neurons. **Acetylcholine** and **substance P** are neurotransmitters of excitatory motor neurons; **dynorphin** and **vasoactive intestinal peptide** are neurotransmitters of inhibitory motor neurons. Parasympathetic stimulation through the vagus nerve is responsible for the slightly constricted smooth muscle tone in a normal resting lung. Parasympathetic fibers also innervate the bronchial glands, and these fibers, when stimulated, increase the synthesis of mucus glycoprotein, which increases the viscosity of mucus. Parasympathetic innervation is greatest in the larger airways and most limited in the smaller conducting airways in the periphery.

Whereas the response of the parasympathetic nervous system is very specific and local, the response of the sympathetic nervous system tends to be more general. Mucous glands and blood vessels are heavily innervated by the sympathetic nervous system; however, airway smooth muscle is not. Neurotransmitters of the adrenergic nerves include **norepinephrine** and dopamine, although dopamine has no influence on the lung. Stimulation of the sympathetic nerves

in mucous glands increases water secretion. This disrupts the balanced response of increased water and increased viscosity between the sympathetic and parasympathetic pathways. Adrenergic fibers are absent in humans. In addition to those in the sympathetic and parasympathetic systems, afferent nerve endings are present in the epithelium and in smooth muscle cells in the lung.

Central Control of Respiration

Breathing is an automatic, rhythmic, and centrally regulated process with voluntary control. The CNS, particularly the **brainstem,** functions as the main control center for respiration (Fig. 20.9). Regulation of respiration requires (1) generation and maintenance of a respiratory rhythm; (2) modulation of this rhythm by sensory feedback loops and reflexes that allow adaptation to various conditions while minimizing energy costs; and (3) recruitment of respiratory muscles that can contract appropriately for gas exchange. Control of respiration is described in greater detail in Chapter 25.

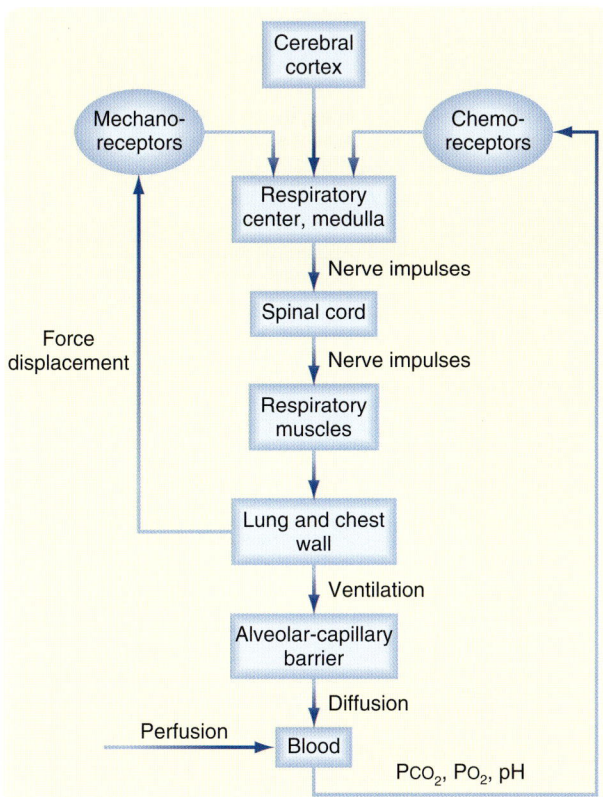

Fig. 20.9 Block Diagram of the Respiratory Control System, Demonstrating Relationships Between the Respiratory Control Center and Muscles of Respiration. The respiratory center neurons, dispersed into several groups in the medulla, demonstrate spontaneous cyclic activity but are strongly influenced by stimuli descending from the cerebral cortex (volitional control) and from two sensory loops: mechanoreceptor and chemoreceptor pathways. Ventilation and perfusion occur together near the end of the cycle, and their output determines partial pressures of arterial and alveolar carbon dioxide (PCO_2) and oxygen (PO_2) and, in part, arterial hydrogen ion concentration (pH). These outputs feed back to the respiratory center via chemoreceptor and mechanoreceptor sensory pathways.

Muscles of Respiration

The major muscles of respiration include the **diaphragm,** the **external intercostal muscles,** and the **scalene muscles,** all of which are skeletal muscles. Skeletal muscles provide the driving force for ventilation; the force of contraction increases when they are stretched and decreases when they are shortened. The force of contraction of respiratory muscles increases at larger lung volumes.

The diaphragm is the major muscle of respiration, and it divides the thoracic cavity from the abdominal cavity (Fig. 20.10). Contraction of the diaphragm forces the abdominal contents downward and forward. This increases the vertical dimension of the chest cavity and creates a pressure difference between the thorax and abdomen. In adults, the diaphragm can generate airway pressures of up to 150 to 200 cm H_2O during maximal inspiratory effort. During quiet breathing (tidal breathing), the diaphragm moves approximately 1 cm; however, during deep-breathing maneuvers (vital capacity), the diaphragm can move as much as 10 cm. The diaphragm

is innervated by the right and left phrenic nerves, whose origins are at the third to fifth cervical segments of the spinal cord (C3 to C5).

The other important muscles of inspiration are the external intercostal muscles, which pull the ribs upward and forward during inspiration (see Fig. 20.10). This causes an increase in both the lateral and anteroposterior diameters of the thorax. Innervation of the external intercostal muscles originates from **intercostal nerves** that arise from the same level of the spinal cord (T1 and T2). Paralysis of these muscles has no significant effect on respiration because respiration is dependent primarily on the diaphragm. This is why individuals with injuries to the lower spinal cord can breathe on their own; it is only when the injury is above C3 that an individual is completely dependent on a ventilator.

Accessory muscles of inspiration (the scalene muscles, which elevate the **sternocleidomastoid muscles;** the **alae nasi,** which cause nasal flaring; and small muscles in the neck and head) do not contract during normal breathing. However, they do contract vigorously during exercise, and when airway obstruction is significant, they actively pull up the rib cage. During normal breathing, they anchor the sternum and upper ribs. Because the upper airway must remain patent during inspiration, the pharyngeal wall muscles (**genioglossus** and **arytenoid**) are also considered muscles of inspiration. All the rib cage muscles are voluntary muscles that are supplied by intercostal arteries and veins and innervated by motor and sensory intercostal nerves.

Exhalation during normal breathing is passive, but it becomes active during exercise and hyperventilation. The most important muscles of exhalation are those of the abdominal wall (**rectus abdominis, internal** and **external oblique,** and **transversus abdominis**) and the **internal intercostal muscles,** which oppose the external intercostal muscles (i.e., they pull the ribs downward and inward). The inspiratory muscles do the work of breathing. During normal breathing, this workload is low, and the inspiratory muscles have significant reserve. Respiratory muscles can be trained to do more work, but there is a limit to the work that they can perform. Respiratory muscle weakness can impair movement of the chest wall, and respiratory muscle fatigue is a major factor in the development of respiratory failure.

Lung Embryology, Development, Aging, and Repair

The epithelium of the lung arises as a pouch from the primitive foregut at approximately 22 to 26 days after fertilization of the ovum. This single lung bud branches into primitive right and left lungs. Over the next 2 to 3 weeks, further branching occurs to create the irregular dichotomous branching pattern. The pathologist Lynne Reid described "three laws of lung development": (1) The bronchial tree has developed by week 16 of intrauterine

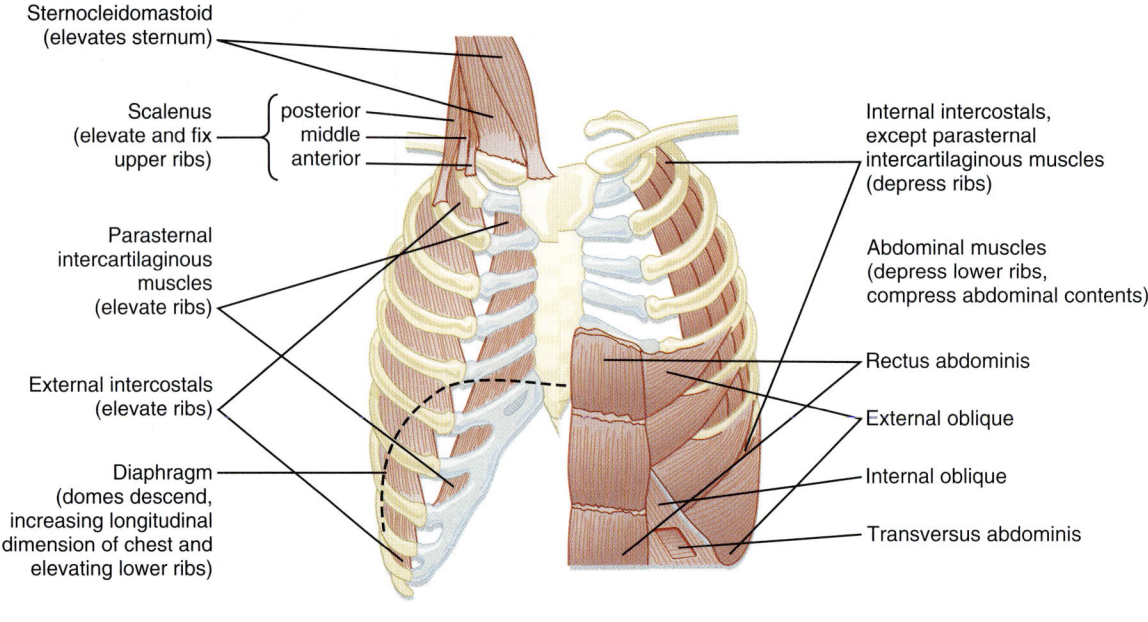

Muscles of inspiration — Muscles of expiration

- Sternocleidomastoid (elevates sternum)
- Scalenus (elevate and fix upper ribs) — posterior, middle, anterior
- Parasternal intercartilaginous muscles (elevate ribs)
- External intercostals (elevate ribs)
- Diaphragm (domes descend, increasing longitudinal dimension of chest and elevating lower ribs)
- Internal intercostals, except parasternal intercartilaginous muscles (depress ribs)
- Abdominal muscles (depress lower ribs, compress abdominal contents)
- Rectus abdominis
- External oblique
- Internal oblique
- Transversus abdominis

A   **Muscles of inspiration**   **Muscles of expiration**

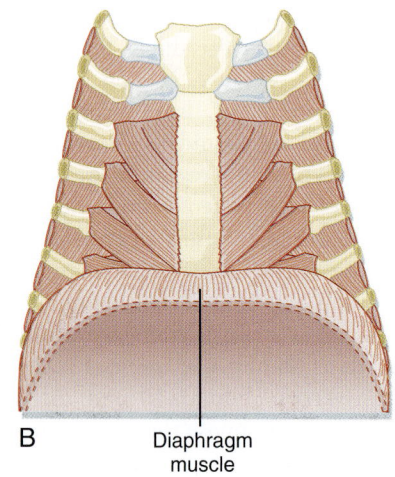

B   Diaphragm muscle

• **Fig. 20.10** Illustrations of the Major Respiratory Muscles. **A,** The inspiratory muscles are depicted on the left side, and the expiratory muscles are depicted on the right side. **B,** The diaphragm muscle in relation to the rib cage. (From Garrity ER, Sharp JT. Respiratory muscles: function and dysfunction. In: American College of Chest Physicians. *Pulmonary and Critical Care Update.* Vol 2. Park Ridge, IL: American College of Chest Physicians; 1986.)

IN THE CLINIC

Because respiratory muscles provide the driving force for ventilation, diseases that affect the mechanical properties of the lung affect the muscles of respiration. For example, in **chronic obstructive pulmonary disease (COPD),** the work of breathing is increased secondary to airflow obstruction. Exhalation no longer is passive but instead requires active, expiratory muscle contraction. In addition, total lung capacity is increased (see Chapter 21). The larger total lung capacity forces the diaphragm downward, shortens the muscle fibers, and decreases the radius of curvature. As a result, the function and efficiency of the diaphragm are decreased. Respiratory muscles can fatigue just as other skeletal muscles do when the workload increases. Respiratory muscles can also weaken in patients with neuromuscular diseases (e.g., **Guillain-Barré syndrome, myasthenia gravis**). In these diseases, sufficient weakness of the respiratory muscles can impair movement of the chest wall and result in respiratory failure, even though the mechanical properties of the lung and chest wall are normal.

life; (2) alveoli develop after birth, the number of alveoli increases until the age of 8 years, and the size of alveoli increases until growth of the chest wall is completed at adulthood; and (3) the development of preacinar vessels (arteries and veins) parallels that of the airways, whereas that of intra-acinar vessels parallels that of the alveoli.

Thus intrauterine events that occur before 16 weeks of gestation will affect the number of airways. A condition known as **congenital diaphragmatic hernia** is an example of a congenital lung disease. It occurs at 6 to 8 weeks of gestation and is due to failure of the pleuroperitoneal canal to close and thereby separate the chest and abdominal cavities; the presence of the abdominal contents in the lung hemithorax results in abnormal lung growth with a decrease in the number of airways and alveoli. Before the birth of an affected infant, the alveolar epithelium is composed solely of type II epithelial cells, and it is not until birth that these cells differentiate into type I epithelial cells.

Growth of the lungs is similar and relatively proportional to growth in body length and stature. The rate of development is fastest in the neonatal and preadolescent periods (≈11 years of age), and girls' lungs mature earlier than boys'. Although the growth rate of the lung slows after adolescence, the body and lung increase in size steadily until adulthood. Improvement in lung function occurs at all stages of growth development; however, once optimal size has been attained in early adulthood (20 to 25 years of age), lung function starts to decline with age. The decrease in lung function with age, estimated at less than 1% per year, appears to begin earlier and proceed faster in individuals who smoke or are exposed to toxic environmental factors. The major physiological insufficiencies caused by aging involve ventilatory capacity and responses, especially during exercise, and they result in abnormal ventilation with normal perfusion. In addition, gas diffusion decreases with age, probably as a result of a decrease in alveolar surface area. Age-related decreases in lung function and altered structure parallel biochemical observations of increased levels of elastin within the lung, which could explain some of the functional abnormalities.

 ## AT THE CELLULAR LEVEL

Type I cells lack free radical scavengers (i.e., superoxide dismutase) and are susceptible to injury and death induced by toxic O_2 compounds and free radical (i.e., H_2O_2, OH^-, and O_2^-). In various inflammatory lung diseases, type I cells die, and the alveolar epithelium thereby becomes denuded, with increased vascular permeability and ensuing fluid accumulation (impaired gas exchange). Type II cells have superoxide dismutase and are thus more resistant to the toxic oxygen radicals. They can survive to proliferate and differentiate into type I cells to restore the normal alveolar architecture. This type of response is dependent on an intact basement membrane to support the proliferation of type II cells and is an example of "phylogeny recapitulating ontogeny." If the basement membrane cannot be repopulated, then the body's recourse for repair is collagen deposition and scar formation, which is not conducive to gas exchange. In lung disease that involves scar formation (i.e., pulmonary fibrosis), the total lung volume decreases as a result of the loss of alveoli and impairment of O_2 diffusion into the capillaries by a thickened, nonpermeable matrix. Historically, idiopathic pulmonary fibrosis has been very difficult to treat because of the lack of specific therapeutics, which can inhibit collagen deposition. Two therapeutic compounds (pirfenidone and nintedanib) have been shown in clinical trials to slow the progression of the disease and improve outcomes in patients with idiopathic pulmonary fibrosis. Pirfenidone is a small low-molecular-weight compound with anti-inflammatory properties (it decreases procollagen types I and II synthesis), and nintedanib is a tyrosine kinase inhibitor (inhibits vascular endothelial growth factor and fibroblast-derived growth factor).

Key Points

1. The lungs demonstrate anatomical and physiological unity; that is, each unit (bronchopulmonary segment) is structurally identical and functions just like every other unit.

2. The upper airways (nose, sinuses, pharynx) condition inspired air for temperature, humidity, and atmospheric pressure, and they control, via the epiglottis, the flow of air into the lungs and food/fluids into the esophagus.

3. Components of the lower airways (trachea, bronchi, bronchioles) are considered conducting airways in which air is transported to the gas-exchanging respiratory units composed of respiratory bronchioles, alveolar ducts, and alveoli.

4. The lungs have unique, dual circulatory systems. The pulmonary circulatory system has the ability to accommodate large volumes of blood at low pressure and brings deoxygenated blood from the right ventricle to the gas-exchanging units in the lung. The bronchial circulation arises from the aorta and provides nourishment (O_2) to the lung parenchyma.

5. Breathing is automatic; the lungs are innervated by the autonomic nervous system of the PNS while under the control of the CNS. Parasympathetic stimulation results in constriction of airway smooth muscles (airway narrowing) whereas sympathetic stimulation results in relaxation of airway smooth muscles (airway opening).

6. Inspiration is the active phase of breathing. The diaphragm is the major muscle of respiration, and its contraction creates a pressure difference (mechanoreceptor response) between the thorax and diaphragm (negative pressure in the chest), which induces inspiration.

7. The respiratory center is located in the medulla and regulates respiration with input from sensory (mechanoreceptor and chemoreceptor) feedback loops.

Additional Readings

Burri PH. Structural aspects of postnatal lung development—alveolar formation and growth. *Biol Neonate.* 2006;89:313-322.

Hameed A, Sherkheli MA, Hussain A, Ul-haq R. Molecular and physiological determinants of pulmonary developmental biology: a review. *Am J Biomed Res.* 2013;1:13-24.

Harding R, Pinkertton KE, eds. *The Lung: Development, Aging and the Environment.* 2nd ed. London: Academic Press; 2014.

Hattrup CL, Gendler SJ. Structure and function of the cell surface (tethered) mucins. *Annu Rev Physiol.* 2007;70:431-457.

Reynolds HY. Lung inflammation and fibrosis: an alveolar macrophage-centered perspective from the 1970s to 1980s. *Am J Respir Crit Care Med.* 2005;171:98-102.

Satir P, Christensen ST. Overview of structure and function of mammalian cilia. *Annu Rev Physiol.* 2007;69:377-400.

Shannon JM, Hyatt BA. Epithelial cell-mesenchymal interactions in the developing lung. *Annu Rev Physiol.* 2004;66:625-645.

Warburton D, El-Hashash A, Carraro G, et al. Lung organogenesis. *Curr Top Devel Bio.* 2010;90:73-158.

21

Static Lung and Chest Wall Mechanics

LEARNING OBJECTIVES

Upon completion of this chapter, the student should be able to answer the following questions:

1. Define the different pressures in the respiratory system.
2. Explain how a pressure gradient is created.
3. Define the different volumes in the lung, and describe how they are measured.
4. Explain how static lung mechanics determines lung volumes.
5. Define lung compliance.
6. Explain how surfactant affects lung compliance, and describe its importance in maintaining unequal alveolar volumes.

To achieve its primary function of gas exchange, air must be moved in and out of the lung. The mechanical properties of the lung and chest wall determine the ease or difficulty of this air movement. Lung mechanics is the study of the mechanical properties of the lung and chest wall (including the **diaphragm**, **abdominal cavity**, and **anterior abdominal muscles**). Lung mechanics is important for how the lungs work both normally and in the presence of disease, inasmuch as most lung diseases affect the mechanical properties of the lungs, chest wall, or both. In addition, death from lung disease is almost always due to respiratory muscle fatigue, which results from an inability of the respiratory muscles to overcome the altered mechanical properties of the lungs, chest wall, or both. Lung mechanics includes static mechanics (the mechanical properties of a lung whose volume is not changing with time) and dynamic mechanics (properties of a lung whose volume is changing with time). Dynamic mechanics of the lung and chest wall are described in Chapter 22.

Pressures in the Respiratory System

In healthy people, the lungs and chest wall move together as a unit. Between these structures is the **pleural space,** which under normal conditions is best thought of as a potential (or virtual) space. Because the lungs and chest wall move together, changes in their respective volumes are equal during inspiration and exhalation. Volume changes in the lungs and chest wall are driven by changes in the surrounding pressure. In accordance with convention,

pressures inside the lungs and chest wall are referenced in relation to atmospheric pressure, which is considered 0. Thus a negative pressure in the pleural space is a pressure that is lower than atmospheric pressure. Also in accordance with convention, pressures across surfaces such as the lungs or chest wall have been defined as the difference between the pressure inside and the pressure outside the surface. The pressure differences across the lung and across the chest wall are defined as the transmural (across a wall or surface) pressures. For the lung, this transmural pressure is called the **transpulmonary** (or translung) **pressure** (P_L), and it is defined as the pressure difference between the air spaces (alveolar pressure [P_A]) and the pressure surrounding the lung (pleural pressure [P_{pl}]):

Equation 21.1

$$P_L = P_A - P_{pl}$$

The **transmural pressure across the chest wall** (P_w) is the difference between pleural (inside) pressure (P_{pl}) and the pressure surrounding the chest wall (P_b), which is the atmospheric pressure or body surface pressure:

Equation 21.2

$$P_w = P_{pl} - P_b$$

The pressure across the respiratory system (P_{rs}) is the sum of the pressure across the lung and the pressure across the chest wall:

Equation 21.3

$$P_{rs} = P_L + P_w$$
$$= (P_A - P_{pl}) + (P_{pl} - P_b)$$
$$= P_A - P_b$$

How a Pressure Gradient Is Created

Air flows into and out of the lungs from areas of higher pressure to areas of lower pressure. In the absence of a pressure gradient, there is no airflow. Thus at end inspiration and at the end of exhalation, which are periods of time when there is no airflow, alveolar pressure (P_A) is the same as atmospheric pressure (P_b), and there is no pressure gradient ($P_b - P_A = 0$). Pleural pressure at these same times, however, is not 0. Before inspiration begins, the pleural pressure in normal individuals is approximately -3 to -5 cm H_2O. Therefore, the pressure in the pleural space is negative in relation to atmospheric pressure. This negative pressure is created by the inward elastic recoil pressure of the lung, and it acts to "pull the lung" away from the chest wall. The lung is not able, however, to pull away from the chest wall, inasmuch as the two function as a unit. Thus the inward elastic recoil pressure of the lung is balanced by the outward recoil of the chest wall.

With the onset of inspiration, the muscles of the diaphragm and chest wall contract, which causes a downward movement of the diaphragm and an outward and upward movement of the rib cage. As a result, pleural pressure decreases during inspiration. This negative pleural pressure is transmitted across the lung tissue and results in a decrease in alveolar pressure. As alveolar pressure decreases below 0 (i.e., from atmospheric pressure to a lower pressure), gas moves into the airways when the glottis is open. As gas flows into the airways to the alveoli, the pressure gradient along the airways decreases, and flow stops when there is no longer a pressure gradient from atmospheric to alveolar pressure. The decrease in pleural pressure at the start of inspiration secondary to inspiratory muscle contraction is greater than the transmitted fall in alveolar pressure, and, as a result, transpulmonary pressure at the start of inspiration is positive (see Eq. 21.1). Positive transpulmonary pressure is necessary to increase lung volume, and lung volume increases with increasing transpulmonary pressure (Fig. 21.1). Similarly, during inspiration, the chest wall expands to a larger volume. Because pleural pressure is negative in relation to atmospheric pressure during quiet breathing, the transmural pressure across the chest wall is negative (see Eq. 21.2).

On exhalation, the diaphragm moves higher into the chest, pleural pressure increases (i.e., becomes less negative), alveolar pressure becomes positive, the glottis opens,

and gas again flows from a higher (alveolar) pressure to a lower (atmospheric) pressure. In the alveoli, the driving force for exhalation is the sum of the elastic recoil of the lungs and pleural pressure (see Chapter 22). This relationship between changes in pressure, changes in airflow, and changes in volume during inspiration and exhalation is displayed in Fig. 21.2. During tidal volume breathing in normal individuals, the decrease in alveolar pressure at the start of inspiration is small (1 to 3 cm H_2O). It is much larger in individuals with airway obstruction because of the larger pressure drop that occurs across obstructed airways. Airflow stops in the absence of a pressure gradient, which occurs whenever alveolar pressure and atmospheric pressure are equal.

Lung Volumes and Their Measurement

Lung volumes (Fig. 21.3) and the factors that determine these volumes are important components of lung mechanics

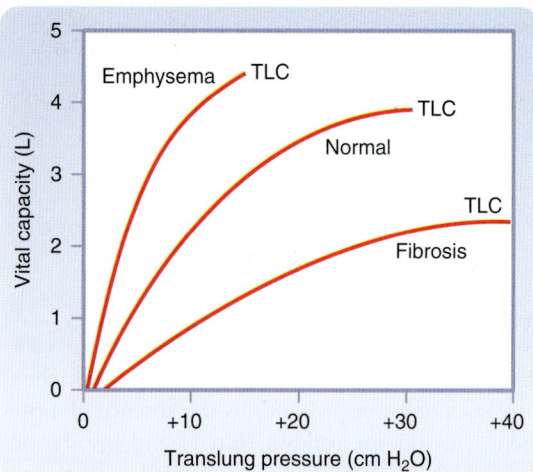

• **Fig. 21.1** Volume of the Lung as a Function of the Transpulmonary Pressure in Health and Disease. As the transpulmonary pressure increases, lung volume increases. Also shown are the changes in lung volume in the presence of emphysema and pulmonary fibrosis. Note that for the same change in transpulmonary pressure, in the presence of either of these types of diseases, the changes in lung volume are different. TLC, total lung capacity (the total volume of gas in the lung).

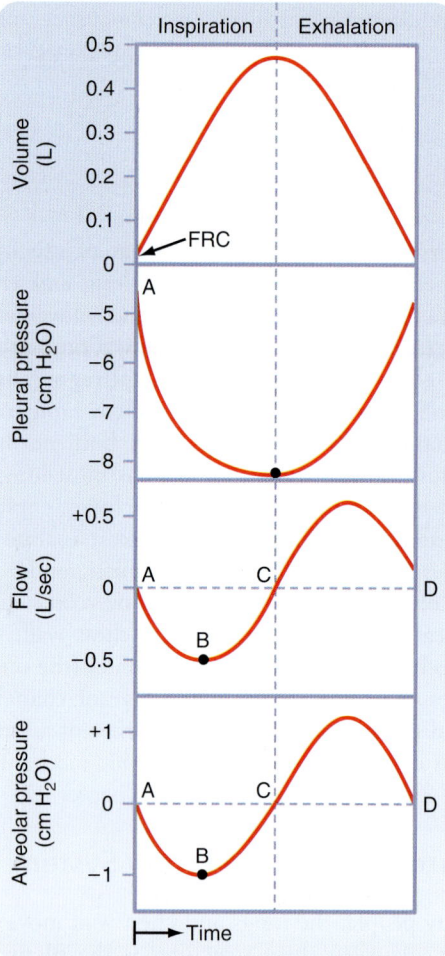

• **Fig. 21.2** Changes in Alveolar and Pleural Pressure During Quiet Breathing (Tidal Volume). Inspiration is represented to the left of the *vertical dotted line,* and exhalation is represented to the right of it. Positive (in relation to atmospheric) pressures are represented above the *horizontal dotted line,* and negative pressures are represented below it. See text for details. At points of no airflow (points A and C), alveolar pressure is 0. FRC, functional residual capacity.

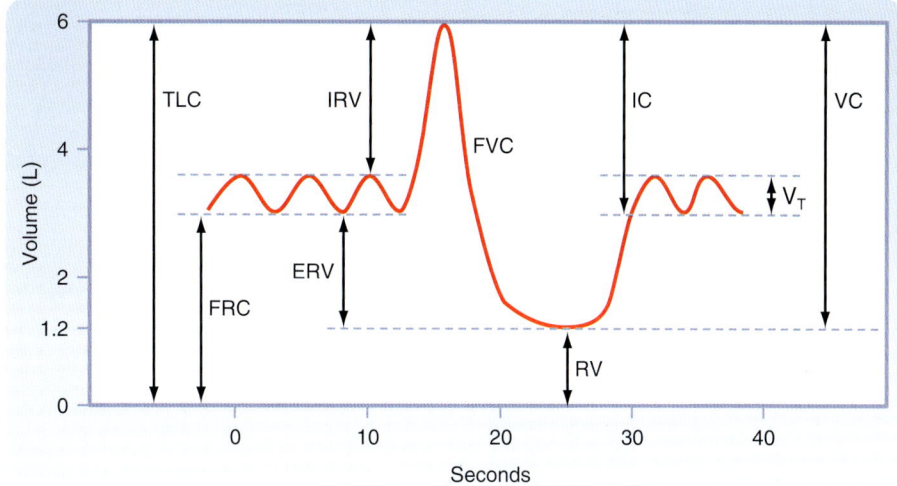

• **Fig. 21.3** The Various Lung Volumes and Capacities. ERV, expiratory reserve volume; FRC, functional residual capacity; FVC, forced vital capacity; IC, inspiratory capacity; IRV, inspiratory reserve volume; RV, residual volume; TLC, total lung capacity; VC, vital capacity; V_T, tidal volume.

and play a major role in the work of breathing (see Chapter 22). All lung volumes are subdivisions of **total lung capacity (TLC),** the total volume of air that is contained in the lung at the point of maximal inspiration. Lung volumes are reported in liters either as volumes or as capacities. A capacity is composed of two or more volumes. Many lung volumes are measured with a **spirometer.** The patient is asked to first breathe normally into the spirometer, and the volume of air (the **tidal volume [V_T]**) that is moved into and out of the lungs with each quiet breath is measured. The patient then inhales maximally and exhales forcefully and completely, and the volume of exhaled air is measured. The total volume of exhaled air, from a maximal inspiration to a maximal exhalation, is the **vital capacity (VC). Residual volume (RV)** is the air remaining in the lungs after a complete exhalation. TLC is the sum of VC and RV; it is the total volume of air contained in the lungs at the end of maximal inspiration, and it includes the volume of air that can be moved (VC) and the volume of air that is always present (trapped) in the lungs (RV). **Functional residual capacity (FRC)** is the volume of air in the lungs at the end of exhalation during quiet breathing and is also called the *resting volume* of the lungs. FRC is composed of RV and the **expiratory reserve volume** (the volume of air that can be exhaled from FRC to RV).

Measurement of Lung Volumes

RV and TLC can be measured in two ways: by helium dilution and by body plethysmography. Both methods are used clinically and provide valuable information about lung function and lung disease. The helium dilution technique is the older and simpler method, but it is often less accurate than body plethysmography, which requires sophisticated and expensive equipment.

TABLE 21.1	Normal Values (Average White Male Adult)
Lung Volumes	
Functional residual capacity (FRC)	2.4 L
Total lung capacity (TLC)	6 L
Tidal volume (V_T)	0.5 L
Breathing frequency (f)	12/min
Static Mechanics	
Pleural pressure (P_{pl}), mean	−5 cm H_2O
Chest wall compliance (C_w) at FRC	0.2 L/cm H_2O
Lung compliance (C_L) at FRC	0.2 L/cm H_2O

 IN THE CLINIC

Pulmonary function tests are often used to diagnose abnormalities in lung function and to assess the progression of lung disease. They can distinguish the two major types of pulmonary pathophysiologic processes: obstructive lung diseases and restrictive lung diseases. For example, in normal individuals, the ratio of RV to TLC is less than 0.25. Thus in a healthy individual, approximately 25% of the total volume of air in the lung is trapped. In **obstructive pulmonary diseases,** an elevation in RV/TLC ratio is secondary to an increase in RV out of proportion to any increase in TLC. In contrast, in **restrictive lung diseases,** the elevation in the RV/TLC ratio is caused by a decrease in TLC.

In normal individuals, the FRC measured by helium dilution and the FRC measured by plethysmography are the same (Table 21.1). This is not true in individuals with lung disease. The FRC measured by helium dilution is the volume of gas in the lung that communicates with the

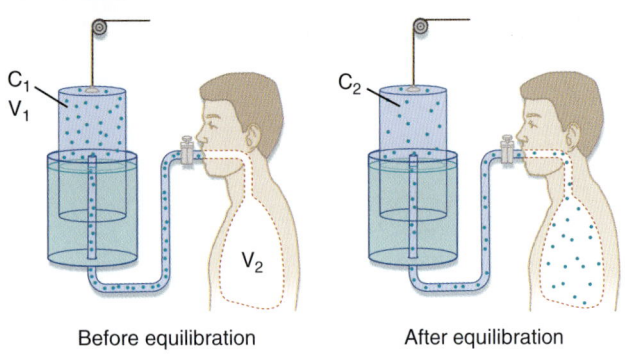

Before equilibration   **After equilibration**

$$C_1 \times V_1 = C_2 \times (V_1 + V_2)$$

• **Fig. 21.4** Measurement of Lung Volume by Helium Dilution. C_1, known concentration of an inert gas; C_2, new (previously unknown) concentration of the gas; V_1, known volume of a box; V_2, lung volume (initially unknown).

airways, whereas the FRC measured by plethysmography is the total volume of gas in the lungs at the end of a normal exhalation. If a significant amount of gas is trapped in the lungs (because of premature airway closure; see Chapter 22), the FRC determined by plethysmography is considerably higher than that determined by helium dilution.

 IN THE CLINIC

In the helium dilution technique, a known concentration (C_1) of an inert gas (such as helium) is added to a box of known volume (V_1). The box is then connected to a volume (V_2) that is unknown (the lung volume to be measured). After adequate time for distribution of the inert gas, the new concentration (C_2) of the inert gas is measured. The change in concentration of the inert gas is then used to determine the new volume in which the inert gas has been distributed (Fig. 21.4). Specifically,

$$C_1 \times V_1 = C_2(V_1 + V_2)$$

In body plethysmograph (body box), Robert Boyle's gas law—that pressure multiplied by volume is constant (at a constant temperature)—is used to measure lung volumes. The patient sits in an airtight box (Fig. 21.5) and breathes through a mouthpiece that is connected to a flow sensor (pneumotach). The patient then makes a panting respiratory effort against a closed mouthpiece. During the expiratory phase of the maneuver, the gas in the lung becomes compressed, lung volume decreases, and the pressure inside the box falls because the gas volume in the box increases. Once the volume of the box and the change in pressure of the box at the mouth are known, the change in volume (ΔV) of the lung can be calculated:

$$P_1 \times V = P_2(V - \Delta V)$$

where P_1 and P_2 are mouth pressures and V is FRC. From the measurement of FRC, inspiratory capacity can be recorded as the volume of air inspired above FRC, and expiratory reserve volume can be determined as the volume of gas exhaled from FRC. These measurements can then be used to determine the other lung volumes.

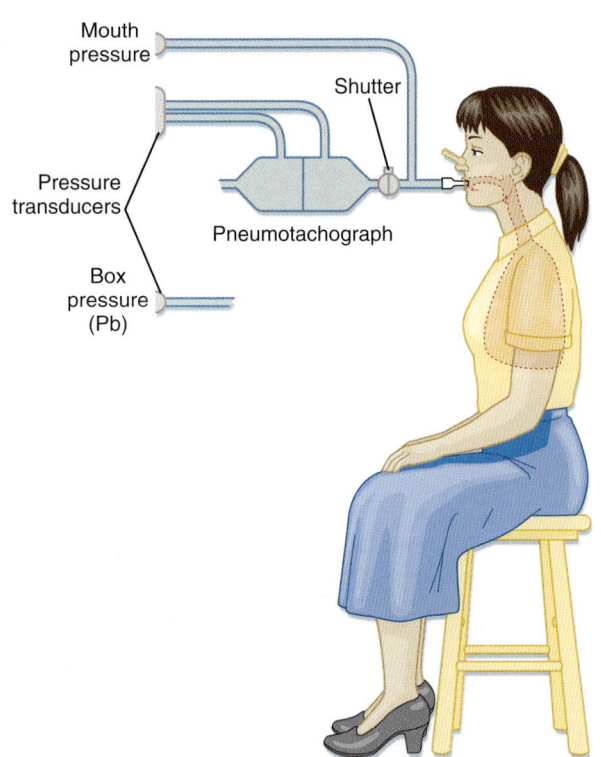

• **Fig. 21.5** The Body Plethysmograph. Note that the box in which the patient sits is not depicted.

Determinants of Lung Volume

What determines the volume of air in the lung at TLC or at RV? The answer lies in the properties of the lung parenchyma and in the interaction between the lungs and the chest wall. The lungs and chest wall always move together as a unit in healthy individuals. The lung contains elastic fibers that (1) stretch when stress is applied, which results in an increase in lung volume, and (2) recoil passively when this stress is released, which results in a decrease in lung volume. The elastic recoil of the lung parenchyma is very high. In the absence of external forces (such as the force generated by the chest wall), the lungs become almost airless (10% of TLC). Similarly, chest wall volume can increase when the respiratory muscles are stretched and decrease when respiratory muscle length is shortened. In the theoretical absence of the lung parenchyma, the resting volume of the chest wall increases and is approximately 60% of TLC.

Lung volumes are determined by the balance between the lung's elastic properties and the properties of the muscles of the chest wall. The maximum volume of air contained within the lung and the chest wall (i.e., TLC) is controlled by the muscles of inspiration (see Chapter 20). With increasing lung volume, the chest wall muscles lengthen progressively. As these muscles lengthen, their ability to generate force decreases. TLC occurs when the inspiratory chest wall muscles are unable to generate the additional force needed to further distend the lung and chest wall. Similarly, the minimal volume of air in the

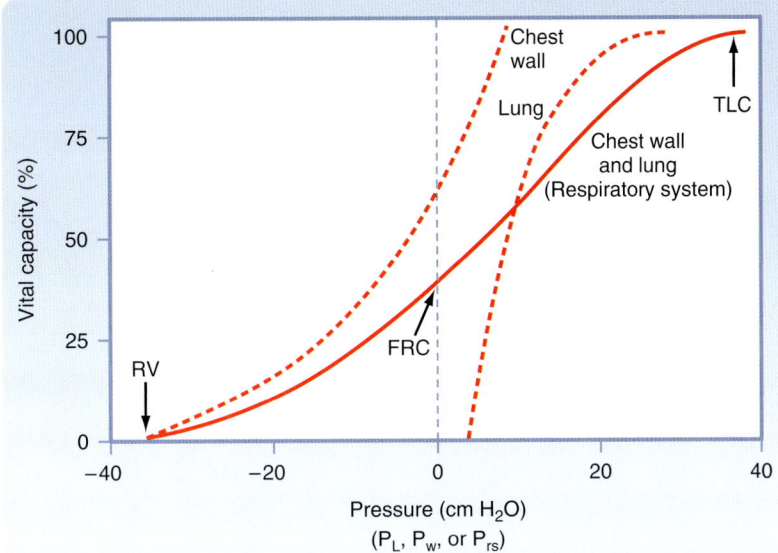

• Fig. 21.6 Relaxation Pressure-Volume Curve of the Lung, Chest Wall, and Respiratory System. The curve for the respiratory system is the sum of the individual curves. The curve for the lung is the same as the one for the normal lung in Fig. 21.1. FVC, forced vital capacity; P_L, transpulmonary (or translung) pressure (P_L); P_{rs}, the pressure across the respiratory system; P_w, the transmural pressure across the chest wall; RV, residual volume; TLC, total lung capacity.

lung (i.e., RV) is controlled by the expiratory muscle force. Decreasing lung volume results in shortening of the expiratory muscles, which, in turn, results in a decrease in muscle force. The decrease in lung volume is also associated with an increase in the outward recoil pressure of the chest wall. RV occurs when expiratory muscle force is insufficient to further reduce chest wall volume.

FRC, or the volume of the lung at the end of a normal exhalation, is determined by the balance between the elastic recoil pressure generated by the lung parenchyma to become smaller (inward recoil) and the pressure generated by the chest wall to become larger (outward recoil). When the chest wall muscles are weak, FRC decreases (lung elastic recoil > chest wall muscle force). In the presence of airway obstruction, FRC increases because of premature airway closure, which traps air in the lung (see Chapter 22).

Pressure-Volume Relationships

A number of important observations can be made from an examination of the pressure-volume curves of the lung, chest wall, and respiratory system (Fig. 21.6). At the resting volume of the lung (FRC), the elastic recoil of the lung acts to decrease lung volume, but this inward recoil is offset by the outward recoil of the chest wall, which acts to increase lung volume. At FRC, these forces are equal and opposite, and the muscles are relaxed. As a result, the transmural pressure across the respiratory system (P_{rs}) at FRC is 0. At TLC, both lung pressure and chest wall pressure are positive, and both require positive transmural distending pressure. The resting volume of the chest wall, in the absence of the lungs, is the volume at which the transmural pressure for the chest wall is 0, and it is approximately 60% of TLC. At volumes

greater than 60% of TLC, the chest wall recoils inward and positive transmural pressure is needed, whereas at volumes below 60% of TLC, the chest wall tends to recoil outward.

The lungs alone are smallest when transpulmonary pressure is 0. The lungs, however, are not totally devoid of air when transpulmonary pressure is 0 because of the surface tension–lowering properties of surfactant (see the section "Surfactant"). The transmural pressure for a healthy lung alone flattens at pressures higher than 20 cm H_2O because the elastic limits of the lung have been reached. Thus further increases in transmural pressure produce little change in volume, and compliance (see the section "Lung Compliance") is low. Further distention is limited by the connective tissue (collagen, elastin) of the lung. If further pressure is applied, the alveoli near the lung surface can rupture, and air can escape into the pleural space. This is called a **pneumothorax.** In a pneumothorax or when the chest is opened, as during thoracic surgery, the lungs and chest wall no longer function as a single unit. The lungs recoil until transpulmonary pressure is 0; the chest wall then increases in size until trans chest wall pressure is 0.

The relationship between transpulmonary pressure and pleural, alveolar, and elastic recoil pressures is depicted in Fig. 21.7. Alveolar pressure is the sum of the pleural pressure and elastic recoil pressure (P_{el}) of the lung:

Equation 21.4
$$P_A = P_{pl} + P_{el}$$

Because transpulmonary pressure (P_L) = $P_A + P_{pl}$,

Equation 21.5
$$P_L = (P_{el} + P_{pl}) - P_{pl}$$
$$P_L = P_{el}$$

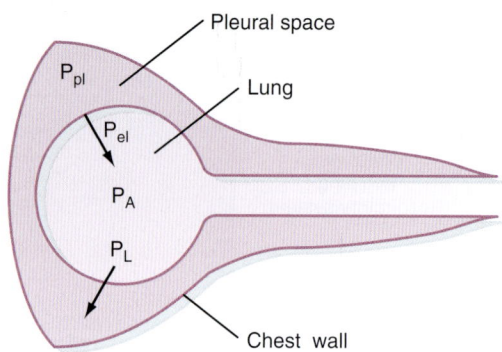

• **Fig. 21.7** Relationship between transpulmonary pressure (P_L) and the pleural (P_{pl}), alveolar (P_A), and elastic recoil (P_{el}) pressures of the lung. Alveolar pressure is the sum of pleural pressure and elastic recoil pressure. Transpulmonary pressure is the difference between alveolar pressure and pleural pressure.

In general, P_L is the pressure distending the lung, whereas P_{el} is the pressure that tends to collapse the lung. Lung elastic recoil increases as the lung inflates.

Lung Compliance

Lung compliance (C_L) is a measure of the elastic properties of the lung. It reflects how easily the lung is distended. Lung compliance is defined as the change in lung volume resulting from a 1–cm H_2O change in the distending pressure of the lung. The units of compliance are in milliliters (or liters) per centimeter of water. When lung compliance is high, the lung is readily distended. When lung compliance is low ("stiff" lung), the lung is not easily distended. The compliance of the lung (C_L) is expressed as

Equation 21.6

$$C_L = \Delta V / \Delta P$$

where ΔV is the change in volume and ΔP is the change in pressure. Graphically, lung compliance is the slope of the line between any two points on the deflation limb of the pressure-volume loop (Fig. 21.8). The compliance of a normal human lung is approximately 0.2 L/cm H_2O, but it varies with lung volume. Note that the lung is less distensible at high lung volumes. For this reason, compliance is corrected for the lung volume at which it is measured (specific compliance; Fig. 21.9). Compliance is not often measured for clinical purposes because it requires placement of an esophageal balloon. The esophageal balloon, which is connected to a pressure transducer, is an excellent surrogate marker for pleural pressure, which is very difficult to measure directly. The change in pleural pressure (P_{pl}) is measured as a function of the change in lung volume; that is, $C_L = \Delta V / \Delta P_{pl}$ or $\Delta P_{pl} = \Delta C_L$.

Surface Tension and Surfactant

Surface Tension

In addition to the elastic properties of the lungs, another major determinant of lung compliance is surfactant and its

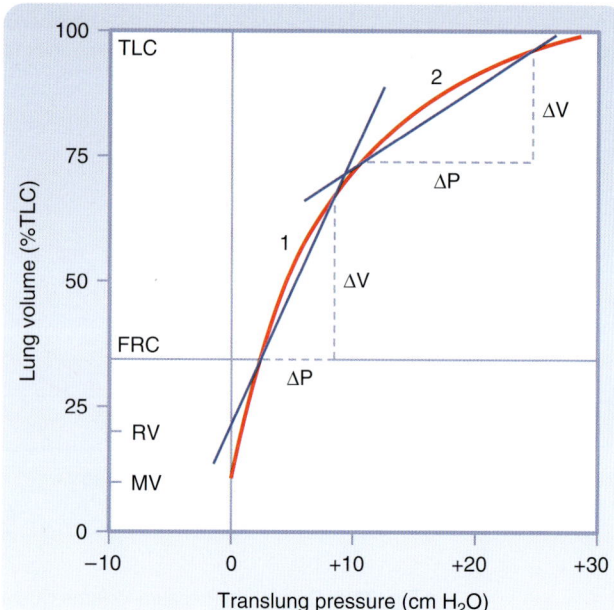

• **Fig. 21.8** Deflation Pressure-Volume Curve. The patient inhales to total lung capacity, and transpulmonary pressure is measured with the use of an esophageal balloon (which measures pleural pressure). The patient then exhales slowly, and pressure is measured at points of no airflow, when the respiratory muscles are relaxed. The pressure-volume curve of the lung is not the same in inspiration (not shown) and exhalation. This difference is called **hysteresis,** and it is caused by the action of surfactant. In accordance with convention, the deflation pressure-volume curve is used for measurements. Compliance at any point along this curve is the change in volume per change in pressure. The curve demonstrates that lung compliance varies with lung volume. A line can be drawn between two different volumes on the curve, and the slope of this line represents the change in volume (ΔV) for a change in pressure (ΔP). Compare the compliance at line 1 versus line 2. The slope of line 2 is less steep than the slope of line 1, and so the compliance is less at this higher lung volume than it is at the lower lung volume. In accordance with convention, lung compliance is the change in pressure from functional residual capacity (FRC) to FRC +1 L. MV, minimal volume; RV, residual volume; TLC, total lung capacity.

🩺 IN THE CLINIC

The compliance of the lungs is affected by several respiratory disorders. In emphysema, an obstructive lung disease that usually occurs in people who smoke and is associated with destruction of the alveolar septa and pulmonary capillary bed, the lung is more compliant; that is, for every 1–cm H_2O increase in pressure, the increase in volume is greater than that in a normal lung (see Fig. 21.1). In contrast, in pulmonary fibrosis, a restrictive lung disease associated with increased collagen fiber deposition in the interstitial space, the lung is noncompliant; that is, for every 1–cm H_2O change in pressure, the change in volume is less. These changes in compliance are of clinical significance because a lung with low compliance requires larger changes in pleural pressure to effect changes in lung volumes; as result, the work of breathing is increased for every breath that the individual takes.

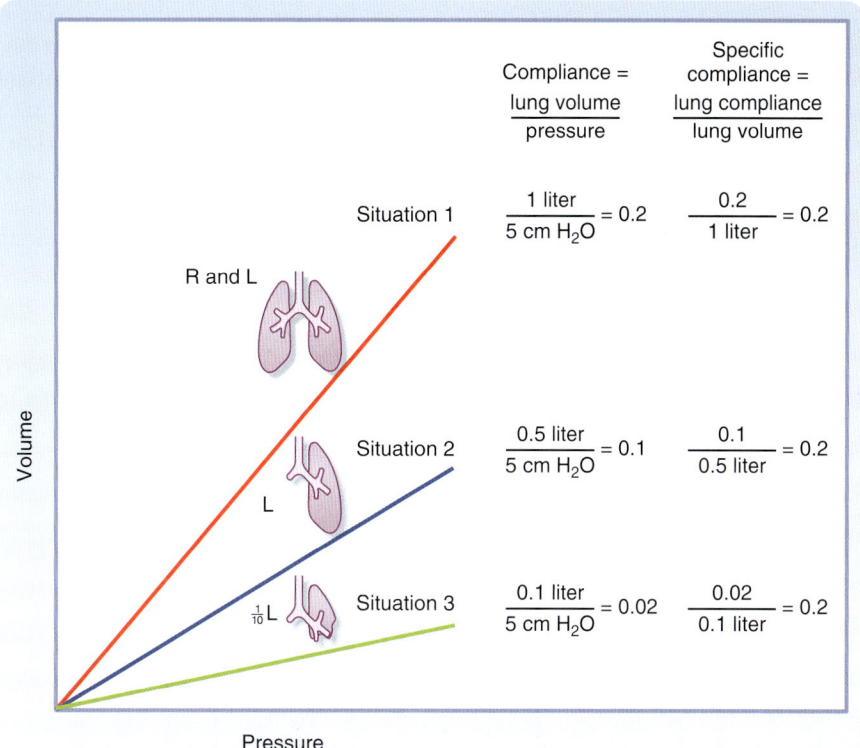

	Compliance = $\dfrac{\text{lung volume}}{\text{pressure}}$	Specific compliance = $\dfrac{\text{lung compliance}}{\text{lung volume}}$
Situation 1	$\dfrac{1\text{ liter}}{5\text{ cm }H_2O} = 0.2$	$\dfrac{0.2}{1\text{ liter}} = 0.2$
Situation 2	$\dfrac{0.5\text{ liter}}{5\text{ cm }H_2O} = 0.1$	$\dfrac{0.1}{0.5\text{ liter}} = 0.2$
Situation 3	$\dfrac{0.1\text{ liter}}{5\text{ cm }H_2O} = 0.02$	$\dfrac{0.02}{0.1\text{ liter}} = 0.2$

• **Fig. 21.9** Relationship Between Compliance and Lung Volume. Imagine lungs in which a 5–cm H₂O change in pressure results in a 1-L change in volume (situation 1). If one lung is removed (situation 2), the compliance decreases, but when corrected for volume of the lung, there is no change (specific compliance). Even when the remaining lung is reduced by 90% (situation 3), the specific compliance is unchanged.

effect on surface tension. **Surface tension** is a force caused by water molecules at the air-liquid interface that tends to minimize surface area, which makes inflating the lungs more difficult. The effect of surface tension on lung inflation is illustrated by a comparison of the volume-pressure curves of a saline-filled lung and of an air-filled lung. Higher pressure is necessary to fully inflate the lung with air than with saline because of the higher surface tension forces in air-filled lungs than in saline-filled lungs. Surface tension is a measure of the attractive force of the surface molecules per unit length of material to which they are attached. The units of surface tension are those of a force applied per unit length. For a sphere (such as an alveolus), the relationship between the pressure within the sphere (P_s) and the tension in the wall is described by the law of Laplace:

Equation 21.7
$$P_s = 2T/r$$

where T is the wall tension (in dynes per centimeter) and r is the radius of the sphere.

The alveoli are lined with a predominantly lipid-based substance called **surfactant.** Pulmonary surfactant serves several physiological roles, including (1) reducing the work of breathing by decreasing surface tension forces; (2) preventing collapse and sticking of alveoli on exhalation;

and (3) stabilizing alveoli, especially those that tend to deflate at low surface tension. In the absence of surfactant, the surface tension at the air-liquid interface would remain constant, and the transalveolar pressure needed to keep it at that volume would be higher when alveolar volumes are lower (Fig. 21.10A). Therefore, higher transalveolar pressure would be necessary to produce a given increase in alveolar volume at lower lung volumes than at higher lung volumes. Surfactant stabilizes the inflation of alveoli because it allows the surface tension to increase as the alveoli become larger (see Fig. 21.10B). As a result, the transalveolar pressure necessary to keep an alveolus inflated increases as lung volume and transpulmonary pressure increases, and it decreases as lung volume decreases. In the presence of surfactant, surface tension is increased at high lung volume and decreased at low lung volume. The result is that the lungs can maintain alveoli at many different volumes. Otherwise, the gas in small alveoli would empty into large alveoli.

Surfactant

Pulmonary surfactant is synthesized by alveolar type II cells, stored in the cell in lamellar bodies, and secreted into the alveolar space in a precursor form (tubular myelin), from where it spreads throughout the entire alveolar surface and attains its ability to decrease surface tension. Surfactant is 85% to 90% lipids, predominantly phospholipids, and 10% to 15% proteins (Table 21.2). The major phospholipid is **phosphatidylcholine**, approximately 75% of which is present as **dipalmitoyl phosphatidylcholine** (DPPC). DPPC decreases surface tension and is the major surface-active component in surfactant. The second most abundant phospholipid is **phosphatidylglycerol,** which accounts for

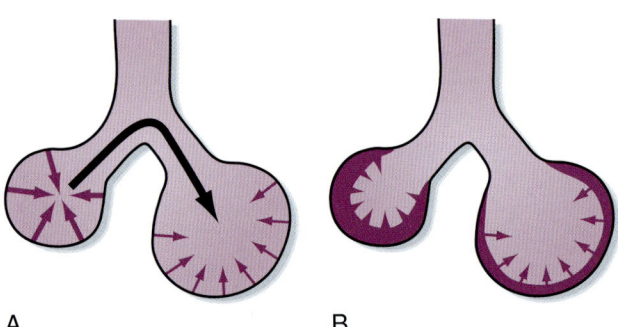

A   B

● **Fig. 21.10** Surface Forces in a Sphere Attempt to Reduce the Area of the Surface and Generate Pressure Within the Sphere. By Laplace's law, the pressure generated is inversely proportional to the radius of the sphere. **A,** In the absence of surfactant, surface forces in the smaller sphere generate higher pressure *(heavier purple arrows)* than do those in the larger sphere *(lighter purple arrows)*. As a result, air moves from the small sphere (higher pressure) to the larger sphere (lower pressure; *black arrow*). This causes the small sphere to collapse and the large sphere to become overdistended. **B,** Surfactant *(shaded layer)* lowers surface tension and lowers it more in the smaller sphere than in the larger sphere. The net results are that the pressures in the small and larger spheres are similar, and the volumes of the spheres are stabilized.

1% to 10% of total surfactant. These lipids are important in formation of the monolayer on the alveolar-air interface, and phosphatidylglycerol is important in the spreading of surfactant over a large surface area. Surfactant protein A, which is the protein most studied, is expressed in alveolar type II cells and in Clara cells in the lungs. Surfactant protein A is involved in the regulation of surfactant turnover, in immunoregulation within the lungs, and in the formation of tubular myelin.

Surfactant is secreted into the airway through exocytosis of the lamellar body by constitutive and regulated mechanisms. Numerous agents, including β-adrenergic agonists, activators of protein kinase C, leukotrienes, and purinergic agonists, stimulate the exocytosis of surfactant. The major routes of clearance of pulmonary surfactant within the lung are reuptake by type II cells, absorption into the lymphatic vessels, and clearance by alveolar macrophages. Surfactant is readily inactivated by hypoxia, infection, and edema fluid, which results in a decrease in lung compliance.

In addition to surfactant, another mechanism, interdependence, contributes to stability of the alveoli. Alveoli, except for those on the pleural surface, are surrounded by other alveoli. The tendency of one alveolus to collapse is opposed by the traction exerted by the surrounding alveoli. Thus collapse of a single alveolus causes stretching and distortion of the surrounding alveoli, which in turn are connected to other alveoli. Small openings **(pores of Kohn)** in the alveolar walls connect adjacent alveoli, whereas the **canals of Lambert** connect the terminal airways to adjacent alveoli. The pores of Kohn and the canals of Lambert provide collateral ventilation and prevent alveolar collapse **(atelectasis).**

TABLE 21.2 Composition and Function of Surfactant Components

Component	% Composition	Function
Phospholipids	80-85	
Phosphatidylcholine	70-80	Decrease surface tension
Phosphatidylglycerol	1-10	Spreading ability
Phosphatidylethanolamine	1-2	Unclear
Phosphatidylserine	1-2	Unclear
Phosphatidylinositol	1-2	Unclear
Neutral Lipids	5-10	
Cholesterol	3-5	Stabilization
Cholesterol Esters	1-3	Stabilization
Free Fatty Acids	1-3	
Proteins	2-5	
Surfactant protein A	2-4	Turn over, immune regulation, tubular myelin formation
Surfactant protein B	2-4	Decrease surface tension, spreading ability, lipid layering
Surfactant protein C	2-4	Decrease surface tension, spreading ability
Surfactant protein D	1-2	Unknown

IN THE CLINIC

In 1959, Avery and Mead discovered that in premature infants who died of hyaline membrane disease (HMD), the lungs were deficient in surfactant. HMD, also known as infant **respiratory distress syndrome,** is characterized by progressive atelectasis and respiratory failure in premature infants. It is a major cause of morbidity and mortality in the neonatal period. The major surfactant deficiency in premature infants is lack of phosphatidylglycerol. In general, as the level of phosphatidylglycerol increases in amniotic fluid, the infant mortality rate decreases. Research in this field has culminated in successful attempts to treat HMD in premature infants with surfactant replacement therapy. Today, surfactant replacement therapy is standard care for premature infants.

Key Points

1. Gas flows from areas of higher pressure to areas of lower pressure. Positive transpulmonary pressure is needed to increase lung volume. The pressure across the respiratory system is 0 at points of no airflow (end inspiration and end exhalation). At functional residual capacity (FRC), the pressure difference across the respiratory system is 0, and lung elastic recoil pressure, which operates to decrease lung volume, and the pressure generated by the chest wall to become larger are equal and opposite.

2. Pressure gradients in the respiratory system are created by the active contraction and subsequent relaxation of the muscles of respiration.

3. Lung volumes are determined by the balance between the lung's elastic recoil properties and the properties of the muscles of the chest wall.

4. Total lung capacity (TLC) is equal to the total volume of air that can be exhaled after a maximal inspiration (vital capacity [VC]) and the air remaining in the lung after a maximal exhalation (residual volume [RV]).

5. Lung compliance is a measure of the elastic properties of the lung. Elastic recoil is lost in patients with emphysema, and this loss is associated with an increase in lung compliance, whereas in diseases associated with pulmonary fibrosis, lung compliance is decreased.

6. The surface tension-reducing and antisticking properties of surfactant increase lung compliance, decrease the work of breathing, and help stabilize alveoli of different size.

Additional Readings

Gibson GJ, Pride NB. Lung distensibility: the static pressure-volume curve of the lungs and its use in clinical assessment. *Br J Dis Chest.* 1976;70:143-184.

Jobe AH. The alveolar lining layer: a review of studies on its role in pulmonary mechanics and in the pathogenesis of atelectasis, by Mary Ellen Avery, MD, Pediatrics, 1962:30:324-330. *Pediatrics.* 1998;102(S1):234-235.

Lumb AB. *Nunn's Applied Respiratory Physiology.* 8th ed. St. Louis: Elsevier; 2016.

Mead J, Macklem PT, vol eds. *American Physiological Society Handbook of Physiology: The Respiratory System, vol. 3: Mechanics.* Bethesda, MD: American Physiological Society; 1986.

Otis AB. A perspective of respiratory mechanics. *J Appl Physiol.* 1983;54:1183-1187.

Otis AB, Fenn WO, Rahn H. Mechanics of breathing in man. *J Appl Physiol.* 1950;2:592-607.

22

Dynamic Lung and Chest Wall Mechanics

LEARNING OBJECTIVES

Upon completion of this chapter the student should be able to answer the following questions:

1. Describe air flow in the airways.
2. Define resistance and its affect upon airflow in the airways.
3. List and describe two categories of factors that contribute to airway resistance.
4. List the features of a spirogram and flow volume curve.
5. Describe how flow limitation occurs at the equal pressure point and the role of dynamic airway compression in flow limitation.
6. Define the components of work of breathing.
7. Understand how dynamic compliance is different from static compliance and its contribution to work of breathing.

Dynamic Lung Mechanics

In this chapter, the principles that control air movement into and out of the lungs are examined. *Dynamic mechanics* is the study of physical systems in motion, and for the respiratory system it is the study of the properties of a lung whose volume is changing with time.

Airflow in Airways

Air flows into and out of an airway when there is a pressure difference at the two ends of the airway. By way of review, during inspiration the diaphragm contracts, pleural pressure becomes more negative, and gas flows into the lung (see Fig. 21.2). To meet the changing metabolic needs of the body, gas exchange depends on the speed at which fresh gas is brought to the alveoli and the rapidity with which the metabolic products of respiration (i.e., CO_2) are removed. Two major factors determine the speed at which gas flows into the airways for a given pressure change: the pattern of gas flow and the resistance to airflow by the airways.

Patterns of Airflow

There are two major patterns of gas flow in the airways—laminar and turbulent. *Laminar flow* is parallel to the airway walls and is present at low flow rates. As the flow rate increases and particularly as the airways divide, the flow stream becomes unsteady and small eddies develop. At higher flow rates the flow stream is disorganized and turbulence occurs.

The pressure-flow characteristics of laminar flow were first described by the French physician Poiseuille and apply to both liquids and air. In straight circular tubes the flow rate ($\dot{V}$) is defined by the following equation:

Equation 22.1

$$\dot{V} = \frac{P\pi r^4}{8\eta l}$$

where *P* is the driving pressure, *r* is the radius of the tube, η is the viscosity of the fluid, and *l* is the length of the tube. It can be seen that driving pressure (P) is proportional to the flow rate ($\dot{V}$); thus the greater the pressure, the greater the flow.

The flow resistance (R) across a set of tubes is defined as the change in driving pressure (ΔP) divided by the flow rate, or:

Equation 22.2

$$R = \frac{\Delta P}{\dot{V}} = \frac{8\eta l}{\pi r^4}$$

The units of resistance are cm $H_2O/L \cdot sec$. This equation is for laminar flow and demonstrates that the radius of the tube is the most important determinant of resistance. If the radius of the tube is reduced by half, the resistance will increase 16-fold. If, however, tube length is increased twofold, the resistance will increase only twofold. Thus the radius of the tube is the principal determinant of resistance. Stated another way, resistance is inversely proportional to the fourth power of the radius, and it is directly proportional to the length of the tube and to the viscosity of the gas.

In *turbulent flow,* gas movement occurs both parallel and perpendicular to the axis of the tube. Pressure is proportional to the flow rate squared. The viscosity of the gas increases with increasing gas density, and therefore the pressure drop increases for a given flow. Overall, gas velocity is blunted because energy is consumed in the process of generating eddies and chaotic movement. As a consequence, higher driving pressure is needed to support a given turbulent flow than to support a similar laminar flow.

Whether flow through a tube is laminar or turbulent depends on the Reynolds number. The *Reynolds number* (R_e) is a dimensionless value that expresses the ratio of two dimensionally equivalent terms (kinematic/viscosity), as seen in the equation:

Equation 22.3

$$R_e = \frac{2rvd}{\eta}$$

where d is the fluid density, v is the average velocity, r is the radius, and η is the viscosity. In straight tubes, turbulence occurs when the Reynolds number is greater than 2000. From this relationship it can be seen that turbulence is most likely to occur when the average velocity of the gas flow is high and the radius is large. In contrast, a low-density gas such as helium is less likely to cause turbulent flow. This is clinically relevant in states of increased airway resistance where a decrease in gas density (e.g., substituting helium for nitrogen in inspired air) can improve airflow.

Although these relationships apply well to smooth cylindrical tubes, application of these principles to a complicated system of tubes such as the airways is difficult. As a result, much of the flow in the airways demonstrates characteristics of both laminar and turbulent flow. In the trachea, for example, even during quiet breathing the Reynolds number is greater than 2000. Hence turbulent flow occurs in the trachea even during quiet breathing. Turbulence is also promoted by the glottis and vocal cords, which produce some irregularity and obstruction in the airways. As gas flows distally the total cross-sectional area increases dramatically, and gas velocities decrease significantly. As a result, gas flow becomes more laminar in the smaller airways even during maximal ventilation. Overall the gas flow in the larger airways (nose, mouth, glottis, and bronchi) is turbulent, whereas the gas flow in the smaller airways is laminar. Breath sounds heard with a stethoscope reflect turbulent airflow. **Laminar flow is silent, which is why it is difficult to "hear" small airway disease with a stethoscope**.

Airway Resistance

Airflow resistance is the second major factor that determines rates of airflow in the airways. Airflow resistance in the airways (R_{aw}) differs in airways of different size. In moving from the trachea toward the alveolus, individual airways become smaller while the number of airway branches increases dramatically. R_{aw} is equal to the sum of the resistance of each of these airways (i.e., $R_{aw} = R_{large} + R_{medium} + R_{small}$). From Poiseuille's equation, one might conclude that the major site of airway resistance is in the smallest airways. In fact, however, the major site of resistance along the bronchial tree is in the first eight generations of airways.

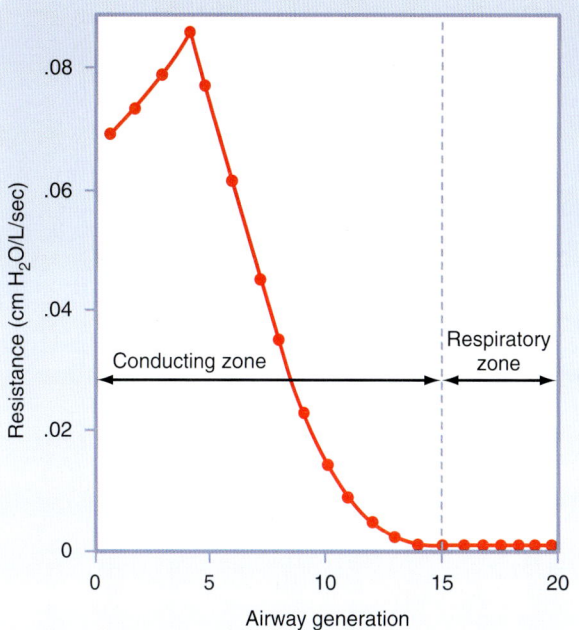

• **Fig. 22.1** Airway resistance as a function of the airway generation. In a normal lung, most of the resistance to airflow occurs in the first eight airway generations.

The smallest airways contribute very little to the overall total resistance of the bronchial tree (Fig. 22.1). The reason for this is twofold: (1) airflow velocity decreases substantially as the effective cross-sectional area increases (i.e., flow becomes laminar), and (2) most importantly, the airway branches in each generation exist in parallel rather than in series. The resistance of airways in parallel is the inverse of the sum of the individual resistances; therefore the overall contribution to resistance of the small airways is very small. As an example, assume that each of three tubes has a resistance of 3 cm H_2O. If the tubes are in series, the total resistance (R_{tot}) is the sum of the individual resistances:

Equation 22.4

$$R_{tot} = R_1 + R_2 + R_3 = 3 + 3 + 3 = 9 \text{ cm } H_2O/L \cdot sec$$

If the tubes are in parallel (as they are in small airways), the total resistance is the sum of the inverse of the individual resistances:

Equation 22.5

$$1/R_{tot} = 1/R_1 + 1/R_2 + 1/R_3 = 1/3 + 1/3 + 1/3$$
$$R_{tot} = 1 \text{ cm } H_2O/L \cdot sec$$

This relationship is in marked contrast to the pulmonary blood vessels, in which most of the resistance is located in the small vessels (see Chapter 23). Thus as airway diameter decreases, the resistance offered by each individual airway increases, but the large increase in the number of parallel pathways and cross-sectional area reduces the resistance at each generation of branching.

During normal breathing, approximately 80% of the resistance to airflow at functional residual capacity (FRC)

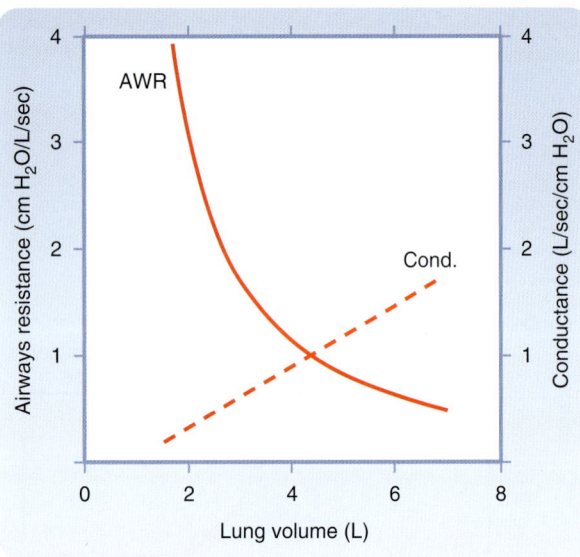

• **Fig. 22.2** Airway resistance (AWR) and conductance (Cond.) as a function of lung volume.

occurs in airways with diameters greater than 2 mm. Because the small airways contribute so little to total lung resistance, measurement of airway resistance is a poor test for detecting small airway obstruction.

Factors That Contribute to Airway Resistance

In healthy individuals, airway resistance is approximately 1 cm H_2O/L • sec. One of the most important factors affecting resistance is lung volume. Increasing lung volume increases the caliber of the airways because it creates a positive transairway pressure. As a result, resistance to airflow decreases with increasing lung volume and increases with decreasing lung volume. If the reciprocal of resistance (i.e., conductance) is plotted against lung volume, the relationship between lung volume and conductance is linear (Fig. 22.2). Other factors that increase airway resistance include airway mucus, edema, and contraction of bronchial smooth muscle, all of which decrease the caliber of the airways.

The density and viscosity of the inspired gas also affect airway resistance. When scuba diving, gas density rises and results in an increase in airway resistance; this increase can cause problems for individuals with asthma and obstructive pulmonary disease. Breathing a low-density gas such as an oxygen-helium mixture results in a decrease in airway resistance and has been exploited in the treatment of **status asthmaticus,** a condition associated with increased airway resistance due to a combination of bronchospasm, airway inflammation, and hypersecretion of mucus.

Neurohumoral Regulation of Airway Resistance

In addition to the effects of disease, airway resistance is regulated by various neural and humoral agents. Stimulation of efferent vagal fibers, either directly or reflexively, increases airway resistance and decreases anatomic dead space (see Chapter 23) secondary to airway constriction (recall that the vagus nerve innervates airway smooth muscle). In contrast, stimulation of sympathetic nerves and release of the postganglionic neurotransmitter norepinephrine inhibits airway constriction. Reflex stimulation of the vagus nerve by inhalation of smoke, dust, cold air, or other irritants can also result in airway constriction and coughing. Agents such as histamine, acetylcholine, thromboxane A_2, prostaglandin F_2, and leukotrienes (LTB_4, LTC_4, and LTD_4) are released by resident cells (e.g., mast cells, airway epithelial cells) and recruited cells (e.g., neutrophils, eosinophils) in response to various triggers such as allergens and viral infections. These agents act directly on airway smooth muscle to cause constriction and an increase in airway resistance. Inhalation of methacholine, a derivative of acetylcholine, is used to diagnose airway hyperresponsiveness, which is one of the cardinal features of certain asthma phenotypes. Although everyone is capable of responding to methacholine, airway obstruction develops in individuals with asthma at much lower concentrations of inhaled methacholine.

Measurement of Expiratory Flow

Measurement of expiratory flow rates and expiratory volumes is an important clinical tool for evaluating and monitoring respiratory diseases. Commonly used clinical tests have the patient inhale maximally to total lung capacity (TLC) and then exhale as rapidly and completely as possible to residual volume (RV). The test results are displayed either as a **spirogram** (Fig. 22.3A) or as a **flow-volume curve/ loop** (Fig. 22.3B). Results from individuals with suspected lung disease are compared with results predicted from normal healthy volunteers. Predicted or normal values vary with age, sex, ethnicity, height, and to a lesser extent, weight (Table 22.1). Abnormalities in values indicate abnormal pulmonary function and can be used to predict abnormalities in gas exchange. These values can detect the presence of abnormal lung function long before respiratory symptoms develop, and they can be used to determine disease severity and the response to therapy.

The Spirogram

A spirogram displays the volume of gas exhaled as a function of time (see Fig. 22.3A) and measures: (1) **forced vital capacity (FVC)**, (2) **forced expiratory volume in 1 second (FEV_1)**, (3) the **ratio of FEV_1 to FVC (FEV_1/ FVC)**, and (4) the **average midmaximal expiratory flow (FEF_{25-75})**.

The total volume of air that is exhaled during a maximal forced exhalation from TLC to RV is called the *FVC*. The volume of air that is exhaled in the first second during the maneuver is called the *FEV_1*. In normal individuals, 70% to 85% (depending on age) of the FVC can be exhaled in the first second. Thus the normal FEV_1/FVC ratio is greater than 70% in healthy adults. A ratio less than 70% suggests

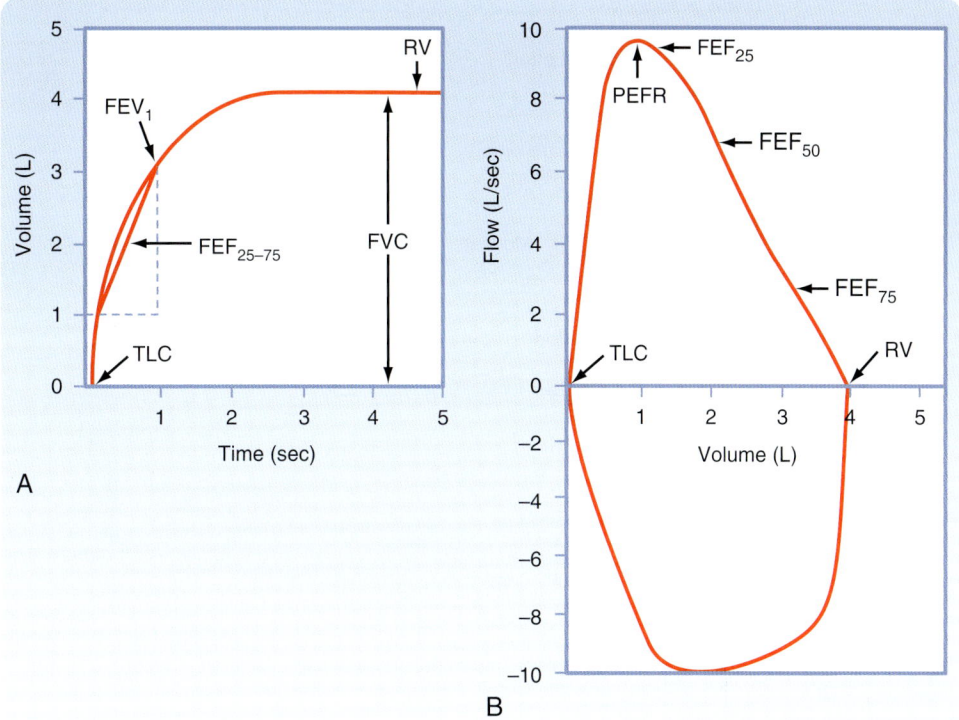

• **Fig. 22.3** The clinical spirogram (A) and flow-volume loop (B). The individual takes a maximal inspiration and then exhales as rapidly, as forcibly, and as maximally as possible. The volume exhaled is plotted as a function of time. In the spirogram that is reported in clinical settings, exhaled volume increases from the bottom of the trace to the top (A). This is in contrast to the physiologist's view of the same maneuver (see Fig. 21.3), in which the exhaled volume increases from the top to the bottom of the trace. In the flow-volume loop (B), exhaled volume is plotted as a function of the instantaneous flow rate, which is measured using a pneumotachometer. The maximal expiratory flow rate achieved during the maneuver is called the *peak expiratory flow rate*. Note the locations of TLC and RV on both tracings.

Pulmonary Function Measurement	Obstructive Pulmonary Disease	Restrictive Pulmonary Disease
FVC (L)	Decreased	Decreased
FEV₁ (L)	Decreased	Decreased
FEV₁/FVC	Decreased	Normal
FEF₂₅₋₇₅ (L/sec)	Decreased	Normal to increased
PEFR (L/sec)	Decreased	Normal
FEF₅₀ (L/sec)	Decreased	Normal
FEF₇₅ (L/sec)	Decreased	Normal
Slope of FV curve	Decreased	Normal to increased

TABLE 22.1 Patterns of Pulmonary Function Test Abnormalities

difficulty exhaling because of obstruction and is a hallmark of obstructive pulmonary disease. One expiratory flow rate—the average flow rate over the middle section of the VC—can be calculated from the spirogram. This expiratory flow rate has several names, including **MMEF (midmaximal expiratory flow)** and **FEF₂₅₋₇₅** (forced expiratory flow from 25%–75% of VC). Although it can be calculated from the spirogram, today's spirometers automatically calculate FEF_{25-75}.

Flow-Volume Loop

Another way of measuring lung function clinically is the flow-volume curve or loop. A flow-volume curve or loop is created by displaying the instantaneous flow rate during a forced maneuver as a function of the volume of gas. This instantaneous flow rate can be displayed both during exhalation (expiratory flow-volume curve) and during inspiration (inspiratory flow-volume curve) (see Fig. 22.3B). Expiratory flow rates are displayed above the horizontal line, and inspiratory flow rates are displayed below the horizontal line. The flow-volume loop measures: (1) the FVC, (2) the greatest flow rate achieved during the expiratory maneuver, called the **peak expiratory flow rate (PEFR),** and (3) multiple expiratory flow rates at various lung volumes. When the expiratory flow-volume curve is divided into quarters, the instantaneous flow rate at which 50% of the VC remains to be exhaled is called the **FEF_{50}** (also known as the $\dot{V}_{max50}$), the instantaneous flow rate at which 75% of the VC has been exhaled is called the **FEF_{75}** ($\dot{V}_{max75}$), and the instantaneous flow rate at which 25% of the VC has been exhaled is called the **FEF_{25}** ($\dot{V}_{max25}$).

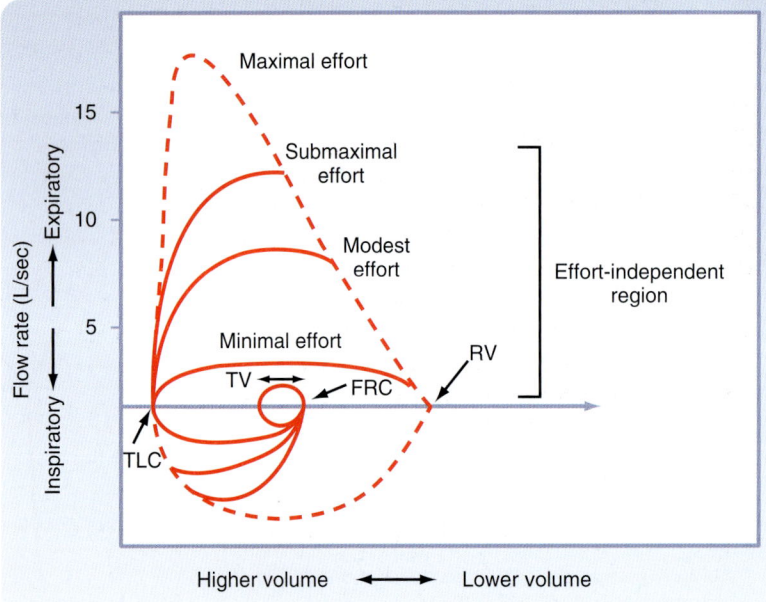

• **Fig. 22.4** Isovolume curves. Three superimposed expiratory flow maneuvers are made with increasing effort. Note that peak inspiratory and expiratory flow rates are dependent on effort, whereas expiratory flow rates later in expiration are independent of effort.

 IN THE CLINIC

In a methacholine challenge test, spirometry measurements are made after the patient inhales increasing concentrations of the muscarinic agonist methacholine. The test is stopped when FEV_1 falls by 20% or more or when a maximum concentration (25 mg/mL) of methacholine has been inhaled. The concentration of methacholine that produces a 20% decrease in FEV_1 is called the *provocation concentration (PC)20*. The lower the PC20, the more sensitive an individual is to methacholine. Most individuals with asthma have a PC20 less than 8 mg/mL of methacholine.

Determinants of Maximal Flow

The shape of the flow-volume loop reveals important information about normal lung physiology that can be altered by disease. Inspection of the flow-volume loop reveals that the maximum inspiratory flow is the same or slightly greater than the maximum expiratory flow. Three factors are responsible for the maximum inspiratory flow. First, the force generated by the inspiratory muscles decreases as lung volume increases above RV. Second, the recoil pressure of the lung increases as the lung volume increases above RV. This opposes the force generated by the inspiratory muscles and reduces maximum inspiratory flow. However, airway resistance decreases with increasing lung volume as the airway caliber increases. The combination of inspiratory muscle force, recoil of the lung, and changes in airway resistance causes maximal inspiratory flow to occur about halfway between TLC and RV.

During exhalation, maximal flow occurs early (in the first 20%) in the maneuver, and flow rates decrease progressively toward RV. Even with increasing effort, maximal flow decreases as RV is approached. This is known as *expiratory flow limitation* and can be demonstrated by asking an individual to perform three forced expiratory maneuvers with increasing effort. Fig. 22.4 shows the results of these three maneuvers. As effort increases, peak expiratory flow increases. However, the flow rates at lower lung volumes converge; this indicates that with modest effort, maximal expiratory flow is achieved. No amount of effort will increase the flow rates as lung volume decreases. For this reason, expiratory flow rates at lower lung volumes are said to be *effort independent* and *flow limited* because maximal flow is achieved with modest effort, and no amount of additional effort can increase the flow rate beyond this limit. In contrast, events early in the expiratory maneuver are said to be *effort dependent;* that is, increasing effort generates increasing flow rates. In general the first 20% of the flow in the expiratory flow-volume loop is effort dependent.

Flow Limitation and the Equal Pressure Point

Why is expiratory flow limited and reasonably effort independent? Factors that limit expiratory flow are important because many lung diseases affect these factors and thus affect the volume and speed with which air is moved into and out of the lung. Flow limitation occurs when the airways, which are intrinsically floppy distensible tubes, become compressed. The airways become compressed when the pressure outside the airway exceeds the pressure inside the airway. How and when this occurs is important to understanding lung disease. Fig. 22.5 shows the events that

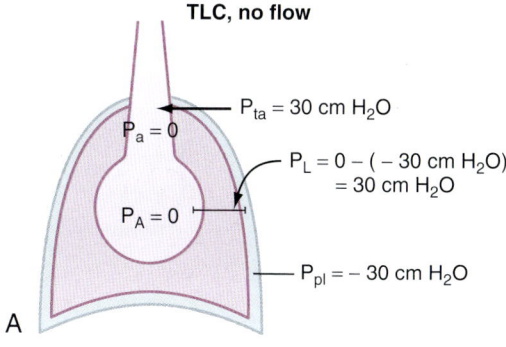

TLC, no flow

$P_{ta} = 30$ cm H_2O

$P_a = 0$

$P_L = 0 - (-30$ cm $H_2O)$
$= 30$ cm H_2O

$P_A = 0$

$P_{pl} = -30$ cm H_2O

A

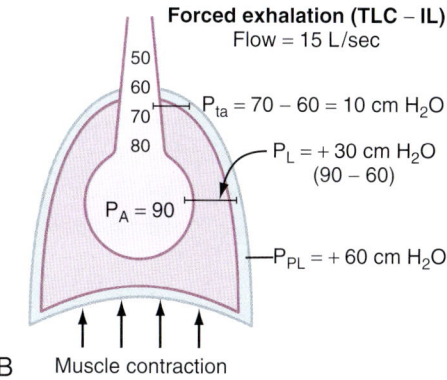

Forced exhalation (TLC – IL)
Flow = 15 L/sec

50
60
70
80

$P_{ta} = 70 - 60 = 10$ cm H_2O

$P_L = +30$ cm H_2O
$(90 - 60)$

$P_A = 90$

$P_{PL} = +60$ cm H_2O

↑ ↑ ↑ ↑
B   Muscle contraction

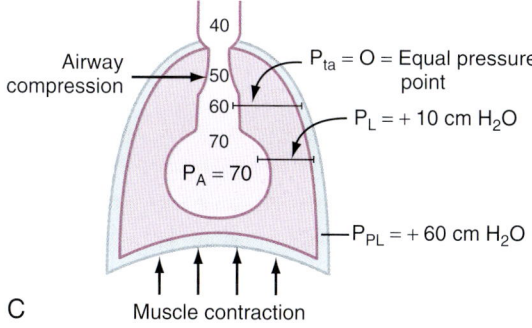

Forced exhalation (TLC – 2L)
Flow = 5 L/sec

40

Airway
compression 50

60

70

$P_{ta} = O = $ Equal pressure
point

$P_L = +10$ cm H_2O

$P_A = 70$

$P_{PL} = +60$ cm H_2O

↑ ↑ ↑ ↑
C   Muscle contraction

• **Fig. 22.5** Flow limitation. A, End inspiration, before the start of exhalation. B, At the start of a forced exhalation. C, Expiratory flow limitation later in a forced exhalation. Expiratory flow limitation occurs at locations where airway diameter is narrowed as a result of negative transmural pressure. See text for details.

occur during expiratory flow limitation at two different lung volumes. The airways and alveoli are surrounded by the pleural space and the chest wall. The airways are shown as tapered tubes because the total or collective airway cross-sectional area decreases from the alveoli to the trachea. At the start of exhalation but before any gas flow occurs, the pressure inside the alveolus (P_A) is zero (no airflow) and pleural pressure (in this example) is -30 cm H_2O. Transpulmonary pressure is thus $+30$ cm H_2O ($P_L = P_A - P_{pl}$). Because there is no flow, the pressure inside the airways is zero and the pressure across the airways (P_{ta}, transairway pressure) is $+30$ cm H_2O [$P_{ta} = P_{airway} - P_{pl}$

$= 0 - (-30$ cm $H_2O)$]. This positive transpulmonary and transairway pressure holds the alveoli and airways open.

When an active exhalation begins and the expiratory muscles contract, pleural pressure rises to $+60$ cm H_2O (in this example). Alveolar pressure also rises, in part because of the increase in pleural pressure ($+60$ cm H_2O) and in part because of the elastic recoil pressure of the lung at that lung volume (which in this case is 30 cm H_2O). Alveolar pressure is the sum of pleural pressure and elastic recoil pressure (i.e., $P_A = P_{el} + P_{pl} = 30$ cm $H_2O + 60$ cm $H_2O = 90$ cm H_2O in this example). This is the driving pressure for expiratory gas flow. Because alveolar pressure exceeds atmospheric pressure, gas begins to flow from the alveolus to the mouth when the glottis opens. As gas flows out of the alveoli, the transmural pressure across the airways decreases (i.e., the pressure head for expiratory gas flow dissipates). This occurs for three reasons: (1) there is a resistive pressure drop caused by the frictional pressure loss associated with flow (expiratory airflow resistance); (2) as the cross-sectional area of the airways decreases toward the trachea, gas velocity increases and this acceleration of gas flow further decreases the pressure; and (3) as lung volume decreases, the elastic recoil pressure decreases.

Thus as air moves out of the lung, the driving pressure for expiratory gas flow decreases. In addition, the mechanical tethering that holds the airways open at high lung volumes diminishes as lung volume decreases. There is a point between the alveoli and the mouth at which the pressure inside the airways equals the pressure that surrounds the airways. This point is called the **equal pressure point.** Airways toward the mouth but still inside the chest wall become compressed because the pressure outside is greater than the pressure inside (**dynamic airway compression**). As a consequence the transairway pressure now becomes negative [$P_{ta} = P_{aw} - P_{pl} = 58 - (+60) = -2$ cm H_2O] just beyond the equal pressure point. No amount of effort will increase the flow further because the higher pleural pressure tends to collapse the airway at the equal pressure point, just as it also tends to increase the gradient for expiratory gas flow. Under these conditions, airflow is independent of the total driving pressure. Hence the expiratory flow is effort independent and flow limited. It is also why airway resistance is greater during exhalation than during inspiration. In the absence of lung disease, the equal pressure point occurs in airways that contain cartilage, and thus they resist collapse. The equal pressure point, however, is not static. As lung volume decreases and as elastic recoil pressure decreases, the equal pressure point moves closer to the alveoli.

Dynamic Compliance

One additional measurement of dynamic lung mechanics should be mentioned, and this is the measurement of dynamic compliance. A dynamic pressure-volume curve can be created by having an individual breathe over a normal lung volume range (usually from FRC to FRC +1 L). The

IN THE CLINIC

What happens in individuals with lung disease? Imagine an individual with airway obstruction secondary to a combination of mucus accumulation and airway inflammation (Fig. 22.6A). At the start of exhalation the driving pressure for expiratory gas flow is the same as in a normal individual; that is, the driving pressure is the sum of the elastic recoil pressure and pleural pressure. As exhalation proceeds, however, the resistive drop in pressure is greater than in the normal individual because of the greater decrease in airway radius secondary to the accumulation of mucus and the inflammation. As a result the equal pressure point now occurs in small airways that are devoid of cartilage. These airways collapse. This collapse is known as **premature airway closure,** which results in a less-than-maximal expiratory volume and produces an increase in lung volume known as *air trapping*. The increase in lung volume initially helps offset the increase in airway resistance caused by the accumulation of mucus and inflammation because it results in an increase in airway caliber and elastic recoil. As the disease progresses, however, inflammation and accumulation of mucus increase further, there is a greater increase in expiratory resistance, and maximal expiratory flow rates decrease.

Now imagine an individual with emphysema and a loss of elastic recoil (see Fig. 22.6B). At the start of exhalation the driving pressure for expiratory gas flow is reduced secondary to a loss of elastic recoil. While the resistive drop in pressure is normal, the smaller initial driving pressure results in an equal pressure point that occurs closer to the alveolus in airways that do not contain cartilage. Premature airway closure again occurs but for a very different reason than the premature airway closure observed in individuals with an increase in airway resistance.

Individuals with premature airway closure frequently have **crackles,** also sometimes called **rales,** a popping sound usually heard during inspiration on auscultation. These crackles are due to the opening of airways during inspiration that closed (i.e., were compressed) during the previous exhalation. Crackles can be due to mucus accumulation, airway inflammation, fluid in the airways, or any mechanism responsible for airway narrowing or compression. They are also heard in individuals with emphysema, in which there is a decrease in lung elastic recoil. In fact, acute and chronic lung diseases can change the expiratory flow-volume relationship by changes in (1) static lung recoil pressure, (2) airway resistance and the distribution of resistance along the airways, (3) loss of mechanical tethering of intraparenchymal airways, (4) changes in the stiffness or mechanical properties of the airways, and (5) differences in severity of the aforementioned changes in various lung regions.

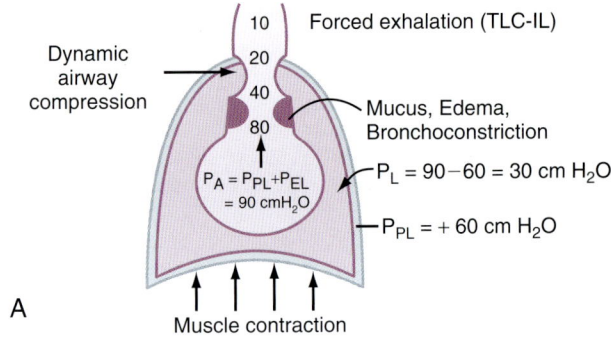

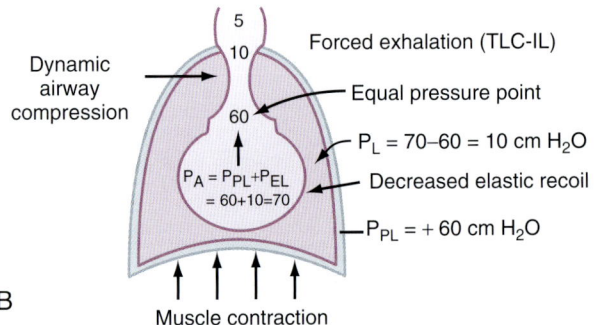

• **Fig. 22.6** A, Flow limitation in the presence of increased airway resistance. B, Flow limitation in the presence of a loss of elastic recoil.

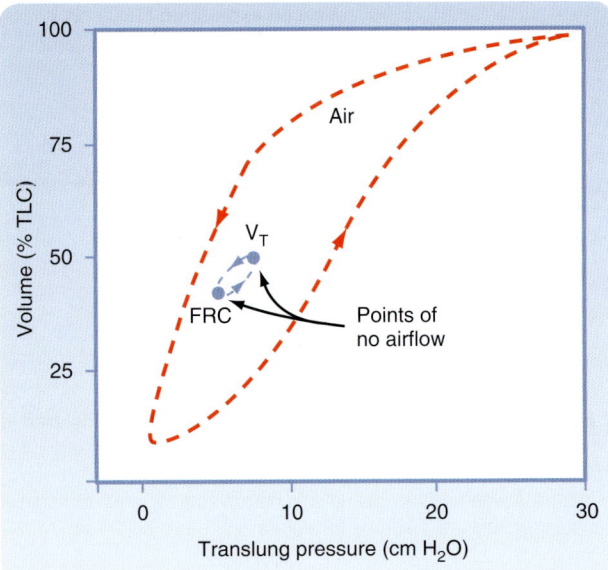

• **Fig. 22.7** Inflation-deflation pressure-volume curve. The direction of inspiration and exhalation is shown by the arrows. The difference between the inflation and deflation pressure-volume curves is due to the variation in surface tension with changes in lung volume. Note the slope of the line joining points of no airflow. This slope is less steep than the slope from the deflation pressure-volume curve at the same lung volume.

mean dynamic compliance of the lung (dyn C_L) is calculated as the slope of the line that joins the end-inspiratory and end-expiratory points of no flow (Fig. 22.7).

Dynamic compliance is always less than static compliance, and it increases during exercise. This is because during tidal volume breathing, a small change in alveolar surface area is insufficient to bring additional surfactant molecules to the surface, and thus the lung is less compliant. During exercise the opposite occurs; there are large changes in

tidal volume, and more surfactant material is incorporated into the air-liquid interface. Therefore the lung is more compliant.

Sighing and yawning increase dynamic compliance by increasing tidal volume and restoring the normal surfactant layer. Both of these respiratory activities are important for maintaining normal lung compliance. In contrast to the lung, the dynamic compliance of the chest wall is not significantly different from its static compliance.

Work of Breathing

Breathing requires the use of respiratory muscles (diaphragm, intercostals, etc.), which expends energy. Work is required to overcome the inherent mechanical properties of the lung (i.e., elastic and flow-resistive forces) and to move both the lungs and the chest wall. This work is known as the **work of breathing.** Changes in the mechanical properties of the lung or chest wall (or both) in the presence of disease result in an increase in the work of breathing. Respiratory muscles can perform increased work over long periods. However, like other skeletal muscles they can fatigue, and respiratory failure may ensue. Respiratory muscle fatigue is the most common cause of **respiratory failure,** a process in which gas exchange is inadequate to meet the metabolic needs of the body. In the respiratory system, the work of breathing is calculated by multiplying the change in volume by the pressure exerted across the respiratory system:

Work of breathing (W) = Pressure (P)
$$\times \text{ Change in volume } (\Delta V)$$

Although methods are not available to measure the total amount of work involved in breathing, one can estimate the mechanical work by measuring the volume and pressure changes during a respiratory cycle. Analysis of pressure-volume curves can be used to illustrate these points. Fig. 22.8A represents a respiratory cycle of a normal lung. The static inflation-deflation curve is represented by line *ABC.* The total mechanical workload is represented by the trapezoidal area *OAECD.*

In restrictive lung diseases, such as pulmonary fibrosis, lung compliance is decreased and the pressure-volume curve is shifted to the right. This results in a significant increase in the work of breathing (see Fig. 22.8B), as indicated by the increase in the trapezoidal area of OAECD. In obstructive lung diseases, such as asthma during an exacerbation or chronic bronchitis, airway resistance is elevated (see Fig. 22.8C) and greater negative pleural pressure is needed to maintain normal inspiratory flow rates. In addition to the increase in total inspiratory work (OAECD), individuals with obstructive lung disease have an increase in positive pleural pressure during exhalation because of the increase in resistance and the increased expiratory workload, which is visualized as area DFO. The stored elastic energy, represented by area ABCF of Fig. 22.8A, is not sufficient, and additional energy is needed for exhalation. With time or disease progression, these respiratory muscles can fatigue and result in respiratory failure. The work of breathing is also increased when deeper breaths are taken (an increase in tidal volume requires more elastic work to overcome) and when the respiratory rate increases (an increase in minute

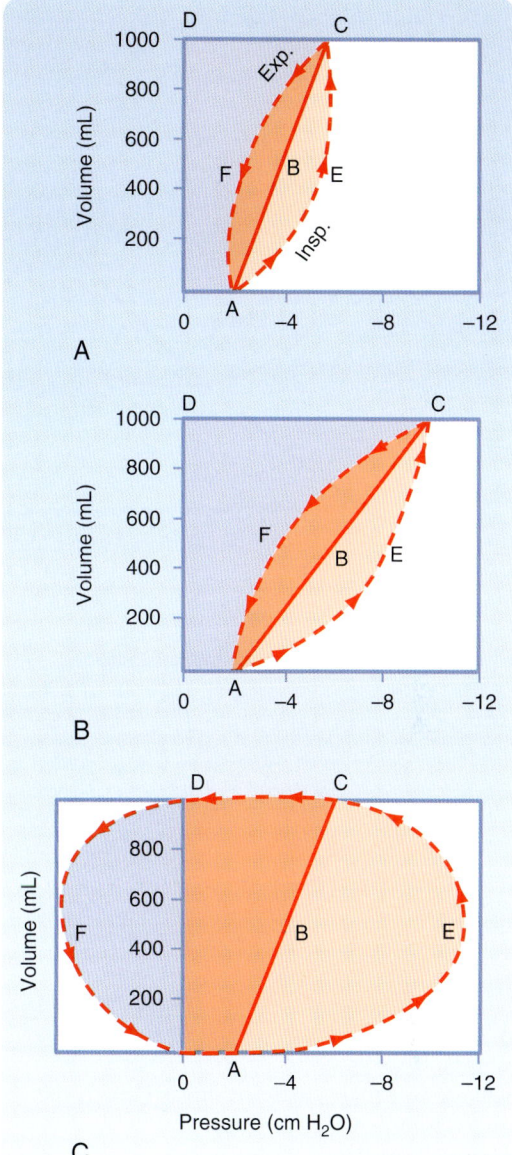

• **Fig. 22.8** Mechanical work done during a respiratory cycle in a normal lung (A), a lung with reduced compliance (B), and a lung with increased airway resistance (C). Breakdown of the trapezoidal areas enables one to appreciate the individual aspects of the mechanical workload, which include the following: OABCD, work necessary to overcome elastic resistance; AECF, work necessary to overcome nonelastic resistance; AECB, work necessary to overcome nonelastic resistance during inspiration; ABCF, work necessary to overcome nonelastic resistance during exhalation (represents stored elastic energy from inspiration).

ventilation requires more flow resistance force to overcome) (Fig. 22.9). Normal individuals and individuals with lung disease adopt respiratory patterns that minimize the work of breathing. For this reason, individuals with pulmonary fibrosis (increased elastic work) breathe more shallowly and rapidly, and those with obstructive lung disease (normal elastic work but increased resistive work) breathe more slowly and deeply.

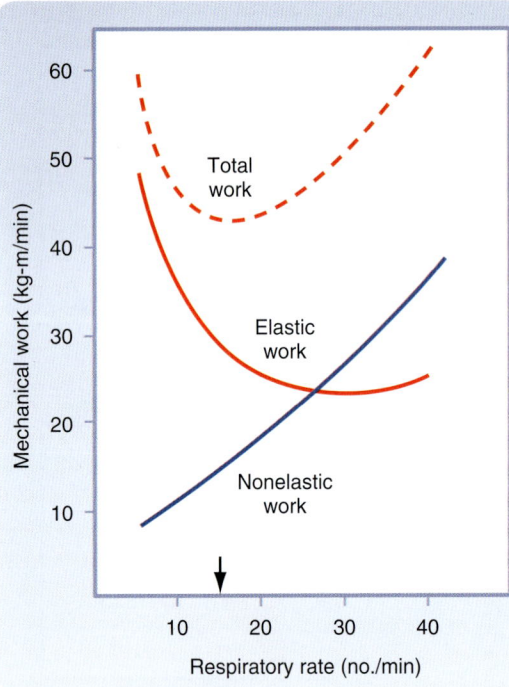

• **Fig. 22.9** Effect of the respiratory rate on the elastic, nonelastic, and total mechanical work of breathing at a given level of alveolar ventilation. Individuals tend to adopt the respiratory rate at which the total work of breathing is minimal *(arrow)* for those without lung disease.

 IN THE CLINIC

Chronic obstructive pulmonary disease (COPD) is a general term that includes diseases such as emphysema and chronic bronchitis. COPD most commonly occurs in individuals who smoke, in whom pathologic changes in the lung consistent with both emphysema and chronic bronchitis can coexist. For individuals with COPD in whom emphysema is a major component, the elastic tissue in the alveolar and capillary walls is progressively destroyed, which results in increased lung compliance and decreased elastic recoil. The decrease in elastic recoil results in movement of the equal pressure point toward the alveolus and premature airway closure. This produces air trapping and increases in RV, FRC, and TLC. Airway resistance is also increased. These increases in lung volumes increase the work of breathing by stretching the respiratory muscles and decreasing their efficiency.

In chronic bronchitis, accumulation of mucus and airway inflammation cause the equal pressure point to move toward the alveolus, which leads to premature airway closure and increases in RV, FRC, and TLC. Airway resistance and the work of breathing are increased, but lung compliance is normal.

In restrictive lung diseases such as pulmonary fibrosis, lung compliance is decreased. Lung volumes are decreased, but flow rates are reasonably normal. Some of the changes in pulmonary function values in obstructive and restrictive pulmonary diseases are shown in Table 22.1.

In the third trimester of pregnancy, the enlarged uterus increases intraabdominal pressure and restricts movement of the diaphragm. The FRC, as a result, decreases. This change in lung volume results in decreased lung compliance and increased airway resistance in otherwise healthy women.

Key Concepts

1. There are two major patterns of gas flow in the airways: turbulent and laminar.
2. *Resistance* to airflow is the change in pressure per unit of flow. Airway resistance varies with the inverse of the fourth power of the radius and is higher in turbulent than in laminar flow. The major site of airway resistance is the first eight airway generations. Airway resistance decreases with increases in lung volume and with decreases in gas density. Airways resistance is also regulated by neural and humoral agents.
3. Pulmonary function tests (spirometry, flow-volume loop, body plethysmography) can detect abnormalities in lung function before individuals become symptomatic. Test results are compared with results obtained in normal individuals and vary with sex, ethnicity, age, and height. COPD is characterized by increases in lung volumes and airway resistance and by decreases in expiratory flow rates. Emphysema, a specific type of COPD, is further characterized by increased lung compliance. Restrictive lung diseases are characterized by decreases in lung volume, normal expiratory flow rates and resistance, and a marked decrease in lung compliance.

4. The *equal pressure point* is the point at which the pressure inside and surrounding the airway is the same. The location of the equal pressure point is dynamic. Specifically, as lung volume and elastic recoil decrease, the equal pressure point moves toward the alveolus in normal individuals. In individuals with chronic obstructive pulmonary disease (COPD), the equal pressure point at any lung volume is closer to the alveolus. Expiratory flow limitation occurs at the equal pressure point.
5. Energy is expended during breathing to overcome the inherent mechanical properties of the lung. Respiratory muscle fatigue is the most common cause of respiratory failure. Individuals breathe at a respiratory rate to minimize work. For individuals with increased airway resistance, work is minimized by breathing at lower frequencies. For individuals with restrictive lung diseases, work is minimized by shallow breathing at high frequencies.
6. The dynamic compliance of the lung is always less than the static compliance and increases during exercise, sighing, and yawning.

Additional Reading

Journal Articles

Calverley PMA, Koulouris NG. Flow limitation and dynamic hyperinflation: key concepts in modern respiratory physiology. *Eur Respir J*. 2005;25:186-199.

Crapo RO, et al. Reference spirometric values using techniques and equipment that meet ATS recommendations. *Am Rev Respir Dis*. 1981;123:659-664.

Otis AB. A perspective of respiratory mechanics. *J Appl Physiol*. 1983;54:1183-1187.

Otis AB, et al. Mechanics of breathing in man. *J Appl Physiol*. 1950;2:592-607.

Books/Book Chapters

Leff AR, Schumacker PT. *Respiratory Physiology: Basics and Applications*. Philadelphia: Saunders; 1993.

Lumb AB. *Nunn's Applied Respiratory Physiology*. 8th ed. St. Louis: Elsevier; 2016.

Mead J, MacKlem PT. Mechanics of breathing. In: *Handbook of Physiology*. Section 3, The Respiratory System. Bethesda, MD: American Physiological Society; 1986. Wiley Online Library doi:10.1002/cphys.cp0303fmo1.

23

Ventilation, Perfusion, and Ventilation/ Perfusion Relationships

LEARNING OBJECTIVES

Upon completion of this chapter, the student should be able to answer the following questions:

1. Define two types of dead space ventilation, and describe how dead space ventilation changes with tidal volume.
2. Describe the composition of gas in ambient air, the trachea, and the alveolus, and understand how this composition changes with changes in oxygen fraction and barometric pressure.
3. Use the alveolar air equation to calculate the alveolar-arterial difference for oxygen ($AaDo_2$).
4. Understand the alveolar carbon dioxide equation and identify how it changes with alterations in alveolar ventilation.
5. Compare the distribution of pulmonary blood flow to the distribution of ventilation.
6. List and define the four categories of hypoxia and the six causes of hypoxic hypoxia.
7. Distinguish the causes of hypoxic hypoxia on the basis of the response to 100% O_2.
8. Describe the two causes of hypercapnia.

The major determinant of normal gas exchange and thus the level of Po_2 and Pco_2 in blood is the relationship between ventilation ($\dot{V}$) and perfusion ($\dot{Q}$). This relationship is called the ventilation/perfusion ($\dot{V}/\dot{Q}$) ratio.

Ventilation

Ventilation is the process by which air moves in and out of the lungs. The incoming air is composed of a volume that fills the conducting airways (dead space ventilation) and a portion that fills the alveoli (alveolar ventilation). Minute (or total) ventilation ($\dot{V}_E$) is the volume of air that enters or leaves the lung per minute:

Equation 23.1
$$\dot{V}_E = f \times V_T$$

where f is the frequency or number of breaths per minute and V_T (also known as TV) is the tidal volume, or volume of air inspired (or exhaled) per breath. Tidal volume varies

with age, sex, body position, and metabolic activity. In an average-sized adult at rest, tidal volume is 500 mL. In children, it is 3 to 5 mL/kg.

Dead Space Ventilation: Anatomical and Physiological

Anatomical Dead Space

Dead space ventilation is ventilation to airways that do not participate in gas exchange. There are two types of dead space: anatomical dead space and physiological dead space. **Anatomical dead space** (V_D) is composed of the volume of gas that fills the conducting airways:

Equation 23.2
$$V_T = V_D + V_A$$

where V refers to volume and the subscripts T, D, and A refer to tidal, dead space, and alveolar. A "dot" above V denotes a volume per unit of time (n):

Equation 23.3
$$V_T \times n = (V_D \times n) + (V_A \times n)$$

or

Equation 23.4
$$\dot{V}_E = \dot{V}_D + \dot{V}_A$$

where $\dot{V}_E$ is the total volume of gas in liters expelled from the lungs per minute (also called exhaled minute volume), $\dot{V}_D$ is the dead space ventilation per minute, and $\dot{V}_A$ is alveolar ventilation per minute.

In a healthy adult, the volume of gas contained in the conducting airways at functional residual capacity (FRC) is approximately 100 to 200 mL, in comparison with the 3 L of gas in an entire lung. The ratio of the volume of the conducting airways (dead space) to tidal volume represents the fraction of each breath that is "wasted" in filling the conducting airways. This volume is related to tidal volume (V_T) and to exhaled minute ventilation ($\dot{V}_E$) in the following way:

Equation 23.5
$$\dot{V}_D = \frac{V_D}{V_T} \times \dot{V}_E$$

IN THE CLINIC

If the dead space volume is 150 mL and tidal volume increases from 500 to 600 mL for the same exhaled minute ventilation, what is the effect on dead space ventilation?

$$V_T = 500 \text{ mL}$$

$$V_D = \frac{150 \text{ mL}}{500 \text{ mL}} \times \dot{V}_E$$

$$= 0.3 \times \dot{V}_E$$

and, similarly,

$$V_T = 600 \text{ mL}$$

$$V_D = \frac{150 \text{ mL}}{600 \text{ mL}} \times \dot{V}_E$$

$$= 0.25 \times \dot{V}_E$$

Increasing tidal volume is an effective way to increase alveolar ventilation (and thus normal blood gas values), as might occur during exercise or periods of stress. As tidal volume increases, the fraction of the dead space ventilation decreases for the same exhaled minute ventilation.

Dead space ventilation (V_D) varies inversely with tidal volume (V_T). The larger the tidal volume, the smaller the proportion of dead space ventilation. Normally, V_D/V_T is 20% to 30% of exhaled minute ventilation. Changes in dead space are important contributors to work of breathing. If the dead space increases, the individual must inspire a larger tidal volume to maintain normal levels of blood gases. This adds to the work of breathing and can contribute to respiratory muscle fatigue and respiratory failure. If metabolic demands increase (e.g., during exercise or with fever), individuals with lung disease may not be able to increase tidal volume sufficiently.

Physiological Dead Space

The second type of dead space is physiological dead space. Often in diseased lungs, some alveoli are perfused but not ventilated. The **total** volume of gas in each breath that does not participate in gas exchange is called the **physiological dead space.** This volume includes the anatomical dead space and the dead space secondary to perfused but unventilated alveoli. The physiological dead space is always at least as large as the anatomical dead space, and in the presence of disease, it may be considerably larger.

Both anatomical and physiological dead space can be measured, but they are not measured routinely in the course of patient care.

Alveolar Ventilation

Composition of Air

Inspiration brings ambient or atmospheric air to the alveoli, where O_2 is taken up and CO_2 is excreted. Ambient air is a

IN THE CLINIC

In individuals with certain types of chronic obstructive pulmonary disease (COPD), such as emphysema, physiological dead space is increased. If dead space doubles, tidal volume must increase in order to maintain the same level of alveolar ventilation. If tidal volume is 500 mL and V_D/V_T is 0.25, then

$$V_T = V_D + V_A$$

$$500 \text{ mL} = 125 \text{ mL} + 375 \text{ mL}$$

If V_D increases to 250 mL in this example, tidal volume (V_T) must increase to 625 mL to maintain a normal alveolar ventilation (i.e., $V_A = 375$ mL):

$$V_T = 250 \text{ mL} + 375 \text{ mL}$$

$$= 375 \text{ mL}$$

---

gas mixture composed of N_2 and O_2, with minute quantities of CO_2, argon, and inert gases. The composition of this gas mixture can be described in terms of either gas fractions or the corresponding partial pressure.

Because ambient air is a gas, the gas laws can be applied, from which two important principles arise. The first is that when the components are viewed in terms of gas fractions (F), the sum of the individual gas fractions must equal one:

Equation 23.6

$$1.0 = FN_2 + FO_2 + F_{argon} \text{ and other gases}$$

It follows, then, that the sum of the **partial pressures** (in millimeters of mercury) of a gas, also known as the gas **tension** (in torr), must be equal to the total pressure. Thus at sea level, where atmospheric pressure (also known as barometric pressure [P_b]) is 760 mm Hg, the partial pressures of the gases in air are as follows:

Equation 23.7

$$P_b = PN_2 + P_{argon} \text{ and other gases}$$

$$760 \text{ mm Hg} = PN_2 + PO_2 + P_{argon} \text{ and other gases}$$

IN THE CLINIC

Three important gas laws govern ambient air and alveolar ventilation. According to **Boyle's law,** when temperature is constant, pressure (P) and volume (V) are inversely related; that is,

$$P_1V_1 = P_2V_2$$

Boyle's law is used in the measurement of lung volumes (see Fig. 21.4). **Dalton's law** is that the partial pressure of a gas in a gas mixture is the pressure that the gas would exert if it occupied the total volume of the mixture in the absence of the other components. Eq. 23.7 is an example of how Dalton's law is used in the lung. According to **Henry's**

law, the concentration of a gas dissolved in a liquid is proportional to its partial pressure.

The second important principle is that the partial pressure of a gas (P_{gas}) is equal to the fraction of that gas in the gas mixture (F_{gas}) multiplied by the atmospheric (barometric) pressure:

Equation 23.8

$$P_{gas} = F_{gas} \times P_b$$

Ambient air is composed of approximately 21% O_2 and 79% N_2. Therefore, the partial pressure of O_2 in inspired ambient air (PO_2) is calculated as follows:

Equation 23.9

$$PO_2 = FiO_2 \times P_b$$
$$PO_2 = 0.21 \times 760 \text{ mm Hg}$$
$$= 159 \text{ mm Hg or } 159 \text{ torr}$$

where (FiO_2) is the fraction of oxygen in inspired air. The partial pressure of O_2, or oxygen tension, in ambient air at the mouth at the start of inspiration is therefore 159 mm Hg, or 159 torr. The O_2 tension at the mouth can be altered in one of two ways: by changing the fraction of O_2 in inspired air (FiO_2) or by changing barometric pressure. Thus ambient O_2 tension can be increased through the administration of supplemental O_2 and is decreased at high altitude.

IN THE CLINIC

The partial pressure of O_2 in ambient air varies with altitude. The highest and lowest points in the contiguous United States are Mount Whitney in Sequoia National Park/ Inyo National Forest (14,505 feet; barometric pressure, 437 mm Hg) and Badwater Basin in Death Valley National Park (282 feet; barometric pressure, 768 mm Hg). On Mount Whitney, the partial pressure of O_2 in ambient air is calculated as follows:

$$PO_2 = 0.21 \times 437 \text{ mm Hg} = 92 \text{ mm Hg}$$

whereas in Death Valley Badwater Basin, the partial pressure of oxygen is calculated as follows:

$$PO_2 = 0.21 \times 768 \text{ mm Hg} = 161 \text{ mm Hg}$$

Note that the FiO_2 does not vary at different altitudes; only the barometric pressure varies. These differences in oxygen tension have profound effects on arterial blood gas values.

As inspiration begins, ambient air is brought into the nasopharynx and laryngopharynx, where it becomes warmed to body temperature and humidified. Inspired air becomes saturated with water vapor by the time it reaches the glottis. Water vapor exerts a partial pressure and dilutes the total pressure in which the other gases are distributed. Water vapor pressure at body temperature is 47 mm Hg. To calculate the partial pressures of O_2 and N_2 in a humidified mixture, the water vapor partial pressure must be subtracted from the total barometric pressure. Thus in the conducting airways, which begin in the trachea, the partial pressure of O_2 is calculated as follows:

Equation 23.10

$$P_{trachea}O_2 = (P_b - PH_2O) \times FiO_2$$
$$= (760 \text{ mm Hg} - 47 \text{ mm Hg}) \times 0.21$$
$$= 150 \text{ mm Hg}$$

and the partial pressure of N_2 is calculated similarly:

Equation 23.11

$$P_{trachea}N_2 = (760 - 47 \text{ mm Hg}) \times 0.79$$
$$= 563 \text{ mm Hg}$$

Note that the total pressure remains constant at 760 mm Hg (150 + 563 + 47 mm Hg) and that the fractions of O_2 and N_2 are unchanged. Water vapor pressure, however, reduces the partial pressures of O_2 and N_2. Note also that in the calculation of the partial pressure of ambient air (Eq. 23.9), water vapor is ignored, and ambient air is considered "dry." The conducting airways do not participate in gas exchange. Therefore, the partial pressures of O_2, N_2, and water vapor remain unchanged in the airways until the air reaches the alveolus.

Alveolar Gas Composition

When the inspired air reaches the alveolus, O_2 is transported across the alveolar membrane into the capillary bed, and CO_2 moves from the capillary bed into the alveolus. The process by which this occurs is described in Chapter 24. At the end of inspiration and with the glottis open, the total pressure in the alveolus is atmospheric; thus, the partial pressures of the gases in the alveolus must equal the total pressure, which in this case is atmospheric. The composition of the gas mixture, however, is changed and can be described as follows:

Equation 23.12

$$1.0 = FO_2 + FN_2 + FH_2O + FCO_2 + F_{argon} \text{ and other gases}$$

where N_2 and argon are inert gases, and therefore the fraction of these gases in the alveolus does not change from ambient fractions. The fraction of water vapor also does not change because the inspired gas is already fully saturated with water vapor and is at body temperature. As a consequence of gas exchange, however, the fraction of O_2 in the alveolus decreases, and the fraction of CO_2 in the alveolus increases. Because of changes in the fractions of O_2 and CO_2, the partial pressures exerted by these gases also change. The partial pressure of O_2 in the alveolus (PAO_2) is given by the **alveolar gas equation,** which is also called the **ideal alveolar oxygen equation:**

Equation 23.13

$$PAO_2 = PiO_2 - \frac{PACO_2}{R}$$
$$= [FiO_2 \times (P_b - PH_2O)] - \frac{PACO_2}{R}$$

TABLE 23.1 Total and Partial Pressures of Respiratory Gases in Ideal Alveolar Gas and Blood at Sea Level (760 mm Hg)

Parameter	Ambient Air (Dry)	Moist Tracheal Air	Alveolar Gas (R = 0.8)	Systemic Arterial Blood	Mixed Venous Blood
P_{O_2}	159	150	102	90	40
P_{CO_2}	0	0	40	40	46
P_{H_2O}, 37°C	0	47	47	47	47
P_{N_2}	601	563	571*	571	571
P_{total}	760	760	760	748	704†

P_{CO_2}, partial pressure of carbon dioxide; P_{H_2O}, partial pressure of water; P_{N_2}, partial pressure of nitrogen; P_{O_2}, partial pressure of oxygen; P_{TOTAL}, partial pressure of all parameters; R, respiratory quotient.

*P_{N_2} is increased in alveolar gas by 1% because R is normally less than 1.

†P_{total} is less in venous than in arterial blood because P_{O_2} has decreased more than P_{CO_2} has increased.

where P_{IO_2} is the partial pressure of inspired O_2, which is equal to the fraction of O_2 (F_{IO_2}) multiplied by the barometric pressure (P_b) minus water vapor pressure (P_{H_2O}); P_{ACO_2} is the partial pressure of alveolar CO_2; and R is the respiratory exchange ratio, or **respiratory quotient.** The respiratory quotient is the ratio of the amount of CO_2 excreted ($\dot{V}_{CO_2}$) to the amount of O_2 taken up ($\dot{V}_{O_2}$) by the lungs. This quotient is the amount of CO_2 produced in relation to the amount of O_2 consumed by metabolism and is dependent on caloric intake. The respiratory quotient varies between 0.7 and 1.0; it is 0.7 in states of exclusive fatty acid metabolism and 1.0 in states of exclusive carbohydrate metabolism. Under normal dietary conditions, the respiratory quotient is assumed to be 0.8. Thus the quantity of O_2 taken up exceeds the quantity of CO_2 that is released in the alveoli. The partial pressures of O_2, CO_2, and N_2 from ambient air to the alveolus at sea level are shown in Table 23.1.

A similar approach can be used to calculate the estimated P_{ACO_2}. The fraction of CO_2 in the alveolus is a function of the rate of CO_2 production by the cells during metabolism and the rate at which the CO_2 is eliminated from the alveolus. This process of elimination of CO_2 is known as **alveolar ventilation.** The relationship between CO_2 production and alveolar ventilation is defined by the **alveolar carbon dioxide equation:**

Equation 23.14

$$\dot{V}_{CO_2} = \dot{V}_A \times F_{ACO_2}$$

or

$$F_{ACO_2} = \dot{V}_{CO_2} / \dot{V}_A$$

where $\dot{V}_{CO_2}$ is the rate of CO_2 production by the body, $\dot{V}_A$ is alveolar ventilation per minute, and F_{ACO_2} is the fraction of CO_2 in dry alveolar gas. This relationship demonstrates that the rate of elimination of CO_2 from the alveolus is related to alveolar ventilation and to the fraction of CO_2 in the alveolus. Like the partial pressure of any other gas (see Eq. 23.8), P_{ACO_2} is defined by the following:

Equation 23.15

$$P_{ACO_2} = F_{ACO_2} \times (P_b - P_{H_2O})$$

Substituting for F_{ACO_2} in the previous equation yields the following relationship:

Equation 23.16

$$P_{ACO_2} = \frac{\dot{V}_{CO_2} \times (P_b - P_{H_2O})}{\dot{V}_A}$$

This equation demonstrates several important relationships. First, there is an inverse relationship between the partial pressure of CO_2 in the alveolus (P_{ACO_2}) and alveolar ventilation per minute ($\dot{V}_A$), regardless of the exhaled CO_2. Specifically, if ventilation is doubled, P_{ACO_2} decreases by 50%. Conversely, if ventilation is decreased by half, the P_{ACO_2} doubles. Second, at a constant alveolar ventilation per minute ($\dot{V}_A$), doubling of the metabolic production of CO_2 ($\dot{V}_{CO_2}$) causes the P_{ACO_2} to double. The relationship between $\dot{V}_A$ and P_{ACO_2} is depicted in Fig. 23.1.

Arterial Gas Composition

In normal lungs, P_{aCO_2} is tightly regulated and maintained at 40 ± 2 mm Hg. Increases or decreases in P_{aCO_2}, particularly when associated with changes in arterial pH, have profound effects on cell function, including enzyme and protein activity. Specialized chemoreceptors monitor P_{aCO_2} in the brainstem (Chapter 25), and exhaled minute ventilation (Eq. 23.1) varies in accordance with the level of P_{aCO_2}.

An acute increase in P_{aCO_2} results in **respiratory acidosis** (pH < 7.35), whereas an acute decrease in P_{aCO_2} results in **respiratory alkalosis** (pH > 7.45). **Hypercapnia** is defined as an elevation in P_{aCO_2}, and it occurs when CO_2 production exceeds alveolar ventilation (hypoventilation). Conversely, hyperventilation occurs when alveolar ventilation exceeds CO_2 production, and it decreases P_{aCO_2} **(hypocapnia).**

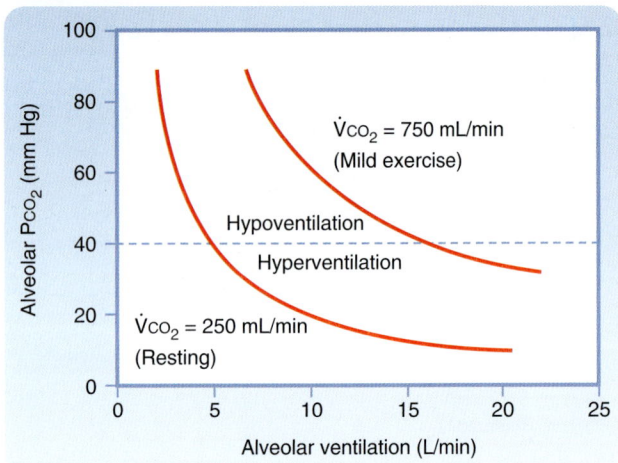

• **Fig. 23.1** The Alveolar Partial Pressure of Carbon Dioxide (PCO$_2$; y-axis) as a Function of Alveolar Ventilation per Minute ($\dot{V}_A$; x-axis) in the Lung. Each line corresponds to a given metabolic rate associated with a constant production of CO$_2$ ($\dot{V}CO_2$ isometabolic line). Normally, alveolar ventilation is controlled to maintain an alveolar PCO$_2$ of approximately 40 mm Hg. Thus at rest, when $\dot{V}CO_2$ is approximately 250 mL/minute, alveolar ventilation of 5 L/minute results in an alveolar PCO$_2$ of 40 mm Hg. A 50% decrease in ventilation at rest (i.e., from 5 to 2.5 L/minute) results in doubling of alveolar PCO$_2$. During exercise, CO$_2$ production is increased ($\dot{V}CO_2 = 750$ mL/min), and to maintain normal alveolar PCO$_2$, ventilation must increase (in this case, to 15 L/minute). Again, however, a 50% reduction in ventilation (from 15 to 7.5 L/minute) results in doubling of the alveolar PCO$_2$.

Distribution of Ventilation

Ventilation is not uniformly distributed in the lung, largely because of the effects of gravity. In the upright position, at most lung volumes, alveoli near the apex of the lung are more expanded than are alveoli at the base. Gravity pulls the lung downward and away from the chest wall. As a result, pleural pressure is lower (i.e., more negative) at the apex than at the base of the lung, and static translung pressure ($P_L = P_A - P_{pl}$) is increased; this results in an increase in alveolar volume at the apex. Because of the difference in alveolar volume at the apex and at the base of the lung (Fig. 23.2), alveoli at the lung base are represented along the steep portion of the pressure-volume curve, and they receive more of the ventilation (i.e., they have greater compliance). In contrast, the alveoli at the apex are represented closer to the top or flat portion of the pressure-volume curve. They have lower compliance and thus receive proportionately less of the tidal volume. The effect of gravity is less pronounced when a person is supine rather than upright, and it is less when a person is supine rather than prone. This is because the diaphragm is pushed in a cephalad direction when a person is supine, and it affects the size of all of the alveoli.

In addition to gravitational effects on the distribution of ventilation, ventilation in alveoli is not uniform. The reason for this is variable airway resistance (R) or compliance (C), and it is described quantitatively by the **time constant** (τ):

Equation 23.17

$$\tau = R \times C$$

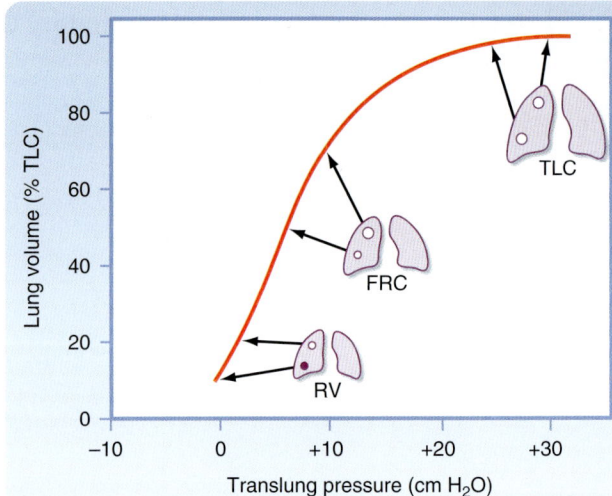

• **Fig. 23.2** Regional Distribution of Lung Volume, Including Alveolar Size *(Circles)* and Location on the Pressure-Volume Curve of the Lung at Different Lung Volumes. Because the lungs are suspended in the upright position, the pleural pressure (P_{pl}) and translung pressure (P_L) of lung units at the apex are greater than those at the base. These lung units are larger at any lung volume than are those at the base. The effect is greatest at residual volume (RV), less so at functional residual capacity (FRC), and absent at total lung capacity (TLC). Note also that because of their "location" on the pressure-volume curve, inspired air is differentially distributed to these lung units; those at the apex are less compliant and receive a smaller proportion of the inspired air than do the lung units at the base, which are more compliant (i.e., are represented at a steeper part of the pressure-volume curve).

Alveolar units with long time constants fill and empty slowly. Thus an alveolar unit with increased airway resistance or increased compliance takes longer to fill and longer to empty. In adults, the normal respiratory rate is approximately 12 breaths per minute, the inspiratory time is approximately 2 seconds, and the expiratory time is approximately 3 seconds. In normal lungs, this time is sufficient to approach volume equilibrium (Fig. 23.3). In the presence of increased resistance or increased compliance, however, volume equilibrium is not reached.

IN THE CLINIC

Adults with COPD have a very long time constant as a result of an increase in resistance and, in the case of individuals with emphysema, an increase in compliance. As a result, such affected adults tend to breathe at a low respiratory rate. Imagine now what happens when individuals with COPD climb a flight of stairs. The increase in respiratory rate does not allow sufficient time for a full exhalation, and a process called *dynamic hyperinflation* occurs (Fig. 23.4); lung volumes, which are already increased, increase further, the lung becomes less compliant, and the work of breathing is very high.

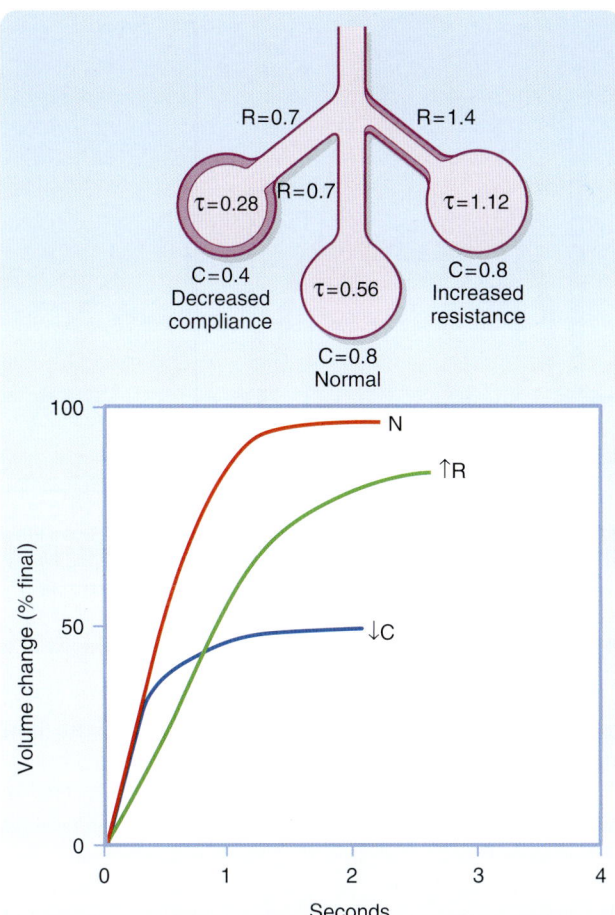

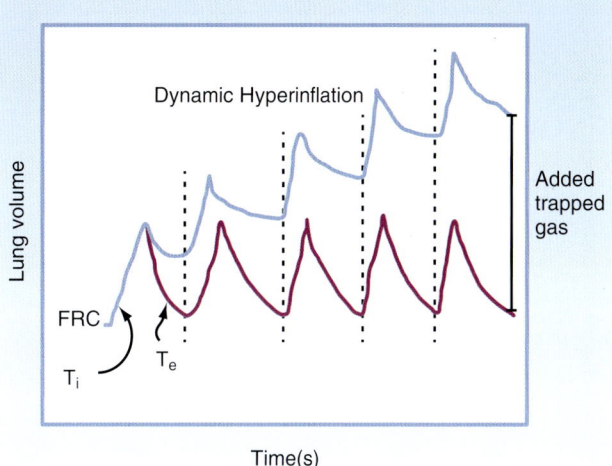

• **Fig. 23.4** Dynamic Hyperinflation. The total time for respiration (T_{tot}) is composed of the time for inspiration (T_i) and the time for exhalation (T_e). When the respiratory rate increases (e.g., during exercise), T_{tot} decreases. In individuals with chronic obstructive pulmonary disease (COPD), the effect of the increase in T_{tot} on Te may not allow for complete emptying of the alveoli with a long time constant, and with each succeeding breath, there is an increase in the lung volume (air trapping). This increase in lung volume eventually results in such a degree of hyperinflation that the affected person is no longer able to do the work needed to overcome the decreased compliance of the lung at this high lung volume. In such individuals, it is a major cause of shortness of breath with activity. FRC, functional residual capacity.

• **Fig. 23.3** Examples of Local Regulation of Ventilation as a Result of Variation in the Resistance (R) or Compliance (C) of Individual Lung Units. *Top,* The individual resistance and compliance values of three different lung units are illustrated. *Bottom,* The graph illustrates the volume of these three lung units as a function of time. In the upper schema, the normal lung has a time constant (τ) of 0.56 second. This lung unit reaches 97% of final volume equilibrium in 2 seconds, which is the normal inspiratory time. The lung unit at the right has a twofold increase in resistance; hence its time constant is doubled. That lung unit fills more slowly and reaches only 80% volume equilibrium during a normal inspiratory time (see graph); thus this lung unit is underventilated. The lung unit on the left has decreased compliance (is "stiff"), which acts to reduce its time constant. This lung unit fills quickly, reaching its maximum volume within 1 second, but receives only half the ventilation of a normal lung unit.

Pulmonary Vascular Resistance

Blood flow in the pulmonary circulation is pulsatile and influenced by pulmonary vascular resistance (PVR), gravity, alveolar pressure, and the arterial-to-venous pressure gradient. PVR is calculated as the change in pressure from the pulmonary artery (P_{PA}) to the left atrium (P_{LA}), divided by the flow (Q_T), which is cardiac output:

Equation 23.18

$$PVR = \frac{P_{PA} - P_{LA}}{Q_T}$$

Under normal circumstances,

Equation 23.19

$$PVR = \frac{14 \text{ mm Hg} - 8 \text{ mm Hg}}{6 \text{ L/minute}} = 1.00 \text{ mm Hg/L/minute}$$

This resistance is about 10 times less than that in the systemic circulation. The pulmonary circulation has two unique features that allow increased blood flow on demand without an increase in pressure: (1) With increased demand, as during exertion or exercise, pulmonary vessels that are normally closed are recruited; and (2) the blood vessels in the pulmonary circulation are highly distensible, and their diameter increases with only a minimal increase in pulmonary arterial pressure.

Lung volume affects PVR through its influence on alveolar capillaries (Fig. 23.5). At end inspiration, the air-filled alveoli compress the alveolar capillaries and increase PVR. In contrast to the capillary beds in the systemic circulation, the capillary beds in the lungs account for approximately 40% of PVR. The diameters of the larger extra-alveolar vessels increase at end inspiration because of radial traction and elastic recoil, and their PVR is lower at higher lung volume. During exhalation, the deflated alveoli apply the least resistance to the alveolar capillaries and their PVR is diminished, whereas the higher pleural pressure during exhalation increases the PVR of extra-alveolar vessels. As a result of these opposite effects of lung volume on PVR, total PVR in the lung is lowest at FRC.

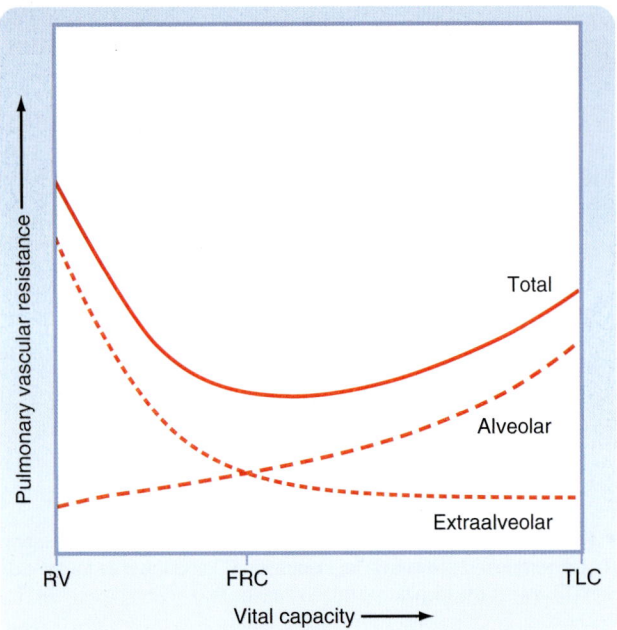

• Fig. 23.5 Schematic Representation of the Effects of Changes in Vital Capacity on Total Pulmonary Vascular Resistance and the Contributions to the Total Afforded by Alveolar and Extra-Alveolar Vessels. During inflation from residual volume (RV) to total lung capacity (TLC), resistance to blood flow through alveolar vessels increases, whereas resistance through extra-alveolar vessels decreases. Thus changes in total pulmonary vascular resistance are plotted as a U-shaped curve during lung inflation, with the nadir at functional residual capacity (FRC).

Distribution of Pulmonary Blood Flow

Because the pulmonary circulation is a low-pressure/low-resistance system, it is influenced by gravity much more dramatically than is the systemic circulation. This gravitational effect contributes to an uneven distribution of blood flow in the lungs. In normal upright persons at rest, the volume of blood flow increases from the apex of the lung to the base of the lung, where it is greatest. Similarly, in a supine individual, blood flow is least in the uppermost (anterior) regions and greatest in the lower (posterior) regions. Under conditions of stress, such as exercise, the difference in blood flow in the apex and base of the lung in upright persons becomes less, mainly because of the increase in arterial pressure.

On leaving the pulmonary artery, blood must travel against gravity to the apex of the lung in upright people. For every 1-cm increase in location of a pulmonary artery segment above the heart, there is a corresponding decrease in hydrostatic pressure equal to 0.74 mm Hg. Thus the pressure in a pulmonary artery segment that is 10 cm above the heart is 7.4 mm Hg less than the pressure in a segment at the level of the heart. Conversely, a pulmonary artery segment 5 cm below the heart has a 3.7–mm Hg increase in pulmonary arterial pressure. This effect of gravity on blood flow affects arteries and veins equally and results in wide variations in arterial and venous pressure from the apex to

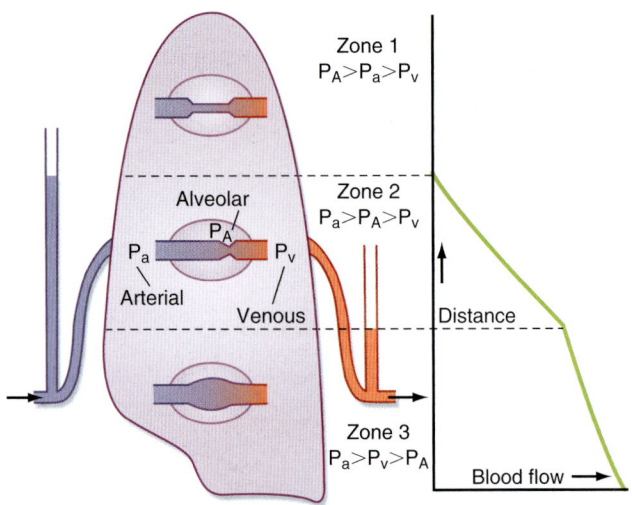

• Fig. 23.6 Model to Explain the Uneven Distribution of Blood Flow in the Lung According to the Pressures Affecting the Capillaries. P_A, pulmonary alveolar pressure; P_a, pulmonary arterial pressure; P_v, pulmonary venous pressure. (From West JB, et al. *J Appl Physiol.* 1964;19:713.)

the base of the lung. These variations influence both flow and ventilation/perfusion relationships.

In addition to the pulmonary arterial pressure (P_a) to pulmonary venous pressure (P_v) gradients, differences in pulmonary alveolar pressure (P_A) also influence blood flow in the lung. Classically, the lung has been thought to be divided into three functional zones (Fig. 23.6). Zone 1 represents the lung apex, where P_a is so low that it can be exceeded by P_A. The capillaries collapse because of the greater external P_A, and blood flow ceases. Under normal conditions, this zone does not exist; however, this state could be reached during positive-pressure mechanical ventilation or if P_a decreases sufficiently (such as might occur with a marked decrease in blood volume). In zone 2, or the upper third of the lung, P_a is greater than P_A, which is in turn greater than P_v. Because P_A is greater than P_v, the greater external P_A partially collapses the capillaries and causes a "damming" effect. This phenomenon is often referred to as the *waterfall effect*. In zone 3, P_a is greater than P_v, which is greater than P_A, and blood flows in this area in accordance with the pressure gradients. Thus, pulmonary blood flow is greater in the base of the lung because the increased transmural pressure distends the vessels and lowers the resistance.

Active Regulation of Blood Flow

Blood flow in the lung is regulated primarily by the passive mechanisms described previously. There are, however, several active mechanisms that regulate blood flow. Although the smooth muscle around pulmonary vessels is much thinner than that around systemic vessels, it is sufficient to affect vessel caliber and thus PVR. Oxygen levels have a major effect on blood flow. **Hypoxic vasoconstriction** occurs in arterioles in response to decreased PAO_2. The response is local, and the result is the shifting of blood flow from

AT THE CELLULAR LEVEL

Endothelin-1 is an amino acid peptide that is produced by the vascular endothelium. Endothelin regulates the tone of pulmonary arteries, and increased expression of endothelin-1 has been found in individuals with pulmonary artery hypertension. Endothelin-1 also decreases endothelial expression of nitric oxide synthase, which reduces levels of nitric oxide, an endothelial vasodilator. Endothelin-1 antagonists (e.g., bosentan, sitaxentan) have been produced and are important drugs in the treatment of pulmonary arterial hypertension.

hypoxic areas to well-perfused areas in an effort to enhance gas exchange. Isolated, local hypoxia does not alter PVR; approximately 20% of the vessels must be hypoxic before a change in PVR can be measured. Low inspired O_2 levels as a result of high altitude have a greater effect on PVR because all vessels are affected. High levels of inspired O_2 can dilate pulmonary vessels and decrease PVR. Other factors and some hormones (Box 23.1) can also influence vessel caliber, but their effects are usually local, brief, and important only in pathological conditions. Pulmonary capillaries lack smooth muscle and are thus not affected by these mechanisms. In some individuals, as a consequence of chronic hypoxia or collagen vascular disease, or for no apparent reason, pulmonary artery vascular resistance and subsequently pulmonary artery pressures rise (**pulmonary artery hypertension**).

Ventilation/Perfusion Relationships

Both ventilation ($\dot{V}$) and lung perfusion ($\dot{Q}$) are essential components of normal gas exchange, but a normal relationship between the two components is insufficient to ensure normal gas exchange. The ventilation/perfusion ratio (also referred to as the $\dot{V}/\dot{Q}$ ratio) is defined as the ratio of ventilation to blood flow. This ratio can be defined for a single alveolus, for a group of alveoli, or for the entire lung. At the level of a single alveolus, the ratio is defined as alveolar ventilation per minute ($\dot{V}_A$) divided by capillary flow ($\dot{Q}_c$). At the level of the lung, the ratio is defined as total alveolar ventilation divided by cardiac output. In normal lungs, alveolar ventilation is approximately 4.0 L/min, whereas pulmonary blood flow is approximately 5.0 L/min. Thus

in a normal lung, the overall ventilation/perfusion ratio is approximately 0.8, but the range of $\dot{V}/\dot{Q}$ ratios varies widely in different lung units. When ventilation exceeds perfusion, the ventilation/perfusion ratio is greater than 1 ($\dot{V}/\dot{Q} > 1$), and when perfusion exceeds ventilation, the ventilation/perfusion ratio is less than 1 ($\dot{V}/\dot{Q} < 1$). Mismatching of pulmonary blood flow and ventilation results in impaired O_2 and CO_2 transfer. In individuals with cardiopulmonary disease, mismatching of pulmonary blood flow and alveolar ventilation is the most frequent cause of systemic arterial **hypoxemia** (reduced Pa_{O_2}). In general, $\dot{V}/\dot{Q}$ ratios greater than 1 are not associated with hypoxemia.

A normal ventilation/perfusion ratio does not mean that ventilation and perfusion of that lung unit are normal; it simply means that the relationship between ventilation and perfusion is normal. For example, in lobar pneumonia, ventilation to the affected lobe is decreased. If perfusion to this area remains unchanged, perfusion would exceed ventilation; that is, the ventilation/perfusion ratio would be less than 1 ($\dot{V}/\dot{Q} < 1$). However, the decrease in ventilation to this area produces hypoxic vasoconstriction in the pulmonary arterial bed supplying this lobe. This results in a decrease in perfusion to the affected area and a more "normal" ventilation/perfusion ratio. Nonetheless, neither the ventilation nor the perfusion to this area is normal (both are decreased), but the relationship between the two could approach the normal range.

Regional Differences in Ventilation/Perfusion Ratios

The ventilation/perfusion ratio varies in different areas of the lung. In an upright individual, although both ventilation and perfusion increase from the apex to the base of the lung, the increase in ventilation is less than the increase in blood flow. As a result, the normal $\dot{V}/\dot{Q}$ ratio at the apex of the lung is much greater than 1 (ventilation exceeds perfusion), whereas the $\dot{V}/\dot{Q}$ ratio at the base of the lung is much less than 1 (perfusion exceeds ventilation). The relationship between ventilation and perfusion from the apex to the base of the lung is depicted in Fig. 23.7.

Ventilation-Perfusion Relationships

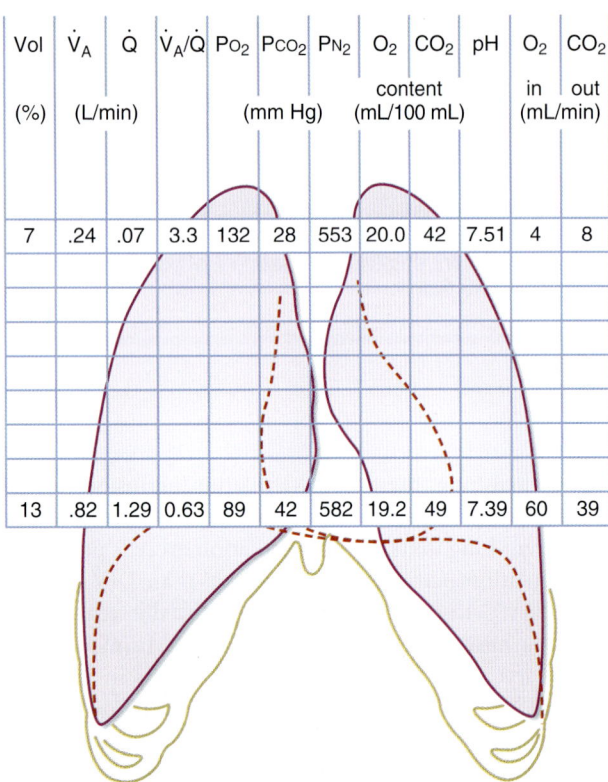

Vol	$\dot{V}_A$	$\dot{Q}$	$\dot{V}_A/\dot{Q}$	P_{O_2}	P_{CO_2}	P_{N_2}	O_2	CO_2	pH	O_2	CO_2
							content			in	out
(%)	(L/min)			(mm Hg)			(mL/100 mL)			(mL/min)	
7	.24	.07	3.3	132	28	553	20.0	42	7.51	4	8
13	.82	1.29	0.63	89	42	582	19.2	49	7.39	60	39

• **Fig. 23.7** Ventilation/Perfusion Relationships in a Normal Lung in the Upright Position. Only the apical and basal values are shown for clarity. In each column, the number on top represents values at the apex of the lung, and the number on the bottom represents values at the base. P_{CO_2}, partial pressure of carbon dioxide; P_{N_2}, partial pressure of nitrogen; P_{O_2}, partial pressure of oxygen; $\dot{Q}$, perfusion per minute; $\dot{V}_A$, alveolar ventilation per minute.

Alveolar-Arterial Difference for Oxygen

PA_{CO_2} and Pa_{CO_2} are equal because of the solubility properties of CO_2 (see Chapter 24). The same is not true for alveolar and arterial O_2. Even in individuals with normal lungs, PA_{O_2} is slightly greater than Pa_{O_2}. The difference between PA_{O_2} and Pa_{O_2} is called *the alveolar-arterial difference for oxygen* (AaDO₂). An increase in the AaDO₂ is a hallmark of abnormal O_2 exchange. This small difference in healthy individuals is not caused by "imperfect" gas exchange, but by the small number of veins that bypass the lung and empty directly into the arterial circulation. The thebesian vessels of the left ventricular myocardium drain directly into the left ventricle (rather than into the coronary sinus in the right atrium), and some bronchial and mediastinal veins drain into the pulmonary veins. This results in venous admixture and a decrease in Pa_{O_2}. (This is an example of an anatomical shunt; see the section "Anatomical Shunts.") Approximately 2% to 3% of the cardiac output is **shunted** in this way.

To measure the clinical effectiveness of gas exchange in the lung, Pa_{O_2} and Pa_{CO_2} are measured. PA_{O_2} is calculated from the alveolar air equation (Eq. 23.13). The difference between the calculated PA_{O_2} and the measured Pa_{O_2} is the AaDO₂. In individuals with normal lungs who are breathing room air, the AaDO₂ is less than 15 mm Hg. The mean value rises approximately 3 mm Hg per decade of life after 30 years of age. Hence, an AaDO₂ lower than 25 mm Hg is considered the upper limit of normal.

Abnormalities in Pa_{O_2} can occur with or without an elevation in AaDO₂. Hence, the relationship between Pa_{O_2} and AaDO₂ is useful in determining the cause of an abnormal Pa_{O_2} and in predicting the response to therapy (particularly to supplemental O_2 administration). Causes of a reduction in Pa_{O_2} (arterial hypoxemia) and their effect on AaDO₂ are listed in Table 23.2. Each of these causes is discussed in greater detail in the following sections.

Arterial Blood Hypoxemia, Hypoxia, and Hypercarbia

Arterial hypoxemia is defined as a Pa_{O_2} lower than 80 mm Hg in an adult who is breathing room air at sea level. **Hypoxia** is defined as insufficient O_2 to carry out normal metabolic functions; hypoxia often occurs when the Pa_{O_2} is less than 60 mm Hg. There are four major categories of hypoxia. The first, *hypoxic hypoxia,* is the most common. The six main pulmonary conditions associated with hypoxic hypoxia—anatomical shunt, physiological shunt, decreased Fi_{O_2}, $\dot{V}/\dot{Q}$ mismatching, diffusion abnormalities, and hypoventilation—are described in the following sections and in Table 23.2. A second category is *anemic hypoxia,* which is caused by a decrease in the amount of functioning hemoglobin as a result of too little hemoglobin, abnormal hemoglobin, or interference with the chemical combination of oxygen and hemoglobin (e.g., carbon monoxide poisoning; see the following "In the Clinic" box). The third category is *hypoperfusion hypoxia,* which results from low

TABLE 23.2	Causes of Hypoxic Hypoxia			
Cause	PaO_2	$AaDO_2$	PaO_2 Response to 100% O_2
Anatomical shunt	Decreased	Increased	No significant change
Physiological shunt	Decreased	Increased	Decreased
Decreased FiO_2	Decreased	Normal	Increased
Low ventilation/perfusion ratio	Decreased	Increased	Increased
Diffusion abnormality	Decreased	Increased	Increased
Hypoventilation	Decreased	Normal	Increased

$AaDO_2$, alveolar-arterial difference for oxygen; FiO_2, fraction of inspired oxygen; PaO_2, partial pressure of arterial oxygen.

blood flow (e.g., decreased cardiac output) and reduced oxygen delivery to the tissues. *Histotoxic hypoxia,* the fourth category of hypoxia, occurs when the cellular machinery that uses oxygen to produce energy is poisoned, as in cyanide poisoning. In this situation, arterial and venous PO_2 are normal or increased because oxygen is not being utilized.

IN THE CLINIC

Carbon monoxide can be generated from a malfunctioning space heater, from car exhaust, or from a burning building. Individuals exposed to carbon monoxide experience headache, nausea, and dizziness, and if it is not recognized, such individuals may die. They often have a cherry-red appearance, and oxygen saturation as measured with an oximeter is high (approaching 100%). Even on an arterial blood gas, the PAO_2 may be normal. Nevertheless, the tissues are depleted of O_2. Thus it is imperative that the clinician recognize a potential case of carbon monoxide poisoning and order an oxygen saturation measurement with the use of a carbon monoxide oximeter. If a patient has carbon monoxide poisoning, there will be a marked difference between the measurement of oxygen saturation by oximetry and that measured with a carbon monoxide oximeter.

Ventilation/Perfusion Abnormalities and Shunts

Anatomical Shunts

A useful way to examine the relationship between ventilation and perfusion is with the two–lung unit model (Fig. 23.8). Two alveoli are ventilated, each of which is supplied by blood from the heart. When ventilation is uniform, half the inspired gas goes to each alveolus, and when perfusion is uniform, half the cardiac output goes to each alveolus. In this normal unit, the ventilation/perfusion ratio in each of the alveoli is the same and is equal to 1. The alveoli are perfused by mixed venous blood that is deoxygenated and contains increased $PaCO_2$. PAO_2 is higher than mixed venous O_2, and this provides a gradient for movement of O_2 into blood. In contrast, mixed venous CO_2 is greater than $PACO_2$, and this provides a gradient for movement

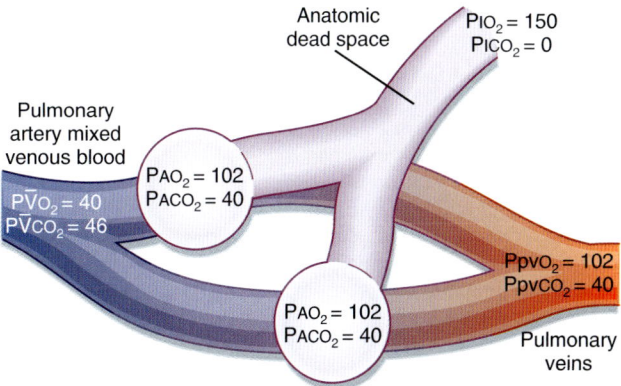

• **Fig. 23.8** Simplified Lung Model of Two Normal Parallel Lung Units. Both units receive equal volumes of air and blood flow for their size. The blood and alveolar gas partial pressures are normal values in a resting person at sea level. $PACO_2$, partial pressure of alveolar carbon dioxide; PAO_2, partial pressure of alveolar oxygen; $PiCO_2$, partial pressure of inspired carbon dioxide; PiO_2, partial pressure of inspired oxygen; $PpvCO_2$, partial pressure of carbon dioxide in portal venous blood; $PpvO_2$, partial pressure of oxygen in portal venous blood; $P\overline{v}CO_2$, partial pressure of carbon dioxide in mixed venous blood; $P\overline{v}O_2$, partial pressure of oxygen in mixed venous blood.

of CO_2 into the alveolus. Note that in this ideal model, alveolar-arterial O_2 values do not differ.

An anatomical shunt occurs when mixed venous blood bypasses the gas-exchange unit and goes directly into the arterial circulation (Fig. 23.9). Alveolar ventilation, the distribution of alveolar gas, and the composition of alveolar gas are normal, but the distribution of cardiac output is changed. Some of the cardiac output goes through the pulmonary capillary bed that supplies the gas-exchange units, but the rest of it bypasses the gas-exchange units and goes directly into the arterial circulation. The blood that bypasses the gas-exchange unit is thus *shunted,* and because the blood is deoxygenated, this type of bypass is called a **right-to-left shunt.** Most anatomical shunts develop within the heart, and they develop when deoxygenated blood from the right atrium or ventricle crosses the septum and mixes with blood from the left atrium or ventricle. The effect of this right-to-left shunt is to mix deoxygenated blood with oxygenated blood, and it results in varying degrees of arterial hypoxemia.

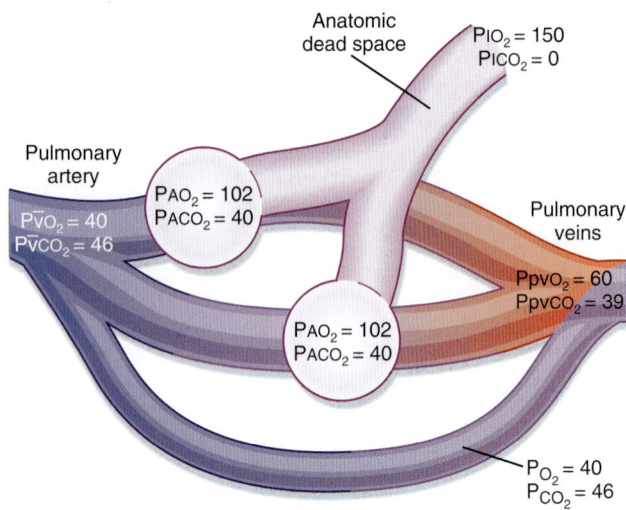

Fig. 23.9 Right-to-Left Shunt. Alveolar ventilation is normal, but a portion of the cardiac output bypasses the lung and mixes with oxygenated blood. PaO_2 varies according to the size of the shunt. $PACO_2$, partial pressure of alveolar carbon dioxide; PAO_2, partial pressure of alveolar oxygen; $PICO_2$, partial pressure of inspired carbon dioxide; PIO_2, partial pressure of inspired oxygen; $PpvCO_2$, partial pressure of carbon dioxide in portal venous blood; $PpvO_2$, partial pressure of oxygen in portal venous blood; $P\overline{v}CO_2$, partial pressure of carbon dioxide in mixed venous blood; $P\overline{v}O_2$, partial pressure of oxygen in mixed venous blood.

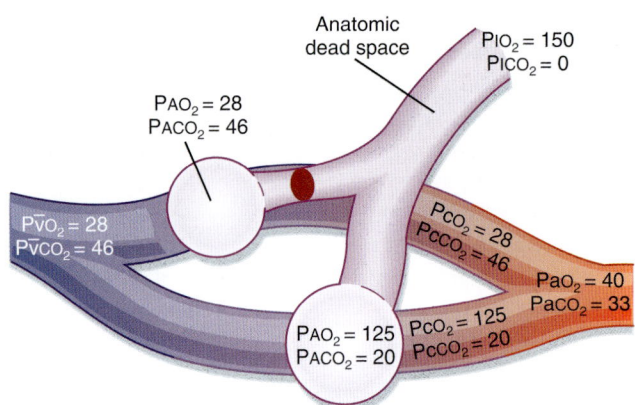

Fig. 23.10 Schema of a Physiological Shunt (Venous Admixture). Notice the marked decrease in PaO_2 in comparison to PcO_2. The alveolar-arterial difference for oxygen ($AaDO_2$) in this example is 85 mm Hg. $PACO_2$, partial pressure of alveolar carbon dioxide; PAO_2, partial pressure of alveolar oxygen; $PICO_2$, partial pressure of inspired carbon dioxide; PIO_2, partial pressure of inspired oxygen; $PpvCO_2$, partial pressure of carbon dioxide in portal venous blood; $PpvO_2$, partial pressure of oxygen in portal venous blood; $P\overline{v}CO_2$, partial pressure of carbon dioxide in mixed venous blood; $P\overline{v}O_2$, partial pressure of oxygen in mixed venous blood.

An important feature of an anatomical shunt is that if an affected individual is given 100% O_2 to breathe, the response is blunted severely. The blood that bypasses the gas-exchanging units is never exposed to the enriched O_2, and thus it continues to be deoxygenated. The PO_2 in the blood that is not being shunted increases and it mixes with the deoxygenated blood. Thus the degree of persistent hypoxemia in response to 100% O_2 varies with the volume of the shunted blood. Normally, the hemoglobin in the blood that perfuses the ventilated alveoli is almost fully saturated. Therefore, most of the added O_2 is in the form of dissolved O_2 (see Chapter 24).

The $PaCO_2$ in an anatomical shunt is not usually increased even though the shunted blood has an elevated level of CO_2. The reason for this is that the central chemoreceptors (see Chapter 25) respond to any elevation in CO_2 with an increase in ventilation and reduce $PaCO_2$ to the normal range. If the hypoxemia is severe, the increased respiratory drive secondary to the hypoxemia increases the ventilation and can decrease $PaCO_2$ to below the normal range.

Physiological Shunts

A physiological shunt (also known as *venous admixture*) can develop when ventilation to lung units is absent in the presence of continuing perfusion (Fig. 23.10). In this situation, in the two–lung unit model, all the ventilation goes to the other lung unit, whereas perfusion is equally distributed between both lung units. The lung unit without ventilation but with perfusion has a $\dot{V}/\dot{Q}$ ratio of 0. The blood perfusing this unit is mixed venous blood; because

there is no ventilation, no gas is exchanged in the unit, and the blood leaving this unit continues to resemble mixed venous blood. The effect of a physiological shunt on oxygenation is similar to the effect of an anatomical shunt; that is, deoxygenated blood bypasses a gas-exchanging unit and admixes with arterial blood. Clinically, **atelectasis** (which is obstruction to ventilation of a gas-exchanging unit with subsequent loss of volume) is an example of a situation in which the lung region has a $\dot{V}/\dot{Q}$ of 0. Causes of atelectasis include mucous plugs, airway edema, foreign bodies, and tumors in the airway.

Low Ventilation/Perfusion

Mismatching between ventilation and perfusion is the most frequent cause of arterial hypoxemia in individuals with respiratory disorders. In the most common example, the composition of mixed venous blood, total blood flow (cardiac output), and the distribution of blood flow are normal. However, when alveolar ventilation is distributed unevenly between the two gas-exchange units (Fig. 23.11) and blood flow is equally distributed, the unit with decreased ventilation has a $\dot{V}/\dot{Q}$ ratio of less than 1, whereas the unit with the increased ventilation has a $\dot{V}/\dot{Q}$ of greater than 1. This causes the alveolar and end-capillary gas compositions to vary. Both the arterial O_2 content and CO_2 content are abnormal in the blood that has come from the unit with the decreased ventilation ($\dot{V}/\dot{Q}$, <1). The unit with the increased ventilation ($\dot{V}/\dot{Q}$, >1) has a lower CO_2 content and a higher O_2 content because it is being overventilated. The actual PaO_2 and $PaCO_2$ vary, depending on the relative contribution of each of these units to arterial blood. The alveolar-arterial O_2 gradient ($AaDO_2$) is increased because the relative overventilation of one unit does not fully

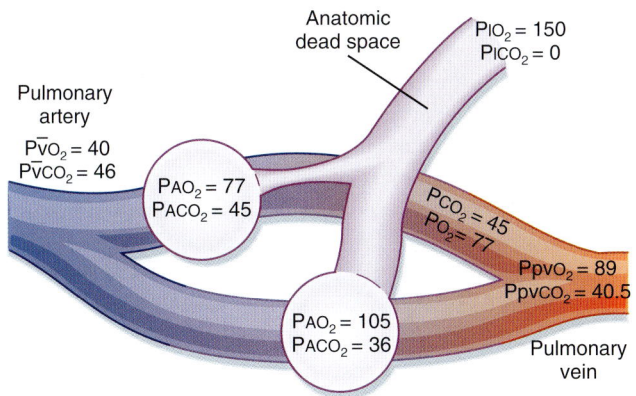

• **Fig. 23.11** Effects of Ventilation/Perfusion Mismatching on Gas Exchange. The decrease in ventilation to the one lung unit could be due to mucus obstruction, airway edema, bronchospasm, a foreign body, or a tumor. PA_{CO_2}, partial pressure of alveolar carbon dioxide; PA_{O_2}, partial pressure of alveolar oxygen; Pi_{CO_2}, partial pressure of inspired carbon dioxide; Pi_{O_2}, partial pressure of inspired oxygen; Ppv_{CO_2}, partial pressure of carbon dioxide in portal venous blood; Ppv_{O_2}, partial pressure of oxygen in portal venous blood; $P\bar{v}_{CO_2}$, partial pressure of carbon dioxide in mixed venous blood; $P\bar{v}_{O_2}$, partial pressure of oxygen in mixed venous blood.

compensate (either by the addition of extra O_2 or by the removal of extra CO_2) for underventilation of the other unit. The failure to compensate is greater for O_2 than for CO_2, as indicated by the flatness of the upper part of the oxyhemoglobin dissociation curve, in contrast to the slope of the CO_2 dissociation curve (see Chapter 24). In other words, increased ventilation increases PA_{O_2}, but it adds little extra O_2 content to the blood because hemoglobin is close to being 100% saturated in the overventilated areas. This is not the case for CO_2, for which the steeper slope of the CO_2 curve indicates removal of more CO_2 when ventilation increases. Thus inasmuch as CO_2 moves by diffusion, then as long as a CO_2 gradient is maintained, CO_2 diffusion will occur.

Alveolar Hypoventilation

The PA_{O_2} is determined by a balance between the rate of O_2 uptake and the rate of O_2 replenishment by ventilation. Oxygen uptake depends on blood flow through the lung and the metabolic demands of the tissues. If ventilation decreases, PA_{O_2} decreases, and Pa_{O_2} subsequently decreases. In addition, V_A and PA_{CO_2} are directly but inversely related. When ventilation is halved, the PA_{CO_2} doubles and thus so does the Pa_{CO_2} (see Eq. 23.16). Ventilation insufficient to maintain normal levels of CO_2 is called **hypoventilation.** Hypoventilation always decreases Pa_{O_2} and increases Pa_{CO_2}.

One of the hallmarks of hypoventilation is a normal AaD_{O_2}. Hypoventilation reduces PA_{O_2}, which in turn results in a decrease in Pa_{O_2}. Because gas exchange is normal, the AaD_{O_2} remains normal. Hypoventilation accompanies diseases associated with muscle weakness and is associated with drugs that reduce the respiratory drive. In the presence of hypoventilation, however, areas of atelectasis develop rapidly; atelectasis creates regions with $\dot{V}/\dot{Q}$ ratios of 0, and when this occurs, the AaD_{O_2} rises.

Diffusion Abnormalities

Abnormalities in diffusion of O_2 across the alveolar-capillary barrier could potentially result in arterial hypoxia. Equilibration between alveolar and capillary O_2 and CO_2 content occurs rapidly: in a fraction of the time that it takes for red blood cells to transit the pulmonary capillary network. Hence, diffusion equilibrium almost always occurs in normal people, even during exercise, when the transit time of red blood cells through the lung increases significantly. An increased AaD_{O_2} attributable to incomplete diffusion **(diffusion disequilibrium)** has been observed in normal persons only during exercise at high altitude ($\geq$10,000 feet). Even in individuals with an abnormal diffusion capacity, diffusion disequilibrium at rest is unusual but can occur during exercise and at altitude. **Alveolar capillary block,** or thickening of the air-blood barrier, is an uncommon cause of hypoxemia. Even when the alveolar wall is thickened, there is usually sufficient time for gas diffusion unless the red blood cell transit time is increased.

Mechanisms of Hypercapnia

Two major mechanisms account for the development of **hypercapnia** (elevated P_{CO_2}): hypoventilation and wasted, or increased, dead space ventilation. As noted previously, alveolar ventilation and alveolar CO_2 are inversely related. When ventilation is halved, PA_{CO_2} and Pa_{CO_2} double. Hypoventilation always decreases Pa_{O_2} and increases Pa_{CO_2} and thereby results in a hypoxemia that responds to an enriched source of O_2. Dead space ventilation is wasted, or increased, when pulmonary blood flow is interrupted in the presence of normal ventilation. This is most often caused by a pulmonary embolus that obstructs blood flow. The embolus halts blood flow to pulmonary areas with normal ventilation ($\dot{V}/\dot{Q} = \infty$). In this situation, the ventilation is wasted because it fails to oxygenate any of the mixed venous blood. The ventilation to the perfused regions of the lung is less than ideal (i.e., there is relative "hypoventilation" to this area because in this situation, it receives all the pulmonary blood flow with "normal" ventilation). If compensation does not occur, Pa_{CO_2} increases and Pa_{O_2} decreases. Compensation after a pulmonary embolus, however, begins almost immediately; local bronchoconstriction occurs, and the distribution of ventilation shifts to the areas being perfused. As a result, changes in arterial CO_2 and O_2 content are minimized.

Effect of 100% Oxygen on Arterial Blood Gas Abnormalities

One of the ways that a right-to-left shunt can be distinguished from other causes of hypoxemia is for the individual to breathe 100% O_2 through a non-rebreathing face mask for approximately 15 minutes. When the individual breathes 100% O_2, all of the N_2 in the alveolus is replaced by O_2. Thus the PAO_2, according to the alveolar air equation (Eq. 23.13), is calculated as follows:

Equation 23.20

$$PAO_2 = [1.0 \times (P_b - PH_2O)] - PaCO_2/0.8$$
$$= [1.0 \times (760 - 47)] - 40/0.8$$
$$= 663 \, mm \, Hg$$

In a normal lung, the PAO_2 rapidly increases, and it provides the gradient for transfer of O_2 into capillary blood. This is associated with a marked increase in PaO_2 (see Table 23.2). Similarly, over the 15-minute period of breathing enriched with O_2, even areas with very low $\dot{V}/\dot{Q}$ ratios develop high alveolar O_2 pressure as the N_2 is replaced by O_2. In the presence of normal perfusion to these areas, there is a gradient for gas exchange, and the end-capillary blood is highly enriched with O_2. In contrast, in the presence of a right-to-left shunt, oxygenation is not corrected because mixed venous blood continues to flow through the shunt and mix with blood that has perfused normal units. The poorly oxygenated blood from the shunt lowers the arterial O_2 content and maintains the $AaDO_2$. An elevated $AaDO_2$ during a properly conducted study with 100% O_2 signifies the presence of a shunt (anatomical or physiological); the magnitude of the $AaDO_2$ can be used to quantify the proportion of the cardiac output that is being shunted.

Regional Differences

The regional differences in ventilation and perfusion and the relationship between ventilation and perfusion were discussed earlier in this chapter. The effects of various physiological abnormalities (e.g., shunt, $\dot{V}/\dot{Q}$ mismatch, and hypoventilation) on arterial O_2 and CO_2 levels were also described. In addition, however, it should be noted that because the $\dot{V}/\dot{Q}$ ratio varies in different regions of the lung, the end-capillary blood coming from these regions has different O_2 and CO_2 levels. These differences are shown in Fig. 23.7, and they demonstrate the complexity of the lung. First, recall that the volume of the lung at the apex is less than the volume at the base. As previously described, ventilation and perfusion are less at the apex than at the base, but the differences in perfusion are greater than the differences in ventilation. Thus the $\dot{V}/\dot{Q}$ ratio is high at the apex and low at the base. This difference in ventilation/perfusion ratios is associated with a difference in alveolar O_2 and CO_2 content between the apex and the base. The PAO_2 is higher and the $PACO_2$ is lower in the apex than in the base. This results in differences in end-capillary contents for these gases. End-capillary PO_2 is lower, and, as a consequence, the O_2 content is lower in end-capillary blood at the lung base than at the apex. In addition, there is significant variation in blood pH in the end capillaries in these regions because of the variation in CO_2 content. During exercise, blood flow to the apex increases and becomes more uniform in the lung; as a result, the difference between the content of gases in the apex and in the base of the lung diminishes with exercise.

Key Points

1. The volume of air in the conducting airways is called the *anatomical dead space*. Dead space ventilation varies inversely with tidal volume. The total volume of gas in each breath that does not participate in gas exchange is called the *physiological dead space*. It includes the anatomical dead space and the dead space secondary to ventilated but unperfused alveoli.

2. The sum of the partial pressures of a gas is equal to the total pressure. The partial pressure of a gas (P_{gas}) is equal to the fraction of the gas in the gas mixture (F_{gas}) multiplied by the total pressure (P_{total}). The conducting airways do not participate in gas exchange. Therefore, the partial pressures of O_2, N_2, and water vapor in humidified air remain unchanged in the airways until the gas reaches the alveolus.

3. The partial pressure of O_2 in the alveolus is given by the alveolar air equation (Eq. 23.13). This equation is used to calculate the $AaDO_2$, a useful measurement of abnormal arterial O_2.

4. The relationship between CO_2 production and alveolar ventilation is defined by the alveolar carbon dioxide equation (Eq. 23.14). There is an inverse relationship between the $PACO_2$ and V_A, regardless of the exhaled quantity of CO_2. In normal lungs, $PaCO_2$ is tightly regulated to remain constant at around 40 mm Hg.

5. Because of the effects of gravity, there are regional differences in ventilation and perfusion. The ventilation/perfusion ($\dot{V}/\dot{Q}$) ratio is defined as the ratio of ventilation to blood flow. In a normal lung, the overall ventilation/perfusion ratio is approximately 0.8. When ventilation exceeds perfusion, the ventilation/perfusion ratio is greater than 1 ($\dot{V}/\dot{Q} > 1$), and when perfusion exceeds ventilation, the ventilation/perfusion ratio is less than 1 ($\dot{V}/\dot{Q} < 1$). The $\dot{V}/\dot{Q}$ ratio at the apex of the lung is high (ventilation is increased in relation to very little blood flow), whereas the $\dot{V}/\dot{Q}$ ratio at the base of the lung is low. In individuals with normal lungs who are breathing room air, the $AaDO_2$ is less than 15 mm Hg; the upper limit of normal is 25 mm Hg.

6. The pulmonary circulation is a low-pressure, low-resistance system. Recruitment of new capillaries and dilation of arterioles without an increase in pressure are unique features of the lung and allow for adjustments during stress, as in the case of exercise. Pulmonary vascular resistance is the change in pressure from the pulmonary artery (P_{PA}) to the left atrium (P_{LA}), divided by cardiac output (Q_T). This resistance is about 10 times less than in the systemic circulation.

7. There are four categories of hypoxia (hypoxic hypoxia, anemic hypoxia, diffusion hypoxia, and histotoxic hypoxia) and six mechanisms of hypoxic hypoxia and hypoxemia: anatomical shunt, physiological shunt, decreased FiO_2, $\dot{V}/\dot{Q}$ mismatching, diffusion abnormalities, and hypoventilation.

8. There are two mechanisms of the development of hypercapnia: increase in dead space ventilation and hypoventilation.

Additional Readings

Leff AR, Schumacker PT. *Respiratory Physiology: Basics and Applications*. Philadelphia: WB Saunders; 1993.

Lumb AB. *Nunn's Applied Respiratory Physiology*. 8th ed. St. Louis: Elsevier; 2016.

Mead J, Macklem PT, vol eds. *American Physiological Society Handbook of Physiology: The Respiratory System*. Vol. 3. Mechanics. Bethesda, MD: American Physiological Society; 1986.

Wasserman K, Beaver WL, Whipp BI. Gas exchange theory and the lactic acidosis (anaerobic) threshold. *Circulation*. 1990;81(1 suppl):1114-1130.

West JB. *Ventilation/Blood Flow and Gas Exchange*. 5th ed. New York: Blackwell Scientific; 1991.

24

Oxygen and Carbon Dioxide Transport

LEARNING OBJECTIVES

Upon completion of this chapter, the student should be able to answer the following questions:

1. Describe the basic gas diffusion principles and how they affect O_2 and CO_2 absorption and expiration.
2. Compare and contrast the chemical transport mechanisms of O_2 and CO_2 in blood.
3. Describe the basic principles and clinical significances of the O_2 and CO_2 dissociation curves.
4. Describe the chemical synthesis of H^+ ions and their role in the regulation of acid-base balance.
5. Explain why the diffusion of O_2 and that of CO_2 are considered perfusion limited and that of CO is considered diffusion limited.
6. Compare and contrast factors that shift the oxyhemoglobin dissociation curve.
7. Explain the clinical significance of the differences in the oxyhemoglobin and carboxyhemoglobin dissociation curves.
8. Compare and contrast the chloride shift and the Haldane effect on CO_2 transport.

The respiratory and circulatory systems function together to transport oxygen (O_2) from the lungs to the tissues to sustain normal cellular activity and to transport carbon dioxide (CO_2) from the tissues to the lungs for expiration. CO_2, a product of active cellular metabolism, is transported from the tissues via systemic veins to the lungs, where it is expired (Fig. 24.1). To enhance uptake and transport of these gases between the lungs and tissues, specialized mechanisms (e.g., binding of O_2 and hemoglobin and HCO_3^- transport of CO_2) have evolved that enable O_2 uptake and CO_2 expiration to occur simultaneously. Moreover, these specialized mechanisms facilitate uptake of O_2 and expiration of CO_2. To understand the mechanisms involved in the transport of these gases, gas diffusion properties, as well as transport and delivery mechanisms, must be considered.

Gas Diffusion

Gas movement throughout the respiratory system occurs predominantly via diffusion. The respiratory and circulatory systems contain several unique anatomical and physiological

features to facilitate gas diffusion: (1) large surface areas for gas exchange (alveolar to capillary and capillary to tissue membrane) with short distances to travel, (2) substantial partial pressure gradient differences, and (3) gases with advantageous diffusion properties. Transport and delivery of O_2 from the lungs to the tissue and vice versa for CO_2 are dependent on basic gas diffusion laws.

Diffusion of Gases From Regions of Higher to Lower Partial Pressure in the Lungs

The process of gas diffusion is passive and similar whether diffusion occurs in a gaseous or liquid state. The rate of diffusion of a gas through a liquid is described by **Graham's law,** which states that the rate is directly proportional to the solubility coefficient of the gas and inversely proportional to the square root of its molecular weight. Calculation of the diffusion properties for O_2 and CO_2 reveals that CO_2 diffuses approximately 20 times faster than O_2. Rates of O_2 diffusion from the lungs into blood and from blood into tissue, and vice versa for CO_2, are predicted by **Fick's law** of gas diffusion (Fig. 24.2). The ratio AD:T represents the conductance of a gas from the alveolus to blood. The diffusing capacity of the lung (D_L) is its conductance (A•D/T) when considered for the entire lung; thus, with Fick's equation, D_L can be calculated as follows:

Equation 24.1

$$\dot{V}_{gas} = A \cdot D \cdot \frac{(P_1 - P_2)}{T}$$

$$\dot{V} = D_L(P_1 - P_2)$$

$$D_L = \frac{\dot{V}}{(P_1 - P_2)}$$

where $\dot{V}_{gas}$ = gas diffusion.

Fick's law of diffusion could be used to assess the diffusion properties of O_2 in the lungs, except that the capillary partial pressure of oxygen cannot be measured. This limitation can be overcome with the use of carbon monoxide (CO) rather than O_2. Because CO has low solubility in the capillary membrane, the rate of CO equilibrium across the capillary is slow, and the partial pressure of CO in capillary blood remains close to 0. In contrast, the solubility of CO in blood is high. Thus the only limitation for diffusion of CO is the alveolar-capillary membrane, and thus CO is a

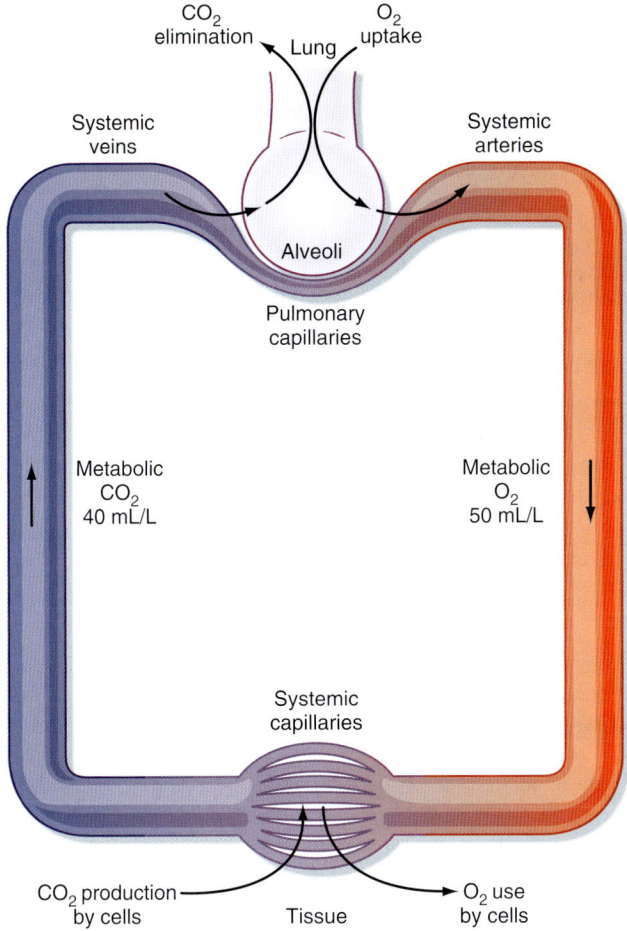

• **Fig. 24.1** Oxygen (O_2) and Carbon Dioxide (CO_2) Transport in Arterial and Venous Blood. Oxygen in arterial blood is transferred from arterial capillaries to tissues. The flow rates for O_2 and CO_2 are shown for 1 L of blood.

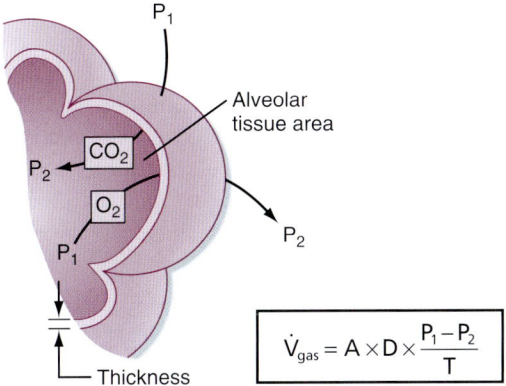

• **Fig. 24.2** According to Fick's law, diffusion of a gas across a sheet of tissue ($\dot{V}_{gas}$) is directly related to the surface area (A) of the tissue, the diffusion constant (D) of the specific gas, and the partial pressure difference ($P_1 - P_2$) of the gas on each side of the tissue, and it is inversely related to tissue thickness (T).

useful gas for calculating D_L. The capillary partial pressure (P_2 in Eq. 24.1) is essentially 0 for CO, and therefore D_L can be measured from the diffusion of carbon monoxide ($\dot{V}CO$) and the average partial pressure of CO in the alveolus; that is,

Equation 24.2

$$\dot{V} = D_L(P_1 - P_2)$$

$$D_LCO = \frac{\dot{V}}{P_1 - P_2} = \frac{\dot{V}CO}{P_ACO}$$

where D_LCO = diffusion capacity of the lung for carbon monoxide.

Assessment of D_LCO has become a classic measurement of the diffusion barrier of the alveolar-capillary membrane. It is useful in the differential diagnosis of certain obstructive lung diseases, such as emphysema.

IN THE CLINIC

A patient with interstitial pulmonary fibrosis (a restrictive lung disease) inhales a single breath of 0.3% CO from residual volume to total lung capacity. He holds his breath for 10 seconds and then exhales. After discarding the exhaled gas from the dead space, a representative sample of alveolar gas from late in exhalation is collected. The average alveolar CO pressure is 0.1 mm Hg, and 0.25 mL of CO has been taken up. The diffusion capacity for CO in this patient is

$$D_L = \frac{\dot{V}CO}{P_ACO}$$

$$= 0.25\,\text{mL}/10\,\text{seconds} \times \frac{60\,\text{seconds/minute}}{0.1\,\text{mm Hg}}$$

$$= 15\,\text{mL/minute/mm Hg}$$

The normal range for D_LCO is 20 to 30 mL/minute/mm Hg. Patients with interstitial pulmonary fibrosis have an initial alveolar inflammatory response with subsequent scar formation within the interstitial space. The inflammation and scar replace the alveoli and decrease the surface area for gas diffusion to occur, which results in decreased D_LCO. This is a classic characteristic of certain types of restrictive lung disease.

Oxygen and Carbon Dioxide Exchange in the Lung Is Perfusion Limited

Different gases have different solubility factors. Gases that are insoluble in blood (i.e., anesthetic gases such as nitrous oxide and ether) do not chemically combine with proteins in blood and equilibrate rapidly between alveolar gas and blood. The equilibration occurs in less time than the 0.75 seconds that the red blood cell spends in the capillary bed (capillary transit time). The diffusion of insoluble gases between alveolar gas and blood is considered **perfusion limited** because the partial pressure of gas in the blood leaving the capillary has reached equilibrium with alveolar gas and is limited only by the amount of blood perfusing the alveolus. In contrast, a gas that is **diffusion limited,** such as CO, has low solubility in the alveolar-capillary membrane but high solubility in blood because of its high affinity for **hemoglobin** (Hgb). These features prevent the equilibration of CO between alveolar gas and blood during the red blood cell transit time.

The high affinity of CO for Hgb enables large amounts of CO to be taken up in blood with little or no appreciable increase in its partial pressure. Gases that are chemically bound to Hgb do not exert a partial pressure in blood. Like CO, both CO_2 and O_2 have relatively low solubility in the alveolar-capillary membrane but high solubility in blood because of their ability to bind to Hgb. However, their rate of equilibration is sufficiently rapid for complete equilibration to occur during the transit time of the red blood cell within the capillary. Equilibration for O_2 and CO_2 usually occurs within 0.25 seconds. Thus O_2 and CO_2 transfer is normally perfusion limited. The partial pressure of a gas that is diffusion limited (i.e., CO) does not reach equilibrium with the alveolar pressure over the time that it spends in the capillary (Fig. 24.3). Although CO_2 has a greater rate of diffusion in blood than O_2 does, it has a lower membrane-blood solubility ratio and consequently takes approximately the same amount of time to reach equilibrium in blood.

Diffusion limitation for O_2 and CO_2 would occur if red blood cells spent less than 0.25 seconds in the capillary bed. This is occasionally the case in very fit athletes during vigorous exercise and in healthy subjects who exercise at high altitude.

Oxygen Transport

Oxygen is carried in blood in two forms: dissolved O_2 and O_2 bound to Hgb. The dissolved form is measured clinically in an arterial blood gas sample as the partial pressure of arterial oxygen (PaO_2). Only a small percentage of O_2 in blood is in the dissolved form, and its contribution to O_2 transport under normal conditions is almost negligible. However, dissolved O_2 can become a significant factor in conditions of severe hypoxemia. Binding of O_2 to Hgb to form **oxyhemoglobin** within red blood cells is the primary transport mechanism of O_2. Hgb not bound to O_2 is referred to as *deoxyhemoglobin* or *reduced Hgb.* The O_2-carrying capacity of blood is enhanced about 65 times by its ability to bind to Hgb.

Hemoglobin

Hgb is the major transport molecule for O_2. The Hgb molecule is a protein with two major components: four nonprotein heme groups, each containing iron in the reduced ferric (Fe^{+++}) form, which is the site of O_2 binding, and a globin portion consisting of four polypeptide chains. Normal adults have two α-globin chains and two β-globin chains (HgbA), whereas children younger than 6 months of age have predominantly fetal Hgb (HgbF), which consists of two α chains and two γ chains. This difference in the structure of HgbF increases its affinity for O_2 and aids in the transport of O_2 across the placenta. In addition, HgbF is not inhibited by 2,3-diphosphoglycerate (2,3-DPG), a product of glycolysis; thus O_2 uptake is further enhanced.

Binding of O_2 to Hgb alters the ability of Hgb to absorb light. This effect of O_2 on Hgb is responsible for the change in color between oxygenated arterial blood (bright red) and deoxygenated venous blood (dark red-bluish). Binding and dissociation of O_2 with Hgb occur in milliseconds, thus facilitating O_2 transport because red blood cells spend only 0.75 seconds in the capillaries. There are approximately 280 million Hgb molecules per red blood cell, which provides an efficient mechanism to transport O_2. Myoglobin, a protein similar in structure and function to Hgb, has only one subunit of the Hgb molecule. It aids in the transfer of O_2 from blood to muscle cells and in the storage of O_2, which is especially critical in O_2-deprived conditions.

Abnormalities of the Hgb molecule occur with mutations in the amino acid sequence (i.e., sickle cell disease) or in the spatial arrangement of the globin polypeptide chains and result in abnormal function. Compounds such as CO, nitrites (nitric oxide), and cyanides can oxidize the

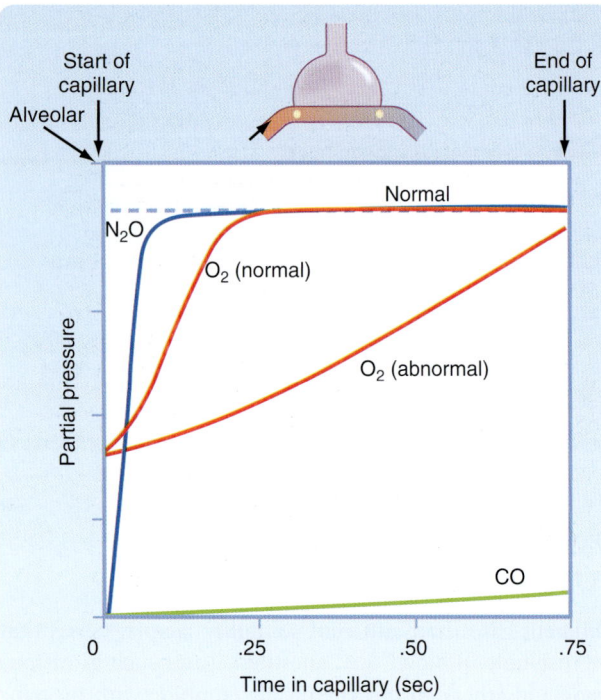

• **Fig. 24.3** Uptake of Nitrous Oxide (N_2O), Carbon Monoxide (CO), and O_2 in Blood in Relation to Their Partial Pressures and the Transit Time of the Red Blood Cell in the Capillary. For gases that are perfusion limited (N_2O and O_2), their partial pressures have equilibrated with alveolar pressure before exiting the capillary. In contrast, the partial pressure of CO, a gas that is diffusion limited, does not reach equilibrium with alveolar pressure. In rare conditions, O_2 uptake can become diffusion limited.

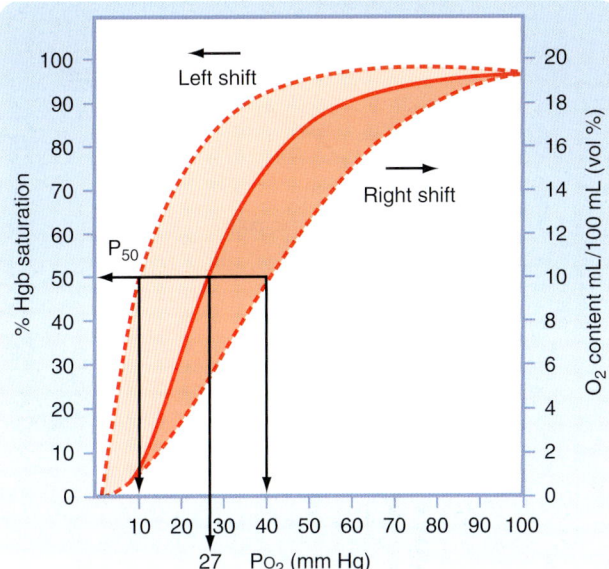

• **Fig. 24.4** Oxyhemoglobin Dissociation Curve Showing the Relationship Between the Partial Pressure of Oxygen (Po₂) in Blood and the Percentage of Hemoglobin (Hgb) Binding Sites That Are Occupied by Oxygen Molecules (Percentage Saturation). Adult hemoglobin (HgbA) is about 50% saturated with oxygen at a Po₂ of 27 mm Hg, 90% saturated at 60 mm Hg, and about 98% saturated at 100 mm Hg. The P₅₀ is the partial pressure at which Hgb is 50% saturated with O₂. When the O₂ dissociation curve shifts to the right, P₅₀ increases. When the curve shifts to the left, P₅₀ decreases.

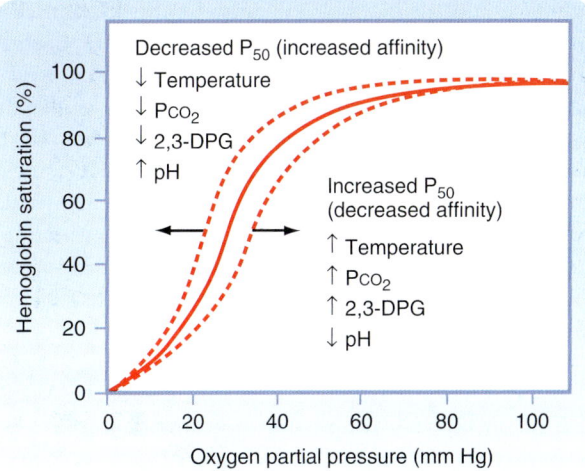

• **Fig. 24.5** Factors that shift the oxyhemoglobin dissociation curve to the right (decreased affinity of Hgb for O₂) or to the left (increased affinity). 2,3-DPG, 2,3-diphosphoglycerate; Pco₂, partial pressure of carbon dioxide.

Physiological Factors That Shift the Oxyhemoglobin Dissociation Curve

The oxyhemoglobin dissociation curve can shift in numerous clinical conditions, either to the right or to the left (Fig. 24.5). The curve is shifted to the right when the affinity of

IN THE CLINIC

In the inherited homozygous condition known as *sickle cell disease,* affected individuals have an amino acid substitution (valine for glutamic acid) on the β chain of the Hgb molecule. This creates a sickle cell Hgb (HgbS), which, when not bound to oxygen (deoxyhemoglobin or desaturated Hgb), can transform into a stiff gelatinous material that distorts the normal biconcave shape of the red blood cell to a crescent, or sickle-shaped, form. This change in appearance from spherical to a sickle shape increases the tendency of the red blood cell to form thrombi or clots that obstruct small vessels and creates a clinical condition known as *acute sickle cell episode.* The symptoms of such an episode vary, depending on the site of the obstruction (e.g., in the brain, stroke; in the lungs, pulmonary infarction) and are commonly associated with intense pain. The spleen is a common site of obstruction/infarction, and the ensuing tissue damage compromises the immune capabilities of affected individuals and renders them susceptible to recurrent infections. In the homozygous form, this condition is life-shortening; however, individuals with the heterozygous form are resistant to malaria. Thus an individual with heterozygous alleles has a survival advantage in regions of the world where malaria is prevalent, which may explain why the sickle cell mutation has been preserved through evolution. The increased affinity of HgbF for O₂ confers advantages to individuals with sickle cell disease in that the cells do not desaturate as much when O₂ is released from Hgb to the tissue and thus are less likely to become deformed in the sickle shape. Sickle cell disease is most prevalent among individuals of African American descent, but it is also observed in Hispanic, Turkish, Asian, and other ethnic groups.

iron molecule in the heme group and change it from the reduced ferrous state (Fe^{++}) to the ferric state (Fe^{+++}), which reduces the ability of O_2 to bind to Hgb.

Oxyhemoglobin Dissociation Curve

In the alveoli, the majority of O_2 in plasma quickly diffuses into red blood cells and chemically binds to Hgb. This process is reversible, so that Hgb quickly gives up its O_2 to tissue through passive diffusion (the concentration of O_2 in Hgb decreases). The oxyhemoglobin dissociation curve illustrates the relationship between Po_2 in blood and the number of O_2 molecules bound to Hgb (Fig. 24.4). The S shape of the curve demonstrates the dependence of Hgb saturation on Po_2, especially at partial pressures lower than 60 mm Hg. The clinical significance of the flat portion of the oxyhemoglobin dissociation curve (>60 mm Hg) is that a drop in Po_2 over a wide range of partial pressures (100 to 60 mm Hg) has a minimal effect on Hgb saturation, which remains at 90% to 100%, a level sufficient for normal O_2 transport and delivery. The clinical significance of the steep portion (<60 mm Hg) of the curve is that a large amount of O_2 is released from Hgb with only a small change in Po_2, which facilitates the release and diffusion of O_2 into tissue. The point on the curve at which Hgb is 50% saturated with O_2 is called the P_{50}, and it is 27 mm Hg in normal adults.

Hgb for O_2 decreases, which enhances O_2 dissociation. This results in decreased Hgb binding to O_2 at a given PO_2, which causes the P_{50} to increase. When the affinity of Hgb for O_2 increases, the curve is shifted to the left, which causes the P_{50} to decrease. In this state, O_2 dissociation and delivery to tissue are inhibited. Shifts to the right or left of the dissociation curve have little effect when they occur at O_2 partial pressures within the normal range (80 to 100 mm Hg). However, at O_2 partial pressures below 60 mm Hg (steep part of the curve), shifts in the oxyhemoglobin dissociation curve can dramatically influence O_2 transport.

Hydrogen Ion Concentration and Carbon Dioxide

Changes in blood hydrogen ion concentration (pH) shift the oxyhemoglobin dissociation curve. An increase in CO_2 production by tissue and its release into blood results in the generation of hydrogen ions (H^+) and a decrease in pH. This shifts the dissociation curve to the right, which has a beneficial effect by aiding in the release of O_2 from Hgb for diffusion into tissues. The shift to the right in the dissociation curve is due to the decrease in pH and to a direct effect of CO_2 on Hgb. Conversely, as blood passes through the lungs, CO_2 is exhaled, which results in an increase in pH, which in turn causes the oxyhemoglobin dissociation curve to shift to the left. This effect of CO_2 on the affinity of Hgb for O_2 is known as the **Bohr effect,** and it serves to enhance O_2 uptake in the lungs and delivery of O_2 to tissues. An increase in body temperature, as occurs during exercise, shifts the oxyhemoglobin dissociation curve to the right and enables more O_2 to be released to tissues, where it is needed because the demand increases. During cold weather, a decrease in body temperature, especially in the extremities (lips, fingers, toes, and ears), shifts the O_2 dissociation curve to the left (higher Hgb affinity). In this instance, PaO_2 may be normal, but release of O_2 in these extremities is not facilitated.

2,3-Diphosphoglycerate

Mature red blood cells do not have mitochondria, and therefore they depend on anaerobic glycolysis. Large quantities of a metabolic intermediary, 2,3-DPG, are formed in red blood cells during glycolysis, and the affinity of Hgb for O_2 decreases as 2,3-DPG levels increase. Thus the oxyhemoglobin dissociation curve shifts to the right. Although the binding sites of 2,3-DPG and O_2 differ on the Hgb molecule, binding of 2,3-DPG creates an allosteric effect that inhibits the binding of O_2. Conditions that increase 2,3-DPG include hypoxia, decreased Hgb, and increased pH. Decreased levels of 2,3-DPG are observed in stored blood samples and thus may present a problem to transfusion recipients because of the greater affinity of Hgb for O_2, which inhibits the unloading of O_2 to tissues.

Fetal Hemoglobin (HgbF)

As discussed previously, HgbF has a greater affinity for O_2 than does adult Hgb, and the oxyhemoglobin dissociation curve thus shifts to the left.

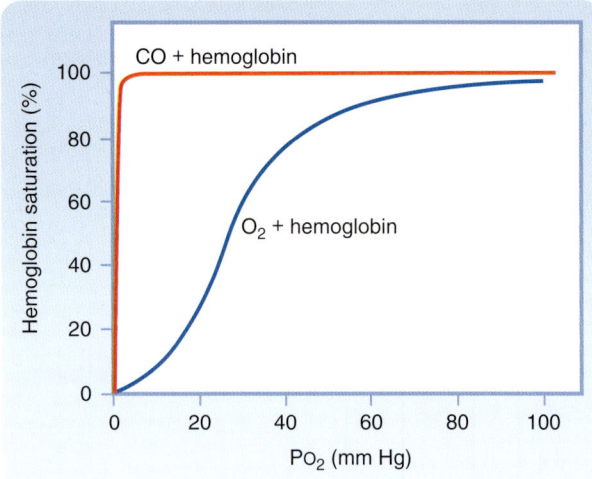

• **Fig. 24.6** The oxyhemoglobin and carboxyhemoglobin dissociation curves clearly illustrate the increased affinity that carbon monoxide (CO) has for Hgb, in comparison with O_2.

Carbon Monoxide

CO binds to the heme group of the Hgb molecule at the same site as O_2 and forms **carboxyhemoglobin** (HgbCO). A major difference between the ability of CO and that of O_2 to bind to Hgb is illustrated by a comparison of the oxyhemoglobin and carboxyhemoglobin dissociation curves. The affinity of CO for Hgb is about 200 times greater than it is for O_2 (Fig. 24.6). Thus small amounts of CO can greatly influence the binding of O_2 to Hgb. In the presence of CO, the affinity of Hgb for O_2 is enhanced. This causes the dissociation curve to shift to the left, which further prevents the unloading and delivery of O_2 to tissues. As the PCO of blood approaches 1.0 mm Hg, all of the Hgb binding sites are occupied by CO, and Hgb is unable to bind to O_2. This situation is not compatible with life and is the cause of death in cases of CO poisoning. In healthy individuals, HgbCO occupies about 1% to 2% of the Hgb binding sites; however, in cigarette smokers and in individuals who reside in high-density urban traffic areas, occupation of Hgb binding sites can be increased to 10%. Levels above 5% to 7% are considered hazardous. Treatment of individuals with high levels of CO, such as those who have inhaled car exhaust or smoke from a burning building, consists of administering high concentrations of O_2 to displace CO from Hgb. Increasing the ambient pressure above atmospheric pressure, through the use of a barometric chamber, substantially increases the O_2 tension, which promotes the dissociation of CO from Hgb. Another gas, nitric oxide, has great affinity (200,000 times greater than O_2) for Hgb, and it binds irreversibly to Hgb at the same site that O_2 does. Endothelial cells synthesize nitric oxide, which has vasodilation properties. Thus nitric oxide is used therapeutically as an inhalant in patients with pulmonary hypertension to reduce pressure. Although nitric oxide poisoning is not common, the clinician should be cautious when administering nitric oxide therapy for long periods. Hgb-bound CO and nitric oxide is referred to as *methemoglobin.* Under normal conditions, about 1% to 2% of Hgb is bound to CO and nitric oxide.

Oxygen Saturation, Content, and Delivery

Each Hgb molecule can bind up to four O_2 atoms, and each gram of Hgb can bind up to 1.34 mL of O_2. The term **O_2 saturation** (SO_2) refers to the amount of O_2 bound to Hgb in relation to the maximal amount of O_2 (100% O_2 capacity) that can bind Hgb. At 100% O_2 capacity, the heme groups of the Hgb molecules are fully saturated with O_2, and at 75% O_2 capacity, three of the four heme groups are occupied. Binding of O_2 to each heme group increases the affinity of the Hgb molecule to bind additional O_2. The O_2 content in blood is the sum of the O_2 bound to Hgb and the dissolved O_2. Oxygen content is decreased in the presence of increased CO_2 and CO and in individuals with anemia (Fig. 24.7).

Oxygen delivery from the lungs to tissues is dependent on several factors, including cardiac output, the Hgb content of blood, and the ability of the lung to oxygenate the blood. Not all of the O_2 carried in blood is unloaded at the tissue level. The actual O_2 extracted from blood by the tissue is the difference between the arterial O_2 content and the venous O_2 content, multiplied by cardiac output. Under normal conditions, Hgb leaves the tissue 75% saturated with O_2, and only about 25% is actually used by tissues. Hypothermia, relaxation of skeletal muscles, and an increase in cardiac output reduce O_2 extraction. Conversely, a decrease in cardiac output, anemia, hyperthermia, and exercise increase O_2 extraction.

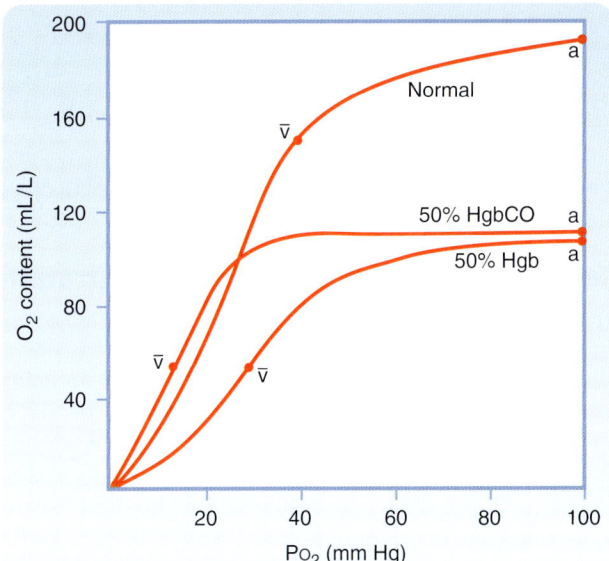

• **Fig. 24.7** A comparison of O_2 content curves under three conditions shows why carboxyhemoglobin (HgbCO) dramatically reduces the O_2 transport system. Fifty percent HgbCO represents the binding of half the circulating Hgb with carbon monoxide (CO). The 50% hemoglobin and 50% HgbCO curves show the same decreased O_2 content in arterial blood. However, CO has a profound effect in lowering venous partial pressure of oxygen. The arterial (a) and mixed venous ($\bar{v}$) points of constant cardiac output are indicated.

Tissue hypoxia is a condition in which O_2 available to cells is insufficient for maintaining adequate aerobic metabolism. Thus anaerobic metabolism is stimulated and results in the increases in levels of lactate and H^+ and the subsequent formation of lactic acid. The net result can lead to a significant decrease in blood pH. In cases of severe hypoxia, the extremities, toes, and fingertips may appear blue-gray **(cyanotic)** because of lack of O_2 and increased deoxyhemoglobin levels. There are four major types of tissue hypoxia (hypoxic hypoxia, circulatory hypoxia, anemic hypoxia, histotoxic hypoxia), discussed in detail in Chapter 23.

Erythropoiesis

Tissue oxygenation depends on the concentration of Hgb and thus on the number of red blood cells available in the circulation. Red blood cell production **(erythropoiesis)** in the bone marrow is controlled by the hormone **erythropoietin,** which is synthesized in the kidneys by cortical interstitial cells. Although Hgb levels are normally very stable, decreased O_2 delivery, low Hgb concentration, and low PaO_2 stimulate the secretion of erythropoietin. This increases the production of red blood cells. Chronic renal disease damages the cortical interstitial cells and thereby suppresses their ability to synthesize erythropoietin. This causes anemia, along with decreased Hgb because of the lack of erythropoietin. Erythropoietin replacement therapy using epoetin alfa (Epogen, Procrit) or darbepoetin alfa (Aranesp) effectively increases red blood cell production.

Carbon Dioxide Transport

Glucose Metabolism and Carbon Dioxide Production

CO_2 is produced at a rate of approximately 200 mL/minute under healthy conditions, and typically, 80 molecules of CO_2 are expired by the lung for every 100 molecules of O_2 that enter the capillary bed. The ratio of expired CO_2 to O_2 uptake is referred to as the **respiratory exchange ratio** and, under normal conditions, is 0.8 (80 molecules of CO_2 to 100 molecules of O_2). This ratio is similar at the tissue to blood compartment, where it is referred to as the **respiratory quotient.**

The body has enhanced storage capabilities for CO_2, in comparison with O_2, and hence PaO_2 is much more sensitive to changes in ventilation than is $PaCO_2$. Whereas PaO_2 is dependent on several factors, in addition to alveolar ventilation, arterial $PaCO_2$ is solely dependent on alveolar ventilation and CO_2 production. There is an inverse relationship between alveolar ventilation and $PaCO_2$.

Bicarbonate and Carbon Dioxide Transport

In blood, CO_2 is transported in red blood cells primarily as bicarbonate (HCO_3^-) but also as dissolved CO_2 and as

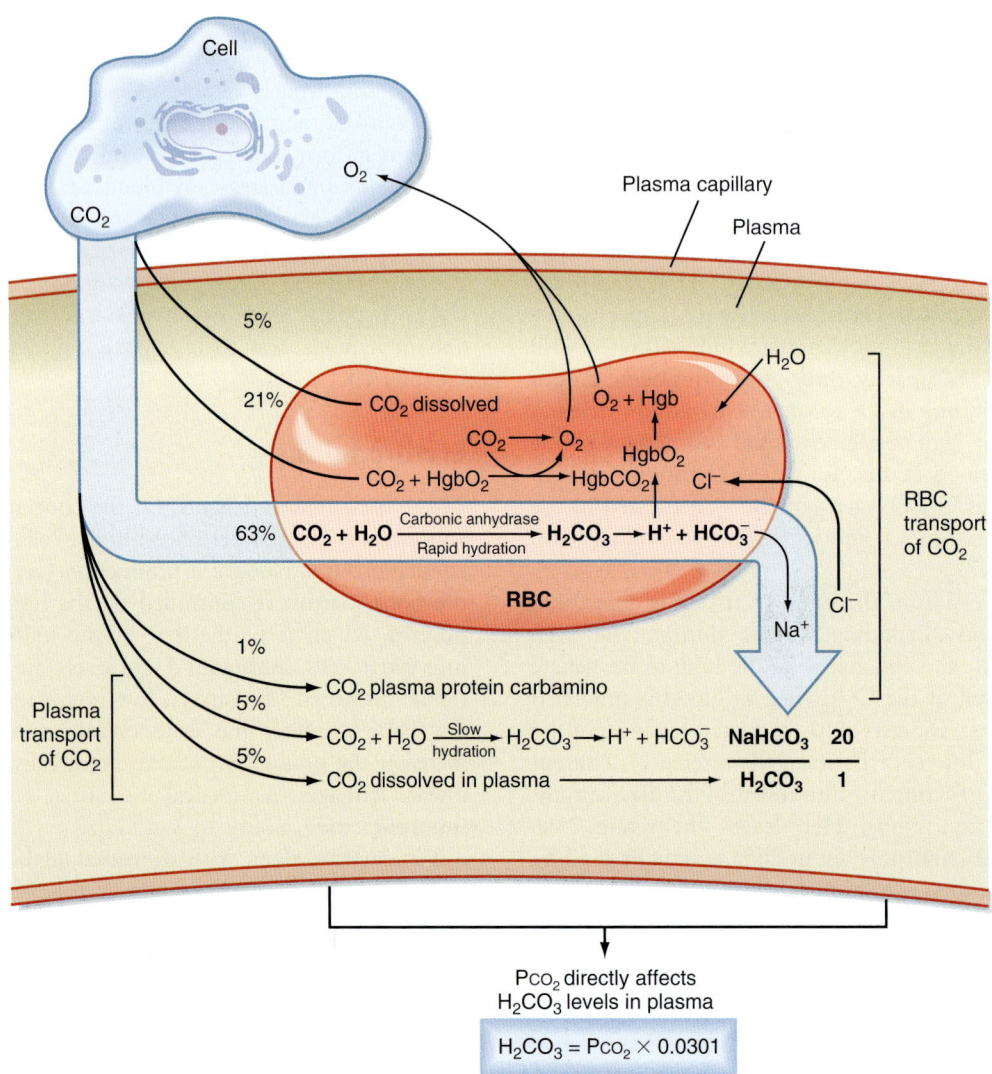

• **Fig. 24.8** Mechanisms of CO_2 Transport in Blood. $\bar{v}$. The predominant mechanism by which CO_2 is transported from tissue cells to the lung is in the form of bicarbonate anion (HCO_3^-). H_2CO_3, carbonic acid; HgbO, oxyhemoglobin; $NaHCO_3$, sodium bicarbonate; RBC, red blood cell.

carbamino protein complexes (i.e., CO_2 binds to plasma proteins and to Hgb; Fig. 24.8). Once CO_2 diffuses through the tissue and enters plasma, it quickly dissolves. The reaction of CO_2 with H_2O to form carbonic acid (H_2CO_3) is the major mechanism for the generation of HCO_3^- in red blood cells:

Equation 24.3

$$CO_2 + H_2O \rightleftharpoons H_2CO_3 \rightleftharpoons H^+ + HCO_3^-$$

The reaction normally proceeds quite slowly; however, it is catalyzed within red blood cells by the enzyme **carbonic anhydrase.** The HCO_3^- diffuses out of the red blood cell in exchange for Cl^-, in a process known as the **chloride shift,** which helps the cell maintain its osmotic equilibrium. This chemical reaction (see Fig. 24.8) is reversible and can be shifted to the right to generate more HCO_3^- in response to more CO_2 entering the blood from tissues, or it can be shifted to the left as CO_2 is exhaled in the lungs, which reduces HCO_3^- levels. The free H^+ is quickly buffered within the red blood cell by binding to Hgb. Buffering of H^+ is critical for keeping the reaction moving toward the synthesis of HCO_3; high levels of free H^+ (low pH) cause the reaction to shift to the left.

Regulation of Hydrogen Ion Concentration and Acid-Base Balance

The H^+ concentration (pH) has a dramatic effect on many metabolic processes within cells, and regulation of pH is essential for normal homeostasis. In the clinical setting, blood pH is measured to assess the concentration of H^+. The normal pH range for adults is 7.35 to 7.45 and is maintained by the lungs, kidneys, and chemical buffer systems. In the respiratory system, conversion of CO_2 to HCO_3^-, illustrated as follows, is a major mechanism of buffering and regulating the H^+ concentration (pH):

Equation 24.4

$$CO_2 + H_2O \rightleftharpoons H_2CO_3 \rightleftharpoons H^+ + HCO_3^-$$

(with hydrogen yielding)

$$H^+ + Hgb \rightleftharpoons H \times Hgb$$

As $PaCO_2$ changes, so does the concentration of HCO_3^- and H_2CO_3, as well as $PaCO_2$.

The **Henderson-Hasselbalch** equation is used to calculate how changes in CO_2 and HCO_3^- affect pH:

Equation 24.5

$$pH = pK' + \frac{\log[HCO_3^-]}{\alpha PCO_2}$$

where α = solubility. Thus

Equation 24.6

$$pH = 6.1 + \frac{\log[HCO_3^-]}{0.03 \times PCO_2}$$

In these equations, the amount of CO_2 is determined from the PCO_2 and its solubility (α) in solution. For plasma at 37°C, α has a value of 0.03. Also, pK' is the negative logarithm of the overall dissociation constant for the reaction and has a logarithmic value of 6.1 for plasma at 37°C.

Acute hyperventilation secondary to exercise or anxiety reduces PCO_2 and thereby causes an increase in pH (respiratory alkalosis). Conversely, if PCO_2 increases because of hypoventilation secondary to an overdose of a respiratory depressant, the pH decreases (respiratory acidosis). Acid-base imbalances are also caused by metabolic disorders such as metabolic acidosis (e.g., lactic acidosis, ketoacidosis, and renal failure) and metabolic alkalosis (e.g., hypokalemia, hypochloremia, vomiting, high doses of steroids).

Carbon Dioxide Dissociation Curve

In contrast to O_2, the dissociation curve for CO_2 in blood is linear and directly related to PCO_2 (Fig. 24.9). The degree of Hgb saturation with O_2 has a major effect on the CO_2 dissociation curve. Although O_2 and CO_2 bind to Hgb

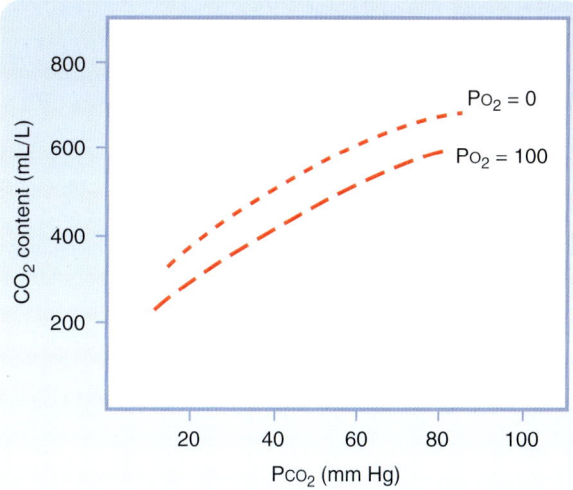

• **Fig. 24.9** Equilibrium Curves for CO_2 in Arterial and Venous Blood. Venous blood can transport more CO_2 than arterial blood can at any given PCO_2. In comparison with the $HgbO_2$ equilibrium curve, the CO_2 curves are essentially straight lines between a PCO_2 of 20 and a PCO_2 of 80 mm Hg. *Long dashes* represent the arterial equilibrium curve; *short dashes* represent the venous equilibrium curve.

at different sites, deoxygenated Hgb has greater affinity for CO_2 than oxygenated Hgb. Thus deoxygenated blood (venous blood) freely takes up and transports more CO_2 than oxygenated arterial blood does. The deoxygenated Hgb more readily forms carbamino compounds and also more readily binds free H^+ released during the formation of HCO_3^-. The effect of changes in oxyhemoglobin saturation on the relationship of CO_2 content to PCO_2 is referred to as the **Haldane effect** and is reversed in the lungs when O_2 is transported from the alveoli to red blood cells. This effect is illustrated by a shift to the left in the CO_2 dissociation curve in venous blood in comparison with arterial blood.

Key Points

1. Gases (nitrous oxide, ether, helium) that have a rapid rate of air-to-blood equilibration are perfusion limited. Gases (CO) that have a slow air-to-blood equilibration rate are diffusion limited. Under normal conditions, O_2 and CO_2 exchange are perfusion limited but can be diffusion limited in some situations.
2. The major transport mechanism of O_2 in blood is within the red blood cell bound to Hgb, and for CO_2, it is within red blood cells in the form of HCO_3^-.
3. The reversible reaction of CO_2 with H_2O to form H_2CO_3, with its subsequent dissociation to HCO_3^- and H^+, is catalyzed by the enzyme carbonic anhydrase within red blood cells and is the major mechanism for generation of HCO_3^-.
4. The O_2 dissociation curve is S shaped. In the plateau area (>60 mm Hg), increasing or decreasing PO_2 has only a minimal effect on Hgb saturation from 100%

to 90%. This ensures adequate Hgb saturation over a large range of PO_2 values.
5. The CO_2 dissociation curve is linear and directly related to PCO_2. PCO_2 is solely dependent on alveolar ventilation and CO_2 production.
6. The CO_2 to HCO_3^- pathway plays a critical role in the regulation of H^+ ions and in maintaining acid-base balance in the body.
7. Tissue oxygenation is dependent on Hgb within red blood cells and subsequently the number (and production) of red blood cells, which is controlled by the hormone erythropoietin. Low O_2 delivery, low Hgb concentration, and low PaO_2 stimulate the secretion of erythropoietin in the kidneys.
8. Tissue hypoxia occurs when insufficient amounts of O_2 are supplied to the tissue to conduct normal levels of aerobic metabolism.

Additional Readings

Butler JP, Tsuda A. Transport of gases between the environment and alveoli—theoretical foundations. *Compr Physiol.* 2011;1:1301-1316.

Calverley PMA, Koulouris NG. Flow limitation and dynamic hyperinflation: key concepts in modern respiratory physiology. *Eur Respir J.* 2005;25:186-199.

Hillman SS, Hancock TV, Hedrick MS. A comparative meta-analysis of maximal aerobic metabolism of vertebrates: implications for respiratory and cardiovascular limits to gas exchange. *J Comp Physiol [B].* 2013;183:167-179.

Hughes JM. Assessing gas exchange. *Chron Respir Dis.* 2007;4:205-214.

Petersson J, Glenny RW. Gas exchange and ventilation-perfusion relationships in the lung. *Eur Respir J.* 2014;44:1023-1041.

Sheel AW, Romer LM. Ventilation and respiratory mechanics. *Compr Physiol.* 2012;2:1093-1142.

Stickland MK, Lindinger MI, Olfert IM, et al. Pulmonary gas exchange and acid-base balance during exercise. *Compr Physiol.* 2013;3:693-739.

Whipp BJ. Physiological mechanisms dissociating pulmonary CO_2 and O_2 exchange dynamics during exercise in humans. *Exp Physiol.* 2007;92:347-355.

25

Control of Respiration

LEARNING OBJECTIVES

Upon completion of this chapter, the student should be able to answer the following questions:

1. Describe the central organization of breathing.
2. Explain the role of central and peripheral chemoreceptors in regulating respiration.
3. Compare and contrast the roles of chemoreceptors and pulmonary mechanoreceptors in regulating respiration.
4. Describe ventilatory control during special circumstances (e.g., exercise and high altitude).
5. Describe the effects of abnormalities in ventilatory control.

People breathe without thinking, and they can willingly modify their breathing pattern and even hold their breath. Control of ventilation includes the generation and regulation of rhythmic breathing by the respiratory center in the brainstem and its modification by the input of information from higher brain centers and from systemic receptors. The goals of breathing are, from a mechanical perspective, to minimize work and, from a physiological perspective, to maintain and regulate positive pressures of arterial blood O_2 (PaO_2) and CO_2 ($PaCO_2$). Another goal of breathing is to maintain acid-base balance in the brain by regulating $PaCO_2$. Automatic respiration begins at birth. In utero, the placenta, not the lung, is the organ of gas exchange in the fetus. Its microvilli interdigitate with the maternal uterine circulation, and PaO_2 transport and $PaCO_2$ removal from the fetus occur by passive diffusion across the maternal circulation.

Ventilatory Control: An Overview

There are four major sites of ventilatory control: (1) the **respiratory control center,** (2) **central chemoreceptors,** (3) **peripheral chemoreceptors,** and (4) **pulmonary mechanoreceptors/sensory nerves.** The respiratory control center is located in the medulla oblongata of the brainstem and is composed of multiple nuclei that generate and modify the basic ventilatory rhythm. This center consists of two main parts: (1) a ventilatory pattern generator, which sets the rhythmic pattern, and (2) an integrator, which controls generation of the pattern, processes input from higher brain centers and chemoreceptors, and controls the

rate and amplitude of the ventilatory pattern. Input to the integrator arises from higher brain centers, including the cerebral cortex, hypothalamus, limbic system including the amygdalae, and cerebellum.

Central chemoreceptors are located in the central nervous system just below the ventrolateral surface of the medulla. These central chemoreceptors detect changes in the $PaCO_2$ and pH of interstitial fluid in the brainstem, and they modulate ventilation. Peripheral chemoreceptors are located on specialized cells in the aortic arch (**aortic bodies**) and at the bifurcation of the internal and external carotid arteries (**carotid bodies**) in the neck. These peripheral chemoreceptors sense the PaO_2, $PaCO_2$, and pH of arterial blood, and they feed information back to the integrator nuclei in the medulla through the vagus nerves and carotid sinus nerves, which are branches of the glossopharyngeal nerves. Pulmonary mechanoreceptors and sensory nerve stimulation, in response to lung inflation or to stimulation by irritants or release of local mediators in the airways, modify the ventilatory pattern.

The collective output of the respiratory control center to motor neurons located in the anterior horn of the spinal column controls the muscles of respiration, and this output determines the automatic rhythmic pattern of respiration. Motor neurons located in the cervical region of the spinal column control the activity of the diaphragm through the **phrenic nerves,** whereas other motor neurons located in the thoracic region of the spine control the intercostal muscles and the accessory muscles of respiration.

In contrast to automatic respiration, voluntary respiration bypasses the respiratory control center in the medulla. The neural activity controlling voluntary respiration originates in the motor cortex, and signaling passes directly to motor neurons in the spine through the **corticospinal tracts.** The motor neurons to the respiratory muscles act as the final site of integration of the voluntary (corticospinal tract) and automatic (ventrolateral tracts) control of ventilation. Voluntary control of these muscles competes with automatic influences at the level of the spinal motor neurons, and this competition can be demonstrated by breath holding. At the start of the breath hold, voluntary control dominates the spinal motor neurons. However, as the breath hold continues, the automatic ventilatory control eventually overpowers the voluntary effort and limits the duration of the breath hold. Motor neurons also innervate muscles

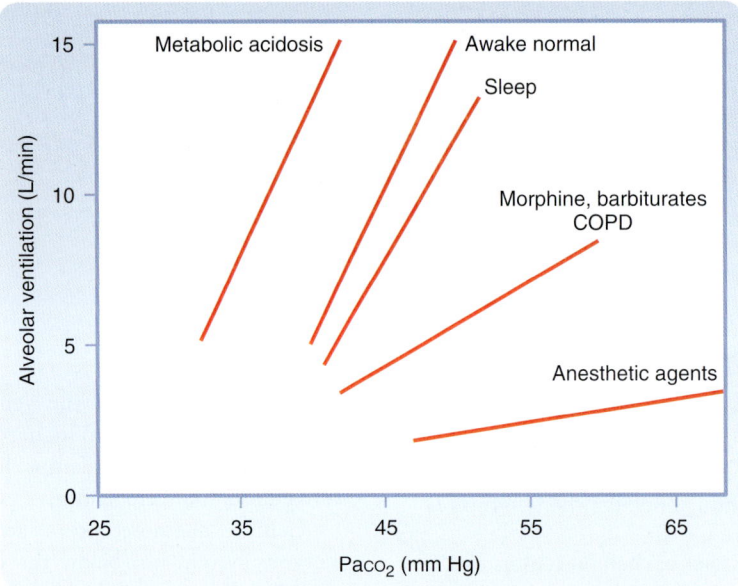

• Fig. 25.1 Relationship between Partial Pressure of Arterial Carbon Dioxide (PaCO₂) and Alveolar Ventilation in Awake Normal States, during Sleep, after Narcotic Ingestion and Deep Anesthesia, and in the Presence of Metabolic Acidosis. Both the slopes of the response (sensitivity) and the position of the response curves (threshold, the point at which the curve crosses the x-axis [not shown]) are changed, which indicates differences in ventilatory responses and response thresholds. COPD, chronic obstructive pulmonary disease.

of the upper airway. These neurons are located within the medulla near the respiratory control center. They innervate muscles in the upper airways through the cranial nerves. When activated, they dilate the pharynx and large airways at the initiation of inspiration.

Response to Carbon Dioxide

Ventilation is regulated by $PaCO_2$, PaO_2, and pH in arterial blood. $PaCO_2$ is the most important of these regulators. Both the rate and depth of breathing are controlled to maintain $PaCO_2$ close to 40 mm Hg. In a normal awake individual, there is a linear rise in ventilation as $PaCO_2$ reaches and exceeds 40 mm Hg (Fig. 25.1). The ventilatory drive or response to changes in $PaCO_2$ can be reduced by hyperventilation and by drugs that depress the respiratory center and decrease the ventilatory response to both CO_2 and O_2, such as morphine, barbiturates, and anesthetic agents. In these instances, the stimulus is inadequate to stimulate the motor neurons that innervate the muscles of respiration. It is also depressed during sleep. In addition, the ventilatory response to changes in $PaCO_2$ is reduced if the work of breathing is increased, which can occur in individuals with chronic obstructive pulmonary disease (COPD). This effect occurs primarily because the neural output of the respiratory center is less effective in promoting ventilation as a result of the mechanical limitation to ventilation.

Changes in $PaCO_2$ are sensed by central and peripheral chemoreceptors, and they transmit this information to the medullary respiratory centers. The respiratory control center then regulates minute ventilation and thereby maintains $PaCO_2$ within the normal range. In the presence of a normal PaO_2, ventilation increases by approximately 3 L/minute for each 1 mm Hg rise in $PaCO_2$. The response to an increase in $PaCO_2$ is further increased when the PaO_2 is low (Fig. 25.2, A). With a low PaO_2, ventilation is greater for any given $PaCO_2$, and the increase in ventilation for a given increment in $PaCO_2$ is enhanced. The slope of the minute ventilation response as a function of the inspired CO_2 is termed the *ventilatory response* to CO_2 and is a test of CO_2 sensitivity. It is important to recognize that this relationship is amplified by low O_2 (see Fig. 25.2, B). The responsiveness to low O_2 is enhanced because different mechanisms are responsible for sensing PaO_2 and $PaCO_2$ in the peripheral chemoreceptors. Thus the presence of both hypercapnia-elevated CO_2 and hypoxemia-low O_2 (often called *asphyxia* when both changes are present) has an additive effect on chemoreceptor output and on the resulting ventilatory stimulation.

Control of Ventilation: The Details

The Respiratory Control Center

When the brain is transected experimentally between the medulla and the pons, periodic breathing is maintained, thus demonstrating that the inherent rhythmicity of breathing originates in the medulla. Although no single group of neurons in the medulla has been found to be the breathing "pacemaker," two distinct nuclei within the medulla are involved in generation of the respiratory pattern (Fig. 25.3). One nucleus is the **dorsal respiratory group** (DRG), which is composed of cells in the **nucleus tractus solitarius** and is located in the dorsomedial region of the medulla. Cells in the DRG receive afferent input from cranial nerves IX

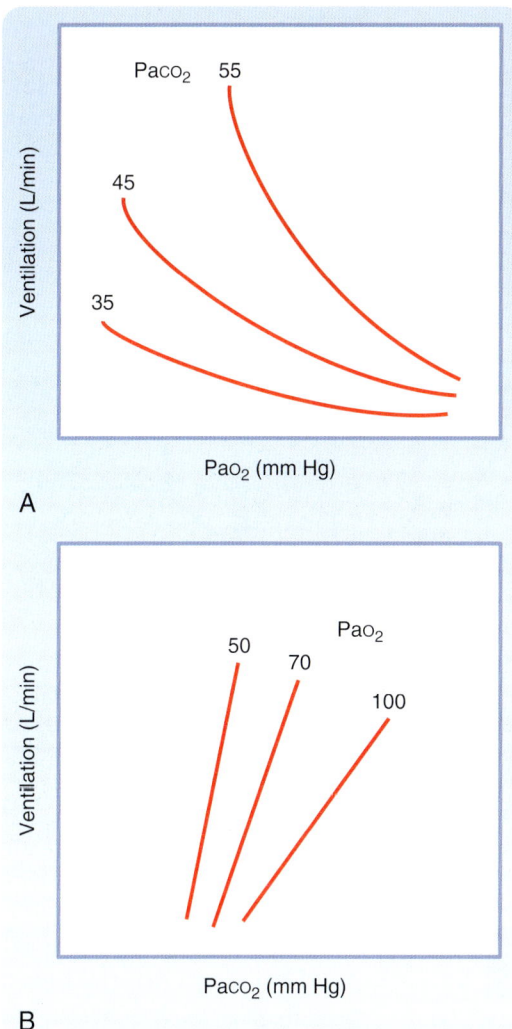

A

B

• **Fig. 25.2** The Effects of Hypoxia and Hypercapnia on Ventilation as the Other Respiratory Gas Partial Pressures Are Varied. **A,** At a given partial pressure of arterial carbon dioxide (Paco$_2$), ventilation increases more and more as partial pressure of arterial oxygen (Pao$_2$) decreases. When Paco$_2$ is allowed to decrease (the normal condition) during hypoxia, there is little stimulation of breathing until Pao$_2$ falls below 60 mm Hg. The hypoxic response is mediated through the carotid body chemoreceptors. **B,** The sensitivity of the ventilatory response to CO$_2$ is enhanced by hypoxia.

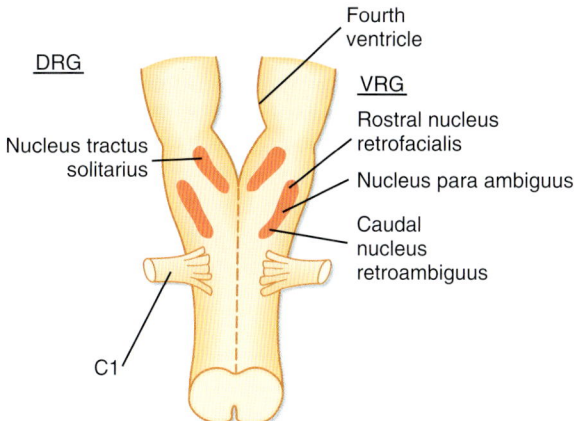

• **Fig. 25.3** The Respiratory Control Center is Located in the Medulla, the Most Primitive Portion of the Brain. The neurons are mainly in two areas: the dorsal respiratory group (DRG), which consists of the nucleus tractus solitarius, and the ventral respiratory group (VRG), which consists of the rostral nucleus retrofacialis, nucleus para-ambiguus, and the caudal nucleus retroambiguus. C1 refers to the first cervical signal segment to the caudal border of the pons. The fourth ventricle of the brain is located below the cerebellum and above, and between, the pons and the medulla.

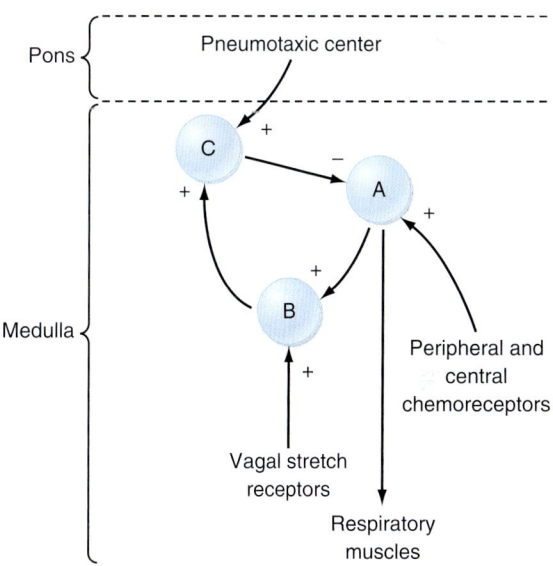

• **Fig. 25.4** Diagram of the Basic Wiring of the Brainstem Ventilatory Controller. The signs of the main output (arrows) of the neuron pools indicate whether the output is excitatory (+) or inhibitory (−). Pool A provides tonic inspiratory stimuli to the muscles of breathing. Pool B is stimulated by pool A and provides additional stimulation to the muscles of breathing, and pool B stimulates pool C. Other brain centers feed into pool C (inspiratory cutoff switch), which sends inhibitory impulses to pool A. Afferent information (feedback) from various sensors acts at different locations: Chemoreceptors act on pool A, and intrapulmonary sensory fibers act via the vagus nerves on pool B. A pneumotaxic center in the anterior pons receives input from the cerebral cortex, and it acts on pool C.

and X, which originate from airways and the lungs and constitute the initial intracranial processing station for this afferent input. The second group of medullary cells is the **ventral respiratory group** (VRG), located in the ventrolateral region of the medulla. The VRG is composed of three cell groups: the **rostral nucleus retrofacialis,** the **caudal nucleus retroambiguus,** and the **nucleus para-ambiguus.** The VRG contains both inspiratory and expiratory neurons. The nucleus retrofacialis and the caudally located cells of the nucleus retroambiguus are active during exhalation, whereas the rostrally located cells of the nucleus retroambiguus are active during inspiration. The nucleus para-ambiguus has inspiratory and expiratory neurons that travel in the vagus nerve to the laryngeal and pharyngeal muscles. Discharges

from cells in these areas excite some cells and inhibit other cells.

At the level of the respiratory control center, inspiration and exhalation involve three phases: one inspiratory and two expiratory (Fig. 25.4). Inspiration begins with an abrupt increase in discharge from cells in the nucleus tractus solitarius, the nucleus retroambiguus, and the nucleus para-ambiguus, followed by a steady ramp-like increase in firing rate throughout inspiration. This leads to progressive contraction of the respiratory muscles during automatic breathing. At the end of inspiration, an "off-switch" event causes neuron firing to decrease markedly, at which point exhalation begins. At the start of exhalation (phase I of expiration), a paradoxical increase in inspiratory neuron firing slows the expiratory phase down by increasing inspiratory muscle tone and expiratory neuron firing. This inspiratory neuron firing decreases and stops during phase II of exhalation. Although many different neurons in the DRG and VRG are involved in ventilation, each cell type appears to have a specific function. For example, the **Hering-Breuer reflex** is an inspiratory-inhibitory reflex that arises from afferent stretch receptors located in the smooth muscles of the airways. Increasing lung inflation stimulates these stretch receptors and results in early exhalation by stimulating the neurons associated with the off-switch phase of inspiratory muscle control. Thus rhythmic breathing depends on a continuous (tonic) inspiratory drive from the DRG and an intermittent (phasic) expiratory drive from the cerebrum, thalamus, cranial nerves, and ascending sensory tracts in the spinal cord.

Central Chemoreceptors

A chemoreceptor is a receptor that responds to a change in the chemical composition of blood or other fluid around it. Central chemoreceptors are specialized cells on the ventrolateral surface of the medulla. Chemoreceptors are sensitive to the pH of the surrounding extracellular fluid. Because this extracellular fluid is in contact with cerebrospinal fluid (CSF), changes in the pH of CSF affect ventilation by acting on these chemoreceptors

CSF is an ultrafiltrate of plasma that is secreted continuously by the **choroid plexus** and is reabsorbed by the arachnoid villi. Because it is in contact with the extracellular fluid in the brain, the composition of CSF is influenced by the metabolic activity of the cells in the surrounding area and the composition of the blood. Although the origin of CSF is plasma, the composition of CSF is not the same as that of plasma because the **blood-brain barrier** exists between the two sites (Fig. 25.5). The blood-brain barrier is composed of endothelial cells, smooth muscle, and the **pial** and **arachnoid membranes,** and it regulates the movement of ions between blood and CSF. In addition, the choroid plexus also determines the ionic composition of CSF by transporting ions into and out of CSF. The blood-brain barrier is relatively impermeable by H^+ and HCO_3^- ions, but it is very permeable by CO_2. Thus the PCO_2 in CSF parallels the arterial PCO_2 tension. CO_2 is also produced by cells of the brain as a product of metabolism. As a consequence, the PCO_2 in CSF is usually a few millimeters of mercury higher than that in arterial blood, and so the pH is slightly more acidic (7.33) in CSF than in plasma (Table 25.1).

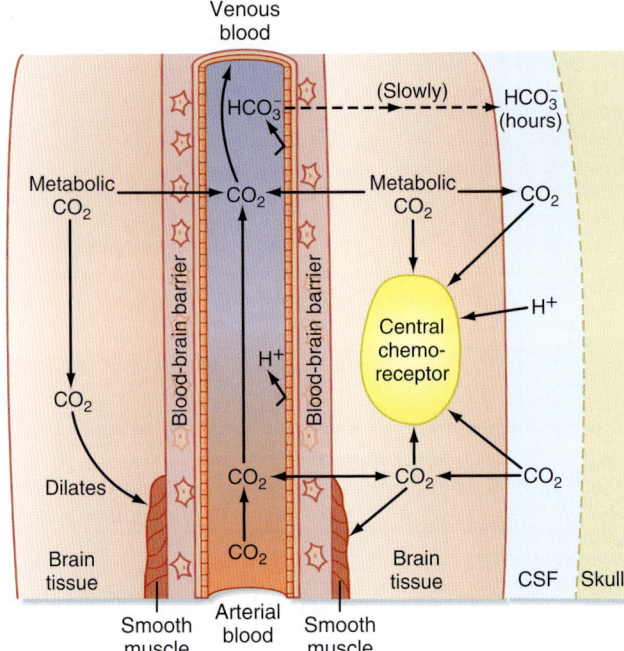

• **Fig. 25.5** Carbon Dioxide and the Blood-Brain Barrier. $PaCO_2$ crosses the blood-brain barrier and rapidly equilibrates with CO_2 in cerebrospinal fluid (CSF). H^+ and HCO_3^- ions cross the barrier slowly. The partial pressure of arterial carbon dioxide ($PaCO_2$) combines with CO_2 generated by metabolism to dilate the smooth muscle. In comparison with arterial blood, the pH of CSF is lower and the PCO_2 is higher, with little protein buffering.

TABLE 25.1	Normal Values for the Composition of Cerebrospinal Fluid and Arterial Blood	
Parameter	Cerebrospinal Fluid	Arterial Blood
pH	7.33	7.40
PCO_2 (mm Hg)	44	40
HCO_3^- (mEq/L)	22	24

PCO_2, partial pressure of carbon dioxide.

Peripheral Chemoreceptors

The **carotid** and **aortic bodies** are peripheral chemoreceptors that respond to changes in PaO_2 (not the O_2 content), $PaCO_2$, and pH, and they transmit afferent information to the central respiratory control center. The peripheral chemoreceptors are the only chemoreceptors that respond to changes in PaO_2. The peripheral chemoreceptors are also responsible for approximately 40% of the ventilatory response to $PaCO_2$. These chemoreceptors are small, highly vascularized structures. They consist of type I **(glomus)** cells that are rich in mitochondria and endoplasmic reticulum. They also have several types of cytoplasmic granules (synaptic vesicles) that contain various neurotransmitters, including dopamine, acetylcholine, norepinephrine, and neuropeptides. Afferent nerve fibers synapse with type I cells, and they transmit information to the brainstem through the carotid sinus nerve (carotid body) and vagus

AT THE CELLULAR LEVEL

The **Henderson-Hasselbalch equation** relates the pH of CSF to the concentration of bicarbonate ([HCO₃⁻]):

$$pH = pK + \frac{\log[HCO_3^-]}{\alpha \times P_{CO_2}}$$

where α is the solubility coefficient (0.03 mmol/L/mm Hg) and pK is the negative logarithm of the dissociation constant for carbonic acid (6.1). The Henderson-Hasselbalch equation demonstrates that an increase in CSF P_{CO_2} causes the pH of CSF to decrease at any given [HCO₃⁻]. The decrease in pH stimulates the central chemoreceptors and thereby increases ventilation. Thus CO_2 in blood regulates ventilation by its effect on the pH of CSF. The resulting hyperventilation reduces the Pa_{CO_2}, and therefore the P_{CO_2} of CSF, and returns the pH of CSF toward a normal value. Furthermore, cerebral vasodilation accompanies an increase in Pa_{CO_2}, and this enhances the diffusion of CO_2 into CSF. In contrast, an increase in CSF [HCO₃⁻] causes an increase in the pH of CSF at any given Pa_{CO_2}.

Changes in Pa_{CO_2} that result from alterations in pH activate homeostatic mechanisms that return the pH back toward a normal value. The blood-brain barrier regulates the pH of CSF by adjusting the ionic composition and [HCO₃⁻] of CSF. These changes in CSF [HCO₃⁻], however, occur slowly, over a period of several hours, whereas changes in CSF P_{CO_2} can occur within minutes. Thus compensation for changes in the pH of CSF requires hours to develop fully.

nerve (aortic body). Type I cells are the cells primarily responsible for sensing Pa_{O_2}, Pa_{CO_2}, and pH. In response to even small decreases in Pa_{O_2}, there is an increase in chemoreceptor discharge, which enhances respiration. The response is robust when Pa_{O_2} decreases below 75 mm Hg. Thus ventilation is regulated by changes in arterial and CSF pH through effects on peripheral and central chemoreceptors (Fig. 25.6).

Pulmonary Mechanoreceptors
Chest Wall and Lung Reflexes

Several reflexes that arise from the chest wall and lungs affect ventilation and ventilatory patterns (Table 25.2). The **Hering-Breuer inspiratory-inhibitory reflex** is stimulated by increases in lung volume, especially those associated with an increase in both ventilatory rate and tidal volume. This stretch reflex is mediated by vagal fibers, and when

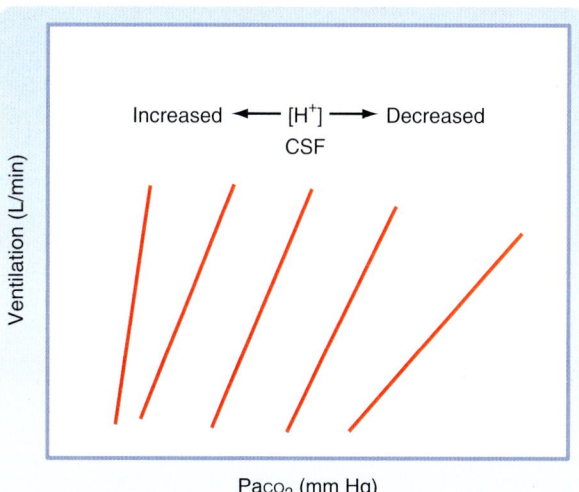

• **Fig. 25.6** The Ventilatory Response to Partial Pressure of Arterial Carbon Dioxide (Pa_{CO_2}) is Affected by the Concentration of Hydrogen ([H⁺]) in Cerebrospinal Fluid (CSF) and Brainstem Interstitial Fluid. During chronic metabolic acidosis (e.g., diabetic ketoacidosis), the [H⁺] in CSF is increased, and the ventilatory response to inspired Pa_{CO_2} is increased (steeper slope). Conversely, during chronic metabolic alkalosis (a relatively uncommon condition), the [H⁺] in CSF is decreased and the ventilatory response to inspired Pa_{CO_2} is decreased (reduced slope). The positions of the response lines are also shifted, which indicates altered thresholds.

TABLE 25.2 Reflexes and Sensory Nerves in the Respiratory Tract

Reflex	Stimuli	End Organ Location)	Receptor Type
Hering-Breuer inflation reflex Hering-Breuer deflation reflex Bronchodilation Tachycardia Hyperpnea	Lung inflation	Airway smooth muscle cells	Myelinated, vagal, slowly adapting receptor
Cough Mucus secretion Bronchoconstriction Hering-Breuer deflation reflex	Lung hyperinflation Exogenous and endogenous agents Histamine Prostaglandins	Airway epithelial cells	Myelinated, vagal, rapidly adapting receptors (irritant receptors)
Apnea, followed by tachypnea Bronchoconstriction Bradycardia Hypotension Mucus secretion	Extensive hyperinflation Exogenous and endogenous agents Capsaicin Phenyl diguanide Histamine Bradykinin Serotonin Prostaglandins	Pulmonary interstitial space Close to the pulmonary circulation Close to the bronchial circulation	Unmyelinated, vagal, C fiber endings (J receptors)

IN THE CLINIC

Imagine flying from New York City to Denver. The barometric pressure in New York is approximately 760 mm Hg, whereas in the mountains surrounding Denver, Colorado, it is 600 mm Hg. At sea level, the PaO_2 is approximately 95 mm Hg and $PAO_2 = [(760 - 47) \times 0.21] - [40/0.8] = 100$ mm Hg (according to the alveolar air equation; see Chapter 23). If the alveolar-arterial PO_2 difference [$AaDO_2$] is 5 mm Hg, $PaO_2 = 100$ mm Hg − 5 mm Hg = 95 mm Hg. In the CSF, pH would be approximately 7.33, $PaCO_2$ would be 44 mm Hg ($PaCO_2 + CO_2$ produced by metabolism of brain cells), and HCO_3^- would be approximately 22 mEq/L.

When you arrive in Denver, there is an abrupt decrease in the partial pressure of inspired O_2 (PiO_2): $PiO_2 = (600 - 47) \times 0.21 = 116$ mm Hg; there are also decreases in the partial pressures of alveolar and arterial O_2: $PAO_2 = 116 - (40/0.8) = 66$ mm Hg, and $PaO_2 = 61$ mm Hg if there is no change in $AaDO_2$). This decrease in arterial O_2 stimulates the peripheral chemoreceptors and thereby increases ventilation. The increase in ventilation decreases $PaCO_2$ and elevates arterial pH. The result of this increase in ventilation is to minimize the hypoxemia by increasing PAO_2. For example, assume that $PACO_2$ decreases to 30 mm Hg. Then $PAO_2 = [(600 - 47) \times 0.21] - [30/0.8] = 78$ mm Hg, a 12–mm Hg increase in PAO_2.

The decrease in $PaCO_2$ also causes a decrease in the PCO_2 of CSF. Because [HCO_3^-] is unchanged, the pH of CSF increases. This increase in the pH of CSF attenuates the rate of discharge of the central chemoreceptors and decreases their contribution to the ventilatory drive. Over the next 12 to 36 hours, [HCO_3^-] in CSF decreases as acid-base transporter proteins in the blood-brain barrier reduce [HCO_3^-]. As a consequence, the pH of CSF returns toward normal. Central chemoreceptor discharge increases, and minute ventilation is further increased. At the same time that [HCO_3^-] in CSF decreases, HCO_3^- is gradually excreted from plasma by the kidneys. This results in a gradual return of arterial pH toward normal values. Peripheral chemoreceptor stimulation increases further as arterial pH becomes normal (peripheral chemoreceptors are inhibited by the elevated arterial pH). Finally, within 36 hours of arriving at high altitude, minute ventilation increases significantly. This delayed response is greater than the immediate effect of the hypoxemia on ventilation. This further increase in ventilation is due to both central and peripheral chemoreceptor stimulation. Thus after 36 hours, both arterial pH and CSF pH are approaching normal values; minute ventilation is increased, PaO_2 is decreased, and $PaCO_2$ is decreased.

You now return home. When you land in New York, the PiO_2 returns to a normal value, and the hypoxic stimulus to ventilation is removed. PaO_2 returns to a normal value, and the peripheral chemoreceptor stimulation to ventilation decreases. This causes an increase in arterial [CO_2] toward normal values, which in turn causes an increase in CSF [CO_2]. This increase is associated with a decrease in the pH of CSF as [HCO_3^-] in CSF is reduced and ventilation is augmented. Over the next 12 to 36 hours, the acid-base transporters in the blood-brain barrier transport HCO_3^- back into CSF, and the pH of CSF gradually returns toward normal values. Similarly, the pH of blood decreases as $PaCO_2$ rises because arterial [HCO_3^-] decreases. This stimulates the peripheral chemoreceptors, and minute ventilation remains augmented. Over the next 12 to 36 hours, blood [HCO_3^-] increases in the kidneys (see Chapter 36), arterial pH returns to a normal value, and minute ventilation returns to a normal level.

elicited, it results in cessation of inspiration by stimulating the off-switch neurons in the medulla. This reflex is inactive during quiet breathing and appears to be most important in newborns. Stimulation of nasal or facial receptors with cold water initiates the **diving reflex.** When this reflex is elicited, **apnea,** or cessation of breathing, and bradycardia occur. This reflex protects individuals from aspirating water in the initial stages of drowning. Activation of receptors in the nose is responsible for the **sneeze reflex.**

The **aspiration** or **sniff reflex** can be elicited by stimulation of mechanical receptors in the nasopharynx and pharynx. This is a strong, short-duration inspiratory effort that brings material from the nasopharynx to the pharynx, where it can be swallowed or expectorated. The mechanical receptors responsible for the sniff reflex are also important in swallowing by inhibiting respiration and causing laryngeal closure. For anatomical reasons, only newborns can breathe and swallow simultaneously, which allows more rapid ingestion of nutrients.

The larynx contains both superficial and deep receptors. Activation of the superficial receptors results in apnea, cough, and expiratory movements that protect the lower respiratory tract from aspirating foreign material. The deep receptors are located in the skeletal muscles of the larynx, and they control muscle fiber activation, as in other skeletal muscles.

Sensory Receptors and Reflexes

Three major types of sensory receptors located in the tracheobronchial tree respond to a variety of different stimuli, and those responses result in changes in the lung's mechanical properties, alterations in the respiratory pattern, and the development of respiratory symptoms. Inhaled dust, noxious gases, and cigarette smoke stimulate **irritant receptors** in the trachea and large airways that transmit information through myelinated vagal afferent fibers. Stimulation of these receptors results in an increase in airway resistance, reflex apnea, and coughing. These receptors are also known as **rapidly adapting pulmonary stretch receptors. Slowly adapting pulmonary stretch receptors** respond to mechanical stimulation, and they are activated by lung inflation. They also transmit information through myelinated, vagal afferent fibers. The increase in lung volume in people with COPD stimulates these pulmonary stretch receptors and delays the onset of the next inspiratory effort. This explains the long, slow expiratory effort in affected individuals, and it is essential to minimize dynamic, expiratory airway compression.

In addition, specialized sensory receptors located in the lung parenchyma respond to chemical or mechanical stimulation in the lung interstitium. These receptors are called **juxta-alveolar,** (or **J**) **receptors.** They transmit their afferent input through unmyelinated, vagal C fibers. They may be responsible for the sensation of **dyspnea** (abnormal shortness of breath) and the rapid, shallow ventilatory patterns that occur in interstitial lung edema and some inflammatory lung states.

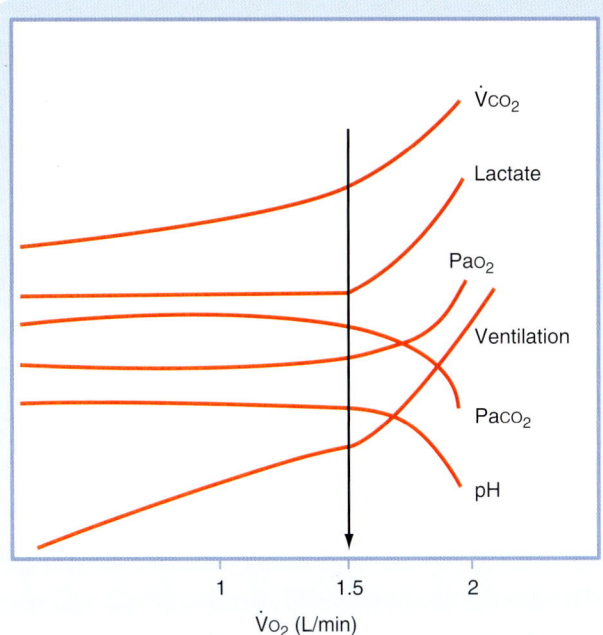

• **Fig. 25.7** Oxygen Consumption ($\dot{V}O_2$) as a Function of the Metabolic Changes That Occur During Exercise. The anaerobic threshold *(arrow)* is the point at which the illustrated variables change and is due to lactic acidosis. $PaCO_2$, partial pressure of arterial carbon dioxide; PaO_2, partial pressure of arterial oxygen; $\dot{V}CO_2$, carbon dioxide consumption.

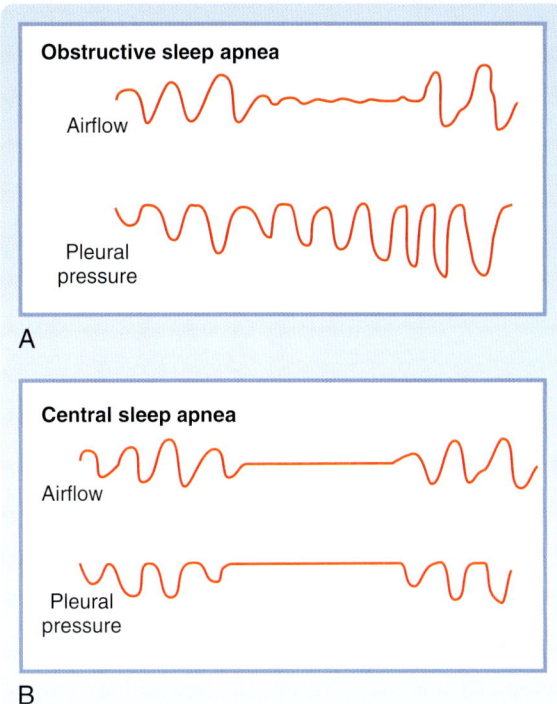

• **Fig. 25.8** The Two Main Categories of Sleep Apnea. **A,** Obstructive sleep apnea, the pleural pressure oscillations increase as CO_2 level rises. This indicates that resistance to airflow is very high as a result of upper airway obstruction. **B,** Central sleep apnea is characterized by no attempt to breathe, as demonstrated by no oscillations in pleural pressure.

Somatic receptors are also located in the intercostal muscles, rib joints, accessory muscles of respiration, and tendons, and they respond to changes in the length and tension of the respiratory muscles. Although they do not directly control respiration, they do provide information about lung volume and play a role in terminating inspiration. They are especially important in individuals with increased airway resistance and decreased pulmonary compliance because they can augment muscle force within the same breath. Somatic receptors also help minimize the chest wall distortion during inspiration in newborns, who have very compliant rib cages.

Exercise

The ability to exercise depends on the capacity of the cardiac and respiratory systems to increase delivery of O_2 to tissues and remove CO_2 from the body. Ventilation increases immediately when exercise begins, and this increase in minute ventilation closely matches the increases in O_2 consumption and CO_2 production that accompany exercise (Fig. 25.7). Ventilation is linearly related to both CO_2 production and O_2 consumption at low to moderate levels (see Fig. 25.7). During maximal exercise, a physically fit individual can achieve an O_2 consumption of 4 L/minute with a minute ventilation volume of 120 L/minute, which is almost 15 times the resting level.

Exercise is remarkable because of the lack of significant changes in blood gases. Except at maximal exertion, changes in $PaCO_2$ and PaO_2 are minimal during exercise. Arterial pH

remains within normal values during moderate exercise. During strenuous exercise, arterial pH begins to fall as lactic acid is liberated from muscles during anaerobic metabolism. This decrease in arterial pH stimulates ventilation that is out of proportion to the level of exercise. The level of exercise at which sustained metabolic (lactic) acidosis begins is called the **anaerobic threshold** (see Fig. 25.7).

Abnormalities in the Control of Breathing

Changes in the ventilatory pattern can occur for both primary and secondary reasons. During sleep, approximately one third of normal individuals have brief episodes of apnea or hypoventilation that have no significant effects on PaO_2 or $PaCO_2$. The apnea usually lasts less than 10 seconds, and it occurs in the lighter stages of slow-wave and rapid eye movement (REM) sleep. In **sleep apnea** syndromes, the duration of apnea is abnormally prolonged, and it changes PaO_2 and $PaCO_2$. There are two major categories of sleep apnea (Fig. 25.8). The first, **obstructive sleep apnea** (OSA), is the most common of the sleep apnea syndromes, and it occurs when the upper airway (generally the hypopharynx) closes during inspiration. Although the process is similar to what happens during snoring, it is more severe, inasmuch as it obstructs the airway and causes cessation of airflow.

The second sleep apnea syndrome is **central sleep apnea.** This variant of apnea occurs when the ventilatory drive to

IN THE CLINIC

The clinical histories of individuals with OSA are very similar. A spouse usually reports that the affected individual snores. The snoring becomes louder and louder and then stops while the individual continues to make vigorous respiratory efforts (see Fig. 25.8). The individual then awakens, falls back to sleep, and continues the same process repetitively throughout the night. Individuals with OSA awaken when the arterial hypoxemia and hypercapnia stimulate both peripheral and central chemoreceptors. Respiration is restored briefly before the next apneic event occurs. Individuals with OSA can have hundreds of these events each night that interrupt sleep. Complications of OSA include sleep deprivation, polycythemia, right-sided cardiac failure (cor pulmonale), and pulmonary hypertension secondary to the recurrent, hypoxic events. OSA is common in individuals with obesity and in those with excessive compliance of the hypopharynx, upper airway edema, and structural abnormalities of the upper airway.

the respiratory motor neurons decreases. Individuals with central sleep apnea have repeated episodes of apnea, during which time they make no respiratory effort, every night (see Fig. 25.8). The degree of hypercapnia and hypoxemia in individuals with central sleep apnea is less than that in individuals with OSA, but the same complications (e.g., polycythemia) can occur when central sleep apnea is recurrent and severe.

Cheyne-Stokes ventilation is another abnormality of breathing that is characterized by varying tidal volume and ventilatory frequency (Fig. 25.9). After a period of apnea, tidal volume and respiratory frequency increase progressively over several breaths, and then they progressively decrease until apnea recurs. This irregular breathing pattern is seen in some individuals with central nervous system diseases, head trauma, and increased intracranial pressure. It is also present on occasion in normal individuals during sleep at high altitude. The mechanism underlying Cheyne-Stokes respiration is not known. In some individuals, it appears to be due to slow blood flow in the brain in association with periods of overshooting and undershooting ventilatory effort in response to changes in P_{CO_2}.

IN THE CLINIC

Central alveolar hypoventilation, also known as *Ondine's curse,* is a rare disease in which voluntary breathing is intact but abnormalities in automaticity exist. It is the most severe of the central sleep apnea syndromes. As a result, people with central alveolar hypoventilation can breathe as long as they do not fall asleep. For these individuals, mechanical ventilation or, more recently, bilateral diaphragmatic pacing (similar to a cardiac pacemaker) can be lifesaving.

Apneustic breathing is another abnormal breathing pattern that is characterized by sustained periods of inspiration separated by brief periods of exhalation (Fig. 25.10, *C*).

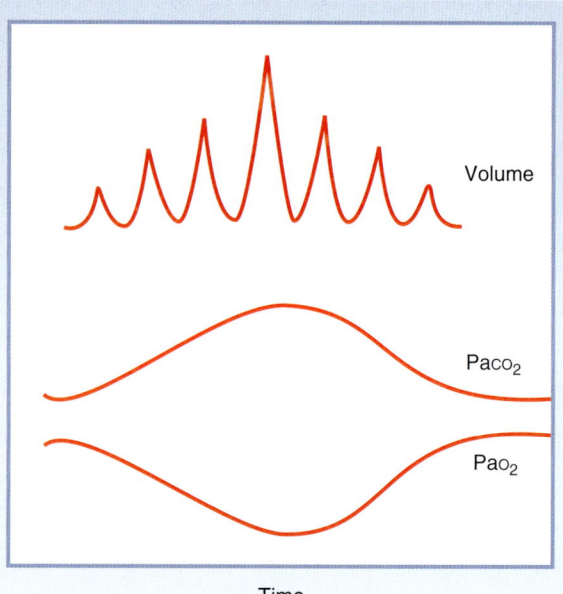

● **Fig. 25.9** In Cheyne-Stokes Breathing, Tidal Volume and, as a Consequence, Arterial Blood Gas Levels Wax and Wane. In general, Cheyne-Stokes breathing is a sign of vasomotor instability, particularly low cardiac output. Pa_{CO_2}, partial pressure of arterial carbon dioxide; Pa_{O_2}, partial pressure of arterial oxygen.

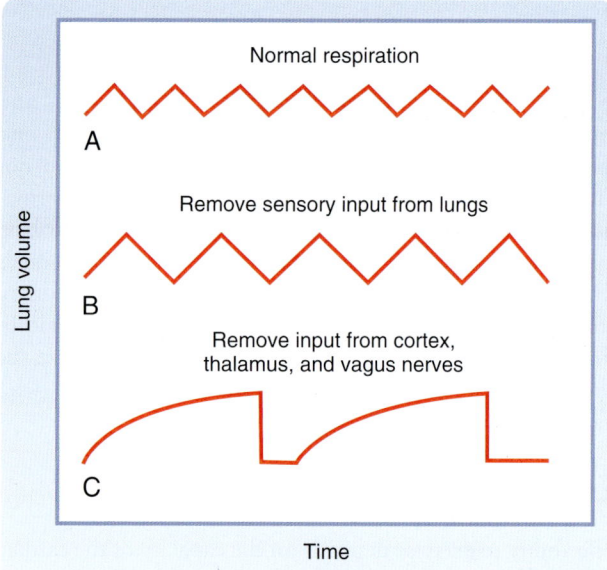

● **Fig. 25.10** Some Patterns of Breathing. **A,** Normal rate of breathing is in the range of 12 to 20 breaths per minute. **B,** When sensory input is removed from various lung receptors (mainly stretch), each breathing cycle is lengthened and tidal volume is increased, so that alveolar ventilation is not significantly affected. **C,** When input from the cerebral cortex and thalamus is also eliminated, together with vagal blockade, the result is prolonged inspiratory activity broken after several seconds by brief expirations (apneusis).

The mechanism underlying this ventilatory pattern appears to be a loss of inspiratory-inhibitory activities that results in augmentation of the inspiratory drive. The pattern sometimes occurs in individuals with central nervous system injury.

IN THE CLINIC

Sudden infant death syndrome (SIDS) is the most common cause of death in infants in the first year of life after the perinatal period. Although the cause of SIDS is not known, abnormalities in ventilatory control, particularly in CO_2 responsiveness, have been implicated. Placing infants on their backs to sleep (which reduces the potential for CO_2 rebreathing) has dramatically decreased (but not eliminated) the rate of death from this syndrome.

Key Points

1. Ventilatory control is composed of the respiratory control center, central chemoreceptors, peripheral chemoreceptors, and pulmonary mechanoreceptors/sensory nerves. $PaCO_2$ is the major factor that influences ventilation.

2. The respiratory control center is composed of the dorsal respiratory group and the ventral respiratory group. Rhythmic breathing depends on a continuous (tonic) inspiratory drive from the dorsal respiratory group and on intermittent (phasic) expiratory input from the cerebrum, thalamus, cranial nerves, and ascending spinal cord sensory tracts. The peripheral and central chemoreceptors respond to changes in $PaCO_2$ and pH. The peripheral chemoreceptors (carotid and aortic bodies) are the only chemoreceptors that respond to changes in PaO_2.

3. Acute hypoxia and chronic hypoxia affect breathing differently because the slow adjustments in CSF [H^+] in chronic hypoxia alter sensitivity to CO_2.

4. Irritant receptors protect the lower respiratory tract from particles, chemical vapors, and physical factors, primarily by inducing cough. C fiber J receptors in the terminal respiratory units are stimulated by distortion of the alveolar walls (by lung congestion or edema).

5. The two most important clinical abnormalities of breathing are obstructive and central sleep apnea.

6. PaO_2, $PaCO_2$, and pH remain within normal limits during moderate exercise; however, during strenuous exercise, pH falls, which stimulates ventilation, whereas PaO_2 and $PaCO_2$ remain relatively normal.

Additional Readings

Calverley PMA, Koulouris NG. Flow limitation and dynamic hyperinflation: key concepts in modern respiratory physiology. *Eur Respir J.* 2005;25:186-199.

Canning BJ. Afferent nerves regulating the cough reflex: mechanisms and mediators of cough in disease. *Otolaryngol Clin N Am.* 2010;43:15-25.

Leff AR, Schumacker PT. *Respiratory Physiology: Basics and Applications.* Philadelphia: WB Saunders; 1993.

Nishino T. Dyspnoea underlying mechanisms and treatment. *Br J Anaesth.* 2011;106:463-474.

Otis AB. A perspective of respiratory mechanics. *J Appl Physiol.* 1983;54:1183-1187.

Sheel AW, Romer LM. Ventilation and respiratory mechanics. *Compr Physiol.* 2012;2:1093-1142.

Wasserman K, Beaver WL, Whipp BI. Gas exchange theory and the lactic acidosis (anaerobic) threshold. *Circulation.* 1990;81(1 suppl):1114-1130.

Wine JJ. Parasympathetic control of airway submucosal glands: central reflexes and the airway intrinsic nervous system. *Auton Neurosci.* 2007;133:35-54.

26

Nonphysiological Functions of the Lung: Host Defense and Metabolism

In addition to their primary function of gas exchange, the lungs act as a primary barrier between the outside world and the inside of the body, with host defense functions. They are also active organs in the metabolism of xenobiotic and endogenous compounds.

Host Defense

To cope with the inhalation of foreign substances, the respiratory system and, in particular, the conducting airways have developed unique structural features: the mucociliary clearance system and specialized adaptive and innate immune response mechanisms.

Mucociliary Clearance System

The mucociliary clearance system protects the conducting airways by trapping and removing inhaled pathogenic viruses and bacteria, in addition to nontoxic and toxic particulates (e.g., pollen, ash, mineral dust, mold spores, and organic particles), from the lungs. These particulates are

inhaled with each breath and must be removed. The three major components of the mucociliary clearance system are two fluid layers, referred to as the *sol* (**periciliary fluid**) and *gel* (**mucus**) layers, and **cilia,** which are positioned on the surface of bronchial epithelial cells (Fig. 26.1). Inhaled material is trapped on the viscoelastic (sticky) mucus layer, whereas the watery periciliary fluid allows the cilia to move freely and establish an upward flow to clear particulates from the lung. Effective clearance requires both ciliary activity and the appropriate balance of periciliary fluid and mucus.

Periciliary Fluid Layer

The periciliary fluid layer is composed of nonviscous serous fluid, which is produced by the pseudostratified ciliated columnar epithelial cells that line the airways. These cells have the ability to either **secrete** fluid, a process that is mediated by activation of cystic fibrosis transmembrane regulator (CFTR) chloride (Cl^-) ion channels (**Na^+ secretion** follows passively between cells across the tight junctions) or **reabsorb** fluid, a process that is mediated by activation of epithelial sodium channels (ENaC; **Cl^- absorption** follows passively between cells across the tight junctions). NaCl secretion or reabsorption temporarily establishes an osmotic gradient across the pseudostratified epithelium, which provides the driving force for passive water movement. The balance between CFTR-mediated Cl^- secretion and ENaC-mediated Na^+ absorption is regulated by a variety of hormones and determines the volume of the periciliary fluid, which in the healthy lung is 5 to 6 µm deep, a level that is optimal for rhythmic beating of the cilia and mucociliary clearance.

Mucus Layer

The mucus layer lies on top of the periciliary fluid layer and is composed of a complex mixture of macromolecules and electrolytes. Because the mucus layer is in direct contact with air, it entraps inhaled substances, including pathogens. The mucus layer is predominantly water (95% to 97%), 5 to 10 µm thick and exists as a discontinuous blanket (i.e., islands of mucus). Mucus has low viscosity and high elasticity and is composed of glycoproteins with groups of oligosaccharides attached to a protein backbone. Healthy

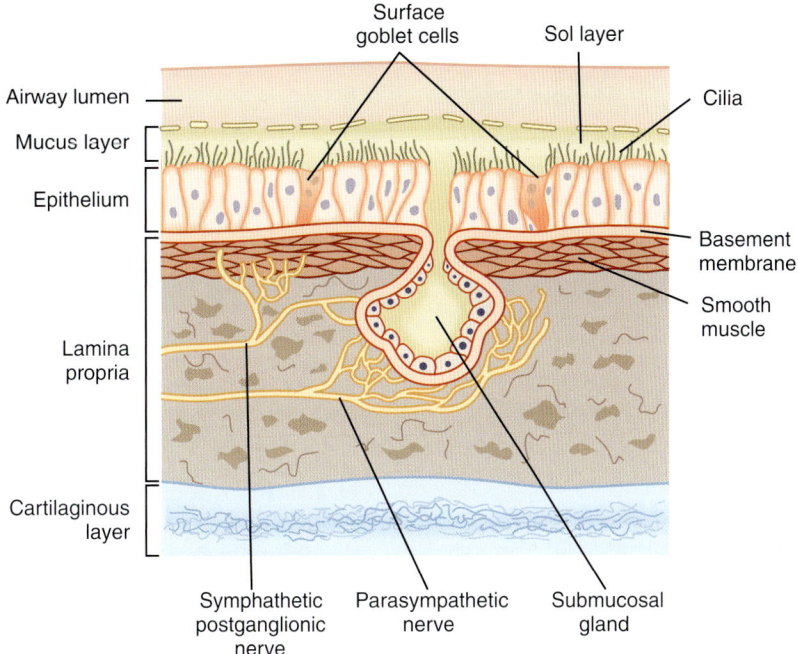

• **Fig. 26.1** Overview of the Epithelial Lining and Innervation of the Tracheobronchial Tree. The cilia of the epithelial cell reside in the periciliary fluid layer, and the mucus layer is on top. Interspersed between the ciliated epithelial cells are surface secretory (goblet) cells and submucosal glands. Sympathetic and parasympathetic nerve fibers descend into the submucosal glands and smooth muscles.

 IN THE CLINIC

Cystic fibrosis (CF) is the most common lethal inherited disease among white people. It is an autosomal recessive disease caused by mutations in the *CFTR* gene. It is characterized by chronic bacterial lung infection, progressive decline in lung function, and premature death at an average age of 38 years. More than 1000 mutations of the *CFTR* gene have been described, but 70% of affected individuals have a deletion of phenylalanine at codon 508 (ΔF508-CFTR) in at least one allele. This mutation results in a lack of Cl^- secretion and an increase in ENaC-mediated Na^+ reabsorption, which in turn results in a reduction in the volume of the periciliary fluid.

Detailed study of the different *CFTR* mutations has resulted in an understanding of various disease-related phenotypes, some of which are associated with milder disease and some with more severe disease. Since the 1980s, research findings have elucidated how many of the most common mutations in the *CFTR* gene cause CF, and this has led to the development of drugs that target specific mutations and reverse the

progressive reduction in lung function. For example, in one mutation (G551D-CFTR, which affects ≈5% of patients with CF), the CFTR Cl^- channel reaches the plasma membrane of airway epithelial cells but does not secrete Cl^-. Through precision medicine, a drug, ivacaftor (Kalydeco), has been found to stimulate Cl^- secretion via the G551D-CFTR, thereby improving lung function and decreasing the rate of disease progression. When the alleles are homozygous for the ΔF508-CFTR mutation (which affects ≈50% of patients with CF), the CFTR Cl^- channel does not reach the plasma membrane of airway epithelial cells.

In 2015, the U.S. Food and Drug Administration approved a combination drug therapy, lumacaftor/ivacaftor (Orkambi), that has been shown to correct the gene defect, increase the amount of ΔF508-CFTR in the plasma membrane, and improve Cl^- transport. Clinically, both ivacaftor and lumacaftor/ivacaftor have been shown to improve lung function significantly and to decrease the rate of decline in lung function.

individuals produce approximately 100 mL of mucus each day. Four cell types contribute to the quantity and composition of the mucus layer: **goblet cells** and **Clara cells** within the tracheobronchial epithelium, and **mucous cells** and **serous cells** within the tracheobronchial submucosal glands. Goblet cells, also referred to as *surface secretory cells,* represent approximately 15% to 20% of the tracheobronchial epithelium, and are found in the tracheobronchial tree up to the 12th division. In many respiratory diseases, goblet cells appear further down the tracheobronchial tree; thus the smaller airways are more susceptible to obstruction

by mucus plugging. Goblet cells secrete neutral and acidic glycoproteins rich in sialic acid in response to chemical stimuli. In the presence of infection or cigarette smoke or in patients with chronic bronchitis, goblet cells can increase in size and number, extend above the 12th division of the tracheobronchial tree, and secrete copious amounts of mucus. Injury and infection increase the viscosity of the mucus secreted by goblet cells, which reduces mucociliary clearance of inhaled particles and pathogens.

Submucosal tracheobronchial glands are present wherever there is cartilage in the upper regions of the conducting

TABLE 26.1	Properties of Serous and Mucous Cells in Submucosal Gland	
Property	**Serous Cells**	**Mucous Cells**
Location	Most distal	Middle to distal
Granules	Small, electron-dense	Large, electron-lucent
Glycoproteins	Neutral Lysozyme, lactoferrin	Acidic
Hormones	α-Adrenergic > β-Adrenergic	β-Adrenergic > α-Adrenergic
Receptors	Muscarinic	Muscarinic
Degranulation	α-Adrenergic Cholinergic Substance P	β-Adrenergic Cholinergic

airways, and they secrete water, ions, and mucus into the airway lumen through a ciliated duct (see Fig. 26.1). Although both mucous and serous cells secret mucus, their cellular structure and mucus composition are distinctly different (Table 26.1). In several lung diseases, including chronic bronchitis, the number and size of submucosal glands are increased, which leads to increases in mucus production, alterations in chemical composition of mucus (i.e., increased viscosity and decreased elasticity), and the formation of mucus plugs that cause airway obstruction. Mucus secretion from submucosal tracheobronchial glands is stimulated by parasympathetic (cholinergic) compounds such as acetylcholine and substance P and inhibited by sympathetic (adrenergic) compounds such as norepinephrine and vasoactive intestinal polypeptide. Local inflammatory mediators such as histamine and arachidonic acid metabolites also stimulate mucus production.

Clara cells, located in the epithelium of bronchioles, also contribute to the composition of mucus through secretion of a nonmucinous material containing carbohydrates and proteins. These cells play a role in bronchial regeneration after injury.

Ciliated Cells and Cilia

As noted previously, the respiratory tract to the level of the bronchioles is lined by a pseudostratified, ciliated columnar epithelium (see Fig. 26.1). These cells maintain the level of the periciliary fluid in which cilia and the mucociliary transport system function. Mucus and inhaled particles are removed from the airways by the rhythmic beating of the cilia. There are approximately 250 cilia per airway epithelial cell, and each is 2 to 5 μm in length. Cilia are composed of nine microtubular doublets that surround two central microtubules held together by dynein arms, nexin links, and spokes. The central microtubule doublet contains an adenosine triposphatase (ATPase) that is responsible for the contractile beat of the cilium. Cilia beat with a coordinated oscillation in a characteristic, biphasic, and wave-like rhythm called **metachronism.** They beat at approximately

AT THE CELLULAR LEVEL

Sputum is expectorated mucus. However, in addition to mucus, sputum contains serum proteins, lipids, electrolytes, Ca^{++}, DNA from degenerated white blood cells (collectively known as *bronchial secretions*), and extrabronchial secretions; including nasal, oral, lingual, pharyngeal, and salivary secretions. The color of sputum is more closely correlated with the amount of time that it has been present in the lower respiratory tract than with the presence of infection. Although not precisely identifiable with disease diagnosis, the color of sputum can be informative in helping lead to a diagnosis and stage of disease. Mucus has many colors: white, yellow, green, red, pink, brown, gray, and black. The coloration is commonly due to the type of cell present in the airways (inflammatory cells, such as neutrophils or eosinophils, or red blood cells) and how long they have been there. Clear or cloudy **white** thin mucus is considered normal; however, if amounts and thickness are increased, it may represent an early sign of infection. Thick white mucus can be the only identifiable feature of gastroesophageal reflux disease caused by gastric acid reflux into the airways. **Yellow** and **green** coloration of mucus is due to the presence and breakdown of neutrophils and eosinophils in infectious and allergic diseases. Yellow is typically associated with more acute disease (infection, allergy) and green usually indicates a more chronic stage with the presence of bacteria (chronic bronchitis, bronchiectasis, cystic fibrosis, and lung abscess). **Red** mucus indicates the presence of red blood cells in the airways and is associated with pneumococcal pneumonia, lung cancer, tuberculosis, and pulmonary emboli. **Pink** mucus is typically associated with the breakdown of eosinophils in individuals with allergies. Gray, brown, and black mucus is often associated with cigarette or marijuana smoking, cocaine use, air pollution (workplace environment, such as coal mines), and old blood.

1000 strokes per minute, with a power forward stroke and a slow return or recovery stroke. During their power forward stroke, the tips of the cilia extend upward into the viscous mucus layer and thereby move it and the entrapped particles. On the reverse beat, the cilia release the mucus and withdraw completely into the sol layer. Cilia in the nasopharynx beat in the direction that propels the mucus into the pharynx, whereas cilia in the trachea propel mucus upward toward the pharynx, where it is swallowed.

Particle Deposition and Clearance

In general, deposition of particles in the lung depends on the particle's size, density, and shape; the distance over which it has to travel; airflow speed; and the relative humidity of the air. The four major mechanisms for deposition are **impaction, sedimentation, interception,** and **brownian movement.** Particle characteristics and properties, which influence the mechanism of deposition, are listed in Table 26.2. In general, particles larger than 10 μm are deposited by **impaction** in the nasal passages and do not penetrate into the lower respiratory tract. Particles 2 to 10 μm in size

TABLE 26.2	Particle Deposition Characteristics			
Method of Deposition	**Particle Size (μm)**	**Deposition Site**	**Airflow**	**Determining Factors**
Impaction	>10	Nasal passages	Fast	Size, density
	2 to 10	Nasal pharynx Trachea Bronchi	Fast	Size, density
Sedimentation	0.2 to 2.0	Distal airways	Slow	Size, density, diameter
Interception	NA	NA	Slow	Shape (elongated)
Brownian movement	<0.2	Smaller airways Alveoli	Slow	Diffusion coefficient (not density)

NA, not applicable.

are deposited in the lower respiratory tract predominantly by inertial impaction at points of turbulent airflow (i.e., nasopharynx, trachea, and bronchi) and at airway bifurcations because their tendency to move in a straight direction prevents them from changing directions rapidly. In more distal areas, where airflow is slower, smaller particles (0.2 to 2 μm) are deposited on the surface by **sedimentation** as a result of gravity. For substances with elongated shapes (i.e., asbestos, silica), the mechanism of deposition is **interception**. The elongated particle's center of gravity is compatible with the flow of air; however, when the distal tip of the particulate comes in contact with a cell or mucus layer, deposition is facilitated. Particles smaller than 0.2 μm are deposited in the smaller airways and alveoli and are influenced mainly by their diffusion coefficient and **brownian motion**. Unlike the deposition of larger particles in the upper airways, particle density does not influence diffusion of these smaller particles, and deposition is enhanced with decreased size. These smaller particles come in contact with the alveolar epithelium, where cilia and the mucociliary transport system do not exist; thus they are removed by the phagocytic activity of alveolar macrophages or absorption into the interstitium with subsequent clearance by lymphatic drainage. Although most alveolar macrophages are adjacent to the epithelium of the alveolus, some are located in the terminal airways and interstitial space.

In the conducting airways, the mucociliary clearance system transports deposited particles from the terminal bronchioles to the major airways, where they are coughed up and either expectorated or swallowed. Deposited particles can be removed in a matter of minutes to hours. In the trachea and main bronchi, the rate of particle clearance is 5 to 20 μm/minute, but it is slower in the bronchioles (0.5 to 1 μm/minute). In general, the longer an inhaled material remains in the airways, the greater is the probability that the material will cause lung damage. The region from the terminal bronchioles to the alveoli is devoid of ciliated cells and is considered the "Achilles heel" in what is otherwise a highly effective system. The relatively slow rate of particle clearance in this area, which is mediated by macrophages, renders it the most common location for many occupational lung diseases.

IN THE CLINIC

In some lung diseases—for example, those caused by inhalation of silica particles **(silicosis)** or coal dust particles **(pneumoconiosis,** the "black lung" disease of coal miners)—alveolar macrophages phagocytize the particles but are unable to destroy them, and the macrophages eventually die. Alveolar macrophages have localized and concentrated the particles in the "Achilles heel" region of the lung. These particles are not removed via mucociliary clearance and eventually enter the lung interstitium. The ensuing inflammatory response leads to a granulomatous-like lesion with fibrosis, a restrictive lung disease. Silicosis and pneumoconiosis are classical examples of diseases originating through environmental workplace exposure. Increased awareness of the cause of these diseases and improved workplace environments have led to reduction in the incidence of these types of lung diseases.

Mucosal Immune System: Adaptive and Innate Immunity

Mucosal Immune System

In nonmucosal tissues (e.g., spleen, liver, kidney), the body's primary defense is the classical proinflammatory adaptive, antigen-specific immune response orchestrated in the local draining lymph nodes with afferent and efferent lymph flow. The major adaptive immune cells in the systemic immune system are T lymphocytes with **αβ T cell receptors (TCRαβ T cells)** for specific antigen recognition and plasma B cells that synthesize immunoglobulin M (IgM) and immunoglobulin G (IgG) complement–binding antibodies, which can induce inflammation. However, mucosal tissues (i.e., those of the respiratory, gastrointestinal, and urinary systems, as well as the eyes, nose, throat, and mouth) must constantly discriminate between what is harmful and what is not, and although inflammation is protective, it usually disrupts the normal physiologic processes and is not desirable unless absolutely necessary. Accordingly, mucosal tissues have developed specialized "noninflammatory" defense mechanisms, which form the basis of the **mucosal immune system** and can function independently of the systemic immune system. The mucosal immune system contains both specialized **innate lymphoid cells** (macrophages, natural killer cells, dendritic cells [DCs]) and **adaptive** T lymphocytes with **γδ T cell receptors (TCRγδ T cells)** and plasma B cells that synthesize immunoglobulin A (IgA), a nonclassical complement-binding antibody. These innate and unique adaptive responses can prevent or limit responses to foreign nonpathological agents while eliminating pathological agents/substances with little or no inflammation. In addition, if this front-line defense system fails or is bypassed, the lungs do have a classical adaptive immune response system in which lymphatic drainage is via the mediastinal lymph node located in the upper region of the thoracic cavity adjacent to the main left-right lung bifurcation.

A distinctive feature of the mucosal immune system is that antigens are processed through **lymphoid aggregates** rather than through a true lymph node. Unlike a true lymph node, which has afferent and efferent lymph flow, lymphoid aggregates have only afferent drainage of material into the aggregate without efferent flow. This lymphoid network is commonly referred to as *mucosa-associated lymphoid tissue* (MALT); in the gastrointestinal tract, it is referred to as *gut-associated lymphoid tissue* (GALT); and in the lungs, it is known as **bronchus-associated lymphoid tissue (BALT;** Fig. 26.2). BALT is present in the conducting airways, where the epithelium is composed mainly of ciliated cells with consistent mucus flow. However, an interesting feature of BALT is that its airway epithelium is not ciliated; it is referred to as a *lymphoepithelium,* which creates a break in the mucus flow (like a drain) and allows the substances/particulates to be processed in the lymphoid aggregate (or follicle).

It appears that there is communication between mucosal tissues and that sensitization via one organ is transposed to all MALT/BALT tissues via a lymphatic-like drainage network. The systemic immune system and MALT/BALT may work independently of each other, and the fact that one is sensitized may not be true of the other. This may serve as a defense mechanism in limiting sensitization only to mucosal tissue. Lymphocytes also cluster in smaller numbers and density in what is referred to as **tertiary ectopic lymphoid tissue (TELT),** in which they can also process antigens (see Fig. 26.2, *A*). Another prominent feature is a diffuse submucosal and intraepithelial network of **solitary lymphocytes** and innate lymphoid cells scattered throughout the respiratory tract. Because inhaled particles are broadly dispersed throughout the respiratory tract, each type of lymphoid cell and tissue (BALT, TELT, and solitary lymphocytes) play important and unique roles in the overall defense of the lungs.

Specialized Adaptive Lymphoid Cells
Plasma Cells Producing Immunoglobulin A

One of the specialized features of MALT, GALT, and BALT is a unique antibody system in which specialized features

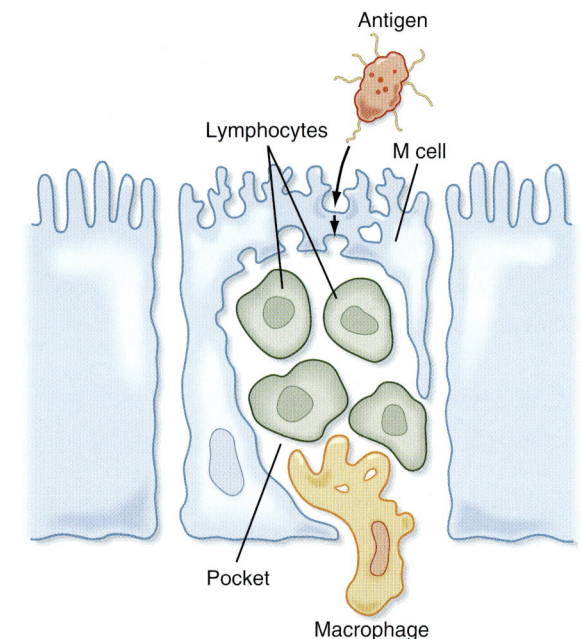

A

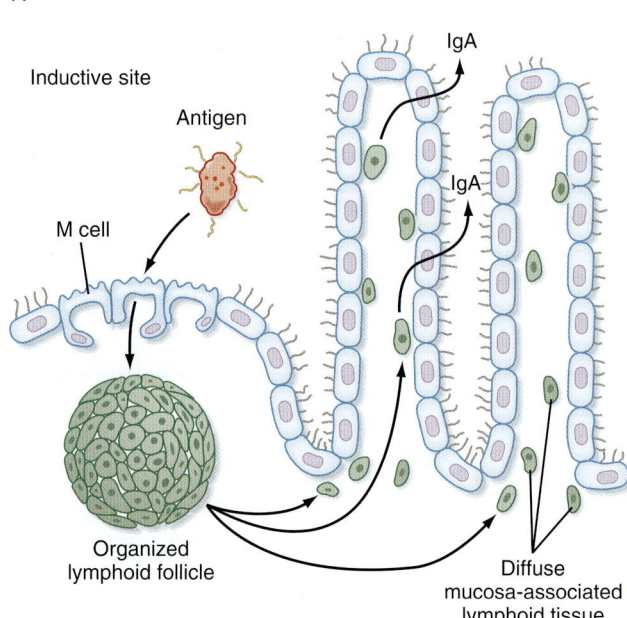

B

• **Fig. 26.2** Representation of Bronchus-Associated Lymphoid Tissue (BALT)/Tertiary Ectopic Lymphoid Tissue (TELT), M Cells, and Immunoglobulin a (IgA) Synthesis. **A,** M cells located in mucosal epithelium endocytose antigen in the lumen and transport it for processing to loosely organized submucosal pockets of lymphoid cells, predominantly lymphocytes and macrophages (TELT). **B,** Diagram of a mucous membrane, showing secretion of IgA antibodies in response to antigen endocytosed by M cells. Activated B cells differentiate into IgA-producing plasma cells (lymphocytes) and migrate from the densely organized lymphoid follicle (in bronchus-associated lymphoid tissue [BALT]) to the nearby submucosa, where they secret IgA.

of the IgA antibody (non–complement-fixing, J chain for transport, and dimeric structure for stability in the airway lumen) are used. In submucosal areas, plasma cells synthesize and secrete IgA, which migrates to the submucosal surface of epithelial cells, where it binds to a surface protein

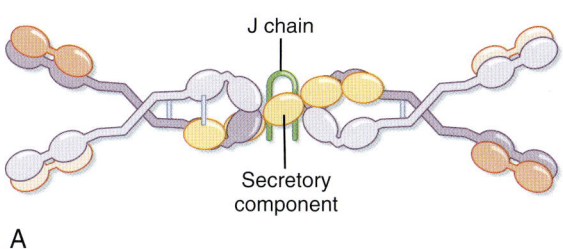

A

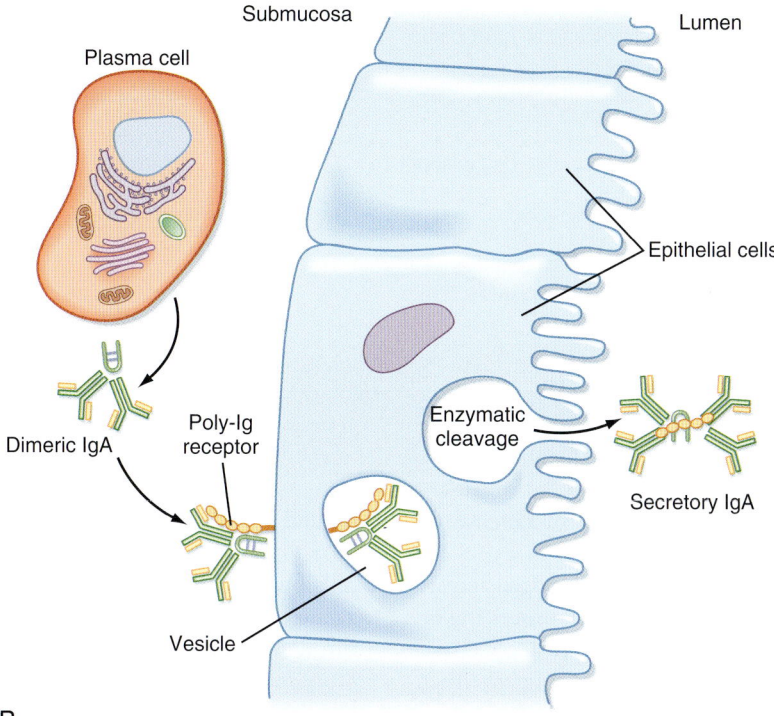

B

• **Fig. 26.3** Structure and Formation of Secretory Immunoglobulin a (IgA). **A,** Secretory IgA consists of at least two IgA molecules that are covalently linked via J chain and covalently associated with the secretory component. The secretory component contains five immunoglobulin-like domains and is linked to dimeric IgA through binding to an IgA heavy chain. **B,** Secretory IgA is formed during transport through epithelial cells. Poly-Ig, polymeric immunoglobulin.

receptor, polymeric immunoglobulin (poly-Ig; Fig. 26.3). The poly-Ig receptor aids in the pinocytosis of IgA into the epithelial cell and its eventual secretion into the airway lumen. During exocytosis of the IgA complex, the poly-Ig is enzymatically cleaved, and a portion of it, the secretory piece, remains associated with the complex. The secretory piece stays attached to the IgA complex in the airway and helps protect it from proteolytic cleavage. The IgA-antigen immune complex does not bind complement in the same classical manner as do other immune complexes, and thus its proinflammatory properties are limited. The IgA-antibody system is very effective in binding particulates and viruses to form a large complex, which promotes its removal via the mucociliary clearance system, before they invade epithelial cells.

T Lymphocytes With γδ T Cell Receptors

Most classical adaptive immune T lymphocytes are CD3⁺ cells with TCRs that are composed of α and β chains (TCRαβ T cells). These cells mature in the thymus and egress mostly to **lymph nodes and the spleen.** The classical activation of TCRαβ cells requires antigen processing/presentation, typically via the major histocompatibility complex, in a DC, usually to induce an inflammatory response. CD3⁺ T lymphocytes with TCRs expressing γ and δ chains (TCRγδ T cells) also mature in the thymus, but the majority of these cells egress to mucosal tissues (i.e., lung, intestines, and skin) and represent only a minority of T cells in the peripheral blood and systemic lymphoid tissues. TCRγδ T cells, often referred to as *intraepithelial lymphocytes,* preferentially localize to submucosal sites and

TABLE 26.3	**Innate and Adaptive Immune Cells in the Respiratory System**	
Cell Type	**Location**	**Function**
Innate Lymphoid Cells		
T lymphocytes with TCRγδ chains	Intraepithelial, submucosa	Selective antigen recognition Immunoregulation (decrease IgE)
Dendritic cells	Diffuse in the lung interstitium	
Conventional		Antigen presentation Immunoregulation (tolerance)
Plasmacytoid		Antiviral
Alveolar macrophages	Alveoli and alveolar ducts	Phagocytosis
Macrophages	Diffuse in the lung interstitium, BALT, and TELT	
M-1 cells		Phagocytosis, proinflammatory cytokines
M-2 cells		Phagocytosis, suppressive cytokines
NK cells	Diffuse in the lung interstitium	Targeted cytotoxicity Immunoregulation (tolerance)
iNKT cells	Diffuse in the lung interstitium	Immunoregulation (cytokines, IL-10, IFN-γ)
Adaptive Immune Cells		
T lymphocytes with TCRαβ chains	Submucosa, BALT, TELT	Specific adaptive immunity Proinflammatory (Th1/Th2 cytokines)
B lymphocytes	Submucosa, BALT, TELT	IgM IgG, IgA, and IgE antibody synthesis

BALT, bronchus-associated lymphoid tissue; IFN-γ, interferon γ; IgA, IgE, IgG, and IgM, immunoglobulins A, E, G, and M; IL-10, interleukin 10; iNKT, invariant natural killer T; NK, natural killer; TCR, T cell receptor; TELT, tertiary ectopic lymphoid tissue; Th1 and Th2, T helper 1 and T helper 2.

epithelium and are considered a first line of defense of epithelial surfaces.

In contrast to TCRαβ T cells, the classical antigen activation of TCRγδ T cells does not require antigen processing or presentation by DCs. Of interest is that TCRγδ T cells have been shown to be capable of responding to antigen either in the manner of an innate cellular response, via pathogen-associated molecular patterns (PAMPs, described later in this chapter), or, in some circumstances, in the manner of a classical TCRαβ T cell response. Whether this variation in responsiveness is due to the presence of subpopulations of TCRγδ T cells or to plasticity (alterations) of the same cell is not clear. Because these cells express this unique duality, they are considered to be a link between adaptive and innate immune responses. Furthermore, TCRγδ cells can respond immediately to antigen and generally do not demonstrate memory. In contrast to the typical TCRαβ T cell proinflammatory response, TCRγδ T cells can provide protection without inducing inflammation. TCRγδ T cells have been shown to expand in response to viral and bacterial infections and synthesize either proinflammatory cytokines (interleukin [IL]–17, interferon [IFN]–γ) or regulatory/suppressive cytokines (transforming growth factor β [TGFβ], lymphocyte-activation gene 3). TCRγδ T cells also suppress the immunoglobulin E (IgE) response to inhaled antigen, further preventing allergen-induced inflammation.

Specialized Innate Lymphoid Cells

Innate lymphoid cells (ILCs) generally fall into four categories, TCRγδ T cells, dendritic cells (DCs), macrophages,

natural killer (NK) cells, and subpopulations of each (Table 26.3). ILCs are distinctively different from adaptive immune cells by their lack of "antigen" specificity for the offending agents and their ability to distinguish self from non-self through recognition of PAMPs present on the pathogens. Since most inhaled substances are nonpathogenic, the body has developed a specialized recognition system to identify harmful pathogenic substances and organisms. Rather than specific antigen recognition, this system enables discrimination of self from non-self material through the recognition of PAMPs on the pathogenic organisms/substances, which are then recognized by a family of receptors on host defense cells (i.e., ILCs) called *pattern recognition receptors* (PRRs). PAMPs are common distinguishing features on many pathogens and are composed of peptidoglycans, lipopolysaccharides, and unmethylated 5'-C-phosphate-G-3' (CpG) DNA, which bind to their corresponding PPRs on ILCs. There are two major classifications for PRRs: **toll-like receptors** (TLRs), which are expressed mainly on cell surfaces, and **nucleotide-oligomerization domain** (NOD) receptors, which are expressed intracellularly in the cytoplasma of ILCs. Both of these pathways activate nuclear factor Kβ (NFKβ), a transcription factor, which stimulates release of proinflammatory cytokines, as well the antiviral type I interferons (IFN-α and IFN-β). The TLRs are a family of transmembrane proteins with different specificities for various pathogens. TLR-2 is specific for lipoteichoic acids associated with gram-positive bacteria, whereas TLR-4 is specific for lipopolysaccharide (endotoxin), a product of gram-negative bacteria. In the lung, bronchial epithelial cells, macrophages, DCs, mast cells, eosinophils, and alveolar type II epithelial cells express both TLR-2 and TLR-4. Other TLRs are specific for viruses: TLR-3 binds to double-stranded RNA viruses, and TLR-7 and TLR-8 bind to single-stranded RNA viruses. NK cells express TLR-3, TLR-7, and TLR-8 and plasmacytoid DCs, eosinophils, and B cells express TLR-7. A variety of phagocytic cells and DCs in the lung and other mucosal tissues also express TLRs. Thus in addition to classical phagocytic cells, bronchial and alveolar epithelial cells play active roles in host defense by means of the PAMP-PRR nexus.

Perhaps the most unique and intriguing aspect of this specialized mucosal defense system is the ability of ILCs to respond to pathogens immediately, as opposed to the days or weeks it takes to mount a classical TCRαβ T cell adaptive immune response with clonal expansion and memory. The fast response, and the lack thereof for innocuous substances, are highly advantageous in the mucosal tissues, which are common sites for parasitic invasion and toxic chemical exposure. In addition, many of the ILCs demonstrate plasticity, meaning that they can be induced by the environment (i.e., cytokines) to alter their functionality to either a proinflammatory or regulatory (suppressive) phentoype.

Macrophages and Dendritic Cells

Both macrophages and DCs originate in the bone marrow but vary somewhat in their differentiation lineage: Macrophages develop through the common myeloid-granulocyte/macrophage progenitor linage with other granulocytic cells (basophils, mast cells, eosinophils, neutrophils), and DCs develop through either the common lymphoid progenitor or the common myeloid progenitor lineages. Macrophages and DCs are the first nonepithelial cells to contact and respond to a foreign substance. In the lungs, there are three major types of macrophages (alveolar, M-1 [proinflammatory], and M-2 [regulatory]) and two types of DCs (conventional and plasmacytoid).

Macrophages

If the inhaled foreign material stays within the airspace in the lower respiratory system (alveolar ducts and alveoli), it will probably be phagocytized by alveolar macrophages. However, if the foreign material/organism penetrates and reaches submucosal areas, it will come into contact with M-1 or M-2 macrophages and conventional or plasmacytoid DCs. Alveolar macrophages are found mostly in the alveolus adjacent to the epithelium and less frequently in the terminal airways and interstitial space. They migrate freely throughout the alveolar spaces and serve as a first line of defense in the terminal bronchioles and alveoli. They phagocytize foreign particles and substances, as well as surfactant and cellular debris from dead cells. For a particle/organism that is phagocytized by a macrophage, the major mechanisms of destruction include formation of O_2 radicals, enzyme activity, and halogen derivatives within lysosomes.

The phagocytic activity of the macrophage inhibits the binding of particulates to the alveolar epithelium and their subsequent penetration into the interstitium. The alveolar macrophage also transports engulfed particles to ciliated regions of the mucociliary transport system for elimination and thus provides an important link between the alveolar spaces, the postterminal bronchiole "Achilles heel" region, and the mucociliary clearance system.

In addition, alveolar macrophages and M-2 macrophages present in the submucosa can also suppress T cell activity by direct contact or by the secretion of soluble factors such as nitric acid, prostaglandin E2, and the immunosuppressive cytokines (IL-10 and TGF-β). M-1 macrophages are also located in submucosal sites and represent the classical "proinflammatory" phagocytic cell, with similar killing capabilities as those of alveolar macrophages. In addition, the T helper 2 (Th2) proinflammatory (and proasthma) cytokines IL-4 and IL-13 have been shown to promote the differentiation of M-1 macrophages into regulatory/suppressive M-2 macrophages, which demonstrates the plasticity of these cells. The ability of these macrophage populations to demonstrate plasticity and dispose of foreign material rapidly with either a proinflammatory or regulatory response considerably enhances the lung defense system and is a unique contributor to the overall mucosal defense system.

Dendritic Cells

Conventional DCs reside in the submucosa of lung and other mucosa tissues and are considered the major antigen-presenting cells. They are usually in a resting immature state in which they function as sentinels to capture and process antigen, after which they mature and migrate to the local draining lymph node (mediastinal node for the lungs). At this node, they present antigen to T cells, which initiates an adaptive immune response, either proinflammatory or suppressive, depending on the antigen and DC. CD103[+] DCs have been shown to induce regulatory/suppressor T cells by means of their synthesis and secretion of TGF-β and indoleamine dioxygenase. Plasmacytoid DCs are located in similar submucosal areas but function more in an antiviral role, as opposed to antigen presentation; upon viral activation, they rapidly secrete the antiviral cytokine IFN-γ. Macrophages and DCs commonly synthesize and secrete many similar cytokines, depending on the stimulus; such cytokines include IL-1β (which activates vascular endothelium and lymphocytes), IL-6 (which activates lymphocytes and enhances antibody production), IL-12 (which activates NK cells, CD4 T cells to Th1 T cells), tumor necrosis factor α (which activates vascular endothelium and increases permeability), IL-8 (which recruits neutrophis, basophils, and T cells), and IFN-γ (which has antiviral effects).

Natural Killer and Invariant Natural Killer T Lymphoid Cells

NK cells originate in the bone marrow and differentiate through the common lymphoid progenitor lineage . They are neither T or B cells, and resident populations of functionally active NK and invariant natural killer T (iNKT) cells are present in mucosal sites and the lung interstitium. NK cells are a major component of the body's innate immune system of defense against invading pathogens such as herpesviruses, various bacterial infections, and tumor cells. NK cells can be activated by interferons (IFN-α and IFN-β) and other macrophage-derived cytokines (IL-12 and IL-18), which enhances their killing capabilities. The mechanism of killing is through the release of granular enzymes (granzymes, perforins, and serine esterases), which create holes or pores within the target cell membranes and thereby cause cell death. In addition, NK cells can synthesize IFN-γ, which is capable of inhibiting viral infections and stimulating CD4 and CD8 T cell responses.

NK cells have a complicated set of receptors to recognize a wide array of infectious bacteria and viruses. Individuals who lack NK cells are very prone to herpesvirus infections and often have recurrent infection. The iNKT cells belong to a category of lymphoid cells termed *innate-like lymphocytes,* reside together with TCRγδ cells in mucosal tissues, and have antigen receptors with limited diversity, unlike the adaptive immune cells. These types of cells are thought to be links between innate and adaptive immunity. A critical aspect of the link is that these cells can respond must faster than cells of the classical adaptive immune response, limiting early damage until the adaptive response can mobilize a stronger defense. The iNKT cells have an invariant T cell receptor α chain and, on activation, secrete the regulatory cytokines IL-10 and IFN-γ.

Epithelial Cells and Commensal Microbiota Protect the Lumen of the Airways

Epithelial Cells

Airway epithelial cells can produce and secret both **antibacterial enzymes** and amphipathic (hydrophilic–positively charged) **antimicrobial peptides**, which generally target and disrupt cell wall components of the bacteria, which leads to cell death. Lysozyme, elastase, hydrolases, and secretory phospholipase A_2 are examples of antibacterial enzymes targeted at various pathogens. Antimicrobial peptides are typically cationic peptides synthesized as propeptides and activated through cleavage of an anionic propiece. Peptides secreted from epithelial cells and alveolar macrophages usually enter the mucosal fluid, whereas peptides secreted from submucosal macrophages typically stay within the local tissue.

There are three classes of antimicrobial peptides: defensins, cathelicidins, and histatins. Defensins, which are categorized as α and β can disrupt cell membranes within minutes and are very effective on bacteria, fungi, and viruses. **β–Defensins** are synthesized in alveolar type II cells, are stored in lamellar bodies, and are a component of surfactant. **α–Defensins** are synthesized and stored within epithelial cells, macrophages, and neutrophils. **Cathelicidins** are stored in secondary granules and activated

intracellularly by fusion of the phagosomes with the granule or also directly by elastase. In the lungs, cathelicidins are secreted from lung type II epithelial cells, as well as from neutrophils and macrophages. **Histatins** are present mainly in the oral cavity and are more specific for pathogenic fungi. Although it is not clear whether they exist in the respiratory tract, the carbodydrate-binding proteins (lectins) in the gut can directly kill Gram-positive bacteria by binding to their cell walls. Furthermore, lactoferrins and lysozyme present in epithelial cells have been shown to have bacteriostatic and bactericidal effects; the precise mechanism by which this occurs is not clearly understood, although it is postulated that they may work synergistically with iron and calcium.

Lung Microbiome-Commensal Microbiota

The human microbiome, especially the gut, has received a lot of attention, and exploration into the role of these commensal microorganisms in health and disease is rapidly developing. Although much is known about the gut microbiome, the lung microbiome has been difficult to explore. Once thought to be a sterile environment, the lung remains a difficult site to sample for a variety of reasons, including subject safety during sampling (invasive procedures), complexity and variety of epithelial surfaces throughout the respiratory tract, and the simple fact that far fewer bacteria are available to sample. Furthermore, the lumen ecosystem and epithelium of the respiratory tract differ considerably along the respiratory tract, from the upper airways to the alveoli; it is likely that the density and types of bacteria residing in these various regions also differ. It is clear, however, that the lung does have a microbiome, and three primary phyla appear to be represented: Bacteroidetes (*Prevotella, Bacteroides*), Firmicutes (*Veillonella, Streptococcus, Staphylococcus*), and Proteobacteria (*Pseudomonas, Haemophilus, Moraxella, Neisseria, Acinetobacter*).

Alterations in the lung microbiome have been shown to be associated with several pulmonary conditions (including asthma, cystic fibrosis, and chronic obstructive pulmonary disease) and with lung transplantation. The increased rate of asthma in children has been linked to the increased rate of antibiotic use, which has a vast effect on the microbiome. Organisms in the Proteobacteria (*Haemophilus, Moraxella*) and Firmicutes (*Streptococcus*) phyla have been shown to correlate with asthma. An experimental study in mice showed an improvement in asthma when Proteobacteria organisms were replaced with Bacteriodetes organisms, which coincided with an increase in regulatory T cells and suppression of the Th2 proinflammatory cytokines prominent in the pathogenesis of asthma.

From a therapeutic standpoint, the use of probiotics to maintain a "normal" microbiome has proved useful in treating infectious diseases in the gut, and the use of probiotic therapy in patients with chronic lung infections (e.g., cystic fibrosis, upper respiratory tract infections) have been promising. The precise mechanism or mechanisms by which the bacteria may "protect" the lungs is still poorly understood but probably varies considerably according to the microorganism.

It has been shown that *Bifidobacterium* can activate DCs and TLRs to secrete IL-10 and retinoic acid, mediators capable of inducing regulatory T cells, which downregulate the Th2 inflammatory response. From a causative perspective, histamine, a major component of mast cell granules and secreted from *Lactobacillus* organisms, has been shown to have an association with priming DCs to stimulate Th2 inflammatory responses, as seen in individuals with asthma. In addition, it has been shown that ILCs can play a role in influencing the microbiome through their ability to bind to receptors on epithelial cells, thus altering the interaction of the epithelium with the microorganisms and possibly promoting colonization of either pathogenic or nonpathogenic microorganisms. Much is still to be learned and absorbed from future studies of the lung microbiome.

Clinical Manifestations Associated With Abnormalities in Mucosal Innate and Adaptive Immunity

By far the most common pathological conditions associated with mucosal tissue are allergic responses (e.g., allergic asthma, allergic rhinitis, and food and skin allergies). Some common human allergens include components of inhaled house dust mites, cockroaches, and cat dander, as well as bee stings and ingestion of peanuts. One of the most common drug allergens is penicillin, which can bind to many endogenous proteins and also alter their antigenicity.

As previously described, the predominant antibody response in MALT is IgA; however, in an allergic response, IgE is the predominant antibody. It is generated by a switchover mechanism induced by the synthesis of IL-4 from Th2-primed CD4 T cells. The IL-4 induces the antibody-producing B cells to switch over from synthesizing IgG antibodies to IgE. IgE binds to the surface of mast cells in the submucosa through its crystallizable fragment (constant) region of the antibody molecule. Upon reexposure to the inhaled antigen and its subsequent migration from the airway lumen to the submucosa, the allergen then binds to the antigen-binding fragment region of the IgE molecule, which forms an immune complex (IgE-Ag) on the surface of the mast cell.

The final step is that IgE-Ag complexes must crosslink with other IgE-Ag complexes on the surface of the mast cell, which induce intracellular signaling pathways to initiate degranulation and immediate release of preformed Th2 mediators (histamine, heparin, prostaglandins, leukotrienes, IL-4, IL-5, IL-13, proteases). These mediators induce the classical signs of asthma: smooth muscle constriction (bronchoconstriction), eosinophil recruitment and activation (inflammation), and connective tissue remodeling. Symptoms of wheezing, coughing, and shortness of breath occur within minutes, followed by a late response of eosinophilia and airway inflammation.

The inflammatory response can resolve spontaneously or as a result of therapy (bronchodilator or anti-inflammatory drugs, such as corticosteroids). Low-grade inflammation may persist and result in a process called *airway remodeling,* manifested by permanent, irreversible structural changes such as submucosal fibrosis and airway smooth muscle hypertrophy. The mechanisms responsible for airway remodeling in allergic diseases are not clearly understood, but chemokines and cytokines such as TGF-β, a potent profibrotic cytokine, play important roles. Thus a very elegant, highly effective defensive system against infectious and parasitic organisms is "tricked" into responding to an innocuous substance, an allergen, as if it were harmful and initiates its defenses; the result is allergic airway disease.

Metabolic Functions of the Lung

The lungs are exposed to and metabolize a wide variety of xenobiotic substances. The endothelial cells within the lung capillary bed have a large surface area that receives very high blood flow, and lung endothelial cells have developed various mechanisms and cell surface receptors to metabolize xenobiotics. Most of the metabolic processing of inhaled or ingested xenobiotic compounds occurs enzymatically within the liver and intestinal tract with members of the cytochrome P-450 (CYP) enzyme families (e.g., CYP1, CYP2, CYP3). The lungs and other organs also selectively participate in the processing of xenobiotics and typically have lower levels of cytochrome P-450 enzymes. Prominent cytochrome P-450 enzymes in the lung include CYP1B1, CYP2B6, CYP2E1, CYP2J2, CYP3A5, and CYP1A1, the last of which is present in high levels in people who smoke cigarettes.

Drugs for the treatment of asthma and chronic obstructive pulmonary disease—such as corticosteroids, long-acting β_2 receptor agonists, leukotriene receptor antagonists, and methylxanthines—are degraded enzymatically in the lungs. In addition, a wide array of endogenous substances are metabolized by endothelial cells within the pulmonary capillary bed, including vasoactive amines, cytokines, lipid mediators, and proteins. Box 26.1 provides a list of compounds metabolized in the lung. Metabolism can occur

---

• BOX 26.1   Compounds Metabolized in the Lungs

Enzymatic Degradation in Pulmonary Circulation

Corticosteroids
Long-acting beta agonists
Methylxanthines

Endothelial Cells in Pulmonary Capillary Bed

Vasoactive amines
Cytokines
Lipid mediators
Proteins

through either intracellular or extracellular processing of endogenous substances that pass though the capillaries or by direct synthesis and secretion by endothelial cells. For example, angiotensin I is activated by angiotensin-converting enzyme, which is on the surface of endothelial cells. Serotonin, a vasoconstrictor, binds to a specific receptor on the surface of endothelial cells and is internalized and metabolized inside cells. Approximately 80% of the serotonin entering the lungs is metabolized in a single pass through the pulmonary capillary bed.

Endothelial cells also have surface receptors for bradykinin, tumor necrosis factor, components of the complement system, immunoglobulin crystallizable fragments, and adhesion molecules. In addition, vascular endothelial cells synthesize and secrete prostacyclin, endothelin, clotting factors, nitric oxide, prostaglandins, and cytokines. Vascular endothelial cells, however, lack 5-lipoxygenase and are not able to synthesize leukotrienes (smooth muscle constrictors). Compounds not metabolized by the pulmonary capillary bed include epinephrine, dopamine, histamine, isoproterenol, angiotensin II, and substance P.

AT THE CELLULAR LEVEL

Angiotensin-converting enzyme (ACE) is present in small indentations (caveolae) on the surface of pulmonary endothelial cells and catalyzes the conversion of the physiological inactive angiotensin I to the active angiotensin II, a potent vasoconstrictor. This is a major mechanism in the body's ability to supply systemic levels of angiotensin II and thus influence blood pressure (see Chapter 36). The therapeutic use of ACE inhibitors is important in the management of patients with high blood pressure.

IN THE CLINIC

The lungs play a key role in the metabolism of many prodrugs, which are inactive, into active drugs. Administration of prodrugs improves the amount of active drug delivered to their targets in the body. In many cases, inactive precursor prodrugs are delivered systemically or locally (through inhalation) to the lungs, where they are activated in situ. An example of this is beclomethasone dipropionate (Qvar, Beconase AQ), a medication inhaled by patients with asthma, which is activated by esterases within the lung to the active form 17-beclamethasone monopropionate.

Key Points

1. The respiratory system has developed unique structural (mucociliary transport system) and immunological (mucosal immune system) features to cope with the constant environmental exposure to foreign substances; these features limit or inhibit inflammation.

2. The three components of the mucociliary transport system are the sol phase (periciliary fluid), the gel phase (mucus), and cilia.

3. The depth of the periciliary fluid layer is maintained by the balance between Cl^- secretion and Na^+ absorption and is essential to normal ciliary beating.

4. Mucus is a complex macromolecule composed of glycoproteins, proteins, electrolytes, and water. It has low viscosity and high elastic mechanical properties.

5. Goblet cells, Clara cells, and the mucous and serous cells residing in the tracheobronchial glands produce mucus.

6. Particle deposition in the lung is dependent on their size, density, and shape; the distance traveled; airflow speed; and relative humidity. The major mechanisms for particle deposition are impaction (particles larger than 10 μm, in nasal passages, and particles 2 to 10 μm, in the nasopharynx, trachea, and bronchi), sedimentation (particles 0.2 to 2 μm in size, in distal airways), interception (particles with elongated shape, in the lower airways), and brownian movement (particles smaller than 0.2 μm, in the alveoli).

7. The respiratory system is part of the mucosal immune system, which is composed of the intestinal (GALT), the respiratory (BALT), and urinary tract systems. These systems do not contain true lymph nodes with afferent and efferent lymph flow; they are composed mainly of nonencapsulated lymph nodules without true lymphatic drainage.

8. The nonciliated lymphoepithelium of BALT establishes a break in the mucociliary blanket that acts as a drain to facilitate the collection and immune processing of foreign particulates throughout the conducting airways.

9. TCRγδ T cells, IgA synthesizing plasma cells, NK cells, and alveolar macrophages are highly specialized innate and adaptive immune cells unique to the anti-inflammatory defense system in the lung and other mucosal tissues.

Additional Readings

Cua DJ, Tato CM. Innate IL-17–producing cells: the sentinels of the immune system. *Nat Rev Immunol.* 2010;10:479-489.

Dickson RP, Erb-Downward JR, Martinez FJ, Huffnagle GB. The microbiome and the respiratory tract. *Annu Rev Physiol.* 2016;78:481-504.

Martin TR, Frevert CW. Innate immunity in the lungs. *Proc Am Thorac Soc.* 2005;2:403-411.

McKenize AN, Spits H, Eberl G. Innate lymphoid cells in inflammation and immunity. *Immunity.* 2014;41:366-374.

Monticelli LA, Sonnenberg GF, Abt MC, et al. Innate lymphoid cells promote lung-tissue homeostasis after infection with influenza virus. *Nat Immunol.* 2011;12:1045-1054.

Olsson B, Bondesson E, Borgström L, et al. Pulmonary drug metabolism, clearance, and adsorption. In: Smyth HDC, Hickey AJ, eds. *Controlled Pulmonary Drug Delivery.* New York: Springer; 2011:21-51.

Spits H, Artis D, Colonna M, et al. Innate lymphoid cells—a proposal for uniform nomenclature. *Nat Rev Immunol.* 2013;13: 145-149.

Vantourout P, Hayday A. The inbetweeners: innate-like lymphocytes. Six of the best: unique contributions of γδ T cells to immunology. *Nat Rev Immunol.* 2014;13:88-100.

Wine JJ. Parasympathetic control of airway submucosal glands: central reflexes and the airway intrinsic nervous system. *Auton Neurosci.* 2007;133:35-54.

SECTION 6

Gastrointestinal Physiology

KIM E. BARRETT AND HELEN E. RAYBOULD

27

Functional Anatomy and General Principles of Regulation in the Gastrointestinal Tract

LEARNING OBJECTIVES

Upon completion of this chapter the student should be able to answer the following questions:

1. What is the neural innervation of the GI tract, and how is GI function regulated?
2. What are some examples of neural, paracrine, and humoral regulation of GI function?

The gastrointestinal (GI) tract consists of the alimentary tract from the mouth to the anus and includes the associated glandular organs that empty their contents into the tract. The overall function of the GI tract is to absorb nutrients and water into the circulation and eliminate waste products. The major physiological processes that occur in the GI tract are **motility, secretion, digestion,** and **absorption.** Most of the nutrients in the diet of mammals are taken in as solids and as macromolecules that are not readily transported across cell membranes to enter the circulation. Thus digestion consists of physical and chemical modification of food such that absorption can occur across intestinal epithelial cells. Digestion and absorption require motility of the muscular wall of the GI tract to move the contents along the tract and to mix the food with secretions. Secretions from the GI tract and associated organs consist of enzymes, biological detergents, and ions that provide an intraluminal environment optimized for digestion and absorption. These physiological processes are highly regulated to maximize digestion and absorption, and the GI tract is endowed with complex regulatory systems to ensure this occurs. In addition the GI tract absorbs drugs administered by the oral or rectal routes.

The GI tract also serves as an important organ for **excretion** of substances. It stores and excretes waste substances from ingested food materials and excretes products from the liver such as cholesterol, steroids, and drug metabolites (all sharing the common property of being lipid-soluble molecules).

When considering the physiology of the GI tract, it is important to remember that it is a long tube that is in contact with the body's external environment. As such, it is vulnerable to infectious microorganisms that can enter along with food and water. To protect itself the GI tract possesses a complex system of defenses consisting of immune cells and other nonspecific defense mechanisms. In fact the GI tract represents the largest immune organ of the body. This chapter provides an overview of the functional anatomy and general principles of regulation in the GI system.

Functional Anatomy

The structure of the GI tract varies greatly from region to region, but there are common features in the overall organization of the tissue. Essentially the GI tract is a **hollow tube** divided into major functional segments; the major structures along the tube are the **mouth, pharynx, esophagus, stomach, duodenum, jejunum, ileum, colon, rectum,** and **anus** (Fig. 27.1). Together the duodenum, jejunum, and ileum make up the small intestine, and the colon is sometimes referred to as the *large intestine.* Associated with the tube are blind-ending glandular structures that are invaginations of the lining of the tube; these glands empty their secretions into the gut lumen (e.g., Brunner's glands in the duodenum, which secrete copious amounts of HCO_3^-). Additionally there are glandular organs attached to the tube via ducts through which secretions empty into the gut lumen—for example, the salivary glands and pancreas.

The major structures along the GI tract have many functions. One important function is storage; the stomach and colon are important storage organs for processed food (also referred to as *chyme*) and exhibit specialization in terms of both their functional anatomy (e.g., shape and size) and control mechanisms (characteristics of smooth muscle to produce tonic contractions) that enable them to perform this function efficiently. The predominant function of the small intestine is digestion and absorption; the major specialization of this region of the GI tract is a large surface area over which absorption can occur. The colon reabsorbs water and ions to ensure they do not get eliminated from the body. Ingested food is moved along the GI tract by the action of muscle in its walls. Separating the regions of the GI tract are also specialized muscle structures called **sphincters.** These function to isolate one region from the next and provide selective retention of contents or prevent backflow, or both.

The blood supply to the intestine is important for carrying absorbed nutrients to the rest of the body. Unlike other organ systems of the body, venous drainage from the GI tract does not return directly to the heart but first enters the **portal circulation** leading to the liver. Thus the liver is unusual in receiving a considerable part of its blood supply from other than the arterial circulation. GI blood flow is also notable for its dynamic regulation. Splanchnic blood flow receives about 25% of cardiac output, an amount disproportionate to the mass of the GI tract it supplies. After a meal, blood can also be diverted from muscle to the GI tract to subserve the metabolic needs of the gut wall and also to remove absorbed nutrients.

The **lymphatic drainage** of the GI tract is important for the transport of lipid-soluble substances that are absorbed across the GI tract wall. As we will see later, lipids and other lipid-soluble molecules (including some vitamins and drugs) are packaged into particles that are too large to pass into the capillaries and instead pass into lymph vessels in the intestinal wall. These lymph vessels drain into larger lymph ducts, which finally drain into the thoracic duct and thus into the systemic circulation on the arterial side. This has major physiological implications in lipid metabolism and also in the ability of drugs to be delivered straight into the systemic circulation.

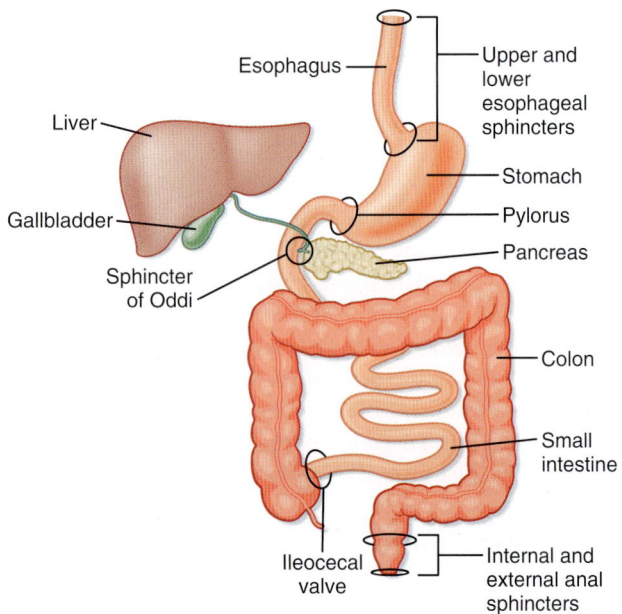

• **Fig. 27.1** General anatomy of the GI system and its division into functional segments.

Cellular Specialization

The wall of the tubular gut is made up of layers consisting of specialized cells (Fig. 27.2).

Mucosa

The **mucosa** is the innermost layer of the GI tract. It consists of the **epithelium,** the **lamina propria,** and the **muscularis mucosae.** The epithelium is a single layer of specialized cells that line the lumen of the GI tract. It forms a continuous layer along the tube and with the glands and organs that drain into the lumen of the tube. Within this cell layer are

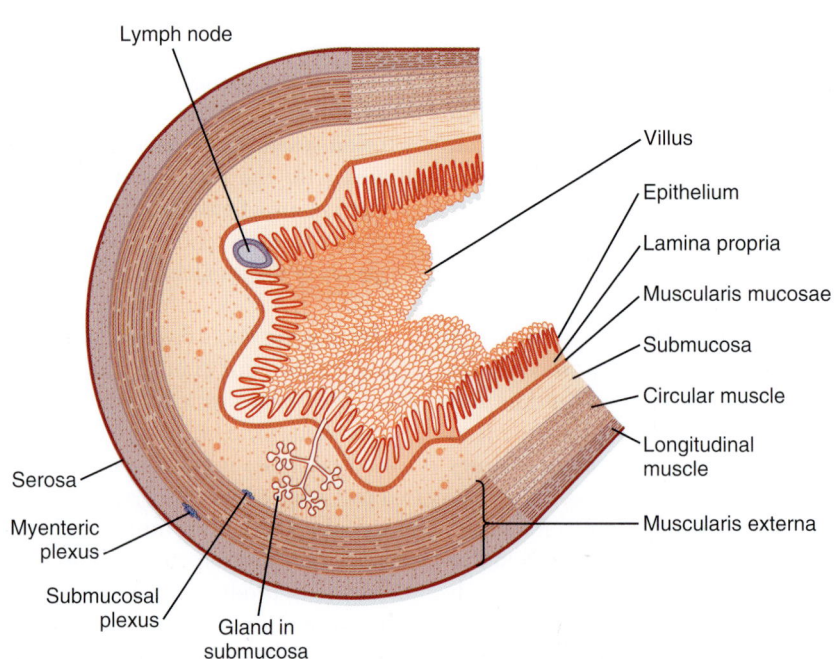

• **Fig. 27.2** General organization of the layers composing the wall of the GI tract.

a number of specialized epithelial cells; the most abundant are cells termed **absorptive enterocytes,** which express many proteins important for digestion and absorption of macronutrients. **Enteroendocrine cells** contain secretory granules that release regulatory peptides and amines to help regulate GI function. In addition, cells in the gastric mucosa are specialized for production of protons, and mucin-producing cells throughout the GI tract produce a glycoprotein (mucin) that helps protect the GI tract and lubricate the luminal contents.

The columnar epithelial cells are linked together by intercellular connections called **tight junctions.** These junctions are complexes of intracellular and transmembrane proteins, and the tightness of these junctions is regulated throughout the postprandial period. The nature of the epithelium varies greatly from one part of the digestive tract to another, depending on the predominant function of that region. For example, the intestinal epithelium is designed for absorption; these cells mediate selective uptake of nutrients, ions, and water. In contrast, the esophagus has a squamous epithelium that has no absorptive role. It is a conduit for transportation of swallowed food and thus needs some protection, provided by the squamous epithelium, from rough food such as fiber.

The surface area of the epithelium is arranged into **villi** and **crypts** (Fig. 27.3). Villi are finger-like projections that serve to increase the surface area of the mucosa. Crypts are invaginations or folds in the epithelium. The epithelium lining the GI tract is continuously renewed and replaced by dividing cells; in humans this process takes about 3 days. These proliferating cells are localized to the crypts, where there is a proliferative zone of intestinal **stem cells.**

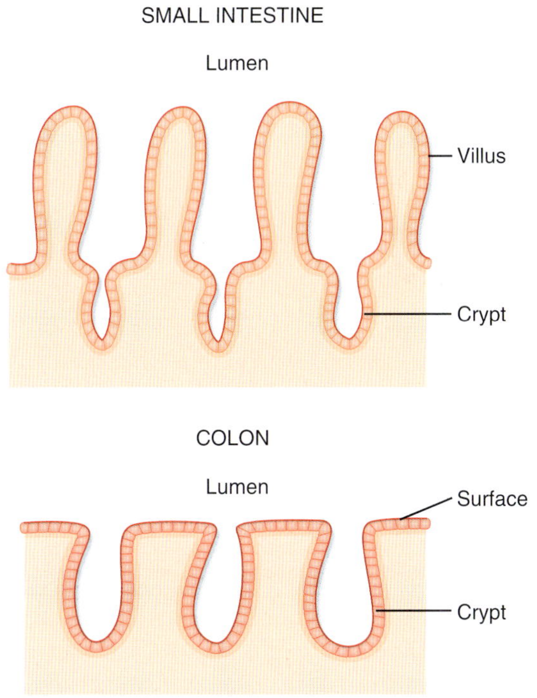

SMALL INTESTINE

• **Fig. 27.3** Comparison of the morphology of the epithelium of the small intestine and colon.

The lamina propria immediately below the epithelium consists largely of loose connective tissue that contains collagen and elastin fibrils (see Fig. 27.2). The lamina propria is rich in several types of glands and contains lymph vessels and nodules, capillaries, and nerve fibers. The muscularis mucosae is the thin innermost layer of intestinal smooth muscle. When seen through an endoscope, the mucosa has folds and ridges that are caused by contractions of the muscularis mucosae.

Submucosa

The next layer is the **submucosa** (see Fig. 27.2), which consists largely of loose connective tissue with collagen and elastin fibrils. In some regions of the GI tract, **glands** (invaginations or folds of the mucosa) are present in the submucosa. The larger nerve trunks, blood vessels, and lymph vessels of the intestinal wall lie in the submucosa, together with one of the plexuses of the enteric nervous system (ENS), the **submucosal plexus.**

Muscle Layers

The **muscularis externa,** or **muscularis propria,** typically consists of two substantial layers of smooth muscle cells: an inner circular layer and an outer longitudinal layer (see Fig. 27.2). Muscle fibers in the **circular muscle layer** are oriented circumferentially, whereas muscle fibers in the **longitudinal muscle layer** are oriented along the longitudinal axis of the tube. In humans and most mammals, the circular muscle layer of the small intestine is subdivided into an inner dense circular layer that consists of smaller, more closely packed cells, and an outer circular layer. Between the circular and longitudinal layers of muscle lies the other plexus of the ENS, the **myenteric plexus.** Contractions of the muscularis externa mix and circulate the contents of the lumen and propel them along the GI tract.

The wall of the GI tract contains many interconnected neurons. The submucosa contains a dense network of nerve cells called the *submucosal plexus* (also referred to as **Meissner's plexus**). The prominent *myenteric plexus* (**Auerbach's plexus**) is located between the circular and longitudinal smooth muscle layers. These intramural plexuses constitute the ENS. The ENS helps integrate the motor and secretory activities of the GI system. If the sympathetic and parasympathetic nerves to the gut are cut, many motor and secretory activities continue because the ENS directly controls these processes.

Serosa

The **serosa,** or **adventitia,** is the outermost layer of the GI tract and consists of a layer of squamous mesothelial cells (see Fig. 27.2). It is part of the **mesentery** that lines the surface of the abdominal wall and suspends the organs within the abdominal cavity. The mesenteric membranes secrete a thin viscous fluid that helps lubricate the abdominal organs so movement of the organs can occur as the muscle layers contract and relax.

Regulatory Mechanisms in the Gastrointestinal Tract

Unlike the cardiovascular or respiratory systems, the GI tract undergoes periods of relative quiescence (intermeal period) and periods of intense activity after the intake of food (postprandial period). Consequently the GI tract has to detect and respond appropriately to food intake. In addition the macronutrient content of a meal can vary considerably, and there have to be mechanisms that can detect this and mount appropriate physiological responses. Thus the GI tract has to communicate with associated organs such as the pancreas. Finally, because the GI tract is essentially a long tube, there have to be mechanisms by which events occurring in the proximal portion of the GI tract are signaled to the more distal parts, and vice versa.

There are three principal control mechanisms involved in the regulation of GI function: **endocrine, paracrine,** and **neurocrine** (Fig. 27.4).

Endocrine Regulation

Endocrine regulation describes the process whereby the sensing cell in the GI tract, an **enteroendocrine cell (EEC),** responds to a stimulus by secreting a regulatory peptide or hormone that travels via the bloodstream to target cells removed from the point of secretion. Cells responding to a GI hormone express specific receptors for the hormone. Hormones released from the GI tract have effects on cells located in other regions of the GI tract and also on glandular structures associated with the GI tract, such as the pancreas. In addition, GI hormones have effects on other tissues that have no direct role in digestion and absorption, including endocrine cells in liver and brain.

EECs are packed with secretory granules, the products of which are released from the cell in response to chemical and mechanical stimuli to the wall of the GI tract (Fig. 27.5). In addition, EECs can be stimulated by neural input or other factors not associated with a meal. The most common EECs in the gut wall are referred to as the "open" type; these cells have an apical membrane that is in contact with the lumen of the GI tract (generally regarded as the location where sensing occurs) and a basolateral membrane through which secretion occurs. There are also "closed"-type EECs that do not have part of their membrane in contact with the luminal surface of the gut; an example is the **enterochromaffin-like** (ECL) cell in the gastric epithelium, which secretes histamine.

There are many examples of hormones secreted by the GI tract (see Table 27.1); it is worth remembering that the first hormone ever identified was the GI hormone **secretin.** One of the most well-characterized GI hormones is **gastrin,** which is released from endocrine cells located in the wall of the distal part of the stomach. Release of gastrin is stimulated by activation of parasympathetic outflow to the GI tract, and gastrin potently stimulates gastric acid secretion in the postprandial period.

ENDOCRINE

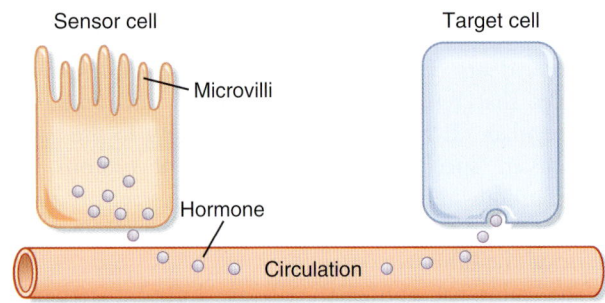

NEUROCRINE

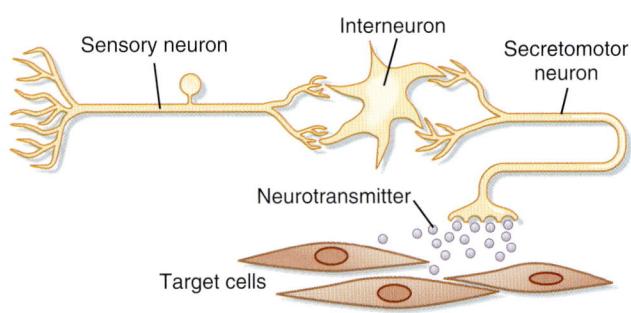

PARACRINE

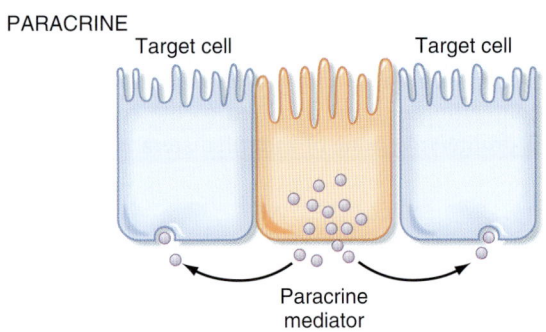

• **Fig. 27.4** The three mechanisms by which function in the GI tract is regulated in the integrated response to a meal.

Paracrine Regulation

Paracrine regulation describes the process whereby a chemical messenger or regulatory peptide is released from a sensing cell (often an EEC) in the intestinal wall that acts on a nearby target cell by diffusion through the interstitial space. Paracrine agents exert their actions on several different cell types in the wall of the GI tract, including smooth muscle cells, absorptive enterocytes, secretory cells in glands, and even on other EECs. There are several important paracrine agents, and they are listed in Table 27.1 along with their site of production, site of action, and function. An important paracrine mediator in the gut wall is histamine. In the stomach, histamine is stored and released by ECL cells located in the gastric glands. Histamine diffuses through the interstitial space in the lamina propria to neighboring

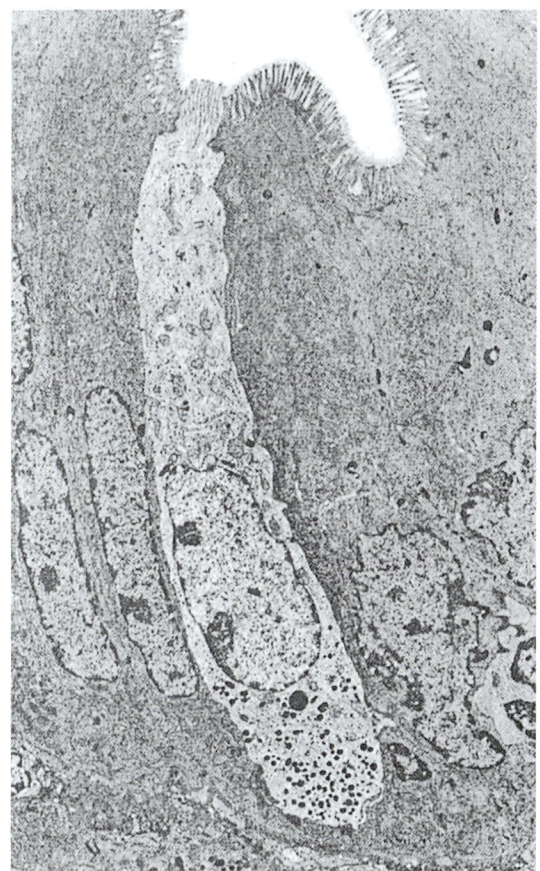

• **Fig. 27.5** Electron micrograph of an open-type endocrine cell in the GI tract. Note the microvilli at the apical projection and the secretory granules in the basolateral portion of the cell. (From Barrett K. *Gastrointestinal Physiology [Lange Physiology Series]*. New York: McGraw-Hill; 2005.) (Courtesy of Leonard R. Johnson, Ph.D.)

parietal cells and stimulates production of acid. **Serotonin** (5-hydroxytryptamine [5-HT]), released from enteric neurons, mucosal mast cells, and specialized EECs called **enterochromaffin cells,** regulates smooth muscle function and water absorption across the intestinal wall. There are other paracrine mediators in the gut wall, including prostaglandins, adenosine, and nitric oxide (NO); the functions of these mediators are not well described, but they are capable of producing changes in GI function.

Many substances can be both paracrine and endocrine regulators of GI function. For example, **cholecystokinin,** which is released from the duodenum in response to dietary protein and lipid, acts locally on nerve terminals in a paracrine fashion and also affects the pancreas. This will be discussed in more detail in Chapter 30.

Neural Regulation of Gastrointestinal Function

Nerves and neurotransmitters play an important role in regulating the function of the GI tract. In its simplest form, neural regulation occurs when a neurotransmitter

is released from a nerve terminal located in the GI tract and the neurotransmitter has an effect on the cell that is innervated. However, in some cases there are no synapses between motor nerves and effector cells in the GI tract. Neural regulation of GI function is very important within an organ, as well as between distant parts of the GI tract.

 AT THE CELLULAR LEVEL

There are multiple receptor subtypes for the regulatory peptide hormones released from endocrine cells in the wall of the gut. The selectivity of receptors to peptide hormones is determined by posttranslational modifications, which then confers receptor selectivity. An example of this is peptide YY (PYY). There are multiple receptor subtypes for PYY, classified as Y1 to Y7. PYY is released from endocrine cells in the wall of the gut, mainly in response to fatty acids. It is released as a 36–amino acid peptide and binds to the Y1, Y2, and Y5 receptors; however, it can be cleaved to PYY3-36 by the enzyme dipeptidyl peptidase IV, a membrane peptidase. This form of the peptide is more selective for the Y2 receptor. Thus the presence of the enzyme that cleaves the peptide can alter the biological response to PYY secretion.

 IN THE CLINIC

Glucagon-like peptide 1 (GLP-1) is a regulatory peptide released from EECs cells in the gut wall in response to the presence of luminal carbohydrate and lipids. GLP-1 arises from differential processing of the glucagon gene, the same gene that is expressed in the pancreas and that gives rise to glucagon. GLP-1 is involved in regulation of the blood glucose level via stimulation of insulin secretion and also insulin biosynthesis. Agonists of the GLP-1 receptor improve insulin sensitivity in diabetic animal models and human subjects. Administration of GLP-1 also reduces appetite and food intake and delays gastric emptying, responses that may contribute to improving glucose tolerance. Long-acting agonists for the GLP-1 receptor (e.g., **exenatide**) have been approved for the treatment of type 2 diabetes.

Neural regulation of the GI tract is surprisingly complex. The gut is innervated by two sets of nerves, the extrinsic and intrinsic nervous systems. The **extrinsic nervous system** is defined as nerves that innervate the gut, with cell bodies located outside the gut wall; these extrinsic nerves are part of the autonomic nervous system (ANS). The **intrinsic nervous system,** also referred to as the **enteric nervous system,** has cell bodies that are contained within the wall of the gut (submucosal and myenteric plexuses). Some GI functions are highly dependent on the extrinsic nervous system, yet others can take place independently of the extrinsic nervous system and are mediated entirely by the ENS. However, extrinsic nerves can often modulate intrinsic nervous system function (Fig. 27.6).

TABLE 27.1	Hormonal and Paracrine Mediators in the GI Tract				
GI Hormone	Source	Stimulus for Release	Pathway of Action	Targets	Effect
Gastrin	Gastric antrum (G cells)	Oligopeptides	Endocrine	ECL cells and parietal cells of the gastric corpus	Stimulation of parietal cells to secrete H⁺ and ECL cells to secrete histamine
Cholecystokinin	Duodenum (I cells)	Fatty acids, hydrolyzed protein	Paracrine, endocrine	Vagal afferent terminals, pancreatic acinar cells	Inhibition of gastric emptying and H⁺ secretion; stimulation of pancreatic enzyme secretion, gallbladder contraction, inhibition of food intake
Secretin	Duodenum (S cells)	Protons	Paracrine, endocrine	Vagal afferent terminals, pancreatic duct cell	Stimulation of pancreatic duct secretion (H_2O and HCO_3^-)
Gluco-insulinotropic peptide (GIP)	Intestine (K cells)	Fatty acids, glucose	Endocrine	Beta cells of the pancreas	Stimulation of insulin secretion
Peptide YY (PYY)	Intestine (L cells)	Fatty acids, glucose, hydrolyzed protein	Endocrine, paracrine	Neurons, smooth muscle	Inhibition of gastric emptying, pancreatic secretion, gastric acid secretion, intestinal motility, food intake
Proglucagon-derived peptides 1/2 (GLP-1/2)	Intestine (L cells)	Fatty acids, glucose, hydrolyzed protein	Endocrine, paracrine	Neurons, epithelial cells	Glucose homeostasis, epithelial cell proliferation

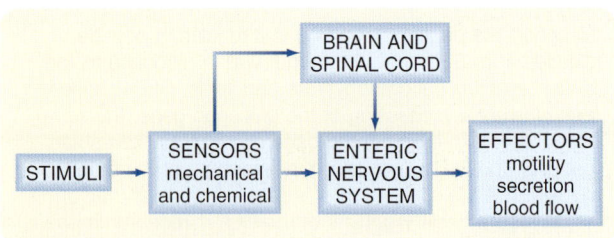

• **Fig. 27.6** Hierarchical neural control of GI function. Stimuli to the GI tract from a meal (e.g., chemical, mechanical, osmotic) will activate both the intrinsic and extrinsic sensory (afferent) pathways, which in turn will activate the extrinsic and intrinsic neural reflex pathways.

Extrinsic Neural Innervation

Extrinsic neural innervation to the gut is via the two major subdivisions of the ANS, namely, parasympathetic and sympathetic innervation (Fig. 27.7). **Parasympathetic innervation** to the gut is via the vagus and pelvic nerves. The **vagus** nerve, the 10th cranial nerve, innervates the esophagus, stomach, gallbladder, pancreas, first part of the intestine, cecum, and the proximal part of the colon. The **pelvic** nerves innervate the distal part of the colon and the anorectal region, in addition to the other pelvic organs that are not part of the GI tract.

Consistent with the typical organization of the parasympathetic nervous system, the **preganglionic** nerve cell bodies lie in the brainstem (vagus) or the sacral spinal cord (pelvic). Axons from these neurons run in the nerves to

the gut (vagus and pelvic nerves, respectively), where they synapse with **postganglionic** neurons in the wall of the organ, which in this case are enteric neurons in the gut wall. There is no direct innervation of these efferent nerves to effector cells within the wall of the gut; the transmission pathway is always via a neuron in the ENS.

Consistent with transmission in the ANS, the synapse between preganglionic and postganglionic neurons is an obligatory nicotinic synapse. That is, the synapse between preganglionic and postganglionic neurons is mediated via acetylcholine released from the nerve terminal and acting at nicotinic receptors localized on the postganglionic neuron, which in this case is an intrinsic neuron.

Sympathetic innervation is supplied by cell bodies in the spinal cord and fibers that terminate in the **prevertebral ganglia** (celiac, superior, and inferior mesenteric ganglia); these are the preganglionic neurons. These nerve fibers synapse with postganglionic neurons in the ganglia, and the fibers leave the ganglia and reach the end organ along the major blood vessels and their branches. Rarely there is a synapse in the **paravertebral** (chain) ganglia, as seen with sympathetic innervation of other organ systems. Some vasoconstrictor sympathetic fibers directly innervate blood vessels of the GI tract, and other sympathetic fibers innervate glandular structures in the wall of the gut.

The ANS, both parasympathetic and sympathetic, also carries the fibers of **afferent** (toward the central nervous system [CNS]) neurons; these are **sensory** in nature. The cell bodies for the **vagal afferents** are in the nodose

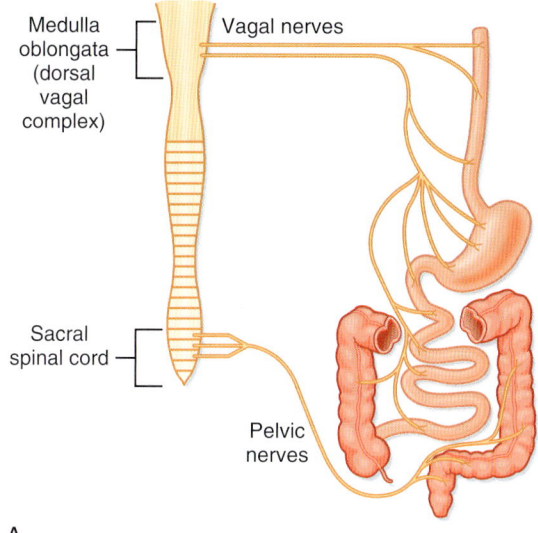

A

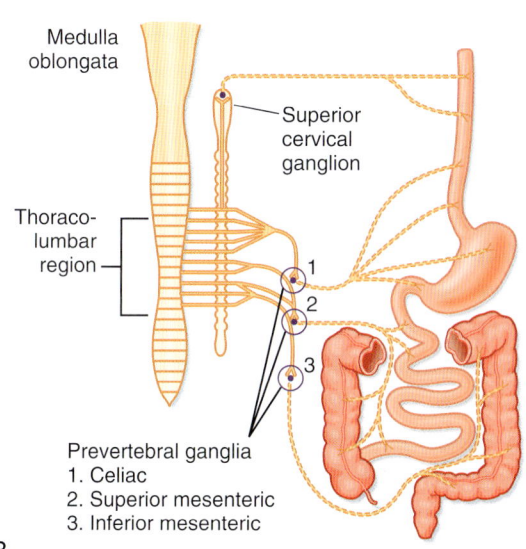

B

• **Fig. 27.7** Extrinsic innervation of the GI tract, consisting of the parasympathetic (A) and sympathetic (B) subdivisions of the autonomic nervous system.

ganglion. These neurons have a central projection terminating in the **nucleus of the tractus solitarius** in the brainstem and the other terminal in the gut wall. The cell bodies of the **spinal afferent** neurons that run with the sympathetic pathway are segmentally organized and are found in the dorsal root ganglia. Peripheral terminals of the spinal and vagal afferents are located in all layers of the gut wall, where they detect information about the state of the gut. Afferent neurons send this information to the CNS. Information sent to the CNS relays the nature of the luminal contents (e.g., acidity, nutrient content, osmolality of the luminal contents), as well as the degree of stretch or contraction in smooth muscle. Afferent innervation is also responsible for transmitting painful stimuli to the CNS.

The components of a **reflex** pathway—afferents, interneurons, and efferent neurons—exist within the extrinsic

innervation to the GI tract. These reflexes can be mediated entirely via the vagus nerve (termed a **vagovagal reflex**), which has both afferent and efferent fibers. The vagal afferents send sensory information to the CNS, where they synapse with an interneuron, which then drives activity in the efferent motor neuron. These extrinsic reflexes are very important in regulating GI function after ingestion of a meal. An example of an important vagovagal reflex is the gastric receptive relaxation reflex, in which distention of the stomach results in relaxation of the smooth muscle in the stomach; this allows filling of the stomach to occur without an increase in intraluminal pressure.

In general, as with other visceral organ systems, the parasympathetic and sympathetic nervous systems tend to work in opposition. However, this is not as simple as in the cardiovascular system, for example. Activation of the parasympathetic nervous system is important in the integrative response to a meal and is discussed in the following chapters. The parasympathetic nervous system generally results in activation of physiological processes in the gut wall, although there are notable exceptions. In contrast, the sympathetic nervous system tends to be inhibitory to GI function and is more frequently activated in pathophysiological circumstances. Overall, sympathetic activation inhibits smooth muscle function. The exception to this is the sympathetic innervation of GI sphincters, in which sympathetic activation tends to induce contraction of smooth muscle. Moreover, the sympathetic nervous system is notably important in regulation of blood flow in the GI tract.

Intrinsic Neural Innervation

The ENS is made up of two major plexuses, which are collections of nerve cell bodies (ganglia) and their fibers, all originating in the wall of the gut (Fig. 27.8). The **myenteric plexus** lies between the longitudinal and circular muscle layers, and the **submucosal plexus** lies in the submucosa. Interganglionic strands link neurons in the two plexuses.

Neurons in the ENS are characterized functionally as afferent neurons, interneurons, or efferent neurons, similar to neurons in the extrinsic part of the ANS. Thus all components of a reflex pathway can be contained within the ENS. Stimuli in the wall of the gut are detected by afferent neurons, which activate interneurons and then efferent neurons to alter function. In this way the ENS can act autonomously from extrinsic innervation. However, neurons in the ENS, as we have already seen, are innervated by extrinsic neurons, and thus the function of these reflex pathways can be modulated by the extrinsic nervous system. Because the ENS is capable of performing its own integrative functions and complex reflex pathways, it is sometimes referred to as the "little brain in the gut" as a result of its importance and complexity. It is estimated that there are as many neurons in the ENS as in the spinal cord. In addition, many GI hormones also act as neurotransmitters in the ENS *and* in the brain in regions involved in autonomic outflow. These mediators and regulatory peptides are thus

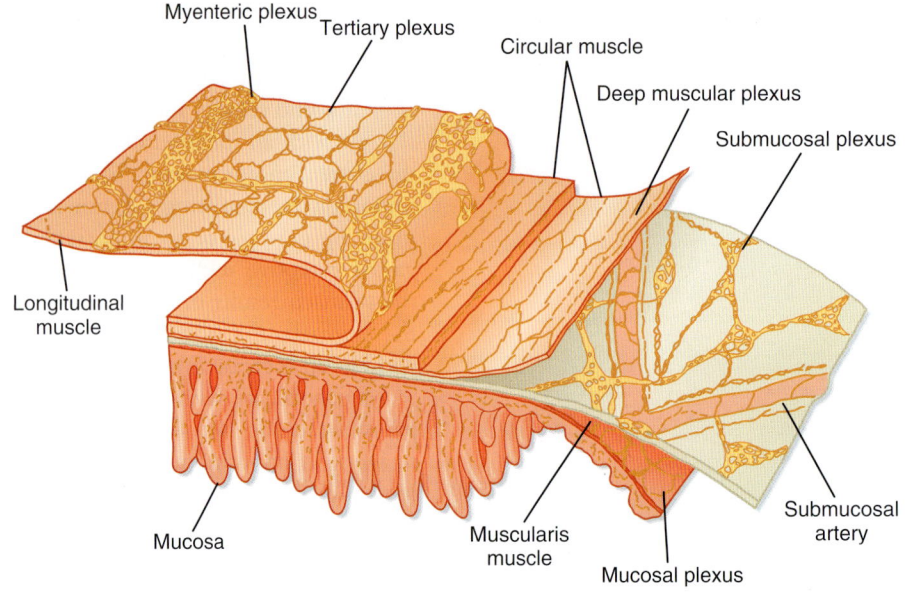

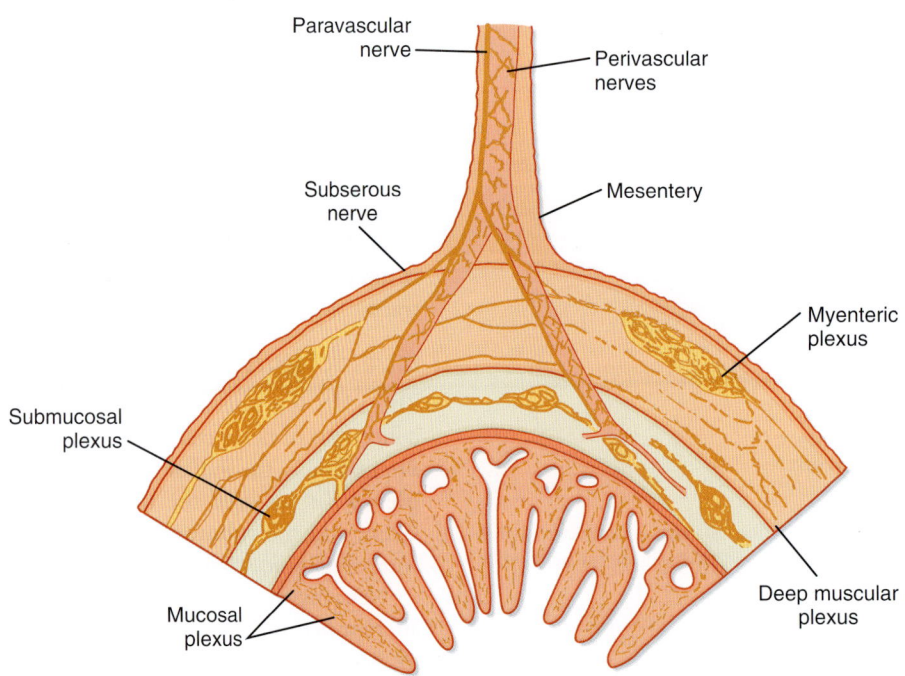

• **Fig. 27.8** The enteric nervous system in the wall of the GI tract.

Response of the GI Tract to a Meal

referred to as **brain-gut peptides,** and the extrinsic and intrinsic components innervating the gut are sometimes referred to as the **brain-gut axis.**

This introductory chapter provides a broad overview of the anatomy and regulatory mechanisms in the GI tract. In the following chapters there will be discussion of the **integrated response to a meal** to provide the details of GI physiology. The response to a meal is classically divided into phases:

cephalic, oral, esophageal, gastric, duodenal, and intestinal. In each phase the meal presents certain **stimuli** (e.g., chemical, mechanical, and osmotic) that activate different **pathways** (e.g., neural, paracrine, and humoral reflexes) that result in changes in **effector function** (e.g., secretion and motility). There is considerable crosstalk between the regulatory mechanisms that have been outlined, and this will be discussed in the next chapters. As with maintenance of homeostasis in other systems of the body, control of GI function requires complex regulatory mechanisms to sense and act in a dynamic fashion.

IN THE CLINIC

Hirschsprung's disease is a congenital disorder of the enteric nervous system characterized by failure to pass meconium at birth or severe chronic constipation in infancy. The typical features are absence of myenteric and submucosal neurons in the distal part of the colon and rectum. It is a polygenic disorder with characteristic mutations in at least three different classes of genes involved in neuronal development and differentiation.

Key Concepts

1. The GI tract is a tube subdivided into regions that subserve different functions associated with digestion and absorption.
2. The lining of the GI tract is subdivided into layers—the mucosal, submucosal, and muscle layers.
3. There are three major control mechanisms: hormonal, paracrine, and neurocrine.
4. The innervation of the GI tract is particularly interesting because it consists of two interacting components, extrinsic and intrinsic.
5. Extrinsic innervation (cell bodies outside the wall of the GI tract) consists of the two subdivisions of the ANS: parasympathetic and sympathetic. Both have an important sensory (afferent) component.
6. The intrinsic or enteric nervous system (cell bodies in the wall of the GI tract) can act independently of extrinsic neural innervation.
7. When a meal is in different regions of the tract, sensory mechanisms detect the presence of the nutrients and mount appropriate physiological responses in that region of the tract, as well as in more distal regions. These responses are mediated by endocrine, paracrine, and neurocrine pathways.

Additional Reading

Baldwin GS. Posttranslational modification of gastrointestinal peptides. In: Johnson LR, ed. *Physiology of the GI Tract*. 5th ed. Waltham, MA: Academic Press; 2012.

Brierley SM, et al. Innervation of the gastrointestinal tract by spinal and vagal afferent nerves. In: Johnson LR, ed. *Physiology of the GI Tract*. 5th ed. Waltham, MA: Academic Press; 2012.

Brookes SJ, et al. Extrinsic primary afferent signalling in the gut. *Nat Rev Gastroenterol Hepatol*. 2013;10:286-296.

Chao C, Hellmich MR. Gastrointestinal peptides: gastrin, cholecystokinin, somatostatin and ghrelin. In: Johnson LR, ed. *Physiology of the GI Tract*. 5th ed. Waltham, MA: Academic Press; 2012.

Furness JB. The enteric nervous system and neurogastroenterology. *Nat Rev Gastroenterol Hepatol*. 2012;9:286-294.

Gomez GA, et al. Postpyloric gastrointestinal peptides. In: Johnson LR, ed. *Physiology of the GI Tract*. 5th ed. Waltham, MA: Academic Press; 2012.

Parker HE, et al. The role of gut endocrine cells in control of metabolism and appetite. *Exp Physiol*. 2014;99:1116-1120.

Sandoval DA, D'Alessio DA. Physiology of proglucagon peptides: role of glucagon and GLP-1 in health and disease. *Physiol Rev*. 2015;95:513-548.

28

The Cephalic, Oral, and Esophageal Phases of the Integrated Response to a Meal

LEARNING OBJECTIVES

Upon completion of this chapter the student should be able to answer the following questions:

1. What are the structures of the functional anatomy of salivary glands, including their secretory elements?
2. What are the cephalic and oral phases (what, why, how it happens) of the response to a meal?
3. What are the general principles of secretion along the gastrointestinal (GI) tract (where do secretions come from, what are the components)?
4. How do the components of secretion vary with the gland or region of the GI tract?
5. What is the correlation between the composition and functions of salivary secretion?
6. How are primary and secondary secretion within salivary glands generated and regulated?
7. What is the sequence of events in swallowing?
8. What are the stimulus and neural pathways generating primary and secondary esophageal peristalsis?
9. What changes in gastric motility take place during swallowing, and what is the significance?
10. What are the major functions of the esophagus and associated structures in terms of protection and propulsion?

This chapter will describe the processes that occur in the gastrointestinal (GI) tract in the early stages of the integrated response to a meal. There are changes in GI tract physiology (1) before food is ingested (the cephalic phase), (2) when ingested food is in the mouth (the oral phase), and (3) when food is transferred from the mouth to the esophagus (the esophageal phase). The responses of the GI tract to the presence of food are mainly associated with preparing the GI tract for digestion and absorption.

Cephalic and Oral Phases

The main feature of the **cephalic phase** is activation of the GI tract in readiness for the meal. The stimuli involved are cognitive and include anticipation or thinking about the consumption of food, olfactory input, visual input (seeing or smelling appetizing food when hungry), and auditory input. The latter may be an unexpected link but was clearly demonstrated in the classic conditioning experiments of Pavlov, in which he paired an auditory stimulus to the presentation of food to dogs; eventually the auditory stimulus alone could stimulate secretion. A real-life analogy is presumably being told that dinner is ready. All these stimuli result in an increase in excitatory parasympathetic neural outflow to the gut. Sensory input (e.g., smell) stimulates sensory nerves that activate parasympathetic outflow from the brainstem. Higher brain sites (e.g., limbic system, hypothalamus, cortex) are also involved in the cognitive components of this response. The response can be both positive and negative; thus anticipation of food and a person's psychological status, such as anxiety, can alter the cognitive response to a meal. However, the final common pathway is activation of the dorsal motor nucleus in the brainstem, the region where the cell bodies of the vagal preganglionic neurons arise. Activation of the nucleus leads to increased activity in efferent fibers passing to the GI tract in the vagus nerve. In turn the efferent fibers activate the postganglionic motor neurons (referred to as *motor* because their activation results in change of function of an effector cell). Increased parasympathetic outflow enhances salivary secretion, gastric acid secretion, pancreatic enzyme secretion, gallbladder contraction, and relaxation of the sphincter of Oddi (the sphincter between the common bile duct and duodenum). All these responses enhance the ability of the GI tract to receive and digest the incoming food. The salivary response is mediated via the ninth cranial nerve; the remaining responses are mediated via the vagus nerve.

Many of the features of the **oral phase** are indistinguishable from the cephalic phase. The only difference is that food is in contact with the surface of the GI tract. Thus there are additional stimuli generated from the mouth, both mechanical and chemical **(taste).** However, many of the responses initiated by the presence of food in the oral cavity are identical to those initiated in the cephalic phase, because the efferent pathway is the same. The responses specifically initiated in the mouth, which consist mainly of the stimulation of salivary secretion, will be discussed next.

The mouth is important for the mechanical disruption of food and for initiation of digestion. Chewing subdivides and mixes the food with the enzymes salivary amylase and lingual lipase and with the glycoprotein mucin, which lubricates food for chewing and swallowing. Minimal absorption occurs in the mouth, although alcohol and some drugs are absorbed from the oral cavity, and this can be clinically important. However, as with the cephalic phase, it is important to realize that stimulation of the oral cavity initiates responses in the more distal GI tract, including increased gastric acid secretion, increased pancreatic enzyme secretion, gallbladder contraction, and relaxation of the sphincter of Oddi, mediated via the efferent vagal pathway.

Properties of Secretion
General Considerations

Secretions in the GI tract come from glands associated with the tract (salivary glands, pancreas, and liver), from glands formed by the gut wall itself (e.g., submucosal glands in esophagus and duodenum), and from the intestinal mucosa itself. The exact nature of the secretory products can vary tremendously, depending on the function of that region of the GI tract. However, these secretions have several characteristics in common. Secretions from the GI tract and associated glands include **water, electrolytes, protein,** and **humoral agents.** Water is essential for generating an aqueous environment for efficient enzyme action. Secretion of electrolytes is important for generation of osmotic gradients to drive the movement of water. Digestive enzymes in secreted fluid catalyze the breakdown of macronutrients in ingested food. Moreover, many additional proteins secreted along the GI tract have specialized functions, some of which are fairly well understood, such as those of mucin and immunoglobulins, and others that are only just beginning to be understood, such as those of trefoil peptides.

Secretion is initiated by multiple signals associated with the meal, including chemical, osmotic, and mechanical components. Secretion is elicited by the action of specific effector substances called **secretagogues** acting on secretory cells. Secretagogues work in one of the three ways that have already been described in Chapter 27—endocrine, paracrine, and neurocrine.

Constituents of Secretions

Inorganic secretory components are region or gland specific, depending on the particular conditions required in that part of the GI tract. The inorganic components are electrolytes, including H^+ and HCO_3^-. Two examples of different secretions include acid (HCl) in the stomach, which is important to activate pepsin and start protein digestion, and HCO_3^- in the duodenum, which neutralizes gastric acid and provides

optimal conditions for the action of digestive enzymes in the small intestine.

Organic secretory components are also gland or organ specific and depend on the function of that region of the gut. The organic constituents are enzymes (for digestion), mucin (for lubrication and mucosal protection), and other factors such as growth factors, immunoglobulins, bile acids, and absorptive factors.

Salivary Secretion

During the cephalic and oral phases of the meal, considerable stimulation of salivary secretion takes place. Saliva has a variety of functions, including those important for the integrative responses to a meal and for other physiological processes (Box 28.1). The main functions of saliva in digestion include lubrication and moistening of food for swallowing, solubilization of material for taste, initiation of carbohydrate digestion, and clearance and neutralization of refluxed gastric secretions in the esophagus. Saliva also has antibacterial actions that are important for overall health of the oral cavity and teeth.

Functional Anatomy of the Salivary Glands

There are three pairs of major salivary glands: parotid, submandibular, and sublingual. In addition, many smaller glands are found on the tongue, lips, and palate. These glands are the typical **tubuloalveolar** structures of glands located in the GI tract (Fig. 28.1). The acinar portion of the gland is classified according to its major secretion: serous ("watery"), mucous, or mixed. The parotid gland produces mainly serous secretion, the sublingual gland secretes mainly mucus, and the submandibular gland produces a mixed secretion.

Cells in the secretory end pieces, or acini, are called *acinar cells* and are characterized by basally located nuclei, abundant rough endoplasmic reticulum, and apically located secretory granules that contain the enzyme amylase and other secreted proteins. There are also mucous cells in the acinus; the granules in these cells are larger and contain the specialized glycoprotein **mucin.** There are three kinds of ducts in the gland that transport secretions from the acinus to the opening in the mouth and also modify the secretion: intercalated ducts drain acinar fluid into larger

---

• BOX 28.1 Functions of Saliva and Chewing

Disruption of food to produce smaller particles
Formation of a bolus for swallowing
Initiation of starch and lipid digestion
Facilitation of taste
Production of intraluminal stimuli in the stomach
Regulation of food intake and ingestive behavior
Cleansing of the mouth and selective antibacterial action
Neutralization of refluxed gastric contents
Mucosal growth and protection in the rest of the GI tract
Aid in speech

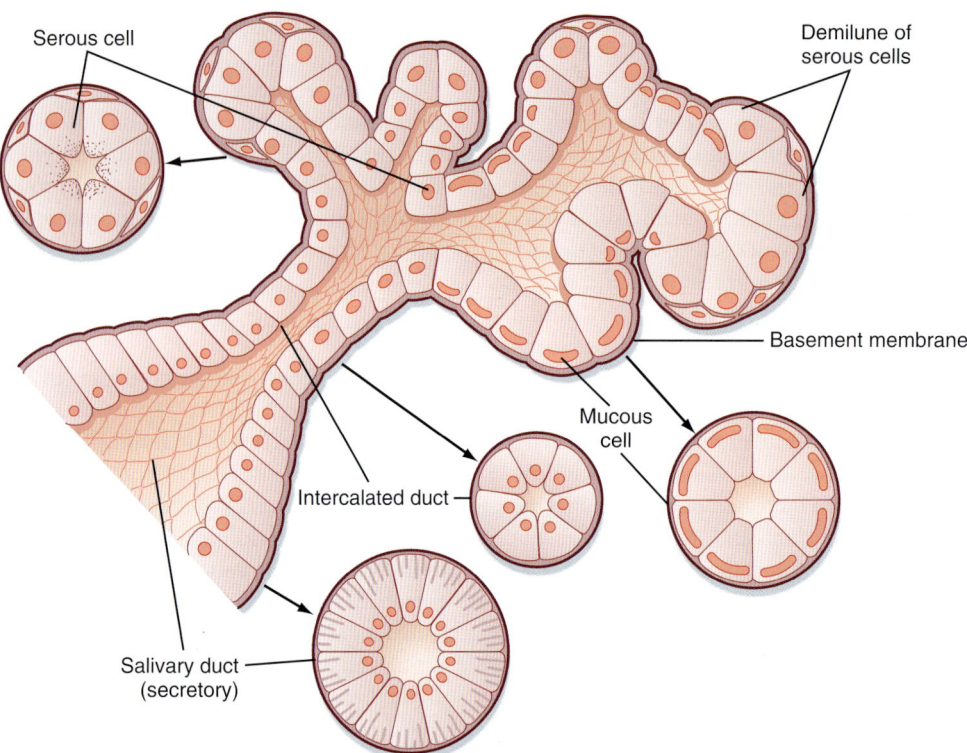

• **Fig. 28.1** General structure of tubuloalveolar secretory glands (e.g., salivary glands, pancreas) associated with the digestive tract.

ducts, the striated ducts, which then empty into even larger excretory ducts. In addition, a single large duct from each gland drains saliva to the mouth. The ductal cells lining the striated ducts, in particular, modify the ionic composition and osmolarity of saliva.

Composition of Saliva

The important properties of saliva are a large flow rate relative to the mass of gland, low osmolarity, high K^+ concentration, and organic constituents, including enzymes (amylase, lipase), mucin, and growth factors. The latter are not important in the integrated response to a meal but are essential for long-term maintenance of the lining of the GI tract.

The inorganic composition is entirely dependent on the stimulus and the rate of salivary flow. In humans, salivary secretion is always hypotonic. The major components are Na^+, K^+, HCO_3^-, Ca^{++}, Mg^{++}, and Cl^-. Fluoride can be secreted in saliva, and fluoride secretion forms the basis of oral fluoride treatment for prevention of dental caries. The concentration of ions varies with the rate of secretion; the flow rate of salivary secretion is stimulated during the postprandial period.

The **primary secretion** is produced by acinar cells in the secretory end pieces (or acini) and is modified by duct cells as saliva passes through the ducts. The primary secretion is isotonic, and the concentration of the major ions is similar to that in plasma. Secretion is driven predominantly by Ca^{++}-dependent signaling, which opens apical Cl^- channels in the acinar cells. Cl^- therefore flows out into the duct

lumen and establishes an osmotic and electrical gradient. Because the epithelium of the acinus is relatively leaky, Na^+ and water then follow across the epithelium via the tight junctions (i.e., via **paracellular transport**). Transcellular water movement may also occur, mediated by aquaporin 5 water channels. The amylase content and rate of fluid secretion vary with the type and level of stimulus. As the fluid passes along the ducts, the excretory and striated duct cells modify the ionic composition of the primary secretion to produce the **secondary secretion.** The duct cells reabsorb Na^+ and Cl^-, and secrete K^+ and HCO_3^- into the lumen. Na^+ is exchanged for protons, but some of the secreted protons are then reabsorbed in exchange for K^+. HCO_3^- on the other hand is secreted only in exchange for Cl^-, thereby alkalinizing salivary secretion.

At rest, final salivary secretion is hypotonic and slightly alkaline. The alkalinity of saliva is important in restricting microbial growth in the mouth, as well as in neutralizing refluxed gastric acid once the saliva is swallowed. When salivary secretion is stimulated, there is a small decrease in the K^+ concentration (but it always remains above plasma concentrations), the Na^+ concentration increases toward plasma levels, and Cl^- and HCO_3^- concentrations increase, thus the secreted fluid becomes even more alkaline (Fig. 28.2). Note that HCO_3^- secretion can be directly stimulated by the action of secretagogues on duct cells. The duct epithelium is relatively tight and lacks expression of aquaporin, and therefore water cannot follow the ions rapidly enough to maintain isotonicity at moderate or high flow rates during stimulated salivary secretion. Thus with

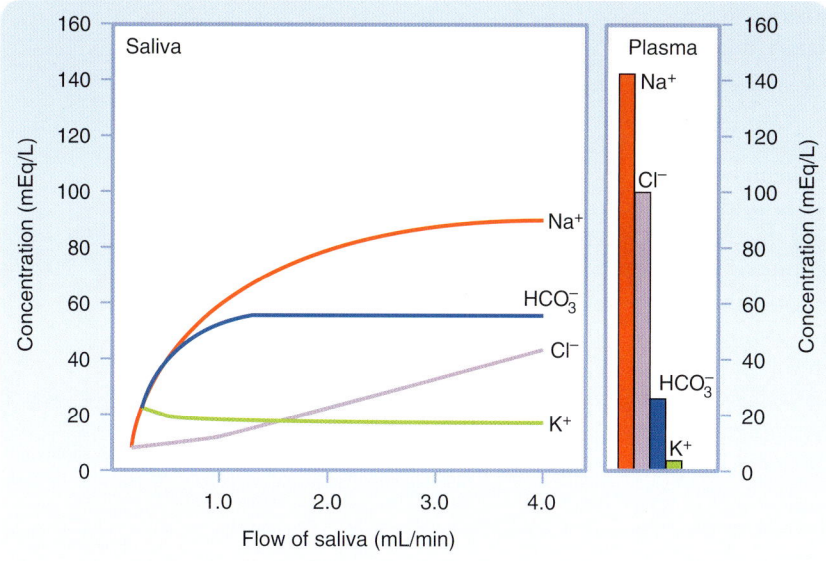

A

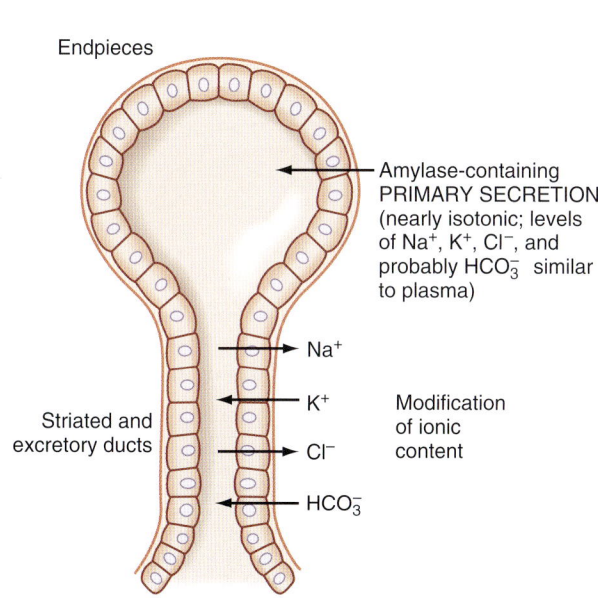

B

• **Fig. 28.2** A, The composition of salivary secretion as a function of the salivary flow rate compared with the concentration of ions in plasma. Saliva is hypotonic to plasma at all flow rates. [HCO_3^-] in saliva exceeds that in plasma except at very low flow rates. B, Schematic representation of the two-stage model of salivary secretion. The primary secretion containing amylase and electrolytes is produced in the acinar cell. The concentration of electrolytes in plasma is similar to that in the primary secretion, but it is modified as it passes through ducts that absorb Na^+ and Cl^- and secrete K^+ and HCO_3^-.

an increase in secretion rate, there is less time for ionic modification by the duct cells, and the resulting saliva more closely resembles the primary secretion and therefore plasma. However, [HCO_3^-] remains high because secretion from duct cells and possibly acinar cells is stimulated (see Fig. 28.2).

The organic constituents of saliva—proteins and glycoproteins—are synthesized, stored, and secreted by the acinar cells. The major products are amylase (an enzyme that initiates starch digestion), lipase (important for lipid digestion), glycoprotein (mucin, which forms mucus when hydrated), and lysozyme (attacks bacterial cell walls to limit colonization of bacteria in the mouth). Although salivary amylase begins the process of digestion of carbohydrates, it is not required in healthy adults because of the excess

of pancreatic amylase. Similarly the importance of lingual lipase is unclear.

Metabolism and Blood Flow of Salivary Glands

The salivary glands produce a prodigious flow of saliva. The maximal rate of saliva production in humans is about 1 mL/min/g of gland; thus at this rate, the glands are producing their own weight in saliva each minute. Salivary glands have a high rate of metabolism and high blood flow; both are proportional to the rate of saliva formation. Blood flow to maximally secreting salivary glands is approximately 10 times that of an equal mass of actively contracting skeletal muscle. Stimulation of the parasympathetic nerves to salivary glands increases blood flow by dilating the vasculature of the glands. Vasoactive intestinal polypeptide (VIP) and acetylcholine are released from parasympathetic nerve terminals in the salivary glands and are vasodilatory during secretion.

Regulation of Salivary Secretion

Control of salivary secretion is exclusively neural. In contrast, control of most other GI secretions is primarily hormonal. Salivary secretion is stimulated by both the sympathetic and parasympathetic subdivisions of the autonomic nervous system. Excitation of either sympathetic or parasympathetic nerves to the salivary glands stimulates salivary secretion. Primary physiological control of the salivary glands during the response to a meal is by the parasympathetic nervous system. If the parasympathetic supply is interrupted, salivation is severely impaired and the salivary glands atrophy.

Sympathetic fibers to the salivary glands stem from the superior cervical ganglion. Preganglionic parasympathetic fibers travel via branches of the facial and glossopharyngeal nerves (cranial nerves VII and IX, respectively). These fibers form synapses with postganglionic neurons in ganglia in or near the salivary glands. The acinar cells and ducts are supplied with parasympathetic nerve endings.

Parasympathetic stimulation increases synthesis and secretion of salivary amylase and mucins, enhances the transport activities of the ductular epithelium, greatly increases blood flow to the glands, and stimulates glandular metabolism and growth.

Ionic Mechanisms of Salivary Secretion

Ion Transport in Acinar Cells

Fig. 28.3 shows a simplified view of the mechanisms of ion secretion by serous acinar cells. The basolateral membrane of the cell contains **Na^+,K^+-ATPase** and an **Na^+-K^+-$2Cl^-$ symporter.** The concentration gradient for Na^+ across the basolateral membrane, which is dependent on Na^+,K^+-ATPase, provides the driving force for entry of Na^+, K^+, and Cl^- into the cell. Cl^- and HCO_3^- leave the acinar cell and enter the lumen via an anion channel located in the apical membrane of the acinar cell. This secretion of anions drives the entry of Na^+ and thus water into the acinar lumen across the relatively leaky tight junctions.

Acinar cell fluid secretion is strongly enhanced in response to elevations in intracellular [Ca^{++}] as a result of activation of the muscarinic receptor for acetylcholine.

Ion Transport in Ductular Cells

Fig. 28.4 shows a simplified model of ion transport processes in epithelial cells of the excretory and striated ducts. Na^+,K^+-ATPase located in the basolateral membrane maintains the electrochemical gradients for Na^+ and K^+ that drive most of the other ionic transport processes of the cell. In the apical membrane the parallel operation of the Na^+/H^+ antiporter, the Cl^-/HCO_3^- antiporter, and the H^+/K^+ antiporter results in absorption of Na^+ and Cl^- from the lumen and secretion of K^+ and HCO_3^- into the lumen. The relative impermeability of the ductular epithelium to

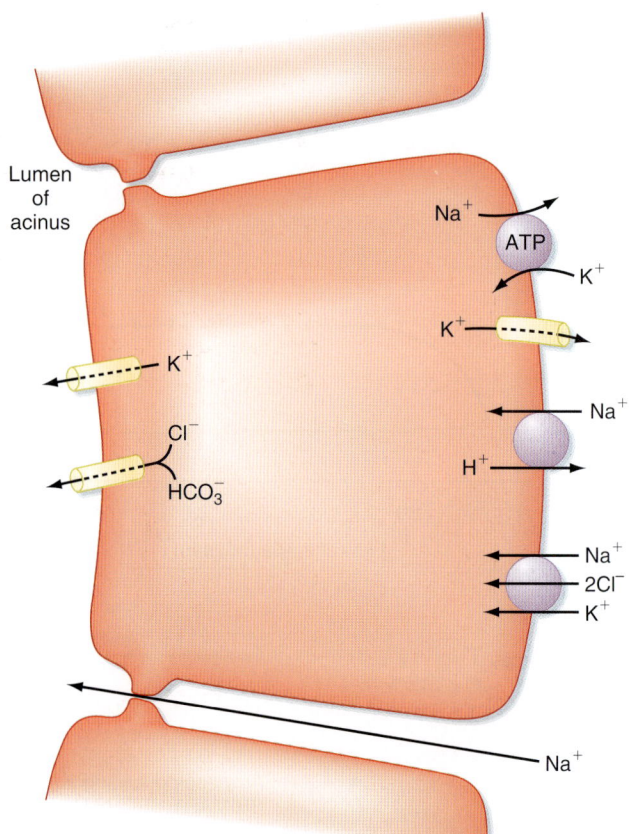

• **Fig. 28.3** Ionic transport mechanism involved in the secretion of amylase and electrolytes in salivary acinar cells.

water prevents the ducts from absorbing too much water by osmosis.

Swallowing

Swallowing can be initiated voluntarily, but thereafter it is almost entirely under reflex control. The **swallowing reflex** is a rigidly ordered sequence of events that propel food from the mouth to the pharynx and from there to the stomach. This reflex also inhibits respiration and prevents entrance of food into the trachea during swallowing. The afferent limb of the swallowing reflex begins when touch receptors, most notably those near the opening of the pharynx, are stimulated. Sensory impulses from these receptors are transmitted to an area in the medulla and lower pons called the *swallowing center.* Motor impulses travel from the swallowing center to the musculature of the pharynx and upper esophagus via various cranial nerves and to the remainder of the esophagus by vagal motor neurons.

The timing of events in swallowing is shown in Fig. 28.5. The voluntary phase of swallowing is initiated when the tip of the tongue separates a bolus of food from the mass of food in the mouth. First the tip of the tongue and later the more posterior portions of the tongue press against the hard palate. The action of the tongue moves the bolus upward and then backward into the mouth. The bolus is forced into the pharynx, where it stimulates the touch receptors

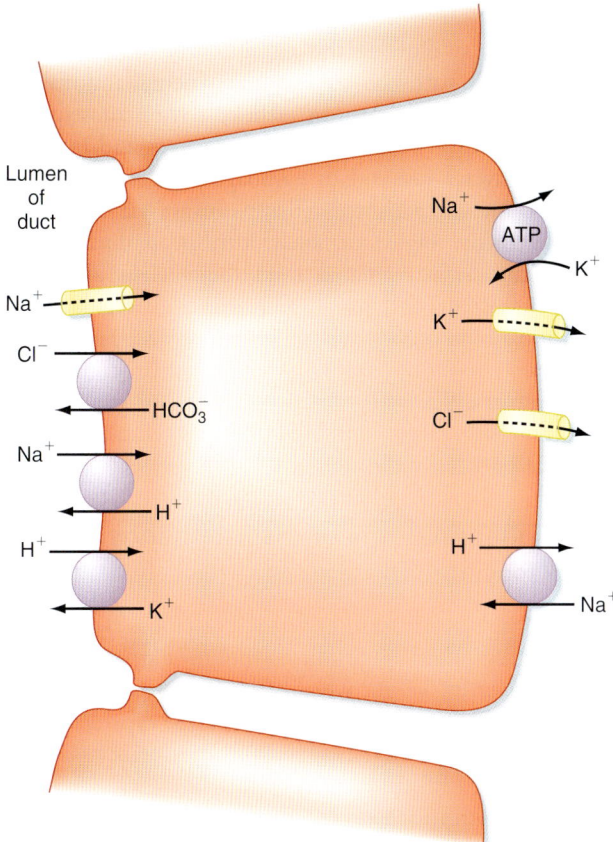

• **Fig. 28.4** Ionic transport mechanism involved in secretion and absorption in epithelial cells of the striated and excretory duct of the salivary gland.

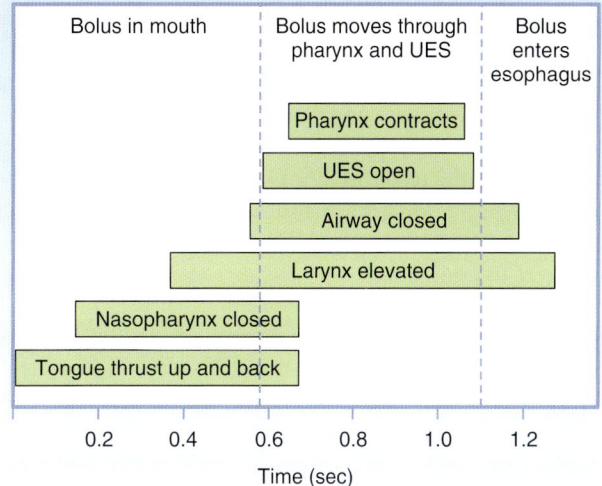

BOLUS TRANSFER FROM THE MOUTH TO THE ESOPHAGUS REQUIRES MULTIPLE EVENTS

 Fig. 28.5 Timing of motor events in the pharynx and UES during a swallow.

AT THE CELLULAR LEVEL

The acinar cells and duct cells of the salivary glands respond to both cholinergic and adrenergic agonists. Nerves stimulate the release of acetylcholine, norepinephrine, substance P, and VIP by salivary glands, and these hormones increase the secretion of amylase and the flow of saliva. These neurotransmitters act mainly by elevating the intracellular concentration of cyclic adenosine monophosphate (cAMP) and by increasing the concentration of Ca^{++} in the cytosol. Acetylcholine and substance P, acting on muscarinic and tachykinin receptors, respectively, increase the cytosolic concentration of Ca^{++} in serous acinar cells. In contrast, norepinephrine, acting on β receptors, and VIP, binding to its receptor, elevate the cAMP concentration in acinar cells. Agonists that elevate the cAMP concentration in serous acinar cells elicit a secretion that is rich in amylase; agonists that mobilize Ca^{++} elicit a secretion that is more voluminous but has a lower concentration of amylase. Ca^{++}-mobilizing agonists may also elevate the concentration of cyclic guanosine monophosphate (cGMP), which may mediate the trophic effects evoked by these agonists.

IN THE CLINIC

Xerostomia, or dry mouth, is caused by impaired salivary secretion. It can be congenital or develop as part of an autoimmune process. The decrease in secretion reduces the pH in the oral cavity, which causes tooth decay and is associated with esophageal erosions. Reduced secretion also causes difficulty swallowing.

IN THE CLINIC

The ability to measure and monitor a wide range of molecular components that are indicative of overall health is useful in diagnosis and monitoring. Saliva is easy to access, and collection of it is noninvasive. It is used to identify individuals with disease (presence of biomarkers) and to monitor the progress of affected individuals under treatment. In endocrinology, levels of steroids can be measured in the free form rather than as the free and bound form, as in plasma (e.g., the stress hormone cortisol and the sex hormones estradiol, progesterone, and testosterone). Viral infections such as human immunodeficiency virus (HIV), herpes, hepatitis C, and Epstein-Barr virus infection can be detected by polymerase chain reaction (PCR) techniques. Bacterial infections, such as **Helicobacter pylori,** can likewise be detected in saliva, and saliva is also used for monitoring drug levels.

that initiate the swallowing reflex. The pharyngeal phase of swallowing involves the following sequence of events, which occur in less than 1 second:
1. The soft palate is pulled upward and the palatopharyngeal folds move inward toward one another; these movements prevent reflux of food into the nasopharynx and open a narrow passage through which food moves into the pharynx.
2. The vocal cords are pulled together and the larynx is moved forward and upward against the epiglottis; these actions prevent food from entering the trachea and help open the upper esophageal sphincter (UES).

3. The UES relaxes to receive the bolus of food.
4. The superior constrictor muscles of the pharynx then contract strongly to force the bolus deeply into the pharynx.

A peristaltic wave is initiated with contraction of the pharyngeal superior constrictor muscles, and the wave moves toward the esophagus. This wave forces the bolus of food through the relaxed UES. During the pharyngeal stage of swallowing, respiration is also reflexively inhibited. After the bolus of food passes the UES, a reflex action causes the sphincter to constrict.

 IN THE CLINIC

Gastroesophageal reflux disease (GERD) is commonly referred to as *heartburn* or *indigestion*. It occurs when the lower esophageal sphincter allows the acidic contents of the stomach to reflux back into the distal part of the esophagus. This region of the esophagus, unlike the stomach, does not have a robust system to protect the mucosal lining. Thus the acid will activate pain fibers and thereby result in discomfort and pain. This is not an uncommon phenomenon, even in healthy individuals. In the long term, continual reflux can result in damage to the esophageal mucosa. In this case, this condition is classed as GERD and can be treated by H_2 receptor antagonists that reduce gastric acid secretion (e.g., ranitidine [Zantac]) or by proton pump inhibitors (e.g., omeprazole [Prilosec]).

Esophageal Phase

The **esophagus,** the **UES,** and the **lower esophageal sphincter (LES)** serve two main functions (Fig. 28.6). First, they propel food from the mouth to the stomach. Second, the sphincters protect the airway during swallowing and protect the esophagus from acidic gastric secretions.

The stimuli that initiate the changes in smooth muscle activity that result in these **propulsive** and **protective functions** are mechanical and consist of pharyngeal stimulation during swallowing and distention of the esophageal wall itself. The pathways are exclusively neural and involve both extrinsic and intrinsic reflexes. Mechanosensitive afferents in both the extrinsic (vagus) nerves and intrinsic neural pathways respond to esophageal distention. These pathways include activated reflex pathways via the brainstem (extrinsic, vagus) or solely intrinsic pathways. The striated muscle is regulated from the nucleus ambiguus in the brainstem, and the smooth muscle is regulated by parasympathetic outflow via the vagus nerve. The changes in function resulting from mechanosensitive stimuli and activation of reflex pathways are peristalsis of striated and smooth muscle, relaxation of the LES, and relaxation of the proximal portion of the stomach.

Functional Anatomy of the Esophagus and Associated Structures

The esophagus, like the rest of the GI tract, has two muscle layers—circular and longitudinal—but the esophagus is one of two places in the gut where striated muscle occurs, the other being the external anal sphincter. The type of muscle (striated or smooth) in the esophagus varies along its length. The UES and LES are formed by thickening of striated or circular smooth muscle, respectively.

Motor Activity During the Esophageal Phase

The UES, esophagus, and LES act in a coordinated manner to propel material from the pharynx to the stomach. At the end of a swallow, a bolus passes through the UES, and the presence of the bolus, via stimulation of mechanoreceptors and reflex pathways, initiates a peristaltic wave (alternating contraction and relaxation of the muscle) along the esophagus that is called **primary peristalsis** (Fig. 28.7). This wave moves down the esophagus slowly (3–5 cm/s). Distention of the esophagus by the moving bolus initiates another

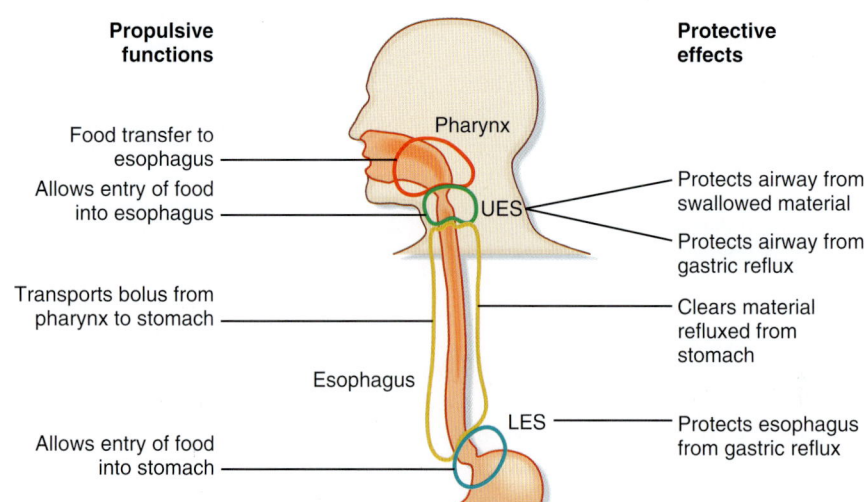

• **Fig. 28.6** The esophagus and associated sphincters have multiple functions involved in movement of food from the mouth to the stomach and also in protection of the airway and esophagus.

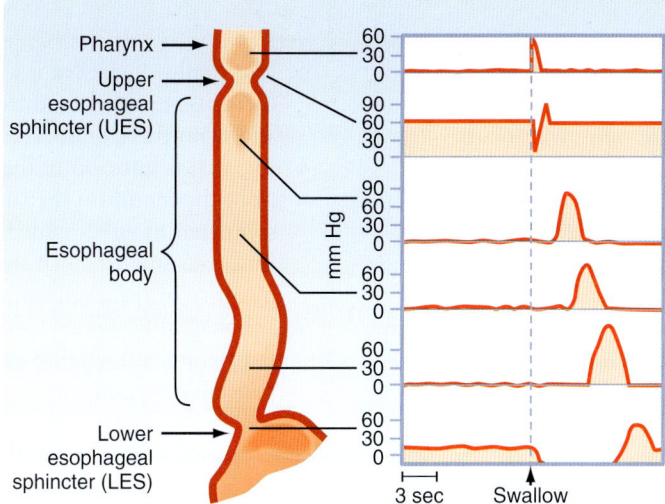

• **Fig. 28.7** Changes in pressure in the different regions of the pharynx, esophagus, and associated sphincters initiated during a swallow. The pressure trace is a diagrammatic representation from that obtained during manometry in an awake human. Stimulation of the pharynx by the presence of a bolus initiates a decrease in pressure (= opening) of the UES and a peristaltic wave of contraction along the esophagus. Stimulation of the pharynx also relaxes the smooth muscle of the LES to prepare for entry of food.

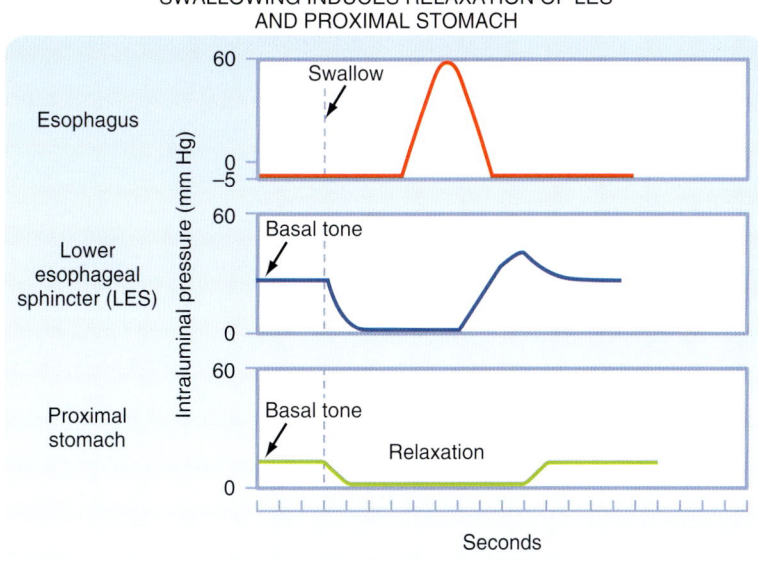

• **Fig. 28.8** Swallowing in the form of pharyngeal stimulation induces neural reflex relaxation of the LES and the proximal part of the stomach to allow entry of food.

wave called **secondary peristalsis.** Frequently, repetitive secondary peristalsis is required to clear the esophagus of the bolus. Stimulation of the pharynx by the swallowed bolus also produces reflex relaxation of the LES and the most proximal region of the stomach. Thus when the bolus reaches the LES, it is already relaxed to allow passage of the bolus into the stomach. Similarly the portion of the stomach that receives the bolus is relaxed. In addition, esophageal distention produces further receptive relaxation of the stomach. The proximal part of the stomach relaxes at the same time as the LES; this occurs with each swallow,

and its function is to allow the stomach to accommodate large volumes with a minimal rise in intragastric pressure. This process is called **receptive relaxation** (Fig. 28.8).

The LES also has important protective functions. It is involved in preventing acid reflux from the stomach back into the esophagus. An insufficient tonic contraction of the LES is associated with reflux disease, a gradual erosion of the esophageal mucosa, which is not as well protected as the gastric and duodenal mucosa. There is also some evidence that peristalsis in the absence of swallowing (secondary peristalsis) is important for clearing refluxed gastric contents.

Key Concepts

1. The cephalic and oral phases of the meal share many characteristics and prepare the remainder of the GI tract for the meal; these responses are neurally mediated, predominantly by the efferent vagus nerve.

2. Salivary secretion has important functions and, together with chewing of the food, allows the formation of a bolus that can be swallowed and passed along the esophagus to the stomach.

3. The ionic composition of salivary secretion varies with the flow rate, which is stimulated during a meal. The primary secretion comes from cells in the acini and is modified by epithelial cells as it passes through the ducts.

4. Regulation of salivary secretion is exclusively neural; parasympathetic innervation is most important in the response to food.

5. The swallowing reflex is a rigidly ordered sequence of events that propel food from the mouth to the pharynx and from there to the stomach.

6. The major function of the esophagus is to propel food from the mouth to the stomach. The esophagus has sphincters at either end that are involved in protective functions important in swallowing and preserving the integrity of the esophageal mucosa.

7. Esophageal peristalsis (primary) is stimulated by mechanical stimulation of the pharynx, and secondary peristalsis is stimulated by distention of the esophageal wall.

8. Esophageal function and the associated sphincters are regulated by extrinsic and intrinsic neural pathways.

Additional Reading

Catalan MA, et al. Salivary gland secretion. In: Johnson LR, ed. *Physiology of the GI Tract*. 5th ed. Waltham, MA: Academic Press; 2012.

Lee MG, et al. Molecular mechanism of pancreatic and salivary gland fluid and HCO3 secretion. *Physiol Rev*. 2012;92:39-74.

Mittal RK. *Motor Function of the Pharynx, Esophagus, and its Sphincters*. San Rafael, CA: Morgan & Claypool Life Sciences; 2011.

29

The Gastric Phase of the Integrated Response to a Meal

LEARNING OBJECTIVES

Upon completion of this chapter the student should be able to answer the following questions:

1. What are the major functions of the monogastric stomach?
2. What are the gross functional regions of the stomach?
3. What is the role of the gastric epithelium in digestion and absorption?
4. What is the role of the proton pump in parietal cell function?
5. What are some examples of how gastric acid secretion is regulated during the postprandial period?
6. What are the differences between gastric mucosal protection and defense?
7. What are the components of the functional anatomy of GI smooth muscle?
8. What is the significance of gap junctions, interstitial cells of Cajal, and pacemaker cells in the functioning of GI smooth muscle?
9. How is the basic electrical rhythm (slow wave) generated, how is it regulated by chemical messengers (hormones, paracrine, neurotransmitters), and what causes contractions associated with the slow wave to occur?
10. What physiological events in gastric motility occur in the gastric phase?

In this chapter, gastrointestinal (GI) tract physiology when food is in the stomach (i.e., the gastric phase of digestion) will be discussed. This includes gastric function and its regulation, in addition to changes in function that occur in more distal regions of the GI tract. The main functions of the stomach are to act as a temporary reservoir for the meal and to initiate protein digestion through secretion of acid and the enzyme precursor pepsinogen. Other functions are listed in Box 29.1.

Food entering the stomach from the esophagus causes mechanical stimulation of the gastric wall via distention and stretching of smooth muscle. Food—predominantly oligopeptides and amino acids—also provides chemical stimulation when present in the gastric lumen. Regulation of gastric function during the gastric phase is dependent on

endocrine, paracrine, and neural pathways. These pathways are activated by mechanical and chemical stimuli, which result in intrinsic and extrinsic neural reflex pathways that are important for regulation of gastric function. Afferent neurons that pass from the GI tract to the central nervous system (and to a lesser extent to the spinal cord) via the vagus nerve respond to these mechanical and chemical stimuli and activate parasympathetic outflow.

The endocrine pathways include the release of **gastrin,** which stimulates gastric acid secretion, and the release of **somatostatin,** which inhibits gastric secretion. Important paracrine pathways include **histamine** release, which stimulates gastric acid secretion. The responses elicited by activation of these pathways include both secretory and motor responses; *secretory responses* include secretion of acid, pepsinogen, mucus, intrinsic factor, gastrin, lipase, and HCO_3^-. Overall these secretions initiate protein digestion and protect the gastric mucosa. *Motor responses* (changes in activity of smooth muscle) include inhibition of motility of the proximal part of the stomach (receptive relaxation) and stimulation of motility of the distal part of the stomach, which causes antral peristalsis. These changes in motility play important roles in storage and mixing of the meal with secretions and are also involved in regulating the flow of contents out of the stomach.

Functional Anatomy of the Stomach

The stomach is divided into three regions: the **cardia,** the **corpus** (also referred to as the *fundus* or *body*), and the **antrum** (Fig. 29.1). However, when discussing the physiology of the stomach, it is helpful to think of it as subdivided into *two functional regions:* the **proximal** and **distal** parts of the stomach. The proximal portion of the stomach (called *proximal* because it is the most cranial) and the distal portion of the stomach (furthest away from the mouth) have quite different functions in the postprandial response to a meal, which will be discussed later.

The lining of the stomach is covered with a columnar epithelium folded into **gastric pits;** each pit is the opening of a duct into which one or more gastric glands empty (Fig. 29.2). The gastric pits account for a significant fraction of the total surface area of the gastric mucosa. The gastric mucosa is divided into three distinct regions based on the structure of the glands. The small cardiac glandular region, located just below the lower esophageal sphincter (LES), primarily contains mucus-secreting gland cells. The remainder of the gastric mucosa is divided into the **oxyntic** or **parietal** (acid-secreting) **gland region,** located above the gastric notch (equivalent to the proximal part of the stomach), and the pyloric gland region, located below the notch (equivalent to the distal part of the stomach).

The structure of a gastric gland from the oxyntic glandular region is illustrated in Fig. 29.2. Surface epithelial cells extend slightly into the duct opening. The opening of the gland is called the **isthmus** and is lined with surface mucous cells and a few parietal cells. Mucous neck cells are located in the narrow **neck** of the gland. Parietal or oxyntic cells, which secrete HCl and intrinsic factor (involved in absorption of vitamin B_{12}), and **chief** or **peptic cells,** which secrete pepsinogens, are located deeper in the gland. Oxyntic glands also contain **enterochromaffin-like** (ECL) cells that secrete histamine, and D cells that secrete somatostatin. Parietal cells are particularly numerous in glands in the fundus, whereas mucus-secreting cells are more numerous in glands

---

• BOX 29.1   Functions of the Stomach

Storage—acts as temporary reservoir for the meal
Secretion of H^+ to kill microorganisms and convert pepsinogen to its active form
Secretion of intrinsic factor to absorb vitamin B_{12} (cobalamin)
Secretion of mucus and HCO_3^- to protect the gastric mucosa
Secretion of water for lubrication and to provide aqueous suspension of nutrients
Motor activity for mixing secretions (H^+ and pepsin) with ingested food
Coordinated motor activity to regulate the emptying of contents into the duodenum

---

of the pyloric (antral) glandular region. In addition the pyloric glands contain G cells that secrete the hormone gastrin. The parietal glands are also divided into regions: the neck (mucous neck cells and parietal cells) and the base (peptic/chief and parietal cells). Endocrine cells are scattered throughout the glands.

Gastric Secretion

The fluid secreted into the stomach is called **gastric juice.** Gastric juice is a mixture of the secretions of the surface epithelial cells and the secretions of gastric glands. One of the most important components of gastric juice is H^+, which is secreted against a very large concentration gradient. Thus H^+ secretion by the parietal mucosa is an energy-intensive process. The cytoplasm of the parietal cell is densely packed with mitochondria, which have been estimated to fill 30% to 40% of the cell's volume. One major function of H^+ is conversion of inactive pepsinogen (the major enzyme product of the stomach) to pepsins, which initiate protein digestion in the stomach. Additionally, H^+ ions are important for preventing invasion and colonization of the gut by bacteria and other pathogens that may be ingested with food. The stomach also secretes significant amounts of HCO_3^- and mucus, which are important for protection of the gastric mucosa against the acidic and peptic luminal environment. The gastric epithelium also secretes intrinsic factor, which is necessary for absorption of vitamin B_{12} **(cobalamin).** The functions of other components of gastric juice are redundant with secretions provided more distally in the GI tract.

Composition of Gastric Secretions

Like other GI secretions, gastric juice consists of inorganic and organic constituents together with water. Among the important components of gastric juice are HCl, salts,

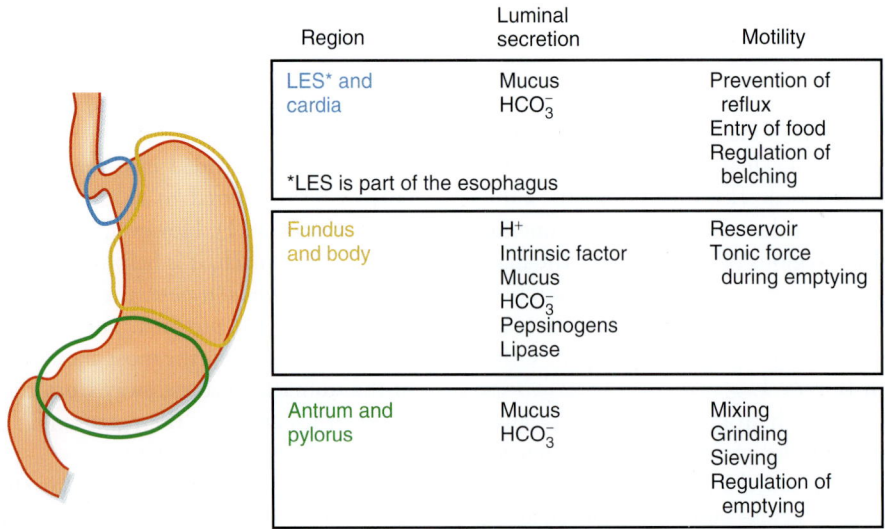

Region	Luminal secretion	Motility
LES* and cardia	Mucus HCO_3^-	Prevention of reflux Entry of food Regulation of belching
*LES is part of the esophagus		
Fundus and body	H^+ Intrinsic factor Mucus HCO_3^- Pepsinogens Lipase	Reservoir Tonic force during emptying
Antrum and pylorus	Mucus HCO_3^-	Mixing Grinding Sieving Regulation of emptying

• Fig. 29.1 The three functional regions of the stomach. The regions have different luminal secretions and patterns of smooth muscle activity indicative of their unique functions in response to food.

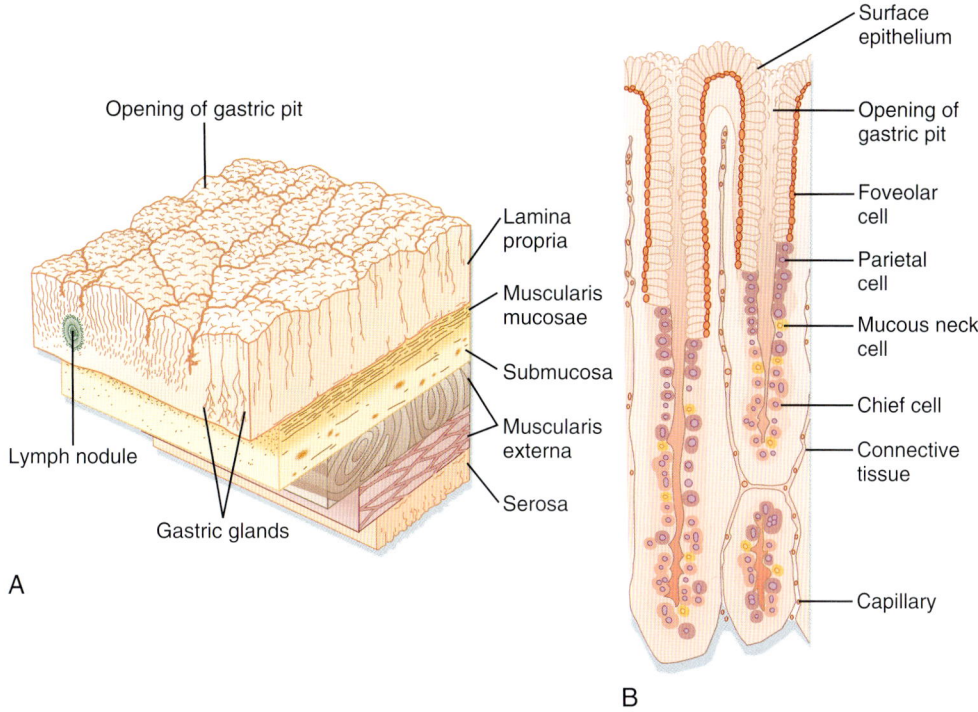

• **Fig. 29.2** Representation of the structure of the gastric mucosa showing a section through the wall of the stomach (A) and detail of the structure of gastric glands and cell types in the mucosa (B).

pepsins, intrinsic factor, mucus, and HCO_3^-. Secretion of all these components increases after a meal.

Inorganic Constituents of Gastric Juice

The ionic composition of gastric juice depends on the rate of secretion. The higher the secretory rate, the higher the concentration of H^+ ions. At lower secretory rates, $[H^+]$ decreases and $[Na^+]$ increases. $[K^+]$ is always higher in gastric juice than in plasma. Consequently, prolonged vomiting may lead to hypokalemia. At all rates of secretion, Cl^- is the major anion of gastric juice. Gastric HCl converts pepsinogens to active pepsins and provides the acid pH at which pepsins are active.

The rate of gastric H^+ secretion varies considerably among individuals. In humans, basal (unstimulated) rates of gastric H^+ production typically range from about 1 to 5 mEq/hr. During maximal stimulation, HCl production rises to 6 to 40 mEq/hr. The basal rate is greater at night and lowest in the early morning. The total number of parietal cells in the stomach of normal individuals varies greatly, and this variation is partly responsible for the wide range in basal and stimulated rates of HCl secretion.

Organic Constituents of Gastric Juice

The predominant organic constituent of gastric juice is **pepsinogen,** the inactive proenzyme of pepsin. Pepsins, often collectively called "pepsin," are a group of proteases secreted by the chief cells of the gastric glands. Pepsinogens are contained in membrane-bound zymogen granules in the chief cells. Zymogen granules release their contents by exocytosis when chief cells are stimulated to secrete

TABLE 29.1	Stimulation of Chief Cells in the Integrated Response to a Meal	
Stimulant		**Source**
Acetylcholine (ACh)		Enteric neurons
Gastrin		G cells in the gastric antrum
Histamine		ECL cells in the gastric corpus
Cholecystokinin (CCK)		I cells in the duodenum
Secretin		S cells in the duodenum

(Table 29.1). Pepsinogens are converted to active pepsins by the cleavage of acid-labile linkages. The lower the pH, the more rapid the conversion. Pepsins also act proteolytically on pepsinogens to form more pepsin. Pepsins are most proteolytically active at pH 3 and below. Pepsins may digest as much as 20% of the protein in a typical meal but are not required for digestion, because their function can be replaced by that of pancreatic proteases. When the pH of the duodenal lumen is neutralized, pepsins are inactivated by the neutral pH.

Intrinsic factor, a glycoprotein secreted by parietal cells of the stomach, is required for normal absorption of vitamin B_{12}. Intrinsic factor is released in response to the same stimuli that elicit secretion of HCl by parietal cells.

Cellular Mechanisms of Gastric Acid Secretion

Parietal cells have a distinctive ultrastructure (Fig. 29.3). Branching secretory canaliculi course through the cytoplasm

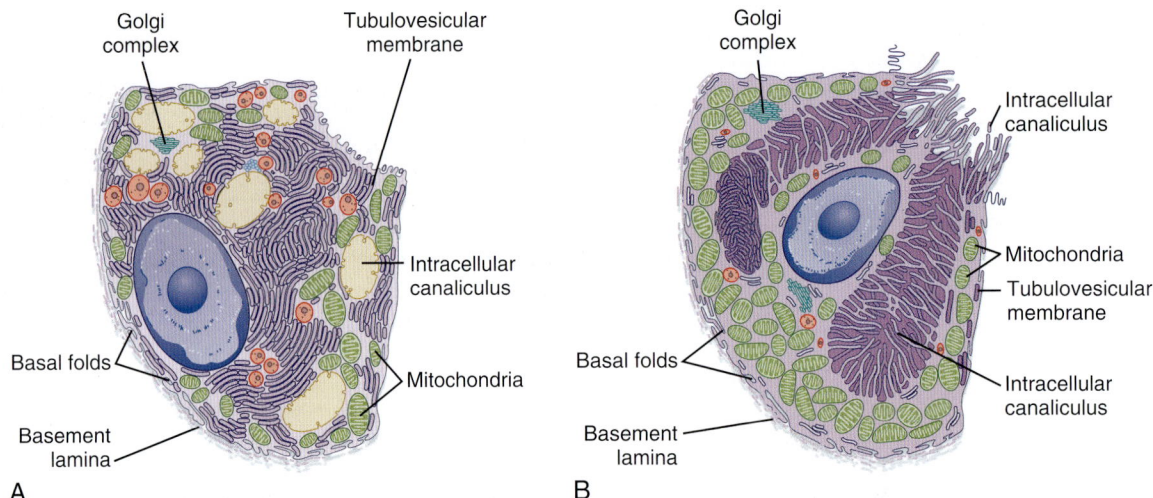

• **Fig. 29.3** Parietal cell ultrastructure. A, A resting parietal cell showing the tubulovesicular apparatus in the cytoplasm and the intracellular canaliculus. B, An activated parietal cell that is secreting acid. The tubulovesicles have fused with the membranes of the intracellular canaliculus, which is now open to the lumen of the gland and lined with abundant long microvilli.

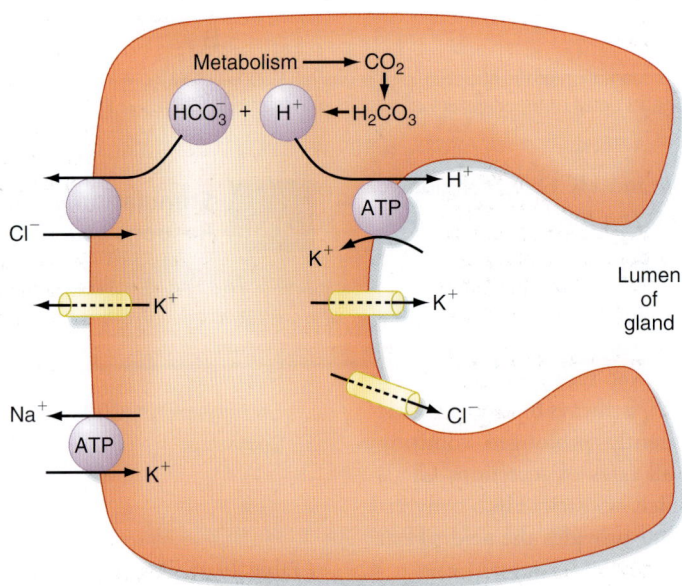

• **Fig. 29.4** Mechanism of H^+ and Cl^- secretion by an activated parietal cell in the gastric mucosa. ATP, adenosine triphosphate.

and are connected by a common outlet to the cell's luminal surface. Microvilli line the surfaces of the **secretory canaliculi.** The cytoplasm of unstimulated parietal cells contains numerous tubules and vesicles called the *tubulovesicular system.* The membranes of tubulovesicles contain the transport proteins responsible for secretion of H^+ and Cl^- into the lumen of the gland. When parietal cells are stimulated to secrete HCl (see Fig. 29.3), **tubulovesicular membranes** fuse with the plasma membrane of the secretory canaliculi. This extensive membrane fusion greatly increases the number of H^+/K^+ antiporters in the plasma membrane of the secretory canaliculi. When parietal cells secrete gastric acid at the maximal rate, H^+ is pumped against a concentration gradient that is about 1 million–fold. Thus the pH

is 7 in the parietal cell cytosol and 1 in the lumen of the gastric gland.

The cellular mechanism of H^+ secretion by the parietal cell is depicted in Fig. 29.4. Cl^- enters the cell across the basolateral membrane in exchange for HCO_3^- generated in the cell by the action of carbonic anhydrase, which produces HCO_3^- and H^+. H^+ is secreted across the luminal membrane by H^+,K^+-ATPase in exchange for K^+. K^+ recycles across the luminal membrane via a K^+ channel. Cl^- enters the lumen via an ion channel (a CLC family Cl^- channel) located in the luminal membrane. Increased intracellular Ca^{++} and cyclic adenosine monophosphate (cAMP) stimulate luminal membrane conduction of Cl^- and K^+. Increased K^+ conductance hyperpolarizes the luminal membrane potential, which increases the driving force for efflux of Cl^- across the luminal membrane. The K^+ channel in the basolateral membrane also mediates the efflux of K^+ that accumulates in the parietal cell via the activity of H^+,K^+-ATPase. In addition, cAMP and Ca^{++} promote trafficking of Cl^- channels into the luminal membrane, as well as fusion of cytosolic tubulovesicles containing H^+,K^+-ATPase with the membrane of the secretory canaliculi (see Figs. 29.3 and 29.4). Parietal cell secretion of H^+ is also accompanied by transport of HCO_3^- into the bloodstream to maintain intracellular pH.

Secretion of HCO_3^-

The surface epithelial cells also secrete a watery fluid that contains Na^+ and Cl^- in concentrations similar to those in plasma but with higher K^+ and HCO_3^- concentrations. HCO_3^- is entrapped by the viscous mucus that coats the surface of the stomach; thus the mucus secreted by the resting mucosa lines the stomach with a sticky alkaline coat. When food is eaten, rates of secretion of both mucus and HCO_3^- increase.

Secretion of Mucus

Secretions that contain **mucins** are viscous and sticky and are collectively termed *mucus*. Mucins are secreted by mucous neck cells located in the necks of gastric glands and by the surface epithelial cells of the stomach. Mucus is stored in large granules in the apical cytoplasm of mucous neck cells and surface epithelial cells and is released by exocytosis.

Gastric mucins are about 80% carbohydrate by weight and consist of four similar monomers of about 500,000 Da each that are linked together by disulfide bonds (Fig. 29.5).

These tetrameric mucins form a sticky gel that adheres to the surface of the stomach. This gel is subject to proteolysis by pepsins that cleave disulfide bonds near the center of the tetramers. Proteolysis releases fragments that do not form gels and thus dissolves the protective mucous layer. Maintenance of the protective mucous layer requires continuous synthesis of new tetrameric mucins to replace the mucins cleaved by pepsins.

Mucus is secreted at a significant rate in the resting stomach. Secretion of mucus is stimulated by some of the same stimuli that enhance acid and pepsinogen secretion, especially acetylcholine released from parasympathetic nerve endings near the gastric glands. If the gastric mucosa is mechanically deformed, neural reflexes are evoked to enhance mucus secretion.

Regulation of Gastric Secretion

Parasympathetic innervation via the vagus nerve is the strongest stimulant of gastric H^+ secretion. Extrinsic efferent fibers terminate on intrinsic neurons that innervate parietal cells, ECL cells that secrete the paracrine mediator histamine, and endocrine cells that secrete the hormone gastrin. In addition, vagal stimulation results in secretion of pepsinogen, mucus, HCO_3^-, and intrinsic factor. Stimulation of the parasympathetic nervous system also occurs during the cephalic and oral phase of the meal. However, the gastric phase produces the largest stimulation of gastric secretion of the postprandial period (Fig. 29.6).

Stimulation of gastric acid secretion is an excellent example of a "feed-forward" (or cascade) response that uses endocrine, paracrine, and neural pathways. Activation of intrinsic neurons by vagal efferent activity results in release

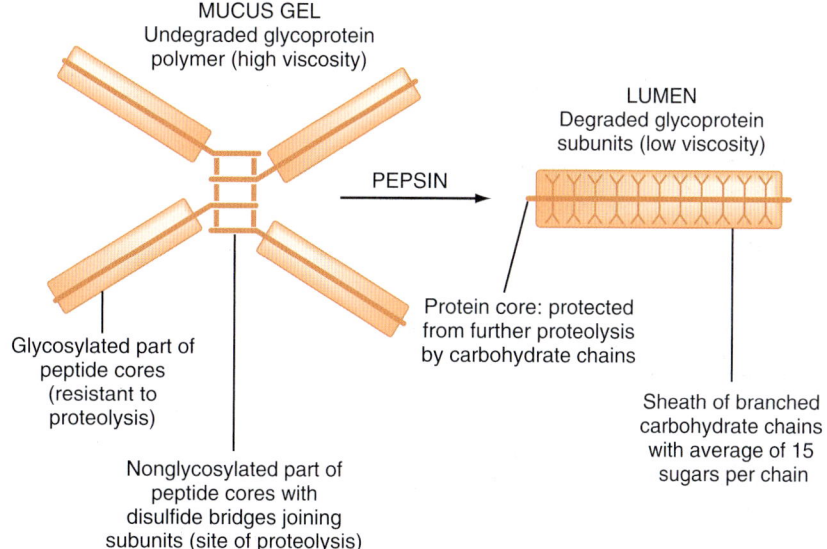

MUCUS GEL
Undegraded glycoprotein
polymer (high viscosity)

PEPSIN

LUMEN
Degraded glycoprotein
subunits (low viscosity)

Glycosylated part of
peptide cores
(resistant to
proteolysis)

Nonglycosylated part of
peptide cores with
disulfide bridges joining
subunits (site of proteolysis)

Protein core: protected
from further proteolysis
by carbohydrate chains

Sheath of branched
carbohydrate chains
with average of 15
sugars per chain

• **Fig. 29.5** Schematic representation of the structure of gastric mucins before and after hydrolysis by pepsin. Intact mucins are tetramers of four similar monomers of about 500,000 Da. Each monomer is largely covered by carbohydrate side chains that protect it from proteolytic degradation. The central portion of the mucin tetramer, near the disulfide cross-links, is more susceptible to proteolytic digestion. Pepsins cleave bonds near the center of the tetramers to release fragments about the size of monomers.

BOTH VAGOVAGAL REFLEX AND ENDOCRINE RELEASE
OF GASTRIN STIMULATE ACID AND PEPSINOGEN
SECRETION DURING THE GASTRIC PHASE

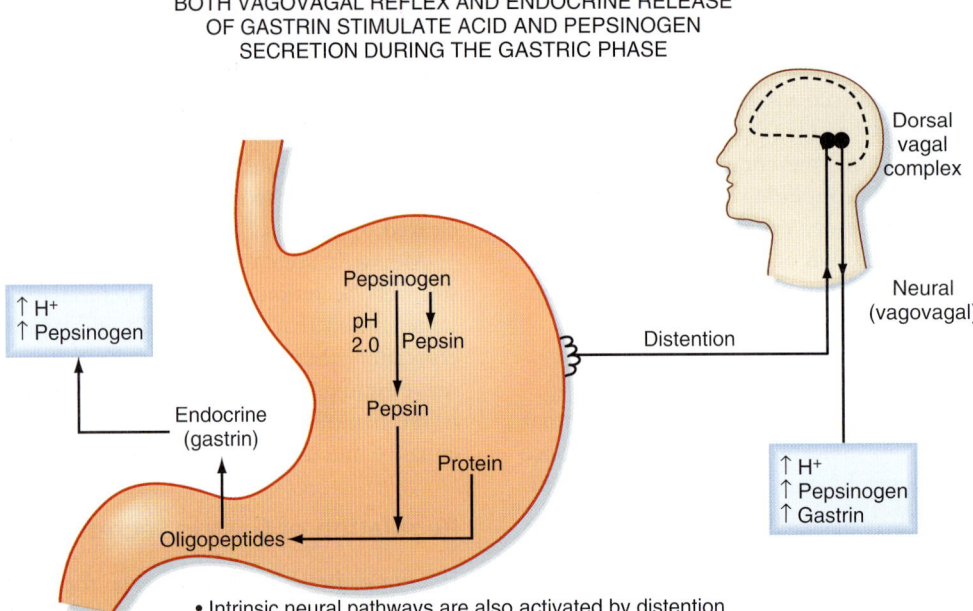

• Intrinsic neural pathways are also activated by distention

• **Fig. 29.6** Neural regulation of gastric acid secretion in the gastric phase of the meal is mediated by the vagus nerve. The stimulation that occurs in the cephalic and oral phases (before food reaches the stomach) results in stimulation of parietal cells to secrete acid and chief cells to secrete pepsinogen. Thus when food reaches the stomach, protein digestion is initiated by generating protein hydrolysate, which further stimulates secretion of gastrin from the mucosa of the gastric antrum. In addition, gastric distention activates a vagovagal reflex that further stimulates gastric acid and pepsinogen secretion.

of acetylcholine from nerve terminals, which activates cells in the gastric epithelium. Parietal cells express muscarinic receptors and are activated to secrete H^+ in response to vagal efferent nerve activity. In addition, parasympathetic activation, via gastrin-releasing peptide from intrinsic neurons, releases gastrin from G cells located in the gastric glands in the gastric antrum (see Fig. 29.6). Gastrin enters the bloodstream and, via an endocrine mechanism, further stimulates the parietal cell to secrete H^+. Parietal cells express cholecystokinin type 2 (CCK2) receptors for gastrin. Histamine is also secreted in response to vagal nerve stimulation, and ECL cells express muscarinic and gastrin receptors. Thus gastrin and vagal efferent activity induce release of histamine, which potentiates the effects of both gastrin and acetylcholine on the parietal cell. Hence activation of parasympathetic (vagal) outflow to the stomach is very efficient at stimulating the parietal cell to secrete acid (Fig. 29.7).

In the gastric phase the presence of food in the stomach is detected and activates **vagovagal reflexes** to stimulate secretion. Food in the stomach results in distention and stretch, which are detected by afferent (or sensory) nerve endings in the gastric wall. These are the peripheral terminals of vagal afferent nerves that transmit information to the brainstem and thereby drive activity in vagal efferent fibers, a vagovagal reflex (see Fig. 29.6). In addition, digestion of proteins increases the concentration of oligopeptides and free amino acids in the lumen, which are detected by **chemosensors** in the gastric mucosa. Oligopeptides and amino acids also stimulate vagal afferent activity. The exact nature of the

ACETYLCHOLINE, GASTRIN, AND HISTAMINE
STIMULATE THE PARIETAL CELL

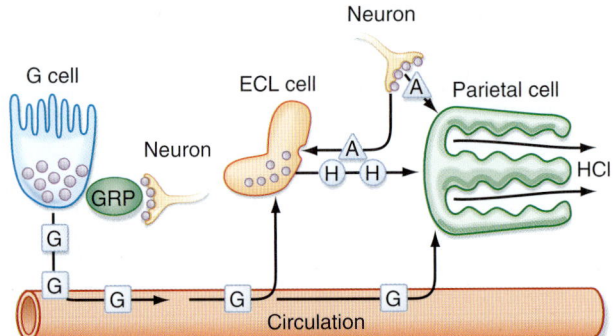

• **Fig. 29.7** The parietal cell is regulated by neural, hormonal, and paracrine pathways. Activation of vagal parasympathetic preganglionic outflow to the stomach acts in three ways to stimulate gastric acid secretion. There is direct neural innervation and activation of the parietal cell via release of acetylcholine (A) from enteric neurons, which acts on the parietal cell via muscarinic receptors. In addition, neural activation of the ECL cell stimulates release of histamine (H), which acts via a paracrine pathway to stimulate the parietal cell. Finally, G cells located in gastric glands in the gastric antrum are activated by release of gastrin-releasing peptide (GRP) from enteric neurons, which acts on the G cell to stimulate release of gastrin (G). Gastrin thereafter acts via a humoral pathway to stimulate the parietal cell.

chemosensors is not clear but may involve endocrine cells that release their contents to activate nerve endings. This topic will be discussed in more detail in Chapter 30.

There is also an important negative feedback mechanism whereby the presence of acid in the distal part of the stomach (antrum) induces a feedback loop to inhibit the parietal cell such that meal-stimulated H^+ secretion does not go unchecked. When the concentration of H^+ in the lumen reaches a certain threshold (<pH 3), somatostatin is released from endocrine cells in the antral mucosa. Somatostatin has a paracrine action on neighboring G cells to decrease the release of gastrin and thereby decrease gastric acid secretion (Fig. 29.8).

The receptors on the parietal cell membrane for acetylcholine, gastrin, and histamine, as well as the intracellular second messengers by which these secretagogues act, are shown in Fig. 29.9. Histamine is the strongest agonist of H^+ secretion, whereas gastrin and acetylcholine are much weaker agonists. However, histamine, acetylcholine, and gastrin potentiate one another's actions on the parietal cell. Antagonists of H_2 histamine receptors (e.g., cimetidine [Tagamet]) block secretagogue-stimulated acid secretion. Thus much of the response to gastrin results from gastrin-stimulated release of histamine. Gastrin also has important trophic effects; elevation of gastrin levels causes ECL cells to increase in size and number. Binding of histamine to H_2 receptors on parietal cell plasma membranes activates adenylyl cyclase and elevates the cytosolic concentration of cAMP. These events stimulate H^+ secretion by activating basolateral K^+ channels and apical Cl^- channels and by causing more H^+,K^+-ATPase molecules and Cl^- channels to be inserted into the apical plasma membrane (see Fig. 29.4). Acetylcholine binds to M_3 muscarinic receptors and opens Ca^{++} channels in the apical plasma membrane. Acetylcholine also elevates intracellular $[Ca^{++}]$ by promoting release of Ca^{++} from intracellular stores, which enhances H^+ secretion by activating basolateral K^+ channels and causing more H^+,K^+-ATPase molecules and Cl^- channels to be inserted into the apical plasma membrane. Gastrin enhances acid secretion by binding to CCK2 receptors (Fig. 29.10).

Digestion in the Stomach

Some digestion of nutrients occurs in the stomach. However, this is not required for full digestion of a meal; intestinal digestion is sufficient. Some amylase-mediated digestion of carbohydrates occurs in the stomach. Amylase is sensitive to pH and inactivated at low pH; however, some amylase is active even in the acidic gastric environment of the stomach because of substrate protection. Thus when carbohydrate occupies the active site of amylase, it protects the enzyme from degradation.

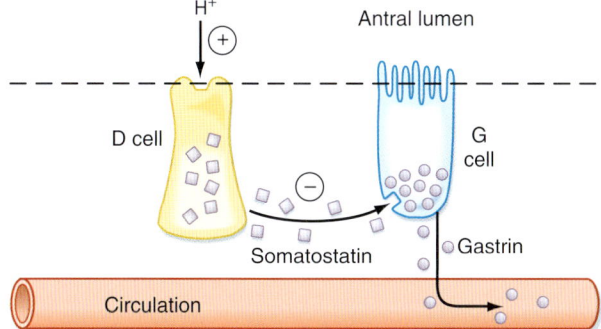

• Fig. 29.8 Feedback regulation of gastric acid secretion by release of somatostatin and its action on G cells in the gastric antrum. Endocrine cells in the mucosa of the gastric antrum sense the presence of H^+ and secrete somatostatin. This in turn acts on specific receptors on G cells to inhibit release of gastrin and thus bring about inhibition of gastric acid secretion.

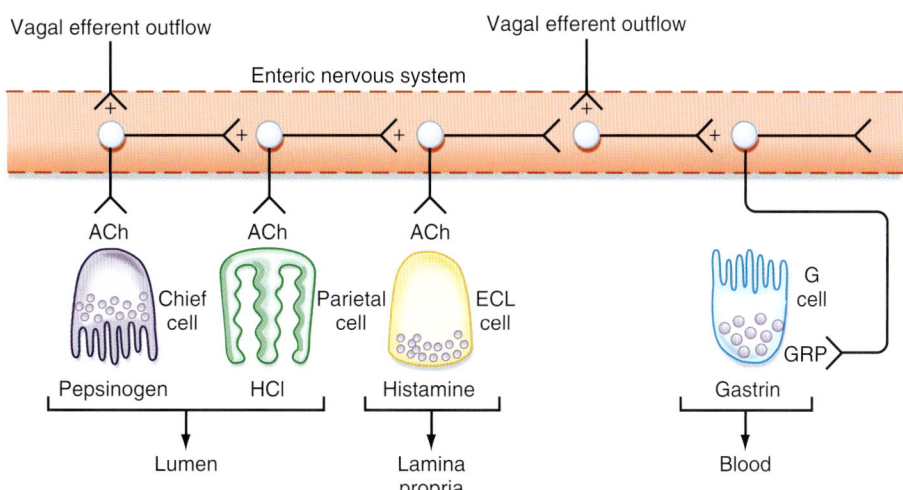

• Fig. 29.9 Vagal parasympathetic stimulation of gastric secretions via enteric neurons. Vagal preganglionic neurons innervate the myenteric and submucosal plexus. The terminals of the vagal preganglionic neurons innervate many enteric neurons and thus bring about changes in function as described in Fig. 29.7. ACh, acetylcholine; GRP, gastrin-releasing peptide.

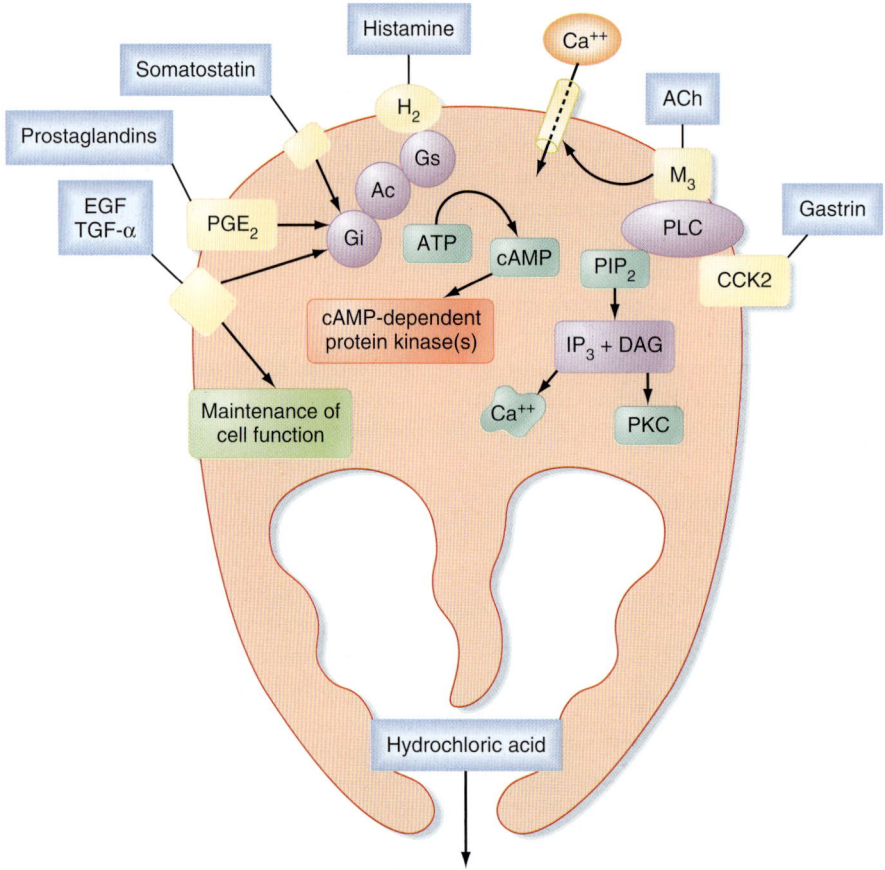

• **Fig. 29.10** Signal transduction mechanisms showing the mechanism of action of agonists (secretagogues) and antagonists that regulate secretion in parietal cells. Acetylcholine (ACh) binds to muscarinic M_3 receptors. Histamine acts via the H_2 receptor. Gastrin binds to the cholecystokinin type 2 (CCK2) receptor. Activation of M_3 and CCK2 receptors results in opening of Ca^{++} channels and release of Ca^{++} from intracellular stores and thus an increase in cytosolic $[Ca^{++}]$. Activation of H_2 receptors activates adenylyl cyclase to increase intracellular levels of cAMP. Ac, adenylyl cyclase; ACh, acetylcholine; CCK, cholecystokinin; DAG, diacylglycerol; EGF, epidermal growth factor; IP_3, inositol triphosphate; PGE_2, prostaglandin E_2; PIP_2, phosphatidylinositol 4,5-diphosphate; PKC, protein kinase C; PLC, protein lipase C; TGF-α, transforming growth factor α.

Digestion of lipids also starts in the stomach. The mixing patterns of gastric motility result in formation of an emulsion of lipids and **gastric lipase,** which attaches to the surface of lipid droplets in the emulsion and generates free fatty acids and monoglyceride from dietary triglyceride. However, the extent of hydrolysis of triglyceride is approximately 10%, and hydrolysis is not essential for normal digestion and absorption of dietary lipids. Moreover, as discussed in the next chapter, the products of lipolysis are not available for absorption in the stomach because of its low luminal pH.

Gastric Mucosal Protection and Defense

Mucus and HCO_3^- protect the surface of the stomach from the effects of H^+ and pepsins. The protective mucus gel that forms on the luminal surface of the stomach, as well as alkaline secretions entrapped within it, constitute a **gastric mucosal barrier** that prevents damage to the mucosa by

gastric contents (Fig. 29.11). The mucus gel layer, which is about 0.2 mm thick, effectively separates the HCO_3^--rich secretions of the surface epithelial cells from the acidic contents of the gastric lumen. The mucus allows the pH of epithelial cells to be maintained at nearly neutral despite a luminal pH of about 2. Mucus also slows the diffusion of acid and pepsins to the epithelial cell surface. Protection of the gastric epithelium depends on both mucus and HCO_3^- secretion.

Gastrointestinal Motility

To understand GI motility it is necessary to review some properties of smooth muscle function. The motion of the gut wall governs the flow of the luminal contents along its length; the main patterns of motility are mixing **(segmentation)** and propulsion **(peristalsis).** In addition, smooth muscle activity in the stomach and colon subserves a storage function.

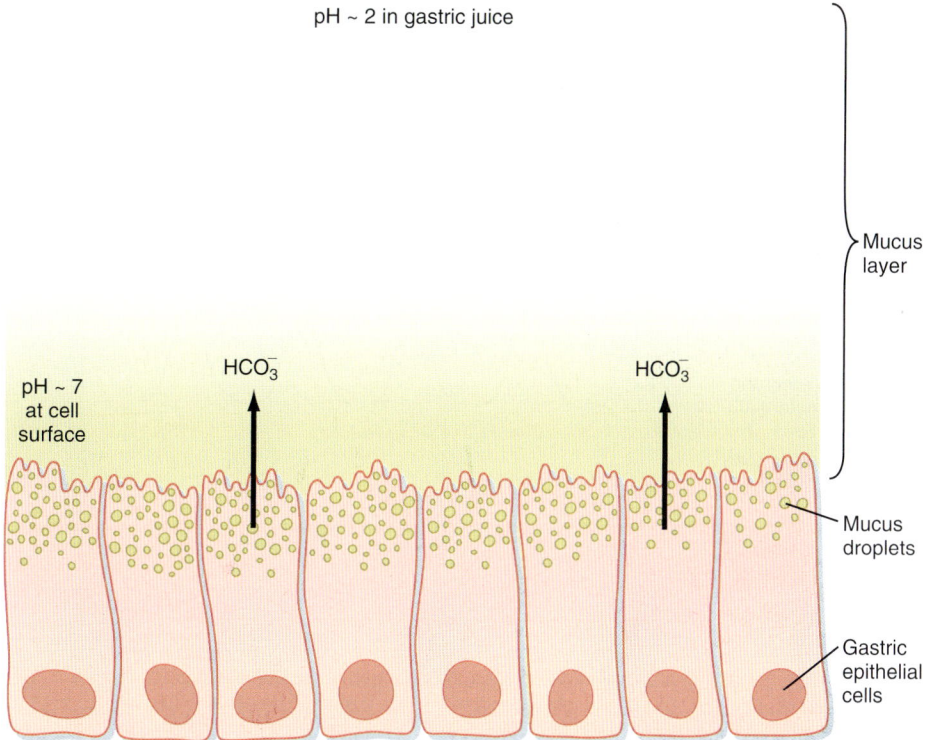

pH ~ 2 in gastric juice

Mucus layer

pH ~ 7 at cell surface

HCO_3^-

HCO_3^-

Mucus droplets

Gastric epithelial cells

• **Fig. 29.11** The surface of the stomach is protected by the gastric mucosal barrier. Buffering by the HCO_3^--rich secretions and the high viscosity of the layer of mucus allow the pH at the cell surface to remain near 7, whereas the pH in the gastric juice in the lumen is 2.

Functional Anatomy of Gastrointestinal Smooth Muscle

The smooth muscle in the GI tract is similar in structure to other smooth muscle found in the body. Fusiform cells are packed together in bundles surrounded by a connective tissue sheath. Gap junctions functionally couple the smooth muscle cells so that contraction of bundles occurs synchronously. The **interstitial cells of Cajal** (ICCs) are a specialized group of cells in the intestinal wall that are involved in transmission of information from enteric neurons to smooth muscle cells (Fig. 29.12). It is also thought that ICCs are **"pacemaker"** cells that have the capacity to generate the basic electrical rhythm, or *slow-wave activity,* that is a consistent feature of GI smooth muscle (Fig. 29.13).

Electrophysiology of Gastrointestinal Smooth Muscle

The cyclic variation in resting membrane potential of GI smooth muscle is called the *basic electrical rhythm or slow wave.* The frequency of slow waves is 3 to 5 per minute in the stomach and about 12 to 20 per minute in the small intestine; it decreases to 6 to 8 per minute in the colon. The frequency of the slow wave is set by a pacemaker region in the different regions of the GI tract (see Fig. 29.13). Slow waves are thought to be generated by ICCs. These cells are located in a thin layer between the longitudinal and

INTERSTITIAL CELLS OF CAJAL (ICC)
ARE THE PACEMAKERS OF THE GUT

Slow waves are generated in interstitial cells of Cajal

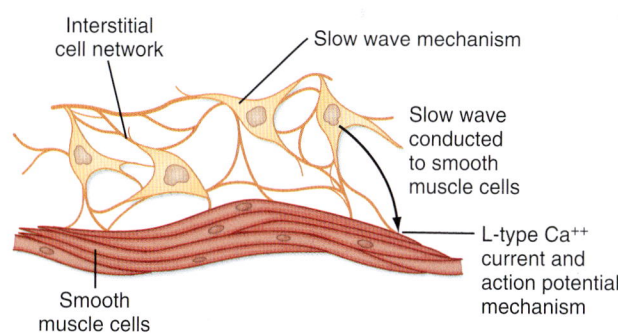

Interstitial cell network

Slow wave mechanism

Slow wave conducted to smooth muscle cells

L-type Ca^{++} current and action potential mechanism

Smooth muscle cells

• **Fig. 29.12** Diagrammatic representation of the interstitial cells of Cajal network in the smooth muscle wall of the GI tract.

circular layers of the muscularis externa and in other places in the wall of the GI tract. Interstitial cells have properties of both fibroblasts and smooth muscle cells. Their long processes form gap junctions with the longitudinal and circular smooth muscle cells; the gap junctions enable the slow waves to be conducted rapidly to both muscle layers. Because gap junctions electrically and chemically couple the smooth muscle cells of both longitudinal and circular layers, the slow wave spreads throughout the smooth muscle of each segment of the GI tract.

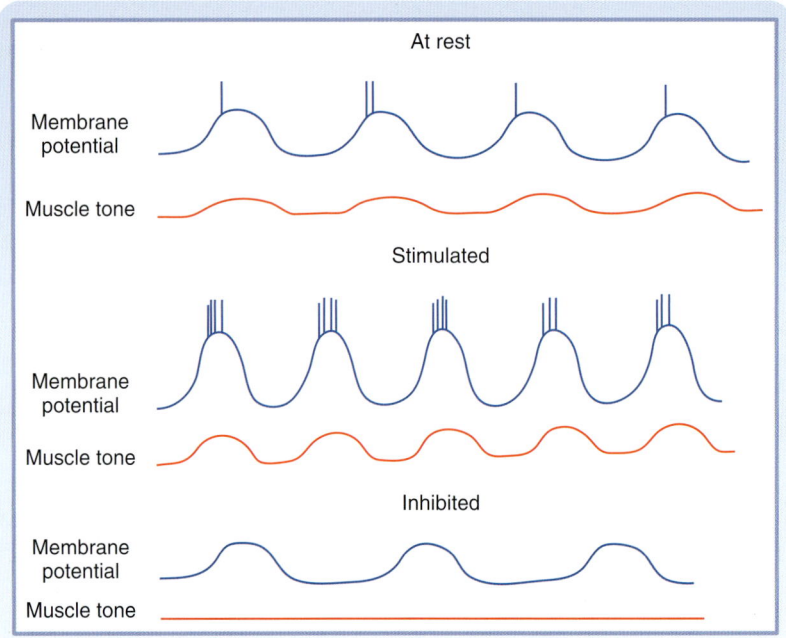

At rest

Membrane potential

Muscle tone

Stimulated

Membrane potential

Muscle tone

Inhibited

Membrane potential

Muscle tone

• **Fig. 29.13** Amplitude of slow wave determines the strength of muscle contraction. The slow wave will initiate a contraction in smooth muscle when it reaches a threshold amplitude. The amplitude of the slow wave is altered by release of neurotransmitters from enteric neurons.

IN THE CLINIC

There are times when the gastric mucosal barrier fails. Superficial breakdowns of the GI lining not involving the submucosa are called *erosions.* They generally heal without intervention. In contrast, breakdowns of the GI lining involving the muscularis and deeper layers are called *ulcers.* Gastric and duodenal erosions and ulcers occur as a result of an imbalance between the mechanisms that protect the mucosa and aggressive factors that can break it down. A healthy stomach/duodenum has ample natural protection against the destructive effects of H^+. Factors that magnify the harmful effect of H^+ on the stomach/duodenum or act separately from H^+ include pepsin, bile, the bacterium *Helicobacter pylori,* and the class of drugs known as *nonsteroidal antiinflammatory drugs* (NSAIDs). Indeed, ulcer disease is becoming more common as the population ages and has more need of NSAIDs for non-GI complaints such as arthritis. Alcohol, tobacco, and caffeine are also risk factors for ulcers. Infectious agents can also cause gastritis (inflammation of the gastric epithelium). *H. pylori* is a spiral bacterium that has now become widely recognized as one factor that can lead to gastritis, ulcer formation, and in humans, gastric carcinoma. *H. pylori* exists in the stomach because it secretes an enzyme, urease, that converts urea to NH_3, which is used to buffer H^+ by forming NH_4^+. An aggressive regimen of antibiotic treatment, sometimes in combination with an H^+,K^+-ATPase inhibitor, can often eliminate the infection, after which the gastritis and ulcer symptoms improve.

The amplitude and, to a lesser extent, the frequency of the slow wave can be modulated by the activity of intrinsic and extrinsic nerves and by hormones and paracrine substances. If the depolarization of the slow wave exceeds the threshold, a train of action potentials may be triggered during the peak of the slow wave.

Action potentials in GI smooth muscle are more prolonged (10–20 msec) than those in skeletal muscle and have little or no overshoot. The rising phase of the action potential is caused by flow of ions through channels that conduct both Ca^{++} and Na^+ and are relatively slow to open. The Ca^{++} that enters the cell during the action potential helps initiate contraction. The extent of depolarization of the cells and the frequency of action potentials are enhanced by some hormones and paracrine agonists and by neurotransmitters from excitatory enteric nerve endings (e.g., acetylcholine and substance P). Inhibitory hormones and neuroeffector substances (e.g., vasoactive intestinal polypeptide and nitric oxide) hyperpolarize the smooth muscle cells and may diminish or abolish action potential spikes.

Slow waves that are not accompanied by action potentials elicit little or no contraction of the smooth muscle cells. Much stronger contractions are evoked by the presence of action potentials. The greater the number of action potentials that occur at the peak of a slow wave, the more intense the contraction of the smooth muscle. Because smooth muscle cells contract rather slowly (about a 10th as fast as skeletal muscle cells), the individual contractions caused by each action potential in a train do not cause distinct twitches; rather they sum temporally to produce a smoothly increasing level of tension.

Between the trains of action potentials, the tension developed by GI smooth muscle falls, but not to zero. This nonzero resting, or baseline, tension of smooth muscle is called **tone.** The tone of GI smooth muscle is altered by neuroeffectors, hormones, paracrine substances, and drugs and is important in the sphincters and also in regions where storage of contents is important, such as the stomach and colon.

Specialized Patterns of Motility

Peristalsis is a moving ring of contraction that propels material along the GI tract. It involves neurally mediated contraction and relaxation of both muscle layers. Peristalsis occurs in the pharynx, esophagus, gastric antrum, and the small and large intestine.

Segmental contractions produce narrow areas of contracted segments between relaxed segments. These movements allow mixing of the luminal contents with GI tract secretions and increase exposure to the mucosal surfaces where absorption occurs. Segmentation occurs predominantly in the small and large intestine.

There are also characteristic pathological patterns of motility. During **spasm,** maximal contractile activity occurs continuously in a dysregulated manner. In **ileus,** contractile activity is markedly decreased or absent; it often results from irritation of the peritoneum, such as occurs in surgery, peritonitis, and pancreatitis.

Gastric Motility

Functional Anatomy of the Stomach

As discussed, the stomach is divided into two functional regions—proximal and distal, with sphincters at either end. The LES and *cardia* (defined as the region of the stomach immediately surrounding the LES) have important functions. Relaxation of the LES and cardia allows entry of food from the esophagus into the stomach and the release of gas, called *belching*. By maintaining tone, reflux of contents from the stomach back into the esophagus is prevented.

The proximal part of the stomach (the fundus together with the corpus or body) produces slow changes in tone compatible with its reservoir function. It is important for receiving and storing food and for mixing the contents with gastric juice (Table 29.2). Generation of tone in the proximal portion of the stomach is also an important driving force in the regulation of gastric emptying. Low tone and consequently low intragastric pressure are associated with delayed or slow gastric emptying, and an increase in tone in this region is required for gastric emptying to occur.

The distal part of the stomach is important in the mixing of gastric contents and for propulsion through the pylorus and into the duodenum. The muscle layers in the region of the gastric antrum are much thicker than in the more proximal regions of the stomach, and thus the antrum is capable of producing strong phasic contractions. Contractions

TABLE 29.2	The Stomach Alters the Physical and Chemical Characteristics of the Meal
Input	**Output**
Bolus	Emulsion, suspension (particles < 2 mm)
Triglyceride	Triglyceride plus small amounts of 2-monoglycerides and free fatty acids
Protein	Protein plus small amounts of peptides and amino acids
Starch	Starch plus oligosaccharides
Water, ions	Addition of large amounts of water and ions of low pH

initiated by the slow wave begin in the midportion of the stomach and move toward the pylorus. The strength of these contractions varies during the postprandial period. In the gastric phase of the meal the pylorus is usually closed, and these antral contractions serve to mix the gastric contents and reduce the size of solid particles (grinding). However, eventually these antral contractions are also important in emptying the stomach of its contents.

The pyloric sphincter is the **gastroduodenal junction** and is defined as an area of thickened circular muscle. This is a region of high pressure generated by tonic smooth muscle contraction. It is important in regulating gastric emptying.

Control of Gastric Motility in the Gastric Phase

Gastric motility is highly regulated and coordinated to perform the functions of storage and mixing. Regulation of emptying of contents into the small intestine, an important part of gastric motor function, will be considered in detail in the discussion of the duodenal phase of the meal, because the controls are generated in the duodenum.

The stimuli regulating gastric motor function that result from the presence of the meal in the stomach are both mechanical and chemical and include distention and the presence of products of protein digestion (amino acids and small peptides). The pathways regulating these processes are predominantly neural and consist of vagovagal reflexes initiated by extrinsic vagal afferent fibers that terminate in the muscle and mucosa. Mucosal afferents respond to chemical stimuli, and mechanosensitive afferents respond to distention and contraction of smooth muscle. This afferent stimulation results in reflex activation of vagal efferent (parasympathetic) outflow and activation of enteric neurons that innervate the smooth muscle. Activation of enteric neurons produces both inhibitory and excitatory effects on gastric smooth muscle; these effects vary depending on the region of the stomach. Thus distention of the gastric wall results in inhibition of smooth muscle in the proximal portion of the stomach and subsequent reflex accommodation, which allows entry and storage of the meal to occur with minimal increase in intragastric pressure.

In contrast, the predominant motor pattern of the distal part of the stomach in the gastric phase of the meal is activation of smooth muscle to produce and strengthen the antral contractions. The rate of antral contractions is set by the gastric pacemaker; however, the magnitude of the contractions is regulated by release of neurotransmitters from enteric neurons, including substance P and acetylcholine, which increase the level of depolarization of the smooth muscle and therefore produce stronger contractions. In this phase of the meal the pylorus is mostly closed. Thus antral contractions will tend to move the contents toward the pylorus; however, because the pylorus is closed, the contents will be returned to the more proximal part of the stomach. In this way the gastric contents will be mixed. In addition, antral contractions can occlude the lumen, and thus larger particles will be dispersed, a process referred to as *grinding* (Fig. 29.14).

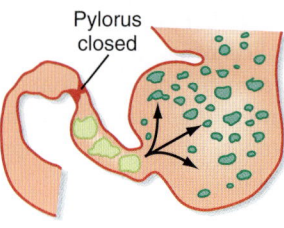

Onset of terminal antral contraction — Pylorus closing

Complete terminal antral contraction — Pylorus closed

• Force for retropulsion is increased pressure in terminal antrum as the antral contraction approaches the closed pylorus.

• **Fig. 29.14** Coordinated activity in the smooth muscle of the proximal and distal portions of the stomach and the pyloric sphincter results in mixing and grinding in the gastric antrum. The peristaltic wave moves down the gastric body and antrum toward the pylorus. If the pylorus is closed, the contents of the gastric antrum are retropulsed back into the more proximal part of the stomach. This pattern of motility results in grinding and mixing of the food with secretions from the gastric wall and eventually leads to a reduction in particle size and the presence of digestive products that will empty into the duodenum.

Key Concepts

1. The main functions of the stomach are storage and initiation of protein digestion.
2. Regulation of gastric function is driven by extrinsic and intrinsic neural pathways together with key humoral (gastrin) and paracrine (histamine) mediators.
3. The key secretions from the stomach are acid and pepsinogen, which together begin protein digestion.
4. H^+ is secreted across the apical plasma membrane of parietal cells via H^+,K^+-ATPase.
5. The stomach also secretes intrinsic factor, which is involved in absorption of vitamin B_{12}.
6. The gastric epithelium secretes HCO_3^- and mucus to form a gel-like mucosal barrier that protects it against the acidic and peptic luminal contents.
7. The smooth muscle of the gut wall undergoes cyclic changes in membrane potential, termed the *basic electrical rhythm* or the *slow wave*.
8. The interstitial cells of Cajal are pacemakers in the gut wall, and they set the frequency of the slow wave.
9. The proximal part of the stomach undergoes a slow change in tone compatible with its storage function.
10. The distal part of the stomach undergoes phasic contractions that can vary considerably in strength.
11. Gastric emptying is regulated by vagovagal reflexes.

Additional Reading

Huizinga JD, Chen JH. Interstitial cells of Cajal: update on basic and clinical science. *Curr Gastroenterol Rep*. 2014;16:363.

Hunt RH, et al. The stomach in health and disease. *Gut*. 2015;64:1650-1668.

Said H, et al. Gastroduodenal mucosal defense mechanisms. *Curr Opin Gastroenterol*. 2015;31:486-491.

Sanders KM, et al. Regulation of gastrointestinal smooth muscle function by interstitial cells. *Physiology (Bethesda)*. 2016;31:316-326.

Schubert ML. Functional anatomy and physiology of gastric secretion. *Curr Opin Gastroenterol*. 2015;31:479-485.

Schubert ML. Regulation of gastric acid secretion. In: Johnson LR, ed. *Physiology of the GI Tract*. 5th ed. Waltham, MA: Academic Press; 2012.

30

The Small Intestinal Phase of the Integrated Response to a Meal

LEARNING OBJECTIVES

Upon completion of this chapter the student should be able to answer the following questions:

1. How are the various components of a mixed meal digested and absorbed in the small intestine?
2. What are the constituents and functions of pancreatic juice, and how is their secretion controlled?
3. How do bile acids assist with digestion and assimilation of lipids?
4. What are the mechanisms that provide for appropriate levels of fluidity of the intestinal contents?
5. What are the motor patterns of the small intestine in the postprandial period as well as during fasting, and what functions do these patterns subserve?

The small intestine is the critical portion of the intestinal tract for assimilation of nutrients. In this site the meal is mixed with a variety of secretions that permit its digestion and absorption, and motility functions ensure adequate mixing and exposure of the intestinal contents **(chyme)** to the absorptive surface. The small intestine has many specializations that enable it to perform its functions efficiently. One of the most obvious specializations is the substantial surface area of the mucosa; this is achieved in a number of different ways. The small intestine is essentially a long tube that is coiled inside the abdominal cavity, there are folds of the full thickness of the mucosa and submucosa, the mucosa has finger-like projections called *villi,* and finally, each epithelial cell has microvilli on its apical surface. Thus a large surface area exists over which digestion and absorption occur.

The main characteristic of the small intestinal phase of the response to a meal is controlled delivery of chyme from the stomach to match the digestive and absorptive capacity of the intestine. In addition there is further stimulation of pancreatic and biliary secretion and emptying of these secretions into the small intestine. Therefore the function of this region is highly regulated by feedback mechanisms that involve hormonal, paracrine, and neural pathways.

The stimuli that regulate these processes are both mechanical and chemical and include distention of the intestinal wall and the presence of increased $[H^+]$, high osmolarity, and nutrients in the intestinal lumen. These stimuli result in a set of changes that represent the **intestinal phase** of the response to the meal: (1) increased pancreatic secretion, (2) increased gallbladder contraction, (3) relaxation of the sphincter of Oddi, (4) regulation of gastric emptying, (5) inhibition of gastric acid secretion, and (6) interruption of the migrating motor complex (MMC). The goal of this chapter is to discuss how such changes are brought about and how they result ultimately in assimilation of nutrients. Changes in small intestinal function that occur after the meal has passed through will also be addressed.

Gastric Emptying in the Small Intestinal Phase

Immediately after a meal the stomach may contain up to a liter of material that will empty slowly into the small intestine. The rate of gastric emptying is dependent on the macronutrient content of the meal and the amount of solids it contains. Thus solids and liquids of similar nutritional composition will empty at different rates. Liquids empty rapidly, but solids do so only after a lag phase, which means that after a solid meal, there is a period of time during which little or no emptying occurs (Fig. 30.1).

Regulation of gastric emptying is achieved by alterations in motility of the proximal part of the stomach (fundus and corpus) and distal part of the stomach (antrum and pylorus) as well as in the duodenum. Motor function in these regions is highly coordinated. Recall that during the esophageal and gastric phase of the meal, the predominant reflex response is receptive relaxation. At the same time, peristaltic movements in the more distal part of the stomach (antrum) mix the gastric contents with gastric secretions. The pyloric sphincter is largely closed. Even if it opens periodically, little emptying will occur, because the proximal portion of the stomach is relaxed and the **antral pump** (antral contractions) is not very strong. Subsequently, gastric emptying is brought about by an increase in tone (intraluminal pressure) in the proximal portion of the stomach, increased strength of antral contractions (increased strength of the

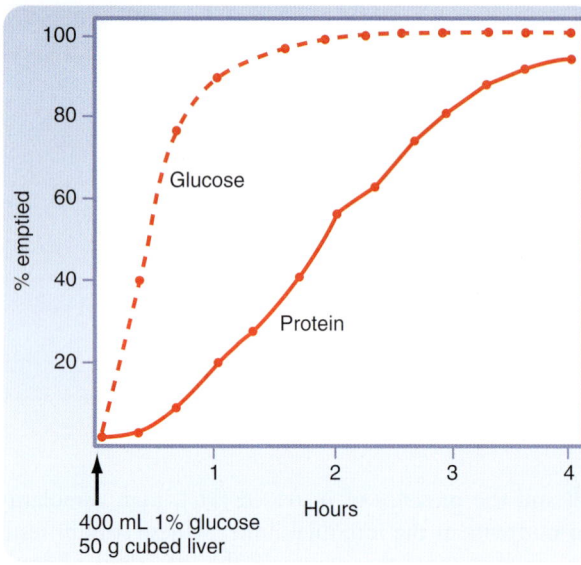

• Fig. 30.1 Rates of emptying of different meals from a dog's stomach. A solution (1% glucose) is emptied faster than a digestible solid (cubed liver). Note the lag phase for emptying of the solids, which is related to the time needed to reduce particles below 2 mm in size. (Adapted from Hinder RA, Kelly KA. *Am J Physiol* 1977;233:E335.)

 IN THE CLINIC

The gastrointestinal (GI) tract plays a major role in the sensing and signaling of ingested nutrients by activating neural and endocrine pathways that connect with other signals, such as fat energy storage and utilization, which together regulate energy homeostasis. Satiety signals from the GI tract are generally involved in short-term regulation of food intake, such as individual meal size and meal duration. For example, the luminal contents activate vagal afferent pathways, leading to suppression of meal size. In addition, several GI hormones released by nutrients also influence food intake. Cholecystokinin (CCK) is a well-described satiety hormone; it is released by nutrients and decreases food intake after exogenous administration. Other GI hormones in this class include glucagon-like peptide 1 (GLP-1) and peptide YY (PYY). In both lean and obese humans, injection of exogenous PYY inhibits food intake. A long-acting analogue of GLP-1, exendin-4, is currently being used as an agent for weight control in humans.

antral pump), opening of the pylorus to allow the contents to pass, and simultaneous inhibition of duodenal segmental contractions. Liquids and the semiliquid chyme flow down the pressure gradient from the stomach to the duodenum.

As the meal enters the small intestine, it feeds back via both neural and hormonal pathways to regulate the rate of gastric emptying based on the chemical and physical composition of the chyme. Afferent neurons, predominantly of vagal origin, respond to nutrients, [H^+], and the hyperosmotic content of chyme as it enters the duodenum. Reflex activation of vagal efferent outflow decreases the strength of antral contractions, contracts the pylorus, and decreases proximal gastric motility (with a decrease in intragastric

pressure), thereby resulting in inhibition (slowing) of gastric emptying. This same pathway is responsible for the inhibition of gastric acid secretion that occurs when nutrients are in the duodenal lumen. Cholecystokinin (CCK) is released from endocrine cells in the duodenal mucosa in response to such nutrients. This hormone is physiologically important, in addition to its role in neural pathways, in the regulation of gastric emptying, gallbladder contraction, relaxation of the sphincter of Oddi, and pancreatic secretion. Recent evidence suggests that CCK acts both directly (inhibits gastric emptying) and indirectly (stimulates vagal afferent fiber discharge to produce an indirect vagovagal reflex–mediated decrease in gastric emptying).

How then can gastric emptying proceed in the face of these inhibitory pathways? The amount of chyme in the duodenum decreases as it passes further down the small intestine into the jejunum; thus the strength of intestinal feedback inhibition fades as there is less activation of duodenal sensory mechanisms by nutrients. At this time, intragastric pressure in the proximal portion of the stomach increases, thereby moving material into the antrum and toward the antral pump. Antral peristaltic contractions again deepen and culminate in opening of the pylorus and release of gastric contents into the duodenum.

 IN THE CLINIC

Surgical treatment of obesity, so-called bariatric surgery, can achieve substantial and lasting weight loss and also help ameliorate associated health problems such as insulin resistance, hyperlipidemia, and elevated blood pressure. Initially, surgery involved *jejunoileal bypass,* the removal of a substantial part of the absorptive small intestine, but this procedure is associated with malabsorption and subsequent undesirable sequelae such as diarrhea. A variety of revised surgical approaches to obesity have been devised, including Roux-en-Y gastric bypass and vertical sleeve gastrectomy. The mechanisms by which these procedures are thought to be successful lie in the small size of the residual gastric pouch, whereby meal size is decreased because of early satiety, and a beneficial effect of the bypass on the profiles of GI hormones. Recent data imply that effects of surgery on bile acids and the microbiome may also contribute to both weight loss and metabolic benefits.

Pancreatic Secretion

Most of the nutrients ingested by humans are in the chemical form of macromolecules. However, such molecules are too large to be assimilated across the epithelial cells that line the intestinal tract and must therefore be broken down into their smaller constituents by processes of chemical and enzymatic digestion. Secretions arising from the pancreas are quantitatively the largest contributors to enzymatic digestion of the meal. The pancreas also provides additional important secretory products that are vital for normal digestive function. Such products include substances that regulate the function or secretion (or both) of other pancreatic products, as well as water and bicarbonate ions. The latter are involved in neutralizing gastric acid so that the small intestinal lumen has a pH approaching 7.0. This is important because pancreatic enzymes are inactivated by high levels of acidity and also because neutralization of gastric acid reduces the likelihood that the small intestinal mucosa will be injured by such acid acting in combination with pepsin. Quantitatively the pancreas is the largest contributor to the supply of bicarbonate ions needed to neutralize the gastric acid load, although the biliary ductules and duodenal epithelial cells also contribute.

As in the salivary glands, the pancreas has a structure that consists of ducts and **acini.** The pancreatic acinar cells line the blind ends of a branching ductular system that eventually empties into the main pancreatic duct and from there into the small intestine under control of the **sphincter of Oddi.** Also in common with salivary glands, a primary secretion arises in the acini, which is subsequently modified as it passes through the pancreatic ducts. In general the acinar cells supply the organic constituents of the pancreatic juice in a primary secretion whose ionic composition is comparable to that of plasma, whereas the ducts dilute and alkalinize the pancreatic juice while reabsorbing chloride ions (Fig. 30.2). The major constituents of pancreatic juice, which amounts to approximately 1.5 L/day in adult humans, are listed in Box 30.1. This list also outlines the functions of pancreatic secretory products. Many of the digestive enzymes produced by the pancreas, particularly the proteolytic enzymes, are produced as inactive precursor forms. Storage in these inactive forms is critically important in preventing the pancreas from digesting itself.

 ## AT THE CELLULAR LEVEL

Pancreatitis can result when enzymes secreted by pancreatic acinar cells become proteolytically activated before they have reached their appropriate site of action in the small intestinal lumen. Indeed, pancreatic juice contains a variety of trypsin inhibitors to reduce the risk of premature activation, because trypsin is the activator of other pro-forms of enzymes secreted in pancreatic juice. A second level of protection lies in the fact that trypsin can be degraded by other trypsin molecules. Despite these defenses, some individuals are susceptible to hereditary pancreatitis that occurs spontaneously in the absence of known risk factors. In some of these patients there is a mutation in trypsin that renders it resistant to degradation by other trypsin molecules. Others harbor mutations in trypsin inhibitors, rendering the inhibitors inactive. In any event, if other defenses have been breached and trypsin becomes active prematurely, a vicious cycle of enzyme activation ensues and bouts of pancreatitis follow.

Characteristics and Control of Ductular Secretion

In this section we consider how the pancreatic ductular cells contribute to the flow and composition of pancreatic juice in the postprandial period. The ducts of the pancreas can be considered the effector arm of a pH regulatory

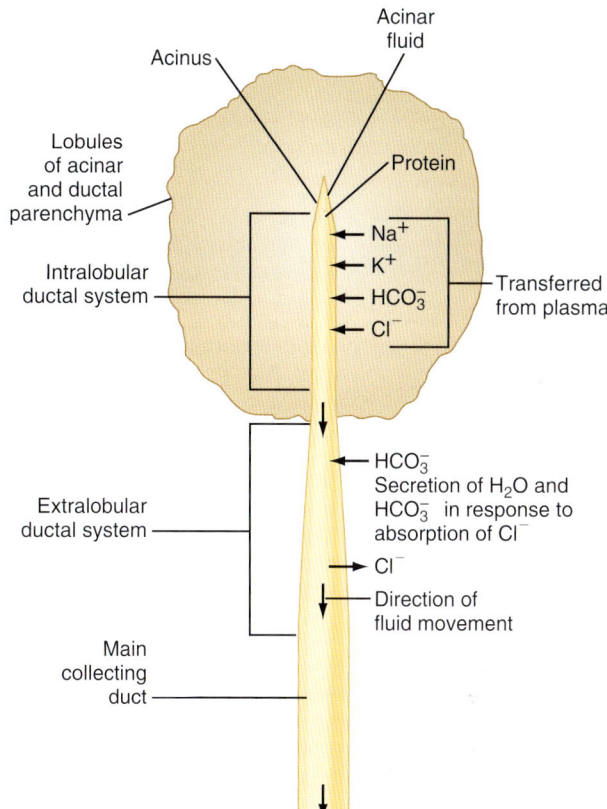

• **Fig. 30.2** Locations of important transport processes involved in the elaboration of pancreatic juice. Acinar fluid is isotonic and resembles plasma in its concentrations of Na^+, K^+, Cl^-, and HCO_3^-. Secretion of acinar fluid and the proteins it contains is stimulated primarily by CCK. The hormone secretin stimulates secretion of water and electrolytes from the cells that line the extralobular ducts. The secretin-stimulated secretion is richer in HCO_3^- than the acinar secretion because of Cl^-/HCO_3^- exchange. (Adapted from Swanson CH, Solomon AK. *J Gen Physiol* 1973;62:407.)

system designed to respond to luminal acid in the small intestine and secrete just enough bicarbonate to restore pH to neutrality (Fig. 30.3). This regulatory function also requires mechanisms to sense luminal pH and convey this information to the pancreas as well as other epithelia (e.g., biliary ductules and the duodenal epithelium itself) capable of secreting bicarbonate. The pH-sensing mechanism is embodied in specialized endocrine cells known as **S cells,** localized within the small intestinal epithelium. When luminal pH falls below approximately 4.5, S cells are triggered to release **secretin** in response to the increase in [H^+]. The components of this regulatory loop constitute a self-limited system. Thus as secretin evokes secretion of bicarbonate, pH in the small intestinal lumen will rise and the signal for release of secretin from S cells will be terminated.

At the cellular level, secretin stimulates epithelial cells to secrete bicarbonate into the ductular lumen, with water following via the paracellular route to maintain osmotic equilibrium. Secretin increases cyclic adenosine monophosphate (cAMP) in the ductular cells and thereby opens cystic fibrosis transmembrane conductance regulator (CFTR)

Precursors of Proteases

Trypsinogen
Chymotrypsinogen
Proelastase
Procarboxypeptidase A
Procarboxypeptidase B

Starch-Digesting Enzymes

Amylase

Lipid-Digesting Enzymes or Precursors

Lipase
Nonspecific esterase
Prophospholipase A_2

Nucleases

Deoxyribonuclease
Ribonuclease

Regulatory Factors

Procolipase
Trypsin inhibitors
Monitor peptide

Cl^- channels (Fig. 30.4) and causes an outflow of Cl^- into the duct lumen. This secondarily drives the activity of an adjacent antiporter that exchanges the chloride ions for bicarbonate. CFTR is also permeable to bicarbonate. Thus the bicarbonate secretory process is dependent on CFTR, which provides an explanation for the defects in pancreatic function seen in the disease **cystic fibrosis,** in which CFTR is mutated. The bicarbonate needed for this secretory process is derived from two sources. Some is taken up across the basolateral membrane of the ductular epithelial cells

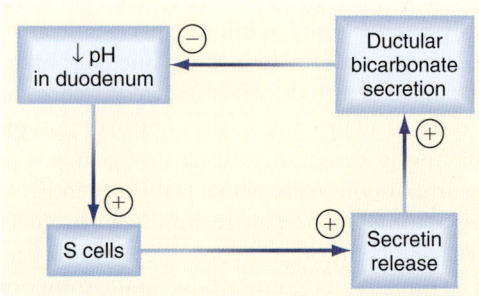

• **Fig. 30.3** Participation of secretin and HCO_3^- secretion in a classic negative-feedback loop that responds to a fall in luminal pH in the duodenum.

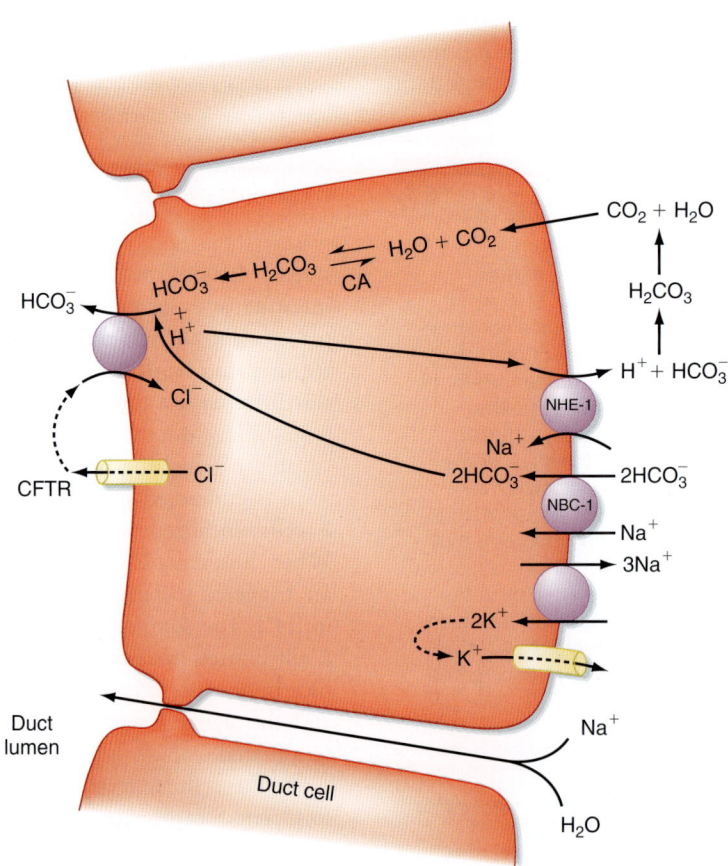

• **Fig. 30.4** Ion transport pathways in pancreatic duct cells. CA, carbonic anhydrase; CFTR, cystic fibrosis transmembrane conductance regulator; NBC-1, sodium/bicarbonate cotransporter (symporter) type 1; NHE-1, sodium-hydrogen exchanger (antiporter) type 1.

via the sodium-bicarbonate cotransporter type 1 (NBC-1) symporter. Recall that the process of gastric acid secretion results in an increase in circulating bicarbonate ions, which can serve as a source of bicarbonate to be secreted by the pancreas. However, bicarbonate can also be generated intracellularly via the activity of the enzyme carbonic anhydrase. The net effect is to move HCO_3^- into the lumen and thereby increase the pH and volume of pancreatic juice.

Characteristics and Control of Acinar Secretion

In contrast to the pancreatic ductules, where secretin is the most important physiological agonist, CCK plays the predominant role at the level of the acinar cells. Thus it is important to understand how release of CCK is controlled during the small intestinal phase of the response to a meal.

 ## IN THE CLINIC

Cystic fibrosis (CF) is a genetic disease that affects the function of a variety of epithelial organs, including the lung, intestine, biliary system, and pancreas. Previously the disease was almost uniformly fatal during adolescence as a result of severe respiratory infections. However, improved antibiotics, drugs that improve clearance of mucus from the lungs, correction of pancreatic insufficiency and undernutrition, and recent US Food and Drug Administration (FDA) approval of drugs such as ivacaftor (Kalydeco) and lumacaftor/ivacaftor (Orkambi) now extend life into the fifth decade or later in some patients. CF is caused by a mutation in *CFTR,* which in the gut impairs the ability to hydrate and alkalinize the luminal contents. In the GI system specifically, this can result in intestinal obstruction, duodenal mucosal injury, and damage to the liver and biliary system as well as the pancreas. In many CF patients the endocrine pancreas is dysfunctional, and they must be given digestive enzyme supplements to maintain adequate nutrient digestion. In other patients with milder mutations, pancreatitis may develop later in life in the absence of other classic CF symptoms, presumably because of retention of digestive enzymes in the pancreas. In either case, improved recognition and treatment of the pulmonary complications of CF mean that GI symptoms such as liver failure, reduced bile flow, pancreatitis, obstruction, and maldigestion/malabsorption of nutrients are acquiring increased importance as facets of the disease that must be managed in adults, often by multidisciplinary teams of physicians and other health care professionals.

CCK is the product of **I cells,** which are also localized in the small intestinal epithelium. These classic enteroendocrine cells release CCK into the interstitial space when specific food components are present in the lumen, particularly free fatty acids and certain amino acids. Release of CCK may occur following direct interaction of fatty acids or amino acids, or both, with the I cells. Release of CCK is also regulated by two luminally acting releasing factors that can stimulate the I cell. The first of these, **CCK-releasing peptide,** is secreted by paracrine cells within the epithelium into the small intestinal lumen in response to products of fat

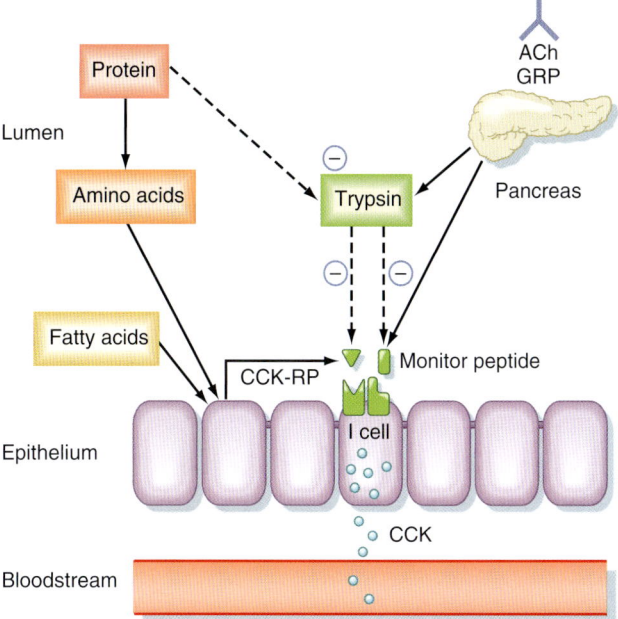

• **Fig. 30.5** Mechanisms responsible for controlling release of CCK from duodenal I cells. ACh, acetylcholine; CCK-RP, CCK-releasing peptide; GRP, gastrin-releasing peptide. Solid arrows represent stimulatory effects, whereas dashed arrows indicate inhibition. (Redrawn from Barrett KE. *Gastrointestinal Physiology.* New York: McGraw-Hill; 2006.)

and protein digestion. The second releasing factor, **monitor peptide,** is released by pancreatic acinar cells into pancreatic juice. Both CCK-releasing peptide and monitor peptide can also be released in response to neural input, which is particularly important in initiating pancreatic secretion during the cephalic and gastric phases, thereby preparing the system to digest the meal as soon as it enters the small intestine.

What is the significance of these CCK-releasing factors? Their primary role is to match CCK release, as well as the resulting availability of pancreatic enzymes, to the need for these enzymes to digest the meal in the small intestinal lumen (Fig. 30.5). Because the releasing factors are peptides, they will be subject to proteolytic degradation by enzymes such as pancreatic trypsin in exactly the same way as dietary protein. However, when dietary protein is ingested, it is present in much greater amounts in the lumen than the releasing factors and thus "competes" with the releasing factors for proteolytic degradation. The net effect is that the releasing factors will be protected from breakdown while the meal is in the small intestine and are therefore available to continue stimulation of CCK release from I cells. However, once the meal has been digested and absorbed, the releasing factors are degraded and the signal for release of CCK is shut off.

CCK evokes secretion by pancreatic acinar cells in two ways. First, it is a classic hormone that travels through the bloodstream to encounter acinar cell CCK1 receptors. However, CCK also stimulates neural reflex pathways that impinge on the pancreas. Vagal afferent nerve endings in

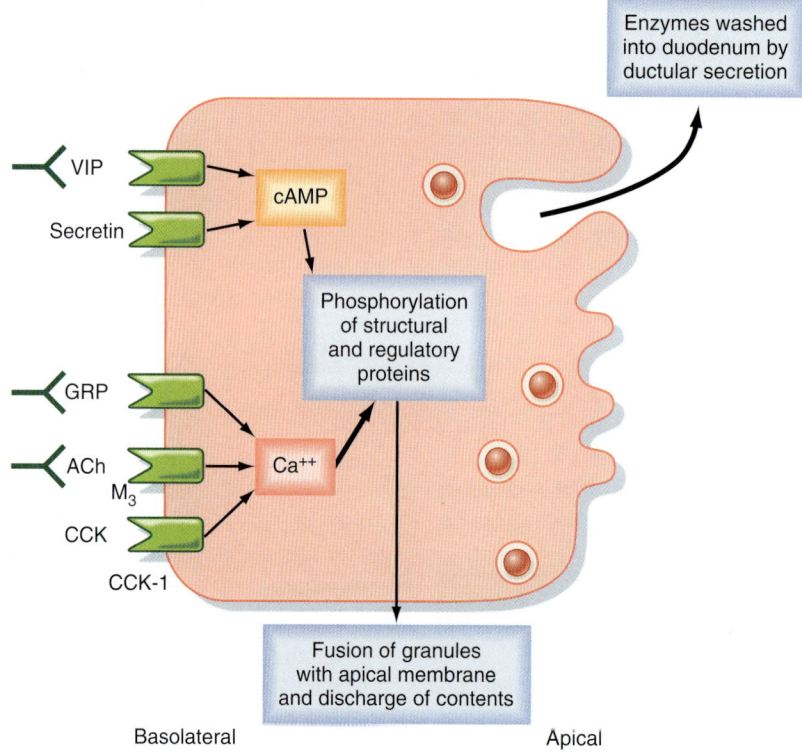

• **Fig. 30.6** Receptors of the pancreatic acinar cell and regulation of secretion. The thick black arrow indicates that Ca⁺⁺-dependent signaling pathways play the most prominent role. ACh, acetylcholine; CCK, cholecystokinin; GRP, gastrin-releasing peptide; VIP, vasoactive intestinal polypeptide; M₃, M₃ muscarinic receptor; CCK-1; CCK receptor type 1. (Redrawn from Barrett KE. *Gastrointestinal Physiology.* New York: McGraw-Hill; 2006.)

the wall of the small intestine are responsive to CCK by virtue of their expression of CCK1 receptors. As described earlier for the effect of CCK on gastric emptying, binding of CCK activates a vagovagal reflex that can further enhance acinar cell secretion via activation of pancreatic enteric neurons and release of a series of neurotransmitters such as acetylcholine, gastrin-releasing peptide, and vasoactive intestinal polypeptide (VIP).

The secretory products of pancreatic acinar cells are largely presynthesized and stored in granules that cluster toward the apical pole of acinar cells (Fig. 30.6). The most potent stimuli of acinar cell secretion, including CCK, acetylcholine, and gastrin-releasing peptide (GRP), act by mobilizing intracellular Ca⁺⁺. Stimulation of acinar cells results in phosphorylation of a series of regulatory and structural proteins within the cell cytosol that move the granules closer to the apical membrane, where the granules fuse with the plasma membrane. The contents of the granule are then discharged into the acinar lumen and washed out by an exudate of plasma crossing the tight junctions, linking the acinar cells together, and subsequently by ductular secretions. In the period between meals the granule constituents are resynthesized by the acinar cells and then stored until needed to digest the next meal. Resynthesis may be stimulated by the same agonists that evoke the initial secretory response.

Biliary Secretion

Another important digestive juice that is mixed with the meal in the small intestinal lumen is **bile.** Bile is produced by the liver, and the mechanisms that are involved, as well as the specific constituents, will be discussed in greater detail in Chapter 32 when we address the transport and metabolic functions of the liver. However, for purposes of the current discussion, bile is a secretion that serves to aid in digestion and absorption of lipids. Bile flowing out of the liver is stored and concentrated in the **gallbladder** until it is released in response to ingestion of a meal. Contraction of the gallbladder, as well as relaxation of the sphincter of Oddi, are evoked predominantly by CCK.

When considering the small intestinal phase of meal assimilation, the bile constituents we are most concerned with are the bile acids. These form structures known as *micelles* that serve to shield the hydrophobic products of lipid digestion from the aqueous environment of the lumen. Bile acids are in essence biological detergents, and large quantities are needed on a daily basis for optimal lipid absorption—as much as 1 to 2 g/day. The majority of the bile acid pool is recycled from the intestine back to the liver after each meal via the **enterohepatic circulation** (Fig. 30.7). Thus bile acids are synthesized in a conjugated form that limits their ability to passively cross the epithelium

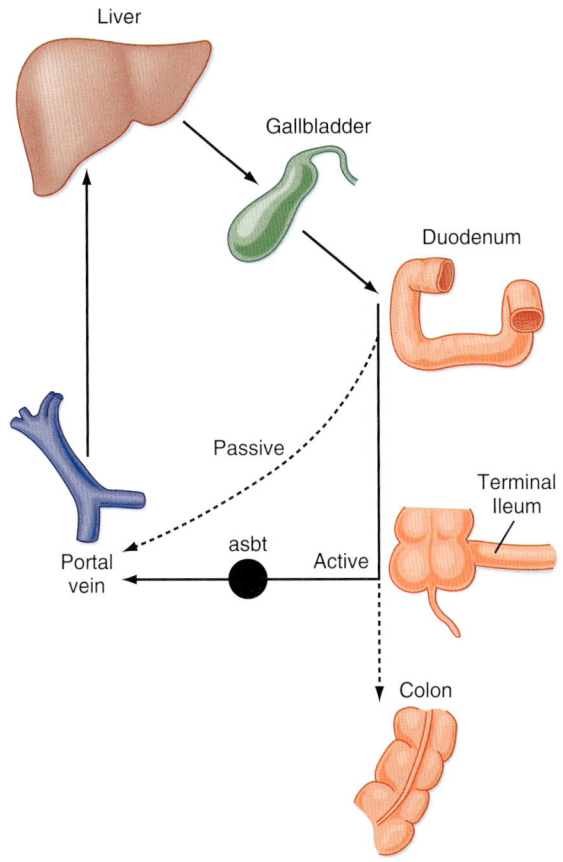

• **Fig. 30.7** Enterohepatic circulation of bile acids. Active uptake of conjugated bile acids occurs via the apical sodium-dependent bile acid transporter (asbt).

lining the intestine so that they are retained in the lumen to participate in lipid assimilation (discussed in detail under that heading). However, when the meal contents reach the terminal ileum, after lipid absorption has been completed, the conjugated bile acids are reabsorbed by a symporter, the **apical Na⁺-dependent bile acid transporter** (asbt), that specifically takes up conjugated bile acids in association with sodium ions. Only a minor portion of the bile acid pool is left to spill over into the colon in health, and here bile acids become deconjugated and subject to passive reabsorption (see Fig. 30.7). The net effect is to cycle the majority of the bile acid pool between the liver and intestine on a daily basis, coincident with signals arising in the postprandial period. Bile acids also exert biological actions beyond their role as detergents by binding to both cell surface and nuclear receptors in a variety of cell types throughout the body. In this way they regulate their own synthesis as well as other metabolic processes.

Carbohydrate Assimilation

The most important physiological function of the small intestine is to take up the products of digestion of ingested nutrients. Quantitatively the most significant nutrients (**macronutrients**) fall into three classes: carbohydrates, proteins, and lipids. The small intestine is critical not only

for absorption of nutrients into the body but also for the final stages of their digestion into molecules that are simple enough to be transported across the intestinal epithelium. We will consider the processes involved in assimilation of each of these nutrients in turn, beginning with carbohydrates. Carbohydrate digestion occurs in two phases: in the lumen of the intestine and then on the surface of enterocytes in a process known as **brush border digestion.** The latter is important in generating simple absorbable sugars only at the point where they can be absorbed. This may therefore limit their exposure to the small number of bacteria present in the small intestinal lumen that might otherwise use these sugars as nutrients.

Digestion of Carbohydrates

Dietary **carbohydrates** are composed of several different molecular classes. Starch, the first of these, is a mixture of both straight- and branched-chain polymers of glucose. The straight-chain polymers are called **amylose** and the branched-chain molecules are called **amylopectin** (Fig. 30.8). Starch is a particularly important source of calories, especially in developing countries, and is found predominantly in cereal products. Disaccharides are a second class of carbohydrate nutrients that includes **sucrose** (consisting of glucose and fructose) and **lactose** (consisting of glucose and galactose), the latter being an important caloric source in infants. It is, however, a key principle that the intestine can absorb only monosaccharides and not larger carbohydrates. Finally, many food items of vegetable origin contain dietary fiber, which consists of carbohydrate polymers that cannot be digested by human enzymes. These polymers are instead digested by bacteria present largely in the colonic lumen (see Chapter 31), thereby allowing salvage of their caloric value.

Dietary disaccharides are hydrolyzed to their component monomers directly on the surface of small intestinal epithelial cells by brush border digestion, mediated by a family of membrane-bound, heavily glycosylated hydrolytic enzymes synthesized by small intestinal epithelial cells. Brush border hydrolases critical to the digestion of dietary carbohydrates include **sucrase, isomaltase, glucoamylase,** and **lactase** (Table 30.1). Glycosylation of these hydrolases is believed to protect them to some extent from degradation by luminal pancreatic proteases. However, between meals the hydrolases are degraded and must therefore be resynthesized by the enterocyte to participate in digesting the next carbohydrate meal. Sucrase/isomaltase and glucoamylase are synthesized in quantities that are in excess of requirements, and assimilation of their products into the body is limited by the availability of specific membrane transporters for these monosaccharides (see Uptake of Carbohydrates). Lactase, in contrast, shows a developmental decline in expression after weaning. The relative paucity of lactase means that digestion of lactose, rather than uptake of the resulting products, is rate limiting for assimilation. If lactase levels fall below a certain threshold, the disease of lactose intolerance results.

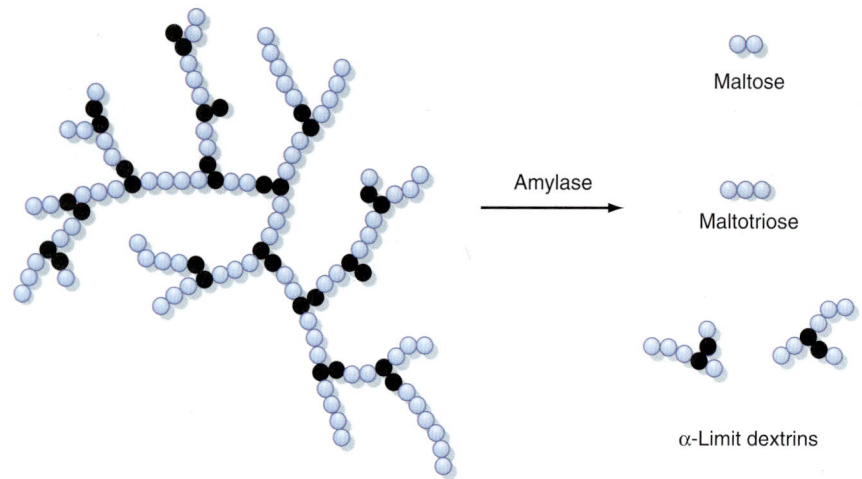

Maltose

Maltotriose

α-Limit dextrins

• **Fig. 30.8** Structure of amylopectin and the action of amylase. The blue circles represent glucose monomers linked by α-1,4 bonds. The black circles represent glucose units linked by α-1,6 bonds at the branch points.

TABLE 30.1	**Brush Border Carbohydrate Hydrolases**		
Enzyme	Specificity/Substrates	Products
Sucrase	α-1,4 bonds of maltose, maltotriose, and sucrose	Glucose, fructose
Isomaltase	α-1,4 bonds of maltose, maltotriose; α-1,6 bonds of α-limit dextrins	Glucose
Glucoamylase	α-1,4 bonds of maltose, maltotriose	Glucose
Lactase	Lactose	Glucose, galactose

 ## IN THE CLINIC

Lactose intolerance is relatively common in adults from specific ethnic groups, such as Asians, African Americans, and Hispanics. The disorder reflects a normal developmental decline in the expression of lactase by enterocytes, particularly when lactose is not a consistent component of the diet. In such individuals, consumption of foods containing large quantities of lactose (e.g., milk, ice cream) can result in abdominal cramping, gas, and diarrhea. These symptoms reflect a relative inability to digest lactose; thus it remains in the lumen, and water is retained. Some lactose-intolerant patients benefit from oral administration of a bacterially derived lactase enzyme before ingesting dairy products.

Digestion of starch occurs in two phases. The first takes place in the lumen and is actually initiated in the oral cavity via the activity of salivary amylase, as discussed in Chapter 28. Salivary amylase, however, is not essential for starch digestion, although it may assume greater importance in neonates or patients in whom the output of pancreatic enzymes is impaired by disease. Quantitatively the most significant contributor to the luminal digestion of starch is pancreatic amylase. Both enzymes hydrolyze internal α-1,4 bonds in both amylose and amylopectin, but not external bonds nor the α-1,6 bonds that form the branch points in the amylopectin molecule (see Fig. 30.8). Thus digestion

of starch by amylase is of necessity incomplete and results in short oligomers of glucose, including dimers (maltose) and trimers (maltotriose), as well as the simplest branching structures, which are called α-limit dextrins. Thus to allow absorption of its constituent monosaccharides, starch must also undergo brush border digestion.

At the brush border, straight-chain glucose oligomers can be digested by the hydrolases glucoamylase, sucrase, or isomaltase (see Table 30.1). All yield free glucose monomers, which can then be absorbed by the mechanisms discussed later. For α-limit dextrins, on the other hand, isomaltase activity is critical because it is the only enzyme that can cleave α-1,6 bonds that make up the branch points as well as α-1,4 bonds.

Uptake of Carbohydrates

Water-soluble monosaccharides resulting from digestion must next be transported across the hydrophobic plasma membrane of the enterocyte. The **sodium/glucose transporter 1** (SGLT1) is a symporter that takes up glucose (and galactose) against its concentration gradient by coupling its transport to that of Na+ (Fig. 30.9). Once inside the cytosol, glucose and galactose can be retained for the epithelium's metabolic needs or can exit the cell across its basolateral pole via a transporter known as *GLUT2*. Fructose, in contrast, is taken up across the apical membrane by GLUT5. However,

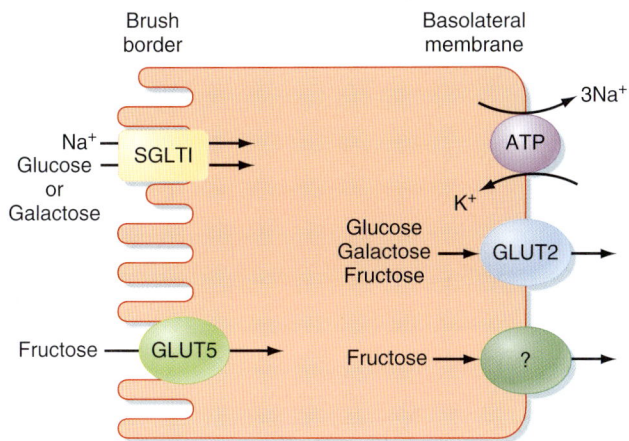

• **Fig. 30.9** Absorption of glucose, galactose, and fructose in the small intestine. GLUT, glucose transporter; SGLT1, sodium/glucose transporter 1.

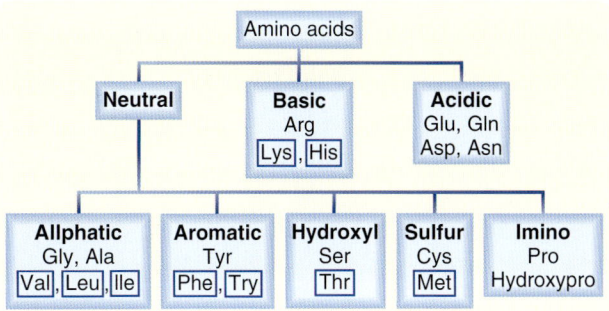

• **Fig. 30.10** Naturally occurring dietary amino acids. Those in boxes are essential amino acids that cannot be synthesized by humans and thus must be obtained from the diet. (Redrawn from Barrett KE. *Gastrointestinal Physiology.* New York: McGraw-Hill; 2006.)

because fructose transport is not coupled to that of Na^+, its uptake is relatively inefficient and can easily be overwhelmed if large quantities of food containing this sugar are ingested. The symptoms that occur from this malabsorption are similar to those experienced by a lactose-intolerant patient who consumes lactose.

Protein Assimilation

Proteins are also water-soluble polymers that must be digested into their smaller constituents before absorption is possible. Their absorption is more complicated than that of carbohydrates because they contain 20 different amino acids and short oligomers of these amino acids (dipeptides, tripeptides, and perhaps even tetrapeptides) can also be transported by enterocytes. The body, particularly the liver (see Chapter 32), has substantial ability to interconvert various amino acids subject to the body's needs. However, some amino acids, termed the **essential amino acids,** cannot be synthesized by the body either de novo or from other amino acids and thus must be obtained from the diet. The amino acids that must be obtained in this way in humans are shown in Fig. 30.10.

🧬 AT THE CELLULAR LEVEL

A rare genetic disorder results in an inability of the intestine to absorb glucose or galactose. This disease has been mapped to a variety of mutations in the *SGLT1* gene that result in a faulty or unexpressed protein or, more commonly, failure of the protein to traffic appropriately to the apical membrane of enterocytes. In patients carrying such mutations, malabsorbed glucose contributes to diarrheal and other symptoms, as discussed earlier for lactose intolerance. Despite the rarity of the disease, it is important in terms of the insight it provided into a critical process of intestinal epithelial transport.

Digestion of Proteins

Proteins can be hydrolyzed to long peptides simply by virtue of the acidic pH that exists in the gastric lumen. However, for assimilation of proteins into the body, three phases of enzymatically mediated digestion are required (Fig. 30.11). Like acid hydrolysis, the first of these phases takes place in the gastric lumen and is mediated by pepsin, the product of chief cells localized to the gastric glands. When gastric secretion is activated by signals coincident with ingestion of a meal, pepsin is released from the chief cells as the inactive precursor pepsinogen. At acidic pH, this precursor is autocatalytically cleaved to yield the active enzyme. Pepsin is highly specialized to act in the stomach, since it is activated by low pH. The enzyme cleaves proteins at sites of neutral amino acids, with a preference for aromatic or large aliphatic side chains. Because such amino acids occur only relatively infrequently in a given protein, pepsin is not capable of digesting protein fully into a form that can be absorbed by the intestine. Instead it yields a mixture of intact protein, large peptides (the majority), and a limited number of free amino acids.

On moving into the small intestine, the partially digested protein encounters the proteases provided in pancreatic juice. Recall that these enzymes are secreted in inactive forms. How then are they activated to begin the process of protein digestion? In fact, activation of proteases is delayed until these enzymes are in the lumen by virtue of the localized presence of an activating enzyme, enterokinase, only on the brush border of small intestinal epithelial cells (Fig. 30.12). Enterokinase cleaves trypsinogen to yield active trypsin. Trypsin in turn cleaves all the other protease precursors secreted by the pancreas, thereby resulting in a mixture of enzymes that can almost completely digest the vast majority of dietary proteins. Trypsin is an **endopeptidase** that cleaves proteins only at internal bonds within the peptide chain rather than releasing individual amino acids from the end of the chain. Trypsin is specific for cleavage at basic amino acids, and such cleavage results in a set of shorter peptides with a basic amino acid at their C-terminus. The two other pancreatic endopeptidases, chymotrypsin and elastase, have a similar mechanism of action but cleave at sites of neutral

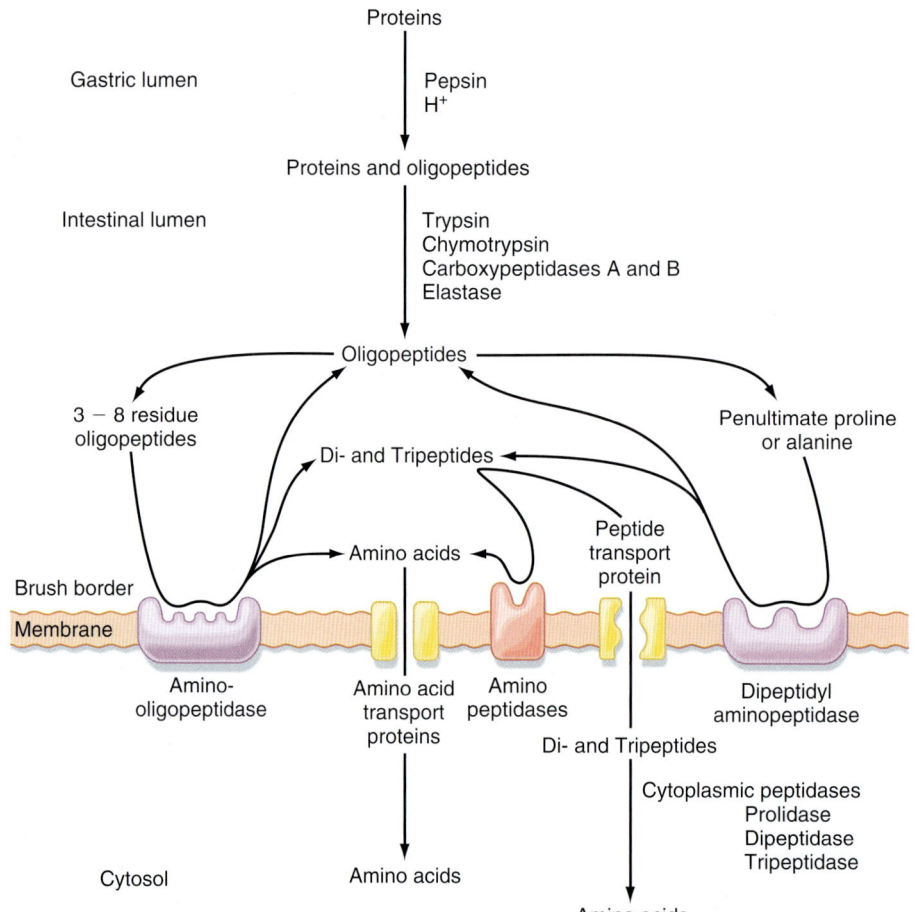

• **Fig. 30.11** Hierarchy of proteases and peptidases that function in the stomach and small intestine to digest dietary protein. Proteins are absorbed as either single amino acids (70%) or short peptides (30%). (Adapted from Van Dyke RW. In: Sleisenger MH, Fordtran JS [eds]. *Gastrointestinal Disease.* 4th ed. Philadelphia: Saunders; 1989.)

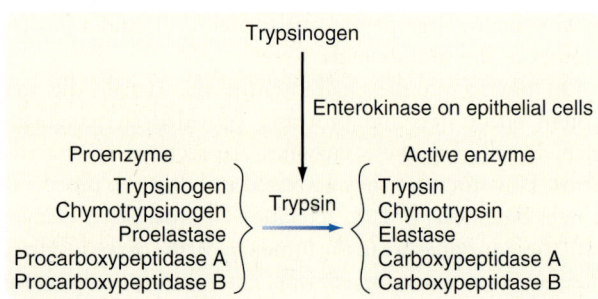

• **Fig. 30.12** Conversion of the inactive proenzymes of pancreatic juice to active enzymes by the action of trypsin. Trypsinogen in pancreatic juice is proteolytically converted to active trypsin by enterokinase expressed on the surface of epithelial cells of the duodenum and jejunum. Trypsin then activates the other proenzymes as shown.

amino acids. The peptides that result from endopeptidase activity are then acted on by pancreatic **ectopeptidases.** These enzymes cleave single amino acids from the end of a peptide chain, and those present in pancreatic juice are specific for either neutral (**carboxypeptidase A**) or basic (**carboxypeptidase B**) amino acids situated at the C-terminus. Thus the products that result after digestion of

a protein meal by gastric and pancreatic secretions include neutral and basic amino acids, as well as short peptides that have acidic amino acids at their C-termini and thus are resistant to carboxypeptidase A or B (Fig. 30.13).

The final phase of protein digestion takes place at the brush border. Mature enterocytes express a variety of peptidases on their brush borders, including both aminopeptidases and carboxypeptidases that generate products suitable for uptake across the apical membrane (see Fig. 30.11). However, it should be noted that even with the substantial complement of active proteolytic enzymes, some dietary peptides are either relatively or totally resistant to hydrolysis. In particular, peptides containing either proline or glycine are digested very slowly. Fortunately the intestine can take up short peptides in addition to single amino acids. The majority of peptides taken up into the enterocyte in their intact form are then subjected to a final stage of digestion in the cytosol of the enterocyte to liberate their constituent amino acids for use in the cell or elsewhere in the body (Fig. 30.14). However, some di- and tripeptides may also be transported into the blood in their intact form.

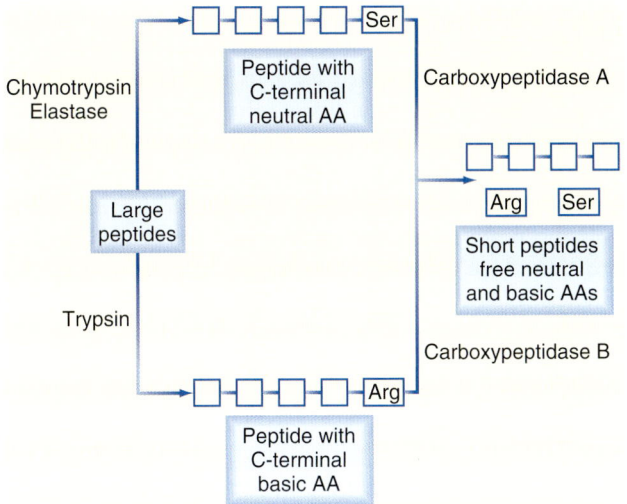

• **Fig. 30.13** Luminal digestion of peptides resulting from partial proteolysis in the stomach. *AA,* amino acid. (Redrawn from Barrett KE. *Gastrointestinal Physiology.* New York: McGraw-Hill; 2006.)

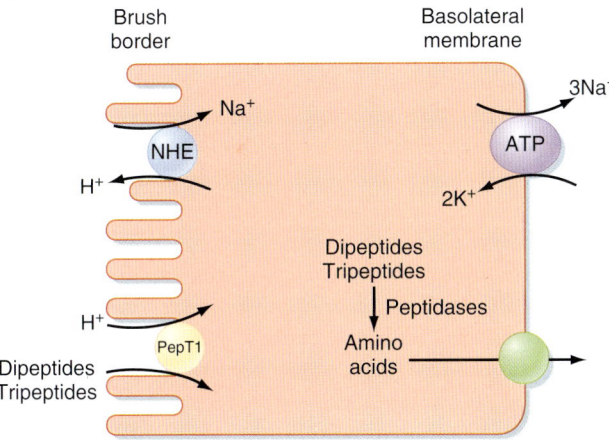

• **Fig. 30.14** A wide variety of dipeptides and tripeptides are taken up across the brush border membrane by the proton-coupled symporter known as *peptide transporter 1* (PepT1). The proton gradient is created by the action of sodium/hydrogen exchangers (NHEs) in the apical membrane. Peptides are largely digested in the cytosol to their constituent amino acids for export to the body, but a small proportion may be exported intact.

Uptake of Peptides and Amino Acids

The body is also endowed with a series of plasma membrane transporters capable of promoting uptake of the water-soluble products of protein digestion. Given the large number of amino acids, there is a relatively large number of specific transporters (see Figs. 30.11 and 30.14). Amino acid transporters are of clinical interest because their absence in a variety of genetic disorders results in diminished ability to transport the relevant amino acid or acids. However, such mutations are often clinically silent, at least from a nutritional standpoint, because the amino acid in question can be assimilated by other transporters with overlapping specificity or in the form of peptides. This does not rule out the possibility of pathology in other organ systems in which the transporter of interest may normally be expressed (e.g., cystinuria). In general, amino acid transporters have reasonably broad specificity and usually transport a subset of amino acids (e.g., neutral, anionic, or cationic) but with some overlap in their affinity for particular amino acids. Furthermore, some (but not all) of the amino acid transporters are symporters that carry their substrate amino acids in conjunction with obligatory uptake of Na^+.

The small intestine is also notable for its ability to take up short peptides (see Fig. 30.14). The primary transporter responsible for such uptake is called **peptide transporter 1 (PepT1)** and is a symporter that transports peptides in conjunction with protons. Amino acids liberated from these peptides that are not required by the enterocyte are in turn exported across the basolateral membrane and enter blood capillaries to be transported to the liver via the portal vein. PepT1 is also of clinical interest because it can mediate the uptake of so-called **peptidomimetic drugs,** which include a variety of antibiotics as well as cancer chemotherapeutic agents. The mechanisms by which amino acids and peptidomimetic drugs exit the enterocyte are not fully understood but are presumed to involve additional transport proteins.

AT THE CELLULAR LEVEL

The redundancy in uptake mechanisms for the products of protein digestion underscores the importance of this process and also means that deficiencies in specific amino acid assimilation across the intestine are relatively rare. However, under certain circumstances, mutations in proteins responsible for specific amino acid transport can lead to pathology in other organs. One example is the disease of cysteinuria, which is a molecularly heterogeneous disease involving mutations in a variety of amino acid transporters capable of transporting cysteine. Because cysteine can also be assimilated across the gut in the form of peptides, nutritional deficiencies do not occur despite a lack of intestinal uptake mechanisms for this particular amino acid. In contrast, cysteine can only be poorly reabsorbed from the urine of patients suffering from cystinuria, and kidney stones can form because this amino acid is relatively insoluble. Pathophysiology can also arise secondary to mutations in *SLC6A19,* a Na^+-independent transporter of neutral amino acids, and result in a condition known as *Hartnup disease,* a relatively rare inborn error of metabolism that causes loss of neutral amino acids in urine, resulting in psychiatric issues, intellectual disability, short stature, headaches, and unsteady gait, although not nutritional deficiencies per se.

Lipid Assimilation

Lipids, defined as substances that are more soluble in organic solvents than in water, are the third major class of macronutrients making up the human diet. Lipids supply more significant calories on a per-gram basis than proteins or carbohydrates do and are thus of major nutritional significance; they also have a propensity to contribute to obesity if consumed in excessive amounts. Lipids also dissolve volatile compounds that contribute to food's taste and aroma.

The predominant form of lipid in the human diet is triglyceride, found in oils and other fats. The majority of these triglycerides have long-chain fatty acids (carbon chains > 12 carbons) esterified to the glycerol backbone. Additional lipid is supplied in the form of phospholipids and cholesterol, mostly arising from cell membranes. It is also important to consider that the intestine is presented daily not only with dietary lipid but also with lipid originating from the liver in biliary secretions, as described in more detail in Chapter 32. Indeed, the cholesterol supplied in bile exceeds that provided in the diet on a daily basis in all but the most egg-loving individuals. Finally, though present in only trace amounts, the fat-soluble vitamins (A, D, E, K) are essential nutrients that should be supplied in the diet to avoid disease. These substances are almost entirely insoluble in water and thus require special handling to promote their uptake into the body.

Emulsification and Solubilization of Lipids

When a fatty meal is ingested, the lipid becomes liquefied at body temperature and floats on the surface of the gastric contents. This would limit the area of the interface between the aqueous and lipid phases of the gastric contents and thus restrict access of enzymes capable of breaking down the lipid to forms that can be absorbed. This is because the lipolytic enzymes, as proteins, reside in the aqueous phase. Therefore an early stage in the assimilation of lipid is its emulsification. The mixing action of the stomach churns the dietary lipid into a suspension of fine droplets, which vastly increases the surface area of the lipid phase.

Lipid absorption is also facilitated by formation of a micellar solution with the aid of bile acids supplied in biliary secretions. Details of this process will be discussed subsequently.

Digestion of Lipids

Lipid digestion begins in the stomach. Gastric lipase is released in large quantities from gastric chief cells; it adsorbs to the surface of fat droplets dispersed in the gastric contents and hydrolyzes component triglycerides to diglycerides and free fatty acids. However, little lipid assimilation can take place in the stomach because of the acidic pH of the lumen, which results in protonation of the free fatty acids released by gastric lipase. Lipolysis is also incomplete in the stomach because gastric lipase, despite its optimum catalytic activity at acidic pH, is not capable of hydrolyzing the second position of the triglyceride ester, which means that the molecule cannot be fully broken down into components that can be absorbed into the body. There is also little if any breakdown of cholesterol esters or the esters of fat-soluble vitamins. Indeed, gastric lipolysis is dispensable in healthy individuals because of the marked excess of pancreatic enzymes.

The majority of lipolysis takes place in the small intestine in health. Pancreatic juice contains three important lipolytic enzymes that are optimized for activity at neutral pH. The first of these is pancreatic lipase. This enzyme differs from the stomach enzyme in that it is capable of hydrolyzing both the 1 and 2 positions of triglyceride to yield a large quantity of free fatty acids and monoglycerides. At neutral pH, the head groups of the free fatty acids are charged, and thus these molecules migrate to the surface of the oil droplets. Lipase also displays an apparent paradox in that it is inhibited by bile acids, which also form part of the small intestinal contents. Bile acids adsorb to the surface of the oil droplets and cause lipase to dissociate. However, lipase activity is sustained by an important cofactor, colipase, which is also supplied in pancreatic juice. Colipase is a bridging molecule that binds to bile acids and to lipase; it anchors lipase to the oil droplet even in the presence of bile acids.

Pancreatic juice also contains two additional enzymes that are important in fat digestion. The first of these is phospholipase A_2, which hydrolyzes phospholipids such as those present in cell membranes. Predictably this enzyme would be quite toxic in the absence of dietary substrates, and thus it is secreted as an inactive pro-form that is activated only when it reaches the small intestine. Furthermore, pancreatic juice contains a relatively nonspecific cholesterol esterase that can break down esters of cholesterol, as its name implies, as well as esters of fat-soluble vitamins and even triglycerides. Interestingly this enzyme *requires* bile acids for activity (contrast with lipase, discussed earlier), and it is related to an enzyme produced in breast milk that plays an important role in lipolysis in neonates.

As lipolysis proceeds, the products are abstracted from the lipid droplet, first into a lamellar (membrane) phase and subsequently into mixed micelles composed of lipolytic products as well as bile acids. The *amphipathic* (meaning they have both a hydrophobic and hydrophilic face) bile acids serve to shield the hydrophobic regions of lipolytic products from water while presenting their own hydrophilic faces to the aqueous environment (Fig. 30.15). Micelles are truly in solution and thus markedly increase lipid solubility in the intestinal contents. This increases the rate at which molecules such as fatty acids diffuse to the absorptive epithelial surface. Nevertheless, given the very large surface area of the small intestine and the appreciable molecular solubility of the products of triglyceride hydrolysis, micelles are not essential for the absorption of triglyceride. Thus patients who have insufficient output of bile acids (caused, for example, by a gallstone that obstructs bile output) do not normally show fat malabsorption. On the other hand, cholesterol and the fat-soluble vitamins are almost totally insoluble in water and accordingly require micelles to be absorbed even after they have been digested. Thus if luminal bile acid concentrations fall below the critical micellar concentration, patients can become deficient in fat-soluble vitamins.

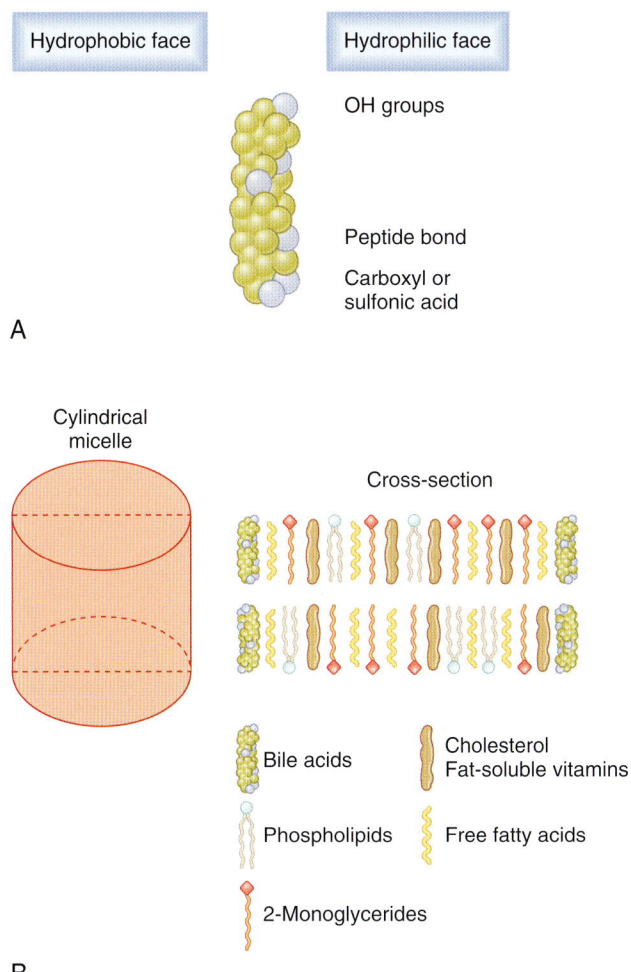

• **Fig. 30.15** Schematic depiction of bile acids (A) and mixed micelles (B). Bile acids in solution are amphipathic. Mixed micelles are cylindrical assemblages of bile acids with other dietary lipids.

IN THE CLINIC

A relatively new treatment for hypercholesterolemia targets absorption of cholesterol, either derived from the diet or in bile, across the small intestinal epithelium. Ezetimibe is a drug that specifically blocks cellular uptake of cholesterol by inhibiting the activity of the NPC1L1 protein expressed in the apical membrane of enterocytes. In conjunction with other drugs designed to counter atherosclerosis, this may be a useful adjunct in that it can interrupt the enterohepatic circulation as well as prevent absorption of dietary cholesterol. Clinical studies suggest that ezetimibe may synergistically improve the efficacy of other strategies designed to reduce circulating levels of low-density lipoprotein cholesterol in those at risk for cardiovascular disease, such as the use of statin drugs.

Lipids also differ from carbohydrates and proteins in terms of their fate after absorption into the enterocyte. Unlike monosaccharides and amino acids, which leave the enterocyte in molecular form and enter the portal circulation, the products of lipolysis are reesterified in the enterocyte to form triglycerides, phospholipids, and cholesterol esters. These metabolic events take place in the smooth endoplasmic reticulum. Concurrently the enterocyte synthesizes a series of proteins known as *apolipoproteins* in the rough endoplasmic reticulum. These proteins are then combined with the resynthesized lipids to form a structure known as a **chylomicron,** which consists of a lipid core (predominantly triglyceride with much less cholesterol, phospholipid, and fat-soluble vitamin esters) coated by the apolipoproteins. The chylomicrons are then exported from the enterocyte by a process of exocytosis. However, on entering the lamina propria, they are too large (≈750–5000 Å in diameter) to permeate through the intercellular spaces of the mucosal capillaries. Instead they are taken up into lymphatics in the lamina propria and therefore bypass the portal circulation and, at least for their first pass, the liver. Eventually, chylomicrons in the lymph enter the bloodstream via the thoracic duct and then serve as the vehicle to transport lipids around the body for use by cells in other organs. The only exception to this chylomicron-mediated transport is for medium-chain fatty acids. These acids are relatively water soluble and can also permeate enterocyte tight junctions appreciably, which means that they bypass the intracellular processing steps described earlier and are not packaged into chylomicrons. They therefore enter the portal circulation and are more readily available to other tissues. A diet rich in medium-chain triglycerides may be of particular benefit in patients with inadequate bile acid pools.

Uptake of Lipids and Subsequent Handling

The products of fat digestion are believed to be capable of crossing cell membranes readily because of their lipophilicity. However, recent evidence suggests that their uptake may alternatively or additionally be regulated via the activity of specific membrane transporters. A microvillus membrane fatty acid–binding protein (MVM-FABP) provides for the uptake of long-chain fatty acids across the brush border. Likewise, Niemann-Pick C1–like 1 (NPC1L1) has recently been identified as an uptake pathway for cholesterol in enterocytes and may be a therapeutic target in patients who suffer from pathological increases in circulating cholesterol (hypercholesterolemia). However, uptake of cholesterol overall is relatively inefficient because this molecule, along with plant sterols, can also be actively effluxed from enterocytes back into the cytosol by a heterodimeric complex of two ATP-binding cassette (ABC) transporters termed *ABC G5* and *G8*. Finally, the glycerol backbone of triglycerides may be transported into intestinal epithelial cells by a number of different aquaglyceroporins.

Water and Electrolyte Secretion and Absorption

The foregoing description of digestion has stressed that these processes take place in the small intestine in an aqueous

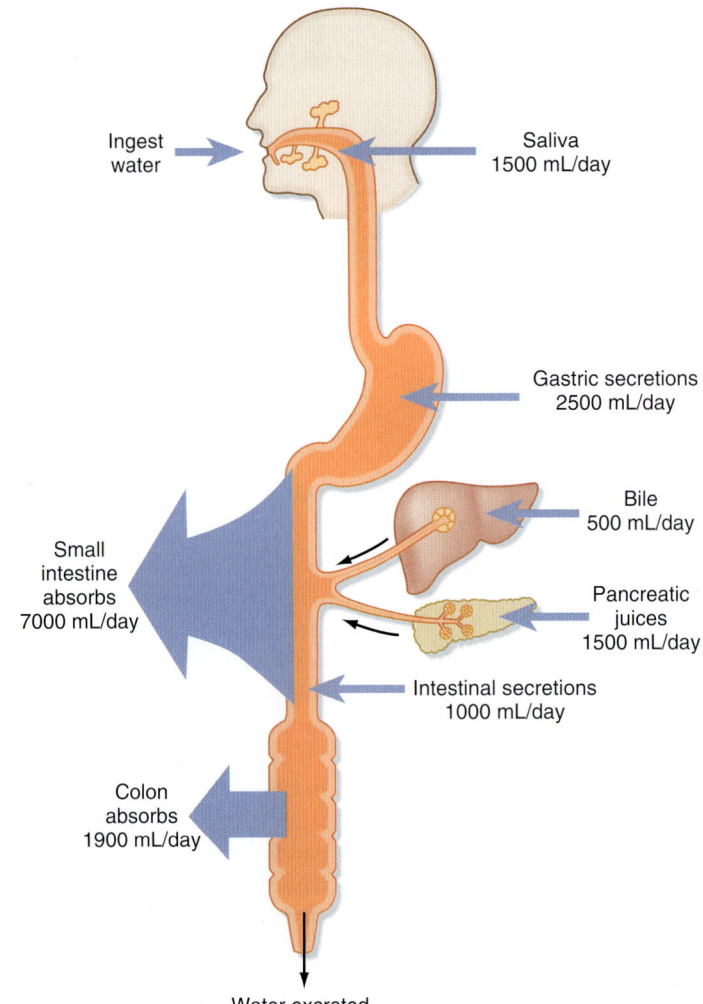

Ingest water

Saliva
1500 mL/day

Gastric secretions
2500 mL/day

Bile
500 mL/day

Small intestine absorbs
7000 mL/day

Pancreatic juices
1500 mL/day

Intestinal secretions
1000 mL/day

Colon absorbs
1900 mL/day

Water excreted

• **Fig. 30.16** Overall fluid balance in the human GI tract. About 1 to 2 L of water is ingested, and 8 L of various secretions enters the GI tract. Of this total, most is absorbed in the small intestine. About 2 L is passed onto the colon, the vast majority of which is absorbed in health. (From Vander AJ et al. *Human Physiology.* 6th ed. New York: McGraw-Hill; 1994.)

milieu. The fluidity of intestinal contents, especially in the small intestine, is important in allowing the meal to be propelled along the length of the intestine and to permit digested nutrients to diffuse to their site of absorption. Part of this fluid is derived from oral intake, but in most adults this consists of only about 1 to 2 L/day derived from both food and drink (Fig. 30.16). Additional fluid is supplied by the stomach and the small intestine themselves, as well as the organs that drain into the GI tract. In total, these secretions add another 8 L, which means that the intestine is presented with approximately 9 to 10 L of fluid on a daily basis. However, in health only about 2 L of this load is passed to the colon for reabsorption, and eventually only 100 to 200 mL exits in stool. Thus the intestine absorbs water overall. During the postprandial period, such absorption is promoted in the small intestine predominantly via the osmotic effects of nutrient absorption. An osmotic gradient is established across the intestinal epithelium that simultaneously drives movement of water across the tight junctions. The generic mechanism for nutrient-driven Na^+

and water absorption in the small intestine is diagrammed in Fig. 30.17. Moreover, in the period between meals, when nutrients are absent, fluid absorption can still occur via the coupled uptake of Na^+ and Cl^- mediated by the cooperative interaction of a Na^+/H^+ antiporter (NHE-3) and a Cl^-/HCO_3^- antiporter (see Fig. 30.17).

Even though net water and electrolyte transport in the small intestine is predominantly absorptive, this does not imply that the tissue fails to participate in electrolyte secretion. Secretion is regulated in response to signals derived from the luminal contents and in response to deformation of the mucosa and intestinal distention. Critical secretagogues include acetylcholine, VIP, prostaglandins, and serotonin. Secretion ensures that the intestinal contents are appropriately fluid while digestion and absorption are ongoing and is important to lubricate the passage of food particles along the length of the intestine. For example, some clinical evidence suggests that constipation and intestinal obstruction, the latter observed in cystic fibrosis, can result when secretion is abnormally low. The majority of the intestinal

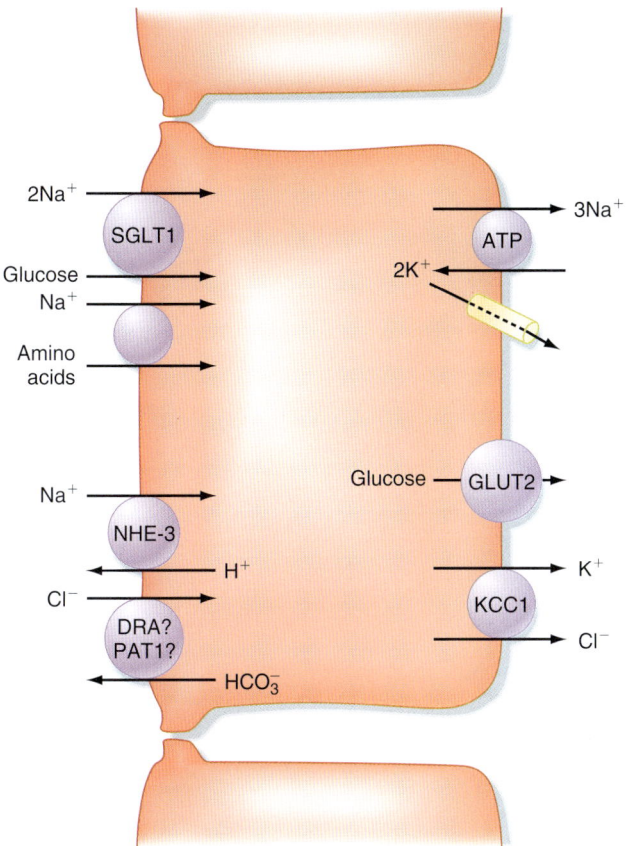

• **Fig. 30.17** Mechanisms of NaCl absorption in the small intestine. DRA, downregulated in adenoma; GLUT2, glucose transporter 2; KCC1, potassium/chloride cotransporter 1; NHE-3, sodium/hydrogen exchanger 3; PAT1, putative anion transporter 1; SGLT1, sodium/glucose transporter 1.

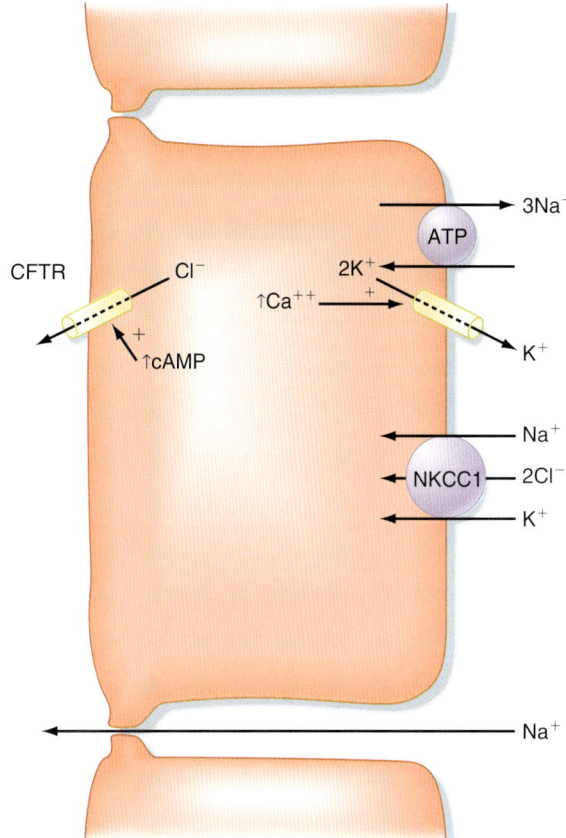

• **Fig. 30.18** Mechanism of Cl⁻ secretion in the small and large intestines. CFTR, cystic fibrosis transmembrane conductance regulator; NKCC1, sodium/potassium/2 chloride cotransporter 1. Note that additional chloride channels (not shown), such as those regulated by intracellular calcium concentration, may also exist in the apical membrane and contribute to overall transport.

secretory flow of fluid into the lumen is driven by active secretion of chloride ions via the mechanism diagrammed in Fig. 30.18. In cystic fibrosis the loss of CFTR Cl channels can partially be compensated for by accessory chloride channels, such as the TMEM16A channel that is activated by elevations in cytoplasmic calcium. Some segments of the intestine may engage in additional secretory mechanisms, such as secretion of bicarbonate ions via the mechanisms shown in Fig. 30.19. Bicarbonate secretion protects the epithelium, particularly in the most proximal portions of the duodenum immediately downstream from the pylorus, from damage caused by acid and pepsin.

Absorption of Minerals and Water-Soluble Vitamins

The small intestine is also an important site for absorption of water-soluble vitamins (e.g., vitamin C, cobalamin) and minerals such as calcium, magnesium, and iron. In general these substances are taken up from the luminal contents via the activity of specific transporters. Their uptake is

also controlled by feedback mechanisms that sense the concentration of the relevant substrate in the circulation and adjust the expression of transporters and accessory molecules accordingly.

Motor Patterns of the Small Intestine

The smooth muscle layers in the small intestine produce motility patterns that mix chyme with the various digestive secretions and propel fluid along the length of the intestine so that nutrients (along with water and electrolytes) can be absorbed. Motor patterns of the small intestine during the postprandial period are directed predominantly toward mixing and consist largely of segmenting and retropulsive contractions that retard the meal while digestion is still ongoing. **Segmentation** is a stereotypical pattern of rhythmic contractions that is displayed in Fig. 30.20 and presumably reflects programmed activity of the enteric nervous system superimposed on the basic electrical rhythm. Hormonal mediators of this fed pattern of motility are poorly defined, although CCK contributes. CCK also plays important roles in slowing gastric emptying when the meal is in the small

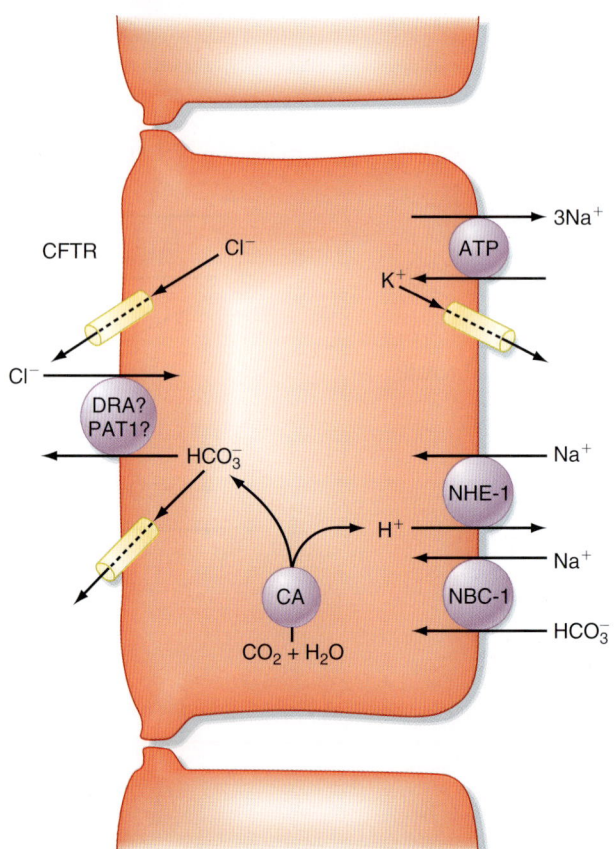

• **Fig. 30.19** Mechanisms of bicarbonate secretion in the duodenum. CA, carbonic anhydrase; CFTR, cystic fibrosis transmembrane conductance regulator; DRA, downregulated in adenoma; NBC-1, sodium/bicarbonate cotransporter 1; NHE-1, sodium/hydrogen exchanger-1; PAT1, putative anion transporter 1.

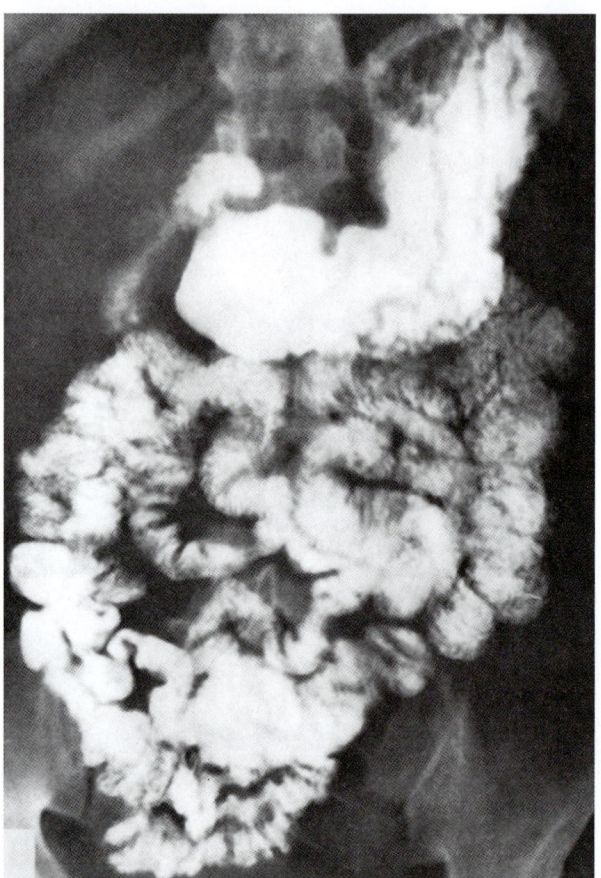

A

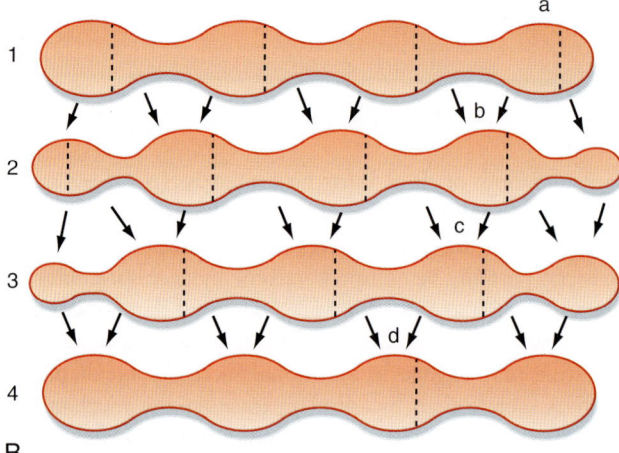

B

• **Fig. 30.20** A, Radiographic view showing the stomach and small intestine filled with barium contrast medium in a normal individual. Note the segmentation of the intestine. B, Sequence of segmental contractions in the small intestine. Lines 1 to 4 represent sequential time points. The dotted lines indicate where contractions will occur next; the arrows depict the direction of movement of the intestinal contents. (A, From Gardener EM et al. *Anatomy: A Regional Study of Human Structure.* 4th ed. Philadelphia: Saunders; 1975; B, redrawn from Cannon WB. *Am J Physiol* 1902;6:251.)

intestine, as described at the beginning of this chapter. This makes sense as a mechanism to match nutrient delivery to the available capacity to digest and absorb the components of the meal.

After the meal has been digested and absorbed, it is desirable to clear any undigested residues from the lumen to prepare the intestine for the next meal. Such clearance is effected by **peristalsis** (Fig. 30.21), a coordinated sequence of contraction occurring above the intestinal contents and relaxation below that permits the contents to be conveyed over considerable distances. Peristalsis reflects the action of acetylcholine and substance P released orad to a site of intestinal distention, which serves to contract the circular muscle, as well as the inhibitory effects of VIP and nitric oxide on the caudad side. Like segmentation, peristalsis originates when action potentials generated by intrinsic innervation are superimposed on sites of cellular depolarization dictated by the basic electrical rhythm. The peristaltic motor patterns that occur during fasting, moreover, are organized into a sequence of phases known as the **migrating motor complex** (Fig. 30.22). Phase I of the MMC is characterized by relative quiescence, whereas small disorganized contractions begin to occur during phase II. During phase III, which lasts about 10 minutes, large contractions that

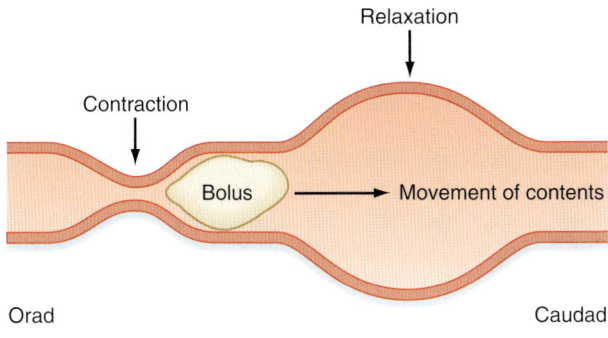

• **Fig. 30.21** Peristaltic motility in the intestine propels intestinal contents along the length of the small intestine.

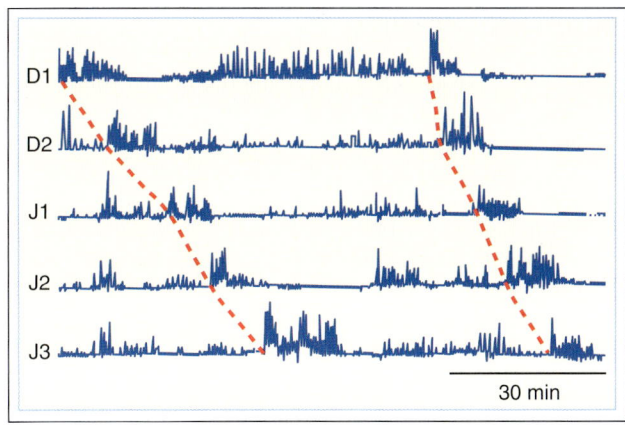

• **Fig. 30.22** Migrating motor complexes in the duodenum and jejunum as recorded from a fasting human subject by manometry. D1, D2, J1, J2, and J3 indicate sequential recording points along the length of the duodenum and jejunum. The intense contractions (phase III) propagate aborally. (Redrawn from Soffer EE et al. *Am J Gastroenterol* 1998;93:1318.)

propagate along the length of the intestine are stimulated by the hormone motilin and sweep any remaining gastric and intestinal contents out into the colon. The pylorus and ileocecal valve open fully during this phase, so even large undigested items can eventually pass from the body. Motility of the intestine then reverts to phase I of the MMC, with the entire cycle taking about 90 minutes in adults unless a meal is ingested, in which case the MMC is suspended. After the meal, motilin levels fall, and the MMC cannot be resumed until they rise again.

Key Concepts

1. On leaving the stomach, the meal enters the small intestine, which consists (sequentially) of the duodenum, jejunum, and ileum. The principal function of the small intestine is to digest and absorb the nutrients contained in the meal.
2. The presence of chyme in the duodenum retards additional gastric emptying, thus helping match nutrient delivery to the ability of the small intestine to digest and absorb such substances.
3. Digestion and absorption in the small intestine are aided by two digestive juices derived from the pancreas (pancreatic juice) and liver (bile). These secretions are triggered by hormonal and neural signals activated by the presence of the meal in the small intestine.
4. Pancreatic secretions arise from the acini and contain various proteins capable of digesting the meal or acting as important cofactors. The secretion is diluted and alkalinized as it passes through the pancreatic ducts.
5. Bile is produced by the liver and stored in the gallbladder until needed in the postprandial period. Bile acids, important components of bile, are biological detergents that solubilize the products of lipid digestion.
6. Carbohydrates and proteins, water-soluble macromolecules, are digested and absorbed by broadly analogous mechanisms. Lipids, the third macronutrient, require special mechanisms to transfer the products of lipolysis to the epithelial surface where they can be absorbed. The small intestine also absorbs fat- and water-soluble vitamins, as well as minerals such as calcium, magnesium, and iron.
7. The small intestine transfers large volumes of fluid into and out of the lumen on a daily basis to facilitate digestion and absorption of nutrients, driven by active transport of ions and other electrolytes.
8. The motor patterns of the small intestine vary depending on whether a meal has been ingested. Immediately after a meal, motility is directed to retaining the meal in the small intestine, mixing it with digestive juices, and providing sufficient time for absorption of nutrients. During fasting, a "housekeeper" complex of intense contractions (the migrating motor complex) sweeps periodically along the length of the stomach and small intestine to clear them of undigested residues.

Additional Reading

Abumrad NA, Davidson NO. Role of the gut in lipid homeostasis. *Physiol Rev*. 2012;92:1061-1085.

Argent BE, et al. Cell physiology of the pancreatic ducts. In: Johnson LR, ed. *Physiology of the GI Tract*. 5th ed. Waltham, MA: Academic Press; 2012.

Barrett KE, Keely SJ. Electrolyte secretion and absorption in the small intestine and colon. In: Podolsky DK, et al., eds. *Yamada's Textbook of Gastroenterology*. 6th ed. Hoboken, NJ: Wiley Blackwell; 2015.

Bharucha AE, Hasler WL. Motility of the small intestine and colon. In: Podolsky DK, et al., eds. *Yamada's Textbook of Gastroenterology*. 6th ed. Hoboken, NJ: Wiley Blackwell; 2015.

Ganapathy V. Protein digestion and absorption. In: Johnson LR, ed. *Physiology of the GI Tract*. 5th ed. Waltham, MA: Academic Press; 2012.

Liddle RA. Regulation of pancreatic secretion. In: Johnson LR, ed. *Physiology of the GI Tract*. 5th ed. Waltham, MA: Academic Press; 2012.

Wright EM, Sala-Rabanal M, Loo DDF, Hirayama BA. Sugar absorption. In: Johnson LR, ed. *Physiology of the GI Tract*. 5th ed. Waltham, MA: Academic Press; 2012.

31

The Colonic Phase of the Integrated Response to a Meal

LEARNING OBJECTIVES

Upon completion of this chapter the student should be able to answer the following questions:

1. What are the structures of the anatomy of the colon and rectum, and what is the role of the large intestine in storing and desiccating the residues of a meal?
2. What are the motility patterns of the colon that provide for its storage function, and what reflexes signal to the colon from more proximal portions of the GI tract?
3. What is the role of intestinal microorganisms in metabolism and host defense?
4. What are the mechanisms that provide for defecation, and how it can be delayed until convenient?

Overview of the Large Intestine

The most distal segment of the gastrointestinal (GI) tract, the **large intestine,** comprises the **cecum;** ascending, transverse, and descending portions of the **colon; rectum;** and **anus** (Fig. 31.1). The primary functions of the large intestine are to digest and absorb components of the meal that cannot be digested or absorbed more proximally, reabsorb the remaining fluid that was used during movement of the meal along the GI tract, and store the waste products of the meal until they can conveniently be eliminated from the body. In fulfilling these functions, the large intestine uses characteristic motility patterns and expresses transport mechanisms that drive the absorption of fluid, electrolytes, and other solutes from the stool. The large intestine also contains a unique biological ecosystem known as the *microbiota,* consisting of many trillions of **commensal bacteria** and other microorganisms that engage in a lifelong symbiotic relationship with their human host. These microorganisms can metabolize components of the meal that are not digested by host enzymes and make their products available to the body via a process known as **fermentation.** Colonic bacteria also metabolize other endogenous substances such as bile acids and bilirubin, thereby influencing their disposition. There is emerging evidence that the colonic microbiota is critically involved in promoting development of the normal colonic epithelium and in stimulating its differentiated functions.

In addition the microbiota can detoxify xenobiotics (substances originating outside the body, such as drugs) and protect the colonic epithelium from infection by invasive pathogens. Finally, the colon is both the recipient and the source of signals that allow it to communicate with other GI segments to optimally integrate function. For example, when the stomach is filled with freshly masticated food, the presence of the meal triggers a long reflex arc that results in increased colonic motility (the **gastrocolic reflex**) and eventually evacuation of the colonic contents to make way for the residues of the next meal. Similarly the presence of luminal contents in the colon causes release of both endocrine and neurocrine mediators that slow propulsive motility and decrease electrolyte secretion in the small intestine. This negative feedback mechanism matches the rate of delivery of colonic contents to the segment's capacity to process and absorb the useful components. Details of the signals that mediate this crosstalk between the colon and other components of the GI system are reviewed in the next section.

Signals That Regulate Colonic Function

The colon is regulated primarily, though not exclusively, by neural pathways. Colonic motility is influenced by local reflexes that are generated by filling of the lumen, thereby initiating distention and the activation of stretch receptors. These regulatory pathways exclusively involve the enteric nervous system. Local reflexes triggered by distortion of the colonic epithelium and produced, for example, by passage of a bolus of fecal material stimulate short bursts of Cl⁻ and fluid secretion mediated by 5-hydroxytryptamine (5-HT) from enteroendocrine cells and acetylcholine from enteric secretomotor nerves. On the other hand, colonic function and motility responses in particular are also regulated by long reflex arcs originating more proximally in the GI tract or in other body systems. One example is the gastrocolic reflex. Distention of the stomach activates a generalized increase in colonic motility and mass movement of fecal material, as described in more detail later. This reflex has both chemosensitive and mechanosensitive components at its site of origin and involves release of 5-HT and

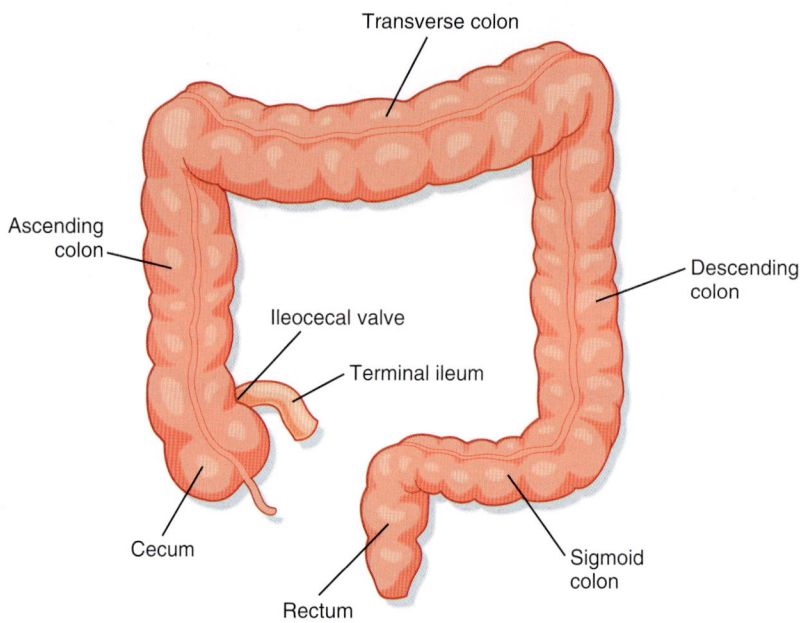

• **Fig. 31.1** Major anatomic subdivisions of the colon.

acetylcholine. Similarly the **orthocolic reflex** is activated on rising from bed and promotes a morning urge to defecate in many individuals.

The colon is relatively poorly supplied with cells that release bioactive peptides and other regulatory factors. Exceptions are **enterochromaffin cells,** which release 5-HT, and cells that synthesize **peptide YY,** so named because its sequence contains two adjacent tyrosine residues. Peptide YY is synthesized by enteroendocrine cells localized in the terminal ileum and colon and is released in response to lipid in the lumen. It decreases gastric emptying and intestinal propulsive motility. Peptide YY also reduces Cl⁻ and thus fluid secretion by intestinal epithelial cells. Thus peptide YY has been characterized as an **"ileal brake"** in that it is released if nutrients, especially fat, are not absorbed by the time the meal reaches the terminal ileum and proximal part of the colon. By reducing propulsion of the intestinal contents, in part by limiting their fluidity and distention-induced motility, peptide YY provides more time for the meal to be retained in the small intestine where its constituent nutrients can be digested and absorbed.

Patterns of Colonic Motility

To appreciate colonic motility the functional anatomy of the colonic musculature will be reviewed first, followed by a discussion of the regulation of colonic motility.

Functional Anatomy of the Colonic Musculature

As in other segments of the intestine, the colon consists of functional layers, with a columnar epithelium most closely apposed to the lumen, which is then underlaid by the lamina propria, serosa, and muscle layers. Similarly the colonic mucosa is surrounded by continuous layers of circular muscle that can occlude the lumen. Indeed, at intervals

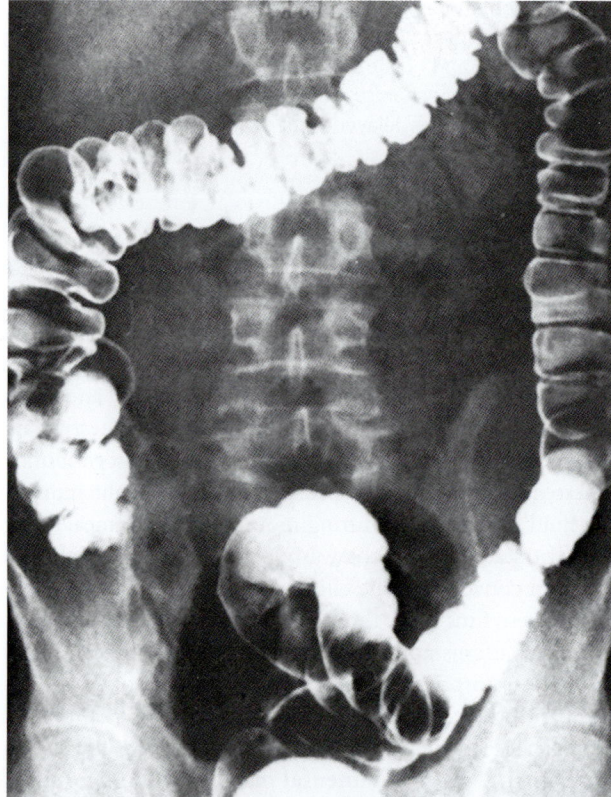

• **Fig. 31.2** Radiograph showing a prominent haustral pattern in the colon of a normal individual. (From Keats TE. *An Atlas of Normal Roentgen Variants.* 2nd ed. St Louis: Mosby–Year Book; 1979.)

the circular muscle contracts to divide the colon into segments called **haustra.** These haustra are readily appreciated if the colon is viewed at laparotomy or by x-ray imaging as shown in Fig. 31.2. The arrangement of the majority of the longitudinal muscle fibers, however, is distinct from

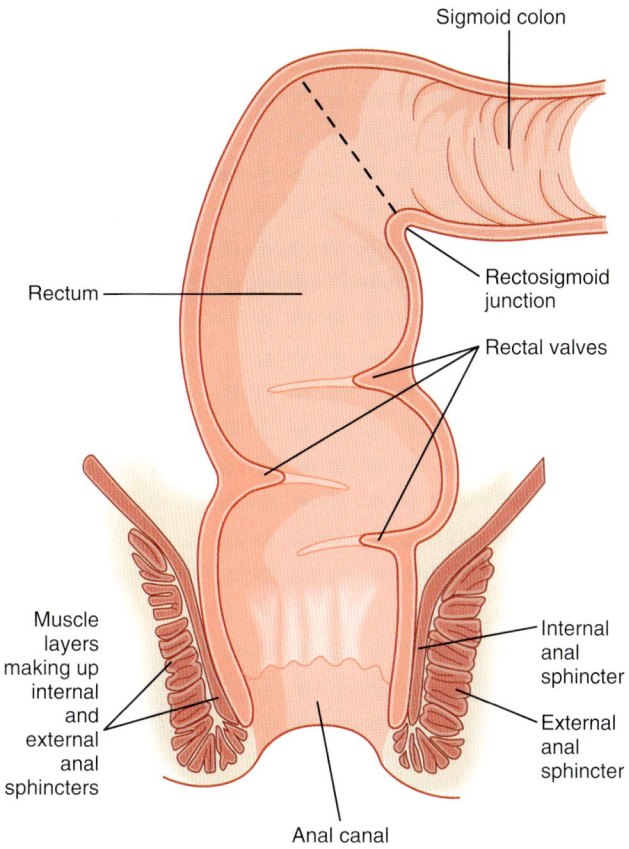

Sigmoid colon

Rectum

Rectosigmoid junction

Rectal valves

Muscle layers making up internal and external anal sphincters

Internal anal sphincter

External anal sphincter

Anal canal

• **Fig. 31.3** Anatomy of the rectum and anal canal.

of circular muscle, whereas the **external anal sphincter** is made up of three different striated muscle structures in the pelvic cavity that wrap around the anal canal. These latter muscles are distinctive because they maintain a significant level of basal tone and can be contracted further either voluntarily or reflexively when abdominal pressure increases abruptly (e.g., when lifting a heavy object).

AT THE CELLULAR LEVEL

Hirschsprung's disease is a condition in which a segment of the colon remains permanently contracted and results in obstruction. It is typically diagnosed in infancy and affects up to 1 in 5000 live births in the United States. The basis of the disease is failure of the enteric nervous system to develop normally during fetal life. During organogenesis, cells destined to become enteric neurons migrate out from the neural crest and populate the gut sequentially from mouth to anus. In some individuals this migration terminates prematurely because of abnormalities in the mechanisms that would otherwise drive this process. Mutations in glial-derived neurotrophic factor and endothelin III, as well as in their receptors, have been described in individuals suffering from this disease, and the affected segment completely lacks the plexuses of the enteric nervous system and associated ganglia. A relative deficiency of interstitial cells of Cajal is also seen in the affected segment, and overall control of motility is markedly impaired. In most individuals the symptoms can be completely alleviated by surgical excision of the affected segment.

that in the small intestine. Three nonoverlapping bands of longitudinal muscle known as the **taeniae coli** extend along the length of the colon.

Although the circular and longitudinal muscle layers of the colon are electrically coupled, this process is less efficient than in the small intestine. Thus propulsive motility in the colon is less effective than in the small intestine. Activity of the enteric nervous system also provides for the segmenting contractions that form the haustra. Contents can be moved back and forward between haustra, which is a means of retarding passage of the colonic contents and maximizing their contact time with the epithelium. In contrast, when rapid propulsion is called for, the contractions forming the haustra relax and the contour of the colon is smoothened.

The colon terminates in the **rectum,** which is joined to the colon at an acute angle (the **rectosigmoid junction**) (Fig. 31.3). The rectum lacks circular muscle and is surrounded only by longitudinal muscle fibers. It is a reservoir wherein feces can be stored before defecation. Muscular contractions also form functional "valves" in the rectum that retard movement of feces and are important in delaying the loss of feces until it is convenient, at least in adults. The rectum in turn joins the anal canal, distinguished by the fact that it is surrounded not only by smooth muscle but also by striated (skeletal) muscle. The combination of these muscle layers functionally accounts for two key sphincters that control evacuation of solid waste and flatus from the body. The **internal anal sphincter** is composed of a thickened band

Contraction of the smooth muscle layers in the proximal part of the colon is stimulated by vagal input as well as by the enteric nervous system. On the other hand, the remainder of the colon is innervated by the pelvic nerves, which also control the caliber of the internal anal sphincter. Voluntary input from the spinal cord via branches of the pudendal nerves regulates contraction of the external anal sphincter and muscles of the pelvic floor. The ability to control these structures is learned during toilet training. This voluntary control distinguishes the anal canal from most of the GI system, with the exception of the striated muscle in the esophagus that regulates swallowing.

Colonic Motility Responses

Consistent with its primary function, the two predominant motility patterns of the large intestine are directed not to propulsion of the colonic contents but rather to mixing of the contents and retarding their movement, thereby providing them with ample time in contact with the epithelium. Two distinctive forms of colonic motility have been identified. The first is referred to as *short-duration contractions,* which are designed to provide for mixing. These contractions originate in the circular muscle and are stationary pressure waves that persist for 8 seconds on average. *Long-duration contractions,* in contrast, are produced by the taeniae coli, last for 20 to 60 seconds, and may propagate over short distances. Notably, however, propagation may move orally as well as aborally, particularly in the more proximal segments of the colon. Both motility patterns are thought to originate largely in response to local conditions such as distention. Note that the basal electrical rhythm that governs the rate and origination sites of smooth muscle contraction in the small intestine does not traverse the ileocecal valve to continue into the colon.

On the other hand, probably as a result of both local influences and long reflex arcs, approximately 10 times per day in healthy individuals the colon engages in a motility pattern that is of high intensity and sweeps along the length of the large intestine from the cecum to the rectum. Such contractions, called ***high-amplitude propagating contractions,*** move exclusively in an aboral direction and are designed to clear the colon of its contents. However, although such a motility pattern can clearly be associated with defecation, it does not necessarily result in defecation for reasons discussed later.

It is also important to note that there is considerable variability among individuals with respect to the rate at which colonic contents are transported from the cecum to the rectum. Although small intestinal transit times are relatively constant in healthy adults, the contents may be retained in the large intestine anywhere from hours to days without signifying dysfunction. This also accounts for significant variation among individuals in their normal patterns of defecation and mandates careful elicitation of a patient's history before diagnosing bowel dysfunction.

Transport Mechanisms in the Colon

The major role of the colonic epithelium is to either absorb or secrete electrolytes and water rather than nutrients. Secretion, which is confined to the crypts, maintains the sterility of the crypts, which might otherwise become stagnant. This is important because the stem cells that renew the epithelium are located at the base of the crypt. The stem cells give rise to daughter cells that migrate out of the crypts and acquire the differentiated properties of surface cells that are responsible for water and electrolyte absorption. The colonic epithelium also absorbs short-chain fatty acids salvaged from nonabsorbed carbohydrates by colonic bacteria. Indeed, one such short-chain fatty acid, butyrate, is a critical energy source for colonocytes. A reduction in butyrate levels in the lumen (as a result of changes in colonic microbiota caused by administration of broad-spectrum antibiotics) may induce epithelial dysfunction.

The colon receives 2 L of fluid each day and absorbs 1.8 L, thus leaving 200 mL of fluid to be lost in stool. The colon has a considerable reserve capacity for fluid absorption and can absorb up to three times its normal fluid load without loss of excessive fluid in the stool. Therefore any illness that results in stimulation of active fluid secretion in the small intestine will cause diarrhea only when the reserve capacity of 4 to 6 L is exceeded.

Absorption and secretion of water by the colon are passive processes driven by absorption or secretion of electrolytes and other solutes. Quantitatively, fluid absorption by the

IN THE CLINIC

Irritable bowel syndrome is the name given to a heterogeneous collection of functional disorders whose sufferers complain of diarrhea, constipation, or alternating patterns of both, often with accompanying pain and distention. The etiology of these disorders is still not fully understood but may involve in part a condition of **visceral hypersensitivity** in which the individual perceives normal signals originating from the bowel (e.g., in response to distention) as painful. This hypersensitivity may be at the level of the enteric or central nervous system (or both) and can be triggered by a variety of factors such as previous infections, childhood abuse, or psychiatric disorders. Most treatments focus on symptomatic relief, but there is the promise of more effective therapies as we learn more about the underlying causes of the condition. Treatment of patients with irritable bowel disorders, which are often refractory to therapy, forms a major part of the practice of many gastroenterologists.

IN THE CLINIC

The colonic epithelium turns over rapidly in health, thus limiting the accumulation of genetic damage. However, this rapid turnover, as well as frequent/prolonged exposure to bacterially synthesized or environmental toxins, or both, makes the large intestine especially vulnerable to malignancy. Colon cancer is second in prevalence only to lung cancer in men in the United States and third behind lung and breast cancer in women. With the decreased incidence of cigarette smoking and therefore lung cancer, colon cancer may assume even greater significance. Colon cancer arises when normal genetic controls on the rate of epithelial proliferation are subverted; initially this leads to growth of a polyp and eventually, if not removed, to an invasive tumor that may metastasize to other parts of the body. Colon cancer can be subdivided according to the basic nature of the underlying molecular defect, which can include overexpression of growth stimulatory factors or a mutation that prevents the cells from responding to factors that would normally be growth suppressive. However, colon cancer mortality can be reduced very substantially by early detection and removal of polyps with malignant potential. This has driven current guidelines for increased screening of even asymptomatic middle-aged individuals for colonic abnormalities via colonoscopy (in which a flexible fiberoptic tube is inserted into the colon to inspect its interior), screening for the presence of so-called occult (or hidden) blood in the stool derived from a bleeding polyp or tumor, or noninvasive imaging techniques such as computed tomography scans.

colon is driven by three transport processes. The first is electroneutral NaCl absorption, which is mediated by the same mechanism that drives NaCl absorption in the intestine (see Chapter 30, Fig. 30.17). NaCl absorption is stimulated by various growth factors, such as epidermal growth factor, and is inhibited by hormones and neurotransmitters that increase levels of cyclic adenosine monophosphate (cAMP) in colonic surface epithelial cells.

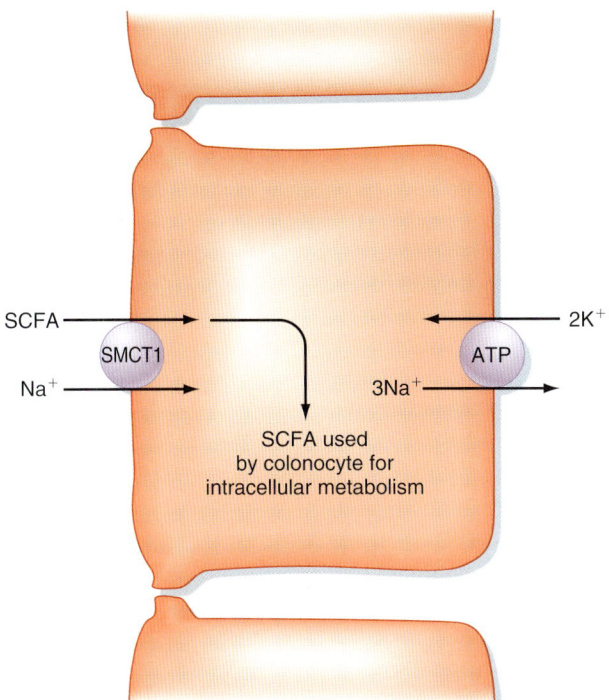

• **Fig. 31.4** Mechanism of short-chain fatty acid (SCFA) uptake by colonocytes. ATP, adenosine triphosphate; SMCT1, sodium/monocarboxylate cotransporter 1.

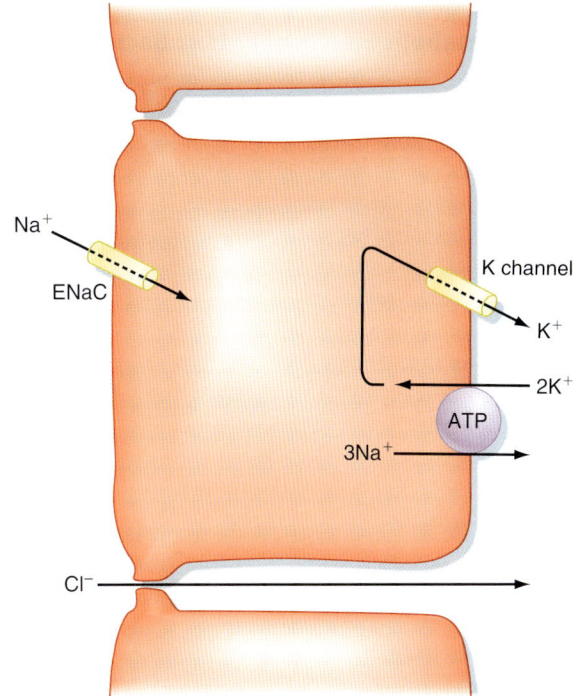

• **Fig. 31.5** Electrogenic Na^+ absorption in the colon. ENaC, epithelial sodium channel.

The second transport process that drives fluid absorption in the colon is absorption of **short-chain fatty acids,** including acetate, propionate, and butyrate. These molecules are absorbed from the lumen by surface (and perhaps crypt) epithelial cells in an Na^+-dependent fashion by a family of symporters related to the Na^+-glucose symporter in the small intestine, and known as **sodium-monocarboxylate transporters** (SMCTs). Uptake of short-chain fatty acids by SMCTs located in the apical plasma membrane is driven by the low intracellular $[Na^+]$ established by the basolateral Na^+,K^+-ATPase (Fig. 31.4). These short-chain fatty acids are used for energy by colonocytes. In addition, butyrate regulates expression of specific genes in colonic epithelial cells and may suppress development of a malignant phenotype. Expression of **SMCT1** (also identified as SLC5A8) is reduced in some colon cancers, thereby leading to a reduction in butyrate uptake, which may contribute to malignant transformation.

The third absorptive process of major significance in the colon is absorption of Na^+ (Fig. 31.5). This transport process is predominantly localized to the distal part of the colon and is driven by the epithelial Na^+ channel (ENaC), which is also involved in reabsorption of Na^+ in the kidney. When the channel is opened in response to activation by neurotransmitters or hormones, or both, Na^+ flows into the colonocyte cytosol and is then transported across the basolateral membrane by Na^+,K^+-ATPase. Cl^- ions follow passively via the intercellular tight junctions to maintain electrical neutrality. Water is absorbed across the tight junctions as a result of the transepithelial osmotic gradient due to solute absorption. This mode of Na^+ absorption is the last line of defense to prevent excessive loss of water in stool, given its strategic location in the distal part of the colon. Indeed, patients suffering from bowel inflammation often show markedly diminished expression of ENaC, perhaps accounting for their diarrheal symptoms. We also know that expression of ENaC can be acutely regulated in response to whole-body Na^+ balance. Thus in situations of reduced Na^+ intake, the hormone aldosterone increases ENaC expression in both the colon and kidney, thereby promoting retention of Na^+.

Adequate hydration of the colonic contents is determined by the balance between water absorption and secretion. Fluid secretion in the colon is driven by Cl^- ion secretion, by the same mechanism driving fluid secretion in the small intestine, and is subject to the same regulation (see Chapter 30, Fig. 30.18). Indeed, some cases of constipation may reflect abnormalities in epithelial transport, and constipation that results from abnormally slow motility can be treated by agents that stimulate Cl^- secretion. Conversely, excessive Cl^- secretion can be one mechanism underlying diarrhea.

Colonic Microbiota

The remnants of the meal entering the colon interact with a vast assortment of bacteria and other microorganisms. This **enteric microbial ecosystem** is established shortly after birth, matures as the child grows, and fluctuates in predictable ways in healthy individuals depending on factors such as diet or circadian rhythms. More drastic perturbations in the microbiota can be provoked by antibiotics or introduction

TABLE 31.1	**Metabolic Effects of Enteric Bacteria**		
Substrate	**Enzymes**	**Products**	**Disposition**
Endogenous Substrates			
Urea	Urease	Ammonia	Passive absorption or excretion as ammonium
Bilirubin	Reductases	Urobilinogen	Passive reabsorption
		Stercobilins	Excreted
Primary bile acids	Dehydroxylases	Secondary bile acids	Passive reabsorption
Conjugated bile acids (primary or secondary)	Deconjugases	Unconjugated bile acids	Passive reabsorption
Exogenous Substrates			
Fiber	Glycosidases	Short-chain fatty acids	Active absorption
		Hydrogen, CO_2, and methane	Excreted in breath or flatus
Amino acids	Decarboxylases and deaminases	Ammonia and bicarbonate	Reabsorbed or excreted (ammonia) as ammonium
Cysteine, methionine	Sulfatases	Hydrogen sulfide	Excreted in flatus

Adapted from Barrett KE. *Gastrointestinal Physiology.* New York: McGraw-Hill; 2006.

of an aggressive pathogen and may then take considerable time to resolve. The enteric bacterial ecosystem contributes to GI physiology in a surprising number of ways. Indeed, the large intestine (and to a lesser extent the distal portion of the small intestine) is an unusual organ in that it maintains a symbiotic relationship with such an extensive microbiota, whereas other body compartments are largely sterile.

 IN THE CLINIC

Diarrheal diseases are a major cause of infant mortality worldwide and are usually the result of inadequate access to clean food and water. Even in developed countries, diarrheal diseases cause substantial suffering and occasional well-publicized deaths and carry a substantial economic burden because of their prevalence. Infectious diarrhea is caused by a number of organisms, with several (e.g., cholera or pathogenic strains of *Escherichia coli*) capable of elaborating toxins that trigger excessive increases in active Cl^- secretion by small and large intestinal epithelial cells. Diarrhea can also result when nutrients are not appropriately digested and absorbed in the small intestine (e.g., lactose intolerance) or as a result of colonic inflammation. In most diarrheal diseases, colonic NaCl and Na^+ absorption are downregulated at the same time Cl^- secretion may be stimulated, thus further worsening fluid loss. On the other hand, nutrient-linked Na^+ absorptive processes typically remain intact. This provides the rationale for the effectiveness of oral rehydration solutions, which are prepackaged mixtures of salt and glucose. Uptake of Na^+ and glucose from these solutions, mediated by SGLT1 (see Chapter 30), restores fluid absorption. These solutions save lives in areas where diarrhea is prevalent and the ability to rehydrate patients with sterile intravenous solutions is limited or absent.

The colonic microbiota is not essential to life because animals raised in germ-free conditions apparently develop normally and are able to reproduce. However, in such

animals the mucosal immune system is immature, and intestinal epithelial cells differentiate more slowly. Importantly the colonic microbiota provides benefits to the host in that the constituent microbes are capable of performing metabolic reactions that do not take place in mammalian cells. Bacterial enzymes act on both endogenous and exogenous substrates. They form secondary bile acids and deconjugate bile acids that have escaped uptake in the terminal ileum, so they can be reabsorbed. They convert bilirubin into urobilinogen (see Chapter 32) and salvage nutrients that are resistant to pancreatic and brush border hydrolases, such as dietary fiber. A summary of the metabolic contributions of the colonic microbiota is provided in Table 31.1. Microbial metabolism can also be exploited for pharmacological purposes. A drug targeted to the colon, for example, can be conjugated in such a way that it will become bioavailable only after it is acted on by bacterial enzymes. Bacterial enzymes may also detoxify some dietary carcinogens, but equally they may generate toxic or carcinogenic compounds from dietary substrates.

Commensal microorganisms also play a critical role in limiting the growth or invasion (or both) of pathogenic microorganisms. They fulfill this antimicrobial role via a number of different mechanisms—by synthesizing and secreting compounds that inhibit the growth of competitor organisms or that are microbicidal, by functioning as a physical barrier to prevent attachment of pathogens and their subsequent entry into colonic epithelial cells, and by triggering patterns of gene expression in the epithelium that counteract the adverse effects of pathogens on epithelial function. These mechanisms provide a basis for understanding why patients who have received broad-spectrum antibiotics, which temporarily disrupt the colonic microbiota, are susceptible to overgrowth of pathogenic organisms and associated intestinal and systemic infections. They may also shed light on the efficacy of **probiotics,** commensal bacteria selected for their resistance to gastric acid and proteolysis that are intentionally ingested to prevent or treat a variety of digestive disorders.

 ## AT THE CELLULAR LEVEL

A toxin known as **heat-stable toxin of _E. coli,_** or STa, is a major causative agent of traveler's diarrhea, which can be contracted by consumption of infected food or water. This toxin binds to a receptor on the apical surface of intestinal epithelial cells known as _guanylyl cyclase C_ (GC-C). In turn this enzyme generates large quantities of intracellular cyclic guanosine monophosphate (cGMP) that trigger increased Cl^- secretion via activation of the cystic fibrosis transmembrane conductance regulator (CFTR) Cl^- channel. However, one could of course question why humans express a receptor for this toxin in a site that would be accessible to luminal bacteria and their products. Indeed, this led to the hypothesis that there is a native ligand for GC-C that could play a physiological role. Researchers subsequently purified and identified guanylin, a hormone synthesized in the intestine. Together with a related molecule, uroguanylin, secreted by the kidney, guanylin is an important regulator of salt and water homeostasis in the body. STa has structural similarities to guanylin, but it has modifications that permit it to persist in the intestinal lumen for prolonged periods. This is an example of molecular mimicry, in which a bacterial product hijacks a receptor and associated signaling for its own purposes (presumably to propagate the toxin-producing bacteria to additional hosts).

The colonic microbiota is also notable for its contribution to formation of **intestinal gas.** Although large volumes of air may be swallowed in conjunction with meals, the majority of this gas returns up the esophagus via belching. However, during fermentation of unabsorbed dietary components, the microbiota generates large volumes of nitrogen, hydrogen, and carbon dioxide. Approximately 1 L of these nonodorous gases is excreted on a daily basis via the anus in all individuals, even those who do not complain of flatulence. Some individuals may generate appreciable concentrations of methane. Trace amounts of odorous compounds are also present, such as hydrogen sulfide, indole, and skatole.

Defecation

The final stage in the journey taken by a meal after its ingestion is expulsion of its indigestible residues from the body in the process known as **defecation.** The feces also contain the remnants of dead bacteria, dead and dying epithelial cells that have been desquamated from the lining of the intestine, biliary metabolites specifically targeted for excretion (e.g., conjugates of xenobiotics [see Chapter 32]), and a small amount of water. In health, the stool contains few if any useful nutrients. The presence of nutrients in stool, particularly lipid (known as **steatorrhea**), signifies maldigestion, malabsorption, or both. Fat in the stool is

a sensitive marker of small intestinal dysfunction because it is poorly used by the colonic microbiota, but loss of carbohydrate and protein in stool can also be seen if the underlying condition worsens.

The process of defecation requires coordinated action of the smooth and striated muscle layers in the rectum and anus, as well as surrounding structures such as the pelvic floor muscles. During the mass movement of feces produced by high-amplitude propagating contractions, the rectum fills with fecal material. Expulsion of this material from the body is controlled by the internal and external anal sphincters, which contribute approximately 70% to 80% and 20% to 30% of anal tone at rest, respectively. Filling of the rectum causes relaxation of the internal anal sphincter via release of vasoactive intestinal polypeptide and generation of nitric oxide. Relaxation of the inner sphincter permits the **anal sampling mechanism,** which can distinguish whether the rectal contents are solid, liquid, or gaseous in nature. After toilet training, sensory nerve endings in the anal mucosa then generate reflexes that initiate appropriate activity of the external sphincter to either retain the rectal contents or permit voluntary expulsion (e.g., of flatus). If defecation is not convenient, the external sphincter contracts to prevent the loss of stool. Then with time the rectum accommodates to its new volume, the internal anal sphincter contracts again, and the external anal sphincter relaxes (Fig. 31.6).

When defecation is desired, on the other hand, adoption of a sitting or squatting position alters the relative orientation of the intestine and surrounding muscular structures by straightening the path for exit of either solid or liquid feces. Relaxation of the puborectalis muscle likewise increases the rectoanal angle. After voluntary relaxation of the external anal sphincter, rectal contractions move the fecal material out of the body, sometimes followed by additional mass movements of feces from more proximal segments of the colon (Fig. 31.7). Evacuation is assisted by simultaneous contraction of muscles that increase abdominal pressure, such as the diaphragm. Voluntary expulsion of flatus, on the other hand, involves a similar sequence of events, except that there is no relaxation of the puborectalis muscle. This permits flatus to be squeezed past the acute rectoanal angle while retaining fecal material.

Cooperative activity of the external anal sphincter, puborectalis muscle, and sensory nerve endings in the anal canal is required to delay defecation until it is appropriate, even if the rectum is acutely distended with stool or intraabdominal pressure rises sharply. This explains why incontinence can develop in individuals in whom the integrity of such structures has been compromised, such as after trauma, surgical or obstetrical injuries, prolapse of the rectum, or neuropathic diseases such as long-standing diabetes. Surgical intervention may be necessary to correct muscle abnormalities in patients with the distressing condition of fecal incontinence, although many can be helped to increase external anal sphincter tone with the use of biofeedback exercises.

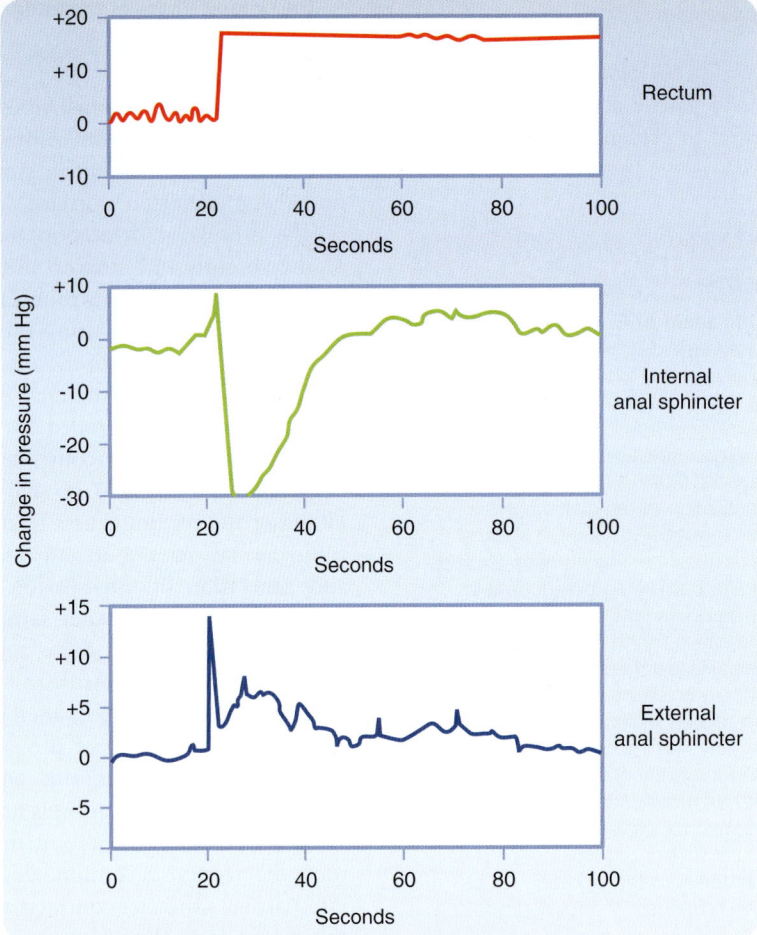

• **Fig. 31.6** Responses of the internal and external anal sphincters to prolonged distention of the rectum. Note that the responses of the sphincters are transient because of accommodation. (Redrawn from Shuster MM et al. *Bull Johns Hopkins Hosp* 1965;116:79.)

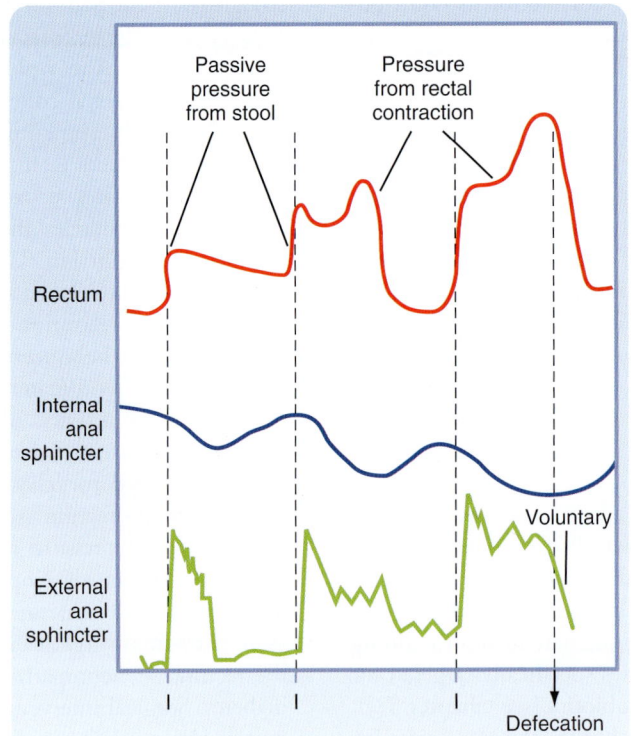

• **Fig. 31.7** Motility of the rectum and anal sphincters in response to rectal filling and during defecation. Note that filling of the rectum causes an initial decrease in internal sphincter tone that is counterbalanced by contraction of the external sphincter. The internal sphincter then accommodates to the new rectal volume, thereby allowing the external sphincter to relax. Finally, defecation occurs when the external anal sphincter is relaxed voluntarily. (Data from Chang EB et al. *Gastrointestinal, Hepatobiliary and Nutritional Physiology.* Philadelphia: Lippincott-Raven; 1996.)

Key Concepts

1. The final segment of the intestine through which the meal traverses is the large intestine, which is composed of the cecum, colon, rectum, and anus. The primary role of the large intestine is to reclaim water used during the process of digestion and absorption and to store the residues of the meal until defecation is socially convenient.

2. Colonic motility primarily serves to mix and delay passage of the luminal contents, other than during periodic large-amplitude contractions that convey fecal material to the rectum.

3. The colon is highly active in transporting water and electrolytes as well as products salvaged from undigested components of the meal by colonic bacteria.

4. The colon maintains a lifelong mutually beneficial relationship with a vast microbial ecosystem that metabolizes endogenous substances, nutrients, and drugs and protects the host from infection with pathogens.

5. Defecation involves both involuntary and voluntary relaxation of muscle structures surrounding the anus and reflex pathways that control these structures.

Additional Reading

Bharucha AE, Brookes SJR. Neurophysiologic mechanisms of human large intestinal motility. In: Johnson LR, ed. *Physiology of the GI Tract*. 5th ed. Waltham, MA: Academic Press; 2012.

Bharucha AE, Hasler WL. Motility of the small intestine and colon. In: Podolsky DK, et al., eds. *Yamada's Textbook of Gastroenterology*. 6th ed. Hoboken, NJ: Wiley Blackwell; 2015.

Kau AL, et al. Human nutrition, the gut microbiome and the immune system. *Nature*. 2011;474:327-336.

Kendig DM, Grider JR. Serotonin and colonic motility. *Neurogastroenterol Motil*. 2015;27:899-905.

Koren A, Ley RE. The human intestinal microbiota and microbiome. In: Podolsky DK, et al., eds. *Yamada's Textbook of Gastroenterology*. 6th ed. Hoboken, NJ: Wiley Blackwell; 2015.

32

Transport and Metabolic Functions of the Liver

LEARNING OBJECTIVES

Upon completion of this chapter the student should be able to answer the following questions:

1. How does the liver contribute metabolic, detoxification, and excretory functions that are vital to life?
2. What cell types make up the liver, what is their unusual arrangement, and what is the liver's unique portal blood supply?
3. What is the function of bile as an excretory fluid, what are the mechanisms that underlie its formation, and what is the role of the enterohepatic circulation?
4. How is bile is stored and concentrated in the gallbladder, and what are the mechanisms that coordinate gallbladder emptying with ingestion of a meal?
5. How are ammonia and bilirubin eliminated from the body?

Overview of the Liver and Its Functions

The liver is a large multilobed organ located in the abdominal cavity whose function is intimately associated with that of the gastrointestinal system. The liver serves as the first site of processing for most absorbed nutrients and also secretes bile acids, which as we learned in Chapter 30, play a critical role in absorption of lipids from the diet. In addition the liver is a metabolic powerhouse, critical for disposing of a variety of metabolic waste products and xenobiotics from the body by converting them to forms that can be excreted. The liver stores or produces numerous substances needed by the body, such as glucose, amino acids, and plasma proteins. In general, key functions of the liver can be divided into three areas: (1) contributions to whole-body metabolism, (2) detoxification, and (3) excretion of protein-bound/lipid-soluble waste products. In this chapter we discuss the structural and molecular features of the liver and biliary system that subserve these functions, as well as their regulation. Although the liver contributes in a pivotal way to the maintenance of whole-body biochemical status, a complete discussion of all the underpinning reactions is beyond the scope of this chapter. We will confine our discussion primarily to hepatic functions that relate to gastrointestinal physiology.

Metabolic Functions of the Liver

Hepatocytes contribute to metabolism of the major nutrients: carbohydrates, lipids, and proteins. Thus the liver plays an important role in glucose metabolism by engaging in **gluconeogenesis,** conversion of other sugars to glucose. The liver also stores glucose as glycogen at times of glucose excess (e.g., postprandial period) and then releases stored glucose into the bloodstream as it is needed. This process is referred to as the **glucose buffer function of the liver.** When hepatic function is impaired, glucose concentrations in blood may rise excessively after ingestion of carbohydrate; conversely, between meals, hypoglycemia may be seen because of an inability of the liver to contribute to carbohydrate metabolism and interconversion of one sugar to another.

Hepatocytes also participate in lipid metabolism. They are a particularly rich source of the metabolic enzymes engaged in **fatty acid oxidation** to supply energy for other body functions. Hepatocytes also convert products of carbohydrate metabolism to lipids that can be stored in adipose tissue and synthesize large quantities of lipoproteins, cholesterol, and phospholipids, the latter two being important in the biogenesis of cell membranes. In addition, hepatocytes convert a considerable portion of synthesized cholesterol to bile acids, which will be discussed in more detail later in this chapter.

The liver also plays a vital role in protein metabolism. The liver synthesizes all the **nonessential amino acids** (see Chapter 30) that do not need to be supplied in the diet, in addition to participating in interconverting and deaminating amino acids so the products can enter biosynthetic pathways for carbohydrate synthesis. With the exception of immunoglobulins, the liver synthesizes almost all the proteins present in plasma (especially **albumin,** which determines plasma oncotic pressure) as well as most of the important **clotting factors.** Patients suffering from liver disease may develop peripheral edema secondary to hypoalbuminemia and are also susceptible to bleeding disorders. Finally, the liver is the critical site for disposal of the ammonia generated from protein catabolism. This is accomplished by converting ammonia to urea, which is

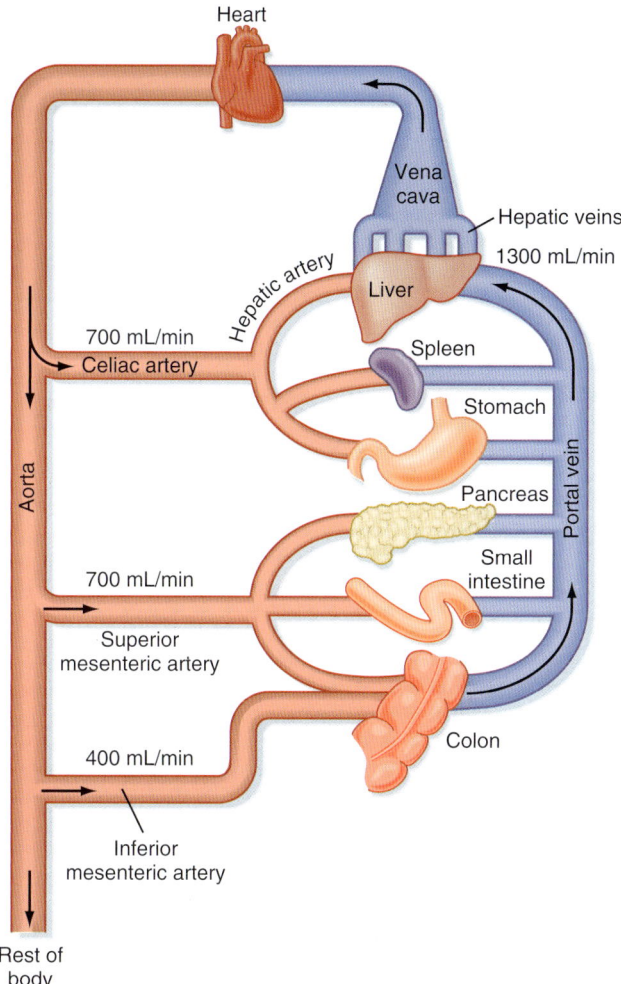

Fig. 32.1 Typical blood flow through the splanchnic circulation in a fasting adult human.

Labels in figure: Heart · Vena cava · Hepatic veins · 1300 mL/min · Hepatic artery · Liver · Spleen · Stomach · Pancreas · Small intestine · Portal vein · Colon · 700 mL/min · Celiac artery · Aorta · 700 mL/min · Superior mesenteric artery · 400 mL/min · Inferior mesenteric artery · Rest of body

The liver has two levels at which it removes and metabolizes/detoxifies substances originating from the portal circulation. The first of these is physical. Blood arriving in the liver percolates among cells of the macrophage lineage, known as **Kupffer cells.** These cells are phagocytic and particularly important in removing particulate material from portal blood, including bacteria that may enter blood from the colon even under normal conditions. The second level of defense is biochemical. Hepatocytes are endowed with a broad array of enzymes that modify both endogenous and exogenous toxins so the products are in general more water soluble and less susceptible to reuptake by the intestine. The metabolic reactions involved are broadly divided into two classes. **Phase I reactions** (oxidation, hydroxylation, and other reactions catalyzed by cytochrome P450 enzymes) are followed by **phase II reactions** that conjugate the resulting products with another molecule (e.g., glucuronic acid, sulfate, amino acids, glutathione) to promote their excretion. The products of these reactions are then excreted into bile or returned to the bloodstream to ultimately be excreted by the kidneys. We will return to the precise mechanisms involved in detoxification of some key metabolic waste products later in this chapter.

Role of the Liver in Excretion

The kidneys play an important role in excretion of water-soluble catabolites, as discussed in the renal section. Only relatively small water-soluble catabolites can be excreted by the process of glomerular filtration. However, larger water-soluble catabolites and molecules bound to plasma proteins, including lipophilic metabolites and xenobiotics, steroid hormones, and heavy metals, cannot be filtered by the glomerulus. All these substances are potentially harmful if allowed to accumulate, so a mechanism must exist for their excretion. The mechanism involves the liver, which excretes these substances in bile. Hepatocytes take up these substances with high affinity by virtue of an array of basolateral membrane transporters, and the substances are subsequently metabolized at the level of microsomes and in the cytosol (Table 32.1). Ultimately, substances destined for excretion in bile are exported across the canalicular membrane of hepatocytes via a different array of transporters. The features of bile allow solubilization of even lipophilic substances, which can then be excreted into the intestine and ultimately leave the body in feces.

Structural Features of the Liver and Biliary System

Hepatocytes, the major cell type in the liver, are arranged in anastomosing cords that form plates around which large volumes of blood circulate (Fig. 32.2). The liver receives a high blood flow that is disproportionate to its mass, which ensures that hepatocytes receive high quantities of both O_2 and nutrients. Hepatocytes receive more than 70% of their

excreted by the kidneys. The details of this process will be discussed later in this chapter.

The Liver and Detoxification

The liver serves both as a gatekeeper, by limiting entry of toxic substances into the bloodstream, and as a garbage disposal, by extracting potentially toxic metabolic products produced elsewhere in the body and converting them to chemical forms that can be excreted. The liver fulfills these functions in part because of its unusual blood supply. Unlike all other organs, the majority of blood arriving at the liver is venous in nature and is supplied via the **portal vein** from the intestine (Fig. 32.1). As such, the liver is strategically located to receive both absorbed nutrients and potentially harmful absorbed molecules such as drugs and bacterial toxins. Depending on the efficiency with which these molecules are extracted by hepatocytes and subjected to **first-pass metabolism,** little or none of the absorbed substance may make it into the systemic circulation. This is a major reason why not all pharmaceutical agents can achieve therapeutic concentrations in the bloodstream if administered orally.

TABLE 32.1	Key Transporters of Hepatocytes		
Name	**Basolateral**	**Canalicular**	**Substrate/Function**
NTCP	Yes	No	Uptake of conjugated bile acids
OATP	Yes	No	Uptake of bile acids and xenobiotics
BSEP	No	Yes	Secretion of conjugated bile acids
MDR3	No	Yes	Secretion of phosphatidylcholine
MDR1	No	Yes	Secretion of cationic xenobiotics
ABC5/ABC8	No	Yes	Secretion of cholesterol
cMOAT/MRP2	No	Yes	Secretion of sulfated lithocholic acid and conjugated bilirubin

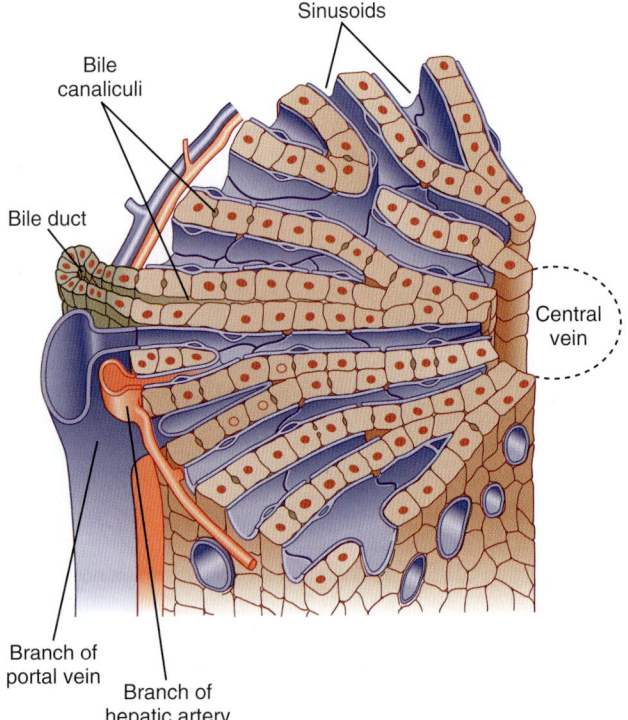

• **Fig. 32.2** Diagrammatic representation of a hepatic lobule. Plates of hepatocytes are arrayed radially around a central vein. Branches of the portal vein and hepatic artery are located on the periphery of the lobule and form the "portal triad" together with the bile duct. Blood from the portal vein and hepatic artery percolates around the hepatocytes via the sinusoids before draining into the central vein. (Modified from Bloom W, Fawcett DW. *A Textbook of Histology.* 10th ed. Philadelphia: Saunders; 1975.)

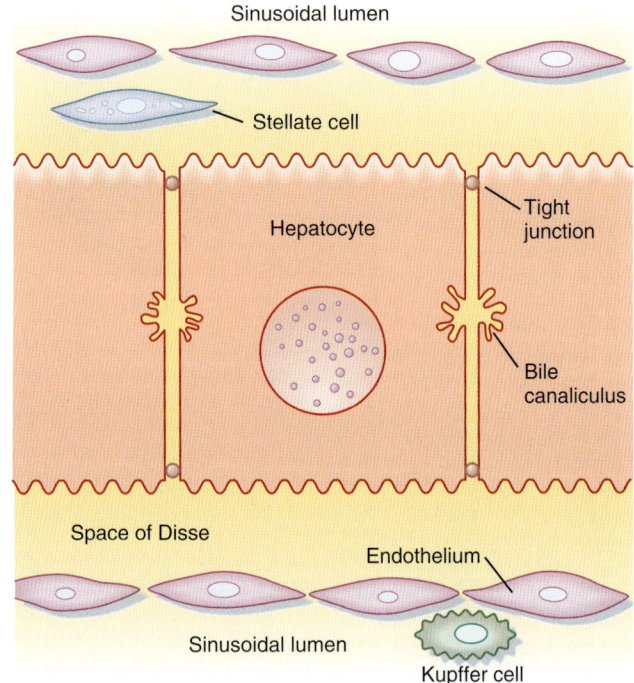

• **Fig. 32.3** Interrelationships of the major cell types in the liver.

blood supply at rest via the portal vein (rising to more than 90% in the postprandial period).

The plates of hepatocytes that constitute the **liver parenchyma** are supplied by a series of **sinusoids,** which are low-resistance cavities supplied by branches of both the portal vein and **hepatic artery.** The sinusoids are unlike the capillaries that perfuse other organs. During fasting, many sinusoids are collapsed, but more can gradually be recruited as portal blood flow increases during the period after a meal when absorbed nutrients are transported to the liver. The low resistance of the sinusoidal cavities means that blood flow through the liver can increase considerably without

a concomitant increase in pressure. Eventually the blood drains into central branches of the **hepatic vein.**

The sinusoids are also unusual in terms of the endothelial cells that line their walls (Fig. 32.3). Hepatic endothelial cells contain specialized openings known as **fenestrations** that are large enough to permit passage of molecules as big as albumin. Sinusoidal endothelial cells also lack a basement membrane, which might otherwise pose a diffusion barrier. These features allow access of albumin-bound substances to the hepatocytes that will eventually take them up. The sinusoids also contain Kupffer cells. Beneath the sinusoidal endothelium and separating the endothelium from the hepatocytes is a thin layer of loose connective tissue called the **space of Disse,** which in health likewise poses little resistance to the movement of molecules even as large as albumin. The space of Disse is also the location of another important hepatic cell type, the **stellate cell.** Stellate cells serve as storage sites for retinoids and in addition are the source of key growth factors for hepatocytes.

Under abnormal conditions, stellate cells are activated to synthesize large quantities of collagen, which contributes to the hepatic dysfunction.

IN THE CLINIC

If the circulation of the liver (particularly its sinusoids) is compressed by fibrosis, the liver loses its ability to accommodate the increases in blood flow that occur after a meal without a concomitant increase in pressure. Because of the fenestrations, albumin escapes from the circulation and albumin-rich fluid weeps from the surface of the liver into the abdominal cavity, where it overwhelms the lymphatic drainage. This condition is known as **ascites** and is reflected in a considerable increase in the girth of many patients with liver disease. As pressure in the liver builds, new collateral blood vessels form in an attempt to circumvent the obstruction and reduce the portal hypertension. Some of these vessels are directed to abdominal structures and, because of their thin weak walls, are prone to rupture. A particular example is formation of high-pressure collaterals to the esophagus, which can then become varices that bleed into the lumen. Bleeding into the esophageal lumen is very hard to control and is thus a medical emergency. Even in the absence of bleeding, moreover, the formation of collateral blood vessels bypasses the remaining metabolic capacity of the liver, and levels of toxins such as ammonia are increased and can exert adverse effects elsewhere in the body.

IN THE CLINIC

Infection of the liver with certain viruses or overexposure to toxic substances such as alcohol kills hepatocytes and activates hepatic stellate cells, which synthesize excessive amounts of collagen that result in the histologic appearance of fibrosis. If the insult is chronic, the fibrosis eventually becomes irreversible, a condition known as **cirrhosis.** The fibrotic scarred areas crowd out the hepatocyte mass, thereby reducing the synthetic, metabolic, and excretory capacity of the liver. Fibrotic masses press on the sinusoids and prevent them from expanding as portal blood flow to the liver increases during the postprandial period. Edema may develop in patients with chronic liver injury as a result of reduced levels of albumin in blood, and ascites may then develop, in which fluid accumulates in the peritoneal cavity secondary to increased portal pressure. Eventually, accumulation of toxic substances in the bloodstream can lead to jaundice, itching, and neurological complications. If hepatic function becomes compromised beyond a certain level, the only effective treatment is liver transplantation.

Hepatocytes are also the origination point for the **biliary system.** Although hepatocytes are considered to be epithelial cells with basolateral and apical membranes, the spatial arrangement of these two cell domains differs from that seen in simple columnar epithelium, such as that lining the gastrointestinal tract. Rather, in the liver the apical surface of the hepatocyte occupies only a small fraction of the cell membrane, and the apical membranes of adjacent cells oppose each other to form a channel between the cells known as the **canaliculus** (see Fig. 32.3). The role of canaliculi is to drain bile from the liver, and these canaliculi drain into biliary ductules that are lined by classic columnar epithelial cells known as **cholangiocytes.** Ultimately the biliary ductules drain into large bile ducts that coalesce into the right and left hepatic ducts to permit exit of bile from the liver. These in turn form the common hepatic duct, from which bile can flow into either the gallbladder (via the cystic duct) or the intestine (via the common bile duct; Fig. 32.4) on the basis of prevailing pressure relationships.

One other feature of the structural organization of the liver bears emphasis because of its clinical significance. Branches of the hepatic vein, hepatic artery, and bile ducts run in parallel in the so-called **hepatic triad.** Hepatocytes lying closest to this triad are referred to as *periportal,* or *zone 1,* and have the greatest supply of oxygen and nutrients. In contrast, hepatocytes lying closest to branches of the hepatic vein are referred to as *pericentral,* or *zone 3.* The latter cells are more sensitive to ischemia, whereas the former are more sensitive to oxidative injury. Thus the location of damaged cells on biopsy may provide clues to the cause of a given case of liver injury. Zone 1 cells are most active in detoxification functions in normal circumstances, but

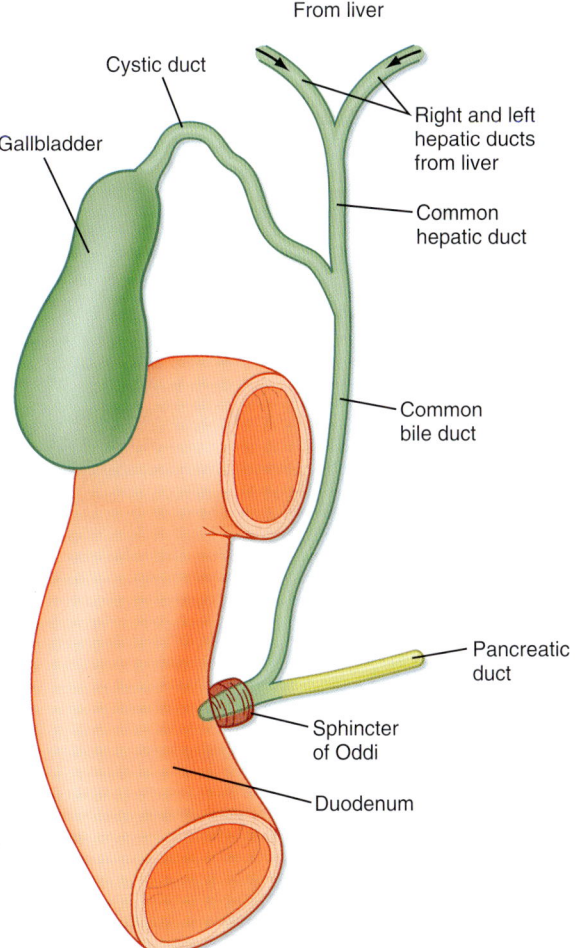

• **Fig. 32.4** Functional anatomy of the biliary system.

zone 2 (intermediate between zones 1 and 3) and zone 3 cells can progressively be recruited in cases of liver disease. Conversely, zone 3 cells are thought to be most active in bile acid synthesis.

Bile Formation and Secretion

Bile is the excretory fluid of the liver that plays an important role in lipid digestion. Bile formation begins in hepatocytes, which actively transport solutes into bile canaliculi across their apical membranes. Bile is a **micellar solution** in which the major solutes are bile acids, phosphatidylcholine, and cholesterol in an approximate ratio of $10:3:1$, respectively. Secretion of these solutes drives concomitant movement of water and electrolytes across the tight junctions that link adjacent hepatocytes to form canalicular bile. The majority of bile flow is driven by the secretion of bile acids across the apical membrane of hepatocytes via an adenosine triphosphatase (ATPase) transporter known as the **bile salt export pump** (BSEP; see Table 32.1). The composition of the resulting fluid can be modified further as it flows through the biliary ductules (resulting in hepatic bile) and still further on storage in the gallbladder (gallbladder bile). Ultimately bile becomes a concentrated solution of biological detergents that aids in solubilization of the products of lipid digestion in the aqueous environment of the intestinal lumen, thereby enhancing the rate at which lipids are transferred to the absorptive epithelial surface. It also serves as a medium in which metabolic waste products are exported from the body.

 ## AT THE CELLULAR LEVEL

Though rare, a variety of familial syndromes that are manifested as progressive cholestasis have taught us a great deal about the molecular nature of the transporters that deliver bile constituents into the canaliculus. For example, **type II progressive familial intrahepatic cholestasis** (PFIC II) has been mapped to a mutation in *BSEP* that results in an almost total absence of bile acids in bile. Cholestasis develops in patients with this disorder, but they have relatively little if any evidence of bile duct injury. Type III PFIC on the other hand is a much more aggressive disease in which cholestasis is accompanied by early increases in circulating γ-glutamyl transpeptidase. The molecular culprit is a mutation that abolishes expression of **multidrug resistance protein 3** (MDR3). In the absence of this transporter, phosphatidylcholine is no longer able to enter bile, thus illustrating the importance of this lipid in protecting cholangiocytes from the injurious effects of bile acids, because mixed micelles cannot form in its absence.

Bile Acid Synthesis

Bile acids are produced by hepatocytes as end products of cholesterol metabolism. Cholesterol is selectively metabolized by a series of enzymes that result in formation of bile acid (Fig. 32.5). The initial and rate-limiting step is addition of a hydroxyl group to the 7 position of the steroid nucleus by the enzyme **cholesterol 7α-hydroxylase.** The side chain of the product of this reaction is then shortened and a carboxylic acid function added by C27 dehydroxylase to yield chenodeoxycholic acid, a dihydroxy bile acid. Alternatively the product is further hydroxylated at the 12 position and then acted on by C27 dehydroxylase to yield cholic acid, a trihydroxy bile acid. Bile acid synthesis can be up- or downregulated depending on the body's requirements (Fig. 32.6). For example, if bile acid levels are reduced in the blood flowing to the liver, synthesis can be increased up to 10-fold. Conversely, feeding of bile acids profoundly suppresses new synthesis of bile acids by hepatocytes. The mechanisms underlying these changes in bile acid synthesis relate to changes in expression of the enzymes involved. Bile acids activate a variety of cell surface and nuclear receptors in hepatocytes, leading ultimately to activation of specific transcription factors that regulate enzyme abundance.

Chenodeoxycholic acid and cholic acid are defined as **primary bile acids** because they are synthesized by the hepatocyte (see Fig. 32.5). However, each can be metabolized in the colonic lumen by bacterial enzymes to yield ursodeoxycholic and deoxycholic acid, respectively. Chenodeoxycholic acid is also converted by bacterial enzymes to form lithocholic acid, which is relatively cytotoxic. Collectively these three products of bacterial metabolism are referred to as **secondary bile acids.** One additional important biochemical modification occurs for both primary and secondary bile acids in the hepatocyte (see Fig. 32.5). These molecules are conjugated with either glycine or taurine, which significantly depresses their pK_a. The result is that conjugated bile acids are almost totally ionized at the pH prevailing in the small intestinal lumen and thus cannot passively traverse cell membranes. Consequently the conjugated bile acids are retained in the intestinal lumen until they are actively absorbed in the terminal ileum via the **apical sodium-dependent bile salt transporter (asbt).** Conjugated bile acids that escape this uptake step are deconjugated by bacterial enzymes in the colon, and the resulting unconjugated forms are passively reabsorbed across the colonic epithelium because they are no longer charged.

Hepatic Aspects of Enterohepatic Circulation of Bile Acids

Bile acids assist in digestion and absorption of lipids by acting as detergents rather than enzymes, and thus a significant mass of these molecules is required to solubilize all dietary lipids. Via the **enterohepatic circulation,** actively reabsorbed conjugated bile acids travel through the portal blood back to the hepatocyte, where they are efficiently taken up by basolateral transporters that may be Na^+ dependent or independent (see Table 32.1). Similarly, bile acids that are deconjugated in the colon also return to the hepatocyte, where they are reconjugated to be secreted into bile. In this way a pool of circulating primary and secondary

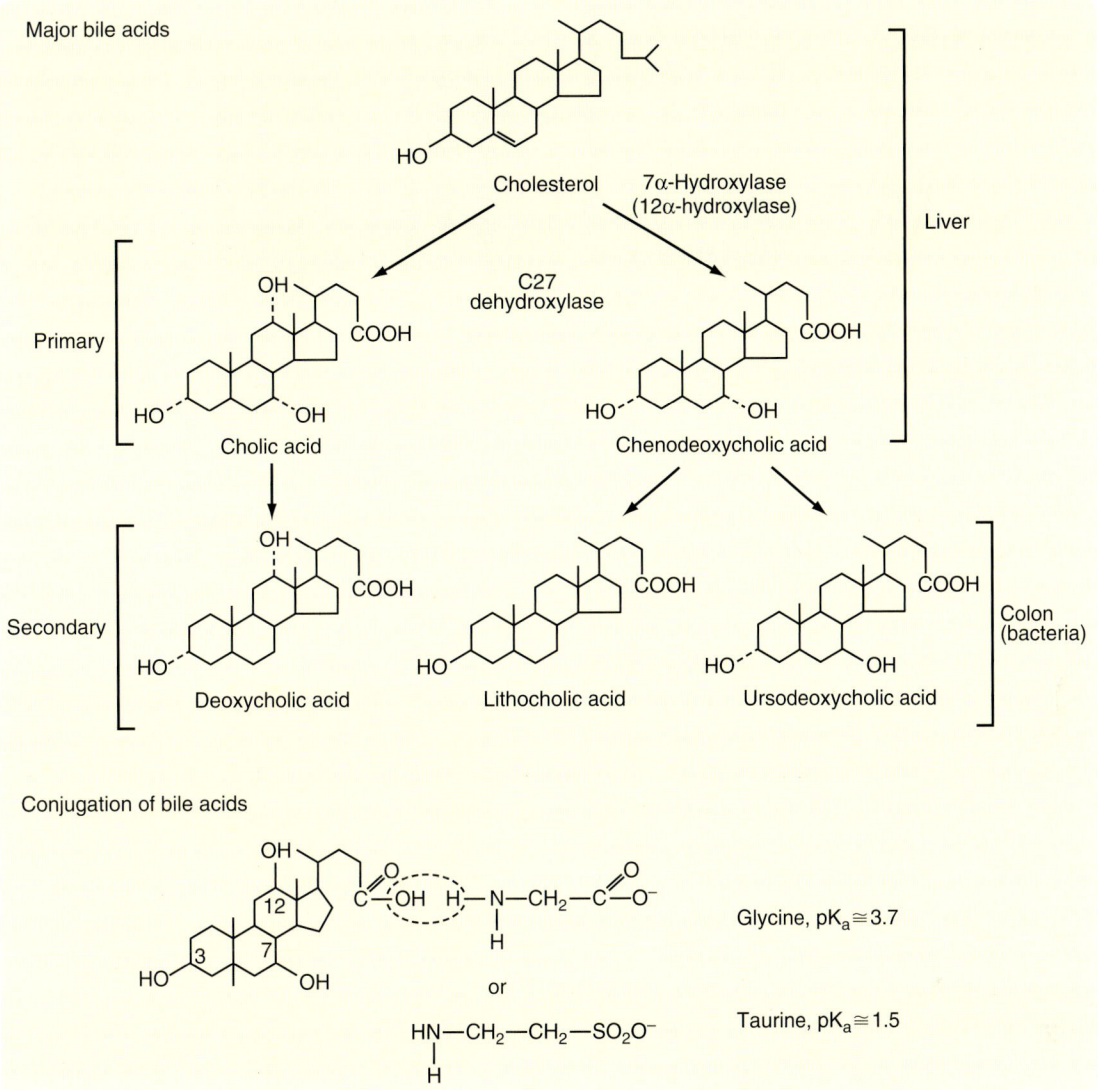

• **Fig. 32.5** Structures of the major primary and secondary bile acids of bile. Primary bile acids are synthesized in the liver. Secondary bile acids are produced when intestinal bacteria act on primary bile acids. At the bottom of the figure the conjugation of cholic acid with glycine or taurine is shown.

bile acids is produced, and daily synthesis is then equal only to the minor fraction ($\approx$10%/day, or 200–400 mg) that escapes uptake and is lost in stool (Fig. 32.7). The only exception to this rule is lithocholic acid, which is preferentially sulfated in the hepatocyte rather than being conjugated with glycine or taurine. The majority of the sulfate conjugates are lost from the body after each meal because they are not substrates for asbt, thereby avoiding accumulation of a potentially toxic molecule.

Some comment should also be made with respect to the role of bile acids in whole-body cholesterol homeostasis. The pool of cholesterol in the body reflects its daily synthesis as well as the relatively minor component derived from inefficient dietary uptake, balanced against loss from the body, which can only occur in health via bile (Fig. 32.8). Cholesterol can be excreted in two forms, either as the native molecule or after its conversion to bile acids. The latter account for up to a third of the cholesterol excreted per day

despite enterohepatic recycling. Thus one strategy for treating hypercholesterolemia is to interrupt the enterohepatic circulation of bile acids, which drives increased conversion of cholesterol to bile acids; the bile acids are then lost from the body in feces.

Other Bile Constituents

As noted earlier, bile also contains cholesterol and phosphatidylcholine. Cholesterol transport across the canalicular membrane is mediated at least in part by a heterodimer of the active transporters we discussed in Chapter 30 as participating in the efflux of cholesterol from the small intestinal epithelial cells, namely, ABC G5 and ABC G8 (see Table 32.1). Phosphatidylcholine derives from the inner leaflet of the canalicular membrane and is specifically "flipped" across the membrane by another ABC family transporter called **multidrug resistance protein 3** (MDR3). Furthermore,

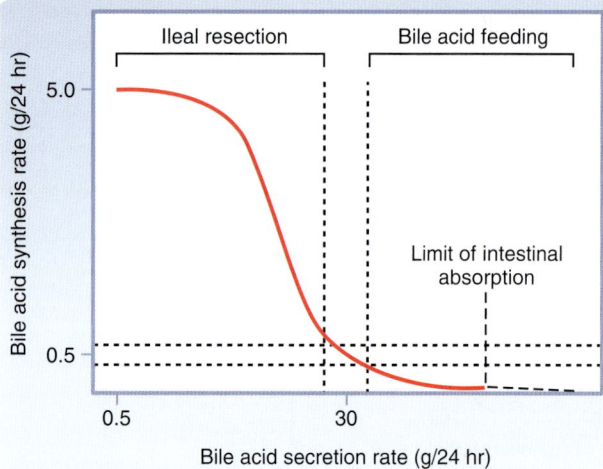

• **Fig. 32.6** Relationship between rates of bile acid synthesis and secretion. The bile acid secretion rate normally averages 30 g/24 h, whereas the synthesis rate averages 0.5 g/24 h. The pairs of vertical and horizontal dotted lines depict the normal range for bile acid secretion and synthesis, respectively. Increased secretion (simulated by bile acid feeding) increases the rate of return of bile acids to the liver via portal blood, which exerts a negative feedback on synthesis. Conversely, interruption of the enterohepatic circulation, such as after ileal resection, can increase synthesis to values more than 10-fold higher than normal. (From Carey MC, Cahalane MJ. In: Arias IM et al. [eds]. *The Liver: Biology and Pathobiology.* 2nd ed. New York: Raven Press; 1988.)

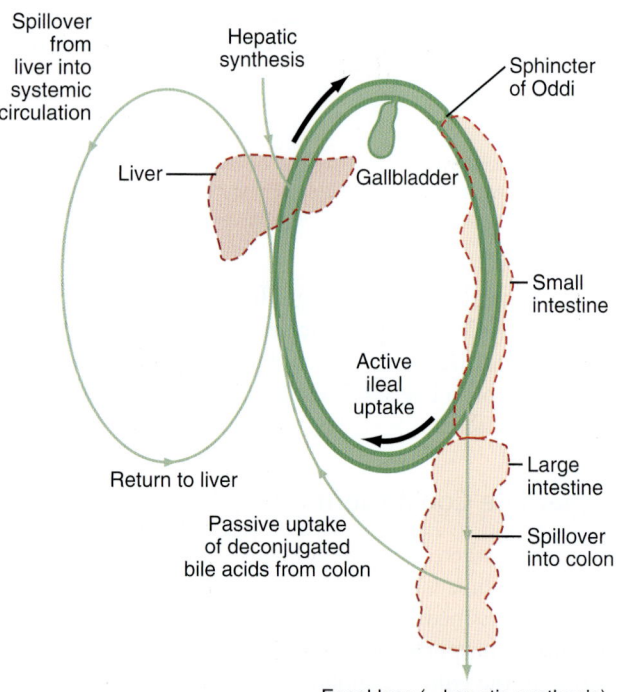

• **Fig. 32.7** Relative amounts of bile acids in different body pools and the enterohepatic circulation. The relative amounts are indicated by the width of the green lines. Thus the figure illustrates that the majority of the bile acid pool circulates between the liver, gallbladder, and small intestine, rather than being in the systemic circulation.

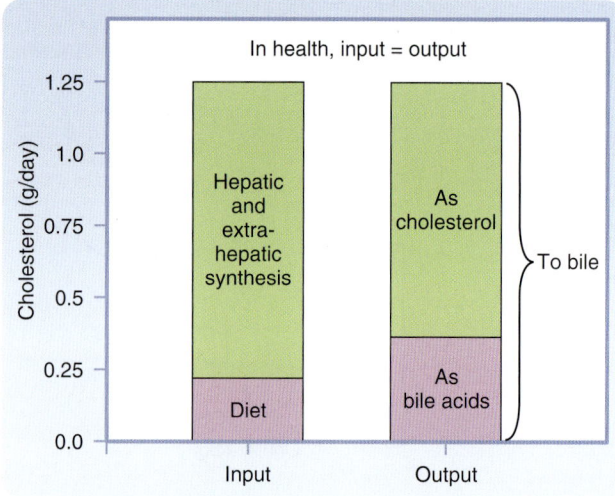

• **Fig. 32.8** Daily cholesterol balance in healthy adult humans.

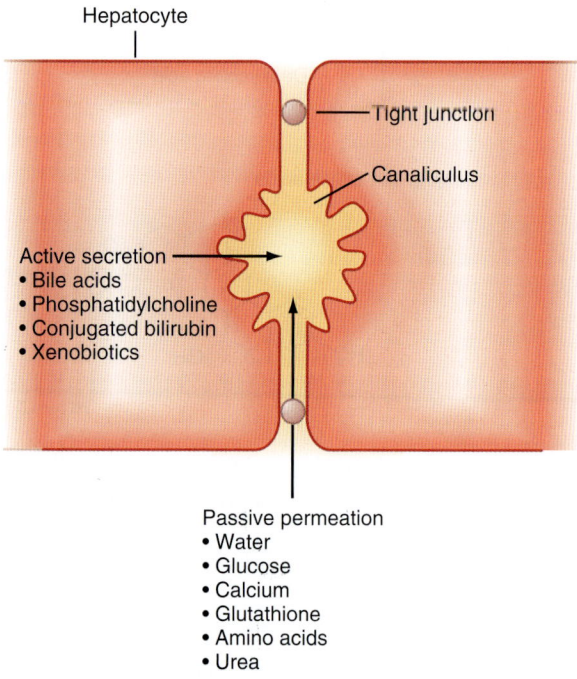

• **Fig. 32.9** Pathways for entry of solutes into bile. (Modified from Barrett KE. *Gastrointestinal Physiology.* New York: McGraw-Hill; 2006.)

because mixed micelles composed of bile acids, phosphatidylcholine, and cholesterol are osmotically active and the tight junctions that link adjacent hepatocytes are relatively leaky, water is drawn into the canalicular lumen, as well as other plasma solutes (e.g., Ca^{++}, glucose, glutathione, amino acids, urea) at concentrations essentially approximating those in plasma (Fig. 32.9). Finally, conjugated bilirubin, which is water soluble, and a variety of additional organic anions and cations formed from endogenous metabolites and xenobiotics are secreted into bile across the apical membrane of the hepatocyte.

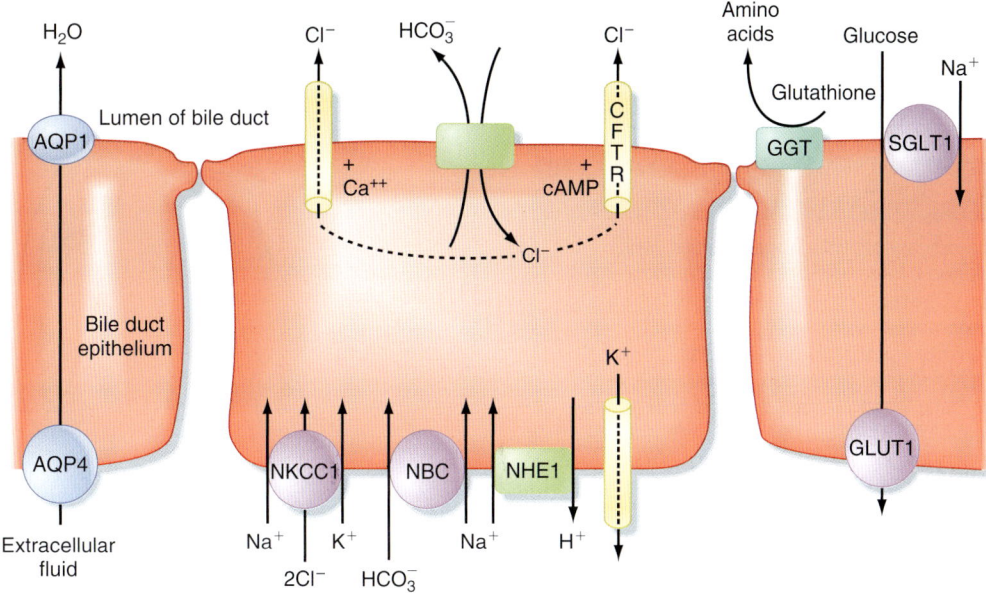

Fig. 32.10 The major transport processes of cholangiocytes that secrete an alkaline-rich fluid and reclaim useful substances. AQP, aquaporin; CFTR, cystic fibrosis transmembrane conductance regulator; GGT, γ-glutamyl transpeptidase; NKCC1, sodium/potassium/2 chloride cotransporter 1; NBC, sodium/bicarbonate cotransporter; NHE1, sodium/hydrogen exchanger 1, SGLT-1, sodium-glucose cotransporter 1; GLUT1, glucose transporter 1.

Bile Modification in Ductules

The cholangiocytes lining the biliary ductules are specifically designed to modify the composition of bile (Fig. 32.10). Useful solutes (e.g., glucose, amino acids) are reclaimed by the activity of specific transporters. Chloride ions in bile are also exchanged for HCO_3^-, rendering the bile slightly alkaline and reducing the risk of precipitation of Ca^{++}. Glutathione is broken down on the surface of cholangiocytes into its constituent amino acids by the enzyme γ-glutamyl transpeptidase (GGT), and the products are reabsorbed. Bile is also diluted at this site, in concert with ingestion of a meal, in response to hormones such as secretin that increase HCO_3^- secretion and stimulate insertion of **aquaporin water channels** into the cholangiocyte's apical membrane. Flow of bile is thereby increased during the postprandial period when bile acids are needed to aid in assimilation of lipid.

Role of the Gallbladder

Finally, bile enters the ducts and is conveyed toward the intestine. However, in the period between meals, outflow is blocked by constriction of the **sphincter of Oddi,** and thus bile is redirected to the **gallbladder.** The gallbladder is a muscular sac lined with high-resistance epithelial cells. During gallbladder storage, bile becomes concentrated because sodium ions are actively absorbed in exchange for protons, and bile acids, as the major anions, are too large to exit across the gallbladder epithelial tight junctions (Fig. 32.11). However, although the concentration of bile acids can rise more than 10-fold, bile remains isotonic because a

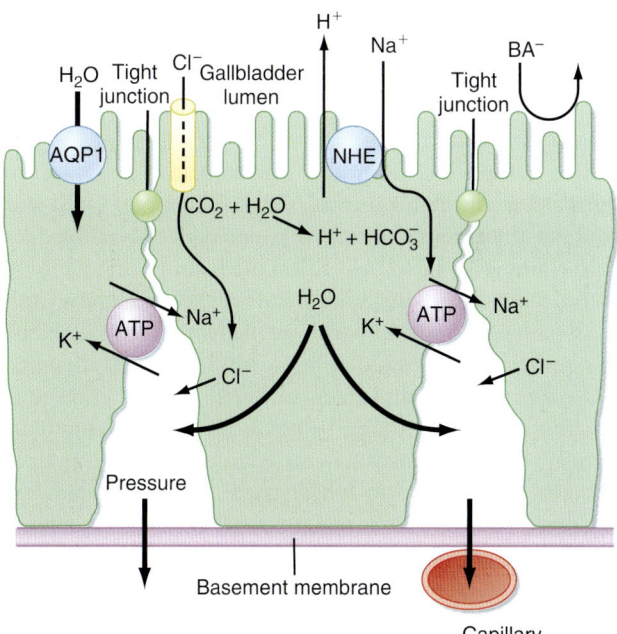

Fig. 32.11 Mechanisms accounting for concentration of bile during storage in the gallbladder. NHE, sodium/hydrogen exchanger; BA⁻, bile acid anion; AQP1, aquaporin 1.

single micelle acts as only one osmotically active particle. Any additional bile acid monomers that become available as a result of concentration are thus immediately incorporated into existing mixed micelles. This also reduces to some extent the risk that cholesterol will precipitate from bile. However, cholesterol is supersaturated in the bile of many adults, with precipitation normally being inhibited by the presence of antinucleating proteins. Prolonged storage of bile increases the chance that nucleation can occur, thus making a good case for never skipping breakfast and perhaps explaining why gallstone disease is relatively prevalent in humans.

🩺 IN THE CLINIC

Humans are unusually susceptible to **gallstones,** which represent precipitated bile constituents that accumulate in the gallbladder or elsewhere in the biliary tree. Gallstones are composed predominantly of cholesterol or Ca^{++} bilirubinate (cholesterol vs. pigment stones, respectively). Their significance lies in their propensity to obstruct biliary flow and thereby result in pain, poor tolerance of large fatty meals, retention of biliary constituents, and (if left untreated) liver injury. In susceptible individuals, mechanisms that normally prevent nucleation of saturated bile are either defective or overcome, and small crystals form and can grow into gallstones. Human bile is often supersaturated in terms of its cholesterol content, thus increasing the risk for stone formation, particularly during prolonged fasting. Cholesterol gallstones are especially common in middle-aged women who are obese, particularly those who have borne children. This increased prevalence is apparently due at least in part to the ability of estrogen to increase hepatic cholesterol secretion. In severe cases of gallstone disease the gallbladder may be removed surgically, which is usually accomplished laparoscopically. Small gallstones that have lodged in the biliary tree can sometimes be retrieved endoscopically by inserting a small snare through the sphincter of Oddi from an endoscope.

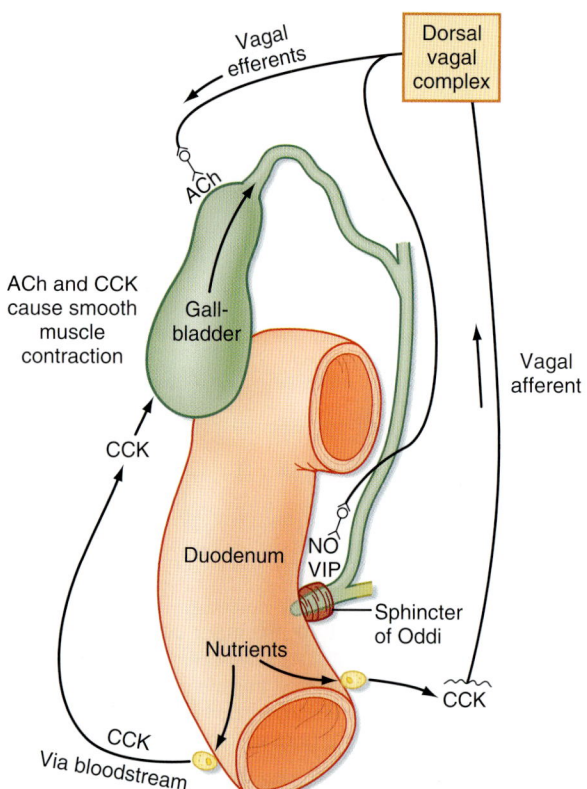

• **Fig. 32.12** Neurohumoral control of gallbladder contraction and biliary secretion. The pathway also involves relaxation of the sphincter of Oddi to permit outflow of bile into the duodenum. ACh, acetylcholine; CCK, cholecystokinin; NO, nitric oxide; VIP, vasoactive intestinal polypeptide.

Bile is secreted from the gallbladder in response to signals that simultaneously relax the sphincter of Oddi and contract the smooth muscle that encircles the gallbladder epithelium (Fig. 32.12). A critical mediator of this response is **cholecystokinin.** In addition, intrinsic neural reflexes and vagal pathways, some of which themselves are stimulated by the ability of cholecystokinin to bind to vagal afferents, also contribute to gallbladder contractility. The net result is ejection of a concentrated bolus of bile into the duodenal lumen, where the constituent mixed micelles can aid in lipid uptake. Then, when no longer needed, the bile acids are reclaimed and reenter the enterohepatic circulation to begin the cycle again. However, the other components of bile are largely lost in stool, thus providing for their excretion from the body.

Bilirubin Formation and Excretion by the Liver

The liver is also important for excretion of **bilirubin,** which is a metabolite of heme that is potentially toxic to the body. Bilirubin is an antioxidant and also serves as a way to eliminate the excess heme released from the hemoglobin of senescent red blood cells. Indeed, red blood cells account for 80% of bilirubin production, with the remainder coming from additional heme-containing proteins in other tissues such as skeletal muscle and the liver. Bilirubin can cross the blood-brain barrier and, if present in excessive levels, results

in brain dysfunction secondary to neuronal cell death and the activation of astrocytes and microglia; it can be fatal if left untreated. Bilirubin and its metabolites are also notable for the fact that they provide color to bile, feces, and to a lesser extent urine. By the same token, when bilirubin accumulates in the circulation as a result of liver disease, it is responsible for the common symptom of **jaundice,** or yellowing of the skin and conjunctiva.

Bilirubin is synthesized from heme by a two-stage reaction that takes place in phagocytic cells of the **reticuloendothelial system,** including Kupffer cells and cells in the spleen (Fig. 32.13). The enzyme heme oxygenase that is present in these cells liberates iron from the heme molecule and produces the green pigment **biliverdin.** This in turn can be reduced to form yellow bilirubin. Because bilirubin is essentially insoluble in aqueous solutions at neutral pH, it is transported through the bloodstream bound to albumin. When this complex reaches the liver, it enters the space of Disse, where bilirubin is selectively taken up across the basolateral membrane of hepatocytes via an **OATP** transporter (see Table 32.1). In the microsomal compartment, bilirubin is then conjugated with one or two molecules of glucuronic acid to enhance its aqueous solubility. The reaction is catalyzed by **UDP glucuronyl transferase** (UGT). This enzyme is synthesized only slowly after birth, which explains why mild jaundice is relatively common in

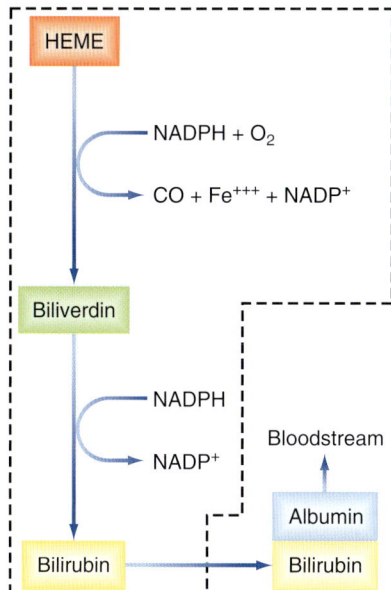

• **Fig. 32.13** Conversion of heme to bilirubin. The reactions inside the dashed box occur in cells of the reticuloendothelial system. NADP$^+$, oxidized form of nicotinamide adenine dinucleotide phosphate; NADPH, reduced form of nicotinamide adenine dinucleotide phosphate.

newborn infants. The bilirubin conjugates are then secreted into bile by a **multidrug resistance–associated protein** (MRP2) located in the canalicular membrane. Notably the conjugated forms of bilirubin cannot be reabsorbed from the intestine, thereby ensuring they can be excreted. However, transport of bilirubin across the hepatocyte (and indeed its initial uptake from the bloodstream) is relatively inefficient, so some conjugated and unconjugated bilirubin is present in plasma even under normal conditions. Both circulate bound to albumin, but the conjugated form is bound more loosely and thus can enter the urine.

In the colon, bilirubin conjugates are deconjugated by bacterial enzymes, whereupon the bilirubin liberated is metabolized by bacteria to yield urobilinogen, which is reabsorbed, and urobilins and stercobilins, which are excreted. Absorbed urobilinogen in turn can be taken up by hepatocytes and reconjugated, thus giving the molecule yet another chance to be excreted.

Measurement of bilirubin in plasma, as well as assessment of whether it is unconjugated or conjugated, is an important tool in the evaluation of liver disease. The presence of unconjugated bilirubin, which is essentially fully albumin bound and cannot be excreted in urine, reflects either loss of UGT (or a normal temporary delay in its maturation in infants) or a sudden oversupply of heme that overwhelms the conjugation mechanism (such as occurs in transfusion reactions or in Rhesus-incompatible newborns). Conjugated bilirubinemia on the other hand is characterized by the presence of bilirubin in urine, to which it imparts a dark coloration. This is indicative of genetic defects in the transporter that mediates bilirubin glucuronide/diglucuronide secretion into the canaliculus, or it may be due to blockage to the flow of bile, perhaps

caused by an obstructing gallstone. In both cases, bilirubin conjugates are formed in the liver, but with no means of exit they regurgitate back into plasma for urinary excretion.

AT THE CELLULAR LEVEL

Crigler-Najjar syndrome is a condition associated with mutations in the hepatocyte enzyme UGT. In type I Crigler-Najjar syndrome, a congenital missense mutation results in complete lack of this enzyme, whereas patients with type II Crigler-Najjar syndrome have a milder mutation that reduces UGT levels to about 10% of those seen in normal individuals. Thus with varying degrees of severity, Crigler-Najjar syndrome impairs the ability of hepatocytes to conjugate bilirubin. The unconjugated bilirubin regurgitates back into the circulation and binds to albumin, with an associated risk of neurological injury if levels rise precipitously. The only effective treatment of type I Crigler-Najjar syndrome at present is liver transplantation, although gene therapy may be a promising option in the near future. Those with type II disease can sometimes be managed effectively with blue light. This converts circulating unconjugated bilirubin to forms that are more water soluble and thus less firmly bound to albumin, which can be excreted in urine.

Ammonia Handling by the Liver

Ammonia (NH$_3$) is a small neutral metabolite that arises from protein catabolism and bacterial activity and is highly membrane permeant. The liver is a critical contributor to prevention of ammonia accumulation in the circulation, which is important because like bilirubin, ammonia is toxic to the central nervous system. The liver eliminates ammonia from the body by converting it to urea via a series of enzymatic reactions known as the **urea**, or **Krebs-Henseleit, cycle** (Fig. 32.14). The liver is the only tissue in the body that can convert ammonia to urea.

Ammonia is derived from two major sources. Approximately 50% is produced in the colon by bacterial ureases. Because the colonic lumen is normally slightly acidic, some of this ammonia is converted to the ammonium ion (NH$_4^+$), which renders it impermeant to the colonic epithelium and therefore allows it to be excreted in stool. However, the remainder of the ammonia generated crosses the colonic epithelium passively and is transported to the liver via the portal circulation. The other major source of ammonia (≈40%) is the kidney (see Chapter 37). A small amount of ammonia (≈10%) is derived from deamination of amino acids in the liver, by metabolic processes in muscle cells, and via release of glutamine from senescent red blood cells.

The "mass balance" for ammonia handling in a healthy adult is presented in Fig. 32.15. As just noted, ammonia is a small neutral molecule that readily crosses cell membranes without the benefit of a specific transporter, although some membrane proteins transport ammonia, including certain aquaporins. Whatever the mechanism for transport, the physicochemical properties of ammonia ensure that it is efficiently extracted from portal and systemic circulation

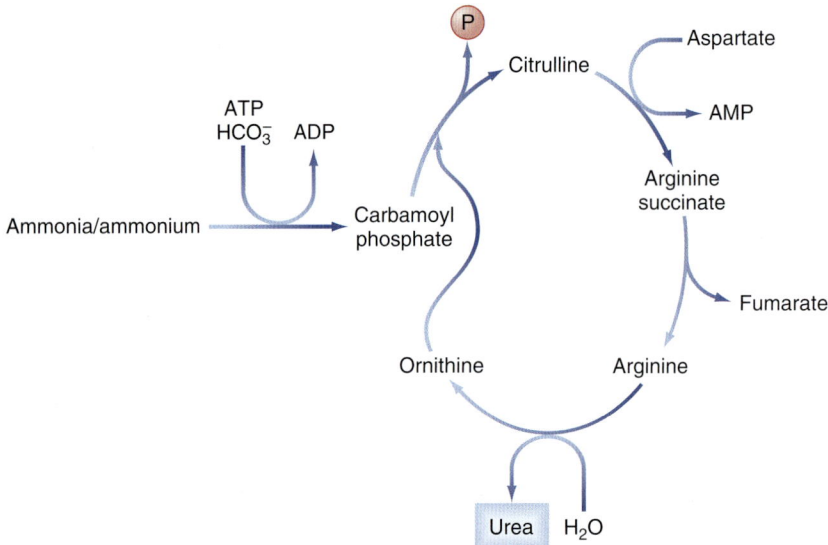

• Fig. 32.14 The urea cycle. ADP, adenosine diphosphate; ATP, adenosine triphosphate.

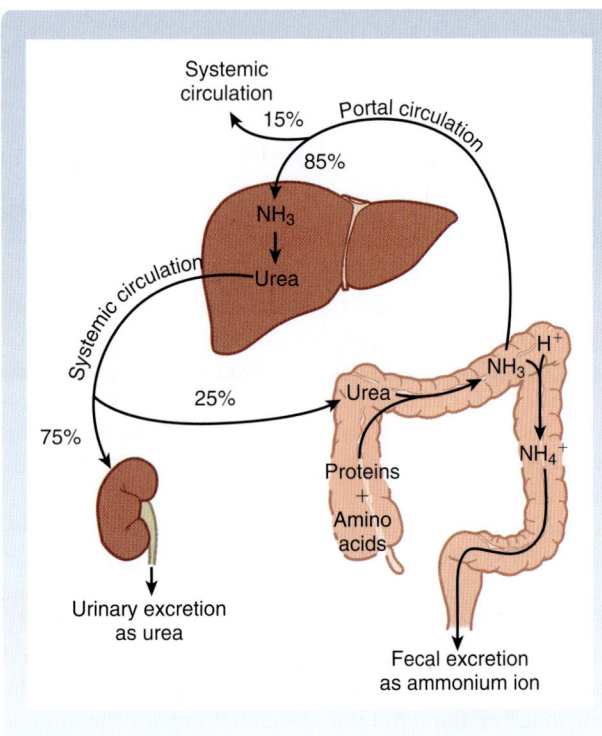

• Fig. 32.15 Ammonia homeostasis in health. (Redrawn from Barrett KE. *Gastrointestinal Physiology.* New York: McGraw-Hill; 2006.)

by hepatocytes, where it then enters the urea cycle to be converted to urea (see Fig. 32.14) and is subsequently transported back into the systemic circulation. Urea is a small neutral molecule that is readily filtered at the glomerulus, and it is reabsorbed by the kidney tubules such that approximately 50% of the filtered urea is excreted in urine (see Chapter 37). Urea that enters the colon is either excreted or metabolized to ammonia via colonic bacteria, with the resulting ammonia being reabsorbed or excreted.

If the metabolic capacity of the liver is compromised acutely, coma and death can rapidly ensue. In chronic liver disease, patients may experience a gradual decline in mental function that reflects the action of both ammonia and other toxins that cannot be cleared by the liver, in a condition known as **hepatic encephalopathy.** Development of confusion, dementia, and eventually coma in a patient with liver disease is evidence of significant progression, and these symptoms can prove fatal if left untreated.

Clinical Assessment of Liver Function

Given the importance of the liver for homeostasis, tests of liver function are a mainstay of clinical diagnosis. Such tests have several goals: (1) to assess whether hepatocytes have been injured or are dysfunctional, (2) to determine whether bile excretion has been interrupted, and (3) to evaluate whether cholangiocytes have been injured or are dysfunctional. **Liver function tests** are also used to monitor responses to therapy or rejection reactions after liver transplantation. However, not all such tests measure function directly. Nevertheless, liver function tests are discussed briefly because of their link to hepatic physiology.

Tests for hepatocyte injury rely on markers that are specific for this cell type. When hepatocytes are killed by necrotic responses to inflammation or infection, for example, they release enzymes that include alanine aminotransferase (ALT) and aspartate aminotransferase (AST). These enzymes, which are essential to interconvert amino acids, are easily measured in serum and indicate hepatocyte injury, although AST may also be released after injury to other tissues, including the heart. Two other tests are markers of injury to the biliary system. Alkaline phosphatase is expressed in the canalicular membrane, and elevations of this enzyme in plasma suggest localized obstruction to bile flow. Similarly, increased levels of GGT are seen when there is damage to cholangiocytes.

Measurement of bilirubin in the circulation or in urine also provides insight into liver function. In addition, measurement of any of the other characteristic secreted products of the liver can be used to diagnose liver disease. Clinically the most common tests are measurements of serum albumin and a blood clotting parameter, the prothrombin time. If results of these tests are abnormal, when considered together with other aspects of the clinical picture, a diagnosis of liver disease may be established. Blood glucose and ammonia levels are frequently monitored in patients with chronic liver disease. Finally, imaging tests and histological examination of biopsy specimens of liver parenchyma, usually obtained percutaneously, are also important in evaluating and monitoring patients with suspected or proven liver disease.

Key Concepts

1. Vital functions of the liver include carbohydrate, lipid, and protein metabolism and synthesis; detoxification of unwanted substances; and excretion of circulating substances that are lipid soluble and carried in the bloodstream bound to albumin. The liver also synthesizes the majority of plasma proteins, including albumin.
2. Liver function depends on its unique anatomy, its constituent cell types (especially hepatocytes), and the unusual arrangement of its blood supply.
3. Substances are excreted from the liver in bile. Bile flow is driven by the presence of bile acids, which are amphipathic end products of cholesterol metabolism that are produced by hepatocytes. Bile acids circulate between the liver and intestine to conserve their mass, and water-insoluble metabolites (e.g., cholesterol) are carried in bile in the form of mixed micelles.
4. Bile is stored in the gallbladder between meals, where it is concentrated and released when hormonal and neural signals simultaneously contract the gallbladder and relax the sphincter of Oddi.
5. The liver is critical for disposing of certain substances that would be toxic if allowed to accumulate in the bloodstream, including bilirubin and ammonia.

Additional Reading

Feranchak AP. Bile secretion and cholestasis. In: Podolsky DK, et al., eds. *Yamada's Textbook of Gastroenterology*. 6th ed. Hoboken, NJ: Wiley Blackwell; 2015.

Kanel GC. Liver: anatomy, microscopic structure, and cell types. In: Podolsky DK, et al., eds. *Yamada's Textbook of Gastroenterology*. 6th ed. Hoboken, NJ: Wiley Blackwell; 2015.

Parekh PJ, Balart LA. Ammonia and its role in the pathogenesis of hepatic encephalopathy. *Clin Liver Dis*. 2015;19:529-537.

Wolkoff AW. Organic anion uptake by hepatocytes. *Compr Physiol*. 2014;4:1715-1735.

Wu T, et al. Gut motility and enteroendocrine secretion. *Curr Opin Pharmacol*. 2013;13:928-934.

The Renal System

BRUCE A. STANTON AND BRUCE M. KOEPPEN

33

Elements of Renal Function

LEARNING OBJECTIVES

Upon completion of this chapter the student should be able to answer the following questions:

1. Which structures in the glomerulus are filtration barriers to plasma proteins?
2. What is the physiological significance of the juxtaglomerular apparatus?
3. What blood vessels supply the kidneys?
4. What nerves innervate the kidneys?
5. What is the location of the kidneys, and what are their gross anatomical features?
6. What are the different parts of the nephron, and what is their locations within the cortex and medulla?
7. What are the major components of the glomerulus, and what are the cell types located in each component?
8. How can the concepts of mass balance be used to measure the glomerular filtration rate (GFR)?
9. Why can inulin clearance and creatinine clearance be used to measure GFR?
10. Why is plasma creatinine concentration used clinically to monitor GFR?
11. What are elements of the glomerular filtration barrier, and how do they determine how much protein enters Bowman's space?
12. What Starling forces are involved in formation of the glomerular ultrafiltrate, and how do changes in each force affect GFR?
13. What is autoregulation of renal blood flow and GFR, and which factors and hormones are responsible for autoregulation?
14. Which hormones regulate renal blood flow?
15. Why do hormones influence renal blood flow despite autoregulation?

Overview of Renal Function

The kidney presents in the highest degree the phenomenon of sensibility, the power of reacting to various stimuli in a direction, which is appropriate for the survival of the organism; a power of adaptation which almost gives one the idea that its component parts must be endowed with intelligence.

E. STARLING—1909

Certainly, mental integrity is a sine qua non of the free and independent life. But let the composition of our internal environment suffer change, let our kidneys fail for even a short time to fulfill their tasks, and our mental integrity, or personality is destroyed.

HOMER W. SMITH—1939

As both Starling and Smith recognized, the kidneys are more appropriately considered regulatory rather than excretory organs. The kidneys regulate (1) body fluid osmolality and volumes, (2) electrolyte balance, and (3) acid-base balance. In addition the kidneys excrete metabolic products and foreign substances and produce and secrete hormones.

Control of body fluid osmolality is important for maintenance of normal cell volume in all tissues of the body. Control of body fluid volume is necessary for normal function of the cardiovascular system. The kidneys are also essential in regulating the amount of several important inorganic ions in the body, including Na^+, K^+, Cl^-, bicarbonate (HCO_3^-), hydrogen (H^+), Ca^{++}, and inorganic phosphate (P_i). Excretion of these electrolytes must be equal to daily intake to maintain appropriate total body balance. If intake of an electrolyte exceeds its excretion, the amount of this electrolyte in the body increases and the individual is in *positive balance* for that electrolyte. Conversely if excretion of an electrolyte exceeds its intake, its amount in the body decreases and the individual is in *negative balance* for that electrolyte. For many electrolytes the kidneys are the sole or principal route for excretion from the body.

Another important function of the kidneys is regulation of acid-base balance. Many metabolic functions of the body are exquisitely sensitive to pH. Thus the pH of body fluids must be maintained within narrow limits. Normal pH is maintained by buffers within body fluids and by the coordinated action of the lungs, liver, and kidneys.

The kidneys excrete a number of the end products of metabolism. These waste products include urea (from amino acids), uric acid (from nucleic acids), creatinine (from muscle creatine), end products of hemoglobin metabolism, and metabolites of hormones. The kidneys eliminate these substances from the body at a rate that matches their production. Thus the kidneys regulate hormone concentrations within body fluids. The kidneys also represent an important route for elimination of foreign substances such as drugs, toxins (e.g., pesticides), and other chemicals from the body.

Finally, the kidneys are important endocrine organs that produce and secrete renin, calcitriol, and erythropoietin. Renin is not a hormone but an enzyme that activates the renin-angiotensin-aldosterone system, which helps regulate blood pressure and Na^+ and K^+ balance. Calcitriol, a metabolite of vitamin D_3, is necessary for normal absorption of Ca^{++} by the gastrointestinal tract and for its deposition in bone (see Chapter 36). In patients with renal disease the kidneys' ability to produce calcitriol is impaired and levels of this hormone are reduced. As a result, Ca^{++} absorption by the intestine is decreased, which over time contributes to the bone formation abnormalities seen in patients with chronic renal disease. Another consequence of many kidney diseases is a reduction in erythropoietin production and secretion. Erythropoietin stimulates red blood cell formation by bone marrow. Decreased erythrocyte production contributes to the anemia that occurs in **chronic kidney disease (CKD),** a progressive loss in kidney function over a period of months or years.

A large variety of diseases impair kidney function and result in renal failure. In some instances the impairment in renal function is transient, but in many cases renal function declines progressively. Patients in whom the **glomerular filtration rate (GFR)** is less than 10% of normal are said to have **end-stage renal disease (ESRD)** and must receive renal replacement therapy in the form of either dialysis or kidney transplantation to survive.

To understand the mechanisms that contribute to renal disease, it is first necessary to understand the normal physiology of renal function. Thus in the following chapters in this section of the book, various aspects of renal function are considered.

Functional Anatomy of the Kidneys

Structure and function are closely linked in the kidneys. Consequently an appreciation of the gross anatomical and histological features of the kidneys is a prerequisite for understanding their functions.

Gross Anatomy

The kidneys are paired organs that lie on the posterior wall of the abdomen behind the peritoneum on either side of the vertebral column. In an adult human, each kidney weighs between 115 and 170 g and is approximately 11 cm long, 6 cm wide, and 3 cm thick.

The gross anatomical features of the human kidney are illustrated in Fig. 33.1. The medial side of each kidney contains an indentation through which pass the renal artery and vein, nerves, and pelvis. If a kidney were cut in half, two regions would be evident: an outer region called the **cortex** and an inner region called the **medulla.** The cortex and medulla are composed of **nephrons** (the functional units of the kidney), blood vessels, lymphatics, and nerves. The

IN THE CLINIC

Kidney disease is a major health problem worldwide. In the United States alone:

- CKD affects over 23 million patients and accounts for more than 90,000 deaths annually.
- Kidney diseases represent the ninth leading cause of death.
- The health care cost for CKD in Medicare patients alone exceeds $44 billion per year.
- Each year, kidney disease is diagnosed in over 3 million new patients.
- Over 500,000 people are treated for **ESRD** every year.
- Approximately 275,000 patients with ESRD are maintained on either hemodialysis or peritoneal dialysis.
- Diabetes and hypertension are the leading causes of ESRD.
- ESRD secondary to diabetes is increasing at an annual rate of more than 11% per year.
- More than 17,000 kidney transplants are performed each year. Unfortunately, in excess of 100,000 patients are awaiting kidney transplants.
- Urinary tract infections (8.3 million visits annually), urolithiasis (i.e., kidney and urinary tract stones; 1.3 million visits annually), interstitial cystitis (i.e., inflammation of the urinary bladder; 700,000 patients), and urinary incontinence (13 million adults affected, mostly older than 65) are also major health care problems.

Individuals with ESRD must undergo renal replacement therapy, which includes peritoneal dialysis, hemodialysis, and kidney transplantation. Both peritoneal dialysis and hemodialysis, as their names suggest, rely on the ability to remove small dialyzable molecules from the blood—including metabolic waste products normally removed by intact kidneys—via diffusion across a selectively permeable membrane into a solution lacking these substances, thereby mitigating both their accumulation and associated adverse health effects. In addition, dialysis helps reestablish both fluid and electrolyte balance via removal of excess fluid, correction of acid-base changes, and normalization of plasma electrolyte concentrations). In **peritoneal dialysis,** the peritoneal membrane lining the abdominal cavity acts as a dialyzing membrane. Several liters of a defined dialysis solution are typically introduced into the abdominal cavity, and small molecules in blood diffuse across the peritoneal membrane into the solution, which can then be iteratively removed, discarded, and replaced. In **hemodialysis,** a patient's blood is pumped through an extracorporeal artificial kidney in which blood is separated from a defined dialysis solution by an artificial semipermeable membrane that allows small molecules to diffuse from the blood down their concentration gradient into the dialysis solution, thereby removing small molecules associated with adverse health effects if allowed to accumulate in patients without functioning kidneys. Patients who are candidates for renal transplantation are often treated with dialysis until an appropriate donor kidney can be procured. Although anemia has historically been a significant problem in ESRD patients owing to severely reduced endogenous erythropoietin production, this problem can now be easily corrected in patients undergoing chronic dialysis via administration of erythropoiesis-stimulating agents (e.g., recombinant human erythropoietin).

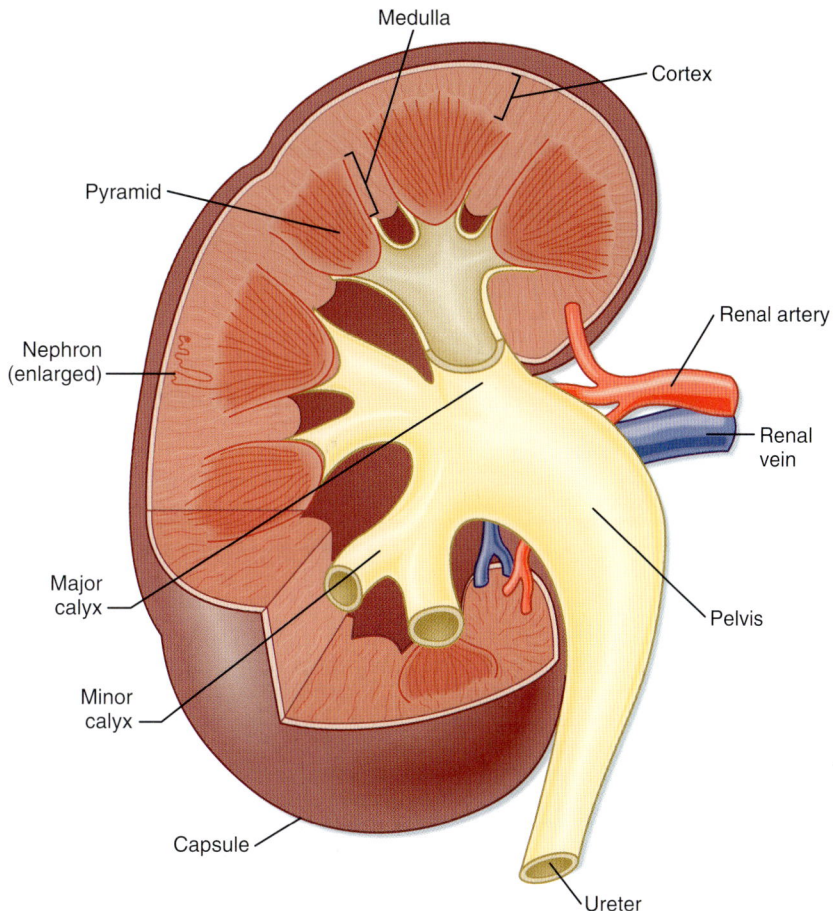

• **Fig. 33.1** Structure of a human kidney, cut open to show the internal structures. (Modified From Boron WF, Boulpaep EL. *Medical Physiology*. 2nd ed. Philadelphia: Saunders Elsevier; 2009.)

medulla in the human kidney is divided into conical masses called **renal pyramids.** The base of each pyramid originates at the corticomedullary border, and the apex terminates in a **papilla,** which lies within a **minor calyx.** Minor calyces collect urine from each papilla. The numerous minor calyces expand into two or three open-ended pouches, the **major calyces.** The major calyces in turn feed into the **pelvis.** The pelvis represents the upper expanded region of the **ureter,** which carries urine from the pelvis to the urinary bladder. The walls of the calyces, pelvis, and ureters contain smooth muscle that contracts to propel the urine toward the **urinary bladder.**

Blood flow to the two kidneys is equivalent to about 25% (1.25 L/min) of the cardiac output in resting individuals. However, the kidneys constitute less than 0.5% of total body weight. As illustrated in Fig. 33.2 *(left),* the **renal artery** branches progressively to form the **interlobar artery,** the **arcuate artery,** the **interlobular artery,** and the **afferent arteriole,** which leads into the **glomerular capillaries**. The glomerular capillaries come together to form the **efferent arteriole,** which leads into a second capillary network, the **peritubular capillaries,** which supply blood to the nephron. The vessels of the venous system run parallel to the arterial vessels and progressively form the **interlobular**

vein, arcuate vein, interlobar vein, and **renal vein,** which courses beside the ureter.

Ultrastructure of the Nephron

The functional unit of the kidneys is the nephron. Each human kidney contains approximately 1.2 million nephrons, which are essentially hollow tubes composed of a single epithelial cell layer. The nephron consists of a **renal corpuscle, proximal tubule, loop of Henle, distal tubule,** and **collecting duct system**[a] (Fig. 33.3; also see Fig. 33.2). The renal corpuscle[b] consists of glomerular capillaries enclosed within **Bowman's capsule.** The proximal tubule exits this structure and initially forms several coils, followed by a straight piece that descends toward the medulla. The next segment is the loop of Henle, which is composed of

[a]The organization of the nephron is actually more complicated than presented here. However, for simplicity and clarity of presentation in subsequent chapters, the nephron is divided into five segments. The collecting duct system is not actually part of the nephron. However, again for simplicity, we consider the collecting duct system part of the nephron.
[b]Although the renal corpuscle is composed of glomerular capillaries and Bowman's capsule, the term *glomerulus* is commonly used to described the renal corpuscle.

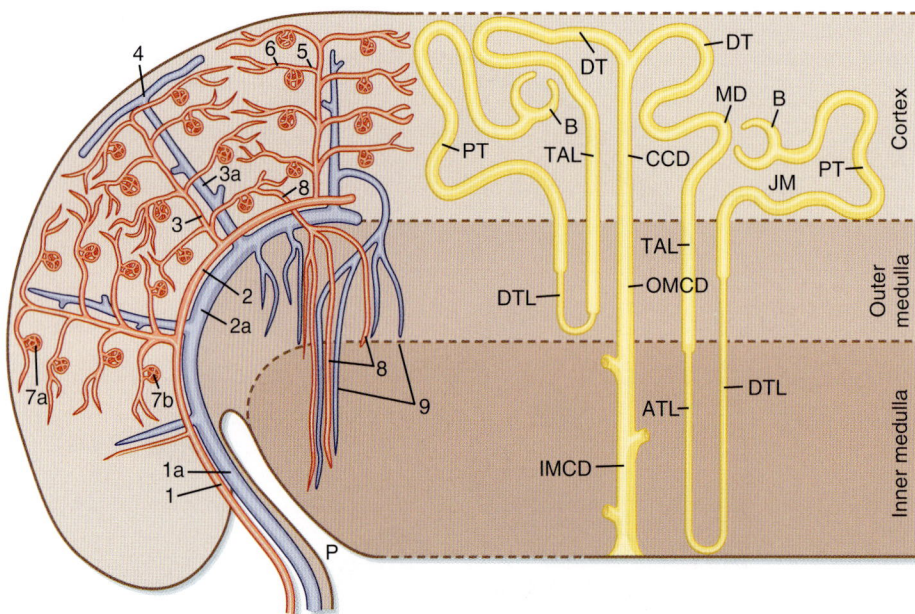

• **Fig. 33.2** *Left,* Organization of the vascular system of the human kidney. 1, interlobar arteries; 1a, interlobar vein; 2, arcuate arteries; 2a, arcuate veins; 3, interlobular arteries; 3a, interlobular veins; 4, stellate vein; 5, afferent arterioles; 6, efferent arterioles; 7a, 7b, glomerular capillary networks; 8, descending vasa recta; 9, ascending vasa recta. *Right,* Organization of the human nephron. A superficial nephron is illustrated on the left and a juxtamedullary (JM) nephron is illustrated on the right. The loop of Henle includes the straight portion of the proximal tubule (PT), descending thin limb (DTL), ascending thin limb (ATL), and thick ascending limb (TAL). B, Bowman's capsule; CCD, cortical collecting duct; DT, distal tubule; IMCD, inner medullary collecting duct; MD, macula densa; OMCD, outer medullary collecting duct; P, pelvis. (Modified from Kriz W, Bankir LA. *Am J Physiol* 1988;254:F1; and Koushanpour E, Kriz W. *Renal Physiology: Principles, Structure, and Function.* 2nd ed. New York: Springer-Verlag; 1986.)

the straight part of the proximal tubule, the descending thin limb (which ends in a hairpin turn), the ascending thin limb (only in nephrons with long loops of Henle), and the thick ascending limb. Near the end of the thick ascending limb, the nephron passes between the afferent and efferent arterioles of the same nephron. This short segment of the thick ascending limb abutting the glomerulus is called the **macula densa** (see Figs. 33.2 and 33.3). The distal tubule begins a short distance beyond the macula densa and extends to the point in the cortex where two or more nephrons join to form a cortical collecting duct. The **cortical collecting duct** enters the medulla and becomes the outer **medullary collecting duct** and then the **inner medullary collecting duct.**

Each nephron segment is made up of cells that are uniquely suited to perform specific transport functions (see Fig. 33.3). Proximal tubule cells have an extensively amplified apical membrane (the ultrafiltrate or urine side of the cell) called the **brush border,** which is present only in the proximal tubule. The basolateral membrane (the interstitial or blood side of the cell) is highly invaginated. These invaginations contain many mitochondria. In contrast, the descending and ascending thin limbs of the loop of Henle have poorly developed apical and basolateral surfaces and few mitochondria. The cells of the thick ascending limb and the distal tubule have abundant mitochondria and extensive infoldings of the basolateral membrane.

The collecting duct is composed of two cell types: principal cells and intercalated cells. **Principal cells** have a moderately invaginated basolateral membrane and contain few mitochondria (see Fig. 33.3). Principal cells play an important role in reabsorption of NaCl (see Chapters 34 and 35) and secretion of K^+ (see Chapter 36). **Intercalated cells,** which play an important role in regulating acid-base balance, have a high density of mitochondria (see Fig. 33.3). One population of intercalated cells secretes H^+ (i.e., reabsorbs HCO_3^-), and a second population secretes HCO_3^- (see Chapter 37). The final segment of the nephron, the inner medullary collecting duct, is composed of inner medullary collecting duct cells, which have poorly developed apical and basolateral surfaces and few mitochondria.

All cells in the nephron except intercalated cells have in their apical plasma membrane a single nonmotile primary cilium that protrudes into the tubule fluid (Fig. 33.4). Primary cilia are mechanosensors (i.e., they sense changes in the rate of flow of tubule fluid) and chemosensors (i.e., they sense or respond to compounds in the surrounding fluid), and they initiate Ca^{++}-dependent signaling pathways, including those that control kidney cell function, proliferation, differentiation, and apoptosis (i.e., programmed cell death).

Nephrons may be subdivided into superficial and juxtamedullary types (see Fig. 33.2), with approximately 10 superficial nephrons for each juxtamedullary nephron. The

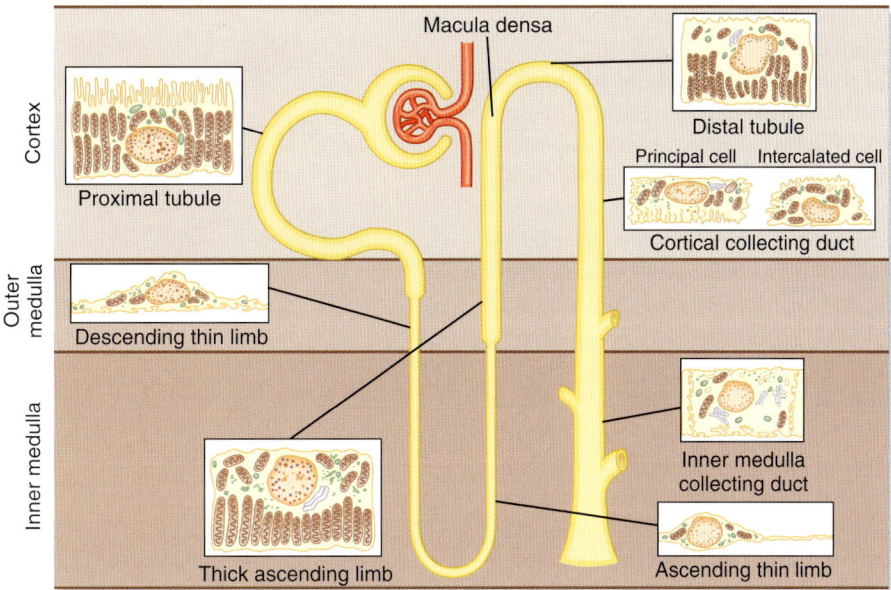

Fig. 33.3 Diagram of a nephron including cellular ultrastructure.

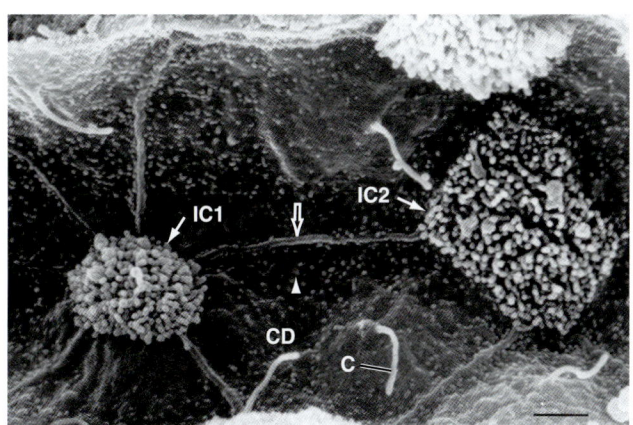

Fig. 33.4 Scanning electron micrograph illustrating primary cilia (C, ≈2 to 30 μm long and 0.5 μm in diameter) in the apical plasma membrane of principal cells in the cortical collecting duct. Note that intercalated cells (IC1 and IC2) do not have cilia but have numerous microvilli. CD, collecting duct principal cells with short microvilli *(arrowhead)*; straight ridges *(open arrow)* represent the cell borders between principal cells; IC1 and IC2, intercalated cells with numerous long microvilli in the apical membrane. (From Kriz W, Kaissling B. Structural organization of the mammalian kidney. In: Seldin DW, Giebisch G [eds]. *The Kidney: Physiology and Pathophysiology.* 3rd ed. Philadelphia: Lippincott Williams & Wilkins; 2000.)

glomerulus of each superficial nephron is located in the outer region of the cortex. The corresponding loops of Henle are short, and associated efferent arterioles branch into peritubular capillaries that surround its associated nephron segments as well as adjacent nephrons. This capillary network conveys oxygen and important nutrients to the nephron segments in the cortex, delivers substances to individual nephron segments for secretion (i.e., movement of a substance from blood into tubular fluid), and serves as a pathway for return of reabsorbed water and solutes to the circulatory system. A few species, including humans,

 AT THE CELLULAR LEVEL

Polycystin 1 (encoded by the *PKD1* gene) and **polycystin 2** (encoded by the *PKD2* gene) are expressed in the membrane of primary cilia and mediate entry of Ca^{++} into cells. PKD1 and PKD2 are thought to play an important role in flow-dependent K^+ secretion by principal cells of the collecting duct. As described in more detail in Chapter 36, increased flow of tubule fluid in the collecting duct is a strong stimulus for secretion of K^+. Increased flow bends the primary cilium in principal cells, which activates the PKD1/PKD2 Ca^{++}-conducting channel complex and allows Ca^{++} to enter the cell and increase intracellular $[Ca^{++}]$. The increase in $[Ca^{++}]$ activates K^+ channels in the apical plasma membrane, which enhances secretion of K^+ from the cell into the tubule fluid.

 IN THE CLINIC

Autosomal dominant polycystic kidney disease (ADPKD) is the most common inherited kidney disease, occurring in about 1 in 1000 people. Approximately 12.5 million people worldwide have ADPKD, which is caused primarily by mutations in *PKD1* (85%–90% of cases) or *PKD2* (≈15% of cases). The major phenotype of ADPKD is enlargement of the kidneys due to the presence of hundreds or thousands of space-occupying renal cysts that can be as large as 20 cm in diameter. Cysts are also seen in the liver and other organs in this condition. About 50% of patients with ADPKD progress to renal failure by the age of 60. Although it is not clear how mutations in *PKD1* and *PKD2* cause ADPKD, renal cyst formation may result from defects in Ca^{++} uptake that alter Ca^{++}-dependent signaling pathways, including those controlling kidney cell proliferation, differentiation, and apoptosis.

also possess very short superficial nephrons whose loops of Henle never enter the medulla.

The glomerulus of each **juxtamedullary nephron** is located in the region of the cortex adjacent to the medulla (see Fig. 33.2, *right*). When compared with superficial nephrons, juxtamedullary nephrons differ anatomically in two important ways: the loop of Henle is longer and extends deeper into the medulla, and the efferent arteriole forms not only a network of peritubular capillaries but also a series of accompanying vascular loops called the **vasa recta.**

As shown in Fig. 33.2 *(left)*, the vasa recta descend into the medulla, where they form capillary networks that surround the collecting ducts and ascending limbs of the loop of Henle. The blood returns to the cortex via the ascending vasa recta. Although less than 0.7% of the renal blood flow (RBF) enters the vasa recta, these vessels serve important functions in the renal medulla that include (1) conveying oxygen and important metabolic substrates to support nephron function, (2) delivering substances to the nephron for secretion, (3) serving as a pathway for return of reabsorbed water and solutes to the circulatory system, and (4) concentrating and diluting urine (urine concentration and dilution are discussed in more detail in Chapter 35).

Ultrastructure of the Glomerulus

The first step in urine formation begins with passive movement of a plasma ultrafiltrate from the glomerular capillaries (i.e., glomerulus) into **Bowman's space.** The term *ultrafiltration* refers to this passive movement of fluid—similar in composition to plasma, except for the fact that the ultrafiltrate protein concentration is much lower than that in the plasma—from the glomerular capillaries into Bowman's space. To appreciate this process, one must understand the anatomy of the glomerulus, which consists of a network of capillaries supplied by the afferent arteriole and drained by the efferent arteriole (Figs. 33.5 and 33.6). During embryological development, the glomerular capillaries press into the closed end of the proximal tubule, forming Bowman's capsule. As the epithelial cells thin on the outside circumference of Bowman's capsule, they form the parietal epithelium (see Fig. 33.5). The epithelial cells in contact with the capillaries thicken and develop into **podocytes**, which form the **visceral layer** of Bowman's capsule (Figs. 33.7–33.9). The visceral cells face outward at the vascular pole (i.e., where afferent and efferent arterioles enter and exit Bowman's capsule) to form the parietal layer of Bowman's capsule. The space between the visceral layer and the parietal layer is Bowman's space, which at the urinary pole (i.e., where the proximal tubule joins Bowman's capsule) of the glomerulus becomes the lumen of the proximal tubule.

The endothelial cells of glomerular capillaries are covered by a basement membrane surrounded by **podocytes.** The capillary endothelium, basement membrane, and foot processes of podocytes form the so-called **filtration barrier** (see Figs. 33.5 and 33.7–33.9). The endothelium is fenestrated

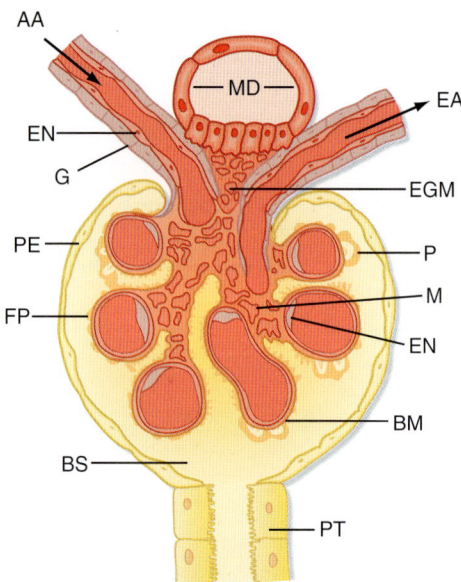

• **Fig. 33.5** Anatomy of the glomerulus and juxtaglomerular apparatus. The juxtaglomerular apparatus is composed of the macula densa (MD) of the thick ascending limb, extraglomerular mesangial cells (EGM), and renin- and angiotensin II–producing granular cells (G) of the afferent arterioles (AA). BM, basement membrane; BS, Bowman's space; EA, efferent arteriole; EN, endothelial cell; FP, foot processes of the podocyte; M, mesangial cells between capillaries; P, podocyte cell body (visceral cell layer); PE, parietal epithelium; PT, proximal tubule cell. (Modified from Kriz W, Kaissling B. Structural organization of the mammalian kidney. In Alpern RJ, Moe OW, Caplan M [eds]: *Seldin and Giebisch's The Kidney,* 5th ed. London, Elsevier, 2013. Figure in that source based on Kriz W, Sakai T, et al. [1988]. Morphological aspects of glomerular function. In *"Nephrology,"* A. M. Davison, Vol. 1, Proceedings of the 10th International Congress of Nephrology, 323. Bailliere Tindall, London.)

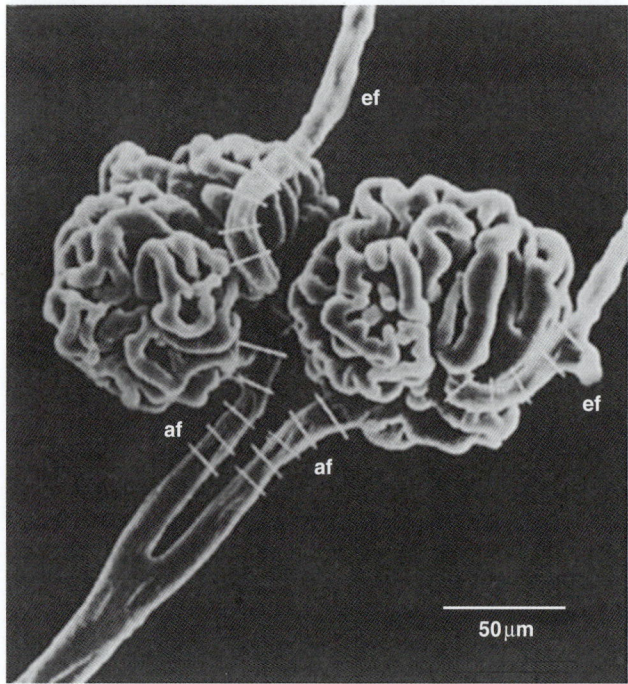

• **Fig. 33.6** Scanning electron micrograph of the interlobular artery, afferent arteriole (af), efferent arteriole (ef), and glomerulus. The white bars on the afferent and efferent arterioles indicate that they are about 15 to 20 μm in diameter. (From Kimura K et al. *Am J Physiol* 1990;259:F936.)

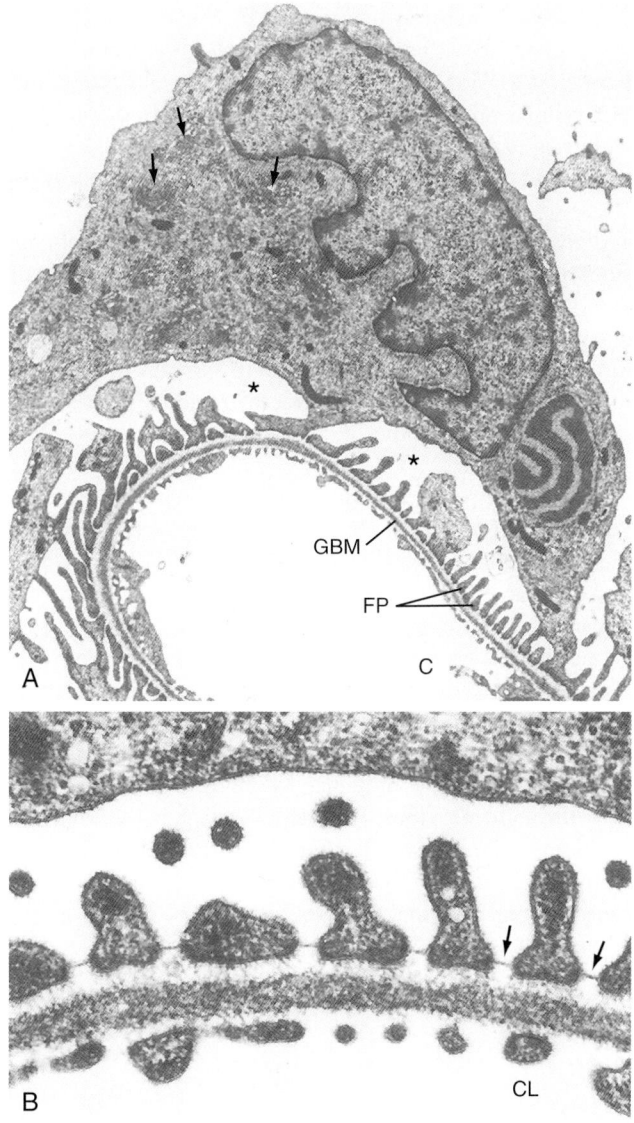

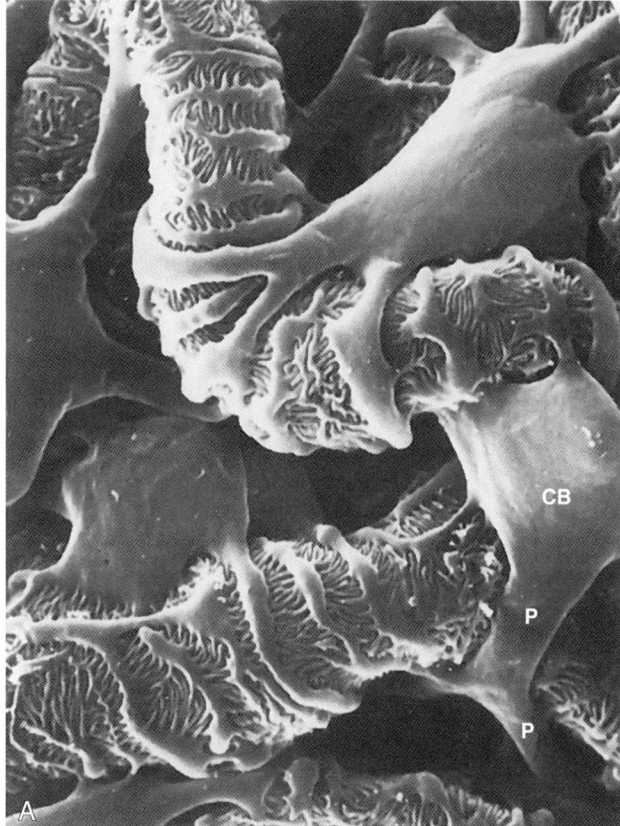

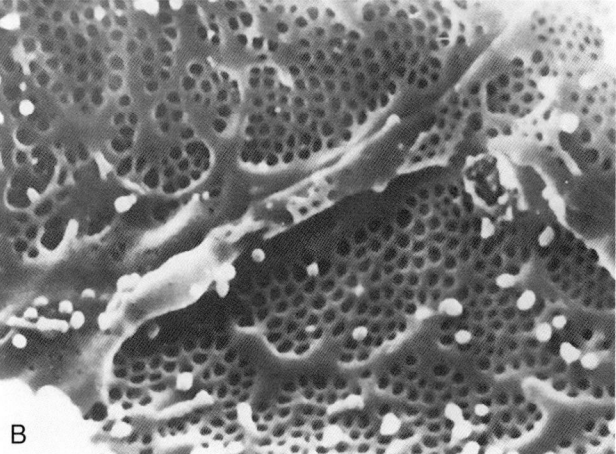

• **Fig. 33.7** A, Electron micrograph of a podocyte surrounding a glomerular capillary. The cell body of the podocyte contains a large nucleus with three indentations. Cell processes of the podocyte form the interdigitating foot processes (FP). The arrows in the cytoplasm of the podocyte indicate the well-developed Golgi apparatus, and the asterisks indicate Bowman's space. C, capillary lumen; GBM, glomerular basement membrane. B, Electron micrograph of the filtration barrier of a glomerular capillary. The filtration barrier is composed of three layers: the endothelium, basement membrane, and foot processes of the podocytes. Note the filtration slit diaphragm bridging the floor of the filtration slits *(arrows)*. CL, capillary lumen. (From Kriz W, Kaissling B: Structural organization of the mammalian kidney. In: Alpern RJ, Moe OW, Caplan M [eds]: *Seldin and Giebisch's The Kidney*, 5th ed. London, Elsevier, 2013.)

• **Fig. 33.8** A, Scanning electron micrograph showing the outer surface of glomerular capillaries. This is the view that would be seen from Bowman's space. Processes (P) of podocytes run from the cell body (CB) toward the capillaries, where they ultimately split into foot processes. Interdigitation of the foot processes creates the filtration slits. B, Scanning electron micrograph of the inner surface (blood side) of a glomerular capillary. This view would be seen from the lumen of the capillary. The fenestrations of the endothelial cells are seen as small 700-Å holes. (From Kriz W, Kaissling B: Structural organization of the mammalian kidney. In: Alpern RJ, Moe OW, Caplan M [eds]: *Seldin and Giebisch's The Kidney*, 5th ed. London, Elsevier, 2013.)

(i.e., contains 700-Å holes, where 1 Å = 10^{-10} m) and freely permeable to water, small solutes (e.g., Na^+, urea, glucose), and most proteins but is not permeable to red blood cells, white blood cells, or platelets. Because endothelial cells express negatively charged glycoproteins on their surface, they minimize the filtration into Bowman's space of albumin, the most abundant plasma protein, and most other plasma proteins. In addition to their role as a barrier to filtration, the endothelial cells synthesize a number of vasoactive substances (e.g., nitric oxide [NO], a vasodilator, and endothelin 1 [ET-1], a vasoconstrictor) that are important in controlling **renal plasma flow (RPF).** The basement

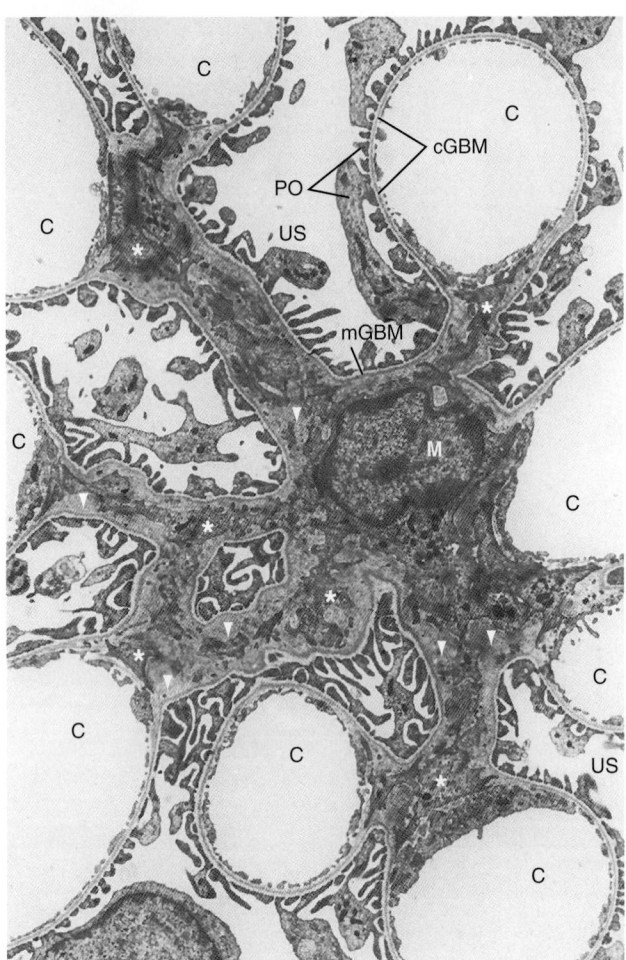

• **Fig. 33.9** Electron micrograph of the mesangium, the area between the glomerular capillaries containing mesangial cells. C, glomerular capillaries; cGBM, capillary glomerular basement membrane surrounded by foot processes of podocytes (PO) and endothelial cells; M, mesangial cell that gives rise to several processes, some marked by stars; mGBM, mesangial glomerular basement membrane surrounded by foot processes of podocytes and mesangial cells; US, urinary space. Note the extensive extracellular matrix surrounded by mesangial cells *(triangles)* (×4100). (From Kriz W, Kaissling B: Structural organization of the mammalian kidney. In: Alpern RJ, Moe OW, Caplan M [eds]: *Seldin and Giebisch's The Kidney,* 5th ed. London, Elsevier, 2013.)

continuous structure when viewed by electron microscopy (see Fig. 33.7B), is composed of several proteins, including **nephrin (NPHS1), NEPH-1,** and **podocin (NPHS2),** and intracellular proteins that associate with the slit diaphragm, including **α-actinin-4 (ACTN4)** and **CD2-AP** (Figs. 33.10 and 33.11). Filtration slits, which function primarily as a size-selective filter, minimize the filtration of proteins and macromolecules that cross the basement membrane from entering Bowman's space. Because both the basement membrane and filtration slits contain negatively charged glycoproteins, some proteins are held back (i.e., not filtered into Bowman's space) on the basis of size and charge. For molecules with an effective molecular radius between 18 and 42 Å, cationic molecules are filtered more readily than anionic molecules.

Another important component of the renal corpuscle is the **mesangium,** which consists of **mesangial cells** and the **mesangial matrix** (see Fig. 33.9). Mesangial cells, which possess many properties of smooth muscle cells, provide structural support for the glomerular capillaries, secrete extracellular matrix, exhibit phagocytic activity by removing macromolecules from the mesangium, and secrete prostaglandins and proinflammatory cytokines. Because they also contract and are adjacent to glomerular capillaries, mesangial cells may influence GFR by regulating blood flow through the glomerular capillaries or by altering the capillary surface area. Mesangial cells located outside the glomerulus (between the afferent and efferent arterioles) are called **extraglomerular mesangial cells.**

 IN THE CLINIC

The **nephrotic syndrome** is produced by a variety of disorders and is characterized by increased permeability of the glomerular capillaries to proteins and by loss of normal podocyte structure, including effacement (i.e., thinning) of foot processes. The augmented permeability to proteins results in increased urinary protein excretion **(proteinuria).** Thus the appearance of proteins in urine can indicate kidney disease. Individuals with this syndrome often develop hypoalbuminemia as a result of the proteinuria. In addition, generalized edema is commonly seen in nephrotic individuals. Mutations in several genes encoding either slit diaphragm proteins (see Figs. 33.10 and 33.11), including **nephrin, NEPH-1,** and **podocin,** or intracellular proteins that functionally interact with slit diaphragm proteins, such as **CD2-AP** and **α-actinin 4** (ACTN4), result in proteinuria and kidney disease. For example, mutations in the nephrin gene *(NPHS1)* in congenital nephrotic syndrome lead to abnormal or absent slit diaphragms, causing massive proteinuria and renal failure. In addition, mutations in the podocin gene *(NPHS2)* cause autosomal recessive, steroid-resistant nephrotic syndrome. These naturally occurring mutations and knockout studies in mice demonstrate that nephrin, NEPH-1, podocin, CD2-AP, and α-actinin-4 play key roles in normal glomerular podocyte structure and function.

membrane, which is a porous matrix of negatively charged proteins (type IV collagen, laminin, the proteoglycans agrin and perlecan, and fibronectin), is an important filtration barrier to plasma proteins. The basement membrane is thought to function primarily as a charge-selective filter in which the ability of proteins to cross the filter is based on charge.

The podocytes, which are endocytic, have long finger-like processes that completely encircle the outer surface of the capillaries (see Figs. 33.7 and 33.8A). The processes of the podocytes interdigitate to cover the basement membrane and are separated by apparent gaps called **filtration slits** (see Figs. 33.7 and 33.8A). Each filtration slit is bridged by a thin diaphragm that contains pores with a dimension of 40 × 140 Å. The **filtration slit diaphragm,** which appears as a

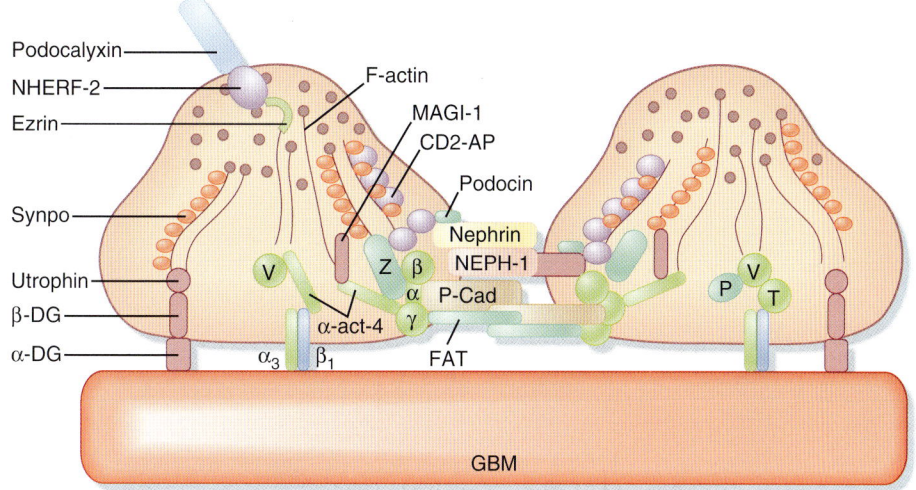

• **Fig. 33.10** Anatomy of podocyte foot processes. This figure illustrates the proteins that make up the slit diaphragm between two adjacent foot processes. Nephrin and NEPH1 are membrane-spanning proteins that have large extracellular domains that interact. Podocin, also a membrane-spanning protein, organizes nephrin and NEPH1 in specific microdomains in the plasma membrane, which is important for signaling events that determine the structural integrity of podocyte foot processes. Many of the proteins that compose the slit diaphragm interact with adapter proteins inside the cell, including CD2-AP. The adapter proteins bind to the filamentous actin (F-actin) cytoskeleton, which in turn binds either directly or indirectly to proteins such as $\alpha3\beta1$ and MAGI-1 that interact with proteins expressed by the glomerular basement membrane (GBM). α-act-4, α-actinin 4; $\alpha_3\beta_1$, $\alpha_3\beta_1$ integrin; α-DG, α-dystroglycan; CD2-AP, an adapter protein that links nephrin and podocin to intracellular proteins; FAT, a protocadherin that organizes actin polymerization; MAGI-1, a membrane-associated guanylate kinase protein; NHERF-2, Na^+-H^+ exchanger regulatory factor 2; P, paxillin; P-Cad, P-cadherin; Synpo, synaptopodin; T, talin; V, vinculin; Z, zona occludens. (Adapted from Mundel P, Shankland SJ. *J Am Soc Nephrol* 2002;13:3005.)

 IN THE CLINIC

Mesangial cells are involved in development of **immune complex–mediated glomerular disease.** Because the glomerular basement membrane does not completely surround all glomerular capillaries (see Fig. 33.9), some immune complexes can escape the blood and enter the mesangium without crossing the glomerular basement membrane. Accumulation of immune complexes induces mesangial infiltration of inflammatory cells and promotes local proinflammatory cytokine and autocoid production. These cytokines and autocoids enhance the immune complex–initiated inflammatory response, which can ultimately lead to mesangial expansion, scarring, and obliteration of the glomerulus.

 IN THE CLINIC

Alport syndrome is characterized by hematuria (i.e., blood in the urine) and progressive glomerulonephritis (i.e., inflammation of the glomerular capillaries), and it accounts for 1% to 2% of all cases of ESRD. Alport syndrome is caused by mutations in type IV collagen, a major component of the glomerular basement membrane. In about 80% of patients with Alport syndrome the disease is X-linked with mutations in the *COL4A5* gene. Some 15% of patients also have mutations in type IV collagen genes (*COL4A3* and *COL4A4*); six have been identified, but the mode of inheritance is autosomal recessive. The remaining 5% of patients with Alport syndrome have autosomal dominant disease that arises from heterozygous mutations in the *COL4A3* or *COL4A4* genes. In Alport syndrome, the glomerular basement membrane becomes irregular in thickness and fails to serve as an effective filtration barrier to blood cells and protein.

Ultrastructure of the Juxtaglomerular Apparatus

The **juxtaglomerular apparatus** is one component of an important feedback mechanism, the tubuloglomerular feedback mechanism, described later in this chapter. Structures that make up the juxtaglomerular apparatus (see Fig. 33.5) include:

1. the **macula densa** of the thick ascending limb
2. extraglomerular mesangial cells
3. renin- and angiotensin II–producing **granular cells** of the afferent arteriole

The cells of the macula densa represent a morphologically distinct region of the thick ascending limb. This region passes through the angle formed by the afferent and efferent arterioles of the same nephron. The cells of the macula densa contact the extraglomerular mesangial cells and the granular cells of the afferent arterioles. The granular cells of the afferent arterioles contain smooth muscle myofilaments and—importantly—manufacture, store, and release **renin** in response to signals associated with decreased effective

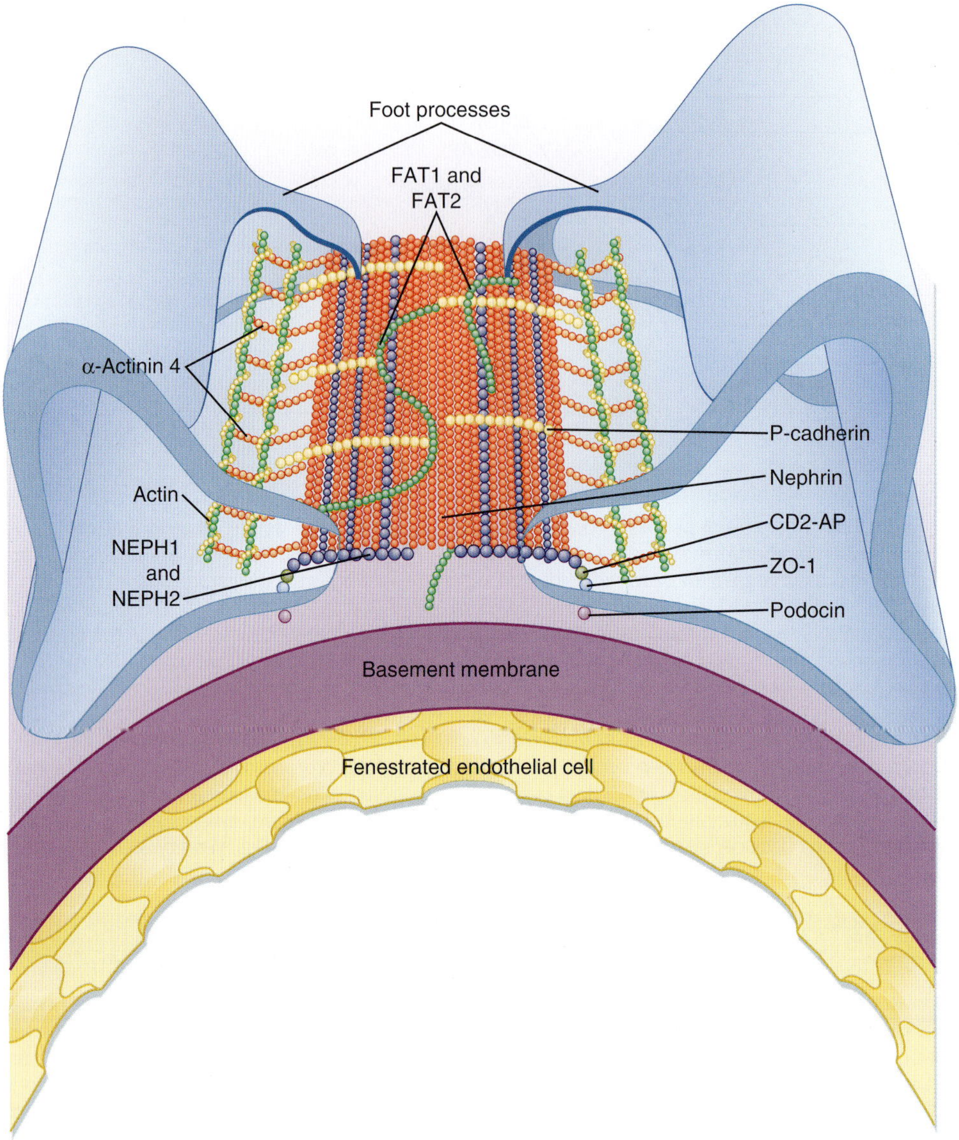

● **Fig. 33.11** Overview of the major proteins that form the slit diaphragm. Nephrins *(red)* from opposite foot processes interdigitate in the center of the slit. In the slit, nephrin interacts with NEPH1 and NEPH2 *(blue)*, FAT1 and FAT2 *(green)*, and P-cadherin. The intracellular domains of nephrin, NEPH1, and NEPH2 interact with podocin and CD2-AP, which connect these slit diaphragm proteins with ZO-1, α-actinin 4, and actin. (Modified from Tryggvason K et al. *N Engl J Med* 2006;354:1387.)

circulating volume and reduced renal perfusion. Renin is involved in proteolytic generation of **angiotensin II** and ultimately in secretion of **aldosterone** (see Chapter 35). The juxtaglomerular apparatus is one component of the tubuloglomerular feedback mechanism involved in autoregulation of RBF and GFR.

Innervation of the Kidneys

Renal nerves regulate RBF, GFR, and salt and water reabsorption by the nephron. The nerve supply to the kidneys consists of sympathetic nerve fibers that originate in the celiac plexus. There is no corresponding parasympathetic innervation. Adrenergic fibers innervating the kidneys release norepinephrine and lie adjacent to the smooth muscle

cells of the major branches of the renal artery (interlobar, arcuate, and interlobular arteries) as well as the afferent and efferent arterioles. In addition, sympathetic nerves innervate the renin-producing granular cells of the afferent arterioles. Renin secretion is stimulated by increased sympathetic activity. Nerve fibers also innervate the proximal tubule, loop of Henle, distal tubule, and collecting duct; activation of these nerves enhances Na^+ reabsorption by these nephron segments.

Assessment of Renal Function

The coordinated actions of the nephron's various segments determine the final amount of a substance that appears in urine. This represents three general processes (1) glomerular

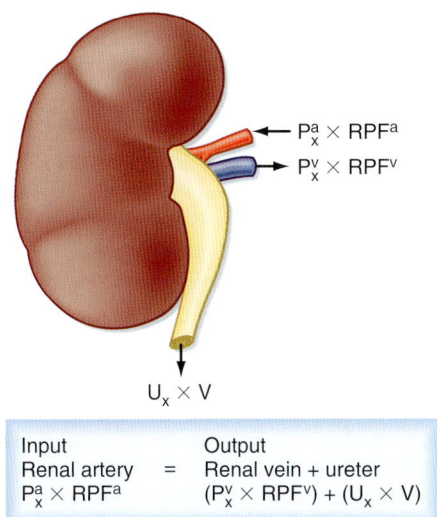

• **Fig. 33.12** Mass balance relationships for the kidney. See text for definition of symbols.

filtration, (2) reabsorption of the substance from tubular fluid back into blood, and (3) (in some cases) secretion of the substance from blood into tubule fluid. The first step in urine formation by the kidneys is production of an ultrafiltrate of plasma across the glomerulus. The process of glomerular filtration and regulation of GFR and RBF are discussed later in this chapter. The concept of renal clearance, which is the theoretical basis for measurement of GFR and RBF, is presented in the following section. Reabsorption and secretion are discussed in subsequent chapters.

Renal Clearance

The concept of **renal clearance** is based on the Fick principle (i.e., mass balance or conservation of mass). Fig. 33.12 illustrates the various factors required to describe the mass balance relationships of a kidney. The renal artery is the single input source to the kidney for substances not synthesized by this organ, whereas the renal vein and ureter constitute the two principal output routes. In other words a nonmetabolized substance entering the renal circulation via the renal artery may only exit this circulation via the renal vein (i.e., the unfiltered fraction plus any filtered amount subsequently reabsorbed back into blood) or the ureter (the combined filtered and secreted fractions less any tubular reabsorption). The following equation defines the mass balance relationship:

Equation 33.1

$$P_x^a \times RPF^a = (P_x^v \times RPF^v) + (U_x \times \dot{V})$$

where

P_x^a and P_x^v are the concentrations of substance x in the renal artery and renal vein plasma, respectively

RPF^a and RPF^v are **renal plasma flow** rates in the artery and vein, respectively

U_x is the concentration of substance x in urine

$\dot{V}$ is the urine flow rate

This relationship permits quantification of the amount of substance x excreted in urine versus the amount returned to the systemic circulation in renal venous blood. Thus for any substance that is neither synthesized nor metabolized, the amount that enters the kidneys is equal to the amount that leaves the kidneys in urine plus the amount that leaves the kidneys in renal venous blood.

The principle of renal clearance emphasizes the excretory function of the kidneys; it considers only the rate at which a substance is excreted into urine and not its rate of return to the systemic circulation in the renal vein. Therefore in terms of mass balance (Eq. 33.1) the urinary excretion rate of substance x ($U_x \times \dot{V}$) is proportional to the plasma concentration of substance x (P_x^a):

Equation 33.2

$$P_x^a \propto U_x \times \dot{V}$$

To equate the urinary excretion rate of substance x to its renal arterial plasma concentration, it is necessary to determine the rate at which it is removed from plasma by the kidneys. This removal rate is the clearance (C_x):

Equation 33.3

$$P_x^a \times C_x = U_x \times \dot{V}$$

If Eq. 33.3 is rearranged and the concentration of substance x in renal artery plasma (P_x^a) is assumed to be identical to its concentration in a plasma sample from any peripheral blood vessel (P_x), the following relationship is obtained:

Equation 33.4

$$C_x = \frac{U_x \times \dot{V}}{P_x}$$

Clearance has the dimensions of volume/time, and it represents a volume of plasma from which all the substance has been removed and excreted into urine per unit time. This last point is best illustrated by considering the following example. If a substance is present in urine at a concentration of 100 mg/mL and the urine flow rate is 1 mL/min, the excretion rate for this substance is calculated as follows:

Equation 33.5

$$\text{Excretion rate} = U_x \times \dot{V} = 100 \text{ mg/mL}$$
$$\times 1 \text{ mL/min} = 100 \text{ mg/min}$$

If this substance is present in plasma at a concentration of 1 mg/mL, its clearance according to Eq. 33.4 is as follows:

Equation 33.6

$$C_x = \frac{U_x \times \dot{V}}{P_x} = \frac{100 \text{ mg/min}}{1 \text{ mg/mL}} = 100 \text{ mL/min}$$

In other words, 100 mL of plasma will be completely cleared of substance x each minute. The definition of *clearance* as a volume of plasma from which all the substance

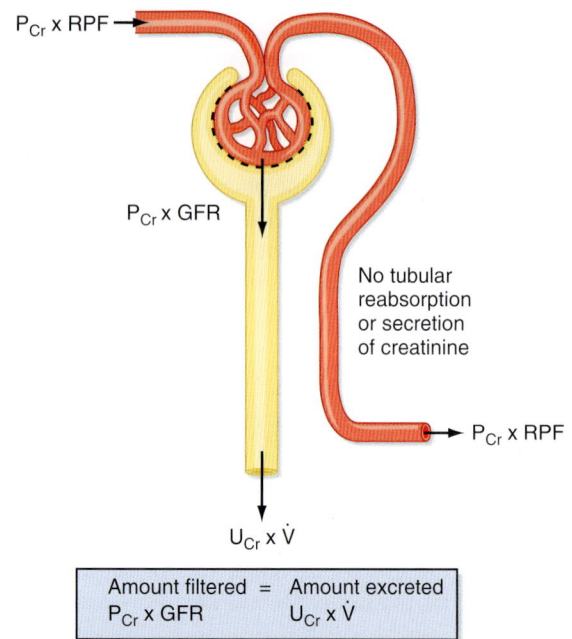

Fig. 33.13 Renal handling of creatinine. Creatinine is freely filtered across the glomerulus and is, to a first approximation, not reabsorbed, secreted, or metabolized by the nephron. Note that all the creatinine coming to the kidney in the renal artery does not get filtered at the glomerulus (normally, 15%–20% of plasma creatinine is filtered). The portion that is not filtered is returned to the systemic circulation in the renal vein. P_{Cr}, plasma creatinine concentration; RPF, renal plasma flow; U_{Cr}, urinary concentration of creatinine; V., urine flow rate.

has been removed and excreted into urine per unit time is somewhat misleading because it is not a real volume of plasma; rather it is a virtual volume.[c] The concept of clearance is important because it can be used to measure GFR and RPF and determine whether a substance is reabsorbed or secreted along the nephron.

Glomerular Filtration Rate

The GFR is equal to the sum of the filtration rates of all functioning nephrons. Thus it is an aggregate index of kidney function. A fall in GFR generally means the kidney disease is progressing, whereas recovery generally suggests recuperation. Thus serial assessment of a patient's GFR is essential to evaluate the severity and course of kidney disease.

Creatinine is a byproduct of normal skeletal muscle creatine metabolism and is freely filtered across the glomerulus into Bowman's space. It is normally generated by the body at a fairly constant rate, and—to a first approximation—it is not appreciably reabsorbed, secreted, or metabolized by the cells of the nephron after its filtration. Accordingly the amount of creatinine excreted in urine per minute is fairly constant at steady state (i.e., when [creatinine] is constant) and equals the amount of creatinine filtered at the glomerulus each minute (Fig. 33.13):

---

[c]For most substances cleared from plasma by the kidneys, only a portion is actually removed and excreted in a single pass through the kidneys.

Equation 33.7
Amount filtered = Amount excreted

$$GFR \times P_{Cr} = U_{Cr} \times \dot{V}$$

where

P_{Cr} = plasma concentration of creatinine
U_{Cr} = urine concentration of creatinine
$\dot{V}$ = urine flow

If Eq. 33.7 is solved for GFR:

Equation 33.8
$$GFR = \frac{U_{Cr} \times \dot{V}}{P_{Cr}}$$

This equation is the same form as that for clearance (see Eq. 33.4). Thus measured creatinine clearance (CrCl) can be used clinically to determine GFR at steady state. Clearance has the dimensions of volume/time, and it represents an equivalent volume of plasma from which all of the substance has been removed and excreted into urine per unit time.

Creatinine is not the only substance that can be used to measure GFR; any substance that meets the following criteria can serve as an appropriate marker. The substance must:

1. achieve a stable plasma concentration
2. be freely filtered across the glomerulus into Bowman's space
3. not be reabsorbed or secreted by the nephron
4. not be metabolized or produced by the kidney
5. not alter GFR

Not all creatinine (or other substances used to measure GFR) that enters the kidney in renal arterial plasma is filtered at the glomerulus. Likewise not all plasma coming into the kidneys is filtered. Although nearly all plasma that enters the kidneys in the renal artery passes through the glomerulus, approximately 10% does not. The *portion of filtered plasma* is termed the **filtration fraction** and is determined as:

Equation 33.9
$$\text{Filtration fraction} = \frac{GFR}{RPF}$$

Under normal conditions the filtration fraction averages 0.15 to 0.20, which means that only 15% to 20% of the plasma that enters the glomerulus is actually filtered. The remaining 80% to 85% continues on through the glomerular capillaries and into the efferent arterioles and peritubular capillaries before finally returning to the systemic circulation via the renal vein.

Glomerular Filtration

The first step in the formation of urine is ultrafiltration of plasma by the glomerulus. In normal adults, GFR ranges from 90 to 140 mL/min in males and from 80 to 125 mL/min in females. Thus in 24 hours as much as 180 L of plasma is filtered by the glomeruli. The plasma ultrafiltrate is devoid of cellular elements (i.e., red and white blood cells and

IN THE CLINIC

Creatinine clearance (CrCl) is used to estimate GFR in clinical practice. It is synthesized at a relatively constant rate, and the amount produced is proportional to the total muscle mass. However, creatinine is not a perfect substance for measuring GFR because it is secreted to a small extent by the organic cation secretory system in the proximal tubule (see Chapter 34). The error introduced by this secretory component is approximately 10%. Thus the amount of creatinine excreted in urine exceeds the amount expected from filtration alone by 10%. However, the method used to measure the plasma creatinine concentration (P_{Cr}) overestimates the true value by 10%. Consequently the two errors cancel each other, and in most clinical situations, CrCl provides a reasonably accurate measure of GFR.

IN THE CLINIC

A fall in GFR may be the first and only clinical sign of kidney disease. Thus measuring GFR is important when kidney disease is suspected. A 50% loss of functioning nephrons reduces GFR only by about 25%. The decline in GFR is not 50% because the remaining nephrons compensate. Because measurements of GFR are cumbersome, kidney function is usually assessed in the clinical setting by measuring the plasma concentration of creatinine (P_{Cr}), which is inversely related to GFR (Fig. 33.14). However, as Fig. 33.14 shows, GFR must decline substantially before an increase in P_{Cr} can be detected in a clinical setting. For example, a fall in GFR from 120 to 100 mL/min is accompanied by an increase in P_{Cr} from 1.0 to 1.2 mg/dL. This does not appear to be a significant change in P_{Cr}, but GFR has actually fallen by almost 20%. In the clinical setting, estimated GFR (eGFR) also includes consideration of several other factors in addition to the plasma concentration of creatinine, including age, sex, body size, and race. A free app to calculate eGFR can be downloaded at: https://www.kidney.org/apps/professionals/egfr-calculator.

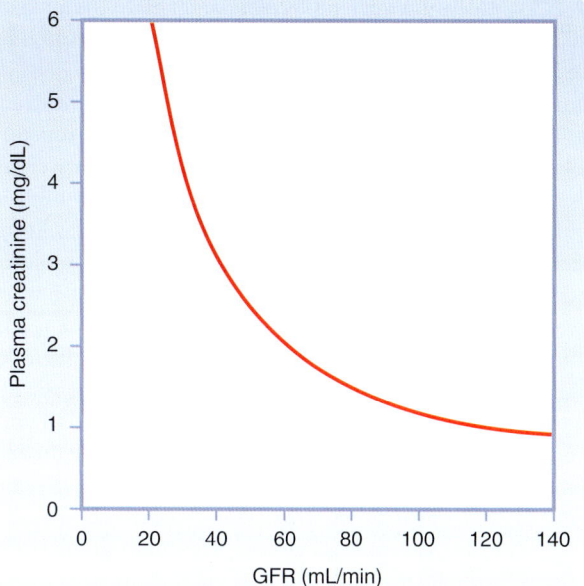

• **Fig. 33.14** Relationship between GFR and plasma [creatinine] (P_{cr}). The amount of creatinine filtered is equal to the amount excreted; thus GFR × P_{Cr} = U_{Cr} × $\dot{V}$. Because the production of creatinine is constant, excretion must be constant to maintain creatinine balance. Therefore if GFR falls from 120 to 60 mL/min, P_{Cr} must increase from 1 to 2 mg/dL to keep filtration of creatinine and its excretion equal to the production rate.

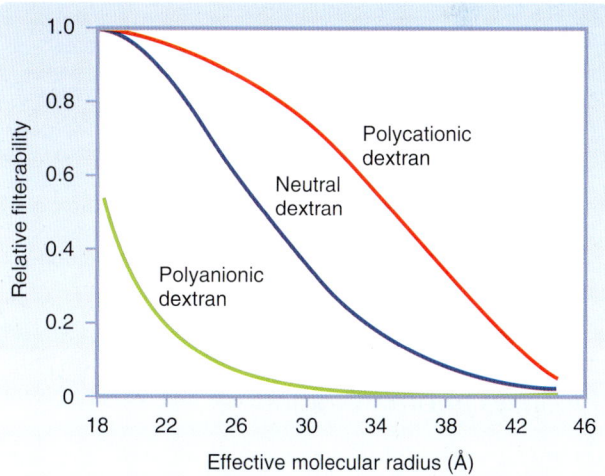

• **Fig. 33.15** Influence of the size and electric charge of dextran on its filterability. A value of 1 indicates that it is filtered freely, whereas a value of zero indicates that it is not filtered. The filterability of dextrans between approximately 18 and 42 Å depends on charge. Dextrans larger than 42 Å are not filtered regardless of charge, and polycationic dextrans and neutral dextrans smaller than 18 Å are freely filtered. The major proteins in plasma are albumin and immunoglobulins. Because the effective molecular radii of immunoglobulin (Ig)G (53 Å) and IgM (>100 Å) are greater than 42 Å, they are not filtered. Although the effective molecular radius of albumin is 35 Å, it is a polyanionic protein, so it does not cross the filtration barrier to a significant degree.

platelets) and is essentially protein free. The concentration of salts and organic molecules (e.g., glucose, amino acids) is similar in plasma and the ultrafiltrate. Starling forces drive ultrafiltration across the glomerular capillaries, and changes in these forces alter GFR. GFR and RPF are normally held within very narrow ranges by a phenomenon called *autoregulation*. The next sections of this chapter review the composition of the glomerular filtrate, the dynamics of its formation, and the relationship between RPF and GFR. In addition, factors that contribute to autoregulation and regulation of GFR and RBF are discussed.

Determinants of Ultrafiltrate Composition

The glomerular filtration barrier determines the composition of the plasma ultrafiltrate. It restricts filtration of molecules on the basis of both size and electric charge (Fig. 33.15). In general, neutral molecules with a radius smaller than about

18 Å are freely filtered, molecules larger than about 42 Å are not filtered, and molecules between about 18 and 42 Å are filtered to varying degrees. Fig. 33.15 shows how electric charge affects filtration of macromolecules (e.g., dextrans) by the glomerulus. Dextrans are a family of exogenous polysaccharides manufactured in various molecular weights. They can be electrically neutral or have either negative (polyanionic) or positive (polycationic) charges. As the size (i.e., effective molecular radius) of a dextran molecule increases, the rate at which it is filtered decreases. For any given molecular radius, cationic molecules are more readily filtered than anionic molecules. The reduced filtration rate for anionic molecules is explained by the presence of negatively charged glycoproteins on the surface of all components of the glomerular filtration barrier. These charged glycoproteins repel similarly charged molecules. Because most plasma proteins are negatively charged, the negative charge on the filtration barrier restricts filtration of anionic proteins more than the filtration of neutral and polyanionic proteins with a molecular radius between approximately 18 to 42 Å. For example, serum albumin, an anionic protein that has an effective molecular radius of 35.5 Å, is poorly filtered. Because the small amount of filtered albumin is normally reabsorbed avidly by the proximal tubule, almost no albumin appears in urine.

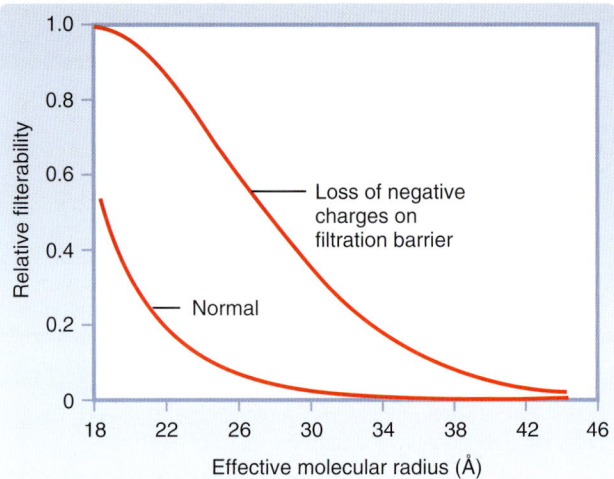

• **Fig. 33.16** Reduction of the negative charges on the glomerular wall results in filtration of proteins on the basis of size only. In this situation the relative filterability of proteins depends only on the molecular radius. Accordingly, excretion of polyanionic proteins (18–42 Å) in urine increases because more proteins of this size are filtered.

IN THE CLINIC

The importance of the negative charges on the filtration barrier in restricting filtration of plasma proteins is shown in Figs. 33.15 and 33.16. Removal of the negative charges from the filtration barrier causes proteins to be filtered solely on the basis of their effective molecular radius (Fig. 33.16). Hence at any molecular radius between approximately 18 and 42 Å, filtration of polyanionic proteins will exceed the filtration that prevails in the normal state (in which the filtration barrier has anionic charges). In a number of glomerular diseases the negative charges on the filtration barrier are reduced because of immunological damage and inflammation. As a result, filtration of anionic proteins between approximately 18 and 42 Å in radius is increased. When the filtered proteins exceed the ability of the proximal tubule to reabsorb and catabolize them, anionic proteins begin to appear in urine (**proteinuria**), which is a marker of kidney disease.

Dynamics of Ultrafiltration

The forces responsible for glomerular filtration of plasma are the same as those in other capillary beds. Ultrafiltration occurs because the Starling forces (i.e., hydraulic and oncotic pressures) combine to drive fluid from the lumen of glomerular capillaries across the filtration barrier and into Bowman's space (Fig. 33.17). The hydraulic pressure inside the glomerular capillary (P_{GC}) is oriented to promote movement of fluid from the glomerular capillary into Bowman's space. Because the glomerular ultrafiltrate is essentially protein free under normal conditions, owing in large part to the paucity of proteins in serum smaller than 18 Å in radius

that can be effectively filtered, the reflection coefficient (σ) for proteins across the glomerular capillary is essentially 1. Thus the oncotic pressure in Bowman's space (π_{BS}) is near zero. Therefore P_{GC} is the principal force favoring filtration. In contrast, the hydraulic pressure in Bowman's space (P_{BS}) and the oncotic pressure in the glomerular capillary (π_{GC}) both oppose filtration.

As shown in Fig. 33.17, a net ultrafiltration pressure (P_{UF}) of 17 mm Hg exists at the afferent end of the glomerulus, whereas at the efferent end it is 8 mm Hg (where $P_{UF} = P_{GC} - P_{BS} - \pi_{GC}$). Two additional points concerning Starling forces and this pressure change are important. First, P_{GC} decreases slightly along the length of the capillary because of the resistance to flow along the length of the capillary. Second, π_{GC} increases as plasma is filtered while protein is retained within the glomerular capillary, thereby progressively increasing the protein concentration along the length of the capillary. GFR is proportional to the sum of the Starling forces that exist across the capillaries [($P_{GC} - P_{BS}$) $- \sigma(\pi_{GC} - \pi_{BS})$] multiplied by the ultrafiltration coefficient (K_f). That is:

Equation 33.10
$$GFR = K_f[(P_{GC} - P_{BS}) - \sigma(\pi_{GC} - \pi_{BS})]$$

K_f is the product of the intrinsic permeability of the glomerular capillary and the glomerular surface area available for filtration. The rate of glomerular filtration is considerably greater in glomerular capillaries than in systemic capillaries, mainly because K_f is approximately 100 times greater in glomerular capillaries. Furthermore P_{GC} is approximately twice as great as the hydraulic pressure in systemic capillaries.

GFR can be altered by changing K_f or by changing any of the Starling forces. In normal individuals, GFR is regulated by alterations in P_{GC} that are mediated mainly by changes

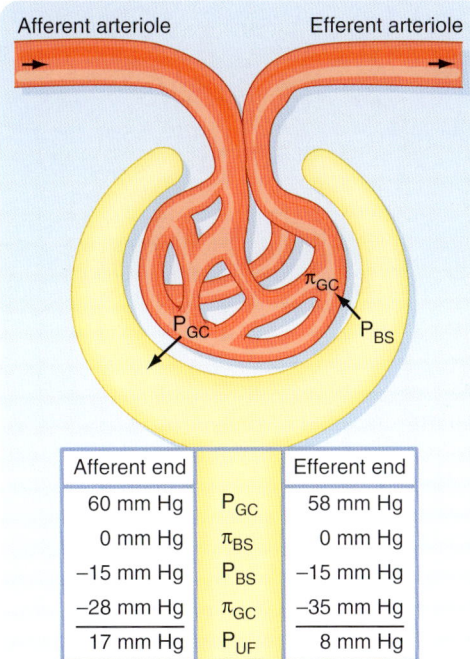

• **Fig. 33.17** Idealized glomerular capillary and the Starling forces across it. The reflection coefficient (σ) for protein across the glomerular capillary is approximately 1. P_{BS}, hydraulic pressure in Bowman's space; P_{GC}, hydraulic pressure in the glomerular capillary; P_{UF}, net ultrafiltration pressure; π_{BS}, oncotic pressure in Bowman's space; π_{GC}, oncotic pressure in the glomerular capillary. The negative signs for P_{BS} and π_{GC} indicate that these forces oppose formation of the glomerular filtrate.

in afferent or efferent arteriolar resistance. P_{GC} is affected in three ways:

1. Changes in afferent arteriolar resistance: A decrease in resistance increases P_{GC} and GFR, whereas an increase in resistance decreases P_{GC} and GFR.
2. Changes in efferent arteriolar resistance: A decrease in resistance reduces P_{GC} and GFR, whereas an increase in resistance elevates P_{GC} and GFR.
3. Changes in renal arteriolar pressure: An increase in blood pressure transiently increases P_{GC} (which enhances GFR), whereas a decrease in blood pressure transiently decreases P_{GC} (which reduces GFR).

Renal Blood Flow

Blood flow through the kidneys serves several important functions:

1. indirectly determines GFR
2. modifies the rate of solute and water reabsorption by the proximal tubule
3. participates in concentration and dilution of urine
4. delivers O_2, nutrients, and hormones to cells along the nephron and returns CO_2, reabsorbed fluid, and solutes to the general circulation
5. delivers substrates for excretion in urine

Blood flow through any organ may be represented by the following equation:

 IN THE CLINIC

A reduction in GFR in disease states is most often due to decreases in K_f because of the loss of filtration surface area. GFR also changes in pathophysiological conditions because of changes in P_{GC}, π_{GC}, and P_{BS}.

1. Changes in K_f: Increased K_f enhances GFR, whereas decreased K_f reduces GFR. Some kidney diseases reduce K_f by decreasing the number of filtering glomeruli (i.e., diminished surface area). Some drugs and hormones that dilate the glomerular arterioles also increase K_f. Similarly, drugs and hormones that constrict the glomerular arterioles also decrease K_f.
2. Changes in P_{GC}: With decreased renal perfusion, GFR declines because P_{GC} falls. As previously discussed, a reduction in P_{GC} is caused by a decline in renal arterial pressure, an increase in afferent arteriolar resistance, or a decrease in efferent arteriolar resistance.
3. Changes in π_{GC}: An inverse relationship exists between π_{GC} and GFR. Alterations in π_{GC} result from changes in protein synthesis outside the kidneys. In addition the protein loss in urine caused by some renal diseases can lead to a decrease in the plasma protein concentration and thus in π_{GC}.
4. Changes in P_{BS}: Increased P_{BS} reduces GFR, whereas decreased P_{BS} enhances GFR. Acute obstruction of the urinary tract (e.g., a kidney stone occluding the ureter) increases P_{BS}.

Equation 33.11

$$Q = \frac{\Delta P}{R}$$

where

Q = blood flow

ΔP = mean arterial pressure minus venous pressure for that organ

R = resistance to flow through that organ

Accordingly, RBF is equal to the pressure difference between the renal artery and the renal vein divided by renal vascular resistance:

Equation 33.12

$$RBF = \frac{\text{Aortic pressure} - \text{Renal venous pressure}}{\text{Renal vascular resistance}}$$

The afferent arteriole, efferent arteriole, and interlobular artery are the major resistance vessels in the kidneys and thereby determine renal vascular resistance. Like most other organs, the kidneys regulate their blood flow by adjusting vascular resistance in response to changes in arterial pressure. As shown in Fig. 33.18 these adjustments are so precise that blood flow remains relatively constant as arterial blood pressure changes between 90 and 180 mm Hg. GFR is also regulated over the same range of arterial pressures. The phenomenon whereby RBF and GFR are maintained relatively constant between blood pressures of 90 and 180 mm Hg, namely **autoregulation,** is achieved by changes in vascular resistance, mainly through the afferent arterioles of the kidneys. Because both RBF and GFR are regulated over the

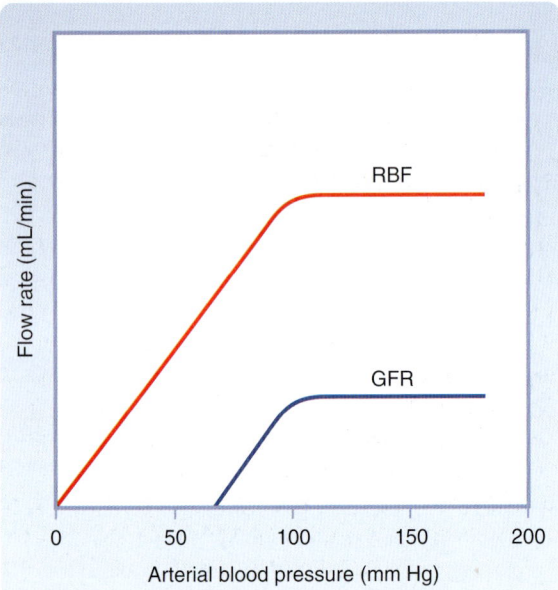

• Fig. 33.18 Relationship between arterial blood pressure and RBF and between arterial blood pressure and GFR. Autoregulation maintains GFR and RBF relatively constant as blood pressure changes from 90 to 180 mm Hg.

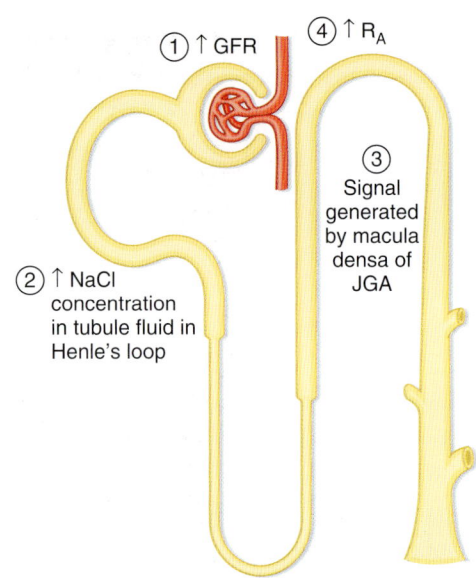

• Fig. 33.19 Tubuloglomerular feedback. An increase in GFR (1) increases [NaCl] in tubule fluid in the loop of Henle (2). The increase in [NaCl] is sensed by the macula densa and converted to a signal (3) that increases the resistance of the afferent arteriole (R_A) (4), which decreases GFR. A decrease in GFR has the opposite effects. (Modified from Cogan MG. *Fluid and Electrolytes: Physiology and Pathophysiology.* Norwalk, CI: Appleton & Lange; 1991.)

same range of pressures and because RBF is an important determinant of GFR, it is not surprising that the same mechanisms regulate both flows.

Two mechanisms are responsible for autoregulation of RBF and GFR: one mechanism that responds to changes in arterial pressure and another that responds to changes in [NaCl] in tubular fluid. Both regulate the tone of the afferent arteriole. The pressure-sensitive mechanism, the so-called **myogenic mechanism,** is related to an intrinsic property of vascular smooth muscle: the tendency to contract when stretched. Accordingly, when arterial pressure rises and the renal afferent arteriole is stretched, the smooth muscle contracts in response. Because the increase in resistance of the arteriole offsets the increase in pressure, RBF, and therefore GFR, remains constant. (That is, RBF is constant if ΔP/R is kept constant [see Eq. 33.11].)

The second mechanism responsible for autoregulation of GFR and RBF is the [NaCl]-dependent mechanism known as **tubuloglomerular feedback.** This mechanism involves a feedback loop in which a change in GFR leads to alteration in the concentration of NaCl in tubular fluid, which is sensed by the macula densa of the **juxtaglomerular apparatus** and converted into signals that affect afferent arteriolar resistance and thus the GFR (Fig. 33.19). For example, when the GFR increases and causes [NaCl] in tubular fluid in the loop of Henle to rise, more NaCl enters the macula densa cells in this segment (Fig. 33.20). This leads to an increase in formation and release of adenosine triphosphate (ATP) and adenosine (a metabolite of ATP) by macula densa cells, which causes vasoconstriction of the afferent arteriole and normalization of GFR. In contrast, when GFR and [NaCl] in tubule fluid decrease, less NaCl enters the macula densa cells, and both ATP and adenosine

production and release decline. The fall in [ATP] and [adenosine] results in afferent arteriolar vasodilation, which returns GFR to normal. NO, a vasodilator produced by the macula densa, attenuates tubuloglomerular feedback, whereas angiotensin II enhances tubuloglomerular feedback. Thus the macula densa may release both vasoconstrictors (e.g., ATP and adenosine) and a vasodilator (e.g., NO) that oppose each other's action at the level of the afferent arteriole. Production plus release of either vasoconstrictors or vasodilators ensures exquisite control over tubuloglomerular feedback.

Fig. 33.20 also illustrates the role of the macula densa in controlling renin secretion by granular cells of the afferent arteriole. This aspect of function of the juxtaglomerular apparatus is considered in detail in Chapter 35.

Because animals engage in many activities that can change arterial blood pressure, mechanisms that maintain RBF and GFR relatively constant despite changes in arterial pressure are highly desirable. If GFR and RBF were to rise or fall suddenly in proportion to changes in blood pressure, urinary excretion of fluid and solute would also change suddenly. Such changes in excretion of water and solutes without comparable changes in intake would alter fluid and electrolyte balance (the reason for which is discussed in Chapter 35). Accordingly, autoregulation of GFR and RBF provides an effective means for uncoupling renal function from arterial pressure, and it ensures that fluid and solute excretion remain relatively constant.

Three points concerning autoregulation should be noted:
1. Autoregulation is absent when arterial pressure is less than 90 mm Hg.

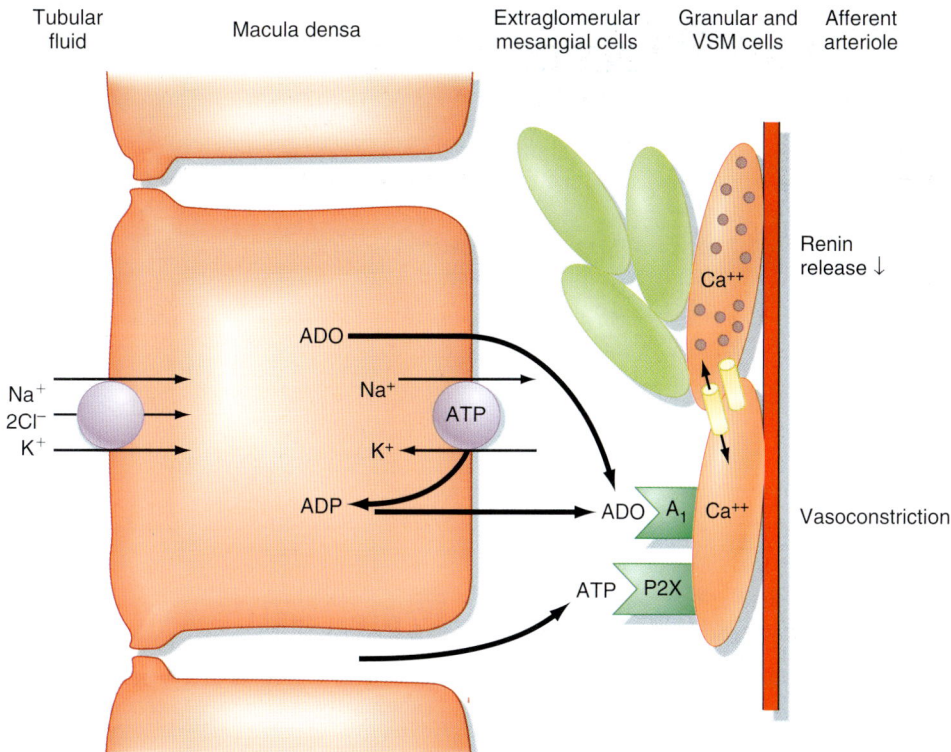

• **Fig. 33.20** Cellular mechanism whereby an increase in delivery of NaCl to the macula densa causes vasoconstriction of the afferent arteriole of the same nephron (i.e., tubuloglomerular feedback). An increase in GFR elevates [NaCl] in tubule fluid at the macula densa. This in turn enhances uptake of NaCl across the apical cell membrane of macula densa cells via the 1Na$^+$-1K$^+$-2Cl$^-$ (NKCC2) symporter, which leads to an increase in [ATP] and [adenosine] (ADO) release. ATP binds to P2X receptors and adenosine binds to adenosine A$_1$ receptors in the plasma membrane of smooth muscle cells surrounding the afferent arteriole, both of which increase intracellular [Ca^{++}]. The rise in [Ca^{++}] induces vasoconstriction of the afferent arteriole, thereby returning GFR to normal levels. Note that ATP and adenosine also inhibit renin release by granular cells in the afferent arteriole. This too results from an increase in intracellular [Ca^{++}] as a reflection of electrical coupling of the granular and vascular smooth muscle (VSM) cells. When GFR is reduced, [NaCl] in tubule fluid falls, as does uptake of NaCl into macula densa cells. This in turn decreases release of ATP and adenosine by the macula densa, which decreases intracellular [Ca^{++}] in smooth muscle cells and thereby increases GFR and stimulates renin release by granular cells. In addition a decrease in entry of NaCl into macula densa cells enhances production of PGE$_2$, which also stimulates renin secretion by granular cells. As discussed in detail in Chapter 34, renin increases plasma [angiotensin II], a hormone that enhances NaCl and water retention by the kidneys. (Modified from Persson AEG et al. *Acta Physiol Scand* 2004;181:471.)

2. Autoregulation is not perfect; RBF and GFR do change slightly as arterial blood pressure varies.
3. Despite autoregulation, RBF and GFR can be changed by several hormones and by alterations in sympathetic nerve activity that change in response to alterations in the extracellular fluid volume (ECFV) (Table 33.1).

Regulation of Renal Blood Flow and Glomerular Filtration Rate

Several factors and hormones affect both RBF and GFR (see Table 33.1). As already discussed, the myogenic mechanism and tubuloglomerular feedback play key roles in maintaining RBF and GFR constant when blood pressure is greater that 90 mm Hg and ECFV is in the normal range. However,

when the ECFV changes sympathetic nerves, angiotensin II, prostaglandins, NO, endothelin, bradykinin, ATP, and adenosine exert major control over RBF and GFR. Fig. 33.21 shows how changes in efferent and afferent arteriolar resistance, mediated by changes in the hormones listed in Table 33.1, modulate both RBF and GFR.

Sympathetic Nerves

The afferent and efferent arterioles are innervated by sympathetic neurons; however, sympathetic tone is minimal when ECFV is normal (see Chapter 35). When ECFV is reduced, sympathetic nerves release norepinephrine and dopamine, and circulating epinephrine (a catecholamine-like norepinephrine and dopamine) is secreted by the adrenal medulla. Norepinephrine and epinephrine cause vasoconstriction by

TABLE 33.1	Major Hormones That Influence Glomerular Filtration Rate and Renal Blood Flow		
	Stimulus	Effect on GFR	Effect on RBF
Vasoconstrictors			
Sympathetic nerves	↓ ECFV	↓	↓
Angiotensin II	↓ ECFV	↓	↓
Endothelin	↑ Stretch, A-II, bradykinin, epinephrine; ↓ ECFV	↓	↓
Vasodilators			
Prostaglandins (PGE₁, PGE₂, PGI₂)	↓ ECFV; ↑ shear stress, A-II	No change/↑	↑
Nitric oxide (NO)	↑ Shear stress, acetylcholine, histamine, bradykinin, ATP	↑	↑
Bradykinin	↑ Prostaglandins, ↓ ACE	↑	↑
Natriuretic peptides (ANP, BNP)	↑ ECFV	↑	No change

A-II, angiotensin II; ACE, angiotensin-converting enzyme; ECFV, extracellular fluid volume.

AT THE CELLULAR LEVEL

Tubuloglomerular feedback (TGF) is absent in mice that do not express the adenosine receptor (A₁). This underscores the importance of adenosine signaling in TGF. Studies have shown that when GFR increases and causes the concentration of NaCl in tubular fluid at the macula densa to rise, more NaCl enters cells via the 1Na⁺-1K⁺-2Cl⁻ symporter (NKCC2) located in the apical plasma membrane (see Fig. 33.20). Increased intracellular [NaCl] in turn stimulates release of ATP via ATP-conducting ion channels located in the basolateral membrane of macula densa cells. In addition, adenosine production is also enhanced. Adenosine binds to A₁ receptors and ATP binds to P2X receptors located on the plasma membrane of smooth muscle cells in the afferent arteriole. Both hormones increase intracellular [Ca⁺⁺], which causes vasoconstriction of the afferent artery and therefore a fall in GFR. Although adenosine is a vasodilator in most other vascular beds, it constricts the afferent arteriole in the kidney.

IN THE CLINIC

Individuals with **renal artery stenosis** (narrowing of the lumen of the artery) caused by atherosclerosis, for example, often have elevated systemic arterial blood pressure mediated by the renin-angiotensin system. Pressure in the renal artery proximal to the stenosis is increased, but pressure distal to the stenosis is normal or reduced. Autoregulation is important in maintaining RBF, P_GC, and GFR in the presence of this stenosis. Administration of drugs to lower systemic blood pressure also lowers the pressure distal to the stenosis; accordingly, RBF, P_GC, and GFR fall.

IN THE CLINIC

Significant **hemorrhage** decreases ECFV and arterial blood pressure and therefore activates sympathetic innervation of the kidneys via the baroreceptor reflex (Fig. 33.22). Norepinephrine causes intense vasoconstriction of the afferent and efferent glomerular arterioles and thereby decreases both RBF and GFR. The rise in sympathetic activity also increases release of epinephrine and angiotensin II, which cause further vasoconstriction and a fall in RBF. The rise in vascular resistance of the kidneys and other vascular beds increases total peripheral resistance. The resulting tendency for blood pressure to increase (blood pressure = cardiac output × total peripheral resistance) offsets the tendency of blood pressure to decrease in response to hemorrhage. Hence this system works to preserve arterial pressure at the expense of maintaining normal RBF and GFR.

binding to α₁-adrenoceptors, which are located mainly in afferent arterioles. Activation of α₁-adrenoceptors decreases RBF and GFR. Dehydration or strong emotional stimuli (e.g., fear, pain) also activate sympathetic nerves and reduce RBF and GFR.

Angiotensin II

Angiotensin II is produced systemically as well as locally within the kidneys. It constricts the afferent and efferent arterioles[d] and decreases both RBF and GFR. Fig. 33.22 shows how norepinephrine, epinephrine, and angiotensin II act together to decrease RBF and GFR and thereby

increase blood pressure and ECFV (e.g., as would occur with hemorrhage).

Prostaglandins

Prostaglandins do not play a major role in regulating RBF in healthy resting individuals. However, during

---

[d]The efferent arteriole is more sensitive than the afferent arteriole to angiotensin II. Therefore with low concentrations of angiotensin II, constriction of the efferent arteriole predominates, and GFR increases and RBF decreases. However, with high concentrations of angiotensin II, constriction of both afferent and efferent arterioles occurs, and GFR and RBF both decrease (see Fig. 33.21).

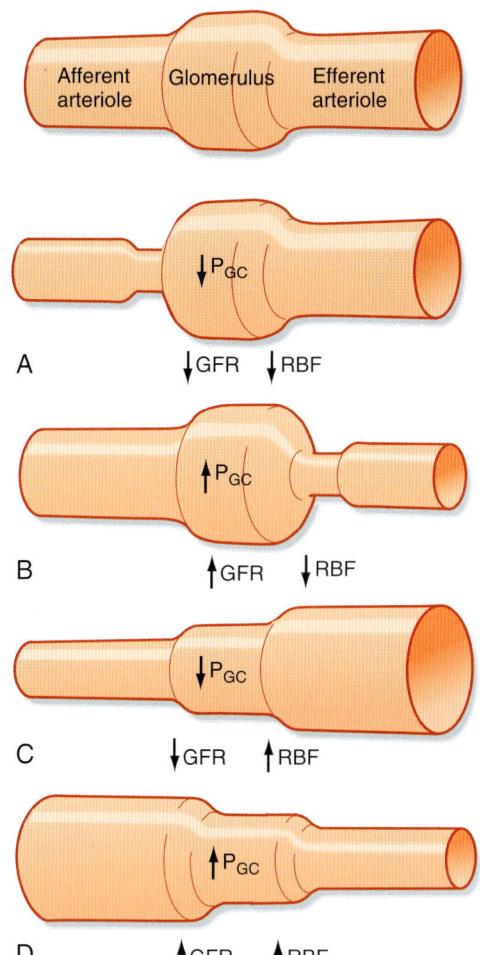

Fig. 33.21 Relationship between selective changes in resistance of either the afferent arteriole or the efferent arteriole on RBF and GFR. Constriction of either the afferent or efferent arteriole increases resistance, and according to Eq. 33.11 ($Q = \Delta P/R$), an increase in resistance (R) decreases flow (Q) (i.e., RBF). Dilation of either the afferent or afferent arteriole increases flow (i.e., RBF). Constriction of the afferent arteriole (A) decreases P_{GC} because less of the arterial pressure is transmitted to the glomerulus, thereby reducing GFR. In contrast, constriction of the efferent arteriole (B) elevates P_{GC} and thus increases GFR. Dilation of the efferent arteriole (C) decreases P_{GC} and thus decreases GFR. Dilation of the afferent arteriole (D) increases P_{GC} because more of the arterial pressure is transmitted to the glomerulus, thereby increasing GFR. (Modified from Rose BD, Rennke KG. *Renal Pathophysiology: The Essentials.* Baltimore: Williams & Wilkins; 1994.)

synthesis. Thus administration of these drugs during renal ischemia and hemorrhagic shock is contraindicated because, by blocking the production of prostaglandins, they decrease RBF and increase renal ischemia. Prostaglandins also play an increasingly important role in maintaining RBF and GFR as individuals age. Accordingly, NSAIDs can significantly reduce RBF and GFR in the elderly.

Nitric Oxide

NO, an endothelium-derived relaxing factor, is an important vasodilator under basal conditions, and it counteracts the vasoconstriction produced by angiotensin II and catecholamines. When blood flow increases, greater shear force acts on endothelial cells in the arterioles and increases production of NO. In addition a number of vasoactive hormones, including acetylcholine, histamine, bradykinin, and ATP, facilitate release of NO from endothelial cells. Increased production of NO causes dilation of the afferent and efferent arterioles in the kidneys. Whereas increased levels of NO decrease total peripheral resistance, inhibition of NO production increases total peripheral resistance.

🩺 IN THE CLINIC

Abnormal production of NO is observed in individuals with **diabetes mellitus** and **hypertension.** Excess renal NO production in diabetes may be responsible for glomerular hyperfiltration (i.e., increased GFR) and damage to the glomerulus, problems characteristic of this disease. Elevated NO levels increase glomerular capillary pressure secondary to a fall in resistance of the afferent arteriole. The ensuing hyperfiltration is thought to cause glomerular damage. The normal response to an increase in dietary salt intake includes stimulation of renal NO production, which prevents an increase in blood pressure. In some individuals, however, NO production may not increase appropriately in response to an elevation in salt intake, so blood pressure rises.

Endothelin

Endothelin is a potent vasoconstrictor secreted by endothelial cells of the renal vessels, mesangial cells, and distal tubular cells in response to angiotensin II, bradykinin, epinephrine, and endothelial shear stress. Endothelin causes profound vasoconstriction of the afferent and efferent arterioles and decreases GFR and RBF. Although this potent vasoconstrictor may not influence GFR and RBF in resting subjects, production of endothelin is elevated in a number of glomerular disease states (e.g., renal disease associated with diabetes mellitus).

Bradykinin

Kallikrein is a proteolytic enzyme produced in the kidneys. Kallikrein cleaves circulating kininogen to bradykinin, which is a vasodilator that acts by stimulating the release

pathophysiological conditions such as hemorrhage and reduced ECFV, prostaglandins (PGI_2, PGE_1, and PGE_2) are produced locally within the kidneys and serve to increase RBF without changing GFR. Prostaglandins increase RBF by dampening the vasoconstrictor effects of both sympathetic activation and angiotensin II. These effects are important because they prevent severe and potentially harmful vasoconstriction and renal ischemia. Synthesis of prostaglandins is stimulated by ECFV depletion and stress (e.g., surgery, anesthesia), angiotensin II, and sympathetic nerves. Nonsteroidal antiinflammatory drugs (NSAIDs), such as ibuprofen and naproxen, potently inhibit prostaglandin

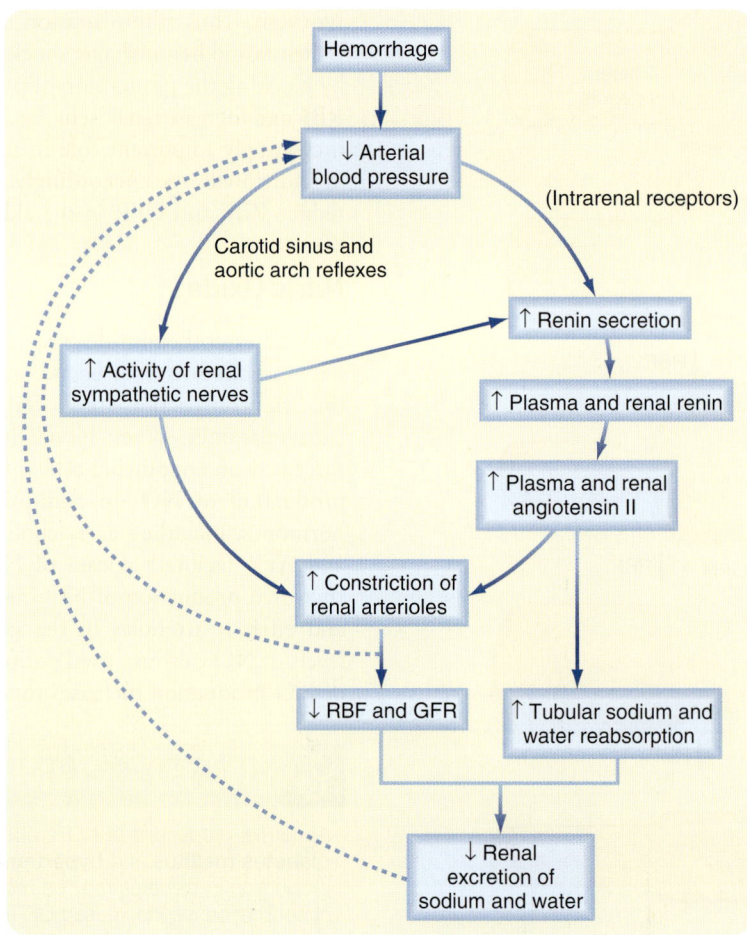

• **Fig. 33.22** Pathway by which hemorrhage activates renal sympathetic nerve activity and stimulates production of angiotensin II. (Modified from Vander AJ. *Renal Physiology.* 2nd ed. New York: McGraw-Hill; 1980.)

of NO and prostaglandins. Bradykinin increases RBF and GFR.

Adenosine

Adenosine is produced within the kidneys and causes vasoconstriction of the afferent arteriole, thereby reducing RBF and GFR. As previously mentioned, adenosine plays an important role in tubuloglomerular feedback.

Natriuretic Peptides

Secretion of atrial natriuretic peptide (ANP) by the cardiac atria and brain natriuretic peptide (BNP) by the cardiac ventricle increases when ECFV is expanded and myocardial wall tension is increased. Both ANP and BNP dilate the afferent arteriole and constrict the efferent arteriole. Therefore ANP and BNP produce a modest increase in GFR with little change in RBF.

Adenosine Triphosphate

Cells release ATP into the renal interstitial fluid. ATP can have bidirectional effects on both RBF and GFR. Under

some conditions, ATP constricts the afferent arteriole, reduces RBF and GFR, and may play a role in tubuloglomerular feedback. Under other conditions, ATP may stimulate NO production and have directionally opposite effects, increasing both RBF and GFR.

Glucocorticoids

Administration of therapeutic doses of glucocorticoids increases GFR and RBF.

Histamine

Local release of histamine modulates RBF during the resting state and during inflammation and injury. Histamine decreases the resistance of the afferent and efferent arterioles and thereby increases RBF without elevating GFR.

Dopamine

The proximal tubule produces the vasodilator substance dopamine. Dopamine has several actions within the kidney, such as increasing RBF and inhibiting renin secretion.

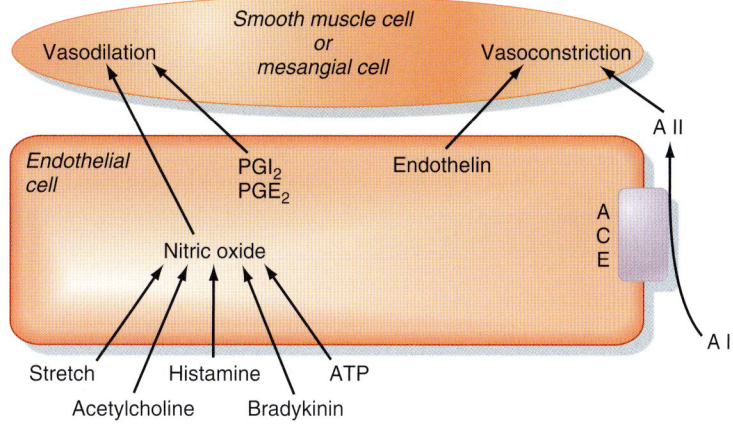

• **Fig. 33.23** Examples of the interactions of endothelial cells with smooth muscle and mesangial cells. ACE, angiotensin-converting enzyme; AI, angiotensin I; AII, angiotensin II. (Modified from Navar LG et al. *Physiol Rev* 1996;76:425.)

Hormones

Finally, as illustrated in Fig. 33.23, endothelial cells play an important role in regulating the resistance of the renal afferent and efferent arterioles by producing a number of paracrine hormones, including NO, prostacyclin (PGI_2), endothelin, and angiotensin II. These hormones regulate contraction or relaxation of smooth muscle cells in afferent and efferent arterioles and mesangial cells. Shear stress, acetylcholine, histamine, bradykinin, and ATP stimulate production of NO, which increases GFR and RBF. **Angiotensin-converting enzyme (ACE),** located on the surface of endothelial cells lining the afferent arteriole and glomerular capillaries, converts angiotensin I to angiotensin II, which decreases GFR and RBF. Angiotensin II is also produced locally by granular cells in the afferent arteriole and by proximal tubular cells. PGI_2 and PGE_2 release by endothelial cells is stimulated by both sympathetic nerve activity and angiotensin II, resulting in increased GFR and RBF. Finally, endothelin release from endothelial cells decreases both RBF and GFR.

IN THE CLINIC

ACE proteolytically inactivates the vasodilatory hormone bradykinin and converts angiotensin I, an inactive hormone, to angiotensin II, an active vasoconstrictive hormone. Thus ACE increases angiotensin II levels and decreases bradykinin levels. **ACE inhibitors** (e.g., lisinopril, enalapril, and captopril) are used clinically to reduce systemic blood pressure in patients with hypertension by decreasing angiotensin II levels and elevating bradykinin levels. Both effects lower systemic vascular resistance, reduce blood pressure, and decrease renal vascular resistance, thereby increasing GFR and RBF. **Angiotensin II receptor antagonists** (e.g., losartan) are also used to treat hypertension. As their name suggests, they block the binding of angiotensin II to the angiotensin II receptor (AT_1). These antagonists block the vasoconstrictor effects of angiotensin II on the afferent arteriole; thus they increase RBF and GFR. In contrast to ACE inhibitors, angiotensin II receptor antagonists do not inhibit kinin metabolism (e.g., bradykinin).

Key Concepts

1. The first step in urine formation is passive movement of a plasma ultrafiltrate from the glomerular capillaries into Bowman's space. The term *ultrafiltration* refers to passive movement of a plasma-like fluid that has a very low concentration of proteins from the glomerular capillaries into Bowman's space. The endothelial cells of glomerular capillaries are covered by a basement membrane that is surrounded by podocytes. The capillary endothelium, basement membrane, and foot processes of podocytes form the so-called filtration barrier.

2. The juxtaglomerular apparatus is one component of an important feedback mechanism (i.e., tubuloglomerular feedback) that regulates RBF and GFR. The structures that make up the juxtaglomerular apparatus include the macula densa, extraglomerular mesangial cells, and renin- and angiotensin II–producing granular cells.

3. Clinically, GFR is frequently estimated using measures of plasma [creatinine] or CrCl.

4. Autoregulation allows GFR and RBF to remain constant despite changes in arterial blood pressure between 90 and 180 mm Hg. When ECFV is altered, sympathetic nerves, catecholamines, angiotensin II, prostaglandins, NO, endothelin, natriuretic peptides, bradykinin, and adenosine exert substantial control over GFR and RBF.

Additional Reading

Journal Articles

Patel A. The primary cilium calcium channels and their role in flow sensing. *Pflugers Arch*. 2015;467:157-165.

Pollak MR, et al. The glomerulus: the sphere of influence. *Clin J Am Soc Nephrol*. 2014;9:1461-1469.

Schlondorff J. How many Achilles' heels does a podocyte have? An update on podocyte biology. *Nephrol Dial Transplant*. 2015;30:1091-1097.

Schnermann J. Concurrent activation of multiple vasoactive signaling pathways in vasoconstriction caused by tubuloglomerular feedback: a quantitative assessment. *Annu Rev Physiol*. 2015;77:301-322.

Taylor AE, Moore TM. Capillary fluid exchange. *Adv Physiol Educ*. 1999;22:s203-s210.

Book Chapters

Arendshorst WJ, Navar LG. Renal circulation and glomerular hemodynamics. In: Schrier RW, et al., eds. *Diseases of the Kidney and Urinary Tract*. 9th ed. Philadelphia: Lippincott Williams & Wilkins; 2012.

Dworkin LD, et al. The renal circulations. In: Taal MW, et al., eds. *Brenner and Rector's The Kidney*. 9th ed. Philadelphia: Saunders; 2012.

Kriz W, Kaissling B. Structural organization of the mammalian kidney. In: Alpern RJ, et al., eds. *Seldin and Giebisch's The Kidney: Physiology and Pathophysiology*. 5th ed. Philadelphia: Academic Press; 2012.

Lafayette RA, et al. Laboratory evaluation of renal function. In: Schrier RW, et al., eds. *Diseases of the Kidney and Urinary Tract*. 9th ed. Philadelphia: Lippincott Williams & Wilkins; 2012.

Madsen K, et al. Anatomy of the kidney. In: Taal MW, et al., eds. *Brenner and Rector's The Kidney*. 9th ed. Philadelphia: Saunders; 2012.

34

Solute and Water Transport Along the Nephron: Tubular Function

LEARNING OBJECTIVES

Upon completion of this chapter the student should be able to answer the following questions:

1. What three processes are involved in the production of urine?
2. What is the composition of "normal" urine?
3. What transport mechanisms are responsible for NaCl reabsorption by the nephron? Where are they located along the nephron?
4. How is water reabsorption "coupled" to NaCl reabsorption in the proximal tubule?
5. Why are solutes but not water reabsorbed by the thick ascending limb of Henle's loop?
6. What transport mechanisms are involved in secretion of organic anions and cations? What is the physiological relevance of these transport processes?
7. What is glomerulotubular balance, and what is its physiological importance?
8. What are the major hormones that regulate NaCl and water reabsorption by the kidneys? What is the nephron site of action of each hormone?
9. What is the aldosterone paradox?

The formation of urine involves three basic processes: (1) **ultrafiltration** of plasma by the glomerulus, (2) **reabsorption** of water and solutes from the ultrafiltrate, and (3) **secretion** of selected solutes into tubular fluid. Although an average of 115 to 180 L/day in women and 130 to 200 L/day in men of essentially protein-free fluid is filtered by the human glomeruli each day,[1] less than 1% of the filtered water and sodium chloride (NaCl) and variable amounts of other solutes are typically excreted in urine (Table 34.1). By the processes of reabsorption and secretion, the renal tubules determine the volume and composition of urine (Table 34.2), which in turn allows the

kidneys to precisely control the volume, osmolality, composition, and pH of the extracellular and intracellular fluid compartments. Transport proteins in cell membranes of the nephron mediate reabsorption and secretion of solutes and water in the kidneys. Approximately 5% to 10% of all human genes code for transport proteins, and genetic and acquired defects in transport proteins are the cause of many kidney diseases (Table 34.3). In addition, numerous transport proteins are important drug targets. This chapter discusses NaCl and water reabsorption, transport of organic anions and cations, the transport proteins involved in solute and water transport, and some of the factors and hormones that regulate NaCl transport. Details on acid-base transport and on K^+, Ca^{++}, and inorganic phosphate (P_i) transport and their regulation are provided in Chapters 35 through 37.

Solute and Water Reabsorption Along the Nephron

The general principles of solute and water transport across epithelial cells were discussed in Chapter 1.

Quantitatively, reabsorption of NaCl and water represent the major function of nephrons. Approximately 25,000 mEq/day of Na^+ and 179 L/day of water are reabsorbed by the renal tubules (see Table 34.1). In addition, renal transport of many other important solutes is linked either directly or indirectly to reabsorption of Na^+. In the following sections, the NaCl and water transport processes of each nephron segment and their regulation by hormones and other factors are presented.

Proximal Tubule

The proximal tubule reabsorbs approximately 67% of filtered water, Na^+, Cl^-, K^+, and most other solutes. In addition the proximal tubule reabsorbs virtually all the glucose and amino acids filtered by the glomerulus, as well as most of the HCO_3^-. The key element in proximal tubule reabsorption is Na^+,K^+-ATPase in the basolateral membrane. Reabsorption of every substance, including water, is linked in some manner to the operation of Na^+,K^+-ATPase.

---

[1]Normal glomerular filtration rate (GFR) averages 115–180 L/day in women and 130–200 L/day in men. Thus the volume of the ultrafiltrate represents a volume that is approximately 10 times that of the extracellular fluid volume (ECFV). For simplicity, we assume throughout the remainder of this section that GFR is 180 L/day.

TABLE
34.1
TABLE 34.1 **Filtration, Excretion, and Reabsorption of Water, Electrolytes, and Solutes by the Kidneys**

Substance	Measure	Filtered[a]	Excreted	Reabsorbed	% Filtered Load Reabsorbed
Water	L/day	180	1.5	178.5	99.2
Na^+	mEq/day	25,200	150	25,050	99.4
K^+	mEq/day	720	100	620	86.1
Ca^{++}	mEq/day	540	10	530	98.2
HCO_3^-	mEq/day	4320	2	4318	99.9+
Cl^-	mEq/day	18,000	150	17,850	99.2
Glucose	mmol/day	800	0	800	100.0
Urea	g/day	56	28	28	50.0

[a]The filtered amount of any substance is calculated by multiplying the concentration of that substance in the ultrafiltrate by the glomerular filtration rate (GFR); for example, the filtered load of Na^+ is calculated as $[Na^+]$ultrafiltrate (140 mEq/L) × GFR (180 L/day) = 25,200 mEq/day.

TABLE 34.2 **Composition of Urine**

Substance	Concentration
Na^+	50–130 mEq/L
K^+	20–70 mEq/L
Ammonium (NH_4^+)	30–50 mEq/L
Ca^{++}	5–12 mEq/L
Mg^{++}	2–18 mEq/L
Cl^-	50–130 mEq/L
Inorganic phosphate (P_i)	20–40 mEq/L
Urea	200–400 mmol/L
Creatinine	6–20 mmol/L
pH	5.0–7.0
Osmolality	500–800 mOsm/kg H_2O
Glucose	0
Amino acids	0
Protein	0
Blood	0
Ketones	0
Leukocytes	0
Bilirubin	0

The composition and volume of urine can vary widely in the healthy state. These values represent average ranges. Normal water excretion typically ranges between 0.5 and 1.5 L/day.
Data from Valtin HV. *Renal Physiology*. 2nd ed. Boston: Little, Brown; 1983.

Na⁺ Reabsorption

Na^+ is reabsorbed by different mechanisms in the first and the second halves of the proximal tubule. In the first half of the proximal tubule, Na^+ is reabsorbed primarily with bicarbonate (HCO_3^-) and a number of other solutes (e.g., glucose, amino acids, P_i, lactate). In contrast, in the second half, Na^+ is reabsorbed mainly with Cl^-. This disparity is mediated by differences in the Na^+ transport systems in the first and second halves of the proximal tubule and by differences in the composition of tubular fluid at these sites. In absolute terms the first half of the proximal tubule reabsorbs significantly more Na^+ than the second half.

In the first half of the proximal tubule, Na^+ uptake into the cell is coupled with either H^+ or organic solutes, including glucose (Fig. 34.1). Specific transport proteins mediate entry of Na^+ into the cell across the apical membrane. For example, the Na^+/H^+ antiporter, NHE3, (see Fig. 34.1A) couples entry of Na^+ with extrusion of H^+ from the cell. H^+ secretion results in reabsorption of sodium bicarbonate ($NaHCO_3$) (see Chapter 37). Na^+ also enters proximal tubule cells via several symporter mechanisms, including Na^+/glucose (SGLT2), Na^+/amino acid, Na^+/P_i, and Na^+/lactate (see Fig. 34.1B). The glucose and other organic solutes that enter the cell with Na^+ leave the cell across the basolateral membrane via passive transport mechanisms. Any Na^+ that enters the cell across the apical membrane leaves the cell and enters the blood via Na^+,K^+-ATPase. Thus reabsorption of Na^+ in the first half of the proximal tubule is coupled to that of HCO_3^- and a number of organic molecules, and this generates a negative transepithelial voltage across the proximal tubule that provides the driving force for the paracellular reabsorption of Cl^-. Reabsorption of many organic molecules, including glucose and lactate, is so avid they are almost completely removed from the tubular fluid in the first half of the proximal tubule (Fig. 34.2). Reabsorption of $NaHCO_3$ and Na^+–organic solutes across the proximal tubule establishes a transtubular osmotic gradient (i.e., the osmolality of the interstitial fluid bathing the basolateral side of the cells is a few mOsm/L higher than the osmolality of tubule fluid) that provides the driving force for the passive reabsorption of water by osmosis. Because more water than Cl^- is reabsorbed in the first half of the proximal tubule, the $[Cl^-]$ in tubular fluid rises along the length of the proximal tubule (see Fig. 34.2).

In the second half of the proximal tubule, Na^+ reabsorption is largely accompanied by Cl^- reabsorption via both transcellular and paracellular pathways (Fig. 34.3). Na^+ is primarily reabsorbed with Cl^- rather than organic solutes or HCO_3^- as the accompanying anion because the Na^+ transport mechanisms in the second half of the proximal tubule

TABLE 34.3 — Selected Monogenic Renal Diseases Involving Transport Proteins

Diseases	Mode of Inheritance	Gene	Transport Protein	Nephron Segment	Phenotype
Cystinuria type I	AR	SLC3A1, SLC7A9	Amino acid symporters	Proximal tubule	Increased excretion of basic amino acids, nephrolithiasis (kidney stones)
Proximal renal tubular acidosis (RTA)	AR	SLC4A4	Na^+/HCO_3^- symporter	Proximal tubule	Hyperchloremic metabolic acidosis
X-linked nephrolithiasis (Dent's disease)	XLR	CLCN, OCRL1	Chloride channel	Distal tubule	Hypercalciuria, nephrolithiasis
Bartter syndrome	AR-type I	SLC12A1	$Na^+/K^+/2Cl^-$ symporter	TAL	Hypokalemia, metabolic alkalosis, hyperaldosteronism
	AR-type II	KCNJ1	ROMK potassium channel	TAL	Hypokalemia, metabolic alkalosis, hyperaldosteronism
	AR-type III	CLCNKB	Chloride channel (basolateral membrane)	TAL	Hypokalemia, metabolic alkalosis, hyperaldosteronism
	AR-type IV	BSND, CLCNKA CLCNKB	Subunit of chloride channel, chloride channels	TAL	Hypokalemia, metabolic alkalosis, hyperaldosteronism
Hypomagnesemia-hypercalciuria syndrome	AR	CLDN16	Claudin-16, also known as *paracellin 1*	TAL	Hypomagnesemia-hypercalciuria, nephrolithiasis
Gitelman syndrome	AR	SLC12A3	Thiazide-sensitive Na^+/Cl^- symporter	Distal tubule	Hypomagnesemia, hypokalemic metabolic alkalosis, hypocalciuria, hypotension
Pseudohypoaldosteronism type I	AR	SCNN1A, SCNN1B, and SCNN1G	α, β, and γ subunit of ENaC	Collecting duct	Increased excretion of Na^+, hyperkalemia, hypotension
Pseudohypoaldosteronism type II	AD	MLR	Mineralocorticoid receptor	Collecting duct	Increased excretion of Na^+ hyperkalemia, hypotension
Liddle syndrome	AD	SCNN1B, SCNN1G	β and γ subunit of ENaC	Collecting duct	Decreased excretion of Na^+, hypertension
Nephrogenic diabetes insipidus (NDI) type II	AR/AD	AQP2	Aquaporin 2 water channel	Collecting duct	Polyuria, polydipsia, plasma hyperosmolality
Distal renal tubular acidosis	AD/AR	SLC4A1	Cl^-/HCO_3^- antiporter	Collecting duct	Metabolic acidosis, hypokalemia, hypercalciuria, nephrolithiasis
Distal renal tubular acidosis	AR	ATP6N1B	Subunit of H^+-ATPase	Collecting duct	Metabolic acidosis, hypokalemia, hypercalciuria, nephrolithiasis

There are over 300 different solute transporter genes that form the so-called SLC (solute carrier) family of genes.
AD, autosomal dominant; AR, autosomal recessive; TAL, thick ascending limb of Henle's loop; XLR, X-linked recessive.
Modified from Nachman RH, Glassock RJ. *NephSAP* 2010;9(3).

differ from those in the first half, and because the tubular fluid that enters the second half contains very little glucose or amino acids. In addition the high $[Cl^-]$ (140 mEq/L) in tubule fluid, which is due to preferential reabsorption of Na^+ with HCO_3^- and organic solutes in the first half of the proximal tubule, facilitates reabsorption of Cl^- with Na^+.

The mechanism of transcellular Na^+ reabsorption in the second half of the proximal tubule is shown in Fig. 34.3. Na^+ enters the cell across the luminal membrane primarily via the parallel operation of a Na^+/H^+ antiporter (NHE3)

IN THE CLINIC

Fanconi syndrome, a renal disease that is either hereditary or acquired, results from an impaired ability of the proximal tubule to reabsorb HCO_3^-, P_i, amino acids, glucose, and low-molecular-weight proteins. Because other downstream nephron segments cannot reabsorb these solutes and protein, Fanconi syndrome results in increased urinary excretion of HCO_3^-, amino acids, glucose, P_i, and low-molecular-weight proteins.

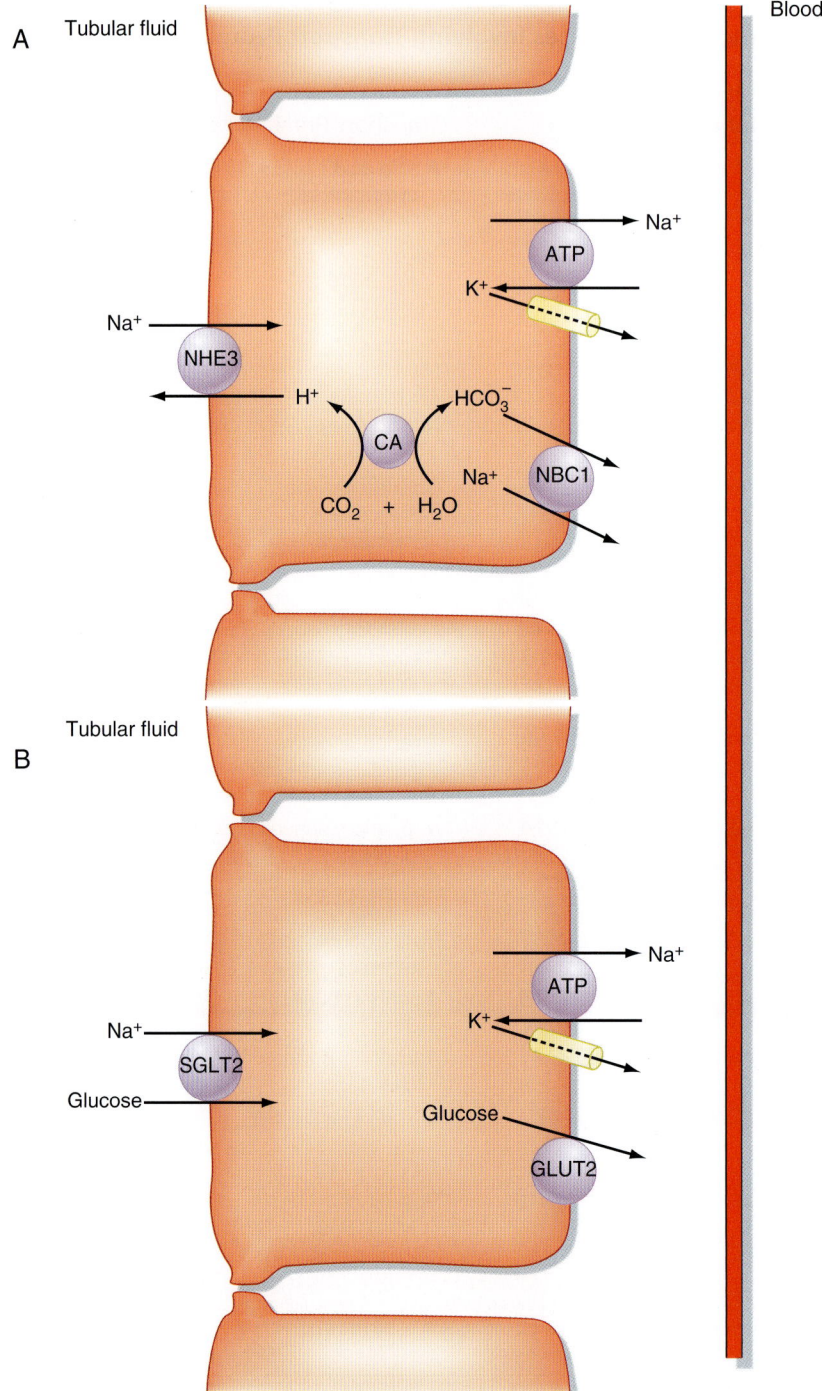

• **Fig. 34.1** Na^+ transport processes in the first half of the proximal tubule. These transport mechanisms are present in all cells in the first half of the proximal tubule but are separated into different cells to simplify the discussion. A, Operation of the Na^+/H^+ antiporter (NHE3) in the apical membrane and the Na^+,K^+-ATPase and HCO_3^- transporters, including the Na^+/HCO_3^- symporter (NBC1; see also Chapter 37) in the basolateral membrane mediates reabsorption of $NaHCO_3$. Carbon dioxide and water combine inside the cells to form H^+ and HCO_3^- in a reaction facilitated by the enzyme carbonic anhydrase (CA). B, Operation of the Na^+/glucose symporter (SGLT2) in the apical membrane, in conjunction with Na^+,K^+-ATPase and the glucose transporter (GLUT2) in the basolateral membrane, mediates Na^+-glucose reabsorption. Inactivating mutations in the *GLUT2* gene lead to decreased glucose reabsorption in the proximal tubule and glucosuria (i.e., glucose in the urine). Though not shown, Na^+ reabsorption is also coupled with other solutes, including amino acids, P_i, and lactate. Reabsorption of these solutes is mediated by the Na^+/amino acid, Na^+/P_i, and Na^+/lactate symporters, respectively, located in the apical membrane and the Na^+,K^+-ATPase, amino acid, P_i, and lactate transporters, respectively, located in the basolateral membrane. Three classes of amino acid transporters have been identified in the proximal tubule: two that transport Na^+ in conjunction with either acidic or basic amino acids and one that does not require Na^+ and transports basic amino acids.

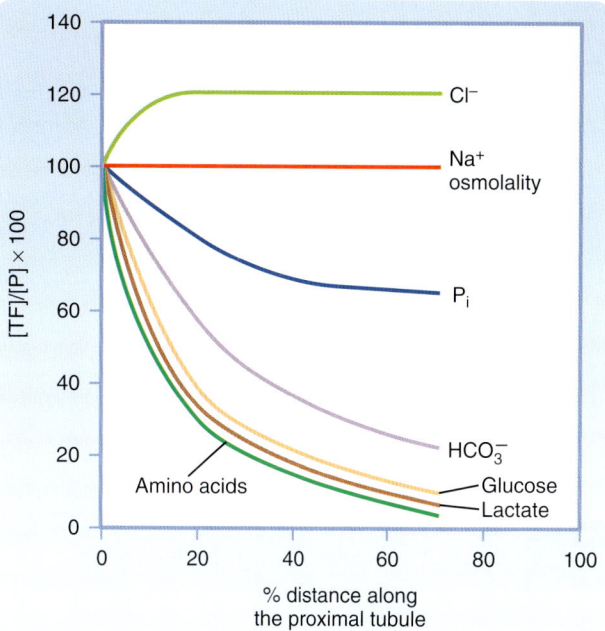

• Fig. 34.2 Concentration of solutes in tubule fluid as a function of length along the proximal tubule. [TF] is the concentration of the substance in tubular fluid; [P] is the concentration of the substance in plasma. Values above 100 indicate that relatively less of the solute than water is reabsorbed, and values below 100 indicate that relatively more of the substance than water is reabsorbed.

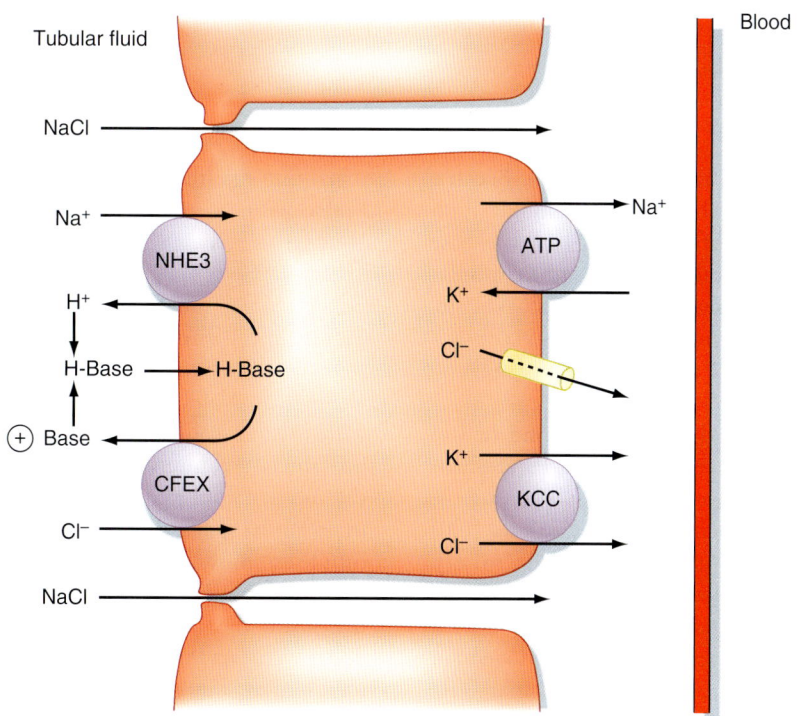

• Fig. 34.3 Na^+ transport processes in the second half of the proximal tubule. Na^+ and Cl^- enter the cell across the apical membrane through the operation of parallel Na^+/H^+ (NHE3) and Cl^--base (e.g., formate, oxalate, and bicarbonate) antiporters (CFEX). More than one Cl^--base antiporter is involved in this process, but only one is depicted. The secreted H^+ and base combine in the tubular fluid to form an H-base complex that can recycle across the plasma membrane. Accumulation of the H-base complex in tubular fluid establishes a H-base concentration gradient that favors H-base recycling across the apical plasma membrane into the cell. Inside the cell, H^+ and the base dissociate and recycle back across the apical plasma membrane. The net result is uptake of NaCl across the apical membrane. The base may be hydroxide ions (OH^-), formate (HCO_2^-), oxalate, HCO_3^-, or sulfate. The positive transepithelial voltage in the lumen, indicated by the plus sign inside the circle in the tubular lumen, is generated by diffusion of Cl^- (lumen to blood) across the tight junction. The high $[Cl^-]$ of tubular fluid provides the driving force for diffusion of Cl^-. Some glucose is also reabsorbed in the second half of the proximal tubule by a mechanism similar to that described in the first half of the proximal tubule, except that the Na^+/glucose symporter (*SGLT1* gene) transports $2Na^+$ with one glucose and has higher affinity and lower capacity than the Na^+/glucose symporter in the first part of the proximal tubule, depicted in Fig. 34.1. In addition, glucose exits the cell across the basolateral membrane via GLUT1 rather than via GLUT2 as in the first part of the proximal tubule (not shown). KCC, KCl symporter.

TABLE 34.4   NaCl Transport Along the Nephron

Segment	% Filtered NaCl Reabsorbed	Mechanism of Na$^+$ Entry Across Apical Membrane	Major Regulatory Hormones
Proximal tubule	67	Na$^+$/H$^+$ antiporter (NHE$_3$), Na$^+$ symporter with amino acids and organic solutes, paracellular	Angiotensin II Norepinephrine Epinephrine Dopamine
Loop of Henle	25	1 Na$^+$/1K$^+$/2Cl$^-$ symporter	Aldosterone Angiotensin II
Distal tubule	≈5	NaCl symporter	Aldosterone Angiotensin II
Late distal tubule and collecting duct	≈3	ENaC Na$^+$ channels	Aldosterone, ANP, BNP, urodilatin, uroguanylin, guanylin, angiotensin II

TABLE 34.5   Water Transport Along the Nephron

Segment	% Filtrate Reabsorbed	Mechanism of Water Reabsorption	Hormones That Regulate Water Permeability
Proximal tubule	67	Passive	None
Loop of Henle	15	Descending thin limb only; passive	None
Distal tubule	0	No water reabsorption	None
Late distal tubule and collecting duct	≈8–17	Passive	AVP, ANP[a], BNP[a]

[a]Atrial natriuretic peptide (ANP) and brain natriuretic peptide (BNP) inhibit vasopressin (AVP)-stimulated water permeability.

and one or more Cl$^-$-base antiporters (e.g., CFEX). Because the secreted H$^+$ and base combine in the tubular fluid and reenter the cell, operation of the Na$^+$/H$^+$ and Cl$^-$-base antiporters is equivalent to uptake of NaCl from tubular fluid into the cell. Na$^+$ leaves the cell via Na$^+$,K$^+$-ATPase, and Cl$^-$ leaves the cell and enters the blood via a K$^+$/Cl$^-$ symporter (KCC) and a Cl$^-$ channel in the basolateral membrane.

Some NaCl is also reabsorbed across the second half of the proximal tubule via a **paracellular route.** Paracellular NaCl reabsorption occurs because the rise in [Cl$^-$] in tubule fluid in the first half of the proximal tubule creates a [Cl$^-$] gradient (140 mEq/L in the tubule lumen and 105 mEq/L in the interstitium). This concentration gradient favors diffusion of Cl$^-$ from the tubular lumen across the tight junctions into the lateral intercellular space. Movement of the negatively charged Cl$^-$ results in the tubular fluid becoming positively charged relative to blood. This positive transepithelial voltage causes diffusion of positively charged Na$^+$ out of the tubular fluid across the tight junction into blood. Thus in the second half of the proximal tubule, some Na$^+$ and Cl$^-$ are reabsorbed across the tight junctions via passive diffusion.

In summary, reabsorption of Na$^+$ and Cl$^-$ in the proximal tubule occurs via both paracellular and transcellular pathways. Approximately 67% of the NaCl filtered each day is reabsorbed in the proximal tubule. Of this amount, two-thirds moves across the transcellular pathway, whereas the remaining one-third moves across the paracellular pathway (Table 34.4).

Water Reabsorption

The proximal tubule reabsorbs 67% of the filtered water (Table 34.5). The driving force for water reabsorption is a transtubular osmotic gradient established by reabsorption of solute (e.g., NaCl, Na$^+$-glucose). Reabsorption of Na$^+$ along with organic solutes, HCO$_3^-$, and Cl$^-$ from tubular fluid into the lateral intercellular spaces reduces the osmolality of the tubular fluid and increases the osmolality of the lateral intercellular space. The osmotic gradient across the proximal tubule established by these transport processes is only a few mOsm/L (Fig. 34.4). Because the proximal tubule is highly permeable to water, primarily owing to expression of aquaporin water channels (AQP1) in the apical and basolateral membranes, water is reabsorbed across cells by osmosis. In addition the tight junctions in the proximal tubule are also water permeable, so some water is also reabsorbed across the paracellular pathway between proximal tubular cells. Accumulation of fluid and solutes within the lateral intercellular space increases hydrostatic pressure in this compartment. The increased hydrostatic pressure forces fluid and solutes into the capillaries.[2] Thus water reabsorption follows solute

---

[2]In addition, protein oncotic pressure in the peritubular capillaries (π_{pc}) is elevated because of the process of glomerular filtration (see Chapter 33). The elevated π_{pc} facilitates uptake of fluid and solute into the capillary.

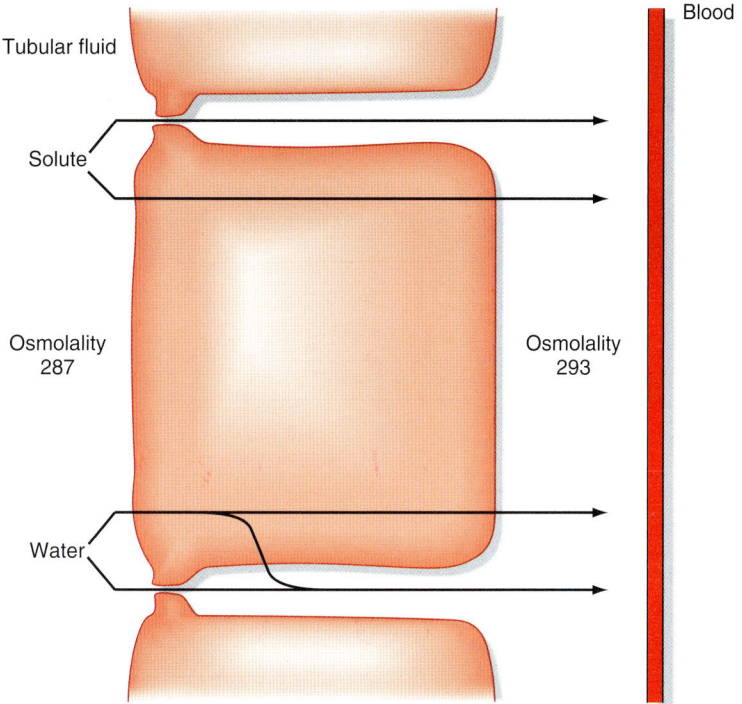

• **Fig. 34.4** Routes of reabsorption of water and solute across the proximal tubule. Transport of solutes, including Na^+, Cl^-, and organic solutes, into the lateral intercellular space increases the osmolality of this compartment, which establishes the driving force for osmotic reabsorption of water across the proximal tubule. This occurs because some Na^+,K^+-ATPase and some transporters of organic solutes, HCO_3^-, and Cl^- are located on the lateral cell membranes and deposit these solutes between cells. Furthermore, some NaCl also enters the lateral intercellular space via diffusion across the tight junction (i.e., paracellular pathway). An important consequence of osmotic water flow across the transcellular and paracellular pathways in the proximal tubule is that some solutes, especially K^+ and Ca^{++}, are entrained in the reabsorbed fluid and thereby reabsorbed by the process of solvent drag.

reabsorption in the proximal tubule. The reabsorbed fluid is slightly hyperosmotic relative to plasma. However, this difference in osmolality is so small it is commonly said that proximal tubule reabsorption is isosmotic (i.e., ≈67% of both the filtered load of solute and water are reabsorbed). Indeed, there is little difference in the osmolality of tubular fluid at the start and end of the proximal tubule. An important consequence of osmotic water flow across the proximal tubule is that some solutes, especially K^+ and Ca^{++}, are entrained in the reabsorbed fluid and thereby reabsorbed by the process of solvent drag (see Fig. 34.4). Reabsorption of virtually all organic solutes, Cl^- and other ions, and water is coupled to Na^+ reabsorption. Therefore changes in Na^+ reabsorption influence reabsorption of water and other solutes by the proximal tubule. This point will be discussed later, notably in Chapter 35, and is especially relevant during volume depletion when increased Na^+ reabsorption by the proximal tubule is accompanied by a parallel increase in HCO_3^- reabsorption, which can contribute to metabolic alkalosis (i.e., volume contraction alkalosis).

Protein Reabsorption

Proteins filtered by the glomerulus are reabsorbed in the proximal tubule. As mentioned previously, peptide hormones, small proteins, and small amounts of larger proteins

such as albumin are filtered by the glomerulus. Overall, only a small percentage of proteins cross the glomerulus and enter Bowman's space (i.e., the concentration of proteins in the glomerular ultrafiltrate is only ≈ 40 mg/L). However, the total amount of protein filtered per day is significant because the glomerular filtration rate (GFR) is so high:

Equation 34.1

$$\text{Filtered protein} = GFR \times [\text{Protein}] \text{ in the ultrafiltrate}$$
$$\text{Filtered protein} = 180 \text{ L/day} \times 40 \text{ mg/L}$$
$$= 7200 \text{ mg/day, or } 7.2 \text{ g/day}$$

Filtered proteins are reabsorbed in the proximal tubule by endocytosis either as intact proteins or after being partially degraded by enzymes on the surface of proximal tubule cells. Once the proteins and peptides are inside the cell, enzymes digest them into their constituent amino acids, which then leave the cell across the basolateral membrane by transport proteins and are returned to the blood. Normally this mechanism reabsorbs virtually all the proteins filtered, and hence the urine is essentially protein free. However, because the mechanism is easily saturated, an increase in filtered proteins can result in **proteinuria** (appearance of protein in urine). Disruption of the glomerular filtration barrier to proteins increases the filtration of proteins and results in proteinuria, which is frequently seen with kidney disease.

Secretion of Organic Anions and Organic Cations

Cells of the proximal tubule also secrete organic anions and organic cations into the tubule fluid. Secretion of organic anions and cations by the proximal tubule plays a key role in regulating the plasma levels of xenobiotics (e.g., a variety of antibiotics, diuretics, statins, antivirals, antineoplastics, immunosuppressants, neurotransmitters, and nonsteroidal antiinflammatory drugs [NSAIDs]) and toxic compounds derived from endogenous and exogenous sources. Many of the organic anions and cations (Boxes 34.1 and 34.2) secreted by the proximal tubule are end products of metabolism that circulate in plasma. Many of these organic compounds are bound to plasma proteins and thus are not readily filtered. Therefore only a small fraction of these potentially toxic substances are eliminated from the body by excretion resulting from filtration alone. Thus secretion of organic anions and cations, including many toxins from the peritubular capillary into the tubular fluid, promote elimination of these compounds from plasma entering the kidneys. Hence these substances are removed from plasma by both filtration and secretion. It is important to note that when kidney function is reduced by disease, urinary excretion of organic anions and cations is severely reduced, which can lead to increased plasma levels of xenobiotics and potentially toxic accumulation of organic anions and cations.

 AT THE CELLULAR LEVEL

Water channels called **aquaporins (AQPs)** mediate transcellular reabsorption of water across many nephron segments. To date, 13 aquaporins have been identified. The AQP family is divided into two groups based on their permeability characteristics. One group (aquaporins) is permeable to water (AQP0, AQP1, AQP2, AQP4, AQP5, AQP6, AQP8, AQP11, and AQP12). The other group (aquaglyceroporins) is permeable to water and small solutes, especially glycerol (AQP3, AQP7, AQP9, AQP10). Aquaporins form tetramers in the plasma membrane of cells, with each subunit forming a water channel. In the kidneys, AQP1 is expressed in the apical and basolateral membranes of the proximal tubule and in portions of the descending thin limb of Henle's loop. The importance of AQP1 in renal water reabsorption is underscored by studies in which the *AQP1* gene was "knocked out" in mice. These mice exhibit increased urine output (polyuria) and reduced ability to concentrate urine. In addition the osmotic water permeability of the proximal tubule is fivefold less in mice lacking APQ1 than in normal mice. AQP7 and AQP8 are also expressed in the proximal tubule. AQP2 is expressed in the apical plasma membrane of principal cells in the collecting duct, and its abundance in the membrane is regulated by arginine vasopressin (AVP) (see Chapter 35). AQP3 and AQP4 are expressed in the basolateral membrane of principal cells in the collecting duct, and mice deficient in these AQPs (i.e., AQP3 and AQP4 knockout mice) have defects in the ability to concentrate urine (see Chapter 35). AQPs are also expressed in many other organs in the body, including the lung, eye, skin, secretory glands, and brain, where they play key physiological roles. For example, AQP4 is expressed in cells that form the blood-brain barrier. Knockout of AQP4 affects the water permeability of the blood-brain barrier such that brain edema is reduced in AQP4-deficient mice after acute water loading and subsequent development of hyponatremia.

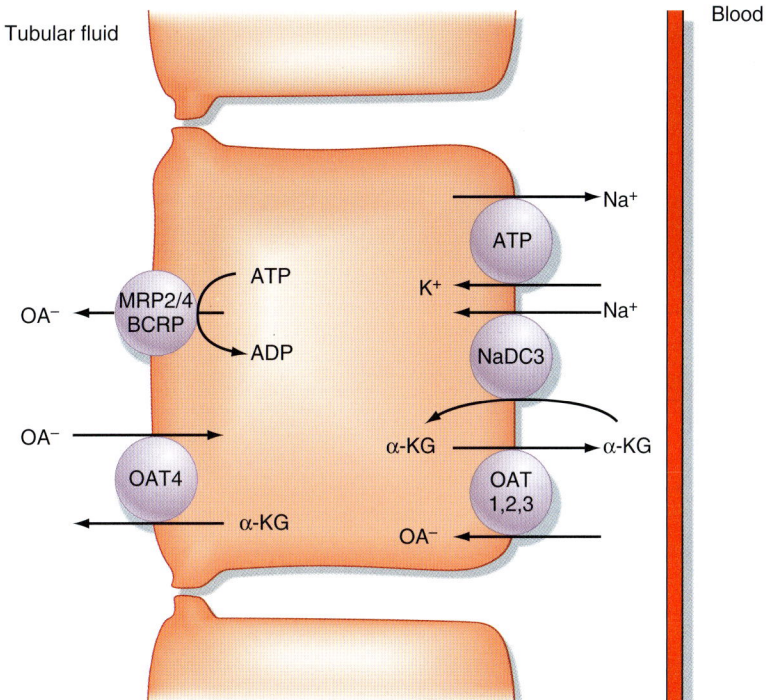

• **Fig. 34.5** Secretion of organic anion (OA^-) across the proximal tubule. OA^-s enter the cell across the basolateral membrane by one of three OA^-/α-ketoglutarate (α-KG) antiporter mechanisms (organic anion transporters, OAT1, OAT2, OAT3). Uptake of α-KG into the cell against its chemical concentration gradient is driven by movement of Na^+ into the cell via the Na^+-dicarboxylate transporter (NaDC3). The $[Na^+]$ inside the cell is low because of the Na^+,K^+-ATPase in the basolateral membrane, which transports Na^+ out of the cell in exchange for K^+. The α-KG recycles across the basolateral membrane on the OATs in exchange for OA^-. OA^-s leave the cell across the apical membrane by multidrug drug resistance proteins (MRP2 and 4), and by breast cancer resistance protein (BCRP), which require ATP. OAT4 in the apical membrane reabsorbs urate, an organic anion.

🧬 AT THE CELLULAR LEVEL

The endocytosis of proteins by the proximal tubule is mediated by apical membrane proteins that specifically bind proteins and peptides in tubule fluid. These receptors, called **multiligand endocytic receptors,** can bind a wide range of peptides and proteins and thereby mediate their endocytosis. **Megalin** and **cubilin** mediate protein and peptide endocytosis in the proximal tubule. Both are glycoproteins, with megalin being a member of the low-density lipoprotein receptor gene family.

🩺 IN THE CLINIC

Urinalysis is an important and routine tool for detection of kidney disease. A thorough analysis of urine includes macroscopic, microscopic, and biochemical assessments. This is performed by visual assessment of the urine, microscopic examination of urinary sediment, and biochemical evaluation of urinary composition using dipstick reagent strips. The dipstick test is both inexpensive and fast (i.e., <5 minutes) and tests urine for both pH and the presence of many substances (e.g., bilirubin, blood, glucose, ketones, protein). It is normal to find trace amounts of protein in urine, particularly concentrated urine. Urinary proteins are derived from two principal sources: (1) filtration exceeding the reabsorptive capacity of the proximal tubule and (2) synthesis and secretion of **Tamm-Horsfall glycoprotein** by the thick ascending limb of Henle's loop. Because the mechanism for protein reabsorption is "upstream" of the thick ascending limb (i.e., in the proximal tubule), the secreted Tamm-Horsfall glycoprotein appears in urine. However, proteinuria in greater than trace amounts is often indicative of renal disease.

Fig. 34.5 illustrates the mechanisms of organic anion (OA^-) transport across the proximal tubule. These secretory pathways have maximum transport rates, low specificity (i.e., they transport many OA^-s), and are responsible for secretion of the OA^-s listed in Box 34.1. OA^-s are taken up into the cell across the basolateral membrane against their chemical gradient in exchange for α-ketoglutarate (α-KG) via several OA^-/α-KG antiporters, including OAT1, OAT2, and OAT3. α-KG accumulates inside the cells via metabolism of glutamate and by a Na^+/α-KG symporter (i.e., the Na^+/dicarboxylate transporter [NaDC3]) also present in the basolateral membrane. Thus uptake of OA^- into the cell against an electrochemical gradient is coupled to the

exit of α-KG out of the cell, down its chemical gradient generated by the Na^+/α-KG symporter mechanism. The exit of OA^-s across the luminal membrane into the tubular fluid are mediated by multidrug resistance proteins 2 and 4 (MRP2/4) and breast cancer resistance protein 1

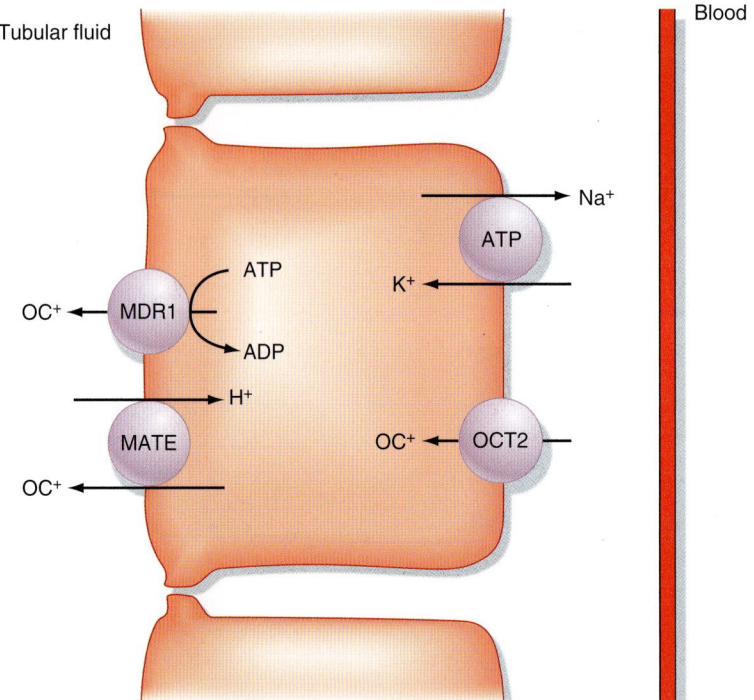

• **Fig. 34.6** Organic cation (OC$^+$) secretion across the proximal tubule. OC$^+$s enter the cell across the basolateral membrane primarily by OCT2. Uptake of OC$^+$s into the cell against their chemical concentration gradient is driven by the cell-negative potential difference. OC$^+$s leave the cell across the apical membrane in exchange with H$^+$ by electrically neutral multidrug and toxin transporters (MATE1 and MATE2-K) and by multidrug resistance protein (MDR1), which requires ATP.

(BCRP), which require adenosine triphosphate (ATP) for their operation. Recent studies reveal that OAT4 mediates reabsorption of the organic anion urate, the end product of purine catabolism, by the proximal tubule (see Fig. 34.5).

Fig. 34.6 illustrates the mechanism of organic cation (OC$^+$) transport across the proximal tubule. Organic cations, including xenobiotics such as the antidiabetic agent metformin, the antiviral agent lamivudine, and the anticancer drug oxaliplatin, and many important mono-amine neurotransmitters including dopamine, epinephrine, histamine, and norepinephrine are secreted by the proximal tubule. Organic cations are taken up into the cell across the basolateral membrane, primarily by the organic cation transporter 2 (OCT2). Uptake of organic cations is driven by the magnitude of the cell-negative potential difference across the basolateral membrane. Organic cation transport across the luminal membrane into the tubular fluid, which is the rate-limiting step in secretion, is mediated primarily by electroneutral multidrug and toxin extrusion transporters (MATEs) and MDR1 (also known as *P-glycoprotein*), which requires ATP for its operation. These transport mechanisms are nonspecific, and several organic cations usually compete for secretion via a given transport pathway.

Henle's Loop

Henle's loop reabsorbs approximately 25% of the filtered NaCl and 15% of the filtered water. Reabsorption of NaCl

IN THE CLINIC

Because many organic anions compete for the same secretory pathways, elevated plasma levels of one transported anion often inhibit secretion of the others. For example, infusing *p*-aminohippuric acid (PAH) can reduce secretion of penicillin by the proximal tubule. Because the kidneys are responsible for eliminating penicillin, infusion of PAH into individuals receiving penicillin reduces penicillin excretion and thereby extends its biological half-life. In World War II, when penicillin was in short supply, hippurates were given with penicillin to extend its therapeutic effect. Similar competition is observed for organic cation secretion by the proximal tubule, and elevated plasma levels of one transported cation species can inhibit secretion of the other competing cations. For example, the histamine H$_2$ antagonist cimetidine used to treat gastric ulcers is secreted via organic cation transport mechanisms in the proximal tubule. If cimetidine is given to patients receiving procainamide (a drug used to treat cardiac arrhythmias), cimetidine reduces urinary excretion of procainamide (also an organic cation) by direct competition for a common secretory pathway. As a consequence, coadministration of cationic drugs competing for the same pathway can increase the plasma concentration of both drugs to levels much higher than those observed when the drugs are given alone. This effect can lead to drug toxicity.

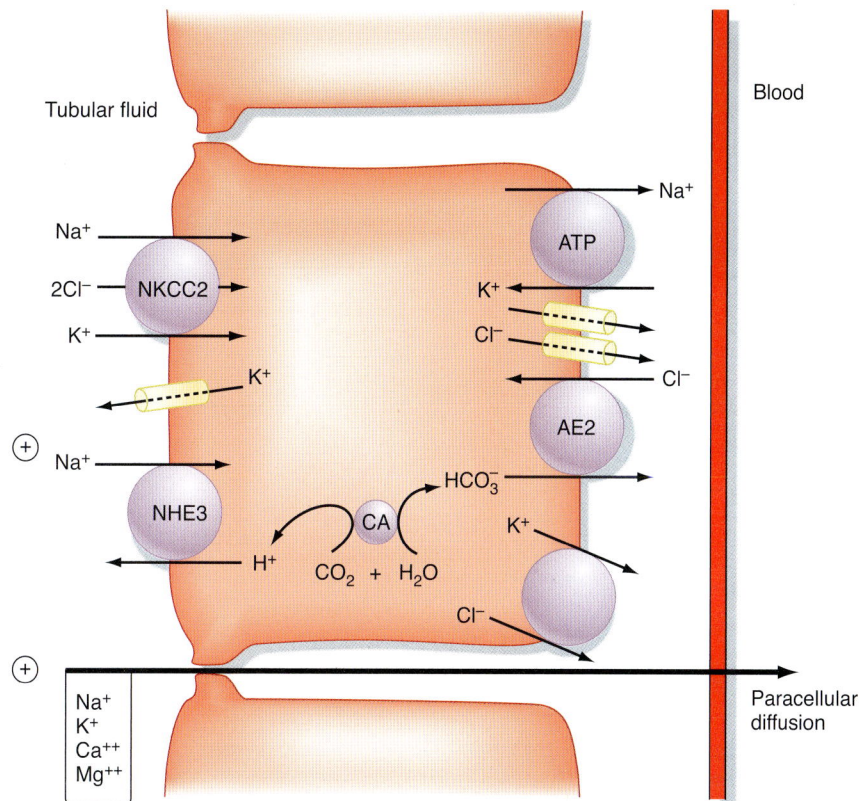

Fig. 34.7 Transport mechanisms for NaCl reabsorption in the thick ascending limb of the loop of Henle. The positive voltage in the lumen plays a major role in driving the passive paracellular reabsorption of cations. Because the apical membrane is conductive primarily to K^+, the apical membrane voltage is more negative than the basolateral membrane voltage, which is conductive to K^+ and Cl^-, thereby resulting in a lumen positive transepithelial potential. Mutations in the apical membrane K^+ channel (ROMK), the apical membrane $1Na^+/1K^+/2Cl^-$ symporter (NKCC2), or the basolateral Cl^- channel (ClCNKB) cause Bartter syndrome (see the clinical box on Bartter syndrome). CA, carbonic anhydrase.

in the loop of Henle occurs in both the thin and thick ascending limbs, whereas the descending thin limb does not reabsorb NaCl. In contrast, water reabsorption mediated by AQP1 water channels is exclusively restricted to the descending thin limb, whereas the ascending limb is impermeable to water. In addition, divalent cations (e.g., Ca^{++} and Mg^{++}) and HCO_3^- are also reabsorbed in the loop of Henle (see Chapters 36 and 37 for more details).

The thin ascending limb reabsorbs NaCl by a passive mechanism. Reabsorption of water, but not NaCl, in the descending thin limb increases [NaCl] in the tubule fluid entering the ascending thin limb. As the NaCl-rich fluid moves toward the cortex, NaCl diffuses out of the tubule lumen across the ascending thin limb and into the medullary interstitial fluid, down a concentration gradient directed from the tubule fluid to the interstitium (see Chapter 35 for details).

The key element in reabsorption of solute by the thick ascending limb is Na^+,K^+-ATPase in the basolateral membrane (Fig. 34.7). As with reabsorption in the proximal tubule, reabsorption of every solute by the thick ascending limb is linked to Na^+,K^+-ATPase activity. This transporter maintains a low intracellular $[Na^+]$, which provides a favorable chemical gradient for the movement of Na^+

from tubular fluid into the cell. This movement of Na^+ across the apical membrane into the cell is mediated by the $1Na^+/1K^+/2Cl^-$ symporter (NKCC2), which couples the movement of $1Na^+$ with $1K^+$ and $2Cl^-$. Using the potential energy released by the downhill movement of Na^+ and Cl^-, this symporter drives the uphill movement of K^+ into the cell. K^+ channels (ROMK and Maxi-K) in the apical plasma membrane play an important role in reabsorption of NaCl by the thick ascending limb. These K^+ channels allow the K^+ transported into the cell via the $1Na^+/1K^+/2Cl^-$ symporter to recycle back into tubule fluid. Because the $[K^+]$ in tubule fluid is relatively low, K^+ recycling is required for continued operation of the $1Na^+/1K^+/2Cl^-$ symporter. A Na^+/H^+ antiporter (NHE3) in the apical cell membrane also mediates Na^+ reabsorption as well as H^+ secretion (HCO_3^- reabsorption) in the thick ascending limb (see also Chapter 37). The operation of the Na^+/H^+ antiporter in the apical membrane results in cellular uptake of Na^+ in exchange for H^+. The production of H^+ inside cells generates HCO_3^-, which exits the cell across the basolateral membrane via a Cl^-/HCO_3^- antiporter (AE2). Na^+ leaves the cell across the basolateral membrane via the Na^+,K^+-ATPase, whereas K^+ and Cl^- leave the cell via separate pathways in the basolateral membrane (i.e., K^+ and Cl^- channels and the K^+/Cl^- symporter).

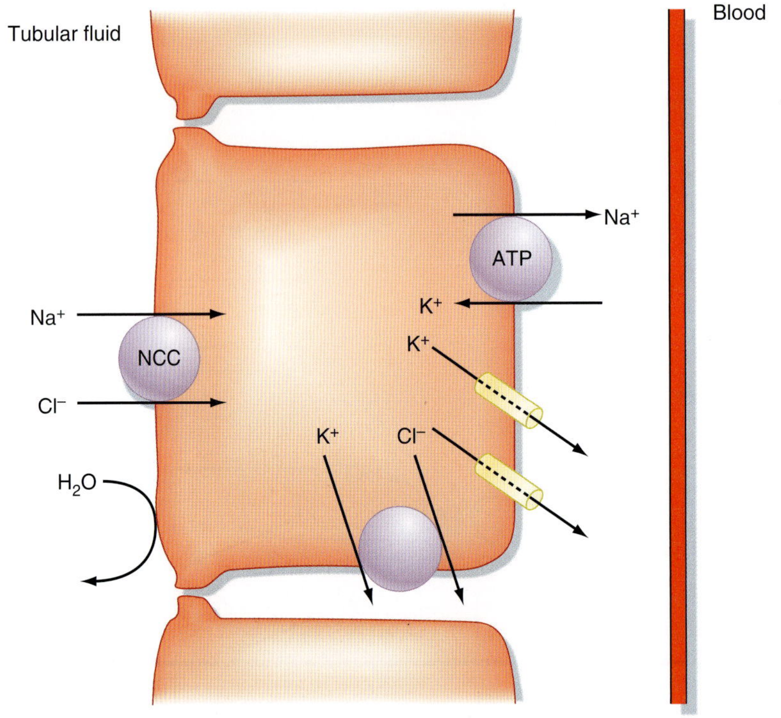

Tubular fluid

Blood

Na⁺

ATP

Na⁺

K⁺

NCC

K⁺

Cl⁻

Cl⁻

H₂O

K⁺

• **Fig. 34.8** Transport mechanism for NaCl reabsorption in the early segment of the distal tubule. This segment is impermeable to water. Mutations in the apical membrane NaCl symporter (NCC) cause Gitelman syndrome.

The voltage across the thick ascending limb is important for reabsorption of several cations. The tubular fluid is positively charged relative to blood because of the unique location of transport proteins in the apical and basolateral membranes. Two points are important: (1) increased NaCl transport by the thick ascending limb increases the magnitude of the positive voltage in the lumen, and (2) this voltage is an important driving force for reabsorption of several cations, including Na⁺, K⁺, Mg⁺⁺, and Ca⁺⁺, across the paracellular pathway (see Fig. 34.7). The importance of the paracellular pathway to solute reabsorption is underscored by the observation that inactivating mutations of the tight junction protein claudin-16 reduce reabsorption of Mg⁺⁺ and Ca⁺⁺ by the ascending thick limb, even in the presence of a lumen positive transepithelial voltage.

In summary, NaCl reabsorption across the thick ascending limb occurs via transcellular and paracellular pathways. Fifty percent of NaCl reabsorption is transcellular, and 50% is paracellular. Because the thick ascending limb does not reabsorb water, owing to a lack of water channels (i.e., AQPs), reabsorption of NaCl and other solutes reduces the osmolality of tubular fluid to less than 150 mOsm/kg H₂O. Thus because the thick ascending limb of Henle's loop produces a fluid that is dilute relative to plasma, this segment and the adjacent distal tubule (as discussed next) are often collectively referred to as the **"diluting segments."**

Distal Tubule and Collecting Duct

The distal tubule and collecting duct reabsorb approximately 8% of the filtered NaCl, secrete variable amounts of K⁺ and H⁺, and reabsorb a variable amount of water (≈8%–17%). The initial segment of the distal tubule (early distal tubule) reabsorbs Na⁺, Cl⁻, and Ca⁺⁺ and is impermeable to water (Fig. 34.8). Entry of NaCl into the cell across the apical membrane is mediated by a Na⁺/Cl⁻ symporter,

AT THE CELLULAR LEVEL

As described in Chapter 2, epithelial cells are joined at their apical surfaces by tight junctions (zonula occludens). A number of proteins have now been identified as components of the tight junction, including proteins that span the membrane of one cell and link to the extracellular portion of the same molecule in the adjacent cell (e.g., occludin and claudins), as well as cytoplasmic linker proteins (e.g., ZO-1, ZO-2, and ZO-3) that link the membrane-spanning proteins to the cytoskeleton of the cell. Of these junctional proteins, claudins appear to be major determinants of the permeability characteristics of tight junctions. For example, claudin-16 and claudin-19 are critical determinants of divalent cation permeability of the tight junctions in the thick ascending limb of Henle's loop. Mutations in human claudin-16 and claudin-19 cause familial hypomagnesemia (i.e., low plasma [Mg⁺⁺]) with hypercalciuria (i.e., increased Ca⁺⁺ in the urine) and nephrocalcinosis (i.e., calcification of the kidney). Claudin-2 is permeable to water and may be responsible for paracellular water reabsorption across the proximal tubule. Claudin-4 has been shown in cultured kidney cells to control the permeability of the tight junction to Na⁺, whereas claudin-15 determines whether a tight junction is permeable to cations or anions. Thus the permeability characteristics of the tight junctions in different nephron segments are determined at least in part by the specific claudins expressed by the cells in that segment.

NCC (see Fig. 34.8). Na$^+$ leaves the cell via the action of Na$^+$,K$^+$-ATPase, and Cl$^-$ leaves the cell via diffusion through Cl$^-$ channels and a K$^+$/Cl$^-$ symporter (KCC4). Thus dilution of tubular fluid begins in the thick ascending limb and continues in the early segment of the distal tubule.

The last segment of the distal tubule (late distal tubule) and the collecting duct are composed of three cell types: **principal cells** and two types of **intercalated cells.** As illustrated in Fig. 34.9, principal cells reabsorb NaCl and water and secrete K$^+$. Both Na$^+$ reabsorption and K$^+$ secretion by these cells depend on the activity of Na$^+$,K$^+$-ATPase in the basolateral membrane. By maintaining a low intracellular [Na$^+$], the Na$^+$,K$^+$-ATPase provides a favorable chemical gradient for movement of Na$^+$ from tubular fluid into the cell. Because Na$^+$ enters the cell across the apical membrane via diffusion through epithelial Na$^+$-selective channels (ENaCs), the negative voltage inside the cell facilitates entry of Na$^+$, which then exits the cell and enters the blood via the basolateral membrane Na$^+$,K$^+$-ATPase. Reabsorption of Na$^+$ generates a negative luminal voltage across the late distal tubule and collecting duct, which provides the driving force for paracellular reabsorption of Cl$^-$. Intercalated cells secrete either H$^+$ or HCO$_3^-$ and play important roles in acid-base homeostasis (see Chapter 37). The α-intercalated cell (see Fig. 34.9, *center*) secretes H$^+$ and reabsorbs both HCO$_3^-$ and K$^+$ and is thus important in regulating acid-base balance (see Chapter 37) and K$^+$ balance (see Chapter 36). α-Intercalated cells reabsorb K$^+$ by the operation of an H$^+$,K$^+$-ATPase (HKA) located in the apical plasma membrane. In contrast, β-intercalated cells (see Fig. 34.9, *bottom*) secrete HCO$_3^-$ and reabsorb both H$^+$ and Cl$^-$. Chloride enters the β-intercalated cell across the apical membrane via a Cl$^-$/HCO$_3^-$ antiporter (pendrin) and leaves the cell across the basolateral membrane via a Cl$^-$ channel. A variable amount of water is reabsorbed across principal cells in the late distal tubule and collecting duct. Water reabsorption in these segments is mediated by the AVP-regulated AQP2 water channel located in the apical plasma membrane and by AQP3 and AQP4 located in the basolateral membrane of principal cells. In the presence of

AVP, water is reabsorbed. By contrast, in the absence of AVP the late distal tubule and collecting duct reabsorb little water (see Chapter 35).

K$^+$ is secreted from blood into tubular fluid by principal cells in two steps (see Fig. 34.9, *top*). First, uptake of K$^+$ across the basolateral membrane is mediated by the action of Na$^+$,K$^+$-ATPase. Second, K$^+$ leaves the cell via passive diffusion. Because [K$^+$] inside principal cells is high ($\approx$150 mEq/L) and [K$^+$] in tubular fluid is low ($\approx$10 mEq/L), K$^+$ diffuses down its concentration gradient through apical cell membrane K$^+$ channels (ROMK and BK) into tubular fluid. Although the negative potential inside these cells favors intracellular K$^+$ retention, the electrochemical gradient across the apical membrane promotes secretion of K$^+$ from the cell into tubular fluid (see Chapter 36). In contrast, K$^+$ reabsorption by α cells is mediated by an H$^+$,K$^+$-ATPase (HKA) located in the apical cell membrane (see Fig. 34.9, *center*). As a consequence these distal nephron segments possess the ability to both secrete and reabsorb K$^+$ via independently regulated mechanisms, which contrasts with the general tendency to reabsorb Na$^+$ along most nephron segments.

Regulation of NaCl and Water Reabsorption

Quantitatively, angiotensin II, aldosterone, catecholamines, natriuretic peptides, and uroguanylin are the most important hormones regulating NaCl reabsorption and thereby urinary NaCl excretion (Table 34.6). However, other hormones (including dopamine and adrenomedullin), Starling forces, and the phenomenon of glomerulotubular balance also influence NaCl reabsorption. AVP is the only major hormone that directly regulates the amount of water excreted by the kidneys.

Angiotensin II has a potent stimulatory effect on the isosmotic reabsorption of NaCl and water in the proximal tubule. It also stimulates reabsorption of Na$^+$ in the thick ascending limb of Henle's loop, as well as the late distal tubule and collecting duct. A decrease in ECFV activates the renin-angiotensin-aldosterone system (see Chapter 35 for more details), thereby increasing the plasma concentration of angiotensin II.

Aldosterone is synthesized by the glomerulosa cells of the adrenal cortex and stimulates reabsorption of NaCl by the thick ascending limb of Henle's loop, the late distal tubule, and the collecting duct. Most of aldosterone's effect on NaCl reabsorption reflects its action on the late distal tubule and collecting duct. Aldosterone enhances reabsorption of NaCl across principal cells in these segments by four mechanisms: (1) increasing the amount of Na$^+$,K$^+$-ATPase in the basolateral membrane; (2) increasing expression of the sodium channel (ENaC) in the apical cell membrane; (3) elevating Sgk1 (*s*erum *g*lucocorticoid-stimulated *k*inase; see the molecular box) levels, which also increases the expression of ENaC in the apical cell membrane; and (4) stimulating

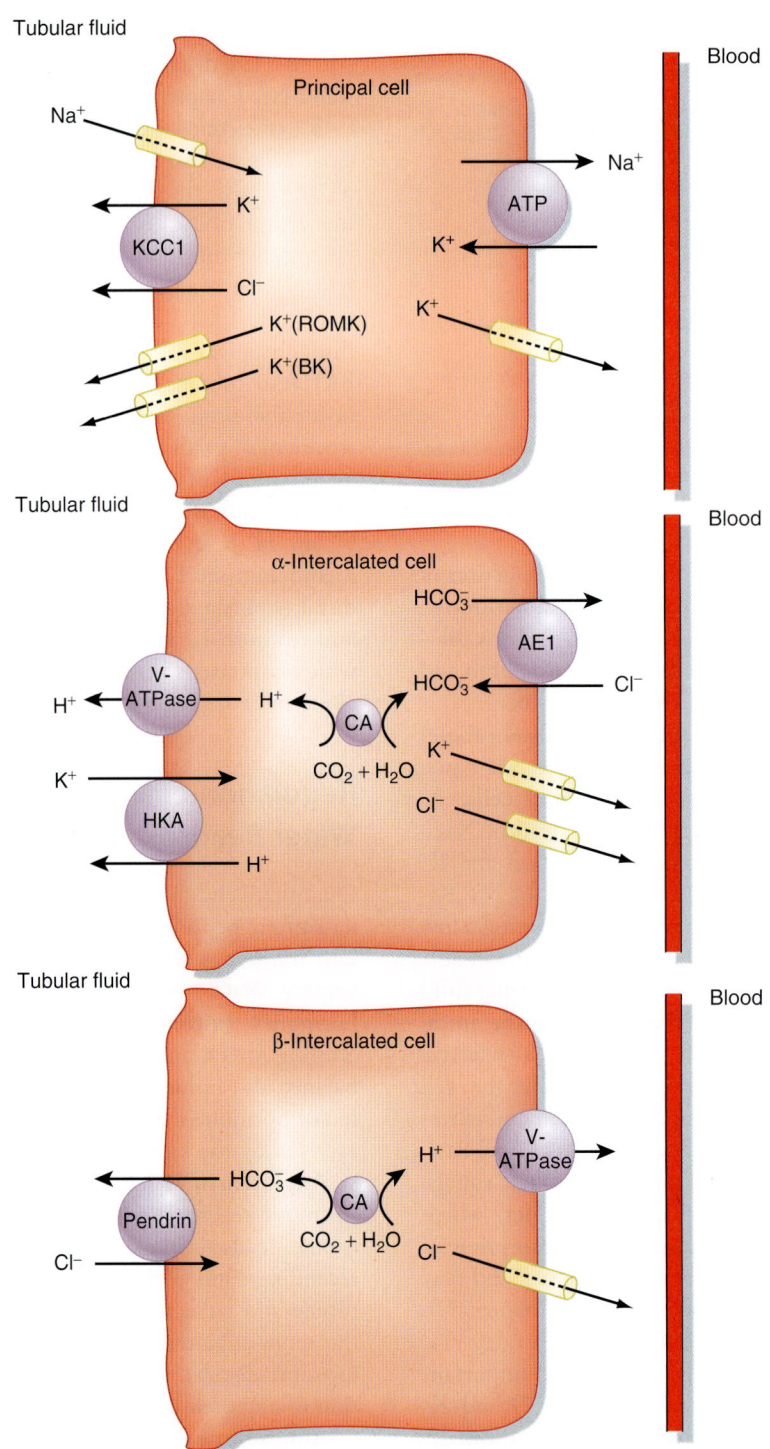

• **Fig. 34.9** Transport pathways in principal cells, α-intercalated cells, and β-intercalated cells of the late segment of the distal tubule and collecting duct. CA, carbonic anhydrase. Principal cells reabsorb Na^+ and secrete K^+. K^+ is secreted by two types of K^+ channels (ROMK and BK) and by a K^+/Cl^- symporter (KCC1). α-Intercalated cells secrete H^+ and reabsorb HCO_3^- and K^+, and β-intercalated cells secrete HCO_3^- and reabsorb H^+ and Cl^-.

CAP1 (*c*hannel-*a*ctivating *p*rotease, also called *prostatin*), a serine protease that directly activates ENaCs by proteolysis. Taken together, these actions increase Na^+ uptake across the apical cell membrane and facilitate Na^+ exit from the cell interior into blood. The increase in reabsorption of Na^+ generates a negative transepithelial luminal voltage across the late distal tubule and the collecting duct. This negative voltage in the lumen provides the electrochemical driving force for reabsorption of Cl^- across the tight junctions. Aldosterone also stimulates secretion of K^+ by the late distal tubule and collecting duct (collectively referred to as the ***aldosterone-sensitive distal nephron*** (**ASDN**)

TABLE 34.6   Hormones That Regulate NaCl and Water Reabsorption

Hormone[a]	Major Stimulus	Nephron Site of Action	Effect on Transport
Angiotensin II	↑Renin	PT, TAL, DT/CD	↑NaCl and H_2O reabsorption
Aldosterone	↑Angiotensin II, ↑$[K^+]_p$	TAL, DT/CD	↑NaCl and H_2O reabsorption[b]
ANP, BNP, urodilatin	↑ECFV	CD	↓H_2O and NaCl reabsorption
Uroguanylin, guanylin	Oral ingestion of NaCl	PT, CD	↓H_2O and NaCl reabsorption
Sympathetic nerves	↓ECFV	PT, TAL, DT/CD	↑NaCl and H_2O reabsorption[b]
Dopamine	↑ECFV	PT	↓H_2O and NaCl reabsorption
AVP	↑P_{osm}, ↓ECFV	DT/CD	↑H_2O reabsorption

[a]All these hormones act within minutes, except aldosterone, which exerts its action on reabsorption of NaCl with a delay of 1 hour. Aldosterone achieves its maximal effect after a few days.
[b]The effect on reabsorption of H_2O does not include the thick ascending limb.
ANP, atrial natriuretic peptide; BNP, brain natriuretic peptide, BP, blood pressure; CD, collecting duct; DT, distal tubule; ECFV, extracellular fluid volume; $[K^+]_p$, plasma K^+ concentration; P_{osm}, plasma osmolality; PT, proximal tubule; TAL, thick ascending limb.

[see Chapter 36]). Aldosterone secretion is increased by hyperkalemia and by hypovolemia (i.e., reduced ECFV via increased angiotensin II following activation of the renin-angiotensin system [RAS]; see Chapter 35 for more details). Aldosterone secretion is decreased by hypokalemia and natriuretic peptides (discussed in more detail next). Through its stimulation of NaCl reabsorption in the collecting duct, aldosterone also indirectly increases water reabsorption by this nephron segment.

The aldosterone paradox. As noted earlier, aldosterone stimulates both NaCl reabsorption and K^+ secretion by the collecting duct. Although both a reduction in the ECFV (i.e., hypovolemia, see Chapter 35) and hyperkalemia (see Chapter 36) increase aldosterone levels, the physiological response of the kidneys differs in these two conditions. In the setting of ECFV depletion, NaCl excretion by the kidneys is reduced to restore ECFV, without an accompanying change in K^+ excretion. By contrast, during hyperkalemia, K^+ excretion by the kidneys is increased to normalize plasma $[K^+]$, albeit without an accompanying change in NaCl excretion. This phenomenon—the apparent independent effects of aldosterone on urinary Na^+ and K^+ excretion—is called the **aldosterone paradox.** The paradox can be explained by the observation that ECFV depletion increases aldosterone release via activation of the RAS, whereas hyperkalemia directly stimulates adrenal release of aldosterone without a requirement for RAS activation. As such, aldosterone increases in both conditions, whereas angiotensin II levels increase only during ECFV depletion and not during hyperkalemia. It is the differential regulation of transport processes in the distal tubule and collecting duct by aldosterone and angiotensin II that accounts for this paradox. The integrated physiological response to a reduction in ECFV is depicted in Fig. 34.10A. During hypovolemia, angiotensin II stimulates NaCl reabsorption by the proximal tubule (not shown in Figure 34.10A), and by the early distal tubule by activating WNK (with no lysine [K] kinase), which enhances NaCl reabsorption by activating the Na^+/Cl^- symporter (NCC). Aldosterone

stimulates ENaC-mediated Na^+ reabsorption in principal cells of the collecting duct by activating SGK1, which increases ENaC abundance in the apical plasma membrane. In parallel, angiotensin II activates WNK in principal cells, which inhibits K^+ secretion via ROMK, thereby preventing increased K^+ excretion despite elevated aldosterone levels, which would be expected to promote K^+ secretion. Angiotensin II stimulation of proximal tubule NaCl and water reabsorption also reduces delivery of NaCl and fluid to the collecting duct, which also suppresses K^+ secretion in this segment (see Chapter 36 for more details). The corresponding integrated physiological response to hyperkalemia is depicted in Fig. 34.10B. During hyperkalemia, aldosterone stimulates ROMK mediated K^+ secretion by principal cells in the collecting duct by activating WNK. Because the early distal tubule is not directly responsive to aldosterone, this hormone does not stimulate NaCl reabsorption in this segment. In fact, because angiotensin II levels are not elevated by hyperkalemia, the basal activity of WNK in the early distal tubule is low, resulting in decreased NaCl reabsorption via NCC (see Fig. 34.10B). This effect of WNK on early distal tubule NaCl reabsorption offsets the stimulatory effect of aldosterone, via SGK1, on ENaC-mediated Na^+ reabsorption in principal cells of the collecting duct.

Atrial natriuretic peptide (ANP) and **brain natriuretic peptide (BNP)** inhibit NaCl and water reabsorption. Secretion of ANP by the cardiac atria and BNP by the cardiac ventricles are both stimulated by increased ECFV and increased myocardial wall pressure. ANP and BNP reduce blood pressure by decreasing total peripheral resistance and enhancing urinary excretion of both NaCl and water, primarily by increasing RBF and GFR. These natriuretic peptides vasodilate the afferent arterioles and vasoconstrict the efferent arterioles, which increases GFR and thus filtration of NaCl, thereby increasing NaCl excretion (see later discussion of glomerulotubular balance for the mechanism). In addition the increase in RBF decreases the concentration of NaCl in the medullary interstitium, which in turn reduces passive NaCl reabsorption by the thin ascending

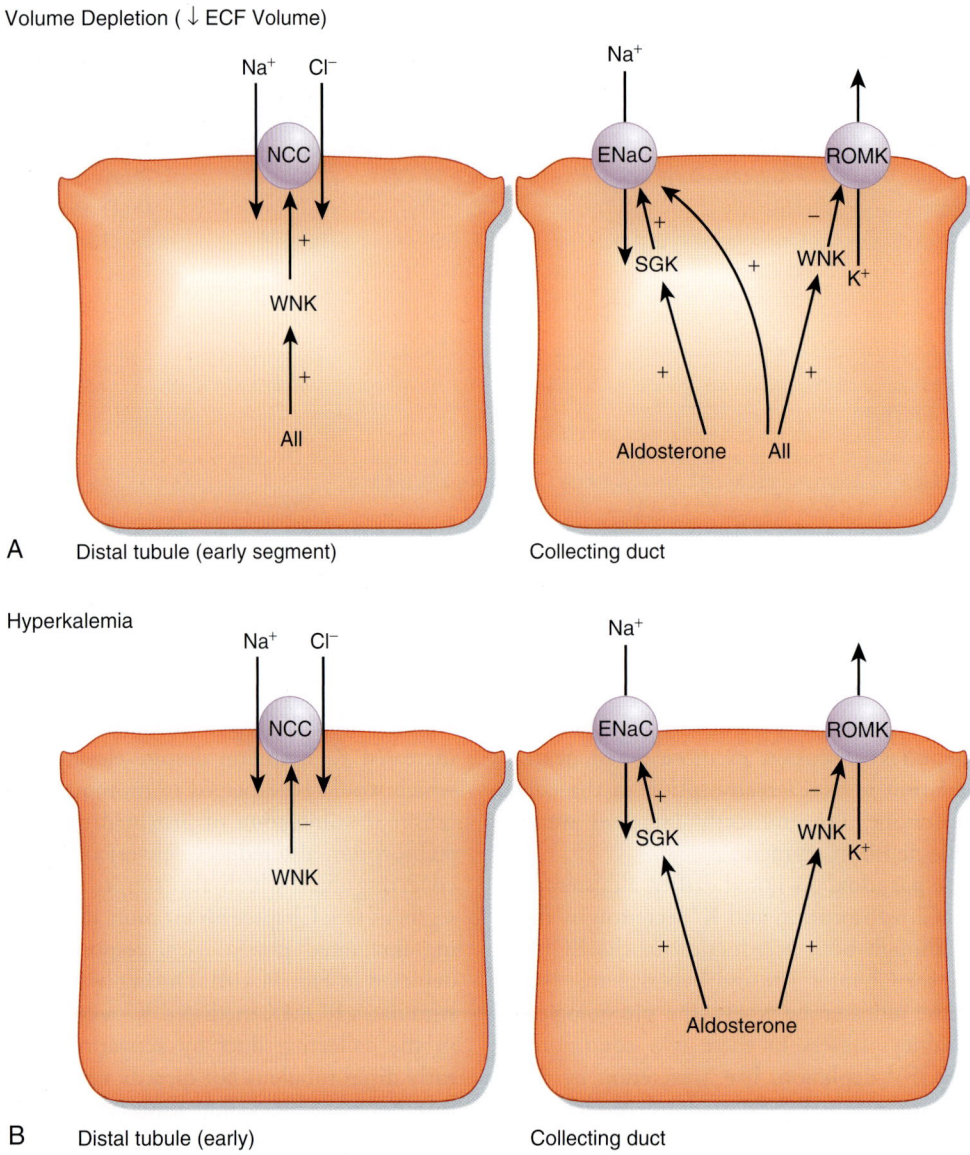

• **Fig. 34.10** Aldosterone paradox (see text for details). +, stimulate; −, inhibit; AII, angiotensin II; NCC, NaCl symporter; ENaC, epithelial sodium channel; ROMK, K⁺ channel; WNK, with no lysine[K] kinases. There are four WNKs in the kidney; details are not given in the figure for clarity and because some of the details of WNK regulation of NCC, ENaC, and ROMK have not been elucidated.

limb of Henle's loop (see earlier discussion for details on NaCl reabsorption by this segment). ANP and BNP also inhibit NaCl reabsorption by the medullary portion of the collecting duct and inhibit AVP-stimulated water reabsorption across the collecting duct. Moreover, ANP and BNP also reduce secretion of AVP from the posterior pituitary. These actions of ANP and BNP are mediated by the activation of membrane-bound guanylyl cyclase receptors, which increases intracellular levels of the second messenger cyclic guanine monophosphate (cGMP). ANP is a more profound natriuretic and diuretic agent than BNP.

Urodilatin and ANP are encoded by the same gene and have similar amino acid sequences. Urodilatin is a 32–amino acid hormone that differs from ANP by the addition of four amino acids to the amino terminus. Urodilatin is secreted by the distal tubule and collecting duct and is not present in the systemic circulation; thus urodilatin influences only the function of the kidneys. Secretion of urodilatin is stimulated by a rise in blood pressure and an increase in ECFV. It inhibits NaCl and water reabsorption across the medullary portion of the collecting duct. Urodilatin is a more potent natriuretic and diuretic hormone than ANP because some of the ANP that enters the kidneys in blood is degraded by a neutral endopeptidase that has no corresponding effect on urodilatin.

Uroguanylin and **guanylin** are produced by neuroendocrine cells in the intestine in response to oral ingestion of NaCl. These hormones enter the circulation and inhibit NaCl and water reabsorption by the kidneys via activation of membrane-bound guanylyl cyclase receptors, which

AT THE CELLULAR LEVEL

Sgk1 (**s**erum **g**lucocorticoid-stimulated **k**inase), a serine/threonine kinase, plays an important role in maintaining NaCl and K^+ homeostasis by regulating excretion of both NaCl and K^+ by the kidneys. Studies in Sgk1 knockout mice reveal that this kinase is required for animals to survive severe NaCl restriction and K^+ loading. NaCl restriction and K^+ loading enhance plasma [aldosterone], which rapidly (in minutes) increases Sgk1 protein expression and phosphorylation. Phosphorylated Sgk1 enhances ENaC-mediated Na^+ reabsorption in the collecting duct, primarily by increasing the number of ENaCs in the apical plasma membrane of principal cells and also by increasing the number of Na^+,K^+-ATPase pumps in the basolateral membrane. Phosphorylated Sgk1 inhibits Nedd4-2, a ubiquitin ligase that monoubiquitinylates ENaC subunits, thereby targeting them for endocytic removal from the plasma membrane and subsequent destruction in lysosomes. Inhibition of Nedd4-2 by Sgk1 reduces the monoubiquitinylation of ENaC, thereby reducing endocytosis and increasing the number of channels in the membrane. Sgk1 induces the translocation of K^+ channels (ROMK) from an intracellular pool to the plasma membrane, and thereby enhances ROMK-mediated K^+ secretion by principal cells. These effects of Sgk1 precede the aldosterone-stimulated increase in ENaC, ROMK, and Na^+,K^+-ATPase abundance, which leads to a delayed (>4 hours) secondary increase in NaCl and K^+ transport by the collecting duct. Activating polymorphisms in Sgk1 cause an increase in blood pressure, presumably by enhancing NaCl reabsorption by the collecting duct, which increases the ECFV and thereby blood pressure.

IN THE CLINIC

Liddle syndrome is a rare genetic disorder characterized by an increase in blood pressure (i.e., hypertension) secondary to an increase in ECFV. Liddle syndrome is caused by activating mutations in either the β or γ subunit of the epithelial Na^+ channel (ENaC). These mutations increase the number of Na^+ channels in the apical cell membrane of principal cells and thereby the amount of Na^+ reabsorbed. In Liddle syndrome, the rate of renal Na^+ reabsorption is inappropriately high, which leads to an increase in ECFV and hypertension. There are two different forms of **pseudohypoaldosteronism (PHA)** (i.e., the kidneys reabsorb NaCl as they do when aldosterone levels are low; however, in PHA, aldosterone levels are elevated). The autosomal recessive form is caused by inactivating mutations in the α, β, or γ subunit of ENaC. The cause of the autosomal dominant form is an inactivating mutation in the mineralocorticoid receptor. Pseudohypoaldosteronism is characterized by an increase in Na^+ excretion, a reduction in ECFV, hyperkalemia, and hypotension. Some individuals with expanded ECFV and elevated blood pressure are treated with drugs that inhibit **angiotensin-converting enzyme (ACE)** (e.g., captopril, enalapril, lisinopril) and thereby lower fluid volume and blood pressure. Inhibition of ACE blocks degradation of angiotensin I to angiotensin II and thereby lowers plasma angiotensin II levels. The decline in plasma angiotensin II concentration has three effects. First, NaCl and water reabsorption by the nephron (especially the proximal tubule) falls. Second, aldosterone secretion decreases, thus reducing NaCl reabsorption in the thick ascending limb, distal tubule, and collecting duct. Third, because angiotensin is a potent vasoconstrictor, a reduction in its concentration permits the systemic arterioles to dilate and thereby lower arterial blood pressure. ACE also degrades the vasodilator hormone bradykinin; thus ACE inhibitors increase the concentration of bradykinin, a vasodilatory hormone. ACE inhibitors decrease ECFV and the arterial blood pressure by promoting renal NaCl and water excretion and by reducing total peripheral resistance.

increase intracellular [cGMP]. The involvement of these gut-derived hormones helps explain why the natriuretic response of the kidneys to an oral NaCl load is more pronounced than when delivered intravenously.

Catecholamines stimulate reabsorption of NaCl. Catecholamines released from the sympathetic nerves (norepinephrine) and the adrenal medulla (epinephrine) stimulate reabsorption of NaCl and water by the proximal tubule, thick ascending limb of the loop of Henle, distal tubule, and collecting duct. Although sympathetic nerves are not active when ECFV is normal, when ECFV declines (e.g., after hemorrhage), sympathetic nerve activity rises and dramatically stimulates reabsorption of NaCl and water by these four nephron segments.

Dopamine, a catecholamine, is released from dopaminergic nerves in the kidneys and is also synthesized by cells of the proximal tubule. The action of dopamine is opposite that of norepinephrine and epinephrine. Secretion of dopamine is stimulated by an increase in ECFV, and its secretion directly inhibits reabsorption of NaCl and water in the proximal tubule.

Adrenomedullin is a 52–amino acid peptide hormone that is produced by a variety of organs, including the kidneys. Adrenomedullin induces a marked diuresis and natriuresis, and its secretion is stimulated by congestive heart failure and hypertension. The major effect of adrenomedullin on the kidneys is to increase GFR and renal blood flow and

thereby indirectly stimulate excretion of NaCl and water (see earlier discussion about ANP and BNP).

Arginine vasopressin (AVP) regulates water reabsorption. It is the most important hormone that regulates reabsorption of water in the kidneys (see Chapter 35). This hormone is secreted by the posterior pituitary gland in response to an increase in plasma osmolality (1% or more) or a decrease in ECFV (>5%–10% of normal). AVP increases the permeability of the collecting duct to water. It increases reabsorption of water by the collecting duct because of the osmotic gradient that exists across the wall of the collecting duct (see Chapter 35). AVP has little effect on urinary NaCl excretion.

Starling forces regulate reabsorption of NaCl and water across the proximal tubule. As previously described, Na^+, Cl^-, HCO_3^-, amino acids, glucose, and water are transported into the intercellular space of the proximal tubule. Starling forces between this space and the peritubular capillaries facilitate movement of the reabsorbed fluid into the capillaries. Starling forces across the wall of peritubular capillaries

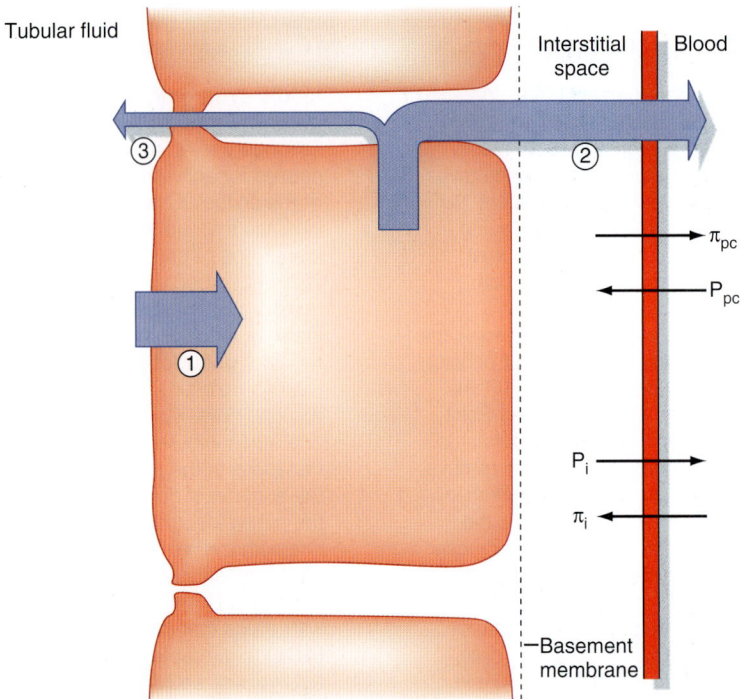

• **Fig. 34.11** Starling forces modify proximal tubule solute and water reabsorption. (1) Solute and water are reabsorbed across the apical membrane. This solute and water then cross the lateral cell membrane. Some solute and water reenters the tubule fluid (3), and the remainder enters the interstitial space and then flows into the capillary (2). The width of the arrows is directly proportional to the amount of solute and water moving by pathways 1 to 3. Starling forces across the capillary wall determine the amount of fluid flowing through pathway 2 versus pathway 3. Transport mechanisms in the apical cell membranes determine the amount of solute and water entering the cell (pathway 1). P_i, interstitial hydrostatic pressure; P_{pc}, peritubular capillary hydrostatic pressure; π_i, interstitial fluid oncotic pressure; π_{pc}, peritubular capillary oncotic pressure. Thin arrows across the capillary wall indicate the direction of water movement in response to each force.

consist of hydrostatic pressure in the peritubular capillary (P_{pc}) and lateral intercellular space (P_i) and oncotic pressure in the peritubular capillary (π_{pc}) and lateral intercellular space (π_i). Thus reabsorption of water as a result of transport of Na^+ from tubular fluid into the lateral intercellular space is modified by the Starling forces. Accordingly:

Equation 34.2

$$J = K_f[(P_i - P_{pc}) + \sigma(\pi_{pc} - \pi_i)]$$

where *J* is flow (positive numbers indicate flow from the intercellular space into blood). Starling forces that favor movement from the interstitium into the peritubular capillaries are π_{pc} and P_i (Fig. 34.11). The opposing Starling forces are π_i and P_{pc}. Normally the sum of the Starling forces favors movement of solute and water from the interstitial space into the capillary. However, some of the solutes and fluid that enter the lateral intercellular space leak back into the proximal tubular fluid. Starling forces do not affect transport by the loop of Henle, distal tubule, and collecting duct because these segments are less permeable to water than the proximal tubule is.

A number of factors can alter the Starling forces across the peritubular capillaries surrounding the proximal tubule. For example, dilation of the efferent arteriole increases P_{pc}, whereas constriction of the efferent arteriole decreases it.

An increase in P_{pc} inhibits solute and water reabsorption by increasing back-leak of NaCl and water across the tight junction, whereas a decrease stimulates reabsorption by decreasing back-leak across the tight junction.

Peritubular capillary oncotic pressure (π_{pc}) is partially determined by the rate of formation of the glomerular ultrafiltrate. For example, if one assumes a constant plasma flow in the afferent arteriole, the plasma proteins become less concentrated in the plasma that enters the efferent arteriole and peritubular capillary as less ultrafiltrate is formed (i.e., as GFR decreases). Hence, π_{pc} decreases. Thus π_{pc} is directly related to the **filtration fraction** (FF = GFR/renal plasma flow [RPF]). A fall in the FF resulting from a decrease in GFR, at constant RPF, decreases π_{pc}. This in turn increases the backflow of NaCl and water from the lateral intercellular space into tubular fluid and thereby decreases net reabsorption of solute and water across the proximal tubule. An increase in FF has the opposite effect.

The importance of Starling forces in regulating solute and water reabsorption by the proximal tubule is underscored by the phenomenon of **glomerulotubular (G-T) balance.** Spontaneous changes in GFR markedly alter the filtered amount of Na^+ (filtered Na^+ = GFR × [Na^+] in the filtered fluid). Without rapid adjustments in Na^+ reabsorption to counter the changes in filtration of Na^+, urinary excretion of Na^+ would fluctuate widely and disturb the Na^+ balance of the body and thus alter ECFV and blood pressure (see Chapter 35 for more details). However, spontaneous changes in GFR do not alter Na^+ excretion in urine or Na^+ balance when ECFV is normal because of the phenomenon of G-T balance. When body Na^+ balance is normal (i.e., ECFV is normal), *G-T balance* refers to the fact that reabsorption of Na^+ and water increases in proportion to the increase in GFR and filtered amount of Na^+. Thus a constant fraction of the filtered Na^+ and water is reabsorbed from the proximal tubule despite variations in GFR. The net result of G-T balance is to reduce the impact of changes in GFR on the amount of Na^+ and water excreted in urine when ECFV is normal.

Two mechanisms are responsible for G-T balance. One is related to the oncotic and hydrostatic pressure differences between the peritubular capillaries and the lateral intercellular space (i.e., Starling forces). For example, an increase in the GFR (at constant RPF) raises the protein concentration in glomerular capillary plasma above normal. This protein-rich plasma leaves the glomerular capillaries, flows through the efferent arterioles, and enters the peritubular capillaries. The increased π_{pc} augments the movement of solute and fluid from the lateral intercellular space into the peritubular capillaries. This action increases net solute and water reabsorption by the proximal tubule.

The second mechanism responsible for G-T balance is initiated by an increase in the filtered amount of glucose and amino acids. As discussed earlier, reabsorption of Na^+ in the first half of the proximal tubule is coupled to that of glucose and amino acids. The rate of Na^+ reabsorption therefore partially depends on the filtered amount of glucose and amino acids. As the GFR and filtered amount of glucose and amino acids increase, reabsorption of Na^+ and water also rises.

In addition to G-T balance, another mechanism minimizes changes in the filtered amount of Na^+. As discussed in Chapter 33, an increase in GFR (and thus in the amount of Na^+ filtered by the glomerulus) activates the *tubuloglomerular feedback mechanism*. This action returns the GFR and filtration of Na^+ to normal values. Thus spontaneous changes in GFR (e.g., caused by changes in posture and blood pressure) increase the amount of Na^+ filtered for only a few minutes. The mechanisms that underlie G-T balance maintain urinary Na^+ excretion constant and thereby maintain Na^+ homeostasis (and ECFV and blood pressure) until the GFR returns to normal.

Key Concepts

1. The four major segments of the nephron (proximal tubule, Henle's loop, distal tubule, and collecting duct) determine the composition and volume of urine by the processes of selective reabsorption of solutes and water and secretion of some solutes.

2. Tubular reabsorption of substances filtered by the glomerulus allows the kidneys to retain substances that are essential and regulate their levels in plasma by altering the degree to which they are reabsorbed. Reabsorption of Na^+, Cl^-, other anions, and organic anions and cations together with water constitutes the major function of the nephron. Approximately 25,200 mEq of Na^+ and 179 L of water are reabsorbed each day. Proximal tubule cells reabsorb 67% of the glomerular ultrafiltrate, and cells of Henle's loop reabsorb about 25% of the NaCl that was filtered and about 15% of the water that was filtered. The distal segments of the nephron (distal tubule and collecting duct system) have a more limited reabsorptive capacity. However, although the proximal tubule reabsorbs the largest fraction of the filtered solutes and water (i.e., 67%), final adjustments in the composition and volume of urine and most of the regulation by hormones and other factors occur primarily in the distal tubule and collecting duct.

3. Secretion of substances from the blood into tubular fluid is a means for excreting various byproducts of metabolism, and it also serves to eliminate exogenous organic anions and cations (e.g., drugs) and toxins from the body. Many organic anions and cations are bound to plasma proteins and are therefore unavailable for ultrafiltration. Thus secretion is their major route of excretion in urine.

4. Various hormones (including angiotensin II, aldosterone, AVP, natriuretic peptides [ANP, BNP, and urodilatin], uroguanylin, and guanylin), sympathetic nerves, dopamine, and Starling forces regulate reabsorption of NaCl by the kidneys. AVP is the major hormone that regulates water reabsorption.

Additional Reading

Journal Articles

Brown D, et al. New insights into the dynamic regulation of water and acid-base balance by renal epithelial cells. *Am J Physiol Cell Physiol.* 2012;302:C1421-C1433.

Dantzler WH, et al. Urine-concentrating mechanism in the inner medulla: function of the thin limbs of the loops of Henle. *Clin J Am Soc Nephrol.* 2014;9:1781-1789.

Divers J, Freedman BI. Genetics in kidney disease in 2013: susceptibility genes for renal and urological disorders. *Nat Rev Nephrol.* 2014;10:69-70.

Hoenig MP, Zeidel ML. Homeostasis, the *milieu interieur,* and the wisdom of the nephron. *Clin J Am Soc Nephrol.* 2014;9: 1272-1281.

Knepper MA, et al. Molecular physiology of water balance. *N Engl J Med.* 2015;372:1349-1358.

Kortenoeven ML, et al. Vasopressin regulation of sodium transport in the distal nephron and collecting duct. *Am J Physiol Renal Physiol.* 2015;309:F280-F299.

McCormick JA, Ellison DH. Distal convoluted tubule. *Compr Physiol.* 2015;5:45-98.

Mount DB. Thick ascending limb of the loop of Henle. *Clin J Am Soc Nephrol.* 2014;9:1974-1986.

Palmer BF. Regulation of potassium homeostasis. *Clin J Am Soc Nephrol.* 2015;10:1050-1060.

Palmer LG, Schnermann J. Integrated control of Na transport along the nephron. *Clin J Am Soc Nephrol.* 2015;10:676-687.

Pearce D, et al. Collecting duct principal cell transport processes and their regulation. *Clin J Am Soc Nephrol.* 2015;10:135-146.

Pelis RM, Wright SH. Renal transport of organic anions and cations. *Compr Physiol.* 2011;1:1795-1835.

Pluznick JL, Caplan MJ. Chemical and physical sensors in the regulation of renal function. *Clin J Am Soc Nephrol.* 2015;10:1626-1635.

Roy A, et al. Collecting duct intercalated cell function and regulation. *Clin J Am Soc Nephrol.* 2015;10:305-324.

Theilig F, Wu Q. ANP-induced signaling cascade and its implications in renal pathophysiology. *Am J Physiol Renal Physiol.* 2015;308:F1047-F1055.

Weiner ID, et al. Urea and ammonia metabolism and the control of renal nitrogen excretion. *Clin J Am Soc Nephrol.* 2015;10:1444-1458.

Book Chapters

Brown D, Nielsen S. The cell biology of vasopressin action. In: Taal MW, et al., eds. *Brenner and Rector's The Kidney.* 9th ed. Philadelphia: Saunders; 2012.

Burckhardt G, Koepsell H. Organic anion and cation transporters in renal elimination of drugs. In: Alpern RJ, et al., eds. *Seldin and Giebisch's The Kidney: Physiology and Pathophysiology.* 5th ed. Philadelphia: Academic Press; 2012.

Christensen EI, et al. Renal filtration, transport, and metabolism of albumin and albuminuria. In: Alpern RJ, et al., eds. *Seldin and Giebisch's The Kidney: Physiology and Pathophysiology.* 5th ed. Philadelphia: Academic Press; 2012.

Gamba G, Schild L. Sodium chloride transport in the loop of Henle, distal convoluted tubule, and collecting duct. In: Alpern RJ, et al., eds. *Seldin and Giebisch's The Kidney: Physiology and Pathophysiology.* 5th ed. Philadelphia: Academic Press; 2012.

Giebisch G, Satlin L. Regulation of potassium excretion. In: Alpern RJ, et al., eds. *Seldin and Giebisch's The Kidney: Physiology and Pathophysiology.* 5th ed. Philadelphia: Academic Press; 2012.

Moe OW, et al. Renal transport of glucose, amino acids, sodium, chloride, and water. In: Taal MW, et al., eds. *Brenner and Rector's The Kidney.* 9th ed. Philadelphia: Saunders; 2012.

Preisig P, et al. Cellular mechanisms of renal tubular acidification. In: Alpern RJ, et al., eds. *Seldin and Giebisch's The Kidney: Physiology and Pathophysiology.* 5th ed. Philadelphia: Academic Press; 2012.

Sands JM, Layton H. The urine concentrating mechanism and urea transporters. In: Alpern RJ, et al., eds. *Seldin and Giebisch's The Kidney: Physiology and Pathophysiology.* 5th ed. Philadelphia: Academic Press; 2012.

35

Control of Body Fluid Osmolality and Volume

LEARNING OBJECTIVES

Upon completion of this chapter the student should be able to answer the following questions:

1. Why do changes in water balance result in alterations in the [Na⁺] of the extracellular fluid (ECF)?
2. How is the secretion of arginine vasopressin (AVP) controlled by changes in the osmolality of body fluids and in blood volume and pressure?
3. What are the cellular events associated with the action of AVP on the collecting duct, and how do they lead to an increase in the water permeability of this segment of the nephron?
4. What is the role of Henle's loop in the production of both dilute and concentrated urine?
5. What is the composition of the medullary interstitial fluid, and how does it participate in the process of producing concentrated urine?
6. What are the roles of the vasa recta in the process of diluting and concentrating urine?
7. How is the diluting and concentrating ability of the kidneys quantitated?
8. Why do changes in Na⁺ balance alter the volume of extracellular fluid?
9. What is the effective circulating volume, how is it influenced by changes in Na⁺ balance, and how does it influence renal Na⁺ excretion?
10. What are the mechanisms by which the body monitors the effective circulating volume?
11. What are the major signals acting on the kidneys to alter their excretion of Na⁺?
12. How do changes in extracellular fluid volume alter Na⁺ transport in the different segments of the nephron, and how do these changes in transport regulate renal Na⁺ excretion?
13. What are the mechanisms involved in formation of edema, and what role do the kidneys play in this process?

T he kidneys maintain the osmolality and volume of the body fluids within a narrow range by regulating excretion of water and NaCl, respectively. This chapter discusses the regulation of renal water excretion (urine concentration and dilution) and NaCl excretion.

The composition and volumes of the various body fluid compartments are reviewed in Chapter 2.

Control of Body Fluid Osmolality: Urine Concentration and Dilution

As described in Chapter 2, water constitutes approximately 60% of the healthy adult human body. Body water is divided into two major compartments—intracellular fluid (ICF) and extracellular fluid (ECF)—that are in osmotic equilibrium because of the high permeability of most cell membranes to water via aquaporins (e.g., AQP1). Water intake into the body generally occurs orally. This may be water contained in beverages as well as water generated during metabolism of ingested foods (e.g., carbohydrates). In many clinical situations, intravenous infusion is an important route of water entry.

The kidneys are responsible for regulating water balance and under most conditions are the major route for elimination of water from the body (Table 35.1). Other routes of water loss from the body include evaporation from cells of the skin and respiratory passages. Collectively, water loss by these routes is termed **insensible water loss** because the individual is unaware of its occurrence. Production of sweat accounts for the loss of additional water. Water loss by this mechanism can increase dramatically in a hot environment, with exercise, or in the presence of fever (Table 35.2). Finally, water can be lost from the gastrointestinal tract. Fecal water loss is normally small (≈100 mL/day) but can increase dramatically with diarrhea (e.g., 20 L/day with cholera). Vomiting can also cause gastrointestinal water losses.

Although water loss from sweating, defecation, and evaporation from the lungs and skin can vary depending on the environmental conditions or during pathological conditions, loss of water by these routes cannot be regulated. In contrast, renal excretion of water is tightly regulated to maintain whole-body water balance. Maintenance of water balance requires that water intake and loss from the body be precisely matched. If intake exceeds losses, **positive water balance** exists. Conversely, when intake is less than losses,

negative water balance exists (see Chapter 2 for review of steady-state balance).

When water intake is low or water losses increase, the kidneys conserve water by producing a small volume of urine that is hyperosmotic with respect to plasma. When water intake is high, a large volume of hypoosmotic urine is produced. In a normal individual, urine osmolality (U_{osm}) can vary from approximately 50 to 1200 mOsm/kg H_2O, and the corresponding urine volume can vary from approximately 18 L/day to 0.5 L/day. Importantly the kidneys can regulate excretion of water separately from excretion of total solute (Fig. 35.1). The ability to regulate water excretion separate from excretion of solutes (e.g., Na^+, K^+, urea, etc.) is necessary for survival because it allows water balance to be achieved without upsetting the other homeostatic functions of the kidneys.

It is important to recognize that disorders of water balance are manifested by alterations in body fluid osmolality, which are usually measured by changes in plasma osmolality (P_{OSM}). Because the major determinant of plasma osmolality is Na^+

(with its anions Cl^- and HCO_3^-), these disorders also result in alterations in plasma or serum $[Na^+]$ (Fig. 35.2). One of the most common fluid and electrolyte disorders seen in clinical practice is an alteration in serum $[Na^+]$. When an abnormal serum $[Na^+]$ is found in an individual, it is tempting to suspect a problem in Na^+ balance. However, the problem most often relates to water balance, not Na^+ balance. As described later, changes in Na^+ balance result in alterations in the volume of ECF, not its osmolality.

The following sections discuss the mechanisms by which the kidneys excrete either **hypoosmotic (dilute)** or **hyperosmotic (concentrated)** urine. Control of arginine vasopressin secretion and its important role in regulating excretion of water by the kidneys are also explained (see also Chapter 41).

TABLE 35.1	Normal Routes of Water Gain and Loss in Adults at Room Temperature (23°C)	
Route		**mL/day**
Water Intake		
Fluid[a]		1200
In food		1000
Metabolically produced from food		300
Total		2500
Water Output		
Insensible		700
Sweat		100
Feces		200
Urine		1500
Total		2500

[a]Fluid intake varies widely for both social and cultural reasons.

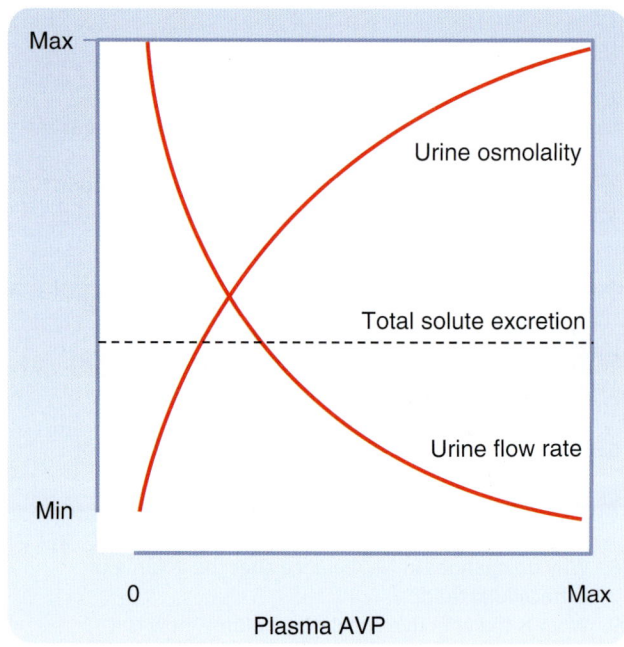

• **Fig. 35.1** Relationships between plasma AVP levels, and urine osmolality, urine flow rate, and total solute excretion. Max, maximum; Min, minimum. (From Koeppen BM, Stanton BA. *Renal Physiology.* 5th ed. Philadelphia: Elsevier; 2013.)

TABLE 35.2	Effect of Environmental Temperature and Exercise on Water Loss and Intake in Adults		
	Normal Temperature	**Hot Weather[a]**	**Prolonged Heavy Exercise[a]**
Water Loss			
Insensible loss			
Skin	350	350	350
Lungs	350	250	650
Sweat	100	1400	5000
Feces	200	200	200
Urine[a]	1500	1200	500
TOTAL LOSS	2500	3400	6700

[a]In hot weather and during prolonged heavy exercise, water balance is maintained by increased water ingestion. Decreased excretion of water by the kidneys alone is insufficient to maintain water balance.

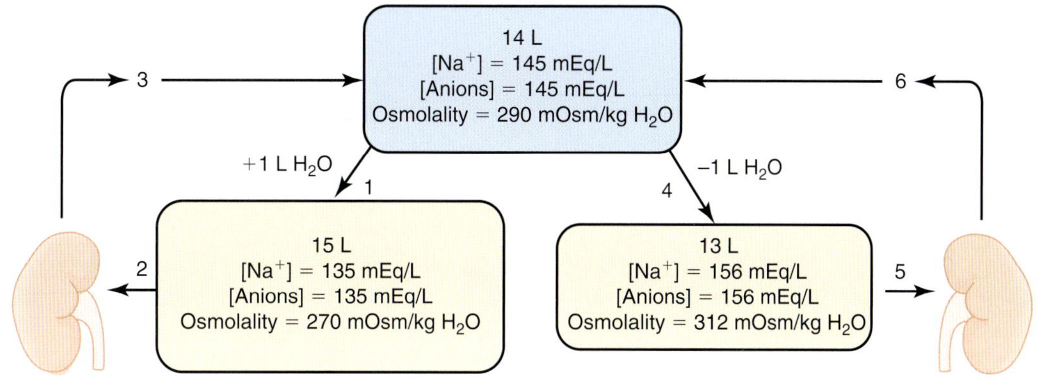

14 L
[Na⁺] = 145 mEq/L
[Anions] = 145 mEq/L
Osmolality = 290 mOsm/kg H₂O

3

6

+1 L H₂O
1

−1 L H₂O
4

15 L
[Na⁺] = 135 mEq/L
[Anions] = 135 mEq/L
Osmolality = 270 mOsm/kg H₂O

2

13 L
[Na⁺] = 156 mEq/L
[Anions] = 156 mEq/L
Osmolality = 312 mOsm/kg H₂O

5

Kidneys excrete 1 L of water
in hypoosmotic urine,
returning volume to 14 L and
restoring [Na⁺] and osmolality
to normal.

Kidneys excrete hyperosmotic
urine as the individual drinks
water, returning volume to 14 L
and restoring [Na⁺] and osmolality
to normal.

• **Fig. 35.2** Response to changes in water balance. Illustrated are the effects of adding or removing 1 L of water from the ECF of a 70-kg individual. **Positive Water Balance:** (1) Addition of 1 L of water increases the ECFV and reduces its osmolality. The [Na⁺] is also decreased (hyponatremia). (2) The normal renal response is to excrete 1 L of water as hypoosmotic urine. (3) As a result of the renal excretion of water, the ECFV, osmolality, and [Na⁺] are returned to normal. **Negative Water Balance:** (4) The loss of 1 L of water from the ECF decreases its volume and increases its osmolality. The [Na⁺] is also increased (hypernatremia). (5) The renal response is to conserve water by excreting a small volume of hyperosmotic urine. (6) With ingestion of water, stimulated by thirst, and conservation of water by the kidneys, the ECFV, osmolality, and [Na⁺] are returned to normal. Size of the boxes indicates relative ECFV. (From Koeppen BM, Stanton BA. *Renal Physiology.* 5th ed. Philadelphia: Elsevier; 2013.)

IN THE CLINIC

In the clinical setting, **hypoosmolality** (a reduction in plasma osmolality) shifts water into cells, and this process results in cell swelling (see Chapter 2). Symptoms associated with hypoosmolality are related primarily to swelling of brain cells. For example, a rapid fall in P_{osm} can alter neurological function and thereby cause nausea, malaise, headache, confusion, lethargy, seizures, and coma. When P_{osm} is increased (i.e., **hyperosmolality**), water is lost from cells. Symptoms of an increase in P_{osm} are also primarily neurological and include lethargy, weakness, seizures, coma, and even death.

Symptoms associated with changes in body fluid osmolality vary depending on how quickly osmolality is changed. Rapid changes in osmolality (i.e., over hours) are less well tolerated than changes that occur more gradually (i.e., over days to weeks). Indeed, individuals who have developed alterations in their body fluid osmolality over an extended period of time may be entirely asymptomatic. This reflects the ability of cells over time to either eliminate intracellular osmoles, as occurs with hypoosmolality, or to generate new intracellular osmoles in response to hyperosmolality and thus minimize changes in cell volume of the neurons (see Chapter 2).

Arginine Vasopressin

The human form of **vasopressin** is **arginine vasopressin (AVP),** which is also known as **antidiuretic hormone (ADH).** AVP, acting though V_1 receptors, causes vascular smooth muscle contraction. As described subsequently,

several nephron segments express a different AVP receptor (V_2) that mediates the kidneys' ability to regulate the volume and osmolality of urine. When plasma AVP levels are low, a large volume of urine is excreted **(diuresis),** and the urine osmolality is less than that of plasma (i.e., dilute).[a] When plasma levels of AVP are high, a small volume of urine is excreted **(antidiuresis),** and the urine osmolality is greater than that of plasma (i.e., concentrated).

AVP is a small peptide that is 9 amino acids in length (arginine is found at position 8). It is synthesized in neuroendocrine cells located within the supraoptic and paraventricular nuclei of the hypothalamus.[b] The synthesized hormone is packaged in granules that are transported down the axon of the cell and stored in nerve terminals located in the neurohypophysis (posterior pituitary). The anatomy of the hypothalamus and pituitary gland are shown in Fig. 35.3 (see also Chapter 41).

Secretion of AVP by the posterior pituitary can be influenced by several factors. The primary physiological regulators of AVP secretion are (1) the osmolality of body fluids (osmotic) and (2) the volume and pressure of the vascular system (hemodynamic, or nonosmotic). Other factors that can alter AVP secretion include nausea (stimulates), atrial

---

[a]*Diuresis* is the term used for excretion of a large volume of urine. This may reflect either excretion of a large volume of water (**water diuresis**), or excretion of a large amount of solute (**solute diuresis**).
[b]Neurons within the supraoptic and paraventricular nuclei synthesize either AVP or the related peptide oxytocin. AVP-secreting cells predominate in the supraoptic nucleus, whereas oxytocin-secreting neurons are primarily found in the paraventricular nucleus.

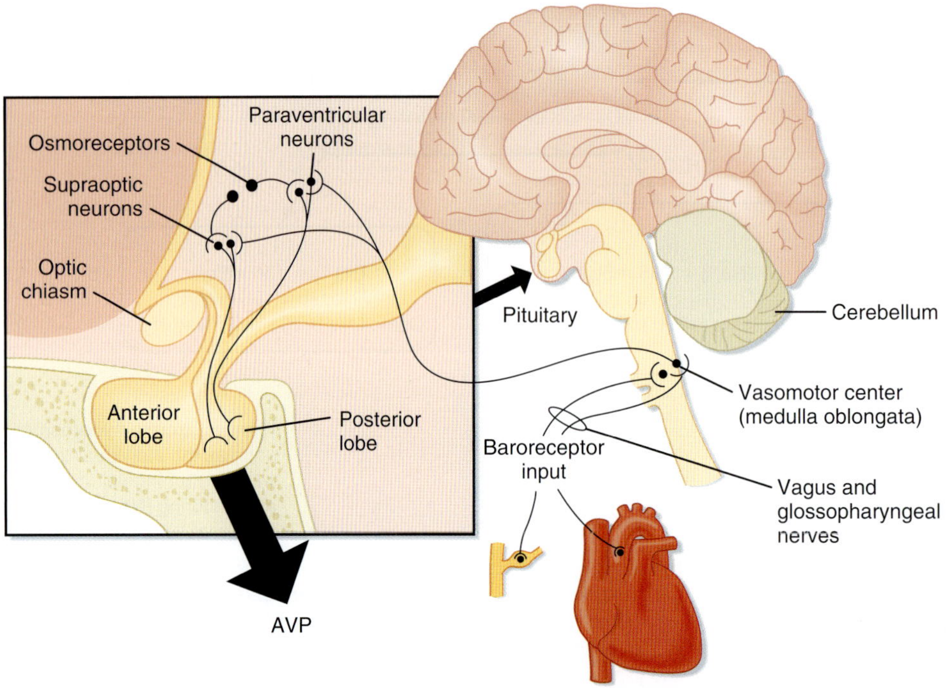

• **Fig. 35.3** Anatomy of the hypothalamus and pituitary gland (midsagittal section) depicting the pathways for AVP section. Also shown are pathways involved in regulating AVP secretion. Afferent fibers from the baroreceptors are carried in the vagus and glossopharyngeal nerves. The inset box illustrates an expanded view of the hypothalamus and pituitary gland.

natriuretic peptide (inhibits), and angiotensin II (stimulates). A number of drugs, prescription and nonprescription, also affect AVP secretion. For example, nicotine stimulates secretion, whereas ethanol inhibits secretion.

Osmotic Control of AVP Secretion

Changes in the osmolality of body fluids (changes as minor as 1% are sufficient) play the most important role in regulating AVP secretion. The receptors that monitor changes in osmolality of body fluids (termed *osmoreceptors*) are distinct from the cells that synthesize and secrete AVP, and are located in the organum vasculosum of the lamina terminalis (OVLT) of the hypothalamus.[c] The osmoreceptors sense changes in body fluid osmolality by either shrinking or swelling. Recent studies have provided evidence that transient receptor potential vanilloid (TRVP) cation channels are involved in the response of the cells to changes in body fluid osmolality. The osmoreceptors respond only to solutes in plasma that are *effective osmoles* (see Chapter 1). For example, urea is an ineffective osmole when the function of osmoreceptors is considered. Thus elevation of the plasma urea concentration alone has little effect on AVP secretion.

When the effective osmolality of the plasma increases, the osmoreceptors send signals to the AVP synthesizing/secreting cells located in the supraoptic and paraventricular nuclei of the hypothalamus, and AVP synthesis and secretion are stimulated. Conversely, when the effective

---

[c]Angiotensin II also stimulates AVP secretion; the cells that mediate this response are located in the subfornical organ (SFO).

🧬 AT THE CELLULAR LEVEL

The gene for AVP is found on chromosome 20. It contains approximately 2000 base pairs with three exons and two introns. The gene codes for a preprohormone that consists of a signal polypeptide, the AVP molecule, neurophysin, and a glycopeptide (copeptin). As the cell processes the preprohormone the signal peptide is cleaved off in the rough endoplasmic reticulum. Once packaged in neurosecretory granules, the preprohormone is further cleaved into AVP, neurophysin, and copeptin molecules. The neurosecretory granules are then transported down the axon to the posterior pituitary and stored in the nerve endings until released. When the neurons are stimulated to secrete AVP, the action potential opens Ca^{++} channels in the nerve terminal, which raises the intracellular $[Ca^{++}]$ and causes exocytosis of the neurosecretory granules. All three peptides are secreted in this process. Neurophysin and copeptin do not have an identified physiological function.

osmolality of the plasma is reduced, secretion is inhibited. Because AVP is rapidly degraded in the plasma, circulating levels can be reduced to zero within minutes after secretion is inhibited. As a result the AVP system can respond rapidly to fluctuations in body fluid osmolality.

Fig. 35.4A illustrates the effect of changes in plasma osmolality on circulating AVP levels. The slope of the relationship is quite steep and accounts for the sensitivity of this system. The set point of the system is the plasma osmolality value at which AVP secretion begins to increase. Below this set point, virtually no AVP is released. The set point varies

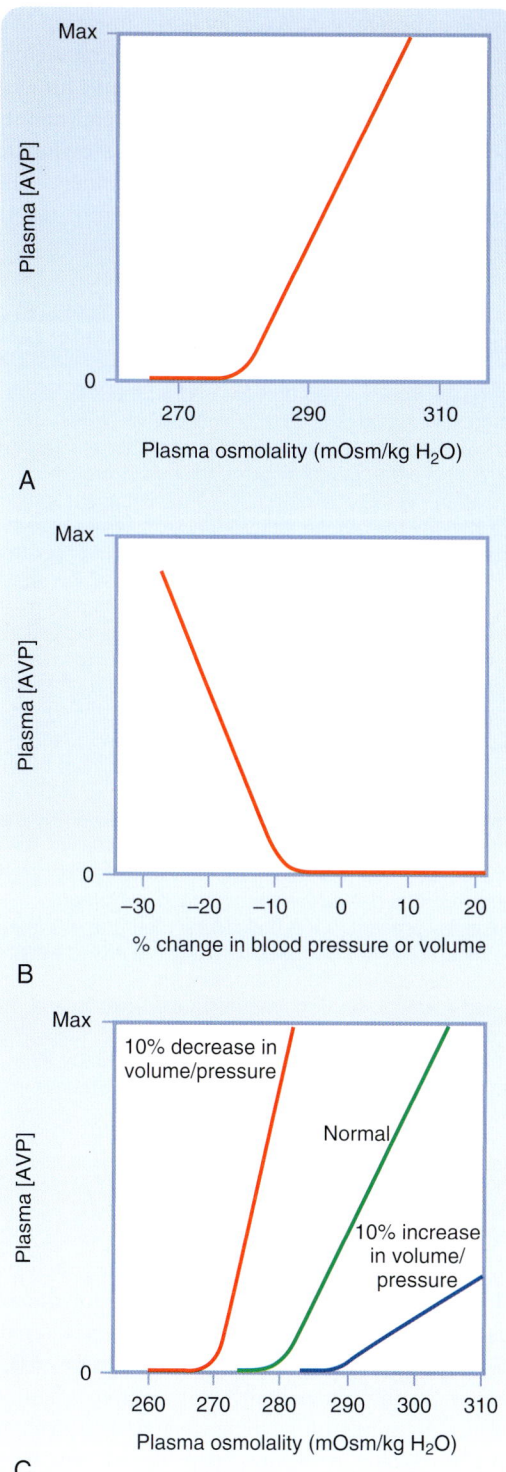

among individuals and is genetically determined. In healthy adults it varies from 275 to 290 mOsm/kg H_2O (average ≈ 280–285 mOsm/kg H_2O). Several physiological factors can also change the set point in a given individual. As discussed later, alterations in blood volume and pressure can shift it. In addition, pregnancy is associated with a decrease in the set point. The mechanism responsible for the set-point shift with pregnancy is not completely understood but is likely due to hormone levels (e.g., relaxin and chorionic gonadotropin) that are elevated during pregnancy.

Hemodynamic (Nonosmotic) Control of AVP Secretion

A decrease in blood volume or pressure also stimulates AVP secretion. The receptors responsible for this response are located in both the low-pressure (left atrium and large pulmonary vessels) and the high-pressure (aortic arch and carotid sinus) sides of the circulatory system. Because the low-pressure receptors are located in the high-compliance side of the circulatory system (i.e., venous), and because the majority of blood is in the venous side of the circulatory system, these low-pressure receptors can be viewed as responding to overall vascular volume. The high-pressure receptors respond to arterial pressure. Both groups of receptors are sensitive to stretch of the wall of the structure in which they are located (e.g., cardiac atrial and aortic arch) and are termed *baroreceptors*. Signals from these receptors are carried in afferent fibers of the vagus and glossopharyngeal nerves to the brainstem (solitary tract nucleus of the medulla oblongata), which is part of the center that regulates heart rate and blood pressure (see also Chapter 18). Signals are then relayed from the brainstem to the AVP secretory cells of the supraoptic and paraventricular hypothalamic nuclei. The sensitivity of the baroreceptor system is less than that of the osmoreceptors, and a 5% to 10% decrease in blood volume or pressure is required before AVP secretion is stimulated. This is illustrated in Fig. 35.4B. A number of substances have been shown to alter the secretion of AVP through their effects on blood pressure. These include bradykinin and histamine, which lower pressure and thus stimulate AVP secretion, and norepinephrine, which increases blood pressure and inhibits AVP secretion.

Alterations in blood volume and pressure also affect the response to changes in body fluid osmolality (see Fig. 35.4C). With a decrease in blood volume or pressure, the set point is shifted to lower osmolality values and the slope of the relationship is steeper. In terms of survival of the individual this means that faced with circulatory collapse, the kidneys will continue to conserve water, even though by doing so they reduce the osmolality of the body fluids. With an increase in blood volume or pressure, the opposite occurs. The set point is shifted to higher osmolality values and the slope is decreased.

AVP Actions on the Kidneys

The primary action of AVP on the kidneys is to enhance absorption of water from the tubular fluid by increasing

• **Fig. 35.4** Osmotic and hemodynamic (nonosmotic) control of AVP secretion. A, Effect of changes in plasma osmolality (constant blood volume and pressure) on plasma AVP levels. B, Effect of changes in blood volume or pressure (constant plasma osmolality) on plasma AVP levels. C, Interactions between osmolar and blood volume and pressure stimuli on plasma AVP levels.

⚕ IN THE CLINIC

Inadequate release of AVP from the posterior pituitary results in excretion of a large volume of dilute urine **(polyuria)**. To compensate for this loss of water the individual must ingest a large volume of water **(polydipsia)** to maintain constant body fluid osmolality. If the individual is deprived of water, body fluids will become hyperosmotic. This condition is called **central diabetes insipidus** or **pituitary diabetes insipidus.** Central diabetes insipidus can be inherited, although this is rare. It occurs more commonly after head trauma and with brain neoplasms or infections. Individuals with central diabetes insipidus have a urine-concentrating defect that can be corrected by administration of exogenous AVP. The inherited (autosomal dominant) form of central diabetes insipidus is due to numerous mutations in all regions of the *AVP* gene (i.e., AVP, copeptin and neurophysin). The human placenta produces a cysteine aminopeptidase that degrades AVP. In some women the levels of this vasopressinase result in diabetes insipidus. The associated polyuria can be treated by administration of the synthetic AVP analogue **desmopressin (DDAVP).**

The **syndrome of inappropriate AVP (ADH) secretion (SIADH)** is a common clinical problem characterized by plasma AVP levels that are elevated above what would be expected on the basis of body fluid osmolality and blood volume and pressure—hence the term *inappropriate AVP (ADH) secretion* (this is alternatively named *syndrome of inappropriate antidiuresis [SIAD]*). In addition the collecting duct overexpresses water channels (see below), thus augmenting the effect of AVP to stimulate water retention by the kidney. Individuals with SIADH retain water, and their body fluids become progressively hypoosmotic. In addition, their urine is more hyperosmotic than expected based on the low body-fluid osmolality. SIADH can be caused by infections and neoplasms of the brain, drugs (e.g., antitumor drugs), pulmonary diseases, and carcinoma of the lung. Many of these conditions stimulate AVP secretion by altering neural input to the AVP secretory cells. By contrast, small cell carcinoma of the lung produces and secretes a number of peptides including AVP. Recently, AVP receptor antagonists (e.g., the nonpeptide antagonists conivaptan [Vaprisol] and tolvaptan [Samsca and Jinarc]) have been developed that can be used to treat SIADH and other conditions in which AVP-dependent water retention by the kidneys occurs (e.g., congestive heart failure and hepatic cirrhosis).

the water permeability of the latter portion of the distal tubule and collecting duct. In addition, and importantly, AVP increases the permeability of the medullary portion of the collecting duct to urea. Finally, AVP stimulates NaCl reabsorption by the thick ascending limb of Henle's loop, distal tubule, and collecting duct.

In the absence of AVP, the apical membrane of principal cells (see Chapter 34), located in the latter portion of the distal tubule and along the collecting duct, is relatively impermeable to water. This reflects the fact that in the absence of AVP the apical membrane of these cells contains few water channels (aquaporins). Thus in the absence of AVP, little water is reabsorbed by these nephron segments. Binding of AVP to the V_2 receptor located in the basolateral

membrane of principal cells results in the insertion of aquaporin (AQP2) water channels into the apical membrane, allowing water to enter the cell from the tubule lumen. This water then exits the cell across the basolateral membrane, which is always freely permeable to water owing to the presence of AQP3 and AQP4 water channels. Thus in the presence of AVP, water is reabsorbed from the tubule lumen.

🧬 AT THE CELLULAR LEVEL

The gene for the V_2 receptor is located on the X chromosome. It codes for a 371–amino acid protein that is in the family of receptors that have seven membrane spanning domains and are coupled to heterotrimeric G proteins. As shown in Fig. 35.5, binding of AVP to its receptor on the basolateral membrane activates adenylcyclase. The increase in intracellular cyclic adenosine monophosphate (cAMP) then activates protein kinase A (PKA), which results in phosphorylation of AQP2 water channels and also results in increased transcription of the *AQP2* gene via activation of a cAMP-response element (CRE). Vesicles containing phosphorylated AQP2 move toward the apical membrane along microtubules driven by the molecular motor dynein. Once near the apical membrane, proteins called *SNAREs* interact with vesicles containing AQP2 and facilitate fusion of these vesicles with the membrane. Addition of AQP2 to the membrane allows water to enter the cell driven by the osmotic gradient (lumen osmolality < cell osmolality). The water then exits the cell across the basolateral membrane through AQP3 and AQP4 water channels, which are constitutively present in the basolateral membrane. When the V_2 receptor is not occupied by AVP, the AQP2 water channels are removed from the apical membrane by clathrin-mediated endocytosis, thus rendering the apical membrane impermeable to water. The endocytosed AQP2 molecules may be either stored in cytoplasmic vesicles, ready for reinsertion into the apical membrane when AVP levels in the plasma increase, or degraded.

AVP also regulates long-term expression of AQP2 (and AQP3). When large volumes of water are ingested over an extended period of time (e.g., psychogenic polydipsia), the abundance of AQP2 and AQP3 in principal cells is reduced. As a consequence, when water ingestion is restricted, these individuals cannot maximally concentrate their urine. Conversely, in states of restricted water ingestion, AQP2 and AQP3 protein expression in principal cells increases, thereby facilitating excretion of maximally concentrated urine.

It is also clear that expression of AQP2 (and in some instances also AQP3) varies in pathological conditions associated with disturbances in urine concentration and dilution. AQP2 expression is reduced in a number of conditions associated with impaired urine concentrating ability (e.g., hypercalcemia, hypokalemia). By contrast, in conditions associated with water retention (e.g., congestive heart failure, hepatic cirrhosis, pregnancy) AQP2 expression is increased.

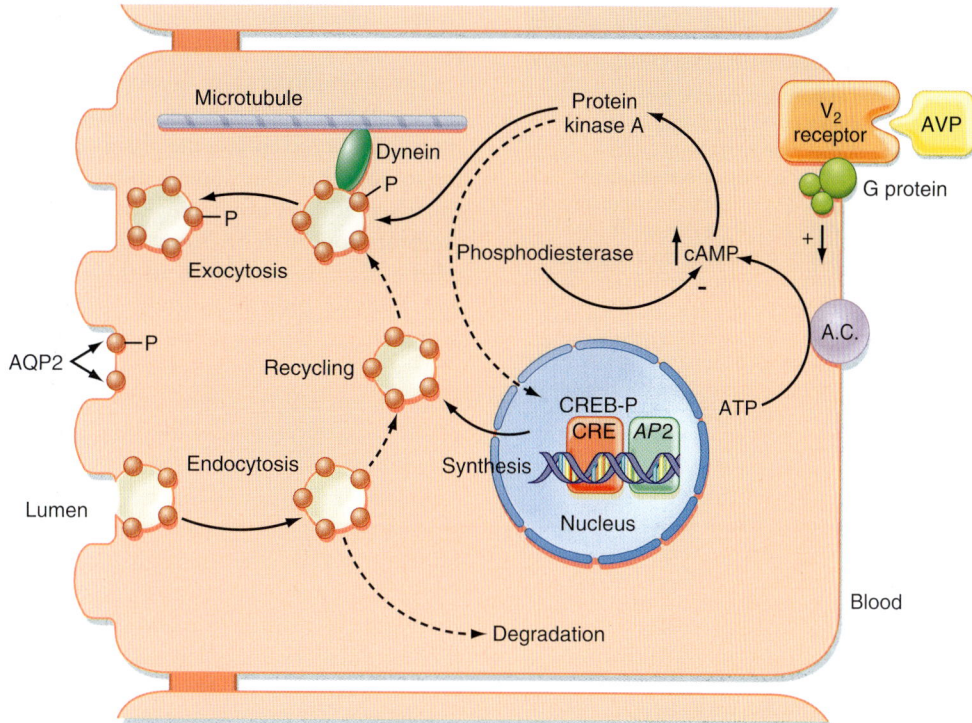

• Fig. 35.5 Action of AVP via the V_2 receptor on the principal cell of the late distal tubule and collecting duct. See text for details. AC, adenylcyclase; cAMP, cyclic adenosine monophosphate; CREB-P, phosphorylated cAMP response element binding protein; CRE, cAMP response element; P, phosphorylated proteins; AQP2, aquaporin 2; AP2, aquaporin 2 gene. (Adapted from Brown D, Nielsen S. The cell biology of vasopressin action. In: Brenner BM [ed]. *The Kidney.* 7th ed. Philadelphia: Saunders; 2004.)

AVP also increases the permeability of the terminal portion of the inner medullary collecting duct to urea. This results in an increase in urea reabsorption and an increase in the osmolality of the medullary interstitial fluid, which as described below is needed for maximal urine concentration. The cells of the collecting duct express two types of urea transporters, UT-A1 and UT-A3. UT-A1 is localized to the apical membrane, and UT-A3 is localized primarily to the basolateral membrane. AVP, acting through the cAMP/PKA cascade, increases expression of both UT-A1 and UT-A3. Increasing the osmolality of the interstitial fluid of the renal medulla also increases the permeability of the inner medullary collecting duct to urea. This effect is mediated by the phospholipase C/protein kinase C (PKC) pathway, which increases UT-A1 and UT-A3 expression.

AVP also stimulates reabsorption of NaCl by the thick ascending limb of Henle's loop and by the distal tubule and cortical segment of the collecting duct. This increase in Na^+ reabsorption is associated with increased abundance of three Na^+ transporters: the $1Na^+/1K^+/2Cl^-$ symporter (thick ascending limb of Henle's loop), the Na^+/Cl^- symporter (distal tubule), and the Na^+ channel (ENaC, in the latter portion of the distal tubule and collecting duct). Stimulation of thick ascending limb NaCl transport may help maintain the hyperosmotic medullary interstitium that is necessary for absorption of water from the medullary portion of the collecting duct (see below).

Thirst

In addition to affecting the secretion of AVP, changes in plasma osmolality and blood volume or pressure lead to alterations in the perception of thirst. When body fluid osmolality is increased or the blood volume or pressure is reduced, the individual perceives thirst. Of these stimuli, hypertonicity is the more potent. An increase in plasma osmolality of only 2% to 3% produces a strong desire to drink, whereas decreases in blood volume and pressure in the range of 10% to 15% are required to produce the same response.

As already discussed, there is a genetically determined threshold for AVP secretion (i.e., a body fluid osmolality above which AVP secretion increases). Similarly there is a genetically determined threshold for triggering the sensation of thirst. However, the thirst threshold is higher than the threshold for AVP secretion. On average the threshold for AVP secretion is approximately 285 mOsm/kg H_2O, whereas the thirst threshold is approximately 295 mOsm/kg H_2O. Because of this difference, thirst is stimulated at a body fluid osmolality where AVP secretion is already maximal.

The neural centers involved in regulating water intake (the thirst center) are located in the same region of the hypothalamus involved with regulating AVP secretion. However, it is not certain whether the same cells serve both functions. Indeed the thirst response, like the regulation of

IN THE CLINIC

The collecting ducts of some individuals do not respond normally to AVP. These individuals cannot maximally concentrate their urine and consequently have polyuria and polydipsia. This clinical entity is termed **nephrogenic diabetes insipidus** to distinguish it from central diabetes insipidus. Nephrogenic diabetes insipidus can result from a number of systemic disorders and more rarely occurs as a result of inherited disorders. Many of the acquired forms of nephrogenic diabetes insipidus are the result of decreased expression of AQP2 in the collecting duct. Decreased expression of AQP2 has been documented in the urine concentrating defects associated with hypokalemia, lithium ingestion (35% of individuals who take lithium for bipolar disorder develop some degree of nephrogenic diabetes insipidus), ureteral obstruction, low-protein diet, and hypercalcemia. The inherited forms of nephrogenic diabetes insipidus reflect mutations in the AVP receptor (V_2 receptor) gene or the *AQP2* gene. Approximately 90% of hereditary forms of nephrogenic diabetes insipidus are the result of mutations in the V_2 receptor gene, with the other 10% the result of mutations in the *AQP2* gene. Since the gene for the V_2 receptor is located on the X chromosome, these inherited forms of nephrogenic diabetes insipidus are X-linked. The gene coding for AQP2 is located on chromosome 12 and is inherited as both an autosomal recessive as well as an autosomal dominant defect. As noted in Chapter 1, aquaporins exist as homotetramers. This homotetramer formation explains the difference between the two forms of AQP2-related nephrogenic diabetes insipidus. In the recessive form, heterozygotes produce both normal AQP2 and defective AQP2 molecules. The defective AQP2 monomer is retained in the endoplasmic reticulum of the cell, and thus the homotetramers that do form only contain normal molecules. Thus mutations in both alleles are required to produce nephrogenic diabetes insipidus. In the autosomal dominant form, the defective monomers can form tetramers with normal monomers as well as defective monomers. However, tetramers containing defective monomers are unable to traffic to the apical membrane.

Recently, individuals have been found that have activating (gain-of-function) mutations in the V_2 receptor gene. Thus the receptor is constitutively activated even in the absence of AVP. These individuals have laboratory findings similar to those seen in SIADH, including reduced plasma osmolality, hyponatremia (reduced plasma [Na^+]), and urine more concentrated than would be expected from the reduced body fluid osmolality. However, unlike SIADH where circulating levels of AVP are elevated and thus responsible for water retention by the kidneys, these individuals have undetectable levels of AVP in their plasma. This new clinical entity has been termed *nephrogenic syndrome of inappropriate antidiuresis.*

AVP secretion, only occurs in response to effective osmoles (e.g., NaCl). Even less is known about the pathways involved in the thirst response to decreased blood volume or pressure, but it is believed that the pathways are the same as those involved in the volume and pressure-related regulation of AVP secretion. Angiotensin II acting on cells of the thirst center also evokes the sensation of thirst. Because angiotensin II levels are increased when blood volume and pressure are reduced, this effect of angiotensin II contributes to the homeostatic response that restores and maintains body fluids at their normal volume.

The sensation of thirst is satisfied by the act of drinking, even before sufficient water is absorbed from the gastrointestinal tract to correct the plasma osmolality. It is interesting to note that cold water is more effective in reducing the thirst sensation. Oropharyngeal and upper gastrointestinal receptors appear to be involved in this response. However, relief of the thirst sensation via these receptors is short lived, and thirst is only completely satisfied when the plasma osmolality or blood volume or pressure is corrected.

It should be apparent that the AVP and thirst systems work in concert to maintain water balance. An increase in plasma osmolality evokes drinking and, via AVP action on the kidneys, conservation of water. Conversely, when plasma osmolality is decreased, thirst is suppressed and, in the absence of AVP, renal water excretion is enhanced. However, most of the time fluid intake is dictated by cultural factors and social situations. This is especially the case when thirst is not stimulated. In this situation, maintaining normal body fluid osmolality relies solely on the ability of the kidneys to excrete water. How the kidney accomplishes this is discussed in detail in the following sections of this chapter.

Renal Mechanisms for Dilution and Concentration of Urine

As already noted, water excretion is regulated separately from solute excretion. For this to occur the kidneys must be able to excrete urine that is either hypoosmotic or hyperosmotic with respect to body fluids. This ability to excrete urine of varying osmolality in turn requires that solute be separated from water at some point along the nephron. As discussed in Chapter 34, reabsorption of solute in the proximal tubule results in reabsorption of a proportional amount of water. Hence solute and water are not separated in this portion of the nephron. Moreover, this proportionality between proximal tubule water and solute reabsorption occurs regardless of whether the kidneys excrete dilute or concentrated urine. Thus the proximal tubule reabsorbs a large portion of the filtered amount of solute and water, but it does not produce dilute or concentrated tubular fluid. The loop of Henle, in particular the thick ascending limb, is the major site where solute and water are separated. Thus excretion of both dilute and concentrated urine requires normal function of the loop of Henle.

Excretion of hypoosmotic urine is relatively easy to understand. The nephron must simply reabsorb solute from the tubular fluid and not allow water reabsorption to also occur. As just noted, and as described in greater detail later, reabsorption of solute without concomitant water reabsorption occurs in the ascending limb of Henle's loop. Under appropriate conditions (i.e., in the absence of AVP) the distal tubule and collecting duct also dilute the tubular fluid by reabsorbing solute but not water.

Excretion of hyperosmotic urine is more complex and thus more difficult to understand. This process in essence

IN THE CLINIC

With adequate access to water, the thirst mechanism can prevent development of hyperosmolality. This mechanism is responsible for the polydipsia seen in response to the polyuria of both central and nephrogenic diabetes insipidus. Most individuals ingest water/beverages even in the absence of the thirst sensation. Normally the kidneys are able to excrete this excess water because they can excrete up to 18 L/day of urine. However, in some instances, the volume of water ingested exceeds the kidneys' capacity to excrete water, especially over short periods of time. When this occurs, body fluids become hypoosmotic.

An example of how water intake can exceed the capacity of the kidneys to excrete water is long-distance running. A study of participants in the Boston Marathon found that 13% of the runners developed hyponatremia during the course of the race.[d]

This reflected the practice of some runners of ingesting water or other hypotonic drinks during the race to remain "well hydrated." In addition, water is produced from the metabolism of glycogen and triglycerides used as fuels by the exercising muscle. Because over the course of the race they ingested (and generated through metabolism) more water than their kidneys were able to excrete or was lost by sweating, hyponatremia developed. In some racers the hyponatremia was severe enough to elicit the neurological symptoms described previously.

The maximum amount of water that can be excreted by the kidneys (e.g., 18 L/day) depends on the amount of solute excreted, which in turn depends on food intake. For example, with maximally dilute urine ($U_{osm} = 50$ mOsm/kg H_2O), the maximum urine output of 18 L/day will be achieved only if the solute excretion rate is 900 mmol/day:

$$U_{osm} = \text{Solute excretion/Volume excreted}$$

$$50 \text{ mOsm/kg } H_2O = 900 \text{ mmol/18 L}$$

If solute excretion is reduced, as commonly occurs in the elderly with reduced food intake, the maximum urine output will decrease. For example, if solute excretion is only 400 mmol/day, a maximum urine output (at $U_{osm} = 50$ mOsm/kg H_2O) of only 8 L/day can be achieved. Thus individuals with reduced food intake have a reduced capacity to excrete water.

[d]Almond CS et al. Hyponatremia among runners in the Boston Marathon. *N Engl J Med* 2005;352:1150-1556.

involves removing water from the tubular fluid without solute. Because water movement is passive, driven by an osmotic gradient, the kidney must generate a hyperosmotic compartment into which water is reabsorbed, without solute, osmotically from the tubular fluid. The hyperosmotic compartment in the kidney that serves this function is the interstitium of the renal medulla. Henle's loop is critical for generating the hyperosmotic medullary interstitium. Once established, this hyperosmotic compartment drives water reabsorption from the collecting duct and thereby concentrates urine.

Fig. 35.6 summarizes tubular fluid osmolality at several points along the nephron, both in the absence and presence

of AVP. Note that tubular fluid entering the loop of Henle from the proximal tubule is isoosmotic with respect to plasma and is so regardless of the absence or presence of AVP. Also, tubular fluid leaving the thick ascending limb is hypoosmotic with respect to plasma, both in the absence and presence of AVP. The osmolality of tubular fluid along the collecting duct is hypoosmotic with respect to plasma in the absence of AVP and becomes progressively hyperosmotic (i.e., from cortex to inner medulla) in the presence of AVP.

Establishment and maintenance of the hyperosmotic medullary interstitium has been a subject of study for more than 50 years. Despite this intense study, the most accepted model for how the medullary osmotic gradient is established, especially within the inner medulla, is incomplete and not consistent with more recent experimental findings regarding the transport properties of the nephron segments in this region of the kidney. With the caveat that the current model needs refinement, it is presented here because it embodies some fundamental concepts that underlie the process.

In the current model the medullary interstitial osmotic gradient is established by a process termed **countercurrent multiplication**. By this process, solute (principally NaCl) is reabsorbed without water from the ascending limb of Henle's loop into the surrounding medullary interstitium. This decreases the osmolality in the tubular fluid and raises the osmolality of the interstitium at this point. The increased osmolality of the interstitium then causes water to be reabsorbed from the descending limb of Henle's loop, thus increasing the tubular fluid osmolality in this segment. Thus at any point along the loop of Henle the fluid in the ascending limb has an osmolality less than fluid in the adjacent descending limb. This osmotic difference was termed the **single effect**. Because of the countercurrent flow of tubular fluid in the descending (fluid flowing into the medulla) and ascending (fluid flow out of the medulla) limbs, this single effect could be multiplied, resulting in an osmotic gradient within the medullary interstitium, where the tip of the papilla has an osmolality of 1200 mOsm/kg H_2O compared to 300 mOsm/kg H_2O at the corticomedullary junction.

Fig. 35.7 schematically depicts the processes for excreting a dilute urine (water diuresis) as well as a concentrated urine (antidiuresis). Three key concepts underlie these processes:

1. Urine is concentrated by AVP-dependent reabsorption of water from the collecting duct.
2. Reabsorption of NaCl from the ascending limb of Henle's loop dilutes the tubular fluid and at the same time generates a high [NaCl] in the medullary interstitium (up to 600 mmol/L at the tip of the papilla), which then drives water reabsorption from the collecting duct.
3. Urea accumulates in the medullary interstitium (up to 600 mmol/L), which allows the kidneys to excrete urine with the same high urea concentration. This allows large amounts of urea to be excreted with relatively little water.

First, how the kidneys excrete dilute urine (**water diuresis**) when AVP levels are low or zero is considered. The following numbers refer to those encircled in Fig. 35.7A:

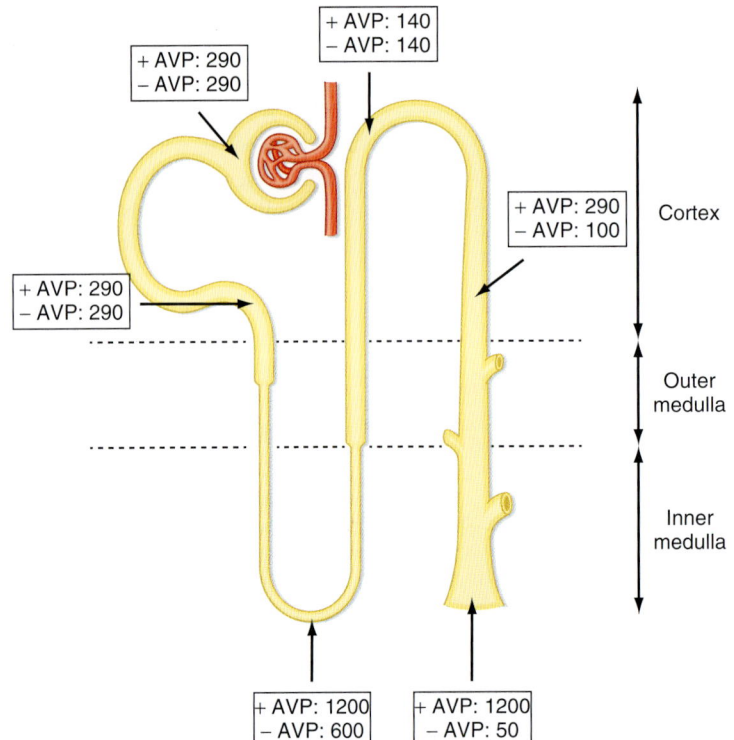

• **Fig. 35.6** Tubular fluid osmolality along the nephron in the presence (+AVP) and in the absence (−AVP) of arginine vasopressin. See text for details. (Adapted from Sands JM et al. Urine concentration and dilution. In: *Brenner and Rector's The Kidney.* 9th ed. Philadelphia: Saunders; 2012.)

1. Fluid entering the descending thin limb of the loop of Henle from the proximal tubule is isosmotic with respect to plasma. This reflects the essentially isosmotic nature of solute and water reabsorption in the proximal tubule (see Chapter 34). (NOTE: Water is reabsorbed from the segments of the proximal tubule via AQP1.)
2. Water is reabsorbed from the thin descending limb of Henle's loop. Most of this water is reabsorbed in the outer medulla, thereby limiting the amount of water added to the deepest part of the inner medullary interstitial space and thus preserving the hyperosmolality of this region of the medulla. (NOTE: Water is reabsorbed via AQP1.)
3. In the inner medulla the terminal portion of the descending thin limb and all of the thin ascending limb is impermeable to water. (NOTE: AQP1 is *not* expressed.) These same nephron segments express the Cl⁻ channel CLC-K1, which mediates Cl⁻ reabsorption, with Na⁺ following via the paracellular pathway. This passive reabsorption of NaCl without concomitant water reabsorption begins the process of diluting the tubular fluid.
4. The thick ascending limb of the loop of Henle is also impermeable to water and actively reabsorbs NaCl from the tubular fluid and thereby dilutes it further (see Chapter 34). Dilution occurs to such a degree that this segment is often referred to as the **diluting segment** of the kidney. Fluid leaving the thick ascending limb is hypoosmotic with respect to plasma (see Fig. 35.6).
5. The distal tubule and cortical portion of the collecting duct actively reabsorb NaCl. In the absence of AVP these segments are not permeable to water (i.e., AQP2 is not present in the apical membrane of the cells). Thus when AVP is absent or present at low levels (i.e., decreased plasma osmolality), the osmolality of tubule fluid in these segments is reduced further because NaCl is reabsorbed without water. Under this condition, fluid leaving the cortical portion of the collecting duct is hypoosmotic with respect to plasma (see Fig. 35.6).
6. The medullary collecting duct actively reabsorbs NaCl. Even in the absence of AVP, this segment is slightly permeable to water and some water is reabsorbed.
7. The urine has an osmolality as low as approximately 50 mOsm/kg H₂O and contains low concentrations of NaCl. The volume of urine excreted can be as much as 18 L/day, or approximately 10% of the glomerular filtration rate (GFR).

Next, how the kidneys excrete concentrated urine **(antidiuresis)** when plasma osmolality and plasma AVP levels are high is considered. The following numbers refer to those encircled in Fig. 35.7B:

1–4. These steps are similar to those for production of dilute urine. An important point in understanding how a concentrated urine is produced is to recognize that while reabsorption of NaCl by the ascending thin and thick limbs of the loop of Henle dilutes the tubular fluid, the reabsorbed NaCl accumulates in the medullary interstitium and raises the osmolality of this compartment.

Water diuresis

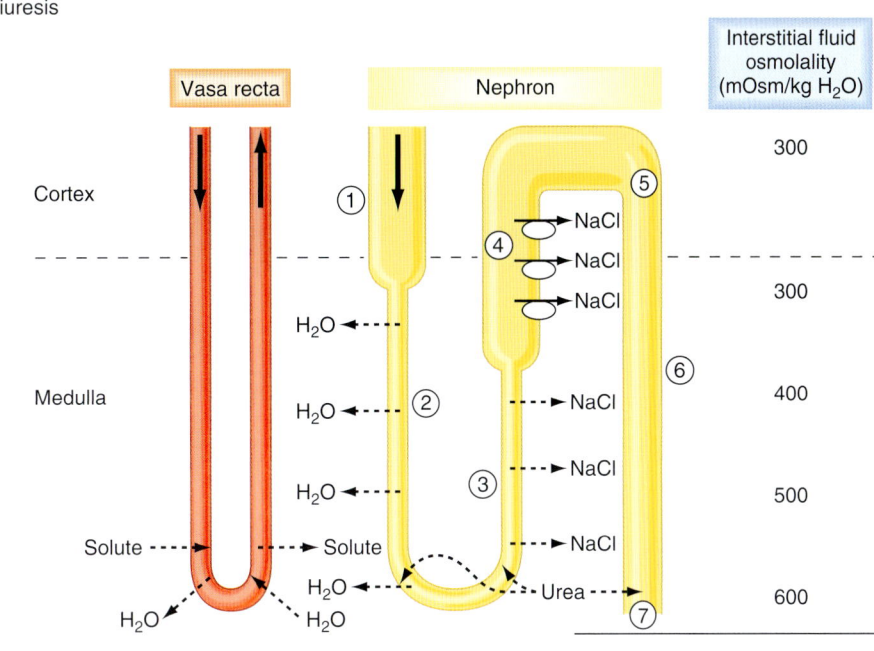

Antidiuresis

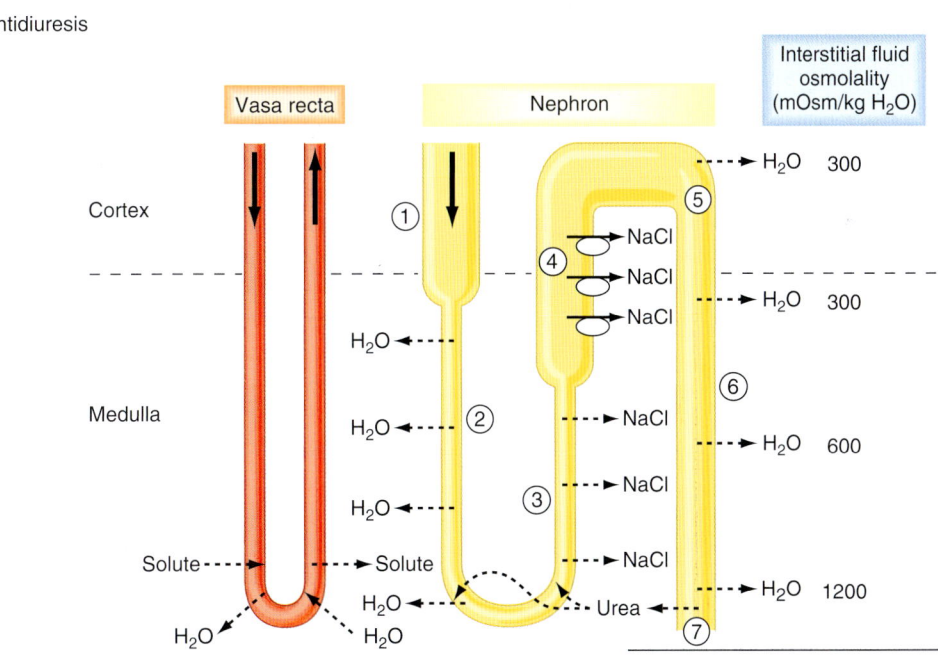

• **Fig. 35.7** Schematic of nephron segments involved in urine dilution and concentration. Henle's loops of juxtamedullary nephrons are shown. A, Mechanism for excretion of dilute urine (water diuresis). AVP) is absent and the collecting duct is essentially impermeable to water. Note also that during a water diuresis the osmolality of the medullary interstitium is reduced as a result of increased vasa recta blood flow and entry of some urea into the medullary collecting duct. B, Mechanism for excretion of a concentrated urine (antidiuresis). Plasma AVP levels are maximal and the collecting duct is highly permeable to water. Under this condition the medullary interstitial gradient is maximal. See text for details.

Accumulation of NaCl in the medullary interstitium is crucial for production of urine hyperosmotic to plasma, because it provides the osmotic driving force for water reabsorption by the medullary collecting duct. As already noted, AVP stimulates NaCl reabsorption by the thick ascending limb of Henle's loop. This is thought to maintain the medullary interstitial gradient at a time when water is being added to this compartment from the medullary collecting duct, which would tend to dissipate the gradient.

5. Because of NaCl reabsorption by the ascending limb of the loop of Henle, the fluid reaching the collecting duct is hypoosmotic with respect to the surrounding interstitial fluid. Thus an osmotic gradient is established across the collecting duct. In the presence of AVP, which increases the water permeability of the latter portion of the distal tubule and the collecting duct by causing insertion of AQP2 into the luminal membrane of the cells, water diffuses out of the tubule lumen and the tubule fluid osmolality increases. This diffusion of water out of the lumen of the collecting duct begins the process of urine concentration. The maximum osmolality the fluid in the distal tubule and cortical portion of the collecting duct can attain is approximately 290 mOsm/kg H_2O (i.e., the same as plasma), which is the osmolality of the interstitial fluid and plasma within the cortex of the kidney.

6. As the tubular fluid descends deeper into the medulla, water continues to be reabsorbed from the collecting duct, increasing the tubular fluid osmolality to 1200 mOsm/kg H_2O at the tip of the papilla.

7. The urine produced when AVP levels are elevated has an osmolality of 1200 mOsm/kg H_2O and contains high concentrations of urea and other nonreabsorbed solutes. Urine volume under this condition can be as low as 0.5 L/day.

In comparing the two conditions just described, it should be apparent that a relatively constant volume of dilute tubular fluid is delivered to the AVP-sensitive portions of the nephron (latter portion of the distal tubule and collecting duct). Plasma AVP levels then determine the amount of water reabsorbed by these segments. When AVP levels are low, a relatively small volume of water is reabsorbed by these segments and a large volume of hypoosmotic urine is excreted (up to 10% of the filtered water). When AVP levels are high, a large volume of water is reabsorbed by these same segments and a small volume of hyperosmotic urine is excreted (<1% of filtered water). During antidiuresis, most of the water is reabsorbed in the distal tubule and cortical and outer medullary portions of the collecting duct. Thus a relatively small volume of fluid reaches the inner medullary collecting duct where it is then reabsorbed. This distribution of water reabsorption along the length of the collecting duct (i.e., cortex > outer medulla > inner medulla) allows for maintenance of a hyperosmotic interstitial environment in the inner medulla by minimizing the amount of water entering this compartment.

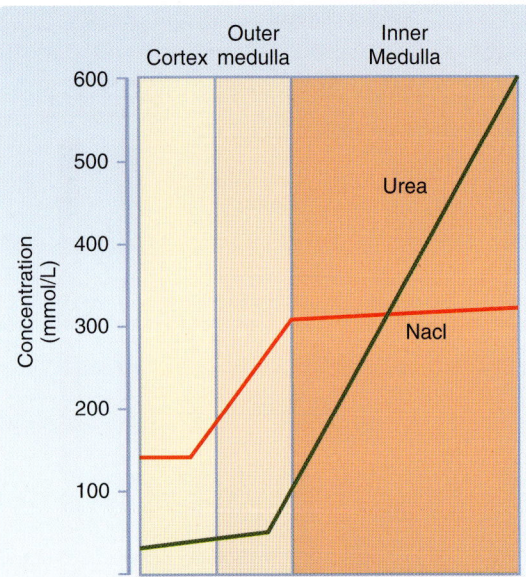

• **Fig. 35.8** The medullary interstitial gradient comprises primarily NaCl and urea. The concentrations for NaCl and urea depicted reflect those found in the antidiuretic state (i.e., excretion of hyperosmotic urine). See text for details. (Adapted from Sands JM et al. Urine concentration and dilution. In: *Brenner and Rector's The Kidney*. 9th ed. Philadelphia: Elsevier; 2012.)

Medullary Interstitium

As noted earlier, the interstitial fluid of the renal medulla is critically important in concentrating urine. The osmotic pressure of the interstitial fluid provides the driving force for reabsorbing water from both the descending thin limb of the loop of Henle and the collecting duct. The principal solutes of the medullary interstitial fluid are NaCl and urea, but the concentration of these solutes is not uniform throughout the medulla (i.e., a gradient exists from cortex to papilla). Other solutes also accumulate in the medullary interstitium (e.g., NH_4^+ and K^+), but the most abundant solutes are NaCl and urea. For simplicity, this discussion assumes that NaCl and urea are the only solutes.

As depicted in Fig. 35.8, NaCl and urea accumulate in the renal medulla, and the interstitial fluid at the tip of the papilla of the inner medulla reaches a maximum osmolality of 1200 mOsm/kg H_2O, with approximately 600 mOsm/kg H_2O attributable to NaCl (300 mmol/L) and 600 mOsm/kg H_2O attributable to urea (600 mmol/L). Establishment of the NaCl gradient is essentially complete at the transition between the outer and inner medulla.

The medullary gradient for NaCl results from accumulation of NaCl reabsorbed by the nephron segments in the medulla during countercurrent multiplication. The most important segment in this regard is the ascending limb of the loop of Henle. Urea accumulation within the medullary interstitium is more complex and occurs most effectively when hyperosmotic urine is excreted (i.e., antidiuresis). When dilute urine is produced, especially over extended periods, the osmolality of the medullary interstitium declines (see Fig. 37.7A). This reduced osmolality is almost entirely caused by a decrease in the concentration of urea. This decrease reflects washout by the vasa recta (discussed later) and diffusion of urea from the interstitium into the tubular fluid within the medullary portion of the collecting duct, which is permeable to urea even in the absence of AVP. (NOTE: The cortical and outer medullary portions of the collecting duct have a low permeability to urea, whereas the inner medullary portion has a relatively high permeability because of the presence of the urea transporters UT-A1 and UT-A3, the expression of which is increased by AVP.) Some of this reabsorbed urea is secreted into the thin descending limbs of Henle's loops via the urea transporter UT-A2, and some enters the vasa recta via the UT-B transporter. The urea that is secreted into the descending thin limbs of Henle's loops is then trapped in the nephron until it again reaches the medullary collecting duct, where it can reenter the medullary interstitium. Thus urea recycles from the interstitium to the nephron and back into the interstitium. This process of urea recycling facilitates accumulation of urea in the medullary interstitium, where it can attain a concentration at the tip of the papilla of 600 mmol/L.

As described, the hyperosmotic medulla is essential for concentrating the tubular fluid within the collecting duct. Because water reabsorption from the collecting duct is driven by the osmotic gradient established in the medullary interstitium, urine can never be more concentrated than that of the interstitial fluid in the papilla. Thus any condition that reduces the medullary interstitial osmolality impairs the ability of the kidneys to maximally concentrate urine. Urea within the medullary interstitium contributes to the total osmolality of the urine. However, because the inner medullary collecting duct is highly permeable to urea, especially in the presence of AVP, urea cannot drive water reabsorption across this nephron segment. Instead, urea in the tubular fluid and medullary interstitium equilibrate, and a small volume of urine with a high concentration of urea is excreted.[c] It is the medullary interstitial NaCl concentration that is responsible for reabsorbing water from the medullary collecting duct and thereby concentrating the nonurea solutes (e.g., NH_4^+ salts, K^+ salts, creatinine) in the urine.

Vasa Recta Function

The **vasa recta,** the capillary networks that supply blood to the medulla, are highly permeable to solute and water. As with the loop of Henle, the vasa recta form a parallel set of hairpin loops within the medulla (see Chapter 33). Not only do the vasa recta bring nutrients and oxygen to the medullary nephron segments, but more importantly they also remove the excess water and solute that is continuously added to the medullary interstitium by these nephron seg-

ments. The ability of the vasa recta to maintain the medullary interstitial gradient is flow dependent. A substantial increase in vasa recta blood flow dissipates the medullary gradient (i.e., washout of osmoles from the medullary interstitium). Alternatively, reduced blood flow reduces oxygen delivery to the nephron segments within the medulla. Because transport of salt and other solutes requires oxygen and ATP, reduced medullary blood flow decreases salt and solute transport by nephron segments in the medulla. As a result the medullary interstitial osmotic gradient cannot be maintained.

Assessment of Renal Diluting and Concentrating Ability

Assessment of renal water handling includes measurements of urine osmolality and the volume of urine excreted. The range of urine osmolality is from 50 to 1200 mOsm/kg H_2O. The corresponding range in urine volume is 18 L to as little as 0.5 L/day. These ranges are not fixed and vary from individual to individual. They can also be affected by disease processes, and as noted previously they are dependent on the amount of solute the kidneys must also excrete.

The ability of the kidneys to dilute or concentrate urine requires the separation of solute and water (i.e., the single effect of the countercurrent multiplication process). This separation of solute and water in essence generates a volume of water that is "free of solute." When urine is dilute, **solute-free water** is excreted from the body. When urine is concentrated, solute-free water is returned to the body (i.e., conserved).

For the kidneys to maximally excrete solute-free water (i.e., 18 L/day) the following conditions must be met:
1. AVP must be absent; without it the collecting duct does not reabsorb a significant amount of water.
2. The tubular structures that separate solute from water (i.e., dilute the tubule fluid) must function normally. In the absence of AVP the following nephron segments can dilute the luminal fluid:
   - ascending thin limb of Henle's loop
   - thick ascending limb of Henle's loop
   - distal tubule
   - collecting duct

   Because of its high transport rate, the thick ascending limb is quantitatively the most important nephron segment involved in the separation of solute and water.
3. An adequate amount of tubular fluid must be delivered to the aforementioned nephron sites for maximal separation of solute and water. Factors that reduce delivery (e.g., decreased GFR or enhanced proximal tubule reabsorption) impair the kidneys' ability to maximally excrete solute-free water.

Similar requirements also apply to conservation of water by the kidneys. For the kidneys to conserve water maximally (6–8 L/day) the following conditions must be met:
1. An adequate amount of tubular fluid must be delivered to those nephron segments that separate solute from water; the most important segment in this regard is the

---

[c]On a typical diet the kidneys must excrete 450 mmol/day of urea. At a maximal urine [urea] of 600 mmol/L this amount of urea can be excreted in less than 1 L of urine. However, if the maximal urine [urea] is reduced because of a decrease in the medullary interstitial fluid [urea], a larger urine volume would be needed to excrete the 450 mmol/day of urea (e.g., 2.25 L of urine would be required if the maximal urine [urea] was only 200 mM).

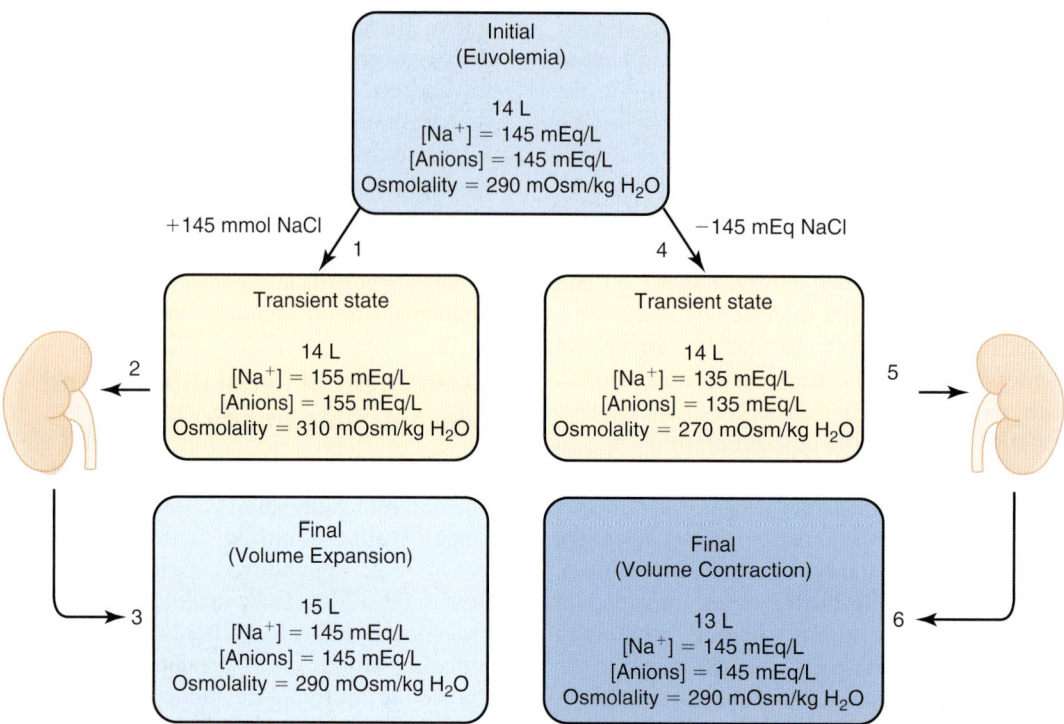

• **Fig. 35.9** Impact of changes in Na+ balance on ECFV. (1) Addition of NaCl (without water) to the ECF increases the [Na+] and osmolality. (2) The increase in ECF osmolality stimulates secretion of AVP from the posterior pituitary, which than acts on the kidneys to conserve water. (3) Decreased renal excretion of water together with water ingestion restore plasma osmolality and plasma [Na+] to normal. However, the ECFV is now increased by 1 L. (4) Removal of NaCl (without water) from the ECF decreases plasma [Na+] and plasma osmolality. (5) The decrease in ECF osmolality inhibits AVP secretion. In response to the decrease in plasma AVP the kidneys excrete water. (6) Increased renal excretion of water returns the plasma [Na+] and plasma osmolality to normal. However, ECFV is now decreased by 1 L. As illustrated, changes in Na+ balance alter ECFV because of the efficiency of the AVP system to maintain normal body fluid osmolality. (Adapted from Koeppen BM, Stanton BA. *Renal Physiology.* 5th ed. Philadelphia: Elsevier; 2013.)

thick ascending limb of Henle's loop. Delivery of tubular fluid to Henle's loop depends on GFR and proximal tubule reabsorption.

2. Reabsorption of NaCl by the nephron segments must be normal; again, the most important segment is the thick ascending limb of Henle's loop.

3. A hyperosmotic medullary interstitium must be present. The interstitial fluid osmolality is maintained by NaCl reabsorption by Henle's loop (conditions 1 and 2) and by effective accumulation of urea. Urea accumulation in turn depends on adequate dietary protein intake.

4. Maximum levels of AVP must be present and the collecting duct must respond normally to AVP.

Control of Extracellular Fluid Volume and Regulation of Renal NaCl Excretion

The major solutes of ECF are the salts of Na+ (see Chapter 2). Of these, NaCl is the most abundant. Because NaCl is also the major determinant of ECF osmolality, alterations in Na+ balance are commonly assumed to disturb ECF osmolality. However, under normal circumstances this is not the case because the AVP and thirst systems maintain body fluid osmolality within a very narrow range (discussed earlier). As illustrated in Fig. 35.9, adding or removing NaCl from ECF changes this body fluid compartment's volume and not the [Na+] (compare initial condition and final conditions). For example, addition of NaCl to the ECF (without water) increases the [Na+] and osmolality of this compartment. (ICF osmolality also increases because of osmotic equilibration with the ECF.) In response, AVP secretion and thirst are stimulated, and as a result water is ingested and renal water loss is reduced. This restores plasma osmolality (and serum [Na+]) to their initial values, but the volume of the ECF is now increased. The opposite occurs when NaCl is lost from the ECF. Changes in ECF volume (ECFV) can be monitored by measuring body weight, because 1 L of ECF equals 1 kg of body weight.

The kidneys are the major route for excretion of NaCl from the body. Only about 10% of the Na^+ lost from the body each day does so by nonrenal routes (e.g., in perspiration and feces). As such, the kidneys are critically important in regulating ECFV. Under normal conditions the kidneys keep ECFV constant (a state termed **euvolemia**) by adjusting the excretion of NaCl to match the amount ingested in the diet. If ingestion exceeds excretion, ECFV increases above normal **(volume expansion)**, whereas the opposite occurs if excretion exceeds ingestion **(volume contraction).**

The typical diet contains approximately 140 mEq/day of Na^+ (8 g of NaCl), and thus daily Na^+ excretion in urine is also about 140 mEq/day. However, the kidneys can vary excretion of Na^+ over a wide range. Excretion rates as low as 10 mEq/day can be attained when individuals are placed on a low-salt diet. Conversely, the kidneys can increase their excretion rate to more than 1000 mEq/day when challenged by ingestion of a high-salt diet. These changes in Na^+ excretion can occur with only modest changes in the ECFV and Na^+ content of the body.

The response of the kidneys to abrupt changes in NaCl intake typically takes several hours to several days, depending on the magnitude of the change. During this transition period the intake and excretion of Na^+ are not matched as they are in the steady state. Thus the individual experiences either **positive Na^+ balance** (intake > excretion) or **negative Na^+ balance** (intake < excretion). However, by the end of the transition period a new steady state is established and intake once again equals excretion.

This section reviews the physiology of the receptors that monitor ECFV and explains the various signals that act on the kidneys to regulate NaCl excretion and thereby ECFV. In addition, responses of the various portions of the nephron to these signals is considered.

Concept of Effective Circulating Volume

As described in Chapter 2, the ECF is subdivided into two compartments: blood plasma and interstitial fluid. Plasma volume is a determinant of vascular volume and thus blood pressure and cardiac output. Maintenance of Na^+ balance, and thus ECFV, involves a complex system of sensors and effector signals that act primarily on the kidneys to regulate NaCl excretion. As can be appreciated from the dependency of vascular volume, blood pressure, and cardiac output on ECFV, this complex system is designed to ensure adequate tissue perfusion. Because the primary sensors of this system are located in the large vessels of the vascular system, changes in vascular volume, blood pressure, and cardiac output are the principal factors regulating renal NaCl excretion (discussed later). In a healthy individual, changes in ECFV result in parallel changes in vascular volume, blood pressure, and cardiac output. Thus a decrease in ECFV results in reduced vascular volume, blood pressure, and cardiac output. Conversely, an increase in ECFV results in increased vascular volume, blood pressure, and cardiac output. The degree to which these cardiovascular parameters change is dependent upon the degree of ECFV contraction or expansion and the effectiveness of cardiovascular reflex mechanisms (see Chapters 18 and 19). When a person is in negative Na^+ balance, ECFV is decreased and renal NaCl excretion is reduced. Conversely, with positive Na^+ balance there is an increase in ECFV, which results in enhanced renal NaCl excretion (i.e., **natriuresis**).

However, in some pathological conditions (e.g., congestive heart failure, hepatic cirrhosis), renal excretion of NaCl does not reflect the ECFV. In both of these situations the volume of the ECF is increased. However, instead of increased renal NaCl excretion, as would be expected, renal excretion of NaCl is reduced. To explain renal Na^+ handling in these situations, it is necessary to understand the concept of **effective circulating volume (ECV)**. Unlike ECF, ECV is *not* a measurable and distinct body fluid compartment. *Effective circulating volume* refers to that portion of the ECF that is contained within the vascular system and is "effectively" perfusing the tissues (*effective blood volume* and *effective arterial blood volume* are other commonly used terms). More specifically the ECV reflects the activity of volume sensors located in the vascular system (discussed later).

IN THE CLINIC

Patients with congestive heart failure frequently have an increase in ECFV that manifests as increased plasma volume and accumulation of interstitial fluid in the lungs **(pulmonary edema)** and peripheral tissues **(generalized edema).** This excess fluid is the result of NaCl and water retention by the kidneys. The kidneys' response (i.e., retention of NaCl and water) is paradoxical because the ECFV is increased. However, this fluid is not in the vascular system but rather in the interstitial fluid compartment. In addition, blood pressure and cardiac output may be reduced because of poor cardiac performance. Therefore the sensors located in the vascular system respond as they do in ECFV contraction and cause NaCl and water retention by the kidneys. In this situation the ECV, as monitored by the volume sensors, is decreased.

In healthy individuals, ECV varies directly with ECFV and in particular the volume of the vascular system (arterial and venous), the arterial blood pressure, and cardiac output. However, as noted this is not the case in certain pathological conditions. In the remaining sections of this chapter the relationship between ECFV and renal NaCl excretion in healthy adults—where changes in ECV and ECFV occur in parallel—are examined.

Volume-Sensing Systems

ECFV (or ECV) is monitored by multiple sensors. A number of these sensors are located in the vascular system, and they monitor its fullness and pressure. These receptors are typically called *vascular volume receptors,* or because they respond to pressure-induced stretch of the walls of the structure in which they are located (e.g., blood vessels or the cardiac atria and ventricles), they are also referred to as *baroreceptors.*[f]

Volume Sensors in the Low-Pressure Cardiopulmonary Circuit

Volume sensors (i.e., baroreceptors) are located within the walls of the left and right atria, right ventricle, and large pulmonary vessels, and they respond to distention of these structures (see Chapters 18 and 19). Because the low-pressure side of the circulatory system has a high compliance, these sensors respond mainly to the "fullness" of the vascular system. These baroreceptors send signals to the brainstem via afferent fibers in the glossopharyngeal and vagus nerves (cranial nerves IX and X). The activity of these sensors modulates both sympathetic nerve outflow and AVP secretion. For example, a decrease in filling of the pulmonary vessels and cardiac atria increases sympathetic nerve activity and stimulates AVP secretion. Conversely, distention of these structures decreases sympathetic nerve activity. In general, 5% to 10% changes in blood volume and pressure are necessary to evoke a response.

The cardiac atria possess an additional mechanism related to control of renal NaCl excretion. The myocytes of the atria synthesize and store a peptide hormone, **atrial natriuretic peptide (ANP).** It is released when the atria are distended and, via mechanisms outlined later in this chapter, reduces blood pressure and increases excretion of NaCl and water by the kidneys. The ventricles of the heart also produce a natriuretic peptide, **brain natriuretic peptide (BNP),** so named because it was first isolated from the brain. Like ANP, BNP is released from myocytes by distention of the ventricles. Its actions are similar to those of ANP.

Volume Sensors in the High-Pressure Arterial Circuit

Baroreceptors are also present in the arterial side of the circulatory system, located in the wall of the aortic arch, carotid sinus, and afferent arterioles of the kidneys. The aortic arch and carotid baroreceptors send input to the brainstem via afferent fibers in the glossopharyngeal and vagus nerves (cranial nerves IX and X). The response to this input alters sympathetic outflow and AVP secretion. Thus a decrease in blood pressure increases sympathetic nerve activity and AVP secretion. An increase in pressure tends to reduce sympathetic nerve activity (and activate parasympathetic nerve activity). The sensitivity of the high-pressure baroreceptors is similar to that in the low-pressure side of the vascular system; 5% to 10% changes in pressure are needed to evoke a response.

The **juxtaglomerular apparatus (JGA)** of the kidneys (see Chapter 33), particularly the afferent arteriole, responds

---

[f]The liver and central nervous system also have sensors that respond to changes in blood pressure and [Na^+] and then signal the kidneys to alter NaCl excretion. These systems do not appear to be as important as vascular receptors in monitoring changes in ECFV and effecting changes in renal NaCl excretion and are not considered here.

directly to changes in pressure. If perfusion pressure in the afferent arteriole is reduced, renin is released from the granular cells. By contrast, renin secretion is suppressed when perfusion pressure is increased. As described later in this chapter, renin determines blood levels of angiotensin II and aldosterone, both of which reduce renal NaCl excretion.

IN THE CLINIC

Constriction of a renal artery (e.g., by an atherosclerotic plaque) reduces perfusion pressure to that kidney. This reduced perfusion pressure is sensed by the afferent arteriole of the JGA and results in renin secretion. The elevated renin levels increase the production of angiotensin II, which in turn increases systemic blood pressure via its vasoconstrictor affect on arterioles throughout the vascular system. The increased systemic blood pressure is sensed by the JGA of the contralateral kidney (i.e., the kidney without stenosis of its renal artery), and renin secretion from that kidney is suppressed. In addition the high levels of angiotensin II also inhibit renin secretion by the contralateral kidney (negative feedback). Treatment of patients with constricted renal arteries includes surgical repair of the stenotic artery, administration of angiotensin II receptor blockers, or administration of an inhibitor of angiotensin-converting enzyme (ACE). The ACE inhibitor blocks conversion of angiotensin I to angiotensin II.

Volume Sensor Signals

When the vascular volume sensors have detected a change in ECFV, they send signals to the kidneys, which results in an appropriate adjustment in NaCl and water excretion. Accordingly, when the ECFV is expanded, renal NaCl and water excretion are increased. Conversely, when the ECFV is contracted, renal NaCl and water excretion are reduced. The signals involved in coupling the volume sensors to the kidneys are both neural and hormonal. These are summarized in Box 35.1, as are their effects on renal NaCl and water excretion.

Renal Sympathetic Nerves

As described in Chapter 33, sympathetic nerve fibers innervate the afferent and efferent arterioles of the glomerulus as well as the cells of the nephron. With ECFV contraction, activation of the low- and high-pressure vascular baroreceptors results in stimulation of sympathetic nerve activity, including those fibers innervating the kidneys. This stimulation has the following effects:

1. The afferent and efferent arterioles are constricted (mediated by α-adrenergic receptors). This vasoconstriction (the effect is greater on the afferent arteriole) decreases the hydrostatic pressure within the glomerular capillary lumen, which results in a decrease in GFR. With this decrease in GFR, the filtered amount of Na^+ is reduced.

2. Renin secretion is stimulated by the cells of the afferent arterioles (mediated by β-adrenergic receptors). As

Signals Involved in Control of Renal NaCl and Water Excretion

Renal Sympathetic Nerves (↑Activity: ↓NaCl Excretion)

↓GFR
↑Renin secretion
↑Na⁺ reabsorption along the nephron

Renin-Angiotensin-Aldosterone (↑Secretion: ↓NaCl Excretion)

↑Angiotensin II stimulates Na⁺ reabsorption along the nephron
↑Aldosterone stimulates Na⁺ reabsorption in the distal tubule and collecting duct and to a lesser degree in the thick ascending limb of Henle's loop
↑Angiotensin II stimulates AVP secretion

Natriuretic Peptides: ANP, BNP & Urodilatin (↑Secretion: ↑NaCl Excretion)

↑GFR
↓Renin secretion
↓Aldosterone secretion (indirect via ↓angiotensin II and direct on adrenal gland)
↓NaCl and water reabsorption by the collecting duct
↓AVP secretion and inhibition of AVP action on the distal tubule and collecting duct

AVP (↑Secretion: ↓H₂O Excretion)

↑H₂O reabsorption by the distal tubule and collecting duct

---

described later, renin ultimately increases the circulating levels of angiotensin II and aldosterone, both of which stimulate Na⁺ reabsorption by the nephron.

3. NaCl reabsorption along the nephron is directly stimulated (mediated by α-adrenergic receptors on the cells of the nephron). Because of the large amount of Na⁺ reabsorbed by the proximal tubule, the effect of increased sympathetic nerve activity is quantitatively most important for this segment.

As a result of these actions, increased renal sympathetic nerve activity decreases NaCl excretion, an adaptive response that works to restore ECFV to normal. With ECFV expansion, renal sympathetic nerve activity is reduced. This generally reverses the effects just described.

Renin-Angiotensin-Aldosterone System

Cells in the afferent arterioles (juxtaglomerular cells, also known as *granular cells*) are the site of synthesis, storage, and release of the proteolytic enzyme renin. Three factors are important in stimulating renin secretion:

1. *Perfusion pressure.* The afferent arteriole behaves as a high-pressure baroreceptor. When perfusion pressure to the kidneys is reduced, renin secretion is stimulated. Conversely, an increase in perfusion pressure inhibits renin release.
2. *Sympathetic nerve activity.* Activation of the sympathetic nerve fibers that innervate the afferent arterioles increases renin secretion (mediated by β-adrenergic receptors).

Renin secretion is decreased as renal sympathetic nerve activity is decreased.

3. *Delivery of NaCl to the macula densa.* Delivery of NaCl to the macula densa regulates GFR by a process termed **tubuloglomerular feedback** (see Chapter 33). In addition the macula densa plays a role in renin secretion. When NaCl delivery to the macula densa is decreased, renin secretion is enhanced. Conversely, an increase in NaCl delivery inhibits renin secretion. It is likely that macula densa–mediated renin secretion helps maintain systemic arterial pressure under conditions of a reduced vascular volume. For example, when vascular volume is reduced, perfusion of body tissues (including the kidneys) decreases. This in turn decreases GFR and the filtered amount of NaCl. The reduced delivery of NaCl to the macula densa stimulates renin secretion, which acts through angiotensin II (a potent vasoconstrictor) to increase blood pressure and thereby maintain tissue perfusion.

AT THE CELLULAR LEVEL

Although many tissues express renin (e.g., brain, heart, adrenal gland), juxtaglomerular (JG) cells are the primary source of circulating renin; they are located in the afferent arterioles of the kidneys. Renin secretion is stimulated by a *decrease* in intracellular [Ca⁺⁺], a response opposite that of most secretory cells, where secretion is stimulated by an increase in intracellular [Ca⁺⁺]. Renin secretion is also stimulated by an increase in intracellular cAMP levels. Conversely, anything that increases intracellular [Ca⁺⁺] will inhibit renin secretion. Renin secretion is also inhibited by increases in intracellular cyclic guanosine monophosphate (cGMP).

Stretch of the afferent arteriole, angiotensin-II, and endothelin increase intracellular [Ca⁺⁺] and thus inhibit renin secretion by JG cells. The stimulatory effect of sympathetic nerve activity on renin secretion is mediated by norepinephrine, which increases intracellular cAMP (via β-adrenergic receptors). Prostaglandin E₂ also increases JG cell cAMP levels and therefore stimulates renin secretion. Natriuretic peptides and nitric oxide (NO) inhibit renin secretion by increasing intracellular cGMP.

Control of renin secretion by the macula densa is complex and appears to involve several paracrine factors, including ATP, adenosine, and prostaglandin E₂ (see Chapter 33).

Fig. 35.10 summarizes the essential components of the renin-angiotensin-aldosterone system (RAAS). Renin alone does not have a physiological function; it functions solely as a proteolytic enzyme. Its substrate is a circulating protein, **angiotensinogen,** which is produced by the liver. Angiotensinogen is cleaved by renin to yield a 10–amino acid peptide, **angiotensin I.** Angiotensin I also has no known physiological function, and it is cleaved to an 8–amino acid peptide, **angiotensin II,** by **angiotensin-converting enzyme (ACE)** found on the surface of vascular endothelial cells. Lung and renal endothelial cells are important sites for the conversion of angiotensin I to angiotensin II. ACE also

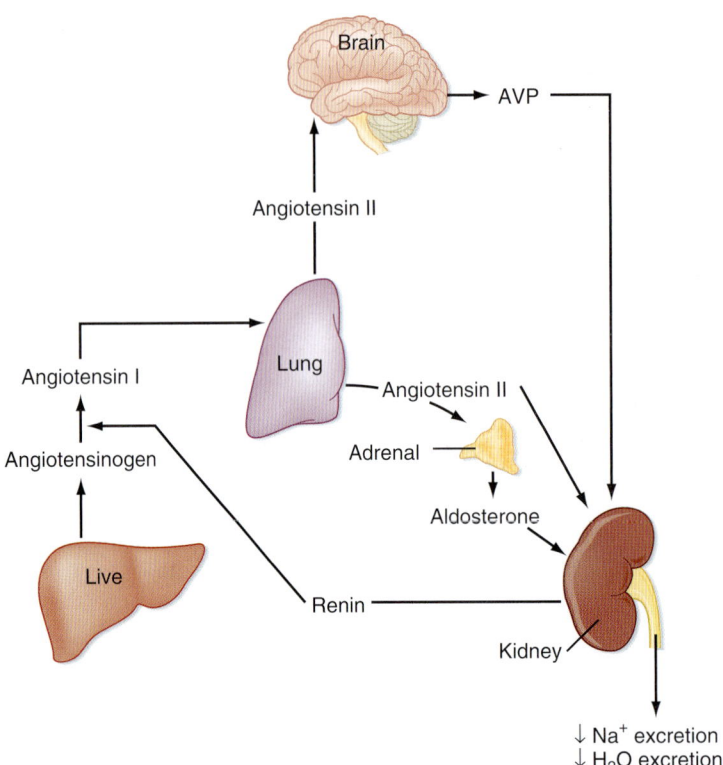

• **Fig. 35.10** Schematic representation of the essential components of the renin-angiotensin-aldosterone system (RAAS). Activation of this system results in a decrease in excretion of Na^+ and water by the kidneys. NOTE: Angiotensin I is converted to angiotensin II by an angiotensin-converting enzyme that is present on all vascular endothelial cells. As shown, the endothelial cells within the lungs play a significant role in this conversion process. See text for details.

degrades bradykinin, a potent vasodilator. Angiotensin II has several important physiological functions:

1. stimulation of aldosterone secretion by the adrenal cortex
2. arteriolar vasoconstriction, which increases blood pressure
3. stimulation of AVP secretion and thirst
4. enhancement of NaCl reabsorption by the proximal tubule, thick ascending limb of Henle's loop, the distal tubule, and the collecting duct (Of these segments the effect on the proximal tubule is quantitatively the largest.)

Angiotensin II is an important secretagogue for **aldosterone.** An increase in plasma K^+ concentration is the other important stimulus for aldosterone secretion (see Chapter 36). Aldosterone is a steroid hormone produced by the glomerulosa cells of the adrenal cortex and acts in a number of ways on the kidneys (see also Chapters 34, 36, and 37). With regard to regulation of the ECFV, aldosterone reduces NaCl excretion by stimulating its reabsorption by several nephron segments. Most importantly it stimulates NaCl reabsorption in the aldosterone-sensitive distal nephron (ASDN), which consists of the latter portion of the distal tubule and the collecting duct. To a lesser extent it also stimulates NaCl reabsorption by the thick ascending limb of Henle's loop and the early portion of the distal tubule.

Aldosterone sensitivity is conferred by the presence of mineralocorticoid receptors as well as the enzyme **11β-hydroxysteroid dehydrogenase 2 (11β-HSD2).** Because the mineralocorticoid receptor also binds glucocorticoids, 11β-HSD2 is required for aldosterone specificity because it metabolizes glucocorticoids and thus prevents them from binding to the mineralocorticoid receptor. The aldosterone-sensitive distal nephron expresses both the mineralocorticoid receptor and 11β-HSD2.

Aldosterone has many cellular actions in responsive cells (see Chapter 34). In the ASDN it increases the abundance and activity of ENaC in the apical membrane of principal cells. This increases Na^+ entry into cells across the apical membrane. Extrusion of Na^+ from cells across the basolateral membrane occurs by Na^+,K^+-ATPase, the abundance of which is also increased by aldosterone. Thus aldosterone increases reabsorption of NaCl from the tubular fluid by distal nephron segments, whereas reduced levels of aldosterone decreases the amount of NaCl reabsorbed by these segments. Although the principal site of action of aldosterone is the ASDN, aldosterone also increases the abundance of the Na^+/Cl^- symporter in the distal tubule and the $Na^+/K^+/2Cl^-$ symporter in the thick ascending limb of Henle's loop.

As summarized in Box 35.1, activation of the RAAS, as occurs with ECFV depletion, decreases excretion of NaCl

IN THE CLINIC

Diseases of the adrenal cortex can alter aldosterone levels and thereby impair the ability of the kidneys to maintain Na^+ balance and euvolemia. With decreased secretion of aldosterone **(hypoaldosteronism)**, the reabsorption of NaCl, mainly by the aldosterone-sensitive distal nephron, is reduced and NaCl is lost in the urine. Because urinary NaCl loss can exceed the amount of NaCl ingested in the diet, negative Na^+ balance ensues and ECFV decreases. In response to ECFV contraction, sympathetic tone is increased and levels of renin, angiotensin II, and AVP are elevated. With increased aldosterone secretion **(hyperaldosteronism)** the effects are the opposite; NaCl reabsorption by the aldosterone-sensitive distal nephron is enhanced and excretion of NaCl is reduced. Consequently, ECFV is increased, sympathetic tone is decreased, and levels of renin, angiotensin II, and AVP are decreased. As described later, ANP and BNP levels are also elevated in this setting.

by the kidneys. The RAAS is suppressed with ECFV expansion, and renal NaCl excretion is therefore enhanced.

Natriuretic Peptides

The body produces a number of substances that act on the kidneys to increase NaCl excretion (see Chapter 34). Of these, natriuretic peptides produced by the heart and kidneys are best understood and will be the focus of the following discussion.

The heart produces two natriuretic peptides. Atrial myocytes produce and store the peptide hormone ANP, and ventricular myocytes produce and store BNP. Both peptides are secreted when the heart dilates (i.e., during volume expansion, with heart failure), and they relax vascular smooth muscle and promote NaCl and water excretion by the kidneys. The kidneys also produce a related natriuretic peptide termed **urodilatin.** Its actions are limited to promoting NaCl excretion by the kidneys. In general the actions of these natriuretic peptides as they relate to renal NaCl and water excretion antagonize those of the RAAS. These actions include:

1. Vasodilation of the afferent and vasoconstriction of the efferent arterioles of the glomerulus. This increases the GFR and the filtered amount of NaCl.
2. Inhibition of renin secretion by the afferent arterioles.
3. Inhibition of aldosterone secretion by the glomerulosa cells of the adrenal cortex. This occurs via two mechanisms: (a) inhibition of renin secretion by the juxtaglomerular cells, thereby reducing angiotensin II–induced aldosterone secretion, and (b) direct inhibition of aldosterone secretion by the glomerulosa cells of the adrenal cortex.
4. Inhibition of NaCl reabsorption by the collecting duct, which is also caused in part by reduced levels of aldosterone. However, natriuretic peptides increase cGMP, which inhibits cation channels in the apical membrane of medullary collecting duct cells and thereby decreases NaCl reabsorption.

5. Inhibition of AVP secretion by the posterior pituitary and AVP action on the collecting duct. These effects decrease water reabsorption by the collecting duct and thus increase excretion of water in the urine.

The net effect of natriuretic peptides is to increase excretion of NaCl and water by the kidneys. Hypothetically a reduction in circulating levels of these peptides would be expected to decrease NaCl and water excretion, but convincing evidence for this has not been reported.

Arginine Vasopressin

As discussed previously, a decreased ECFV stimulates AVP secretion by the posterior pituitary. Elevated levels of AVP decrease water excretion by the kidneys, which serves to reestablish euvolemia.

Control of NaCl Excretion During Euvolemia

Maintenance of Na^+ balance and therefore euvolemia requires precise matching of the amount of NaCl ingested with the amount excreted from the body. As already noted the kidneys are the major route for NaCl excretion. Accordingly, in a euvolemic individual we can equate daily urine NaCl excretion with daily NaCl intake.

As noted, the amount of NaCl excreted by the kidneys can vary widely. Under conditions of salt restriction (i.e., low-NaCl diet), virtually no NaCl appears in the urine. Conversely, in individuals who ingest large quantities of NaCl, renal NaCl excretion can exceed 1000 mEq/day. The time course for adjustment of renal NaCl excretion varies (hours to days) and depends on the magnitude of the change in NaCl intake. Acclimation to large changes in NaCl intake requires a longer time than acclimation to small changes in intake.

The general features of NaCl transport along the nephron are illustrated in Fig. 35.11. Most (67%) of the Na^+ filtered by the glomerulus is reabsorbed by the proximal tubule. An additional 25% is reabsorbed by the thick ascending limb of the loop of Henle, and the remainder by the distal tubule and collecting duct.

In a normal adult the filtered amount (load) of Na^+ is approximately 25,000 mEq/day:

Equation 35.1

$$\text{Filtered load of } Na^+ = (GFR) \times (\text{Plasma } [Na^+])$$
$$= (180 \text{ L/day}) \times (140 \text{ mEq/L})$$
$$= 25,200 \text{ mEq/day}$$

With a typical diet, less than 1% of this filtered load is excreted in urine ($\approx$140 mEq/day).[g] Because of the large filtered load of Na^+, small changes in its reabsorption by the nephron can profoundly affect Na^+ balance and thus ECFV. For example, an increase in Na^+ excretion from 1%

---

[g]The percentage of the filtered load excreted in urine is termed **fractional excretion.** In this example the fractional excretion of Na^+ is 140 mEq/day ÷ 25,200 mEq/day = 0.005, or 0.5%.

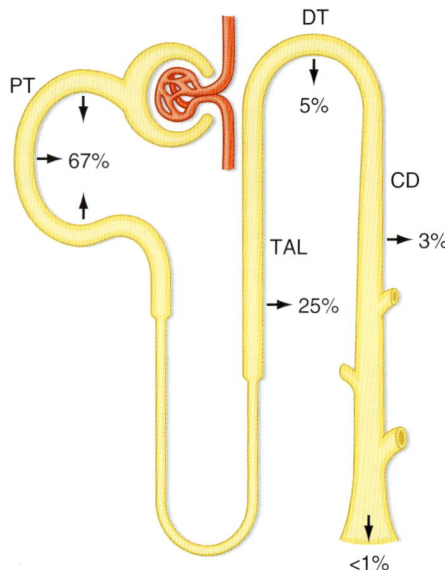

• **Fig. 35.11** Segmental Na$^+$ reabsorption. The percentage of the filtered load of Na$^+$ reabsorbed by each nephron segment is indicated. PT, proximal tubule; TAL, thick ascending limb; DT, distal tubule; CD, cortical collecting duct.

to 3% of the filtered load represents an additional loss of approximately 500 mEq/day of Na$^+$. Because the ECF Na$^+$ concentration is 140 mEq/L, such a Na$^+$ loss would decrease the ECFV by more than 3 L (i.e., water excretion would parallel the loss of Na$^+$ to maintain body fluid osmolality constant: 500 mEq/day ÷ 140 mEq/L = 3.6 L/day of fluid loss). Such fluid loss in a 70-kg individual would represent a 26% decrease in ECFV.

In euvolemic individuals the nephron segments distal to the loop of Henle (distal tubule and collecting duct) are the main nephron segments where Na$^+$ reabsorption is adjusted to maintain excretion at a level appropriate for dietary intake. However, this does not mean the other portions of the nephron are not involved in this process. Because the reabsorptive capacity of the distal tubule and collecting duct is limited, these other portions of the nephron (i.e., proximal tubule and loop of Henle) must reabsorb the bulk of the filtered load of Na$^+$. Thus during euvolemia, Na$^+$ handling by the nephron can be explained by two general processes:

1. Na$^+$ reabsorption by the proximal tubule and loop of Henle is regulated so that a relatively constant portion of the filtered load of Na$^+$ is delivered to the distal tubule. The combined action of the proximal tubule and loop of Henle reabsorbs approximately 92% of the filtered load of Na$^+$, and thus 8% of the filtered load is delivered to the distal tubule.
2. Reabsorption of this remaining portion of the filtered load of Na$^+$ by the distal tubule and collecting duct is regulated so that the amount of Na$^+$ excreted in the urine matches the amount ingested in the diet. Thus these later nephron segments make final adjustments in Na$^+$ excretion to maintain the euvolemic state.

Mechanisms for Maintaining Constant Delivery of NaCl to the Distal Tubule

A number of mechanisms maintain a constant delivery of Na$^+$ to the beginning of the distal tubule. These processes are autoregulation of the GFR (and thus the filtered load of Na$^+$), glomerulotubular balance, and load dependency of Na$^+$ reabsorption by the loop of Henle.

Autoregulation of the GFR (see Chapter 33) allows maintenance of a relatively constant filtration rate over a wide range of perfusion pressures. Because the filtration rate is constant, the filtered load of Na$^+$ is also constant.

Despite the autoregulatory control of GFR, small variations occur. If these changes are not compensated for by an appropriate adjustment in Na$^+$ reabsorption by the nephron, Na$^+$ excretion would change markedly. Fortunately, Na$^+$ reabsorption in the euvolemic state, especially by the proximal tubule, changes in parallel with changes in GFR. This phenomenon is termed **glomerulotubular balance.** Thus if GFR increases, the amount of Na$^+$ reabsorbed by the proximal tubule also increases. The opposite occurs if GFR decreases (see Chapter 34 for a more detailed description of glomerulotubular balance).

The final mechanism that helps maintain constant delivery of Na$^+$ to the distal tubule and collecting duct involves the ability of the loop of Henle to increase its reabsorptive rate in response to increased delivery of Na$^+$.

Regulation of Distal Tubule and Collecting Duct NaCl Reabsorption

When delivery of Na$^+$ is constant, small adjustments in Na$^+$ reabsorption, primarily by the ASDN, are sufficient to balance excretion with intake. Aldosterone is the primary regulator of Na$^+$ reabsorption by the ASDN and thus of NaCl excretion. When aldosterone levels are elevated, Na$^+$ reabsorption by these segments is increased (Na$^+$ excretion is decreased). When aldosterone levels are decreased, Na$^+$ reabsorption is decreased (NaCl excretion is increased). Other factors have been shown to alter Na$^+$ reabsorption by the ASDN (e.g., angiotensin II and natriuretic peptides), but their role during euvolemia is unclear.

As long as variations in dietary intake of NaCl are minor, the mechanisms previously described can regulate renal Na$^+$ excretion appropriately and thereby maintain euvolemia. However, these mechanisms cannot effectively handle significant changes in NaCl intake. When NaCl intake changes significantly, ECFV expansion or contraction occurs. In such cases, additional factors act on the kidneys to adjust Na$^+$ excretion and thereby reestablish the euvolemic state.

Control of NaCl Excretion With Volume Expansion

During ECFV expansion, the high-pressure and low-pressure vascular volume sensors send signals to the kidneys that result in increased excretion of NaCl and water. The signals acting on the kidneys include:

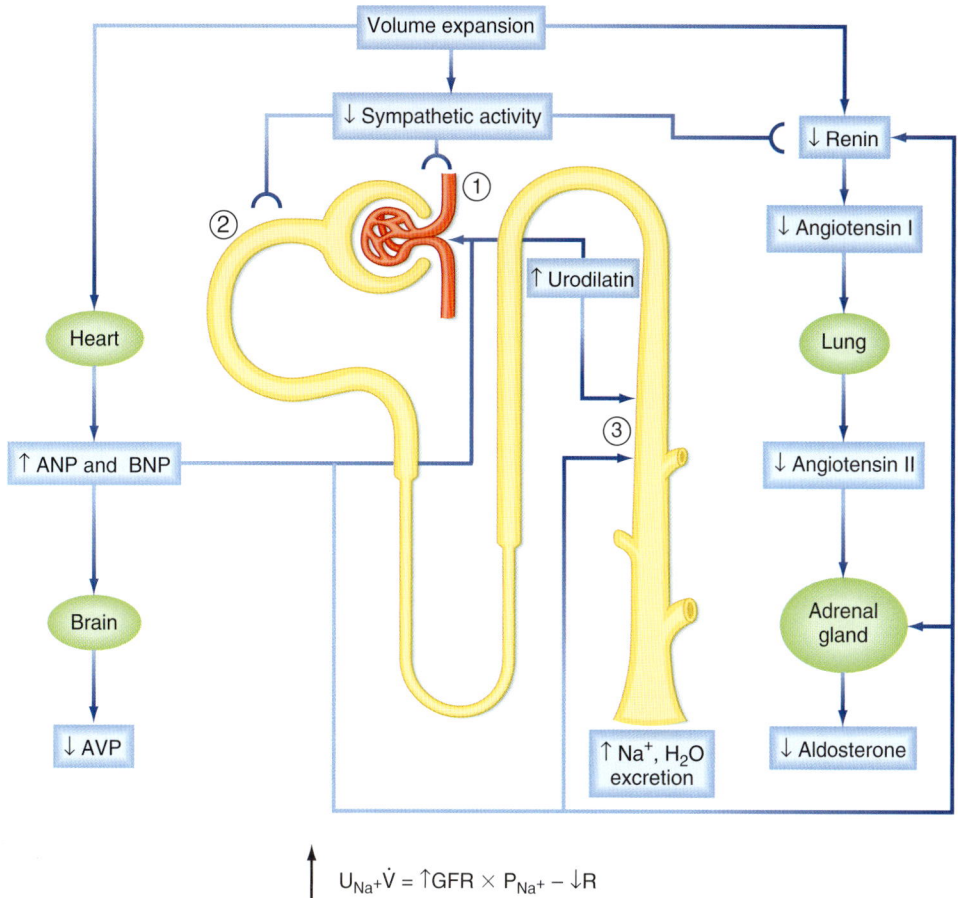

Fig. 35.12 Integrated response to ECFV expansion. Numbers refer to the description of the response in the text. ANP, atrial natriuretic peptide; BNP, brain natriuretic peptide; GFR, glomerular filtration rate; P_{Na+}, plasma [Na$^+$]; R, tubular reabsorption of Na$^+$; $U_{Na+}\dot{V}$, Na$^+$ excretion rate.

1. decreased activity of renal sympathetic nerves
2. release of ANP and BNP from the heart and urodilatin in the kidneys
3. inhibition of AVP secretion from the posterior pituitary and decreased AVP action on the collecting duct
4. decreased renin secretion and thus decreased production of angiotensin II
5. decreased aldosterone secretion, which is caused by reduced angiotensin II levels, and elevated natriuretic peptide levels

The integrated response of the nephron to these signals is illustrated in Fig. 35.12. Three general responses to ECFV expansion occur (the numbers match those encircled in Fig. 35.12):

1. *GFR increases.* GFR increases mainly as a result of the decrease in sympathetic nerve activity. Sympathetic fibers innervate the afferent and efferent arterioles of the glomerulus and control their diameter. Decreased sympathetic nerve activity leads to arteriolar dilation and an increase in renal plasma flow (RPF). Because the effect appears to be greater on the afferent arterioles, the hydrostatic pressure within the glomerular capillary is increased, thereby increasing the GFR. Because RPF increases to a greater degree than GFR, the filtration

fraction (GFR/RPF) decreases. Natriuretic peptides also increase GFR by dilating the afferent arterioles and constricting the efferent arterioles. Thus the increased natriuretic peptide levels that occur during ECFV expansion contribute to this response. With the increase in GFR, the filtered load of Na$^+$ increases.

2. *The reabsorption of Na$^+$ decreases in the proximal tubule and loop of Henle.* Several mechanisms reduce Na$^+$ reabsorption by the proximal tubule. Because activation of the sympathetic nerve fibers that innervate this nephron segment stimulates proximal tubule Na$^+$ reabsorption, the decreased sympathetic nerve activity that results from ECFV expansion decreases Na$^+$ reabsorption. In addition, angiotensin II directly stimulates Na$^+$ reabsorption by the proximal tubule. Because angiotensin II levels are also reduced under this condition, proximal tubule Na$^+$ reabsorption decreases as a result. Starling forces across the proximal tubule also change. The elevated hydrostatic pressure within the glomerular capillaries in ECFV expansion also tends to increase the hydrostatic pressure within the peritubular capillaries. In addition the decrease in filtration fraction reduces the peritubular oncotic pressure. These alterations in the capillary Starling forces (i.e., hydrostatic and oncotic) reduce

the absorption of solute (e.g., NaCl) and water from the lateral intercellular space and thus reduce tubular reabsorption of NaCl (see Chapter 34 for a complete description of this mechanism). Both the increase in the filtered load and the decrease in NaCl reabsorption by the proximal tubule result in delivery of more NaCl to the loop of Henle. Because activation of sympathetic nerves and angiotensin II and aldosterone stimulate NaCl reabsorption by the thick ascending limb of the loop of Henle, the reduced nerve activity and low angiotensin II and aldosterone levels that occur with ECFV expansion reduce NaCl reabsorption by the thick ascending limb. Thus the fraction of the filtered load delivered to the distal tubule is increased.

3. *Na$^+$ reabsorption decreases in the distal tubule and collecting duct.* As noted, the amount of Na$^+$ delivered to the distal tubule exceeds that observed in the euvolemic state (i.e., the amount of Na$^+$ delivered to the distal tubule varies in proportion to the degree of ECFV expansion). This increased load of Na$^+$ overwhelms the reabsorptive capacity of the distal tubule and the collecting duct, and this capacity is also impaired by reduced levels of angiotensin II and aldosterone, as well as increased levels of natriuretic peptides.

The final component in the response to ECFV expansion is increased excretion of water. As Na$^+$ excretion increases, plasma osmolality begins to fall. This decreases secretion of AVP. AVP secretion is also decreased in response to the elevated levels of natriuretic peptides. In addition, natriuretic peptides inhibit the action of AVP on the collecting duct. Together, these effects decrease water reabsorption by the collecting duct and thereby increase water excretion by the kidneys. Thus excretion of NaCl and water occurs in concert; euvolemia is restored and body fluid osmolality remains constant. The time course of this response (hours to days) depends on the magnitude of the ECFV expansion. Thus if the degree of ECFV expansion is small, the mechanisms just described generally restore euvolemia within 24 hours. However, with large degrees of ECFV expansion, the response can take several days.

In brief, the renal response to ECFV expansion involves the integrated action of all parts of the nephron; (1) the amount of Na$^+$ filtered is increased, (2) Na$^+$ reabsorption by the proximal tubule and loop of Henle is reduced (glomerulotubular balance does not occur under this condition), and (3) reabsorption of Na$^+$ by the distal tubule and collecting duct is decreased (while the delivery of Na$^+$ is increased), primarily secondary to a reduction in aldosterone. In summary, ECFV expansion leads to a reduction in Na$^+$ reabsorption by both the proximal tubule and the ASDN, which results in excretion of a larger fraction of the filtered Na$^+$, thereby restoring euvolemia.

Control of NaCl Excretion With Volume Contraction

During ECFV contraction, the high-pressure and low-pressure vascular volume sensors send signals to the kidneys to reduce NaCl and water excretion and thereby restore ECFV. The signals that act on the kidneys include:

1. increased renal sympathetic nerve activity
2. increased secretion of renin, which results in elevated angiotensin II levels and thus increased secretion of aldosterone by the adrenal cortex
3. inhibition of ANP and BNP secretion by the heart and urodilatin production by the kidneys
4. stimulation of AVP secretion by the posterior pituitary

The integrated response of the nephron to these signals is illustrated in Fig. 35.13. The general response is as follows (the numbers correlate with those encircled in Fig. 35.13):

1. *GFR decreases.* Afferent and efferent arteriolar constriction occurs as a result of increased renal sympathetic nerve activity. The effect appears to be greater on the afferent than on the efferent arteriole, because the hydrostatic pressure in the glomerular capillaries fall, which thereby decreases GFR. Because RPF decreases more than GFR, filtration fraction increases. The decrease in GFR reduces the filtered amount of Na$^+$.

2. *Na$^+$ reabsorption by the proximal tubule and loop of Henle is increased.* Several mechanisms augment Na$^+$ reabsorption in the proximal tubule. For example, increased sympathetic nerve activity and angiotensin II levels directly stimulate Na$^+$ reabsorption. The decreased hydrostatic pressure within the glomerular capillaries also leads to a decrease in the hydrostatic pressure within the peritubular capillaries. In addition, as just noted, the increased filtration fraction results in an increase in the peritubular oncotic pressure. These alterations in the capillary Starling forces facilitate movement of fluid from the lateral intercellular space into the capillary and thereby stimulate reabsorption of NaCl and water by the proximal tubule (see Chapter 34 for a complete description of this mechanism). Increased sympathetic nerve activity as well as elevated levels of angiotensin II and aldosterone stimulate Na$^+$ reabsorption by the thick ascending limb.

3. *Na$^+$ reabsorption by the distal tubule and collecting duct is enhanced.* The small amount of Na$^+$ that is delivered to the ASDN, owing to decreased filtration and increased reabsorption by the proximal tubule and loop of Henle, is almost completely reabsorbed because transport in the segments of the ASDN is enhanced. Stimulation of Na$^+$ reabsorption by the ASDN is primarily the result of increased aldosterone levels, although increased sympathetic nerve activity and increased angiotensin II levels may also contribute to this response. Lastly, levels of natriuretic peptides, which inhibit collecting duct reabsorption, are reduced.

Finally, water reabsorption by the latter portion of the distal tubule and the collecting duct is enhanced by AVP, the levels of which are elevated through activation of the low- and high-pressure vascular volume sensors as well as by the elevated levels of angiotensin II. As a result, water excretion is reduced. Because both water and Na$^+$ are retained by the kidneys in equal proportions, euvolemia is reestablished and

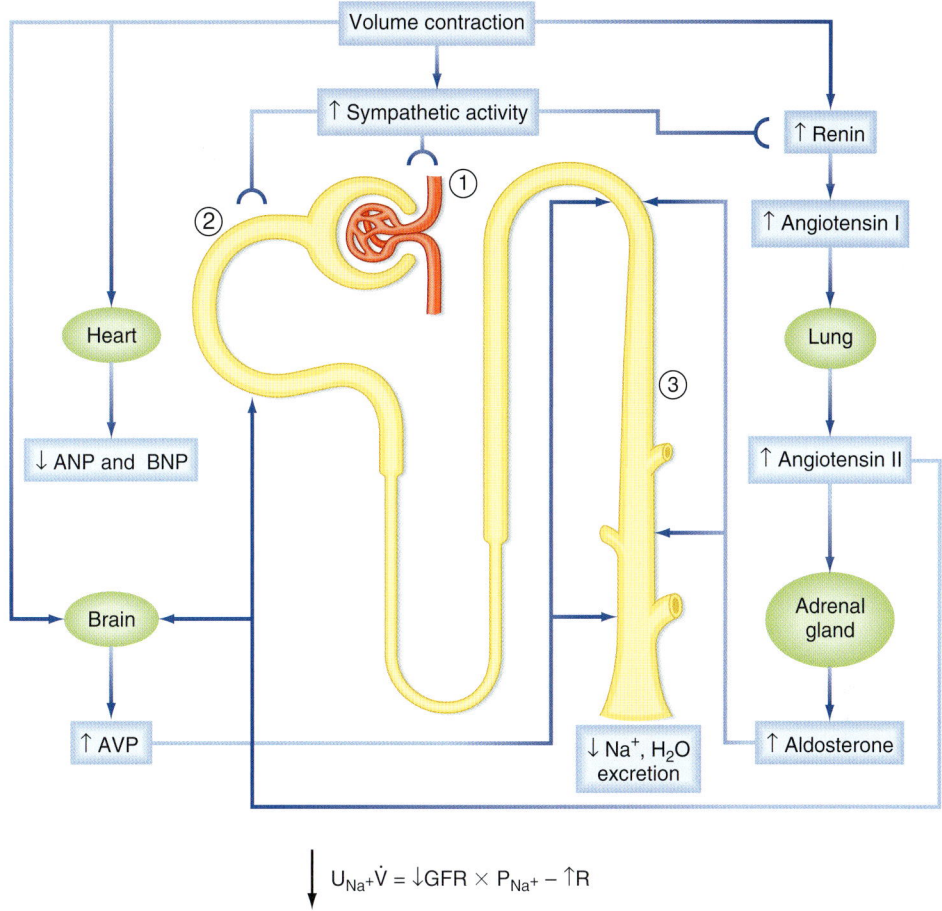

• **Fig. 35.13** Integrated response to ECFV contraction. Numbers refer to the description of the response in the text. ANP, atrial natriuretic peptide; BNP, brain natriuretic peptide; GFR, glomerular filtration rate; P_{Na+}, plasma [Na^+]; R, tubular reabsorption of Na^+; $U_{Na+}\dot{V}$, Na^+ excretion rate.

$$U_{Na+}\dot{V} = \downarrow GFR \times P_{Na+} - \uparrow R$$

body fluid osmolality remains constant. The time course of this restoration of ECFV (hours to days) and the degree to which euvolemia is attained depend on the magnitude of ECFV contraction as well as the dietary intake of Na^+. Euvolemia can be restored more rapidly if additional NaCl is ingested in the diet.

In brief, the nephron's response to ECFV contraction involves the integrated action of all its segments: (1) the filtered amount of Na^+ is decreased, (2) reabsorption in the proximal tubule and loop of Henle is enhanced (GFR is decreased, whereas proximal reabsorption is increased, and thus glomerulotubular balance does not occur under this condition), and (3) delivery of Na^+ to the ASDN is reduced. This decreased delivery, together with enhanced Na^+ and water reabsorption by the ASDN, virtually eliminates Na^+ from the urine and reduces urine volume.

Key Concepts

1. Regulation of body fluid osmolality (i.e., steady-state balance) requires that the amount of water added to the body exactly matches the amount lost from the body. Water is lost from the body by several routes (e.g., during respiration, with sweating, and in feces). The kidneys are the only regulated route of water excretion. Excretion of water by the kidneys is regulated by AVP secreted from the posterior pituitary. When AVP levels are high, the kidneys excrete a small volume of hyperosmotic urine. When AVP levels are low, a large volume of hypoosmotic urine is excreted.

2. Disorders of water balance alter body fluid osmolality. Because Na^+ with its anions are the major determinant of ECF osmolality, disorders of water balance manifest as changes in ECF [Na^+]. Positive water balance (intake > excretion) results in a decrease in body fluid osmolality and hyponatremia. Negative water balance (intake < excretion) results in an increase in body fluid osmolality and hypernatremia.

3. The volume of the ECF is determined by the amount of Na^+ in this compartment. To maintain a constant ECFV (i.e., euvolemia) NaCl excretion must match NaCl intake. The kidneys are the major route for regulating excretion of NaCl from the body. Volume sensors located primarily in the vascular system monitor volume and pressure. When ECFV expansion

occurs, neural and hormonal signals are sent to the kidneys to increase excretion of NaCl and water and thereby restore euvolemia. When ECFV contraction occurs, neural and hormonal signals are sent to the kidneys to decrease NaCl and water excretion and

thereby restore euvolemia. The sympathetic nervous system, the RAAS, and natriuretic peptides are important components of the system that maintain steady-state Na^+ balance.

Additional Reading

Water Balance

Berl T. Vasopressin antagonists. *N Engl J Med*. 2015;372:2207-2216.

Bichet DG. Polyuria and diabetes insipidus. In: Alpern RJ, et al., eds. *Seldin and Giebisch's The Kidney: Physiology and Pathophysiology*. 5th ed. Philadelphia: Academic Press; 2013.

Brown D, Fenton RA. The cell biology of vasopressin action. In: Taal MW, et al., eds. *Brenner and Rector's The Kidney*. 9th ed. Philadelphia: Saunders; 2012.

Dantzler WH, et al. Urine-concentrating mechanism in the inner medulla: function of the thin limbs of the loops of Henle. *Clin J Am Soc Nephrol*. 2014;9:1781-1789.

Danziger J, Zeidel M. Osmotic homeostasis. *J Am Soc Nephrol*. 2015;10:852-862.

Knepper MA, et al. Molecular physiology of water balance. *N Engl J Med*. 2015;372:1349-1358.

Nielsen S, et al. Aquaporin water channels in mammalian kidney. In: Alpern RJ, et al., eds. *Seldin and Giebisch's The Kidney: Physiology and Pathophysiology*. 5th ed. Philadelphia: Academic Press; 2013.

Robertson GL. Thirst and vasopressin. In: Alpern RJ, et al., eds. *Seldin and Giebisch's The Kidney: Physiology and Pathophysiology*. 5th ed. Philadelphia: Academic Press; 2013.

Sands JM, et al. Urine concentration and dilution. In: Taal MW, et al., eds. *Brenner and Rector's The Kidney*. 9th ed. Philadelphia: Saunders; 2012.

Sands JM, Layton HE. The urine concentrating mechanism and urea transporters. In: Alpern RJ, et al., eds. *Seldin and Giebisch's The Kidney: Physiology and Pathophysiology*. 5th ed. Philadelphia: Academic Press; 2013.

Sands JM, Layton HE. Advances in understanding the urine-concentrating mechanism. *Annu Rev Physiol*. 2014;76:387-409.

Schrier RW. Systemic arterial vasodilation, vasopressin, and vasopressinase in pregnancy. *J Am Soc Nephrol*. 2010;21:570-572.

Sterns RH. Disorders of plasma sodium–causes, consequences, and correction. *N Engl J Med*. 2015;372:55-65.

Verbalis JG. Disorders of water balance. In: Taal MW, et al., eds. *Brenner and Rector's The Kidney*. 9th ed. Philadelphia: Saunders; 2012.

NaCl Balance

Eladari D, et al. A new look at electrolyte transport in the distal tubule. *Annu Rev Physiol*. 2012;74:325-349.

Gamba G, et al. Sodium chloride transport in the loop of Henle, distal convoluted tubule, and collecting duct. In: Alpern RJ, et al., eds. *Seldin and Giebisch's The Kidney: Physiology and Pathophysiology*. 5th ed. Philadelphia: Academic Press; 2013.

Mount DB. Transport of sodium, chloride, and potassium. In: Taal MW, et al., eds. *Brenner and Rector's The Kidney*. 9th ed. Philadelphia: Saunders; 2012.

Mount DB. Thick ascending limb of the loop of Henle. *Clin J Am Soc Nephrol*. 2014;9:1974-1986.

Palmer LG, Schnermann J. Integrated control of Na transport along the nephron. *Clin J Am Soc Nephrol*. 2015;10:676-687.

Pearce D, et al. Aldosterone regulation of transport. In: Taal MW, et al., eds. *Brenner and Rector's The Kidney*. 9th ed. Philadelphia: Saunders; 2012.

Pearce D, et al. Collecting duct principal cell transport processes and their regulation. *Clin J Am Soc Nephrol*. 2015;10:135-146.

Slotki IN, Skorecki KL. Disorders of sodium balance. In: Taal MW, et al., eds. *Brenner and Rector's The Kidney*. 9th ed. Philadelphia: Saunders; 2012.

Staub O, Loffing J. Mineralocorticoid action in aldosterone sensitive distal nephron. In: Alpern RJ, et al., eds. *Seldin and Giebisch's The Kidney: Physiology and Pathophysiology*. 5th ed. Philadelphia: Academic Press; 2013.

Subramanya AR, Ellison DH. Distal convoluted tubule. *Clin J Am Soc Nephrol*. 2014;9:2147-2163.

Vesley DL. Natriuretic hormones. In: Alpern RJ, et al., eds. *Seldin and Giebisch's The Kidney: Physiology and Pathophysiology*. 5th ed. Philadelphia: Academic Press; 2013.

Weinstein AM. Sodium and chloride transport: proximal nephron. In: Alpern RJ, et al., eds. *Seldin and Giebisch's The Kidney: Physiology and Pathophysiology*. 5th ed. Philadelphia: Academic Press; 2013.

36

Potassium, Calcium, and Phosphate Homeostasis

LEARNING OBJECTIVES

Upon completion of this chapter the student should be able to answer the following questions:

1. How does the body maintain K⁺ homeostasis?
2. What is the distribution of K⁺ within the body compartments? Why is this distribution important?
3. What are the hormones and factors that regulate plasma K⁺ levels? Why is this regulation important?
4. How do the various segments of the nephron transport K⁺, and how does the mechanism of K⁺ transport by these segments determine how much K⁺ is excreted in the urine?
5. Why are the distal tubule and collecting duct so important in regulating K⁺ excretion?
6. How do plasma K⁺ levels, aldosterone, vasopressin, tubular fluid flow rate, and acid-base balance influence K⁺ excretion?
7. What is the physiological importance of calcium (Ca^{++}) and inorganic phosphate (P_i)?
8. How does the body maintain Ca^{++} and P_i homeostasis?
9. What roles do the kidneys, intestinal tract, and bone play in maintaining plasma Ca^{++} and P_i levels?
10. What hormones and factors regulate plasma Ca^{++} and P_i levels?
11. What are the cellular mechanisms responsible for Ca^{++} and P_i reabsorption along the nephron?
12. What hormones regulate renal Ca^{++} and P_i excretion by the kidneys?
13. What is the role of the calcium-sensing receptor (CaSR)?
14. What are some of the more common clinical disorders of Ca^{++} and P_i homeostasis?
15. What is the role of the kidneys in the production of calcitriol (active form of vitamin D)?
16. What effects do loop and thiazide diuretics have on Ca^{++} excretion?

K⁺ Homeostasis

Potassium (K⁺) is one of the most abundant cations in the body, and it is critical for many cell functions, including regulation of cell volume, regulation of intracellular pH, synthesis of DNA and protein, growth, enzyme function, resting membrane potential, and cardiac and neuromuscular activity. Despite wide fluctuations in dietary K⁺ intake, [K⁺] in cells and extracellular fluid (ECF) remains remarkably constant. Two sets of regulatory mechanisms safeguard K⁺ homeostasis. First, several mechanisms regulate [K⁺] in the ECF. Second, other mechanisms maintain the amount of K⁺ in the body constant by adjusting renal K⁺ excretion to match dietary K⁺ intake. It is the kidneys that regulate excretion of K⁺.

Total body [K⁺] is 50 mEq/kg of body weight, or 3600 mEq for a 70-kg individual. Ninety-eight percent of the K⁺ in the body is located within cells, where the average [K⁺] is 150 mEq/L. High intracellular [K⁺] is required for many cell functions, including cell growth and division and volume regulation. Only 2% of total body [K⁺] is located in the ECF, where its normal concentration is approximately 4 mEq/L. A [K⁺] in ECF that exceeds 5.0 mEq/L constitutes **hyperkalemia.** Conversely, a [K⁺] in ECF of less than 3.5 mEq/L constitutes **hypokalemia.**

 ## IN THE CLINIC

Hypokalemia is one of the most common electrolyte disorders in clinical practice and can be observed in as many as 20% of hospitalized patients. The most frequent causes of hypokalemia include administration of diuretic drugs, surreptitious vomiting (e.g., bulimia), and severe diarrhea. Gitelman syndrome (a genetic defect in the Na^+/Cl^- symporter in the apical membrane of distal tubule cells) also causes hypokalemia (see Chapter 36). **Hyperkalemia** is also a common electrolyte disorder and is seen in 1% to 10% of hospitalized patients. Hyperkalemia often occurs in patients with renal failure, in patients taking drugs such as angiotensin-converting enzyme (ACE) inhibitors and K⁺-sparing diuretics, in patients with hyperglycemia (i.e., high blood sugar), and in the elderly. **Pseudohyperkalemia,** a falsely high plasma [K⁺], is caused by traumatic lysis of red blood cells during blood drawing. Red blood cells, like all cells, contain K⁺, and lysis of red blood cells releases K⁺ into plasma, thereby artificially elevating plasma [K⁺].

The large concentration difference of K⁺ across cell membranes (≈146 mEq/L) is maintained by the operation of Na^+,K^+-ATPase. This [K⁺] gradient is important in

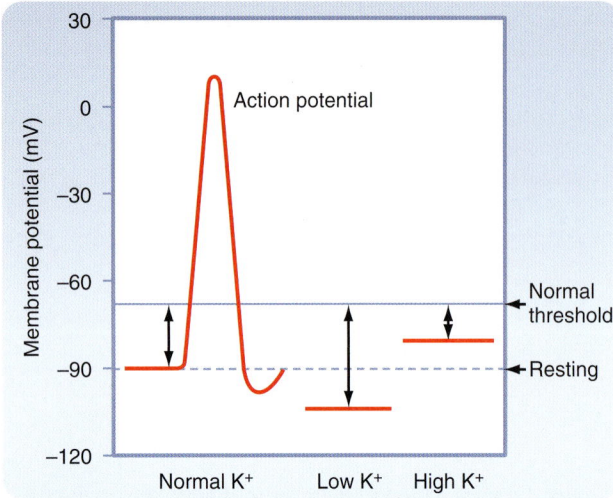

• **Fig. 36.1** Effects of variations in plasma [K$^+$] on resting membrane potential of skeletal muscle. Hyperkalemia causes membrane potential to become less negative, which decreases excitability by inactivating the fast Na$^+$ channels responsible for the depolarizing phase of the action potential. Hypokalemia hyperpolarizes the membrane potential and thereby reduces excitability because a larger stimulus is required to depolarize the membrane potential to the threshold potential. *Resting* indicates "normal" resting membrane potential. *Normal threshold* indicates the membrane threshold potential.

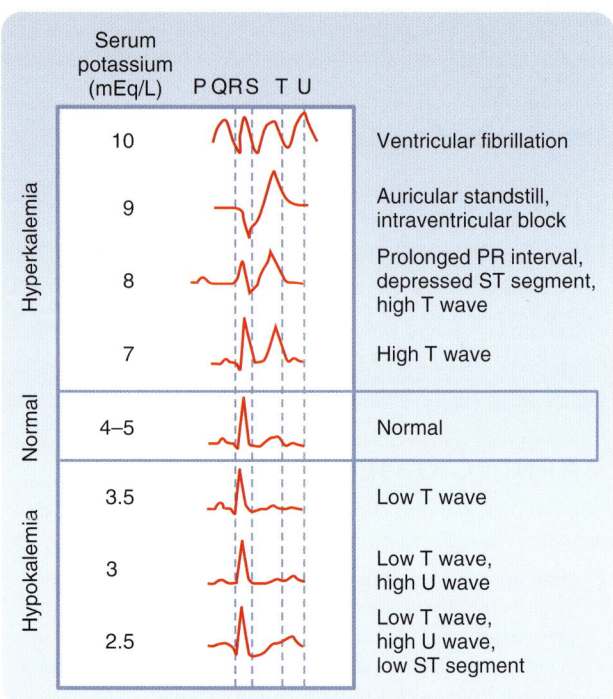

• **Fig. 36.2** Electrocardiographs from individuals with varying plasma [K$^+$]. See text for details. (Modified from Barker L et al. *Principles of Ambulatory Medicine.* 5th ed. Baltimore: Williams & Wilkins; 1999.)

maintaining the potential difference across cell membranes. Thus K$^+$ is critical for the excitability of nerve and muscle cells, as well as for the contractility of cardiac, skeletal, and smooth muscle cells (Fig. 36.1).

IN THE CLINIC

Cardiac arrhythmias are produced by both hypokalemia and hyperkalemia. The electrocardiogram (ECG; Fig. 36.2) (also see Chapter 16) monitors the electrical activity of the heart and is a fast and easy way to determine whether changes in plasma [K$^+$] influence the heart and other excitable cells. In contrast, measurement of plasma [K$^+$] by the clinical laboratory requires a blood sample, and values are often not immediately available. The first sign of hyperkalemia is the appearance of tall, thin T waves on the ECG. Further increases in plasma [K$^+$] prolong the PR interval, depress the ST segment, and lengthen the QRS interval of the ECG. Finally, as plasma [K$^+$] approaches 10 mEq/L, the P wave disappears, the QRS interval broadens, the ECG appears as a sine wave, and the ventricles fibrillate (i.e., manifest rapid, uncoordinated contractions of muscle fibers). Hypokalemia prolongs the QT interval, inverts the T wave, and lowers the ST segment of the ECG.

After a meal the K$^+$ absorbed by the gastrointestinal (GI) tract enters the ECF within minutes (Fig. 36.3). If the K$^+$ ingested during a normal meal ($\approx$33 mEq) were to remain in the ECF compartment (14 L), plasma [K$^+$] would increase by 2.4 mEq/L (33 mEq added to 14 L of ECF):

Equation 36.1

$$33 \text{ mEq}/14 \text{ L} = 2.4 \text{ mEq}/\text{L}$$

This rise in plasma [K$^+$], which could have deleterious effects on the electrical activity of the heart and other excitable tissues, is prevented by the rapid (minutes) uptake of K$^+$ into cells. Because excretion of K$^+$ by the kidneys after a meal is relatively slow (hours), uptake of K$^+$ by cells is essential to prevent life-threatening hyperkalemia. Maintaining total body [K$^+$] constant requires that all the K$^+$ absorbed by the GI tract eventually be excreted by the kidneys. This process requires about 6 hours.

Regulation of Plasma [K$^+$]

Several hormones, including epinephrine, insulin, and aldosterone, increase uptake of K$^+$ into skeletal muscle, liver, bone, and red blood cells (Box 36.1; see Fig. 36.3) by stimulating Na$^+$,K$^+$-ATPase, the 1Na$^+$/1K$^+$/2Cl$^-$ symporter, and the Na$^+$/Cl$^-$ symporter in these cells. Acute stimulation of K$^+$ uptake (i.e., within minutes) is mediated by increased activity of existing Na$^+$,K$^+$-ATPase, 1Na$^+$/1K$^+$/2Cl$^-$, and Na$^+$/Cl$^-$ transporters, whereas a chronic increase in K$^+$ uptake (i.e., within hours to days) is mediated by an increase in the quantity of Na$^+$,K$^+$-ATPase. The rise in plasma [K$^+$] that follows K$^+$ absorption by the GI tract stimulates secretion of insulin from the pancreas, release of aldosterone from the adrenal cortex, and secretion of epinephrine from the adrenal medulla (see Fig. 36.3). In contrast, a decrease in plasma [K$^+$] inhibits release of these hormones. Whereas insulin and epinephrine act within a few minutes, aldosterone requires about an hour to stimulate uptake of K$^+$ into cells.

Epinephrine

Catecholamines affect the distribution of K^+ across cell membranes by activating α- and β_2-adrenergic receptors. Stimulation of α-adrenoceptors releases K^+ from cells, especially in the liver, whereas stimulation of β_2-adrenoceptors promotes K^+ uptake by cells.

For example, activation of β_2-adrenoceptors after exercise is important in preventing hyperkalemia. The rise in plasma $[K^+]$ after a K^+-rich meal is greater if the patient has been pretreated with propranolol, a β-adrenoceptor antagonist. Furthermore, release of epinephrine during stress (e.g., myocardial ischemia) can rapidly lower plasma $[K^+]$.

Insulin

Insulin also stimulates uptake of K^+ into cells. The importance of insulin is illustrated by two observations. First, the rise in plasma $[K^+]$ after a K^+-rich meal is greater in patients with diabetes mellitus (i.e., insulin deficiency) than in healthy people. Second, insulin (and glucose to prevent insulin-induced hypoglycemia) can be infused to correct hyperkalemia. Insulin is the most important hormone that shifts K^+ into cells after ingestion of K^+ in a meal. In patients with chronic kidney disease, although insulin-stimulated

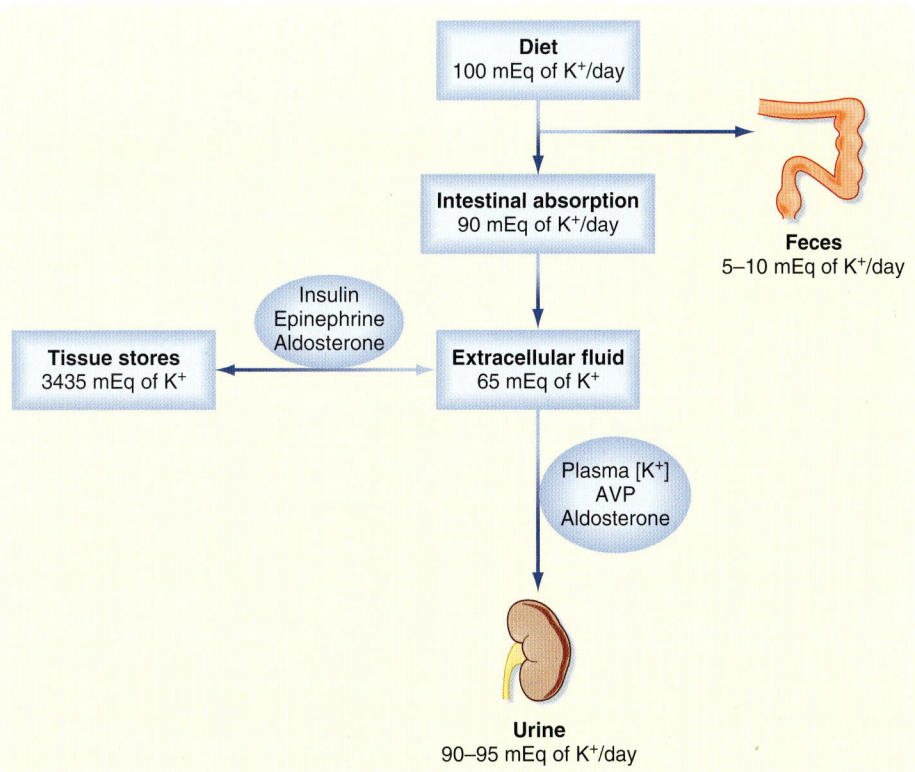

• **Fig. 36.3** Overview of K⁺ homeostasis. An increase in plasma insulin, epinephrine, or aldosterone stimulates movement of K⁺ into cells and decreases plasma [K⁺], whereas a fall in plasma concentration of these hormones has the opposite effect and increases plasma [K⁺]. The amount of K⁺ in the body is determined by the kidneys. An individual is in K⁺ balance when dietary intake and urinary output (plus output by the GI tract) are equal. Excretion of K⁺ by the kidneys is regulated by plasma [K⁺], aldosterone, and AVP.

glucose uptake into cells is impaired, insulin stimulation of K+ uptake into cells is normal.

Aldosterone

Aldosterone, like catecholamines and insulin, also promotes uptake of K+ into cells. A rise in aldosterone levels (e.g., primary aldosteronism) causes hypokalemia, whereas a fall in aldosterone levels (e.g., Addison's disease) causes hyperkalemia. As discussed later and as illustrated in Fig. 36.3, aldosterone also stimulates urinary K+ excretion. Thus aldosterone alters plasma [K+] by acting on uptake of K+ into cells and altering urinary K+ excretion.

Alterations in Plasma [K+]

Several factors can alter plasma [K+] (see Box 36.1). These factors are not involved in regulation of plasma [K+] but rather alter the movement of K+ between the intracellular fluid (ICF) and ECF and thus cause development of hypokalemia or hyperkalemia.

Acid-Base Balance

Metabolic acidosis increases plasma [K+], whereas metabolic alkalosis decreases it. Respiratory alkalosis causes hypokalemia. In contrast, respiratory acidosis has little or no effect on plasma [K+]. Metabolic acidosis produced by addition of inorganic acids (e.g., HCl, H_2SO_4) increases plasma [K+] much more than an equivalent acidosis produced by accumulation of organic acids (e.g., lactic acid, acetic acid, ketoacids). The reduced pH (i.e., increased [H+]) promotes movement of H+ into cells and the reciprocal movement of K+ out of cells to maintain electroneutrality. This effect of acidosis occurs in part because acidosis inhibits the transporters that accumulate K+ inside cells, including Na^+,K^+-ATPase and the $1Na^+/1K^+/2Cl^-$ symporter. In addition, movement of H+ into cells occurs as the cells buffer changes in [H+] of the ECF (see Chapter 37). As H+ moves across cell membranes, K+ moves in the opposite direction, and thus cations are neither gained nor lost across cell membranes. Metabolic alkalosis has the opposite effect; plasma [K+] decreases as K+ moves into cells and H+ exits.

Although organic acids produce a metabolic acidosis, they do not cause significant hyperkalemia. Two explanations have been suggested for the reduced ability of organic acids to cause hyperkalemia. First, the organic anion may enter the cell with H+ and thereby eliminate the need for K+-H+ exchange across the membrane. Second, organic anions may stimulate insulin secretion, which moves K+ into cells. This movement may counteract the direct effect of the acidosis, which moves K+ out of cells.

Plasma Osmolality

The osmolality of plasma also influences the distribution of K+ across cell membranes. An increase in the osmolality of

ECF enhances the release of K+ by cells and thus increases extracellular [K+]. Plasma [K+] may increase by 0.4 to 0.8 mEq/L with a 10 mOsm/kg H_2O elevation in plasma osmolality. In patients with diabetes mellitus who do not take insulin, plasma [K+] is often elevated, in part because of the lack of insulin and in part because of the increase in plasma [glucose] (i.e., from a normal value of ≈ 100 mg/dL to as high as ≈ 1200 mg/dL in some cases), which increases plasma osmolality. Hypoosmolality has the opposite action. The alterations in plasma [K+] associated with changes in osmolality are related to changes in cell volume. For example, as plasma osmolality increases, water leaves cells because of the osmotic gradient across the plasma membrane (see Chapter 1). Water leaves cells until the intracellular osmolality equals that of ECF. This loss of water shrinks cells and causes [K+] in cells to rise. The rise in intracellular [K+] provides a driving force for the exit of K+ from cells. This sequence increases plasma [K+]. A fall in plasma osmolality has the opposite effect.

Cell Lysis

Cell lysis causes hyperkalemia as a result of addition of intracellular K+ to ECF. Severe trauma (e.g., burns) and some conditions such as **tumor lysis syndrome** (i.e., chemotherapy-induced destruction of tumor cells) and **rhabdomyolysis** (i.e., destruction of skeletal muscle) destroy cells and release K+ and other cell solutes into ECF. In addition, gastric ulcers may cause seepage of red blood cells into the GI tract. The blood cells are digested, and the K+ released from the cells is absorbed and can cause hyperkalemia.

Exercise

More K+ is released from skeletal muscle cells during exercise than during rest. The ensuing hyperkalemia depends on the degree of exercise. In people walking slowly, plasma [K+] increases by 0.3 mEq/L. With vigorous exercise, plasma [K+] may increase by 2.0 or more mEq/L.

K+ Excretion by the Kidneys

The kidneys play a major role in maintaining K+ balance. As illustrated in Fig. 36.3 the kidneys excrete 90% to 95% of the K+ ingested in the diet. Excretion equals intake even when intake increases by as much as 10-fold. This balance in urinary excretion and dietary intake underscores the importance of the kidneys in maintaining K+ homeostasis. Although small amounts of K+ are lost each day in feces and sweat (≈5%–10% of the K+ ingested in the diet), except during severe diarrhea this amount is essentially constant, is not regulated, and therefore is relatively less important than the K+ excreted by the kidneys. K+ secretion from blood into tubular fluid by cells of the distal tubule and collecting duct system is the key factor in determining urinary K+ excretion (Fig. 36.4).

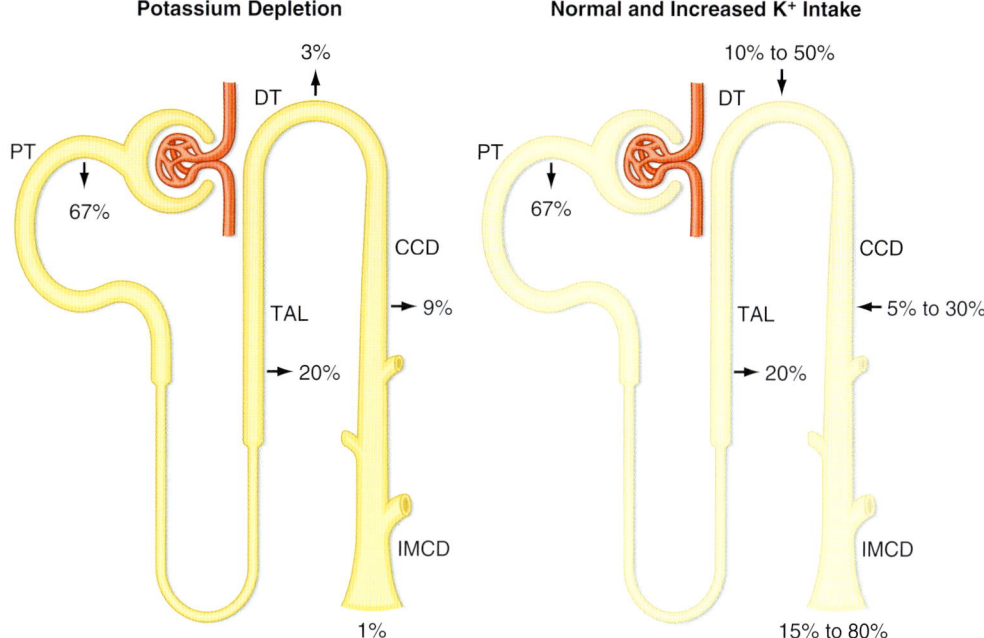

• Fig. 36.4 K⁺ transport along the nephron. Excretion of K⁺ depends on the rate and direction of K⁺ transport by the late segment of the distal tubule and collecting duct. Percentages refer to the amount of filtered K⁺ reabsorbed or secreted by each nephron segment. Arrows indicate direction of transport. *Left,* Dietary K⁺ depletion. An amount of K⁺ equal to 1% of the filtered load of K⁺ is excreted. *Right,* Normal and increased dietary K⁺ intake. An amount of K⁺ equal to 15% to 80% of the filtered load is excreted. *CCD,* cortical collecting duct; *DT,* distal tubule; *IMCD,* inner medullary collecting duct; *PT,* proximal tubule; *TAL,* thick ascending limb.

 IN THE CLINIC

Exercise-induced changes in plasma [K⁺] do not usually produce symptoms and are reversed after several minutes of rest. However, vigorous exercise can lead to life-threatening hyperkalemia in individuals (1) who have endocrine disorders that affect release of insulin, epinephrine (a β-adrenergic agonist), or aldosterone; (2) whose ability to excrete K⁺ is impaired (e.g., renal failure); or (3) who take certain medications, such as β₁-adrenergic blockers. For example, during vigorous exercise, plasma [K⁺] may increase by at least 2 to 4 mEq/L in individuals who take β₁-adrenergic receptor antagonists for hypertension. Because acid-base balance, plasma osmolality, cell lysis, and exercise do not maintain plasma [K⁺] at a normal value, they do not contribute to K⁺ homeostasis (see Box 36.1). The extent to which these pathophysiological states alter plasma [K⁺] depends on the integrity of the homeostatic mechanisms that regulate plasma [K⁺] (e.g., secretion of epinephrine, insulin, and aldosterone).

Because K⁺ is not bound to plasma proteins, it is freely filtered by the glomerulus. When individuals ingest 100 mEq of K⁺ per day, urinary K⁺ excretion is about 15% of the amount filtered. Accordingly K⁺ must be reabsorbed along the nephron. When dietary K⁺ intake increases, however, K⁺ excretion can exceed the amount filtered. Thus K⁺ can also be secreted.

The proximal tubule reabsorbs about 67% of the filtered K⁺ under most conditions. Approximately 20% of the filtered K⁺ is reabsorbed by the loop of Henle, and as with the proximal tubule, the amount reabsorbed is a constant fraction of the amount filtered. In contrast to these segments, which can only reabsorb K⁺, the distal tubule and collecting duct are able to reabsorb or secrete K⁺. The rate of K⁺ reabsorption or secretion by the distal tubule and collecting duct depends on a variety of hormones and factors. When 100 mEq/day of K⁺ is ingested, it is secreted by these nephron segments. A rise in dietary K⁺ intake increases K⁺ secretion. K⁺ secretion can increase the amount of K⁺ that appears in urine so that it approaches 80% of the amount filtered (see Fig. 36.4, right panel). In contrast, a low-K⁺ diet activates K⁺ reabsorption along the distal tubule and collecting duct so that urinary excretion falls to about 1% of the K⁺ filtered by the glomerulus (see Fig. 36.4, left panel). The kidneys cannot reduce K⁺ excretion to the same low levels as they can for Na⁺ (i.e., 0.2%). Therefore hypokalemia can develop in individuals placed on a K⁺-deficient diet. Because the magnitude and direction of K⁺ transport by the distal tubule and collecting duct are variable, the overall rate of urinary K⁺ excretion is determined by these tubular segments.

Cellular Mechanism of K⁺ Secretion by Principal Cells and Intercalated Cells

Fig. 36.5A illustrates the cellular mechanisms of K⁺ secretion by principal cells in the late segment of the distal

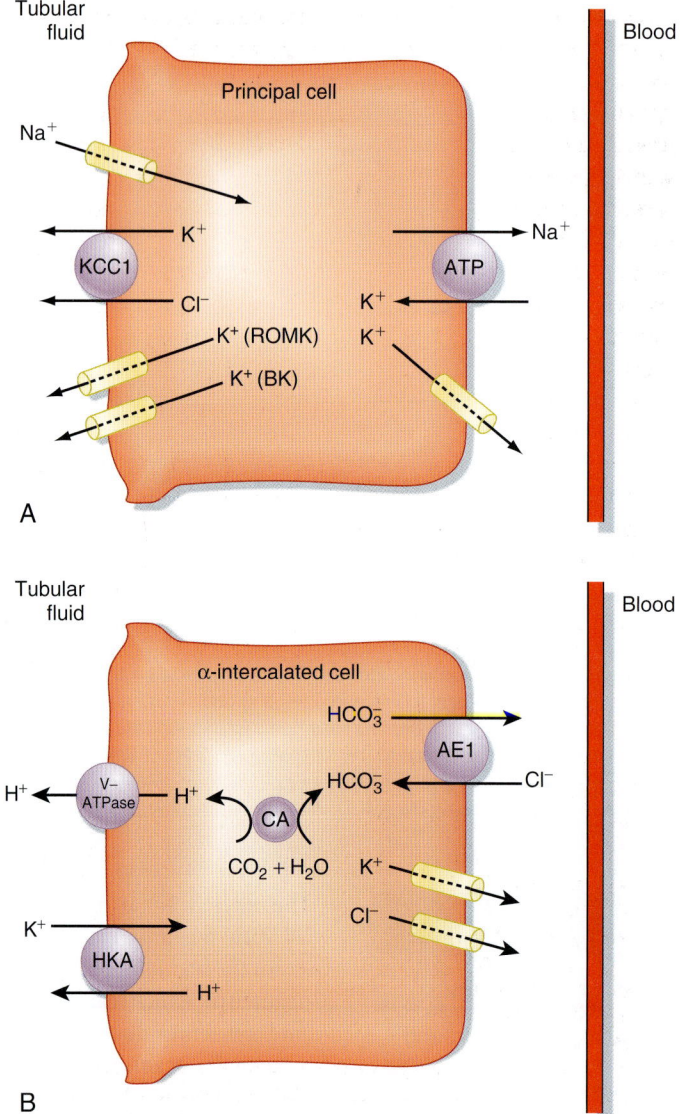

• **Fig. 36.5** Cellular mechanism of K$^+$ secretion by principal cells (A) and α-intercalated cells (B) in the late segment of the distal tubule and collecting duct. α-Intercalated cells contain very low levels of Na$^+$,K$^+$-ATPase in the basolateral membrane (not shown). K$^+$ depletion increases K$^+$ reabsorption by α-intercalated cells by stimulating the H$^+$,K$^+$-ATPase (HKA).

IN THE CLINIC

In individuals with **advanced renal disease,** the kidneys are unable to eliminate ingested K$^+$ from the body. Therefore plasma [K$^+$] rises. The resulting hyperkalemia reduces the resting membrane potential (i.e., the voltage becomes less negative), and this reduced potential decreases the excitability of neurons, cardiac cells, and muscle cells by inactivating fast Na$^+$ channels, which are critical for the depolarization phase of the action potential (see Fig. 36.1). Severe rapid increases in plasma [K$^+$] can lead to cardiac arrest and death. In contrast, in patients taking diuretic drugs for hypertension, urinary K$^+$ excretion often exceeds dietary K$^+$ intake. Accordingly K$^+$ balance is negative and hypokalemia develops. This decline in extracellular [K$^+$] hyperpolarizes the resting cell membrane (i.e., the voltage becomes more negative) and reduces the excitability of neurons, cardiac cells, and muscle cells. Severe hypokalemia can lead to paralysis, cardiac arrhythmias, and death. Hypokalemia can also impair the ability of the kidneys to concentrate urine and can stimulate renal production of NH$_4^+$, which affects acid-base balance (see Chapter 37). Therefore maintenance of high intracellular [K$^+$], low extracellular [K$^+$], and a high [K$^+$] gradient across cell membranes is essential for a number of cellular functions.

tubule and the collecting duct. Secretion from blood into the tubule lumen is a two-step process: (1) uptake of K⁺ from blood across the basolateral membrane by Na⁺,K⁺-ATPase and (2) diffusion of K⁺ from the cell into tubular fluid via K⁺ channels (ROMK and BK). A K⁺/Cl⁻ symporter (KCC1) in the apical plasma membrane also secretes K⁺. Na⁺,K⁺-ATPase creates a high intracellular [K⁺] that provides the chemical driving force for exit of K⁺ across the apical membrane through K⁺ channels. Although K⁺ channels are also present in the basolateral membrane, K⁺ preferentially leaves the cell across the apical membrane and enters the tubular fluid. K⁺ transport follows this route for two reasons. First, the electrochemical gradient of K⁺ across the apical membrane favors its downhill movement into tubular fluid. Second, the permeability of the apical membrane to K⁺ is greater than that of the basolateral membrane. Therefore K⁺ preferentially diffuses across the apical membrane into tubular fluid. K⁺ secretion across the apical membrane via the K⁺/Cl⁻ symporter is driven by the favorable concentration gradient of K⁺ between the cell and tubular fluid. The three major factors that control the rate of K⁺ secretion by the late segment of the distal tubule and the collecting duct are:

1. the activity of Na⁺,K⁺-ATPase
2. the driving force (electrochemical gradient for K⁺ channels and the chemical concentration gradient for the K⁺/Cl⁻ symporter) for movement of K⁺ across the apical membrane
3. the permeability of apical membrane K⁺ channels to K⁺

Every change in K⁺ secretion by principal cells results from an alteration in one or more of these factors (Fig. 36.6).

α-Intercalated cells reabsorb K⁺ by a H⁺,K⁺-ATPase transport mechanism (HKA) located in the apical membrane (see Fig. 36.5B). This transporter mediates K⁺ uptake across the apical plasma membrane in exchange for H⁺. K⁺ exit from intercalated cells into the blood is mediated by a K⁺ channel. Reabsorption of K⁺ is activated by a low-K⁺ diet.

Regulation of K⁺ Secretion by the Distal Tubule and Collecting Duct

Regulation of K⁺ excretion is achieved mainly by alterations in K⁺ secretion by principal cells of the late segment of the distal tubule and collecting duct. Plasma [K⁺] and aldosterone are the major physiological regulators of K⁺ secretion. Ingestion of a K⁺-rich meal also stimulates renal K⁺ excretion by a mechanism involving an unknown gut-dependent mechanism. Arginine vasopressin (AVP), also stimulates K⁺ secretion; however, it is less important than plasma [K⁺] and aldosterone. Other factors, including the flow rate of tubular fluid and acid-base balance, influence secretion of K⁺ by the distal tubule and collecting duct. However, they are not homeostatic mechanisms, because they disturb K⁺ balance (Box 36.2).

Plasma [K⁺]

Plasma [K⁺] is an important determinant of K⁺ secretion by the distal tubule and collecting duct. Hyperkalemia (e.g., resulting from a high-K⁺ diet or from rhabdomyolysis)

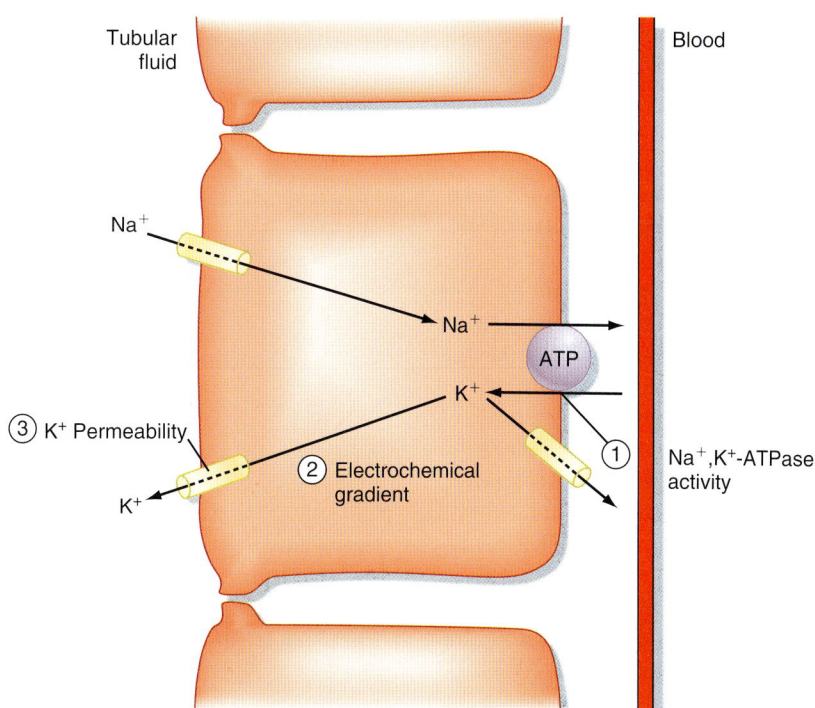

• **Fig. 36.6** Cellular mechanism of K⁺ secretion by principal cells. The numbers indicate where K⁺ secretion is regulated. 1, Na⁺,K⁺-ATPase; 2, electrochemical gradient of K⁺ across the apical membrane; 3, permeability of the apical membrane to K⁺.

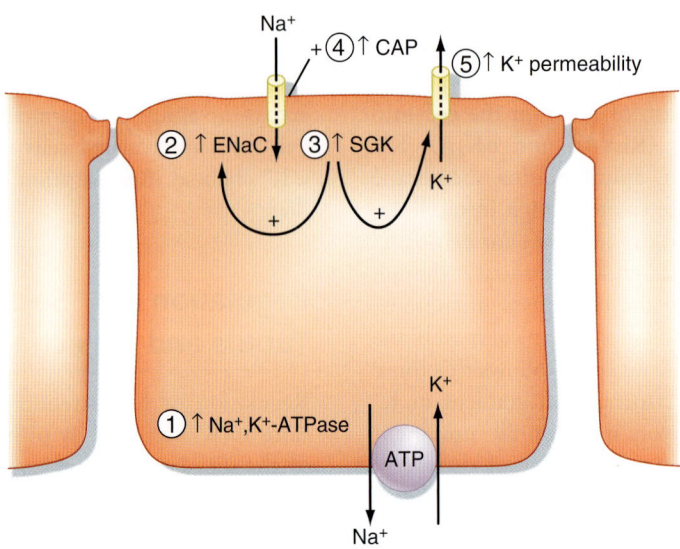

• **Fig. 36.7** Effects of aldosterone on secretion of K^+ by principal cells in the late segment of the distal tubule and collecting duct. Numbers refer to the five effects of aldosterone discussed in text.

• **BOX 36.2** **Major Factors and Hormones Influencing K^+ Excretion**

Physiological: Keep K^+ Balance Constant

Plasma $[K^+]$
Aldosterone
AVP

Pathophysiological: Displace K^+ Balance

Flow rate of tubule fluid
Acid-base disorders
Glucocorticoids

---

stimulates secretion of K^+ within minutes. Several mechanisms are involved. First, hyperkalemia stimulates Na^+, K^+-ATPase and thereby increases K^+ uptake across the basolateral membrane. This uptake raises intracellular $[K^+]$ and increases the electrochemical driving force for exit of K^+ across the apical membrane. Second, hyperkalemia also increases the permeability of the apical membrane to K^+. Third, hyperkalemia stimulates secretion of aldosterone by the adrenal cortex, which as discussed later, acts synergistically with plasma $[K^+]$ to stimulate secretion of K^+. Fourth, hyperkalemia also increases the flow rate of tubular fluid, which as discussed later, stimulates secretion of K^+ by the distal tubule and collecting duct.

Hypokalemia (e.g., caused by a low-K^+ diet or loss of K^+ in diarrhea fluid) decreases K^+ secretion via actions opposite those described for hyperkalemia. Hence hypokalemia inhibits Na^+,K^+-ATPase, decreases the electrochemical driving force for efflux of K^+ across the apical membrane, reduces permeability of the apical membrane to K^+, and decreases plasma aldosterone levels.

IN THE CLINIC

Chronic hypokalemia (plasma $[K^+]$ < 3.5 mEq/L) occurs most often in patients who receive diuretics for hypertension. Hypokalemia also occurs in patients who vomit, undergo nasogastric suction, have diarrhea, abuse laxatives, or have hyperaldosteronism. Hypokalemia occurs because excretion of K^+ by the kidneys exceeds dietary intake of K^+. Vomiting, nasogastric suction, diuretics, and diarrhea can all decrease ECF volume (ECFV), which in turn stimulates secretion of aldosterone (see Chapter 35). Because aldosterone stimulates excretion of K^+ by the kidneys, its action contributes to development of hypokalemia.

Chronic hyperkalemia (plasma $[K^+]$ > 5.0 mEq/L) occurs most frequently in individuals with reduced urine flow, low plasma aldosterone levels, and renal disease in which the glomerular filtration rate (GFR) falls below 20% of normal. In these individuals, hyperkalemia occurs because excretion of K^+ by the kidneys is less than dietary intake of K^+. Less common causes of hyperkalemia occur in people with deficiencies in insulin, epinephrine, and aldosterone secretion or in people with metabolic acidosis caused by inorganic acids.

Aldosterone

Chronically elevated (i.e., ≥24 hours) plasma aldosterone levels enhance secretion of K^+ across principal cells in the late segment of the distal tubule and collecting duct (i.e., aldosterone-sensitive distal nephron [**ASDN**]) via five mechanisms (Fig. 36.7): (1) by increasing the amount of Na^+,K^+-ATPase in the basolateral membrane; (2) by increasing expression of the epithelial sodium channel (ENaC) in the apical cell membrane; (3) by elevating SGK1 (*s*erum *g*lucocorticoid-stimulated *k*inase) levels, which also increases expression of ENaC in the apical membrane and activates K⁺ channels; (4) by stimulating CAP1 (*channel-activating protease*, also called **prostatin**), which directly activates ENaC; and (5) by stimulating the permeability of the apical membrane to K⁺. Aldosterone increases the permeability of the apical membrane to K⁺ by increasing the number of K⁺ channels in the membrane. However, the cellular mechanisms involved in this response are not completely known. Increased expression of Na⁺,K⁺-ATPase facilitates uptake of K⁺ across the basolateral membrane into cells and thereby elevates intracellular [K⁺]. The increase in the number and activity of Na⁺ channels enhances entry of Na⁺ into the cell from tubular fluid, an effect that depolarizes the apical membrane voltage. Depolarization of the apical membrane and increased intracellular [K⁺] enhance the electrochemical driving force for secretion of K⁺ from the cell into the tubule fluid. Taken together, these actions increase uptake of K⁺ into the cell across the basolateral membrane and enhance exit of K⁺ from the cell across the apical membrane. Secretion of aldosterone is increased by hyperkalemia and by angiotensin II (after activation of the renin-angiotensin system). Secretion of aldosterone is decreased by hypokalemia and natriuretic peptides released from the heart.

Although an acute (e.g., within hours) increase in aldosterone levels enhances the activity of Na⁺,K⁺-ATPase, K⁺ excretion does not increase immediately. The reason for this relates to the effect of aldosterone on Na⁺ reabsorption and tubular flow. Aldosterone stimulates reabsorption of Na⁺ and water and thus decreases tubular flow. Reduction in flow in turn decreases K⁺ secretion (as discussed in more detail later). However, chronic stimulation of Na⁺ reabsorption increases the ECFV and thereby returns tubular flow to normal. These actions allow a direct stimulatory effect of aldosterone on the ASDN to enhance K⁺ excretion.

AVP

Although AVP does not affect urinary K⁺ excretion, this hormone does promote secretion of K⁺ by the ASDN (Fig. 36.8). AVP increases the electrochemical driving force for exit of K⁺ across the apical membrane of principal cells by stimulating uptake of Na⁺ across the apical membrane of these cells. The increased Na⁺ uptake reduces the electrical potential difference across the apical membrane (i.e., the interior of the cell becomes less negatively charged). Despite this effect, AVP does not change K⁺ secretion by these nephron segments. The reason for this relates to the effect of AVP on tubular fluid flow. AVP decreases flow of tubular fluid by stimulating water reabsorption. The decrease in tubular flow in turn reduces secretion of K⁺ (explained later). The inhibitory effect of decreased flow of tubular fluid offsets the stimulatory effect of AVP on the electrochemical driving force for exit of K⁺ across the apical membrane (see Fig. 36.8). If AVP did not increase the electrochemical gradient favoring K⁺ secretion, urinary K⁺ excretion would fall as AVP levels increased and urinary flow rates decreased. Hence K⁺ balance would change in

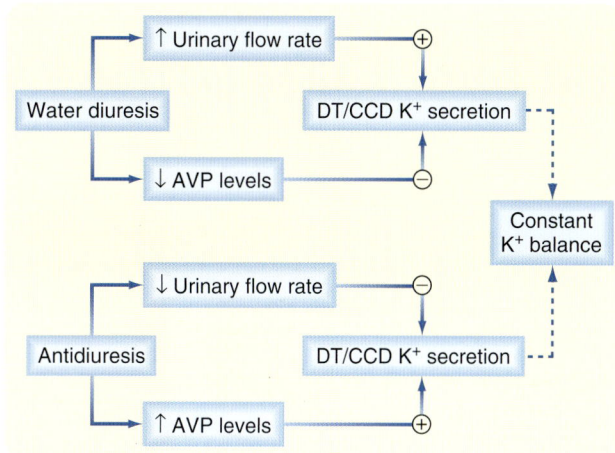

• **Fig. 36.8** Opposing effects of AVP and urine flow rate on secretion of K⁺ by the ASDN. K⁺ secretion is stimulated by an increase in urinary flow rate and reduced by a fall in AVP levels. In contrast, K⁺ secretion is reduced by a decrease in urinary flow rate and increased by a rise in AVP levels. Because the effects of flow and AVP oppose each other, net K⁺ secretion is not affected by water diuresis or antidiuretics.

response to alterations in water balance. Thus the effects of AVP on the electrochemical driving force for exit of K⁺ across the apical membrane and on tubule flow enable urinary K⁺ excretion to be maintained constant despite wide fluctuations in water excretion.

Factors That Perturb K⁺ Excretion

Plasma [K⁺], aldosterone, and AVP play important roles in regulating K⁺ balance; however, the factors and hormones discussed next perturb K⁺ balance (see Box 36.2).

Flow of Tubular Fluid

A rise in the flow of tubular fluid (e.g., with diuretic treatment, ECFV expansion) stimulates secretion of K⁺ within minutes, whereas a fall (e.g., ECFV contraction caused by hemorrhage, severe vomiting, or diarrhea) reduces secretion of K⁺ by the late segment of the ASDN. Increments in tubular fluid flow are more effective in stimulating secretion of K⁺ as dietary K⁺ intake is increased. Recent studies on the primary cilium in principal cells have elucidated some of the mechanisms whereby increased flow stimulates secretion of K⁺ (Fig. 36.9). Increased flow bends the primary cilium in principal cells, which activates the PKD1/PKD2 Ca⁺⁺-conducting channel complex. This allows more Ca⁺⁺ to enter principal cells and increases intracellular [Ca⁺⁺]. The increase in [Ca⁺⁺] activates BK K⁺ channels in the apical plasma membrane, which enhances K⁺ secretion from the cell into the tubule fluid. Increased flow may also stimulate secretion of K⁺ by other mechanisms. As flow increases, such as after administration of diuretics or as the result of an increase in ECFV, so does the [Na⁺] of tubule fluid. This increase in [Na⁺] facilitates entry of Na⁺ across the apical membrane of ASDN cells, thereby decreasing the cells'

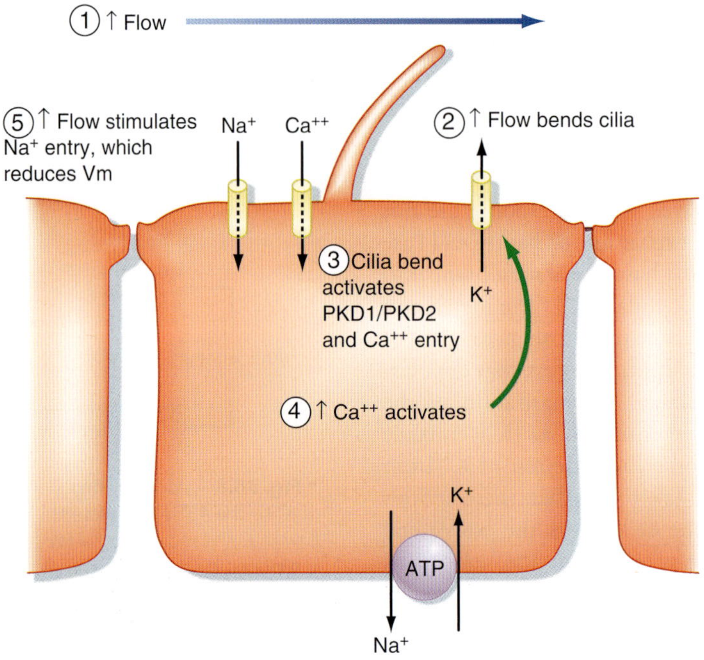

① ↑ Flow

⑤ ↑ Flow stimulates Na⁺ entry, which reduces Vm

Na⁺   Ca⁺⁺

② ↑ Flow bends cilia

③ Cilia bend activates PKD1/PKD2 and Ca⁺⁺ entry

K⁺

④ ↑ Ca⁺⁺ activates

K⁺

ATP

Na⁺

• **Fig. 36.9** Cellular mechanism whereby an increased flow rate of tubule fluid stimulates secretion of K⁺ by principal cells. See text for details.

interior negative membrane potential. This depolarization of the cell membrane potential increases the electrochemical driving force that promotes secretion of K⁺ across the apical cell membrane into tubule fluid. In addition, increased uptake of Na⁺ into cells activates the Na⁺,K⁺-ATPase in the basolateral membrane, thereby increasing uptake of K⁺ across the basolateral membrane and consequently elevating [K⁺]. However, it is important to note that an increase in flow rate during a water diuresis does *not* have a significant effect on excretion of K⁺, most likely because during a water diuresis the [Na⁺] of tubule fluid does not increase as flow rises.

Acid-Base Balance

Another factor that modulates secretion of K⁺ is the [H⁺] of ECF. Acute alterations (within minutes to hours) in the pH of plasma influences secretion of K⁺ by the ASDN. Alkalosis (i.e., plasma pH above normal) increases secretion of K⁺, whereas acidosis (i.e., plasma pH below normal) decreases it. An acute acidosis reduces K⁺ secretion via two mechanisms: (1) it inhibits Na⁺,K⁺-ATPase and thereby reduces cell [K⁺] and the electrochemical driving force for exit of K⁺ across the apical membrane, and (2) it reduces the permeability of the apical membrane to K⁺. Alkalosis has the opposite effects.

The effect of metabolic acidosis on excretion of K⁺ is time dependent. As already noted, and illustrated in Fig. 36.10, acute metabolic acidosis reduces K⁺ excretion. However, when metabolic acidosis lasts for several days, urinary K⁺ excretion is stimulated (see Fig. 36.10). This occurs because chronic metabolic acidosis decreases reabsorption of water and solutes (e.g., NaCl) by the proximal tubule by

inhibiting Na⁺,K⁺-ATPase. Hence the flow of tubular fluid is augmented along the ASDN. Inhibition of water and NaCl reabsorption by the proximal tubule also decreases ECFV and thereby stimulates secretion of aldosterone. In addition, chronic acidosis caused by inorganic acids increases

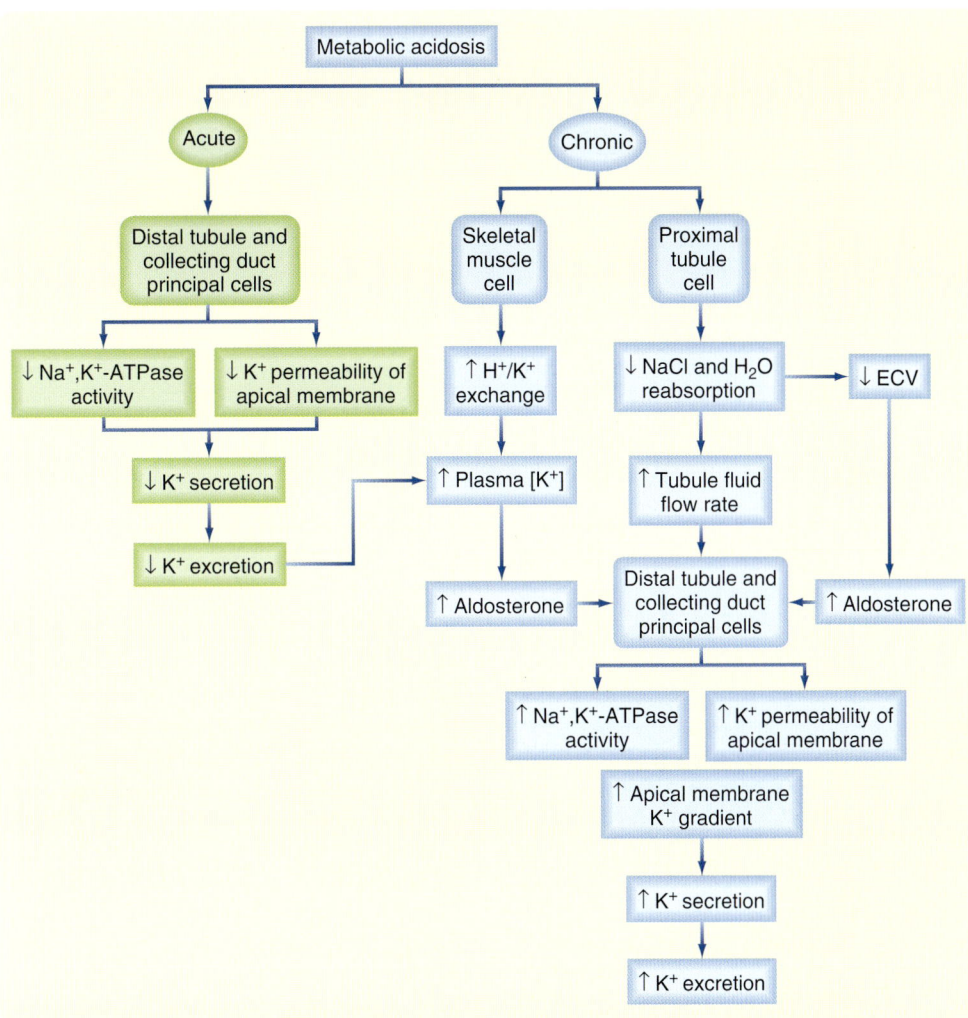

• **Fig. 36.10** Acute versus chronic effect of metabolic acidosis on excretion of K⁺. See text for details. ECV, effective circulating volume.

plasma [K⁺], which stimulates secretion of aldosterone. The rise in tubular fluid flow, plasma [K⁺], and aldosterone levels offsets the effects of acidosis on cell [K⁺] and apical membrane permeability, and K⁺ secretion rises. Thus metabolic acidosis may either inhibit or stimulate excretion of K⁺, depending on the duration of the disturbance. As noted, acute metabolic alkalosis stimulates excretion of K⁺. Chronic metabolic alkalosis, especially in association with ECFV contraction, significantly increases renal K⁺ excretion because of the associated increased levels of aldosterone.

Glucocorticoids

Glucocorticoids increase urinary K⁺ excretion. This effect is mediated in part by an increase in GFR, which enhances the urinary flow rate, a potent stimulus of K⁺ excretion, and by stimulation of SGK1 activity (see earlier).

As discussed, the rate of urinary K⁺ excretion is frequently determined by simultaneous changes in hormone levels, acid-base balance, or the flow rate of tubule fluid (Table 36.1). The powerful effect of flow often enhances or opposes the response of the ASDN to hormones and changes in

acid-base balance. This interaction can be beneficial in the case of hyperkalemia, in which the increase in flow enhances excretion of K⁺ and thereby restores K⁺ homeostasis. However, this interaction can also be detrimental, as in the case of alkalosis, in which changes in flow and acid-base status alter K⁺ homeostasis.

Overview of Calcium and Inorganic Phosphate Homeostasis

Ca⁺⁺ and inorganic phosphate $(P_i)^a$ are multivalent ions that subserve many complex and vital functions. Ca⁺⁺ is an important cofactor in many enzymatic reactions; it is a key second messenger in numerous signaling pathways; it plays an important role in neural transduction, blood clotting, and muscle contraction; and it is a critical component of the extracellular matrix, cartilage, teeth, and bone. P_i, like Ca⁺⁺, is a key component of bone. P_i is essential for metabolic

---

[a]At physiological pH, inorganic phosphate exists as HPO_4^- and $H_2PO_4^-$ (pK = 6.8). For simplicity, we collectively refer to these ion species as P_i.

TABLE 36.1	Effects of Hormones and Other Factors on K⁺ Secretion by Distal Tubule and Collecting Duct, and on Urinary K⁺ Excretion			

Condition	Effect	Tubule Fluid Flow	Urinary Excretion
Hyperkalemia	Increase	Increase	Increase
Aldosterone			
Acute	Increase	Decrease	No change
Chronic	Increase	No change	Increase
Glucocorticoids	Increase	Increase	Increase
AVP	Increase	Decrease	No change
Acidosis			
Acute	Decrease	No change	Decrease
Chronic	Decrease	Large increase	Increase
Alkalosis	Increase	Increase	Large increase

Modified from Field MJ et al. In Narins R (ed). *Textbook of Nephrology: Clinical Disorders of Fluid and Electrolyte Metabolism.* 5th ed. New York: McGraw-Hill; 1994.

AT THE CELLULAR LEVEL

The cellular mechanisms whereby changes in the K⁺ content of the diet and acid-base balance regulate secretion of K⁺ by the early segment of the ASDN have recently been elucidated. Elevated K⁺ intake increases secretion of K⁺ by several mechanisms, all related to increased serum [K⁺]. Hyperkalemia increases the activity of the ROMK channel in the apical plasma membrane of principal cells. Moreover, hyperkalemia inhibits reabsorption of NaCl and water by the proximal tubule, thereby increasing the ASDN flow rate, a potent stimulus to secretion of K⁺. Hyperkalemia also enhances [aldosterone], which increases K⁺ secretion by three mechanisms. First, aldosterone increases the number of K⁺ channels in the apical plasma membrane. Second, aldosterone stimulates uptake of K⁺ across the basolateral membrane by increasing the number of Na⁺,K⁺-ATPase pumps, thereby enhancing the electrochemical gradient driving secretion of K⁺ across the apical membrane. Third, aldosterone increases movement of Na⁺ across the apical membrane, which depolarizes the apical plasma membrane voltage and thus increases the electrochemical gradient, promoting secretion of K⁺. A low-K⁺ diet dramatically reduces secretion of K⁺ by the ASDN by increasing the activity of protein tyrosine kinase, which causes ROMK channels to be endocytosed from the apical plasma membrane, thereby reducing K⁺ secretion. Acidosis decreases secretion of K⁺ by inhibiting the activity of ROMK channels, whereas alkalosis stimulates secretion of K⁺ by enhancing ROMK channel activity.

processes, including formation of adenosine triphosphate (ATP), and it is an important component of nucleotides, nucleosides, and phospholipids. Phosphorylation of proteins is an important mechanism of cellular signaling, and P_i is an important buffer in cells, plasma, and urine.

In adults the kidneys play important roles in regulating total body Ca^{++} and P_i by excreting the amount of Ca^{++} and P_i that is absorbed by the intestinal tract (normal bone remodeling results in no net addition of Ca^{++} and P_i to, or Ca^{++} and P_i release from, bone). If plasma concentrations of

Ca^{++} and P_i decline substantially, intestinal absorption, bone resorption (i.e., loss of Ca^{++} and P_i from bone), and renal tubular reabsorption increase and return plasma concentrations of Ca^{++} and P_i to normal levels. During growth and pregnancy, intestinal absorption exceeds urinary excretion, and these ions accumulate in newly formed fetal tissue and bone. In contrast, bone disease (e.g., osteoporosis) or a decline in lean body mass increases urinary Ca^{++} and P_i loss without a change in intestinal absorption. These conditions produce a net loss of Ca^{++} and P_i from the body. Finally, during chronic renal failure, P_i accumulates in the body because absorption by the intestinal tract exceeds excretion in the urine. This can lead to accumulation of P_i in the body and changes in bone (see the In The Clinic box discussion of chronic renal failure).

This brief introduction reveals that the kidneys, in conjunction with the GI tract and bone, play a major role in maintaining plasma Ca^{++} and P_i levels as well as Ca^{++} and P_i balance (see Chapter 40). Accordingly this section of the chapter discusses Ca^{++} and P_i handling by the kidneys, with an emphasis on the hormones and factors that regulate urinary excretion.

Calcium

Cellular processes in which Ca^{++} plays an important role include bone formation, cell division and growth, blood coagulation, hormone-response coupling, and electrical stimulus-response coupling (e.g., muscle contraction, neurotransmitter release). Nearly 99% of Ca^{++} is stored in bone and teeth, approximately 1% is found in ICF, and 0.1% in ECF. The total Ca^{++} concentration ($[Ca^{++}]$) in plasma is 10 mg/dL (2.5 mM or 5 mEq/L), and its concentration is normally maintained within very narrow limits. Approximately 50% of the Ca^{++} in plasma is ionized, 40% is bound to plasma proteins (mainly albumin), and 10% is complexed to several anions, including P_i, HCO_3^-, citrate, and SO_4^{2-} (Fig. 36.11). The pH of plasma influences this distribution (Fig. 36.12). Acidemia increases the

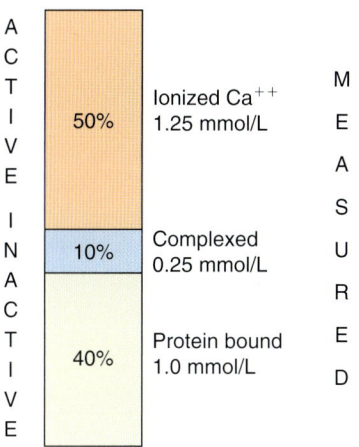

• **Fig. 36.11** Distribution of Ca⁺⁺ in plasma. (From Koeppen BM, Stanton BA. *Renal Physiology.* 5th ed. Philadelphia: Elsevier; 2013.)

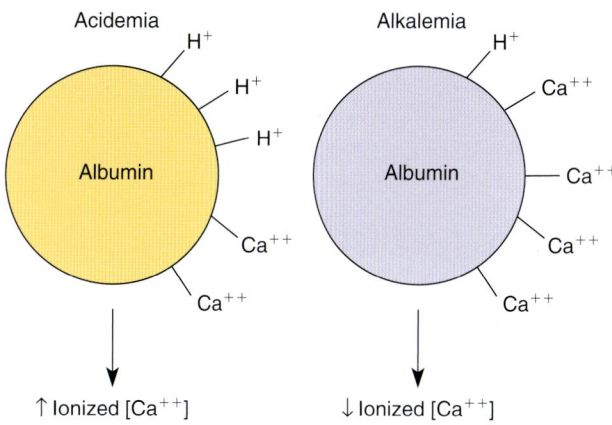

• **Fig. 36.12** Effect of pH on plasma [Ca⁺⁺]. (From Koeppen BM, Stanton BA. *Renal Physiology.* 5th ed. Philadelphia: Elsevier; 2013.)

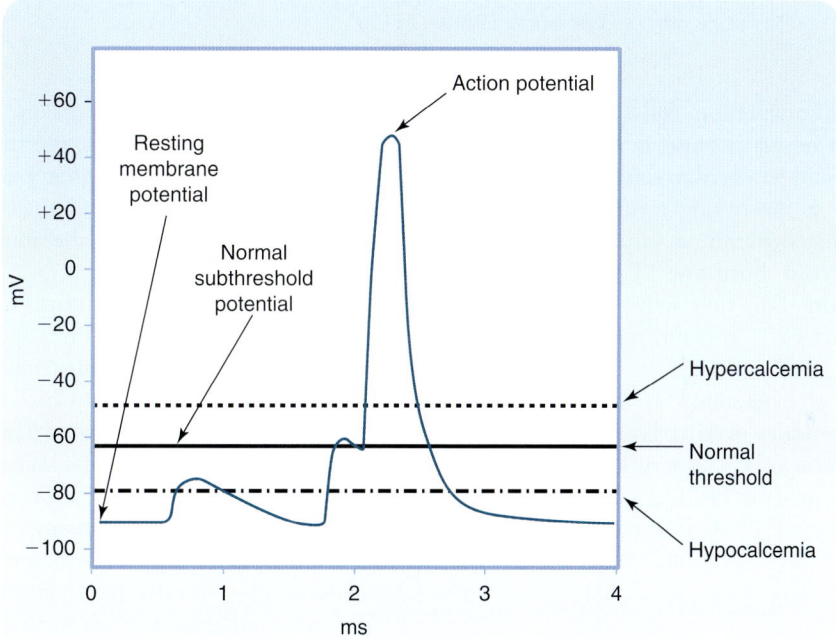

• **Fig. 36.13** Effect of Ca⁺⁺ on nerve and muscle excitability. (From Koeppen BM, Stanton BA. *Renal Physiology.* 5th ed. Philadelphia: Elsevier; 2013.)

percentage of ionized Ca^{++} at the expense of Ca^{++} bound to proteins, whereas alkalemia decreases the percentage of ionized Ca^{++}, again by altering the Ca^{++} bound to proteins. Individuals with alkalemia are susceptible to **tetany** (tonic muscular spasms), whereas individuals with acidemia are less susceptible to tetany, even when total plasma Ca^{++} levels are reduced. The increase in $[H^+]$ in patients with metabolic acidosis causes more H^+ to bind to plasma proteins, P_i, HCO_3^-, citrate, and SO_4^{2-}, thereby displacing Ca^{++}. This displacement increases the plasma concentration of ionized Ca^{++}. In alkalemia the $[H^+]$ of plasma decreases. Some H^+ ions dissociate from plasma proteins, P_i, HCO_3^-, citrate, and SO_4^{2-} in exchange for Ca^{++}, thereby decreasing the plasma concentration of ionized Ca^{++}. In addition, the

plasma albumin concentration also affects ionized plasma $[Ca^{++}]$.

Hypoalbuminemia increases the ionized $[Ca^{++}]$, whereas **hyperalbuminemia** decreases ionized plasma $[Ca^{++}]$. The total measured plasma $[Ca^{++}]$ does not reflect the total ionized $[Ca^{++}]$, which is the physiologically relevant measure of plasma $[Ca^{++}]$. A low ionized plasma $[Ca^{++}]$ **(hypocalcemia)** increases the excitability of nerve and muscle cells and can lead to hypocalcemic tetany. Tetany associated with hypocalcemia occurs because hypocalcemia causes the threshold potential to shift to more negative values (i.e., closer to the resting membrane voltage) (Fig. 36.13). An elevated ionized plasma $[Ca^{++}]$ **(hypercalcemia)** may decrease neuromuscular excitability or produce cardiac

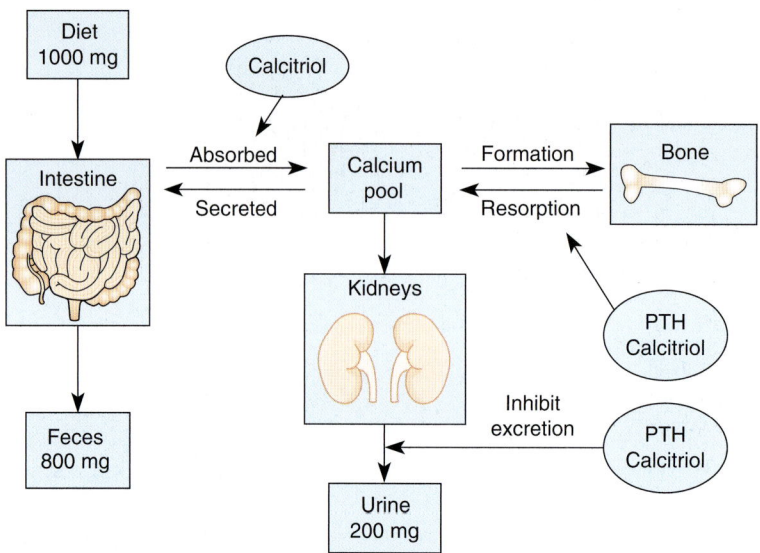

• **Fig. 36.14** Overview of Ca++ homeostasis. PTH, parathyroid hormone. (From Koeppen BM, Stanton BA. *Renal Physiology.* 5th ed. Philadelphia: Elsevier; 2013.)

arrhythmias, lethargy, disorientation, and even death.[b] This effect of hypercalcemia occurs because an elevated plasma [Ca++] causes the threshold potential to shift to less negative values (i.e., farther from the resting membrane voltage). Plasma [Ca++] is regulated within a very narrow range, primarily by **parathyroid hormone (PTH)**, **calcitriol (1,25-dihydroxyvitamin D)**, the active metabolite of vitamin D₃, and plasma Ca++, and this regulation will be discussed next.

Within cells, Ca++ is sequestered in the endoplasmic reticulum and mitochondria, or it is bound to proteins. Thus the free intracellular [Ca++] is very low (≈100 nM). The large concentration gradient for [Ca++] across cell membranes is maintained by a Ca++-ATPase pump (PMCa1b) in all cells and by a 3Na+/Ca++ exchanger (NCX1) in some cells.

Overview of Calcium Homeostasis

Ca++ homeostasis depends on two factors: (1) the total amount of Ca++ in the body and (2) the distribution of Ca++ between bone and ECF. The total body Ca++ level is determined by the relative amounts of Ca++ absorbed by the intestinal tract and excreted by the kidneys (Fig. 36.14). The intestinal tract absorbs Ca++ through an active carrier-mediated transport mechanism that is stimulated by calcitriol, the active metabolite of vitamin D₃ that is produced in the proximal tubule of the kidneys. Net Ca++ absorption by the intestine is normally 200 mg/day, but it can increase to 600 mg/day when calcitriol levels rise. In adults, Ca++ excretion by the kidneys equals the amount absorbed by the GI tract (200 mg/day), and it changes in

parallel with intestinal absorption. Thus in adults, Ca++ balance is maintained because the amount of Ca++ ingested in an average diet (1000 mg/day) equals the amount lost in feces (800 mg/day, the amount that escapes absorption by the intestinal tract) plus the amount excreted in urine (200 mg/day).

The second factor that controls Ca++ homeostasis is the distribution of Ca++ between bone and ECF (see Fig. 36.14). Two hormones (PTH and calcitriol[c]) regulate the distribution of Ca++ between bone and ECF and thereby, in concert with the kidneys, regulate plasma [Ca++]. PTH is secreted by the parathyroid glands, and its secretion is stimulated by a decline in plasma [Ca++] (i.e., hypocalcemia). Plasma Ca++ is an agonist of the **calcium sensing receptor (CaSR)**, which is located in the plasma membrane of chief cells in parathyroid glands (discussed later). Hypercalcemia activates the CaSR, which decreases PTH release, whereas hypocalcemia reduces CaSR activity that in turn increases PTH release. PTH increases plasma [Ca++] by: (1) stimulating bone resorption, (2) increasing Ca++ reabsorption by the distal tubule of the kidney, and (3) stimulating the production of calcitriol, which in turn increases Ca++ absorption by the intestinal tract and facilitates PTH-mediated bone resorption. Production of calcitriol in the kidney is stimulated by hypocalcemia and hypophosphatemia. Calcitriol increases plasma [Ca++], primarily by stimulating Ca++ absorption from the intestinal tract. It also facilitates the action of PTH on bone and enhances Ca++ reabsorption in the kidneys by increasing expression of key Ca++ transport and binding proteins in the kidneys (details discussed later). In addition,

---

[b]In clinical practice the terms *hypercalcemia* and *hypocalcemia* are often used to describe a high or low total plasma [Ca++], respectively, even though this usage is not the physiologically correct usage of *hypercalcemia* and *hypocalcemia*.

[c]Calcitonin is secreted by thyroid C cells (parafollicular cells), and its secretion is stimulated by hypercalcemia. Calcitonin decreases plasma [Ca++], mainly by stimulating bone formation (i.e., deposition of Ca++ in bone). Although it plays an important role in Ca++ homeostasis in lower vertebrates, calcitonin plays only a minor role in Ca++ homeostasis in humans, so it will not be discussed further.

hypercalcemia activates the CaSR in the TAL of Henle's loop, inhibiting Ca++ reabsorption in this segment, resulting in an increase in urinary Ca++ excretion and thereby a reduction in plasma [Ca++]. Hypocalcemia has the opposite effect. Importantly, regulation of Ca++ excretion by the kidneys is one of the major ways the body regulates plasma [Ca++].

IN THE CLINIC

Conditions that lower PTH levels (i.e., hypoparathyroidism after parathyroidectomy for an adenoma) reduce plasma [Ca++], which can cause **hypocalcemic tetany** (intermittent muscular contractions). In severe cases, hypocalcemic tetany can cause death by asphyxiation. Hypercalcemia can also cause lethal cardiac arrhythmias and decreased neuromuscular excitability. Clinically the most common causes of hypercalcemia are primary hyperparathyroidism and malignancy-associated hypercalcemia. Primary hyperparathyroidism results most often from overproduction of PTH caused by a benign tumor of the parathyroid glands. In contrast, malignancy-associated hypercalcemia, which occurs in 10% to 20% of all patients with cancer, is caused by secretion of **parathyroid hormone–related peptide (PTHrP),** a PTH-like hormone secreted by carcinomas in various organs. Increased levels of PTH and PTHrP cause hypercalcemia and hypercalciuria.

Calcium Transport Along the Nephron

The Ca++ available for glomerular filtration consists of the ionized fraction and the amount complexed with anions. Thus about 60% of the Ca++ in plasma is available for glomerular filtration. Normally 99% of filtered Ca++ is reabsorbed by the nephron (Fig. 36.15). The proximal tubule reabsorbs about 50% to 60% of the filtered Ca++. Another 15% is reabsorbed in the loop of Henle (mainly the cortical portion of the TAL), about 10% to 15% is reabsorbed by the distal tubule, and less than 1% is reabsorbed by the collecting duct. About 1% (200 mg/day) is excreted in urine. This fraction is equal to the net amount absorbed daily by the intestinal tract.

Ca++ reabsorption by the proximal tubule occurs primarily via the paracellular pathway. This passive paracellular reabsorption of Ca++ is driven by the lumen-positive transepithelial voltage (V_{te}) across the second half of the proximal tubule and by a favorable concentration gradient of Ca++, both of which are established by transcellular sodium and water reabsorption in the first half of the proximal tubule (see Chapter 34).

Ca++ reabsorption by the loop of Henle also occurs primarily via the paracellular pathway. Like the proximal tubule, Ca++ and Na+ reabsorption in the TAL parallel each other. These processes are parallel because of the significant component of Ca++ reabsorption that occurs via passive paracellular reabsorption secondary to Na+ reabsorption that generates a lumen-positive V_{te}. Loop diuretics inhibit Na+ reabsorption by the TAL of the loop of Henle, and in so doing reduce the magnitude of the lumen-positive V_{te} (see Chapter 34). This action in turn inhibits reabsorption of Ca++ via the paracellular pathway. Thus loop diuretics are used to increase renal Ca++ excretion in patients with hypercalcemia.

AT THE CELLULAR LEVEL

Mutations in the tight junction protein **claudin-16 (CLDN16)** reduce the permeability of the paracellular pathway to Ca++ and Mg++ and thereby reduce the diffusive reabsorptive movement of Ca++ and Mg++ across tight junctions in the TAL of Henle's loop. **Familial hypomagnesemic hypercalciuria** is caused by mutations in claudin-16, which is a component of the tight junctions in TAL cells. This disorder is characterized by enhanced excretion of Ca++ and Mg++ due to a fall in passive reabsorption of these ions across the paracellular pathway in the TAL. Affected individuals have high levels of Ca++ in their urine, which leads to stone formation (nephrolithiasis).

In the distal tubule where the voltage in the tubule lumen is electrically negative with respect to blood, reabsorption of Ca++ is entirely active because Ca++ is reabsorbed against its electrochemical gradient (Fig. 36.16). Thus Ca++ reabsorption by the distal tubule is exclusively transcellular. Calcium enters the cell across the apical membrane through Ca++-permeable epithelial ion channels (TRPV5). Inside the cell, Ca++ binds to calbindin-D28K. The calbindin-Ca++ complex carries Ca++ across the cell and delivers Ca++ to the basolateral membrane, where it is extruded from the cell primarily by the 3Na+/1Ca++ antiporter (NCX1); however, plasma membrane Ca++-ATPase isoform 1b (PMCA1b) may also contribute. Urinary Na+ and Ca++ excretion usually change in parallel. However, excretion of these ions does not always change in parallel, because reabsorption of Ca++ and Na+ by the distal tubule is independent and differentially

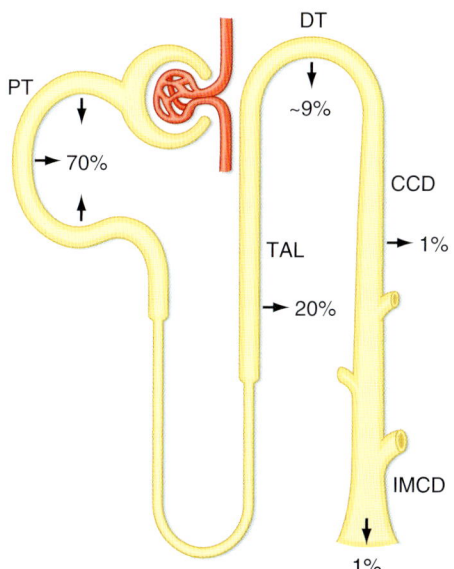

• **Fig. 36.15** Overview of Ca++ transport along the nephron. Percentages refer to amount of filtered Ca++ reabsorbed by each segment. CCD, cortical collecting duct; DT, distal tubule; IMCD, inner medullary collecting duct; PT, proximal tubule; TAL, thick ascending limb.

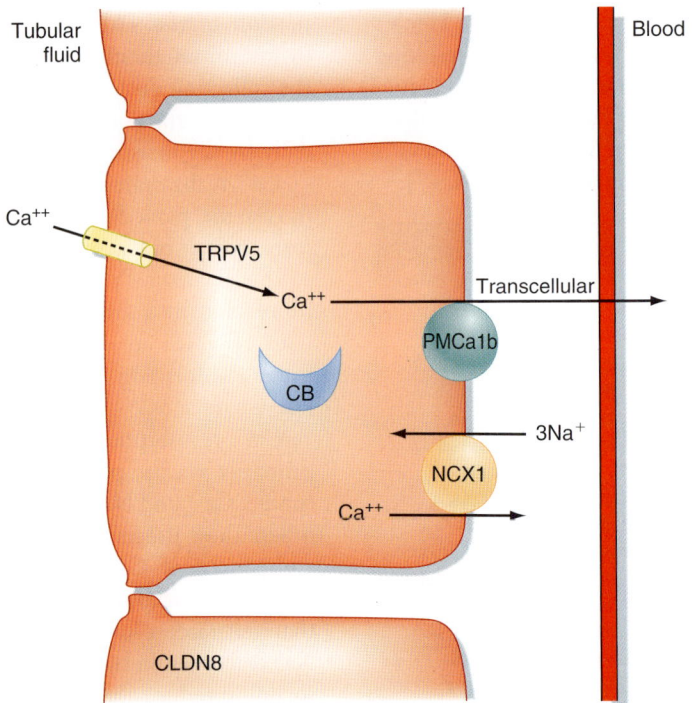

• **Fig. 36.16** Cellular mechanism of Ca^{++} reabsorption by the distal tubule. Ca^{++} is reabsorbed exclusively by a cellular pathway. Ca^{++} enters the cell across the apical membrane via a Ca^{++}-permeable ion channel (TRPV5). Inside cells, Ca^{++} binds to calbindin (calbindin-D_{28K}), and the Ca^{++}-calbindin complex diffuses across the cell to deliver Ca^{++} to the basolateral membrane. Ca^{++} is transported across the basolateral membrane primarily by a 3 (or 4) Na^+/Ca^{++} antiporter (NCX1) and also by a Ca^{++}-H^+-ATPase (PMCa1b). Claudin 8 (CLDN8) is a tight junction protein that is impermeable to Ca^{++} and thereby prevents the back diffusion of Ca^{++} across the tight junction into the tubule lumen, which is electrically negative compared to the blood side of the cell.

regulated. For example, **thiazide diuretics** inhibit Na^+ reabsorption by the distal tubule and stimulate Ca^{++} reabsorption by this segment. Accordingly the net effects of thiazide diuretics are to increase urinary Na^+ excretion and reduce urinary Ca^{++} excretion. Because thiazide diuretics reduce urinary Ca^{++} excretion, they are often given to reduce urinary [Ca^{++}] in individuals who produce Ca^{++}-containing kidney stones.

Regulation of Urinary Calcium Excretion

Several hormones and factors influence urinary Ca^{++} excretion (Table 36.2). Of these, PTH exerts the most powerful control on renal Ca^{++} excretion; it is the primary hormone/factor responsible for maintaining Ca^{++} homeostasis. Overall this hormone stimulates Ca^{++} reabsorption by the kidneys (i.e., reduces Ca^{++} excretion). Although PTH inhibits reabsorption of NaCl and fluid, and therefore Ca^{++} reabsorption by the proximal tubule, PTH stimulates Ca^{++} reabsorption by the TAL of the loop of Henle and the distal tubule. Thus the net effect of PTH is to enhance renal Ca^{++} reabsorption. Changes in plasma [Ca^{++}] also regulate urinary Ca^{++} excretion, with hypercalcemia increasing excretion and hypocalcemia decreasing excretion. Hypercalcemia increases urinary Ca^{++} excretion by: (1) reducing proximal tubule Ca^{++} reabsorption (reduced paracellular reabsorption due to increased interstitial fluid [Ca^{++}]); (2) inhibiting Ca^{++}

reabsorption by the TAL of the loop of Henle via activation of the CaSR located in the basolateral membrane of these cells (NaCl reabsorption is decreased, thereby reducing the magnitude of the lumen-positive V_{te}); and (3) suppressing Ca^{++} reabsorption by the distal tubule by reducing PTH levels. As a result, urinary Ca^{++} excretion increases. Hypocalcemia has the opposite effect on urinary Ca^{++} excretion, primarily by increasing Ca^{++} reabsorption by the proximal tubule and TAL. Calcitriol enhances Ca^{++} reabsorption by the distal tubule, but it is less effective than PTH.

Several factors disturb Ca^{++} excretion. An increase in plasma [P_i] concentration (e.g., caused by a dramatic increase in dietary intake of P_i or by reduced kidney function) elevates PTH levels both directly and by decreasing the ionized plasma [Ca^{++}] and thereby decreases Ca^{++} excretion. A decline in plasma [P_i] (e.g., caused by dietary P_i depletion) has the opposite effect (NOTE: with normal kidney function, changes in dietary P_i intake over a sevenfold range has no effect on plasma [P_i]). Changes in ECFV alter Ca^{++} excretion mainly by affecting NaCl and fluid reabsorption in the proximal tubule. Volume contraction increases NaCl and water reabsorption by the proximal tubule and thereby enhances Ca^{++} reabsorption. Accordingly, urinary Ca^{++} excretion declines. Volume expansion has the opposite effect. Acidemia increases Ca^{++} excretion, whereas alkalemia decreases excretion. Regulation of Ca^{++} reabsorption by pH

Summary of Hormones, Factors, and Diuretics Affecting Ca^{++} Reabsorption

TABLE 36.2

Factor/Hormone	Nephron Location		
	Proximal Tubule	**TAL**	**Distal Tubule**
PTH (PTHrP)[a]	Decrease	Increase	Increase
Calcitriol			Increase
Volume expansion	Decrease	No Change	Decrease
Hypercalcemia	Decrease	Decrease (via CaSR)	Decrease (via PTH)
Hypocalcemia	Increase	Increase	
Phosphate loading (hyperphosphatemia)			Increase (via PTH)
Phosphate depletion (hypophosphatemia)	Decrease		Decrease (via PTH)
Acidemia			Decrease
Alkalemia			Increase
Loop diuretics		Decrease	
Thiazide diuretics			Increase

[a]PTH inhibits Ca^{++} reabsorption by the proximal tubule but stimulates reabsorption by the TAL and distal tubule. Overall the net effect is to increase Ca^{++} reabsorption and thereby reduce urinary Ca^{++} excretion.

CaSR, calcium-sensing receptor.

Modified from Mount DB, Yu A. Transport of inorganic solutes: sodium, chloride, potassium, magnesium, calcium and phosphate. In: Brenner BM (ed). *Brenner and Rector's The Kidney.* 8th ed. Philadelphia: Saunders; 2008.

occurs primarily in the distal tubule. Alkalosis stimulates the apical membrane Ca^{++} channel (TRPV5), thereby increasing Ca^{++} reabsorption. By contrast, acidosis inhibits the same channel, thereby reducing Ca^{++} reabsorption. Finally, as noted earlier, loop diuretics inhibit Ca^{++} reabsorption by the TAL, and thiazide diuretics stimulate Ca^{++} reabsorption by the distal tubule.

Calcium-Sensing Receptor

The calcium-sensing receptor (CaSR) is a receptor expressed in the plasma membrane of cells involved in regulating Ca^{++} homeostasis; it senses small changes in extracellular [Ca^{++}]. Ca^{++} binds to CaSR receptors in PTH-secreting cells of the parathyroid gland and calcitriol-producing cells of the proximal tubule. Activation of the receptor by an increase in plasma [Ca^{++}] results in inhibition of PTH secretion and production of calcitriol by the proximal tubule. Moreover, reduction in PTH secretion also contributes to decreased production of calcitriol because PTH is a potent stimulus of calcitriol synthesis. By contrast, a fall in plasma [Ca^{++}] has the opposite effect on PTH and calcitriol secretion.

The CaSR also maintains Ca^{++} homeostasis by directly regulating Ca^{++} excretion by the kidneys. CaSRs in the TAL respond directly to changes in plasma [Ca^{++}] and regulate Ca^{++} absorption. An increase in plasma [Ca^{++}] activates CaSR in the TAL and inhibits Ca^{++} absorption, thereby stimulating urinary Ca^{++} excretion. By contrast, a fall in plasma [Ca^{++}] leads to an increase in Ca^{++} absorption by the TAL and a corresponding decrease in urinary Ca^{++} excretion. Thus the direct effect of plasma [Ca^{++}] on CaSRs in the TAL acts in concert with changes in PTH, which regulates Ca^{++} reabsorption by the distal tubule to regulate urinary Ca^{++} excretion and thereby maintain Ca^{++} homeostasis.

IN THE CLINIC

Mutations in the gene coding for **CaSR** cause disorders in Ca^{++} homeostasis. **Familial hypocalciuric hypercalcemia (FHH)** is a haploinsufficient state caused by an inactivating mutation of CaSR. The hypercalcemia is caused by deranged Ca^{++}-regulated PTH secretion (i.e., the set point for Ca^{++}-regulated PTH secretion is shifted such that PTH levels are elevated at any level of plasma [Ca^{++}], and are not suppressed in the setting of hypercalcemia). The hypocalciuria is caused by enhanced Ca^{++} reabsorption in the TAL and distal tubule owing to elevated PTH levels and defective CaSR regulation of Ca^{++} transport in the kidneys. **Autosomal-dominant hypoparathyroidism** is caused by an activating mutation in CaSR. Activation of CaSRs causes deranged Ca^{++}-regulated PTH secretion (i.e., the set point for Ca^{++}-regulated PTH secretion is shifted such that PTH levels are decreased at any level of plasma [Ca^{++}]). Hypercalciuria results and is caused by decreased PTH levels and defective CaSR-regulated Ca^{++} transport in the kidneys.

Phosphate

P$_i$ is an important component of many organic molecules, including DNA, RNA, ATP, nucleotides, nucleosides, and phospholipids and intermediates of metabolic pathways. Like Ca^{++} it is a major constituent of bone. Its concentration in plasma is an important determinant of bone formation and resorption. In addition, urinary P$_i$ is an important buffer (i.e., it is one of many titratable acids) involved in the maintenance of acid-base balance (see Chapter 37). Some 85% of P$_i$ is located in bone and teeth, 14% is located in the ICF, and 1% is located in the ECF. Normal plasma

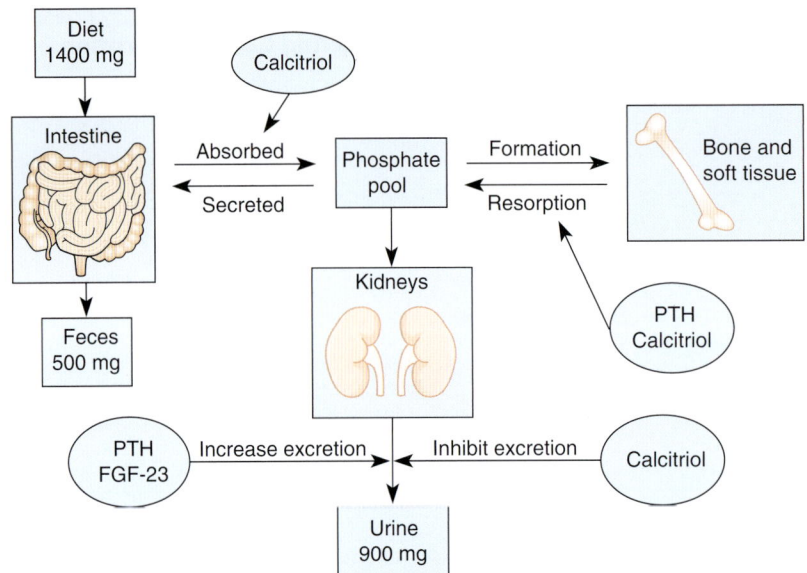

● **Fig. 36.17** Overview of P_i homeostasis. (From Koeppen BM, Stanton BA. *Renal Physiology.* 5th ed. Philadelphia: Elsevier; 2013.)

$[P_i]$ is 3 to 4 mg/dL (1–1.5 mM). P_i in plasma is ionized (45%), complexed (30%), and bound to protein (25%). Phosphate deficiency causes muscle weakness, rhabdomyolysis, and reduced bone mineralization resulting in **rickets** (in children) and **osteomalacia** (in adults).

Overview of Phosphate Homeostasis

A general scheme of P_i homeostasis is shown in Fig. 36.17. Maintenance of P_i homeostasis depends on two factors: (1) the amount of P_i in the body and (2) the distribution of P_i between the ICF and ECF compartments. Total body P_i levels are determined by the relative amount of P_i absorbed by the intestinal tract versus the amount excreted by the kidneys. P_i absorption by the intestinal tract occurs via active and passive mechanisms; P_i absorption increases as dietary P_i rises, and it is stimulated by calcitriol. Despite variations in P_i intake between 800 and 1500 mg/day, in adults (i.e., the steady state) the kidneys maintain total body P_i balance constant by excreting an amount of P_i in the urine equal to the amount absorbed by the intestinal tract (normal bone remodeling results in no net addition of P_i to, or P_i release from, bone). By contrast, during growth, P_i is accumulated in the body. Renal P_i excretion is the primary mechanism by which the body regulates P_i balance and thereby P_i homeostasis.

The second factor that maintains P_i homeostasis is distribution of P_i among bone and the ICF and ECF compartments. Regulation of plasma $[P_i]$ is controlled by two hormones, PTH and calcitriol, in concert with the kidneys (see Fig. 36.17). Release of P_i from bone is stimulated by the same hormones (i.e., PTH and calcitriol) that release Ca^{++} from this pool. Thus release of P_i from bone is always accompanied by release of Ca^{++}.

The kidneys also make an important contribution to maintaining plasma $[P_i]$ within a narrow range (1–1.5 mM).

P_i excretion by the kidneys is regulated by PTH and calcitriol (see Fig. 36.17). PTH increases P_i excretion, whereas calcitriol inhibits P_i excretion. Plasma $[P_i]$ is determined by: (1) intestinal absorption, (2) storage in bone, and (3) P_i excretion by the kidneys. Maintenance of plasma $[P_i]$ is essential for optimal Ca^{++}-P_i complex formation required for bone mineralization without deposition of Ca^{++}-P_i in vascular and other soft tissues.

A rise in plasma $[P_i]$ directly stimulates PTH synthesis and release and also decreases the ionized $[Ca^{++}]$, which stimulates PTH release by its interaction with the CaSR. PTH enhances urinary P_i excretion by inhibiting proximal tubule P_i reabsorption. Hyperphosphatemia also decreases calcitriol production by the proximal tubule, which leads to a reduction in P_i absorption by the intestine. Both the increase in PTH and the decrease in calcitriol reduce plasma $[P_i]$.

Phosphate Transport Along the Nephron

Fig. 36.18 summarizes P_i transport by the various portions of the nephron. The proximal tubule reabsorbs 80% of the P_i filtered by the glomerulus; the loop of Henle, distal tubule, and collecting duct reabsorb negligible amounts of P_i. Therefore approximately 20% of the P_i filtered across the glomerular capillaries is excreted in urine.

P_i reabsorption by the proximal tubule occurs by a transcellular route (Fig. 36.19). P_i uptake across the apical membrane of the proximal tubule occurs via two Na^+/P_i symporters (IIa and IIc). Type IIa transports $3Na^+$ with one divalent P_i (HPO_4^{2-}), and carries positive charge into the cell. Type IIc transports $2Na^+$ with one monovalent P_i ($H_2PO_4^-$) and is electrically neutral. P_i exits across the basolateral membrane by a P_i-inorganic anion antiporter that has not been characterized.

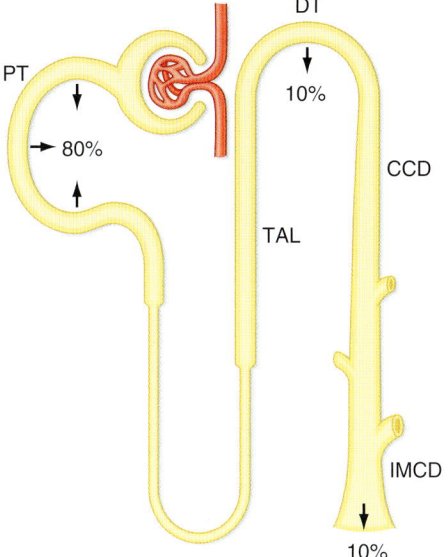

• **Fig. 36.18** P_i transport along the nephron. P_i is reabsorbed primarily by the proximal tubule. Percentages refer to the amount of the filtered P_i reabsorbed by each nephron segment. Approximately 20% of the filtered P_i is excreted. CCD, cortical collecting duct; DT, distal tubule; IMCD, inner medullary collecting duct; PT, proximal tubule; TAL, thick ascending limb.

IN THE CLINIC

In patients with **chronic renal failure,** the kidneys cannot excrete P_i. Because of continued P_i absorption by the intestinal tract, P_i accumulates in the body and plasma $[P_i]$ rises. The excess P_i complexes with Ca^{++} and reduces the ionized plasma $[Ca^{++}]$. P_i accumulation also decreases production of calcitriol. This response reduces Ca^{++} absorption by the intestine, an effect that further reduces plasma $[Ca^{++}]$. This reduction in plasma $[Ca^{++}]$ increases PTH secretion and Ca^{++} release from bone. These actions result in **renal osteodystrophy** (i.e., increased bone resorption with replacement by fibrous tissue, which renders bone more susceptible to fracture). Chronic hyperparathyroidism (i.e., elevated PTH levels due the fall in plasma $[Ca^{++}]$) during chronic renal failure can lead to metastatic calcifications in which Ca^{++} and P_i precipitate in arteries, soft tissues, and viscera. Deposition of Ca^{++} and P_i in the heart may cause myocardial failure. Prevention and treatment of hyperparathyroidism and P_i retention include a low-P_i diet or administration of a "phosphate binder" (i.e., an agent that forms insoluble P_i salts and thereby renders P_i unavailable for absorption by the intestinal tract) in the diet. Supplemental Ca^{++} and calcitriol are also prescribed to increase plasma $[Ca^{++}]$.

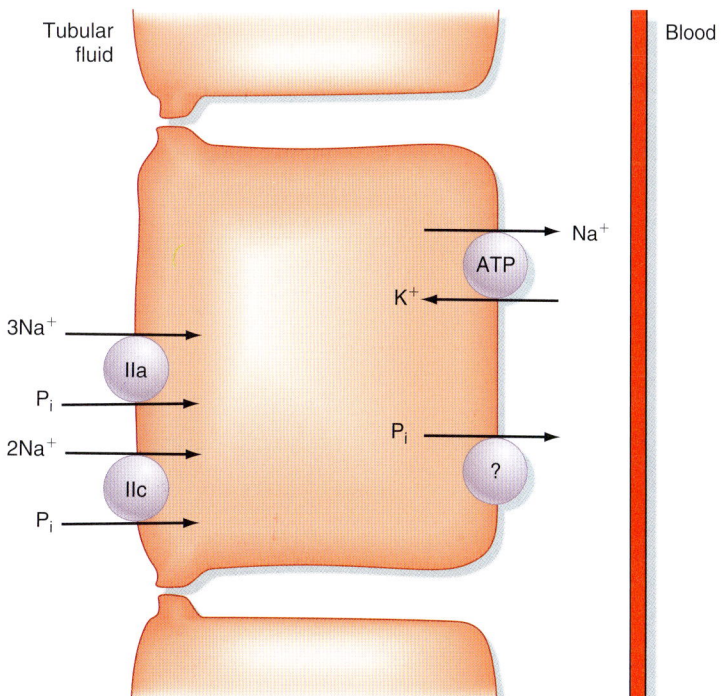

• **Fig. 36.19** Cellular mechanisms of P_i reabsorption by the proximal tubule. The apical transport pathway contains two Na^+/P_i symporters, one that transports three Na^+ for each P_i (IIa) and one that transports two Na^+ for each P_i (IIc). P_i leaves the cell across the basolateral membrane by an unknown mechanism. ATP, adenosine triphosphate.

Fibroblast growth factor 23 (FGF-23) increases renal P_i excretion and thereby contributes to regulation of plasma $[P_i]$ (see Fig. 36.17). FGF-23 is secreted by osteocytes and osteoblasts and inhibits P_i reabsorption and calcitriol production by the proximal tubule. Secretion of FGF-23 is stimulated by sustained hyperphosphatemia, PTH, and calcitriol. Activating mutations in the *FGF23* gene cause hypophosphatemia, low plasma calcitriol, and rickets/osteomalacia, whereas inactivating mutations cause hyperphosphatemia, high serum calcitriol, and calcification of soft tissue.

Regulation of Urinary Phosphate Excretion

Several hormones and factors regulate urinary P_i excretion (Table 36.3 and Fig. 36.20). Increased plasma P_i reduces

TABLE 36.3	Summary of Hormones and Factors Affecting P_i Reabsorption by Proximal Tubule	
Factor/Hormone	**Proximal Tubule Reabsorption**
PTH	Decrease
FGF-23	Decrease
Phosphate loading	Decrease
Phosphate depletion	Increase
Metabolic acidosis: chronic	Decrease
Metabolic alkalosis: chronic	Increase
ECFV expansion	Decrease
Growth hormone	Increase
Glucocorticoids	Decrease

plasma $[Ca^{++}]$ and therefore increases plasma PTH, which increases P_i excretion by the kidneys. PTH, the most important hormone that controls P_i excretion, inhibits P_i reabsorption by the proximal tubule and thereby increases P_i excretion. PTH reduces P_i reabsorption by stimulating endocytic removal of Na^+/P_i transporters from the brush border membrane of the proximal tubule. Increased plasma P_i also increases FGF-23, which inhibits P_i reabsorption and calcitriol production by the proximal tubule. Elevated plasma P_i also suppresses calcitriol production, which results in a decrease in intestinal P_i reabsorption. Dietary P_i intake also regulates P_i excretion by mechanisms unrelated to changes in PTH levels. P_i loading increases excretion, whereas P_i depletion decreases it. Changes in dietary P_i intake modulate P_i transport by altering the transport rate of each Na^+/P_i symporter and the number of symporters in the apical membrane of the proximal tubule.

ECFV also affects P_i excretion. Expansion of the ECF enhances P_i excretion by: (1) increasing GFR and thus the filtered load of P_i; (2) decreasing Na/P_i coupled reabsorption, which reduces ECFV, and (3) reducing plasma $[Ca^{++}]$, thereby increasing PTH, which inhibits P_i reabsorption in the proximal tubule. Acid-base balance also influences P_i excretion; chronic acidosis increases P_i excretion, and chronic alkalosis decreases it. These effects of acid-base balance, like the effect of PTH, are mediated by changes of expression of the Na /P_i symporters in the apical membrane. Systemic acidosis increases glucocorticoid secretion, and glucocorticoids increase excretion of P_i by inhibiting P_i reabsorption by the proximal tubule. This inhibition, together with the direct effect of acidosis on P_i reabsorption by the proximal tubule, enables the distal tubule and collecting duct to secrete more H^+ as titratable acid and to generate more HCO_3^- because P_i is an important urinary buffer. Growth hormone decreases P_i excretion.

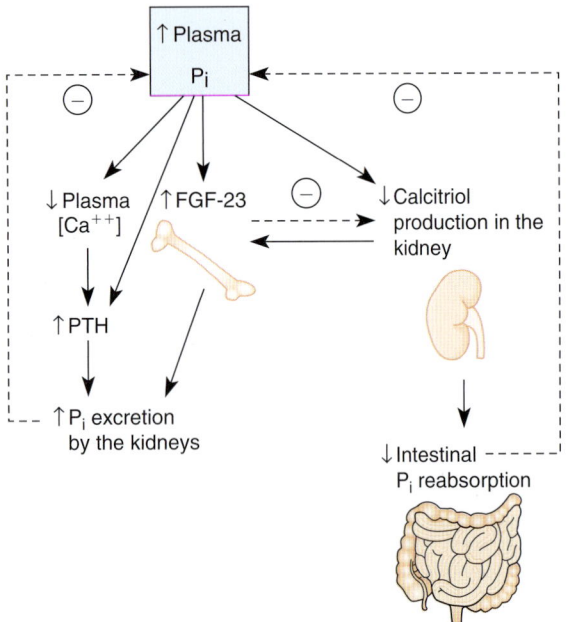

• **Fig. 36.20** Overview of the major hormones regulating plasma [P_i]. FGF-23, fibroblast growth factor 23; PTH, parathyroid hormone. Dashed lines indicate negative feedback. (From Koeppen BM, Stanton BA. *Renal Physiology.* 5th ed. Philadelphia: Elsevier; 2013.)

 IN THE CLINIC

Klotho, which was identified in 1997, is highly expressed in the early distal tubule of the kidney. Klotho knockout mice have a phenotype that resembles chronic kidney disease (CKD), including soft tissue calcification, hyperphosphatemia, and elevated plasma FGF-32. Klotho exists as both a membrane-bound and a soluble protein. The membrane-bound form is a coreceptor for FGF-23, thus Klotho promotes P_i excretion by the kidneys and reduces serum levels of 1,25-dihydroxyvitamin D_3. Soluble circulating Klotho has a number of additional functions, including modulating ion transport and Wnt signal transduction, inhibiting renin-angiotensin signaling, and modulating FGF-23 regulation of PTH production. A considerable body of experimental data suggests that Klotho may be a biomarker for CKD, and that Klotho deficiency may contribute to development of CKD. Moreover, experimental data also suggest that Klotho therapy may slow the progression of CKD.

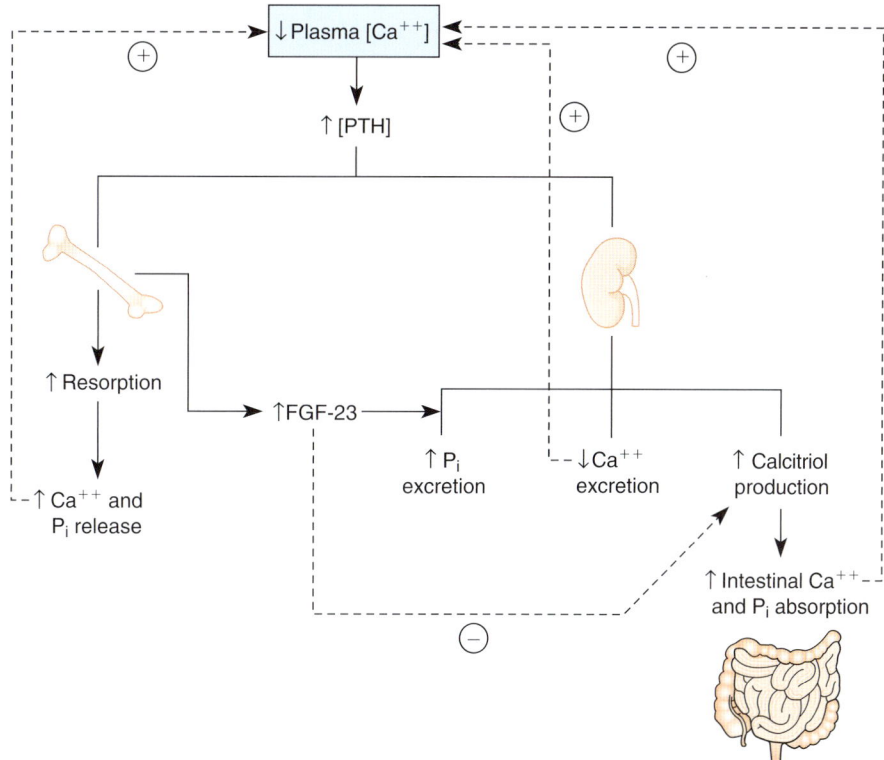

• **Fig. 36.21** Overview of the major hormones regulating plasma [Ca^{++}]. Dashed lines indicate negative feedback. *FGF-23*, fibroblast growth factor 23; *PTH*, parathyroid hormone. (From Koeppen BM, Stanton BA. *Renal Physiology.* 5th ed. Philadelphia: Elsevier; 2013.)

Integrative Review of Parathyroid Hormone and Calcitriol on Ca^{++} and P$_i$ Homeostasis

As summarized in Fig. 36.21, PTH has numerous effects on Ca^{++} and P$_i$ homeostasis. Hypocalcemia is the major stimulus of PTH secretion. PTH stimulates bone resorption, increases urinary P$_i$ excretion, decreases urinary Ca^{++} excretion, and stimulates production of calcitriol, which stimulates Ca^{++} and P$_i$ absorption by the intestine. Because changes in P$_i$ handling in bone, intestines, and kidneys tend to balance out, PTH increases plasma [Ca^{++}] while having little effect on plasma [P$_i$]. Overall a rise in plasma PTH levels in response to hypocalcemia returns plasma [Ca^{++}] to the normal range. A decline in plasma [Ca^{++}] has the opposite effect.

Calcitriol (the active form of vitamin D) also plays an important role in Ca^{++} and P$_i$ homeostasis (Fig. 36.22).

The primary action of calcitriol is to stimulate Ca^{++} and P$_i$ absorption by the intestine. To a lesser degree it acts with PTH to release Ca^{++} and P$_i$ from bone and to decrease Ca^{++} excretion by the kidneys. The net effect of calcitriol is to increase plasma [Ca^{++}] and [P$_i$]. Thus the major stimuli of calcitriol production are hypocalcemia via PTH and hypophosphatemia (i.e., a low plasma [P$_i$]).

IN THE CLINIC

In the absence of glucocorticoids (e.g., in **Addison's disease**), excretion of P$_i$ is depressed, as is the ability of the kidneys to excrete titratable acid and to generate new HCO$_3^-$. Growth hormone increases reabsorption of P$_i$ by the proximal tubule. As a result, growing children are in positive P$_i$ balance and have a higher plasma [P$_i$] than adults, and this elevated [P$_i$] is important for bone formation.

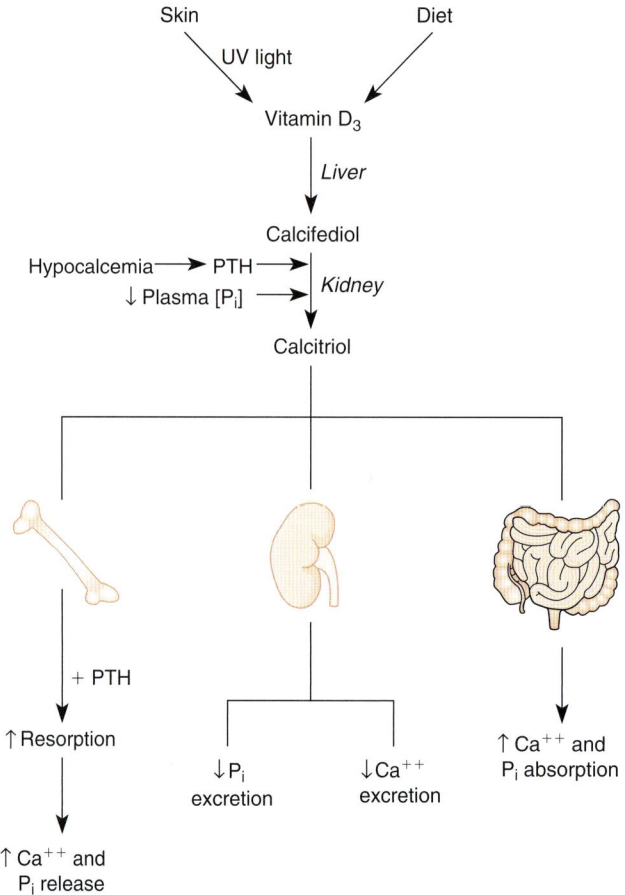

• **Fig. 36.22** Activation of calcitriol (vitamin D$_3$) and its effect on Ca^{++} and P$_i$ homeostasis. Hypocalcemia (via PTH) and hypophosphatemia are the major stimuli of the metabolism of calcifediol to calcitriol in the kidneys. The net effect of calcitriol is to increase plasma [Ca^{++}] and [P$_i$]. (From Koeppen BM, Stanton BA. *Renal Physiology.* 5th ed. Philadelphia: Elsevier; 2013.)

Key Concepts

1. K$^+$ homeostasis is maintained by the kidneys, which adjust K$^+$ excretion to match dietary K$^+$ intake, and by the hormones insulin, epinephrine, and aldosterone, which regulate the distribution of K$^+$ between the ICF and ECF compartments. Other events, such as cell lysis, exercise, and changes in acid-base balance and plasma osmolality, disturb K$^+$ homeostasis and plasma [K$^+$].

2. Excretion of K$^+$ by the kidneys is determined by the rate and direction of K$^+$ transport by the distal tubule and collecting duct. Secretion of K$^+$ by these tubular segments is regulated by plasma [K$^+$], aldosterone, and AVP. In contrast, changes in tubular fluid flow and acid-base disturbances perturb K$^+$ excretion by the kidneys. In K$^+$-depleted states, K$^+$ secretion is inhibited and the distal tubule and collecting duct reabsorb K$^+$.

3. The kidneys, in conjunction with the intestinal tract and bone, play a vital role in regulating plasma [Ca^{++}] and [P$_i$].

4. Plasma [Ca^{++}] is regulated by PTH and calcitriol. Calcitonin is not a major regulatory hormone in humans. Ca^{++} excretion by the kidneys is regulated by PTH, plasma [Ca^{++}], and calcitriol and is altered by changes in acid-base status, ECFV, and plasma P$_i$.

5. Ca^{++} reabsorption by the TAL and distal tubule are regulated by PTH and calcitriol, both of which stimulate Ca^{++} reabsorption, and by plasma [Ca^{++}].

6. Plasma [P$_i$] is regulated by PTH, FGF-23, and calcitriol. P$_i$ excretion is regulated by PTH, FGF-23, dietary phosphate, and growth hormone and is altered by acid base balance, ECFV expansion, and glucocorticoids. Bone tumors secrete FGF-23, which enhances renal P$_i$ excretion and thereby causes hypophosphatemia, hyperphosphatemia, and a defect in bone mineralization (i.e., osteomalacia).

Additional Reading

Journal Articles

Biber J, et al. Phosphate transporters and their function. *Annu Rev Physiol*. 2013;75:535-550.

Christov M, Juppner H. Insights from genetic disorders of phosphate homeostasis. *Semin Nephrol*. 2013;33:143-157.

Miller T. Control of renal calcium, phosphate, electrolyte, and water excretion by the calcium-sensing receptor. *Best Pract Res Clin Endocrinol Metab*. 2013;27:345-358.

Palmer BF. Regulation of potassium homeostasis. *Clin J Am Soc Nephrol*. 2015;10:1050-1060.

Patel A. The primary cilium calcium channels and their role in flow sensing. *Pflugers Arch*. 2015;467:157-165.

Pearce D, et al. Collecting duct principal cell transport processes and their regulation. *Clin J Am Soc Nephrol*. 2015;10:135-146.

Book Chapters

Bernardo JF, Friedman PA. Renal calcium metabolism. In: In: Alpern RJ, et al., eds. *Seldin and Giebisch's The Kidney: Physiology and Pathophysiology*. 5th ed. Waltham, MA: Elsevier; 2013:2225.

Berndt TJ, Kumar R. Clinical disturbances of phosphate homeostasis. In: Alpern RJ, et al., eds. *Seldin and Giebisch's The Kidney: Physiology and Pathophysiology*. 5th ed. Waltham, MA: Elsevier; 2013:2369.

Malnic G, et al. Regulation of K⁺ excretion. In: Alpern RJ, et al., eds. *Seldin and Giebisch's The Kidney: Physiology and Pathophysiology*. 5th ed. Waltham, MA: Elsevier; 2013:1659.

Murer H, et al. Proximal tubular handling of phosphate. In: Alpern RJ, et al., eds. *Seldin and Giebisch's The Kidney: Physiology and Pathophysiology*. 5th ed. Waltham, MA: Elsevier; 2013:2351.

Smogorzewski MJ, et al. Disorders of calcium, magnesium, and phosphate balance. In: Taal MW, et al., eds. *Brenner and Rector's The Kidney*. 9th ed. Philadelphia: Saunders; 2012:689.

Wang W, Huang CL. The molecular biology of renal K⁺ channels. In: Alpern RJ, et al., eds. *Seldin and Giebisch's The Kidney: Physiology and Pathophysiology*. 5th ed. Waltham, MA: Elsevier; 2013: 1601.

37

Role of the Kidneys in the Regulation of Acid-Base Balance

LEARNING OBJECTIVES

Upon completion of this chapter the student should be able to answer the following questions:

1. How does HCO_3^- operate as a buffer, and why is it an important buffer of the extracellular fluid?
2. How does metabolism of food produce acid and alkali, and what effect does the composition of the diet have on systemic acid-base balance?
3. What is the difference between volatile and nonvolatile acids, and what is net endogenous acid production (NEAP)?
4. How do the kidneys and lungs contribute to systemic acid-base balance, and what is renal net acid excretion (RNAE)?
5. Why are urinary buffers necessary for excretion of acid by the kidneys?
6. What are the mechanisms for H^+ transport in the various segments of the nephron, and how are these mechanisms regulated?
7. How do the various segments of the nephron contribute to the process of reabsorbing the filtered HCO_3^-?
8. How do the kidneys produce new HCO_3^-?
9. How is ammonium produced by the kidneys, and how does its excretion contribute to renal acid excretion?
10. What are the major mechanisms by which the body defends itself against changes in acid-base balance?
11. What are the differences between simple metabolic and respiratory acid-base disorders, and how are they differentiated by arterial blood gas measurements?

The concentration of H^+ in body fluids is low compared with that of other ions. For example, Na^+ is present at a concentration some three million times greater than that of H^+ ($[Na^+] = 140$ mEq/L; $[H^+] = 40$ nEq/L). Because of the low $[H^+]$ of the body fluids, it is commonly expressed as the negative logarithm, or pH.

Virtually all cellular, tissue, and organ processes are sensitive to pH. Indeed, life cannot exist outside of a range of extracellular fluid (ECF) pH from 6.8 to 7.8 (160–16 nEq/L of H^+). Normally the pH of ECF is maintained between 7.35 and 7.45. As described in Chapter 2 the pH of intracellular fluid (ICF) is slightly lower (7.1–7.2) but also tightly regulated.

Each day, acid and alkali are ingested in the diet. Also, cellular metabolism produces a number of substances that have an impact on the pH of body fluids. Without appropriate mechanisms to deal with this daily acid and alkali load and thereby maintain acid-base balance, many processes necessary for life could not occur. This chapter reviews the maintenance of whole-body acid-base balance. Although the emphasis is on the role of the kidneys in this process, the roles of the lungs and liver are also considered. In addition the impact of diet and cellular metabolism on acid-base balance is presented. Finally, disorders of acid-base balance are considered, primarily to illustrate the physiological processes involved. Throughout this chapter, **acid** is defined as any substance that adds H^+ to body fluids, whereas **alkali** is defined as a substance that removes H^+ from body fluids.

The HCO_3^- Buffer System

Bicarbonate (HCO_3^-) is an important buffer of the ECF. With a normal plasma $[HCO_3^-]$ of 23 to 25 mEq/L and a volume of 14 L (for a 70-kg individual) the ECF can potentially buffer 350 mEq of H^+. The HCO_3^- buffer system differs from the other buffer systems of the body (e.g., phosphate) because it is regulated by both the lungs and kidneys. This is best appreciated by considering the following reaction:

Equation 37.1

$$CO_2 + H_2O \overset{Slow}{\leftrightarrow} H_2CO_3 \overset{Fast}{\leftrightarrow} H^+ + HCO_3^-$$

As indicated the first reaction (hydration/dehydration of CO_2) is the rate-limiting step. This normally slow reaction is greatly accelerated in the presence of carbonic anhydrase.[a] The second reaction, ionization of H_2CO_3 to H^+ and HCO_3^-, is virtually instantaneous.

---

[a]Carbonic anhydrase (CA) actually catalyzes the following reaction:

$$H_2O \overset{CA}{\rightarrow} H^+ + OH^- + CO_2 \rightarrow HCO_3^- + H^+ \rightarrow H_2CO_3$$

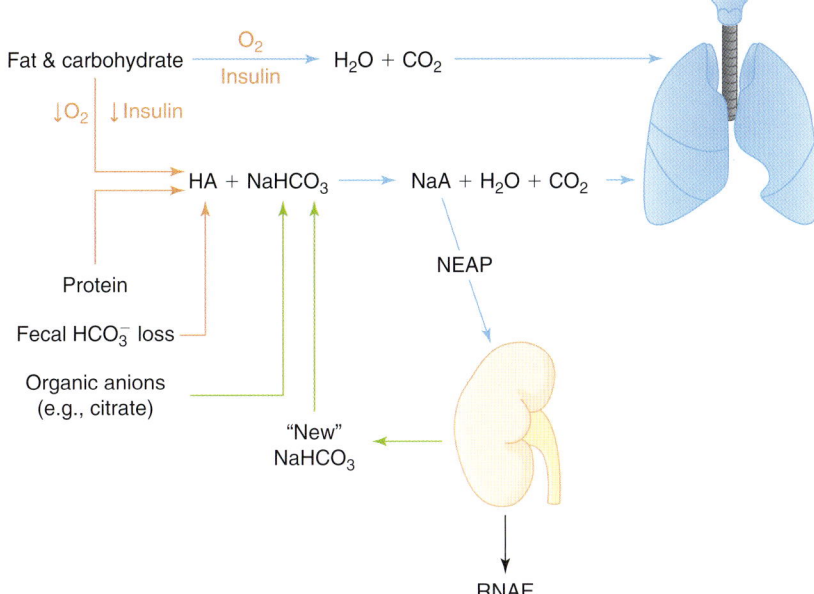

• **Fig. 37.1** Overview of acid-base balance. The lungs and kidneys work together to maintain acid-base balance. The lungs excrete CO_2 (volatile acid), and the kidneys excrete acid (renal net acid excretion [RNAE]) equal to net endogenous acid production (NEAP), which reflects dietary intake, cellular metabolism, and loss of acid and alkali (e.g., HCO_3^- loss in feces) from the body. See text for details. (From Koeppen BM, Stanton BA. *Renal Physiology*. 5th ed. Philadelphia: Elsevier; 2013.)

The **Henderson-Hasselbalch equation** is used to quantitate how changes in CO_2 and HCO_3^- effect pH:

Equation 37.2

$$pH = pK' + \log\frac{[HCO_3^-]}{\alpha P_{CO_2}}$$

or

Equation 37.3

$$pH = 6.1 + \log\frac{[HCO_3^-]}{0.03 P_{CO_2}}$$

In these equations the amount of CO_2 is determined from the partial pressure of CO_2 (P_{CO_2}) and its solubility (α) in solution. For plasma at 37°C, α has a value of 0.03. Also, pK′ is the negative logarithm of the overall dissociation constant for the reaction in Eq. 37.1 and has a value for plasma at 37°C of 6.1. Alternatively the relationship between HCO_3^- and CO_2 on the [H^+] can be determined as follows:

Equation 37.4

$$[H^+] = \frac{24 \times P_{CO_2}}{[HCO_3^-]}$$

Inspection of Eqs. 36.3 and 36.4 show that pH and [H^+] vary when either [HCO_3^-] or P_{CO_2} are altered. Disturbances of acid-base balance that result from a change in [HCO_3^-] are termed *metabolic acid-base disorders,* whereas those resulting from a change in P_{CO_2} are termed *respiratory acid-base disorders.* These disorders are considered in more detail in a subsequent section. The kidneys are primarily responsible for regulating the [HCO_3^-] of ECF, whereas the lungs control the P_{CO_2}.

Overview of Acid-Base Balance

The diet of humans contains many constituents that are either acid or alkali. In addition, cellular metabolism produces acid and alkali. Finally, alkali is normally lost each day in feces. As described later, although diet dependent, the net effect of these processes is the addition of acid to body fluids. For acid-base balance to be maintained, acid must be excreted from the body at a rate equivalent to its addition. If acid addition exceeds excretion, **acidosis** results. Conversely, if acid excretion exceeds addition, **alkalosis** results.

As summarized in Fig. 37.1, the major constituents of the diet are carbohydrates and fats. When tissue perfusion is adequate, O_2 is available to tissues, and insulin is present at normal levels, carbohydrates and fats are metabolized to CO_2 and H_2O. On a daily basis, 15 to 20 moles of CO_2 are generated through this process. Normally this large quantity of CO_2 is effectively eliminated from the body by the lungs. Therefore this metabolically derived CO_2 has no impact on acid-base balance. CO_2 is usually termed **volatile acid** because it has the potential to generate H^+ after hydration with H_2O (see Eq. 36.1). Acid not derived directly from hydration of CO_2 is termed **nonvolatile acid** (e.g., lactic acid).

The cellular metabolism of other dietary constituents also has an impact on acid-base balance. For example, cysteine and methionine, sulfur-containing amino acids, yield sulfuric acid when metabolized, whereas hydrochloric acid results from metabolism of lysine, arginine, and histidine. A portion of this nonvolatile acid load is offset by production of HCO_3^- through metabolism of the amino acids aspartate and glutamate. On average the metabolism of dietary amino acids yields net nonvolatile acid production. Metabolism of certain organic anions (e.g., citrate) results in production of HCO_3^-, which offsets nonvolatile acid production to some degree. Overall in individuals ingesting a meat-containing diet, acid production exceeds HCO_3^- production. In contrast, a vegetarian diet produces less nonvolatile acid. In addition to the metabolically derived acids and alkalis, the foods ingested contain acid and alkali. For example, the presence of phosphate ($H_2PO_4^-$) in ingested food increases the dietary acid load. Finally, during digestion, some HCO_3^- is normally lost in feces. This loss is equivalent to the addition of nonvolatile acid to the body. In an individual ingesting a meat-containing diet, dietary intake, cellular metabolism, and fecal HCO_3^- loss result in addition of approximately 0.7 to 1.0 mEq/kg body weight of nonvolatile acid to the body each day (50–100 mEq/day for most adults). This acid, referred to as **net endogenous acid production (NEAP),** results in an equivalent loss of HCO_3^- from the body that must be replaced.

 IN THE CLINIC

When insulin levels are normal, carbohydrates and fats are completely metabolized to $CO_2 + H_2O$. However, if insulin levels are abnormally low (e.g., diabetes mellitus), cellular metabolism leads to production of several organic ketoacids (e.g., β-hydroxybutyric acid and acetoacetic acid from fatty acids).

In the absence of adequate levels of O_2 (hypoxia), anaerobic metabolism by cells can also lead to production of organic acids (e.g., lactic acid) rather than $CO_2 + H_2O$. This frequently occurs in normal individuals during vigorous exercise. Poor tissue perfusion, such as occurs with reduced cardiac output, can also lead to anaerobic metabolism by cells and thus to acidosis. In these conditions the organic acids accumulate and the pH of body fluids decreases (acidosis). Treatment (e.g., administration of insulin in the case of diabetes) or improved delivery of adequate levels of O_2 to tissues (e.g., in the case of poor tissue perfusion) results in the metabolism of these organic acids to $CO_2 + H_2O$, which consumes H^+ and thereby helps correct the acid-base disorder.

Nonvolatile acids do not circulate throughout the body but are immediately neutralized by the HCO_3^- in ECF.

Equation 37.5

$$H_2SO_4 + 2NaHCO_3 \leftrightarrow Na_2SO_4 + 2CO_2 + 2H_2O$$

Equation 37.6

$$HCl + NaHCO_3 \leftrightarrow NaCl + CO_2 + H_2O$$

This neutralization process yields the Na^+ salts of the strong acids and removes HCO_3^- from the ECF. Thus HCO_3^- minimizes the effect of these strong acids on the pH of ECF. As noted previously, ECF contains approximately 350 mEq of HCO_3^-. If this HCO_3^- were not replenished, the daily production of nonvolatile acids ($\approx$70 mEq/day) would deplete the ECF of HCO_3^- within 5 days. To maintain acid-base balance the kidneys must replenish the HCO_3^- lost by neutralization of the nonvolatile acids, a process termed **renal net acid excretion (RNAE).**

Net Acid Excretion by the Kidneys

Under steady-state conditions, NEAP must equal RNAE to maintain acid-base balance. Although NEAP varies from individual to individual and from day to day in anyone individual, it is not regulated. Instead the kidneys regulate RNAE to match NEAP and in so doing replenish the HCO_3^- (**new HCO_3^-**) lost by neutralization of nonvolatile acids. In addition the kidneys must prevent the loss of HCO_3^- in urine. This latter task is quantitatively more important because the filtered load of HCO_3^- is approximately 4320 mEq/day (24 mEq/L × 180 L/day = 4320 mEq/day), compared with only 50 to 100 mEq/day needed to balance NEAP.

Both reabsorption of filtered HCO_3^- and excretion of acid are accomplished via H^+ secretion by the nephrons. Thus in a single day the nephrons must secrete approximately 4390 mEq of H^+ into the tubular fluid. Most of the secreted H^+ serves to reabsorb the filtered load of HCO_3^-. Only 50 to 100 mEq of H^+, an amount equivalent to NEAP, is excreted in urine. As a result of this acid excretion, urine is normally acidic.

The kidneys cannot excrete urine more acidic than pH 4.0 to 4.5. Even at a pH of 4.0, only 0.1 mEq/L of H^+ can be excreted. Therefore to excrete sufficient acid, the kidneys excrete H^+ with urinary buffers such as phosphate (P_i).[b] Other constituents of urine can also serve as buffers (e.g., creatinine), although their role is less important than P_i. Collectively the various urinary buffers are termed **titratable acid.** This term is derived from the method by which these buffers are quantitated in the laboratory. Typically, alkali (OH^-) is added to a urine sample to titrate its pH to that of plasma (i.e., 7.4). The amount of alkali added is equal to the H^+ titrated by these urine buffers and is termed *titratable acid.*

Excretion of H^+ as a titratable acid is insufficient to balance NEAP. An additional and important mechanism by which the kidneys contribute to maintenance of acid-base balance is through synthesis and excretion of **ammonium (NH_4^+).** The mechanisms involved in this process are discussed in more detail later in this chapter. With regard to renal regulation of acid-base balance, each NH_4^+ excreted in urine results in the return of one HCO_3^- to the systemic circulation, which replenishes the HCO_3^- lost during neutralization of the nonvolatile acids. Thus production and excretion of NH_4^+, like excretion of titratable acid, is equivalent to excretion of acid by the kidneys.

In brief the kidneys contribute to acid-base homeostasis by reabsorbing the filtered load of HCO_3^- and excreting an

---

[b]The titration reaction is: $HPO_4^{2-} + H^+ \leftrightarrow H_2PO_4^-$. This reaction has a pK of approximately 6.8.

amount of acid equivalent to NEAP. This process can be quantitated as follows:

Equation 37.7

$$RNAE = (U_{NH_4^+} \times \dot{V}) + (U_{TA} \times \dot{V}) - (U_{HCO_3^-} \times \dot{V})$$

where $(U_{NH_4^+} \times \dot{V})$ and $(U_{TA} \times \dot{V})$ are the rates of excretion (mEq/day) of NH_4^+ and titratable acid (TA), and $(U_{HCO_3^-} \times \dot{V})$ is the amount of HCO_3^- lost in urine (equivalent to adding H^+ to the body).[c] Again, maintenance of acid-base balance means that net acid excretion must equal nonvolatile acid production. Under most conditions, very little HCO_3^- is excreted in urine. Thus net acid excretion essentially reflects titratable acid and NH_4^+ excretion. Quantitatively, titratable acid accounts for approximately one-third and NH_4^+ for two-thirds of RNAE.

HCO_3^- Reabsorption Along the Nephron

As indicated by Eq. 37.7, net acid excretion is maximized when little or no HCO_3^- is excreted in urine. Indeed, under most circumstances, very little HCO_3^- appears in urine. Because HCO_3^- is freely filtered at the glomerulus, approximately 4320 mEq/day is delivered to the nephrons and is then reabsorbed. Fig. 37.2 summarizes the contribution of each nephron segment to reabsorption of filtered HCO_3^-.

The proximal tubule reabsorbs the largest portion of the filtered load of HCO_3^-. Fig. 37.3 summarizes the primary

[c]This equation ignores the small amount of free H^+ excreted in urine. As already noted, urine with a pH = 4.0 contains only 0.1 mEq/L of H^+.

transport processes involved. H^+ secretion across the apical membrane of the cell occurs by both a Na^+/H^+ antiporter and H^+-ATPase (V-type). The Na^+/H^+ antiporter (NHE3) is the predominant pathway for H^+ secretion (accounts for ≈ two-thirds of HCO_3^- reabsorption) and uses the lumen-to-cell $[Na^+]$ gradient to drive this process (i.e.,

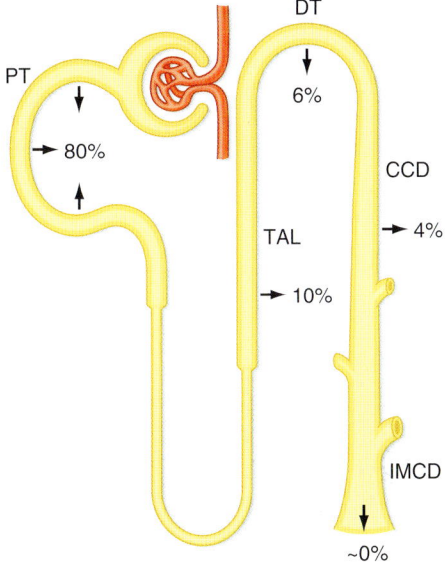

• **Fig. 37.2** Segmental reabsorption of HCO_3^-. The fraction of the filtered load of HCO_3^- reabsorbed by the various segments of the nephron is shown. Normally the entire filtered load of HCO_3^- is reabsorbed and little or no HCO_3^- appears in the urine. CCD, cortical collecting duct; DT, distal tubule; IMCD, inner medullary collecting duct; PT, proximal tubule; TAL, thick ascending limb.

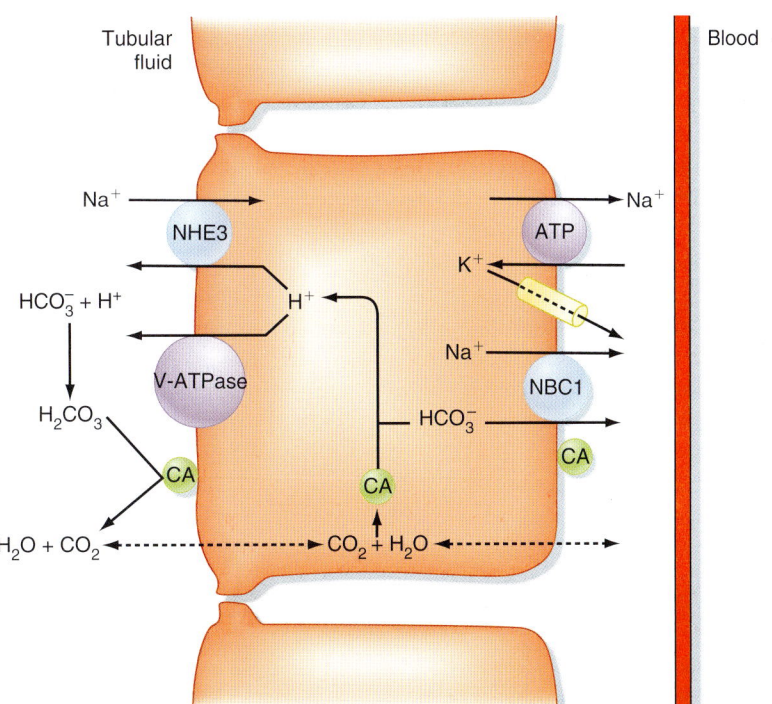

• **Fig. 37.3** Cellular mechanism for reabsorption of filtered HCO_3^- by cells of the proximal tubule. Only the primary H^+ and HCO_3^- transporters are shown. ATP, adenosine triphosphate; CA, carbonic anhydrase.

secondary active secretion of H⁺). Within the cell, H⁺ and HCO₃⁻ are produced in a reaction catalyzed by carbonic anhydrase (CA-II). The H⁺ is secreted into the tubular fluid, whereas the HCO₃⁻ exits the cell across the basolateral membrane and returns to the peritubular blood. HCO₃⁻ movement out of the cell across the basolateral membrane is coupled to other ions. The majority of HCO₃⁻ exits via a symporter that couples the efflux of Na⁺ with HCO₃⁻ (sodium bicarbonate symporter, NBC1). Some HCO₃⁻ exits the cell by other transporters, but they are not as important as the Na⁺/HCO₃⁻ symporter. As noted in Fig. 37.3, carbonic anhydrase (CA-IV) is also present in the brush border and basolateral membrane of the cell. The brush border enzyme catalyzes dehydration of H_2CO_3 in the luminal fluid, whereas the enzyme localized to basolateral membrane facilitates HCO₃⁻ exit from the cell. The movement of CO_2 into and out of the cell occurs via AQP1, which is localized to both the luminal and basolateral membranes.

The cellular mechanism for HCO₃⁻ reabsorption by the thick ascending limb (TAL) of the loop of Henle is very similar to that in the proximal tubule. H⁺ is secreted by a Na⁺/H⁺ antiporter and H⁺-ATPase. Like in the proximal tubule, the Na⁺/H⁺ antiporter (NHE3) is the predominant pathway for H⁺ secretion. HCO₃⁻ exit from the cell involves both a Na⁺/HCO₃⁻ symporter (NBC1) and a Cl⁻/HCO₃⁻ antiporter (anion exchanger, AE-2). Some HCO₃⁻ may also exit the cell through Cl⁻ channels present in the basolateral membrane.

The distal tubule[d] and collecting duct reabsorb the small amount of HCO₃⁻ that escapes reabsorption by the proximal tubule and loop of Henle. Fig. 37.4 shows the cellular mechanism of H⁺/HCO₃⁻ transport by the intercalated cells located within these segments (see also Chapter 33).

One type of intercalated cell secretes H⁺ (reabsorbs HCO₃⁻) and is called the *A-* or *α-intercalated cell*. Within this cell, H⁺ and HCO₃⁻ are produced by hydration of CO_2; this reaction is catalyzed by carbonic anhydrase (CA-II). H⁺ is secreted into the tubular fluid via two mechanisms. The first involves an apical membrane H⁺-ATPase (V-type). The second couples secretion of H⁺ with reabsorption of K⁺ via an H⁺-K⁺-ATPase similar to those found in the stomach and colon (HKα1 and HKα2). HCO₃⁻ exits the cell across the basolateral membrane in exchange for Cl⁻ (via a Cl⁻/HCO₃⁻ antiporter, AE-1) and enters the peritubular capillary blood.

A second population of intercalated cells secretes HCO₃⁻ rather than H⁺ into the tubular fluid (also called *B-* or

β-*intercalated cells*).[e] In these cells the H⁺-ATPase (V-type) is located in the basolateral membrane and the Cl⁻/HCO₃⁻ antiporter is located in the apical membrane (see Fig. 37.4). However, the apical membrane Cl⁻/HCO₃⁻ antiporter is different from the one found in the basolateral membrane of the H⁺-secreting intercalated cells and has been identified as **pendrin.** The activity of the HCO₃⁻-secreting intercalated cell is increased during metabolic alkalosis, when the kidneys must excrete excess HCO₃⁻. However, under most conditions (e.g., ingestion of a meat-containing diet) H⁺ secretion predominates in these segments.[f]

The apical membrane of collecting duct cells is not very permeable to H⁺, and thus the pH of tubular fluid can become quite acidic. Indeed, the most acidic tubular fluid along the nephron (pH = 4.0–4.5) is produced there. In comparison the permeability of the proximal tubule to H⁺ and HCO₃⁻ is much higher, and the tubular fluid pH falls to only 6.5 in this segment. As explained later the ability of the collecting duct to lower the pH of the tubular fluid is critically important for excretion of urinary titratable acids and NH₄⁺.

Regulation of H⁺ Secretion

A number of factors influence secretion of H⁺ and thus reabsorption of filtered HCO₃⁻ by the cells of the nephron. From a physiological perspective the primary factor that regulates H⁺ secretion by the nephron is a change in systemic acid-base balance. Thus acidosis stimulates RNAE, whereas RNAE is reduced during alkalosis.

The response of the kidneys to changes in acid-base balance includes both immediate changes in the activity and/or number of transporters in the membrane and longer-term changes in the synthesis of transporters. For example, with metabolic acidosis, H⁺ secretion is stimulated by multiple mechanisms, depending on the particular nephron segment. First, the decrease in intracellular pH that occurs with acidosis will create a more favorable cell-to–tubular fluid H⁺ gradient and thereby make secretion of H⁺ across the apical membrane more energetically favorable. Second, the decrease in pH may lead to allosteric changes in transport proteins, thereby altering their kinetics. Lastly, transporters may be shuttled to the membrane from intracellular vesicles. With long-term acidosis the abundance of transporters increases, either by increased transcription of appropriate transporter genes or by increased translation of transporter mRNA.

---

[d]Here and in the remainder of the chapter we focus on the function of intercalated cells. The early portion of the distal tubule, which does not contain intercalated cells, also reabsorbs HCO₃⁻. The cellular mechanism appears to involve an apical membrane Na⁺/H⁺ antiporter (NHE2) and a basolateral Cl⁻/HCO₃⁻ antiporter (AE2).

[e]A third group of intercalated cells shares features of both H⁺-secreting and HCO₃⁻-secreting intercalated cells. The precise function of this cell type in acid-base transport is not fully understood.
[f]Traditionally it was believed that intercalated cells were only involved in acid-base transport. There is now good evidence that NaCl reabsorption is also carried out by intercalated cells (B type). Reabsorption of NaCl occurs by the tandem operation of an apical membrane Cl⁻/HCO₃⁻ antiporter (pendrin) and an apical membrane Na⁺/HCO₃⁻/2Cl⁻ antiporter (NDCBE). This mechanism of NaCl reabsorption is inhibited by thiazide diuretics.

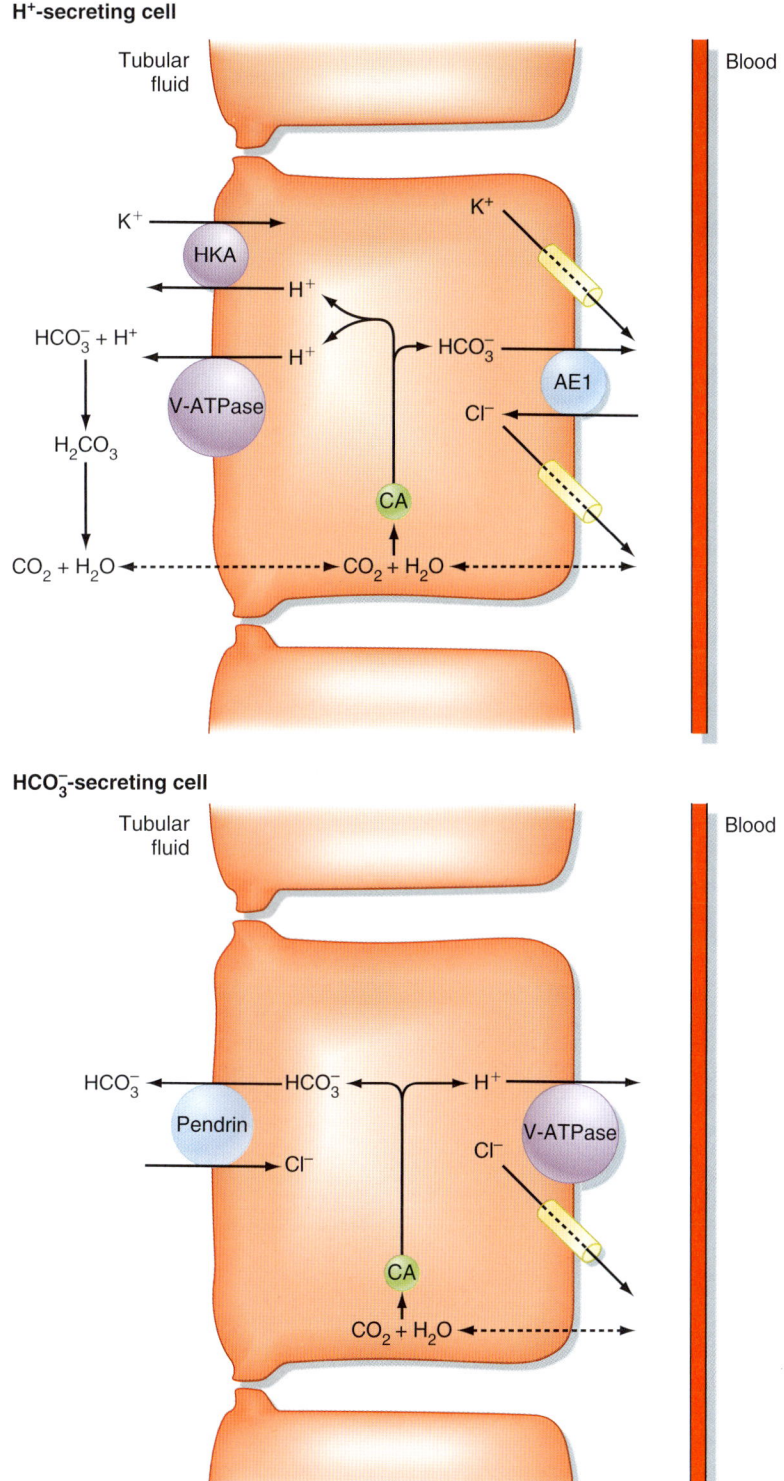

H⁺-secreting cell

• **Fig. 37.4** Cellular mechanisms for reabsorption and secretion of HCO_3^- by intercalated cells of the distal tubule and collecting duct. Only the primary H^+ and HCO_3^- transporters are shown. ATP, adenosine triphosphate; CA, carbonic anhydrase.

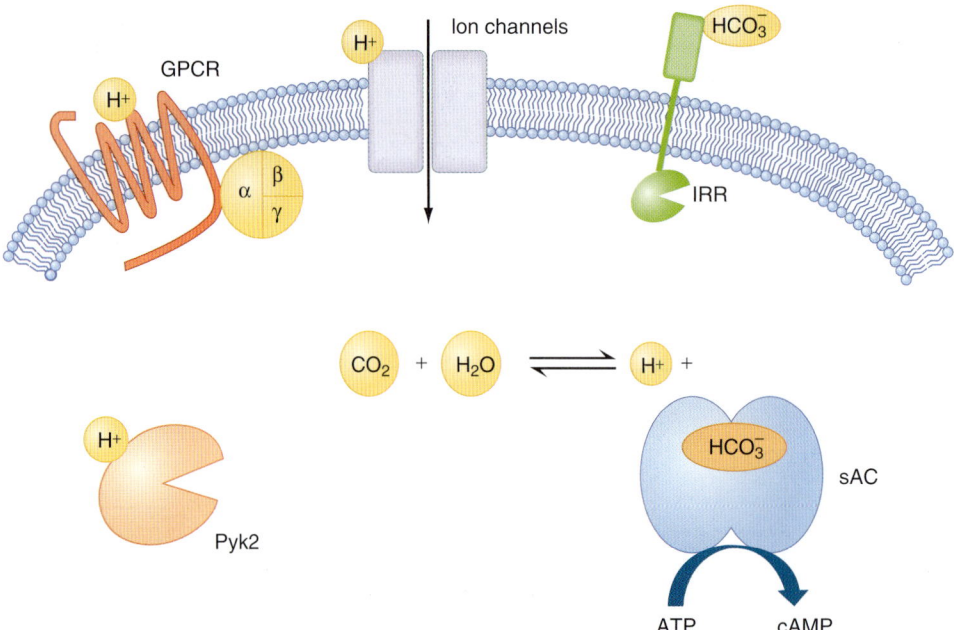

• **Fig. 37.5** Examples of cellular H^+ and HCO_3^- sensors. ATP, adenosine triphosphate; cAMP, cyclic adenosine monophosphate; GPCR, G protein–coupled receptor; IRR, insulin receptor–related receptor; Pyk2, nonreceptor tyrosine kinase; sAC, soluble adenylyl cyclase. (Adapted from: Levin LR, Buck J. *Annu Rev Physiol* 2015;77:347.)

AT THE CELLULAR LEVEL

Cells of the kidney express receptors that monitor acid-base status and therefore play a critical role in regulating H^+ and HCO_3^- transporters along the nephron (Fig. 37.5). For example, a G protein–coupled H^+ receptor (GPCR–GPR4) has been localized to the collecting duct. Activation of this receptor by an increase in ECF $[H^+]$ stimulates H^+ secretion. Also in the collecting duct, HCO_3^--secreting intercalated cells (B- or β-ICs) express a basolateral insulin-related receptor (IRR) that is a tyrosine kinase. It is activated by an increase in ECF $[HCO_3^-]$ and stimulates HCO_3^- secretion by the cell. A soluble adenylyl cyclase (sAC) regulated by intracellular HCO_3^- appears to also play a roll in regulating collecting duct H^+ secretion. In the proximal tubule, basolateral membrane receptor tyrosine kinases (ErbB1 and ErbB2) sense changes in ECF P_{CO_2}. Activation of these receptors by an increase in P_{CO_2} results in generation of angiotensin II, which acting from the lumen via AT-1A receptors, stimulates H^+ secretion/ HCO_3^- reabsorption. Also in the proximal tubule, the nonreceptor tyrosine kinase (Pyk2) senses intracellular $[H^+]$. When it is activated by an increase in intracellular $[H^+]$, H^+ secretion/HCO_3^- reabsorption is stimulated. Finally, the gating of several ion channels (e.g., the renal outer medullary K^+ channel [ROMK]) is effected by changes in either ECF or ICF pH. These too have the potential to serve as cellular acid-base sensors.

AT THE CELLULAR LEVEL

In the proximal tubule, metabolic acidosis increases the transport kinetics of the Na^+/H^+ antiporter (NHE3) and increases apical membrane expression of the Na^+/H^+ antiporter, H^+-ATPase, and the basolateral $Na^+/3HCO_3^-$ symporter (NBCe1). In the collecting duct, acidosis leads to exocytic insertion of H^+-ATPase into the apical membrane of intercalated cells. With long-term acidosis the abundance of key acid-base transporters is increased in the proximal tubule (NHE3 and NBCe1) and in collecting duct intercalated cells (H^+-ATPase and AE1). Lastly, acidosis decreases expression of the Cl^-/HCO_3^- antiporter pendrin in HCO_3^--secreting intercalated cells.

Although some of the effects just described may be attributable directly to acidosis, many of these changes in cellular H^+ transport are mediated by hormones or other factors. Three known mediators of the renal response to acidosis are endothelin, cortisol, and angiotensin II. **Endothelin (ET-1)**

is produced by endothelial and proximal tubule cells. With acidosis, ET-1 secretion is enhanced. In the proximal tubule, ET-1 stimulates phosphorylation and subsequent insertion of the Na^+/H^+ antiporter into the apical membrane, and insertion of the $Na^+/3HCO_3^-$ symporter into the basolateral membrane. ET-1 may mediate the response to acidosis in other nephron segments as well. Acidosis also stimulates secretion of the glucocorticoid hormone **cortisol** by the adrenal cortex. Cortisol increases the abundance of the Na^+/H^+ antiporter and $Na^+/3HCO_3^-$ symporter in the proximal tubule. Angiotensin II is produced in proximal tubule cells in response to acidosis. It is secreted into the tubular fluid, where it binds to the angiotensin I receptor and thereby stimulates H^+ secretion/HCO_3^- reabsorption by the proximal tubule. Both cortisol and angiotensin II also stimulate production and secretion of NH_4^+ by the proximal tubule, which as described later is an important component of the kidney's response to acidosis.

Acidosis also stimulates secretion of parathyroid hormone (PTH). PTH inhibits phosphate (P_i) reabsorption by the proximal tubule (see Chapter 36). In so doing, more P_i is delivered to the distal nephron, where it serves as a urinary buffer and thus increases the capacity of the kidneys to excrete titratable acid.

The response of the kidneys to alkalosis is less well characterized. RNAE is decreased because of increased urinary HCO_3^- excretion and because excretion of titratable acid and NH_4^+ are reduced. The factors that regulate this response are not well characterized.

Other factors not necessarily related to maintaining acid-base balance can influence secretion of H^+ by the cells of the nephron. Because a significant H^+ transporter in the nephron is the Na^+/H^+ antiporter, factors that alter Na^+ reabsorption can secondarily affect H^+ secretion. For example, with volume contraction (negative Na^+ balance), Na^+ reabsorption by the nephron is increased (see Chapter 35), including reabsorption of Na^+ via the Na^+/H^+ antiporter. As a result, H^+ secretion is enhanced. This occurs by several mechanisms. One mechanism involves the renin-angiotensin-aldosterone system, which is activated by volume contraction. As noted earlier, angiotensin II acts on the proximal tubule to stimulate the apical membrane Na^+/H^+ antiporter as well as the basolateral $Na^+/3HCO_3^-$ symporter. To a lesser degree, angiotensin II stimulates H^+ secretion in the TAL of Henle's loop and the early portion of the distal tubule, a process also mediated by the Na^+/H^+ antiporter. Aldosterone's primary action on the distal tubule and collecting duct is to stimulate Na^+ reabsorption by principal cells (see Chapter 34). However, it also stimulates intercalated cells in these segments to secrete H^+. This effect is both indirect and direct. By stimulating Na^+ reabsorption by principal cells, aldosterone hyperpolarizes the transepithelial voltage (i.e., the lumen becomes more electrically negative). This change in transepithelial voltage then facilitates secretion of H^+ by intercalated cells. In addition to this indirect effect, aldosterone (and angiotensin II) act directly on intercalated cells to stimulate H^+ secretion via H^+-ATPase and H^+,K^+-ATPase.

Another mechanism by which ECF volume (ECFV) contraction enhances H^+ secretion (HCO_3^- reabsorption) is through changes in peritubular capillary Starling forces. As described in Chapters 34 and 35, ECFV contraction alters the peritubular capillary Starling forces such that overall proximal tubule reabsorption is enhanced. With this enhanced reabsorption, more of the filtered load of HCO_3^- is reabsorbed.

Potassium balance influences secretion of H^+ by the proximal tubule. Hypokalemia stimulates and hyperkalemia inhibits H^+ secretion. It is thought that K^+-induced changes in intracellular pH are responsible at least in part for this effect, with hypokalemia acidifying and hyperkalemia alkalinizing the cells. Hypokalemia also stimulates H^+ secretion by the collecting duct. This occurs as a result of increased expression of the H^+,K^+-ATPase in intercalated cells.

Formation of New HCO_3^-

As discussed previously, reabsorption of the filtered load of HCO_3^- is important for maximizing RNAE. However, HCO_3^- reabsorption alone does not replenish the HCO_3^- lost during neutralization of the nonvolatile acids produced during metabolism. To maintain acid-base balance, the kidneys must replace this lost HCO_3^- with new HCO_3^-. Generation of new HCO_3^- occurs by excretion of titratable acid and by synthesis and excretion of NH_4^+.

Production of new HCO_3^- as a result of titratable acid excretion is depicted in Fig. 37.6. Because of HCO_3^- reabsorption by the proximal tubule and loop of Henle, fluid reaching the distal tubule and collecting duct normally contains little HCO_3^-. Thus when H^+ is secreted it will combine with non-HCO_3^- buffers (primarily P_i) and be excreted as titratable acid. Because the H^+ was produced inside the cell from hydration of CO_2, a HCO_3^- is also produced. This HCO_3^- is returned to the ECF as new HCO_3^-. As noted, P_i excretion increases with acidosis. However, even with increased P_i available for titratable acid formation, this response is insufficient to generate the required amount of new HCO_3^-. The remainder of new HCO_3^- generation occurs as a result of NH_4^+ generation and excretion.

NH_4^+ is produced by the kidneys, and its synthesis and subsequent excretion adds HCO_3^- to ECF. Importantly, this process is regulated in response to the acid-base requirements of the body.

NH_4^+ is produced in the kidneys via metabolism of **glutamine.** Essentially the kidneys metabolize glutamine, excrete NH_4^+, and add HCO_3^- to the body. However, formation of new HCO_3^- via this process depends on the kidneys' ability to excrete NH_4^+ in urine. If NH_4^+ is not excreted in urine but instead enters the systemic circulation, it is converted into urea by the liver. This conversion process generates H^+, which is then buffered by HCO_3^-. Thus production of urea from renal-generated NH_4^+ consumes HCO_3^- and negates formation of HCO_3^- through synthesis and excretion of NH_4^+ by the kidneys. However, normally the kidneys excrete NH_4^+ in urine and thereby produce new HCO_3^-.

The process by which the kidneys excrete NH_4^+ is complex. Fig. 37.7 illustrates the essential features of this process. NH_4^+ is produced from glutamine in the cells of the proximal tubule, a process termed **ammoniagenesis.** Each glutamine molecule produces two molecules of NH_4^+ and the divalent anion 2-oxoglutarate^{2-}. Metabolism of this anion ultimately provides two molecules of HCO_3^-. HCO_3^- exits the cell across the basolateral membrane and enters the peritubular blood as new HCO_3^-. NH_4^+ exits the cell across the apical membrane and enters the tubular fluid. The primary mechanism for NH_4^+ secretion into the tubular fluid involves the Na^+/H^+ antiporter, with NH_4^+ substituting for H^+. In addition, some NH_3 can diffuse out of the cell into the tubular fluid, where it is protonated to NH_4^+.

A significant portion of the NH_4^+ secreted by the proximal tubule is reabsorbed by the loop of Henle. The TAL

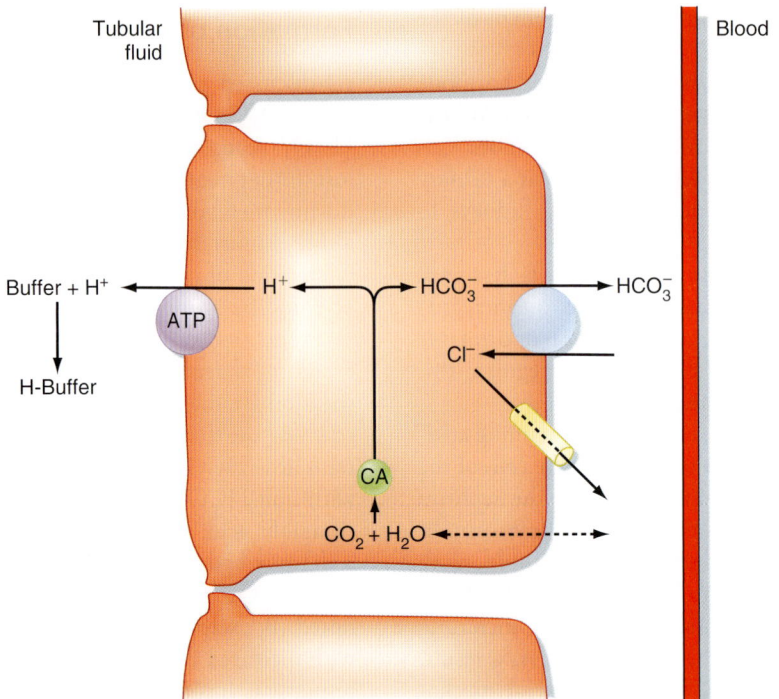

• **Fig. 37.6** General scheme for excretion of H^+ with non-HCO_3^- urinary buffers (titratable acid). The primary urinary buffer is phosphate (HPO_4^{2-}). A H^+-secreting intercalated cell is shown. For simplicity, only the H^+-ATPase is depicted. H^+ secretion by the H^+,K^+-ATPase also titrates luminal buffers. ATP, adenosine triphosphate; CA, carbonic anhydrase.

is the primary site of this NH_4^+ reabsorption, with NH_4^+ substituting for K^+ on the $1Na^+/1K^+/2Cl^-$ symporter. In addition the lumen-positive transepithelial voltage in this segment drives paracellular reabsorption of NH_4^+.

The NH_4^+ reabsorbed by the TAL of the loop of Henle accumulates in the medullary interstitium. From there it is then secreted into the tubular fluid by the collecting duct.

The cells of the collecting duct express two NH_3 membrane transporters known as *Rhesus (Rh) glycoproteins* (RhBG and RhCG).[g] RhBG is present in the basolateral membrane of H^+-secreting intercalated cells and principal cells, and RhCG is present in both the apical and basolateral membranes of these cells. As depicted in Fig. 37.7, NH_3 is transported across the collecting duct, a process traditionally termed *nonionic diffusion*. The secreted NH_3 is protonated in the tubule lumen as a result of intercalated H^+ secretion. Because the apical membrane has a low permeability to NH_4^+ it is effectively trapped in the tubular lumen, a process traditionally termed *diffusion trapping*.

H^+ secretion by the collecting duct is critical for excretion of NH_4^+. If collecting duct H^+ secretion is inhibited, the NH_4^+ reabsorbed by the TAL of Henle's loop will not be excreted in the urine. Instead it will be returned to the systemic circulation, where as described previously, it will be converted to urea by the liver and consume HCO_3^- in

the process. Thus new HCO_3^- is produced during the metabolism of glutamine by cells of the proximal tubule. However, the overall process is not complete until the NH_4^+ is excreted (i.e., production of urea from NH_4^+ by the liver is prevented). Thus NH_4^+ excretion in urine can be used as a marker of glutamine metabolism in the proximal tubule.

🩺 IN THE CLINIC

Assessing NH_4^+ excretion by the kidneys is done indirectly because assays of urine NH_4^+ are not routinely available. Consider for example the situation of metabolic acidosis, wherein the appropriate renal response is to increase net acid excretion. Accordingly, little or no HCO_3^- will appear in urine, urine will be acidic, and NH_4^+ excretion will be increased. To assess this, and especially the amount of NH_4^+ excreted, the "urinary net charge" or "urine anion gap" can be calculated by measuring urinary concentrations of Na^+, K^+, and Cl^-:

$$\text{Urine anion gap} = ([Na^+] + [K^+]) - [Cl^-]$$

The concept of urine anion gap during a metabolic acidosis assumes that the major cations in urine are Na^+, K^+, and NH_4^+ and that the major anion is Cl^- (with urine pH < 6.5, virtually no HCO_3^- is present). As a result the urine anion gap will yield a negative value when NH_4^+ is being excreted. Indeed, the absence of a urine anion gap or the existence of a positive value indicates a renal defect in NH_4^+ production and excretion.

---

[g]Both RhBG and RhCG transport NH3. There is some evidence that RhBG may also transport some NH_4^+.

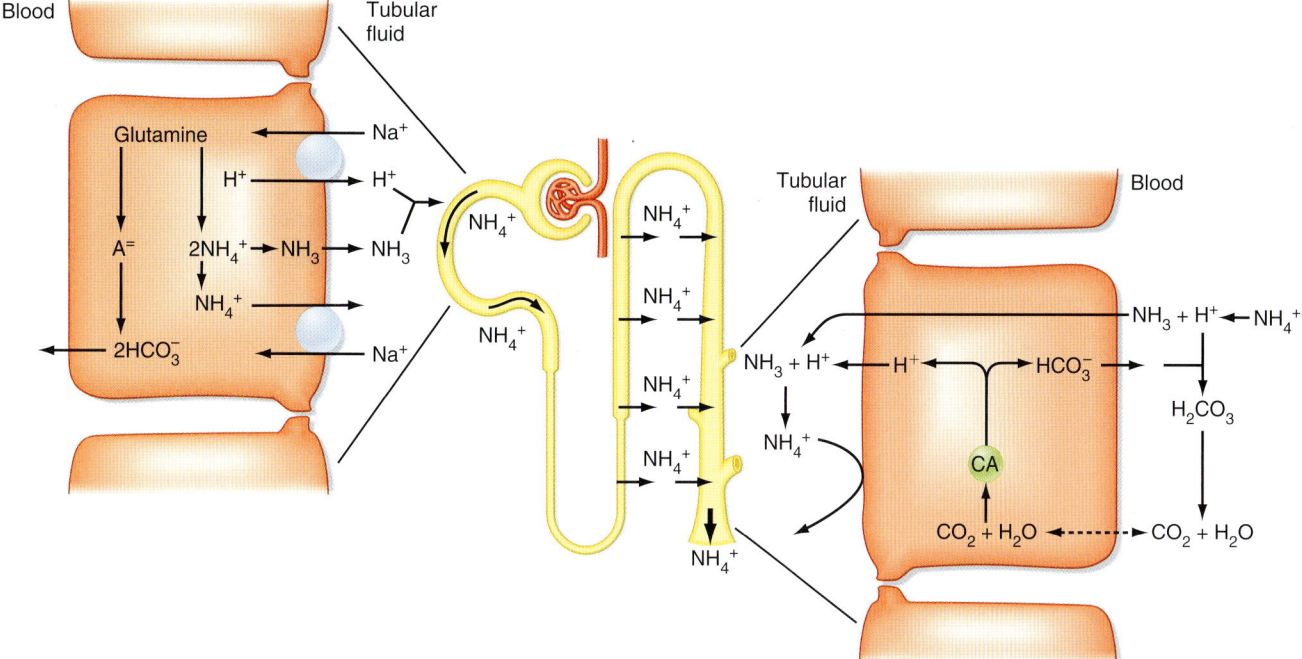

• **Fig. 37.7** Production, transport, and excretion of NH_4^+ by the nephron. Glutamine is metabolized to NH_4^+ and HCO_3^- in the proximal tubule. The NH_4^+ is secreted into the lumen, and the HCO_3^- enters the blood. The secreted NH_4^+ is reabsorbed in Henle's loop, primarily by the thick ascending limb, and accumulates in the medullary interstitium. NH_3 is secreted by the collecting duct via rhesus glycoproteins, and H^+ secretion traps NH_4^+ in the lumen. For each molecule of NH_4^+ excreted in the urine, a molecule of "new" HCO_3^- is added back to the ECF. CA, carbonic anhydrase.

In the net, one new HCO_3^- is returned to the systemic circulation for each NH_4^+ excreted in the urine.

An important feature of the renal NH_4^+ system is that it is regulated by systemic acid-base balance. As already described, cortisol levels increase with acidosis, as does angiotensin II secretion into the lumen of the proximal tubule. Both cortisol and angiotensin II stimulate ammoniagenesis (i.e., NH_4^+ production from glutamine). During systemic acidosis, the enzymes in the proximal tubule cell responsible for metabolism of glutamine are stimulated. This involves synthesis of new enzyme and requires several days for complete adaptation. With increased levels of these enzymes, NH_4^+ production is increased, allowing enhanced production of new HCO_3^-. Conversely, glutamine metabolism is reduced with alkalosis. Acidosis also increases the abundance of RhBG and RhCG in the collecting duct. Thus the ability to secrete NH_4^+ is also enhanced.

Response to Acid-Base Disorders

The pH of the ECF is maintained within a very narrow range (7.35–7.45).[h] Inspection of Eq. 37.3 shows that the pH of

ECF varies when either $[HCO_3^-]$ or PCO_2 are altered. As already noted, disturbances of acid-base balance that result from a change in $[HCO_3^-]$ of ECF are termed **metabolic acid-base disorders,** whereas those resulting from a change in PCO_2 are termed **respiratory acid-base disorders.** The kidneys are primarily responsible for regulating $[HCO_3^-]$, whereas the lungs regulate PCO_2.

When an acid-base disturbance develops, the body uses several mechanisms to defend against the change in ECF pH. These defense mechanisms do not correct the acid-base disturbance but merely minimize the change in pH imposed by the disturbance. Restoration of the blood pH to its normal value requires correction of the underlying process or processes that produced the acid-base disorder. The body has three general mechanisms to defend against changes in body fluid pH produced by acid-base disturbances: (1) extracellular and intracellular buffering, (2) adjustments in blood PCO_2 via alterations in the ventilatory rate of the lungs, and (3) adjustments in RNAE.

Extracellular and Intracellular Buffers

The first line of defense against acid-base disorders is extracellular and intracellular buffering. The response of the extracellular buffers is virtually instantaneous, whereas the response to intracellular buffering is slower and can take several minutes.

Metabolic disorders that result from addition of non-volatile acid or alkali to body fluids are buffered in both

---

[h]For simplicity of presentation in this chapter, the value of 7.40 for body fluid pH is used as normal, even though the normal range is from 7.35 to 7.45. Similarly the normal range for PCO_2 is 35 to 45 mm Hg. However, a PCO_2 of 40 mm Hg is used as the normal value. Finally, a value of 24 mEq/L is considered a normal ECF $[HCO_3^-]$, even though the normal range is 22 to 28 mEq/L.

the extracellular and intracellular compartments. The HCO_3^- buffer system is the principal ECF buffer. When nonvolatile acid is added to body fluids (or alkali is lost from the body), HCO_3^- is consumed during the process of neutralizing the acid load, and the $[HCO_3^-]$ of ECF is reduced. Conversely, when nonvolatile alkali is added to body fluids (or acid is lost from the body), H^+ is consumed, causing more HCO_3^- to be produced from the dissociation of H_2CO_3. Consequently, $[HCO_3^-]$ increases.

Although the HCO_3^- buffer system is the principal ECF buffer, P_i and plasma proteins provide additional extracellular buffering. The combined action of the buffering processes for HCO_3^-, P_i, and plasma protein accounts for approximately 50% of the buffering of a nonvolatile acid load and 70% of a nonvolatile alkali load. The remainder of the buffering under these two conditions occurs intracellularly. Intracellular buffering involves movement of H^+ into cells (during buffering of nonvolatile acid) or movement of H^+ out of cells (during buffering of nonvolatile alkali). H^+ is titrated inside the cell by HCO_3^-, P_i, and the histidine groups on proteins.

Bone represents an additional source of extracellular buffering. However, with acidosis, buffering by bone results in its demineralization.

When respiratory acid-base disorders occur, the pH of body fluid changes as a result of alterations in PCO_2. Virtually all buffering in respiratory acid-base disorders occurs intracellularly. When PCO_2 rises (respiratory acidosis), CO_2 moves into the cell, where it combines with H_2O to form H_2CO_3. H_2CO_3 then dissociates to H^+ and HCO_3^-. Some of the H^+ is buffered by cellular proteins, and HCO_3^- exits the cell and raises the ECF $[HCO_3^-]$. This process is reversed when PCO_2 is reduced (respiratory alkalosis). Under this condition the hydration reaction ($H_2O + CO_2 \leftrightarrow H_2CO_3$) is shifted to the left by the decrease in PCO_2. As a result the dissociation reaction ($H_2CO_3 \leftrightarrow H^+ + HCO_3^-$) also shifts to the left, thereby reducing the ECF $[HCO_3^-]$.

Respiratory Compensation

The lungs are the second line of defense against acid-base disorders. As indicated by the Henderson-Hasselbalch equation (see Eq. 37.2), changes in PCO_2 alter the blood pH: a rise decreases the pH, and a reduction increases the pH.

The ventilatory rate determines the PCO_2. Increased ventilation decreases PCO_2, whereas decreased ventilation increases it (Fig. 37.8). The blood PCO_2 and pH are important regulators of the ventilatory rate. Chemoreceptors located in the brainstem (ventral surface of the medulla) and periphery (carotid and aortic bodies) sense changes in PCO_2 and $[H^+]$ and alter the ventilatory rate appropriately. Thus when metabolic acidosis occurs, a rise in the $[H^+]$ (decrease in pH) stimulates the ventilatory rate. Conversely, during metabolic alkalosis, a decreased $[H^+]$ (increase in pH) reduces the ventilatory rate. With maximal hyperventilation, the PCO_2 can be reduced to approximately

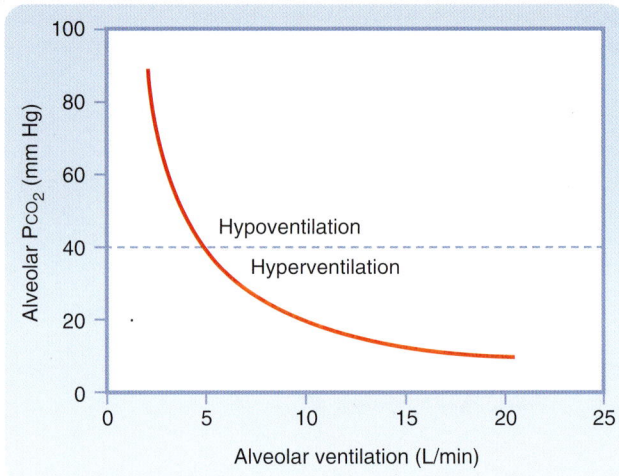

• **Fig. 37.8** Effect of ventilatory rate on alveolar PCO_2 and thus arterial blood PCO_2.

10 mm Hg. Because hypoxia, a potent stimulator of ventilation, also develops with hypoventilation, the degree to which the PCO_2 can be increased is limited. In an otherwise normal individual, hypoventilation cannot raise the PCO_2 above 60 mm Hg. The respiratory response to metabolic acid-base disturbances may be initiated within minutes but may require several hours to complete.

Renal Compensation

The third and final line of defense against acid-base disorders involves the kidneys. In response to an alteration in plasma pH and PCO_2, the kidneys make appropriate adjustments in the excretion of RNAE. The renal response may require several days to reach completion because it takes hours to days to increase the synthesis and activity of the proximal tubule enzymes involved in NH_4^+ production. In the case of acidosis (increased $[H^+]$ or PCO_2), the secretion of H^+ by the nephron is stimulated and the entire filtered load of HCO_3^- is reabsorbed. Titratable acid excretion is increased and production and excretion of NH_4^+ is also stimulated, and thus RNAE is increased (Fig. 37.9). The new HCO_3^- generated during the process of net acid excretion is added to the body, and the plasma $[HCO_3^-]$ increases.

When metabolic alkalosis exists (decreased $[H^+]$), the filtered load of HCO_3^- is increased (plasma $[HCO_3^-]$ is elevated). With respiratory alkalosis (decreased PCO_2), plasma $[HCO_3^-]$ is decreased and thus the filtered load is decreased. In both conditions, secretion of H^+ by the nephron is inhibited. As a result, HCO_3^- excretion is increased. At the same time, excretion of both titratable acid and NH_4^+ is decreased. Thus RNAE is decreased and HCO_3^- appears in urine. Also, some HCO_3^- is secreted into urine by the HCO_3^--secreting intercalated cells of the distal tubule and collecting duct. With enhanced excretion of HCO_3^-, plasma $[HCO_3^-]$ decreases.

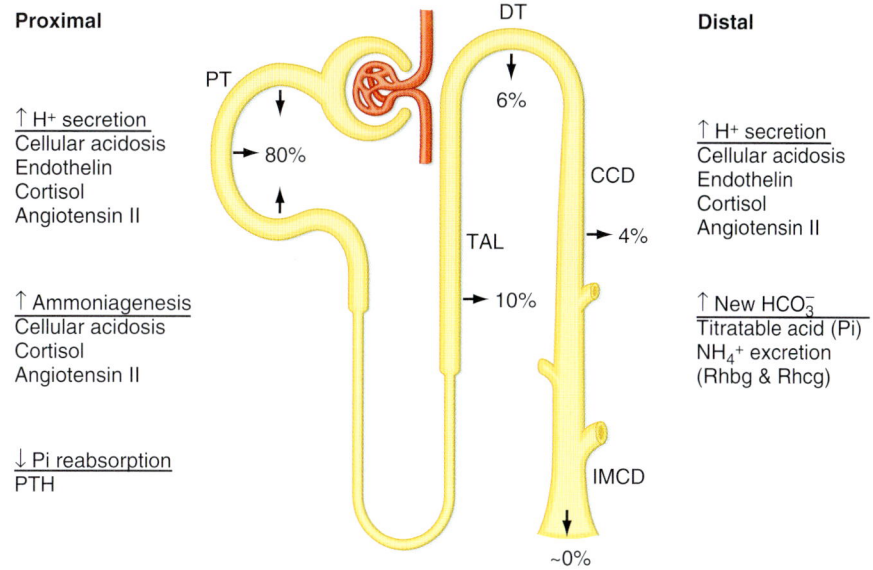

Proximal

PT

↑ H⁺ secretion
Cellular acidosis
Endothelin
Cortisol
Angiotensin II

↑ Ammoniagenesis
Cellular acidosis
Cortisol
Angiotensin II

↓ Pi reabsorption
PTH

80%

DT

6%

TAL

10%

Distal

CCD

4%

↑ H⁺ secretion
Cellular acidosis
Endothelin
Cortisol
Angiotensin II

↑ New HCO₃⁻
Titratable acid (Pi)
NH₄⁺ excretion
(Rhbg & Rhcg)

IMCD

~0%

$$\uparrow\uparrow RNAE = \uparrow (U_{NH_4} + x\ \dot{V}) + \uparrow (U_{TA} * \dot{V}) - \downarrow (U_{HCO_3^-} x\ \dot{V})$$

• **Fig. 37.9** Response of the nephron to acidosis. P_i, phosphate; *PTH*, parathyroid hormone; *Rhbg & Rhcg*, rhesus glycoproteins; *RNAE*, renal net acid excretion; *TA*, titratable acid; $\dot{V}$, urine flow rate.

IN THE CLINIC

Loss of gastric contents from the body (e.g., vomiting, nasogastric suctioning) produces metabolic alkalosis secondary to the loss of HCl. If the loss of gastric fluid is significant, ECFV contraction occurs. Under this condition the kidneys cannot excrete sufficient quantities of HCO_3^- to compensate for the metabolic alkalosis. The inability of the kidneys to excrete HCO_3^- is a result of the need to reduce Na^+ excretion to correct the ECFV contraction. As described previously (see Chapter 35 for details) the response of the kidneys to volume contraction is to reduce the glomerular filtration rate, which reduces the filtered load of HCO_3^-, and to increase Na^+ reabsorption along the nephron. Because a large amount of Na^+ reabsorption occurs via the Na^+/H^+ antiporter, this results in an increase in H^+ secretion (HCO_3^- reabsorption) by the proximal tubule. In this setting the entire filtered load of HCO_3^- is reabsorbed and new HCO_3^- generation may even be enhanced. The latter response occurs because aldosterone, the levels of which are elevated in volume contraction, not only stimulates distal Na^+ reabsorption but also H^+ secretion by intercalated cells. This stimulation of H^+ secretion generates new HCO_3^- by enhancing titratable acid and NH_4^+ excretion. Thus in individuals who lose gastric contents, metabolic alkalosis and paradoxically acidic urine characteristically occur. Correction of the alkalosis occurs only when euvolemia is reestablished.

Simple Acid-Base Disorders

Table 37.1 summarizes the primary alterations and the subsequent compensatory or defense mechanisms of the various simple acid-base disorders. In all acid-base disorders the compensatory response does not correct the underlying

disorder but simply reduces the magnitude of the change in pH. Correction of the acid-base disorder requires treatment of its cause.

Types of Acid-Base Disorders

Metabolic Acidosis

Metabolic acidosis is characterized by a decreased ECF $[HCO_3^-]$ and pH. It can develop via the addition of non-volatile acid to the body (e.g., diabetic ketoacidosis), the loss of nonvolatile base (e.g., HCO_3^- loss caused by diarrhea), or failure of the kidneys to excrete titratable acid and NH_4^+ (e.g., renal failure). As previously described, the buffering of H^+ occurs in both the ECF and ICF compartments. When the pH falls, the respiratory centers are stimulated and the ventilatory rate is increased (respiratory compensation).

Finally, in metabolic acidosis, RNAE is increased. This occurs via elimination of all HCO_3^- from urine (enhanced reabsorption of filtered HCO_3^-) and via increased titratable acid and NH_4^+ excretion (enhanced production of new HCO_3^-). If the process that initiated the acid-base disturbance is corrected, the enhanced RNAE will ultimately return the pH and $[HCO_3^-]$ to normal. After correction of the pH, the ventilatory rate also returns to normal.

Metabolic Alkalosis

Metabolic alkalosis is characterized by an increased ECF $[HCO_3^-]$ and pH. It can occur via the addition of non-volatile base to the body (e.g., ingestion of antacids), as a result of volume contraction (e.g., hemorrhage), or more commonly from loss of nonvolatile acid (e.g., loss of gastric HCl because of prolonged vomiting). Buffering occurs predominantly in the ECF and to a lesser degree in the

TABLE 37.1	Characteristics of Simple Acid-Base Disorders			
Disorder	**Plasma pH**	**Primary Alteration**	**Defense Mechanisms**	
Metabolic acidosis	↓	↓ECF [HCO_3^-]	ICF and ECF buffers Hyperventilation (↓P_{CO_2}) ↑RNAE	
Metabolic alkalosis	↑	↑ECF [HCO_3^-]	ICF and ECF buffers Hypoventilation (↑P_{CO_2}) ↓RNAE	
Respiratory acidosis	↓	↑P_{CO_2}	ICF buffers ↑RNAE	
Respiratory alkalosis	↑	↓P_{CO_2}	ICF buffers ↓RNAE	

ECF, extracellular fluid; ICF, intracellular fluid; RNAE, renal net acid excretion

IN THE CLINIC

When nonvolatile acid is added to body fluids, as in **diabetic ketoacidosis**, the [H^+] increases (pH decreases), and the [HCO_3^-] decreases. In addition the concentration of the anion associated with the nonvolatile acid increases. This change in anion concentration provides a convenient way of analyzing the cause of a metabolic acidosis by calculating what is termed the **anion gap.** The anion gap represents the difference between the concentration of the major ECF cation (Na^+) and the major ECF anions (Cl^- and HCO_3^-):

$$\text{Anion gap} = [Na^+] - ([Cl^-] + [HCO_3^-])$$

Under normal conditions the anion gap ranges from 8 to 16 mEq/L. It is important to recognize that an anion gap does not actually exist. All cations are balanced by anions. The gap simply reflects the parameters that are measured. In reality:

$$[Na^+] + [\text{Unmeasured cations}]$$
$$= [Cl^-] + [HCO_3^-] + [\text{Unmeasured anions}]$$

If the anion of the nonvolatile acid is Cl^-, the anion gap will be normal (i.e., the decrease in [HCO_3^-] is matched by an increase in [Cl^-]). Metabolic acidosis associated with diarrhea or renal tubular acidosis (i.e., defect in renal H^+ secretion) has a normal anion gap. In contrast, if the anion of the nonvolatile acid is not Cl^- (e.g., lactate, β-hydroxybutyrate), the anion gap will increase (i.e., the decrease in [HCO_3^-] is not matched by an increase in [Cl^-] but rather by an increase in the concentration of the unmeasured anion). The anion gap is increased in metabolic acidosis–associated ketoacidosis (e.g., diabetes mellitus) with renal failure, lactic acidosis, or ingestion of toxins or certain drugs (e.g., large quantities of aspirin). Thus calculation of the anion gap is a useful way of identifying the etiology of metabolic acidosis in the clinical setting.

Albumin is a negatively charged macromolecule, and it makes a considerable contribution to "unmeasured anions." As a result the anion gap must be adjusted in patients who have an abnormal serum [albumin]. For each 1 g/dL change in serum [albumin], the anion gap needs to be adjusted in the same direction by 2.5 mEq/L.

ICF. The increase in the pH inhibits the respiratory centers, which reduces the ventilatory rate, thus the P_{CO_2} is elevated (respiratory compensation).

The renal compensatory response to metabolic alkalosis is to increase excretion of HCO_3^- by reducing its reabsorption along the nephron. Normally this occurs quite rapidly (minutes to hours) and effectively. Enhanced renal excretion of HCO_3^- eventually returns the pH and [HCO_3^-] to normal, provided the underlying cause of the initial acid-base disturbance is corrected. When the pH is corrected, the ventilatory rate also returns to normal.

Respiratory Acidosis

Respiratory acidosis is characterized by an elevated P_{CO_2} and reduced ECF pH. It results from decreased gas exchange across the alveoli as a result of either inadequate ventilation (e.g., drug-induced depression of the respiratory centers) or impaired gas diffusion (e.g., pulmonary edema, such as occurs in cardiovascular and lung disease). In contrast to the metabolic disorders, buffering during respiratory acidosis occurs almost entirely in the ICF compartment. The increase in P_{CO_2} and the decrease in pH stimulate both HCO_3^- reabsorption by the nephron and titratable acid and NH_4^+ excretion (renal compensation). Together these responses increase RNAE and generate new HCO_3^-. The renal compensatory response takes several days to occur. Consequently, respiratory acid-base disorders are commonly divided into acute and chronic phases. In the acute phase the time for the renal compensatory response is not sufficient, and the body relies on ICF buffering to minimize the change in pH. In the chronic phase, renal compensation occurs. Correction of the underlying disorder returns the P_{CO_2} to normal and RNAE decreases to its initial level.

Respiratory Alkalosis

Respiratory alkalosis is characterized by a reduced P_{CO_2} and an increased ECF pH. It results from increased gas exchange in the lungs, usually caused by increased ventilation from stimulation of the respiratory centers (e.g., via drugs or disorders of the central nervous system). Hyperventilation also occurs at high altitude and as a result of anxiety, pain, or fear. Buffering is primarily in the ICF compartment. As with respiratory acidosis, respiratory alkalosis has both acute and chronic phases reflecting the time required for renal compensation to occur. The acute phase of respiratory alkalosis reflects intracellular buffering, whereas the chronic phase reflects renal compensation. With renal compensation, the elevated pH and reduced PCO_2 inhibit HCO_3^- reabsorption by the nephron and reduce titratable acid and NH_4^+ excretion. As a result of these effects, RNAE is reduced. Correction of the underlying disorder returns the PCO_2 to normal, and renal excretion of acid then increases to its initial level.

Analysis of Acid-Base Disorders

Analysis of an acid-base disorder is directed at identifying the underlying cause so appropriate therapy can be initiated. The patient's medical history and associated physical findings often provide valuable clues about the nature and origin of an acid-base disorder. In addition, analysis of an arterial blood sample is frequently required. Such an analysis is straightforward if approached systematically. For example, consider the following data:

$$pH = 7.35$$
$$[HCO_3^-] = 16 \text{ mEq/L}$$
$$PCO_2 = 30 \text{ mmHg}$$

The acid-base disorder represented by these values or any other set of values can be determined using the following three-step approach:

1. *Examination of the pH.* When pH is considered first, the underlying disorder can be classified as either an acidosis or an alkalosis. The defense mechanisms of the body cannot correct an acid-base disorder by themselves. Thus even if the defense mechanisms are completely operative, the change in pH indicates the acid-base disorder. In the example provided, the pH of 7.35 indicates acidosis.
2. *Determination of metabolic versus respiratory disorder.* Simple acid-base disorders are either metabolic or

respiratory. To determine which disorder is present the clinician must next examine the ECF $[HCO_3^-]$ and PCO_2. As previously discussed, acidosis could be the result of a decrease in $[HCO_3^-]$ (metabolic) or an increase in PCO_2 (respiratory). Alternatively, alkalosis could be the result of an increase in ECF $[HCO_3^-]$ (metabolic) or a decrease in PCO_2 (respiratory). For the example provided, the ECF $[HCO_3^-]$ is reduced from normal (normal = 24 mEq/L), as is the PCO_2 (normal = 40 mm Hg). The disorder must therefore be metabolic acidosis; it cannot be a respiratory acidosis because the PCO_2 is reduced.

3. *Analysis of a compensatory response.* Metabolic disorders result in compensatory changes in ventilation and thus in PCO_2, whereas respiratory disorders result in compensatory changes in RNAE and thus in ECF $[HCO_3^-]$ (Table 37.2). In an appropriately compensated metabolic acidosis, the PCO_2 is decreased, whereas it is elevated in compensated metabolic alkalosis. With respiratory acidosis, compensation results in an elevation of the $[HCO_3^-]$. Conversely, ECF $[HCO_3^-]$ is reduced in response to respiratory alkalosis. In this example, the PCO_2 is reduced from normal, and the magnitude of this reduction (10 mm Hg decrease in PCO_2 for an 8 mEq/L decrease in ECF $[HCO_3^-]$) is as expected (see Table 37.2). Therefore the acid-base disorder is a simple metabolic acidosis with appropriate respiratory compensation.

A **mixed acid-base disorder** reflects the presence of two or more underlying causes for the acid-base disturbance. For example, consider the following data:

$$pH = 6.96$$
$$[HCO_3^-] = 12 \text{ mEq/L}$$
$$PCO_2 = 55 \text{ mm Hg}$$

When the three-step approach is followed, it is evident that the disturbance is an acidosis that has both a metabolic component (ECF $[HCO_3^-] < 24$ mEq/L) and a respiratory component ($PCO_2 > 40$ mm Hg). Thus this disorder is mixed. Mixed acid-base disorders can occur, for example, in an individual who has a history of a chronic pulmonary disease such as emphysema (i.e., chronic respiratory

TABLE 37.2	**Simple Acid-Base Disorders With Compensation**			
Primary Disturbance	pH	HCO_3^-	PCO_2	Compensation
Metabolic acidosis	<7.4	Primary ↓	Compensatory ↓	1–2 mm Hg ↓PCO_2 for every 1 mEq/L ↓ in $[HCO_3^-]$ *or* $PCO_2 = (1.5 \times [HCO_3^-]) + 8 \pm 2$ *or* $PCO_2 = [HCO_3^-] + 15$
Metabolic alkalosis	>7.4	Primary ↑	Compensatory ↑	0.6–0.75 mm Hg ↑PCO_2 for every 1 mEq/L ↑$[HCO_3^-]$
Respiratory acidosis	<7.4	Compensatory ↑	Primary ↑	*Acute*: 1–2 mEq/L ↑$[HCO_3^-]$ for every 10 mm Hg ↑PCO_2 *Chronic*: 3–4 mEq/L ↑$[HCO_3^-]$ for every 10 mm Hg ↑PCO_2
Respiratory alkalosis	>7.4	Compensatory ↓	Primary ↓	*Acute*: 1–2 mEq/L ↓$[HCO_3^-]$ for every 10 mm Hg ↓PCO_2 *Chronic*: 3–4 mEq/L ↓$[HCO_3^-]$ for every 10 mm Hg ↓PCO_2

acidosis) and who develops an acute gastrointestinal illness with diarrhea. Because diarrhea fluid contains HCO_3^-, its loss from the body results in the development of metabolic acidosis.

A mixed acid-base disorder is also indicated when a patient has abnormal PCO_2 and ECF [HCO_3^-] values but the pH is normal. Such a condition can develop in a patient who has ingested a large quantity of aspirin. Salicylic acid (active ingredient in aspirin) produces metabolic acidosis and at the same time stimulates the respiratory centers, causing hyperventilation and respiratory alkalosis. Thus the patient has a reduced ECF [HCO_3^-] and a reduced PCO_2. (NOTE: The PCO_2 is lower than would occur with normal respiratory compensation of a metabolic acidosis.)

Key Concepts

1. The kidneys maintain acid-base balance through excretion of an amount of acid equal to the amount of nonvolatile acid produced by metabolism and the quantity ingested in the diet (termed *renal net acid excretion* [RNAE]). The kidneys also prevent loss of HCO_3^- in urine by reabsorbing virtually all the HCO_3^- filtered at the glomeruli. Both reabsorption of filtered HCO_3^- and excretion of acid are accomplished by secretion of H^+ by the nephrons. Acid is excreted by the kidneys in the form of titratable acid (primarily as P_i) and NH_4^+. Both titratable acid and NH_4^+ excretion result in generation of new HCO_3^-, which replenishes the ECF HCO_3^- lost during neutralization of nonvolatile acids.

2. The body uses three lines of defense to minimize the impact of acid-base disorders on body fluid pH: (1) ECF and ICF buffering, (2) respiratory compensation, and (3) renal compensation.

3. Metabolic acid-base disorders result from primary alterations in ECF [HCO_3^-], which in turn results from addition of acid to or loss of alkali from the body. In response to metabolic acidosis, pulmonary ventilation is increased, which decreases PCO_2. The pulmonary response to metabolic acid-base disorders occurs in a matter of minutes. RNAE is also increased, but this takes several days. An increase in ECF [HCO_3^-] causes alkalosis. This rapidly (minutes to hours) decreases pulmonary ventilation, which elevates PCO_2 as a compensatory response. RNAE is also decreased, but this takes several days.

4. Respiratory acid-base disorders result from primary alterations in PCO_2. Elevation of PCO_2 produces acidosis, and the kidneys respond with an increase in RNAE. Conversely, a reduction of PCO_2 produces alkalosis, and RNAE is reduced. The kidneys respond to respiratory acid-base disorders over several hours to days.

Additional Reading

Book Chapters

Curthoys NP. Renal ammonium ion production and excretion. In: Alpern RJ, et al., eds. *Seldin and Giebisch's The Kidney: Physiology and Pathophysiology*. 5th ed. Waltham, MA: Elsevier; 2013.

Dubose TD Jr. Disorders of acid-base balance. In: Taal MW, et al., eds. *Brenner and Rector's The Kidney*. 9th ed. Philadelphia: Saunders; 2012.

Hamm LL, et al. Cellular mechanisms of renal tubule acidification. In: Alpern RJ, et al., eds. *Seldin and Giebisch's The Kidney: Physiology and Pathophysiology*. 5th ed. Waltham, MA: Elsevier; 2013.

Weiner ID, Verlander JW. Renal acidification mechanisms. In: Taal MW, et al., eds. *Brenner and Rector's The Kidney*. 9th ed. Philadelphia: Saunders; 2012.

Journal Articles

Berend K, et al. Physiological approach to assessment of acid-base disturbances. *N Engl J Med*. 2014;371:1434-1445.

Breton S, Brown D. Regulation of liminal acidification by the V-ATPase. *Physiology (Bethesda)*. 2013;28:318-329.

Brown D, Wagner CA. Molecular mechanisms of acid-base sensing in the kidney. *J Am Soc Nephrol*. 2012;23:774-780.

Curthoys NP, Moe OW. Proximal tubule function and response to acidosis. *Clin J Am Soc Nephrol*. 2014;9:1627-1638.

Hamm LL, et al. Acid-base homeostasis. *Clin J Am Soc Nephrol*. 2015;10:2232-2242.

Levin LR, Buck J. Physiological roles of acid-base sensors. *Annu Rev Physiol*. 2015;77:347-362.

Roy A, et al. Collecting duct intercalated cell function and regulation. *Clin J Am Soc Nephrol*. 2015;10:305-324.

Seifter JL. Integration of acid-base and electrolyte disorders. *N Engl J Med*. 2014;371:1821-1831.

Weiner ID, Verlander JW. Ammonia transport in the kidney by rhesus glycoproteins. *Am J Physiol Renal Physiol*. 2014;306:F1107-F1120.

The Endocrine and Reproductive Systems

BRUCE A. WHITE AND JOHN R. HARRISON

38

Introduction to the Endocrine System

Upon completion of this chapter, the student should be able to answer the following questions:

1. Name the major endocrine glands and their hormonal product or products.
2. Map out and differentiate a simple endocrine negative feedback loop and one involving the hypothalamus, anterior pituitary and peripheral endocrine gland, and list the major endocrine glands under each type of feedback loop.
3. Define a releasing hormone and a tropic hormone.
4. Explain the chemical nature and the characteristics of protein/peptide hormones, catecholamine hormones, steroid hormones, and iodothyronines (thyroid hormones). Include such characteristics as site of regulation (synthesis or secretion), circulating form of hormone, subcellular localization of hormone receptor, and metabolic clearance.
5. Integrate the concept of peripheral conversion with the function/action of a secreted hormone.
6. Integrate the intracellular steps associated with a hormone response in a target cell.

The ability of cells to communicate with each other is an underpinning of human biology. As discussed in Chapter 3, cell-to-cell communication exists at various levels of complexity and distance. **Endocrine signaling** involves (1) the **regulated secretion** of an extracellular signaling molecule, called a **hormone,** into the extracellular fluid; (2) diffusion of the hormone into the **vasculature** and its circulation throughout the body; and (3) diffusion of the hormone out of the vascular compartment into the extracellular space and binding to a **specific receptor** within cells of a **target organ.** Because of the spread of hormones throughout the body, one hormone often regulates the activity of several target organs. Conversely, cells frequently express receptors for multiple hormones.

The **endocrine system** is a collection of glands whose function is to regulate multiple organs within the body to (1) meet the growth and reproductive needs of the organism and (2) respond to fluctuations within the internal environment, including various types of stress. The endocrine system comprises the following major glands (Fig. 38.1):

Endocrine tissues of the pancreas
Parathyroid glands

Pituitary gland (in association with hypothalamic nuclei)
Thyroid gland
Adrenal glands
Gonads (testes or ovaries)

These endocrine glands synthesize and secrete bioactive hormones and, with the exception of gonads, which perform both endocrine and gametogenic functions, are dedicated to hormone production (Table 38.1). A transitory organ, the **placenta,** also performs a major endocrine function.

In addition to dedicated endocrine glands, there are endocrine cells within organs whose primary function is not endocrine (see Table 38.1). These include cells within the heart that produce **atrial natriuretic peptide,** liver cells that produce **insulin-like growth factor type 1 (IGF-1),** cells within the kidney that produce **erythropoietin,** and numerous cell types within the gastrointestinal tract that produce gastrointestinal hormones. There also exist collections of cell bodies (called *nuclei*) within the hypothalamus that secrete peptides, called *neurohormones,* into capillaries associated with the pituitary gland.

A third subset of the endocrine system is represented by numerous cell types that express intracellular enzymes, ectoenzymes, or secreted enzymes that modify inactive precursors or less active hormones into highly active hormones (see Table 38.1). An example is the generation of **angiotensin II** from the inactive polypeptide angiotensinogen by two subsequent proteolytic cleavages (see Chapter 43). Another example is activation of **vitamin D** by two subsequent hydroxylation reactions in the liver and kidneys to produce the highly bioactive hormone 1,25-dihydroxyvitamin D (vitamin D).

Configuration of Feedback Loops Within the Endocrine System

The predominant mode of a closed feedback loop among endocrine glands is **negative feedback.** In a negative feedback loop, a hormone acts on one or more target organs to induce a change (either a decrease or increase) in circulating levels of a specific component, and the change in this component then inhibits secretion of the hormone. Negative feedback loops confer stability by keeping a physiological parameter (e.g., blood glucose level) within a normal range. There are also a few examples of **positive feedback**

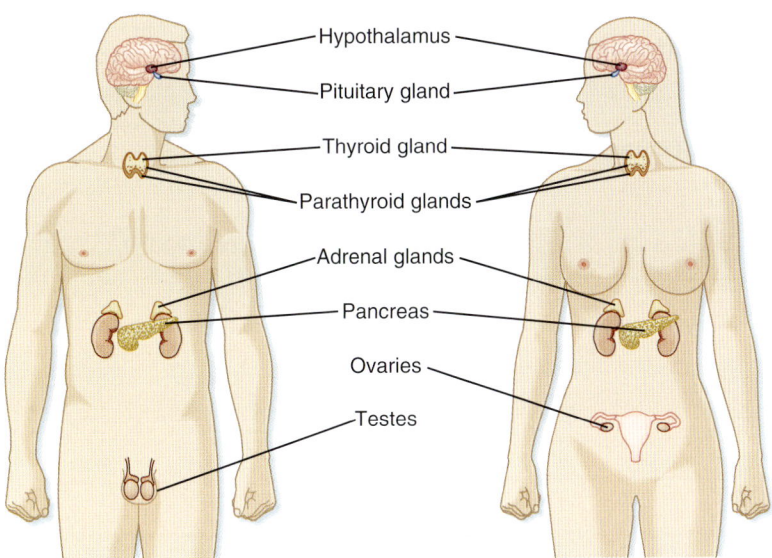

• **Fig. 38.1** Glands of the endocrine system.

in endocrine regulation. A closed positive feedback loop, in which a hormone increases levels of a specific component and this component stimulates secretion of the hormone, confers instability. Under the control of positive feedback loops, something has got to give; for example, positive feedback loops control processes that lead to rupture of a follicle through the ovarian wall or expulsion of a fetus from the uterus.

There are two basic configurations of negative feedback loops within the endocrine system: a **physiological response–driven** feedback loop (referred to simply as a *response-driven feedback loop*) and an **endocrine axis–driven** feedback loop (Fig. 38.2). The response-driven feedback loop is observed in endocrine glands that control blood glucose levels (pancreatic islet cells), blood Ca^{++} and P_i levels (parathyroid glands, kidneys), blood osmolarity and volume (hypothalamus/posterior pituitary gland), and blood Na^+, K^+, and H^+ levels (zona glomerulosa of the adrenal cortex and atrial cells). In the response-driven configuration, secretion of a hormone is stimulated or inhibited by a change in the level of a specific extracellular parameter (e.g., an increase in blood glucose level stimulates insulin secretion). Alterations in hormone levels lead to changes in the physiological characteristics of target organs (e.g., decreased hepatic gluconeogenesis, increased uptake of glucose by muscle) that directly regulate the parameter (in this case, blood glucose level) in question. The change in the parameter (decreased blood glucose level) then inhibits further secretion of the hormone (i.e., insulin secretion drops as blood glucose level falls).

Much of the endocrine system is organized into **endocrine axes;** each axis consists of the hypothalamus, the pituitary gland, and the peripheral endocrine glands (see Fig. 38.2). Thus the endocrine axis–driven feedback loop involves a three-tiered configuration. The first tier is represented by **hypothalamic neuroendocrine neurons** that secrete **releasing hormones.** Releasing hormones stimulate

(or, in a few cases, inhibit) the production and secretion of **tropic hormones** from the **pituitary gland** (second tier). Tropic hormones stimulate the production and secretion of hormones from **peripheral endocrine glands** (third tier). The peripherally produced hormones—namely, thyroid hormone, cortisol, sex steroids, and IGF-1—typically have **pleiotropic** actions (e.g., multiple phenotypic effects) on numerous cell types. However, in endocrine axis–driven feedback, the primary feedback loop involves feedback inhibition of pituitary tropic hormones and hypothalamic releasing hormones by the peripherally produced hormone. In contrast to response-driven feedback, the physiological responses to the peripherally produced hormone play only a minor role in regulation of feedback within endocrine axis–driven feedback loops. From a clinical perspective, endocrine diseases are described as **primary, secondary,** or **tertiary diseases** (e.g., secondary hyperthyroidism, tertiary hypogonadism). **Primary** disease is a lesion in the **peripheral endocrine gland; secondary** disease is a lesion in the **anterior pituitary gland;** and **tertiary** disease is a lesion in the **hypothalamus.**

An important aspect of the endocrine axes is the ability of descending and ascending neuronal signals to modulate release of the hypothalamic releasing hormones and thereby control the activity of the axis. A major neuronal input to releasing hormone–secreting neurons comes from another region of the hypothalamus called the **suprachiasmatic nucleus (SCN).** SCN neurons impose a daily rhythm, called a **circadian rhythm,** on the secretion of hypothalamic releasing hormones and the endocrine axes that they control (Fig. 38.3). SCN neurons represent an intrinsic circadian clock, as evidenced by the fact that they demonstrate a spontaneous peak of electrical activity at the same time every 24 to 25 hours. The 24- to 25-hour cycle can be **"entrained"** by the normal environmental light-dark cycle created by the earth's rotation, so that the periodicity of the clock appears to be environmentally controlled (Fig. 38.4).

TABLE 38.1	Hormones and Their Sites of Production in Nonpregnant Adults
Gland	**Hormone**
Hormones Synthesized and Secreted by Dedicated Endocrine Glands	
Pituitary gland	Growth hormone (GH)
	Prolactin
	Adrenocorticotropic hormone (ACTH)
	Thyroid-stimulating hormone (TSH)
	Follicle-stimulating hormone (FSH)
	Luteinizing hormone (LH)
Thyroid gland	Thyroxine
	Triiodothyronine
	Calcitonin
Parathyroid glands	Parathyroid hormone (PTH)
Islets of Langerhans (endocrine tissues of the pancreas)	Insulin
	Glucagon
	Somatostatin
Adrenal gland	Epinephrine
	Norepinephrine
	Cortisol
	Aldosterone
	Dehydroepiandrosterone sulfate (DHEAS)
Ovaries	Estradiol-17β
	Progesterone
	Inhibin
Testes	Testosterone
	Antimüllerian hormone (AMH)
	Inhibin
Hormones Synthesized in Organs With a Primary Function Other Than Endocrine	
Brain (hypothalamus)	Antidiuretic hormone (ADH; vasopressin)
	Oxytocin
	Corticotropin-releasing hormone (CRH)
	Thyrotropin-releasing hormone (TRH)
	Gonadotropin-releasing hormone (GnRH)
	Growth hormone–releasing hormone (GHRH)
	Somatostatin
	Dopamine
Brain (pineal gland)	Melatonin
Heart	Atrial natriuretic peptide (ANP)
Kidneys	Erythropoietin
Adipose tissue	Leptin
	Adiponectin
Stomach	Gastrin
	Somatostatin
	Ghrelin
Intestines	Secretin
	Cholecystokinin
	Glucagon-like peptide-1 (GLP-1)
	Glucagon-like peptide-2 (GLP-2)
	Glucose-dependent insulinotropic peptide (gastrin inhibitory peptide [GIP])
	Motilin
Liver	Insulin-like growth factor type 1 (IGF-1)
Hormones Produced to a Significant Degree by Peripheral Conversion	
Lungs	Angiotensin II
Kidney	1,25-Dihydroxyvitamin D (vitamin D)
Adipose, mammary glands, other organs	Estradiol-17β
Liver, sebaceous gland, other organs	Testosterone
Genital skin, prostate, other organs	5-Dihydrotestosterone (DHT)
Many organs	Triiodothyronine

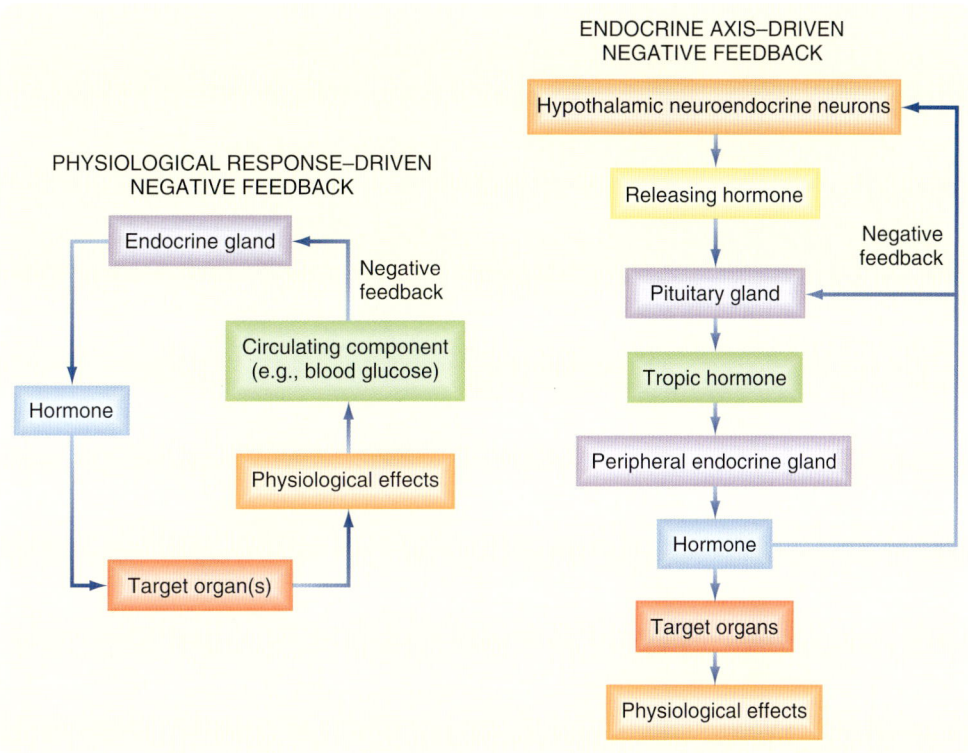

ENDOCRINE AXIS–DRIVEN
NEGATIVE FEEDBACK

Hypothalamic neuroendocrine neurons

Releasing hormone

Negative feedback

Pituitary gland

Tropic hormone

Peripheral endocrine gland

Hormone

Target organs

Physiological effects

PHYSIOLOGICAL RESPONSE–DRIVEN
NEGATIVE FEEDBACK

Endocrine gland

Negative feedback

Circulating component
(e.g., blood glucose)

Hormone

Physiological effects

Target organ(s)

• **Fig. 38.2** Physiological response–driven and endocrine axis–driven negative feedback loops.

Neural input is generated from specialized light-sensitive retinal cells that are distinct from rods and cones and from signals to the SCN via the retinohypothalamic tract. Under constant conditions of light or dark, however, the SCN clock becomes "free running" and slightly drifts away from a 24-hour cycle each day.

The **pineal gland** forms a neuroendocrine link between the SCN and various physiological processes that require circadian control. This tiny gland, close to the hypothalamus, synthesizes the hormone **melatonin** from the neurotransmitter **serotonin,** of which tryptophan is the precursor. The rate-limiting enzyme for melatonin synthesis is N-acetyltransferase. The amount and activity of this enzyme in the pineal gland vary markedly in a cyclic manner, which accounts for the cycling of melatonin secretion and its plasma levels. Synthesis of melatonin is inhibited by light and markedly stimulated by darkness (see Fig. 38.4). Thus melatonin may transmit the information that nighttime has arrived, and body functions are regulated accordingly. Melatonin feedback to the SCN at dawn or dusk may also help evoke day-night entrainment of the SCN 24- to 25-hour clock. Melatonin has numerous other actions, including induction of sleep.

Another important input to hypothalamic neurons and the pituitary gland is stress, either as **systemic stress** (e.g., hemorrhage, inflammation) or as **processive stress** (e.g., fear, anxiety). Major medical or surgical stress overrides the circadian clock and causes a pattern of persistent and exaggerated hormone release and metabolism that mobilizes endogenous fuels, such as glucose and free fatty acids, and augments their delivery to critical organs. Growth

and reproductive processes, in contrast, are suppressed. In addition, cytokines released during inflammatory or immune responses, or both, directly regulate the release of hypothalamic releasing hormones and pituitary hormones.

Chemical Nature of Hormones

Hormones are classified biochemically as **proteins/peptides, catecholamines, steroid hormones,** or **iodothyronines.** The chemical nature of a hormone determines (1) how it is synthesized, stored, and released; (2) how it is transported in blood; (3) its biological half-life and mode of clearance; and (4) its cellular mechanism of action.

Proteins/Peptides

Protein and peptide hormones can be grouped into structurally related molecules that are encoded by gene families. Protein/peptide hormones obtain their specificity from their primary amino acid sequence and from posttranslational modifications, especially glycosylation.

Because protein/peptide hormones are destined for secretion outside the cell, their synthesis and processing are differently from those of proteins destined to remain within the cell or to be continuously added to the membrane (Fig. 38.5). These hormones are synthesized on the polyribosome as larger preprohormones or prehormones. The nascent peptides have at their N-terminus a group of 15 to 30 amino acids called the **signal peptide.** The signal peptide interacts with a ribonucleoprotein particle, which ultimately directs the growing peptide chain through a pore

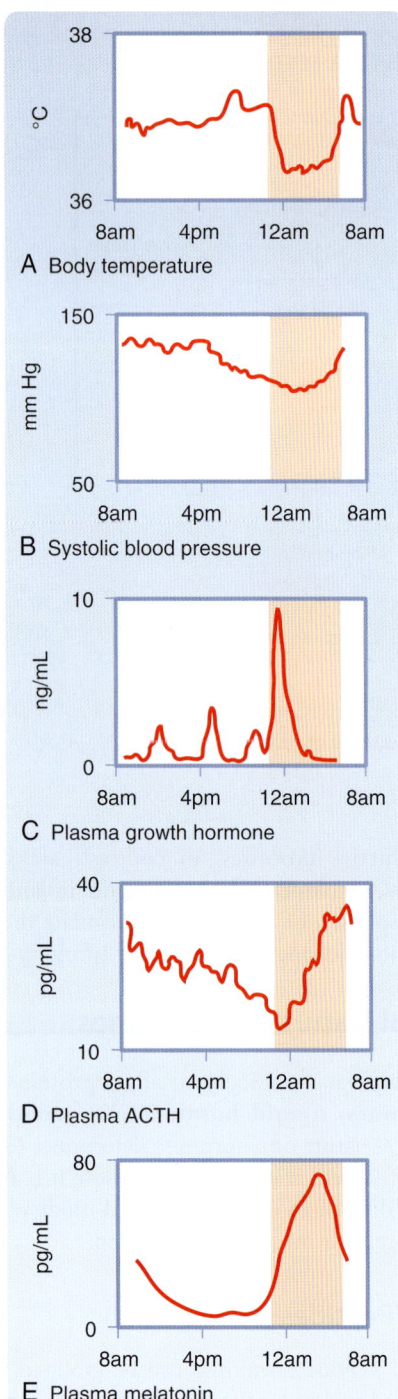

A Body temperature

B Systolic blood pressure

C Plasma growth hormone

D Plasma ACTH

E Plasma melatonin

• **Fig. 38.3** A circadian pacemaker directs numerous endocrine and body functions, each with its own daily schedule. The nighttime rise in plasma melatonin may mediate certain other circadian patterns. ACTH, adrenocorticotropic hormone. (Data from Schwartz WJ. *Adv Intern Med.* 1994;38:81.)

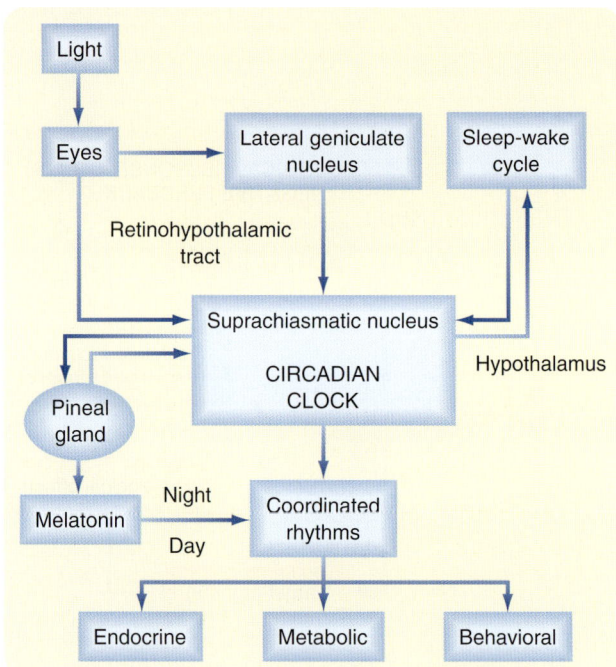

• **Fig. 38.4** Origin of circadian rhythms in endocrine gland secretion, metabolic processes, and behavioral activity. (Modified from Turok FW. *Recent Prog Horm Res.* 1994;49:43.)

a membrane-bound secretory vesicle that is subsequently released into the cytoplasm. The carbohydrate moiety of glycoproteins is added in the Golgi apparatus.

Most hormones are produced as **prohormones.** Prohormones harbor the peptide sequence of the active hormone within their primary sequence. However, prohormones are inactive or less active and require the action of endopeptidases to trim away the neighboring inactive sequences.

Protein/peptide hormones are stored in the gland as membrane-bound secretory vesicles and are released by **exocytosis** through the **regulated secretory pathway.** Thus these hormones are not continually secreted. Rather, they are secreted in response to a stimulus through a mechanism of **stimulus-secretion coupling.** Regulated exocytosis requires energy, Ca^{++}, an intact cytoskeleton (microtubules, microfilaments), and the presence of coat proteins that specifically deliver secretory vesicles to the cell membrane. The ultrastructure of protein hormone–producing cells is characterized by abundant rough endoplasmic reticulum and Golgi membranes and the presence of secretory vesicles (Fig. 38.6).

Protein/peptide hormones are soluble in body fluids and, with the notable exceptions of IGFs and growth hormone, circulate in blood predominantly in an unbound form and therefore have short biological half-lives. Protein hormones are removed from blood primarily by endocytosis and lysosomal degradation of hormone-receptor complexes (see the section "Cellular Responses to Hormones"). Many protein hormones are small enough to appear in urine in a physiologically active form. For example, follicle-stimulating hormone and luteinizing hormone are present in urine.

in the membrane of the endoplasmic reticulum located on the cisternal (i.e., inner) surface of the endoplasmic reticular membrane. Removal of the signal peptide by a **signal peptidase** generates a hormone or prohormone, which is then transported from the cisternae of the endoplasmic reticulum to the Golgi apparatus, where it is packaged into

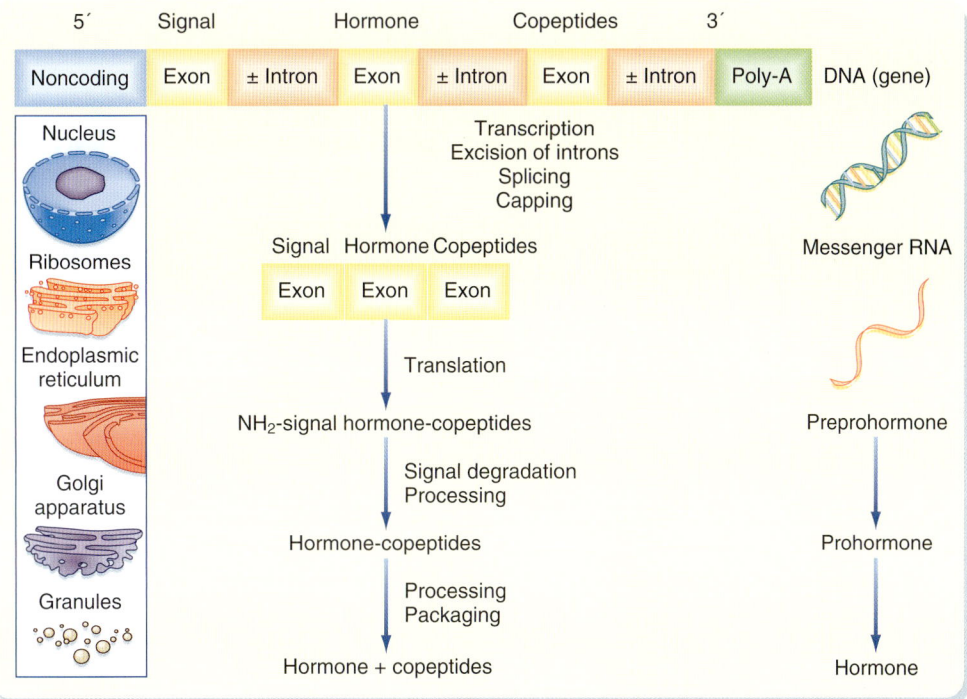

• **Fig. 38.5 Schematic Representation of Peptide Hormone Synthesis.** In the nucleus, the primary gene transcript, a messenger RNA precursor molecule, undergoes excision of introns, splicing of exons, capping of the 5′ end, and addition of polyadenylation (poly-A) at the 3′ end. The resultant mature messenger RNA enters the cytoplasm, where it directs the synthesis of a preprohormone peptide sequence on ribosomes. In this process, the N-terminus signal is removed, and the resultant prohormone is transferred vectorially into the endoplasmic reticulum. The prohormone undergoes further processing and packaging in the Golgi apparatus. After final cleavage of the prohormone within the granules, they contain the hormone and copeptides ready for secretion by exocytosis. NH_2, amidogen.

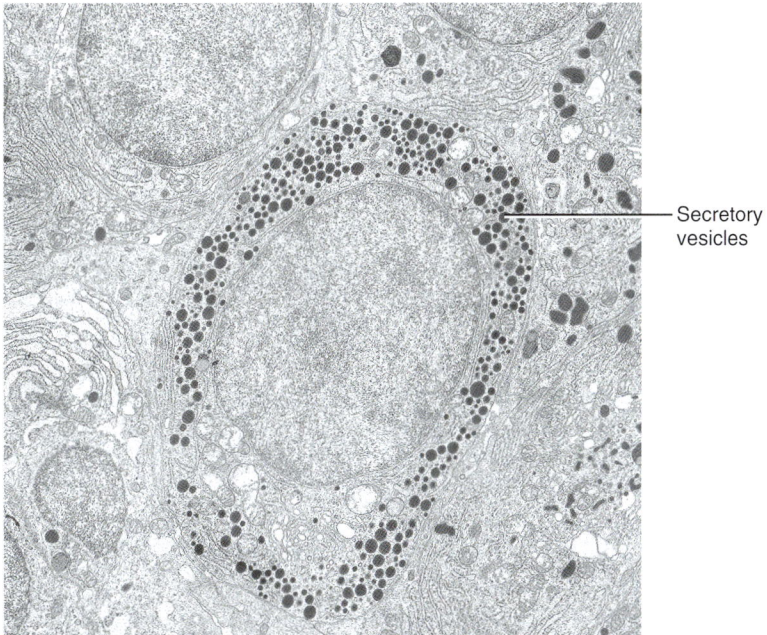

• **Fig. 38.6 Ultrastructure of a Protein Hormone–Producing Cell.** Note the presence of secretory vesicles and rough endoplasmic reticulum in the protein hormone–secreting cell. (From Kierszenbaum AL. *Histology and Cell Biology: An Introduction to Pathology.* 2nd ed. Philadelphia: Mosby; 2007.)

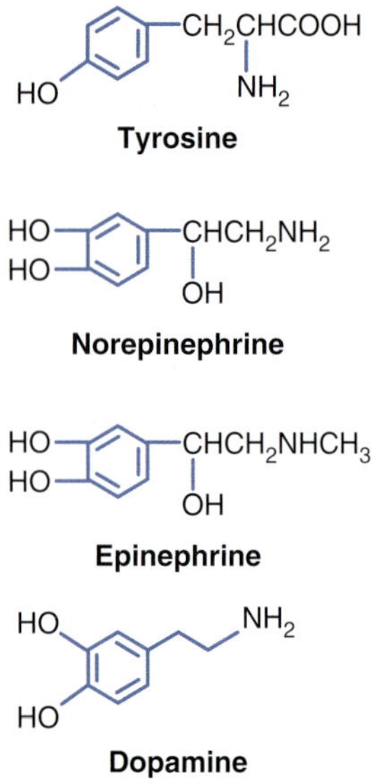

Tyrosine

Norepinephrine

Epinephrine

Dopamine

• **Fig. 38.7** Chemical structures of catecholamines.

Proteins/peptides are readily digested in the gastro-intestinal tract if administered orally. Hence they must be administered by injection or, in the case of small peptides, through a mucous membrane (sublingually or intranasally). Because proteins/peptides do not cross cell membranes readily, they signal through membrane receptors (see Chapter 3).

Catecholamines

Catecholamines are synthesized by the adrenal medulla and neurons and include **norepinephrine, epinephrine,** and **dopamine** (Fig. 38.7). The primary hormonal products of the adrenal medulla are epinephrine and, to a lesser extent, norepinephrine. Catecholamines obtain their specificity through enzymatic modifications of the amino acid tyrosine. Catecholamines are stored in secretory vesicles that are part of the regulated secretory pathway. They are copackaged with adenosine triphosphate, Ca^{++}, and proteins called **chromogranins.** Chromogranins play a role in the biogenesis of secretory vesicles and in the organization of components within the vesicles. Catecholamines are soluble in blood and circulate either unbound or loosely bound to albumin. They are similar to protein/peptide hormones in that they do not cross cell membranes readily and hence produce their actions through cell membrane receptors. Catecholamines have short biological half-lives (1 to 2 minutes) and are removed from blood primarily by cell uptake and enzymatic modification.

Steroid Hormones

Steroid hormones are made by the **adrenal cortex, ovaries, testes,** and **placenta.** Steroid hormones from these glands belong to five categories: **progestins, mineralocorticoids, glucocorticoids, androgens,** and **estrogens.** Progestins,

mineralocoricoids, and glucocorticoids are 21-carbon steroids, whereas androgens are 19-carbon steroids and estrogens are 18-carbon steroids (Table 38.2). Steroid hormones also include the active metabolite of **vitamin D** (see Chapter 40), which is a secosteroid (i.e., one of the rings has an open conformation).

TABLE 38.2	**Steroid Hormones**			
Family	**Number of Carbons**	**Specific Hormone**	**Primary Site of Synthesis**	**Primary Receptor**
Progestin	21	Progesterone	Ovary Placenta	Progesterone receptor
Glucocorticoid	21	Cortisol Corticosterone	Adrenal cortex	Glucocorticoid receptor
Mineralocorticoid	21	Aldosterone 11-Deoxycorticosterone	Adrenal cortex	Mineralocorticoid receptor
Androgen	19	Testosterone Dihydrotestosterone	Testis	Androgen receptor
Estrogen	18	Estradiol-17β Estriol	Ovary Placenta	Estrogen receptor

Steroid hormones are synthesized by a series of enzymatic modifications of cholesterol, and a cyclopentanoperhydrophenanthrene ring (or a derivative thereof) serves as their core (Fig. 38.8). The enzymatic modifications of cholesterol are of three general types: hydroxylation, dehydrogenation/reduction, and lyase reactions. The purpose of these modifications is to produce a cholesterol derivative that is sufficiently unique to be recognized by a specific receptor. Thus progestins bind to the **progesterone receptor,** mineralocorticoids bind to the **mineralocorticoid receptor,** glucocorticoids bind to the **glucocorticoid receptor,** androgens bind to the **androgen receptor,** estrogens bind to the **estrogen receptor,** and the active vitamin D metabolite binds to the **vitamin D receptor.** The complexity of steroid hormone action is increased by the expression of multiple forms of each receptor. In addition, there is some degree of nonspecificity between steroid hormones and the receptors to which they bind. For example, glucocorticoids bind to the mineralocorticoid receptor with high affinity, and progestins, glucocorticoids, and androgens can all interact with the progesterone, glucocorticoid, and androgen receptors to some degree. As discussed later, steroid hormones are hydrophobic and pass through cell membranes easily. Accordingly, classic steroid hormone receptors are localized intracellularly and act by regulating gene expression. There is mounting evidence of the presence of plasma membrane and juxtamembrane steroid hormone receptors that mediate rapid, nongenomic actions of steroid hormones.

Steroidogenic cell types are defined as cells that can convert cholesterol to pregnenolone, which is the first reaction common to all steroidogenic pathways. Steroidogenic cells have some capacity for cholesterol synthesis but often obtain cholesterol from cholesterol-rich lipoproteins (low-density lipoproteins and high-density lipoproteins). Pregnenolone is then further modified by several enzymatic reactions. Because of their hydrophobic nature, steroid hormones and precursors can leave the steroidogenic cell easily and thus are not stored. Therefore, steroidogenesis is regulated at the level of uptake, storage, and mobilization of cholesterol and at the level of steroidogenic enzyme gene expression and activity. Steroids are not regulated at the

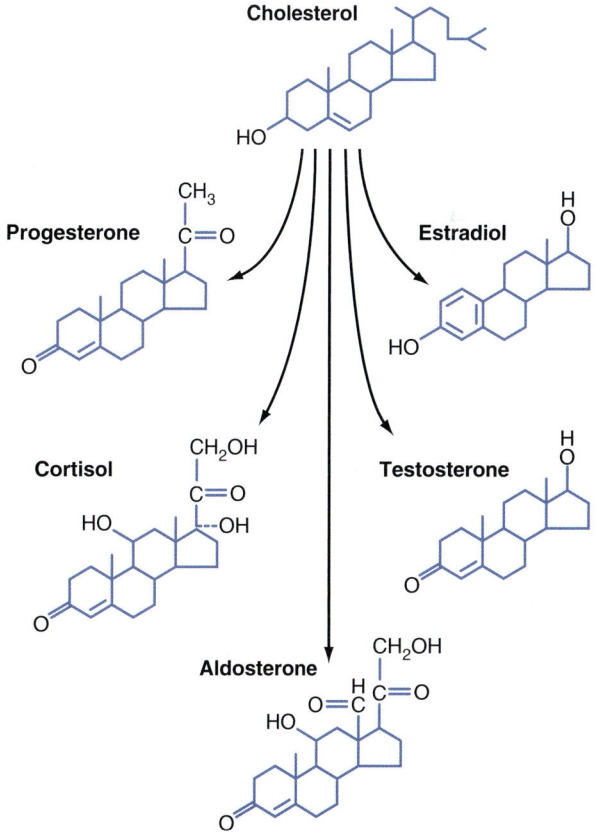

A

B

• **Fig. 38.8 A,** Structure of cholesterol, the precursor of steroid hormones. **B,** Structure of steroid hormones.

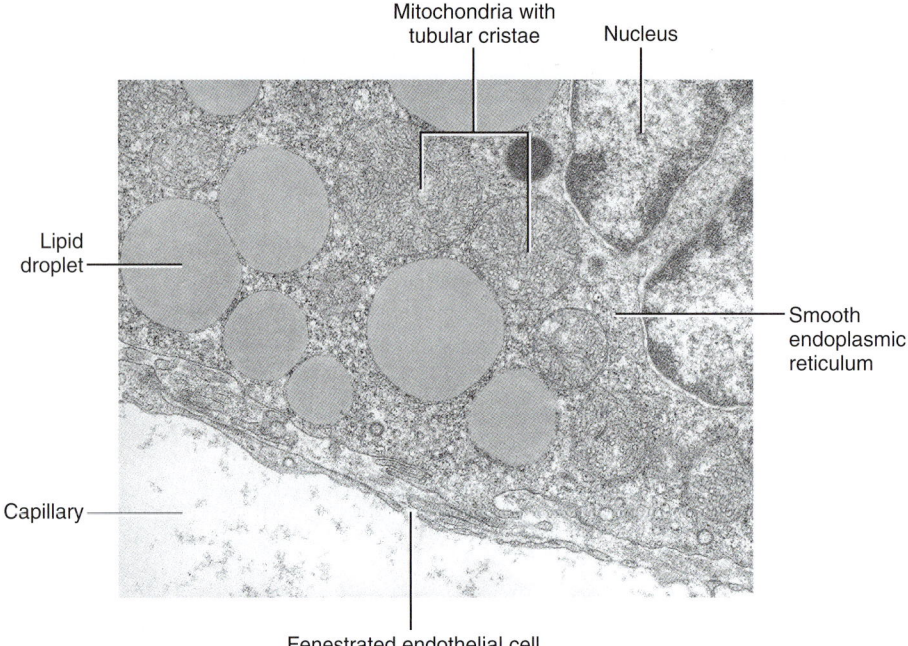

Mitochondria with tubular cristae

Nucleus

Lipid droplet

Smooth endoplasmic reticulum

Capillary

Fenestrated endothelial cell

● **Fig. 38.9** Ultrastructure of a Steroidogenic Cell.　Note the abundance of lipid droplets, smooth endoplasmic reticulum, and mitochondria with tubular cristae. (From Kierszenbaum AL. *Histology and Cell Biology: An Introduction to Pathology.* 2nd ed. Philadelphia: Mosby; 2007.)

level of secretion of the preformed hormone. A clinical implication of this mode of secretion is that high levels of steroid hormone precursors are easily released into blood when a steroidogenic enzyme within a given pathway is inactive or absent. The ultrastructure of steroidogenic cells is distinct from protein- and catecholamine-secreting cells. Steroidogenic enzymes reside within the inner mitochondrial membrane or the membrane of the smooth endoplasmic reticulum. Thus steroidogenic cells typically contain extensive mitochondria and smooth endoplasmic reticulum (Fig. 38.9). These cells also contain lipid droplets, which represent a store of cholesterol esters.

An important feature of steroidogenesis is that steroid hormones often undergo further modifications (apart from those involved in deactivation and excretion) after their release from the original steroidogenic cell. For example, estrogen synthesis by the ovary and placenta requires at least two cell types to complete the conversion of cholesterol to estrogen. This means that one cell secretes a precursor and a second cell converts the precursor to estrogen. There is also considerable **peripheral conversion** of active steroid hormones. For example, the testes secrete little estrogen. However, adipose, muscle, and other tissues express the enzyme for converting testosterone (a potent androgen) to estradiol-17β (a potent estrogen). Thus the overall production of a particular steroid hormone is equivalent to the sum of the secretion of this steroid hormone from a steroidogenic cell type and peripheral conversion of other steroids to the particular steroid hormone (Fig. 38.10). Peripheral conversion can produce (1) a more active but similar class of hormone (e.g., conversion of 25-hydroxyvitamin D to

1,25-dihydroxyvitamin D); (2) a less active hormone that can be reversibly activated by another tissue (e.g., conversion of cortisol to cortisone in the kidneys, followed by conversion of cortisone to cortisol in abdominal adipose tissue); or (3) a different class of hormone (e.g., conversion of testosterone to estrogen). Peripheral conversion of steroids plays an important role in several endocrine disorders (see Chapters 43 and 44).

Because of their nonpolar nature, steroid hormones are not readily soluble in blood. Therefore, steroid hormones circulate bound to **transport proteins,** including albumin, but also the specific transport proteins **sex hormone–binding globulin** and **corticosteroid-binding globulin** (see the section "Transport of Hormones in the Circulation"). Excretion of hormones from the body typically involves inactivating modifications, followed by **glucuronide or sulfate conjugation** in the liver, which is often coupled to biliary excretion. These modifications also increase the water solubility of the steroid and decrease its affinity for transport proteins, thereby allowing the inactivated steroid hormone to be excreted by the kidneys. Steroid compounds are absorbed fairly readily in the gastrointestinal tract and may therefore be administered orally.

Iodothyronines

Thyroid hormones are iodothyronines (Fig. 38.11) that are made by the coupling of iodinated tyrosine residues through an ether linkage. Their specificity is determined by the thyronine structure, as well as by where the thyronine is iodinated. Thyroid hormones cross cell membranes by

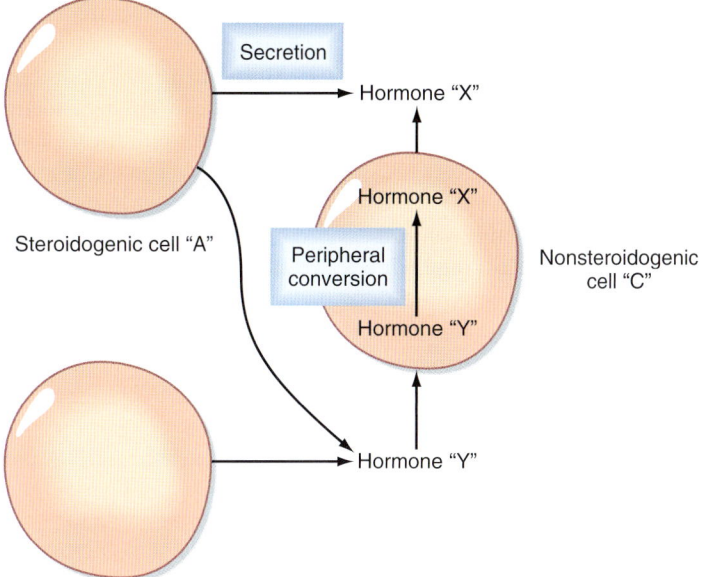

Total production of hormone "X" =
secretion of hormone "X" + peripheral conversion of hormone "Y"
into hormone "X"

• **Fig. 38.10** Peripheral conversion of steroid hormones.

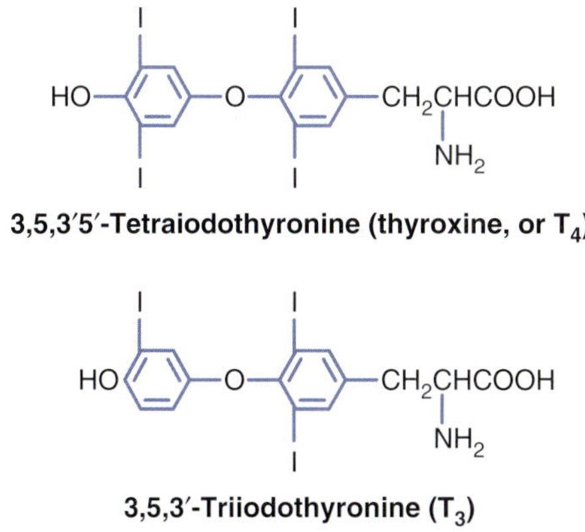

3,5,3′5′-Tetraiodothyronine (thyroxine, or T₄)

3,5,3′-Triiodothyronine (T₃)

• **Fig. 38.11** Structure of thyroid hormones, which are iodothyronines.

transport systems. They are stored extracellularly in the thyroid as an integral part of the glycoprotein molecule thyroglobulin. Thyroid hormones are sparingly soluble in blood and aqueous fluids and are transported in blood that is bound (>99%) to serum-binding proteins. A major transport protein is **thyroid hormone–binding globulin.** Thyroid hormones have long half-lives (7 days for thyroxine; 18 hours for triiodothyronine). Thyroid hormones are similar to steroid hormones in that the **thyroid hormone receptor** is intracellular and acts as a transcription factor. In fact, the thyroid hormone receptor belongs to the same gene family that includes steroid hormone receptors and vitamin D receptor. Thyroid hormones can be administered orally; the amount absorbed intact is sufficient for this to be an effective mode of therapy.

Transport of Hormones in the Circulation

A significant fraction of steroid and thyroid hormones is transported in blood that is bound to plasma proteins that are produced in a regulated manner by the liver. Protein and polypeptide hormones are generally transported free in blood. The concentrations of bound hormone, free hormone, and plasma transport protein are in equilibrium. If free hormone levels drop, hormone will be released from the transport proteins. This relationship may be expressed as follows:

Equation 38.1

$$[H] \times [P] = [HP] \: or \: K = [H] \times [P]/[HP]$$

where [H] = concentration of free hormone, [P] = concentration of plasma transport protein, [HP] = concentration of bound hormone, and K = the dissociation constant.

Free hormone is the biologically active form for action on the target organ, feedback control, and clearance by cellular uptake and metabolism. As a consequence, when hormonal status is evaluated, sometimes free hormone levels must be determined in addition to total hormone levels. This is particularly important because hormone transport proteins themselves are regulated by altered endocrine and disease states.

Protein binding serves several purposes. It prolongs the circulating half-life of the hormone. Many hormones cross cell membranes readily and would either enter cells or be excreted by the kidneys if they were not protein bound. The bound hormone represents a reservoir of hormone and, as such, can serve to buffer acute changes in hormone secretion. Some hormones, such as steroids, are sparingly soluble in blood, and protein binding facilitates their transport.

Cellular Responses to Hormones

Hormones are also referred to as **ligands,** in the context of ligand-receptor binding, and as **agonists,** in that their binding to the receptor is transduced into a cellular response. Receptor **antagonists** typically bind to a receptor and lock it in an inactive state, in which the receptor is unable to induce a cellular response. Loss or inactivation of a receptor results in **hormonal resistance.** Constitutive activation of a receptor leads to unregulated, hormone-independent activation of cellular processes.

Hormones regulate essentially every major aspect of cellular function in every organ system. Hormones control the growth of cells, ultimately determining their size and competency for cell division. Hormones regulate the differentiation of cells and their ability to survive or to undergo programmed cell death. They influence cellular metabolism, the ionic composition of body fluids, and cell membrane potential. Hormones orchestrate several complex cytoskeleton-associated events, including cell shape, migration, division, exocytosis, recycling/endocytosis, and cell-cell and cell-matrix adhesion. Hormones regulate the expression and function of cytosolic and membrane proteins, and a specific hormone may determine the level of its own receptor or the receptors for other hormones.

Although hormones can exert coordinated, pleiotropic control on multiple aspects of cell function, any given hormone does not regulate every function in every cell type. Rather, a single hormone controls a subset of cellular functions in only the cell types that express receptors for that hormone. Thus selective receptor expression determines which cells respond to a given hormone. Moreover, the differentiated state of a cell determines how it responds to a hormone. Thus the specificity of hormonal responses resides in the structure of the hormone itself, the receptor for the hormone, and the cell type in which the receptor is expressed. Serum hormone concentrations are typically extremely low (10^{-11} to 10^{-9} mol/L). Therefore, a receptor must have high affinity, as well as specificity, for its cognate hormone.

How does hormone-receptor binding get transduced into a cellular response? Hormone binding to a receptor induces conformational changes in the receptor. These changes are collectively referred to as a **signal.** The signal is transduced into the activation of one or more **intracellular messengers.** Messenger molecules then bind to **effector proteins,** which in turn modify specific cellular functions. The combination of hormone-receptor binding (signal), activation of messengers (transduction), and regulation of one or more effector proteins is referred to as a **signal transduction pathway** (also called simply a **signaling pathway**), and the final outcome is referred to as the **cellular response.**

Signaling pathways are usually characterized by the following properties:

1. Multiple, hierarchical steps in which "downstream" effector proteins are dependent on and driven by "upstream" receptors, transducers, and effector proteins. This means that loss or inactivation of one or more components within the pathway leads to general resistance to the hormone, whereas constitutive activation or overexpression of components can drive a pathway in an unregulated manner.
2. Amplification of the initial hormone-receptor binding. Amplification can be so great that maximal response to a hormone is achieved when the hormone binds to a small percentage of receptors.
3. Activation of multiple pathways, or at least regulation of multiple cell functions, from one hormone-receptor binding event. For example, binding of insulin to its receptor activates three separate signaling pathways. Even in fairly simple pathways (e.g., glucagon activation of adenylate cyclase), divergent downstream events allow the regulation of multiple functions (e.g., posttranslational activation of glycogen phosphorylase and increased phosphoenolpyruvate carboxykinase gene transcription).
4. Antagonism by constitutive and regulated negative feedback reactions. This means that a signal is dampened or terminated (or both) by opposing reactions and that loss or gain of function of opposing components can cause hormone-independent activation of a specific pathway or hormone resistance.

As discussed in Chapter 3, hormones signal to cells through membrane or intracellular receptors. Membrane receptors have rapid effects on cellular processes (e.g., enzyme activity, cytoskeletal arrangement) that are independent of the synthesis of new protein. Membrane receptors can also rapidly regulate gene expression through either mobile kinases (e.g., cyclic adenosine monophosphate–dependent protein kinase [PKA], mitogen-activated protein kinases [MAPKs]) or mobile transcription factors (e.g., signal transducer and activator of transcription proteins [STATs], Mothers against decapentaplegic homologs [Smads]). Steroid hormones have slower, longer term effects that involve chromatin remodeling and changes in gene expression. Increasing evidence indicates that steroid hormones have rapid, nongenomic effects as well, but these pathways are still being elucidated.

The presence of a functional receptor is an absolute requirement for hormone action, and loss of a receptor produces essentially the same symptoms as loss of hormone. In addition to the receptor, there are fairly complex pathways involving numerous intracellular messengers and effector proteins. Accordingly, endocrine diseases can arise from abnormal expression or abnormal activity, or both, of any of these signal transduction pathway components. Finally, hormonal signals can be terminated in several ways, including hormone/receptor internalization, phosphorylation/dephosphorylation, proteosomal destruction of receptor, and generation of feedback inhibitors.

IN THE CLINIC

Endocrine diseases can be broadly categorized as hyperfunction or hypofunction of a specific hormonal pathway. Hypofunction can be caused by lack of active hormone or by **hormone resistance** as a result of inactivation of hormone receptors or postreceptor defects. **Testicular feminization syndrome** is a dramatic form of hormone resistance in which the androgen receptor is mutated and cannot be activated by androgens. In patients in whom the diagnosis is not made before puberty, the testis becomes hyperstimulated because of abrogation of the negative feedback between the testis and the pituitary gland. The increased androgen levels have no direct biological effect as a result of the receptor defect. However, the androgens are peripherally converted to estrogens. Thus affected individuals are genetically male (i.e., 46,XY) but have a strongly feminized external phenotype, a female sexual identity, and usually a sexual preference for men (i.e., heterosexual in relation to sexual identity). Treatment involves removal of the hyperstimulated testes (which reside in the abdomen and pose a risk for cancer), estrogen replacement therapy, and counseling for the patient and, if one exists, the partner/spouse to address infertility and social/psychological distress.

Key Points

1. Endocrine signaling involves (1) regulated secretion of an extracellular signaling molecule, called a *hormone,* into the extracellular fluid; (2) diffusion of the hormone into the vasculature and circulation throughout the body; and (3) diffusion of the hormone out of the vascular compartment into the extracellular space and binding to a specific receptor within cells of a target organ.
2. The endocrine system is composed of the endocrine tissue of the pancreas, the parathyroid glands, the pituitary gland, the thyroid gland, the adrenal glands, and the gonads (testes or ovaries).
3. Negative feedback represents an important control mechanism that confers stability on endocrine systems. Hormonal rhythms are imposed on negative feedback loops.
4. Protein/peptide hormones are produced on ribosomes and stored in endocrine cells in membrane-bound secretory granules. They typically do not cross cell membranes readily and act through cell membrane–associated receptors.
5. Catecholamines are synthesized in the cytosol and secretory granules and do not readily cross cell membranes. They act through cell membrane–associated receptors.
6. Steroid hormones are not stored in tissues and generally cross cell membranes relatively readily. They act through intracellular receptors.
7. Thyroid hormones are synthesized in follicular cells and stored in follicular colloid as thyroglobulin. They cross cell membranes and associate with nuclear receptors.
8. Some hormones act through membrane receptors, and their responses are mediated by rapid intracellular signaling pathways.
9. Other hormones bind to nuclear receptors and act by directly regulating gene transcription.

Additional Readings

Bae YJ, Kratzsch J. Corticosteroid-binding globulin: modulating mechanisms of bioavailability of cortisol and its clinical implications. *Best Pract Res Clin Endocrinol Metab.* 2015;29:761-772.

Vija L, Ferlicot S, Paun D. Testicular histological and immunohistochemical aspects in a post-pubertal patient with 5 alpha-reductase type 2 deficiency: case report and review of the literature in a perspective of evaluation of potential fertility of these patients. *BMC Endocr Disord.* 2014;14:43.

39

Hormonal Regulation of Energy Metabolism

LEARNING OBJECTIVES

Upon completion of this chapter the student should be able to answer the following questions:

1. Explain the different requirements for and utilization by different cells of fuels during the digestive phase as opposed to the interdigestive and fasting phases.
2. Integrate the structure, synthesis, and secretion of insulin with circulating fuel levels, especially glucose.
3. Utilize the different signaling pathways regulated by insulin to link insulin to itscellular effects at the molecular level.
4. Integrate the structure, synthesis, and secretion of glucagon with the levels of circulating fuels, insulin, and catecholamines.
5. Map out and integrate the actions of insulin on the utilization and storage of glucose, free fatty acids (FFAs), and amino acids (AAs) by hepatocytes, skeletal muscle, and adipocytes during the digestive phase.
6. Map out and integrate the actions of counterregulatory hormones (glucagon, catecholamines) on the utilization of glucose, the sparing of glucose, and the utilization of FFAs and AAs by hepatocytes, skeletal muscle, and adipocytes during the interdigestive and fasting phases.
7. Integrate the changes in fuel utilization and hormonal signaling in hepatocytes during the interdigestive and fasting phases that allow for and promote hepatic glucose production and ketogenesis.
8. Compare signaling pathways that have orexigenic and anorexigenic actions via the hypothalamus.
9. Link several pathologies related to metabolism, especially those caused by the absolute or relative absence of insulin and by obesity.

Continual Energy Supply and Demand: The Challenge

There are an estimated 40 trillion cells in the human body, not including the approximately 40 trillion nonhuman cells that comprise the human microbiome. All these cells must continually perform **work** to stay alive. This work includes maintenance of cellular composition and structural integrity, along with the integrated synthesis and breakdown (i.e., turnover) of macromolecules and organelles. This work also involves the functions of cells that contribute to the human body as a whole (e.g., contraction of the muscle fibers of the diaphragm). Additional work is required of cells when the human body is engaged in a variety of activities, including (but not limited to) manual labor, exercise, and outdoor play; body growth spurt and maturation of the reproductive systems at puberty; pregnancy and breastfeeding; combating a serious infection or cancer; and the healing of damaged tissues/organs (e.g., a broken bone or healing from surgery). On average the **resting metabolic rate** of a relaxed, awake, stationary, healthy adult human accounts for about 70% of their total energy expenditure each day (Fig. 39.1).

To perform this work, cells need **fuels,** along with the capability to convert fuels into potential chemical energy in the form of **adenosine triphosphate (ATP).** Cells then convert the energy within ATP into chemical and mechanical work (see Fig. 39.1). This means that the need for ATP is immediate and unending, and consequently all living cells must continually synthesize ATP. In fact, humans produce about the equivalent of their body weight in ATP daily. This places a demand on the body to continually supply fuel in some form to all cells. All fuel originates from the **diet,** but humans do not eat in a nonstop manner all day long. *Thus the constant cellular demand for fuels to make ATP and perform work is paired with an intermittent intake of fuels.* Diet-derived fuels are oxidized for ATP, but some are also stored for future use, and some are converted to other fuels that can be used by other cell types, depending on the metabolic phase, type of fuel, and cell type in question.

In trying to make sense of energy metabolism, it is important to organize one's thinking around the following:

1. *Metabolic phases.* Metabolic phases refer to the hourly and daily differences in fuel usage and energy metabolism, which are dictated largely by the abundance or scarcity of certain fuels and orchestrated by phase-specific hormones. In general there are three metabolic phases (Fig. 39.2): (1) the **digestive** or **absorptive phase,** which occurs during the 2 to 3 hours it takes to digest a meal; (2) the **interdigestive** or **postabsorptive phase,** which normally occurs between meals; and (3) the

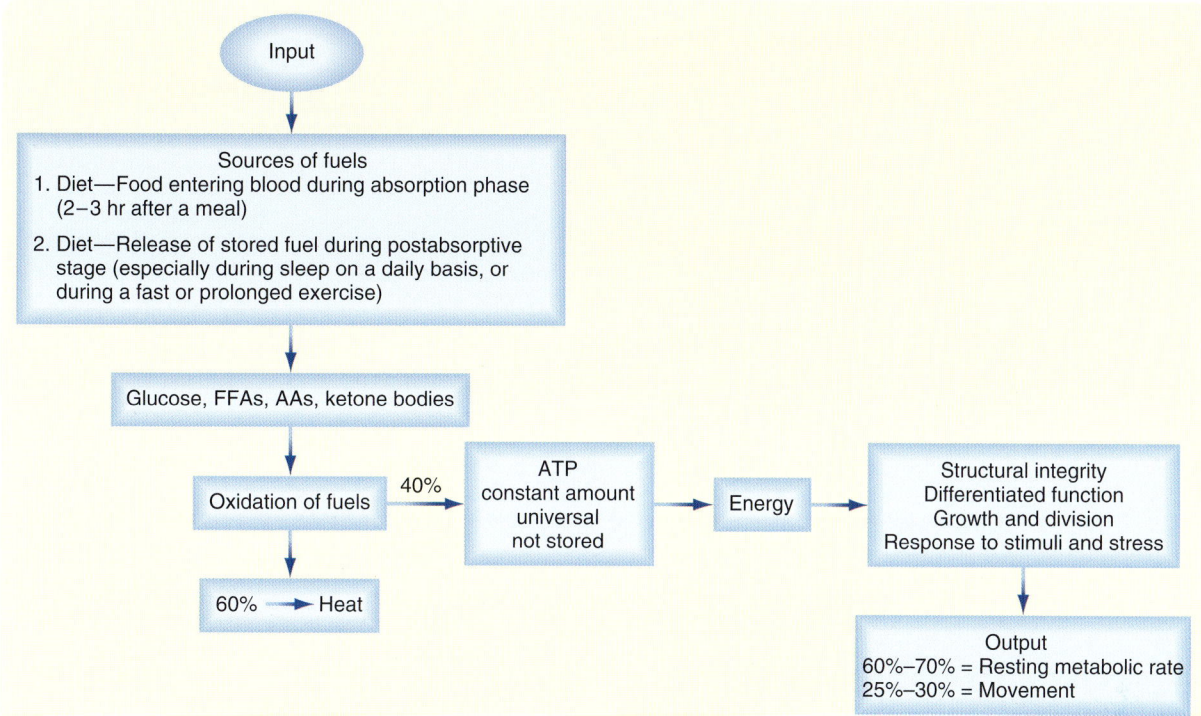

• **Fig. 39.1** Overview of energy metabolism.

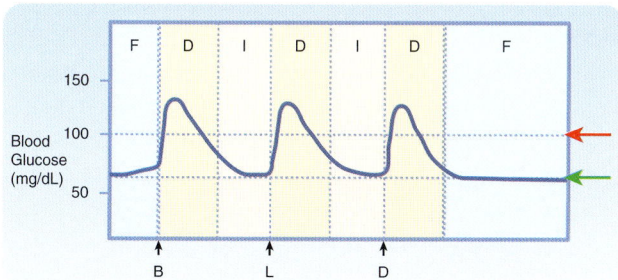

• **Fig. 39.2** Blood glucose levels during the three metabolic phases: D, digestive phase; F, fasting phase; I, interdigestive phase. B, breakfast; D, dinner; L, lunch. *Red arrow* indicates the upper limit for normal fasting glucose; *green arrow* indicates the lower limit for normal fasting glucose.

fasting phase, which most commonly occurs between the last snack before bedtime and breakfast. (In fact, physicians refer to a blood value as "fasting," e.g., "fasting blood glucose," if the patient abstains from eating after midnight and has blood drawn about 8 AM; prolonged fasting and starvation are more extreme forms of fasting.) **Physical exertion,** which imposes a heightened energy demand, is another type of metabolic phase that occurs with some frequency and regularity for some individuals. This chapter primarily compares how metabolism differs between the digestive phase and the fasting phase, and how different hormones orchestrate these metabolic differences.

2, *Metabolic actions of hepatocytes, adipocytes, and skeletal myocytes.* All cells are involved in energy metabolism,

but these three cell types have a profound impact on whole-body metabolism. Key features with respect to metabolism of these three cell types are listed in Table 39.1.

3, *Insulin and the counterregulatory hormones.* Metabolism during the digestive phase is orchestrated almost entirely by **insulin.** During the fasting phase, insulin drops to low levels, and this alone allows for some of the metabolic adaptions associated with fasting. In addition, **glucagon** and **catecholamines (epinephrine, norepinephrine)** stimulate metabolic pathways that integrate the body's response to an absence of ingested and absorbed fuels. These hormones are referred to as *counterregulatory hormones* based on their opposition to insulin. Growth hormone (see Chapter 41) and cortisol (see Chapter 43) also contribute somewhat to fasting-phase metabolism.

Integrated Overview of Energy Metabolism

Digestive Phase

Fuels enter the body from the diet during the digestive phase. Our diet includes both monomeric and polymeric forms (the latter are converted into monomeric forms during digestion and absorption) of the following fuels: (1) **monosaccharides,** including **glucose,** fructose, and galactose; (2) **long-chain free fatty acids** (referred to in this chapter as simply **FFAs**); and (3) **amino acids (AAs).** The diet may also include other fuels such as **ethanol.**

TABLE 39.1	Fate of Fuels During Digestive and Fasting Phases		
	Hepatocytes	**Adipocytes**	**Skeletal Myocytes**
Glucose, fed phase	• Utilization for ATP • Storage as glycogen • DNL	• Increased uptake by GLUT4 • Utilization for ATP • Utilization for G3P	• Increased uptake by GLUT4 (largest impact on glucose tolerance) • Utilization for ATP • Storage as glycogen
Glucose, fasting phase	• Breakdown of glycogen and release of glucose into blood • New synthesis of glucose from small precursors and release into blood • Use of alternative fuel for ATP	• Decreased uptake by GLUT4 • Use of alternative fuels	• Decreased uptake by GLUT4 • Use of alternative fuels • Breakdown of glycogen and use of glucose intracellularly (no export), especially during exercise
FFA/TG, fed phase	• Make FFAs from glucose by DNL • Esterify FFAs into intrahepatic TG • Uptake of chylomicron remnants	• Lipolysis of chylomicrons and uptake of FFAs • Esterification of FFAs into storage TG • Inhibition of lipolysis of stored TG	• Minimal involvement
FFA/TG, fasting phase	• Utilization for ATP • Utilization to produce KBs • Assembly of TG into VLDL • Secretion of VLDL	• Release of FFAs from TG stores • Utilization for ATP	• Utilization for ATP
AAs, fed phase	• Utilization for multiple anabolic pathways	• Utilization for multiple anabolic pathways	• Utilization for multiple anabolic pathways
AAs, fasting phase	• Utilization for gluconeogenesis • Utilization for ketogenesis	• Proteolysis and release of AAs	• Proteolysis and release of AAs
KBs, fed phase	• Should be absent	• Should be absent	• Should be absent
KBs, fasting phase	• Synthesis from FFAs and some amino acids • Cannot be utilized for ATP	• Utilization for ATP	• Utilization for ATP

AA, amino acid; ATP, adenosine triphosphate; DNL, de novo lipogenesis; FFA, free fatty acid; G3P, glycerol-3-phosphate; KB, ketone body; TG, triglyceride; VLDL, very low-density lipoprotein.

During the digestive phase, absorbed fuels are partitioned and used for different purposes. **Insulin** (discussed in detail later) regulates essentially every aspect of metabolism during the digestive phase. **Glucose** is the primary fuel used for energy (i.e., ATP production) during the digestive phase (Fig. 39.3A). Glucose is considered a universal fuel in that most cells can perform the following:

1. import glucose via bidirectional facilitative GLUT transporters
2. "trap" and "activate" imported glucose by converting glucose into glucose-6-phosphate (G6P) through the activity of one or more hexokinases. G6P cannot pass through GLUT transporters ("trapping") and is now a substrate for several enzymatic pathways ("activation").
3. metabolize G6P to pyruvate via the glycolytic pathway, which yields a small amount of ATP without requiring mitochondria or O_2. Cells without mitochondria ferment pyruvate to lactate and export lactate as waste. Most cells import pyruvate into mitochondria, convert it to acetyl CoA by pyruvate dehydrogenase, and then condense acetyl CoA with oxaloacetate to form citrate. Citrate is cycled through the tricarboxylic acid (TCA) cycle back

to oxaloacetate. This metabolism of pyruvate through the TCA cycle releases CO_2 as waste and generates guanosine triphosphate (GTP) along with flavin adenine dinucleotide hydride ($FADH_2$) and nicotine adenine dinucleotide hydride (NADH). $FADH_2$ and NADH are used by the electron transport system and oxidative phosphorylation to ultimately generate relatively large amounts of ATP through a process that is absolutely dependent on O_2.

Cells with no or very few mitochondria (e.g., erythrocytes, lens cells of the eye) are absolutely dependent on glucose for energy. Additionally the brain can only use glucose under normal conditions, in that it has essentially no capacity for oxidation of FFAs and only marginal capacity for the use of lactate, ketone bodies (KBs), and AAs as fuels. Thus in all metabolic phases, maintenance of blood glucose above a certain minimal threshold is absolutely necessary to avoid central nervous system (CNS)-related symptoms, beginning with those caused by a hypoglycemia-activated autonomic response (e.g., nausea, sweating, cardiac arrhythmias). If blood glucose continues to fall, progression to symptoms caused by neuroglycopenia (e.g., cognitive dysfunction, loss of coordinated motor function, and ultimately even coma

A. Digestive Phase

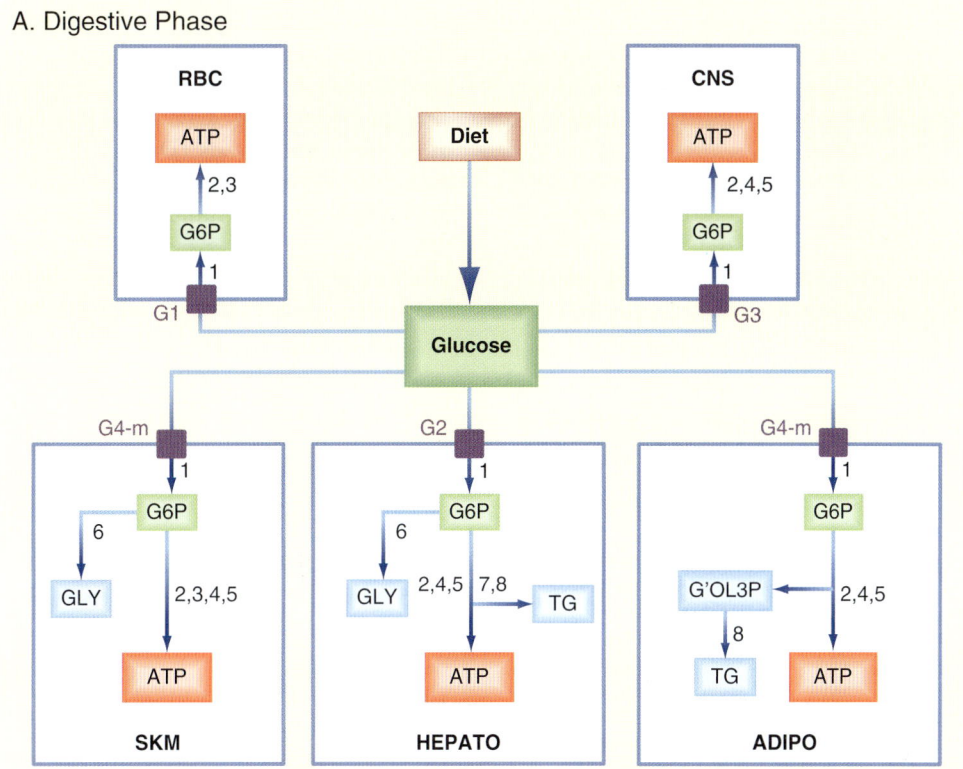

B. Fasting Phase

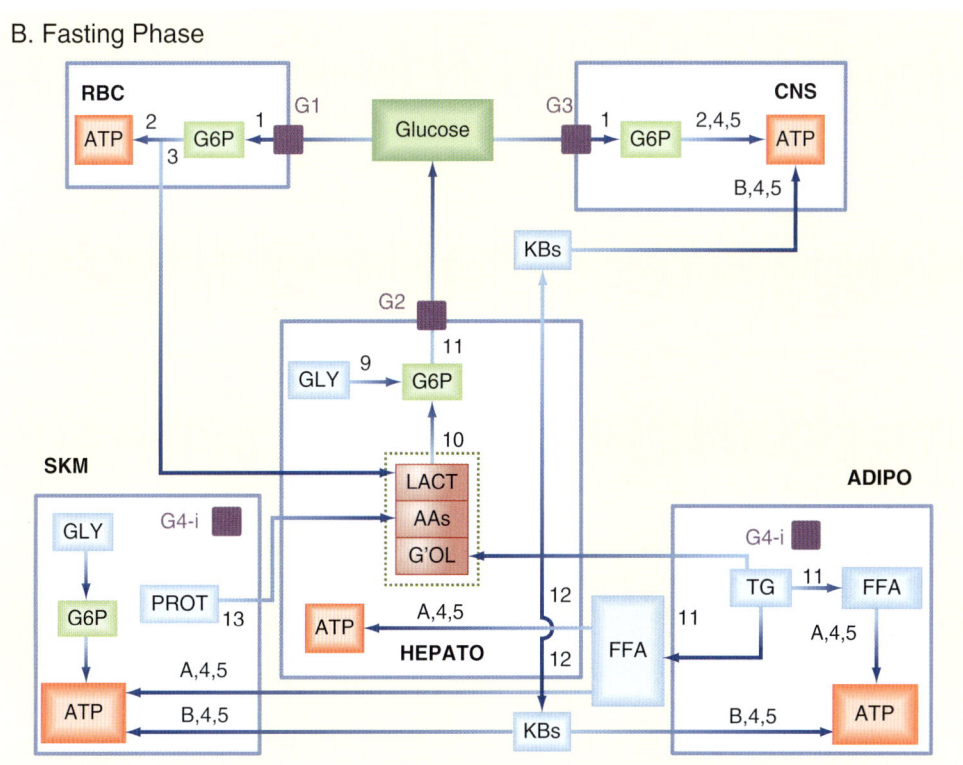

• **Fig. 39.3 A, Overview of glucose utilization during the digestive phase.** GLUT transporters: G1, GLUT1; G2, GLUT2; G3, GLUT3; G4-m, functional GLUT4 localized to cell membrane. Cell types: ADIPO, adipocyte; CNS, central nervous system neurons and glia; HEPATO, hepatocyte; RBC, red blood cell; SKM, skeletal myocyte. Metabolites: G6P, glucose-6-phosphate; GLY, glycogen; G'OL3P, glycerol-3-phosphate; TG, triglyceride. Metabolic reactions/pathways: 1, hexokinase/glucokinase; 2, glycolysis; 3, lactate dehydrogenase; 4, TCA cycle ± pyruvate dehydrogenase; 5, oxidative phosphorylation; 6, glycogen synthesis; 7, de novo lipogenesis; 8, esterification of FFAs to G'OL3P to form TG. **B, Overview of energy metabolism during the fasting phase.** GLUT transporters: see legend for A, plus: G4-I, inactive GLUT4 with intracellular localization. Cell types: see legend for A Metabolites: see legend for A, plus: AAs, amino acids; FFA, free fatty acid; G'OL, glycerol; KBs, ketone bodies; LACT, lactate; PROT, protein. Metabolic reactions/pathways: see legend for A, plus: 9, glycogenolysis; 10, gluconeogenesis; 11, G6Pase; 12, ketogenesis; 13, proteolysis; A, β-oxidation; B, ketolysis.

and death) can occur. This means that whole-body metabolism during the interdigestive and fasting phases must meet the challenge of maintaining blood glucose above 60 mg/dL (see Fig. 39.2, *green arrow*).

On the other hand, glucose levels must be maintained below a certain maximal threshold (i.e., fasting blood glucose < 100 mg/dL) to avoid glucose-induced side reactions (both enzymatic and nonenzymatic) as well as osmotic imbalances that lead to multiple forms of damage to cell function (see Fig. 39.2, *red arrow*).

🩺 IN THE CLINIC

Glucotoxicity Within Microvasculature

The endothelium of the microvasculature of the kidney and retina, as well as the endothelium of the vasa nervosum of the autonomic nervous system, are particularly sensitive to hyperglycemia. Chronically high blood glucose results in pathologically high intracellular levels of glucose in these endothelial cells, resulting in altered protein and lipid structure, oxidative stress, and altered signaling pathways. These insults, collectively referred to as **glucotoxicity,** cause pathological changes in intracellular and membrane components as well as in secreted molecules that either signal and/or make up the extracellular matrix. Indeed, glucotoxicity is the root cause of the **nephropathy, retinopathy, and peripheral neuropathy** that occur in poorly controlled diabetes mellitus. Therefore whole-body metabolism during all metabolic phases must meet the challenge of minimizing the magnitude and duration of the rise in blood glucose associated with ingestion of a meal and must maintain blood glucose below a safe maximal threshold of 100 mg/dL during all other times. Fasting blood glucose between 100 and 124 mg/dL is indicative of **impaired glucose tolerance,** and values at 125 mg/dL and above are evidence of **diabetes mellitus.**

Glucose is consumed by erythrocytes and the brain continually throughout all metabolic phases. In contrast, **hepatocytes**, **skeletal myocytes,** and **adipocytes** primarily use glucose during the digestive phase only. Insulin stimulates glycolysis and entry of pyruvate (end product of glycolysis) into the TCA cycle and oxidative phosphorylation for ATP production in hepatocytes, skeletal myocytes, and adipocytes (see Table 39.1).

Hepatocytes express the **GLUT2** isoform of the glucose transporter, which is not regulated by insulin for its insertion into the cell membrane. In contrast, skeletal myocytes and adipocytes express the **GLUT4** isoform. Newly synthesized GLUT4 exists in an intracellular inactive state with GLUT4 storage vesicles (G4-i in Fig. 39.3B). Insulin induces translocation of GLUT4 to the cell membrane, where it can function as an active glucose transporter (G4-m in Fig. 39.3A).

After its phosphorylation to G6P by glucokinase, hepatocytes convert some of the imported glucose into the storage form, **glycogen,** during the digestive phase (see Fig. 39.3A). Similarly, skeletal muscle converts some of the G6P

from imported glucose into glycogen. Hepatocytes can only store a finite amount of glucose as glycogen. Hepatocytes also convert excess glucose into FFAs through the process of **de novo lipogenesis (DNL).** These FFAs are typically esterified to **glycerol-3-phosphate** (G3P) to form **triglyceride (TG),** which accumulates as **intrahepatic TG** during the digestive phase (see Fig. 39.3A). As discussed later for insulin signaling, an excessive accumulation of intrahepatic TG (i.e., fatty liver, hepatic steatosis) can result in insulin resistance.

During the digestive phase, AAs are used in multiple anabolic pathways to regenerate degraded molecules, including other AAs, proteins, nucleotides and nucleic acids, glutathione, and complex lipids.

FFAs represent the most efficient fuel type in terms of ATP molecules made per carbon of fuel. However, utilization of FFAs competes effectively with glucose utilization in the mitochondria. High FFA levels during the digestive phase would promote a greater magnitude and duration of the glucose surge, thereby contributing to hyperglycemia. Thus most of the FFAs in an average meal are prevented from entering the circulation by their reesterification into TG and packaging into **chylomicrons** within the intestinal enterocyte. Chylomicrons are secreted, enter lymphatic vessels, and ultimately enter the blood. Once in the general circulation, chylomicrons convey the FFAs (as TG) to the adipose tissue where they are released from chylomicron-associated TG through lipolysis, imported by adipocytes, and reesterified as intracellular TG for storage (discussed in detail later).

Fasting Phase

During the digestive phase, **hepatocytes, skeletal myocytes,** and **adipocytes** function largely independently of each other. In contrast the actions of these three cell types become highly integrated during the **fasting phase** to maintain adequate blood glucose levels while providing alternative energy substrates for each cell type (see Fig. 39.3B and Table 39.1).

As stated earlier, cells without mitochondria (e.g., erythrocytes) and cells of the CNS require glucose for ATP production during all metabolic phases, and thus the body must maintain blood glucose above 60 mg/dL even days or weeks after the last ingestion of food (an exception to this rule is that the brain can use **KBs** [discussed later] during a prolonged fasting phase). Two general processes contribute to maintenance of blood glucose during the fasting phase: **hepatic glucose production** and **glucose sparing.** Hepatic glucose production is in turn based on two metabolic pathways. The first is the rapid catabolic process of **glycogenolysis** (pathway 9 in Fig. 39.3B). Hepatocytes express the enzyme **glucose-6-phosphatase (G6Pase)** (pathway 11 in Fig. 39.3B), allowing them to convert G6P back to glucose, which can then exit the cell through a bidirectional GLUT transporter. Release of glucose derived from glycogenolysis is relatively short

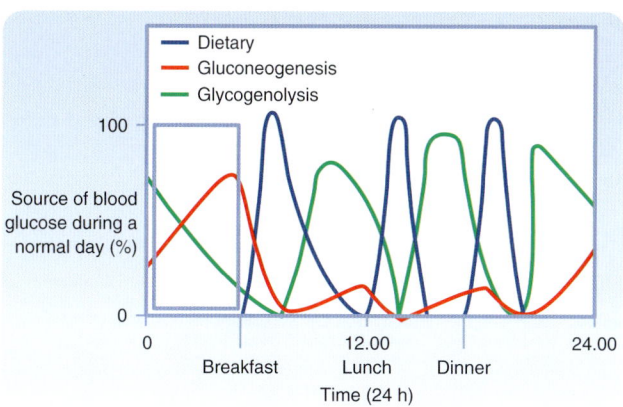

Fig. 39.4 Relative contributions of the three sources of blood glucose relative to meals and time of day. The inset box stresses replacement of glycogenolysis with gluconeogenesis during the fasting phase (i.e., sleep). (Adapted from Baynes JW, Dominiczak JH [eds]. *Medical Biochemistry.* 3rd ed. Philadelphia: Mosby/Elsevier; 2009.)

lived because the liver glycogen supply becomes exhausted by about 8 hours. The second metabolic contribution to hepatic glucose production during the fasting phase is the gradual pathway of **gluconeogenesis** (pathway 10 in Fig. 39.3B). The onset of gluconeogenesis during fasting is slower than glycogenolysis, but gluconeogenesis continues essentially nonstop throughout a fasting phase (Fig. 39.4). Gluconeogenesis requires precursors, especially **lactate, "gluconeogenic" AAs,** and **glycerol.** How are these precursors supplied during the fasting phase? Lactate is continually produced by erythrocytes. Lactate is also produced by glycolytic skeletal muscle fibers during exercise (exercise tends to occur more frequently during the interdigestive and fasting phases as opposed to "on a full stomach"), although much of this lactate is utilized by aerobic skeletal muscle and cardiac muscle during exercise. But additionally the overall anabolism of the digestive phase switches over to a general **catabolism** during the fasting phase (see Fig. 39.3B). TGs within adipocytes undergo **lipolysis** to FFAs and glycerol, and there is a general net **proteolysis** with the release of AAs during the fasting state. The glycerol and gluconeogenic AAs are released from cells and circulate to the liver, where they are subsequently used for gluconeogenesis. Thus gluconeogenesis requires an integration of catabolic pathways in adipocytes and skeletal myocytes with anabolic gluconeogenesis in hepatocytes. Gluconeogenesis eventually supplants glycogenolysis and can continue as long as precursors flow into the liver.

Glucose sparing represents the other general process that contributes to maintenance of adequate circulating glucose levels during the fasting phase. *Glucose sparing* means the switching of fuel utilization from glucose to a **nongluconeogenic fuel** in most cell types, but especially in skeletal muscle, which represents the potentially largest single consumer of glucose. First, the uptake of glucose by skeletal muscle and adipocytes is greatly reduced because the GLUT4 transporter isoform exists in an intracellular

inactive state (G4-i in Fig. 39.3B) during the fasting phase. Thus alternative fuels need to be delivered to skeletal muscle and adipocytes.

The nongluconeogenic fuels (i.e., cannot be used for gluconeogenesis by the liver) are **FFAs** and **KBs.** FFAs are primarily released from adipocytes (see Fig. 39.3B) but are also released after packaging of intrahepatic TGs into very low-density lipoproteins (VLDLs) by hepatocytes (discussed later). KBs are produced by hepatocytes from acetyl CoA, which in turn originates from FFAs and **ketogenic AAs,** both of which become abundant during the fasting phase. Thus glucose sparing depends on catabolic adipocyte metabolism, which results in lipolysis of stored TGs and release of FFAs. FFAs are imported by hepatocytes, which use FFAs to produce acetyl CoA. Protein degradation in skeletal muscle and other tissues also makes certain AAs available for ketogenesis. High levels of intramitochondrial acetyl CoA in the hepatocyte not only provides ample carbons for ATP synthesis but serves to: (1) inhibit conversion of pyruvate to acetyl CoA, (2) promote conversion of pyruvate to oxaloacetate for gluconeogenesis, and (3) promote synthesis of KBs (see Fig. 39.3B). After several days of fasting, the CNS can start using KBs for energy, thereby further sparing glucose for erythrocytes. Many other cell types with mitochondria use KBs along with FFAs for ATP production, especially skeletal muscle.

The hormones that drive glycogenolysis, gluconeogenesis, lipogenesis, and hepatic ketogenesis as well as VLDL production by the liver during the fasting phase are glucagon and catecholamines. In the presence of low glucose, insulin levels fall, and that removes the inhibition by insulin of the secretion of another pancreatic hormone, **glucagon.** Thus diminished blood glucose causes a rise in the circulating **glucagon-to-insulin ratio.** Hepatocytes are the primary target organ of glucagon, which directly drives glycogenolysis, gluconeogenesis, FFA oxidation, and ketogenesis by the hepatocyte (Fig. 39.5B). Hepatocytes also express β_2- and α_1-adrenergic receptors so that norepinephrine from sympathetic innervation and epinephrine from the adrenal medulla (see Chapter 43) can reinforce the actions of glucagon. Adipocytes also express the glucagon receptor, as well as the β_2- and β_3-adrenergic receptors that respond to catecholamines in response to hypoglycemia, exertion, or certain stresses. Skeletal muscle is not a target of glucagon but does respond to catecholamines stimulation through β_2-adrenergic receptors. Skeletal muscle is very responsive to **intracellular Ca^{++},** which increases during physical exertion/movement, and to an increase in the **intracellular adenosine monophosphate (AMP):ATP ratio,** which activates **AMP kinase.**

Finally, it is important to understand that the pathways upregulated during the fasting phase are opposed by insulin-dependent pathways that are most active during the digestive phase (discussed later). Thus **attenuation of insulin signaling** also contributes to the ability of hepatocytes, skeletal myocytes, and adipocytes to display an integrated response to the metabolic challenges of the fasting phase.

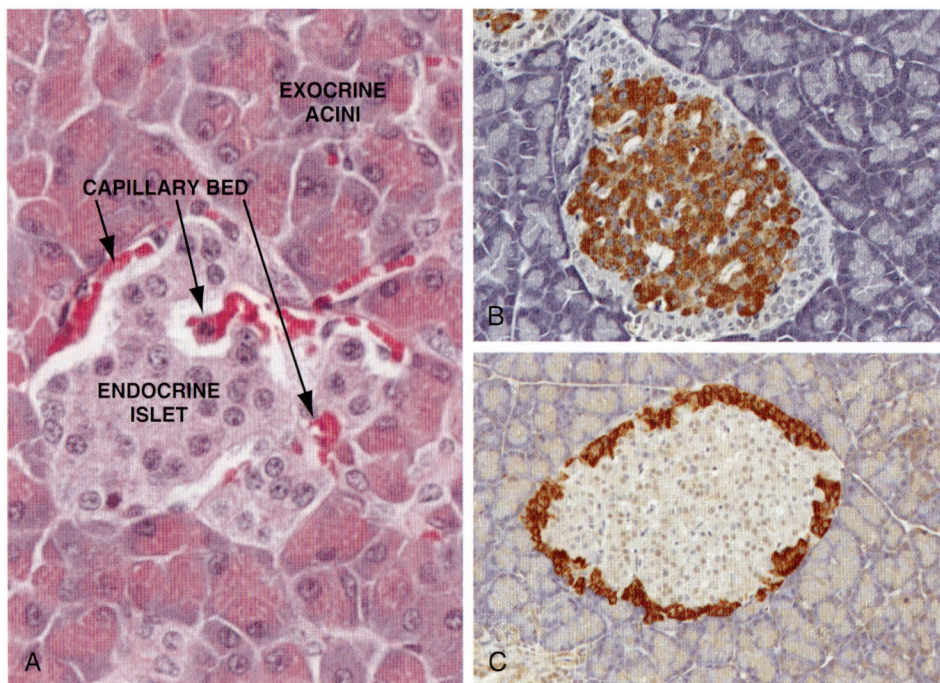

• **Fig. 39.5** The islets of Langerhans (endocrine pancreas) from rat. **A,** Pancreas histology showing exocrine acini where digestive enzymes are produced to be delivered to the duodenum via the pancreatic duct, and an endocrine islet where insulin and glucagon are produced and delivered to the circulation upon uptake by a rich capillary bed. **B,** Staining of endocrine islet for insulin within beta cells; these are the most numerous cell type and are primarily located centrally within the islet. **C,** Staining of endocrine islet for glucagon with alpha cells; these are much less numerous than beta cells and are primarily located along the periphery of the islet.

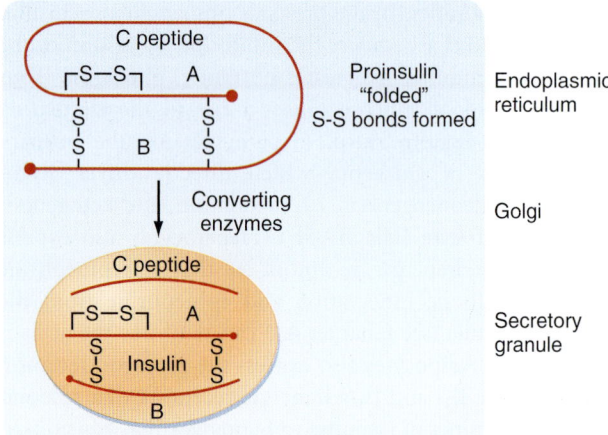

• **Fig. 39.6** Proinsulin is processed by prohormone convertases into a mature insulin molecule with two peptide strands linked by H-bonds and a C peptide. Both are secreted in equimolar ratios. (From White BA, Porterfield SP [eds]. *Endocrine and Reproductive Physiology.* 4th ed. Philadelphia: Mosby/Elsevier; 2013.)

Pancreatic Hormones Involved in Metabolic Homeostasis During Different Metabolic Phases

The islets of Langerhans constitute the **endocrine pancreas** (Fig. 39.6). Approximately 1 million islets making up about 1% to 2% of the pancreatic mass are spread throughout the **exocrine pancreas** (see Chapter 27). The islets are composed of several cell types, each producing a different hormone. Beta cells make up about three-fourths of the cells of the islets and produce the hormone **insulin** (see Fig. 39.5B). **Alpha cells** account for about 10% of islet cells and secrete **glucagon** (see Fig. 39.5C). Other endocrine cell types reside within islets, but their respective hormone products are of marginal or unclear importance and thus will not be discussed further.

Blood flow to the islets is somewhat autonomous from blood flow to the surrounding exocrine pancreatic tissue. Blood flow through the islets passes from beta cells, which predominate in the center of the islet, to alpha and delta cells, which predominate in the periphery (see Fig. 38.5B-C). Consequently the first cells affected by circulating insulin are the alpha cells, in which insulin inhibits glucagon secretion.

Insulin

Insulin is the primary anabolic hormone that dominates regulation of metabolism during the digestive phase. Insulin is a protein hormone that belongs to the gene family that includes **insulinlike growth factors I and II (IGF-I, IGF-II)** and **relaxin.** The insulin gene encodes preproinsulin. Insulin is synthesized as preproinsulin, which is converted to proinsulin as the hormone enters the endoplasmic reticulum. **Proinsulin** is packaged in the

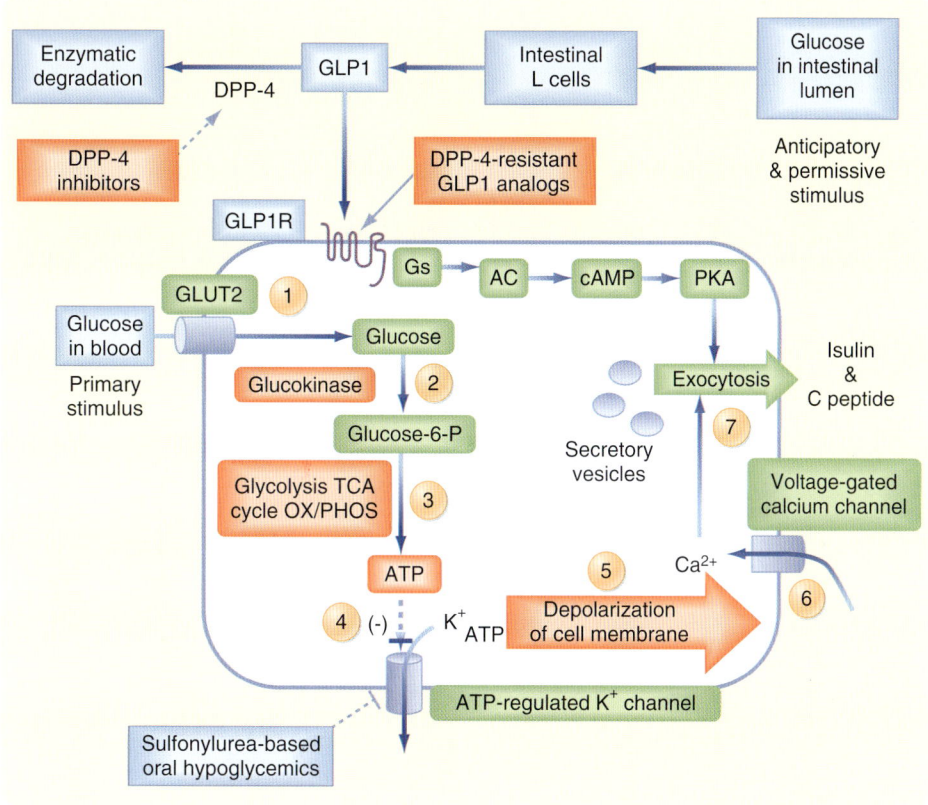

• **Fig. 39.7** Glucose is the primary stimulus of insulin secretion and is enhanced by sulfonylurea drugs as well as GLP-1 analogues/DPP-4 inhibitors. See text for explanation of numbered steps in glucose-stimulated insulin secretion (GSIS).

Golgi apparatus into membrane-bound secretory granules. Proinsulin contains the AA sequence of insulin plus the **C (connecting) peptide.** The proteases that cleave proinsulin (proprotein convertases) are packaged with proinsulin within secretory vesicles. Proteolytic processing clips out the C peptide and generates the mature hormone, which consists of two chains, an α chain and a β chain, connected by two disulfide bridges (Fig. 39.7). A third disulfide bridge is contained within the α chain. Insulin is stored in secretory granules in zinc-bound crystals. On stimulation, the granule's contents are released to the outside of the cell by exocytosis. Equimolar amounts of mature insulin and C peptide are released, along with small amounts of proinsulin. C peptide has no known biological activity but is useful in assessing endogenous insulin production. C peptide is more stable in blood than insulin (making it easier to assay) and helps distinguish endogenous insulin production from injected insulin, insofar as the latter has been purified from C peptide.

Insulin has a short half-life of about 5 minutes and is cleared rapidly from the circulation. It is degraded by **insulin-degrading enzyme (IDE;** also called *insulinase)* in the liver, kidney, and other tissues. Because insulin is secreted into the **hepatic portal vein,** it is exposed to liver IDE before it enters the peripheral circulation. About half the insulin is degraded before leaving the liver. Thus peripheral tissues are exposed to significantly less serum

insulin concentrations than the liver. **Recombinant human insulin** and **insulin analogues** with different characteristics of speed of onset and duration of action and peak activity are now available. Serum insulin levels normally begin to rise within 10 minutes after ingestion of food and reach a peak in 30 to 45 minutes. The higher serum insulin level rapidly lowers blood glucose to baseline values.

Glucose is the primary stimulus of insulin secretion ("steps" in glucose-stimulated insulin secretion described in the discussion that follows refer to Fig. 39.8). Entry of glucose into beta cells is facilitated by the **GLUT2 transporter** (step 1). Once glucose enters the beta cell, it is phosphorylated to **G6P** by the low-affinity hexokinase **glucokinase** (step 2). Glucokinase is referred to as the **"glucose sensor"** of the beta cell because the rate of glucose entry is correlated to the rate of glucose phosphorylation, which in turn is directly related to insulin secretion. Metabolism of G6P through glycolysis, the TCA cycle, and oxidative phosphorylation by beta cells increases the intracellular ATP:ADP ratio (step 3) and closes an **ATP-sensitive K$^+$ channel** (step 4). This results in depolarization of the beta cell membrane (step 5), which opens **voltage-gated Ca^{++} channels** (step 6). Increased intracellular [Ca^{++}] activates microtubule-mediated exocytosis of insulin/proinsulin-containing secretory granules (step 7).

Ingested glucose has a greater effect on insulin secretion than *injected glucose.* This phenomenon, called the **incretin**

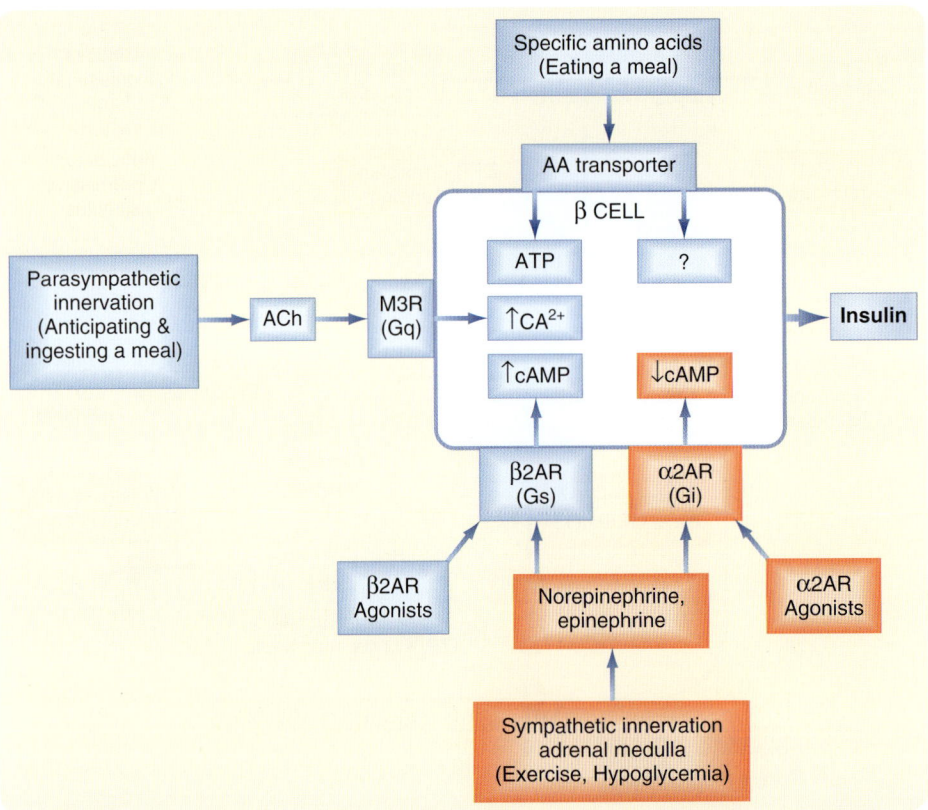

• **Fig. 39.8** Secondary regulators of insulin secretion. See text for explanation of abbreviations.

effect, is due to stimulation by glucose of **incretin hormones** from the gastrointestinal tract. One clinically relevant incretin hormone is **glucagon-like peptide 1 (GLP-1),** which is released by **L cells** of the ileum in response to glucose in the ileal lumen (Fig. 39.9). As a hormone, GLP-1 enters the circulation and ultimately binds to the Gs-coupled **GLP1 receptor** (GLP1R) on beta cells. This GLP1R/Gs/adenylyl cyclase/protein kinase A (PKA) signaling pathway amplifies the intracellular effects of Ca^{++} on insulin secretion. GLP-1 is rapidly degraded in the circulation by **dipeptidyl peptidase 4 (DPP-4).**

Several AAs and vagal (parasympathetic) cholinergic innervation via muscarinic receptor 3 (MR3) also stimulate insulin through increasing intracellular $[Ca^{++}]$ (Fig. 39.10). Insulin secretion is primarily dampened by sympathetic autonomic regulation through α_2-**adrenergic receptors**. Binding of **norepinephrine** or **epinephrine** to α_2-adrenergic receptors decreases cyclic adenosine monophosphate (cAMP), possibly by closing Ca^{++} channels (see Fig. 39.6). Adrenergic inhibition of insulin serves to protect against hypoglycemia, especially during exercise. Beta cells also express Gs-coupled β_2-adrenergic receptors that normally play a minor role in promoting insulin secretion. Note, however, that β-adrenergic receptor agonists oppose the actions of insulin on overall metabolism and may antagonize the actions of administered insulin in diabetics. Conversely, β-blockers (i.e., β-adrenergic receptor antagonists) may increase the severity of hypoglycemic episodes in patients receiving exogenous insulin.

IN THE CLINIC

Oral and Injectable Hypoglycemic Drugs

The ATP-sensitive K^+ channel is an octameric protein complex that contains four ATP-binding subunits called **SUR subunits.** These subunits are bound by **sulfonylurea drugs,** which also close the K^+ channel and are widely used as oral hypoglycemics to treat hyperglycemia in patients with partially impaired beta cell function. Hypoglycemia is a significant side effect of sulfonylurea drugs if used in excess or incorrectly in combination with other drugs, owing to inappropriately high release of insulin.

Both **DPP-4–resistant analogues of GLP-1** and **inhibitors of DPP-4** are currently approved for treatment of patients with type 2 DM with some beta cell function. Importantly these drugs are **permissive** to the actions of glucose on the beta cell and thus only weakly increase insulin secretion in the absence of glucose. Thus GLP-1 analogues induce hypoglycemia much less frequently than sulfonylurea drugs.

Insulin Receptor

The **insulin receptor (InsR)** is a member of the **receptor tyrosine kinase *(RTK)* gene family** (see Chapter 3). Most of the actions of insulin on metabolism involve activation of the protein kinase Akt, which in turn has pleiotropic actions on cell metabolism.

The InsR is expressed on the cell membrane as a homodimer, with each monomer containing a tyrosine kinase

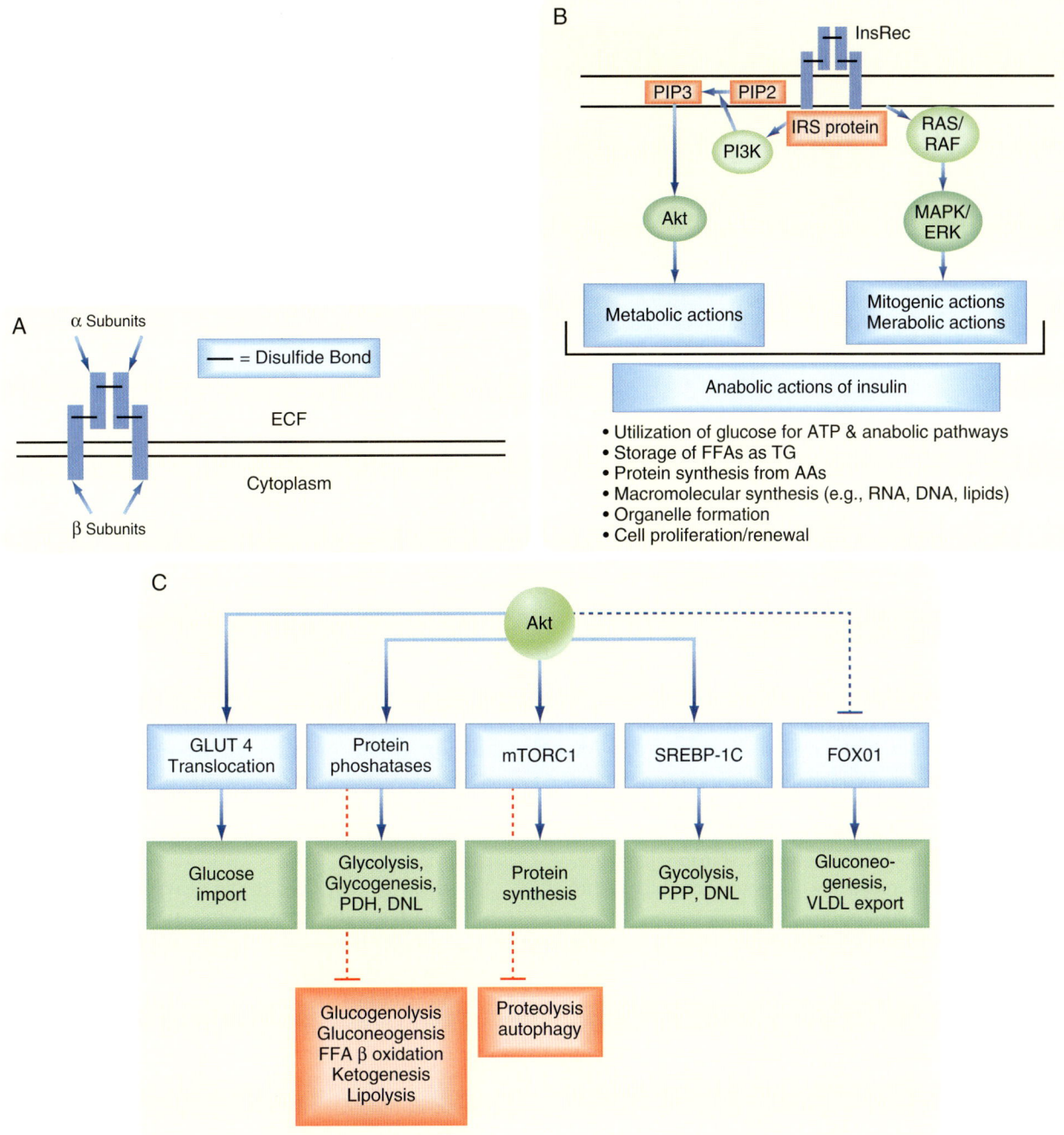

• **Fig. 39.9** A, Structure of dimerized insulin receptor in cell membrane. B, Simplified diagram of the Akt kinase and MAPK pathways downstream of the InsR. C, Summarized actions of insulin/InsR-activated Akt kinase.

IN THE CLINIC

MODY and Beta Cell Transcription Factors

Insulin gene expression and islet cell biogenesis are dependent on several transcription factors specific to the pancreas, liver, and kidney. These transcription factors include **hepatocyte nuclear factor 4α (HNF-4α), HNF-1α, insulin promoter factor 1 (IPF-1), HNF-1β, and neurogenic differentiation 1/beta cell E-box trans-activator 2 (NeuroD1/β₂).** A heterozygous null mutation of one of these factors results in progressively inadequate production of insulin and **maturity-onset diabetes of the young (MODY)** before the age of 25. MODY is characterized by nonketotic hyperglycemia, often asymptomatic, that begins in childhood or adolescence. In addition to the five transcription factors, mutations in **glucokinase** also give rise to MODY.

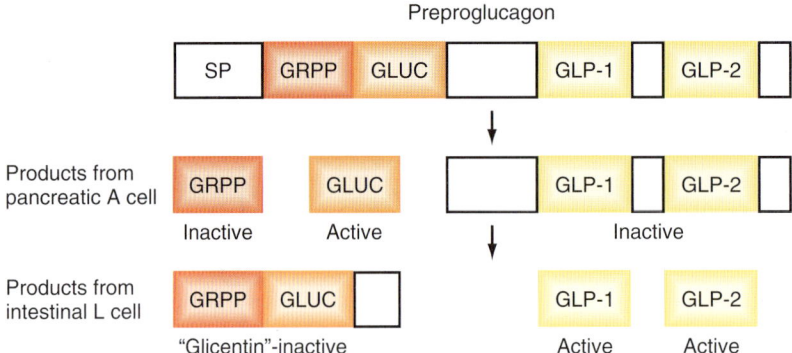

Fig. 39.10 Divergent proteolytic cleavage patterns of the proglucagon molecule. GLUC, glucagon; GLP, glucagon-like peptide; GRPP, glucagon-related polypeptide.

domain on the cytosolic side (see Fig. 39.10A). Binding of insulin to the receptor induces cross-phosphorylation of the subunits. These phosphotyrosine residues are then bound by the **insulin receptor substrate (IRS) proteins** (i.e., IRS proteins are "recruited" to the InsR). The IRS proteins themselves are phosphorylated by the InsR on specific tyrosines, which then recruits **phosphoinositide-3-kinase (PI3K)** to the IRS protein bound to the InsR (see Fig. 39.10B). PI3K converts phosphoinositol-4,5-bisphosphate (PIP2) to **phosphoinositol-3,4,5-trisphosphate (PIP3)**. PIP3 is an informational lipid that recruits proteins to the membrane. In this pathway, PIP3 recruits **Akt protein kinase** to the cell membrane where it becomes activated. This pleiotropic Akt protein kinase signaling pathway orchestrates the numerous metabolic actions of insulin in hepatocytes, skeletal muscle, and adipocytes, including (see Fig. 39.10C):

1. *translocation* of the **GLUT4 glucose transporter** to the cell membrane, thereby allowing import of glucose into **skeletal myocytes** and **adipocytes;**
2. *activation* of **protein phosphatases,** which in turn regulate the activity of multiple metabolic enzymes in all insulin target cells;
3. *activation* of the protein complex **mechanistic target of rapamycin complex 1 (mTORC1),** which promotes protein synthesis and may inhibit proteosomal-mediated protein degradation in insulin target cells;
4. *activation* of the transcription factor **sterol response element binding protein 1 (SREBP1).**

    SREBP1 is especially important for insulin effects on the liver, where it orchestrates glycolysis and **de novo lipogenesis (DNL)** for production of phospholipids, FAs, and TGs from excessive ingested glucose and fructose. InsR/Akt signaling stimulates SREBP1 directly as well as indirectly through activation of mTORC1, which also activates SREBP1. SREBP1 also induces the enzyme that catalyzes the first reaction in the oxidative arm of the **pentose phosphate pathway (PPP).** This reaction generates the coenzyme NADPH, which is required for DNL.

5. *inactivation* of the transcription factor **FOXO1.** Akt-mediated phosphorylation of FOXO1 promotes nuclear exclusion of FOXO1. In the absence of insulin/Akt signaling, FOXO1 induces expression of genes encoding gluconeogenic enzymes and proteins involved in hepatic VLDL assembly and export.

All these actions of Akt will be discussed in more detail later. The InsR also promotes **proliferation/renewal** of some target cells through the **Ras/Raf/mitogen-activated protein kinase (MAPK) pathway** (see Fig. 39.10B). The MAPK pathway also participates in some metabolic regulation.

Glucagon

Glucagon is the primary **counterregulatory hormone** that increases blood glucose levels, primarily through its effects on liver glucose output. Glucagon also enhances intramitochondrial fatty acid oxidation and ketogenesis in hepatocytes.

    Glucagon is a member of the secretin gene family. The precursor **preproglucagon** harbors the AA sequences for glucagon, **GLP-1,** and **GLP-2** (Fig. 39.11). Preproglucagon is proteolytically cleaved in the alpha cell in a cell-specific manner to produce the peptide glucagon. Glucagon circulates in an unbound form and has a short half-life of about 6 minutes. The predominant site of glucagon degradation is the liver, which degrades as much as 80% of circulating glucagon in one pass. Because glucagon enters the hepatic portal vein and is carried to the liver before reaching the systemic circulation, a large portion of the hormone never reaches the systemic circulation. The liver is the primary target organ of glucagon, with lesser effects on adipocytes. Skeletal muscle does not express the glucagon receptor.

    The glucagon receptor is is a Gs-linked G protein–coupled receptor that increases adenylyl cyclase activity and thus cAMP levels. Glucagon exerts many rapid actions through PKA signaling. Glucagon also exerts some transcription effects through phosphorylation and activation of transcription factors such as CREB (cAMP response element binding) protein.

    The **insulin-glucagon ratio** determines the net effect of metabolic pathways on blood glucose. A major stimulus for secretion of glucagon is a decline in blood glucose. Insulin inhibits glucagon secretion, so low blood glucose has an

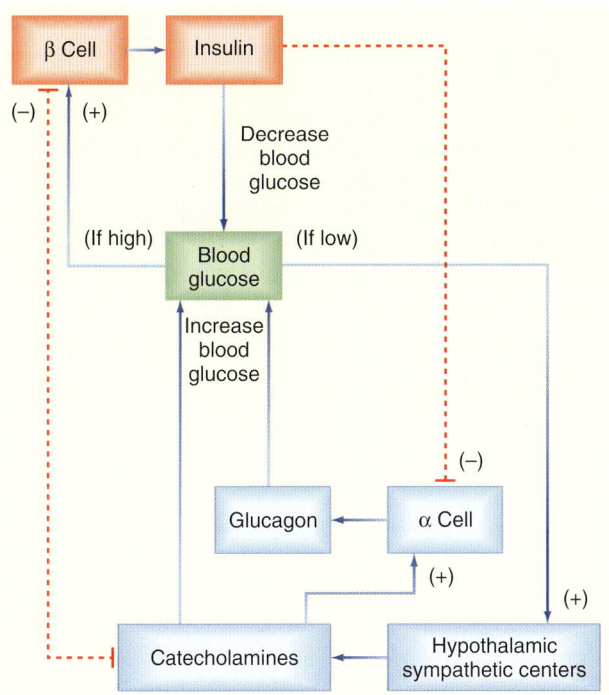

Fig. 39.11 Integrated regulation of blood glucose by insulin and the counterregulatory factors glucagon and catecholamines (norepinephrine, epinephrine).

indirect effect on glucagon secretion through removal of inhibition by insulin (Fig. 39.12). Some recent evidence also indicates that low glucose has a direct effect on alpha cells to increase glucagon secretion.

Circulating catecholamines, which inhibit secretion of insulin via α_2-**adrenergic** receptors, stimulate secretion of glucagon via β_2-**adrenergic receptors** (see Fig. 39.12). Serum AAs promote secretion of glucagon. This means a protein meal will increase postprandial levels of both insulin and glucagon (which protects against hypoglycemia), whereas a carbohydrate meal stimulates only insulin.

Catecholamines: Epinephrine and Norepinephrine

The other major counterregulatory factors are the catecholamines **epinephrine** and **norepinephrine.** Epinephrine is the primary product of the **adrenal medulla** (see Chapter 43), whereas norepinephrine is released from **postganglionic sympathetic nerve endings** (see Chapter 11). Catecholamines are released in response to decreased glucose concentrations, various forms of stress, and exercise. Decreased glucose levels (i.e., hypoglycemia) are primarily sensed by neurons in the CNS, which initiate an integrated

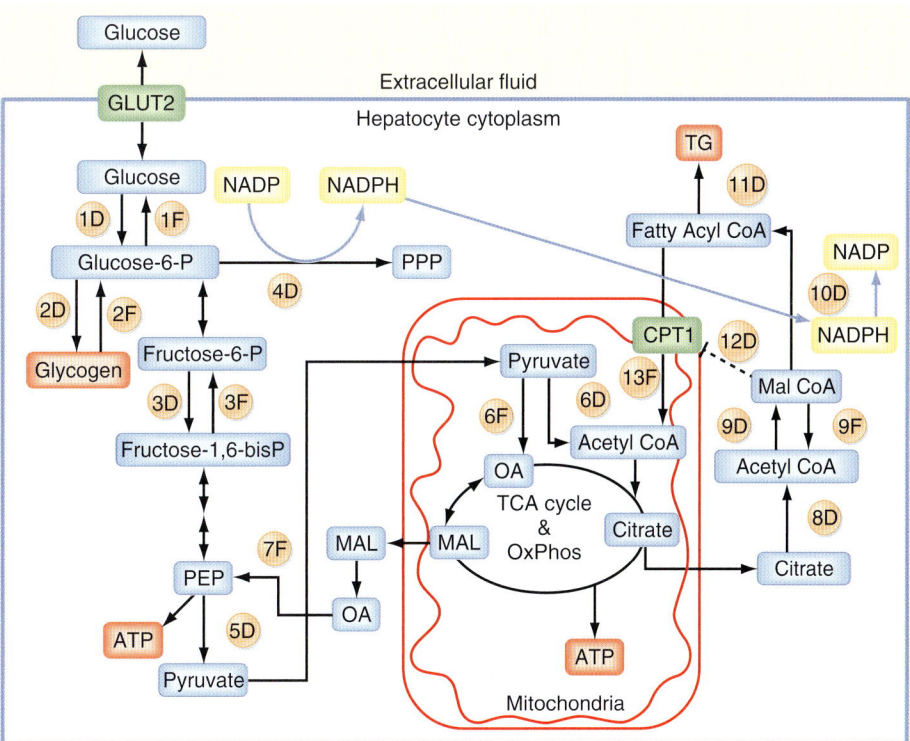

Fig. 39.12 Metabolic pathways in hepatocytes during digestive ("D" numbers) and fasting ("F" numbers) phases. Reactions/pathways: 1D, glucokinase; 1F; G6Pase; 2D, glycogen synthesis; 2F, glycogenolysis; 3D, phosphofructokinase 1; 3F, fructose-1,6-bisphosphatase; 4D, glucose-6-phosphate dehydrogenase; 5D, pyruvate kinase; 6D, pyruvate dehydrogenase; 6F, pyruvate carboxylase; 7F, phosphoenolpyruvate carboxykinase; 8D, ATP-citrate lyase; 9D, acetyl CoA carboxylase; 9F, malonyl CoA decarboxylase; 10D, fatty acid synthase; 11D, glycerol phosphate/fatty acyl CoA transferase and other enzymes involved in esterification and formation of TG; 12D, inhibition by malonyl CoA (Mal CoA) of fatty acyl CoA transporter, carnitine/palmitoyl transporter 1 (CPT1) on outer mitochondrial membrane; 13F, movement of fatty acyl CoA into mitochondrion through CPT1 and beta-oxidation to acetyl CoA.

sympathetic response through the hypothalamus. The direct metabolic actions of catecholamines are mediated primarily by α_1-, β_2-, and β_3-**adrenergic receptors** located on muscle, adipose, and liver tissue (see later). Like the glucagon receptor, β-adrenergic receptors (β_2 and β_3) increase intracellular cAMP. The α_1-adrenergic receptor is also expressed by some organs, especially the liver.

Hormonal Regulation of Specific Metabolic Reactions and Pathways

This section discusses the main pathways in hepatocytes, skeletal myocytes, and adipocytes that contribute to integrated metabolism. For even more detailed description, the student is referred to biochemistry textbooks.

Hepatocyte Metabolism: Digestive vs. Fasting Phases

Some of the key metabolic steps regulated by **insulin and glucagon (and catecholamines)** in the **liver** are as follows (refer to Fig. 39.13 for numbered reactions; D denotes digestive phase, F denotes fasting phase):

1. *Trapping vs. releasing intracellular glucose.* Although glucose enters hepatocytes through insulin-independent GLUT2 transporters, insulin increases hepatic retention and utilization of glucose by increasing expression of **glucokinase** (reaction 1D). Insulin increases glucokinase gene expression through increased expression and activation of the transcription factor **sterol regulatory element–binding protein 1C (SREBP1C),** which acts

as a "master switch" in the fed state to coordinately increase levels of several enzymes involved in glucose utilization and TG synthesis. Hepatocytes also express the enzyme G6Pase (reaction 1F), which converts G6P back to glucose, which can then exit the hepatocyte via the GLUT2 transporter. Insulin prevents the futile cycle of glucose phosphorylation-dephosphorylation by repressing gene expression of the enzyme **G6Pase.** The transcription factor FOXO1 stimulates gene expression of G6Pase. Insulin-activated Akt kinase phosphorylates and inactivates FOXO1. During the fasting phase, FOXO1 is active and promotes G6Pase expression, whereas SREBP1C is inactive and does not stimulate glucokinase expression. The reciprocal regulation of SREBP1C and FOXO1 are thus regulated primarily by the presence or absence of insulin.

2. *Glycogen synthesis vs. breakdown.* Insulin indirectly increases glycogen synthesis through increased expression of glucokinase because high levels of G6P allosterically increase **glycogen synthase** activity. Through stimulation of specific protein phosphatases, insulin promotes dephosphorylation and thereby activation of **glycogen synthase** (reaction 2D). Insulin also prevents the futile cycle of glycogen synthesis to glycogenolysis through inhibition of **glycogen phosphorylase** (reaction 2F). Glucagon-activated PKA phosphorylates phosphorylase kinase, which in turn phosphorylates and activates glycogen phosphorylase.

3. *Increasing glycolysis.*

   A. **Activating phosphofructokinase 1 (PFK1) and inhibiting fructose-2,6-bisphosphatase.** Insulin increases the activity of **PFK1**, which phosphorylates

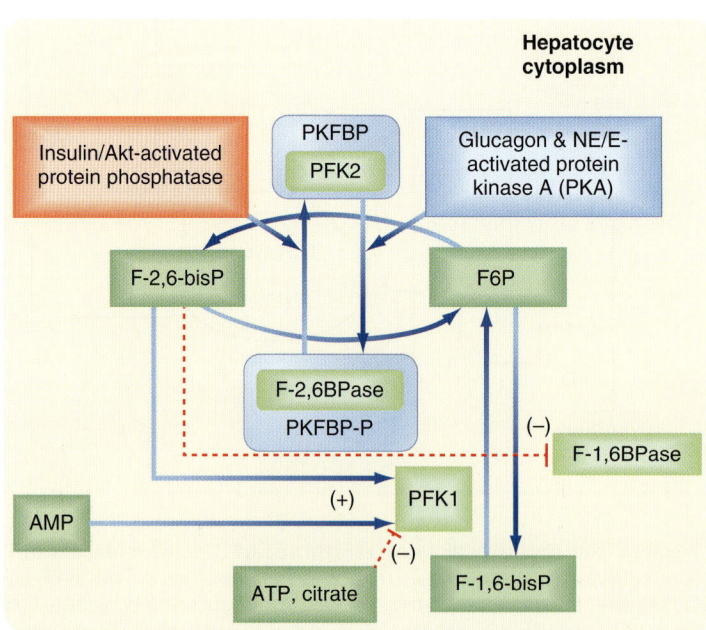

● **Fig. 39.13** Insulin and counterregulatory hormone regulation of phosphofructokinase 1 (PFK1; reaction 3D in Fig. 39.12) and fructose-1,6-bisphosphatase (F1,6BPase; reaction 3F in Fig. 39.12) through changing the activity of the bifunctional enzyme phosphofructokinase 2/fructose-2,6-bisphosphatase (PKFBP) and thus the levels of the allosteric regulatory metabolite fructose-2,6-bisphosphate (F-2,6-bisP).

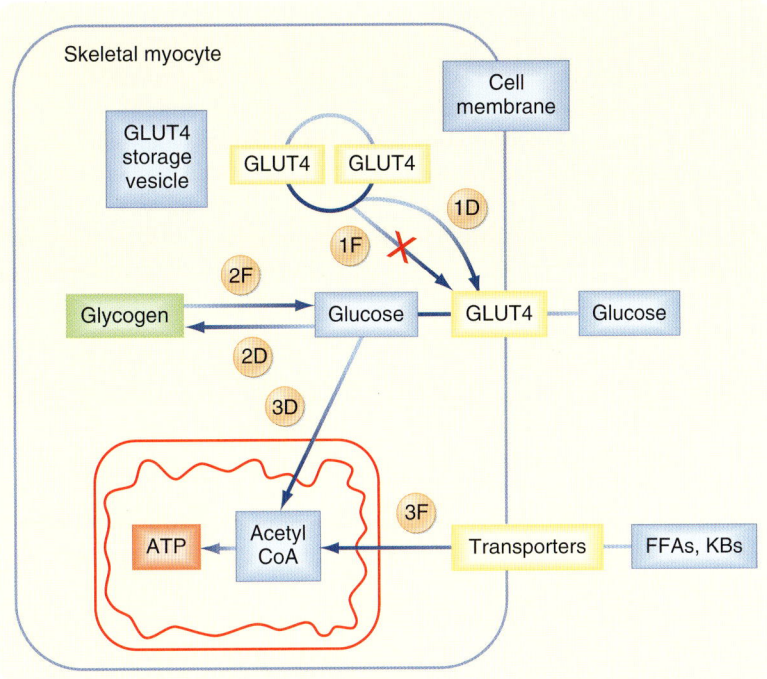

• **Fig. 39.14** Metabolism in skeletal muscle during digestive ("D" reactions) vs. fasting ("F" reactions) phases. Reactions/pathways: 1D, translocation of GLUT4 transporter to cell membrane; 1F, loss of translocation of GLUT4 transporter to cell membrane; 2D, glycogen synthesis; 2F, glycogenolysis; 3D, glycolysis and lactate dehydrogenase, or pyruvate dehydrogenase/TCA cycle/oxidative phosphorylation (OxPhos), depending on muscle fiber type; 3F, β-oxidation of FFAs or ketolysis followed by the TCA cycle and OxPhos.

fructose-6-phosphate (F6P) to fructose-1,6-bisphosphate (reaction 3D). This reaction is referred to as the "commitment" reaction for glycolysis. Insulin also inhibits the reverse reaction, as catalyzed by the gluconeogenic enzyme **fructose-1,6-bisphosphatase.** Insulin regulates these two enzymes through an indirect two-step mechanism that is diagramed in Fig. 39.14. This mechanism involves the bifunctional enzyme **phosphofructokinase-2/ fructose bisphosphatase (PKFBP;** see Fig. 39.14). Insulin/Akt-activated protein phosphatases promote dephosphorylation of PKFBP, thereby activating the kinase function and lessening the phosphatase function. This phosphorylates F6P to **fructose-2,6-bisphosphate (F-2,6-bisP).** F-2,6-bisP in turn allosterically activates PFK1, thereby driving glycolysis. F-2,6-bisP also competitively inhibits fructose-1,6-bisphosphatase, thereby blocking the futile cycle of F6P to fructose-1,6-bisphosphate to F6P.

B. Activating pyruvate kinase (PK). PK catalyzes the irreversible conversion of phosphoenolpyruvate (PEP) to pyruvate (see reaction 5D in Fig. 39.12). Again, insulin/Akt kinase activation of a protein phosphatase dephosphorylates PK, which activates the enzyme. Insulin also increases *PK* gene expression through SREBP1C. Finally, fructose-1,6-bisphosphate (product of reaction 3D) allosterically activates PK.

4. *Activating pyruvate dehydrogenase (PDH) complex.* PDH converts pyruvate to acetyl CoA, which can then enter the TCA cycle upon condensation with oxaloacetate (OA) to form citrate. Insulin increases PDH activity through Akt kinase activation of PDH phosphatase, which in turn dephosphorylates and activates PDH (reaction 6D).

5. *Increasing synthesis of intrahepatic TG.* During the digestive phase, some **acetyl CoA** is transferred from the mitochondria to the cytosol in the form of **citrate,** which is then converted back to acetyl CoA and oxaloacetate by the cytosolic enzyme **ATP-citrate lyase** (reaction 8D). Insulin increases ATP-citrate lyase gene expression through transcription factor SREBP1C. Once in the cytoplasm, acetyl CoA can enter fatty acid synthesis. The first step involves conversion of **acetyl CoA** to **malonyl CoA** by the enzyme **acetyl-CoA carboxylase** (reaction 9D). Insulin stimulates acetyl-CoA carboxylase gene expression through the transcription factor SREBP1C. Insulin also promotes dephosphorylation of acetyl-CoA carboxylase, which activates the enzyme. Malonyl CoA is converted to the 16-carbon fatty acid **palmitoyl CoA** by repetitive additions of acetyl groups (contributed by malonyl CoA) by the **fatty acid synthase (FASN) complex** (reaction 10D). *FASN* gene expression is enhanced by insulin through the transcription factor SREBP1C. Insulin also stimulates **palmitoyl-CoA desaturase,** which produces unsaturated fatty acids, and

glycerol phosphate–fatty acyl transferases that esterify FFAs to G3P to form intrahepatic TG.

Palmitate synthesis requires the coenzyme **NADPH.** A major source of NADPH is the **pentose phosphate pathway** (see Fig. 39.12). The first reaction converts G6P to 6-phosphogluconolactone by the enzyme **glucose-6-phosphate dehydrogenase (G6PD; reaction 4D).** Insulin increases *G6PD* gene expression through the transcription factor SREBP1C.

By activating steps that lead to generation of malonyl CoA, insulin indirectly inhibits oxidation of FFAs. Malonyl CoA inhibits the activity of CPT-I, which transports FFAs from the cytosol into the mitochondria (reaction 12D). As a result, FFAs that are synthesized by DNL cannot be transported into mitochondria, where they undergo β-oxidation (reaction 13F). Thus increased malonyl CoA prevents the futile cycle of FFA synthesis to FFA oxidation.

FFAs are converted to TGs by the liver (reaction 11D) and are either stored in the liver or transported to adipose tissue and muscle in the form of VLDL (see later). Insulin acutely promotes degradation of the VLDL apoprotein apoB-100. This keeps the liver from secreting VLDL during a meal when the blood is rich with chylomicrons. Thus the lipid made in response to insulin during a meal is released as VLDL during the interdigestive and fasting phases and provides an important source of energy to skeletal and cardiac muscle.

6. *Activation vs. inhibition of the gluconeogenic enzymes pyruvate carboxylase (PC) and phosphoenolpyruvate carboxykinase (PEPCK).* Pyruvate can also be converted to OA by PC (reaction 6F). However, this reaction is indirectly inhibited by insulin in several ways. First, insulin activates PDH as just discussed, thereby diverting pyruvate away from the PC reaction. Additionally, PC is allosterically activated by high levels of intramitochondrial acetyl CoA. Insulin keeps intramitochondrial levels of acetyl CoA low by activation of cytosolic DNL, which promotes removal of acetyl CoA via citrate from the mitochondria. Another key mechanism is to prevent β-oxidation of FFAs within the mitochondria, which generates high levels of acetyl CoA. By stimulating DNL, insulin also increases levels of cytosolic malonyl CoA, which inhibits transport of FFAs into the mitochondria (reaction 12D). Also, inhibitory actions of insulin on glucagon secretion and on lipolysis of TG within adipocytes prevents release of FFAs by adipose tissue and their import into hepatocytes.

In contrast, during the fasting phase, low insulin coupled with high glucagon and/or catecholamines stimulate release of FFAs from adipocytes (see later), which increases the flow of FFAs into hepatocytes. Glucagon also phosphorylates and activates the enzyme malonyl decarboxylase, which converts malonyl CoA back to acetyl CoA (reaction 9F). Enhanced malonyl CoA decarboxylase, along with generally low DNL due to low insulin, reduces malonyl CoA levels and thus removes the inhibition on the CPT1 transporter. This allows FFAs to enter the mitochondria and undergo β-oxidation (reaction 13F), generating high levels of intramitochondrial acetyl CoA, activating PC (reaction 6F) and also allosterically inhibiting PDH (reaction 6D).). The enzymes involved in β-oxidation are activated by PKA signaling. Glucagon also activates the transcription factor PPARα, which further induces expression of enzymes involved in β-oxidation. Fibrate drugs activate PPARα, promoting oxidation of intrahepatic TG and ameliorating insulin resistance.

Insulin also represses gene expression of the gluconeogenic enzyme **PEPCK** (reaction 7F), which converts pyruvate (by way of malate transferase out of the mitochondria) to phosphoenolpyruvate. PEPCK is primarily regulated at the level of transcription. Similar to its actions on G6Pase, FOXO1 stimulates transcription of PEPCK during the fasting phase, and insulin/Akt kinase signaling inactivates FOXO1 during the digestive phase. Glucagon and catecholamines also increase *PEPCK* gene expression through PKA-CREB signaling during the fasting phase.

Skeletal Muscle and Adipose Tissue Metabolism: Digestive vs. Fasting Phases

1. *Skeletal muscle* (Fig. 39.15). **Glucose tolerance** refers to the ability of an individual to minimize the increase in blood glucose concentration after a meal. A primary way by which insulin promotes glucose tolerance is activation of glucose transporters in skeletal muscle. Insulin stimulates translocation of preexisting **GLUT4 transporters** to the cell membrane (reaction 1D). Insulin also promotes storage of glucose in muscle as glycogen (reaction 2D) and promotes oxidation of glucose through glycolysis (reaction 3D). During the fasting phase, low insulin results in a low number of GLUT4 transporters at the membrane (reaction 1F), so these cells consume less glucose (**glucose sparing**). Skeletal muscle fibers with mitochondria switch to the use of FFAs from adipocytes and KBs from hepatocytes (reaction 3F). Skeletal myocytes do not express glucagon receptors. Uptake of FFAs and KBs and their oxidation for ATP is largely upregulated by intracellular Ca^{++} levels and a high AMP:ATP ratio. Exercise also activates these pathways, as does glycogenolysis (reaction 2F), through adrenergic receptor stimulation.

2. *Adipocytes → glucose* (Fig. 39.16A). Insulin also stimulates GLUT4-dependent uptake of glucose and subsequent glycolysis in adipose tissue (reactions 1D and 2D). Adipose tissue uses glycolysis for energy needs but also for generating G3P (reaction 3D), which is required for esterification of FFAs into TGs (reaction 4D). During the fasting phase, insulin is low, so GLUT4 movement to the cell membrane is blocked (reaction 1F).

2. *Adipocytes → FFAs and TG* (see Fig. 39.16B). Insulin stimulates expression of **lipoprotein lipase (LPL)**

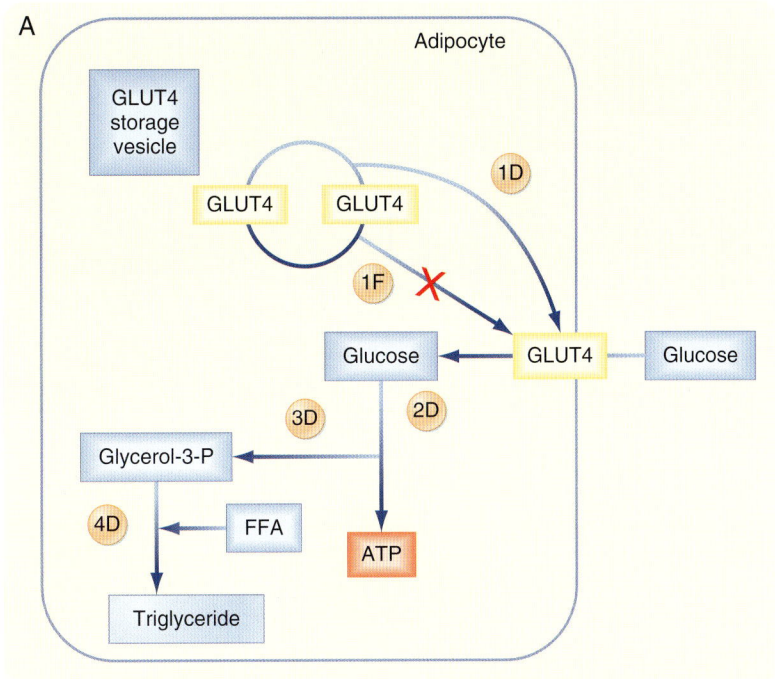

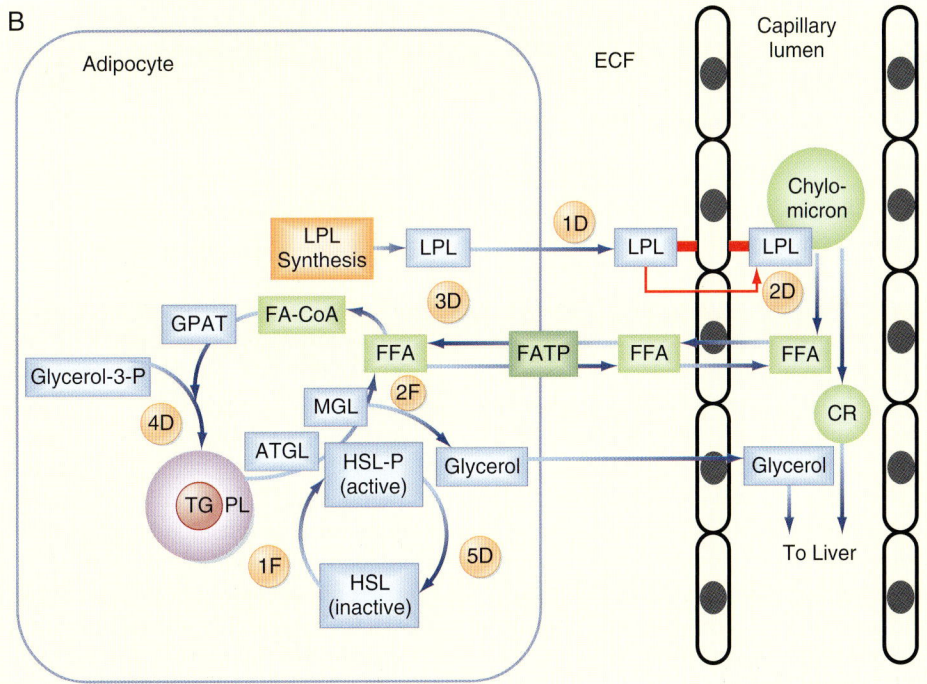

• **Fig. 39.15 A, Glucose metabolism in an adipocyte during digestive ("D" reactions) and fasting ("F" reactions) phases.** Reactions/pathways: 1D and 1F (also see Fig. 39.14 legend); 2D, glycolysis, pyruvate dehydrogenase/TCA cycle/OxPhos; 3D, glycerol-3-phosphate dehydrogenase; 4D, esterification of FFAs to G3P to form triglyceride. **B, Lipid metabolism in adipocyte during digestive ("D" reactions) and fasting ("F" reactions) phases.** Reactions/pathways: 1D, synthesis of lipoprotein lipase (LPL) and LPL secretion into subcapillary space, binding to GPI-anchored protein *(red box),* and migration to luminal side of capillary endothelial cell; 2D, lipolysis of chylomicron TG and releasing free FFA (after digestion, chylomicron remnant is cleared from circulation by liver); 3D, activation of imported FFAs by transfer to acetyl CoA to form fatty acyl CoA; 4D, esterification of fatty acyl CoAs to G3P to form TG (TG droplets are covered and stabilized by perilipins [PL]); 5D, dephosphorylation and inactivation of hormone-sensitive lipase (HSL), which promotes storage of TG; 1F, phosphorylation and activation of HSL, which contributes to complete lipolysis of TG (ATGL, adipocyte TG lipase; HSL, diacylglyceride lipase; MGL, monoacylglyceride lipase); 2F, final step in TG lipase by MGL releases FFA and glycerol.

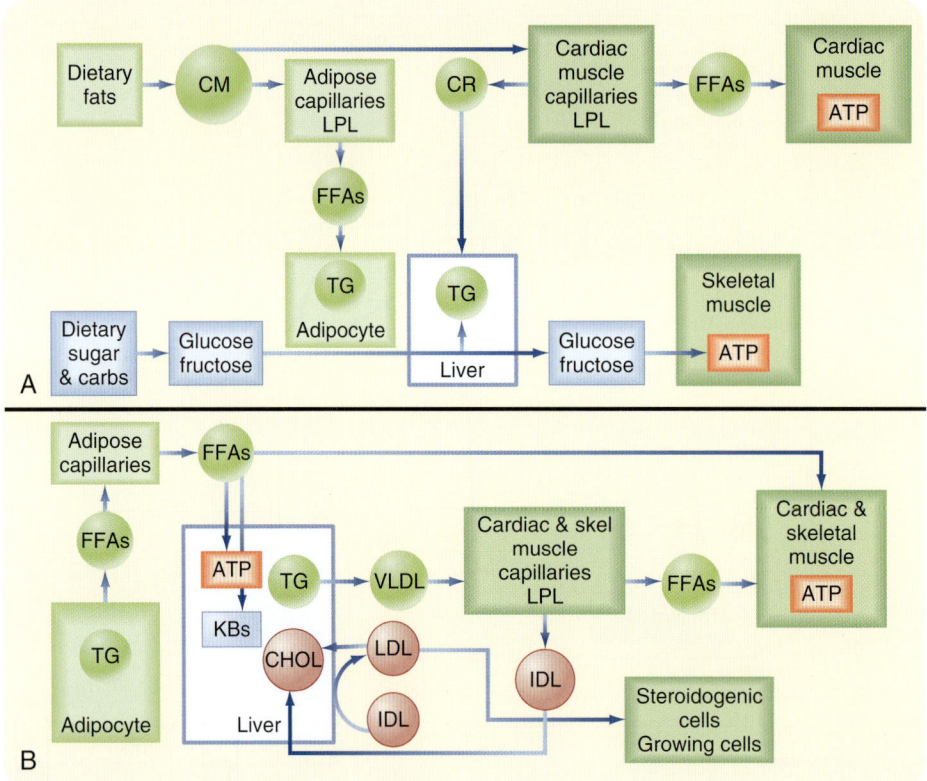

• **Fig. 39.16** Role of lipoproteins in energy metabolism. A, Digestive phase. B, Fasting phase.

within adipocytes and its migration to the apical side of endothelia in adipose capillaries (reaction 1D). This action of insulin allows LPL to extract FFAs from chylomicrons within adipose tissue capillary beds (reaction 2D). The chylomicron remnants (discussed later) are removed by the liver. Insulin also stimulates activation of imported FFAs by their conversion to fatty acyl CoAs (reaction 3D). Insulin stimulates glycolysis in adipocytes, which generates the G3P required for reesterification of FFAs into TGs (reaction 4D). Insulin directly inhibits **hormone-sensitive lipase** (**HSL**; reaction 5D), thereby promoting storage of FFAs as opposed to their release. During the fasting phase, glucagon and catecholamines phosphorylate and activate HSL (reaction 1F), thereby promoting release of FFAs and glycerol from stored TG (reaction 2F). In the absence of insulin these two products of lipolysis are exported into the blood.

Protein Metabolism in All Hormone Target Cells: Digestive vs. Fasting Phases

Insulin promotes protein synthesis in muscle and adipose tissue by stimulating AA uptake and mRNA translation. Insulin also inhibits proteolysis. Although the liver uses AAs for ATP synthesis, insulin also promotes synthesis of proteins during the digestive phase and attenuates the activity of urea cycle enzymes in the liver. Glucagon and catecholamines activate proteosomal degradation of proteins and release of AAs during the fasting phase.

Metabolic Roles of Lipoproteins: Digestive vs. Fasting Phases

This section deals with lipoproteins in some detail. For more details, consult a biochemistry textbook.

FFAs circulate in the blood primarily bound to **albumin.** However, TG, free cholesterol, cholesterol esters, phospholipids and some lipid-soluble vitamins, all of which are hydrophobic and would partition into the membranes of endothelial cells instead of circulating, are transported through blood within lipid aggregations (i.e., a mix of the above) bound by specific apoproteins. These lipid-protein complexes are referred to as **lipoproteins.** The **TG-rich lipoproteins** are **chylomicrons** and **VLDL** and primarily function to deliver FFAs (as TG) to skeletal and cardiac muscle for energy and to adipocytes for storage. The **cholesterol-rich lipoproteins** include **low-density lipoprotein (LDL)** and **high-density lipoprotein (HDL),** which deliver cholesterol to proliferating cells, steroidogenic cells, and bile-producing hepatocytes. HDL also removes excess cholesterol (i.e., from macrophage-engulfed dead cells) from the periphery. There are also **"remnants"** of lipoproteins that have their lipid cargo partially digested and then cleared from the circulation by the liver.

Digestive Phase: Chylomicrons and Chylomicron Remnants

TGs in a meal are enzymatically digested to FFAs and 2-monoglycerides within the lumen of the intestine.

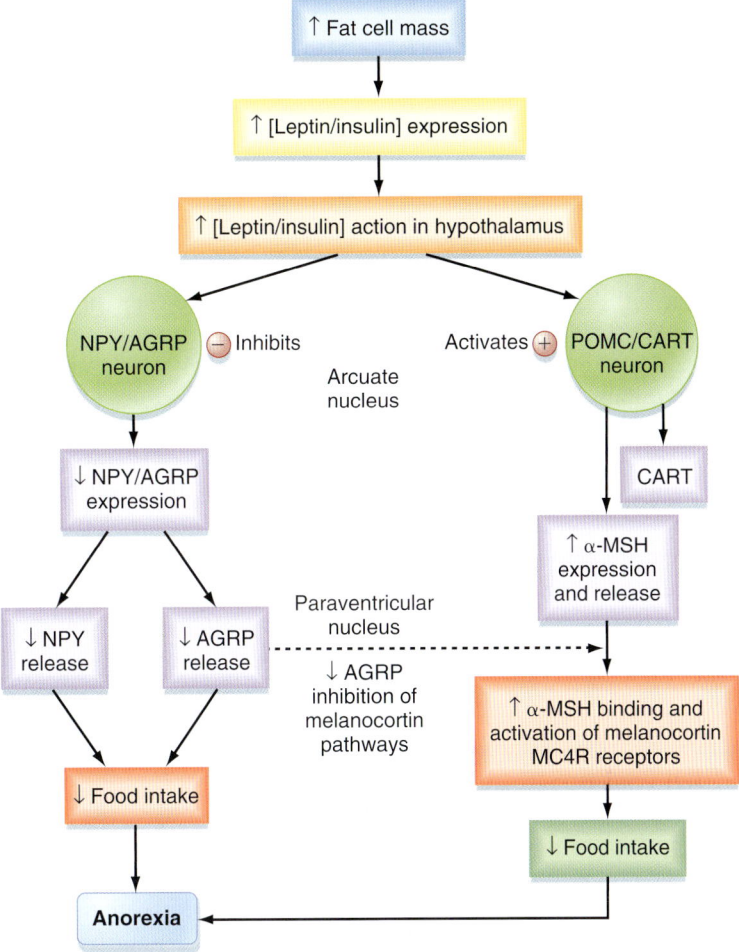

• **Fig. 39.17** Leptin and hypothalamic centers involved in regulation of appetite. See text for explanations of abbreviations.

Intestinal enterocytes import both of these lipids and reesterify them to form TGs. TGs, along with fat-soluble vitamins, cholesterol, cholesterol esters, and phospholipids, are complexed with the protein **ApoB48** to form **chylomicrons.** Chylomicrons are secreted, move into lymphatics, and then ultimately enter the circulation. While in the blood, other apoproteins such as **ApoE** and **ApoC2** are transferred to the chylomicrons from HDL particles (one function of HDL is to provide a circulating reservoir of various apoproteins). This converts nascent chylomicrons into mature chylomicrons.

When chylomicrons enter the capillaries of adipose tissue during the digestive phase, they are partially digested by **lipoprotein lipase (LPL).** LPL is synthesized by adipocytes and secreted into the subendothelial space. LPL then binds to the endothelial membrane GPI-anchored protein GPIHBP1, which transports LPL to the luminal (apical) surface of the capillary endothelial cell. Once in this position, LPL molecules come into contact with chylomicrons. ApoC2 within chylomicrons is an activator of LPL dimerization and activity. FFAs are released from chylomicrons by LPL-mediated lipolysis of TG. (See earlier discussion and Fig. 39.16B for an

explanation of the processing of FFAs to stored TG within adipocytes.

LPL is also expressed in cardiac and skeletal muscle. Cardiac muscle preferentially uses FFA for energy and obtains most FFAs from lipoprotein particles (Fig. 39.17). Thus cardiac muscle also extracts FFA from chylomicrons during the digestive phase. The activity of LPL in cardiomyocytes is highly regulated by local factors such as the local concentration of FFAs within the coronary capillary beds. LPL activity in skeletal muscle is relatively low during the digestive phase.

After lipolytic digestion within adipose tissue capillary beds, chylomicrons are converted to smaller, denser **chylomicron remnants (CR)** that now have reduced TG content. CR particles are able to penetrate the tunica intima of blood vessels at sites with endothelial dysfunction and thus are atherogenic. Because they still have ApoE protein associated with them, they can bind to one of several membrane receptors that recognize ApoE. CRs can also bind to specific glycolipids. Bound CRs are then endocytosed by hepatocytes (see Fig. 39.17). Remaining FFAs that are released after endocytosis of CRs are reesterified into intrahepatic TG.

Fasting Phase: VLDL, IDL, and LDL

The source of circulating TG during the fasting phase is primarily the liver (see Fig. 39.17). During the digestive phase, intrahepatic TGs accumulate from de novo lipogenesis using the carbons in cytosolic citrate and from endocytosed CRs. Intrahepatic TG, along with other lipids including cholesterol and cholesterol esters, is exported by hepatocytes as **VLDL.** VLDL particles are assembled as lipids complexed to the **ApoB100 protein.** Expression of ApoB100, along with other components involved in VLDL assembly, is stimulated by transcription factor **FOXO1.** FOXO1 in turn is inhibited by the insulin signaling pathway. This means that hepatic VLDL production is minimal during the time when blood is rich in chylomicrons. During the fasting phase, insulin levels are low, so FOXO1 activity is high, and VLDL assembly and secretion resumes. Once VLDL particles enter the circulation, they accept other apoproteins (e.g., ApoE, ApoC2) and become mature VLDL.

Adipocytes display low LPL activity during the fasting phase, in part owing to low insulin levels. However, cardiomyocytes and skeletal myocytes express LPL, which digests VLDL and provides FFA to these muscle cell types during the fasting phase. Lipolytic extraction of some FFAs from VLDL generates a remnant particle called **intermediate-density lipoprotein (IDL).** IDL circulates to the liver, where it s processed in one of two ways (see Fig. 39.17). About half of the IDL binds to one of several **ApoE-recognizing receptors,** undergoes receptor-mediated endocytosis, and is digested in endolysosomes. Released lipids can be reassembled into VLDL particles and returned to the circulation to provide fuel for cardiac and skeletal muscle as the fasting phase progresses. The other half of the IDL undergoes further digestion from the hepatocyte-specific LPL-related enzyme **hepatic lipase.** Hepatic lipase extracts most of the remaining TG in the IDL, forming the final remnant of VLDL, **LDL.** LDL is TG poor but cholesterol rich. It should be noted that both mature chylomicrons and VLDL can receive additional cholesterol from HDL while in the circulation through the action of **cholesterol ester transport protein (CETP),** so cholesterol content of the remnant particles (ChyR, IDL, and LDL) can vary. In any case the LDL particle is a small, dense, cholesterol-rich particle that is potentially very atherogenic in the face of endothelial damage. LDL particles are safely imported into cells through the **LDL receptor.** It should be noted that in the conversion of IDL to LDL, the **ApoE protein** disassociates from the particle. This means that only receptors for which ApoB100 is a ligand can remove LDL from the blood. In contrast to the multiple ApoE receptors, only one receptor, the LDL receptor, can recognize and bind ApoB100. Thus loss or decrease of a functional LDL receptor has significant clinical consequences (see At the Cellular Level box). LDL receptor is expressed on **proliferating cells,** including some cancer cells, which need to synthesize new cell membranes. LDL receptor is also expressed on **steroidogenic cells,** which use cholesterol to make steroid hormones. The major site of LDL uptake is the **liver,** which secretes cholesterol as well as cholesterol-based **bile acids,** as bile into the biliary tree. Some cholesterol is excreted by the intestines. Other cholesterol byproducts (e.g., steroid hormones) are excreted primarily at the kidney.

Leptin and Energy Balance

White adipose tissue (WAT) is composed of several cell types. The TG-storing cell is called the **adipocyte.** These cells develop from preadipocytes during gestation in humans. This process of adipocyte differentiation, which may continue throughout life, is promoted by several transcription factors. One of these factors is SREBP1C, which is activated by lipids as well as insulin and several growth factors and cytokines. Another important transcription factor in WAT is **PPARγ.** Activated PPARγ promotes expression of genes involved in TG storage. Thus an increase in food consumption leads to activation of SREBP1C and PPARγ, which increase the differentiation of preadipocytes into small adipocytes and upregulation of enzymes within these cells to allow storage of excess fat.

Adipose tissue produces paracrine and endocrine factors, including adiponectin, TNF-α, resistin, interleukin-6, angiotensinogen, and acylation-stimulating protein.

Leptin

Leptin is an adipocyte-derived protein that signals information to the hypothalamus about the degree of adiposity and nutrition, which in turn controls eating behavior and energy expenditure. Leptin-deficient mice and humans become morbidly obese. These findings originally raised hope that leptin therapy could be used to combat morbid obesity. However, administration of leptin to individuals who suffer from diet-induced obesity does not have a significant anorectic or energy-consuming effect. In fact, obese individuals already have elevated endogenous circulating levels of leptin and appear to have developed **leptin resistance.**

Leptin has an important role in liporegulation in peripheral tissues. Leptin protects peripheral tissues (e.g., liver, skeletal muscle, cardiac muscle, beta cells) from accumulation of too much lipid by directing storage of excess caloric intake into adipose tissue. This action of leptin, though opposing the lipogenic actions of insulin, contributes significantly to maintenance of insulin sensitivity (as defined by insulin-dependent glucose uptake) in peripheral tissues. Leptin also acts as a signal that the body has sufficient energy stores to allow reproduction and to enhance erythropoiesis, lymphopoiesis, and myelopoiesis. For example, in women suffering from anorexia nervosa, leptin levels are extremely low and result in low ovarian steroids, amenorrhea (lack of menstrual bleeding), anemia from low red blood cell production, and immune dysfunction.

Structure, Synthesis, and Secretion

Leptin, a 16-kDa protein secreted by mature adipocytes, is structurally related to cytokines. Thus it is sometimes

AT THE CELLULAR LEVEL

SREBP2 was discovered as a transcription factor that resides in the membrane of the endoplasmic reticulum (ER). In the presence of high intracellular cholesterol, SREBP2 is held in the ER by a lipid-sensing protein called *SCAP* (SREBP cleavage–activating protein). In response to depleted sterols, SCAP escorts SREBP2 to the Golgi, where SREBP is cleaved sequentially by proteases and released into the cytoplasm. SREBP2 then translocates to the nucleus and increases the transcription of genes involved in synthesis and uptake of cholesterol. A more recently discovered member of this transcription factor family is **SREBP1C,** which is highly expressed in adipose and liver. In contrast to SREBP2, SREBP1C stimulates genes involved in synthesis of FA and TG. Regulation of SREBP1C occurs at the transcriptional level of the *SREBP1C* gene, with cleavage induced by polyunsaturated fatty acids and activation by the MAPK pathway.

   Peroxisome proliferation activator receptors (PPARs) belong to the nuclear hormone receptor superfamily that also includes steroid hormone receptors and thyroid hormone receptors. PPARs heterodimerize with the **retinoid X receptors (RXRs).** Unlike steroid and thyroid hormone receptors, PPARs bind to ligands in the micromolar range (i.e., with lower affinity). PPARs bind saturated and unsaturated fatty acids as well as natural and synthetic prostanoids. **PPARγ** is highly expressed in adipose tissue and at a lower level in skeletal muscle and liver. Its natural ligands include several polyunsaturated fatty acids. PPARγ regulates genes that promote fat storage. It also synergizes with SREBP1C to promote differentiation of adipocytes from preadipocytes. Tissue-specific knockout of PPARγ in mice and PPARγ-dominant negative mutations in humans give rise to **lipodystrophy** (i.e., lack of white adipose tissue), which leads to deposits of TG in muscle and liver (called *steatosis*), insulin resistance, diabetes, and hypertension.

The **thiazolidinediones** are exogenous ligands for PPARγ. Although they promote weight gain, moderate levels of thiazolidinediones significantly improve insulin sensitivity. PPARγ also stimulates secretion of **adiponectin,** which promotes oxidation of lipids in muscle and fat and thereby improves insulin sensitivity. **PPARα** is abundantly expressed in liver and to a lesser extent in skeletal and cardiac muscle and kidney. PPARα promotes uptake and oxidation of FFAs. Thus PPARα is an antisteatotic molecule. The **fibrates** are exogenous ligands of PPARα and are used to reduce TG deposits in muscle and liver, thereby improving insulin sensitivity. A third member, **PPARδ,** similarly promotes fatty acid oxidation in adipose and muscle tissue. PPARδ promotes development of slow-twitch oxidative muscle fibers and increases muscle stamina. PPARδ has a positive effect on lipoprotein metabolism by increasing production of ApoA apoproteins and the number of HDL particles.

   Another family of lipid-sensing transcription factors is the **liver X receptor (LXR)** family, which is composed of LXRα and LXRβ. LXRα is expressed primarily in adipose tissue, liver, intestine, and kidney, whereas LXRβ is ubiquitously expressed. LXRs are related to PPARs in that they are members of the nuclear hormone receptor family and heterodimerize with RXR. LXRs are cholesterol sensors. In high-cholesterol conditions, LXRs upregulate expression of ATP-binding cassette (ABC) proteins. In the face of excess cholesterol, LXRs also increase ABC protein expression in the gastrointestinal tract, which promotes efflux of cholesterol from enterocytes to the lumen for excretion. Mutations in these transporters (ABCG5 and ABCG8) cause **sitosterolemia,** characterized by excessive absorption of cholesterol and plant sterols. In the liver, LXRs promote conversion of cholesterol to bile acids for excretion or to cholesterol esters for storage. In the latter action, LXRs increase SREBP1C expression, thereby increasing the fatty acyl CoAs needed for esterification.

---

referred to as an **adipocytokine.** Circulating levels of leptin have a direct relationship with adiposity and nutritional status. Leptin output is increased by insulin, which prepares the body for correct partitioning of incoming nutrients. Leptin is inhibited by fasting and weight loss and by lipolytic signals (e.g., increased cAMP and β₃-agonists).

Diet-induced obesity, advanced age, and T2DM are associated with leptin resistance. Thus mechanisms that turn off leptin signaling are potential therapeutic targets.

Energy Storage

The amount of energy stored by an individual is determined by caloric intake and calories expended as energy per day. In many individuals, input and output are in balance, so weight remains relatively constant. However, the abundance of inexpensive high-fat, high-carbohydrate food, along with more sedentary lifestyles, is currently contributing to a pandemic of obesity and the pathological sequelae of obesity, including T2DM and cardiovascular disease.

The preponderance of stored energy consists of fat, and individuals vary greatly in the amount and percentage

of body weight that is accounted for by adipose tissue. About 25% of the variance in total body fat appears to be due to genetic factors. A genetic influence on fat mass is supported by (1) the tendency for the body mass of adopted children to correlate better with that of their biological parents than with that of their adoptive parents; (2) the greater similarity of adipose stores in identical (monozygotic) twins, whether reared together or apart, than in fraternal (dizygotic) twins; (3) the greater correlation between gains in body weight and abdominal fat in identical twins than in fraternal twins when they are fed a caloric excess; and (4) the discovery of several genes that cause obesity.

In addition, the gestational environment has a profound effect on body mass of the adult. The effect of maternal diet on the weight and body composition of offspring is called **fetal programming.** Low birth weights correlate with increased risk for obesity, cardiovascular disease, and diabetes. These findings suggest that the efficiency of fetal metabolism has plasticity and can be altered by the in utero environment. The development of a "thrifty" metabolism would be advantageous to an individual born to a mother

IN THE CLINIC

Diabetes mellitus is a disease in which insulin levels or responsiveness of tissues to insulin (or both) is insufficient to maintain normal levels of plasma glucose. Although the diagnosis of diabetes is based primarily on plasma glucose, diabetes also promotes imbalances in circulating levels of lipids and lipoproteins (i.e., **dyslipidemia**). Major symptoms of diabetes mellitus include hyperglycemia, polyuria, polydipsia, polyphagia, muscle wasting, electrolyte depletion, and ketoacidosis (in T1DM). With normal fasting (i.e., no caloric intake for at least 8 hours), plasma glucose levels should be below 110 mg/dL. A patient is considered to have impaired glucose control if fasting plasma glucose levels are between 110 and 126 mg/dL, and the diagnosis of diabetes is made if fasting plasma glucose exceeds 126 mg/dL on 2 successive days. Another approach to the diagnosis of diabetes is the oral glucose tolerance test. After overnight fasting the patient is given a bolus of glucose (usually 75 g) orally, and blood glucose levels are measured at 2 hours. A 2-hour plasma glucose concentration greater than 200 mg/dL on 2 consecutive days is sufficient to make the diagnosis of diabetes. The diagnosis of diabetes is also indicated if the patient has symptoms associated with diabetes and has a nonfasting plasma glucose level greater than 200 mg/dL.

Diabetes mellitus is currently classified as **type 1 (T1DM)** or **type 2 (T2DM).** T2DM is by far the more common form and accounts for 90% of diagnosed cases. However, T2DM is usually a progressive disease that remains undiagnosed in a significant percentage of patients for several years. T2DM is often associated with visceral obesity and lack of exercise—indeed, obesity-related T2DM is reaching epidemic proportions worldwide. Usually there are multiple causes for the development of T2DM in a given individual that are associated with defects in the ability of target organs to respond to insulin (i.e., **insulin resistance**), along with some degree of **beta cell deficiency.** Insulin sensitivity can be compromised at the level of the InsR or at the level of postreceptor signaling. T2DM appears to be the consequence of insulin resistance, followed by reactive hyperinsulinemia, but ultimately by **relative hypoinsulinemia** (i.e., inadequate release of insulin to compensate for the end-organ resistance) and **beta cell failure.**

The underlying causes of insulin resistance differ among patients. Three major underlying causes of obesity-induced insulin resistance are:

1. *Decreased ability of insulin to increase GLUT4-mediated uptake of glucose, especially by skeletal muscle.* This function, which is specifically a part of **glucometabolic regulation by insulin,** may be due to excessive accumulation of TG in muscle in obese individuals. Excessive caloric intake induces hyperinsulinemia. Initially this leads to excessive glucose uptake into skeletal muscle. Just as in the liver, excessive calories in the form of glucose promote lipogenesis and, through generation of malonyl CoA, repression of fatty acyl CoA oxidation. Byproducts of fatty acid and TG synthesis (e.g., diacylglycerol, ceramide) may accumulate and stimulate signaling pathways (e.g., protein kinase C–dependent pathways) that antagonize signaling from the InsR or IRS proteins, or both. Thus insulin resistance in the skeletal muscle of obese individuals may be due to **lipotoxicity.**

2. *Decreased ability of insulin to repress hepatic glucose production.* The liver makes glucose by glycogenolysis in the short term and by gluconeogenesis in the long term. The ability of insulin to repress key hepatic enzymes in both these pathways is attenuated in insulin-resistant individuals. Insulin resistance in the liver may also be due to lipotoxicity in obese individuals (e.g., **fatty liver** or **hepatic steatosis**). Visceral adipose tissue is likely to affect insulin signaling at the liver in several ways, in addition to the effects of lipotoxicity. For example, visceral adipose tissue releases the cytokine **tumor necrosis factor (TNF)-α,** which has been shown to antagonize insulin signaling pathways. Also, TG in visceral adipose tissue has a high rate of turnover (possibly because of rich sympathetic innervation), so the liver is exposed to high levels of FFAs, which further exacerbates hepatic lipotoxicity.

3. *Inability of insulin to repress hormone-sensitive lipase or increase LPL in adipose tissue (or both).* High HSL and low LPL are major factors in the dyslipidemia associated with insulin resistance and diabetes. Although the factors that resist the actions of insulin on HSL and LPL are not completely understood, there is evidence for increased production of paracrine diabetogenic factors in adipose tissue, such as TNF-α. The dyslipidemia is characterized as hypertriglyceridemia with large TG-rich VLDL particles produced by the liver. Because of their high TG content, large VLDLs and IDLs are digested very efficiently, thereby giving rise to small, dense LDL particles that are very atherogenic. In addition, HDL takes on excess TG in exchange for cholesterol esters, which appears to shorten the circulating half-life of HDL and ApoA proteins. Thus there are lower levels of HDL particles, which normally play a protective role against vascular disease.

T1DM is characterized by destruction of beta cells, almost always by an autoimmune mechanism. T1DM is also termed *insulin-dependent diabetes mellitus.* Characteristics of T1DM are:

1. People with T1DM need exogenous insulin to maintain life and prevent ketosis; virtually no pancreatic insulin is produced.

2. There is pathological damage to the pancreatic beta cells. Insulinitis with pancreatic mononuclear cell infiltration is a characteristic feature at the onset of the disorder. Cytokines may be involved in the early destruction of the pancreas.

3. People with T1DM are prone to ketoacidosis.

4. Ninety percent of cases begin in childhood, mostly between 10 and 14 years of age. This common observation led to application of the term *juvenile diabetes* to the disorder. This term is no longer used because T1DM can arise at any time of life, although juvenile onset is the typical pattern.

5. Islet cell autoantibodies are frequently present around the time of onset. If T1DM is induced by a virus, the autoantibodies are transient. Occasionally antibodies will persist long term, particularly if they are associated with other autoimmune disorders. About 50% of T1DM is related to problems with the major histocompatibility complex on chromosome 6. It is correlated with an increased frequency of certain human leukocyte antigen (HLA) alleles. The HLA types DR3 and DR4 are most commonly associated with diabetes.

who received poor nutrition and into a life that meant chronic undernourishment.

Body Mass Index

A measure of adiposity is the body mass index (BMI). The BMI of an individual is calculated as:

Equation 38.1

$$BMI = Weight\ (kg)/Height\ (m)^2$$

The BMI of healthy lean individuals ranges from 20 to 25. A BMI greater than 25 indicates that the individual is overweight, whereas a BMI higher than 30 indicates obesity. The condition of being overweight or obese is a risk factor for multiple pathologies, including insulin resistance, dyslipidemia, diabetes, cardiovascular disease, and hypertension.

WAT tissue is divided into subcutaneous and intraabdominal (visceral) depots. Intraabdominal *WAT* refers primarily to omental and mesenteric fat and is the smaller of the two depots. These depots receive different blood supplies that are drained in a fundamentally different way in that venous return from intraabdominal fat leads into the hepatic portal system. Thus intraabdominally derived FFAs are mostly cleared by the liver, whereas subcutaneous fat is the primary site for providing FFAs to muscle during exercise or fasting. Regulation of intraabdominal and subcutaneous adipose tissue also differs. Abdominal fat is highly innervated by autonomic neurons and has a greater turnover rate. Furthermore, these two depots display differences in hormone production and enzyme activity.

Men tend to gain fat in the intraabdominal depot **(android [apple-shaped] adiposity),** whereas women tend to gain fat in the subcutaneous depot, particularly in the thighs and buttocks **(gynecoid [pear-shaped] adiposity).** Clearly an excess of abdominal fat poses a greater risk factor for the pathologies mentioned earlier. Thus another indicator of body composition is circumference of the waist (measured in inches around the narrowest point between the ribs and hips when viewed from the front after exhaling) divided by the circumference of the hips (measured at the point where the buttocks are largest when viewed from the side). This **waist-hip ratio** may be a better indicator of body fat than BMI, especially as it relates to risk for development of diseases. A waist-hip ratio of greater than 0.95 in men or 0.85 in women is linked to a significantly higher risk for development of diabetes and cardiovascular disease.

Central Mechanisms Involved in Energy Balance

In recent years, numerous hormones and neuropeptides have been implicated in both chronic and acute regulation of appetite, satiety, and energy expenditure in humans. One simplified model involves two peptide hormones, **leptin** and **insulin** (see Fig. 39.17), already discussed. Leptin acts on at least two neuron types in the arcuate nucleus of the hypothalamus. In the first, leptin represses production of **neuropeptide Y (NPY),** a very potent stimulator of food-seeking behavior (energy intake) and an inhibitor of energy expenditure. Norepinephrine, another appetite stimulator, co-localizes with NPY in some of these neurons. At the same time, leptin represses production of agouti-related peptide (AGRP), an endogenous antagonist that acts on MC4R, a hypothalamic receptor for the anorexigenic peptide α-melanocyte–stimulating hormone (α-MSH), which inhibits food intake. In another type of arcuate neuron, leptin stimulates production of proopiomelano-cortin (POMC) products, one of which is α-MSH, and production of cocaine-amphetamine–regulated transcript (CART), both of which inhibit food intake. Thus leptin decreases food consumption and increases energy expenditure by simultaneously inhibiting NPY and the α-MSH antagonist AGRP and by stimulating α-MSH and CART (see Fig. 39.17). These second-order neuropeptides are transmitted to and interact with receptors in neurons of the paraventricular hypothalamic nucleus ("satiety" neurons) and lateral hypothalamic nucleus ("hunger" neurons). In turn these hypothalamic neurons generate signals that coordinate feeding behavior and autonomic nervous system activity (especially sympathetic outflow) with diverse endocrine actions on thyroid gland function, reproduction, and growth.

Another regulator of food intake and body energy stores is **melanin-concentrating hormone (MCH).** This neuropeptide increases food seeking and adipose tissue by antagonizing the satiety effect of α-MSH downstream from the interaction of α-MSH with its MC4R receptor. The probable importance of this molecule is demonstrated by the fact that it is the only regulator whose ablation by gene knockout actually results in leanness.

To maintain overall energy homeostasis, the system must also balance specific nutrient intake and expenditure—for example, carbohydrate intake with carbohydrate oxidation. This may account for some specificity in neuropeptide and neurotransmitter responses to meals. Serotonin produces satiety after ingestion of glucose. Gastrointestinal hormones such as cholecystokinin and GLP-1 produce satiety by humoral effects, but their local production in the brain may participate in nutrient and caloric regulation. The recently discovered hormone **ghrelin** is an acylated peptide with potent orexigenic activity that arises in cells of the oxyntic glands in the stomach. Plasma levels of ghrelin rise in humans in the 1 to 2 hours that precede their normal meals. Plasma levels of ghrelin fall drastically to minimum values about 1 hour after eating. Ghrelin appears to stimulate food intake by reacting with its receptor in hypothalamic neurons that express NPY.

Key Concepts

1. Cells make ATP to meet their energy needs. ATP is made by glycolysis and by the TCA cycle coupled to oxidative phosphorylation.

2. Cells can oxidize carbohydrate (primarily in the form of glucose), AAs, and FFAs to make ATP. Additionally the liver makes KBs for other tissues to oxidize for energy in times of fasting.

3. Some cell types are limited in the energy substrates they can oxidize for energy. The brain is normally exclusively dependent on glucose for energy. Thus blood glucose must be maintained above 60 mg/dL for normal autonomic and CNS function. Conversely, inappropriately high levels of glucose (i.e., fasting glucose > 100 mg/dL) promote glucotoxicity and thereby lead to the long-term complications of diabetes.

4. The endocrine pancreas produces the hormones insulin, glucagon, somatostatin, gastrin, and pancreatic polypeptide.

5. Insulin is an anabolic hormone that is secreted in times of excess nutrient availability. It allows the body to use carbohydrates as energy sources and store nutrients.

6. Major stimuli for insulin secretion include increased serum glucose and some AAs. Activation of cholinergic (muscarinic) receptors also increases insulin secretion, whereas activation of α_2-adrenergic receptors inhibits insulin secretion. The gastrointestinal tract releases incretin hormones that stimulate pancreatic insulin secretion. GLP-1 is particularly potent in augmenting glucose-dependent stimulation of insulin secretion (GSIS). GLP-1 is degraded by dipeptidyl peptidase (DPP)-4. DPP-4–resistant GLP-1 analogues and inhibitors of DPP-4 are currently used to increase GSIS in patients with type 2 diabetes.

7. Insulin binds to the insulin receptor (InsR), which is linked to multiple pathways that mediate the metabolic (Akt kinase) and growth effects (MAPK) of insulin.

8. During the digestive phase, insulin acts on the liver to promote trapping of glucose as G6P. Insulin also increases glycogenesis, glycolysis, and de novo lipogenesis (DNL) in the liver. Insulin inhibits gluconeogenesis, glycogenolysis, and assembly of lipids into VLDL.

9. Insulin increases GLUT4-mediated glucose uptake in muscle and adipose tissue.

10. Insulin increases glycogenesis, glycolysis, and in the presence of caloric excess, lipogenesis in muscle.

11. Insulin increases glycolysis and generation of G3P in adipocytes. Insulin induces expression of LPL and its transport to the luminal side of capillary endothelial cells. Insulin promotes uptake and activation of FFAs and esterification of fatty acyl CoAs to G3P to form TG, and it decreases hormone-sensitive lipase activity in adipocytes.

12. Insulin increases AA uptake and protein synthesis in skeletal muscle but also essentially all insulin target cells. Insulin/Akt kinase signaling activates mTORC1 and S6K to promote synthesis of ribosomal proteins and proteins involved in mRNA translation, as well as other types of proteins. Insulin inhibits proteosomal degradation of protein.

13. Glucagon is a catabolic counterregulatory hormone. Its secretion increases during periods of food deprivation, and it acts to mobilize nutrient reserves. It also mobilizes glycogen, fat, and even protein.

14. Glucagon is released in response to decreased serum glucose (and therefore decreased insulin) and increased serum AA levels and β-adrenergic signaling.

15. Glucagon binds to the glucagon receptor, which is linked to PKA-dependent pathways. The primary target organ for glucagon is the liver. Glucagon increases liver glucose output by increasing glycogenolysis and gluconeogenesis. It increases β-oxidation of fatty acids and ketogenesis.

16. Glucagon regulates hepatic metabolism both by regulation of gene expression and through posttranslational PKA-dependent pathways.

17. The major counterregulatory factors in muscle and adipose tissue are the adrenal hormone epinephrine and the sympathetic neurotransmitter norepinephrine. These two factors act through β_2- and β_3-adrenergic receptors to increase cAMP levels. Epinephrine and norepinephrine enhance glycogenolysis and fatty acyl oxidation in muscle and increase hormone-sensitive lipase in adipose tissue.

18. Diabetes mellitus is classified as type 1 (T1DM) and type 2 (T2DM). T1DM is characterized by destruction of pancreatic beta cells, and exogenous insulin is required for treatment. T2DM can be due to numerous factors but is usually characterized as insulin resistance coupled to some degree of beta cell deficiency. Patients with T2DM may require exogenous insulin at some point to maintain blood glucose levels.

19. Obesity-associated T2DM is currently at epidemic proportions worldwide and is characterized by insulin resistance due to lipotoxicity, hyperinsulinemia, and inflammatory cytokines produced by adipose tissue. T2DM is often associated with obesity, insulin resistance, hypertension, and coronary artery disease. This constellation of risk factors is referred to as the *metabolic syndrome.*

20. Major symptoms of diabetes mellitus include hyperglycemia, polyuria, polydipsia, polyphagia, muscle wasting, electrolyte depletion, and ketoacidosis (in T1DM).

21. The long-term complications of poorly controlled diabetes are due to excess intracellular glucose (glucotoxicity), especially in the retina, kidney, and peripheral nerves. This leads to retinopathy, nephropathy, and neuropathy.

22. Adipose tissue has an endocrine function, especially in terms of energy homeostasis. Hormones produced by adipose tissue include leptin and adiponectin. Leptin acts on the hypothalamus to promote satiety.

Additional Reading

Bailey CJ, et al. Future glucose-lowering drugs for type 2 diabetes. *Lancet Diabetes Endocrinol.* 2016;4:350-359.

Davidson JA, et al. Glucagon therapeutics: dawn of a new era for diabetes care. *Diabetes Metab Res Rev.* 2016;doi:10.1002/dmrr.2773; [Epub ahead of print].

Font-Burgada J, et al. Obesity and cancer: the oil that feeds the flame. *Cell Metab.* 2016;23:48-62.

Klein MS, Shearer J. Metabolomics and type 2 diabetes: translating basic research into clinical application. *J Diabetes Res.* 2016;2016:3898502. doi:10.1155/2016/3898502; [Epub 2015 Nov 9].

Marion-Letellier R, et al. Fatty acids, eicosanoids and PPAR gamma. *Eur J Pharmacol.* 2016;785:44-49.

Pawlak M, et al. Molecular mechanism of PPARα action and its impact on lipid metabolism, inflammation and fibrosis in nonalcoholic fatty liver disease. *J Hepatol.* 2015;62:720-733.

40

Hormonal Regulation of Calcium and Phosphate Metabolism

LEARNING OBJECTIVES

Upon completion of this chapter the student should be able to answer the following questions:

1. Describe the pool of serum calcium and phosphate, including ionized, complexed, and protein bound. Describe the normal concentration ranges of these ions and the major routes of influx and efflux.
2. Discuss the role of the parathyroid gland in the regulation of serum calcium and explain the role of the calcium-sensing receptor in the regulation of parathyroid hormone (PTH) secretion.
3. Describe the production of 1,25-dihydroxyvitamin D, including sources of vitamin D precursor, sites and key regulators of vitamin D hydroxylation, and transport of vitamin D metabolites in the blood.
4. List the target organs of PTH and describe its effects on calcium and phosphate mobilization or handling at each of these sites.
5. List the target organs and key actions of 1,25-dihydroxyvitamin D.
6. Discuss the regulation of phosphate metabolism by FGF23.
7. Predict the hormone responses that would be triggered by perturbations of serum calcium and phosphate or by vitamin D deficiency, and discuss the consequences of these compensatory hormone actions.

Calcium (Ca) and phosphate are essential to human life because they play important structural roles in hard tissues (i.e., bones and teeth) and important regulatory roles in metabolic and signaling pathways. In biological systems, **inorganic phosphate (P_i)** consists of a mixture of dihydrogen phosphate ($H_2PO_4^-$) and hydrogen phosphate (HPO_4^-). The two primary sources of circulating Ca and P_i are the diet and the skeleton (Fig. 40.1). Two hormones, **1,25-dihydroxyvitamin D** (also called **calcitriol**) and **parathyroid hormone (PTH),** regulate intestinal absorption of Ca and P_i and release of Ca and P_i into the circulation after bone resorption. The primary processes for removal of Ca and P_i from blood are renal excretion and bone mineralization (see Fig. 40.1). 1,25-Dihydroxyvitamin

D and PTH regulate both processes. Other hormones and paracrine growth factors also regulate Ca and P_i homeostasis.

Crucial Roles of Calcium and Phosphate in Cellular Physiology

Ca is an essential dietary element. In addition to obtaining Ca from the diet, humans contain a vast store (i.e., >1 kg) of Ca in bone mineral, which can be called upon to maintain normal circulating levels of Ca in times of dietary restriction and during the increased demands of pregnancy and nursing. Circulating Ca exists in three forms (Table 40.1): free ionized Ca^{++}, protein-bound Ca, and Ca complexed with anions (e.g., phosphates, HCO_3^-, citrate). The ionized form represents about 50% of circulating Ca. Since it is critical to so many cellular functions, $[Ca^{++}]$ in both the extracellular and intracellular compartments is tightly controlled. Circulating Ca^{++} is under direct hormonal control and normally maintained within a relatively narrow range. Either too little calcium (**hypocalcemia;** total serum calcium < 8.7 mg/dL [2.2 mM]) or too much Ca (**hypercalcemia;** total serum Ca > 10.4 mg/dL [2.6 mM]) in blood can lead to a broad range of pathophysiological changes, including neuromuscular dysfunction, central nervous system dysfunction, renal insufficiency, calcification of soft tissue, and skeletal pathology.

P_i is also an essential dietary element, and it is stored in large quantities in mineral. Most circulating P_i is in the free ionized form, but some P_i (<20%) circulates as a protein-bound form or complexed with cations (see Table 40.1). Because soft tissues contain 10-fold more P_i than Ca, tissue damage (e.g., crush injury with massive muscle cell death) can result in **hyperphosphatemia,** whereupon the increased P_i complexes with Ca^{++} to cause acute hypocalcemia.

P_i is a key intracellular component. Indeed, it forms the high-energy phosphate bonds of adenosine triphosphate (ATP) that maintain life. Phosphorylation and dephosphorylation of proteins, lipids, second messengers, and cofactors represent key regulatory steps in numerous metabolic and signaling pathways, and phosphate also serves as the backbone for nucleic acids.

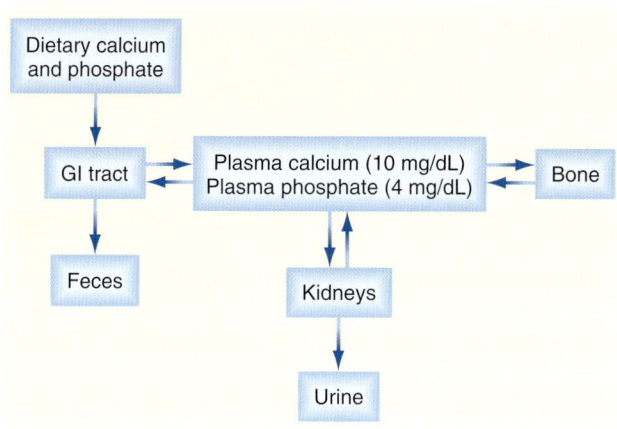

• **Fig. 40.1** Daily Ca⁺⁺ and P$_i$ flux.

TABLE 40.1	**Forms of Ca and P$_i$ in Plasma**			
Ion	mg/dL	Ionized	Protein Bound	Complexed
Ca	8.5–10.2	50%	45%	5%
P$_i$	3–4.5	84%	10%	6%

Ca⁺⁺ is bound (i.e., complexed) to various anions in plasma, including HCO$_3^-$, citrate, and SO$_4^{2-}$. P$_i$ is complexed to various cations, including Na⁺ and K⁺.

From Koeppen BM, Stanton BA. *Renal Physiology.* 4th ed. Philadelphia: Mosby; 2007.

Physiological Regulation of Calcium and Phosphate: Parathyroid Hormone and 1,25-Dihydroxyvitamin D

PTH and **1,25-dihydroxyvitamin D** are the two physiologically most important hormones dedicated to maintenance of normal blood Ca and P$_i$ in humans. As such they are referred to as **calciotropic hormones.** The structure, synthesis, and secretion of these two hormones and their receptors will be discussed first. In the following section, the detailed actions of PTH and 1,25-dihydroxyvitamin D on the three key sites of Ca/P$_i$ homeostasis (i.e., gut, bone, and kidney) are discussed.

Parathyroid Glands

The predominant parenchymal cell type in the parathyroid gland is the **principal** (also called **chief**) **cell** (Fig. 40.2).

Parathyroid Hormone

PTH is the primary hormone that protects against hypocalcemia. The primary targets of PTH are bone and the kidneys. PTH also functions in a positive feed-forward loop by stimulating production of 1,25-dihydroxyvitamin D.

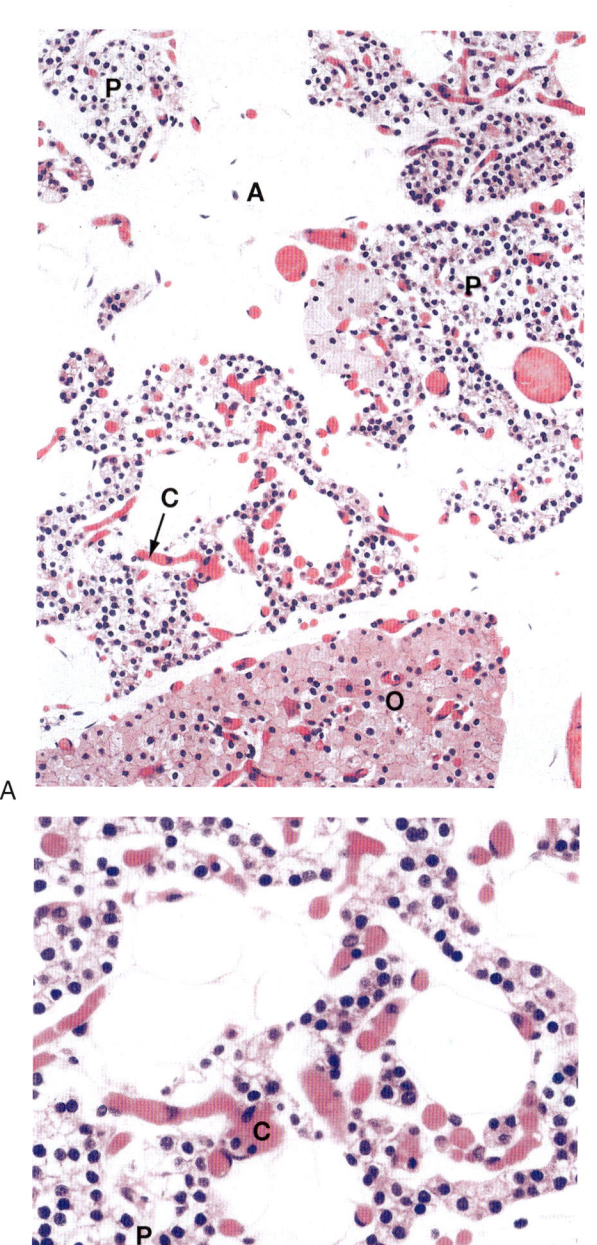

• **Fig. 40.2** A and B, Histology of parathyroid glands. A, adipose tissue within parathyroid glands; C, capillaries; O, oxyphil cells; P, principal or chief cells. (From Young B et al. *Wheater's Functional Histology.* 5th ed. Philadelphia: Churchill Livingstone; 2006.)

Structure, Synthesis, and Secretion

PTH is secreted as an 84–amino acid polypeptide and is synthesized as a **prepro-PTH,** which is proteolytically processed to **pro-PTH** in the endoplasmic reticulum and then to PTH in the Golgi and secretory vesicles. PTH has a short half-life in the circulation (2 minutes), consistent with its role in minute-to-minute regulation of plasma calcium.

Parathyroid Hormone Receptor

Because the PTH receptor also binds PTH-related peptide (PTHrP), it is usually referred to as the **PTH/PTHrP receptor.** The PTH/PTHrP receptor is expressed on osteoblasts in bone and in the proximal and distal tubules of the kidney, and it is the receptor that mediates the systemic actions of PTH. However, the PTH/PTHrP receptor is also expressed in many developing organs in which PTHrP has important paracrine functions. One such example is regulation of chondrocyte proliferation in the growth plate during endochondral bone growth.

Vitamin D

Vitamin D is a prohormone that must undergo two successive hydroxylation reactions to become the active form known as **1,25-dihydroxyvitamin D** or **calcitriol** (Fig. 40.5). This hormone plays a critical role in Ca absorption

 ## AT THE CELLULAR LEVEL

Extracellular [Ca^{++}] is sensed by the parathyroid chief cell through a plasma membrane **calcium-sensing receptor (CaSR).** The primary signal that stimulates PTH secretion is a decrease in circulating [Ca^{++}] (Fig. 40.3). Conversely, increasing amounts of extracellular Ca^{++} bind to the CaSR and stimulate signaling pathways that repress PTH secretion. Although the CaSR binds to extracellular Ca^{++} with relatively low affinity, the CaSR is extremely sensitive to minute changes in extracellular [Ca^{++}]. The relationship between [Ca^{++}] and the rate of PTH secretion is described by a steep inverse sigmoidal curve. A 0.2-mM difference in blood [Ca^{++}] spans the full range of the curve, altering PTH secretion from basal (5% of maximum) to maximum levels (Fig. 40.4). The steady-state "set point" will vary between individuals but typically resides below the midpoint of the curve (i.e., half-maximal PTH secretion). Thus the CaSR is a rapid, robust, and continuous regulator of PTH output in response to subtle [Ca^{++}] fluctuations.

In addition to inhibiting PTH secretion, activation of the CaSR also promotes degradation of stored PTH in the parathyroid chief cell. As a result, biologically inactive carboxy-terminal PTH fragments are secreted from the parathyroid gland and are also produced by peripheral metabolism of PTH by the liver and kidney. Therefore current PTH assays use two antibodies that recognize epitopes from both ends of the molecule to accurately measure intact PTH[1-84].

Over a longer time frame, PTH production is also regulated at the level of mRNA stability and gene transcription (see Fig. 40.3). Decreased [Ca^{++}] leads to production of proteins that bind the 3-untranslated region of PTH mRNA and stabilize it, leading to increased PTH translation. *PTH* gene transcription is repressed by 1,25-dihydroxyvitamin D in a negative feedback loop (acting through vitamin D response elements—see later). The ability of 1,25-dihydroxyvitamin D to hold *PTH* gene expression in check is reinforced by the coordinated upregulation of *CASR* gene expression by positive vitamin D response elements in the promoter region of the *CASR* gene (see Fig. 40.3). It should be noted, however, that during a hypocalcemic challenge, the decrease in [Ca^{++}] overrides the inhibitory effect of 1,25-dihydroxyvitamin D on PTH transcription, allowing both of these hormones to be elevated simultaneously.

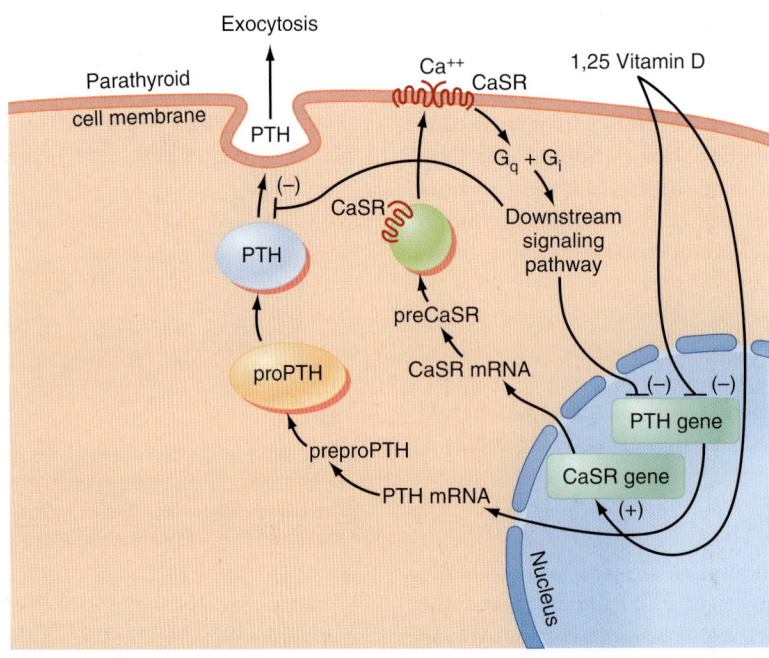

• **Fig. 40.3** Regulation of *PTH* gene expression and secretion. (Modified from Porterfield SP, White BA. *Endocrine Physiology.* 3rd ed. Philadelphia: Mosby; 2007.)

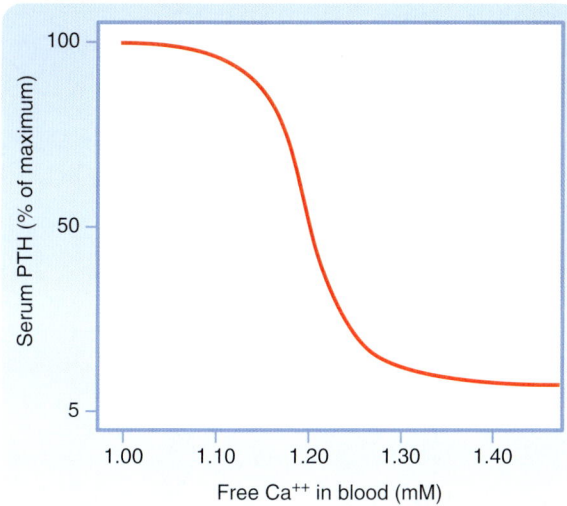

• **Fig. 40.4** Sigmoidal relationship between serum [Ca⁺⁺] and serum PTH, which reflects the rate of PTH secretion. (Modified from Porterfield SP, White BA. *Endocrine Physiology*. 3rd ed. Philadelphia: Mosby; 2007.)

IN THE CLINIC

Patients with **benign familial hypocalciuric hypercalcemia (FHH)** are heterozygous for inactivating mutations of the CaSR. In these patients, because of complete or partial loss of one CaSR allele, higher levels of [Ca⁺⁺] are required to suppress PTH secretion. This results in an elevated [Ca⁺⁺] set point for PTH secretion, accounting for the hypercalcemia. The CaSR is also expressed in the thick ascending limb of the renal tubule, where it normally inhibits Ca⁺⁺ reabsorption when blood Ca⁺⁺ rises. The hypocalciuria in the face of hypercalcemia in FHH is due to the reduced ability of the CaSR in the kidney to sense and respond to elevated blood [Ca⁺⁺] by increasing Ca excretion.

AT THE CELLULAR LEVEL

Parathyroid hormone–related peptide (PTHrP) is a peptide paracrine hormone produced by several adult tissues (skin, hair, breast), where it may regulate proliferation and differentiation. It also plays a role in relaxation of smooth muscle in response to stretch in blood vessels, uterus, and bladder. During lactation, PTHrP promotes maternal bone resorption and the transport of calcium into milk. During development, PTHrP regulates calcium transport across the placenta and is a key regulator of chondrocyte proliferation and differentiation in the growth plate of long bones. The 30 amino acids at the N-terminus of PTHrP have significant structural homology with PTH. PTHrP is not regulated by circulating Ca⁺⁺ and normally does not play a role in Ca/Pᵢ homeostasis in adults. However, certain tumors secrete high levels of PTHrP, which causes **hypercalcemia of malignancy** and symptoms that resemble hyperparathyroidism.

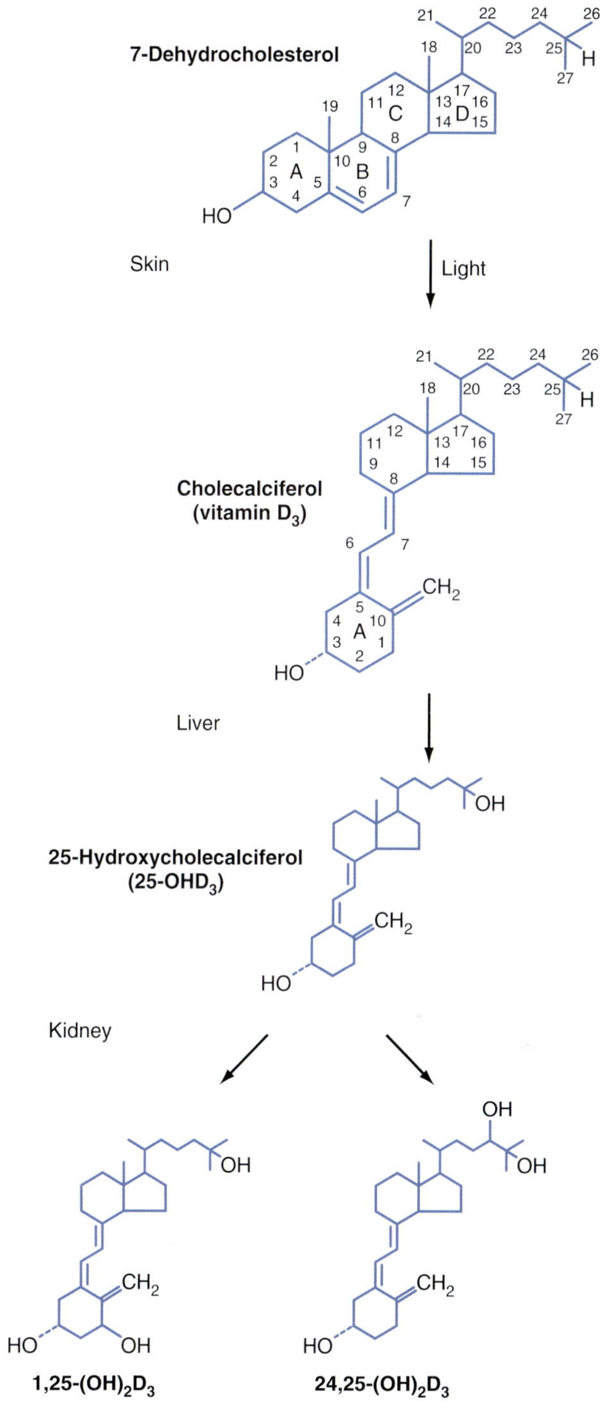

7-Dehydrocholesterol

Skin

Light

Cholecalciferol (vitamin D₃)

Liver

25-Hydroxycholecalciferol (25-OHD₃)

Kidney

1,25-(OH)₂D₃   24,25-(OH)₂D₃

• **Fig. 40.5** Biosynthesis of 1,25-dihydroxyvitamin D. (Modified from Porterfield SP, White BA. *Endocrine Physiology*. 3rd ed. Philadelphia: Mosby; 2007.)

and to a lesser extent P_i absorption by the small intestine. It also regulates bone remodeling and renal reabsorption of Ca and P_i.

Structure, Synthesis, and Transport of Active Vitamin D Metabolites

Vitamin D₃ (also called **cholecalciferol**) is synthesized via conversion of 7-dehydrocholesterol by ultraviolet B

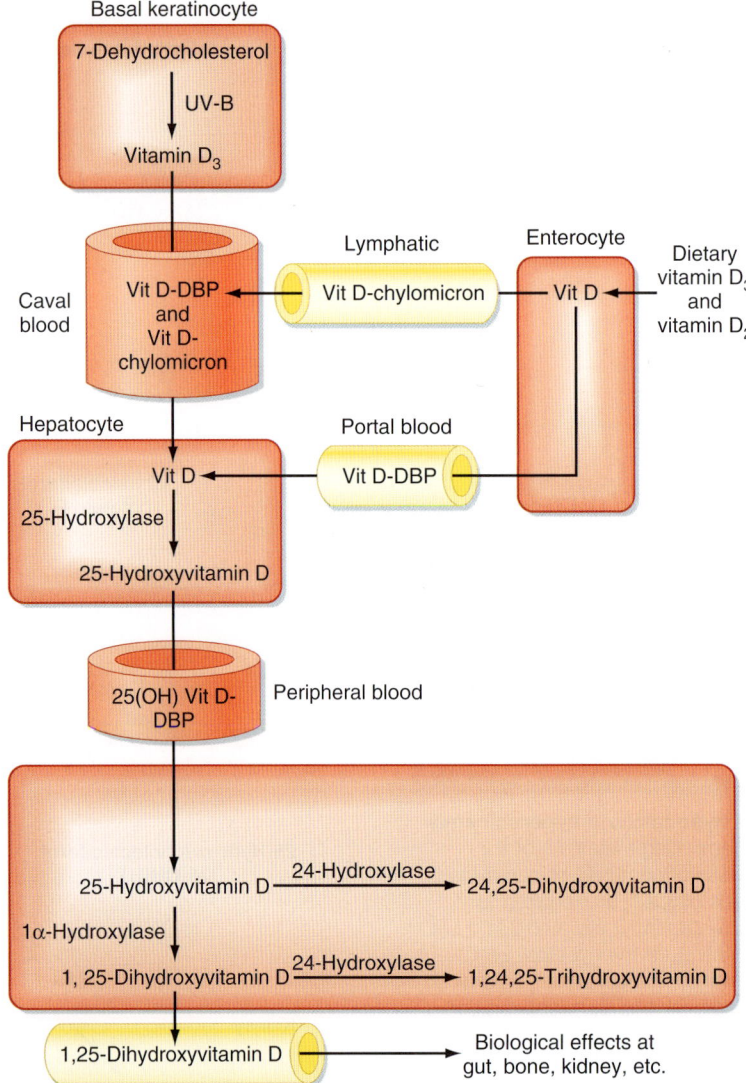

• **Fig. 40.6** Vitamin D metabolism. (Modified from Porterfield SP, White BA. *Endocrine Physiology.* 3rd ed. Philadelphia: Mosby; 2007.)

(UVB) light in the more basal layers of the skin (Fig. 40.6). Chemically, vitamin D_3 is a **secosteroid** in which one of the cholesterol rings is opened (see Fig. 40.5). **Vitamin D_2 (ergocalciferol)** is produced in plants. Vitamin D_3 and to a lesser extent vitamin D_2 are absorbed from the diet and are equally effective after conversion to active hydroxylated forms. The balance between UVB-dependent endogenously synthesized vitamin D_3 and absorption of the dietary forms of vitamin D becomes important in certain situations. Individuals with higher melanin content in skin who live at higher latitudes convert less 7-dehydrocholesterol to vitamin D_3 and thus are more dependent on vitamin supplements or dietary sources of vitamin D (natural or fortified, e.g., milk). Institutionalized elderly patients who stay indoors and avoid dairy products are particularly at risk for development of **vitamin D deficiency.**

Vitamin D is transported in blood from the skin to the liver. Dietary vitamin D reaches the liver directly via transport in the portal circulation and indirectly via chylomicrons (see Fig. 40.6). In the liver, vitamin D is hydroxylated at the 25-carbon position to yield **25-hydroxyvitamin D.** The hepatic 25-hydroxylase is constitutively expressed and unregulated, so circulating levels of 25-hydroxyvitamin D reflect the amount of precursor available for 25-hydroxylation. For this reason, and because of its relatively long half-life in the circulation (2–3 weeks), measurement of 25-hydroxyvitamin D levels is used to assess vitamin D status.

25-Hydroxyvitamin D undergoes further hydroxylation in the proximal tubule of the kidney (see Figs. 40.5 and 40.6). Hydroxylation at the 1α position generates **1,25-dihydroxyvitamin D,** the most active form of vitamin D. Hydroxylation at the 24 position generates **24,25-dihydroxyvitamin D,** which does not play a major biological role and serves as an inactivation pathway.

Renal 1α-hydroxylase is tightly regulated by a number of factors (Fig. 40.7). PTH and hypophosphatemia are the primary inducers of 1α-hydroxylase activity, resulting in

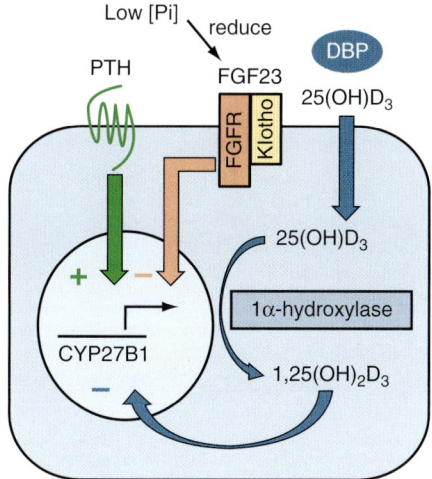

• **Fig. 40.7** Regulation of 1α-hydroxylase gene *(CYP27B1)* expression in the proximal tubule, showing stimulation by PTH and inhibition by FGF23 and 1,25-dihydroxyvitamin D. Hypophosphatemia probably stimulates 1α-hydroxylase by reducing FGF23 levels at least in part.

increased levels of 1,25-dihydroxyvitamin D. Conversely, [Ca^{++}] and 1,25-dihydroxyvitamin D, the enzyme product, inhibit it. Fibroblast growth factor (FGF)23, a major regulator of P$_i$ metabolism (see later), also represses 1α-hydroxylase activity; a reduction of FGF23 levels likely mediates the effect of hypophosphatemia on 1,25-dihydroxyvitamin D production at least in part.

Vitamin D and its metabolites circulate in blood primarily bound to **vitamin D–binding protein (DBP).** DBP is a serum glycoprotein that is synthesized by the liver. DBP binds more than 85% of 1,25-hydroxyvitamin D and 24,25-dihydroxyvitamin D. Because of binding to other proteins, only 0.4% of 1,25-dihydroxyvitamin D circulates as free hormone. DBP transports the highly lipophilic vitamin D in blood and provides a reservoir of vitamin D that protects against vitamin D deficiency.

1,25-Dihydroxyvitamin D Receptor

1,25-Dihydroxyvitamin D exerts its actions primarily through binding to the nuclear **vitamin D receptor (VDR),** which is a member of the nuclear hormone receptor family. The VDR is a ligand-dependent transcription factor that binds to cognate DNA sequences (**vitamin D response elements**) as a heterodimer with the **retinoid X receptor** (RXR). Thus the primary action of 1,25-dihydroxyvitamin D is to regulate gene expression in its target tissues, including the small intestine, bone, kidneys, and parathyroid gland.

The genomic actions of 1,25-dihydroxyvitamin D mediated by the VDR occur over a period of hours to days. 1,25-Dihydroxyvitamin D also has rapid effects (seconds to minutes). For example, 1,25-dihydroxyvitamin D rapidly induces absorption of Ca^{++} by the duodenum. The VDR is also expressed in the plasma membrane of cells and is linked to rapid signaling pathways (e.g., G proteins, phosphatidylinositol-3′-kinase). Current molecular modeling has led to development of ligands that specifically bind

to the nuclear- versus the membrane-localized VDR, paving the way for selective treatment of disorders related to the rapid versus slow actions of 1,25-dihydroxyvitamin D with synthetic vitamin D analogues.

Regulation of [Ca^{++}] and [P$_i$] by Small Intestine and Bone

An overview of the regulation of [Ca^{++}] and P$_i$ by the action of PTH and 1,25-dihydroxyvitamin D on the small intestine, bone, and parathyroid glands is summarized in Table 40.2 and in the following paragraphs. For details on renal handling of Ca^{++}, consult Chapter 36.

AT THE CELLULAR LEVEL

Calcitonin is a peptide hormone produced by the medullary cells, or C-cells, of the thyroid gland. Calcitonin secretion is positively regulated by serum [Ca^{++}] via the CaSR. The calcitonin receptor is expressed in osteoclasts, where calcitonin acts rapidly and directly to inhibit bone resorption. However, in humans, calcitonin does not appear to play a major role in regulating serum Ca. In support of this view, production of excess calcitonin or complete absence of calcitonin (e.g., following thyroidectomy) does not perturb serum Ca levels. More potent forms of the hormone (e.g., salmon calcitonin) have been used therapeutically as an antiresorptive in the treatment of **Paget's disease** (characterized by excessive osteoclastic bone resorption) and in osteoporosis. Calcitonin is also a useful histochemical marker of medullary thyroid cancer.

Ca^{++} and P$_i$ Transport by Small Intestine

Dietary intake of Ca can vary widely among individuals and from day to day. Assuming an intake of 1000 mg (the RDA for ages 19–50), 350 mg would typically be absorbed, counterbalanced by 150 mg secreted by the intestine, for a net intake of 200 mg. Most Ca^{++} absorption takes place in the proximal small intestine. Importantly, absorption of Ca^{++} is stimulated by 1,25-dihydroxyvitamin D, so absorption is more efficient in the face of declining dietary Ca^{++}.

Ca^{++} is absorbed from the duodenum and jejunum by both a Ca^{++}-regulated and a hormonally regulated transcellular route and by a passive paracellular route. The transcellular route of Ca^{++} absorption is summarized in Fig. 40.8. Movement of Ca^{++} from the gastrointestinal lumen into the enterocyte, which is driven by both chemical and electrical gradients, occurs via apical calcium channels called **TRPV5** and **TRPV6.** Once inside the cell, Ca^{++} ions bind to **calbindin-D$_{9K}$,** which maintains a low cytoplasmic [Ca^{++}], preserving the favorable transluminal membrane Ca^{++} gradient. Calbindin-D$_{9K}$ also plays a role in apical-to-basolateral shuttling of Ca^{++}, which is transported across the basolateral membrane against an electrochemical gradient by **plasma membrane calcium ATPase (PMCA).** The **Na$^+$/Ca^{++} exchanger (NCX)** also contributes to the

TABLE 40.2	Actions of PTH and 1,25-Dihydroxyvitamin D on Ca++/Pi Homeostasis			
	Small Intestine	**Bone**	**Kidney**	**Parathyroid Gland**
PTH	No direct action	Intermittent PTH promotes osteoblastic bone formation. Regulates M-CSF, RANKL, OPG in osteoblasts Chronic high levels promote osteoclastic bone resorption, Ca++ and Pi release from bone.	Stimulates 1α-hydroxylase activity Stimulates Ca++ reabsorption by thick ascending limb of Henle's loop and distal tubule Inhibits Pi reabsorption in proximal tubule (inhibits NPT2a)	No direct action
1,25-Dihydroxyvitamin D	Increases Ca++ absorption by increasing TRPV, calbindin, and PMCA expression Modestly increases Pi absorption	Regulates osteoclast differentiation via RANKL expression in osteoblasts Maintains [Ca++] and [Pi] to support bone mineralization	Supports actions on Ca++ reabsorption through calbindin expression Promotes Pi reabsorption by proximal nephrons (stimulates NPT2a expression)	Directly inhibits *PTH* gene expression (negative feedback) Directly stimulates *CASR* gene expression

CASR, calcium-sensing receptor; M-CSF, monocyte colony-stimulating factor; NPT2, Na+/Pi cotransporter; OPG, osteoprotegerin; PTH, parathyroid hormone; RANKL, receptor activator of nuclear factor κ-B.

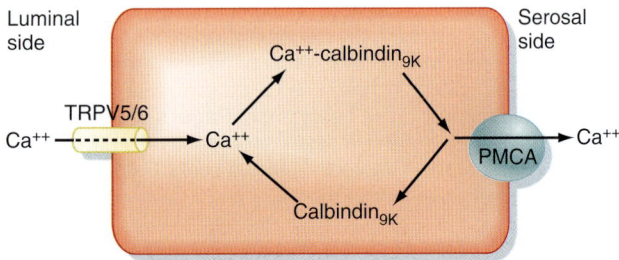

• **Fig. 40.8** Intestinal absorption of Ca++ via the transcellular route. (Modified from Porterfield SP, White BA. *Endocrine Physiology.* 3rd ed. Philadelphia: Mosby; 2007.)

transport of Ca++ out of enterocyte. 1,25-Dihydroxyvitamin D stimulates expression of all the components involved in absorption of Ca++ by the small intestine.

The fraction of dietary Pi absorbed by the jejunum remains relatively constant at about 70% and is under minor hormonal control by 1,25-dihydroxyvitamin D. The limiting process in transcellular Pi absorption is transport across the apical brush border, which is mediated by the **Na+/Pi cotransporter (NPT2).**

Ca++ and Pi in Bone

Bone stores vast amounts of Ca and Pi. Once peak bone mass has been achieved in an adult, the skeleton is constantly remodeled through the concerted activities of bone cells. The processes of **bone formation** and **bone resorption** are in balance in a healthy, physically active, and well-nourished individual. Of the 1 kg of Ca immobilized in bone, about 500 mg (i.e., 0.5%) is mobilized from and deposited into bone each day. However, the process of bone remodeling

can be modulated to provide a net gain or loss of Ca++ and Pi into blood and is responsive to physical activity (loading), diet, age, and hormonal regulation. Because the integrity of bone is absolutely dependent on Ca and Pi, chronic dysregulation of these ions or the hormones that regulate them lead to pathological changes in bone.

Physiology of Bone

The processes of pattern formation, growth, and remodeling of the skeleton is complex and beyond the scope of this chapter. The key elements required for understanding the role of adult bone in the hormonal regulation of Ca and Pi metabolism are discussed.

In adults, bone remodeling involves (1) destruction of fatigued or microdamaged bone with the release of Ca++, Pi, and hydrolyzed fragments of bone matrix into blood and (2) synthesis of **osteoid** (yet to be mineralized bone matrix) at the site of resorption, followed by controlled mineralization of the osteoid by Ca++ and Pi to form new bone. Bone remodeling occurs continually at about 2 million discrete sites throughout the skeleton by packets of bone cells referred to as the **basic multicellular units (BMU).**

The cells involved in bone remodeling fall into two major classes: cells that form bone (**osteoblasts**) and cells that destroy or resorb bone (**osteoclasts**). The process of bone remodeling is a highly integrated process (Fig. 40.9). Osteoblast-lineage cells express factors that induce differentiation of osteoclasts from progenitors of the monocyte/macrophage lineage and also promote mature osteoclast function. Osteoblasts release **monocyte colony-stimulating factor (M-CSF),** which expands and differentiates early hematopoietic progenitors (CFU-GM) into preosteoclasts that express a cell surface receptor called **RANK (receptor**

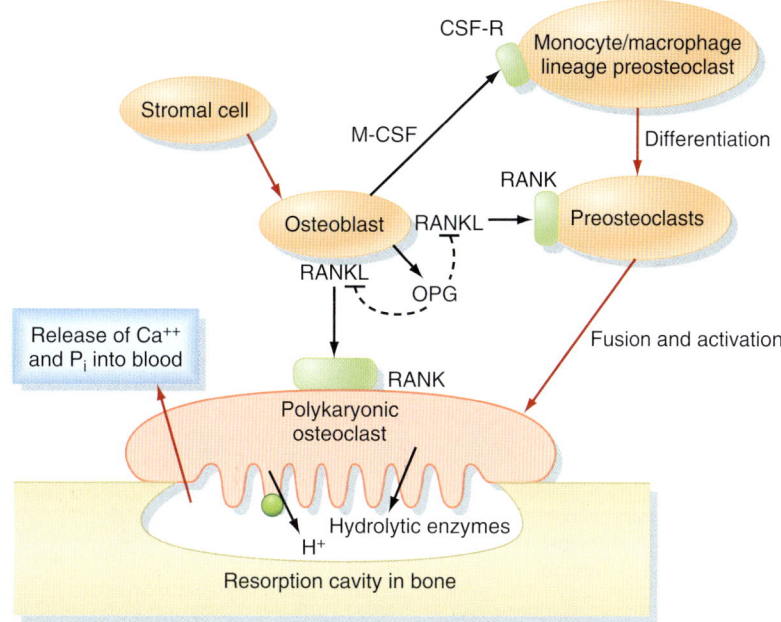

● **Fig. 40.9** Osteoblast regulation of osteoclast differentiation and function. (Modified from Porterfield SP, White BA. *Endocrine Physiology.* 3rd ed. Philadelphia: Mosby; 2007.)

activator of nuclear factor [NF]-κB). Osteoblast-lineage cells display **RANK ligand (RANKL)** on their cell surface. **RANKL** then binds to **RANK** on preosteoclasts and induces osteoclastogenesis. This process involves fusion of several osteoclast precursors, giving rise to a large multinucleated osteoclast. The perimeter of the osteoclast membrane facing mineralized bone adheres tightly to the bone and seals off the area of osteoclast-bone contact (see Fig. 40.9). The region within the sealed zone forms a highly invaginated membrane called the *ruffled border,* from which HCl and hydrolytic lysosomal enzymes are secreted. The acidic enzyme-rich microenvironment beneath the osteoclast dissolves the bone mineral, thereby releasing Ca^{++} and P_i into blood, and also degrades the bone matrix. There is an additional inhibitory component of the RANK/RANKL system. Osteoblast-lineage cells can also produce a soluble factor called **osteoprotegerin (OPG),** which acts as a decoy receptor for RANKL and inhibits osteoclast differentiation and function (see Fig. 40.9). Therefore the balance between RANKL and OPG expression by osteoblasts determines how much osteoclast differentiation and bone resorption will occur.

Following bone resorption in the BMU, there is a brief reversal phase, then adjacent osteoblasts migrate into the resorbed area and begin to lay down osteoid. Several components within osteoid (pyrophosphate, alkaline phosphatase, specific glycoproteins) promote slow, controlled mineralization, a process that removes Ca^{++} and P_i from blood. As the osteoblasts become surrounded by and entrapped within bone, they become **osteocytes** that sit within small spaces called *lacunae.* Osteocytes remain interconnected through cell processes that run within canaliculi and form communicating junctions with adjacent cell processes. The new

concentric layers of bone, along with the interconnected osteocytes and the central canal, are referred to collectively as a **Haversian system or osteon.** Emerging evidence indicates that osteocytes are able to sense mechanical stress in bone and signal that additional local bone formation is needed. They can also detect microdamage in bone that serves to initiate remodeling at that location.

🧬 AT THE CELLULAR LEVEL

As a calciotropic hormone, PTH is a potent regulator of bone resorption in adults. The PTH/PTHrP receptor is expressed on osteoblasts but not on osteoclasts. Therefore PTH acts on osteoblasts to increase expression of osteoblast paracrine factors (i.e., M-CSF, RANKL) that upregulate osteoclast differentiation and bone resorption. 1,25-Dihydroxyvitamin D also stimulates bone resorption by upregulating RANKL expression in osteoblasts.

It is important to recognize that PTH (along with 1,25-dihydroxyvitamin D) will promote bone resorption when PTH levels are high (i.e., during a hypocalcemic challenge). When PTH levels are normal, however, bone remodeling is a locally controlled process by which old damaged bone is replaced. Interestingly it has been shown that intermittent administration of low-dose PTH promotes osteoblast survival and bone anabolic functions, increases bone density, and reduces the risk of fracture in humans.

Discovery of the RANK/RANKL/OPG system has presented new therapeutic opportunities for treating osteoporosis. A biological antiresorptive drug based on a humanized antibody directed against RANKL is now available for treatment of postmenopausal osteoporosis. This has proven to be an effective treatment that improves bone density and reduces the risk of fracture.

IN THE CLINIC

Vitamin D deficiency (Fig. 40.10C) produces a hypocalcemic challenge by decreasing gastrointestinal absorption of Ca^{++} and P_i. A drop in serum $[Ca^{++}]$ increases compensatory *PTH* gene expression, PTH secretion, parathyroid cell proliferation, and PTH-mediated upregulation of renal 1-hydroxylase. In the absence of sufficient 25-hydroxyvitamin D precursor, however, 1,25-dihydroxyvitamin D levels fall. The secondary elevation of PTH mobilizes Ca^{++} from bone and kidney but promotes renal excretion of P_i, causing hypophosphatemia. Because the $Ca^{++} \times P_i$ product in serum is low, bone mineralization is impaired. In children this leads to **rickets,** in which the growth of long bones is abnormal and impaired. The rib cage, wrists, and ankles show characteristic bone deformities, and the impaired mineralization causes bowing of the legs. In adults, vitamin D deficiency leads to **osteomalacia,** which is characterized by poor mineralization of newly formed osteoid, visible on radiographs as pseudofractures. In severe cases, osteomalacia results in weakness, bone pain, and increased risk of fracture.

Regulation of Serum Phosphate by FGF23

Study of hypophosphatemic disorders has led to the discovery that FGF23, a peptide hormone produced by osteocytes, is a regulator of P_i metabolism. FGF23 binds to a receptor complex in proximal tubule cells and, like PTH, inhibits NPT2 to promote P_i excretion. Several diseases are associated with excess production of FGF23, including rickets in children and osteomalacia secondary to hypophosphatemia in adults. In autosomal dominant hypophosphatemic rickets (ADHR), a mutation in FGF23 prevents its cleavage and inactivation. X-linked hypophosphatemic rickets (XLHR) is caused by a mutation of the *PHEX* gene (**p**rotein with **h**omology to **e**ndopeptidases on the **X** chromosome), which also causes overproduction of FGF23. Finally, FGF23 is sometimes ectopically produced by slow-growing occult mesenchymal tumors, causing a hypophosphatemic paraneoplastic syndrome. In addition to inhibition of P_i reabsorption, FGF23 also inhibits expression of 1α-hydroxylase in the proximal tubule, thereby inhibiting production of 1,25-dihydroxyvitamin D and exacerbating hypophosphatemia. The physiological role of this pathway is not completely understood, and many questions remain,

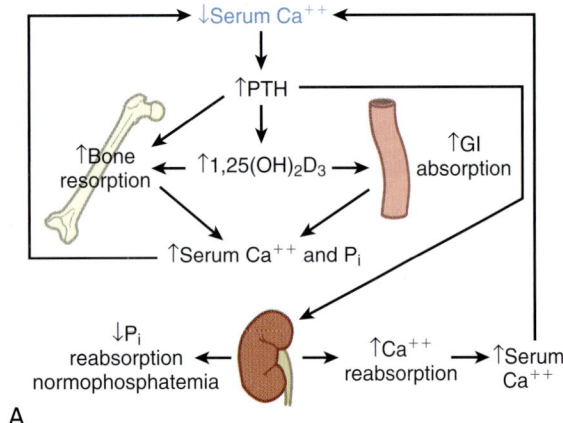

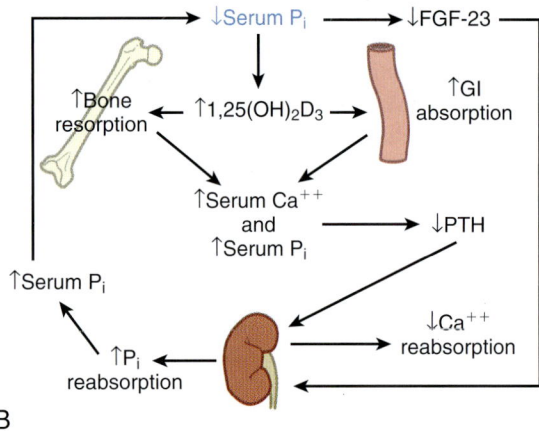

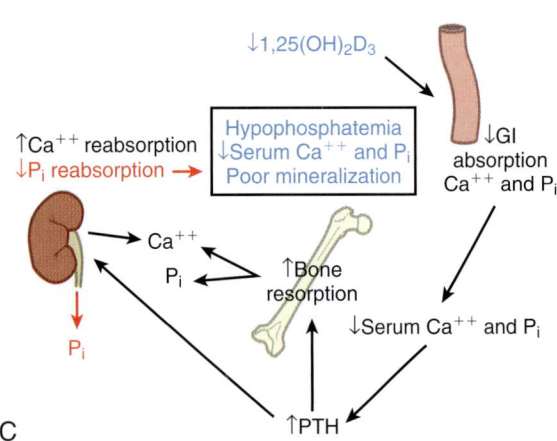

• **Fig. 40.10** Integrated hormone responses to perturbations of Ca^{++} (A), P_i (B), and vitamin D (C).

including how and where P_i levels are sensed. P_i is not as tightly regulated as calcium, either temporally or with respect to concentration range, but recent evidence suggest that long-term elevation of P_i is associated with increased production of FGF23. In what appears to be an emerging negative feedback loop, 1,25-dihydroxyvitamin D decreases production of FGF23 by osteocytes.

Regulation by Gonadal and Adrenal Steroid Hormones

Gonadal and **adrenal steroid hormones** have profound effects on bone. **17β-Estradiol** (**E_2;** see Chapter 44) has important anabolic effects on bone and is a potent regulator of osteoblast and osteoclast function. Estrogen promotes survival of osteoblasts and apoptosis of osteoclasts, thereby favoring bone formation over resorption. **Androgens** also have bone anabolic effects, although some of these effects are due to local conversion of testosterone to E_2 in men (see Chapter 44). The combined effects of testosterone and E_2 account for the higher peak bone mass observed in men. In postmenopausal women, estrogen deficiency results in an initial phase of rapid bone loss that lasts about 5 years, followed by a second phase of slower age-related bone loss that is similar in both sexes. For this reason, women are susceptible to **postmenopausal osteoporosis.**

Glucocorticoids at high therapeutic doses promote bone resorption and inhibit intestinal Ca absorption. However, the most critical adverse effect is inhibition of osteoblast differentiation, which impairs bone formation. Therefore patients treated with high levels of a glucocorticoid as an antiinflammatory or immunosuppressive drug are at risk for **glucocorticoid-induced osteoporosis** and should be monitored carefully.

Integrated Physiological Regulation of Ca^{++}/P_i Metabolism

Hypocalcemic Challenge

The integrated response of PTH and 1,25-dihydroxyvitamin D to a hypocalcemic challenge is shown in Fig. 40.10A. A decrease in serum $[Ca^{++}]$ detected by the CaSR on parathyroid chief cells stimulates secretion of PTH. In the kidney, PTH increases Ca^{++} reabsorption in the distal tubule and to a lesser extent in the distal thick ascending limb of the loop of Henle. In bone, elevated PTH stimulates osteoblast lineage cells to express RANKL, which increases osteoclast activity and leads to increased bone resorption and release

of Ca^{++} and P_i into blood. PTH stimulates 1α-hydroxylase expression in the proximal renal tubule, thereby increasing 1,25-dihydroxyvitamin D levels. 1,25-Dihydroxyvitamin D stimulates absorption of Ca and P_i in the small intestine and upregulates osteoblast expression of RANKL, thereby amplifying the effect of PTH on bone resorption. In the kidney, PTH inhibits NPT2 in the proximal tubule to lower P_i reabsorption and increase P_i clearance, thereby counterbalancing P_i mobilized from the bone and gut.

Hypophosphatemic Challenge

Although not as tightly regulated as Ca^{++}, perturbations in serum P_i will also elicit hormonal responses (see Fig. 40.10B). Low serum P_i stimulates production of 1,25-dihydroxyvitamin D in the kidney, which in turn will mobilize Ca and P_i from the intestine. The rise in Ca^{++} will suppress PTH secretion to prevent hypercalcemia. This drop in PTH will enhance P_i reabsorption in the proximal tubule to help restore serum P_i. Over a longer time course, a decrease in serum P_i will inhibit FGF23 production, which will favor P_i reabsorption in the proximal tubule. These integrated responses will allow correction of hypophosphatemia while maintaining normocalcemia. For hormonal responses to vitamin D deficiency, see the In the Clinic box and Fig. 40.10C.

IN THE CLINIC

Primary hyperparathyroidism is caused by excessive production of PTH by the parathyroid glands. It is most frequently caused by a single **adenoma** confined to one of the parathyroids. Owing to elevated PTH, patients with primary hyperparathyroidism have high serum $[Ca^{++}]$ and, in most cases, low serum $[P_i]$. **Hypercalcemia** is a result of bone resorption, increased gastrointestinal Ca absorption (mediated by 1,25-dihydroxyvitamin D), and increased renal Ca^{++} reabsorption. The major symptoms of the disorder are related to increased bone resorption, hypercalcemia, and **hypercalciuria.** These include radiographic manifestations of excessive bone resorption and psychological disorders, particularly depression. Progressive neurological symptoms include fatigue, mental confusion, and at very high levels (>15 mg/dL), coma. Kidney stones **(nephrolithiasis)** composed of calcium phosphate are common because hypercalcemia leads to hypercalciuria, and increased P_i clearance causes **phosphaturia.** Fortunately, routine blood chemistry screening over the past several decades has resulted in earlier detection of primary hyperparathyroidism, precluding development of severe symptoms in most cases.

Key Concepts

1. Serum $[Ca^{++}]$ is determined by the rate of Ca absorption by the gastrointestinal tract, bone formation and resorption, and renal excretion. Serum $[Ca^{++}]$ is normally maintained within a very narrow range.

2. Serum $[P_i]$ is determined by the rate of P_i absorption by the gastrointestinal tract, soft tissue influx and efflux, bone formation and resorption, and renal excretion. Serum $[P_i]$ normally fluctuates over a relatively wider range.

3. The major physiological hormones regulating serum $[Ca^{++}]$ and $[P_i]$ are PTH, 1,25-dihydroxyvitamin D (calcitriol), and FGF23.

4. Vitamin D is synthesized from 7-dehydrocholesterol in skin in the presence of UVB light or acquired in the diet. It is hydroxylated to 25-hydroxycholecalciferol in the liver and activated by renal 1α-hydroxylase to 1,25-dihydroxyvitamin D.

5. 1,25-Dihydroxyvitamin D promotes intestinal Ca^{++} absorption and modestly increases P_i absorption.

6. The flux of Ca^{++} and P_i into and out of bone is determined by the relative rates of osteoblastic bone formation and osteoclastic bone resorption.

7. The PTH/PTHrP receptor is expressed on osteoblasts, not on osteoclasts. PTH has both anabolic and catabolic actions in bone depending on the dose and timing of administration. PTH promotes bone resorption by upregulation of M-CSF and RANKL in osteoblasts.

8. 1,25-Dihydroxyvitamin D binds to the VDR in osteoblasts to support osteoclast differentiation via RANKL and promotes bone mineralization by maintaining appropriate serum $[Ca^{++}]$ and $[P_i]$.

Additional Reading

Bhattacharyya N, et al. Fibroblast growth factor 23: state of the field and future directions. *Trends Endocrinol Metab*. 2012;23:610-618.

Boyce BF. Advances in the regulation of osteoclasts and osteoclast functions. *J Dent Res*. 2013;92:860-867.

Christakos S, et al. Vitamin D: metabolism, molecular mechanism of action, and pleiotropic effects. *Physiol Rev*. 2016;96:365-408.

41

The Hypothalamus and Pituitary Gland

LEARNING OBJECTIVES

Upon completion of this chapter the student should be able to answer the following questions:

1. Describe the structure and composition of the pituitary gland and its structural and functional relationship to magnocellular and parvicellular hypothalamic neurons.
2. Discuss the mechanisms by which the neurohormones antidiuretic hormone (ADH) and oxytocin are synthesized, transported, and released by magnocellular neurons.
3. Diagram a basic scheme illustrating the components and feedback loops of a typical endocrine axis, including central input, hypothalamic releasing factors, pituitary hormones, and a peripheral endocrine gland. Explain the concept of a set point.
4. List the endocrine cell types of the adenohypophysis and the tropic hormones they produce, noting hormones that share a common subunit.
5. Contrast the axes of somatotropes and lactotropes with the classic endocrine axes and explain how they differ.
6. Discuss the actions of growth hormone (GH) and insulinlike growth factor I (IGF-I) in the regulation of growth, and the role of growth hormone in the fasted state.
7. Describe the role of prolactin in the initiation and maintenance of lactation.

The **pituitary gland** (also called the **hypophysis**) is a small (≈0.5 g in weight) yet complex endocrine structure at the base of the forebrain (Fig. 41.1). It is composed of an epithelial component called the **adenohypophysis** and a neural structure called the **neurohypophysis.** The adenohypophysis is composed of five cell types that secrete six hormones. The neurohypophysis releases several neurohormones. All endocrine functions of the pituitary gland are regulated by the hypothalamus and by negative- and positive-feedback loops.

Anatomy

Microscopic examination of the pituitary reveals two distinct types of tissue: epithelial and neural (Fig. 41.2). The epithelial portion of the human pituitary gland is called the **adenohypophysis.** The adenohypophysis makes up the anterior portion of the pituitary and is often referred to as the **anterior lobe of the pituitary,** and its hormones are referred to as **anterior pituitary hormones.** The adenohypophysis is composed of three parts: (1) the **pars distalis,** which makes up about 90% of the adenohypophysis, (2) the pars tuberalis, which wraps around the stalk, and (3) the pars intermedia, which regresses and is absent in adult humans.

The neural portion of the pituitary is called the **neurohypophysis** and represents a downgrowth of the hypothalamus. The most inferior portion of the neurohypophysis is called the **pars nervosa,** also called the **posterior lobe of the pituitary** (or simply **posterior pituitary**). At the superior end of the neurohypophysis, a funnel-shaped swelling called the **median eminence** develops. The portion of the neurohypophysis that extends from the median eminence down to the pars nervosa is called the **infundibulum.** The infundibulum and the pars tuberalis make up the pituitary stalk—a physical connection between the hypothalamus and pituitary gland (see Fig. 41.2).

The pituitary gland (anterior and posterior lobes) is situated within a depression of the sphenoid bone called the **sella turcica.** Generally, cancers of the pituitary have only one way to expand, which is up into the brain and against the optic chiasma. Thus any increase in size of the pituitary is commonly associated with visual field or visual acuity abnormalities and headaches. The sella turcica is sealed off from the brain by a membrane called the **diaphragma sellae.**

The Neurohypophysis

The pars nervosa is a **neurovascular** structure that is the site of neurohormone release adjacent to a rich capillary bed. The peptide hormones that are released are **antidiuretic hormone (ADH, or arginine vasopressin)** and **oxytocin.** The cell bodies of the neurons that project to the pars nervosa are located in the **supraoptic nuclei (SON)** and **paraventricular nuclei (PVN)** of the **hypothalamus** (a *nucleus* refers to a collection of neuronal cell bodies residing within the central nervous system (CNS); a *ganglion* is a collection of neuronal cell bodies residing outside the CNS). The large cell bodies of these neurons are described as **magnocellular,** and they project axons down the infundibular stalk as the **hypothalamohypophyseal tracts.** Individual magnocellular neurons are hormone specific, producing

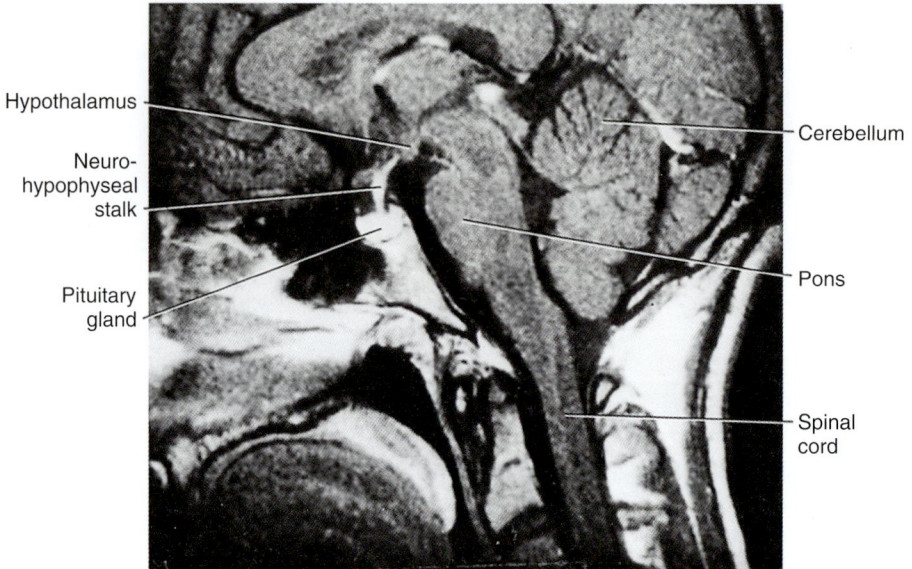

• **Fig. 41.1** Cross-sectional image of the head demonstrating the proximity of the hypothalamus and pituitary gland and their connection by a neurohypophyseal (pituitary) stalk.

either ADH or oxytocin. These axons terminate in the pars nervosa (Fig. 41.3). In addition to axonal processes and termini from the SON and PVN, there are glial-like support cells called **pituicytes.** The posterior pituitary is extensively vascularized and the capillaries are fenestrated, thereby facilitating diffusion of hormones into the systemic circulation.

Synthesis of ADH and Oxytocin

ADH and oxytocin are small peptides (nine amino acids) that differ in only two amino acids, yet they have limited overlapping activity. ADH and oxytocin are synthesized as preprohormones (Fig. 41.4). Each prohormone harbors the structure of oxytocin or ADH and a co-secreted peptide, either **neurophysin I** (associated with ADH) or **neurophysin II** (associated with oxytocin). These preprohormones are called **preprovasophysin** and **preprooxyphysin.** The N-terminal signal peptide is cleaved as the peptide is transported into the endoplasmic reticulum. In cell bodies within the SON and PVN, the prohormones are packaged in the endoplasmic reticulum and Golgi apparatus in membrane-bound secretory granules (Fig. 41.5). The secretory granules are conveyed through a "fast" (i.e., millimeters per hour) adenosine triphosphate (ATP)-dependent axonal transport mechanism down the infundibular stalk to axonal termini in the pars nervosa. During transit of the secretory granule, the prohormones are proteolytically cleaved to produce equimolar amounts of hormone and neurophysin. Secretory granules containing fully processed peptides are stored in the axonal termini. Expansions of the termini due to the presence of stored secretory granules can be observed by light microscopy and are termed **Herring bodies.**

ADH and oxytocin are released from the pars nervosa in response to stimuli that are primarily detected at the cell body and its dendrites in the SON and PVN. These stimuli are mainly in the form of neurotransmitters released from hypothalamic interneurons. With sufficient stimulus the neurons will depolarize and propagate an action potential down the axon. At the axonal termini the action potential increases intracellular [Ca^{++}] and results in a stimulus-secretion response, with exocytosis of ADH or oxytocin along with neurophysins into the extracellular fluid of the pars nervosa (see Fig. 41.5). Hormones and neurophysins enter the peripheral circulation, and both can be measured in blood.

Actions and Regulation of ADH and Oxytocin

ADH acts primarily at the kidney to retain water (antidiuresis). The actions of ADH and regulation of ADH secretion were described in Chapter 35. Oxytocin primarily acts on the pregnant uterus to induce labor and on myoepithelial cells of the breast to promote milk letdown during nursing. The actions and regulation of oxytocin are discussed in Chapter 44.

The Adenohypophysis

The pars distalis is composed of five endocrine cell types that produce six hormones (Table 41.1). Because of the histological staining properties of the cell types, the corticotropes, thyrotropes, and gonadotropes are referred to as pituitary **basophils,** whereas the somatotropes and lactotropes are referred to as pituitary **acidophils** (see Fig. 41.2B).

Endocrine Axes

Before discussing the individual hormones of the adenohypophysis, it is important to understand the structural

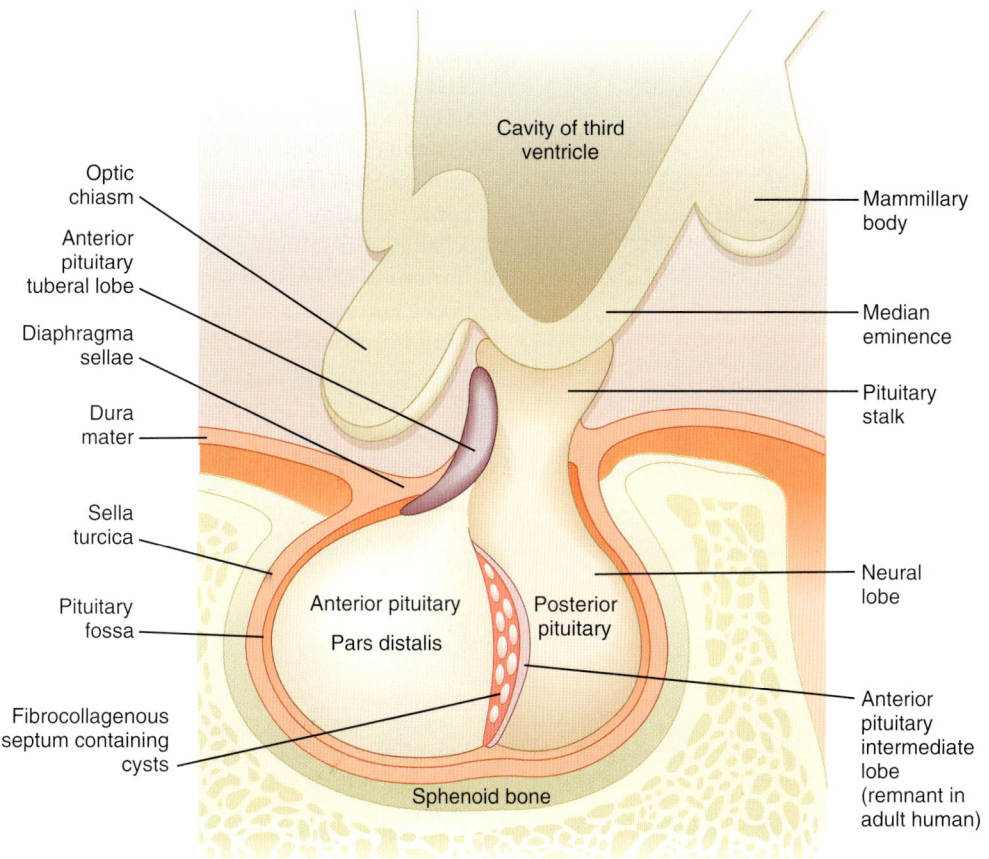

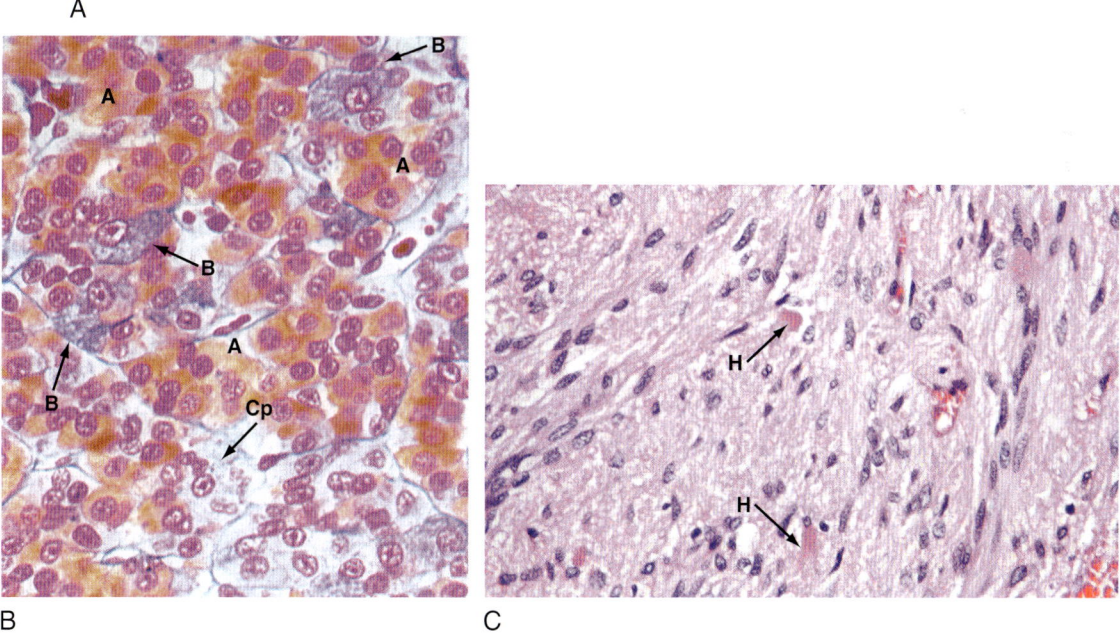

• **Fig. 41.2** A, Gross structure of the pituitary gland. The pituitary gland is below the hypothalamus and is connected to it by the pituitary stalk. The gland sits within the sella turcica, a fossa within the sphenoid bone, and is covered by a dural reflection, the diaphragma sellae. The pars distalis makes up most of the anterior pituitary. B, The pars distalis is derived from epithelial tissue that is composed of acidophils (A) (somatotropes and lactotropes) and basophils (B) (thyrotropes, gonadotropes, and corticotropes). C, The posterior pituitary is derived from neural tissue and has a histological appearance of nonmyelinated nerves. Cp, chromophobes; H, Herring bodies. (A, Modified from Stevens A. In: Lowe JS [ed]. *Human Histology.* 3rd ed. Philadelphia: Elsevier; 2005. B and C, From Young B et al [eds]. *Wheater's Functional Histology.* 5th ed. Philadelphia: Churchill Livingstone; 2006.)

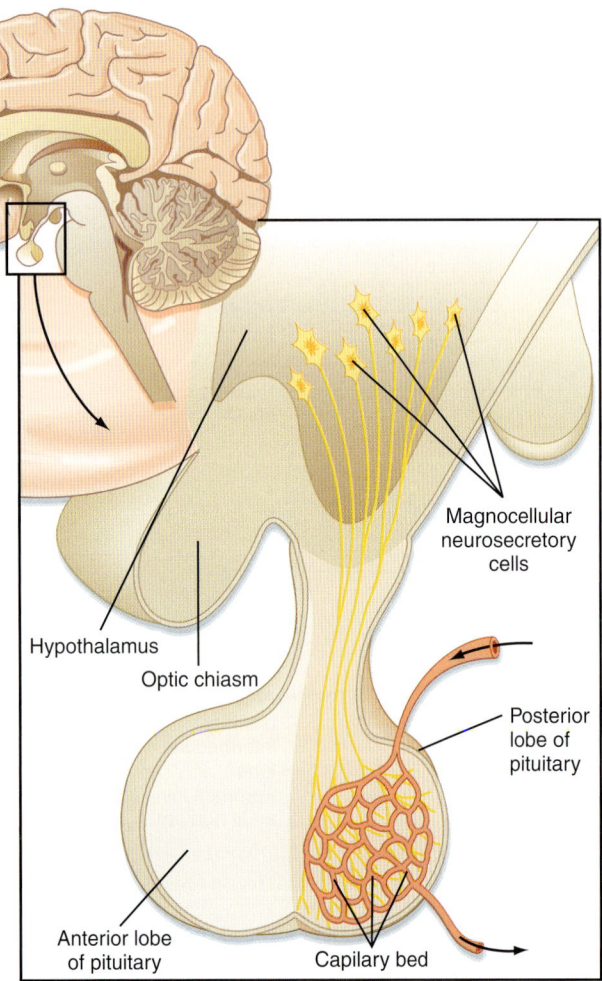

and functional organization of the adenohypophysis in the context of the **endocrine axes** (Fig. 41.6; also see Table 41.1 and Chapter 38). Each endocrine axis is composed of three levels of endocrine cells: (1) hypothalamic neurons, (2) anterior pituitary cells, and (3) peripheral endocrine glands. Hypothalamic neurons release specific **hypothalamic releasing hormones** (designated XRH in this generic scheme) that stimulate secretion of specific **pituitary tropic hormones** (XTH). In some cases, production of a pituitary tropic hormone is secondarily regulated by a **release-inhibiting hormone** (XIH). Pituitary tropic hormones then act on specific peripheral target endocrine glands and stimulate them to release peripheral hormones (X). The peripheral hormone X has two general functions: it regulates several aspects of human physiology, and it negatively feeds back on the pituitary gland and hypothalamus to inhibit production and secretion of tropic hormones and releasing hormones, respectively (see Fig. 41.6).

• **Fig. 41.3** Magnocellular neurons of the hypothalamus (paraventricular and supraoptic nuclei) project their axons down the infundibular process and terminate in the pars nervosa (posterior lobe), where they release their hormones (either ADH or oxytocin) into a capillary bed. (Modified from Larsen PR et al [eds]. *Williams Textbook of Endocrinology.* 10th ed. Philadelphia: Saunders; 2003.)

 ## AT THE CELLULAR LEVEL

Significant progress has been made in understanding the differentiation of the five endocrine cells of the pars distalis from one precursor cell. The homeodomain transcription factor **PROP-1** is expressed soon after Rathke's pouch (embryological precursor of the adenohypophysis) forms and promotes the cell lineages of somatotropes, lactotropes, thyrotropes, and gonadotropes. In humans, rare mutations in the *PROP1* gene result in a type of **combined pituitary hormone deficiency.** These individuals display dwarfism due to lack of GH, cognitive deficits secondary to hypothyroidism, and infertility due to lack of gonadotropins. A subsequently expressed pituitary-specific homeodomain transcription factor called **POU1F1** (formerly known as *Pit-1*) is required for differentiation of thyrotropes, somatotropes, and lactotropes, and it directly stimulates transcription and expression of TSH, GH, and prolactin. Affected individuals with *POU1F1* mutations have dwarfism and intellectual disability. The nuclear hormone receptor–related transcription factor **steroidogenic factor-1 (SF-1)** was originally identified in the adrenal cortex and gonads as a regulator of steroidogenic enzyme gene expression. SF-1 is also expressed in GnRH neurons in the hypothalamus and in pituitary gonadotropes, where it regulates transcription of LH and FSH. Mutations in the *SF1* gene disrupt adrenal and gonadal function, including loss of gonadotropes in the pituitary gland. **TPIT** is a transcription factor involved in the differentiation of corticotropes. TPIT, acting with other transcription factors, promotes differentiation of corticotropes and expression of the *POMC* gene (see Corticotropes section). Mutations in the human *TPIT* gene result in **isolated ACTH deficiency.** This results in a form of **secondary adrenal insufficiency** that requires lifelong replacement with glucocorticoids (see Chapter 43).

IN THE CLINIC

Because posterior pituitary hormones are synthesized in the hypothalamus rather than the pituitary, **hypophysectomy** (pituitary removal) does not necessarily permanently disrupt synthesis and secretion of these hormones. Immediately after hypophysectomy, secretion of the hormones decreases. However, over a period of weeks the severed proximal end of the tract will show histological modification and pituicytes will form around the neuron terminals. Secretory vacuoles are seen, and secretion of hormone resumes from this proximal end. Secretion of hormone can even potentially return to normal levels. In contrast, a lesion higher up on the pituitary stalk can lead to loss of neuronal cell bodies in the PVN and SON.

The hypothalamic regulation of anterior pituitary function is neurohormonal. An area of the hypothalamus collectively referred to as the **hypophysiotropic** (i.e., stimulatory to the hypophysis) **region** contains nuclei composed of small, or **parvocellular,** cell bodies that project axons to the

GENE

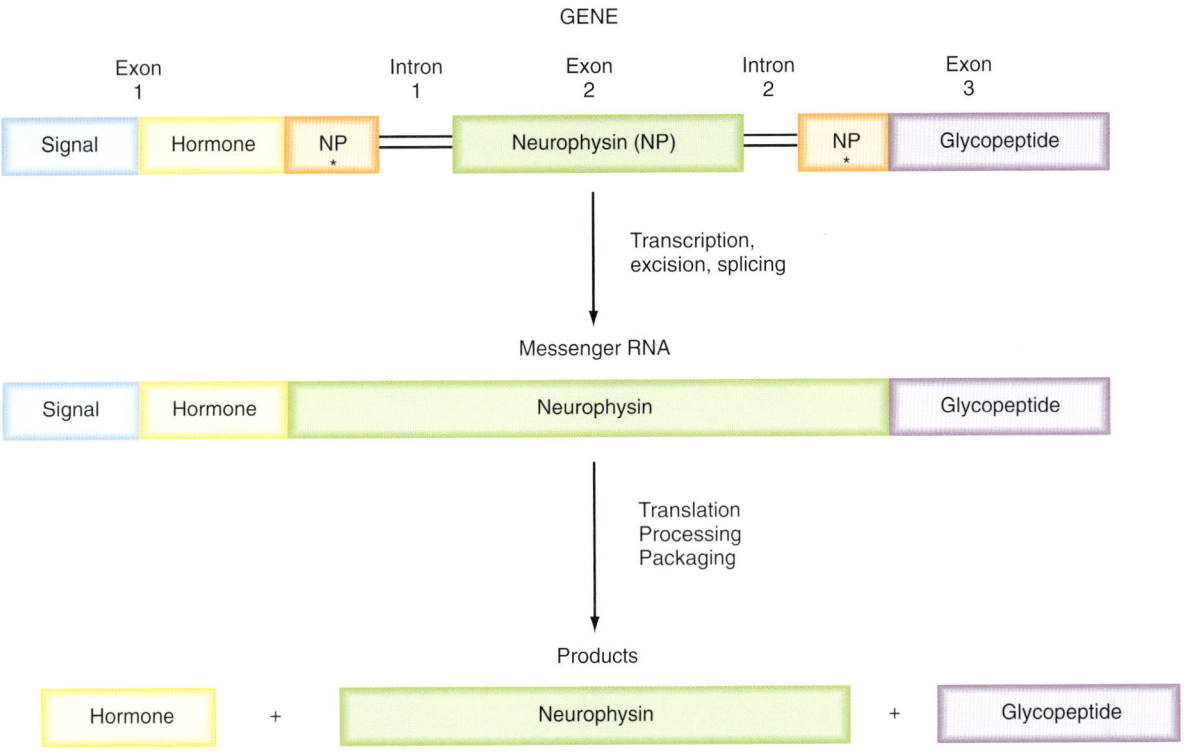

• **Fig. 41.4** Synthesis and processing of preprovasopressin or preprooxytocin.

median eminence. They are distinct from the magnocellular neurons of the PVN and SON that project to the pars nervosa. Parvocellular neurons secrete **releasing hormones** from their axonal termini at the median eminence (Fig. 41.7). The releasing hormones enter a primary plexus of fenestrated capillaries and are then conveyed to a second capillary plexus located in the pars distalis by the **hypothalamohypophyseal portal vessels** (a *portal vessel* is defined as a vessel that begins and ends in capillaries without going through the heart). At the secondary capillary plexus, the releasing hormones diffuse out of the vasculature and bind to their cognate receptors on specific cell types within the pars distalis. The neurovascular link (i.e., pituitary stalk) between the hypothalamus and pituitary is somewhat fragile and can be disrupted by physical trauma, surgery, or hypothalamic disease. Damage to the stalk and subsequent functional isolation of the anterior pituitary result in a decline in all anterior pituitary tropic hormones except prolactin (discussed later).

The cells of the adenohypophysis make up the intermediate level of the endocrine axes. The adenohypophysis secretes protein hormones that are referred to as **tropic hormones**—adrenocorticotropic hormone (ACTH, also called *corticotropin*), thyroid-stimulating hormone (TSH), follicle-stimulating hormone (FSH), luteinizing hormone (LH), growth hormone (GH), and prolactin (PRL) (see Table 41.1). With a few exceptions, tropic hormones bind to their cognate receptors on peripheral endocrine glands. Because of this arrangement, pituitary tropic hormones generally do not directly regulate physiological responses (see Chapter 38).

The endocrine axes have the following important features:

1. The activity of a specific axis is normally maintained at a **set point,** which varies from individual to individual, usually within a normal range. The set point is determined by the integration of hypothalamic stimulation and peripheral hormone negative feedback. Importantly, negative feedback generally is not exerted by the physiological responses regulated by a specific endocrine axis but by the peripheral hormone itself acting on the pituitary and hypothalamus (see Fig. 41.6). Thus if the level of a peripheral hormone drops, secretion of hypothalamic releasing hormones and pituitary tropic hormones will increase. As the level of peripheral hormone rises, the hypothalamus and pituitary will decrease secretion because of negative feedback. Although certain nonendocrine physiological parameters (e.g., acute hypoglycemia) can regulate some endocrine axes, the axes function semiautonomously with respect to the physiological changes they produce. This configuration means a peripheral hormone (e.g., thyroid hormone) can regulate multiple organ systems without these organ systems exerting competing negative-feedback regulation of the hormone. Clinically this partial autonomy means that multiple aspects of a patient's physiology are at the mercy of whatever derangements might exist within a specific axis.

2. Hypothalamic hypophysiotropic neurons are often secreted in a **pulsatile** manner and are entrained to daily and seasonal rhythms through CNS input. Additionally, hypothalamic nuclei receive a variety of neuronal inputs from higher and lower levels of the brain. These can be

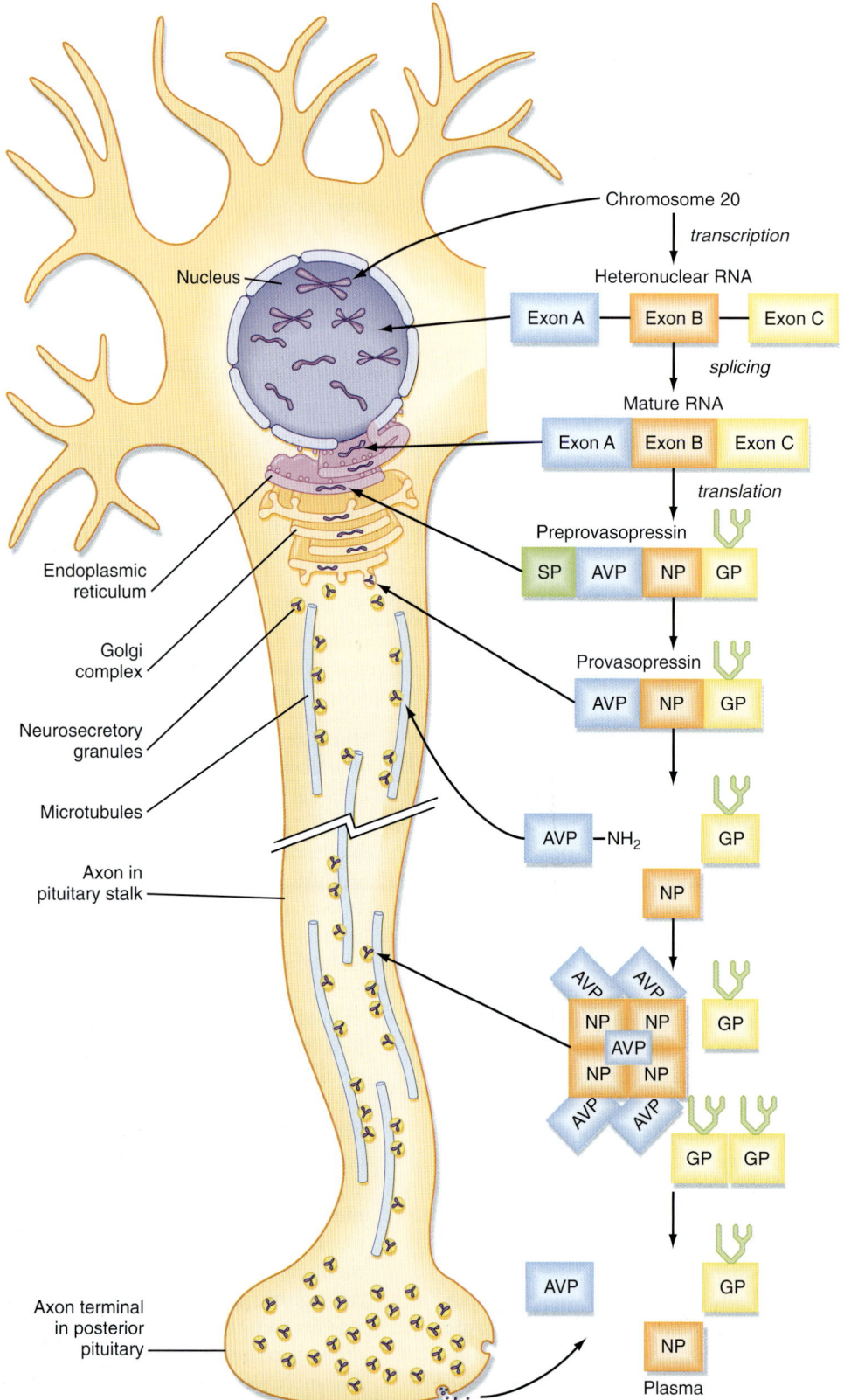

• **Fig. 41.5** Synthesis, processing, and transport of preprovasopressin. Human ADH (also called *arginine vasopressin* [AVP]) is synthesized in the hypothalamic magnocellular cell bodies and packaged into neurosecretory granules. During intraaxonal transport of the granules down the infundibular process to the pars nervosa, provasopressin is proteolytically cleaved into the active hormone (AVP = ADH), neurophysin (NP), and a C-terminal glycoprotein (GP). NP arranges into tetramers that bind five AVP molecules. All three fragments are secreted from axonal termini in the pars nervosa (posterior pituitary) and enter the systemic blood. Only AVP (ADH) is biologically active. (Modified from Larsen PR et al [eds]. *Williams Textbook of Endocrinology.* 10th ed. Philadelphia: Saunders; 2003.)

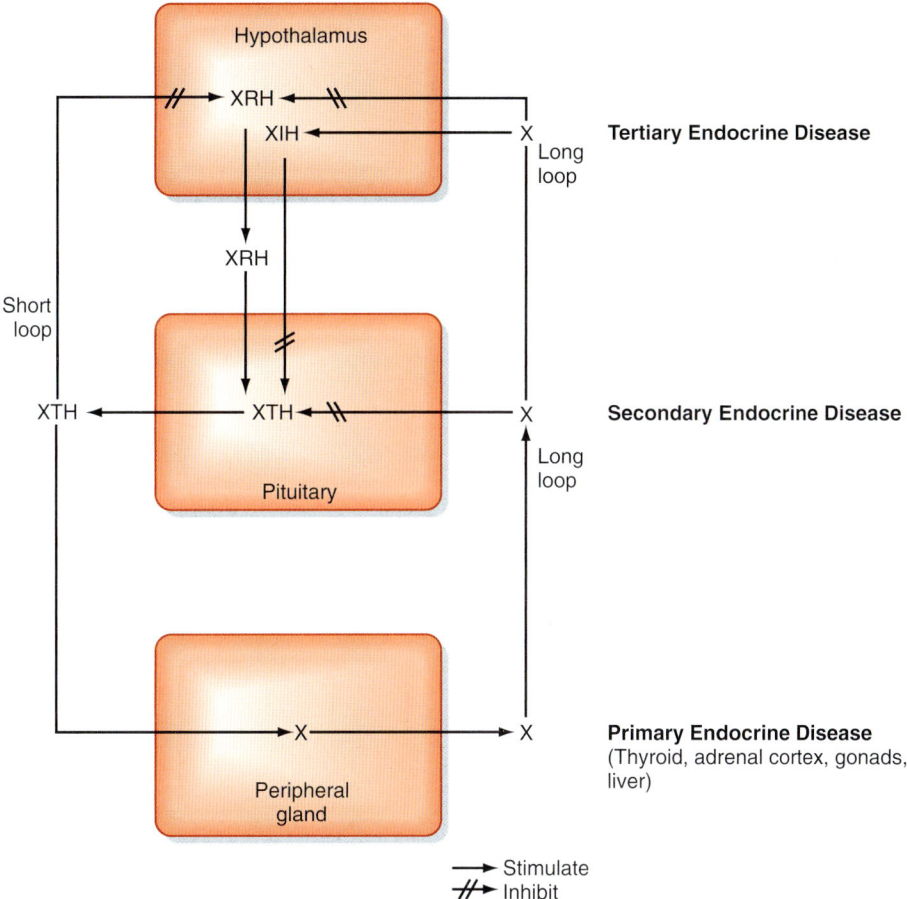

• **Fig. 41.6** Negative-feedback loops regulating hormone secretion in a typical hypothalamus-pituitary-peripheral gland axis. X, peripheral gland hormone; XIH, hypothalamic-inhibiting hormone; XRH, hypothalamic-releasing hormone; XTH, pituitary tropic hormone.

short-term (e.g., various stress/infections) or long-term (e.g., onset of reproductive function at puberty) inputs. Thus inclusion of the hypothalamus in an endocrine axis allows integration of a considerable amount of information for setting or changing the set point of that axis. Clinically this means a broad range of complex neurogenic states can alter pituitary function. **Psychosocial dwarfism** is a striking example in which children subject to abuse or intense emotional stress have lower growth rates as a result of decreased growth hormone secretion by the pituitary gland.

3. Abnormally low or high levels of a peripheral hormone (e.g., thyroid hormone) may be due to a defect at the level of the peripheral endocrine gland (e.g., thyroid), the pituitary gland, or the hypothalamus. Such lesions are referred to as **primary, secondary, and tertiary endocrine disorders,** respectively (see Fig. 41.6). A thorough understanding of the feedback relationships within an axis allows the physician to determine where the defect lies. Primary endocrine deficiencies tend to be the most severe because they often involve complete absence of the peripheral hormone.

Endocrine Function of the Adenohypophysis

The adenohypophysis consists of the following endocrine cell types: **corticotropes, thyrotropes, gonadotropes, somatotropes,** and **lactotropes** (see Table 41.1).

Corticotropes

Corticotropes stimulate the adrenal cortex as part of the **hypothalamic-pituitary-adrenal (HPA) axis.** Corticotropes produce the hormone **ACTH (corticotropin),** which stimulates two zones of the adrenal cortex (see Chapter 43). ACTH is a 39–amino acid peptide that is synthesized as part of a larger prohormone called **proopiomelanocortin (POMC).** Thus corticotropes are also referred to as **POMC cells.** POMC harbors the peptide sequence for ACTH, two isoforms of melanocyte-stimulating hormone (MSH), endorphins (endogenous opioids), and enkephalins (Fig. 41.8). However, the human corticotrope expresses only prohormone convertase-1, which produces ACTH as the sole active hormone secreted by these cells. The other fragments that are cleaved from POMC are the N-terminal fragment and β-lipotropic hormone (β-LPH), neither of which plays a physiological role in humans.

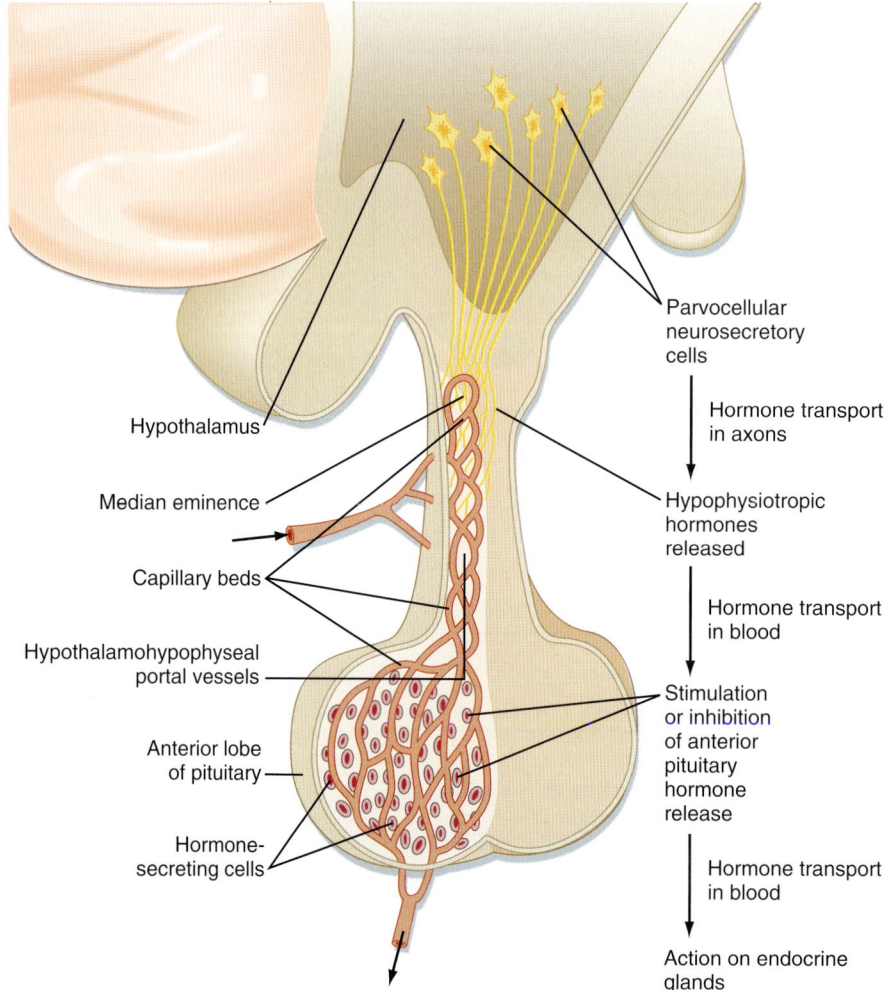

- **Fig. 41.7** Neurovascular link between the hypothalamus and the anterior lobe (pars distalis) of the pituitary. Parvicellular "hypophysiotropic" neurosecretory neurons within various hypothalamic nuclei project axons to the median eminence, where they secrete releasing hormones (RHs). RHs flow down the pituitary stalk in the hypothalamohypophyseal portal vessels to the anterior pituitary. RHs (and release-inhibiting hormones [see text]) regulate secretion of tropic hormones from the five cell types of the anterior pituitary. (From Larsen PR et al [eds]. *Williams Textbook of Endocrinology.* 10th ed. Philadelphia: Saunders; 2003.)

ACTH circulates as an unbound hormone and has a short half-life of about 10 minutes. It binds to the **melanocortin-2 receptor (MC2R)** on cells in the adrenal cortex (Fig. 41.9). ACTH acutely increases cortisol and adrenal androgen production by increasing expression of steroidogenic enzyme genes. In the long term, ACTH promotes growth and survival of two zones within the adrenal cortex (see Chapter 43).

ACTH is under stimulatory control by the hypothalamus. A subset of parvocellular hypothalamic neurons expresses the peptide **procorticotropin-releasing hormone (pro-CRH)** (see Table 41.1). Pro-CRH is processed to **CRH,** an amidated 41–amino acid peptide. CRH acutely stimulates ACTH secretion and increases transcription of the *POMC* gene. The parvocellular neurons that express CRH also express ADH, which potentiates the action of CRH on corticotropes. ACTH secretion has a pronounced diurnal pattern, with a peak in early morning and a nadir

AT THE CELLULAR LEVEL

At supraphysiological levels, **ACTH** causes darkening of the skin (e.g., in Cushing's disease). Keratinocytes in the basal layer of the epidermis also express the POMC gene but process it to **α-MSH** instead of ACTH. Keratinocytes secrete α-MSH in response to ultraviolet light, and α-MSH acts as a paracrine factor on neighboring melanocytes to darken the skin. α-MSH binds to the **MC1R** on melanocytes. At very high levels, ACTH can cross-react with the MC1R receptor on skin melanocytes (see Fig. 41.9). Thus increased skin pigmentation is one indicator of excess circulating ACTH.

in late afternoon (Fig. 41.10). In addition, secretion of CRH—and hence secretion of ACTH—is pulsatile.

There are multiple regulators of the HPA axis, and many of them are mediated through the CNS (Fig. 41.11). Many types of stress, both neurogenic (e.g., fear) and systemic

TABLE 41.1 Adenohypophysis Cell Types: Hormonal Production and Action, Hypothalamic Regulation, and Feedback Regulation

	Basophils			Acidophils	
	Corticotrope	Thyrotrope	Gonadotrope	Somatotrope	Lactotrope
Primary hypothalamic regulation	Corticotropin-releasing hormone (CRH): 41–amino acid peptide, stimulatory	Thyrotropin-releasing hormone (TRH): tripeptide, stimulatory	Gonadotropin-releasing hormone (GnRH): decapeptide, stimulatory	Growth hormone–releasing hormone (GHRH): 44–amino acid peptide, stimulatory. Somatostatin: tetradecapeptide, inhibitory	Dopamine (catecholamine): inhibitory. PRL-releasing factor?: stimulatory
Tropic hormone secreted	Adrenocorticotropic hormone (ACTH): 4.5-kDa protein	Thyroid-stimulating hormone (TSH): 28-kDa glycoprotein hormone	Follicle-stimulating hormone and luteinizing hormone (FSH, LH): 28- and 33-kDa glycoprotein hormones	Growth hormone (GH): ≈22-kDa protein	Prolactin (PRL): ≈23-kDa protein)
Receptor	MC2R (Gs-linked GPCR)	TSH receptor (Gs-linked GPCR)	FSH and LH receptors (Gs-linked GPCRs)	GH receptor (JAK/STAT-linked cytokine receptor)	PRL receptor (JAK/STAT-linked cytokine receptor)
Target endocrine gland	Zona fasciculata and zona reticularis of adrenal cortex	Thyroid epithelium	Ovary (theca and granulosa[a]). Testis (Leydig and Sertoli cells)	Liver (but also direct actions—especially in terms of metabolic effects)	No endocrine target organ—not part of an endocrine axis
Peripheral hormone involved in negative feedback	Cortisol	Triiodothyronine	Estrogen,[b] progesterone, testosterone, and inhibin[c]	IGF-I. GH (short loop)	None

[a]Both follicular and luteinized thecal and granulosa cells.
[b]Estrogen can also have a positive feedback in women.
[c]Inhibin selectively inhibits release of FSH from the gonadotrope.

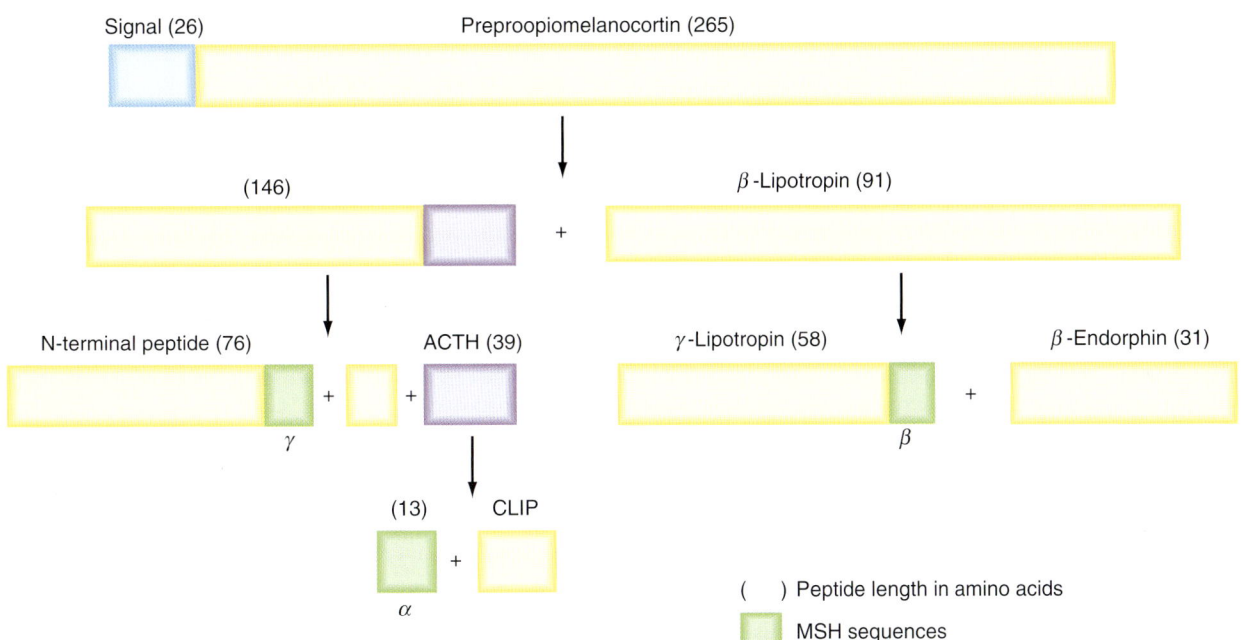

• **Fig. 41.8** The original gene transcript of proopiomelanocortin contains structures of multiple bioactive compounds. ACTH, adrenocorticotropic hormone; CLIP, corticotropin-like intermediate peptide; MSH, melanocyte-stimulating hormone. Note that ACTH is the only bioactive peptide released by the human corticotrope.

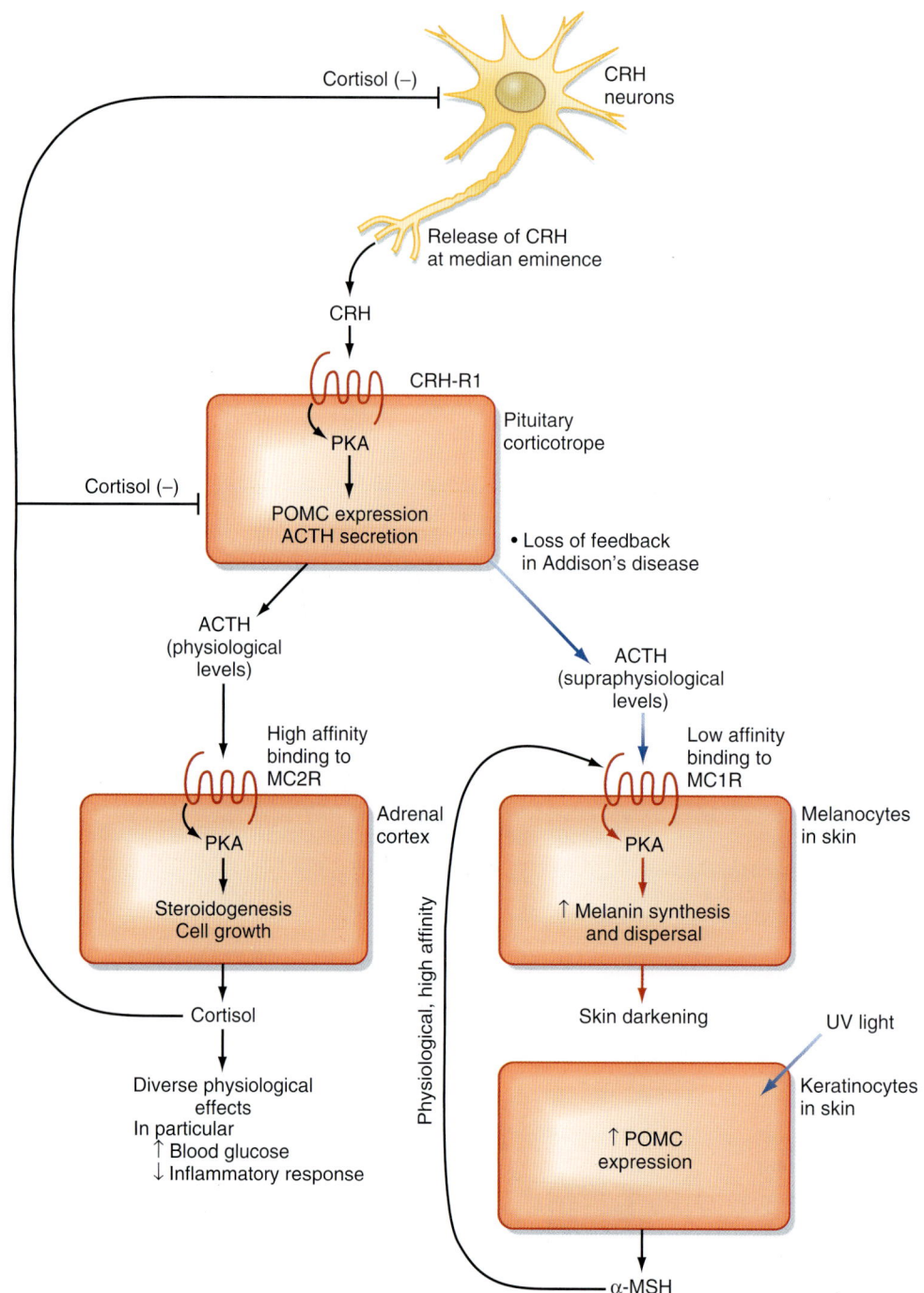

- **Fig. 41.9** Normal levels of ACTH act on the MC2R to increase cortisol. Supraphysiological levels of ACTH due to decreased cortisol production act on both the MC2R and the MC1R on melanocytes and cause skin darkening. (Modified from Porterfield SP, White BA. *Endocrine Physiology.* 3rd ed. Philadelphia: Mosby; 2007.)

(e.g., infection), stimulate ACTH. The stress effects are mediated through CRH and ADH via the CNS. The response to many forms of severe stress can persist despite negative feedback from high cortisol levels. This means the hypothalamus has the ability to alter the set point of the HPA axis in response to stress. Severe chronic depression can reset the HPA axis as a result of hypersecretion of CRH and is a factor in the development of **tertiary hypercortisolism.**

Cortisol exerts negative feedback on the pituitary, where it suppresses *POMC* gene expression and ACTH secretion, and on the hypothalamus, where it decreases pro-CRH gene expression and release of CRH. Because cortisol has profound effects on the immune system (see Chapter 43), the HPA axis and the immune system are closely coupled. Moreover, cytokines—particularly interleukin (IL)-1, IL-2, and IL-6—stimulate the HPA axis.

Thyrotropes

Thyrotropes regulate thyroid function by secreting the hormone **TSH (thyrotropin)** as part of the **hypothalamic-pituitary-thyroid axis.** TSH is one of three **pituitary glycoprotein hormones** (see Table 41.1) that also include **FSH** and **LH** (discussed later). TSH is a heterodimer composed of an α subunit, called the **α-glycoprotein subunit (α-GSU),** and a β subunit **(β-TSH)** (Fig. 41.12). The α-GSU is common to TSH, FSH, and LH, whereas the β subunit is hormone specific (i.e., β-TSH, β-FSH, and β-LH are all unique). Glycosylation of the subunits increases their stability in circulation and enhances the affinity and specificity of the hormones for their receptors. The half-lives of TSH, FSH, and LH (and an LH-like placental glycoprotein hormone, **human chorionic gonadotropin [hCG]**) are relatively long, ranging from tens of minutes to several hours.

TSH binds to the TSH receptor on thyroid follicle cells (see Chapter 42). As discussed in Chapter 42, production of thyroid hormones is a complex multistep process, and TSH stimulates essentially every aspect of thyroid function. TSH also has a strong tropic effect and stimulates hypertrophy, hyperplasia, and survival of thyroid epithelial cells. In geographical regions where the availability of iodide is limited (iodide is required for the synthesis of thyroid hormone),

TSH levels are elevated because of reduced negative feedback. Elevated TSH levels can produce striking growth of the thyroid, producing a bulge in the neck called a **goiter.**

The pituitary thyrotrope is stimulated by the releasing hormone **thyrotropin-releasing hormone (TRH)** (see Table 41.1). TRH, produced by a subset of parvocellular

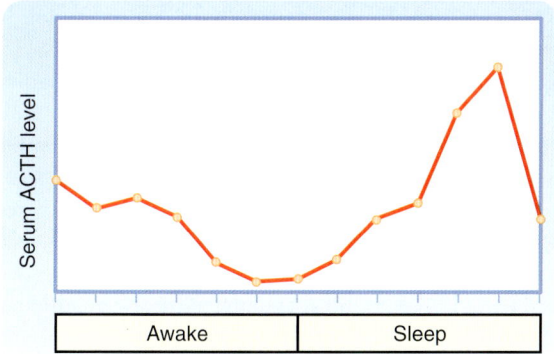

• **Fig. 41.10** Diurnal pattern of serum ACTH. (Modified from Porterfield SP, White BA. *Endocrine Physiology.* 3rd ed. Philadelphia: Mosby; 2007.)

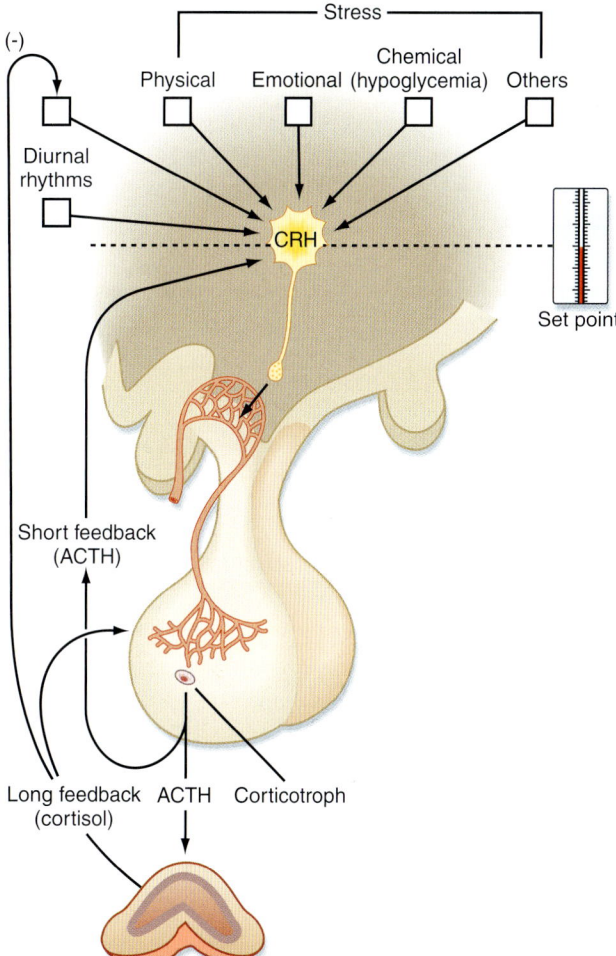

• **Fig. 41.11** Hypothalamic-pituitary-adrenal axis illustrating factors regulating secretion of corticotropin-releasing hormone (CRH). ACTH, adrenocorticotropic hormone. (Modified from Porterfield SP, White BA. *Endocrine Physiology.* 3rd ed. Philadelphia: Mosby; 2007.)

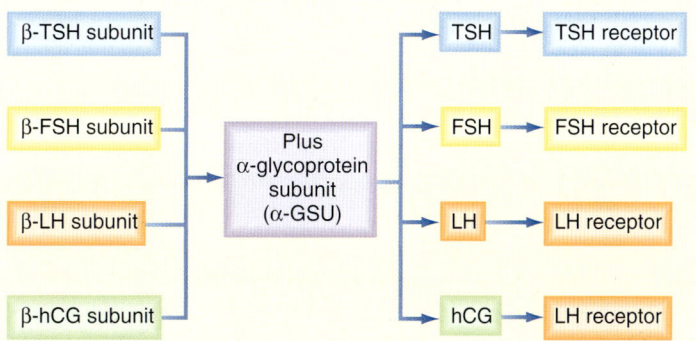

• **Fig. 41.12** Pituitary glycoprotein hormones. hCG is made by the placenta (see Chapter 44) and binds to the LH receptor. FSH, follicle-stimulating hormone; hCG, human chorionic gonadotropin; LH, luteinizing hormone; TSH, thyroid-stimulating hormone.

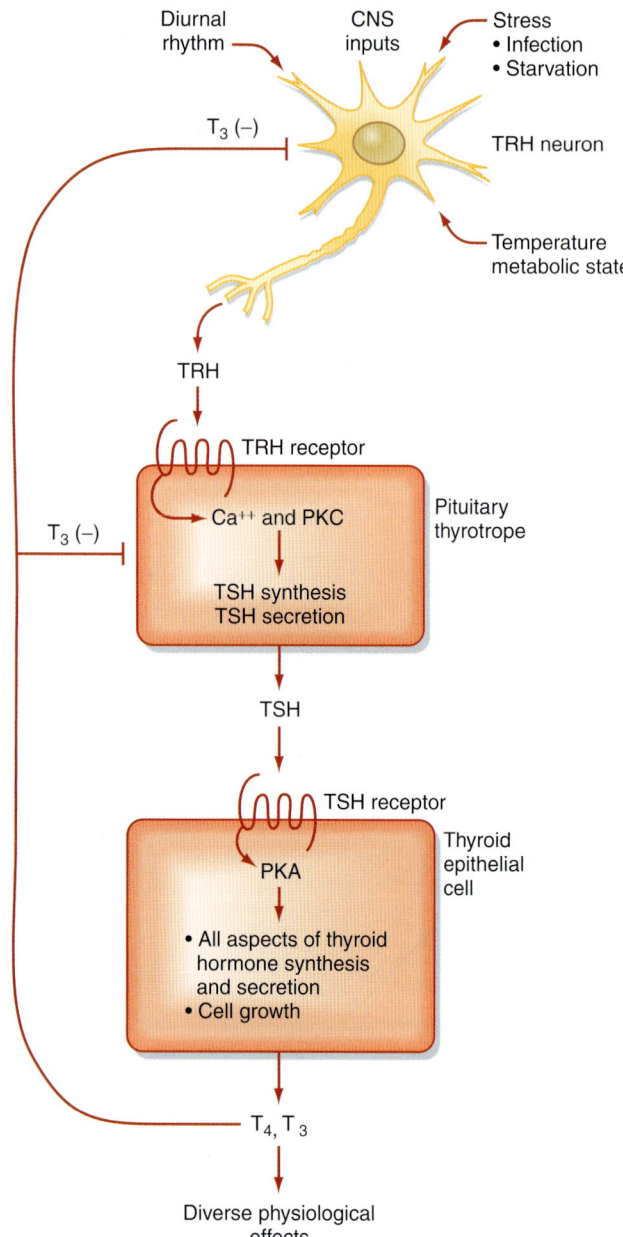

• **Fig. 41.13** Hypothalamic-pituitary-thyroid axis. PKA, protein kinase A; PKC, protein kinase C; T₃, triiodothyronine (active form of thyroid hormone); T₄, thyroxine; TRH, thyrotropin-releasing hormone; TSH, thyroid-stimulating hormone. (Modified from Porterfield SP, White BA. *Endocrine Physiology.* 3rd ed. Philadelphia: Mosby; 2007.)

hypothalamic neurons, is a tripeptide with cyclization of a glutamine at its N-terminus (pyro-Glu) and an amidated C-terminus. TRH is synthesized as a larger prohormone that contains six copies of TRH within its sequence. It binds to the TRH receptor on thyrotropes (Fig. 41.13). TRH neurons are regulated by numerous CNS-mediated stimuli, and TRH is released according to a diurnal rhythm (highest during overnight hours, lowest around dinner time). TRH secretion is also regulated by stress, but in contrast to CRH, stress inhibits secretion of TRH. This includes physical stress, starvation, and infection. Triiodothyronine (T₃) and thyroxine (T₄) (the latter via type 2 deiodinase-mediated

conversion to T₃; see Chapter 42) negatively feed back on both pituitary thyrotropes and TRH-producing neurons. Thyroid hormone represses both β-TSH expression and the sensitivity of pituitary thyrotropes to TRH while also inhibiting TRH production and secretion by parvocellular neurons.

IN THE CLINIC

During embryonic development, GnRH neurons migrate to the mediobasal hypothalamus from the nasal placode. Patients with **Kallmann syndrome** have **tertiary hypogonadotropic hypogonadism,** often associated with loss of the sense of smell (anosmia). This is due to a mutation in the **KAL gene,** which results in failure of the GnRH neuronal precursors to properly migrate to the hypothalamus and establish a neurovascular link to the pars distalis.

The Gonadotrope

The gonadotrope secretes FSH and LH (collectively called *gonadotropins*) and regulates the function of gonads in both sexes. As such the gonadotrope plays an integral role in the **hypothalamic-pituitary-testis axis** and the **hypothalamic-pituitary-ovarian axis** (Fig. 41.14).

FSH and LH are segregated into different secretory granules and are not co-secreted in equimolar amounts (in contrast to ADH and neurophysin, for example). This allows independent regulation and secretion of FSH/LH by gonadotropes. The actions of FSH and LH on gonadal function are complex, especially in women, and will be discussed in detail in Chapter 44. In general, gonadotropins promote testosterone secretion in men and estrogen and progesterone secretion in women. FSH also increases secretion of a transforming growth factor (TGF)-β–related protein hormone called **inhibin** in both sexes.

FSH and LH secretion are regulated by one hypothalamic releasing hormone, **gonadotropin-releasing hormone (GnRH;** formerly called **LHRH).** GnRH is a 10–amino acid peptide produced by a subset of parvocellular hypothalamic GnRH neurons (see Fig. 41.14). GnRH is produced as a larger prohormone and, as part of its processing to a decapeptide, is modified with a cyclized glutamine (pyro-Glu) at its N-terminus and an amidated C-terminus.

GnRH is released in a pulsatile manner (Fig. 41.15), and both the pulsatile secretion and the frequency of the pulses have important effects on the gonadotrope. Continuous infusion of GnRH downregulates the GnRH receptor, thereby resulting in a decrease in FSH and LH secretion. In contrast, pulsatile secretion does not desensitize the gonadotrope to GnRH, and FSH and LH secretion is normal. At a frequency of one pulse per hour, GnRH preferentially increases LH secretion (Fig. 41.16). At a slower frequency of one pulse per 3 hours, GnRH preferentially increases FSH secretion. Gonadotropins increase sex steroid synthesis (see Fig. 41.14). In men, testosterone and estrogen negatively feed back at the level of the pituitary and the hypothalamus.

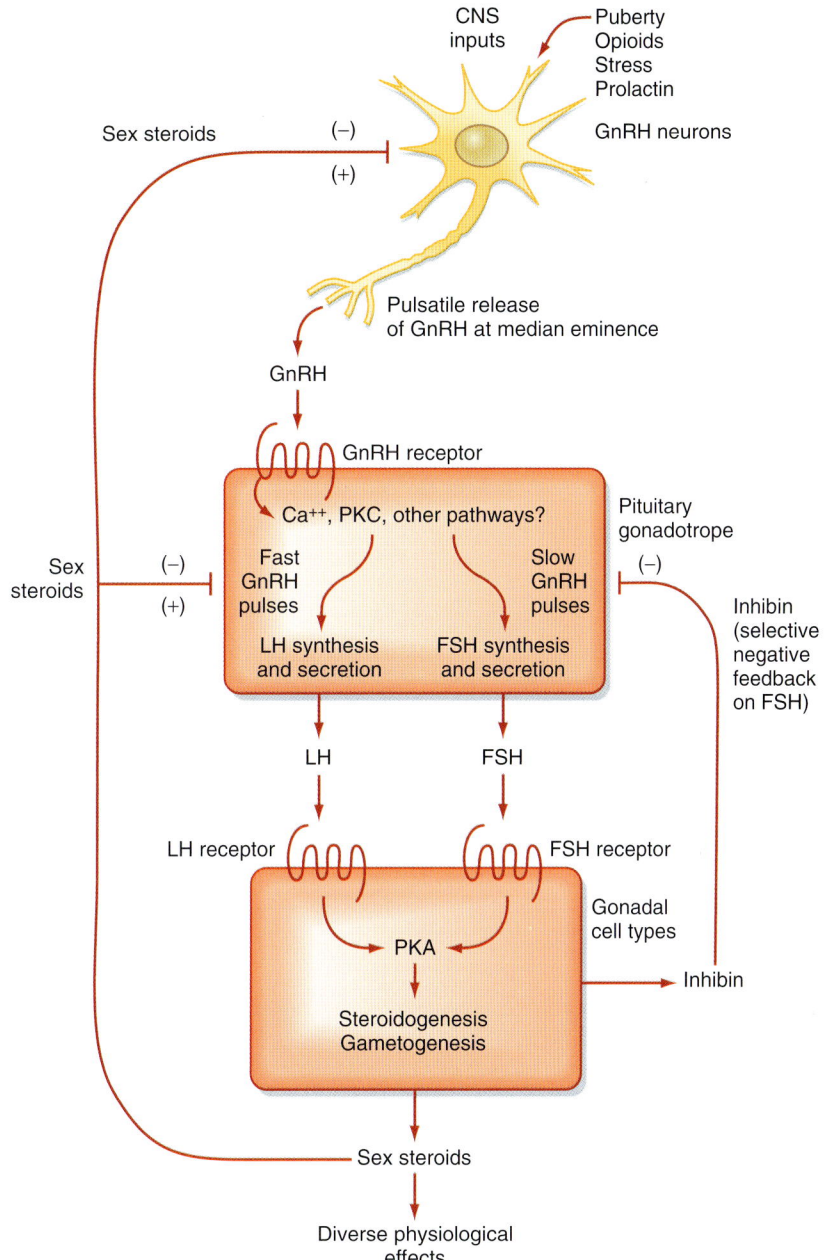

• **Fig. 41.14** Hypothalamic-pituitary-gonadal axis. FSH, follicle-stimulating hormone; GnRH, gonado-tropin-releasing hormone; LH, luteinizing hormone. (Modified from Porterfield SP, White BA. *Endocrine Physiology*. 3rd ed. Philadelphia: Mosby; 2007.)

Exogenous progesterone also inhibits gonadotropin function in men and has been considered as a possible component of a male contraceptive pill. Additionally, inhibin negatively feeds back selectively on FSH secretion in men and women. In women, progesterone and testosterone negatively feed back on gonadotropic function at the level of the hypothalamus and pituitary. At low doses, estrogen also exerts negative feedback on FSH and LH secretion. However, high estrogen levels maintained for 3 days cause a surge in LH and to a lesser extent FSH secretion. This positive feedback, which is critical in promoting ovulation, is observed at the hypothalamus and pituitary. At the hypothalamus, GnRH pulse amplitude and frequency increase. At the pituitary,

high estrogen levels greatly increase the sensitivity of the gonadotrope to GnRH, both by increasing GnRH receptor levels and by enhancing postreceptor signaling (see Chapter 44).

The Somatotrope

The somatotrope produces **GH (somatotropin)** and is part of the hypothalamic-pituitary-liver axis (Fig. 41.17). A major target of GH is the liver, where it stimulates production of **insulinlike growth factor (IGF)-I.** GH is a 191–amino acid protein that is similar to **PRL** and **human placental lactogen (hPL);** accordingly there is some overlap in activity among these hormones. Multiple forms of GH

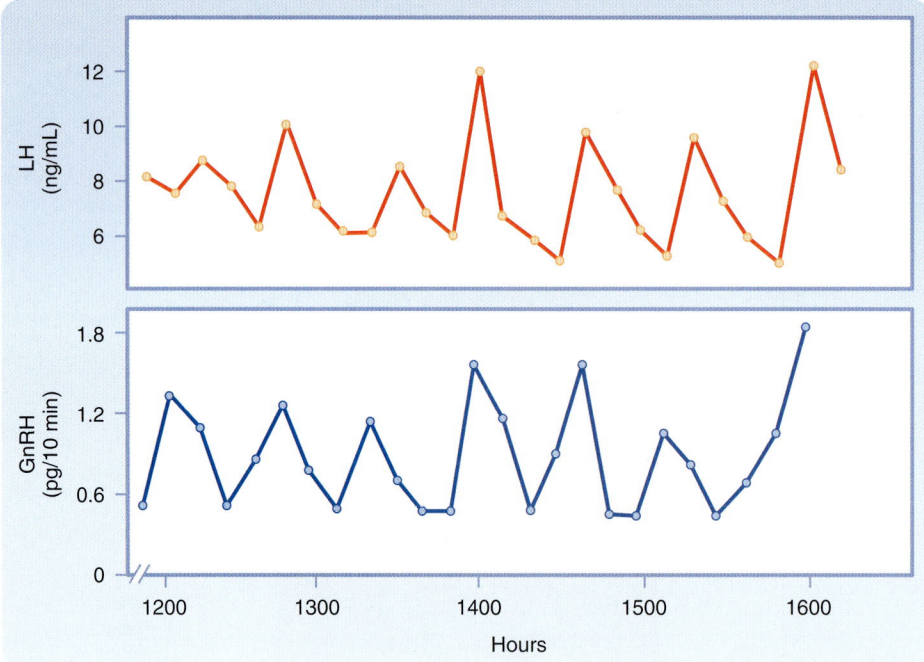

• **Fig. 41.15** Fluctuation of peripheral vein plasma LH levels and portal vein plasma GnRH levels in unanesthetized, ovariectomized female sheep. Each pulse of LH is coordinated with a pulse of GnRH. This supports the view that pulsatility of LH release is dependent on pulsatile stimulation of the pituitary by GnRH. (From Levine J et al. *Endocrinology* 1982;111:1449.)

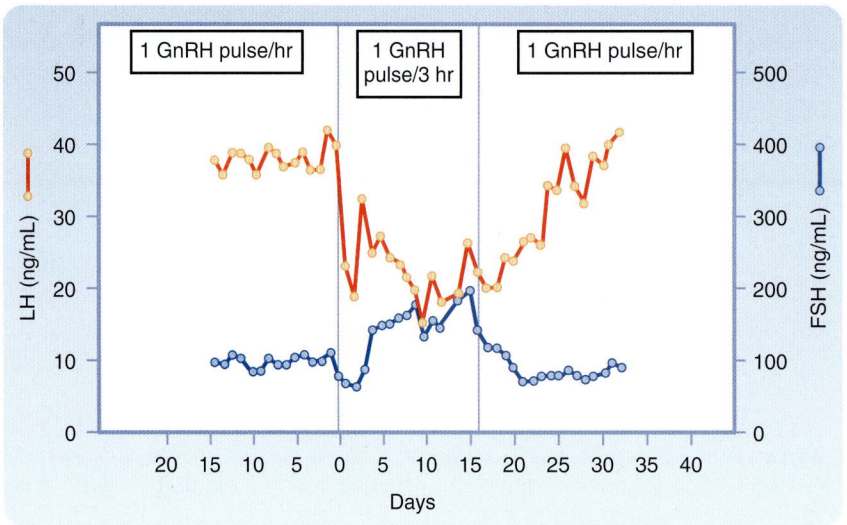

• **Fig. 41.16** Frequency-encoded regulation of FSH and LH secretion from gonadotropes. A high frequency of GnRH (1 pulse/hr) preferentially stimulates LH secretion, whereas a slower frequency of GnRH promotes FSH secretion. (From Larsen PR et al [eds]. *Williams Textbook of Endocrinology.* 10th ed. Philadelphia: Saunders; 2003.)

are present in serum, with the 191–amino acid (22-kDa) form representing approximately 75% of circulating GH. The GH receptor is a member of the cytokine/GH/PRL/erythropoietin receptor family and as such is linked to the JAK/STAT signaling pathway (see Chapter 3). Human GH can also act as an agonist of the PRL receptor. About 50% of the 22-kDa form of GH in serum is bound to **GH-binding protein (GHBP)**, which is derived from the N-terminal portion (the extracellular domain) of the

GH receptor. Individuals with Laron syndrome, who lack normal GH receptors but have normal GH secretion, do not have detectable GHBP in their serum. GHBP reduces renal clearance and thus increases the biological half-life of GH, which is about 20 minutes. The liver and kidney are major sites of GH degradation.

GH secretion is under dual positive/negative control by the hypothalamus (see Fig. 41.17). The hypothalamus predominantly stimulates GH secretion via the peptide **growth**

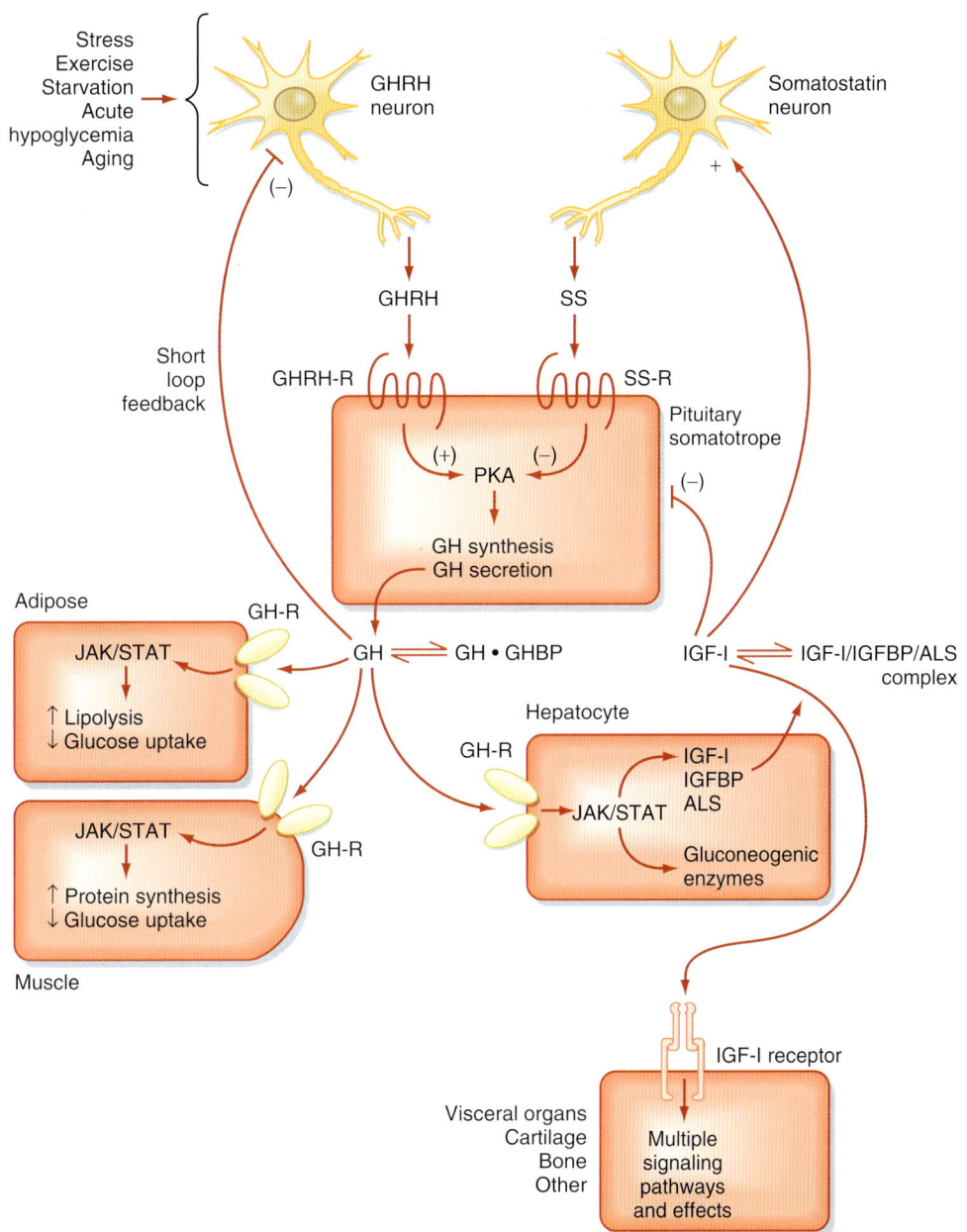

• Fig. 41.17 Hypothalamic-pituitary-liver axis. ALS, acid labile subunit; GHBP, growth hormone–binding protein; GHRH, growth hormone–releasing hormone; IGFBP, insulinlike growth factor–binding protein; IGF-I, insulinlike growth factor I; SS, somatostatin. (From Porterfield SP, White BA. *Endocrine Physiology.* 3rd ed. Philadelphia: Mosby; 2007.)

hormone–releasing hormone (GHRH). This hormone is a member of the vasoactive intestinal polypeptide (VIP)/secretin/glucagon family and is processed from a larger prohormone into a 44–amino acid peptide with an amidated C-terminus. GHRH enhances GH secretion and GH gene expression. The hypothalamus inhibits pituitary GH synthesis and release via the peptide **somatostatin.** In the anterior pituitary, somatostatin inhibits release of GH and TSH. GH secretion is also stimulated by **ghrelin,** which acts through the GH secretagogue receptor on somatotropes. Ghrelin is primarily produced by the stomach but is also expressed in the hypothalamus. Ghrelin increases appetite

and may serve as a signal to coordinate nutrient acquisition with growth.

The primary negative feedback on the somatotrope is exerted by IGF-I (see Fig. 41.17). GH stimulates IGF-I production by the liver, and IGF-I then inhibits GH synthesis and secretion by the pituitary and hypothalamus in a classic "long feedback" loop. In addition, GH itself exerts negative feedback on release of GHRH through a "short feedback" loop. GH also increases somatostatin release.

GH secretion, like ACTH, shows prominent diurnal rhythms, with peak secretion occurring in the early morning just before awakening. Its secretion is stimulated during deep slow-wave sleep (stages III and IV). GH secretion is lowest during the day. This rhythm is entrained to sleep-wakefulness patterns rather than light-dark patterns, so a phase shift occurs in people who work night shifts. As is typical of anterior pituitary hormones, GH secretion is pulsatile. Levels of GH in serum vary widely (0–30 ng/mL, with most values usually falling between 0 and 3). Because of this marked variation and the heterogeneity of circulating GH, measurement of serum GH levels is of limited clinical utility. Since IGF-I secretion is regulated by GH and possesses a longer half-life that buffers pulsatile and diurnal changes in GH secretion, it may be used to assess the status of the GH axis, especially in young patients.

GH secretion is differentially regulated depending on the physiological state. GH is classified as one of the **"stress hormones"** and is increased by neurogenic and physical stress. It promotes lipolysis, increases protein synthesis, and antagonizes the ability of insulin to reduce blood glucose levels. It is not surprising, therefore, that acute hypoglycemia is a stimulus for GH secretion and GH is classified as a **hyperglycemic hormone.** A rise in the serum concentration of some amino acids also stimulates GH secretion; administration of arginine is used for provocative testing of GH secretion. In contrast, an increase in blood glucose or free fatty acids inhibits secretion of GH. Obesity also inhibits GH secretion, in part because of insulin resistance (relative hyperglycemia) and increased circulating free fatty acids. Conversely, exercise and starvation stimulate GH secretion.

The lifetime pattern of GH secretion is shown in Fig. 41.18. GH secretion increases in the neonatal period as growth becomes GH and IGF-I dependent. It remains high throughout childhood and peaks during puberty, when estrogen (in females and also males via aromatization) promotes even higher rates of GH secretion. Thyroid hormone also enhances GH and IGF-I secretion to support bone growth and maturation. Adults continue to produce GH, consistent with its role in metabolism, before levels fall during senescence.

IGFs are multifunctional hormones that regulate cellular proliferation, differentiation, and metabolism. These protein hormones resemble insulin in structure and function. The two hormones in this family, IGF-I and IGF-II, are produced in many tissues and have autocrine, paracrine, and endocrine actions. IGF-I is the major form produced in most adult tissues. IGF-II is the major form produced in the fetus, where it regulates growth of both the fetus and the placenta in a GH-independent manner. Both hormones are structurally similar to proinsulin, with IGF-I exhibiting 42% structural homology with proinsulin. IGFs and insulin show receptor cross-reactivity; IGFs in high concentration mimic the metabolic actions of insulin. Both IGF-I and IGF-II act through type I IGF receptors, which are similar to insulin and epidermal growth factor (EGF) receptors and contain intrinsic tyrosine kinase activity. However, IGF-II also binds to the type II IGF/mannose-6-phosphate receptor. This receptor does not resemble the insulin receptor, does not have intrinsic tyrosine kinase activity, and probably functions to limit IGF-II signaling through the type I receptor. IGFs stimulate glucose and amino acid uptake and protein and DNA synthesis. They were initially called **somatomedins** because of their growth-mediating actions on cartilage, bone, and other organs. It was originally proposed that IGF-I is produced exclusively in the liver upon GH stimulation. During puberty, when GH levels increase (Fig. 41.19), IGF-I levels increase in parallel. However, it is now known that IGFs are produced in many extrahepatic tissues, exhibiting both autocrine and paracrine actions. Some of these are under the control of GH, whereas others are not. In bone, for example, IGF-I has both endocrine and paracrine effects on linear growth, some of which are GH independent. Hormones such as parathyroid hormone (PTH) and estradiol are also effective stimuli for IGF-I production by osteoblasts. At the same time, GH exerts stimulatory effects on the growth plate that are independent of IGF-I. The liver appears to be the predominant source of the circulating pool of IGF-I (see Fig. 41.19).

Essentially all circulating IGFs are transported in serum bound to **IGF binding proteins (IGFBP).** IGFBP-3 binds to IGF then associates with another protein called the **acid labile subunit (ALS)** (see Fig. 41.19). GH stimulates hepatic production of IGF-I, IGFBP-3, and ALS. The IGFBP-3/ALS/IGF-I complex mediates transport and bioavailability of IGF-I. Although IGFBPs generally inhibit IGF action, they greatly increase the biological half-life of IGFs (up to 12 hours). **IGFBP proteases** degrade IGFBP and play a role in locally generating free (i.e., active) IGFs. This is of interest in the context of IGF-responsive cancers (e.g., prostate cancer), which may overexpress one or more IGFBP proteases.

Growth Hormone Actions

GH plays a dual role in metabolism that is highly dependent on physiological context. At the risk of oversimplification, its dual roles are to: (1) promote growth and protein anabolism when nutritional status is favorable, and (2) switch fuel consumption to lipids, sparing glucose in the fasted state.

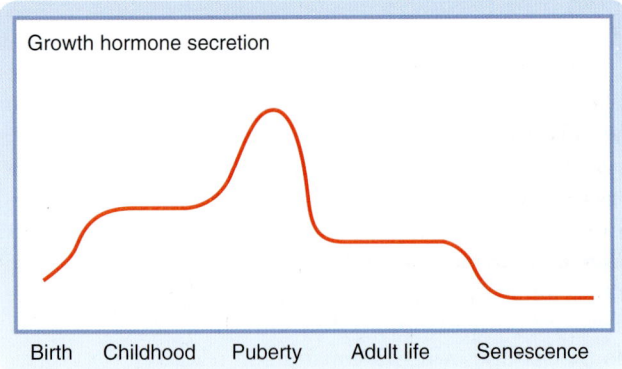

Growth hormone secretion

Birth   Childhood   Puberty   Adult life   Senescence

• **Fig. 41.18** Lifetime pattern of GH secretion. GH levels are higher in children than in adults, with a peak period during puberty. GH secretion declines with aging.

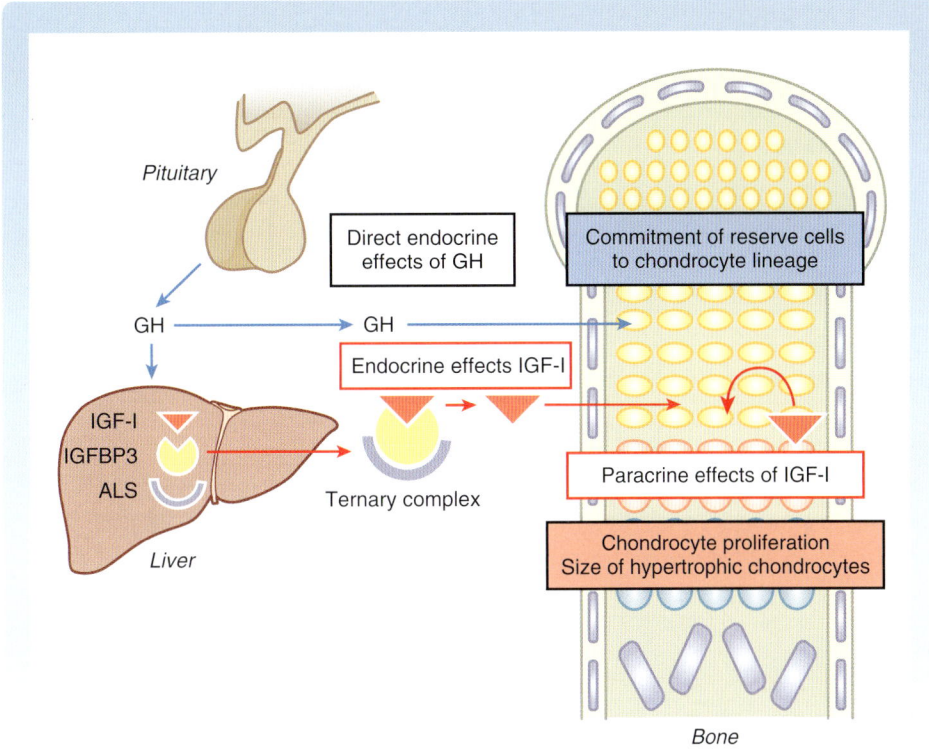

• Fig. 41.19 Relationship of GH and IGF-I. GH has direct endocrine actions on growth and stimulates the production of IGF-I, IGFBP-3 and ALS in the liver. Circulating IGF-I exerts endocrine actions on target organs. IGF-1 is also produced locally in bone, where it exerts paracrine effects. Part, but not all of this local IGF-I production is GH-dependent.

GH acts through a specific GH receptor **(GHR)** that is a member of the cytokine receptor family. A GHR dimer binds to GH, which triggers activation of the JAK/STAT signaling pathway (see Fig. 41.17). This results in phosphorylation of STAT5b, which translocates to the nucleus to stimulate transcription of GH-responsive genes. Additional signaling pathways activated by GH include MAPK and PI3K among others.

In the **fed state,** GH is a **protein anabolic hormone** that increases cellular amino acid uptake and incorporation into protein. Consequently it produces nitrogen retention (positive nitrogen balance) and decreases urea production. The muscle wasting that occurs concomitant with aging has been proposed to be caused at least in part by the decrease in GH secretion that occurs during senescence. In children, GH increases skeletal, muscular, and visceral growth; children without GH show growth stunting or dwarfism. GH promotes cartilage growth and both linear and appositional growth of long bones (Fig. 41.20, *green arrowheads*).

Although GH is an effective stimulator of IGF production, this response requires insulin, which supports GH receptor expression and signaling in hepatocytes. When a balanced supply of nutrients is available, high serum glucose levels stimulate insulin secretion and high serum amino acid levels promote GH secretion (Fig. 41.21, *top*). These conditions are appropriate for growth, and GH in turn stimulates IGF-I production by the liver. IGFs are mitogenic and have profound anabolic effects on many organs and tissues, including muscle, cartilage, and bone. Together, GH and IGF-I promote chondrocyte proliferation, differentiation, and hypertrophy during the process of endochondral ossification (see Fig. 41.20, *green arrowheads*). After closure of the epiphyses, longitudinal growth ceases but appositional growth of long bones continues. IGF-I stimulates osteoblast replication and synthesis of collagen and bone matrix. Not surprisingly, serum IGF levels correlate well with growth in children. The role of GH changes with alteration of nutritional status. If the diet is high in calories but low in amino acids, for example, high carbohydrate availability promotes insulin secretion but low serum amino acid levels inhibit GH and IGF production (see Fig. 41.21, *middle*). These responses allow dietary carbohydrates and fats to be stored, but conditions are unfavorable for growth.

In the **fasted state,** on the other hand, when nutrient availability wanes, serum GH levels rise and serum insulin levels fall in response to hypoglycemia (see Fig. 41.21, *bottom*). In the absence of insulin, peripheral glucose utilization decreases, thereby conserving glucose for essential tissues such as the brain. In these circumstances the rise in

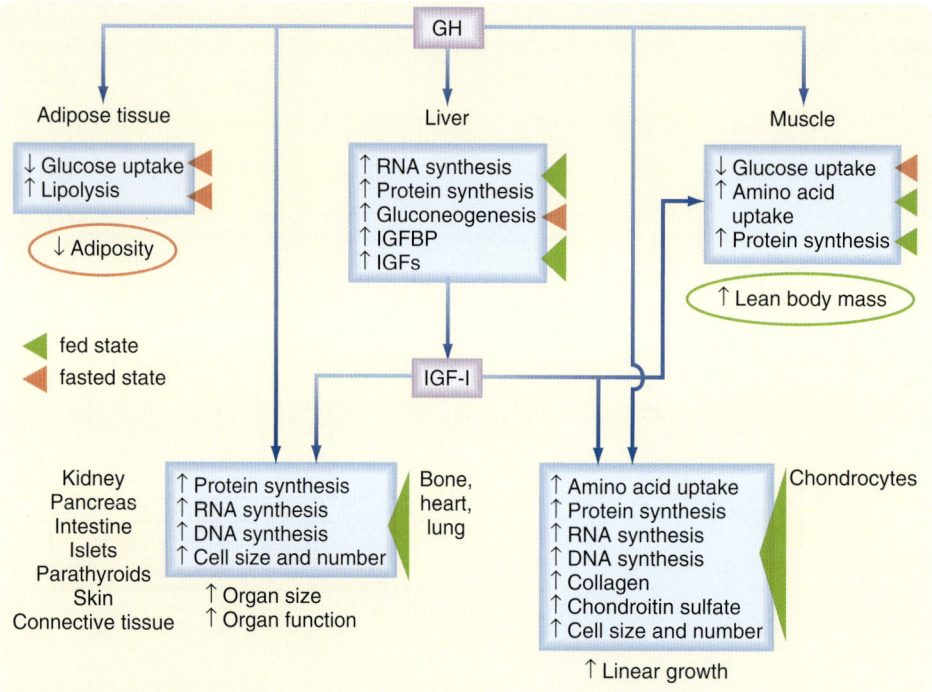

• **Fig. 41.20** Biological effects of GH and IGF-I. Anabolic growth-promoting effects that occur when nutritional status is favorable are indicated by green arrowheads. Metabolic effects of GH that mobilize fat while sparing glucose and protein during fasting are denoted by red arrowheads. IGFBP, insulinlike growth factor–binding protein.

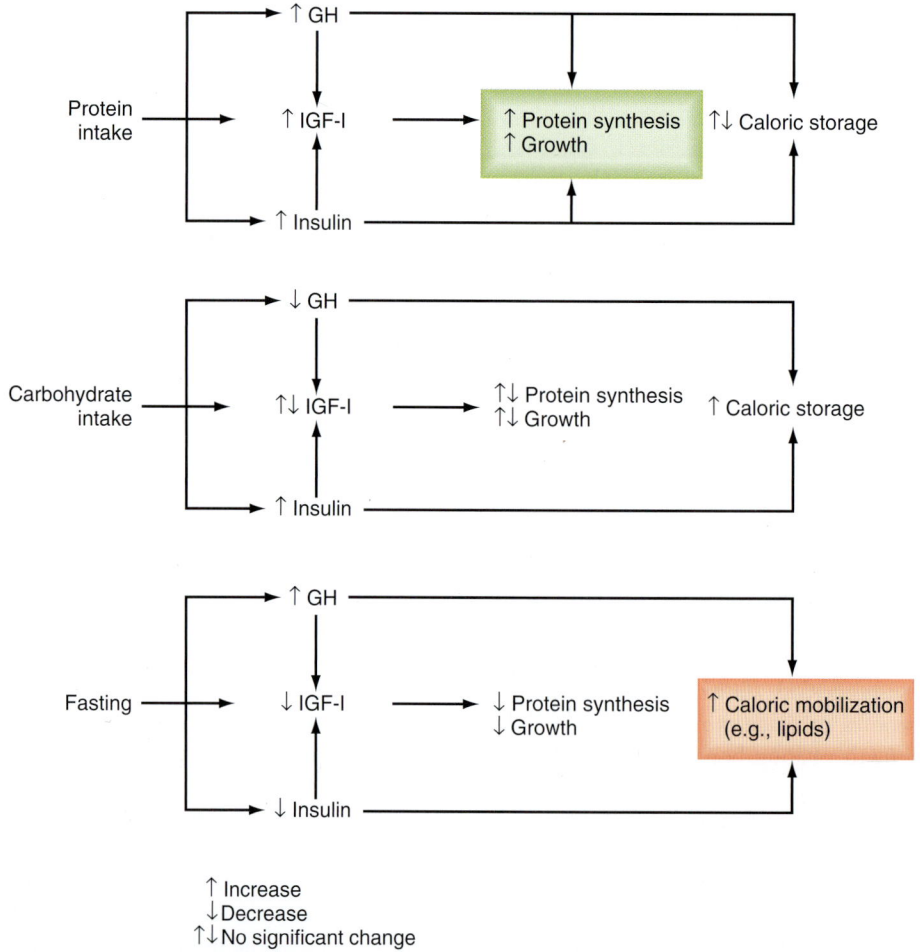

• **Fig. 41.21** Differential regulation of GH, insulin, and IGF-I secretion coordinates availability of nutrients with growth and protein anabolism, caloric storage, or caloric mobilization (primarily lipids).

GH secretion is beneficial because it shifts metabolism to lipid as an energy source, thereby conserving carbohydrate and protein. This involves coordinated direct actions of GH on the liver, muscle, and adipose tissue (see Fig. 41.20, *red arrowheads*).

GH is a **lipolytic** hormone. In adipocytes it mobilizes fatty acids and glycerol from triacylglycerol by combined direct and indirect activation of adipocyte lipases. An important indirect action of GH is sensitization of adipocytes to the lipolytic actions of catecholamines, which are also elevated during fasting. Serum fatty acid levels rise as a result of GH action, and more fats are used for energy production. Fatty acid uptake and β-oxidation increase in skeletal muscle and liver. GH can be ketogenic as a result of the increase in fatty acid oxidation when insulin is absent. GH also alters carbohydrate metabolism, causing blood glucose levels to rise. Many of its actions may be secondary to increased fat mobilization and oxidation. For example, an increase in serum free fatty acids inhibits uptake of glucose in skeletal muscle and adipose tissue. The hyperglycemic effects of GH are mild and slower than those of glucagon and epinephrine. Liver glucose output increases, but this is not an effect of GH on glycogenolysis. The increase in fatty acid oxidation and hence the rise in liver acetyl CoA stimulate gluconeogenesis. GH also directly stimulates expression of the gluconeogenic enzyme PEPCK through activation of STAT5b. These actions increase glucose production by the liver from substrates such as lactate and glycerol. The latter is released into the circulation as a result of GH-induced lipolysis in adipocytes.

GH antagonizes the action of insulin at the postreceptor level in skeletal muscle and adipose tissue (but not the liver). **Hypophysectomy** (removal of the pituitary gland) can improve diabetic management because GH, like cortisol, decreases insulin sensitivity. Because GH produces **insulin insensitivity,** it is considered a **diabetogenic hormone.** Therefore when secreted in excess (e.g., in acromegaly), GH can cause diabetes mellitus, and the insulin levels necessary to maintain normal metabolism increase. Excessive insulin secretion resulting from an excess of GH can cause damage to pancreatic beta cells. In the absence of GH, insulin secretion declines. Thus normal levels of GH are required for normal pancreatic function and insulin secretion.

GH deficiency in adults is becoming recognized as a pathological syndrome. If GH deficiency occurs after the epiphyses close, growth is not impaired. GH deficiency is one of many possible causes of hypoglycemia. Recent studies have shown that extended deficiencies of GH lead to changes in body composition. Fat as a percentage of body weight increases, whereas lean body mass declines. In addition, muscle weakness and early exhaustion are symptoms of GH deficiency. There has been interest in using GH in elderly populations to reverse age-related physical decline and body composition, but studies to date have shown small changes in body composition, no functional benefits, and increased risk of adverse events.

IN THE CLINIC

GH is necessary for growth before adulthood. GH deficiencies result in severe growth deficits, and excesses result in gigantism. Excess GH in adulthood after epiphyseal closure causes **acromegaly,** characterized by insidious enlargement of the hands and feet, coarsening of facial features, insulin resistance, and diabetes. Genetic disorders of the GH–IGF-I axis cause severe growth impairment. Identified mutations causing isolated GH deficiency most commonly occur in the GH and GHRH receptor genes. These patients can be treated with recombinant hGH to restore function of the downstream axis. In Laron syndrome, a mutation of the GH receptor causes GH resistance. In this instance the liver does not produce IGF-I; this is due to a lack of responsiveness to GH. These patients can be treated with IGF-I, but without the direct actions of GH, the effectiveness of the treatment is limited. Other downstream genetic mutations that have been reported include those in STAT5B, IGF-I, and ALS.

The Lactotrope

The lactotrope produces the hormone **prolactin,** which is a 199–amino acid single-chain protein. PRL is structurally related to GH and hPL (see Chapter 44). Like GH, the PRL receptor is a member of the cytokine family coupled to the JAK/STAT signaling pathways. Because the primary action of PRL in humans is related to breast development and function during pregnancy and lactation, the regulation and actions of prolactin will be discussed in detail in Chapter 44.

In the context of the pituitary gland, it should be appreciated that the lactotrope differs from the other endocrine cell types of the adenohypophysis in two major ways:

1. The lactotrope is not part of an endocrine axis. This means PRL acts directly on nonendocrine cells (primarily of the breast) to induce physiological changes.
2. Production and secretion of PRL are predominantly under inhibitory control by the hypothalamus. Thus disruption of the pituitary stalk and the hypothalamo-hypophyseal portal vessels (e.g., secondary to surgery or physical trauma) results in an increase in PRL levels but a decrease in ACTH, TSH, FSH, LH, and GH.

PRL circulates unbound to serum proteins and thus has a relatively short half-life of about 20 minutes. Normal basal serum concentrations are similar in men and women. Release of PRL is normally under tonic inhibition by the hypothalamus. This is exerted by dopaminergic tracts that secrete **dopamine** in the median eminence. There is also evidence for existence of a **prolactin-releasing factor (PRF).** The exact nature of this compound is not known, although many factors, including TRH and hormones in the glucagon family (secretin, glucagon, VIP, and gastric inhibitory polypeptide [GIP]) can stimulate release of PRL.

PRL is one of the many hormones released in response to **stress.** Surgery, fear, stimuli causing arousal, and exercise are all effective stimuli. As is the case with GH, sleep increases PRL secretion, and PRL has a pronounced sleep-associated diurnal rhythm. However, unlike GH, the rise in sleep-associated PRL is not associated with a specific sleep phase. Drugs that interfere with the synthesis or action of dopamine increase PRL secretion. Many commonly prescribed antihypertensive drugs and tricyclic antidepressants are dopamine inhibitors. Bromocriptine is a dopamine agonist that can be used to inhibit PRL secretion. Somatostatin, TSH, and GH also inhibit PRL secretion.

Key Concepts

1. The pituitary gland (also called the *hypophysis*) is composed of epithelial tissue (adenohypophysis [anterior lobe]) and neural tissue (neurohypophysis [posterior lobe]).

2. Magnocellular hypothalamic neurons in the paraventricular and supraoptic nuclei project axons down the infundibular stalk and terminate in the pars nervosa. The pars nervosa is a neurovascular organ from which neurohormones are released into the vasculature.

3. Two neurohormones, ADH and oxytocin, are synthesized in the hypothalamus in the magnocellular neuronal cell bodies. ADH and oxytocin are transported intraaxonally down the hypothalamohypophyseal tracts to the pars nervosa. Stimuli received at the cell bodies and dendrites in the hypothalamus control release of ADH and oxytocin at the pars nervosa.

4. The adenohypophysis secretes several tropic hormones that are part of endocrine axes. An endocrine axis includes the hypothalamus, the pituitary, and a peripheral endocrine gland. The set point of an axis is largely controlled by central input and negative feedback by the peripheral hormone on the pituitary and hypothalamus.

5. The adenohypophysis contains five endocrine cell types: corticotropes, thyrotropes, gonadotropes, somatotropes, and lactotropes. Corticotropes secrete ACTH, thyrotropes secrete TSH, gonadotropes secrete FSH and LH, somatotropes secrete GH, and lactotropes secrete PRL.

6. The hypothalamus regulates the anterior pituitary by secreting releasing hormones. These small peptides are carried via the hypophyseal portal system to the anterior pituitary, where they control synthesis and release of the pituitary hormones ACTH, TSH, FSH, LH, and GH. PRL secretion is inhibited by the hypothalamus through the catecholamine dopamine.

7. GH stimulates growth directly and via regulation of the growth-promoting hormone IGF-I. When nutritional status is favorable, GH promotes anabolic protein synthesis and growth. During fasting, GH stimulates lipolysis to mobilize fatty acids as an energy source, sparing glucose and protein. GH raises blood glucose by decreasing peripheral glucose uptake and stimulating hepatic gluconeogenesis.

8. PRL initiates and maintains lactation.

Additional Reading

Murray PG, et al. 60 YEARS OF NEUROENDOCRINOLOGY: The hypothalamo-GH axis: the past 60 years. *J Endocrinol*. 2015;226: T123-T140.

42

The Thyroid Gland

LEARNING OBJECTIVES

Upon completion of this chapter the student should be able to answer the following questions:

1. Describe the anatomy and histology of the thyroid gland, including the structure of the thyroid follicle.
2. Explain how thyroid hormones are synthesized within the thyroid gland, including the processes of iodine uptake, iodination of tyrosine residues in thyroglobulin by thyroid peroxidase/dual oxidase, and coupling to form T_4 and T_3.
3. Describe the process of endocytosis by which thyroglobulin is retrieved from the follicle lumen and processed to yield T_3 and T_4, which are secreted into the circulation.
4. Diagram the hypothalamic-pituitary-thyroid axis to show how TSH regulates thyroid function and how thyroid hormones feed back to regulate the axis. List examples of how central input can alter the set point of the axis.
5. Discuss the role of thyroid-binding proteins in the transport and stability of thyroid hormones, and the role of peripheral deiodinases in the activation of T_4 to T_3 or inactivation to reverse T_3. Contrast the cellular location and function of the D1 and D2 deiodinases.
6. Describe the mechanisms of thyroid hormone action, including the nature and location of the thyroid hormone receptor and its ability to either repress or activate target gene transcription.
7. Discuss the actions of thyroid hormone during development, especially on the central nervous system (CNS) and skeleton, including the consequences of severe hypothyroidism.
8. Describe the effects of thyroid hormone on basal metabolic rate and thermogenesis, on the cardiovascular system (heart rate, cardiac output, systemic vascular resistance), and on other organ systems (skin, skeletal muscle, digestive tract).

The thyroid gland produces the prohormone tetraiodothyronine (T_4, also called *thyroxine*) and the active hormone triiodothyronine (T_3). Synthesis of T_4 and T_3 requires iodine, which can be a limiting factor in some parts of the world. Much of T_3 is also made by peripheral conversion of T_4 to T_3. Thyroid hormone acts primarily through a nuclear receptor that regulates gene transcription. T_3 is critical for normal brain and bone development and has broad effects on metabolism and cardiovascular function in adults.

Anatomy and Histology of the Thyroid Gland

The thyroid gland is composed of right and left lobes that sit anterolateral to the trachea (Fig. 42.1). Typically the two lobes are connected by a midventral isthmus. The thyroid gland receives a rich blood supply. It is drained by three sets of veins on each side: the superior, middle, and inferior thyroid veins. The thyroid gland receives sympathetic innervation that is vasomotor but not secretomotor.

The functional unit of the thyroid gland is the **thyroid follicle,** a spherical structure about 200 to 300 μm in diameter that is surrounded by a single layer of thyroid epithelial cells (Fig. 42.2). The epithelium sits on a basal lamina, the outermost structure of the follicle, and is surrounded by a rich capillary supply. The apical side of the follicular epithelium faces the lumen of the follicle. The follicular lumen itself is filled with **colloid,** which is composed of **thyroglobulin.** This large (660 kDa) protein is secreted into the lumen and iodinated by the thyroid epithelial cells, serving as a scaffold for production of thyroid hormones. The size of the epithelial cells and the amount of colloid are dynamic features that change with activity of the gland. The thyroid gland contains another type of cell in addition to follicular cells. Scattered within the gland are **parafollicular cells,** or **C cells,** which are the source of the polypeptide hormone **calcitonin** (see Chapter 40).

Production of Thyroid Hormones

The secretory products of the thyroid gland are **iodothyronines** (Fig. 42.3), a class of hormones formed by the coupling of two iodinated tyrosine molecules. Approximately 90% of the thyroid output is **3,5,3′,5′-tetraiodothyronine (thyroxine,** or **T_4).** T_4 is primarily a prohormone. About 10% is **3,5,3′-triiodothyronine (T_3),** which is the active form of thyroid hormone. Less than 1% of thyroid output is **3,3′,5′-triiodothyronine (reverse T_3,** or **rT_3),** which is inactive. Normally these three products are secreted in the same proportions at which they are stored in the gland.

Because the primary product of the thyroid gland is T_4, yet the active form of thyroid hormone is T_3, the thyroid axis relies heavily on **peripheral conversion** through the action of **thyronine-specific deiodinases** (see Fig. 42.3). Most

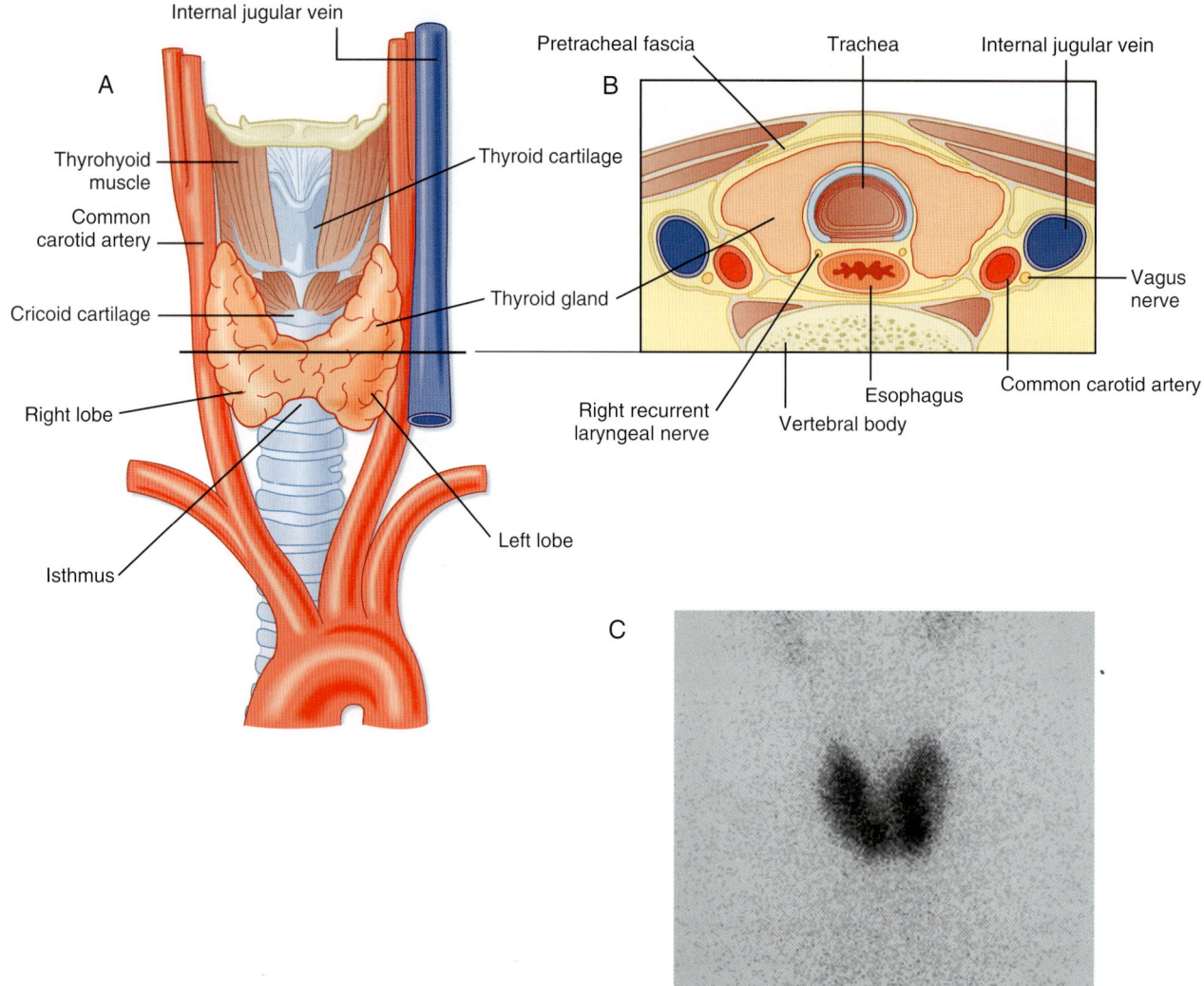

• **Fig. 42.1** A and B, Anatomy of the thyroid gland. C, Image of pertechnetate uptake by a normal thyroid gland. (Modified from Drake RL et al. *Gray's Anatomy for Students.* Philadelphia: Churchill Livingstone; 2005.)

conversion of T_4 to T_3 by **type 1 deiodinase (D1)** occurs in tissues with high blood flow and rapid exchange with plasma, such as the liver and kidneys. This process supplies basal circulating T_3 for uptake by other tissues in which local T_3 generation is low or absent. D1 is also expressed in the thyroid (again, where T_4 is abundant) and has relatively low affinity (i.e., a K_m of 1 μM) for T_4. Levels of D1 are paradoxically increased in hyperthyroidism and contribute to the elevated circulating T_3 levels in this disease.

The brain maintains constant intracellular levels of T_3 by a high-affinity deiodinase called **type 2 deiodinase (D2)** that is expressed in glial cells of the CNS. D2 has a K_m of 1 nM and maintains intracellular concentrations of T_3 even when circulating T_4 falls to low levels. D2 is also present in pituitary thyrotropes. There, D2 acts as a "thyroid axis sensor" that mediates the ability of circulating T_4 to feed back on secretion of **thyroid stimulating hormone (TSH)**. Expression of D2 is increased during hypothyroidism, which helps maintain constant T_3 levels in the brain.

There is also an "inactivating" deiodinase called **type 3 deiodinase (D3)**. D3 is a high-affinity inner ring deiodinase that converts T_4 to the inactive rT_3. Type 3 deiodinase is increased during hyperthyroidism, which helps blunt overproduction of T_4. All forms of iodothyronines are eventually further deiodinated to noniodinated thyronine.

Iodide Balance

Because iodide plays a unique role in thyroid physiology, a description of thyroid hormone synthesis requires some understanding of iodide turnover (Fig. 42.4). An average of 400 μg of iodide per person is ingested daily in the United States, versus a minimum daily requirement of 150 μg for adults, 90 to 120 μg for children, and 200 μg for pregnant women. In the steady state, the same amount, 400 μg, is excreted in urine. Iodide is actively concentrated in the thyroid gland, salivary glands, gastric glands, lacrimal glands, mammary glands, and choroid plexus. About 70

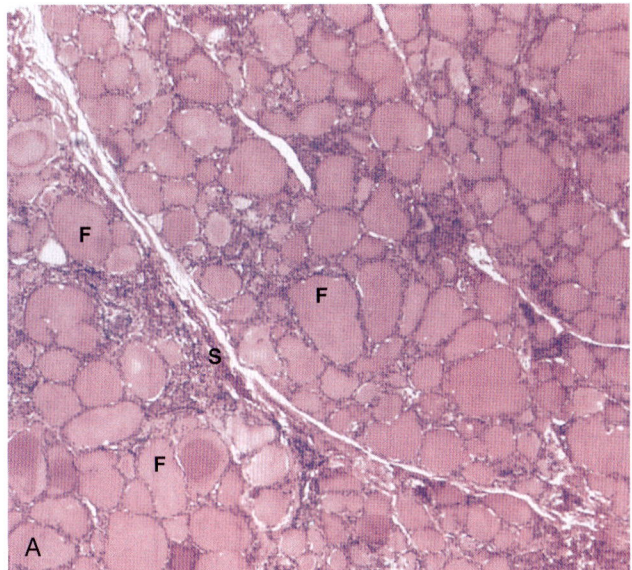

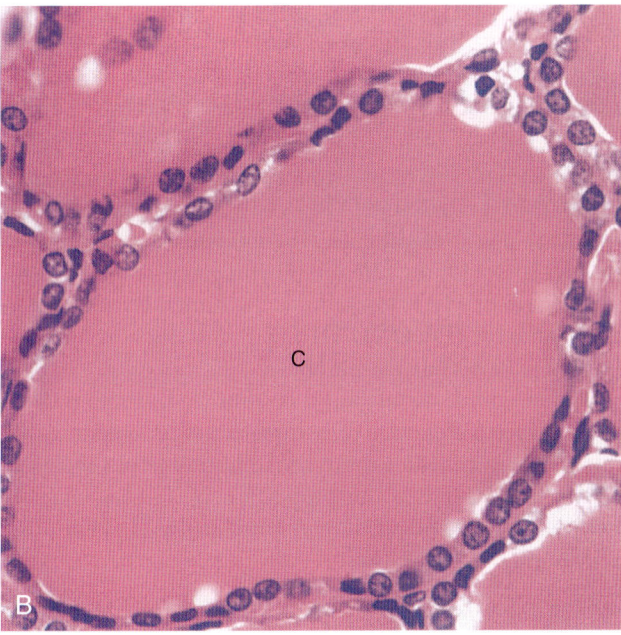

• **Fig. 42.2** Histology of the thyroid gland at low *(upper panel)* and high *(lower panel)* magnification. C, colloid; F, thyroid follicles; S, connective tissue septa. (From Young B et al. *Wheater's Functional Histology.* 5th ed. Philadelphia: Churchill Livingstone; 2006.)

to 80 μg of iodide is taken up daily by the thyroid gland from a circulating pool that contains approximately 250 to 750 μg of iodide. The total iodide content of the thyroid gland averages 7500 μg, virtually all of which is in the form of stored iodothyronine in colloid thyroglobulin. In the steady state, 70 to 80 μg of iodide, or about 1% of the total, is released from the gland daily. Of this amount, 75% is secreted as thyroid hormone and the remainder as free iodide. The large ratio (100:1) of iodide stored in the form of hormone to the amount turned over daily protects against iodide deficiency for about 2 months. Iodide is also conserved by a marked reduction in renal excretion of iodide as its concentration in serum falls.

Overview of Thyroid Hormone Synthesis

To understand thyroid hormone synthesis and secretion, one must appreciate the directionality of each process as it relates to the polarized thyroid epithelial cell (Fig. 42.5). Synthesis of thyroid hormone requires two precursors: iodide and thyroglobulin. Iodide is transported across cells from the basal (vascular) side to the apical (follicular luminal) side of the thyroid epithelium. Thyroglobulin is synthesized and secreted across the apical membrane into the follicular lumen. Thus synthesis involves a basal-to-apical movement of these precursors into the follicular lumen (see Fig. 42.5). Actual synthesis of iodothyronines occurs enzymatically within the follicular lumen close to the apical membrane of the epithelial cells (see Synthesis of Iodothyronines on a Thyroglobulin Backbone). Secretion of thyroid hormone involves endocytosis of iodinated thyroglobulin and apical-to-basal movement of the endocytotic vesicles, which fuse with lysosomes. Thyroglobulin is enzymatically degraded by lysosomal enzymes, resulting in release of thyroid hormones from the thyroglobulin backbone. Finally, thyroid hormones move across the basolateral membrane, probably through a specific transporter, and ultimately into the blood. Thus secretion involves apical-to-basal movement (see Fig. 42.5).

Synthesis of Iodothyronines on a Thyroglobulin Backbone

Iodide is actively transported into the gland against chemical and electrical gradients by a **sodium-iodide symporter (NIS)** located in the basolateral membrane of thyroid epithelial cells (see Fig. 42.5). NIS is highly expressed in the thyroid gland, but it is also expressed at lower levels in the placenta, salivary glands, and actively lactating breast. One iodide ion is transported uphill against an iodide gradient while two sodium ions move down their electrochemical gradient from extracellular fluid into the thyroid cell. The driving force for this secondary active transporter is provided by plasma membrane Na^+,K^+-ATPase. Expression of the *NIS* gene is inhibited by iodide and stimulated by TSH. A reduction in dietary iodide intake depletes the circulating iodide pool and greatly enhances the activity of the iodide trap. When dietary iodide intake is low, the percentage of thyroid uptake of iodide can reach 80% to 90%.

The steps in thyroid hormone synthesis are shown in Fig. 42.6. After entering the gland, iodide rapidly moves to the apical plasma membrane of epithelial cells. From there, iodide is transported into the lumen of the follicles by a sodium-independent iodide/chloride transporter called **pendrin**. Iodide is immediately oxidized and incorporated into tyrosine residues within **thyroglobulin** (see Fig. 42.5). A single **iodination** forms a **monoiodotyrosine (MIT)**; a second iodination of the same residue produces **diiodotyrosine (DIT)** (see Fig. 42.6). After iodination, two DIT molecules are **coupled** to form T_4; one MIT and one DIT are coupled to form T_3. Coupling occurs between iodinated tyrosines that remain part of the primary structure of

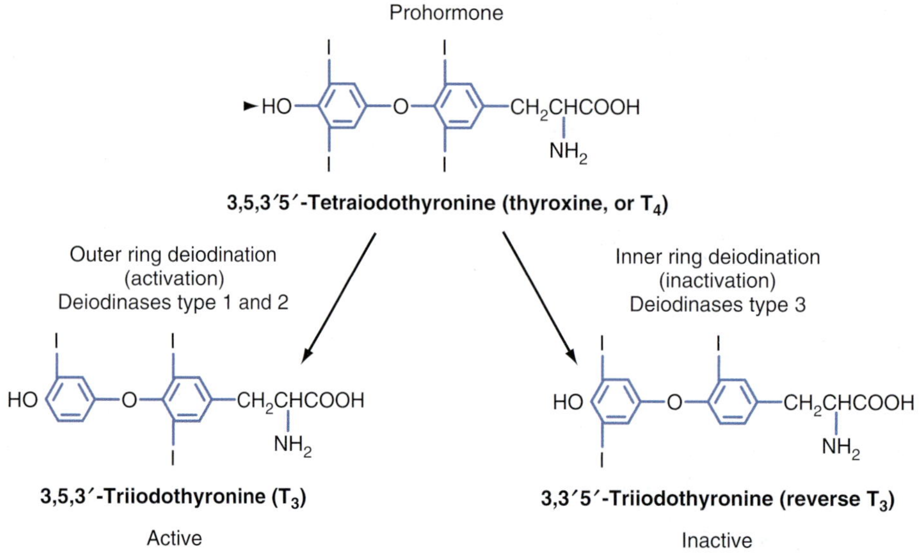

Prohormone

HO—⬡—O—⬡—CH₂CHCOOH
                        |
                       NH₂

3,5,3′5′-Tetraiodothyronine (thyroxine, or T₄)

Outer ring deiodination
(activation)
Deiodinases type 1 and 2

Inner ring deiodination
(inactivation)
Deiodinases type 3

HO—⬡—O—⬡—CH₂CHCOOH
                        |
                       NH₂

3,5,3′-Triiodothyronine (T₃)

Active

HO—⬡—O—⬡—CH₂CHCOOH
                        |
                       NH₂

3,3′5′-Triiodothyronine (reverse T₃)

Inactive

• **Fig. 42.3** Structure of the iodothyronines T_4, T_3, and reverse T_3.

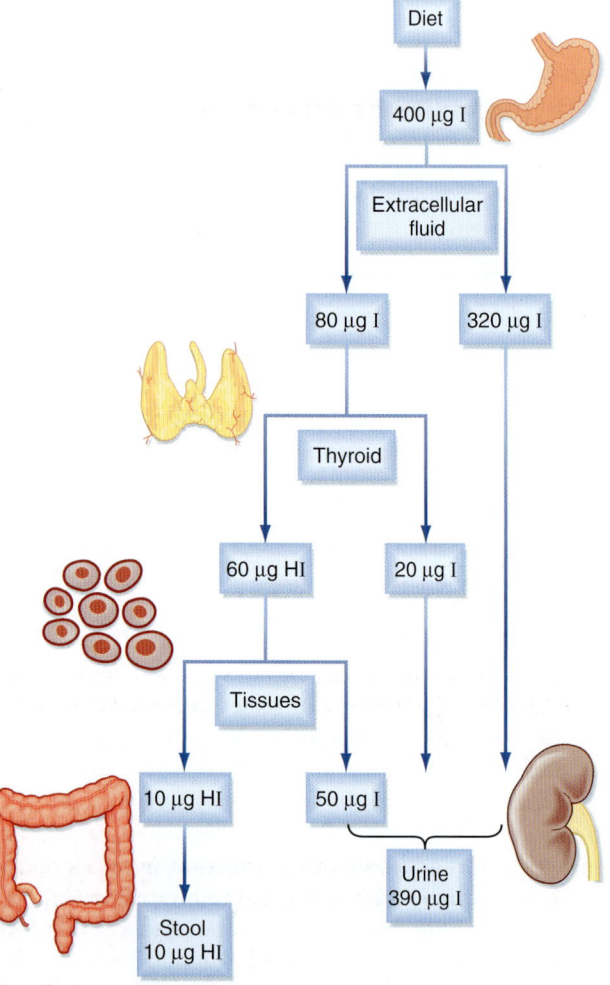

• **Fig. 42.4** Iodine distribution and turnover in humans. HI, hormone-associated iodine.

thyroglobulin. This entire sequence of reactions is catalyzed by **thyroid peroxidase (TPO),** an enzyme complex that spans the apical membrane. The immediate oxidant (electron acceptor) for the reaction is hydrogen peroxide (H_2O_2). Generation of H_2O_2 in the follicular lumen is catalyzed by **dual oxidases (DUOX1, DUOX2)** that are also localized in the apical plasma membrane.

When iodide availability is restricted, formation of T_3 is favored. Because T_3 is three times as potent as T_4, this response provides more active hormone per molecule of organified iodide. The proportion of T_3 also increases when the thyroid gland is hyperstimulated by TSH or other activators.

Secretion of Thyroid Hormones

Once thyroglobulin has been iodinated, it is stored in the lumen of the follicle as colloid (see Fig. 42.2). Release of T_4 and T_3 into the bloodstream is initiated by endocytosis of colloid from the follicular lumen by the processes of macro- and micropinocytosis. Endocytotic vesicles then fuse with lysosomes and thyroglobulin is degraded (Fig. 42.7; also see Fig. 42.5). MIT and DIT molecules, which also are released during proteolysis of thyroglobulin, are rapidly deiodinated within the follicular cell by the enzyme **iodotyrosine deiodinase** (see Fig. 42.5). This deiodinase is specific for MIT and DIT and cannot use T_4 and T_3 as substrates. The iodide is then recycled into synthesis of T_4 and T_3. Amino acids from the digestion of thyroglobulin reenter the intrathyroidal amino acid pool and can be reused for protein synthesis. Only minor amounts of intact thyroglobulin leave the follicular cell under normal circumstances. Enzymatically released T_4 and T_3 are transported across the basal side of the cell and enter the blood.

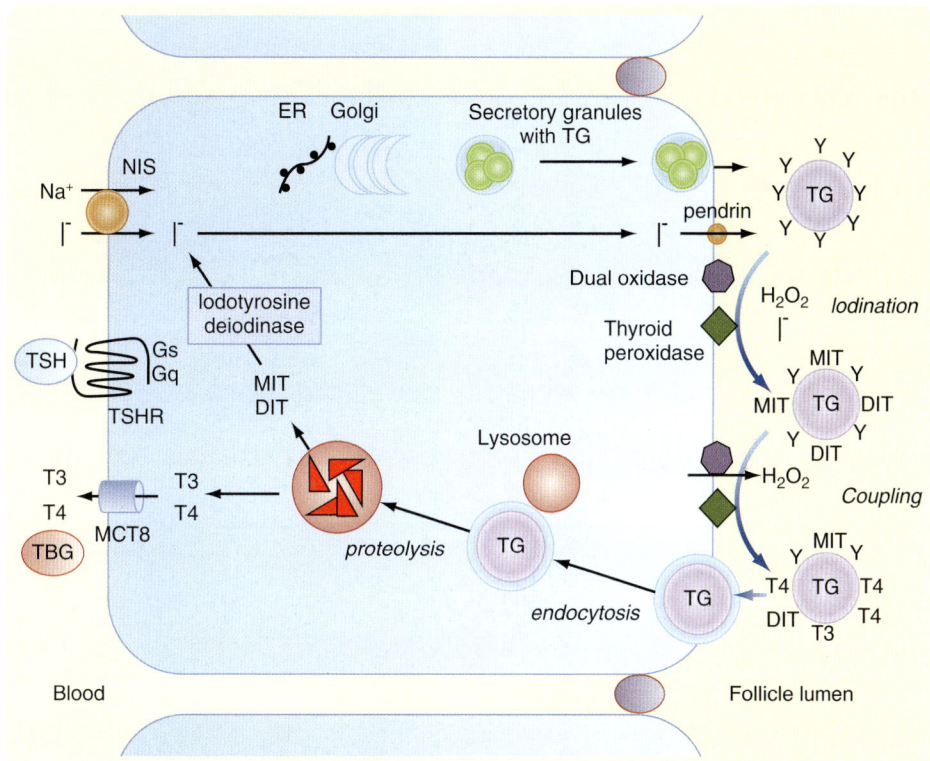

• **Fig. 42.5** Synthesis and secretion of thyroid hormones by the thyroid epithelial cell.

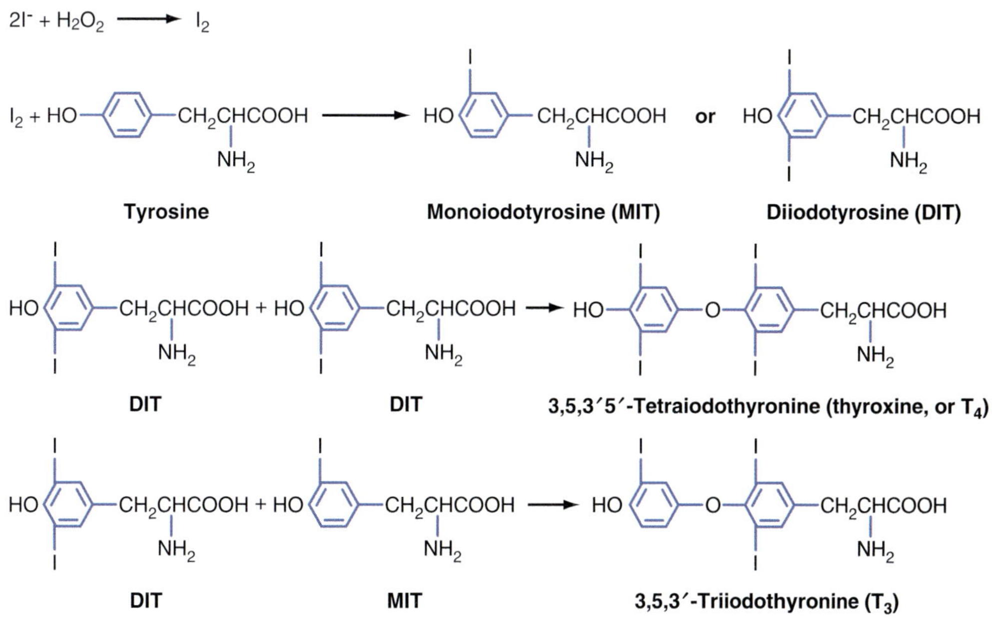

$$2I^- + H_2O_2 \longrightarrow I_2$$

$I_2 + HO$—〈 〉—$CH_2CHCOOH \longrightarrow HO$—〈 〉—$CH_2CHCOOH$ or HO—〈 〉—$CH_2CHCOOH$

with NH_2 below each structure.

| Tyrosine | Monoiodotyrosine (MIT) | Diiodotyrosine (DIT) |

HO—〈 〉—$CH_2CHCOOH + HO$—〈 〉—$CH_2CHCOOH \longrightarrow HO$—〈 〉—$O$—〈 〉—$CH_2CHCOOH$

with NH_2 below each.

| DIT | DIT | 3,5,3′5′-Tetraiodothyronine (thyroxine, or T_4) |

HO—〈 〉—$CH_2CHCOOH + HO$—〈 〉—$CH_2CHCOOH \longrightarrow HO$—〈 〉—$O$—〈 〉—$CH_2CHCOOH$

with NH_2 below each.

| DIT | MIT | 3,5,3′-Triiodothyronine (T_3) |

• **Fig. 42.6** Reactions involved in the generation of iodide, MIT, DIT, T_3, and T_4.

Transport and Metabolism of Thyroid Hormones

Secreted T_4 and T_3 circulate in the bloodstream almost entirely bound to proteins. Normally only about 0.03% of total plasma T_4 and 0.3% of total plasma T_3 exist in the free state (Fig. 42.8). Free T_3 is biologically active and mediates the effects of thyroid hormone on peripheral tissues in addition to exerting negative feedback on the pituitary and hypothalamus. The major binding protein is **thyroxine-binding globulin (TBG),** which is synthesized in the liver and binds one molecule of T_4 or T_3. About 70% of circulating T_4 and T_3 is bound to TBG; 10% to 15% is bound to another specific thyroid-binding protein called **transthyretin (TTR). Albumin** binds 15% to 20%, and 3% is bound to lipoproteins. Ordinarily only alterations in

AT THE CELLULAR LEVEL

Regulation of thyroid hormone secretion by TSH is under exquisite negative-feedback control (see Chapter 41). Circulating thyroid hormones feed back on the pituitary gland to decrease TSH secretion, primarily by repressing TSHβ subunit gene expression. The pituitary gland expresses the high-affinity D2, which converts T_4 entering these cells to T_3. Thus feedback in thyrotropes mediated by intracellular T_3 represents an integrated measure of circulating free T_4 and T_3 (see Transport and Metabolism of Thyroid Hormones). Because the diurnal variation in TSH secretion is small, thyroid hormone secretion and plasma concentrations are relatively constant. Only small nocturnal increases in secretion of TSH and release of T_4 occur. Thyroid hormones also feed back on hypothalamic thyroid-releasing hormone (TRH)-secreting neurons. In these neurons, T_3 inhibits expression of the prepro-TRH gene.

Autoregulation of thyroid gland function is caused by iodide itself, which has a biphasic action. At relatively low levels of iodide intake, the rate of thyroid hormone synthesis is directly related to the availability of iodide. However, if the intake of iodide exceeds 2 mg/day, the intraglandular concentration of iodide reaches a level that paradoxically suppresses TPO activity, blocking hormone biosynthesis. This phenomenon is known as the **Wolff-Chaikoff effect.** Adaptation to high iodide intake normally occurs by reducing expression of NIS, which causes the intrathyroidal iodide level to fall. TPO activity then returns to normal, and thyroid hormone synthesis resumes within days to weeks. In unusual instances, failure of NIS to downregulate leads to prolonged inhibition of hormone synthesis by iodide and resultant hypothyroidism. The temporary reduction in hormone synthesis by excess iodide has also been used therapeutically in hyperthyroidism.

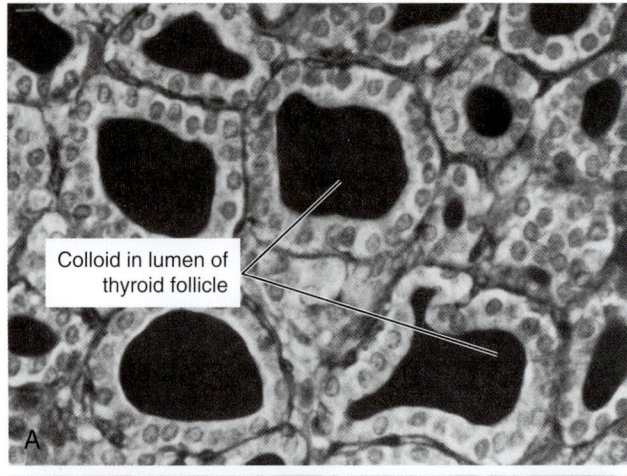

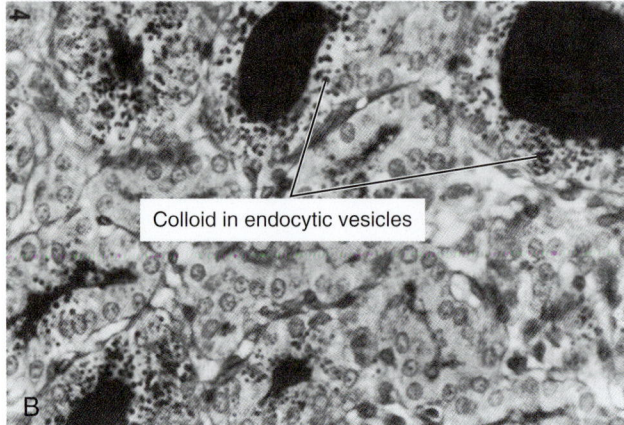

Colloid in lumen of thyroid follicle

Colloid in endocytic vesicles

• **Fig. 42.7** Before (A) and minutes after (B) rapid induction of thyroglobulin endocytosis by TSH. (From Wollman SH et al. *J Cell Biol* 1964;21:191.)

IN THE CLINIC

Because of its ability to **trap** and incorporate iodine into thyroglobulin (called **organification**), the activity of the thyroid can be assessed by **radioactive iodine uptake (RAIU).** In this test a tracer dose of ^{123}I is administered and RAIU is measured by placing a gamma detector on the neck at 4 to 6 hours and at 24 hours. In the United States, where the diet is relatively rich in iodine, RAIU is typically around 15% after 6 hours and 25% after 24 hours (Fig. 42.9). Abnormally high RAIU (>60%) after 24 hours indicates hyperthyroidism. Abnormally low RAIU (<5%) after 24 hours indicates hypothyroidism. In individuals with extreme chronic stimulation of the thyroid (Graves disease–associated thyrotoxicosis), iodide is trapped, organified, and released as hormone very rapidly. In these cases of elevated turnover, 6-hour RAIU will be very high but 24-hour RAIU will be lower (see Fig. 42.8). A number of anions, such as thiocyanate (CNS⁻), perchlorate ($HClO_4^-$), and pertechnetate (TcO_4^-), are competitive or noncompetitive inhibitors of iodide transport via NIS. If iodide cannot be rapidly incorporated into tyrosine (**organification defect**) after its uptake by the cell, administration of one of these anions will, by blocking further iodide uptake, cause rapid release of iodide from the gland (see Fig. 42.9).

This release occurs as a result of the high thyroid-plasma concentration gradient.

The thyroid can be imaged with a rectilinear scanner or gamma camera after administration of a tracer, ^{123}I, ^{131}I, or the iodine-mimic pertechnetate (^{99m}Tc). Imaging can display the size and shape of the thyroid (see Fig. 42.1C) as well as heterogeneities of active versus inactive tissue within the thyroid gland. Such heterogeneities are often due to the development of **thyroid nodules,** which are regions of enlarged follicles with evidence of regressive changes due to cycles of stimulation and involution. **"Hot" nodules** (i.e., nodules that display high RAIU on imaging) are not usually cancerous but may lead to thyrotoxicosis (hyperthyroidism). **"Cold" nodules** are 10 times more likely to be cancerous. Such nodules can be sampled for pathological analysis by **fine-needle aspiration biopsy.**

The thyroid can also be imaged by **ultrasonography,** which is superior in resolution to RAIU imaging. Ultrasonography is used to guide the physician during fine-needle aspiration biopsy of a nodule. The highest resolution of the thyroid is achieved with **magnetic resonance imaging (MRI).**

TBG concentration significantly affect total plasma T_4 and T_3 levels. Two important biological functions have been ascribed to TBG. First, it maintains a large circulating reservoir of T_4 that buffers any acute changes in thyroid gland function. Second, binding of plasma T_4 and T_3 to proteins prevents loss of these relatively small hormone molecules in urine and thereby helps conserve iodide. TTR transports T_4 in cerebrospinal fluid and provides thyroid hormones to the CNS.

Regulation of Thyroid Function

The most important regulator of thyroid gland function and growth is the hypothalamic-pituitary **thyroid releasing**

hormone–thyroid stimulating hormone axis (see Chapter 41). TSH stimulates every aspect of thyroid function. TSH has immediate, intermediate, and long-term actions on the thyroid epithelium. Rapid actions of TSH include pinocytosis of colloid droplets in the cytoplasm, which represent thyroglobulin within endocytic vesicles (see Fig. 42.7). TSH stimulates the proteolysis of thyroglobulin and release of T_4 and T_3 from the gland. Iodide uptake and TPO activity increase. TSH also stimulates entry of glucose into the hexose monophosphate shunt pathway, which generates the reduced nicotinamide adenine dinucleotide phosphate (NADPH) needed for the peroxidase reaction. Intermediate effects of TSH on the thyroid gland occur after hours to days and involve protein synthesis and expression of numerous genes, including those encoding NIS, thyroglobulin, and TPO. Sustained TSH stimulation leads to the long-term effects of hypertrophy and hyperplasia of follicular cells. Capillaries proliferate and thyroid blood flow increases. These actions, which underlie the growth-promoting effects of TSH on the gland, are supported by local production of growth factors. A noticeably enlarged thyroid gland is called a *goiter* (Fig. 42.10). Endemic goiter is due to lack of adequate iodine in the diet, which results in low thyroid hormone and elevated TSH levels.

Physiological Effects of Thyroid Hormone

Thyroid hormone acts on essentially all cells and tissues, and imbalances in thyroid function constitute some of the most common endocrine diseases. Thyroid hormone has many direct actions, but it also acts in more subtle ways to optimize the actions of several other hormones and neurotransmitters.

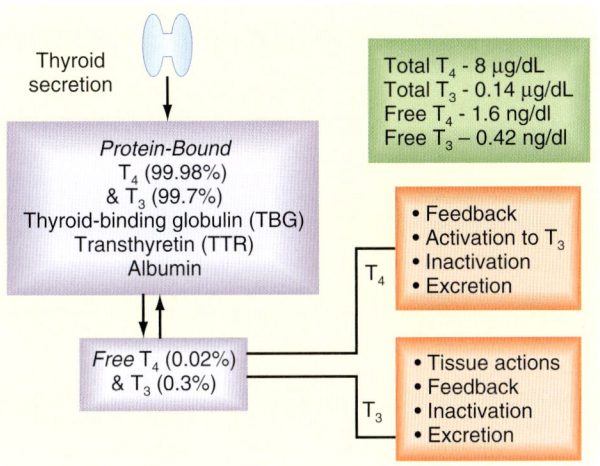

• **Fig. 42.8** Transport of T_4 and T_3 in serum by transport proteins and percentages of bound and free hormone.

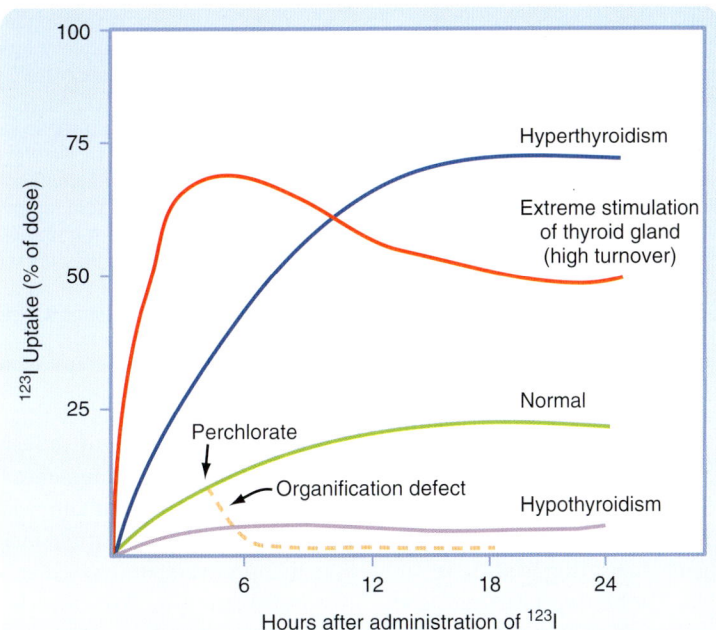

• **Fig. 42.9** Thyroid gland iodothyronine uptake curves for normal, hypothyroid, hyperthyroid, and defective organification states.

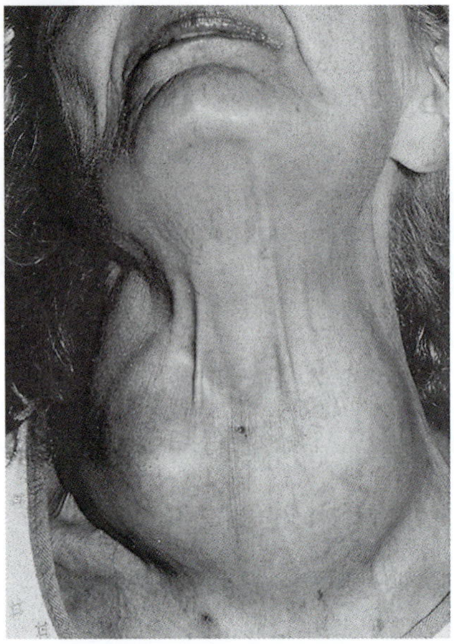

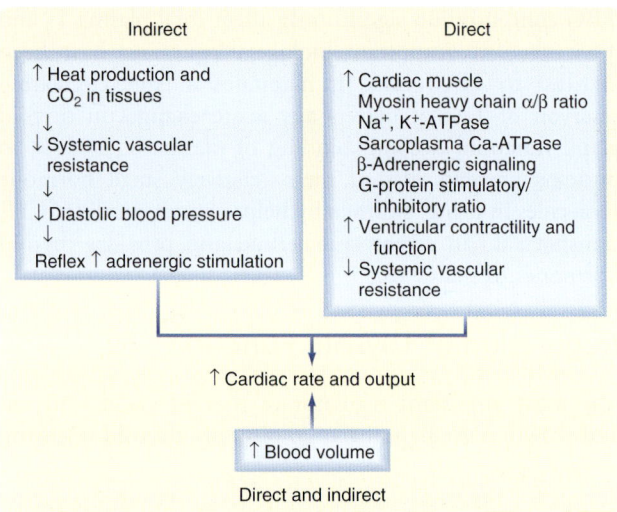

• Fig. 42.10 The thyroid gland is located in the anterior aspect of the neck, where it is easily visualized when enlarged (goiter).

• Fig. 42.11 Mechanisms by which thyroid hormone increases cardiac output. The indirect mechanisms are probably quantitatively more important.

IN THE CLINIC

Graves disease is the most common form of **hyperthyroidism.** It occurs most frequently between the ages of 20 and 50 and is 10 times more common in women than in men. Graves disease is an autoimmune disorder in which activating autoantibodies are produced against the TSH receptor. Hyperthyroidism driven by the antibody is often accompanied by a diffuse goiter as a result of hyperplasia and hypertrophy of the gland. The follicular epithelial cells become tall columnar cells, and the colloid shows a scalloped periphery indicative of rapid turnover.

The primary clinical state found in Graves disease is **thyrotoxicosis**—the state of excessive thyroid hormone in blood and tissues. A patient with thyrotoxicosis presents one of the most striking pictures in clinical medicine. The large increase in metabolic rate is manifested as weight loss despite increased food intake. Excess heat production causes discomfort in warm environments, sweating, and greater intake of water. The increase in adrenergic activity produces a rapid heart rate, hyperkinesis, tremor, nervousness, and a wide-eyed stare. Weakness is caused by a loss of muscle mass as well as impaired muscle function. Other symptoms include a labile emotional state, breathlessness during exercise, and difficulty swallowing or breathing because of compression of the esophagus or trachea by the enlarged thyroid gland. The most common cardiovascular sign is sinus tachycardia. There

is increased cardiac output associated with a widened pulse pressure secondary to a positive inotropic effect coupled with decreased systemic vascular resistance. A major clinical sign in Graves disease is **exophthalmos** (abnormal protrusion of the eyeball) and **periorbital edema.** This is caused by autoantibody binding to the TSH receptor expressed on orbital fibrocytes, leading to production of inflammatory cytokines.

Graves disease is usually diagnosed by elevated serum free and total T_4 and T_3 and the clinical signs of diffuse goiter and ophthalmopathy. Serum TSH levels are low because the hypothalamus and pituitary are inhibited by the high levels of T_4 and T_3. In most cases, radioiodine uptake by the thyroid is excessive and diffuse. Assay of TSH levels and the presence of circulating thyroid stimulating immunoglobulin will distinguish Graves disease (a primary disorder) from a rare adenoma of pituitary thyrotrophs (a secondary disorder) that produces high levels of TSH.

Treatment of Graves disease is usually removal of the thyroid tissue, followed by lifelong replacement therapy with T_4. Thyroid tissue can be removed either by radioablation with ^{131}I or by surgery. With surgical removal of the gland, precautions must be taken to avoid a massive, potentially life-threatening release of thyroid hormones known as **thyroid storm.** An alternative to removal of thyroid tissue is administration of **antithyroid drugs** that inhibit TPO activity.

Cardiovascular Effects

Perhaps the most clinically important actions of thyroid hormone are those on cardiovascular physiology. T_3 increases cardiac output, thereby ensuring sufficient O_2 delivery to tissues (Fig. 42.11). The resting heart rate and stroke volume are increased. The speed and force of myocardial contractions are enhanced (positive chronotropic and inotropic effects, respectively), and the diastolic relaxation

time is shortened (positive lusitropic effect). Systolic blood pressure is modestly augmented and diastolic blood pressure is decreased. The resultant widened pulse pressure reflects the combined effects of the increased stroke volume and the reduction in systemic vascular resistance secondary to blood vessel dilation in skin, muscle, and heart. These effects in turn are partly due to the increase in tissue production of heat and CO_2 that thyroid hormone induces (see Effects on Basal Metabolic Rate and Thermogenesis). In addition, however, thyroid hormone decreases systemic resistance by dilating arterioles in the peripheral circulation. Total blood volume is increased by activation of the renin-angiotensin-aldosterone axis, thereby increasing renal tubular sodium reabsorption (see Chapter 34).

The cardiac inotropic effects of T_3 are both direct and indirect. The latter are due primarily to enhanced responsiveness to catecholamines (see Chapter 43). Direct inotropic effects (see Fig. 42.11) involve regulation of multiple proteins that enhance contractility, including increased **α-myosin heavy chain** expression and inhibition of the plasma membrane **Na^+/Ca^{++} exchanger.** **Sarcoplasmic reticulum Ca^{++}-ATPase (SERCA)** is increased by T_3, whereas phospholamban is decreased. As a result, sequestration of calcium during diastole is enhanced and the relaxation time is shortened. Increased **ryanodine Ca^{++} channels** in the sarcoplasmic reticulum promote release of Ca^{++} from the sarcoplasmic reticulum during systole.

IN THE CLINIC

Thyroid hormone levels in the normal range are necessary for optimum cardiac performance. A deficiency of thyroid hormone in humans reduces stroke volume, left ventricular ejection fraction, cardiac output, and the efficiency of cardiac function. The latter defect is shown by the fact that the stroke work index [(stroke volume/left ventricular mass) × peak systolic blood pressure] is decreased to a greater extent than myocardial oxidative metabolism. The rise in systemic vascular resistance may contribute to this cardiac debility. In contrast, excess thyroid hormone enhances cardiac output by increasing both heart rate and stroke volume. Pulse pressure is widened by increased systolic pressure and decreased diastolic pressure due to decreased systemic vascular resistance. Thyrotoxicosis is associated with palpitations, atrial fibrillation, and mitral valve prolapse (see Chapter 15).

Effects on Basal Metabolic Rate and Thermogenesis

Increased O_2 use ultimately depends on an increased supply of substrates for oxidation. T_3 augments glucose absorption from the gastrointestinal tract and increases glucose turnover (glucose uptake, oxidation, and synthesis). In adipose tissue, thyroid hormone induces enzymes for the synthesis of fatty acids, including acetyl-CoA carboxylase and fatty acid synthase, and enhances lipolysis by increasing the number of β-adrenergic receptors (see Effects on the Autonomic Nervous System). Thyroid hormone also enhances the clearance of chylomicrons. Thus lipid turnover (free fatty acid release from adipose tissue and oxidation) is augmented.

Protein turnover (release of muscle amino acids, protein degradation, and to a lesser extent protein synthesis and urea formation) is also increased. T_3 potentiates the respective stimulatory effects of epinephrine, norepinephrine, glucagon, cortisol, and growth hormone on gluconeogenesis,

lipolysis, ketogenesis, and proteolysis of the labile protein pool. The overall metabolic effect of thyroid hormone has been aptly described as accelerating the physiological response to starvation. In addition, thyroid hormone stimulates synthesis of bile acids from cholesterol and promotes biliary secretion. The net effect is a decrease in the body pool and plasma levels of total and low-density lipoprotein cholesterol. Metabolic clearance of adrenal and gonadal steroid hormones, some B vitamins, and certain administered drugs is also increased by thyroid hormone.

Thyroid hormones stimulate **thermogenesis** by affecting both adenosine triphosphate (ATP) utilization and the efficiency of ATP synthesis. ATP utilization is enhanced by upregulation of several energy-dependent processes, including Na^+,K^+-ATPase and SERCA, particularly in skeletal muscle, where calcium cycling between the cytoplasm and sarcoplasmic reticulum uses ATP and generates heat. Recently it has been demonstrated that brown fat in humans, once thought to be important only in neonates, appears to play a role in facultative thermogenesis in adults. Imaging studies have demonstrated the presence of brown fat in the mediastinum, particularly in lean individuals, and metabolic activity in brown fat is enhanced by exposure to cold. Brown fat expresses **uncoupling protein-1 (UCP1)**, also called *thermogenin,* which causes the proton gradient across the inner mitochondrial membrane to be dissipated as heat, which is then disseminated to the rest of the body by the circulation. UCP1 is regulated by thyroid hormone, and brown fat expresses D2, providing intracellular conversion of T_4 to T_3. Brown fat thermogenesis involves a synergistic interaction between thyroid hormones and the sympathetic nervous system. Catecholamines promote lipolysis and upregulate expression of D2. T_3 in turn upregulates adrenergic receptors and enhances catecholamine responsiveness. Hyperthyroidism is accompanied by heat intolerance, whereas hypothyroidism is accompanied by cold intolerance.

Respiratory Effects

Thyroid hormone stimulates O_2 utilization and enhances O_2 delivery. Appropriately, T_3 increases the **resting respiratory rate, minute ventilation,** and the **ventilatory response** to hypercapnia and hypoxia. These actions maintain a normal arterial PO_2 when O_2 utilization is increased and a normal PCO_2 when CO_2 production is increased. Additionally the hematocrit increases slightly to enhance O_2-carrying capacity. This increase results from stimulation of **erythropoietin** production by the kidney.

Skeletal Muscle Effects

Normal function of skeletal muscles also requires optimal amounts of thyroid hormone. This requirement may be related to regulation of energy production and storage. Glycolysis and glycogenolysis are increased, whereas glycogen and creatine phosphate are reduced by thyroid hormone excess. The inability of muscle to take up and phosphorylate creatine leads to its increased urinary excretion.

Effects on the Autonomic Nervous System and Catecholamine Action

As already mentioned, there is important synergism between catecholamines and thyroid hormones. Thyroid hormones are synergistic with catecholamines in increasing the metabolic rate, heat production, heart rate, motor activity, and excitation of the CNS. T_3 may enhance sympathetic nervous system activity by increasing the number of β-adrenergic receptors in heart muscle and the generation of intracellular second messengers such as cyclic adenosine monophosphate (cAMP).

Effects on Growth and Maturation

A major effect of thyroid hormone is to promote growth and maturation. A small but crucial amount of thyroid hormone crosses the placenta, and the fetal thyroid axis becomes functional at midgestation. Thyroid hormone is extremely important for normal neurological development and proper bone formation in the fetus. In infants, insufficient fetal thyroid hormone causes congenital hypothyroidism, characterized by irreversible intellectual disability and short stature (see In the Clinic box).

Effects on Bone, Hard Tissue, and Dermis

Thyroid hormone promotes endochondral ossification, linear bone growth, and maturation of the epiphyseal bone centers. T_3 enhances maturation and activity of chondrocytes in the cartilage growth plate, in part by increasing local growth factor production and action. During linear postnatal growth, T_3 supports the actions of growth hormone, insulinlike growth factor (IGF)-I, and other growth factors. T_3 also supports normal adult bone remodeling.

The progression of tooth development and eruption depends on thyroid hormone, as does the normal cycle of growth and maturation of the epidermis, its hair follicles, and nails. The normal degradative processes in these structural and integumentary tissues are stimulated by thyroid hormone. Thus either too much or too little thyroid hormone can lead to hair loss and abnormal nail formation. Thyroid hormone regulates the structure of subcutaneous tissue by inhibiting synthesis and increasing degradation of mucopolysaccharides (glycosaminoglycans) and fibronectin in the extracellular connective tissue (see later description of myxedema).

Effects on the Nervous System

Thyroid hormone regulates the timing and pace of development of the CNS. Thyroid hormone deficiency in utero and in early infancy inhibits growth of the cerebral and cerebellar cortex, proliferation of axons and branching of dendrites, synaptogenesis, myelination, and cell migration. Irreversible CNS impairment results when neonatal thyroid hormone deficiency is not recognized and treated promptly. These morphological defects are paralleled by biochemical abnormalities. Decreased thyroid hormone levels reduce cell size, RNA and protein content, tubulin- and microtubule-associated protein, protein and lipid content of myelin, local production of critical growth factors, and rates of protein synthesis.

Thyroid hormone also enhances wakefulness, alertness, responsiveness to various stimuli, auditory sense, awareness of hunger, memory, and learning capacity. In addition, normal emotional tone depends on proper thyroid hormone availability. Furthermore the speed and amplitude of peripheral nerve reflexes are increased by thyroid hormone, as is motility of the gastrointestinal tract.

Effects on Reproductive Organs and Endocrine Glands

In both women and men, thyroid hormone plays an important permissive role in regulation of reproductive function. The normal ovarian cycle of follicular development, maturation, and ovulation, the homologous testicular process of spermatogenesis, and maintenance of the healthy pregnant state are all disrupted by significant deviations in thyroid hormone levels from the normal range. In part these deleterious effects may be caused by alterations in the metabolism or availability of steroid hormones. For example, thyroid hormone stimulates hepatic synthesis and release of sex steroid–binding globulin.

 IN THE CLINIC

Hypothyroidism refers to insufficient production of thyroid hormones and can occur as primary, secondary, or tertiary endocrine disease (see Chapter 41). In primary hypothyroidism, T_4 and T_3 levels are abnormally low and TSH is high. In secondary and tertiary hypothyroidism, both thyroid hormones and TSH are low. The response of TSH levels to synthetic TRH can be used to distinguish between pituitary and hypothalamic disease.

Hypothyroidism in the fetus or early childhood leads to **congenital hypothyroidism** (formerly called *cretinism* [Fig. 42.12]). Affected individuals have severe intellectual disability, short stature with incomplete skeletal development, coarse facial features, and a protruding tongue. The most common cause of hypothyroidism in children worldwide is iodide deficiency. Historically, iodide deficiency was viewed as a major cause of hypothyroidism in certain mountainous regions of South America, Africa, and Asia, but recent evidence suggests that the problem is even more widespread. This tragic form of **endemic hypothyroidism** can be prevented by public health programs that add iodide to table salt or provide yearly injections of a slowly absorbed iodide preparation. **Congenital defects** are a less common cause of neonatal/child hypothyroidism. In most cases the thyroid gland simply does not develop **(thyroid gland dysgenesis)**. Less frequent causes of childhood

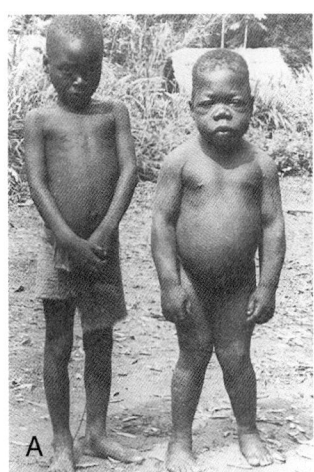
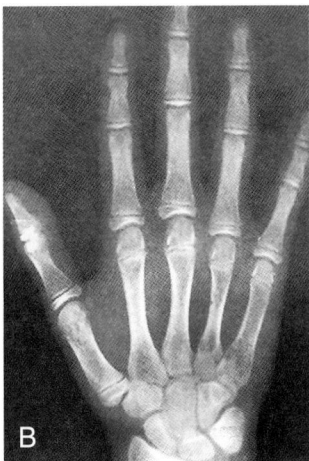
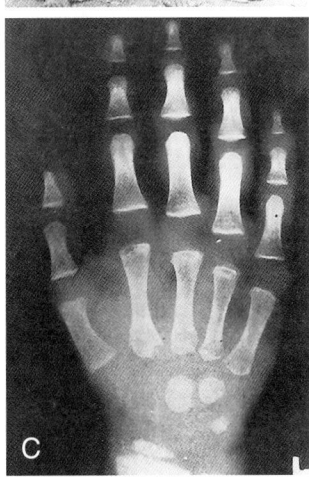

• **Fig. 42.12** A, Normal 6-year-old child *(left)* and a congenitally hypo-thyroid 17-year-old *(right)* from the same village in an area of endemic hypothyroidism. Note the short stature, obesity, malformed legs, and dull expression of the intellectually disabled hypothyroid child. Other features are a prominent abdomen, a flat broad nose, a hypoplastic mandible, dry scaly skin, delayed puberty, and muscle weakness. Radiographs of the hand comparing a normal 13-year-old (B) to that of a 13-year-old suffering from hypothyroidism (C). Note that the patient with hypothyroidism has a marked delay in development of the small bones of the hands, in growth plates at either end of the fingers, and in the growth plate of the distal radius. (A, From Delange FM. In: Braverman LE, Utiger RD [eds]. *Werner and Ingbar's The Thyroid.* 7th ed. Philadelphia: Lippincott-Raven; 1996. B, From Tanner JM et al. *Assessment of Skeletal Maturity and Prediction of Adult Height (TW2 Method).* New York: Academic Press; 1975. C, From Andersen HJ. In: Gardner LI [ed]. *Endocrine and Genetic Diseases of Childhood and Adolescence.* Philadelphia: Saunders; 1975.)

hypothyroidism are mutations in genes involved in thyroid hormone production (e.g. NIS, TPO, thyroglobulin, pendrin) or blocking antibodies to the TSH receptor. The severity of the neurological and skeletal defects is closely linked to the timing of diagnosis and thyroid hormone (T_4) replacement, with early treatment resulting in a normal cognitive ability and subtle neurological deficits. On the other hand, if hypothyroidism at birth remains untreated for only 2 to 4 weeks, the CNS will not mature normally in the first year of life. Developmental milestones such as sitting, standing, and walking will be late, and severe irreversible cognitive deficits can result. Hypothyroid babies usually appear normal at birth because of protection by maternal thyroid hormones. **Neonatal screening** (T_4 and TSH levels) has therefore

played a critical role in diagnosis and prevention of congenital hypothyroidism.

Hypothyroidism in adults who are not iodide deficient most often results from another autoimmune disorder known as **Hashimoto's disease** (formerly called *lymphocytic thyroiditis*). In contrast to the stimulatory effect of autoantibodies seen in Graves disease, thyroid autoantibodies in Hashimoto's disease (against TPO, thyroglobulin, or TSH receptor) cause apoptosis of thyroid cells and destruction of thyroid follicles. These antibodies fix complement and promote lysis of thyroid cells, causing release of thyroglobulin into the circulation. The thyroid gland becomes infiltrated by both B and T lymphocytes, which may cause enlargement of the gland.

Other causes of hypothyroidism include iatrogenic causes (e.g., radiochemical damage or surgical removal for treatment of hyperthyroidism), nodular goiters, and pituitary or hypothalamic disease. Treatment of patients with the antiarrhythmic drug amiodarone, which contains a large amount of iodine, may cause either hypo- or hyperthyroidism. Thyroid function must be carefully monitored in patients taking this medication.

The clinical picture of hypothyroidism in adults is in many respects the exact opposite of that seen in hyperthyroidism. The lower-than-normal metabolic rate leads to weight gain without an appreciable increase in caloric intake. The decreased thermogenesis lowers body temperature and causes intolerance to cold, decreased sweating, and dry skin. Adrenergic activity is decreased, and therefore bradycardia may occur. Movement, speech, and thought are all slowed, and lethargy, sleepiness, and lowering of the upper eyelids (ptosis) occur. Accumulation of negatively charged mucopolysaccharides in connective tissues attracts sodium and fluid. The resulting nonpitting **myxedema** produces puffy features, an enlarged tongue, hoarseness, joint stiffness, effusions in the pleural, pericardial, and peritoneal spaces, and pressure on peripheral and cranial nerves entrapped by excess ground substance. Constipation, loss of hair, menstrual dysfunction, and anemia are other signs. In adults lacking thyroid hormone, positron emission tomography demonstrates a generalized reduction in cerebral blood flow and glucose metabolism. This abnormality may explain the psychomotor impairment and depressed affect of hypothyroid individuals.

Replacement therapy with a daily dose of T_4 that normalizes TSH levels is usually curative in adults. In most patients, T_3 is not needed because it is generated as needed by peripheral D1 and D2. Furthermore, administration of T_3 is complicated by its high potency and short half-life, requiring frequent dosing and causing difficulty in maintaining consistent physiological levels of T_3.

Thyroid hormone also has significant effects on other parts of the endocrine system. Pituitary production of growth hormone is increased by thyroid hormone, whereas that of prolactin is decreased. Adrenocortical secretion of cortisol (see Chapter 43) as well as metabolic clearance of this hormone is stimulated, but plasma free cortisol levels remain normal. The ratio of estrogens to androgens (see Chapter 44) is increased in men (in whom breast enlargement may occur with hyperthyroidism). Decreases in both parathyroid hormone and 1,25-$(OH)_2$-vitamin D production are compensatory consequences of the effects of thyroid hormone on bone resorption (see Chapter 40). Kidney size, renal plasma flow, glomerular filtration rate, and transport rates for a number of substances are also increased by thyroid hormone.

Non-thyroidal illness syndrome (NTIS), also known as *euthyroid sick syndrome,* occurs frequently in severely ill patients who require hospitalization. NTIS is characterized by decreased levels of both circulating thyroid hormone and TSH caused by CNS-mediated suppression of the hypothalamic-pituitary-thyroid axis. In addition, peripheral metabolism of T_4 to inactive rT_3 is also increased. A similar pattern is seen upon prolonged fasting. Although it remains incompletely understood, NTIS has been proposed to represent a physiological energy-sparing adaptation to chronic illness or starvation.

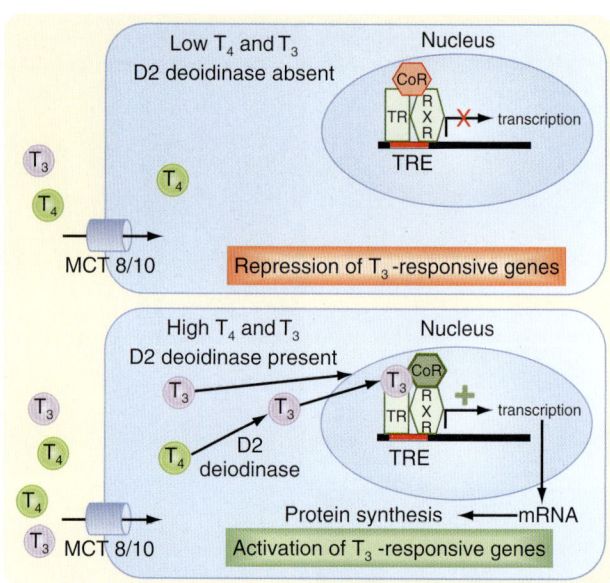

• **Fig. 42.13** Mechanisms of thyroid hormone action, including the role of MCT transporters, D2 deiodinase, and TR-RXR heterodimers. CoA, coactivator; CoR, co-repressor; RXR, retinoid X receptor.

AT THE CELLULAR LEVEL

Mechanism of Thyroid Hormone Action

For many years it was thought that thyroid hormones diffuse passively across cell membranes, but this is now known to require transport proteins. These include the monocarboxylate transporters MCT8 and MCT10, which are capable of transporting both T_4 and T_3 across the plasma membrane (Fig. 42.13; also see Fig. 42.5). Recently, mutations in MCT8 have been shown to cause an X-linked developmental syndrome in humans characterized by elevated T_3 levels, muscle hypoplasia, and severe neurological impairment. Another transporter, OATP1C1, appears to play a role in transport of T_4 across the blood-brain barrier.

Many but not all T_3 actions are mediated through its binding to one of the members of the **thyroid hormone receptor (TR) family.** The TR family belongs to the nuclear hormone receptor superfamily of transcription factors (see also Chapters 3 and 40). In humans there are two thyroid hormone receptor genes, **THRA** and **THRB,** located on chromosomes 17 and 3, respectively, that encode the nuclear thyroid hormone receptors. THRA encodes **TR_α,** which is alternatively spliced to form two main isoforms. $TR_{\alpha 1}$ is a bonafide TR, whereas the other isoform does not bind T_3. THRB encodes **$TR_{\beta 1}$** and **$TR_{\beta 2}$,** which are high-affinity receptors for T_3. The tissue distribution of $TR_{\alpha 1}$ and $TR_{\beta 1}$ is widespread. $TR_{\alpha 1}$ is strongly expressed in cardiac and skeletal muscle. $TR_{\alpha 1}$ is the primary mediator of thyroid hormone action on the heart. In contrast, $TR_{\beta 1}$ is expressed mainly in brain, liver, and kidney. $TR_{\beta 2}$ expression is restricted to the pituitary and critical areas of the hypothalamus, as well as the cochlea and retina. T_3 acting via $TR_{\beta 2}$ is responsible for inhibiting expression of the prepro-TRH gene in the paraventricular neurons of the hypothalamus and the β subunit TSH gene in pituitary thyrotropes. Thus the negative-feedback effects of thyroid hormone on both TRH and TSH secretion bere largely mediated by $TR_{\beta 2}$.

TR forms heterodimers with RXR (see Fig. 42.13). Unliganded TR-RXR binds to thyroid response elements in target genes and recruits co-repressors that inhibit gene transcription. Upon T_3 binding, the co-repressors are released, and coactivators are recruited to the hormone-receptor complex, inducing gene transcription.

An understanding of TR subtypes is important because inactivating TR mutations have been found to cause **thyroid hormone resistance** syndromes. The most common mutations occur in the $TR_{\beta 2}$ subtype, resulting in incomplete negative feedback at the hypothalamic-pituitary level. Thus T_4 levels are elevated, but TSH is not suppressed. When resistance is predominantly at the hypothalamic-pituitary level, the patient may exhibit signs of hyperthyroidism due to effects of elevated thyroid hormone levels on peripheral tissues, particularly on the heart, mediated by $TR_{\alpha 1}$. TR isoforms may also offer potential therapeutic targets. For example, research is underway to develop TR_β-specific agonists that have beneficial effects on lipid and cholesterol metabolism without the risk of adverse cardiovascular side effects.

There is emerging evidence for nongenomic actions of T_3 and T_4 that are mediated by receptors acting in the plasma membrane, mitochondria, or cytoplasm. In some cases these are modified versions of the nuclear thyroid receptors. For example, truncated isoforms of $TR_{\alpha 1}$ have been reported that bind T_3 in the plasma membrane to mediate nongenomic effects in bone or that bind T_4 in the cytoplasm to regulate microfilament organization. It has also been reported that an integrin, $\alpha_v\beta_3$, can act as a T_4 receptor at the cell surface to regulate cellular proliferation and angiogenesis by a nongenomic mechanism. The interplay between the classical genomic and nongenomic actions of thyroid hormones is likely to be another active area of future research.

Key Concepts

1. The thyroid gland is situated in the ventral aspect of the neck and is composed of right and left lobes anterolateral to the trachea and connected by an isthmus.

2. The thyroid gland is the source of tetraiodothyronine (thyroxine, T_4) and triiodothyronine (T_3).

3. The basic endocrine unit in the gland is a follicle that consists of a single spherical layer of epithelial cells surrounding a central lumen that contains colloid or stored hormone.

4. Iodide is taken up into thyroid cells by a sodium-iodide symporter in the basolateral plasma membrane.

5. T_4 and T_3 are synthesized from tyrosine and iodide by the enzyme complex of dual oxidase and thyroid peroxidase. Tyrosine residues in thyroglobulin undergo iodination, after which two iodotyrosine molecules are coupled to yield the iodothyronines.

6. Secretion of stored T_4 and T_3 requires retrieval of thyroglobulin from the follicle lumen by endocytosis. Thyroglobulin is then degraded in endolysosomes to liberate T_4 and T_3. Iodide is conserved by recycling any iodotyrosine molecules that did not undergo coupling within thyroglobulin.

7. TSH acts on the thyroid gland via its plasma membrane receptor to stimulate all steps in the production of T_4 and T_3. These steps include iodide uptake, iodination and coupling, and retrieval from thyroglobulin. TSH also stimulates glucose oxidation, protein synthesis, and growth of epithelial cells.

8. More than 99.5% of T_4 and T_3 circulates bound to the following proteins: thyroid-binding globulin, transthyretin, and albumin. Only the free fractions of T_4 and T_3 are biologically active.

9. T_4 functions largely as a prohormone whose disposition is regulated by three types of deiodinases. Monodeiodination of the outer ring yields 75% of the daily production of T_3, which is the principal active hormone. Alternatively, monodeiodination of the inner ring yields reverse T_3, which is biologically inactive. Proportioning of T_4 between T_3 and reverse T_3 regulates the availability of active thyroid hormone.

10. Thyroid hormone is a major positive regulator of the basal metabolic rate and thermogenesis. Other important actions of thyroid hormone are increased heart rate, cardiac output, and ventilation and decreased systemic vascular resistance. Substrate mobilization and disposal of metabolic products are enhanced.

11. Thyroid hormone action on the CNS and skeleton are crucial for normal growth and development. Absence of the hormone causes congenital hypothyroidism, characterized by poor brain development, short stature, and immature skeletal development. In adults, thyroid hormone supports bone remodeling and degradation of skin and hair.

12. T_3 binds to thyroid hormone receptor subtypes responsible for the various actions of thyroid hormone. The thyroid hormone receptor heterodimerizes with RXR to regulate thyroid response elements on target genes, resulting in induction or repression in the presence or absence of T_3, respectively.

Additional Reading

Bernal J, et al. Thyroid hormone transporters–functions and clinical implications. *Nat Rev Endocrinol.* 2015;11:406-417.

Bianco AC. Minireview: cracking the metabolic code for thyroid hormone signaling. *Endocrinology.* 2011;152:3306-3311.

Brent GA. Mechanisms of thyroid hormone action. *J Clin Invest.* 2012;122:3035-3043.

de Vries EM, et al. The molecular basis of the non-thyroidal illness syndrome. *J Endocrinol.* 2015;225:R67-R81.

Krude H, et al. Treatment of congenital thyroid dysfunction: achievements and challenges. *Best Pract Res Clin Endocrinol Metab.* 2015;29:399-413.

43

The Adrenal Gland

LEARNING OBJECTIVES

Upon completion of this chapter the student should be able to answer the following questions:

1. Describe the anatomy and microscopic anatomy of the adrenal gland, including the chromaffin cells of the adrenal medulla and the three zones of the adrenal cortex.
2. Explain the enzymatic reactions involved in generating norepinephrine and epinephrine and integrate those reactions with the regulation of epinephrine synthesis and secretion by the adrenal medulla.
3. Utilize the specific actions of catecholamines to explain an overall sympathetic response to a stress imposed on the body.
4. Describe the first two common reactions of the steroidogenic pathway, and their subcellular locations, and the function of StAR protein in the first reaction.
5. Compare the steroidogenic pathways within the zona glomerulosa, zona fasciculata, and zona reticularis with respect to common and zona-specific reactions.
6. Describe the mechanism of action of glucocorticoids and mineralocorticoids, including the cross-reactivity of cortisol with the mineralocorticoid receptor, and the mechanism to prevent this.
7. Integrate the multiple actions of cortisol throughout the body to explain the hormone's role during normal development and physiology, and to describe the multiple aspects of the pathophysiology of Addison's disease and Cushing's syndrome.
8. Map out the hypothalamic-pituitary-adrenal axis, including the "loophole" in the feedback mechanisms that leads to excessive androgen production (e.g., in congenital adrenal hyperplasia) in the face of an enzyme deficiency specific to the zona fasciculata and cortisol synthesis.
9. Review the regulation and actions of aldosterone.

In adults the adrenal glands emerge as fairly complex endocrine structures that produce two structurally distinct classes of hormones: steroids and catecholamines. The catecholamine hormone **epinephrine** acts as a rapid responder to stresses such as hypoglycemia and exercise to regulate multiple parameters of physiology, including energy metabolism and cardiac output. Stress is also a major secretagogue of the longer-acting steroid hormone **cortisol,** which regulates glucose utilization, immune and inflammatory homeostasis, and numerous other processes. In addition the adrenal glands regulate salt and volume homeostasis through the steroid hormone **aldosterone.** Finally, the adrenal gland secretes large amounts of the androgen precursor **dehydroepiandrosterone sulfate (DHEAS),** which plays a major role in fetoplacental estrogen synthesis and as a substrate for peripheral androgen synthesis in women.

Anatomy

The **adrenal glands** are bilateral structures located immediately above the kidneys (*ad,* near; *renal,* kidney) (Fig. 43.1). In humans they are also referred to as the **suprarenal glands** because they sit on the superior pole of each kidney. The adrenal glands are similar to the pituitary in that they are derived from both neuronal tissue and epithelial (or epithelial-like) tissue. The outer portion of the adrenal gland, called the **adrenal cortex** (Fig. 43.2), develops from mesodermal cells in the vicinity of the superior pole of the developing kidney. These cells form cords of epithelial endocrine cells. The cells of the cortex develop into steroidogenic cells (see Chapter 38). In adults the adrenal cortex is composed of three zones—the **zona glomerulosa,** the **zona fasciculata,** and the **zona reticularis**—that produce mineralocorticoids, glucocorticoids, and adrenal androgens, respectively (see Fig. 43.2B).

Soon after the cortex forms, neural crest–derived cells associated with the sympathetic ganglia, called **chromaffin cells,** migrate into the cortex and become encapsulated by cortical cells. Thus the chromaffin cells establish the inner portion of the adrenal gland, which is called the **adrenal medulla** (see Fig. 43.2). The chromaffin cells of the adrenal medulla have the potential to develop into postganglionic sympathetic neurons. They are innervated by cholinergic preganglionic sympathetic neurons and can synthesize the catecholamine neurotransmitter **norepinephrine** from tyrosine. However, high levels of cortisol that drain into the medulla from the adrenal cortex induce expression of the enzyme **phenylethanolamine *N*-methyl transferase (PNMT),** which transfers a methyl group onto norepinephrine to produce the catecholamine hormone **epinephrine,** the primary hormonal product of the adrenal medulla (see Fig. 43.2B).

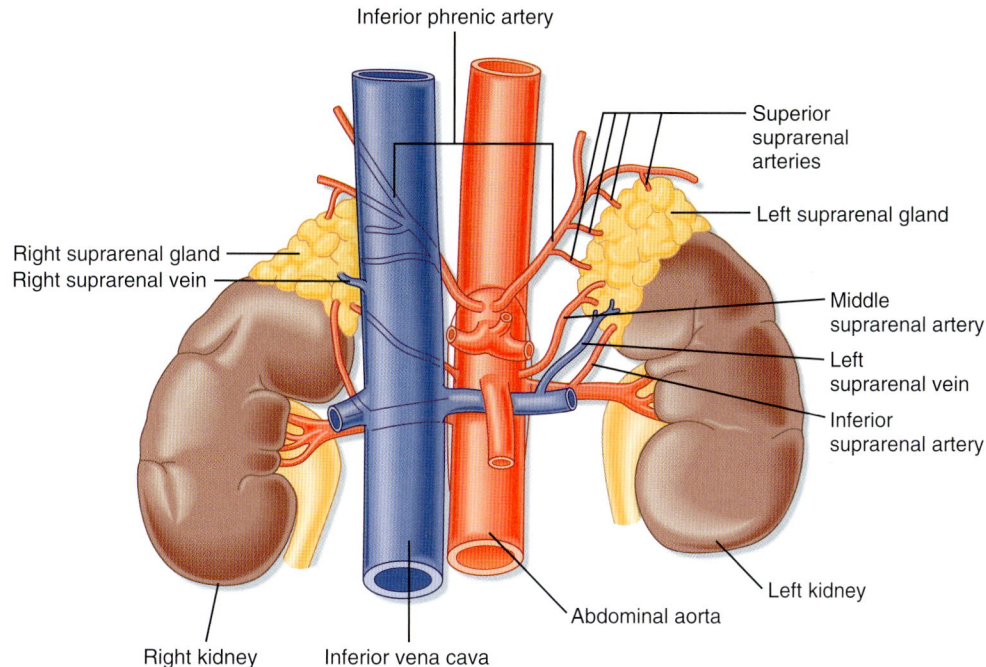

Inferior phrenic artery

Superior suprarenal arteries

Left suprarenal gland

Right suprarenal gland
Right suprarenal vein

Middle suprarenal artery

Left suprarenal vein

Inferior suprarenal artery

Left kidney

Abdominal aorta

Right kidney   Inferior vena cava

 Fig. 43.1 The adrenal glands sit on the superior poles of the kidneys and receive a rich arterial supply from the inferior, middle, and superior suprarenal arteries. The adrenals are drained by a single suprarenal vein. (Modified from Drake RL et al. *Gray's Anatomy for Students.* Philadelphia: Churchill Livingstone; 2005.)

Adrenal Medulla

Instead of being secreted near a target organ and acting as neurotransmitters, adrenomedullary catecholamines are secreted into blood and act as hormones. About 80% of the cells of the adrenal medulla secrete **epinephrine,** and the remaining 20% secrete **norepinephrine.** Although circulating epinephrine is derived entirely from the adrenal medulla, only about 30% of the circulating norepinephrine comes from the medulla. The remaining 70% is released from postganglionic sympathetic nerve terminals and diffuses into the vascular system. Because the adrenal medulla is not the sole source of catecholamine production, this tissue is not essential for life.

Synthesis of Epinephrine

The enzymatic steps in epinephrine synthesis are shown in Fig. 43.4. Synthesis begins with transport of the amino acid **tyrosine** into the chromaffin cell cytoplasm and subsequent hydroxylation of tyrosine by the rate-limiting enzyme **tyrosine hydroxylase** to produce **dihydroxyphenylalanine (DOPA).** DOPA is converted to **dopamine** by a cytoplasmic enzyme, aromatic amino acid decarboxylase, and is then transported into the secretory vesicle (also called the **chromaffin granule**). Within the granule, all dopamine is completely converted to **norepinephrine** by the enzyme dopamine β-hydroxylase. In most adrenomedullary cells, essentially all of the norepinephrine diffuses out of the chromaffin granule by facilitated transport and is methylated by the cytoplasmic enzyme **PNMT** to form epinephrine.

AT THE CELLULAR LEVEL

The high local concentration of cortisol in the medulla is maintained by the vascular configuration within the adrenal gland. The outer connective tissue capsule of the adrenal gland is penetrated by a rich arterial supply coming from three main arterial branches: the inferior, middle, and superior suprarenal arteries (see Fig. 43.1). These give rise to two types of blood vessels that carry blood from the cortex to the medulla (Fig. 43.3): (1) relatively few medullary arterioles, which provide high oxygen- and nutrient-laden blood directly to the medullary chromaffin cells, and (2) relatively numerous cortical sinusoids into which cortical cells secrete steroid hormones (including cortisol). Both vessel types fuse to give rise to the medullary plexus of vessels that ultimately drains into a single suprarenal vein. Thus secretions of the adrenal cortex percolate through the chromaffin cells and bathe them in high concentrations of cortisol before leaving the gland and entering the inferior vena cava. Cortisol inhibits neuronal differentiation of the medullary cells, so they fail to form dendrites and axons. Additionally, cortisol induces expression of the enzyme **PNMT,** which converts norepinephrine to epinephrine (Fig. 43.4). Glucocorticoid receptor–knockout mice have an enlarged cortex, but the size of the medulla is decreased and PNMT activity is undetectable.

Epinephrine is then transported back into the granule for storage and to undergo regulated exocytosis.

Secretion of epinephrine and norepinephrine from the adrenal medulla is regulated primarily by descending sympathetic signals in response to various forms of stress, including exercise, hypoglycemia, and hemorrhagic hypovolemia

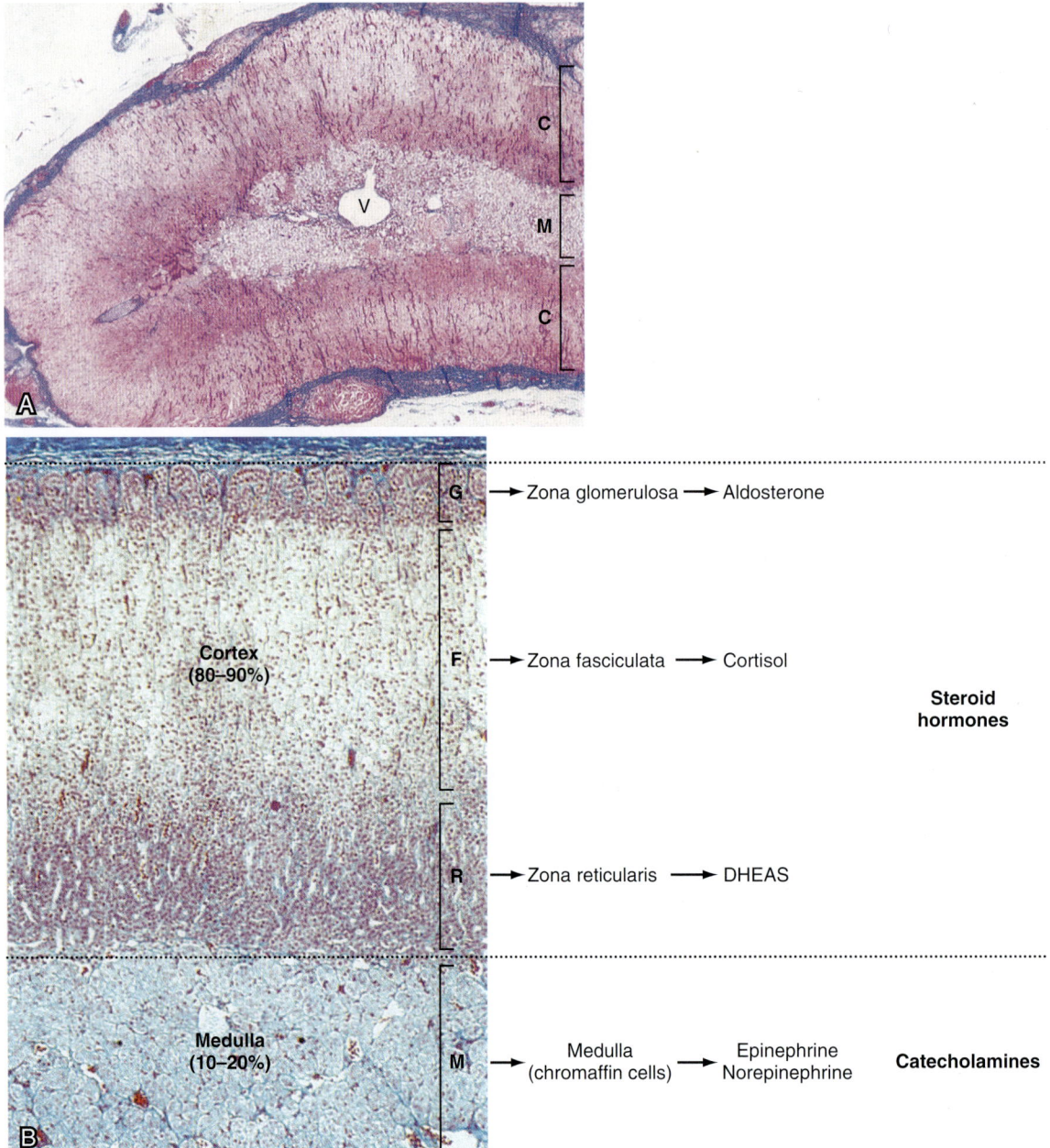

• **Fig. 43.2** Histology of the adrenal gland. A, Low magnification illustrating the outer cortex (C) and inner medulla (M; note the central vein [V]). B, Higher magnification clearly illustrating the zonation of the cortex. The corresponding endocrine function and the different zones of the cortex and the medulla are noted. (From Young B et al. *Wheater's Functional Histology.* 5th ed. Philadelphia: Churchill Livingstone; 2006.)

(Fig. 43.5). The primary autonomic centers that initiate sympathetic responses reside in the hypothalamus and brainstem, and they receive input from the cerebral cortex, the limbic system, and other regions of the hypothalamus and brainstem.

The chemical signal for secretion of catecholamine from the adrenal medulla is **acetylcholine (ACh),** which is secreted from **preganglionic sympathetic neurons** and binds to **nicotinic receptors** on chromaffin cells (see Fig. 43.5). ACh increases the activity of the rate-limiting enzyme tyrosine hydroxylase in chromaffin cells (see Fig. 43.4). It

also increases the activity of dopamine β-hydroxylase and stimulates exocytosis of the chromaffin granules. Synthesis of epinephrine and norepinephrine is closely coupled to secretion so that levels of intracellular catecholamines do not change significantly even in the face of changing sympathetic activity.

Mechanism of Action of Catecholamines

Adrenergic receptors are generally classified as α- and **β-adrenergic receptors,** with the α-adrenergic receptors

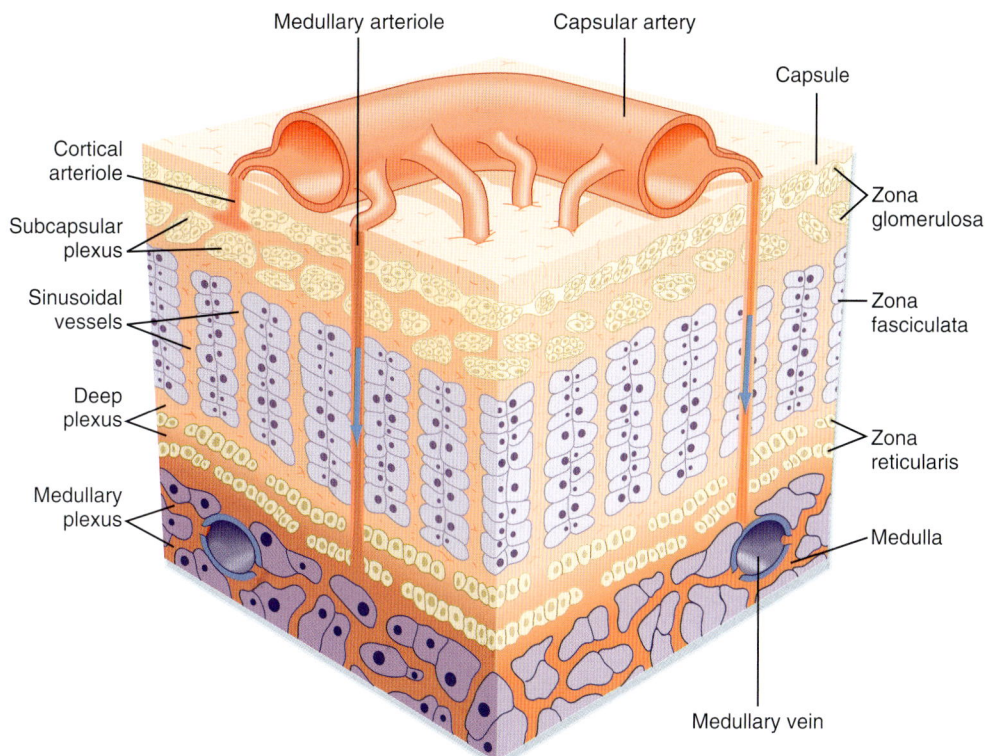

Medullary arteriole   Capsular artery

Capsule

Cortical arteriole

Subcapsular plexus

Sinusoidal vessels

Deep plexus

Medullary plexus

Zona glomerulosa

Zona fasciculata

Zona reticularis

Medulla

Medullary vein

• **Fig. 43.3** Blood flow through the adrenal gland. Capsular arteries give rise to sinusoidal vessels that carry blood centripetally through the cortex to the medulla. (Modified from Young B et al. *Wheater's Functional Histology.* 5th ed. Philadelphia: Churchill Livingstone; 2006.)

further divided into α_1 and α_2 **receptors** and the β-adrenergic receptors divided into β_1, β_2, and β_3 **receptors** (Table 43.1). These receptors can be characterized according to:

1. Relative potency of endogenous and pharmacological agonists and antagonists. Epinephrine and norepinephrine are potent agonists for α receptors and for β_1 and β_3 receptors, whereas epinephrine is more potent than norepinephrine for β_2 receptors. A large number of synthetic selective and nonselective adrenergic agonists and antagonists now exist.
2. Downstream signaling pathways. Table 43.1 shows the primary pathways that are coupled to the different adrenergic receptors. This is an oversimplification, because differences in signaling pathways for a given receptor have been linked to the duration of agonist exposure and cell type.
3. Location and relative density of receptors. Importantly, different receptor types predominate in different tissues. For example, although both α and β receptors are expressed by pancreatic islet beta cells, the predominant response to a sympathetic discharge is mediated by α_2 receptors.

Physiological Actions of Adrenomedullary Catecholamines

Because the adrenal medulla is directly innervated by the autonomic nervous system, adrenomedullary responses are very rapid. Furthermore, because of the involvement of several centers in the central nervous system (CNS), most notably the cerebral cortex, adrenomedullary responses can precede onset of the actual stress (i.e., they can be anticipated) (see Fig. 43.5). In many cases the adrenomedullary output, which is primarily epinephrine, is coordinated with sympathetic nervous activity as determined by the release of norepinephrine from postganglionic sympathetic neurons. However, some stimuli (e.g., hypoglycemia) evoke a stronger adrenomedullary than sympathetic nervous response and vice versa.

Many organs and tissues are affected by a sympathoadrenal response (Table 43.2). An informative example of the major physiological roles of catecholamines is the sympathoadrenal response to exercise. Exercise is similar to the **"fight-or-flight" response** but without the subjective element of fear, and it involves a greater adrenomedullary response (i.e., endocrine role of epinephrine) than a sympathetic nervous response (i.e., neurotransmitter role of norepinephrine). The overall goal of the sympathoadrenal system during exercise is to meet the increased energy demands of skeletal and cardiac muscle while maintaining sufficient oxygen and glucose supply to the brain. The response to exercise includes the following major physiological actions of epinephrine (Fig. 43.6):

1. Increased blood flow to muscles is achieved by the integrated action of norepinephrine and epinephrine on the heart, veins and lymphatics, and nonmuscular (e.g., splanchnic) and muscular arteriolar beds.

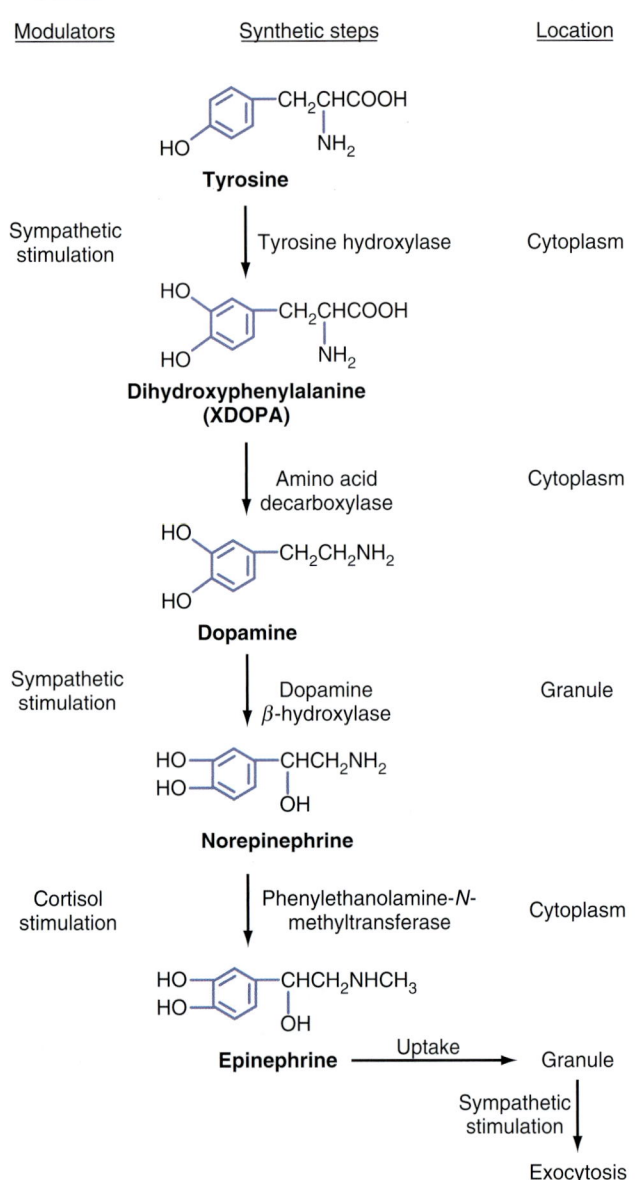

Modulators | Synthetic steps | Location

Fig. 43.4 Steps in synthesis and secretion of catecholamines from adrenal medullary chromaffin cells.

2. Epinephrine promotes glycogenolysis in muscle. Exercising muscle can also utilize free fatty acids (FFAs), and epinephrine and norepinephrine promote lipolysis in adipose tissue. Epinephrine increases blood glucose by increasing hepatic glycogenolysis and gluconeogenesis. The promotion of lipolysis in adipose tissue is also coordinated with an epinephrine-induced increase in hepatic ketogenesis. Finally, the effects of catecholamines on metabolism are reinforced by the fact that they stimulate glucagon secretion (β_2 receptors) and inhibit insulin secretion (α_2 receptors). Efficient production of adenosine triphosphate (ATP) during normal exercise

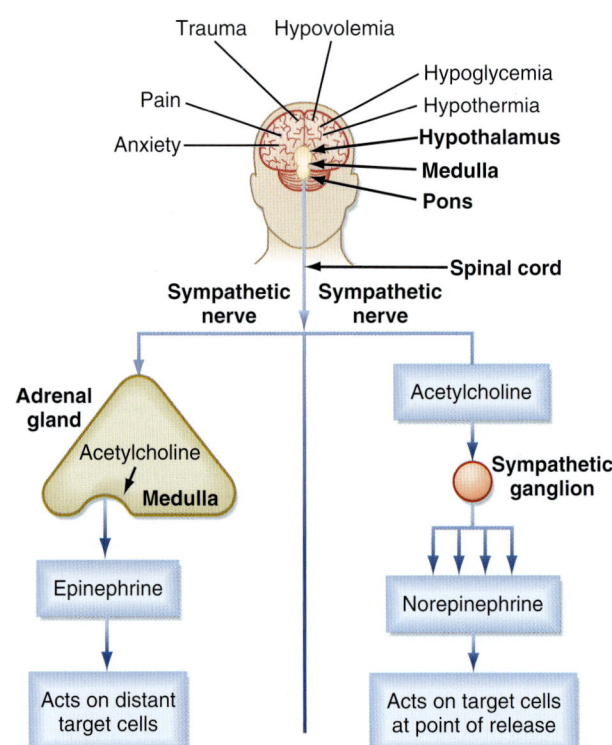

Fig. 43.5 Stimuli that enhance catecholamine secretion.

TABLE 43.1	**Adrenergic Receptors**		
Receptor Type	Primary Mechanism of Action	Examples of Tissue Distribution	Examples of Action
α_1	↑ IP3 and Ca++, DAG	Sympathetic postsynaptic nerve terminals	Increase vascular smooth muscle contraction
α_2	↓ cAMP	Sympathetic presynaptic nerve terminals, beta cell of pancreatic islets	Inhibit norepinephrine release, inhibit insulin release
β_1	↑ cAMP	Heart	Increase cardiac output
β_2	↑ cAMP	Liver; smooth muscle of vasculature, bronchioles, and uterus	Increase hepatic glucose output; decrease contraction of blood vessels, bronchioles, and uterus
β_3	↑ cAMP	Liver, adipose tissue	Increase hepatic glucose output, increase lipolysis

cAMP, cyclic adenosine monophosphate; DAG, diacylglycerol.

TABLE 43.2	Some Actions of Catecholamine Hormones	
β: Epinephrine > Norepinephrine	α: Norepinephrine > Epinephrine
↑ Glycogenolysis	↑ Gluconeogenesis (α_1)
↑ Gluconeogenesis (β_2)	↑ Glycogenolysis (α_1)
↑ Lipolysis (β_3) (β_2)	
↑ Calorigenesis (β_1)	
↓ Glucose utilization	
↑ Insulin secretion (β_2)	↓ Insulin secretion (α_2)
↑ Glucagon secretion (β_2)	
↑ Muscle K⁺ uptake (β_2)	↑ Cardiac contractility (α_1)
↑ Cardiac contractility (β_1)	
↑ Heart rate (β_1)	
↑ Conduction velocity (β_1)	
↑ Arteriolar dilation: ↓ BP (β_2) (muscle)	↑ Arteriolar vasoconstriction; ↑ BP (α_1) (splanchnic, renal, cutaneous, genital)
↑ Muscle relaxation (β_2)	↑ Sphincter contraction (α_1)
Gastrointestinal	Gastrointestinal
Urinary	Urinary
Bronchial	↑ Platelet aggregation (α_2)
	↑ Sweating ("adrenergic")
	↑ Dilation of pupils (α_1)

BP, blood pressure.

(i.e., a 1-hour workout) also requires efficient exchange of gases with an adequate supply of oxygen to exercising muscle. Catecholamines promote this by relaxation of bronchiolar smooth muscle.

3. Catecholamines decrease energy demand by visceral smooth muscle. In general a sympathoadrenal response decreases overall motility of the smooth muscle in the gastrointestinal (GI) and urinary tracts, thereby conserving energy where it is not immediately needed.

Metabolism of Catecholamines

Two primary enzymes are involved in the degradation of catecholamines: **monoamine oxidase (MAO)** and **catechol-*O*-methyltransferase (COMT)**. The neurotransmitter norepinephrine is degraded by MAO and COMT after uptake into the presynaptic terminal. This mechanism is also involved in the catabolism of circulating adrenal catecholamines. However, the predominant fate of adrenal catecholamines is methylation by COMT in nonneuronal tissues such as the liver and kidney. Urinary **vanillylmandelic acid (VMA)** and **metanephrine** are sometimes used clinically to assess the level of catecholamine production in a patient. Much of the urinary VMA and metanephrine is derived from neuronal rather than adrenal catecholamines.

IN THE CLINIC

Pheochromocytoma is a tumor of chromaffin tissue that produces excessive quantities of catecholamines. These are commonly adrenal medullary tumors, but they can occur in other chromaffin cells of the autonomic nervous system. Although pheochromocytomas are not common tumors, they are the most common cause of hyperfunctioning of the adrenal medullary. The catecholamine most frequently elevated in pheochromocytoma is norepinephrine. For unknown reasons the symptoms of excessive catecholamine secretion are often sporadic rather than continuous. Symptoms include hypertension, headaches (from hypertension), sweating, anxiety, palpitations, and chest pain. In addition, patients with this disorder may show orthostatic hypotension (despite the tendency for hypertension). This occurs because hypersecretion of catecholamines can decrease the postsynaptic response to norepinephrine as a result of downregulation of the receptors (see Chapter 3). Consequently the baroreceptor response to blood shifts that occurs on standing is blunted.

Adrenal Cortex

Zona Fasciculata

The zona fasciculata produces the glucocorticoid hormone **cortisol.** This zone is an actively steroidogenic tissue composed of straight cords of large cells. These cells have a "foamy" cytoplasm because they are filled with lipid droplets that represent stored cholesterol esters (CEs). These cells make some cholesterol de novo but import a significant amount of cholesterol from the blood in the form of low-density lipoprotein (LDL). LDL particles bind to their receptor (LDLR) and are endocytosed. Within endolysosomes, free cholesterol (FC) is released from CEs by a lysosomal lipase, and the FC is transported out of the endolysosome by Niemann-Pick C (NPC) proteins. Free cholesterol is stored in lipid droplets in the cytoplasm after esterification by acyl CoA-cholesterol acyltransferase (ACAT) (Fig. 43.7). The stored cholesterol is continually turned back into free cholesterol by **hormone-sensitive lipase (HSL),** a process that is increased in response to adrenocorticotropic hormone (ACTH; see Regulation of Cortisol Production).

All steroid hormone synthesis begins in the mitochondria, where the first enzyme, CYP11A1, is attached to the inner mitochondrial membrane. Although several proteins appear to be involved in the transfer of FC into the inner mitochondrial matrix, one protein called **steroidogenic acute regulatory protein (StAR protein)** is indispensable in this process (see Fig. 43.7). StAR protein is short-lived and rapidly activated posttranslationally (phosphorylation) and transcriptionally by pituitary tropic hormones. In patients with inactivating mutations in StAR protein, cells of the zona fasciculata become excessively laden with lipid ("lipoid") because cholesterol cannot be accessed by CYP11A1 within the mitochondria and used for cortisol

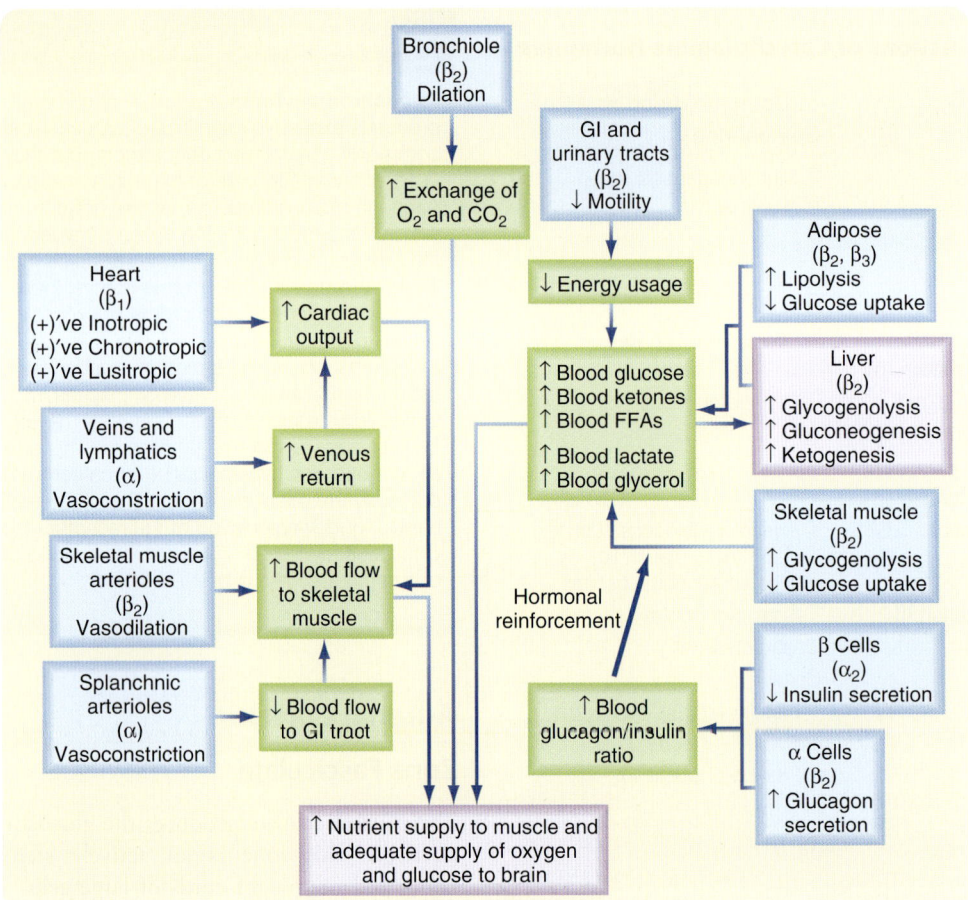

• **Fig. 43.6** Some of the individual actions of catecholamines that contribute to the integrated sympathoadrenal response to exercise. (Modified from White BA, Porterfield SP. *Endocrine and Reproductive Physiology.* 4th ed. Philadelphia: Mosby; 2013.)

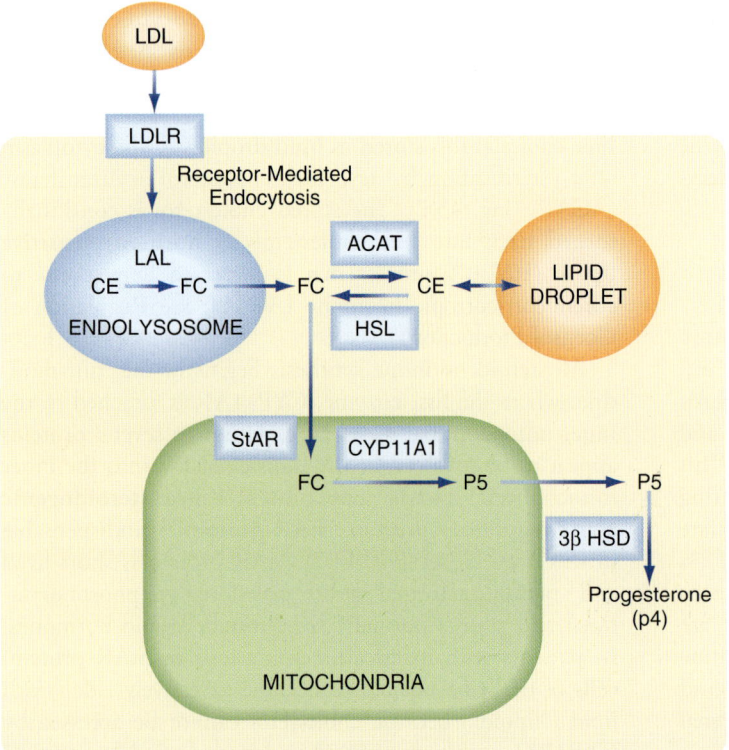

• **Fig. 43.7** Events involved in the first two reactions in the steroidogenic pathway: conversion of cholesterol to pregnenolone; conversion of pregnenolone (P5) to progesterone (P4) in zona fasciculata cells. ACAT, acyl CoA:cholesterol acyltransferase; 3β HSD, 3β hydroxysteroid dehydrogenase; CE, cholesterol esters; CYP11A1 also called *P450 side-chain cleavage enzyme;* FC, free cholesterol; HSL, hormone-sensitive lipase; LAL, lysosomal acid hydrolase; LDL, low-density lipoprotein; LDLR, low-density lipoprotein receptor; P5, pregnenolone; StAR, steroidogenic acute regulatory protein. (Modified from White BA, Porterfield SP. *Endocrine and Reproductive Physiology.* 4th ed. Philadelphia: Mosby; 2013.)

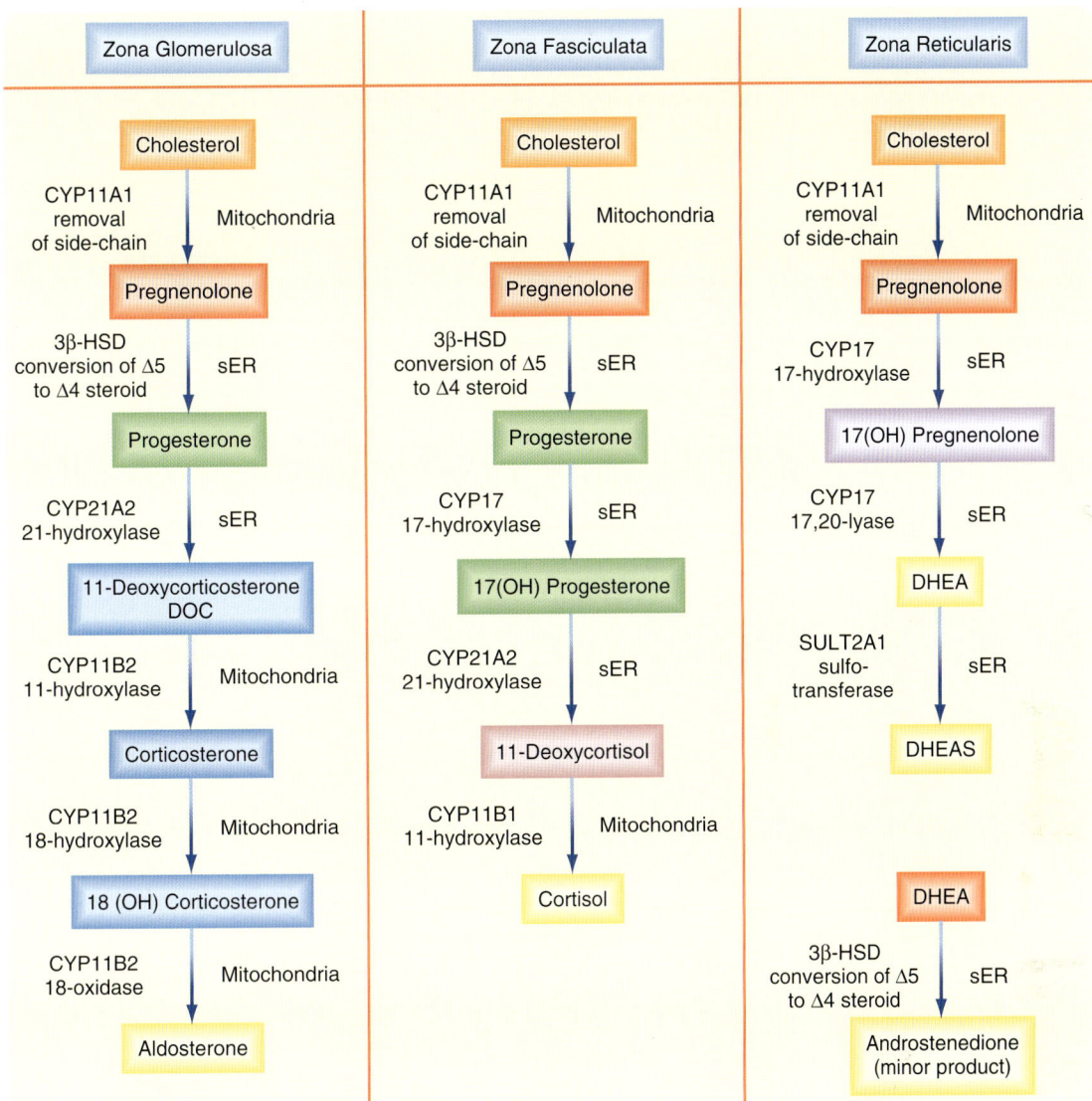

Fig. 43.8 Summary of the steroidogenic pathways for each of the three zones of the adrenal cortex. The enzymatic reactions are color coded across zones. sER, smooth endoplasmic reticulum. (Modified from White BA, Porterfield SP. *Endocrine and Reproductive Physiology.* 4th ed. Philadelphia: Mosby; 2013.)

synthesis. Moreover, these individuals cannot synthesize gonadal steroid hormones. The placenta does not express StAR, so these individuals have normal placental steroid production in utero.

In the zona fasciculata, cholesterol is converted sequentially to pregnenolone, progesterone, 17-hydroxyprogesterone, 11-deoxycortisol, and cortisol (Figs. 43.8 and 43.9). A parallel pathway in the zona fasciculata involves a pathway that bypasses 17-hydroxylation, in which progesterone is converted to 11-deoxycorticosterone (DOC) and then to corticosterone (see Fig. 43.9C). This pathway is minor in humans, but in the absence of active CYP11B1 (11-hydroxylase activity), the production of DOC is significant. Because DOC acts as a weak mineralocorticoid (Table 43.3), elevated levels of DOC cause hypertension.

Transport and Metabolism of Cortisol

Cortisol is transported in blood predominantly bound to **corticosteroid-binding globulin [CBG]** (also called **transcortin**), which binds about 90% of cortisol, and albumin, which binds 5% to 7% of cortisol. The liver is the predominant site of steroid inactivation. It both inactivates cortisol and conjugates active and inactive steroids with glucuronide or sulfate so that they can be excreted more readily by the kidney. The circulating half-life of cortisol is about 70 minutes.

Cortisol is reversibly inactivated by conversion to **cortisone.** This action is catalyzed by the enzyme **11β-hydroxysteroid dehydrogenase type 2 (11β-HSD2).** Inactivation of cortisol by 11β-HSD2 occurs in cells that also express the mineralocorticoid receptor (MR) and are

AT THE CELLULAR LEVEL

Steroidogenic enzymes fall into two superfamilies. Most belong to the **cytochrome P-450 monooxidase gene family** and are thus referred to as **CYPs.** These enzymes are located either in the inner mitochondrial matrix, where they use molecular oxygen and a flavoprotein electron donor, or in the smooth endoplasmic reticulum, where they use a different flavoprotein for electron transfer. Different CYP enzymes act as hydroxylases, lyases (desmolases), oxidases, or aromatases. Two of these enzymes have multiple functions. CYP17 has both a 17-hydroxylase function and a 17,20-lyase (desmolase) function. CYP11B2, also called *aldosterone synthase,* has three functions: 11-hydroxylase, 18-hydroxylase, and 18-oxidase.

The other enzymes involved in steroidogenesis belong to three **hydroxysteroid dehydrogenase** (HSD) families. **3β-HSDs** have two isoforms that convert the hydroxyl group on carbon 3 of the cholesterol ring to a ketone and shift the double bond from the 5-6 (**Δ5**) position to the 4-5 (**Δ4**) position. All active steroid hormones must be converted to Δ4 structures by 3β-HSD. The **17β-HSDs** have at least five members and can act as either oxidases or reductases. 17β-HSDs primarily act on sex steroids and can be activating or deactivating. Finally, the **11β-HSDs** have two isoforms that catalyze the interchange between cortisol (active) and cortisone (inactive).

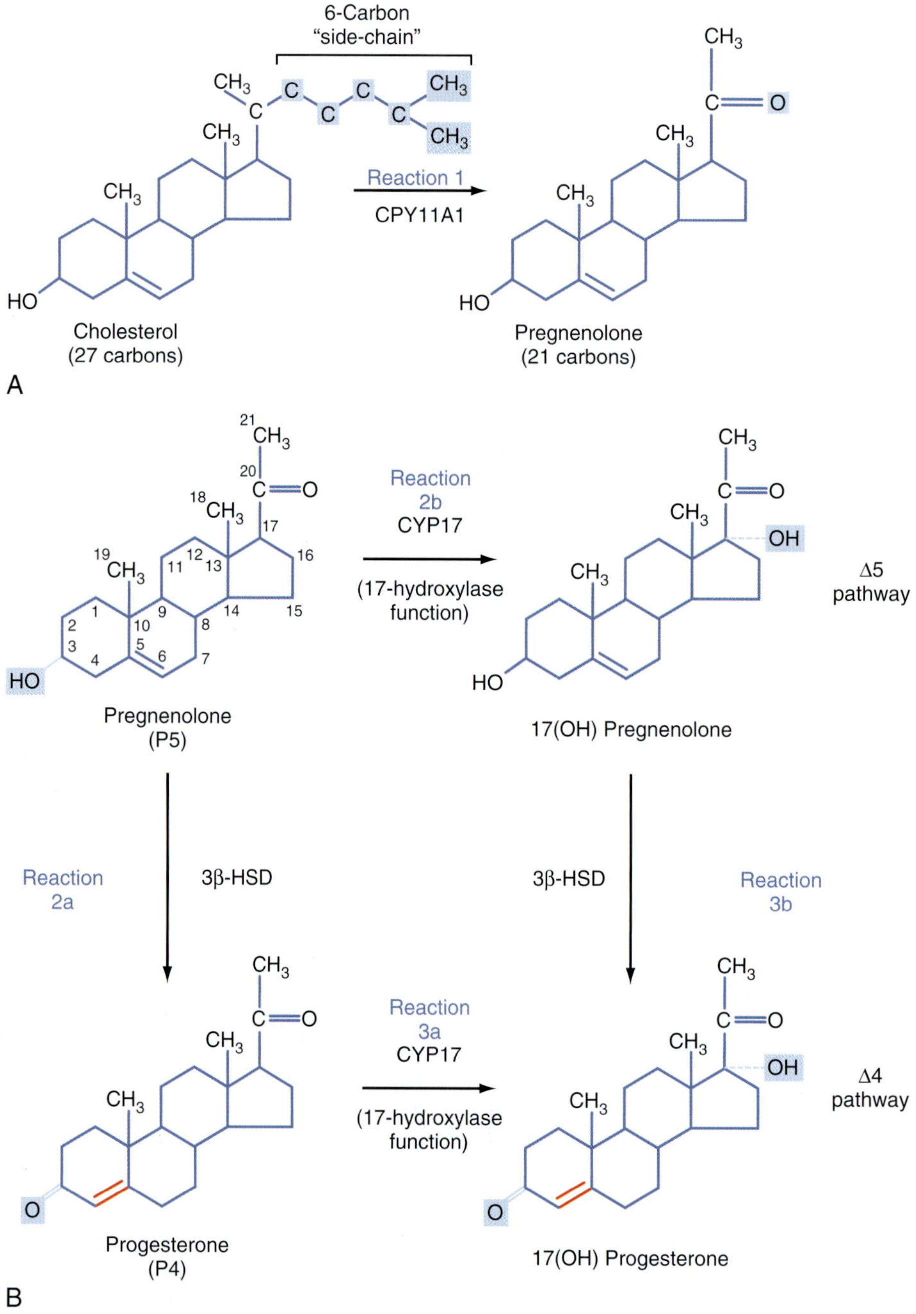

• **Fig. 43.9** A, Reaction 1, catalyzed by CYP11A1, in making cortisol. B, Reactions 2a/b and reactions 3a/b, involving CYP17 (17-hydroxylase function) and 3β-hydroxysteroid dehydrogenase (3β-HSD), in making cortisol. This figure shows the Δ5 versus Δ4 pathway.

• **Fig. 43.9, cont'd** C, Reactions 4 and 5, involving CYP21B and CYP11B1, in which the last two steps in cortisol synthesis are carried out. Also shown is the minor pathway leading to corticosterone synthesis in the zona fasciculata. (Modified from White BA, Porterfield SP. *Endocrine and Reproductive Physiology.* 4th ed. Philadelphia: Mosby; 2013.)

target cells of aldosterone (see below). The conversion of cortisol to cortisone prevents the binding of cortisol to the MR and having inappropriate mineralocorticoid actions on these cells. Inactivation of cortisol by 11β-HSD2 is reversible in that another enzyme, **11β-HSD1,** converts cortisone back to cortisol. This conversion occurs in tissues expressing the glucocorticoid receptor (GR), including liver, adipose tissue, and the CNS, as well as in skin.

Mechanism of Action of Cortisol

Cortisol acts primarily through the **glucocorticoid receptor,** which regulates gene transcription (see Chapter 3). In the absence of hormone, the GR resides in the cytoplasm in a stable complex with several **molecular chaperones,** including heat shock proteins and cyclophilins. Cortisol-GR binding promotes dissociation of the chaperone proteins, followed by:

TABLE 43.3 Relative Glucocorticoid and Mineralocorticoid Potency of Natural Corticosteroids and Some Synthetic Analogues in Clinical Use[a]

	Glucocorticoid	Mineralocorticoid
Corticosterone	0.5	1.5
Prednisone (1.2 double bond)	4	<0.1
6α-Methylprednisone (Medrol)	5	<0.1
9α-Fluoro-16α-hydroxyprednisolone (triamcinolone)	5	<0.1
9α-Fluoro-16α-methylprednisolone (dexamethasone)	30	<0.1
Aldosterone	0.25	500
Deoxycorticosterone	0.01	30
9α-Fluorocortisol	10	500

[a]All values are relative to the glucocorticoid and mineralocorticoid potencies of cortisol, which have each been arbitrarily set at 1.0. Cortisol actually has only 1/500 the potency of the natural mineralocorticoid aldosterone.

1. rapid translocation of the cortisol-GR complex into the nucleus,
2. dimerization and binding to **glucocorticoid response elements** (GREs, both "positive" GREs and "negative" GREs) near the basal promoters of cortisol-regulated genes, and
3. recruitment of **coactivator proteins** and assembly of general transcription factors leading to increased or decreased transcription of the targeted genes.

In some cases the GR interacts with other transcription factors, such as the proinflammatory nuclear factor (NF)-κB transcription factor, and interferes with their ability to activate gene expression.

Physiological Actions of Cortisol

Cortisol has a broad range of actions and is often characterized as a "stress hormone." In general, cortisol maintains blood glucose levels, CNS function, and cardiovascular function during fasting and increases blood glucose during stress at the expense of muscle protein. Cortisol protects the body against the self-injurious effects of unbridled inflammatory and immune responses. Cortisol also partitions energy to cope with stress by inhibiting reproductive function. As stated later, cortisol has several other effects on bone, skin, connective tissue, the GI tract, and the developing fetus that are independent of its stress-related functions.

Metabolic Actions

As the term *glucocorticoid* implies, cortisol is a steroid hormone from the adrenal cortex that regulates **blood glucose.** It increases blood glucose by stimulating **gluconeogenesis** (Fig. 43.10). Cortisol enhances gene expression of the key hepatic gluconeogenic enzyme **phosphoenolpyruvate carboxykinase (PEPCK).** Cortisol also decreases GLUT4-mediated glucose uptake in skeletal muscle and adipose tissue. During the interdigestive period (low insulin-glucagon ratio), cortisol promotes glucose sparing by potentiating the effects of catecholamines on lipolysis, thereby making FFAs available as energy sources. Cortisol inhibits protein synthesis and increases proteolysis, especially in skeletal muscle, thereby providing a rich source of carbon for hepatic gluconeogenesis.

Fig. 43.10 also contrasts the normal role of cortisol in response to stress and the effects of **chronically elevated cortisol** as a result of pathological conditions. As discussed later, there are important differences in the overall metabolic effects of cortisol between these two states, particularly with respect to lipid metabolism. During stress, cortisol synergizes with catecholamines and glucagon to promote a lipolytic, gluconeogenic, ketogenic, and glycogenolytic metabolic response while synergizing with catecholamines to promote an appropriate cardiovascular response. During chronically elevated cortisol secondary to pathological overproduction, cortisol synergizes with insulin in the context of elevated levels of glucose (from increased appetite) and hyperinsulinemia (from elevated glucose and glucose intolerance) to promote lipogenesis and truncal (abdominal/visceral) adiposity.

Cardiovascular Actions

Cortisol reinforces its effects on blood glucose by its positive effects on the cardiovascular system. Cortisol has permissive actions on catecholamines by increasing **adrenergic receptor** expression and thereby contributes to cardiac output and blood pressure. Cortisol stimulates **erythropoietin** synthesis and hence increases red blood cell production. **Anemia** occurs when cortisol is deficient, and **polycythemia** occurs when cortisol levels are excessive.

Antiinflammatory and Immunosuppressive Actions

Inflammation and immune responses are often part of the response to stress. However, inflammation and immune responses have the potential for significant harm and may cause death if they are not held in homeostatic balance. As a stress hormone, cortisol plays an important role in maintaining immune homeostasis. Cortisol, along with epinephrine and norepinephrine, represses production of proinflammatory cytokines and stimulates production of antiinflammatory cytokines.

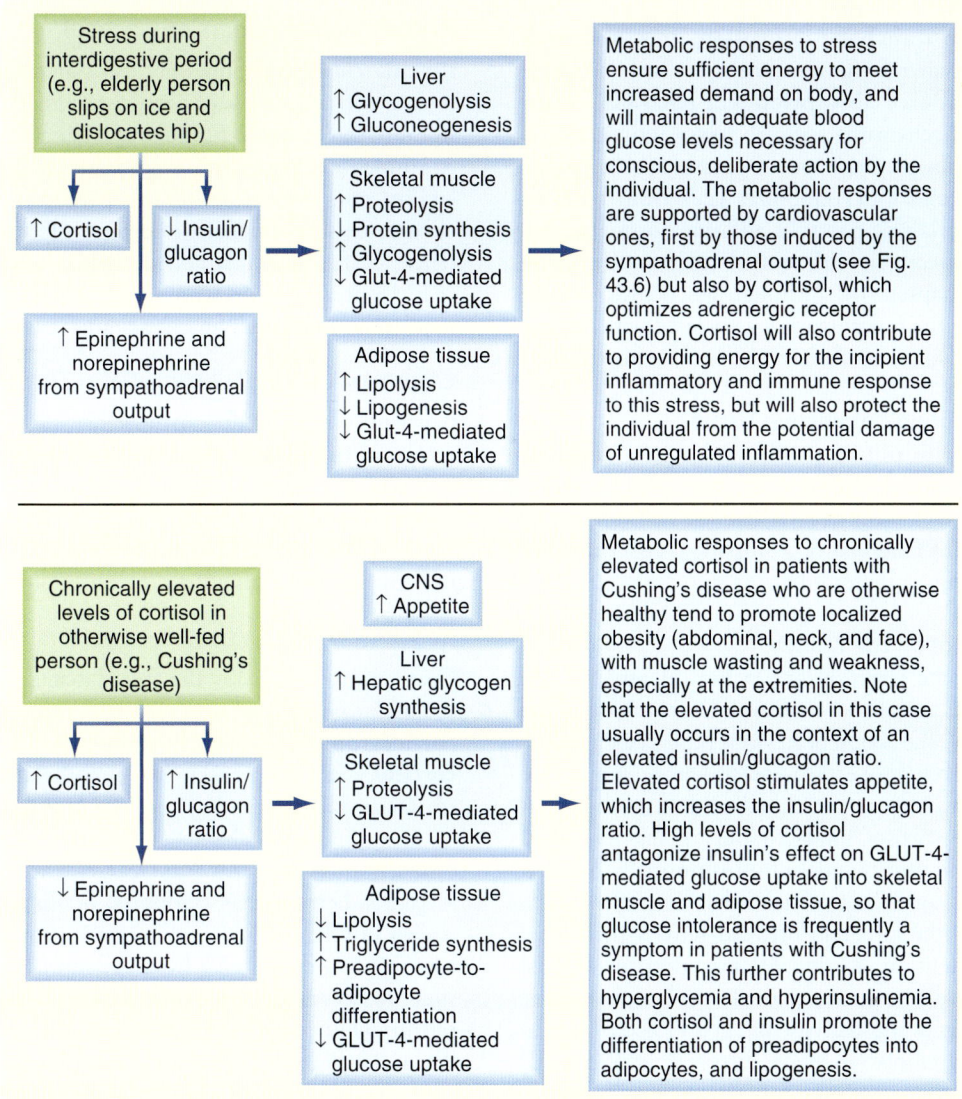

• Fig. 43.10 Metabolic actions of cortisol (integrated with catecholamines and glucagon) in response to stress *(upper panel)* and contrasted to the actions of chronically elevated cortisol (integrated with insulin) in an otherwise healthy individual *(lower panel)*. (Modified from White BA, Porterfield SP. *Endocrine and Reproductive Physiology.* 4th ed. Philadelphia: Mosby; 2013.)

The inflammatory response to injury consists of local dilation of capillaries and increased capillary permeability, with resultant local edema and accumulation of white blood cells. These steps are mediated by **prostaglandins, thromboxanes, and leukotrienes**. Cortisol inhibits **phospholipase A₂**, a key enzyme in prostaglandin, leukotriene, and thromboxane synthesis. Cortisol also stabilizes lysosomal membranes, thereby decreasing release of the proteolytic enzymes that augment local swelling. In response to injury, leukocytes normally leave the vascular system and migrate to the site of injury. This complex process is generally inhibited by cortisol, as is the phagocytic activity of neutrophils, although release of neutrophils from bone marrow is stimulated. Analogues of glucocorticoid are frequently used pharmacologically because of their antiinflammatory properties.

Cortisol inhibits the immune response, and for this reason glucocorticoid analogues have been used as **immunosuppressants** in organ transplants. High cortisol levels decrease the number of circulating T lymphocytes (particularly helper T lymphocytes) and reduce their ability to migrate to the site of antigenic stimulation. Glucocorticoids promote atrophy of the thymus and other lymphoid tissue. Although corticosteroids inhibit cellular-mediated immunity, antibody production by B lymphocytes is not impaired.

Effects of Cortisol on the Reproductive Systems

Reproduction exacts a considerable anabolic cost on the organism. In humans, reproductive behavior and function are dampened in response to stress. Cortisol decreases the function of the **reproductive axis** at the hypothalamic, pituitary, and gonadal levels.

Effects of Cortisol on Bone

Glucocorticoids increase **bone resorption.** They have multiple actions that alter bone metabolism. Glucocorticoids decrease **intestinal Ca^{++} absorption** and **renal Ca^{++} reabsorption.** Both mechanisms serve to lower serum [Ca^{++}]. As serum [Ca^{++}] drops, secretion of parathyroid hormone (PTH) increases, and PTH mobilizes Ca^{++} from bone by stimulating resorption of bone. In addition to this action, glucocorticoids directly inhibit **osteoblast bone-forming functions** (see Chapter 40). Although glucocorticoids are useful for treating the inflammation associated with **arthritis,** excessive use will result in bone loss **(osteoporosis).**

Actions of Cortisol on Connective Tissue

Cortisol inhibits **fibroblast proliferation** and **collagen formation.** In the presence of excessive amounts of cortisol, the **skin** thins and is more readily damaged. The connective tissue support of capillaries is impaired, and **capillary injury,** or **bruising,** is increased.

Actions of Cortisol on the Kidney

Cortisol inhibits the secretion and action of **antidiuretic hormone (ADH),** and thus it is an ADH antagonist. In the absence of cortisol, the action of ADH is potentiated, which makes it difficult to increase free water clearance in response to a water load and increases the likelihood of water intoxication. Although cortisol binds to the mineralocorticoid receptor with high affinity, this action is normally blocked by inactivation of cortisol to cortisone by the enzyme 11β-HSD2. However, the mineralocorticoid activity (i.e., renal Na$^+$ and H$_2$O retention, K$^+$ and H$^+$ excretion) of cortisol depends on the relative amount of cortisol (or synthetic glucocorticoids) and the activity of 11β-HSD2. Certain agents (e.g., compounds in black licorice) inhibit 11β-HSD2 and thereby increase the mineralocorticoid activity of cortisol. Cortisol increases the glomerular filtration rate by both increasing cardiac output and acting directly on the kidney.

Actions of Cortisol on Muscle

When cortisol levels are excessive, **muscle weakness and pain** are common symptoms. The weakness has multiple origins. In part it is a result of the excessive **proteolysis** cortisol produces. High cortisol levels can result in **hypokalemia** (via mineralocorticoid actions), which can produce muscle weakness because it hyperpolarizes and stabilizes the muscle cell membrane and thus makes stimulation more difficult.

Actions of Cortisol on the Gastrointestinal Tract

Cortisol exerts a trophic effect on the **GI mucosa.** In the absence of cortisol, GI motility decreases, GI mucosa degenerates, and GI acid and enzyme production decreases. Because cortisol stimulates **appetite,** hypercortisolism is frequently associated with weight gain. The cortisol-mediated stimulation of gastric acid and pepsin secretion increases the risk for development of **ulcers.**

Psychological Effects of Cortisol

Psychiatric disturbances are associated with either excessive or deficient levels of corticosteroids. Excessive corticosteroids can initially produce a feeling of well-being, but continued excessive exposure eventually leads to emotional lability and depression. Frank psychosis can occur with either excessive or deficient hormone. Cortisol increases the tendency for insomnia and decreases rapid eye movement (REM) sleep. People who are deficient in corticosteroids tend to be depressed, apathetic, and irritable.

Effects of Cortisol During Fetal Development

Cortisol is required for normal development of the **CNS, retina, skin, GI tract, and lungs.** The best studied system is the lungs, in which cortisol induces differentiation and maturation of type II alveolar cells. During late gestation these cells produce **surfactant,** which reduces surface tension in the lungs and thus allows the onset of breathing at birth.

Regulation of Cortisol Production

Cortisol production by the zona fasciculata is regulated by a standard hypothalamic-pituitary-adrenal axis involving **corticotropin-releasing hormone (CRH), ACTH, and cortisol** (see Chapter 41). The hypothalamus and pituitary stimulate cortisol production, and cortisol negatively feeds back on the hypothalamus and pituitary to maintain its set point. Both **neurogenic** (e.g., fear) and **systemic** (e.g., hypoglycemia, hemorrhage, cytokines) **forms of stress** stimulate release of CRH. CRH is also under strong **diurnal rhythmic** regulation emerging from the suprachiasmatic nucleus, such that cortisol levels surge during the early predawn and morning hours and then continually decline throughout the day and evening. CRH acutely stimulates release of ACTH and chronically increases proopiomelanocortin (POMC) gene expression and corticotrope hypertrophy and proliferation. Some parvicellular neurons coexpress CRH and ADH, which potentiates the actions of CRH.

ACTH binds to the **melanocortin 2 receptor (MC2R)** located on cells in the zona fasciculata (Fig. 43.11). The effects of ACTH can be subdivided into three phases:

1. The **acute effects** of ACTH occur within minutes. Cholesterol is rapidly mobilized from lipid droplets by posttranslational activation of cholesterol ester hydrolase and transported to the outer mitochondrial membrane. ACTH both rapidly increases StAR protein gene expression and activates StAR protein through protein kinase A (PKA)-dependent phosphorylation. Collectively, these acute actions of ACTH increase pregnenolone levels.

2. The **chronic effects** of ACTH occur over a period of several hours. These effects involve increasing transcription of the genes encoding the steroidogenic enzymes and their coenzymes. ACTH also increases expression of the LDL receptor and scavenger receptor BI (SR-BI; the HDL receptor).

3. The **trophic actions** of ACTH on the zona fasciculata and zona reticularis occur over a period of weeks and

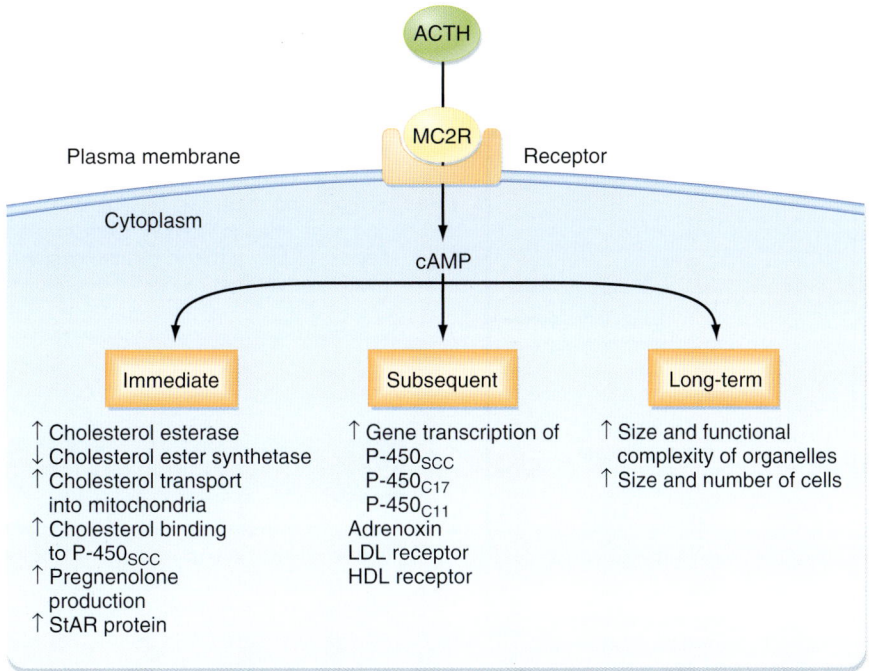

• **Fig. 43.11** Overview of the actions of ACTH on target adrenocortical cells. Note that the major second messenger, cAMP, activates immediate protein mediators and also induces production of later protein mediators. HDL, high-density lipoprotein; LDL, low-density lipoprotein.

months. This effect is exemplified by atrophy of the zona fasciculata in patients receiving therapeutic (i.e., supraphysiological) levels of glucocorticoid analogues for at least 3 weeks. Under these conditions the exogenous corticosteroids completely repress CRH and ACTH production, thereby resulting in atrophy of the zona fasciculata and a decline in endogenous cortisol production (Fig. 43.12). At the end of therapy, these patients need to be slowly weaned off exogenous glucocorticoids to allow the hypothalamic-pituitary-adrenal axis to reestablish itself and the zona fasciculata to enlarge and produce adequate amounts of cortisol.

Cortisol inhibits both *POMC* gene expression at the corticotropes and pro-CRH gene expression at the hypothalamus. However, intense stress can override the negative-feedback effects of cortisol at the hypothalamus and reset the "set point" at a higher level.

Zona Reticularis

The innermost zone, the zona reticularis, begins to appear after birth at about 5 years of age. Adrenal androgens, especially DHEAS, the main product of the zona reticularis, become detectable in the circulation at about 6 years of age. This onset of adrenal androgen production is called **adrenarche,** and it contributes to the appearance of axillary and pubic hair at about age 8. DHEAS levels continue to increase, peak during the mid-20s, and then progressively decline with age.

Androgen Synthesis by the Zona Reticularis

The zona reticularis differs from the zona fasciculata in several important ways with respect to steroidogenic enzyme activity (see Fig. 43.8). First, 3β-HSD is expressed at very low levels in the zona reticularis; thus the Δ5 pathway predominates in the zona reticularis. Second, the zona reticularis expresses cofactors or conditions that enhance the 17,20-lyase function of CYP17, thereby generating the 19-carbon androgen precursor molecule **dehydroepiandrosterone (DHEA)** from 17-hydroxypregnenolone. Additionally the zona reticularis expresses **DHEA sulfotransferase (*SULT2A1* gene),** which converts DHEA into **DHEAS** (Fig. 43.13). A limited amount of the Δ4 androgen **androstenedione** is also made in the zona reticularis. Although small amounts of potent androgens (e.g., testosterone) or 18-carbon estrogens are normally produced by the human adrenal cortex, most active sex steroids are produced primarily from **peripheral conversion** of DHEAS and androstenedione.

Metabolism and Fate of DHEAS and DHEA

DHEAS can be converted back to DHEA by peripheral **sulfatases,** and DHEA and androstenedione can be converted to active androgens (testosterone, dihydrotestosterone) peripherally in both sexes. DHEA binds to albumin and other globulins in blood with low affinity, so it is excreted efficiently by the kidney. The half-life of DHEA is 15 to 30 minutes. In contrast, DHEAS binds to albumin with very high affinity and has a half-life of 7 to 10 hours.

Normal HPA axis

Quiescent HPA axis in patient undergoing long-term (>3 wk) treatment with glucocorticoid

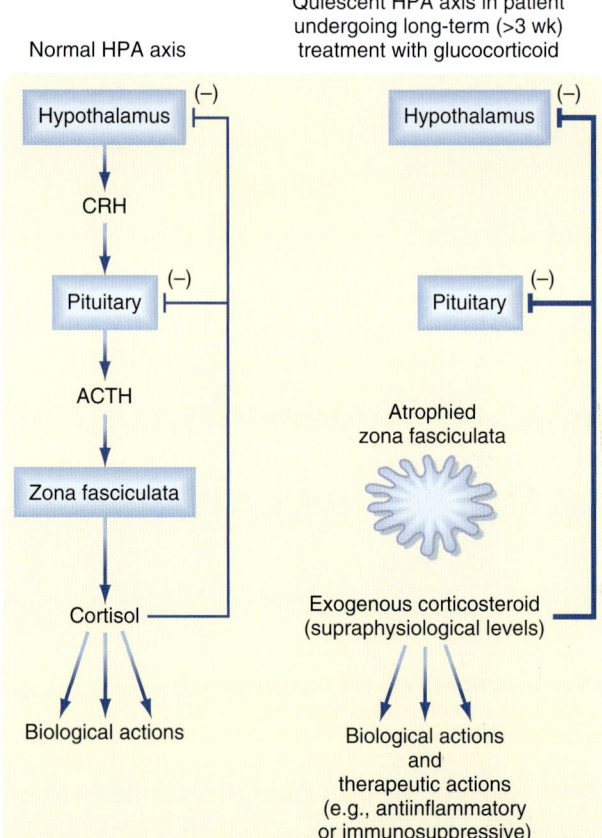

• **Fig. 43.12** Comparison of a normal hypothalamic-pituitary-adrenal (HPA) axis to a quiescent HPA axis in individual receiving exogenous glucocorticoid therapy. The latter causes the zona fasciculata to atrophy after 3 weeks, thus requiring a careful withdrawal regimen to allow rebuilding of the adrenal tissue before total cessation of exogenous corticosteroid administration. (Modified from White BA, Porterfield SP. *Endocrine and Reproductive Physiology.* 4th ed. Philadelphia: Mosby; 2013.)

Physiological Actions of Adrenal Androgens

In men the contribution of adrenal androgens to active androgens is negligible. However, in women the adrenal contributes to about 50% of circulating active androgens, which are required for the growth of axillary and pubic hair and for libido.

Apart from providing androgen precursors, it is not clear what other role or roles if any the zona reticularis plays in adult humans. DHEAS is the most abundant circulating hormone in young adults. It increases steadily until it peaks in the mid-20s and then steadily declines thereafter. Thus there has been considerable interest in the possible role of DHEAS in the aging process. However, the function of this abundant steroid in young adults and the potential impact of its gradual disappearance on aging are still poorly understood. It should be noted that the age-related decline in DHEA and DHEAS has led to the popular use of these steroids as dietary supplements, even though recent studies indicate no beneficial effects.

IN THE CLINIC

During adrenal androgen excess (e.g., adrenal tumor, Cushing's syndrome, congenital adrenal hyperplasia), **masculinization of women** can occur. This involves masculinization of the external genitalia (e.g., enlarged clitoris) in utero and excessive facial and body hair (called **hirsutism**) and acne in adult women. Excessive adrenal androgens also appear to play a role in ovarian dysovulation (i.e., polycystic ovarian syndrome).

IN THE CLINIC

A crucial clinical aspect of regulation of the zona reticularis is that neither adrenal androgens nor their more potent metabolites (e.g., testosterone, dihydrotestosterone, estradiol-17β) negatively feed back on ACTH or CRH (Fig. 43.14). This means that an enzymatic defect associated with the synthesis of cortisol (e.g., CYP21B deficiency) is associated with a dramatic increase in both ACTH (no negative feedback from cortisol) and adrenal androgens (because of the elevated ACTH). It is this "loophole" in the hypothalamic-pituitary-adrenal axis that gives rise to **congenital adrenal hyperplasia (CAH).**

Regulation of Zona Reticularis Function

ACTH is the primary regulator of the zona reticularis. Both DHEA and androstenedione display the same diurnal rhythm as cortisol (DHEAS does not because of its long circulating half-life). Moreover, the zona reticularis shows the same atrophic changes as the zona fasciculata in conditions typified by little or no ACTH. However, other factors must regulate adrenal androgen function. Adrenarche occurs in the face of constant ACTH and cortisol levels, and the rise and decline of DHEAS is not associated with a similar pattern of ACTH or cortisol production. However, the other factors, whether extraadrenal or intraadrenal, remain unknown.

Zona Glomerulosa

The thin outermost zone of the adrenal, the zona glomerulosa, produces the mineralocorticoid aldosterone, which regulates salt and volume homeostasis (see Chapters 35 and 36). The zona glomerulosa is minimally influenced by ACTH. Rather it is regulated primarily by the renin-angiotensin system, plasma [K+], and atrial natriuretic peptide (ANP).

An important feature in the steroidogenic capacity of the zona glomerulosa is that it does not express CYP17. Therefore zona glomerulosa cells never make cortisol, nor do they make adrenal androgens in any form. Pregnenolone is converted to progesterone and DOC by 3β-HSD and

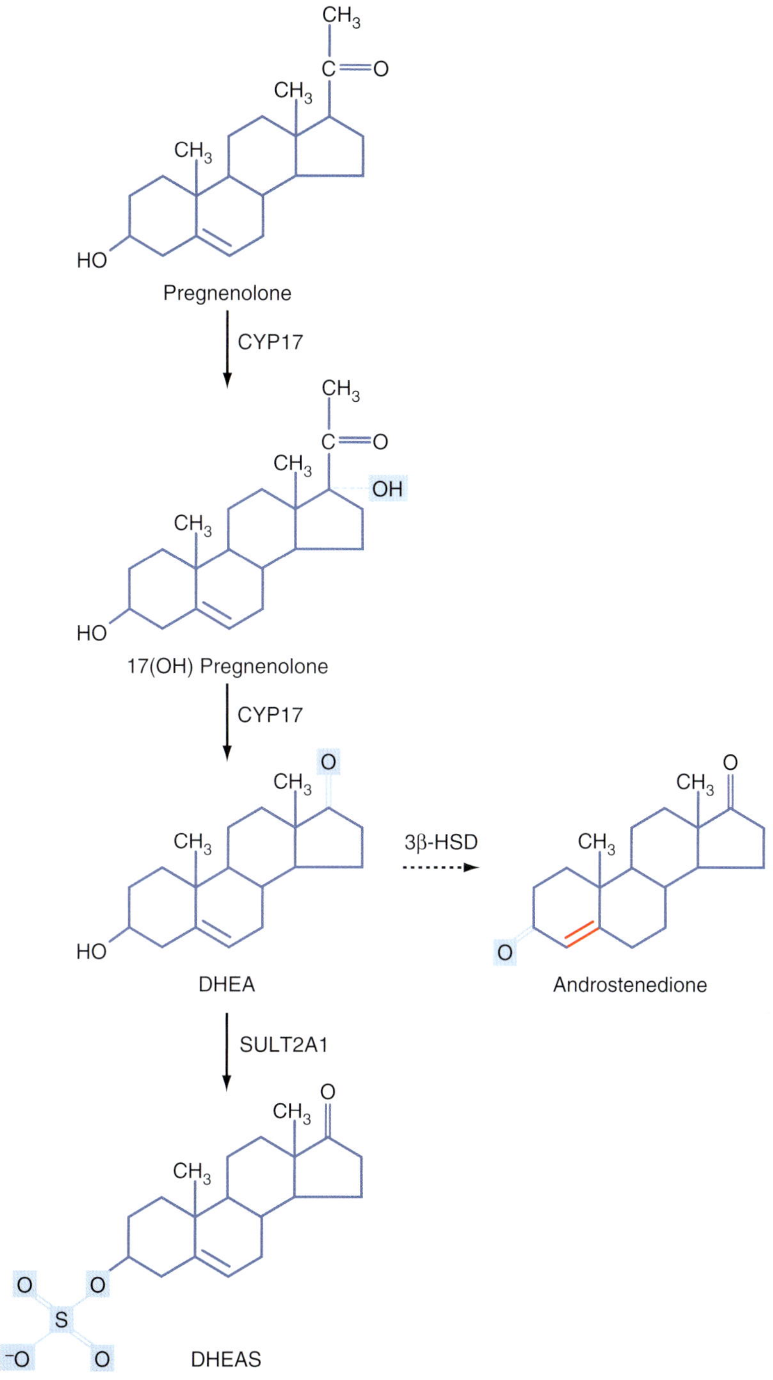

• **Fig. 43.13** Steroidogenic pathways in the zona reticularis. The first common reaction in the pathway, conversion of cholesterol to pregnenolone by CYP11A1, is not shown. Expression of 3β-hydroxysteroid dehydrogenase (3β-HSD) is relatively low in the zona reticularis, so androstenedione is a minor product in comparison to DHEA and DHEAS. The zona reticularis also makes a small amount of testosterone and estrogens (not shown). (Modified from White BA, Porterfield SP. *Endocrine and Reproductive Physiology.* 4th ed. Philadelphia: Mosby; 2013.)

CYP21, respectively (Fig. 43.15). A completely unique feature of the zona glomerulosa among the steroidogenic glands is its expression of CYP11B2, which is regulated by different signaling pathways. Furthermore the enzyme coded by CYP11B2, **aldosterone synthase,** catalyzes the last three reactions from DOC to aldosterone within the zona glomerulosa. These reactions are 11-hydroxylation of DOC to form corticosterone, 18-hydroxylation to form 18-hydroxycorticosterone, and 18-oxidation to form aldosterone (see Figs. 43.8 and 43.15).

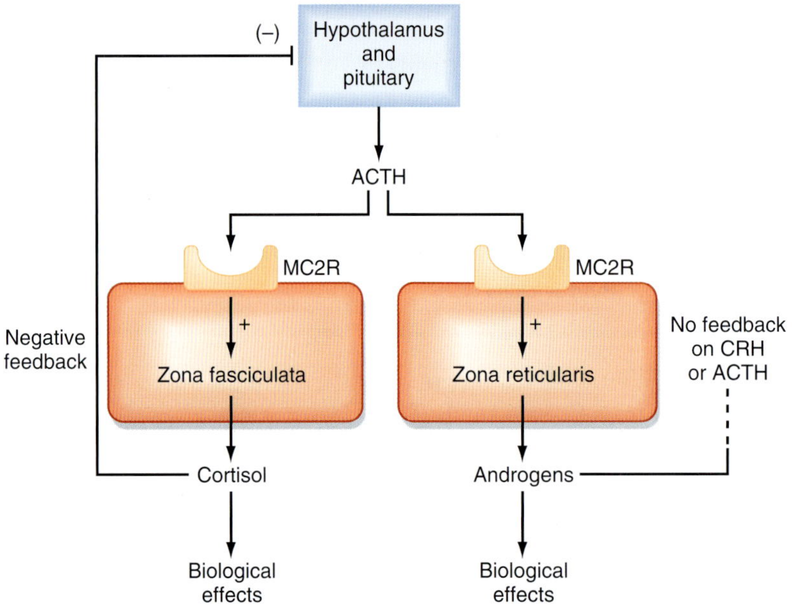

• **Fig. 43.14** The "loophole" in the hypothalamic-pituitary-adrenal axis. ACTH stimulates production of both cortisol and adrenal androgens, but only cortisol negatively feeds back on ACTH and CRH. Thus if cortisol production is blocked (i.e., CYP11B1 deficiency), ACTH levels increase along with adrenal androgens. (Modified from White BA, Porterfield SP. *Endocrine and Reproductive Physiology.* 4th ed. Philadelphia: Mosby; 2013.)

AT THE CELLULAR LEVEL

CYP11B1 (expressed only in the zona fasciculata) and CYP11B2 (expressed only in the zona glomerulosa) are located on chromosome 8 in humans, display 95% similarity, and are separated from each other by only about 50 kilobases. This increases the possibility of uneven crossing over during gametogenesis, with the formation of hybrid genes. In one case the promoter region and 5′ end of the *CYP11B1* gene is fused to the 3′ end of the *CYP11B2* gene. This arrangement leads to aldosterone synthase being expressed in the zonae fasciculata and reticularis under the control of ACTH. Because aldosterone is no longer under feedback control by the renin-angiotensin system (see Chapter 35), aldosterone levels are high and hypertension ensues. This form of primary aldosteronism is called **glucocorticoid-remediable aldosteronism,** and it is inherited in an autosomal dominant manner. This disease can be confirmed by the polymerase chain reaction technique and by measurement of 18-hydroxycortisol and 18-oxicortisol in a 24-hour urine sample. The disease is treated by the **administration of glucocorticoid,** which suppresses ACTH and thus expression of the hybrid gene.

Transport and Metabolism of Aldosterone

Aldosterone binds to albumin and corticosteroid-binding protein in blood with low affinity and therefore has a biological half-life of about 20 minutes. Almost all aldosterone is inactivated by the liver in one pass, conjugated to a glucuronide group, and excreted by the kidney.

Mechanism of Aldosterone Action

Aldosterone acts much like cortisol (and other steroid hormones) in that its primary mechanism of action is mediated by binding to a specific intracellular receptor (i.e., **mineralocorticoid receptor [MR]**). After dissociation of chaperone proteins, nuclear translocation, dimerization, and binding to the mineralocorticoid response element (MRE), the aldosterone-MR complex regulates expression of specific genes (see Chapter 3). As discussed earlier, cortisol binds to the MR with significant affinity. However, as also discussed, cells that express MR also express 11β-HSD2, which converts cortisol to the inactive steroid cortisone (Fig. 43.16). Cortisone can be converted back to cortisol by 11β-HSD1, which is expressed in several glucocorticoid-responsive tissues, including the liver and skin.

IN THE CLINIC

Clinical studies in humans have revealed a deleterious effect of aldosterone on cardiovascular function independent of its effects on renal sodium and water reabsorption. Aldosterone has a **proinflammatory, profibrotic effect** on the cardiovascular system and causes left ventricular hypertrophy and remodeling. This effect of aldosterone is associated with increased morbidity and mortality in patients with essential hypertension.

Physiological Actions of Aldosterone

The actions and regulation of aldosterone are discussed in Chapter 35.

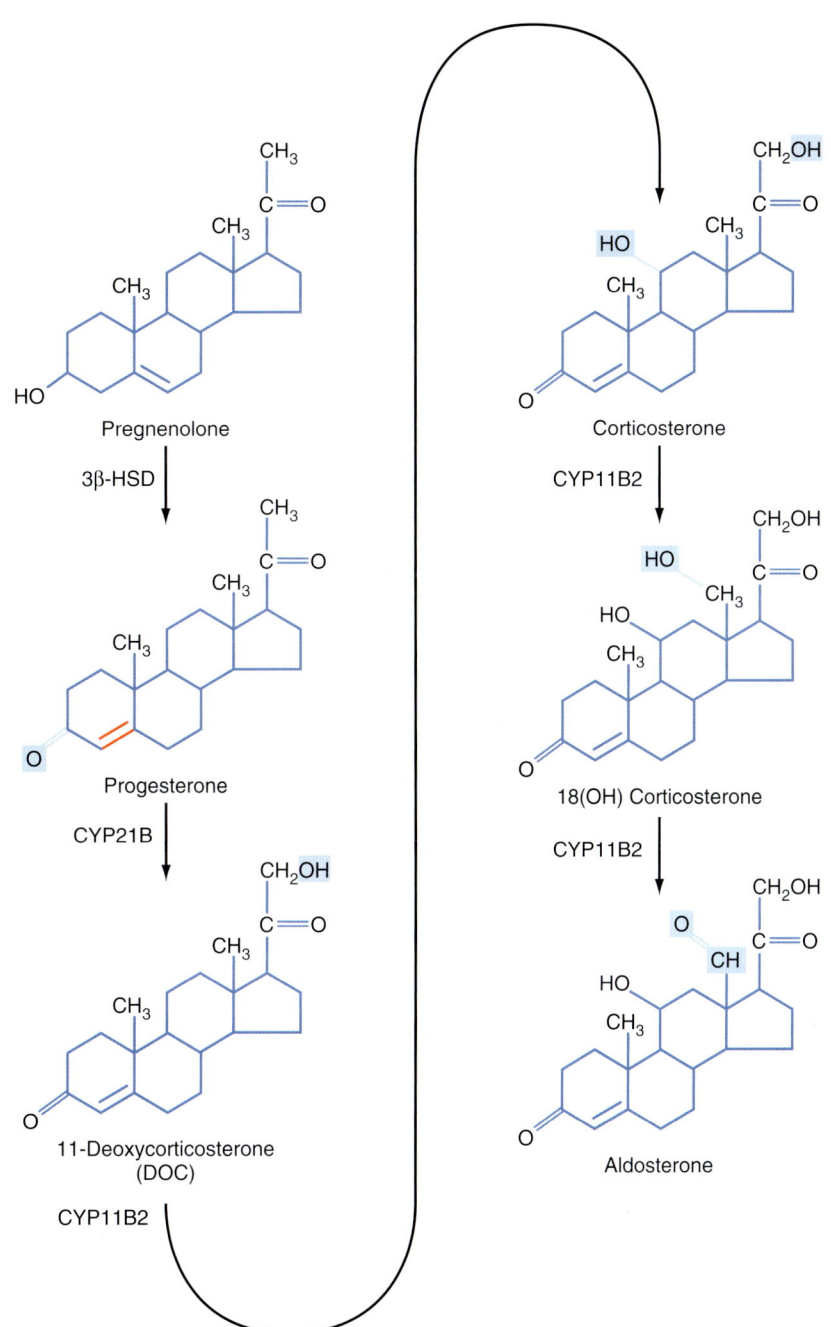

• **Fig. 43.15** Steroidogenic pathways in the zona glomerulosa. The first common reaction in the pathway, conversion of cholesterol to pregnenolone by CYP11A1, is not shown. Note that the last three reactions are catalyzed by CYP11B2. (Modified from White BA, Porterfield SP. Endocrine and Reproductive Physiology. 4th ed. Philadelphia: Mosby; 2013.)

⚕ IN THE CLINIC

Addison's disease is defined by **primary adrenal insufficiency,** with both mineralocorticoids and glucocorticoids usually being deficient. In North America and Europe, the most prevalent cause of Addison's disease is **autoimmune destruction of the adrenal cortex.** Because of the cortisol deficiency, ACTH secretion increases. Elevated levels of ACTH can compete for MC1R in melanocytes and cause an increase in skin pigmentation, particularly in skin creases, scars, and gums (see Fig. 43.14). The loss of mineralocorticoids results in contraction of extracellular volume, which produces circulatory hypovolemia and therefore a drop in blood pressure. Because loss of cortisol decreases the vasopressive response to catecholamines, peripheral vascular resistance drops, thereby facilitating the development of hypotension. Individuals with Addison's disease are also prone to hypoglycemia when stressed or fasting, and water

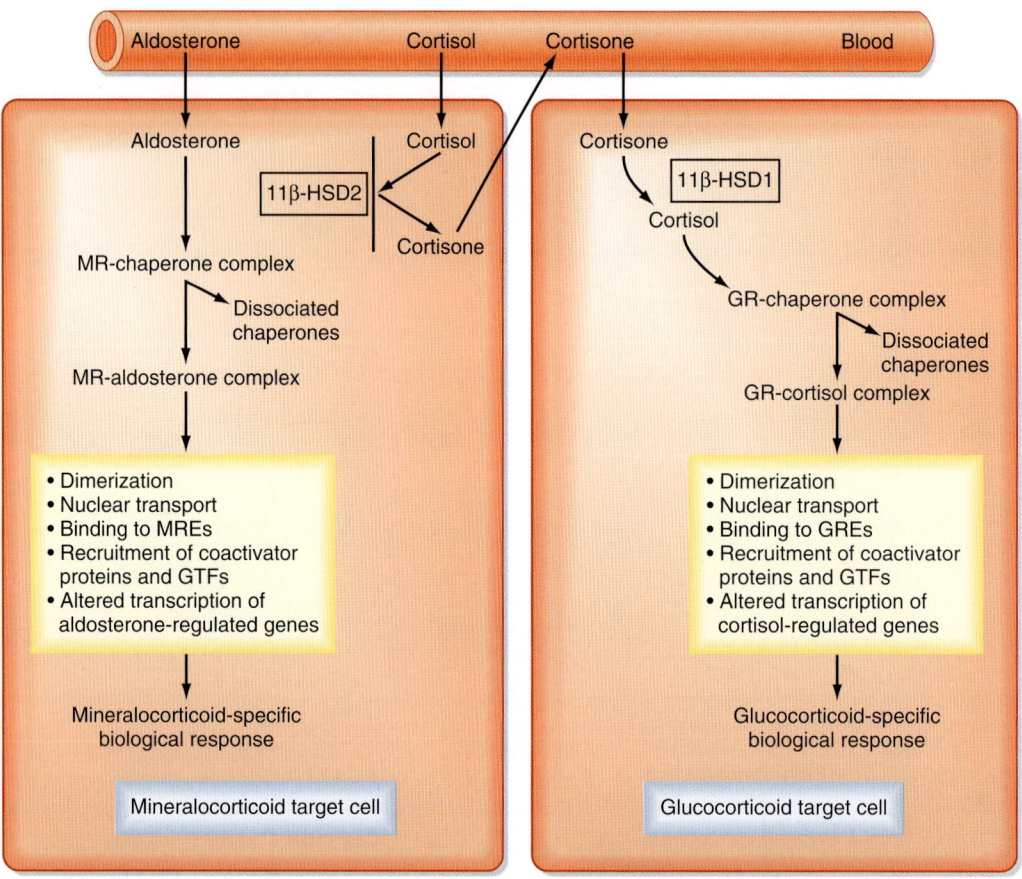

• **Fig. 43.16** The mineralocorticoid receptor (MR) is protected from activation by cortisol by the enzyme 11β-hydroxysteroid dehydrogenase type 2 (11β-HSD2), which converts cortisol to inactive cortisone. Cortisone can be converted back to cortisol in glucocorticoid target cells by the enzyme 11β-HSD type 1. GRE, glucocorticoid response element; GTF, general transcription factors; MRE, mineralocorticoid response element. (Modified from White BA, Porterfield SP. *Endocrine and Reproductive Physiology.* 4th ed. Philadelphia: Mosby; 2013.)

intoxication can develop if excess water is ingested. Because cortisol is important for muscle function, muscle weakness also occurs in cortisol deficiency. Loss of cortisol results in anemia, decreased GI motility and secretion, and reduced iron and vitamin B_{12} absorption. Appetite decreases with cortisol deficiency, and this decreased appetite coupled with the GI dysfunction predisposes these individuals to weight loss. These patients often have disturbances in mood and behavior and are more susceptible to depression.

Adrenocortical hormone excess is termed **Cushing's syndrome. Pharmacological** use of exogenous corticosteroids is now the most common cause of Cushing's syndrome. The next most prevalent cause is **ACTH-secreting tumors.** The form of Cushing's syndrome caused by a functional pituitary adenoma is called *Cushing's disease.* The fourth most common cause of Cushing's syndrome is **primary hypercortisolism** resulting from a functional adrenal tumor. If the disorder is primary or if it is a result of corticosteroid treatment, secretion of ACTH will be suppressed and increased skin pigmentation will not occur. However, if hypersecretion of the adrenal is the result of an ACTH-secreting nonpituitary tumor, ACTH levels sometimes become high enough to increase skin pigmentation.

Increased cortisol secretion causes weight gain with a characteristic centripetal fat distribution and a "buffalo hump."

The face appears rounded (fat deposition), and the cheeks may be reddened, in part because of the polycythemia. The limbs are thin owing to skeletal muscle wasting (from increased proteolysis), and muscle weakness is evident (from muscle proteolysis and hypokalemia). Proximal muscle weakness is apparent, so the patient may have difficulty climbing stairs or rising from a sitting position. The abdominal fat accumulation coupled with atrophy of the abdominal muscles and thinning of the skin produces a large protruding abdomen. Purple abdominal striae are seen as a result of damage to the skin by the prolonged proteolysis, increased intraabdominal fat, and loss of abdominal muscle tone. Capillary fragility occurs because of damage to the connective tissue supporting the capillaries. Patients are likely to show signs of osteoporosis and poor wound healing. They have metabolic disturbances that include glucose intolerance, hyperglycemia, and insulin resistance (see Fig. 43.10). Prolonged hypercortisolism can lead to manifestations of diabetes mellitus. Because of suppression of the immune system caused by glucocorticoids, patients are more susceptible to infection. The mineralocorticoid activities of glucocorticoids and the possible increase in aldosterone secretion produce salt retention and subsequent water retention that result in hypertension. Excessive androgen secretion in women can produce hirsutism, male pattern baldness, and clitoral enlargement (adrenogenital syndrome).

IN THE CLINIC

Any enzyme blockage that decreases cortisol synthesis will increase ACTH secretion and produce adrenal hyperplasia. The most common form of congenital adrenal hyperplasia is due to deficiency of the enzyme **21-hydroxylase (CYP21).** These individuals cannot produce normal quantities of cortisol, **deoxycortisol,** DOC, corticosterone, or aldosterone (see Figs. 43.8 and 43.10C). Because of impaired cortisol production and resultant elevated ACTH levels, steroidogenesis is stimulated, thereby increasing the synthesis products "upstream" of the missing enzyme, as well as products of the zona reticularis. Because the latter include the adrenal androgens, a female

fetus will be masculinized. Because they are unable to produce the mineralocorticoids, aldosterone, DOC, and corticosterone, patients with this disorder have difficulty retaining salt and maintaining extracellular volume. Consequently they are likely to be hypotensive. If the blockage is at the next step, **11β-hydroxylase** (CYP11B1), **DOC** will be formed and levels of DOC will accumulate (see Figs. 43.8 and 43.9C). Because DOC has significant **mineralocorticoid activity** and its levels become high, these individuals tend to retain salt and water and become hypertensive.

Key Concepts

1. The adrenal gland is composed of a cortex that is of mesodermal origin and a medulla that is of neuroectodermal origin. The cortex produces steroid hormones, and the medulla produces catecholamines.

2. The rate-limiting enzymes in medullary catecholamine synthesis are tyrosine hydroxylase and dopamine β-hydroxylase, which are induced by sympathetic stimulation, and phenylethanolamine-*N*-methyltransferase, which is induced by cortisol.

3. Catecholamines increase serum glucose and fatty acid levels. They stimulate gluconeogenesis, glycogenolysis, and lipolysis. Catecholamines increase cardiac output but have selective effects on blood flow to different organs.

4. Pheochromocytoma is a tumor of chromaffin tissue that produces excessive quantities of catecholamines. Symptoms of pheochromocytoma are often sporadic and include hypertension, headaches, sweating, anxiety, palpitations, chest pain, and orthostatic hypotension.

5. The adrenal cortex displays clear structural and functional zonation: the zona glomerulosa produces the mineralocorticoid aldosterone, the zona fasciculata produces the glucocorticoid cortisol, and the zona reticularis produces the weak androgens DHEA and DHEAS.

6. Cortisol binds to the glucocorticoid receptor. During stress, cortisol increases blood glucose by increasing gluconeogenesis in the liver and breaking muscle protein down to supply gluconeogenic precursors. Cortisol also decreases glucose uptake by muscle and adipose tissue and has permissive actions on glucagon and catecholamines. Cortisol has multiple effects on other tissue. From a pharmacological point of view, the most important is the immunosuppressive/antiinflammatory effect.

7. Cortisol is regulated by the CRH-ACTH-cortisol axis. Cortisol negatively feeds back at the hypothalamus on both CRH-producing neurons and pituitary corticotropes. CRH is regulated by several

forms of stress, including proinflammatory cytokines, hypoglycemia, neurogenic stress, and hemorrhage, and by diurnal input.

8. The adrenal androgens DHEA, DHEAS, and androstenedione are androgen precursors. They can be converted to active androgens peripherally and provide about 50% of circulating androgens in women. In men the role of adrenal androgens if any remains obscure. In women, adrenal androgens promote pubic and axillary hair growth and libido. Excessive adrenal androgens in women can lead to various degrees of virilization and ovarian dysfunction.

9. The zona glomerulosa of the adrenal cortex is the site of aldosterone production. Aldosterone is the strongest naturally occurring mineralocorticoid in humans. It promotes Na^+ and water reabsorption by the distal tubule and collecting duct while promoting renal K^+ and H^+ secretion. Aldosterone promotes Na^+ and water absorption in the colon and salivary glands. It also has a proinflammatory, profibrotic effect on the cardiovascular system and causes left ventricular hypertrophy and remodeling.

10. Major actions of angiotensin II on the adrenal cortex are increased growth and vascularity of the zona glomerulosa, increased StAR and CYP11B2 enzyme activity, and increased aldosterone synthesis.

11. Major stimuli for aldosterone production are a rise in angiotensin II and a rise in serum $[K^+]$. The major inhibitory signal is ANP.

12. Addison's disease is adrenocortical insufficiency. Common symptoms include hypotension, hyperpigmentation, muscle weakness, anorexia, hypoglycemia, and hyperkalemic acidosis.

13. Cushing's syndrome results from hypercortisolemia. If the basis of the disorder is increased pituitary adrenocorticotropin secretion, the disorder is called *Cushing's disease.* Common symptoms of Cushing's syndrome include centripetal fat distribution, muscle wasting, proximal muscle weakness, thin skin with

abdominal striae, capillary fragility, insulin resistance, and polycythemia.

14. Congenital adrenal hyperplasia is caused by a congenital enzyme deficiency that blocks production of cortisol. The enzyme blockage results in elevated ACTH secretion, which stimulates adrenal cortical growth and secretion of precursors produced before the block. 21-Hydroxylase (CYP21B) deficiency is the most common form.

Additional Reading

De Bosscher K, et al. Activation of the glucocorticoid receptor in acute inflammation: the SEDIGRAM concept. *Trends Pharmacol Sci.* 2016;37:4-16.

Guerineau NC, et al. Functional chromaffin cell plasticity in response to stress: focus on nicotinic, gap junction, and voltage-gated Ca^{2+} channels. *J Mol Neurosci.* 2012;48:368-386.

Namsolleck P, Unger T. Aldosterone synthase inhibitors in cardiovascular and renal diseases. *Nephrol Dial Transplant.* 2014;29(suppl 1): i62-i68.

Rafacho A, et al. Glucocorticoid treatment and endocrine pancreas function: implications for glucose homeostasis, insulin resistance and diabetes. *J Endocrinol.* 2014;223:R49-R62.

Sundahl N, et al. Selective glucocorticoid receptor modulation: new directions with non-steroidal scaffolds. *Pharmacol Ther.* 2015;152:28-41.

Turcu AF, Auchus RJ. Adrenal steroidogenesis and congenital adrenal hyperplasia. *Endocrinol Metab Clin North Am.* 2015;44: 275-296.

Whittier X, Saag KG. Glucocorticoid-induced osteoporosis. *Rheum Dis Clin North Am.* 2016;42:177-189.

44

The Male and Female Reproductive Systems

LEARNING OBJECTIVES

Upon completion of this chapter the student should be able to answer the following questions:

1. Describe the general anatomical components of the male and female reproductive system.
2. Map out the organization of the testis, with the Sertoli cells and developing sperm cells within the intralobular compartment, and the Leydig cells and capillary plexus within the interlobular/interstitial compartment.
3. Describe the processes of spermatogenesis and spermiogenesis.
4. List the functions of the Sertoli cell.
5. Diagram the process of testosterone synthesis within the Leydig cells, and the peripheral conversion of testosterone to estradiol or dihydrotestosterone.
6. Diagram the male hypothalamus/pituitary/testis axis, including all cell types and hormones involved.
7. Map out the organization of the ovary, and describe the various stages of follicular development, ovulation and corpus luteum formation.
8. List the stages and control of female germ cell progression from oogonia to egg.
9. Map out the steroidogenic pathways in the corresponding cell types that lead to androgen, estrogen and progesterone synthesis.
10. Diagram the female hypothalamus/pituitary/ovarian axis during the menstrual cycle, including all cell types and hormones involved.
11. Explain changes in the female tract, with emphasis on the uterine endometrium, during the menstrual cycle.
12. List the events involved in fertilization.
13. Describe the development and function of the placenta.
14. Describe the development and function of the mammary glands

The two most basic components of the reproductive system are the **gonads** and the **reproductive tract.** The gonads (**testes** and **ovaries**) perform an **endocrine function** that is regulated within a **hypothalamic-pituitary-gonadal axis.** The gonads are distinct from other endocrine glands in that they also perform **gametogenesis.** The reproductive tract is involved in several aspects of gamete development, function, and transport and in women allows fertilization, implantation, and gestation. Normal gametogenesis in the gonads and development and physiology of the reproductive tract are absolutely dependent on the endocrine function of the gonads. The clinical ramifications of this hormonal dependence include infertility in the face of low sex hormone production, ambiguous genitalia in dysregulated hormone or receptor expression, and hormone-responsive cancers, especially uterine and breast cancer in women and prostate cancer in men.

THE MALE REPRODUCTIVE SYSTEM

The male reproductive system has evolved for **continuous lifelong gametogenesis** coupled with occasional **internal insemination** with a **high density of sperm** ($>60 \times 10^6$/mL in 3–5 mL of semen). In adult men the basic roles of gonadal hormones are (1) support of gametogenesis (**spermatogenesis**), (2) maintenance of the male reproductive tract and production of semen, and (3) maintenance of secondary sex characteristics and libido. There is no overall cyclicity of this activity in men.

The Testis

Histophysiology

Unlike the ovaries, the testes reside outside the abdominal cavity in the **scrotum** (Fig. 44.1). This location maintains testicular temperature at about 2°C lower than body temperature, which is crucial for optimal sperm development. The human **testis** is covered by a connective tissue capsule and divided into about 300 **lobules** by fibrous septa (Fig. 44.2). Within each lobule are two to four loops of **seminiferous tubules.** Each loop empties into an anastomosing network of tubules called the **rete testis.** The rete testis is continuous with small ducts, the **efferent ductules,** that lead the sperm out of the testis into the head of the **epididymis** on the superior pole of the testis (see Fig. 44.2). Once in the epididymis, the sperm pass from the **head,** to the **body,** to the **tail** of the epididymis and then to the **vas (ductus) deferens.** Viable **sperm** can be stored in the tail of the epididymis and vas deferens for several months.

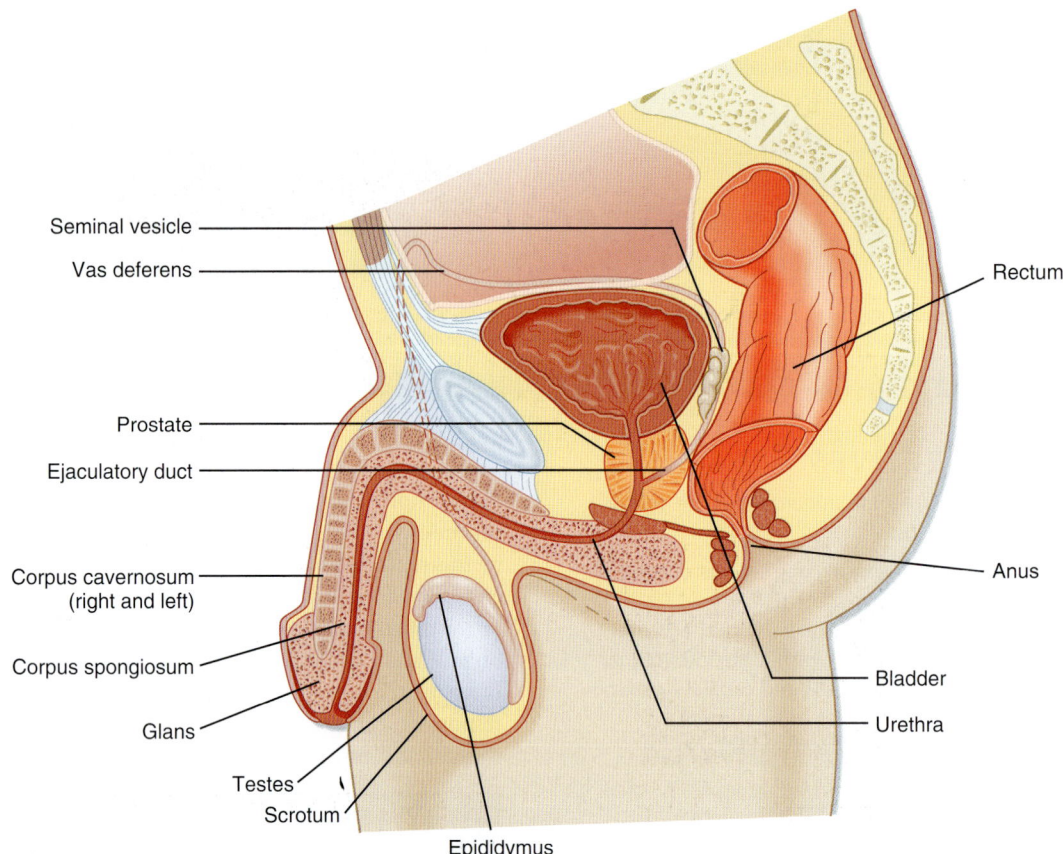

• **Fig. 44.1** Anatomy of the male reproductive system. (Modified from Drake RL et al. *Gray's Anatomy for Students.* Philadelphia: Churchill Livingstone; 2005.)

The presence of the seminiferous tubules creates two compartments within each lobule: an intratubular compartment, which is composed of the avascular **seminiferous epithelium** of the seminiferous tubule, and a peritubular compartment, which is composed of neurovascular elements, connective tissue cells, immune cells, and the **interstitial cells of Leydig,** whose main function is to produce **testosterone** (Fig. 44.3).

Intratubular Compartment

The seminiferous tubule is lined by a complex **seminiferous epithelium** composed of two cell types: **sperm cells** in various stages of **spermatogenesis** and the **Sertoli cell,** which is a "nurse cell" in intimate contact with all sperm cells (Fig. 44.4).

Developing Sperm Cells

Spermatogenesis involves the processes of **mitosis** and **meiosis.** Stem cells called **spermatogonia** reside at the basal level of the seminiferous epithelium (see Fig. 44.4, S_A and S_B). Spermatogonia divide mitotically to generate daughter spermatogonia **(spermatocytogenesis).** One or more spermatogonia remain within the stem cell population, firmly adherent to the basal lamina. However, the majority of these daughter spermatogonia enter meiotic division, which

results in haploid spermatozoa on completion of meiosis. These divisions are accompanied by **incomplete cytokinesis** such that all daughter cells remain interconnected by a cytoplasmic bridge. This configuration contributes to the synchrony of development of a clonal population of sperm cells. Spermatogonia migrate apically away from the basal lamina as they enter the first meiotic prophase. At this time they are called **primary spermatocytes** (see Fig. 44.4, S_1). During the first meiotic prophase the hallmark processes of sexual reproduction involving chromosomal reduplication, synapsis, crossing over, and homologous recombination take place. Completion of the first meiotic division gives rise to **secondary spermatocytes,** which quickly (i.e., within 20 minutes) complete the second meiotic division. The initial products of meiosis are haploid **spermatids** (see Fig. 44.4, S_3).

Spermatids are small round cells that undergo a remarkable metamorphosis called **spermiogenesis** (Fig. 44.5). The products of spermiogenesis are the streamlined **spermatozoa** see Fig. 44.4, S_4). As the spermatid matures into a spermatozoon the size of the nucleus decreases and a prominent tail is formed. The tail contains microtubular structures that propel sperm, similar to a flagellum. The chromatin material in the sperm nucleus condenses, and most of the cytoplasm is lost. The **acrosome** is a membrane-enclosed structure on the head of the sperm that acts as a lysosome

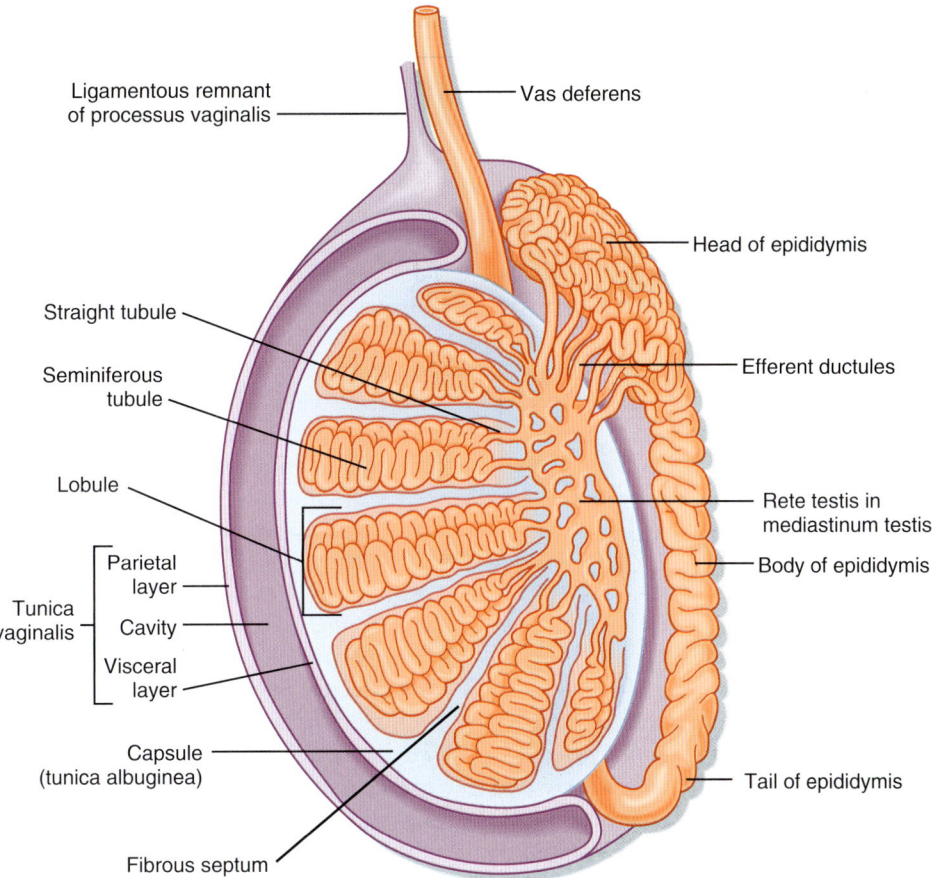

• **Fig. 44.2** Anatomy and organization of the testis. (Modified from Drake RL et al. *Gray's Anatomy for Students.* Philadelphia: Churchill Livingstone; 2005.)

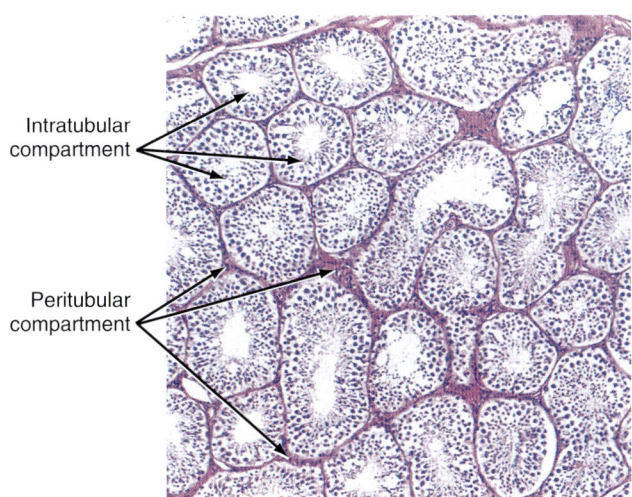

• **Fig. 44.3** Histology of a testicular lobule. (From Young B et al. *Wheater's Functional Histology. A Text and Colour Atlas.* 5th ed. London: Churchill Livingstone; 2006.)

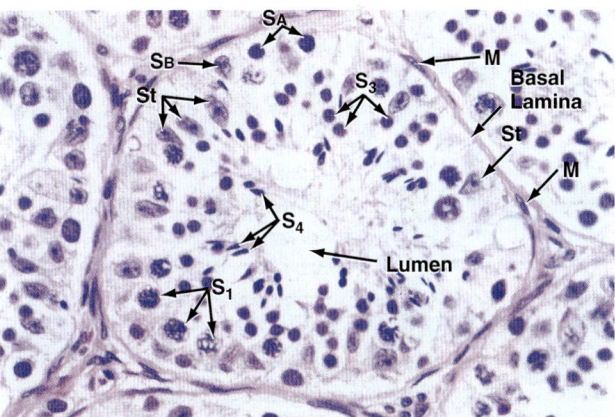

• **Fig. 44.4** Histology of a seminiferous tubule. M, myoid cell just outside the basal lamina; S_1, primary spermatocyte; S_3, spermatid; S_4, mature spermatid or spermatozoon; S_B and S_A, spermatogonia; St, Sertoli cell. (From Young B et al. *Wheater's Functional Histology. A Text and Colour Atlas.* 5th ed. London: Churchill Livingstone; 2006.)

and contains hydrolytic enzymes that are important for fertilization. These enzymes remain inactive until the acrosomal reaction occurs (see Fertilization).

Spermatozoa (see Fig. 44.4, S_4) are found at the luminal surface of the seminiferous tubule. Release of sperm, or **spermiation,** is controlled by Sertoli cells. The process of

spermatogenesis takes about 72 days. A cohort of adjacent spermatogonia enter the process every 16 days so that the process is staggered at one point along a seminiferous tubule. In addition the process is staggered along the length of a seminiferous tubule (i.e., not all spermatogonia enter the process of spermatogenesis at the same time along the

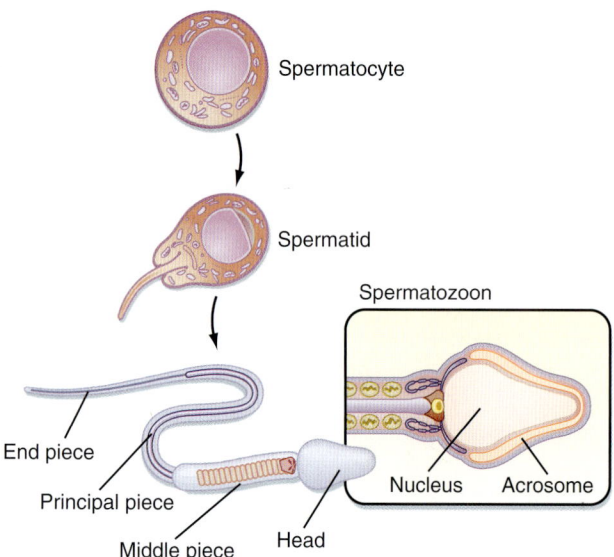

• **Fig. 44.5** Structure of sperm cells during the process of spermatogenesis and spermiogenesis.

entire length of the tubule or in synchrony with every other tubule; there are about 500 seminiferous tubules per testis; see later). Because the seminiferous tubules within one testis are about 400 m in length, spermatozoa are continually being generated at many sites within the testis at any given time.

The Sertoli Cell

Sertoli cells are the true epithelial cells of the seminiferous epithelium and extend from the basal lamina to the lumen (see Fig. 44.4, St). Sertoli cells surround sperm cells and provide structural support within the epithelium, and they form adhering and gap junctions with all stages of sperm cells. Through formation and breakdown of these junctions, Sertoli cells guide sperm cells toward the lumen as they advance to later stages of spermatogenesis. Spermiation requires the final breakdown of Sertoli–sperm cell junctions.

Another important structural feature of Sertoli cells is the formation of tight junctions between adjacent Sertoli cells (Fig. 44.6). These Sertoli-Sertoli cell occluding junctions divide the seminiferous epithelium into a **basal compartment** containing the spermatogonia and early-stage primary spermatocytes and an **adluminal compartment** containing later-stage primary spermatocytes and all subsequent stages of sperm cells. As early primary spermatocytes move apically from the basal to the adluminal compartment, the tight junctions need to be disassembled and reassembled. These tight junctions form the physical basis for the **blood-testis barrier** (see Fig. 44.6), which creates a specialized immunologically safe microenvironment for developing sperm. By blocking paracellular diffusion the tight junctions restrict movement of substances between blood and the developing germ cells through a trans–Sertoli cell transport pathway and in this manner allow the Sertoli cell to control the availability of nutrients to germ cells.

Healthy Sertoli cell function is essential for sperm cell viability and development. In addition, spermatogenesis is absolutely dependent on testosterone produced by peritubular Leydig cells (see The Leydig Cell), yet it is the Sertoli cells that express the **androgen receptor** and respond to testosterone, not the developing sperm cells. Similarly the pituitary hormone follicle-stimulating hormone (FSH) is also required for maximal sperm production, and again it is the Sertoli cell that expresses the **FSH receptor,** not the developing sperm. Thus these hormones support spermatogenesis indirectly through stimulation of Sertoli cell function.

Sertoli cells have multiple additional functions. They express the enzyme CYP19 (also called *aromatase*), which converts Leydig cell–derived testosterone to the potent estrogen estradiol-17β (see Intratesticular Androgen). This local production of estrogen may enhance spermatogenesis in humans. Sertoli cells also produce **androgen-binding protein** (ABP), which maintains a high androgen level within the adluminal compartment, the lumens of the seminiferous tubules, and the proximal part of the male reproductive tract. Sertoli cells also produce a large amount of fluid. This fluid provides an appropriate bathing medium for the sperm and assists in moving the immotile spermatozoa from the seminiferous tubule into the epididymis. Sertoli cells perform an important phagocytic function by engulfing **residual bodies,** which represent cytoplasm shed by spermatozoa during spermiogenesis.

Finally, the Sertoli cell has an important endocrine role. During development, Sertoli cells produce **antimüllerian hormone (AMH;** also called **müllerian inhibitory substance**), which induces regression of the embryonic müllerian duct that is programmed to give rise to the female reproductive tract (discussed later). The Sertoli cells also produce the hormone **inhibin.** Inhibin is a heterodimer protein hormone related to the transforming growth factor-β family. FSH stimulates inhibin production, which then negatively feeds back on gonadotropes to inhibit FSH production. Thus inhibin keeps FSH levels within a set point.

Peritubular Compartment

The peritubular compartment contains the primary endocrine cell of the testis, the **Leydig cell** (Fig. 44.7). This compartment also contains the common cell types of loose connective tissue and an extremely rich peritubular capillary network that provides nutrients to the seminiferous tubules (by way of Sertoli cells) while conveying testosterone away from the testes to the peripheral circulation.

The Leydig Cell

Leydig cells are steroidogenic stromal cells. These cells synthesize cholesterol de novo, as well as acquire it through low-density lipoprotein (LDL) receptors and high-density lipoprotein (HDL) receptors (also called *scavenger receptor BI* [SR-BI]), and store cholesterol as cholesterol esters as

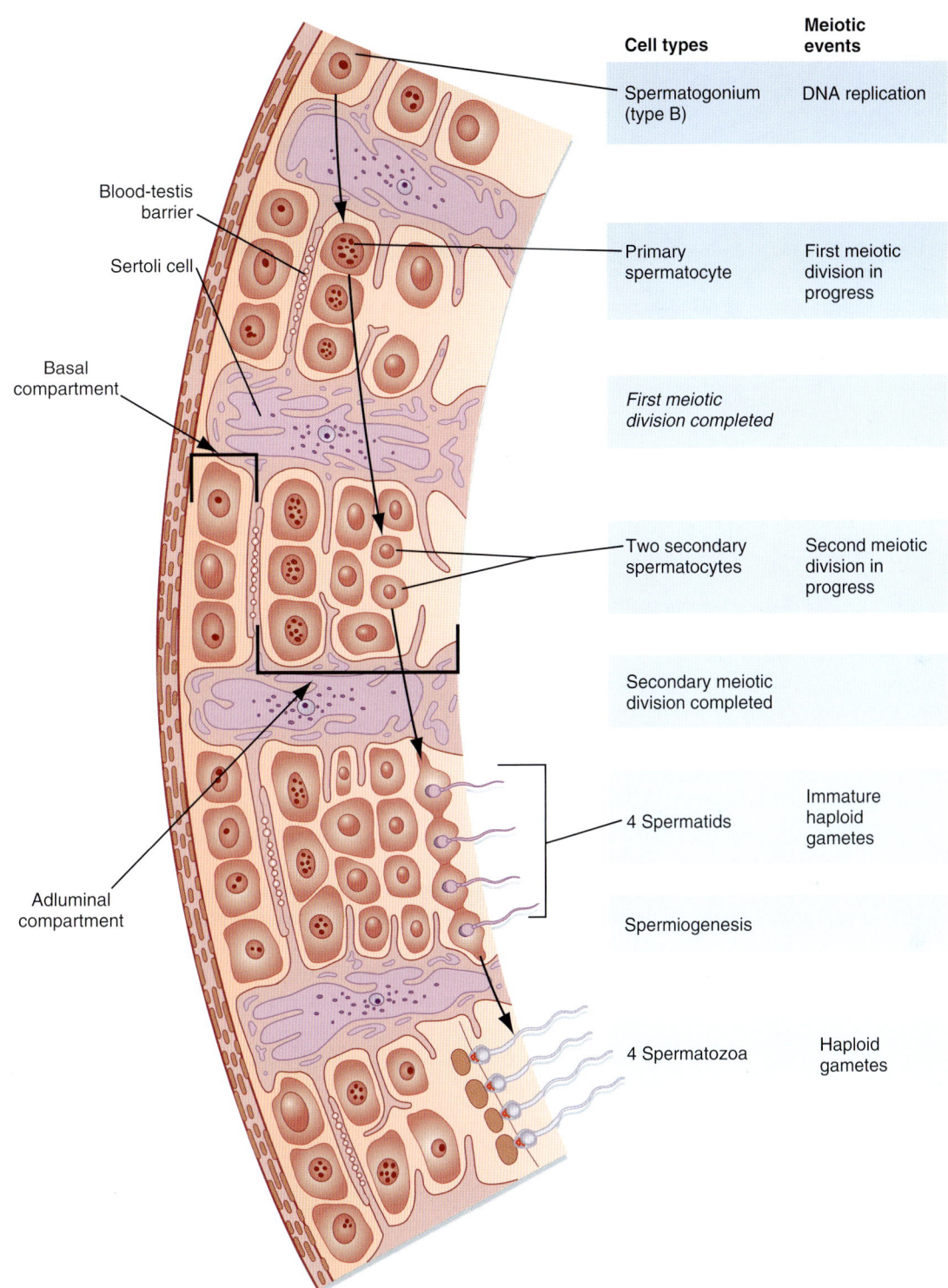

	Cell types	Meiotic events
	Spermatogonium (type B)	DNA replication
	Primary spermatocyte	First meiotic division in progress
		First meiotic division completed
	Two secondary spermatocytes	Second meiotic division in progress
		Secondary meiotic division completed
	4 Spermatids	Immature haploid gametes
		Spermiogenesis
	4 Spermatozoa	Haploid gametes

Blood-testis barrier

Sertoli cell

Basal compartment

Adluminal compartment

• **Fig. 44.6** Placement of germ cells within seminiferous tubule as they progress through spermatogenesis (From Carlson BM. *Human Embryology and Developmental Biology.* Philadelphia: Mosby; 2004.)

described for adrenocortical cells (see Chapter 43). Free cholesterol is generated by a cholesterol ester hydrolase and transferred to the outer mitochondrial membrane and then to the inner mitochondrial membrane in a **steroidogenic acute regulatory (StAR) protein**–dependent manner. As in all steroidogenic cells, cholesterol is converted to pregnenolone by CYP11A1. Pregnenolone is then processed to progesterone, 17-hydroxyprogesterone, and androstenedione by 3β-hydroxysteroid dehydrogenase (**3β-HSD**) and **CYP17** (Fig. 44.8). Recall from Chapter 43 that CYP17 is a bifunctional enzyme with **17-hydroxylase activity** and **17,20-lyase activity.** CYP17 displays a robust level of both activities in the Leydig cell. In this respect the Leydig cell is similar to the zona reticularis cell, except that

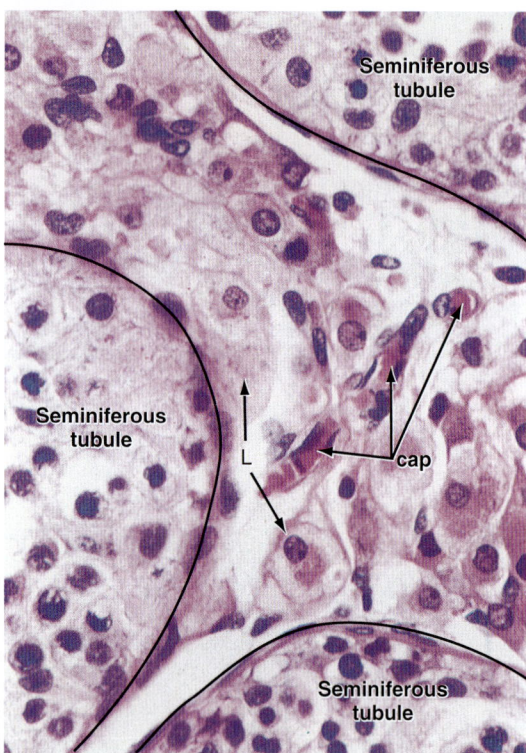

• **Fig. 44.7** Histology of the peritubular space (between three semi-niferous tubules) containing Leydig cells (L) and richly vascularized by peritubular capillaries (cap). (Modified from Young B et al. *Wheater's Functional Histology. A Text and Colour Atlas.* 5th ed. London: Churchill Livingstone; 2006.)

it expresses a higher level of 3β-HSD, so the **Δ4 pathway** is ultimately favored. Another major difference is that the Leydig cell expresses a Leydig cell–specific isoform of **17β-hydroxysteroid dehydrogenase (17β-HSD type 3),** which efficiently converts **androstenedione** to **testosterone** (see Fig. 44.8).

Fates and Actions of Androgens

Intratesticular Androgen

The testosterone produced by Leydig cells has several fates and multiple actions. Because of the proximity of Leydig cells to the seminiferous tubules, significant amounts of testosterone diffuse into the seminiferous tubules and become concentrated within the adluminal compartment by ABP (see Fig. 44.8). Testosterone levels within the seminiferous tubules that are greater than 100 times more concentrated than circulating testosterone levels are absolutely required for normal spermatogenesis. As mentioned earlier, Sertoli cells express the enzyme **CYP19 (aromatase),** which converts a small amount of testosterone into the highly potent estrogen **estradiol-17β.** Human sperm cells express at least one isoform of the **estrogen receptor,** and there is some evidence from aromatase-deficient men that this locally produced estrogen optimizes spermatogenesis in humans.

Peripheral Conversion to Estrogen

In several tissues (especially adipose tissue), testosterone is converted to estrogen (see Fig. 44.8). Studies involving men with aromatase deficiency have shown that an inability to produce estrogen results in tall stature because of the lack of epiphyseal closure in long bones, as well as osteoporosis. Thus peripheral estrogen plays an important role in bone maturation and biology in men. These studies also implicated estrogen in promoting insulin sensitivity, improving lipoprotein profiles (i.e., increasing HDL, decreasing triglycerides and LDL), and exerting negative feedback on gonadotropins at the pituitary and hypothalamus.

Peripheral Conversion to Dihydrotestosterone

Testosterone can also be converted into a potent **nonaromatizable androgen, 5αα-dihydrotestosterone (DHT),** by the enzyme **5α-reductase** (see Fig. 44.8). There are two isoforms of 5α-reductase, type 1 and type 2. Major sites of 5α-reductase 2 expression are the male urogenital tract, genital skin, hair follicles, and liver. 5α-Reductase 2 generates DHT, which is required for masculinization of the external genitalia in utero and for many of the changes associated with puberty, including growth and activity of the prostate gland (see Male Reproductive Tract), growth of the penis, darkening and folding of the scrotum, growth of pubic and axillary hair, growth of facial and body hair, and increased muscle mass (Fig. 44.9). The onset of 5α-reductase 1 expression occurs at puberty. This isozyme is expressed primarily in the skin and contributes to sebaceous gland activity and the acne associated with puberty. Because DHT has strong growth-promoting (i.e., trophic) effects on its target organs, development of **selective 5α-reductase 2 inhibitors** has benefited the treatment of prostatic hypertrophy and prostatic cancer.

Peripheral Testosterone Actions

Testosterone has a direct action (i.e., without conversion to DHT) in several cell types (see Fig. 44.9). As mentioned earlier, testosterone regulates Sertoli cell function. It induces development of the male tract from the mesonephric duct in the absence of 5α-reductase. Testosterone has several metabolic effects, including increasing very low-density lipoprotein (VLDL) and LDL while decreasing HDL, promoting deposition of abdominal adipose tissue, increasing red blood cell production, promoting bone growth and health, and exerting a protein anabolic effect on muscle. Testosterone is sufficient to maintain erectile function and libido.

Mechanism of Androgen Action

Testosterone and DHT act through the same androgen receptor (AR). The AR resides in the cytoplasm bound to chaperone proteins in the absence of ligand. Testosterone-AR

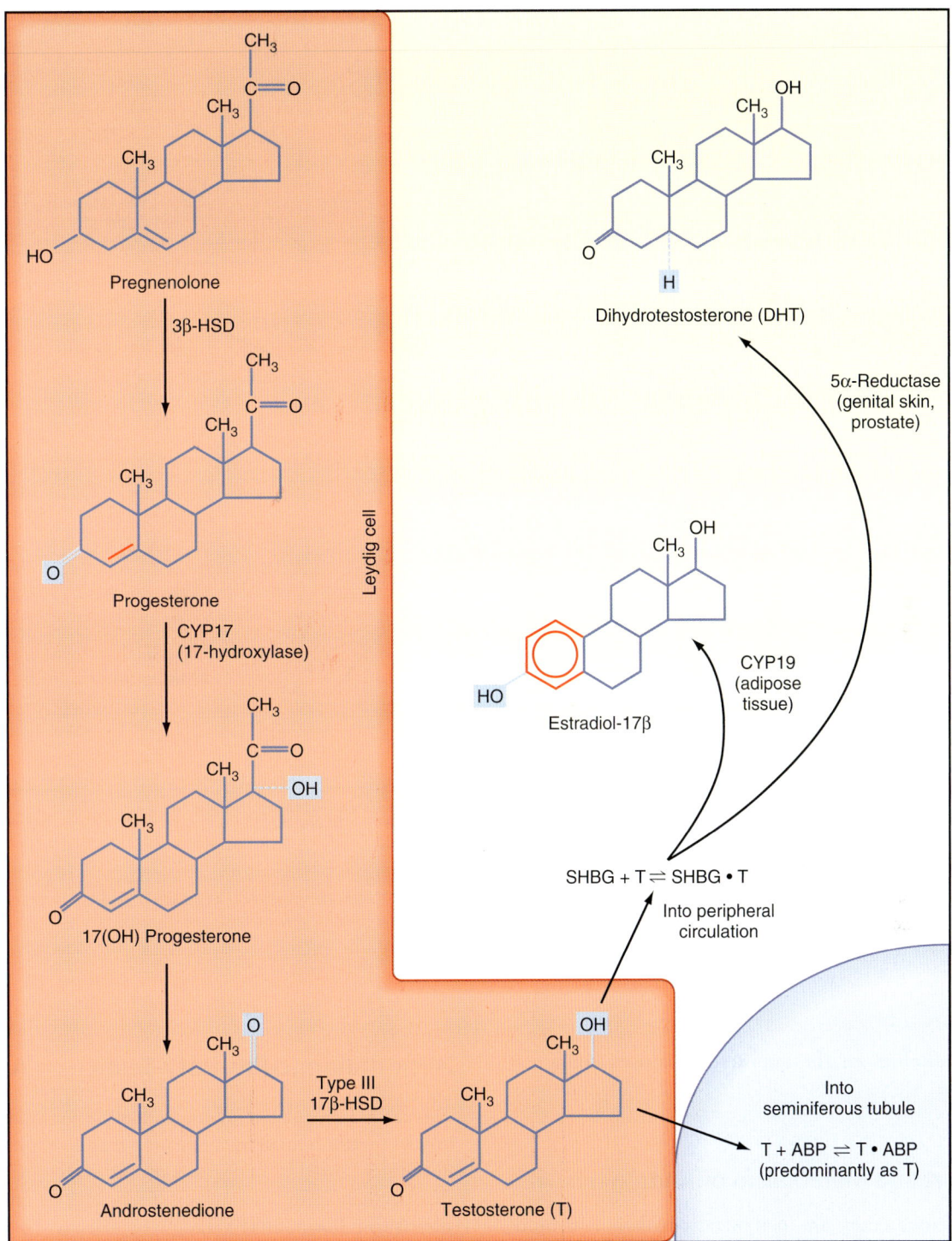

• **Fig. 44.8** Steroidogenic pathway in Leydig cells (the first step of converting cholesterol to pregnenolone is omitted). Testosterone is sequestered by binding to androgen-binding protein (ABP) within the seminiferous tubules or circulates within the peripheral circulation bound to sex hormone–binding globulin (SHBG) and can be peripherally converted to dihydrotestosterone (DHT) or estradiol-17β (E₂). (Modified from White BA, Porterfield SP. *Endocrine Physiology.* 4th ed. Philadelphia: Mosby; 2013.)

binding or DHT-AR binding causes dissociation of the chaperone proteins, followed by nuclear translocation of the androgen-AR complex, dimerization, binding to an **androgen response element (ARE),** and recruitment of coactivator proteins and general transcription factors to the vicinity of a specific gene's promoter. It remains unclear how testosterone and DHT differ in their ability to activate the AR in the context of different cell types, although the presence of different coactivator proteins in different cell types is probably involved. Such coactivator proteins would have

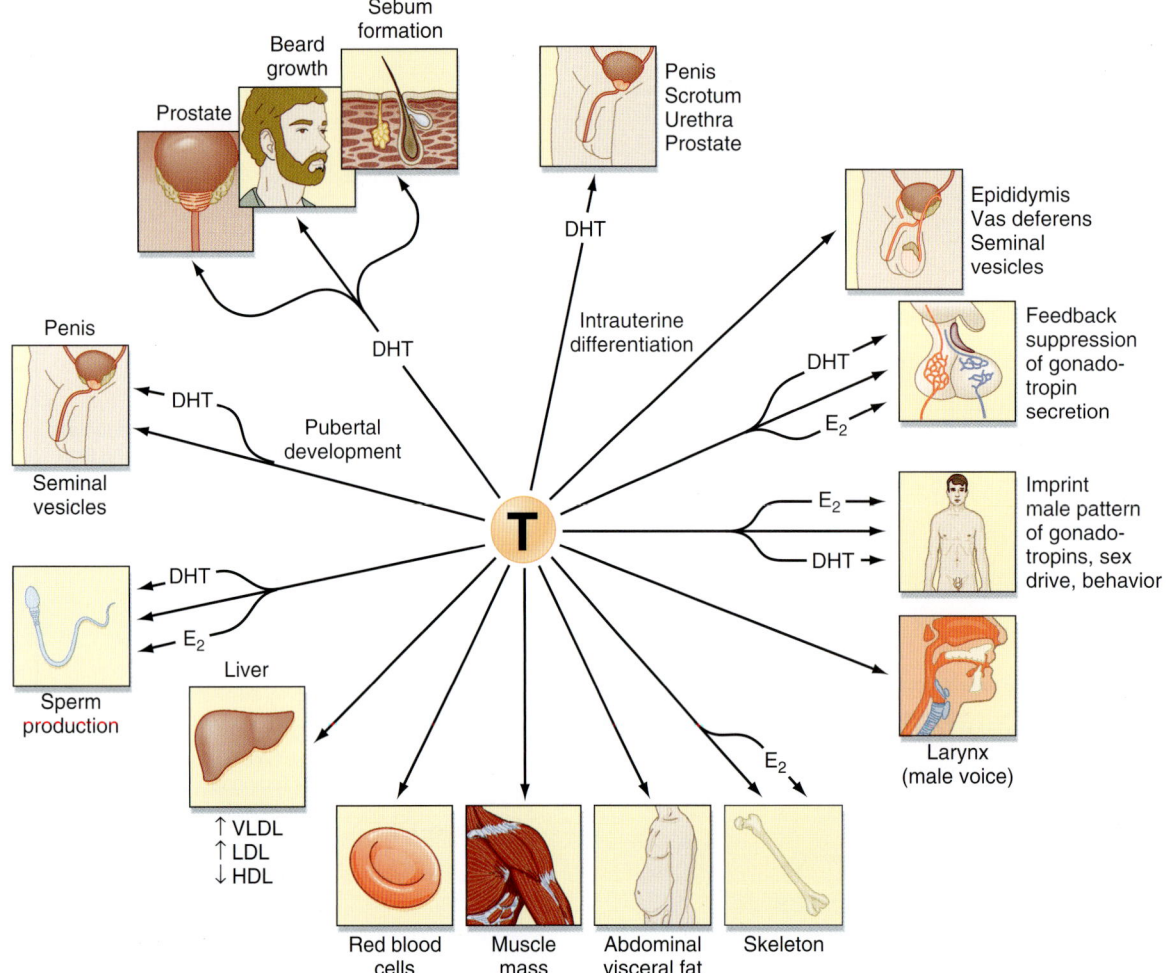

• **Fig. 44.9** Spectrum of effects of testosterone (T). Note that some effects result from the action of testosterone itself, whereas others are mediated by dihydrotestosterone (DHT) and estradiol (E_2) after they are produced from testosterone. VLDL, LDL, HDL, very-low-density, low-density, and high-density lipoproteins, respectively.

different affinities for the testosterone-induced configuration of the AR versus the DHT-induced configuration of the AR.

Transport and Metabolism of Androgens

As testosterone enters the peripheral circulation, it binds to and quickly reaches equilibrium with serum proteins. About 60% of circulating testosterone is bound to sex hormone–binding globulin (SHBG), 38% is bound to albumin, and about 2% remains as "free" hormone. Testosterone and its metabolites are primarily excreted in urine. Approximately 50% of excreted androgens are found as **urinary 17-ketosteroids,** with most of the remainder being conjugated androgens or diol or triol derivatives. Only about 30% of the 17-ketosteroids in urine are from the testis; the rest are produced from adrenal androgens. Androgens are conjugated with glucuronate or sulfate in the liver, and these **conjugated steroids** are excreted in urine.

Hypothalamic-Pituitary-Testicular Axis

The testis is regulated by an endocrine axis (Fig. 44.10) involving parvicellular hypothalamic gonadotropin-releasing hormone (GnRH) neurons and pituitary gonadotropes that produce both luteinizing hormone (LH) and follicle-stimulating hormone (FSH).

Regulation of Leydig Cell Function

The Leydig cell expresses the **LH receptor,** which acts on Leydig cells much like adrenocorticotropic hormone (ACTH, or corticotropin) does on zona fasciculata cells in the adrenal cortex (see Chapter 43). Rapid effects include hydrolysis of cholesterol esters and new expression of StAR protein. Less acute effects include an increase in steroidogenic enzyme gene expression and expression of the LDL. Over the long term, LH promotes Leydig cell growth and proliferation.

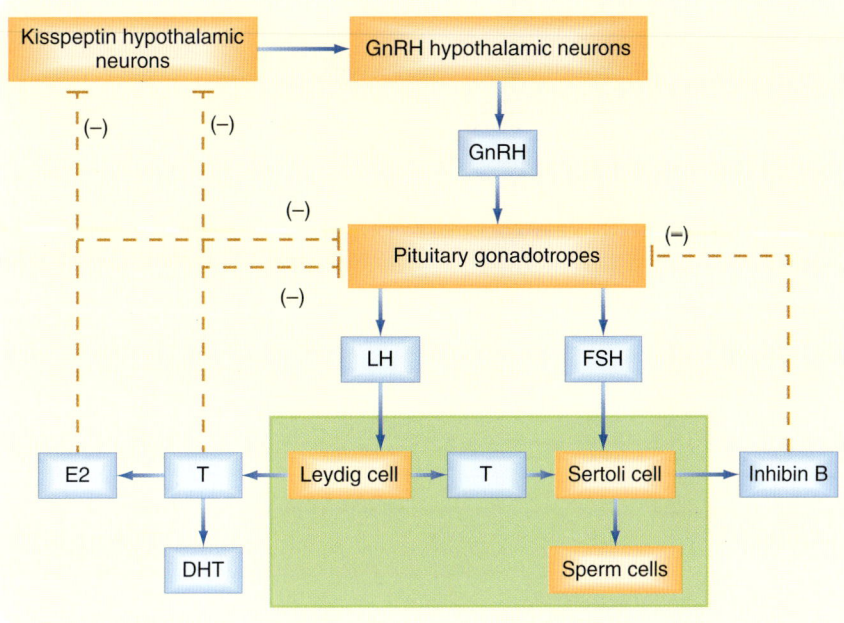

• **Fig. 44.10** The hypothalamic-pituitary-testicular axis. Abbreviations as in other figures.

Testosterone, and estradiol produced from peripheral conversion of testosterone, negatively feed back on GnRH hypothalamic neurons indirectly through inhibition of kisspeptin-producing neurons (see Fig. 44.10). Testosterone and estradiol also negatively feed back on pituitary gonadotropes. DHT, the other major product of peripheral conversion of testosterone, has little effect on LH or FSH levels.

Regulation of Sertoli Cell Function

The Sertoli cell is stimulated by both testosterone and FSH. In addition to stimulating synthesis of proteins involved in the "nurse cell" aspect of Sertoli cell function (e.g., ABP), FSH stimulates synthesis of the dimeric protein **inhibin.** Inhibin is induced by FSH and negatively feeds back on the gonadotrope to selectively inhibit FSH production (see Fig. 44.10).

Male Reproductive Tract

Once spermatozoa emerge from the efferent ductules, they leave the gonad and enter the male reproductive tract (see Fig. 44.1). The segments of the tract are the: **epididymis** (**head, body,** and **tail**), **vas deferens, ejaculatory duct, prostatic urethra**, **membranous urethra**, and **penile urethra.** Unlike the female tract, there is a **contiguous lumen** from the seminiferous tubule to the end of the male tract (i.e., the tip of the penile urethra), and the male reproductive tract connects to the **distal urinary tract** (i.e., **male urethra**). In addition to conveying sperm, the primary functions of the male reproductive tract are:
1. *Sperm maturation.* Sperm spend about a month in the **epididymis**, where they undergo further maturation.

AT THE CELLULAR LEVEL

There is an important "loophole" in the male reproductive axis that is based on the fact that **intratesticular levels of testosterone** need to be greater than 100-fold higher than circulating levels of the hormone to maintain normal rates of spermatogenesis, yet it is the **circulating levels of testosterone (and estradiol)** that provide the negative feedback to the pituitary and hypothalamus. This means that exogenous administration of testosterone can raise circulating levels of testosterone and estradiol sufficient to inhibit LH but not sufficient to accumulate in the testis at the required concentration for normal spermatogenesis. However, the decreased LH levels will diminish intratesticular production of testosterone by Leydig cells, which results in reduced levels of spermatogenesis (Fig. 44.11). This "loophole" is currently being investigated as a possible strategy for developing a **male oral contraceptive.** It is also the basis for **sterility** in some cases of **steroid abuse** in men.

The epithelium of the epididymis is secretory and adds numerous components to the seminal fluid. Spermatozoa that enter the head of the epididymis are weakly motile but are strongly unidirectionally motile by the time they exit the tail. Spermatozoa also undergo the process of **decapacitation,** which involves changes in the cell membrane to prevent spermatozoa from undergoing the acrosome reaction before contact with an egg (see later). Sperm become capacitated by the female reproductive tract within the oviduct. The function of the epididymis is dependent on **luminal testosterone-ABP complexes** that come from the seminiferous tubules and on testosterone from blood.

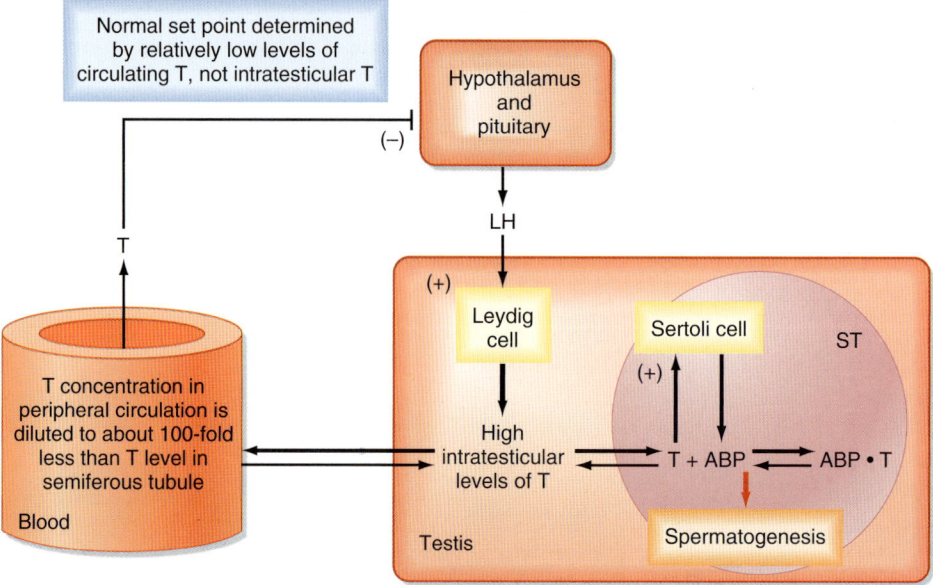

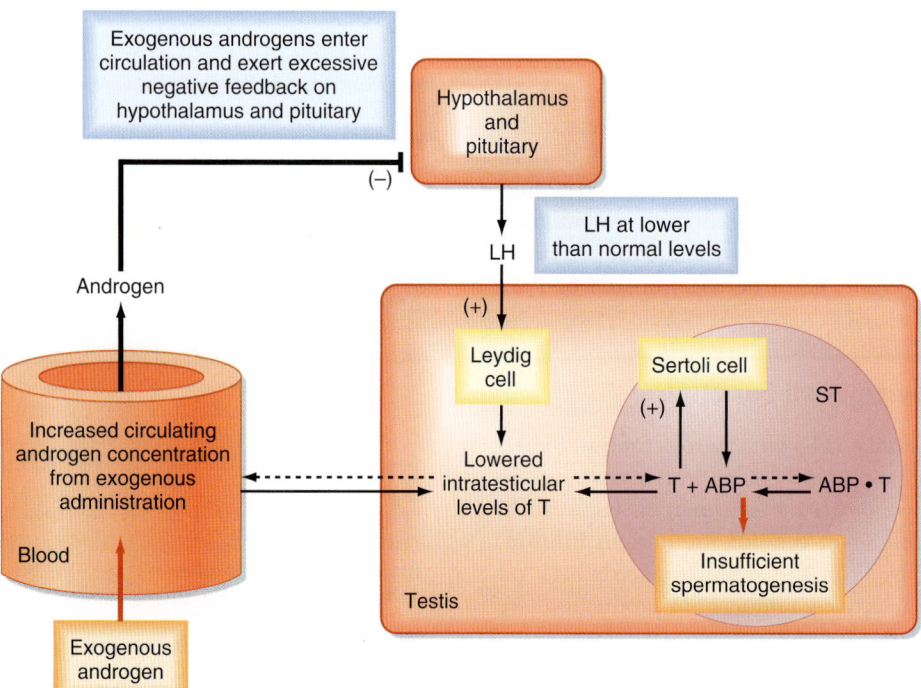

• **Fig. 44.11** The difference in intratesticular testosterone versus circulating testosterone concentrations and its importance in the hypothalamic-pituitary-testis axis. *Upper panel,* Feedback loop in a normal adult man. *Lower panel,* Administration of testosterone (or an androgenic analogue) increases circulating testosterone (androgen) levels, which in turn increase negative feedback on release of LH. Decreased LH levels diminish Leydig cell activity and intratesticular production of androgen. Lowered intratesticular testosterone levels result in reduced sperm production and can cause infertility. (The inhibin feedback loop has been omitted from this diagram.) (Modified from White BA, Porterfield SP. *Endocrine Physiology.* 4th ed. Philadelphia: Mosby; 2013.)

2. *Sperm storage and emission.* Sperm are stored in the **tail of the epididymis** and **vas deferens** for several months without loss of viability. The primary function of the vas deferens, besides providing a storage site, is to propel sperm during sexual intercourse into the male urethra.

The vas deferens has a very thick muscularis that is richly innervated by sympathetic nerves. Normally in response to repeated tactile stimulation of the penis during coitus, the muscularis of the vas deferens receives bursts of sympathetic stimulation that cause peristaltic contractions.

Emptying of the contents of the vas deferens into the prostatic urethra is called **emission.** Emission immediately precedes **ejaculation,** which is the propulsion of semen out of the male urethra.

3. *Production and mixing of sperm with seminal contents.* During emission, contraction of the vas deferens coincides with contraction of the muscular coats of the two accessory sex glands, the **seminal vesicles** (right and left) and the **prostate gland** (which surrounds the prostatic urethra). At this point, sperm become mixed with all the components of **semen.** The seminal vesicles secrete approximately 60% of the volume. These glands are the primary source of **fructose,** a critical nutrient for sperm. The seminal vesicles also secrete **semenogelins,** which induce coagulation of semen immediately after ejaculation. The alkaline secretions of the prostate, which make up about 30% of the volume, are high in **citrate, zinc, spermine,** and **acid phosphatase. Prostate-specific antigen (PSA)** is a serine protease that liquefies coagulated semen after a few minutes. PSA can be detected in blood under conditions of prostatic infection, benign prostatic hypertrophy, and prostatic carcinoma and is currently used as one indicator of prostatic health. The predominant buffers in semen are phosphate and bicarbonate. A third accessory gland, the **bulbourethral glands** (also called *Cowper's glands*), empty into the penile urethra in response to sexual excitement before emission and ejaculation. This secretion is high in mucus, which lubricates, cleanses, and buffers the urethra. Average sperm counts are between 60 and 100 million/mL semen. Men with sperm counts below 20 million/mL, less than 50% motile sperm, or less than 60% normally conformed sperm are usually infertile.

4. *Erection, penetration, and ejaculation.* Emission and ejaculation occur during coitus in response to a reflex arc that involves sensory stimulation from the penis (via the pudendal nerve) followed by sympathetic motor stimulation to the smooth muscle of the male tract and somatic motor stimulation to the musculature associated with the base of the penis. However, for sexual intercourse to occur in the first place, the man has to achieve and maintain an **erection** of the **penis.** The penis has evolved as an intromittent organ designed to separate the walls of the vagina, pass through the potential space of the vaginal lumen, and deposit semen at the distal end of the vaginal lumen near the cervix. This process of **internal insemination** can be performed only if the penis is stiffened from the process of erection.

Erection is a neurovascular event. The penis is composed of three erectile bodies: two **corpora cavernosa** and one **corpus spongiosum** (Fig. 44.12A). The penile urethra runs through the corpus spongiosum. These three bodies are composed of **erectile tissue**—an anastomosing network of potential **cavernous vascular spaces** lined with continuous endothelia within a loose connective tissue support. During the **flaccid state,** blood flow to the cavernous spaces is minimal (see Fig. 44.12A). This is due to vasoconstriction

of the vasculature (called the *helicine arteries*) and shunting of blood flow away from the cavernous spaces. In response to sexual arousal the parasympathetic cavernous nerves innervating the vascular smooth muscle of the helicine arteries release nitric oxide (NO). NO activates guanylyl cyclase, thereby increasing cyclic guanosine monophosphate (cGMP), which decreases intracellular $[Ca^{++}]$ and causes muscular relaxation (see Fig. 44.12B). Vasodilation allows blood to flow into the cavernous spaces to induce engorgement and erection. It also presses on veins in the penis and reduces venous drainage (see Fig. 44.12B).

IN THE CLINIC

Inability to achieve or maintain an erection is termed **erectile dysfunction (ED)** and is one cause of infertility in men. Multiple factors can lead to ED, including insufficient androgen production; neurovascular damage (e.g., from diabetes mellitus, spinal cord injury); structural damage to the penis, perineum, or pelvis; psychogenic factors (e.g., depression, performance anxiety); and prescribed medications and recreational drugs, including alcohol and tobacco. A major development in the treatment of some forms of erectile dysfunction is use of selective cGMP phosphodiesterase inhibitors that assist in maintaining an erection (see Fig. 44.12B).

Andropause

There is no distinct **andropause** in men. However, as men age, gonadal sensitivity to LH decreases and androgen production drops. As this occurs, serum LH and FSH levels rise. Although sperm production typically begins to decline after age 50, many men can maintain reproductive function and spermatogenesis throughout life.

THE FEMALE REPRODUCTIVE SYSTEM

The female reproductive system is composed of the gonads, called **ovaries,** and the female reproductive tract, which includes the **oviducts, uterus, cervix, vagina,** and **external genitalia.**

The Ovary

The ovary is located within a fold of peritoneum called the **broad ligament,** usually close to the lateral wall of the pelvic cavity (Fig. 44.13). Because the ovary extends into the peritoneal cavity, ovulated eggs briefly reside within the peritoneal cavity before they are captured by the oviducts.

The ovary is divided into an outer cortex and inner medulla (Fig. 44.14). Neurovascular elements innervate the medulla of the ovary. The cortex of the ovary is composed of a densely cellular stroma. Within this stroma reside the **ovarian follicles** (see Fig. 44.14, F), which contain a primary oocyte surrounded by follicle cells. The cortex is covered by a connective tissue capsule, the tunica albuginea,

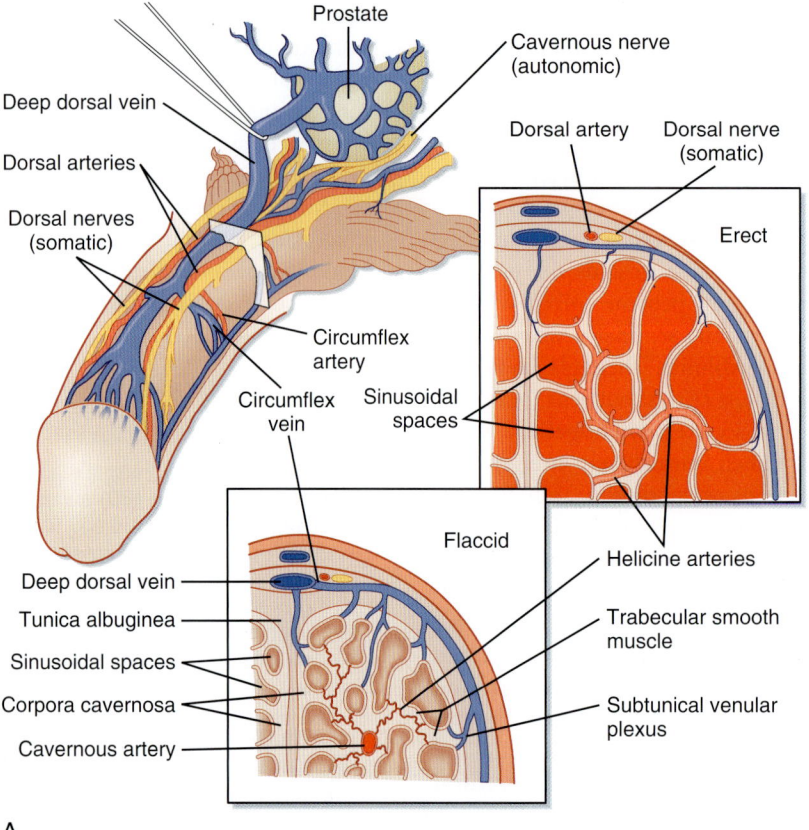

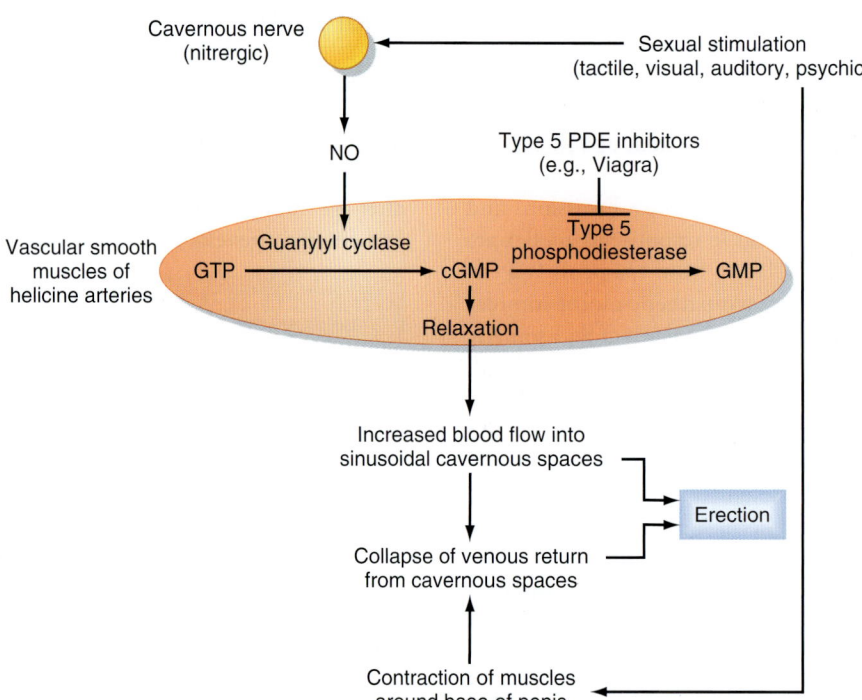

• **Fig. 44.12** A, Arrangement of the vasculature and cavernous tissue within the penis. During the flaccid state, blood flow into the cavernous spaces is limited by contraction of the helicine arteries. B, Outline of neurovascular events leading to penile erection. (A, From Bhasun S et al. In: Larsen P et al [eds]. *Williams Textbook of Endocrinology.* 10th ed. Philadelphia: Saunders; 2003.)

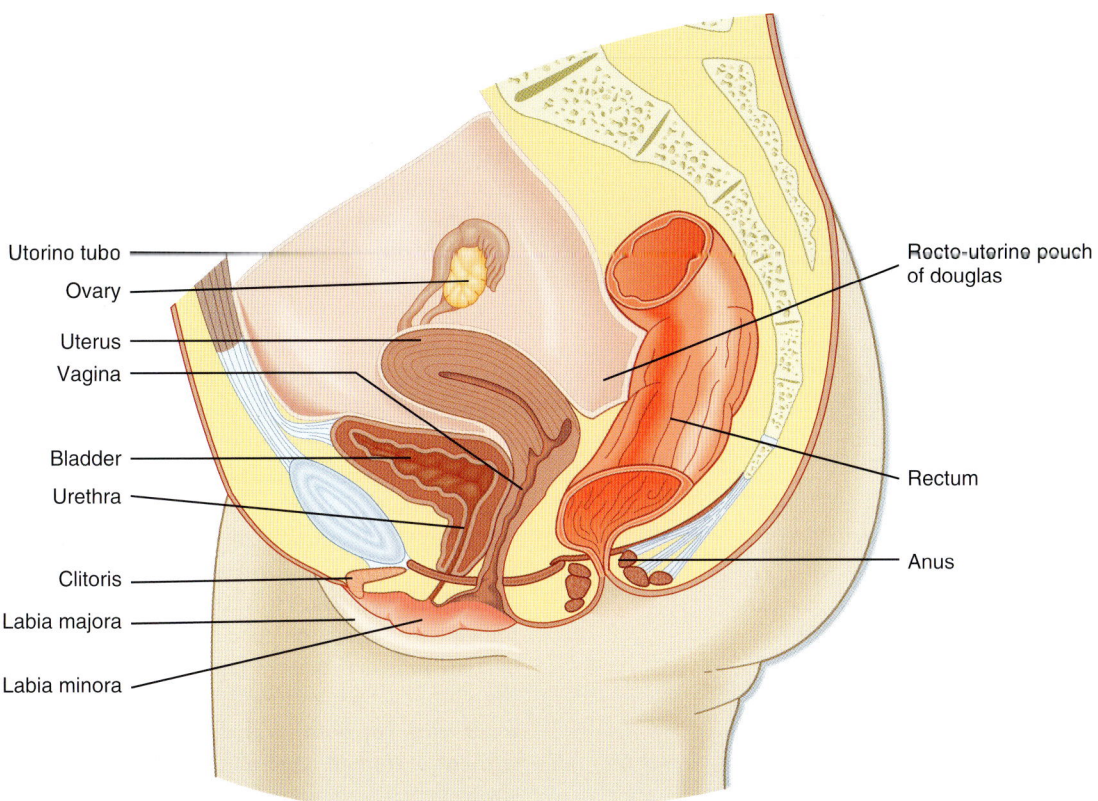

• **Fig. 44.13** Anatomy of the female reproductive system. (Modified from Drake RL et al. *Gray's Anatomy for Students.* Philadelphia: Churchill Livingstone; 2005.)

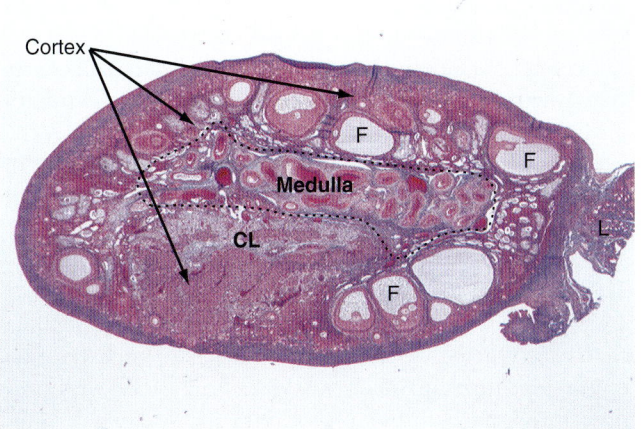

• **Fig. 44.14** Histology of the ovary. CL, corpus luteum; F, follicle. (Modified from Young B et al. *Wheater's Functional Histology. A Text and Colour Atlas.* 5th ed. London: Churchill Livingstone; 2006.)

Growth, Development, and Function of the Ovarian Follicle

The ovarian follicle is the functional unit of the ovary, and it performs both gametogenic and endocrine functions. A histological section of the ovary from a premenopausal cycling woman contains follicular structures at many different stages of development. The life history of a follicle can be divided into the following stages:

1. resting primordial follicle
2. growing preantral (primary and secondary) follicle
3. growing antral (tertiary) follicle
4. dominant (preovulatory, graafian) follicle
5. dominant follicle within the periovulatory period
6. corpus luteum (of menstruation or of pregnancy)
7. atretic follicles

Resting Primordial Follicle

Growth and Structure

Resting **primordial follicles** (Fig. 44.15) represent the earliest and simplest follicular structure in the ovary. Primordial follicles appear during midgestation through the interaction of gametes and somatic cells. Primordial germ cells that have migrated to the gonad continue to divide mitotically as oogonia until the fifth month of gestation in humans. At this point the approximately 7 million oogonia enter the process of meiosis and become **primary oocytes.** During this time the primary oocytes become surrounded by a

and a layer of simple epithelium consisting of **ovarian surface epithelial cells.** There are no ducts emerging from the ovary to convey its gametes to the reproductive tract. Thus the process of ovulation involves an inflammatory event that erodes the wall of the ovary. After ovulation the ovarian surface epithelial cells rapidly divide to repair the wall. The majority of ovarian cancer originates from this highly proliferative epithelium.

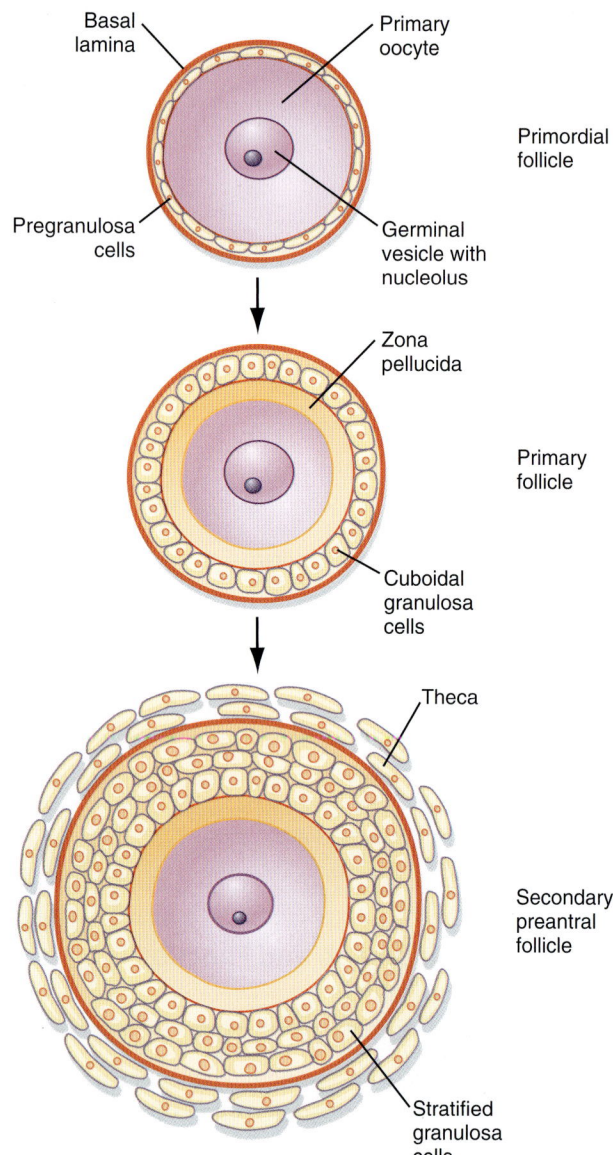

Basal lamina — Primary oocyte

Primordial follicle

Pregranulosa cells — Germinal vesicle with nucleolus

Zona pellucida

Primary follicle

Cuboidal granulosa cells

Theca

Secondary preantral follicle

Stratified granulosa cells

● **Fig. 44.15** Development of a primordial follicle up to a secondary preantral follicle. (Modified from White BA, Porterfield SP. *Endocrine Physiology.* 4th ed. Philadelphia: Mosby; 2013.)

simple epithelium of somatic **follicle cells,** thereby creating the primordial follicles (see Fig. 44.15). The follicle cells establish **gap junctions** with each other and the oocyte. The follicle cells themselves represent a true avascular epithelium surrounded by a basal lamina. Similar to Sertoli cell–sperm interactions, a subpopulation of granulosa cells remains intimately attached to the oocytes throughout their development. Granulosa cells provide nutrients such as amino acids, nucleic acids, and pyruvate to support oocyte maturation.

The primordial follicles represent the **ovarian reserve** of follicles (Fig. 44.16). This reserve is reduced from a starting number of about 7 million to less than 300,000 follicles at reproductive maturity. Of these, a woman will ovulate about 450 between **menarche** (first menstrual cycle) and **menopause** (cessation of menstrual cycles). At menopause, less than 1000 primordial follicles are left in the ovary. Primordial follicles are lost primarily from death as a result of **follicular atresia.** However, a small subset of primordial follicles will enter follicular growth in waves. Because the ovarian follicular reserve represents a fixed finite number, the rate at which resting primordial follicles die or begin to develop (or both) will determine the reproductive life span of a woman. Age at the onset of menopause has a strong genetic component but is also influenced by environmental factors. For example, cigarette smoking significantly depletes the ovarian reserve. An overly rapid rate of atresia or development will deplete the reserve and give rise to **premature ovarian insufficiency.**

Pituitary gonadotropins maintain a normal ovarian reserve by promoting the general health of the ovary. However, the rate at which resting primordial follicles enter the growth process appears to be independent of pituitary gonadotropins. The decision of a resting follicle to enter the early growth phase is primarily dependent on intraovarian paracrine factors produced by both the follicle cells and oocytes.

The Gamete

In primordial follicles the gamete is derived from oogonia that have entered the first meiotic division; such oogonia are referred to as **primary oocytes.** Primary oocytes progress through most of prophase of the first meiotic division

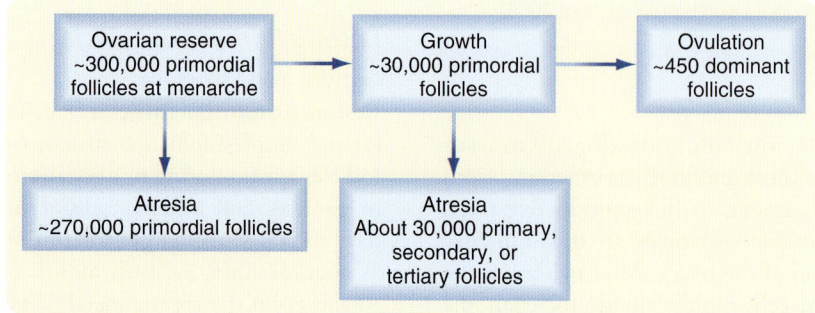

Ovarian reserve ~300,000 primordial follicles at menarche → Growth ~30,000 primordial follicles → Ovulation ~450 dominant follicles

Atresia ~270,000 primordial follicles

Atresia About 30,000 primary, secondary, or tertiary follicles

● **Fig. 44.16** Fate of ovarian follicles. (Modified from White BA, Porterfield SP. *Endocrine Physiology.* 4th ed. Philadelphia: Mosby; 2013.)

(termed *prophase I*) over a 2-week period and then arrest in the **diplotene stage.** This stage is characterized by the decondensation of chromatin, which supports the transcription needed for oocyte maturation. Meiotic arrest at this stage, which may last for up to 50 years, appears to be due to "maturational incompetence," or lack of the cell cycle proteins needed to support the completion of meiosis. The nucleus of the oocyte, called the **germinal vesicle,** remains intact at this stage.

Growing Preantral Follicles

Growth and Structure

The first stage of follicular growth is **preantral,** which refers to the development that occurs before the formation of a fluid-filled **antral cavity.** One of the first visible signs of follicle growth is the appearance of **cuboidal granulosa cells.** At this point the follicle is referred to as a **primary follicle** (see Fig. 44.15). As granulosa cells proliferate, they form a multilayered (i.e., stratified) epithelium around the oocyte. At this stage the follicle is referred to as a **secondary follicle** (see Fig. 44.15).

Once a secondary follicle acquires three to six layers of granulosa cells, it secretes paracrine factors that induce nearby stromal cells to differentiate into epithelioid **thecal cells.** Thecal cells form a flattened layer of cells around the follicle. Once a thecal layer forms, the follicle is referred to as a **mature preantral follicle** (see Fig. 44.15). In humans it takes several months for a primary follicle to reach the mature preantral stage.

Follicular development is associated with an inward movement of the follicle from the outer cortex to the inner cortex, closer to the vasculature of the ovarian medulla. Follicles release **angiogenic factors** that induce development of one to two arterioles that form a vascular wreath around the follicle.

The Gamete

During the preantral stage, the oocyte begins to grow and produce cellular and secreted proteins. The oocyte initiates secretion of extracellular matrix glycoproteins called **ZP1, ZP2,** and **ZP3** that form the **zona pellucida** (see Fig. 44.15). The zona pellucida increases in thickness and provides a species-specific binding site for sperm during fertilization (see Pregnancy). Importantly, granulosa cells and the oocyte maintain gap junctional contact via cellular projections through the zona pellucida. The oocyte also continues to secrete paracrine factors that regulate follicle cell growth and differentiation.

Endocrine Function

Granulosa cells express the **FSH receptor** during this period, but they are primarily dependent on factors from the oocyte to grow. They do not produce ovarian hormones at this early stage of follicular development. The newly acquired thecal cells are analogous to testicular Leydig cells in that they reside outside the epithelial "nurse" cells, express the **LH receptor,** and produce **androgens.** The main difference

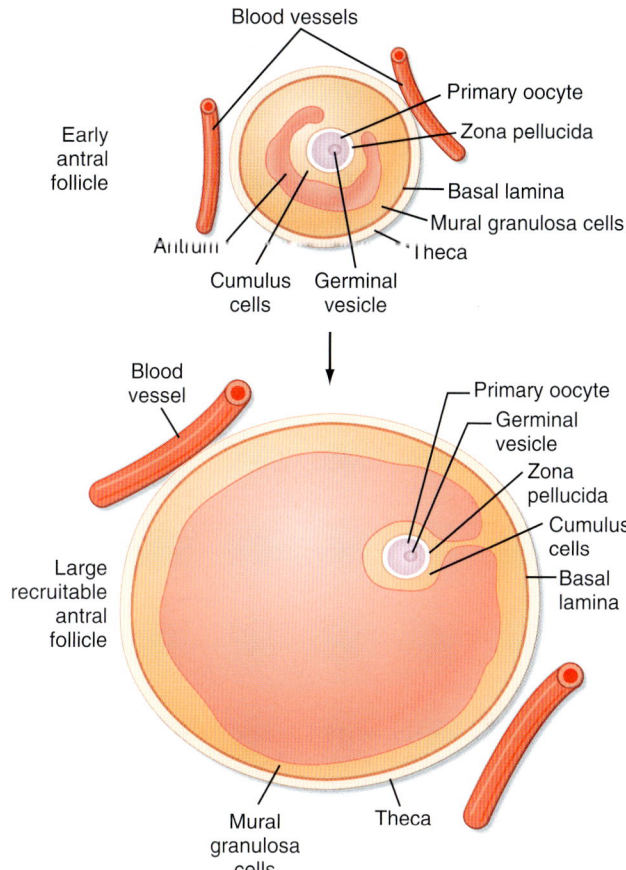

• **Fig. 44.17** Development of an early antral follicle to a mature preovulatory follicle. (Modified from White BA, Porterfield SP. *Endocrine Physiology.* 4th ed. Philadelphia: Mosby; 2013.)

between Leydig cells and thecal cells is that thecal cells do not express high levels of 17β-HSD. Thus the major product of theca cells is androstenedione as opposed to testosterone. Androstenedione production at this stage is minimal.

Growing Antral Follicles

Growth and Structure

Mature preantral follicles develop into **early antral follicles** (Fig. 44.17) over a period of about 25 days, during which they grow from a diameter of about 0.1 mm to a diameter of 0.2 mm. Once the granulosa epithelium increases to six to seven layers, fluid-filled spaces appear between cells and coalesce into the **antrum.** Over a period of about 45 days, this wave of small antral follicles will continue to grow to **large recruitable antral follicles** that are 2 to 5 mm in diameter. This period of growth is characterized by about a 100-fold increase in granulosa cells (from about 10,000 to 1,000,000 cells). It is also characterized by swelling of the antral cavity, which increasingly divides the granulosa cells into two discrete populations: mural granulosa cells and cumulus cells (see Fig. 44.17).

Mural granulosa cells (also called the **stratum granulosum**) form the outer wall of the follicle. The basal layer is adherent to the basal lamina and in close proximity to

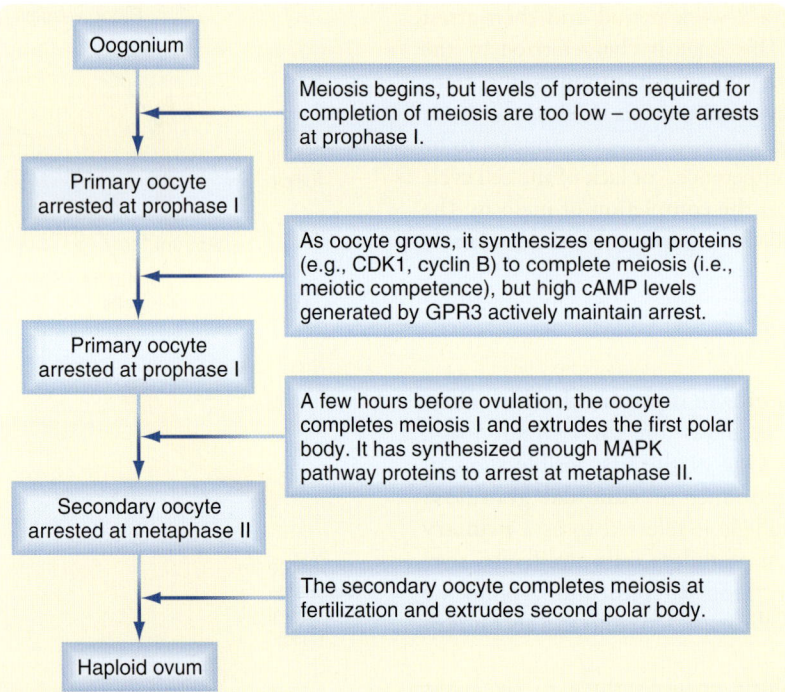

• **Fig. 44.18** Events involved in meiotic arrest and maturation of the oocyte. MAPK, mitogen-activated protein kinase. (Modified from White BA, Porterfield SP. *Endocrine Physiology*. 4th ed. Philadelphia: Mosby; 2013.)

the outer-lying thecal layers. Mural granulosa cells become highly steroidogenic and remain in the ovary after ovulation to differentiate into the corpus luteum.

Cumulus cells are the inner cells that surround the oocyte (they are also referred to as **the cumulus oophorus** and **corona radiata**). The innermost layer of cumulus cells maintains gap and adhesion junctions with the oocyte. Cumulus cells are released with the oocyte (collectively referred to as the **cumulus-oocyte complex**) during the process of ovulation. Cumulus cells are crucial for the ability of the fimbriated end of the oviduct to "capture" and move the oocyte by a ciliary transport mechanism along the length of the oviduct to the site of fertilization (see Pregnancy).

Early antral follicles are dependent on pituitary FSH for normal growth. Large antral follicles become highly dependent on pituitary FSH for their growth and sustained viability. As discussed later, 2- to 5-mm follicles are recruited to enter a rapid growth phase via the transient increase in FSH that occurs toward the end of the previous menstrual cycle.

The Gamete

The oocyte grows rapidly in the early stages of antral follicles; growth then slows in larger follicles. During the antral stage the oocyte synthesizes sufficient amounts of cell cycle components so it becomes competent to complete meiosis I at ovulation. (Note that the human egg arrests after ovulation at a second point, metaphase II, until it is fertilized by sperm.) Thus in early primary and secondary follicles, the oocyte fails to complete meiosis I because of a dearth

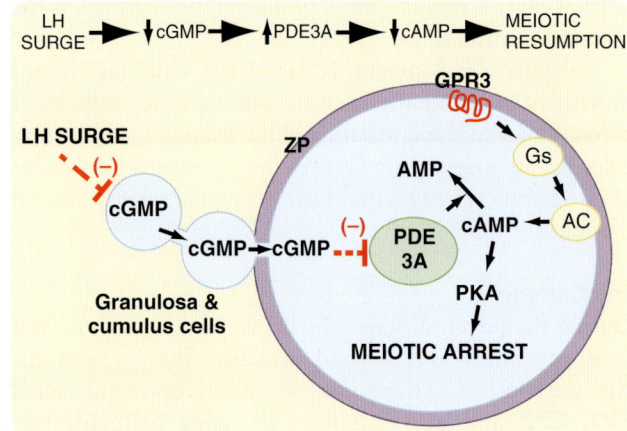

• **Fig. 44.19** Model of how LH surge leads to resumption of meiosis. PDE, phosphodiesterase.

of specific meiosis-associated proteins. However, larger antral follicles gain **meiotic competence** but still maintain **meiotic arrest** until the midcycle LH surge. Meiotic arrest is achieved by maintenance of **elevated cyclic adenosine monophosphate (cAMP) levels** in the mature oocyte (Figs. 44.18 and 44.19). The constitutively active (i.e., not requiring a ligand) Gs protein-coupled receptor **GPR3** maintains high cAMP. The oocyte-specific phosphodiesterase PDE3A degrades cAMP to inactive AMP. Before the LH surge, PDE3A is inhibited by cGMP, which is produced within cumulus and granulosa cells and enters the oocyte via gap junctions.

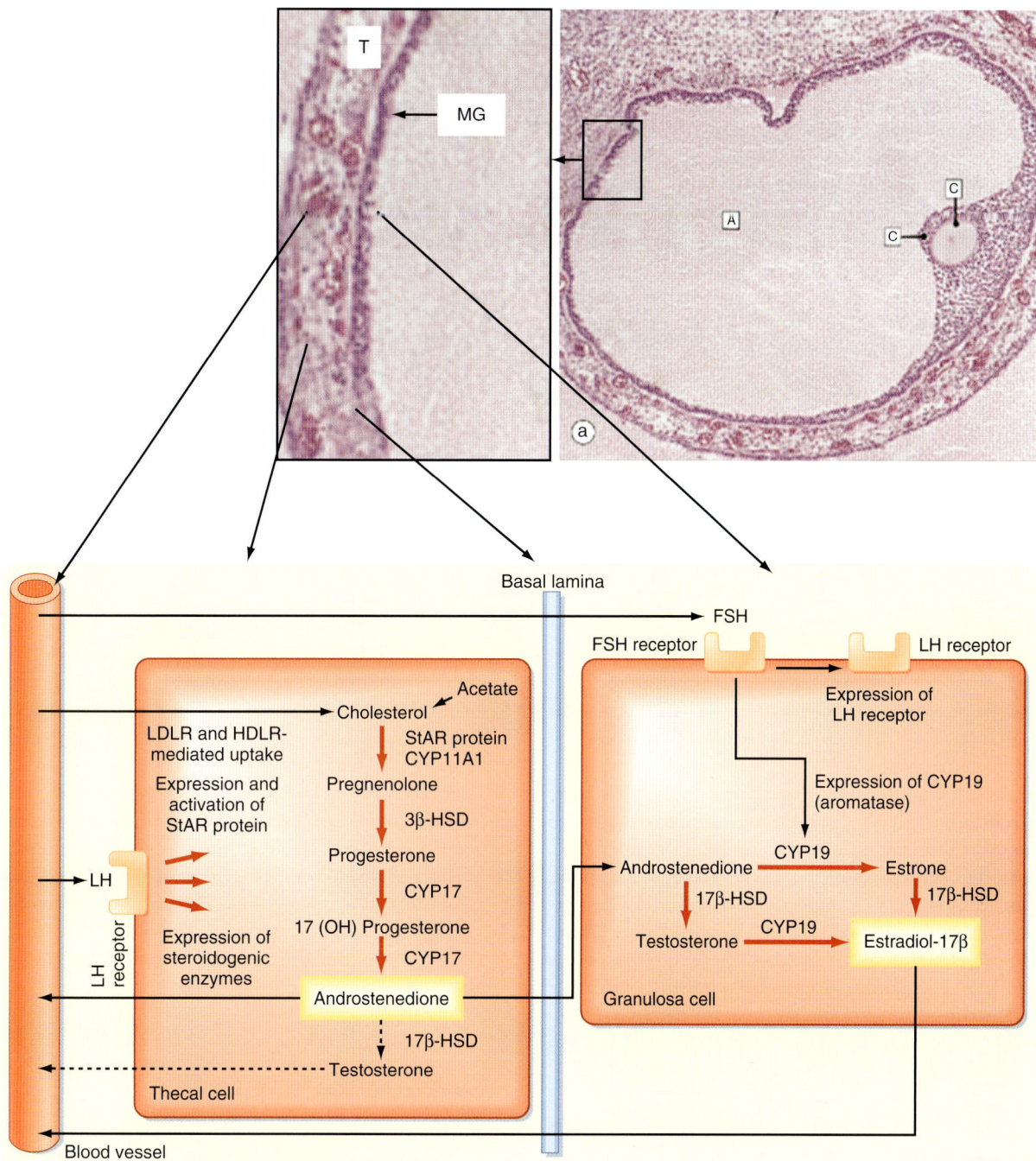

• Fig. 44.20 Two-cell model for steroidogenesis in the dominant follicle. *Top panel:* MG, mural granulosa; T, theca. (Modified from White BA, Porterfield SP. *Endocrine Physiology.* 4th ed. Philadelphia: Mosby; 2013.)

Endocrine Function

The thecal cells of large antral follicles produce significant amounts of **androstenedione** and less testosterone. Androgens are converted to **estradiol-17β** by the granulosa cells (Fig. 44.20). At this stage, FSH stimulates proliferation of granulosa cells and induces the expression of **CYP19 (aromatase)** required for estrogen synthesis. Additionally the mural granulosa cells of the large antral follicles produce increasing amounts of **inhibin** during the early follicular

phase. Low levels of estrogen and inhibin negatively feed back on FSH secretion, thereby contributing to selection of the follicle with the most FSH-responsive cells.

Dominant Follicle

Growth and Structure

As already discussed, at the end of a previous menstrual cycle, a crop of large (2–5 mm) antral follicles (see Fig. 44.17) are **recruited by a rise in FSH** to begin rapid gonadotropin-dependent development. The total number of recruited follicles in both ovaries can be as high as 20 in a younger woman (<33 years old) but rapidly declines at older ages. The number of recruited follicles is reduced to the **ovulation quota** (one in humans) by the process of **selection.** As FSH levels decline the rapidly growing follicles progressively undergo atresia until one follicle is left. Generally the largest follicle with the most FSH receptors of the recruited crop becomes the **dominant follicle.** Selection occurs during the early follicular phase. By midcycle the dominant follicle becomes a large **preovulatory follicle** that is 20 mm in diameter and contains about 50 million granulosa cells by the midcycle gonadotropin surge.

The Gamete

The oocyte is competent to complete meiosis I but remains arrested in the dominant follicle until the LH surge. Growth of the oocyte continues but at a slower rate until the oocyte reaches a diameter of about 140 µm by ovulation (i.e., ≈20 times the diameter of an erythrocyte).

Endocrine Function

The newly selected follicle emerges as a significant steroidogenic "gland." Ovarian steroidogenesis requires both theca and granulosa cells. As discussed earlier, **thecal cells** (see Fig. 44.20, T) express **LH receptors** and produce primarily androstenedione. Basal levels of LH stimulate expression of steroidogenic enzymes in thecal cells. The thecal cells are richly vascularized and thus have access to cholesterol within the lipoprotein articles LDL and HDL. LH promotes expression of the **LDL receptor and HDL receptor (SR-B1),** which import cholesterol. LH also promotes robust expression of **CYP11A1** (side chain cleavage enzyme), **3β-HSD, and CYP17** with both 17-hydroxylase activity and 17,20-lyase activity. Androgens (primarily **androstenedione** but also some **testosterone**) released from the theca diffuse into the **mural granulosa cells** or enter the vasculature surrounding the follicle.

The **mural granulosa cells** (see Fig. 44.20, MG) of the selected follicle have a high number of **FSH receptors** and are very sensitive to FSH, which upregulates **CYP19 (aromatase)** gene expression and activity. CYP19 converts androstenedione to the weak estrogen **estrone** and converts testosterone to the potent estrogen **estradiol-17β.** Granulosa cells express activating isoforms of **17β-HSD,** which converts the less active estrone to highly active estradiol-17β. In addition, FSH induces expression of **inhibin B** during the follicular phase.

Importantly, FSH also induces expression of **LH receptors** in mural granulosa cells during the second half of the follicular phase (see Fig. 44.20). Thus mural granulosa cells acquire the ability to respond to LH, which allows these cells to maintain high levels of CYP19 in the face of declining FSH levels. Acquisition of LH receptors also ensures that mural granulosa cells respond to the LH surge.

The Dominant Follicle During the Periovulatory Period

The **periovulatory period** is defined as the time from the onset of the LH surge to expulsion of the cumulus-oocyte complex out of the ovary (i.e., ovulation). This process lasts for 32 to 36 hours in women. Starting at the same time and superimposed on the process of ovulation is a change in the steroidogenic function of theca and mural granulosa cells. This process is called **luteinization** and culminates in formation of a **corpus luteum** that is capable of producing high amounts of **progesterone,** along with estrogen, within a few days after ovulation. Thus the LH surge induces the onset of complex processes during the periovulatory period that complete the gametogenic function of the ovary for a given month and switches the endocrine function to prepare the female reproductive tract for implantation and gestation.

Growth and Structure

The LH surge induces dramatic structural changes in the dominant follicle that involves its rupture, ovulation of the cumulus-oocyte complex, and biogenesis of a new structure called the **corpus luteum** from the remaining thecal and mural granulosa cells. Major structural changes occur during this transition:

1. Before ovulation the large preovulatory follicle presses against the ovarian surface and generates a poorly vascularized bulge of the ovarian wall called the **stigma.** The LH surge induces release of inflammatory cytokines and hydrolytic enzymes from the theca and granulosa cells. These secreted components lead to breakdown of the follicle wall, tunica albuginea, and surface epithelium in the vicinity of the stigma (Fig. 44.21). At the end of this process the antral cavity becomes continuous with the peritoneal cavity.

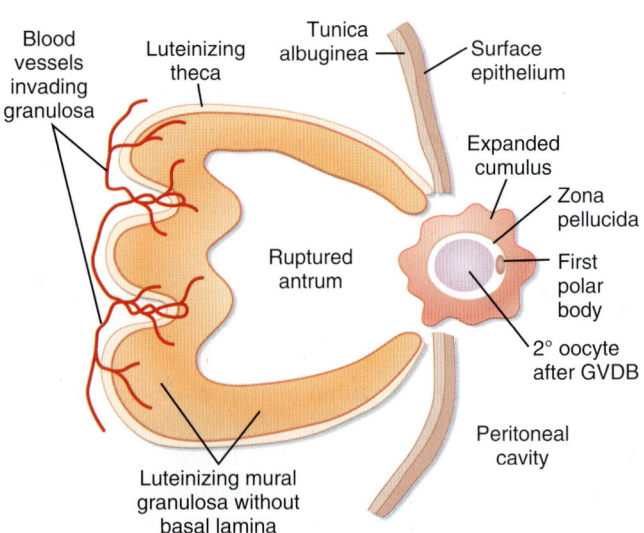

• **Fig. 44.21** Ovulation. GVBD, germinal vesicle breakdown. (Modified from White BA, Porterfield SP. *Endocrine Physiology.* 4th ed. Philadelphia: Mosby; 2013.)

2. The attachment of the cumulus cells to the mural granulosa cells degenerates, and the cumulus-oocyte complex becomes free-floating within the antral cavity (see Fig. 44.21). Cumulus cells also respond to the LH surge by secreting hyaluronic acid and other extracellular matrix components. These substances enlarge the entire cumulus-oocyte complex, a process called **cumulus expansion** (see Fig. 44.21). This enlarged cumulus-oocyte complex is more easily captured and transported by the oviduct. The expanded cumulus also makes the cumulus-oocyte complex easier for spermatozoa to find. Sperm express a **membrane hyaluronidase** that allows them to penetrate the expanded cumulus. The cumulus-oocyte complex is released through the ruptured stigma through a relatively slow process.

3. The basal lamina of mural granulosa cells is broken down so that blood vessels and outer-lying theca cells can push into the granulosa cells. Granulosa cells secrete **angiogenic factors** such as vascular endothelial growth factor (VEGF), angiopoietin 2, and basic fibroblast growth factor (bFGF), which significantly increase the blood supply to the new corpus luteum.

The Gamete

Before ovulation the primary oocyte is competent to complete meiosis, but it is arrested in prophase I (see Fig. 44.18). The LH surge inhibits production of cGMP by granulosa and cumulus cells, thereby removing inhibition of the oocyte-specific PDE3A. PDE3A proceeds to degrade cAMP to inactive AMP, thereby removing the brake on meiotic progression. The oocyte then progresses to metaphase II and subsequently arrests at metaphase II until fertilization.

Endocrine Function

Both theca and mural granulosa cells express LH receptors at the time of the LH surge. The LH surge induces differentiation of the granulosa cells—a process that continues for several days after ovulation. During the periovulatory period, the LH surge induces the following shifts in steroidogenic activity of the mural granulosa cells (now turning into granulosa lutein cells).

1. *Transient inhibition of CYP19 expression and consequently estrogen production.* The rapid decline in estrogen helps turn off the positive feedback on LH secretion.

2. *Breakdown of the basal lamina and vascularization of the granulosa cells.* This makes LDL and HDL cholesterol accessible to these cells for steroidogenesis. The LH surge also increases expression of the LDL receptor and HDL receptor (SR-BI) in granulosa cells.

3. *Onset of expression of StAR protein, CYP11A1 (side chain cleavage enzyme), and 3β-HSD* (Fig. 44.22). Expression of these enzymes are key to the onset of production of high levels of progesterone by these cells. As discussed later, high progesterone synthesis is absolutely necessary for maintenance of pregnancy. Because CYP17 activity, especially its 17,20-lyase function, is largely absent in granulosa lutein cells, progesterone is not further metabolized to another steroid but instead exits the cells and enters the circulation.

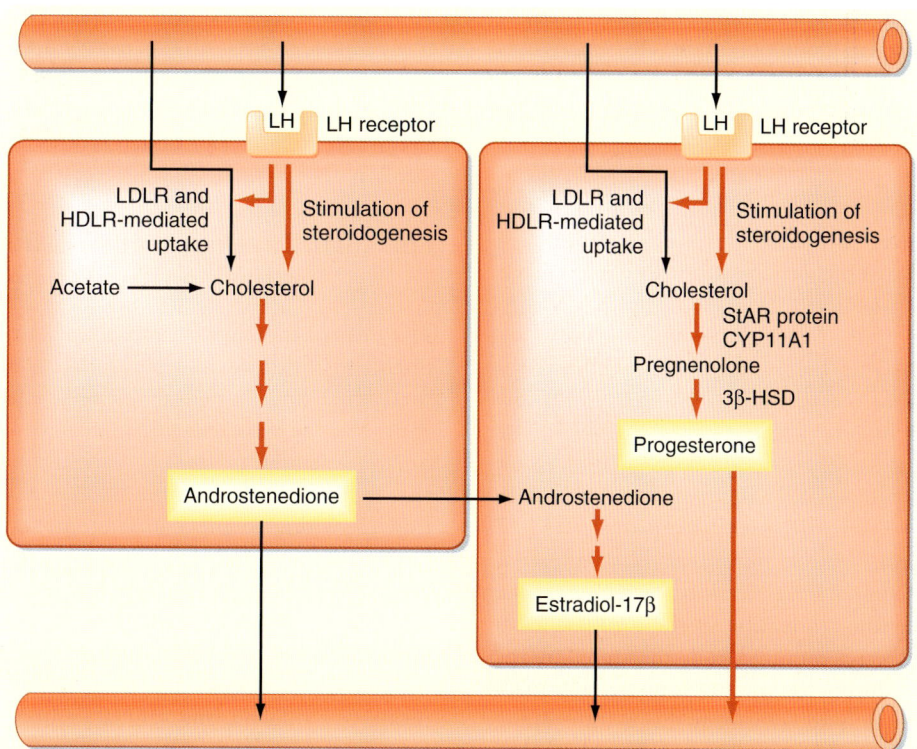

• **Fig. 44.22** Steroidogenic pathways in the corpus luteum (Modified from White BA, Porterfield SP. *Endocrine Physiology.* 4th ed. Philadelphia: Mosby; 2013.)

The Corpus Luteum

Growth and Structure

After ovulation the remnant of the antral cavity fills with blood from damaged blood vessels in the vicinity of the stigma. This gives rise to a **corpus hemorrhagicum.** Within a few days, red blood cells and debris are removed by macrophages, and fibroblasts fill in the antral cavity with a hyaline-like extracellular matrix. In the mature corpus luteum the granulosa cells, now called **granulosa lutein cells,** enlarge and become filled with lipid (cholesterol esters). The enlarged granulosa lutein cells collapse into and partially fill in the old antral cavity. The theca along with blood vessels, mast cells, macrophages, leukocytes, and other resident connective tissue cells infiltrate the granulosa layer at multiple sites.

The human corpus luteum is programmed to live for 14 days, plus or minus 2 days (**corpus luteum of menstruation),** unless "rescued" by the LH-like hormone **human chorionic gonadotropin (hCG),** which originates from an implanting embryo. If rescued, the **corpus luteum of pregnancy** will remain viable during the pregnancy (usually $\approx$ 9 months). The mechanism by which the corpus luteum of menstruation regresses in 14 days is not fully understood. The corpus luteum appears to become progressively less sensitive to basal levels of LH, so that hCG binding to the LH receptor is needed for continued health and function of the corpus luteum. Regression appears to involve release of the **prostaglandin PGF$_{2\alpha}$** from both granulosa lutein cells and the uterus in response to declining levels of progesterone during the second week of the luteal phase. Several paracrine factors (endothelin, monocyte chemotactic protein-1) from immune and vascular cells are likely to play a role in the demise and removal of granulosa lutein cells. The corpus luteum is ultimately turned into a scarlike body called the **corpus albicans,** which sinks into the medulla of the ovary and is slowly absorbed.

The Gamete

The LH surge induces two parallel events, ovulation and luteinization. If ovulation occurs normally, the corpus luteum is devoid of a gamete.

Endocrine Function

Before the LH surge, the granulosa cells have very low capacity to convert cholesterol into a steroid hormone. The LH surge induces the onset of expression of CYP11A1, 3β-HSD and StAR protein, allowing granulosa lutein cells to convert cholesterol into progesterone. Because CYP17 expression is extremely low, progesterone accumulates and moves out of the granulosa lutein cells and enters the vasculature. Progesterone production by the corpus luteum (see Fig. 44.22) increases steadily from the onset of the LH surge and peaks during the midluteal phase. The main purpose of this timing is to transform the uterine lining into an adhesive and supportive structure for implantation and early pregnancy. As discussed later the midluteal phase is synchronized with early embryogenesis so the uterus is

optimally primed when a blastocyst tumbles into the uterus around day 22 of the menstrual cycle. Estradiol continues to be produced by theca lutein cells and granulosa lutein cells. Estrogen production transiently decreases in response to the LH surge but then rebounds and peaks at the midluteal phase. Estradiol induces the progesterone receptor in progesterone target cells, such as the uterine endometrium, and ensures a full response to progesterone.

Luteal hormonal output is absolutely dependent on basal LH levels (see Fig. 44.22). In fact, progesterone output is closely correlated with the pulsatile pattern of LH release in women. Both FSH and LH are reduced to basal levels during the luteal phase by negative feedback from progesterone and estrogen. In addition, granulosa lutein cells secrete **inhibin,** which selectively represses FSH secretion.

The corpus luteum must generate large amounts of progesterone to support implantation and early pregnancy. Accordingly the life of the corpus luteum is very regular, and a shortened luteal phase typically leads to infertility. The quality of the corpus luteum is largely dependent on the size and health of the dominant follicle from which it developed, which in turn is dependent on normal hypothalamic and pituitary stimulation during the follicular phase. Numerous factors that perturb hypothalamic and pituitary output during the follicular phase, including heavy exercise, starvation, high prolactin levels, and abnormal thyroid function, can lead to **luteal phase deficiency** and **infertility.**

Atretic Follicles

Follicular atresia refers to the demise of an ovarian follicle. During atresia the granulosa cells and oocytes undergo **apoptosis.** The thecal cells typically persist and repopulate the cellular stroma of the ovary. The thecal cells retain LH receptors and the ability to produce androgens and are collectively referred to as the **"interstitial gland"** of the ovary. Follicles can undergo atresia at any time during development.

Follicular Development With Respect to the Monthly Menstrual Cycle

The **human menstrual cycle** strictly refers to the monthly discharge of discarded uterine lining as **menstrual blood** or **menstrual flow** (a period referred to as **menses**) through the process of **menstruation** (see later). In fact it is the onset or lack thereof of menses, as detected by the woman herself, that is the primary evidence for the cessation of menstruation (e.g., due to pregnancy or menopause) or a change in the duration and/or frequency of the menstrual cycle. However, it is useful from an endocrinological perspective to consider the human menstrual cycle as having an ovarian cycle and a uterine cycle, with the latter being driven by the former. As discussed later, there are also hypothalamic, pituitary, oviductal, and vaginal components of the human menstrual cycle. The reproductive function of the menstrual cycle is the collective orchestration by ovarian hormones of the functions of the hypothalamus, pituitary, uterus, oviduct, cervix, and vagina—and even the ovary itself—to: (1) produce a fertilizable gamete (egg), (2) provide a supportive environment for intercourse, reception of sperm, fertilization of egg and early embryogenesis, (3) prepare the uterine lining for implantation, placentation and pregnancy, and (4) minimize the possibility of a superimplantation (i.e., a second implantation) from occurring, and/or prevent an ascending infection from moving up to the uterus from the vagina.

The first half of the monthly menstrual cycle is referred to as the **follicular phase** of the **ovary** and is characterized by recruitment and growth of a large antral follicle, selection of the dominant follicle, and growth of the dominant follicle until ovulation. The dominant follicle must contain a fully developed oocyte and somatic follicle cells that secrete high levels of estrogen.

The second half of the monthly menstrual cycle is referred to as the **luteal phase** of the ovary and is dominated by hormonal secretions of the **corpus luteum**. The corpus luteum must secrete both **progesterone** and **estradiol** for progression of the normal cycle.

Regulation of Late Stages of Follicular Development, Ovulation, and Luteinization: The Human Menstrual Cycle

As stated earlier, late stages of follicular development and luteal function are absolutely dependent on normal hypothalamic and pituitary function. As in the male, hypothalamic neurons secrete **GnRH** in a **pulsatile** manner. GnRH in turn stimulates LH and FSH production by pituitary gonadotropes. A high frequency of GnRH pulses (1 pulse per 60–90 minutes) selectively promotes LH production, whereas a slow frequency promotes FSH production. A major difference between the male and female reproductive axes is the midcycle gonadotropin surge, which is dependent on a constant high level of estrogen coming from the dominant follicle.

A highly dynamic "conversation" occurs among the ovary, pituitary, and hypothalamus in which the events of the menstrual cycle are orchestrated, beginning with the ovary at the end of the luteal phase of a previous nonfertile cycle (Fig. 44.23). Events that follow are numbered according to Fig. 44.24:

Event 1: In the absence of fertilization and implantation, the **corpus luteum regresses and dies** (called **luteolysis**). This leads to a dramatic **decline** in levels of **progesterone, estrogen, and inhibin** by day 24 of the menstrual cycle.

Event 2: The **pituitary gonadotrope** perceives the end of luteal function as a release from negative feedback (see Fig. 44.23B, Late Luteal Phase). This leads to an **increase** in **FSH** about 2 days before the onset of menstruation. The basis for the selective increase in FSH is incompletely understood, but it may be due to the slow frequency of GnRH pulses during the luteal phase, which in turn is due to high progesterone levels.

Event 3: The rise in FSH levels recruits a crop of **large (2–5 mm in diameter) antral follicles** to begin rapid, highly gonadotropin-dependent growth. These follicles produce **low levels of estrogen and inhibin B.**

Event 4: The gonadotrope responds to the slowly rising levels of estrogen and inhibin B by **decreasing FSH secretion** (see Fig. 44.23A, Early Follicular Phase). The absence of progesterone promotes an **increase** in the **frequency of GnRH pulses,** thereby selectively **increasing LH** synthesis and secretion by the gonadotrope. Thus the **LH:FSH ratio slowly increases** throughout the follicular phase.

Event 5: The ovary's response to declining FSH levels is **follicular atresia of all of the recruited follicles** except for one dominant follicle (see Fig. 44.23A, Early Follicular Phase). Thus the process of **selection** is driven by an extreme dependency of follicles on FSH in the face of declining FSH secretion. Usually only the largest follicle with the most FSH receptors and best blood supply can survive. This follicle produces **increasing amounts of estradiol-17β and inhibin B**. A critical action of FSH at this time is induction of the expression of **LH receptors** in the **mural granulosa cells** of the dominant follicle (see Fig. 44.23A, Late Follicular Phase).

Event 6: Once the dominant follicle causes **circulating estrogen levels to exceed 200 pg/mL for about 50 hours** in women, estrogen exerts a **positive feedback** on the gonadotrope to produce the **midcycle LH surge**. This is enhanced by the small amount of progesterone secreted at midcycle. The exact mechanism of the positive feedback is unknown, but it occurs largely at the level of the pituitary. **GnRH receptors** and the sensitivity to GnRH signaling increase dramatically in the gonadotropes. The hypothalamus contributes to the gonadotropin surge by increasing the frequency of GnRH pulses.

Event 7: The LH surge drives **meiotic maturation, ovulation**, and **differentiation of granulosa cells** into progesterone-producing cells (see Fig. 44.23A, Late Follicular Phase).

Event 8: **Rising levels of progesterone, estrogen, and inhibin A** by the mature **corpus luteum** negatively feed back on pituitary gonadotropes. Even though estradiol levels exceed the 200-pg/mL threshold for positive feedback, the high progesterone levels now produced by the corpus luteum block any positive feedback of estradiol. Consequently both **FSH and LH levels decline** to basal levels (see Fig. 44.23B, Mid-Luteal Phase).

Event 9: **Basal levels of LH** (but not FSH) are absolutely required for **normal corpus luteum function.** However, the corpus luteum becomes progressively insensitive to LH signaling and will die unless LH-like activity (i.e., hCG from an implanted embryo) increases. In a nonfertile cycle the corpus luteum of menstruation will regress in 14 days, and progesterone and estrogen levels will start to decline by about 10 days, thereby cycling back to event 1 (see Fig. 44.23B, Late Luteal Phase).

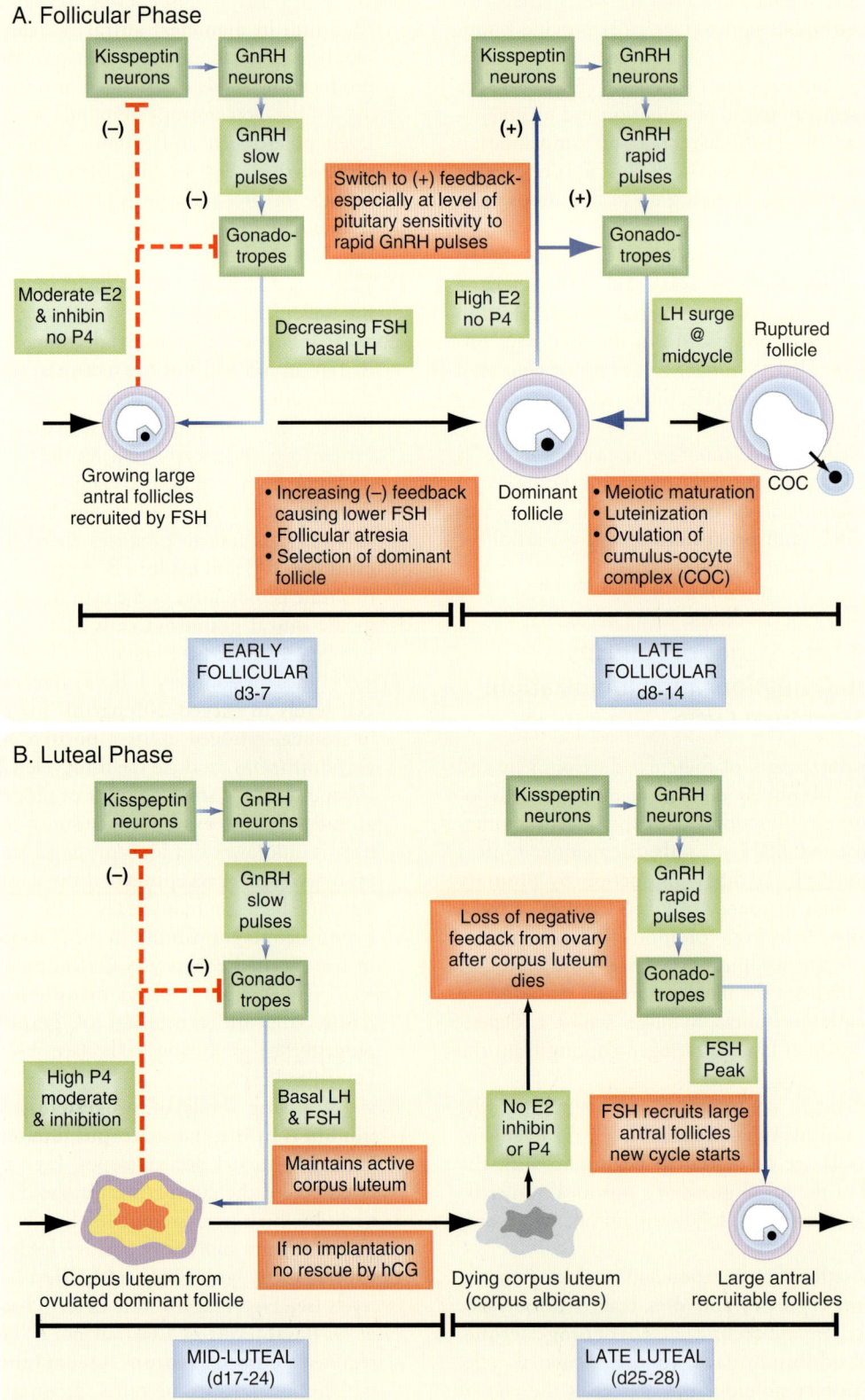

• Fig. 44.23 A, Endocrine signaling leading to the ovulation of a dominant follicle at the end of the follicular phase of the menstrual cycle. B, Endocrine signaling during the luteal phase of a non-pregnant menstrual cycle leading to the death of the corpus luteum and recruitment of follicles to begin next cycle.

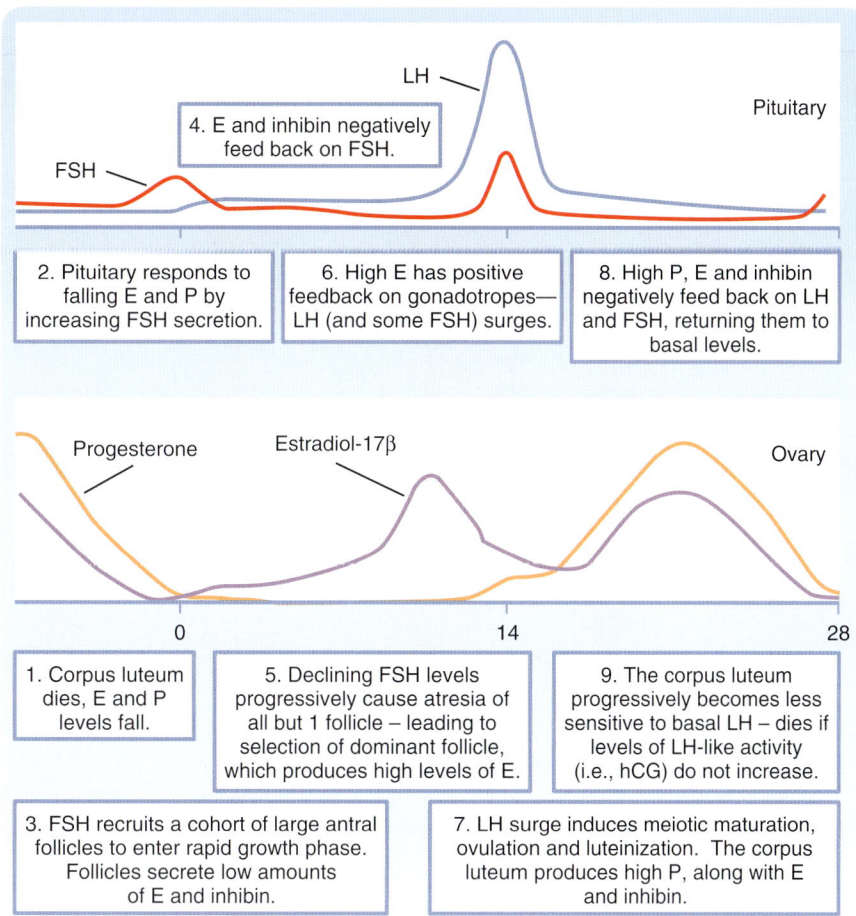

2. Pituitary responds to falling E and P by increasing FSH secretion.

4. E and inhibin negatively feed back on FSH.

6. High E has positive feedback on gonadotropes— LH (and some FSH) surges.

8. High P, E and inhibin negatively feed back on LH and FSH, returning them to basal levels.

1. Corpus luteum dies, E and P levels fall.

5. Declining FSH levels progressively cause atresia of all but 1 follicle – leading to selection of dominant follicle, which produces high levels of E.

9. The corpus luteum progressively becomes less sensitive to basal LH – dies if levels of LH-like activity (i.e., hCG) do not increase.

3. FSH recruits a cohort of large antral follicles to enter rapid growth phase. Follicles secrete low amounts of E and inhibin.

7. LH surge induces meiotic maturation, ovulation and luteinization. The corpus luteum produces high P, along with E and inhibin.

• **Fig. 44.24** The human menstrual cycle, with emphasis on the "dialogue" between ovary and pituitary gonadotropes. Note that the relative changes in the levels of E_2 and inhibin are shown by the same line.

From this sequence of events it is evident that the **ovary is the primary clock for the menstrual cycle.** The timing of the two main pituitary-based events—the transient rise in FSH that recruits large antral follicles and the LH surge that induces ovulation—is determined by two ovarian events. These are respectively the highly regular life span of the corpus luteum and its demise after 14 days, and growth of the dominant follicle to the point at which it can maintain a sustained high production of estrogen that induces a switch to positive feedback at the pituitary. In essence the dominant follicle tells the pituitary it is ready to undergo ovulation and luteinization.

The Oviduct

Structure and Function

The **oviducts** (also called the **uterine tubes** and the **fallopian tubes**) are muscular tubes with the distal ends close to the surface of each ovary and the proximal ends traversing the wall of the uterus. The oviducts are divided into four sections (going from distal to proximal): the **infundibulum,** or open end of the oviduct, which has fingerlike projections called **fimbriae** that sweep over the surface of the ovary; the

ampulla, which has a relatively wide lumen and extensive folding of the mucosa; the **isthmus,** which has a relatively narrow lumen and less mucosal folding; and the **intramural** or **uterine segment,** which extends through the uterine wall at the superior corners of the uterus (Fig. 44.25).

The main functions of the oviducts are to:

1. Capture the **cumulus-oocyte complex** at ovulation and transfer the complex to a midway point (the **ampullary-isthmus junction**), where **fertilization** takes place. Oviductal secretions coat and infuse the cumulus-oocyte complex and are likely required for viability and fertilizability.

2. Provide a site for **sperm storage.** Women who ovulate up to 5 days after sexual intercourse can get pregnant. Sperm remain viable by adhering to the epithelial cells lining the isthmus. The secretions of the oviduct also induce **capacitation** and **hyperactivity of sperm.**

3. Secrete fluids that provide nutritional support to the **preimplantation embryo.**

The timing of movement of the embryo into the uterus is critical because the uterus has an implantation window of about 3 days. The oviduct needs to hold the early embryo until it reaches the blastocyst stage (5 days after fertilization) and then allow it to pass into the uterine cavity.

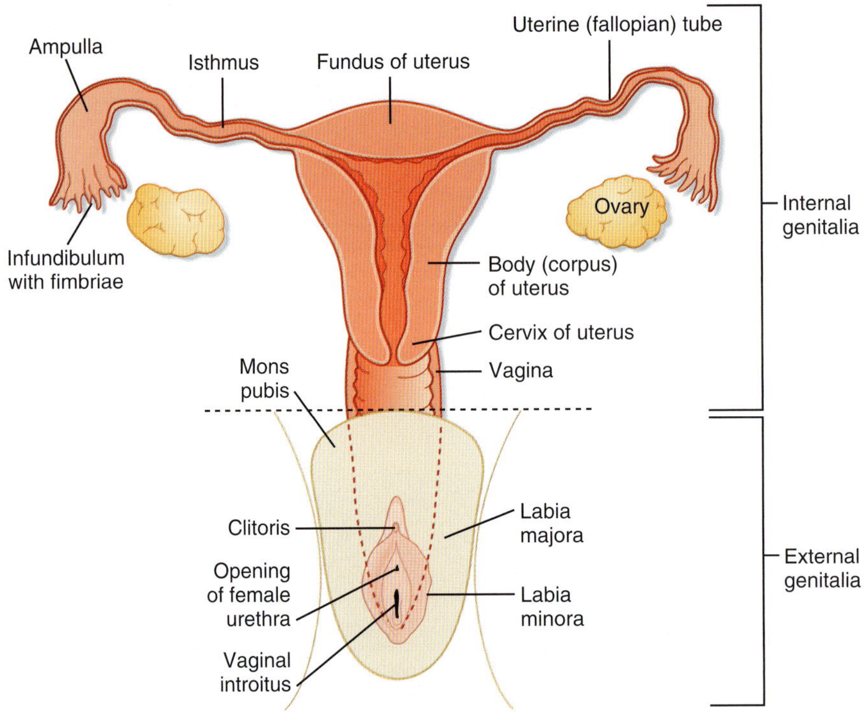

• **Fig. 44.25** Schematic of the female reproductive system. (Modified from White BA, Porterfield SP. *Endocrine Physiology.* 4th ed. Philadelphia: Mosby; 2013.)

The wall of the oviduct is composed of a mucosa (called the **endosalpinx**), a two-layered muscularis (called the **myosalpinx**), and an outer-lying connective tissue (the **perisalpinx**). The endosalpinx is thrown into many folds, almost to the extent that the lumen is obliterated, and is lined by a simple epithelium made up of two cell types: **ciliated cells** and **secretory cells.** The cilia are most numerous at the infundibular end and propel the cumulus-oocyte complex toward the uterus. The cilia on the fimbriae are the sole mechanism for transport of the ovulated cumulus-oocyte complex. Once the complex passes through the ostium of the oviduct and enters the ampulla, it is moved by both cilia and peristaltic contractions of the muscularis.

The secretory cells produce a protein-rich mucus that is conveyed along the oviduct to the uterus by the cilia. This ciliary-mucus escalator maintains a healthy epithelium, moves the cumulus-oocyte complex toward the uterus, and may provide directional cues for swimming sperm. Movement of the cumulus-oocyte complex slows at the ampullary-isthmus junction, where fertilization normally takes place. This appears to be due in part to thick mucus produced by the isthmus and increased tone of the muscularis of the isthmus. The composition of oviductal secretions is complex and includes growth factors, enzymes, and oviduct-specific glycoproteins. Note that the clinical process of in vitro fertilization has shown that secretions of the oviduct are not absolutely necessary for fertility. However, normal oviductal function is absolutely required for both fertilization and implantation after in vivo insemination (i.e., natural sexual intercourse). Normal oviductal function also minimizes the risk of **ectopic implantation** and **ectopic pregnancy,** which occurs most often within the oviduct.

Hormonal Regulation During the Menstrual Cycle

In general, estrogen secreted during the follicular phase increases epithelial cell size and height in the endosalpinx. Estrogen increases blood flow to the lamina propria of the oviducts, promotes production of oviduct-specific glycoproteins (whose functions are poorly understood), and increases ciliogenesis throughout the oviduct. Estrogen promotes secretion of thick mucus in the isthmus and increases the tone of the muscularis of the isthmus, thereby keeping the cumulus-oocyte complex at the ampullary-isthmus junction for fertilization. High progesterone, along with estrogen, during the early luteal to midluteal phase decreases epithelial cell size and function. Progesterone promotes deciliation. It also decreases the secretion of thick mucus and relaxes the tone in the isthmus. In addition it should be noted that oviductal epithelial cells express the LH receptor, which may synergize with estrogen to optimize oviductal function during the periovulatory period.

The Uterus

Structure and Function

The **uterus** is a single organ that sits in the midline of the pelvic cavity between the bladder and the rectum. The mucosa of the uterus is called the **endometrium,** the three-layered thick muscularis is called the **myometrium,** and the outer connective tissue and serosa are called the **perimetrium.** The parts of the uterus are (1) the **fundus,** which is the portion that rises superiorly from the entrance of the oviducts, (2) the **body** of the uterus, which makes up most of the uterus, (3) the **isthmus,** a short narrowed part of the body at its inferior end, and (4) the **cervix,** which extends into the **vagina** (see Figs. 44.13 and 43.25). Because the cervical mucosa is distinct from the rest of the uterus and does not undergo the process of menstruation, it will be discussed separately later.

The established functions of the uterus are all related to fertilization and pregnancy (discussed later). The main functions of the uterus are to:

1. assist movement of sperm from the vagina to the oviducts
2. provide a suitable site for attachment and implantation of the blastocyst, including a thick nutrient-rich stroma
3. limit the invasiveness of the implanting embryo so it stays in the endometrium and does not reach the myometrium

4. provide a maternal side of the mature placental architecture, including the basal plate to which the fetal side attaches, and large intervillous spaces that become filled with maternal blood after the first trimester
5. grow and expand with the growing fetus so it develops within an aqueous nonadhesive environment
6. provide strong muscular contractions to expel the fetus and placenta at term

To understand the function of the uterus and uterine changes during nonfertile menstrual cycles, the fine structure of the endometrium and the relationship of uterine blood supply to the endometrium will be reviewed (Fig. 44.26). The luminal surface of the endometrium is covered with a simple cuboidal/columnar epithelium. The epithelium is

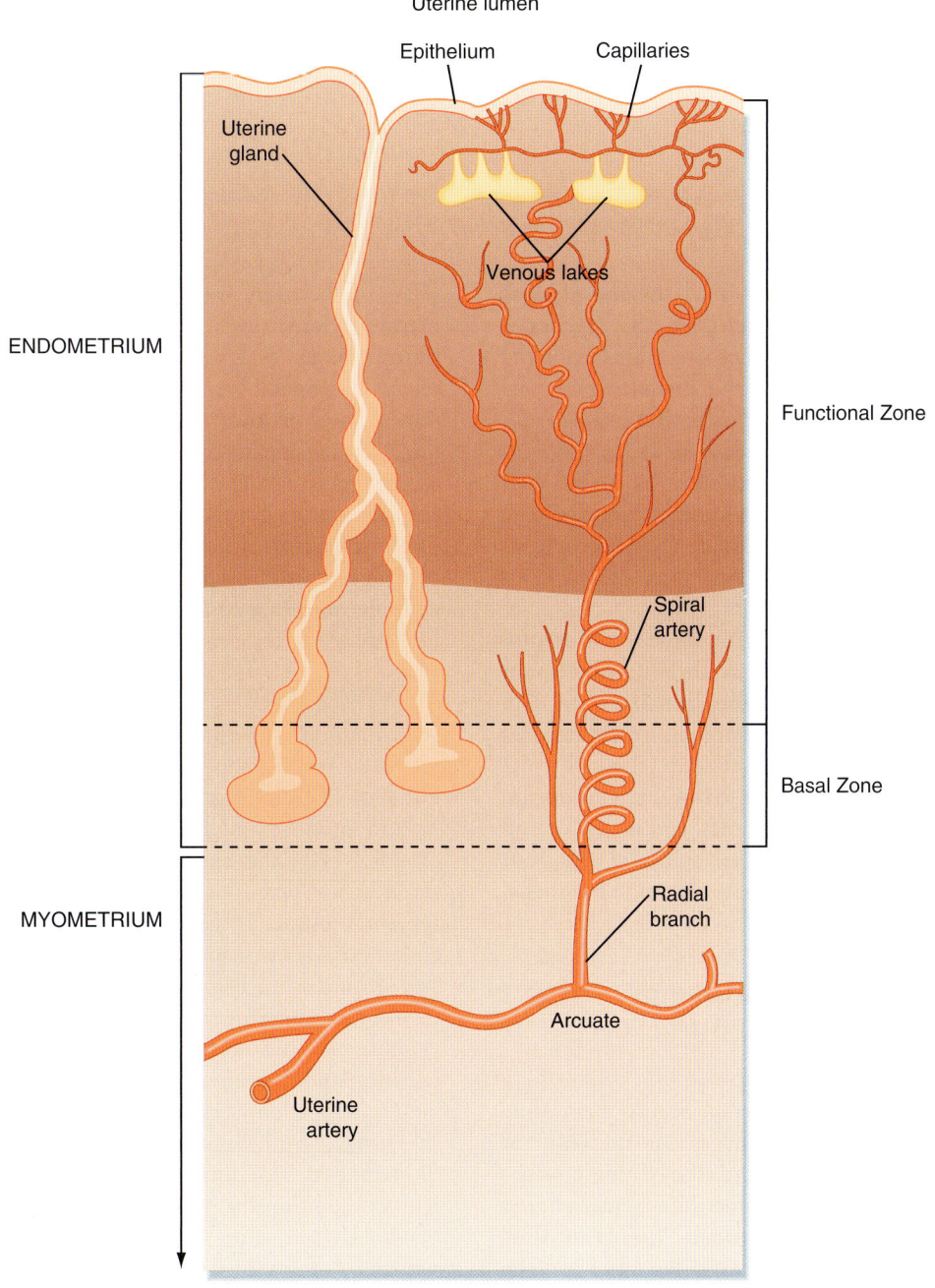

• **Fig. 44.26** Diagram of the organization of glands and blood flow within the uterine endometrium. (From Straus III. In: Yen SSC et al [eds]. *Reproductive Endocrinology.* 4th ed. Philadelphia: Saunders; 1999.)

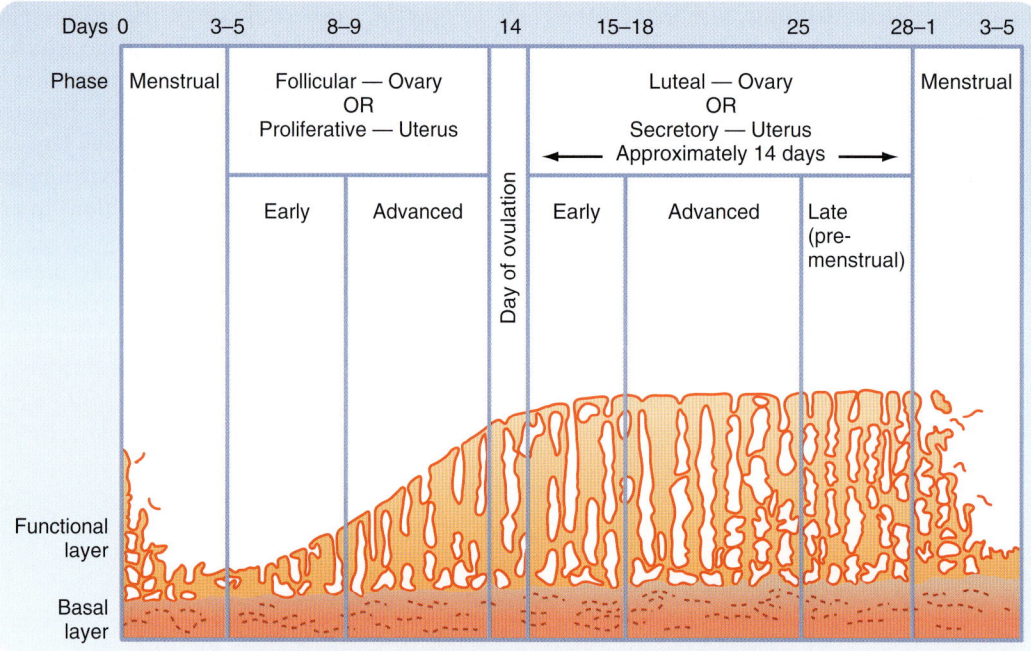

• **Fig. 44.27** The menstrual cycle of the uterine endometrium. (Modified from White BA, Porterfield SP. *Endocrine Physiology.* 4th ed. Philadelphia: Mosby; 2013.)

continuous with mucosal glands (called **uterine glands**) that extend deep into the endometrium. The mucosa is vascularized by **spiral arteries,** which are branches of the **uterine artery** that run through the myometrium. The terminal arterioles of the spiral arteries project just beneath the surface epithelium. These arterioles give rise to a subepithelial plexus of capillaries and venules that have ballooned thin-walled segments called **venous lakes** or **lacunae.** The lamina propria itself is densely cellular. The stromal cells of the lamina propria play important roles during both pregnancy and menstruation.

About two-thirds of the luminal side of the endometrium is lost during menstruation and is called the **functional zone** (also called the **stratum functionalis**) (see Fig. 44.26). The basal third of the endometrium that remains after menstruation is called the **basal zone** (also called the **stratum basale**). The basal zone is fed by straight arteries that are separate from the spiral arteries, and it contains all the cell types of the endometrium (i.e., epithelial cells from the remaining tips of glands, stromal cells, and endothelial cells).

Hormonal Regulation of the Uterine Endometrium During the Menstrual Cycle

Proliferative Phase

Monthly oscillations in ovarian steroids induce the uterine endometrium to enter different stages. At the time of selection of the dominant follicle and its elevating production of estradiol, the uterine endometrium is just ending menstruation. The stratum functionalis has been shed and only the stratum basale remains (Fig. 44.27). The rising levels of estrogen during the mid to late follicular phase of the ovary induce the **proliferative phase** of the uterine endometrium. Estrogen induces all cell types in the stratum basale to grow and divide. In fact the definition of an **"estrogenic"** compound has historically been one that is **"uterotropic."** Estrogen increases cell proliferation directly through its cognate receptors (ER-α and ER-β), which regulate gene expression (Fig. 44.28). Estrogen also controls uterine growth indirectly through local production of growth factors. In addition, estrogen induces expression of **progesterone receptors,** thereby "priming" the uterine endometrium so it can respond to progesterone during the luteal phase of the ovary.

Secretory Phase

By ovulation, the thickness of the stratum functionalis has been reestablished under the proliferative actions of estradiol-17β (see Fig. 44.27). After ovulation the corpus luteum produces high levels of progesterone along with estradiol-17β. The luteal phase of the ovary switches the proliferative phase of the uterine endometrium to the **secretory phase.** In general, progesterone inhibits further endometrial growth and induces differentiation of epithelial and stromal cells. Progesterone induces the uterine glands to secrete a nutrient-rich product that supports blastocyst viability. As the secretory phase proceeds the mucosal uterine glands become corkscrewed and sacculated (see Fig. 44.27). Progesterone also induces changes in adhesivity of the surface epithelium, thereby generating the "window of receptivity" for implantation of an embryo (see Pregnancy). Additionally, progesterone promotes differentiation of stromal cells into **"predecidual cells,"** which must be prepared to form the **decidua** of pregnancy or to orchestrate menstruation in the absence of pregnancy.

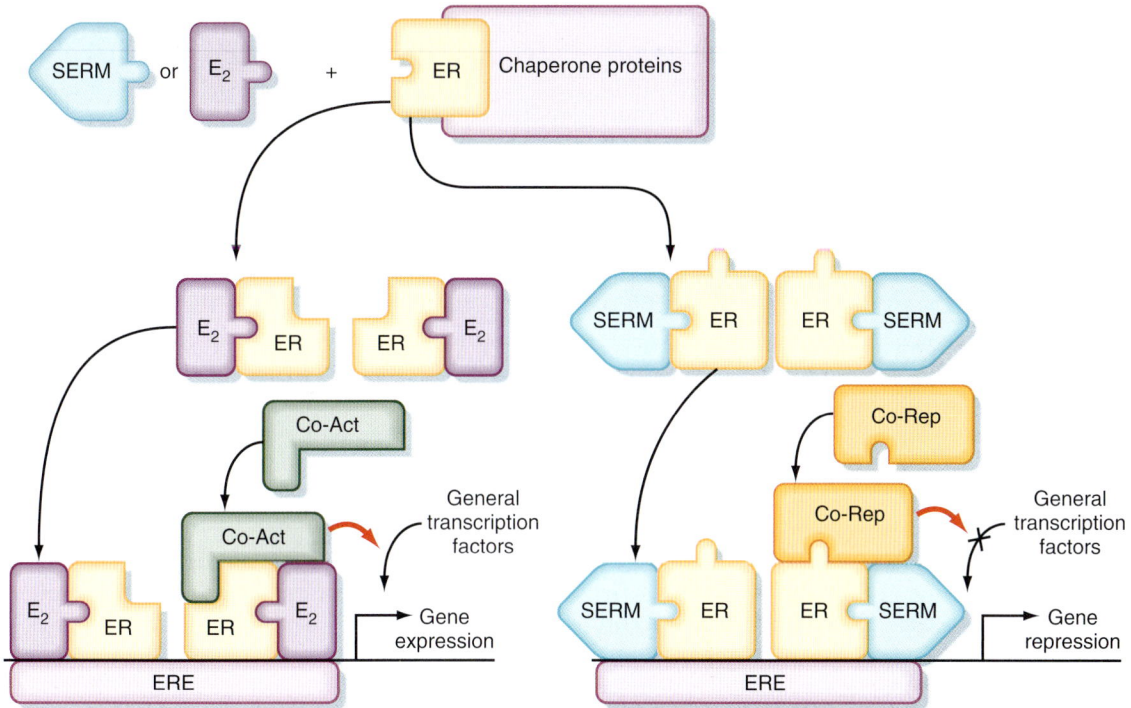

• **Fig. 44.28** Molecular mechanism by which the estrogen receptor (ER) regulates gene expression. *Left,* Estradiol-17β binds to the ER and changes its conformation so that it binds as a dimer to estrogen-response element (ERE) and recruits coactivator proteins (Co-Act), which leads to stimulation of gene expression. *Right,* Selective estrogen receptor modulators (SERMs), such as tamoxifen in the breast, alter ER conformation so that it recruits co-repressor proteins (Co-Rep), thereby inhibiting gene expression. In this case the SERM acts as an ER antagonist, but in some tissues the same SERM can act as an ER agonist. (Modified from White BA, Porterfield SP. *Endocrine Physiology.* 4th ed. Philadelphia: Mosby; 2013.)

AT THE CELLULAR LEVEL

Progesterone opposes the proliferative actions of estradiol-17β and downregulates the estrogen receptor (ER). Progesterone also induces **inactivating isoforms of 17β-HSD,** thereby converting the active estradiol-17β into the inactive estrone. This opposition of the mitogenic actions of estradiol-17β by progesterone is important to protect the uterine endometrium from estrogen-induced uterine cancer. In contrast, administration of **"unopposed estrogen"** to women significantly increases the risk for uterine cancer.

Drugs called **selective estrogen receptor modulators (SERMs)** have been developed that inhibit ER function in a tissue-specific manner (see Fig. 44.28). For example, the SERM **tamoxifen** is used as an ER antagonist for the treatment of breast cancer (whose early progression is promoted by estrogen). Binding of SERM to the ER induces conformational changes that allow co-repressors to bind to the ER or promote degradation of the ER (or both; see Fig. 44.28). Because tamoxifen has some uterotropic activity (i.e., makes uterine endometrial tissue grow), newer SERMs such as **raloxifene** have been developed to have ER antagonist activity on the breast, beneficial ER agonist activity on bone (see later), and no activity or ER antagonistic activity on the uterine endometrium.

Menstrual Phase

In a nonfertile cycle, death of the corpus luteum results in sudden withdrawal of progesterone, which leads to changes in the uterine endometrium that result in loss of the lamina functionalis (see Fig. 43.27). **Menstruation** normally lasts for 4 to 5 days (called a **period**), and the volume of blood loss ranges from 25 to 35 mL. Menstruation coincides with the early follicular phase of the ovary.

Hormonal Regulation of the Myometrium

The smooth muscle cells of the myometrium are also responsive to changes in steroid hormones. Peristaltic contractions of the myometrium favor movement of the luminal contents from the cervix to the fundus at ovulation, and these contractions probably play a role in rapid bulk transport of ejaculated sperm from the cervix to the oviducts. During menstruation, contractions propagate from the fundus to the cervix, thereby promoting expulsion of sloughed stratum functionalis. The size and number of smooth muscle cells are determined by estrogen and progesterone. Healthy cycling women maintain a robust myometrium, whereas the myometrium progressively thins in postmenopausal women. The most drastic changes are seen during pregnancy, when the smooth muscle cells increase from 50 to 500 μm in length. The pregnant myometrium also has a greater number of smooth muscle cells and more extracellular matrix.

 IN THE CLINIC

Disorders of menstruation are relatively common and include **menorrhagia** (heavy menstrual flow leading to loss of more than 80 mL of blood), **metrorrhagia** irregular and sometimes prolonged menstrual flow between normal periods), and **dysmenorrhea** (painful periods). The existence of a few irregular periods, called **oligomenorrhea,** and the absence of periods, called **amenorrhea,** are often due to dysfunction of the hypothalamic-pituitary-ovarian axis as opposed to local pelvic pathophysiology.

Because endometrial tissue is naturally sloughed in fragments that contain viable cells, endometrial tissue occasionally gains access to other parts of the female tract (e.g., oviducts, ovary), as well as the lower part of the abdomen and associated structures (e.g., rectouterine pouch of Douglas, as shown in Fig. 44.13). These implants give rise to **endometriosis**—a foci of hormonally responsive endometrial tissue outside the uterus. The spread of endometriosis may be due to reflux of menstrual tissue into the oviducts or movement of tissue through lymphatics, or both. Endometriosis frequently exhibits cyclic bleeding and is associated with infertility, pain on defecation, pain on urination, pain with sexual intercourse, or generalized pelvic pain.

The Cervix

Structure and Function

The cervix is the inferior extension of the uterus that projects into the vagina (see Figs. 44.13 and 44.25). It has a mucosa that lines the **endocervical canal,** which has a highly elastic lamina propria and a muscularis that is continuous with the myometrium. The part of the cervix that extends into the vaginal vault is called the **ectocervix,** whereas the part surrounding the endocervical canal is called the **endocervix.** The openings of the endocervical canal at the uterus and vagina are called the **internal cervical os** and the **external cervical os,** respectively. The cervix acts as a gateway to the upper female tract—at midcycle the endocervical canal facilitates sperm viability and entry. During the luteal phase the endocervical canal impedes passage of sperm and microbes, thereby inhibiting **superimplantation** of a second embryo or ascending infection into the placenta, fetal membranes, and fetus. The cervix physically supports the weight of the growing fetus. At term, **cervical softening and dilation** allow passage of the newborn and placenta from the uterus into the vagina.

Hormonal Regulation of Cervical Mucus During the Menstrual Cycle

The endocervical canal is lined by simple columnar epithelium that secretes **cervical mucus** in a hormonally responsive manner. Estrogen stimulates production of a copious quantity of thin, watery, slightly alkaline mucus that is an ideal environment for sperm. Progesterone stimulates production of a scant, viscous, slightly acidic mucus that is hostile to sperm. During the normal menstrual cycle the conditions of the cervical mucus are ideal for sperm penetration and viability at the time of ovulation.

The Vagina

Structure and Function

The vagina is one of the copulatory structures in women and acts as the birth canal (see Figs. 44.13 and 44.25). Its mucosa is lined by a nonkeratinized stratified squamous epithelium. The mucosa has a thick lamina propria enriched with elastic fibers and is well vascularized. There are no glands in the vagina, so lubrication during intercourse comes from (1) cervical mucus (especially with intercourse that occurs midcycle), (2) a transudate (i.e., ultrafiltrate) from the blood vessels of the lamina propria, and (3) the vestibular glands. The mucosa is surrounded by a relatively thin (i.e., relative to the uterus and cervix) two-layered muscularis and an outer connective tissue. The vaginal wall is innervated by branches of the pudendal nerve, which contribute to sexual pleasure and orgasm during intercourse.

Hormonal Regulation During the Menstrual Cycle

The superficial cells of the vaginal epithelium are continually desquamating, and the nature of these cells is influenced by the hormonal environment. Estrogen stimulates proliferation of the vaginal epithelium and increases its glycogen content (referred to as **"cornification"**—but in humans, true cornification or keratinization does not occur). The glycogen is metabolized to lactic acid by commensal lactobacilli, thereby maintaining an acidic environment. This inhibits infection by noncommensal bacteria and fungi. Progesterone increases the desquamation of epithelial cells.

The External Genitalia

Structure and Function

The female external genitalia are surrounded by the **labia majora** (homologues of the scrotum) laterally and the **mons pubis** anteriorly (see Fig. 44.25). The **vulva** collectively refers to an area that includes the labia majora and mons pubis plus the **labia minora,** the **clitoris,** the **vestibule of the vagina,** the **vestibular bulbs** (glands), and the **external urethral orifice.** The vulva is also referred to as the **pudendum** by clinicians. The structures of the vulva serve the functions of sexual arousal and climax, directing the flow of urine, and partially covering the opening of the vagina, thereby inhibiting entry of pathogens.

The clitoris is the embryological homologue of the penis and is composed of two **corpora cavernosa,** which attach the clitoris to the ischiopubic rami, and a **glans.** These structures are composed of erectile tissue and undergo the process of erection in essentially the same manner as the penis. Unlike the penis, clitoral tissue is completely separate from the urethra. Thus the clitoris is involved in sexual arousal and climax at orgasm. The vagina is likewise involved in sexual satisfaction but also serves as the copulatory organ and birth canal.

Hormonal Regulation During the Menstrual Cycle

The structures of the vulva do not show marked changes during the menstrual cycle. However, the health and function of these structures are dependent on hormonal support. The external genitalia and vagina are responsive to androgens (testosterone and dihydrotestosterone) and estrogen. Androgens also act on the central nervous system (CNS) to increase libido in women.

Biology of Estradiol-17β and Progesterone

Biological Effects of Estrogen and Progesterone

Estradiol-17β and progesterone fluctuate during the menstrual cycle, and they have multiple effects that can be categorized according to whether they are directly related to the reproductive system or not. Both hormones have profound effects on the ovary, oviduct, uterus, cervix, vagina, and external genitalia and on the hypothalamus and pituitary. Estrogen and progesterone also have important effects on nonreproductive tissues:

Bone: Estrogen is required for closure of the epiphyseal plates of long bones in both sexes. Estradiol-17β has a **bone anabolic** and **calciotropic effect** (see Chapter 40). It stimulates intestinal Ca^{++} absorption. Estradiol-17β is also one of the most potent regulators of osteoblast and osteoclast function. Estrogen promotes survival of osteoblasts and apoptosis of osteoclasts, thereby favoring bone formation over resorption. Low estrogen levels associated with menopause leads to bone loss and **osteoporosis.**

Liver: The overall effect of estradiol-17β on the liver is to improve circulating lipoprotein profiles. Estrogen increases expression of the **LDL receptor,** thereby increasing clearance of cholesterol-rich **LDL** particles by the liver. Estrogen also increases circulating levels of **HDL.** Estrogen regulates hepatic production of several transport proteins, including cortisol-binding protein, thyroid hormone–binding protein, and SHBG.

Cardiovascular organs: Premenopausal women have significantly less cardiovascular disease than men or postmenopausal women do. Estrogen promotes vasodilation through increased production of **nitric oxide,** which relaxes vascular smooth muscle and inhibits platelet activation. Single-nucleotide polymorphisms in the estrogen receptor have been associated with increased cardiovascular disease.

Integument: Estrogen and progesterone maintain healthy smooth skin with normal epidermal and dermal thickness. Estrogen stimulates proliferation and inhibits apoptosis of keratinocytes. In the dermis, estrogen and progesterone increase collagen synthesis and inhibit breakdown of collagen by suppressing matrix metalloproteinases. Estrogen also increases glycosaminoglycan production and deposition in the dermis and promotes wound healing.

CNS: Estrogen is neuroprotective—that is, it inhibits neuronal cell death in response to hypoxia or other insults. Estrogen's positive effects on angiogenesis may account for some of the beneficial and stimulant-like actions of estrogen on the CNS. Progesterone acts on the hypothalamus to increase the set point for thermoregulation, thereby elevating body temperature approximately 0.5°F. This is the basis for using body temperature measurements to determine whether ovulation has occurred. Progesterone is a CNS depressant. Loss of progesterone on demise of the corpus luteum of menstruation is the basis for **premenstrual dysphoria (premenstrual syndrome [PMS]).** Progesterone also acts on the brainstem to sensitize the ventilatory response to PCO_2 so that ventilation increases and PCO_2 decreases.

Adipose tissue: Estrogen decreases adipose tissue by decreasing lipoprotein lipase activity and increasing hormone-sensitive lipase (i.e., it has a lipolytic effect). Loss of estrogen results in accumulation of adipose tissue, especially in the abdomen.

Transport and Metabolism of Ovarian Steroids

Steroid hormones are slightly soluble in blood and are bound to plasma proteins. Approximately 60% of the estrogen is transported bound to **SHBG,** 20% is bound to albumin, and 20% is in the free form. Progesterone binds primarily to **cortisol-binding globulin (transcortin)** and albumin. Because it has relatively low binding affinity for these proteins, its circulating half-life is about 5 minutes.

Although the ovary is the primary site of estrogen production, peripheral aromatization of androgens to estrogens can generate locally high levels of estradiol-17β in some tissues. Peripheral conversion of adrenal and ovarian androgens serves as an important source of estrogen after menopause (discussed later). The fact that CYP19 (aromatase) is expressed in the breast is the basis for the use of **aromatase inhibitors** in the treatment of estrogen-dependent breast cancer in postmenopausal women.

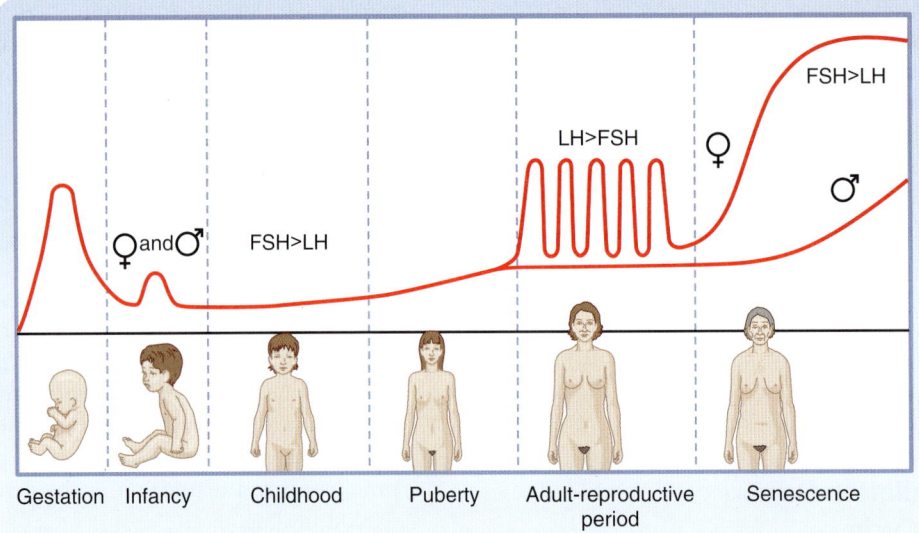

• **Fig. 44.29** Pattern of gonadotropin secretion throughout life. Note the transient peaks during gestation and early infancy and the low levels thereafter in childhood. Women subsequently have monthly cyclic bursts, with luteinizing hormone (LH) exceeding follicle-stimulating hormone (FSH); men do not. Both genders show increased gonadotropin production after age 50, with FSH exceeding LH.

Estrogens and progestins are degraded in the liver to inactive metabolites, conjugated with sulfate or glucuronide, and excreted in urine. Major metabolites of estradiol include estrone, estriol, and catecholestrogens (2-hydroxyestrone and 2-methoxyestrone). The major metabolite of progesterone is pregnanediol, which is conjugated with glucuronide and excreted in urine.

Ontogeny of the Reproductive Systems

Unlike most other organ systems, the reproductive systems undergo significant changes in their activity during the life span of a man or woman (Fig. 44.29). Development of the reproductive systems occurs in utero and results in female or male fetuses. After birth and during infancy, the reproductive systems are largely quiescent. At puberty the hypothalamic-pituitary-gonadal axes "wake up" and the gonads begin producing sex steroids, which in turn induce the sexually dimorphic changes in appearance and behavior associated with men and women. The reproductive life span of women is set by their ovarian reserve and degree of follicular development (see earlier) and ends at menopause, usually in the fifth decade of life. Loss of estrogen production by the ovaries has a clear clinical impact on many postmenopausal women. Men continue to produce sperm throughout life but can experience a decline in androgen production (andropause), which is associated with its own clinical sequelae.

Pregnancy

The reproductive system of women undergoes dramatic changes during pregnancy. Production of gonadotropin and gonadal steroids is switched from the maternal hypothalamic-pituitary-ovarian axis, which is strongly repressed during pregnancy, to the **fetal placenta.** Indeed, it is the endocrine function of fetal placental tissue that (1) maintains a quiescent gravid uterus, (2) alters maternal physiology to ensure fetal nutrition in utero, (3) alters maternal pituitary function and mammary gland development to ensure ongoing fetal nutrition after birth, and (4) determines the time of labor and delivery (also called **parturition**). The placenta also plays an important role in fetal testosterone production and male differentiation of the reproductive system before the fetal hypothalamus and pituitary develop into a functional axis.

Fertilization, Early Embryogenesis, Implantation, and Placentation

Synchronization With Maternal Ovarian and Reproductive Tract Function

Fertilization, early embryogenesis, implantation, and early gestation are all synchronized with the human menstrual cycle (Fig. 44.30). Just before ovulation the ovary is in the late follicular stage and produces high levels of estrogen. Estrogen promotes growth of the uterine endometrium and induces expression of the progesterone receptor. Estrogen ultimately induces the LH surge, which in turn induces meiotic maturation of the oocyte and ovulation of the cumulus-oocyte complex.

The events between fertilization and implantation take about 6 days to complete, so implantation occurs at about day 22 of the menstrual cycle. At this time the ovary is in the midluteal phase and secreting large amounts of progesterone. Progesterone stimulates secretion from the uterine glands, which provide nutrients to the embryo. This is referred to as *histiotropic nutrition* and is an important

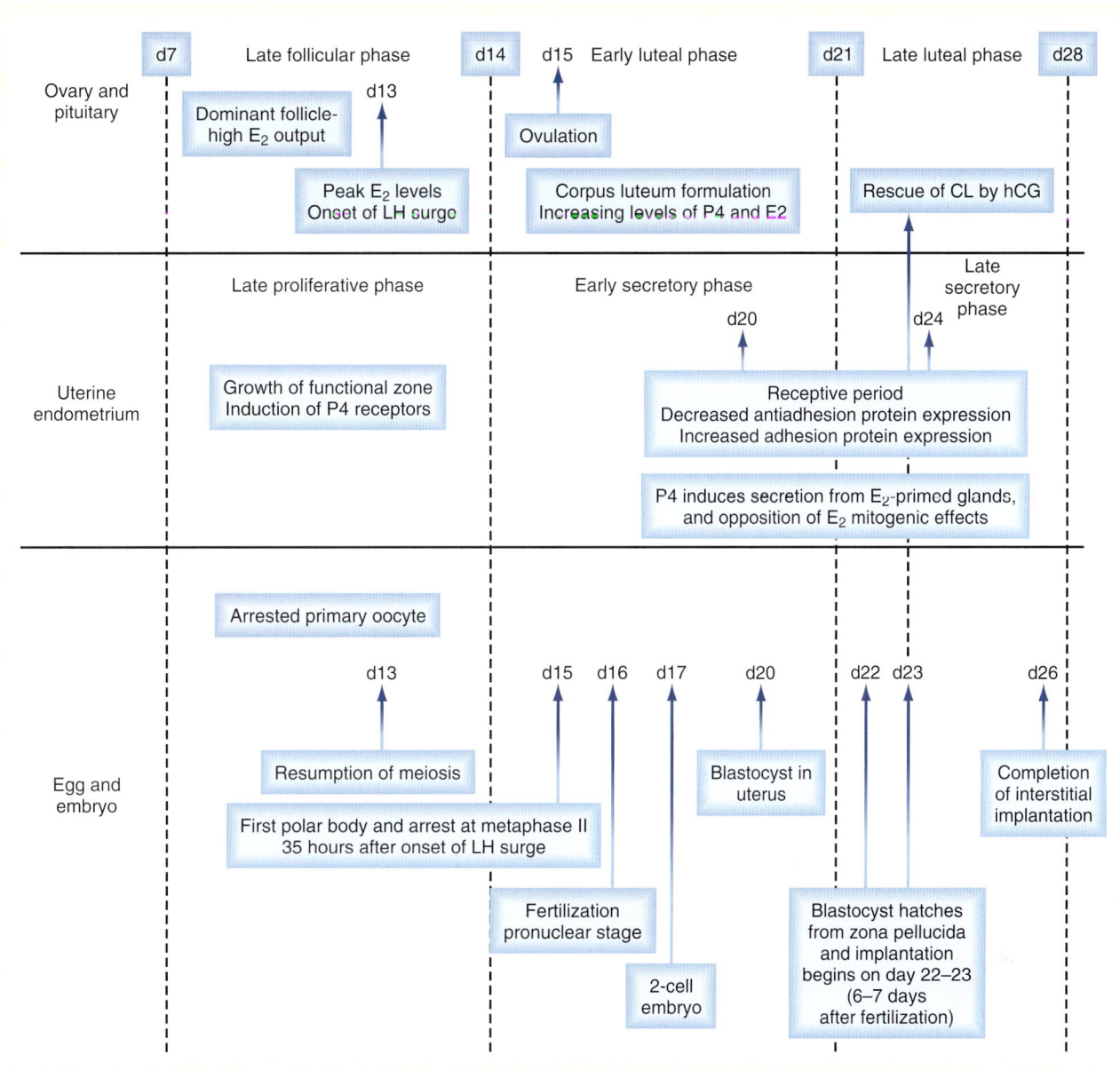

• **Fig. 44.30** Synchronization of events of the menstrual cycle (ovary and endometrium) with fertilization, early embryonic development within the oviduct, and implantation of embryo (blastocyst) into the uterine endometrium. E_2, estradiol; P4, progesterone. (Modified from White BA, Porterfield SP. *Endocrine Physiology.* 4th ed. Philadelphia: Mosby; 2013.)

mode of maternal-to-fetal transfer of nutrients for about the first trimester of pregnancy, after which it is replaced by hemotropic nutrition (see later). Progesterone inhibits myometrial contraction and prevents release of paracrine factors (e.g., cytokines, prostaglandins, chemokines, and vasoconstrictors) that lead to menstruation. Progesterone induces the **"window of receptivity"** in the uterine endometrium, which exists from about day 20 to day 24 of the menstrual cycle. This receptive phase is associated with increased adhesivity of the endometrial epithelium and involves formation of cellular extensions called **pinopodes** on the apical surface of endometrial epithelia, along with increased expression of adhesive proteins (e.g., integrins,

cadherins) and decreased expression of antiadhesive proteins (e.g., mucins) in the apical cell membrane.

When a fertilized egg implants in the uterus, the uterine endometrium is at its full thickness, is actively secreting, and is capable of tightly adhering to the implanting embryo.

Fertilization

Fertilization accomplishes both recombination of genetic material to form a new genetically distinct organism and initiation of events that begin embryonic development. Several steps must occur to achieve successful (unassisted) fertilization (Fig. 44.31):

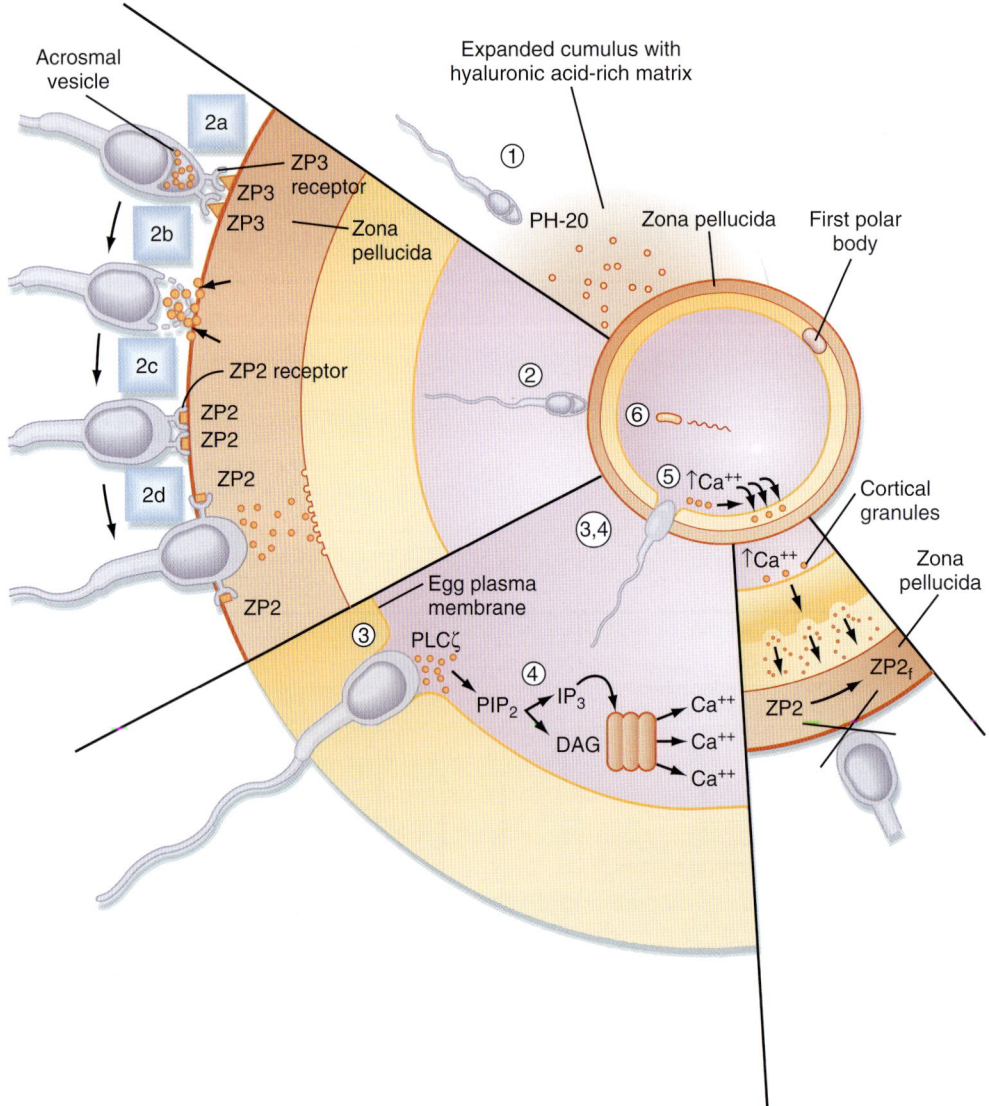

• **Fig. 44.31** Events involved in fertilization (see text for details). (Modified from White BA, Porterfield SP. *Endocrine Physiology.* 4th ed. Philadelphia: Mosby; 2013.)

Step 1: Penetration of the expanded cumulus by the sperm. This involves digestion of the extracellular matrix of the cumulus by a membrane hyaluronidase, PH-20.

Step 2: Penetration of the zona pellucida by the sperm. This involves binding of the sperm to the zona protein ZP3 (step 2a), which induces release of acrosomal enzymes (called the **acrosomal reaction** (step 2b). The sperm secondarily bind to another zona protein, ZP2 (step 2c), as the zona pellucida is digested and the sperm swims through to the egg (step 2d).

Step 3: Fusion of the sperm and egg membrane takes place.

Step 4: A Ca^{++} signaling cascade (see Chapter 3) occurs.

Step 5: The signaling cascade activates the exocytosis of enzyme-filled vesicles called **cortical granules** that reside in the outermost, or cortical, region of the unfertilized egg. The enzymes contained in the cortical granules are released to the outside of the egg upon exocytosis. These enzymes modify both ZP2 and ZP3 of the zona

pellucida such that ZP2 can no longer bind acrosome-reacted sperm, and ZP3 can no longer bind capacitated acrosome-intact sperm. Thus only one sperm usually enters the egg. Occasionally, more than one sperm does enter the egg. This results in a **triploid** cell that is unable to develop further. Therefore prevention of polyspermy is critical for normal development of the fertilized egg.

Step 6: The entire sperm enters the egg during fusion. The flagellum and mitochondria disintegrate, so most of the mitochondrial DNA in cells is maternally derived. Once inside the egg, decondensation of the sperm DNA occurs. A membrane called the *pronucleus* forms around the sperm DNA as the newly activated egg completes the second meiotic division.

In mammalian eggs a large initial release of Ca^{++} is followed by a series of subsequent smaller Ca^{++} oscillations that can last for hours. A major consequence of this signaling pathway is that it "wakes up" the metabolically quiescent

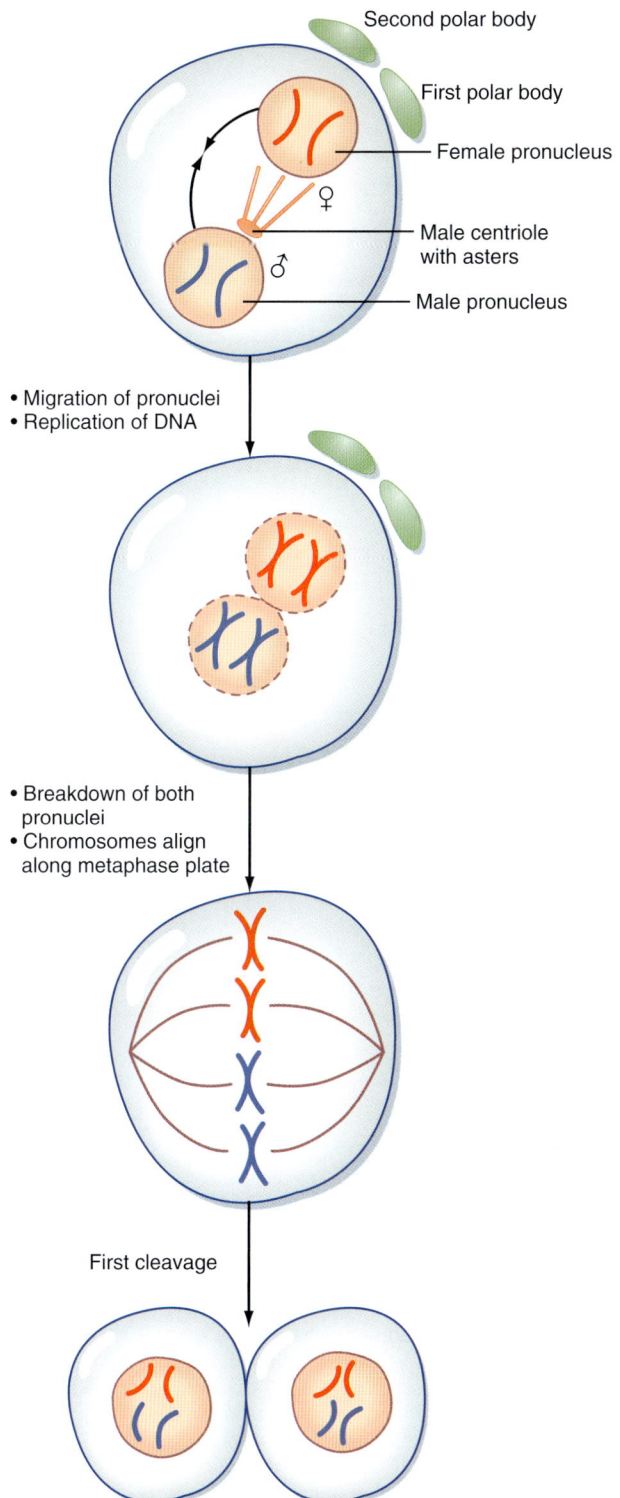

Second polar body

First polar body

Female pronucleus

♀

Male centriole
with asters

♂

Male pronucleus

• Migration of pronuclei
• Replication of DNA

• Breakdown of both
  pronuclei
• Chromosomes align
  along metaphase plate

First cleavage

• **Fig. 44.32** Overview of genetic events after fertilization up to the first embryonic cleavage. (Modified from White BA, Porterfield SP. *Endocrine Physiology.* 4th ed. Philadelphia: Mosby; 2013.)

egg so it can resume meiosis and begin embryonic development. This process is called **egg activation.**

The activated egg completes the second meiotic division as the sperm DNA decondenses and a pronucleus forms around it (Fig. 44.32). Once the egg has completed meiosis,

a pronucleus forms around the female chromosomes as well. A **centrosome** contributed by the sperm becomes a microtubule organizing center from which microtubules extend until they contact the female pronucleus. The male and female DNA replicate as the two pronuclei are pulled together. Once the pronuclei contact each other, the nuclear membranes break down, the chromosomes align on a common metaphase plate, and the first cleavage occurs.

Early Embryogenesis and Implantation

Fertilization typically occurs on day 16 to 17 of the menstrual cycle, and implantation occurs about 6 days later. Thus the first week of embryogenesis takes place within the lumens of the oviduct and uterus. For most of this time the embryo remains encapsulated by the zona pellucida. The first two cleavages take about 2 days, and the embryo reaches a 16-cell **morula** by 3 days. The outer cells of the morula become tightly adhesive with each other and begin transporting fluid into the embryonic mass. During days 4 and 5 the transport of fluid generates a cavity called the *blastocyst cavity,* and the embryo is now called a **blastocyst** (Fig. 44.33). The blastocyst is composed of two subpopulations of cells: an eccentric **inner cell mass** and an outer epithelial-like layer of **trophoblasts.** The region of the trophoblast layer immediately adjacent to the inner cell mass is referred to as the **embryonic pole,** and it is this region that attaches to the uterine endometrium at implantation (see Fig. 44.33).

The embryo resides within the oviduct during the first 3 days and then enters the uterus. By 5 to 6 days of development the trophoblasts of the blastocyst secrete proteases that digest the outer-lying zona pellucida. At this point, corresponding to about day 22 of the menstrual cycle, the **"hatched" blastocyst** is able to adhere to and implant into the receptive uterine endometrium (see Fig. 44.33).

At the time of attachment and implantation the trophoblasts differentiate into two cell types: an inner layer of **cytotrophoblasts** and an outer layer of multinuclear/multicellular **syncytiotrophoblasts** (see Fig. 44.33). The cytotrophoblasts initially provide a feeder layer of continuously dividing cells. Syncytiotrophoblasts initially perform three general types of function: adhesive, invasive, and endocrine. Syncytiotrophoblasts express adhesive surface proteins (i.e., cadherins and integrins) that bind to uterine surface epithelia and, as the embryo implants, to components of the uterine extracellular matrix. In humans the embryo completely burrows into the superficial layer of the endometrium (see Fig. 44.33). This mode of implantation, called **interstitial implantation,** is the most invasive among placental mammals. Invasive implantation involves adhesion-supported migration of syncytiotrophoblasts into the endometrium, along with the breakdown of extracellular matrix by secretion of matrix metalloproteinases and other hydrolytic enzymes.

The endocrine function begins with the onset of implantation, when syncytiotrophoblasts start secreting the LH-like protein **hCG,** which maintains the viability of

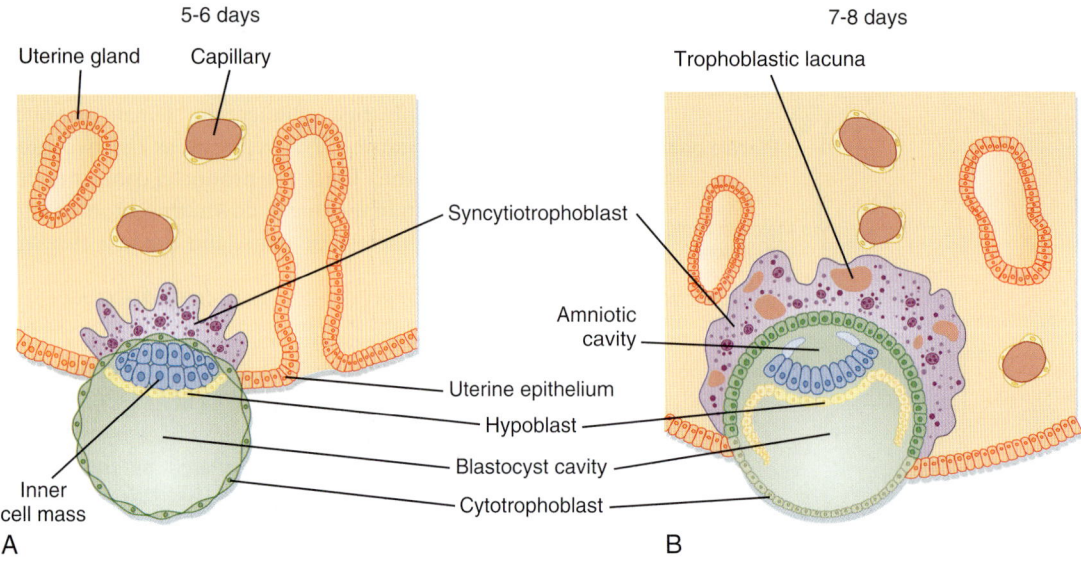

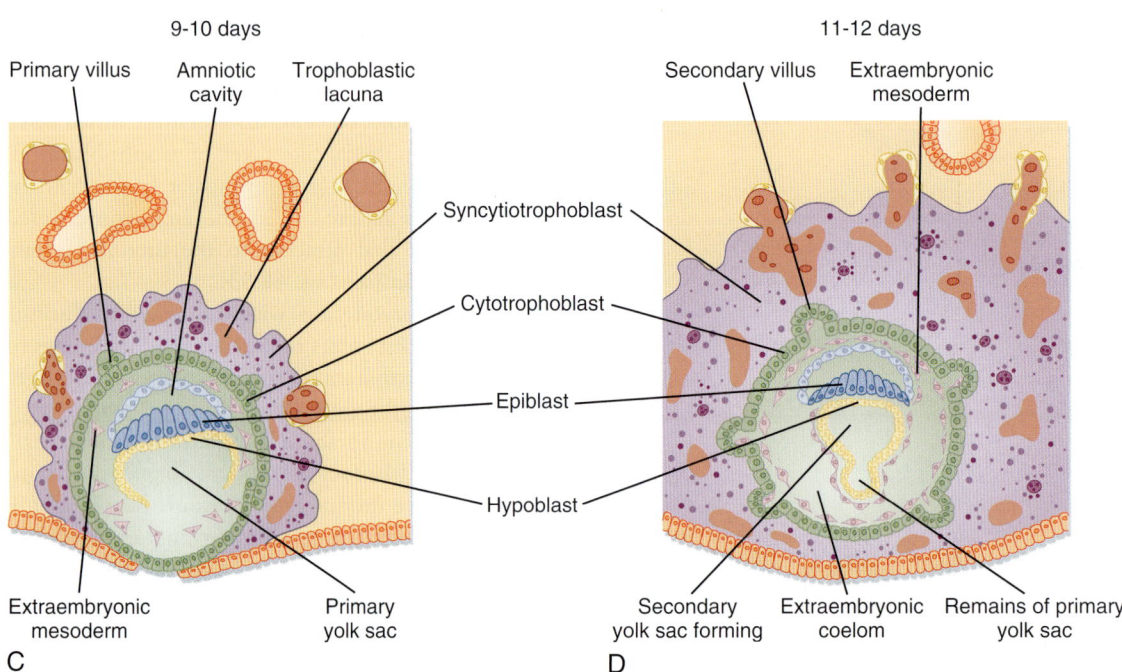

• **Fig. 44.33** A, Beginning of implantation. The trophoblast has differentiated into cytotrophoblast and syncytiotrophoblast layers. B, As the syncytiotrophoblast layer increases in size and invades deeper, this layer begins to surround and erode maternal vessels, forming lacunae filled with maternal blood. C, Interstitial implantation is almost complete. Extensions of cytotrophoblasts have formed that will become covered by a layer of syncytiotrophoblast. At this point , they are called "primary villi". D, Interstitial implantation is complete. Extraembryonic mesodermal has developed from the epithelial layers (amnion, primary yolk sac), and will form an inner layer of the villi, forming "secondary villi". Ultimately, the mesoderm will give rise to umbilical blood vessels within the core of the villus, thereby forming tertiary villi. (From Carlson BM. *Human Embryology and Developmental Biology.* Philadelphia: Mosby; 2004.)

the corpus luteum and thus progesterone secretion. Syncytiotrophoblasts also become highly steroidogenic. By 10 weeks the syncytiotrophoblasts acquire the ability to make progesterone at sufficient levels to maintain pregnancy independently of a corpus luteum. Syncytiotrophoblasts produce several other hormones as well as enzymes that modify hormones.

As implantation and placentation progress, syncytiotrophoblasts take on the important functions of phagocytosis (during histiotropic nutrition) and bidirectional placental transfer of gases, nutrients, and wastes. Exchange across the syncytiotrophoblasts involves diffusion (e.g., gases), facilitated transport (e.g., GLUT1-mediated transfer of glucose), active transport (e.g., amino acids by specific transporters), and pinocytosis/transcytosis (e.g., of iron-transferrin complexes).

There is also a maternal response to implantation that involves transformation of the endometrial stroma. This response, called **decidualization,** involves an enlargement of stromal cells as they become lipid- and glycogen-filled decidual cells (at this time the endometrium is referred to as the **decidua**). The decidua forms an epithelial-like sheet with adhesive junctions that inhibit migration of the implanting embryo. The decidua also secretes factors such as **tissue inhibitors of metalloproteinases (TIMPs)** that moderate the activity of syncytiotrophoblast-derived hydrolytic enzymes in the endometrial matrix. Consequently, decidualization allows regulated invasion during implantation. Normally the implanting embryo and placenta do not extend to and involve the myometrium.

IN THE CLINIC

Placenta accreta is the burrowing of the embryo completely through the endometrium and adhesion of the placenta to the myometrium, a condition associated with potentially life-threatening **postpartum hemorrhage.** The decidual response occurs only in the uterus. Thus the highly invasive nature of the human embryo poses considerable risk to the mother in the case of **ectopic implantation.** *Ectopic implantation* refers to implantation of an embryo at a site other than the uterus, and *ectopic pregnancy* refers to a developing embryo at a site of ectopic implantation. Most ectopic pregnancies (>90%) occur within the oviducts (called **tubal pregnancies**), but they can also occur in the ovary and abdominal cavity. Implantation in the oviducts is often associated with long-term infection and inflammation (called **pelvic inflammatory disease**) and obstruction of the tube. In a tubal pregnancy the highly invasive nature of the human syncytiotrophoblast, which is normally moderated by the uterine decidual response, usually leads to burrowing of the implanted embryo through the wall of the oviduct. Although abdominal pregnancies can proceed to term, undetected oviductal pregnancies usually lead to rupture of the oviductal wall. The resulting internal hemorrhage can be catastrophic to the mother and requires immediate surgical intervention.

Placental Endocrinology

Human Chorionic Gonadotropin

The first hormone produced by syncytiotrophoblasts is hCG, which is structurally related to the pituitary glycoprotein hormones (see Chapter 41). As such, hCG is composed of a common **α-glycoprotein subunit (α-GSU)** and a **hormone-specific β subunit (β-hCG).** Antibodies used to detect hCG (i.e., in laboratory assays and over-the-counter pregnancy tests) are designed to specifically detect the β subunit. hCG is most similar to LH and binds with high affinity to the LH receptor. The β subunit of hCG is longer than that of LH and contains more sites for **glycosylation,** which greatly increases the half-life of hCG to 24 to 30 hours. The stability of hCG allows it to rapidly accumulate in the maternal circulation such that hCG is detectable

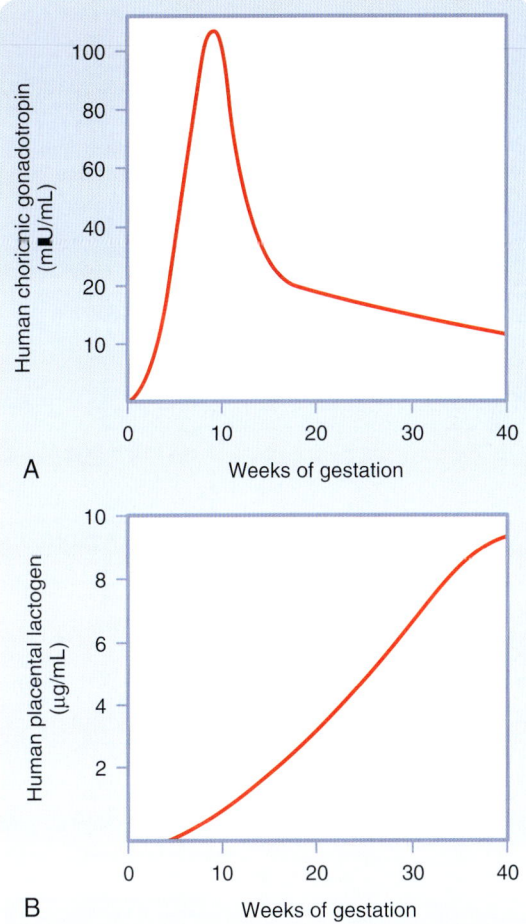

• **Fig. 44.34** Circulating levels of hCG and hPL in maternal blood during pregnancy. (Modified from White BA, Porterfield SP. *Endocrine Physiology.* 4th ed. Philadelphia: Mosby; 2013.)

within maternal serum within 24 hours of implantation. Serum hCG levels double every 2 days for the first 6 weeks and peak at about 10 weeks. Serum hCG then declines to a constant level at about 50% of the peak value (Fig. 44.34A).

The primary action of hCG is to stimulate LH receptors on the corpus luteum. This prevents luteolysis and maintains a high level of luteal-derived progesterone production during the first 10 weeks. The rapid increase in hCG is responsible for the nausea of **"morning sickness"** associated with early pregnancy. A small amount (i.e., 1%–10%) of hCG enters the fetal circulation. hCG stimulates fetal Leydig cells to produce testosterone before the fetal gonadotropic axis is fully mature. hCG also stimulates the fetal adrenal cortex (see later) during the first trimester.

Progesterone

The placenta produces a high amount of progesterone, which is absolutely required to maintain a quiescent myometrium and a pregnant uterus. Progesterone production by the placenta is largely unregulated—the placenta produces as much progesterone as the supply of cholesterol and the levels of CYP11A1 and 3β-HSD allow (Fig. 44.35). Notably, placental steroidogenesis differs from that in the

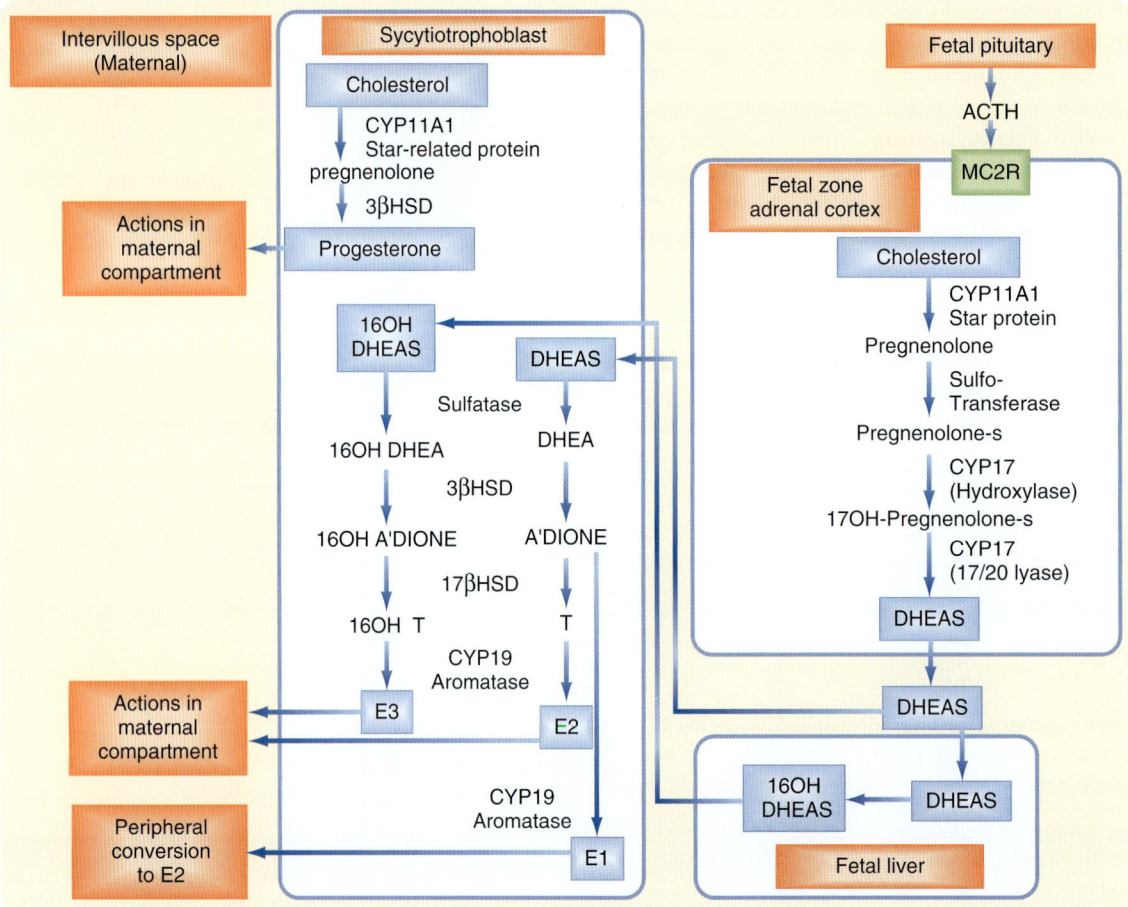

• **Fig. 44.35** Production of progesterone by syncytiotrophoblast and estrogens by fetoplacental unit.

adrenal cortex, ovaries, and testis in that cholesterol is transported into the placental mitochondria by a mechanism that is independent of **StAR protein.** Thus this first step in steroidogenesis is not a regulated rate-limiting step in the placenta as it is in other steroidogenic glands. This means that fetuses with an inactivating mutation in StAR protein will develop **lipoid congenital adrenal hyperplasia** (see Chapter 43) and **hypogonadism** but will have normal progesterone levels produced by their placenta. Progesterone production by the placenta does not require fetal tissue. Consequently, progesterone levels are largely independent of fetal health and cannot be used as a measure of fetal health. Maternal progesterone levels continue to increase throughout pregnancy.

Progesterone is released primarily into the maternal circulation and is required for implantation and maintenance of pregnancy. Progesterone also has several effects on maternal physiology and induces breast growth and differentiation. The switch from corpus luteum–derived progesterone to placental-derived progesterone (referred to as the **luteal-placental shift**) is complete at about the eighth week of pregnancy. Progesterone (and pregnenolone) are used by the transitional zone of the fetal cortex to make cortisol late in pregnancy.

Estrogen

Estrogens are also produced by syncytiotrophoblasts. Syncytiotrophoblasts are similar to ovarian granulosa cells in that they lack CYP17 and are dependent on another cell type to provide 19-carbon androgens for aromatization (see Fig. 44.35). The ancillary androgen-producing cells reside in the **fetal adrenal cortex.**

The fetal adrenal cortex contains an outer **definitive zone,** a middle **transitional zone,** and an inner **fetal zone.** The definitive and transitional zones give rise to the zona glomerulosa and zona fasciculata, respectively. Aldosterone synthesis is initiated close to parturition. Synthesis of cortisol begins at about 6 months and increases during late gestation. The fetal zone is the predominant portion of the adrenal cortex in the fetus; it constitutes as much as 80% of the bulk of the large fetal adrenal and is the site of most fetal adrenal steroidogenesis. The fetal zone strongly resembles the zona reticularis in that it expresses little or no 3β-HSD (see Fig. 44.35). The fetal zone primarily releases the sulfated form of the inactive androgen **dehydroepiandrosterone sulfate** (DHEAS) throughout most of gestation. Production of DHEAS from the fetal adrenal is absolutely dependent on fetal ACTH from the fetal pituitary by the end of the first trimester.

The DHEAS released from the fetal zone has two fates. First, DHEAS can go directly to the syncytiotrophoblast, where it is desulfated by a placental **steroid sulfatase** and used as a 19-carbon substrate for the synthesis of estradiol-17β and estrone (see Fig. 44.35). The second fate of DHEAS is **16-hydroxylation** in the fetal liver by the enzyme CYP3A7. 16-Hydroxyl-DHEAS is then converted by syncytiotrophoblasts to the major estrogen of pregnancy, **estriol** (see Fig. 44.35). In X-linked ichthyosis, the steroid sulfatase is low or missing, resulting in loss of active (i.e., desulfated) estrogen production by the fetoplacental unit. Pregnancy is normal, but because estrogens promote parturition, the pregnancy is prolonged and usually ends with physician-induced labor. The baby boy has a skin disorder of varying degrees of severity that is called *ichthyosis* (scaly skin), owing to buildup of layers of shed cells within the stratum corneum. This form of ichthyosis is readily treatable by topical creams.

Maternal estrogen levels increase throughout pregnancy. Because estrogen production is dependent on a healthy fetus, estriol levels can be used as one measure of fetal health. The collective term used for the placental syncytiotrophoblasts and fetal organs in the context of estrogen production is the **fetoplacental unit.** Estrogens increase uteroplacental blood flow, enhance LDL receptor expression in syncytiotrophoblasts, and induce several components (e.g., prostaglandins, oxytocin receptors) involved in parturition. Estrogens increase breast growth directly and indirectly through stimulation of maternal pituitary prolactin production. Estrogens also increase lactotrope size and number, thereby increasing overall pituitary mass by more than twofold by term. Estrogens also affect several other aspects of maternal physiology.

Human Placental Lactogen

Human placental lactogen (hPL), also called **human chorionic somatomammotropin (hCS),** is a 191–amino acid protein hormone produced in the syncytiotrophoblast that is structurally similar to growth hormone (GH) and prolactin (PRL). Its function overlaps those of both GH and PRL. It can be detected within the syncytiotrophoblast by 10 days after conception and in maternal serum by 3 weeks' gestation (see Fig. 44.34). Maternal serum levels rise progressively throughout the remainder of the pregnancy. The quantity of hormone produced is directly related to the size of the placenta, such that as the placenta grows during gestation, hPL secretion increases. As much as 1 g/day of hPL can be secreted late in gestation.

Like GH, hPL is protein anabolic and lipolytic. Its antagonistic action to insulin is the major basis for the diabetogenicity of pregnancy. Like PRL, it stimulates mammary gland growth and development. Mammary gland development in pregnancy results from the actions of hPL, PRL, estrogens, and progestins. hPL inhibits maternal glucose uptake and use, thereby increasing serum glucose levels. Glucose is a major energy substrate for the fetus, and hPL increases fetal glucose availability.

As with hCG, far less hPL is found in the fetal circulation than in the maternal circulation. This suggests that the hormones may play a more important role in the mother than in the fetus. hPL is not essential for the pregnancy.

Both hPL and PRL act as fetal growth hormones and stimulate production of the fetal growth-promoting hormones insulinlike growth factor (IGF)-I and IGF-II. Ironically, fetal GH does not appear to regulate growth, and anencephalic infants and GH-deficient children typically have normal birth weight.

Diabetogenicity of Pregnancy

Pregnancy represents an **insulin-resistant state** (Fig. 44.36). During the last half of pregnancy when hPL levels are highest, maternal energy metabolism shifts from an anabolic state in which nutrients are stored to a catabolic state sometimes described as **accelerated starvation,** in which maternal energy metabolism shifts toward fat utilization with sparing of glucose. As maternal use of glucose for energy decreases, lipolysis increases and fatty acids become major energy sources. Peripheral responsiveness to insulin decreases and pancreatic insulin secretion increases. Beta cell hyperplasia occurs in pregnancy. Although this does not usually lead to a clinical condition, pregnancy aggravates existing diabetes mellitus, and diabetes can develop for the first time in pregnancy. If the diabetes resolves spontaneously with delivery, the condition is referred to as **gestational diabetes.** Other hormones contributing to the diabetogenicity of pregnancy are estrogens and progestins, because both these hormones decrease insulin sensitivity.

Parturition

Human pregnancy lasts an average of 40 weeks from the beginning of the last menstrual period (gestational age). This corresponds to an average fetal age of 38 weeks. **Parturition** is the process whereby uterine contractions lead to childbirth. **Labor** consists of three stages: strong uterine contractions that force the fetus against the cervix, with dilation and thinning of the cervix (several hours); delivery of the fetus (<1 hour); and delivery of the placenta, along with contractions of the myometrium to halt bleeding (<10 minutes).

Control of parturition in humans is complex, and the exact mechanisms underlying its control are not well understood.

Placental CRH and the Fetal Adrenal Axis

The placenta produces **corticotropin-releasing hormone (CRH),** which is identical to the 41–amino acid peptide produced by the hypothalamus. Placental CRH production and maternal serum CRH levels increase rapidly during late pregnancy and labor. Moreover, circulating CRH is either in the form of free CRH, which is bioactive, or complexed to a CRH-binding protein. Maternal levels of CRH-binding protein plummet during late pregnancy and labor, so free

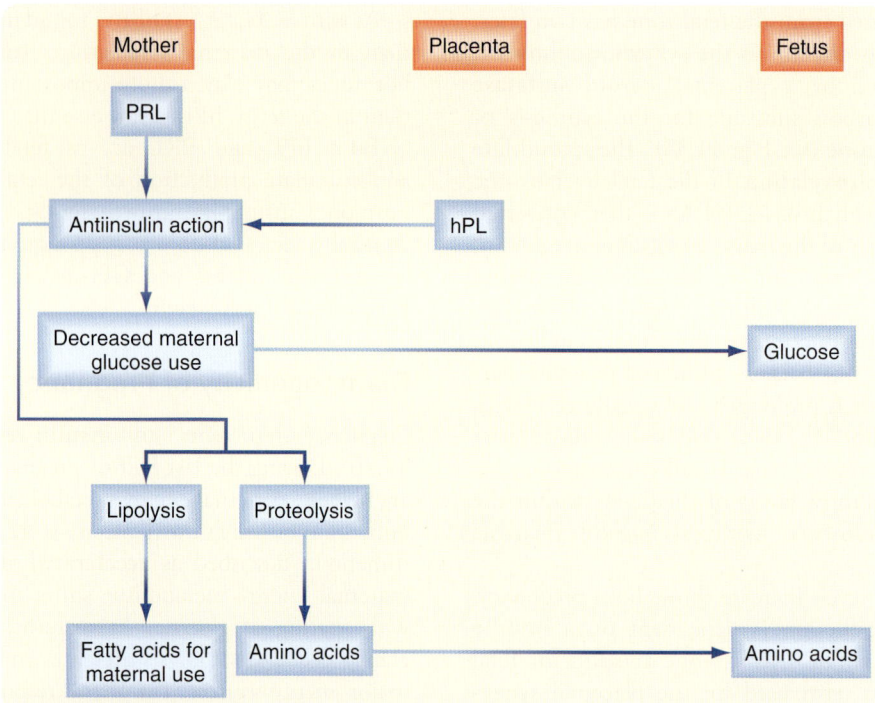

• **Fig. 44.36** Overview of energy use by the maternal and fetal compartments. (Modified from White BA, Porterfield SP. *Endocrine Physiology.* 4th ed. Philadelphia: Mosby; 2013.)

CRH levels increase. Placental CRH also accumulates in the fetal circulation and stimulates fetal ACTH secretion. ACTH stimulates both fetal adrenal cortisol production and fetoplacental estrogen production. In contrast to the inhibitory effect of cortisol on hypothalamic CRH production, cortisol stimulates placental CRH production. This establishes a self-amplifying positive feedback. CRH itself promotes myometrial contractions by sensitizing the uterus to oxytocin and prostaglandins (see Oxytocin, Prostaglandins). Estrogens also directly and indirectly stimulate myometrial contractility. Moreover, this model correlates the onset of parturition with cortisol-induced maturation of fetal systems, including the lungs and gastrointestinal tract.

Estrogen and Progesterone Secretion

Although a rise in maternal serum estrogen and a drop in progesterone levels occur late in gestation in some species, no change in the ratio of the two hormones is seen in human serum. However, "functional" progesterone withdrawal involving changes in uterine progesterone receptor and progesterone metabolism has been proposed.

Oxytocin

Oxytocin is secreted from the **pars nervosa (posterior pituitary)** (see Chapter 41). Oxytocin, which stimulates powerful uterine contractions, plays a major role in progression and completion of parturition. Oxytocin is released in response to stretch of the cervix through a **neuroendocrine reflex;** it stimulates uterine contractions and thereby facilitates delivery. Oxytocin can be used to induce parturition,

and uterine sensitivity to oxytocin increases before parturition. Because maternal serum oxytocin levels do not increase until after parturition has begun, oxytocin is not thought to initiate parturition. However, progesterone inhibits and estrogen stimulates synthesis of oxytocin receptors, and although maternal serum progesterone levels do not decrease immediately before human parturition, estrogen levels rise and oxytocin receptor synthesis increases.

Prostaglandins

Prostaglandins and other cytokines increase uterine motility, and levels of these compounds increase during parturition, thereby facilitating delivery. Their exact role in initiating parturition is not known. Prostaglandin levels in amniotic fluid, fetal membranes, and uterine decidua increase before the onset of labor. **Prostaglandin $F_{2\alpha}$** and **prostaglandin E_2** increase uterine motility. Large doses of these compounds have been used to induce labor. Because estrogens stimulate prostaglandin synthesis in the uterus, amnion, and chorion, the rising estrogen levels late in gestation can increase uterine prostaglandin formation before parturition.

Uterine Size

Uterine size is thought to be a factor regulating parturition because stretch of smooth muscle, including the uterus, increases muscle contraction. In addition, uterine stretch stimulates uterine prostaglandin production. Multiple births generally occur prematurely. The tendency for early delivery can be a result of increased uterine size, increased fetal production of chemicals stimulating delivery, or both.

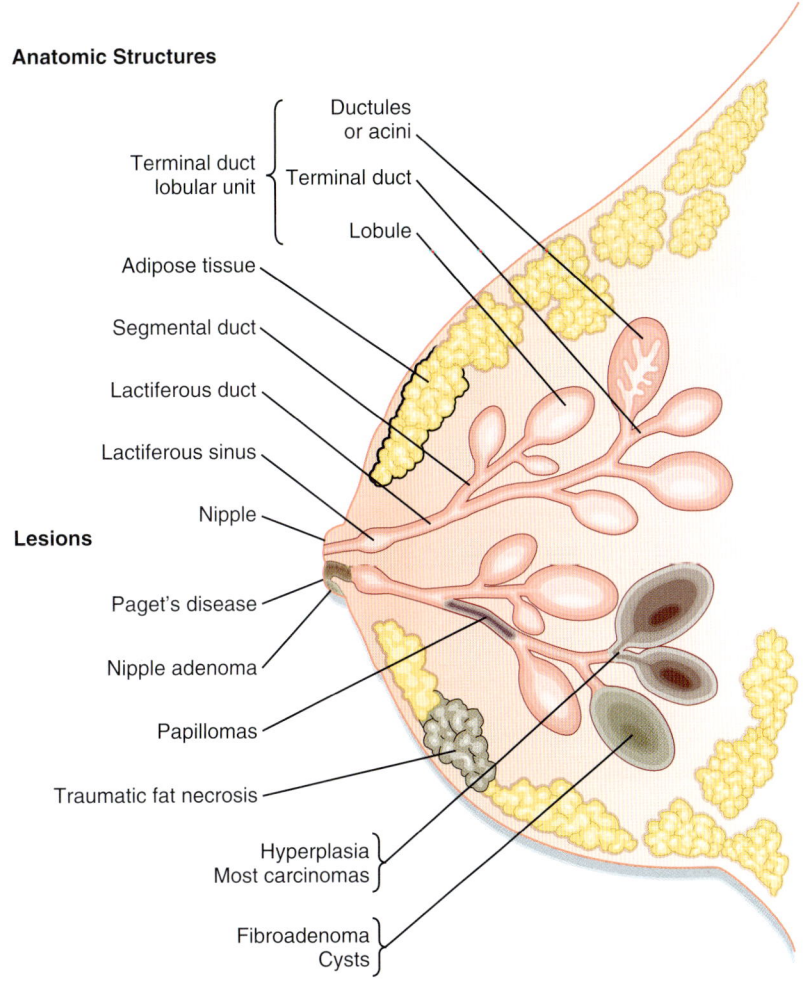

Anatomic Structures

Terminal duct lobular unit
- Ductules or acini
- Terminal duct
- Lobule

Adipose tissue

Segmental duct

Lactiferous duct

Lactiferous sinus

Nipple

Lesions

Paget's disease

Nipple adenoma

Papillomas

Traumatic fat necrosis

Hyperplasia
Most carcinomas

Fibroadenoma
Cysts

• **Fig. 44.37** Diagram of the structure of the breast, along with some pathological conditions of the breast and where they occur. (From Crum CP et al. In: Kumar V et al [eds]. *Robbins Basic Pathology.* 7th ed. Philadelphia: Saunders; 2003.)

Mammogenesis and Lactation

Structure of the Mammary Gland

The **mammary gland** is composed of 15 to 20 lobes, each with an excretory **lactiferous duct** that opens at the nipple (Fig. 44.37). The lobes in turn are composed of several lobules that contain secretory structures called **alveoli** and the terminal portions of the **ducts.** The epithelium of the alveoli and ducts is composed of two cell layers: apical **luminal epithelial cells** and basal **myoepithelial cells.** There is strong evidence for the presence of adult mammary stem cells within this epithelium. The luminal epithelial cells of the alveoli are the producers of milk, and the luminal cells of the ducts convey and modify the secreted milk. Myoepithelial cells are stellate smooth muscle–like cells, and contraction of these cells in response to a stimulus (i.e., milk let-down) expels milk from the lumens of the alveoli and ducts. Lobes and lobules are supported within a connective tissue matrix. The other major tissue component of the breast is adipose tissue. The lactiferous ducts empty at the **nipple,** a highly innervated hairless protrusion of

the breast designed for suckling by an infant. The nipple is surrounded by a pigmented hairless areola that is lubricated by sebaceous glands. Protrusion of the nipple, called **erection,** is mediated by sympathetic stimulation of smooth muscle fibers in response to suckling and other mechanical stimulation, erotic stimulation, and cold.

Hormonal Regulation of Mammary Gland Development

At **puberty, estrogen** increases ductal growth and branching. With onset of the luteal phases of the ovary, **progesterone and estrogen** induce ductal growth and formation of rudimentary alveoli. During nonpregnant cycles the breasts develop somewhat and then regress. Estrogen also increases deposition of **adipose tissue,** which makes a major contribution to breast size and overall form. Adipose tissue expresses **CYP19/aromatase,** so accumulation of this tissue in the breast increases local production of estrogens from circulating androgens.

Breast development is facilitated by pregnancy, during which extensive ductal growth and branching

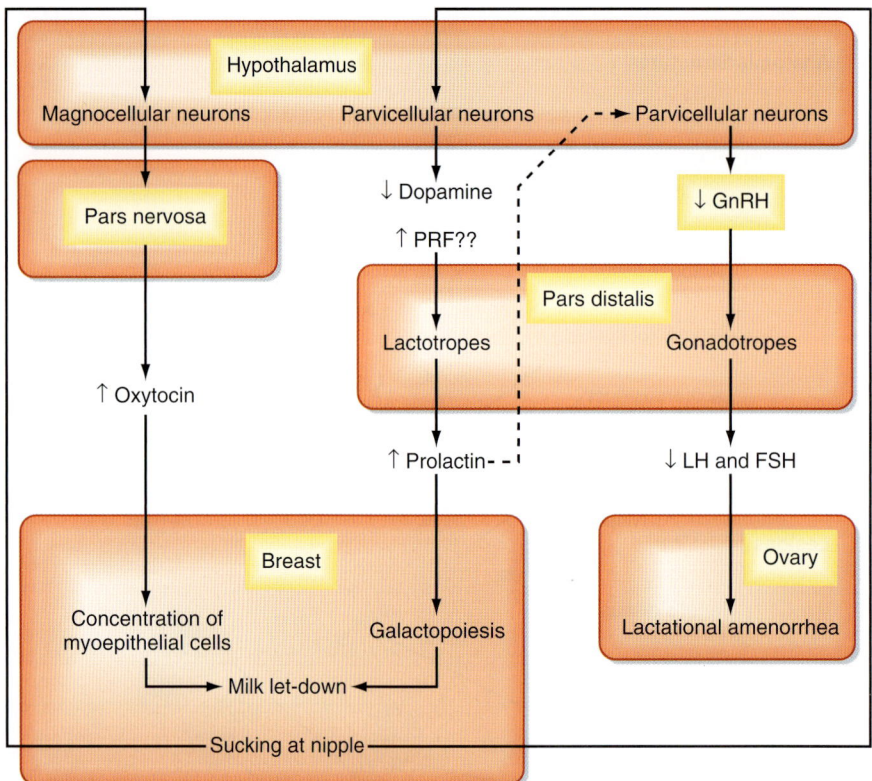

• **Fig. 44.38** Neuroendocrine reflex caused by suckling at the nipple and leading to secretion of oxytocin and prolactin. In turn these hormones induce continued milk production (galactopoiesis) and milk let-down. Prolactin also induces lactational amenorrhea. (Modified from White BA, Porterfield SP. *Endocrine Physiology.* 4th ed. Philadelphia: Mosby; 2013.)

and lobuloalveolar development occur. The parenchymal growth of the breast during development occurs at the expense of stroma, which is degraded to make room for enlarging lobuloalveolar structures. Several placental hormones stimulate breast development, including **estrogen, progesterone, placental lactogen,** and a **growth hormone variant (GH-V).** Estrogen acts on the breast both directly and indirectly through increasing maternal **pituitary PRL.** Estrogen increases PRL secretion from pituitary lactotropes. Estrogen also stimulates **lactotrope hypertrophy and proliferation,** which accounts for the twofold increase in pituitary volume during pregnancy in humans. Although epithelial cells express genes encoding milk protein and enzymes involved in milk production, progesterone inhibits the onset of milk production and secretion **(lactogenesis).**

After parturition, the human breast produces **colostrum,** which is enriched with antimicrobial and antiinflammatory proteins. In the absence of placental progesterone, normal breast milk production occurs within a few days. The lobuloalveolar structures produce milk, which is subsequently modified by the ductal epithelium. **Lactogenesis** and maintenance of milk production **(galactopoiesis)** require stimulation by pituitary PRL in the presence of normal levels of other hormones, including insulin, cortisol, and thyroid hormone. Although placental estrogen stimulates PRL secretion during pregnancy, the stimulus for PRL

secretion during the nursing period is suckling by the infant (Fig. 44.38). Levels of PRL are directly correlated with frequency and duration of suckling at the nipple. The link between suckling at the nipple and PRL secretion involves a neuroendocrine reflex in which dopamine secretion at the median eminence is inhibited (the PRL release inhibitory factor; see Chapter 41). It is also possible suckling increases secretion of unidentified PRL-releasing hormones.

PRL also inhibits release of GnRH, and consequently, nursing can be associated with **lactational amenorrhea** (see Fig. 44.38). This effect of prolactin has been called "nature's contraceptive," and it may play a role in spacing out pregnancies. However, only regular nursing over a 24-hour period is sufficient to induce a PRL-induced anovulatory state in the mother. Thus lactational amenorrhea is not an effective or reliable form of birth control for most women. Inhibition of GnRH by high levels of PRL is important clinically. A **prolactinoma** is the most common form of hormone-secreting pituitary tumor, and **hyperprolactinemia** is a significant cause of infertility in both sexes. Hyperprolactinemia can likewise be associated with **galactorrhea** (inappropriate flow of breast milk) in men and women.

Suckling at the nipple also stimulates release of **oxytocin** from the pars nervosa (see Chapter 41) through a neuroendocrine reflex (see Fig. 44.38). Contraction of myoepithelial cells induces **milk let-down,** or expulsion of milk from the

IN THE CLINIC

Invasive breast cancer (IBC) is a major cancer in women and can be classified into several categories. Most newly diagnosed IBC is classified as **luminal A,** which is usually derived from luminal cells of terminal ducts or alveoli. Luminal A IBC displays some epithelial organization, including E cadherin–mediated cell-cell contacts, and is poorly motile and poorly aggressive. This form also expresses the **estrogen receptor α**

(ERα) and is dependent on estrogenic stimulation for growth. Early diagnosis of luminal A IBC confers a good prognosis. Treatment for early small tumors that are "node-negative" (i.e., have not spread to nearby lymph nodes) typically involves surgical removal ("lumpectomy"), followed by radiation treatment, followed by 5 years of daily tamoxifen treatment. Tamoxifen is a SERM that opposes estrogen in the breast.

IN THE CLINIC

There are multiple behavioral methods of **contraception.** Total abstinence is the best way to avoid getting pregnant. Other methods include the rhythm method, which relies on abstinence from sexual intercourse during fertile periods around the time of ovulation. The fertile period extends from 3 to 4 days before the time of ovulation until 3 to 4 days afterward. A second method is withdrawal before ejaculation, **coitus interruptus.** Both these methods have higher failure rates (20%–30%) than **barrier methods** (2%–12%), **intrauterine devices (IUDs)** (<2%), and **oral contraceptives** (<1%) do. Barriers such as **condoms** or **diaphragms** are more effective when used with **spermicidal jellies.** Of all methods, only condoms provide effective protection from sexually transmitted diseases in sexually active individuals. IUDs are relatively effective. They prevent implantation by locally producing an inflammatory response in the endometrium. Some forms of IUDs contain copper, zinc, or progestins, which inhibit sperm transport or viability in the female reproductive tract.

Oral contraceptives have been marketed in the United States since the early 1960s. The doses of steroids used today are significantly lower than those used 35 years ago. Properly used, oral contraceptives have a low failure rate. Many forms of oral contraceptives are marketed today. The trend over the years has been to decrease the dosage of steroids used because the side effects are dose dependent. All oral steroidal contraceptives contain either a combination of an estrogen and a progestin or a progestin alone. Oral

contraceptives work through multiple mechanisms. Most block the LH surge that triggers ovulation. However, some pills (e.g., progestin-only minipill) do not prevent LH surges. Fertility is also blocked by changing the nature of cervical mucus, altering endometrial development, or regulating fallopian tube motility. Because these contraceptives suppress FSH, they impair early follicular development.

Emergency contraception involves hormonal treatment designed to inhibit or delay ovulation, inhibit corpus luteum function, disrupt the function of the oviducts and uterus, or any combination of these mechanisms. For example, candidates for emergency contraception include women who are sexually assaulted or who experienced failure of a barrier method (e.g., ruptured condom). There are more than 20 types of commercially available "morning-after" pills. The currently preferred medication is **levonorgestrel (Plan B),** which is a synthetic progestin-only pill. The efficacy of the pill is inversely correlated with the time it is taken after intercourse. The exact mechanism of action is not known. Treatment has no effect if implantation has occurred.

Medical **(hormonal) termination** of pregnancy (abortion) can be achieved up to 49 days' gestation by administration of **mifepristone (RU-486),** a progesterone receptor antagonist that induces collapse of the pregnant endometrium. Mifepristone is followed 48 hours later by ingestion or vaginal insertion of a **synthetic prostaglandin E** (e.g., misoprostol), which induces myometrial contractions.

---

alveolar and ductal lumens. Thus the nursing infant does not gain milk by applying negative pressure to the breast from suckling. Rather, milk is actively ejected through a neuroendocrine reflex. Oxytocin release and milk let-down can be induced by psychogenic stimuli such as the mother hearing a baby crying on television or thinking about her baby. Such psychogenic stimuli do not affect PRL release.

Menopause

Though related to depletion of ovarian follicles, the causes and process of **menopause** are poorly understood. Age-related changes in the CNS, including critical patterns of GnRH secretion, precede follicular depletion and may play an important role in menopause. Because follicles do not develop in response to LH and FSH secretion, estrogen and progesterone levels drop. Loss of the negative-feedback inhibition of estrogen on GnRH and LH/FSH results in a

marked rise in serum LH and FSH. FSH levels rise more than LH levels. This could result from loss of ovarian inhibin.

Menopause typically occurs between 45 and 55 years of age. It extends over a period of several years. Initially the cycles become irregular and are periodically anovulatory. The cycles tend to shorten, primarily in the follicular phase. Eventually the woman ceases to cycle altogether. Serum estradiol levels drop to about a sixth the mean levels for younger cycling women, and progesterone levels drop to about a third those in the follicular phase of younger women. Production of these hormones does not cease entirely, but the primary source of these hormones in postmenopausal women becomes the adrenal, although interstitial cells of the ovarian stroma continue to produce some steroids. Most circulating estrogens are now produced peripherally from androgens. Because estrone is the primary estrogen produced in adipose tissue, it becomes the predominant estrogen in postmenopausal women.

Most symptoms associated with menopause result from **estrogen deficiency.** The vaginal epithelium atrophies and becomes dry, and bone loss is accelerated and may lead to osteoporosis. The incidence of coronary artery disease increases markedly after menopause. **Hot flashes** result from periodic increases in core temperature, which produces peripheral vasodilation and sweating. Hot flashes are thought to be linked to increases in LH release and are probably associated not with the pulsatile rise in LH secretion but rather with central mechanisms controlling GnRH release. Hot flashes typically subside within 1 to 5 years of the onset of menopausal symptoms.

Key Concepts

1. The reproductive systems are composed of gonads, an internal reproductive tract with associated glands, and external genitalia. Mammary glands are accessory reproductive glands in women.

2. Gonads have two main functions: production of gametes and production of hormones. Hormones (primarily sex steroids) are absolutely necessary for normal function of the reproductive system, and their production is regulated by a hypothalamic-pituitary-gonadal axis.

3. Seminiferous tubules in the testis contain Sertoli cells and developing sperm cells.

4. *Spermatogenesis* refers to the progression of sperm cells from spermatogonia through the processes of meiosis and spermiogenesis to form mature spermatozoa.

5. Testosterone and pituitary FSH are required for normal sperm production. Only Sertoli cells express the androgen receptor and the FSH receptor, so these hormones regulate spermatogenesis indirectly through their actions on Sertoli cells. Sertoli cells produce the hormone inhibin, which negatively feeds back on pituitary FSH production.

6. Sertoli cells have many functions, including production of androgen-binding protein (ABP) and fluid and creation of the blood-testis barrier.

7. Leydig cells are stromal cells that reside outside the seminiferous tubules. They respond to LH by producing testosterone.

8. Testosterone is an active androgen. It can be converted peripherally to DHT, which is more active in certain tissues (e.g., prostate), or to estradiol.

9. Leydig cells are regulated within a hypothalamic-pituitary-testicular axis. The hypothalamus produces GnRH, which stimulates pituitary gonadotropes to secrete LH and FSH. Testosterone, DHT, and estradiol negatively feed back at the pituitary and hypothalamus and inhibit LH more than FSH secretion. Inhibin from the Sertoli cells selectively inhibits FSH.

10. Testosterone, DHT, and estradiol have numerous actions on the male reproductive tract, external genitalia, and male secondary sex characteristics, as well as on other organ systems (e.g., blood cell production, lipoprotein production, bone maturation).

11. The male tract includes tubal structures (epididymis, ductus deferens, and male urethra), accessory sex glands (seminal vesicles, prostate), and the penis. The seminal vesicles and the prostate produce most of the ejaculate, which nourishes, buffers, and protects sperm.

12. Penile erection involves a complex neurovascular response leading to engorgement of the erectile tissue within the penis base and shaft with blood.

13. The follicle is the functional unit of the ovary. Follicles contain epithelial cells (granulosa and cumulus) and outer stromal cells (thecal). All these cells surround a primary oocyte that remains arrested in the first meiotic prophase until just before ovulation.

14. Follicles develop from the smallest (primordial) to a large antral follicle over a period of months. The latter part of follicular development requires gonadotropins.

15. The *menstrual cycle* refers to an approximately 28-day cycle that is driven by the following ovarian events: development of one large antral follicle to a preovulatory follicle (follicular phase), ovulation, and formation and death of a corpus luteum of menstruation (luteal phase).

16. The follicular phase of the ovary corresponds to the menstrual and proliferative phases of the uterine endometrium. The luteal phase of the ovary corresponds to the secretory phase of the uterine endometrium.

17. One dominant follicle is selected per menstrual cycle—usually the largest follicle with the most FSH receptors.

18. High levels of estradiol occur around midcycle and exert positive feedback on gonadotropin secretion. This induces the LH (and a smaller FSH) surge. The midcycle gonadotropin surge induces (a) meiotic maturation of the primary oocyte so that it progresses to a secondary oocyte (with one polar body) arrested at metaphase of the second meiotic division, (b) breakdown of the ovarian and follicular wall so that the oocyte-cumulus complex is extruded (called *ovulation*), and (c) differentiation of the remaining follicular cells into a corpus luteum. The corpus luteum produces high levels of progesterone, estradiol, and inhibin.

19. If pregnancy does not occur, the corpus luteum will die in 14 days. This constitutes the luteal phase of the menstrual cycle.

20. The oviducts capture the ovulated cumulus-oocyte complex and transport it medially into the oviduct and toward the uterus. Estrogen promotes ciliation and transport; progesterone inhibits transport.

21. The uterine mucosa, called the *endometrium,* is the normal site of embryonic implantation. The mucosa is increased in thickness in preparation for implantation and is sloughed away if no pregnancy occurs.

22. During the mid to late follicular phase (days 6–14 of the menstrual cycle), the ovary produces estradiol, which induces all cells of the endometrium to proliferate (called the *proliferative phase* of the uterus).

23. After ovulation the ovary enters the luteal phase (days 16–28) and produces progesterone. Progesterone stimulates secretion from the uterine glands (called the *secretory phase* of the uterus).

24. In the absence of an implanting embryo the corpus luteum dies, progesterone production ceases, and the uterine endometrium is sloughed (called the *menstrual phase,* or *period,* of the uterus—this corresponds to days 1 to 5 of the follicular phase of the ovary).

25. The cervix is the lower portion of the uterus. Cervical mucus is hormonally regulated so that at midcycle in response to estrogen, cervical mucus promotes entry of sperm into the uterus from the vagina. During the luteal phase in response to progesterone, cervical mucus becomes thick and poses a barrier to entry of sperm and microbes into the uterus.

26. Fertilization is a complex series of events that occur in the oviduct and lead to penetration of the oocyte by sperm.

27. Early embryogenesis (up to day 6 after fertilization) occurs in the oviduct and gives rise to a blastocyst that hatches from the zona pellucida.

28. The placenta develops from the outer extraembryonic trophoblast. The endocrine function of the placenta includes production of hCG, progesterone, estrogens, and placental lactogen. Estrogen production requires placental cells (syncytiotrophoblasts) as well as the fetal adrenal and liver—collectively called the *fetoplacental unit.*

29. Pregnancy and the hormones of pregnancy induce major changes in maternal physiology, including an increase in insulin resistance, an increase in the use of free fatty acids by the mother, and development of the mammary glands. Mammary gland development (but not lactation) is promoted by estrogen, progesterone, and placental lactogen but also by maternal pituitary prolactin, whose secretion is stimulated by placental estrogens.

30. Oxytocin is a pituitary hormone that promotes contraction of certain smooth muscles, including myometrial contractions during labor and myoepithelial contractions in the breasts that lead to let-down of milk in response to suckling.

31. Menopause results from exhaustion of the ovarian reserve and is characterized by low ovarian hormone and elevated gonadotropin levels.

Acknowledgement

We wish to thank Dr. Lisa Mehlmann for her advice on this chapter, and especially for help in drawing Figures 44.31 and 44.32.

Additional Reading

Achar S, et al. Cardiac and metabolic effects of anabolic-androgenic steroid abuse on lipids, blood pressure, left ventricular dimensions, and rhythm. *Am J Cardiol.* 2010;106:893-901.

Coticchio G, et al. Oocyte maturation: gamete-somatic cells interactions, meiotic resumption, cytoskeletal dynamics and cytoplasmic reorganization. *Hum Reprod Update.* 2015;21:427-454.

Defeudis G, et al. Erectile dysfunction and its management in patients with diabetes mellitus. *Rev Endocr Metab Disord.* 2015; 16:213-231.

Franca LR, et al. The Sertoli cell: one hundred fifty years of beauty and plasticity. *Andrology.* 2016;4:189-212.

Kosaka T, et al. Is DHT production by 5alpha-reductase friend or foe in prostate cancer? *Front Oncol.* 2014;4:article 247.

Larney C, et al. Switching on sex: transcriptional regulation of the testis-determining gene Sry. *Development.* 2014;141:2195-2205.

Mehlmann LM. Stops and starts in mammalian oocytes: recent advances in understanding the regulation of meiotic arrest and oocyte maturation. *Reproduction.* 2005;130:791-799.

Pasqualini JR. Enzymes involved in the formation and transformation of steroid hormones in the fetal and placental compartments. *J Steroid Biochem Mol Biol.* 2005;97:401-415.

Sreekumar A, et al. The mammary stem cell hierarchy: a looking glass into heterogeneous breast cancer landscapes. *Endocr Relat Cancer.* 2015;22:T161-T176.

Umetani M, Shaul PW. 27-Hydroxycholesterol: the first identified endogenous SERM. *Trends Endocrinol Metab.* 2011;22:130-135.

Index

Page numbers followed by "*f*" indicate figures, "*t*" indicate tables, and "*b*" indicate boxes.

A

A band, 243, 244*f*
A kinase anchoring protein (AKAP), 274*b*
A wave, 337
Abdominal cavity, 447
Abducens nerve, 127–128
Absolute refractory period, 74
Absorption, 511
  in vectorial transport, 29–30
Absorptive enterocytes, 512–513
Absorptive phase, of energy metabolism, 698–699
Accelerated starvation, 823
Accessory optic nuclei (AON), 202
Accommodation, in action potential, 73–74
Accommodation response, of pupil, 234
Acetyl coenzyme A (acetyl CoA), 711–712
Acetylcholine (ACh), 98, 387
  in capillary endothelium, 361
  in lung function, 441–442
  myocardial performance and, 397*f*, 398
  and pancreatic function, 546
  receptors, 105
    nicotinic, 768
    in skeletal muscle, 261
  in salivary glands, 523–524
  and secretion of catecholamine, 768
Acetylcholinesterase, 387
Acetyl-CoA carboxylase, 711–712
Achromatopsia, 142*b*
Acid(s), 670
Acid labile subunit (ALS), 747*f*, 748
Acid-base balance, 486–487, 671–672, 671*f*
  and HCO₃- buffer system, 670–671
  and phosphate excretion, 666–667
  and potassium, 650
    and excretion of, 656–657, 657*f*
  renal regulation of, 581
  role of kidneys in regulation of, 670–684, 672*b*, 676*b*, 678*b*, 681*b*–682*b*

Acid-base disorders. *see also* Acidosis.
  analysis of, 683–684
  mixed, 683
  response to, 679–680
    extracellular and intracellular buffers, 679–680
    renal compensation, 680, 681*f*
    respiratory compensation, 680, 680*f*
  simple, 681–684, 682*t*
    with compensation, 683*t*
  types of, 681–683
    metabolic acidosis, 681
    metabolic alkalosis, 681–682
    respiratory acidosis, 682
    respiratory alkalosis, 682–683
Acidemia, and calcium, 658–659
Acidophils, 734
Acidosis, 671. *see also* Metabolic acidosis.
Acinar cells, 521–522, 525*b*
  ionic transport in, 524, 524*f*
  pancreatic, products of, 544*b*
Acinar secretion, 545–546
Acoustic neuroma, 149*b*
Acrosomal reaction, 818
Acrosome, of spermatozoa, 788–789, 790*f*
Actin, of skeletal muscle, 243
Actin filaments
  in cardiac muscle, 270
  in microvilli, 28–29
  in stereocilia, 28–29
α-Actinin-4 (ACTN4), 588, 588*b*, 589*f*
Actin-myosin interaction, in skeletal muscle, 251–252, 252*f*
Action potential(s)
  cardiac, 304–306, 305*f*, 305*t*
    fast-response, 306–311, 306*b*
    slow-response, 311–312, 312*f*
  coding of information by, 80–81
  components of, 69–70, 69*f*
  conductance changes during, 70–73, 72*f*–73*f*
  conduction of, 58, 75–82
  conduction velocity, axon diameter correlated with, 75–76, 76*f*

Action potential(s) (*Continued*)
  definition of, 65
  generation of, 24–25, 24*f*, 65–83
  in GI, 538
  ionic basis of, 69–70
  nerve, 24–25, 24*f*
  propagation of, 70
  refractory periods of, 74–75
  as self-reinforcing signal, 75–76, 75*f*
  versus subthreshold and passive responses, 69
  suprathreshold response in, 69–75
Active transport, 12, 13*f*
Active zones, 85*f*, 86, 90
Acute heart failure, 414*b*
Acute left ventricular failure, 418*b*
Adaptation, 132
  visual, 132–134
Adaptin, 8–9, 9*f*
Adaptive immune cells, 501–502, 504*t*
Adaptive immunity, clinical manifestations associated with, 507
Addisonian crisis, 399*b*
Addison's disease, 399*b*, 667*b*, 783*b*–784*b*
Adenohypophysis, 733–752
  cell types of, 734, 739, 741*t*
  endocrine function of, 739–752
  pars distalis, 733
  pars intermedia, 733
  pars tuberalis, 733
Adenosine
  as candidate vasodilator substances, 401
  and cerebral blood flow, 379
  effect on GFR and RBF, 600
  in gastrointestinal tract, 514–515
Adenosine triphosphatase (ATPase), 500
Adenosine triphosphate (ATP)
  for contraction, of skeletal muscle, 258–259
  effect on GFR and RBF, 600
  as energy source, 698
  hydrolysis, in active transport, 12
  in myocardial performance, 397*f*
  synthesis of, 3*t*, 698

Premature airway closure, 462

Premature atrial depolarization, 328, 328*f*

Premature beat
ventricular contraction in, 395, 396*f*
weakness of, 395

Premature depolarizations, 326–328, 328*f*

Premature ventricular depolarization, 328, 328*f*

Premenstrual dysphoria (premenstrual syndrome [PMS]), 815

Premotor area, 183

Preoptic area, autonomic control of, 235–236

Prepiriform cortex, in olfactory pathway, 159

Preprooxyphysin, 734

Preprovasophysin, 734

Presbycusis, 143, 146*b*

Presbyopia, 129*b*

Pressoreceptors, 403–405

Pressure, flow and, relationship between, 346–350
laminar flow in, 349
Poiseuille's law in, 346–347, 346*f*
resistance to flow and, 347–348, 347*f*
series and parallel, resistance in, 348–349
turbulent flow in, 349, 349*b*
vessel wall, shear stress on, 349–350, 349*b*

Pressure gradient, creation of, 447–448

Pressure-volume relationships, 338–339, 338*f*–339*f*, 451–452, 451*f*
deflation, 452, 452*f*

Presynaptic cell, 53*f*

Presynaptic inhibition, 97, 97*f*

Presynaptic receptors, transmitter release modulated by, 97

Pretectum, 143

Prevertebral ganglia, 516

Primary afferents, 118–119

Primary ending, 166
responses of, 167*f*

Primary hyperkalemic paralysis, 74*b*

Primary motor cortex, 181

Primary peristalsis, 526–527, 527*f*

Primary secretion, 522

Primary spermatocytes, 788, 789*f*

Primary visual cortex, 139

Principal cells
of collecting duct, 615, 616*f*
of collecting duct system, 584, 585*f*
of parathyroid glands, 723, 723*f*

Principal cells *(Continued)*
secretion of potassium by, cellular mechanism of, 651–653, 652*f*–653*f*

Procedural memory, 222

Procorticotropin-releasing hormone (pro-CRH), 740

Progesterone, 693*t*, 773, 793*f*
biology of, 815–816
and mammary gland development, 825–826
metabolism of, 815–816
placental, 821–822, 822*f*
production of, corpus luteum, 804
secretion of, 824
structure of, 693*f*
transport of, 815–816
"unopposed estrogen" and, 813*b*

Progesterone receptor, 693, 693*t*

Progestin(s), chemistry of, 692, 693*t*

Proglucagon-derived peptides 1/2 (GLP-1/2), 516*t*

Progressive familial intrahepatic cholestasis, type II, 572*b*

Prohormone(s), 690, 692*b*

Prohormone convertase, 692*b*

Proinsulin, 704–705

Prolactin (PRL), 737, 751
secretion of, 751–752

Prolactinoma, 826, 826*f*

Prolactin-releasing factor (PRF), 751

Proliferating cells, 716

Proopiomelanocortin (POMC), 692*b*, 739, 741*f*, 719

Prop-1, 736*b*

Propranolol, 393*b*
heart rate and, 386, 387*f*

Proprotein convertases, 704–705

Prosopagnosia, 142*b*

Prostacyclin, 47, 360–361

Prostaglandin(s), 47
effect on GFR and RBF, 598–599, 598*t*
in gastrointestinal tract, 514–515
and parturition, 824

Prostaglandin I2 (PGI2), 360–361

Prostaglandin PGF2α, 806

Prostate gland, 788*f*, 797

Prostate-specific antigen (PSA), 797

Prostatin, 654–655

Protein(s), 689–692
assimilation, 549–550
digestion of, 549–550
proteases and peptidases, 550*f*
intracellular, signaling, 42, 43*f*
membrane, 5–6

Protein(s) *(Continued)*
metabolism of
digestive *vs.* fasting phase of, 714
in liver, 568–569
plasma, 19
glomerular filtration of, 594*b*, 594*f*
oncotic pressure, 15, 15*f*
relationship between osmotic pressure, 15*f*
pressure, 15, 15*f*
reabsorption of, in proximal tubule of kidneys, 609–610
secretion of, 521

Protein anabolic hormone, 749

Protein kinase A, 45
stimulation of , by G protein-coupled receptors, 46*f*

Protein kinase C, 47

Protein phosphatase(s), activation of, 708

Protein turnover, thyroid hormones and, 761

Proteinuria, 588*b*, 594*b*, 609–610

Proteolysis, 702–703

Proteosome, 3*t*

Prothrombin time, 579

Proximal renal tubular acidosis (RTA), 605*t*

Proximal tubule(s), 583–584, 584*f*
HCO₃- reabsorption in, 673–674, 673*f*
organic anions secreted by, 610–612, 610*b*, 611*f*
organic cations secreted by, 610*b*, 612*b*, 612*f*
phosphate reabsorption by, 664, 665*f*
and potassium absorption, 651
solute and water reabsorption in, 603–612, 608*t*

Pseudohyperkalemia, 647*b*

Pseudohypoaldosteronism, 605*t*, 619*b*

Psychomotor epilepsy, 217*b*

Psychosocial dwarfism, 737–739

Pudendum, 814–815

Pulmonary artery hypertension, 472–473

Pulmonary circulation, 301, 438–441
blood volume in, 303*t*

Pulmonary edema, 418*b*, 637*b*

Pulmonary fibrosis, 452*b*

Pulmonary function test abnormalities, patterns of, 459*t*

Pulmonary mechanoreceptors, 493–495, 493*t*

Pulmonary reflex(es), peripheral blood flow and, 406